AUG   2009

641.3374 GRIVETTI
GRIVETTI, LOUIS EVAN (E
CHOCOLATE :HISTORY,
    CULTURE, AND HERITAGE

P9-DVE-599

WITHDRAWN

641.3374 GRIVETTI
GRIVETTI, LOUIS EVAN (E
CHOCOLATE :HISTORY,
    CULTURE, AND HERITAGE

Chocolate

Tlapolcacanatl.

Chocolate

HISTORY, CULTURE, AND HERITAGE

Edited by

Louis Evan Grivetti

University of California
Davis, California

Howard-Yana Shapiro

Mars Incorporated,
and
University of California
Davis, California

Alameda Free Library
1550 Oak Street
Alameda, CA 94501

WILEY

A John Wiley & Sons, Inc., Publication

Copyright © 2009 by John Wiley & Sons, Inc. All right reserved

Published by John Wiley & Sons, Inc., Hoboken, New Jersey
Published simultaneously in Canada

No part of this publication may be reproduced, stored in a retrieval system, or transmitted in any form or by any means, electronic, mechanical, photocopying, recording, scanning, or otherwise, except as permitted under Section 107 or 108 of the 1976 United States Copyright Act, without either the prior written permission of the Publisher, or authorization through payment of the appropriate per-copy fee to the Copyright Clearance Center, Inc., 222 Rosewood Drive, Danvers, MA 01923, (978) 750-8400, fax (978) 750-4470, or on the web at www.copyright.com. Requests to the Publisher for permission should be addressed to the Permissions Department, John Wiley & Sons, Inc., 111 River Street, Hoboken, NJ 07030, (201) 748-6011, fax (201) 748-6008, or online at http://www.wiley.com/go/permission.

Limit of Liability/Disclaimer of Warranty: While the publisher and author have used their best efforts in preparing this book, they make no representations or warranties with respect to the accuracy or completeness of the contents of this book and specifically disclaim any implied warranties of merchantability or fitness for a particular purpose. No warranty may be created or extended by sales representatives or written sales materials. The advice and strategies contained herein may not be suitable for your situation. You should consult with a professional where appropriate. Neither the publisher nor author shall be liable for any loss of profit or any other commercial damages, including but not limited to special, incidental, consequential, or other damages.

For general information on our other products and services or for technical support, please contact our Customer Care Department within the United States at (800) 762-2974, outside the United States at (317) 572-3993 or fax (317) 572-4002.

Wiley also publishes its books in a variety of electronic formats. Some content that appears in print may not be available in electronic formats. For more information about Wiley products, visit our web site at www.wiley.com.

Library of Congress Cataloging-in-Publication Data:
Grivetti, Louis.
   Chocolate : history, culture, and heritage / Louis Evan Grivetti, Howard-Yana Shapiro.
     p. cm.
   Includes index.
   ISBN 978-0-470-12165-8 (cloth)
   1. Chocolate–History.   I. Shapiro, Howard-Yana.   II. Title.
   TX767.C5G747 2009
   641.3'374–dc22

                                                        2008041834

Printed in the United States of America
10 9 8 7 6 5 4 3 2 1

To our parents and wives:

Blanche Irene Carpenter Grivetti
Rex Michael Grivetti
Georgette Stylanos Mayerakis Grivetti

Pesche Minke Shapiro
Yankel Shapiro
Nancy J. Shapiro

Contents

Foreword

The Portuguese poet Fernando Pessoa had the right idea when he wrote early in the last century "Look, there's no metaphysics on earth like chocolate." Chocolate is a substance long regarded as magical, even supernatural, not to mention salubrious, today for its heart-healthy properties, yesterday because of a solid medicinal reputation as well as an aphrodisiacal one. Chocolate begins as seeds in a pod, that pod the fruit of the cacao tree *Theobroma cacao*. Not incidentally, the scientific name means "drink of the gods," by way of continuing the metaphysical.

Until relatively recently nobody gave much thought to eating chocolate. Drink was its original use and, despite evidence of an Amazonian origin, Mesoamericans were probably its original users. Cacao was employed in ancient Maya ceremonies and rituals and later used in religious rites to keep alive the memory of Quezalcoatl, the god of the air who made earthly visits from time to time dispensing instructions on how to grow various foods, cacao among them. In addition cacao nibs (the almond-shaped seeds) were put to work as coins so that by the time the Europeans sailed into the New World cacao was well entrenched in all facets of Mesoamerican life: spiritual, nutritional, and financial.

The European phase of cacao's history dates from 1502 when Columbus, then in the Gulf of Honduras on his forth voyage encountered natives who gave him the drink *xocoatl* made of cacao, honey, spices, and vanilla. The Explorer carried some nibs back to Spain, where they were viewed as curiosities only and it took another introduction in 1528 by Hernando Cortés (the conqueror of Mexico) to establish the plant in Iberia. Before long the Spaniards had figured out how to turn the nibs into an agreeable drink and by 1580 cocoa had achieved widespread popularity among Spain's elite and its cacao plantations became sources of considerable wealth. As sugar grew cheaper and more readily available in the seventeenth century, chocolate spread across Europe, chocolate houses sprang up and cocoa, although expensive, was charming everyone who could afford it. Doubtless, part of that charm resided in its alleged aphrodisiac properties, and chocolate found its way into confections and was tinkered with as candy.

An international phase of chocolate history was launched in 1819, when the first eating chocolate was produced in Switzerland. In the following decade Cadbury's Chocolate Company opened in England, the Baker Chocolate Company in the United States, a Dutch chocolate maker produced the world's first chocolate candy, and an instant cocoa powder was invented. The commercial chocolate industry was born.

If there is little passion in my nutshell early history of chocolate, the same is not true of the pages that follow. They reflect the energy and enthusiasm of the chocolate history research group established at the University of California at Davis a decade ago with the backing of Mars, Incorporated. Led by Professor Louis Grivetti, its members have investigated myriad aspects of chocolate history and have generated mountains of materials. Nonetheless, the editors explain that their intention has not been to produce a full history of chocolate, which would have taken many more years to complete. Instead, what they have done is to assemble a veritable archive of the subject in 56 chapters and 10 appendices for which food historians will be forever grateful.

The chapters are wide ranging and head in whatever directions their authors' expertise and curiosity dictate. Within this work they are organized roughly chronologically as well as geographically and topically, so that they begin with pre-Maya cacao use and contain in the penultimate chapter searches for chocolate references made during the American Civil War. Medicinal application is a recurring theme and one chapter examines twenty-first century attitudes about such uses. Chocolate pots for serving are given considerable space and five chapters are devoted to cacao and chocolate in the Caribbean with another to cacao production in Brazil and West Africa.

The final chapter scouts new terrain for future chocolate research with the appendices intended to help in this regard by disclosing archives, libraries, museums, other institutions, and digitized resources consulted in this effort. Some 99 chocolate-associated quotations are provided, as is a chocolate timeline and an important discussion of early written works on chocolate. Finally, there is a brief discussion of the nutritional properties of cocoa.

All of this may not constitute a full history of chocolate but it comes close. This work is both a major contribution to the field and to a growing body of food-history literature.

KENNETH F. KIPLE

Preface

To study the history of chocolate is to embark upon an extraordinary journey through time and geographical space. The chocolate story spans a vast period from remote antiquity through the 21st century. Historical evidence for chocolate use appears on all continents and in all climes, from tropical rain forests to the icy reaches of the Arctic and Antarctic. The story of chocolate is associated with millions of persons, most unknown, but some notables including economists, explorers, kings, politicians, and scientists. Perhaps no other food, with the exception of wine, has evoked such curiosity regarding its beginnings, development, and global distribution. But there is a striking difference: wine is forbidden food to millions globally because of its alcohol content but chocolate can be enjoyed and savored by all.

The chocolate history group at the University of California, Davis, was formed in 1998 at the request of Mars, Incorporated. The purpose of this association was to identify chocolate-associated artifacts, documents, and manuscripts from pre-Columbian America and to trace the development and evolution of culinary and medical uses of chocolate into Europe and back to North America. Our initial activities (1998–2001) were characterized by archive/library research and on-site field work observations and interviews conducted in Cuba, the Dominican Republic, Guatemala, Mexico, Panama, and the United States. The primary objectives during this research period were the following:

1. Identify early medical and culinary data associated with cacao and chocolate use in the Americas and Europe.

2. Interview traditional healers and chocolate vendors in the Americas to better understand contemporary, 20th and 21st century, cultural uses of chocolate.

3. Identify indigenous, historical, and early 20th century chocolate recipes.

In 2004, the chocolate history research group was expanded after a second generous gift from Mars, Incorporated. Our team of scholars during 2004–2007 included colleagues and independent scholars affiliated with the following institutions: Albany Institute of History and Art, Albany, New York; Museum of Fine Arts, Boston, Massachusetts; Department of History, Boston University, Boston, Massachusetts; California Parks System, Sacramento, California; Department of Ethnic Studies, California State University, San Luis Obispo, California; Colonial Deerfield, Deerfield, Massachusetts; Colonial Williamsburg, Williamsburg, Virginia; Center for Anthropology, University of Havana, Cuba; Department of Art History, East Los Angeles Community College, Los Angeles, California; Florida Institute for Hieroglyphic Research, Palmetto, Florida; Fort Ticonderoga, Ticonderoga, New York; Fortress Louisbourg, Nova Scotia, Canada; Harokopio University, Athens, Greece; Mars, Incorporated, Elizabethtown, Pennsylvania, Hackettstown, New Jersey, and McLean, Virginia; The McCord Museum, Montreal, Canada; Mills College, Oakland, California; Division of Social Sciences, New College of Florida, Sarasota, Florida; Oxford University, Oxford, England; Oxford Brookes University, Oxford, England; Parks Canada, Ottawa, Canada; Department of Community Development, Department of Engineering, Department of Food Science, Department of Native American Studies, Department of Nutrition, Graduate Group in Geography, and Peter J. Shields Library, University of California, Davis, California; and University of Massachusetts, North Dartmouth, Massachusetts.

Activities during the second research period (2004–2007) continued to identify chocolate-related documents available in archives, libraries, and museums located in Central America, Mexico, and the Caribbean, and efforts were expanded into additional countries of South and North America, western and southeastern Europe, western Africa, and south Asia. Our primary objectives were the following:

1. Determine historical patterns of introduction and dispersal of chocolate products throughout North America.

2. Identify the development and evolution of chocolate-related technology in North America.

3. Identify and trace the culinary, cultural, economic, dietary/medical, military, political, and social uses of chocolate in North America from the Colonial Era through the early 20th century.

4. Develop a state-of-the-art database and web portal for the history of chocolate, to be used by students, scholars, and scientists.

5. Publish chocolate-related findings via the popular press and scholarly journals, and relate findings via local, national, and international symposia and professional meetings.

The present book contains 56 chapters written by members of our chocolate history team. The story of chocolate is traced from earliest pre-Columbian times, through uses by Central American societies prior to European arrival, through the global spread of cacao trees to Africa and Asia, through Caribbean and South American trade, and ultimately the culinary and medical uses of chocolate in Europe, North America, and globally.

While much of the chocolate story has been told elsewhere, it is characteristic of chocolate-associated research that new documents can be identified and brought to light daily. Historical research on chocolate-associated topics has been facilitated in recent years by important, easily available on-line services through university and governmental subscriptions, whether the Library of Congress, Paper of Record, NewsBank/Readex, or other services. These sites (and others) have made it relatively easy to search millions of newspaper and journal/magazine advertisements and articles and other documents that cover historical North America (United States and Canada) from the 16th through early 20th centuries. These on-line services provide users with topical, keyword search engines that permit easy identification, retrieval, and cataloging of tens of thousands of documents within a short period in sharp contrast to the more laborious and time-consuming use of microfilm and microfiche services of previous decades. Still, it has been the slow, detailed tasks associated with archive and library research that has characterized much of our current efforts, and that has revealed many of the most exciting findings chronicled within the present book.

Our vision was to recruit a team of scholars with diversified training and research methods who would apply their special talents and skills to investigate chocolate history. Our team consisted of 115 colleagues and represented a broad range of professional fields: agronomy, anthropology, archaeology, archive science, art history, biochemistry, business management and product development, computer science, culinary arts, curatorial arts, dietetics, economics, engineering, ethnic studies, food science, gender studies, genetics and plant breeding, geography, history, legal studies (both historical and contemporary), library science, linguistics, marketing, museum administration, nutrition, paleography, and statistics. Team members also were skilled in a variety of languages, an important consideration given that chocolate-related documents regularly have appeared in Dutch, French, German, Greek, Italian, Latin, Portuguese, Russian, Spanish (Castiliano and contemporary national dialects), and Swedish, as well as ancient and contemporary Mesoamerican languages.

Topics investigated by members of our team also reflected diversified research interests: agriculture and agronomy (cacao cultivation and ecology), collectables (chocolate-associated posters, ephemera, toys, and trading cards), culinary arts (recipes and serving equipment), culture in its broadest sense (art, linguistics, literature, music, religion, and theater), diet and health (chocolate in preventive and curative medicine), economics (advertising, import/export, manufacturing, marketing, product design, and sales), education (18th century North American school and library books), ethics (issues associated with 17th to 19th century child labor and slavery), gender (division of labor and women's roles in chocolate production), legal issues (chocolate-associated crime and trial accounts, copyright, and patent law), military (chocolate as rations and as hospital/medical supplies), and politics (chocolate-associated legislation at local, state, regional, national, and international levels).

Team members selected historical eras for their chocolate-related research that suited their interests, talents, and previous experience. These conceptual eras included: Pre-Columbian America; Colonial Era North, Central, South America, and the Caribbean; American Revolutionary War Era; America and Canada in the Post-Revolutionary War Era; Early American Federal Period; Continental Exploration and Westward Expansion (both Canadian and American); Spanish and Mexican Periods (American Southeast, Southwest borderlands, and West Coast regions of North America); California Gold Rush Era; American Civil War Era; Postwar Reconstruction; Early Industrial North America; and Early Modern Era.

The types of information available for inspection by team members included advertisements (magazines and newspapers, advertising posters, signs, and trade cards); archaeological materials (murals, paintings, pottery, statues, and actual chocolate residues from ancient containers); art (lithographs, paintings, prints, and sculpture); commonplace books, diaries, and handwritten travel accounts; expedition records; government documents; hospital records; personal correspondence; literature (diaries, novels, and poetry); magazine articles; menus; military documents; newspaper accounts; obituaries; probate records; religious documents; and shipping manifests.

During the early stages of our work, we elected not to produce an integrated global history of chocolate. In our view, such an effort would have exceeded several thousand pages in print and would have been out of date upon publication due to continued evidence uncovered almost daily during our archive, library, and museum research. Instead, the thematic chapters presented in the present book reflect in-depth snapshots that illustrate specific themes within the breadth and scope of chocolate history. As a collection, the chapters presented herein present a common thread that reveals the sustained importance of chocolate through the millennia. The chapters also reveal where additional scholarship and future activities might be productive. It is our hope that readers of our work, those interested in expanding and furthering archive, library, and museum research on chocolate, will themselves embark upon their own voyage of discovery and make additional contributions to chocolate research.

LOUIS EVAN GRIVETTI
HOWARD-YANA SHAPIRO

Davis, California
January 2009

Acknowledgments

This book reflects the efforts of many persons and organizations. We wish to thank Deborah and Forrest Mars, Jr. for their deep interest in history and for their vision that led to the founding of the Chocolate History Group. We also extend our thanks to Dr. Harold Schmitz, Chief Scientist, Mars, Incorporated, for his valuable support throughout the years. To our editors at Wiley, Jonathan Rose and Lisa Van Horn, we thank you for your skills and dedication to produce a volume that is beautiful and content rich. We thank Lee Goldstein of Lee Goldstein Design for the design of the text and insert. We thank Dr. Teresa Dillinger for her support in the early days of our research and Dr. Deanna Pucciarelli who helped manage this enormous undertaking during the last three years. We thank Steven Oerding, Senior Artist/Supervisor, and Samuel Woo, Principal Photographer, both from IET-Academic Technology Services, Mediaworks, at the University of California, Davis, for photography and map production included within the present book. We also wish to thank the Administration and Librarians of the Peter J. Shields Library, University of California, Davis, especially Daryl Morrison and Axel Borg, for their assistance in locating key volumes and manuscripts during our research.

We sincerely extend our personal thanks to Mars, Incorporated for their generous support and their enduring respect and appreciation for all things chocolate, allowing us to document the enormous breadth of chocolate's role throughout history.

Finally, we thank each of the chocolate history researchers who worked as part of our team throughout the last 10 years. Our lives have been enriched by each of them!

L. E. G.
H.-Y. S.

Chocolate Team (1998–2009)

Shelly Allen
Undergraduate Researcher
University of California
Davis, California

Brent Anderson
Process Development Engineer
Historic Division of Mars, Incorporated
Hackettstown, New Jersey

Jennifer Anderson
Professor
Department of Anthropology
California State University
San Jose, California

Margaret Asselin
Marketing Director
Mars, Incorporated
Hackettstown, New Jersey

Richard Bailey
Captain
Ocean Classrooms Foundation
Watch Hill, Rhode Island

Diane Barker
Undergraduate Researcher
University of California
Davis, California

James Barrett
Postgraduate Researcher
University of California
Davis, California

Patricia Barriga
Archivist and Paleographer
Mexico City, Mexico

Steve Beck
California State Parks Service
Sutter's Fort
Sacramento, California

William Bellody
Research and Development Officer
Mars, Incorporated
Hackettstown, New Jersey

Carmen Bernett
Mars, Incorporated
McLean, Virginia

Anne Blaschke
Postgraduate Researcher
Boston University
Boston, Massachusetts

Axel Borg
Librarian
Biological and Agricultural Sciences Department
Shields Library
University of California
Davis, California

Fred Bowers
National Sales Director
Mars, Incorporated
Jasper, Georgia

Laura Pallas Brindle
Postgraduate Researcher
University of Georgia
Athens, Georgia

Eileen Brown
Senior Franchise Manager
Mars, Incorporated
McLean, Virginia

BEATRIZ CABEZON
Paleographer and Independent Scholar
Davis, California

HALLEY CARLQUIST
Undergraduate Researcher
University of California
Davis, California

KATI CHEVAUX
Mars, Incorporated
Hackettstown, New Jersey

FRANK CLARK
Supervisor, Historic Foodways
Colonial Williamsburg Foundation
Williamsburg, Virginia

CHRISTOPHER CLAYTON
Undergraduate Researcher
University of California
Davis, California

JEAN COLVIN
Director
University of California Research Expeditions
   (UREP)
Davis, California

KARL CRANNELL
Public Programs Coordinator
Fort Ticonderoga
Ticonderoga, New York

BRANDON DAVIS
Undergraduate Researcher
University of California
Davis, California

VICTORIA DICKINSON
Director
McCord Museum
Montreal, Canada

TERESA DILLINGER
Postgraduate Researcher
University of California
Davis, California

CLEO DIMITRIADOU
Undergraduate Researcher
Harokopio University
Athens, Greece

VASSILIKI DRAGOUMANIOTI
Undergraduate Researcher
Harokopio University
Athens, Greece

PHIL DUNNING
Material Culture Researcher
Parks Canada
Ottawa, Canada

SYLVIA ESCARCEGA
Assistant Professor
Department of Anthropology
DePaul University
Chicago, Illinois

JENNIFER FOLLETT
Postgraduate Researcher
University of California
Davis, California

RUBY FOUGÈRE
Curatorial Collections Specialist
Fortress of Louisbourg
National Historic Site of Canada
Nova Scotia, Canada

CHRISTOPHER D. FOX
The Anthony D. Pell Curator of Collections
Fort Ticonderoga
Ticonderoga, New York

MARJORIE FREEDMAN
Professor
Department of Nutrition
California State University
San Jose, California

ESTHER FRIEDMAN
Independent Researcher
Boston, Massachusetts

ENRIQUE GARCÍA-GALIANO
Professor
Department of Food Science
National University of Mexico
Mexico City, Mexico

VANESSA GARDIA-BRITO
MPM/Counsel, Americas
Mars, Incorporated
McLean, Virginia

JAMES F. GAY
Journeyman
Historic Foodways
Colonial Williamsburg Foundation
Williamsburg, Virginia

NICOLE GEURIN
Undergraduate Researcher,
University of California
Davis, California

ROSE GIORDANO
Postgraduate Researcher
University of California
Davis, California

ESTRELLA GONZÁLEZ NORIEGA
Investigadora Auxiliar and
Adjunct Professor
Center for Anthropology
University of Havana
Havana, Cuba

BERTRAM M. GORDON
Professor
Department of History
Mills College
Oakland, California

JIM GRIESHIP
Extension Specialist
Department of Community Development
University of California
Davis, California

LOUIS EVAN GRIVETTI
Professor Emeritos
Department of Nutrition
University of California
Davis, California

JUDY HAMWAY
Mars, Incorporated
Hackettstown, New Jersey

E. JEANNE HARNOIS
Independent Researcher
Boston, Massachusetts

LISA HARTMAN
Historic Division of Mars, Incorporated
Bel Air, Maryland

KATY HECKENDORN
Undergraduate Researcher
University of California
Davis, California

JEYA HENRY
Professor
Department of Nutrition and Molecular Biology
Oxford Brookes University
Oxford, England

MARTHA JIMENEZ
Postgraduate Researcher
University of California
Davis, California

ALIZA JOHNSON
Undergraduate Researcher
University of California
Davis, California

ANNE MARIE LANE JONAH
Historian
Fortress of Louisbourg
National Historic Site of Canada
Nova Scotia, Canada

LOIS KAMPINSKI
Independent Scholar
Washington, DC

ALEXANDRA KAZAKS
Postgraduate Researcher
University of California
Davis, California

CHRISTOPHER KELLY
Postgraduate Researcher
Department of History
University of Massachusetts
North Dartmouth, Massachusetts

GALE KEOGH-DWYER
Mars, Incorporated
Hackettstown, New Jersey

AMANDA LANGE
Curatorial Department
Chair and Curator of Historic Interiors
Historic Deerfield
Deerfield, Massachusetts

MATTHEW LANGE
Postgraduate Researcher
University of California
Davis, California

KRISTINE LEE
Undergraduate Researcher
University of California
Davis, California

JULIO LOPEZ
Postgraduate Researcher
University of California
Davis, California

CATHERINE MACPHERSON
Independent Researcher
McCord Museum
Montreal, Canada

MARTHA J. MACRI
Professor
Department of Native American Studies
University of California
Davis, California

SILVIU MAGARIT
Undergraduate Researcher
University of California
Davis, California

DEBORAH MARS
President, Advisory Board
Historic Division of Mars, Incorporated
McLean, Virginia

ANTONIA-LEDA MATALA
Assistant Professor
Department of Nutrition
Harokopio University
Athens, Greece

ANNE MCCARTY
Director of Membership and Special Initiatives
Fort Ticonderoga
Ticonderoga, New York

W. DOUGLAS MCCOMBS
Curator of History
Albany Institute of History and Art
Albany, New York

TIMOTEO MENDOZA
Advisor to the California Department of Education
Madera, California

CATLIN MERLO
Undergraduate Researcher
University of California
Davis, California

JANET HENSHALL MOMSEN
Professor Emerita
Department of Community Development
University of California
Davis, California

VICTOR MONTEJO
Professor
Department of Native American Studies
University of California
Davis, California

HEIDI MOSES
Archaeology Collections Manager
Fortress of Louisbourg
National Historic Site of Canada
Nova Scotia, Canada

JUAN CARLOS MOTAMAYOR
Senior Scientist
Mars, Incorporated
Miami, Florida

MARY MYERS
Group Research Manager of
    Chocolate, Cocoa, Dairy
Mars, Incorporated
Elizabethtown, Pennsylvania

NATARAJ NAIDU
Undergraduate Researcher
University of California
Davis, California

EZRA NEALE
Postgraduate Researcher
University of California
Davis, California

MADEILINE NGUYEN
Undergraduate Researcher
University of California
Davis, California

BENJAMIN NOWICKI
Postgraduate Researcher
DePaul University
Chicago, Illinois

NIURKA NUÑEZ GONZÁLEZ
Investigadora Agregada
Center for Anthropology
University of Havana
Havana, Cuba

BRADLEY FOLIART OLSEN
Postgraduate Researcher
University of California
Davis, California

CHRISTIAN OSTROSKY
Postgraduate Researcher
University of California
Davis, California

ADRIANA PARRA
Postgraduate Researcher
University of California
Davis, California

SUZANNE PERKINS
Art Historian and Independent Scholar
Berkeley, California

SUE PROVENZALE
American Heritage Chocolate
Mars, Incorporated
Hackettstown, New Jersey

DEANNA PUCCIARELLI
Assistant Professor
Food and Consumer Sciences
Ball State University
Muncie, Indiana

SEZIN RAJANDRAN
Archivist and Independent Scholar
Seville, Spain

PAMELA RICHARDSON
Postgraduate Researcher
Oxford University
Oxford, England

KURT RICHTER
Postgraduate Researcher
University of California
Davis, California

Peter G. Rose
Independent Researcher
South Salem, New York

Robert Rucker
Biochemist and Nutritionist
Department of Nutrition
University of California
Davis, California

Diana Salazar
Independent Researcher and Translator
Davis, California

Brianna Schmid
Undergraduate Researcher and
University of California
Davis, California

Harold Schmitz
Mars, Incorporated
McLean, Virginia

Rebecca Shacker
Undergraduate Researcher
University of California
Davis, California

Celia D. Shapiro
Archivist and Independent Scholar
Washington, DC

Howard-Yana Shapiro
Director of Plant Science
Mars, Incorporated
Mclean, Virginia, and
University of California
Davis, California

Adam Siegal
Librarian
Humanities/Social Sciences Department
Shields Library
University of California
Davis, California

Rodney Snyder
Mars, Incorporated
Senior Research Engineer
Elizabethtown, Pennsylvania

Eduardo Somarriba
Professor, Tropical Agroforestry
Leader, Cocoa Thematic Group
CATIE (Centro Agronómico Tropical de
    Investigación y Enseñanza)
Turrialba, Costa Rica

Ward Speirs
Mars, Incorporated
Hackettstown, New Jersey

Margaret Swisher
Postgraduate Researcher
University of California
Davis, California

Nghiem Ta
Undergraduate Researcher
University of California
Davis, California

Josef Toledano
Agriculture and Agroforestry Consultant
Tel Aviv, Israel

Gabrielle Vail
Research Scholar and Director
Florida Institute for Hieroglyphic Research
Division of Social Sciences
New College of Florida
Sarasota, Florida

Lucinda Valle
Instructor
Department of Art History
East Los Angeles Community College
Los Angeles, California

Victor Valle
Professor
Chair, Department of Ethnic Studies
California State University
San Luis Obispo, California

Eric VanDeWal
Marketing Director
Mars, Incorporated
Hackettstown, New Jersey

Marilyn Villalobos
Regional Coordinator
Central America Cacao Project
CATIE (Centro Agronómico Tropical de Investigación
    y Enseñanza)
Turrialba, Costa Rica

Timothy Walker
Assistant Professor
Department of History
University of Massachusetts
North Dartmouth, Massachusetts

Gerald W. R. Ward
The Katharine Lane Weems Senior Curator of
    Decorative Arts and Sculpture
Art of the Americas
Museum of Fine Arts
Boston Massachusetts

Nicholas Westbrook
Director
Fort Ticonderoga
Ticonderoga, New York

Virginia Westbrook
Public Historian
Ticonderoga, New York

Eric Whitacre
Applied Food Science and Product Design
Elizabethtown, Pennsylvania

Amanda Zompetti
Undergraduate Researcher
University of Massachusetts
North Dartmouth, Massachusetts

PART

I

Beginnings and Religion

*C*hapter 1 (Vail) considers chocolate use by Mayan cultures in the pre-Hispanic Yucatán Peninsula, as evidenced through cacao- and chocolate-associated texts and information from actual residues of chocolate beverages discovered in ceremonial pots excavated at archaeological sites. Her chapter traces the role of cacao in Mayan religion and explores its function both as food and as a ceremonial item. Chapter 2 (Macri) examines theories on the origins of the word *cacao* (originally given as *kakaw*) and traces the scholarly debates regarding the linguistic origins of *cacao* and how the word diffused throughout Mesoamerica. Her chapter also reviews the controversial suggestion that first use of the word *chocola-tl* in the Nahua/Aztec language appeared only *after* the Spanish conquest of Mexico in the 1520s. Chapter 3 (Grivetti and Cabezon) identifies the several Mayan, Mixeca/Aztec, and contemporary Native American religious texts that report how the first cacao tree was given to humans by the gods. Their chapter considers how chocolate found a niche within Catholic ritual and social uses during religious holidays in New Spain/Mexico. Chapter 4 (Cabezon and Grivetti) translates and comments on a suite of extraordinary New World, Spanish texts written during that tragic period known as the Inquisition. The documents reveal how chocolate sometimes was associated with behaviors considered by the Church at this time to be heretical, among them blasphemy, extortion, seduction, and witchcraft, as well as accusations and denouncement for being observant Jews. Their chapter casts bright light on these dark actions practiced during this terrible period of Mesoamerican history. Chapter 5 (Shapiro) documents the intriguing and rich history of Jewish merchants influential in the 18th century cacao trade between the Caribbean islands of Aruba and Curaçao, New Amsterdam/New York, and elsewhere in New England. Her chapter reveals and describes for the first time the important role played by Jewish merchants who developed and expanded cacao trade in North America.

Cacao Use in Yucatán Among the Pre-Hispanic Maya

Gabrielle Vail

Introduction

Much of the discussion of cacao in ancient Mesoamerica centers on Classic Maya culture, especially the period between 500 and 800 CE, because of the abundance of ceramics that reference cacao (*kakaw*) in their texts, and painted scenes that depict its use. Chemical analyses of residues from the bottoms of Classic period vessels reveal that cacao was an ingredient of several different drinks and gruels and that it was served in a wide variety of vessel types. The best known of these is the lidded vessel from Río Azul (Fig. 1.1), where the chemical signature of cacao was first identified by scientists from Hershey Corporation in 1990 [1, 2].[1]

Cacao played an important role in the economic, ritual, and political life of the pre-Hispanic Maya, and it continues to be significant to many contemporary Maya communities [3–5]. The different drinks and foods containing cacao that were served are described in hieroglyphic texts [6–9], the most common being a frothy drink that is depicted in scenes showing life in royal courts and as a beverage consumed by couples being married (Fig. 1.2).[2] Field work among contemporary Maya groups suggests that it continues to be used as part of marriage rituals in some areas, where it is most commonly given as a gift—in either seed or beverage form—from the family of the prospective bridegroom to the bride's family [11–13].[3] Cacao also represents an important offering in other ritual contexts, including those of an agricultural nature [15–17].

During the pre-Hispanic period, cacao seeds (also called "beans") served as a unit of currency, and they were an important item of tribute [18, 19]. Because the tree can only be grown in certain regions (those that are especially humid), the seeds had to be imported to various parts of the Maya area that were not productive for growing cacao. It was considered one of the most important trade items, along with items like jade beads and feathers from the quetzal bird, which were worn by rulers to adorn their headdresses and capes.[4]

Cacao consumption was not important in and of itself but rather as part of elite feasting rituals that cemented social and political alliances. In addition to being one of the chief components of the feasts, cacao was one of the items gifted to those participating in the ritual, as were the delicate vases with their finely painted scenes of court life from which the beverage was consumed [20]. The scientific name of cacao, *Theobroma cacao*, meaning "food of the gods," is an apt

Chocolate: History, Culture, and Heritage. Edited by Grivetti and Shapiro
Copyright © 2009 John Wiley & Sons, Inc.

FIGURE 1.1. Vessel from Tomb 19, Río Azul, Guatemala. Hieroglyphic caption includes the glyph for cacao (at left on vessel lid). *Source*: Museo Nacional de Arqueología y Etnología in Guatemala City. (Used with permission.) (See color insert.)

FIGURE 1.2. Cacao exchange during wedding ceremony. *Source*: *The Codex Nuttall. A Picture Manuscript from Ancient Mexico*, edited by Zelia Nuttall. Courtesy of Dover Publications, New York. (Used with permission.)

description of the role it played within ancient Maya culture.

Cacao use in the Maya area can be traced back to somewhere between 600 and 400 BCE, based on chemical remains recovered from spouted vessels from Belize [21, 22].[5] But it has a much longer history that extends back, according to Maya beliefs, to the time before humans were created. In the creation story recorded in the text known as the *Popol Vuh*,[6] cacao is one of the precious substances that is released from "Sustenance Mountain" (called Paxil in the *Popol Vuh*, meaning "broken, split, cleft"), along with maize (corn), from which humans are made, at the time just prior to the fourth or present creation of the world [29].[7] Before this, it belonged to the realm of the Underworld lords, where it grew from the body of the sacrificed god of maize, who was defeated by the lords of the Underworld in an earlier era [31].

At a later time, the maize god was resurrected by his sons the Hero Twins, who were able to overcome the Underworld gods [32, 33]. Thereafter, human life became possible, once the location where the grains and maize were hidden was located and its bounty released by the rain god Chaak and the deity K'awil, who represents both the lightning bolt and the embodiment of sustenance and abundance [34, 35].[8]

The deities of importance to this story include the maize god Nal, Chaak, K'awil, and various Underworld lords, including the paramount death god (named Kimil in some sources) and a deity known by the designation God L. God L played various roles in Maya mythology, representing a merchant deity, an Underworld lord, and a Venus god.

The Role of Cacao in the Northern Maya Area

Previous discussions of cacao in Maya culture emphasize its use and depiction in Classic period contexts (ca. 250–900 CE) from the southern Maya lowlands. The focus of this chapter is on cacao from the northern Maya lowlands, the area known today as the Yucatán Peninsula. In this region, cacao becomes visible in the archaeological record during the Late Classic period, where it is mentioned in glyphic texts inscribed on pottery in the style known as Chocholá [37, 38]. Additionally, recent investigations suggest that it was grown in damp areas such as collapsed caves and cenotes (sinkholes) [39, 40].

Following the conquest of the Yucatán Peninsula by the Spanish in the early 16th century, we learn from European chronicles that cacao was important in various rituals, and that it was grown both locally on plantations and imported from Tabasco and Honduras [41–43]. Additionally, Bishop Diego de Landa, writing in ca. 1566, noted that cacao beans formed a unit of currency for exchange and were given in tribute. The owners of cacao plantations celebrated a ritual in the month of Muwan (corresponding to late April and early May) to several deities, including Ek' Chuwah, who was the god of merchants [44]. Cacao was also mixed with sacred water and crushed flowers and used to anoint children and adolescents participating in an initiation ceremony, or what Landa termed a "baptism" ritual [45]. Landa further noted that:

They make of ground maize and cacao a kind of foaming drink which is very savory, and with which they celebrate their feasts. And they get from the cacao a grease which resembles butter, and from this and maize they make another beverage which is very savory and highly thought of. [46]

Archaeological data from sites in Yucatán, in combination with images and hieroglyphic texts painted in screenfold books (codices) and on the stones used to bridge the vaults of buildings (called capstones), are useful in reconstructing the ritual and other uses of cacao in ancient Maya society.

CACAO OFFERINGS TO THE DEITIES

We learn, for instance, of its use as an offering to the gods in several scenes from the pre-Hispanic Maya codices, which are believed to have been painted in Yucatán during the Postclassic period.[9] A scene on Dresden Codex p. 12a (Fig. 1.3), showing the god of sustenance K'awil seated with a bowl of cacao beans in his outstretched hand, recalls an offering found at the site of Ek' Balam dating to the Late Classic period, of

FIGURE 1.3. K'awil, god of sustenance with cacao on Dresden Codex p. 120. *Source: Die Maya Handschrift der Königlichen öffentlichen Bibliothek zu Dresden* (Dresden Codex), by Ernst Förstemann. Mit 74 Tafeln in Chromo-Lightdruck. Verlag der A. Naumannschen Lichtdruckeret, Leipzig, 1880. (Used with permission.)

a pottery bowl filled with carved shells replicating cacao beans. Several other almanacs from the Dresden Codex picture the plant or its beans being held by various Maya deities, including the rain god Chaak, K'awil, the death god Kimil, and the Underworld god Kisin.[10] The texts associated with these almanacs describe cacao (spelled hieroglyphically as *kakaw*) as the god's sustenance, or *o'och* [49]. We have already seen the importance of these deities in relation to cacao, and we will see further examples of the role played by K'awil below.

Another scene, this one from the Madrid Codex p. 95a (Fig. 1.4), pictures four deities piercing their ears with obsidian blades in order to draw blood. Bloodletting rituals such as these are described by Landa, who noted that piercings were made in the ear, tongue, lips, cheek, and penis [50].[11] As a devout 16th century Catholic cleric, he was repulsed by these rituals, but they were an essential part of Maya religion and focused on the reciprocity between the gods (who created people out of maize and water) and people (who, in turn, were required to feed the gods with their blood) [51, 52]. It is interesting to note that Landa described women as being specifically excluded from bloodletting rituals, although one of the figures in the Madrid almanac is a female deity, as is another figure pictured letting blood from her tongue on Madrid Codex p. 40c [53]. This mirrors scenes from the Classic Maya area, where an important wife of the Yaxchilán ruler Shield Jaguar is shown drawing a thorny rope through her tongue to conjure ancestral spirits [54].

Cacao is mentioned as one of the offerings in the text on Madrid Codex p. 95a, in conjunction with incense (both *pom* from the copal tree and *k'ik'* made from the sap of the rubber tree). Bar-and-dot numbers associated with these offerings in the text indicate the number of offerings that were given. The deities pictured in the almanac's four frames include the creator Itzamna, a female deity associated with the earth, the wind and flower god, and a second depiction of Itzamna. Unfortunately, the glyph beginning each text caption is largely eroded, making it difficult to determine the action being referenced.[12]

Sophie and Michael Coe have suggested that the blood in the almanac is pictured falling from the deities' ears onto cacao pods in the four frames illustrated [55]. It is difficult to tell what this object is, since its appearance is fairly nondescript. It is possible that, rather than depicting cacao, the elliptical objects represent some sort of paper-like material, since blood was usually collected on such strips and then burned in order to reach the gods in the Upperworld [56]. The Lacandón Maya of Chiapas still practice rituals similar to those depicted in the codices; a red dye called *annatto* (also known as achiote) is used in place of human blood, but it has the same general significance [57].

FIGURE 1.4. Cacao and incense as offerings during blood letting ritual. *Source*: Codex Tro-Cortesianus (Madrid Codex). Akademische Druck-und Verlagsanstalt, Graz, 1967. Courtesy of the Museo de América, Madrid, Spain. (Used with permission.)

CACAO IN ASSOCIATION WITH OTHER PRECIOUS OBJECTS

Another interesting correspondence in both the codices and other painted scenes is the association between cacao and the birds known as quetzals (*Pharomachrus mocino*). In the Madrid Codex, for instance, the deity Nik, whose associations include wind, life, and flowers, leans with his back against a cacao tree, grasping a second cacao plant with his hand (Fig. 1.5). A quetzal bird appears above the scene, with a leaf in its beak. The text above the frame includes a reference to the god Nik, who is described as a day-keeper or priest; this is followed by the glyphic collocation for cacao.

This scene recalls one of the mural paintings in the Red Temple at the Late Classic site of Cacaxtla in Tlaxcala (Fig. 1.6). Although located in the Mexican highlands, the murals reflect the Maya style and were probably painted by a Maya artist [58]. The scene in question pictures the Classic Maya merchant god, God L, standing in front of a blue-painted cacao tree; a quetzal is perched on (or about to land on) the tree. Simon Martin suggests that this scene corresponds to the Underworld realm [59]. Although they come from very different environments, it is not difficult to imagine what served to link cacao and quetzals in the minds of the Maya who painted these images—both were exotic trade items and had sacred associations.

In later times, as discussed previously, the merchant deity Ek' Chuwah—a Maya god with clear links to Yacatecuhtli, the Aztec deity of merchants and travelers—was also one of the patrons of cacao. During the month Muwan (corresponding to late April and early May), the owners of cacao plantations in Yucatán celebrated a festival in honor of this deity. Additionally, travelers and merchants prayed to Ek' Chuwah and burned incense in his honor when they stopped to camp for the night. Landa noted that:

Wherever they came they erected three little stones, and placed on each several grains of the incense; and in front they placed three other flat stones, on which they threw incense, as they offered prayers to God whom they called Ek Chuah [Ek' Chuwah] that he would bring them back home again in safety. [60]

Three stones were also used to build one's hearth, in replication of the original three stones set in the sky by the gods at the time of creation. This is illustrated in a scene from the Madrid Codex p. 71a (Fig. 1.7), which shows the turtle (corresponding to the constellation Orion) as the celestial location where the stones were set [61].[13]

CACAO IN MARRIAGE RITUALS

One of the most interesting of the almanacs featuring cacao in the Maya codices occurs on p. 52c of the Madrid Codex (Fig. 1.8). Its text reads: *tz'á'ab' u kakaw cháak ix kàab'* "Chaak [the rain god] and Ixik Kaab' [the earth goddess] were given their cacao."[14] Recently, Martha Macri pointed out that the verb *ts'ab'a* is defined as "payment of the marriage debt [from the wife to the husband or between marriage partners]" in the Cordemex dictionary [65, 66]. This calls to mind rituals pictured in the codices from the Mixtec area of highland Mexico, where a frothy cacao beverage is exchanged by those being married (see Fig. 1.2) [67]. Contemporary Mayan speakers in the Guatemalan highlands still

FIGURE 1.5. Flower god Nik with cacao tree. *Source*: Codex Tro-Cortesianus (Madrid Codex). Akademische Druck-und Verlagsanstalt, Graz, 1967. Courtesy of the Museo de América, Madrid, Spain. (Used with permission.)

FIGURE 1.6. God L and cacao tree. From the Red Temple at Cacaxtla, Tlaxcala, Mexico. *Source*: Photograph by David R. Hixson. (Used with permission.) (See color insert.)

follow this tradition. In a folktale from the Alta Verapaz region of Guatemala, the marriage vows are sealed with a vessel full of "foamy chocolate" [68].

In the Madrid Codex scene, the deity pair stands across from each other, holding what may be honeycomb in their hands, as suggested by the scene on Madrid Codex p. 104c [69].[15] The cacao mentioned in the text is shown hieroglyphically on the side of the vessel (an abbreviated spelling of /ka-ka/). Two elements that remain to be identified (with another two above the deities' outstretched hands) appear on top of the vessel. Scholars have suggested that similar elements found in a different context (see later discussion) represent jade earspools or jeweled flowers, a symbol of preciousness [70]. Another possibility may be suggested, as pointed out by Simon Martin [71], based on similarities between the elements depicted

and their use in the glyph that has been read as *inah*, or "seed" by Floyd Lounsbury [72]. Among both the contemporary and pre-Hispanic Maya, cacao beverages were often flavored with honey and crushed seeds [73, 74].

An element of the composition that remains unexplained is the rattlesnake that forms a U-shape around the vessel. Rattlesnakes commonly appear in Maya art in association with the rain god Chaak, who is mentioned in the almanac's text. Additionally, the U-shape it assumes is reminiscent of the shape formed by cenotes (and glyphs representing cenotes) in Maya art.[16] In this regard, it is of interest to note that recent research in the Yucatán Peninsula indicates that cacao trees grow wild in cenotes, which provide the humid microenvironment necessary for their survival [75].[17] Scholars have proposed that groves of cacao trees were planted in dry sinkholes (*rejolladas*) by the Maya of Yucatán in pre-Hispanic times to supplement cacao obtained from cacao-rich areas in Belize, Tabasco, and Honduras, a possibility first suggested by native chroniclers [78]. Recent archaeological research in the vicinity of Ek' Balam and several other sites in the northern Maya lowlands is suggestive of cacao production in those regions during pre-Hispanic times [79–81]. The rattlesnake in the picture, therefore, may symbolize a cenote where cacao and other precious substances (including virgin water for making the cacao beverage) were gathered. It may also indicate that the ceremony in question took place in a location below ground, such as a cave or dry cenote. Recent studies suggest that caves were the loci of many different types of rituals of importance to the pre-Hispanic Maya, and they continue to function in this manner [82].

FIGURE 1.7. Turtle constellation with three hearthstones from the Madrid Codex. *Source*: Codex Tro-Cortesianus (Madrid Codex). Akademische Druck-und Verlagsanstalt, Graz, 1967. Courtesy of the Museo de América, Madrid, Spain. (Used with permission.)

FIGURE 1.8. Wedding ceremony with cacao and honeycomb from Madrid Codex p. 500. *Source*: *Codex Tro-Cortesianus* (Madrid Codex). Akademische Druck-und Verlagsanstalt, Graz, 1967. Courtesy of the Museo de América, Madrid, Spain. (Used with permission.)

FIGURE 1.9. God K'awil with cacao and precious foods. From the Temple of the Owls, Chichén Itzá, Yucatán, Mexico. Courtesy of the Peabody Museum of Archaeology and Ethnology, Harvard University, Cambridge, MA. (Used with permission.)

RESCUE OF CACAO FROM THE UNDERWORLD

A capstone recovered from a structure called the Temple of the Owls at Chichén Itzá (Fig. 1.9) has a painted image that is similar in certain ways to that on Madrid Codex p. 52c. The similarities include a serpent, depicted here in a curled position with an open mouth, from which the god of sustenance K'awil emerges; an emphasis on cacao, here represented by the hanging pods; and globular elements like those occurring in the madrid scene. The lower part of the picture is enclosed within a U-shaped element symbolizing a cenote, whereas the upper part includes a glyphic text along the top and glyphs indicating celestial elements along the two sides. The outer edges of the upper

FIGURE 1.10. God K'awil as "burden" of year. *Source: Die Maya Handschrift der Königlichen öffentlichen Bibliothek zu Dresden* (Dresden Codex), by Ernst Förstemann. Mit 74 Tafeln in Chromo-Lightdruck. Verlag der A. Naumannschen Lichtdruckeret, Leipzig, 1880. (Used with permission.)

area are marked by bands and rays, perhaps representing the sun or the scales of a caiman [83, 84].

As we have seen, K'awil is associated with cacao on several occasions in the codices, including Dresden Codex pp. 12a and 25a (Figs. 1.3 and 1.10). In the latter scene, he corresponds to the patron of the year (he is depicted as the deity being carried on the back of the richly attired figure), and cacao is mentioned as one of the offerings he receives. These images differ significantly from the capstone from Chichén Itzá, which scholars interpret as showing the emergence of K'awil from the Underworld [85, 86]. He is shown with grains and plants important to the survival of humans, which were originally stored within the earth (mentioned earlier), until they were revealed and brought to the earth's surface. Among these, according to Maya mythology, were maize (corn), cacao, honey, and various tropical fruits [87]. In light of this, it is possible that the elements resembling jade earspools might instead represent gourds or one of the tropical fruits mentioned in the creation story.[18] If this were the case, then the scene on Madrid Codex p. 52c, like the Chichén capstone, might refer not only to a wedding ceremony, but perhaps also to one of the paramount moments in the story of creation—that taking place just prior to the formation of the first human couple out of maize and water. The substitution of Itzamna in the picture for Chaak in the text, therefore, takes on greater meaning, as the rain god was associated with releasing maize and the other foods from their home within the earth, whereas Itzamna is

FIGURE 1.11. God Chaak emerging from serpent representing a cenote. *Source: Die Maya Handschrift der Königlichen öffentlichen Bibliothek zu Dresden* (Dresden Codex), by Ernst Förstemann. Mit 74 Tafeln in Chromo-Lightdruck. Verlag der A. Naumannschen Lichtdruckeret, Leipzig, 1880. (Used with permission.)

the primordial creator to whom humans owe their existence [89, 90].[19] The earth goddess may here be signifying the forces of fertility and creation that are one of her primary attributes in the Maya codices [91].[20]

An almanac in the Dresden Codex, running across the middle (b) register of p. 31 through 35 (Fig. 1.11), pictures a scene very similar to that on the capstone—a deity emerging from the gaping mouth of a serpent. In this instance, however, it is Chaak, rather than K'awil, who is pictured, and he holds his lightning axe aloft (rather than a vessel with precious substances).

FIGURE 1.12. Emergence of creation and death deities. *Source: Codex Tro-Cortesianus* (Madrid Codex). Akademische Druck- und Verlagsanstalt, Graz, 1967. Courtesy of the Museo de América, Madrid, Spain. (Used with permission.)

This almanac likely pictures the role played by Chaak in bringing water from cenotes to the sky, where it falls to the earth as rain. A textual reference to rain (or *chaak*) as the sustenance of the year provides support for this identification [92]. The scenes in question may also, however, be referencing myths concerning the role of the rain deity in providing maize, cacao, and other plants to humans. Imagery picturing the rain god with an upraised axe has been suggested to symbolize the moment before he struck the mountain of sustenance to make its bounty available [93]. In that case, the images on Dresden Codex pp. 31b–35b would also be a reference to the origin of life on earth.

A further scene, from an almanac on Madrid Codex p. 19a (Fig. 1.12), also is relevant to this discussion. It contains two frames, the first picturing the creator Itzamna emerging from the open jaws of a rattlesnake, and the second the death god Kimil in a similar position. If these scenes are interpreted in the same way as those previously discussed, they appear to show the creator and the death god emerging from the Underworld. The former brings *ha' waah* "food and drink," while the latter has his hand raised to his face in a gesture signifying death. This almanac clearly shows that, with the emergence of Itzamna and Kimil onto the earthly plane, humans were given both the means to survive and the reality of death at the end of their life's journey.

Conclusion

The rich mythology surrounding the production and use of cacao by Mayan speakers in different eras and regions points to its enduring quality as a gift from the gods. Although cacao no longer is used as frequently in Maya rituals today, it continues to function as a marker of important life events such as birth and marriage, and in agricultural ceremonies such as those documented among the Ch'orti' Maya of Honduras [94, 95]. It is also an important component of the ceremonies involving offerings to the Lacandón Maya "god pots," where it is characterized as a melding of the female (represented by cacao, because it is prepared by women) and the male (represented by a fermented drink called *balché*, made by the men) [96]. This mirrors its use in feasts 1500 years ago, when it was also prepared and frothed by women, as illustrated by Classic period artists portraying life at royal Maya courts on ceramic vessels [97].

Acknowledgments

This chapter has benefited from discussions with many of my colleagues in anthropology, as well as the inspiration and guidance of the History of Chocolate Project directors and team members. I would especially like to thank Louis Grivetti for the invitation to participate in the project initially, when I was based at the University of California, Davis (1998–1999), and for his continued generous support and encouragement—and that of Harold Shapiro and Kati Chevaux of Mars—after I relocated to New College of Florida.

While at the University of California, Davis, I enjoyed the interaction and intellectual stimulation of project meetings and gained especially from working closely with Martha Macri, Theresa Dillinger, and

Patricia Barriga. I also benefited from conversations and correspondence with Robert Rucker and Alyson Mitchell of the Department of Nutrition.

More recently, I have been funded by a collaborative research grant from the National Endowment for the Humanities to develop an on-line database of the Maya hieroglyphic codices (grant RZ-50311-04, from July 2004 to April 2007). My discussion of cacao-related imagery in this volume owes much to their generous support, but I must note that any views, findings, conclusions, or recommendations expressed in this chapter do not necessarily represent those of the National Endowment for the Humanities.

I would also like to thank the many colleagues who have contributed to this research over the years, including Anthony Andrews, Traci Ardren, Ellen Bell, Christine Hernández, Martha Macri, Dorie Reents-Budet, Bob Sharer, and Loa Traxler.

Additionally, I appreciate the editorial assistance of Mary Grove and the support of Ty Giltinan, the members of the New College Glyph Study Group, and my research assistant, New College student Jessica Wheeler. *Muchas gracias* to all!

Endnotes

1. For a discussion of the methodology of analyzing residues, see Jeffrey Hurst [2].

2. A fermented drink made from the pulp of the plant may also have been consumed [10], although it assumes little importance in Mayan art and writing.

3. But see Elin Danien for an example of chocolate (in beverage form) being given by the bride to her husband to consummate their wedding [14].

4. The quetzal makes its home in the highland Maya area. Its brilliant blue and green tail feathers were prized by peoples throughout Mesoamerica during the pre-Hispanic period and represented an important trade item. Today, the quetzal is the national bird of Guatemala.

5. The history of its cultivation and the linguistic etymology of the term *kakaw* are considered in a recent book edited by Cameron McNeil, and previously by Lyle Campbell and Terrence Kaufman, and Karen Dakin and Søren Wichmann [23–25].

6. The *Popol Vuh* is an alphabetic text that dates to the 16th century from the highland kingdom of the K'iche' Maya. Its anonymous authors note that it had pre-Hispanic antecedents; it was probably recorded hieroglyphically in a screenfold book, or codex [26, 27]. The stories it contains—and others that were not written down—may be seen as early as the second century BCE from various parts of the Maya area and beyond [28]. Clearly, the mythology represents a tradition that spanned the region called Mesoamerica.

7. The concept of Sustenance Mountain is still found among highland Maya cultures, where it represents the realm of the honored dead and the place where rain clouds are created. Its surface overflows with life, including game and fruit trees [30].

8. His name consists of two parts: *k'aa* "surplus, abundance" and *wi'il* "sustenance" [36].

9. Although the almanacs and tables in the codices refer to events in the ninth through 15th centuries, they are thought to be 14th or 15th century copies of much earlier manuscripts [47, 48].

10. In the Maya codices, deities are pictured engaged in the activities that should be performed in their honor by those worshipping them. Deities had generally positive or negative associations. Of those named above, Chaak and K'awil were associated with positive auguries (abundance of food being the most common), and the death and underworld gods were associated with auguries such as "evil thing" and "dead person."

11. Scenes from the Madrid Codex show deities drawing blood from their ears, tongue, and genitals, whereas other almanacs depict the implements used (including obsidian blades and stingray spines).

12. Maya texts are usually in the form verb–object (optional)–subject–title or augury, a feature shared with the Mayan languages today.

13. There is disagreement among scholars as to the Western correlate to the Maya turtle constellation. I follow Harvey and Victoria Bricker in their identification [62], although Gemini has been suggested as an alternative [63].

14. The passive form of the verb was first recognized by Alfonso Lacadena [64].

15. Although the male deity is named as the rain god Chaak in the text (see Figure 1.8), the creator deity Itzamna is pictured, painted a brilliant blue color used to signify rainfall and fertility.

16. A cenote is a sinkhole filled with water. Sinkholes, caverns, and cenotes are common features of the topography of the Yucatán Peninsula, which is composed of limestone crisscrossed by underground rivers.

17. The soils of Yucatán are dry and rocky and not suited to trees such as cacao, nor does the climate provide enough water in the form of rain. For many years, scholars were therefore confused by statements by Spanish chroniclers who commented on cacao plantations in Yucatán [76, 77].

18. In his analysis of the capstone image, Karl Taube provides evidence suggesting that these motifs are flowers, specifically flowers from the cacao tree [88].

19. In the Maya codices, as in the *Popol Vuh*, creation is the work of a paired male and female couple. According to Maya beliefs, men and women play complementary roles in all aspects of daily life.

20. An older female deity, pictured with wrinkles, is more commonly shown as Itzamna's counterpart in scenes relating to creation.

References

1. Hall, G. D., Tarka, S. M. Jr., Hurst, W. J., Stuart, D., and Adams, R. E. W. Cacao residues in ancient maya vessels from Río Azul, Guatemala. *American Antiquity* 1990;55(1):138–143.

2. Hurst, W. J. The determination of cacao in samples of archaeological interest. In: McNeil, C. L., editor. *Chocolate in Mesoamerica: A Cultural History of Cacao*. Gainesville: University Press of Florida, 2006; pp. 105–113.

3. McNeil, C. L. Introduction: the biology, antiquity, and modern uses of the chocolate tree (*Theobroma cacao* L.). In: McNeil, C. L., editor. *Chocolate in Mesoamerica: A Cultural History of Cacao*. Gainesville: University Press of Florida, 2006; pp. 1–28.

4. McNeil, C. L. Traditional cacao use in modern Mesoamerica. In: McNeil, C. L., editor. *Chocolate in Mesoamerica: A Cultural History of Cacao*. Gainesville: University Press of Florida, 2006; pp. 341–366.

5. Thompson, J. E. S. Notes on the use of cacao in Middle America. *Notes on Middle American Archaeology and Ethnology*, Publication No. 128. Washington, DC: Carnegie Institution of Washington, 1956.

6. Coe, S. D., and Coe, M. D. *The True History of Chocolate*. London, England: Thames and Hudson, 1996.

7. Macri, M. Nahua loan words from the Early Classic period: words for cacao preparation on a Río Azul ceramic vessel. *Ancient Mesoamerica* 2006;16(2): 321–326.

8. Reents-Budet, D. *Painting the Maya Universe: Royal Ceramics of the Classic Period*. Durham, NC: Duke University Press, 1994.

9. Stuart, D. The language of chocolate: references to cacao on Classic Maya drinking vessels. In: McNeil, C. L., editor. *Chocolate in Mesoamerica: A Cultural History of Ccacao*. Gainesville: University Press of Florida, 2006; pp. 184–201.

10. Henderson, J. S., and Joyce, R. A. Brewing distinction: the development of cacao beverages in formative Mesoamerica. In: McNeil, C. L., editor. *Chocolate in Mesoamerica: A Cultural History of Cacao*. Gainesville: University Press of Florida, 2006; pp. 140–153.

11. Bunzel, R. L. *Chichicastenango: A Guatemalan Village*. Ethnological Society Publications, Volume 22. Locust Valley, NY: J. J. Augustin, 1952.

12. Tedlock, D. *Popol Vuh: The Definitive Edition of the Mayan Book of the Dawn of Life and the Glories of Gods and Kings*, revised edition. New York: Simon and Schuster, 1996; p. 322, note 184.

13. Redfield, R., and Villa Rojas, A. *Chan Kom, A Maya Village*. Washington, DC: Carnegie Institution of Washington, Publication 448, 1934; p. 40.

14. Danien, E., editor. *Maya Folktales from the Alta Verapaz*. Philadelphia: University of Pennsylvania Museum, 2005; pp. 57–59.

15. McNeil, C. L. Introduction: the biology, antiquity, and modern uses of the chocolate tree (*Theobroma cacao* L.). In: McNeil, C. L., editor. *Chocolate in Mesoamerica: A Cultural History of Cacao*. Gainesville: University Press of Florida, 2006; pp. 1–28.

16. McNeil, C. L. Traditional cacao use in modern Mesoamerica. In: McNeil, C. L., editor. *Chocolate in Mesoamerica: A Cultural History of Cacao*. Gainesville: University Press of Florida, 2006; pp. 341–366.

17. Kufer, J., and Heinrich, M. Food for the rain gods: cacao in Ch'orti' ritual. In: McNeil, C. L., editor. *Chocolate in Mesoamerica: A Cultural History of Cacao*. Gainesville: University Press of Florida, 2006; pp. 384–407.

18. Millon, R. F. *When Money Grew On Trees: A Study of Cacao in Ancient Mesoamerica*. Unpublished Ph.D. dissertation. New York: Department of Political Science, Columbia University, 1955.

19. Thompson, J. E. S. Notes on the use of cacao in Middle America. *Notes on Middle American Archaeology and Ethnology*, Publication No. 128. Washington, DC: Carnegie Institution of Washington, 1956.

20. Reents-Budet, D. The social context of *kakaw* drinking among the ancient Maya. In: McNeil, C. L., editor. *Chocolate in Mesoamerica: A Cultural History of Cacao*. Gainesville: University Press of Florida, 2006; pp. 202–223.

21. Hurst, W. J., Tarka, S. M. Jr., Powis, T. G., Valdez, F., and Hester, T. R. Cacao usage by the earliest Maya civilization. *Nature* 2002;418:289–290.

22. Powis, T. G., Valdez, F. Jr., Hester, T. R., Hurst, W. J., and Tarka, S. M. Jr., Spouted vessels and cacao use among the Preclassic Maya. *Latin American Antiquity* 2002;13(1):85–106.

23. McNeil, C. L., editor. *Chocolate in Mesoamerica: A Cultural History of Cacao*. Gainesville: University Press of Florida, 2006.

24. Campbell, L., and Kaufman, T. A linguistic look at the Olmecs. *American Antiquity* 1976;41(1):80–89.

25. Dakin, K., and Wichmann, S. Cacao and chocolate: a Uto-Aztecan perspective. *Ancient Mesoamerica* 2000; 11(1):55–75.

26. Christenson, A. J. *Popol Vuh: The Sacred Book of the Maya*. New York: O Books, 2003.

27. Tedlock, D. *Popol Vuh: The Definitive Edition of the Mayan Book of the Dawn of Life and the Glories of Gods and Kings*, revised edition. New York: Simon and Schuster, 1996.

28. Saturno, W. A., Taube, K. A., and Stuart, D. *The Murals of San Bartolo, El Petén, Guatemala. Part 1: The North Wall*. Ancient America 7. Barnardsville, NC: Center for Ancient American Studies, 2005.

29. Christenson, A. J. *Popol Vuh: The Sacred Book of the Maya*. New York: O Books, 2003; pp. 193–194.

30. Martin, S. Cacao in ancient Maya religion: first fruit from the maize tree and other tales from the underworld. In: McNeil, C. L., editor. *Chocolate in Mesoamerica: A Cultural History of Cacao*. Gainesville: University Press of Florida, 2006; pp. 154–183 (information cited from p. 158).

31. Martin, S. Cacao in ancient Maya religion: first fruit from the maize tree and other tales from the underworld. In: McNeil, C. L., editor. *Chocolate in Mesoamerica: A Cultural History of Cacao*. Gainesville: University Press of Florida, 2006; pp. 154–183.

32. Christenson, A. J. *Popol Vuh: The Sacred Book of the Maya*. New York: O Books, 2003.

33. Tedlock, D. *Popol Vuh: The Definitive Edition of the Mayan Book of the Dawn of Life and the Glories of Gods and Kings*, revised edition. New York: Simon and Schuster, 1996.

34. Martin, S. Cacao in ancient Maya religion: first fruit from the maize tree and other tales from the underworld. In: McNeil, C. L., editor. *Chocolate in Mesoamerica: A Cultural History of Cacao*. Gainesville: University Press of Florida, 2006; pp. 154–183 (information cited from p. 172).

35. Taube, K. A. The temple of Quetzalcoatl and the cult of sacred war at Teotihuacan. *RES* 1992;21:53–87 (information cited from pp. 55–58).

36. Stuart, D. *Ten Phonetic Syllables*. Research reports on ancient Maya writing 14. Washington, DC: Center for Maya Research, 1987; pp. 15–16.

37. Grube, N. The primary standard sequence on Chocholá style ceramics. In: Kerr, J., editor. *The Maya Vase Book*, Volume 2, New York: Kerr Associates, 1990; pp. 320–330.

38. Stuart, D. The language of chocolate: references to cacao on Classic Maya drinking vessels. In: McNeil, C. L., editor. *Chocolate in Mesoamerica: A Cultural History of Cacao*. Gainesville: University Press of Florida, 2006; pp. 184–201 (information cited from p. 190).

39. Kepecs, S., and Boucher, S. The pre-Hispanic cultivation of *rejolladas* and stone-lands: new evidence from northeast Yucatán. In: Fedick, S. L., editor. *The Managed Mosaic: Ancient Maya Agriculture and Resource Use*. Salt Lake City: University of Utah Press, 1996; pp. 69–91.

40. Houck, C. W. *The Rural Survey of Ek Balam, Yucatan, Mexico*. Ph.D. dissertation. New Orleans, LA: Department of Anthropology, Tulane University, 2004.

41. Landa, D. de. *Yucatan Before and After the Conquest*. Translated and with notes by William Gates. New York: Dover, 1978 (original ca. 1566).

42. Tozzer, A. M. *Landa's relación de las cosas de Yucatan*. Papers of the Peabody Museum of American Archaeology and Ethnology, Volume 18. Cambridge, MA: Harvard University, 1941 (original ca. 1566, Landa, D. de); pp. 94–95, 96, note 429.

43. Tozzer, A. M. *Landa's relación de las cosas de Yucatan*. Papers of the Peabody Museum of American Archaeology and Ethnology, Volume 18. Cambridge, MA: Harvard University, 1941 (original ca. 1566, Landa, D. de); p. 230.

44. Tozzer, A. M. *Landa's relación de las cosas de Yucatan*. Papers of the Peabody Museum of American Archaeology and Ethnology, Volume 18. Cambridge, MA: Harvard University, 1941 (original ca. 1566, Landa, D. de); p. 164.

45. Tozzer, A. M. *Landa's relación de las cosas de Yucatan*. Papers of the Peabody Museum of American Archaeology and Ethnology, Volume 18. Cambridge, MA: Harvard University, 1941 (original ca. 1566, Landa, D. de); p. 105.

46. Tozzer, A. M. *Landa's relación de las cosas de Yucatan*. Papers of the Peabody Museum of American Archaeology and Ethnology, Volume 18. Cambridge, MA: Harvard University, 1941 (original ca. 1566, Landa, D. de); p. 90.

47. Bricker, V. R., and Bricker, H. M. A method for cross-dating almanacs with tables in the Dresden codex. In: Aveni, A. F., editor. *The Sky in Mayan Literature*. New York: Oxford University Press, 1992; pp. 43–86.

48. Bricker, V. R., and Vail, G., editors. *Papers on the Madrid Codex*, Publication No. 64. New Orleans, LA: Middle American Research Institute, Tulane University, 1997.

49. Vail, G., and Hernández, C. The Maya hieroglyphic codices, version 2.0; 2005. A website and database available online at: www.mayacodices.org.

50. Landa, D. de. *Yucatan Before and After the Conquest*. Translated and with notes by William Gates. New York: Dover, 1978 (original ca. 1566); pp. 47–48.

51. Christenson, A. J. *Popol Vuh: The Sacred Book of the Maya*. New York: O Books, 2003.

52. Tedlock, D. *Popol Vuh: The Definitive Edition of the Mayan Book of the Dawn of Life and the Glories of Gods and Kings*, revised edition. New York: Simon and Schuster, 1996.

53. Vail, G., and Hernández, C. The Maya hieroglyphic codices, version 2.0; 2005. Available at http://mayacodices.org/codex2/frameDetail.asp?almNum=72&frameNum=2. (Accessed February 14, 2007.)

54. Miller, M. E., and Martin, S. *Courtly Art of the Ancient Maya*. New York: Thames and Hudson, 2004; p. 106, Pl. 49.

55. Coe, S. D., and Coe, M. D. *The True History of Chocolate*. London, England: Thames and Hudson, 1996; p. 45.

56. Miller, M. E., and Martin, S. *Courtly Art of the Ancient Maya*. New York: Thames and Hudson, 2004; p. 106.

57. McGee, R. J. *Life, ritual, and religion among the Lacandon Maya*. Belmont, CA: Wadsworth, 1990; p. 85.

58. Miller, M. E. *Maya Art and Architecture*. New York: Thames and Hudson, 1999; p. 179.

59. Martin, S. Cacao in ancient Maya religion: first fruit from the maize tree and other tales from the underworld. In: McNeil, C. L., editor. *Chocolate in Mesoamerica: A Cultural History of Cacao.* Gainesville: University Press of Florida, 2006; pp. 154–183 (information cited from p. 170).

60. Tozzer, A. M. *Landa's relación de las cosas de Yucatan.* Papers of the Peabody Museum of American Archaeology and Ethnology, Volume 18. Cambridge, MA: Harvard University, 1941 (original ca. 1566, Landa, D. de); p. 107.

61. Schele, L. *Notebook for the XVIth Maya Hieroglyphic Workshop.* Austin: Department of Art and Art History, University of Texas at Austin, 1992; p. 140.

62. Bricker, H. M., and Bricker, V. R. Zodiacal references in the Maya codices. In: Aveni, A. F., editor. *The Sky in Mayan Literature.* New York: Oxford University Press, 1992; pp. 148–183 (information cited from p. 171, Table 6.5).

63. Paxton, M. *The Cosmos of the Yucatec Maya: Cycles and Steps from the Madrid Codex.* Albuquerque: University of New Mexico Press, 2001; pp. 153–181, Appendix D.

64. Lacadena, A. Bilingüismo en el Códice de Madrid. In: *Los investigadores de la cultura maya,* No. 5. Campeche, Mexico: Publicaciones de la Universidad Autónoma de Campeche, 1997; pp. 184–204.

65. Macri, M. Department of Native American Studies, University of California, Davis. Personal communication. July 20, 2005.

66. Barrera Vásquez, A., Bastarrachea Manzano, J. R., Brito Sansores, W., Vermont Salas, R., Dzul Góngora, D., and Dzul Poot, D., editors. *Diccionario maya cordemex: maya–español, español–maya.* Merida: Ediciones Cordemex, 1980; p. 871.

67. Coe, S. D., and Coe, M. D. *The True History of Chocolate.* London, England: Thames and Hudson, 1996; p. 95.

68. Danien, E., editor. *Maya Folktales from the Alta Verapaz.* Philadelphia: University of Pennsylvania Museum, 2005; p. 59.

69. Vail, G., and Hernandez, C. The Maya hieroglyphic codices, version 2.0; 2005. Available at: http://mayacodices.org/codex2/frameDetail.asp?almNum=229&frameNum=1. (Accessed February 14, 2007.)

70. Martin, S. Cacao in ancient Maya religion: first fruit from the maize tree and other tales from the underworld. In: McNeil, C. L., editor. *Chocolate in Mesoamerica: A Cultural History of Cacao.* Gainesville: University Press of Florida, 2006; pp. 154–183 (information cited from p. 175).

71. Martin, S. Cacao in ancient Maya religion: first fruit from the maize tree and other tales from the underworld. In: McNeil, C. L., editor. *Chocolate in Mesoamerica: A Cultural History of Cacao.* Gainesville: University Press of Florida, 2006; pp. 154–183 (information cited from p. 183, note 32).

72. Lounsbury, F. G. Glyphic substitutions: homophonic and synonymic. In: Justeson, J. S., and Campbell, L., editors. *Phoneticism in Mayan Hieroglyphic Writing.* Albany, NY: Institute for Mesoamerican Studies, State University of New York, 1984; pp. 167–184 (information cited from p. 177).

73. Reents-Budet, D. *Painting the Maya Universe: Royal Ceramics of the Classic Period.* Durham, NC: Duke University Press, 1994.

74. Stuart, D. The language of chocolate: references to cacao on Classic Maya drinking vessels. In: McNeil, C. L., editor. *Chocolate in Mesoamerica: A Cultural History of Cacao.* Gainesville: University Press of Florida, 2006; pp. 184–201.

75. Gómez-Pompa, A., Salvador Flores, J., and Aliphat Fernández, M. The sacred cacao groves of the Maya. *Latin American Antiquity* 1990;1(3):247–257.

76. Tozzer, A. M., editor. *Landa's relación de las cosas de Yucatan.* Papers of the Peabody Museum of American Archaeology and Ethnology, Volume 18. Cambridge, MA: Harvard University, 1941 (original ca. 1566, Landa, D. de); p. 164.

77. Chi, G. A. Relación. In: Tozzer, A. M., editor. *Landa's relación de las cosas de Yucatan.* Papers of the Peabody Museum of American Archaeology and Ethnology, Volume 18. Cambridge, MA: Harvard University, 1941 (original 1582); pp. 230–232, Appendix C.

78. Chi, G. A. Relación. In: Tozzer, A. M., editor. *Landa's relación de las cosas de Yucatan.* Papers of the Peabody Museum of American Archaeology and Ethnology, Volume 18. Cambridge, MA: Harvard University, 1941 (original 1582); pp. 230–232, Appendix C.

79. Houck, C. W. *The Rural Survey of Ek Balam, Yucatan, Mexico.* Ph.D. dissertation. New Orleans, LA: Department of Anthropology, Tulane University, 2004.

80. Kepecs, S., and Boucher, S. The pre-Hispanic cultivation of *rejolladas* and stone-lands: new evidence from northeast Yucatán. In: Fedick, S. L., editor. *The Managed Mosaic: Ancient Maya Agriculture and Resource Use.* Salt Lake City: University of Utah Press, 1996; pp. 69–91.

81. Ardren, T., Burgos Villanueva, R., Manahan, T. K., Dzul Góngora, S., and Estrada Faisal, J. Recent investigations at Xuenkal, Yucatán. *Mexicon* 2005; 27:92–97.

82. Prufer, K., and Brady, J. E., editors. *Stone Houses and Earth Lords: Maya Religion in the Cave Context.* Boulder: University Press of Colorado, 2005.

83. Hernández, C. Middle American Research Institute, Tulane University, New Orleans. Personal communication. February 3, 2007.

84. Martin, S. Cacao in ancient Maya religion: first fruit from the maize tree and other tales from the underworld. In: McNeil, C. L., editor. *Chocolate in Mesoamerica: A Cultural History of Cacao.* Gainesville:

University Press of Florida, 2006; pp. 154–183 (information cited from p. 174).

85. Martin, S. Cacao in ancient Maya religion: first fruit from the maize tree and other tales from the underworld. In: McNeil, C. L., editor. *Chocolate in Mesoamerica: A Cultural History of Cacao*. Gainesville: University Press of Florida, 2006; pp. 154–183 (information cited from p. 175).

86. Miller, M. E., and Martin, S. *Courtly Art of the Ancient Maya*. New York: Thames and Hudson, 2004; p. 63.

87. Christenson, A. J. *Popol Vuh: The Sacred Book of the Maya*. New York: O Books, 2003; pp. 193–194.

88. Taube, K. A. The iconography of Toltec period Chichen Itza. In: Prem, H. J., editor. *Hidden Among the Hills: Maya Archaeology of the Northwestern Yucatan Peninsula. Acta Mesoamericana*, Volume 7. Maskt Schwaben, Germany: Verlag. Anton Saurwein, 1999; pp. 212–246 (information cited from p. 227).

89. Martin, S. Cacao in ancient Maya religion: first fruit from the maize tree and other tales from the underworld. In: McNeil, C. L., editor. *Chocolate in Mesoamerica: A Cultural History of Cacao*. Gainesville: University Press of Florida, 2006; pp. 154–183 (information cited from p. 172).

90. Taube, K. A. Itzamná. In: Carrasco, D., editor. *The Oxford Encyclopedia of Mesoamerican Cultures: The Civilizations of Mexico and Central America*, Volume 2. New York: Oxford University Press, 2001; pp. 56–57.

91. Vail, G., and Stone, A. Representations of women in Postclassic and Colonial Maya literature and art. In: Ardren, T., editor. *Ancient Maya Women*. Walnut Creek, CA: Alta Mira Press, 2002; pp. 203–228.

92. Vail, G., and Hernandez, C. The Maya hieroglyphic codices, version 2.0; 2005. Available at http://mayacodices.org/codex2/text.asp?almNum=365&frameNum=4&clause=1&dispOrd=1. (Accessed February 14, 2007.)

93. Urcid, J. Department of Anthropology, Brandeis University, Waltham, MA. Personal communication. February 1, 2007.

94. McNeil, C. L. Introduction: the biology, antiquity, and modern uses of the chocolate tree (*Theobroma cacao* L.). In: McNeil, C. L., editor. *Chocolate in Mesoamerica: A Cultural History of Cacao*. Gainesville: University Press of Florida, 2006; pp. 1–28.

95. Kufer, J., and Heinrich, M. Food for the rain gods: cacao in Ch'orti' ritual. In: McNeil, C. L., editor. *Chocolate in Mesoamerica: A Cultural History of Cacao*. Gainesville: University Press of Florida, 2006; pp. 384–407.

96. Pugh, T. W. Cacao, gender, and the northern Lacandon god house. In: McNeil, C. L., editor. *Chocolate in Mesoamerica: A Cultural History of Cacao*. Gainesville: University Press of Florida, 2006; pp. 367–383.

97. Coe, S. D., and Coe, M. D. *The True History of Chocolate*. London, England: Thames and Hudson, 1996; p. 50.

Tempest in a Chocolate Pot

Origin of the Word Cacao

Martha J. Macri

Introduction

The production, distribution, preparation, and ceremonial consumption of cacao were activities that touched the lives of all ancient Mesoamerican societies (Table 2.1 and Fig. 2.1). A majority of the languages spoken and written by the peoples of ancient Mesoamerica have the word *cacao* or a form that is clearly derived from it. The origin of the word *cacao* has become a topic of debate between two groups of historical linguists. One group argues for an origin in the Mixe-Zoquean family, a group of languages they believe are related to the language spoken by the ancient Olmec [1, 2], the builders of archaeological sites such as San Lorenzo, La Venta, and Tres Zapotes. A second group offers an etymology in a Uto-Aztecan language related to contemporary Nahuatl. They believe that Nahuatl (or languages closely related to it, usually referred to as Nahua)[1] has been present in ancient Mesoamerica since the Early Classic period,[2] at least by about 400 CE [3].

This chapter outlines their respective arguments and explains how this controversy relates to the larger question of the development of Mesoamerican civilization and the roles played at various times by certain linguistic and ethnic groups. In particular, the arguments relate to another controversy about how long any languages of the Uto-Aztecan family have been present in Mesoamerica. Several linguists based in the United States have maintained that Uto-Aztecan speakers did not enter the Valley of Mexico until the fall of Teotihuacan, and did not move south into Central America until after 900 CE [4–6]. Recently, Jane Hill has presented data supporting a very early Uto-Aztecan presence in Mesoamerica, believing such speakers were participants in the development of agriculture [7]. Other scholars have suggested that many features typically associated with Mesoamerica, such as semantic couplets—in Spanish, *difrasismos*—can be seen in widespread Uto-Aztecan languages, also supporting an early presence for those language in the region [8]. Since chocolate was unknown to Europeans before contact with the Americas, words for it would be expected to have originated in the languages of the Americas. In fact, there is general agreement that the words came into European languages by way of Nahuatl, which functioned as the primary language of contact between the Spanish conquistadors and the indigenous peoples of Mexico. The "tempest in a chocolate pot" has to do not with how the word spread to Europe and throughout the world, but with which language invented the word *cacao*.

Origin of the Word Cacao

One of the murkier concepts of this debate is what it means to "invent" a word. No one doubts that cacao

Chocolate: History, Culture, and Heritage. Edited by Grivetti and Shapiro
Copyright © 2009 John Wiley & Sons, Inc.

Table 2.1 Words for Cacao and Chocolate in Mesoamerican Languages

Word	Language	Geographical Location	Reference
Cacao *Theobroma cacao* L.			
*kakaw	Proto-Mayan		Campbell (1985), Kaufman (2003)
*kakaw(a)	Proto-Mixe-Zoquean		Campbell (1985), Kaufman (2003)
*xVwa	Proto-Amuzgo-Mixtecan		Campbell (1985)
kakawa-t	Nahua (Pipil)	El Salvador, Guatemala	Campbell (1985)
tëxua	Amuzgo	San Pedro, Oaxaca	Stewart and Stewart (2000)
du^{4}ndu^{3}cha^3 (possibly < dun^3 "ground" + du^{2}cha^2 "counted")	Cuicateco	Santa María Pápalo, Oaxaca	Anderson and Roque (1983)
mɨɨ'dxah^3 (mɨɨ^1 'round')	Chinanteco	San Juan Lealao, Oaxaca	Rupp and Rupp (1996)
na'iasi	Chiapaneca	Chiapa de Corzo, Chiapas	Aguilar Penagos (1992)
na'iasi cambó (cacaíto, *Ardisia paschalis D. Sm.*, fruit and plant)	Chiapaneca	Chiapa de Corzo, Chiapas	Aguilar Penagos (1992)
nuusí (fruit and plant)	Chiapaneca	Chiapa de Corzo, Chiapas	Aguilar Penagos (1992)
nuusí cambó (cacao patashete, *Theobroma bicolor H. & B.*, fruit and plant)	Chiapaneca	Chiapa de Corzo, Chiapas	Aguilar Penagos (1992)
nuusí yambusi (cacaíto, chocolatillo, *Ardisiam paschalis D. Sm.*, fruit and plant)	Chiapaneca	Chiapa de Corzo, Chiapas	Aguilar Penagos (1992)
zgüia'	Zapotec	San Bartolomé Zoogocho, Oaxaca	Long C. and Cruz M. (1999)
biziaa	Zapotec	Isthmus	Pickett et al. (1979)
Chocolate			
chucula't	Nahua	Pochutla, Oaxaca	Boas (1917)
chukulat	Nahua	Guatemala	Campbell (1985)
chicolüt	Huave	San Mateo del Mar, Oaxaca	Stairs and Scharfe de Stairs (1981)
cho^{3}co^{3}la^{2}te^4	Cuicateco	Santa María Pápalo, Oaxaca	Anderson and Roque (1983)
chocolate	Chinanteco	San Juan Lealao, Oaxaca	Rupp and Rupp (1996)
deju	Otomí	Valle del Mezquital, Hidalgo	Herández Cruz et al. (2004)
na'nbiusi	Chiapaneca	Chiapa de Corzo, Chiapas	Aguilar Penagos (1992)
xua	Amuzgo	San Pedro, Oaxaca	Stewart and Stewart (2000)
shcwlat, shicwlat	Zapotec	San Bartolomé Zoogocho, Oaxaca	Long C. and Cruz M. (1999)
dxuladi	Zapotec	Isthmus	Pickett et al. (1979)
dun^3	Cuicateco	Santa María Pápalo, Oaxaca	Anderson and Roque (1983)

References:
Aguilar Penagos, M. (1992) *Diccionario de la Lengua Chiapaneca*. Miguel Angel Porrúa, México.
Anderson, E. R., and Roque, H. C. (1983) *Diccionario Cuicateco*. Instituto Lingüístico de Verano. México.
Boas, F. (1917) El dialecto Mexicano de Pochutla, Oaxaca. *International Journal of American Linguistics* 1:9–44.
Campbell, L. (1985) *The Pipil Language of El Salvador*. Berlin: Mouton.
Hernández Cruz, L., Torquemada, M. V., and Crawford, D. S. *Diccionario del Hñähñu (Otomí)*. Instituto Lingüístico de Verano, Tlalpan.
Kaufman, T. (2003) *A Preliminary Mayan Etymological Dictionary*. http://famsi.org/reports/01051/index.html
Long, C. R., and Cruz, M. S. *Diccionario Zapoteco de San Bartolomé Zoogocho, Oaxaca*. Instituto Lingüístico de Verano, Coyoacán.
Pickett, V., et al. (1979) *Vocabulario Zapoteco del Istmo*. Instituto Lingüístico de Verano, México.
Rupp, J., and Rupp, N. (1996) *Diccionario Chianteco de San Juan Lealao, Oaxaca*. Summer Institute of Linguistics, Tucson, AZ.
Stewart, C., Stewart, R. D., and Amuzgo Collaborators. (2000) *Diccionrio Amuzgo de San Pedro Amuzgos, Oaxaca*. Instituto Lingüístico de Verano, Coyoacán.
Stairs, K., Albert, G., and Scharfe de Stairs, E. F. (1981) *Diccionario Huave de San Mateo del Mar*. Instituto Lingüístico de Verano, México.

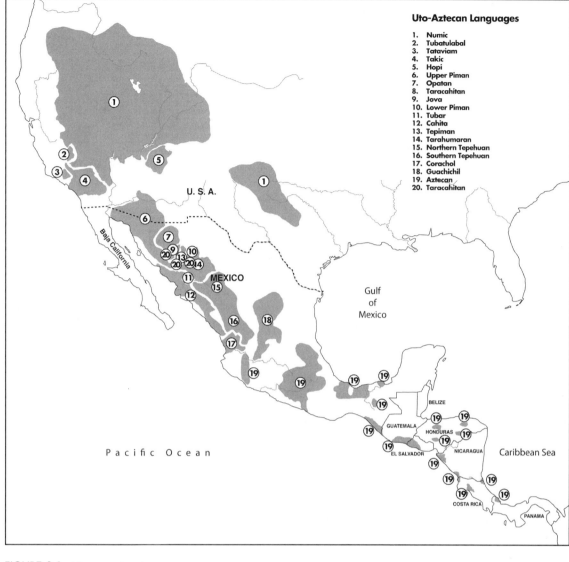

Uto-Aztecan Languages

1. Numic
2. Tubatulabal
3. Tataviam
4. Takic
5. Hopi
6. Upper Piman
7. Opatan
8. Taracahitan
9. Jova
10. Lower Piman
11. Tubar
12. Cahita
13. Tepiman
14. Tarahumaran
15. Northern Tepehuan
16. Southern Tepehuan
17. Corachol
18. Guachichil
19. Aztecan
20. Taracahitan

FIGURE 2.1. Mesoamerican language map.

was first grown and first consumed in South America. As a cultigen, it was originally a foreign item to all Mesoamerican peoples. The earliest date at which cacao plants reached Guatemala and Mexico is uncertain and may be significantly different from the date at which it began to be consumed. Could a word for cacao in a South American language accompany the introduction of its use in the north? Did this coincide with its cultivation, with its consumption, with its ceremonial use, and with its economic importance? Each of these developments occurred at different times, before spreading throughout the region. Since the first peoples to control commercial distribution of cacao undoubtedly played pivotal roles in the history of Mesoamerican cultures, the origins of these and other words of cultural importance may assist in the linguistic identification of such peoples, and suggest which modern groups are their most direct descendants.

The indigenous language most familiar to the first Spanish conquerors and colonizers was Nahuatl, the language of the Mexica empire.[1] Consequently, it was widely assumed that Nahuatl was the source of two words for chocolate found, not just in Europe, but in many languages throughout the world today—in English, *cacao* and *chocolate*. Nahuatl is part of the Uto-Aztecan language stock and is most closely related to several other language varieties including Pipil (Guatemala), Pochutec (Oaxaca, Mexico), and the languages Cora and Huichol (Nayarit, Mexico). Other southern Uto-Aztecan languages of northern Mexico and the bordering United States include Tohono O'odham, Tepehuan, Tarahumara, Yaqui, and Mayo [9]. More recently, however, other groups such as the Mayan languages of Guatemala, Belize, and Mexico (the Yucatán Peninsula, Chiapas, Tabasco), and the Mixe-Zoquean languages of the Mexican states of Veracruz,

Tabasco, Oaxaca, and Chiapas, have been proposed as ultimate sources of these words [1, 2, 10]. Nahuatl was the immediate source from which the Europeans borrowed the words *cacao* and *chocolate*, just as they also borrowed words for other native flora and fauna: for example, *avocado*, *chili*, *coyote*, *ocelot*, and *tomato*. Current debate, however, centers not on which language gave the word *cacao* to the Spanish, but on whether Nahuatl speakers invented the word themselves or borrowed it from some other language. This study does not provide direct evidence for either a Mixe-Zoquean or a Uto-Aztecan source for the word *cacao*. However, it challenges several of the reasons that have been invoked in rejecting a Uto-Aztecan language as a possible source.

Lyle Campbell, in *Quichean Linguistic Prehistory*, listed the proto-Mayan form *kakaw* as one of 13 Pan-Mesoamerican loans (i.e., loan words) that are found in so many languages that it would be difficult to determine the language of origin [4]. In addition to the proto-Mayan word, he provided examples in three other languages: Aztec *cacaua-tl*; Lenca *cua*; and Xinca *tuwa*. In a co-authored article entitled *A Linguistic Look at the Olmecs*, published earlier than *Quichean Linguistic Prehistory*, but evidently written subsequently to it, Campbell and Kaufman [1] list "cacao" among 13 cultigens for which words were borrowed throughout Mesoamerica, from, according to their hypothesis, Mixe-Zoquean languages. They state further that because the pan-Mayan *kakaw* (Chol *käkäw,* Tzotzil *kokow*) violates the single syllable form that is typical of Mayan languages, Mayan languages should be eliminated as a likely source. They reject Nahuatl on the basis that the Nahua word *kakawa* lacks cognates in other Uto-Aztecan languages, and that cacao itself is "not found in the Uto-Aztecan homeland." They give examples of the word in six other Mexican and Central American languages: Totonac *kakaw*; Jicaque *kʰaw*; Paya *kaku*; Huave *kakaw*; Lenca *kaw*; and Tarascan *kahékua*. Hypothesizing that the language of the ancient Olmec culture of the Gulf region was Mixe-Zoquean, the authors suggest that the source of the word is proto-Mixe-Zoquean *kakawa*, and that it spread to other Mesoamerican languages along with other features of the ancient Olmec culture.

Kaufman later mentions *cacao* as a loan from Mixe-Zoque into Huastec, a Mayan language spoken in the Mexican states of San Luis Postosí and Veracruz [11]. He asserts that *kakaw*[+] "cacao" later becomes "peanut" as in the local Aztec word *kakawatl* from proto-Mixe-Zoque *kakawa* "cacao." In *Foreign Impact on Lowland Mayan Language and Script*, Kaufman and Campbell, along with John Justeson and William Norman, invoke proto-Mixe-Zoque *kakawa* "cacao" as a source of the Greater Lowland Mayan word *kakaw* [12]. Kaufman [6] considers the Nahua to be intrusive into Central Mexico and suggests that they may have been among the destroyers of the great city of Teoti-

huacan (at his time of writing, believed to have dated to the seventh or eighth centuries; the fall of Teotihuacan is now dated to the mid-sixth century). He asserts that at 600 CE Nahua speakers were only on the northwestern borders of Mesoamerica. Suggestions that Nahua speakers were present earlier are idle speculation, "done in ignorance of basic linguistic facts about Nahua." Listing 34 words he claims to be Mixe-Zoquean loans into other Mesoamerican languages, he includes the following as linguistic facts: According to glottochronological calculations (based on a rate of replacement of 14 percent of an arbitrary set of so-called basic words over one millennium), Nahua did not diversify into Western, Central, Northern, Eastern Nahuatl, and Pipil until about 900 CE; an assertion that all Nahua loans resemble contemporary Nahuatl enough so that they must be relatively recent; and only languages that were Toltec-influenced but not Mexico-influenced had Nahua loans (he does not provide the names of these languages).

Søren Wichmann, in his volume on Mixe-Zoquean languages [13], argues that irregularities among Mixe-Zoque reflexes of proto-Mixe-Zoque *kakawa* "cacao/cacao tree" "point to the possibility that a complicated flow of borrowings between Mixe-Zoquean languages, which may or may not have involved non-Mixe-Zoquean sources, took place." The realization that Mixe-Zoquean forms exhibit variations not easily explained by a linear development from proto-Mixe-Zoquean to contemporary languages was an impetus for him to look elsewhere for the origin of the word *cacao*. In *Cacao and Chocolate: A Uto-Aztecan Perspective*, Karen Dakin and Søren Wichmann include an appendix listing 87 languages with words for cacao and chocolate [3]. Most of these are from Mesoamerica, though Dutch and some South American languages relevant to their argument are included. The authors review the problems with a Mixe-Zoquean reconstruction and demonstrate that, contrary to the claim by Campbell and Kaufman, Uto-Aztecan languages do provide a satisfactory etymology for the origin of *cacao*. Rather than seeing *tlalkakawa-tl*, the Nahuatl word for "peanut,"—literally "earth cacao"—as an extension of *kakawa*, they argue that *kawa* derives from proto-Southern-Uto-Aztecan *ka*[N] "hard, brittle" plus *pa*[N] "pod" combining to form the word *ka*[N]-*pa*[N] "hard pod, shell." Tables provide examples of Uto-Aztecan languages for each of the monosyllabic protoforms (for "hard, brittle" and for "pod"), as well as for the combined form. The reduplication of the first syllable yields the Nahuatl form, *kakawa*. Reduplication is a characteristic feature of these languages, indicating imitation or diminutive forms. The authors cite such Nahuatl words as *nanaka-tl* "mushroom" from *nakatl* "meat" and *kakahli* "shelter, awning" from *kahli/-kal* "house."

Campbell and Kaufman's second objection to a Uto-Aztecan origin for *cacao* is that the plant is not found in the Uto-Aztecan homeland [1]. This statement is built on three questionable assumptions. First, if the origin of the word is in a Uto-Aztecan language it must have been present in the protolanguage. Second, the homeland of proto-Uto-Aztecan language is known. Third, languages can only name local items. What Dakin and Wichmann have demonstrated is not that the word *kakawa*—"cacao"—itself had to have been present in proto-Uto-Aztecan, but that the lexical roots from which it derives are found in the protolanguage. The homeland of proto-Uto-Aztecan has been the subject of much debate, with the earlier unanimity of a northern Mexico or southern California homeland called into question by Hill [7]. However, more relevant to *cacao* is not the location of the proto-Uto-Aztecan homeland thousands of years ago, but the geographic distribution of languages related to Nahuatl at the time of the first attestations of the word.

A clear illustration of the problem between these two arguments is the early presence of the word *cacao* in the Mayan region. Lists of words for cacao and for the preparation of the chocolate drink made from it in the Yukatek Mayan language are presented here (Table 2.2). The variety of forms suggests that cacao has a long history of use by Yukatek speakers. Although the earliest use of these words cannot be associated with an absolute chronological date, there is one source that shows that *cacao* was known to Mayan speakers at least by the Early Classic period.

The syllabic spelling ka-ka-wa occurs in a hieroglyphic text on the rim of a beautifully decorated ceramic vessel, found in a burial of a high ranking person, that dates from the middle of the fifth century CE [14–16]. Both the syllabically spelled words and the arrangement of the words (the syntax) indicate that the text itself is written in a Mayan language of the Ch'olan language family and follows a well-known pattern of texts found on ceramic pots called the primary standard sequence. So we can be certain that by this date not only was the word *cacao* known to Mayan speakers, but they were also knowledgeable of its complex preparation as a drink and its ceremonial use. This evidence of the word's early presence in the Mayan area becomes a critical point in the argument of its origin, since according to those who argue for a Mixe-Zoquean origin, a Uto-Aztecan presence in the region at this time should have been impossible.

In the 15th century the Spanish found Nahuatl speakers not only in the Valley of Mexico, but also living throughout the Gulf region, as well as in parts of Chiapas, Guatemala, El Salvador, Honduras, and scattered even farther south into Central America [9]. It has been assumed that these outlying Nahua speakers represented the frontiers of Mexica political and economic influence. True, when the Spanish first arrived, Tenochtitlan, the city of the Mexica, dominated much of Mesoamerica, but this was not always the case. The Gulf Coast region was also an important economic and cultural center, even during the florescence of Teoti-

huacan. It has even been suggested as the source for some of the "Mexican" influences seen in Guatemala at Kaminaljuyu and Tikal, and in Honduras at Copán. Since this region has a long tradition of cacao cultivation, words for it in local languages would certainly have existed. If the Gulf regions were already home to Nahua speaking populations, this could well have been the source for the word *cacao*.

To add another dimension to this investigation, one of the hallmarks of ancient Mesoamerica was an abundance of long-distance trade. Salt, obsidian, and jade are among items that archaeologists have shown to have been transported over great distances. Many of the major archaeological sites are strategically located at seaports, headwaters, confluences, and mouths of major rivers, and in valleys and mountain passes crucial to the transportation of goods. Economic control over such trade goods does not necessarily rest with the people living where the goods originate. Words for items in an economically or politically dominant language may be introduced to the language or languages in the originating locations and may even replace the local term. It would be a mistake to assume that any of the peoples who participated in the complex trade networks of ancient Mesoamerica were unfamiliar with certain animals or food items simply because they were not native to their region.

The most recent installment in the Mixe-Zoque/Nahuatl debate is a rebuttal of Dakin and Wichmann by Kaufman and Justeson [2]. I leave a summary and an evaluation of their interpretation of the development of *cacao* in Mixe-Zoque languages to others with more extensive knowledge of these languages. What follows is an assessment of their reasons for dismissing Nahuatl as a possibility, and the problem of circular argumentation when it comes to an examination of areal linguistic features. Kaufman and Justeson list six reasons why *cacao* could not have originated in Nahua. First, they examine the direction of borrowing: They claim that both Mixe-Zoque and the northern Mayan language of Huastec have few borrowings from Nahua. Second, Nahua has substantial lexical and grammatical borrowings from Mixe-Zoquean languages. Third, proto-Mixe-Zoquean and all genetic subgroups show no evidence of borrowing from Nahua or Uto-Aztecan—only individual languages have some borrowed words. Fourth, no Nahua loans in any Mesoamerican language predate the Late Classic period (ca. 750–900 CE). Since all loans reflect Nahua phonology as known in the 16th century, they cannot be earlier than about 1000 CE. Fifth, Nahua words are usually borrowed in their unpossessed form with the absolutive suffix *-tl*, *-tli*, *-li*. Sixth, cacao does not grow in the Basin of Mexico or farther north.

The first three points having to do with the direction of borrowing must be examined more carefully to see whether some words presumed to be Mixe-Zoquean might have been present in Uto-Aztecan

Table 2.2 **Words Related to Cacao and Chocolate Preparation in Yukatek Mayan [18]**

Word	Gloss
kakaw	cacao, tree and fruit
b'alamte'	cacao
b'alam	jaguar
-te'	tree; plant (Ch'olan from)
kokox	tree-ripe cacao
chakaw haa'	hot water; chocolate
chakaw	something warm; hot
haa'	grind (haa'); water (ha')
(u) ta' chakaw ha'	chocolate rolls; chocolate cookies
u-	its
ta'	excrement
patb'il chakaw ha'	chocolate rolls; chocolate cookies
patb'il	made of dough
t'oh haa'	chocolate
t'oh chakaw haa'	to make chocolate pouring rapidly from one vessel to another
t'oh	pour; throw
u kaa'il chakaw haa'	chocolate mill; grinder
u-	its
kaa'	mill; grinder
-il	noun suffix
haa'	scrape; grind
(u) k'ab' kakao	cacao pod
u-	its
k'ab'	hand
tulis kakao	cacao pod
tulis	whole
nixtul	earflower (*Cymbopetalum penduliflorum*) chocolate flavor
tew (te u) xuchit	earflower (*Cymbopetalum penduliflorum*) chocolate flavor (plant?)
-te'	tree; plant (Ch'olan from)
u-	its
xuchit	sip; suck
yom kakaw	chocolate foam
y- (u-)	its
om	foam
p'iste'	small cacao fruit; a variety of small chile
p'is	to measure
-te'	tree; plant (Ch'olan from)
sak'a	atole, maize drink sometimes with cacao
sak'	white
ha'	water
(ix) chukwa'	chocolate maker
ix-	female
chuk wa' < chakaw haa'	hot water, chocolate
ximte'	quota, contribution of cacao or money; the coins given by the bridegroom to the bride
xim	corn ear
-te'	tree; plant (Ch'olan from)
-xim	numeral classifier for counting cacao, corn ears, threads
ximila'	present or give cacao or money as an offering
(ah) haa'	chocolate maker
ah	person who
(ah) uk' chuk wa'	chocolate drinker
uk'	to drink
chuk wa' < chakaw haa'	hot water; chocolate

languages outside of Nahua—an important task, but one that is beyond the present investigation. The fourth point, that the only Nahua loans are so similar to Nahuatl of the 16th century that they cannot be early, may arise from the fact that the authors only recognize those words as loans from Nahuatl that resemble the more recent form. It may well be that some of the sound changes that took place in early loans render them difficult to recognize, especially by those who are not expecting to find them. The fifth point, that Nahuatl words are always borrowed in their possessed form, can be argued on the basis of attested Nahua loanwords spelled syllabically in Mayan hieroglyphic texts that date prior to the Late Classic period, that appear to have been borrowed without the absolute suffix. It is these fourth and fifth points that are addressed here.

To counter these arguments, I offer two possible loans that imperfectly resemble 16th century Nahuatl, and that were not borrowed with the absolutive suffix. I then cite three Nahua words spelled syllabically in the Mayan script, again borrowed without -tl, -tli, or -li. Finally, two Nahua words that appear to have motivated the value of certain Mayan syllabic signs are discussed here.

The first examples are Mayan words that have similar sounds and exact or related meanings to Uto-Aztecan, and more specifically Nahua forms, in which some of the Nahuatl consonants correspond to glottalized consonants in the Mayan words. The first example is *k'oha'n*, reconstructed by Kaufman [17] for proto-Yukatekan as *k'oja'n* meaning "sick." The word cannot be reconstructed for the entire Mayan family. If it were a borrowing, it seems to have been borrowed only into Yukatekan languages, either borrowed before they diversified or spread across only Yukatekan languages after they had split. In Yukatek Mayan there are nominal, adjectival, and verbal forms, respectively: *k'oha'n* "sickness," *k'oha'nil* "sick," and *k'oha'ntal* "to be sick" [18]. The word for "sick; hurt" has been reconstructed for proto-Uto-Aztecan as *ko*, *koko* by Miller and Taube [19] and as *ko'o(ko)-* by Dakin [20]. Dakin reconstructs proto-Nahua forms as *koko-wa* "to be sick," *koko-ya* "to become sick," and *koko:-k* "painful." Since this word for "sick" has such a limited distribution in Mayan languages, and since it otherwise resembles proto-Uto-Aztecan forms with the same meaning, if it were not for the fact that the initial consonant in the Mayan word is glottalized, it would appear to be a loan from some language in that family.

The words *tz'i* "small," *tz'itz'* "to be small," and *tz'eek* "a little" occur in Mopan. In Yukatek *tz'itz'i* occurs in a phrase from an early dictionary *tz'itz'i ich* "he that has small eyes" [18] and with a vowel change in the verb *tz'etz'hal* "to make something small" [18]. Kaufman and Norman [21] reconstruct *tz'it-a'* "little" for proto-Ch'olan, and Kaufman reconstructs *tz'i* as "little" in proto-Lowland Mayan (Yukatekan and Ch'olan families). Chontal has *tz'ita'* "little" [17], and

again with the vowel /e/, *tz'etz'* "thin." [22]. I propose that this word was borrowed into Mayan languages from Nahua, and that the sound /tz/ was interpreted by Mayan speakers as /tz'/. The word is one among several examples of Nahua loans in Mayan languages in which consonants, especially affricates, have become glottalized in the Mayan word. Dakin reconstructs *-tzi/-tzi* as the diminutive in proto-Uto-Aztecan. In Nahuatl there are several related words: *tziqui* "a bit of something," and the reduplicated form *tzi:tziqui* "something very small" [23]. Again, with the limited distribution in the Mayan language family contrasted with a wide distribution in Uto-Aztecan, it would appear, except for the glottalized consonant, that the word was borrowed from a Uto-Aztecan language into Yukatekan, and either directly or through Yukatekan, into Ch'olan.

I have recently enumerated several examples of Nahuatl words spelled in the Classic Mayan hieroglyphic texts [14, 24, 25]. A word not previously published is *ikatz* "bundle, cargo," reconstructed for proto-Common Mayan by Kaufman [17]. Its distribution in Mayan languages is limited to Tzeltalan, Tojolab'al, and Mamean languages. It is spelled in Classic Mayan texts with the syllabic signs i-ka-tzi (see Table 2.1) [26], and in several cases it occurs as a label on an image of a bundle (Fig. 2.2). Although I have not been able to find a similar form in any Nahua language, the word *'ikat* "net bag" has been reconstructed for proto-Cupan, a subfamily of the Takic language family, a group of Uto-Aztecan languages found in Southern California [27]. Here both the form of the words and their meanings suggest that the two have a common origin. One of them, almost certainly, is a borrowing. A problem lies in the distance between the languages in which the two words occur, one in the Mayan region, the other beyond the northern limits of Mesoamerica. A relationship between these two words may be found with the Nahuatl form *meca-tl* "cord, rope" [23] (from which net bags are made), and the many words derived from it. If the Mayan word were a borrowing from a Uto-Aztecan language, it would not have been a recent one. This brings up again the question of the amount of time Uto-Aztecan languages have been in Mesoamerica and, more specifically, in contact with Mayan speakers.

Hill's hypothesis placing the Uto-Aztecan homeland within Mesoamerica [7] opens the door for early contact between Uto-Aztecan and Mayan languages. It would be a circular argument to cite the fifth century spelling of *cacao* on the Río Azul pot [15, 16]. A Nahua origin for three other words spelled syllabically in that text (*koxom, witik, mul*) has been presented elsewhere [14]. Early contact with Uto-Aztecan languages may account for the syllabic values of several Mayan graphemes. Motivation for the syllabic values of these signs appears to come from Nahua words, but it is not certain whether there might already have been

(a)

(b)

FIGURE 2.2. (a) I-ka-tzi ikatz "bundle" and figures holding bundles. From Yaxchilan Lintels 1 and 5. (b) Detail of ziza ka and tzi glyphs. *Source*: Drawings by Matthew G. Looper and Ian Graham. (Used with permission.)

a Nahua script from which these syllabic signs and their syllabic values were taken, or whether Mayan scribes familiar with certain Nahua terms invented the signs, or whether certain words themselves had been borrowed into Mayan languages and were subsequently lost. The first is the sign **xo** from Nahua "flower;" the second is **xi** from Nahua "peel, flay, shave" [23].

The grapheme known as the "decorated ajaw" is composed of the sign for the Mayan day Ajaw with

FIGURE 2.3. Xo decorated Ajaw. *Source*: Drawing by Matthew G. Looper. (Used with permission.)

FIGURE 2.4. Xi flayed skull. *Source*: Drawing by Matthew G. Looper. (Used with permission.)

decorative curls around it, and sometimes with a split in the top (Fig. 2.3). The earliest example occurs on the Río Azul chocolate pot dated to the mid-5th century [14–16]. Other examples of it are not known until nearly two hundred years later [26]. I have previously suggested that the syllabic value of *xo* (the *x* here sounds like the English *sh*) for the decorated Ajaw derives acrophonically, that is, from the first syllable of the Nahuatl day name *Xochitli*, meaning "flower" [24]. The day Xochitli, like Ajaw, occurs in the same place in the day name sequence, following the day for "storm; rain." The Mexican day is represented visually by a stylized image of a flower. By adding curls or U-shaped elements to the sides of the Mayan sign, the scribe invokes the idea of a flower, and the Nahuatl word for the day Flower, Xochitli, from which is derived the syllabic value **xo**. If this analysis is correct, it confirms that Mayan scribes had knowledge of Nahua names for the 20 days of the calendar of 260 days well before the fall of Teotihuacan. That is, it implies that Nahua speakers were already players in the Mayan region by the beginning of the Classic period, and gives further credence to the argument that the word *kakawa* may have been brought to the Mayan directly from Nahua or some other Uto-Aztecan language. Dakin reconstructs **siyo* as the proto-Uto-Aztecan form [20]. Miller gives cognates found throughout the Uto-Aztecan family.

A second grapheme is an image of a flayed skull with a series of dots around the eyehole, with teeth showing, and usually with a bared lower jawbone (Fig. 2.4). The earliest known examples are from lintels, specifically, from Lintels 18 and 22 from Structure 22 at Yaxchilan. They are considered to be among the oldest known texts from that site, though neither text contains a date [28]. They are probably no later than the early sixth century. The motivation for the syllabic value may come from the Nahua form *xi:p-* "to flay, peel, or shave" [23]. Cognates of the protoform **sipa* "scrape" occur in languages throughout the Uto-Aztecan family [19]. Xipe Totec, the name of a deity from Highland Mexico, is known also from Oaxaca and the Gulf regions. Most depictions show a person wearing the skin of a sacrificial victim [29]. The Mayan sign, however, shows a skull that has had the skin and scalp removed. Whether or not the sign may be directly related to the Mexican deity is not known, but it would seem that the syllabic value may well derive from the Uto-Aztecan word.[3]

Conclusion

As this chapter demonstrates, an understanding of words related to cacao and to chocolate preparation has the potential for providing unique insight into the history of trade and cultural relationships among the diverse peoples of Mesoamerica. That is why determining the ultimate source of words such as *cacao* becomes important. If the history of that word can be successfully sorted out, either the early dominance of Mixe-Zoquean speaking peoples will be confirmed, or an early influential presence of Uto-Aztecan speakers in Mesoamerica will have to be accepted. However, neither of these conclusions rules out the other. Both of them could be correct, regardless of any conclusion about the word *cacao*. Data presented here calls into question several of the reasons given for eliminating a Uto-Aztecan language as the source of the word *cacao*. An early presence of Nahua speakers even as far south as the Mayan lowlands must be considered.

Acknowledgments

I would like to acknowledge Diane M. Barker who helped compile a database of Mesoamerican words for chocolate. Tables 2.1 and 2.2 are derived, in part, from her efforts.

Endnotes

1. In this chapter, language varieties, past and present, closely related to Nahuatl are referred to as a group by the name Nahua.

2. Approximate dates for Mayan archaeological periods relevant to this discussion are as follows: Early Preclassic, 1700–1100 BCE; Middle Preclassic, 1100–500 BCE; Late Preclassic, 200 BCE–200 CE; Classic, 200–900 CE; Postclassic, 900–1500 CE.

3. Gabrielle Vail first suggested to me that the syllabic spelling *xi-b'i* on p. 22c of the Mayan Dresden Codex might refer to the Aztec deity Xipe Totec (personal communication, June 3, 2006).

References

1. Campbell, L., and Kaufman, T. A linguistic look at the Olmecs. *American Antiquity* 1976;41:80–89.

2. Kaufman, T., and Justeson, J. S. History of the word for "cacao" and related terms in ancient Mesoamerica. In McNeil, C. L., editor. *Chocolate in Mesoamerica: A Cultural History of Cacao*. Gainesville: University Press of Florida, 2006; pp. 117–139.

3. Dakin, K., and Wichmann, S. Cacao and chocolate: an Uto-Aztecan perspective. *Ancient Mesoamerica* 2000;11: 55–75.

4. Campbell, L. *American Indian Languages: The Historical Linguistics of Native America*. Oxford, England: Oxford University Press, 1997.

5. Kaufman, T. Areal linguistics and Middle America. *Current Trends in Linguistics* 1973;11:459–483.

6. Kaufman, T. Archaeological and linguistic correlations in Mayaland and associated areas of Meso-America. *World Archaeology* 1976;8:101–118.

7. Hill, J. Proto-Uto-Aztecan: a community of cultivators in Central Mexico? *American Anthropologist* 2001;103: 913–934.

8. Montes de Oca Vega, M. Los difrasismos: ¿Núcleos conceptuales Mesoamericanos? In: Montes de Oca Vega, M., editor. *La Metáfora en Mesoamérica*, pp. 225–251. Estudios Sobre Lenguas Americanas 3. Universidad Nacional Autónoma de México, Instituto de Investigaciones Filológicas, Seminario de Lenguas Indígenas, México, 2004.

9. Miller, W. Uto-Aztecan languages. In Ortiz, S. editor. *Handbook of North American Indians, Volume 10, Southwest*. Washington, DC: Smithsonian Institution, 1984; pp. 113–124.

10. Campbell, L. *Quichean Linguistic Prehistory*. University of California Publications In Linguistics No. 81. Berkeley: University of California Press, 1977.

11. Kaufman, T. Pre-Columbian borrowing involving Huastec. In: Klar, K., et al., editors. *American Indian and Indoeuropean Studies: Papers in Honor of Madison S. Beeler*. The Hague: Mouton, 1980; pp. 101–122.

12. Justeson, J. S., Norman, W. M., Campbell, L., and Kaufman, T. The foreign impact on lowland Mayan language and scripts. *Middle American Research Institute*, Volume 53, New Orleans, LA: Tulane University, 1985.

13. Wichmann, S. *The Relationship Among the Mixe-Zoquean Languages of Mexico*. Salt Lake City: University of Utah Press, 1995.

14. Macri, M. J. Nahua loanwords from the Early Classic: words for cacao preparation on a Río Azul ceramic vessel. *Ancient Mesoamerica* 2005;16:321–326.

15. Stuart, D. The Río Azul cacao pot: epigraphic observations on the function of a Maya ceramic vessel. *Antiquity* 1988;62:153–157.

16. Stuart, D. The language of chocolate: references to cacao on Classic Maya drinking vessels. In: McNeil, C. L., editor. *Chocolate in Mesoamerica: A Cultural History of Cacao*. Gainesville: University Press of Florida, 2006; pp. 184–201.

17. Kaufman, T. *A Preliminary Mayan Etymological Dictionary*; 2003. Available at http://www.famsi.org/reports/ 01051/index.html. (Accessed March 10, 2007.)

18. Barrera Vásquez, A., Ramón, J., Manzano, B., and Sansores, W. B. *Diccionario maya Cordemex: maya–español, español–maya*. Mérida, Yucatán: Ediciones Cordemex, 1980.

19. Miller, W. R. *Uto-Aztecan Cognate Sets*. University of California Publications in Linguistics No. 48. Berkeley: University of California Press, 1967.

20. Dakin, K. *La evolución fonológica del protonáhuatl*. Mexico City: Universidad Nacional Autónoma de México, 1982.

21. Kaufman, T., and Norman, W. An outline of proto-Cholan phonology, morphology and vocabulary. In: Justeson, J. S., and Campbell, L., editors. *Phoneticism in Mayan Hieroglyphic Writing*. Institute for Mesoamerican Studies Publication No. 9. Albany: State University of New York, 1984; pp. 77–166.

22. Keller, K. C., and Luciano, G. P. *Diccionario chontal de Tabasco*. Tucson, AZ: Instituto Lingüístico de Verano, 1997.

23. Karttunen, F. *An Analytical Dictionary of Nahuatl*. Norman: University of Oklahoma Press, 1983.

24. Macri, M. J. T536 Xo, from Nahuatl Xochitli "Flower." *Glyph Dwellers Report 11*. Maya Hieroglyphic Database Project, University of California, Davis, 2000. Available at http://nas.ucdavis.edu/NALC/glyphdwellers.html. (Accessed March 10, 2007.)

25. Macri, M. J., and Looper, M. G. Nahua in Ancient Mesoamerica: evidence from Maya inscriptions. *Ancient Mesoamerica* 2003;14:285–297.

26. Macri, M. J. *Maya Hieroglyphic Database*. Unpublished database on file at the Native American Language Center, University of California, Davis.

27. Bright, W., and Hill, J. The linguistic history of the Cupeño. In: Hymes, D. H., and Bittle, W. E., editors. *Studies in Southwestern Ethnolinguistics: Meaning and History in the Languages of the American Southwest*. The Hague: Mouton, 1967; pp. 351–371.

28. Tate, C. E. *Yaxchilan: The Design of a Maya Ceremonial City*. Austin: University of Texas Press, 1992.

29. Miller, M., and Taube, K. *The Gods and Symbols of Ancient Mexico and the Maya*. London, England: Thames and Hudson, 1993.

30. Graham, I. *Corpus of Maya Hieroglyphic Inscriptions*, Volume III, Part 1. Peabody Museum of Archaeology and Ethnology. Cambridge, MA: Harvard University, 1977.

3

Ancient Gods and Christian Celebrations

Chocolate and Religion

Louis Evan Grivetti and Beatriz Cabezon

Introduction

All religions celebrate their festivals and spiritual events with food. Some faiths associate specific foods with celebrations; others are more general. Christian celebrations of Christmas and Easter, for example, commonly include hard-boiled eggs, ham, and lamb, but are highly variable and dependent on geographical region, country, family traditions, and individual taste preferences. Dietary codes of permitted and forbidden foods are characteristic of a number of faiths including Judaism, Christianity, Islam, and Hinduism [1–5].

Chocolate, being manufactured from *Theobroma cacao* beans—a New World domesticate—would not have played any role in Old World religions until after 1492 and would only have been involved from the early to mid-16th century CE. Nevertheless, the religious importance of chocolate is easily documented by New World, pre-Columbian artifacts, religious objects, and texts. Furthermore, religious traditions practiced in contemporary Central America in the 20th and 21st centuries have their origins in antiquity and represent the blending of ancient religious beliefs and practices with Christian rituals (Fig. 3.1). Some of these practices are associated with the rites of passage (i.e., birth, coming of age, marriage, and death) or exemplified by the well-known Mexican rituals associated with *Día de la Muertos*, or Day of the Dead celebrations (Figs. 3.2 and 3.3) [6, 7].

The theme of cacao/chocolate with religion cuts a large path through time and geographical space. Early cosmological texts produced by Native Americans during the pre-Columbian era are rich with symbolism and content. They relate the importance of cacao in daily life of different societies, and reveal that such an important tree must have had divine origins.

With the arrival of the Spanish in the early 16th century, the Mesoamerican cultures were exposed to Christianity for the first time. Many individuals converted and adopted Christian rituals and trappings characteristic of the Western Church; others professed outward conversions but in their hearts maintained their older belief systems as well. Still others rejected Christianity and followed the old ways as long as possible.

The multiple layering and blending of pre-Columbian and Christian faiths commonly included cocoa and chocolate. Some of these themes have long histories, whereas others can be dated more recently. The cocoa and chocolate religious themes tell contem-

Chocolate: History, Culture, and Heritage. Edited by Grivetti and Shapiro
Copyright © 2009 John Wiley & Sons, Inc.

FIGURE 3.1. Religious procession in Oaxaca, Mexico. Participants carrying cacao pods. *Source*: Copyright 2008, Howard Shapiro. (Used with permission.) (See color insert.)

FIGURE 3.2. Image of Saint decorated with cacao beans. *Source*: Copyright 2008. Guillermo Aldana. (Used with permission.) (See color insert.)

porary readers much about peoples living in earlier times throughout Central America and Mexico. But what should these themes be called: legends, myths, or stories? Were they not more, and in fact, were they not central to the religious faiths of the people who recorded them? Just as it is not proper to attribute the central themes of Judaism, Christianity, or Islam as "legends, myths, and stories," it is also not proper to

FIGURE 3.3. Altar constructed in celebration of Day of the Dead in Mexico. *Source*: Copyright 2008, Howard Shapiro. (Used with permission.) (See color insert.)

address the religious practices of past and contemporary Maya, Mixtec, Nahua, Otomi, Zapotec, and other indigenous peoples of the Americas as "mythology." The chocolate-related themes that formed part of these ancient people's cosmologies, therefore, are presented here with respect.

How Cacao Came to Humans

Ethnobotanists, food historians, and geneticists search for the botanical origins and earliest geographical distribution patterns for *Theobroma cacao* and related species using contemporary, 21st century methods. Balancing the scientific approach to how cacao came to humans are religious texts and oral traditions that explain the first appearance of the noble tree. In these texts the gods presented *Theobroma cacao*, the cacao tree, or products of the tree to humans as gifts. Several of these documents are well known, others less so; some have survived from antiquity in the form of written texts, while others have been preserved through oral traditions. Representative examples are presented here.

POPUL VUH

The *Popul Vuh* manuscript is the Sacred Book of the Quiché Maya. The document—along with Quiché/ K'iche oral traditions—was discovered and summarized in Latin during the 16th century by Father Francisco Ximénez, a Dominican priest, who served in the Guatemala highlands at Santo Tomas Chichicastenango [8]. The document reflects the cosmology, religious practices, and history of the Quiché/ K'iche Mayan.

> These are the names of the animals which brought the food: yac [the mountain cat], utiú [the coyote], quel [a small parrot], and hoh [the crow]. These four animals gave tidings of the yellow ears of corn and the white ears of corn, they told them that they should go to Paxil and they showed them the road to Paxil. And thus they found the food, and this was what went into the flesh of created man, the made man; this was his blood; of this the blood of man was made. So the corn entered [into the formation of man] by the work of the Forefathers. And in this way they were filled with joy, because they had found a beautiful land, full of pleasures, abundant in ears of yellow corn and ears of white corn, and abundant also in pataxte and cacao, and in innumerable zapotes, anonas, jocotes, nantzes, matasanos, and honey. There was an abundance of delicious food in those villages called Paxil and Cayalá. There were foods of every kind, small and large foods, small plants and large plants. [9]

According to Mayan religious beliefs, cacao had a divine origin. Xmucane, one of the creation gods, invented nine beverages; of these nine, three were prepared from cacao and corn. The Mayan god of cacao, Ek' Chuwah, was honored in April with animal sacrifices and offerings of feathers, incense, and cacao, followed by gift exchanges. During the month of Muan, the 15th month of the Mayan year, owners of cacao plantations and those going to plant cacao celebrated a festival in honor of their gods Chac, Hobnil, and Ek' Chuwah. In order to solemnize the day, all family members went to the plantations, where they sacrificed a dog that had cacao-colored spots on its skin. They also burned incense before images of their gods, after which they offered blue iguanas and feathers from certain types of birds, and presented cacao tree branches to each official. The day ended with a ceremonial meal where cacao was served and the number of cups (*xicaras*) drunk was limited to three [10, 11].

TREE OF KNOWLEDGE–TREE OF SIN

Contemporary Quiché/ K'iche Mayans relate that their ancestors considered the cacao tree especially revered and still call the tree today the tree of sin and knowledge. While conducting field work in northern Guatemala during August 2000, Dr. Victor Montejo met with Quiché/ K'iche elders who told him how cacao had come to humans.

> The cacao tree originally was just like any other tree in the forest. Then, Christ appeared to us and was persecuted by His

> enemies. Christ fled and took refuge in the forest. To escape His enemies Christ hid under a cacao tree. He touched the trunk and the tree immediately blossomed and covered Him with white flowers, so that His enemies would not find Him. Because of the tree's help, Christ told the tree that it would be blessed, and that its cacao fruits would be used for beverages in ceremonies. This is why it is called tree of knowledge. [12]

Quiché/ K'iche respondents told Montejo that the cacao tree also is called *awas che'* (tree of sin). The reason for this designation stems from the dual function of the tree as being sacred and a source for food, and because of its former use as currency among Quiché/ K'iche peoples when cacao beans were used as a medium of exchange and ultimately became equated with money (see also Chapter 56). Quiché/ K'iche respondents related that humans initially had received the gift of cacao from Christ and in order to perpetuate Christ's memory the ancestors were instructed that cacao was to be drunk only during religious ceremonies. The Quiché/ K'iche elders then related: "There came a time when we did not know that the original gift and meaning of cacao was to serve as a tree of knowledge, and [through the centuries] mankind sinned and we did not drink cacao to honor Christ, but used the beans as money. Therefore, it became a tree of sin [*awas che'*]." According to Montejo, other Quiché/ K'iche he interviewed claimed the tree that protected Christ was a different, unnamed tree [13].

A variation of the Quiché/ K'iche cacao tree/ Christ tradition initially was published in 1967 and reviewed recently in a monograph on chocolate in Mesoamerica, as told by the contemporary Quiché/ K'iche Mayan of Chichicastenango.

> When Christ first asked for help, the cacao tree initially responded, with some trepidation, that it was not capable of the task and that "our lord Manuel Lorenzo [the whirlwind]" will shake the tree. Christ, in response, told the tree that if it helped him, it would be remembered forever, but that if it did not he would destroy it at Judgment day. Christ also offered to bless the tree in heaven, but the tree replied that it could not go to heaven because its roots were stuck down into the ground. Christ told the tree that he would put it "into the clouds and mists of heaven," bringing these to the tree if he could not bring the tree to them and said that the tree would thrive along the coasts. Then the cacao tree covered Christ in white blossoms, sheltering him from the sun and his enemies. [14, 15]

QUETZALCÓATL AND THE GOOD PEOPLE OF TOLLAN (TULA)

The Toltec peoples settled in the Central Valley of Mexico during the 10th century of the Common Era. Monuments to their greatness still stand at their religious and commercial center Teotihuacan, approximately 60 kilometers northeast of modern Mexico City [16, 17]. Toltec cosmology is contained in the *Tonalamatl*, the Book of Prophecies from the priests of the Goddess Xochiquetzal.

The gods felt sorry for all the work that the Toltec people had to perform and decided that one of themselves should go down to the Earth to help humans by teaching the sciences and the arts. Thus it was that Quetzalcóatl took the form of a man and descended on Tollan, the city of workers and good men. And it was done. Quetzalcóatl loved all mankind and gave them the gift of a plant that he had taken from his brothers the gods, who jealously guarded it. He took the small bush which had red flowers, hung on branches which had long leaves, and were sloped towards the ground, which offered up its dark fruit. Thus it was that Quetzalcóatl planted the tree in the fields of Tollan, and asked Tlaloc to nourish it with rain, and Xochiquetzal to decorate it with flowers, and the little tree gave off fruit and Quetzalcóatl picked the pods, had the fruit toasted, and taught the women who followed the man's work to grind it, and to mix it with water in pots, and in this manner chocolate was obtained. [18]

THE FAITHFUL MEXICA PRINCESS

The Mexica/Aztecs were relative latecomers to the Central Valley of Mexico, arriving during the mid-13th century CE from their mythical home of Aztlan, an undefined region to the north. Their arrival precipitated great upheaval within the valley and the people ultimately settled ca. 1325 on an island in the midst of Lake Texcoco, where they would establish the great city of Tenochtitlan [19]. The Mexica relate how cacao came to humans.

A Mexica princess guarded the treasure that belonged to her husband, who had left home to defend the empire. While he was away she was assaulted by her husband's enemies who attempted to make her reveal the secret place where the royal treasure was hidden. She remained silent and out of revenge for her silence they killed her. From the blood shed by the faithful wife and princess, the cacao plant was born, whose fruit hides the real treasure of seeds—that are bitter like the suffering of love; seeds—that are strong like virtue; seeds— that are lightly pink like the blood of the faithful wife. And it was that Quetzalcóatl gave to humans the gift of the cacao tree for the faithfulness of the princess. [20]

SIBÖ AND THE CACAO TREE

The Bribri are Native American peoples that currently occupy the geographical region of southeastern Costa Rica and northern Panama. They are members of the Chibchan language family and linked linguistically to the Cuna/Kuna of Panama (Fig. 3.4) [21, 22]. Field work conducted by one of us (Grivetti) among the Bribri of Costa Rica in September 2006 explored cacao- and chocolate-associated traditions. After appropriate permissions had been obtained, and introductions concluded, interviews on the origins of cacao were held with a Bribri elder, Mr. Samuel López, and his daughter, Marina López Morales, who reported the following:

FIGURE 3.4. Cacao pods. Cuna textile from Panama. *Source:* Private Collection, Grivetti Family Trust. (Used with permission.) (See color insert.)

All things God made. We do not know from where the seeds [of cacao] have come from. People do not make anything. God brings all things to us. Sibö made the earth and said: I made the seeds of all the foods. I made the human people. Sibö thought how would the humans have food? Sibö had a sister named Tsura and a niece Iricia. He called a meeting and asked the sister how to transform the earth. He asked his sister to help him look for a special beverage. He, Sibö, knew that the niece, Iricia, was the key. Sibö knew that the niece would help to create the earth. Sibö convinced his niece to participate in the ceremony and Iricia brought people [humans] to help. Sorbon and Iricia danced and they married and formed the earth. Sibö said that the cacao is a fruit that I will leave to people. Sibö made different tests with different seeds and he saw that they grew and could be good for the temporal or physical world. Sibö then transformed the trees and the girls and slept with the girls. [23]

A second version related to Sibö also was reported regarding how the cacao tree came to humans.

Long ago Sibö was looking for a girl to marry. He went on a walk to find a girl. He came to a farm and found four sisters. Three of the sisters were beautiful, one was homely. Sibö transformed himself into an ugly old man. He went to the house of the sisters and asked for help since he needed food and something to drink. The three beautiful girls did not want anything to do with him. The homely girl, named Tsura, said: "Come, I will help you." After eating and drinking, Sibö wanted to bathe in the river. He asked Tsura to come with him. He asked Tsura to marry and as he bathed in the river, he changed into a handsome man. Tsura was surprised. Sibö and Tsura got married and he said to her: "When you go to the physical world you will become a special tree." And she became the cacao tree. The three beautiful sisters came to the physical world as other plants similar to the cacao tree but these plants were not good to eat—they smelled and tasted bad. Sibö visited the physical world and told the people: "This is a good tree—it is my wife—and must never be burned and the tree must always be cared for. When you go to prune the tree, be careful, and then it will be good for you always." [24]

Contemporary Mayan Ritual Uses of Cacao

Professor Victor Montejo, Department of Native American Studies, University of California at Davis, conducted field work among Mayan communities in western and northern Guatemala during the summers of 1998, 1999, and 2000. The following excerpts are summarized from his field reports.

Cacao is used in each stage of the life cycle, especially during baptisms, marriages, and funerals. Starting with pregnancy and birth, cacao is used in ceremonies, such as petitioning for the hand of a bride, during religious festivities, and during the wake, following a death. Most importantly, cacao and cacao–corn drinks are used during communal work projects and ceremonies related to the Mayan calendar [25].

Before a child is born, the mother is constantly visited by her neighbors who bring *pozole* (corn–cacao beverage). This drink is considered very healthy and provides energy to the mother who must be very strong in order to deliver a healthy baby. The corn–cacao drink is brought by the visitor in a gourd (*jícara*) filled with the beverage. Usually, the *pozole* is served with sweet bread for the expecting mother. During pregnancy, the expecting mother is well treated and fed, since the neighbors visit almost every day bringing *pozole* or some other foods [26].

After delivery, the mother is also given a bath with medicinal plants. Then she is wrapped in a blanket and then her mother brings her hot *pozole* to drink. Drinking quantities of *pozole con cacao* is believed to help the mother become strong and healthy so that she will not have problems breast-feeding her baby. After delivery, the mother must comply with a 20-day after-birth prohibition, during which she will always be wrapped in a warm cloth. During this period she does not work and the women in the village/town visit her, bringing foods (e.g., eggs with herbs) and *pozole* to drink. After the 20 days of postpartum rest, she resumes her regular tasks as mother and head of the household, and the visits with offerings/gifts of *pozole* stop at this time [27].

During childhood, children usually do not drink *pozole con cacao*. The main reason is that cacao is almost an elite item and is expensive. The *pozole con cacao*, therefore, is drunk only on special occasions, the most common being pregnancy, postpartum, petitioning for the hand of a bride, marriage, and during communal work activities such as construction of a house or temple in the community. *Pozole con cacao* is a special food because it is also an offering to the supernatural beings. Simple *pozole* prepared from white maize flour without cacao is a common beverage and is consumed almost every day. But the *kakaw yal ixim* (cacao–corn drink) is reserved for special occasions [28].

In ancient times corn and cacao together were central to incantations spoken to placate anger, or conversely, to produce anger or project evil desires onto someone who was an enemy. Incantations and prayers are said over seeds of corn and cacao to infuse the grains with the power of human desire. In modern times, cacao and corn are still used to make beverages and are used for divination. Cacao primarily is used as a basis into which other natural medicines are mixed. Cacao is used in this region of Guatemala for attaining strength and energy, especially by hard-working farmers. Cacao is used to rehabilitate the weak and sick as during postpartum or an illness, or if suffering from malnutrition [29].

Use of cacao becomes more relevant during the Day of the Dead, since all families make *pozole con cacao* (*kakaw yal ixim*) and place the cacao-filled gourds (*jícaras*) of this sacred drink on their home altar. The cacao-filled *jícaras* are placed beside bowls or plates of food, mostly poultry, which are the favorite food for the spirits when they visit their home and families during the day of All Saints, November 1st [30].

Interviews conducted by Professor Montejo during 2000 also explored symbolic and ritual uses of cacao among the Jakaltak and Mam Mayan communities as noted in these excerpts summarized from his field report:

Some respondents replied that while cacao was still being used within their communities, it was losing its importance since much of the land once used to cultivate cacao was being turned into coffee plantations. Still, people considered cacao as part of their diet and used cacao for beverages. Cacao symbolized both power and sacredness; cacao or chocolate remains the sacred food for the gods and it is considered a sacred beverage for humans. It symbolizes a perfect offering to placate the thirst of rulers [i.e., officials] and is a sacred offering to spirits. (See Fig. 3.5.) [31]

Arrival of Christianity

Priests who accompanied Hernán Cortés during the initial period of the conquest of Mexico had little or no impact on Mexica/Aztec conversions to Christianity. Shortly after the conquest, the Franciscan priest Pedro de Gante arrived in Mexico in 1523, followed by the first wave of Franciscans—known as The Twelve[1]—who arrived in Mexico during May 1524 [32]. Under the direction of Friar Martín de Valencia, the Franciscans initiated their work and opened schools for children. Slowly, adults began to accept Christianity and by the late 1520s many had been baptized. The events of 1531, when the Virgin Mary appeared as Our Lady of Guadalupe to the Indian, Juan Diego, increased the numbers of conversions to Christianity. Over time, geographical distribution of missions, churches, and administrative rules began to formalize, with the Franciscans working in what are now the states of Coahuila, Nuevo León, Tamaulipas, and Texas, whereas the

FIGURE 3.5. Mayan cacao god. *Source*: Drawing by Mari Montejo, 2002. (Used with permission.)

Jesuit "territory" included what is now the states of Chihuahua, Durango, Sinaloa, Sonora, and *Baja* California. Subsequently, the Jesuits were expelled from Mexico (and other New World Spanish colonies) by Charles III, whereupon the Franciscans emerged with more responsibilities, staffing abandoned missions in *Baja* California, and founding others in *Alta* California (see Chapter 33) [33].

CONVENT AND MONASTERY DOCUMENTS

A document attributed to the Jesús María Convent in Mexico City identified the annual expenditures of the convent in 1645 and included notations of costs associated with assisting the sick, expenses for religious celebrations, and expenses for the sacristy. Chocolate is mentioned as an extraordinary expense to be given as gifts to visitors during vigil days of Lent. Items identified included flour for food and for the consecrated wafers (host), salt, firewood, candles, butter, meat, soap, wine, and chocolate.

> *And during Lent we give 5 pesos for the three meals of the Vigil and as an extraordinary expense we offer chocolate as a gift for the persons that come to visit the convent.* [34]

From the Jesuit College in Mexico City a letter from Julio Ortiz to the Proctor, Juan Nicolas of the Society of Jesus (Jesuits), at the Royal College of San Ignacio de Loyola in Mexico City dated to 1688, asked payment for Easter-related provisions and expenses:

> *May my Brother have a very happy Easter beginning and a very happy Easter end, and may you find consolation of the absence of the Brother Bartolomé Gonzales. I am writing to you, my Brother, asking your kindness and benevolence, to send me 150 pesos to pay the Sunday's people and 2 arrobas of chocolate, one arroba of one and another arroba of the other and one arroba of sugar. May God guard my Brother for many years. April 1688, Julio Ortiz.* [35]

In response, the good Father's requests were honored by Juan Nicolas with an invoice that contained the following amounts and quantities:

> *I remit 17 arrobas and 14 pounds [i.e., 439 pounds] of fine chocolate [chocolate fino] produced from 8 arrobas of cocoa from Maracaibo and 14 pounds of sugar; I remit 20 arrobas and 19 pounds [519 pounds] of ordinary chocolate produced from 9 arrobas and 4.5 pounds of cocoa from Caracas and 14 arrobas and 4 pounds of sugar, asking the Colegio [College] to reserve 14 pounds in order to be able to complete the grinding. . . . It remains in my power [that I kept for my personal use] 8 arrobas and 4 pounds [164 pounds] of cocoa from one and the other.* [36]

Another Jesuit document associated with vigils held ten years later during Easter week 1698 at the Real Colegio de San Ignacio de Loyola in Mexico City itemized payments for foods and preparation expenses associated with the Jesuit brotherhood (*Cofradía de Nuestra Señora de Aránzanzu*). The events of Holy Thursday specifically identified included the following:

> *9 pesos that I paid to the musicians who sang on the Holy Thursday; 5 pesos for the sculptor that prepared and removed the adorned altar that we used for the evening mass on Holy Thursday; 1 peso and 3 reales spent for the mistela [beverage made of wine, water, sugar, and cinnamon] and chocolate for those that remained awake all night.* [37]

An extraordinary document covered the dates June 1, 1697 to the end of April 1702, and identified expenses associated with the Santa Clara Convent in Mexico City. In addition to stated costs for items such as bread, meat, fish, beans, various legumes (chickpeas and lentils), rice, sugar and cinnamon, saffron, tomatoes, and green peppers, the document listed five specific accounts where chocolate was used for specific celebratory purposes.

> *[For] extraordinary expenses of the Mother Abbess [i.e., Superior] and of the Mother Contadora [in charge of expenses] for celebrations made when the wife of the Viceroy and the Archbishop visited the convent: sweets, waters [distilled liquids from fruits or flowers], and chocolate 326 pesos;* [38]

> *For burials . . . chocolate for the religious [priests] that were present and assisted with the burials of nuns that died during this period. Also dinners for them—80 pesos;* [39]

Chocolate for the nuns that is given once a year during Lent and on Palm Sunday; 36 arrobas ~$\frac{1}{2}$ chocolate [925 pounds] and 36 arrobas of sugar [900 pounds]—543 pesos and 3 tomines [1 tomín = 1 real]; [40]

Chocolate for the cooks of the religious [nuns], four times each month from July 1st to October 1697—16 pesos and 0 tomines; [41]

Chocolate for Father Vicente and our Chaplain, 4 times a month, each month—483 pesos. [42]

A companion document within the same file suggested that the nuns were under the "watchful eye" of the Inquisition, and this consideration may account for the extraordinary detail given to the chocolate-associated expenditures so that the legitimacy of the expenses could not be questioned. This companion document, signed individually by each of the 66 nuns residing at the Convent, also revealed the strict structural and managerial controls over convent life during this period of Mexican history (see Chapter 4) [43].

Through the centuries, the Santa Clara Convent, Mexico City, continued to document its expenses. A second set of documents associated with this convent covered the period July 1795 to November 1798 and contained descriptions of expenses on a monthly basis. Two types of chocolate were identified in these monthly documents: one type for the nuns (consumed on a regular basis), and the second identified for the mass (i.e., served after the mass) [44].

A document dated to June 1, 1799, from the Convent of the Religious Ladies of the Very Sweet Name of Maria and Our Lord the Manor San Bernardo Abad in Mexico City, described expenditures during Passion Week:

> 101 *pesos for the people who sing the Passions [religious chanting];*
> 81 *pesos for the services [several masses] of this week;*
> 69 *pesos for the chocolate [for the priests]. [45]*

An early 19th century account by Diego de Agreda, Superintendent of the Girls School of Our Lady of Charity in Mexico City presented a detailed list of expenses for the sustenance of schoolgirls during Lent:

> 28 *pesos for the chocolate and biscuits for the Sundays, Thursdays, Fridays, and Saturdays during Lent . . . with other things of regular consumption. [46]*

Conclusion

In Central America, chocolate as a beverage has had many associations: a drink offered to men about to be sacrificed, a medicine, a beverage drunk as part of religious ceremonies, even drunk in advance of anticipated sexual pleasure. The association of chocolate with religion, however, is one that continues to connect both the ancient and early modern history of the New World.

Before the arrival of the Spanish, the indigenous peoples of Central America had a long, rich history that blended cacao-related rituals and their respective faiths. Central to these traditions was how the cacao tree, *Theobroma cacao*, came to humankind as a gift to humans from the gods. And while it is widely held today in the 21st century that the earliest cultivation of cacao occurred in northeastern Ecuador, along the headwaters of the western Amazon basin [47], the strength of these early religious associations with cacao in Mexico offers the suggestion that cacao as a beverage originated in lands north of the equator. While cacao trees were domesticated initially in South America, the beans most likely were not used there to make chocolate by indigenous peoples, who instead focused on the delightful, sweet-tasting pulp, and we support the view that has been advanced that beverages prepared from cacao beans were first developed in Mesoamerica [48]. And if this is the case, then the suggestion can be advanced further that cacao—in the sense of chocolate—truly was a gift from the gods, and the rich religious heritage associated with Mayan and Mexica beliefs support this view.

Acknowledgments

We would like to acknowledge Victor Montejo for his field work efforts, and Sylvia Escarcega and Patricia Barriga for locating some of the archival documents used when preparing this chapter.

Endnote

1. The twelve Franciscans who arrived in New Spain in 1524 known as "the twelve" were: Friar Martín de Valencia, O.F.M., Custos or Superior; Friar Francisco de Soto, O.F.M.; Friar Martín de la Coruña, O.F.M.; Friar Juan Xuarez, O.F.M.; Friar Antonio de Ciudad Rodrigo, O.F.M.; Friar Toribio de Benavente, O.F.M.; Friar García de Cisneros, O.F.M.; Friar Luis de Fuensalida, O.F.M.; Friar Juan de Ribas, O.F.M.; Friar Francisco Ximenes, O.F.M.; Friar Andrés de Córdoba, O.F.M.; and Friar Juan de Palos, O.F.M. Two Friars, Bernardino de la Torre, O.F.M. and José de la Coruña, O.F.M., originally had been named to make the voyage but were replaced prior to sailing for Mexico.

References

1. Grivetti, L. E., and Pangborn, R. M. Origin of selected Old Testament dietary prohibitions. *Journal of the American Dietetic Association*. 1974;65:634–638.

2. Korff, S. L. The Jewish dietary code. *Food Technology* 1966;20:76–78.

3. Sakr, A. H. Dietary regulation and food habits of Muslims. *Journal of the American Dietetic Association* 1975;58:123–126.

4. Müller, F. M. *Manu*. In: The Laws of Manu. Translated with Extracts from Seven Commentaries. Volume 25, *The Sacred Books of the East*. Translated by G. Buhler. Oxford, England: Clarendon Press, 1886.

5. Grivetti, L. E. Food prejudices and taboos. In: Kiple, K. F., and Ornelas, C., editors. *The Cambridge World History of Food*, Volume 2. New York: Cambridge University Press, 2000; pp. 1495–1513.

6. Carmichael, E., and Sayer, C. *The Skeleton at the Feast. The Day of the Dead in Mexico*. Austin: University of Texas Press, 1992.

7. Garciagodoy, J. *Digging the Days of the Dead. A Reading of Mexico's Dias de Muertos*. Boulder: University Press of Colorado, 1998.

8. Ximénez, F. *Historia de la Provincia de San Vicente de Chiapas y Guatemala, de la Orden de Predicadores. Compuesta por el R.P. Pred. Gen. Fray Francisco Ximenez, Hijo de la Misma Provincia*, Three Volumes. Guatemala, Centro América: Biblioteca "Goathemala" de la Sociedad de Geografía e Historia, 1929 (Volume I); 1930 (Volume II); 1931 (Volume III).

9. Goetz, D., and Morley, S. G. *The Book of the People. Popul Vuh*. Translated into English by D. Goetz and S. G. Morley, from Adrián Recino's Translation from Quiché into Spanish. Los Angeles, CA: Plantin Press, 1954; Part 3, Chapter 1, p. 105.

10. Grivetti, L. E. From aphrodisiac to health food; a cultural history of chocolate. *Karger Gazette*. Issue 68 (Chocolate). Basel, Switzerland: S. Karger, 2005.

11. Tozzer, A. M. *Diego de Landa. Relación de las cosas de Yucatán. Manuscrito en la Real Academia de la Historia, Madrid. Copia fotostática completa con tres mapas en 135 hojas, de una por cara*. Papers of the Peabody Museum of Archaeology and Ethnology, Volume 18. Cambridge, MA: Harvard University Press, 1941 (original ca. 1566). Cited in: Gómez-Pompa, A., Flores, J. S., and Fernandez, M. A. The sacred cacao groves of the Maya. *Latin American Antiquity* 1990;1:247–257.

12. Montejo, V. Unpublished field report dated 2000. Cacao: The Tree of Sin and Knowledge. Ethnographic Research on Cacao Among the Mayas of Guatemala; pp. 4–5.

13. Montejo, V. Unpublished field report dated 2000. Cacao: The Tree of Sin and Knowledge. Ethnographic Research on Cacao Among the Mayas of Guatemala; p. 5.

14. Bunzel, R. L. *Chichicastenango: A Guatemalan Village*. Ethnological Society Publications, Volume 33. Locust Valley, NY: J. J. Augustin, 1967; pp. 239–241.

15. McNeil, C. L., editor. *Chocolate in Mesoamerica. A Cultural History of Cacao*. Gainesville: University Press of Florida, 2006.

16. Davies, N. *The Toltec Heritage. From the Fall of Tula to the Rise of Tenochtitlán*. Norman: University of Oklahoma Press, 1980.

17. Diehl, R. A. *Tula, the Toltec Capital of Ancient Mexico*. London, England: Thames and Hudson, 1983.

18. Vayssade, M. R., editor. *Cacao. Historia, Economía y Cultura*. Mexico City: Compañía Nestle, S.A. de C.V., 1992; p. 20.

19. Davies, N. *The Aztecs. A History*. Norman: University of Oklahoma Press, 1973.

20. Vayssade, M. R., editor. *Cacao. Historia, Economía y Cultura*. Mexico City: Compañía Nestle, S.A. de C.V., 1992; p. 22.

21. http://www.native-languages.org/famchi.htm. (Accessed January 20, 2007.)

22. Sánchez, J. *Mi Libro de Historias Bribrís*. San José, Costa Rica: Lara Segura, 2001.

23. Grivetti, L. Unpublished fieldnotes. September 2, 2006, Bribri Reserve, Costa Rica.

24. Grivetti, L. Unpublished fieldnotes. September 2, 2006, Puerto Viejo, Costa Rica.

25. Montejo, V. Unpublished field report dated July–August–September 1999; pp. 2–3.

26. Montejo, V. Unpublished field report dated July–August–September 1999; p. 3.

27. Montejo, V. Unpublished field report dated July–August–September 1999; p. 3.

28. Montejo, V. Unpublished field report dated July–August–September 1999; p. 3.

29. Montejo, V. Unpublished field report dated July–August–September 1999; p. 4.

30. Montejo, V. Unpublished field report dated July–August–September 1999; p. 4.

31. Montejo, V. Unpublished field report dated 2000. Cacao. The Tree of Sin and Knowledge. Ethnographic Research on Cacao Among the Mayas of Guatemala; pp. 1–7.

32. Arjona, D. K. "The Twelve" meet a language requirement. *Hispania*. 1952;35:259–266.

33. *New Advent, Catholic Encyclopedia*. Mexico. http://www.newadvent.org/cathen/10250b.htm. (Accessed February 25, 2007.)

34. Archivo General de la Nación [AGN]. Templos y Conventos, Vol. 158, exp. 92, f. 1003f.–1004f.

35. AGN. Jesuitas Vol. III-10, Year 1688.

36. AGN. Jesuitas Vol. III-10, Year 1688.

37. Archivo Histórico del Real Colegio de San Ignacio de Loyola (AHRCIL), estante 6, Tabla 1, Vol. 3, fol. 119v.

38. AGN. Archivo Histórico de Hacienda (AHH), Vol. 1403, exp. 3, fs 100–107v, pp. 102v–103r.

39. AGN. AHH, Vol. 1403, exp. 3, fs 100–107v, pp. 102v–103r.

40. AGN. AHH, Vol. 1403, exp. 3, fs 100–107v, p. 103v.

41. AGN. AHH, Vol. 1403, exp. 3, fs 100–107v, p. 104r.

42. AGN. AHH, Vol. 1403, exp. 3, fs 100–107v, pp. 106r–106v.

43. AGN. AHH, Vol. 1403, exp. 3, fs 40v–42v, pp. 41v–42v.

44. AGN. Templos y Conventos, Vol. 130, exp. 1. Libro de Gastos Generales del Convento de Santa Clara de México 1795–1798.

45. AGN. Templos y Conventos, Vol. 50, exp. 1, f. 39f.

46. AHRCIL. Estante 14, Tabla IV, Vol. 5, 1806, fs 154–154v.

47. Motamayor, J. C. Germplasm resources and geographical origins of *T. cacao*. In: Bennett, A. B., Keen, C., and Shapiro, H.-Y., editors. *Theobroma Cacao: Biology, Chemistry And Human Health*. Hoboken, NJ: Wiley, 2009; in preparation.

48. McNeil, C. L., editor. *Chocolate in Mesoamerica. A Cultural History of Cacao*. Gainesville: University Press of Florida, 2006.

4

Chocolate and Sinful Behaviors

Inquisition Testimonies

Beatriz Cabezon, Patricia Barriga, and Louis Evan Grivetti

Introduction

Mention of the word Inquisition evokes strong mental images of religious excess, intolerance, torture, and zeal. The world of art is filled with hundreds of graphic depictions of torture and execution in the name of religious faith. It would be incorrect to believe that the events associated with "The Inquisition" were conducted by amorphous nameless others at distant times and places far removed from 21st century North America. But they were not anonymous: the inquisitors had names, and the locations where Inquisition officials took testimony and determined the life or death of thousands of persons are known. Such church-sponsored events even continued into the 19th century—in New Spain—and fear engendered by such activities extended northward into Texas, New Mexico, Arizona, and California.

To understand how and why the Inquisition came to existence in the New World—and how chocolate was involved—we should begin with a short background to the political and social situation of 15th century Spain. The unification of Spain into a single kingdom began in 1469, when Isabella—heiress of the Crown of Castile—married Ferdinand of Aragon, heir of the Crown of Aragon. This marriage led to the unification and emergence of a new Kingdom (Aragon and Castile); with its appearance a series of historical events were set in motion, as the Moors (Muslims) and Jews were expelled from Spain, and the American continents were "discovered" with their subsequent conquest and colonization.[1]

The Inquisition as a religious institution preceded these events and already had been instituted by Pope Gregory IX in 1227. By 1254 Pope Innocent IV had conferred Inquisition rites and responsibilities on the Dominican Friars of Spain. During the 13th and 14th century the Inquisition grew with the power and authority to imprison all who were denounced for heresy. The Spanish Inquisition—as it would be known—was formally instituted November 1, 1480. The last Muslim bastion in Spain, the city of Granada, fell to the Catholic Monarchs on January 2, 1492, whereupon Christianity was imposed as a state religion.

But Spain and Portugal competed fiercely, and these two nations posed a threat to Christian unity. This led to the decision by Pope Alexander VI in 1493 to divide the known world into two spheres: exploration and commercial rights to the west of the

Chocolate: History, Culture, and Heritage. Edited by Grivetti and Shapiro
Copyright © 2009 John Wiley & Sons, Inc.

Pope's line[2] were ceded to Spain, while exploration and commercial rights to the east of the line were given to Portugal. Complicating Christian discord was the event two decades later when on October 31, 1517, Martin Luther, the German monk and theologian, posted his 95 theses on the door of the castle church in Wittenberg, thereby initiating the Christian Reformation movement. This event, known as the Lutheran heresy, led to increased levels of prosecution by the Inquisition not only in Europe but in the newly "discovered" Indies [1].

The Spanish Crown feared the economic activities and proselytizing of Lutheran Reformers in the American colonies, and the Inquisition increased efforts by the Dominicans to seek out and punish such individuals. In 1527 inquisitorial powers were delegated to bishops in New Spain and the first *auto de fé*[3] was conducted in 1536 in Mexico City [2]. A formal Inquisition Tribunal, however, was not appointed until Moya de Contreras was selected Inquisitor-General, or Senior Inquisitor, for Mexico on January 3, 1570 and took office in September 1571 [3]. His first *auto de fé*—one of extraordinary "grandeur"—was conducted in Mexico City on February 28, 1574 and included between 70 and 80 accused individuals, mostly Lutherans and English sailors [4, 5].

Chocolate and the Inquisition

What actions and behaviors connect such terrible Inquisition events in Europe and the New World with chocolate? Sometimes it was the mere description of drinking chocolate while attending an event. It is sometimes told, for example, that Charles II, King of Spain, sipped chocolate while observing Inquisition victims being put to death [6]. It is also held that the Inquisition in Spain and Portugal drove Jewish chocolate makers into France, where they established businesses at Bayonne (see Chapter 42) [7]. More specifically, Inquisition documents from New Spain sometimes identify the activities of chocolate merchants participating in presumed anti-Christian behavior and the use of chocolate in seduction and in witchcraft.

The documents presented and reviewed in this chapter reveal the workings of the Inquisition. Some of the documents are denunciations, as when neighbors turned against neighbors; at other times the respondents apparently appeared before the Tribunal on a voluntary basis. The variety of charges that could lead to an "audience" with the officials, and how and where testimony was recorded, provide insights on social mores and cultural traditions during this troubled era in the Americas.

Our search through the archives in Mexico City identified 23 Inquisition documents—not previously published—where individuals were accused of using chocolate in "non-Christian" ways, or where specific chocolate makers were denounced and deposed by Inquisition officials. The documents examined and presented here contained only the testimonies/recorded depositions, and only two files inspected provided outcomes regarding innocence, guilt, or sentencing if convicted. The testimonies we located are presented here in chronological sequence.[4]

Inquisition Case Records

May 30, 1604. In this document Francisco Sanchez Enriquez volunteered to appear before the Tribunal in order to confess the sin that he drank chocolate prior to receiving the Holy Sacrament:

[Deposition location not identified]: *On May 30 1604 . . . Francisco Sanchez Enriquez came on his own account to make a confession. He is an honest man. I know him. Your Lordship will see what to do with him, but I do not think we need to examine his wife. He is a sick and old man [and he said] that he had received the sacrament of the Holy Eucharist without remembering [that] he had drunk a cup of chocolate.*

SIGNED EL CANÓNIGO SANTIAGO. [8][5]

January 25, 1620. A brief but intriguing document is this letter signed by Friar Andrés de Carrera, a Franciscan, dated January 25, 1620, addressed to an unidentified high officer of the Inquisition. The document mentioned that when the Father Guardian[5] departed—to continue his journey to Sacatecas (Zacatecas, Mexico)—he had left behind 60 tablets (*tablillas*) of chocolate and some "*exquisitas letras*" (exquisite writings).[6] These writings seem to have made him uncomfortable and probably consisted of some type of literature he thought was dangerous to possess, since an important role of the Inquisition was censorship and book-burning. Possession of such forbidden texts could lead to an accusation of heresy and prosecution:

. . . in all the occasions I am writing to Your Lordship I am always praying for your health and it had always rejoiced my soul to know that you are in good health; Father Guardian was ready to continue his journey to Sacatecas [Zacatecas, Mexico] and after mass he left, leaving [behind] 60 tablets [tablillas] of chocolate and a small cask of vinegar; also [he left] very exquisite writings and I kept them just to show them to Your Lordship, so you would decide accordingly [to your wish] what to do with them, because neither do I understand them nor they are for me. Your humble servant:

FRAY ANDRÉS DE CARRERA [9].

September 17, 1620. The following document records a "family affair" whereby chocolate was used to seduce an important individual:

In the city of Guadalajara on September 17th, 1620 . . . The Commissary [7] of the Holy Office of the Inquisition, Don Juan Martin de Argatimendia, ordered me the Notary and the undersigned to go to the houses of the maid-servants of Doña Antonia Dejaz, wife of the Captain Juan de Mesa, to take a deposition in the house of the above mentioned Doña Antonia Dejaz. [The Notary arrived and started to depose Juana de Bracamonte.] In order to have a clean conscience, and because it was established in the edicts that have been proclaimed at the Cathedral Church of this city of the Inquisition, Juana de Bracamonte declares: about a year and a half ago when she was in the house of Doña Francisca Sezor, wife of Juan Gonçales also called Juan el Mayor, of this city. Doña Francisca, already dead, prepared a tecomate [8] [large bowl] of cacao for Don Antonio de Figueroa [who Juana de Bracamonte was in love with] and she said that her cousin, already dead, had given her magical powders but Doña Francisca said that was Doña Ynes de Tapia [who had given the powders], in order that Don Antonio de Figueroa would fall in love with her. She reported that she threw those powders into the tecomate in her presence and this is the truth. She did not say this [give her deposition] out of hatred or malevolence. She looked to be a woman of 23 years of age. She did not sign because she said she did not know how to write. This, I attest,

RICARDO DE SOLEDA, NOTARY. [10]

Month and Date Missing, 1621. This document represents testimony given before the Tribunal regarding a corrupt priest accused of extortion and seeking illegal bribes/payments from his church membership. The indigenous population served by the accused, Father Bernabé (?) de Cárcamo, stopped confessing their sins to him and refused to be coerced by the priest to pay for their confessions with cacao beans and guavas.

In the city of San Miguel de Totonicapa [Totonicapan, Guatemala], on [date and month not recorded], 1620 . . . The first thing he [Father Bernabé (?) de Cárcamo] did during Lent he told to all the Indians [Naturales], the men as well as the women, that came to him for confession, that if they wanted to be confessed they had to give him 40 grains [i.e., beans] of cacao and at least two guavas [guayabas], each person, and that he would not administer the Sacrament of the Confession until they brought the cacao grains. . . . This minister or rather a sorcerer [mago] was selling the Holy Sacrament. The result was that not a single Indian received confession.

SIGNED FATHER BONILLA. [11]

Although corrupt priests always have existed—in all faiths—one of the more popular activities that inquisitors were always eager to punish was the ancient art of witchcraft that very often was conducted by women of all social layers. As the following documents

illustrate, these women often were Black, Indian, Mestizo, or Mulatto, the ones who knew "how" to make "magic potions." Desperate Spanish women of the "upper social layers"—either wives or lovers, mothers-in-law or widows—could obtain those potions from such "knowledgeable" women who were deemed to be masters of the ancient art of sorcery.

But witchcraft was not only the domain of women, as men also utilized love potions to obtain favors from the opposite sex. Chocolate as a beverage was the perfect vehicle for these potions and concoctions were used in various ways, for example, to attract or repel lovers, to tame men, even to make a man/woman forget a rival. Chocolate was an excellent medium for such concoctions because as a liquid that commonly was bitter, or semi-sweet, chocolate could mask the flavor of unusual additives, among them crow's heart, excrement, human flesh, and menstrual blood. The offering of a fragrant cup of chocolate was—and still is—a common social custom in certain regions of Spain and Latin America today in the 21st century, particularly during cold weather. Such a hospitable gift of a cup of chocolate was even more important during the 17th and 18th centuries. Through the following documents we can follow stories of despair, intrigue, love, and deceit. In each it is not too difficult to imagine someone with an irresistible smile saying, *"Por favor, aquí está su chocolate. Espero que lo disfrute"* (Please, here is your chocolate. I hope you enjoy it)!

April 6, 1620. This case reported a quarrel between a mother and her son-in-law with a further confession regarding witchcraft, where a slave asked her to prepare a love elixir from *puyomate* [8] to assist a friend.

In the city of Guadalajara, on April 6, 1620, at 4 in the afternoon, came before the Commissary [of the Holy Office of the Inquisition] Doña Luisa de Barrera and say that to alleviate her conscience—she [she wanted to say that] had remembered that 4 months ago and because she did not have peace with her son-in-law, asked a woman—a widow—[what she could do] and she told her to take some toads and burn them on a pan, and bring them to his house—but she finally decided not to do it. She also confessed about [something she did herself] and declared that 10 years ago a woman, slave of Don Vergara, begged her to grind [for her] a root of puyomate [8] because she wanted to give it to one of her friends. [12]

April 6, 1620. In this document the flesh of a man, a presumed criminal (since he had been hanged, then drawn and quartered) was mixed with chocolate. The document does not, however, specify for whom the bizarre cup of chocolate was destined.

In the city of Guadalajara, on the same day [on April 6, 1620] came before the Commissary [of the Holy Office of the Inquisition] Don Baltasar Peña and he said and denounced that he had found under the bed of one of the mulatto woman [that served in his household] a piece of flesh taken from the quarter of a man that had been hanged, and that she has roasted it, and then mixed with chocolate. [13]

The following documents offer insights regarding how women of 17th century colonial New Spain dealt with and coped in a society where they had no voice. Whether within Native American or Judeo/Christian traditions, a woman's menstruation cycle has been a source of continued mythology and misunderstandings, commonly with negative results. Conversely, the use of menstrual blood disguised in chocolate also reveals a positive side, since menstrual blood could be seen as an element to entice prospective lovers. Although today in the 21st century the effects of smell and taste in attraction have been described in many scholarly articles on animal and human behavior and chemical communication through pheromones, the woman of these documents did not know the "chemistry of love." Their actions, however, seem from our vantage point of the 21st century to be intuitive, since they thought their actions would attract the men they loved. And in one of the cases listed here, a woman did what she deemed to be necessary to protect herself from a violent husband, a case where chocolate, worms, and witchcraft collided.

April 7, 1621.

The example presented here documents use of menstrual blood to attract a man and the use of worms to tame a husband who might become violent, because his wife had become pregnant in his absence. Both "confessions" by Doña Mariana de Aguilar are good examples to ponder from various viewpoints: that of the Inquisition, social behavior, and fear due to "unexpected" consequences.

In the city of Guadalajara on April 7, 1621 in front of me, Licenciate[9] [Licenciado] Juan Martinez—Commissary of the Holy Office of the Inquisition, Prelate of the cathedral of this city, ordered to me Ricardo Soleda notary of the Holy Office to obtain a declaration from Doña Mariana de Aguilar, widow and neighbor of this city, and obeying I went to her house. A woman of 37 years who says her name is Doña Mariana de Aguilar, said that because she wants to have peace of mind and also to clear her from a charge of guilt, she says and denounces that: about three years ago, she heard Doña Petronila de Baro, already dead, say that a woman, a mulatto woman whose name is Isabel Portillo, known also as La Montana [i.e., a person who lives in the woods] and a black woman are both witches and that they bewitch people [with their sorceries]; she also says that the slave [the black woman mentioned above] had heard that María López widow and neighbor of this city, would add her menstrual blood to the chocolate before to serve it to [one of] her male friends. She also says that Doña Juana de Bracamonte, already dead, said that while her husband, also dead, was absent, she called an Indian woman because she was pregnant and was afraid of her husband [her husband's reaction since he had been absent for a long period of time] and asked her [the Indian woman] if she could see her husband and tell him about it [about the

pregnancy]. The Indian woman made a wax figure and then left; a few days ago she came back and told her: "Do not be afraid, your husband is already sick." Also she [the pregnant wife] asked the Indian woman if she could give her something [a potion] for her husband. The Indian woman gave her some powders, wrapped in piece of paper, and also gave her some little worms, that she called esticulinos, and she explained to her that those worms were good for taming a man and that she should give them to her husband mixed with chocolate. [14]

November 24, 1622.

Women who "interact" with sailors and who accompany the fleet commonly are known by other terms. This Inquisition document is interesting for other reasons as well, since one of the prostitutes also operated an illegal chocolate-selling business without a government license.

Havana, Cuba, on the 24th of November, 1622 . . . The Commissary of New Spain, Don Julio Maszias y Lorenzo informs about Doña Isabel de la Parra [of] Habana. . . . In a previous letter to your Lordship I had written about detaining Juana de Valenzuela and also about how Doña Isabel de la Parra that left [Havana] with the fleet. . . . Some persons who have come from Sabana[10] know that she has been selling chocolate without telling the Governor's house. . . . Your Lordship will tell me if I have to inform the Commissary of that city [Sabana], so she would be arrested. In waiting for your letter, May God give your Lordship many more years [of life]. [15]

March 30, 1626.

Persons in love sometimes resort to unorthodox methods to make themselves attractive in order to seduce the opposite sex. The following testimony presented to the Tribunal revealed that the accused, María Bravo, had obtained the recipe for a magical spell from a local Indian woman in order to secure the affection of a man she loved. Bravo complained that she had prepared the ingredients, but that the spell did not work as planned and confessed to having practiced witchcraft. The outcome of her confession was rather modest—considering the alternative verdicts sometimes assigned—but led to her being ostracized by her community. Because of social pressure she was forced to attend church elsewhere. The document also reveals issues of local poverty and social mores, and documents lack of education of women.

In the town of Teotitlán, on March 30, 1626 at 2 o-clock in the afternoon, before the Attorney General in charge of the Commission of the Holy Office of this town appeared María Bravo, without being summoned, who swore to tell the truth and stated that she was a mestizo, daughter of Francisco Bravo and that she does not know the name of her mother. Said declarer stated that she was single and was raised in the town of Teotitlán and is 30 years old, more or less. She came to unburden her conscience and stated that approximately three years ago, while in her house in said town of Teotitlán, she complained to an Indian woman of the same town named Ana María, who the declarer did not know about, who said that she understood the cause, because a man that she loved very much did not love her. The Indian woman told her to take some of her menstrual blood and mix it with chocolate and give it to him, and that with this he would love her. The declarer, believing in this, did so and gave it to him to drink once and seeing that it did not achieve what she wanted decided that it was a lie and repented for having done it. And within about a year, during Lent, she confessed and this town did not want to absolve her and she no longer saw the confessor of this jurisdiction. Later she confessed and declared to the closest Inquisition Tribunal and this is the truth according to the vow that she has made and said that this is well written [i.e., her statement was recorded correctly] and that she did not sign it because she does not know how to write [Signature and seal of the Commissioner follow]. [16]

June 15, 1626. Whereas the previous documents presented evidence of the use of menstrual blood to seduce a lover, this document represents more of a sad case, where a wife experienced loss of affection from her husband and attempted to correct the situation using a magical spell. The document, as written, became convoluted as the testimony unfolded, and keeping track of the persons identified requires concentration. The declaration took place in Tepeaca (today, the state of Puebla). A person named Ana Perdomo—widow of Ardiga—declared that her husband was in love with a woman named Agustina de Vergara. A second person, Francisca Hernandez, told Ana Perdomo that while her husband was alive he had never loved her. Agustina then gave Ana's husband chocolate mixed with her own menstrual blood. The document also reports that the Inquisitor ordered Ana to think again and make certain she knew the name of the person who said that Agustina had gave Ardiga chocolate laced with menstrual blood. The Inquisitor also asked Ana if she knew if Agustina had done this with other men, and she answered that she did not know.

In the city of Tepeaca on the 15th day of the month of June of 1626, between eight and nine o'clock at night, before Mr. Domingo Carvajal y Sousa, Commissioner of the Holy Office of this city, appeared, without being called and swearing to tell the truth, a woman who said that her name was Ana Perdomo, widow of Ardiga, of this city, who is 28 years old. In order to unload her conscience she states Hernandez who the declaring party understands to be now deceased, that her husband did not love Agustina de Vergara very much but [so] she gave him menstrual blood to drink in his chocolate. The declaring party states that being in the city of Los Angeles[11] she went to make this stated declaration before the gentleman Commissioner of the Holy Office of the town and having told the Commissioner the above-mentioned, he told her to go and search her memory well and think of who was the person and the name of the person who had said that the stated Agustina de Vergara gave the menstrual blood to her husband. The declaring party went and learned that it was the stated

Francisca Hernandez who was mentioned and that she did not go to declare it as she was ordered to because on that occasion her stated husband had gone out of the city and that because she was alone, so on this occasion she declares. When she was asked if anyone knows that the stated Agustina de Vergara had given her husband or other men said menstrual blood in order to get them to like her, she responded that she did not know if other people might know better than they do, that this is the truth by the oath that she has made and it having been read she said it was well written [i.e., her statement was recorded correctly] and does not sign it, because she does not know how to write it is signed for her [Signature and seal of the Commissioner follow]. [17]

April 4, 1629. Sometimes love affairs required more than one magic spell, each with different elements according to the purpose. This is the case of a woman who, being in love with a married man, needed a potion to attract him and a second concoction to make him forget his wife.

On April 4, 1629 Magdalena Mendes, the concubine [media mujer: half woman] of Francisco Palacios—a Spanish man—confessed to the priest [Catholic ritual of confession], Father Juan del Val, and because they [Inquisition officers?] allowed me to say I [the priest ?] declare that: A mulatto woman whose name she [Magdalena] does not know, gave her a herb so she could give it to the married man [she was with] so he would forget his wife. The mulatto woman told her that she has to give it [to him] four times a day mixed with his chocolate and that she also should put her menstrual blood. If she followed her advice, the mulatto woman said, he will love her well. And, to make him forget his wife, because that he still loves her, she must obtain the heart of a crow and together with the excrements of his wife, and add these [items] to his chocolate [i.e., the chocolate he customarily drinks]. Signed by José de Herrera and Artega. Notary of the Inquisition Office. [18]

February 15, 1647. The Inquisition as a highly regulated institution had well-defined and precise procedures. One of these was the confiscation of all possessions of persons declared guilty of heresy. These activities were controlled by the *Fisco de la Inquisición* (Inquisition Treasury). Once the sentences were issued—be it the terrible sentence of *relaxation*[12] or any other punishment—all property and assets of the guilty person became property of the Inquisition Tribunal. It was a characteristic of the Inquisition administrators to register assets and property in great detail even when circumstances arose where ownership was disputed. The following documents reveal in unexpected detail how ownership procedures were accomplished as in the case where the three persons identified were sentenced to death as heretics. Provided here is information regarding the confiscation of their respective properties, the auction mandate whose revenues would enrich the Inquisition coffers, and a precise list of the assets/properties associated with chocolate.

In the city of Mexico, on February 15, 1647, in the presence of the doctors [judges] Don Francisco de Estrada y Escobedo, Don Juan Saenz de Mayorca, and the licenciate [licenciado] Don Bernabé de la Higuera Jaramilla, the petition presented to the General Treasurer [Receptor General] of this Holy Office of the Inquisition was read. And it was mandated that the notary-present here attached to this petition a memorandum with all the chocolate cakes [panes de chocolate] that the Commissary of Veracruz sent [from Veracruz] and that are part of the properties and possessions of Manuel de Acosta, and that Don Juan Bautista de Avila and Don Francisco Martinez Guadina, familiares[13] of this Holy Office, and also merchants of this city, must appraise the above mentioned chocolate cakes; and that they [the familiares] have sworn under oath that there will not be fraud nor deceit in their appraisal. And that [i.e., after the appraisal is completed] those chocolate cakes have to be sold in public auction. Signed by Miguel de Almonacir. The Receptor [Treasurer] of this Inquisition mandates the sale of Mantas de Campeche and the ordinary chocolate cakes, remitted by the Commissary of Veracruz, that belongs to Manuel Acosta[14] and to sell 13 or 14 crude cakes of wax of Campeche [marquetas de cera] that belongs to Fernando Rodriguez. It must be noted that there has been considerable confusion: the properties that belonged to the above mentioned Miguel Acosta and Fernando Rodriguez were mixed, and that badly. [The property] also was mixed with that belonging to Diego Lopez Coronel who even though he was taken prisoner by this Inquisition, was released without any punishment or condemnation.

Notices of the result of the diligences made to adjust what belonged to each of them. . . .

—Manuel Acosta—

It belonged to him 3 bundles with wax [cakes] from Campeche, that were appraised at 9 reales each piece; 108 dozen chocolate cakes [appraised at] 12 reales each. It seems that the chocolate cakes were auctioned at 9 reales each cake to Don Esteban Beltrán who also bought a bundle that belonged to Diego de Lopez; also 448 mantas [short ponchos worn by men] with a value of 504 reales. From the chocolate cakes that belonged to Acosta, it still must be determined the amount that was returned to Diego Lopez when he was released from prison. Despite all diligence it should be noticed that the sale of 108 dozen chocolate cakes was not registered, and it was necessary [that I] reviewed the accounts registered by the Receptor of the Inquisition.

—Fernando Rodriguez—

It belonged to him 17 tercios of white wax from Campeche that was estimated to be 15 pesos per arroba, and another 17 tercios [of white wax] whose weight was 149 arrobas and 6 pounds, that were auctioned to Don Fernando del Bosque, neighbor of this city of Mexico.

—Diego Lopez Coronel—

It belonged to him a small bundle of mantas, that was appraised and sold at the same price than the mantas of Acosta. [19]

March (Date Unclear) 1650.

The Spanish wealthy classes in Mexico did not escape from the vigilant eyes and ears of the Inquisition. Sometimes important, wealthy, Spanish women also were victims of persecution and their social status did not protect them. The following document identifies a group of exceptional women (recluses) who had been arrested and were being held by the Inquisition in order to extract further information. It is not clear whether their incarceration was to explore further their own involvement in heresy or to extract additional information that would lead the Inquisitors to pursue lines of evidence against others of high status known to the women.[15] Curiously, the document here considers women, a lizard (iguana), and chocolate offerings to Saint Anthony of Padua, patron saint of lost causes.

I, Fray Marcos de Rivera, religious, priest, confessor and preacher of the order of Our Lady of Mercy [Nuestra Señora de la Merced] in the Convent of Concepción [Convento de la Concepción], born in Mexico and of 51 years old, I present myself in front of Your Lordship and declare that in the past few days Doña Francisca Velazquez, legitimate wife of Joan de Rivera notary of the city of Mexico who lives in the houses [sic] of Pedro Callejas Panadero, in the street that goes from the new church to the College of San Pablo [Colegio de San Pablo] told me: that one day when she [Doña Francisca] was talking with Fray Alosnso Sanches, [a] religious of the order of Santo Domingo, he[the friar] told her about certain women that this saint tribunal of the Holy Office [of the Inquisition] has as recluses. [One day] other women that had stolen a large quantity, I do not remember if it was a large quantity of money or a large quantity of clothes, brought those things to the recluses house [i.e., house where all the women had been held] and when they were there with the recluses they saw that under one of the beds there was an animal. I do not know if it was a lizard or a dragon.[16] At the sight of the lizard the women were struck with panic and one of the recluses told them to calm down because it was a tamed animal, and that the animal had been raised in the house. [While she was saying this] the lizard crawled out from under the bed and walked [across the room] and crawled into a large earthenware jar that was filled with water in one of the corners of the bedroom; the recluses told the women that the animal was going to take a bath. Also, one of the recluses told Saint Anthony [i.e., prayed to the statue of Saint Anthony], "Oh! San Antonio! You are angry with me because you did not have yet your chocolate." And she beat [frothed] the chocolate in a small glass and gave it to the glorious saint, so he would drink it. Also, the recluses gave cigars to Saint Nicolas Bonorio [?]. Also, they had a young woman, I do not know if she was a mestizo woman, that when the recluses would order her, she would put her hands over her head and will rise from the earth [i.e., levitation]. Then when the recluses send her to the Inquisitor so she would do it in front of them, she could not do it. Also the recluses will not be allowed to leave this year, nor next year, because there was still much to do with them, and the worst [of all] was that these recluses were in communication with the most illustrious persons of Mexico.

[Signature] Marcos de Rivera. [20]

August 22, 1650. In this document a female parishioner presented testimony before the Tribunal accusing her priest of breaking a religious taboo—and that he had drunk chocolate before mass and also had consorted with an unidentified woman over a cup(s) of chocolate.

In the town of Tampico on the 22nd day of August, 1650. . . . A woman whose name is [illegible] came and said that in order to have a clear conscious she had to say that Fray Francisco Ortiz [a priest of the order of Saint Francis, 29 years of age] has drunk chocolate before the mass, and that he left the altar for half an hour to drink chocolate with a woman. [Signed] Joseph Borel [?], Notary Bartholeme del [last name illegible]. [21]

December 6, 1689. In the Spanish colonies of the New World, the number of Inquisition victims burnt at the stake was much lower than in Spain [22]. Still, the declared heretics in the Spanish colonies suffered persecution and in many cases were put to death. One of the last crypto-Jews, or hidden Jews, identified, accused, convicted, and sentenced to death by the Inquisition in Mexico was Don Diego Munõz de Alvarado. Don Diego Munõz had been mayor of the city of Puebla and had accumulated substantial wealth. He died while undergoing torture. Although dead, his body was convicted by the Inquisition on June 1684 and his effigy was burned in an *auto de fé* celebrated in Mexico City at the end of October 1688 [23]. The following Inquisition document describes the confiscation of Don Diego Muñoz's possessions after he was "relaxed," a dreadful euphemism applied to those condemned to death.

The Holy Office of the Inquisition of Mexico–December 6th, 1689 On the account of the possessions that have been confiscated to Diego Muñoz de Albarado [Alvarado], relaxed in this city of Mexico Against Lucas Romero-neighbor of Zacatecas for 424 pesos and 4 reales that he owed for cacao. As a consequence of what has been mandated by the Apostolic Inquisitors of the Inquisition of this New Spain in an Auto on October 23, of the past year [1688], the possessions of Diego de Albarado [Alvarado], relaxed, were disclosed; And in concordance with the above mentioned Auto I certified and attest that having recognize and seeing the list of debtors of the forenamed Albarado, the part that correspond to Lucas Romero, neighbor of Zacatecas has been registered in the books by this tribunal, in the following manner: Lucas

Romero, neighbor of Zacatecas is a debtor of 425 pesos and 4 tomines for the 3 tercios of cacao Caracas with 647 pounds of net weight at 5 reales and one cuartillo per pound that was delivered to Alvarado on December 5th, 1687, as it shows in the declaration that the aforementioned [Lucas Romero] presented in this tribunal, and that is registered in a book covered with a red sheep skin.

[Signature] Diego Pardo y Aguila. [24]

May 16, 1690. Sometimes under psychological pressure and fear of punishment, individuals even would denounce themselves to the Inquisition. In the following case, Simón Hernán, under pressure from his parish priest confessor, reported to the Inquisition officials in order to obtain absolution for his sins. Hernán was married and his sin was seduction. His was an especially dangerous act given that his wife was a mulatto, and the woman he attempted to seduce was Spanish.

In the city of Mexico, on the 16th of May, 1690, the Inquisitor ordered a man to appear [in front of the tribunal] that has come on his own will; the Reverend Father [name illegible; probably his confessor] was present and according with his ministry he will hold secret [i.e., sanctity of the confessional]. A man of 35 years, that said his name is Simón Hernán, born in this city [of Mexico] and married with a mulatto woman, named María [last name illegible] appeared before this Office by mandate of his confessor who refused to absolve him from his sins unless he would appear in front of this Office; and he [Simón Hernán] says and denounces that: During the mass at the church of Nuestra Señora de la Merced [Our Lady of Mercy] in this province of Mexico, he gave some powders to a Spanish woman named Micaela, dissolved in her chocolate [cup]. Eight days later, and without telling anything to his wife, in another mass he gave her [Micaela] those powders again, and as before mixed with chocolate. He declares that he gave her those powders because he wanted to seduce her [Micaela] and in doing so, he would able to enjoy her favors. But he said those powders had no effect on her. He also said that the powders were green.[17] This, he [Simón Hernán] declares under oath.

Signed by Diego de Vergara. [25]

(Month and Date Not Provided) 1697. Sometimes strong-willed women could be seducers. In the following document a man related to the Holy Office of the Inquisition that he had been forcefully seduced—at least initially—by a woman who came to his cacao and sugar shop for something more than just purchasing chocolate. He reported that after he had allowed himself to be seduced, he subsequently became weak and ill. His testimony also alludes to the probability that he had become impotent.

Don Francisco de las Casas, neighbor of this city [of Mexico] and owners of a cacao and sugar store in the corner of the Casas de los Sirvientes [House of the Servants] says that a Spanish woman whose name is Michaela de Orbea, who maybe because she was thankful or because I had bought something from her own store, or because she was pressed by the demon, she brought him [i.e., to his cacao-sugar store] stuffed manchet.[18] *And he said that as soon he ate them he felt suddenly ill. Some of my neighbors who were there [seeing me so ill] brought me to a small bed [I had at the back of the store]. At that moment she began to close windows and doors [in the cacao shop] and asked the people [the customers] who where there to leave; and then she sat over me and did many dishonest actions; and seeing myself a single man and with such an occasion so near my reach, I said to her loving words to show her that I had understood her affection, and she condescended. And since that day of weakness, I cannot have any action as a man, big or small and this is why I decided to tell this to the Inquisition.*

[SIGNATURE] FRANCISCO DE LAS CASAS. [26]

December 13, 1717. The following case presented before the Inquisition Tribunal represented not only sorcery and witchcraft associated with chocolate, but presented details where a sister-in-law attempted to seduce her brother-in-law (but with the support of his wife—her sister).

The Inquisitor Don Antonio del Rincón y Mendoza, with the ecclesiastic benefits of Vicar and Ecclesiastic Judge, bestowed by His Majesty, in the name [representing] of His Lordship, Monseñor Joseph de Lanziego y Eguilas, Monk of the Great Patriarch San Benito, Archbishop of the saint and glorious city of Mexico, and Commissary of the Holy Office of the Inquisition, declares: At 8:00 in the morning came Doña Petronila del Rincón y Mendoza and said that Juan Barrial had come [to her] asking to be cured. . . . He said that he had drunk chocolate in his house and after drinking he felt sick. . . . A boy named Miguel Martin helped him and brought him to his house [i.e., not his house, but the house of the parish priest where he worked] and gave him turnip oil [aceite de nabo] so he could vomit the chocolate. After vomiting he felt better. A few days after he came again feeling sick, and feeling a tingling in the stomach. Another day he came back with a very serious stomach ache and asked for water. He said that he had drunk chocolate. Hearing this the boy ran out to look for a priest to confess him, and the priest gave him a consecrated wafer [hostia] because he was suspecting an act of evil sorcery. When they asked who had given him the chocolate he said that his sister-in-law had prepared it, but that his wife was the one who beat it [i.e., stirred the chocolate]. His wife said that her sister had paid an Indian woman for preparing a potion [love potion?] to give to her husband. [27]

It was customary in Inquisition interrogations that witnesses were held in separate rooms, and the proceedings were conducted in secret. Related to the deposition just cited were subsequent testimonies in the same case. Only one of these, however, considered the issue of the presumed poisoned chocolate. This deposition was the declaration of young Miguel Martin, the boy who had helped Juan Barrial to vomit the chocolate that seemed to be causing the pain.

In this city [of Mexico] on the same day [December 13, 1717], by virtue of the Auto—de-Fe of the Ecclesiastic Judge, appears in front of me the notary, [the boy] Miguel Martín. He said he was a mestizo, and he made the sign of the cross and promise to say the truth in all that he would be asked. When asked if he knew about the illness that was affecting Juan Barrial, he said that when he was a servant in the house of the Parish Priest [name not identified], that Juan Barrial came one day at 9 in the morning, saying he had a pain so strong in his stomach that he felt he was about to die, and asked for some water to drink. So, I gave water to him and asked him if he had eaten something that could be causing the pain, and he said that he has not eaten anything that morning but [he] had only drank the chocolate that was served to him in his house this morning. [He also said that] as soon as he drank it he felt a very strong pain all over his body. Then, I told him to put his fingers inside of his mouth so he would be able to empty his stomach [i.e., vomit]. [He said that] Doña Petronila gave him [also] some oil to drink, and he emptied his stomach well. While I was holding his head because he was shaking and vomiting he told me that yesterday night he had a quarrel with his wife . . . [testimony breaks off]. [28]

October 12, 1758. The following letter, written to the Inquisition Office in Mexico City, denounced the production and sale of crude religious images sold in chocolate shops in Mexico City and alerted officials to the "sacrilege."

Mexico, October 12th, 1758. It has come to my knowledge that in some candle as well as chocolate shops and peanut stores they are used to give as pilones [solid brown sugar cones] figures of the Holy Sacrament [monstrance: custodias] as well as images of Holy Maria and other saints to attract children and they have copious sales. This results in a lack of appropriate veneration, not only for having them piled up in a basket among grease and butter in corners [in the store's corners] but also for degrading them by giving them with such low regard [inferioridad] as if it was worth less than the goods being sold as it is given in bunches. Furthermore, it is also like a kind of commerce being done with other images and crosses that are printed in the mentioned custodias. If this were to be known by the heretics it would give them an opportunity to start an uprising [levantar una polvareda] and make more fun and ridicule of the cult we tribute to the Holy Cross, Holy Mary and the Saints, revolted by the indecencies that by other causes are executed and the vilification and insult with which the boys treat them because they [the shopkeepers] give the boys the sacred images in sugar instead of giving them panocha.[19] *What is more, as these are made by Indians and it is their profit and that of the shop-keepers [compradores] it is their interest in selling many of them, so they are made and carved [in ways] so deformed and imperfect that only the diadems and other insignias represent the images and nothing more. For all this it is proper that Your Holiness totally prohibit in the mentioned places to sell these pieces and give severe punishment to those who abuse the images by selling them for cacao grains or giving them for free thus lowering the damage and hurt that the adoration, cult and reverence that God and Your Holiness deserve.* [29]

September 19, 1780. The following Inquisition documents represent a three-part set that is extremely unusual for their content: They detail events surrounding identification and denouncement of a lesbian chocolate-maker. The first document presents the accusation, where testimony is given that the woman in question was known only as Maria *La Chocolatera* (Maria the chocolate-maker). The text was unusually frank for the period and emphasizes through the context that homosexuality was considered a sin in the eyes of the Catholic Church in 18th century New Spain.

In the Inquisitional Convent of Our Father Santo Domingo in the City of Mexico, on September 19th, 1780, in the afternoon, a woman that said her name is Maria Josefa de Ita, Spanish, maiden, daughter of Francisco Ita, deceased, and of Anna Ignacia Murcia, or 21 years of age, born and neighbor of this city, appeared in front of the Reverend Father Apostolic Inquisitor Francisco Larrea, qualifier and Commissary of the Holy Office. She came without being called to achieve peace of mind [descargo de conciencia] and she said and denounced that: Maria, a woman whose last name she ignores, owner of the chocolate shop on the San Lorenzo Street, Spanish, unmarried, that she had asked God to send her a woman so she could sin with her [have a sexual relationship], and that several times she asked her to sin with her, she the declarant [sic], as a woman, and she [Maria La Chocolatera][Maria, the chocolate maker] as a man. The last time she asked this was on a Saturday, eight nights ago, and that when she told her that she should pray and give her a rosary, she threw it on the floor and said that she would not pray nor to make the sign of the cross [persignarse] until God would send her the woman she desired. There being no witnesses [of this conversation] because although there were two more persons [in the chocolate shop] they did not hear the conversation. When we asked the declarant [sic] if the so called Maria La Chocolatera had said that these actions [torpezas] were a sin, she answered that it should not so bad to do it because her Confession Father had told her that she could do it. The declarant [sic] added that she [Maria La Chocolatera] has two children but that they do not live with her.

SIGNED, FRANCISCO LARREA AND
MARIA JOSETA DE ITA. [30]

The second document in the series presented an order by the Inquisitor that Maria La Chocolatera be sought and her full name identified. The Tribunal also required that the address of her chocolate shop be verified [31].

The third document reported to the Inquisitor that upon further inquiry those who had searched for the woman had determined that the true name of Maria La Chocolatera was Maria Gertrudis de la Zerda, and that her chocolate shop was located on Calle del Leon (near) San Lorenzo Street across from the school (school not named) [32].

November 24, 1781. In contrast with Maria "La Chocolatera"—the daring woman that openly defied the moral traditions of her time—the last document reviewed in this chapter reveals another type of woman who also had a chocolate shop in Mexico City. The woman identified, a María Dolores Fernández, was especially pious and after confession had followed her priest's advice and presented herself to the Holy Office to denounce a male customer for blasphemies that he had uttered when he was a customer inside her chocolate shop.

In the city of Mexico, on November 24, 1781 the Inquisitor Don Juan Antonio Bergosa y Jordán, ordered to appear in front of him, a woman that came without being asked. We received her oath, and she made the sign of the cross, and promised to say the truth and all she knew, and that she has to maintain in secrecy whatever she would say and hear. She said that her name was María Dolores Fernández, born in this city [of Mexico], daughter of Don Gabriel Fernández, old Castilian, and [she said] that she was a maiden of 25 years old and that she lived with her mother María Rita Hernández, in the street of Tabqueros. She said that she came because her confessor had advised her to come and denounce for her peace of mind that 15 days ago when she was at her house, a chocolate store [chocolatería], Sivestre Diaz born in Galicia [Spain] a musician at the Military Corps [of this city] and he began to talk to her about a woman that was living illicitly with him, and said that "Not even the God Almighty can separate me from that woman." Upon hearing this, she reprimanded him and told him that God was not deaf and that He [God] might punish him. He began to blaspheme against the Holy Sacrament of the Eucharist, and continued using foul language to address Our Lady of Carmen, Our Lady of Mercy, and Our Lady of the Rosary. [At this point] she changed the conversation; all this happened in the presence of her niece Juana Jaques. [33]

Conclusion

The several chocolate-associated Inquisition documents presented here appear as relics from a past era, where distrust, fear, and suspicion influenced human behaviors and practices. Prying eyes and ever-vigilant Dominicans characterized the era and one never knew when there might be a knock on the door with an order to appear before the Tribunal (or worse). The chocolate-related documents we reviewed included Tribunals that sat in Cuba, Guatemala, and Mexico.

Most of the individuals accused and deposed were poor, with the exception of a crypto-Jew, tortured and murdered in secret, then burned at the stake in effigy. One was an accused priest. The interesting case of the "recluse" wealthy women reveals the absolute power of the Inquisition to hold/detain the accused for indefinite periods of time, especially for the purpose of eliciting further testimony and denouncements of others. The theme that linked most of the testimonies presented here was witchcraft, where presumed magical ingredients were mixed with chocolate in order to seduce or to restore affection of one's mate.

Given the severe outcomes of many Inquisition depositions in terms of heresy or "rooting-out" nonbelievers or followers of other religions, the chocolate-associated "crimes" of the accused seem minor when viewed from the perspective of 21st century tolerance. Readers of the documents, however, cannot be assured that the crimes of the accused were minor in the eyes of the Tribunal officials since dispositions of the various cases were not attached to the testimonies. What would be the punishment, for example, of a priest who extorted his church membership to pay him with cacao beans, otherwise, he would not hear their confessions? What would be the punishment of two women, one married, who helped her sister to seduce her husband using chocolate? One suspects, however, that the punishment of Maria Gertrudis de la Zerda, known initially to the Inquisitors as only Maria La Chocolatera, would have been severe. But how severe, one only may speculate. Perhaps the outcome of these interesting chocolate-associated cases can be revealed during further research.

We are left, however, with a curious question: Why was chocolate used as a vehicle for witchcraft? Why not other foods, perhaps soups or other beverages such as wine? First, chocolate was readily available and was a good medium for mixing all types/kinds of ingredients. The fine taste of chocolate would mask ill-flavors that resulted from the addition of crow's heart, excrement, human flesh, menstrual blood, and other substances identified in the documents presented here. Second, chocolate as a thick beverage would also have a thick texture that could hide ground powders or substances. Third, one offered chocolate to acquaintances, friends, and even enemies as tokens of hospitality, thus defusing suspicion and offsetting potential fears that something consumed might bring about illness or death. Regarding the use of chocolate by women—or women being offered chocolate by men—such hospitalities were commonplace in New Spain and elsewhere. In contrast, it would have been improper for a man to offer wine to a woman (and vice versa) ... with the intent of seduction. But why not use soup in witchcraft? Perhaps the answer is simple and obvious: Who would ever offer soup to a person one wanted to seduce?

Acknowledgments

We wish to thank our colleagues at the various archives and libraries in Mexico for their assistance in identifying the documents translated and presented here.

Endnotes

1. Although there had been sporadic periods of violence against Jews in Spain, Spanish Jews—known as Sefardi (Sephardics)—had found a haven of tolerance in Muslim Spain. Medieval Spanish Catholic society was far more tolerant toward religious beliefs of others than the rest of European states. In the year 929, with the proclamation of the Califato de Córdoba by Abderramán III, there began a period where Christian, Jewish, and Muslim scholars worked together in an invigorating intellectual environment. In contrast with the Dark Ages of the rest of Medieval Europe, Cordoba flourished as a center of medicine, science, and language, where Greek and Roman manuscripts were translated and thus preserved for history. But this remarkable historical period of tolerance and cooperation among individuals of three different religions ended when Granada—the last Moorish bastion—fell on January 2, 1492, to the Catholic Monarchs. The reasons for the expulsion of Jews—as well as for instituting the Inquisition in Spain—were due not only to religious zeal but also to economic conflicts of interest between the feudalistic nobility and the strong, and growing, Jewish middle class.

2. Pope Alexander VI decreed all new lands west of a meridian 100 leagues west of the Cape Verde Islands should belong to Spain; all new lands east of the line would belong to Portugal. The treaty of Tordesillas, signed June 7, 1494, moved the line to 370 leagues west of Cape Verde. The line intersected the eastern portions of Brazil, which is why Portuguese is spoken there. Once the line was extended completely around the globe, portions of the Philippines extended into the Spanish sphere of influence; this is why the Spaniards were able/allowed to colonize these islands.

3. Public procession of those found guilty by the Inquisition Tribunals; executions were held after the processions, not during the auto de fé.

4. The documents presented in this chapter do not duplicate the interesting and valuable examples presented by María Águeda Méndez in her 2001 publication, *Secretos del Oficio. Avatares de la Inquisicion Novohispana*. Mexico City: El Colegio de Mexico Universidad Nacional Autonoma de Mexico, and the examples provided by Martha Few in her publication, "Chocolate, Sex, and Disorderly Women in Late-

Seventeenth- and Early-Eighteenth-Century Guatemala" (*Ethnohistory* 2005;52:673–687).

5. Canónigo is a member of the clergy belonging to a cathedral or collegiate church with ecclesiastic benefits.

6. Such *exquisitas letras* in the sense used here could have been poetry or other texts dealing with humanistic knowledge (not mathematics or natural sciences), for which possession could have led to an Inquisition summons.

7. Commissary of the Inquisition or Commissary of the Holy Office was the name given to each of the priests that represented the ecclesiastic tribunal in the main cities of the Kingdom of Spain (including all Spanish possessions in the Americas).

8. A plant indigenous to Mexico widely used for purposes both of attraction and repulsion in colonial times.

9. In the context used here, a university graduate.

10. Sabana-Camaguey or Jardines del Rey; a coral archipelago of approximately 400 islets north of Cuba.

11. Not Los Angeles, California, but Puebla de los Angeles (Puebla, Mexico).

12. Relaxation was an Inquisition procedure to deliver a condemned heretic to the hands of the secular authorities for execution. Being an ecclesiastical organization, the Inquisition could not itself carry out the execution since one of the Ten Commandments was "thou shall not kill."

13. Secular police of the Inquisition who conducted secret investigations.

14. The name Manuel de Acosta of this document appears in Liebman's 1974 list of Jews brought before the Inquisition in New Spain (Mexico) and convicted (Liebman, S. B. *The Inquisitors and the Jews in the New World. Summaries of Procesos, 1500–1810 and Bibliographical Guide.* Coral Gables, FL: University of Miami Press, 1974; p. 38). He was the son of Antonio de Acosta, born in Orense, Galicia, and of Juana Lopez, Portuguese, and both were Jewish. He arrived in Mexico in 1638, married Isabel Tinoco, and fathered a daughter, Micaela. He was accused and publicly reconciled in an auto de fé of 1648; his property was confiscated, he was forced to wear the *sanbenito* (dunce cap and cape announcing the crime of the accused), sentenced to 200 lashes, and five years service in the galleys (most sentenced to galley service died within three months), but in case he survived, he also was sentenced to life imprisonment.

15. Otherwise why would it be proposed that they continue to be held for two years?

16. Perhaps this large lizard, thought by the deposed to be a dragon, was in fact an iguana.

17. Actually green, or were the powders from unripe "magical" herbs?

18. Loaf of bread made from fine, white flour.

19. Panocha: a brown sugar fudge-like confection made with butter, milk, and nuts.

References

1. Lea, H. C. *The Inquisition in the Spanish Dependencies. Sicily, Naples, Sardinia, Milan, The Canaries, Mexico, Peru, New Granada.* New York: Macmillan Company, 1908; p. 191.

2. Lea, H. C. *The Inquisition in the Spanish Dependencies. Sicily, Naples, Sardinia, Milan, The Canaries, Mexico, Peru, New Granada,* New York: Macmillan Company, 1908; p. 196.

3. Plaidy, J. *The Spanish Inquisition. Its Rise, Growth, and End.* New York: The Citadel Press, 1969; pp. 65–66.

4. Lea, H. C. *The Inquisition in the Spanish Dependencies. Sicily, Naples, Sardinia, Milan, The Canaries, Mexico, Peru, New Granada.* New York: Macmillan Company, 1908; pp. 204–208.

5. Plaidy, J. *The Spanish Inquisition. Its Rise, Growth, and End.* New York: The Citadel Press, 1969; pp. 65–66.

6. Coe, S. D., and Coe, M. D. *The True History of Chocolate.* London: Thames and Hodson, 1996; p. 138.

7. Terrio, S. J. *Crafting the Culture and History of French Chocolate.* Berkeley: University of California Press, 2000; pp. 74–75.

8. Archivo General de la Nación, México (AGN). Inquisición, Vol. 368, exp. 141, f. 5725.

9. AGN. Inquisición, Vol. 220, exp. 1, f. 13.

10. AGN. Inquisición, Vol. 339, exp. 89, f. 623r.

11. AGN. Inquisición, Vol. 339, exp. 17, f. 174r–174v.

12. AGN. Inquisición, Vol. 339, exp. 84, f. 580r–580v.

13. AGN. Inquisición, Vol. 339, exp. 84, f. 586r–586v.

14. AGN. Inquisición, Vol. 339, exp. 84, f. 611r–642r; 561r–v 561v.

15. AGN. Inquisición, Vol. 335, exp. 64, f. 261r.

16. AGN. Inquisición, Vol. 356, exp. 78, f. 115r–115v.

17. AGN. Inquisición, Vol. 353, exp. 46, f. 78r.

18. AGN. Inquisición, Vol. 363, exp. 30, f. 240v.

19. AGN. Inquisición, Vol. 435, exp. 104, f. 139r–140r.

20. AGN. Inquisición, Vol. 436, exp. 28, f. 129–129v.

21. AGN. Inquisición, Vol. 435, exp. 104, f. 139r–140v.

22. Lea, H. *The Inquisition in the Spanish Dependencies. Sicily, Naples, Sardinia, Milan, The Canaries, Mexico, Peru, New Granada.* New York: Macmillan Company, 1908; pp. 209.

23. Liebman, S. B. *The Inquisitors and the Jews in the New World. Summaries of Procesos, 1500–1810 and Bibliographical Guide.* Coral Gables, FL: University of Miami Press, 1974; p. 40.

24. AGN. Inquisición, Vol. 1474, exp. 14, f. 105r–106r.

25. AGN. Inquisición, Vol. 520, exp. 168, f. 266v–267v.

26. AGN. Inquisición, Vol. 1312, exp. 1, f. 1r–2v.

27. AGN. Inquisición, Vol. 1169, (exp. missing), f. 241r–252r.

28. AGN. Inquisición, Vol. 1169, f. pp. 242r–242v.

29. AGN. Inquisición, Vol. 952, exp. 15 (recto/verso data missing, but with handwritten page numbers, 183–184).

30. AGN. Inquisición, Vol. 1203, exp. 16, f. 122r–123v.

31. AGN. Inquisición, Vol. 1203, exp. 16, f. 124r.

32. AGN. Inquisición, Vol. 1203, exp. 16, f. 125r–125v.

33. AGN. Inquisición, Vol. 1355, exp. 12, f. 11r–12v.

5

Nation of Nowhere

Jewish Role in Colonial American Chocolate History

Celia D. Shapiro

Introduction

It is hard to say when the story really begins. Was it during the Crusades when pilgrims terrorized non-Christians? Was it in 1242 in Rome when the Inquisition condemned the Talmud—the compilation of commentaries on the Torah and the Old Testament—or in France in 1288 with the first mass burning-at-the-stake of Jews? More likely, the story began in 1391 when the Inquisition moved to Spain and brought with it the mass slaughter of 50,000 Jews in Barcelona, Cordova, Mallorca, Seville, and Toledo (see Chapter 4). The story then moved forward in 1478 when Pope Sixto IV approved the establishment of a Holy Inquisition to discover and punish hidden Jews. And the story became a familiar one with the Spanish Inquisition that Isabella and Ferdinand authorized in 1480, that started in earnest in 1481, and resulted in the great expulsion, a time when blind hatred was institutionalized and the Jews had to leave a country where they had flourished, a country that itself flourished in part because of the Jews.

The year 1492 was particularly important for the Jews of Spain. On March 31, 1492, Isabella and Ferdinand issued the Expulsion Decree that stipulated that within four months all Jews and Jewesses had to leave the kingdom and lands of Spain. On the same day—April 30, 1492—that the decree was publicly announced, Columbus was ordered to equip ships for his voyage to the Indies. Three months later, on July 30, 1492, Inquisitor General Father Tomas de Torquemada, expelled the Jews. Three days after that, August 2, 1492, it was reported that 300,000 Jews left Spain, hoping to settle somewhere they could live in peace. And the following day—August 3, 1492—Columbus set sail. Because of the large number of Jews attempting to leave Spain by ship, he departed from the port of Palos rather than Cadiz.

Known Jews who sailed with Columbus included Roderigo de Triana, a sailor; Luis de Torres, an interpreter fluent in Hebrew, Chaldaic, and Arabic as well as Spanish; Maestre Bernal, a physician; Roderigo Sanchez de Segovia, Queen Isabella's inspector; Marco, a surgeon; and Alfonso de La Calle; a sailor [1]. On October 12, 1492, Columbus reached the West Indies. Some sources report that Roderigo de Triana was the first to sight land [2, 3]; some sources document that Luis de Torres was the first to step ashore. What is known is that de Torres lived in Cuba for a portion of the year 1492, perhaps the first Jew to live in the West Indies, and also was responsible for introducing tobacco to Europe [4, 5].

The two stories were set: the Jews were under constant duress in Spain and Portugal, and at the same time the West Indies were becoming colonized. Eight Jews were expelled from Portugal in 1496, and in 1497 tens of thousands were forcibly converted to Christianity. Amerigo Vespucci wrote that in 1499 Vicente Yáñez

Chocolate: History, Culture, and Heritage. Edited by Grivetti and Shapiro
Copyright © 2009 John Wiley & Sons, Inc.

Pinzón, a Spaniard, "discovered" Brazil; in 1500, Pedro Alvarez de Cabral, a Portuguese, made the same "discovery" [6]. Voyaging with de Cabral was a Jewish mariner, Gaspar da Gama.

Chocolate enters this story of Jewish expulsion in 1527 in Spain, when Hernando Cortés arrived home from his voyage bringing cacao, and the recipes and tools for preparing it [7]. Jews continued to leave the Iberian Peninsula and they took their knowledge of cacao and chocolate production with them to Amsterdam and the West Indies. Some historians report that Jews were responsible for introducing chocolate to France (see Chapter 42) [8].

The story of Jews and chocolate is one not previously explored in depth. The search has been intriguing and sometimes difficult. Documentation is complicated due to a large number of terms, some with dual meanings that have also changed through time (see Appendix 1). Further confounding the search have been spelling variations for even common terms, and more so with family surnames. It commonly is accepted, for example, that the spelling system of the English language did not normalize until the mid-1600s, and it was not until the 1800s that most words had standardized spelling. For the period under consideration in this chapter, it was common to find variant spellings of common and proper nouns. To complicate matters even more, the people who are the focus of this chapter were comfortable using a multiplicity of languages: Dutch, Hebrew, Ladino, Portuguese, Spanish, and Yiddish. To simplify reading, therefore, the most common spelling variant is used unless the word is set within a direct quotation.

The spelling of personal names becomes even more complicated. In his book *500 Years in the Jewish Caribbean*, Harry Ezratty describes the multiple names that Jewish persons might have [9]. Sephardim, for example, often had Hebrew first names (perhaps Elihu or Solomon); their family names frequently were place names (i.e., Toledo = city in Spain), trade names (i.e., Mercado = merchant), or Arabic names (i.e., Abudiente = father of *diente*). To emphasize a family's commitment to Judaism, a Hebrew name might be added to the family name (i.e., Cohen-Toledo). To tease historians even more, an alias often was assumed when traveling or in business matters to prevent retaliation if discovered by the Inquisition or its agents. The following is an example.

> The legendary Amsterdam ship owner, Manuel Rodrigues, elsewhere was known as Jacob Tirado. Tirado or Tyrado may also have assumed the names Guimes Lopez da Costa or Simon Lopez da Costa, [10, 11]

The variable-names puzzle was made even more difficult by lack of uniform spellings. Arnold Wiznitzer in his important text, "The Members of the Brazilian Jewish Community (1648–1653)," provides the following example.

> Benjamin Sarfatti de Pinah signed his name sometimes as Binjamy Sarfatti and sometimes as Binjamy de Pinah. The signature of Arao depina appears as Aron and Aharon Serfatty. The name of Isque Montesinos also appears as Isque Montissinos Mesquita. Jehosuah Jessurn de Haro signed his name as Josua de Haro. Three men had identical names: David Senior Coronel. [12]

In the present chapter, the most common spelling of the name is used.

The Nation of Nowhere

The Jewish people do not have a history of using chocolate in food or in religious observance. Chocolate does not play a role in the storytelling of the nation. Rather, the Jewish involvement is all about supporting oneself in a new location—again and again and again. The story of the Jewish role in the history of chocolate in Colonial America is really the story of the nation of Jews—a nation of nowhere. All of the groups involved in the history of chocolate in Colonial America have common motivations—often economic growth, frequently territorial expansion, sometimes a search for new personal liberties. Jews, however, unlike the Spaniards, Portuguese, French, and Dutch, do not have a geographic boundary to their cohort.

THE 1500–1600S: A DIVERGING STORY

The 1500s and early 1600s brought more events—both good and terrible—that are important for understanding the story behind the history of chocolate in Colonial America. In 1530 Jews already expelled from Spain landed in Jamaica and made Spanish Town (also known as St. Jago de la Vega) their home. In 1531, the Inquisition was extended to Portugal by Pope Leo X. In 1532, sugar cane was imported to Brazil from Madeira by a marrano. In 1540, the Jews were expelled from the Kingdom of Naples, and ten years after that from Genoa and Venice. On February 7, 1569, Philip the Second of Spain ordered by royal patent that the Inquisition be established in the Americas [13]. In 1536, the first *auto de fé* took place in Mexico (see Chapter 4). Next came an event that would have a direct and very powerful effect on the history of chocolate in Colonial America: Holland achieved independence from Spain in 1581. In 1585, Joachim Gaunse (or Ganz) landed on Roanoke Island and became the first known Jew in what is today the United States [14]. In 1588, England defeated the Spanish Armada, weakening Spain and thus decreasing the reach of the Inquisition in Europe, especially in the Netherlands. In 1591, however, the

Inquisition arrived in Brazil via the Portuguese; and in 1593, Pope Clement VIII, through the *Caeca et obdurate*, expelled the Jews from all the Papal states except Rome and Ancona.

In 1596, Yom Kippur services were held for the first time in Amsterdam. The services, though controversial, were held openly. This privilege of free worship, however, remained uncommon. In Lisbon in 1605, Freo Diogo Da Assumpacao, a partly Jewish friar who embraced Judaism, was burned alive. In Hamburg, as early as 1612, the city council permitted Sephardic Jews from Portugal to take up residence. It was recorded that these Jews

were tradespeople [sic] who specialized in the wholesale trade of exotic goods such as tobacco, sugar, coffee, cocoa, calico, and spices. [15]

In France in 1615, King Louis XIII decreed that all Jews must leave the country within one month or face death. And in Persia in 1619, Shah Abbasi of the Persian Sufi dynasty increased persecution of the Jews and forced many to outwardly practice Islam. Around the world, the Jews became wary and sought refuge.

In 1620, the Jewish connection to the history of chocolate in Colonial America and the Jewish pursuit of safety took a major step when the Dutch West Indies Company was formed with Jewish stockholders. In that same year, Christian Puritans began to immigrate to America. Also in 1620, in Brazil, Joseph Nunes de Fonseca—known later as David Nassy or David the Leader—was born, probably in Recife, Brazil. In 1624, the Jews in Recife, some of whom had been living quietly as conversos, united and organized a colony. Six hundred of the leading Jews of Holland joined them [16]. The Dutch declared religious tolerance in 1624 in Brazil and the situation was peaceful. In 1630 the Dutch captured Pernambuco and transferred the capital to Recife.

During the same period in central Europe, in Vienna in 1625, the Jews were forced to move into Leopoldstadt, a ghetto [17]. In Spain in 1632, Miguel and Isabel Rodreguese (and five others) were burned alive in front of the King and Queen after being discovered holding Jewish rites. Hatred and chocolate came together in 1691 in Bayonne, France, when Christian chocolate makers obtained an injunction to prevent Jewish chocolate makers from selling to private customers within the walls of the city [18]. It was not until 1767 that Jewish chocolate makers in Bordeaux received permission from Parliament to sell chocolate in Bayonne (see Chapter 42).

Given this tumult in Europe, the mid-1600s were a period when Jews willingly came to many of the Caribbean nations as well to Brazil, and they soon arrived in Barbados, Curaçao, and Surinam [19]. In 1634, Samuel Coheno became the first Jew to set foot on and establish himself in Curaçao. He was an interpreter, pilot, and Indian guide for Johan van Walbeeck, the Dutch naval commander who took Curaçao from the Spanish. The Jews were comfortable in the West Indies, but as they moved to the Spanish- and Portuguese-held areas surrounding the West Indies, they reencountered the Inquisition. On April 13, 1638, in Mexico City, Tomas Tremino de Sobremonte was tried. The evidence about him included the fact that he had offered chocolate to men accusing him of not being a Christian [20]. In 1639, in Lima, Peru, more than 80 New Christians were burned at the stake after the Inquisition caught them holding regular Jewish services [21]. Mexico City was the site of repeated Inquisition horror. Gabriel de Granada, a young man, was accused in October 1642, January 1643, and May 1643. On all three occasions part of his guilt was related to chocolate: his relatives sent eggs and chocolate to a mourner, his close relative put chocolate aside for him to eat, and his close relative did not serve chocolate to other people because they were observing the Yom Kippur fast and could not eat [22]. There is some thought that the refusal to drink chocolate during the Yom Kippur fast was "a kind of code" among secret Jews in Mexico [23]. The Inquisition in Mexico peaked in 1649 when, in the largest *auto de fé* ever held in the New World, 109 crypto-Jews were accused of Judaizing and several were burned alive.

It was years before the Inquisition would end. In Madrid in 1680, Inquisition officials were serving biscuits and chocolate to the guests of an *auto de fé* (Fig. 5.1) [24]. A few years earlier, Roger Williams, a Puritan minister, was banished from Salem, Massachusetts, when he proposed that Jews, Turks, and all others be granted the right to worship in their own ways. Williams established Rhode Island and instituted religious tolerance and separation of church and state. Those precedents led to the colony of Rhode Island granting religious liberty to Jews in 1658. The stage was set for the eventual immigration of Jews to and participation in the life of Colonial America. But before then, they found at least partial sanctuary in Amsterdam and the West Indies.

EMANCIPATION: AMSTERDAM

The Dutch decided as early as 1588 to recognize and provide governmental protection to Jews. As David Liss wrote in *The Coffee Trader: A Novel*:

To those of us who had lived under the thumb of the Inquisition, or in lands such as England where our religion was outlawed, or in places such as the cities of the Turks where it was barely tolerated, to dwell in Amsterdam seemed a small taste of the World to Come. We were free to congregate and observe our holidays and our rituals, to study our texts in the

FIGURE 5.1. *Auto de fé* of June 30, 1680, Plaza Mayor, Madrid, by Francisco Ricci. Date: 1683. *Source*: Courtesy of the Museo Nacional del Prado, Madrid, Spain. (Used with permission.) (See color insert.)

light of day. For us who belonged to a small nation, cursed with having no land to call our own, the simple freedom to live as we chose was a kind of bliss. [25]

Simply put: "the racial stigma the 'Men of the Nation' had encountered in the [Iberian] Peninsula ... was absent in the Netherlands" [26].

The Dutch West Indies Company was formed in 1620 with a primary goal of rapidly acquiring wealth [27]. Jews immediately became stockholders and were able to exert influence at times that were critical for the survival of their co-religionists. The list of "main-participants" for 1656 includes seven Jewish names out of 167; the list for 1658 includes 11 Jewish names. An unpublished list of 1674 includes the names Antonio Lopes Suasso, Johacob de Pinto, Simon and Louis Rodriques de Sousa, Jeronimus Nunes da Costa, and Jacob and Moses Nunes Rodrigues—all names that appear time and time again in the history of chocolate in Colonial America [28].

NEW LANDS: THE WEST INDIES, VENEZUELA, AND BRAZIL

Colonization of the West Indies offered the nation of Jews a chance to save themselves and their religion. In 1641 in Recife, Brazil, the first synagogue in the New World, Kahal Zur Israel, was built on Rua dos Judeus (Street of Jews). Also in that year, Isaac Aboab da

Fonseca became the first Haham in the New World. Soon after, the Jewish community in Recife numbered between 2000 and 3000 persons. In 1644, the first Jews arrived in Paramaribo; and in 1652, a second group of Jewish settlers came to Surinam with Lord Willoughby of Parham. In 1654, Jewish settlers arrived in Guadeloupe and Haiti [29].

Even as the Jews settled in the colonies and contributed to the well-being of the West Indian islands, there were problems. The Jewish history in Curaçao exemplifies this struggle. Peter Stuyvesant, the Governor of Curaçao, objected to the equal treatment of the Jews. Fortunately for the Jews, the Directors of the Dutch West Indies Company believed in the value that the Jews were adding to the community. On March 21, 1651, the Directors wrote to Stuyvesant and upheld their contract with Jaoa d'Yllan (also known as Jan de Illan) to bring Jews to Curaçao [30].

The Jews were comfortable in many of the West Indian colonies. In 1654, the Dutch granted civil rights to the Jews of Guiana. That document began, "To the People of the Hebrew nation that are to goe [sic] to the Wilde Cust [sic]" [31]. In 1658, Josua Nunez Netto and Joseph Pereira, Jews specializing in different Arawak dialects and serving as translators between the English and Dutch authorities and the native Indian tribes, arrived in Pomeroon, Guiana. The native Indians had their own system for processing cocoa and the Jews learned it [32]. Also in 1658, Jews arrived in Tobago. On September 12, 1659, Jews from Recife under the leadership of David Nassy arrived in Cayenne [33]. In 1660, 152 Jews from Livorno settled in Cayenne [33]. In May 1664, after the fall of Cayenne, some of these Dutch–Portuguese–Brazilian Jews expanded the community in Surinam when they moved there under Nassy's leadership [34]. On April 8, 1661, Benjamin de Caseres, Henry de Caseres, and Jacob Fraso petitioned the King of England for permission to "live and trade in Barbados and Surinam," and their request was granted [35].

The Jews' serious involvement with cocoa in the New World began in 1654 when Benjamin d'Acosta de Andrade arrived in Martinique. With knowledge acquired from Native Indians, it is thought that he established the first cocoa-producing plant in the New World. The monk Jean-Baptiste Labat wrote:

Un Juif nommé Benjamin y planta la première vers l'année 1660 . . . [A Jew named Benjamin planted there (Martinique) the first (cocoa tree) in approximately the year 1660. . . .] [36]

It also is written of d'Acosta de Andrade that:

One special target of Jesuits was Benjamin d'Acosta de Andrade. He owned two sugar refineries and was the first to process and manufacture cocoa in a French territory. He exported his product to France, calling it chocolate. [37–39]

D'Acosta de Andrade based his work on the Mexican production of chocolate. Then, using international connections, the Jews entered the worldwide cocoa business. What were these international connections? They appear to have been of two sorts: the business connections best exemplified by the Dutch West Indies Company, and the family relationships all over Europe and, starting in 1654, in New Amsterdam in North America [40]. Herbert Bloom has written: "Meanwhile more Jewish settlers had been coming from Holland [to New Amsterdam], among them, in 1655, Joseph da Costa, a large shareholder in the West India Company" [41]. It is hard not to speculate that Joseph *da Costa* and Benjamin *d'Acosta* de Andrade belonged to the same family of Sephardic Jews who came from Portugal to England and the West Indies. With da Costa's arrival in New Amsterdam, the trade routes from the West Indies to Europe as well as the trade route from the West Indies to New Amsterdam were in place.

By 1655, just five years after d'Acosta de Andrade arrived, the Jews of Amsterdam were buying cocoa from Curaçao and that cocoa business continued to grow [42]. In 1661, the Jews of Curaçao exchanged manufactured goods for tobacco, hides, coffee, corn, powdered gold, and cocoa with the Jews of Coro, Venezuela [43]. In the late 1600s, de Andrade and others were shipping cocoa grown in Venezuela to Amsterdam for processing. "In 1688, the English trea-sury allowed Peter Henriques of London . . . to import twenty tons of cocoa from Holland. . . . It originated in the Spanish West Indies, reached Holland through Curaçao and was then transferred to England" [44]. In 1693, Jews traveled from Curaçao to Tucacas (Venezuela) to make large-scale purchases of cocoa to send back to Curaçao [45]. In 1711, Juan Jacobo Montero de Espinos, the mayor of Coro, wrote that 12,000 bales of cocoa were exported from the Tucacas valley by the Jews [46]. In 1722, it was reported that Jews were present in Tucacas in January and July at the fairs when the cocoa beans were collected [47]. Montero de Espinos also arranged attacks on several mule trains carrying cocoa beans to the Jews of Tucacas [48].

The Jews in the New World also were involved in the related businesses of sugar and vanilla production. In 1663, in Surinam, David Mercato (also known as David de Mercado) found a new way to build sugar mills [49]. On March 31, 1684, Abraham Beekman, an official in Essequibo, Guiana, wrote to the Dutch West Indies Company that Salomon de la Roche had died and no one else could be found to process vanilla—which Beekman had used in chocolate [50].

Cocoa production—despite its attendant problems—was a more and more important activity in the West Indies. On December 17, 1671, Sir Thomas Lynch, the Governor of Jamaica, wrote:

What falls heaviest on them [the Jews] is the blasting of their cocoas; fear most of the old trees will die, as in San Domingo and Cuba; yet hopes to pick up a few nuts for the King and his Lordship, with a bunch or two of vanillas; and hopes to send his Majesty some off his own land, for he is sending a Jew to the inland provinces where the vine grows to see whether he can cure any. [51]

In 1700, Samuel Beekman wrote to the Directors of the Dutch West Indies Company that he was trading salt fish for cocoa and sending them:

. . . an anker of cocoa, together with twelve cakes of chocolate, as a sample, hoping the same will please well. [52]

By 1708, cocoa production was the principal reason that settlers came to Jamaica [53–55].

In 1682, cocoa was imported to Curaçao at 22–25 pesos per hundredweight and the cocoa export duty was set at 2.5 percent [56]. In 1684, chocolate was the most important export of Martinique, but that changed in 1685 with the publication of the law known as the *Code Noir*. The law stated, in part:

We command all our officers to chase out of our islands all Jews who have established their residence there whom as declared enemies of the Christian faith we command to get out in three months counting from the day of the publication of these results upon penalty of confiscation of their persons and property" [57]. *More simply, the* Code Noir *called for the expulsion of the Jews from all French islands.*

D'Acosta de Andrade, father of the cocoa business in Martinique, left for Curaçao, the Dutch haven. During the years 1704 through 1731, Jacob Andrade da Costa, another member of the extended family, operated there as a broker. By some estimates there were 200 Jewish brokers operating in Curaçao between 1660 and 1871 [58]. Marcus writes that Rachel Luis brought in cocoa from Curaçao during this time period [59]. While it would be valuable to include Jewish women in the Colonial American chocolate history, there is no other evidence—save this singular account—of Rachel Luis.

Surinam was especially important to the Jews—both for religious freedom and business opportunities. In 1669, Governor Capitän Phillipe Julius Lichtenberg formally promised that the Jews would be allowed free exercise of religion and all the other privileges granted by the English [60]. The congregation in Surinam included many traders, among them David Nassy, Isaac Pereira, Abraham de Fonseca, Jacob Nunez, Isaac da Costa, and Benjamin da Costa. By 1700, the Jewish community had grown to 90 Sephardic families. In 1706, Surinam exported 900 pounds of cocoa, and by 1730, Surinam, with its plantation-based economy—sugar cane, coffee, and chocolate—was the leading community of the Americas, far surpassing Boston, New York, and Philadelphia. Lavaux's map of 1737 shows 436 plantations of sugar, coffee, cocoa, wood, and cotton [61].

Jewish-related cocoa production and export continued to grow in the West Indies. In Curaçao in 1734, there were 40 Jewish marine insurance brokers [62]. On May 7, 1741, in the Bill "An act for raising several sums of money and applying the same to several uses," the Jamaican House of Assembly decided that Jews should pay extra tax on cocoa because they "are a very wealthy body . . . if they ever do import cocoa . . . they generally contrive to avoid paying any duty on it" [63]. This act was passed after Governor Edward Trelawny rescinded his earlier promise of April 16, 1739 not to levy a special tax on the Jews [64]. In 1744, Luis Diaz Navarro wrote that the Jamaican Jews "erected tents on the land and kept their merchandise in them, and the citizens of Cartago [Costa Rica] would come to the fair twice a year to sell cacao" [65]. Also in 1774, Edward Long, the secretary of the governor, wrote from Spanish Town, Jamaica:

> The Jews here are remarkably healthy and long lived . . . [they] may owe their good health and longevity, as well as their fertility, to a sparring use of strong liquers, early rising, indulgence in garlic and fish, Mosaic laws, sugar, chocolate, and fast. [66]

In Surinam between 1743 and 1781, there were more than 450 shipments of cocoa; almost all went to Amsterdam [67]. In 1745 alone, 674,749 pounds of cocoa were exported [68]. In 1787, 802,724 pounds of cocoa were exported [69]. In 1779, in St.

Eustatius, the Jews exported 15,220 pounds of cocoa to North America and 422,770 to the Netherlands [70].

Colonial America

NEW AMSTERDAM

In 1654, the Jews who were prospering in Brazil again met the Portuguese—and the Inquisition. The Portuguese conquered the Dutch Colony of Recife on January 25, 1654, and the Jews were forced to leave. Some settled in safer places in the West Indies, some returned to Amsterdam, and 23 arrived in New Amsterdam on the bark *St. Catarina*. These Brazilian Jews are mentioned in the record of the Burgomasters and Schepens on September 7, 1654 [71]; just five days later, on September 12, 1654, Rosh Hashonah, the Jewish New Year (year 5415), was celebrated for the first time in New Amsterdam [72].

Peter Stuyvesant once again lobbied against the Jews, and once again the Dutch West Indies Company insisted that the Jews could settle in that Colonial American colony. On April 26, 1655, Stuyvesant was instructed that the Jews could reside permanently and trade in New Amsterdam. This decision was influenced by the Dutch history of tolerance, the trading history of the Jews in the West Indies, and the seven (of 167) stockholders who were Sephardic Jews [73]. The Jews were not, however, granted religious freedom at the same time [74]. In 1657, Jews arrived in Rhode Island and for the first time Jews obtained the rights of citizens [75]. In 1661, Asser Levy of Albany became the first Jew to own real estate in New York and, in 1665, became the first Jew to become denized—a status between an alien and a naturalized person—in New York [76].

NEW YORK

In 1664, the English conquered New Amsterdam and the Jews became English subjects of New York. In 1665, Benjamin Bueno de Mesquita was banished from Jamaica and came to New York. Approximately 15 years later, his cousin, Joseph Bueno de Mesquita (known as Joseph Bueno), arrived in New York. He had established trading ties with Surinam by 1685, and as early as 1702 was involved in the cocoa trade [77]. In 1685, the Jews of New York were granted the right to participate in wholesale trade. They became major businessmen who traded with both their co-religionists and the general community, and traded locally, with England, and with the West Indies. Most importantly, they traded in cocoa and chocolate and used chocolate themselves. In their study of New York commerce from 1701 through 1709, Bloch and co-workers found that 12.3 percent of the customs records involved Jewish merchants, 60 percent of the trade went to

Barbados, and these cargoes were mainly cocoa, rum, wine, fur, and fabrics. Isaac Marquez, Moses Levy, and Joseph Nunes were especially active in these transactions [78].

Nathan Simson was one of the earliest and most prominent Colonial Jewish merchants. While the exact date of his arrival in North America is unknown, it most likely was between 1701 and 1704 [79]. Archival documents found in the Nathan Simson Collection, Public Records Office, Kew, England, provide a sense of the Jewish involvement in the early Colonial chocolate trade. How prominent was Simson? Between 1715 and 1722, he was the major Curaçao trader in New York. During those years, "he organized twenty-three voyages to [Curaçao], the equivalent of 13% of all voyages from Manhattan in that period" [80]. Between February 22 and June 1, 1722, a total of 150,000 pounds of cocoa (sold at 13.5 pesos hundredweight) arrived in Curaçao from the Venezuela coast on his behalf [81, 82]. This three-month period was especially notable because, as Abraham Ulloa and Isaac Levy Maduro explained in their letters to Simson, the governor of Caracas was creating trade difficulties for these merchants [83, 84].

Who was Nathan Simson? He seems to have appeared in New York in the early 1700s; he returned to London in 1722 and died there in 1725 [85]. He was granted letters patent in England in 1713 that were recorded years later in New York in 1728 [86]. He was elected constable of the South Ward, New York in 1718 [87]. His correspondence in the Public Record Office collection appears in multiple languages including Dutch, Hebrew, Portuguese, and Yiddish [88]. Records indicate that he participated in international trade in many ports, including Amsterdam, Barbados, Curaçao, Jamaica, and London; he also had kinsmen in many places including Amsterdam, Bonn, and London [89]. In addition to his successes as an international merchant, he was a man who cared about his friends. It is clear from business correspondence that these correspondents took care to personally address each other with flowery salutations before discussing business matters.

How involved in the cocoa trade were other Sephardic Jews? On May 10, 1711, Moses Levy wrote to Samuel Levy that Gomez "can Draw noe [sic] more being he can have noe [sic] Cocoa" [90]. On May 6, 1715, Benjamin Pereira wrote to Nathan Simson about "foure [sic] Pipes of Cocoa" [91]. On May 27, 1719, a single ledger page shows Simson was conducting cocoa dealings with Isaac Levy, Samuel Levy, Meir Wagg, and Jacob Keiser (of London) [92]. On September 24, 1719, Samuel Levy wrote to Isaac Levy (in London) about cocoa trade with Moros Mitchells, Aron Cohen, and Benjamin Lo(?)os [93]. Aaron Louzada entered the picture in 1721 when he was noted for manufacturing chocolate for Simson [94]. On May 12, 1725, Judah Baruch was trading cocoa with Simson [95]. In sum,

the Simson letters reveal many Jewish merchants and many cocoa transactions.

Rodrigo Pacheco's activities provide more evidence of the bidirectional nature of the trading routes and relationships. On February 11 and 14, 1732, James Alexander in Curaçao shipped cocoa to Rodrigo Pacheco in New York [96]. In 1734, a sloop of Pacheco's carried a cargo to Curaçao, where it was loaded with cocoa and lime to be exchanged for rice in Charleston and then sailed to Falmouth [97]. Many trades, many destinations.

Trade among these co-religionists and relatives was especially valuable. By trading with family and friends, merchants could be sure about the other party's reputation and trust that bad practices would be resolved [98]. These connections included all of the West Indies, North America, England, Holland, and those countries that were home to the Jews expelled from the Iberian Peninsula. "These connections were not generally a prearranged geographical or formal economic organization but rather a loose bond of cooperative merchants ... As economic centers shifted, so did Jewish merchants, creating an informal, self-energizing fraternity usually willing to extend assistance to one another" [99]. A shipping receipt that describes "two pipes and four whole bar of cocoa for my proper account and risqué and goes consigned to Mr. Joseph Levy merchant in London," on behalf of Moses Levy, is just one example of intrafamily trade [100].

Hasia Diner describes the following example of merchant family ties.

> In the late seventeenth century . . . Louis Moses Gomez, born in Madrid about 1655, emigrated by way of France to New York. There he married Esther Marques, a New Yorker with family ties to Barbados. . . . The sons of Louis Moses and Esther eventually married Rebecca Torres, daughter and sister of Jamaican merchants; the Curaçaoan Esther Levy; and Esther Nunes of Barbados. . . . Daniel's [a son] business associates also included one Miguel Gomez, of the Spanish island of Madeira, perhaps also a relative. [101]

Through their private correspondence with family members and friends, Jewish merchants often were able to learn of market fluctuations before others. Richard Janeway, among others, frequently apprised Nathan Simson of prices. Janeway wrote that "sugar will fall, cocoa will rise," then he predicted that the price of cocoa would fall.

As late as 1772, a rabbi preached in Spanish to the Newport congregation, more evidence of the influence of the Sephardim, especially Aaron Lopez, recently arrived from Spain and the wealthiest Jew in the American colonies [102].

In addition to being loyal kinsmen and excellent businessmen, the chocolate traders and merchants were part of the larger community of Jews. Frequently, they gave charity (tzedakah—צדקה in Hebrew). It was not uncommon to see an entry, such as for "the poor

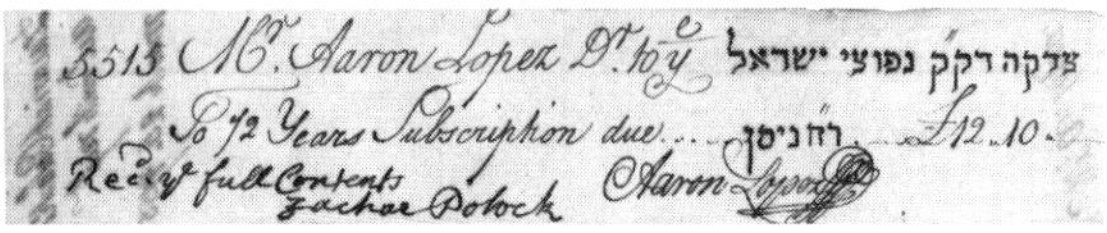
Mr. Aaron Lopez Dr. to y צדקה דקק נפוצי ישראל
To 12 Years Subscription due ... £12.10.
Receiv'd full Contents
Zachar Polock Aaron Lopez

FIGURE 5.2. Tzedakah receipt (pledge of charity to synagogue) signed by Aaron Lopez. *Source*: Aaron Lopez Collection (P-11), Box 12, Miscellaneous. Accounts–Household. Courtesy of the Americana Jewish Historical Society, New York. (Used with permission.)

of the nation" in merchants' account books or for pledges of charity made to the synagogue (Fig. 5.2).

An interesting example of chocolate tzedakah is the story of Judah Abraham. When he suffered financial reverses and left the Newport community for Surinam, the Newport synagogue provided him with

> *an iron cooking pot, 40 pounds of kosher beef, 28 pounds of bread, one half pound of tea, six pounds of sugar, and two pounds of chocolate*. [103]

Leo Hershkowitz, writing in *American Jewish History,* provided key economic information for 13 Colonial New York Jews. These inventories "list all property owned by the deceased, including personal possessions, real estate, cash on hand, debts owed to others, and debts owed to the estate." Data derived primarily from these inventories (Table 5.1) show chocolate-related items in the estates of Joseph Tores Nunes, Joseph Bueno de Mesquita (known as Joseph Bueno, the same Bueno who unloaded English-bound cocoa in New York while he waited for a better price), Isaac Pinheiro, Abraham de Lucena, Isaac Levy, and Mordecai Gomez. These accounts provide additional evidence of the important role that Jewish businessmen played in the Colonial chocolate business [104, 105].

Data presented in Table 5.1 also reveal that the executors and assessors were either Jewish or well-known local personalities. The currency abbreviations of the day included: "£" for English pounds, "s" for Dutch *stivers,* a coin commonly used within the Caribbean and coastal South America, and "f" for Dutch florins (alternatively called guilders). The Isaac Levy inventory was taken on September 18, 1745, but was not recorded until 1748. By 1751, Daniel Gomez, who was an overseer of the accounting for Modecai Gomez's estate, "did business with New York, Boston, New Haven, New London, Norwalk, Newport, Pennsylvania, Amboy, Raritan, Princeton, South Carolina, and Maryland. He had business connections with Madeira, Barbados, Curaçao, London, Dublin, and Liverpool" [106].

How did it happen that these Jews of New York became traders? When the Dutch West Indies Company granted the Jews the right to settle in New Amsterdam, the Jews did not receive the right to participate in the retail trade. Under English administration, the exclusion continued. Thus, being unable to "sell," the Jews became importers. When the Jews were expelled from Brazil and dispersed among the West Indies islands, the trading network was established. Although the distance between the West Indies and North America was much less than that between New York and Europe, the island crops were different enough to be valuable [107]. Cocoa was one of these trade-worthy crops.

One of the interesting anecdotes about the relationships between and among the Sephardic Jews in the chocolate trade in New York dates to 1722. Moses Levy sued Nathan Simson and Jacob Franks, executors of Samuel Levy's estate. Moses Levy claimed that Samuel Levy sold him spoiled cocoa—cocoa that had salt thrown into it [108]. This is especially interesting because Samuel Levy had been described as "the honestest [*sic*] Jew . . . and a man of the most Easy Temper" [109]. Equally interesting is the fact that Samuel Levy was accused by his own brother. James Alexander, the attorney who represented the estate of Samuel Levy, wrote "probably that other Jews in London knew the trick of putting salt water into cocoa in order to get allowance for damage" [110]. Perhaps this technique explains how the cocoa in the Bueno (de Mesquita) estate came to be spoiled.

At the same time that the chocolate trade in New York City was so vibrant, other men in other areas were involved in smaller ways. In Annapolis, Maryland, Isaac Navarro placed an advertisement in the November 8, 1748, *Maryland Gazette* that read:

> *Notice is hereby given, that at the House of John Campbell, Taylor, in the City of Annapolis, the Subscriber makes and sells as good Chocolate as was every made in England, at 4s. 6d. per Pound*. [111]

Uriah Hendricks, a Dutch Jew and the forebear of the Colonial Jewish family that became so important to the copper trade, came to North America in 1755. He opened a dry goods store in lower Manhattan, where he traded chocolate [112]. Samuel Jacobs had a small store in St. Denis, Quebec. In 1764, his store inventory was 3.7 percent foodstuffs, including rice, cheese, and chocolate [113]. Abraham Wagg was in the grocery and chocolate business in New York in 1770, and Michael Gratz used the Globe Mill in Philadelphia to make mustard and chocolate in 1773 [114, 115].

A Jewish woman appears in the chocolate business around this time. On December 2, 1780, Rebecca Gomez (née de Lucena) placed the following advertisement in New York City's *Royal Gazette*: "Has for sale at the Chocolate Manufactory . . . Superfine warranted Chocolate, wholesale and retail" [116] (Fig. 5.3)."

Table 5.1 Chocolate Inventories: Prominent New York Jewish Households

Name	Date of Birth	Date of Death	Date of Inventory	Assessors and Executors	Chocolate-Related Entries
Joseph Tores Nunes[a]	Unknown	10/2/1704	10/8/1700	Lewis Gomes	An old Coat & small parcel cocoa, [value] 3s;
Joseph Bueno de Mesquita (Joseph Bueno)[a]	1611	11/1/1708	11/12/1708	Luis Gomes, Abraham de Lucena, Rachel (Dovale) Bueno	25 Baggs of Cocoa wt. 37 lb. [a bag at 4.3?] [value] £185; 1 Chocolate Stone to Justus Buss; [value] 11s; 1 Chocolate Stone & Router to Giles Shelly, [value] 9s 6f; No. 11 Bag cocoa 1-2-21 wt to Ab. De Lucena, 5 lbs [missing], [value] £8 8s 9f; No. 21 [ditto] 1-3-9 to [ditto] 4 lbs. 16, [value] £8 15s 8f; No. 31 [ditto] 1-2-3 to Moses Levy but was returned on March 28, 1709; 22 Bags [ditto] 18-2-27 to Ab. De Lucena, 5 lb., [value] £93 14s 2f; 10 Bagg [ditto] 15-3-6 [ditto] 3 lb. 19, [value] £62 8s 6f; 1 Bag damaged Cocoa returned by Moses Levy after sold 1-1-11 to Justus Buss @12/1 d, 16s; No. 1 Bag Cocoa that was formerly sold to Mr. Levey 29 lb.[?] [value] 6s; 1710 (payment to) Ab. De Peyster for attachment of cocoa, [value] 9s 15f.
Isaac Pinheiro[a]	1636	2/17/1710	2/22/1710	Will Anderson Wm Chambers	$\frac{1}{2}$ Bar[ell] Coca wt 24 lbs, [value] 18s; 3 Course earthen Dishes, 3 Bassins, 10 plates, 2 porringers, one of y[he]m broke, 1 Mugg, 1 Mustard Pot, 10 Chocolate Cups of wh[i]ch 3 are broken & 1 Cann, [value] 14s; 1 Pr. Money Scales & about 6 lb. of Chocolate, [value] 15s 9f.
Abraham de Lucena[b]	Unknown	8/4/1725	9/22/1725	Robt. Lustig, Fra. Harison	1 Brass Candlestick, 1 Gridiron, 1 pr. Tongs, 1 Copper chocalet Pott, [value] 6s 6f.
Isaac Levy[b]	Unknown	8/27/1745	9/18/1745	Not identified	4 Duss Shokelate [Knifs] @6/, [value] £1 4s; 1 Duss Shokelate knifes @7/6d, [value] 7s 6f; $4\frac{1}{2}$ lbs. of Shokelate @2/, [value] 9s.
Mordecai Gomez[b,c]	1688	11/1/1750	11/12/1750	Daniel Gomez, David Gomez, Isaac Gomez, Benjamin Gomez, Isaac Gomez, Jr.	1 Chocolate [pott], no value given; 16 Chocolate Cupps, [broken & whole], no value given; 2 boxes Chocolate—50 lbs. Each, no value given; 6 Surunis Coco [Surinam], no value given.

Sources:
[a]Hershkowitz, L. Original inventories of early New York Jews (1682–1763). *American Jewish History* 2002;90(3)239–321.
[b]Hershkowitz, L. Original inventories of early New York Jews (1682–1763) (Concluded). *American Jewish History* 2002;90(4)385–448.
[c]Gomez Family (P-62), Box 1, Inventory of Estate, American Jewish Historical Society, New York.

Howard Sachar writes in *A History of the Jews in America* that "By mid-century, Jews accounted for possibly 15 percent of the colonies' import–export firms, dealing largely in cocoa, run, wine, fur, and textiles" [117]. That trade had been important in New Amsterdam/New York since the early 1700s; by the later 1700s, it also was important in Newport, Rhode Island.

NEWPORT, RHODE ISLAND

It is not clear from where or when the first Jews arrived in Newport, Rhode Island. Some early Jewish families arrived from Holland in 1658 [118]; other accounts report that some of the Jews of Newport came from Jamaica in 1658, and still others have written that the

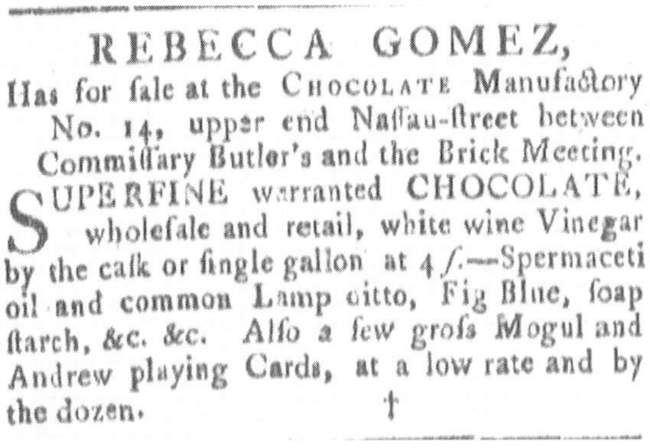

REBECCA GOMEZ,

Has for fale at the CHOCOLATE Manufactory No. 14, upper end Naffau-ftreet between Commiffary Butler's and the Brick Meeting.

SUPERFINE warranted CHOCOLATE, wholefale and retail, white wine Vinegar by the cafk or fingle gallon at 4 f.—Spermaceti oil and common Lamp ditto, Fig Blue, foap ftarch, &c. &c. Alfo a few grofs Mogul and Andrew playing Cards, at a low rate and by the dozen.   †

FIGURE 5.3. Chocolate advertisement by Rebecca Gomez. *Source: Royal Gazette,* February 12, 1780, Issue 436, p. 3. Courtesy of Readex/Newsbank, Naples, FL. (Used with permission.)

Newport Jews emigrated from Barbados in 1678 [119]. What is certain, however, is that "on August 24th, 1694, a ship arrived at Newport . . . from one of the West India islands, with a number of Jewish families of wealth and respectability on board" [120].

Joseph Bueno was the first Newport Jewish trader to buy large amounts of Curaçao cocoa. This is the same Joseph Bueno who received a license to trade and traffic within the City of New York on February 8, 1683 [121]. A November 19, 1703 document in the Rhode Island State Archives reveals the following:

Mr Joseph Bueno . . . having a bond of yours in my hand . . . for three thousand three hundred sixty . . . of . . . Cocoa . . . Corraso payable at Corraso. [122]

On April 28, 1704, the records show that Bueno purchased "twenty four thousand waight [sic] of Coko" [123]. Also in 1704, Bueno was accused of trying to keep cocoa in New York instead of sending it on to London because the market in London was depressed [124]. Despite this early trading, however, the chocolate business in Newport did not develop and expand until the arrival of Aaron Lopez.

Aaron Lopez was born Duarte Lopez in 1731 in Portugal, where his parents were members of the Converso community. As a young adult, he was interested in practicing Judaism and he immigrated directly to North America, arriving in Newport in October 1752 [125]. Aaron Lopez was Colonial Newport's most important merchant—without regard to religion—and also was one of Colonial Jewry's most important citizens. On October 15, 1762, he became the first Jew to be naturalized in America, an event that took place in Massachusetts where Lopez had a summer home [126].

Aaron Lopez was widely respected. As Charles P. Daly, Chief Justice, Court of Common Pleas of New York, wrote in his 1853 monographs, *The Settlement of the Jews in North America:* "At the breaking out of the American Revolution he [Lopez] was himself the owner of thirty vessels engaged in European and West India trade and the whale fisheries, and was then and for some years previously looked upon as the most eminent and successful merchant in New England" [127]. At Lopez's death, Ezra Stiles, the President of Yale, composed his epitaph.

He was a merchant of eminence, of polite and amiable manners. Hospitality, liberality, and benevolence were his true characteristics. An ornament and valuable pillar to the Jewish society, of which he was a member. His knowledge in commerce was unbounded and his integrity irreproachable; thus he lived and died, much regretted, esteemed and loved by all. [128]

Not all of Lopez's business, however, was with his co-religionists in the West Indies, New York, and Europe. As an example, in December 1760, Joseph and William Rotch of Nantucket reported to Lopez about the chocolate, molasses, and rum they sold on his account [129]. Most interestingly, in Lopez's Newport ledger books from 1766 through 1769, there are many instances of his doing business with a Negro named Prince Updike. Lopez delivered raw cocoa to Updike and Updike returned ground chocolate for which he received five shillings for every pound prepared [130, 131]. Between 1766 and 1767, Updike produced 2000 pounds of chocolate from 2500 pounds of cocoa, and between 1768 and 1769, Updike produced 4000 pounds of ground chocolate from 5000 pounds of cocoa (Fig. 5.4) [132].

As Frances Fitzgerald wrote in *The New Yorker:* "Slaves were a part of the fabric of life in pre-Revolutionary New England, and in the 1770s Rhode Island had more slaves per capita than any other New England colony. According to the 1774 [Rhode Island] census, 3,761 people—6.3 per cent of Rhode Island's population—were 'Negroes,' and it can be assumed that most of them were enslaved" [133]. Lopez's records contain transactions with many Negro customers, both free and slave, and the status of Prince Updike between 1766 and 1769 is not clear.

Lopez conducted his business in English, French, Portuguese, and Spanish, with additional use of Hebrew and Yiddish. His Sephardic roots stayed with him. The Colonial Jews continued observing their religious practices—many remained kosher. The Revolution brought food shortages for everyone; especially kosher Jews. In 1779, Lopez wrote to Joseph Anthony that the Jews were "forced to subsist on chocolate and coffee" [134]. The Revolution also cost Lopez his fortune: he died insolvent because of his support of the American cause [135]. Aaron Lopez had been an innovative man: he came straight to America without stopping in England, he was the patriarch of a large and successful family, he was prosperous to the point of being one of the richest men in Newport, and he was a major importer of cocoa and producer of chocolate.

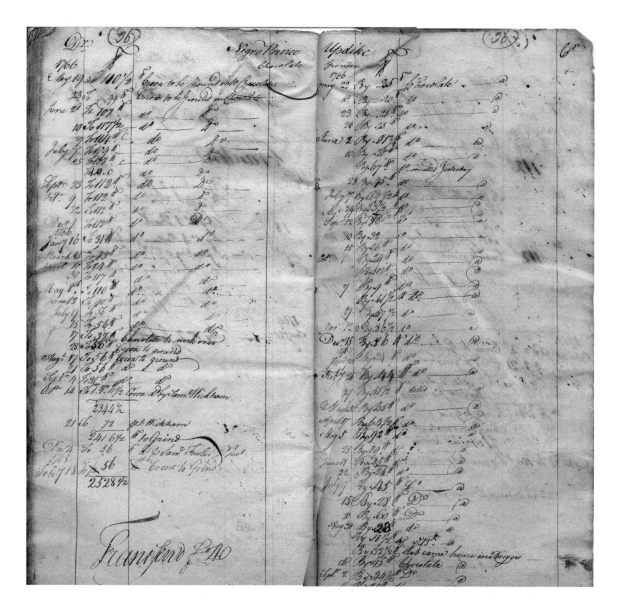

FIGURE 5.4. Aaron Lopez's ledger Negro Prince Updike for the year 1766. Notation for Updike grinding cacao. *Source*: Aaron Lopez Collection, Lopez Account Book 715, p. 37, left. Courtesy of the Newport Historical Society, Newport, RI. (Used with permission.)

Conclusion

How does the story of the Jewish involvement in chocolate play out? The relationship continued, but the intensity decreased. Levy Solomons was one of the first to manufacture chocolate in Albany. On March 29, 1829, he extended his lease at Rennslaerville, where he manufactured chocolate, for one year at $200 per annum [136]. Daniel L. M. Peixotto, a medical doctor, published a circular on August 1, 1832, in New York City, that suggested that Jews not observe the Tisha B'av fast during cholera season. He wrote that instead of fasting, "a slight meal, say of coffee, tea, or cocoa, with dry toast, be allowed" [137]. In 1910, *Haibri*, a Hebrew periodical founded in Berlin that circulated in America, carried advertisements for Suchard chocolate

[138]. In 1931, Carl M. Loeb & Company acquired a seat on the New York Stock Exchange and provided merchant banking for houses trading in rubber, hides, and cocoa [139]. In 1941, chocolate matzo bars were introduced for Passover, and chocolate *alef-bet* bars and boxes were introduced for Hanukkah [140].

The number of Sephardic Jews in North America—the Jews who originally were interwoven with the story and business of chocolate in Colonial America—has grown proportionally smaller, but their story remains rich. Today, there are many Jews involved in the chocolate life cycle, everything from sustainable agriculture through distribution of a wide variety of products.

History is not just names and dates; it is stories. Strife and poetry: that is the story of the Jewish role in the history of chocolate in Colonial America. It also is the story of the Spanish Inquisition, which was abolished only on July 15, 1834, by Queen Mother Maria Christina. The Spanish government, however, did not declare the expulsion order void until December 17, 1968—476 years after Queen Isabella and King Ferdinand ordered the Jews expelled. It is impossible not to wonder how the history of chocolate in Colonial America was affected by that long-ago order.

Acknowledgments

I wish to acknowledge the following individuals and institutions for their help and assistance during the research for this chapter: Philip Dikland, Jodensavanne Foundation, Suriname; Sharon Horowitz, African and Middle Eastern Reading Room, Library of Congress; Lois Kampinsky, independent film producer; Joann Lesser, Adas Israel Synagogue Library; Bertram Lippincott III, Newport Historical Society; Jane Mandelbaum, Library of Congress; Jennifer Manuel, American Jewish Archives; Irwin J. Miller, The Jewish Historical Society of Lower Fairfield County, Connecticut; Dan Pine, *The Jewish News Weekly of Northern California*, San Francisco, California; Pete Schlosser, Cocoa Pete's Chocolate Adventures; Deena Schwimmer, American Jewish Historical Society; Amy F. Stempler, Edward Kiev Judaica Collection, Gelman Library, The George Washington University; and Mark Wizniter.

References

1. Feinman, B. *The Jews Who Sailed with Columbus*. Available at http://www.jewishvirtuallibrary.org/jsource/Judaism/columbus.html. (Accessed August 31, 2006.)

2. Emmanuel, I. S. *The Jews of Coro, Venezuela*. Cincinnati, OH: American Jewish Archives, Hebrew Union College–Jewish Institute of Religion, 1973.

3. Feingold, H. L. *Zion In America: The Jewish Experience from Colonial Times to the Present*. Mineola, NY: Dover Publications, 1974.

4. Kayserling, M. *Christopher Columbus and the Participation of Jews in the Spanish and Portuguese Discoveries*. Translated by Charles Gross. New York: Longmans, Green, 1894.

5. Feingold, H. L. *Zion in America: The Jewish Experience from Colonial Times to the Present*. Mineola, NY: Dover Publications, 1974.

6. Kayserling, M. *Christopher Columbus and the Participation of Jews in the Spanish and Portuguese Discoveries*. Translated by Charles Gross. New York: Longmans, Green, 1894.

7. Rosenblum, M. *Chocolate: A Bittersweet Saga of Dark and Light*. New York: Northpoint Press, 2005.

8. Terrio, S. J. *Crafting the Culture and History of French Chocolate*. Berkeley: University of California Press, 2000.

9. Ezratty, H. A. *500 Years in the Jewish Caribbean: The Spanish and Portuguese Jews in the West Indies*. Baltimore, MD: Omni Arts, 1977.

10. Dubiez, F. J. *De Portugees Israëlitische Gemeente te te Amsterdam*. Amsterdam, Holland: Privately published, undated.

11. Ezratty, H. A. *500 Years in the Jewish Caribbean: The Spanish and Portuguese Jews in the West Indies*. Baltimore, MD: Omni Arts, 1977.

12. Wiznitzer, A. The members of the Brazilian Jewish community (1648–1653). *Publications of the American Jewish Historical Society* 1952;42:387–397.

13. Kohut, A. Jewish martyrs of the Inquisition in South America. *Publications of the American Jewish Historical Society* 1896;4:101–187.

14. American Jewish Historical Society. *American Jewish Desk Reference: The Ultimate One-Volume Reference to the Jewish Experience in America*. New York: Random House, 1999.

15. Gilman, S. *Jewish Frontiers: Essays on Bodies, Histories, and Identities*. New York: Palgrave Macmillan, 2003.

16. Becker, M. *Jews in Suriname*. Chapter 4 Historical Timeline. Available at http://www.angelfire.com/mb2/jodensavanne/. (Accessed September 11, 2006.)

17. Weiner, R. *The Virtual Jewish History Tour: Vienna*. Available at http://www.jewishvirtuallibrary.org/jsource/vjw/Vienna.html. (Accessed September 11, 2006.)

18. Terrio, S. J. *Crafting the Culture and History of French Chocolate*. Berkeley: University of California Press, 2000.

19. Arbell, M. *The Jewish Nation of the Caribbean: The Spanish–Portuguese Jewish Settlements in the Caribbean and the Guianas*. Jerusalem, Israel: Gefen Publishing House, 2002.

20. Adler, C. Original unpublished documents relating to Thomas Tremino de Sobremonte (1638). *Publications of the American Jewish Historical Society* 1909;17:27–31 (Inquisition Folio 270).

21. *Dominance of Ottoman Muslim Empire in Turkey: 1500–1920*. Available at http://www.jewishvirtuallibrary.org/jsource/History/ottotime.html. (Accessed September 12, 2006.)

22. Anonymous. Epoca Colonial; Mexico Viejo, Noticias Históricas, tradiciones, leyendas y costumbres. Cited in: Process [or trial] of Gabriel de Granada (13 years old), observer of the Law of Moses, 1642 to 1645. *Publications of the American Jewish Historical Society* 1899;7:1–127.

23. Ferry, R. J. *Don't Drink the Chocolate: Domestic Slavery and the Exigencies of Fasting for Crypto-Jews in Seventeenth Century Mexico*. Available at http://nuevomundo.revues.org/document934.html. (Accessed August 31, 2006.)

24. del Olmo, J. *Relacion historia del auto general de fe que ce celebro en Madrid en 30 de junio de 1680.* Madrid: Rocque rico de Miranco, 1680.

25. Liss, D. *The Coffee Trader: A Novel.* New York: Random House, 2003.

26. Bodian, M. *Hebrews of the Portuguese Nation: Conversos and Community in Early Modern Amsterdam.* Bloomington: Indiana University Press, 1997.

27. Bloom, H. I. *The Economic Activities of the Jews of Amsterdam in the Seventeenth and Eighteenth Centuries.* Port Washington, NY: Kennikat Press, 1937.

28. Fortune, S. A. *Merchants and Jews: The Struggle for British West Indian Commerce, 1650–1750.* Gainesville: University Press of Florida, 1984.

29. Adler, C., and Brunner, A. W. *America.* Available at http://www.jewishencyclopedia.com/view.jsp?artid=1384&letter=A&search=paramaribo. (Accessed September 14, 2006.)

30. Documents Relating to Colonial History of the State of New York, Volume 14, p. 135. Cited by Cone, G. H. The Jews in Curacao: According to Documents from the Archives of the State of New York. *Publications of the American Jewish Historical Society* 1902;10:141–157.

31. Egerton Manuscript. British Museum, Volume 2395, f. 46. Cited by Oppenheim S. An Early Jewish Colony in Western Guiana, 1658–1666: and its Relation to the Jews in Surinam, Cayenne, and Tobago. *Publications of the American Jewish Historical Society* 1907;17:95–186.

32. Meijer, J. Pioneers of Pauroma (Pomeroon), Paramaribo 1954, Based on the Compilation by R. Bjilsma. *Archief der Nederlandische Portageesh-Israelitische Gemeente in Suriname* pp. 23–24.

33. Arbell, M. *The Jewish Nation of the Caribbean: The Spanish–Portuguese Jewish Settlements in the Caribbean and the Guianas.* Jerusalem, Israel: Gefen Publishing House, 2002.

34. Adler, C., and Kohut, G. A. *Cayenne.* Available at http://www.jewishencyclopedia.com/view.jsp?artid=285&letter=C. (Accessed September 14, 2006.)

35. Arbell, M. *The Jewish Nation of the Caribbean: The Spanish–Portuguese Jewish Settlements in the Caribbean and the Guianas.* Jerusalem, Israel: Gefen Publishing House, 2002.

36. Diderot D. *Encyclopédie, ou, Dictionnaire raisonné des sciences, des arts et des métiers, par une Société de Gens de lettres.* Paris, France: André Le Breton, 1751.

37. Ezratty, H. A. *500 Years in the Jewish Caribbean: The Spanish and Portuguese. Jews in the West Indies.* Baltimore, MD: Omni Arts, 1977.

38. Diderot D. *Encyclopédie, ou, Dictionnaire raisonné des sciences, des arts et des métiers, par une Société de Gens de lettres.* Paris, France: André Le Breton, 1751.

39. Labat Pere, J.-B. *Nouveau voyage aux isle Françoise de l'Amérique.* Paris, France: G. Cavalier, 1722.

40. Birmingham, S. *The Grandees. America's Sephardic Elite.* Syracuse, NY: Syracuse University Press, 1971.

41. Bloom, H. I. *The Economic Activities of the Jews of Amsterdam in the Seventeenth and Eighteenth Centuries.* Port Washington, NY: Kennikat Press, 1937.

42. Emmanuel, I. S., and Emmanuel, S. A. *History of the Jews of the Netherlands Antilles.* Cincinnati, OH: American Jewish Archives, 1970.

43. Emmanuel, I. S. *The Jews of Coro, Venezuela.* Cincinnati, OH: American Jewish Archives, Hebrew Union College–Jewish Institute of Religion, 1973.

44. Fortune, S. A. *Merchants and Jews: The Struggle for British West Indian Commerce, 1650–1750.* Gainesville: University Press of Florida, 1984.

45. Corcos, J. A *Synopsis of the History of the Jews of Curaçao.* Curaçao: Imprenta de la Liberia, 1897.

46. Archivo General de Indias (AGI). Letter: Montero to the King, 9 April 1711 (Santo Domingo), Document 697.

47. Archivo General de Indias (AGI), Portales to the King, 29 January 1722 (Santo Domingo), Document 759.

48. Arbell, M. 2002. *Portuguese Jews—Pioneers of Cocoa and Vanilla Production in South America and the Caribbean: Sixteenth to Eighteenth Centuries.* Available at http://www.sefarad.org/publication/lm/046/7.html. (Accessed August 31, 2006.)

49. Becker, M. 2006. *Jews in Suriname.* Chapter 4 Historical Timeline. Available at http://www.angelfire.com/mb2/jodensavanne/. (Accessed September 11, 2006.)

50. *Letter of Abraham Beekman, Commandeur at Essequibo, to the West India Company.* Cited by Oppenheim S. An early Jewish colony in Western Guiana, 1658–1666: and its relation to the Jews in Surinam, Cayenne, and Tobago. *Publications of the American Jewish Historical Society* 1907;17:95–186.

51. Calendar of British State Papers, Colonial 1669–1674, No. 570, pp. 298–300. Cited by Friedenwald, H. Material for the history of the Jews in the British West Indies. *Publications of the American Jewish Historical Society* 1897;5:45–106.

52. Calendar of British State Papers, CO 116/19, Nos. 4, 4i–xii, Aug 9/20, 1700; Fort Kyck Overal, Rio Issquibe. Samuel Beekman to Directors of the Dutch West Indies Company, Middleburgh. (Original letter in obscure Dutch; translated in the Calendar of State Papers.)

53. Arbell, M. *The Jewish Nation of the Caribbean: The Spanish–Portuguese Jewish Settlements in the Caribbean and the Guianas.* Jerusalem, Israel: Gefen Publishing House, 2002.

54. Arbell, M. 2002. *Portuguese Jews—Pioneers of Cocoa and Vanilla Production in South America and the Caribbean: Sixteenth to Eighteenth Centuries.* Available at http://www.sefarad.org/publication/lm/046/7.html. (Accessed August 31, 2006.)

55. Friedman, L. M. *Jewish Pioneers and Patriots*. Philadelphia: Jewish Publication Society, 1948.

56. Emmanuel, I. S., and Emmanuel, S. A. *History of the Jews of the Netherlands Antilles*. Cincinnati, OH: American Jewish Archives, 1970.

57. Friedman, L. M. *Jewish Pioneers and Patriots*. Philadelphia: Jewish Publication Society, 1948.

58. Emmanuel, I. S., and Emmanuel, S. A. *History of the Jews of the Netherlands Antilles*. Cincinnati, OH: American Jewish Archives, 1970.

59. Marcus, J. R. *The Colonial American Jew, 1492–1776*. Detroit, MI: Wayne State University Press, 1970.

60. Arbell, M. *The Jewish Nation of the Caribbean: The Spanish–Portuguese Jewish Settlements in the Caribbean and the Guianas*. Jerusalem, Israel: Gefen Publishing House, 2002.

61. Becker M. *Jews in Suriname*. Chapter 4 Historical Timeline. Available at http://www.angelfire.com/mb2/jodensavanne/. (Accessed September 11, 2006.)

62. Emmanuel, I. S., and Emmanuel, S. A. *History of the Jews of the Netherlands Antilles*. Cincinnati, OH: American Jewish Archives, 1970.

63. Judah, G. F. The Jews' tribute in Jamaica. Extracted from the Journals of the House of Assembly of Jamaica. *Publications of the American Jewish Historical Society* 1909;18:149–177.

64. Jamaica Jewish Community Collection (1–82), Box 1, Document Relating to Jews in Jamaica, American Jewish Historical Society, New York.

65. Navarro, L. D. Informes sobre la Provincia de Costa Rica, presentado al Capitan General de Guatemala en 1744. *Revista de Archivos Nacionales (Costa Rica)* 1939;3(11–12):583.

66. Long, E. *The History of Jamaica*. New York: Arno Press, 1972.

67. Dikland, P. Jodensavanne Foundation (Suriname). Scheepsreizen 1729–1794. Personal communication, dated June 6, 2005.

68. Becker, M. *Barter to Production*. Available at http://www.angelfire.com/mb2/jodensavanne/. (Accessed September 19, 2006.)

69. Marcus, J. R., and Chyet, S. F. *Historical Essay on the Colony of Surinam, 1788*. Translated by Simon Cohen. Cincinnati, OH: American Jewish Archives, 1974.

70. Arbell, M. *The Jewish Nation of the Caribbean: The Spanish–Portuguese Jewish Settlements in the Caribbean and the Guianas*. Jerusalem, Israel: Gefen Publishing House, 2002.

71. Wiznitzer, A. The exodus from Brazil and arrival in New Amsterdam of the Jewish Pilgrim fathers, 1654. *Publications of the American Jewish Historical Society* 1954;44:80–99.

72. Birmingham, S. *The Grandees. America's Sephardic Elite*. Syracuse, NY: Syracuse University Press, 1971.

73. Feingold, H. L. *Zion in American: The Jewish Experience from Colonial Times to the Present*. Mineola, NY: Dover Publications, 1974.

74. Documents Related to the Colonial History of New York. Vol. XIV, p. 315. Cited in Daly, C. P. *The Settlement of the Jews in North America*. New York: P. Cowen, 1893.

75. Peterson, E. *History of Rhode Island*. New York: J. S. Taylor, 1853.

76. Hershkowitz, L. Some aspects of the New York Jewish merchant and community, 1654–1820. *American Jewish Historical Quarterly* 1976;66:10–34.

77. Hershkowitz, L. Original inventories of early New York Jews (1682–1763) (Concluded). *American Jewish History* 2002;90(4):385–448.

78. Bloch, J. M. *An Account of Her Majesty's Revenue in the Province of New York, 1701–09: The Custom Records of Early Colonial New York*. Ridgewood, NJ: Gregg Press, 1966.

79. Ben-Jacob, M. Nathan Simson: a biographical sketch of a Colonial Jewish merchant. *American Jewish Archives Journal* 1999;51(1–2):11–37.

80. Klooster, W. *Illicit Riches: Dutch Trade in the Caribbean, 1648–1795*. Leiden, the Netherlands: KITLV Press, 1998.

81. Emmanuel, I. S. *The Jews of Coro, Venezuela*. Cincinnati, OH: American Jewish Archives, Hebrew Union College–Jewish Institute of Religion, 1973.

82. Nathan Simson. Shipper. The National Archives, Kew, England. Record Reference: C104/13-14. GB/NNAF/B22778. Microfilm copies at Nathan Simson, Correspondence, bills of lading, receipts and ledgers 1710–1725. American Jewish Archive, Cincinnati, OH, Reels 528, 528A, and 528B.

83. Emmanuel, I. S., and Emmanuel, S. A. *History of the Jews of the Netherlands Antilles*. Cincinnati, OH: American Jewish Archives, 1970.

84. Nathan Simson. Shipper. The National Archives, Kew, England. Record Reference: C104/13-14. GB/NNAF/B22778. Microfilm copies at Nathan Simson, Correspondence, bills of lading, receipts and ledgers 1710–1725. American Jewish Archive, Cincinnati, OH, Reels 528, 528A, and 528B.

85. Marcus, J. R. *Studies in American Jewish History: Studies and Addresses by Jacob R. Marcus*. Cincinnati, OH: Hebrew Union College Press, 1969.

86. Hershkowitz, L. Some aspects of the New York Jewish merchant and community, 1654–1820. *American Jewish Historical Quarterly* 1976;66:10–34.

87. Liber 36 of Conveyances, p. 50. Register's Office, New York County. Cited in Oppenheim S. Will of Nathan Simson, a Jewish merchant in New York before 1722, and genealogical note concerning him and Joseph Simson. *Publications of the American Jewish Historical Society* 1917;25:87–91.

88. Nathan Simson. Shipper. The National Archives, Kew, England. Record Reference: C104/13-14. GB/NNAF/B22778. Microfilm copies at Nathan Simson, Correspondence, bills of lading, receipts and ledgers 1710–1725. American Jewish Archive, Cincinnati, OH, Reels 528, 528A, and 528B.

89. Ben-Jacob M. Nathan Simson: a biographical sketch of a Colonial Jewish merchant. *American Jewish Archives Journal* 1999;51(1–2):11–37.

90. Nathan Simson. Shipper. The National Archives, Kew, England. Record Reference: C104/13-14. GB/NNAF/B22778. Microfilm copies at Nathan Simson, Correspondence, bills of lading, receipts and ledgers 1710–1725. American Jewish Archive, Cincinnati, OH, Reels 528, 528A, and 528B.

91. Nathan Simson. Shipper. The National Archives, Kew, England. Record Reference: C104/13-14. GB/NNAF/B22778. Microfilm copies at Nathan Simson, Correspondence, bills of lading, receipts and ledgers 1710–1725. American Jewish Archive, Cincinnati, OH, Reels 528, 528A, and 528B.

92. Nathan Simson. Shipper. The National Archives, Kew, England. Record Reference: C104/13-14. GB/NNAF/B22778. Microfilm copies at Nathan Simson, Correspondence, bills of lading, receipts and ledgers 1710–1725. American Jewish Archive, Cincinnati, OH, Reels 528, 528A, and 528B.

93. Nathan Simson. Shipper. The National Archives, Kew, England. Record Reference: C104/13-14. GB/NNAF/B22778. Microfilm copies at Nathan Simson, Correspondence, bills of lading, receipts and ledgers 1710–1725. American Jewish Archive, Cincinnati, OH, Reels 528, 528A, and 528B.

94. Marcus, J. R. *The Colonial American Jew, 1492–1776*. Detroit, MI: Wayne State University Press, 1970.

95. Nathan Simson. Shipper. The National Archives, Kew, England. Record Reference: C104/13-14. GB/NNAF/B22778. Microfilm copies at Nathan Simson, Correspondence, bills of lading, receipts and ledgers 1710–1725. American Jewish Archive, Cincinnati, OH, Reels 528, 528A, and 528B.

96. Papers of James Alexander, New York Historical Society.

97. Naval Office List of Ships at Charleston, SC 1717–1767.

98. Diner, H. R., and Benderly, B. L. *Her Works Praise Her: Jewish Women in America from Colonial Times to the Present*. New York: Basic Books, 2002.

99. Fortune, S. A. *Merchants and Jews: The Struggle for British West Indian Commerce, 1650–1750*. Gainesville: University Press of Florida, 1984.

100. Nathan Simson. Shipper. The National Archives, Kew, England. Record Reference: C104/13-14. GB/NNAF/B22778. Microfilm capies at Nathan Simson, Correspondence, bills of lading, receipts and ledgers 1710–1725. American Jewish Archive, Cincinnati, OH, Reels 528, 528A, and 528B.

101. Diner, H. R. and Benderly, B. L. *Her Works Praise Her: Jewish Women in American from Colonial Times to the Present*. New York: Basic Books, 2002.

102. Ezratty, H. A. *500 Years in the Jewish Caribbean: The Spanish and Portuguese Jews in the West Indies*. Baltimore, MD: Omni Arts, 1977.

103. Marcus, J. R. *The Colonial American Jew, 1492–1776*. Detroit, MI: Wayne State University Press, 1970.

104. Hershkowitz, L. Original inventories of early New York Jews (1682–1763). *American Jewish History* 2002;90(3):239–321.

105. Hershkowitz L. Original inventories of early New York Jews (1682–1763) (Concluded). *American Jewish History* 2002;90(4):385–448.

106. Fortune, S. A. *Merchants and Jews: The Struggle for British West Indian Commerce, 1650–1750*. Gainesville: University Press of Florida, 1984.

107. Kohler, M. Phases of Jewish life in New York before 1800. *Publications of the American Jewish Historical Society* 1894;2:77–100.

108. *Minute Books, Court of Chancery*, Volume 1720–1748, New York County Clerk's Office, Hall of Records.

109. *Minute Books, Court of Chancery*, Volume 1720–1748, New York County Clerk's Office, Hall of Records.

110. Miller, G. J. James Alexander and the Jews, especially Isaac Emmanuel. *Publications of the American Jewish Historical Society* 1939;35:171–188.

111. *Maryland Gazette*, November 8, 1748. p. 4.

112. Nathan, J. *Jewish Cooking in America*. New York: Alfred A. Knopf, 1994.

113. Greer, A. *Peasant, Lord, and Merchant: Rural Society in Three Quebec Parishes 1740–1840*. Toronto, Canada: University of Toronto Press, 1985.

114. Roth, C. A Jewish voice for peace in the War of American Independence. The life and writings of Abraham Wagg, 1719–1803. *Publications of the American Jewish Historical Society* 1928;31:33–75.

115. Roth, C. *Essays and Portraits in Anglo-Jewish History*. Philadelphia, PA: The Jewish Publication Society of America, 1968.

116. *Royal Gazette*, February 12, 1780, Issue 436, p. 3.

117. Sachar, H. *A History of the Jews in America*. New York: Alfred A. Knopf, 1992.

118. Freund, M. K. *Jewish Merchants in Colonial America: Their Achievements and Their Contributions to the Development of America*. New York: Behrman's Jewish Book House, 1939.

119. Oppenheim, S. The first settlement of the Jews in Newport: some new matter on the subject. *Publications of the American Jewish Historical Society* 1937;34:1–10.

120. Kohler, M. The Jews in Newport. *Publications of the American Jewish Historical Society* 1898;2:61–80.

121. O'Callaghan, E. B. *Calendar of Historical Manuscripts in the Office of the Secretary of State, Albany, New York*. Albany, NY: Weed, Parsons, and Company, 1865.

122. Rhode Island Land Records (Public Notary), (1694–1708), Volume 2, 1671–1708, Providence, Rhode Island, p. 208.

123. Rhode Island Land Records (Public Notary), (1694–1708), Volume 2, 1671–1708, Providence, Rhode Island, p. 210.

124. Letter: Governor Edward Cornbury to the Lords of Trade [June 1705?] Doc. Rel., 1144. Peter Force Collection, Library of Congress, Washington, DC.

125. Kohler, M. The Jews in Newport. *Publications of the American Jewish Historical Society* 1898;2:61–80.

126. Birmingham, S. *The Grandees. American's Sephardic Elite*. Syracuse, NY: Syracuse University Press, 1971.

127. Kohler, M. The Jews in Newport. *Publications of the American Jewish Historical Society* 1898;2:61–80.

128. Snyder, H. *Guide to the Papers of Aaron Lopez (1731–1782), 1752–94, 1846, 1852, 1953*. American Jewish Historical Society, Center for Jewish History. Available at http://www.cjh.org/nhprc/AaronLopez.html. (Accessed September 28, 2006.)

129. Commerce of Rhode Island, 1726–1800. CRI 1:66–72.

130. Aaron Lopez ledger books (555 and 715), Newport Historical Society, Newport, RI.

131. Marcus, J. R. *The Colonial American Jew, 1492–1776*. Detroit, MI: Wayne State University Press, 1970.

132. Marcus, J. R. *The Colonial American Jew, 1492–1776*. Detroit, MI: Wayne State University Press, 1970.

133. Fitzgerald, F. Brown University Looks at the Slave Traders in Its Past. *The New Yorker*, September 12, 2005, pp. 68–77.

134. Chyet, S. F. *Lopez of Newport: Colonial American Merchant Prince*. Detroit, MI: Wayne State University Press, 1970.

135. Korn, H. Documents relative to the state of Aaron Lopez. *Publications of the American Jewish Historical Society* 1939;35:139–143.

136. Solomons, L. Sr. Items relating to the Solomons family, New York. *Publications of the American Jewish Historical Society* 1920;27:278–280.

137. Anonymous. Miscellaneous items relating to Jews in New York. *Publications of the American Jewish Historical Society* 1920;27:150–160.

138. Brown, M. G. All, all alone: the Hebrew Press in America from 1914–1922. *American Jewish Historical Quarterly* 1969;59(2):139–178.

139. Carosso, V. P. A financial elite: New York's German–Jewish investment bankers. *American Jewish Historical Quarterly* 1976;66(1):67–88.

140. Duker, A. G. Emerging cultural patterns in American Jewish life: the psycho-cultural approach to the study of Jewish life in America. *Publications of the American Jewish Historical Society* 1949;39:351–388.

PART

II

Medicine and Recipes

Chapter 6 (Grivetti) explores the history and use of chocolate as medicine from early pre-Columbian times in Mesoamerica, through the early Spanish colonial period in Mexico, and ultimately into traditional medicine as practiced in Europe and North America during the 16th to 19th centuries. Chapter 7 (Grivetti) examines the Boston smallpox epidemic of 1764 and how fear gripped city elders and merchants, causing many to flee and locate to Boston suburbs with their chocolate inventory in attempts to avoid the pox. His chapter reveals how doctors and pharmacists of the era used chocolate during treatment of smallpox cases. Chapter 8 (Grivetti) identifies a suite of chocolate recipes that date from pre-Columbian times through early 19th century European and North American culinary traditions. His chapter draws sharp distinctions between ingredients used to manufacture chocolate before and after 1492, to separate indigenous New World recipes from those developed during the colonial era and later. Chapter 9 (Pucciarelli) continues the theme of culinary chocolate and discusses chocolate recipes published in North American cookbooks during the 19th century. Her chapter documents for the first time the importance of these early cookbooks and how recipes and advice contained therein imparted dietary advice and moral values to women of the era.

Medicinal Chocolate in New Spain, Western Europe, and North America

Louis Evan Grivetti

Introduction

The history and scientific investigation of chocolate as medicine is seductive. Researchers are drawn by sirens into dusty archives, along paths that lead to remote traditional villages and markets in the highlands and lowlands of Mesoamerica, ultimately into laboratories where scrapings from the bottoms of ancient Aztec and Mayan pots that once contained chocolate are analyzed. Chocolate and medical-associated themes link ancient peoples and places and draw upon information from the fields of archaeology, art, biochemistry and chemistry, ethnobotany, ethnography, geography, linguistics, and nutrition. Ancient manuscripts reveal the numerous social and cultural uses of chocolate, and variable medicinal uses through the ages, while recent advances in flavonoid and polyphenolic research has revealed the health-promoting properties of chocolate as well [1].

The present chapter considers chocolate-associated medical texts from the pre-Columbian period through the early decades of the 20th century. The documents identified reflect both indigenous and European attitudes toward chocolate as a healing or preventive medicine for a wide range of diseases and infirmities. Some of the information contained in this chapter has been published previously and has been redacted and resummarized here.[1] Additional historical sources on medicinal chocolate identified during archival research conducted since these previous publications have been added and integrated here.

Medicinal Chocolate: Indigenous and Early New World Accounts

Indigenous Mexica/Aztec medical views of cacao-chocolate are recorded in several 16th century documents, among them the *Codex Barberini, Latin 241* published in 1552, commonly known as the *Badianus Manuscript* [4] and the *Florentine Codex* published in 1590 [5]. While both manuscripts postdate Spanish colonial contact, they were compiled by Spanish priests who obtained the information from Mexica/Aztec respondents, so the information contained in these documents reflects earlier, pre-European contact considerations regarding medical uses of chocolate.

Chocolate: History, Culture, and Heritage. Edited by Grivetti and Shapiro
Copyright © 2009 John Wiley & Sons, Inc.

Tlapolcacauatl.

FIGURE 6.1. *Badianus Codex.* Manuscript page 38 verso, top register. *Source:* National Institute of Anthropology and History, Mexico City. Used with permission.

The *Badianus Manuscript*, written in both Nahuatl/Aztec and Latin, is a Mexica herbal document that identified more than 100 medical conditions common to the Central Valley of Mexico and their treatments. The author of the *Badianus Manuscript* was a teacher at the College of Santa Cruz, founded in 1536 by the Spanish in Mexico City. The manuscript contains a striking color painting of a cacao tree among the healing plants identified in the text (Fig. 6.1). The manuscript also contains a passage that describes how cacao flowers strewn in perfumed baths reduced fatigue experienced by Mexica government administrators [6].

The *Florentine Codex*, published toward the end of the 16th century, was compiled by the priest Bernardino de Sahagún, who arrived in New Spain in 1529. While many Catholic priests associated with the early decades of Spanish colonial rule in Mexico (or New Spain) viewed local inhabitants as savages, their

customs and traditions and literary documents as "ungodly," Sahagún held an opposing view and exhibited a deep interest and curiosity about Mexica/Aztec medical knowledge. Because he took time to learn their social traditions and history, Sahagún's Mexica/Aztec informants reported a vast array of knowledge that he dutifully recorded and preserved for posterity. Without his labor and efforts to preserve this important cultural information, 21st century scholars would have relatively few documents to work with and interpret when attempting to understand and reconstruct the precolonial era of Mexico.

The *Florentine Codex* is a critical document as it contains some of the earliest known medical-related passages associated with chocolate. In one passage it is recorded that drinking large quantities of green cacao made consumers confused and deranged, but if used reasonably the beverage both invigorated and refreshed.

[Green cacao] makes one drunk, takes effect on one, makes one dizzy, confuses one, makes one sick, deranges one. When an ordinary amount is drunk, it gladdens one, refreshes one, consoles one, invigorates one. Thus it is said: I take cacao. I wet my lips. I refresh myself. [7]

Chocolate was drunk by the Mexica/Aztecs to treat stomach and intestinal complaints, and when the cacao was combined with liquid from the bark of the silk cotton tree (*Castilla elastica*), it reportedly cured infections [8]. Childhood diarrhea was treated with a prescription that used five cacao beans. These were ground and blended with the root of *tlayapoloni xiuitl* (unknown plant) and then drunk [9]. To relieve fever and faintness the prescription called for 8 to 10 cacao beans to be ground with dried maize kernels and blended with *tlacoxochitl* (*Calliandra anomala*); then the mixture was drunk [10]. Patients stricken with cough who expressed phlegm drank an infusion prepared from opossum tail, followed by a second medicinal chocolate beverage into which had been blended three herbs: *mecaxochitl* (*Piper sanctum*), *uey nacaztli* (*Chiranthodendron pentadactylon*), and *tlilixochitl* (*Vanilla planifolia*) [11].

Another passage from the *Florentine Codex* reveals that cacao was mixed with various medicinal products and used to offset or mask the flavor of ill-tasting drugs. Preparations of *tlatlapaltic* root (unknown plant) used to control fever were made palatable when mixed with cacao [12]. Sahagún's informants also reported that a local mineral product known as *quinametli* (identified as "bones of ancient people called giants") was blended with chocolate and used to treat bloody dysentery [13].

The text known as the *Ritual of the Bacabs* (*Princeton Codex*) is a Mayan-language codex discovered in 1914 in Yucatán. This manuscript contains a number of medical incantations used by priests to treat medical conditions. The various illnesses were provided names and causal origins were presumed, sometimes attributed to specific birds or trees. At the conclusion of one chant used to cure skin eruptions, fever, and seizures, a bowl of *chacah* (i.e., medicinal chocolate) was drunk by patients. Such medicinal chocolate was prepared using two peppers, honey, and juice from the tobacco plant [14].

EARLY SPANISH AND OTHER EUROPEAN ACCOUNTS (16TH–17TH CENTURIES)

In the decades that followed Spanish conquest of the Americas, cacao was introduced as a beverage to the Spanish court, and within a century both culinary and medical uses of cacao spread to France and to England. Increased demand for chocolate led the French to establish the first Caribbean cacao plantations, while Spain developed cacao as a cash crop in their Philippine colony. By 1822, the English had established their cacao plantations at the Gold Coast colony in western Africa, a geographical area today known as Ghana. From the 16th through the early 19th centuries, numerous European travel accounts and medical texts documented the presumed merits and medicinal value of chocolate. It is to these interesting accounts that we now turn.

Friar Agustín Dávila Padilla wrote a manuscript dated to the second half of the 16th century and related how Friar Jordán de Santa Catalina was treated for kidney disease:

> At the end of his days, his urine was afflicted, and the doctors ordered him to use a drink that in the Indies they call chocolate. It is a little bit of hot water in which they dissolve something like almonds that they call cacaos, and it is made with some spices and sugar. [15]

Francisco Hernández published his botanical text in 1577, which contained one of the first detailed descriptions of the natural history of the cacao tree. Hernández commented on the attributes of drinking chocolate:

> The cacahoaquahuitl [cacao tree] is a tree of a size and leaves like the citron-tree, but the leaves are much bigger and wider, with an oblong fruit similar to a large melon, but striated and of a red color, called cacahoacentli, which is full of the seed cacahoatl, which, as we have said, served the Mexicans as coin and to make a very agreeable beverage. It is formed of a blackish substance divided into unequal particles, but very tightly fit among themselves, tender, of much nutrition, somewhat bitter, a bit sweet and of a temperate nature or a bit cold and humid. [16]

Hernández identified the varieties of cacao trees and types of cacao beans differentiated by the Mexica/Aztecs whether as use for currency or for beverages:

> There are, that I am aware of, four varieties of this tree: the first, called quauhcacahoatl, is the biggest of all and gives the biggest fruits; the second is mecacacahoatl, that is of a medium size, extended and with a fruit that follows in size the former;

> the third, called xochicacahoatl, is smaller, and gives a smaller fruit, and a seed that is reddish on the outside and like the rest on the inside; the fourth, which is the smallest of all and, for this reason, is called tlalcacahoatl or small, gives a fruit that is smaller than the others but of the same color. All the varieties are of the same nature and serve for the same uses, although the latter serves more for beverages while the others are more appropriate for coins. [17]

Hernández then turned his attention to the medical attributes of cacao and its uses in treating specific medical complaints. He mentioned that a simple preparation of cacao, not mixed with other ingredients, was administered to patients suffering from fever and infirmities of the liver [18]. He wrote that if four cacao seeds and a quantity of gum (*holli*) were toasted, ground, and mixed, the preparation "contained dysentery" and that a medicine called *atextli*, identified as a thin paste made of cacao seeds and maize, was "compounded" by adding fruits of *mecaxochitl* and *tlilxochitl* that would excite the "venereal appetite" [19]. Hernández concluded his description of cacao by identifying a beverage called *chocolatl*, made by mixing grains of *pochotl* and *cacahoatl* in equal quantities, that had the properties of making the consumer "extraordinarily fat" when used frequently. This beverage, therefore, was prescribed to "thin and weak" patients [20].

The physician Juan de Cárdenas completed his book, *Problems y Secretos Maravillosos de las Indias* (*The Problems and Secret Marvels of the Indies*) in 1591. Using the humoral medical system of his time,[2] Cárdenas described three substances found in chocolate. The first, identified as cold, dry, thick, earthy, and melancholy, had the potential of harming consumers by engendering anxiety and faltering of the heart, as if "the soul of the person who has eaten it, has left him." The second component, corresponding to the nature of air, Cárdenas described as a substance like yellow grease that floated on the chocolate; consumers of this element found it "warm and humid in constitution, and oily and buttery in taste." Cárdenas claimed that because of the buttery component, "the more cacao or chocolate one takes, the more sustenance it gives, and the person will experience said fattening." This so-called fattening attributed to cacao was thought to be increased if the beans were toasted, whereby the "fire [converted] the earthy [component] into oil." The third element of cacao identified by Cárdenas was described as warm and penetrating, a substance that provided fire to the consumer's body. He wrote that this third element contributed the bitterness to raw, untoasted cacao beans, and since with digestion it "[rose] quickly to the brain," this third element commonly led to headaches. This third element, too, was said to penetrate the body, cause perspiration, and accelerate "expulsion of excrement" [25].

Perhaps the most interesting aspect of the Cárdenas text, however, was his identification and description of European and Native American spices added to chocolate. In his manuscript, Cárdenas differentiated between Castilian spices (e.g., anise, black pepper, cinnamon, and sesame), then provided a list of local New World additives that comprised different chocolate recipes [26]:

Achiote: Although sometimes considered a spice in early manuscripts, its primary use was to provide red color to the cacao, and to "fatten" the consumer through its fatty composition. Cárdenas identified *achiote* as very healthy and recommended its addition to chocolate to "remove obstructions, to provoke menstrual periods, and to cause sweats."

Gueyncaztle: This was said to strengthen and comfort the "vital virtues" of the consumer and to engender "life spirits." This combination with chocolate was said to comfort the liver, aid digestion, and "[remove] gases and bad humors from the stomach."

Mecasuchil: Identified as aromatic thin threads or thin sticks, these were mixed with cacao in order to warm and consume "phlegmatic dampness . . . comfort the liver . . . and increase appetite."

Tlixochil: Identified as aromatic vanilla seeds or good aroma, this was considered "friendly to the heart." Such seeds also warmed the stomach and brought about the fermentation of "thick humors."

After identifying these four main chocolate additives, Cárdenas recommended two additional products: toasted chili and dry cilantro seed, the latter sometimes identified as "earth pepper" [27]. Cárdenas wrote that in ancient times chocolate preparations were stirred or beaten so that the froth arose from it, and the "frothier" the chocolate, the better. He argued that the reason for "frothing" was not olfactory or for taste properties, but that "frothing" thinned or eliminated the "thickness and raw elements" of the cacao. Cárdenas advised against drinking only the froth from chocolate beverages because such practices "put gas in the stomach, impeded digestion . . . [weighed] on the heart, and caused terrible sadness" [28].

Cárdenas concluded his chocolate observations with two important statements on both dietary and medicinal uses of this important food. The first summarized the basic, prevailing humoral medical theory of the time and how chocolate fit into the system:

Aside from giving good sustenance to the body, [chocolate] helps to spend bad humors, evacuating them through sweat, excrement, and urine. Again, I say that in no land of the world is chocolate more necessary than in that of the Indies. Because the atmosphere [here] is humid and listless, bodies and stomachs are full of phlegm and excess humidity, which with the heat of the chocolate is fermented, and converts into blood. [29]

The second identified how and when chocolate should be drunk by consumers:

The most appropriate time to [drink chocolate] is in the morning at seven, or at eight, and before breakfast, because then the heat of this substantial drink helps to spend all that phlegm which has remained in the stomach from the [previous] dinner and supper. . . . The second time for using it is at five or six in the afternoon, when it is presumed that digestion is completed of what was eaten at noon, and then it helps him to distribute [that] heat throughout the whole body. [30]

Cárdenas wrote that untoasted cacao, prepared without other ingredients, constipated the stomach, drained menstruation, closed the urinary tracts, blocked the liver and spleen, reduced facial color, weakened digestion within the stomach, caused shortness of breath, and led to fatigue and fainting. But he also argued that if cacao was toasted, ground, and mixed with *atole* (ground maize and water), the preparation caused weight gain in consumers, sustained people, and provided a healthy, laudable substance [31].

José de Acosta prepared his treatise, *The Naturall and Morall Historie of the East and West Indies*, in 1604. He wrote:

The Spaniards, both men and women, are very greedy of this chocolaté. They say they make diverse sortes [sic] of it, some hote [sic], some colde [sic], and put therein much chili . . . that the chocolate paste is good for the stomach. [32]

Santiago de Valverde Turices wrote an extensive treatise on chocolate in 1624 and argued that cacao was suitable for patients suffering from "cold" or "wet" illnesses. Valverde Turices argued that chocolate should be called a medicine, since it changed the patient's constitution. He concluded that chocolate drunk in great quantities was beneficial for the ailments of the chest and was good for the stomach if drunk in small quantities [33].

Antonio Colmenero de Ledesma published his treatise on chocolate in 1631 and mentioned that cacao preserved health, and made consumers fat, corpulent, fair, and amiable (Fig. 6.2). He stated that chocolate:

Vehemently incites to Venus, and causeth conception in women, hastens and facilitates their delivery; it is an excellent help to digestion, it cures consumptions, and the cough of the lungs, the New Disease, or plague of the guts, and other fluxes, the green sickness, jaundice, and all manner of inflammations and obstructions. It quite takes away the Morpheus, cleaneth the teeth, and sweetneth the breath, provokes urine, cures the stone, and expels poison, and preserves from all infectious diseases. [34]

Tomas Hurtado wrote an ecclesiastical treatise on chocolate and tobacco in 1645. In this work he explored the issue of whether or not drinking chocolate was permitted during Christian fasting periods. He concluded that basic chocolate, if consumed as a beverage, would not break the fast, but if the paste was mixed with milk and eggs it became a food and the faithful should abstain. Elsewhere in this religious text

FIGURE 6.2. Native American presenting gift of cacao to Europe. *Source:* Colmenero de Ledesma. *Curioso Tratado de la Naturaleza y Calidad del Chocolate.* Madrid, Spain: Francisco Martinez, 1631. Courtesy of the Peter J. Shields Library, University of California, Davis. (Used with permission.) (See color insert.)

Hurtado noted that cacao was an important remedy commonly used to treat "illness or thinness of [the] stomach" [35]. He also expressed the opinion the basic chocolate beverages did not sustain the body or "take away hunger," but that when drunk they provided comfort and, since they burned up undigested foods, then served as digestive aides [36].

Thomas Gage published his account of travels in the New World in 1648 and devoted the bulk of his Chapter XVI to the description of cacao and chocolate. At the time of his visit, chocolate commonly was drunk throughout the West Indies, as well as in European Flanders, Italy, and Spain. Gage was aware of Colmenero de Ledesma's text published in 1631, as he included mention of this work. Gage described a type of medicinal chocolate used in New Spain, a product blended with black pepper and administered to patients with "cold livers" [37]. He reported that medicinal chocolate mixed with cinnamon promoted urine flow and should be administered to patients with kidney disorders [38]. Gage noted that *achiote* (*Bixa orellana*), a red plant-based coloring agent, was sometimes added to chocolate and provided an "attenuating quality." This preparation then was administered to patients who suffered from shortness of breath and reduced urine flow [39]. He also wrote that persons who drank chocolate grew fat and corpulent [40].

Henry Stubbe wrote his monograph, *The Indian Nectar, or a Discourse Concerning Chocolata* (sic), in 1662. His objective was to advise readers on chocolate-related misconceptions that were widely held during this period. His text represents a compilation of the works of numerous authors, botanists, physicians, and travelers and provides 20th and 21st century readers with a wealth of chocolate-related authorship. He noted that in the West Indies, chocolate was drunk on the advice of physicians once or twice each day and was especially helpful to restore energy if patients were tired because of business pursuits and wanted speedy refreshment [41]. Stubbe wrote that in Mexico chocolate was used:

In acute diseases [associated with] heat and fervour [sic], and in hot distempers of the liver, [they] give the cacao nut, punned [?] and dissolved in water, without any other mixture. In case of the bloody flux, they mixed the said nuts with a guman [tree gum] called olli, and so curred [them] miraculously. [42]

Elsewhere he wrote that when the flower *xochinacaztli* (*Cymbopetalum penduliflorum*) was added to chocolate, the preparation was used to treat weak "phlegmatique and windy stomachs;" that mixtures of chocolate and *tlilxochitl* (vanilla) were medicines that strengthened the brain and womb; that *achiotl* (*achiote*) mixed with chocolate strengthened debilitated stomachs and treated diarrhea; and that *tepeyantli* (unknown plant) was added to chocolate to treat cough [43].

Stubbe reported that cacao nuts were "very nourishing," that they were filled with nutrients (*multi nutrimenti*), and when chocolate was drunk frequently it "doth fatten" consumers [44]. He reported that English soldiers stationed in Jamaica lived for long periods on a cacao-based paste prepared with sugar that was dissolved in water. He said that soldiers sustained themselves eating only this paste and that Indian women ate chocolate so often that they scarcely consumed any solid meat (i.e., food meant) and still did not decline in strength [45]. Stubbe suggested that if the oily fraction of cacao was removed, the remainder had special curative functions, especially against general inflammations, and specifically was useful when curing the fire of St. Anthony, or ergot poisoning [46]. He wrote that when chocolate was mixed with Jamaica pepper the combination provoked urine and menstrual flow, strengthened the brain, comforted the womb, and dissipated excessive wind, whereas if vanilla was added to chocolate this combination strengthened the heart, "beget strong spirits," and promoted digestion in the stomach [47].

Stubbe expanded upon his views relative to the curing properties of chocolate and wrote that *achiote* mixed with chocolate:

Allayed feverish distempers, it helpeth the bloody-flux, and repels praeternatural tumors . . . it is mixed with chocolate [sic] to . . . helpeth the tooth-ach [sic] arising from hot causes, it strengthens the gums, it provokes urine, it quencheth thirst . . . and being mixed with rosin [sic], it cureth the itch and ulcers; it strengthens the stomach, stoppeth the fluxes of the belly, it encreaseth milk [in lactating women]. [48]

Further elaborating on medicinal forms of chocolate, he wrote that different varieties of peppers, specifically *mecaxochitl* or *piso,* when mixed with cacao paste, "opens obstructions, cures colds, and distempers arising from cold causes; it attenuates gross humours, it strengthens the stomach, and it amends the breath" [49]. Several varieties of ear flowers (*xochinacaztlis* or *orichelas*) when mixed with chocolate provided a pleasing scent and taste to the medicine that was used to strengthen the stomach, revive the spirit, "beget good blood," and "provoke monthly evacuations in women." The same mixtures, however, were thought by other physicians to be stronger, more effective medicines and were used with caution to strengthen the heart and vital parts [50].

Stubbe also observed that "the Indians used [chocolate] as food, and daily aliment; upon occasion of fevers and other hot distempers, they made some little alteration of it" [51]. He cited a Dr. Franciscae Ferdinandez, the Principal Physician in Mexico during the reign of Philip II, who wrote the following:

> [Chocolate] is one of the most wholesome and pretious [sic] drinks, that have been discovered to this day: because in the whole drink there is not one ingredient put in, which is either hurtful in it self, or by commixtion [sic]; but all are cordial, and very beneficial to our bodies, whether we be old, or young, great with child, or others acustomed [sic] to a sedenary [sic] life. And we aught not to drink or eat after the taking chocolata [sic]; no, nor to use any exercise after it: but to rest for a while after it without stiring [sic]. It must be taken very hot. [52]

Stubbe included an account by an unnamed Spanish physician from Seville who compared wine and chocolate and noted that "none hath been known to live above seven dayes [sic] by drinking wine alone, [however] one may live moneths [sic], and years using nothing but chocolate" [53]. He quoted the same Spanish physician who testified that he, himself, "saw a childe weaned, which could not be brought by any artifice to take any food, and for four moneths [sic] space he was preserved alive by giving him *chocolata* [sic] only, mixing now and then some crumbs of bread therewith" [54].

In 1672, William Hughes published his ethnobotanical monograph on plants growing in English plantations in the Americas. Appended to his general text was a specific discussion entitled *Discourse of the Cacao-Nut-Tree, and the Use of Its Fruit: With All the Ways of Making of Chocolate: The Like Never Extant Before.* In this appendix Hughes identified several varieties of medicinal chocolates and their respective uses:

> To strengthen the stomack [sic] much debilitated, there is put in achiote, or rather saffron: [to treat] fluxes, cinnamon, nutmegs, or a little steel-powder: for coughs, almonds, and the oyl [sic] of almonds, sugar, or sugar candied: for a phlegmatick stomack [sic], they put in pepper, cloves, etc. [55]

Having identified the ingredients of so-called medicinal chocolates, Hughes then elaborated on the use of chocolate in medical care:

> Chocolate is most excellent, it nourishing and preserving health entire, purging by expectorations, and especially the sweat-vents of the body, preventing unnatural fumes ascending to the head, yet causing a pleasant and natural sleep and rest . . . eaten twice a day, a man very well [may] subsist therewith, not taking any thing else at all. [56]

Most interesting from a nutritional perspective, however, was Hughes's view that chocolate could cure the "pustules, tumours, or swellings" experienced by "hardy sea-men long kept from a fresh diet." He wrote that, once ashore, sailors should drink chocolate since it:

> Is excellent to drive forth such offensive humours, opening the pores, and causing moderate sweats. [57]

Hughes also wrote that chocolate nourished consumers who required "speedy refreshment after travel, hard labour, or violent exercise," that the beverage was "exhilerating [sic] and corroborating [to] all parts and faculties of the body" [58]. He urged readers living in England to drink chocolate, especially patients with:

> Weak constitutions, and have thin attenuate bodies, or are troubled with sharp rheums, catarrhs, and such as consumption . . . and all aged people may safely take it, especially in the heat of summer, when the skin and pores are relaxed by great expence [sic] of spirit, causing a faintness. [59]

Hughes also offered a ringing endorsement to the medical merits of chocolate:

> Chocolate is the only drink in the Indies, and I am fully perswaded [sic] is instrumental to the preservation and prolonging of many an Europeans life that travels there . . . for my own part, I think I was ever fatter in all my life, then when I was in that praise-worthy Island of Jamaica, partily [sic] by the frequent use there-of, neither had I one sick day during the time I was there, which was more than half a year. [60]

Hughes cited two physicians, identified only as Dr. Juanes and Dr. Ferdinandez, on other medical aspects of chocolate and reported:

> It is the most wholesome and most excellent drink that is yet found out . . . it is good alone to make up a breakfast, needing no other food, either bread or drink, is beneficial to the body, and without exception, may be drunk by people of all ages, young as well as old, of what sex or what constitution so ever and is very good for women with childe, nourishing the embryo, and preventing fainting fits, which some breeding women are subject unto: it helpeth nature to concoct phlegme and superfluous moisture in the stomack [sic]; it voideth the excrements by urine and sweat abundantly, and breedeth store of very good blood, thereby supplying the expence [sic] of spirits, it expels gravel, and keepth the body fat and plump, and also preserveth the countenance fresh and fair: it strengthens the vitals, and is good against fevers, cattarrhs, asthmaes, and consumptions of all sorts. [61]

Sylvestre Dufour published his monograph, *The Manner of Making of Coffee, Tea, and Chocolate,* in 1685.

The value of the Dufour text, however, lies in his acknowledgment and credit that much of what he presented in his book on chocolate had been previously published by the Spanish physician from Ecija in Andalusia, Antonio Colmenere de Ledesma. Dufour noted that medicinal preparations of chocolate commonly contained anise seed, and when mixed with chocolate, the result was effective for patients suffering from diseased or infected bladders, kidneys, as well as others with throat or womb problems [62]. He noted that a blend of *achiote* with chocolate, when applied externally, allayed the "fever of love," and reduced dysentery, which he called the "griping of the guts" [63]. Dufour also wrote that *mecaxuchil* (vanilla) mixed with chocolate had a positive effect on both heart and stomach, attenuated thick and slow humours, and was an excellent antidote and medicine against undefined types of poison [64].

The last portion of the Dufour text considered associations among climate, seasons, and timing when consumers should drink chocolate. He wrote that chocolate prepared with endive water should be drunk during winter and that chocolate prepared with rhubarb water treated hot distempers of the liver. Patients were cautioned not to drink chocolate during the "dog days," that is, early July through early September [65]. Dufour debated why chocolate fattened those who drank it. The prevailing theory of the time postulated that the various ingredients in medicinal cacao—because of their "hot–cold" valence—should make the body lean. Dufour, however, believed the cause was something inherent within cacao itself, and in his opinion this was the fatty-oil fractions contained in medicinal and everyday chocolate beverages:

> [It is due to the] buttery parts [of the cacao] . . . which fatten [because] the hot ingredients of medicinal chocolate serve as a type of pipe or conduit . . . and make it pass by the liver, and the other parts till they arrive at the fleshy parts, where finding a substance which is like and comfortable to them, to wit hot and moist . . . convert themselves into the substance of the subject they augment and fatten it. [66]

Nicolas de Blégny published his treatise on chocolate, coffee, and tea in 1687. The third part of his interesting text considered chocolate, its preparation, composition, and various properties:

> Taken with the vanilla syrup at different times of the day and especially in the evening, at least two doses, it [chocolate] has an effect equally . . . to suspend the violent cause of rheumatoids and inflammation of the lungs, and to dull the irritation and ferocity which incites cough [and] to put out the inflammations of the throat and lungs [pleure], to calm the different courses of insomnia and to restore the fatigue of preachers and other persons who frequently engage in public activities. Prepared the same way, it is a great help to deaden the spleen [bile] overflow which provokes vomiting and which makes the stomach bilious, [leading to] death-producing diarrhea and dysentery [le colera morbus]. It is also a very effective remedy [to reduce] ethic fever [éthique fiévre], and I

want to say [is effective in relieving] dryness of the chest which leads to pulmonary disease, which we can [use chocolate] to stop the advancement to soften the infirmity, especially in the place of water we prepare [the chocolate] with milk which we must skim before boiling. If we prepare it with the "syrup of coins" [sirop de coins] to which we have added some drops of tincture of gold, or oil of amber, it [becomes a] very efficient [medicine to relieve] indigestion and heart palpitations so well that in need, it might serve all together as a sufficient nourishment and as a remedy in [treating] more familiar illnesses. [67]

LATER SPANISH AND OTHER EUROPEAN ACCOUNTS (18TH TO EARLY 20TH CENTURIES)

In his *Natural History of Chocolate* originally published in 1718 (the English translation of 1730 is cited here), de Quélus considered that chocolate was a temperate food, nourishing, easy to digest, and essential to good health. He noted that women living in the Americas, subject to "the whites" (i.e., leukorea), were "cured of this distemper, by eating a dozen cocao [sic] kernels for breakfast every morning" [68]. De Quélus remarked that drinking chocolate repaired "exhausted spirits" and "decayed strength," that the beverage preserved health and prolonged the lives of old men [69]. He noted that drinking cacao quenched thirst, was refreshing and "feeding," and procured "easy quiet sleep" [70]. De Quélus described the case of an unfortunate woman who could not chew because an accident had injured her jaw and therefore "did not know how to subsist." She was encouraged by her physician to take:

> Three dishes of chocolate, prepared after the manner of the country, one in the morning, one at noon, and one at night . . . [only] cocao [sic] kernels dissolved in hot water, with sugar, and seasoned with a bit of cinnamon . . . [and] lived a long while since, more lively and robust than before [her] accident. [71]

Elsewhere, he wrote that an ounce of chocolate "contained as much nourishment as a pound of beef" [72].

De Quélus summarized the prevailing controversy of his time regarding the digestibility of chocolate and concluded:

> Digestion of chocolate is soon brought about without trouble, without difficulty, and without any sensible rising of the pulse; the stomach very far from making use of its strength, acquires new force. . . . I have seen several persons who had but weak digestion, if not quite spoiled, who have been entirely recovered by the frequent use of chocolate. [73]

He mentioned that should agitated persons consume chocolate they would perceive an effect nearly instantly, that faintness would cease, and strength would be recovered even before digestion had been initiated [74]. Waxing enthusiastic about the positive medical properties of chocolate, de Quélus wrote:

Before chocolate was known in Europe, good old wine was called the milk of old men; but this title is now applied with greater reason to chocolate, hence its use has become so common that it has been perceived that chocolate is with respect to them, what milk is to infants. [75]

Expanding upon a popular view that chocolate could be a "sole" food, de Quélus provided testimonial evidence that chocolate was more than beneficial for health, that it extended longevity:

There lately died at Martinico [sic, Caribbean Island of Martinique meant] a councilor about a hundred years old, who, for thirty years past, lived on nothing but chocolate and biscuit. He sometimes, indeed, had a little soup at dinner, but never any fish, flesh, or other victuals: he was, nevertheless, so vigorous and nimble, that at fourscore and five, he could get on horseback without stirrups. [76]

De Quélus concluded his medical observations on chocolate with a phrase that has rung down through the centuries: *In multis eseis erit infirmitas, propter crapulam multi obierunt: Qui autem abstinens est, adjiet vitam* (Plentiful feeding brings diseases, and excesses has killed numbers; but the temperate person prolongs his days) [77]. But lest one consider Quélus a seer, a prophet of medical-nutritional theory, it should be noted that he also concluded that chocolate could be used as a vehicle when it became necessary to cure patients with "powders of millipedes, earthworms, vipers, and the livers and galls of eels" [78]. He also recorded an instance when, during Lent, there was insufficient availability of olive oil, whereupon chocolate "oil" was substituted and was well received [79]. De Quélus noted further that "chocolate oil" served as:

[An] easer of pain, it is excellent, taken inwardly, to cure hoarsenes [sic], and to blunt the sharpness of the salts that irritate the lungs . . . [when] taken reasonably, may be a wonderful antidote against corrosive poisons. [80]

De Quélus further described uses of "chocolate oil" and wrote that when applied externally to the body such products could:

Clear and plump the skin when it [was] dry, rough . . . without making it appear either fat or shining . . . [and] there is nothing so proper as this to keep [one's] arms from rusting, because it contains less water, than any other oil made use of for that purpose. [81]

Elsewhere De Quélus wrote that "chocolate oil" was used to cure piles, sometimes as a sole ingredient, in other instances mixed with lead dross reduced to a fine powder, and in other instances the "chocolate oil" was blended with millipede powder, "sugar of lead," and laudanum. Beyond hemorrhoids, however, oil of chocolate also was used to ease the pain of gout [82].

The famous naturalist Carl von Linné, or Linnaeus, examined the medicinal uses of chocolate in his 1741 monograph entitled *Om Chokladdryken*. He wrote that chocolate was an excellent source of nourishment

and that it cured many ills. He identified three categories of illness that responded well to chocolate therapy: (1) wasting or thinness brought on by lung and muscle diseases; (2) hypochondria; and (3) hemorrhoids. Linnaeus also wrote that chocolate was an effective aphrodisiac, a statement based presumably upon personal experience [83].

Philip Dormer Stanhope, the Earl of Chesterfield, wrote a letter in 1775 to his son, Philip Stanhope, who was about to embark upon a trip to Paris. The father knew of his son's obesity and cautioned the younger Stanhope on the proper way to lose weight:

Mr. Tollot says, that you are inclined to be fat: but I hope you will decline it as much as you can; not by taking any thing corrosive to make you lean, but by taking as little as you can of those things that would make you fat. Drink no chocolate, take your coffee without cream; you cannot possibly avoid suppers at Paris, unless you avoid company too, which I would by [no] means have you do; but eat as little at supper as you can and make even an allowance for that little at your dinners. Take, occasionally a double dose of riding and fencing; and now, that the summer is come, walk a good deal in the Tuilleries; it is a real inconveniency to any body to be fat; and besides, it is ungraceful for a young fellow. [84]

Vincente Lardizabel published his treatise on chocolate in 1788 and discussed how chocolate drinking countered the ill effects of mineral water, and to control vomiting. He wrote that stagnant humors were cleared after drinking chocolate, and it could be used to treat severe belching and flatulence [85].

Alexander Peter Buchan published a medical monograph in 1792 and recommended that women in labor should be served chocolate. He identified additional medicinal uses of chocolate, specifically chocolate's role in preventing fainting after severe blood loss. Buchan recommended that infirm patients eat frequently and that convalescent diets be composed of light, nutritive foods such as chocolate, but served in small portions [86].

Antonio Lavedan wrote his influential treatise, *Tratado de los Usos, Abusos, Propiendades y Virtudes del Tabaco, Café, Te, y Chocolate*, in 1796. This important work contained a wealth of medical-related information regarding the use of chocolate. He claimed that chocolate was beneficial only if drunk in the morning, and he urged prohibition of chocolate drinking in the afternoon:

Chocolate is most beneficial when it is taken only in the morning, and its use in the afternoon should be prohibited. This is because it is a drink that should not be mixed with other food, since it ferments with them easily, causing a precipitated and corruptible digestion. This is much worse for those who tend to drink chocolate after having eaten and drunk well, or a short time after eating, for the same reason we have just stated, that it becomes corrupted when mixed with other dishes. Even though many fans and excessive users of chocolate say that it aids digestion, this is not verified in practice, and if it works well for one, for a thousand more it is bad, causing as we have said, indigestion. You can almost be sure that the greater part of obstructions and flatulence proceed from lunch and chocolate that are taken in the afternoon, or from not waiting long enough after eating. This regards the use of chocolate for those who are healthy. [87]

Lavedan wrote extensively on "health chocolate" (*chocolats de santé* or *chocolats thérapeutiques du médicinaux*) and concluded:

Health Chocolate made without aromas is preferable and has the properties to awaken the appetite in those who do not usually drink it. Chocolate is good sustenance for those who typically drink it in the morning. ... The chocolate drink made with lightly toasted cacao with little or no aromas, is very healthy for those who are suffering from tuberculosis and consumption. It protects against obstructions, and if they are able to recover, cures sufferers of tuberculosis who seek this remedy on time, by replacing the loss of nutrient balsams that have stolen the consumptive warmth, dominating and sweetening the feverish acid that the spirits absorb. ... Chocolate is a food that repairs and fortifies quickly and therefore it is better for phlegmatic persons that need stimulation. ... It is possible for chocolate alone to keep a man robust and healthy for many years, if he takes it three times a day, that is, in the morning, at noon and at night, and there are examples of this. ... Without help from other food, chocolate can prolong life through the great nutrients that it supplies to the body and it restores strength, especially when one mixes an egg yolk with some spoonfuls of meat broth. It is a good stomach remedy, repairing all weaknesses, afflictions, indigestion, vomiting and heart pain, freeing the intestines of flatulence and colic. Those who have weakness of the stomach because of diarrhea or because of some purging substance will experience relief with the chocolate drink. It strengthens those suffering from tuberculosis, who are without hope, and its daily use reestablishes their health more than what could have been expected. For gout or podagra it is of great use—those suffering from gout should drink this nectar of the gods without worrying about any ill effects, for it will be very beneficial to them ... it is a universal medicine ... not only for preserving health, but to undo many ills, and for this reason it strengthens and increases natural warmth, generating more spirituous blood. It vivifies the substance of the heart, diminishes flatulence, takes away obstructions, helps the stomach, and awakens the appetite, which is a sign of health for those that drink it. It increases virility, slows the growth of white hair, and extends life until decrepitude. To people of any age, including the youngest, it can be given. [88]

Anthelme Brillat-Savarin died in 1825, whereupon, his writings were assembled and published under the title *Handbook of Gastronomy*, sometimes with the title, *The Physiology of Taste*. While best known for his aphorism "Tell me what you eat, and I will tell you what you are," Brillat-Savarin also wrote that:

Chocolate, when properly prepared, is a food as wholesome as it is agreeable ... it is nourishing, easily digested ... and

is an antidote to the [inconveniences] ... ascribed to coffee. [89]

Brillat-Savarin wrote that chocolate was most suitable for those who had much brain work in their occupation, identified as clergymen, lawyers, and travelers [90]. He recommended a medicinal form of chocolate, a preparation mixed with ground amber dust, as a remedy for hangover, or when the "faculties [are] temporarily dulled," such as periods of "tormented thinking" [91]. Brillat-Savarin also identified a form of medicated chocolate mixed with orange-blossom water given to patients with delicate nerves, whereas irritated persons were prescribed chocolate mixed with almond milk [92].

Thomas J. Graham wrote his medical treatise in 1828 and included several popular treatments that sometimes included cacao as an ingredient. To treat various disorders he recommended the following:

For Asthma: This diet should be uniformly light and easy to digest, consisting mainly of a fresh food of animal origin, such as eggs, as well as bread, tea and chocolate. [93]

For Indigestion or Dyspepsia: For breakfast and in the afternoon, one should drink tea, cacao or light chocolate, with biscuits, bread and butter, or dry toast. Rolls, and any other type of spongy bread are bad, and it is important to refrain completely from coffee. One fresh egg passed lightly through water can be taken with breakfast, if you feel well. With any liquid that is taken in the morning and the afternoon, one should not exceed the size of one common lunch cup, each time. [94]

Milne Edwards and P. Vavasseur wrote their medical treatise in 1835 and briefly commented on cacao butter and both the pleasurable and medicinal uses of chocolate:

The cacao, after having been toasted, serves to make chocolate, which has wide use as a food. With regard to the oil, it is used as an emollient, in the flegmasias [sic] of the digestive, respiratory and urinary organs. It is often useful in cases of cancer of the stomach. Externally, it is applied to hemorrhoid tumors, and on chapped lips and nipples. [95]

A book by Francisco X. Estrada identified instructions taken to counter the effects of a measles epidemic in Mexico. The author wrote his recommendations to help the people of northeastern Mexico, especially the state of San Luis Potosi, who did not have access to physicians. He described the disease, its symptoms, progression, and a plan of healing:

Put the sick person in a dry room. Not too cold, not too hot, but warm. Serve a healing diet consisting of: a small amount of atole, bread without butter, soup, and encourage the patient to sweat. During the convalescence: Slowly replace the healing diet for a more abundant diet that includes: champurrado with some chocolate; very ripe apples (boiled), toasted tortillas; chick-peas, and chicken. [96]

Auguste Saint-Arroman wrote his influential work *Coffee, Tea and Chocolate: Their Influence upon the Health, the Intellect, and the Moral Nature of Man* in 1846. Throughout his manuscript, Saint-Arroman encouraged the use of chocolate as part of medical treatment and healing. After drinking chocolate, he also recommended drinking a glass of water [97]. He argued in his treatise that chocolate was suited to the aged, to the weak, and to worn out persons, but that it was injurious to the young and to those with liver conditions [98]. Saint-Arroman identified several varieties of medical chocolate; one, ferruginous chocolate, was considered:

Beneficial to women who are out of order, or have the green sickness, is prepared by adding to the paste of chocolate iron in the state of filings, oxide or carbonate. [99]

He also held that chocolate exerted an effect on the moral nature of consumers and suggested that chocolate paste could produce changes in the brain, if the stomach digested it easily, and that chocolate was a nourishing "aliment," but if given at an improper time, might cause poor vision [100].

That chocolate drinking contributed to longevity was widely held throughout the 18th and 19th centuries. The journal *Scientific American* published an interview with Andrew Loucks in 1849, a gentleman who had reached the age of 97, who responded to the following:

[Question] "Why were people of your day healthier than those born at a later period?" Loucks replied, "We ate lighter food when I was a boy than at present—such as soups; used a great deal of milk, and but little tea and coffee. We sometimes made chocolate by roasting wheat flour in a pot, though not often." "But, ah!" added the old man, "Young people are now up late at night—to run about evenings is not good, but to take the morning air is good." [101]

Auguste Debay wrote *Les influences du Chocolat du thé et du café sur l'économie humaine* in 1864. He provided recipes for medicinal, healing chocolates (*chocolats de sante*) that combined cacao seeds of different geographical locations with refined powdered sugar [102]. For patients suffering from general debilities, weak stomach, and nervous-gastrointestinal distress, Debay recommended a formula that consisted of different varieties of cacao seeds, blended with wheat gluten [103]. Debay identified several *analeptique* (restorative) forms of chocolate prepared using cacao seeds ground and mixed with a variety of ingredients, including cinnamon, gum, sugar, and tincture of vanilla [104]. Other medicinal chocolates included various combinations of cinnamon, iron hydrate, iodine, ground lichen, quinine extract, starch, and sugar [105]. Chocolate also was a primary ingredient in an antihelminthic (i.e., antiworm or vermifuge) prescription that was combined with calomel, cinnamon, oil of croton, and sugar [106]. Debay identified a medical chocolate used to treat syphilis that called for balm of Pérou (*sic*), aromatic cacao, sugar, and an unidentified sublimated corrosive, to be dissolved in alcohol [107].

Debay concluded his treatise by providing opinions and testimonial evidence from distinguished physicians and scientists of his era regarding the positive effects of chocolate as a nutritious food. The physician to the King of France, identified as a Dr. Alibert, reported that chocolate was "*très-salutaire*" for persons suffering from weakness and exhaustion. The physician to the King of Prussia, identified as a Dr. Huffeland, stated that chocolate was useful to treat persons who were excitable, nervous, or violent, and that medical chocolate could combat fatigue and debilitation and could improve the life of invalids. Dr. Huffeland also recommended chocolate for patients with chronic intestinal distress and praised its use by lactating women. Debay also quoted "*le grand naturaliste*" (i.e., Baron Georges Lépold Chrétien Frédéric Dagobert Cuvier) who stated that drinking chocolate could help emaciated persons gain weight [108].

Jose Panadés y Poblet wrote his *La Educacion de la Mujer (Education of Women)* in 1878. While directed toward the education of women, his treatise reviewed the history of cacao, identified cacao as the national food of Mexico, and presented a formula for *chocolate de la salud* (health chocolate) that consisted of cacao, sugar, and several secret "aromas" [109]. While he generally extolled the merits of chocolate, Panadés y Poblet commented on the dichotomy that some previous authors had affirmed that chocolate was digestible, while others reported chocolate caused indigestion. He summarized this puzzle in an amusing manner by stating that the differences were due simply to the fact that "not all chocolates and not all stomachs are the same" [110]. He concluded that chocolate served to convalescents or to patients with "delicate stomachs" should be prepared with water; that chocolate was nutritious when consumed in small quantities; and that chocolate repaired the loss of energy due to excessive work, pleasure, and staying up late at night [111].

The *Manual del Farmacéutico* published in Mexico in 1881 provided several recipes for *chocolate de salud* (health chocolate). The *Manual* identified other mixtures where chocolate served as the administrative vehicle or masked unpleasant medicinal flavors. Further recipes were identified that served as purgatives where ground cacao was combined with *scammony (Convolvulus scammonia)* and *jalapa (Convolvulus jalapa)*, while others consisted of calcinated magnesia and powdered iron filings as active ingredients [112]. The manuscript also identified how to prepare antihemorrhoid suppositories using cacao butter, cocaine hydrochlorate, and ergot. Another suppository, identified as "calming," consisted of cacao butter, belladonna extract, and laudanum [113].

In 1888, Gustavo Reboles y Campos translated into Spanish the French treatise *Higiene Terapéutica; La*

Higiene Alimentica by Georges Octave Dujardin-Beaumetz. This work offered a range of positive and negative views regarding the therapeutic role of chocolate. One passage provided instructions on how to feed patients to cure different illnesses:

> *One can use mixes of flour, lentils and meat powders in the form of soups. But it is preferable to mix this powder with chocolate or with liqueurs, making a mix that is known as meat powder grog.* [114]

Mariano Villanueva y Francesconi wrote the *Arte de Hacer Fortuna: 5000 Recetas de Artes, Oficios, Ciencias y de Familia* in 1890. This popular book included specific recipes/discussions on health and home medical treatments that commonly included cacao as ingredients, for example:

> *Recipe 1340: [If suffering from] anemia, convalescence, chlorosis, [or] cancerous diseases [do the following] Soups without fat, red meats, roasts, beefsteak and roasts on the broiler . . . wild game and fowl, like partridge, duck, pheasant, woodcock, etc. Avoid coffee and tea . . . chocolate is preferable . . . no salads . . . avoid acids and alcoholic beverages.* [115]

In Recipe 1345, chocolate was included as an ingredient along with ground melon/pumpkin seeds, ground almonds, and milk of sweet almonds in order to treat diarrhea [116].

Mariano Villanueva y Francesconi wrote another popular book, *El Medico y La Botica en Casa*, in 1897 and touted the value of chocolate as a primary food source. He wrote that chocolate should be a breakfast food for children because of its nutritious qualities and that the ideal breakfast would consist of "a little chocolate and a little glass of milk, or a custard or also toasted bread with butter" [117].

Dr. J. Millam Ponce wrote his *Pequeno Tratado de Medicina Domestica* in 1902, a work to educate the general public with homemade preparations of "indispensable medicines" to cure illnesses. He explained that when nursing infants became ill, wet nurses should initiate a light diet of coffee with milk or chocolate [118]. Ponce also recommended cacao butter suppositories to treat hemorrhoids [119].

At the transition between the 19th and 20th centuries, medicinal uses for chocolate products such as cacao butter were widely recommended by physicians. Juan Bardina published his *Modern Hygiene, a Manual for Hispanoamericans,* in 1905 and discussed a range of healthful foods. He identified cacao as a fruit and cautioned that candy bars of his era commonly were wrapped in toxic, silvered paper. Bardina recommended that an ointment prepared from one part cacao butter, one part white wax, and four parts oil of sweet almonds be applied to the breasts of nursing women who developed sores or abscesses [120].

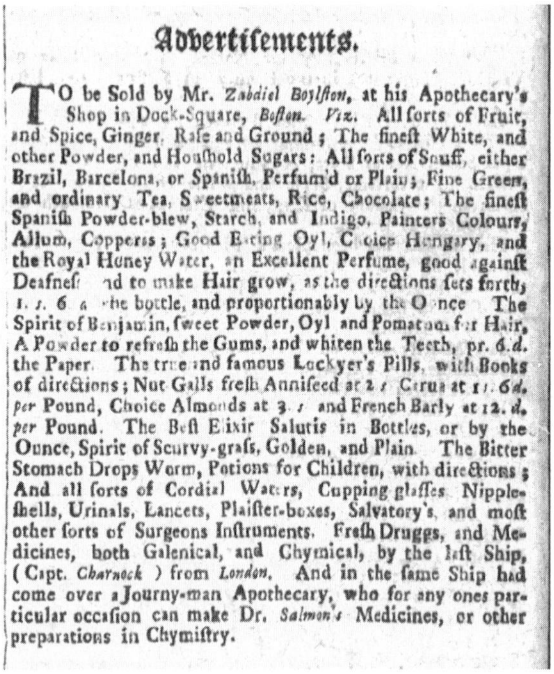

Advertisements.

TO be Sold by Mr. *Zabdiel Boylston*, at his Apothecary's Shop in Dock-Square, *Boston. Viz.* All sorts of Fruit, and Spice, Ginger, Rase and Ground; The finest White, and other Powder, and Houshold Sugars: All sorts of Snuff, either Brazil, Barcelona, or Spanish, Perfum'd or Plain; Fine Green, and ordinary Tea, Sweetmeats, Rice, Chocolate; The finest Spanish Powder-blew, Starch, and Indigo, Painters Colours, Allum, Copperss; Good Eating Oyl, Choice Hungary, and the Royal Honey Water, an Excellent Perfume, good against Deafness and to make Hair grow, as the directions sets forth, 1. s. 6 d. the bottle, and proportionably by the Ounce The Spirit of Benjamin, sweet Powder, Oyl and Pomatum for Hair, A Powder to refresh the Gums, and whiten the Teeth, pr. 6. *d.* the Paper. The true and famous Lockyer's Pills, with Books of directions; Nut Galls fresh Anniseed at 2 *s* Carus at 1 *s*. 6 *d*. *per* Pound, Choice Almonds at 3. *s* and French Barly at 12. *d*. *per* Pound. The Best Elixir Salutis in Bottles, or by the Ounce, Spirit of Scurvy-grass, Golden, and Plain. The Bitter Stomach Drops Worm, Potions for Children, with directions; And all sorts of Cordial Waters, Cupping glasses Nipple-shells, Urinals, Lancets, Plaister-boxes, Salvatory's, and most other sorts of Surgeons Instruments. Fresh Druggs, and Medicines, both Galenical, and Chymical, by the last Ship, (Capt. *Charnock*) from *London*. And in the same Ship had come over a Journy-man Apothecary, who for any ones particular occasion can make Dr. *Salmon's* Medicines, or other preparations in Chymistry.

FIGURE 6.3. Chocolate in a medicinal context. Zebediah Boylston advertisement. *Source: The Boston News-Letter*, March 17, 1711, p. 2. Courtesy of Readex/Newsbank, Naples, FL. (Used with permission.)

Chocolate and Medicine in North America: A Sampling of Themes

While a more detailed review of medical-associated chocolate in relation to diet and health, as evidenced in 18th and 19th century cookbooks is discussed elsewhere (see Chapter 9), it is appropriate here to review several additional medical-related themes.

SALE OF CHOCOLATE IN APOTHECARY SHOPS

Some early 18th century newspaper advertisements by Boston apothecaries and physicians reveal that chocolate was listed among the balms, medicines, salves, and tinctures available at stores managed by these professionals.

1711: Zabdiel Boylston (Boston) was a well-known physician of Boston who advertised the sale of drugs, foods, and medicines at his Apothecary Shop located in Dock-Square (Fig. 6.3). Among the items available to customers were the following:

> *Household Sugars, All sorts of Snuff; Fine Green and ordinary Tea, Rice, Chocolate; an Excellent Perfume good against Deafness and to make Hair grow; Powder to refresh the Gums, and whiten the Teeth; The true and famous Lockyer's Pills; Nut Galls, fresh Anniseed [sic]; Choice Almonds; Scurvy-grass, Golden and Plain; The Bitter Stomach Drops Worm*

Potions for Children; Cupping glasses; Urinals, Lancets, Fresh Druggs [sic], and Medicines, both Galenical, and Chymical [sic]. [121]

1774: Seth Lee (Boston) advertised drugs and medicines that included the following:

Turlington's Balsam of Life, Eaton's Styptic, James' Universal Fever Powder, Spirits of Scurvy Grass, also Mace, Cinnamon, Cloves, Nutmegs, Ginger, Pepper, Allspice, Raisins, Figgs [sic], and Chocolate. [122]

1775: Francis Daymon (Philadelphia) reported stocks of the following:

A very good assortment of MEDICINES, and the very best chocolate, already sweetened fit for the gentlemen of the army. [123]

1785: Aaron Hastings (Bennington, Vermont):

Sells chocolate, drugs, and medicines. [124]

1788: Ashur Shephard (Bennington, Vermont):

Sells chocolate at his medicinal store. [125]

CHOCOLATE, CHURCH, AND DIET

Cotton Mather wrote his *Manuductio ad Ministerium* (*Directions for a Candidate of the Ministry*) in 1726. In this religious document he provided both dietary and health recommendations:

To feed much on Salt-Meats, won't be for your Safety. Indeed, if less Flesh were eaten, and more of the Vegetable and Farinaceous Food were used, it were better. The Milk-Diet is for the most part some of the wholesomest [sic] in the World! And not the less wholesome, for the Cocoa-Nutt giving a little Tincture to it. [126]

CHOCOLATE, CLIMATE, AND SEASONALITY

Thomas Cadwalader published his essay on the West Indies in 1745 and included discussion of the illness he called the West-India dry-gripes and offered a "method of preventing and curing that cruel distemper." He remarked that if residents were to stay well while living in the West Indies, they should abstain from strong punch (see Chapters 36 and 37) and highly seasoned meats, and avoid immoderate exercise, which he believed was improper since it caused much perspiration. He noted that it was important to "rise early in the Morning . . . to take Chocolate for Breakfast and Supper" [127].

An anonymous publication dated to 1751 entitled *A Friendly Caution against Drinking Tea, Coffee, Chocolate, & Very Hot,* advanced the thesis that beverages should be served at temperatures below normal body heat, lest the higher degree of heat "coagulate and thicken the Blood to such a Degree as to endanger Life" [128].

A school of thought also emerged in North America, advanced by T. Philomath, that drinking chocolate during springtime should be discouraged (Fig. 6.4).

Of the four quarters of the YEAR, and first of the SPRING,

I shall conclude this quarter, however, with Virgil's advice to young people, that while they are gathering wild strawberries and nosegays they should have a care of the *snake in the grass.* And here I must fall in with the prescriptions of our ast-rological physicians, such as a spare and simple diet, with the moderate use of phlebotomy, during this season.

And I must beg leave to advise my readers who are of the fair sex to be in a particular manner careful how they meddle with romances, chocolate, novels, and the like inflamers, which I look upon as very dangerous to be made use of during this great carnival of nature.

FIGURE 6.4. Chocolate use discouraged during spring season. *Source*: Philomath, T. *The Virginia Almanack* [sic] *for the Year of Our Lord God 1770 . . . Calculated According to Art, and Referred to the Horizon of 38 Degrees North Latitude, and a Meridian of Five Hours West from the City of London; Fitting Virginia, Maryland, North Carolina, & c.* Williamsburg, VA: Purdie and Dixon, 1769; p. 16. Courtesy of Readex/Newsbank, Naples, FL. (Used with permission.)

Close reading of his text is amusing because of the correlation that he drew between springtime—when human thoughts turn to those of love and lust—and chocolate. Since it was widely recognized/thought that chocolate was stimulating, Philomath's admonition was an attempt to dampen the ardor of young couples [129].

CHOCOLATE AND HOSPITAL SUPPLIES

Inspection of hospital ledgers frequently reveals the purchase and disposition of cacao beans, cocoa powder, and prepared chocolate. Cacao beans or processed chocolate regularly was served to patients in hospitals in both Central and North America. From the Archives in Mexico City come the following selected examples that document this use:

[College of Sancti-Spirit, city of Puebla, Mexico] *And the chocolate that was consumed in the 12 days of this month [month not identified] year 1769 totals 22 pounds, 2 ounces, because 4 Fathers drank 3 ounces daily; the two insane [Fathers], the Commissioner, the Pantry Administrator, the [male] Nurse, and 18 Servants = 7 pounds ½ ounces.* [130]

[College of Sancti-Spirit, city of Puebla, Mexico] *Expenses for January 23rd . . . 12 ½ pounds meat; 4 hens; 1 chicken; 1 pound beans; [?] jitomates [tomatoes]; [?] chick peas; 1 cigar [each] for the insane fathers; [?] milk for Father Salvador; [?] sugar; 30 pounds [of the large; i.e., large tablets?] chocolate; 2 fruits for Father Miranda; 5 eggs.* [131]

[Chocolate Distribution: Hospital de Jesús, 1748–1751] *During this time [September, 1748] we had 103 patients that consumed 611 pounds of bread, 102 pounds of meat, 10,863 tabillas [chocolate bars], and 296 hens. Ten people died this month and 26 were discharged. [Signature] Antonio Avila, Principal Nurse, October 1748.* [132]

On January 30, 1771, Joseph Laporta, surgeon at the Hospital de la Religion Betlemitica in the city of Guadalajara, wrote to the Hospital Prefect asking for an increase in salary of 250 pesos (the normal salary being 400–500 pesos), plus an additional daily allowance for bread, meat, and chocolate. While Laporta was granted the salary increase, the request also generated an investigation into work patterns at the Betlemitica (Bethlehem) hospital. The Inspector asked the Hospital Prefect if Laporta deserved the salary increase and received the following reply:

Surgeon Laporte only comes to the hospital in the morning from 7:00 to 8:00, and while he keeps his surgical instruments at the hospital, this does not mean that he uses them. [133]

Chocolate expenditures at the Xalapa Hospital, southeast of Mexico City, were carefully recorded for the year 1775:

May. One chocolate grinding and general chocolate confections: 12 pounds of cocoa and 126 pounds of sugar;

September. Chocolate making: 127 pounds of cocoa, 27 pounds of sugar for chocolate and another 13 pounds of sugar for the making of chocolate;

October. One chocolate grinding and chocolate making: 9 pounds of cocoa and 9 pounds of sugar;

November. One chocolate grinding and chocolate making: 9 pounds of cocoa and 9 pounds of sugar;

December: One chocolate grinding and chocolate making: 9 pounds of cocoa and 9 pounds of sugar. [134]

After a serious typhoid epidemic in 1778, treatment at the Royal Hospital for Indians was reformed under a document entitled *Constituciones y ordenanzas para el régimen y gobierno del Hospital Real y General des los Indios de esta Nueva España.* This reorganization was ordered by the King of Spain. The hospital was important in the history of Mexico as it originally was founded in 1551 and was the location where Francisco Hernández wrote his important work, *Historia Natural de las cosas naturales de la Nueva España* [135]. The hospital reorganization document challenges the hospital staff to work for the benefit of the Indians:

The cooks, the atole makers [atoleras] and the chocolate makers [chocolateras] that are in charge to produce the food and other substances for the sick, they should apply great care in the fulfillment of their obligations in the care of the poor Indians, that even though they are very poor Indians and when they are healthy they eat poorly, when they are sick we should take good care of them. [136]

A second manuscript from the same hospital dated to 1778 contained a detailed description of the salaries received by all employees. The Apothecary, Don Francisco Pasapera, is mentioned in a chocolate-related context, that besides his salary, he also received two loaves of bread, two quarts of *atole*, six tortillas, and a chocolate tablet daily. On Fridays during Lent, Don

Pasapera also received 4 *reales* in compensation for the two pounds of mutton (*carnero* = tough meat from male sheep) that he was unable to consume due to fasting, and one small box of sweets each week [137].

Still other documents provide hospital-related information relative to chocolate:

[Contract to Supply Chocolate to Hospitals: 1782] *In the town of San Blas, Veracruz, on the 8th month of 1782, appeared before me Don Francisco Trillo y Hernandez, Commissary of this Department, with witnesses and he says: I will provide the food, medicines and all the other necessary things for the sick people that are in the hospital. I will assist the sick with the diets indicated by the Surgeon of the Department of the city, Don Juan García. These diets are: Morning: ordinary champurrado made with one ounce of chocolate and once ounce of sugar and one cup of atole. Noon time: a cup of stock without solids, fresh meat, ham, chicken [hen], and garbanzos [chick peas]. In the case of extreme prostration a wine soup or biscuits with eggs should be given. [138]*

An interesting document dated to December 31, 1784, signed by the Prior Father, Joseph Maxtinez (Martinez) considered expenditures at Convent-Hospital, San Roque, founded in 1592 in the city of Los Angeles (located in Oaxaca-Mexico; not California). The manuscript, *Expenditures for the Sustenance of the Poor and for their Care,* contained the following chocolate-associated comments:

We spent annually 1688 pesos to provide bread, meat, chicken, chocolate, atole for the poor sick and demented people, about 26 to 30 persons depending upon the month. [139]

We have spent 953 pesos annually to assist the brothers and the priests with bread, meat, chocolate and all the other things that were necessary to assist them. [140]

A medical document dated to 1808 recorded all donations made to the Hospital San Lazaro in Mexico City, with chocolate being among the items received:

It is recorded that 750 pesos have been paid to Felipe de Mendoze for 120 arrobas [=3000 pounds] for chocolate to be given to the sick; It is recorded that 60 pesos have been donated by Don Juan Echevereste to be used for chocolate for the sick; It is recorded that 60 arrobas of chocolate [=1500 pounds] are ordered to be ground by the Trustee of the Hospital San Lazaro, Manual García, for the sick and the servants of that hospital, at the cost of 375 pesos. [141]

Turning to hospital records for North America, these also reveal the extensive use of cacao beans, cocoa, and chocolate for treating the infirm, as this representative document reveals for the Salem Hospital, Salem, Massachusetts, in 1773 (Fig. 6.5):

Section II. Of the Patients [sic] Diet.

As it will be much for the Credit of the Hospital and Physician, and the Health of the Patients, to observe a suitable Regimen: It is Ordered—

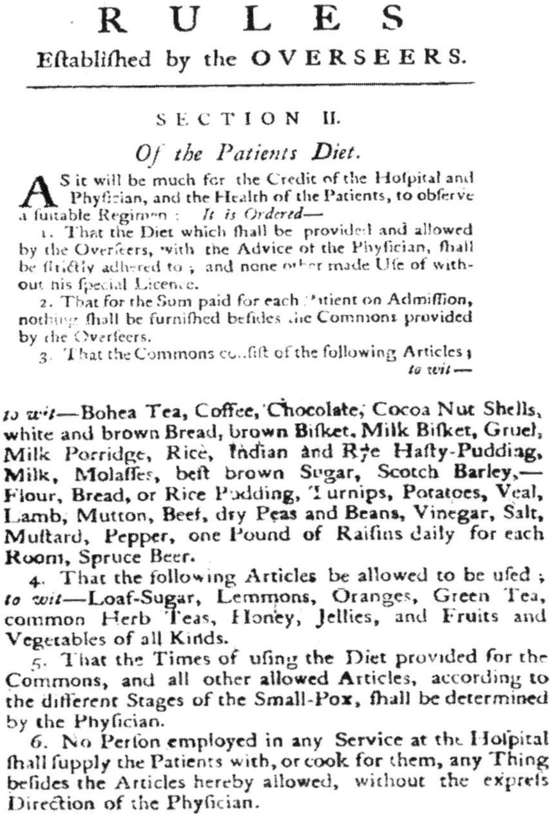

RULES

Established by the OVERSEERS.

SECTION II.

Of the Patients Diet.

AS it will be much for the Credit of the Hospital and Physician, and the Health of the Patients, to observe a suitable Regimen : *It is Ordered—*

1. That the Diet which shall be provided and allowed by the Overseers, with the Advice of the Physician, shall be strictly adhered to ; and none other made Use of without his special Licence.

2. That for the Sum paid for each Patient on Admission, nothing shall be furnished besides the Commons provided by the Overseers.

3. That the Commons consist of the following Articles ;
to wit—

*to wit—*Bohea Tea, Coffee, Chocolate, Cocoa Nut Shells, white and brown Bread, brown Bisket, Milk Bisket, Gruel, Milk Porridge, Rice, Indian and Rye Hasty-Pudding, Milk, Molasses, best brown Sugar, Scotch Barley,—Flour, Bread, or Rice Pudding, Turnips, Potatoes, Veal, Lamb, Mutton, Beef, dry Peas and Beans, Vinegar, Salt, Mustard, Pepper, one Pound of Raisins daily for each Room, Spruce Beer.

4. That the following Articles be allowed to be used ; *to wit—*Loaf-Sugar, Lemmons, Oranges, Green Tea, common Herb Teas, Honey, Jellies, and Fruits and Vegetables of all Kinds.

5. That the Times of using the Diet provided for the Commons, and all other allowed Articles, according to the different Stages of the Small-Pox, shall be determined by the Physician.

6. No Person employed in any Service at the Hospital shall supply the Patients with, or cook for them, any Thing besides the Articles hereby allowed, without the express Direction of the Physician.

FIGURE 6.5. Chocolate use in a hospital context. *Source:* Salem Hospital. Rules for Regulating Salem Hospital. Salem, MA: Samuel and Ebenezer Hall, 1773; pp. 4–5. Courtesy of Readex/Newsbank, Naples, FL. (Used with permission.)

1. That the Diet which shall be provided and allowed by the Overseers, with the Advice of the Physician, shall be strictly adhered to; and none other made Use of without his special License.

2. That for the Sum paid for each Patient on Admission, nothing shall be furnished besides the Commons [i.e., general food] provided by the Overseers.

3. That the Commons consist of the following Articles; to wit—Bohea Tea, Coffee, Chocolate, Cocoa Nut Shells, white and brown Bread, brown Bisket [sic], Milk Bisket [sic], Gruel, Milk Porridge, Rice, Indian and Rye Hasty-Pudding, Milk, Molasses, best brown Sugar, Scotch Barley, Flour, Bread, or Rice Pudding, Turnips, Potatoes, Veal, Lamb, Mutton, Beef, dry Peas and Beans, Vinegar, Salt, Mustard, Pepper, one Pound of Raisins daily for each Room, Spruce Beer.

4. That the following Articles be allowed to be used, to wit—Loaf-Sugar, Lemmons [sic], Oranges, Green Tea, common Herb Teas, Honey, Jellies, and Fruits and Vegetables of all Kinds.

5. That the Times of using the Diet provided for the Commons, and all other allowed Articles, according to the different States of the Small-Pox, shall be determined by the Physician.

6. No Person employed in any Service at the Hospital shall supply the Patients with, or cook, for them, any Thing besides the Articles hereby allowed, without the express Direction of the Physician. [142]

MISCELLANEOUS CHOCOLATE-RELATED ACCOUNTS

Suicide, commonly attributed as a mental disorder, has characterized all human societies. While we identified no specific instances of self-inflicted death using poisoned chocolate as the medium, in contrast to several accounts where murder was attempted or achieved through use of chocolate (see Chapter 21), accounts of suicide that appeared in Colonial era newspapers frequently provided the occupation/profession of the deceased, as in this sad case dated to 1772:

> Last Monday night one Dunn, chocolate grinder, journeyman to Mr. Wallace, at the north-end [of Boston] being delirious, threw himself out of a garret window into the street, where he was found the next morning. [143]

Fear of battle, malingering, and general unrest of soldiers during the Revolutionary War are themes not readily reported or discussed in American history books. All too often the images portrayed in history books are those akin to the staunch Minutemen who held the ground at Lexington and Concord. The negatives that have been reported generally have been expressed as disgruntlement over lack of food and quality thereof, or irregular or lack of payment for the risks and services rendered. The following document, a letter from American General Schuyler to Jonathan Trumbull, Supply Officer for the Continental Army and subsequently Governor of Connecticut, dated to 1775, reveals that not all military volunteers were pleased with their service and found ways for "early discharge" from the ranks of the Continental Army:

> It is certain, however, that some [of the recruits] have feigned sickness; for Dr. Stringer informs me, that on his way up here, about the 6th of September [1775], he met many men that looked very well; and, upon inquiry, some acknowledged they had procured their discharges by swallowing tobacco juice, to make them sick. Others had scorched their tongues with hot chocolate, to induce a believe [sic] that they had a fever, &c. [144]

CONTEMPORARY FIELD WORK: MEDICAL USES OF CHOCOLATE IN TRADITIONAL CULTURES

Between 1998 and 2006, members of our team conducted field work interviews Brazil, Colombia, Costa Rica, Cuba, Dominican Republic, Guatemala, Mexico, Panama, and Mixtec communities of the Central Valley of California to learn more about contemporary 20th and 21st century medicinal uses of chocolate.

Rebecca Schacker worked in the Dominican Republic and reported that cacao/chocolate beverages have a long history of use in traditional medicine, to improve kidney function, reduce anemia, halt diarrhea, cure sore throat, ease the brain when overexerted, and soothe stomachache. In the Dominican Republic, chocolate beverages blended with coconut milk and onion were drunk to reduce symptoms caused by the common cold. Respondents also informed her that chocolate beverages were encouraged to strengthen the lungs and to energize consumers [145].

Sylvia Escarcega conducted field work in the Mixteca Alta region of Oaxaca, Mexico, where she identified *curanderos* (traditional healers) who used cacao to treat the medical conditions known as *espanto* or *susto*, conditions thought to result when persons are startled or frightened. Both patients and healers returned to the exact location where the fright occurred, and the *curanderos* brought with them tobacco, bowls of fermented beverages, herbs, and cacao beans. The healers then fed the earth by planting cacao beans into the ground, as a form of payment in order to heal the patient. By restoring wealth to the earth, the evil that caused the fright was distracted, and persons suffering from *espanto* could then be healed and returned to health. Other healers interviewed in Oaxaca used chocolate to treat bronchitis, and to protect consumers against the stings of bees, wasps, and scorpions [146].

Jim Greishop and Temotao Mendoza surveyed the Mixtec Indian community at Madera, in the California San Joaquin Valley north of Fresno, to better understand chocolate and healing in contemporary times. Members of the Mixtec community in Madera had their cultural origins in Oaxaca and exhibited a wide range of medicinal uses for chocolate. Respondents reported that chocolate was healthy: a blend of chocolate and fresh beaten eggs added to hot water was used to combat fatigue; chocolate added to *manzanilla* tea was used to counter pain; and chocolate mixed with cinnamon and *ruda* (rue) alleviated stomachache. The Mixtec living in Madera reported that eating and drinking chocolate lowered high blood pressure; alternatively, chocolate was drunk in attempts to raise low blood pressure. Chocolate beverages also were drunk to alleviate symptoms of the common cold [147].

MEDICINAL USES OF OTHER PARTS OF THE CACAO TREE

The continued use of cacao as medicine by indigenous peoples of the Americas has been reviewed by a number of scholars, among them Eric Thompson and Julia Morton. Efforts by these scholars, coupled with information provided to our team members during field work interviews, reveals sustained medicinal uses throughout Central and South America of different components of the cacao tree, for example:

Cacao bark—used to treat bloody stools and to reduce abdominal pain.

Cacao fat—used to disinfect cuts; applied to burns; given to patients with liver and lung disorders; applied to the face to reduce chapping.

Cacao flowers—used to treat cuts specifically on the feet; mixed with water and drunk to improve mental apathy and to reduce timidity.

Cacao fruit pulp—prepared as a decoction and drunk by pregnant women to facilitate delivery.

Cacao leaves, especially young leaves—applied as antiseptics to wounds; astringent properties of young cacao leaves used to stanch excessive bleeding.

Cacao oil/butter—applied to wounds, burns, cracked lips, sore breasts; also applied to the anus to treat hemorrhoids and used vaginally to relieve irritation [148, 149].

Carmen Aguilera, in her *Flora y Fauna Mexicana, Mitologia y Tradiciones,* related that in rural Mexico chocolate is used to alleviate abdominal pains and poisonings and is considered "refreshing and invigorating." She noted widespread beliefs that chocolate preparations excited consumers in advance of sexual pleasure, but if the chocolate beverages are prepared using green seeds, this led to drunkenness and insanity. Aguilera noted that in ancient Mayan tradition seeds of wild cacao were burned as incense to represent the heart, liver, and kidney of the god *Ek Chuac*, patron deity of merchants, and to *Chaak* and *Hobnill*, deities who protected merchants (see Chapter 1) [150].

Conclusion

Cacao and chocolate have had extensive medicinal roles in the Americas and Europe. In some manuscripts, chocolate was touted and praised for a broad sweep of medical complaints; in other documents, chocolate beverages were recommended for narrower ranges of specific ills; in still other accounts, chocolate was not the central, healing element but was used instead to mask strong or irritating flavors or served as "binders" or "mixers," whereby other ingredients could readily be combined and easily administered to patients.

There are intriguing historical accounts that suggest eating or drinking chocolate could/would have had a positive effect on patients beyond merely the placebo effect and pleasure of consuming this food. Hughes wrote in 1672 that drinking chocolate alleviated asthma spasms [151]; Stubbe wrote in 1662 that drinking chocolate increased breast milk production [152]; and Colmenro de Ledesma suggested in 1631 that drinking chocolate could expel kidney stones [153]. Modern 21st century science has identified the vasodilatation and diuretic effects that follow chocolate consumption, and with chocolate's high energy value, the concept that chocolate could be a galactagogue (milk producer) at least should be considered. That chocolate is a diuretic is commonly recognized today in the 21st century, and through increased urine flow, it would be logical that some kidney stones would be expelled.

Throughout the historical period of the past 450 years, five consistent medical-related uses for chocolate can be identified through archive and library documents:

First: Numerous accounts suggest that chocolate caused weight gain in consumers. Ancient, colonial, and early modern physicians regularly recommended or prescribed chocolate drinking for the specific purpose of adding or restoring "flesh" to emaciated patients. The high caloric energy and potential weight gain would be a positive outcome of chocolate consumption especially in emaciated patients.

Second: Chocolate commonly was prescribed to patients to stimulate the nervous system, especially in patients identified as feeble, who lacked energy, or who suffered from apathy, exhaustion, lassitude, or symptoms that possibly could have been depression.

Third: At the same time, chocolate was said to calm overstimulated persons and patients who suffered from hyperactivity, due to its calming, soothing, almost tranquilizing effect.

Fourth: Numerous documents support the contention that drinking chocolate on a regular basis improved digestion and elimination; that chocolate was an effective prescription that countered weak or stagnant stomachs; that chocolate stimulated the kidneys and hastened urine flow; or that chocolate (because of its fat content) improved bowel function, softened stools, and thereby reduced or cured hemorrhoids.

Fifth: Chocolate regularly served as a pharmacological binder, whereby 16th through early 20th century physicians could administer drugs within chocolate beverages or pastes [154].

From earliest times into the 21st century, humans have enjoyed the product of this remarkable tree. How cacao beans pass from tree to farmer, to broker/merchant, to manufacturer, to ultimately become chocolate—has been retold in numerous publications [155, 156], and technological changes through the century are reviewed elsewhere in this book (see Chapter 46). Chocolate as food—chocolate as medicine: Some modern chocolate consumers linger and drink their beverage at the counters or tables of upscale shops in American urban centers en route to the office, while others, more traditional, prepare their daily

chocolate from scratch. Still other consumers treat cacao and chocolate as a gift from the gods, prepare cups or bowls of cacao, and take them as gifts into mountain areas where the bowls of chocolate are set upon the damp, soft, cushioned floors of tropical forests. At these times prayers are offered, as with the following words of the Jakaltek Maya of Guatemala, collected by Victor Montejo during summer field work in 2000 [157]:

Jakaltek	Spanish	English
Tawet, xhwa' xahanb'al,	*Para ti, mi ofrenda de posol,*	*For you, my gift of posol,*
Yal ixim b'oj kakaw, mamin,	*Posol con cacao, señor,*	*Posol with cocoa, milord,*
Yunhe mach xhtaj ha ti',	*Para aplacar tu sed,*	*To placate your thirst.*
Yulb'al tz'ayik sunil,	*Por nuchos dias,*	*For many days,*
Yub'al sunib'al hab'il, mamin.	*Y por muchos años señor.*	*And for many years, milord.*

So it is that the next time you purchase chocolate . . . unwrap the package carefully, examine the color (it will range from reddish-brown to sleek, dark black), enjoy the odor and attempt to describe it in words (what adjectives can be used?), anticipate, then experience the multiplicity of textures and flavors as the chocolate melts slowly inside your mouth. At such times, pause, reflect, and appreciate the unnamed peoples whose ancestors first domesticated and cultivated the cacao tree; appreciate the unnamed persons who created the first recipes for solid and liquid chocolate; appreciate the unnamed traditional healers who first used chocolate as medicine; and appreciate all who brought chocolate into the realm of human experience.

Pause, reflect, and appreciate—
Chocolate—Mesoamerica's gift to the world!

Acknowledgments

I wish to acknowledge the research efforts of the following colleagues, who contributed to the archival/field research or data summary at early stages in our medicinal chocolate project: Patricia Barriga, Teresa L. Dillinger, Sylvia Escárcega, James Grieshop, Martha Jimenez, Temoteo Mendoza, Victor Montejo, Diana Salazar Lowe, and Rebecca Schacker.

Endnotes

1. Portions of this chapter have been published previously [2–3]. The present chapter reflects a redaction and revision of these two papers and incorporates new materials identified during recent archive and library research.

2. Allopathic medicine, sometimes referred to as humoral medicine, originated in India, ca. 1500 BCE. The core of the ancient Indian, Ayurvedic medical system is based upon a balance among air, earth, fire, and water. The system spread westward from India into the eastern Mediterranean lands and influenced ancient Greek and Roman medical concepts. Subsequently, the system spread eastward from India into China. During the era of global exploration by the Spanish and Portuguese, their physicians took the humoral system and spread it globally. There remains the possibility that a hot–cold/we–dry system of allopathic medicine existed in pre-European contact Meso-America and developed independently from that introduced to Mexico by the Spanish and to Brazil by the Portuguese [21–24].

References

1. Bennett, A., Keen, C. L., and Shapiro, H.-Y. *Theobroma cacao*. In: *Biology, Chemistry, and Human Health*. Hoboken, NJ: Wiley Interscience, 2008.

2. Dillinger, T. L., Barriga, P., Escarcega, S., Jimenez, M., Salazar Lowe, D., and Grivetti, L. E. Food of the gods: cure for humanity? A cultural history of the medicinal and ritual use of chocolate. *Journal of Nutrition* 2000; 130 (Supplement):2057S–2072S.

3. Grivetti, L. E. From aphrodisiac to health food; a cultural history of chocolate. In: *Karger Gazette*, Issue 68 (Chocolate). Basel, Switzerland: S. Karger, 2005.

4. De la Cruz, M. (1552) *The Badianus Manuscript (Codex Barberini, Latin 241) Vatican Library; an Aztec Herbal of 1552*. Baltimore, MD: The Johns Hopkins University Press, 1940.

5. Sahagún, B. (1590) *Florentine Codex*. In: *General History of the Things of New Spain*. Santa Fe, NM: The School of American Research, 1981; and the University of Utah Monographs of the School of American Research and The Museum of New Mexico.

6. De la Cruz, M. (1552) *The Badianus Manuscript (Codex Barberini, Latin 241) Vatican Library; an Aztec Herbal of 1552*. Plate 70, 1940.

7. Sahagún, B. (1590). *Florentine Codex*. In: *General History of the Things of New Spain*. Santa Fe, NM: The School of American Research, 1981; and the University of Utah Monographs of the School of American Research and The Museum of New Mexico. Part 12: pp. 119–120.

8. Sahagún, B. (1590). *Florentine Codex*. In: *General History of the Things of New Spain*. Santa Fe, NM: The School of American Research, 1981; and the University of Utah Monographs of the School of American Research and The Museum of New Mexico. Part 12: pp. 119–120.

9. Sahagún, B. (1590). *Florentine Codex*. In: *General History of the Things of New Spain*. Santa Fe, NM: The School of American Research, 1981; and the University of Utah Monographs of the School of American Research and The Museum of New Mexico. Part 12: p. 170.

10. Sahagún, B. (1590). *Florentine Codex*. In: *General History of the Things of New Spain*. Santa Fe, NM: The School of American Research, 1981; and the University of Utah Monographs of the School of American Research and The Museum of New Mexico. Part 12: p. 176.

11. Sahagún, B. (1590). *Florentine Codex*. In: *General History of the Things of New Spain*. Santa Fe, NM: The School of American Research, 1981; and the University of Utah Monographs of the School of American Research and The Museum of New Mexico. Part 12: p. 12.

12. Sahagún, B. (1590). *Florentine Codex*. In: *General History of the Things of New Spain*. Santa Fe, NM: The School of American Research, 1981; and the University of Utah Monographs of the School of American Research and The Museum of New Mexico. Part 12: p. 178.

13. Sahagún, B. (1590). *Florentine Codex*. In: *General History of the Things of New Spain*. Santa Fe, NM: The School of American Research, 1981; and the University of Utah Monographs of the School of American Research and The Museum of New Mexico. Part 12: p. 189.

14. Roys, R. L. *Ritual of the Bacabs*. Norman: University of Oklahoma Press, 1965; pp. 35–37.

15. Torres, E. M. *La Relación Fraile-Indígena en Méxiso a Través de las Plantas. Tres Ejemplos: el Tabaco, el Cacao y las Nopaleras*. In: *Institutum Historicum Fratrum Praedictatorum, Romae. Archivum Fratrum Praedictorum*, Volume 61. Rome, Italy: Instituto Storico Domenicano, 1997; p. 244.

16. Hernández, F. *Historia de las Plantas de la Neuva Espana*. Mexico City, Mexico: Imprenta Universitaria, 1577; p. 304.

17. Hernández, F. *Historia de las Plantas de la Neuva Espana*. Mexico City, Mexico: Imprenta Universitaria, 1577; p. 304.

18. Hernández, F. *Historia de las Plantas de la Neuva Espana*. Mexico City, Mexico: Imprenta Universitaria, 1577; p. 305.

19. Hernández, F. *Historia de las Plantas de la Neuva Espana*. Mexico City, Mexico: Imprenta Universitaria, 1577; p. 305.

20. Hernández, F. *Historia de las Plantas de la Neuva Espana*. Mexico City, Mexico: Imprenta Universitaria, 1577; p. 305.

21. Grivetti, L. E. Nutrition past—nutrition today. Prescientific origins of nutrition and dietetics. Part 1. Legacy of India. *Nutrition Today* 1991;26(1):13–24.

22. Grivetti, L. E. Nutrition past—nutrition today. Prescientific origins of nutrition and dietetics. Part 2. Legacy of the Mediterranean. *Nutrition Today* 1991; 26(4):18–29.

23. Grivetti, L. E. Nutrition past—nutrition today. Prescientific origins of nutrition and dietetics. Part 3. Legacy of China. *Nutrition Today* 1991;26(6):6–17.

24. Grivetti, L. E. Nutrition past—nutrition today. Prescientific origins of nutrition and dietetics. Part 4. Aztec patterns and Spanish legacy. *Nutrition Today* 1992;27(3):13–25.

25. Cárdenas, Juan de. *Problemas y Secretos Maravillosos de las Indias.* Madrid, Spain: Ediciones Cultura Hispanica, 1591; Chapter 7, pp. 108f–108v.

26. Cárdenas, Juan de. *Problemas y Secretos Maravillosos de las Indias.* Madrid, Spain: Ediciones Cultura Hispanica, 1591; Chapter 7, pp. 110f–112v.

27. Cárdenas, Juan de. *Problemas y Secretos Maravillosos de las Indias.* Madrid, Spain: Ediciones Cultura Hispanica, 1591; Chapter 7, p. 112v.

28. Cárdenas, Juan de. *Problemas y Secretos Maravillosos de las Indias.* Madrid, Spain: Ediciones Cultura Hispanica, 1591; Chapter 8, p. 115f.

29. Cárdenas, Juan de. *Problemas y Secretos Maravillosos de las Indias.* Madrid, Spain: Ediciones Cultura Hispanica, 1591; Chapter 8, p. 117v.

30. Cárdenas, Juan de. *Problemas y Secretos Maravillosos de las Indias.* Madrid, Spain: Ediciones Cultura Hispanica, 1591; Chapter 8, pp. 117v–118f.

31. Cárdenas, Juan de. *Problemas y Secretos Maravillosos de las Indias.* Madrid, Spain: Ediciones Cultura Hispanica, 1591; Chapter 8. p. 117v.

32. Acosta, J. de. *The Naturall and Morall Historie of the East and West Indies. Intreating of the Remarkable Things of Heaven, of the Elements, Metals, Plants and Beasts.* . . . London, England: Val. Sims for Edward Blount and William Aspley, 1604; p. 271.

33. Valverde Turices, S. de. *Un Discurso del Chocolate.* Seville, Spain: J. Cabrera, 1624; pp. D1–2.

34. Colmenero de Ledesma, A. *Curioso Tratado de la Naturaleza y Calidad del Chocolate.* Madrid, Spain: Francisco Martinez, 1631; p. A4.

35. Hurtado, T. *Chocolate y Tabaco Ayuno Eclesiastico y Natural: si este le Quebrante el Chocolate: y el Tabaco al Natural, para la Sagrada Comunion. Por Francisco Garcia, Impressor del Reyno. A Costa de Manuel Lopez.* Madrid, Spain: Mercador de Libros, 1645; Volume 1, 1:1, 3:21.

36. Hurtado, T. *Chocolate y Tabaco Ayuno Eclesiastico y Natural: si este le Quebrante el Chocolate: y el Tabaco al Natural, para la Sagrada Comunion. Por Francisco Garcia, Impressor del Reyno. A Costa de Mauel Lopez.* Madrid, Spain: Mercador de Libros, 1645; Volume 1, 2:13.

37. Gage, T. *The English American: His Travail by Sea; or a New Survey of the West Indies Containing a Journall [sic] of Three Thousand Three Hundred Miles within the Main Land of America.* . . . London, England: R. Cotes, 1648; pp. 107–108.

38. Gage, T. *The English American: His Travail by Sea; or a New Survey of the West Indies Containing a Journall [sic] of Three Thousand Three Hundred Miles within the Main Land of America.* . . . London, England: R. Cotes, 1648; p. 108.

39. Gage, T. *The English American: His Travail by Sea; or a New Survey of the West Indies Containing a Journall [sic] of Three Thousand Three Hundred Miles within the Main Land of America.* . . . London, England: R. Cotes, 1648; p. 108.

40. Gage, T. *The English American: His Travail by Sea; or a New Survey of the West Indies Containing a Journall [sic] of Three Thousand Three Hundred Miles within the Main Land of America.* . . . London, England: R. Cotes, 1648; p. 110.

41. Stubbe, H. *The Indian Nectar, or, a Discourse Concerning Chocolata [sic]: the Nature of the Cacao-Nut and the Other Ingredients of that Composition is Examined and Stated According to the Judgment and Experience of Indian and Spanish Writers.* London, England: J. C. for Andrew Crook, 1662; p. 3.

42. Stubbe, H. *The Indian Nectar, or, a Discourse Concerning Chocolata [sic]: the Nature of the Cacao-Nut and the Other Ingredients of that Composition is Examined and Stated According to the Judgment and Experience of Indian and Spanish Writers.* London, England: J. C. for Andrew Crook, 1662; p. 8.

43. Stubbe, H. *The Indian Nectar, or, a Discourse Concerning Chocolata [sic]: the Nature of the Cacao-Nut and the Other Ingredients of that Composition is Examined and Stated According to the Judgment and Experience of Indian and Spanish Writers.* London, England: J. C. for Andrew Crook, 1662; p. 11.

44. Stubbe, H. *The Indian Nectar, or, a Discourse Concerning Chocolata [sic]: the Nature of the Cacao-Nut and the Other Ingredients of that Composition is Examined and Stated According to the Judgment and Experience of Indian and Spanish Writers.* London, England: J. C. for Andrew Crook, 1662; p. 30.

45. Stubbe, H. *The Indian Nectar, or, a Discourse Concerning Chocolata [sic]: the Nature of the Cacao-Nut and the Other Ingredients of that Composition is Examined and Stated According to the Judgment and Experience of Indian and Spanish Writers.* London, England: J. C. for Andrew Crook, 1662; p. 30.

46. Stubbe, H. *The Indian Nectar, or, a Discourse Concerning Chocolata [sic]: the Nature of the Cacao-Nut and the Other Ingredients of that Composition is Examined and Stated According to the Judgment and Experience of Indian and Spanish Writers.* London, England: J. C. for Andrew Crook, 1662; p. 43.

47. Stubbe, H. *The Indian Nectar, or, a Discourse Concerning Chocolata [sic]: the Nature of the Cacao-Nut and the Other Ingredients of that Composition is Examined and Stated According to the Judgment and Experience of Indian and Spanish Writers.* London, England: J. C. for Andrew Crook, 1662; pp. 53–54.

48. Stubbe, H. *The Indian Nectar, or, a Discourse Concerning Chocolata [sic]: the Nature of the Cacao-Nut and the Other Ingredients of that Composition is Examined and Stated According to the Judgment and Experience of Indian and

Spanish Writers. London, England: J. C. for Andrew Crook, 1662; pp. 58–60.

49. Stubbe, H. *The Indian Nectar, or, a Discourse Concerning Chocolata [sic]: the Nature of the Cacao-Nut and the Other Ingredients of that Composition is Examined and Stated According to the Judgment and Experience of Indian and Spanish Writers*. London, England: J. C. for Andrew Crook, 1662; p. 67.

50. Stubbe, H. *The Indian Nectar, or, a Discourse Concerning Chocolata [sic]: the Nature of the Cacao-Nut and the Other Ingredients of that Composition is Examined and Stated According to the Judgment and Experience of Indian and Spanish Writers*. London, England: J. C. for Andrew Crook, 1662; pp. 68–69.

51. Stubbe, H. *The Indian Nectar, or, a Discourse Concerning Chocolata [sic]: the Nature of the Cacao-Nut and the Other Ingredients of that Composition is Examined and Stated According to the Judgment and Experience of Indian and Spanish Writers*. London, England: J. C. for Andrew Crook, 1662; p. 79.

52. Stubbe, H. *The Indian Nectar, or, a Discourse Concerning Chocolata [sic]: the Nature of the Cacao-Nut and the Other Ingredients of that Composition is Examined and Stated According to the Judgment and Experience of Indian and Spanish Writers*. London, England: J. C. for Andrew Crook, 1662; pp. 83–84.

53. Stubbe, H. *The Indian Nectar, or, a Discourse Concerning Chocolata [sic]: the Nature of the Cacao-Nut and the Other Ingredients of that Composition is Examined and Stated According to the Judgment and Experience of Indian and Spanish Writers*. London, England: J. C. for Andrew Crook, 1662; pp. 97–98.

54. Stubbe, H. *The Indian Nectar, or, a Discourse Concerning Chocolata [sic]: the Nature of the Cacao-Nut and the Other Ingredients of that Composition is Examined and Stated According to the Judgment and Experience of Indian and Spanish Writers*. London, England: J. C. for Andrew Crook, 1662; p. 98.

55. Hughes, W. *The American Physitian [sic], or A Treatise of the Roots, Plants, Trees, Shrubs, Fruit, Herbs etc. Growing in the English Plantations in America . . . whereunto is Added a Discourse of the Cacao-Nut-Tree, and the Use of its Fruit; with All the Ways of Making of Chocolate. The Like Never Extant Before*. London, England: J. C. for William Crook the Green Dragon without Temple-Bar, 1672; p. 124.

56. Hughes, W. *The American Physitian [sic], or A Treatise of the Roots, Plants, Trees, Shrubs, Fruit, Herbs etc. Growing in the English Plantations in America . . . whereunto is Added a Discourse of the Cacao-Nut-Tree, and the Use of its Fruit; with All the Ways of Making of Chocolate. The Like Never Extant Before*. London, England: J. C. for William Crook the Green Dragon without Temple-Bar, 1672; p. 143.

57. Hughes, W. *The American Physitian [sic], or A Treatise of the Roots, Plants, Trees, Shrubs, Fruit, Herbs etc. Growing in the English Plantations in America . . . whereunto is Added a Discourse of the Cacao-Nut-Tree, and the Use of its Fruit; with All the Ways of Making of Chocolate. The Like Never Extant Before*. London, England: J. C. for William Crook the Green Dragon without Temple-Bar, 1672; p. 144.

58. Hughes, W. *The American Physitian [sic], or A Treatise of the Roots, Plants, Trees, Shrubs, Fruit, Herbs etc. Growing in the English Plantations in America . . . whereunto is Added a Discourse of the Cacao-Nut-Tree, and the Use of its Fruit; with All the Ways of Making of Chocolate. The Like Never Extant Before*. London, England: J. C. for William Crook the Green Dragon without Temple-Bar, 1672; p. 145.

59. Hughes, W. *The American Physitian [sic], or A Treatise of the Roots, Plants, Trees, Shrubs, Fruit, Herbs etc. Growing in the English Plantations in America . . . whereunto is Added a Discourse of the Cacao-Nut-Tree, and the Use of its Fruit; with All the Ways of Making of Chocolate. The Like Never Extant Before*. London, England: J. C. for William Crook the Green Dragon without Temple-Bar, 1672; p. 146.

60. Hughes, W. *The American Physitian [sic], or A Treatise of the Roots, Plants, Trees, Shrubs, Fruit, Herbs etc. Growing in the English Plantations in America . . . whereunto is Added a Discourse of the Cacao-Nut-Tree, and the Use of its Fruit; with All the Ways of Making of Chocolate. The Like Never Extant Before*. London, England: J. C. for William Crook the Green Dragon without Temple-Bar, 1672; pp. 147–148.

61. Hughes, W. *The American Physitian [sic], or A Treatise of the Roots, Plants, Trees, Shrubs, Fruit, Herbs etc. Growing in the English Plantations in America . . . whereunto is Added a Discourse of the Cacao-Nut-Tree, and the Use of its Fruit; with All the Ways of Making of Chocolate. The Like Never Extant Before*. London, England: J. C. for William Crook the Green Dragon without Temple-Bar, 1672; pp. 153–154.

62. Dufour, P. S. *The Manner of Making of Coffee, Tea, and Chocolate as it is Used in Most Parts of Europe, Asia, Africa and America, with Their Vertues [sic]*. London, England: William Crook, 1685; pp. 75–76.

63. Dufour, P. S. *The Manner of Making of Coffee, Tea, and Chocolate as it is Used in Most Parts of Europe, Asia, Africa and America, with Their Vertues [sic]*. London, England: William Crook, 1685; p. 77.

64. Dufour, P. S. *The Manner of Making of Coffee, Tea, and Chocolate as it is Used in Most Parts of Europe, Asia, Africa and America, with Their Vertues [sic]*. London, England: William Crook, 1685; p. 99.

65. Dufour, P. S. *The Manner of Making of Coffee, Tea, and Chocolate as it is Used in Most Parts of Europe, Asia, Africa and America, with Their Vertues [sic]*. London, England: William Crook, 1685; pp. 111–113.

66. Dufour, P. S. *The Manner of Making of Coffee, Tea, and Chocolate as it is Used in Most Parts of Europe, Asia, Africa and America, with Their Vertues [sic]*. London, England: William Crook, 1685; pp. 115–116.

67. Blégny, N. *Le Bon Useage du Thé, du Caffé, et du Chocolat pour la Préservation & pour la Guérison des Maladies.* Paris, France: E. Michallet, 1687; pp. 282–285.

68. de Quélus. *The Natural History of Chocolate: Being a Distinct and Particular Account of the Cocoa-Tree . . . the Best Way of Making Chocolate is Explain'd; and Several Uncommon Medicines Drawn from It, Are Communicated.* London, England: J. Roberts, 1730; p. 44.

69. de Quélus. *The Natural History of Chocolate: Being a Distinct and Particular Account of the Cocoa-Tree . . . the Best Way of Making Chocolate is Explain'd; and Several Uncommon Medicines Drawn from It, Are Communicated.* London, England: J. Roberts, 1730; p. 45.

70. de Quélus. *The Natural History of Chocolate: Being a Distinct and Particular Account of the Cocoa-Tree . . . the Best Way of Making Chocolate is Explain'd; and Several Uncommon Medicines Drawn from It, Are Communicated.* London, England: J. Roberts, 1730; p. 46.

71. de Quélus. *The Natural History of Chocolate: Being a Distinct and Particular Account of the Cocoa-Tree . . . the Best Way of Making Chocolate is Explain'd; and Several Uncommon Medicines Drawn from It, Are Communicated.* London, England: J. Roberts, 1730; p. 46.

72. de Quélus. *The Natural History of Chocolate: Being a Distinct and Particular Account of the Cocoa-Tree . . . the Best Way of Making Chocolate is Explain'd; and Several Uncommon Medicines Drawn from It, Are Communicated.* London, England: J. Roberts, 1730; p. 48.

73. de Quélus. *The Natural History of Chocolate: Being a Distinct and Particular Account of the Cocoa-Tree . . . the Best Way of Making Chocolate is Explain'd; and Several Uncommon Medicines Drawn from It, Are Communicated.* London, England: J. Roberts, 1730; p. 50.

74. de Quélus. *The Natural History of Chocolate: Being a Distinct and Particular Account of the Cocoa-Tree . . . the Best Way of Making Chocolate is Explain'd; and Several Uncommon Medicines Drawn from It, Are Communicated.* London, England: J. Roberts, 1730; p. 51.

75. de Quélus. *The Natural History of Chocolate: Being a Distinct and Particular Account of the Cocoa-Tree . . . the Best Way of Making Chocolate is Explain'd; and Several Uncommon Medicines Drawn from It, Are Communicated.* London, England: J. Roberts, 1730; p. 56.

76. de Quélus. *The Natural History of Chocolate: Being a Distinct and Particular Account of the Cocoa-Tree . . . the Best Way of Making Chocolate is Explain'd; and Several Uncommon Medicines Drawn from It, Are Communicated.* London, England: J. Roberts, 1730; p. 58.

77. de Quélus. *The Natural History of Chocolate: Being a Distinct and Particular Account of the Cocoa-Tree . . . the Best Way of Making Chocolate is Explain'd; and Several Uncommon Medicines Drawn from It, Are Communicated.* London, England: J. Roberts, 1730; p. 59.

78. de Quélus. *The Natural History of Chocolate: Being a Distinct and Particular Account of the Cocoa-Tree . . . the Best Way of Making Chocolate is Explain'd; and Several Uncommon Medicines Drawn from It, Are Communicated.* London, England: J. Roberts, 1730; p. 73.

79. de Quélus. *The Natural History of Chocolate: Being a Distinct and Particular Account of the Cocoa-Tree . . . the Best Way of Making Chocolate is Explain'd; and Several Uncommon Medicines Drawn from It, Are Communicated.* London, England: J. Roberts, 1730; p. 76.

80. de Quélus. *The Natural History of Chocolate: Being a Distinct and Particular Account of the Cocoa-Tree . . . the Best Way of Making Chocolate is Explain'd; and Several Uncommon Medicines Drawn from It, Are Communicated.* London, England: J. Roberts, 1730; pp. 76–77.

81. de Quélus. *The Natural History of Chocolate: Being a Distinct and Particular Account of the Cocoa-Tree . . . the Best Way of Making Chocolate is Explain'd; and Several Uncommon Medicines Drawn from It, Are Communicated.* London, England: J. Roberts, 1730; pp. 77–78.

82. de Quélus. *The Natural History of Chocolate: Being a Distinct and Particular Account of the Cocoa-Tree . . . the Best Way of Making Chocolate is Explain'd; and Several Uncommon Medicines Drawn from It, Are Communicated.* London, England: J. Roberts, 1730; p. 78.

83. Linné (Linneaus), C. von. *Om Chokladdryken.* Stockholm, Sweden: Fabel, 1965.

84. Stanhope, P. D. *Letters Written by the Late Right Honourable Philip Dormer Stanhope, Earl of Chesterfield, to His Son, Philip Stanhope, Esq; Late Envoy Extraordinary at the Court of Dresden.* Edited by Eugenia Stanhope. New York: J. Rivington and H. Caine, 1775; Volume 3, pp. 81–82.

85. Lardizabal, V. *Memoria Sobre las utilidades de el Chocolate para Precaber las Incomodidades, que resultan del uso de las Aguas minerales, y promover sus buenos efectos, como los de los Purgantes, y otros remedios; y para curar diertas dolencias.* Pamplona, Spain: Antonio Castilla, 1788; pp. 16–18.

86. Buchan, A. P. *Medicina Domestica ó Tratado de las Enfermadades Quirurgicas y Cirugia en General del Celebre Buchan M.d. del Real Colegio Medico de Edimburgo.* Madrid, Spain: La Imprenta Real, 1792; p. 224.

87. Lavedan, A. *Tratado de Los Usos, Abusos, Propiendades y Virtudes del Tabaco, Café, Te y Chocolate.* Madrid, Spain: Imprenta Real, 1796; Chapter V, p. 223.

88. Lavedan, A. *Tratado de Los Usos, Abusos, Propiendades y Virtudes del Tabaco, Café, Te y Chocolate.* Madrid, Spain: Imprenta Real, 1796; Chapter V, pp. 221–237.

89. Brillat-Savarin, J. A. (1825?) *The Physiology of Taste.* New York: Houghton Mifflin, 1915; p. 95.

90. Brillat-Savarin, J. A. (1825?) *The Physiology of Taste.* New York: Houghton Mifflin, 1915; pp. 95–96.

91. Brillat-Savarin, J. A. (1825?) *The Physiology of Taste.* New York: Houghton Mifflin, 1915; p. 97.

92. Brillat-Savarin, J. A. (1825?) *The Physiology of Taste.* New York: Houghton Mifflin, 1915; p. 100.

93. Graham, T. J. *Medicina Moderna Casera. O Tratado Popular, en el que se ilustran el caracter, sintomas, causas, distincion, y plan curativo currecto, de todas las enfermedades incidentales al cuerpo humano.* London, England: Juan Davy, 1828; p. 231.

94. Graham, T. J. *Medicina Moderna Casera. O Tratado Popular, en el que se ilustran el caracter, sintomas, causas, distincion, y plan curativo currecto, de todas las enfermedades incidentales al cuerpo humano.* London, England: Juan Davy, 1828; pp. 412–413.

95. Milne Edwards, X., and Vavasseur, P. *Manual de Materia Medica ó de los Medicamentos, por los Doctores en Medicina,* 2nd edition, Volume 2. Barcelona, Spain: R. R. M. Indar, 1835; pp. 339–340.

96. General Archive of the Nation, Mexico City (AGN). Gobernacion. Legajos. Num. 165. Caja 250, Exped. 1, San Luis Potosi, 1836.

97. Saint-Arroman, A. *Coffee, Tea and Chocolate: Their Influence upon the Health, the Intellect, and the Moral Nature of Man.* Philadelphia, PA: Townsend Ward, 1846; p. 84.

98. Saint-Arroman, A. *Coffee, Tea and Chocolate: Their Influence upon the Health, the Intellect, and the Moral Nature of Man.* Philadelphia, PA: Townsend Ward, 1846; p. 85.

99. Saint-Arroman, A. *Coffee, Tea and Chocolate: Their Influence upon the Health, the Intellect, and the Moral Nature of Man.* Philadelphia, PA: Townsend Ward, 1846; p. 86.

100. Saint-Arroman, A. *Coffee, Tea and Chocolate: Their Influence upon the Health, the Intellect, and the Moral Nature of Man.* Philadelphia, PA: Townsend Ward, 1846; p. 87.

101. *Scientific American*, Volume 4, No. 31, April 21, 1849.

102. Debay, A. *Les Influences du Chocolat du Thé et du Café sur l'Économie Humaine. Leur Analyse Chimique, leurs Falsifications, leur Role Important dans l'Alimentation. Ouvrage Faisant Suite a l'Hygiène Alimentaire.* Paris, France: Dentu, 1864; p. 58.

103. Debay, A. *Les Influences du Chocolat du Thé et du Café sur l'Économie Humaine. Leur Analyse Chimique, leurs Falsifications, leur Role Important dans l'Alimentation. Ouvrage Faisant Suite a l'Hygiène Alimentaire.* Paris, France: Dentu, 1864; p. 60.

104. Debay, A. *Les Influences du Chocolat du Thé et du Café sur l'Économie Humaine. Leur Analyse Chimique, leurs Falsifications, leur Role Important dans l'Alimentation. Ouvrage Faisant Suite a l'Hygiène Alimentaire.* Paris, France: Dentu, 1864; pp. 86–87.

105. Debay, A. *Les Influences du Chocolat du Thé et du Café sur l'Économie Humaine. Leur Analyse Chimique, leurs Falsifications, leur Role Important dans l'Alimentation. Ouvrage Faisant Suite a l'Hygiène Alimentaire.* Paris, France: Dentu, 1864; pp. 88–89.

106. Debay, A. *Les Influences du Chocolat du Thé et du Café sur l'Économie Humaine. Leur Analyse Chimique, leurs Falsifica-tions, leur Role Important dans l'Alimentation. Ouvrage Faisant Suite a l'Hygiène Alimentaire.* Paris, France: Dentu, 1864; p. 90.

107. Debay, A. *Les Influences du Chocolat du Thé et du Café sur l'Économie Humaine. Leur Analyse Chimique, leurs Falsifications, leur Role Important dans l'Alimentation. Ouvrage Faisant Suite a l'Hygiène Alimentaire.* Paris, France: Dentu, 1864; p. 91.

108. Debay, A. *Les Influences du Chocolat du Thé et du Café sur l'Économie Humaine. Leur Analyse Chimique, leurs Falsifications, leur Role Important dans l'Alimentation. Ouvrage Faisant Suite a l'Hygiène Alimentaire.* Paris, France: Dentu, 1864; pp. 101–108.

109. Pandés y Poblet, J. *La Educacion de la Mujer. Segun los mas Ilustres Moralistas é Higienistas de Ambos Sexos,* Volume 2. Barcelona, Spain: Jaime Seix y Compañia, 1878; pp. 187–188.

110. Pandés y Poblet, J. *La Educacion de la Mujer. Segun los mas Ilustres Moralistas é Higienistas de Ambos Sexos,* Volume 2. Barcelona, Spain: Jaime Seix y Compañia, 1878; p. 191.

111. Pandés y Poblet, J. *La Educacion de la Mujer. Segun los mas Ilustres Moralistas é Higienistas de Ambos Sexos,* Volume 2. Barcelona, Spain: Jaime Seix y Compañia, 1878; p. 192.

112. *Manual del Farmacéutico,* 2nd edition. Mexico City, Mexico: No publisher given, 1881–1882; pp. 202–203.

113. *Manual del Farmacéutico,* 2nd edition. Mexico City, Mexico: No publisher, 1881–1882; pp. 354–356.

114. Reboles y Campos, G. *Higiene Therapéutica. La Higiene Alimenticia.* Madrid, Spain: Carlos Bailly-Bailliere, 1888; p. 183.

115. Villanueva y Francesconi, M. *El Medico y La Botica en Casa. Manual de Medicina Domestica.* Mexico City, Mexico: Imprenta de Aguilar é Hijos, 1890; p. 329.

116. Villanueva y Francesconi, M. *El Medico y La Botica en Casa. Manual de Medicina Domestica.* Mexico City, Mexico: Imprenta de Aguilar é Hijos, 1890; p. 333.

117. Villanueva y Francesconi, M. *El Médico y la Botica en Casa: Manuyal de Medicina Domestica.* Mexico City, Mexico: Imprenta de Aguilar e Hijos, 1897; p. 23.

118. Ponce, J. M. *La Medicina en el Hogar. Pequeno Tratado de Medicina Domestica.* Mexico City, Mexico: Editorial Católica, 1902; p. 91.

119. Ponce, J. M. *La Medicina en el Hogar. Pequeno Tratado de Medicina Domestica.* Mexico City, Mexico: Editorial Católica, 1902; p. 123.

120. Bardina, J. *Higiene Moderna. Manual Hispanoamericano.* Madrid, Spain: Sociedad General de Publicaciones, 1905; p. 307.

121. *The Boston News-Letter.* March 17, 1711, p. 2.

122. *Connecticut Courant.* July 19, 1774, p. 2.

123. *The Pennsylvania Packet*. January 17, 1777, no page.

124. *Vermont Gazette*. November 7, 1785, p. 3.

125. *Vermont Gazette*. February 11, 1788, p. 2.

126. Mather, C. *Manudictio ad ministerium. Directions for a Candidate of the Ministry. Wherein, First, a Right Foundation is Laid for His Future Improvement; and, then, Rules Are Offered for Such a Management of His Academical & Preparatory Studies; and Thereupon, for Such a Conduct After His Appearance in the World; as May Render Him a Skilful and Useful Minister of the Gospel*. Boston, MA: Thomas Hancock, 1726; p. 132.

127. Cadwalader, T. *An Essay on the West-India Dry-Gripes; With the Methods of Preventing and Curing that Cruel Distemper. To Which is Added, an Extraordinary Case in Physick*. Philadelphia, PA: Benjamin Franklin, 1745; p. 6.

128. *The Boston Gazette, or Weekly Journal*. March 5, 1751, p. 1.

129. Philomath, T. *The Virginia Almanack [sic] for the Year of Our Lord God 1770 ... Calculated According to Art, and Referred to the Horizon of 38 Degrees North Latitude, and a Meridian of Five Hours West from the City of London; Fitting Virginia, Maryland, North Carolina, & c.* Williamsburg, VA: Purdie and Dixon, 1769; p. 16.

130. Archivo General de la Nación (AGN) (General Archive of the Nation), Mexico City. AGN. Colegios, Vol. 1, exp. 1, fs. 1–8, 1769, 12 fs. p. 4.

131. AGN. Colegios, Vol. 1, exp. 1, fs. 1–8, 1769, 12fs. p. 9.

132. AGN. Hospital de Jesús, Vol. 30, exp. 1, fs. 212f–221f.

133. Medrano, F. A. Historia de los Hospitales Coloniales de Hispanoamérica. New York: Editorial Arenas, 1992; Vol. 3, Sec. 4, p. 201.

134. AGN. Templos y Conventos, Vol. 31, exp.5, fs. 145f.

135. Hernández, F. *Historia de las Plantas de la Neuva Espana*. Mexico City, Mexico: Imprenta Universitaria, 1577; p. 304.

136. Constituciones y Ordenanzas para el Régimen, y Gobierno del Hospital Real, y General de los Indios de esta Nueva España, mandadas guarder por S. M. en Real Cédula de 27 de octubre del año de 1776. Edición Facsimilar con Licencia del Superior Gobierno. Impresas en México, en la neuva Oficina Madrileña de D. Felipe de Zuñiga y Ontiveros, calle de la Palma, año 1778. Mexico City, Mexico: Rolston-Bain, 1982; p. 49.

137. *Historia de los Hospitales Coloniales de Hispanoamerica*. Siglos XVI–XIX, Volume 2, p. 220.

138. AGN. Hospitales, Vol. 67, exp. 3, fs. 87f–90v.

139. AGN. Templos y Conventos, Vol. 23, exp. 17, fs. 138f–149f, Year 1784, manuscript page 140.

140. AGN. Templos y Conventos, Vol. 23, exp. 17, fs. 138f–149f, Year 1784, manuscript page 140.

141. AGN. Hospitales, Vol. 5, exp. 4, fs. 251f–253f.

142. Salem Hospital. *Rules for Regulating Salem Hospital*. Salem, MA: Samuel and Ebenezer Hall, 1773; pp. 4–5.

143. *The Massachusetts Spy*. February 27, 1772, p. 207.

144. Letter: Philip Schuyler to Governor Jonathan Trumbull, written at Fort Ticonderoga, October 12, 1775. (American Archives, Series 4, Vol. 3, p. 1035).

145. Schacker, R. 2001. Field Report: Dominican Republic. Typed ms.

146. Escarcega, Sylvia. 2000. Field Report: Oaxaca, Mexico. Typed ms.

147. Greishop, J., and Mendoza, T. 2001. Field Report: Madera, California. Typed ms.

148. Thompson, J. E. S. Notes on the use of cacao in Middle America. Notes on Middle American Archaeology. *Ethnology* 1956;128:95–116.

149. Morton, J. F. *Atlas of Medicinal Plants of Middle America*. Springfield, IL: Charles C. Thomas, 1981.

150. Aguilera, C. *Flora y Fauna Mexicana. Mitologia y tradiciones*. Mexico City, Mexico: Editorial Everest Mexicana, 1985; p. 120.

151. Hughes, W. *The American Physitian [sic], or A Treatise of the Roots, Plants, Trees, Shrubs, Fruit, Herbs etc. Growing in the English Plantations in America ... whereunto is Added a Discourse of the Cacao-Nut-Tree, and the Use of its Fruit; with All the Ways of Making of Chocolate. The Like Never Extant Before*. London, England: J. C. for William Crook the Green Dragon without Temple-Bar, 1672; pp. 153–154.

152. Stubbe, H. *The Indian Nectar, or, a Discourse Concerning Chocolata [sic]: the Nature of the Cacao-Nut and the Other Ingredients of that Composition is Examined and Stated According to the Judgment and Experience of Indian and Spanish Writers*. London, England: J. C. for Andrew Crook, 1662; pp. 58–60.

153. Colmenero de Ledesma, A. *Curioso Tratado de la Naturaleza y Calidad del Chocolate*. Madrid, Spain: Francisco Martinez, 1631; p. A4.

154. Dillinger, T. L., Barriga, P., Escarcega, S., Jimenez, M., Salazar Lowe, D., and Grivetti, L. E. Food of the gods: cure for humanity? A cultural history of the medicinal and ritual use of chocolate. *Journal of Nutrition* 2000;130 (Supplement):2057S–2072S.

155. Coe, S. D., and Coe, M. D. *The True History of Chocolate*. London, England: Thames and Hudson, 1996.

156. Young, A. *The Chocolate Tree. A Natural History of Cacao*. Washington, DC: Smithsonian Institute Press, 1994.

157. Montejo, V. 2000. Field Report: Guatemala. Typed ms.

7

Chocolate and the Boston Smallpox Epidemic of 1764

Louis Evan Grivetti

Introduction

Smallpox has been a human scourge from remote antiquity until the late 20th century of the Common Era, when the World Health Organization in Geneva, Switzerland, reported that the disease had been eliminated [1].[1] Evidence of smallpox can be seen in Egyptian mummies dated to the 18th to 20th Dynasties (16th to 11th centuries BCE), and it is likely that the ancient Egyptian king Ramses V died of smallpox [2, 3]. From earliest times until the 16th century, smallpox was restricted in geographical distribution only to the Old World (Africa, Asia, and Europe) but the era of global exploration by the English, French, Portuguese, and Spanish dramatically changed this distribution.

It is not certain when the first documented cases of smallpox reached the New World (North, Central, South America, and the Caribbean) but it may have been in Haiti (island of Hispaniola) by 1518, and then spread to Cuba in 1519 [4]. The first case of smallpox on the American mainland, however, can be identified easily, since the disease was described by Bernal Diaz del Castillo in his account of the conquest of Mexico. Bernal Diaz related how Hernán Cortés ventured to the east coast of Mexico to put down a revolt and in his commentary reported that a Black man (slave?), Francisco de Eguía [5], had arrived from Cuba with smallpox:

> Narváez [Pánfilo de Narváez] brought with him a Negro who was in the small pox; an unfortunate importation for that country, for the disease spread with inconceivable rapidity, and the Indians died by thousands; for not knowing the nature of it, they brought it to a fatal issue by throwing themselves into cold water in the heat of the disorder. [6]

Cases of smallpox then rapidly spread among indigenous peoples of Mexico, Central America, and South America and devastated native populations [7]. A plate in the Florentine Codex painted ca. 1575 shows the Mexica/Aztecs suffering from the disease [8]. Evidence suggests that smallpox epidemics struck the Yucatán Peninsula shortly after arrival of the Europeans and may have killed half the population there, with similar death levels suffered later by the Inca in Peru [9].

Between the years 1520 and 1581 Mexico suffered epidemics that caused the deaths of untold hundreds of thousands of Native Americans who had no immunity to Old World diseases. The first smallpox outbreak in 1520 was called *hueyzahuatl* by the Mexica/Aztecs and was followed 11 years later in 1531 by a second smallpox outbreak, this time perhaps associated with measles. Subsequent epidemics in 1545 and 1576–1581 of a disease called *cocoliztli* by the Mexica/Aztecs killed an estimated one-third of the remaining inhabitants of New Spain and led to widespread depopulation [10, 11].

Chocolate: History, Culture, and Heritage. Edited by Grivetti and Shapiro
Copyright © 2009 John Wiley & Sons, Inc.

Smallpox knew no geographical boundaries and followed exploration and commercial routes into North America as well. By one estimate, 90 percent of the Native Americans living in Massachusetts between the years 1617 to 1619 died of smallpox, and another epidemic struck Plymouth Colony in 1633, killing both settlers and regional Native Americans alike [12]. Smallpox epidemics ravaged geographical areas of Louisiana (1865–1873), Massachusetts (1721–1722, 1764, and 1775–1782), Pennsylvania (1865–1873), South Carolina (1738), and the Black Hills of modern South Dakota (1876) [13].

During early centuries, smallpox as a disease was poorly understood and there was essentially no defense. Once smallpox was contracted, medical care primarily consisted of bed rest and palliative care. Physicians of the time did not understand how smallpox was transmitted but they recognized it was contagious. Patients were quarantined, commonly with warning flags posted outside homes, and treatment was designed to keep patients comfortable. Whether one lived or died was thought to be in the "hands of God."

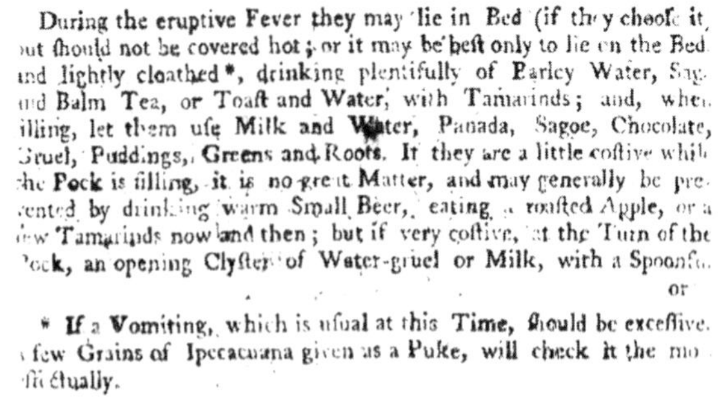
During the eruptive Fever they may lie in Bed (if they choose it, but should not be covered hot; or it may be best only to lie on the Bed and lightly cloathed *, drinking plentifully of Barley Water, Sage and Balm Tea, or Toast and Water, with Tamarinds; and, when filling, let them use Milk and Water, Panada, Sagoe, Chocolate, Gruel, Puddings, Greens and Roots. If they are a little costive while the Pock is filling, it is no great Matter, and may generally be prevented by drinking warm Small Beer, eating a roasted Apple, or a few Tamarinds now and then; but if very costive, at the Turn of the Pock, an opening Clyster of Water-gruel or Milk, with a Spoonfu... or

* If a Vomiting, which is usual at this Time, should be excessive, a few Grains of Ipecacuana given as a Puke, will check it the most effectually.

FIGURE 7.1. Chocolate as an ingredient in a recipe to avoid smallpox. *Source*: Saunders, R. (Benjamin Franklin). *Poor Richard improved: being an almanack and ephemeris . . . for the year of our Lord 1761 . . . Fitted to the latitude of forty degrees, and a meridian of near five hours west from London; but may, without sensible error, serve all the northern colonies*. Philadelphia: B. Franklin and D. Hall, 1760. Courtesy: Readex/Newsbank, Naples, FL. (Used with permission.)

Chocolate and Smallpox Treatment in 18th Century North America

By 1759, certain North American physicians had advanced the view that chocolate had a role to play during the treatment of smallpox. Doctors prescribed what in the 21st century would be called high caloric diets to patients that, if tolerated, would have lessened potential weight loss associated with the disease. Such diets primarily were palliative, with little influence on the disease course.[2] Dietary recommendations for the treatment of smallpox commonly included chocolate. Chocolate was recommended for smallpox patients by Richard Saunders (i.e., Benjamin Franklin) in *Poor Richard's Almanack* (*sic*) edition for the year 1761 (Fig. 7.1) [14]. The author of *Hutchin's Improved Almanack* (*sic*) also identified chocolate among the foods recommended to be consumed prior to sitting for smallpox inoculations [15]. Thomas Dimsdale republished the Hutchin's account without acknowledgment in 1771 [16]. The well-known and respected Philadelphia physician, Benjamin Rush, reaffirmed in 1781 that "tea, coffee, and even weak chocolate with biscuit" would help those residents who had decided to be inoculated against smallpox [17]. Throughout the decade of the 1780s, Rush continued to recommend chocolate in the diet for those suffering from smallpox [18] and restated the same recommendation in his 1794 medical text:

> *Tea, coffee, and even weak chocolate, with biscuit or dry toast, may be used as usual.* [19]

Others advised readers what to eat prior to receiving their smallpox inoculations:

> *I order such of my patients, as constitute the first class, and who are by much the majority, to live in the following manner: to abstain from all animal food, including broths, also butter and cheese, and from all fermented liquors, excepting small beer, which is allowed sparingly, and from all spices, and whatever possesses a manifest heating quality. The diet is to consist of pudding, gruel, sago, milk, rice-milk, fruit pyes [sic], greens, roots, and vegetables of all the kinds in season, prepared or raw. Eggs, tho' not to be eaten alone, are allowed in puddings, and butter in pye-crust [sic]; the patients are to be careful that they do not eat such a quantity as to overload their stomachs, even of this kind of food. Tea, coffee, or chocolate are permitted for breakfast, to those who choose or are accustomed to them. In this manner they are to proceed about nine or ten days before the operation [i.e., inoculation].* [20]

Once patients had contracted smallpox and the fever and pustules were manifest, they were advised to do the following:

> *During the eruptive Fever they may lie in Bed (if they choose it) but should not be covered hot; or it may be best only to lie on the Bed and lightly cloathed [sic], drinking plentifully of Barley Water, Sage and Balm Tea, or Toast and Water, with Tamarinds; and, when filling, let them use Milk and Water, Panda, Sagoe [sic], Chocolate, Gruel, Puddings, Greens and Roots. If they are a little costive while the Pock is filling, it is no-great Matter, and may generally be prevented by drinking warm Small Beer, eating a roasted Apple, or a few Tamarinds now and then; but if very costive, at the Turn of the Pock, an opening Clyster of Water-gruel or Milk, with a Spoonfull [sic] or two of Melasses [sic] and Oil in it, will be proper every Day or two.* [21]

Smallpox in Boston

Boston was one of the primary ports in North America, and sailors and merchants mingled freely in the give-and-take of Colonial enterprise. It was in this atmosphere of trade and commerce that the population of Boston periodically was exposed to myriad diseases that sometimes blossomed into fully developed epidemics. Perhaps the most violent and destructive of these occurred in 1721 when smallpox ravaged Boston. At this time the estimated population of Boston was 11,000 and approximately 6000 residents became infected; of these, 844 died or about 1 person for every 6 or 7 cases contracted [22]. The 1721 Boston smallpox epidemic also is important because of the influences of two individuals, Cotton Mather, the famed jurist, and Zabdiel Boylston, a prominent Boston physician, who both championed the cause of inoculation when other physicians doubted and the general population feared the process [23].

When the 1721 smallpox epidemic abated, Boston aldermen, physicians, and residents continued to debate the efficacy of smallpox inoculations; there was much talk but little action and resolution. The stage then was set for another disaster and it is to the Boston smallpox epidemic of 1764 that we now turn. This outbreak remains important to history because of its extensive newspaper coverage and documentation, whereas the 1721 outbreak received relatively little newspaper attention for reasons that are not clear.[3] The 1764 outbreak is interesting, too, because of direct and tangential references to chocolate.

SMALLPOX IN BOSTON, 1764

The Boston Post Boy issue for January 2, 1764 carried the dreaded news that smallpox not only had returned to Boston but that Sea Captain Joseph Buckley had died of the disease (Fig. 7.2). The newspaper provided the terse announcement of Buckley's death [24].

But the report of no other cases proved premature. By January 16 the epidemic had spread further within the city (Fig. 7.3) *The Boston Post Boy* carried the news that three families were infected; a cluster of four families that lived along Fish Street near the Old North Meeting House, and another family identified at a second site near the Reverend Eliot's Meeting House. Each of these families was identified by name:

> *THE Public are hereby Notified, That [sic] the SMALL-POX IS NOW BUT IN Five Families in this Town, viz. At the Widow Demett's, Mr. Benjamin Adams's, Mr. Thomas Anderson's, and Mr. Daniel Warten's, all in Fish-Street, in the Neighbourhood of the Old North Meeting House; and at Mr. Benjamin Labree's, near the Rev. Mr. Eliot's Meeting House.*
>
> *By Order of the Select Men,*
>
> *WILLIAM COOPER, Town-Clerk. [25]*

One week later, on January 23, *The Boston Post Boy* published even worse news and informed the public that as of 11:00 a.m. (on the date of publication) 13 families were infected. The announcement was crafted with carefully chosen words as if to calm the readership. The public was alerted that *only* 13 families were infected, as if identification of this low number would limit the spread of false information. The announcement, ordered by William Cooper, Town-Clerk, also noted that "smallpox flags" were to be displayed outside all homes where the ill were treated (Fig. 7.4) [26].

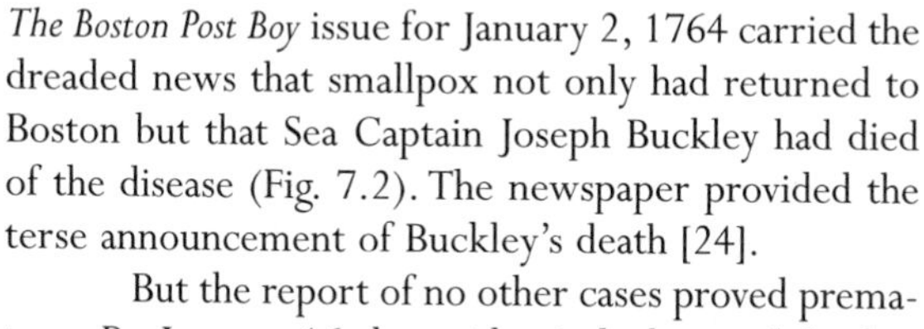

Boſton, January 2, 1764.
Capt. *Joſeph Buckley*, who has been ſick of the Small-Pox for ſome Days at the North Part of this Town, died this Morning. No other Perſon in Town has that Diſtemper.

FIGURE 7.2. Smallpox epidemic announcement, Boston, Massachusetts. Identification of first case: death of Sea Captain Joseph Buckley. *Source: The Boston Post Boy*, January 2, 1764. Courtesy of Readex/Newsbank, Naples, FL. (Used with permission.)

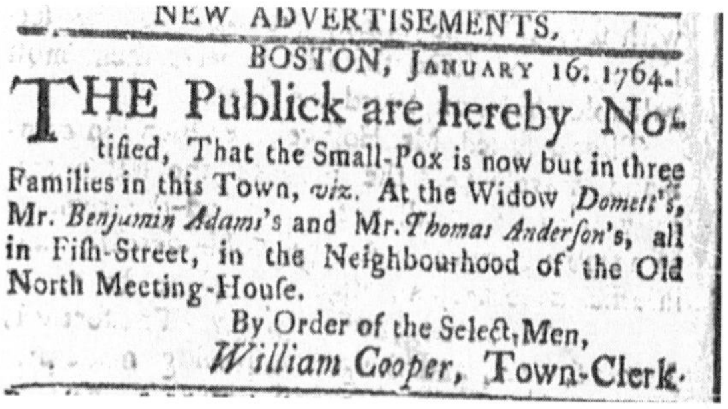

NEW ADVERTISEMENTS.
BOSTON, JANUARY 16. 1764.
THE Publick are hereby Notified, That the Small-Pox is now but in three Families in this Town, *viz.* At the Widow *Domett's*, Mr. *Benjamin Adams's* and Mr. *Thomas Anderſon's*, all in Fiſh-Street, in the Neighbourhood of the Old North Meeting-Houſe.
By Order of the Selećt Men,
William Cooper, Town-Clerk.

FIGURE 7.3. Smallpox epidemic announcement, Boston, Massachusetts. Disease has spread within the city. *Source: The Boston Post Boy*, January 16, 1764. Courtesy of Readex/Newsbank, Naples, FL. (Used with permission.)

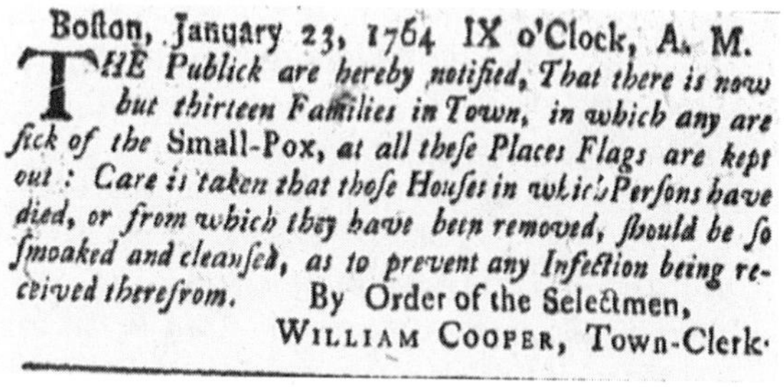

Boſton, January 23, 1764 IX o'Clock, A. M.
THE Publick are hereby notified, That there is now but thirteen Families in Town, in which any are ſick of the Small-Pox, *at all theſe Places Flags are kept out : Care is taken that thoſe Houſes in which Perſons have died, or from which they have been removed, ſhould be ſo ſmoaked and cleanſed, as to prevent any Infection being received therefrom.* By Order of the Selećtmen,
WILLIAM COOPER, Town-Clerk.

FIGURE 7.4. Smallpox epidemic announcement, Boston, Massachusetts. Houses of those infected to be marked with flags. *Source: The Boston Post Boy*, January 23, 1764. Courtesy of Readex/Newsbank, Naples, FL. (Used with permission.)

Also on January 23, *The Boston Post Boy* published the text of an Act passed three days previously on January 20 by the Selectmen (Councilmen) of Boston. The intent of the Act was to restrict the spread of smallpox and it established a baseline number of cases—set at 30 families—before the Boston Selectmen could implement specific measures that restricted individual rights. The January 20 Act informed the public that it was criminal to conceal family members who had smallpox, and readers were informed of their personal responsibility and obligation to report all cases. A second component of the Act deemed it illegal for individuals to take it upon their own responsibility to inoculate themselves (or family) members *before* the 30-family threshold had been identified. The text continued, however, that should specific residents of Boston wish to undergo inoculation prior to identification of the 30 families, they could petition the Selectmen in writing for a review of their request:

The following ACT passed the Great and General Court of this Province on Friday last: and is here inserted [in the newspaper] by Order of the Select-Men of the Town of Boston, for the Information of all Persons whom it may concern. An ACT to prevent, if possible, the further spreading of the Small-Pox in the Town of Boston. WHEREAS it is represented to this Court, that there is still hope that the spreading the Small-Pox in the Town of Boston may be prevented, if due Care be taken: inasmuch as the Families visited with that Distemper, generally live in the same Neighborhood: Be it therefore Enacted by the Governor, Council and House of Representatives: That all Persons in the said Town shall be held and bound to observe all the Directions of the Law of this Province, made in the Sixteenth Year of his late Majesty's Reign, intitled [sic] An Act to prevent the spreading of the Small-Pox and other infections Sickness, and to prevent the concealing the same, and under all the Penalties in the said Act contained, until that thirty Families are known to be visited in the said Town at one Time with that Distemper. AND no Person shall presume to inoculate or be Innoculated [sic] in the said Town, without the Leave of the major Part of the Selectmen in Writing, at their Meeting for such Purpose, until that thirty Families are known to be visited with the said Distemper at one Time, unless before that Time the Selectmen of the said Town shall give public Notice, that they have no hope to stop the Progress of the said Distemper, on the Penalty of the Sum of Fifty Pounds, to be recovered and applied as in said Act is mentioned: And the Selectmen of said Town are hereby required, so soon as the Number of thirty Families shall be visited with that Distemper, to give Notice thereof in the several Boston News-Papers, for the Information and Satisfaction of such as are minded to be Innoculated [sic]. This Act to continue and be in Force for two Months from this Day, and no longer. [Published January 20, 1764]. [27]

Individual noncompliance to any part of the Act resulted in a fine of 50 pounds (sterling), an astonishing sum

given that daily wages of Bostonian unskilled laborers at this time would not have exceeded two shillings per day [28].

Fear of the pox continued and families were especially concerned for their children. *The Boston Evening Post*, January 23, 1764, published a front-page statement about how to protect newborn infants from contracting smallpox (and ever exhibiting this disease later in life). While published with good intent and based on presumed knowledge of the time, such recommendations would not have been effective:

To the Publishers of the BOSTON EVENING-POST. Please to publish in your next [issue] for the Benefit of Posterity, the following approved Method to prevent, at the Birth of a Child, its having the Small Pox or any Diseases of Putrefaction ever after, viz. WHEN a Child is new-born, and the Midwife going to tye [sic] and cut the navel-string, let the thread that is to tye [sic] it not be drawn close immediately: but when it is about the navel-string, and ready to draw and knit close, drive up with the finger and thumb, the blood that is at the root of the navel, that so you may drive away out of the child's body, the loose blood that is newly come in by the spring [sic—string meant] and when all is strained out, draw presently the thread close, and knit it fast, and cut off the navel-string: this will cause that this child will never at any age, have the SMALL.POX, tho' [sic] he or she should converse daily with those which are infected with it. This hath been often tried. [29]

Throughout the remainder of January into March 1764, the incidence of smallpox in Boston increased. Individuals who feared the disease were forced to decide what actions were best for themselves and their families. Some merchants elected to stay inside the city; others feared their neighbors and relocated. Newspapers published early in February 1764 documented that Boston merchants moved from locations associated geographically with smallpox outbreaks to sites presumably far from the diseased areas.

Bartholemew Stavers, merchant, published an advertisement in the *Boston News-Letter and New-England Chronicle* issue for February 2 and informed his customers that he had moved himself and his goods from Boston's North End to a new location on King Street in central Boston for the purpose of once more attracting customers. In this case, it is interesting that Stavers moved from the periphery into the core of Boston:

Bartholemew Stavers, Carrier from Boston to Piscataqua, GIVES Notice That he has removed from the North-End, to Mrs. Bean's in King-street, being at a great Distance from the Houses infected with the Small-Pox: Whoever has any Business for him to transact, may see him from Wednesday evenings to Thursday Evenings. [30]

Rebecca Walker, merchant, relocated her inventory that included chocolate to Roxbury, south

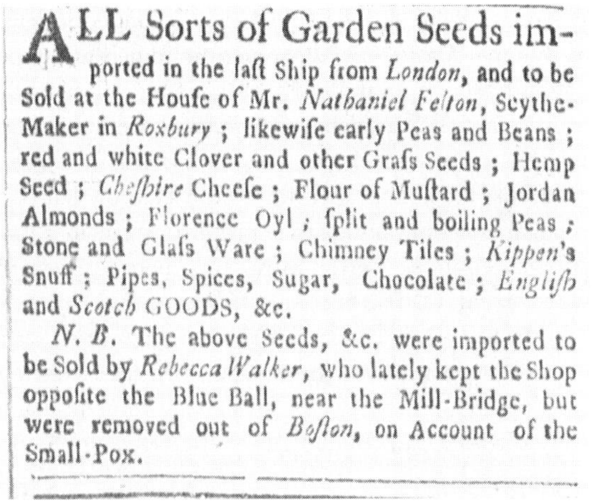

ALL Sorts of Garden Seeds imported in the laſt Ship from *London*, and to be Sold at the Houſe of Mr. *Nathaniel Felton*, Scythe-Maker in *Roxbury* ; likewiſe early Peas and Beans ; red and white Clover and other Graſs Seeds ; Hemp Seed ; *Cheſhire* Cheeſe ; Flour of Muſtard ; Jordan Almonds ; Florence Oyl ; ſplit and boiling Peas ; Stone and Glaſs Ware ; Chimney Tiles ; *Kippen's* Snuff ; Pipes, Spices, Sugar, Chocolate ; *Engliſh* and *Scotch* GOODS, &c.

N. B. The above Seeds, &c. were imported to be Sold by *Rebecca Walker*, who lately kept the Shop oppoſite the Blue Ball, near the Mill-Bridge, but were removed out of *Boſton*, on Account of the Small-Pox.

FIGURE 7.5. Smallpox epidemic announcement, Boston, Massachusetts. Rebecca Walker, chocolate merchant, has relocated to avoid the pox. *Source: The Boston Gazette and Country Journal*, February 20, 1764. Courtesy of Readex/ Newsbank, Naples, FL. (Used with permission.)

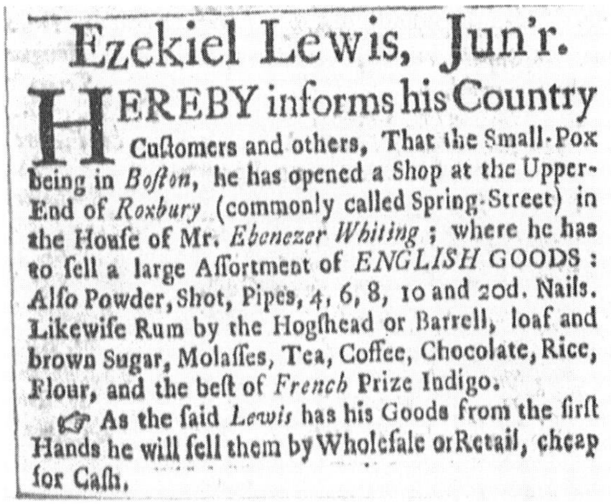

Ezekiel Lewis, Jun'r.

HEREBY informs his Country Cuſtomers and others, That the Small-Pox being in *Boſton*, he has opened a Shop at the Upper-End of *Roxbury* (commonly called Spring-Street) in the Houſe of Mr. *Ebenezer Whiting* ; where he has to ſell a large Aſſortment of *ENGLISH* GOODS : Alſo Powder, Shot, Pipes, 4, 6, 8, 10 and 20d. Nails. Likewiſe Rum by the Hogſhead or Barrell, loaf and brown Sugar, Molaſſes, Tea, Coffee, Chocolate, Rice, Flour, and the beſt of *French* Prize Indigo.

☞ As the ſaid *Lewis* has his Goods from the firſt Hands he will ſell them by Wholeſale or Retail, cheap for Caſh.

FIGURE 7.6. Smallpox epidemic announcement, Boston, Massachusetts. Ezekiel Lewis, Junior, chocolate merchant, has relocated to avoid the pox. *Source: The Boston Post Boy*, March 3, 1764. Courtesy of Readex/Newsbank, Naples, FL. (Used with permission.)

of Boston proper. A friend/relative (?) of Walker, identified as a Mr. Nathaniel Felton, scythe-maker, published an advertisement in *The Boston Gazette and Country Journal* on February 20 that informed Walker's customers that she had relocated to his house because of the smallpox epidemic. Walker's advertisement informed her usual customers that they could purchase a broad range of imported goods—among them chocolate—at her new location (Fig. 7.5) [31].

Ezekiel Lewis, merchant, also left Boston for Roxbury and announced his departure in *The Boston Post Boy* on March 3, 1764, (Fig. 7.6). Lewis then sold his Boston merchandise, including chocolate, from the house of his relative/friend (?), Mr. Ebenezer Whiting, located on Spring Street, in Roxbury [32].

The Jackson brothers, William and James, in partnership for many years, placed their relocation announcement in the March 12, 1764, issue of *The Boston Gazette and Country Journal* (Fig. 7.7). Troubled by the smallpox epidemic, the brothers relocated to Waltham (approximately 10 miles west of Boston) and asked their customers for consideration [33].

Inoculations against the smallpox began in late January or early February 1764 in Boston but the precise date cannot be determined through newspaper publications. The March 31 issue of *The Providence Gazette and Country Journal* carried an announcement from the Governor and His Majesty's Council (approved five days earlier on March 26) that the inoculation program was proving successful:

BOSTON, March 26th. His Excellency the Governor, with the Advice of His Majesty's Council, has been pleased to appoint Thursday the 12th Day of April next, to be observed as a Day of Fasting and Prayer throughout this Province. The Practice of Inoculation for the Small-Pox in Town, goes on successfully; upwards of a Thousand Persons are already

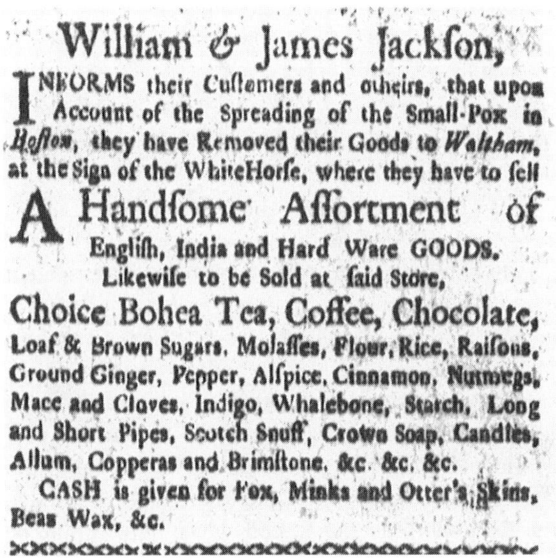

William & James Jackſon,

INFORMS their Cuſtomers and others, that upon Account of the Spreading of the Small-Pox in *Boſton*, they have Removed their Goods to *Waltham*, at the Sign of the White Horſe, where they have to ſell A Handſome Aſſortment of Engliſh, India and Hard Ware GOODS. Likewiſe to be Sold at ſaid Store, Choice Bohea Tea, Coffee, Chocolate, Loaf & Brown Sugars, Molaſſes, Flour, Rice, Raiſons, Ground Ginger, Pepper, Alſpice, Cinnamon, Nutmegs, Mace and Cloves, Indigo, Whalebone, Starch, Long and Short Pipes, Scotch Snuff, Crown Soap, Candles, Allum, Copperas and Brimſtone, &c. &c. &c.

CASH is given for Fox, Minks and Otter's Skins, Bees Wax, &c.

FIGURE 7.7. Smallpox epidemic announcement, Boston, Massachusetts. William and James Jackson, chocolate merchants, have relocated to avoid the pox. *Source: The Boston Gazette and Country Journal*, March 12, 1764. Courtesy of Readex/Newsbank, Naples, FL. (Used with permission.)

recovered of that Distemper, which makes many who opposed the Practice now embrace it. [34]

Evidence suggests, however, that the smallpox epidemic continued since inoculations proceeded into the spring and early summer of 1764. Evidence for continuation also is confirmed by a letter from a citizen known only by the initials "A. B." published in the June 18, 1764 issue of *The Boston Gazette and Country Journal*. This letter requested that the editors, Enes and Gill, publish a prayer said to have the properties of protecting those about to be inoculated. The Editors of the

Journal complied and published A.B.'s lengthy account that requested God's protection, of which a segment is presented here:

> O Strengthen and support me during this alarming Trial: soften the Pains, and abate the Violence of the Disorder: let thy good Spirit suggest the most proper Means for my Preservation and Recovery: and let thy gracious Providence give those Means their best and most beneficial Effects! And, O! be pleased to hearken [sic] to the Prayers of my Friends for me every where [sic], that if it be thy good Pleasure, I may be restored to them with renewed Life, and Health, and Vigour [sic]. But, O Lord, while I am preparing my mortal Part for this dreaded Trial, let me not neglect to prepare my Soul for Eternity. The utmost I can hope from Success in this Pursuit, is to prolong my Live [sic], perhaps, for a few transient Years: let me not fail then to make Provision for that immortal State, which will continue when Time shall be no more, beyond the Reach of Disaster or Casualty. O pardon all my Frailties, Negligences [sic] and Sins, wheresoever [sic] committed, or of whatsoever Nature: erase them from the Annals of my Life, and grant they may never rise in Judgment against me. And hrwever [sic] thou art pleased to dispose of Life here, or of this still more brittle and changeful Form: grant I may be finally happy with thee hereafter, when my Soul shall be cloathed [sic] with eternal Health, and Youth, and Beauty. O grant all this, and whatever else is needful for me, through Jesus Christ our Lord! AMEN. [35]

During the spring and into the summer of 1764, some Boston merchants, among them John and Thomas Stevenson, prepared additional advertisement copy in their attempts to attract customers. The Stevenson brothers published their advertisement on June 25 in *The Boston Evening Post* and suggested that the smallpox epidemic was due to mercantile goods being landed at the port of Boston. In the Stevenson's perception, if goods were unloaded elsewhere—and subsequently transported overland to their shop located in the Cornhill district of Boston—such items would be smallpox-free, safe for purchase, and customers should thus be assured:

> TO BE SOLD By John and Thomas Stevenson, At the Sign of the Three Nuns in Cornhill, BOSTON, and at the London Coffee-House, opposite the Custom-House in SALEM,—an Assortment of Scotch and English GOODS—such as they usually trade in;—which they will sell very reasonable by Wholesale. N.B. As the Goods at Salem were never landed at Boston; they are free of the Infection of the Small-Pox. [36]

CONCLUDING COMMENTS ON CHOCOLATE AND 1764 SMALLPOX EPIDEMIC

The year 1764 was a smallpox disaster in Boston, but not as terrible as the epidemic of 1721. As the epidemic of 1764 subsided, it became clear that city and health officials had begun to manage the number of patients through a quarantine system. We searched for but found no information regarding incidence of

persons infected or death rates. We found evidence that home quarantine edicts were modified and some patients were removed to the Point Shirley hospital, located outside the densely populated area of central Boston northwest of the city on a narrow, isolated peninsula at the entrance to Boston Harbor. Here, in relative isolation, authorities presumed that smallpox patients could be better managed. Announcement of patient removal also included the statement—important for its time in history—that increased global trade exposed merchants to more types of diseases such as smallpox, and that city officials needed to be prepared well in advance to counter more effectively any future outbreaks:

> The Growth of the British Colonies, and their increasing Intercourse with one another, and with Europe, must render the Inhabitants more and more exposed to the Small-Pox; a Distemper, fatal to a large Proportion of such as are seized with it accidentally and without Preparation. From this Consideration, the General Court has wisely encouraged the Establishment of an Hospital at Point Shirley, where such as are inclined may at any time and with a moderate Charge be conducted thro' this Disease in the safe and easy Method of Inoculation. An Institution of this Kind has long been desired by the most sensible and considerate Part of the Community: And whoever visits the Hospital will at once perceive that no Situation could be more happy for the Design and afford so many Accommodations with equal Security from Communicating the Infection abroad: — The Inhabitants of Chelsea are so satisfied of the Care taken at the Hospital in this Respect, that they readily allow such as are recovered to pass through the Town in [sic] their Way home. [37]

Other examples of "good news" touted the high quality of care at the Point Shirley hospital (Fig. 7.8) [38]. These notices emphasized the availability of personal accommodations nearby where family members could stay so as to be able to visit and assist their infected relatives more readily:

> Physicians of the Town of Boston who are engaged in carrying on the inoculating Hospital at Point-Shirley, being prevented giving their constant Attendance there during the continuance of the Small-Pox in Town, hereby notify the public, that they are join'd by Doctor Barnett of New Jersey, who will constantly attend at said Hospital with one or other of said Physicians whose Business will permit, and employ the utmost Diligence and Attention for the relief of those that put themselves under their care: They further notify, that Point-Shirley contains as many comfortable and decent Houses as will be sufficient to accommodate as many Persons as will probably ever offer for Inoculation at one Time, from this or the neighbouring [sic] Governments, and is will furnished with every requisite Convenience both for Sickness and Health. [39]

Then, as quickly as the smallpox virus had infected the citizens of Boston in 1721 and 1764, it disappeared and lay dormant until the next outbreak

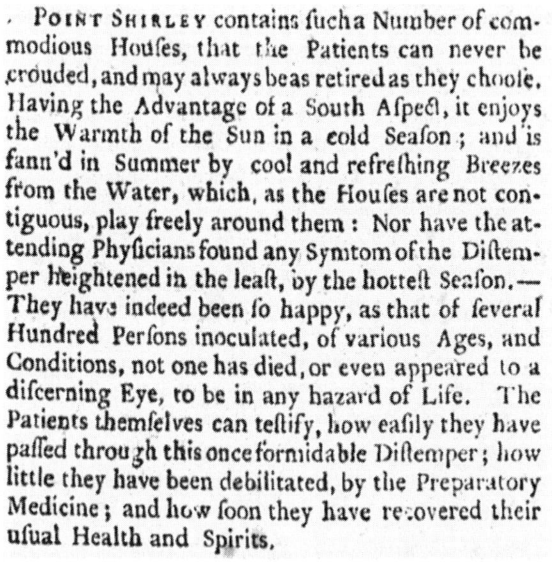

POINT SHIRLEY contains fucha Number of com-
modious Houfes, that the Patients can never be
crouded, and may always beas retired as they choofe.
Having the Advantage of a South Afpect, it enjoys
the Warmth of the Sun in a cold Seafon ; and is
fann'd in Summer by cool and refrefhing Breezes
from the Water, which, as the Houfes are not contiguous, play freely around them : Nor have the attending Phyficians found any Symtom of the Diftemper heightened in the leaft, by the hotteft Seafon. —
They have indeed been fo happy, as that of feveral
Hundred Perfons inoculated, of various Ages, and
Conditions, not one has died, or even appeared to a
difcerning Eye, to be in any hazard of Life. The
Patients themfelves can teftify, how eafily they have
paffed through this once formidable Diftemper ; how
little they have been debilitated, by the Preparatory
Medicine ; and how foon they have recovered their
ufual Health and Spirits.

FIGURE 7.8. Smallpox epidemic announcement, Boston, Massachusetts. Inoculations and patient accommodations available at Point Shirley. *Source: The Boston Evening Post*, August 6, 1764, p. 3. Courtesy of Readex/Newsbank, Naples, FL. (Used with permission.)

in 1775–1776 [40]. The practice of inoculation as a prevention against smallpox continued until the pioneering research of Edward Jenner, who discovered in 1796 that inoculation of patients with cow pox (vaccination) would prevent smallpox [41].

Elsewhere beyond Boston, smallpox and other Old World diseases killed Native Americans in the New World at an extraordinary rate and by some estimates between 8 and 15 million, perhaps more, died as a result of the accidental introduction of diseases during the era of global exploration [42, 43]. A series of smallpox epidemics among the Cherokee nation during 1729, 1738, and 1753 killed approximately 50 percent of the population [44].

Across the North American continent on the West Coast in California other Native Americans and Spanish colonists feared the pox as well (see Chapter 33). A letter dated 1782 written at the Santa Barbara Mission in *Alta* California by Father Fermin Lasuen reported to Father Junipero Serra on the difficulties faced by another mission. Father Lasuen wrote that the San Diego Mission suffered from a very poor location related to "sterility of the land, epidemics, and hostility from the Native Americans" and he closed his letter to Father Serra with the following words:

Thanks God we do not suffer now, but we had some epidemics [like] those of Antiqua California [Baja] . . . we are always fearing smallpox. [45]

Approximately 200 years after these terrible events along both eastern and western coasts of North America, the last known human case of smallpox, any-

where globally, was identified on April 14, 1978, in the town of Merka, Somalia [46].

When smallpox infected residents of colonial North America, many died but others survived. The survivors achieved protection against subsequent outbreaks—for reasons unknown at the time. As these "protected" Native Americans, Anglos, Hispanics, and others married and had children, the new generation did not receive the same protective benefit as their parents. And thus it would continue that smallpox would be cyclical in nature and primarily attack children and young adults—those without immunity, as we know the process today in the 21st century. As time passed, inoculations and then vaccinations against smallpox became acceptable to most individuals and cultures.[4] With vaccinations, smallpox incidence dropped, leaving only the faces and scarred bodies of adults and the elderly to attest to the horrors and ravages of the disease.

Finally, what about the chocolate merchants who relocated from Boston to the suburbs in 1764 because of the smallpox epidemic fear? Perhaps experts in Bostonian genealogy one day will determine how they fared after the 1764 smallpox outbreak.

Acknowledgments

I wish to thank my mentor, William J. Darby, former Chair, Department of Biochemistry, Division of Nutrition, Vanderbilt University, and my friend and colleague for encouraging me to pursue an academic career that blended geography and history with science.

Endnotes

1. Colonies of the smallpox virus still are maintained at high-security locations in North America and elsewhere. Smallpox vaccinations are no longer required as an adjunct to international travel.

2. It could be, however, that if chocolate beverages were prepared after water had been heated to a high degree, perhaps boiled, then the patients would have had a safe beverage to drink, one that would have been superior to ordinary drinking water that commonly was contaminated.

3. Using smallpox and pox as keywords, a Readex/Newsbank search of Boston newspapers for the year 1721 revealed only passing mention.

4. The author worked in the eastern Kalahari Desert of the Republic of Botswana between the years 1973 and 1975 and was attached to the Botswana Ministry of Health, Office of Medical Services (Maternal and Child Health/Family Planning unit). During this period there was a strong effort on the part of the government of Botswana to eradicate smallpox, and the World Health Organization had a representative stationed in Gaborone, the

national capital. Immigrants to Botswana who fled Rhodesia's racist government included the maZezuru religious communities, whose elders refused to allow members to be vaccinated. Refusal led to a religious/personal choice/community health impasse until such time as the government issued the maZezuru an ultimatum: either vaccinate your membership or return to Rhodesia. As the author recalls, shortly thereafter, the senior maZezuru elder had a "vision" that changed the antivaccination tenet and the community submitted to vaccination to meet the Botswana government's requirement (personal observations and discussions held in Botswana, 1974).

References

1. Deria, A., Jezek, Z., and Markvart, K. The world's last endemic case of smallpox: surveillance and containment measures. *Bulletin, World Health Organization* 1980;**58**:279–283.

2. Nunn, J. F. *Ancient Egyptian Medicine*. Norman: University of Oklahoma Press, 1996; p. 77.

3. Hopkins, D. R. *Princes and Peasants: Smallpox in History*. Chicago, IL: University of Chicago Press, 1983; pp. 14–15.

4. Thomas, H. *The Conquest of Mexico*. London, England: Hutchinson, 1993; p. 443.

5. Thomas, H. *The Conquest of Mexico*. London, England: Hutchinson, 1993; p. 444.

6. Díaz del Castillo, B. *The True History of the Conquest of Mexico by Captain Bernal Diaz del Castillo, One of the Conquerors, Written in the Year 1568*. Translated by M. Keatinge. London: John Dean, 1800, p. 206.

7. Thomas, H. *The Conquest of Mexico*. London, England: Hutchinson, 1993; pp. 443–446.

8. Florentine Codex, ca. 1575. In: *Historia De Las Cosas de Nueva Espana*, Volume **4**, Book 12, Lam. cliii, plate 114. Peabody Museum of Archaeology and Ethnology, Harvard University. *Biblioteca Medicea Laurenziana, Florence, IT. Alberto Scardigli photo.*

9. Koplow, D. A. *Smallpox. The Fight to Eradicate a Global Scourge*. University of California e-edition. Available at http://www.ucpress.edu/books/pages/9968/9968. ch01.html. (Accessed January 10, 2007.)

10. Risse, G. B. Medicine In New Spain. In: Numbers, R. L., editor. *Medicine in the New World. New Spain, New France, and New England*. Knoxville: University of Tennessee Press, 1987; pp. 25–27.

11. Grivetti, L. E. Nutrition past—nutrition today. Prescientific origins of nutrition and dietetics. Part 4. Aztec patterns and Spanish legacy. *Nutrition Today* 1992; **27**(3):13–25.

12. http://www.ucpress.edu/books/pages/9968/9968. ch01.html. (Accessed January 10, 2007.)

13. http://www.cagenweb.com/eldorado/research/epidemics.htm. (Accessed January 10, 2007.)

14. Saunders, R. (Benjamin Franklin). *Poor Richard improved: being an almanack and ephemeris . . . for the year of our Lord 1761 . . . Fitted to the latitude of forty degrees, and a meridian of near five hours west from London; but may, without sensible error, serve all the northern colonies.* Philadelphia: B. Franklin and D. Hall, 1760; page not numbered (4th page after title page).

15. Hutchins, J. N. *Hutchin's Improved. Being an Almanack and Ephemeris . . . for the Year of Our Lord 1768*. New York: Hugh Gaine, 1768; page not numbered (12th page after title page).

16. Dimsdale, T. *The Present Method of Inoculating for the Small-Pox. To Which Are Added, Some Experiments, Instituted With a View to Discover the Effects of a Similar Treatment in the Natural Small-pox*. Philadelphia: John Dunlap for John Sparhawk, 1771; p. 13.

17. Rush, B. *The New Method of Inoculating for the Small Pox*. Delivered in a Lecture in the University of Philadelphia, February 20, 1781. Philadelphia: Charles Cist, 1781; p. 11.

18. Rush, B. *Medical Inquiries and Observations*, Volume **1**. Philadelphia: Prichard and Hall, 1789; p. 210.

19. Rush, B. *Medical Inquiries and Observations*, 2nd edition, Volume **1**. Philadelphia: Thomas Dobson, 1794; p. 298.

20. Dimsdale, T. *The Present Method of Inoculating for the Small-Pox. To Which Are Added, Some Experiments, Instituted With a View to Discover the Effects of a Similar Treatment in the Natural Small-Pox*. Philadelphia: John Dunlap for John Sparhawk, 1771; p. 13.

21. Anonymous [By a Person Properly Qualified]. Directions Concerning Inoculation, Chiefly Collected from the Late Pieces on That Subject. With Instructions How to Prepare Those Who are Soonest Likely to Take the Small-pox in the Natural Way. The Whole being Carefully Adapted to Town and Country. Philadelphia: Franklin and Hall, 1760, See also: *New Hampshire Gazette*, March 2, 1764.

22. Winslow, O. W. *A Destroying Angel. The Conquest of Smallpox in Colonial Boston*. Boston: Houghton Mifflin Company, 1974; pp. 45, 58.

23. Winslow, O. W. *A Destroying Angel. The Conquest of Smallpox in Colonial Boston*. Boston: Houghton Mifflin Company, 1974, pp. 49–50.

24. *The Boston Post Boy*. January 2, 1764.

25. *The Boston Post Boy*. January 16, 1764.

26. *The Boston Post Boy*. January 23, 1764.

27. *The Boston Post Boy*. January 23, 1764.

28. United States Bureau of Labor Statistics, E. M. Stewart, and J. C. Bowen. *History of Wages in the United States from Colonial Times to 1928. Revision of Bulletin No. 499 with Supplement, 1929–1933*. Detroit, MI: Gale Research Co., 1966.

29. *The Boston Post Boy*. January 23, 1764.

30. *Boston News-Letter and New-England Chronicle.* February 2, 1764.

31. *The Boston Gazette and Country Journal.* February 20, 1764.

32. *The Boston Post Boy.* March 3, 1764.

33. *The Boston Gazette and Country Journal.* March 12, 1764.

34. *The Providence Gazette and Country Journal.* March 31, 1764.

35. *The Boston Gazette and Country Journal.* June 18, 1764.

36. *The Boston Evening Post.* June 25, 1764.

37. *The Boston Evening Post.* August 6, 1764, p. 3.

38. *The Boston Evening Post.* August 6, 1764, p. 3.

39. *The Boston Evening Post.* March 19, 1764, p. 3.

40. Winslow, O. E. *A Destroying Angel. The Conquest of Smallpox in Colonial Boston.* Boston: Houghton Mifflin Company, 1974; p. 89.

41. Jenner, E. *An Inquiry into the Causes and Effects of Variolae Vaccinae or Cow-Pox.* London, England: Sampson Lowe, 1798. (First edition reprinted—London, England: Dawson, 1966.)

42. Ramenofsky, A. F. Death by disease. *Archaeology* 1992;**45**(2):47–49.

43. Thornton, R. *American Indian Holocaust and Survival: A Population History Since 1492.* Norman: University of Oklahoma Press, 1987.

44. http://www.cagenweb.com/eldorado/research/epidemics.htm. (Accessed January 10, 2007.)

45. Santa Barbara Mission Archives. Serra Collection, Document # 885.

46. Deria, A., Jezek, Z., and Markvart, K. The world's last endemic case of smallpox: surveillance and containment measures. *Bulletin World Health Organization* 1980;**58**:279–283.

CHAPTER

8

From Bean to Beverage

Historical Chocolate Recipes

Louis Evan Grivetti

Introduction

The cacao tree (*Theobroma cacao*), source of the beans used to prepare chocolate, was first domesticated in the western headwaters of the Amazon basin, perhaps in northeastern Ecuador, approximately 4000 years before the Common Era [1]. The natural ethnobotanical distribution of *Theobroma cacao* prior to the 16th century was within a relatively narrow geographical range within tropical Central and South America. The suggestion has been made that Native Peoples of western Amazonia did not develop or produce chocolate in the strict sense and consumed only the sweet, viscous, white pulp that surrounded the beans inside the cacao pod. This theory holds that "chocolate" in the strict sense first was produced by Central American peoples in the geographical region of Oaxaca and Chiapas in southern Mexico or in northern Guatemala [2]. Various cosmologies associated with Mayan, Mexica/Aztec, and Bribri cultures in Central America relate how the cacao tree, and eventually chocolate, were introduced to humankind by the gods [3]. Accompanying these cosmologies are pre-Columbian texts and current traditional beliefs that speak to the human desire to prepare and to drink chocolate (see Chapter 3).[1]

While the white pulp inside cacao pods is sweet and easily recognized as food, it is difficult to perceive how humans initially turned to the terribly bitter cacao beans embedded in the pulp and ultimately developed a specific taste for such an ill-tasting food. This dichotomy of sweet pulp and bitter cacao beans, therefore, poses a dilemma, perhaps best approached if one accepts the proposition that the initial use of the bitter cacao beans was not for food, but as medicine. Even today in the 21st century, numerous traditional societies hold the contention that the more bitter the taste, the stronger the medicine. This belief holds that medicines should be bitter (perhaps even painful) in order to be an effective cure.[2] Initial preparations of cacao beans by traditional healers would have produced extremely bitter medicines. At the same time, these medicines also would have produced a noticeable effect on consumers, as such concoctions would have stimulated and energized patients. The next step in the development of chocolate as food, therefore, could have been the intention to maintain the stimulant and energy effects of drinking unadulterated chocolate—without sweeteners added—and to modulate bitterness through addition of various other products. It could have been in this way that the first recipes for making chocolate were developed.

During the course of research, our project team members examined pre-Colonial era archaeological remains and written documents, whether diaries, monographs/treatises, or travel reports, dating from antiquity through the early 20th century.

Chocolate: History, Culture, and Heritage. Edited by Grivetti and Shapiro
Copyright © 2009 John Wiley & Sons, Inc.

Written documents located in archives and libraries in North, Central, and South America, the Caribbean, and Western Europe were systematically reviewed for information related to chocolate preparation and recipe development (see Appendix 2). The chocolate recipes presented in this chapter reflect pre-Columbian through early 20th century culinary traditions. These recipes would have been developed and prepared initially as solid chocolate balls, bars, chunks, or tablets for easy storage and subsequent use. When chocolate was needed for drinking, this solid chocolate was grated and the particles mixed with water (heated or boiled), then drunk. Furthermore, until the mid-16th century, chocolate was only consumed as a beverage by adult males, since Mayan and Mexica/Aztec traditions held that chocolate was too "stimulating" for adult females and children [4].

Chocolate Recipes

The chocolate recipes identified during research are grouped here in four clusters based on relative chronology and geographical region: (1) pre-Columbian era, (2) Early New Spain, (3) 18th–19th century New Spain/Mexico and Europe, and (4) 18th century North America. Identification and analysis of 19th and early 20th century recipes is treated elsewhere in this book (see Chapter 9). Following this presentation of historical recipes, there is appended a representative sample of contemporary chocolate recipes from southern Mexico (state of Oaxaca) and from northern Guatemala collected by project team members during field work in 1998–2000.

PRE-COLUMBIAN ERA

The chocolate-related documents that survive from the pre-Columbian era provide information only on medicinal recipes; we found no primary documentation that identified ingredients used to prepare chocolate for personal consumption. Recipes for pre-Columbian era medicinal chocolate are uncommon, but the following examples may be identified.

Chocolate (unmixed with other products; very bitter) was drunk by the Mexica/Aztecs to treat stomach and intestinal complaints; when combined with liquid extruded from the bark of the silk cotton tree (*Castilla elastica*), this beverage was used by traditional healers to cure infections [5]. In another recipe prescribed to reduce fever and prevent fainting, 8–10 cacao beans were ground along with dried maize kernels; this powder then was mixed with *tlacoxochitl* (*Calliandra anomala*) and the resulting beverage was drunk [6]. Patients with severe cough who expressed much phlegm were advised to drink infusions prepared from opossum tails, followed by a second medicinal beverage where chocolate was mixed with three

herbs—*mecaxochitl* (*Piper sanctum*), *uey nacaztli* (*Chiranthodendron pentadactylon*), and *tlilixochitl* (*Vanilla planifolia*) [7]. Preparations of *tlatlapaltic* root (an unknown, ill-tasting plant) were made palatable by mixing with cacao, and the beverage was given to patients with fever [8]. In another medical recipe, bloody dysentery was treated using a mixture of chocolate blended with *quinametli*, identified as "bones of ancient people called giants" [9].

EARLY NEW SPAIN

The Spanish landed on the east coast of what is now Mexico just north of modern Veracruz on April 21, 1519. The march overland and the ultimate conquest of the Mexica/Aztecs was detailed in manuscripts written by Hernán Cortés [10] and several of his literate officers, among them Bernal Díaz del Castillo [11] and a soldier known to historians as the Anonymous Conquistador [12]. The Mexica/Aztecs used cacao beans as a form of currency and, as such, cacao was highly valued by their culture. In what would become a classic case of cultural misunderstanding, the Mexica/Aztecs offered the Spanish the most valuable commodity within their realm—gifts of cacao. But the Spanish wanted gold and looked, initially, upon cacao beans as something inconsequential—not different in fact than almonds—and puzzled why they were given these "almonds" when gold was desired?

Eyewitness accounts describe a dinner held in 1520 at Tenochtitlan, the Mexica/Aztec capitol, when Montezuma dined with Cortés and his Spanish officers. At this meal the Mexica/Aztec king reportedly drank chocolate from cups of pure gold. The texts that describe the meal are similar in general content, but each reflects subtle differences. One was authored by Bernal Díaz del Castillo:

> From time to time the men of Montezuma's guard brought him, in cups of pure gold a drink made from the cacao-plant, which they said he took before visiting his wives. I saw them bring in a good fifty large jugs of chocolate, all frothed up, of which he would drink a little. I think more than a thousand plates of food must have been brought in for them, and more than two thousand jugs of chocolate frothed up in the Mexican style. [13]

The Spanish historian Antonio de Solis y Rivadeneyra published his *History of the Conquest of Mexico* in 1684. He recounted the same meal with Montezuma as described by Bernal Díaz. He was not an eyewitness to the event, however, and it is not clear what sources he may have used for his compilation:

> Before sitting down to eat, the Emperor inspected the plates so as to recognize the different pleasures that they contained. Tablecloths were of white and thin cotton and the napkins were of the same materials and somewhat large. Up to twenty beautifully dressed women served the food and the drink with the same type of reverences that they used in their temples.

The plates were of very fine clay and were only used once. Gold glasses were placed on trays of the same material. They used, with moderation, wines or to be more exact, beers that those Indians made, making a liquid of the grains of maize through infusion and cooking. At the end of the meal, he ordinarily drank some type of chocolate in his manner, in which one added the substance of the cacao, beating with a molinillo [stirring stick] until the jicara [gourd cup for serving chocolate] was filled with more foam than liquid. And after that, they used the smoke of tobacco. [14]

An early account of chocolate dated to 1524 was written by Father Toribio de Benavente, a Franciscan priest and one of "the twelve" (i.e., priests) who arrived in New Spain shortly after the defeat of the Mexica/Aztecs by Cortés. He mentioned that cacao was mixed with corn and other "ground" seeds:

This cocoa is a very general drink; they grind it and mix it with corn and other ground seeds; and this is also a major use of the cocoa seeds. It is good, it is good and it is considered as a nutritious drink. [15]

One of Cortés's literate officers, known only as the Anonymous Conquistador, published a firsthand account of the conquest in 1556 after his return to Spain. His manuscript contained the following general description of chocolate preparation by the Mexica/Aztecs. While generalized regarding the names of ingredients, the document also represents one of the earliest chocolate recipes written by a European.

These seeds, which they call almonds or cacao, are ground and made into powder, and some other small seeds they have are also ground, and the powder put into certain vessels that have a spout. Then they add water and stir it with a spoon, and after it is well mixed they pour it back and forth from one vessel to another until it is foamy. The foam is gathered and put in a cup, and when they are ready to drink the beverage they beat it with some small spoons made of gold, or silver or wood. To drink it one must open the mouth wide, for since it has a froth it is necessary to make room for it to dissolve and go in gradually. This drink is the most wholesome and substantial of any food or beverage in the world, because whoever drinks a cup of this liquor can go through the day without taking anything else even if he is on a journey, and it is better in warm weather than in cold, since it is a cold drink. [16]

Juan de Cárdenas published his account of New Spain in 1591 entitled *Problemas y Secretos Maravillosos de las Indias* (*Marvelous Problems and Secrets of the Indies*). In Chapter VII of his treatise he offered a brief, general statement on the preparation of chocolate and identified ingredients that could be blended with it. Of interest is the fact that Cárdenas differentiated between spices and flavorings from Castile and those indigenous to the New World.

Cacao by itself, largely being eaten raw, causes all this harm of which we spoke, but that toasted and incorporated with

warm spices, as it is mixed in chocolate, it has great benefits for everything. In this precious and medicinal drink [chocolate] there are spices called "from Castile [sic]," and others that here we call "from the earth." The Castillian spices [added to chocolate] are cinnamon, pepper, anise, sesame. . . . I will refer to the Indian spices and the qualities, constitution, properties and effects they have, both by themselves and when mixed into chocolate. [17]

Elsewhere in his Chapter VII, Cárdenas listed the local ingredients from Mexico added to chocolate and he provided information on the medicinal attributes of the various chocolate combinations:

The first spice added to said drink is called gueyncaztle by the Indians, and ear-flower by the Spanish . . . because of its good aroma . . . it gives the pleasure of its fragrance to this drink. . . . Therefore, this drink strengthens and comforts the vital virtues, helping to engender life spirits. It likewise has a very pleasant taste, which makes what is drunk even more beneficial. . . . Gueynacaztle is followed second by mecasuchil, which is nothing more than some sticks or threads, brown and thin, which because they have this shape of thin threads, are called by this name, which means "rose in the form of a thread." . . . The third in order, and the first in soft and delicate aromas, is the so-called tlixochil, which in our language is called aromatic vanilla, because they are really long and brown vanilla pods with insides full of black seeds, but fewer than those of mustard . . . they compete with musk and amber. . . . They add to said chocolate a very soft and mild odor and therefore have an advantage over all other spices in being cordial and friendly to the heart. . . . Achiote is also counted as a spice, since it is no less valued in this drink than cardamom in medicinal and aromatic compounds. . . . Achiote [is added] to this drink in order to give it a pretty red color, and in order to give sustenance and to fatten the person who drinks it. . . . Some people tend to add a little of everything [to chocolate] if they feel cold in the stomach, or the belly, and also add some toasted chilies to the chocolate, and some large seeds of dry cilantro, called earth pepper. [18]

Cárdenas continued his description of chocolate and recorded his recommendations for toasting cacao beans to assure quality of the ultimate chocolate beverages. In his view chocolate tablets could be kept for two years if made properly. Of special interest in his account is reference to a stirring rod, now called a *molinillo*, commonly assumed to be a Spanish invention and not a pre-Colonial era tool:

Spices that are added to this appetizing Indian drink should be obtained fresh . . . they should not be aged, expired, moldy, or decayed, but be the best that can be found of their genus. Only the cacao should be aged, because the older it is, the more oily and buttery it will be . . . when said materials are together, this will be the dosage or ordinary amount: to 100 cacaos is added one half ounce of spices, of any spices . . . all spices should be put together with the cacao and toasted, but I warn that because cacao bears more fire than the spices, it should be toasted separately from the spices. Cacao is toasted until it starts to blister and becomes the color of brown turning to black, and the spices are toasted until they acquire a red color turning to black . . . once toasted they are ground together very well . . . after having ground everything very well, there are differences in how it is shaped. Those who want to store it for a long time make it into the shape of tablets, and these can be kept for at least two years. . . . It is now an ancient custom that when chocolate is made, it is stirred and beaten to such a degree that froth arises from it, and the frothier the chocolate, the better it is considered to be. [19]

The Jesuit priest, Jose de Acosta, lived and worked in the Americas from 1572 until 1587. Upon his return to Spain he published a book in 1587 that described the geography and history of New Spain. His work received considerable attention in Europe and subsequently was translated into English and published in London in 1604 under the title *The Naturall [sic] and Morall Historie[sic] of the East and West Indies. Intreating of the Remarkeable [sic] Things of Heaven, of the Elements, Metals, Plants and Beasts.* He included a brief recipe for chocolate and noted that chili peppers were added to chocolate beverages:

They say they make diverse sorts of it [chocolate], some hote [sic], some colde [sic] and some temperate, and put therein much of that chili; yea they make paste thereof. [20]

Santiago de Valverde Turices published his treatise on chocolate, *Un Discurso del Chocolate*, in 1624. In part 1 of his manuscript he explored the medical uses of chocolate and provided two recipes that could be prepared at home:

Chocolate for the home: 6 pounds cacao; 1 pound anise; 1/2 pound cinnamon; 5 pounds sugar; 2 ounces little ears (ear flower); 1 ounce black pepper; 1/2 ounce cloves; and some chili [quantity not defined]. [21]

Chocolate for the home: 6 pounds cacao; 1 pound anise; 1/2 pound cinnamon; 5 pounds sugar; 2 ounces black pepper; 1/5 ounce cloves; 4 peppers from the Caribbean. [22]

Regarding further preparation and mixing of ingredients, Valverde Turices wrote:

Some put a lot of chocolate in a little water, so it will be thick, and others less chocolate, so it will be more liquid. Some are content with the amount of sugar and others add more. Some grind more cinnamon, cloves, pepper. [23]

Antonio Colmenero de Ledesma wrote his monograph in 1631, entitled *Curioso Tratado de la Naturaleza y Calidad del Chocolate*, a much-cited and much-translated publication [24]. Colmenero's text was highly circulated throughout Europe, so much so that different editions and translations of his work have been difficult to attribute to specific authors. Arthur Knapp in his 1920 text, *Cocoa and Chocolate. Their History from Plantation to Consumer*, clarified much of this

confusion surrounding Colmenero's work and subsequent translations [25]. The essence of Colmenero's work, as noted by Knapp, appeared in translations prepared in English, by Don Diego de Vadesforte (1640) [26]; in French, by Rene Moreau (1643) [27]; in Latin, by J. G. Volckamer (1644) [28]; in English, by J. Wadsworth (1652) [29]; and in Italian, by A. Vitrioli (1667) [30]; a third English translation was prepared by J. Chamberlaine (1685) [31]. While much of Colmenero's manuscript considers medical attributes of chocolate, Chamberlaine provided a recipe that commonly was copied (without attribute) by subsequent authors:

To every 100 cacao [beans] you must put two cods of the long red pepper [chili] that are called in the Indian tongue, chilparlangua. One handful of aniseed, orejnelas, which are otherwise called vinacaxlidos; add two of the flowers called mechasuchil [alternatively add] 6 roses of Alexandria; beat to powder. [Add to this] one cod [i.e., pod] of Campeche or Logwood. Two drams of cinnamon; almonds and hazel-nuts; of each one dozen; of white sugar, half-a-pound; of achiote enough to give it [a red] color. [32]

Thomas Hurtado, an obscure Catholic priest, examined the use of tobacco and chocolate within the context of religious fasting and whether or not drinking chocolate would break Christian fasting obligations during Lent and at other times. His volume, *Chocolate y Tabaco Ayuno Eclesiastico y Natural: si este le Quebrante el Chocolate: y el Tabaco al Natural, para la Sagrada Comunion*, was published in Madrid in 1645. Hurtado concluded that chocolate was not a "food" unless mixed with milk and eggs. He commented on medicinal aspects of chocolate and included one recipe for its preparation:

How to make a chocolate confection:
Blend chocolate with almonds, dates, and physic-nuts [i.e., piñon]. [33]

The Englishman Thomas Gage (the theologian not the American Revolutionary War era military general), was a Dominican priest trained in Spain. Due to travel restrictions imposed on the English, he was smuggled out of Spain and ultimately reached the Americas in 1627. Gage remained in Mexico and Guatemala until 1637, whereupon he returned to England. His account of travels and work in New Spain was published in 1648 under the title *The English American: His Travail by Sea.* Imbedded in his text were observations and recordings on the preparation of chocolate. Gage certainly knew of other chocolate-related manuscripts/texts since he included notations to Antonio Colmenero de Ledesma and, in fact, copied much from Colmenero's chocolate recipe when writing his own (compare with Ledesma's cited earlier):

[They] mix chocolate paste with black pepper; [They] mix chocolate paste with chili pepper. How to make chocolate: for every 100 cacao beans add . . . 2 pods chili (long red ones);

1 handful [of] anise seed [and] orejuela/ear flower [Cymbopetalum penduliflorum]; 2 mexacochitl flowers [vanilla: Piper amalago]—alternatively add rose petals; beat to powder and mix [the following]: 2 drams cinnamon; 12 almonds; 12 hazel nuts; 1/2 pound white sugar; achote [Bixa orellana] add enough to color mixture red. [34]

The English physician Henry Stubbe (sometimes spelled Stubbes) wrote a treatise on chocolate in 1662 entitled *The Indian Nectar, or, a Discourse Concerning Chocolata [sic]; the Nature of the Cacao-Nut and the Other Ingredients of that Composition is Examined and Stated According to the Judgment and Experience of Indian and Spanish Writers.* Stubbe cited the works of previous chocolate-related researchers and filled the pages of his treatise with information on the medicinal roles played by chocolate in the Americas. He reported that *achiotl* (achiote: *Bixa orellana*), *tepeyantli* (unknown), *tlilxochitl* (vanilla), and *xochinacaztli* (*Cymbopetalum penduliflorum*) were added to chocolate to alleviate/cure stomach problems, strengthen the brain and womb, or to treat coughs [35].

Stubbe lauded medicinal chocolate and provided a recipe for its preparation—copied primarily from Colmenero, but with minor alterations:

To every hundred nuts of cacao, put two cods of chile [sic] called long red pepper, one handful of anise-seeds, and orichelas [orejaelas], and two of the flowers called mecasuchill, one vaynilla [sic] or instead thereof fix Alexandrian roses beaten to powder, two drams of cinnamon, twelve almonds and as many hasel-nuts [sic], half a pound of sugar, and as much achiote as would color it. [36]

Elsewhere in his treatise Stubbe wrote that "chocolate mixed with Jamaica pepper provoked urine and menstrual flow, strengthened the brain, comforted the womb and dissipated excessive winde [sic]"; if vanilla was substituted such chocolate "strengthened the heart, beget strong spirits, and promoted digestion" [37]. He described how different varieties of chili pepper were mixed with chocolate, especially those called *mecaxochitl* and *piso,* and how the resulting chocolate paste when prepared as a beverage would "open obstructions, cure cold and distempers arising from cold causes; strengthen the stomach, and amend the breath" [38]. He also wrote that varieties of ear flowers, *orichelas* or *xochinacaztlis* (*Cymbopetalum penduliflorum*) when mixed with chocolate provided nice odor and taste to the medicinal chocolate; the mixture then was drunk by patients to "beget good blood and to provoke monthly evacuations n women" [39].

William Hughes worked in Jamaica and published his cacao and chocolate monograph in London in 1672 under the title *The American Physitian [sic] . . . Where-unto is Added a Discourse on the Cacao-Nut-Tree, and the Use of Its Fruit, With All the Ways of Making Chocolate.* In contrast to earlier manuscripts that offered information on chocolate recipes, Hughes's provided both recipes and substantially more detail on how the beans should be prepared, cleanliness of the equipment, and need to regulate flame heat:

Take as many of the cacao's as you have a desire to make up at one time, and put as many of them at once into a frying-pan (being very clean scoured) as will cover the bottom thereof, and hold them over a moderate fire, shaking them so, that they may not burn (for you must have a very great care of that) until they are dry enough to peel off the outward crust skin; and after they are dried and peeled then beat them in an iron mortar, until it will rowl [sic] up into great balls or rows and be sure you beat it not over-much neither, for then it will become too much oyly [sic]. [40]

Hughes presented the argument (erroneous) that, prior to the arrival of the Spanish, Native Americans drank chocolate without addition of other items. He wrote that English-origin residents of Jamaica had adopted Spanish traditions of preparing chocolate:

The Native Indians seldom or never use any compounds [add anything to their chocolate] desiring rather to preserve their health, than to gratify and please their palats [sic], until the Spaniards coming amongst them, made several mixtures and compounds, which instead of making the former better (as they supposed) have made it much worse. And many of the English (especially those that know not the nature of the cacao) do not imitate them: for in Jamaica, as well as other places in making it into lumps, balls, cakes, they add to the cacao-paste, chili, or red pepper; achiote, sweet pepper, commonly known by the name of Jamaica-pepper, or some or one of them; as also such other ingredients as the place affordeth [sic], or as most pleaseth [sic] the maker thereof or else as the more skilful [sic] persons may think it to agree with this or that individual person, adding thereto as much sugar only as will sweeten it first of all drying and beating every ingredient apart, and then at the last all mixing them together, as it is wrought up into a mass. [41]

Elsewhere in his text Hughes identified several New World products added to chocolate based on the whims or desires of persons preparing the beverage. In the citation that follows it should be noted that while *achiote* (*Bixa orellana*) at first glance may appear akin to a chili pepper, it is not a sweet-scented pepper and the fruit is basically inedible. The seeds contained in the fruit are crushed and added to chocolate to provide the blood-red color that in the final mixture with chocolate turns the mass a dark reddish-brown. Several ingredients added to chocolate identified by Hughes in the following passage are difficult to identify.

Some of the ingredients put in are chili, or red pepper, achiote [or] sweet-scented pepper, orejuela [ear-flower], boil [?], pocolt [?] or prnise [?], atolls [sic] or maize-flower, sugar, and more of less of these, or any one of them are put in, as the makers thereof see good. . . . But if there be any addition made to the cacao . . . [some] are much more properly used than other spices which in lieu thereof, in Spain, and other countries, are often put in: such are aniseeds, fennel-seeds, sweet-almonds, nutmegs, cloves, black, white, and long pepper, cinnamon, saffron, musk, amber-greece [sic], orange-flower-water, lemon, and citron-pill [sic, peel meant], cardamonet [sic for cardamom], oyl [sic] of nutmegs, cinnamon, and many other ingredients are usually put in, as is thought fit by the physicians or others; either when it is made up in the mass, or else more or less be added when the drink is made. [42]

The Frenchman Philippe Sylvestre Dufour was interested in chocolate monographs and has been accredited with publishing a revision of René Moreau's translation of Antonio Colmenero de Ledesma's treatise in 1671. It has been suggested, too, that Dufour was not in fact a Frenchman but the pseudonym of Jacob Spon who reportedly died in December 1685 [43]. Regardless of the authorship/name controversy, Dufour (perhaps Spon?) translated into French and published a new edition of Colmenero's work in 1685 (same year as his death?) under the title *Traitez Nouveaux et Curieux du Café, du Thé et du Chocolat*. While it may readily be seen that the work primarily was "lifted" from Colmenero, additional recipes or variations thereof appeared in the new edition:

As for the other ingredients which go to the making of your confection of chocolate, I find many different sorts, some put therein black pepper ... [others say] that the pepper of Mexico, called chili, is far the better. [44]

For family and everyday use, Dufour recommended the following recipe:

For every 700 cacao beans add: 1 1/2 pound sugar (white); 2 ounces cinnamon; 14 grains chili pepper; 1/2 ounce clove; 3 straws vanilla (alternatively anise seed equal to weight of a shilling); achote [Bixa orellana] enough to color mixture red. Blend [these and] add sufficient amounts of almonds, filberts, and orange flower water. [45]

To prepare medicinal chocolate upon recommendation of physicians, Dufour wrote that a slightly different recipe was required:

The receipt [recipe] of our physician to make chocolate is thus: Take 700 cacao nuts, a pound and a half of white sugar, two ounces of cinnamon, fourteen grains of Mexico pepper, call'd [sic] chili or pimento, half an ounce of cloves, three little straws or vanilla's de Campeche, so for want thereof, as much anise-seed as will equal the weight of a shilling, of achiot [sic] a smal [sic] quantity as big as a filbeard [sic for filbert], and the water of orange flowers. [46]

Elsewhere in the manuscript Dufour identified a broad range of New and Old World spices and flavoring agents that could be added to recipes when preparing medicinal chocolate:

There be other ingredients that they put into this composition [chocolate] the chief of which they call mecasuchil ... [it goes by] name mecaxuchitl creeping upon the earth, whose leaves are great, thick and almost round, sweet-smelling, and of a sharp taste, it bears a fruit like long pepper, the which they mix with the drink of the cacao, call'd [sic] chocolate, to

which it gives an agreeable flavour [sic]. ... Another ingredient is the vinacaxtli [Dufour preferred the spelling huclimacutzli] which is a tree the flower whereof is called by the Spaniards, flor de la orejas, or flower of the ear, because of its near resemblance with the ear ... it is of a very sweet and pleasant smell. ... The mecasuchil is purgative, and the Indians make thereof a purging syrup. Those that live in Europe for want of mecasuchil may put therein powder of roses of Alexandria. ... There be two other ingredients [added to chocolate] ... the one is the flower of a certain pitchy or rosi'ny [sic resin] tree, which yields a gum like that of the Siorax, but of a finer colour [sic], its flower is like that of the orange tree, of a good smell, which they mix with the chocolate ... the other ingredient is the shale [sic shell] or cod of the tlixochitl, which is a creeping herb having leaves like the plantane [sic], but longer, and thick, it climbs up to the top of the trees, and intwines [sic] itself with them, and bears a shale [sic shell] long, strait, and as it were round, which smells of the balm of New Spain, they mix this shale [sic shell] with their drink of cacao; their pith is black full of little seeds, like that of the poppy, they say that two of these in water provoke urine wonderfully. [47]

According to Arthur Knapp's review and assessment of translations and first editions of chocolate-related monographs and texts, John Chamberlaine translated Sylvestre Dufour's treatise on chocolate in London in 1685, which appeared with the title *The Manner of Making Coffee, Tea, and Chocolate* [48]. As with several of the previous translations of Colmenero's work, additional chocolate-related recipes and variations were included, commonly without acknowledgment. The following recipe contains the comment that rhubarb added to chocolate was effective for treating "young green ladies" (chlorosis), a malady now recognized as iron deficiency anemia whereupon dermal changes produced a characteristic, curious greenish hue:

In the common sort [of chocolate] the cacaw nuts [sic] may take up half the composition, in the worst as third part only. As to the other ingredients for making up chocolate, they may be varied according to the constitutions of those that are to drink it; in cold constitutions Jamaica pepper, cinnamon, nutmegs, cloves may be mixed with the cacao nut; some add musk, ambergrease [sic], citron, lemmon-peels [sic], and odoriferous aromatick [sic] oyls [sic]; in hot consumptive tempers you may mix almonds, pistachios, sometimes china [?], serfa [?], and saunders [?]; and sometimes steel [?] and rhubarb may be added for young green Ladies. [49]

In a subsequent passage Chamberlaine instructed his readers how to prepare chocolate:

I will give you a short direction ... as for the managing the cacao nut ... you must peel, dry, beat and searce [sift] it very carefully, before you beat it up into a mass with other simples [term applied to a category of medicines]; as for the great quantity of sugar which is commonly put in, it may destroy the native and genuine temper of the chocolate ...

*for preparing the drink of chocolate, you may observe the
following measures. Take of the mass of chocolate, cut into
small pieces, one ounce, of milk and water well boyl'd [sic]
together, of each half a pint, one yolk of an egg well beaten,
mix them together, let them boyl [sic] but gently, till all is
dissolved, stirring them often together with your mollinet [sic]
or chocolate mill.* [50]

James Lightbody wrote his treatise, *Every Man
His Own Gauger ... Together with the Compleat [sic]
Coffee-Pan, Teaching how to Make Coffee, Tea, Chocolate*, in
1695. One recipe Lightbody recommended was the
following:

*To make Chocolate Cakes and Rowles [sic] pot, and peel off
the husks, then powder them very small, so that they may be
sifted through a fine Searce [i.e., sieve]; then to every pound
of the said Powder add seven Ounces of fine white Sugar, half
an Ounce of Nutmegs, one Ounce of Cinnamon, Ambergreace
[sic] and Musk each four Graines [sic], but these two may be
ommitted [sic], unless it be for extraordinary use.* [51]

A second recipe by Lightbody was more
complex:

*To make Chocolate. Take of Milk one Pint, and of Water,
half as much and Boil it a while over a gentle Fire; then
grate the quantity of one Ounce of the best Chocolate; and
put therein; then take a small quantity of the Liquor out, and
beat it with six Eggs; and when it is well beat, pour it into
the whole quantity of Liquor, and let it boil half an Hour
gently, stiring [sic] it often with your Mollinet; then take it
off the Fire, and set it by the Fire to keep it hot; and when
you serve it up, stir it well with your Mollinet. If you Toast a
thin slice of white Bread, and put there-in, it will eat
extraordinary well. It will not be amiss to Insert the manner
of making the Chocolate Cake; or Rowles [sic].* [52]

Nicolas de Blégny, physician to the French King
Louis XIV, published a treatise on coffee and chocolate
in 1687. The title of his work was *Le bon usage du thé
du caffé et du chocolat pour la préservation & pour la guérison
des maladies.* His text has remained a rich source of
information regarding medical attributes of chocolate
and he recommended mixing with vanilla syrup to
"suspend the violent cause of rheumatoids and inflammation of the lungs" [53]. He also recommended that
milk be added to medicinal chocolate to lessen chest
dryness [54]. De Blégny also reported that adding
"syrup of coins," tincture of gold, or oil of amber to
the chocolalate would relieve indigestion and heart
palpitations [55].

The French physician de Quélus (sometimes
spelled de Chélus; his first initial is problematical) published his monograph on cacao and chocolate in 1719
[56]. An English translation of this work prepared by
R. Brookes subsequently appeared in 1724 under the
title *Natural History of Chocolate* [57]. The English
edition of de Quélus (also a Brookes translation) examined during the course of our research, however, was
published in 1730 [58]. Major sections of his work

consider the medical uses of drinking chocolate, for
example, to cure distemper, improve exhausted spirits,
induce quiet sleep, and promote long life [59]. Elsewhere in his manuscript de Quélus identified the
preparation of chocolate and ingredients that were
added:

*They chuse [sic] cocoa-nuts that are half ripe, and take out
the kernels one by one, for fear of spoiling them; they then
lay them to soak for 40 days in spring water, which they care
to change morning and evening; afterwards, having taken
them out and wiped them, they lard them with little bits of
citron-bark and cinnamon, almost as they make the nuts of
Rouen [located in French Normandy]. In the meantime they
prepare a syrup of the finest sugar, but very clear and that is
to say, wherein there is but little sugar: and after it has been
clarified and purified, they take it boiling-hot off the fire,
and put in the cocoa-kernels and let them lie 24 hours. Then
repeat this operation fix or seven times increasing every time
the quantity of sugar, without putting it on the fire, or doing
anything else to it; last of all they boil another syrup to the
consistency of sugar, and pour it on the kernels well wiped
and put in a clean earthen pot; and when the syrup is almost
cold, they mix with it some drops of the essence of amber.
When they would have these in a dry form, they take them
out of the syrup, and after it, is well drained from them, they
put them into a basin full of a very strong clarified syrup,
then they immediately put it in a stove, or hot-house, where
they candy it. This confection, which nearly resembles the
nuts of Rouen, is excellent to strengthen the stomach without
heating it too much. ... Cinnamon is the only spice which
has had general approbation, and remains in the composition
of chocolate. ... Vanilla is a cod [pod] of a brown colour
[sic] and delicate smell; it is flatter and longer than our
French beans, it contains a luscious substance, full of little
black shining grains. They must be chosen fresh, full, and
well grown, and care must be taken that they are not smeared
with balsam, nor put in a moist place. The agreeable smell,
and exquisite taste that they communicate to chocolate, have
prodigiously recommended it. ... The sugar being well mix'd
[sic] with the paste, they add a very fine powder made of
vanilla and cinnamon powdered and searced [sifted] together.
They mix all over again upon the stone very well, and then
put it in tin molds, of what form you please, where it grows
as hard as before. Those that love perfumes, pour a little
essence of amber on it before they put it in the molds. ...
Those that make chocolate for sale, that they may be thought
to have put in a good deal of vanilla, put in pepper, ginger,
etc.* [60]

In 1741, the Swedish naturalist Carl von Linné
(Linnaeus) wrote a monograph on chocolate (*Om
Chokladdryken*) that was printed in 1778 and republished
in 1965 [61]. Linné provided the scientific binomial
nomenclature for the cacao tree (*Theobroma cacao*) that
he derived from a blend of Greek and New World
languages: *Theobroma* translates as "food of the gods."
Much of his text considered the medical aspects
of chocolate, where he praised its use to combat

pulmonary diseases and hypochondria. He also mentioned, candidly, that through drinking chocolate he cured himself of hemorrhoids! Also included in his treatise were three recipes used to prepare chocolate:

[Recipe 1] *1 pound cocoa beans (roasted); 1 / 2 pound sugar, salt and rosewater (combined); 1 / 2 pound corn flour. Crush, cook over fire all the time stirring so it does not burn; form past into a dough.* [62]

[Recipe 2] *6 pounds cocoa beans (roasted); 3.5 pounds sugar; 7 straws vanilla 1.5 pounds corn flour; 0.5 pound cinnamon; 6 cloves; 1 dracma [i.e., Swedish equivalent for dram] Spanish pepper; 2 dracmas oleana color in rose water. Crush ingredients in a pot, stir all the time over a very slow fire until all mixed; treat it and kneed to a dough, then add amber and musk, according to taste.* [63]

[Recipe 3] *17 pounds roasted cocoa beans; 10 pounds sugar; 28 units (?) vanilla; 1 dracma amber; 6 pounds cinnamon.* [64]

Elsewhere in the manuscript Linné alerted his readers how to prepare chocolate tablets:

Cocoa beans are roasted, then grated on a stone mortar which is heated over a fire; the remainder of ingredients are added and mixed by more grating and working into a dough, which is cut into cakes, then dried for 15–20 days—the longer the better. [65]

Once the chocolate tablets had been prepared, he described how to make chocolate beverages:

Dissolve 1 ounce of chocolate cake [above] in 6 ounces of water or luke-warm milk; heat until it comes to a slow boil, then stir over hot ashes for 15 minutes; whip it until it is skimmy [i.e., foamy], pour skim into a cup; whip again so it forms new skim. [66]

Antonio Lavedan published his book in 1796, *Tratado de Los Usos, Abusos, Propiendades y Virtudes del Tabaco, Café, Te y Chocolate* [67]. Lavedan made use of a broad range of documents available to him and provided his readers with perhaps the most detailed description how to prepare and manufacture chocolate as reflected by Spanish practices during the late 18th century:

In order to prepare and make chocolate paste, the first step is to toast the cacao. Place a small amount of river sand, which is white, fine and has been sifted, to make it even finer, in an iron toaster in the form of a pan, (that in Madrid this is called a paila), and put over a fire. Heat the sand and stir it with a wooden stick so that it heats evenly, then add the cacao. Continue to stir it in with the sand until the heat penetrates the shell, which separates it from the bean without burning it. It is better to toast it this way instead of without the sand for the following reasons: First, the chocolate becomes more intense in color; second, it separates from the shell without burning the bean; and third and most importantly, the fatty, oily and volatile parts of the cacao are not dissipated. Therefore, it keeps all of its medicinal and stomachic qualities, which will be described below. Once the

worker or miller knows that it is well-toasted, he puts it in a sack or a bag and leaves it until the next day when he pours it through a sieve in order to separate the sand from the cacao. Then it is hulled and cleaned from its shell as follows: it is placed on a quadrilateral, somewhat arched rock with a thickness of three fingers. A small brazier is placed underneath the rock with a little bit of carbon fire. The miller smashes and grinds the cacao with another round cylindrical rock and works it into a roll, which he takes in his hands by either end, thereby making a paste which is then mixed with one-half or third parts sugar. Once this mixture it made, it is ground again. Then, while still hot it is placed in molds of tin or wood, according to individual custom. Sometimes it is placed on paper and made into rolls or blocks, where it is fixed and becomes solid quickly. Others mash it and make it into a paste called ground chocolate. Prepared in this manner, it is called "Health Chocolate." Some people claim that it is good to mix in a small amount of vanilla in order to facilitate digestion because of its stomachic and tonic qualities. [68]

When you want a chocolate that further delights the senses, you add a very fine powder made from vanilla and cinnamon. This is added after mixing in the sugar, and then it is ground a third time lightly, in order to make a good mixture. It is claimed that with these aromas it is digested better. Some people who like aromas add a little essence of amber. [69]

—————

When the chocolate is made without vanilla, the amount of cinnamon added is usually two drachmas for every pound of cacao [eight drachmas is one medicinal ounce]. But when vanilla is used the amount of cinnamon is usually reduced by at least half. When using vanilla, one adds one or two small pods to each pound of cacao. Some chocolate makers add pepper and ginger, and wise people should be careful to not take chocolate without first knowing its composition. In Spain, people do not usually add either of these two spices, but in other countries, they tend to use one or another, although this is not very common because it is so disagreeable. [70]

—————

On the French Islands, breads [?] of pure chocolate are made without anything added. When they want to drink chocolate they reduce the squares to powder and add greater or lesser amounts of cinnamon, sugar and orange flower, already ground into a powder. Chocolate prepared in this way is dark but very agreeable with an exquisite aroma. Even though vanilla is very common on the Islands, it is not widely used in this confection. [71]

—————

The way to make and drink chocolate so that it has all of its properties, is the following: First the squares or rolls of chocolate are cut into pieces, and placed in the chocolate pot with cold water, not hot. They are placed over a slow fire and stirred well with the chocolate beater until well dissolved. Chocolate should never be heated at a high temperature because it does not need to cook or boil when it is well dissolved. This would cause it to coagulate, and the oily or fatty part separates when it boils. Those with weak stomachs cannot digest it in this form, and therefore it does not have the desired effects. It is even worse when already dissolved chocolate is kept and later boiled again when someone wants to drink it. In order to fully enjoy the benefits of chocolate, it should be drunk immediately after it is dissolved and stirred. Others reduce the squares of chocolate to powder and add it to boiling water in the chocolate pot. They dissolve it, stirring it well with the chocolate beater and immediately pour it into the serving cups. This is the best way to make chocolate because there is no time for the oil to separate. [72]

Another passage within Lavedan's text identified a chocolate-related medical publication authored by Geronimo Piperi entitled *De Potione Chocolate*, date and publisher unknown. According to Lavedan, Piperi wrote that chocolate was protective, served as an overall cure for many diseases, extended the lives of those who drank the beverage, and suggested that the best medicinal chocolate should be mixed with egg yolk and meat broth:

I cannot refrain from transcribing the praises and uses for chocolate described by Geronimo Piperi in his document titled "de potione chocolate" fol. mihi 6: "Chocolate is celestial, a divine drink, the glistening of the stars, vital seed, divine nectar, drink of the gods, panacea and universal medicine. In addition, it is the best nutrient and remedy, not only for those that usually drink it and are fans of this drink, but it also protects against and cures many illnesses. It is very beneficial especially for the old and decrepit, because it lengthens their lives, and all those that serve themselves of this good nectar enjoy firm health. Without help from other food, chocolate can prolong life through the great nutrients that it supplies to the body and it restores strength, especially when one mixes an egg yolk with some spoonfuls of meat broth." [73]

17TH AND 18TH CENTURY CHOCOLATE RECIPES: NORTH AMERICA

The earliest cookbooks to appear in North America were printed in England and contained texts based on common English foods with ingredients and preparations. Many recipes contained in English cookbooks were unsuitable given the differences in available foodstuffs between the Old and New Worlds; for example, originally there was no wheat in early northern or southern Virginia and the principal grain available was Indian corn or maize [74]. Furthermore, it may be noted that most of the European immigrant women to New England could not read or write [74]. It was not until after the American Revolution that Amelia Simmons published the first cookbook in North America in 1796 based on local ingredients and foods; her cookbook, however, did not contain any mention of cacao or chocolate [75].

While many Colonial era American families could not afford to purchase imported cookbooks, that

did not mean they lacked the ability to cook or to create and share recipes. The mechanism used in many such households for keeping and/or preserving recipes became known as *Commonplace Books*. These were essentially scrapbooks where adults who could read and write documented their daily activities of interest, pasted or otherwise assembled information deemed important, whether home remedies, religious quotations, poems of interest, or recipes.

During the course of our team research on the history of chocolate in North America, the earliest recipe identified was by Jim Gay, Colonial Willamsburg Foundation. He found the following recipe—*To Make Chocolate Almonds*—in an anonymous *Commonplace Book* dated to 1700:

Take your Sugar & beat it & Serch [sift] if then: great youre [sic] Chocolatt [sic]; take to 1 lb Sugar: 5 oz of Chocolatt [sic] mix well together put in 2 Spoonful Gundragon [?] soaked in rosewater & a grm [gram] musk & ambergrease [sic] & beat all well together in mortar. Rowl [sic] out and markwt [sic] ye molds & lay on tin plates to dry. turn everyday. [76]

As the 18th century progressed, chocolate recipes began to be diversified and included more items than merely variations of beverages. Chocolate specialized items that appeared included meringues, puffs, and tarts as represented by the following examples.

A letter from Esther Edwards Burr to Sarah Prince Gill, dated January 14, 1755 identified another early divergence from chocolate as a beverage when she provided the recipe for cocoa-nut tarts:

1755. Teusday [sic] To morrow [sic] the Presbytyry [sic] are to meet here, so you may think I am prety [sic] busy in the Citchin [sic], making Mince-pyes [sic] and Cocoa-nut Tarts etc. [77]

In his *New Art of Cookery, According to the Present Practice*, Richard Briggs identified the following recipe for chocolate puffs:

1792 Chocolate Puffs: Take half a pound of double-refined sugar, beat and sift it fine, scrape into it one ounce of chocolate very fine, and mix them together; beat up the white of an egg to a very high froth, then put in your chocolate and sugar, and beat it till it is as stiff as a paste; then strew sugar on some writing-paper, drop them on about the size of a sixpence, and bake them in a very slow oven; when they are done take them off the paper and put them in plates. [78]

Another late 18th century cookbook, *Valuable Secrets Concerning Arts and Trades 1795*, included the following recipe:

The meringues, with chocolate, or cinnamon, are made as follows. Pound and sift into subtile [sic] powder and distinctly each by itself the cinnamon, and a quantity of the above described paste, after a thorough drying. Then mix these two powders and a discretionable [sic] quantity of sugar together in the same mortar, by means of whites of eggs beaten, *continuing to pound the whole till the paste be firm and however flexible. Now spread it with the rolling pin to the thickness you like, and cut it in the shape and form you please, then bake and ice it as usual.* [79]

It is perhaps appropriate to end this chapter on chocolate recipes with several selected examples from the 19th century—one for chocolate ice cream and another for chocolate custard that appeared in popular magazines of the day, *The Saturday Evening Post* and *Godey's Lady's Book*:

Chocolate Ice Cream—Chocolate used for this purpose must have neither sugar nor spice in it. Baker's prepared Cocoa is the best. For each quart of cream, scrape down three large ounces of cocoa or chocolate, put it into a saucepan with half a pint of hot water for each ounce, and mix it into a smooth paste with a spoon. Place it over hot coals, and when it has come to a boil, take it off the fire and set it to cool. When it has become lukewarm, stir the chocolate into the cream, and strain it. When straining, add gradually, for each quart of cream, three-quarters of a pound of powdered loaf-sugar, and give the whole a boil up. Then put it into a freezer. [80]

Chocolate Custards. Dissolve gently by the side of the fire an ounce and a half of the best chocolate in rather more than a wineglassful of water, and then boil it until it is perfectly smooth; mix with it a pint of milk well flavored with lemon-peel or vanilla, and two ounces of fine sugar, and when the whole boils, stir to it five well-beaten eggs that have been strained. Put the custard into a jar or jug, set it into a pan of boiling water, and stir it without ceasing until it is thick. Do not put it into glasses or a dish till nearly or quite cold. These, as well as all other custards, are infinitely finer when made with the yolks only of the eggs. [81]

Conclusion

When discussing historical chocolate recipes it is important to differentiate ingredients and form two categories: before and after 1519 and Spanish contact with the mainland of Central America. Ingredients native to the Americas and identified as ingredients in chocolate recipes/preparations within a decade or two after initial contact are listed in Table 8.1. As noted elsewhere in this book (see Chapter 6), various Mexica/Aztec and Mayan medical texts identify a broad range of items added to chocolate, among them bark of the silk cotton tree (*Castilla elastica*), *mecaxochitl* (*Piper sanctum*), *tlacoxochitl* (*Calliandra anomala*), *tlatlapaltic* root (unknown), *tlayaploni xiuitl* (unknown), *uey nacaztli* (*Chiranthodendron pentadactylon*), and *tlilixochitl* (*Vanilla planifolia*). While these were medicinal chocolate beverages, it is not clear from the historical record what items were added to chocolate that was drunk for pleasure by adult males. As identified in the present chapter, other ingredients such as *achote* (*Bixa orellana*) were added to provide color, while ground corn (*Zea*

Table 8.1   Chocolate Ingredients Native to the Americas

Ingredients
Achote (*Bixa orellana*)
Allspice (*Pimienta dioica*)
Chichiualxochitl (breast flower?)
Chili pepper (many varieties)
Eloxochiquauitl (?)
Gie'cacai (flower?)
Gie'suba (flower?)
Hueinacazztli (*Cymbopetalum penduliflorum*)
Izquixochitl (*Bourreria* spp.)
Maize (*Zea mais*)
Marigold (*Tagetes lucida*)
Mexacochitl (string flower)
Piñon nuts (*Pinus edulis*)
Piztle (*Calorcarpum mammosum*)
Pochoctl (*Ceiba* spp.)
Quauhpatlachtli
Tlilxochitl (flower of *Vanilla planifolia*)
Tlilxochitl (pod/beans of *Vanilla planifolia*)
Xochinacaztli (*Cymbopetalum penduliflorum*)
Yolloxochitl (*Magnolia mexicana*)
Zapayal (*Calorcarpum mammosum*)
Zapote (*Achradelpha mammosa*)
Sweeteners
Syrup of maguey
Bee honey

mais) and piñon nuts (*Pinus edulis*) would have added texture and thickening. Allspice (*Pimienta dioica*) and numerous varieties of chili peppers (*Capiscum* spp.) would have added both flavor and "zing" to traditional pre-Columbian chocolate beverages. Flavors contributed by the leaves of various flowers, among them marigold (*Tagetes lucida*) and *Tlilxochitl* (flower of *Vanilla planifolia*), would have pleased many palates. The bitter chocolate bean base could have been sweetened using honey as suggested in the *Ritual of the Bacabs* (see Chapter 3) or the boiled-down syrups of various tropical fruits whether *Zapote* (*Achradelpha mammosa*) or others. Sugar—from sugar cane—was not available and would be introduced to the New World only during the Colonial era.

But the question still remains: What did pre-Columbian chocolate taste like?

The arrival of the Europeans initiated what has been called the Columbian Exchange, and the trans-Atlantic trade/import of European and African goods to Central America, and the reciprocal trade/import of New World foods and commodities to Africa, Asia, and Europe [82]. This trade/exchange of culinary items brought about major changes in chocolate recipes. Old World products subsequently became part of chocolate recipes from the mid-16th century onward (Table 8.2). The key product on this list certainly was the addition of sugar (from sugar cane), which was used to sweeten the chocolate products that were now consumed by all—young and old—and by both genders.

Table 8.2 Chocolate Ingredients Introduced to the Americas After 1519

Ingredients
  Ambergris
  Almond
  Anise
  Cardamom
  Cinnamon (bark or oil)
  Citron (bark or peel)
  Cloves (nail or oil)
  Dates
  Fennel
  Filbert
  Garlic
  Ginger
  Hazel nut
  Lemon (juice or peel)
  Milk
  Musk
  Nutmeg (ground or oil)
  Orange (blossom water, juice, or peel)
  Pepper (black)
  Pepper (white)
  Pistachio
  Rhurbarb
  Rose (petals)
  Rose (water)
  Saffron
  Salt
  Sesame (seeds)
Sweetener: Sugar (cane—raw or refined)

In many cases the importation of new ingredients would not have resulted in major recipe changes. Would there have been noticeable differences, for example, between the inclusion of ground almonds, filberts, hazel nuts, pistachios, or finely ground sesame seeds (Old World) versus piñon nuts (New World); or with the inclusion/substitution of black pepper (Old World) for chili peppers (New World)? Other items, however, would have radically changed the flavors of traditional New World chocolate, such as the addition of products like anise, cardamom, cinnamon, cloves, fennel, mace, and nutmeg. Additions such as lemon (juice or peel) and rhubarb would have added tart flavors to chocolate preparations that may or may not have existed previously.

Common recipes for Mexican chocolate obtained during field work in Oaxaca, Mexico, conducted between the years 1998 and 2000, included cinnamon and finely ground almonds (both Old World plants). During field work we did not encounter recipes that included the addition of vanilla, which would have represented a "true" traditional chocolate. Given the adoption of cinnamon into contemporary recipes in Mexico and Central America, it would be reasonable to conclude that chocolate and cinnamon went hand-in-hand through the centuries and would have been a desired consumer flavor—but to make such an assumption would be an error.

Finally, in 2006 Häagen-Daz® launched a new ice cream product called Mayan Chocolate, an effort that provided a wonderful taste and flavor, but one that also represented a curious misunderstanding of historical ingredients and their use among the ancient Mayans. The advertising sites claimed the following:

Inspired by the original chocolate first created by ancient Mayans in 500 BC . . . [a product that] blends the finest cocoa beans with cinnamon to create [a] magical Mayan offering: rich, chocolate ice cream with a fudge swirl and a hint of cinnamon. It's truly made like no other. [83]

The only difficulty with this advertising campaign is that cinnamon trees (*Cinnamomum zeylanicum*), cinnamon in the strict sense, did not grow in the New World during the Mayan era and this spice was not available to the ancient Mayans 500 years before the Common era, as claimed by the Häagen-Daz® advertisement. [84]

Throughout most of history, chocolate was consumed only as a beverage, and most of the historical recipes presented in this chapter reflect preparation of chocolate to be served in this form. The chocolate "tablets" identified in some of the recipes represented the base ingredient that subsequently was shaved and "dissolved" in water. The water used to prepare chocolate through the centuries commonly was heated to a boil, and this process, even without the addition of chocolate, would have been salutary. When analyzing different chocolate recipes and their reputed medical effects it is necessary, therefore, to separate presumed healthful effects of drinking chocolate prepared using boiled water from reputed positive medical effects inherent in chocolate itself. Separating the two, however, is difficult. What is clear is that chocolate beverages were widely used to treat diseases and, at the very least, administrating a variety of chocolate beverages produced a palliative, pleasurable effect on patients. The recipes for "health chocolate" identified as *chocolats de santé* or *chocolats thérapeutiques du médicinaux* reflect the importance of understanding the dual nature of chocolate: as food and as medicine.

We conclude this chapter with a presentation of contemporary chocolate recipes.

The following were collected by Dr. Sylvia Escarcega during project-related field work conducted in Oaxaca, Mexico, during 1998–2000.

RECIPE 1: Chocolate Oaxaquero: Oaxaca, Mexico

1 kg	Toasted cacao beans
1 kg	Granulated sugar
250 gr	Mexican cinnamon
250 gr	Almonds
Not defined	Vanilla and coffee beans (to taste)

Wash the cacao and almonds before toasting on a clay *comal* (griddle). Toast the almonds and the Mexican cinnamon. Once the cacao beans cool, carefully remove the skin. Grind the cacao beans on a *metate* (heated, slanted grinding stone), then add the almonds and cinnamon. When well ground, add sugar and mix. Take the hot paste and form bars of chocolate. Let them cool. These can be stored outside until used.

RECIPE 2: Chocolate de agua al estilo Oaxaquero: Oaxaca, Mexico

Melt the bars of chocolate in water or milk. Whip the beverage with a *molinillo* (long wooden stick with rings at the bottom that spin when the stick is rolled between the palms) until the *espuma* (froth) comes out. Serve immediately.

RECIPE 3: Chocolate Oaxaquero Amargo (Bitter): Oaxaca, Mexico

1 kg	Toasted cacao beans
1 kg	Granulated sugar
2 oz	Mexican cinnamon
2 oz	Almonds

RECIPE 4: Chocolate Oaxaquero Semi-amargo (Semi-sweet): Oaxaca, Mexico

2 kg	Toasted cacao beans
1 kg	Granulated sugar
25 gr	Mexican cinnamon
150 gr	Almonds

RECIPE 5: Chocolate Oaxaquero Dulce (Sweet): Oaxaca, Mexico

2 kg	Toasted cacao beans
2 kg	Granulated sugar
25 gr	Mexican cinnamon
25 gr	Almonds

RECIPE 6: Chocolate Atole: Oaxaca, Mexico

To prepare the *espuma* (froth)

500 gr	Toasted cacao beans
50–250 gr	*Petaxtli* [white cacao beans]
1 kg	Toasted corn or wheat
25–100 gr	Mexican cinnamon

To prepare the *atole* (cooked corn flour)

500 gr	Corn cooked in water
	Sugar [to taste]

Toast and peel the cacao beans. Peel the black cover off the white cacao beans and toast. Grind the cacao beans and cinnamon on the *metate*. Grind the corn/wheat separately, then add to the chocolate. Re-grind to powder. Add water to humidify. Soak the paste all night. In the morning, stir the paste quickly so it will not be solid or liquid. Pour into a *chocolatera* (ceramic pitcher) and add cold water. Whip using a *molinillo* until the *espuma* forms. Separate the froth with a wooden spoon. To make the *atole*, add water to the corn. Heat until soft. Grind and sieve the soft corn. Boil the powder and add sugar. Mix with a *chiquihuite* (thin wooden stick), pour into a cup, and then add the *espuma*.

RECIPE 7: Chocolate Tejate Beverage: Oaxaca, Mexico

1 kg	Cooked corn with ash (sift out any charcoal)
3	Mamey seeds
50 gr	Toasted *rosita de cacao*[3] (rose of cacao)
50 gr	Toasted cacao beans

Wash the corn, cook and remove skin. Cook corn in enough water but not too much. Add mamey seeds and make a paste. Toast the paste on the clay *comal* (griddle). Grind the corn, *rosita de cacao*, and the mamey. Add a little bit of water and blend by hand everything until all is integrated. Continue blending and whipping with the hand until the *espuma* (froth) comes out. Pour into *jicaras de calabaza* (pumpkin serving bowls) from high above to form more froth.

RECIPE 8: Champurrado: Oaxaca, Mexico, A Recipe Supplied by Vendors at the Benito Juarez Market, in Oaxaca City

3	Chocolate bars
1 kg	Corn
1 large piece	*Piloncillo* (raw sugar) or granulated sugar to taste
1 liter	Water

Cook the corn in the water for half an hour or until it softens. Grind the corn on the *metate* and pass through a sieve. Cook again, moving constantly with a wooden spoon, until it thickens. When it boils, add the chocolate and the *piloncillo* or sugar. Pour into the *chocolatera* (ceramic pitcher) and whip with a *molinillo* (wooden stick).

RECIPE 9: Espuma de Zaachila: Oaxaca, Mexico, A Recipe Supplied by Vendors at the Zaachila Market, Oaxaca City

1 kg	Rice, *comal* toasted
50 gr	Cacao beans (1st class); yellowish to add color; washed and toasted
1 oz	Mexican cinnamon
2 kg	*Petastli* (white cacao beans, almost rotten)
2 kg	White *atole* (cooked corn flour)
2 kg	Sugar

Grind all the ingredients on a *metate* or at the mill. The powder that comes out is called *pinole* and can be stored in this way. To prepare the *espuma*, mix the *pinole* with a little bit of cold water until it becomes a dough. Put in refrigerator. Add water as the paste is whipped with a *molinillo para espuma* (*molinillo* without rings) until the froth comes out. Keep doing it until enough froth has come out. The froth is taken out with a *jicara* (pumpkin serving bowl) without getting any water in it. The froth is then added to white *atole* previously prepared. Add sugar to taste. Use a *chiquihuite* (wooden stick) to mix.

RECIPE 10: Molé Negro Oaxaquero: Oaxaca, Mexico (Another from the Market)

Chili guajillo	5 pieces
Chili chilhuacle negro	5 pieces
Chili pasilla mexicano	5 pieces
Chili mulato or ancho negro	5 pieces
Chili chilhuacle rojo	2 pieces
Tomatillos	125 grams
Tomatoes	250 grams
Cloves	3 pieces
Allspice berries	3 pieces
Marjoram	3 sprigs
Thyme	3 sprigs
Avocado leaf (dried)	1 leaf
Oregano (dried)	1 tbs.
Lard or vegetable shortening	2 tbs.
Sesame seeds	1 cup
Peanuts (with skin)	10 pieces
Almonds (unpeeled)	10 pieces
Raisins	3 tbs.
Pecans	6 pieces
Onion	1 medium
Garlic (unpeeled)	6 cloves
Cinnamon (Mexican)	1 large stick
Plantain (peeled; sliced)	1 large
Corn tortillas	2 large
French bread	Some pieces
Mexican chocolate	60 grams (or more to taste)
Sugar	60 grams
Oil	60 grams
Salt	To taste
Chicken broth	As necessary
Chicken	10 pieces
Onion	1 (medium)
Garlic (peeled)	2 cloves

PREPARATION: PART 1

With a damp cloth clean the chilies and remove the stems, seeds, and veins. Reserve the seeds. Toast the chilies until black but not burnt. Cover them with hot water and let them soak for 10–20 min. On a skillet toast the seeds, medium heat, until golden. Increase the heat and toast them until black. Cover with cold water and let them soak for 5 minutes. Transfer the chilies to the blender and enough of the liquid to make it pass through a sieve. Save; set aside. Roast the tomatoes, *tomatillos*, onion, unpeeled garlic cloves for 10 minutes. Peel the garlic cloves. Save; set aside. Heat 2 tablespoons of oil and use it to fry the following ingredients but separately, save them separately: the raisins, the bread until browned, the tortillas, the plantain until golden (add more oil if needed), the sesame seeds.

Pass through a sieve to remove the excess oil, and in the reserved oil, fry at the same time peanuts, pecans, and almonds. Grind the seeds on the *metate* (alternatively use a food processor) adding water if needed.

PREPARATION: PART 2

Blend the tomatoes, *tomatillos*, garlic, onion, and spices. Separately blend the seeds, nuts, banana, raisins, bread, and tortillas, adding chicken broth as needed, until well blended. Heat 2 tablespoons of vegetable shortening on a large kettle, fry the chili paste, until it dries. Then fry the tomato mixture. Let it simmer for about 10 min or until it changes color. Add the rest of the blended ingredients except the chocolate and the avocado leaf.

PREPARATION: PART 3

Let it boil for about half an hour and add the chocolate. Toast slightly the avocado leaf over the flame adding it to the *molé*. Leave it simmering for a time, then taste and check for the flavors of chocolate and sugar. Add chicken broth as much as needed; the *molé* should have the consistency to cover the back of a spoon. On a large pan cook the chicken pieces with garlic, onion, and salt. Place a piece of the chicken on the serving dish, cover with the *molé*, and serve it with rice and hot tortillas.

The following recipes were collected by Dr. Victor Montejo during project-related field work conducted in Guatemala in 2000.

RECIPE 1: K'echi Maya Chocolate Beverage

Cacao (toasted)	5 pounds
Cinnamon	Add as necessary
Sugar	15 pounds

Place ingredients on grinding stone; grind until you have a paste. Make cakes of the paste by hand or by pressing into round molds. Let stand for one day. On the next day place 4 tablets of chocolate into hot water; stir it; you have a chocolate drink.

RECIPE 2: K'echi Maya Chocolate Beverage

Cacao	Quantity not given
Black pepper	"
Anise	"
Cinnamon	"
Muk (unidentified spice)	"
Sugar	"

Place ingredients on grinding stone; grind until you have a paste. Make cakes of the paste by hand or by pressing into round molds. Let stand for one day. On the next day place 4 tablets of chocolate into hot water; stir it; you have a chocolate drink.

Drink and enjoy!

Acknowledgments

I wish to acknowledge the research efforts of the following colleagues who contributed to the archival/field research or data summary at early stages of data collection: Patricia Barriga, Teresa L. Dillinger, Sylvia Escárcega, Victor Montejo, and Rebecca Schacker.

Endnotes

1. Professor Victor Montejo, a member of our chocolate-history team, conducted project-related field work in northern Guatemala, where he identified previously unpublished information on how cacao came to humankind. My own field work conducted in Costa Rica among indigenous Bribri Native Americans identified a different explanation of how humans came to use cacao (see Chapter 3).

2. The author's field work experience in rural Egypt (1964–1967) and in Botswana (1973–1975) confirmed the important role of bitter substances and pain when evaluating the "power" of traditional medicines: the more it hurt, the more bitter the taste, the stronger the medicine.

3. Rosita de cacao is the small white flower called *cacahuaxochitl* in Nahuatl (Aztec language). The Latin name is *Quararibea funebris*. It is a member of the Bombax family.

References

1. Motamayor, J. C. Germplasm resources and geogiaphic of J. Cacao. (submitted). Bennett, A., Keen, C., and Shapiro, H.-Y. *Theobroma cacao: Biology, Chemistry, and Woman Health*, Hoboken, NJ: Wiley, 2008.

2. Motamayor, J. C. Germplasm resources and geogiaphic of J. Cacao. (submitted). Bennett, A., Keen, C., and Shapiro, H.-Y. *Theobroma cacao: Biology, Chemistry, and Woman Health*, Hoboken, NJ: Wiley, 2008.

3. Dillinger, T. L., Barriga, P., Escarcega, S., Jimenez, M., Salazar Lowe, D., and Grivetti, L. E. Food of the gods: cure for humanity? A cultural history of the medicinal and ritual use of chocolate. *Journal of Nutrition* 2000;130(Supplement):2057S–2072S.

4. Grivetti, L. E. From aphrodisiac to health food; a cultural history of chocolate. In: *Karger Gazette*. Issue 68 (Chocolate). Basel, Switzerland: S. Karger, 2005.

5. Sahagún, B. (1590). (Florentine Codex). *General History of the Things of New Spain*. Santa Fe, NM: The School of American Research, 1981; and the University of Utah Monographs of The School of American Research and The Museum of New Mexico. Part 12, pp. 119–120.

6. Sahagún, B. (1590). (Florentine Codex). *General History of the Things of New Spain*. Santa Fe, NM: The School of American Research, 1981; and the University of Utah Monographs of The School of American Research and the Museum of New Mexico. Part 12, p. 176.

7. Sahagún, B. (1590). (Florentine Codex). *General History of the Things of New Spain*. Santa Fe, NM: The School of American Research, 1981; and the University of Utah Monographs of The School of American Research and the Museum of New Mexico. Part 12, p. 12.

8. Sahagún, B. (1590). (Florentine Codex). *General History of the Things of New Spain*. Santa Fe, NM: The School of American Research, 1981; and the University of Utah Monographs of The School of American Research and the Museum of New Mexico. Part 12, p. 178.

9. Sahagún, B. (1590). (Florentine Codex). *General History of the Things of New Spain*. Santa Fe, NM: The School of American Research, 1981; and the University of Utah Monographs of The School of American Research and the Museum of New Mexico. Part 12, p. 189.

10. Cortés, H. (1519). *Hernan Cortés: Letters from Mexico*. New Haven, CT: Yale University Press, 1986.

11. Díaz del Castillo, B. (1560–1568). *The Conquest of New Spain*. New York: Penguin Books, 1983.

12. Anonymous Conquistador. The Chronicle of the Anonymous Conquistador. In: *The Conquistadors. First-Person Accounts of the Conquest of Mexico*. Edited and translated by P. de Fuentes. New York: Orion Press, 1963; pp. 165–181.

13. Díaz del Castillo, B. (1560–1568). *The Conquest of New Spain*. New York: Penguin Books, 1983; pp. 226–227.

14. de Solis, A. *Historia De La Conquista De Mexico, Poblacion, y Progresos de la America Septentrional, Conocida por el Nombre de Nueva España. (History of the Conquest of Mexico, Population and Progress of the Northern America, Known by the Name of New Spain). By Antonio de Solis, Secretary of His Majesty, and His Major Chronicler of the Indias [sic]*. Madrid, Spain: de B. de Villa-Diego, 1766; p. 238.

15. Fray Toribio de Benavente, 1524, p. IBID.

16. Anonymous Conquistador. The Chronicle of the Anonymous Conquistador. In: *The Conquistadors: First-Person Accounts of the Conquest of Mexico*. Edited and translated by P. de Fuentes. New York: Orion Press, 1963; p. 173.

17. Cárdenas, Juan de. *Problemas y Secretos Maravillosos de las Indias*. Madrid, Spain: Ediciones Cultura Hispanica, 1591; Chapter VII, p. 109v.

18. Cárdenas, Juan de. *Problemas y Secretos Maravillosos de las Indias*. Madrid, Spain: Ediciones Cultura Hispanica, 1591; Chapter VII, pp. 110v–111v.

19. Cárdenas, Juan de. *Problemas y Secretos Maravillosos de las Indias*. Madrid, Spain: Ediciones Cultura Hispanica, 1591; Chapter VII, pp. 114f–114v.

20. de Acosta, J. *The Naturall [sic] and Morall Historie[sic] of the East and West Indies. Intreating of the Remarkeable [sic] Things of Heaven, of the Elements, Metals, Plants and Beasts*. London, England: Val. Sims for Edward Blount and William Aspley, 1604; p. 271.

21. Valverde Turices, S. *Un Discurso del Chocolate*. Seville, Spain: J. Cabrera, 1624; Part 1; no page numbers.

22. Valverde Turices, S. *Un Discurso del Chocolate*. Seville, Spain: J. Cabrera, 1624; Part 1, no page numbers.

23. Valverde Turices, S. *Un Discurso del Chocolate*. Seville, Spain: J. Cabrera, 1624; Part 1, no page numbers.

24. Colmenero de Ledesma, A. *Curioso Tratado de la Naturaleza y Calidad del Chocolate*. Madrid, Spain: Francisco Martinez, 1631.

25. Knapp, A. *Cocoa and Chocolate. Their History from Plantation to Consumer*. London, England: Chapman and Hall, 1920; pp. 191–203.

26. de Vades-Forte, D. *A Curious Treatise of the Nature and Quality of Chocolate by Antonio de Ledesma Colmenero*. London, England: I. Okes, 1640.

27. Moreau, R. *Du Chocolate Discours Curieux by Antonio de Ledesma Colmenero*. Paris, France: Sebastien Cramoisy, 1643.

28. Volckamer, J. G. *Chocolate Inda Opusculum de Qualitate et Natura Chocolatae by Antonio de Ledesma Colmenero*. Norimbergae (Nuremburg), Germany: Typis Wolfgang Enderi, 1644.

29. Wadsworth, J. *Chocolate: Or an Indian Drinke [sic] by Antonio Ledesma Colmenero*. London, England: J. G. for Iohn Dakins, 1652.

30. Vitrioli, A. *Della Cioccolata Discorso*. Rome, Italy: R. C. A., 1667.

31. Chamberlaine, J. *The Manner of Making Coffee, Tea and Chocolate*. London, England: Christopher Wilkinson, 1685.

32. Colmenero de Ledesma, A. *Curioso Tratado de la Naturaleza y Calidad del Chocolate*. Madrid, Spain: Francisco Martinez, 1631.

33. Hurtado, T. *Chocolate y Tabaco Ayuno Eclesiastico y Natural: si este le Quebrante el Chocolate: y el Tabaco al Natural, para la Sagrada Comunion. Por Francisco Garcia, Impressor del Reyno. A Costa de Mauel Lopez*. Madrid, Spain: Mercador de Libros, 1645; Chap. 3, Part 17, no page numbers.

34. Gage, T. *The English American: His Travail by Sea; or a New Survey of the West Indies Containing a Journall [sic] of Three Thousand Three Hundred Miles within the Main Land of America. . . .* London, England: R. Cotes, 1648; pp. 153–154.

35. Stubbe, H. 1662. *The Indian Nectar, or, a Discourse Concerning Chocolata [sic]: the Nature of the Cacao-Nut and the Other Ingredients of that Composition is Examined and Stated According to the Judgment and Experience of Indian and Spanish Writers*. London, England: J. C. for Andrew Crook, 1662; p. 11.

36. Stubbe, H. *The Indian Nectar, or, a Discourse Concerning Chocolata [sic]: the Nature of the Cacao-Nut and the Other Ingredients of that Composition is Examined and Stated According to the Judgment and Experience of Indian and Spanish Writers*. London, England: J. C. for Andrew Crook, 1662; p. 13.

37. Stubbe, H. *The Indian Nectar, or, a Discourse Concerning Chocolata [sic]: the Nature of the Cacao-Nut and the Other Ingredients of that Composition is Examined and Stated According to the Judgment and Experience of Indian and Spanish Writers*. London, England: J. C. for Andrew Crook, 1662; pp. 53–54.

38. Stubbe, H. *The Indian Nectar, or, a Discourse Concerning Chocolata [sic]: the Nature of the Cacao-Nut and the Other Ingredients of that Composition is Examined and Stated According to the Judgment and Experience of Indian and Spanish Writers*. London, England: J. C. for Andrew Crook, 1662; p. 67.

39. Stubbe, H. *The Indian Nectar, or, a Discourse Concerning Chocolata [sic]: the Nature of the Cacao-Nut and the Other Ingredients of that Composition is Examined and Stated According to the Judgment and Experience of Indian and Spanish Writers*. London, England: J. C. for Andrew Crook, 1662; pp. 68–69.

40. Hughes, W. *The American Physitian [sic], or A Treatise of the Roots, Plants, Trees, Shrubs, Fruit, Herbs etc. Growing in the English Plantations in America . . . whereunto is Added a Discourse of the Cacao-Nut-Tree, and the Use of its Fruit; with All the Ways of Making of Chocolate. The Like Never Extant Before*. London, England: J. C. for William Crook the Green Dragon without Temple-Bar, 1672; p. 118.

41. Hughes, W. *The American Physitian [sic], or A Treatise of the Roots, Plants, Trees, Shrubs, Fruit, Herbs etc. Growing n the English Plantations in America . . . whereunto is Added a Discourse of the Cacao-Nut-Tree, and the Use of its Fruit; with All the Ways of Making of Chocolate. The Like Never Extant Before*. London, England: J. C. for William Crook the Green Dragon without Temple-Bar, 1672; pp. 119–120.

42. Hughes, W. *The American Physitian [sic], or A Treatise of the Roots, Plants, Trees, Shrubs, Fruit, Herbs etc. Growing n the English Plantations in America . . . whereunto is Added a Discourse of the Cacao-Nut-Tree, and the Use of its Fruit; with All the Ways of Making of Chocolate. The Like Never Extant Before*. London, England: J. C. for William Crook the Green Dragon without Temple-Bar, 1672; pp. 122–123.

43. http://www.daileyrarebooks.com/0902rarebefore 1700.htm. (Accessed January 10, 2007.)

44. Dufour, P. S. *The Manner of Making of Coffee, Tea, and Chocolate as it is Used in Most Parts of Europe, Asia, Africa and America, with Their Vertues [sic]*. London, England: William Crook, 1685; p. 71.

45. Dufour, P. S. *The Manner of Making of Coffee, Tea, and Chocolate as it is Used in Most Parts of Europe, Asia, Africa and America, with Their Vertues [sic]*. London, England: William Crook, 1685; p. 72.

46. Dufour, P. S. *The Manner of Making of Coffee, Tea, and Chocolate as it is Used in Most Parts of Europe, Asia, Africa and America, with Their Vertues [sic]*. London, England: William Crook, 1685; pp. 72–73.

47. Dufour, P. S. *The Manner of Making of Coffee, Tea, and Chocolate as it is Used in Most Parts of Europe, Asia, Africa and America, with Their Vertues [sic]*. London, England: William Crook, 1685; pp. 90–92.

48. Knapp, A. *Cocoa and Chocolate. Their History from Plantation to Consumer*. London, England: Chapman and Hall, 1920; p. 194.

49. Chamberlaine, J. *The Manner of Making Coffee, Tea and Chocolate*. London, England: Christopher Wilkinson, 1685; p. 15.

50. Chamberlaine, J. *The Manner of Making Coffee, Tea and Chocolate*. London, England: Christopher Wilkinson, 1685; pp. 16–17.

51. Lightbody, J. *Every Man His Own Gauger. Wherein Not only the Artist is shown . . . The Art of Brewing Beer, Ale . . . The Vintners Art of Fining, Curing, Preserving and Rectifying all Sorts of Wines . . . Together with the Compleat Coffee-Pan, Teaching how to Make Coffee, Tea, Chocolate*. London, England: The Ring, 1695; p. 62.

52. Lightbody, J. *Every Man His Own Gauger. Wherein Not only the Artist is shown . . . The Art of Brewing Beer, Ale . . . The Vintners Art of Fining, Curing, Preserving and Rectifying all Sorts of Wines . . . Together with the Compleat Coffee-Pan, Teaching how to Make Coffee, Tea, Chocolate*. London, England: The Ring, 1695; p. 62.

53. Blégny, N. de *Le bon usage du thé, Du caffé, et Du chocolat pour La préservation & Pour La guérison des maladies*. Paris, France: E. Michallet, 1687; pp. 282–285.

54. Blégny, N. de *Le bon usage du thé, Du caffé, et Du chocolat pour La préservation & Pour La guérison des maladies*. Paris, France: E. Michallet, 1687; pp. 282–285.

55. Blégny, N. de *Le bon usage du thé, Du caffé, et Du chocolat pour La préservation & Pour La guérison des maladies*. Paris, France: E. Michallet, 1687; pp. 282–285.

56. de Quélus (first name unknown). *Histoire naturelle du Cacao et du Sucre*. Paris, France: L. d'Houry, 1719.

57. de Quélus. *Natural History of Chocolate*. Translated by R. Brookes. London, England: J. Roberts, 1724.

58. de Quélus. *Natural History of Chocolate*. Translated by R. Brookes. London, England: J. Roberts, 1730.

59. de Quélus. *Natural History of Chocolate*. Translated by R. Brookes. London, England: J. Roberts, 1730; p. 45.

60. de Quélus. *Natural History of Chocolate*. Translated by R. Brookes. London, England: J. Roberts, 1730; pp. 61–62, 64, 66.

61. Linné (Linneaus), C. von. *Om Chokladdryken*. Stockholm, Sweden: Fabel, 1965.

62. Linné (Linneaus), C. von. *Om Chokladdryken*. Stockholm, Sweden: Fabel, 1965.

63. Linné (Linneaus). C. von. *Om Chokladdryken*. Stockholm, Sweden: Fabel, 1965.

64. Linné (Linneaus). C. von. *Om Chokladdryken*. Stockholm, Sweden: Fabel, 1965.

65. Linné (Linneaus). C. von. *Om Chokladdryken*. Stockholm, Sweden: Fabel, 1965.

66. Linné (Linneaus). C. von. *Om Chokladdryken*. Stockholm, Sweden: Fabel, 1965.

67. Lavedan, A. *Tratado de Los Usos, Abusos, Propiendades y Virtudes del Tabaco, Café, Te y Chocolate*. Madrid, Spain: Imprenta Real, 1796.

68. Lavedan, A. *Tratado de Los Usos, Abusos, Propiendades y Virtudes del Tabaco, Café, Te y Chocolate*. Madrid, Spain: Imprenta Real, 1796; Chap. IV, unpaginated.

69. Lavedan, A. *Tratado de Los Usos, Abusos, Propiendades y Virtudes del Tabaco, Café, Te y Chocolate*. Madrid, Spain: Imprenta Real, 1796; Chap. IV, unpaginated.

70. Lavedan, A. *Tratado de Los Usos, Abusos, Propiendades y Virtudes del Tabaco, Café, Te y Chocolate*. Madrid, Spain: Imprenta Real, 1796; Chap. IV, unpaginated.

71. Lavedan, A. *Tratado de Los Usos, Abusos, Propiendades y Virtudes del Tabaco, Café, Te y Chocolate*. Madrid, Spain: Imprenta Real, 1796; Chap. IV, unpaginated.

72. Lavedan, A. *Tratado de Los Usos, Abusos, Propiendades y Virtudes del Tabaco, Café, Te y Chocolate*. Madrid, Spain: Imprenta Real, 1796; Chap. IV, unpaginated.

73. Lavedan, A. *Tratado de Los Usos, Abusos, Propiendades y Virtudes del Tabaco, Café, Te y Chocolate*. Madrid, Spain: Imprenta Real, 1796; Chap. V, unpaginated.

74. Grivetti, L. E., Corlett, J. L., and Lockett, C. T. Food in American history. Part 1. Maize. Bountiful gifts: America on the eve of European colonization (antiquity to 1565). *Nutrition Today* 2001;36(1):19–28.

75. Simmons, A. *American Cookery. A Facsimile of the First Edition, 1796*. New York: Oxford University Press, 1958; p. 34.

76. Harbury, K. E. *Colonial Virginians Cooking Dynasty*. University of South Carolina Press, 1700; p. 189.

77. Esther Edwards Burr, E. E. 1732–1758. *The Journal of Esther Edwards Burr, 1754–1757*. New Haven, CT: Yale University Press, 1984. Letter from Esther Edwards Burr to Sarah Prince Gill, dated January 14, 1755; p. 318.

78. Briggs, R. *The New Art of Cookery, According to the Present Practice*. Philadelphia: W. Spotswood, R. Campbell, and B. Johnson, 1792; p. 375.

79. Anonymous. *One thousand valuable secrets, in the elegant and useful arts collected from the practice of the best artists, and containing an account of the various methods of engraving on brass, copper and steel . . . Of making vinegars. Of liquors, essential oils, &c. Of confectionary. Of preparing various kinds of snuffs. Of taking out spots and stains. Of fishing, angling, bird-catching, and a variety of other curious, entertaining and useful articles*. First American edition. Philadelphia: B. Davies, 1795; p. 314.

80. *Saturday Evening Post*. Volume 30, September 21, 1850, p. 4.

81. *Godey's Lady's Book*. February Issue, 1854, p. 186.

82. Grivetti, L. E. Clash of cuisines. Notes on Christoforo Colombo 1492–1503. *Nutrition Today* 1992;26(2): 13–15.

83. http://www.nestle.ca/Haagen_Dazs/en/Products/Ice_Cream/mayan_chocolate.htm. (Accessed January 10, 2007.)

84. Dalby, A. *Dangerous Tastes. The Story of Spices*. Berkeley: University of California Press, 2000; pp. 36–41.

Chocolate as Medicine

Imparting Dietary Advice and Moral Values Through 19th Century North American Cookbooks

Deanna Pucciarelli

Introduction

It would be possible to generalize that every maturing civilization—the Chinese in the fifth century, the Muslim in the eleventh and twelfth, and the Western in the Italy of the fifteenth—gives its cuisine a formal structure and decorum. But can an author and his book be considered only as a symptom of something outside them? [1]

Cooking and presenting food is more than merely itemizing basic ingredients and preparation procedures in standard form. Furthermore, even the word "cookbook" is problematic. It can be argued that cookbook authors reflect snapshots of society and the historical eras and events therein. It is the premise of this chapter that through the study of cookbooks, scholars can understand cultural norms of a given era. In reviewing famous gastronomic texts and cookbooks from the ancient past, including Athenaeus's *The Deipnosophists* (*Sophists at Dinner*) dated to the early third century CE, *De re Coquinaria* (*The Art of Cooking*) written by Apicius, during the late fourth and early fifth centuries CE, and *Liber de Coquina* (*The Book of Cookery*), written in separate codices by two unknown authors from France and Italy during the late 13th and early 14th centuries CE, it can be documented that these texts not only contained information on ingredients and directions for meal preparation but also served as directives on how to live economic, healthful, and righteous lives.

Descriptions of food, culinary preparation, recipes, meals, banquets, and social tradition associated with food can be traced to antiquity. Food offerings and in some instances complete meals were left for the deceased to consume in ancient Egypt [2]. Behavior at meals, rations for soldiers and workers, and food-related proverbs have been documented for more than 5000 years in Egyptian tomb and temple art, as well as in surviving papyrus documents [3]. The preparation, serving, and social behavior associated with foods also have been documented for ancient India [4], China [5], Mesopotamia [6], and Greece and Rome [7].

Perhaps the most important food-related manuscript to survive from antiquity, however, is dated to the early third century CE [8]. This manuscript, *The Deipnosophists* (*Sophists at Dinner*) was written in Rome by the Greek-Egyptian Athenaeus, who had access to hundreds of ancient manuscripts. *The Deipnosophists* contains a wealth of information on hundreds of foods within the Mediterranean region and beyond, their geographical origins, effect on the human body, relative popularity, preparation, and social use [9].

The first manuscript recognized by scholars as a formal cookbook, however, is attributed to Apicius

Chocolate: History, Culture, and Heritage. Edited by Grivetti and Shapiro
Copyright © 2009 John Wiley & Sons, Inc.

of Rome, entitled *De re Coquinaria*, or *The Art of Cooking*. Although several Romans named Apicius were known for their appetite, the author of *De re Coquinaria* most likely lived during the reign of Caesar Augustus and Caesar Tiberius [10]. Unlike *The Deipnosophists*, which gives a more broad view on the social use of food, *De re Coquinaria* strictly contains recipes with associated ingredient lists and preparatory methodology. While readers obtain an understanding of cooking methods employed, social norms are conspicuously absent.

It may be argued that Bartolomeo Platina's cookbook, *De Honesta Voluptate* (*On Honest Indulgence and Good Health*) published in 1475, or 25 years after Guttenberg's invention of the printing press in 1450, is the first dated, published collection of recipes, or what scholars in the 21st century would define as a cookbook [11]. Although widely lauded, Platina, Pope Sixtus IV's librarian, copied most of his recipes from a manuscript entitled *Libro de Arte Coquinara* written by (Maestro) Martino during this same time period. The pilfering of previously written recipes continued to be customary well into the 19th century.

Platina's cookbook written during the 15th century serves as a prologue to modern 21st century technically written, recipe-focused cookbooks, yet he does not escape altogether the following centuries' cookbook genre of dispensing "healthful living" advice to the populace. In the early pages of his work, Platina identified desirable behavioral norms of his time, specifically: "what to do upon arising in the morning, what should be done to further the enjoyment of life and, on tarrying with a woman" [12]. Given that the writer of the first printed cookbook integrated social and behavioral norms as adjuncts to his recipes, it is intriguing to note that this pattern of blending culinary and social aspects of life continued to characterize cookbooks well into the 19th and early 20th centuries. The earliest cookbooks were written by men—for men. But by the 19th century, authorship gender had reversed so that cookbook writers primarily were women and the literary focus of cookbooks shifted from that of primarily being instructional food preparation to one where virtuous living was coached, an approach that also happened to include recipes.

Cookbooks

INFLUENTIAL ENGLISH COOKBOOKS IN THE COLONIES

Eliza Smith's *The Compleat [sic] Housewife: Or Accomplished Gentlewomen's Companion*, first issued in London in 1727, was very popular in 18th century England [13]. Reprinted along with other popular English cookbooks in Boston, New York, and Philadelphia, this work is the first recorded cookbook printed in 18th century North America. William Parks, a printer in 18th century Colonial Williamsburg, reviewed Smith's work and omitted recipes where he believed ingredients could not be found in early America.

For daily cooking activities prior to the printing of Smith's book, Colonial housewives and kitchen help used cookbooks printed in England then shipped to the Colonies. Imported books, expensive for most families, complemented what have become known as "commonplace books." Commonplace books were a type of diary or scrapbook wherein recipes, lines of poetry, proverbs, and prayers were collected and recorded. By reviewing compilations of these manuscripts, scholars can weave together general themes particular to a given geographic region and time. Moreover, repetitive foodstuff entries can highlight the product's usage—or commonplace—among the published and literate.

Scholars credit the first North American chocolate recipe "to make chocolat [sic] almonds" to a Colonial Williamsburg, handwritten commonplace book probably compiled at the end of the 17th or early 18th century [14]. The same titled recipe appeared four decades later in Parks's edition of *The Compleat [sic] Housewife*, 1742, with only minor changes to the text. Whereas the unidentified author from the earlier period used much less chocolate per batch, and rosewater, Smith's recipe required a pound of chocolate and orange flower water [15]. The significance is not that the ingredient quantities differed; rather, it is the inclusion of chocolate at all—that chocolate, an ingredient dependent on importation, had become a household commodity in early America. Chocolate's availability is further evidenced in a book written by an Englishman sent to the United States to report back to England on the country's climate, cultivation of the land, grocery prices, and housekeeping practices. He states:

> Groceries, as they are called, are, upon an average, at far less than half the English price. Tea, sugar, coffee, spices, chocolate, cocoa . . . are so cheap as to be within the reach of everyone. Chocolate, which is a treat to the rich, in England, is here used even by the negroes. [16]

English writers continued throughout the 18th century to influence American cookbook authors. Susannah Carter's *The Frugal Housewife*, became a pivotal English contribution to American cookery because Amelia Simmons directly copied recipes from this work and placed them into her own cookbook, *American Cookery, or The Art of Dressing Viands, Fish, Poultry and Vegetables, and the Best Modes of Making*

Pastes, Puffs, Pies, Tarts, Puddings, Custards and Preserves, and All Kinds of Cakes . . . Adapted to This Country, and All Grades of Life [17, 18]. Simmons' cookbook commonly is credited as being the first known to be authored and published in North America with recipes that utilized ingredients indigenous to the Colonies. This argument is supported by inclusion of recipes such as Indian Slapjacks and Indian Pudding, two recipes that required corn as a central ingredient [19]. Throughout the early 19th century English cookbooks were reprinted and widely used by American women; however, increasingly, Americans began to write and publish cookbooks for the home market [20].

NINETEENTH CENTURY AMERICAN COOKBOOKS

American cookbooks of the 19th century can be clustered into four distinct periods: (1) early century (1800–1849), (2) pre-Civil War (1850–1860), (3) Civil War (1861–1865), and (4) post-Civil War (1866–1899). Each period presents a distinct reflection of historical phenomena inherent to American life. The phenomena are complex and multifaceted, including but not limited to culture, economics, politics, religion, socioeconomic factors, and technology. These factors influenced the American population in a variety of ways, as evidenced by both ingredient lists and household management advice contained in 19th century cookbooks.

Culinary historians identify Mary Randolph's *The Virginia Housewife* cookbook, first printed in 1824 (reprinted 19 times prior to the American Civil War), as reflective of dietary norms, at least in the South, during the early part of the century [21]. Randolph's utilitarian narrative mirrors both her cookbook's subheading, *Methodical Cook* [22], as well as her livelihood as a boardinghouse proprietor. Raised and married into privileged environments, Randolph was formally educated in reading, writing, and arithmetic. In addition, she was trained in the household arts. After many years of comfortable living, her husband, Davis Randolph, openly criticized Thomas Jefferson's governmental policies and consequently the family fell from high social and economic status [23]. Randolph's many years of hosting parties enabled her to operate a boardinghouse kitchen. The recipes found in *The Virginia Housewife* are operative of serving groups—working-class groups.

The abundance of ingredients listed in the recipes indicates that Virginians consumed a variety of foodstuffs, including over 35 different vegetables and 22 different ice cream flavors [24]. The ice cream recipes almost always included the addition of sugar, the exception being that for chocolate [25]. One can speculate that this could be an editorial mistake; however, the only other recipe in her collection to include chocolate, that for *chocolate cakes*, oddly does not use chocolate (Fig. 9.1) [26]. In this recipe collection *chocolate cakes* are plain cakes served with a chocolate beverage. It is difficult to determine, therefore, whether or not chocolate recipes with sugar were common during this era.

A contemporary of Randolph, Eliza Leslie, published two cookbooks during this period. The first, *Seventy-five Receipts for Pastry, Cakes, and Sweetmeats,* lacked a single recipe that included chocolate as a listed ingredient [27]. In spite of that, as Coe and Coe point out in their seminal book on chocolate, "the first family of recipes containing cacao (apart from recipes for the hot drink) appeared between 1680 and 1684 [in Italy], and were common in the next century" [28]. A recipe found in a manuscript in Naples for chocolate sorbet dates to 1794 and utilizes a kilogram of chocolate and 680 g of sugar [29]. At least in Europe, chocolate paired with sugar was commonly included in dessert recipes for well over a century prior to Leslie's cookbook. Her second volume, *Directions for Cookery, in Its Various Branches* [30], printed in 1840, or sixteen years after *The Virginia Housewife,* included three recipes that mention chocolate [31]. Of these chocolate-related recipes, two are clustered under the subheading "Preparations for the Sick" (Fig. 9.2).

In both beverage recipes "Cocoa" and "Cocoa Shells," sugar is missing. Sugar, during the 19th century, was thought by many to contain magical properties [32]. Indeed, sugar used for the infirm depended on the cookbook author's philosophical view toward sugar's magical powers. Likewise, chocolate was associated with magical properties throughout the 17th and 18th centuries in Mesoamerican countries, specifically Guatemala. After Europeans colonized Guatemala, Inquisition records (see Chapter 4) describe how chocolate was used by women to bewitch men into sexual submission:

> De Varaona gave her some powders and told Maneuela to wash her genitals with water, then beat the powders and the water into a hot chocolate drink and give it to a man she desired. [33]

In spite of Leslie's omission, the use of sugar with chocolate either for medicinal purposes or in confectionary can be found elsewhere. Thomas Cooper, a physician, wrote a medical text on caring for the sick that contained an appendix on domestic cookery [34]. His general advice followed Galen's allopathic system of medicine based on the four humours (black bile, yellow bile, blood, and phlegm) complemented by the four elements (air, earth, fire, and water) and integrated with qualities of hot/cold and moist/dry [35]. Good health and cures for illnesses

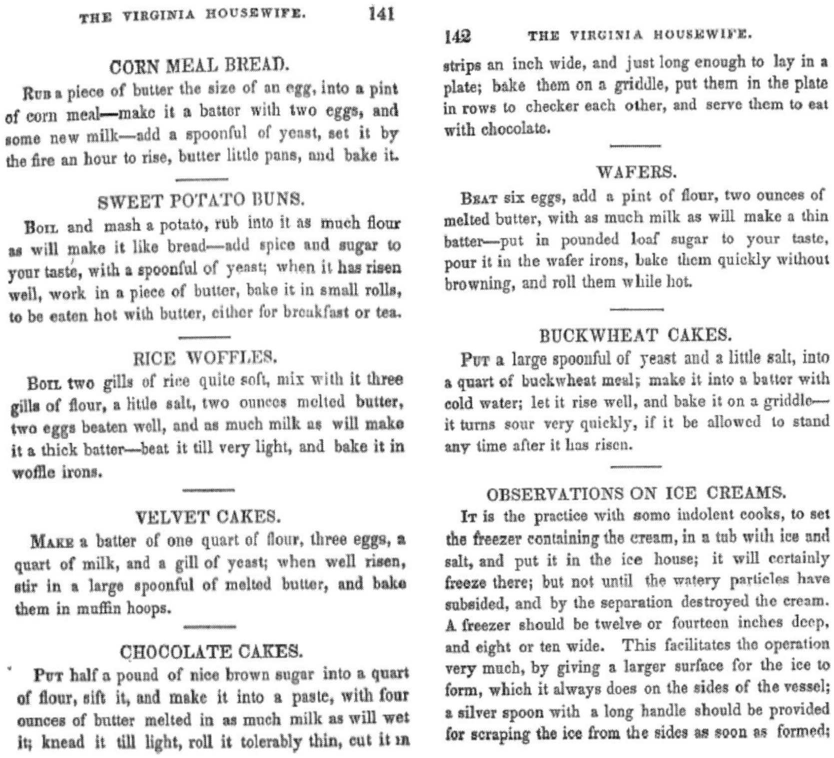

CORN MEAL BREAD.

Rub a piece of butter the size of an egg, into a pint of corn meal—make it a batter with two eggs, and some new milk—add a spoonful of yeast, set it by the fire an hour to rise, butter little pans, and bake it.

SWEET POTATO BUNS.

Boil and mash a potato, rub into it as much flour as will make it like bread—add spice and sugar to your taste, with a spoonful of yeast; when it has risen well, work in a piece of butter, bake it in small rolls, to be eaten hot with butter, either for breakfast or tea.

RICE WOFFLES.

Boil two gills of rice quite soft, mix with it three gills of flour, a little salt, two ounces melted butter, two eggs beaten well, and as much milk as will make it a thick batter—beat it till very light, and bake it in woffle irons.

VELVET CAKES.

Make a batter of one quart of flour, three eggs, a quart of milk, and a gill of yeast; when well risen, stir in a large spoonful of melted butter, and bake them in muffin hoops.

CHOCOLATE CAKES.

Put half a pound of nice brown sugar into a quart of flour, sift it, and make it into a paste, with four ounces of butter melted in as much milk as will wet it; knead it till light, roll it tolerably thin, cut it in

strips an inch wide, and just long enough to lay in a plate; bake them on a griddle, put them in the plate in rows to checker each other, and serve them to eat with chocolate.

WAFERS.

Beat six eggs, add a pint of flour, two ounces of melted butter, with as much milk as will make a thin batter—put in pounded loaf sugar to your taste, pour it in the wafer irons, bake them quickly without browning, and roll them while hot.

BUCKWHEAT CAKES.

Put a large spoonful of yeast and a little salt, into a quart of buckwheat meal; make it into a batter with cold water; let it rise well, and bake it on a griddle—it turns sour very quickly, if it be allowed to stand any time after it has risen.

OBSERVATIONS ON ICE CREAMS.

It is the practice with some indolent cooks, to set the freezer containing the cream, in a tub with ice and salt, and put it in the ice house; it will certainly freeze there; but not until the watery particles have subsided, and by the separation destroyed the cream. A freezer should be twelve or fourteen inches deep, and eight or ten wide. This facilitates the operation very much, by giving a larger surface for the ice to form, which it always does on the sides of the vessel; a silver spoon with a long handle should be provided for scraping the ice from the sides as soon as formed;

FIGURE 9.1. Unusual recipe for chocolate cake. Recipe lacks any chocolate or cocoa as an ingredient. *Source*: Randolph, M. *Virginia Housewife; or, Methodical Cook*. Baltimore, Md: Plaskitt & Cugle, 1838; p. 142. Courtesy of Peter J. Shields Library, University of California, Davis. (Used with permission.)

depended on the balance between the aforementioned qualities—these qualities were manifested in foodstuffs. Therefore, if a patient had a melancholic (cold and dry) condition, a warm and moist food would be prescribed.

Depending on the writer's perspective, chocolate was either classified within this system as *moist and cold* due it its high fat content, or as *dry* owing to its astringent tannins that puckered the mouth. *Hot* was associated with fiery chili peppers, which were commonly added to chocolate, as reported by the 16th century Spanish physician Dr. Juan de Cárdenas, who claimed chocolate was neutral—the hot chili-peppers balanced chocolate's cold nature [36]. As early as 1591, the lack of consensus on chocolate's medicinal attributes is apparent in Cárdenas' writings:

> In terms of the harms and benefits, I hear every person with his own opinion: some despise chocolate, considering it the inventor of numerous sicknesses; others say that there is nothing comparable in the world. [37]

The practice of allopathic medicine continued until the middle of the 20th century, well after Louis Pasteur posited germ theory to the scientific community. Many cultures continued to adhere to this style of medicine regardless of Western medicine's ubiquitous influence [38]. It was the balancing of

these properties through foodstuffs, espoused by Cooper and his female cookbook writing contemporaries, where the juxtaposition of sugar and chocolate usage constantly shifted from cold to hot, and moist to dry.

In part twelve, "Cookery For the Sick and For the Poor," Cooper advised the addition of sugar to chocolate: "put a teaspoon [of chocolate] or two into milk, boil it with sugar, and mill it well" [39]. Yet in reviewing 19th century American cookbooks, most chocolate recipes for invalids lacked the addition of sugar. The question may be posed: Why this omission of sugar in chocolate beverages served to the infirm?

By the mid-19th century sugar could be purchased easily. Economic barriers to consumption were eliminated as sugar supplies continued to increase. Transportation networks expanded exponentially throughout the mid to late 19th century with the exception of the Civil War years. Consequently, transportation costs of sugar decreased, lowering the sale price [40]. Moreover, the industrial revolution in the United States had been underway for more than 20 years, resulting in technological advances in sugar refining. Correspondingly, sugar yields increased, further decreasing the difficulty and expense of sugar procurement. The availability of sugar was so prevalent that refineries organized to stabilize the markets.

FIGURE 9.3. Chocolate and seduction: woman stirring cocoa. *Source*: Private Collection. Courtesy of The John Grossman Collection of Antique Images, Tucson, AZ. (Used with permission.) (See color insert.)

A WELCOME VISITOR.
A BEDROOM IN HINDELOOPEN (HOLLAND).

FIGURE 9.2. Nurse serving cocoa to the infirm. *Source*: Trade Card Collection. Ephemera: Groceries, Cocoa-Maple Syrup. Courtesy of the American Antiquarian Society, Worcester, MA. (Used with permission.) (See color insert.)

The Sugar Trust, an American organization of eight major refining companies formed in 1887, often kept production low in order to prevent market saturation [41]. Sugar's accessibility and low cost placed it within reach of middle-class Americans. Not using sugar in medicinal chocolate was not an economic decision. It was a choice, one influenced by culture.

The answer to the omission of sugar in medicinal chocolate, therefore, may be found in what sociologists term "cultural context." Cultural context is defined as "an assembly of cultural traits [including] social structures, customary behaviors, ideas, words, and material objects that [make] sense to contemporaries as elements to their world" [42]. North America during the mid to late 19th century can be regarded as an aggregation of overlapping cultural contexts as an influx of immigrants flooded the eastern seaboard, bringing with them their European beliefs, ideas, and values. This perceived "outside" threat was reflected in an editorial in *The New York Times*:

To the immense number of foreigners yearly coming to our shores, bringing with them social sympathies entirely antagonistic to those formerly accepted here, it would be difficult to determine; but whatever the cause may be, the fact

is patent that restaurants and boarding-houses are fast multiplying, and threaten at no distant day to usurp the place of the family dinner table. [43]

Social norms metamorphosed as the various immigrant groups imposed their respective countries' customs. Whether in regard to dining in a restaurant or a home, or caring for the infirm, the competing cultures' practices influenced the collective American societal norms. Some taboos, however, transcended national boundaries.

Adding sugar to medicinal chocolate was viewed by many, especially 19th century female cookbook writers, as decadent, sinful, and feminine [44]. Chocolate was first introduced to Colonial America via Spain and retained some of its pre-Colonial cultural origins. Thus it was linked not only with medicine, but also with sexuality, a long-held view dating to the early 16th century account of chocolate-drinking depicted at Montezuma's Court as recorded by Bernal Díaz del Castillo (see Chapter 6):

From time to time the men of Montezuma's guard brought him, in cups of pure gold, a drink made from the cacao-plant, which they said he took before visiting his wives. [45]

To the present day, chocolate remains linked to aphrodisiacs, and moral colloquialisms still abound, as in "sinfully" rich chocolate cake.

Nineteenth century advertisers furthered the sexual image of chocolate through evocative graphics in posters and other print media, of frilly boxes, suggestive portraits, and romantic packaging (Fig. 9.3). According to Woloson:

By 1900, there was no doubt that chocolate's reputation as "the apex of the seductive edifice of confectionary" has been cemented, perpetuated through generations of Europeans and Americans since Montezuma's time. [46]

Public morality was the cornerstone of Victorian respectability. The incongruity between the ascribed-to Victorian female ideal and late 19th century commercial advertising can be explained by the division between male and female economic hierarchy.

Men controlled the production, distribution, and consumption of chocolate; chocolates primarily were given as gifts from men to women during this era [47]. As Jackson Lears argues: "the gift of chocolates . . . operated within the bounds of accepted cultural rituals but privately represented elements of sexual drive" [48]. In other words, public exoticism displayed in packaging was permissible within Victorian refinement provided that established male–female courtship protocols were adhered to. Tangential to overt advertising would be technological advances in chocolate-related technology that created a sensual product.

In 1828, Conrad Van Houten patented a process whereby chocolate was *defatted*, enabling the mixture of cocoa solids (minus the hydrophobic cocoa butter and hydrophilic sugar) to emulsify into a creamy mass. At the next production stage, the previously extracted cocoa butter was added back into the mixture. The end result produced a creamy product that melted at body temperature. The mid-19th century brought other technological advances to chocolate production via Swiss inventors Henri Nestle and Daniel Peter, who discovered the processes to make powdered milk and milk chocolate bars, respectively [49]. By 1879, conching—or the process of transforming chocolate from a semigritty product into a smooth mass—was utilized in confectionary production (see Chapter 46). Manufacturers took advantage of these technological advances by taking the process one step further and produced bonbons, or cream-filled chocolates. Bonbons, as opposed to solid-chocolate penny bars, were purchased by men as gifts for women as part of courtship rituals, rather than for immediate consumption. Women, conversely, could not purchase and subsequently consume bonbons—since such an action would be considered "promiscuous" [50].

Through advertising, bonbons became associated with sexuality as exemplified in 19th century print media including women's magazines such as *Godey's*, *International Journal of Ethnics*, and *Popular Science Monthly* [51]. Woloson has suggested that women's morality, including "their temptation of luxurious things" and sinfulness of "self-indulgence," was expressed in their desire for bonbon consumption [52]. Furthermore, the *Boston Cooking School Magazine* echoed this sentiment by declaring food a moral agent [53]. Reinforcing chocolate's sexually charged stereotype are the words of James Wadsworth's jingle: "Twill make old Women young and Fresh; Create New Motions of the Flesh. And cause them long for you know what, If they but taste chocolate" [54]. Consequently, it is not surprising that female cookbook writers created medicinal recipes for invalids that purposely excluded sugar in cocoa drink recipes, while at the same time added sugar to cake recipes, confections, and pastries. Chocolate beverages, too, echoed this dissection between the medicinal—without sugar—and recreational—with sugar (Table 9.1).

Table 9.1 List of Chocolate Drink Preparations

Beverage Name	Ingredients
French Chocolate	Vanilla, sugar, nibs
Spanish Chocolate	Curacoa (sic), sugar, cinnamon, cloves, sometimes almonds
Vanilla Chocolate	Caracas, Mexican vanilla, cinnamon and cloves
Cocoa Nibs	Bruised roasted seeds deprived of their covering
Flake Cocoa	Whole seed nib and husk ground up together until a flake-like appearance
Rock Cocoa	Same as Flake, but with arrowroot and sugar
Homeopathic Cocoa	Same as Rock, but without sugar
Maravilla (sic) Cocoa	Contains sugar and sago flour
Cocoa Essence, Cocoa Extract and Cocoatina (sic)	Consists of pure cocoa deprived of 60–70 percent of its fat
Pressed Cocoa	Cocoa nibs, with only 30 percent cocoa fat
Vi-Cocoa	Preparation in which an extract of kola nut is said to be added

Source: Adapted from Edwards, W. N. *The Beverages We Drink*. London: Ideal Publishing Union Ltd., 1898; pp. 100–111.

Attention needs to be paid, however, to the use of the term chocolate versus cocoa. Mary Henderson wrote in her cookbook *Diet For The Sick*: "When it [chocolate] is flavored with vanilla and mixed with sugar it is called chocolate" [55]. Henderson differentiated between cocoa, a therapeutic beverage lacking high levels of fat and caffeine, and chocolate products served for their gustatory properties. Standards of identity for chocolate would not be established and adopted for another 50 years (see Chapter 47).

A contemporary and fellow cookbook writer of Cooper was Prudence Smith. Whereas Cooper's prose was rather dry and scientific, Smith's preface extolled and lamented the virtues of domestic cooking: "for were it not for the art of cookery, it would forever remain impossible to give a just definition of man. He is emphatically a cooking animal, or he is nothing" [56]. Simultaneously, Smith adhered to the hot/cold, wet/dry approach to medicine:

The headache might be cured by a proper application of calves' head soup . . . and obstinate love-fit, by a regimen of ox heart stuffed with heartsease[sic], and served up with truffle-sauce, which vegetable, growing entirely underground, is naturally cooling and consolatory. [57]

Like Cooper, Smith added sugar to her chocolate drink recipe for the sick. Furthermore, albeit an uncommon addition, Smith instructs that "if the stomach is weak, make some gruel as thick as the chocolate, strain it, and mix them together" [58]. The addition of gruel to chocolate for the infirm is yet another variation in recipe format. Medicinal recipes differed depending on the author's observance of humoral medicine practices or adherence to medical advice passed down through the generations—usually through the maternal side of the family: from grandmother to mother to daughter. This practice of changing pharmacological recipes, particularly as it relates to medicinal chocolate, can be traced back to Spanish physicians in Mexico [59].

The addition of spices to 19th century chocolate recipes served to the infirm was also variable. At the bottom of a cocoa recipe that omitted sugar—written by a wife of an American homeopathic physician—a footnote advertised: ". . . pure cocoa and [h]omoeopathic chocolate, without any admixture of spices" [60]. According to the system of humoral physiology, foods were classified according to their sensory/gustatory properties. Consequently, hot or spicy foods represented fire, also hot and dry. In treating the infirm, therefore, it would have been ill-advised to provide foods that excited the body.

Concurrent with the rise and incorporation of sugar usage in medicinal chocolate at this time, the United States experienced rapid population growth with all the associated influences that economic expansion demanded. During the 40 years preceding the American Civil War, the United States experienced in relative terms the most rapid urbanization in its history [61]. As divergent cultures, norms, political values, and religious affiliations converged, cookbook authors asserted influence on defining moral living through increasingly lengthy prefaces, and specific chapters in their publications with titles such as "On the Health of [the] Mind" and "On the Management of Young Children" [62]. A central moral topic commonly discussed in these prefaces was alcohol consumption. Similar to sugar consumption, alcohol's dual roles as medicine and as an impetus to moral decay was problematic to 19th century cookbook writers.

Alcohol, like chocolate at this time, was used medicinally and extolled by some and condemned by others. There are cookbooks specific to the Temperance Movement, and these editions, like other cookbooks printed during this time, contained sections dedicated to moral living. An early 19th century example attests to the dangers of alcohol and women's place in guarding the family from such influences. The author warns that "[f]emale influence, though powerful, is unobtrusive, and when a woman forgets her proper sphere, she loses that influence" [63]. Moreover, Neal instructed women to leave the room when alcohol was served rather than confronting their husbands verbally.

Another female cookbook author placed the responsibility of men's alcohol consumption clearly with women. In warning against the dangers associated with consumption of spicy pickles and alcohol, Dodds wrote that "the wives and mothers of this country are themselves responsible for much of the ruin wrought in their own households" [64]. The duality of being responsible for every family member's moral behavior concurrent with self-deprecation in the arena of learned, scientific knowledge is a theme repeated throughout several centuries of cookbooks, as female authors prefaced their medical advice with the caveat that when administering food as medicine, "[it is] not intended as a substitute for the family physician" [65]. Family physicians during this era primarily were male. Even after the American Civil War, cookbook authors continued to preach moral code set within the respected genre of scientific discovery as illustrated in the Beecher sisters' much-revered cookbook—*The American Woman's Home* (Fig. 9.4), whereby they pontificate

FIGURE 9.4. Chocolate and righteous living: post-Civil War cookbook. *Source:* Beecher, C. E. and Beecher Stowe, H. *The American Woman's Home.* New York: J. B. Ford and Company, 1869. Courtesy of the Peter J. Shields Library, University of California, Davis. (Used with permission.) (See color insert.)

"[t]here is no direction in which a women more needs both scientific knowledge and moral force than in using her influence to control her family in regard to simulating beverages" [66].

Whether male or female, gender defines hierarchy in having a public voice. During the 19th century it was the patriarchal voice that dominated. Equally important to power in the public arena, however, was one's social status. A cookbook author known only as "A Housekeeper" proclaimed: "The condition of society in America is peculiar and very distinct from that of the Old World" [67]. The author continued by describing the various classes (low, middle, and upper) with the middle class being the majority, and noted:

> [We have] general spirit of enterprise, the eager ambition, and the habit of self-respect and confidence, lead our people to pretend to and to seek all that is desirable in comfort, delightful in taste, showy in fashion, or in any way agreeable to ourselves or producing an effect upon others. Hence our social ambition over-taxes our abilities, and our absolute social necessities require, in most cases, an economy careful and judicious, to enable us to satisfy our wants and wishes. [68]

The author ended her preface by instructing American housewife readers that household order, along with "well-prepared and acceptable fare," would bring happiness and "great moral as well as physical blessings" to their families and nation [69].

North American cookbook authors during this period were cognizant that Europeans had available to them a wider selection of ingredients. Nevertheless, they wrote that American women could and should create a *proper* table. In fact, the proper table—like the proper housewife—created respectability for her husband and by extension her country. Simultaneously, frugality was stressed: "frugality and order must be the corner-stones of our republican edifice" and decadence was viewed with distain [70]. Europeans of the *genteel society* could afford to be wanton. In contrast to this view, indoctrinated in the early 19th century, Americans learned that thrift and industry were virtues, positions argued by Richard Bushman in his book *The Refinement of America: Persons, Houses and Cities* [71]. Consequently, advice stated in prefaces to 19th century cookbooks promoted American housewives to set an abundant table, while substitutions for more economical ingredients were listed in the recipe sections.

In the 18th century the unique needs of frontier life, which required women to work side-by-side with men, allowed women to circumvent restrictive laws and exert more public influence. But the 19th century brought women back into the home where their activities became more restrictive, and quieted women's public voice [72].

At the same time, beginning mid-century and climaxing at the turn of the 20th century, scientific knowledge began to be extolled as a virtue. The sub-title to the cookbook *The American Matron; or, Practical and Scientific Cookery*, published in the mid-19th century, was indicative of science's esteem; the author advised homemakers to obtain scientific knowledge associated with cooking methodologies [73]. Later in the century Mary Boland reiterated the value of science in her cookbook preface and quoted a Harvard professor: "no man could be a gentleman without knowledge of chemistry" [74]. Moreover, in subsequent pages Boland declared that "no woman could be a lady without a knowledge of the chemistry of the household—what a glorious prospect would there be opened for the future health of the nation" [75]. The earlier unknown author of *The American Matron*, published in 1851, would not have known about germ theory, but Boland, whose work was published after Louis Pasteur's experiments during the 1860s when germ theory was demonstrated, not only knew about the theory but goes to great length describing in her introduction the dangers of unseen microscopic forms of life.

By the mid-19th century, evidence of chemistry's influence over health advice is found in other sources: "wherever oxygen seizes upon carbon, whether in the shape of coals in a stove or fat in our bodies, the result of the struggle is heat" [76]. Hales argued that summer meals should be lighter fare, thereby reducing oxygen load to the consumer. This myth of consuming lighter foods in the summer has continued into the 21st century, albeit the argument today is that heavy winter meals increase digestion, consequently heating the individual.

By the late 19th century, science, particularly chemistry, dominated the discussion and advice given in cookbook prefaces. Furthermore, cookbook authors incorporated chemical analysis directly into the recipes. An example of this genre is Fannie Farmer's cookbook, *Food and Cookery for the Sick and Convalescent*, published at the beginning of the 20th century [77] (Fig. 9.5). Listed under beverages for the sick are "breakfast cocoa", defined as a beverage where half of the cocoa butter had been removed, and "chocolate," listed as a full-fat beverage alternative. Unusual for the time is the fact that the Farmer cookbook presented chemical analysis for the various recipes that included three macronutrients, mineral content, and caloric density (Fig. 9.6) [78]. Interestingly, although the analysis—in this case for chocolate—is 100 years old, it is relatively consistent with 21st century analytical values. Farmer also advised patients who suffered from digestive problems to consume cocoa, the lower fat variety [79].

The scientific approach to health, whereby laboratory experiments are conducted, reviewed, and published, dominated the literature, and one result was the decline of chocolate's medicinal use. Medical practitioners recognized chocolate's weight-inducing qualities while recognizing that chocolate was not symptom relieving. Nutrition advice at the dawn of the

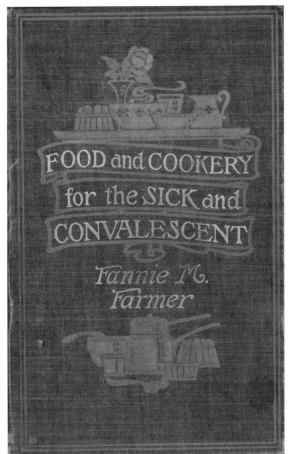

FIGURE 9.5. Chocolate and health. *Source*: Farmer, F. M. *Food and Cookery for the Sick and Convalescent* (title page). Boston: Little, Brown, and Company, 1905. Courtesy of Peter J. Shields Library, University of California, Davis. (Used with permission.) (See color insert.)

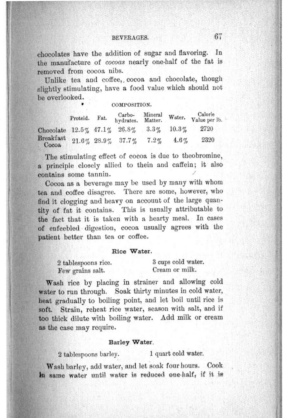

FIGURE 9.6. Early nutritional analysis of cocoa. *Source*: Farmer, F. M. *Food and Cookery for the Sick and Convalescent* (title page). Boston: Little, Brown, and Company, 1905. Courtesy of the Peter J. Shields Library, University of California, Davis. (Used with permission.)

21st century originated from the laboratory, rather than from cookbook authors. Cookbooks began to focus strictly on food preparation, while women's magazines took the lead in dispensing domestic and ethical advice.

Conclusion

Nineteenth century American women, as primary caretakers of family health practices, adhered to the medicinal theory of hot/cold, wet/dry popular during this period, and the evidence is present in North American cookbooks of the era. Moreover, within sections sometimes titled "Cookery for the Sick and Convalescent," cookbooks were a means for filtering dietary advice to the general public while giving voice to "Others," specifically women. With chapter titles such as "Health of the Mind" and "Running a Household," these books provided the context whereby scholars today may understand women's, and as an extension society's, perceived cultural norms. Nine-

teenth century cookbook prefaces are rich sources of data demonstrating societal norms. Advertisements placed in cookbooks as well as printed in 19th century newspapers, likewise, reflect attitudes of the era (Fig. 9.7). The trajectory of the medicinal use of chocolate can be traced through 19th century cookbooks and other ephemera of this time period.

Interestingly, health information in the 21st century is transferred predominantly through women's popular literature (see Chapter 50). By researching and analyzing 19th century cookbooks as a means of giving voice to women, scholars can then compare the dispersion of homeopathic, dietary advice over a century and a half by comparing the magnitude of influence that both channels have exerted over North American society's dietary behaviors.

If cookbooks were a resource for women during the 19th century in caring for their family, and health magazines capture today's female audience, the 21st century medical community can use analyses such as this one to understand the cultural context wherein health messages were, and are, sought after and

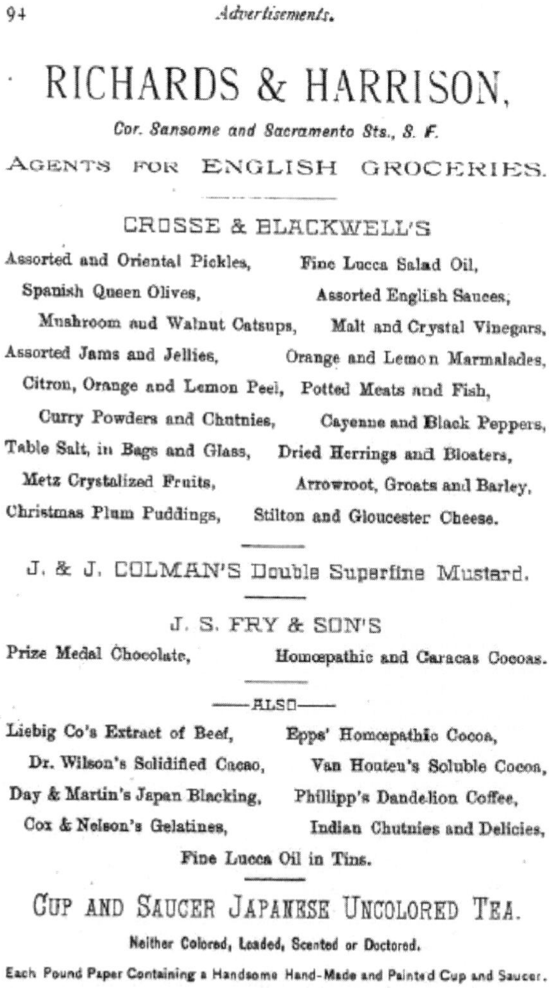

RICHARDS & HARRISON,

Cor. Sansome and Sacramento Sts., S. F.

AGENTS FOR ENGLISH GROCERIES.

CROSSE & BLACKWELL'S

Assorted and Oriental Pickles,   Fine Lucca Salad Oil,

Spanish Queen Olives,   Assorted English Sauces,

Mushroom and Walnut Catsups,   Malt and Crystal Vinegars,

Assorted Jams and Jellies,   Orange and Lemon Marmalades,

Citron, Orange and Lemon Peel,   Potted Meats and Fish,

Curry Powders and Chutnies,   Cayenne and Black Peppers,

Table Salt, in Bags and Glass,   Dried Herrings and Bloaters,

Metz Crystalized Fruits,   Arrowroot, Groats and Barley,

Christmas Plum Puddings,   Stilton and Gloucester Cheese.

J. & J. COLMAN'S Double Superfine Mustard.

J. S. FRY & SON'S

Prize Medal Chocolate,   Homoeopathic and Caracas Cocoas.

—————ALSO—————

Liebig Co's Extract of Beef,   Epps' Homoeopathic Cocoa,

Dr. Wilson's Solidified Cacao,   Van Houten's Soluble Cocoa,

Day & Martin's Japan Blacking,   Phillipp's Dandelion Coffee,

Cox & Nelson's Gelatines,   Indian Chutnies and Delicies,

Fine Lucca Oil in Tins.

CUP AND SAUCER JAPANESE UNCOLORED TEA.

Neither Colored, Leaded, Scented or Doctored.

Each Pound Paper Containing a Handsome Hand-Made and Painted Cup and Saucer.

FIGURE 9.7. Mystery advertisement that predates 1883. Document pasted in the back of a cookbook listing homeopathic cocoa. *Source:* Clayton, H. J. *Claytons Quaker Cookbook: Being a Practical Treatise.* San Francisco: Womens Co., 1883 (after p. 94). Courtesy of the Huntington Library, San Marino, CA. (Used with permission.)

integrated into the lives of women and their families. Furthermore, scholars that research societal norms in either historical or contemporary periods will find cookbooks an insightful resource to their work.

Acknowledgments

I wish to thank the following individuals for their assistance during the research and preparation of this chapter: Romaine Ahlstrom, The Huntington Library, San Marino, California; Charlotte Biltekoff, University of California, Davis, California; Axel Borg, Shields Library, University of California, Davis, California; Jim Gay, Colonial Williamsburg, Williamsburg, Virginia; John Grossman, The John Grossman Collection of Antique Images, Richmond, California; and Dan Strehl, Los Angles Public Library, Los Angeles, California.

References

1. Beck, L. N. *Two "Loaf-Givers."* Washington DC: Library of Congress, 1984.

2. Emery, W. B. *A Funerary Repast in an Egyptian Tomb of the Archaic Period.* Leiden: Nederlands Institut voor het Nabije Osten, 1962.

3. Darby, W. J., Ghalioungui, P., and Grivetti, L. E. *Food. The Gift of Osiris.* London, England: Academic Press, 1977.

4. Prakish, O. *Food and Drinks In Ancient India.* Delhi, India: Munshi Ram Manohar Lal, 1961.

5. Simoons, F. J. *Food in China. A Cultural and Historical Inquiry.* Boca Raton, FL: CRC Press, 1991.

6. Bottero, J. *The Oldest Cuisine in the World. Cooking in Mesopotamia.* Chicago: University of Chicago Press, 2004.

7. Grivetti, L. E. Mediterranean food patterns. In: Matalas A. L., Antonis Z., Vassilis S., and Wolinsky, I., editors. *The Mediterranean Diet.* Boca Raton, FL: CRC Press, 2001; pp. 3–30.

8. Athenaeus. *The Deipnosophists,* VII Volumes. Translated by Charles Burton Gulick. Cambridge, MA: Harvard University Press, 1927.

9. Braund, D., and Wilkins, J. *Athenaeus and His World. Reading Greek Culture in the Roman Empire.* Exeter, England: University of Exeter Press, 2000.

10. Lempriere, J., and Wright, F. A. *Lempriere's Classical Dictionary of Proper Names Mentioned in Ancient Authors. With a Chronological Table.* London, England: Routledge & Kegan Paul, 1963.

11. Platina, B. *De Honesta Voluptate.* Venice, Italy: Biblioteca in Vaticano, 1475. Not paginated.

12. Platina, B. *De Honesta Voluptate.* Venice, Italy: Biblioteca in Vaticano, 1475. Not paginated.

13. Longone, J. B., and Longone, D. T. *American Cookbooks and Wine Books 1797–1950.* Ann Arbor, MI: The Clements Library, 1984.

14. Anonymous. *Handwritten day book, as known as commonplace book.* Williamsburg, VA, ca. 1700; p. 252.

15. Smith, E. *The Compleat [sic] Housewife.* Williamsburg, VA, 1742; p. 224.

16. Cobbett, W. *A Year's Residence in the United States of America.* New York: Clayton and Kingsland, 1819; p. 335.

17. Carter, S. *The Frugal Housewife.* London, England: F. Newbery, 1772.

18. Simmons, A. *American Cookery, or The Art of Dressing Viands, Fish, Poultry and Vegetables, and the Best Modes of Making Pastes, Puffs, Pies, Tarts, Puddings, Custards and Preserves, and All Kinds of Cakes … Adapted to This Country, and All Grades of Life.* Hartford, CT: Printed by Hudson & Goodwin for the author, 1796.

19. Simmons, A. *American Cookery, or The Art of Dressing Viands, Fish, Poultry and Vegetables, and the Best Modes of*

Making Pastes, Puffs, Pies, Tarts, Puddings, Custards and Preserves, and All Kinds of Cakes ... Adapted to This Country, and All Grades of Life. Hartford, CT: Printed by Hudson & Goodwin for the author, 1796; p. 34.

20. Longone, J. B., and Longone D. T. *American Cookbooks and Wine Books 1797–1950*. Ann Arbor, MI: The Clements Library, 1984; p. 1.

21. *The Historic American Cookbook Project: Feeding America.* East Lansing: Michigan State University, 2004. Available at http://digital.lib.msu.edu/projects/cookbooks/html/books/book_10.cfm. (Accessed June 2, 2006.)

22. Randolph, M. *Virginia Housewife; or, Methodical Cook.* Baltimore, MD: Plaskitt & Cugle, 1838.

23. *The Historic American Cookbook Project: Feeding America.* East Lansing: Michigan State University, 2004. Available at http://digital.lib.msu.edu/projects/cookbooks/html/books/book_10.cfm. (Accessed June 3, 2006.)

24. Randolph, M. *Virginia Housewife; or, Methodical Cook.* Baltimore, MD: Plaskitt & Cugle, 1838; pp. 143–146.

25. Randolph, M. *Virginia Housewife; or, Methodical Cook.* Baltimore, MD: Plaskitt & Cugle, 1838; p. 141.

26. Randolph, M. *Virginia Housewife; or, Methodical Cook.* Baltimore, MD: Plaskitt & Cugle, 1838; p. 144.

27. Leslie, E. *Seventy-five Receipts for Pastry, Cakes and Sweetmeats.* Boston: Munroe and Francis, 1832.

28. Coe, S. D., and Coe, M. D. *The True History of Chocolate.* New York: Thames and Hudson, 1996; p. 218.

29. Corrado, V. F. *La Manovra della Cioccolata e del Caffe,* 2nd edition. Naples, Italy. Quoted in: Coe, S. D., and Coe, M. D. *The True History of Chocolate.* New York: Thames and Hudson, 1996; p. 220.

30. Leslie, E. *Directions for Cookery, in Its Various Branches,* 10th edition. Philadelphia: E. L. Cary & A. Hart, 1840.

31. Leslie, E. *Directions for Cookery, in Its Various Branches,* 10th edition. Philadelphia: E. L. Cary & A. Hart, 1840; p. 418.

32. Woloson, W. *Refined Tastes: Sugar, Confectionary, and Consumers in Nineteenth-century America.* Baltimore, MD: The Johns Hopkins University Press, 2002.

33. Few, M. Chocolate, sex, and disorderly women in late-seventeenth- and early-eighteenth-century Guatemala. *Ethnohistory* 2005;52(4):673–687.

34. Cooper, T. *A Treastie [sic] of Domestic Medicine, Intended for Families, in Which the Treatment of Common Disorders are Alphabetically Enumerated.* Reading, PA: George Getz, 1824.

35. Grivetti, L. E. *Cultural Aspects of Nutrition. The Intergration of Art and Science.* Oxford, England: Oxford Brookes University, 2003; pp. 1–258.

36. Cárdenas, J. *Problemas y Secretos Maravillosos de Las Indias.* Madrid, Spain: E. A. Durán, 1591.

37. Cárdenas, J. *Problemas y Secretos Maravillosos de Las Indias.* Madrid, Spain: E. A. Durán, 1591, p. 115.

38. Grivetti, L. E. *Cultural Aspects of Nutrition. The Integration of Art and Science.* Oxford, England: Oxford Brookes University, 2003; pp. 175–188.

39. Cooper, T. *A Treastie [sic] of Domestic Medicine, Intended for Families, in Which the Treatment of Common Disorders are Alphabetically Enumerated.* Reading, PA: George Getz, 1824; p. 108.

40. Taylor, G. R. *The Transportation Revolution, 1815–1860.* New York: Harper & Row, 1951.

41. Vogt, P. L. *The Sugar Refining Industry in the United States: Its Development and Present Condition.* Philadelphia: University of Pennsylvania, 1908.

42. Smith, W. D. Complications of the commonplace: tea, sugar, and imperialism. *Journal of Interdisciplinary History* 1992;23(2):261.

43. Anonymous. Cheap restaurants. *The New York Times.* August 6, 1871, p. 5, col. 5.

44. Woloson, W. *Refined Tastes: Sugar, Confectionary, and Consumers in Nineteenth-century America.* Baltimore, MD: The Johns Hopkins University Press, 2002; pp. 109–154.

45. Díaz del Castillo, B. *The Conquest of New Spain.* New York: Penguin Books, 1560; pp. 226–227.

46. Woloson, W. *Refined Tastes: Sugar, Confectionary, and Consumers in Nineteenth-century America.* Baltimore, MD: The Johns Hopkins University Press, 2002; p. 133.

47. Copper, G. Love, war, and chocolate. In: Horowitz, R., and Mohun, A., editors. *His and Hers.* Charlottesville: University Press of Virgina, 1998; pp. 68–94.

48. Lears, T. J. *Fables of Abundance: A Cultural History of Advertising in America.* New York: Basic Books, 1994; p. 135.

49. Coe, S. D., and Coe, M. D. *The True History of Chocolate.* New York: Thames and Hudson, 1996; p. 50.

50. Thomas, N. *Entangled Objects: Exchange, Material Culture, and Colonialism in the Pacific.* Cambridge, MA: Harvard University Press, 1991; p. 26.

51. Woloson, W. *Refined Tastes: Sugar, Confectionary, and Consumers in Nineteenth-century America.* Baltimore, MD: The Johns Hopkins University Press, 2002; pp. 109–154.

52. Woloson, W. *Refined Tastes: Sugar, Confectionary, and Consumers in Nineteenth-century America.* Baltimore, MD: The Johns Hopkins University Press, 2002; p. 137.

53. Anonymous. *Boston Cooking School Magazine I.* Boston: Boston Cooking School, 1896; pp. 88–89.

54. Wadsworth, J. A curious history of the nature and quality of chocolate. Quoted in: Fuller, L. K. *Chocolate Fads, Folklore, and Fantasies.* New York: Haworth Press, 1994; p. 137.

55. Henderson, M. F. *Diet For The Sick.* New York: Harper & Brothers, 1885; p. 5.

56. Smith, P. Preface. *Modern American Cookery*. New York: J&J Harper, 1831.

57. Smith, P. *Modern American Cookery*. New York: J&J Harper, 1831; p. 7.

58. Smith, P. *Modern American Cookery*. New York: J&J Harper, 1831; p. 173.

59. Graziano, M. M. Food of the gods as mortals' medicine: the uses of chocolate and cacao products. *Pharmacy in History* 1998;40(4):121–168.

60. Anonymous. *A Manual of Homoeopathic Cookery, Designed Chiefly for the Use of Such Persons as Are Under Homoeopathic Treatment, by the Wife of a Homoeopathic Physician; With Additions by the Wife of An American Homoeopathic Physician*. New York: W. Radde, 1846; p. 142.

61. Blumin, S. M. *The Emergence of the Middle Class: Social Experience in the American City, 1760–1900*. Cambridge, MA: Cambridge University Press, 1989.

62. Beecher, C. *Miss Beecher's Domestic Receipt Book: Designed as a Supplement to the Treatise on Domestic Economy*, 3rd edition. New York: Harper & Brothers, 1850.

63. Neal, C. A. *The Temperance Cookbook*. Philadelphia: Temperance and Tract Depository, 1835; p. 6.

64. Dodds, S. W. *Health in the Household; or, Hygienic Cookery*, 2nd edition. New York: Fowler & Wells, 1885; p. 82.

65. Howland, E. A. Preface. *American Economical Housekeeper & Family Receipt Book*. Worcester, MA: William Allen, 1845; not paginated.

66. Beecher, C. E., and Beecher Stowe, H. *The American Woman's Home*. New York: J. B. Ford and Company, 1869; p. 138.

67. Anonymous. *The American Matron; or Practical and Scientific Cookery*. Boston: James Munroe and Company, 1851; p. 5.

68. Anonymous. *The American Matron; or Practical and Scientific Cookery*. Boston: James Munroe and Company, 1851; p. 5.

69. Anonymous. *The American Matron; or Practical and Scientific Cookery*. Boston: James Munroe and Company, 1851; p. 7.

70. Anonymous. *The American Matron; or Practical and Scientific Cookery*. Boston: James Munroe and Company, 1851; p. 7.

71. Bushman, R. L. *The Refinement of America: Persons, Houses, Cities*. New York: Alfred A. Knopf, 1992.

72. Bushman, R. L. *The Refinement of America: Persons, Houses, Cities*. New York: Alfred A. Knopf, 1992; pp. 50–79.

73. Anonymous. *The American Matron; or Practical and Scientific Cookery*. Boston: James Munroe and Company, 1851; p. 5.

74. Boland, M. A. *A Handbook of Invalid Cooking: For the Use of Nurses in Training Schools, Nurses in Private Practice and Others Who Care For the Sick*. New York: The Century Company, 1893; p. 2.

75. Boland, M. A. *A Handbook of Invalid Cooking: For the Use of Nurses in Training Schools, Nurses in Private Practice and Others Who Care For the Sick*. New York: The Century Company, 1893; p. 3.

76. Hale, M. *Mrs. Hales New Cookbook: A Practical System for Private Families in Town and Country*. Philadelphia: TB Peterson & Bros, 1857; p. xxxix.

77. Farmer, F. M. *Food and Cookery for the Sick and Convalescent*. Boston: Little, Brown, and Company, 1905.

78. Farmer, F. M. *Food and Cookery for the Sick and Convalescent*. Boston: Little, Brown, and Company, 1905; p. 67.

79. Farmer, F. M. *Food and Cookery for the Sick and Convalescent*. Boston: Little, Brown, and Company, 1905; p. 67.

PART

III

Serving and Advertising

Chapter 10 (Lange) presents a historical overview of chocolate preparation and serving equipment based on analysis of silver, pewter, china, and porcelain chocolate pots and cups held in museum collections in the American Northeast. Chapter 11 (Ward) builds on this topic and provides a detailed exploration and description of silver chocolate pots and related items in Boston museum collections. Chapter 12 (Perkins) examines the history of French chocolate pots and analyzes the distinguishing features which identify them. Chapter 13 (Swisher) expands the study of historical, Colonial era, and foreign chocolate pots using representative images obtained from 21st century web-based search engines. Chapter 14 (Westbrook) examines the evolution of chocolate advertising through "Trade Cards" used by merchants in their shops to provide customers with information on a wide array of topics in order to attract customers. Chapter 15 (Swisher) considers the development and evolution of themes in chocolate advertising posters used by companies to attract customers. Chapter 16 (Westbrook) expands the theme of chocolate advertising and documents how different manufacturing companies advertised their products at the several World's Fairs during the period 1851–1964.

Chocolate Preparation and Serving Vessels in Early North America

Amanda Lange

Introduction

The introduction of tea, coffee, and chocolate into the western European diet during the mid-17th century revolutionized drinking habits. Chocolate and its fellow stimulating beverages gradually replaced traditional ale, beer, and hard cider as the breakfast beverages of choice in early America. Chocolate was first introduced into colonial regions of northeastern North America in the late 17th century; having been introduced earlier in southeastern Spanish Florida (see Chapter 33). Chocolate often was used as an exotic and expensive medicine to nourish the sick, cure hangovers, or aid digestion. According to John Worlidge:

Chocolate is a very great Restorative, comforting and cherishing the inward parts, and reviving natural strength, and hath a wonderful effect upon Consumptive and antient [sic] people, being drank hot in a morning. [1]

Even later authors continued to preach the virtues of chocolate for health. Rees's *The Cyclopaedia*, published in Philadelphia between 1810 and 1824, described chocolate as a drink:

Esteemed not only as an excellent food, as being very nourishing, but also as a good medicine; at least a diet, for keeping up the warmth of the stomach, and assisting digestion. [2]

The preparation of chocolate was a complicated and time-consuming process, quite unlike today's ready-

mix of nonfat powdered milk, sugar, and cocoa. In addition to the expense, preparation of chocolate beverages required space, time, equipment, and patience. Period recipes from cookbooks or medical manuals of the day underscore the many ingredients needed:

Chocolate is made with Chocolate, Milk, Eggs, White-wine, Rose-water, and Mace or Cinnamon, which the party fancies, they being all boiled together over a gentle fire, two ounces of Chocolate, eight Eggs, half pound of Sugar, a pint of White-wine, an ounce of Mace or Cinnamon, and half a pound of Sugar answering in this case a Gallon of Milk. [3]

Other writers focused on the complexity of preparation and equipment needed to produce outstanding chocolate:

A portion of one of the cakes must be scraped fine, added to a sufficient quantity of water, and simmered for a quarter of an hour; but milling is necessary to make it completely smooth. For this purposes [chocolate pots have] a circular wheel of

Chocolate: History, Culture, and Heritage. Edited by Grivetti and Shapiro
Copyright © 2009 John Wiley & Sons, Inc.

wood or metal within, fixed to a stem that passes through the lid, and which, being whirled about rapidly by the palms of the hand, bruises and mixes the chocolate with the water. The chocolate must be milled off the fire, then put on again to simmer some time, then milled again until it is quite smooth. From the fineness there should be no sediment, and the whole should be drunk; cream is generally used with it . . . Sugar may be put in with the scraped chocolate, or added afterward. [4]

Because of the amount of time needed to prepare chocolate as a breakfast beverage, some cookbook authors advised starting the process the evening before:

To make chocolate. Scrape four ounces of chocolate and pour a quart of boiling water upon it; mill it well and sweeten it to your taste; give it a boil and let it stand all night; then mill it again very well; boil it in two minutes, then mill it till it will leave the froth upon the tops of your cups. [5]

Early recipes start with the chocolate itself. The manufacture of chocolate from imported cacao nibs may have occurred soon after its introduction into the American colonies. Most chocolate consumed by Americans during the Colonial era was purchased ready-made for use, with roasting and grinding of the beans already completed. Commercial chocolate makers manufactured their product using a variety of milling methods. Due to the uncertainty and expense of the cocoa trade, many chocolate makers also manufactured other commodities besides chocolate—often grinding mustard and tobacco too [6, 7].

One of the earliest references to Boston chocolate mills appears in a 1751 newspaper advertisement for Joseph Palmer and Richard Cranch who "make very good chocolate at 12 shillings per pound, and the superfine sort for 14 shillings per pound" [8]. Although cheaper than tea on a per-pound basis, on a per-cup basis, chocolate was more expensive than tea or coffee (see Chapter 18), and was a luxury only the wealthy and aspiring middle classes could afford.[1] Easy access to shipping, water mills, and large amounts of cacao imports combined to make New England manufacturers the major source of chocolate in eastern North America throughout the 18th and 19th centuries.

Very few cooks purchased cacao nibs for roasting and grinding on their own.[2] Most chocolate available to consumers in early America was in the form of ground lozenges or cakes of chocolate, varying in weight from two to four ounces. In April 1687, for example, Samuel Sewall of Boston, Massachusetts, bought "21 balls of chokolatto"—also called rowls, lumps, lozenges, cakes, or tablets in the contemporary literature [9, 10]. These purchased cakes were often already combined with white sugar and flavored with spices such as nutmeg, cinnamon, red peppers, cloves, or orange flower water.

Chocolate cakes were most commonly sold and stored in linen bags or boxes. According to Henry Stubbe, "enclosed in boxes" was one method of selling the material [11]. Although boxes specifically used for storing chocolate have yet to be identified, a 1710 English broadside reported that "to preserve it a good long time from its great Enemy moisture, keep it close in Tin-Boxes, or in old thick Oaken Chests in a dry place" [12]. Through time, chocolate develops a white haze or gray streaks on the surface called bloom, due to exposure to humidity, changes in temperature, or improper storage. Bloom does not adversely affect the taste, but it is unsightly.

Boxes kept the chocolate safe from vermin in addition to protecting it from losing its flavor or absorbing odors. Cocoa butter does not become rancid as quickly as other fats, so properly stored chocolate kept for up to a year. Yet in 1785, Thomas Jefferson complained that chocolate had little favor in the American diet because of an impression that it went rancid quickly. In a letter to John Adams, he wrote:

Chocolate. This article when ready made, and also the Cacao becomes so soon rancid, and the difficulties of getting it fresh have been so great in America that its use has spread little. [13]

Preparing Chocolate

Cakes of chocolate needed to be scraped or grated in preparation for the addition of liquid, which often could be a combination of water, wine, or milk. According to period recipes, chocolate cakes could be cut, sliced, scraped, or grated before adding to a mixture of hot liquid and white sugar (see Chapters 8 and 23). Evidence of the equipment used to prepare chocolate appears in probate inventories such as that of Samuel Hanson of Charles County, Maryland, who owned "1 old chocolate grater" valued at 3 pence in 1741 [14, 15]. Maria Eliza Rundell's *A New System of Domestic Cookery*, included in Thomas Jefferson's library, instructed:

Cut a cake of chocolate into very small bits; put a pint of water into the pot, and, when it boils put in the above. . . . When wanted, put a spoonful or two of milk, boil it with the sugar, and mill it well. [16]

Eggs were commonly added to chocolate to increase nutrition as well as encourage frothing. James Lightbody's recipe advocated the following:

Take a small quantity of the Liquor out, and beat with six Eggs; and when it is well beat, pour it into the whole quantity of Liquor. [17]

The high percentage of cocoa butter or fat present in the cakes added to the complexity of chocolate's preparation. With at least 50 percent of the beverage being cocoa butter, chocolate was extremely nourishing, but the fat content also became one of its detracting characteristics. Rees's *The Cyclopaedia* referred to chocolate as "of a dusky color, soft and oily"

[18]. To counteract this thick layer of cocoa butter, one could either skim off the fat or add starches—bread, flour, toast, or wafer[3]—to absorb excess fat and to make the drink more palatable [19]. John Nott in *The Cook's and Confectioner's Dictionary* added "fine flour or starch half a quarter of an ounce" as part of his recipe for chocolate [20].

Being thick with cocoa butter, the chocolate must also be mixed with a stirring rod, "chocolate mill," or *molinet*, prior to pouring.[4] According to John Worlidge's 1675 description:

> The mill is only a knop [sic] at the end of a slender handle or stick, turned in a turner's lathe, and cut in notches, or rough at the end. They are sold at turners for that purpose. This being whirled between your hands, whilst the pot is over the fire, and the rough end in the liquor causes an equal mixture of the liquor with the chocolate and raises a head of froth over it. [21]

Commonly called chocolate mills, these stirring rods sometimes cause confusion today since the word "mill" is more identified with a type of grinder or "hand mill" used to pulverize the chocolate [22, 23]. The chocolate mill, typically made of wood and measuring 10–12 inches in length, was the product of wood turners in the 18th century.[5] When immersed in chocolate and whirled between the hands, the notched or roughened *knop* (i.e., knob) at its lower end produced a uniform consistency in the chocolate and raised the desired froth. An early image of these essential stirring rods occurs in Nicolas de Blégny's *Le Bon Usage du Thé, du Caffé, et Chocolat Pour la Preservation & Pour la Guerison des Maladies* (Lyon: Thomas Amaulry, 1687) with the caption "*Moulinets de diverses formes: pour faire moussior le chocolat,*" which roughly translates as "Stirring rods of various forms, for the frothing of chocolate" (Fig. 10.1). No one has adequately explained the many variations on the notched ends of the stirring rods, but this same diversity continues in both antique and modern Mexican *molinillos*.

Few chocolate pots survive with their original mills, as this part easily was lost or became separated over time. The Farrer Collection of the Ashmolean Museum, Oxford University, contains a rare example of an English silver chocolate pot made by London silversmith Paul Crespin (1694–1770) in 1738–1739 that retains its original silver and wood mill [24]. The most frequently encountered mills are those found with miniature English silver chocolate pots; in such instances they are made entirely of silver.

Chocolate mills served a variety of culinary purposes. When Mrs. Elizabeth Raffald wrote her *Experienced English Housekeeper* in 1769, she recommended using a chocolate mill to make frothy desserts such as syllabubs and whips, and encouraged cooks to "raise your Froth with a Chocolate Mill" [25]. As the 19th century progressed, chocolate mills remained in use even as the formulation for drinking chocolate

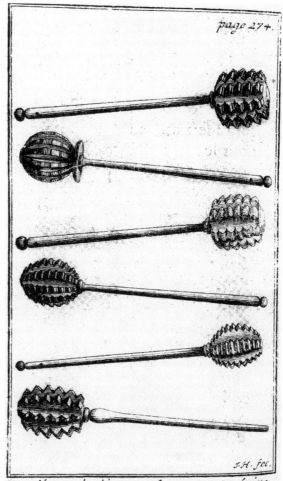

Moulinets de diverses formes pour faire moußer le Chocolat.

FIGURE 10.1. Moulinets de diverses formes (chocolate mills of diverse forms). *Source*: Blégny, N. de. *Le Bon Usage du Thé, du Caffé, et Chocolat Pour la Preservation & Pour la Guerison des Maladies*. Lyon, France: Thomas Amaulry, 1687; p. 274. Courtesy of the Peter J. Shields Library, University of California, Davis. (Used with permission.)

changed. Isabella Beeton's *The Book of Household Management* [26] illustrated a conventional chocolate mill with a turned handle and pierced flanges and instructed:

> Chocolate, prepared with a mill as shown in the engraving, is made by putting in the scraped chocolate, pouring over it the boiling milk and water, and milling it over the fire until hot and frothy. [26]

Chocolate Pots

SILVER CHOCOLATE POTS

French silversmiths are credited with the invention of the chocolate pot or *chocolatière* in the late 17th century (see Chapters 12 and 42), which English silversmiths quickly imitated [27].[6] A chocolate pot's most distinctive feature is an opening in the lid that allows for the insertion of the stirring rod or chocolate mill. This was achieved by a removable cap or finial, or a small hole with a swinging or hinged cover. Otherwise, chocolate pots and coffeepots are virtually indistinguishable.

One of the earliest references to the ownership of silver chocolate pots is found in the New York probate inventory of William Pleay, who owned a silver "jocolato pot" in 1690 [28, 29]. In 1701, William Fitzhugh of Virginia wrote a codicil to his will that bequeathed "to my son Thomas Fitzhugh my Silver Chocolate Pott which I brought out of England" [30].

Some of the most exquisite objects for the consumption of chocolate in early America were created in silver. Some of these silver chocolate pots were made by American craftsmen, while others were imports from England. A Maryland family owned a pear-shaped chocolate jug made in London during 1733–1734 by the Huguenot silversmith Peter Archambo I (d. 1767) that now resides in the collection of the Clark Art Institute, Williamstown, Massachusetts [31]. Samuel Chew (1704–1737) of Annapolis, Maryland, and his wife, Henrietta Maria (Lloyd) Chew, commissioned a chocolate jug designed with a beak-shaped spout and ornamented with their engraved coat of arms. This form of chocolate jug was fashionable in English silver in the 1730s and 1740s. Typically, the aperture for a chocolate mill was concealed by a sliding disk onto which the finial was soldered. A small clasp that hooked around a pin or rivet located beside the hole secured the disk [32, 33].

Silver chocolate pots are extremely rare forms in American silver, with only eight known Boston examples.[7] Despite the paucity of survivors, American manufactured silver chocolate pots provide a glimpse of the importance and status of chocolate in early American life. Extraordinarily stylish and costly, the pots were faddish in their response to the elite's adoption of this new beverage.

While few chocolate pots survive, those by Boston silversmiths Zachariah Brigden, John Coney, Peter Oliver, Edward Webb, and Edward Winslow show an impressive array of styles (see Chapter 11). Even though these forms were manufactured in the same city, and in some cases by the same artisan, this group exhibits diversity in shapes from vase forms to pear-shaped examples. Some have their handles opposite their spouts; others have handles at a right angle to their spout. Some of the pots have pierced, interior strainers, while others have none.[8]

Historic Deerfield owns an extraordinary and rare object, an elegant chocolate pot made by the Boston silversmith Zachariah Brigden (1734–1787), crafted ca. 1760 (Fig. 10.2). Like many an aspiring craftsman, Brigden apprenticed with the prominent Boston silversmith Thomas Edwards and later married his master's daughter, Sarah Edwards (often a convenient pathway to prosperity). Designed like a coffeepot, his chocolate pot's tall, cylindrical form has a

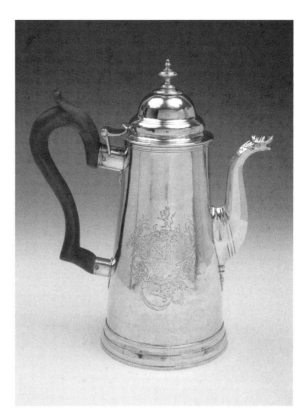

FIGURE 10.2. Chocolate pot, manufactured ca. 1760 by Zachariah Brigden (1734–1787), Boston, Massachusetts. Courtesy of the Historic Deerfield, Deerfield, MA. (Used with permission.) (See color insert.)

domed lid with a removable finial for the insertion of a chocolate mill or stirring rod. The attachment of the spout to the body of the pot appears to be slightly higher than in a corresponding coffeepot, possibly to prevent the heavy chocolate solids from pouring out. Brigden's piece is a late example of the form: Most of the other surviving Boston silver pots were created in the first two decades of the 18th century.

The arms engraved on the Brigden pot's body are those of the Thompson family impaling the arms of De(e)ring (Fig. 10.3). This type of heraldic device commemorated a marriage—the merger of two families. A similar Brigden chocolate pot, owned by the Museum of Fine Arts, Boston, is engraved with the arms of the Storer family and was owned by Ebenezer Storer. Brigden and Storer were brothers-in-law: Brigden married Sarah Edwards, while Storer married her sister, Mary Edwards [34].

NEW YORK SILVER CHOCOLATE POTS

New York-made examples of chocolate pots are also rare. The Mead Art Museum of Amherst College owns an example marked by the New York silversmith Ephraim Brasher (1744–1810), but unfortunately it does not retain any provenance [35]. The most notable New York silver chocolate pot was commissioned by

FIGURE 10.3. Chocolate pot, Manufactured ca. 1760 by Zachariah Brigden. Object shows detail of the Thompson family arms impaling the De(e)ring family. Courtesy of Historic Deerfield, Deerfield, MA. (Used with permission.)

York from 1824 to 1838, produced a pear-shaped chocolate pot or jug for Dr. Valentine Mott, a prominent New York surgeon known for his bold and original surgical procedures (Fig. 10.4) [37]. In appreciation for the surgeon's skills, the piece is engraved, "Presented to/ Dr Valentine Mott/ by Mary Williams at/ the request of her late sister/ Jane, as a token of their/ mutual regard/ 1827." Fashioned as a large and elegant jug on three claw and ball feet, the pot is decorated with elegant chased and floral *repoussé* work on the body, lid, and spout. In 1827, Mott successfully tied the common iliac artery for an aneurism of the external iliac, and the patient survived [38].

FOREIGN VISITORS, THOMAS JEFFERSON, AND CHOCOLATE

While Boston and New York silver chocolate pots are known, examples made in other regional centers have received less attention. Chocolate drinking in the American South proved extremely popular, as evidenced by comments from visitors. The journal of William Byrd (1674–1744) of Westover, Virginia, is filled with numerous references to drinking chocolate; he recorded that he drank chocolate for breakfast about once every three weeks from February 1709 through September 1712 [39]. London traveler John F. D. Smyth Stuart (1745–1814) visited Williamsburg, Virginia, in 1770 and recorded what passed for a popular Virginia breakfast. He observed that:

> cold turkey, cold meat, fried hominy, toast and cyder, ham, bread, and butter, tea, coffee, or chocolate, which last, however, is seldom tasted but by the Woman [sic]. [40]

Thomas Jefferson, who was a strong proponent of the beverage in America, owned an extraordinary example of a chocolate pot. Jefferson preached the virtues of chocolate in a letter to his rival John Adams, writing that:

> The superiority of chocolate, both for health and nourishment, will soon give it the preference over tea and coffee in America, which it has in Spain. [41]

Jefferson probably experienced chocolate first in his native Virginia, but his tenure as American minister to France undoubtedly shaped his culinary tastes.

While in France, Thomas Jefferson traveled south examining the architecture, vineyards, and agricultural methods of the local region. Jefferson was taken with the Maison Carrée, a Roman temple, as well as an excavated antiquity from the site—a bronze ewer or *askos* on view in the Cabinet of Antiquities at Nîmes (Nismes). The vessel had been found at the Maison Carrée site and was inspired by wine jugs or wine skins created from animal bladders. Jefferson so admired the form that he had a mahogany model created in 1787. After returning to the United States,

Robert Livingston, Jr. (1708–1790), third and last Lord of Livingston Manor, and his wife, Mary Thong Livingston (1711–1765), whom he married in 1731. Robert Livingston was a member of the New York House of Assembly from 1737 to 1759. In 1749, he succeeded his father as lord of Livingston Manor, a grant of 160,000 acres in the area of Dutchess and Columbia counties, New York, along the east bank of the Hudson River. The Livingstons moved in the most elite circles in colonial New York society. Robert's brother William served as Governor of New Jersey, and his brother Philip was a signer of the Declaration of Independence. Made around 1760 by Thomas Hammersley (active 1756–1769) of New York, the piece features a tucked-in base, a more conservative and old-fashioned design than that of contemporaneous examples made by Zachariah Brigden. Hammersley's chocolate pot was lavishly engraved with the arms, crest, and motto of the Livingston family, *Spero Meliora* (*I Hope for Better Things*) [36].⁹

In the early 19th century, silver pear-shaped jugs—now commonly referred to as hot-water jugs—may have served double duty as chocolate pots. The earlier form of chocolate pots with an aperture for the insertion of a chocolate mill disappeared from manufacture in American silver. Frederick Marquand (1799–1882), a silversmith actively working in New

FIGURE 10.4. Chocolate pot in the form of a Roman *askos*, manufactured in 1801 by Anthony Simmons and Samuel Alexander, Philadelphia, Pennsylvania. *Source*: Courtesy of Thomas Jefferson Foundation, Charlottesville, VA. (Used with permission.) (See color insert.)

FIGURE 10.5. Chocolate pot, manufacturer unknown, made in either the Netherlands or possibly New York in 1703. Courtesy of the Winterthur Library and Museum, Winterthur, DE. (Used with permission.)

Jefferson commissioned Philadelphia silversmiths Anthony Simmons (d. 1808) and Samuel Alexander (d. 1847) to create a silver version of the *askos* in 1801 (Fig. 10.4) [42–44]. But instead of using it to serve wine, Jefferson's silver *askos* was used by the family for serving chocolate. A family letter suggests that they called this unusually shaped vessel "the duck," and that it functioned as the family's "silver chocolate-pot" [45, 46].

Usually chocolate was stirred up until the point of serving into cups. Because Jefferson's chocolate pot did not permit the insertion of a chocolate mill for stirring, that process was probably done in the kitchen immediately before pouring into the *askos*. The unusual shape of the *askos,* while fashionable and exotic, may have meant that the chocolate served at Monticello was cool and oily from separation that occurred on the long trip from the kitchen to the Tea Room.

BASE METAL CHOCOLATE POTS

The majority of Americans prepared and served their morning draught of chocolate in a pot composed of a more humble material. In 1675, John Worlidge advised that chocolate must be "mixed in a deep pot of Tin, copper, or stone, with a cover with a hole in the middle of it, for the handle of the mill to come out at, or without a cover" [47]. Specialized forms for the preparation of chocolate often appear in early American documents throughout the 18th century. For example, the 1775 probate inventory of architect Peter Harrison listed "1 Copper chocolate pot" valued at 16 shillings in his estate [48, 49]. Copper, brass, and tinned sheet iron chocolate pots appear far more frequently than their silver counterparts, although marked examples or ones with American histories of owner-

ship are still extremely rare. Unlike silver, copper and brass objects are often unmarked and anonymous (see Chapter 54).

The Winterthur Museum owns an initialed and dated "VR/ 1703" copper pot, which probably hails from Holland but may also have been made in New York (Fig. 10.5) [50]. The cylindrical, straight-sided form with its folded foot has a removable stepped lid with a swinging cover hiding the hole for the chocolate mill. The handle of wrought iron and turned wood is attached to the copper pot by means of several brass rivets. Historic Deerfield recently acquired a later, unmarked copper example with its original lid and attached iron handle. All of these pots are lined with a layer of molten tin, intended to reduce any metallic taste to the chocolate and to prevent toxins leaching from the copper (Fig. 10.6). Without family histories or makers' marks, these copper and brass chocolate pots are difficult to date; but according to Christopher Fox, Curator of Fort Ticonderoga, most of these pear-shaped or bellied examples with flared bases date to the third quarter of the 18th century.[10] In addition, determining the country of manufacture for these pieces presents challenges as many of them could be Dutch, English, or French imports.

One known American manufacturer of chocolate pots was Benjamin Harbeson (d. 1809), a Philadelphia coppersmith. Evidence from his 1764 broadside indicates that he made a wide variety of copper wares "at the Golden Tea Kettle in Market Street," including tea kettles, coffee pots, stills, saucepans, and chocolate pots (Fig. 10.7). His elaborately engraved broadside features a chocolate pot on the upper left-hand side above a still, but no marked examples of chocolate pots are known to survive from his shop.

FIGURE 10.6. Chocolate pot, manufacturer unknown, made possibly in the Netherlands ca. 1750–1775. Courtesy of Historic Deerfield, Deerfield, MA. (Used with permission.) (See color insert.)

FIGURE 10.7. Broadside engraved by Henry Dawkins (ca. 1764) for Benjamin Harbeson, coppersmith of Philadelphia, Pennsylvania. Courtesy of the Winterthur Library and Museum, Winterthur, DE. (Used with permission.)

Unlike their silver counterparts, base metal chocolate pots intended for use with chocolate mills continued to be produced well into the 19th century. F. A. Walker and Company's *Illustrated Catalogue of Useful and Ornamental Goods*, published in Boston, ca. 1872–1873, depicts tinned sheet iron "Chocolate Pots, to use with a Muddler" in six different sizes. The catalogue also sold similar chocolate pots without an aperture [51].

The ability to purchase chocolate did not necessarily signify that the purchaser used specialized equipment to prepare it. Several historical references document the use of other kitchenware for making chocolate. An article in the *Pennsylvania Gazette* about an attempted murder case (see Chapter 21) revealed that a "skillet" was used to prepare chocolate. In 1735, Mr. Humphrey Scarlet, his wife, and two children of Boston accused their two slaves, "a Man named Yaw, and a Boy named Caesar," of poisoning them. Arsenic or "ratsbane" had been placed in their "Skillet of Chocolate which they eat for Breakfast" [52]. Additionally, a 1755 French cookery manual refers to chocolate made in a coffeepot:

> *Vous mettez autant tablettes de chocolat que de tasses d'eau dans une caffetiere . . . [Put as many tablets of chocolate as cups of water in a coffeepot].* [53]

Skillets and coffee pots for preparing chocolate were both mentioned in Lydia Childs's *The American Frugal Housewife*, published in Boston in 1832:

> *Many people boil chocolate in a coffee-pot; but I think it is better to boil it in a skillet, or something open.* [54]

CERAMIC CHOCOLATE VESSELS

Ceramic chocolate pots are not as easily identifiable or as frequently encountered as their metalware counterparts. These ceramic vessels may have been less successful than metal ones for a variety of reasons. The ceramic body did not keep the chocolate from cooling and thickening; the chocolate pot might have been too big for it to be kept over a spirit lamp; and pottery or porcelain bodies may not have been sturdy enough for vigorous stirring and frothing of the chocolate [55]. Very few ceramic examples are fitted with a hole for the insertion of a chocolate mill, stirring rod, or *molinet* [56, 57]. Undoubtedly ceramic coffeepots served double duty for chocolate, although without a lid, stirring the chocolate must have made a mess.

One possible American-owned ceramic chocolate pot is located at Montpelier, the reconstructed home of General Henry Knox (1750–1806), built in Thomaston, Maine. Knox—like several of his military comrades—received gifts of Chinese export porcelain from Samuel Shaw (1754–1794). During the American Revolution, Shaw had been the aide-de-camp to Knox and was a close friend of the family. In 1789, Shaw, then American Consul to Canton, purchased sets of china for fellow members of the Society of the Cincinnati, a fraternal organization of French and American Revolutionary War officers. Each piece was decorated with an insignia of the Society and the entwined initials of the recipient. Knox and his wife, Lucy, received porcelain complete for the service of tea, coffee, and chocolate. An inventory of the contents of the box that accompanied Shaw's letter to Knox includes a listing for a "chocolate pot [and stand]" as well as "6 chocolate bowls covers and dishes" [58]. Of the entire service only two serving vessels—the chocolate pot and the coffeepot—along with two double-handled cups survive [59].[11] The chocolate pot at Montpelier is cylindrical in form with a domed lid and a pistol grip handle perpendicular to its spout. This pot does not possess a large hole in the center of its lid, but instead has two sets of paired holes. One set is drilled along the edge of the lid and another at the neck of the pot. These holes may have enabled a hinged metal mount to be added to lift the lid [60].[12]

CERAMIC CHOCOLATE CUPS AND SAUCERS

When tea, coffee, and chocolate were introduced into Europe, there were no serving vessels specifically associated with their use. European earthenware and metal jugs, tankards, and mugs proved unsuitable for sipping these hot and costly beverages brewed in small quantities [61]. Given that metals conduct heat efficiently and often taint the taste of beverages, ceramic vessels were the preferred forms for drinking tea, coffee, and chocolate. Porcelain vessels, and eventually their less expensive pottery counterparts, soon filled the void. Porcelain had the advantage of being lightweight, thinly potted, translucent, and heat resistant, making it the perfect material for brewing and serving these exotic beverages [62].

In the early 18th century, teacups were usually small, broad forms without handles. By the 1730s teacups began to have the option of added handles, although handleless cups continued to be popular well into the 19th century. Coffee cups, on the other hand, developed as U-shaped beaker forms and by the 1730s most had handles [63].[13] As tea and coffee became cheaper in the early 19th century, both forms became larger in size.

Identifying a specific ceramic vessel for drinking chocolate is difficult, since new fashions for drinking the beverage continually evolved over time. Chocolate was first served in tall, handless beakers with slightly flared lips.[14] Chocolate was supposed to be served with a frothy, foamy head: this may explain why these ceramic cups were taller. Early 18th century chocolate cups also had the option of a saucer and lid, used to keep the beverage warm [64]. English sale records of private trade goods from 1706 mention "12 Japan chocolat [sic] cups, covers and saucers" [65]. These early chocolate cups did not have handles.

Beyond fashion, handleless cups benefited merchants who would profit from compact packing and less breakage. But as early as 1712, the shipping documents of the English East India Company record the addition of chocolate cups with handles. The supercargo or business agent of the ship *Loyal Blisse* was instructed to purchase "Handle Chocolatetts [sic], upright of these two sorts Twenty Thousand Vizt. of the biggest blue and white of different Flowers Ten thousand, Ditto the smaller Sort with brown edges Ten Thousand" [66]. In 1729, the Dutch East India Company archives indicate that a full tea set should consist of a teapot and stand, milk jug and stand, slop bowl, sugar bowl with lid and stand, twelve tea bowls and saucers, six chocolate cups with handles and six without. In the China trade, chocolate cups and saucers were sold as parts of tea services—never separately [67]. The Dutch East India Company sent drawings to Canton, China, in 1758, to instruct the Chinese potters on how to design a chocolate cup. These rough sketches depict four different versions of chocolate cups with single handles. All have a basic U-shaped outline with variations in the style of handles featuring either a loop or scallop design [68].[15]

While it was acceptable to drink chocolate from cups both with and without handles until the mid-18th century, Christiaan Jörg's research has shown that after the 1760s, Chinese export porcelain chocolate cups were more likely to have handles [69]. By the late 18th and early 19th centuries, British pottery manufacturers used pattern books to sell their range of wares to an international clientele of merchants. The pattern books of the Castleford Pottery (1796), James and Charles Whitehead Pottery (1798), the Don Pottery (1807), and the Leeds Pottery (1814) all record the availability of chocolate cups.[16] These forms consistently have a single handle attached to a cylindrical can-shaped cup or a U-shaped cup. Strangely, none of these pattern books identify a separate coffee cup. Despite the existence of a specific chocolate cup, certainly tea and coffee cups could have served chocolate as well.

References to sale of chocolate-related ceramics abound in early American newspapers and account books. Josiah Blakeley of Hartford, Connecticut, advertised a wide range of ceramics including "cream color'd Coffee and Chocolate, Do [cups and saucers]" in the September 1, 1778, issue of the *Connecticut Courant* [70]. "Cream color'd" referred to English lead-glazed creamware, a type of durable pottery popular in the late 18th century. These chocolate cups may have resembled an example in the Historic Deerfield collection (Fig. 10.8). Connecticut River Valley merchants' invoice books of the 1770s and 1780s include numerous references to "chocolate bowls," and "chocolate bowls and saucers" [71]. The range of pottery types being sold to Connecticut River Valley residents included creamware, agate ware, and blackware examples.

FIGURE 10.8. Chocolate cup and saucer. Staffordshire or possibly Yorkshire, England, ca. 1770. Courtesy of Historic Deerfield, Deerfield, MA. (Used with permission.) (See color insert.)

George and Martha Washington were well-documented lovers of chocolate. Martha Washington apparently enjoyed chocolate, but also drank an infusion of cocoa shells in water similar in color and flavor to coffee. In 1789, President Washington wrote to his agent, Clement Biddle: "She will . . . thank you to get 20 lb. of the shells of Cocoa nuts, if they can be had of the Chocolate makers" [72]. The use of cocoa shells at Mount Vernon further appears in a letter by Burges Ball, who wrote: "I wd. take the liberty of requesting you'll be so good as to procure & send me 2 or 3 Bush: of the Chocolate Shell, such as we're frequently drank Chocolate of at Mt. Vernon, as my Wife thinks it agreed with her better than any other Breakfast" [73].

The Washingtons enjoyed tea, coffee, and chocolate in an imported gold and white "Save" (Sèvres) service purchased for them by the Count de Moustier in France in 1778 and 1790 (Fig. 10.9). All of the china was white with gold rims, though not all of the pieces were made at the Sèvres Factory. Surviving examples reveal that Moustier supplemented a basic Sèvres dinner and dessert service with pieces from other French factories. The invoice from the 1790 bill of sale survives and lists "12 Chocolate cups & saucers" [74]. The chocolate cups are barely different from the coffee cups—just slightly taller and larger in capacity.

Mount Vernon also retains a set of Chinese export porcelain teaware, a gift from the Dutch merchant Andreas Everardus van Braam Houckgeest (1739–1801) in 1796. Van Braam (as he was called) created a mercantile fortune as an agent in the Dutch East India Company. Intending to make America his new home, Van Braam brought with him a variety of Chinese objects including "A Box of China for Lady Washington." The porcelain service was designed especially for Martha Washington, ornamented with her initials within a central gold sunburst design. A border of a chain of unbroken links containing the names of all of the then 15 states decorates the outside rim, and the motto *DECUS ET TUTAMEN AB ILLO* (Our

FIGURE 10.9. Chocolate cup (not mug), hard-paste porcelain with gilding. Possibly Sèvres Factory, Sèvres, France, ca. 1775. *Source*: Mount Vernon Porcelain Collection. Courtesy of Mount Vernon Ladies' Association, Mount Vernon, VA. (Used with permission.)

FIGURE 10.10. Chocolate cup with cover and saucer, manufacturer unknown, made in Jingdezhen, China, ca. 1795. *Source*: Mount Vernon Porcelain Collection. Courtesy of Mount Vernon Ladies' Association, Mount Vernon, VA. (Used with permission.) (See color insert.)

Union is our Glory, and our Defense against Him) is emblazoned on a ribbon [75]. Very few pieces of this service survive to the present day; those that do include two-handled covered cups with saucers, plates, and a sugar bowl (Fig. 10.10). It has long been debated whether this vessel served tea, hot caudle (a drink made with "ale" or "wine" thickened with bread), or chocolate [76, 77]. Given the reference to Henry Knox receiving "six chocolate bowls covers and dishes" as a gift from Samuel Shaw, it is entirely possible that Martha Washington's example represents that form.

From documented sources we know that the British and their American colonists were far less rigid and specialized when serving chocolate, using everything from porringers to mugs. The same 1735 *Pennsylvania Gazette* article about the attempted poisoning of Humphrey Scarlet and his family included a reference on how the chocolate containing the poison was served. Another slave in the family also fell ill when she "lick'd her Master's Porringer after he had been to Breakfast" [78]. Porringers were multipurpose serving vessels, usually composed of pewter, silver, or pottery and fashioned as a small bowl with a tab handle. While porringers quickly went out of style in England, Americans retained the useful form to eat soups, stews, and pottages well into the 19th century. The Reverend Thomas Prince (1687–1758), minister of Old South Church in Boston, also once mentioned drinking chocolate using a porringer. He wrote: "At $6\frac{1}{2}$ go to Family Prayer & only the porringer of Chocolate for Breakfast" [79].

Ceramic vessels may have been the most common, but they were not the only way to serve chocolate. Peter Faniel (1700–1742/3), reputed to be the wealthiest man in Boston at the time of his death in 1742/3, served the beverage in "6 lignum vitae chocolate cups lin'd with silver" [80].

An undated Carrington Bowles print illustrates another alternative way to consume chocolate. Kitty is a young, beautiful woman—perhaps a prostitute—who has managed to find a wealthy Jewish man to support her in style. Dressed in her loose robe and he in his banyan and cap, they receive an early morning visit from a jeweler. Kitty points toward her new earrings and her lover reaches into his pocket to pay for them. A servant completes the scene of early morning wealth and luxury by bearing a tray with a chocolate pot with a mill and a pair of pottery mugs. The mugs look like they are painted English delftware or tin-glazed earthenware.

CHOCOLATE STANDS

The chocolate or trembleuse stand, also called a *mancerina* in Mexico, is shaped in the form of a large shell or leaf with an attached cup holder that prevents trembling hands from spilling the contents (Fig. 10.11). This form related to the use of chocolate as nourishment for the aged or the ill, and provided an additional surface for holding a roll or wafer. Although available in silver and porcelain, chocolate stands appear frequently in English pottery pattern books of the late 18th and early 19th centuries. The form seems to be rare or nonexistent in the American market and may not have been imported, as no examples have appeared

FIGURE 10.11. Chocolate stand (trembluese), manufacturer unknown, hard-paste porcelain with overglaze enamels and gilding, made in Jingdezhen, China, ca. 1740–1750. Courtesy of Historic Deerfield, Deerfield, MA. (Used with permission.) (See color insert.)

with American histories of ownership or within documented shipments of ceramics to the North American continent.

Chocolate stands achieved greatest popularity in the predominantly Roman Catholic countries of Spain, Portugal, Italy, and Mexico—all of which indulged in the consumption of chocolate, especially during religious fasts. The term *mancerina* derives from the 23rd viceroy of Mexico, Antonio Sebastian de Toledo, the Marqués de Mancera (served in office 1664–1673), who reportedly had palsy and popularized the all-in-one shape. The earliest known chocolate stands were manufactured in Spain at the Alcora faience (tin-glazed earthenware) factory in the second quarter of the 18th century [81].[17]

Chocolate Drinking in the 19th century

The year 1828 marked the beginning of the modern era of chocolate making. The Dutch chemist Coenraad van Houten patented the process to manufacture a new kind of powdered chocolate with a low fat content. He had been experimenting in his factory to find a better way to make chocolate rather than boiling and skimming off the cocoa butter. His machine reduced the cocoa fat content from 53 percent to 28 percent fat; the resulting cake could be ground into a powder. He also added alkaline salts to make the chocolate darker and milder—often called Dutched chocolate. Van Houten's invention of the defatting process made drinking chocolate lighter and easier to mix with liquids. The full effect of Van Houten's innovations took many years to be fully utilized, one problem being the lack of a market for the resulting cocoa butter [82]. Eventually this commodity found great use in confectionery chocolate as well as in the cosmetics industry (see Chapter 46).

In the first half of the 19th century, American consumers bypassed chocolate in favor of cheaper teas and coffee for their morning beverage. Chocolate was often relegated to children or as a nutritional supplement for the infirm. But in the late 19th century, chocolate experienced a resurgence in popularity. As sales of lighter and easy to mix chocolate powders expanded, the demand for fashionable ceramics also increased. In the late 19th and early 20th centuries, imported and domestic-made chocolate pots were frequently designed as tall, cylindrical forms with a short spout or snip. Given the new formulation for chocolate, no longer was there any need for a stirring rod or chocolate mill. Trade catalogues testify to the large numbers of porcelain chocolate pots and cups and saucers for sale; often companies marketed chocolate sets as gifts. Higgins & Seiter of New York sold chocolate sets of Austrian china for $23.45. Each piece was decorated with a "celebrated court beauty," and the set included one tray, one chocolate jug, and six chocolate cups and saucers, all displayed in a white leatherette, satin-lined case [83]. Although the price of chocolate plummeted, making it affordable to a mass audience, drinking chocolate still remained a fashionable and elite social custom well into the 20th century.

Acknowledgments

The author would like to thank the following individuals for all their help in preparing this chapter for publication: David Bosse and Martha Noblick, Historic Deerfield, Incorporated; Carol Borchert Cadou, Mount Vernon Ladies' Association; Ellen Dyer, Montpelier, The General Henry Knox Museum; Christopher Fox and Nicholas Westbrook, Fort Ticonderoga Museum; Jim Gay, Colonial Williamsburg Foundation; Margi Hofer, New York Historical Society; Patricia Kane and John Stuart Gordon, Yale University Art Gallery; Donald Fennimore and Leslie Grigsby, Winterthur Museum; Anne Marie Lane Jonah and Ruby Fougere, Fortress Louisbourg, Parks Canada; Anna Gruber Koester, Charlottesville, Virginia; Susan McLellan Plaisted, Heart to Hearth Cookery; Jane Spillman, Corning Museum of Glass; Susan Stein, Monticello, The Thomas Jefferson Foundation; Gerald W. R. Ward, Museum of Fine Arts, Boston; Beth Carver Wees, Metropolitan Museum of Art; and Timothy Wilson, Ashmolean Museum, Oxford University.

Endnotes

1. In this period a day's wage in Massachusetts currency varied from two shillings (lawful currency) to three shillings, nine pence (new tenor); therefore making a pound of chocolate slightly less than one week's wages for a laborer.

2. A notable exception was the kitchen staff of the Royal Governor Lord Botetourt in Williamsburg, Virginia, who roasted and ground cacao.

3. A wafer can be seen in the chocolate cups pictured in William Hogarth's *Marriage à la Mode IV: The Countess's Levée*, ca. 1743/44, The National Gallery, London.

4. Various terms are given to this form. They include chocolate mill, stirring rod, muddler, whisk, *molinet, moulinet, moussoir*, and *molinillo*.

5. A variety of chocolate mills are illustrated in Pinto [23], p. 294, plate 312. Although G. Bernard Hughes in his article "Silver Pots for Chocolate," refers to green bottle glass chocolate mills, Jane Spillman, Curator of American Glass at the Corning Museum of Glass, has never encountered one. Correspondence with the author, February 12, 2007.

6. Virtually no French examples of *chocolatiéres* before 1700 have survived, but the form appears in drawings of Louis XIV's silver commissioned by the King of Sweden at the beginning of the 18th century.

7. The Museum of Fine Arts, Boston currently owns four examples of Boston-made, silver chocolate pots: one by Edward Webb (museum number 1993.61), two by John Coney (museum numbers 29.1091 and 1976.771), and one by Zachariah Brigden (museum number 56.676). Yale University Art Gallery owns one example by Edward Winslow (1944.71). Historic Deerfield, Incorporated owns one example by Zachariah Brigden (museum number 75.463). The Metropolitan Museum in New York owns one example by Edward Winslow (museum number 33.120.221). An example by Peter Oliver owned by Mr. and Mrs. Walter M. Jeffords was sold at Sotheby's New York on October 29, 2004 and is now in a private collection.

8. The lack of interior strainers and handles perpendicular to their spouts do not distinguish coffeepots from chocolate pots.

9. Now in a private collection.

10. Correspondence with Christopher Fox, September 15, 2005.

11. The coffeepot is located in the Yale University Art Gallery. The two-handled cups are more likely remnants of the "24 coffee [cups], 2 handles, do [saucers]" listed on the inventory.

12. Another similarly shaped chocolate pot exists at the New York Historical Society and was part of a service that Samuel Shaw ordered for himself (1972.11a). This pot also shares the same unexplained two sets of paired holes as Knox's example.

13. According to a lecture by Carl Crossman, "Taste for the Exotic: Tea, Coffee, and Chocolate Wares," New York Ceramics Fair, January 19, 2007, coffee was also served in handleless shallow bowls in the first half of the 18th century.

14. Tall, handleless beakers are shown in Robert Bonnard's *Un Cavelier et une Dame Beuvant du Chocolat* (Cavalier and Lady Drinking Chocolate), ca. 1690–1710; *Still Life of Chocolate Service* by Luis Melendez (1716–1780),

Spanish, oil on canvas, collection of the Museo del Prado, Madrid, Spain.

15. The seven sheets of drawings are located in the State Archives of The Hague.

16. For more information, see *The Castleford Pottery Pattern Book* (Wakefield, England: EP Publishing, 1973); James & Chas. Whitehead Manufacturers, *Designs of Sundry Articles of Earthenware* (Bletchley, England: D. B. Drakard, 1971); *Don Pottery Pattern Book*, reprint of 1807 edition (Doncaster, England: Doncaster Library, 1983); and Donald C. Towner, *The Leeds Pottery* (London, England: Cory, Adams & Mackay, 1963).

17. A silver chocolate stand in the shape of a leaf is in the collection of Isaac Backal, Mexico City, Mexico.

References

1. Worlidge, J. *Vinetum Britannicum*. London, England: Printed by J. C. for Tho. Dring and Tho. Burrel, 1676; p. 153.

2. Rees, A. *The Cyclopaedia; or, Universal Dictionary of Arts, Sciences, and Literature*, 47 volumes. Philadelphia: Samuel F. Bradford, 1810–1824; Volume 9: s.v. "Chocolate."

3. Shirley, J. *The Accomplished Ladies Rich Closet of Rarities, Or, the Ingenious Gentlewoman and Servant Maids Delightful Companion*. London, England: n.p., 1690; p. 20.

4. Webster, T. *Encyclopedia of Domestic Economy*. New York: Harper and Brothers, 1845; p. 717.

5. Raffald, E. *The Experienced English Housekeeper*. Manchester, England: n.p., 1769; p. 316. Quoted in: Walton, P. *Creamware and Other English Pottery at Temple Newsom House, Leeds*. Bradford, England: Manningham Press, 1976; p. 158.

6. *Pennsylvania Gazette*. October 12, 1769.

7. *Pennsylvania Gazette*. March 8, 1764.

8. *Boston Gazette*. March 12, 1751.

9. Ward, G. W. R. "The silver chocolate pots of Colonial Boston." In: Falino, J., and Ward, G. W. R., editors. *New England Silver & Silversmithing, 1620–1815*. Boston: The Colonial Society of Massachusetts, 2001; p. 62.

10. Stubbe, H. *The Indian Nectar, or, A Discourse Concerning Chocolata*. London, England: Printed by J.C. for Andrew Cook, 1662; p. 5.

11. Stubbe, H. *The Indian Nectar, or, A Discourse Concerning Chocolata*. London, England: Printed by J.C. for Andrew Cook, 1662; p. 5.

12. Anonymous. "Imperial chocolate made by a German lately come into England," ca. 1710. Quoted in: Brown, P. *In Praise of Hot Liquors: The Study of Chocolate, Coffee and Tea-Drinking, 1600–1850*. York, England: York Civic Trust, 1995; p. 33.

13. Letter: Thomas Jefferson to John Adams, November 27, 1785. In: Boyd, J. P., editor. *The Papers of Thomas Jefferson*. Princeton, NJ: Princeton University Press, 1954; Volume 9, p. 63.

14. Probate Inventory: Samuel Hanson, Charles County, Maryland, Charles County Inventories, 1735–1752, pp. 147–151, taken May 11, 1741, recorded July 10, 1741. Available at http://www.gunstonhall.org/probate/HANSON41.PDF. (Accessed January 11, 2007.)

15. Hughes, W. *The American Physitian; or A Treatise of the Roots, Shrubs, Plants, Fruit, Trees, Herbs, & c. Growing in the English Plantations in America*. London, England: Printed by J.C. for William Crook, at the Green Dragon without Temple-Bar, 1672; p. 128.

16. Rundell, M. E. *A New System of Domestic Cookery*, 3rd edition. Exeter, New Hampshire, 1808. 5 Quoted in: Fowler, D. E., editor. *Dining at Monticello: In Good Taste and Abundance*. Charlottesville, VA: Thomas Jefferson Foundation, 2005; p. 185.

17. Lightbody, J. *Every Man His Own Gauger*. London, England: Printed for Hugh Newman at the Grasshopper near the Rose Tavern in the Poultrey, 1695; p. 104.

18. Rees, A. *The Cyclopaedia; or, Universal Dictionary of Arts, Sciences, and Literature*, 47 volumes. Philadelphia: Samuel F. Bradford, 1810–1824; Volume 9: s.v. "chocolate."

19. Walvin, J. *Fruits of Empire: Exotic Produce and British Taste: 1660–1800*. New York: New York University Press, 1997; p. 97.

20. Nott, J. *The Cook's and Confectioner's Dictionary*. London, England: Printed for C. Rivington, 1723; pp. 183–184.

21. Hughes, G. B. "Silver pots for chocolate." *Country Life* 1960;128(Oct. 20):856.

22. Davis, J. D. *English Silver at Williamsburg*. Williamsburg, VA: Colonial Williamsburg Foundation, 1976; p. 84.

23. Pinto, E. H. *Treen and Other Wooden Bygones: An Encyclopedia and Social History*. London, England: G. Bell and Sons, 1969; p. 292.

24. Schroder, T. *The National Trust Book of English Domestic Silver, 1500–1900*. New York: Viking Press in association with the National Trust, 1988; p. 184.

25. Pinto, E. *Treen and Other Wooden Bygones: An Encyclopedia and Social History*. London, England: G. Bell and Sons, 1969; p. 292.

26. Beeton, I. *The Book of Household Management*. London, England: Ward, Lock, & Tyler, 1869; p. 912.

27. Hartop, C. *The Huguenot Legacy: English Silver 1680–1760, from the Alan and Simone Hartman Collection*. London, England: Thomas Heneage, 1996; p. 290.

28. Mudge, J. M. *Chinese Export Porcelain in North America*. New York: Riverside Book Company, 1986; p. 102.

29. Singleton, E. *Dutch New York*. New York: Dodd, Mead, and Company, 1909; pp. 109 and 133.

30. Davis, J. D. *English Silver at Willamsburg*. Williamsburg, VA: Colonial Williamsburg Foundation, 1976; p. 85.

31. Wees, B. C. *English, Irish, & Scottish Silver at the Sterling and Francine Clark Art Institute*. New York: Hudson Hills Press, 1997; pp. 310–313.

32. Wees, B. C. *English, Irish, & Scottish Silver at the Sterling and Francine Clark Art Institute*. New York: Hudson Hills Press, 1997; p. 311.

33. Wenham, E. "Silver chocolate-pots." *The Antique Collector* 1946;17(5):170.

34. Museum of Fine Arts, Boston, Accession No. 1956.676.

35. Mead Art Museum, Amherst College, Amherst, Massachusetts, Accession Number 1945.168.

36. Sotheby's New York, January 21, 2005, Lot number 213 (Sotheby's Images).

37. Metropolitan Museum of Art, New York, Accession No. 1970.31.

38. *Dictionary of American Biography*, Volume 3. New York: Charles Scribner's Sons, 1964; pp. 290–291.

39. Ward, G. W. R. "The Silver chocolate pots of Colonial Boston." In: Falino, J., and Ward, G. W. R., editors. *New England Silver & Silversmithing, 1620–1815*. Boston: The Colonial Society of Massachusetts, 2001; p. 85, footnote 10.

40. Smyth Stuart, J. F. D. *A Tour in the United States of America*. London, England: Printed for G. Robinson, J. Robson, & J. Sewell, 1784; p. 142.

41. Letter: Thomas Jefferson to John Adams, dated November 27, 1785. In: Boyd, J. P., editor. *The Papers of Thomas Jefferson*. Princeton, NJ: Princeton University Press, 1954; Volume 9, p. 63.

42. Boyd, J. P. "Thomas Jefferson and the Roman Askos of Nimes." *The Magazine Antiques* 1973;104(1):116–124.

43. Stein, S. R. *The Worlds of Thomas Jefferson at Monticello*. New York: Harry N. Abrams, Inc. in association with the Thomas Jefferson Memorial Foundation, 1993; pp. 328–329.

44. Garvan, B. B. *Federal Philadelphia, 1785–1825: The Athens of the Western World*. Philadelphia: Philadelphia Museum of Art, 1987; pp. 78–79.

45. Fowler, D. L., editor. *Dining at Monticello: In Good Taste and Abundance*. Charlottesville, VA: Thomas Jefferson Foundation, 2005; pp. 78 and 184.

46. Letter: Joseph Coolidge to Nicholas Trist, dated January 5, 1827, Library of Congress. Letter: Joseph Coolidge to Thomas Jefferson Randolph, dated December 16, 1826, Edgehill-Randolph Papers. File Number 1397, University of Virginia.

47. Hughes, G. B. "Silver pots for chocolate." *Country Life* 1960;128(Oct. 20):856.

48. Ward, G. W. R. "The Silver chocolate pots of Colonial Boston." In: Falino, J., and Ward, G. W. R., editors. *New England Silver & Silversmithing, 1620–1815*. Boston: The Colonial Society of Massachusetts, 2001; p. 88, footnote.

49. Bridenbaugh, C. *Peter Harrison: First American Architect*. Chapel Hill: University of North Carolina Press, 1949; p. 176.

50. Fennimore, D. L. *Metalwork in Early America: Copper and Its Alloys from the Winterthur Collection*. Winterthur, DE: Henry Francis du Pont Winterthur Museum, 1996; p. 105, Item No. 33.

51. F. A. Walker & Co. *Illustrated Catalogue of Useful and Ornamental Goods* (Boston, ca. 1872–1873), p. 3, No. 17; p. 10, No. 67. Boston: Baker Library, Harvard University.

52. *Pennsylvania Gazette*. August 21, 1735.

53. Anonymous. *Les Soupes de la Cour, ou l'Art de Travailler Toutes Sortes d'Alimens*. Tome III. Paris, France: n.p., 1755. Reprinted in Geneva, Switzerland: Slatkine Reprints, 1978; pp. 332–333.

54. Child, L. M. F. *The American Frugal Housewife*, reprint of the 12th edition. Boston: Applewood Books, 1986; p. 83.

55. Sheaf, C., and Kilburn, R., *The Hatcher Porcelain Cargoes: The Complete Record*. Oxford, England: Phaidon and Christies, 1988; p. 104.

56. Emerson, J. *Coffee, Tea, and Chocolate Wares in the Collection of the Seattle Art Museum*. Seattle, WA: Seattle Art Museum, 1991; p. 22, No. 10.

57. Schiedlausky, G. *Tee, Kaffee, Schokolade: Ihr Eintritt in Die Europäische Gesellschaft*. Munich, Germany: Prestel Verlag, 1961; Figures 16 and 43.

58. Dyer, E. S. *Montpelier, This Spot So Sacred to A Name So Great: Collections Catalogue of the General Henry Knox Museum*. Thomaston, ME: Friends of Montpelier, 2004; p. 17.

59. Mudge, J. M. *Chinese Export Porcelain in North America*. New York: Riverside Book Company, 1986; p. 213, Figure 345.

60. Howard, D. S. *New York and the China Trade*. New York: The New York Historical Society, 1984; pp. 78–79, B17.

61. Emerson, J., Chen, J., and Gates, M. G. *Porcelain Stories: From China to Europe*. Seattle, WA: Seattle Art Museum in association with the University of Washington Press, 2000; p. 108.

62. Emerson, J., Chen, J., and Gates, M. G. *Porcelain Stories: From China to Europe*. Seattle, WA: Seattle Art Museum in association with the University of Washington Press, 2000; p. 108.

63. Emerson, J., Chen, J., and Gates, M. G. *Porcelain Stories: From China to Europe*. Seattle, WA: Seattle Art Museum in association with the University of Washington Press, 2000; p. 23, no. 11.

64. Jörg, C. J. A. *Fine & Curious: Japanese Export Porcelain in Dutch Collections*. Amsterdam, Holland: Hotei, 2003; pp. 201–202, No. 256–258.

65. Godden, G. A. *Oriental Export Market Porcelain and Its Influence on European Wares*. London, England: Granada Publishing, 1979; p. 320.

66. Mudge, J. M. *Chinese Export Porcelain in North America*. New York: Riverside Book Company, 1986; p.144.

67. Jörg, C. J. A. *Porcelain and the Dutch China Trade*. The Hague, Holland: M. Nijhoff, 1982; pp. 110, 112, and 115.

68. Jörg, C. J. A. *Interaction In Ceramics: Oriental Porcelain & Delftware*. Hong Kong: Hong Kong Museum of Art, 1984; p. 28, Figure 7.

69. Jörg, C. J. A. *Interaction in Ceramics: Oriental Porcelain & Delftware, Exhibition Catalogue*. Hong Kong: Hong Kong Museum of Art, 1984; p. 164.

70. *Connecticut Courant*. September 1, 1778.

71. Samuel Boardman, Invoice Book, 1772–1774, Wethersfield Historical Society, Wethersfield, CT; Frederick Rhinelander, "Book of Bad Debts," 1772–1773, New York Historical Society.

72. Letter: George Washington to Clement Biddle, Mount Vernon, February 11, 1789. Available at http://rotunda.upress.virginia.edu:8080/pgwde/dflt.xqy?keys=search-Pre01d189&hi=chocolate. (Accessed January 11, 2007.)

73. Letter: Burgess Ball to George Washington, dated February 13, 1794, George Washington Papers at the Library of Congress, 1741–1799: Series 4. General Correspondence. 1697–1799. Available at http://memory.loc.gov/cgiin/ampage?collId=mgw4&fileName=gwpage105.db&recNum=241&tempFile=./temp/~ammem_Fy2Z&filecode=mgw&next_filecode=mgw&prev_filecode=mgw&itemnum=21&ndocs=100. (Accessed January 11, 2007.)

74. Detweiler, S. G. *George Washington's Chinaware*. New York: Harry N. Abrams, 1982; p. 123, Figure 100.

75. Cadou, C. B. *The George Washington Collection: Fine and Decorative Arts at Mount Vernon*. New York: Mount Vernon Ladies' Association and Hudson Hills Press, 2006; p. 149, cat. 43.

76. Cadou, C. B. *The George Washington Collection: Fine and Decorative Arts at Mount Vernon*. New York: Mount Vernon Ladies' Association and Hudson Hills Press, 2006; p. 149, cat. 43.

77. Detweiler, S. G. *George Washington's Chinaware*. New York: Harry N. Abrams, 1982; p. 156, No. 135.

78. *Pennsylvania Gazette*. August 21, 1735.

79. Ward, G. W. R., "The Silver chocolate pots of Colonial Boston." In: Falino, J., and Ward, G. W. R., editors. *New England Silver & Silversmithing, 1620–1815*. Boston: The Colonial Society of Massachusetts, 2001; p. 63.

80. Ward, G. W. R., "The Silver chocolate pots of Colonial Boston." In: Falino, J., and Ward, G. W. R., editors. *New England Silver & Silversmithing, 1620–1815*. Boston: The Colonial Society of Massachusetts, 2001; p. 80.

81. Chinese Porcelain Company. *Chinese Glass Paintings and Export Porcelain, Exhibition and Sale*, October 8–November 9, 1996. New York: Chinese Porcelain Company, 1996; p. 79.

82. Clarence-Smith, W. G. *Cocoa & Chocolate 1765–1914*. London, England: Routledge, 2000; p. 24.

83. Higgins & Seiter, Trade catalogue, New York, New York, ca. 1905, p. 247. Winterthur, DE: Downs Printed Books and Manuscript Collection, Winterthur Library.

Silver Chocolate Pots of Colonial Boston[1]

Gerald W. R. Ward

Introduction

Tea, coffee, and chocolate—three drinks we now take for granted—were all introduced to England and North America in the 17th century. Initially considered exotic beverages, each arrived to the Anglo-American world from a remote part of the globe. Tea came from the Orient, coffee from North Africa and the Middle East. Chocolate was the gift of Mexico, Central America, and South America; it first arrived in Spain in the 16th century and then migrated to France, England, and the rest of Europe, before making the trip to North America and to our area of concern here, New England, through the West Indies trade. Flavorful, rich, and nutritious, chocolate was indeed a boon to the 17th and 18th century diet. Usually consumed as a beverage in those days, chocolate was "of a dusky colour, soft, and oily; usually drank hot, and esteemed not only an excellent food, as being very nourishing, but also a good medicine; at least a diet, for keeping up the warmth of the stomach, and assisting digestion." Wealthy consumers and silversmiths responded to this expensive novelty in the Anglo-American world by demanding and creating a specialized form—the chocolate pot—to be used for serving this luxurious new drink [1–7].

Most scholars give 1657 as the date of the first documented instance of chocolate's appearance in London [8–10]. By the 1660s, Samuel Pepys took chocolate frequently there, often at breakfast, as on May 3, 1664, when he "went to Mr. Blands and there drinking my morning draught in good chocolatte, and slabbering my band sent home for another" [11]. It seems likely that chocolate appeared in Boston not long after it arrived in England. In the winter of 1667–1668, the merchant and goldsmith John Hull was trading in cocoa and tobacco, and in 1670, chocolate was common enough in town that the Boston selectmen approved the separate petitions of two women, Dorothy Jones and Jane Barnard, "to keepe [*sic*] a house of publique [*sic*] Entertainment for the sellinge [*sic*] of Coffee and Chucalettoe." These women and other individuals were granted similar licenses in the years following, as chocolate became a more standard item on the bill of fare in public establishments [12, 13].

In typical New England fashion, chocolate was too enjoyable to be considered entirely respectable. In 1676, Benjamin Tompson (1642–1714), a largely forgotten New England poet, wrote a lengthy poem entitled "New-Englands Crisis," published in the middle of the bloody conflict known to the English colonists as King Philip's War. Tompson laments that the "golden times" of New England have passed—that the strength of the Puritan had been "quickly sin'd away for love of gold." Citizens forebearers once noted for their strong character

Chocolate: History, Culture, and Heritage. Edited by Grivetti and Shapiro
Copyright © 2009 John Wiley & Sons, Inc.

and religious piety have been corrupted by licentiousness and idle pleasures. Chocolate figures into his elegy. Remembering the past, he recalled the days before outside influences worked their evil ways. This age, he recalled,

Twas ere the Islands sent their Presents in,
Which but to use was counted next to sin.
Twas ere a Barge had made so rich a fraight
as Chocholatte, dust-gold and bitts of eight.

He further blames "fruits and dilicacies" from "western Isles" that "Did rot maids teeth and spoil their hansome faces."

Despite such concerns, chocolate soon became accepted, even among the Puritan elite [14]. Samuel Sewall (1652–1730) of Boston (Fig. 11.1) mentions chocolate several times in his famous diary and in so doing gives us our first glimpses of chocolate drinking as a form of social interaction in New England. On October 20, 1697, he visited Lieutenant Governor William Stoughton in Dorchester, and they had "breakfast together on Venison and Chockalatte." Sewall observed that "Massachusetts and Mexico met at his Honour's Table" [15]. On October 1, 1709, Sewall went to Mr. Belcher's in Dedham, where in the morning he drank "warm chockelat, and no Beer; find my self much refresh'd by it after great Sweating

to day, and yesterday" [16]. On other occasions, as revealed by entries in his diary and letter-book, Sewall gave gifts of chocolate to friends. In April 1687, for example, he bought "21 balls [of] chokolatto," probably to be shipped in linen bags in chunks known as balls, rowls, lumps, cakes, or tablets in the contemporary literature [17]. These balls may have resembled the irregular pieces of chocolate depicted in the lower right side of a still-life painting of 1770 by the Spanish artist Luis Melendez (Fig. 11.2). In 1707, Sewall "gave Mr. Solomon Stoddard two half pounds of Chockalat, instead of Commencement Cake," and on another occasion in 1723 he presented a new mother with a gift of "two pounds of Chockalet." Sewall also gave silver objects as gifts; his use of chocolate for similar presentation purposes suggests its high social status [18].

Other Bostonians no doubt shared Sewall's enjoyment of chocolate. The Reverend Thomas Prince (1687–1758), minister of Old South Church and historian of early New England, made it part of his daily routine. Following graduation from Harvard, Prince spent a decade abroad before returning to Boston in 1717 and assuming his position at Old South the next year. After his marriage, he drew up his plans in 1719 for carrying out each day. His "proposed order" called for rising at 5:00 a.m. and spending an hour in prayer

FIGURE 11.1. Judge Samuel Sewall, portrait by John Smibert (1688–1751); dated 1729. *Source*: Courtesy of the Museum of Fine Arts, Boston, MA. (Used with permission.) (See color insert.)

FIGURE 11.2. Still life chocolate service by Luis Melendez (1716–1780); dated ca. 1760. *Source*: Courtesy of the Museo Nacional del Prado, Madrid, Spain. (Used with permission.) (See color insert.)

and reading the Bible in his study, after which he would wake up the rest of the family. At 6:30, they would have family prayers, and then "only the Porringer of Chocolat for Breakfast," before setting out on an arduous day [19–21].[2]

Prince's porringer of "Chocolat" and the "warm chockelat" Sewall enjoyed were probably prepared with equipment and utensils of the type depicted in the Melendez painting, following directions of the kind described in an English publication of 1675 by John Worlidge. Some people "boil [the chocolate] in water and sugar; others mix half water and half milk and boil it, then add powdered chocolate to it and boil them together; others add wine and water." The author continues:

> Be sure whilst it is boiling to keep it stirring, and when it is off the fire, whir it with your hand mill. . . . That is, it must be mixed in a deep pot of Tin, copper [as in the Melendez painting] or stone [stoneware], with a cover with a hole in the middle of it, for the handle of the mill to come out at, or without a cover. The mill is only a knop [sic] at the end of a slender handle or stick, turned in a turner's lathe, and cut in notches, or rough at the end. They are sold at turners for that purpose. This being whirled between your hands, whilst the pot is over the fire, and the rough end in the liquor causes an equal mixture of the liquor with the chocolate and raises a head of froth over it. Then pour it out for use in small dishes for that purpose. You must add a convenient quantity of sugar to the mixture. [22]

In this passage the author has identified the distinguishing characteristic of a chocolate pot—the opening in the lid, which allows for the insertion of the essential stirring rod. The opening can be achieved in a variety of ways—by a removable cap or finial, or the presence of a small hole accessed by a sliding cover—but it must be present in order to consider an object a chocolate (as opposed to a coffee or tea) pot. The sediment in hot chocolate—unlike that in tea and coffee—is desirable, and thus the mixture needs to be stirred continually. Otherwise, chocolate pots and coffeepots—both usually taller and generally larger than teapots—are virtually indistinguishable. Both forms were made with the spout in line with the handle, or occasionally with the spout at a right angle to the handle.

While there must have been many individual methods of preparation, the basic recipe outlined by Worlidge changed little for more than a century. Chocolate was probably prepared in a copper or brass pot in large quantities and perhaps served in a ceramic or base metal pouring vessel. But in a few of the most fashionable New England homes, it was served in a silver chocolate pot. The earliest reference to such an object in Boston comes in 1690, when William Pleay owned a "jocolato pot" [23]. Only eight examples made in colonial Boston are known to have survived;

a potential ninth example, whereabouts unknown today, has been published as both a chocolate and a coffeepot. Their rarity is underlined by the fact that chocolate pots represent only an infinitesimal percentage of the more than six thousand pieces of extant silver made by colonial Massachusetts silversmiths [24, 25].[3,4]

Despite their low numbers, the Boston silver chocolate pots—especially the six made before 1720 by John Coney, Edward Winslow, Edward Webb, and Peter Oliver—provide a glimpse of life in Boston during a period of florescence in the decorative arts. Extraordinarily stylish and costly, the pots were faddish in their response to a new custom. Used in the process of consuming a luxurious beverage in a custom that migrated from Catholic Spain and southern Europe, silver chocolate pots seem almost antithetic in Protestant Boston, yet their existence—when taken into account with other stylish forms of silver, furniture, and architecture—is a small slice of material evidence of the changes, ultimately dramatic in their extent, that were moving Boston from its origins as a Puritan enclave in the 17th century to its place as a cosmopolitan, sophisticated, commercial colonial city in the very earliest years of the 18th century.

The Boston Chocolate Pots

Of this group of eight, the earliest example is probably the one made by John Coney about 1701 (Fig. 11.3). It is engraved on the bottom "The gift of Wm Stoughton Esquire / to Mrs. Sarah Tailor : 701," presumably in error for 1701. In his will, executed on July 6, 1701, Lieutenant Governor Stoughton left his niece, Mrs. Sarah Byfield Tailer, twelve pounds to buy a piece of plate as a "particular remembrance" of him, and presumably this vessel is the result of that bequest. It may have been a fitting choice, for Stoughton is known to have enjoyed a cup of chocolate at breakfast with Samuel Sewall, and he may have had a fondness for the drink. In form, this Coney pot resembles an Oriental vase; it would have been virtually at the height of fashion in London and smaller English towns when it was made. It has its handles at right angles to the spout, as many (but not all) examples do, perhaps to facilitate stirring while pouring. The turned-ball finial covers the hole necessary for insertion of the stirring rod. The curved spout originally had a hinged lid at the end, now lost, that perhaps helped to keep the contents warm. The Coney pot is further distinguished by its cut-card ornament around the base [26].

Sarah Tailer was the wife of Lieutenant Governor William Tailer (1677–1732). Their household included many stylish goods, including two English cane chairs that have survived (one in the Massachusetts Historical Society and the other in the Bostonian

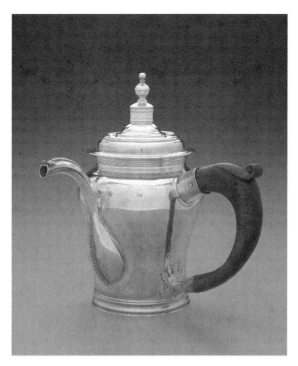

FIGURE 11.3. Chocolate pot, made by John Coney (1665–1722); dated 1701. *Source*: Courtesy of the Museum of Fine Arts, Boston, MA. (Used with permission.) (See color insert.)

Society); these were probably among 18 such chairs in the fashionable Tailer household [27, 28].

An English example of a chocolate pot (Fig. 11.4), made by Isaac Dighton of London in 1697, provides a high-style analogue to the Coney pot, with its more elaborate gadrooning and other ornament, and its gilt surface. A London pot made by George Garthorne in 1686 is more similar to the Coney example, and some provincial English examples, like one made by R. Williamson of Leeds about 1695, are also close in feeling. In other words, Coney's work, while not at the highest level of court silver, was nevertheless stylish and up-to-date in the Anglo-American community [29–31].

Coney, who was probably apprenticed to Jeremiah Dummer, was the leading Boston goldsmith of his day. His work included rare forms such as sugar boxes, punch bowls, and monteith bowls, as well as the earliest known American silver teapot, and thus it is not surprising that a second chocolate pot by him is known (Fig. 11.5). This example was probably made between 1715 and 1720, as indicated by its curvilinear pear-shaped body in the incipient late baroque taste. Its most decorative feature is its serpent-headed spout. This pot originally had a sliding or pivoted finial for the insertion of the stirring rod, and the underside of its base shows a circular pattern of wear that suggests the

FIGURE 11.4. Chocolate pot, made by Isaac Dighton, London, England; dated 1697. *Source*: Courtesy of the Metropolitan Museum of Art, New York. (Used with permission.) (See color insert.)

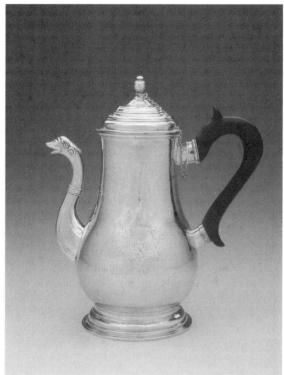

FIGURE 11.5. Chocolate pot, made by John Coney; dated ca. 1715–1720. *Source*: Private collection. (Used with permission.) (See color insert.)

pot was often placed on a small brazier or chafing dish to keep it warm. The original owners of this pot, unfortunately, are not known—their initials have been erased and replaced with those of William Downes and his wife, Elizabeth Edwards Cheever, who were not married until 1749. This Coney example also has its counterparts in English work [32–34].

Perhaps the most beautiful American chocolate pot was made by Edward Winslow of Boston between 1700 and 1710 (Fig. 11.6). It is richly decorated with gadrooned ornament that is played off of plain areas, creating what the silver scholar John Marshall Phillips of Yale University called the "beautiful equipoise of baroque silver." This pot, now in the Metropolitan Museum of Art, has an acorn finial that unscrews and is then held in place by a chain. It was owned originally by Thomas Hutchinson (1675–1739), a wealthy merchant and husband of Sarah Foster, the daughter of Colonel John Foster. They were the parents of the notorious Tory Governor Thomas Hutchinson [35, 36].

Another Winslow pot, very similar to the one in the Metropolitan Museum, is now in the collection of the Yale University Art Gallery. Also made in the first decade of the 18th century, the Yale Winslow pot has had its spout moved to a higher location on the body, perhaps as early as 1740—the hole where the original spout was located was patched with a decorative plaque.

This alteration probably had its origins in the function of the object. A higher location for the spout allowed for sediment to settle more easily in the pot; thus the change may have allowed the chocolate to flow more smoothly. The Yale Winslow pot is engraved with what are thought to be the Auchmuty family arms; if such is the case, the pot's original owner may have been lawyer Robert Auchmuty (1687–1750) of Roxbury [37, 38].

A fifth Boston silver chocolate pot was made by Peter Oliver, who died at the age of thirty and left only about seven examples of his work behind (Fig. 11.7). Oliver probably was apprenticed to Coney, "his trusty friend." Oliver's chocolate pot must have been made in the short period of time between the completion of his training in about 1703 and his death in 1712. Visually, his pot is similar to the Winslow examples, sharing the use of gadrooned ornament; the finial, originally removable, has been soldered in place. The pot was probably made for Beulah Jacquett, who married Thomas Coates; it subsequently descended in their family in the female line through many generations [39].

Edward Webb was the maker of a sixth Boston silver chocolate pot (Fig. 11.8). While it bears an overall resemblance to the Winslow and Oliver examples, it has some significant differences. It has an

FIGURE 11.6. Chocolate pot, made by Edward Winslow (1669–1753), Boston, Massachusetts; dated ca. 1705. *Source*: Courtesy of the Metropolitan Museum of Art, New York. (Used with permission.) (See color insert.)

FIGURE 11.7. Chocolate pot, made by Peter Oliver (ca. 1682–1712), Boston, Massachusetts; dated ca. 1705. *Source*: Private collection. (See color insert.)

FIGURE 11.9. Monteith (large silver punch bowl), made by William Denny, London, England; dated 1702. *Source*: Christie's Images. (Used with permission.) (See color insert.)

FIGURE 11.8. Chocolate pot, made by Edward Webb (1666–1718), Boston, Massachusetts; dated ca. 1710. *Source*: Courtesy of the Museum of Fine Arts, Boston, MA. (Used with permission.) (See color insert.)

unusual finial assembly, consisting essentially of a screw-on cap, and it has fluted, rather than gadrooned, ornament on its body and cover. Webb, born in England about 1666, learned his craft in London during an apprenticeship to William Denny. Denny had opened shop in London in 1679 and worked there until his death in 1709. He had just entered into business when Webb started his training in 1680, which lasted until about 1687. Denny's prominent shop filled many important commissions, and Webb was thus exposed to sophisticated English silver styles during his formative years there. It is possible that Webb stayed on in Denny's shop as a journeyman. It is not known when Webb arrived in this country, although evidence places him here as early as 1704.

Objects that have Denny's mark, like a monteith dated 1702 (Fig. 11.9) and others dated about the same time, often bear the type of fluting—tight, somewhat pinched, irregular, almost nervous—that Webb must have learned in Denny's shop and brought to America. Most other objects bearing Webb's mark, including tankards, porringers, and spoons, are simpler than the chocolate pot. Some of them have well-executed cast ornament, but none is as ambitious as the pot. However, Webb was a wealthy man when he died

in 1718, and his estate inventory indicates that he owned all the tools necessary to create such a complex pot. As an English immigrant and one with specialized skills, Webb may have worked primarily as a jobber or journeyman for native-born silversmiths. Although the writer is not aware of any chocolate pots bearing William Denny's mark that resemble the Webb pot, some contemporaneous English work, such as an example of 1703 by William Charnelhouse of London, is closely related. Unfortunately, the original owner of the Webb pot is not known. The applied cartouche on the side of the pot is a later addition of uncertain, although possibly 18th century, date, and the engraved initials it bears, P / TA, remain unidentified [40–50].

The last two Boston colonial chocolate pots known were made by Zachariah Brigden, one in the Museum of Fine Arts (Fig. 11.10) and the other at Historic Deerfield (see Fig. 10.3), both dated to about 1755–1760 and thus made very early in Brigden's career. The MFA Brigden pot was made for Ebenezer and Mary (Edwards) Storer, who were married in 1723, and is engraved with the coat of arms used by Ebenezer [51]. The Deerfield Brigden example, which is slightly larger but otherwise very similar, was made for a member of the De(e)ring family [52]. Both have removable finials; the MFA example has an open-link chain to secure the finial when it is removed. Brigden's chocolate pots, last in the Boston series, document the survival of their kind for more than half a century, into the era when tea and coffeepots were the more common forms.

PROVINCIAL LUXURY

This group of eight objects—perhaps statistically insignificant even when one considers that undoubtedly a few examples have been lost to the melting pot in the last two centuries—is of extraordinary significance when considered in light of developments in Boston in

FIGURE 11.10. Chocolate pot, made by Zachariah Brigden (1734–1787), Boston, Massachusetts; dated ca. 1755–1760. *Source*: Courtesy of the Museum of Fine Arts, Boston, MA. (Used with permission.) (See color insert.)

sophistication that characterized Boston in this period [54].

The Foster–Hutchinson house, built on Garden Court Street in the North End about 1692 (and destroyed in 1833), is representative of this new wave of change. As Abbott Lowell Cummings has observed, the Foster–Hutchinson house is the first "recorded example of the English Renaissance in Boston," with its Ionic pilasters topped by capitals of imported Portland stone [55, 56]. The house is central to our theme because it was built by John and Abigail Foster, and descended to Mrs. Foster's nephew, the merchant-captain Thomas Hutchinson, the owner of the Winslow chocolate pot now at the Metropolitan (Fig. 11.7). The Foster–Hutchinson house stood next door to the Clark–Frankland house, built by the merchant William Clark about 1712 (sadly, it too was taken down in 1832 or 1833). This three-story brick mansion was an even more fully developed example of the Georgian style, with its string courses, rhythmic facade, and alternating segmental and triangular pediments. Many less ambitious houses, like the Moses Pierce–Hichborn house of ca. 1711, which still stands in the North End, came from the same mold, as did collegiate buildings at Harvard and churches and civic buildings in Boston, such as the landmark building now known as the Old State House of 1712–1713 [57].

These developments in architecture, in which the Georgian style supplanted the timber-frame first-period buildings of the postmedieval style, started to transform the landscape. They were mirrored in changes in furniture and furnishings that similarly began to affect domestic space. Such objects as the Warland family chest-on-chest (Fig. 11.11) and the Reverend John Avery's desk and bookcase (Fig. 11.12) are keys to understanding Boston furniture of possibly as early as 1715. Both are very English in style, and advanced stylistically. Edward S. Cooke, Jr. has suggested that they are the cabinetmaking equivalent of the elaborate Boston silver made by Coney, Winslow, and others—silver that includes the sophisticated group of chocolate pots, as well as other notable examples of conspicuous consumption, such as the wrought candlesticks of 1690–1710 by Coney and gadrooned candlesticks by John Noyes of about 1695–1700, lighting devices rarely made in silver in this country. Other rare forms made in this period include John Coney's spectacular monteith (made about 1705–1710 for rinsing and cooling wineglasses) and his simple yet elegant punch bowl of about 1710 [58, 59].

The chocolate pots bear perhaps the most significant relationship to Boston silver sugar boxes, a similarly small group of ten known examples, including ones by John Coney of ca. 1685 (Fig. 11.13) and Edward Winslow of ca. 1700 (Fig. 11.14). Both forms are rare, costly, stylish, and linked to relatively short-lived customs. The chocolate pots lack the overt snake symbolism and other embellishments that are

the Colonial era, especially in the years 1700 to 1715 or 1720, when all but the Brigden pots were made. As the 17th century came to a close, changes began to be felt in the small seaport town of about ten thousand people. "The worldwide commercial interests of the town," according to the historian G. B. Warden, "did provide luxuries and diversions from Augustan England which seriously affected the town's way of life" [53]. Many of these "luxuries and diversions" were brought to Boston by the succession of royal governors and other officials who administered the province's affairs after 1692. Changes in architecture, furniture, silver, and other furnishings took place, and the little cluster of chocolate pots is perhaps best seen as part of a constellation of material objects, large and small, that introduced new styles, customs, and behaviors to Boston.

The house built in 1679 by Peter Sargeant was probably the grandest house in Massachusetts in its time and for many years thereafter. In the 1710s, it was remodeled to become the fashionable city home of the colonial governors and became known as the Province House. Portraits of kings, queens, and governors hung in its council chambers, and it symbolized the existing power structure in Boston. It also was the epicenter of the increased urbanization and

FIGURE 11.11. Chest-on-chest, black walnut, burl walnut veneer, and Eastern white pine; maker unknown; Boston, Massachusetts; dated ca. 1715–1725. *Source*: Courtesy of the Museum of Fine Arts, Boston, MA. (Used with permission.)

FIGURE 11.12. Desk and bookcase, walnut, walnut veneer, Eastern white pine; maker unknown; Boston, Massachusetts; dated ca. 1715–1720. *Source*: Courtesy of the Museum of Fine Arts, Boston, MA. (Used with permission.)

FIGURE 11.14. Silver sugar box, made by Edward Winslow (1669–1753), Boston, Massachusetts; dated ca. 1700. *Source*: Courtesy of the Museum of Fine Arts, Boston, MA. (Used with permission.) (See color insert.)

FIGURE 11.13. Silver sugar box, made by John Coney, Boston, Massachusetts; dated 1680–1690. *Source*: Courtesy of the Museum of Fine Arts, Boston, MA. (Used with permission.) (See color insert.)

practically unique to Winslow's sugar boxes, as explained by Ed Nygren, who explored the complex iconography and sexual connotations of these objects, which are related to issues of marriage, fertility, and fecundity. Although sugar boxes were used primarily in the service of wine, one can imagine a wealthy family using sugar boxes and chocolate pots in the same household, as sugar was also important as a sweetener for the otherwise bitter chocolate [60].

Both forms also were produced by only a few craftsmen. Of the combined total of 18 sugar boxes and chocolate pots, Coney (four sugar and two chocolate) and Winslow (five sugar and two chocolate) produced 13, or 72 percent; another chocolate pot was made by Peter Oliver, who was probably apprenticed to Coney. It is possible that Coney and Winslow were able to employ a journeyman with experience in London—perhaps Edward Webb or Henry Hurst—to assist in the fashioning of such stylish, sophisticated forms [61].

The "Excellent Nectar"

Drinking chocolate was both a private and a public custom. Chocolate was taken at coffee houses, or houses of public entertainment, in a public setting, although because of its higher price due to high duties, chocolate initially took a secondary role to coffee and was soon almost completely eclipsed by the more caffeine-laden drink. A silver chocolate pot was not necessary, of course, for even though silver's purity and thermal conductivity made it suitable for hot beverages, base metal vessels would do as well and were commonly used in these commercial establishments (see Chapter 54).

In the home, the principal context for silver articles, chocolate was often taken in the morning, at least during the first years after its introduction, but it was undoubtedly drunk at many times during the day [62–67]. For example, Madam Sarah Kemble Knight (1666–1727) arrived in the evening to a clean, comfortable house in the Narragansett country while en route in her famous trip from Boston to New Haven in October 1704. The owner of the home asked what Madam Knight would like to eat, and she replied, "I told her I had some Chocolett, if shee [sic] would prepare it; which with the help of some Milk, and a little clean brass Kettle, she soon effected to my satisfaction" [68]. But the longstanding, fashionable tradition among the European moneyed class—one that migrated from the Catholic countries of southern Europe, Italy, and Spain—was to take chocolate in the morning, as Samuel Sewall did at least once and undoubtedly did more frequently. The chocolate was poured, in most cases, from silver pots into ceramic chocolate cups, often small, handleless beakers imported from abroad, although smaller teacups would serve nearly as well. No doubt a variety of drinking vessels were used. Peter Fanueil (1700–1743), who owned some 1400 ounces of silver and was reputed to be the wealthiest man in Boston at his death in 1743, served the beverage in "6 lignum vitae chocolate cups lin'd with silver," and, as we saw earlier, the Reverend Thomas Prince used a porringer [69].

Bostonians obtained their chocolate from local merchants, such as Mrs. Hannah Boydell, who offered "Tea, Coffee, Chocolate, Loaf and Muscovado Sugars of all Sorts" at her shop on King Street in the 1730s, or the grocer John Merrett, at the sign of the Three Sugar Loaves and Canister, also on King Street, who sold "Super-fine Chocolate, Coffee, raw and roasted, [and] very choice Teas," along with all manner of sugars, spices, and other delicacies. Another merchant offered "Italian chocolate ready prepared with sugar" in 1739, and in 1769, John Goldsmith advertised "Choice Chocolate made and Sold" at his corner shop leading down John Hancock's wharf (see Chapter 27) [70, 71].

Although chocolate contains theobromine, a mild stimulant, it has more nutritional than stimulating value. In its early days, chocolate was also thought to have medicinal value and to be an aphrodisiac [72, 73]. An English verse of 1652 observes that chocolate:

'Twill make old women Young and Fresh;
Create new notions of the flesh,
And cause them long for you know what,
If they but taste of chocolate. [74]

But within a few years, chocolate, like tea, which had similarly been accorded many magical and mystical powers when first introduced, became primarily a fashionable, expensive drink that did not carry a great deal of attendant symbolic meaning. Thus, the French cavalier and lady seen taking chocolate in a French print published between 1690 and 1710 (Fig. 11.15), were primarily simply enjoying themselves in a cabaret [75, 76].

This theme of enjoyment and relaxation typifies many 18th century images of chocolate drinking. A French depiction of a lady's boudoir—or possibly a brothel—features a chocolate service at its center, for example, with the mill (or molinet) clearly visible (Fig. 11.16). A painting by the French artist Jean-Baptiste Le Prince, entitled La Crainte (The Fear) (1769) depicts a woman in bed leaning toward her departing lover; the chocolate cups and pot, with its mill clearly visible, also are seen (Fig. 11.17). The method of drinking chocolate was to adopt a "fluid, lazy, lanquid [sic] motion," well typified by this young lady. Dress for taking chocolate was meant to be casual (carried to an extreme in this image) [77–79].

Most everything about chocolate drinking suggested ample amounts of leisure time, which is perhaps the greatest symbol of power. The Englishman William Hughes, writing in 1672, cautioned his readers that:

it is not convenient, as experience hath sufficiently taught us,
to eat or drink any thing else quickly after the drinking of it;
or presently to use any immoderate exercise; but rather to rest
awhile, whether it be taken hot or cold: because it is apt to
open the Pores, and thereby it causeth the greater expence [sic]
of Spirits by transpiration, and so consequently nourisheth the
less. [80]

FIGURE 11.15. *Un Cavalier et une Dame beuvant du chocolat* (*Cavalier and Lady Drinking Chocolate*), engraving on paper by Robert Bonnart; dated ca. 1690–1710. *Source*: Courtesy of the Pierpont Morgan Library and Museum, New York. (Used with permission.) (See color insert.)

In other words, take it easy.

As the scholar Wolfgang Schivelbusch has observed, a long breakfast involving chocolate "does not start off a workday—rather it marks the start of a day's carefully cultivated idleness," a "morning-long awakening to the rigors of studied leisure" [81].

Such a style of life—of slow, languorous mornings—was at odds with the Puritan and Yankee modes of behavior, and this divergence may account for the small number of silver chocolate pots made in Massachusetts Bay. Certainly the Reverend Prince was not an idle aristocrat—his morning draft of chocolate inaugurated a long day of scholarship and meditation. But coffee was more in tune with New England Protestantism. The silver chocolate pots of Coney, Winslow, Webb, and Oliver allow us to see a few Bostonians, in the first few years of the 18th century, trying on a Continental style of life that, apparently, did not fit. Although chocolate remained a part of the culinary landscape throughout the 18th century, it was not afforded the exalted status of silver pouring vessels, designed and reserved just for chocolate, with the exception of the two Brigden pots made at midcentury.

Instead, other forms became more popular in silver. Based on surviving examples, for example, Massachusetts silversmiths are known to have made at least 95 silver teapots before about 1775. Most of these, however, were made relatively late. The earliest is one

FIGURE 11.16. *La Jolie Visiteuse* (*The Pleasant Houseguest*) by Jean-Baptiste Mallet (1759–1835); dated ca. 1750. *Source*: Courtesy of the Museum of Fine Arts, Boston, MA. (Used with permission.) (See color insert.)

FIGURE 11.17. Jean-Baptiste Le Prince (French, 1734–1781), *The Fear (La Crainte)*, 1769, oil on canvas, 50 × 64 cm, Toledo Museum of Art, purchased with funds from the Libbey Endowment, Gift of Edward Drummond Libbey, 1970.444. *Source*: Courtesy of The Toledo Museum of Art, Toledo, OH. (Used with permission.) (See color insert.)

by John Coney of about 1710; only five others dating before 1730 are known. By the middle of the 1730s, there was a spike in the production of teapots, with at least 25 known examples, including many by Jacob Hurd, that can be dated to ca. 1735–1745; the numbers continued to increase during the following decades.

Silver coffeepots were crafted rarely by Massachusetts silversmiths in the first half of the 18th century. Only about six made before 1750 survive, although at least 21 are known dating to the period ca. 1750–1775. Silver coffeepots or pots of other materials may have been used for chocolate after 1720, but the era when it demanded specialization had largely passed [24, 82, 83].[5]

The historian Cary Carson, in a lengthy study of the style of life in the 18th century, has observed that there was something in the air in the years just before and after 1700. The world was changing and, among the many consequences of those changes, he suggests that "none was more novel or conspicuous than the pleasure that men and women took in their physical well-being and the value they placed on material things" [84]. Why this was so remains a matter of debate among historians in the academy, but it is likely that the story of Boston's silver chocolate pots might add a footnote to the ultimate resolution. Certainly people drank chocolate because it tasted good, and wealthy people apparently took pleasure in using silver vessels to do so. The surviving chocolate pots indicate that the period of 1700–1715 was especially significant for these changes in one small area of "cultivated gentility." Thus, aside from their intrinsic qualities as works of art, these pots have something to add to the current academic concerns about the rise of gentility and

genteel behavior in America, about changes in etiquette and manners, and about the ongoing concern in American life over the pernicious effects of luxury [85, 86]. They offer information about the consumer revolution, and no doubt, like the story of sugar as told by Sidney Mintz, they say something about power relationships based on wealth and exploitation [87].

The thoughts expressed by William Hughes in 1672, however, probably come closer to the true meaning of chocolate to people of the 18th century than do our latter-day theoretical speculations. His volume was entitled *The American Physitian; or, A Treatise of the Roots, Plants, Trees, Shrubs, Fruit, Herbes, &c. Growing in the English Plantations in America . . . whereunto is added a Discourse of the Cacao-Nut-Tree, And the use of its Fruit; with all the ways of making Chocolate, the like never extant before*. The words Hughes wrote ring true today in the 21st century:

> But what shall I say more of this excellent Nectar? It is a very good aliment, a clear Pabulum multi nutriment: that it doth fatten . . . is undeniable; and that it nouriseth [sic] . . . is without dispute. . . . [I]t revives the drooping spirits, and chears [sic] those that are ready to faint; expelling sorrow, trouble, care, and all perturbations [sic] of the minde [sic]: it is an Ambrosia: And finally, in a word, it cannot be too much praised. [88]

Acknowledgments

I am grateful to many people for their assistance with this chapter, including Jeannine Falino and Jane L. Port. In all my work, I am indebted to Jonathan L. Fairbanks and to my wife, Barbara McLean Ward, for their ongoing support.

Endnotes

1. This chapter first appeared in slightly different form in: Falino, J., and Ward, G.W.R., editors. *New England Silver and Silversmithing, 1620–1815.* Boston: Colonial Society of Massachusetts, 2001, pp. 61–88. It is reprinted here by permission of the Society and its editor of publications, John W. Tyler.

2. The Virginian William Byrd (1674–1744) recorded that he drank chocolate for breakfast about once every three weeks from February 1709 through September 1712 as recorded in: Wright, L.B., and Tinling, M., editors. *The Secret Diary of William Byrd of Westover, 1709–1712.* Richmond, VA: Dietz Press, 1941.

3. An object by Edward Winslow is listed by Patricia E. Kane [Ref. 24, p. 977] as a chocolate pot. However, in the entry on the Yale Winslow chocolate pot, Kathryn C. Buhler and Graham Hood discuss the various chocolate pots known at that time, and note that "Winslow made a later coffee pot (privately owned) in the tapered-cylindrical, Queen Anne style, which is apparently engraved with the Saltonstall arms," citing the Yale University Art Gallery files as their source.

4. Its present whereabouts is unknown to this writer. Inquiries to current members of the Saltonstall family have not been successful in locating the object. For the purposes of the present essay, it will be regarded as a coffeepot.

5. Specialized chocolate forms continue to appear in documents throughout the 18th century, often in base metals (see also Chapter 54). The architect Peter Harrison had a copper chocolate pot valued at 16s. in his estate when he died in New Haven, Connecticut, in 1775; outside New England, Lord Botetourt of Williamsburg, Virginia, owned 24 pounds of chocolate and had "3 chocolate pots with four mills" in his 1770 estate.

References

1. Coe, S. D. *America's First Cuisines.* Austin: University of Texas Press, 1994; pp. 50–58.

2. Coe S. D., and Coe, M. D. *The True History of Chocolate.* New York: Thames and Hudson, 1996.

3. Toussaint-Samat, M. *A History of Food. Translated by Anthea Bell.* Cambridge, MA: Blackwell Publishers, 1992; Chapter 18.

4. Braudel, F. Civilization and Capitalism, 15th–18th Century, Volume 1, *The Structures of Everyday Life: The Limits of the Possible.* Translated by Siân Reynolds. New York: Harper & Row, 1981; pp. 249–260.

5. Rees, A. *The Cyclopaedia; or, Universal Dictionary of Arts, Sciences, and Literature,* 47 volumes. Philadelphia: Samuel F. Bradford and Murray, Fairman and Co., 1810–1824; s.v. "Chocolate."

6. Wenham, E. Silver chocolate pots. *Antiques* 1942; 42(5):248.

7. Hughes, G. B. Silver pots for chocolate. *Country Life* 1960;128(3320):856–857.

8. *Cocoa and Chocolate: A Short History of Their Production and Use.* Dorchester, MA: Walter Baker & Co., 1886; p. 39.

9. Morton, F., and Morton, M. *Chocolate: An Illustrated History,* New York: Crown, 1986; p. 59.

10. Wees, B. C. *English, Irish, and Scottish Silver in the Sterling and Francine Clark Art Institute.* New York: Hudson Hills Press, 1997; pp. 267–272.

11. Latham, R., and Matthews, W., editors. *The Diary of Samuel Pepys,* 11 Volumes. Berkeley: University of California Press, 1970–1983; Volume 5, p. 139. See also references for June 19, 1660 (Volume 1, p. 178), October 17, 1662 (Volume 3, pp. 226–227), January 6, 1663 (Volume 4, p. 5), February 26, 1664 (Volume 5, p. 64), and November 24, 1664 (Volume 5, p. 329).

12. *The Diaries of John Hull, Mint-Master and Treasurer of the Colony of Massachusetts Bay.* Boston: John Wilson and Son, 1857; p. 158.

13. *A Report of the Record Commissioners of the City of Boston Containing the Boston Records from 1660 to 1701.* Boston: Rockwell and Churchill, 1881; pp. 58, 60, 64, 68, and 73.

14. Harrison, T., and Meserole, H. T. *Seventeenth-Century American Poetry.* New York: W. W. Norton, 1968; pp. 225–229.

15. Thomas, M. H. *The Diary of Samuel Sewall, 1674–1729,* 2 Volumes. New York: Farrar, Straus and Giroux, 1973; Volume 1, p. 380.

16. Thomas, M. H. *The Diary of Samuel Sewall, 1674–1729,* 2 Volumes. New York: Farrar, Straus and Giroux, 1973; Volume 2, p. 626.

17. Sewall, S. *Letter-Book,* Collections of the Massachusetts Historical Society, 6th Series, Volumes 1 and 2. Boston: Massachusetts Historical Society, 1886; Volume 1, pp. 46 and 157.

18. Thomas, M. H. *The Diary of Samuel Sewall, 1674–1729,* 2 Volumes. New York: Farrar, Straus, and Giroux, 1973; Volume 1, pp. 476, 563–564, and 570.

19. Hill, H. A. *History of the Old South Church (Third Church), Boston, 1669–1684,* 2 Volumes. Boston: Houghton Mifflin, 1890; Volume 1, p. 398.

20. Blake, F. E. *History of Princeton.* Princeton, MA: Published by the Town, 1915; p. 110.

21. Wright, L. B., and Tinling, M. *The Secret Diary of William Byrd of Westover, 1709–1712.* Richmond, VA: Dietz Press, 1941; Volume 1, passim.

22. Davis, J. D. *English Silver at Williamsburg.* Williamsburg, VA: Colonial Williamsburg Foundation, 1976; p. 85.

23. Clayton, M. *The Collector's Dictionary of the Silver and Gold of Great Britain and North America.* New York: World Publishing Company, 1971; s.v. "chocolate pot."

24. Kane, P. E. *Colonial Massachusetts Silversmiths and Jewelers: A Biographical Dictionary*. New Haven, CT: Yale University Art Gallery, 1998.

25. Buhler, K. C., and Hood, G. *American Silver: Garvan and Other Collections in the Yale University Art Gallery*, 2 Volumes. New Haven, CT: Yale University Press for the Yale University Art Gallery, 1970; Volume 1, p. 56, Catalogue No. 49.

26. Buhler, K. C. *American Silver, 1655–1825, in the Museum of Fine Arts, Boston*, 2 Volumes. Boston: Museum of Fine Arts, 1972; Catalogue No. 50.

27. Kane, P. E. Furniture owned by the Massachusetts Historical Society. *Antiques* 1976;109(5):960, Figure 3.

28. Cabot, H. R. *Handbook of the Bostonian Society*. Boston: Old State House, 1979; p. 10 (left).

29. Ward B. M. The Edwards family and the silver smithing trade in Boston, 1692–1762. In: Puig, F. J., and Conforti, M., editors. *The American Craftsman and the European Tradition, 1620 to 1820*. Minneapolis, MN: Minneapolis Institute of Arts, 1989; pp. 77–79.

30. New Orleans Auction Galleries Inc. *Antiques* 1999; 155(1):104.

31. Clayton, M. *The Collector's Dictionary of the Silver and Gold of Great Britain and North America*. New York: World Publishing Company, 1971; Figure 124.

32. Fairbanks, J. L. A decade of collecting decorative arts and sculpture at the Museum of Fine Arts, Boston. *Antiques* 1981;120(3)627, plate 42 (right).

33. Fairbanks, Cooke, Jr, E. S., Falino, J. J., Foss, L. L., Monfredo, R. J., and Pulsone, M. *Collecting American Decorative Arts and Sculpture, 1971–1991*. Boston: Museum of Fine Arts, 1991; p. 70.

34. Clayton, M. *The Collector's Dictionary of the Silver and Gold of Great Britain and North America*. New York: World Publishing Company, 1971; Figure 127.

35. Avery, C. L. *American Silver of the Seventeenth and Eighteenth Centuries: A Study Based on the Clearwater Collection*. New York: Metropolitan Museum of Art, 1920; Catalogue Nos. 11, 21–24.

36. Phillips, J. M. *American Silver*. New York: Chanticleer Press, 1949; pp. 57–58.

37. Buhler, K. C., and Hood, G. *American Silver: Garvan and Other Collections in the Yale University Art Gallery*, 2 Volumes. New Haven, CT: Yale University Press for the Yale University Art Gallery, 1970; Volume 1, Catalogue No. 49.

38. Townsend, A. *The Auchmuty Family of Scotland and America*. New York: Grafton Press, 1932.

39. Buhler, K. C. *Colonial Silversmiths: Masters and Apprentices*. Boston: Museum of Fine Arts, 1956; Catalogue No. 111, Figure 45.

40. Ward, B. M. Edward Wabb, C. 1666–1718 In: Kane, P. E., editor. *Colonial Massachusetts Silversmiths and Jewelers*. New Haven, CT: Yale University Art Gallery, 1998; pp. 951–957.

41. Grimwade, A. G. *London Goldsmiths, 1697–1837: Their Marks and Lives*, 2nd edition. London, England: Faber and Faber, 1982; p. 490.

42. Wyler, S. Incorporated. *Antiques* 1998;154(4):459.

43. Lee, G. E. with assistance from Lee, R. A. *British Silver Monteith Bowls Including American and European Examples*. Byfleet, England: Manor House Press, 1978; Figure 32 and appropriate entries in Appendix II.

44. *Catalogue of Fine English Silver Plate of the 17th and 18th Centuries, The Property of Captain H. C. S. Ward*. London, England: Christie, Manson, and Wood, 1914; lot 44.

45. McFadden, D. R., and Clark, M. A. *Treasures for the Table: Silver from the Chrysler Museum*. New York: Hudson Hills Press in association with The American Federation of Arts, 1989; Catalogue No. 28.

46. Jackson, C. J. *An Illustrated History of English Plate*, 2 Volumes, 1911. Reprinted New York: Dover Publications, 1969; Volume 1, pp. 272–273.

47. Clayton, M. *Christie's Pictorial History of English and American Silver*. Oxford, England: Phaidon/Christie's, 1985; p. 122, Figure 1.

48. Clayton, M. *The Collector's Dictionary of the Silver and Gold of Great Britain and North America*. New York: World Publishing Company, 1971; Figure 129.

49. Wees, B. C. *English, Irish, and Scottish Silver in the Sterling and Francine Clark Art Institute*. New York: Hudson Hills Press, 1997; catalogue No. 182.

50. Waldron, P. *The Price Guide to Antique Silver*, 2nd edition. Woodbridge, England: Antique Collectors' Club, 1982; p. 257, Figure 812.

51. Buhler, K. C. *American Silver, 1655–1825, in the Museum of Fine Arts, Boston*, 2 Volumes. Boston: Museum of Fine Arts, 1972; Catalogue No. 326.

52. Flynt, H. N., and Fales, M. G. *The Heritage Foundation Collection of Silver, with Biographical Sketches of New England Silversmiths, 1625–1825*. Old Deerfield, MA: Heritage Foundation, 1968; pp. 100–101, Figure 79.

53. Warden, G. B. *Boston, 1689–1776*. Boston: Little, Brown and Company, 1970; p. 24.

54. *Old-Time New England* 1972; pp. 85–123.

55. Cummings, A. L. The domestic architecture of Boston, 1660–1725. *Archives of American Art Journal* 1971;9(4):1–16; quotation from p. 7.

56. Cummings, A. L. The Foster–Hutchinson house. *Old-Time New England* 1964;54(3):59–76.

57. Cummings, A. L. The domestic architecture of Boston, 1660–1725. *Archives of American Art Journal* 1971;9(4):1–16, Figures 18–11.

58. Cooke, E. S. Jr., 15, chest-on-chest. In: Fairbanks, J. L., editor. *Collecting American Decorative Arts and Sculpture, 1971–1991*. Boston: Museum of Fine Arts, 1991; p. 32.

59. Forman, B. M. *American Seating Furniture, 1630–1730: An Interpretive Catalogue*. New York: W. W. Norton, 1988.

60. Nygren, E. J. Edward Winslow's sugar boxes: colonial echoes of courtly love. *Yale University Art Gallery Bulletin* 1971;33(2):38–52.

61. Ward, B. M. Boston goldsmiths, 1690–17830. In: Quimby, I. M. G., editor. *The Craftsman in Early America.* New York: W. W. Norton for the Henry Francis du Pont Winterthur Museum, 1984; pp. 126–157 (especially pp. 144–150).

62. Wheaton, B. K., and Kelly, P. *Bibliography of Culinary History: Food Resources in Eastern Massachusetts.* Boston: G. K. Hall & Company.

63. Brown, P. B. *In Praise of Hot Liquors: The Study of Chocolate, Coffee, and Tea-Drinking, 1600–1850.* York, England: Fairfax House, 1995.

64. Emerson, J. *Coffee, Tea, and Chocolate Wares in the Collection of the Seattle Art Museum.* Seattle, WA: Seattle Art Museum, 1991.

65. Pinto, E. H. *Treen and Other Wooden Bygones: An Encyclopedia and Social History.* London, England: G. Bell and Sons, 1969; p. 291, Plate 312.

66. Newman, H. *An Illustrated Dictionary of Silverware.* New York: Thames and Hudson, 1987; p. 214.

67. Fairbanks, J. L., Dicken, R. F., Dockstader, F. J., Ewers, J. C., Farnam, A., Sussman, E., Truettner, W. H., and Wohlauer, G. S. *Frontier America: The Far West.* Boston: Museum of Fine Arts, 1975; p. 148.

68. *The Journal of Madam Knight.* Boston: David R. Godine, 1972; pp. 9, 97.

69. Fairbanks, J. L., Whitehill, W. M., Cooper, W. A., Farnam, A., Jobe, B. W., Katz-Hyman, M. B., *Paul Revere's Boston: 1735–1818.* Boston: Museum of Fine Arts, 1975; p. 29, Catalogue No. 26.

70. Dow, G. F. *The Arts and Crafts in New England, 1704–1775.* Topsfield, MA: Wayside Press, 1927; pp. 209, 263–264, 292–294.

71. *Boston Gazette.* January 2, 1769.

72. Ackerman, D. *A Natural History of the Senses.* New York: Random House, 1990; pp. 153–157.

73. Schivelbusch, W. *Tastes of Paradise: A Social History of Spices, Stimulants, and Intoxicants.* New York: Pantheon Books, 1992; especially Chapter 3.

74. Wadsworth, J. *A Curious History of the Nature and Quality of Chocolate* (1652). Quoted in: Brown, P. B. *In Praise of Hot Liquors: The Study of Chocolate, Coffee, and Tea-Drinking, 1600–1850.* York, England: Fairfax House, 1995; p. 18.

75. Maccubbin, R. P., and Hamilton-Phillips, M. *The Age of William III and Mary II: Power, Politics, and Patronage, 1688–1702.* Williamsburg, VA: College of William and Mary, 1989; p. xxxii.

76. Roth, R. *Tea Drinking In Eighteenth-Century America: Its Etiquette and Equipage.* United States Museum Bulletin 225, Contributions from the Museum of History and Technology, Paper 14, 1–30. Washington, DC: Smithsonian Institution, 1961.

77. Schivelbusch, W. *Tastes of Paradise: A Social History of Spices, Stimulants, and Intoxicants.* New York: Pantheon Books, 1992; p. 91.

78. Munger, J. H., Zafram, E. M., Poulet, A. L., Mowry, R. D., Alcorm, E. M., Hawes, U. S., and Secondo, J., *The Forsyth Wickes Collection in the Museum of Fine Arts, Boston.* Boston: Museum of Fine Arts, 1992; Catalogue No. 73.

79. Rosenberg, P. *The Age of Louis XV: French Painting, 1710–1774.* Toledo, OH: Toledo Museum of Art, 1975; Catalogue No. 66.

80. Hughes, W. *The American Physitian; or, A Treatise of the Roots, Plants, Trees, Shrubs, Fruit, Herbes, &c. Growing in the English Plantations in America . . . whereunto is added a Discourse of the Cacao-Nut-Tree, And the use of its Fruit; with all the ways of making Chocolate, the like never extant before.* London, England: Printed by J.C. for William Crook, at the Green Dragon Without Temple Bar, 1672; p. 141.

81. Schivelbusch, W. *Tastes of Paradise: A Social History of Spices, Stimulants, and Intoxicants.* New York: Pantheon Books, 1992; p. 91.

82. Bridenbaugh, C. *Peter Harrison: First American Architect.* Chapel Hill, NC: University of North Carolina Press, 1949; p. 176.

83. Hood, G. *The Governor's Palace in Williamsburg: A Cultural Study.* Williamsburg, VA: Colonial Williamsburg Foundation, 1991; pp. 287–289.

84. Carson, C. The consumer revolution in Colonial America. In: Carson, C., Hoffman, R., and Albert, P. J., editors. *Of Consuming Interests: The Style of Life in the Eighteenth Century.* Charlottesville, VA: University Press of Virginia for the United States Capitol Historical Society, 1994; pp. 483–697 (quotation on p. 494).

85. Sweeney, K. M. High-style vernacular: lifestyles of the Colonial elite. In: Carson, C., Hoffman, R., and Albert, P. J., editors. *Of Consuming Interests: The Style of Life in the Eighteenth Century.* Charlottesville, VA: University Press of Virginia for the United States Capitol Historical Society, 1994; pp. 1–58.

86. Conroy, D. W. *In Public Houses: Drink and the Revolution of Authority in Colonial Massachusetts.* Chapel Hill, NC: University of North Carolina Press for the Institute of Early American History and Culture, 1995; pp. 22–23.

87. Mintz, S. W. *Sweetness and Power: The Place of Sugar in Modern History.* New York: Penguin Books, 1986; especially pp. 106–139.

88. Hughes, W. *The American Physitian; or, A Treatise of the Roots, Plants, Trees, Shrubs, Fruit, Herbes, &c. Growing in the English Plantations in America . . . whereunto is added a Discourse of the Cacao-Nut-Tree, And the use of its Fruit; with all the ways of making Chocolate, the like never extant before.* London, England: Printed by J.C. for William Crook, at the Green Dragon Without Temple Bar, 1672; p. 148.

CHAPTER

Is It A Chocolate Pot?

Chocolate and Its Accoutrements in France from Cookbook to Collectible

Suzanne Perkins

Introduction: The Chocolatière and the Refinement of Aristocratic Manners in Early Modern France

European social life and dining styles were characterized in the Early Modern period by an increasing refinement that can be seen in many of the fine and decorative arts of the era. Nowhere is this more evident than in France, which emerged after the Thirty Years' War (1618–1648) as the dominant political and military power on the Continent and one of the wealthiest countries in Europe, if not in the world, becoming noted for the production of luxury products. The era of Louis XIV, the "Sun King," was marked by the construction of the Versailles palace, admired and copied by royalty and aristocracy throughout Europe. An increased delicacy in aristocratic and bourgeois dining styles included the arrival of specialized dining rooms and dining tables with increasingly elaborate tablecloths and table settings, not the least of which was the popularization of the fork. This growing refinement of dining styles included the adoption of beverages such as tea, coffee, and chocolate.

Just as new ceramic (faience and porcelain) vessels for preparing and serving beverages were created for the new drinks, new terms were invented or adapted to describe both individual pots as well as sets or services of several pieces for serving these beverages. According to the French etymological diction-

ary *Le Trésor de la Langue Française,* by or during the 18th century, terms for chocolate pot (*chocolatière*), teapot (*théière*), and coffeepot (*cafetière*), in a variety of different spellings, had appeared by 1671, 1698, and 1690, respectively. The term *déjeuner,* referring to a breakfast service or meal, and *tête-à-tête,* a beverage service for two persons, appeared in 1728 and 1780, respectively [1]. A *déjeuner* service typically included a tray, a chocolate pot, a milk-pot or creamer, a sugar-basin (bowl), and cups and saucers [2]. Later, the term *solitaire* referred to a beverage service for one person.

Chocolate: History, Culture, and Heritage. Edited by Grivetti and Shapiro
Copyright © 2009 John Wiley & Sons, Inc.

The introduction of chocolate and its spread, largely as a beverage, in royal and aristocratic circles brought with it the development of the *chocolatière*, or chocolate pot, as a specialized item to facilitate the stirring, frothing, and serving of hot chocolate, beginning in the late 17th century. It remained in use until the introduction of more mechanized ways of making chocolate beverages in the mid-19th century rendered it obsolete. In the 20th century, the *chocolatière* assumed a second life as an object of historical and purely aesthetic interest, and as a collectible [3].

Chocolatière history is complicated by the lack of agreement on what exactly constitutes a *chocolatière*, as they can be similar or, on occasion, identical in form to a coffeepot. In general, the *chocolatière* is marked by a hole in its unattached cover (*couvercle*), or hinged lid, into which a wooden stirrer (*moulinet* or *moussoir*) is inserted to stir and froth the hot chocolate in the pot below. Other terms for a *moulinet* were *bâton* or even *baguette* (both mean "stick"). Because chocolate was a mixture in which the heavier particles fell to the bottom, 17th and 18th century chocolate had to be stirred and frothed just before pouring. Froth was considered an appealing attribute of the chocolate beverage, requiring the *moulinet*. In some cases, an item is identified as a *chocolatière* because of written documentation accompanying it. Porcelain and faience (earthenware) examples without a hole in the cover, whether *chocolatières* or works of art depicting chocolate use, can be said to be for chocolate use rather than coffee by the existence of written records that identify them precisely as such.

Chocolate pots or vessels made of earthenware and of metal appear in Mesoamerica, where use of chocolate preceded the arrival of the Europeans. Some Spanish Colonial vessels were a combination chocolate-maker and chocolate-cup, with one handle, such that the hot chocolate could be drunk from the same vessel in which it was made. An example of this type of *jarro* (mug) or *jicara de chocolate* from ca. 1780–1820, of tin-lined copper, is on exhibition in the La Nueva Casa (Spanish Colonial) kitchen exhibit in the Albuquerque Museum of Art and History in New Mexico [4]. In their *True History of Chocolate*, Sophie and Michael Coe wrote that Spaniards introduced the *molinillo*, or stirrer, which they twirled with two hands, and added the closed top, through which the stirrer was inserted [5]. From Spain, chocolate, the chocolate pot, and its related cups and saucers, together with the chocolate beverage itself, were diffused to other European countries.

The common argument is that the French developed and introduced the *chocolatière*, in metal, with a hinged lid and hole for the wooden *moulinet* and a handle placed at a right angle to the spout. The handle unscrewed clockwise so that it would remain tight while pouring from the spout in a counterclockwise direction [6]. Philippe du Chouchet has also seen "left-handed" *chocolatières*, including a silver one with a wooden handle [7]. In the collection of the Musée des Arts Décoratifs in Paris is a three-footed, pear-shaped copper *chocolatière*, ca. 1750–1770, with a wooden handle attached at ninety degrees from the spout, a thumb piece for ease in opening the lid, a domed cover with a pierced hole for the *moulinet*, and a second small hole to fasten the small round lid that pivots to cover the hole [8]. Some covers have pull-out finials, others pivot or unscrew. Turned handles were often made of rare woods such as ebony. This is a typical form for 18th century chocolate pots made of various metals, found not only in France but also throughout Europe. Examples made of silver and/or gold have hallmarks attesting to the quality of the metal, a longstanding practice of goldsmiths and silversmiths since the mid-14th century [9]. They also have maker's marks, identifying the artisans.

The use of metal versus ceramic material influenced the *chocolatières*' lids; porcelain covers are generally not hinged and most often do not have a hole for the *moulinet* due to ceramics' greater fragility in the face of vigorous frothing, although we will discuss several notable exceptions among royal or aristocratic collections and a *faïence fine* example later. Guillemé-Brulon defines *faïence fine* as an opaque paste, white or ivory, with a fine texture, dense and resonant when struck, and covered by a transparent lead glaze. In contrast, the term *faïence* refers to pottery made of a colored paste, permeable to water, with a porous texture, and covered by an opaque white tin glaze that conceals the color of the clay [10].

The Function of the Chocolatière

An early user of chocolate, for medicinal purposes, was Cardinal Alphonse de Richelieu of Lyon (brother of the more famous Richelieu), to whom René Moreau dedicated his *Du chocolat, discours curieux divisé en quatre parties*, an annotated translation of an earlier Spanish text, published in France in 1643. Moreau refers to the widespread use of chocolate in the West Indies, where Indians and Negroes used *apastles*, vessels similar to *terrines*. He also mentions a *mollinet*, which he described as similar to a stick twisted to make rope from threads in Spain. Twisting this stick aerated the chocolate and cooled the froth or foam gathered at the top [11].

The first French *chocolatière* reference is an often-cited letter by the Marquise de Sévigné to her adult daughter, Madame de Grignan, whom she advised in 1671 to take chocolate as a medical restorative for fatigue. But, the Marquise virtually exclaimed in her letter to her daughter:

> You have no *chocolatière*—I've thought about it a thousand times! How will you make it [the chocolate]? [12]

Implied in Sévigné's letter was a familiarity with a *chocolatière* as a specialty item, not widely available in all parts of France. Pierre Richelet's *Dictionnaire François* of 1680, which was the first French dictionary to include the word *chocolat*, also defined *chocolatière* as a "metal vase in which one keeps chocolate until one wishes to take it" [13].

References to *chocolatières* appear in treatises on chocolate as well as in cookbooks and culinary literature for confectioners. In his 1685, *Traitéz Nouveaux & Curieux du café, du thé et du chocolat*, Sylvestre Dufour cited the English explorer Thomas Gage and described the Mexicans' use of hot chocolate as a beverage:

> They dissolve a tablet [tablette] in hot water [he writes], and stir it with a moulinet in the cup from which it will be drunk. When froth [écume—i.e., espume] has developed at the top, they fill the rest of the cup with hot atolle [a liquid prepared from maize] and they gulp it down hot. [14]

Two years later, in 1687, Nicolas Blégny was more explicit in his discussion of *chocolatières*. His *Le bon usage du thé, du caffé et du chocolat* states that the most common and best way to have chocolate was as a beverage, necessitating the use of a *chocolatière* [15]. In his instructions for preparation he writes that *chocolatières* were different from *caffetières* (coffeepots) only in the hole in their lids for the *moulinet* [16]. The *moulinet*, a wooden stick that passed through the hole in the lid of the *chocolatière*, had a ball-like head at the lower end that agitated the liquid chocolate when turned or twirled with the hands. According to Blégny, the head must not come into contact with the bottom of the pot, otherwise the stirring effect would be lost [17]. In order to keep the chocolate frothy, one needed to pour it by turns sequentially into several cups, stirring it all the while. Keeping the *chocolatière* closed during this process would prevent air from getting in to help generate the foam, or *mousse*, but the necessary opening and closing of the pot with the handle of the *moulinet* passed through a hole in the lid would be cumbersome [18]. Accordingly, Blégny concluded that *chocolatière* lids should not be pierced and they should be no different from *caffetières* [19].

François Massaliot, whose *Cuisinier Roial et Bourgeois* first appeared in 1691 and contained some of the earliest chocolate recipes in France, subsequently published in 1734 his *Nouvelle Instruction pour les Confitures, les Liqueurs, et les Fruits*, a confectionary cookbook that included a discussion of *chocolatières*. Like Blégny, Massaliot made little distinction between *chocolatières* and *caffetières*, arguing that either could serve for the making of chocolate. His description of the *moulinet* also followed that of Blégny. The frothy chocolate then was poured into a goblet (defined as a cup without a handle by Geneviève Le Duc, a specialist in the history of French ceramics), or into a jigger [20, 21]. The *chocolatière* was to be refilled and stirred again until a new froth was made, and the remaining cups filled in

the same way until it was empty. As with Blégny, Massaliot emphasized the importance of allowing the chocolate to come into contact with air to create the froth. Chocolate with milk, he added, could be made the same way, except that milk was substituted rather than water [22]. A similar discussion of the *chocolatière* and its use appears in Joseph Gillier's dictionary of terms for the specialist in sweets and confectionary, *Le Cannameliste français*, published in 1751 [23]. In his definition of *chocolatière*, Gillier seemed to oppose Blégny's argument against the hole in the lid. To Gillier, a *chocolatière* was the:

> Equipment of the Office [the term for the various specialties and subspecialties of chefs in the kitchen] in which one prepares chocolate, is a type of cafetière, in which the lid is not attached with a hinge [charnière] and which has a hole in the middle for passing the handle of the moulinet. [24]

Presumably, being able to remove the lid (a separate part) from the main body of the chocolatière permitted the insertion of the *moulinet* with greater ease. In his 1755 *Traité des Alimens* (*Treatise on Foods*) Louis Lémery also addressed the subject of *chocolatières* and how to use them. He, too, referred to boiling the chocolate in the *chocolatières* but made specific reference to the chocolate *pâte*, or paste, which must be broken into small pieces prior to boiling. Again, the *moulinet* was used to create a froth, which was poured into cups. The remaining chocolate was then poured into the cups as well in what, Lémery added, was the method of the Spanish and the Italians [25]. Lémery concluded with what seemed the essential difference between chocolate pots and coffeepots:

> The chocolatières, which are in a truncated cone [cône tronqué], are more suited than the caffetières, to make the froth. [26]

Shortly before the French Revolution, Pierre-Joseph Buc'hoz, a leading naturalist and private physician for the King's brother, in a treatise on tobacco, coffee, cacao, and tea, also described the function of the *chocolatière*, which, he wrote, was to give the consistency of honey to the hot beverage. Noting that water or milk could be used, Buc'hoz described the heating and pouring processes in terms similar to the earlier writers. Ingredients included chocolate, sugar, cinnamon, fresh eggs for textural consistency, and possibly some orange flower water or essence of amber (ambergris) to produce a highly perfumed chocolate drink leaving no sediment in the *chocolatière* or the cups [27].

Chocolatières in Precious and Other Metals

The earliest *chocolatières* in France appear to have been made from metal, and those that survive are largely in precious metals, specifically silver. The 18th century *Encyclopédie* of Diderot and d'Alembert noted in 1776 that *chocolatières* were made of silver, tin-lined copper, pewter, and clay (ceramic) and that clay was not as suitable a material as metal. Ceramic was said to be more vulnerable to breakage from heating and strong boiling [28]. Although there are none made of pewter, there is a small 19th century *chocolatière*, a rare survivor, made of iron on exhibit at the Musée du Chocolat in Biarritz, implying middle or lower social class use [29]. In his dictionary of home furnishings, Henry Havard defined *chocolatière* similarly as a "type of vase . . . made of metal, silver, or silver-plated copper or brass." He cited the "Inventory of the Furnishings of the Crown" (Louis XIV), dated April 22, 1697, as listing "one *chocolatière* of gold, with its re-heater" [30].

The earliest reference to specific *chocolatières* appears to come from Bayonne in the Basque country, which is not surprising because chocolate made its way into France via that route. Jews fleeing the Inquisition in Spain and Portugal had been given refuge in Saint-Esprit, a suburb of Bayonne, by the French authorities. The Jewish communities included makers and sellers of chocolate [31]. Accordingly, one of the earliest references to *chocolatières* comes from an estate in Bayonne, which listed one made of copper in 1684 [32].

One of the more striking events in the history of *chocolatières* in France occurred with a gift of them made by King Narai of Siam to Louis XIV and some of his close associates in 1686. At the time, the Siamese were involved in a complex set of diplomatic negotiations involving the French, Dutch, and other Western powers in Southeast Asia. Constantin Phaulkon, a Greek merchant, had become chief minister of King Narai and appears to have counseled him in his choice of gifts to the French. Recorded in the *Mercure Galant* (issue of July 1686), a newspaper of the day that recorded royal activities, the Siamese gift included gold and silver *chocolatières* with their covers to Louis XIV, the Queen, and several of their officials. The *Mercure Galant* reported that Louis XIV received two silver *chocolatières* with their covers, *du Japon* (of Japan) [33, 34]. Raphaël Masson, Curator of Heritage at the Research Center in the Château of Versailles, kindly consulted the inventory of the gifts sent by the King of Siam to Louis XIV and his entourage. According to the inventory, the King was given two *pairs* of silver *chocolatières* [35]. The Dauphin was given two silver *chocolatières*, according to the *Mercure Galant*, with gold flowers, finely worked (i.e., *relevé* or *repoussé*) of Japan [36]. Once again, the inventory shows a discrepancy, listing the gift as one silver *chocolatière* rather than the two of the journalistic account [37]. Additional gifts included "a small gold *chocolatière* of Japan" for the Queen [38]. The Queen of Siam also gave the Duke of Burgundy a small gold *chocolatière* with an accompanying small silver plate, of Japan [39]. Phaulkon, cited in

the *Mercure Galant* as Mr. Constance, presented Louis XIV with five *chocolatières*, of which three were of silver, also of Japan [40], and to Jean-Baptiste Colbert, Louis XIV's influential finance minister, also known as the Marquis de Seignelay, he gave three silver *chocolatières*, of which one was described as larger than the other two [41].

It is unlikely that the *chocolatières* were actually made in Japan. Sophie and Michael Coe suggest that these chocolatières were "japanned," implying that they were not necessarily made in Japan [42]. Japanning or japanned ware, however, are terms applied to ceramics decorated with a black ground and decoration in gold and silver in imitation of Japanese lacquer ware [43]. Wares made of silver and gold were not japanned. After consulting the inventory, Raphaël Masson confirmed with us that none of the *chocolatières* in the Siamese gift were ceramic—all were metal—so none of them was japanned. Masson also advised caution, noting that for Alexandre, Chevalier de Chaumont, Louis XIV's Ambassador to Siam at the time, the word "Japan" could mean a geographic area extending from as far as the Japanese islands to Macau [44]. Bertrand Rondot, Curator at the Musée des Arts Décoratifs in Paris, agreed that the descriptions "*ouvrage du Japon*" and "*du Japon*" do not imply that the *chocolatières* were made in Japan [45]. The French historian Alfred Franklin noted that the 1686 Siamese gift was the first mention of specific precious metal *chocolatières* that he found in France [46]. It should also be noted that in England, George Garthorne made the earliest known silver chocolate pot in 1685, just one year before the Siamese gift [47].

Chocolatières reappeared in the *Mercure Galant* in the July 1689 issue with a report that the Duc d'Orléans had staged a lottery at Saint-Cloud to give away "jewels for women." It was reported that one woman, a Madame de Maré, won two *chocolatières*, one of silver and one of "porcelain," together with seven chocolate beating sticks (*batons*), and a tea box [48, 49].

An engraving by Antoine Trouvain (Fig. 12.1) probably completed in the years between 1694 and 1698, "represents the 'sixth drawing-room' in the great State apartments at Versailles, and shows a dresser adorned with large gold and silver dishes" [50]. In the center of the engraving, positioned on the dresser, is a pot with three feet, a pear-shaped body with a bulbous lower section, a cover, and a handle that comes out straight on the left side of the pot. It appears to be made of metal, but is it a *chocolatière*? It has some features in common with many metal *chocolatières*, but does it have a hole in the center of the cover or is it a coffeepot or some other kind of pot? From the drawing alone one cannot tell.

In addition to *chocolatières* for the aristocracy, evidence indicates that by the early 18th century, use of the pots had filtered to middle and lower social classes, at least in southwestern France, shown by a

FIGURE 12.1. Engraving by Antoine Trouvain, dated 1694–1698. *Source*: © Réunion des Musées Nationaux / Art Resource, New York. (Used with permission.)

variety of different kinds of *chocolatières* in French Basque inventories at Pau in 1713 [51]. In contrast, one of the more striking *chocolatières* still extant is one made in 1729–1730 of silver-gilt (gilded silver), in rococo style, by Henry-Nicolas Cousinet. It is part of a complete tea, coffee, and chocolate service of well over twenty pieces, including its re-heater, in a fitted case suitable for travel, called a *nécessaire de voyage* (a portable case holding implements to make tea, coffee, and chocolate). This case was given by King Louis XV to his queen, Maria Leczinska, on the occasion of the birth of their son, the Dauphin, and is in the permanent collection of the Musée du Louvre in Paris. The spout and feet of the *chocolatière*'s stand are in the form of dolphins, a device used to represent the Dauphin. A bouquet of seven small roses crowns its pivoting finial. A subsequent owner removed the royal arms in two ovals (Fig. 12.2). The service includes Meissen, Chinese, and Japanese porcelain cups, as it was created before Sèvres porcelain was being made. The authors of *French Master Goldsmiths and Silversmiths* have stated: "It is one of the extremely rare examples of Paris silverware of royal origin from the time of Louis XV that has survived" [52].

Elaborate *chocolatières* were made not only in Paris, but also in cities such as Strasbourg and

FIGURE 12.2. *Chocolatière*, Louis XV silver-gilt rococo style, manufactured by Henry-Nicolas Cousinet; dated ca. 1729–1730. *Source*: © Réunion des Musées Nationaux / Art Resource, New York. (Used with permission.) (See color insert.)

Bordeaux. A second rococo example, with asymmetrical design elements, is a silver-gilt *chocolatière* from Strasbourg (which was annexed to France in 1678), by Jean-Jacques Ehrlen, dated variously to the period between 1736 and 1750. It bears the arms of the Landgrave (Count) of Hesse Darmstadt and his wife, Countess Marie-Louise Albertine de Leiningen-Dagsburg-Heidesheim, and is adorned with a flower bud or pistil finial. There is a matching coffeepot (*cafétière*), similar except for the base, which does not have three feet as seen on the *chocolatière*. Both pots are also in the collection of the Musée du Louvre. They were originally part of the Countess's *nécessaire de voyage* of 29 pieces, which became separated at an auction at the Hôtel Drouot in Paris in 1963. The rest of the boxed set is now in the Toledo Museum of Art.

Two additional silver *chocolatières*, identifiable as such by their pear-shaped bodies—the truncated cone (*cône tronqué*), to use the words of Lémery—are now in the collections of New York's Metropolitan Museum of Art. Both have marks identifying the silversmiths who crafted them. One has a spirally fluted pear-shaped body and was made by François-Thomas Germain in Paris in 1765–1766. Both stand on three feet, have a lid with a central opening for a stirrer, a thumb-piece for ease in opening the cover, a wooden handle screwed into a socket, and a prominent spout, with several hallmarks. Pierre Vallières in Paris made the second one in 1781. The two *chocolatières* resemble the copper example from 1750–1770, but are more gracefully proportioned and have more decorative elements (Figs. 12.3 and 12.4). Although these and most marked *chocolatières* were made by men, there were some made by women, such as an example by the widow Veuve Louis Waustrud who assumed her late husband's profession in Valenciennes around 1745 [53].

An exceptional order for sixteen silver *chocolatières* in different sizes ranging from two to ten cups was part of the famous Orloff Service, originally ordered by Catherine the Great of Russia in 1770 as a present for Count Gregorii Orlov (Orloff). The entire dining service of approximately 3000 pieces, one of the most enormous ever ordered, took over three years to complete. The contract that Catherine the Great signed on June 14, 1770 stipulated, among other terms, "that the pieces would be made of the usual standard of silver (. . . slightly higher than sterling) and would be marked with the marks of Paris—the prestige of Parisian silver outside France in the 18th century was so great the marks themselves became a symbol of fashion, the guarantee of craftsmanship" [54]. In September 1771, the Tsarina placed a supplementary order that included *"réchauds, cafétières, chocolatières et pots à lait"* (re-heaters, coffeepots, chocolate pots and milk pitchers) [55].

The sixteen *chocolatières*, by Jacques-Nicolas Roettiers of Paris, have a tapering cylindrical form, in neoclassical style, with the Russian Imperial Coat of

FIGURE 12.3. *Chocolatière*, and *cafetière*, silver-gilt, asymmetrical form, manufactured by Jean-Jacques Ehrlen, Strasbourg, France; dated ca. 1736–1750. *Source*: Courtesy of the Musée du Louvre, Paris, France. © Réunion des Musées Nationaux/Art Resource, New York. (Used with permission.) (See color insert.)

FIGURE 12.4. *Chocolatière*, silver three-footed, pear-shaped form, manufactured by Pierre Vallières, Paris, France; dated 1781. *Source*: Accession Number 28.156. © The Metropolitan Museum of Art, New York. (Used with permission.)

FIGURE 12.5. *Chocolatière*, silver, commissioned by Catherine the Great of Russia as a gift for Count Gregorii Orlov. Manufactured by Jacques-Nicolas Roettiers, Paris, France; dated early 1770s. *Source*: Christie's Sales Catalogue, May 25, 1993, Object Number 296. Courtesy of Christie's Auction House, New York. (Used with permission.) (See color insert.)

Arms engraved on one side and a berry motif finial on the small pivoting lid (Fig. 12.5). This appears to be the single largest royal order of French *chocolatières*. The Orloff service was largely dispersed after 1784, the year after Orloff's death. Parts of it returned to the Imperial collections, where the Russian Imperial Coat of Arms was added to the *chocolatières*, and many objects disappeared between subsequent inventories. Some of these *chocolatières* were sold by the Soviet government by 1931, were subsequently dispersed through auction house and other sales, and are now in private hands or museum collections.

Chocolatières in Porcelain

Production of *chocolatières* in hard-paste porcelain (*pâte dure*) appeared nearly seventy years after the introduction of those in silver. Curiously, the July 1689 *Mercure Galant* account of the Duc d'Orléans's lottery, lists a "porcelain" *chocolatière* as having been given away [56, 57], but it was probably made of faience, which was produced at a faience manufactory at Saint-Cloud beginning in 1666 [58, 59]. The Saint-Cloud factory began producing soft-paste porcelain (*pâte tendre*) only in the early 1690s, thus after the 1689 lottery, with an official patent to make porcelain in 1702 [60]. There

is little evidence, however, that many *chocolatières* were produced there. A rare example pictured in Winifred Williams's *Eighteenth Century French Porcelain, Sales Exhibition Catalogue* is a Saint-Cloud chocolate pot and cover in "white glazed with applied *prunus* sprigs" from approximately 1730 [61]. Chinese porcelain had been prized for centuries, but hard-paste porcelain had not been developed yet in France and the word was used earlier to describe faience. A ceramics factory, established in 1738 at Vincennes, became the Royal Manufactory in 1752 and, in 1756, moved to Sèvres. After 1767, the Royal Manufactory imitated the hard-paste porcelain developed in early 18th century Saxony and branched out into its own forms, decorative styles, and colors. Sèvres produced porcelain to order for the royal family, and porcelain *chocolatières* joined those in silver as high prestige items.

The middle of the 18th century saw the establishment of France as a center of luxury production in Europe and an increase in the production of upscale *chocolatières* there. In addition to limitations on demand resulting from the high cost of silver and, later, porcelain *chocolatières*, the state imposed or at least sanctioned limitations on their production. Guilds, supported by the state, set standards for silversmiths, and royal patents were later required for porcelain factories, most of which were owned either by royalty or aristocrats.

Most early porcelain *chocolatières* were made as individual specialty items or in small services with related cups, saucers, and trays, called *déjeuners*. From 1743 to 1749, the "Rue de Charenton" factory in Paris, and from 1749 through 1788, its successor, "Pont-aux-Choux," produced high-quality ceramics. An example was a *chocolatière* of *faïence fine*, or fine earthenware, that used finer quality clays and glazes different from the then standard faience, and was similar to English creamware, called *façon angleterre* (Fig. 12.6) [62]. Presently in the collection of the Musée National de Céramique at Sèvres, this *chocolatière* is composed of several parts: a pear-shaped pot whose bulbous lower portion accommodates the action of a *moulinet*, as in metal *chocolatières*; an unusual divided cover with two hinges; an unusual *moulinet* made of the same faience with a smooth and pointed cone-shaped mass at the end, in contrast to the more common wooden *moulinets* with carved and faceted ends; and a two-piece container to hold the *chocolatière*. The container could serve as a *bain-marie* (double boiler) to heat or keep warm the contents of the chocolate pot [63]. Geneviève Le Duc describes what appears to be another example of the same model *chocolatière* in a private collection [64]. A variation on the *moulinet* is an example with a wooden handle but having a blue and white block-like ceramic end, somewhat like a reamer for citrus fruits, in a private collection [65].

Of the ceramic *chocolatières* illustrated in the present chapter, only the one made of *faïence fine* is

FIGURE 12.6. *Chocolatière, faïence fine* (earthenware), manufactured by the Rue de Charenton factory, Paris, France; dated ca. 1750–1760. *Source*: Accession Number MNC 4007. © Réunion des Musées Nationaux/Art Resource, New York. (Used with permission.) (See color insert.)

pear-shaped. The others are based on a cylindrical, or tapering cylindrical, form and, due to their relative fragility, do not have the bulbous lower portion found in metal *chocolatières*, to accommodate the frothing action of the *moulinet*. It may be that these porcelain *chocolatières* were used for serving chocolate drinks with only mild stirring, after they had been prepared in a metal pot, such as a saucepot, in the kitchen.

By the 1760s, France was producing hard-paste porcelain *chocolatières*. An inventory of 1762 shows the Royal Porcelain Manufactory at Sèvres having produced goblets for chocolate with saucers, possibly for sale to Madame de Pompadour [66]. France's reputation in England as a producer of luxury goods was evidenced by the purchases of luxury products made in Paris by affluent English travelers, such as Horace Walpole, an influential arbiter of English fashion, also said to be the father of the Gothic novel. The Reverend William Cole, a close friend of Walpole's who accompanied him on a trip in 1765, wrote of the two of them being entranced by Sèvres porcelain in the Parisian shops. At Madame Dulac's shop on the Rue Saint-Honoré—then and now a fashionable shopping street in Paris—Walpole "bought a chocolate cup and saucer for his nephew, the Bishop of Exeter, which cost between three and four guineas," less expensive than some of the coffee cups he had purchased elsewhere in Paris [67].

The prestige of the Versailles court under Louis XIV and his successors in the 18th century influenced taste in many of the arts throughout court and aristocratic circles elsewhere in Europe. Walpole is but one example of the desire among those who could afford it to acquire French dining services including silverware and Sèvres porcelain objects of the highest quality. Today, many of these objects are in royal or state col-

lections. In addition to Catherine the Great, other Russian aristocrats sought out French porcelain. Prince Nikolai Borisovich Yusupov, in particular, formed a collection that was "unique in its quantity, variety and elaborate selection" [68]. During frequent trips to Paris between 1782 and 1809, Prince Yusupov purchased and ordered Sèvres porcelain, which was seldom sold in Russia at the time. He personally was acquainted with and received presents of Sèvres porcelain from Louis XVI and Marie Antoinette, as well as, later, from Napoleon. Sèvres porcelain, including *chocolatières*, thereby appeared at the apex of Tsarist society in late 18th and early 19th century Russia.

Following the Russian revolution of October 1917, many of the museum-quality collections and palaces of the wealthiest families were nationalized. Some works of art and decorative art were dispersed, with many going, for example, to the Finance Ministry. Gradually, porcelain objects were transferred to the Hermitage's Porcelain Gallery. Two outstanding Sèvres *chocolatières*, purchased by Russian aristocrats while traveling in France, are now in the Hermitage Museum in Saint Petersburg. One is from the collection of Prince Yusupov (Fig. 12.7), which had been owned by the son of Carl Fabergé, and the other is from the collection of Count Pavel-Petrovitch Shuvalov (Chouvaloff), who inherited it from Countess Katarina Petrovna Shuvalova, who formed the collection on her frequent trips to Europe (Fig. 12.8).

These two cylindrical-shaped *chocolatières*, and a third in the Louvre of the same model, have a high level of decoration. *Chocolatières* of the same basic form but with simpler decoration have been dispersed and are seen in auction house sales records from time to time. The two in the Hermitage have an interesting model of a lid with a pierced hole. On this model the

FIGURE 12.7. *Chocolatière*, Sèvres porcelain, owned by Prince Yusupov. Chinoiserie decor (black background to imitate lacquer work) inspired by Oriental sources; dated 1781. *Source*: Accession Number 20549. Courtesy of the State Hermitage Museum, Saint Petersburg, Russia. (Used with permission.) (See color insert.)

FIGURE 12.9. Detail of *chocolatière* in Figure 12.7: close-up of lid top. *Source*: Accession Number 20549. Courtesy of the State Hermitage Museum, Saint Petersburg, Russia. (Used with permission.) (See color insert.)

FIGURE 12.8. *Chocolatière*, Sèvres porcelain with white background. Part of a breakfast service (*déjeuner*) belonging to Count Shuvalov; dated to the 1780s. *Source*: Accession Number 24984. Courtesy of the State Hermitage Museum, Saint Petersburg, Russia. (Used with permission.) (See color insert.)

FIGURE 12.10. Detail of *chocolatière* in Figure 12.7: close-up of lid bottom. *Source*: Accession Number 20549. Courtesy of the State Hermitage Museum, Saint Petersburg, Russia. (Used with permission.) (See color insert.)

porcelain lid pulls out and has a silver, shaped handle that can be positioned upright, or can fold to the side when the *moulinet* is inserted. The single visible hole has a silver hinged cap. The underside of the cover has several metal parts, including two or three prongs that hold the cover on the pot while chocolate is being poured, and three fasteners, attached to a metal disk similar to a washer, with which the silver handle and cap are anchored (Figs. 12.9–12.11).

The first *chocolatière* discussed here, made in 1781 of hard-paste porcelain, is unique [69]. Part of a table service, now dispersed, it has a rare black ground, in imitation of lacquer, with gold decoration and with *chinoiserie* (European decoration inspired by Oriental

FIGURE 12.11. Detail of *chocolatière* in Figure 12.7: close-up of *chocolatière* bottom. *Source:* Accession Number 20549. Courtesy of the State Hermitage Museum, Saint Petersburg, Russia. (Used with permission.) (See color insert.)

FIGURE 12.12. *Chocolatière*, Sèvres porcelain off-white with gold decoration, manufactured for Madame Victoire (daughter of King Louis XV) in 1786. *Source:* © Réunion des Musées Nationaux/Art Resource, New York. (Used with permission.) (See color insert.)

sources, particularly Chinese), scenes in various shades of gold, including reddish-gold, as well as silver or platinum [70]. Hard-paste porcelain allowed a greater range of decorating methods and this *chocolatière* was one of the earliest porcelain objects made with this decorative scheme [71]. Motifs on *chocolatières*, as on other decorative art objects, reflect the prevailing styles of their times. *Chinoiserie* scenes, and black lacquer effects, as on this *chocolatière*, were used extensively on textiles and furniture by artists and artisans producing at the highest level. Sources for scenes included engravings after paintings by well-known artists, such as François Boucher and Jean Pillement, the latter also a designer, which were interpreted by the porcelain painters at Sèvres [72].

The other Sèvres *chocolatière* from the Hermitage Museum is part of a breakfast service, a *déjeuner*, that includes a tray, cream pitcher, sugar bowl, and cup and saucer. It has no date mark but is thought to have been made in the 1780s. It is decorated with sprigs of pink, lilac, red, and gold flowers as well as bands of blue and lilac edging on a white ground. Unlike many other *chocolatière* forms, the lid does not cover the spout. Both of the above *chocolatières* have the Sèvres trademark on the bottom. Marking objects with trademarks (for which a precedent existed in the markings of objects from some faience factories) began early on in the factories of the completely new industry of porcelain production in Europe. The black ground pot also has a year date mark for 1781 and the symbol of painter and gilder Louis-François L'Écot (Lécot), a master of *chinoiserie* [73, 74].

One of the more interesting *chocolatières*, usually on exhibition in the permanent collection, is displayed at the Musée National Adrien-Bubouché, a major ceramics museum in Limoges, France (Fig. 12.12) [75]. Made in 1786, this *chocolatière* is of Sèvres porcelain, with gold décor of strands of garlands on an off-white ground, with a wooden handle. A larger hole in the lid is for a *moulinet*, a smaller one for the cover fastener. Its gilder was Étienne-Henry Le Guay. In an unusual set of circumstances, Pierre Verlet, Chief Curator of Decorative Arts at the Louvre, found the original drawings for it in the Sèvres Royal Porcelain Manufactory archives and was able to link the drawings to the object. It is rare to have detailed documentation that can be linked to a specific *chocolatière*; written records are usually insufficiently precise to be tied to a specific object. In my view, the disposition of the *chocolatière* in the drawing that seems to be a cross-sectional view shows the spout or cup and the handle separated by 180 degrees, whereas the porcelain *chocolatière* itself has the spout and handle separated by 90 degrees, so they do not seem to match up exactly. After determining that the object corresponded to the drawings, Verlet showed that this *chocolatière* had been made for Madame Victoire, a daughter of Louis XV [76]. Madame Victoire's coat of arms applied on the *chocolatière* helped to identify the object as hers. Parts of the *chocolatière*, the funnel and a small chain attached to the handle and ending in a small lid that fitted the rounded spout, are missing. Verlet called this pot a coffeepot, as one of the drawings refers to *café* [77]. Illustrating some of the complexities involved in identifying these antique objects, some contemporary specialists, however, deem it to be a *chocolatière* because of the hole in its lid [78], while others are skeptical of its purpose being for either chocolate or coffee due to the unusual cup. Madame Victoire ordered three such pots, no others were made, and only one known example has survived.

Chocolatières and Portraiture, 18th Century France

Eighteenth century portraiture confirmed the upper class use of the *chocolatière* and related cups, although given the cautionary accounts of Blégny and Massaliot, it is often difficult to be certain. Paintings depicted scenes in which *chocolatières* were featured and *chocolatières* themselves were decorated with a variety of landscapes and related scenes. François Boucher's 1739 *Déjeuner*, now in the Musée du Louvre in Paris, shows what appears to be a chocolatière (Fig. 12.13). A chocolate cup is depicted in Jean-Etienne Liotard's *La Belle Chocolatière* of 1744–1745, a pastel on parchment, now in the Dresden Gemäldegalerie. This picture, of a young woman serving chocolate, was later used as the emblem for Baker's Chocolate Company in the United States. Jean-Baptiste Charpentier's *La Tasse de chocolat* (*The Cup of Chocolate*), now at the Versailles Musée National du Château et des Trianons, was painted in 1768. In the painting, well-dressed women and men are shown together with cups and what appear to be saucers. The inclusion of both sexes hints that the scene may have been later in the day rather than breakfast, as often women took breakfast in their private apartments. Frequently, the only indication that cups in the paintings are for chocolate rather than some other beverage is the title of the painting, as in the case of Charpentier's tableau (Fig. 12.14) [79].

Another portrait featuring chocolate is that of Madame du Barry at her dressing table by Jean-Baptiste André Gautier d'Agoty, completed sometime in the 1770s or early 1780s and now also in the Musée National du Château et des Trianons at Versailles. A young black servant known as Zamore is seen serving chocolate to Madame du Barry. In paintings of this period, chocolate is sometimes shown as being presented by a servant, often a young black boy.

FIGURE 12.13. *Déjeuner* (*Breakfast*), by François Boucher: informal scene of drinking chocolate with children; dated 1739. *Source*: © Réunion des Musées Nationaux / Art Resource, New York. (Used with permission.) (See color insert.)

FIGURE 12.14. *La Tasse de chocolat* (*The Cup of Chocolate*), by Jean-Baptiste Charpentier: interior scene of men and women drinking chocolate; dated 1768. *Source*: Accession Number MV 7716. © Réunion des Musées Nationaux / Art Resource, New York. (Used with permission.) (See color insert.)

Chocolate was frequently served to women in their private chambers and as Madame du Barry is shown at her dressing table, the scene is probably one of chocolate being brought as part of breakfast. The relatively small cup appears to have one handle (Fig. 12.15).

A hand-colored engraving "dressed" with pieces of fabric, dating from the period between 1690 and 1710, designed by Robert Bonnart, and published by Nicolas Bonnart, portrays a delightful scene. The work, *Un cavalier, et une dame beuvant du chocolat* (*A gentleman and a lady drinking chocolate*), depicts the couple seated on an outdoor terrace wearing elegant outfits and headgear (see Fig. 11.15) [80, 81]. The gentleman and lady are being served by a young black male servant, holding a tray on which appears to be a jug of water and two goblets, and a maid, who is rolling the *moulinet* between her hands to stir and froth the chocolate. The *chocolatière* is pear-shaped with three feet and has a hole in its cover for the *moulinet*. The tray, two cups (without handles) with saucers, and a spoon complete the *déjeuner* service.

Another group portrait is a detail from the oil painting by Nicholas Lancret, ca. 1740, titled *La Tasse de Thé* (*The Cup of Tea*) in The National Gallery, London. Peter Brown, however, in his *In Praise of Hot Liquors*, a 1995 exhibition catalogue from Fairfax House, York (England), interprets the beverage being offered in a spoon to a young girl to be not tea, but either chocolate or coffee. A servant is seen pouring the beverage from a silver, three-legged pear-shaped pot (Fig. 12.16). Is it a *chocolatière*?

Turning to the decorations on the *chocolatières* themselves, there are several interesting examples of

FIGURE 12.15. Color engraving of *Madame Du Barry à sa Toilette (Madame du Barry at her Dressing Table)*, by Jean-Baptiste André Gautier d'Agoty: Zamore serving chocolate to Madame du Barry; dated ca. 1769–1786. *Source*: Accession Number MV 8649; Inv.gravures6677. Courtesy of the Musée National du Château et des Trianons, Versailles, France. © Réunion des Musées Nationaux/Art Resource, New York. (Used with permission.) (See color insert.)

FIGURE 12.16. *La Tasse de Thé* (*The Cup of Tea*), by Nicholas Lancret: detail showing a family taking chocolate outdoors; dated ca. 1740. *Source*: © The National Gallery, London, England. (Used with permission.) (See color insert.)

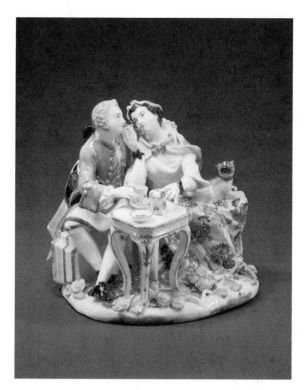

FIGURE 12.17. Couple drinking chocolate: hand-painted Meissen porcelain figure, by Johann Joachim Kändler; dated ca. 1744. *Source*: Accession Number 1982.60.326. © The Metropolitan Museum of Art, New York. (Used with permission.)

scenes painted on the pots. One otherwise undated 18th century Sèvres *chocolatière* shows, on the side opposite an ebony handle, an oval medallion with a gold frame surrounded by a garland of roses, containing the portrait of a woman holding a chocolatière, with its *moulinet*. The portrait and the garland are painted in a naturalistic style. The *chocolatière* itself is decorated in gold with laurel garlands, shells and palmette motifs on a white ground; its cover is pierced with a circular opening for a *moulinet* [82].

A German example is a charming three-dimensional figural scene of a couple drinking chocolate, modeled in porcelain by Johann Joachim Kändler, in approximately 1744, in Meissen in Saxony. A miniature *chocolatière* and two cups and saucers are depicted on the table (Fig. 12.17). Porcelain manufactories sprang up throughout Europe once knowledge of how to make porcelain was disseminated; Meissen was the most important competitor to the Sèvres factory in France. Meissen was the first to make true porcelain and Meissen *chocolatières* are dispersed around the world in private and public collections [83].

Another Sèvres *chocolatière*, dated to 1777, also is painted on the side opposite its handle. There is an oval reserve surrounded with a gold frame, portraying four figures, two women and two men, seated in a pastoral scene, drinking chocolate. The central figure holds a ceramic mug and a *chocolatière* is pictured set on the ground. The *chocolatière* itself is made of soft-paste porcelain rather than hard-paste porcelain. Soft-paste was an "imitation of true porcelain" made from clay and powdered glass, whereas "hard-paste or true porcelain is made from kaolin (white china clay) and a feldspathic rock," the process developed in Meissen in the early 18th century [84]. Due to greater breakage during the firing process in the kiln, soft-paste is far less common than hard-paste, so this *chocolatière* is rarer than hard-paste *chocolatières*.

Detailed inspection of this *chocolatière* reveals a primarily deep blue ground, richly gilded. The separate lid, gilded and painted with a garland of various flowers on a white ground, has a central hole for a *moulinet*. A separate blue and gold cover can be placed over the hole. Marks on the bottom of the *chocolatière* reveal that the painter was Antoine Caton and the gilder was Michel-Barnabé Chauvaux, the Elder. The object now is part of the Royal Collection in London. In the years following the French Revolution, The Prince Regent (later King George IV) of England was able to purchase and form one of the most important collections of Sèvres porcelain, including services with French royal provenance.

Chocolatières are found occasionally in 18th century French literature as well as paintings. In *Paul et Virginie*, a novel critiquing French aristocratic sophistication—as well as slavery—written just before the French Revolution, the botanist Jacques-Henri Bernardin de Saint-Pierre has one of his protagonists roll a small branch between his hands as "one rolls a *moulinet* with which one wishes to make chocolate frothy" [85].

Decline and Renewal: The French Revolution and After

The French Revolution swept away the monarchy, the aristocracy, and the guilds. Marie Antoinette, hoping to flee the Revolution with her family in 1791, ordered a *nécessaire de voyage,* which contained a silver *chocolatière*. This service, made in 1787–1788, originally contained over one hundred items made of silver, crystal, porcelain, ivory, ebony, and steel. The *chocolatière,* like most of the silver items, was made by Jean-Pierre Charpenat and was marked with the Queen's monogram "MA," as are many of the other pieces. The form of her *chocolatière* is cylindrical, in neoclassical style; the *moulinet* is of ebony and ivory. This *chocolatière* had an interesting travel history after Marie Antoinette's death. It bears an official city customs mark indicating that it was imported to Milan in 1794, a year after she died. According to tradition, it was subsequently acquired by French General Jean-Baptiste Cervoni during the 1796–1797 French military campaign in northern Italy. Cervoni is then said to have sold it, possibly to Napoleon, who gave it to his wife, Josephine. In any event, a receipt dated December 21, 1804 shows it to have been sold by an agent of Josephine to Felice Origoni, an Italian merchant, in whose family the *chocolatière* remained until 1942. Ultimately, in 1955, the object was acquired for the permanent collection of the Musée du Louvre, where it is today (Fig. 12.18) [86].

FIGURE 12.18. *Chocolatière*, silver, owned by Marie Antoinette and manufactured by Jean-Pierre Charpenat, Paris; dated 1787–1788. *Source*: Museum Accession Number OA 9594. © Réunion des Musées Nationaux/Art Resource, New York. (Used with permission.) (See color insert.)

FIGURE 12.19. *Verseuse* (pitcher), silver, owned by Napoleon, converted into a *chocolatière*. Manufactured by Martin-Guillaume Biennais; dated 1798–1809. *Source*: Museum Accession Number: OA 10270. © Réunion des Musées Nationaux/Art Resource, New York. (Used with permission.) (See color insert.)

Generic vessels or pitchers used for pouring liquids, called *verseuses*, sometimes were converted into *chocolatières*. One example is a silver-gilt *chocolatière* with ebony handle crafted by Napoleon's silversmith, Martin-Guillaume Biennais, with hallmarks certifying the date between 1798 and 1809. This item now in the Louvre in Paris may have been a pitcher or a coffeepot (Fig. 12.19). Small, serving only one to two cups, this vessel is important because it is decorated with the imperial arms of Napoleon in his title of King of Italy, which is extremely rare for any piece of silver or gold (Fig. 12.20). Other decorations on the piece include two profiles in the antique style, one a woman's head, the other a man's head, bordered by two rows of pearls and laurel leaves, a device favored by Napoleon, recalling the laurel wreaths worn by ancient Roman emperors. An eagle is posed on a rosette on the pivoting small lid, but the addition of this redone lid and the hole for the *moulinet* are not believed to be the work of Biennais because the relatively poor workmanship is considered to be beneath the quality of his work [87].

Like the social order in postrevolutionary France, the world of chocolate was changing in the 19th century. In 1828, in Amsterdam, Coenraad Johannes Van Houten invented a process to more efficiently separate cocoa powder from cocoa butter, intensifying the industrialized production of chocolate, already underway. By defatting cacao, Van Houten's process produced a liquid that no longer required constant stirring in a *chocolatière* to make it palatable [88], making the specialized pots less necessary. In 1875, Daniel Peter developed milk chocolate in Switzerland and, four years later, the first milk chocolate bar was produced [89]. The effects of technology shifted chocolate consumption in large measure from chocolate as a hot beverage to solid chocolate in bar form. The chocolate product itself had

FIGURE 12.20. Close-up of the *chocolatière* in Figure 12.20 showing the Imperial Arms of Napoleon as King of Italy. *Source*: Museum Accession Number: OA 10270. © Réunion des Musées Nationaux/Art Resource, New York. (Used with permission.) (See color insert.)

changed as well as the custom of drinking it, and *choco-latières* became more dispensable.

In contrast, industrialization promoted the rise of the chocolate mold—which, together with the *chocolatière*, would become a collectible in the 20th century. Chocolate candies or bonbons existed in 18th century France, but they were rarely made from molds. The 1763 *L'Encyclopédie Diderot et d'Alembert* shows molds [90] but until the first half of the 19th century, with the development of enhanced technology including Van Houten's process, chocolate molds for candies were exceptional. Used to shape *chocolat ouvragé* (worked or sculpted hard chocolate), molds became more prominent beginning with the 1855 Paris Exposition (see Chapter 16) [91].

Chocolatières continued to be produced in the middle and later 19th century, but their function shifted to curiosities and collectibles, sometimes as parts of sets. A partial coffee or *déjeuner* service, made of Sèvres porcelain and decorated by Charles Develly in 1836, had as its theme the cultivation and harvesting of cacao. Seven scenes painted on different pieces in the service depict a somewhat romanticized view of cacao production. A scene on the tray shows three seated women separating the cacao pods from the beans while another woman moves a *moulinet,* in a large pot, rolling it between her hands (Figs. 12.21–23). This partial coffee service, now in the collection of the Metropolitan Museum of Art in New York, originally may have included a *chocolatière* [92].

The *chocolatière* model also became an interesting one to use for exhibition pieces at international expositions, where artisans in many media created masterpieces that would never, or perhaps could not ever, be actually used. Both silver and porcelain *chocolatières* were produced; examples of each in art nouveau

style from the beginning of the 20th century are at the Musée des Arts Décoratifs in Paris. A silver and ivory art nouveau *chocolatière* created by designer Lucien Bonvallet and silversmith Ernest Cardeilhac was made by 1900 and exhibited in both the 1900 Paris Exposition and the St. Louis World's Fair (The Louisiana Purchase International Exposition) four years later. The form of this *chocolatière* is primarily cylindrical, curving inward at its middle, with a gently curved wooden handle. It sits on a round base without the three feet found on earlier *chocolatières*. Its interior is silver-gilt and its exterior is engraved with undulating wave-like designs. Its unusual *moulinet* and matching small, pivoting cover are both made of turned and carved ivory highlighted in chestnut brown, each with flower pistil finials. Cardeilhac made both identical and similar *chocolatières,* of which four other examples are now in museums in Amsterdam, Berlin, Cologne, and Copenhagen, again highlighting the changing role of *chocolatières* as they became collectibles.

A porcelain art nouveau model designed by Georges de Feure and manufactured by the Gérard, Dufraisseix, and Abbot (GDA) Company in Limoges, France, was made sometime between 1900 and 1902. The object is cylindrical in form, modified toward the base by four bulbous nodes formed by four stylized flowers tinted pale rose and joined together with undulating pale green leaves on a white ground. The bulbous lower portion suggests the pear shape of metal *chocolatières,* but could not accommodate a *moulinet,* and there is no hole in the cover where one could be inserted. A curved handle is placed at 180 degrees from the spout, rather than the 90 degrees often found on both ceramic and metal *chocolatières*. This example entered the collection of the Musée des Arts Decoratifs in 1908, becoming a collectible less than ten years after

FIGURE 12.21. Painting on Sèvres porcelain tray showing preparation of hot chocolate, by Charles Develly; dated 1836. *Source*: Accession Number 1986.281.1ab-4. © (2000–2006) The Metropolitan Museum of Art, New York. (Used with permission.) (See color insert.)

FIGURE 12.22. Detail of Figure 12.21. Painting showing three seated women separating beans from cacao pods and another woman preparing hot chocolate. *Source*: Accession Number 1986.281.1ab-4. © (2000–2006) The Metropolitan Museum of Art, New York. (Used with permission.) (See color insert.)

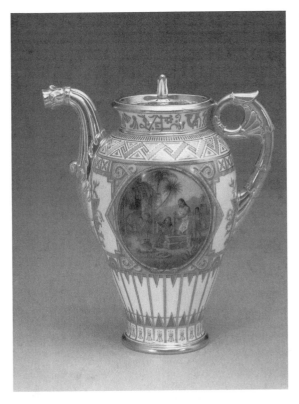

FIGURE 12.23. Coffeepot, one of four pieces from a partial coffee set made at the Sevres Manufactory, 1836. New York The Metropolitan Museum of Art, *Source*: Accession Number 1986.281.1ab-4. © (2000–2006) (Used with permission.) (See color insert.)

it was made. Factory, painters', and gilders' marks, which served business control purposes in preceding centuries, became increasingly important to collectors and museum curators who sought objects created and decorated by specific artisans, which also determined the current market demand and market value of each object.

Conclusion

Even if no longer used to make chocolate drinks, *choco-latières* have continued to be the subjects of paintings, however complex their designation and even if they were no longer depicted in their original uses. In 1902, Henri Matisse painted an oil on canvas entitled *Bouquet of Flowers in a Chocolate Pot* [93]. Matisse, short on funds, sold it soon thereafter to the art dealer Ambroise Vollard who, shortly before his death in 1939, sold it to Pablo Picasso. Curiously, the painting appeared as *Still Life Silver Coffee Pot* in a 1904 exhibit of Matisse's paintings in Vollard's Paris gallery. Following Picasso's acquisition of the painting in 1939, Matisse wrote to him about it, saying that he had confused it with another of his paintings and asked Picasso to describe it for him. Picasso complied, sending Matisse a sketch and a photograph of the painting . . . it now is identified as *Bouquet of Flowers in a Chocolate Pot* and is located in the permanent collection of the Musée Picasso in Paris [94].

After joining museum and private collections, the surviving French *chocolatières* now live the cosseted life of museum objects: preserved in an appropriate environment, conserved and/or restored as necessary, researched by curators and scholars, occasionally traveling to other museums around the world for exhibitions, and spending time on display as well as in storage. They continue to delight with their beauty and to inform us about the history and culture that produced them.

Acknowledgments

I wish to thank the following who assisted me during the preparation of this chapter: Frédéric Ballon, Director and Professor, Christie's Education, Paris; Sir Geoffrey de Bellaigue, Surveyor Emeritus of the Queen's Works of Art and Director of the Royal Collection, London; Chantal and Philippe du Chouchet, authors and collectors, Versailles, France; Aileen Dawson, Curator, Department of Prehistory and Europe, The British Museum, London; Bernard Dragesco and Didier Cramoisan, dealers, Galerie Dragesco-Cramoisan, Paris; Antoinette Faÿ-Hallé, Conservateur général du Patrimoine et Directeur du Musée National de Céramique de Sèvres, France; Deborah Gage (descendent of Thomas Gage), dealer, French Porcelain Society, London; Bertram M. Gordon, Professor of History, Mills College, Oakland, California; Joanna Gwilt, Royal Collection, French Porcelain Society, London; Philip and Mary Hyman, historians, Paris; Thierry de Lachaise, Directeur, Département Orfèvrerie, Sotheby's, Paris; Ulrich Leben, Associate Curator, Waddesdon Manor, The Rothschild Collection, Buckinghamshire, England; Raphaël Masson, Conservateur du Patrimoine, Centre de Recherche du Château de Versailles, France; Margaret Connors McQuade, Curator of Ceramics, Hispanic Society of America, New York; Chantal Meslin-Perrier, Conservateur en Chef du Musée, Musée National de Porcelaine Adrien Dubouché, Limoges, France; Jeffrey Munger, Associate Curator, European Sculpture and Decorative Arts, The Metropolitan Museum of Art, New York; David Peters, author and collector, London; Tamara Préaud, archivist, Manufacture Nationale de Sèvres, France; Letitia Roberts, independent scholar, formerly Senior Vice President and Director of European Ceramics and Chinese Export Porcelain, Sotheby's, New York; Marie-Laure de Rochebrune, Conservateur, Départment des Objets d'Art, Musée du Louvre, Paris; Bertrand Rondot, Conservateur, Département XVIIe-XVIIIe siècles, Musée des Arts Décoratifs, Paris; William R. Sargent, Curator of Asian Export Art, Peabody Essex Museum, Salem, Massachusetts; Laura Stalker, Avery Head of Collection Development and Reader Services, The Huntington Library, San Marino, California; Jan Vilensky, Curator of French porcelain, State Hermitage Museum, Saint Petersburg, Russia; John Whitehead, author and dealer, London; Ghenete Zelleke, Samuel and M. Patricia Grober Curator of European Decorative Arts, The Art Institute of Chicago.

References

1. Trésor de la Langue Française Informatisée. Available at http://atilf.atilf.fr/dendien/scripts/tlfiv5/advanced.exe?8;s=3826981095. (Accessed February 10, 2007.)

2. Savage, G., and Newman, H. *An Illustrated Dictionary of Ceramics.* London, England: Thames and Hudson, 1992 (original edition 1974); pp. 60 and 94.

3. Interview with Chantal du Chouchet, Versailles, France, July 23, 2005.

4. The Albuquerque Museum of Art and History, Albuquerque, New Mexico.

5. Coe, S. D., and Coe, M. D. *The True History of Chocolate.* London, England: Thames and Hudson, 1996; p. 160.

6. Coe, S. D., and Coe, M. D. *The True History of Chocolate.* London, England: Thames and Hudson, 1996; pp. 160–161.

7. Interview with Philippe du Chouchet, Versailles, France, July 23, 2005.

8. Interview with Bertrand Rondot, Conservateur (Curator), Département XVIIe-XVIIIe siècles, Musée des Arts Décoratifs, Paris, July 5, 2006.

9. "Poinçons" in Trésor de la Langue Française Informatisée. Available at http://atilf.atilf.fr/dendien/scripts/tlfiv5/advanced.exe?8;s=3826981095. (Accessed February 10, 2007.)

10. Guillemé-Brulon, D. *La faïence fine française, 1750–1867.* Paris, France: C. Massin, 1995; p. 13.

11. Moreau, R. *Du chocolate: Discours Curieux divisé en quatre parties. Par Antoine Colmenero de Ledesma Medecin & Chirurgien de la ville de Ecija de l'Andalouzie. Traduit d'Espagnol en François sur l'impression faite à Madrid l'an 1631, & éclaircy de quelques annotations. Par René Moreau Professeur du Roy en Medicin à Paris. Plus est ajouté un Dialogue touchant le même chocolate.* Paris, France: Sébastien Cramoisy, 1643; pp. 12 and 52.

12. Sévigné, M. de Rabutin-Chantal. *Lettres de Madame de Sévigné, de sa famille et de ses amis recueillies et annotées par M. Monmerqué,* 14 Volumes. Paris, France: L. Hachette, 1862–1868; Sévigné, to Madame de Grignan, II, No. 133, p. 60.

13. Richelet, P. *Dictionnaire françois: contenant les mots et le choses, plusieurs nouvelles remarques sur la langue françoise.* Geneva, Switzerland: Slatkine Reprints, 1994 (original, Geneva, Switzerland: J. H. Widerhold, 1680); I, p. 136.

14. Dufour, S. *Traitéz Nouveaux & Curieux du café, du thé et du chocolat.* Lyon, France: Jean Girin and B. Rivière, 1685; pp. 363–364.

15. Blégny, N. de. *Le bon usage du thé, du caffé et du chocolat pour la préservation et pour la guérison des maladies.* Paris, France: E. Michallet, 1687; p. 265.

16. Blégny, N. de. *Le bon usage du thé, du caffé et du chocolat pour la préservation et pour la guérison des maladies,* Paris, France: E. Michallet, 1687; pp. 266–267.

17. Blégny, N. de. *Le bon usage du thé, du caffé et du chocolat pour la préservation et pour la guérison des maladies*, Paris, France: E. Michallet, 1687; pp. 269–270.

18. Blégny, N. de. *Le bon usage du thé, du caffé et du chocolat pour la préservation et pour la guérison des maladies*, Paris, France: E. Michallet, 1687; pp. 270–271.

19. Blégny, N. de. *Le bon usage du thé, du caffé et du chocolat pour la préservation et pour la guérison des maladies*, Paris, France: E. Michallet, 1687; pp. 272–273.

20. Massaliot, F. *Nouvelle Instruction pour les Confitures, les Liqueurs, et les Fruits*. Amsterdam, Holland: Aux Depens de la Compagnie, 1734; Part 3, p. 226.

21. Le Duc, G. *Porcelain tendre de Chantilly au XVIIIe siècle*. Paris, France: Hazan, 1996; p. 288.

22. Massaliot, F. *Nouvelle Instruction pour les Confitures, les Liqueurs, et les Fruits*. Amsterdam, Holland: Aux Depens de la Compagnie, 1734; Part 3, p. 226.

23. Gillier, J. *Le Cannameliste français, ou Nouvelle instruction pour ceux qui désirent d'apprendre l'office*. Nancy, France: Abel-Denis Cusson, 1751; p. 38.

24. Gillier, J. *Le Cannameliste français, ou Nouvelle instruction pour ceux qui désirent d'apprendre l'office*. Nancy, France: Abel-Denis Cusson, 1751; p. 38.

25. Lémery, L. *Traité des Alimens*, 3rd edition. Paris, France: Durand, 1755; Volume I, p. 102.

26. Lémery, L. *Traité des Alimens*, 3rd edition. Paris, France: Durand, 1755; Volume I, p. 103.

27. Buc'hoz, P.-J. *Dissertations sur l'utilité, et les bons et mauvais effets du Tabac, du Café, du Cacao et du Thé, ornées de Quatre planches en taille-douce*, 2nd edition. Paris, France: Chez l'Auteur, De Burc l'aîné, et la Veuve Tilliard et Fils, 1788; pp. 124–125.

28. Diderot, D. *Nouveau dictionnaire pour server de supplement aux Dictionnaires des sciences, des arts et des métiers*, 4 Volumes. Paris, France: Panckoucke, 1776–1777, Volume 2.

29. Interview with Madame Dany Pelong, Musée du Chocolat, Biarritz, France, July 14, 2005.

30. Havard, H. *Dictionnaire de l'ameublement et de la décoration depuis le XIIIe siècle jusqu'à nos jours*. Paris, France: Quantin, 1894; I, p. 816.

31. Gordon, B. M. Chocolate in France: the evolution of a luxury product. In: Grivetti, L. E., and Shapiro, H.-Y, editors. *Chocolate: History Culture, and Heritage*. Hoboken, NJ: Wiley, 2009; pp. 569–582.

32. Inventaire des biens et effets de feu Messire Salvat, Seigneur Vicomte Durturbie, du 9e janvier 1684. *Société des Sciences, Lettres et Arts de Bayonne*. Nouvelle Série, No. 11 (janvier–juin 1933).

33. Presens du Roy de Siam au Roy. *Mercure Galant*, July 1686, pp. 299–300 in Bibliothèque de l'Histoire de la Ville de Paris. Per 8° 4C (No. 131).

34. Franklin, A. *La Vie Privée d'autrefois. Le Café, le Thé et le Chocolat*. Paris, France: Plon, 1893; p. 170.

35. Masson, R. Conservateur du Patrimoine, Centre de Recherche du Château de Versailles. E-mail dated July 12, 2006.

36. Presens du Roy de Siam à Monseigneur le Dauphin. *Mercure Galant*, July 1686, p. 307 in Bibliothèque de l'Histoire de la Ville de Paris. Per 8° 4C (No. 131).

37. Masson, R. Conservateur du Patrimoine, Centre de Recherche du Château de Versailles. E-mail dated July 12, 2006.

38. Presens de la Princess Reyne à la Dauphine. *Mercure Galant*, July 1686, p. 311 in Bibliothèque de l'Histoire de la Ville de Paris. Per 8° 4C (No. 131).

39. Presens de la Princess Reyne à Monseigneur le Duc de Bourgogne. *Mercure Galant*, July 1686, pp. 315–316 in Bibliothèque de l'Histoire de la Ville de Paris. Per 8° 4C (No. 131).

40. Presens de Mr. Constance au Roy. *Mercure Galant*, July 1686, p. 318 in Bibliothèque de l'Histoire de la Ville de Paris. Per 8° 4C (No. 131).

41. Presens de Mr. Constance à Mr. le Marquis de Seignelay. *Mercure Galant*, July 1686, p. 323 in Bibliothèque de l'Histoire de la Ville de Paris. Per 8° 4C (No. 131).

42. Coe, S. D., and Coe, M. D. *The True History of Chocolate*, London, England: Thames and Hudson, 1996; p. 162.

43. Savage, G., and Newman, H. *An Illustrated Dictionary of Ceramics*. London, England: Thames and Hudson, 1992 (original edition 1974); p. 161.

44. Masson, R. Conservateur du Patrimoine, Centre de Recherche du Château de Versailles. E-mail dated July 12, 2006.

45. Interview with Bertrand Rondot, Conservateur (Curator), Département XVIIe-XVIIIe siècles, Musée des Arts Décoratifs, Paris, July 5, 2006.

46. Franklin, A. *La Vie Privée d'autrefois. Le Café, le Thé et le Chocolat*, Paris, France: Plon, 1893; p. 170.

47. Deitz, P. Antiques/chocolate pots brewed ingenuity. *The New York Times*, February 19, 1989, Section 2, p. 38, Column 1.

48. *Mercure Galant*. 1689, p. 180.

49. Franklin, A. *La Vie Privée d'autrefois. Le Café, le Thé et le Chocolat*, Paris, France: Plon, 1893; p. 170.

50. *French Master Goldsmiths and Silversmiths from the Seventeenth to the Nineteenth Century*. New York: French & European Publications, Inc. and The Connaissance des Arts Collection, 1966; p. 45.

51. Duhart, Frédéric. *Habiter et consommer à Bayonne au XVIIIe siècle, éléments d'une culture matérielle urbaine*. Paris, France: L'Harmattan, 2001, p. 170.

52. *French Master Goldsmiths and Silversmiths from the Seventeenth to the Nineteenth Century*. New York: French & European Publications, Inc. and The Connaissance des Arts Collection, 1966; p. 132.

53. Interview with Thierry de Lachaise, Directeur, Département Orfèvrerie, Sotheby's, Paris, August 3, 2005.

54. Le Corbeiller, C. Grace and Favor. *The Metropolitan Museum of Art Bulletin* 1969; 27(6):289–298.

55. Le Corbeiller, C., Kuodriavceca, A., and Lopato, M. Le Service Orloff. In Versailles et les Tables Royales en Europe, XVIIème–XIXème siècles, *Exhibition Catalogue*. Paris, France: Editions de la Réunion des Musées Nationaux, 1993, p. 316.

56. *Mercure Galant*. 1689, p. 180.

57. Franklin, A. *La Vie Privée d'autrefois. Le Café, le Thé et la Chocolat*, Paris, France: Plon, 1893; p. 170.

58. Ledes, A. E. French soft-paste porcelain. Porcelain from Saint-Cloud, France. Antiques, August 1, 1999, Bard Graduate Center for Studies in the Decorative Arts, New York. Available at http://www.encyclopedia.com/doc/1G1-55487304.html. (Accessed February 11, 2007.)

59. Rondot, B., editor. *Exhibition Catalogue: Discovering the Secrets of Soft-Paste Porcelain at the Saint-Cloud Manufactory, ca. 1690–1766, July 15–Oct. 24, 1999*. New Haven, CT: Yale University Press, 1999.

60. Saint-Cloud Porcelain Manufactory. The Getty. Available at http://www.getty.edu/art/gettyguide/artMakerDetails?maker=1224. (Accessed February 11, 2007).

61. Williams, W. *Eighteenth Century French Porcelain, Sales Exhibition Catalogue "An exhibition of 18th century French Porcelain," London, 3rd–20th July 1978*, p. 21, Figure 64, "St. Cloud Chocolate Pot and Cover."

62. Guillemé-Brulon, D. *La faïence fine française, 1750–1867*, Paris, France: C. Massin, 1995; pp. 24–25.

63. Guillemé-Brulon, D. *La faïence fine française, 1750–1867*, Paris, France: C. Massin, 1995; p. 24.

64. Le Duc, G. *Catalogue: The International Ceramics Fair and Seminar*, "Parisian White Earth in the English Manner 1743–1749." London, England: n.p., 1998; p. 22.

65. Interview with Chantal and Philippe du Chouchet, Versailles, France, August 11, 2005.

66. Sèvres. Inventory of objects produced, October 1, 1762. Royal Porcelain Manufactory, Sèvres. Microfilm. Vy Folio 115.

67. Cole, W. *A Journal of My Journey to Paris in the Year 1765*. London, England: Constable, 1931: Cited in: Sargentson, C. *Merchants and Luxury Markets, The Marchands Merciers of Eighteenth-Century Paris*. London, England: Victoria and Albert Museum and The J. Paul Getty Museum, Malibu, California, 1996; p. 141, note 144.

68. Vilensky, J. Educated Fancy. *The Collection of Nikolai Borisovich Yusopov, Exhibition Catalogue*, "Sèvres Porcelain from the Collection of Nikolai Borisovich Yusopov in the Hermitage," Moscow, 2001, pp. 273–274 (Russian text).

69. Kazakevitch, N. Porcelaine de Sèvres du XVIIIe siècle à l'Ermitage. *L'Estampille/L'Objet d'Art*, No. 382, juillet-août 2003, p. 70.

70. Savage, G., and Newman, H., *An Illustrated Dictionary of Ceramics*, London, England: Thames and Hudson, 1992; p. 75.

71. Birioukova, N., and Kazakevitch, N. *La Porcelaine de Sèvres du XVIII Siècle; Catalogue de la Collection*. Saint Petersburg, Russia: Editions du Musée de l'Ermitage, 2005, p. 323, see also pp. 305–308 (Russian and French text).

72. Préaud, T. Sèvres, la Chine et les "chinoiseries" au XVIIIe siècle. *The Journal of the Walters Art Gallery (Baltimore)* 1989;47:48–50.

73. Peters, D. *Decorator and date marks on C18TH [sic] Vincennes and Sèvres Porcelain*. London, England: The Author, 1997.

74. Danckert, L. *Directory of European Porcelain*. Translated by Christine Bainbridge and Rita Kipling. London, England: N.A.G. Press, Robert Hale, 2004 (original German edition 1992); p. 722.

75. Musée National Adrien-Bubouché (Ceramics), Limoges. Collection Casnault, 1881.

76. Verlet, P. Notes on eighteenth century French objets d'art. *The Art Quarterly* 1968;31(4):368–371.

77. Verlet, P. Notes on eighteenth century French objets d'art. *The Art Quarterly* 1968;31(4):371.

78. Meslin-Perrier, C. Letter, October 31, 2006.

79. Le Duc, G. *Porcelain tendre de Chantilly au XVIIIe siècle*. Paris, France: Hazan, 1996; p. 290.

80. Gioffré, R. *Cioccolato Nuove Armonie*. Milan, Italy: Giunti Editore, 2003; p. 41.

81. Corsair, The Online Catalog of The Pierpont Morgan Library. Available at http://corsair.morganlibrary.org/cgibin/Pwebrecon.cgi?v1=4&ti=1,4&SEQ=20070306031957&Search_Arg=French+Costume+Prints+Box+1&SL=Submit%26LOCA%3DDept.+of+Drawings+and+Prints%7C4&Search_Code=CALL&PID=527&CNT=50&SID=4. (Accessed August 15, 2006).

82. Musée des Arts Décoratifs, Paris.

83. Metropolitan Museum of Art, New York.

84. Fleming, J., and Honour, H. *The Penguin Dictionary of Decorative Arts*. London, England: Viking, 1989 (original edition 1977); pp. 644–645.

85. Bernardin de Saint-Pierre, J.-Henri. *Paul et Virginie*. Paris, France: Garnier, 1788; p. 1243.

86. de Plinval de Guillebon, R. *Catalogue des porcelaines françaises: Musée du Louvre, Département des objets d'art*. Paris, France: Réunion des Musées Nationaux, 1992; I, p. 166.

87. Dion-Tenenbaum, A. *L'orfèvre de Napoléon, Martin-Guillaume Biennais*. Paris, France: Réunion des Musées Nationaux, 2003; pp. 72–73.

88. Coe, S. D., and Coe, M. D. *The True History of Chocolate*. London, England: Thames and Hudson, 1996; p. 162.

89. Coe, S. D., and Coe, M. D. *The True History of Chocolate*. London, England: Thames and Hudson, 1996; pp. 242 and 250.

90. "Confiseurs," in "Artisanats au 18ème Siècle." *L'Encyclopédie Diderot et d'Alembert*. Paris, France: Inter-Livres, 1986; pp. 3–4.

91. Dorchy, H. *Le Moule à chocolat. Un nouvel objet de collection*. Paris, France: Éditions de l'Amateur, 1987; p. 12.

92. Cultivation and Harvesting of Cacao, 1836, Sèvres porcelain, decorated by Jean-Charles Develly, in Metropolitan Museum of Art, New York. Available at http://www.metmuseum.org/toah/hd/sevr/hob_1986.281.1-4_av8.htm. (Accessed February 25, 2007).

93. Rabinow, R. A. Vollard and Matisse. In: Rabinow, R. A., editor. *Cézanne to Picasso. Ambroise Vollard, Patron of the Avant-Garde*. New York: Metropolitan Museum of Art; and New Haven, CT: Yale University Press, 2006; p. 135, Figure 151.

94. Rabinow, R. A. Vollard and Matisse. In: Rabinow, R. A., editor. *Cézanne to Picasso. Ambroise Vollard, Patron of the Avant-Garde*. New York: Metropolitan Museum of Art; and New Haven, CT: Yale University Press, 2006; pp. 139 and 383.

Commercial Chocolate Pots

Reflections of Cultures, Values, and Times

Margaret Swisher

Introduction

Between December 2004 and August 2006, members of the chocolate history research group at the University of California, Davis, searched the World Wide Web for images of chocolate pots using Google and Netscape. Collectively, we identified more than 100 commercial auction houses and sites developed by private owners of chocolate-related pots and ephemera. These efforts led to the identification of 1134 images of chocolate pots.

Thumbnail images of the pots were captured and transferred to an Excel data sheet, along with accompanying Meta data when available: specifically, country of origin, date of fabrication/manufacture, composition material, decorative motifs and colors, and image sources. These small-scale images (without commercial value) were uploaded for description and analysis to the Chocolate Research Portal (see Chapter 53) using the *Fair Use* doctrine. Selected examples that illustrate the diversity of chocolate pots we identified are presented here, following this brief synopsis of our work, and are reproduced here with permission.

Findings

Ancient Mayan and other Central American chocolate pots may be dated well before the Common Era (see Chapters 1 and 2; Appendix 8). So-called "chocolate pots" produced before the Common Era by indige-

nous Native Americans in Costa Rica are "chocolate pots" only in color, and this designation may confuse unsophisticated researchers during Web-based searches. Although not chocolate pots in the strict sense, nevertheless, they are intriguing in shape and style; some portray exotic animal-associated designs.

EUROPEAN-SOURCE CHOCOLATE POTS

Logic holds that in Europe the practice of using chocolate pots would have spread from Spain into nearby France (see Chapter 42). The earliest European chocolate pots we identified during our Web-based search, however, were English, not French, and dated to the mid-17th to mid-18th century. Examples of English silver chocolate pots commonly were decorated around the top and base with raised metal designs, or imprinted on the sides with family crests. Handles of these pots commonly were made of wood and set at a 90 degree angle from the pouring spout, a distinction that sometimes differentiated them from silver teapots (Fig. 13.1).

European chocolate pots changed in material and fabrication sites during the 19th century as decorative pots made of silver, and unadorned everyday pots

Chocolate: History, Culture, and Heritage. Edited by Grivetti and Shapiro
Copyright © 2009 John Wiley & Sons, Inc.

FIGURE 13.1. Chocolate pot (English), mid-17th to mid-18th century. Private Collection. Previous owner Robert Elliott & Charles Wamsley Jr. (Used with permission.)

made of copper, were augmented by china and porcelain imports [1]. Porcelain manufacturing was developed in Germany independently of Asian techniques that had been in use before 1800 [2]. Echoing earlier Asian chocolate pots, floral decorations with roses became common, with roses on European porcelain chocolate pots being larger and more detailed than Asiatic designs. Changes in chocolate pot form that occurred in Germany during the 19th century were echoed in France, although French chocolate pots tended to be taller and more slender, usually with a less bulbous base than those from Germany, providing an overall form that seemed more slender and delicate. French chocolate pots, like those manufactured in Germany, were decorated, predominantly, with floral designs (Fig. 13.2). English factories began making porcelain about thirty years after Germany and France, but still maintained a strong silver tradition [3]. One unusual silver English chocolate pot we identified carried the motif of everyday working people and was etched with a scene of a woman mopping a floor, an unusually mundane scene to decorate an expensive item.

ASIAN-SOURCE CHOCOLATE POTS

During the 19th century, Japan produced a wide array of porcelain chocolate pots for use in North America, but the main market was export to Europe [4]. Styles, shapes, and patterns were relatively similar. Floral designs seemed to be most popular. The flowers depicted are less detailed than those of European designs; however, Japanese roses were commonly more fully open with only five petals and the interior fully exposed (Fig. 13.3.). Japanese chocolate pots tended

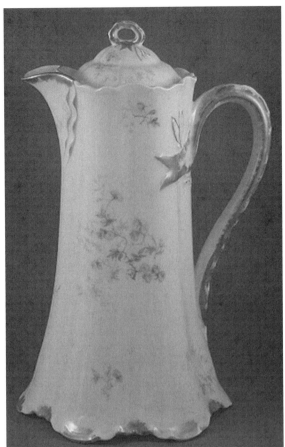

FIGURE 13.2. Chocolate pot (French), 19th century. Private Collection. Previous owner Michael Tese. (Used with permission.)

FIGURE 13.3. Chocolate pot (Japanese), 19th century. Private Collection. Previous owner Wen Chen. (Used with permission.)

to be tall and slender with handles less elaborate than European manufactured types. Decorations typically covered the surface of the pot, and small clay dots, known as *moriage*, were used as decoration and added a three-dimensional look to the design.

NORTH AMERICAN-SOURCE CHOCOLATE POTS

Silver and base metal American chocolate pots are reviewed elsewhere (see Chapters 10, 11, and 54). Our Web-based search located only fourteen chocolate pots, with most dated after 1850. Metal chocolate pots remained popular in America, as in Europe, but porcelain chocolate pots were more common and most did not have the characteristic bulbous base of European pots of the time. We identified an exceptional example of a chocolate pot whose entire surface was engraved with leaves, with the pouring spout modeled after a bust of Zeus.

Porcelain chocolate pots used in North America were decorated less elaborately. White with pink roses arranged in bunches over the surface of the pot or in a linear distribution predominated. The flowers tended to be less detailed and most examples examined were light pink in color, giving the pots an overall muted characteristic.

TWENTIETH CENTURY CHOCOLATE POTS: EUROPE AND NORTH AMERICA

The majority of the images collected during our Web-based survey were from 20th century chocolate pots dated 1900–1949. New trends in material use and decoration appeared during this time period; floral decorations remained popular but more flower types were represented. Scenes of people or animals commonly were incorporated and Asiatic geisha designs became popular. Furthermore, chocolate pots were decorated with a wider variety of colors.

German chocolate pots during this period tended to be more bulbous at the top rather than at the base and had an overall rounded appearance (Fig. 13.4). The handles became more ornate and delicate and were commonly gilded. Porcelain remained the material of choice, as high-quality silver pots became rare. The most popular design remained floral motifs, and roses were the most popular flower.

French chocolate pots during the 20th century commonly remained bulbous at the base, in contrast to German pots, which were now usually bulbous at the top. French chocolate pots of this period commonly used less color than German pots. The body of many French chocolate pots remained largely white, with sprays of small delicate flowers scattered over the surface, or exhibited large designs in the center of the pot, surrounded by white. Gold was used mainly in edging and for small details in the designs.

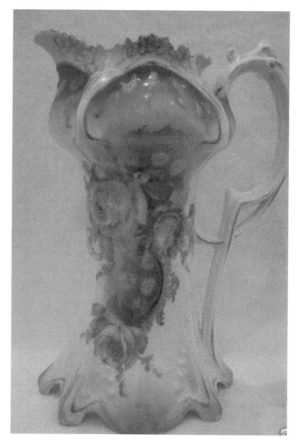

FIGURE 13.4. Chocolate pot (German), 20th century. Private Collection. Previous owner Gareth and Amy Jones. (Used with permission.) (See color insert.)

Chocolate pots manufactured in England during the 20th century commonly were copper, pewter, porcelain, or silver. Short and stout chocolate pot forms remained popular, but examples of tall slender forms also appeared. Base-metal pots usually had no decoration (see Chapter 54). English porcelain chocolate pots frequently exhibited floral designs, sometimes animals, and English landscape scenes were common (Fig. 13.5). Japanese chocolate pots produced from 1900 to 1950 and imported to North America and Europe were similar in shape to French chocolate pots of this time period. They tended to be tall and slender, usually tapered in the middle and wider at both bottom and top. Japanese chocolate pots commonly were decorated with floral motifs, animals (i.e., butterflies and dragons), landscapes, and people (geishas) (Fig. 13.6 and 13.7).

American chocolate pots manufactured from 1900 to 1950 varied widely in shape: some tall and slender, others short and stout. Our Web-based survey identified few porcelain American chocolate pots during this period. The metal pots identified usually were undecorated or decorated only sparingly. Some of the metal American chocolate pots of this era placed

FIGURE 13.5. Chocolate pot (English), 20th century Wedgewood. Private Collection. Previous owner Wendy and Tim Burri. (Used with permission.) (See color insert.)

FIGURE 13.7. Chocolate pot (Japanese), early to middle 20th century. Private Collection. Previous owner Mike Moore. (Used with permission.)

FIGURE 13.6. Chocolate pot (Japanese), early to middle 20th century. Private Collection. Previous owner Bob Brenner. (Used with permission.)

the handle at 90 degrees from the pouring spout, a design that was common in the early 1800s in Europe but generally had disappeared in Europe by this time. Several of the American metal chocolate pots we identified were manufactured specifically for use by hotels or railroads. This could explain the metal trend in chocolate pots, which would be more durable for public use than fragile porcelain.

Conclusion

Chocolate pots represent a unique art form that records symbols important to the culture of each country of manufacture. They represent a prism through which the aesthetic and cultural values of countries and regions may be compared and contrasted. Chocolate pots, like any art form, tell a historical story. Our most important finding, however, was not related to the changing designs, forms, and shapes of chocolate pots—it was that the history of this art form is disappearing. The chocolate pots examined by our survey were privately owned and were offered for sale at auction houses, merchant sites, and on-line bidding companies such as eBay. Once these items were sold, the images disappeared from public view as the pots were returned to private collections—and to new, anonymous owners. The images of the chocolate pots initially placed for sale on the World Wide Web were copyrighted by the original owners—not eBay or Google. After sales took place, the images vanished, and no longer were available for researchers to examine, describe, or analyze. The legal departments of both eBay and Google were contacted for permission to use images in our research publications: in both instances permission was denied and we were referred to the original owner/seller (whose name and contact information also had vanished). The images reproduced in the present chapter, however, were those that had been posted recently (between January 1 and August 31, 2007). They are reproduced here, with permission of the original owners. In the weeks that passed between the preparation and final publication of this chapter, these too, have disappeared from the several auction sites.

Chocolate pots manufactured after 1950 were uncommon, perhaps because drinking chocolate became passé and out of fashion as coffee and tea became more popular beverages globally. The era of elaborately decorated and ornate porcelain and silver chocolate pots has given way to basic, functional coffee and tea pots/urns. The 1950s in the United States also was an era of coffee in mugs, where strong, simple, black coffee could be purchased for $0.05, a nickel-per-cup. Paralleling this basic over-the-counter coffee were more expensive, exotic coffees known by Italian names, served in upscale, specialty coffee houses—dark, interesting venues for "beat" poetry and the folk-music revival. And who could have foreseen a future time when customers would be willing to pay $3.50 or more for specialty coffees in order to receive their morning or afternoon "fix" of caffeine?

Acknowledgments

I would like to acknowledge the efforts of previous workers at the University of California, Davis, affiliated with the Chocolate History Project, who also assisted with identification and compilation of chocolate pot images, especially, Ezra Neale. I also would like to acknowledge my mother, Barbara Swisher, for editorial support during the preparation of this chapter.

References

1. Impey, O. Japanese porcelain. In: Atterbury, P., editor. *The History of Porcelain*. London, England: Orbis Publishing, 1982; pp. 43–54.

2. Raffo, P. The development of European porcelain. In: Atterbury, P., editor. *The History of Porcelain*. London, England: Orbis Publishing, 1982; pp. 79–126.

3. Schleger, M. Eighteenth-century English porcelain. In: Atterbury, P., editor. *The History of Porcelain*. London, England: Orbis Publishing, 1982; pp. 127–138.

4. Coe, S., and Coe, M. *The True History of Chocolate*. London, England: Thames and Hudson, 1996.

14

Role of Trade Cards in Marketing Chocolate During the Late 19th Century

Virginia Westbrook

Introduction

Samuel Johnson characterized "the soul of an advertisement" in one of his "Idler" commentaries in 1759 [1].[1] Dr. Johnson lamented the extravagant language of print advertisements whose tools for getting attention were limited to hyperbole and typography. Over the course of the next century, printers would use decorative rules, boxes, and small images (called "dingbats" in the trade) to gain attention for their advertisers. Copperplate engravings added fancy script and line art to the list of possibilities, though their services were expensive. A century after Dr. Johnson's observation, the new technique of lithography had developed into a full-color process, allowing advertisers to increase their "promise" power a thousandfold with the use of chromolithography. Eager consumers embraced these exciting images and soon gave them the nickname "chromos."

As an introduction to the world of "chromos," consider the impact this delightful image might have had on a middle-class 19th century housewife as she went about her daily shopping (Fig. 14.1). Hers was a world of black and white print. Popular magazines and gift books had illustrations, but any color was the result of hand-tinting. Only people of means could afford to hang portraits and landscape paintings in their homes. Figure 14.1 represents a brightly colored image of a beautiful woman anticipating a morning cup of cocoa. She is well dressed, seated comfortably on a stylish sofa, awaiting the service of a charming cherub. Who would not want to replicate this experience in their own life? Best of all, by purchasing a package of Runkel Brothers cocoa, the housewife would be able to take the image home with her for free! Such was the promise of the trade card, as it evolved over the second half of the 19th century.

Trade Cards

The term "trade card" encompasses a broad range of ephemeral material produced over a period of several centuries. The earliest examples, printed on paper, simply stated the tradesman's name and location, with references to important intersections and public buildings in the years before the institution of street numbers. After British law prohibited the use of hanging signs as a hazard to public safety in the 1760s, use of trade cards increased, often reproducing the image of the former shop sign [2].[2] Card stock, copperplate

Chocolate: History, Culture, and Heritage. Edited by Grivetti and Shapiro
Copyright © 2009 John Wiley & Sons, Inc.

FIGURE 14.1. Chocolate trade card (Runkel). *Source*: Private Collection. Courtesy of The John Grossman Collection of Antique Images, Tucson, AZ. (Used with permission.) (See color insert.)

engraving, textured paper, and embossing added interest to the medium over the next century in an evolutionary process that continues with present-day business cards.

Trade cards changed dramatically with the introduction of color printing. The art of lithography—literally "writing on stone"—started in Germany at the end of the 18th century. By the 1820s, lithography had largely replaced engraving because a stone was easier and more economical to prepare, produced unlimited impressions, and offered larger dimensions than copper plates. While British printers worked mainly in black and white, European lithographers, especially in Germany, Moravia, and Switzerland, experimented with multiple impressions using different colors of ink [3]. Success of the final product depended on precise "registration"—positioning the paper so that each successive impression retained the same alignment as the

previous one. European lithographers migrating to America found a ready market for their skills.

The term chromolithography originated with a French patent in 1837 [4].[3] Strictly speaking, it referred to a print made using at least three separate colors, as contrasted with those colored by hand or tinted with a wash that flooded the stone with color to produce an atmospheric effect. For nearly half a century, chromolithographers concentrated on producing inexpensive prints of fine art, illustrations for books and educational teaching aids [5].[4] Louis Prang, a German immigrant, receives credit for the idea of mass-produced advertising cards that could be customized for individual advertisers, but he was able to do so only after building a reputation for extraordinary color reproductions. He used steam-driven presses to produce remarkable likenesses that required ten or more separate stones to match the nuanced colors of an original painting. Following upon the success of John James Audubon's *Birds of America* [6],[5] Prang launched a series of bird prints designed to fit in custom-made albums manufactured at his factory in Dorchester, Massachusetts. Along the way, he appropriated the term "chromo" as a trademark for his best color prints [7].

Louis Prang produced his first full-color trade card to advertise his own products at the Vienna International Exposition in 1873. Until that time, most colored advertising cards had been produced by lithographers to promote their own services [8]. The opening of the Centennial Exposition in Philadelphia in 1876[6] offered the ideal opportunity to launch a new advertising medium. Exhibitors at the fair wanted to hand out an *aide de memoire* to visitors, who in turn sought to collect cards from each of the exhibits they had seen [9].[7] The stage was set for an explosion in trade card production complemented by a concurrent public craze for collecting the cards [10].[8]

CHOCOLATE CHROMOS

Trade cards promoting chocolate products seldom have been mentioned or discussed by scholars of trade card history. Robert Jay traced the emergence of great American brands of tools and machinery (Fairbanks, McCormick, Remington), patent medicines (Mrs. Pinkham's Vegetable Compound, Dr. Morse's Indian Root Pills), household furnishings (Singer sewing machines, Glenwood stoves, Florence lamp stoves), soap, tobacco, alcohol, and thread companies and personal grooming tools [11, 12]. William McGlaughlin explored the pictorial melodrama of humorous cards and the use of patriotic and factory imagery [13]. But the examples consulted for this chapter testify to the enthusiasm with which chocolate makers embraced the new medium and the reciprocal response from a consuming public whose collecting impulses ensured that the specimens survive today [14, 15].

Chocolate companies produced their chromos in various shapes and sizes common to the medium. Each form played a particular role in the business of marketing. "Showcards" were designed for store display. In the days before fancy packaging, manufacturers would supply retail merchants with large chromos to stimulate sales. These often came as "hangers" with a prepunched hole at center top, or as dimensional cards with wings that folded out so they could stand up on a counter (Fig. 14.2).

Cards for grocers were much more businesslike. One printed by Josiah Webb & Company dispensed with imagery altogether in favor of a simple inventory of products. The reverse carried a grocer's price list. A card distributed by the Dutch company Blookers retained the imagery—one of many romantic seascapes—backed by an informative explanation of their product aimed at helping the grocer select their wares for sale. The Franco-American Food Company, *"Sole Agents for the U.S.,"* distributed the Blookers card to retailers. Boston agents for Maillards distributed the manufacturer's standard card after printing their

name and address along edges of the reverse side (Fig. 14.3).

A great array of novelty cards employed some of the dimensional tricks of the larger showcards. Rountree produced a card with a cocoa pod that opened to reveal a cup of "Rountree's Elect Cocoa." A Schaal card featured a boy trundling an oversize can of cocoa on a wheelbarrow, one of several such cards that could expand into a three-dimensional scene. A broader survey might reveal whether these were common offerings or high-end productions for special events such as one of the many fairs and expositions that exposed consumers to the wide world of possibilities. This may be the case for a Van Houten's card picturing the famous windmill whose screw press removed cocoa butter starting in 1828. Although now lost, the vanes of the paper "mill" turned, creating a visual phenomenon (unknown) in the center. On the back, instructions to "bend here" indicated the fold line for a tab that would prop the card upright (Fig. 14.4) [16].[9]

Most trade cards targeted customers. Manufacturers supplied them to shopkeepers to hand out at the counter or enclosed them in the package. The cards followed a standard format, already well established by

FIGURE 14.2. Chocolate trade card (Walter H. Baker). *Source*: Private Collection. Courtesy of The John Grossman Collection of Antique Images, Tucson, AZ. (Used with permission.) (See color insert.)

FIGURE 14.3. Chocolate trade card (Maillard). *Source*: Courtesy of the American Antiquarian Society, Worcester, MA. (Used with permission.) (See color insert.)

FIGURE 14.4. Chocolate trade card (Van Houten). *Source*: Courtesy of the American Antiquarian Society, Worcester, MA. (Used with permission.) (See color insert.)

the 1870s, with an appealing image on the front to catch the viewer's attention and product information printed on the back. Often, the product information on the reverse reminded customers of the packaging details—the pink wrapper on Huyler's Vanilla Chocolate or the blue wrapper on Runkel's cooking and baking chocolate—so they could easily find the desired product another time (Fig. 14.5).

Producers of trade cards responded promptly to the public's enthusiasm for collecting. Van Houten's Royal Dutch Chocolate produced a series of bird prints with the species description printed on the back, along with a testimonial from Dr. Braithwaite's *Retrospect of Medicine*. These came in a protective envelope. Scrafft and Cadbury also produced nature scenes. Nearly all the specimens in the John Grossman archive have evidence of glue on the back, or scars where paste tore free from collectors' albums. Surprisingly, there are only a few examples of the kind of generic advertising for which Louis Prang gets credit: cheap, generic cards illustrating humorous vignettes that were distributed to merchants for overprinting [17].[10] A single example of this type was distributed by C. D. Brooks Chocolate in Chicago (Fig. 14.6). The dearth of this type may result from an overabundance of fancy chromos being supplied by established manufacturers, or it may simply derive from the selectivity of collectors.

Postcards came into use at the end of the 19th century. When postal regulations expanded to permit privately printed postcards in 1898, merchants moved quickly to get their trade card images into the post. In 1906, someone with the initials G. M. J. sent a Bensdorp's postcard to Mr. Samuel Wares in Townsend, Massachusetts, with the message, "We want all we can have" scribbled in the space left blank for a local merchant (Fig. 14.7).

Huyler's
"If you only knew how nice it is!"
Huyler's Vanilla Chocolate New York
½ lb
VANILLA
Chocolate.
(PINK WRAPPER)
COPYRIGHT, 1891

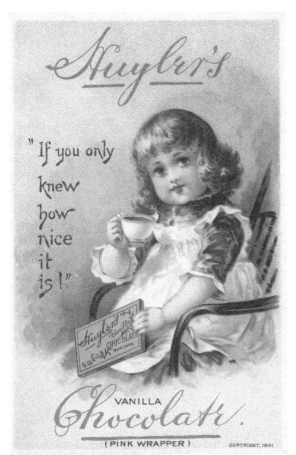

FIGURE 14.5. Chocolate trade card (Huyler). *Source*: Courtesy of the American Antiquarian Society, Worcester, MA. (Used with permission.) (See color insert.)

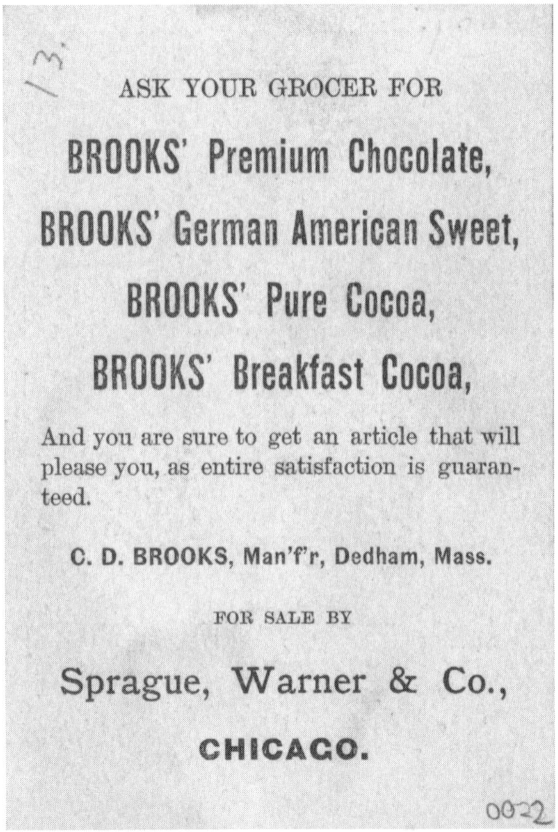
ASK YOUR GROCER FOR
BROOKS' Premium Chocolate,
BROOKS' German American Sweet,
BROOKS' Pure Cocoa,
BROOKS' Breakfast Cocoa,
And you are sure to get an article that will please you, as entire satisfaction is guaranteed.
C. D. BROOKS, Man'f'r, Dedham, Mass.
FOR SALE BY
Sprague, Warner & Co.,
CHICAGO.

FIGURE 14.6. Chocolate trade card (Brooks). *Source*: Courtesy of the American Antiquarian Society, Worcester, MA. (Used with permission.)

Potent Pictures

Throughout the heyday of trade card advertising, Victorian sentimentality prevailed. Chocolate companies relied as heavily on the cult status of childhood as other advertisers. Sweet images of young children dominate the medium, whether they are crawling babies, strolling siblings, or youngsters sipping cocoa, occasionally accompanied by a pet or some dolls (Fig. 14.8). Only one card, issued by Runkel Brothers, strikes the popular patriotic note with an image of Uncle Sam and an eagle carrying a box of cocoa. A few adventurous companies—Cherub, Princess, and Stollwerck—pictured women in exotic clothing (Fig. 14.9). Domestic scenes of well-dressed couples at ease vied with cherubs stirring steamy cocoa as the perfect image of chocolate enjoyment.

Remarkably, the Dutch chocolate companies enjoyed an unexpected advantage in the contest for customers. Concurrent with the rise in popularity of chromo trade cards was a surge of interest in Dutch culture and the Dutch chapter in early American history [18, 19].[11] Companies such as Bensdorp, Blooker, and Van Houten struck a resonant chord in the popular imagination with advertising images of Dutch seascapes and country folk dressed in wooden shoes and traditional clothing.

The Promise

Trade cards delivered complex messages to customers. The images came freighted with meaning, only some of which can be deciphered today in the 21st century. The pictures carried more power than the words, especially for children, then, as now, often the primary targets of the advertisement. Images of children suggested innocence, delicacy, and, by extrapolation, purity. The pictures of Dutch country folk mentioned earlier evoked all the connections associated with the culture: cleanliness, industriousness, and an appreciation for good food and drink. Domestic scenes parse out more easily. The image of a happy family sitting in a well-appointed breakfast room while a maid pours morning cocoa clearly promises prosperity and domestic tranquility as well as satisfaction from the Van Houten's brand of cocoa (Fig. 14.10). The image of "a welcome visitor" to the bedroom of a woman lying-in with her baby in a nearby cradle assures that cocoa is, indeed, good food for a nursing mother (see Fig. 9.2).

FIGURE 14.7. Chocolate postcard (Bensdorp). *Source*: Courtesy of the American Antiquarian Society, Worcester, MA. (Used with permission.) (See color insert.)

FIGURE 14.8. Chocolate trade card (Phillips). *Source*: Courtesy of the American Antiquarian Society, Worcester, MA. (Used with permission.) (See color insert.)

Perhaps the most obscure visual cue comes from a Cadbury card depicting a serene Antarctic seascape with a lone ship in among the icebergs. Contemporary viewers may need to read the fine print to understand the message in this image, but to anyone who followed British exploratory expeditions (and there were many), the picture cried out—"energy food!"

The essential purpose of a trade card was twofold: to make a sale and to help the customer feel comfortable using the product. Headlines and slogans accomplished the first goal. Words like "pure," "digestible," "soluble," and "dietetic" appear most often, followed closely by "nutritious," "healthy," "invigorating," and "delicious." Slogans often appeared integrated with the graphic. Van Houten's slogan on the face of their card—"Best & Goes Farthest"—was combined with a marginal notation on the back, "for rich & poor—made instantly—easier than tea." Cadbury's proclaimed "Absolutely Pure Therefore Best." Mack's Milk Chocolate was one of only a handful of generic cards overprinted with the company message that announced, "The Best! It's Pure!" Runkel's slogan, "Right in it!"

made sense only if you had seen the full version of the message, a sequence of three captioned pictures of the same child, saying "I Want It!" I'll Get it!" Right in it!" An insidious message appears on one card that depicts two very young children in a basket, with the caption: "Always contented when we have plenty of Clark & Morgan's Chocolates & and Hand-Made Cream Candies."

Descriptive language on the back of the cards instructed consumers in the use of chocolate. Maillard's advertised a chocolate school with "Free lessons & illustrations given on Monday, Wednesday and Friday afternoons between 3 and 5 o'clock." Stolwerck's card gave directions for preparing cocoa. Bensdorp did the same, along with a recipe for frosting, and recommended chocolate for use in ice cream, custards, and jellies. Other Bensdorp cards featured recipes as incentive to collect the cards. One featured a recipe for cocoa sponge drops "by Mrs. D.A. Lincoln, author of the Boston Cook Book."[12] Another, with a recipe for cocoa soufflé, admonished that "good cocoa can always be used with better results than cake chocolate." This company also recommended that their product was "unequalled on board ships, in railway stations, the camp and hospitals."

FIGURE 14.9. Chocolate trade card (Cherub). *Source*: Private Collection. Courtesy of The John Grossman Collection of Antique Images, Tucson, AZ. (Used with permission.) (See color insert.)

FIGURE 14.10. Chocolate trade card (Van Houten). *Source*: Private Collection. Courtesy of The John Grossman Collection of Antique Images, Tucson, AZ. (Used with permission.) (See color insert.)

Only a few cards addressed medicinal use of chocolate. The image of the nursing mother mentioned earlier supports the implication articulated in a Phillips' card whose descriptive literature stated that their cocoa was a "superior drink for growing children and for nervous and delicate women." Another medical context appeared on the Van Houten aviary series as a recommendation from a Dr. Braithwaite.

The era of the chromolithographic trade card drew to a close with the 19th century. Changes in postal regulations in 1885 reduced the cost of sending second-class mail to a penny a pound. Liberated from weight constraints, periodical publishers could accept more advertising. Although it took some time for magazines to duplicate the lively color quality of trade cards, they soon prevailed as the primary medium for printed advertisements.[13] In the 20th century, the trade card form shifted to collectibles and premiums like cigarette and bubble gum cards.

Trade cards may have been a short-lived phenomenon, but the surviving examples permit understanding of how chocolate was presented to the burgeoning consumer marketplace in the last decades of the 19th century.

At the time chromo trade cards flooded the market, the familiar forms of chocolate were available: cocoa (thanks to Van Houton's extraction of cocoa butter, 1828), edible chocolate bars (from Fry's of

England, 1848), and super smooth chocolate for covering candies (courtesy of Rodolphe Lindt's conching machine, 1879). But these commodities did not receive equal billing on trade cards. The vast majority of cards touted cocoa. The word "chocolate" appears nearly as often, but the term refers to several different products, including tablets that dissolve in water to make a beverage, bars or blocks for baking and confectionary use, and bite-sized candies sold in boxes. Only a few cards advertise "bonbons" or "cream candies." Clearly, cocoa for use as a hot drink dominated the market during the era of trade card advertising.

A substantial body of chocolate trade cards survives today, documenting the assertive marketing strategy of foreign and domestic producers alike. These cards may have survived because of their high production numbers, or because the popular sentiments they expressed appealed to customers. But they may have been saved simply because people liked chocolate.

Acknowledgments

I wish to acknowledge Georgia Barnhill, Curator of Graphic Arts at the American Antiquarian Society, who offered counsel concerning her own collection and also recommended the John Grossman Collection as a fruitful avenue for research. I also thank John Grossman, owner of the John Grossman Collection, Tucson, Arizona, who personally helped with the search of his collections.

Endnotes

1. "The art of advertising exemplified" appeared in the *London Weekly, the Universal Chronicle*, on January 20, 1759. A total of 103 essays, of which this is #40, appeared under the "Idler" byline between April 1758 and April 1760. Dr. Johnson wrote all but 12 of them. The essays carried no title until compiled into book form under the title *The Idler*.

2. Maurice Rickards defined ephemera as "minor transient documents of everyday life," while fully recognizing that all the examples collected over the years have resisted that very transience.

3. The color printing process began in Germany, but the French word translated easily into English and quickly took over.

4. Currier & Ives, the most recognizable name in American print making, concentrated on producing inexpensive prints for a broad market, making them available on easy terms to independent salesmen. Most of their work consisted of hand-colored images until after the founder, Nathaniel Currier, retired in 1880.

5. John James Audubon's *Birds of America* was first published in England in 1826. The life-size images were printed from aquatint engravings on huge plates that then were hand colored. Audubon published a popular edition in America in 1842. A chromolithographic edition was initiated in 1858 but only one volume was ever published.

6. The official title of the event was "International Exhibit of Arts, Manufactures and Products of Soil and Mine," but even *Frank Leslie's Illustrated Historical Register of the Centennial Exposition, 1876* referred to it in the vernacular.

7. A visit to the 1893 World's Fair inspired John Springer to begin a collection of trade cards and other ephemera. He donated his collection to the University of Iowa Libraries in 1937.

8. Rickards recommends that confusion in use of the term "trade card" could be avoided by adopting more specific terminology, for example, trade cards, chromo trade cards, and commercial collecting cards.

9. Van Houten's made possible the mass-marketing of dry cocoa powder with the development (in 1828) of two important techniques: extraction of cocoa butter from liquid chocolate by means of a screw press and "dutching," the treatment of the resulting powder with potash or sodium carbonate to reduce acidity and increase miscibility.

10. Rickards cites the Union Card Company of Montpelier, Vermont, and the Calvert Lithography Company, Chicago, Illinois, as leaders in this form, among others.

11. The extraordinary popularity of *Hans Brinker, or the Silver Skates*, published in 1865, exposed a generation of young readers to a romantic notion of Holland. As those readers came of age, they traveled to the Netherlands, studied the Dutch masters, copied Dutch architecture, and hung pictures of Dutch people in traditional dress on their walls. Wallace Nutting, the taste-maker and advocate of "old America" reserved a substantial section of his catalogue for views of Holland. The fascination with Dutch culture continued into the 1920s.

12. Mrs. Lincoln taught at the Boston Cooking School from 1879 to 1885. One of her students, Fannie Farmer, administered the school from 1894 to 1902, during which time she published a revised edition of the *Boston Cooking School Cook Book*, famous for the first use of standardized measurements.

13. During the transition, some manufacturers, beginning with Ivory soap, continued to print their own chromos and sent them to publishers to be bound into the magazines.

References

1. Johnson, S. The art of advertising exemplified (#40). *Universal Chronicle,* January 20, 1759. Available at http://en.wikipedia.org/wiki/The_Idler_%281758-1760%29. (Accessed February 24, 2007.)

2. Rickards, M. *The Encolopedia of Ephemera*. London, England: The British Library, 2000; p. 334.

3. Rickards, M. *The Encyopedia of Ephemera*. London, England: The British Library, 2000; p. 335.

4. Marzio, P. C. *The Democratic Art: Pictures for a 19-Century America*. Boston: David Godine, 1979; p. 6.

5. Marzio, P. C. *The Democratic Art: Pictures for a 19-Century America*. Boston: David Godine, 1979; pp. 59–63.

6. Marzio, P. C. *The Democratic Art: Pictures for a 19-Century America*. Boston: David Godine, 1979; pp. 55–59.

7. Marzio, P. C. *The Democratic Art: Pictures for a 19-Century America*. Boston: David Godine, 1979; p. 11.

8. Jay, R. *The Trade Card in Nineteenth-century America*. Columbia: University of Missouri Press, 1987; p. 25.

9. The John Springer Printing Ephemera Collection, Special Collections, University of Iowa Libraries. Available at http://www.lib.uiowa.edu/speccoll/Msc/ToMsc250/MsC202/MsC202.htm. (Accessed February 25, 2007.)

10. Rickards, M. *The Encyclopedia of Ephemera*. London, England: The British Library, 2000; p. 335.

11. Jay, R. *The Trade Card in Nineteenth-century America*. Columbia: University of Missouri Press, 1987; pp. 61–98.

12. John, W. Hartman Center for Sales, Advertising & Marketing History, Duke University. Available at http://scriptorium.lib.duke.edu/eaa/ephemera.html. (Accessed February 9, 2007.)

13. McGlaughlin, W. Trade cards. *American Heritage* 1967;XVII(2):48.

14. Trade Card Collection, American Antiquarian Society, Worcester, MA. Available at http://www.americanantiquarian.org/tradecards.htm.

15. John Grossman Collection of Antique Images. Available at http://johngrossmancollection.com/_wsn/page2.html.

16. Coe, S. D., and Coe, M. D. *The True History of Chocolate*. New York: Thames and Hudson, 2000; p. 242.

17. Rickards, M. *The Encylopedia of Ephemera*. London, England: The British Library, 2000; p. 335.

18. Denenberg, T. A. *Wallace Nutting and the Invention of Old America*. New Haven, CT: Yale University Press, 2003; pp. 80–83.

19. Stott, A. *Holland Mania*: The Unknown Dutch Period in American Art & Culture. Woodstock, NY: Overlook Press, 1998.

Commercial Chocolate Posters

Reflections of Cultures, Values, and Times

Margaret Swisher

Introduction

During the 19th century, liquid chocolate became unfashionable in many parts of the world, especially in the United States, and solid chocolate—sold as candy—became the most popular form. Before that time, chocolate had been an expensive drink favored by the aristocracy, but new technologies created during the Industrial Revolution transformed it into an inexpensive food, available to the general public [1]. As more confectioners specialized in making and selling chocolate, they competed directly with one another for customers. During the 19th century, advertising took the form of posters, and chocolate confectioners took full advantage of new developments in graphic arts, lithography, and commercial advertising, as companies boasted that their chocolate was the most pure, was the most filling or satisfying, and gave consumers strength. In addition, chocolate advertising posters depicted scenes meant to elicit different emotions from consumers, such as adventure, comfort, and sensuality. Just as chocolate pots provide researchers with insights into the cultures that created them (see Chapter 13), chocolate advertising posters illustrate values and emotions tied to chocolate and provide 21st century researchers insights not only into chocolate, but into the cultures that created the poster art.

From December 2004 through August 2006, members of the chocolate history research group at the University of California, Davis, searched the World Wide Web for images of chocolate advertising posters using Google and Netscape. In the process, we identified more than 100 commercial auction houses, poster gallery sites, and sites developed by private owners of chocolate-related posters and ephemera. During our search we identified more than 500 images of commercial chocolate advertising posters from 11 countries: Austria, England, France, Germany, Holland, Italy, Monaco, Russia, Spain, Switzerland, and the United States. The majority of the poster images, however, were from France and Switzerland.

Thumbnail images of the posters were captured and transferred to an Excel data sheet, along with accompanying Meta data when available: specifically, country of origin, date of fabrication/manufacture, artist, content motifs and colors, and image sources. These small-scale images (without commercial value) were uploaded for description and analysis to the Chocolate Research Portal (see Chapter 53) using the *Fair Use* doctrine. Selected examples that illustrate the diversity of chocolate advertising we identified are

Chocolate: History, Culture, and Heritage. Edited by Grivetti and Shapiro
Copyright © 2009 John Wiley & Sons, Inc.

presented here, following this brief synopsis of the work, and are reproduced here with permission.

Findings

FRANCE

The most striking characteristic of French chocolate advertising posters is their use of color. Yellow, red, and blue were present in nearly every poster, while green and black were common but secondary. Most French posters were drawn in a realistic style and relatively few were highly romanticized. We encountered only two examples of French posters drawn as cartoons.

The dominant theme of French chocolate advertising posters was incorporation of children. Young children, usually girls and infants (of both genders), commonly were depicted with chocolate bars or cups of hot chocolate. Sometimes children were portrayed playing games such as tug-of-war or rolling hoops. Children also were depicted accompanied by adult women, either mothers or caretakers/nurses. Several examples showed child–mother bonding, as when mothers fed their infants/children chocolate—or the reverse—when mothers were shown receiving chocolate from their children. One poster showed a mother sitting down to a cup of drinking chocolate with her daughter, while a cat attempts to get a taste from the daughter's cup (Fig. 15.1). Another theme common to French chocolate posters was interactions among children: fighting over chocolate, stealing chocolate from one another, or sharing beverages or bars. These scenes of children often were coupled with mischief and commonly depicted them spilling hot chocolate or dropping boxes of chocolate. One poster we identified showed a toddler dangling from a shelf stacked with chocolate that was in the process of collapsing. Other child/chocolate motifs depicted children standing in front of candy stores, the most interesting (to us) being a girl writing on the shop window seen from the perspective of the owner (inside the shop) so that the girl's writing is reversed. Other examples blended children, chocolate, and fairy tales, as with the well-known tale of Little Red Riding Hood, where the young girl holds a bar of chocolate and the wolf attempts to steal it.

Adult women also were common themes used in French chocolate posters. Mothers were depicted serving chocolate to their delighted children, conveying the message to consumers that good mothers should serve chocolate as a wholesome treat. Women also were shown as servants bringing chocolate on serving trays to their employers. In one rare poster a matronly woman is seen giving chocolate bars to others from around the world; the recipients of the chocolate were dressed in traditional clothing from their countries and included representatives from China, the Middle East,

FIGURE 15.1. Chocolate poster (Compagnie Francaise des Chocolats et des Thés), color lithography by Theophile Alexandre Steinlen; dated ca. 1898. *Source*: Bibliotheque des Arts Decoratifs, Paris, France. Courtesy of The Bridgeman Art Library, London, England. (Used with permission.) (See color insert.)

and Russia; most interesting, a Native American Indian was included in this group.

Women also were associated with elegance in French chocolate posters. They were often depicted in beautiful dresses and hats, and not uncommonly—but more intriguingly—displayed subtle or overt sexuality. Our Web-based survey also identified overtly sexual themes in two French chocolate posters. One depicted a woman dancing with a glass of liqueur in one hand and the bottle in the other, while her skirt was raised to show her calves and her top cut low to reveal an adequate décolleté. The other example, even more risqué, depicted a woman dressed in lingerie, wearing thigh-high black stockings with a garter belt, and black stiletto heels: she is shown kneeling and winking at the (presumed) chocolate customer! Women were also depicted in more sedate scenes of romance or courtship, where chocolate commonly would have been given or received as gifts. In another instance, young girls are dressed as cupids with butterfly wings, armed with arrows—but instead of shooting the arrows, the cupids are sending the adult lovers a box of chocolate.

Curiously, adult men seldom were depicted as the primary theme on French posters, but when they appeared they usually were shown as chocolate mer-

FIGURE 15.2. Chocolate poster (Gala Peter), color lithograph, French School. *Source*: Courtesy of The Bridgeman Art Library, London, England. (Used with permission.) (See color insert.)

FIGURE 15.3. Chocolate poster (Suchard), color lithograph, Swiss School; dated late 19th century. *Source*: Courtesy of The Bridgeman Art Library, London, England. (Used with permission.) (See color insert.)

chants. Occasionally, men and chocolate were paired with the theme of adventure. One poster showed men in a desert, sitting near their camel, which is laden with boxes of chocolate (Fig. 15.2). Another poster we examined depicted a man dressed as a Mexican cowboy, holding the reins of a black stallion. Another French poster, one commonly reproduced, showed an African man (obviously a servant) carrying a cup of hot chocolate on a tray.

Animals also played an important role in French chocolate posters. Most common representations were of cats, dogs, and horses. Cats, while appearing in the posters, were not the central theme and usually were depicted in the background and commonly seen drinking spilled hot chocolate. Dogs were sometimes shown as the only character in the poster, or were depicted as attempting to steal chocolate from children. Horses often were depicted with women, and one unusual poster showed a horse stealing a chocolate bar from a young girl. Other animals appearing in French chocolate posters included bears, bees, birds, and wolves, variously intermeshed with fantasy themes. We also identified a number of French posters that depicted chocolate and clowns as symbolic of chocolate being part of an enjoyable event, such as the circus.

SWITZERLAND

Swiss chocolate posters also were characterized by bright, vivid colors, especially blue, red, and yellow, with green and orange used by artists less frequently. Swiss chocolate posters usually featured romantic or realistic styles, but cartoon styles were common as well, more so than in French posters.

Children—usually girls—also were a dominant theme in Swiss chocolate advertising posters (Fig. 15.3). When boys were shown, they were portrayed playing sports such as football/soccer, swimming, or tennis. The images sometimes were accompanied with slogans, such as "a healthy heart in a healthy body." Posters that featured girls usually portrayed just one girl, perhaps holding a bar of chocolate, drinking a cup of hot chocolate, or pouring cacao mix into a cup, even finding chocolate bars inside a bird's nest.

Adult women in Swiss chocolate posters commonly were depicted in scenes of elegance, commonly with a side view of the face in a romantic style, and shown holding flowers or surrounded by "nature." In another instance, an elegant woman was painted wearing a formal gown and riding sidesaddle on a red horse. We identified an unusual example from Switzerland that showed two women in flowing robes reminiscent of classical Greece—holding a chocolate bar.

In contrast with French posters, we identified only two examples of romance in Swiss chocolate advertising posters. One depicted a woman in formal dress sitting in a chair watching a man who appears to be directing an orchestra with a chocolate bar. The other poster revealed a couple dressed in costume and mask, with the man holding a chocolate bar out of reach of the woman who is stretching her hand out to receive it.

Several Swiss chocolate advertising posters featured images only of candy bars, an approach not seen as often in French posters. Like French posters, however, Swiss examples also featured animals, especially cats, dogs, and horses. Several interesting examples show dogs as the central character, whether St. Bernards or perhaps Bernese mountain dogs, trekking in the snow with alpine towns and mountains in the background. In one instance, the dog had a bar of chocolate suspended around its neck (rather than the traditional keg of brandy), implying that chocolate had equal or better reviving qualities. A unique theme, not seen in the chocolate posters of other countries, was the integration of chocolate bars with green alpine pastures and barns in the foreground with snow capped Alps in the background, reminiscent of Switzerland.

AUSTRIA

We found only four examples of chocolate advertising posters from Austria. One was similar to the Swiss romantic style posters, painted in pale hues and depicting a mother surrounded by flowers, giving chocolate to her two children. The other three posters were cartoon style, drawn with clean lines and bright colors. All these posters depicted humans in a similar way, with a smile reduced by the artist to a simple white splash and eyes that were merely dark marks on the faces. Two of these posters depicted children, whether leaning out of a train window with chocolate in their hands, or a boy dressed in traditional Austrian clothing carrying a suitcase and chocolate bars. The third (and perhaps most curious) Austrian poster showed a Mexican man wearing a poncho and sombrero, holding an oversized chocolate bar. All these cartoon-style posters evoked thoughts of travel.

GERMANY

German chocolate advertising posters also incorporated travel and energy themes. One poster we examined showed a scene of a vigorous woman full of energy atop a mountain holding a chocolate bar above her head. Her male companion, in contrast, was seated

FIGURE 15.4. Chocolate poster (Kasseler), color lithograph, German School; dated early 20th century. *Source*: Bibliotheque des Arts Decoratifs, Paris, France, Archives Charmet. Courtesy of The Bridgeman Art Library, London, England. (Used with permission.) (See color insert.)

behind her in shadows, implying it was a hard climb to the top. Another poster shows a woman carrying a medical staff serving liquid chocolate to an invalid (Fig. 15.4). Other German posters depicted individuals from Africa or Native Americans serving chocolate; both images evoked thoughts of travel and, curiously, reinforced chocolate's geographical origin in the Americas. Similar to French posters, one German poster shows children sharing chocolate with each other.

HOLLAND

Chocolate advertising posters from Holland displayed a sense of nationalism with traditional dress. Although the posters depicted different scenes—whether children, courting couples, or mothers—the characters usually were dressed in traditional clothing that included bonnets and clogs. The style of Dutch posters usually was realistic and very few were drawn in a romantic style.

ITALY

Unlike those from Holland, most of the Italian chocolate posters we identified were drawn in an abstract style with clean lines and solid colors, with no details such as clothing or facial features. One poster shows a

FIGURE 15.5. Chocolate poster (Vinay), color lithograph, French School. *Source*: Courtesy of The Bridgeman Art Library, London, England. (Used with permission.) (See color insert.)

simple drawing of a cow against a solid red background (Fig. 15.5). Another popular figure was a clown whose body resembled a cacao/coffee bean: round brown face with simple white lines for facial features and arms and legs like ribbons. Women and children also were popular themes in Italian posters. One depicted a sophisticated woman sitting at a table, writing and sipping a cup of hot chocolate, while another was of a woman dressed in candy wrappers, sprinkling candy from her hand. The posters with children depicted toddlers with chocolate: one a young girl eating a bar of chocolate sitting on an old man's lap and another showing a young girl about to drink from an oversized cup of hot chocolate. Some interesting posters from Italy depicted exotic scenes, such as a geisha by a lake on a moonlit night drinking hot chocolate, or a family of elephants, smiling, playing, and wearing hats.

MONACO AND SPAIN

We identified only one chocolate advertising poster from Monaco, but this item deserves mention simply because of its rarity. The poster depicted the relatively common child and cat theme, with a young girl dressed in a bright blue jumpsuit holding bars of chocolate above her head and smiling, while a black cat looks up at the chocolate with an open mouth. The colors in the poster are vibrant, with a bright red background and bright blue clothes on the child.

Several of the chocolate advertising posters from Spain pictured women in highly romantic styles, dressed in flowing robes with flowers in their hair and also surrounded by flowers. Some of the more realistic-style Spanish posters depicted children sharing chocolate with each other. Others showed children, women sitting at desks writing letters (perhaps to a distant loved one?), and African men serving hot chocolate.

RUSSIA

Chocolate advertising posters from Russia commonly depicted women in traditional clothing with head-pieces and extravagant jewelry, or courting couples, set within religious motifs. Others were more fanciful, as one where a courting couple was seated on an enormous chocolate egg. One poster commonly reproduced depicted a baby with chocolate smeared over his face and clothing, notable for its use of bright green, red, and yellow colors. Another Russian chocolate poster depicted a black child drawn in a cartoon style that resembled blackface: the child in cartoon form holds a box of chocolate while the box itself was illustrated with a realistic portrait of a white woman.

ENGLAND

Chocolate advertising posters from England used more muted colors than other countries, although blue, red, and yellow remained the most popular. Women were often depicted in leisure situations, sometimes elegantly dressed in formal Victorian era clothing and sipping hot chocolate. Often the women were paired, and one poster had two women walking together with a girl through a pasture with cows in the background. Early Cadbury posters depicted different scenes and added clever phrases, as with a bulldog guarding chocolate and the word "unapproachable" printed underneath. Other posters had scenes of men playing rugby and a text that boasted Cadbury's cacao is "absolutely pure" and would give the drinker "strength and staying power." One poster pictured a young boy in a field dressed in a cricket uniform, with the words "not beaten" written underneath. Two English posters we identified depicted servants, one being "assaulted" by a young boy and girl with toys in hands, each pulling on the servant's bonnet. The words "mother's cacao in danger" appear below in the poster text.

UNITED STATES

One characteristic that defined American chocolate advertising posters and set them apart from their European counterparts was their focus on chocolate bars as the primary point of interest (Fig. 15.6). In addition, American chocolate bar posters were the only ones we identified that contained information on ingredients or fillings, whether coconut, marshmallow, nuts, or peanut butter. We identified several examples of elegant American posters, picturing well-dressed men and women at parties serving hot chocolate. In contrast to European themes, the American posters we identified presented few examples of children. One interesting example, however, showed a baby emerging from a cacao nut grasping a bar of chocolate, while another more traditionally showed children drinking out of an oversized cup.

FIGURE 15.6. Chocolate poster (Bunte), Christmas advertisement; dated 1930. *Source*: Courtesy of The Bridgeman Art Library, London, England. (Used with permission.) (See color insert.)

Conclusion

Chocolate advertising posters can be viewed as a form of art history. Although posters available on the World Wide Web seldom had associated dates (when created), the various styles, form, and content could be reviewed and analyzed. The colors used in the chocolate posters most commonly were blue, red, and yellow; some companies seemed to prefer bolder shades while others decorated their posters in pale tones. Themes common to the countries we identified regularly included children, courtship/romance, elegance, and mothers. Posters that featured human images dressed in traditional clothing were important icons of national culture. As with any form of advertising, the posters were designed to associate chocolate as a positive food in the mind of consumers.

Acknowledgments

I would like to acknowledge the efforts of previous workers at the University of California, Davis, affiliated with the Chocolate History Project, who also assisted with identification and compilation of chocolate poster images, especially Ezra Neale. I also would like to acknowledge my mother, Barbara Swisher, for editorial support during the preparation of this chapter.

Reference

1. Coe, S. D., and Coe, M. D. *The True History of Chocolate.* London, England: Thames and Hudson, 1996.

CHAPTER

Chocolate at the World's Fairs, 1851–1964

Nicholas Westbrook

Introduction

In 1876, Americans celebrated their nation's coming of age with a grand "International Exhibition of Arts, Manufactures, and Products of the Soil and Mines" (or more familiarly, the "Centennial Exposition") in Philadelphia. During the 159 days of the fair, some 10 million people, a quarter of the nation's population, came to see the products of 13,720 exhibitors and 38 countries. Centerpiece and icon of the Exposition was the world's largest steam engine, the giant 40-foot-high Corliss, which was the prime mover for the 14 acres of machines in Machinery Hall. President Grant challenged visitors: "I hope a careful examination of what is now about to be exhibited to you will not only inspire you with a profound respect for the skill and taste of our friends from other nations, but also satisfy you with the attainments made by our own people during the past one hundred years" [1, 2]. America was taking her place among the major industrial powers of the world.

Many exhibitors presented their products in artful arrays suggesting plenitude and variety. Twenty exhibitors focused on candy production. Chocolate manufacturers created freestanding pavilions within Agricultural Hall to exhibit their work. So at location number 53a, for example, one found a dark-brown, richly polished wooden display case surmounted by flags. The exhibit case bore the gilt-lettered boast of Joseph Storrs Fry & Sons, London and Bristol, that they were suppliers to Queen Victoria and the Prince of Wales (Fig. 16.1).[1] Nearby, at location number 183, stood the proud showcase of *Chocolat Menier* , Paris, just then at the beginning of its dramatic corporate rise [4].[2] At location number 23 stood the exhibit of Emile Menier, London. Matias Lopez of Madrid exhibited his work at location number 1806, and next to him was the chocolate exhibit of Lopez and Vaquez, also from Madrid [8]. Philadelphia manufacturer Stephen F. Whitman and Son had exhibits in both Agricultural Hall and in Machinery Hall—which his trade cards announced as a "branch manufactory" of his main plant downtown on Market Street. Whitman's exhibit in Machinery Hall included a panning machine for chocolate-coating nuts, and marble-topped production tables (Fig. 16.2). Whitman proudly noted that the Viennese bakery at the Exposition used his chocolates [9, 10]. The enthusiasm for seeing the manufacturing processes in motion continued at the J.M. Lehmann exhibit of German-made machines at the 1893 Columbian Exposition in Chicago [11]. At the 1904 Louisiana Purchase Exposition in St. Louis, Walter Baker & Company exhibited German-made miniature machines, powered "by its own little motor," demonstrating the steps in transforming the raw beans into bolted cocoa powder [12].

Chocolate: History, Culture, and Heritage. Edited by Grivetti and Shapiro
Copyright © 2009 John Wiley & Sons, Inc.

FIGURE 16.1. Chocolate exhibit: J. S. Fry of Bristol and London, Agricultural Building, Centennial Exposition, Philadelphia, 1876. *Source:* Image C020428. Courtesy of the Free Library of Philadelphia, Philadelphia, PA. (Used with permission.) (See color insert.)

Most impressive in Philadelphia were the pair of exhibits produced by Henry Maillard of New York. In Machinery Hall, Maillard organized a comprehensive manufacturing exhibit of "new machines . . . in working order" (Fig. 16.3) [13]. In Agricultural Hall, Maillard presented a confectionary exhibit in a tall walnut case (Fig. 16.4). Towering in the center of the case was a 15-foot "monument of white sugar and chocolate" depicting episodes in American history, including America's first victory of the Revolution— the capture of Fort Ticonderoga by Ethan Allen, Benedict Arnold, and the Green Mountain Boys, an event that occurred exactly a century to the day before the Exposition opened [14, 15].

These chocolate exhibits (and similar ones in cutlery, machine tools, papermaking, and hundreds of other trades) at the Centennial Exposition established a pattern that would hold for the next century of world's fairs: attention to both product and manufacturing process, emphasis on dramatic size or presentation in order to create a memorable corporate image, and, for chocolate, the simple pleasures of taste. The consistent strategy of exhibitors during the past 150 years of world's fairs has been More! Bigger! Better!

Discovering the Botany and Agriculture of Cacao

In an era when scholars and laypersons alike were devoted to collecting, grouping, and categorizing the flora and fauna of the world, the great Victorian

FIGURE 16.2. Chocolate exhibit: Stephen F. Whitman and Son of Philadelphia, Machinery Hall, Centennial Exposition, Philadelphia, 1876. *Source:* Image C021232. Courtesy of the Free Library of Philadelphia, Philadelphia, PA. (Used with permission.) (See color insert.)

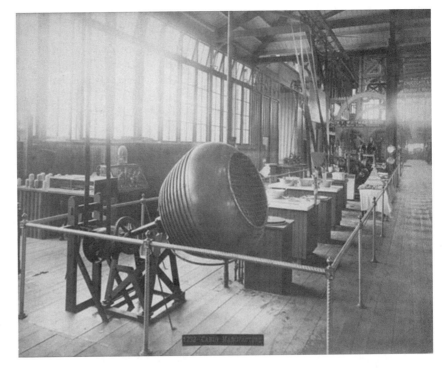

FIGURE 16.3. Chocolate exhibit: Henry Maillard, New York, Machinery Hall, Centennial Exposition, Philadelphia, 1876. *Source:* Image C02182. Courtesy of the Free Library of Philadelphia, Philadelphia, PA. (Used with permission.) (See color insert.)

FIGURE 16.4. Chocolate exhibit: Henry Maillard, New York, sugar and chocolate sculpture depicting major events in American history; Philadelphia Centennial Exposition, 1876. *Source:* Frank Leslie's Illustrated Historical Register of the Centennial Exposition 1876. p. 162. Courtesy of Peter J. Shields Library, University of California, Davis. (Used with permission.)

enterprise, fair organizers during the 19th century attempted to do likewise for the "Arts, Manufactures, and Products of the Soil and Mines," and even the anthropological aspects of world cultures.

Guests at the first true world's fair, London's 1851 "The Great Exhibition of the Works of Industry of all Nations" (more familiarly the "Crystal Palace Exhibition"), visited the section categorized as interpretive exhibits of "substances used as food," including chocolate, coffee, dates, and honey [16–18]. In 1851, this exhibition area remained a showcase of foods from faraway places, rather than a food court full of familiar concessions. Schweppes paid £5500 for concession rights to provide refreshments at this fair. Sale of intoxicating beverages was forbidden: vendors offered tea, coffee, chocolate, cocoa, lemonade, ginger beer, and soda water [19, 20]. As late as 1904, visitors to the Louisiana Purchase Exposition could visit the Haiti pavilion in the Palace of Forestry, Fish, and Game, where they viewed the "Grand Prize winners for chocolate and cocoa in the pod" [21]. They could study the raw materials for chocolate under large glass jars: cacao pods, beans, shells, cocoa powder, and cocoa butter, and finished chocolate at the Walter Baker pavilion [22].

Discovering the Taste of Chocolate

Chocolate manufacturers immediately realized that offering taste samples could expand business. The Walter Baker & Company pavilion at the 1893

Columbian Exposition offered samples of chocolate and cocoa—and provided complimentary copies of celebrity cook Maria Parloa's "Choice Receipts," specially prepared for Baker (Figs. 16.5 and 16.6) [23, 24]. Chocolat Menier introduced an automatic chocolate dispenser to visitors at the 1900 Paris *Exposition Universelle* [25]. At the 1901 Pan-American Exposition in Buffalo, both Baker and Walter M. Lowney offered cups of cocoa for 5 cents, dispensed from their lavish pavilions [26]. The Haitian pavilion at the 1904 St. Louis Fair provided samples of its native coffee and cocoa [27]. The Baker pavilion at the St. Louis Fair offered "Miss Burr of the Boston School of Domestic Science" lecturing daily on the uses of chocolate in scientific cookery [28]. As late as the 1939 World's Fair in New York City, refreshment stands offered chocolate milk and cocoa. The Belgian Restaurant at that World's Fair offered chocolate or cocoa on its menu [29].

The 1893 Columbian Exposition introduced two new chocolate products to the world. Boston chocolatière Walter M. Lowney introduced America's first viable chocolate bars [31–33].[3] According to some sources, Mrs. Potter (Bertha) Palmer introduced a new way of enjoying chocolate via her husband's Palmer House Hotel. She asked the chef to make a "ladies' dessert" to be used in the box lunches at the Women's Building at the Fair, edible without getting a lady's gloved fingers dirty. The "brownie" is still served at the Hilton Palmer House today [35].[4] "Brownie"-shaped chocolates were offered in the 1897 Sears, Roebuck catalogue. The earliest brownie recipe was published in the 1897 Sears, Roebuck catalogue [36]. Fannie Farmer's revised edition of her *Boston Cooking-School Cook Book* (1905) included a recipe for brownies. Maria Willett Howard followed her mentor with a 1907 recipe, enriched by an extra egg, in *Lowney's Cook Book.* During the following decades of "brownie"-mania, a period that spanned 1887 through the 1920s, Farmer, Howard, and their competitors promoted "brownie" recipes, with slight variations as to quantities of eggs and chocolate [37–39].

Creating Corporate Identity

At the 1876 Centennial Exposition, Maillard had created the fair's most memorable chocolate exhibit— not with its new machines but with its 15-foot monument encapsulating dramatic moments in American history. At the 1889 Paris Exposition, Maillard challenged the host country by bringing back "home" one of France's most cherished antiquities: a reproduction of the *Venus de Milo* in chocolate, weighing some 3000 pounds. Parisians and even Americans were derisive, readily visualizing the authentic *Venus* on exhibit nearby at the museum in the Louvre. A French writer lamented ironically, after a hot summer, that "the poor Venus had lost some of her elegance of form. No, chocolate

FIGURE 16.5. Chocolate trade card (Walter Baker & Company, Exhibition Building), Columbian Exposition, Chicago, 1893. *Source:* Private Collection. Courtesy of Nicholas Westbrook. (Used with permission.) (See color insert.)

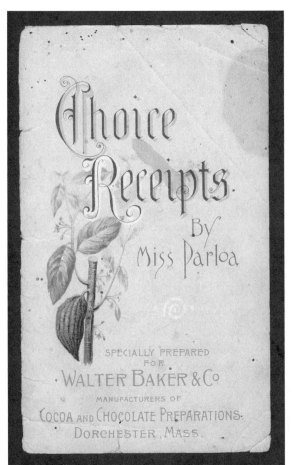

FIGURE 16.6. Chocolate trade card (Walter Baker & Company, Choice Receipts by Miss Parloa), Columbian Exposition, Chicago, 1893. *Source:* Private Collection. Courtesy of Nicholas Westbrook. (Used with permission.) (See color insert.)

FIGURE 16.7. Chocolate trade card (Stollwerck), obverse, Columbian Exposition, Chicago, 1893. *Source:* Private Collection. Courtesy of Nicholas Westbrook. (Used with permission.) (See color insert.)

is decidedly not the sculptural material of the future" [40]. The official American observer, Professor Arthur J. Stace of Notre Dame University, was even less charitable:

> *Of Maillard's chocolate Venus, we do not know what to say. Venus is presumably sweet in any form, but in chocolate she may be thought too sweet for any use. At the time of this writing, however, she shows the results of various nibblings, although the Exposition is not half over.* [42]

Exhibitors at future fairs would have to try to outdo the Maillard precedents. At the 1893 World Columbian Exposition, the Stollwerck brothers of Cologne, Germany, responded (Fig. 16.7). In the Agricultural Building, Stollwerck presented a reproduction of the famous *Germania* statue at Niederwald—here 11 feet high and weighing 3000 pounds. "Thus does civilization progress from one triumph to another." The entire Stollwerck pavilion was made of chocolate and towered 38 feet high in the form of a Renaissance temple, made of 15 tons of chocolate and cocoa butter. On the base of the pavilion are "reliefs, more than life

size," of the emperors William I, Frederick III, William II, Bismarck, Von Moltke, and other historic characters [45–48].

The Architecture of Chocolate

Companies created architectural fantasies to establish a memorable corporate image. At the 1889 Paris *Exposition Universelle*, Chocolat Menier erected an *Arc de Triomphe* of 250,000 chocolate bars wrapped in their trademark gold wrappers—representing just one day's production [49].[5] At the 1893 Columbian Exposition, Chocolat Menier and Lowney's of Boston created ornate *Beaux-Arts* pavilions [51].[6] The C. J. van Houten

FIGURE 16.9. The Dutch House, Brookline, Massachusetts. Formerly the C. J. Van Houten & Zoon Pavilion, Columbian Exposition, Chicago, 1893. *Source:* Courtesy of Amy Houghton. (Used with permission.)

FIGURE 16.8. Postcard: The Dutch House, Brookline, Massachusetts. Formerly the C.J. Van Houten & Zoon Pavilion, Columbian Exposition, Chicago, 1893. *Source:* Private Collection. Courtesy of Nicholas Westbrook. (Used with permission.) (See color insert.)

& Zoon Cocoa Company of Weesp, the Netherlands, recreated the 1591 brick town hall of Franker, Holland (Figs. 16.8 and 16.9) [56].[7] Another company, J. & C. Blooker, built a Dutch cocoa windmill and a 17th century traditional Dutch house from which to serve their samples. The mill was a replica of the one used by the Blooker brothers since the beginning of the century to grind their cocoa: "Adjoining the mill, or rather forming a portion of it, is a Dutch house of the old style, built and furnished to illustrate the surroundings of a burgher of moderate means and refined tastes. . . . There are girls who serve the cocoa dressed in the national costumes of the Netherlands as worn in early days" (Fig. 16.10) [59–61].

The Walter Baker & Company concocted *Beaux-Arts* confections as their "colonial" structures for the 1893 Columbian Exposition, the 1901 Pan-American Exposition in Buffalo, and a two-story faux-marble structure inside the Agriculture Palace at the 1904 Louisiana Purchase Exposition in St. Louis (Fig. 16.11) [62–64]. For the 1907 Jamestown Exposition, the company managed to create a reasonable facsimile

of a wood-framed middle-class New England house of 1780, then promoted as the year of the company's founding. The "Chocolate House" still stands on Dillingham Boulevard ("Admiral's Row") at the Naval Station Norfolk, the world's largest naval base, erected on the site of the Exposition.[8] During the fair, the Baker pavilion housed an exhibit of the company's chocolate-making machinery and dispensed samples [65–67].

Conclusion

The most dramatic chocolate transformation story focuses on Pennsylvania caramel-manufacturer Milton W. Hershey. Thirty-six-year-old Hershey visited the 1893 Chicago fair, curious about products and possibilities. Encountering the J. M. Lehmann exhibit of milk-chocolate manufacturing equipment from Dresden, Germany, Hershey recognized an opportunity and bought the entire assembly of machines. When the fair closed, he ordered the machines shipped to Lancaster, Pennsylvania. Hershey soon quit his caramel business and focused on the chocolate business, made possible by the 1893 Chicago Columbian Exposition equipment [69]. By 1900, Hershey had introduced the first Hershey bars (plain and almonds). In 1903, Hershey broke ground for a new chocolate factory—and town—at Derry Church (renamed Hershey in 1906), Pennsylvania [70–72]. The rest of the Hershey story is history!

FIGURE 16.10. Chocolate trade card (Blooker), obverse, Columbian Exposition, Chicago, 1893. *Source:* Courtesy of the American Antiquarian Society, Worcester, MA. (Used with permission.) (See color insert.)

FIGURE 16.11. Chocolate exhibit: Walter Baker & Company pavilion, Pan-American Exposition, Buffalo, 1901. *Source:* Image Number 475. Courtesy of Dorcester Atheneum. (Used with permission.)

At the 1964 New York World's Fair, Hershey offered a comprehensive history of chocolate in the Better Living Center, the third-largest pavilion at the fair [73].

Acknowledgments

The author wishes to thank Georgia B. Barnhill, American Antiquarian Society; Stan Daniloski whose Web sites on World's Fairs are invaluable; the Dorchester (Massachusetts) Athenaeum; Amy Houghton, Sasaki Associates; the Free Library of Philadelphia; and the John Springer Collection, the University of Iowa.

Endnotes

1. In 1847, the great-grandson of the founder of Fry mixed sugar and cocoa butter back into the defatted cocoa powder to create edible bars. Cadbury (established 1831) absorbed Fry in 1916; the firms officially merged in 1919. Earlier attempts at producing chocolate bars were made by Callier (1819), van Houten (1828), and Menier (1836). Lowney created a true chocolate-bar product in the United States in 1893 [3, 5].

2. Launched in 1821, Chocolat Menier grew out of a pharmaceutical grinding business located at an ancient water-power site at Noisiel on the Marne. The word *Confection* originally meant a medicinal preparation bound either in honey or a sugar syrup. As part of an expanding medicinal line, company founder Jean-Antoine Brutus Menier put on the market in 1836 the first chocolate bars in France, wrapped in gold paper "in order to distinguish, in a complete manner, the products of this factory, with a wrapper and a label echoing the [gold]-medal awards hitherto unknown in commerce." The golden packaging became Menier's trademark for more than a century. Menier was also celebrated at successive World's Fairs at the end of the 19th century for the company's paternalistic role with its employees. The deaths of the last Menier brothers in 1959 and 1967 led to the absorption of Chocolat Menier by Rowntree Mackintosh in 1973–1976, and in turn by Nestlé in 1988. In 1990, Nestlé announced the closing of the chocolate manufacturing business at Noisiel. In 1992, the old mill, le Moulin Saulnier, was declared a historical landmark [6, 7].

3. Fry had made an earlier attempt at producing chocolate bars in 1847. Chocolat Menier had created a monument of chocolate bars at the 1889 Paris Exposition. Lowney's was the first successful American attempt [30].

4. Palmer Cox had begun a series of stories about his "brownies" in the St. Nicholas children's magazine in 1883. The first collection of Cox's "brownie" stories was published in 1887. Cox licensed the "brownie" name to many other manufacturers, most notably Eastman Kodak [34].

5. The Eiffel Tower was the great icon of the 1889 Paris fair. Menier boasted that one day's production of chocolate bars would be 17 times taller than the Eiffel Tower. Six months' production was equivalent to the weight of the Tower. Chocolat Menier repeated the *Arc de Triomphe* trope at the Chicago fair four years later [50].

6. The Chocolat Menier pavilion was designed by Peter Joseph Weber [52]. Chocolat Menier was declared the world's finest chocolate at the Chicago fair [53]. Menier boasted that its chocolate was "the only chocolate dispensed at all the restaurants of the Vienna Bakery" [54].

7. Conrad J. van Houten & Zoon was established in 1828 (some sources say 1814, but van Houten's own advertising in the 19th century provided the date as 1828). Van Houten patented a method for pressing cocoa butter from the roasted beans. Van Houten was awarded a royal warrant for its cocoa in 1889. The company's pavilion for the 1893 Columbian Exposition was designed by Dutch architect G. Weyman. The award-winning "Dutch House," as it is known today, is one of just three buildings surviving from the 1893 Columbian Exposition. The building was prefabricated in Holland and Belgium, then dismantled for reerection in Chicago. Captain Charles Brooks Appleton became so enamored of the building when he saw it in Chicago that he purchased it at auction when the fair closed and had it shipped to Brookline, Massachusetts, where it stands today at 20 Netherlands Road. The Dutch House has been on the National Register of Historic Places since January 24, 1986 [55]. Van Houten & Zoon was acquired by Leonhard Monheim AG, Aachen, Germany, in 1971 [57, 58].

8. The Chocolate House and other surviving buildings from the Exposition were named an historic district on the National Register of Historic Places in 1975 [68].

References

1. Kenin, R. Introduction. In: *Frank Leslie's Illustrated Historical Register of the Centennial Exposition 1876*. New York: Paddington Press Ltd., 1974.

2. Post, R. C. *1876: A Centennial Exhibition*. Washington, DC: Smithsonian Institution, 1976.

3. Kimmerle, B. *Chocolate: The Sweet History*. Portland, OR: The Collector's Press, 2005; pp. 22, 99–102.

4. Centennial Exhibition Digital Collection, Free Library of Philadelphia. Available at http://www.library.phila. gov/CenCol/Details.cfm?ItemNo=c021642&source URL=cedcpubsrch9.cfm and image c021643. (Accessed February 24, 2007.)

5. Trager, J. *The Food Chronology*. New York: Henry Holt, 1995; pp. 208, 216, 300.

6. Collection Saga Menier, Noisiel. Available at http://perso.orange.fr/pone.lateb/chronologie.htm. (Accessed February 24, 2007.)

7. Turpin, N. L. Hot chocolate: the social question in the chocolate exhibits at the 1900 Paris Universal Exposition. *Journal for the Study of Food and Society* 2003;**6**(2):72–77.

8. Centennial Exhibition Digital Collection, Free Library of Philadelphia. Available at http://libwww.library. phila.gov/CenCol/Details.cfm?ItemNo=c021642& sourceURL=cedcpubsrch9.cfm and image c021643. (Accessed February 24, 2007.)

9. Centennial Exhibition Digital Collection, Free Library of Philadelphia. Available at http://libwww.library. phila.gov/CenCol/Details.cfm?ItemNo=c110150& sourceURL=cedcpubsrch9.cfm and image c021232. (Accessed February 24, 2007.)

10. Kimmerle, B. *Chocolate: The Sweet History*. Portland, OR: The Collector's Press, 2005; p. 135.

11. Bancroft, H. H. *The Book of the Fair: An Historical and Descriptive Presentation*. Chicago: Bancroft Company, 1893; p. 332.

12. 1904 World's Fair Agriculture Palace. Available at http://www.lyndonirwin.com/04chocol.htm. (Accessed October 1, 2006.)

13. Centennial Exhibition Digital Collection, Free Library of Philadelphia. Available at http://libwww.library. phila.gov/CenCol/Details.cfm?ItemNo=c021382& sourceURL=cedcpubsrch9.cfm. (Accessed February 24, 2007.)

14. Kenin, R. *Frank Leslie's Illustrated Historical Register of the Centennial Exposition 1876*. New York: Paddington Press Ltd., 1974; pp. 162 and 212.

15. Maillard's Chocolate for Breakfast, Lunch, and Travellers (trade card), 1876, collection of American Antiquarian Society.

16. Ffrench, Y. *The Great Exhibition: 1851*. London, England: Harvill Press, 1950; p. 23.

17. Gibbs-Smith, C. H. *The Great Exhibition of 1851*. London, England: Her Majesty's Stationery Office, 1981.

18. Gloag, J. Introduction. *The Crystal Palace Exhibition Illustrated Catalogue*. New York: Dover Publications, 1970. p. xxv. An unabridged republication of the Art-Journal Catalogue special issue.

19. Daniloski, S. The World's Fair and Exposition Information and Reference Guide. Available at www. earthstation9.com/1851_lon.htm. (Accessed February 12, 2007.)

20. Trager, J. *The Food Chronology*. New York: Henry Holt, 1995; p. 247.

21. A Brief Sketch of Haiti for the Visitors of the World's Fair at St. Louis, MO. St. Louis, MO: Perrin & Smith Printing, 1904. Available at http://www.webster. edu/~corbetre/haiti/misctopic/texts/stlouis.htm. (Accessed October 1, 2006.)

22. 1904 World's Fair Agriculture Palace. Available at http://www.lyndonirwin.com/04chocol.htm. (Accessed October 1, 2006.) Derived from The World's Work Advertiser, August 1904.

23. Walter Baker & Co. trade cards, 1893. Author's collection.

24. Shapiro, L. *Perfection Salad: Women and Cooking at the Turn of the Century*. New York: Modern Library, 2001; pp. 44–66.

25. Quantin, A. *L'Exposition du Siècle 1900*. Paris, France: Le Monde Moderne, (ca. 1901); p. 231.

26. 1901 Pan-American Exposition. Available at http://panam1901.bfn.org/visiting/food/foodbeverages.htm. (Accessed February 15, 2007.)

27. http://www.scripophily.net/balgarinc.html. (Accessed October 1, 2006.)

28. 1904 World's Fair Agriculture Palace. Available at http://wwwlyndonirwin.com/04chocol.htm. (Accessed October 1, 2006.)

29. http://www.armchair.com/recipe/fair39a.html. (Accessed October 1, 2006.)

30. Trager, J. *The Food Chronology*. New York: Henry Holt, 1995; pp. 208 and 216.

31. Overfelt, M. A world (fair) of invention. *Fortune Small Business*, April 1, 2003.

32. Kimmerle, B. *Chocolate: The Sweet History*. Portland, OR: The Collector's Press, 2005; pp. 22, 68–69, 99–102.

33. Howard, M. W. *Lowney's Cookbook . . . A New Guide for the Housekeeper . . . for Any Well-to-do Family*. Boston: W. M. Lowney Company, 1907.

34. http://www.townshipsheritage.com/Eng/Hist/Arts/cox.html. (Accessed February 18, 2007.)

35. Chicago's Palmer House Chocolate Fudge Recipe. Available at http://www.recipezaar.com/recipe.print?id=207091. (Accessed February 18, 2007.)

36. Trager, J. *The Food Chronology*. New York: Henry Holt, 1995; p. 354.

37. Zanger, M. H. Brownies. In: Smith, A. F., editor. *Oxford Encyclopedia of Food and Drink in America*, Volume I. New York: Oxford University Press, 2004; pp. 136–138.

38. Farmer, F. M. *The Boston Cooking-School Cook Book*. Boston: Little, Brown & Company, 1924; pp. 613–614 and 631.

39. Kimmerle, B. *Chocolate: The Sweet History*. Portland, OR: The Collector's Press, 2005; p. 166.

40. Huard, L. *Livre d'Or de l'Exposition*. Paris, France: L. Boulanger, 1889; Volume 2, pp. 485–486.

41. Chandler, A. Revolution: The Paris *Exposition Universelle*, 1889. World's Fair 1986;7(1). Available at http://charonsfsu.edu/publications/PARISEXPOSITIONS/1889EXPO.html. (Accessed October 1, 2006.)

42. Stace, A. J. *Education and the Liberal Arts. Report of the United States Commissioners to the Universal Exposition of 1889 . . .* Washington, DC: Government Printing Office, 1890–1891; Volume II, pp. 115–191. Cited in: Havlik, R. J. University of Notre Dame and the Napoleon III Telescope. Available at http://transitofvenus.org/napoleon3.htm. (Accessed February 15, 2007.)

43. Walton, W. *[World's Columbian Exposition] Art and Architecture, Architecture Volume*. Philadelphia: George Barrie, 1893; p. xxxvi.

44. Boime, A. The chocolate Venus, "tainted" pork, the wine blight, and the tariff: Franco-American stew at the Fair. In: Blaugrund, A., editor. *Paris 1889: American Artists at the Universal Exposition*. New York: Harry Abrams, Inc., 1989.

45. Walton, W. *[World's Columbian Exposition] Art and Architecture, Architecture Volume*. Philadelphia: George Barrie, 1893; p. xxxvi.

46. Bancroft, H. H. *The Book of the Fair: An Historical and Descriptive Presentation*. Chicago: Bancroft Company, 1893; pp. 371–372.

47. Galvin, R. M. *Sybaritic to some, sinful to others, but how sweet it is! Editor's Choice Smithsonian: An Anthology of the First Two Decades of Smithsonian Magazine*. Washington, DC: Smithsonian Books, 1990; pp. 42–51.

48. Kimmerle, B. *Chocolate: The Sweet History*. Portland, OR: The Collector's Press, 2005; pp. 84–87.

49. *Les Réflexions d'un Visiteur Curieux devant L'Exposition du Chocolat-Menier*. Available at http://perso.orange.fr/pone.lateb/exposition.htm. (Accessed February 24, 2007.)

50. Bancroft, H. H. *The Book of the Fair: An Historical and Descriptive Presentation*. Chicago: Bancroft Company, 1893; p. 370.

51. Bancroft, H. H. *The Book of the Fair: An Historical and Descriptive Presentation*. Chicago: Bancroft Company, 1893; p. 397.

52. Blum, B. J. Interview with Bertram A. Weber, 1983. In: *Chicago Architects Oral History Project*. Chicago: Art Institute of Chicago, 1995; Revised edition, 2004. www.artic.edu.

53. http://perso.orange.fr/pone.lateb/exposition.htm. (Accessed February 24, 2007.)

54. Garden & Forest 1893;6(277) Jun 14. Advertisement in miscellaneous unnumbered back pages.

55. http://www.nationalregisterofhistoricplaces.com/MA/Norfolk/state2.html. (Accessed February 15, 2007.)

56. Update. Department of Planning and Community Development, Town of Brookline (MA), Nov–Dec 2002; p. 5.

57. John Springer Collection, University of Iowa; Box 10: booklet describing van Houten presence at the 1893 Columbian Exposition; Box 12: van Houten trade card collected during Springer's visit to the fair.

58. Trager, J. *The Food Chronology*. New York: Henry Holt, 1995; pp. 204, 216, and 607.

59. Blooker's Dutch Cocoa trade cards, 1893, collection of the American Antiquarian Society.

60. Blooker advertisement. *Cosmopolitan*, December 1893, p. 93.

61. Bancroft, H. H. *The Book of the Fair: An Historical and Descriptive Presentation*. Chicago: Bancroft Company, 1893; p. 397.

62. Walter Baker & Co. trade card, 1893. Author's collection. Available at http://www.dorchesteratheneum.org/page.php?id=565. (Accessed February 15, 2007.)

63. 1901 Pan-American Exposition: Eck, S. J. Doing the Pan. Available at http://panam1901.bfn.org/miscbuildings/bakerbuilding.html. (Accessed February 15, 2007.)

64. Missouri State University, Agricultural History Series: 1904 World's Fair Agriculture Palace. Available at http://www.lyndonirwin.com/04chocol.htm. (Accessed October 1, 2006.)

65. http://www.navstanorva.navy.mil/expo.htm. (Accessed October 1, 2006.)

66. http://www.earthstation9.com/index.html?1907_jam.htm. (Accessed October 1, 2006.)

67. http://www.noahent.com/portfolio-historical.html#3. (Accessed February 25, 2007.)

68. http://www.nationalregisterofhistoricplaces.com/va/Norfolk/state.html. (Accessed February 15, 2007.)

69. Brenner, J. G. *The Emperors of Chocolate: Inside the Secret World of Hershey and Mars*. New York: Random House, 1999; pp. 85, 88, and 166.

70. Trager, J. *The Food Chronology*. New York: Henry Holt, 1995; pp. 342, 366, 376.

71. D'Antonio, M. *Hershey: Milton S. Hershey's Extraordinary Life of Wealth, Empire, and Utopian Dreams*. New York: Simon & Schuster, 2007.

72. http://www.hersheys.com/discover/milton/columbian_exp.asp. (Accessed October 1, 2006.)

73. Stanton, J. Better Living Center, 1964 World's Fair, 1997. Available at http://72.14.209.104/search?q=cache:Kr5vt-n0Li4J:naid.sppsr.ucla.edu. (Accessed October 1, 2006.)

PART

IV

Economics, Education, and Crime

Chapter 17 (Richter and Ta) considers reports of cacao and chocolate in 18th and early 19th century *Shipping News* documents. Their analysis reveals how cacao and chocolate were transported from points of debarkation in South and Central America and the Caribbean to primary ports along the eastern seaboard of North America, especially Baltimore, Boston, New York, and Philadelphia. Chapter 18 (Richter and Ta) presents an economic analysis of 18th and early 19th century *Price Current* documents that established prices for cacao and chocolate in the major European and North American East Coast cities. They report how the prices for these commodities fluctuated during periods of peace versus hostilities and by transportation distance. Chapter 19 (Grivetti) summarizes information from 18th century almanacs, religious tracts, and primary school textbooks that included cacao- and chocolate-related homespun medical advice, recommendations for mental and personal improvement, morality tales, and spiritual instruction. Chapter 20 (Grivetti) analyzes a suite of English documents where defendants were tried for cacao- and chocolate-associated crimes that ranged from assault, to grand theft, to murder, and reviews the verdicts for relative fairness/consistency or lack thereof. Chapter 21 (Grivetti) continues the "dark side" theme of chocolate and presents selected criminal cases that involved cacao/chocolate-associated arson, blackmail, counterfeiting, murder, smuggling, and theft.

Pirates, Prizes, and Profits

Cocoa and Early American East Coast Trade

Kurt Richter and Nghiem Ta

Our wants are various, and nobody has been found able to acquire even the necessaries without the aid of other people, and there is scarcely any Nation that has not stood in need of others. The Almighty himself has made our race such that we should help one another. Should this mutual aid be checked within or without the Nation, it is contrary to Nature. [1]

Introduction

Cocoa and chocolate have played profound roles in American history. Regardless of the form in which chocolate was consumed, a complex web of commerce and trade linked the production of cocoa to its consumption in the United States. In this chapter we document the commercial web that brought cocoa beans out of the production regions of the Caribbean and Central and South America into the busy ports along the eastern seaboard of North America. The history of cocoa trade in the 18th and 19th centuries is a story of the high seas, where great tall ships sailed from tropical Caribbean ports of call to the chaotic ports along the eastern seaboard of North America. Along the way, some ships were lost to storms, to pirates, and to other countries during times of war.

In this chapter we document the importation of cocoa into North America, a process that begins the commercial business of transforming the beans into an increasingly popular beverage and ultimately to a confection. We identify the sea captains and the merchants

who were the leading businessmen of their time, and we document how the chocolate manufacturers of Baltimore, Boston, New York, Philadelphia, and elsewhere were able to source the raw material of cocoa beans to produce their products for sale. Chocolate has a rich history in the Americas. In this chapter we also will show that this history starts far from North American shores and travels on the decks of ships commanded by seasoned seamen who sailed sometimes perilous waters to deliver raw materials from far-flung ports like Caracas, Cayenne, and Martinique to the docks of Baltimore, Boston, or New York.

Shipping News Documents

In the 18th and 19th centuries, seas were the highways of commerce and ports were hubs of commercial activity. Because of the importance of shipping commerce to port cities, ship arrivals regularly were reported in local newspapers. Such announcements were commonly called *Shipping News*. An example of a *Shipping News* document is presented here, as published in the *New-York Commercial Advertiser*, June 12, 1808 (Fig. 17.1). It lists the arrival of the Schooner *Pegasus* with a cargo of cotton, rocoa,[1] and cocoa [2].

This *Shipping News* example contains the basic information needed to link production to consumption

Chocolate: History, Culture, and Heritage. Edited by Grivetti and Shapiro
Copyright © 2009 John Wiley & Sons, Inc.

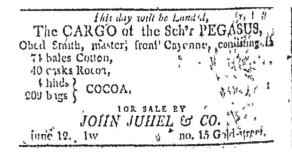

FIGURE 17.1. *Shipping News* announcement: John Juhel & Company, New York City. *Source:* Commercial Advertiser, dated June 12, 1808. Courtesy of Readex/Newsbank, Naples, FL. (Used with permission.)

during 1808. First, it gives the name of the ship, *Pegasus*, and its description as a schooner.[2] The report states that *Pegasus* was captained by Obed Smith and that the port of debarkation was Cayenne, capital of former French Guiana located in northeastern South America. This *Shipping News* document also lists the quantity of cocoa transported, along with other cargo. In this example the document also includes the name of the merchant who imported the cocoa—John Juhel—and provides the address of his office, 15 Gold Street, New York City.[3]

Research Methodology

The source we used to identify historical *Shipping News* documents was *Readex*®, a division of NewsBank, Incorporated (Early American Newspapers, Series I (1690–1876).[4] The Readex database allows researchers to search Early American newspapers and limit their queries through multiple search options. We used the internal Readex options to limit our search results to documents that Readex classified as *Shipping News* that contained the word *cocoa*. Each of the Readex results was reviewed for relevancy. Documents that contained the word *cocoa* without shipping information were removed from the results.

Readex presents researchers with specific tools that allow documents to be "captured" and exported from their company database. In our case, relevant results were exported as PDFs and saved to our laptops using a basic document storage program. We also exported citations directly from Readex into Endnote®, a reference management software program. For our analysis all relevant information from each historical *Shipping News* document was entered into *Access*®: document date, source, name of ship, type of ship, name of captain/master, associated cargo, quantities of cacao, port of debarkation, date of debarkation, port of embarkation, date of arrival, and any additional pertinent information. Using *Access*® tools, we

developed the tables and charts presented in this chapter.

Results

We captured 528 *Shipping News* documents that recorded cocoa imports to North American East Coast port cities. The years covered in our sample were 1721–1809, with the majority of documents published in the late 1790s and the early 1800s. The quantity of *Shipping News* documents available and the variable frequency of cocoa imports both worked to limit the number of reports we captured.

PRODUCTION AREAS

During the 18th and 19th centuries, cocoa production in the Americas centered in geographical areas that either bordered or were located within the Caribbean Sea. We identified 62 ports throughout northern South America; Mexico and Central America; and the Caribbean islands that had the climate necessary for cocoa production and export. While cocoa was native to tropical regions of South and Central America, because of the usefulness of cocoa beans, the distribution eventually expanded and quickly spread through neighboring humid tropical regions with good soil, plentiful rainfall, and sufficient foliage canopy for shade. The top 10 cocoa-exporting ports we identified are presented by their modern name and national affiliation (Table 17.1).

The island of Curaçao in the Lesser Antilles was blessed with a large natural harbor that developed into the port of Willemstad, a site that became the base for the Dutch West Indies Trading Company. Although too small for the development of significant cacao plantations, Curaçao became a leading hub for import/export and trans-shipment of cacao during the late 1700s through the early 1800s.

South of Curaçao, the Spanish colony of Venezuela was a primary cocoa production region because of its vast tropical lands suitable for cacao tree growth. Early adoption of cocoa by Spain [3], coupled with Venezuela as a major Spanish colony, increased the economic role of the port of La Guaira (Venezuela) during the 1700s. By the early 19th century the number of shipments of cacao from La Guaira nearly matched those that originated in Curaçao.

Cayenne in French Guiana and St. Pierres on the island of Martinique were primary French ports where cacao was either produced or exported. The tropical environmental conditions surrounding the port of Cayenne were ideal and supported cocoa production within French Guiana. Ports associated with English Caribbean colonies, especially St. Thomas (now the U.S. Virgin Islands) and the port

Table 17.1 Top 10 Cocoa Production Areas by Number of Shipments and Years (1794–1809)

Port	Number of Shipments	Beginning	Ending
Curaçao, Lesser Antilles	48	1806	1808
La Guaira, Venezuela	41	1806	1807
St. Thomas, U.S. Virgin Islands	28	1806	1807
Cayenne, French Guiana	26	1794	1809
La Guira, Dominican Republic	24	1806	1809
St. Pierres, Martinique	23	1806	1808
Kingston, Jamaica	16	1806	1809
St. Domingo, Dominican Republic	12	1797	1809
Saint-Barthélemy	11	1806	1809

of Kingston, Jamaica, also were important exporters of cacao.

SHIPS, CAPTAINS, AND TRADE ROUTES

In addition to identifying ship names, Captains/ Masters, and trade routes, *Shipping News* documents present clear evidence of cocoa importation and sometimes reveal the names of cacao importers. We collected this information to determine whether or not relationships could be established between types of ship, captains, ports of call, and wholesalers in the major North American port cities.

Hollywood pirate movies and the occasional sailing of the *U.S.S. Constitution* have firmly imbedded the three-masted frigate in the minds of most Americans as the image of the typical 18th and 19th century sailing ship. But our research revealed that no frigates were used in the commercial transportation of cocoa to Colonial America; trade was conducted using much smaller ships. Early in our research we also discovered that ships with identical names were involved in the cocoa trade, and it became critical, therefore, to document dates, captains/masters, and ports in order to clarify which specific ships were meant in the documents.

Schooners and brigantines (usually shortened to brigs) clearly hauled the majority of cocoa from the production areas in Central and South America and the Caribbean into North American ports (Table 17.2). While both ship types were much smaller than frigates, they were well adapted for carrying cargo and designed for speed. Schooners, with their characteristic double masts and shallow draft, weighed approximately 100 tons and because of their design gained ready access to less-well-developed shallow harbors. Brigs, characterized by three masts, were heavier and therefore were restricted to well-established deep-water ports. Barques (three masts) and sloops (one mast) were less commonly used to transport cacao.

Table 17.2 Ship Types Used in Cocoa Trade with Number of Imports

Ship Type	Number of Imports
Schooner	156
Brig	150
Ship (unidentified)	46
Barque	41
Sloop	15
Spanish schooner	2

Captains/masters of merchant ships were responsible for the cargo, crew, navigation, and overall safety. Commonly, captains/masters were more than basic nautical persons; they were businessmen who received their pay as a percentage of the market price for cacao (or other goods) upon delivery at the conclusion of the voyage. Most captains/masters did not own the ships they commanded, but sailed on behalf of specific merchants or teams of merchants who owned the ships and supplied the goods for barter/ trade or funds to purchase goods at distant ports. Relationships between captains/masters and merchants varied depending on whether or not the ships returned to their North American ports with the cargos intact and in a timely manner. Merchants typically hired captains/masters based on previous abilities to meet contractual agreements and through time a high degree of trust (in each other) emerged that sometimes lasted for long periods of time. Since captains/masters rather than the respective merchants were commonly listed by name in *Shipping News* documents, they represented the first named links to cocoa importation into the emerging United States (Table 17.3).

It also was likely that merchants who owned ships would employ several captains/masters during the lifetime of a specific ship active in the cacao trade.

The ships *Mary Ann, Argus*, and *Hope* head the list of those commonly used by cocoa importers (Table 17.4). A review of this table, however, still raises the question of whether or not there was more than one ship with the same name with cocoa as cargo. Consider the case of the name *Mary Ann*, as applied to ships of two types (Table 17.5). It is clear from the *Shipping News* evidence that the schooner *Mary Ann* and the brig *Mary Ann* were two different ships. It is interesting, however, that the brig *Mary Ann* was captained by two different

men in 1807, identified as Captains Reed and Lawson. The brig *Mary Ann* (1806–1807) completed six voyages but as not under the ownership/direction of a single merchant. *Shipping News* documents reveal that the Company of Saltus and Son was listed as the merchant on all but one of the brig's voyages, but the ship was commanded by two different individuals, Reed and Lawson, both hired by Saltus and Son. Furthermore, linkages between captains/masters and specific ships also may be tallied. Captain Bryan, for example, sailed the *Alfred* on five cocoa-laden voyages from 1806 to 1807, while Captain Lawson, on the *Mary Ann*, completed four cocoa voyages during the same time period (Table 17.6).

Our initial research into ship and captain combinations identified specific individuals responsible for physically moving cocoa from production to North American ports, but did not reveal who was responsible for selling cocoa on the open market at each port. To determine this vital link between importation and the cocoa entering the American economy, we turned our focus to specific port cities and merchants.

PORTS AND MERCHANTS

Examination of available *Shipping News* documents (1721–1809) revealed that Philadelphia and New York

Table 17.3 **Ship Captains: Number of Voyages**

Captain	Voyages
Lawson	6
Turner	6
Brown	5
Bryan	5
Congleton	5
Dodge	5
Gardner	5
Jones	5
Clough	4
Davis	4

Table 17.4 **Most Frequent Cocoa Trading Ships by Port, Number of Voyages, and Years (1806–1809)**

Ship Name	Port	Voyages	Beginning	Ending
Mary Ann	New York	6	1806	1807
Argus	Philadelphia	5	1789	1809
Hope	Baltimore	5	1806	1807
Brutus	Philadelphia	4	1806	1806
Eliza and Mary	New York	4	1806	1809
Farmer	Philadelphia	4	1806	1807
Jane	Philadelphia	4	1807	1807
Neptune	Philadelphia	4	1806	1808
Sukey and Peggy	Philadelphia	4	1806	1806
Alert	Philadelphia	3	1789	1807

Table 17.5 **Voyages of the Brig and Schooner *Mary Ann* by Captain and Years (1806–1808)**

Ship Name	Ship Type	Captain	Voyages	Beginning	Ending
Mary Ann	Brig	Lawson	2	1806	1806
Mary Ann	Brig	Reed	1	1807	1807
Mary Ann	Brig	Reed	1	1807	1807
Mary Ann	Brig	Lawson	1	1807	1807
Mary Ann	Brig	Lawson	1	1807	1807
Mary Ann	Schooner	Bowers	1	1807	1807
Mary Ann	Schooner	Morse	1	1808	1808

Table 17.6 Captain and Ship Frequency (1806–1807)

Captain	Ship	Voyages	Beginning	Ending
Bryan	Alfred	5	1806	1807
Congleton	Sukey and Peggy	5	1806	1806
Clough	Northern Liberties	4	1806	1809
Lawson	Mary Ann	4	1806	1807
Turner	Resolution	4	1806	1806
Wheeler	Valona	4	1806	1807
Bell	Neptune	3	1806	1807
Conklin	Fame	3	1806	1806
Dawson	Hope	3	1806	1807
Gale of Amboy	Ceres	3	1807	1807

Table 17.7 Cocoa Trade: Port Frequency (1721–1809)

Ports	Number of Imports	Beginning	Ending
Philadelphia, PA	144	1721	1809
New York, NY	141	1806	1809
Baltimore, MD	55	1794	1809
Boston, MA	24	1729	1809
Newburyport, MA	13	1806	1808
Norfolk, VA	5	1806	1808
Newport, RI	4	1763	1808
Charleston, SC	4	1794	1809
Salem, MA	3	1807	1808
Savannah, GA	2	1764	1767
Providence, RI	1	1807	1807
Portland, ME	1	1807	1807
New London, CT	1	1786	1786
New Haven, CT	1	1798	1798

Table 17.8 Leading Cocoa Importing Merchants (1806–1807)

Merchant	Port	Number of Imports	Beginning	Ending
H. Cherriot	New York	5	1807	1807
J. Howland and Son	New York	5	1806	1806
Sultus, Son and Co.	New York	5	1806	1807
Snell, Stagg, and Co.	New York	4	1806	1807
W. & A.M. Buckly	Philadelphia	4	1806	1806

were the leading ports where cocoa was discharged. Their high percentage of the cocoa imports may be reflective of the higher volume of goods flowing into these two ports. However, it is also important to consider that these two cities may have had better (or more consistent?) newspaper reporting of ship arrivals (Table 17.7).

As cocoa was being imported into these North American cities, there were numerous chocolate merchants in the urban areas serviced by the ports (see Chapters 27, 28, 30, and 54). Using *Shipping News* documents, we identified the top five merchants and an additional 154 others involved with the cacao trade; of these 159, an astonishing number (117) imported cocoa only one time (Table 17.8).

COCOA AND OTHER GOODS

After review of the 528 *Shipping News* documents surveyed, we found that barely 10 percent (54 ships) of the vessels arrived in North American ports laden with cocoa as the only reported cargo. The fact that merchants sometimes were willing to load an entire ship with one commodity illustrates economic volatility, great risk, and how highly cocoa was prized (and profitable). Experienced importers would never allocate all stowage space aboard ship to one type of cargo unless they could expect to recoup all of the expenses incurred plus achieve an adequate profit level. Given that a standard brig had a cargo capacity of 150 tons, while that of a schooner was 100 tons, the quantity of cocoa aboard these 54 ships that entered North American ports can be calculated. And is it not astonishing to visualize the off-loading of 150 tons of chocolate—300,000 pounds—in North American ports?

Most shipments, however, involved the importation of cocoa with other goods (Table 17.9). *Shipping News* documents reveal that coffee, followed by sugar, were the most common consumable commodities shipped in conjunction with cocoa. Since cocoa, coffee, and sugar were grown in similar tropical climates, it is understandable that such merchant ships would carry these items on board. The presence of cotton, indigo, and rum are also understandable. The frequent importation of hides with cocoa, however, raises an interesting question since hide production (i.e., cattle ranching) and cacao production both require large tracks of land.

The top three cacao merchants, identified as H. Cherriot, J. Howland and Son, and Sultus, Son and Company, all sent ships to the Caribbean that returned with only coffee, cocoa, and sugar (see Table 17.8). The ships of H. Cherriot along with Sultus, Son and Company, listed St. Pierres, Martinique, as the port of call for all of their shipments. The connection between Cherriot and Sultus with St. Pierres was the most frequent one we found.

While coffee, sugar, and cocoa were frequently shipped at the same time aboard ship, cocoa sometimes was shipped with lesser-known goods. The company of Snell and Stagg imported both cocoa and beef jerky from Montevideo, Uruguay, into New York in 1807. Several documents also list annatto or rocoa[5] being

shipped with cocoa. Wood was also a frequent shipmate with cocoa. Wood was commonly listed as either timbers or logwood, but there were two geographically named wood products, Nicaraguan wood and Peruvian bark (a medicinal product), listed as cargo. The Nicaraguan wood was shipped from Curacoa (Curaçao) while the Peruvian bark was sourced only from Bermuda. Sprinkled throughout the *Shipping News* documents were mention of the importation of cacao along with rum and brandy into North American ports. While these spirits were not imported with cocoa at the same rate as coffee and sugar, spirits were a frequent cargo that accompanied cocoa.

LINKAGE: IMPORTS → SALES

One of our research goals was met when we discovered a document that reported the importation of cocoa to New York City on June 12, 1808, by John Juhel and Company (see Fig. 17.1) [2]. This notice is unique because it combines information typically found on a *Shipping News* article with information announcing when the specific cacao—made into chocolate—could be purchased. This document revealed that John Juhel and Company were located at 15 Gold Street in New York City, an address still used today—but without its former cacao/chocolate mercantile connection (the location is now a Holiday Inn!).

John Juhel and Company documents are part of the special collections #2176 at the Division of Rare and Manuscript Collections, Cornell University Library. The documents include the following:

Trans-Atlantic and coastal bills of lading, insurance policies, and other shipping documents issued in New York City, Philadelphia, Wilmington, Savannah, Bordeaux, Liverpool, Hamburg, and Guadeloupe to John Juhel & Company and other firms, 1799–1807. Other items include a letter, 1815, to Ichabod Parsons of Denmark, New York, from a brother in Boston telling of plans to sail on privateer schooner, and items from Lowville and Denmark, New York, including a miscellaneous manuscript. [4]

One of our future activities will be to search this collection for additional invoices, ledgers, and references to cocoa importation and sales made by John Juhel and Company. We know that John Juhel imported and sold multiple items, with cocoa being a portion of his activities. It remains our objective to examine this special collection and attempt to calculate the percentage of the business that came from cocoa sales, and the economic profitability of importing and marketing cocoa in New York.

Links to Modern Commodity Research

Numerous scholars in the social sciences study modern commodity chains that link production areas with

Table 17.9 Goods Shipped in Conjunction with Cocoa

Good	Number of Shipments
Coffee	198
Sugar	118
Hides	75
Cotton	42
Indigo	28
Rum	27
Wood/logwood	21
Rocoa	20
Molasses	19
Salt	15

consumption. Such chains may span a few miles for local food networks or may extend across the globe. Regardless of the length of a specific commodity chain, modern economic principles can be applied to the study and analysis of the historical cocoa trade as well.

GLOBAL COMMODITY CHAINS

Global commodity chain research examines how geographical areas of commodity production are connected with other geographical areas of consumption. The use of global in the title does not limit the use of this theory as a means to examine cocoa commerce in the 18th and 19th centuries. Global commodity chain research is a world system theory [5] initially developed by Hopkins and Wallerstein [6] along with Marsden and Little [7]. While these researchers are credited with first working the idea, many scholars credit Gereffi and Korzeniewicz's work [8] as the most comprehensive address of this theory. Global commodity chain theory when applied to modern commerce systems uses detailed trade data to record the flow of commodities from points of production to consumption. Models of goods with a large number of inputs create very complex chains. The commodity chain for cocoa in the 18th and 19th centuries is very simple because very few inputs are used in the production and cocoa was produced in very few areas.

The ports of debarkation at the supply end (i.e., export ports) were either primary sources of cocoa where local production was sold and shipped directly to North America on ships from a local port, or secondary ports. Evidence from our examination of *Shipping News* documents revealed that Venezuela and French Guiana were primary production regions and the majority of cocoa shipped to North America originated in the geographical hinterlands of these ports. Therefore, Venezuelan and French Guianan ports would be classified as primary ports. In contrast, the port of Curaçao in the Lesser Antilles is an example of a secondary port, one that shipped large quantities of cocoa that was produced in another area. Given the geographical location of Venezuela, French Guiana, and Curaçao and the factor of distance, it makes economic sense that cocoa would be shipped from these ports primarily to the major port cities on the Eastern Coast of North America during the time period covered.

CLUSTERS

Michael Porter first popularized the term *clusters* to explain how specific geographic regions become centers for certain industries [9]. Common examples of clusters are the entertainment cluster in Los Angeles, the technology and computer cluster in Silicon Valley, and the wine industry cluster in Napa County, California. Different academic disciplines have debated the validity

of clusters [10–14]. Reasons for clustering are variable and much debated in academic literature. The basic idea behind the concept of *cluster*, however, is that similar businesses located close to each other experience lowered infrastructure and labor search costs. In our research we used the concept of *cluster* to search for evidence that one or more North American ports had a higher proportion of cocoa importation than others. Higher levels of importation should point toward higher consumption rates and a higher number of chocolate-associated merchants in that city. The data show strong support for this claim, with most chocolate imports arriving to Baltimore, Boston, New York, and Philadelphia—all cities with high numbers of chocolate-associated merchants.

Conclusion

This chapter documents cocoa trade patterns from the original production regions within the Americas into port cities of eastern North America during the 1700s and early 1800s. We identified the major cacao exporting and importing locations, trade routes, ship types, and captains/masters and identified the key merchants who imported cocoa. The Readex Early American Newspapers on-line database provided our research with more cocoa *Shipping News* documents than could be processed during our short 12-month project. While we identified and examined 528 documents, many more can be brought into our project database in the years to come.

We would like to further explore how changes in governance of production areas impacted trade patterns, for example:

1. Did changes in administration and economic system after U.S. independence change cocoa trading patterns? If so, how?

2. Given there were Dutch, French, English, Portuguese, Spanish, as well as North American interests in the cacao trade, how did nationalism and political relationships between and among these nations remain constant or change through the decades of the 18th and 19th centuries?

Shipping News documents represent a greatly underutilized source of historical information on cocoa production and commerce. Besides the basic information on ship type, cargo, personnel, and ports, they commonly contain references to pirates, military hostilities, nautical prizes seized on the open seas, even information on shipwrecks. Further research through these documents would be valuable and could cast additional light on relationships between specific trade routes and risk—whether economic or personal. The captains/masters and their crews that comprised the North American cacao trade are more than just names. These men risked more than just economic disaster if their ships foundered in storms or were captured. Further study of *Shipping News* documents can reveal insights on life and activities on the high seas, and point to further connections between cocoa production regions and North America during the Colonial and Federal eras of American history.

Acknowledgments

We wish to acknowledge the assistance of Matthew Lange and Deanna Pucciarelli during the preparation of this chapter.

Endnotes

1. The orange-colored pulp covering the seeds of the tropical plant *Bixa orellana*, from which annatto is prepared.

2. A schooner is a type of sailing ship characterized by the use of two or more masts set in line with the keel.

3. A Holiday Inn now occupies the address 15 Gold Street, New York, New York.

4. http://infoweb.newsbank.com/. (Accessed throughout the research period, September 2005 to August 2006.)

5. The red dye produced from the pulp surrounding the seed of the *Bixa orellana* tree.

References

1. Chydenius, A. In: Schaumann, G., editor. *The National Gain*. London, England: Ernest Benn Limited, 1931.

2. *Shipping News*. In: *New-York Commercial Advertiser* 1808: New York, NY.

3. Walvin, J. *Fruits of Empire: Exotic Produce and British Taste, 1660–1800*. New York: New York University Press, 1997; pp. xiii, 219.

4. Badgley, L. *Lester Badgley, collector papers*. In *#2176. Division of Rare and Manuscript Collections, Cornell University Library*, L. Badgley, editor. 1799–1919.

5. Hughes, A., and Reimer, S. *Geographies of Commodity Chains*. London, England: Routledge, 2004; pp. x, 276.

6. Hopkins, T. K. W. I. Commodity chains in the world-economy prior to 1800. *Review* 1986;10:157–170.

7. Marsden, T., and Little, J. *Political, Social, and Economic Perspectives on the International Food System*. Aldershot, Hants, England: Avebury, 1990; pp. xii, 259.

8. Gereffi, G., and Korzeniewicz, M. Commodity chains and global capitalism. In: *Contributions in Economics and Economic History*. Westport, CT: Greenwood Press, 1994; pp. xix, 149, 334.

9. Porter, M. Clusters and the new economics of competition. *Harvard Business Review* 1998;76(Nov.–Dec.): 77–90.

10. Schmitz, H., and Nadvi, K. Clustering and industrialization: introduction. *World Development* 1999;27(9): 1503.

11. Visser, E.-J. A comparison of clustered and dispersed firms in the small-scale clothing industry of Lima. *World Development* 1999;27(9):1553.

12. Clark, G. L., Feldman, M. P., and Gertler, M. S. *The Oxford Handbook of Economic Geography*. New York: Oxford University Press, 2001.

13. Martin, R., and Sunley, P. Deconstructing clusters: chaotic concept or policy panacea? *Journal of Economic Geography* 2003;3(1):5–35.

14. Mueller, R. A. E., and Sumner, D. A. Clusters of grapes and wine. In: *3rd International Wine Business Research Conference*, 2006, Montpellier, France.

How Much Is That Cocoa in the Window?

Cocoa's Position in the Early American Marketplace

Kurt Richter and Nghiem Ta

Introduction

One aspect of life that is the same in modern times as in the 18th and 19th centuries is that consumers spend money. The type of currency may have changed through the centuries and goods available and items desired may be different, but the issue has remained: consumers use earned income to purchase goods and services. This similarity in human behavior through time makes the analysis of historical price information a unique window into the lives of everyday people at specific time periods.

In this chapter we address the basic question of what cocoa and chocolate cost during the late 1700s and early 1800s. We focused on historical data to provide an economic context to the work of other scholars collaborating with this project. As historical information was collected by project researchers, the basic question of how expensive these items were became a constant recurring theme. In response, we initiated the collection and analysis of price information for cocoa and chocolate.

Prices can be used to reveal what was affordable to average persons during the Colonial and Federal eras of North American history. Examination of prices allows researchers to identify and separate goods

affordable to common laborers from those beyond their economic reach. During our research, we were able to couple price and wage data to show what was affordable to specific classes/groups of North Americans and how expensive different items were in relation to each other. With enough historical price and wage data, a shopping trip to a market could be simulated and possible purchasing/affordability options presented.

This chapter considers historical cocoa price information in three sections. First, we present descriptive statistics on price data collected. The time period covered, frequency of price quotes, and geographical distribution of the data are summarized and analyzed. Second, we present the temporal and spatial distribution of cocoa prices in North America and at select international ports. Third, we explore the affordability

Chocolate: History, Culture, and Heritage. Edited by Grivetti and Shapiro
Copyright © 2009 John Wiley & Sons, Inc.

of cocoa by simulating the activities of a "typical" assistant carpenter as a "representative adult male" who would have lived in Boston during the year 1811. In this model the price of cocoa is compared to goods typically available and consumed with or without cocoa during this time period.

Price Information: Cautions and Caveats

It is difficult to convert prices listed in the early 1800s to current 21st century (2006–2009) prices. Historical price and economic information that has survived into the 21st century is not directly comparable since modern economists compute currency equivalencies and inflation using a variety of methods unavailable in the past. Modern currency equivalency, for example, is based on the free market price for a currency as determined by the international money market: this market did not exist in the 1800s. Furthermore, inflation must be considered.

Inflation in the 21st century is computed by tracking changes in the price of a standard basket of "consumer goods" through time. As the type of products change, the goods used to calculate inflation change as well: as items change, accuracy of inflation estimates decreases. If, for instance, researchers wanted to track the inflation rate of selected consumable goods between the years 2006 and 2009, it would be obvious that many of these commodities would be identical in type, size, and composition during each of these three years, making an inflation calculation easy. If researchers wanted to estimate inflation rates during the past 20 years, it would be more challenging: the items could be identical, but it may be difficult to locate identical product market prices for each of the years 1988–2009. Computing and comparing inflation rates between 1700–1800 and 2007–2009 then becomes even more complex since few products found today in the marketplace were available 200 years ago. Because of limitations of converting historical data to different time periods, we present here only price data and do not attempt currency conversions or attempt to adjust historical prices for inflation.

Research Methodology

Our source for historical price information was *Readex*®, a division of NewsBank, Incorporated, Early American Newspapers, Series I (1690–1876) database: (http://infoweb.newsbank.com/). The Readex database allows researchers to search Early American newspapers beginning in 1690 through 1876 using key words and to limit queries through multiple search options. We used the internal options to limit our search results to documents classified/identified as

being *Price Currents*[1] that contained the word cocoa. Each retrieved result was reviewed for relevancy. Documents that contained the word cocoa without price information were removed from the data set. Readex also presents researchers with numerous tools to export results from their database as PDFs, into personal computers. Accordingly, we exported price information, citation data, and newspaper images directly into *Access*®, a reference management software program. Information from each historical document was entered into spreadsheets that listed all relevant information; we then created data tables and charts used to develop the present chapter.

Descriptive Statistics

Our data set spanned the years 1736–1820. Although some Early American newspapers provided *Price Current* data prior to 1800, the information was inconsistent, with considerable gaps in publication. Boston and New York newspapers consistently reported local prices starting ca. 1800. Many *Price Currents* also quoted prices for chocolate. In our data set cocoa is defined as fermented and dried whole or ground cocoa beans, while the term chocolate refers to processed cocoa ready for consumption [1]. During a 6 month research period we collected 3751 *Price Currents* that reported local prices for cocoa at 47 different cities in North America and Europe. The fifteen ports with the highest number of published *Price Currents* are identified in Table 18.1.

Our collection of *Price Currents* information also was influenced by two factors: (1) the frequency that newspapers reported the information (whether weekly, monthly, or irregularly), and (2) the rate at which Readex scanned the Early American newspapers (an ongoing process). Sometimes newspapers located in one port city published *Price Currents* for other Colonial

Table 18.1 *Price Currents* Captured by Year (1736–1820)

Port	Count	Beginning	Ending
New York	2687	1736	1820
Boston	496	1789	1817
Liverpool	133	1799	1820
London	108	1785	1818
Philadelphia	85	1785	1817
Bordeaux	48	1798	1819
Baltimore	21	1804	1815
La Havre	18	1795	1819
Amsterdam	13	1805	1817
Rotterdam	13	1797	1808
Bermuda	12	1788	1810
Hamburg	12	1795	1818
Dutch	11	1802	1817
Marseilles	10	1803	1817
New York and Boston	10	1795	1796

or Federal era ports (i.e., a Boston newspaper reporting prices for New York). The newspapers that published the highest number of *Price Currents* are identified by name (with range of dates) in Table 18.2. Boston, New York, and Philadelphia publications presented the greatest number of *Price Currents* (Table 18.3). Numerous foreign ports were identified that also provided *Price Current* information. Those with the highest number of retrieved examples are presented in Table 18.4. We did not convert different local currencies into United States dollars because of potential conversion inaccuracies prior to the establishment of the Gold Standard [2].

Table 18.2 Newspapers Reporting Cocoa Price Information (1786–1820)

Paper	Count	Beginning	Ending
Ming's New-York Price-Current	386	1805	1817
Republican Watch-Tower	213	1803	1809
Chronicle Express	156	1802	1804
The Spectator	145	1797	1804
New-York Price-Current	128	1799	1805
Oram's New-York Price-Current	117	1802	1804
Columbian Gazetteer	113	1793	1794
The Gazette of the United States	80	1789	1792
The Daily Advertiser	76	1786	1807
New-York Spectator	65	1805	1820

Table 18.3 Most Frequent North American Ports with Advertised Cocoa Prices (1736–1820)

Port	Count	Beginning	Ending
New York	2687	1736	1820
Boston	496	1789	1817
Philadelphia	85	1785	1817
New York and Boston	10	1795	1796
Charleston	3	1803	1805

Table 18.4 Most Frequent Foreign Ports with Advertised Cocoa Prices (1785–1820)

Port	Count	Beginning	Ending
Liverpool	133	1799	1820
London	108	1785	1818
Bordeaux	48	1798	1819
La Havre	18	1795	1819
Amsterdam	13	1805	1817
Rotterdam	13	1797	1808
Bermuda	12	1788	1810
Hamburg	12	1795	1818
Dutch	11	1802	1817
Marseilles	10	1803	1817

Results

IMPACT OF MILITARY HOSTILITIES ON THE PRICE OF COCOA

Our research focused on the late 1700s to 1820 because of more complete sets of *Price Current* information from Boston and New York newspapers. Due to the completeness of the data set, we examined the effect of the War of 1812 on the price of cocoa. In 1800, we retrieved 74 *Price Currents* published in eight domestic and foreign ports: Bilboa (*sic*) (Spain), Bremen (Germany), Curaçao (Dutch Colony in the Caribbean), Hamburg (Germany), Lisbon (Portugal), Liverpool (England), London (England), and New York. The majority of *Price Current* information retrieved was for ports in the United Kingdom, which is not surprising. London and Liverpool accounted for over 50 percent of the foreign *Price Currents* collected. Changes in *Price Current* values reported during the years 1800, 1813, and 1820 are reported in Table 18.5. The first hint that the United States of America adjusted prices based on the political winds was noticed with the 1813 data. Since all East Coast foreign goods entered 19th century North America by ship, the locations reported in the *Price Currents* reflected the locations where American merchants "sourced" their goods.

Military blockades during the War of 1812 forced American merchants to change their shipping routes from British to French ports during 1812. It would be expected that a change in shipping routes would occur because of hostilities, and the historical data published as *Price Currents* reflects the expected change. British ports were the primary sources of goods prior to the War of 1812—and after cessation of hostilities in 1813—but during the War, American newspapers reported prices from French ports (see Table 18.5). By 1820, however, American newspapers once again reported *Price Currents* from British ports, revealing American merchant adjustments to shipping patterns.

PRICE DIFFERENCES AMONG AMERICAN CITIES

New York clearly dominated the number of *Price Currents* retrieved and provided the clearest picture of changes in cocoa prices through time (Fig. 18.1). This figure clearly reveals an economic spike in cocoa prices

Table 18.5 Number of *Price Currents* by Port: 1800, 1813, and 1820

Port	1800	1813	1820
New York	153	46	59
London, England	8	0	0
Liverpool, England	6	0	1
Bordeaux, France	0	3	0

FIGURE 18.1. New York City cocoa prices, 1797–1820.

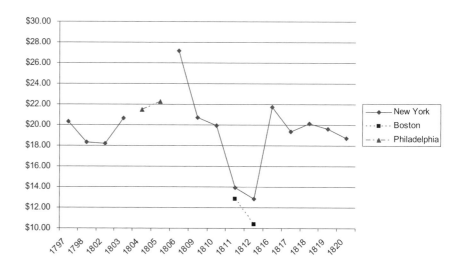

FIGURE 18.2. New York City, Boston, and Philadelphia cocoa prices, 1797–1820.

during 1806, whereas prices fell repeatedly in subsequent years until the War of 1812, when once again they rose dramatically from ca. $13.00 per cwt (hundred pounds) to almost $22 per cwt. While it is not possible to conclude that this rise in cocoa price specifically was due to the War of 1812, the supposition can be advanced that hostilities between America and England resulted in an interruption of the shipping lanes and cocoa supply sources. Fig. 18.1 also reveals that after the conclusion of hostilities (in 1813), cocoa prices settled at approximately $20.00 per cwt.

Price information also was collected for Boston and Philadelphia, then compared with New York (Fig. 18.2). The small price difference between the three port cities could be accounted for by differences in transportation costs. It is curious, however, that Boston showed a lower price for cocoa during the period of hostilities associated with the War of 1812, as lower prices seemingly would run counter to the idea that cocoa price differences could be accounted for by differences in transportation costs (Boston being considerably north of New York).

LONDON

The London markets for cocoa were as volatile as American markets (Fig. 18.3). Cocoa produced in Caracas and shipped to London received premium prices in the late 1780s, but lost its market advantage by 1817. The London market for cocoa prior to and during the War of 1812 dipped sharply and mirrored the decline in New York markets. This would suggest that the world market price for cocoa (in reality, North American and western European markets) decreased substantially in the years leading up to the onset of hostilities. The London data as revealed through *Price Currents* also rebounded after 1812. It is impossible to tell from these data if the war caused world cocoa prices to decline; nevertheless, they rebounded after 1812 (see Fig. 18.3).

The data set revealed five geographical sources for cocoa brokered on the London market: Caracas, Grenada, Martinique, Trinidad, and a generic term, West Indies. Of these destinations, Caracas, Venezuela, was the preferred source because its cocoa

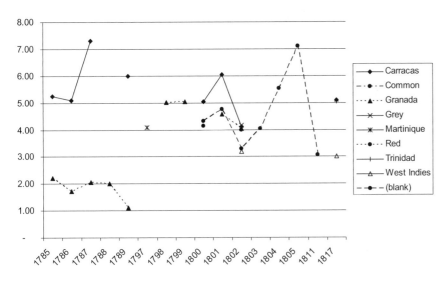

FIGURE 18.3. London cocoa prices by variety, 1785–1817.

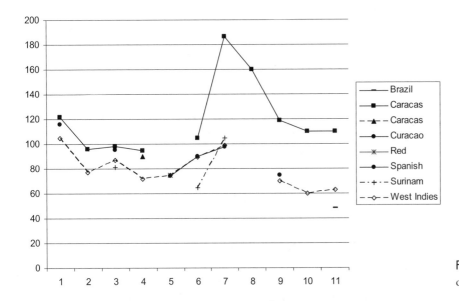

FIGURE 18.4. Liverpool cocoa prices, 1799–1818.

demanded the highest market value. Cocoa also was imported to London at this time from the island of Grenada.[2] It is uncertain, however, whether Grenada was the source of the cacao beans or acted as a transshipment port.

LIVERPOOL

Price Current cocoa data from Liverpool (England) also were captured during our research. Fig. 18.4 presents the average yearly price for the different types of cocoa reported in Liverpool. Prices here also reflected a similar upward trend in cocoa until 1805, then revealed a sharp decrease through 1807. The data suggest that there was a global decrease in cocoa prices in 1805; prices reported in the United States and England both show similar slumps from which cocoa did not recover during the years covered in our research.

Purchase Choices Available to a "Representative Adult Male"

Price Currents are a window to view what goods and items cost at specific points in time. The number of specified commodities reported in the various *Price Currents* reviewed ranged from 10 or less to over 100 specified items. While collecting and summarizing the individual prices for all goods on each *Price Current* was beyond the scope of the present project, we collected prices for selected, representative food items for several North American cities for the period 1800–1815. Our objective was to determine how expensive cocoa and chocolate were in comparison to other goods.

We focused this specific *Price Current* collection on four cities in geographical sequence from south to north: Charleston, Philadelphia, New York, and Boston. We eliminated all *Price Currents* from 1800–1815 that did not include a price for cocoa so as to compare these values against prices for other goods at a specific time periods and locations.

The next step was to determine wage rates for typical workers in North America during 1800–1815. Due to the complexity of comparing the value of 19th century wage rates with 21st century wages, we made comparisons only within the same year. Economic historians have called the quest to determine the current value of historical data to be highly challenging [3]. Reasons for this complexity include the issue that wage data available during this period of American history are not the types used by 21st century economists.

We created, therefore, a "representative" adult, male worker, who lived in Boston during 1811. Our creation was a young carpenter's apprentice. We determined that his daily wage of ca. $1.00 was received for a full day of work, perhaps more than 10 hours of effort. Once this daily wage had been determined, we then established a series of parameters (calculated on a daily basis) for his expenses. From his daily wage, he had to pay for an average accommodation, which would be ca. $0.40. Once this primary expense was covered, leaving ca. $0.60 to spend, he would be free to make purchases at a local marketplace. Entering a store/shop/market with his remaining $0.60, the carpenter had numerous options or choices. Consider his options for the week of August 23, 1811. Our "representative man" had options to make individual purchases or combinations using his available income. For example, if he elected to purchase basic cocoa, it would be likely that he also would purchase sugar and spices used in flavoring cocoa (what economists call complementary goods) [4].

During the year in question, 1811, cocoa was consumed primarily as a beverage. Accordingly, other beverages, among them coffee and tea, wine, gin, rum, and other alcoholic spirits, would compete against cocoa as consumer choices, whether as thirst quenchers, pick-me-ups, or for their "mind-altering" properties. When the "representative man" entered the store/market in Boston, his beverage options were variably priced (Table 18.6).

In regard to beverage choices, the first decision was whether or not to make an alcoholic or nonalcoholic choice. Considering nonalcoholic beverages such as coffee and tea, the economic data are clear as expressed on a purely price basis: tea was the most expensive, followed by cocoa, then coffee. If each of the three nonalcoholic beverage products produced exactly the same benefit, then the "representative man" would purchase coffee because it is the least expensive on a unit basis and therefore he would receive a cup of coffee at the lowest price compared to cocoa and tea. Human nature and other economic issues, however, require further examination into the recipes and methods of beverage preparation that influenced consumer decisions.

COFFEE

Coffee beans must be roasted in order for them to develop the rich, deep aroma and flavors consumers enjoy. Converting unroasted coffee beans into ground coffee, however, is not a complicated process. The beans can be roasted in a pan over an open fire and then ground in a mortar and pestle. This process would require that our "representative man" invest time and energy to complete this process. Time and energy also are economic issues, as both factors influence the decision-making process whether or not to purchase coffee beans. The next step would be to determine the quantity of coffee used to prepare a "reference cup" of coffee, then relate this quantity to price.[3] We determined the average weight for a tablespoon of ground coffee. Based on this calculation and knowing the price of coffee purchased in Boston in 1811 ($0.162/pound), this quantity would have produced ca. 38 servings of coffee at a cost of $0.004/cup [5].

TEA

If it would be possible to transport our "representative man" to the 21st century and into a Starbucks® (coffee shop), he would not recognize the beverage known as espresso macchiato, but the process of steeping tea would be identifiable. Using *Price Current* information and ingenuity, it also is possible to determine what tea cost (per serving) in 1811.[4] Based on an average weight that we determined for a tablespoon of tea, a pound of tea purchased in Boston in 1811 that cost $0.62/pound would have produced ca. 145 servings of tea at a cost of $0.004/cup (the same equivalent price for coffee [6]). Since both coffee and tea require time for steeping, but tea does not involve time for roasting, it would be economically cheaper for our "representative man" to produce a cup of tea—over coffee.

Table 18.6 **Boston Beverage Prices: 1811**

Gin (gallon)	Rum (gallon)	Brandy (gallon)	Coffee (pound)	Tea (pound)	Cocoa (pound)	Chocolate (pound)	Sugar (pound)
1.87	1.24	2.05	0.162	0.62	0.22	0.275	0.12

Table 18.7 **Boston Foodstuff Prices: 1811**

Corn (bushel)	Oats (bushel)	Rye (bushel)	Rice (cwt)	Beans (bushel)	Beef (pound)	Butter (pound)
0.975	0.475	1.025	4.06	1.33	0.055	0.165

COCOA

As participants in the Chocolate History Research Project at the University of California, Davis, one of us (Richter) assisted Jim Gay to process cocoa beans and prepare chocolate using techniques developed at Colonial Williamsburg (see Chapter 48). The process was long, laborious, and sometimes unpleasant (i.e., hard, tedious work). Experiencing 18th century chocolate-making made us realize that consuming chocolate was far more complicated and difficult than merely roasting coffee beans or steeping tea leaves. We experienced a substantial economic cost in terms of time, energy, and resources to convert cocoa beans into a delicious and nutritious cup of hot chocolate.[5] Based on an average weight that we determined for a tablespoon of cocoa (prepared from shaving a block of chocolate), a pound of block chocolate purchased in Boston in 1811 that cost $0.22/pound would have produced a little more than 4 servings of chocolate beverage at a cost of $0.05/cup. Expressed in this manner, the price of chocolate in 1812 would have been more than ten times that of either coffee or tea—and that would not include the time or energy used in processing the cocoa into chocolate.

Without computing the energy and time used to process cocoa into chocolate, the conclusion may be drawn that cocoa and chocolate were expensive, compared to either coffee or tea on a per serving basis. The fact that coffee and tea were almost exactly the same price is interesting. But the fact that chocolate was almost 10 times the price of these alternative nonalcoholic beverages positioned cocoa and chocolate as luxury products when compared to coffee and teas—especially since the three were drunk and enjoyed for similar reasons and in similar manners.

ALCOHOLIC BEVERAGES

When the price of cocoa is compared to alcoholic beverages, the relative cost of cocoa is even greater than the price difference between coffee and tea. The 1811 *Price Current* for brandy, gin, and rum revealed that brandy was the most expensive at $2.05 per gallon. Each standard North American gallon contained 128 fluid ounces. A typical "shot" of a high proof alcoholic beverage would be approximately one ounce. On a per serving basis, a gallon of brandy at $2.05 would provide the 128 servings at $0.016 per serving; rum would cost $0.0096; and gin would be slightly less at a cost of $0.014 per serving.

Chocolate was consumed for very different reasons when compared to alcoholic beverages. This difference makes it difficult to compare options/choices between chocolate and alcoholic beverages except for the issue of price. And while chocolate has been credited with near "miraculous powers" by different individuals and social groups through the ages, no one to our knowledge has ever suggested that cocoa had the equivalent power of brandy, gin, or rum over body and mind.

FOODSTUFFS

Our research also recorded prices for selective foodstuffs in Boston in 1811, specifically, corn (maize), oats, rice, rye, beans, beef, and butter (Table 18.7). These data are presented here for comparative purposes only, since as a beverage cocoa did not compete directly against these goods. Still, it is interesting to note the relatively high price for rice in 1811.

Conclusion

In the time period covered by our data set, cocoa and chocolate were used primarily as beverages. This places the decision of purchasing cocoa or chocolate as a choice between chocolate, tea, or coffee and possibly beer, wine, and spirits. Given the raw cocoa price and the amount required to produce a cup of chocolate, cocoa was the most expensive beverage choice in 1811 Boston.

Still, based on wage and price information collected, cocoa does not appear to be so expensive that it would have been out of reach for average individuals in the working class. However, cocoa was clearly more expensive than the other available beverages. It is not possible to determine from *Price Current* documents how often an average person in 1812 chose to purchase chocolate . . . but a survey of early 19th century diaries and store ledgers (where customers were identified by name) could help identify chocolate purchasers per week/month.

Our data clearly show that the price of cocoa varied depending on location and year. Possible future avenues of research would be to explore the causes of the apparent crash in world cocoa prices that occurred during 1806 as demonstrated here (Figs. 18.1–18.4). It would be interesting to determine whether or not the price collapse could be traced back to a specific catastrophic weather event or simple oversupply.

Historical economic price data provided a background to the research other historians were completing as part of this international project. Because the act of purchasing goods is a common activity for people who lived during Early America and today in the 21st century, examination of prices allows researchers to better understand how an everyday "representative reference adult male" viewed cocoa. The research in this chapter has provided an economic foundation that allows a better understanding of the role chocolate played in society during the early years of American history.

Acknowledgments

We wish to acknowledge the assistance of Matthew Lange and Deanna Pucciarelli during the preparation of this chapter.

Endnotes

1. *Price Currents* are documents published in Early American newspapers that listed the current price for a range of commodities. *Price Currents* were published in all major North American East Coast cities, commonly on a weekly or monthly basis. The purpose was to alert potential consumers of what would be called "fair market prices" so as to avoid price gouging.

2. The island of Grenada ceded by the French to the British in 1763 as part of the Treaty of Paris.

3. In the office kitchen of one of us (Richter) a bag of coffee displayed printed recommended instructions: use two teaspoons of ground coffee for every 6 fluid ounces of water [5]. The coffee was weighed using a standard kitchen scale; 10 tablespoons of medium ground coffee had an average weight of 0.42 ounces.

4. The directions on the back of a standard tea tin provide the recommended amount of 1 level tablespoon of loose tea for every 12 fluid ounces of water [6]. One of us (Richter) weighed 10 individual tablespoons on a kitchen scale and determined the average weight to be 0.11 ounces per tablespoon.

5. While not a perfect substitute for an 1811 recipe for hot drinking chocolate, one of us (Richter) used a recipe printed on the back of a box of Mayordomo Chocolate produced in Oaxaca, Mexico. The recipe called for 4 ounces of chocolate to be used with two cups of milk. Based on this recipe, each pound of chocolate—ground or shaved from the block—would produce four servings of chocolate.

References

1. Clarence-Smith, W. G. *Cocoa and Chocolate, 1765–1914*. New York: Routledge, 2000; pp. xiv and 319.

2. McCusker, J. J. *How Much Is That in Real Money? A Historical Commodity Price Index for Use as a Deflator of Money Values in the Economy of the United States*, 2nd edition. Worcester, MA: American Antiquarian Society, 2001; p. 142.

3. United States Bureau of Labor Statistics, Stewart, E. M., and Bowen, J.C. *History of Wages in the United States from Colonial Times to 1928. Revision of Bulletin No. 499 with Supplement, 1929–1933 (page 523)*. Detroit, MI: Republished by Gale Research Co., 1966; pp. ix, 574.

4. *New-Bedford Mercury*. August 23, 1811, p. 4.

5. Starbucks. *Tips to Make Great Coffee—Arabian Mocha Sanani*. Product label and packaging information, 2007.

6. World Market. *Dragon Well Green Tea*. Product label and packaging information, 2007.

CHAPTER

19

"C" Is for Chocolate

Chocolate and Cacao as Educational Themes in 18th Century North America

Louis Evan Grivetti

Introduction

Recent publications related to chocolate history commonly have focused on culinary and medical–nutritional themes [1–3]. Less well recognized as a theme in chocolate history is the role that cacao/chocolate-related information has played in educating children, young adults, and families in North America during the Colonial and Early Federal periods. The present chapter illustrates the extent to which cacao and chocolate had become commonplace as both food and medicine, and one measure of this extent is its use as an educational device in early North America.

Electronic, digitized databases of North American books, pamphlets, and ephemera dated to the Colonial and Early Federal periods were accessed through The Peter J. Shields Library, University of California, Davis. The most productive databases utilized were: *Early American Imprints, Series I: Evans, 1639–1800* [4]; *Early American Imprints, Series II: Shaw–Shoemaker, 1801–1819* [5]; and *American Periodicals Series Online 1740–1900* [6]. Key words searched were cacao, chocolate, and cocoa. This approach retrieved more than 5000 "hits" that represented chocolate-associated information contained in books, pamphlets, and ephemera. Titles obtained through the word searches were reviewed for education-related topics and for obvious references to education or to teaching. As this search produced many duplicates, the decision was made to incorpo-

rate only North American first editions and printings during the preparation of this chapter. The search revealed no obvious 17th century education-related documents, while 18th century materials were extraordinarily rich and diffuse. The decision therefore was made to concentrate only on the period between 1700 and 1800. Chocolate-associated educational-related materials dated to the 19th century, of course, remain topics for future activities.

Education Materials

Materials retrieved represented a very broad range of educational materials used to train school children and instill family, moral, and religious values in North America during the 18th century. The documents are presented here by theme in alphabetical order: Arithmetic and Accounting/Bookkeeping; Essays; Ethical, Moral, and Religious Tracts; Geography and History; Health, Home Remedies, and Medical Advice; Linguis-

Chocolate: History, Culture, and Heritage. Edited by Grivetti and Shapiro
Copyright © 2009 John Wiley & Sons, Inc.

tics; Poetry, Proverbs, and Saws; Religion; and Spelling/Grammar Dialogues.

ARITHMETIC, ACCOUNTING, AND BOOKKEEPING

Arithmetic schoolbooks in 18th century America primarily trained students for experiences they might expect to encounter in everyday life, whether as farmers or merchants. Such books commonly were filled with barter, percentage, and profit–loss problems that used commodities and examples familiar to both students and prospective mercantile employers. Complicating the calculations from the perspective of 21st century readers, however, were problems associated with conversions for British weights (presented in ounces, pounds, and cwt or hundredweight) and commercial monetary values for goods expressed as pounds (Sterling), shillings, and pence. Shortly after the Revolution, Americans abandoned the British monetary system in favor of dollars, cents, and mills $\left(\frac{1}{10}\text{ of a cent}\right)$, thus requiring revisions and modifications of earlier textbook problems to conform to the new types of currency.

At least twelve authors published arithmetic books in North America during the survey period: Anonymous (1794) [7], John Bonnycastle (1786) [8], Nathan Daboll (1800) [9], Thomas Dilworth (1781, 1798) [10, 11], John Gough (1788, 1800) [12, 13], Zachariah Jess (1799) [14], Gordon Johnson (1792) [15], William Milns (1797) [16], Nicolas Pike (1788) [17], Thomas Sarjeant (1789) [18], Edward Shepherd (1800) [19], and Consider Sterry (1795) [20]. Some of the more popular texts went through several printings and editions, with essentially no or only minor textual changes. Only information from the earliest edition or printing of each author's works is summarized here. When the documents identified are reviewed in chronological sequence, one obvious finding is how some authors copied from the works of others without acknowledgment as would be required today with 21st century contemporary scholarship. Differences in answers to the arithmetic problems may reflect currency value adjustments within the American colonies at various time periods.

Barter Problems

What quantity of tea, at 12 shillings per lb. must be given in barter for 1 1/2 cwt. of chocolate, at 5 shillings per lb. Answer: 2 qrs [?] 14 lb. [21]

_____

What quantity of tea, at [___?] per lb. must be given for 1 C. wt. of chocolate, at 4 shillings per lb? Answer: 44 lb. 12 oz. [22]

_____

What Quantity of Tea, at 10 shillings per lb. must be given in Barter for 1 C. of Chocolate at 4 shillings per lb? Answer: 44 pounds 12 oz 5/10. [23]

_____

What quantity of chocolate, at 4 shillings per lb. must be delivered in barter for 2 cwt. of tea, at 9 shillings per lb. Answer: 50 lb. of chocolate. [24]

_____

What quantity of tea, at 10 s per lb., must be given for 1 Cwt. of chocolate, at 4 s. per lb? Answer: 44 lb. 12 oz. [25]

_____

How much tea at 9 shillings per pound can I have in barter for 4 cwt 2 qrs of chocolate at 4 shillings per pound? Answer: 2 cwt. [26]

_____

What Quantity of Tea, at 10 shillings per pound must be given in Barter for 1 C of Chocolate at 4 shillings per pound? Answer: 44 pounds 12 ounces, 3/5 [?]. [27]

_____

How much tea at $1\frac{1}{2}$ dol. [dollars] per lb. can I have in barter for 4 cwt. 2 qr. of chocolate at 66.6 ct. per lb? Answer: 2 cwt. [28]

Commodity Problems

What is the value of 5 cwt. of chocolate at 1£ 3 s. per lb? [Answer not provided]. [29]

_____

What quantity of Coffee at 5 s. 3d a lb. must be given for 72 lb of Chocolate, at 4 s 6d a lb.? Answer: 61 lb. 11 1/4 oz. [30]

Conversion Problems

If 27 lb. of chocolate cost 6£ 15 shillings what is that per cwt? Answer: 28£. [31]

_____

If the price be 2 shillings and an aliquot or aliquant part thereof; first find the price for 2 shillings and then take the proper parts out of said price and add thereto [and then answer]. How much cost 3782 lb. of chocolate, at 2 shillings 8 1/2 d. per lb? Answer: 512 £ 2 shillings 11 d. [32]

_____

When the first term is an aliquot part of the second or third; divide by the first that of which it is an aliquot part, and multiply the other by the quotient [to determine the answer]. If 27 lb. of chocolate cost 6 pounds 15 shillings, what is that per C.wt. Answer: 28 pounds. [33]

Loss and Gain Problems

Brought 137 lb. of Chocolate, at 4 s 1 1/2 d a lb. and sold it at 4 s. 9 d a lb. What was the Gain? Answer: £4 57 1/2 s. [34]

_____

If 40 lbs. of chocolate be sold at 1 shilling 6 pence per lb and I gain 9 per Cent [sic] what did the whole cost me? [Answer not provided]. [35]

Percentage Problem

If 40 lb. of chocolate be sold at 25 cts. per lb. and I gain 9 per cent; what did the whole cost me? Answer: 9 dollars 17 cents + 4 mills. [36]

Numerous chocolate-related entries appear in the 1796 accounting schoolbook authored by William Mitchell. The following examples are reflective of his text.

June 9th. Sold Robert Grier, 2 Boxes Chocolate, 160 lb. at 22 c. [37]

June 9th. Cash Dr. to Chocolate, for 1 Box 80 lb. 16 dollars, 80 cents. [38]

ESSAYS

Many books and pamphlets printed during the survey period contained essays on diverse topics with the intent of educating and amusing readers. Given the tone and content of many of the essays, a secondary purpose clearly was to inflate the ego and demonstrate the author's scholarship. The essays reviewed here sometimes contain straightforward, factual information, but other essays are pedantic and labored. The essay subjects ranged from the presumed "longings" of a long-dead Roman gourmand; to identification of luxury items transported across the Pacific to North America; to description of a mental exercise whereby indecisive individuals could be taught how to make decisions when forced to make a single choice from a group of objects of equal value. Also illuminating is an essay of the period about how to maintain health and fitness after taking "the waters" at the Saratoga spa, New York:

APICIUS [Roman gourmand speaking]. *I cannot indeed but exceedingly lament my ill fate, that America was not discovered, before I was born. It tortures me when I hear of chocolate, pineapples, and a number of other fine fruits, or delicious meats, produced there, which I have never tasted.* [39]

At last the ship arrived from Japan, bringing much [sic] goods, as rich Persian silks, cotton, linen, spices, fruit, sugar, tea, chocolate, liquors, live fowls of several kinds for breed [sic], tame beasts, and all things wanting. Tanganor, with these treated and made presents to his guests of what they wanted; and the ship being to return to Japan he proposed to them what to do. They resolved to go for Mexico with the ship, which being now unloaded might easily go thither before it returned to Japan. [40]

There are many things, which appear fit and agreeable, which are never chosen; nor yet proper, refused; i.e., the mind feels no abhorrence of them; nor does it will the absence of them. They

pass through the mind as agreeable things; and that is all that can be said of them. We may observe further, that when two different things appear equally eligible to the mind, we find no difficulty in taking one, and leaving the other. Mr. Edwards allows that two things may appear equally eligible to the mind: but, in this case, he supposes, that we are determined by accident—one is nigher [sic] to us than the other, or is last in the eye or mind. Let us suppose, that I am at a Gentleman's table—he asks me which I choose, tea, coffee, or chocolate: They are neither of them present to affect my eye; and I am told, that either of them can be had, with equal ease; and they all of them appear equally eligible to my mind. I determine to take coffee. Was the reason of determining to take coffee, because coffee was last in my mind? Chocolate was mentioned last. Surely, then, it was not the sound that caused the coffee to be last in my mind. What, then, was the reason, or accident, which caused the coffee to be last in my mind? I believe, that it will be impossible, in this, and a multitude of similar instances, to assign any accident or circumstance, which determines the mind to its choice among things which appear equally fit and eligible. [41]*

Samuel Stearns published *The American Oracle* in 1791. One purpose of his essay was to identify proper behaviors that would promote "felicity and future happiness." One selection within his tract was directed to those traveling to Saratoga, New York, to enjoy the mineral baths and what to eat and drink accordingly. He crafted his recommendation using poetry:

[Prescription 5]:
And when you thus have took [sic] the air,
Unto your house again repair;
Drink coffee, chocolate, or tea,
Or such things as best suiteth [sic] thee. [42]

Nature and taming the wilderness were themes widely associated with exploration of North America. The common view held by many Americans of the time was that the vast central lands of the continent were uninhabited (except by tribal peoples regarded as savages), that such territories basically were God-given for Americans to use and exploit. The first American edition of *Studies of Nature* by James Henry Bernardin de Saint Pierre published in 1797 in Worcester, Massachusetts, contained the following provocative lines that also mention the cacao tree:

Nature invited, into the wildernesses of America, the overflowings [sic] of the European Nations: She had there disposed every thing, with an attention truly maternal to indemnify the Europeans for the loss of their country . . . She [Nature] has there deposited milk and butter in the nuts of the cocoa tree; perfumed creams in the apples of the atte [?]; table linen and provision in the large sattiny [sic] leaves, and in the delicious figs of the banana; loaves ready for the fire in the potatoes, and the roots of the manioc . . . She [Nature] had scattered about, for the purposes at once of delight and of commerce, along the rivers, in the bosom of the rocks, and in the very bed of torrents, the maize, the sugar cane, the chocolate nut, the tobacco plant, with a multitude of other useful vegetables. [43]

ETHICAL, MORAL, AND RELIGIOUS TRACTS

Some 18th century Americans, belonging to a nation based on Christian values, were torn between the poles of behavior that accepted and tolerated slavery versus the alternative of speaking out and condemning the practice. While certainly an evil and immoral practice, slavery was not introduced to North America by European colonists: indeed, slavery was practiced throughout North, Central, and South America at the time of initial European contact [44, 45]. The Northwestern and Iberian Peninsula Europeans who colonized North America did not invent slavery: slavery was practiced in Mediterranean antiquity [46, 47], the Middle East [48, 49], China [50], and also in Europe well before 1492 [51].

The year 1441 commonly is accepted as the beginning of the Portuguese slave trade between Europe and Africa, with Spanish merchants joining soon after during the 1470s [52, 53]. Systematic trading of African slaves between the Old and New Worlds began in 1510, less than 20 years after initial contact between Columbus and Native peoples in the Caribbean [54]. Documentation of first arrival of indentured Africans (i.e., slaves) to North America occurred in 1619 at Jamestown, Virginia [55].

One of the earliest North American antislavery tracts, *The Selling of Joseph*, was published by Samuel Sewall in 1700 [54]. That slaves were used to harvest cacao beans in the Caribbean was of little consequence to most 18th century North Americans. The ethical tract by Louis-Sebastien Mercier, *Memoires of the Year Two Thousand Five Hundred* (English translation by W. Hooper) brought the issue of slavery to a broader American audience:

It may be very agreeable to sip chocolate, to breathe the odour [sic] of spices, to eat sugar and ananas [pineapple], to drink Barbados water, and to be clothed in the gaudy stuffs of India. But are these sensations sufficiently voluptuous to close our eyes against the crowd of unheard of evils that your luxury engendered in the two hemispheres? You violated the most sacred ties of blood and nature on the coast of Guinea. You armed the father against the son, while you pretended to the name of Christians and of men. Blind barbarians! . . . A thirst for gold extolled by every heart; amiable moderation banished by avidity; justice and virtue regarded as chimeras; avarice pale, and restless, plowing the waves, and peopling with carcasses the depths of the ocean; a whole race of men bought and sold, treated as the vilest animals; kings become merchants, covering the seas with blood for the flag of a frigate. [56]

Other moral tracts urged readers to become more Christian in spirit and heart as with John Brown's essay (not John Brown of antislavery, Harper's Ferry fame), one that appeared in the *Christian Journal* published by Robert Campbell in 1796:

Can I partake of pottage, coffee, or chocolate without lifting up my heart to him [sic] who, as the fruit of the earth, excellent and [__mely?], was roasted in flames, and grinded in the [__ill?] of unbounded wrath; that being, as it were, mingled with the full flood of everlasting love, with the Spirit of all grace, he [sic] might be delicate provision for me! [57]

Some 18th century writers viewed the world through rose-colored glasses and found wonderment even while experiencing life's misfortunes. Mary Wollstonecraft translated Christian Salzmann's text into English, and it appeared in 1795 as *Elements of Morality, For the Use of Children, With an Introductory Address to Parents*. Salzmann presented the thesis that sorrowful events in life were to be cherished, since without sorrow and despair, individuals would not otherwise come to appreciate, enjoy, or value the good things of life. Salzmann (via Wollstonecraft) thus argued that joy and sorrow balanced one's existence:

If we had no sorrow, we should be so accustomed to the comforts of life, that we should no more think of their real value, or enjoy the days of health and peace; but when, now and then, a day or week of anguish and distress comes, we really rejoice in the pleasant days that follow. Yes, how good does this chocolate taste in the company of such dear friends, and with my beloved family, who seem to be snatched from the grave. I should not have felt the pleasure I now do, if I had never known sorrow. [58]

Some British social reformers viewed coffee and chocolate houses as hotbeds of evil, promiscuity, and sedition—vile places patronized by debauchees, drunkards, and spendthrifts. Vicesimus Knox, the English essayist, touched on this issue in his literary efforts, *Essays Moral and Library*. The 12th edition of this work published in 1793 (first edition not inspected) received wide attention in North America. Knox argued that chocolate drinking led to moral decay:

When we read the list of dukes, marquises, earls, viscounts, barons, and baronets, exhibited in the Court Calendar, we cannot help wondering at the great number of those who are sunk in obscurity, or branded with infamy; and at the extreme paucity of characters to which may be applied with justice the epithets of decent, virtuous, learned, and devout. Here we see a long list of titled shadows, whose names are seldom heard, and whose persons are seldom seen but at Newmarket and the chocolate-house. There we mark a tribe who fame has celebrated for those feats of gallantry called, in an old-fashioned book, adultery. Here we point out a wretch

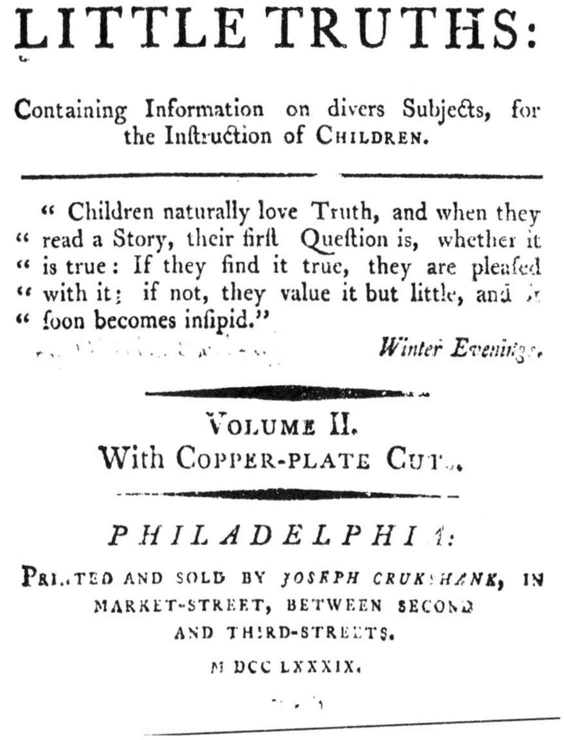

LITTLE TRUTHS:

Containing Information on divers Subjects, for the Instruction of CHILDREN.

――――――――   ――――――――

" Children naturally love Truth, and when they
" read a Story, their first Question is, whether it
" is true: If they find it true, they are pleased
" with it; if not, they value it but little, and it
" soon becomes insipid."

Winter Evenings.

VOLUME II.
With COPPER-PLATE CUTS.

――――――――――――――――

PHILADELPHIA:

PRINTED AND SOLD BY *JOSEPH CRUKSHANK*, IN
MARKET-STREET, BETWEEN SECOND
AND THIRD-STREETS.

M DCC LXXXIX.

FIGURE 19.1. Chocolate: educating children. *Source:* Darton, W. *Little Truths Better than Great Fables. In Variety* [sic] *of Instruction for Children from Four to Eight Years Old,* 2 Volumes. Philadelphia: Joseph Crukshank, 1789; Vol. 1, p. 18. Courtesy of Readex/Newsbank, Naples, FL. (Used with permission.)

stigmatized for unnatural crimes, there a blood-thirsty duelist [sic]. *Debauchees, drunkards, spendthrifts, gamesters, tyrannical neighbours* [sic], *and bad masters of families, occur to the mind of the reader so frequently, that they almost cease by familiarity to excite his animal version.* [59]

William Darton, Sr., an English writer and publisher of children's books, was well known and highly regarded throughout Europe. His volume *Little Truths Better than Great Fables* was first published in England in 1787 with the intent of instructing parents how to teach moral values to 4—8 year old children. The 1789 North American edition of this important volume, published in Philadelphia by Joseph Crukshank, contained one of the more famous chocolate-associated stories from the 18th century, a tale that discussed Hannah and her cup of chocolate (Fig. 19.1):

Hannah, ring the bell for Molly; desire her to bring a cup of chocolate as soon as she can; for it looks so pleasant, I think a walk would contribute to health. Pray, does chocolate grow? Or is it made from any thing? Chocolate is made chiefly from the kernels of cocoa nuts, and is a very nourishing drink. [60]

Elizabeth Rowe, a highly respected and admired 18th century English writer, was perhaps best known

for her book with the unusual title: *Friendship in Death. In Twenty Letters from the Dead to the Living. To Which Are Added, Moral & Entertaining Letters, in Prose and Verse. In Three Parts.* The 1782 American edition of *Friendship* published in Boston contained the following chocolate-associated exchange:

It was now about ten o'clock. Aurelia ordered tea and chocolate to be brought. All her attendance was a fresh-coloured [sic] country lass; who withdrew as soon as we had breakfasted. I was impatient to hear a relation of Aurelia's misfortunes; but durst not ask any question, for fear it would look like insulting her distress; only renewed my excuses for interrupting her privacy. [61]

GEOGRAPHY AND HISTORY

Numerous geographical and historical textbooks printed in North America during the 18th century identified cacao beans and chocolate as commodities associated with specific tropical countries. Some of these passages reported on cacao/chocolate-related production, while others commented on social behavior associated with drinking chocolate. In some instances erroneous information was uncritically included by authors, as with the statement:

1520—Chocolate [was] first brought from Mexico [to Spain] by the Spaniards. [62]

This comment, copied from an earlier text without citation, is without foundation. There is no evidence suggesting that within one year after contact between the troops of Cortés and the Mexica/Aztecs that chocolate suddenly appeared in Europe. It is correct, however, to report that by 1520 Cortés and his literate officers had dined with Montezuma and had been served chocolate beverages [63].

The following selections from standard 18th century geography and history textbooks illustrate the richness of the cacao- and chocolate-related information contained therein. As with the arithmetic textbooks cited earlier, the authors of 18th century geography and history books regularly copied from earlier editions without providing acknowledgment or citations.

[Spain] *Question. What are the customs of the Spanish? Answer. The ladies paint themselves very much. Both sexes live very temperately, drinking but little wine or tea. They usually drink coffee and chocolate, morning and evening, in bed, and eat flesh at noon. Both men and women commonly sleep after eating.* [64]

[Spain] *The temperance of the Spaniards in eating and drinking is remarkable. They frequently breakfast, as well as sup, in bed: their breakfast is usually chocolate; tea being very seldom drank. They live much upon garlic, salad, and radishes. The men drink very little wine; and the women use water or chocolate. Both sexes usually sleep after dinner, and take the air in the cool of the evening.* [65]

[Sweden] *The use of tea, coffee, and chocolate, is every day extending more and more.* [66]

[Mexico] *Mexico, Old, or New Spain. . . . The chief articles produced here, which supply the different branches of domestic and foreign trade, are hides, tallow, cotton, indigo, sugar, chocolate, cochineal, tobacco, honey, feathers, balsams, dying woods, gold, silver, precious stones, ginger, amber, and salt.* [67]

[Mexico] *Old Mexico, or New Spain. Boundaries and Extent. The cotton and cedar trees, and those which bear the cocoa, of which chocolate is made, abound here.* [68]

[Mexico] *To these, which are chiefly the productions of Spanish-America, may be added a great number of other commodities, which, though of less price, are of much grater use. Of these are the plentiful supplies of cochineal, indigo, anatto, logwood, brazil, sustic, pimento, lignum vitae, rice, ginger, cocoa, or the chocolate-nut, sugar, cotton, tobacco, banillas [sic], red-wood, the balsams of Tolu, Peru, and Chili, that valuable article in medicine, the Jesuit's bark, mechoacan, sassafras, sarsaparilla, cassia, tamarinds, hides, furs, ambergris, and a great variety of woods, roots, and plants; to which, before the discovery of America, the Europeans were either entire strangers, or which they were forced to buy at an extravagant rate from Asia and Africa, through the hands of the Venetians and Genoese, who then [sic] engrossed the trade of the Eastern-World.* [69]

[Mexico] *To these, which are chiefly the productions of Spanish-America, may be added a great number of other commodities, which, though of less price, are of much greater use; and many of them make the ornament and wealth of the British Empire in this part of the world. Of these are the plentiful supples [sic] of cochineal, indigo, ananto, log-wood, brazil, fustic, pimento, lignum vitae, rice, ginger, cocoa, or the chocolate nut, sugar, cotton, tobacco, banillas [sic], red-wood, the balsams of Tolu, Peru, and Chili; that valuable article in medicine the Jesuit's bark, mechoacan, sassafras, sarsaparilla [sic] cassia, tamarinds, hides, furs, ambergrease, and a great variety of woods, roots, and plants; to which, before the Discovery of America, we were either strangers, or forced to buy at an extravagant rate from Asia and Africa, through the hands of the Venetians and Genoese, who crossed the trade of the eastern world.* [70]

[Mexico] *Mexico, lying principally in the torrid zone, is excessively hot. The cotton and cedar trees, and those which bear the cocoa, of which chocolate is made, abound here.* [71]

[Guatemala] *Guatimala [sic], a province of New Spain, 750 miles long, and 450 broad. It is bounded E. by Honduras, N. by Chiapa [sic], and Vera Paz, S. and W. by the S. sea. It includes 12 provinces. A large chain of mountains intersect it from E. to W. The valleys between these are fertile, producing cotton, indigo, chocolate, cochineal, honey, wood and some balsam. The inhabitants, partly Spaniards and partly Indians, make use of chocolate as money.* [72]

[Nicaragua] *It has a fruitful soil, [___da?] temperature [sic] climate. Sugar, chocolate, and cochineal, are its chief productions.* [73]

[Tropical America] *There is another sort of a nut called the cocoa [sic]; this grows in the West-Indies and South America. The tree which produces it is something like our cherry-tree, and the nut about the size of an almond. There are seeds withinside [sic], which are made into chocolate, with the addition of some other ingredients. The best sort of this nut is imported from Caraca [i.e., Caracas, Venezuela].* [74]

[Jamaica] *The produce is sugar, rum, cotton, ginger, pymento [sic], indigo, chocolate, medicinal dregs [sic], and several kinds of wood.* [75]

[North America; New England] *The principal manufacturers of the town [Dorchester] are paper, chocolate, snuff, leather, and shoes of various sorts. [Regarding] Mills: There are belonging to Dorchester, 10 mills; viz. 1 paper-mill; 2 chocolate-mills; 1 snuff-mill; 2 fulling-mills; and 4 grist-mills; all which are situated on [the] Naponset River.* [76]

HEALTH, HOME REMEDIES, AND MEDICAL ADVICE

The long history and development of chocolate use in preventive and curative medicine was treated in medical-related monographs of the 16th through 19th centuries. This history is reviewed elsewhere in the present book (see Chapter 6). Presented here are selected examples of chocolate-associated health and medical examples that would have been readily available to North Americans via almanacs, pamphlets, and general household-related books.

CHOCOLATE: Which is the cocoa-nut mixed with either flour or sugar, is very wholesome when not boiled too much; it should only be properly dissolved, as heat, when too great, coagulates it, and consequently makes it harder of digestion. [77]

CHOCOLATE: is nourishing, and proper for the debilitated hypochondriacs, and those who labour [sic] under the piles, or costiveness. [78]

Other [Medical] Cases which Require Immediate Assistance: To prevent the return of the fits [fainting-fits], he ought to take often, but in small quantities, some light yet strengthening nourishment, as pandao made with soup instead of water, new laid eggs lightly poached, chocolate, light roasted meats, jellies, and such like. These fainting-fits, which are the effect of bleeding, or of the violent operation of purges, belong to this class. [79]

A Receipt for Curing the Bite of a Mad-dog . . . And in case a cupping-glass is not to be had, the operation may be performed by a cyder-glass [sic], chocolate cup, pepper box, or other tight domestic implement. However probable the success of this may be, it would be imprudent to trust to it alone; application should be made to some skillful physician, lest the whole of the poison should not be extracted. [80]

Breakfasts are as necessary as suppers; only those who are troubled with phlegm should eat less at this meal than others. A cup of chocolate, not made too strong, is a good breakfast. Coffee I cannot advise generally: But the exceptions against tea are in a great measure groundless. [81]

Chocolate is nourishing and balsamic, if fresh and good; but is disagreeable to the stomach if the nut is badly prepared. [82]

[Caution regarding mixing chocolate and romance; see Fig. 19.2] Of the Four Quarters of the Year, and first of the SPRING. . . . And I must beg leave to advise my readers who are of the fair sex to be in a particular manner careful how they meddle with romances, chocolate, novels, and the like inflamers, which I look upon as very dangerous to be made use of during this great carnival of nature [i.e., Spring]. [83]

[Use of chocolate in treatment of gout] Drink no malt liquor on any account. Let your beverage at dinner consist of two glasses of wine diluted with three half-pints of water. On

Of the four quarters of the YEAR, and first of the SPRING,

our to devotion.

I shall conclude this quarter, however, with Virgil's advice to young people, that while they are gathering wild strawberries and nosegays they should have a care of the snake in the grass. And here I must fall in with the prescriptions of our astrological physicians, such as a spare and simple diet, with the moderate use of phlebotomy, during this season.

And I must beg leave to advise my readers who are of the fair sex to be in a particular manner careful how they meddle with romances, chocolate, novels, and the like inflamers, which I look upon as very dangerous to be made use of during this great carnival of nature.

FIGURE 19.2. Chocolate: educating young adults. *Source:* Philomath, T. *The Virginia Almanack* [sic] *for the Year of Our Lord God 1770 . . . Calculated According to Art, and Referred to the Horizon of 38 Degrees North Latitude, and a Meridian of Five Hours West from the City of London; Fitting Virginia, Maryland, North Carolina, &c.* Williamsburg, VA: Purdie and Dixon, 1769; p. 16. Courtesy of Readex/Newsbank, Naples, FL. (Used with permission.)

no account drink any more wine or spirituous liquors in the course of the day; but, if you want more liquid, take cream and water, or milk and water, or lemonade, with tea, coffee, chocolate. Use the warm bath twice a week for half an hour before going to bed, at the degree of heat which is most grateful to your sensations. Eat meat constantly at dinner and with it any kind of tender vegetables you please. Keep the body open by two evacuations daily, if possible without medicine, if not take the size of a nutmeg of lenitive electuary occasionally, or five grains of rhubarb every night. Use no violent exercise, which may subject yourself to sudden changes from heat to cold; but as much moderate exercise as may be, without being much fatigued or starved with cold. Take some supper every night; a small quantity of animal food is preferred; but if your palate refuses this, take vegetable food, as fruit pie, or milk; something should be eaten, as it might be injurious to you to fast too long. [84]

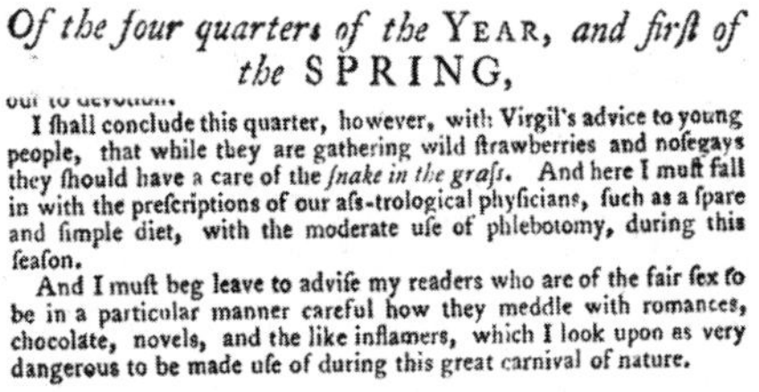

[Chocolate used to reduce the incidence of fainting] *Of Swoonings [sic] Occasioned by Weakness:* . . . Lastly, during the whole time that all other precautions are taken to oppose the cause of the swooning, care must be had for some days to prevent any deliquiu [sic] or fainting, by giving them often, and but little a time, some light yet strengthening nourishment, such as panada made with soup instead of water, new laid eggs very lightly poached, light roast meats with sweet sauce, chocolate, soups of the most nourishing meats, jellies, milk, &c. [85]

Liquid aliments, are milk, eggs, chocolate, soups, broths, &c. Milk, requiring but little preparation in the stomach, is a good aliment for persons whose stomachs are weak, and children; new-laid eggs are very nourishing, and easy of digestion, therefore agree with exhausted and old persons. Chocolate nourishes greatly, strengthens the stomach, helps digestion, and softens sharp humours [sic]; whence it is proper for weak stomachs and consumptive persons. [86]

Of special interest are recommendations that appear in home remedies that contained information to be followed when preparing patients to receive smallpox inoculations (see Chapter 7). Those about to be treated were prepared both spiritually and physically for the inoculation ordeal, that was not without considerable risk in the 18th century:

[Preparing patients to receive smallpox inoculations] *The Diet for Breakfast. Tea, coffee, or chocolate, with dry toast or ordinary cake, rice, milk-gruel, skimmed milk, honey and bread, &c.* [87]

[Advice for treating smallpox] *During the eruptive Fever they may lie in Bed (if they choose it) but should not be covered hot; or it may be best only to lie on the Bed and lightly cloathed [sic], drinking plentifully of Barley Water, Sage and Balm Tea, or Toast and Water, with Tamarinds; and, when filling, let them use Milk and Water, Panda, Sagoe [sic], Chocolate, Gruel, Puddings, Greens and Roots. If they are a little costive while the Pock is filling, it is no-great Matter, and may generally be prevented by drinking warm Small Beer, eating a roasted Apple, or a few Tamarinds now and then; but if very costive, at the Turn of the Pock, an opening Clyster of Water-gruel or Milk, with a Spoonfull [sic] or two of Melasses [sic] and Oil in it, will be proper every Day or two.* [88]

[On preparing patients to receive smallpox inoculations] *I order such of my patients . . . to abstain from all animal food, including broths, also butter and cheese, and from all fermented liquors, excepting small beer, which is allowed sparingly, and from all spices, and whatever possesses a manifest heating quality. The diet is to consist of pudding, gruel, sago, milk, rice-milk, fruit pyes [sic], greens, roots, and vegetables of all the kinds in season, prepared or raw. Eggs, tho' not to be eaten alone, are allowed in puddings, and butter in pye-crust [sic]; the patients are to be careful that they do not eat such a quantity as to overload their stomachs, even of this kind of food. Tea, coffee, or chocolate are permitted for breakfast, to those who choose or are accustomed to them. In this manner they are to proceed about nine or ten days before the operation [i.e., inoculation].* [89]

LINGUISTICS

Dictionaries and foreign language textbooks, whether in French, Spanish, or German, were widely available in North America during the 18th century. These books commonly contained examples of chocolate-associated passages as revealed by these representative selections:

Chocolate, tchokul-et, n. A preparation of the Indian cocoa nut, the liquor made of it. [90]

Xicara, a dish for chocolate; Leche? Tostadas? Chocolate? [Milk? Toasts? Chocolate?]. [91]

Je ne suis pas amateur de caffe, je pretere du chocolat. [I am no lover of coffee; I prefer chocolate.] [92]

To Mill, mil. v. to grind, to comminute; to beat up chocolate; to stamp letters or other work around the edges of coin in the mint. [93]

Words of Three Syllables. Table 1: Note, the Accent is on the first Syllable . . . Cho co late. [94]

The following Nouns, taken in a limited Sense, are to be declined through all their Cases: chocolate. [95]

The Sounds of the Consonants: Ch, 2 sh (English) [as in] chocolate. [96]

O—this vowel has three variations in its sound . . . O is short and closed in the following words: chocolat [sic]. [97]

The Sounds of the Consonants: Ch, 2 sh (English) [as in] chocolate. [98]

Hier ist Wein, Bier, Wasser, Punch, Brandwein, Thee, Caffee, und Chocolate [Here is Wine, Beer, Water, Punch, Brandy, Tea, Coffee, and Chocolate] . . . Hier ist Coffee und Schocolade [Here is Coffee and Chocolate]. [99]

POETRY, PROVERBS, AND SAWS

Chocolate-related poetry abounds and should be considered a separate topic for investigation. Chocolate-related poetry appeared in a variety of 18th and 19th century North American magazines. In 1746, Fulke Greville published a long poem entitled "The Miserable Glutton," which contained the following lines:

Next day obedient to his word,
'The dish appear'd at course the third;
But matters now were alter'd quite,
In bed till noon he'd stretch'd the night.
Took chocolate at ev'ry dose,
And just at twelve his worship rose.
Then eat a tost and sip'd bohea [tea],
Till one, and fat to dine at three. [100]

Several additional examples also could serve as inspiration for further search efforts in archives and libraries, given the unusual nature and themes of the poetic compositions. Richard Saunders, the pseudonym used by Benjamin Franklin when writing *Poor Richard's Almanac* (*Poor Richard Improved: Being an Almanack [sic] and Ephemeris . . . for the Year of Our Lord 1759 . . . Fitted to the Latitude of Forty Degrees, and a Meridian of Near Five Hours West from London; but May, Without Sensible Error, Serve All the Northern Colonies*), used the following lines to fill the bottom of the page dedicated to the month of May:

May hath xxxi days
Ye Nymphs that pine, o'er Chocolate and Rolls,
Hence take fresh Bloom, fresh Vigour [sic] to your Souls.
[101]

It would not be appreciated widely that Franklin's insertion of these two lines as a "page-filler" required truncation of the full poem that had been penned some years earlier (author obscure). The poem remained popular through subsequent decades as "space allowed" further publication of additional lines that struck true to the hearts of frugal Colonial era residents and in contrast to Franklin's truncated lines and carried considerable moral weight:

Ye Nymphs, that pine o'er Chocolate and Rolls,
Hence take fresh Bloom, fresh Vigour [sic] to your Souls.
Fast and fear not—you'll need no Drop or Pill;
Hunger may starve, Excess is sure to kill. [102]

The poem with a full representation of lines, however, was published in 1791 by Joseph Leland and appeared in his *Farmer's Diary* for use in the State of Connecticut:

> Hear now a rhyming Quack, who spurns your wealth,
> Ant gratis, gives a sure Receipt for Health.
> Foul Luxury's the Cause of all your Pain;
> To scoure th' [sic] obstructed Glands, Abstain! Abstain!
> Fast and be fat, thou starveling in a Gown,
> Ye bloated fast, I'll warrant it brings you down,
> Ye Nymphs that pine our Chocolate and Rolls,
> Hence take fresh Bloom, fresh Vigour 'sic' to your Souls,
> Fast, and fear not, you'll need no Drop, nor Pill,
> Hunger may starve, Excess is sure to kill. [103]

A collection of poetry entitled *The Poetical Flower Basket, Being a Selection of Approved and Entertaining Pieces of Poetry, Calculated for the Improvement of Young Minds*, was published in Worcester, Massachusetts, in 1799. The author has remained anonymous, perhaps for reasons that will be obvious shortly. Readers, whether in the 18th or 21st centuries, would have expected that poetry "written and approved" for young minds would have been prim and proper. Imagine, then, the setting: Consider sedate mothers and fathers reading the poem "A Modern Morning" aloud to family members and children gathered around the hearth. Now smile as you contemplate the actions and conversation that probably followed the reading:

> At four on monday [sic] morn, 'tis said,
> The Dawn sprung from his truckle bed,
> And in a passion with old Night,
> Unbarr'd the rosy gates of Light . . .
> Well now I'll sleep again, begone,
> And get my chocolate by one:
> No, bring my gown, I'l [sic] put it on,
> For see the paltry sunbeams come . . .
> And Ma'am, if you approve the night,
> The day'd be out of fashion quite.
> Hussy, you flatter me, begone,
> And send my chocolate by John.
> Nay, Ma'am. Then curt'seying, exit Nell;
> My lady laughs, and all is well—
> When enters John, and bows his head,
> And brings the chocolate to bed. [104]

Another literary collection of prose and poetry, by a second "anonymous," was published in 1798 in Philadelphia. This volume, *The Pocket Miscellany. In Prose and Verse*, contained the following lines:

> Let us survey the Old Bachelor in all his glory. He gets up in the morning, and rings his bell; his servant attends to know what he would be pleased to have—because he is paid for it—The Old Bachelor orders breakfast of coffee, or tea, or chocolate, and his housekeeper makes the tea, or coffee, or chocolate—because she is paid for it. [105]

The following selection published in 1800—on the cusp of the 19th century—appeared in a text on grammar authored by Lindley Murray. The tone and content, however, reflect more of a sense of poetry set within a series of proverbs and life recommendations:

> Souchbong-tea and Turkey-coffee were his [unclear who the author means] favourite [sic] beverage: chocolate he seldom drank. If we injur [sic] others, we must expect retaliation. Peace and honour [sic] are the sheaves of virtue's harvest. High seasoned food vitiates the palate, and disgusts it with plain fare. Alexander, the conqueror of the world, was, in fact, a robber and a murderer. Honest endeavours [sic], if persevered in, will finally be successful. [106]

RELIGION

An unusual religious tract by the English author Daniel Defoe, written in 1793 and entitled *Religious Courtship*, became widely available in a North American edition. The volume related how fathers and daughters should be encouraged to sit together at home over cups of chocolate to discuss topics such as courtship, marriage, and how to behave if/when a wife finds that her husband may have "strayed." The text presented in some detail how fathers are obliged to tell their daughters about the evils of "promiscuous conversations [that take place] at chocolate houses," and how men of virtue should not indulge such lusts or practices. In Defoe's text the father suggested to his daughter that husbands who find themselves in such conversational difficulties at chocolate houses should speak up in moral tones and turn the conversation to positive subjects, such as religious news or international affairs [107].

George Riley's 1792 religious tract presented a wonderful, descriptive, and complex title: *The Beauties of the Creation; or, A New Moral System of Natural History; Displayed in the Most Singular, Curious, and Beautiful, Quadrupeds, Birds, Insects, Trees, and Flowers: Designed to Inspire Youth with Humanity Towards the Brute Creation, and Bring Them Early Acquainted with the Wonderful Works of the Divine Creator*. In one portion of his manuscript Riley described the cacao tree as naturally wonderful, something that resulted from Divine creation:

> This tree, bearing the cocoa or chocolate nut, resembles our heart cherry tree; except that, when full grown, it is much higher and broader. . . . Each fruit contains from between 15 and 25 small nuts, or almonds, covered with a thin yellow skin; which being separated, a tender substance appears, divided into several unequal particles, that, although sharp to the palate, are nourishing to the constitutior [sic = constitution]. These trees grow in all the Spanish West-Indies, Jamaica, &c, where they commonly produce fruit every seven years at most, after the first planting . . . The cocoa-nuts, attending [?] to the Indians and Spaniards food, raiment, riches, and delight, are received in payment, as currency. It is unnecessary to add, that from this extraordinary tree, that [___some?] beverage chocolate is made, in such

quantities as to supply the greater part of the world with a liquor distinguished for its [___ive?] and restorative qualities. [108]

In 1636, a moral question was posed within the Spanish Catholic Church and continued to challenge clerics for several centuries. The issues were whether or not chocolate was a food or beverage, and whether or not drinking chocolate during Lent constituted breaking religious tradition [109, 110]. Within Colonial era Christianity, fasting also was a source of much concern and comment. This topic served as the focus of a dialogue held among Christian ministers, delegates of the Methodist Episcopal Church in America, at their Baltimore convention held December 27, 1784. A long slate of religious issues was debated and one of these was the topic of whether or not religious fasting compromised one's health:

Question 53. But how can I fast since it hurts my Health?

Answer. There are several Degrees of Fasting, which can not hurt your Health. We will instance in one. Let us every Friday (beginning on the next) avow: this Duty throughout the Continent, by touching no Tea, Coffee or Chocolate in the Morning, but (if we want it) half a Pint of Milk or Water-Gruel. Let us dine on Vegetables, and (if we need it) eat three or four Ounces of Flesh in the Evening. At other Times let us eat no Flesh-suppers. These exceedingly tend to breed nervous Disorders. [111]

On December 17, 1727, the well-respected North American jurist Cotton Mather delivered an oration at the funeral service of Pastor Peter Thatcher (Thacher) held at Milton, Massachusetts. Unknown to those attending and listening to the great orator was the fact that this presentation would be Mather's last sermon prior to his own death in 1728. It is interesting that Mather's 1727 funeral oration subsequently was published nearly 70 years later in 1796 and while it eulogizes Thatcher (Thacher), the document also

contains a geography lesson. The remarks, however, seem out of place from the vantage point of the 21st century, since Mather used the funeral pulpit to describe the economic base of Milton, well known in New England for the chocolate mill that characterized the settlement:

There are seven mills upon Naponset River, in three of which the manufacture of Paper is carried on; a chocolate, slitting, saw and grist mill make up the rest. The inhabitants subsist chiefly by agriculture. [112]

Religious themes of temperance, exercise, and mental purity commonly recommended certain culinary behaviors, whereby frugal meals were deemed better suited for those who were "godly":

But methinks I hear some object and say, that eating a hearty supper makes them restless in the night, and prevents their sleeping. To such I would with joy. It is a proof that nature has not yet sunk under the weight of two hearty meals a day; for I never heard any one make this complaint who did not likewise eat a hearty dinner. Leave off dining in your usual manner, and, instead of eating half a pound, or a pound of flesh, with vegetables proportioned to it, allay your appetite with a little bread and cheese, a bowl of light soup, a cup of coffee or chocolate, or, after the French custom, with a few raisins, or an apple, and I am persuaded you will feel no inconvenience from eating a moderate supper. [113]

SPELLING/GRAMMAR DIALOGUES

Noah Webster, the 18th century grammarian and wordsmith, wrote several books on education that included sections on diction and pronunciation, grammar, and spelling. It was common during the 18th century to merely repeat passages from one volume to another, without citation, credit, or reference to earlier versions or editions. Presented here (Table 19.1) are two texts written by Webster, who merely copied the

Table 19.1 **Comparison of Daniel Webster Chocolate-Related Passages**

Webster, 1783 [114]	Webster, 1790 [115]
Is it true that you have heard good news? *It is true indeed.* *Do you believe what you have heard?* *I am very certain, it is true.* *I think I may rely on your word.* *I would not tell a lie for all America.* *Will you drink a dish of tea [sic].* *Sir, I am much obliged to you;* *I chuse [sic] not to drink any.* *What! do not chuse [sic] to drink any?* *No Sir, I am not fond of it.* *Perhaps you like coffee better.* *No Sir [sic]. I like chocolate.* *At what o'clock shall you prefer it?* *Just at what hour you please.*	*Is it true that you have heard good news?* *It is true indeed.* *Do you believe what you have heard?* *I am very certain it is true.* *I think I may rely on your word.* *I would not tell a lie for all America.* *Will you drink a dish of tea?* *Sir, I am much obliged to you.* *I choose not to drink any.* *What! do not choose to drink any!* *No, Sir, I am not fond of it.* *Perhaps you like coffee better.* *No, Sir, I like chocolate.* *At what o'clock shall you prefer it?* *At eight.*

essence of one passage to another, perhaps an example of what scholars today might call "autoplagiarism" [114–115]. When presented side-by-side it can be seen that both passages contain reference to chocolate, set within a child's dialogue used to teach good manners and how to converse with one's elders. Inspection reveals that the two passages are not identical and that Webster had "polished" and improved upon his work. The incorrect punctuation mark (period) that appeared on line 7 of the 1783 version has been changed, appropriately, to a question mark; the typographical error that appeared on line 13 has been corrected; the spelling of "*chuse*" has been changed to the more familiar "choose"; and Webster's addition to the last line added precision to the overall context of the passage. It also may be noted that another prominent educator, William Biglow, "lifted" verbatim Webster's 1783 passage without attribute of acknowledgment—something considered unacceptable today [116].

Conclusion

The chocolate-associated passages identified and reviewed in this chapter reflect a broad range of educational materials used to train school-children and instill family, moral, and religious values in North America during the 18th century. The documents that contain these chocolate-associated passages can be clustered conveniently into five basic categories:

1. Almanacs (regional and local).

2. Primary school textbooks (arithmetic, accounting/ bookkeeping; English grammar and spelling books; foreign language training books in French, German, and Spanish; and geography and history books).

3. Ethical/moral tracts, religious essays, and sermons.

4. Health-related home remedy documents.

5. Collections of poems, proverbs, and saws/ sayings.

Each of the passages presented here was drawn from documents that would have had wide public distribution in North America within the survey period, 1700–1800. The passages presented reveal insights on teaching methods and provide a prism through which to view everyday life in Colonial and Early Federal era America. Sorting the documents by date also revealed that 18th century American authors commonly did not provided credit to other authors and frequently "lifted" or copied directly previously published passages (or textbook examples) and "inserted" the copied texts into their own work—thus claiming the text as their own. Most of all, the passages cited herein reveal the extent to which cacao and chocolate had become an accepted, integral part of daily life in North America during the 18th century.

Acknowledgments

I wish to acknowledge the efforts of Aliza Johnson and Kristine Lee, who, at an early stage of research on this topic in 2005, identified and collected a portion of the documents used to prepare this chapter.

References

1. Dillinger, T. L., Barriga, P., Escarcega, S., Jimenez, M., Salazar Lowe, D., and Grivetti, L. E. Food of the gods: cure for humanity? A cultural history of the medicinal and ritual use of chocolate. *Journal of Nutrition* 2000; 130(Supplement): 2057S–2072S.

2. Grivetti, L. E. From aphrodisiac to health food; a cultural history of chocolate. In: *Karger Gazette*. Issue 68 (Chocolate). Basel, Switzerland: S. Karger, 2005.

3. Rosenblum, M. *Chocolate. A Bittersweet Saga of Dark and Light*. New York: North Point Press, 2005.

4. http://infoweb.newsbank.com/iw-search/we/ Evans/?p_action=help. (Accessed January 10, 2007.)

5. http://www.readex.com/readex/product. cfm?product=5. (Accessed January 10, 2007.)

6. http://www.proquest.com/products_pq/ descriptions/aps.shtml. (Accessed January 10, 2007.)

7. Anonymous. *The American Tutor's Assistant. Or, A Compendious System of Practical Arithmetic; Containing, the Several Rules of That Useful Science, Concisely Defined, Methodically Arranged, and Fully Exemplified. The Whole Particularly Adapted to the Easy and Regular Instruction of Youth in Our American Schools*, 2nd edition. Philadelphia: Zacaharia Poulson, Jr., 1794.

8. Bonnycastle, J. *The Scholar's Guide to Arithmetic. Or, A Complete Exercise-book, for the Use of Schools . . . 2nd Edition Corrected*. Boston: John West Folsom, 1786.

9. Daboll, N. *Daboll's Schoolmaster's Assistant. Being a Plain Practical System of Arithmetic; Adapted to the United States*. New London, CT: Samuel Green, 1800.

10. Dilworth, T. *The Schoolmaster's Assistant. Being a Compendium of Arithmetic, Both Practical and Theoretical . . . To Which is Perfixt [sic] an Essay on the Education of Youth; Humbly Offered to the Consideration of Parents*. Philadelphia: Joseph Crukshank, 1781.

11. Dilworth, T. *Dilworth's Assistant. Adapted to the Commerce of the Citizens of the United States. Being a Compendium of Arithmetic both Practical and Theoretical. In Five Parts . . . Carefully Revised and Adapted to the Commerce of the Citizens of the United States, with Many Additions in the Various Rules*. New York: Hurtin & M'Farlane, 1798.

12. Gough, J. *A Treatise of Arithmetic in Theory and Practice. Containing Every Thing Important in the Study of Abstract and Applicate Numbers . . . To Which are Added, Many Valuable Additions and Amendments; more Particularly Fitting the Work for the Improvement of the American Youth*. Philadelphia: J. M'Culloch, 1788.

13. Gough, J. *Practical Arithmetick [sic] in Four Books ... Revised by Thomas Telfair Philomath. And Now Fitted to the Commerce of America.With an Appendix of Algebra.* Wilimgton, DE: Peter Brynberg, 1800.

14. Jess, Z. *The American Tutor's Assistant, Improved: Or, A Compendious System of Decimal, Practical Arithmetic; Comprising the Usual Methods of Calculation, With the Addition of Federal Money, and Other Decimals, Dispersed Through the Several Rules of That Useful Science. Adapted for the Easy and Regular Instruction of Youth in the United States.* Wilmington, DE: Bonsal and Niles, 1799.

15. Johnson, G. *An Introduction to Arithmetic. Designed for the Use of Schools.* Springfield, MA: Ezra W. Weld, 1792.

16. Milns, W. *The American Accountant. Or, A Complete System of Practical Arithmetic ... The Whole Calculated to Ease the Teacher and Assist the Pupil; It Will be Found Likewise Extremely Useful to American Merchants, &c.* New York: J. S. Mott, 1797.

17. Pike, N. *A New and Complete System of Arithmetic, Composed for the Use of the Citizens of the United States.* Newbury-Port, MA: John Mycall, 1788.

18. Sarjeant, T. *Elementary Principles of Arithmetic; With Their Application to the Trade and Commerce of the United States of America ... For the Use of Schools and Private Education.* Philadelphia: Dobson and Lang, 1789.

19. Shepherd, E. *The Columbian Accountant; Or, A Complete System of Practical Arithmetic; Particularly Adapted to the Commerce of the United States of America. To Which is Annexed, a Short Sketch of Mensuration.* New York: T. & J. Swords, 1800.

20. Sterry, C. *A Complete Exercise Book in Arithmetic: Designed for the Use of Schools in the United States. By Consider & John Sterry, Authors of The American Youth.* Norwich, CT: John Sterry and Company, 1795.

21. Shepherd, E. *The Columbian Accountant; Or, A Complete System of Practical Arithmetic; Particularly Adapted to the Commerce of the United States of America. To Which is Annexed, a Short Sketch of Mensuration.* New York: T. & J. Swords, 1800; p. 90.

22. Jess, Z. *The American Tutor's Assistant, Improved: or, A Compendious System of Decimal, Practical Arithmetic; Comprising the Usual Methods of Calculation, With the Addition of Federal Money, and Other Decimals, Dispersed Through the Several Rules of That Useful Science. Adapted for the Easy and Regular Instruction of Youth in the United States.* Wilmington, DE: Bonsal and Niles, 1799; p. 116.

23. Dilworth, T. *Dilworth's Assistant. Adapted to the Commerce of the Citizens of the United States. Being a Compendium of Arithmetic both Practical and Theoretical. In Five Parts ... Carefully Revised and Adapted to the Commerce of the Citizens of the United States, with Many Additions in the Various Rules.* New York: Hurtin & M'Farlane, 1798; p. 104.

24. Milns, W. *The American Accountant. Or, A Complete System of Practical Arithmetic ... The Whole Calculated to Ease the Teacher and Assist the Pupil; It Will be Found Likewise Extremely Useful to American Merchants, &c.* New York: J. S. Mott, 1797; p. 165.

25. Anonymous. *The American Tutor's Assistant. Or, A Compendious System of Practical Arithmetic; Containing, the Several Rules of That Useful Science, Concisely Defined, Methodically Arranged, and Fully Exemplified. The Whole Particularly Adapted to the Easy and Regular Instruction of Youth in Our American Schools,* 2nd edition. Philadelphia: Zacaharia Poulson, Jr., 1794; p. 110.

26. Bonnycastle, J. *The Scholar's Guide to Arithmetic. Or, A Complete Exercise-book, for the Use of Schools ... 2nd Edition Corrected.* Boston: John West Folsom, 1786; p. 75.

27. Dilworth, T. *The Schoolmaster's Assistant. Being a Compendium of Arithmetic, Both Practical and Theoretical ... To Which is Perfixt [sic] an Essay on the Education of Youth; Humbly Offered to the Consideration of Parents.* Philadelphia: Joseph Crukshank, 1781; p. 96.

28. Sterry, C. *A Complete Exercise Book in Arithmetic: Designed for the Use of Schools in the United States. By Consider & John Sterry, Authors of The American Youth.* Norwich, CT: John Sterry and Company, 1795; p. 84.

29. Johnson, G. *An Introduction to Arithmetic. Designed for the Use of Schools.* Springfield, MA: Ezra W. Weld, 1792; p. 25.

30. Sarjeant, T. *Elementary Principles of Arithmetic; With Their Application to the Trade and Commerce of the United States of America ... For the Use of Schools and Private Education.* Philadelphia: Dobson and Lang, 1789; p. 63.

31. Gough, J. *Practical Arithmetick [sic] in Four Books ... Revised by Thomas Telfair Philomath. And Now Fitted to the Commerce of America. With an Appendix of Algebra.* Wilimgton, DE: Peter Brynberg, 1800; Volume 1, p. 92.

32. Gough, J. *Practical Arithmetick [sic] in Four Books ... Revised by Thomas Telfair Philomath. And Now Fitted to the Commerce of America. With an Appendix of Algebra.* Wilimgton, DE: Peter Brynberg, 1800; Volume 3, p. 168.

33. Gough, J. *A Treatise of Arithmetic in Theory and Practice. Containing Every Thing Important in the Study of Abstract and Applicate Numbers ... To Which are Added, Many Valuable Additions and Amendments; more Particularly Fitting the Work for the Improvement of the American Youth.* Philadelphia: J. M'Culloch, 1788; p. 101.

34. Sarjeant, T. *Elementary Principles of Arithmetic; With Their Application to the Trade and Commerce of the United States of America ... For the Use of Schools and Private Education.* Philadelphia: Dobson and Lang, 1789; p. 69.

35. Pike, N. *A New and Complete System of Arithmetic, Composed for the Use of the Citizens of the United States.* Newbury-Port, MA: John Mycall, 1788; p. 279.

36. Daboll, N. *Daboll's Schoolmaster's Assistant. Being a Plain Practical System of Arithmetic; Adapted to the United States.* New London, CT: Samuel Green, 1800; p. 141.

37. Mitchell, W. *A new and complete system of book-keeping, by an improved method of double entry; adapted to retail, domestic and foreign trade: exhibiting a variety of transactions which usually occur in business. The whole comprised in three sets of books; the last set, being a copy of the second according to those systems most generally in use, is given in order to exhibit, by a comparative view, the advantages of the system now laid down. To which is added, a table of the duties payable on goods, wares and merchandise, imported into the United States of America. The whole in dollars and cents*s. Philadelphia: Bioren & Madan, 1796; p. 271.

38. Mitchell, W. *A new and complete system of book-keeping, by an improved method of double entry; adapted to retail, domestic and foreign trade: exhibiting a variety of transactions which usually occur in business. The whole comprised in three sets of books; the last set, being a copy of the second according to those systems most generally in use, is given in order to exhibit, by a comparative view, the advantages of the system now laid down. To which is added, a table of the duties payable on goods, wares and merchandise, imported into the United States of America. The whole in dollars and cents*. Philadelphia: Bioren & Madan, 1796; p. 323.

39. Lyttelton, G. *Dialogues of the Dead. By the Late Lord (George) Lyttleton*, 1st American edition. Worcester, MA: Thomas, Son and Thomas, 1797; p. 155.

40. Aubin, P. *The Noble Slaves. Being an Entertaining History of the Surprising Adventures, and Remarkable Deliverances, from Algerine [sic] Slavery, of Several Spanish Noblemen and Ladies of Quality*. Danbury, CT: Douglas and Nichols, 1797; p. 46.

41. West, S. *Essays on Liberty and Necessity; In Which the True Nature of Liberty is Stated and Defended; and the Principal Arguments Used by Mr. Edwards, and Others, for Necessity, Are Considered*. Boston: Samuel Hall, 1793; p. 18.

42. Stearns, S. *The American Oracle. Comprehending an Account of Recent Discoveries in the Arts and Sciences, with a Variety of Religious, Political, Physical, and Philosophical Subjects, Necessary to be Known in All Families, for the Promotion of Their Present Felicity and Future Happiness*. New York: Hodge and Campbell, Berry and Rogers, and T. Allen, 1791; pp. 308–309.

43. Saint Pierre, J. H. B. de. *Studies of Nature. By James Henry Bernardin de Saint Pierre*. Translated by H. Hunter. 4 Volumes 1, American edition. Worcester, MA: J. Nancrede, 1797; Volume 2, pp. 471–472.

44. http://www.slaveryinamerica.org/history/hs_es_indians_slavery.htm. (Accessed January 10, 2007.)

45. http://tarlton.law.utexas.edu/rare/aztec/Slavery.htm. (Accessed January 10, 2007.)

46. Garlan, Y. *Esclaves en Grèce ancienne [Slavery in Ancient Greece]*. Translated by J. Lloyd. Revised and expanded edition. Ithaca, NY: Cornell University Press, 1988.

47. Hezser, C. *Jewish Slavery in Antiquity*. Oxford, England: Oxford University Press, 2005.

48. Gordon, M. *Slavery in the Arab World*, 1st American edition. New York: New Amsterdam Press, 1989.

49. Tora, M., and Philips, J. E. *Slave Elites in the Middle East and Africa. A Comparative Study*. London, England: Kegan Paul International, 2000.

50. Wilbur, C. M. *Slavery in China during the Former Han Dynasty, 206 B.C.–A.D. 25*. Field Museum of Natural History, Chicago, Anthropological Series, Volume 34. Chicago: Field Museum of Natural History, 1943.

51. Bonnassie, P. *From Slavery to Feudalism in South-Western Europe*. Cambridge, England: Cambridge University Press, 1991.

52. Rout, L. B., Jr. *The African Experience in Spanish America, 1502 to the Present Day*. Cambridge, England: Cambridge University Press, 1976.

53. Landers, J. G., and Robinson, B. M. *Slaves, Subjects, and Subversives. Blacks in Colonial Latin America*. Albuquerque: University of New Mexico Press, 2006.

54. Galenson, D. W. *Traders, Planters, and Slaves. Market Behavior in Early English America*. Cambridge, England: Cambridge University Press, 1986.

55. http://www.nps.gov/archive/colo/Jthanout/AFRICANS.html. (Accessed January 10, 2007.)

56. Mercier, L.-S. *Memoirs of the Year Two Thousand Five Hundred*. Translated from the French by W. Hooper. Richmond, VA: N. Pritchard, 1799; p. 313.

57. Brown, J. *The Christian Journal; Or, Common Incidents, Spiritual Instructors. Being a Series of Meditations on a Spring, Summer, Harvest, Winter, and Sabbath-Day*, 1st American edition. Philadelphia: Robert Campbell, 1796; p. 91.

58. Salzmann, C. *Elements of Morality, For the Use Of Children; With An Introductory Address To Parents*. Translated from the German of the Rev. C. G. Salzmann (by M. Wollstonecraft). 1st American edition. Providence, RI: Carter and Wilkinson, 1795; p. 259.

59. Knox, V. *Essays Moral and Literary*, 2 Volumes, 12th edition. New York: T. Allen, 1793; Volume 2, p. 301.

60. Darton, W. *Little Truths Better than Great Fables. In Variety [sic] of Instruction for Children from Four to Eight Years Old*, 2 Volumes. Philadelphia: Joseph Crukshank, 1789; Volume 1, p. 18.

61. Rowe, E. *Friendship in Death. In Twenty Letters from the Dead to the Living. To Which Are Added, Moral & Entertaining Letters, in Prose and Verse. In Three Parts*. Boston: R. Hodge, W. Green, and J. Norman, 1782; p. 238.

62. Anonymous. *The Gentleman's Pocket Library. Containing, 1. The Principles of Politeness. 2. The Economy of Human Life. 3. Rochefoucauld's Moral Reflections. 4. Lavater's Aphorisms on Man. 5. The Polite Philosopher. 6. The Way to Wealth, by Dr. Franklin. 7. Select Sentences. 8. Detached Sentences. 9. Old Italian, Spanish and English Proverbs. 10. A Tablet of Memory*. Boston: W. Spotswood, 1794; p. 233.

63. Díaz del Castillo, B. (1560–1568). *The Conquest of New Spain*. New York: Penguin Books, 1983.

64. Dwight, N. *A Short but Comprehensive System of the Geography of the World. By Way of Question and Answer. Principally Designed for Children, and Common Schools.* Hartford, CT: Elisha Babcock, 1795; p. 78.

65. Guthrie, W. *A New System of Modern Geography: Or, A Geographical, Historical, and Commercial Grammar; and Present State of the Several Nations of the World*, 2 Volumes. Philadelphia: Matthew Carey, 1794; Volume 2, p. 507.

66. Guthrie, W. *A New System of Modern Geography: Or, A Geographical, Historical, and Commercial Grammar; and Present State of the Several Nations of the World*, 2 Volumes. Philadelphia: Matthew Carey, 1794; Volume 2, p. 119.

67. Scott, J. *The New and Universal Gazetteer. Or, Modern Geographical Dictionary. Containing a Full and Authentic Description of the Different Empires, Kingdoms, Republics, States, Provinces, Islands, Cities, Towns, Forts, Mountains, Caves, Capes, Canals, Rivers, Lakes, Oceans, Seas, Bays, Harbours, &c. in the Known World.* Philadelphia: Francis and Robert Bailey, 1799; no pagination.

68. Morse, J. *Geography Made Easy. Being a Short but Comprehensive System of That Very Useful and Agreeable Science . . . Calculated Particularly for the Use and Improvement of Schools in the United States.* New Haven, CT: Peigs, Bowen, and Dana, 1784; p. 101.

69. Morse, J. *The History of America, in Two Books. Containing, 1. A General History of America. II. A Concise History of the Late Revolution. Extracted from the American Edition of the Encyclopaedia.* Philadelphia: Thomas Dobson, 1790; p. 110.

70. Winterbotham, W. *A Geographical, Commercial, and Philosophical View of the Present Situation of the United States of America. Comprehending a Description of the United States . . . To Which is Prefixed, a General Account of the Discovery of America, by Columbus; General Description of the Whole Continent of America, and the Numerous Tribes of American Indians, Their Manners, Customs.* New York: Tiebout & O'Brien, 1795; p. 156.

71. Morse, J. *The American Geography. Or, A View of the Present Situation of the United States of America . . . To Which is Added, a Concise Abridgment of the Geography of the British, Spanish, French and Dutch Dominions in America, and the West Indies . . . of Europe, Asia and Africa.* Elizabethtown, NJ: Shepard Kollock, 1789; p. 480.

72. Scott, J. *The New and Universal Gazetteer. Or, Modern Geographical Dictionary. Containing a Full and Authentic Description of the Different Empires, Kingdoms, Republics, States, Provinces, Islands, Cities, Towns, Forts, Mountains, Caves, Capes, Canals, Rivers, Lakes, Oceans, Seas, Bays, Harbours, &c. in the Known World . . . Philadelphia*, Pennsylvania: Francis and Robert Bailey, 1799; no pagination.

73. Scott, J. *The New and Universal Gazetteer. Or, Modern Geographical Dictionary. Containing a Full and Authentic Description of the Different Empires, Kingdoms, Republics, States, Provinces, Islands, Cities, Towns, Forts, Mountains, Caves, Capes, Canals, Rivers, Lakes, Oceans, Seas, Bays, Harbours, &c. in the Known World . . . Philadelphia*, Pennsylvania: Francis and Robert Bailey, 1799; no pagination.

74. Trimmer, [S]. *An Easy Introduction to the Knowledge of Nature. Adapted to the Capacities of Children. By Mrs. Trimmer. Revised, Corrected, and Greatly Augmented; and Adapted to the United States of America.* Boston: Manning and Loring, 1796; p. 18.

75. Scott, J. *The New and Universal Gazetteer. Or, Modern Geographical Dictionary. Containing a Full and Authentic Description of the Different Empires, Kingdoms, Republics, States, Provinces, Islands, Cities, Towns, Forts, Mountains, Caves, Capes, Canals, Rivers, Lakes, Oceans, Seas, Bays, Harbours, &c. in the Known World . . . Philadelphia*, Pennsylvania: Francis and Robert Bailey, 1799; no pagination.

76. Anonymous. *Geographical Gazetteer of the Towns in the Commonwealth of Massachusetts.* Boston: Greenleaf and Freeman, 1784; p. 35.

77. Anonymous [By a Gentleman of the Faculty]. *Concise Observations on the Nature of Our Common Food, So Far as It Tends to Promote or Injure Health; with Remarks on Water, Bread, Meat, Cheese, Butter, Milk, Wine, Punch, Beer, Coffee, Tea, Sugar, &c. &c. To Which Are Prefixed General Rules for a Course of Diet.* New York: T. & J. Swords, 1790; p. 31.

78. Burrell, W. *Medical Advice. Chiefly for the Consideration of Seamen and Adapted for the Use of Travellers [sic], or Domestic Life. Containing Practical Essays on Diseases in General—with Plain and Full Directions for Their Prevention and Cure: Gun-shot wounds, Fractures, Dislocations— and on the Venereal Disease.* New York: R. Wilson, 1798; p. 21.

79. Buchan, W. *Domestic Medicine. Or, the Family Physician. Being an Attempt to Render the Medical Art More Generally Useful, by Shewing [sic] People What is in Their Own Power Both With Respect to the Prevention and Cure of Diseases*, 3rd American edition. Norwich, CT: John Trumbull, 1778; p. 433.

80. Goddard, M. *Mary K. Goddard's Pennsylvania, Delaware, Maryland, and Virginia Almanack [sic] for 1779.* Baltimore, MD: Mary K. Goddard, 1778; p. 14.

81. Hill, J. *The Old Man's Guide to Health and Longer Life. With Rules for Diet, Exercise, and Physic. For Preserving a Good Constitution, and Preventing Disorders in a Bad One.* Philadelphia: John Dunlap, 1775; p. 12.

82. Weatherwise, A. *Father Abraham's Almanack [sic] for the Year of Our Lord, 1771 . . . Fitted to the Latitude of Forty Degrees, and a Meridian of Near Five Hours West from London.* Philadelphia: John Dunlap, 1770; p. 13.

83. Philomath, T. *The Virginia Almanack [sic] for the Year of Our Lord God 1770 ... Calculated According to Art, and Referred to the Horizon of 38 Degrees North Latitude, and a Meridian of Five Hours West from the City of London; Fitting Virginia, Maryland, North Carolina, &c.* Williamsburg, VA: Purdie and Dixon, 1769; p. 16.

84. Darwin, E. *Zoonomia. Or, the Laws of Organic Life, Part Second*, 2 Volumes. Philadelphia: T. Dobson, 1797; Volume 2, pp. 86–87.

85. Mackenzie, J. *Advice to the People in General, With Regard to Their Health: But Particularly Calculated for Those, Who Are the Most Unlikely to be Provided in Time with the Best Assistance, in Acute Diseases, or Upon any Inward or Outward Accident*, 2 Volumes. Boston: Mein and Fleming, 1767; Volume 2, pp. 403–404.

86. Theobald, J. *Every Man His Own Physician. Being a Complete Collection of Efficacious and Approved Remedies for Every Disease Incident to the Human Body. With Plain Instructions for Their Common Use*, 10th edition. Boston: W. Griffin, 1765; p. 43

87. Nathan, J. *Hutchin's Improved. Being an Almanac and Ephemeris ... For the Year of Our Lord, 1768*. New York: Hugh Gaine, 1768; p. 13.

88. Saunders, R. (Benjamin Franklin). *Poor Richard Improved. Being an Almanack [sic] and Ephemeris ... for the Year of Our Lord 1761 ... Fitted to the Latitude of Forty Degrees, and a Meridian of Near Five Hours West from London; but May, Without Sensible Error, Serve all the Northern Colonies*. Philadelphia: B. Franklin and D. Hall, 1760; p. 5.

89. Dimsdale, T. *The Present Method of Inoculating for the Small-pox. To Which Are Added, Some Experiments, Instituted with a View to Discover the Effects of a Similar Treatment in the Natural Small-pox*. Philadelphia: John Dunlap, 1771; pp. 12–13.

90. Alexander, C. *The Columbian Dictionary of the English Language; In Which Many New Words Peculiar to the United States, and Many Words of General Use, Not Found in Any Other English Dictionary, are Inserted*. Boston: Isaiah Thomas and Ebenezer T. Andrews, 1800; p. 105.

91. Giral del Pino, H. S. J. *A New Spanish Grammar. Or, The Elements of the Spanish Language. Containing an Easy and Compendious Method to Speak and Write It Correctly*, First American edition. Philadelphia: Colerick and Hunter, 1795; p. 294.

92. Perrin, J. *The Elements of French Conversation, with New, Familiar and Easy Dialogues, Each Preceded by a Suitable Vocabulary in French and English. Designed Particularly for the Use of Schools*. Philadelphia: Thomas Bradford, 1794; p. 90.

93. Sheridan, T. *A Complete Dictionary of the English Language, Both With Regard to Sound and Meaning: One Main Object of Which Is, To Establish a Plain and Permanent Standard of Pronunciation*, 4th edition. Revised and corrected by J. Andrews. Philadelphia: William Young, 1789; no pagination.

94. Dilworth, T. *A New Guide to the English Tongue. In Five Parts ... The Whole, Being Recommended by Several Clergymen and Eminent Schoolmasters, as the Most Useful Performance, for the Instruction of Youth, Is Designed for the Use of Schools in Great-Britain, Ireland, and in the Several Colonies and Plantations Abroad*. Hartford, CT: Nathaniel Patten, 1788; p. 41.

95. Mary, J. *A New French and English Grammar. Wherein the Principles are Methodically Digested, with Useful Notes and Observations, Explaining the Terms of Grammar, and Further Improving Its Rules*. Boston: J. Norman, 1784; p. 164.

96. Perrin, J. *The Practice of French Pronunciation Alphabetically Exhibited. Wherein the Several Sounds of the Letters are Distinguished, and the Words Which Have the Same Sound are Placed in One Class*. Philadelphia: Styner and Cist, 1780; p. 3.

97. Perrin, J. *The Practice of French Pronunciation Alphabetically Exhibited. Wherein the several Sounds of the Letters are Distinguished, and the Words Which Have the Same Sound are Placed in One Class*. Philadelphia: Styner and Cist, 1780; p. 15.

98. Perrin, J. *A Grammar of the French Tongue, Grounded Upon the Decisions of the French Academy, Wherein all the Necessary Rules, Observations, and Examples are Exhibited in a Manner Entirely New*, 2nd edition. Philadelphia: Styner and Cist, 1779; p. 16.

99. Bachmair, J. *A Complete German Grammar, in Two Parts. The First Part Containing the Theory of the Language Through All the Parts of Speech; the Second Part is the Practice in as Ample a Manner as Can be Desired*, 3rd edition. Philadelphia: Henry Miller, 1772; pp. 142 and 160.

100. Greville, H. *The Miserable Glutton, Or, The Pleasures of Sense, Dependent on Virtue. The American Magazine and Historical Chronicle* 1746; July Issue: 326.

101. Saunders, R. (Benjamin Franklin). *Poor Richard Improved: Being an Almanack [sic] and Ephemeris ... for the Year of Our Lord 1759 ... Fitted to the Latitude of Forty Degrees, and a Meridian of Near Five Hours West from London; but May, Without Sensible Error, Serve All the Northern Colonies*. Philadelphia: B. Franklin and D. Hall, 1758; p. 8.

102. Weatherwise, A. *The New-England Town and Country Almanack [sic] ... For the Year of Our Lord 1769 ... Fitted to the Latitude of Providence, in New-England; but May, Without Sensible Error, Serve All the Northern Colonies*. Providence, RI: Sarah Goddard and John Carter, 1768; p. 24.

103. Leland, J. *The Farmer's Diary: Or, The United States Almanack [sic] for the Year of Our Lord Christ, 1792 ... Calculated for the Meridian of Danbury, in the State of Connecticut*. Danbury, CT: Douglas and Ely, 1791; no pagination.

104. Anonymous. *The Poetical Flower Basket, Being a Selection of Approved and Entertaining Pieces of Poetry, Calculated for*

the *Improvement of Young Minds*. Worcester, MA: Isaiah Thomas, Jr., 1799; pp. 42–43.

105. Anonymous. *The Pocket Miscellany. In Prose and Verse*. Philadelphia: Mathew Carey, 1798; p. 36.

106. Murray, L. *Exercises Adapted to Murray's English Grammar Designed for the Benefit of Private Learners, as Well as for the Use of Schools; With a Key*. Philadelphia: Asbury Dickins, 1800; p. 15.

107. Defoe, F. *Religious Courtship. Being Historical Discourses on the Necessity of Marrying Religious Husbands and Wives Only; As Also, of Husbands and Wives Being of the Same Opinions in Religion with One Another*. New York: W. Durrell, 1793; pp. 9 and 47.

108. Riley, G. *The Beauties of the Creation; Or, A New Moral System of Natural History; Displayed in the Most Singular, Curious, and Beautiful, Quadrupeds, Birds, Insects, Trees, and Flowers: Designed to Inspire Youth with Humanity Towards the Brute Creation, and Bring Them Early Acquainted with the Wonderful Works of the Divine Creator*. Philadelphia: William Young, 1792; pp. 259, 260.

109. Pinelo, A. de L. *Question Moral. Si el Chocolate Quebranta el Ayuno Eclesiastico. Facsímile de la primera edición (Madrid, 1636)*. Mexico City, Mexico: Centro de Estudios de Historia de México, 1636.

110. Hurtado, T. *Chocolate y Tabaco Ayuno Eclesiastico y Natural: si este le Quebrante el Chocolate: y el Tabaco al Natural, para la Sagrada Comunion. Por Francisco Garcia, Impressor del Reyno. A Costa de Mauel Lopez*. Madrid, Spain: Mercador de Libros, 1645; Volume 1, 1:1, 3:21.

111. Coke, T., Francis, A. (and others not named). *Minutes of Several Conversations between the Rev. Thomas Coke, LLD, the Rev. Francis Asbury and Others, at a Conference, Begun in Baltimore in the State of Maryland, on Monday, the 27th of December, in the Year 1784. Composing a Form of Discipline for the Ministers, Preachers and Other Members of the Methodist Episcopal Church in America*. Philadelphia: Charles Cist, 1785; p. 20.

112. Mather, C. *The Comfortable Chambers, Opened and Visited, Upon the Departure of that Aged and Faithful Servant of God, Mr. Peter Thatcher [Thacher], the Never to be Forgotten Pastor of Milton, Who Made His Flight Thither, on December 17, 1727. By Cotton Mather, D.D. & F. R. S.* Boston: Thomas Fleet, 1796; p. 28.

113. Anonymous (Possibly Benjamin Rush?) *Sermons to the Rich and Studious, on Temperance and Exercise*. Litchfield, CT: T. Collier, 1791; p. 18.

114. Webster, N. *A Grammatical Institute, of the English Language, Comprising, an Easy, Concise, and Systematic Method of Education, Designed for the Use of English Schools in America. In Three Parts. Part 1. Containing a New and Accurate Standard of Pronunciation*. Hartford, CT: Hudson and Goodwin, 1783; Volume 1, p. 108.

115. Webster, N. *The American Spelling Book. Containing an Easy Standard of Pronunciation. Being the First Part of a Grammatical Institute of the English Language*. Boston: Isaiah Thomas and Ebenezer T. Andrews, 1790; p. 140.

116. Biglow, W. *The Child's Library. Part Second. Containing a Selection of Lessons for Spelling, Reading and Speaking*. Salem, MA: Joshua Cushing, 1800; pp. 21–22.

CHAPTER 20

Chocolate, Crime, and the Courts

Selected English Trial Documents, 1693–1834

Louis Evan Grivetti

Introduction

Today, members of the general public are unlikely to consider cacao beans, cocoa, or even manufactured chocolate to be items of high value, especially when compared or contrasted with the value of precious metals or gems. But in centuries past, chocolate was a rare, expensive commodity, one not widely separated in economic value from gold or silver. Cacao beans, cocoa, and manufactured chocolate presented criminals with easy opportunities. Raw ingredients (beans) or the manufactured goods (cocoa powder; finished chocolate) were relatively "easy targets." While it might be possible to identify the rightful owners of elegantly crafted silver chocolate pots, how could previous owners of recovered bags of cacao beans, tins of ground cacao, or clumps/tablets of manufactured chocolate be identified with certainty? The problem was serious: The chocolate manufacturing profession carried considerable economic risk due to criminal behavior.

Merchants and manufacturers of chocolate were aware that even "branded" blocks of stolen chocolate easily could be "fenced" after identification marks had been melted off or cut away by thieves. Unless criminals were observed "red-handed," chances were that the perpetrators would not be caught or prosecuted. For thieves, therefore, the risk of capture and a court trial was slight—and even if the stolen beans, cacao, or chocolate could not be "fenced," at the end of the day the products could be prepared and drunk as a refreshing, stimulating beverage!

The present chapter identifies and analyzes a rich cache of chocolate-associated crime documents from England, the Old Bailey Trial Archive. The Old Bailey Courthouse was constructed in western London, not far from St. Paul's Cathedral, and served as London's primary criminal court during the 17th, 18th, and 19th centuries (Fig. 20.1). The name Old Bailey stems from the English term *bailey* meaning "fortified wall" and refers to its location adjacent to Newgate Street and the western wall of medieval London. The court's proximal location to Newgate Prison also was convenient for court officials, police, and prisoners (and their legal representatives), since transportation distances between docket and prison could be reduced [1, 2].

Chocolate: History, Culture, and Heritage. Edited by Grivetti and Shapiro
Copyright © 2009 John Wiley & Sons, Inc.

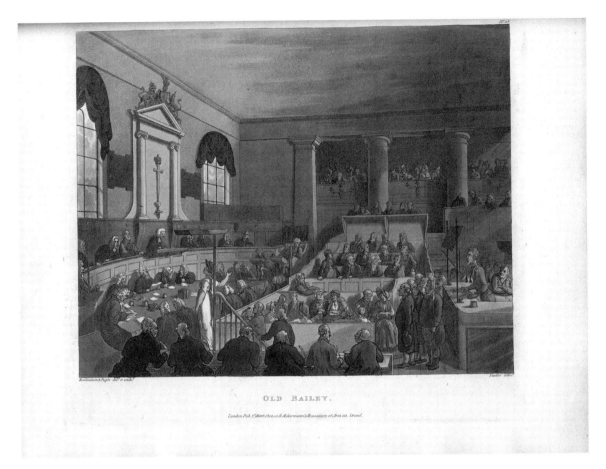

FIGURE 20.1. *Old Bailey Criminal Court*, by Thomas Rowlandson and Augustus Pugin. *Source*: Microcosm of London, 1808. Courtesy of the Peter J. Shields Library, University of California, Davis. (Used with permission.) (See color insert.)

The Old Bailey Trial Archive (London)

English trial documents recently digitized and made available on-line reflect a rich archive that can be searched using key words [3]. A total of 82 chocolate-associated trials for crimes perpetrated by 113 defendants were identified using three key words: cacao, cocoa, and chocolate. The earliest trial document identified was dated to May 31, 1693, [4], and the last case that considered a chocolate-associated crime was dated September 4, 1834 [5]. Defendants were accused of 15 specific crimes (Table 20.1), the most common being simple grand larceny (63 instances) followed by burglary (19 instances).

The documents reveal that violence rarely accompanied chocolate-associated or chocolate-related crime: only two of 82 cases. One of the violent cases occurred in 1693: The defendant, identified only by his initials (D.P.), was accused of murder. The Court's witness related that he and the defendant had conspired the day before the murder while drinking cups of chocolate taken at the Bridges Street Chocolate-house located in Covent Garden. The defendant was con-

Table 20.1 Chocolate-Related Crime: Old Bailey Court Records[a]

Defendants Accused of the Following Crimes	Number of Cases
Theft: simple grand larceny	63
Theft: burglary	19
Theft: receiving stolen goods	6
Theft: shoplifting	3
Theft: at specified place	3
Theft: petty larceny	1
Theft with violence: highway robbery	1
Theft: not specified	9
Breaking the peace: assault	2
Deception: forgery	1
Deception: not specified	1
Housebreaking: theft specified	1
Murder	1
Offence against the King: counterfeiting	1
Offence against the King: tax fraud	1

[a]The Old Bailey Court records can be accessed electronically at http://www.oldbaileyonline.org/. (Accessed January 24, 2007.)

Table 20.2 Chocolate-Related Crime: Old Bailey Court Records,*a* Outcomes (113 Defendants)

Number of Cases	Sentence
42	Transported 　7 years (to America) (1) 　7 years (location?) (8) 　14 years (location?) (3) 　Term not identified (30)
13	Executed
9	Fined 　1 shilling (4) 　39 shillings (1) 　4 shillings (1) 　6 d (pence) (1) 　Not specified (2)
6	Sent to House of Corrections 　3 Months (4) 　6 Months (1) 　9 Months (1)
5	Whipped 　Private (2) 　Public (1) 　Not differentiated (2)
4	Sent to Newgate Prison 　1 month (2) 　3 months (1) 　2 years (1)
3	Branded
1	Printed apology required
2	Other 　"Ten marks"—(lashes?) (1) 　Punishment not specified (1)

*a*The Old Bailey Court records can be accessed electronically at http://www.oldbaileyonline.org/. (Accessed January 24, 2007.)

victed but trial documents do not reveal his punishment [6]. The second instance of chocolate-associated violence was linked with highway robbery. On December 6, 1721, stagecoach passenger Sir Edward Lawrence was beaten and robbed by Butler Fox and relieved of his purse and the 6 pounds of chocolate that he carried. Trial testimony revealed that after the robbery Fox took shelter with an accomplice where they prepared and drank the chocolate stolen from Lawrence [7].

Of the 113 defendants brought to trial in these 82 cases, 79 defendants (62 percent) were convicted. Old Bailey Court records also provide information on sentences (Table 20.2). In nine instances guilty parties were fined amounts that ranged from 6 pence (pennies) to 39 shillings. Monetary fines, however, were not in accord with the value of goods stolen: sometimes higher fines were levied against low-value thefts, and sometimes the reverse. Other punishments ranged from public humiliation (printed apology, branding, or whipping) to incarceration for periods of one to three

years either in an unspecified "House of Corrections" or at Newgate Prison. The majority of the convicted felons (42 of 79 cases, or 53 percent), however, received the sentence known as "transportation," a seemingly innocuous euphemism that in reality meant deportation from England, either to the Americas or to Australia (i.e., to Botany Bay) for terms that ranged between 7 and 14 years. Execution awaited 13 of the guilty convicted of chocolate-associated crime. Execution methods were not identified in court documents, but the most likely mechanism was hanging.

Chocolate-associated thefts documented in the Old Bailey Archive can be sorted into five categories based on the type of product or equipment stolen:

1. Manufactured chocolate
2. Ground cocoa
3. Cocoa nuts (i.e., beans)
4. Cocoa shells (shelled beans; used to make a type of tea)
5. Chocolate-related manufacturing or serving equipment and utensils

Case numbers and selected data for chocolate-associated crimes are presented in Table 20.3.

MANUFACTURED CHOCOLATE

A total of 25 cases involved theft of manufactured chocolate. Two representative cases presented here reveal a side of human nature associated with this crime:

> *Frances Jacobs, of the Parish of Christ Church, was indicted for feloniously stealing 26 Ounces of Gum Elamy [?], value 10 s. in the Shop of John Antrim, the 20th of May last [1718]. The Prosecutor's Servant deposed, That the Prisoner came into his Master's Shop under pretence of buying some Coco-Nuts, and took the Opportunity to steal the Gum Elamy. Another Evidence [sic] deposed, that the Prisoner brought it to his Shop to sell, and he suspecting she had stollen [sic] it, remembring [sic] he had seen such Gum lie in his Neighbour [sic] a Druggist's Window, and sending to enquire, the Prosecutor's Servant came and owned it. The Prisoner deny'd [sic] the Fact, pleading she used to deal in such things; and that finding it in her Handkerchief among some Chocolate, she did not know but that she had brought it from home. This did not avail but that the Jury found her Guilty to the Value of 4 s. 10 d. Transportation for 7 Years. [8]*

> *William Claxton and Frances Claxton, of Christ-church, were indicted, the former for feloniously stealing 1 Ounce of Mace, 1 Ounce of Nutmegs, half a Pound of Bohea Tea, half a Pound of Chocolate and half a Pound of Citron, the Goods of John' Pledger, on the 19th of January last [1721]: and*

Table 20.3 Chocolate-Related Crime: Old Bailey Court Records

Case Number	Trial Date	Accused	Gender	Crime	Verdict	Punishment
T-16930531-25	1693, May 31	D. P.	Male	Murder	Guilty	Not identified
T-17070903-4	1707, September 3	Levy, Philip	Male	Theft: burglary	Guilty	Branding
T-17111205-29	1711, December 5	Goodale, Charles	Male	Housebreaking: theft; specified place	Guilty	Death
T-17180530-1	1718, May 30	Jacobs, Frances	Female	Theft: shoplifting	Guilty	Transported: 7 years
T-17190225-29	1719, February 25	Moor, Susanna	Female	Theft: simple grand larceny	Guilty	Transported: term not identified
T-17190514-3	1719, May 14	Davis, Gothard	Female	Theft: shoplifting	Acquitted	Not identified
T-17211206-41	1721, December 6	Fox, Butler	Male	Theft with violence: highway robbery	Acquitted	Acquitted in this case, but because of two additional robbery charges, ordered to remain in custody
T-17210113-45	1721, January 13	Tompkins, Ann	Female	Theft: specified place	Acquitted	Not identified
T-17210301-1	1721, March 1	Claxton, William	Male	Theft	Guilty	Transported: term not identified
T-17210301-1	1721, March 1	Claxton, Frances	Female	Receiving stolen goods	Acquitted	Not identified
T-17210525-15	1721, May 25	Chapman, Edward	Male	Theft: simple grand larceny (theft under 1 shilling)	Guilty	Whipped
T-17210525-15	1721, May 25	Edgerly, Thomas	Male	Theft: simple grand larceny (theft under 1 shilling)	Acquitted	Not identified
T-17240415	1724, April 15	Stevenson, Henry	Male	Breaking the peace: assault	Guilty	Fine of 10 marks (lashes?)
T-17240812-16	1724, August 12	Wetherel, Francis	Male	Theft: simple grand larceny	Guilty	Transported: term not identified
T-17281204-27	1728, December 4	Moor, John (alias Holland)	Male	Theft: shoplifting	Guilty	Transported: term not identified
T-17281204-38	1728, December 4	White, Robert	Male	Theft: simple grand larceny	Acquitted	Not identified
T-17321011-30	1732, October 11	Marshall, Richard	Male	Theft: burglary	Guilty	Death
T-17321011-30	1732, October 11	Horsenail, Mary (alias Marshall)	Female	Theft: burglary	Acquitted	Not identified
T-17321011-30	1732, October 11	Mason, Amy (alias Griffin)	Female	Theft: burglary	Acquitted	Not identified
T-17330112-29	1733, January 12	Webb, Edmund	Male	Theft	Acquitted	Not identified
T-17341204-67	1734, December 4	Taylor, Elizabeth	Female	Theft	Acquitted	Not identified
T-17360225-6	1736, February 25	Hughes, Mary	Female	Theft: simple grand larceny	Guilty	Transported: term not identified
T-17360225-6	1736, February 25	Hughes, Grace	Female	Theft: shoplifting	Guilty	Transported: term not identified
T-17360225-6	1736, February 25	Williams, Edward	Male	Theft: receiving stolen goods	Acquitted	Not identified
T-17360225-6	1736, February 25	Williams, Diana Edward	Female	Theft: receiving stolen goods	Acquitted	Not identified
T-17360115-2	1736, January 15	Wood, Martha	Female	Theft: simple grand larceny	Guilty	Not identified
T-17360115-37	1736, January 15	Starling, Elizabeth	Female	Theft: simple grand larceny	Guilty	Whipped
T-17380628-30	1738, June 28	Hawkins, Joseph	Male	Theft: simple grand larceny	Guilty	Transported: term not identified

Reference	Date	Name	Sex	Offence	Verdict	Punishment
T-17400227-3	1740, February 27	Hill, Samuel	Male	Theft: burglary	Guilty	Death
T-17410828-35	1741, August 28	Middleton, Thomas	Male	Theft: no type specified	Acquitted	Not identified
T-17410116-27	1741, January 16	Jones, John	Male	Theft: simple grand larceny	Guilty	Transported: term not identified
T-17410116-27	1741, January 16	Pennington, John	Male	Theft: simple grand larceny	Guilty	Transported: term not identified
T-17410116-27	1741, January 16	Collins, John	Male	Theft: simple grand larceny	Guilty	Transported: term not identified
T-17430413-14	1743, April 13	Read, John	Male	Theft: burglary	Guilty	Transported: term not identified
T-17430519-18	1743, May 19	Johnson, Jane	Female	Theft: receiving stolen goods	Acquitted	Not identified
T-17440223-8	1744, February 23	Durose, Richard	Male	Theft: simple grand larceny	Acquitted	Not identified
T-17440113-18	1744, January 13	Palmer, Rachel	Female	Theft: simple grand larceny	Guilty	Transported: term not identified
T-17450424-16	1745, April 24	Rankin, Jane	Female	Theft: simple grand larceny	Guilty	Branded
T-17450424-34	1745, April 24	Parsons, Stephen	Male	Theft: specified place	Guilty	Death
T-17451204-16	1745, December 4	Barton, William	Male	Theft	Guilty	Transported: term not identified
T-17461205	1746, December 5	Identified only as 66 (?)	??	Theft	Acquitted	Not identified
T-17461015-6	1746, October 15	Goodman, Elizabeth	Female	Theft: simple grand larceny	Guilty	Transported: term not identified
T-17480420-48	1748, April 20	Keys, John (alias Thornton)	Male	Theft: simple grand larceny	Guilty	Transported: term not identified
T-17481207-27	1748, December 7	Cable, Daniel	Male	Breaking the peace: assault	Guilty	Required to print an apology in a public paper
T-17480706-40	1748, July 6	Eames, Sarah	Female	Theft: burglary	Acquitted	Not identified
T-17480706-40	1748, July 6	Higby, Mary	Female	Theft: burglary	Guilty	Transported: term not identified
T-17480706-40	1748, July 6	Myers, Ann	Female	Theft: burglary	Guilty	Transported: term not identified
T-17480706-47	1748, July 6	Cable, Daniel	Male	Deception	Acquitted	Not identified
T-17481012-9	1748, October 12	Coates, Benjamin	Male	Theft: simple grand larceny	Guilty	Public whipping
T-17500711-5	1750, July 11	King, William	Male	Theft: simple grand larceny	Guilty	Transported: term not identified
T-17501017-41	1750, October 17	Godd, Francis	Male	Theft: simple grand larceny	??	Not identified
T-17520219-20	1752, February 19	Thorne, John	Male	Theft: simple grand larceny	Guilty	Transported: term not identified
T-17520219-3	1752, February 19	Whitmore, Ann	Female	Offences against the King: coining	Acquitted	Not identified
T-17520116-13	1752, January 16	Titten, Richard	Male	Theft: burglary	Guilty	Transported: term not identified
T-17580222-44	1758, February 22	Jones, William	Male	Theft: simple grand larceny	Guilty	Transported: term not identified
T-17590117-31	1759, January 17	Bowden, Mary	Female	Simple grand larceny	Guilty	Transported: term not identified
T-17590711-10	1759, July 11	Ridgway, John	Male	Theft: simple grand larceny	Guilty	Transported: term not identified
T-17610625	1761, June 25	Smith, Mary	Female	Theft: simple grand larceny	Acquitted	Not identified
T-17650116-36	1765, January 16	Williams, Edward	Male	Theft: simple grand larceny	Guilty	Death
T-17650918-33	1765, September 18	Milward, Susannah	Female	Theft: simple grand larceny	Acquitted	Not identified
T-17671209-78	1767, December 9	Flint, Richard	Male	Theft (not otherwise specified)	Guilty	Transported: term not identified
T-17670715-52	1767, July 15	Blanch, John	Male	Theft: simple grand larceny	Guilty	Branded

Table 20.3 Continued

Case Number	Trial Date	Accused	Gender	Crime	Verdict	Punishment
T-17711204-53	1771, December 4	Gulley, Richard	Male	Theft: burglary	Guilty	Death
T-17711204-53	1771, December 4	Hurdley, John	Male	Theft: burglary	Guilty	Death
T-17710911-9	1771, September 11	White, John	Male	Theft: simple grand larceny; receiving stolen goods	Guilty	Transported: term not identified
T-17710911-9	1771, September 11	Summerayse, Ann	Female	Theft: simple grand larceny; receiving stolen goods	Guilty	Transported: 14 years
T-17720603-61	1772, June 3	Stack, Edward	Male	Theft: simple grand larceny	Acquitted	Not identified
T-17720603-61	1772, June 3	Wood, Constantia	Female	Theft: simple grand larceny	Acquitted	Not identified
T-17720909-60	1772, September 9	Smith, William (alias Johnson)	Male	Theft: simple grand larceny	Acquitted	Not identified
T-17720909-60	1772, September 9	Wood, Elizabeth (alias Smith)	Female	Theft: simple grand larceny	Guilty	Transported: term not identified
T-17731208-52	1773, December 8	Robinson, Benjamin	Male	Theft: simple grand larceny	Guilty	Transported: term not identified
T-17731208-52	1773, December 8	Kirby, Thomas	Male	Theft: simple grand larceny	Guilty	Transported: term not identified
T-17730707-40	1773, July 7	Cotton, William	Male	Theft: simple grand larceny	Acquitted	Not identified
T-17730626-59	1773, June 26	Lewis, Abraham	Male	Theft: petty larceny	Guilty	Transported: term not identified
T-17820911-89	1782, September 11	Anderson, Joseph	Male	Theft: simple grand larceny	Guilty	Transported to America for 7 years
T-17820911-89	1782, September 11	Woodson, John	Male	Theft: simple grand larceny	Acquitted	Not identified
T-17840526-10	1784, May 26	Wigmore, Christina (also known as Catherine)	Female	Theft: simple grand larceny	Guilty	Privately whipped; confined to hard labor for 6 months in House of Corrections
T-17840526-10	1784, May 26	Blake, Francis	Male	Theft: burglary	Guilty	Transported: term not identified
T-17870912-114	1787, September 12	Stokes, Charles	Male	Theft: simple grand larceny	Guilty	Transported: 7 years
T-17870912-114	1787, September 12	Bramsley, William	Male	Theft: simple grand larceny	Guilty	Transported: 7 years
T-17870912-114	1787, September 12	Nadan, George	Male	Theft: simple grand larceny	Guilty	Transported: 7 years
T-17870912-114	1787, September 12	Lamb, William	Male	Theft: simple grand larceny	Guilty	Discharged
T-17880109-47	1788, January 9	Mackey, John	Male	Offence against the King (assault on an Excise Tax officer)	Acquitted	Not identified
T-17891028-38	1789, October 28	Cave, John	Male	Theft: burglary	Guilty	Death
T-17891028-38	1789, October 28	Partington, John	Male	Theft: burglary	Guilty	Death
T-17930410-42	1793, April 10	Corless, Henry	Male	Theft: simple grand larceny	Guilty	Confined in House of Corrections for 6 months
T-17930911-89	1793, September 11	Seville, John (alias Wright)	Male	Theft: specified place	Guilty	Not identified
T-17930911-89	1793, September 11	Lazarus, Eleazor	Male	Theft: receiving stolen goods	Acquitted	Not identified

Trial ID	Date	Name	Gender	Offence	Verdict	Punishment
T-17941111-10	1794, November 11	Wiley, William	Male	Theft: burglary	Guilty	Transported: 7 years
T-18001203-19	1800, December 3	Mead, Mary	Female	Theft: simple grand larceny	Acquitted	Not identified
T-18010520-52	1801, May 20	Wright, Richard	Male	Theft: burglary	Guilty	Death
T-18010520-52	1801, May 20	Smith, John (alias Parker)	Male	Theft: burglary	Guilty	Death
T-18010520-52	1801, May 20	Johnstone, Thomas	Male	Theft: burglary	Guilty	Death
T-18011028-14	1801, October 28	Grantham, Francis	Male	Theft: simple grand larceny	Guilty	Confined at Newgate Prison for 2 years; publicly whipped
T-18010916-61	1801, September 16	Myers, Robert	Male	Theft: simple grand larceny	Guilty	Confined for 3 months in Newgate; public whipping on Cox's Quay
T-18010916-61	1801, September 16	Kent, John	Male	Theft: simple grand larceny	Acquitted	Not identified
T-18010916-120	1801, September 16	Gittons, Edward	Male	Theft: simple grand larceny	Guilty	Confined for 1 month in Newgate, delivered to his Sargeant
T-18040111-13	1804, January 11	Hamilton, Robert	Male	Theft: simple grand larceny	Acquitted	Not identified
T-18040111-18	1804, January 11	Townsend, Thomas	Male	Theft: simple grand larceny	Guilty	Transported: 7 years
T-18050424-17	1805, April 24	Kilbeaumont, William	Male	Theft: simple grand larceny	Guilty	Confined in House of Corrections for 12 months
T-18070701-55	1807, July 1	Thomas, David	Male	Theft: simple grand larceny	Guilty	Confined for 1 month at Newgate Prison
T-18100110-16	1810, January 10	Everett, Anthony	Male	Theft: simple grand larceny	Acquitted	Not identified
T-18160403-154	1816, April 3	Jones, John	Male	Theft: simple grand larceny	Guilty	Confined (House of Corrections) for 6 months
T-18210411-72	1821, April 11	Snape, John	Male	Deception: forgery	Guilty	Death
T-18210718-104	1821, July 18	Brown, William	Male	Theft: simple grand larceny	Guilty	Confined (House of Corrections) for 1 month
T-18230409-209	1823, April 9	Dunkin, Catherine	Female	Theft: simple grand larceny	Guilty	Confined for 6 months
T-18230409-209	1823, April 9	Riley, Catherine	Female	Theft: receiving stolen goods	Guilty	Transported: 14 years
T-18231022-49	1823, October 22	Adams, William	Male	Theft: simple grand larceny	Guilty	Confined for 9 months
T-18240114-9	1824, January 14	Willis, Robert	Male	Theft: simple grand larceny	Guilty	Transported: 7 years
T-18280221-76	1828, February 21	Brookes, Thomas	Male	Theft: simple grand larceny	Guilty	Recommended for mercy: whipped and discharged
T-18320405-126	1832, April 5	Padwick, Letitia	Female	Theft: simple grand larceny	Guilty	Transported: 14 years
T-18320906-214	1832, September 6	Lyons, Mary	Female	Theft: simple grand larceny	Acquitted	Not identified
T-18340904-79	1834, September 4	Vickers, Ann	Female	Theft: simple grand larceny	Guilty	Transported: 7 years

Frances Claxton for receiving the same knowing them to be stole [sic]. It appeared that the Prisoner William was the Prosecutor's Apprentice, and took the Mace out of his Master's Drawer while he was at Dinner, and gave it to his Brother, who sold some for a Penny to a Woman who told the Prosecutor of it. The Prisoner William owned the Fact, before the Justice, though he denied it on his Tryal [sic] and his Confession was Read in Court. There being no Evidence to charge his Mother [Frances] the Jury acquitted her; but the Fact being plain against him the Jury found him Guilty. [Verdict] Transportation. [9]

COCOA NUTS

The theft of cacao beans (commonly called cocoa nuts in the trial documents) posed unusual difficulties for conviction, but when individuals were caught with the "goods," verdicts often were harsh, as revealed in these representative examples:

Ann Tompkins, alias Webb, of St. Sepulchres, was indicted for feloniously stealing 14 lb of Cocoa-Nuts, 20 Gallons of Ale and 40 s. in Money, in the Dwelling House of Mary Benning on the 20th of October last [1721]. She was a second time indicted for feloniously stealing 5 Guineas, 80 Pound of Cocoa-Nuts, 10 Pound of Chocolate, some Coffee and 12 Flasks of Florence Wine. But the Evidence not being sufficient, the Jury Acquitted her of both Indictments. [10]

Henry Stevenson, was indicted for an Assault on Joseph Woolley, the 11th of this Instant April [1724]. Joseph Woolley depos'd, That he carrying a Burden, pitch'd it to rest him at Middle Fleet Bridge, and the Prisoner laid his Hand upon the Bag, and pull'd out 4 or 5 Cocoa Nuts, and asked him where he was going with them? That he answered him, what was that to him; that he demanded of him the Bill of Parcels; that he told him he had none; whereupon the Prisoner said, then he would stop them, and call'd two Porters to carry them away, but neither of them would do it. And that then the Prisoner mark'd the Bags with the King's Mark, the Broad Arrow; and then a Gentleman came by, and seiz'd the Prisoner, and sent for a Constable. The Person that was coming by depos'd, That he taking notice of the Difference between the Porter and the Prisoner, and seeing the Broad Arrow upon the Bags, and a Porter coming to take the Goods, he asked the Prisoner for his Deputation, but he refus'd to show it him, saying, he had shown it once already, and he was not oblig'd to show it to 50 People; but at last, he insisting upon seeing it, he pretended to consent to it, but would not do it there, but took him up 2 or 3 Alleys, and he supposing that he intended to give him the Drop, he secur'd him, and charg'd a Constable with him. The Prisoner pleaded he was in Drink, and the Porter call'd for Assistance to help him down, and there was a Hole in the Bag, and 2 or 3 of the Nuts came out, and that he did not know that he set the Broad Arrow on the Bags. He acknowledg'd in Court, he had no Authority so to do. The Jury found him guilty, and the Court set on him a Fine of 10 Marks [i.e., 10 lashes]. [11]

LOCATIONS OF CHOCOLATE-ASSOCIATED BUSINESSES

Trial documents identified several London-based chocolate makers, chocolate shops, and pier/warehouse locations where cacao was off-loaded or stored, as documented by the following examples:

1693. Chocolate shop located on Bridges Street in Covent Garden. [12]

1750. White's Chocolate House on St. James Street, London, where the accused stole a nutmeg-grater. [13]

1761. William Hastings chocolate-maker living in the Carnaby (sic) market area solicited a Ms. Smith on the street for a "nightly stroll," then claimed that she picked his pocket. [14]

1767. Richard Flint, chocolate-maker, was employed for four years by Hugh James at his grocery shop located in Fleet Street. Flint was accused of stealing from his employer and was caught with an undefined amount of cocoa in the pocket of his great coat. In another pocket of his great coat he had secreted half a pound of chocolate and two lumps of sugar weighing one pound. Upon apprehension a warrant was secured to search Flint's lodgings located at Stretton Ground, where authorities found additional stolen items identified as: $9\frac{3}{4}$ pounds of chocolate, some cocoa shells, coffee, hard soap, sugar, candles, and three linen bags. The prisoner related in his defense:

I have a relation that belongs to an Indiaman [seaman aboard a trading ship] who gave me some of the nuts [cocoa beans], and the rest I took home as I roasted them; I did take sugar and candle; when God leaves a person, the Devil gets into him. [15]

1773. Prisoner Abraham Lewis—accused of stealing a linen handkerchief, the property of John Nicholson with a value of 6 pence—reported to the court that his father, Hyam Lazarus, was a chocolate-maker although they had different last names. [16]

GEOGRAPHY LESSONS IN THE TRIAL DOCUMENTS

In one cocoa-related trial dated to 1801, the court interjected questions during testimony regarding the value and origin of cacao beans:

Sworn testimony of Henry Major: I saw Mr. Brassett with a parcel in his hand which turned out to be more cocoa; the prisoner [Francis Grantham] put it under his coat, and in that situation took it to the accompting-house [accounting office in a business]; a constable was sent for; the cocoa in the silk stocking was delivered to the constable. Sworn testimony of Joseph Reeves: I am a coachman; on the 19th of September, I heard the cry of stop thief in Scott's yard; I saw the prisoner running; I ran up to him, and laid hold of him; he had a handkerchief with cocoa-nuts in it; and as soon as I laid hold of him, he threw it away. Sworn testimony of Benjamin Gray: Question from the Court: what is the value of a pound of cocoa-nuts? Answer [from Gray]: from eleven-pence to one shilling. Gray cross-examined by Mr. Gurney: Question: This is the cocoa from which chocolate is made? Answer: Yes. Question from the Court: do you mean to say these are cocoa-nuts? Answer: Yes, they are. When it is ground, it is called cocoa; in its present state, it is cocoa-nut. [17]

Of special interest to the history of advertising and marketing are references in several accounts in the Old Bailey Archive that date between 1743 and 1772 that provide information on branding and product identification:

1743. Testimony of Josias Taylor: I had 120 pound weight of chocolate brought from the office, which had been sent there to be stamped [marked]. It was brought in on the Saturday and the Monday before the Tuesday Night on which this [the theft] happened. It was put in the window to dry, because it is damp when it comes from the office. Sworn testimony of David Shields: the prisoner [John Read], and one who goes by the Nickname of Cutos, Emanual Hubbard that is dead, and myself, went to Mr. Taylor's in Bishopsgate-street [sic], shoved up the window, and took out 26 pound weight of chocolate, and sold it for a shilling a pound in Rag-Fair. We all of us helped to take the chocolate; every one had a hand in it. [18]

1743. Sworn testimony of David Shields: the chocolate was carried to Jane Johnson's house; she [the prisoner Jane Johnson] asked the price and I [David Shields] told her 18d/pound was the common price; she said she would give me but a shilling; I saw her deliver the chocolate in a hand-basket to one Isaac Aldridge, to carry it out, in order to sell it; as I was sitting by the fire, she showed me a piece of gold which she had for it, and said it was safe, for she had seen some part of it [the chocolate] melted down. It was melted down because it was stamped. This was the same day we stole it. [19]

1744. Sworn testimony of Charles Gataker, apprentice to Mr. Barton: Question: how do you know this to be your master's chocolate? Gateker: It was marked with our mark; I believe she was a little in liquor [i.e., intoxicated when she stole the chocolate] when she did it. [20]

1759. Three pounds weight of chocolate with a value of 12 shillings was stolen chocolate from the shop of John Wilson and Thomas Thornhill. The shop owners testified that the chocolate was marked with the two initial letters of their names, W.T. [21]

1771. Sworn testimony of John Saunders: I received the warrant and I searched Mrs. Summerayse's house, and there I found the goods mentioned in the indictment. The woman said the chocolate was smuggled chocolate. It has a stamp and our-own private mark on it. I don't know who marked it. Cross Examination: Question: You sell a great many hundred pounds of these [chocolates] I suppose? John Saunders. Yes. Question: You can't sware [sic] it is not chocolate sold out of your shop. John Saunders. I can't sware [sic] that. [22]

1772. Deposition of Mrs. Farrell: She gave the prosecutor this chocolate, and knows the mark upon it. [23]

Other court documents identified the names of different sailing ships used to import chocolate/cacao to London:

1750. William Slocomb: I am steward on board the ship Chester that came into the river [Thames] on the 30th of August from Antigua; one John Key and I bought half a hundred weight of cocoa between us in that country, some we made use of in our passage home, and the rest I gave [to] Jenkins and Wright to dispose of for me. [24]

1788. Thomas Groves sworn: I am an inspector on the Thames; I remember the ship Mary laying at St. Catherine's stairs; that is within the port of London; I had been on board that ship, in consequence of an information from the last witness, on the 28th of December, in the morning, my assistant seized 340 lb. weight of chocolate. [25]

Analysis and Commentary

These English cocoa/chocolate-associated trial accounts provide insights on London daily life and the various vices associated with human nature: envy, greed, lust, and temptation. The documents confirm that cocoa beans, shells, and prepared chocolate were valuable commodities; that chocolate-making equipment could be "fenced" easily; and that many cases of chocolate-related theft were not well planned but were instead crimes of opportunity or "spur of the moment" decisions.

The documents also reveal that some of the accused were in fact guilty but acquitted due to court-related errors or lack of supporting testimony. Death was not an unusual sentence for repeat criminals: 13 of the 76 convictions for chocolate-associated crimes resulted in sentences for execution. Expulsion verdicts from England and subsequent transportation of criminals upon conviction to the Americas or to Botany Bay, Australia, commonly were assigned.

Most often the thieves were simple servants or employees who stole from their place of work. The defendant's state of mind or financial situation played little to no role in judgments. Several of the accused were intoxicated when the crimes were committed and such behavior had no bearing on the verdicts or sentences. Court documents revealed the accused generally were poor or subsisted at a low poverty level. While most of the accused were adult males, a significant number were adult women (30 percent). Ages of defendants rarely were provided in court documents. In one instance, however, a William Wiley was identified as being 12 years of age in 1794 when he stole 1/2 pound of cocoa from William Phelps's store (commercial value 1 shilling 3 pence). Wiley was convicted of the theft and transported for 7 years to an unknown destination outside England [26]. Religion of the accused was reported only if the defendant was Jewish. Only in one specific instance in our data set, the 1719 trial of Gothard Davis, was the accused specifically identified as Jewish. [27]. Still, the 1773 case of Abraham Lewis (above), whose father Hyam (Chaim) Lazarus had a different name, almost certainly was Jewish and this was not noted in the court record.

Some of the convicted obviously were professional thieves, perhaps better described as repeat "flim-flam" or con artists who had attempted on more than one occasion to defraud civilians, customers, or merchants. In other instances, criminal teams of husbands and wives worked together as thieves. The most common opportunity for chocolate-associated crime was breaking and entering homes or business establishments. In contrast to citizen behavior of the 21st century, several of the accused were apprehended by good citizens who responded to the call "stop thief"—who then performed their civic duty and chased down and captured the perpetrators (especially one who could not run quickly), and waited for the police to arrive [28, 29].

Commercial values for amounts of stolen chocolate do not appear in court documents prior to 1736. After this date, court records provide the values of goods stolen as itemized trial exhibits. These itemized documents, however, revealed considerable inconsistencies. This problem (outlined later) may have several explanations and reflect differences in the quality of chocolate/cocoa stolen, recording errors, perhaps even economic inflation. But the inconsistencies also could be due to victims lying to police investigators to inflate the value of stolen goods, or otherwise providing false information to the authorities. The inconsistencies and anomalies in weights versus attributed values can be seen upon review of the following cases:

1736   Stolen 2.00 pounds of chocolate—value 9 shillings, 6 pence [30]

1738   Stolen 0.75 pound of chocolate—value 2 shillings [31]

1743   Stolen 26.00 pounds of chocolate—value 4 pounds [32]

1748   Stolen 0.50 pound of chocolate—value 2 pence [33]

1752   Stolen 2.00 pounds of chocolate—value 5 shillings [34]

1758   Stolen 6.00 pounds of chocolate—value 13 shillings [35]

1759   Stolen 3.00 pounds of chocolate—value 12 shillings [36]

1771   Stolen 1.00 pound of chocolate—value 3 shillings [37]

1784   Stolen 0.50 pound of chocolate—value 2 pence [38]

1801   Stolen 4.00 pounds of chocolate—value 8 shillings [39]

1805   Stolen 1.00 pound of chocolate—value 2 shillings [40]

1816   Stolen 8.00 ounces of chocolate—value 3 shillings [41]

1821   Stolen 0.25 ounce of chocolate—value 6 pence [42]

1824   Stolen 0.50 pound of chocolate—value 2 shillings [43]

1832   Stolen 0.50 pound of chocolate—value 1 shilling [44]

Reported values of chocolate-related preparation/serving equipment also were highly variable and inconsistent through time, as noted here. Values of stolen silver chocolate pots ranged from 3 to 12 pounds Sterling (£); copper chocolate pots between 1 and 4 shillings; a court record also exists for theft of a single china chocolate cup valued by the owner at 1 pence.

1711   Stolen 1 silver chocolate pot—value 12 pounds [45]

1745   Stolen 1 silver chocolate pot—value 7 pounds, 10 shillings [46]

1789   Stolen 1 silver chocolate pot—value 3 pounds [47]

1793   Stolen 1 silver chocolate pot—value 3 pounds [48]

1740   Stolen 1 copper chocolate pot—value 2 shillings [49]

1745   Stolen 1 copper chocolate pot—value 1 shilling [50]

1771   Stolen 1 copper chocolate pot—value 4 shillings [51]

1772   Stolen 1 copper chocolate pot—value 2 shillings [52]

1773   Stolen 1 copper chocolate pot—value 4 shillings [53]

1773 Stolen 1 copper chocolate pot—value 4 shillings [54]

1782 Stolen 1 copper chocolate pot—value 2 shillings [55]

1784 Stolen 1 copper chocolate pot—value 3 shillings [56]

1810 Stolen 1 copper chocolate pot—value 4 shillings [57]

1765 Stolen 1 china chocolate cup—value one pence [58]

Execution verdicts were independent of the reported value of goods stolen. Death was the verdict assigned for each of the following thefts:

1711 One silver chocolate pot: value given as 12£ [59]

1732 One copper chocolate pot: value not given (estimated at between 1 and 4 shillings) [60]

1740 One copper chocolate pot: value given as 2 shillings [61]

1745 One silver chocolate pot: value given as 7£ 10 shillings [62]

1765 Three silver chocolate stands (on which to set chocolate pots): value given as 30 shillings [63]

1771 One copper chocolate pot: value given as 4 shillings [64]

1789 One silver chocolate pot: value given as 3£ [65]

1801 Four pounds of chocolate and 3 pounds of cocoa: value given as 8 and 3 shillings, respectively [66]

Executions for these crimes—in contrast with 20th and 21st century justice in England and North America—were swift and commonly conducted by hanging!

The Old Bailey criminal cases presented here suggest that chocolate-associated crimes in 17–19th century England commonly were spur-of-the-moment decisions, crimes of opportunity, and only rarely associated with violence. Only slightly more than 50 percent of the chocolate-associated crimes occurred during late fall or winter (cold months)—outside temperature appears not to have been an issue when the crimes were perpetrated. Poverty and opportunity were the driving forces associated with these crimes.

Acknowledgments

I wish to thank my colleagues at Oxford Brooks University, especially Professor Jeya Henry, for encouraging me to explore the extensive records from Old Bailey Courthouse.

References

1. Hooper, W. E. *The History of Newgate and the Old Bailey and a Survey of the Fleet Prison and Fleet Marriages*. London, England: Underwood Press, 1935.

2. http://www.oldbaileyonline.org/history/the-old-bailey/. (Accessed January 24, 2007.)

3. http://www.oldbaileyonline.org/. (Accessed January 24, 2007.)

4. Old Bailey Proceedings (OBP) On-line (www.oldbaileyonline.org). (Accessed January 24, 2007.) May 31, 1693, trial of D.P. (not otherwise identified). Case #T-16930531-25.

5. OBP. September 4, 1834, trial of Ann Vickers. Case #T18340904-79.

6. OBP. May 31, 1693, trial of D.P. (not otherwise identified). Case #T-16930531-25.

7. OBP. December 6, 1721, trial of Butler Fox. Case #T-17211206-41.

8. OBP. May 30, 1718, trial of Frances Jacobs. Case #T-17180530-1.

9. OBP. March 1, 1721, trial of William and Frances Claxton. Case #T-17210301-1.

10. OBP. January 13, 1721, trial of Ann Tompkins (alias Webb). Case #T-17210113-45.

11. OBP. April 15, 1724, trial of Henry Stevenson. Case #T-17240415.

12. OBP. May 31, 1693, trial of D.P. (not otherwise identified). Case #T-16930531-25.

13. OBP. July 11, 1750, trial of William King. Case #T-17500711-5.

14. OBP. June 25, 1761, trial of Mary Smith. Case #T-17610625-10.

15. OBP. December 9, 1767, trial of Richard Flint. Case #T17671209-78.

16. OBP. June 26, 1773, trial of Abraham Lewis. Case #T-17730626-59.

17. OBP. October 28, 1801, trial of Francis Grantham. Case #T-18011028-14.

18. OBP. April 13, 1743, trial of John Read. Case #T-1743-0413-14.

19. OBP. May 19, 1743, trial of Jane Johnson. Case #T-17430519-18.

20. OBP. January 13, 1744, trial of Rachel Palmer. Case #T-17440113-18.

21. OBP. July 11, 1759, trial of John Ridgway. Case #T-17590711-10.

22. OBP. September 11, 1771, trial of Ann Summerayse. Case #T-17710911-9.

23. OBP. September 9, 1772, trial of Elizabeth Wood (alias Smith). Case #T-17720909-60.

24. OBP. October 17, 1750, trial of Francis Godd. Case #T-17501017-41.

25. OBP. January 9, 1788, trial of John Mackey. Case #T-17880109-47.

26. OBP. November 11, 1794, trial of William Wiley. Case #T-17941111-10.

27. OBP. May 14, 1719, trial of Gothard Davis. Case #T-17190514-3.

28. OBP. October 28, 1801, trial of Francis Grantham. Case #T-18011028-14.

29. OBP. April 3, 1816, trial of John Jones. Case #T-18160403-154.

30. OBP. January 15, 1736, trial of Martha Wood. Case #T-17360115-2.

31. OBP. June 28, 1738, trial of Joseph Hawkins. Case #T-17380628-30.

32. OBP. May 19th 1743, trial of Jane Johnson. Case #T-17430519-18.

33. OBP. October 12, 1748, trial of Benjamin Coates. Case #T-17481012-9.

34. OBP. February 19, 1752, trial of John Thorn. Case #T-17520219-20.

35. OBP. February 22, 1758, trial of William Jones. Case #T-17580222-44.

36. OBP. July 11, 1759, trial of John Ridgway. Case #T-17590711-10.

37. OBP. September 11, 1771, trial of John White. Case #T-17710911-9.

38. OBP. May 26, 1784, trial of Francis Blake. Case #T-17840526-10.

39. OBP. May 20, 1801, trial of Thomas Johnstone, John Smith (alias Parker), and Richard Wright. Case #T-18010520-52.

40. OBP. April 24, 1805, trial of William Kilbeaumont. Case #T-18050424-17.

41. OBP. April 3, 1816, trial of John Jones. Case #T-18160403-154.

42. OBP. July 18, 1821, trial of William Brown. Case #T-18210718-104.

43. OBP. January 14, 1824, trial of Robert Willis. Case #T-18240114-9.

44. OBP. September 6, 1832, trial of Mary Lyons. Case #T-18320906.

45. OBP. December 5, 1711, trial of Charles Goodale. Case #T-17111205-29.

46. OBP. April 24, 1745, trial of Stephen Parsons. Case #T-17450424-34.

47. OBP. October 28, 1789, trial of John Cave and John Partington. Case #T-17891028-38.

48. OBP. September 11, 1793, trial of John Seville (alias Wright) and Eleazor Lazarus. Case #T-17930911-89.

49. OBP. February 27, 1740, trial of Samuel Hill. Case #T-17400227-3.

50. OBP. April 24, 1745, trial of Jane Rankin. Case #T-17450424-16.

51. OBP. December 4, 1771, trial of Richard Gulley and John Hurdley. Case #T-17711204-53.

52. OBP. June 3, 1772, trial of Edward Stack and Constantia Wood. Case #T-17720603-61.

53. OBP. July 7, 1773, trial of William Cotton. Case #T-17730707-40.

54. OBP. December 8, 1773, trial of Benjamin Robinson and Thomas Kirby. Case #T-17731208-52.

55. OBP. September 11, 1782, trial of Joseph Anderson and John Woodson. Case #T-17820911-89.

56. OBP. May 26, 1784, trial of Christina Wigmore [aka Catherine Wigmore] and Francis Blake. Case #T-17840526-10.

57. OBP. January 10, 1810, trial of Anthony Everett. Case #T-18100110-16.

58. OBP. September 18, 1765, trial of Susannah Milward. Case #T-17650918-33.

59. OBP. December 5, 1711, trial of Charles Goodale. Case #T-17111205-29.

60. OBP. October 11, 1732, trial of Mary Horsenail (alias Marshall), Richard Marshall, and Amy Mason. Case #T-17321011-30.

61. OBP. February 27, 1740, trial of Samuel Hill. Case #T-17400227-3.

62. OBP. April 24, 1745, trial of Stephen Parsons. Case #T-17450424-34.

63. OBP. January 16, 1765, trial of Edward Williams. Case #T-17650116-36.

64. OBP. December 4, 1771, trial of Richard Gulley and John Hurdley. Case #T-17711204-53.

65. OBP. October 28, 1789, trial of John Cave and John Partington. Case #T-17891028-38.

66. OBP. May 20, 1801, trial of Thomas Johnstone, John Smith (alias Parker), and Richard Wright. Case #T-18010520-52.

Dark Chocolate

Chocolate and Crime in North America and Elsewhere

Louis Evan Grivetti

Introduction

How can it be that a food as good as chocolate arouses passions associated with both love and criminal behavior? During the course of our historical chocolate research, we encountered chocolate-associated crime as a recurring theme. The documents were diverse and interesting and the cases involved a wide range of crimes, among them arson, assault, burglary and theft, counterfeiting, fraud, homicide/murder, negligence, smuggling, and tampering.

Throughout the centuries, cacao beans, cocoa, and manufactured chocolate have presented criminals with targets of opportunity. Raw ingredients (beans) or the manufactured goods (cocoa powder; finished chocolate) have been "easy targets" for thieves. But besides thieves, other chocolate-associated criminals also emerged. Sometimes they were merchants themselves who adulterated their products by adding unappetizing substances such as brick dust, chalk, clay, dirt, paraffin, talc, and other items to extend the ingredients and to defraud customers (see Chapter 47). Indeed, midway through the 18th century chocolate manufacturers in both Europe and the Americas alert to such product misrepresentation started "branding" their manufactured chocolate with specific marks, for example, "I. B." for John Brewster, chocolate manufacturer in Boston, so to distance themselves from unscrupulous competitors (Fig. 21.1) [1]. Chocolate-associated problems continued during the 19th century and forced legitimate merchants to certify their products as good,

pure, or unadulterated so as to maintain their customer base [2–4]. Even the late 19th and early 20th centuries saw continued sales of debased chocolate, activities that ultimately resulted in magazine articles that cautioned consumers regarding the dangers of eating impure chocolate [5–7]. Crimes associated with chocolate purity represented but one facet of the chocolate/criminal spectrum.

Chocolate-Associated Crimes

The criminal cases examined during the course of preparing this chapter suggest that chocolate-associated crimes through the centuries commonly have been spur-of-the-moment decisions, crimes of opportunity, and with the exception of homicide and murder (by chocolate), only rarely associated with violence. The present chapter complements the information presented in Chapter 20, where English chocolate crimes are identified and analyzed. In the present chapter we identify and discuss representative examples of chocolate-associated crimes based

Chocolate: History, Culture, and Heritage. Edited by Grivetti and Shapiro
Copyright © 2009 John Wiley & Sons, Inc.

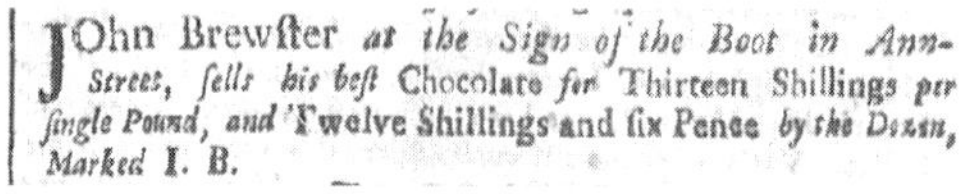

JOhn Brewſter *at the Sign of the Boot in Ann-Street, ſells his beſt Chocolate for Thirteen Shillings per ſingle Pound, and Twelve Shillings and ſix Pence by the Dozen,* Marked I. B.

FIGURE 21.1. Early evidence for branding chocolate (John Brewster of Boston). *Source: The Boston Weekly News-Letter,* February 16, 1738, p. 2. Courtesy of Readex/Newsbank, Naples, FL. (Used with permission.)

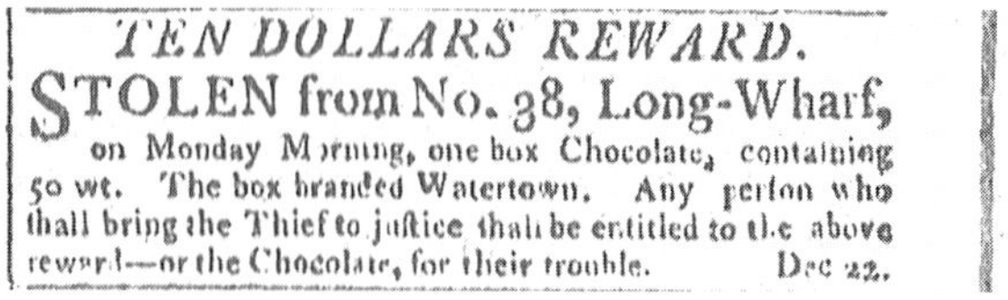

TEN DOLLARS REWARD. STOLEN from No. 38, Long-Wharf, on Monday Morning, one box Chocolate, containing 50 wt. The box branded Watertown. Any perſon who ſhall bring the Thief to juſtice ſhall be entitled to the above reward—or the Chocolate, for their trouble. Dec 22.

FIGURE 21.2. Reward for stolen chocolate. *Source: Columbian Centinel (sic),* January 5, 1799, p. 4. Courtesy of Readex/Newsbank, Naples, FL. (Used with permission.)

on information published in 18th through 19th century North American newspapers and magazines. While the focus of this chapter remains on crime-related examples from North America, chocolate-elicited criminal behavior elsewhere has been sampled as well.

BLACKMAIL/EXTORTION

Chocolate and blackmail seem a curious combination of activities, but in 1881 the prominent Canadian chocolate manufacturer John Mott was indeed blackmailed. The misguided extortionist demanded $600 in gold coins from Mott, otherwise he would be murdered. After receiving his written death threat Mott contacted the police and together they set a trap and the blackmailer was apprehended:

Halifax, Nova Scotia, January 25th [1881]. John P. Mott, the soap, candle, and chocolate manufacturer, and one of the wealthiest men in Nova Scotia, received last Friday an unsigned letter, threatening him with death unless he placed a bag containing $600 in gold behind a door at the smoking-room on the Dartmouth ferry steamer, at 8 o'clock last night. At the hour named, when the boat was on the Dartmouth side, Mr. Mott, having prepared a bag containing pennies, deposited it behind the door and departed. Just before she [i.e., the ship] reached the wharf a young man approached the hiding place of the money, grasped the bag, and was walking rapidly away, when a detective, who had been lounging around the smoking room in disguise, on the watch, arrested him, and he was taken to the Police station. The young man proved to be Warren P. Herman, who had, until lately, been following the sea, but is at present unemployed. [8]

BURGLARY AND THEFT

A large number of advertisements appeared in Colonial era newspapers that announced the theft of chocolate or chocolate-associated equipment or serving items. The following examples are representative (see Fig. 21.2 [9]):

1768. *Taken out of a House at the South End [of Boston], some Time last Week, one Two-Quart Copper Tunnel [sic, funnel meant], also one Silver Chocolate Ladle,*

that weights about 3 Ounces and an [sic] Half, mark'd M.S. on the bottom, it's desired if either of them are offered for sale the Printer hereof may be informed; if the present Possessor will Return them, or either of them, they shall receive One DOLLAR Reward for each and no Questions asked. [10]

1769. *A THEFT. WHEAREAS Some evil-minded Person or Persons have again broke open the Store of the Subscribers, on Spear's Wharf [Boston], and last Night took from thence a Number of Articles amongst which was Two Firkins Butter, about 50 or 60 lb. Of Chocolate mark'd W. Call, and S. Snow, half a Barrel of Coffee, nine or ten Pair Lynn Shoes, some Cocoa and Sugar. Any Person that will discover the Thief or Thieves, so that he or they may be brought to Justice, shall be handsomely rewarded by Bossenor &William Foster. [11]*

Two burglaries committed in March and August of 1808, also contained information that the stolen chocolate had been branded and could be identified by manufacturers' marks:

STOLEN yesterday, from Store No. 22, Long Wharf [Boston], 2 boxes of Chocolate, one of them Branded E. Baker of Dorchester, Mass. No. 1 [i.e. first quality chocolate], the other not branded—one contained 24 pounds, the other 23. Whoever will give information, so that the Goods may be recovered, and the Thief or Thieves brought to punishment, shall receive FIVE DOLLARS Reward, from EDWARD HOLDEN. [12]

Fifty Dollars Reward. The Store of the subscriber, was broken open on Monday night last and the following articles, with a number of others, were Stolen there from—1 chest Hyson Tea, marked I. P. inclosed [sic] in triangles; 2 bags Coffee; 1 small bale Cassia, marked B. I.; about 40 wt. Chocolate, marked A. Child's No 1; some Cod [fishing] Lines; 1 doz. Red morocco Pocket Books; 1 ream Writing Paper; 1 doz horn Combs, &c &c. The above reward will be paid on recovery of the articles or detection of the Thieves by JNO. KENNEDY. [13]

Two robberies the same year with rewards of $5.00 and $50.00 for theft of approximately the same weight of chocolate! The Kennedy robbery, however, also included coffee and tea, items that perhaps warranted the higher reward.

The life and activities of Thomas Mount, a well-known New England thief, were characterized by a long history of crime. Eventually Mount was captured, tried, and executed by the state of Rhode Island on Friday, May 27, 1791. As he stood on the gallows awaiting the hangman's noose, Mount was offered the opportunity to speak a few last words. His eloquent, albeit curious confession and "guilt-easing" was recorded for posterity, subsequently published, and attracted widespread attention during the late 18th century:

At a dance in Norwich Landing I took a silk cloak from a young lady, and sold it about two or three miles from Pockatanock bridge, and then set off for Providence. . . . Then leaving Providence I set off for Boston, where I fell in company with James Williams for the first time of our connexion [sic]. I stole some buckles, and then, in company with Williams, set out for Newbury-Port, there, I broke open a schooner and took out of her some clothes, some chocolate, some tobacco, and a bottle of rum: Williams stood upon the wharf and helped to take the things; we then set off for Portsmouth, and on the way I stole an axe and sold it for a pair of shoes. [14]

CIVIL DISOBEDIENCE

Henry David Thoreau was arrested, tried, and convicted in 1846 for refusing to pay his poll tax. In his well-known *Resistance to Civil Government*, Thoreau described his incarceration and his feelings as he looked out onto the green Massachusetts landscape awaiting his breakfast:

It was to see my native village in the light of the Middle Ages, and our Concord was turned into a Rhine stream, and visions of knights and castles passed before me. They were the voices of old burghers that I heard in the streets. I was an involuntary spectator and auditor of whatever was done and said in the kitchen of the adjacent village inn—a wholly new and rare experience to me. It was a closer view of my native town. I was fairly inside of it. I never had seen its institutions before. This is one of its peculiar institutions; for it is a shire town. I began to comprehend what its inhabitants were about. In the morning, our breakfasts were put through the hole in the door, in small oblong-square tin pans, made to fit, and holding a pint of chocolate, with brown bread, and an iron spoon. When they called for the vessels again, I was green enough to return what bread I had left, but my comrade seized it, and said that I should lay that up for lunch or dinner. [15]

COUNTERFEITING

Two New York children, Powell Lopell and Ellis Elting, were arrested in 1861 and tried before Judge Nelson of the United States Circuit Court. They were accused of two crimes, the first for passing and the second for attempting to pass counterfeit coins with the intent to purchase chocolate and tobacco-related products. The evidence presented at trial revealed that, on March 6, 1861, Elting successfully presented and passed a counterfeit quarter of a dollar in his attempt to purchase chocolate at a value of three cents; the coin was accepted by and he received change from the shop proprietor. It was further charged that Elting and Lopell entered a second store, handed a second counterfeit quarter to the merchant in an attempt to fraudulently purchase papers used to roll tobacco cigarettes. In this case, however, the merchant refused to accept the coin and called for the police. Since witnesses for the prosecution did not appear in court on the trial date to support the prosecution's case, and there being no further evidence presented, both children were acquitted [16].

FRAUD

Prominent merchants also needed to protect themselves from "perceived" fraudulent representation that resulted from use of damaged or incorrect scales that provided customers with short weights. We documented early cases of chocolate-associated fraud in Mexico, as in the case where port officials added lead to scale weight bars but were caught by the Royal Inspector in charge of taxes:

I Juan Lorenzo Saavedra notary in charge of the inspection of the Royal Taxes [Cargas Reales] . . . inspected the cocoa load brought from Maracaibo to Veracruz in the sloop Fancisco Ramirez, that anchored in this port on December 18th, 1776. The coaoa [sic] load was inspected by Bartolome de Arriaga and Joseph Calderón. They weighed the cocoa [sic] sacks with the Royal Purveyor's scales and they found a difference of 24 pounds per sack. . . . When they inspected the scales they found out that they [port workers] had added pieces of lead to the steelyard drop . . . [the criminals] were fined 2,000 pesos. [17]

An interesting apology offered by Gideon Foster was published in the *Boston Gazette* in 1806. Foster, chocolate manufacturer at Salem, Massachusetts, desired to alert his customers to the following information:

Notice. WHEREAS a small quantity of Chocolate, manufactured and sold by the subscriber, has been deficient in weight, in consequence of an accident in my manufactory, viz. the loss of a wire from a Seal Beam—I request every person who has purchased such Chocolate per box, to call upon LUKE BALDWIN & CO. at No. 44, Long-Wharf, Boston, who will make good to them any loss they may have suffered, in consequence of such deficiency in weight. [18]

The curious aspect of Foster's published apology is how he expected his customers to comply with his request. The only customers who would have access to scales were merchants, as everyday citizens would not. After some of the chocolate had been grated to prepare beverages . . . how would it be possible for clients to determine whether or not the original quantity purchased had been "light"?

MURDER AND ATTEMPTED MURDER

Since antiquity, chocolate preparations have been used medicinally to mask the taste of bitter medicines [19]. Chocolate's strong flavor and own inherent bitterness have also been perceived by some criminals as a means whereby ill-tasting poisons could be administered and not easily detected by the unsuspecting consumer/victim. Cases of murder or attempted murder where chocolate was the vehicle for administering the supposedly lethal dose have been reported through the centuries. Among the earliest suggestion of "death-by-chocolate" is the oft-reported description of the death of the English King Charles II in 1685, while staying at the home of the Duchess of Portsmouth when she was residing in England. The Dutchess "told Lord Chancellor Cowper that King Charles was poisoned at her house by one of her footmen," and that the King had been served a "dish" of chocolate into which poison had been added [20]. The reality is, however, that King Charles II died of complications from kidney disease, and while the symptoms associated with his death included nausea, vomiting, headache, convulsions and coma (common to poisoning from various agents), the more likely cause of the King's death was kidney-related toxicity from uremia [21].

An account published August 21, 1735, in the *Pennsylvania Gazette*, recounted a case where arsenic was used in an attempt to murder members of a Boston family:

> Last Tuesday [August 4th, 1735] a horrid Attempt was made here to poison Mr. Humphrey Scarlet of this Town, Victuller [sic], his Wife and 2 Children; some Arsenick [sic] or Ratsbane having been put into a Skillet of Chocolate when they eat for Breakfast; Finding themselves soon after all disordered thereby, immediately applied to a Physician, and upon using proper Means, we hear, they are in a likely Way to do well. [22]

An account published in 1750 in *The New York Evening Post* summarized an attempt the previous year by the Turkish Pasha of Rhodes to massacre the Knights of Malta by poisoning the water supply used to prepare the Knights' coffee and chocolate. According to the report the plot was uncovered, and the article concluded that it would be likely that the Pasha would receive a sentence of life in prison [23].

As several attempts were made to assassinate Fredrick the Great, King of Prussia, it was reported that the King was in frequent danger for his life. Between 1790 and 1792, several North American newspapers carried the same report of the means Fredrick had developed to "test" the safety of his chocolate:

> The late King of Prussia was frequently in danger of being poisoned, but never sentenced those to death who made an attempt upon his life. One of his valets-de-chambre mediated the perpetration of this abominable act. The wretch, one morning, carried the King his chocolate as usual; but, in presenting it, his resolution failed him, and the King remarked his extraordinary confusion. "What is the matter with you?" observes Frederick, looking steadfastly at him, "I believe you mean to poison me." At these words, the villain's agitation augments; he throws himself at the feet of the monarch, avows his crime, and begs his pardon. "Quit my presence! Miscreant!" answered the King; and this was all his punishment. From that [time on], Frederick, before he took his chocolate, constantly gave a little of it to his dogs. [24]

One of the more interesting chocolate-related reports of attempted poisoning considered an assassination attempt on the life of General Napoleon Bonaparte. This intriguing story, published in 1807, has captured the interest of scholars through the centuries and even today seems like a 21st century American soap opera, since the various elements of the story include a spurned lover, opportunity, revenge, last-minute discovery, the presumed assassin being forced to taste the chocolate—her death—and continued political intrigue through the centuries. The story involves a Pauline Riotti, reportedly a former mistress of Bonaparte, who fell upon hard times and became destitute. She then was hired as a monastery kitchen inspector by a sympathetic priest. The opportunity for Riotti's mischief came about when the monastery where she worked was visited by Napoleon. The sordid tale, as told by a British propagandist and published several times in North American newspaper and magazine accounts, related that Bonaparte commonly drank chocolate during the late morning hour, and on the day of his departure from the monastery, Riotti took it upon herself to prepare the chocolate beverage for Napoleon—but fortunately:

> One of the cooks observed that she mixed with it something from her pocket, but without saying a word to her that indicated suspicion, he warned Bonaparte. In a note delivered to a page, to be on his guard. When the chamberlain carried in the chocolate, Napoleone [sic] ordered the person who had prepared it to be brought before him. This being said Pauline, she fainted away, after having first drank the remaining contents of the chocolate pot. Her convulsions soon indicated that she was poisoned, and notwithstanding the endeavors of Bonaparte's physician, Corvisart [? spelling garbled], she expired within an hour, protesting that her crime was an act of revenge against Napoleon, who had seduced her when young, under a promise of marriage, but who since his elevation, had not only neglected her to despair, by refusing an honest support for herself and child, sufficient to preserve her from the degradation of servitude. . . . The cook was,

with the reward of a pension, made a member of the Legion of Honor, and it was given out by Corvisart that Pauline died insane. [25]

Employing the concept of best evidence, the suggestion may be advanced that the Bonaparte assassination attempt by poisoned chocolate was fictitious, a clever piece of English propaganda, a story probably conjured by Lewis Goldsmith. Goldsmith originally published the assassination story in his book, *The Secret History of the Court and Cabinet of St. Cloud, written during the months of August, September, and October, 1805*. The edition examined was published in Philadelphia and its intent most likely was to discourage American support for France [26]. Whether true or not, the story of Pauline Riotti and Napoleon Bonaparte has continued to ring out across the centuries as a classic example of a spurned lover's attempted vengeance—using poisoned chocolate as her weapon.

A criminal case from Carlisle, Pennsylvania, dated to 1879, reported that a Mr. Wynkoop, an officer of the court, and a Mrs. Zell conspired to murder Mary Kiehl by poisoned chocolate:

On May 28th, of the present year, Mrs. Mary Kiehl, aged eighty-one years . . . was taken ill and after three days of great suffering she died. . . . In the house were two tin pots which had contained coffee and chocolate, of which she is known to have passion. What had remained in the vessels was subjected to an analytical examination, which resulted in the discovery of large quantities of arsenic. Subsequently the body of Mrs. Kiehl was exhumed, and the contents of her stomach, with the liver and other organs of the body, were subjected to analysis with a similar result. A man named Wynkook, a Justice of the Peace, and no relative of the deceased, was her sole [heir]. He was suspected of having procured her murder. . . . Another point against Wynkook was the expressed determination of the deceased to change her will in favor of a relative. Mrs. Zell was arrested and committed for trial and Wynkook was released on entering into $5,000 bail. [27]

Poisoning by chocolate, however, also could be accidental, as in 19th century cases where consumers became ill after eating chocolates that had been wrapped in green paper. Green pigments used during this period contained arsenic as a basic ingredient to accentuate the color:

Arsenic has been found in a number of colors, but it is generally confined to a light green. . . . It is very popular on account of its brilliancy. Its use is not confined to staining house-paper (i.e., wallpaper). We find it wrapped around our cake of chocolate, and in the fancy boxes that contain dried fruits and confectionery. [28]

NEGLIGENCE

Neglect commonly is associated with or related to bad manners, lack of attention to detail, sometimes egotistical behavior, or willful misbehavior. But what if

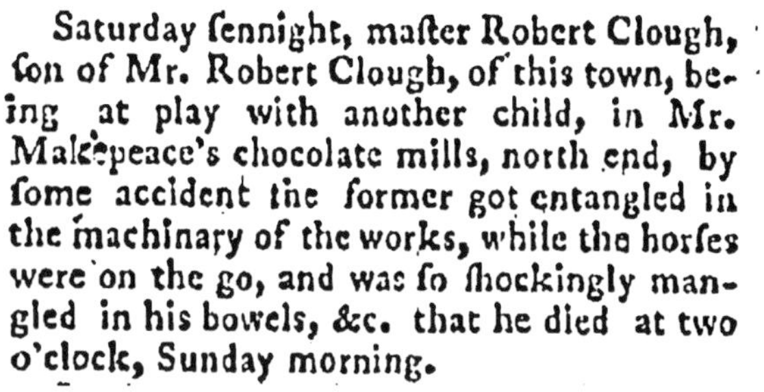

Saturday fennight, mafter Robert Clough, fon of Mr. Robert Clough, of this town, being at play with another child, in Mr. Makepeace's chocolate mills, north end, by fome accident the former got entangled in the machinary of the works, while the horfes were on the go, and was fo fhockingly mangled in his bowels, &c. that he died at two o'clock, Sunday morning.

FIGURE 21.3. Child killed by chocolate-manufacturing equipment. *Source: Worcester Magazine*, September 1, 1787, p. 307. Courtesy of Readex/Newsbank, Naples, FL. (Used with permission.)

responsible persons make poor judgments and their error results in the death of a second or third party? In such instances neglect may or may not be treated as a crime (Fig. 21.3 [29]).

Another example of negligence was reported in 1816 and described the tragic deaths of the sons of John Wait, Boston chocolate-maker, who were scalded to death while aboard a steamship whose boiler exploded:

In New-York, two sons of Mr. John Wait, jun. Chocolate-maker, late of Boston, one aged 8, the other 6 years. They had been at home on a visit from school, at Elizabeth-town, and had taken their passage on their return in a new Steam-boat, when the boiler burst, and so scalded them as to occasion their death in a short time. Several others were injured. Those interested in steam-boats, say, the boat was not finished; and that had there been a partition between the fire room and cabin, no great injury would have occurred. But why did the owners take passengers before the boat was finished? [30]

The following account of negligence, however, provided insights on the value of human life in France during the mid-19th century:

A girl seven years of age, affected with worms, was ordered by a physician [to take] some chocolate lozenges with santonin [i.e., wormwood: Artemisia pauciflora]. The apothecary's assistant who made up the prescription [mistakenly] took the bottle containing strychnine, which the principal had neglected to keep locked up. The lozenges were taken and death resulted. The apothecary and his assistant were both tried, the latter for involuntary homicide by imprudence and inattention; the former as responsible to the act of his assistant. The latter was declared guilty, and condemned to a year's imprisonment, and a fine of ten dollars and the costs, amounting to about sixty dollars. [31]

When individuals ignore their responsibilities to tend campfires and "live" embers subsequently flare up and ignite forest fires, are they not to blame? When individuals discard cigarette butts carelessly and the burning stubs ignite grass fires along highways, are these individuals not guilty of neglect as well?

Differences between the words "accident" and "neglect" commonly center upon the issue of intent. How is it possible—without a confession—to certify or measure intent or lack thereof? Is it not possible that certain instances of "presumed neglect," coupled with expressions of "sorrow," in fact are lies, where sobbing and facial anguish merely reflect crocodile tears? Accident, neglect, or intent: What is to be made of the following account?

> The chocolate factory of Runkel Brothers, Nos. 328 and 330 Seventh-avenue [New York] was badly damaged by fire early yesterday morning. The origin of the fire is a mystery and there are suspicions of incendiarism. The building is a four-story brick structure on the west side of Seventh-avenue, between Twenty-eighth and Twenty-ninth streets. The neighborhood is a very bad one and is infested with young thieves and corner loafers who give the police a great deal of trouble. . . . Their machinery and stock was valued at $60,000 and their loss will be about $25,000. . . . Although the police believe that the building was fired by thieves, it is not at all impossible that it may have been accidentally started by the policemen engaged in searching the building incautiously throwing away burning matches. [32]

It should also be mentioned here under the general theme of negligence that troops commanded by General George Washington, while encamped at Trenton after the famous battle, accidentally set fire to the home and out-buildings of William Richards, and his chocolate mill located on the site was destroyed [33].

SMUGGLING

We discovered an early case of smuggling of chocolate and cacao dated to December 20, 1622. The account consists of correspondence between officials of the *Case de Contratacion* in Seville, Spain, that identified fraud and misrepresentation of ship manifests. The letters commented on problems associated with several ships and summarized the items smuggled into Spain. The primary document addressed to Don Alonso Hurtado was signed by Don Gabriel Ocaña y Alarcón and stated:

> 29 loads [i.e., shipments] of sugar, chocolate, cacao, pearls, gold, silver, and other precious things valued in millions [of reales?] that have entered [Spain] without registration. [34]

In July 1773, Pope Clement XIV ordered the expulsion of the Jesuits from all Catholic countries. The effect of the expulsion order was that Jesuit priests were forced from Spanish-controlled regions of the New World and ultimately replaced by Franciscan friars throughout much of New Spain. One of these Franciscans, Junípero Serra, ultimately was instrumental in the foundation of the California mission system and would play a part in the introduction of chocolate into California as well (see Chapter 34). Most of the Jesuits departed Mexico from East Coast ports and set sail for Cadiz, Spain, taking aboard their material goods and possessions associated with their Order. The evidence suggests, however, that Jesuit priests sometimes returned to Spain with more than simple possessions as revealed in this document dated to 1792:

> A small fleet arrived at Cadiz; it contained upwards of sixty millions of livers [i.e., silver pound/coin] in gold and silver, and twelve millions in merchandise, besides smuggled goods. In unloading the vessels, eight large cases of chocolate were said to have been found. . . . These cases threatening to break the backs of the porters. . . . [Customs] officers became curious to know the cause. They opened one amongst themselves, and found nothing but very large cakes of chocolate, piled on each other. They were all equally heavy, and the weight of each surprising [sic]. Attempting to break one, the cake resisted, but the chocolate shivering off, [revealed] an inside of gold covered round with chocolate to the thickness of an inch. [35]

This finding, not unexpectedly, caused a considerable "difficulty." The customs officials confiscated the Jesuit's cases containing the gold and sought advice from officials in Madrid. The Jesuits refused to acknowledge that the cases were their property—although the containers had been transported aboard their ship—and professed ignorance regarding how the gold had become smothered in chocolate. They continued protesting that the cases were not theirs, and as a result the gold ultimately was sent to the Spanish King [36].

Smugglers commonly "doctored" invoices for goods imported to North America and commonly labeled sea-freight containers as "chocolate," as in the following account dated to 1817:

> In an invoice of imported goods which lately underwent the scrutiny of examination, as prescribed by the late order of the Secretary of the Treasury, we learn, there were twenty-six packages found of a fraudulent character, being invoiced and entered as Chocolate, Confected Citron, Gentian Root, Gum Lac, Orange Peel, &c. They were found to contain, besides these articles, upwards of seventy small boxes secreted therein, containing Sewing Silks, Ribbons, Silk Stockings, Silk Velvets, Prussian Blue, Boots, Shoes, &c. to the amount of five thousand dollars. [37]

Smuggling always has been a gender-equality crime as young women through the centuries have contributed their share of such illegal behavior while attempting to avoid paying import duties. The 1885 smuggling case of Emilie Sichel attracted considerable attention for several reasons. First, accounts of the day describe Ms. Sichel as a young, attractive lady. Second, customs officials identified Ms. Sichel as a smuggler, clearly identified contraband in her luggage, and she certainly was guilty. But how was it possible that such a proper, attractive young lady could do such a thing? Sichel's personal belongings and the smuggled goods were confiscated by Port Authorities as evidence and stored in a bonded warehouse while she awaited trial. During the intervening period, however, events took an unusual turn:

> Two trim and pretty young ladies smiled at the jury and Judge Wheeler in the United States Circuit Court yesterday in such a way as to win their sympathy and attention. One of the young ladies, Miss Emilie Sichel, came over from Europe in the Steamer Egypt in 1883 and had her baggage examined by Customs Inspector Peterson on January 31st of that year. The Inspector found in her trunk—12 corsets, 10 pounds of chocolate, and some lace curtains, which he declared were dutiable, and he held the trunk despite her demand for it. He ordered it sent to the public stores. That night a fire destroyed the Inman Line pier and Miss Sichel's baggage was consumed. . . . The jury promptly came in with a verdict for Miss Sichel for $459, interest being added by the jury. The case was the first of the kind [that it was the] Collector's responsibility for trunks held under such conditions. [38]

TAMPERING

Tampering with food products with the intent of causing harm is another crime that has seemed to linger through the centuries. Sowing darnel (tares) into the wheat fields of one's enemies was an oft-used military technique to contaminate the enemies' food supply [39]. Mixing poisons with foods or beverages also were common practices through the ages. Tampering in the strict sense used in the following example, however, is unusual because of its sense of familiarity. Although reported more than a century ago, this New York City case is reminiscent of 20th and 21st century North American examples of tampering with Halloween candy, or attempts by extortionists to extract funds by blackmail by threatening to harm consumers:

> Mrs. Abbie Smart, the wife of William Smart, a ship fastener, took to Sanitary Headquarters yesterday some crushed chocolate creams in which were a number of bent pins, and said that on two occasions they had been found at the foot of the basement steps of her home, No. 306 East Third-street. She suspected no one, and could not comprehend the motive of the person who put them there. Sanitary Detective Kennedy made an investigation, but could not arrive at any other conclusion than that some mischievous child had put the pins in the candies to injure Mrs. Smart's children. [40]

Conclusion

Rene Millon completed his Ph.D. dissertation on chocolate in 1955 [41]. The title of his dissertation, *When Money Grew on Trees: A Study of Cacao in Ancient Mesoamerica*, related the long, historical use of cacao beans as a form of currency during the pre-Columbian era in Central America. Once economic value became associated with cacao beans, the prepared products of ground cocoa and manufactured chocolate would increase dramatically in economic value, and chocolate-related crime would rise as well. From the earliest manufacture of "fake" cacao beans used to defraud merchants in ancient Central America, to the rise and proliferation of cases associated with adulterated chocolate, to burglary and theft of chocolate and silver chocolate-related serving utensils, it can be seen that chocolate-associated crime has a long history from antiquity into the 21st century. The crime of cacao-related fraud even was reported in February 2006, at a meeting held at the National Academy of Sciences in Washington, DC. A speaker mentioned the case that some cacao bean suppliers regularly defrauded cacao brokers (and ultimately chocolate manufacturing companies and consumers) by adding excessive amounts of dirt, rocks, shells, and other detritus to increase the weight of commercial bags of raw cacao beans [42]. It would seem that little has changed through the centuries.

It is perhaps appropriate to end this chapter on chocolate-associated crime with a relatively recent incident that took place in August 2004. A large, beautiful mural depicting cacao being harvested was stolen from the lobby of Cadbury chocolate headquarters in Birmingham, England. Theft of the chocolate mural had to be well planned in advance since the target object was large. It was likely, too, that several persons had to be involved since a vehicle was required to transport the mural. Cadbury offered a reward for the recovery of the painting, and the police asked for assistance in solving the crime [43].

Eight years have passed since the theft: Where is the Cadbury mural?

Acknowledgments

I wish to acknowledge the following students from the University of California, Davis, who assisted in identifying a broad range of chocolate-associated crimes: Silviu Margarit, Nataraj Naidu, Stacie Sakauye, Xun Su, and Naason Toledo.

References

1. *The Boston Weekly News-Letter*. February 16, 1738, p. 2.

2. *Boston Post Boy*. March 6, 1769, p. 2.

3. *The Boston News-Letter*. March 8, 1770, p. 2.

4. *Boston Post Boy*. May 27, 1771, p. 3.

5. *The New York Times*. May 6, 1852, p. 1.

6. *The Farmers' Cabinet*. Volume 77, Issue 34, 1879, p. 1.

7. *Peterson's Magazine*. Volume XCIX, March 1891, p. 269.

8. *The New York Times*. January 26, 1881, p. 2.

9. *Columbian Centinel [sic]*. January 5, 1799, p. 4.

10. *The Boston News-Letter and New-England Chronicle*. October 27, 1768, p. 2.

11. *The Boston News-Letter and New-England Chronicle*. February 9, 1769, p. 2.

12. *Columbian Centinel [sic]*. March 2, 1808, p. 4.

13. *Columbian Centinel [sic]*. August 3, 1808, p. 4.

14. Mount, T. *The confession, &c of Thomas Mount, who was executed at Little-Rest, in the state of Rhode-Island, on Friday the 27th of May, 1791, for burglary*. p. 11.

15. Thoreau, H. D. *On the Duty of Civil Disobedience*. (Project Gutenberg. Release Date, June 12, 2004. E-Text #71, Updated, March 1, 2005).

16. *The New York Times*. May 17, 1861, p. 3.

17. Archivo General de la Nación (AGN), Mexico City, Mexico. Criminal. Volume 77, exp. 3, fs. 47f–58f.

18. *Boston Gazette*. March 17, 1806, p. 4.

19. Dillinger, T. L., Barriga, P., Escarcega, S., Jimenez, M., Salazar Lowe, D., and Grivetti, L. E. Food of the gods: cure for humanity? A cultural history of the medicinal and ritual use of chocolate. *Journal of Nutrition* 2000;130(Supplement):2057S–2072S.

20. *The Antheneum, or Spirit of the English Magazines*. April 15, 1820, p. 67.

21. http://en.wikipedia.org/wiki/Charles_II_of_England. (Accessed January 10, 2007.)

22. *Pennsylvania Gazette*. August 21, 1735. Available at http://www.accessible.com/accessible/text/gaz1/00000019/00001962.htm. (Accessed January 10, 2007.)

23. *The New York Evening Post*. February 12, 1750, p. 2.

24. *The Mail, or Claypoole's Daily Advertiser*. January 26, 1792, p. 2.

25. *Massachusetts Spy, or Worcester Gazette*. May 13, 1807, p. 1.

26. Goldsmith, L. Memoires of the Court of St. Cloud. Being Secret Letters from a Gentleman at Paris to a Nobleman in London. Letter XXXIV, dated Paris, August, 1805. (Project Gutenberg. Release Date, September 11, 2006. E-Book 3899.)

27. *The National Police Gazette*. Volume 35, No. 114, November 29, 1879, p. 10.

28. *Potter's American Monthly*. March 1877, p. 235.

29. *Worcester Magazine*. September 1, 1787, p. 307.

30. *Columbian Centinel [sic]*. December 21, 1816, p. 2.

31. *Medical News*. May 1851, p. 40.

32. *The New York Times*. May 1, 1885, p. 8.

33. *Pennsylvania Evening Post*. January 14, 1777, p. 20.

34. Archivo General de Indias (AGI) (Archive of the Indies, Seville, Spain). Indiferente. 435, L. 11/1/615.

35. *Massachusetts Magazine/Monthly Museum*. 1792, p. 416.

36. *Massachusetts Magazine/Monthly Museum*. 1792, p. 416.

37. *Boston Gazette*. June 16, 1817, p. 2.

38. *The New York Times*. January 31, 1885, p. 8.

39. *Holy Bible*. Matthew, 13:24–30.

40. *The New York Times*. February 12, 1895, p. 8.

41. Millon, R. F. *When Money Grew on Trees: A Study of Cacao in Ancient Mesoamerica*. Ph.D. dissertation. New York: Columbia University, 1955.

42. http://www.cocoasymposium.com/. (Accessed January 24, 2007.)

43. http://news.bbc.co.uk/2/hi/uk_news/england/west_midlands/3962591.stm. (Accessed January 24, 2007.)

PART

V

Colonial and Federal Eras (Part 1)

Chapter 22 (Clark) considers the wide variety of beverages available in the American colonies during the 18th century, both alcoholic and nonalcoholic, examines their comparative cost and economic importance, and considers how political change influenced popularity and consumer purchase decisions. Chapter 23 (Gay) challenges the commonly held assumption that North American chocolate preferences were dependent on European traditions. His chapter also describes American manufacturing methods and how producers maximized their competitive advantage using homemade equipment, local labor, and innovative marketing methods. Chapter 24 (Macpherson) traces the introduction and spread of cacao and chocolate in Canada from the early 17th century into the 20th century and identifies relationships among chocolate and health, marketing, manufacture, cooking, importation and taxation, and popular customs. Chapter 25 (Jonah, Fougère, and Moses) examines the 18th century chocolate history of Fortress Louisbourg, capital of the French colony of Île Royale, Cape Breton, Canada. Using archaeological and curatorial collections, the authors illustrate the complex patterns of exchange during this historic period defined by military conflicts and changing political ideas. Chapter 26 (Blaschke) considers the development of the cocoa and chocolate industry in New England from the early 18th century through early 20th century, especially the growth of the chocolate industry as related to labor and gender issues, and use of chocolate and cocoa by Union forces during the American Civil War. Chapter 27 (Grivetti) considers production, sale, and social uses of chocolate in Boston during the 18th and early 19th centuries based on Colonial and Federal era newspaper articles and advertisements that described events occurring in this important New England port.

22

Chocolate and Other Colonial Beverages

Frank Clark

Introduction

We begin the discussion of cocoa and its role among beverages of the 18th century by classifying the types of beverages available at the time. Eighteenth century beverages in North America—as elsewhere—can be divided into two main categories: alcoholic and nonalcoholic. Alcoholic beverages were more important and—while consumers did not know about or understand microbial organisms—provided a safe drinking alternative in a time before pasteurization and modern water supplies. As a category, alcoholic beverages can be further subdivided into various fermented varieties—beer, cider, or wine—and distilled beverages or, as they were called then, "strong drink" or "*aqua vita*." The nonalcoholic drinks also can be subdivided into two categories: those served cold and those served hot. The present chapter examines each of these categories to obtain a basic understanding of the various libations, how they were made, how they were viewed by society, and the role of chocolate as one of several options available.

We also start with the basic premise that consuming liquids is necessary for life and survival. In the long distant past when humans formed societies, procurement of safe beverages was a basic survival need. While this may seem like common sense, it was not easy to achieve before public water and sewage systems had been developed. Although early medical understanding was unable to adequately explain why, it was perhaps basic common knowledge that alcoholic beverages—in moderation—were safer and led to fewer illnesses than "ordinary" water. It may also have been held that alcoholic beverages were more "nutritious" than water. The ancient Romans had their varieties of wine, and before them the Egyptians as well. The primary beverage among both the Egyptians and

Sumerians was beer—as was also initially characteristic of the English empire.

Alcoholic Beverages

MALT LIQUOR: ALES AND BEERS

When discussing beer, it is better to use the 18th century term "malt liquor." The reason for this distinction is that while both beverages were based on grain, definitions for beer and ale have changed over time. Malt liquors were natural choices for the English because of local climate. England's northern climes are such that grains grow better than grapes. Indeed, there is evidence that the English have brewed beer for nearly 4000 years [1].

Malt beverages, of course, are made from grain. Grain is sprouted then placed in a kiln to stop the growth process. This process, called malting,

Chocolate: History, Culture, and Heritage. Edited by Grivetti and Shapiro
Copyright © 2009 John Wiley & Sons, Inc.

allows the natural starchy components in grains to begin their conversion into sugars. Malted grains are crushed, then mashed in hot water, a process that continues this conversion. The mashed, water-soaked grain then is removed and the resulting liquid is called wort. The wort then is boiled, yeast added, fermentation occurs, and malt liquor is produced.

A wide variety of beverages can be produced by this basic process. These beverages differ only by type of grain used, amount of time and temperature at which the grains are kiln-dried, and various flavor components added to the wort. The early English process of brewing often mashed the same grain two or three times: with each subsequent mashing, fewer sugars could be extracted, and beers of three descending strengths would be produced.

The oldest type of English malt beverage was called ale. Ale usually was made from malted barley, but might also be prepared from either oats or wheat. Without preservatives or refrigeration, ale soured quickly. To compensate for such "spoilage," medieval brewers added various spices and herbs to help flavor the brew. Mixtures of spices added to ale were known as *gruit*. Stephen Harrod Buhner has provided a nice description of gruit in his book, *Sacred and Herbal Healing Beers*:

> Gruit was, primarily, a combination of three mild to moderately narcotic herbs; sweet gale (Myrica gale), also called bog myrtle, yarrow (Achillea millefolium), and wild rosemary (Ledum palustre), also called marsh rosemary. Gruit varied somewhat, each gruit producer adding additional herbs to produce unique tastes, flavors, and effects. Other adjunct herbs were juniper berries, ginger, caraway seed, aniseed, nutmeg, and cinnamon. The exact formula for each gruit was like that for Coca Cola, proprietary—a closely guarded secret. [2]

In Germany during the 9th century, brewers began to use hops (*Humulus lupulus*) as one of the herbs in brewing and soon realized that in addition to adding a bitter flavor, hops also helped to preserve the beverage. The use of hops spread slowly throughout northern Europe and eventually reached England by way of Holland sometime around 1450 [3]. After some early opposition, hops became fully accepted for use in brewing in England by the late 17th century. At this point there is some confusion about the terms beer and ale. Originally, the word ale meant a malt beverage made without hops, whereas beer was a malt beverage made with hops. By the late 17th century, however, all recipes for ale also contained hops. Some brewers then began to use the term ale to indicate the stronger beverage made from the first mashing of the grain, and the term beer to indicate the weaker second and third mashing, the last being called "small beer" [4].

Malt liquors were the traditional drink of England in addition to being the most popular. This is demonstrated by a quote from a French traveler who remarked in a letter:

> Would you believe it, although water is to be had in abundance in London and of fairly good quality, absolutely none of it is drunk? The lower classes, even the paupers, do not know what it is to quench their thirst with water. In this country nothing but beer is drunk and it is made in several qualities. Small beer is what everyone drinks when thirsty: it is used even in the best houses and costs only a penny a pot. Another kind of beer is called porter . . . because the greater quantity is consumed by the working classes. It is a thick strong beverage and the effect that it produces if drunk in excess, is the same as wine. . . . It is said that more grain is consumed in England for making beer then bread. [5]

When the English began to colonize the New World, their preference for malt drinks went with them (Fig. 22.1). This preference was confirmed in an early letter from the Governor of the Virginia colony to the other members of the Virginia charter company. Francis Wyatt wrote in 1623:

> To plant a colony by water drinkers was an inexcusable error in those who laid the first foundations. And have made it a received custom, which until it is laid down again there is little hope of health. [6]

Eventually brewers and better water supplies were located and more colonies were established, but it is clear that malt liquors were still quite popular. We can see this in a quote from Robert Beverly in his *The History and Present State of Virginia* when he wrote:

> Their small drink is either wine and water, beer, milk and water or water alone. The richer sort generally brew their beer with malt, which they have from England, though they have as good barley of their own, as any in the world; but for want of malt houses, the inhabitants take no care to sow it. The poorer sort brews their beer with molasses and bran; with Indian corn malted by drying in a stove; with persimmons dried in cakes and baked; with potatoes; with the green stalks of Indian corn dried and cut small and bruised; with the Bates canedencis, or Jerusalem artichoke. [7]

WINES

The evidence is clear that beer was king of the fermented beverages, but wine was a close second, especially among the gentry. The climate of England made growing good wine grapes difficult. Wine, therefore, was imported and often faced stiff taxes and duties in addition to the basic economic cost of importation. The English of the 18th century preferred the strong sweet fortified wines of Portugal and the Madeira Islands. Such fortified wines were made by adding a distilled spirit, usually brandy, to a fermenting barrel of wine. The addition of alcohol stopped the fermentation process by killing the yeast. This gave such wines a sweeter flavor because the yeast otherwise would have continued to consume the sugars left in the wine, resulting in a "dry" wine. Such fortified wines included varieties known as Canary, Madeira, Port, and Sack

TOBY FILLPOT.

FIGURE 22.1. *Toby Philpot*.
Source: Courtesy of The
Colonial Williamsburg
Foundation, Williamsburg,
VA. (Used with permission.)
(See color insert.)

(Sherry). Within the nonfortified wine category the English preferred the sweeter German Rein (i.e., Rhein) wines or Rennish as they were known, to the drier French wines, even though the wines of Burgundy already were famous. Wine terminology during this period referred to wines not based on the grape variety but upon their place of origin. Wines imported to England or North America were often relatively expensive because of taxes placed for political reasons and to protect the English brewing and distilling industries.

In the 21st century it is common to think of wine as being exclusively produced from grapes, but because grapes did not do well in England an examination of period cookbooks reveals that the English made "wine" from a wide variety of fruits, the more popular being blackberry, cherry, currant, peach, plum, raspberry, and strawberry. But even less well-known fruits such as elderberry and gooseberry were used as well.

These fruit juices often had sugar added to them to increase their alcoholic strength [8].

In addition to fruit wines, the English were hopeful that the New World would prove to be a good place to grow grapes. There were numerous early attempts to grow wine grapes in colonial Virginia. John Custis, for example, attempted to establish a vineyard in the vicinity of Williamsburg. Thomas Jefferson, a colonial Virginian very enamored of wine, viewed Virginia as a perfect place for its production, a view that led him to make a gift of 193 acres to an Italian winemaker, Phillip Mazzei. Mazzei, however, never was financially successful at winemaking in Virginia and most of his grapes died within the first two years of his farming operation. Still, Jefferson continued to believe that the New World was a good place to grow grapes. Eventually, after the conquest of *phylloxera* (*Daktulosphaira vitifoliae*) in the 19th century, and understanding of the role of sulfur in controlling fungus that attacked vines in eastern North America, Jefferson's vision proved to be correct—especially in California [9].

In addition to drinking wine "straight" (i.e., without mixing with water), the English often used it in various mixed drinks. The most popular were spiced or mulled wines created by adding sugar and spices then heating. There were also wine-based punches popular at the time, especially a type imported from Spain called *sangria*.

Because of costs associated with importation and taxes, wine was the most expensive alcoholic beverage by volume available in North America, as revealed in the Alexandria Virginia Hastings Court Records of 1786. The records list the cost of a quart of Madeira wine at 6 shillings. For the same price consumers could purchase a gallon of peach or apple brandy, whereas a quart of porter beer sold for 2 shillings and a quart of cider for 77 pence [10].

CIDER

Technically, cider can be considered a form of wine because its sugar comes from a fruit, but we will treat it here as a separate entity because of its economic and social importance. Cider is produced by squeezing apples using a press. The extracted juice begins as a nonalcoholic beverage but ultimately ferments due to yeast on the apple skin (or added yeasts) and ultimately can yield up to 10 percent alcohol. Historically, cider has been a popular beverage in England, especially in the apple growing areas of Hertford and Kent. Because it was produced locally and cheaply, cider commonly has been the drink of farmers, allowing a reasonable distinction to be drawn: beer in urban areas—cider in rural areas.

When the English colonized the New World they brought varieties of apples with them. The New World proved to be an excellent place to grow apples and many new cider varieties actually were developed here, including the Hews crab apple and the white crab apple. Cider, in turn, often is composed of several apple varieties combined to produce distinctive blends that range in flavor between sweet and sour. Economically, cider was very important in North America because it provided early farmers a means to preserve and transport apples in a more convenient form [11].

DISTILLED BEVERAGES IN GENERAL (*Aqua Vita*)

Distillation requires both fermentation and specialized equipment. The making of *aqua vita*—"water of life"—starts with fermentation. The fermented beverage is heated in a still, sometimes called an alembic. As alcohol has a lower boiling point than water, it evaporates sooner than water. By keeping their stills at the "correct" constant heat, producers of *aqua vita* boiled-off the alcohol, which then condensed in long tubes or coils and was collected. Alcohol thus obtained could then be used to make a range of beverages. The type of distilled drink produced depends on the type of fermented beverage being processed: distilling malted barley results in Scotch whiskey; distilling a mix of sour, fermented corn and sugar produces bourbon; distilling molasses and water yields rum; distilled wine results in brandy. Adding various flavorings to the basic fermented beverage being distilled produces gin and schnapps. The distillation process prevents the formation of vinegar, and beverages can be produced that can last for many years—economic implications not lost on our North American ancestors.

The origins of distilling are unclear but may represent a discovery that happened independently in a number of cultures throughout history. Distillation clearly was a process developed by Muslim cultures in the Middle East, and it is likely through the Muslims that distillation initially was introduced to Europe. It is sometimes claimed that the distillation process was introduced to England via the Dutch, when English soldiers returning from war in Holland brought the process of gin-making home to England. While the exact route and date of the introduction to England may not be known, by the 17th and 18th centuries distilled beverages had eroded the popularity of traditional English fermented drinks.

PUNCH

Distilled beverages can be consumed plain, but as such they tend to be harsh on the mouth and they quickly intoxicate drinkers. To soften the "taste" and to slow intoxication, distilled spirits often were mixed with other liquids. These mixtures of distilled spirits, water, and spices became known as punch. Peter Brown has provided an explanation for the origin of the word *punch*:

> Punch as it became known, was a pugilistic mixture of arrack (distilled from coconut sap), water, lemon juice, sugar and spices. Its name probably derived from the Hindu panch meaning five, presumably referring to the number of ingredients. [12]

Punch became extremely popular in England and her colonies. Punch drinking could be a rather raucous affair. The famous 18th century painter William Hogarth provided an excellent social portrait of punch drinking in his painting, *A Modern Midnight Conversation*, where a number of obviously drunk gentlemen are depicted lounging around an extremely large punch bowl (Fig. 22.2). Punch, properly made, consisted of five ingredients (citrus, spices, spirits, sugar, and water) but cheaper or "lesser punches" also were widely available. Bumbo, for instance, was prepared from rum, sugar, and water. Grog, in contrast, was made by mixing only rum with water (and of course became famous as the beverage of choice of the

FIGURE 22.2. *A Midnight Modern Conversation*, by William Hogarth, dated 1732–1733. *Source:* Accession Number 1966-498. Courtesy of The Colonial Williamsburg Foundation, Williamsburg, VA. (Used with permission.)

royal navy). Milk punch was prepared using rum, milk, and various mixtures of spices. At some point milk punch began to be called nog (hence eggnog would be a "lesser punch" prepared from rum, milk, and nutmeg). Basically, punch could be constructed/prepared from a broad range of diverse ingredients on hand by merely combining them with whatever distilled spirits were available.

Punch also made the journey to the North American English colonies, where it became as popular as it was in the homeland. In the Virginia Colony the term punch included beverages prepared not only from arrack and rum, but also from local brandies. Brown pointed out in his book, *Come Drink the Bowl Dry*, that "many of these colonial concoctions also included peach brandy; Lord Botetourt had large quantities of this in his cellars in Williamsburg, together with copious supplies of the other ingredients used to formulate a social bowl" [13].

Punch became very important in social rituals of 18th century Virginians and was served at almost all social occasions including weddings, financial deals, elections, and even everyday gatherings of gentlemen. Punch bowls appear in the inventories of almost all gentry and even many middle-class citizens [14]. Archaeological excavations at Shields Tavern in Williamsburg revealed the broken parts to as many as 80 different punch bowls, indicating the popularity of punch in taverns [15].

With all this punch consumption taking place in the North American colonies, it would be easy to conclude that drunkenness regularly resulted (and was widespread). Such conclusions would be correct (Fig. 22.3). There was a fair amount of lip service paid to the importance of not getting too drunk, but drunkenness certainly occurred frequently in all levels of 18th century English and, by inference, North American

"high society" as well. This view is reinforced by descriptions from a wide number of sources and evidenced from the following quote from a French traveler who described the aftermath of a gentry's dinner:

> At this point all the servants disappear. The ladies drink a glass or two of wine and at the end of a half an hour all go out together. It is then that the real enjoyment begins—there is not an Englishman who is not supremely happy at this particular moment. One proceeds to drink—sometimes in alarming measure. Everyone has to drink his turn, for the bottles make a continuous circuit of the table and the host takes note that everyone is drinking in his turn. After this has gone on for some time and when thirst has become inadequate reason for drinking, a fresh stimulus is supplied by the drinking of toasts, that is to say the host begins by giving the name of a lady; he drinks to her health and everyone is obliged to do likewise. After the host someone else gives a toast and everyone drinks to the health of everyone else's lady. Then each member of the party names some man and the whole ceremony begins again. If more drinking is required fresh toasts are always on hand; politics can supply plenty. [16]

GIN

The wealthy consumed their distilled beverages in the form of punch because they could afford the citrus, spices, and sugar to make it. For the poor, however, their distilled drink was gin—often drunk straight and by the dram. Gin became a serous social and health problem in 18th century London. This was the direct result of the passage of laws that raised the tax on malt beverages and wines and reduced taxes on distilled spirits. Passage of these laws benefited the wealthy by giving them a cheap way to convert their surplus grain into the more durable and profitable commodity, *aqua*

FIGURE 22.3. *Christmas in the Country*, by Barlow, dated January 7, 1791. *Source*: Accession Number 1975-286. Courtesy of The Colonial Williamsburg Foundation, Williamsburg, VA. (Used with permission.)

vita. The laws for retailing strong drink also were reduced so that individuals of different economic strata could open gin houses—and a great many did. All that was required for a cheap license was ownership of a still and access to spices and herbs. In some geographical districts of London three out of every four buildings retailed gin. The resulting flood of drunkenness shocked England and caused divisions in attitudes toward and perceptions of alcoholic beverages (Fig. 22.4) [17].

Socially, distilled spirits began to be viewed as dangerous and deadly, whereas fermented beverages were seen as healthy and nutritious. This view is reflected in a quotation from the Bishop of Oxford:

> Strong liquors produce in everyman a high opinion of their own merit; that they blow the latent spark of pride into flower and there fore, destroy all voluntary submissions, they put an end to subordination and raise every man on equal with his master. [18]

Another version of the same idea was espoused by Sir Joseph Jekyll (Figs. 22.5 and 22.6):

> The fear of a house of correction, imprisonment, or danger of the gallows made little impression on the rabble who drank gin. . . . People can not so soon nor so easily get drunk with beer as with gin. [19]

WHISKEY

North Americans during the Colonial era also held a similar dual view of alcoholic beverages. But in North America, in contrast to England, the problem beverage was not gin but whiskey. This attitude was summed up by Thomas Jefferson when he wrote in a letter:

> I wish to see this beverage [beer] become the common instead of whiskey which kills one third of our citizens an ruins their families. [20]

And just like gin in England there were interesting economic reasons why whiskey became an important product in North America. There is an excellent description of the economic argument for whiskey in *The Alcoholic Republic, an American Tradition*, by William Rorabach:

> The process of distillation interested Americans because it preformed a vital economic function by transforming fragile, perishable, bulky, surplus fruits and grains into non-perishable spirits that could easily be stored, shipped, and sold. [21]

There also were political reasons for the popularity of whiskey. During the American Revolution era, America's source of cheap rum dried up when trade with the Caribbean colonies became illegal as a result of nonimportation agreements. Americans needed a new source for distilled beverages: American corn provided it. It then became patriotic to drink locally produced spirits rather than imported ones. Americans had created their own distilled spirit in the form of bourbon whiskey and were drinking more and more of it as the 19th century approached.

Despite the increased consumption of distilled spirits, or perhaps because of it, attitudes toward drinking continued to change in America. Rorbach has provided another description of these factors:

FIGURE 22.4. *An Election Entertainment*, by William Hogarth, dated February 24, 1755. *Source*: Accession Number 1966-499. Courtesy of The Colonial Williamsburg Foundation, Williamsburg, VA. (Used with permission.)

FIGURE 22.5. *Beer Street*, by William Hogarth, dated 1761. *Source*: Accession Number 1972-409,93. Courtesy of The Colonial Williamsburg Foundation, Williamsburg, Virginia. (Used with permission.)

GIN LANE.

FIGURE 22.6. *Gin Lane*, by William Hogarth, dated 1761. *Source*: Accession Number 1972-409,94. Courtesy of The Colonial Williamsburg Foundation, Williamsburg, Virginia. (Used with permission.)

This change in mind was stimulated by a number of impulses among which were the spread of rationalist philosophy, the rise of mercantile capitalism, advances in science, particularly the science of medicine, and an all pervasive rejection of custom and tradition. [22]

The change in medical opinion regarding alcohol in America came from a source none other than the prominent Philadelphia physician, Doctor Benjamin Rush. He published a pamphlet in 1784 titled *An Inquiry into the effects of Ardent spirits on the human mind and body*. In this essay Rush blamed heavy alcohol consumption for such medical problems as vomiting, uncontrollable tremors, liver disease, and madness. The doctor recommended that alcohol consumption be limited to fermented beverages in moderate amounts. He also created a visual aid to make and reinforce this point, a "measuring device" that he called The Moral and Physical Thermometer, or Scale of Temperance (Fig. 22.7). Rush's "thermometer"—consisting of two

sides with written text/comments—listed the beverages from water to whiskey on one side and the vices, diseases, and punishment that resulted from consumption on the other. This "reference diagram" and other publications became popular with many citizens, who believed that alcoholic consumption was out of control, and were prominent pieces of information used frequently by the Temperance Movement, which began in the 1780s. Benjamin Rush, in fact, has been referred to as the Father of the American Temperance Movement.

There was strong support for the Temperance Movement in religious circles, and throughout the 18th century waves of religious revivalism swept through America, a phenomenon known as the Great Awakening. Religious tolerance was becoming more common and new freedoms allowed nontraditional religions like the Baptists and Methodists to gain large followings. These groups often took their religion on the road, traveling and holding revival meetings.

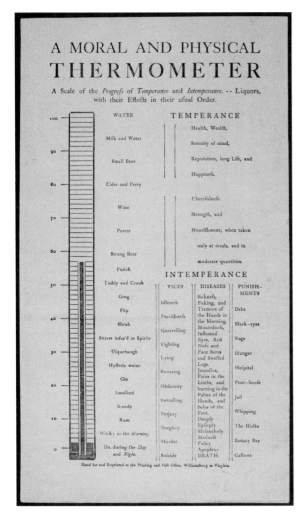

FIGURE 22.7. *A Moral and Physical Thermometer*, by Benjamin Rush. *Source*: Courtesy of The Colonial Williamsburg Foundation, Williamsburg, VA. (Used with permission.)

Quakers, and later Methodists and Baptists, began speaking out against alcohol and strong drink, and religious elements would remain an important part of the Temperance Movement throughout its lifetime [23].

Nonalcoholic Beverages

To this point we have examined the basic types of alcoholic drinks available in Colonial era North America, and we now turn to an exploration of non-alcoholic varieties. Previously we had divided this class of beverages into two categories: cold drinks and hot liquors. We will begin with cold drinks, although the term is misleading because these beverages were not served all that cold and never would have been served with ice as some are today.[1] Commonly "cold" drinks in early North America were served at cellar temperature, around 55 degrees Fahrenheit.

WATER

The most obvious nonalcoholic cold beverage was water. Water, as previously mentioned, often was viewed with mistrust. The 16th century physician Andrew Borde summed up the early viewpoint of water with his statement: "Water is not wholesome, sole by its self for an Englishman" [24]. This view of water was especially prevalent in the cities where public sewage and unsanitary conditions meant water could be quite dangerous. Early settlers in Jamestown also had a good reason to distrust local water supplies. In New England, however, the situation was different, as described by William Bradford: "Our landing party discovered some spring water which in their great thirst was as pleasant to them as wine and beer had been in foretimes." Eventually Governor Bradford decided that while the water (just described) was not as wholesome as the beer and wine of London, it was as good as any in the world and "wholesome enough to us that we may be contented therewith" [25].

Water originating from wells, especially those in large settlements or cities, was not as safe as spring water. River water quality and safety, of course, also varied depending on what was happening upstream. Spring water and rain water, therefore, were the safest forms, but water still was distrusted into the 19th century, as exemplified by a comment from John Bickerdyke, dated 1889: "Water is the purveyor of epidemics and hardly ever obtained pure by the working class" [26]. Water also formed the basis of the hot liquors, but with one important difference: in the manufacture/production of hot liquors water first was boiled. Why boiled water was safe was not well understood until the late 19th century.

MILK AND FRUIT JUICES

Milk also was viewed with suspicion and only became an important beverage after development of pasteurization during the mid-19th century. Peter Brown has provided a good description of how milk was perceived in early America in his book, *In Praise of Hot Liquors*:

> Milk, a substance we take so much for granted today was seen as moderately hot and moist[2] but there persisted a basic mistrust of this product in the seventeenth century which prevented its widespread use. Ian Bersten [and others have] argued that many people of the Near East and elsewhere were unable to digest milk.[3] . . . But whether or not individuals had digestive tract distress, the question of hygiene and freshness was foremost in their minds. [27]

Milk more often was used for cooking than drunk as a beverage.

In addition to milk and water, early North Americans also drank various fruit juices, whether prepared as lemonade or apple juice. Depending on freshness, these juices would have been nonalcoholic or at least low-alcoholic alternatives.

Hot Liquors

This group of beverages consists of tea, coffee, and chocolate and each arrived in England during the middle part of the 17th century. When these drinks were introduced to England they were viewed initially with suspicion because of the practice of consuming them hot. The prevailing medical theory of the time (see Chapter 6) generally held that consuming foods hotter than blood or milk from a cow could unbalance the body. Another factor that made these drinks unique is that each was bitter. The popularity of these hot drinks, therefore, depended on the introduction of sugar to make them more palatable to English tastes [28].

Sugar production had became one of the primary sources of income for the British empire, a situation made possible through the acquisition of sugarcane-growing colonies in the Caribbean coupled with British Navigation Acts that helped to control trade and shipping of goods from these English colonies to the European homeland. England also increased the size and skill of its navy to help establish further control of the seas. The expansion of the British Empire made imported goods that were once rare and costly cheaper and available to a wider segment of the population. The cost of many imported goods began to slowly decline as the 17th and 18th centuries progressed. The most important of these imported commodities was sugar; the gradual price decline affected the cost of chocolate, coffee, tea, and various spices as well [29]. Hot liquids were considered luxuries and therefore subject to taxation, and the tax rates often were very high, thus increasing their cost and encouraging smuggling (Fig. 22.8).

COFFEE

The Ottoman Empire was primarily responsible for the introduction of coffee to Europe. The Muslim world had been introduced to coffee via original production areas in Yemen (or from an alternative region across the Red Sea to the west in the Horn of Africa). The consumption of coffee was the source of debate in the Muslim world: banned by some and encouraged by others. One description of the arrival and spread of coffee in the urban Muslim world is attributed to Ralph Hattox:

> The central fact in the introduction of coffee into a common urban context, and its spread to other areas of the Muslim world was its adoption by Sufi groups in the larger cities of Yemen. This from all available evidence can be traced to the mid–fifteenth century. [30]

FIGURE 22.8. *Un Caffetier*, by Martin Englebrecht, Germany, ca. 1720. *Source*: Courtesy of The Colonial Williamsburg Foundation, Williamsburg, VA. (Used with permission.) (See color insert.)

Coffee arrived in England around 1650. One of the ways that coffee merchants diffused the initial distrust of their product and hot drinks in general was to claim healthful attributes. One such propaganda tract claimed:

> It does the orifices of the stomach good, it fortifies the heart within, helpeth digestion, quickens the spirits . . . is good against eyesores, cough, or colds, rhumes, consumption, head ache, dropsy, gout, scurvy, Kings evil and many others. [31]

Certainly not all consumers would accept such statements, and coffee had its early detractors as well. Coffee, in such tracts, was blamed for many ailments chiefly among them impotence. A consistent thread in coffee statements, however, was coffee's role as a stimulant, its sleep-preventing attribute, and how it caused consumers to be alert. This consideration of coffee as a stimulant would become important in how coffee was consumed [32].

Coffee was first available in specialized coffee houses. The first coffee house was established in Oxford in 1651 (see Chapters 37 and 43) [33]. Since coffee initially was available only in public settings, and its stimulant nature encouraged conversation, it quickly became viewed as a "social beverage." John Burnett has written:

From its beginnings coffee-drinking was a social activity, associated with lively dissemination of news, local and national. Coffeehouses therefore attracted men of letters, scholars, poets, wits and men of affairs. [34]

London coffee houses were artistic and commerce centers of the city's intellectual life. Dignitaries like Christopher Wren and Robert Boyle met at the Oxford coffee house. The use of coffee as a social beverage extended to North America as well. In important cities like Boston, New York, and Philadelphia, coffee houses were as much birthplaces of the Revolution as taverns. Even Williamsburg, Virginia, a town with a small population, had three coffee houses, one located just down the street from the Capitol building. Coffee eventually moved out of the coffee house and began to be consumed in homes of the gentry as well. Coffee drinking in North America received a significant boost from the Revolution, when it was seen as a patriotic alternative to tea.

TEA

Tea came to England from the Far East. It first arrived in tiny quantities brought by travelers, but soon began to be sold in the London coffee houses. Early on tea, like coffee, was given so many virtues that it was perceived more as a drug than as a beverage and, again like coffee, had its supporters and detractors [35]. Early coffee houses sold not only coffee but often tea and chocolate as well. Eventually they also sold beer, wine, and spirits. One of the first advertisements for tea published in London linked its sale with a coffee house:

That excellent and by all Physicians approved, China drink, called by the Chinese, Tcha, by other nations Tay alias Tee, is sold at the Sultaness-head a Cophee-house [sic] in sweetings rents by the royal exchange. [36]

Tea—like coffee—began in the male-dominated world of coffee houses but soon its acceptance and use spread to women. It is reported that a woman, Catherine of Braganza, first introduced tea to the court of Charles the Second. It is said that she brought a chest of tea as part of her dowry with her to England along with the right to trade it in certain Portuguese ports.

Once tea had found a place in and had been accepted by the English King's court, tea drinking moved slowly down the social ladder to all of England. Since coffee houses were unsuitable venues for respectable women, specialty tea shops—for ladies—were opened. The first of these was founded by Thomas Twining, in Devereux Court, London, in 1717 [37]. Despite the creation of tea houses (and later tea gardens), tea soon became popular as a home-based beverage served by women. This change resulted in the creation of entire industries to make tea equipage, as teapots, teacups, plates, tables, and caddies all become commonplace items in wealthy households. Anne Wilson has described the actions and implications of this change:

The hostess could thus brew and serve tea herself in the presence of her guests, a ceremony which had great appeal. The wives of the nobility and gentry eagerly acquired the new possessions, and served tea to their friends, who reciprocated. The very costliness of the tea itself gave an extra cachet to the person who offered it at her entertainments . . . a psychological advantage which in due course helped to spread tea drinking further and further down the social scale. [38]

It was probably only the high cost and high tax rates, coupled with the East India Company monopoly, that slowed tea's rise in England, but even these issues could not stop its growing popularity. The rise in tea consumption was nothing short of astounding. Although accurate figures are difficult to determine, due to a presumed high rate of tea smuggling, but by one estimate in 1701 approximately 100,000 pounds of tea were sold in England; by 1799, the quantity of tea sold had increased to 23 million pounds [39]. By the end of the 18th century tax rates had been lowered to a much more reasonable rate, and political acts basically eliminated tea smuggling. Peter Brown described the situation before that time in the following terms:

Customs and excise were levying higher and higher taxes on theses commodities [tea and chocolate] knowing that wholesale smuggling was causing a huge loss in revenue and they kept trying to balance this out by penalizing the legitimate trade. [40]

In addition to smuggling, in the 18th century tea also was subject to adulteration and recycling. Roy Moxham has described these processes in his book, *Tea*:

The favorite leaves used for adulteration were hawthorn for green teas and sloe for black teas, but birch, ash, and elder were also used. Of course, the leaves of these trees did not make convincing liquor, so it was necessary to add various coloring agents. In addition to the Terra japonica mentioned above, additives included verdigris, ferrous sulphate, Prussian blue, Dutch pink, copper carbonate, and even sheep's dung. Of these the sheep's dung was probably the least harmful. . . . Used tea leaves were purchased from servants or from the poor or even as a regular business from the coffeehouses. They were then dried on hotplates and in the case of green teas augmented with copper compounds. [41]

Regardless of purity or price, tea continued to gain in popularity in England. By 1730 tea imports began to surpass those of coffee. By the end of the 18th century tea imports had increased so that tea overtook beer as the most commonly consumed beverage in parts of England. The primary cause for this shift in beverage choice has been described by Anne Wilson:

In 1784 high grain prices were affecting the cost and also the quality of beer, and the duty on tea was at last lowered to a nominal sum based on the grounds that "tea has become an economical substitute to the middle and lower classes of society for malt liquor, the price of which renders it impossible of them to procure the quantity sufficient for them as their only drink." [42]

This time period then marked the beginning of England's transition from a primarily fermented alcoholic beverage-consuming country, to one where non-alcoholic beverages dominated. This transition also would be mirrored in North America, but would take longer in the Americas because of continental size and the slower rate of industrialization.

Tea in North America did not have the same success as a beverage as it did in Britain. The primary cause, of course, was politics. Tea was not the only beverage to be affected by the politics of the American Revolution—beer and rum also became viewed as politically improper. But it was tea that was most affected by this perception. Liza Gusler described this phenomenon in her paper, *More the Water Bewitched:*

Within a century of its introduction, tea was an established symbol of creature comfort in England, but by the late 1760s it was becoming a rallying point for political protest in the American colonies. Tea became the emblem of British oppression forsworn by patriotic associations who gave less than hospitable "tea parties" in Boston and Yorktown for merchants who continued to sell the politically incorrect brew. Coffee and chocolate became the drinks of choice for Americans who opposed tyranny imposed by a distant parliament. [43]

We see this aversive attitude toward tea in the inventory of Peyton Randolph in Williamsburg, Virginia. Randolph died in 1775 and an inventory was taken of his possessions. The inventory lists 40 pounds of coffee, but no tea. Certainly Randolph was wealthy enough to afford tea, but as one of the primary authors of the nonimportation agreements, he would not be "caught dead" with it in his house [44].

CHOCOLATE

Today, chocolate is viewed primarily as a food, but this is because of the invention of milk chocolate during the 19th century. Prior to the invention of processes to remove the fat and blend the cocoa with milk and sugar, culinary uses of chocolate were fairly limited. Anne Wilson, author of *Food and Drink in Britain, from the Stone Age to the 19th century*, has offered appropriate observations regarding the use of chocolate:

Chocolate Almonds could be purchased from French confectioners in the 1670s, and . . . these were soon copied by English housewives.[4] Flavored first with musk and ambergris, and later with orange flower water, the almonds were served as dessert along with crystallized fruits and candies. Chocolate itself flavored other confectionaries, notably chocolate puffs (little light cakes made of sugar and beaten eggs). Chocolate creams were a second course dish. [45]

As seen in the numerous chapters of this book, chocolate originated in the New World and the cacao tree probably originated in the eastern slopes of the Andes Mountains in the western Amazon basin [46]. The Olmec civilization was likely the first to use the plant as a beverage. The early Mesoamerican drinks made with cocoa beans bear little resemblance, however, to today's hot cocoa. Early cacao was unsweetened, thick, spicy, and consumed at room temperature. The Spanish transformed the *cacahuatl* they learned from the Aztecs into chocolate by adding sugar, often vanilla beans, and European spices. The Spanish also began the practice of serving cacao as a hot beverage (see Chapter 8).

Chocolate in an economic sense did not reach England as a viable commodity until after the English captured Jamaica from the Spanish in 1655 [47]. The early consumption of chocolate in England followed a similar pattern to coffee: cacao or chocolate beverages were initially consumed in coffee houses and later specific chocolate houses by wealthy gentlemen, and eventually reached individual households. Once in the house, it commonly moved into the bedroom, where many drank chocolate for its perceived "special" attributes. This use reflected a popular view that chocolate was an aphrodisiac, as summed up by this 17th century poem:

T'will make old women young and fresh;
create new notions of the flesh;
And cause them to long for you know what,
if they but taste of chocolate. [48]

Chocolate, in time, was taken as a breakfast beverage, sometimes with cold meats and breads. A version of chocolate made from cocoa shells was a favorite breakfast for Martha Washington [49]. In addition to these uses, chocolate also was viewed as a stimulant and as being a wholesome and filling beverage.

During the 18th century, there were three main ways of preparing chocolate as a beverage in England and the North American colonies. These methods were explained by William Salmon in his *Dictionary:*

That made with water alone (but fortified with a little brandy); that made with milk; and wine chocolate in which the liquid element was supplied by three quarters of a pint of water and half a pint of choice red port or rather choice sherry. [50]

Chocolate was different from the other hot liquors in that the cocoa beans were processed into chocolate bars or tablets, and then grated and prepared as a drink. Coffee used the natural beans roasted and then crushed, and tea was simply the dried leaves. Chocolate required more preprocessing before it could be made into a beverage. The processing first involved fermenting the beans after harvest, then roasting the beans, shelling them, grinding them on a hot stone, and sometimes adding spices and sugar during the processing. The chocolate mass—in a semiliquid state—then was poured into molds and allowed to harden. When desired, these bars or "chunks" of chocolate were grated/powdered and mixed with the desired liquid, heated, and finally frothed with a chocolate mill or *molinillo*.

Crushing cacao beans—on a small scale—was completed using curved chocolate stones, and workers could control how much spices and other ingredients were added (see Chapter 48). When large quantities were required, chocolate was made in mills. Large-scale commercial chocolate operations also were open to the same types of fraud and adulteration associated with tea: one of the most common adulterants added to chocolate was brick dust (see Chapter 47). During most of the Colonial era, most North Americans did not take the time and effort to make their own chocolate and bought pre-made chocolate from commercial chocolate makers.

Chocolate was also considered a luxury by the British government and subject to taxation. The first taxes were on the cocoa beans or nuts as they were called, but merchants evaded these taxes by importing pre-made chocolate bars. To counter this practice, the English government changed the law and this loophole was removed. In addition, the English also regulated the production of chocolate by British tradesmen and restricted the import of chocolate manufactured in North America, with the result that chocolate remained relatively expensive and less available in England, compared to North America [51].

In North America, chocolate was easy to obtain, either legitimately or as smuggled goods. With the exception of some small production in English colonies in the Caribbean, the majority of the cocoa coming to North American ports originated in non-English colonies. Cacao beans were sent to major seaport cities in the Northeast, where they were converted into chocolate. The primary chocolate producing cities in colonial America were Boston, Newport, New York, and Philadelphia. Chocolate was manufactured in mills (both large and small), that also ground products as diverse as ginger, mustard, snuff, and wheat. Millers needed a variety of products to prepare since chocolate-making was seasonal and quantities were not manufactured during the hot weather of summer. Chocolate-making also required considerable quantities of unskilled labor to participate in the processing of cacao beans (i.e., shelling and grinding). Often slaves were employed in these tasks in North America [52].

COMPARATIVE PRICES OF HOT LIQUORS

Comparing prices of the hot liquors is more difficult than for fermented liquors. This is because tea leaves can be reused, whereas chocolate and coffee prepared from beans are one-time-use beverages. The situation is further complicated by taxes imposed at different times in North American history, and the ever-existing problem of inflation through the decades. The law of supply and demand also affected prices as production increased. Still, one may make comparisons by examining the rates charged by different coffee houses, keeping in mind that buying these drinks in coffee houses would be considerably more expensive than preparing the beverages at home.

There is an account from York, England, dated to 1733, where Anne Wilson (Proprietor) sold coffee at 12 pence a pot; tea at 18 pence a pot; and chocolate at 3 pence a dish. The prices in York, however, were high since in Oxford the price of a pot of coffee was 4 pence [53]. Prices varied throughout the middle part of the 18th century and rose quickly, especially during times of war. The Commutation Act of 1784 reduced the taxes on tea and the other hot liquors to 12.5 percent, lowering prices of these products significantly as well as reducing the black market. Chocolate at this time sold for up to 5 shillings 6 pence a pound; after the Act it sold for between 2 shillings 3 pence a pound and 4 shillings a pound for the finest type [54]. In Williamsburg, Virginia, chocolate sold for 2 shillings 6 pence per pound, while coffee prices were 1 shilling 6 pence a pound, with tea fluctuating between 12 and 15 shillings a pound depending on the quality [55].

Conclusion

During the 18th century in England and her colonies, the availability and use of different beverages were in transition. The traditional fermented beverages faced competition from distilled spirits and the hot liquors. Attitudes toward consuming alcohol on a daily basis started to transform because of changing medical opinion, political factors, and religious zeal. The Industrial Revolution also changed public attitudes toward drinking. The use of machinery in factories helped to encourage a transition away from alcoholic drinks toward other types. Other social factors that influenced attitudes away from alcohol and towards non-alcoholic beverages—and in some instances encouraged cacao/chocolate drinking—included the religious fervor brought by the Great Awakening and the rise of the Temperance Movement in both England and North America. A transition away from alcohol was taking place in North America—linked to the rapid rise in popularity of hot liquors—and it was in this time period that North Americans commonly turned their attention to chocolate.

Acknowledgments

I would like to thank Jim Gay, whose enthusiasm for chocolate is contagious. I would also like to thank Master Foods U.S.A., a division of Mars Company, particularly Deborah and Forest Mars, for their support and their company. Eric and Judy Whitacre, and Rodney Snyder taught me about the world of chocolate and lighted my passion for this wonderful beverage/food. I also wish to thank the staff at Colonial Williamsburg who have helped and supported this project during the course of my work on chocolate and other Colonial era beverages.

Endnotes

1. The preference for serving drinks very cold or with ice is a relatively new one that began to take root in the United States after World War II.

2. This is in reference to the humoral food/medical system that predated the concept of germ theory.

3. Numerous reviews of human genetic differences in the ability to digest fresh milk have appeared during the past 35 years.

4. Indeed, the first recipe for using chocolate that the chocolate history research group has found for North America was for chocolate almonds.

References

1. Jackson, M. *Michael Jackson's Beer Companion*. London, England: Duncan Baird Publishers, 1993.

2. Buhner, S. H. *Sacred and Herbal Healing Beers*. Bolder, CO: Siris Books, 1998; p. 169.

3. Bickerdyke, J. *The Curiosities of Ale and Beer*. London, England: Spring Books, 1889 (reprinted 1965); p. 68.

4. Sambrook, P. *Country House Brewing in England, 1500 to 1900*. London, England: Hambledon and London, 1996.

5. Corran, H. S. *The History of Brewing*. London, England: David and Charles, 1974; p. 11.

6. Baron, S. *Brewed in America*. Cleveland, OH: Beerbooks, 1962; p. 6.

7. Beverly, R. *The History and Present State of Virginia*. Raleigh, NC: University of North Carolina Press, 1947; p. 293.

8. Bradley, R. *The Country housewife and Ladys director in the management of a house and the delights and profits of the farm*, 6th edition. London, England: D. Brown, 1736; distributed by University of Virginia Press, 1980.

9. Lawrence, R. DeTreville. *Jefferson and Wine*. The Planes, VA: The Vinifera Wine Growers Association, Inc., 1989.

10. Microfilm, Reel 1 #00035, Alexandria Public Library, special collections.

11. Rorabach, W. L. *The Alcoholic Republic, an American Tradition*. Oxford, England: Oxford University Press, 1974; p. 74.

12. Brown, P. *Come Drink the Bowl Dry*. York, England: York Civic Trust, 1996; p. 45.

13. Brown, P. *Come Drink the Bowl Dry*. York, England: York Civic Trust, 1996; p. 53.

14. Stephenson, M., and Carson, J. *Peyton Randolph House Historical Report*. Williamsburg, VA: Colonial Williamsburg Foundation, 1952; revised 1967, p. 155.

15. Brown, G., Higgins, J., Thomas, F., Muraca, D. F., Pepper, K. S., and Polk, R. H. *Archeological Investigation of Shields Tavern Site*. Colonial Williamsburg Foundation. 1990; p. 90.

16. Brown, P. *Come Drink the Bowl Dry*. York, England: York Civic Trust, 1996; p. 99.

17. Dillon, P. *The Much Lamented Death of Madam Geneva, The Eighteenth-Century Gin Craze*. London, England: Review, Headline Book Publishing, 2002.

18. Warner, J. *Craze, Gin Debauchery in an Age of Reason*. New York: Four Walls Eight Windows Publishing, 2002; p. 186.

19. Jekyll, J. *The Trial of the Spirits*, 2nd edition, London, England: T Cooper, 1736; p. 13.

20. Baron, S. *Brewed in America*. Cleveland, OH: Beerbooks, 1962; p. 146.

21. Rorabach, W. L. *The Alcoholic Republic, an American Tradition*. Oxford, England: Oxford University Press, 1974; p. 74.

22. Rorabach, W. L. *The Alcoholic Republic, an American Tradition*. Oxford, England: Oxford University Press, 1974; p. 35.

23. Lender, M. E., and Martin, K. J. *Drinking in America, A History*. New York: Freepress, 1982; p. 37.

24. Bored, A. *A Compendious Regyment or a Dyetary of Health* London, England: Robert Wyer, 1547; p. 67.

25. Hooker, R. *Food and Drink in America*. New York: Bobs-Merrill Company, 1981; p. 35.

26. Bickerdyke, J. *The Curiosities of Ale and Beer*. London, England: Spring Books, 1889 (reprinted 1965); p. 373.

27. Brown, P. *In Praise of Hot Liquors*. York, England: York Civic Trust, 1995; p. 26.

28. Brown, P. *In Praise of Hot Liquors*. York, England: York Civic Trust, 1995; p. 10.

29. Brown, P. *In Praise of Hot Liquors*. York, England: York Civic Trust, 1995; p. 11.

30. Hattox, R. S. *Coffee and Coffeehouses*. Seattle: University of Washington Press, 1985; p. 26.

31. Brown, P. *In Praise of Hot Liquors*. York, England: York Civic Trust, 1995; p. 37.

32. Brown, P. *In Praise of Hot Liquors*. York, England: York Civic Trust, 1995; p. 37.

33. Brown, P. *In Praise of Hot Liquors*. York, England: York Civic Trust, 1995; p. 36.

34. Brown, P. *In Praise of Hot Liquors*. York, England: York Civic Trust, 1995; p. 12.

35. Moxham, R. *Tea, Addiction, Exploitation and Empire*. London, England: Constable and Robinson, 2003; p. 34.

36. Moxham, R. *Tea, Addiction, Exploitation and Empire*. London, England: Constable and Robinson, 2003; p. 18.

37. Wilson, A. C. *Food and Drink in Britain, from the Stone Age to the 19th Century*. Chicago, IL: Academy Chicago Publishers, 1973; p. 413.

38. Wilson, A. C. *Food and Drink in Britain, from the Stone Age to the 19th Century*. Chicago, IL: Academy Chicago Publishers, 1973; p. 413.

39. Moxham, R. *Tea, Addiction, Exploitation and Empire*. London, England: Constable and Robinson, 2003; p. 50.

40. Brown, P. *In Praise of Hot Liquors*. York, England: York Civic Trust, 1995; p. 36.

41. Moxham, R. *Tea, Addiction, Exploitation and Empire*. London, England: Constable and Robinson, 2003; p. 29.

42. Wilson, A. C. *Food and Drink in Britain, from the Stone Age to the 19th Century*. Chicago, IL: Academy Chicago Publishers, 1973; p. 416.

43. Gusler, E. *More the Water Bewitched*. Williamsburg, VA: Colonial Williamsburg Foundation, 1992; p. 1.

44. Stephenson, M. A., and Carson, J. *Payton Randolph House Historical Report*. Williamsburg, VA: Colonial Williamsburg Foundation, 1952 (revised 1967); p. 21.

45. Wilson, A. C. *Food and Drink in Britain, from the Stone Age to the 19th Century*. Chicago, IL: Academy Chicago Publishers, 1973; p. 410.

46. Maricel, P. *The New Taste of Chocolate: A Cultural and Natural History of Cocoa with Recipes*. Berkley, CA: Ten Speed Press, 2001; p. 10.

47. Wilson, A. C. *Food and Drink in Britain, from the Stone Age to the 19th Century*. Chicago, IL: Academy Chicago Publishers, 1973; p. 408.

48. Gay, J. Chocolate Production and Consumption in North America in the 17th and 18th Centuries. *Chocolate: History, Culture and Heritage*. Hoboken, NJ: Wiley, 2009; Chapter 23.

49. Brown, P. *In Praise of Hot Liquors*. York, England: York Civic Trust, 1995; p. 18.

50. Wilson, A. C. *Food and Drink in Britain, from the Stone Age to the 19th Century*. Chicago, IL: Academy Chicago Publishers, 1973; p. 410.

51. Brown, P. *In Praise of Hot Liquors*. York, England: York Civic Trust, 1995; p. 36.

52. Gay, J., and Clark, F. Making Colonial Era Chocolate. The Colonial Williamsburg Experience. *Chocolate: History, Culture and Heritage*. Hoboken, NJ: Wiley, 2009; Chapter 48.

53. Brown, P. *In Praise of Hot Liquors*. York, England: York Civic Trust, 1995; p. 20.

54. Brown, P. *In Praise of Hot Liquors*. York, England: York Civic Trust, 1995; pp. 36–37.

55. Carramia, J. Unpublished paper Colonial Williamsburg Foundation, *A Research Study Comparing Costs of Consumer Goods Sold in Williamsburg Stores, Private Accounts, and Newspaters*. 1992.

Chocolate Production and Uses in 17th and 18th Century North America

James F. Gay

Introduction

Legends when told by historians become history. The success and longevity of Baker's Chocolate® has spawned an enduring folklore repeated again and again by historians, food writers, and marketing departments. Namely, that the first chocolate in North America was produced in Dorchester, Massachusetts, by John Hannon and Dr. James Baker in 1765 (even though the current Baker's Chocolate logo seems to refute this date by printing the year "1780" on every box). As we shall see, Americans had been making chocolate for at least a century prior to 1765. But, because of the dominance of the Walter Baker Chocolate Company throughout the 19th century, and the iconic nature of Baker's Chocolate brand today, this claim has gone largely unchallenged. The 17th and 18th century North American chocolate experience has been ignored by historians who have focused their attention almost exclusively on Latin America and Europe. And, the other North American chocolate manufacturers (Table 23.1) besides Baker and Hannon, have remained anonymous to historians—until now.

The history of cocoa beans and chocolate in North America is more than 300 years old. The oldest record of North Americans trading in cocoa beans is found in the diary of Massachusetts Bay's mint-master John Hull. In the winter of 1667–1668, he noted the loss of "our ship *Providence* . . . cast away on the French shore . . . [carrying] . . . cocoa" [1]. One of the earliest records of chocolate in North America (New England region) dates to 1670 when Dorothy Jones and Jane Barnard were given approval to serve "Coffee and Chucaletto" in houses of "publique Entertainment" by the selectmen of Boston [2]. Did Jones and Barnard

manufacture the chocolate themselves or did they import it? The answer is unclear. The oldest British customs record showing cocoa arriving in America reads: "1682 . . . Jamaica . . . to . . . Boston" [3]. Was this the first shipment? Perhaps. There may be earlier examples yet to be discovered. Shipping records from the Colonial period are filled with gaps measured in years. Surviving Boston import records, for example, only begin in 1686 [4]. Was the first parcel of cocoa grown in Jamaica or was it pirated from the Spanish Main? Was the first cocoa even delivered on a British ship? The probability is that we will never know.

What is known, however, is the following.

Chocolate is more American than apple pie. It was created in the New World after the Spanish conquest of Mexico. While it combined ingredients from

Chocolate: History, Culture, and Heritage. Edited by Grivetti and Shapiro
Copyright © 2009 John Wiley & Sons, Inc.

Table 23.1 **Eighteenth Century North American Chocolate Makers**

Location	Name	Religion	Date	Source
New Haven, CT	Nathaniel Jocelin		October 26, 1770	*Connecticut Journal*
Danvers, MA	Amos Trask		September 18, 1771	Walter Baker, *Cocoa and Chocolate*, 1917, p. 35
Boston, MA	Jonas Welch		January 1, 1770	Crooked & Narrow St. of Boston
Boston, MA	Caleb Davis		January 1, 1776	John W., Tyler, *Smugglers and Patriots*, p. 263
Boston, MA	Edward Romney		January 1, 1789	Crooked & Narrow St. of Boston
Boston, MA	John Brown/Browne		1702–1771	Crooked & Narrow St. of Boston
Boston, MA	John Goldsmith		1751–1781	Crooked & Narrow St. of Boston
Milton, MA	John Hannon		1768–1777	James Baker Business Records
Boston, MA	Joseph Palmer and Richard Cranch		March 12, 1751	*Boston Gazette*
Boston, MA	James Lubbock		April 25, 1728	Crooked & Narrow St. of Boston
Boston, MA	Joseph Mann		March 6, 1769	*Boston Post Boy*
Dorchester, MA	Dr. James Baker		May 17, 1771	James Baker Business Records
Boston, MA	Newark Jackson		February 26, 1740	*New England Weekly Journal*
Boston, MA	John Merrett		February 9, 1736	*Boston Evening Post*
Boston, MA	Samuel Watts		April 23, 1751	*Boston Gazette*
Boston, MA	Israel Eaton		December 20, 1750	*Boston News-Letter*
Boston, MA	Sam		June 2, 1763	*Boston News-Letter*
Charlestown, MA	Samuel Dowse		October 10, 1763	*Boston Gazette, and Country Journal*
Boston, MA	Richard Floyd		April 7, 1763	*Boston News-Letter and New-England Chronicle*
Boston, MA	Henry Snow		November 30, 1767	*Boston Gazette, and Country Journal*
Boston, MA	Richard Catton		March 27, 1769	*Boston Evening Post*
Boston, MA	James Cooke		April 79, 1770	*Boston Gazette, and Country Journal*
Boston, MA	Mr. Feacham		March 25, 1771	*Boston Evening Post*
Boston, MA	George Leonard		March 11, 1771	*Boston Gazette, and Country Journal*
Boston, MA	Benjamin Leigh		May 20, 1771	*Boston Post Boy*
Boston, MA	Deacon Davis		December 21, 1772	*Boston Evening Post*
Boston, MA	George Fechem		September 2, 1776	*Boston Gazette, and Country Journal*
Boston, MA	John Brewster		February 20, 1749	*Boston Evening Post*
Salem, MA	A Gentleman		September 5, 1737	*Boston Gazette*
Annapolis, MD	Issac Narvarro	Jewish	1748–1749	J. R. Marcus, *The American Colonial Jew*, 1970, p. 538
New York, NY	Moses Gomez	Jewish	January 1, 1702	J. R. Marcus, *The American Colonial Jew*, 1970, p. 673
New York, NY	Peter Low		November 23, 1769	*New York Gazette and Weekly Post Boy*
New York, NY	Abraham Wagg	Jewish	January 1, 1770	J. R. Marcus, *The American Colonial Jew*, 1970, p. 1296
New York, NY	Peter Swigart		November 18, 1758	*New York Gazette*
New York, NY	Jacob Louzada	Jewish	1710–1729	myohiovoyager.net/~wkfisher/ Luzadder.html
New York, NY	Joseph Pinto	Jewish	1750–1760	J. R. Marcus, *The American Colonial Jew*, 1970, p. 673
New York, NY	Aaron Louzada	Jewish	1710–1764	J. R. Marcus, *The American Colonial Jew*, 1970, p. 673
Philadelphia, PA	John Dorsey	Quaker	January 1, 1765	*Pennsylvania Gazette/Quaker Encyclopedia*
Philadelphia, PA	Edward Loudon		January 1, 1785	*Philadelphia City Directory*
Philadelphia, PA	William Young	Quaker?	January 7, 1762	*Pennsylvania Gazette/Quaker Encyclopedia*
Philadelphia, PA	William Smith	Quaker	November 17, 1763	*Pennsylvania Gazette/Quaker Encyclopedia*
Philadelphia, PA	Dr. John Bard		1743–1745	*Pennsylvania Gazette*
Philadelphia, PA	James Humphreys	Quaker?	1748–1750	*Pennsylvania Gazette/Quaker Encyclopedia*

Table 23.1 Continued

Location	Name	Religion	Date	Source
Philadelphia, PA	Glover Hunt		1752–1760	*Pennsylvania Gazette*
Philadelphia, PA	Benjamin Jackson		1756–1769	*Pennsylvania Gazette*
Philadelphia, Pa	Capt Jonathon Crathorne		1760–1767	*Pennsylvania Gazette*
Philadelphia, PA	George Rankin		1763–1764	*Pennsylvania Gazette*
Philadelphia, PA	Mary Crathorn		1767–1769	*Pennsylvania Gazette*
Philadelphia, PA	William Norton	Quaker	1769–1776	*Pennsylvania Gazette / Quaker Encyclopedia*
Philadelphia, PA	Thomas White	Quaker?	February 12, 1752	*Pennsylvania Gazette / Quaker Encyclopedia*
Philadelphia, PA	Thomas Roker		February 14, 1776	*February 14, Pennsylvania Gazette*
Philadelphia, PA	Thomas Hart	Quaker	February 19, 1745	*Pennsylvania Gazette / Quaker Encyclopedia*
Philadelphia, PA	A likely Negro		February 19, 1752	*Pennsylvania Gazette*
Philadelphia, PA	Samuel Garrigues	Quaker	February 8, 1775	*Pennsylvania Gazette / Quaker Encyclopedia*
Philadelphia, PA	Roger Hiffernan		March 25, 1755	*Pennsylvania Gazette*
Philadelphia, PA	Robert Iberson	Quaker	March 26, 1754	*Pennsylvania Gazette / Quaker Encyclopedia*
Philadelphia, PA	George Justice	Quaker	March 3, 1768	*Pennsylvania Gazette / Quaker Encyclopedia*
Philadelphia, PA	John Browne		March 8, 1764	*Pennsylvania Gazette*
Philadelphia, PA	Dick		April 2, 1752	*Pennsylvania Gazette*
Philadelphia, PA	John Maes		April 28, 1743	*Pennsylvania Gazette*
Philadelphia, PA	Jacob Cox		June 9, 1763	*Pennsylvania Gazette*
Philadelphia, PA	William Richards		August 29, 1771	*Pennsylvania Gazette*
Newport, RI	Casey Family / Aaron Lopez		1750–1782	J. R. Marcus, *The American Colonial Jew*, 1970, p. 673
Newport, RI	Prince Updike / Aaron Lopez	Jewish	1750–1782	J. R. Marcus, *The American Colonial Jew*, 1970, p. 673
Providence, RI	Tillinghast and Holroyd		April 8, 1775	*Providence Gazette and Country Journal*
Providence, RI	Daniel Torres	Jewish	January 1, 1750	J. R. Marcus, *The American Colonial Jew*, 1970, p. 673
Providence, RI	Humphrey Palmer and William Wheat		July 9, 1774	*Providence Gazette and Country Journal*
Providence, RI	Obadiah Brown		January 1, 1752	M. Thompson, *Moses Brown*, 1962, p. 11–12
Charleston, SC	Angelo Santi		February 4, 1795	*City Gazette and Daily Advertiser*

the Old World and the New, it was new to both. Neither exclusively Spanish nor Aztec, chocolate was a Creole (i.e., American, in the larger sense) invention. It was available to a greater variety of people and at cheaper prices than in Europe. This was because the raw ingredient, cocoa, was available without the high transportation costs or import duties experienced by the Europeans. Furthermore, most chocolate in North America was mass produced for an American market on machines that were invented and built in America. There were no monopolies or exclusive manufacturing agreements in North America as there were in Europe. American chocolate manufacturers were concentrated in four major production centers: Boston, Newport (Rhode Island), New York City, and Philadelphia. Chocolate makers in these cities were the exclusive suppliers of the North American market. Practically no chocolate came from Britain or anywhere else. We also know that chocolate was accepted along with coffee and tea as a staple and was considered a necessity. While the recipes were ostensibly European, American chocolate practices reflected a different experience. After the American Revolution, the surviving Boston chocolate makers consolidated their dominance of the infant industry. Their surnames became iconic and these icons evolved into brands. Likewise, chocolate as a confection became more available as the confectioners trade was established in the cities.

Food of the Gods and Turmoil in the Caribbean

Everything chocolate begins with the seeds of cacao. Historical literature is expansive on the origins of cacao and its Mesoamerican roots. Humans have been exploiting the cacao fruit and its seeds for food for at least 3000 years. Carl Linnaeus, the 18th century scientist labeled it *Theobroma cacao—food of the gods*. Once fermented and dried, the seeds from the fruit are referred to as *cocoa*. Eighteenth century scientists called it the "chocolate nut tree" [5]. Cocoa beans were called "chocolate nuts" or "cocoa nuts" or simply "cocoa" until the 19th century.

Eighteenth century cocoa was typically referred to by its port of origin. Maricel Presilla has written:

> European and Latin [and North] American connoisseurs knew a lot about the origin of the chocolate. They did not generally use the names criollo or forestero, but they identified most cacao by the name of the area in which it was grown. Soconosco and Caracas beans were considered the cream de la cream. The Soconosco came from an area of Chiapas that used to send tribute to Aztec emperors; Caracas came from Venezuelan regions that sent cacao to the port of La Guaira, near the city of Caracas. Equally valued was Maracaibo cacao brought from the foothills of the Venezuelan Andes and the southern environs of Lake Maracaibo and shipped from Maracaibo port. [6]

American consumers were probably savvier about their chocolate in the 18th century than they are in the modern world. Colonial chocolate makers routinely advertised the geographic sources of their cocoa, much like modern coffee vendors do for their coffee beans. The physician R. Brookes wrote in 1730:

> The Kernels that come to us from the Coast of Caraqua, are more oily, and less bitter, than those of the French Islands, and in France and Spain they prefer them to these latter: But in Germany, and the North they have a quite opposite Taste. Several People mix that of Caraqua with that of the Islands, half in half, and pretend by this Mixture to make Chocolate better. I believe in the bottom, the difference of Chocolates is not considerable, since they are only obliged to increase or diminish the Proportion of Sugar, according as the Bitterness of the Kernels require it. [7]

Brookes continued:

> The Kernels of Caraqua are flattish, and for Bulk and Figure not unlike our large Beans [probably fava]. Those of St. Domingo, Jamaica, and Cuba are generally larger than those of the Antilloes. The more bulky the Kernals are, and better they have been nourished, the less Waste there is after they have been roasted and cleaned. [8]

Since most cacao was grown in the Spanish colonies, smuggling by the Americans, Dutch, English, and French was a major component of the supply network. With Curaçao, a mere 30 miles off the coast of Venezuela, the Dutch were so successful that Caracas cocoa was cheaper in Amsterdam than in Madrid [9]. Throughout the Colonial period, the French supplied their cocoa needs from the islands of Martinique and Guadeloupe, also augmented by smuggling Venezuelan cocoa [10]. Dutch interests were not confined to the West Indies. In 1625, to supplement their fur trading activities in the Hudson Valley, they established the colony of New Amsterdam on Manhattan Island with settlers from Curaçao and Surinam. This planted a seed for what became a flourishing trade between New York and the West Indies and the Spanish Main (see Appendix 1) even after the British took New Amsterdam and renamed it New York.

One circular from the British Council of Trade and Plantations concerning Curaçao states the following:

> From the English Plantations upon the mainland the inhabitants of Curaçao have all sort of provisions, as bread, flower, butter cheese pease [sic], rice, beef, pork and corn; from Pennsylvania and New York strong and small beer; from Carolina and New England pitch and tarr [sic]; from the Charibbee Islands [sic] and Jamaica rum, sugar, cotton, ginger, indigo, and tobacco; in return of which our plantations have chiefly cocoa, linens, muslins, silks and other goods for wearing apparel, with great quantity of riggings, sail, canvas, anchors, and other sorts of iron-work, powder and shott, which is never taken notice of by our men at war, when they meet with any sloop from thence. [11]

> Curaçao loads home for Holland in one year about 50 sail of ships, and most of them are richly laden, and a great part of their loading comes out of the English Plantations, chiefly sugar, cotton, tobacco, indigo and ginger; of their own produce by trade with the Spaniards, cocoa, hides, tobacco, logwood, stockfish wood and mony [sic]. [12]

Beginning in 1651, the English passed a series of "Navigation Acts" requiring that goods brought to England had to be transported aboard English ships, manned by English sailors, and no foreign (i.e., Dutch) vessels were to conduct coastal trade with England. The effect was to triple English shipping in the last half of the 17th century. The English intended to monopolize the shipping of the major bulk cargoes, such as tobacco, sugar, molasses, and cocoa. Marcus Rediker has written:

> With the passage of the Navigation Act of 1696 and the Board of Trade it established, the English state enlarged its role in the direction of commerce, seeking to tighten imperial controls by regulating and rationalizing trade. These efforts simultaneously helped to defeat the Dutch and to launch England's "Commercial Revolution." [13]

Ranking second only to London in shipbuilding, Boston produced ships at half the cost of London shipyards due to the abundance of high quality wood, and "built more ships than the rest of the British colonies combined." By the 1680s, New England vessels represented about half the ships that served the English Caribbean [14]. This New England dominance continued throughout the 18th century with significant impact on where cocoa was delivered and where chocolate was produced. In 1760, a Virginia merchant seeking to export wheat could not find a single Chesapeake sea captain familiar with the New England coast in the winter [15].

Beginning in 1689 and continuing for eight decades, the British and French fought over control of the sea lanes, colonial territory, and predominance in Europe. The British built a navy equal to the size of both the French and Spanish navies combined [16]. The end of the Seven Years War in 1763 resulted in British control of Quebec, the Ohio Valley, and Spanish Florida and acquisition of frontier islands in the Lesser Antilles. The cocoa growing islands of Granada, St. Vincent, Tobago, and Dominica were now in British hands. These islands gave the British new colonies nearly as close to the coast of Venezuela as Dutch Curaçao. As a result, the American chocolate experience changed as cocoa became available from a variety of sources, both legal and illegal.

Because of high transportation costs and excessive import duties on cocoa, European chocolate was both expensive and exclusive. It was a beverage for the elite and demand was relatively low. In 1776, there was only one British chocolate maker [17]. In order to recapture some of the costs of the war, Parliament passed the 1767 Townsend Duties. While most students of American history might be familiar with this law because of an annoying tax on tea, the Act also included "drawbacks" of importation duties for cocoa if re-exported [18]. Parliament also hoped to stimulate production of West Indian cocoa for export to Britain. If that cocoa were re-exported from Britain, then the heavy import duties would be returned (i.e., a "drawback") thus making it more profitable for *both* importer and exporter. The British export target was Spain, the main market for cocoa in Europe [19]. In 1728, a London court sentenced John Moor (alias Holland) to transportation to America instead of hanging, for the theft of one pound of chocolate (see Chapter 20) [20].

In North America, by contrast, chocolate was more available at cheaper prices and consumed by a wider variety of people. The quantity of domestically produced chocolate was sufficient enough to give it away to the poor. The Almshouses of Philadelphia and New York regularly provided chocolate and sugar to its needy residents, something that did not happen in England for the fear of indulging the poor [21]. Like any commodity, however, its quality was reflected by price. In 1744, Dr. Alexander Hamilton, traveling in Maryland, noted in his diary:

I breakfasted upon some dirty chocolate, but the best that the house could afford. [22]

When discussing the breakfasts of the "lower and middling classes," the writer of *Smyth's Travels in Virginia, In 1773*, noted:

A man . . . breakfasts about ten o'clock, on cold turkey, cold meat, fried hominy, toast and cider, ham, bread and butter, tea, coffee, or chocolate, which last, however, is seldom tasted but by the women. [23]

American chocolate makers routinely advertised chocolate for sale in newspapers throughout the 18th century. Approximately 70 commercial chocolate makers have been identified from these sources (see Table 23.1). Additionally, probate inventories and purchasing accounts listed chocolate making equipment and quantities of chocolate nuts for personal use in some elite households. In 1776, Joseph Fry was the only British chocolate manufacturer with 56 agents throughout the island [24]. Not so in North America.

American Production and Manufacturing

American chocolate manufacturers were concentrated in four major production centers: Boston, Philadelphia, New York City, and Newport (Rhode Island). Since these locations regularly were engaged in the trade with the West Indies, it is logical that the domestic chocolate production also occurred here. British customs records for the five year period from January 1768 to January 1773 are the most complete accounting of the American import and export activity for 40 ports along the Atlantic coast. Called "Ledger of Imports and Exports (America); January 5th, 1768 to January 5th, 1773," or "PRO/Customs 16/1," these records also include Newfoundland, Bermuda, the Bahamas, and Florida (Tables 23.2 and 23.3).

From these data, it is clear that there were four distinct import commodities. Cocoa was differentiated by its origin whether "British," "Foreign," or "Coastwise." Since Great Britain was not listed as a source for the cocoa, the "British" designation refers to cocoa from the British possessions in the West Indies. Likewise, the "Foreign" designation refers to cocoa from any other West Indian or mainland source. "Coastwise" obviously refers to Atlantic ports. The other distinct commodity is the chocolate itself, which was exclusively North American. Examination of Tables 23.2 and 23.3 reveals that Americans relied heavily on "Foreign" cocoa to sustain their chocolate production. Nearly 70 percent of the imported cocoa was from these sources, whereas only 21 percent was British. American "Coastwise" traffic comprised the remaining 10 percent of imports. However, since this cocoa was actually trans-shipped as opposed to domestically produced, the actual percentage of Foreign and British

Table 23.2 American Imports and Exports: January 5, 1768 to January 5, 1773

Port	Imports				Exports						
	British Cocoa (lb)	Foreign Cocoa (lb)	Coastwise Chocolate (lb)	Coastwise Cocoa (lb)	To Southern Parts of Europe and Wine Islands Chocolate (lb)	Coastwise Chocolate (lb)	Coastwise Cocoa (lb)	To Britain Foreign Cocoa (lb)	To Africa Chocolate (lb)	To Foreign and British West Indies Chocolate (lb)	To Ireland Foreign Cocoa (lb)
Newfoundland	0	0	115,861	100	0	1,190	0	0	0	0	
St. John's Island	0	0	950	0	0	0	0	0	0	0	
Quebec	330	0	31,072	0	0	710	30	0	0	0	
Halifax	0	0	28,712	300	0	1,500	100	0	0	0	
Piscataqua	58,166	130,465	5,509	0	0	21,816	33,074	0	0	0	
Falmouth	24,344	103,671	990	0	0	2,051	0	183	0	250	
Salem/Marblehead	35,605	430,523	1,677	2,036	352	70,976	46,794	0	0	1,550	
Boston	140,536	357,757	13,589	47,672	4,210	220,968	32,208	0	0	1,840	
Rhode Island	15,906	246,104	11,223	10,942	300	23,270	18,099	0	800	50	
New Haven	4,050	177,687	972	800	0	200	51,107	8,492	0	0	
New London	11,156	158,785	28,393	392	0	2,900	15,467	1,976	50	0	
New York	70,409	206,322	2,830	121,159	12,856	31,918	5,612	150	900	120	
Salem	0	7,500	0	0	0	0	0	0	0	0	
Perth Amboy	17,400	6,200	400	0	0	400	0	2,500	0	0	
Philadelphia	243,229	242,996	2,626	113,716	0	73,528	26,676	400	0	1,920	1,000
New Castle	0	0	0	0	0	50	0	0	0	0	
Lewis (Lewes?)	0	0	1,300	0	0	1,300	0	0	0	0	

Pocomoke	0	0	3,562	0	0	210	0	0	0	0
Patuxent	0	0	22,624	100	0	1,010	0	0	0	0
Chester	0	0	970	0	0	0	0	0	0	0
Northern Potomack	0	0	14,420	0	0	606	0	0	0	0
Accomack	0	0	3,556	0	0	180	0	0	0	0
Southern Potomack	0	0	6,130	0	0	0	0	0	0	0
Rappahannock	0	0	8,303	0	0	395	0	0	0	0
York River	201	173	7,905	0	0	350	0	0	0	0
Lower James River	22,730	0	26,037	0	0	270	12,000	0	0	0
Upper James River	0	0	11,978	0	0	4,000	0	0	200	0
Currituck	0	0	1,390	0	0	0	1,800	0	0	0
Roanoke	647	0	18,423	0	0	750	0	0	0	0
Bathtown	0	530	4,995	0	0	250	0	0	0	0
Beaufort	4,000	100	3,104	400	0	225	2,695	0	0	0
Brunswick	600	2,344	4,885	0	0	0	1,200	0	0	0
Wynyaw	0	0	4,326	0	0	601	0	0	0	0
Port Royal	0	0	253	0	0	0	0	0	0	0
Charlestown	10,751	50	9,237	0	0	1,023	7,150	6,743	0	0
Sanburg	0	0	230	0	0	0	0	0	0	0
Savannah	300	0	8,215	0	0	0	3,200	0	0	150
St. Augustine	0	0	810	0	0	100	0	0	0	0
Pensacola	0	0	860	0	0	0	0	0	0	0
Bahamas	0	0	1,153	0	0	0	0	0	0	0
Bermuda	250	0	100	0	0	0	0	0	0	0

Table 23.3 Ledger of Cocoa and Chocolate Imports and Exports (American): January 5, 1768 to January 5, 1773

| | Total Imports | | | | Total Exports | | | | | | |
| | British Foreign Cocoa (lb) | Cocoa (lb) | Coastwise Chocolate (lb) | Coastwise Cocoa (lb) | To Southern Parts of Europe and the Wine Islands Chocolate (lb) | Coastwise Chocolate (lb) | Coastwise Cocoa | To Britain Foreign Cocoa (lb) | To Africa Chocolate (lb) | To Foreign and British West Indies Chocolate (lb) | To Ireland Foreign Cocoa (lb) |
Port											
Canada	330	0	176,595	400	0	3,400	130	0	0	0	0
New England	289,763	1,604,992	62,353	61,842	4,862	342,181	196,749	10,651	850	3,690	0
New York/East New Jersey	87,809	220,022	3,230	121,159	12,856	32,318	5,612	2,650	900	120	1,000
Delaware Valley	243,229	242,996	3,926	113,716	0	74,878	26,676	400	0	1,920	0
Chesapeake	22,931	173	105,485	100	0	7,021	12,000	0	200	0	0
Carolinas	15,998	3,024	46,613	400	0	2,849	12,845	6,743	0	150	0
Georgia/Florida	300	0	10,115	0	0	100	3,200	0	0	0	0
Bermuda/Bahamas	250	0	1,253	0	0	0	0	0	0	0	0
Total all ports	*660,610*	*2,071,207*	*409,570*	*297,617*	*17,718*	*462,747*	*257,212*	*20,444*	*1,950*	*5,880*	*1,000*

cocoa imports was 76 and 24 percent, respectively. New England, the greatest producing area, imported over 81 percent of the cocoa from non-British sources; a little over half of the cocoa used in New York was from these sources; and 40 percent was imported into Philadelphia. Both New York and Philadelphia imported 28 and 20 percent, respectively, from "Coastwise" sources, presumably from New England. The southern colonies were almost exclusively importers of North American chocolate with only a minor role in the cocoa trade. It is interesting that the cocoa that found its way to the Chesapeake, the Carolinas, and Georgia was almost exclusively from British sources.

NEW ENGLAND

New England imported nearly two million of the three million pounds of British and "Foreign" cocoa, and exported 74 percent of the chocolate produced in North America to its North American neighbors (Table 23.2). During the five year period covered in the tables, commercial chocolate makers advertised in Boston, New Haven, and Newport, Rhode Island, newspapers. Additionally, there was a rich tradition of commercial chocolate making in New England that covered the entire 18th century. At least 36 18th century New England chocolate makers may be identified (see Table 23.1). While most worked in and around Boston, there also was chocolate manufacturing activity in Newport and Providence (Rhode Island) and New Haven, Connecticut.

On November 3, 1718, Samuel Payton from New York guided the *Royal Prince* into Boston Harbor carrying "eighteen bags of cocoa and chocolate, one bag of cotton, twenty small casks of molasses along with four casks of pork, salt and skins" [25]. The Boston customs official who recorded this routine entry transcribed information into the log that may be the oldest surviving record of "Coastwise" chocolate exchanged between North American colonies. Colonial shipping records were not contiguous and have gaping holes in coverage. Boston coffeehouse owners had been banking their survival on cocoa imports and satisfying chocolate consumers for half a century before Payton's arrival. Colonial North American chocolate history has "New England" stamped all over the industry.

Rhode Island, because of its trading relationships with the West Indies, was also important in the cocoa trade and chocolate production. Two traders were prominent in the 18th century: Obadiah Brown from Providence and Aaron Lopez from Newport. Obadiah Brown, of the Rhode Island merchant family, had a variety of interests including the African slave trade, West Indian trade (legal and illegal), and spermaceti candlemaking. Brown University is named after the family, of which Obadiah was the patriarch. He owned a watermill that made chocolate [26]. This watermill might have been the first of its kind in North

America although there were others. Aaron Lopez arrived in Newport in 1750 to join a Jewish community that had been present for almost a century. Lopez became the "merchant prince" of New England, owning 30 of the 130 Newport ships engaged in the West Indies trade [27]. His interests included African slave trading, whaling, West Indian and European trading, and chocolate manufacture. Jacob Marcus has written of Aaron Lopez:

> [He] saw food-processing as ancillary to his involvement in the coastal and West Indian traffic. Since provisions were the prime staples sent by the New Englanders to the West Indies, it was with the islands in mind that Lopez contracted with various processors for thousands of pounds of cheese, while the chocolate he secured through outwork was destined for local and North American consumption. . . . In Newport relatively large quantities of chocolate were prepared for Lopez by Negroes whose Jewish masters may have taught them the art of making the confection. Prince Updike, one of Lopez' Negro workers, ground thousands of pounds for the Newport merchant. He received five shillings for every pound he prepared, and one batch of cocoa which he turned into chocolate weighed over 5000 pounds. Another of the Negro craftsmen, a member of the cocoa-grinding Casey family, was so useful that when the man was thrown in jail for drunkenness, Lopez paid his fine and put him to work. [28]

PHILADELPHIA

After Boston, Philadelphia was the second-largest chocolate manufacturing center for North America. Examination of Table 23.1 reveals the names of 24 chocolate makers, and most of these advertised in Benjamin Franklin's *Pennsylvania Gazette*. Franklin himself advertised chocolate for sale in his print shop [29]. From 1768 to 1773, the Delaware Valley imported nearly 600,000 pounds of cocoa, but only shipped a little less than 75,000 pounds of chocolate by sea to other colonies in the flourishing "coastwise" traffic. In the same period, the region received only 3926 pounds of chocolate in the intercolonial trade, either from New England or New York (see Table 23.3). Clearly, Philadelphia chocolate makers produced most of their chocolate for the local population and sent the rest overland.

One of the pitfalls in suggesting that coastwise shipping was the only means of conveyance between colonies is that it disregards overland and river traffic. Arthur Middleton has written:

> There was considerable trade at the head of the Chesapeake Bay between Maryland and Pennsylvania, and at the southern end of the Bay between Virginia and North Carolina. The trade between Pennsylvania which sprang up late in the 17th century consisted of foodstuffs (flour and bread), West Indian commodities, and horses in return for cash, bills of exchange, and European goods. Overland transportation, which accounted for most of this trade, took place across eight

different portages ranging from five to thirteen miles in length between the headwaters of the Eastern Shore tributaries of Chesapeake and Delaware Bays. [30]

Likewise, New York overland and river traffic could reach New Jersey and Connecticut and vice versa. In 1772, New London, Connecticut, received the most cocoa of any seaport in North America. In that year, colonial customs officials recorded 103,367 pounds of "foreign cocoa" coming off the decks of arriving ships. This amount was more than the quantity received by New York and Rhode Island combined [31]. Exporting very little chocolate or cocoa, New London probably was a trading port for larger cities whose goods were delivered overland or by small watercraft. Likewise, New Haven, Connecticut, was also an active cocoa trading port. From January 1768 to January 1773, New Haven imported 182,537 pounds of cocoa and exported 59,599 pounds of cocoa, but only 200 pounds of chocolate. Boston, Salem, Providence, and Newport were all within a relatively easy distance overland. Records from one Massachusetts chocolate maker show expenses paid for cocoa carted overland from Providence and Newport, and other carting fees [32].

NEW YORK

When the British took New Amsterdam from the Dutch in 1664, they found the most ethnically diverse, religiously tolerant colony in North America. Besides the Dutch, there were English Puritans, Portuguese Jews from Brazil and the West Indies, French Huguenots, Germans, Flemings and Walloons from Belgium, Native Americans, and African slaves. Nearly half the white colonists were not Dutch [33]. Twenty years earlier, a French priest claimed to have heard 18 different languages spoken in New Amsterdam. By 1686, New York both imported and exported Venezuelan cocoa. Commercial chocolate making developed to a scale to allow chocolate exports by the early 18th century. New York's chocolate makers were heavily concentrated in the Jewish community. Jacob Marcus has written:

Among the industries in which colonial Jewish businessmen interested themselves was food processing. Jewish shopkeepers specialized in cocoa and chocolate which they secured in large quantities from their coreligionists in Curaçao. [34]

The surviving records show New York shipped chocolate by water in 1718 [35]. Active in the cocoa and chocolate trade throughout the period, New York imported nearly 430,000 pounds of cocoa from 1768 to 1773. Only 20 percent came from the British West Indies. In that period, New York shipped 32,318 pounds of chocolate to other North American ports along with 900 pounds to Africa.[1] It imported only 3230 pounds of chocolate, an indication that New York manufacturers were able to satisfy local demand.

CHOCOLATE EXPORTS

It is clear that both British and "foreign" cocoa shipments were routinely received and made into chocolate, then primarily exported to other American ports in the "coastwise" traffic. There were only modest transfers of cocoa to Britain and no shipments of chocolate to Britain. By contrast, chocolate was exported to Africa and southern Europe, presumably for use in the slave and wine trades. The most curious data from these records are the extensive chocolate shipments to Newfoundland. Not much is written about New England's trade with Newfoundland, so other reasons why there would have been so much chocolate exported there remains a mystery. A New England merchant's interest in Newfoundland primarily concerned whaling and cod fishing. There was a small British garrison on St. Johns to protect British fishing interests and Newfoundland's administration fell under the British Admiralty [36]. A hot cup of chocolate might have been the stimulant of choice among the fishermen and British sailors on the Grand Banks. But the British Admiralty was not known for concerning themselves with the creature comforts of their sailors until later in history. One possibility was that the chocolate was smuggled to Europe, with Newfoundland being a place of exchange. British writers complained of illegal transactions between Americans and the French along the coasts of Newfoundland and Labrador. In 1765, British Commodore Palliser was dispatched with a squadron of small "ships of war" to Newfoundland to disrupt smuggling between the Americans and the French [37]. Did Americans trade chocolate for French brandy? Was some of the chocolate consumed in Europe made in America? In 1776, Joseph Fry, England's only commercial chocolate maker, complained to British customs authorities about "trading vessels from Ireland" bringing "very large quantities of Chocolate, which is a quality equal to much that is made in England." He also noted that for "about two Years past, Smuggling is vastly increased . . . by a desperate gang of Villains" [38].

Where was the chocolate smuggled by these "villains" manufactured? Who besides the Dutch and the Spanish had excesses of cocoa and chocolate? American chocolate illegally smuggled to England with Newfoundland as the place of exchange is an intriguing possibility, although this idea remains to be confirmed. Smugglers who were captured ultimately wound up in the Admiralty courts and eventually in the history books. So who knows? Nevertheless, in the five year period from 1768 to 1773, Newfoundland imported more chocolate by sea than the Chesapeake (see Table 23.2).

THE CHESAPEAKE, CAROLINAS, AND GEORGIA

Whereas the mid-Atlantic and New England colonies were producers and exporters of chocolate, Virginia, Maryland, and the other southern colonies were primarily consumers. One of the earliest accounts of chocolate consumed in Virginia comes from Governor Francis Nicholson. Nicholson, the city planner for two colonial capitals, Annapolis and Williamsburg, was never popular with his early 18th century contemporaries. Although in Williamsburg today two streets are named for him (Francis Street and Nicholson Street), in his time he had to confront his opponents in person and in writing. Among other things, Nicholson was accused of not being a gentleman: the accusation was that he only served one meat for dinner. In 1705, Nicholson, complaining about his chief rival Reverend James Blair, who founded the College of William and Mary, wrote:

> Mr. President Blair used . . . to invite . . . the Burgesses . . . to his lodgings in the Colledge [sic] to drink chocolate in the morning, and maybe sometimes in the afternoon a glass of wine; and this I think he used [to] for about 2 years. [39]

Bottom line: Blair served chocolate for political purposes while Nicholson did not serve enough meat . . . match point to Blair.

A few years later on February 16, 1709, Councilor William Byrd wrote:

> I rose at 6 o'clock this morning and read a chapter in Hebrew and 200 verses in Homer's Odyssey. I said my [prayers] and ate chocolate for breakfast with Mr. Isham Randolph, who went away immediately after. [40]

Certainly not every woman or man who could afford chocolate for breakfast chose to drink it. Some wanted the stimulating effects of theobromine, cocoa's main alkaloid, but not the weighty feeling from the fat. Their alternative was a hot beverage made from steeping cocoa shells (the membrane from the roasted cocoa beans) in hot water. The result is an infusion similar in color, flavor, and bitterness to coffee. When sweetened, more chocolate flavors emerge. Advertisements for chocolate makers routinely offered cocoa shells for sale. While this might seem like a chocolate substitute for the poor or "lower sort," this was not the case, as it was a known product to the very wealthy. Martha Washington apparently enjoyed this beverage. In 1789, President George Washington noted in a letter to his agent:

> She will . . . thank you to get 20 lb. of the shells of Cocoa nuts, if they can be had of the Chocolate makers. [41]

In 1794, George Washington received a letter from his cousin that also referred to chocolate shells:

> I wd. [sic] take the liberty of requesting you'll be so good as to procure and send me 2 or 3 Bush[els]: of the Chocolate Shells such as we're [sic] frequently drank Chocolate of at Mt.

Vernon, as my Wife thinks it agreed with her better than any other Breakfast. [42]

Chocolate was also eaten in puddings, creams, and as ice cream among the very wealthy. The first cookbook printed in North America was a copy of Eliza Smith's *The Compleat Housewife*. The printer, William Parks of Williamsburg, Virginia, culled Smith's 1732 British cookbook and in 1742 published the recipes that he felt suitable for Virginia dining tables [43]. The only chocolate recipe that Parks printed was one called "chocolate almonds." The recipe called for scraped chocolate, sugar, gum tragacanth as a binder, and orange flower water. These would be shaped "into what form you please" and allowed to dry [44]. Served as dessert, chocolate almonds were part of Virginia cuisine from the beginning of the 18th century. Published cookbooks were too expensive to be in a smoky 18th century kitchen. Grease smudges were the enemies of paper. Instead, housewives copied their special recipes into what became known as "commonplace" books. The common, everyday recipes were not transcribed, as a housewife knew them by memory. Her "commonplace" book was a gift passed to a daughter and on to a future granddaughter that she might not ever see, given that it was not uncommon for women to die during childbirth. Commonplace books were a "compilation of recipes from family and friends."

One of the earliest manuscript cookbooks—or commonplace book—in North America that included a chocolate recipe was handwritten by an anonymous Virginia mistress dated to ca.1700. Katherine Harbury has described this text in detail and offered insights into the writer's person and social status:

> Anonymous (1700) appears to have been a well-educated woman of respectable standing who interacted with various members of the Randolph family, among others. She may well have been a Randolph or connected to them by marriage or social network. [45]

Although the manuscript binding dates to the late 17th century, most of the recipes were copied from British cookbooks that appeared in 1705–1714 [46]. Anonymous wrote:

> To Make Chocolate Almonds: Take your Sugar & beat it & Serch [sift] it then: great youre [sic] Chocolatt [sic]: take to 1 lb Sugar: 5 oz. of Chocolatt mix well together put in 2 Spoonfull Gumdragon [sic] Soaked in rosewater & a grn musk & ambergrease & beat all well together in mortar rowl [sic] out and mark wt: ye molds & lay on tin plats [sic] to dry turn everyday. [47]

Another unsigned Virginia manuscript dated to 1744 described "Almonds in Chocolat," a recipe that might be the 18th century ancestor of the red, yellow, blue, and green M&M™:

> Almonds in Chocolat: Take 3 quarters of a pound of Sugar and half a pound of Chocolat and make it a high Candy then put in two pound of right Jordan Almonds and keep them

*Sturing tell they are almost Cold then lay them out to dry on
Sives and coulier [color] them thus for the couller Red
Scutcheneel [cochineal] for Yellow Termermick o[r] Saffron
for blew Stone blew for Green the Juice of Spinage and steep
your Gum in ye Juice of your Green.* [48]

Thirty-one percent of 325 Chesapeake probate inventories compiled by Gunston Hall Plantation yielded chocolate-related items including chocolate pots, cups, bowls, graders, mullers, and stones [49]. In pre-Revolutionary Williamsburg, chocolate imported from New England, Philadelphia, and New York retailed for 2 shillings 6 pence per pound (Virginia currency). Coffee was less expensive at 1s.6d. per pound. Tea, by comparison cost 12–15s per pound. A free unskilled laborer or a sailor earning approximately 2 shillings a day might not have tasted chocolate very often, if at all. Likewise, the amount of money spent for a pound of chocolate could have also purchased 15 pounds of salted fish, so a slave's chance of tasting it was even less [50].

While Williamsburg merchants sold chocolate in their stores, there were no commercial chocolate makers who advertised in *The Virginia Gazette*. Commercial chocolate making required an economic infrastructure, including cocoa merchants, experienced chocolate makers in the labor supply, skilled metal workers to make or repair equipment, printers, paper makers who made wrapping paper, coffeehouses and other outlets, and a steady flow of retail and wholesale consumers such as in large cities. Warm weather would have made chocolate manufacturing difficult on a commercial scale. In 1766, *The Virginia Gazette* reported that manufacturing had begun in Savannah, Georgia, to export chocolate "from cocoa of their own produce" [51]. This venture was probably an experiment that failed. In the five year period from 1768 to 1773, only 150 pounds of chocolate was exported from Savannah (see Table 23.2 and Chapter 52).

South Carolina, on the other hand, imported cocoa but very little chocolate throughout the Colonial period. While it is tempting to imagine a commercial chocolate maker using a large mill in wealthy Charleston, the tropical weather would have made this a heroic effort at best. Another hurdle was lack of free customers to purchase the chocolate. The 1790 Federal Census shows Charleston County with a population of nearly 67,000 people. Of these, 50,000 were enslaved [52]. However, the possibility exists that a confectioner might have made chocolate on consignment but if so, it was probably hand-ground. However, even with stones, chocolate could be produced in excess of local needs. In 1764, a quantity of 700 pounds of chocolate was exported from South Carolina to Gosport, England (in Hampshire opposite Portsmouth) [53]. In 1772, only two of 19 seaports south of the Mason–Dixon Line imported any cocoa: one in Virginia (Lower James River) and the other in Charleston, South Carolina.

Their combined total was slightly over 10,400 pounds of cocoa. Those same 19 seaports imported nearly 52,000 pounds of chocolate from New England, New York, and Philadelphia. In the same year, 200 pounds of chocolate was (re)exported from the Lower James River to West Africa for barter in the slave trade [54]. The chocolate that was made in Virginia and perhaps other southern provinces was made by black hands on chocolate stones or mortars and pestles.

MAKING OF AMERICAN CHOCOLATE

While there is debate about the religious motivations of American colonists, the founding of both Pennsylvania and New England was distinctively religious in character. Likewise, Jewish chocolate makers and traders in New York and Newport, Rhode Island, were refugees from the Spanish and Portuguese Inquisition. Many of the American chocolate makers and cocoa traders were tied by family and religion. Religious affiliations of those involved in the chocolate trade also can be identified (see Table 23.1). It is no surprise that the Quakers in Philadelphia and Jews in New York were prominent. Likewise, it would be logical that most New England chocolate makers were Protestant. Chocolate, like coffee and tea, was synonymous with middling sort industriousness in Protestant Europe and North America. Like all nonalcoholic stimulants, all three were consumed at the beginning of the day to energize the inner spirits for the world of work. In Catholic Europe, by contrast, chocolate was associated with aristocratic decadence. Portraiture from the period often showed half-clad aristocratic women sitting in bed being served chocolate for breakfast or prior to a late-night seduction [55]. In 1774, Pope Clement XIV was murdered. The turmoil in the Catholic countries was because of the dissolution of the Jesuit Order, a major player in the cocoa trade. The initial rumor in Britain was that he was poisoned with a cup of chocolate, although the *Virginia Gazette* reported that he was poisoned with the sacrament [56].

Since New England was the leading chocolate-producing area in the Colonial period, one of the earliest reports of chocolate actually consumed in North America was from a Massachusetts minister. On October 20, 1697, Reverend Samuel Sewall wrote:

*I wait on the Lieut. Governor at Dorchester, and there meet
with Mr. Torry, breakfast together on Venison and
Chockalatte: I said Massachutset [sic] and Mexico met at his
Honour's Table.* [57]

Reverend Sewall also made the rounds distributing chocolate to a sick Samuel Whiting: "I gave him 2 Balls of Chockalett [sic] and a pound of figs" [58]. His chocolate also came with a price. On March 31, 1707, Sewall wrote: "Visited Mr. Gibbs, presented him with a pound a Chockalett [sic], and 3 of Cousin Moody's sermons, gave one to Mrs. Bond, who came in while I was there" [59]. From this last entry, one cannot tell if Mrs. Bond received "chockalett" or a sermon. On another visit, Sewall gave "two half pounds of Chocka-lat [sic], instead of Commencement Cake; and a Thesis" [60]. He does not mention the chocolate maker or how much it cost, but clearly, he thought of it as special, at least as special as sermons and a thesis.

Reverend Sewall also noted on one occasion that he drank "warm chockelat [sic] and no Beer; find my self much refresh'd by it after great Sweating to day, and yesterday" [61]. He might have also meant that his spirit was stimulated to do more good work ... a Protestant ethic. The reverend would have found agreement with his Quaker and Jewish counter-parts. Chocolate was nonalcoholic, stimulating, whole-some, nourishing, and even medicinal. All three religions stressed moderation of spirits. While neither the Quakers nor the Puritans were teetotalers, they punished drunkenness, the Quakers even more severely than the Puritans [62]. Quaker and Jewish traditions have purity as one of their tenets while Protestantism stressed doing good deeds with the emphasis on the "doing."

When it came to ready-made foodstuffs, 18th century consumers were naturally suspicious about adulteration (see Chapter 47). There were no laws regulating purity of foodstuffs so it was buyer beware when it came to putting something into your mouth. Trust was the coin of the realm when it came to chocolate or other purchased foods. Once sick, a doctor's cure might have killed faster than the disease. Jewish butchers were renowned for their care taken in slaughter and cleanliness [63]. Quakers were known for their integrity. Philadelphia chocolate makers were Friends, friendly, or just plain neighbors as much as manufacturers. Many chocolate makers stamped or printed their names on the wrapped packages of choc-olate, emphasized purity in their advertisements, and offered money back guarantees. Sometimes shopkeep-ers like Boston merchant Joseph Barrell would do the same thing. In February 1773, his advertisement read that his chocolate was not "'Hannon's much approv'd,' 'Palmer's superior,' nor 'made by Deacon Davis,'" never-theless it was "warranted pure, and at least, equal to either" [64]. The significance of this advertisement is that the chocolate maker's name is associated with the product. "Hannon" referred to John Hannon and "Palmer" referred to Joseph Palmer. Of the two, Joseph Palmer had advertised chocolate for at least 20 years prior to Barrell's notice [65]. John Hannon's name is famous in American chocolate lore because of his association with Dr. James Baker. Baker was the patriarch who began dabbling in chocolate in the 1770s and whose son and grandson greatly expanded the chocolate company and the Baker brand in the 19th century [66]. Baker's Chocolate remains one of the oldest brand names on American grocery shelves to this day. Whether or not "Deacon Davis" was really a deacon is unknown. But the name itself certainly denotes images of trust and faith in the manufacturer.

Another feature of American chocolate was that it was primarily machine-made and purchased in stores. Chocolate histories written from a European perspective generally ignore American manufacturing methods. American newspaper advertisements, however, provide insight regarding chocolate-making equipment and the chocolate makers themselves. Since there were no monopolies or manufacturing guilds, there were no barriers to entry into the chocolate trade other than capital formation and access to cocoa. American manufacturing equipment was generally homemade and varied from foot-powered mills capable of producing small quantities to watermills capable of producing several thousand pounds a day. Likewise, there were no patent restrictions or monopolies on the types of mills, so there were many different methods of production. Manufacturers used some sort of rotary machine, powered by horse, water, or human feet. Some chocolate makers also produced other commodi-ties at the same time. The cocoa trade was tenuous, at best, especially during wartime. Chocolate makers could ill afford disruption in a steady supply of cocoa unless they were diversified into other commodities. Besides chocolate, chocolate makers commonly ground coffee, oats, spices, mustard, and even tobacco [67]. One chocolate maker advertised a horse-powered mill that could grind chocolate and tobacco at the same time [68].

Chocolate making was hard work. The labor was at times intense and at other times tedious. Often it was both. Whether roasting and shelling hundreds of pounds of cocoa at a time, or walking on a treadmill for hours, or hand-grinding ten pounds of chocolate a day for the Master, the work was mind numbing. And those working in large watermills also had their trials. If the order was for a ton of chocolate for a ship sailing on the next high tide, then well over a ton of cocoa would have had to have been manhandled onto carts, roasted, shelled, winnowed, taken to the hopper, ground up, mixed and molded, wrapped in paper, packaged into perhaps 50 pound boxes, and loaded onto carts. This in an age where most of that labor would have been by hand, sun up to sun down. Choco-late generally was not manufactured in the summer because higher temperatures did not allow the choco-late to harden. It could stay "wet" for days. Therefore, chocolate-making activities started in the fall and ended in late spring. New England chocolate workers also had to contend with freezing temperatures inside the mill as well as the difficulties of moving heavy wooden containers over ice or snow.

To my knowledge, American 18th century chocolate has not be identified in archeological excava-tions: it is anyone's guess what it tasted like. We may surmise, however, that with the variety of manufacturing methods, the different locations in which it was made, and the year-to-year variations in the quality, the cocoa yielded results that were highly variable and all over the map. Water mills, because of complex and multiple gearing systems, simultaneously could grind chocolate, pound rags to make paper, full or finish wool, and grind tobacco into snuff, and wheat or mustard into flour. Some chocolate mills were horse powered. In other words, the environment in which chocolate makers performed their work probably affected the flavors of their product.[2] We also can surmise that 18th century chocolate was not wrapped in cellophane or tin foil. Lacking a vapor barrier, the chocolate, therefore, would eventually absorb and incorporate flavors or odors from whatever items were stored near it (perhaps even dried fish!). This explains why consumers might have preferred locally made chocolate over that imported from another colony. Like other food purchases in the face-to-face society of the 18th century, trusting the chocolate meant knowing the producer. Simply stated, the product might have tasted better than the one packed in pine boxes made of green wood shipped over water in leaking vessels that also carried salted cod. This might explain why some recipes used chocolate sparingly while greatly increasing the quantity of sugar. This also could explain why some of the colonial elite had their chocolate ground and prepared by hand.

The spices added are a feature of hand-ground chocolate. The degree of spiciness was purely a matter of personal preference of whoever wanted the chocolate ground in the first place. It distinguished hand-ground from machine-manufactured chocolate that was more likely than not unsweetened and unspiced. The vast majority of 18th century chocolate recipes called for sugar to be added. Once the whole mass was spiced and sweetened, the chocolate was removed from the stone, put into tin molds, and left to dry. Brooks wrote of the process:

> Now the Kernels being sufficiently rubb'd and ground upon the Stone . . . if you would compleat [sic]the Composition in the Mass, there is nothing more to be done, than to add to this Paste a Powder sifted thro [sic] a fine Searce [sieve], composed of Sugar, Cinnamon, and, if it be desired, of Vanillas . . . mix it well upon the Stone, the better to blend it and incorporate it together, and then fashion it in Moulds made of Tin in the form of Lozenges of about 4 ounces each, of desired, half a pound. [69]

> The Spaniards taught by the Mexicans, and convinced by their own Experience, that this Drink, as rustick [sic] as it appeared to them, nevertheless yielded very wholesome Nourishment; try'd to make it more agreeable by the Addition of Sugar, some Oriental Spices, and Things. . . . Because there are an infinite Variety of Tastes, and everyone expects that we should have regard to his, and one Person is for

> adding what the other rejects. Besides, when it is agreed upon what things to put in, it not possible to hit upon Proportions that will be universally approved. [70]

Some of the African or Oriental spices called for in 18th century recipes include anise, cardamom, cinnamon, cloves, and nutmeg. Additionally, some recipes called for ambergris, ground almonds or pistachios, musk, or orange-flower (blossom) water. The New World additions found in published recipes include achiote, chili peppers, and vanilla (see Chapter 8). British cookbooks were sold in stores and most of these ingredients were available during the Colonial period, with vanilla being the possible exception. Various authors made comments, derogatory or complimentary, about the preferences of the English, French, and Spanish [71]. As usual, North American preferences were ignored, so researchers can only speculate what might have been preferred by the colonists. Some clues, however, exist, as at least one Boston chocolate maker offered "Italian Chocolate" for sale [72].

REVOLUTION

The effect of the Townshend Duties of 1767 on American food practices was electric. The Act made tea politically incorrect while stimulating American demand for coffee and chocolate. In 1767, consumption of chocolate became a patriotic act. Between 1737 and 1775, *The Virginia Gazette* announced chocolate-laden ships arriving in Virginia 54 times. Of those, 41 arrived between 1767 and 1775. In this same period, four shipments of cocoa also arrived [73]. Clearly, the duties on tea got the colonists' attention and probably stimulated chocolate production over what it had been previously. So it was that coffee and chocolate replaced tea in the parlors and coffeehouses of North America. One can imagine the women of North America happier, too, since both coffee and chocolate were cheaper than tea and household food budgets were given some relief. One shudders to think of the pain inflicted by the women on the men of Boston had they instead thrown 342 boxes of chocolate into the harbor instead of tea on that night in December 1773!

If women determined the degree of chocolate consumption in peacetime, soldiers and sailors were the major determinants in war. Since the time of the Aztecs up to present day, warriors from the New World have carried chocolate in their pockets or on their backs. Light, nutritious, and highly caloric, chocolate was and has remained the perfect travel food. Whether Spanish conquistador, French Canadian, British, or American, soldiers of different nations and times made room for chocolate. Benjamin Franklin organized shipments of provisions, including chocolate, to a Pennsylvania regiment with General Braddock's army during the French and Indian War [74]. One war earlier, in 1747, an American officer at

Saratoga captured chocolate among the provisions carried by a French and Indian war party [75]. In 1776, Virginia rifleman Charles Portfield recorded in his diary that American prisoners at Quebec complained to a British officer that they were having trouble getting "necessities," including chocolate [76]. In that same year, Ethan Allen, while a prisoner of the British, received "a hamper of wine, sugar, fruit, chocolate, &c." [77]. At the beginning of the Revolution, the State of Virginia included chocolate among the provisions for its Continental troops [78].

While chocolate might have been readily available and considered a necessity in the early years of the Revolutionary War, commissary officers and eventually civilians began to find it scarce as the war progressed. This might be discounted as effects of rampant inflation, but there were some other reasons as well: (1) British occupation of cocoa distribution and chocolate making centers, (2) cocoa shortages due to declining imports, and (3) trade embargoes imposed by the states themselves on each other.

Boston, Newport, New York, and Philadelphia all fell under British occupation at least for a time during the Revolutionary War. The only city that would survive the war as a center of chocolate production was Boston. While cocoa and chocolate trading might have occurred between ports in New England, the British control of New York and Newport effectively disrupted the New England and West Indian cocoa trade, the coastwise flow of cocoa between Philadelphia and Boston, and chocolate between Massachusetts and the southern colonies. Only when the British left Newport in 1779 and focused their attention on the South did Massachusetts begin to see cocoa again and then at greatly inflated prices.

Cocoa imports were reduced from prewar levels by loss of American shipping because of capture or diversion to wartime purposes. Almost from the beginning of the war, ships that might have carried cocoa in peacetime now carried military equipment, gunpowder, and saltpeter to make more gunpowder. Also, many of these vessels, particularly schooners, were converted into privateers used to disrupt British shipping in the Atlantic and Caribbean [79].

In the previous colonial wars, New York was the "privateering capitol of the American colonies" [80]. With its harbor being a British lake, New York now supported loyalist privateers. With the British contesting Long Island Sound, loyalist privateers were able to cruise from Newport and New York, further isolating New England chocolate makers from their cocoa and their customers. Likewise, loyalist privateers cruised up and down the entire coast of North America and the West Indies. With the advent of war, French, Dutch, and Spanish cocoa would have been at least legal for the Americans to import, if ships and sailors were available to carry on the trade. However, there were sporadic entries of cocoa from the French

and Dutch West Indies and captures of cocoa from British vessels. Bottom line: Cocoa became expensive in real terms. It was paid for with money that was becoming worthless on land and with blood from sailors at sea.

In February 1777, the Massachusetts Assembly passed an embargo on exports beyond the boundaries of the state for "Rum, Molasses and sundry other Articles" including "salt, coffee, cocoa, chocolate" [81]. In Williamsburg, the last advertisement in *The Virginia Gazette* for the sale of chocolate—for the duration of the war—appeared the following November [82].

The economies of the mid-Atlantic states also were in shambles. With British occupation and battles in 1777 and 1778 occurring in the Pennsylvania and New Jersey countryside, by June 1779 a pound of chocolate cost the Philadelphia consumer 30 times more than three years earlier [83]. To put it in perspective, butter was not considered worthy of mentioning in price-control schemes in dairy-rich Pennsylvania in 1776. By 1779, complaints about the price of butter were what drove the Pennsylvania Legislature to set new fixed prices [84]. In the fall of 1779, the inflationary spiral caused the Massachusetts Assembly to put another embargo on exports from the state of various foodstuffs including salt, coffee, cocoa, chocolate, and rum "either by land or water" [85].

When the hostilities shifted southward in 1779, Virginia towns set fixed prices on commodities. Chocolate was not included, perhaps because of its scarcity. Looking north from the Chesapeake, the outlook for chocolate was bleak: New York and Newport were completely isolated; Massachusetts embargoed chocolate from exportation; and two years of warfare in the Pennsylvania and New Jersey countryside had decimated the economy. Chocolate would have been rare beyond a day's cart-ride from the manufacturer . . . if he or she could obtain cocoa. Additionally, every "chocolate nut" had to compete for cargo space with gunpowder, rum, and salt on every northbound ship running the gauntlet from the West Indies. Furthermore, the value of Virginia currency had inflated enormously. Prior to the war, a Virginia pound traded at about one and a half times the value of a British pound sterling. By July 31, 1779, Dixon's *Virginia Gazette* noted that the Virginia pound was valued at 50 to one [86]. By 1780, a Virginia commissary officer applied to Governor Jefferson for "Rum, Tea, Sugar, Coffee, Chocolate &c of which there was none except the first mentioned article" [87].

THE FEDERAL PERIOD

Massachusetts never heard a shot fired in anger on its soil after March 1776, so the few North American chocolate makers remaining at war's end were in and around Boston. Previously, for at least a century, quality had been assured by knowing and trusting the local manufacturer and the origin of the cocoa. After the Revolution, it was more important to know that the chocolate was "Boston-made." Chocolate makers from surrounding cities advertised that their product was as "low as in Boston" [88]. Additionally, consumers in New York or Providence (places where chocolate had been locally produced before the war) extended their trust to Boston chocolate makers they never met. The chocolate maker's surnames like Baker or Welsh became protobrands or icons of the product itself. For the distant chocolate consumer, knowing the cocoa's origin was less important than knowing that the origin of the chocolate was Boston ... and the name of the chocolate maker. The chocolate maker's name had always been important to local customers. However, after the war, this name recognition extended to other cities as well.

In the late 18th century, the surnames of the Boston chocolate makers themselves started appearing in advertisements in other cities. In 1794, New York newspaper advertisements referred to "Welsh's first quality chocolate" [89] or "Welsh's Boston Chocolate" [90]. Likewise, Welsh's chocolate was advertised in Philadelphia [91] and Charleston, South Carolina [92]. An 1806 abstract of the goods entered into the Baltimore Custom-House listed Boston as the only source for the chocolate on hand [93].

Those Boston manufacturers who had survived the war had directed their business activities away from chocolate making and focused instead on producing other goods. When the time came to start making chocolate again, the people who had not diversified were wiped out. Take the case of Jonas Welsh of Boston. Between 1770 and 1798, Welsh was referred to as a "miller," "merchant," and a "chocolate grinder" [94]. His watermill was capable of producing 2500 pounds of chocolate a day [95]. However, close reading of his dealings reveal that he was a speculator, buying, selling, or renting interests in mills, wharfs, houses, buildings, stores, and even part of a tomb. The Suffolk County records list 34 real estate transactions where Welsh owned or mortgaged mills in partnership with others, plus surrounding land and passages or right-of-way to ponds and the sea [96]. Welsh was similar to other tradesmen and entrepreneurs of the period: either diversify or be history. Chocolate grinders could not rely on just chocolate to make a living because of the uncertainties of the cocoa trade. It would not have been unreasonable for a "chocolate maker" to consider himself a "miller" or a miller to think of his or herself as "merchant." Dr. James Baker owned a dry goods store and made chocolate and finished woolens in a fulling mill, but called himself "Doctor," although he never practiced medicine.

Another feature of chocolate in the late 18th century is that expressions of quality became rational. Beginning in late 18th century and continuing into the 19th century, manufacturers sold chocolate categorized by numeric grades of quality. Words like "superior" or "much approved" became "No. 1," "No. 2," or "No. 3." An 1817 Boston newspaper advertised "Baker's Chocolate and Shells of the first quality ... No. 1 Chocolate and Shells are warranted to be of as good a quality as any manufactured in the United States" and that "No. 3 Chocolate, much approved of for the Southern Markets." This same advertisement listed No. 1 chocolate for family use sold at 25 cts. per pound. Likewise, No. 3 chocolate sold for 16 cts. per pound [97]. The last reference alluded to a growing regional preference that was distinctly related to chocolate. We can only surmise from a distance of the 21st century what the real differences were. What were the particular qualities of the chocolate called "No. 1" that made it better than "No. 2"? Why was "No. 3" the inferior grade, preferred in the southern markets? Were shipping costs the reason? Grittiness? Sweetness (more or less)? Roasting techniques? Packaging? Who knows?

Another change in the American chocolate experience was the introduction of vanilla. Whereas vanilla was generally associated with chocolate in Europe, hardly any was imported into North America until the 19th century. Searching for "vanilla" in Colonial advertisements yields few results. As late as 1792, newspaper advertisements referring to vanilla discuss it as being rare and "scarcely known" to American consumers [98]. But it was actually advertisements from confectioners that began appearing in the mid-1790s that introduced "vanilla chocolate" to the American palette. By the turn of the century, vanilla beans were generally available but the chocolate made with the flavoring still carried the titles "vanilla chocolate," "chocolate a la Vanille," or "No. 1 with vanilla" and being available from confectioners rather than chocolate makers [99]. These same confectioners also offered other spices in their chocolate and would use labels like "with cinnamon" or "plain prepared" [100, 101].

Conclusion

One might make a general assumption that chocolate making in North America was a trade learned in England or the West Indies and that the technology was imported. As we have seen, however, the European labor supply of commercial chocolate makers was quite small. Those who emigrated to America would have joined a native-born labor pool using machines and techniques that were cutting edge for their day. More akin to the milling trades than confectionary, American chocolate makers were diversified businessmen who had to withstand cocoa shortages, wars, and the ups and downs of overseas or "coastwise" markets. Large producers had the advantage of waterfalls near the coast to site their mills along with access to large amounts of cocoa at an affordable price. They also had the advantage of having a "Plan B" if the cocoa market was disrupted or their mill burned down, which happened frequently. But, they were the most successful when they expanded their selling horizons and received the trust of customers beyond their own city limits or state borders.

Acknowledgments

I would like to acknowledge the following persons whose cumulative guidance and support have been instrumental in the writing of this chapter. From the Colonial Williamsburg Foundation: Barbera Scherer, Historic Foodways; Robert Brantley, Historic Foodways; Dennis Cotner, Historic Foodways; Susan Holler, Historic Foodways; Frank Clark, Historic Foodways; Jay Gaynor, Director of Trades; Gayle Greve, Curator of Special Collections Library; Juleigh Clarke, Rockefeller Library; Pat Gibbs, Research; Martha Katz-Hyman, Research; Donna Seale, Historic Foodways; Nicole Henning, Historic Foodways; and Kristi Engle, Historic Foodways. From the University of California, Davis: Deanna Pucciarelli. From Fort Ticonderoga, New York: Nick Westbrook and Virginia Westbrook. From Masterfoods USA/Mars, Incorporated: Deborah Mars, Eric Whitacre, Judy Whitacre, Rodney Synder, Doug Valkenburg, and Mary Myers. From Oxford University: Pamela Richardson. And I wish to thank the following members of my family: Jan Gay, Andrea West, and Bill Gay for their support and encouragement.

Endnotes

1. This chocolate was for use as food and as an exchange item during slave-trade dealings.

2. Imagine chocolate today in the 21st century having minute particles of mustard seed, wool lint, pieces of cotton rags, tobacco, even traces of horse manure!

References

1. Bercovitch, S., editor. *The Diaries of John Hull, Mint-master and Treasurer of the Colony of Massachusetts Bay.* Boston: J. Wilson and Son, 1857. Reprinted in *Puritan Personal Writings: Diaries.* New York: AMS Press, Inc., 1982, p. 158.

2. Ward, G., The silver chocolate pots of Colonial Boston. In: Ward, G., and Falino, J., editors. *New England Silver and Silversmithing 1620–1815.* Boston: The Colonial Society of Massachusetts; distributed by the University of Virginia Press, 2001; p. 61.

3. Colonial Port Commodity Import and Export Records for Virginia (Accomac, Rappahannock, South Potomac), Jamaica, New York, and South Carolina (1680–1769). M-1892 1–9 at the Rockefeller Library, Colonial Williamsburg Foundation.

4. Massachusetts Naval Officer Shipping List, 1686–1765, PRO HE 752.

5. *Encyclopaedia Britannica; or A Dictionary of Arts and Sciences, Compiled Upon a New Plan. In which The different Sciences and Arts are digested into distinct Treatises or Systems; and The various Technical Terms, &c. are explained as they occur in the order of the Alphabet. By a Society of Gentlemen in Scotland in Three Volumes, Volume III.* Edinburgh, Scotland: A. Bell and C. MacFarquhar, 1771; p. 895.

6. Presilla, M. *The New Taste of Chocolate: A Cultural and Natural History of Cacao with Recipes.* Berkeley, CA: Ten Speed Press, 2001; p. 27.

7. Brookes, R. *The Natural History of Chocolate.* London; England: J. Roberts (printer), 1730; p. 30.

8. Brookes, R. *The Natural History of Chocolate*, London, England: J. Roberts (printer), 1730; p. 32.

9. Emmanuel, I. S., and Emmanuel, S. A. *History of the Jews of the Netherlands Antilles, First Volume, Royal Vangorcum, LTD.* Cincinnati, OH: American Jewish Archives, 1970; p. 71.

10. Clarence-Smith, W. G. *Cocoa and Chocolate, 1765–1914.* New York: Routledge, 2000; p. 40.

11. Emmanuel, I. S., and Emmanuel, S. A. *History of the Jews of the Netherlands Antilles, First Volume, Royal Vangorcum, LTD.* Cincinnati, OH: American Jewish Archives, 1970; p. 71.

12. Emmanuel, I. S., and Emmanuel, S. A. *History of the Jews of the Netherlands Antilles, First Volume, Royal Vangorcum, LTD.* Cincinnati, OH: American Jewish Archives, 1970; p. 72.

13. Rediker, M. *Between the Devil and the Deep Blue Sea: Merchant Seamen, Pirates, and the Anglo-American World, 1700–1750.* New York: Cambridge University Press, 1987; p. 20.

14. Taylor, A. *American Colonies.* New York: Penguin Books, 2001; p. 176.

15. Middleton, A. P. *The Tobacco Coast.* Baltimore, MD: Johns Hopkins University Press, 1953; p. 219.

16. Taylor, A. *American Colonies.* New York: Penguin Books, 2001; p. 298.

17. Clarence-Smith, W. G. *Cocoa and Chocolate, 1765–1914.* New York: Routledge, 2000; p. 41.

18. Patriot's Resource: Townshend Acts. Available at http://www.patriotresource.com/documents/townshend.html. (Accessed January 11, 2007.)

19. Clarence-Smith, W. *Cocoa and Chocolate, 1765–1914.* New York: Routledge, 2000; p. 68.

20. The Proceedings of the Old Bailey. Available at http://www.oldbaileyonline.org/html_units/1720s/t17281204-27.html. (Accessed January 11, 2007.)

21. Huey, P. R. The Almshouse in Dutch and English Colonial North America and its precedent in the Old World: historical and archaeological evidence. *International Journal of Historical Archaeology* 2001;5(2):149.

22. Hamilton, Dr. A. Hamiltons Itinerarium; being a narrative of a journey from Annapolis, Maryland, through Delaware, Pennsylvania, New York: New Jersey, Connecticut, Rhode Island, Massachusetts and New Hampshire, from May to September, 1744. Edited by Albert Bushnell Hart. p. 16.

23. *The Virginia Historical Register*. Smyth's Travels in Virginia, 1773, p. 77.

24. Clarence-Smith, W. *Cocoa and Chocolate, 1765–1914*. New York: Routledge, 2000; p. 41.

25. Massachusetts Naval Office Shipping Lists, 1686–1765; PRO, Reel 1, 1979.

26. Thompson, M. *Moses Brown: Reluctant Reformer*. Published for the Institute of Early American History and Culture at Williamsburg, VA. Chapel Hill: University of North Carolina Press, 1962; pp. 11–12.

27. Smith, H. W. A Note to Longfellow's "The Jewish Cemetery at Newport." *College English* 1956;18(2): 103–104. Available at http://links.jstor.org/sici?sici= 00100994%28195611%2918%3A2%3C103%3AAN TL%22J%3E2.0.CO%3B2-M. (Accessed January 11, 2007.)

28. Marcus, J. R. *The Colonial American Jew, 1492–1776*. Detroit, MI: Wayne State University Press, 1970; p. 673.

29. *The Pennsylvania Gazette*, May 22, 1735. The Accessible Archives CD-ROM *Pennsylvania Gazette* Folio I 1728– 1750. Malvern, PA: Accessible Archives, Inc., 1993; #1881.

30. Middleton, A. *The Tobacco Coast*, Baltimore, MD: Johns Hopkins University Press, 1953; p. 221.

31. Ledger of Imports and Exports (American); January 5,1768–January 5,1773; PRO/Customs 16/1.

32. James Baker Records, Box 1, Folder 11, Jan–Jun 1771 held at the American Antiquarian Society, Worcester, MA. Jan 3, 1771, Baker pays Enoch Brown for cart of 601cwt of cocoa from Providence.

33. Taylor, A. *American Colonies*. New York: Penguin Books, 2001; p. 255.

34. Marcus, J. *The Colonial American Jew, 1492–1776*. Detroit, MI: Wayne State University Press, 1970; p. 607.

35. Colonial Port Commodity Import and Export Records for Virginia (Accomac, Rappahannock, South Potomac), Jamaica, New York: and South Carolina (1680–1769). M-1892 1–9 at the Rockefeller Library, Colonial Williamsburg Foundation.

36. Janzen, O. *Garrison Life in the 18th Century: Newfoundland and Labrador, 1991*. Available at http://www. heritage.nf.ca/exploration/glife_18c_1.html. (Accessed January 11, 2007.)

37. *Boston Post Boy*. April 8, 1765; Issue 299, p. 3.

38. Joseph Fry Letter to British Customs, June 27, 1776; PRO T1 S23/332–336.

39. Governor Nicholson Letter to the Council of Trade and Plantations, March 6, 1705. Calendar of State Papers, Colonial: North America and the West Indies CD-ROM. New York: Routledge, 2000.

40. Wright, L. B., and Tinling, M., editors. *The Secret Diary of William Byrd of Westover, 1709–1712*. Richmond, VA: Dietz Press, 1941; p. 4.

41. Fitzpatrick, J. C. The writings of George Washington from the original manuscript sources. Washington, George, 1732–1799. Available at http://etext. virginia.edu/etcbin/ot2wwwwashington?specfile=/ texts/english/washington/fitzpatrick/search/gw.o2 w&act=surround&offset=37666381&tag=Writings+ of+Washington,+Vol.+30:+To+CLEMENT+BIDD LE&query=chocolate&id=gw300170. (Accessed January 12, 2007.)

42. Fitzpatrick, J. C. The writings of George Washington from the original manuscript sources. Washington, George, 1732–1799. Available at http://etext. virginia.edu/etcbin/ot2wwwwashington?specfile=/ texts/english/washington/fitzpatrick/search/gw.o2 w&act=surround&offset=41733107&tag=Writings+ of+Washington,+Vol.+33:+*To+BURGES+BALL &query=Chocolate&id=gw330223. (Accessed January 12, 2007.)

43. Carson, J. *Colonial Virginia Cookery*. Williamsburg, VA: Colonial Williamsburg Foundation, 1985; p. xiii.

44. Smith, E. *The Compleat Housewife*. Reprinted by William Parks, Williamsburg, VA, 1742; p. 224.

45. Harbury, K. E. *Colonial Virginians Cooking Dynasty*. Columbia: University of South Carolina Press, 2004; p. xiv.

46. Harbury, K. *Colonial Virginians Cooking Dynasty*. Columbia: University of South Carolina Press, 2004; p. xv.

47. Harbury, K. *Colonial Virginians Cooking Dynasty*. Columbia: University of South Carolina Press, 2004; p. 189.

48. Harbury, K. *Colonial Virginians Cooking Dynasty*. Columbia: University of South Carolina Press, 2004; p. 189.

49. Gunston Hall Plantation Probate Inventory Database. Available at http://gunstonhall.org/probate/inventory.htm. (Accessed January 12, 2007.)

50. Colonial Williamsburg Foundation unpublished research study comparing costs of consumer goods sold in Williamsburg stores, private accounts, and newspaper advertisements. See also Walsh, L. S. *Provisioning Early American Towns. The Chesapeake: A Multidisciplinary Case Study*. Williamsburg, VA: Colonial Williamsburg Foundation, 1997. Available at http://research.history.org/Historical_ Research/Technical_Reports/DownloadPDF.

cfm?ReportID=HistRes02. (Accessed January 12, 2007.)

51. *The Virginia Gazette*. December 18, 1766, 2:2.

52. Geospatial and Statistical Data Center, Historic Census Browser, University of Virginia Library. Available at http://fisher.lib.virginia.edu/. (Accessed January 12, 2007.)

53. Colonial Port Commodity Import and Export Records for Virginia (Accomac, Rappahannock, South Potomac), Jamaica, New York: and South Carolina (1680–1769) M-1892 1–9 at the Rockefeller Library, Colonial Williamsburg Foundation.

54. Ledger of Imports and Exports (American); January 5, 1768–January 5, 1773; PRO/Customs 16/1.

55. Clarence-Smith, W. *Cocoa and Chocolate, 1765–1914*. New York: Routledge, 2000; p. 13.

56. Coe, S. D., and Coe, M. D. *The True History of Chocolate*. New York: Thames and Hudson, 1996; p. 215 for description of chocolate used in murder. See also *The Virginia Gazette*. December 15, 1774; p. 2 for the accusation of the Jesuits.

57. Thomas, M. H. *The Diary of Samuel Sewall, Volume 1, 1674–1708*. New York: Farrar, Straus and Giroux, Inc., 1973; p. 380.

58. Thomas, M. *The Diary of Samuel Sewall, Volume 1, 1674–1708*. New York: Farrar, Straus, and Giroux, Inc., 1973; p.476.

59. Thomas, M. *The Diary of Samuel Sewall, Volume 1, 1674–1708*. New York: Farrar, Straus, and Giroux, Inc., 1973; pp. 563–564.

60. Thomas, M. *The Diary of Samuel Sewall, Volume 1, 1674–1708*. New York: Farrar, Straus, and Giroux, Inc., 1973; p. 570.

61. Thomas, M. *The Diary of Samuel Sewall, Volume 1, 1674–1708*. New York: Farrar, Straus, and Giroux, Inc., 1973; p. 626.

62. Fischer, D. H. *Albion's Seed*. New York: Oxford University Press, 1989; p. 540.

63. Randolph, M. *The Virginia House-wife, 1824*. Columbia: University of South Carolina Press, 1984; p. 17.

64. *Boston Evening Post*. February 8, 1773; Issue 1950; p. 3.

65. *Boston Post Boy*. January 28, 1751; p. 2.

66. Dorchester Athenaeum. Available at http://www.dorchesteratheneum.org/page.php?id=67. (Accessed January 12, 2007.)

67. *The Pennsylvania Gazette*. January 28, 1762; The Accessible Archives CD-ROM *Pennsylvania Gazette* Folio II 1751–1765. Malvern, PA: Accessible Archives, Inc., 1993; #27995.

68. *The Pennsylvania Gazette*. March 8, 1764. The Accessible Archives CD-ROM *Pennsylvania Gazette* Folio II 1751–1765. Malvern, PA: Accessible Archives, Inc., 1993; #32819.

69. Brookes, R. *The Natural History of Chocolate*, London, England: ••, 1730; p. 37.

70. Brookes, R. *The Natural History of Chocolate*, London, England: ••, 1730; p. 64.

71. *Encyclopaedia Britannica, Volume 2*. Edinburgh, Scotland: Colin MacFarquhar, 1771; p. 193.

72. Dow, G. F. *The Arts and Crafts of New England, 1704–1775*. New York: Da Capo Press, 1967; p. 294.

73. *The Virginia Gazette*. Accessible through the Colonial Williamsburg Foundation, http://research.history.org/JDRLibrary/Online_Resources/Virginia Gazette/VGPPIndex.cfm?firstltr=C. (Accessed January 12, 2007.)

74. Franklin, B. Autobiography. In: Eliot, C. W., editor. The Autobiography of Benjamin Franklin. *The Journal of John Woolman, the Fruits of Solitude William Penn*. New York: P. F. Collier and Son, 1909. Available at http://mith2.umd.edu/eada/html/display.php?docs=franklin_autobiography.xml. (Accessed January 12, 2007.)

75. The Accessible Archives CD-ROM *Pennsylvania Gazette* Folio I 1728–1750. Malvern, PA: Accessible Archives, Inc., 1993; #8490.

76. *The Virginia Magazine of History and Biography, Volume IX, House of the Society*. Richmond, VA: The Virginia Historical Society, 1902; p. 147.

77. The Accessible Archives CD-ROM *Pennsylvania Gazette* Folio II 1751–1765. Malvern, PA: Accessible Archives, Inc., 1993; #59354.

78. *The Virginia Magazine of History and Biography, Volume XXI, House of the Society*. Richmond, VA: The Virginia Historical Society, 1913; p. 156. (Reprinted with permission of the original publisher Kraus Reprint Corporation, New York: 1968.)

79. Frayler, J. "Privateers in the American Revolution." National Park Service. Available at http://www.nps.gov/revwar/about_the_revolution/privateers.html. (Accessed January 12, 2007.)

80. Rediker, M. *Between the Devil and the Deep Blue Sea: Merchant Seamen, Pirates, and the Anglo-American World, 1700–1750*. New York: Cambridge University Press, 1987.

81. Article Bookmark (OpenURL Compliant): Early American Imprints, 1st series, no. 15427 (filmed). Available at http://docs.newsbank.com/openurl?ctx_ver=z39.88-2004&rft_id=info:sid/iw.newsbank.com:EVANL&rft_val_format=info:ofi/fmt:kev:mtx:ctx&rft_dat=0F2F82E70740B080&svc_dat=Evans:eaidoc&req_dat=0D0CB57AEDE52A75. (Accessed January 12, 2007.)

82. *The Virginia Gazette*. November 14, 1777, 3:1.

83. The Accessible Archives CD-ROM *Pennsylvania Gazette* Folio III 1766–1783. Malvern, PA: Accessible Archives, Inc., 1993; #64559.

84. The Accessible Archives CD-ROM *Pennsylvania Gazette Folio III 1766–1783*. Malvern, PA: Accessible Archives, Inc., 1993; #64559.

85. The Accessible Archives CD-ROM *Pennsylvania Gazette Folio III 1766–1783*. Malvern, PA: Accessible Archives, Inc., 1993; #64841.

86. *The Virginia Gazette*. July, 31, 1779, 3:1.

87. Palmer, Wm. P., M.D., editor. *Calender of Virginia State Papers: 1652–1781; Volume 1*. Richmond, VA: Virginia State Library, 1875, p. 374. (Reprinted with the permission of the Virginia State Library by Kraus Reprint Corp., New York, 1968.)

88. *Massachusetts Spy: Or, Worcester Gazette*. January 20, 1785, p. 3.

89. *The Daily Advertiser* (New York). January 3, 1794, p. 1.

90. *New-York Daily Gazette*. May 19, 1794, p. 2.

91. *Gazette of the United States* (Philadelphia). October 4, 1796, p. 4.

92. *City Gazette And Daily Advertiser* (Charleston, SC). January 23, 1797, p. 1.

93. *Baltimore Weekly Price Current*. February 27, 1806, p. 3.

94. *Thwing Index of Early Boston Inhabitants*. Boston: Massachusetts Historical Society, 2005.

95. Baker, W. *Cocoa and Chocolate*. Dorchester, MA: Walter Baker & Company, 1917.

96. *Thwing Index of Early Boston Inhabitants*. Boston Massachusetts Historical Society, 2005.

97. *Boston Daily Advertiser*. January 18, 1817, p. 4.

98. *The Federal Gazette, and Philadelphia Evening Post*. October 9, 1792, p. 2.

99. *Federal Gazette and Baltimore Advertiser*. January 1, 1796, p. 4

100. *City Gazette and Daily Advertiser* (Charleston, SC). February 4, 1795, p. 4.

101. *American Citizen and General Advertiser* (New York). June 10, 1801, p. 3.

Chocolate's Early History in Canada

Catherine Macpherson

Introduction

The McCord Museum has undertaken the sweet yet sweeping task of documenting the history of chocolate in Canada from the Colonial era to the early 20th century.[1] Very little has been written on the subject thus far. Many of the home-grown chocolate companies familiar to Canadians—Ganong, Moir's, Laura Secord, Purdy's, and Roger's—have been around for generations, some since the early 20th century and others from the late 1800s. Beyond some more commonly shared contemporary experiences with chocolate—Laura Secord eggs at Easter, Maple Buds, Smarties, or Cherry Blossoms, and the de rigueur hot cocoa for a childhood's worth of Canadian winters—what was the daily role, if any, that chocolate played in the social or cultural life of Canadians in the past? Was chocolate a part of the Canadian diet in the 19th, 18th, or even 17th century? Was it always a popular confection or did it play another alimentary role? How easy or even desirable was it to obtain chocolate, and was there indigenous manufacturing of the sweet?

Some of these questions might be answered through import–export records, when and where available. Other answers might be revealed through scans of newspaper databanks or archived trade publications. Searches through journals, diaries, letter books, and medical or educational texts might further construct a portrait of chocolate's use and value in Canada's past. Still other questions might never derive a conclusive answer. For every bit of evidence as to chocolate's early life in Canada, other mysteries reveal themselves simultaneously; among them, the biggest question asked by any food historian: What did chocolate *taste* like back then?

L'arbre qui s'apelle cacau: Chocolate's Earliest Connection to Canada

The earliest Canadian references to chocolate occur among French Colonial explorers and settlers. Research into archival documentation relating to New France and, in particular, Île Royale (site of the Fortress of Louisbourg National Historic Site of Canada) netted some of the earliest references to chocolate use among Canadians. In 1713, the French first came to Louisbourg and by 1719 construction had begun on the fortified town. The town fell from the French to the British on two separate occasions, the British ultimately razing the fortifications after the second capture in 1758.

Chocolate: History, Culture, and Heritage. Edited by Grivetti and Shapiro
Copyright © 2009 John Wiley & Sons, Inc.

CHOCOLATE AT LOUISBOURG

The Fortress of Louisbourg is rich archaeological terrain. Among the various ceramics unearthed, there is a preponderance of brown faience ware, similar to excavated sites of early towns in Quebec. While certain jugs have been cited as chocolate pots, they could just as easily have held milk, water, or other liquids. The same could be said for a two-handled ceramic glazed cup unearthed at Louisbourg, frequently labeled as a "chocolate cup." Better archaeological proof of chocolate use at Louisbourg lies in some of the metal vessels discovered at the site. The lids and partial lids have holes pierced through the middle of them. This allowed a chocolate whisk or "mill" to pass through. The lid could remain closed while the mill was twirled between the palms of the hands to mix grated chocolate with hot water, milk, or cream to create a pleasing froth. Small swivel lips covered the hole on such pots. The lip was, presumably, to keep the hole covered and the beverage hot when the mill was not in place (Fig. 24.1).[2]

Correspondence and contacts with collections technicians and other researchers at the fortress led to further documentary evidence and conclusive proof of chocolate's use by certain inhabitants of the town, most notably one of Louisbourg's governors, Governor Duquesnel. A study of inventories, provisioning lists, and additional textual evidence has determined chocolate's use by 18th century French militia in Canada and has also examined the social customs associated with chocolate use among inhabitants of the fortified town of Louisbourg (see Chapter 25).[3]

Was there chocolate in Canada earlier than at Île Royale? It is doubtful that the Acadians had chocolate, but similar chocolate consumption patterns would have existed throughout New France. Chocolate was a luxury and not everyone would have consumed it. The nobility and senior officers would have had chocolate. The first mention we have located, thus far, of chocolate in Quebec is from the probate inventory of M. Charles Aubert de la Chesnaye; his goods were inventoried upon his death in the fall of 1702.[4] Aubert de la Chesnaye, successful merchant and fur trader, was possibly the most important businessman of 17th century New France.

CHAMPLAIN'S CURIOUS RELATION TO CACAO

To speak generally in terms of chocolate's earliest connections to Canada, Samuel de Champlain, immediately prior to his exploration of Port Royal, New France, and the founding of Quebec City in 1608, allegedly traveled to Mexico and the West Indies, where he supposedly made extensive notes on the local flora and fauna, documented in rather exquisitely illustrated journals. The journal of 1602 contains an illustration with the label *l'arbre qui s'apelle cacou* accompanied by a lengthy description of the tree and its uses. Curiously, the description conflates the attributes of the cacao tree and those of the maguey or agave plant [1]. The text reads, in part:

> There is [a] tree, which is called cacao, the fruit of which is very good and useful for many things, and even serves for money among the Indians. . . . When this fruit is desired to be made use of, it is reduced to powder, then a paste is made, which is steeped in hot water, in which honey,[5] which comes from the same tree, is mixed, and a little spice; then the whole being boiled together, it is drunk in the morning, warm . . . and they find themselves so well after having drunk of it, that they can pass a whole day without eating or having great appetite. [2]

Whether or not Champlain had first-hand knowledge of the plant is a subject of debate. The text goes on to describe the fiber of the plant and its use by native tribes. The illustration portrays characteristics of each plant—agave and cacao—combined into one image. This confusing depiction feeds preexisting questions of authenticity as regards this text in general.[6] Could Champlain have been familiar with cacao and chocolate at the time? Given the timeline of chocolate's introduction to Europe, it seems unlikely that he would have encountered it there prior to his exploration of the New World. It is certain he did not have chocolate with him during his time in New France (Figs. 24.2 and 24.3).

Another historic figure, Pierre Le Moyne D'Iberville, renowned son of New France and hero in defending French interests in Hudson's Bay, had purchased a lucrative cocoa plantation in Saint-Domingue in 1701, but this was after his time of service in Canada. It is unlikely that he possessed cocoa or chocolate while in Canada, although its value would have been recognized at the time.[7] Chocolate had only recently been introduced to the French upper classes and was hardly

FIGURE 24.1. Reproduction of metal chocolate pot, used in the interpretive program at Fortress Louisbourg. *Source:* Courtesy of the Fortress of Louisbourg National Historic Site of Canada, Parks Canada. (Used with permission.) (See color insert.)

PLANCHE XXXII.

PLANCHE XXXIII.

Original in the John Carter Brown Library at Brown University

FIGURE 24.2. *L'arbre qui s'appelle cacao* (The Tree That is Called Cacao), manuscript version. *Source: Brief discours des choses plus remarquables que Samuel Champlain de Brouage a reconnues aux Indes occidentals.* Courtesy of The John Carter Brown Library at Brown University. (Used with permission.) (See color insert.)

FIGURE 24.3. *L'arbre qui s'appelle cacao* (The Tree That is Called Cacao), printed version. *Source: Brief discours des choses plus remarquables que Samuel Champlain de Brouage a reconnues aux Indes occidentals.* Courtesy of The John Carter Brown Library at Brown University. (Used with permission.)

a common commodity among Europeans at this point in history. With French holdings in the West Indies, however, its popularity and use would continue to grow.

"Chocolate—Here Made"—Early English Settlement in Canada

What were English-Canadian consumption patterns of chocolate like during the early Colonial period? According to research completed thus far into British Admiralty and Colonial import–export records, trade statistics, and *Blue Book* listings, Halifax, largest of the British colonial towns in Canada, did not receive much in the way of British cocoa shipments from territories in the Caribbean—most of it went directly to England (See Chapters 26 and 27). Raw cocoa would not be imported in any quantity or with any regularity into Halifax until the early 1800s. But chocolate as a pre-pared product would have made its way from England

to Canada as an import and through the military. Halifax also received chocolate from the New England colonies.

Chocolate, measured by boxes, pounds, or occasionally by cask, keg, or barrel, made its way to Halifax via heavy trade traffic in this important port (Table 24.1). Most ships bringing chocolate to Halifax in the latter half of the 18th century arrived from New York, Boston, or Philadelphia. The chocolate may or may not have been from indigenous manufacture in these regions; it could have been of English provenance as revealed by export statistics for the principal Eastern Seaboard Atlantic ports (see Chapters 23 and 52).

Chocolate outward bound from Halifax is listed as early as 1750. Given the absence of cocoa imports, one could assume that most of this chocolate was of English or New England manufacture and was simply making its way up or down the coast via Halifax. The same could be said for the few instances of cocoa to leave Halifax, likely en route from the West Indies, overseas. But on April 14, 1752, the schooner *Jolly Robin* left Halifax bound for Newbury with cargo that

Table 24.1 Chocolate: Inward Bound to Halifax

Year	Date	Cocoa	Chocolate	From	Ship Name	Ship Type	Master	Tons	Crew Number	Owner
1749	August 10	None	50lb	Salem	*Salisbury*	Schooner	Benjamin Ober?	25	3	John Grigg & Co.
	September 18	None	2 boxes	New York	*Two Sisters*	Sloop	John Grigg	26	6	Jo. Whipple & Co.
	October 4	None	20lb	Rhode Island	*John*	Sloop	John Clark	35	5	Amos Goudy & Co.
1750	January 3	None	100lb	Boston	*Chancey*	Schooner	Edward Sohier	60	6	Daniel Siles & Co.
	April 2	None	12 boxes	New York	*Deborah*	Sloop	Thomas Coger	30	7	Jacob Hart
	June 25	None	6 boxes	St. John's	*Little Moses*	Sloop	Israel Boardman	25	3	Benjamin Weaver & Co.
	August 20	None	2 boxes	Rhode Island	*Endeavour*	Sloop	Benjamin Weaver	20	4	James Nichols & Co.
	October 1	None	8 boxes	Rhode Island	*Endeavour*	Sloop	Hugh M. Clein	85	5	Henry Creighton
	October 3	None	2 boxes	New York	*Henry*	Sloop	James Creighton	15	4	
1751	March 25	None	6 boxes	New York	*Huza*	Sloop	Thomas Barnes	30	6	Thomas Barnes
	April 14	None	6 boxes	Philadelphia	*Mifflin*	Ship	John Peele	80	8	J. Mfflin & Co.
	May 28	None	2 boxes	Boston	*St. John*	Sloop	Lewis Turner	40	4	Giles Tidmarth & Co.
	August 3	None	4 boxes	New York	*Huza*	Sloop	Thomas Barnes	30	4	Thomas Barnes
	October 17	None	4 boxes	Rhode Island	*Pelican*	Sloop	Benjamin Coffin	24	4	Benjamin Thurston
1752	March 11	None	? boxes, 1 brl	Philadelphia	*Hamilton*	Brigantine	George Rankin	50	6	John Pote & Co.
	April 4	None	1 box	Philadelphia	*Diamond*	Sloop	Jonah Titcomb	45	5	Jonah Titcomb & Co.
	April 10	None	18 boxes	New York	*Hannah*	Sloop	Saul Berrien	15	5	James Campbell & Co.
	April 28	None	1 cask	Philadelphia	*Dolphin*	Sloop	Sam Tubbs	15	3	Sam Tubbs & Co.
	June 8	None	6 boxes	Philadelphia	*Boyn*	Ship	Henry Ash	130	7	Irwin & Co.
	October 6	None	1 box, 1 keg	Philadelphia	*Katherine & Ann*	Sloop	Zeph. Holwell	30	5	Z. Holwell & Co.
1753	April 9	None	20 boxes	Rhode Island	*Hannah*	Sloop	Jeremiah Eddy	25	4	William Bensen & Co.
	April 14	None	12 boxes	Philadelphia	*Endeavour*	Brigantine	George Houston	40	6	Townsend White & Co.
	June 19	None	4 boxes	Philadelphia	*Endeavour*	Brigantine	George Houston	40	6	Townsend White & Co.
	September 4	None	4 boxes	Philadelphia	*Prospect*	Schooner	William Condy	40	5	John Pole
	September 21	None	1 cask	New York	*Two Sisters*	Sloop	David Fry	15	4	John Gifford
	October 9	None	1 box	Philadelphia	*Carpenter*	Ship	Hugh Bowes	90	8	Reese Meredith
	October 14	None	8 boxes	Philadelphia	*Patience*	Schooner	Caleb Turner	30	3	Nathan Miller & Co.
1754	July 8	None	1 box	New York	*Nelly*	Ship	James Gibbons	60	8	John Bell & Co.

Year	Date		Amount	Origin	Type	Vessel	Master			Consignee
1755	March 28	None	6 boxes	New York	Sloop	*Martha & Hannah*	Henry Roomes	25	4	John Butler
	August 4	None	1 box	Rhode Island	Sloop	*Sea Flower*	Simon Rhodes	40	6	Simon Rhodes
	August 4	None	3 boxes	Boston	Schooner	*Sarah*	William Trefry	40	4	Thomas Cushing & Co.
	August 18	None	1 box	Newport	Sloop	*Kinnicut*	John Updike	30	7	Sal Mitchell & Co.
	August 18	None	1 weight	Boston	Brig	*Devonshire*	Geoge Willmot	75	6	John Gould
	September 3	None	1 box	Philadelphia	Schooner	*Swallow*	David Stewart	20	4	Robert Ragg
	September 27	None	4 boxes	New York	Sloop	*Anne*	Peter Dobson	15	4	Luke Kiersted & Co.
	October 7	None	3 boxes	Philadelphia	Brigantine	*Esther*	Joseph House	30	7	Joseph House
1756	March 25	None	Not given	New York	Sloop	*John*	Henry Bogart	25	6	Sam Tingley & Co.
	April 8	None	4 boxes	Boston	Schooner	*Sarah*	William Trefry	40	4	Thomas Cushing & Co.
	April 12	None	6 boxes	Philadelphia	Brig	*Sarah & Kath*	William Condy	40	5	John Howell & Co.
	June 21	None	4 boxes	Philadelphia	Brig	*Sarah & Kath*	William Condy	40	6	John Howell & Co.
	August 9	None	3 boxes	Boston	Sloop	*Swallow*	William Trefry	45	4	Israel Stone & Co.
	October 7	None	14 boxes	Philadelphia	Sloop	*Patty*	Sam House	20	5	William Ritchie & Co.
	November 9	None	2 boxes	Philadelphia	Ship	*Isabella Maria*	William Cuzzins	80	11	John Bell & Co.
1757	January 24	None	50 boxes	Philadelphia	Sloop	*Patty*	Sam House	20	4	William Ritchie & Co.
	July 2	None	6 boxes	Philadelphia	Sloop	*Sally*	Alex Sage	25	6	Alex Ritchie & Co.
	July 2	None	10 boxes	New York	Sloop	*Sally*	William Montgomery	15	4	Law & Kortwright
	July 5	None	4 boxes	New York	Sloop	*Elizabeth*	James Creighton	20	4	William Bayard
	July 5	None	4 boxes	New York	Schooner	*Mary Magdalene*	Anthony Water	10	4	Thomas Dungan
	July 25	None	1 box	Boston	Schooner	*Hawke*	Joseph Frost	35	5	James Pitt
	July 30	None	6 boxes	Philadelphia	Brig	*Sarah & Kath*	William Condy	40	6	John Howell & Co.
	August 2	None	6 boxes	Boston	Sloop	*Pursue*	William MacKay	32	4	William MacKay & Co.
	August 17	None	5 boxes	New York	Sloop	*Mary*	Samuel Tudor	25	5	Samuel Tudor & Co.
	August 29	None	6 boxes	Philadelphia	Sloop	*Harlequin*	Augustine Hicks	25	5	Barnard Badger & Co.
	September 16	None	2 boxes	Philadelphia	Sloop	*Patty*	James Haydon	45	6	John Collins & Co.
	November 10	None	10 boxes	Philadelphia	Schooner	*Blakely (sp?)*	Thomas Martin	60	5	Thomas Martin & Co.
1758	June 2	None	100lb	Rhode Island	Sloop	*Dolphin*	Theo. Topsham	35	5	In . . . Tweedy
	June 3	None	10 boxes	Boston	Sloop	*Phoenix*	Benjamin Homer	50	4	Anthony Thomas & Co.
	November 1	None	10 boxes	Boston	Sloop	*Ruth & Mary*	Jacob Wilds	35	4	Jacob Wilds & Co.
1759	March 1	None	4 boxes	Boston	Schooner	*Susanah*	Thomas Wood	45	4	Thomas James Gruchy
	September 17	None	24 boxes	Boston	Schooner	*Desire*	Solomon Bangs	30	5	Benjamin Bangs
	September 29	None	10 boxes	Boston	Schooner	*Barnstaple*	William Trefry	40	4	Samuel Sturgis & Co.
1760	February 8	None	8 boxes	Boston	Sloop	*Rachael*	William Lord	70	5	Jonathon Chandler & Co.
	March 15	None	10 boxes, 3 cases	Philadelphia	Schooner	*Jolly Robin*	William Kidd	28	4	William Kidd & Co.
	April 3	None	4 boxes	New York	Sloop	*John Robert*	John Smith	20	5	John Smith
	September 8	None	1 weight	Rhode Island	Sloop	*Char.g Elizabeth*	William Holmes	30	5	Thomas Green & Co.

Table 24.1 Continued

Year	Date	Cocoa	Chocolate	From	Ship Type	Ship Name	Master	Tons	Crew Number	Owner
1761	March 11	None	7 boxes	New York	Sloop	*Molly*	Dan Sprout	50	5	James Miller
	December 1	None	13 boxes	Boston	Schooner	*Betsey*	John Atwood	30	5	James Callender & Co.
1762	June 25	None	6 boxes	Boston	Sloop	*Swallow*	Samuel Doggett	50	5	Silvanus Drew & Co.
	July 6	None	10 boxes	Philadelphia	Schooner	*Leopard*	Thomas Church	80	6	Thomas Church & Co.
	July 16	None	2 boxes	Newport	Sloop	*Mary*	William Gardner	35	5	Joshua Amry
	November 1	None	6 boxes	Boston	Sloop	*Swallow*	Samuel Doggett	50	5	Silvanus Drew & Co.
	November 1	None	1 box	Boston	Sloop	*Jolly Roger*	Jesse Hall	40	5	Zenas Drew & Co.
	December 30	None	14 boxes	Philadelphia	Schooner	*Two Brothers*	Jacob Parker	60	4	Alexander Campbell & Co.
1763	July 14	None	4 boxes	Boston	Schooner	*Murray*	Ephraim Deane	50	4	William Deane & Co.
	July 30	None	4 boxes	Boston	Schooner	*Humbird*	Rufus Ripley	25	4	Hezchiah Ripley
1764	May 7	None	3 boxes	Philadelphia	Sloop	*Nabby*	William Maxwell	55	6	Alexander Grant
	May 10	None	8 boxes	Philadelphia	Brig	*Peggy*	James Brokinson	120	8	William Funn
	May 15	None	70lb	Boston	Sloop	*Two Sisters*	Thomas Mitchell	30	4	Thomas Mitchell & Co.
	May 17	None	2 boxes	Boston	Sloop	*Two Brothers*	Ephraim Deane	60	5	Ephraim Deane & Co.
	July 9	None	6 boxes	Boston	Sloop	*Kingston*	William Ruggles	50	4	Robert Pierpont
	November 17	None	3 boxes	Philadelphia	Ship	*Delaware*	Peter Creighton	80	9	Thomas Wharton
	December 15	None	4 boxes	Boston	Sloop	*Swallow*	Nathaniel Atwood	50	5	Nathaniel Atwood
	December 17	None	2 boxes	Philadelphia	Schooner	*Adventure*	John Wheldon	18	4	In . . . McLean
1765	February 23	None	8 boxes	Boston	Sloop	*Swallow*	Nathaniel Atwood	50	5	Nathaniel Atwood & Co.
	April 15	None	2 boxes	New York	Schooner	*Polly*	Jasper Griffin	16	4	Jasper Griffin
	April 28	None	1 box	Philadelphia	Schooner	*Margery*	Edmond Butler	20	6	Edmond Butler
	May 11	None	6 boxes	Philadelphia	Sloop	*Union*	Sam Harlow	30	5	Nathaniel Delaware
	August 19	None	2 boxes	Boston	Schooner	*Elizabeth*	Nehem . . . Somes	60	4	Nehem . . . Somes & Co.
	September 21	None	10 boxes	Philadelphia	Schooner	*Mary*	Edmond Butler	20	5	Edmond Butler & Co.
	September 24	None	2 boxes	Boston	Sloop	*Sarah*	James Ford	70	5	Cornelius White
	October 5	None	2 boxes	Piscatagua	Sloop	*Good Intent*	Samuel Harris	60	5	Nathaniel Dennell
	November 29	None	5 boxes	Philadelphia	Schooner	*Mary*	Edmond Butler	20	6	Edmond Butler & Co.

Sources: CO221 Colonial Office, Nova Scotia and Cape Breton Miscellanea, B-3228, Shipping Returns—Nova Socotia—No. 28; 1758–1761, and No. 31; 1762–1765.

included two boxes of chocolate. The export record noted that the boxes of chocolate were "here made" [3]. Whether the designation "here" referred to Halifax, specifically, or to the North American colonies in general, is not clear. Thus far, no other evidence has been found of commercial chocolate manufacture in Nova Scotia for this period, and there are no significant cocoa imports to imply as much. Still, chocolate was an increasingly common commodity and some small-scale local manufacture may have existed (Table 24.2).

PRIVATEERING AND SMUGGLING

Simeon Perkins, 18th century businessman, politician, and diarist from Liverpool, Nova Scotia, frequently referenced chocolate and cocoa in his journals. Perkins was keenly interested in matters of trade. On December 16, 1786, he noted the *Price Current* of sundry items listed in the New York paper, the *Morning Post* of November 2; among the items noted: "Cinnamon, 28/ per lb. Cloves, 16/. Chocolate, $\frac{1}{4}$" [4].

Perkins also was involved in privateering, particularly during times of warfare. On September 10, 1797, Perkins wrote of a Spanish prize brigantine, the *Nuestro Senior de la Carmen*, taken by the privateer ship *Charles Mary Wentworth* on September 4:

Loaded, as is Said, with Cocoa & Cotton, from Havana, bound to Spain. . . . The Prize Brig comes into the River, and is laid at Parker's wharf. . . . The Prize appears to be very Clearly Spanish property, both Vessel and Cargo. . . . The Cocoa in the Rum is warm. We also find it warm forward. [5]

This last observation is curious. Was Perkins referring to both the cocoa and a cargo of rum as being warm from storage near the bow of the ship? There is also the possibility of fire damage, as the brig would likely have suffered some blows from the privateer ship. Perkins was, in fact, a co-owner of the *Charles Mary Wentworth*.

Not long after the taking of the *Nuestro Senior de la Carmen*, Perkins further increased his fortune through investments in privateer ships. Four more Spanish prizes—the schooners *Casualidad, Diligence,* and *Fortuna,* and the brig *La Liebre*—were captured by his ship *Charles Mary Wentworth* in the spring of 1799. On June 3, 1799, shortly after the ships' arrival in Liverpool, the cargoes were auctioned and Perkins saw a decent profit from the sale. Two of the schooners held valuable cargoes of cocoa that totaled over 70 tons and sold for an average price of 96 shillings per unit (not identified). Two merchants from Halifax purchased the cocoa, and while Perkins awaited word of a ship to deliver the cocoa from Liverpool to Halifax, he provided details regarding the unloading of the cocoa from the schooners to stores onshore; he and his men packed 50 hogsheads of cocoa in all.[8]

Perkins mentioned numerous, subsequent prizes of cocoa captured and brought to Liverpool and frequently commented on whether or not the ships sold for fair prices. Privateering was a very lucrative undertaking for him. His accounts of cocoa being auctioned and shipped around the province and coastwise provide insights regarding the nature and magnitude of cocoa's value in British North America for this period of commerce.

Smuggling also played a role in the chocolate consumption of British North America. From Mather Byles's letter books, ca. 1785, part of *The Winslow Papers*, a collection of Loyalist era journals and personal correspondence, there is an interesting account of smuggling chocolate:

Waddington had been detected running brandy and chocolate and was under prosecution for it at the time of the election which is the foundation of all the squibs upon him. [6]

Both chocolate and brandy were prized goods and the English sometimes took their hot chocolate with a little brandy (Fig. 24.4).

CHOCOLATE IN THE NAVY

The Winslow Papers (Fig. 24.5) also contain accounts of cocoa being bought and sold between captains and merchants, likely as a result of privateering.[9] As in Perkins's diary, a fair price for cocoa remained a preoccupation.[10] There also were accounts of soldiers drawing on the military for chocolate supplies.[11] By the mid-19th century, approximately half of all raw cocoa shipped to Britain was destined for use by the Royal Navy.[12] Some of the earlier duty and taxation records, as listed in the *Blue Book* statistics for Nova Scotia and Cape Breton in the first quarter of the 19th century, consistently note that cocoa may be imported free of duty "when imported for the use of His Majesty's Navy and Army" [7].

During the War of 1812, correspondence between officers in the Royal Navy and the Provincial Marine in British North America indicated that chocolate was not simply a privilege of the officers, as it was with the French militia in the previous century; it had become an expected ration for sailors and soldiers. In a letter dated May 9, 1813, R. H. Barclay, Captain and Senior Naval Officer, stationed in Kingston on Lake Ontario (then a region known as Upper Canada) wrote to Quebec to request supplies and approval for the construction of a new ship, and to request additional ration items for the crews from the Royal Navy, who were due to arrive in Kingston to join members of the Upper Canadian force:

In the Royal Navy, the sea men are allowed Butter, and cheese, or in cases where these are not to be obtained Cocoa, and Sugar; I perceive that neither is allowed in the Provincial Marine. I beg leave to suggest to his Excellency the propriety of sending up from Quebec a sufficient supply of these

Table 24.2 Chocolate: Outward Bound from Halifax

Year	Date	Cocoa	Chocolate	Destination	Ship Type	Ship Name	Master	Tons	Crew Number	Owner
1750	June 27		6 boxes	Rhode Island	Sloop	*Little Moses*	Israel Boardman	25	3	Jacob Hart
1752	March 21		10 boxes	Chignecto	Schooner	*Anna*	Zech Foss	30	4	Zech Foss & Co.
	April 14		2 boxes	Newbury	Schooner	*Jolly Robin*	Thomas Follingsby	50	4	John Knight & Co.
	April 22		200 (?)	Newbury	Sloop	*Phoebe*	Jonathon Buck	50	4	Jonathon Buck & Co.
1757	August 29		11 boxes	Newport	Sloop	*Lydia*	John Ollesbe	30	5	John Sweeney & Co.
Missing data	xx	xx	xx	xx	xx	xx	xx	xx	xx	xx
1811	July 13	10 bags		Quebec	Schooner	*Margaret*	A. Cameron	78	5	J. Broian (sp?)
	July 20		40 boxes	Newfoundland	Schooner	*Ann*	D. Kelly	70	4	D. Kelly
1812	April 22		25 bags	Quebec	Schooner	*Industry*	F. Demuel	84	4	F. Demuel
	April 25		8 boxes	New Brunswick	Ship	*Neptune*	J. Smith	483	23	Fon?gth Black & Co.
	July 18		25 boxes	Quebec	Schooner	*Industry*	F. Demuel	84	4	F. Demuel
	September 12		25 boxes	Quebec	Schooner	*Linnet*	J. Koch	95	5	L. Doyle
1813	October 5		21 tierces	Quebec	Schooner	*Nathan*	W. Baine	108	5	E. Perkins
Missing data	xx	xx	xx	xx	xx	xx	xx	xx	xx	xx
1816	January 17	xx	25 boxes	Newfoundland	xx	xx	xx	xx	xx	xx
	June 25	10 bls		New Brunswick	xx	*King*	xx	xx	xx	xx
	July 9	xx	5 boxes	BWI (British West Indies?)	xx	*Bellona*	xx	xx	xx	xx
1819	January 27	xx	20 boxes	Newfoundland	xx	*John*	xx	xx	xx	xx
	May 4	xx	100 boxes	Quebec	xx	*Three Sisters*	xx	xx	xx	xx
	June 3	xx	20 boxes	Newfoundland	xx	*Mary Ann*	xx	xx	xx	xx
	August 19	xx	20 boxes	Quebec	xx	*Alligator*	xx	xx	xx	xx
	September 1	xx	42 boxes	Quebec	xx	*Garland*	xx	xx	xx	xx
	October 7	xx	43 boxes	Quebec	xx	*William*	xx	xx	xx	xx
1820	August 23	xx	80 boxes	Quebec	xx	*Glasgow*	xx	xx	xx	xx
	October 7	xx	7 boxes	Newfoundland	xx	*Four Sons*	xx	xx	xx	xx

Sources: CO221 Colonial Office, Nova Scotia and Cape Breton Miscellanea, B-3228, Shipping Returns—Nova Scotia—No. 28; 1753–1757; No. 30; 1758–1761; No. 31; 1762–1767; No. 32; 1811–1815; and No. 22; 1816–1820.

In my letter by the express I gave you the secret history of the St. Johns election and I now inclose a budget of the Halifax puffs which Belcher was industrious enough to collect for me – they will serve to amuse especially as you know the persons. Waddington had been detected running brandy & chocolate, & was under prosecution for it at the time of the election which is the foundation of all the squibs upon him.

Inclose a number of Letters which [illegible] in Dalzel

Sally will for[illegible] one from Mr. Watson w[illegible] is really many strains higher than I thought so costive a man could ever reach, and I'm sure it must gratify you vastly.

The appointment of Sir Guy (now Lord) Carleton Governor General Mr. Watson Commissary General Mr. Coffin Secretary Mr. Smith Chief Justice in Canada & Mr. Achmuty Chief Justice in Nova Scotia will hardly be news at the time this letter reaches you – but

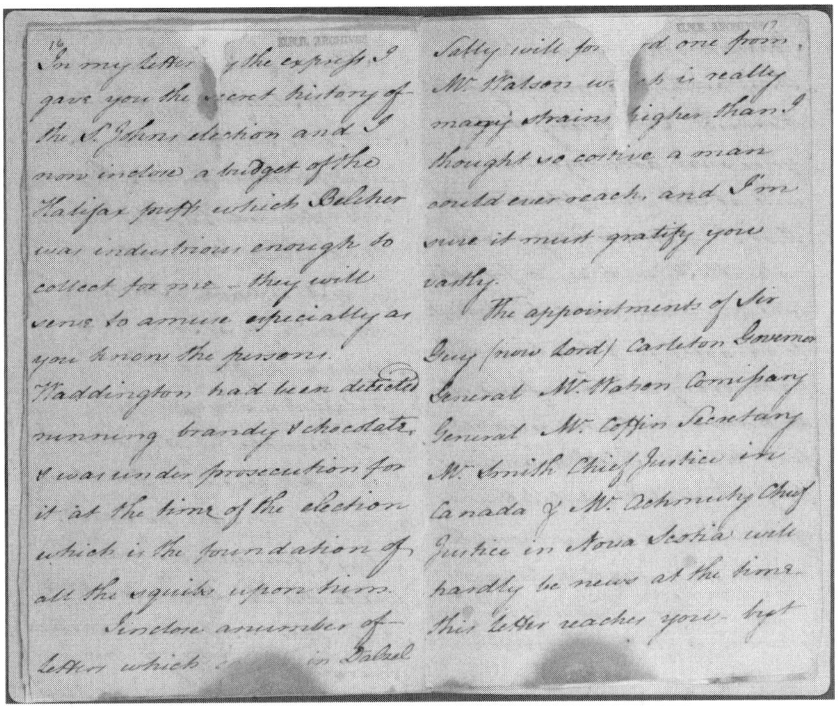

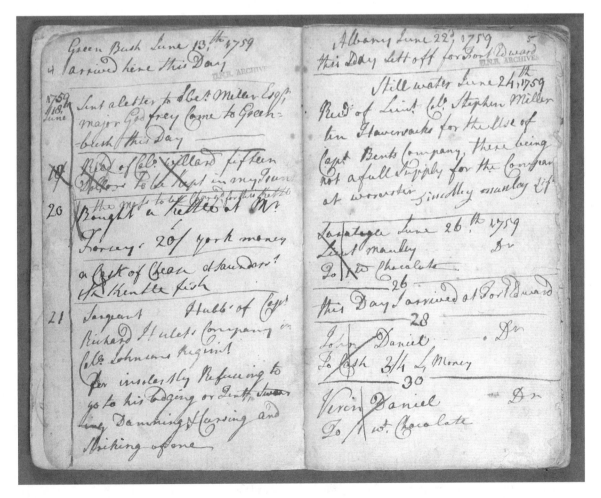

FIGURE 24.4. Chocolate smuggling. *Source: Mather Byles' Letter Books*, No. 4, pp. 16–17, dated 1785. Courtesy of The Winslow Papers, University of New Brunswick. (Used with permission.)

Green Bush June 13th 1759
4 arrived here this Day
1759 18th June Sent a letter to Lieut. Miller Esqr
major Godfrey Come to Green-
bush this Day
19 Recd of Col. Willard fifteen
Dollars to be kept in my [illegible]
the mess to be [illegible]
20 Bought a kettle at Mr.
Lorey 20/ york money
a Cask of Chees at Saunders
[illegible] kettle fish
21 Sargeant Stubbs of Capt
Richard Hulets Company in
Coll. Johnsons Regimt
for insolently Refusing to
go to his Lodging or Duty swear
-ing Damning & Cursing and
Striking of one

Albany June 22d 1759 5
this Day left off for Fort Edward
Still water June 24th 1759
Recd of Lieut. Col. Stephen Miller
ten Haversacks for the Use of
Capt. Bents Company, there being
not a full Supply for the Company
at worcester Timothy Manley Lt
Saratoga June 26th 1759
Lieut. Manley Dr
To 1 wt Chocolate
26
this Day I arrived at Fort Edward
28
John Daniel Dr
To Cash 3/4 L Money
30
Vercin Daniel Dr
To 1 wt. Chocolate

FIGURE 24.5. Chocolate as a military ration. *Source: Stephen Miller's Letter Books*, No. 1, p. 5. Dated 1759. Courtesy of The Winslow Papers, University of New Brunswick. (Used with permission.)

provisions to be ready to issue to the crews when they arrive as part of their accustomed Rations. [8]

Barclay also suggests sending up tobacco with the cocoa and sugar.

By June 2, 1813, a scale of provisions for the men engaged in the Lake Service had been established. His Excellency the Commander of the Forces allowed that all seamen should receive the same rations, be they members of the Royal Navy or the Provincial Marine. The daily rations for each man were to be distributed as follows:

One pound of Flour or Biscuit, Half a pint of Rum, 9 & 1/7 Ounces of Beef, Four and 4/7 Ounces of Pork, 2/7 Pint of Pease [sic], 3/14 Pint of Oatmeal, 6/7 Ounces of Sugar, 6/7 Ounces of Butter or Sugar, One and 5/7 Ounces of Cheese, Rice or Cocoa, 1/14 of a Pint of Vinegar. [9]

As a military foodstuff, cocoa did not spoil, was easy to transport and to prepare, and would warm and fortify the body. It would continue to be an important ration for British, Canadian, and American soldiers of all ranks in the 20th century.

"To Be Sold Cheap for Ready Money"—Early Canadian Chocolate Advertisements

While commercial chocolate manufacture would not establish itself in Canada until the early 19th century, chocolate was a readily available commodity in Colonial British North America. The earliest advertisement found for chocolate in Canada appeared in *The Halifax Gazette* on March 30, 1752, the second issue of the paper, which contained two listings of chocolate for sale.[13] *The Halifax Gazette* was the first newspaper published in Canada. The merchant John Codman carried "Bohea Tea, Loaf and Brown Sugar, Chocolate, Coffee, Starch, Mustard, Chandlery Ware" among numerous other sundry items for sale [10]. Cornelius Durant advertised chocolate "imported in the last Vessels from Boston and to be sold cheap for ready money" at the store of Samuel Shipton [11].

It is exciting to see chocolate represented so early on in Canada's publishing history. The advertisements are typical for their time as regards the type and variety of goods sold by one merchant; Codman and Shipton both had a number of subsequent identical advertisements in later editions of the paper. Advertisements for coffeehouses, "where the best coffee, tea, chocolate and other Refreshments will be provided" appear as early as the April 25, 1752 edition of the paper [12]. An advertisement for wood chocolate mills (presumably wooden mills used for frothing chocolate) appeared in *The Halifax Gazette* in 1785 (Table 24.3).

"Our Only Spark of Comfort": Chocolate in Canadian Exploration, Fur Trade, and Early Arctic Voyages

There is an amusing but unfortunate tendency for writers of popular food histories to use a single-food subject text as a means to exaggerate the importance of a given food's role and place in history. That said, while chocolate certainly did not assure the success of Canadian explorers and voyagers in conquering this vast and unforgiving wilderness, it was an important, at times even vital provision to carry into the woods and snow. Hardened chocolate cakes from earlier centuries differed considerably from the "eating chocolate" consumers enjoy today. Still, chocolate continues to be a desirable staple for hikers and modern-day adventurers. It was, and remains, a valuable source of carbohydrates, fats, and calories; the chemical properties of cacao, combined with the sweeteners added to the prepared paste, provided important sources of energy for explorers in the Canadian wilderness. It was easy to transport and could be made into a beverage if water and fire were on hand, or could be nibbled plain as a "boost" to the system. Prepared hard chocolate cakes kept well and survived a range of temperature fluctuations; chocolate was often left in caches for fur traders and arctic exploration parties—welcome provisions after long, cold, and tiring treks.

THE FUR TRADE AND EXPLORATION OF THE INTERIOR

Chocolate was a common item in inventories from fur forts. While not an everyday staple among fur traders, it was prized as a food that kept well, traveled well, and provided energy. Early Canadian exploration accounts contain inventories listing chocolate as a supply item, but it was likely a foodstuff that only certain classes, ranks, or wealthier individuals would have possessed.[14] A recipe for *biscuits au chocolat* (dated to 1750) is cited as a possible menu item for a winter's meal at a Canadian fur fort ca. 1799, although its inclusion in the hypothetical menu may only be because the recipe was traditional for the time and chocolate was a desirable ingredient among traders and explorers.[15] Chocolate and flour would have been shipped into interior posts by canoe; the eggs called for in the recipe could have been obtained from domestic fowl raised on site.

The North West Company (NWC), a fur trading company based in Montreal, operated in Canada from 1779 until 1821. The NWC was one of the two principal fur trading companies in Colonial Canada, the other being the Hudson's Bay Company

Table 24.3 Atlantic Canada Newspaper Survey: Data for Nova Scotia

Year	Date	Newspaper	Merchant or Auctioneer	Product	Vessel and Provenance	Advertisement Text and Notes
1752	March 30, 1752	*The Halifax Gazette*	Cornelius Durant; Mr. Shipton	Chocolate		Imported in the last vessel from Boston and to be sold cheap for ready money, by Cornelius Durant, at Mr. Shipton's near the North gate. . . . cheap for ready money; n.b. said Codman buys oyl, blubber.
	March 30, 1752	*The Halifax Gazette*	John Codman	Chocolate		
	April 6, 1752	*The Halifax Gazette*	Nathans and Hart	Chocolate		Just imported and to be sold by Nathans and Hart, at their dwelling house in Hollis Street, opposite to his Excellency's for ready money. (*This ad repeats once. His Excellency refers to the Lieutenant Governor of Nova Scotia.*)
	April 18, 1752	*The Halifax Gazette*	Samuel Shipton	Chocolate		Shipton is selling his goods near the Northgate. (*This ad repeats once.*)
	April 18, 1752	*The Halifax Gazette*	Nathans and Hart	Chocolate		Nathans and Hart will also buy oyl, blubber at their house on Hollis Street.
	May 9, 1752	*The Halifax Gazette*	Joseph Hillborn; Captain Salter	Chocolate		
	December 23, 1752	*The Halifax Gazette*	John Codman	Chocolate		
1753	April 14, 1753	*The Halifax Gazette*	Malachy Salter; Jeremiah Eddy	Chocolate, boxes	*Hannah.* Sloop	Nova Scotia Vice Admiralty. To be sold by public auction by order of said court, on Tuesday next at eleven o'clock a.m. at Mr. Malachy Salter's store, sundry damaged goods, being part of the cargo of the sloop *Hannah*, Jeremiah Eddy, Master.
	April 28, 1753	*The Halifax Gazette*	Joseph Yard; Joseph Fairbanks	Chocolate		Just imported from Philadelphia and to be sold by Joseph Yard at the store of Mr. Joseph Fairbank's . . . all at reasonable rates for ready money. (*This ad repeats once.*)
	July 7, 1753	*The Halifax Gazette*	John Anderson; George Taylor	Chocolate		To be sold by John Anderson at his store, where Mr. George Taylor, baker, lately lived, sundry goods . . . (*This ad repeats once.*)
1755	February 15, 1755	*The Halifax Gazette*	Paul Prichard	Chocolate		
1761	May 21, 1761	*The Halifax Gazette*	Charles Morris, June	Chocolate, single box or large quantity		(*This ad repeats once.*)
1770	June 19, 1770	*Nova Scotia Gazette and Weekly Chronicle*	Andrew Cuenod	Chocolate		Cuenod intends to leave the province and wants all accounts settled. (*This ad repeats seven times.*)
1773	April 27, 1773	*Nova Scotia Gazette and Weekly Chronicle*	John Fillis	Chocolate, good; boxes		Items for sale at Fillis's store. (*This ad repeats nine times.*)
1781	November 27, 1781	*Nova Scotia Gazette and Weekly Chronicle*	John George Pike	Chocolate	*St. Laurence*	

Table 24.3 Continued

Year	Date	Newspaper	Merchant or Auctioneer	Product	Vessel and Provenance	Advertisement Text and Notes
1782	July 9, 1782	*Nova Scotia Gazette and Weekly Chronicle*	Robert Fletcher	Chocolate		*(This ad repeats twice.)*
1783	May 27, 1783	*The Nova Scotia Gazette*	Moore & Tuttle	Chocolate		Auction May 27, 1783 at 1 pm by Moore and Tuttle at Fillis's Wharf
1785	?	*Port Roseway Gazetteer and Shelburne Advertiser*	James Donaldson & Company	Chocolate		
	January 4, 1785	*Nova Scotia Gazette and Weekly Chronicle*	Hugh Chalmers	Wood chocolate mills	*Neptune* (Ship); London	
	April 5, 1785	*Nova Scotia Gazette and Weekly Chronicle*	Samuel Buttle (Buttler?)	Chocolate, by the pound		To be sold at Samuel Buttle's shop next door to Mr. Welch's. *(This ad repeats twice. First instance of price: "chocolate, at 1s.2d. Per pound.")*
	August 1, 1785	*The Royal American Gazette*	McLean & Bogle	Chocolate	*New Hope; Friendship.* London; Glasgow	*(This ad repeats once.)*
1786	January 3, 1786	*Nova Scotia Gazette and Weekly Chronicle*	John Newton, Arthur Goold	Chocolate		Notice is given that duties will be placed on rum, liquors[. . .]chocolate, [etc.] Anyone possessing such articles are directed to "enter" the goods at the Impost & Excise office or be prosecuted. Notice given by Newton and Goold, collectors, through the Impost & Excise office, Halifax. *(This ad repeats twenty-nine times.)*
	February 7, 1786	*Nova Scotia Gazette and Weekly Chronicle*	John Lawson	Chocolate, superior quality		
1790	January 12, 1790	*Royal Gazette and Nova Scotia Advertiser*	John Stealing	Chocolate		Stealing has goods for sale at his store on Water Street. *(This ad repeats eleven times.)*
	July 22, 1790	*The Halifax Journal*	James Moody	Chocolate	*Rashleigh*	*(This ad repeats twice.)*
1791	February 19, 1791	*Weekly Chronicle— Halifax*	P. McAllum	Chocolate	*Neptune.* (Ship); Greenock, UK	McAllum has goods for sale at store formerly occupied by Tritten in Water Street. *(Captain Cambridge, of the "Neptune" is noted as advertiser. This ad repeats seven times.)*
	February 19, 1791	*Weekly Chronicle— Halifax*	Mullowny & Stealing	Chocolate	*Catherine.* (Brig); Philadelphia	"Mullowny and Stealing" have imported a cargo of goods. Stealing has goods for sale at his Water Street Store. *(This ad repeats twenty-two times.)*
1792	August 30, 1792	*The Halifax Journal*	James Moody; George Grant	Chocolate	*Rashleigh;* London	The goods are for sale a little northward of Mr. George Grant at Moody's shop

(HBC). In a collection of documents relating to the NWC, guidelines were laid out by company management for the provisioning of posts and the supplying of proprietors and clerks to these posts. As a winter supply, proprietors were to receive 6 lb of tea, 4 lb coffee, and 4 lb of chocolate. Principal clerks in charge of posts would receive 2 lb tea, 1 lb coffee, and 1 lb of chocolate. Inferior (i.e., lower level) clerks received only 1 lb of tea [13].

The Hudson's Bay Company, likewise, outfitted its posts with chocolate. York Factory was one of the company's principal posts, falling under the jurisdiction of the Council of the Northern Department of Rupert Land (northern Manitoba, on the west coast of Hudson's Bay). In 1824, the Council renewed its request with the Company to provide a vessel on Hudson's Bay to assist with the unloading and loading of a ship from England. A requisition had been made for "fifty rolls of North West twist tobacco at ten to fifteen percent below London prices and 600 lb of chocolate at fifty percent below London prices" [14].

Upper Canada, the region known today as Muskoka and Haliburton (i.e., the landmass between the Ottawa River and Georgian Bay), was explored comprehensively during the late 18th and early 19th centuries. David Thompson, in his *Journal of Occurrences from Lake Huron to the Ottawa River* (1837), raised a sentiment frequently mentioned in personal logs and diaries from the fur trade and exploration era: "For want of something fresh find myself weak, took Chocolate for Dinner with Crackers and [felt] much better" [15]. Chocolate was a welcome source of quick and digestible energy for explorers and traders in the Canadian wilderness. It was a familiar and tasty foodstuff, an essential item that could be counted on to keep well and to revive the body and spirit. It was a welcome comestible in an often strange and harsh environment.

In 1885–1886, Joseph Burr (J. B.) Tyrrell, a hydrographer and meteorological observer with the Geological Survey of Canada, was stationed at Ashe Inlet, Nunavut. Tyrrell's career saw him travel throughout northern Manitoba, the North Saskatchewan River region, and central Alberta, and, most notably, he explored the Barren Lands and Hudson's Bay regions. A provision list from the Ashe Inlet Station in 1886 listed items packed at the end of the season when the station was to be dismantled. The list included one case of cocoa and another notation for 20 lb of cocoa. The case of cocoa was listed among other items from the larder such as butter, currants, and flour and possibly was used in baking. The 20 lb of cocoa was listed along with 60 lb of coffee and one chest of tea.[16]

The Tyrrell brothers (Joseph Burr and James) later explored the territory referred to as the Barren Lands. James wrote of their work and travels in the account *Across the sub-Arctics of Canada*, where conditions often were harsh and storms persisted for days on end. During a particularly cold storm endured while paddling around Dubawnt Lake in Nunavut, James Tyrrell wrote:

This storm continued with fury for two days, and during this time, wet and shivering in the tents, we found our only spark of comfort in the brewing and imbibing of hot chocolate prepared over the spirit lamp. [16]

Chocolate was listed as a provision for the brothers' 3200 mile trek through the Barren Lands in a report issued by James Tyrrell that was presented before the Ontario Land Surveyors Association on February 27, 1896. Tea was the only other beverage listed; three caddies of tobacco were brought along "for use chiefly in securing the goodwill of the natives" [17].

ARCTIC EXPLORATION

There exist numerous accounts of chocolate as a staple in caches on Arctic expeditions. Arrowroot, dried meat, pemmican, dried soups, tea, and chocolate are listed on countless provision lists for Arctic parties throughout the course of the 19th century. Occasionally, condensed milk also is listed with instructions that it be used in the preparation of chocolate, but mostly the chocolate was taken simply mixed with heated water, or eaten plain.[17] In extensive British Admiralty Correspondence records relating to Arctic parties, suggestions were made to improve the provisioning of these parties. Scurvy was of concern, as was the quantity of various items provided. In one such report, a letter entitled "General Remarks connected with Travelling," written to the Admiralty by George F. McDougall, Second Master, commanding Sledge *Endeavour*, McDougall noted that:

The pemmican was made more palatable when mixed with soup. The only alteration we could have wished was the substitution of an additional quarter pound of bread for the same quantity of pemmican, which could easily have been dispensed with and chocolate every morning, instead of tea, which makes but a light meal to travel on. [18]

Chocolate was a common provision for Arctic explorers, among them Sir John Franklin, who brought chocolate from England to the Canadian Arctic. In the initial provisioning lists, Franklin requested 60 lb of coffee essence, 110 lb of tea, and 180 lb of chocolate.[18] Franklin noted chocolate's excellent restorative and keeping properties. He remarked on taking chocolate for breakfast and, on occasion, administering chocolate to help stave off hypothermia among members of the exploration party:

Our friend Augustus was seized with a shivering fit . . . [and] was put between blankets and provided with warm chocolate, and the only inconvenience that he felt the next morning was pain in his limbs. [19]

In his *Narrative of a Journey to the Shores of the Polar Sea, in the Years 1819, 20, 21, and 22*, Franklin's personal account of the first organized trek under his command to find a northwest passage, he mentioned packing up stores before leaving the mouth of the Yellowknife River:

> Our provision was two casks of flour, two hundred dried rein-deer [sic] tongues, some dried moose meat, portable soup, and arrow-root, sufficient in the whole for ten days' consumption, besides two cases of chocolate, and two canisters of tea. [20]

Upon surveying the stores for an excursion on July 15, 1820, he noted:

> Our stock of provision unfortunately did not amount to more than sufficient for one day's consumption, exclusive of two barrels of flour, three cases of preserved meats, some chocolate, arrowroot, and portable soup, which we had brought from England, and intended to reserve for our journey to the coast next season. [21]

It is poignant to note that, following the death of Franklin and his men during their last expedition to the Canadian Arctic in 1845 (aboard the ships *Erebus* and *Terror*), subsequent search parties who went on the quest to solve the mystery behind Franklin's disappearance provisioned themselves similarly to the ill-fated expedition. Between 1847 and 1859, over 30 expeditions were launched. In 1859, the expedition of Francis Leopold (F. L.) McClintock provided the last pieces of the puzzle in the disappearance of Franklin's party. On King William Island, the McClintock party found relics, human remains, and the few written records that survive from the last Franklin expedition. Among items found by members of the search team were "a little tea, forty pounds of chocolate, and a small quantity of pemmican" [22].

Aboriginal Encounters with Chocolate

We did not find evidence that chocolate was a significant item of trade between Canadian fur traders and Northern Aboriginal tribes. Several personal exploration accounts and settlers' diaries, however, reference a substance identified as "wild chocolate" or "Indian chocolate." Wild chocolate was made by boiling dried, crushed bloodroot in water, then adding sugar and milk. The color of the root imbued the liquid with a pinkish hue, thereby suggesting the same color as a cup of chocolate. This practice was common throughout Maine[19] and there are mentions of Indian chocolate in New Brunswick, including a recipe for such by Doctor James Odell, in 1816:

> Boil a good handful of the roots in one quart of water to a pint. Strain this infusion and sweeten it with loaf sugar. A moderate tea-cup will hold as much as the stomach can well bear at a time. [23]

Dr. Odell advises that this concoction may "incite a considerable nausea" as bloodroot is indeed a powerful substance [24].[20]

Yet settlers hungry for a taste of chocolate and wanting the "real thing" might easily have turned to this starchy drink to appease their craving. In the second quarter of the 19th century, Robina and Kathleen MacFarlane Lizars wrote of the social customs, the joys, the hardships, and the innovations and adaptations required of new settlers in Upper Canada:

> Tea, fried pork, and bread and butter were great luxuries; but a tramp through the woods . . . cultivated appetites which looked askance at nothing. Crust coffee[21] without milk or sugar, heavy Indian meal bread, wild chocolate, beef tallow, made their appearance on the menu cards of 1828 at these primeval inns. [25]

Women frequently tended to the fires and the cooking in shanties set up along the lines and rough roads through the new territory. Supply teams might bring in fresh pork. During one description of a cozy fireside scene, shanty-cakes were cooked on the spider (skillet)[22] in the hearth, water boiled, and "wild chocolate was made, sweetened with sugar but innocent of milk" [26].

Pioneering life in the Canadian wilderness was full of extreme hardships and frustrations. A profile given by the MacFarlane Lizars of an especially stoic woman states that despite having survived so many of the heartaches of life in the woods, including poverty, sickness, and the death of several children, the one item she could not do without was her tea. "Wild chocolate and the makeshifts of the country tea-table were her last straw" [27]. When her husband had trouble selling a sheep in the nearby town, he traded the animal for a pound of tea.

Did settlers learn this bloodroot preparation from Aboriginal tribes? Did the Aboriginals themselves ever imbue this substance in any great quantity? It is unclear if "Indian chocolate" is indeed an "Indian" invention or simply the result of the ingenuity required of European inhabitants, looking for trappings of home or "civility" in foreign, unexplored land. What is interesting is that chocolate had become such a fixture in the daily diet that some substitute, even one that only bore scant resemblance to the real item, was desirable.

"Strength to the Whole System": Chocolate and Cocoa Manufacture in Canada

During the French Colonial period, little is known thus far regarding possible indigenous commercial manufacture of chocolate. Chocolate, both "prepared" and "unprepared," that is, either already mixed with sweeteners and aromatics, or simply a plain, hardened cocoa paste, could be had and settlers may have mixed the hot beverage with local products such as maple sugar or honey on occasions when imported sugar was in short supply.

The tools required for home preparation of chocolate also were available, such as chocolate pots and chocolate mills. A dozen chocolate pots were listed for auction in a St. John's, Newfoundland, paper, *The Royal Gazette and Newfoundland Advertiser*, in 1815.[23] A silver pot, labeled as a chocolate pot and crafted by Quebec silversmith Laurent Amiot, is housed in the Archbishop's Palace in Quebec City. Born in Quebec in 1764, Amiot apprenticed briefly in France from 1782 to 1786, and then returned to Quebec. The silver vessel is likely from the late 18th or early 19th century.[24] It is unclear from photographic evidence if the pot has the characteristic hole in the lid that would conclusively distinguish it as a chocolate pot. Amiot was a premier craftsman of ecclesiastical silverware and the clergy frequently consumed chocolate, especially during the Lenten fast. The Catholic Church, after much debate, decreed that the consumption of chocolate did not violate the abstinence from food called for during the fast; chocolate, therefore, was a much-welcomed source of energy and calories.

The 19th century brought major developments in chocolate processing and manufacturing that led to its increased position as a more affordable and common commodity; it also became a more refined product, akin to the "eating chocolate" recognized today. Chocolate continued to be consumed as a beverage, but it also developed as both a confectionery and as an ingredient in baking.

PROTECTING A BLOSSOMING DOMESTIC INDUSTRY

Mr. John F. Ferguson established in Halifax what is possibly Canada's earliest chocolate mill and factory, ca. 1809. It is around this time that cocoa imports to Halifax begin to increase significantly (Table 24.4). By 1833, the Nova Scotia House of Assembly reported on a petition on behalf of Mr. Ferguson who sought a bounty on the manufacture of chocolate and a tax on foreign (imported) chocolate. The final report on John Ferguson's petition, in 1839, reads in part as follows:

That the petitioner has carried on the manufacture of Chocolate in Halifax for upward of thirty years; and last year manufactured about 100,000 lbs, weight, yielding upwards of £2000. A large proportion of this quantity was exported by the petitioner to Canada, NB [New Brunswick], Newfoundland, and the West Indies; and the cheaper of the two sorts manufactured by him, which he can afford to sell at 7d. a pound, is used extensively by the Fishermen of our own Province. [28]

The document goes on to highlight the importance of taking such protectionist measures.

If he is obliged to pay the duty of 5 s [shillings] Sterling on Cocoa, besides the Imperial Duties on Lard and Flour, which are largely used in his manufacture, and cannot be protected either by a drawback or bounty, he is exposed to a most unequal competition with the American Manufacturer, who gets all the Raw Material he uses, free of duty; and the consequence will be the surrender of the Newfoundland, West India, New Brunswick and Canada market, to the Foreign supplier and the destruction of our own Manufacture and Export. [29]

In the end, Ferguson won his petition and was granted a bounty on chocolate, paid quarterly by the province. It is interesting to note the possible use of flour and lard in Ferguson's chocolate manufacture. Perhaps this yielded a cheaper product, affordable to fishermen as a portable energy supplement. Was Ferguson's intention to add additional, cheaper fats and carbohydrates to the product? As the domestic manufacture of chocolate and other products grew throughout Canada and Nova Scotia, so too did concerns over adulteration (see Chapter 47). Parliamentary papers from the mid-19th century detail the introduction of regular food inspections for chocolate and cocoa, among other comestibles such as alcohol, bread, coffee, milk, and spices.[25]

"TRANQUILITY TO THE NERVES" — CANADIAN COCOA MANUFACTURE AND MARKETING

With his invention of the cacao screw-press in 1828, the Dutchman Conrad van Houten introduced a method for processing the ground cocoa beans and nibs that effectively removed about two-thirds of the cacao butter from the chocolate paste, leaving the powder that became known as cocoa. Cocoa's solubility in water and improved digestibility (compared with the full-fat chocolate) led to a new, expanded industry of cocoa manufacturing, and powdered cocoa eventually would take over from grated chocolate pastes as the preferred way to prepare drinking chocolate.

Powdered cocoa was marketed heavily from the middle of the 19th century onward. Billed as a healthful drink, beneficial to the sick and frail, more easily digested by the old and young, and more healthful than the fatty chocolate pastes of the previous era, cocoa was twinned with such adjectives as "hygienic," "soluble," "homeopathic," and "absolutely pure" (see Chapter 9).[26] An advertisement for Baker's cocoa, listed in *Canadian Grocer* toward the end of the 19th century, stated that their breakfast cocoa:

From which the excess of oil has been removed, is absolutely pure and it is soluble. . . . It has more than three times the strength of cocoa mixed with starch, arrowroot or sugar, and is therefore far more economical, costing less than one cent a cup. It is delicious, nourishing, strengthening, easily digested, and admirably adapted for invalids as well as for persons in health. [30]

Baker's, started in 1780, was one of the earliest chocolate and cocoa mills in the United States and remains a popular baking brand (see Chapter 26). The mid-19th century saw two major domestic cocoa and

Table 24.4 Cocoa: Inward Bound to Halifax

Year	Date	Cocoa	Chocolate	From	Ship Type	Ship Name	Master	Tons	Crew Number	Owner
1811	September 28	3 barrels	xx	Martinique	Schooner	*Mary*	J. Bouratte	89	6	J. Mahon
	October 7	37 bags	xx	Miramichi	Schooner	*Providence*	F. Langlois	73	3	B. Bouche
	November 7	74 bags	xx	Jamaica	Brig	*Arabella*	P.Burley	103	8	J.M. Morris
	December 21	10 bags	xx	Jamaica	Schooner	*Lark*	J. Moch	90	7	J. Woodward
1812	February 1	12 tierces	xx	Jamaica	Brig	*Margaret*	J. Ayres	198	10	A. Belcher
	May 12	3 casks	xx	Guadeloupe	Brig	*Dasher*	A.G. Kish	132	7	E. Collins
	August 28	3 casks	xx	Martinique	Ship	*Berkeley*	W. Jackson	201	13	A. Belcher
	November 18	4 tierces, 180 bags	xx	Surinam	Schooner	*Mary*	M. Donaldson	114	7	E. Connelly
Break in statistics due to unreadable microfiched documents										
1814	December 2	4 bags	xx	Trinidad	Brig	*E.L. Sherbrooke*	E. Rumford	186	8	G. Miller
1815	Unreadable	12 bags	xx	Boston	Unreadable	Unreadable	Unreadable	Unreadable	Unreadable	Unreadable
1816	April 26	6 bags	xx	Surinam	Brig	*Duke of Kent*	In. Ames	132	6	L. Doyle
	April 29	2 bags	xx	Liverpool		*Ajax*				
	May 15	214 bags	xx	Jamaica		*Parker*				
	May 22	50 bags	xx	St. Lucia		*George McIntosh*				
	October 4	83 barrels	xx	Grenada		*Paragon*				
1818	May 8	38 bags	xx	Grenada		*Sarah Jane*				
	August 18	40 bags	xx	Boston		*Four Sons*				
1819	April 3	55 lb (?)	xx	Grenada		*Douglas*				
	May 20	18 bags	xx	Grenada		*Thalia*				
	September 21	100 barrels	xx	Grenada		*Helen*				
	November 18	18 lb (?)	xx	Jamaica		*Margaret*				
1820	June 2	5 bags	xx	Berbice		*Lyon*				

Sources: CO221 Colonial Office, Nova Scotia and Cape Breton Miscellanea, B-3228, Shipping Returns; Nova Scotia, No. 32; 1811–1815; and No. 33; 1816–1820.

chocolate manufacturers begin operating in Canada. John P. Mott started Mott's Spice & Cocoa Company in Dartmouth, Nova Scotia, around 1853.[27] He advertised heavily in maritime newspapers in the 1850s, extolling the virtues of his cocoa (Table 24.5). One such advertisement reads as follows:

Motts Broma is one of the most innocent and nutritious beverages that can be obtained from the cocoa nut—it imparts tranquility to the nerves and gives strength to the whole system; Mott's soluble cocoa is instantly soluble and a cup of boiling water added results in hot chocolate; cocoa is highly nutritious and agreeable; from the facility with which it may be prepared, it is particularly convenient for travelers and is sold for 6d/package. [31]

Evidence suggests Mott did quite well by his cocoa and chocolate manufacture. A William Notman photograph of Mott in a carriage, in front of what is likely Niagara Falls, suggests that Mott had the leisure time and the means to travel.[28] Professional photography was a rather expensive proposition at the time, generally reserved for members of high society. Hazelhurst, Mott's residence in Dartmouth, was a lavish estate.[29] By the end of the 19th century, Mott's company was producing upward of 15 different cocoa and chocolate products, ranging from "homeopathic" cocoa powder, broma, and breakfast chocolate, to cocoa nibs and shells, various sorts of cooking chocolate, and chocolate liquors.[30]

Mott makes an appearance in the legal background of another prominent Nova Scotian in the chocolate business—James Moir, son of William Church Moir. James Moir received permission from his father in 1873 to shift the concentration of the Moir Steam Bakery and Flour Mill from biscuits to chocolate. Court summons suggest that Mr. Moir was, at times, lax in paying his invoices to Mr. Mott.[31] Presumably, Mott sold some form of cocoa or raw chocolate to Moir.

Another Canadian cocoa manufacturer who started business in the 19th century was the Cowan Company of Toronto. John Warren Cowan, like Mott, was an adept self-promoter. Tramcar advertisements, bookmarks, and collectible trading cards—popular promotional items of the time (see Chapter 14)—were all used by Cowan to promote his cocoa. One set of collectible cards depicts Birds of Canada.[32] Trade cookbooks were becoming a popular marketing tool for manufacturers. Cookbooks published by the Walter Baker & Co. feature their distinctive emblem *La Belle Chocolatiere* on the cover.[33] The Cowan Company published a number of cookbooks promoting their cocoa and ideas for its culinary use. *Cowan's Dainty Recipes* was first published in 1915 and underwent many subsequent editions. The book gives tips on measuring and melting chocolate and lists many recipes for chocolate beverages, both hot and cold. The dessert recipes range from custards, puddings, and ice creams, to sauces, cakes, cookies, and candies (Fig. 24.6).

Cowan also produced a wide range of chocolate and cocoa products at the turn of the century. Mott, Cowan, Walter Baker, and another Canadian company, Todhunter, Mitchell, & Co., all sold cocoa and chocolate for baking and eating to the Canadian market. The *Price Current* section of the January 2, 1891 edition of *Canadian Grocer* listed approximately 20 items in Cowan's repertoire.

THE CHOCOLATE MAKERS

The year 1880 heralded a major development in chocolate processing. Rodolphe Lindt developed the process of conching. The creation of a palatable eating chocolate involves adding some of the extracted cocoa butter back to the ground cocoa. Lindt increased the amount of cocoa butter added back to the chocolate mass and with a slow, steady process of heating and mixing, created a smoother product with a pleasant mouth feel. Lindt's innovation paved the way for the growing popularity of chocolate as a confection. Combined with falling prices of the raw commodity and an increase in the disposable income of the middle class, the era of the beloved foodstuff as we know it today was ushered in (see Chapter 18).

Improved eating chocolate, in turn, meant the launch and growth of chocolate candy companies, and many of the Canadian chocolate makers mentioned in the introduction to this chapter launched their businesses in the last quarter of the 19th century or in the first quarter of the 20th century. Both Moir and Ganong started business in 1873, although chocolate confectionery was not their immediate focus. Ganong started as a general merchant, and Moir sold biscuits. Chocolate, however, soon proved to be a viable business for both families.

Ganong Bros. Ltd., family-owned and operated and still based in St. Stephen, New Brunswick, lays claim to many Canadian "firsts" in the chocolate candy business. In 1910, Arthur Ganong invented and introduced the first 5-cent chocolate nut bar in North America. They were the first confectioner in Canada to use cellophane in packaging and in 1887 the Ganongs installed the first lozenge-making machine in Canada. Still in use today, it may possibly be the world's oldest candy-making machine in operation. Ganong also introduced the heart-shaped box to the Canadian marketplace; it originally was issued at Christmastime.[34]

Early in the 20th century, competition grew. The Walter M. Lowney Company, William Neilson Ltd., and Willard's Chocolates Ltd. all established factories in Canada by 1915. In Toronto, the first Laura Secord shop or "studio," as originally called, opened in 1913. It is interesting that both Ganong and Laura Secord chose to use Canadian heroines as the face of their chocolates. Laura Secord braved danger to warn British forces of an impending American attack during the War of 1812. As the masthead for the *chocolatier*

Table 24.5 Atlantic Canada Newspaper Survey: Data for Prince Edward Island

Year	Date	Newspaper	Merchant or Auctioneer	Product	Vessel and Provenance	Advertisement Text and Notes
1818	January 12, 1818	*PEI Gazette*	Paul Mabey	Chocolate	*Success* (Schooner)	Just received this morning.
1830	October 19, 1830	*Royal Gazette*	John Morris	Chocolate		
1832	January 3, 1832	*Royal Gazette*	John Morris	Chocolate		
	December 4, 1832	*Royal Gazette*	A. MacDonald	Chocolate		A. MacDonald has just received, by last arrivals from Halifax, his winter supply, which he is selling at reduced prices for cash or barter.
1838	January 6, 1838	*Colonial Herald*	Josiah Parkin	Chocolate		
1842	December 17, 1842	*Colonial Herald*	Donald MacDonald; Mr. Brenan; Irivng & K'Kay	Chocolate, 6 boxes	*Walthron* (Schooner); Halifax	Auction; to be sold, at the warehouse of Mr. Brenan, Sydnet St., on Tueday, the 20th inst. At 12 o'clock; and immediately after, at the store of Messrs. Irving & K'Kay, Queen St.; landed from the schooner *Walthron*, from Halifax, the same being damaged, and ordered to be sold for the benefit of all concerned.
1851	July 18, 1851	*Islander*	M.W. Skinner	Chocolate: Mott's, Howard's, Thomas's, other		Mr. Skinner advertises he has everything needed by the sick and invalids. (*This ad repeats seven times.*)
	November 7, 1851	*Islander*	John Rigg	Chocolate		New Cheap Cash Store. (*This ad repeats five times.*)
	November 14, 1851	*Islander*	S.C. Holman	Chocolate		The subscriber hourly expects the arrival of the following goods . . . (*This ad repeats seven times.*)
1852	April 30, 1852	*Islander*	S.C. Holman	Chocolate		(*This ad repeats seven times. This ad was also inserted in "Haszard's Gazette".*)
	May 14, 1852	*Islander*	A.H. Yates	Chocolate, Preston's patent, 56 boxes	*Hope* (Schooner); Boston	The auction will take place on board the schooner *Hope* at 10 o'clock, on Saturday, May 15, 1852. Terms liberal, and made known at sale
	May 28, 1852	*Islander*	A.H. Yates	Chocolate, Preston's patent		The auction will take place on Tuesday, June 1, 1852, at 11 o'clock
	June 18, 1852	*Islander*	John Andrew MacDonald	Chocolate		(*This ad repeats five times.*)
	July 2, 1852	*Islander*	S.C. Holman	Chocolate		(*This ad repeats seven times.*)
	October 10, 1852	*Islander*	William Elliot & Company; Henry Palmer	Chocolate		Constantly on hand; large supply sold as low as they can be imported. (*This ad repeats seven times.*)
	October 29, 1852	*Islander*	Frederick Norton; S.C. Holman; George Greig	Chocolate		Holman died and Norton was given the power of attorney by Greig to settle all accounts and to sell the following goods. (*This ad was carried by "all the papers" (no names given) and the Islander ran the ad three times.*)

Year	Date	Newspaper	Advertiser	Product	Vessel / Origin	Notes
1853	April 15, 1853	*Islander*	James Morris	Chocolate		Auction on Monday, April 25, 1853 at 11 o'clock . . . (*This ad repeats once.*)
	June 10, 1853	*Islander*	James Morris; Charles Dempsey	Chocolate, 40 boxes		Executive and unreserved auction, auction to take place Thursday, June 30 at 12:00. (*This ad repeats twice.*)
	November 11, 1853	*Islander*	H. Haszard	Chocolate	*Sir Alexander and Helen* (Brigs); England	Just received. (*This ad repeats twice.*)
	November 25, 1853	*Islander*	John W. Morrison	Chocolate		Morrison has completed his stock of Fall and Winter goods; low prices for cash. (*This ad repeats seven times.*)
	December 23, 1853	*Islander*	Henry Palmer; William Elliot & Company	Chocolate		Boston merchants have the following constantly on hand in Charlottetown; to be disposed of as low as they can be imported.
1854	February 24, 1854	*Islander*	Edward Castell	Chocolate		Castell keeps constantly a large assortment of confectionary including confectionary of his own manufacture. Orders from the provinces are respectfully solicited, and all goods are carefully packed fro transportation without extra charge. (*This ad repeats seven times.*)
	November 17, 1854	*Islander*	H. Haszard	Chocolate	*Cicely, Peeping Tom, and Anne Hall* (Schooners) London; Liverpool	Just received; extensive supply of British and foreign new Fall goods. (*This ad repeats twice.*)
1855	December 8, 1854	*Islander*	E. Parker	Chocolate	*Rapid and Abigail*	Fall stock on sale low at new store.
	January 15, 1855	*Examiner*	Benjamin Chappell	Chocolate	*Rapid and Abigail*; London	Mr. Chappell also has land for sale. (*This ad repeats once.*)
	January 15, 1855	*Examiner*	E. Parker	Chocolate		
	May 11, 1855	*Islander*	M.Y. Mott and Sons	Chocolate, 809 boxes, "warranted superior"		(*Spices listed as well. Mott's of Halifax have taken out a local ad.*)
	June 8, 1855	*Islander*	Beer and Son	Chocolate		Free trade! Goods in excellent condition and will be sold for small advance.
	June 11, 1855	*Examiner*	Beer and Son	Chocolate	*Friends* (Schooner); Boston	Beer and Son are advertising the arrival of "American Goods"
	August 31, 1855	*Islander*	William Smardon	Chocolate, sweet		Notice of a business opening; groceries, Liquors, etc. wholesale and retail.
	September 3, 1855	*Examiner*	Hugh Fraser; William Crabb; William Smardon	Chocolate, sweet		Frazer has taken over the shop that had been occupied by Wiliam Crabb. (*This ad repeats seven times.*)
	September 14, 1855	*Islander*	W.R. Watson	Chocolate		Fall supplies beginning to arive at the City Drug Store Watson advertising Fall supplies. (*This ad repeats six times.*)
	September 17, 1855	*Examiner*	W.R. Watson	Chocolate		
	October 5, 1855	*Islander*	John W. Morrison	Chocolate	*Emily*; Boston	Morrison is selling his fall importation chosen himself from the best houses in Boston; will be sold wholesale and retail; British dry goods expected daily.
	October 5, 1855	*Islander*	Walter Baker & Company	Chocolate, pure		Their chocolate, cocoa, and broma won the first premium at the World's Fair in New York and are recommended by more physicians as the more soothing and nourishing than the more stimulating infusions of tea and coffee and are sold by all principal grocers.

Table 24.5 Continued

Year	Date	Newspaper	Merchant or Auctioneer	Product	Vessel and Provenance	Advertisement Text and Notes
	October 19, 1855	*Islander*	Thomas Williams	Chocolate	*Monte Christo* (Brig)	Fall supplies; does work at reasonable terms . . .
	November 5, 1855	*Examiner*	W.R. Watson	Chocolate		Fall supplies are arriving at the store. (*This ad repeats seven times.*)
	December 14, 1855	*Islander*	M.W. Skinner	Chocolate, No.1		Skinner's drug store is advertising some health foods; Mott's broma is one of the most innocent and nutritious beverages that can be obtained from the cocoa nut—it imparts tranquility to the nerves and gives strength to the whole system; Mott's soluble cocoa is instantly soluble and a cup of boiling water added results in hot chocolate' cocoa is highly nutritious and agreeable, from the facility with which it may be prepared, it is particularly convenient for travelers and is sold for 6d/package.
	December 17, 1855	*Examiner*	M. W. Skinner	Chocolate, No.1		Food for invalids and children. (*Preparations and benefits included in the ad. This ad repeats four times.*)
1856	February 11, 1856	*Examiner*	K. Elderidge	Chocolate	*Lydian Polly* (Schooner)	The auction will take place on the schooner at 11 am. on Feb. 25. (*This ad repeats in the Islander on 02/15/1856.*)
	May 30, 1856	*Islander*	John W. Morrison	Chocolate	*Ann* (Brig)	Sold low for cash. (*This ad repeats June 6, 1856.*)
	October 24, 1856	*Islander*	W.R. Watson	Chocolate	*Elizabeth*; Halifax	
1857	February 6, 1857	*Islander*	W.D. Waters	Chocolate		(*Unclear as to whether these goods are for sale in PEI by a Boston merchant, or if the merchant is simple placing an ad in PEI newspapers, that someone might wish to distribute his goods locally.*)
	March 27, 1857	*Islander*	J. William Morrison	Chocolate		
	December 28, 1857	*Examiner*	Benjamin Chappell	Chocolate		For sale for cash. (*This as repeats six times.*)
1859	June 6, 1859	*Examiner*	Rogers Dodd,	Chocolate drops	*Alma* (Schooner); Boston	
	August 21, 1859	*Examiner*	William Dodd, Hugh Fraser	Chocolate, Chocolate burnt		Auction, Thursday, August 30th, 11 a.m.; Selling the whole of Fraser's stock . . .
1861	August 12, 1861	*Examiner*	Mrs. A. Margaret McKenzie	Almonds, 300 pounds		(*See her ad in Ross's Weekly, August 29, 1861.*)
	August 29, 1861	*Ross's Weekly*	Mrs. A. McKenzie	Chocolate, 100 & 200 lb		Adviser thanks public for support she has received in her wholesale and retail confectionary business. Besides stating goods, she offers a liberal discount to country purchasers. (*This ad repeats once.*)
	October 17, 1861	*Ross's Weekly*	S. W. McKenzie	Chocolate	*Ariel* (Brigantine); Boston	Besides these products, advertiser announces arrival soon of English goods from "Isabel" from Liverpool. (*This ad repeats five times.*)
	December 16, 1861	*Examiner*	W.R. Watson	Chocolate		(*Also runs ad in Ross's Weekly.*)
	December 19, 1861	*Ross's Weekly*	W.R. Watson	Chocolate		(*This ad repeats three times.*)
	December 23, 1861	*Broad-Axe*	Apothecaries' Hall	Chocolate, prepared		
	December 23, 1861	*Examiner*	M.W. Skinner	Chocolate, "prime family article"		Nice things for Christmas. (*Note how chocolate advertised.*)

Year	Date	Newspaper	Merchant	Product	Origin	Notes
1863	June 11, 1863	*Ross's Weekly*	M.W. Skinner	Chocolate, superior, good assortment		
	August 14, 1863	*Semi-Weekly Advertiser*	W.R. Watson	Chocolate, No.1		
1864	May 26, 1864	*Ross's Weekly*	H.J. Richardson	Chocolate	*Atlantic* (Schooner); New York	
1868	July 8, 1868	*Herald*	W.H. MacEachern Wilson	Chocolate		Just What the Doctor Ordered
	September 30, 1868	*Herald*	P. Foley	Chocolate		New provision grocery and sugar store; wholesale, retail, provision and liquor store; low prices
1870	November 8, 1870	*Island Argus*	The Confectionery	Chocolate creams		Assorted "good things" for sale at the confectionery on Queen Street. (*No merchant name given.*)
	December 7, 1870	*Herald*	Alexr. McKenzie	Chocolate		Confectionery; sell cheaper than can be imported for cash. (*Related to Mrs. A McKenzie?*)
1871	January 24, 1871	*Island Argus*	The Confectionery	Chocolate creams		Everton Taffy chocolate drops, fig cream candy, maple candy, and a variety of other confections are always to be found at the confectionery. (*This ad appears in Island Argus, May 9, 1871.*)
	April 25, 1871	*Island Argus*	The Confectionery	Chocolate drops		
	May 10, 1871	*Herald*	Fenton T. Newberry	Chocolate	*Alhambra* (Steamer); Halifax	Received on consignment from the manufactory of J.P. Mott, Halifax; Warranted.
	May 30, 1871	*Island Argus*		Chocolate creams; eating chocolate		
	July 4, 1871	*Island Argus*	Fenton T. Newberry	Chocolate, Mott's Chocolate		
	July 5, 1871	*Herald*	Fenton T. Newberry	Chocolate		
	August 1, 1871	*Island Argus*		Chocolate cream drops		
1877	September 11, 1877	*Examiner*	Fenton T. Newberry and Company	Chocolate, Mott's breakfast, 7 boxes		
	September 17, 1877	*Examiner*		Chocolate creams		
1880	December 17, 1880	*Weekly Examiner*	W.F. Carter	Chocolate caramels, very nice		
1881	January 7, 1881	*Weekly Examiner*	B. Balderston	Chocolate caramels; spiced chocolate sticks		Balderston the confectioner stocks the best confectionary in the Dominion
1883	December 12, 1883	*Herald*	W.R. Watson	Chocolate		A Merry Christmas, Happy New year; Christmas presents, requisites at Watson's Drug Store
	December 21, 1883	*Herald*	B. Balderston	Chocolate drops		
1892	December 14, 1892	*Herald*	A. Quirk, C. Quirk	Chocolate candy cream, mixed		Good things for X-mas and New Year now in stock

FIGURE 24.6. *Cowan's Dainty Recipes* (Cover). *Source:* Cowan Company of Toronto. *Cowan's Dainty Recipes—Dainty & Delicious Dishes Prepared from Cowan's Cocoa and Chocolate.* Toronto, Canada: The Cowan Company of Toronto, 1915. Courtesy of Nestle Canada, North York, Ontario. (Used with permission.) (See color insert.)

she has undergone various "beauty treatments" in her time, morphing from *dour doyenne* to the rosy-cheeked lass who adorned chocolate boxes at the end of the 20th century. In 1904, the Ganongs introduced the figure of Evangeline on their chocolate boxes, the romantic heroine of Acadia, encapsulated in the poem *A Tale of Acadie* by Henry Wadsworth Longfellow. She embodied the qualities of purity, excellence and constancy, romance, sentiment, and sweetness—in short, the qualities Ganong wished to associate with their chocolates. Evangeline was a constant with Ganong until 1978 (Fig. 24.7).

Conclusion

Because Canada's colonial history is defined first by French inhabitants, and then by British settlement, chocolate consumption patterns in Canada served as reflections of chocolate's cultural status in France and England. The Fortress of Louisbourg maintained, for the most part, the social hierarchy associated with chocolate drinking in France—it was an indulgence reserved primarily for the nobility (see Chapter 25).

FIGURE 24.7. Laura Secord chocolate, 20th century. Style of box produced between 1920 and 1950. *Source:* Courtesy of McCord Museum, Montreal, Canada. (Used with permission.)

British consumption patterns, on the other hand, represented the more democratic status of chocolate in England. It was served in coffeehouses and was available to anyone who could afford it. The enterprising spirit of England in the 17th and 18th centuries came to the North American colonies, where chocolate was readily advertised, retailed, imported, and exported between the colonies. Local industries eventually were encouraged, spurred along by raw materials from English cacao plantations, the Industrial Revolution, favorable taxation, and the rise of the middle class. Although tea eventually surpassed chocolate as the beverage of choice among the English, chocolate manufacture was supported and promoted as a valuable and practical provision for explorers, traders, and members of the military.

Chocolate in Canada also reflected the American evolution of chocolate consumption and the value placed on cacao. Cocoa was a target for smuggling and was always a desirable prize for a privateer, something that could fetch good prices at auction. In reading the pricing preoccupations of someone like Simeon Perkins, or scanning the weekly *Prices Currents* for cocoa and chocolate in merchant's publications such as *Canadian Grocer* or *Maritime Merchant*, it is intriguing to consider how the early, sometimes volatile, trade in cocoa and the commercial manufacture of chocolate eventually led to the founding of the Cocoa Exchange.

Chocolate fortified Canada's earliest explorers and settlers. Providing energy in a durable, portable, digestible form, hardened chocolates pastes were carried by soldiers and fur traders alike. Arctic adventurers could come across chocolate in provisions caches, months later, and it would provide welcome relief. While the thought of chocolate stored in the Arctic wilderness for months on end might not tempt modern palates, it is unlikely that the chocolate of the 18th and 19th centuries would have appealed in the first place; a coarse, fatty, gritty, and unevenly bitter substance, it was a far cry from the eating chocolate enjoyed today in the 21st century. And yet chocolate has never lost its ability to simultaneously boost the system and comfort the soul.

Chocolate consumption eventually became concentrated in the home. Hot cocoa was given to the ill or frail, to children and to the elderly. It was a popular breakfast beverage. It also gained favor as an ingredient in baking; housewives familiarized themselves with cocoa cookery through the promotional cookbooks distributed by the blossoming cocoa and chocolate industry. Advertising for cocoa and chocolate increasingly focused on homemakers and their families, and chocolate candy eventually would be targeted almost exclusively toward children. In 1947, when the price of a chocolate bar rose from five to eight cents, Canadian children, in what has been dubbed the "Five Cent War," took to the streets in protest, determined to fight for the candy they loved. Such has been the enduring power of chocolate in Canada.

Acknowledgments

This research could never have covered as much ground as it did without the assistance of a great many people. Working from East to West: In Newfoundland thanks go to Paul Smith, Department of Folklore, Memorial University of Newfoundland. In Nova Scotia I would like to thank Anita Campbell and Janet Stoddard with the Atlantic Service Centre, Parks Canada; Gary Shutlek, Nova Scotia Archives and Records Management; Mary Guildford, Museum of Industry; Scott Robson, History Collection, Nova Scotia Museum; Stephanie Marshall and Eric Ruff, Yarmouth County Museum; and special thanks to the staff of the Fortress of Louisbourg National Historic Site of Canada, especially Ruby Powell, Collections Specialist; Sandy Balcom, Curator of Furnishings; Heidi Moses, Archaeology Unit; and Anne Marie Lane Jonah, Historian. In Prince Edward Island, thanks to Linda Berko and Boyde Beck of the PEI Museum and Heritage Foundation. In New Brunswick, many thanks to David Folster, Independent Historian; Daryl E. Johnson, New Brunswick Museum Archives; Cheryl Hewitt, Ganong Bros. Limited; Irene Ritch, Charlotte County Museum; Margaret Conrad, University of New Brunswick; Patti Auld Johnson, Harriet Irving Library, University of New Brunswick; and Darrell Butler, King's Landing Historical Settlement. In Quebec, I would like to thank Chris Lyons, Osler Library, McGill University; Nathalie Cooke, McGill Institute for the Study of Canada; Brian Cowan, Department of History, McGill University; Sherry Olson, Geography Department, McGill University; Jordan Le Bel, Concordia University (past), and Cornell University; Marie Marquis, Université de Montréal; Marc Lacasse, Archives des Prêtres de Saint-Sulpice de Montréal; Yvon Desloges, Parks Canada, Quebec City; and Éric Normand, Choco-Musée Érico. In Ontario, many thanks to Elizabeth Driver, Montgomery's Inn Museum; Fiona Lucas, Spadina Museum; Mary Williamson, York University; Richard Feltoe, Redpath Sugar Museum; the staff of the Archival and Special Collections, University of Guelph Library; and Sheldon Posen, Canadian Museum of Civilization. In Manitoba, I would like to thank Anne Lindsay, The Centre for Rupert's Land Studies, University of Winnipeg. In British Colombia, thanks to Kate Phoenix, Rogers' Chocolates Ltd. Many, many thanks to all members of the Mars and University of California–Davis chocolate research team. I am indebted to the staff of the McCord Museum, including Moira McCaffrey, Director, Research and Exhibitions; Cynthia Cooper and Conrad Graham, Curators; Dolorès Contré-Migwans, Native Programs; Francois Cartier, archivist; and Christian Vachon, Head, Collections Management. Finally, I express my thanks to Dr. Victoria Dickenson, Executive Director, McCord Museum, without whose enthusiasm and guidance this research might never have been realized.

Endnotes

1. The McCord Museum has been charged with the task of documenting the Canadian segment in the history of chocolate as part of the collective research effort to uncover chocolate's definitive history in North America. To date, the research has been fairly general in scope, as efforts have moved from recording early import–export data, to examining newspaper advertisements for chocolate and cocoa, to sifting through recipe books and medical texts, to reading early exploration journals and accounts. Through the McCord's resources, attempts are being made to record a summarized, overarching history of chocolate in Canada. The research strives to document a portrait of chocolate's "life," such as it was, in Canada, from the Colonial era through the early 20th century. While it is exciting to engage in a research project as broad in scope as this, for it permits one to gain a sense of chocolate's place in Canadian life and of the evolution of a foodstuff that is very familiar to us today, the sheer magnitude of the task means that it is a research project forever evolving and growing, constantly adding to our understanding of the role of cocoa and chocolate in Canada's early history and ultimately telling us much about the social history of its eaters, in this case, Canadians.

2. As per an on-site interview and archaeological collection tour with Ruby Powell, Collections Technician, Fortress of Louisbourg, Parks Canada.

3. Colleagues from the Fortress of Louisbourg have subsequently joined research efforts and have undertaken an analysis of documentation relating to chocolate's use among early French colonies in Canada including

Louisbourg and Quebec. They have focused on discussing the history of chocolate in New France and among early French militia. A detailed examination of chocolate among early French inhabitants of Canada can be found in Chapter 25.

4. Charles Aubert de la Chesnaye's probate inventory taken from the files of notary Florent de la Cetière, dated October 27, 1702. Archives nationales du Québec à Québec. Thanks to Yvon Desloges, Parks Canada for this reference.

5. The honey substance referenced by Champlain is mentioned twice in this description; it is pressed from the pith at the heart of the tree. He describes a juice made by pressing the leaves of the tree and describes a sort of thread from this tree, used by the Indians. "This tree bears numbers of thorns, which are very pointed; and when they are torn off, a thread comes from the bark of the said tree, which they spin as fine as they please; and with this thorn, and the thread which is attached to it, they can sew as well as with a needle and other thread. The Indians make very good, fine and delicate thread of it" [2] These latter traits, including the honey substance, do not apply to *Theobroma cacao*; they do, however, describe *Agave Americana*. In the illustration that accompanies the description (see Fig. 24.2), once again, the two trees are combined: we see the very distinct pods of the cacao tree, suspended directly from the trunk. Yet the bark is incredibly spiny and a figure, a Native American, is pictured beneath the tree, sewing with thorn and thread. What might explain these inaccuracies? The description is taken from the above-cited volume under the section titled *Brief discours des choses plus remarquables que Samuel Champlain de Brouage a reconnues aux Indes occidentals*. While often attributed to him, Champlain never published this particular text. Champlain did travel in the West Indies but whether or not he did so according to the time and itinerary stated in this text is not clear. Champlain may have written the text later in life, or someone else to whom Champlain recounted the details may have written it. This might account for the confused description. The illustrations are generally credited to and accepted as the work of Champlain; they are originals whereas the text is a copy. Champlain was a painter and master draftsman. Because the illustration listed as "cacou" reflects the combined physical attributes of both the cacao and agave plants, it is unclear if Champlain was basing the drawing on the text's confused description, on information drawn from multiple external sources, or on his own possibly muddied memory. For further discussion of the authenticity of this text, see Claude de Bonnault, "Encore le Brief discours: Champlain a-t-il été à Blavet en 1598?" *BRH* 1954; LX: 59–69; Jean Bruchési, "Champlain a-t-il menti?" *Cahiers des Dix* 1950; XV: 39–53; Marcel Delafosse, "L'oncle de Champlain," *RHAF* 1958–59; XII: 208–216; Jacques Rousseau, "Samuel de Champlain, botaniste mexican et antillais," *Cahiers des Dix* 1951;

XVI: 39–61; L.-A. Vigneras, "Le voyage de Samuel Champlain aux Indes occidentales," *RHAF* 1957–58; XI: 163–200; "Encore le capitaine provençal," *RHAF* 1959–60; XIII: 544–49; and Marcel Trudel's entry in the *Dictionary of Canadian Biography Online* (http://www.biographi.ca/EN/ShowBio.asp?BioId=34237&query=Samuel%20AND%20de%20AND%20Champlain).

6. It is important to cite the full passage:

   There is another tree which is called cacou, the fruit of which is very good and useful for many things, and even serves for money among the Indians, who give sixty for one real; each fruit is of the size of a pine-seed, and of the same shape; but the shell is not so hard; the older it is the better; and to buy provisions; such as bread, meat, fish, or herbs, this money my serve, for five or six objects. Merchandise for provision can only be procured with it from the Indians, as it is not current among the Spaniards, nor to buy often other merchandise than fruits. When this fruit is desired to be made use of, it is reduced to powder, then a paste is made, which is steeped in hot water, in which honey, which comes from the same tree, is mixed, and a little spice; then the whole being boiled together, it is drunk in the morning, warm, as our sailors drink brandy, and they find themselves so well after having drunk of it, that they can pass a whole day without eating or having great appetite.

7. See Bernard Pothier, *Dictionary of Canadian Biography Online* (http://www.biographi.ca/EN/ShowBio.asp?BioId=35062&query=Pierre%20AND%20Le%20AND%20Moyne%20AND%20D'Iberville).

8. Simeon Perkins, *The diary of Simeon Perkins, 1797–1803*, edited with an introduction and notes by Charles Bruce Fergusson (Toronto: Champlain Society, 1967), pp. 169–173.

9. Benjamin Marston, *Benjamin Marston's Diary, 1782–1787* (New Brunswick: c. 1782), pp. 20–21. In The Winslow Papers (http://www.lib.unb.ca/winslow/winslowunb.html).

10. Benjamin Marston, *Benjamin Marston's Diary, 1782–1787*, p. 66.

11. Stephen Miller, *Stephen Miller's Letter Books #1, 1759–1782* (New Brunswick: ca. 1759), p. 5. In The Winslow Papers (http://www.lib.unb.ca/winslow/winslowunb.html).

12. John A. West. "A brief history and botany of cacao." In: Foster, N., and Cordell, L.S., editors, *Chilies to Chocolate: Food the Americas Gave the World* Tucson: The University of Arizona Press, 1992: pp. 113–114.

13. *The Halifax Gazette*. March 30, 1752, p. 2 in Paper of Record, on-line full-text database of historic newspapers (www.paperofrecord.com).

14. For a complete table of items and provisions found in fur forts, see Jeff and Angela Gottfred, "A compendium of material culture; or, what we dug up." Copyright 1994–2002 *Northwest Journal* (ISSN 1206-4203) online: http://www.northwestjournal.ca/X2.htm. Gottfred and Gottfred have subsequently rated or graded their

sources used to compile this table, indicating when the sources were concrete (e.g., archaeological remains, inventories) or slightly less definitive, such as personal diaries or journals. They also indicate how common or uncommon the item was, and any site-specifics of the objects (i.e., if it was only found at major forts).

15. See A. Gottfred. "Fur fort food—receipts for the winter." Copyright 1994–2002 *Northwest Journal* (ISSN 1206-4203) online: http://www.northwestjournal. ca/114.htm. Gottfred discusses why and how she has pieced together this theoretical menu from a typical fur fort, and elaborates on the provisioning of the forts, discussing what foods were brought in by supply parties and what foods may have been cultivated, raised, or gathered on site.

16. See James Williams Tyrrell, "Station Provision List" (Nunavut: 1886), *Barren Lands* collection, University of Toronto Libraries, Ms. Coll. 310, Box 4, folder 1 Available at http://www.library.utoronto.ca/tyrrell/writings/W10004/0001-2-0.jpg and http://www.library. utoronto.ca/tyrrell/writings/W10004/0003-2-0.jpg

17. The brand of chocolate frequently mentioned on Arctic expeditions is Moore's chocolate. For additional provisioning lists and commentary on supplies, see also: *Further Correspondence connected with the Arctic Expedition* (London: Eyre & Spottiswoode, 1852); *Report of the Committee appointed by the Lord Commissioners of the Admiralty, to enquire into the Causes of the Outbreak of Scurvy in the Recent Arctic Expedition; the Adequacy of the Provision made by the Admiralty in the way of Food, Medicine and Medical Comforts; and the propriety of the Orders given by the Commander of the Expedition for Provisioning the Sledge Parties* (London: Harrison & Sons, 1877); *Papers Relative to the Recent Arctic Expeditions in search of Sir John Franklin and the crews of H.M.S. "Erebus" and "Terror"* (London: Eyre & Spottiswoode, 1854).

18. Sir John Franklin, *Sir John Franklin's journals and correspondence: the first Arcticland expedition, 1819–1822* (Toronto: Champlain Society, 1995), p. 302.

19. See "Blood-root 'Chocolate.'" *The Journal of American Folklore* 1906; 19(75): 347–348.

20. Thanks to Daryl Johnson of the New Brunswick Museum for assistance with this letter as an "expanded" chocolate search. Thanks also to Dolores Contre-Migwans, who advised me on the uses of bloodroot among Northeastern North American Aboriginal tribes.

21. Crust coffee, sometimes described as a "food for the sick," was prepared by toasting bread until quite brown, pouring boiling water over the well-done toast, straining the water, then adding cream and sugar, and sometimes nutmeg to the drink. See *Buckeye Cookery and Practical Housekeeping,* edited by Estelle Woods Wilcox, with an introduction by Virginia M. Westbrook. First published by Buckeye Publishing Co., 1880 (St. Paul: Borealis Books–Minnesota Historical Society Press, 1988), p. 477.

22. A cast-iron pan, common in fireplace cookery. It resembles a shallow frying pan or griddle with legs so that it might stand in the embers.

23. See *The Royal Gazette and Newfoundland Advertiser,* October 26, 1815, in Atlantic Canada Newspaper Survey, record #005773. Canadian Heritage on-line reference library: http://daryl.chin.gc.ca:8000/BASIS/acns/user/www/sf. The Atlantic Canada Newspaper Survey does not provide a scan or pdf of the original document; rather, it is a fully transcribed, searchable database of newspapers from Newfoundland, Nova Scotia, Prince Edward Island, and New Brunswick. The papers date as far back as *The Halifax Gazette,* Canada's first newspaper, which launched publication in March 1752.

24. John E. Langdon. *Canadian Silversmiths: 1700–1900* (Toronto: Stinehour Press, 1966), plate 29.

25. Library and Archives Canada Government Publications Collection. *Sessional papers of the Dominion of Canada: Volume 3, fourth session of the third Parliament, session 1877* (Ottawa: MacLean, Roger, 1877), 4-1. The next two decades show a concern for adulteration with respect to certain foodstuffs. Subsequent Parliamentary session appendices contain food inspection reports.

26. See the Atlantic Canada Newspaper Survey for cocoa and chocolate advertisements from numerous newspapers throughout Newfoundland, Nova Scotia, Prince Edward Island, and New Brunswick. Through the Canadian Heritage on-line reference library: http://daryl. chin.gc.ca:8000/BASIS/acns/user/www/sf.

27. Mary Jane Katzman Lawson. *History of the Townships of Dartmouth, Preston, and Lawrencetown; Halifax Country, N.S.* Edited by Harry Piers (Halifax: Morton & Co., 1893), p. 97. No business records pertaining to John P. Mott or Mott's Spice & Cocoa Company could be found in the Nova Scotia Archives and Records Management. Likewise, there exists scant material pertaining to Moir, a chocolate manufacturer in Halifax, who started operations not long after Mott. There exist personal fonds for Moir but most business records were destroyed in one of the numerous factory fires suffered by Moir's Chocolates. It is possible that the same fate may have met Mott's Spice & Cocoa Company. A picture of this early cocoa company has been pieced together through photographic archives, newspaper surveys, marketing materials, and social histories of the area.

28. See the Nova Scotia Archives and Records Management (NSARM), Notman Studio Collection, reference No. 1983-310/91250. William Notman operated the largest photographic enterprise in North America, with seven studios throughout Canada and 19 in the northeastern United States, including seasonal studios. The superior quality of the vast collection of Notman photographs, as found throughout archives across Canada, makes them valuable from a research perspective, but they also provide an important photographic record of Canadian social history.

29. See Nova Scotia Archives and Records Management (NSARM), Notman Studio Collection, reference No. 1983-310/5367. Thanks to Scott Robson of the Nova Scotia Museum for his assistance in tracking down photographic materials for Nova Scotia chocolate factories.

30. See *Maritime Merchant—Maritime Grocer and Commercial Review* (January 12, 1893). Nova Scotia Archives and Records Management (microfiche). *Maritime Merchant* and *Canadian Grocer* were popular Canadian trade publications of the late 19th and early 20th centuries. Published by and for the merchant and manufacturer, these catalogues were the business magazines of their time, and give invaluable references to goods available and their prices.

31. See Nova Scotia Archives and Records Management (NSARM), Moirs Limited Fonds, 1866–1966. 1-91 and 1-71.

32. Thanks to the staff of the PEI Museum and Heritage Foundation for their assistance in viewing their collection.

33. *Choice Recipes by Miss Maria Parloa and other noted Teachers, Lecturers and Writers* (Dorchester, MA: Walter Baker & Co. Ltd., 1902). *La Belle Chocolatière* was painted by Swiss artist Jean-Etienne Liotard ca. 1743 and became the emblem for Walter Baker & Co. Ltd. See also West, "A brief history and botany of cacao," p. 115.

34. David Folster, *Ganong—A Sweet History of Chocolate.* (Fredericton, NB: Goose Lane Editions, 2006), pp. 52–70. See also Folster, *The Chocolate Ganongs of St. Stephen, New Brunswick* (Fredericton, NB: Goose Lane Editions, 1990) for a complete history of the Ganong family and their chocolate company. As an interesting connection, William Francis Ganong, son of company founder James Ganong, was an accomplished writer, scientist, and historian. A number of the Canadian exploration texts consulted in the research for this chapter and published by the Champlain Society were edited in part by William Ganong. The Ganong family was descended from Loyalists.

References

1. Champlain, S. de. *Oeuvres De Champlain, Tome 1.* Published under the patronage of Charles-Honoré Laverdière. Quebec, Canada: Université Laval, 1870; p. 160, plate xxxiii. From the digital reprint at Early Canadiana Online: http://www.canadiana.org/ECO/PageView/26833/0160?id=0fb85d17d4b1c51e.

2. Champlain, S. de. *Narrative of a voyage to the West Indiens and Mexico in the years 1599–1602.* Translated from the original, unpublished manuscript by Alice Wilmere. Edited by Norton Shaw. London, England: Printed for the Hakluyt Society, 1859: pp. 26–27. Scanned from a CIHM microfiche of the original publication held by the Scott Library, York University. From the digital reprint at Early Canadiana Online: http://www.canadiana.org/ECO/PageView/33073/0156?id=0fb85d17d4b1c51e.

3. CO221 Colonial Office—Nova Scotia and Cape Breton Miscellanea B-3228, No. 28. Shipping Returns—Halifax 1752 (microfiche).

4. Perkins, S. *The Diary of Simeon Perkins, 1797–1803;* edited with an introduction and notes by Charles Bruce Fergusson. Toronto, Canada: Champlain Society, 1967; p. 348.

5. Perkins, S. *The Diary of Simeon Perkins, 1797–1803;* edited with an introduction and notes by Charles Bruce Fergusson, Toronto, Canada: Champlain Society, 1967; p. 120.

6. Byles, M. *Mather Byles' Letter Books No. 4* (New Brunswick: c. 1785), p. 16. In The Winslow Papers (http://www.lib.unb.ca/winslow/winslowunb.html).

7. National Archives of Canada, CO 221 (Series part of Colonial Office fonds—R10976-0-4-E). Nova Scotia and Cape Breton, Miscellanea, microfilm reel B-1525 (No. 42, NS & CB—Misc. Blue Book of statistics, etc. 1827).

8. Wood, W. *Select British Documents of the Canadian War of 1812, Volume II.* Toronto, Canada: Champlain Society, 1920–1928; p. 116.

9. Wood, W. *Select British Documents of the Canadian War of 1812, Volume III, Part 2.* Toronto, Canada: Champlain Society, 1920–1928; p. 756.

10. *The Halifax Gazette.* March 30, 1752, p. 2.

11. *The Halifax Gazette.* March 30, 1752, p, 2.

12. *The Halifax Gazette.* April 25, 1752, p, 2.

13. Wallace, W. S. *Documents Relating to the North West Company.* Toronto, Canada: Champlain Society, 1934; p. 218.

14. *Minutes of Council, Northern Department Of Rupert Land, 1821–31.* Rupert's Land. Northern Dept. Council; edited by R. Harvey Fleming, with an introduction by H. A. Innis. General editor, E. E. Rich. Toronto, Canada: Champlain Society, 1940; p. xxxviii.

15. Excerpts from David Thompson, "Journal of Occurrences from Lake Huron to the Ottawa River," 1837. In *Muskoka and Haliburton, 1615–1875: A Collection of Documents.* Edited with an introduction by Florence B. Murray. Toronto, Canada: Champlain Society, 1963; p. 95.

16. Tyrrell, J. W. *Across the Sub-Arctics of Canada : A Journey of 3,200 Miles by Canoe and Snowshoe Through the Hudson Bay Region.* Toronto, Canada: William Briggs, 1897; p. 103.

17. Tyrrell, J. W. *Through the Barren Lands: An Exploration Line of 3,200 Miles.* Toronto, Canada: C. Blackett Robinson printer, 1896? Barren Lands collection, University of Toronto Libraries Ms. Coll. 310, Box 11, folder 6. Available at http://www.library.utoronto.ca/tyrrell/text/T10001/0006-3-0.jpg.

18. *Additional Papers Relative to the Arctic Expedition Under the Orders of Captain Austin and Mr. William Perry.* London, England: Eyre & Spottiswoode, 1852; p. 20.

19. Franklin, Sir John. *Thirty Years in the Arctic Regions or the Adventures of Sir John Franklin.* New York: George Cooper, 1859; p. 463.

20. Franklin, Sir John. *Narrative of a Journey to the Shores of the Polar Sea, in the Years 1819-20-21-22*, 3rd edition, two volumes. London, England: J. Murray, 1824; p. 324.

21. Franklin, Sir John. *Thirty Years in the Arctic Regions or the Adventures of Sir John Franklin.* New York: George Cooper, 1859; p. 95.

22. Mudge, Z. A. *Arctic Heroes: Facts and Incidents of Arctic Explorations from the Earliest Voyages to the Discovery of the Fate of Sir John Franklin, Embracing Sketches of Commercial and Religious Results.* New York: Nelson & Phillips, 1875? p. 303.

23. Instructional letter entitled "Indian Chocolate," Dr. James Odell (New Brunswick: January 29, 1816). Odell Papers Collection, New Brunswick Museum Archives and Research Library.

24. Instructional letter entitled "Indian Chocolate," Dr. James Odell (New Brunswick: January 29, 1816). Odell Papers Collection, New Brunswick Museum Archives and Research Library.

25. MacFarlane Lizars, R., and MacFarlane Lizars, K. *In the Days of the Canada Company: The Story of the Settlement of the Huron Tract and a View of the Social Life of the Period, 1825–1850.* With an introduction by G. M. Grant. Toronto, Canada: William Briggs, 1896; p. 90.

26. MacFarlane Lizars, R., and MacFarlane Lizars, K. *In the Days of the Canada Company: The Story of the Settlement of the Huron Tract and a View of the Social Life of the Period, 1825–1850.* With an introduction by G. M. Grant. Toronto, Canada: William Briggs, 1896; p. 420.

27. MacFarlane Lizars, R., and MacFarlane Lizars, K. *In the Days of the Canada Company: The Story of the Settlement of the Huron Tract and a View of the Social Life of the Period, 1825–1850.* With an introduction by G. M. Grant. Toronto, Canada: William Briggs, 1896; p. 340.

28. *Journal and Proceedings of the House of Assembly of the Province of Nova Scotia* (Halifax: 1839), Appendix 53, pp. x–xi.

29. *Journal and Proceedings of the House of Assembly of the Province of Nova Scotia* (Halifax: 1839), Appendix 53, pp. x–xi.

30. *Canadian Grocer* (Toronto: March 6, 1891) 16 (microfiche: reel 1–295, Vol. 5 reel 1). Collection the Thomas Fisher Rare Books Library, University of Toronto.

31. Islander. December 14, 1855. Atlantic Canada Newspaper Survey record No. 871478PS through the Canadian Heritage on-line reference library: http://daryl.chin.gc.ca:8000/BASIS/acns/user/www/sf.

CHAPTER

A Necessary Luxury

Chocolate in Louisbourg and New France

Anne Marie Lane Jonah, Ruby Fougère, and Heidi Moses

Introduction

In the 1700s, the term New France referred to all French possessions in North America, a diverse collection of separate colonies that varied according to different treaties. Canada, the name given to the French-settled area along the St. Lawrence River, was the heartland of New France and theoretically its administrative center. In fact, the far-flung colonies usually communicated directly with France. New France after 1713 was Canada, Louisiana, and Île Royale (Cape Breton), which included Île St. Jean (Prince Edward Island), as well as the *pays d'en haut*, the vast territories beyond the Great Lakes and along interior river systems, traded in by the French but still primarily the lands of North America's Native peoples.

This chapter focuses on the military uses of chocolate in areas that became part of modern Canada—the Saint Lawrence River Valley, the *pays d'en haut*, and Île Royale—as well as the development of a trade in chocolate in the French North Atlantic empire and a taste for chocolate in New France's society. Chocolate in French Louisiana is a topic that needs a full exposition and discussion and is not included in depth in the present book. This study of chocolate trade and use in the present chapter, therefore, is centered on Louisbourg, the capital of Île Royale, a pivotal shipping port in France's North Atlantic trade.

In 1717, when the French chose Louisbourg to become the capital and military stronghold of Île Royale, chocolate was a rare and unfamiliar commodity in New France. Introduced to France in the 17th century, it was only becoming common for the aristocracy during the Regency of Louis XV [1]. During

Louisbourg's short history as an important fortified port in New France, the use of chocolate increased and spread in the French colonies. Who in New France used chocolate, why they had it, and how they acquired it provide a history of evolving ideas about chocolate, its appeal, and its powers. Louisbourg's records provide details of how this tropical product came to be traded in the North Atlantic and of how the taste for chocolate and ideas about chocolate were spread in this colonial society. Official records of New France provide evidence of its use by the military for its food value.

Louisbourg in New France

Louisbourg became important within New France after the signing of the Treaty of Utrecht in 1713, which ceded France's North Atlantic colonies of Acadia (mainland Nova Scotia) and Plaisance (in Newfoundland) to the British. Eager to stay involved in the valuable cod fishery, France reestablished itself in the Atlantic region by moving its administrative and commercial operations to Île Royale. France also retained

---

Chocolate: History, Culture, and Heritage. Edited by Grivetti and Shapiro
Copyright © 2009 John Wiley & Sons, Inc.

FIGURE 25.1. Modern depiction of 1744 Louisbourg showing commercial activity on the busy quay. *Source: View from a Warship*, by Lewis Parker. Photo by Ruby Fougère. Image Accession Number E-00-29. Courtesy of the Fortress of Louisbourg National Historic Site of Canada, Parks Canada. (Used with permission.) (See color insert.)

Île St. Jean, which it developed slightly later and less extensively. France chose Louisbourg, a harbor previously used for seasonal fishery, to be the administrative and commercial base for lucrative North Atlantic cod fishery (Fig. 25.1). Impressive fortifications were begun around the new town in the Vauban style, like fortifications built on France's southern and western borders in the late 17th century (Fig. 25.2). Although remote by today's standards, this colonial capital was well situated on the North Atlantic shipping lanes and functioned as a valuable entrepôt between France, the West Indies, and the North American interior. In fact, the St. Lawrence River Valley and the Great Lakes region were more remote than Louisbourg, in the context of the 18th century. By the mid-18th century, Louisbourg was one of the most heavily fortified towns in North America, and one of the five busiest ports on the North Atlantic coast [2–5].

Military officers and soldiers, administrators and clerks, merchants, artisans and fishermen, and their families came to Louisbourg from the former colonies, from France, and from other parts of Europe. Merchants and fishermen came from northern France, Brittany and Normandy, as well as the Basque region in southwestern France. Soldiers of the *compagnies franches de la Marine* (regular colonial troops) came from the cities of France, as well as a detachment from a Swiss mercenary regiment. Louisbourg's social hierarchy was similar to that of France, with the military and merchant elite at the head. Classes within Louisbourg society, however, were not as strictly divided as in France, because of the power of new wealth created by trade. The consumption of chocolate in Louisbourg was linked to both class and culture; the patterns of consumption illustrate the relative proximity of groups within these categories in this colonial port [6, 7].

Louisbourg, as a key French naval and commercial base in North America, was an important target in times of war between the British and French. Twice it was besieged and taken by British forces. Louisbourg fell in 1745 to a force from New England supported by the British navy. The British then occupied the fortress town until the terms of the Treaty of Aix-La-Chapelle in 1749 returned it to the French, in exchange for the Austrian Netherlands [8]. The British besieged Louisbourg again in 1758 with a combined land and naval force of more than 27,000 supported by 39 British naval vessels [9–11]. This force took seven weeks to defeat Louisbourg, after which the French population was deported to France. After the fall of Louisbourg in 1758, came the fall of Québec in 1759 and the end of New France by treaty in 1763. British engineers destroyed the walls of Louisbourg in 1760 and left the town to fall into ruin.

Louisbourg is a valuable vantage point from which to study Colonial era material culture because after the fall of the fortress to the British in 1758 no further development took place on the site. To date, archaeologists have studied 25 percent of Louisbourg, generating one of the largest collections of artifacts in the world for an 18th century town (Figs. 25.3–25.12,

FIGURE 25.2. Modern reconstruction of the Fortress of Louisbourg, operated by Parks Canada, represents approximately one-quarter of the original fortifications and one-fifth of the 18th century town of Louisbourg. *Source*: Atlantic Service Centre, Parks Canada. Photo by Ron Garnett. Image Accession Number NS-FL-1999-14. Courtesy of the Fortress of Louisbourg National Historic Site of Canada, Parks Canada. (Used with permission.) (See color insert.)

14, 16). This collection complements the carefully kept administrative records of the period and the French and British records of the sieges. These sources were the basis for the largest historical reconstruction in North America, the Fortress of Louisbourg National Historic Site of Canada, where today, in homes reconstructed following the detailed records of Louisbourg still in the French archives, servants prepare chocolate for their wealthy masters as it was prepared 250 years ago (Figs. 25.13 and 25.15).

Les Compagnies Franches de la Marine and Chocolate

The first reference to the use of chocolate in Louisbourg attests to the 18th century French belief that chocolate had a medicinal value—that it warmed and fortified—rather than to its taste or luxury value [12]. In 1725, a royal transport vessel, the *Chameau*, ran aground and sank during a storm on Île Royale's coast and the vessel and all 316 persons on board, including

FIGURE 25.3. Assortment of chocolate pot lids and swivels (18th century), excavated at the Fortress of Louisbourg. *Source*: Archaeological Collection of the Fortress of Louisbourg. Accession Numbers: Lids 1B18K7-28, 1B16G1-1764, 16L94H13-5. Swivels 1B4E2-32, 4L54K9-4. Photograph by Heidi Moses. Image Accession Number 6796E. Courtesy of the Fortress of Louisbourg National Historic Site of Canada, Parks Canada. (Used with permission.)

FIGURE 25.4. Chocolate pot lid (18th century), excavated at the Fortress of Louisbourg. *Source*: Archaeological Collection of the Fortress of Louisbourg. Accession Number: 1B18K7-28. Photograph by Heidi Moses. Image Accession Number 6652E. Courtesy of the Fortress of Louisbourg National Historic Site of Canada, Parks Canada. (Used with permission.)

the new *Intendant* of New France, the governor of Trois Rivières, and a son of the governor of New France, were lost. After the wreck, local fishermen and soldiers found victims and debris scattered along the coast, but no one had found the chests containing 176,000 *livres* (the equivalent of £8800 sterling) in gold, silver, and copper coins loaded by the treasurer in France to fund the colony for the year. In September of 1726, *compagnies franches* soldiers specially trained as divers came from Québec to assist the Louisbourg garrison and port captain with the difficult salvage of the *Chameau*. As the divers prepared to dive into the North Atlantic to seek the lost treasury, they greased their bodies for protection from the cold water, and they fortified themselves with a diet of fresh meat and chocolate. The commander of the divers listed chocolate

see page 334

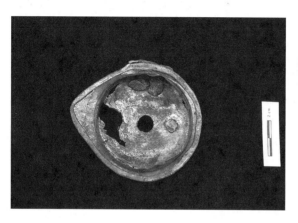

FIGURE 25.5. Construction of 18th century chocolate pot lid. Underside shows different pieces used in construction. Excavated at the Fortress of Louisbourg. *Source*: Archaeological Collection of the Fortress of Louisbourg. Accession Number 1B18K7-28. Photograph by Heidi Moses. Image Accession Number 6652E. Courtesy of the Fortress of Louisbourg National Historic Site of Canada, Parks Canada. (Used with permission.)

FIGURE 25.6. Construction of 18th century chocolate pot lid. A copper alloy band forms part of the lid style (see Fig. 25.5). Excavated at the Fortress of Louisbourg. *Source*: Archaeological Collection of the Fortress of Louisbourg. Accession Number 1B16G1-912. Photograph by Heidi Moses. Image Accession Number 6629E. Courtesy of the Fortress of Louisbourg National Historic Site of Canada, Parks Canada. (Used with permission.)

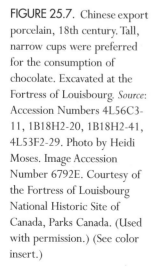

FIGURE 25.7. Chinese export porcelain, 18th century. Tall, narrow cups were preferred for the consumption of chocolate. Excavated at the Fortress of Louisbourg. *Source*: Accession Numbers 4L56C3-11, 1B18H2-20, 1B18H2-41, 4L53F2-29. Photo by Heidi Moses. Image Accession Number 6792E. Courtesy of the Fortress of Louisbourg National Historic Site of Canada, Parks Canada. (Used with permission.) (See color insert.)

FIGURE 25.8. Chocolate/ coffee Cups, French faience style. Excavated at the Fortress of Louisbourg. *Source*: Accession Numbers: 3L19A3-10, 4L52M21-1, 1B5A7-4. Photo by Heidi Moses. Image Accession Number 6791E. Courtesy of the Fortress of Louisbourg National Historic Site of Canada, Parks Canada. (Used with permission.) (See color insert.)

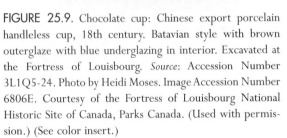

FIGURE 25.9. Chocolate cup: Chinese export porcelain handleless cup, 18th century. Batavian style with brown outerglaze with blue underglazing in interior. Excavated at the Fortress of Louisbourg. *Source*: Accession Number 3L1Q5-24. Photo by Heidi Moses. Image Accession Number 6806E. Courtesy of the Fortress of Louisbourg National Historic Site of Canada, Parks Canada. (Used with permission.) (See color insert.)

FIGURE 25.10. Chocolate cup: Chinese export porcelain handleless cup, 18th century. Chinese Imari style with blue underglaze with overpainting of gold and red. Excavated at the Fortress of Louisbourg. *Source*: Accession Number 1B18H2-41. Photo by Heidi Moses. Image Accession Number 6613E. Courtesy of the Fortress of Louisbourg National Historic Site of Canada, Parks Canada. (Used with permission.) (See color insert.)

FIGURE 25.11. Chocolate cup: English manufactured agateware handled cup, 18th century. Agateware style named for use of multicolored clay that mimics agate. John Astbury, Staffordshire Potter, credited with innovation of agateware, also produced white salt glaze stoneware. Excavated at the Fortress of Louisbourg. *Source*: Accession Number 17L31C3-1. Photo by Heidi Moses. Image Accession Number 6799E. Courtesy of the Fortress of Louisbourg National Historic Site of Canada, Parks Canada. (Used with permission.) (See color insert.)

FIGURE 25.12. Chocolate cup: English manufactured Astbury style handled cup. Example of a fine red earthenware body with white slip decoration around cup lip rim and on top of handle. Ware type is named after John Astbury. Excavated at the Fortress of Louisbourg. *Source*: Accession Number 6L92N12-3. Photo by Heidi Moses. Image Accession Number 6798E. Courtesy of the Fortress of Louisbourg National Historic Site of Canada, Parks Canada. (Used with permission.)

FIGURE 25.14. Chocolate/coffeepot and handled cup: French Rouen style. Excavated at the Fortress of Louisbourg. *Source*: Accession Numbers 4L52M14-4 and 3L1Q5-4. Photo by Heidi Moses. Image Accession Number 6794E. Courtesy of the Fortress of Louisbourg National Historic Site of Canada, Parks Canada. (Used with permission.)

FIGURE 25.13. Chocolate pot: copper, 18th century. *Source*: Fortress of Louisbourg Curatorial Collection. Photograph by Ruby Fougère. Image Accession Number BL.68.1.117. Courtesy of the Fortress of Louisbourg National Historic Site of Canada, Parks Canada. (Used with permission.)

FIGURE 25.15. Chocolate/coffeepot: silver, 18th century. *Source*: Fortress of Louisbourg Curatorial Collection. Accession Number BL.68.1.122. Photo by Ruby Fougère. Image Accession Number BL.68.1.122. Courtesy of the Fortress of Louisbourg National Historic Site of Canada, Parks Canada. (Used with permission.) (See color insert.)

among the provisions consumed during the first 15 days of diving and requested it again for them as a *gratification*, a bonus. In his request, he mentioned that they had plenty of milk to prepare the chocolate [13, 14]. The divers consumed their chocolate as a warm drink, as it was consumed in Europe at that time. The people of Louisbourg may have preferred milk with their chocolate to water because the quality of their drinking water was poor, as it came from shallow wells that risked contamination by human and animal waste.

Chocolate was not part of the normal rations of 18th century French soldiers. Its inclusion among the provisions for these divers indicates their officers' belief in its powers to nourish and to strengthen. A *livre* (roughly a pound) of prepared chocolate cost more

FIGURE 25.16. Chocolate/coffeepot: plain, white, French faience. Excavated at the Fortress of Louisbourg. *Source:* Accession Number 4L52M12-19. Photo by Heidi Moses. Image Accession Number 6793E. Courtesy of the Fortress of Louisbourg National Historic Site of Canada, Parks Canada. (Used with permission.) (See color insert.)

than a French soldier was paid in a month, after his deductions for barracks and rations, so these divers probably did not taste chocolate except in preparation for their grueling work. Although chocolate is not one of the official rations required or allowed for French soldiers, the *compagnies franches'* growing confidence in the nourishing power of chocolate was demonstrated in its inclusion among the supplies for raiding parties leaving Québec for overland missions in the 1740s. This was the case with the party of French and Natives lead by Paul Marin de la Malgue, a lieutenant with long experience in the *pays d'en haut*. At the beginning of the War of the Austrian Succession in 1744, experienced officers such as Marin were chosen to lead raiding parties against the English. Marin's party set out in January of 1745 to participate in an attack on Annapolis Royal, in the former French territory of Acadia, and then to continue on to reinforce Louisbourg. The party of 120 Frenchmen and about 200 Natives carried among their provisions 83 *livres* of chocolate [15, 16]. This quantity of chocolate was not enough for all members of the force to consume it regularly; it may have been reserved for the officers and cadets in the force.

The supplies issued to another expedition the following year prove the existence of a class and rank distinction in the distribution of chocolate provisions in the French military. A large French and Native force on a shorter duration expedition to New England also had chocolate among its supplies. In this instance, the chocolate was issued to French officers and cadets, to officer level volunteers, and to the militia officers, but not to the common militia soldiers or the Native fighters [17]. A document from Montréal detailed provisions given to Native and combined Native and French raiding parties and garrisons for borderland forts in the last three months of 1746. Chocolate was issued to officers only when they were in raiding parties and was never given to regular or militia troops, or Native fighters. Chocolate was issued based on rank, but it was not given to officers in established forts; possibly it was not seen as necessary for men who had secure protection from the weather [18].

Another such document recorded provisions for raiding parties, garrisons, and expeditions in 1747 [19]. In this year some garrison officers received chocolate provisions. Sugar was also issued with the chocolate, unlike the previous year. In 1746 and 1747, officers in raiding parties usually received one *livre* of chocolate for each excursion. A larger ration of chocolate, three *livres*, was given to officers in a French and Native party commanded by the *Chevalier* Louis La Corne, voyaging into the *pays d'en haut* for the winter to intercept Mohawk/British raiding parties [20, 21]. This list also recorded that two Jesuit priests voyaging to Détroit received 20 *livres* of chocolate with their provisions for the winter [22]. Chocolate was the only commodity on the lists of supplies that was distributed consistently on the basis of officer rank or membership in the noble class. Officers were almost always of the noble class, as were members of the educated Jesuit order. Prerogatives were commonly reserved for nobles in French society, and even though nobility was not as strictly respected in the colonies, it was important and signaled by distinctions, which included the provision of chocolate [23]. The distribution of chocolate may have also depended on the perception of a need for it and the available supply. Wine, normally distributed to officers, was given to soldiers, Natives, and *habitants* who were ill, whereas chocolate was not. In the documents for 1746 and 1747, wine was allocated for the sick but not chocolate. In the last year there was chocolate as well as wine given to a sick officer cadet [24].

Gilles Hocquart, the *Intendant* of New France, produced these detailed provision lists to answer the Minister of the Marine's inquiry as to why expenditures on the war effort were so high [25]. Hocquart recorded in detail all of the provisions for raiding parties in these years, providing an insight into the use of chocolate by the *compagnies franches*. The purchase of 130 *livres* of chocolate for the military was also documented in this period [26]. These documents show a pattern of usage that may have continued but was not documented in later lists of payments, *bordereaux*, which were less detailed. In Québec in 1759, during the Seven Years' War, the *Intendant* François Bigot wrote the Minister of the Marine to again justify high expenditures. He referred to the expenses on raiding parties and frontier posts, admitted that they were higher, and promised to send a detailed *bordereau*, which this researcher has not yet found. He added in his letter that he felt the need to draw on his own funds to daily offer a fine table to the officers of the colonial and regular army troops serving together in Québec, in order to help them cope with the difficulty of that time and maintain their unity [27]. The offerings at his "fine tables" remain a matter of speculation; however, we know that Bigot in the early 1750s had offered his guests chocolate [28].

A French officer in the *compagnies franches*, known only as J.-C. B., wrote a book describing details of his life during his ten years of service in Canada on inland campaigns during 1751–1761. He did not mention having chocolate, but when describing the unfamiliar maple sugar he compared it to chocolate: "*On fait de ce sucre de petites tablettes comme du chocolat afin de le transporter plus facilement en voyage . . .* [They make small bars with this sugar like with chocolate so that they can carry it more easily on trips]" [29].

For him, chocolate was a well-known reference point, which would easily bring to mind the appearance of the unfamiliar bars of maple sugar. He, as an officer in the *compagnies franches*, was familiar with both of these made into small flat portions that could easily be packed and carried.

For the French military, chocolate was a necessity reserved in all cases, except for the *Chameau* divers in 1726, for the officers. Although it was assigned medicinal or nutritional value, as demonstrated when it was given to a sick cadet, and in its being reserved for officers on overland missions rather than in posts, its use was exclusive to those identified with the noble class. Thus, although a practical food to carry in winter voyages, chocolate had a luxury association for these French officers in the mid-18th century.

Louisbourg Society and Chocolate

The civilian consumers of chocolate in New France may have also believed that it would strengthen and restore them, but they were less likely to test it against the North Atlantic or the Canadian forest in winter. This section studies the ownership and exchange of chocolate in Louisbourg society, representing one of the urban areas of New France, and considers examples of the use of chocolate throughout New France. In Louisbourg the consumers of chocolate were almost all from the upper class, noble or bourgeois. Class, as demonstrated by rank in the military and by social status, noble or nonnoble, and wealth in civil society, ruled every aspect of life in New France. As in Europe, it was expressed in every material aspect of life, the clothing worn and the food consumed. One illustration of this class distinction is the difference between the

monotonous list of provisions for the crews of vessels that sailed from Louisbourg and the galley list for the captain of the *Hermione*, a supply ship of the French navy that left Louisbourg in 1758. Sailors normally received flour or biscuits, lard, dried peas, rum or perhaps "common" wine, dried fish or cured pork, water, and a bit of salt. The *Hermione*'s captain had linen cloths, silver, crystal, and porcelain at his table, and his chef had a three-page list of wines and luxury food supplies to offer his patron, including chocolate [30].

The governor of Louisbourg in 1744, Jean-Baptiste-Louis LePrévost Duquesnel, was among Louisbourg's consumers of chocolate. Duquesnel's naval career had taken him from northern to southern France, to Martinique, where he met his wife, and to the North Atlantic. He was an ill and elderly man when he died at Louisbourg in 1744, leaving behind his possessions to be inventoried by colonial officials [31]. Duquesnel had a total of 29.5 *livres* of chocolate, some of which was from Manila and some of which was in the form of unprepared chocolate from the French West Indies, a region closely linked to Louisbourg by the seaborne trade that was the source of wealth for both colonies (Figs. 25.17 and 25.18). Duquesnel's chocolate was kept in his *cabinet*, a small room off his public bedchamber, rather than in the main pantry of his apartments, implying it was for his personal use or that he wanted to keep a close watch on it. Like the divers for the *Chameau*, and the officers in the Canadian forest, he probably believed that chocolate had powers to warm the internal organs and to give energy. One of the purchasers of his chocolate at the auction of his estate, Gabriel Schonherr, a Swiss Karrer officer stationed at Louisbourg, was also older and ill, and perhaps also had faith in the restorative powers of chocolate [32].

As an expensive and portable commodity, chocolate also made an appealing target for thieves (see Chapters 20 and 21). In 1743, Bernard Muiron, an *entrepreneur* of the fortifications, realized that someone had been robbing his home and storehouse. He apprehended Valerien Louis *dit* Bourguignon, a stonecutter and former soldier he had hired in Dijon the year before, hiding among the barrels in his yard. In the investigation of the crime, the accused had explanations for his incriminating activities, selling building supplies like glass panes and nails and large amounts of powder and shot in the local taverns. He had a trunk containing stolen goods, which he said he had been given, and another sack of goods, including the stolen chocolate, which he denied owning. He repeatedly denied robbing his employer even though many witnesses incriminated him. Valerien's personal possessions provide a clue to his possible motivation: the desire for profit and the wealth he saw around him. He had several good shirts, some trimmed with lace, at a laundress. This clothing was more like that of a bourgeois than a laborer. As well, he had hair or

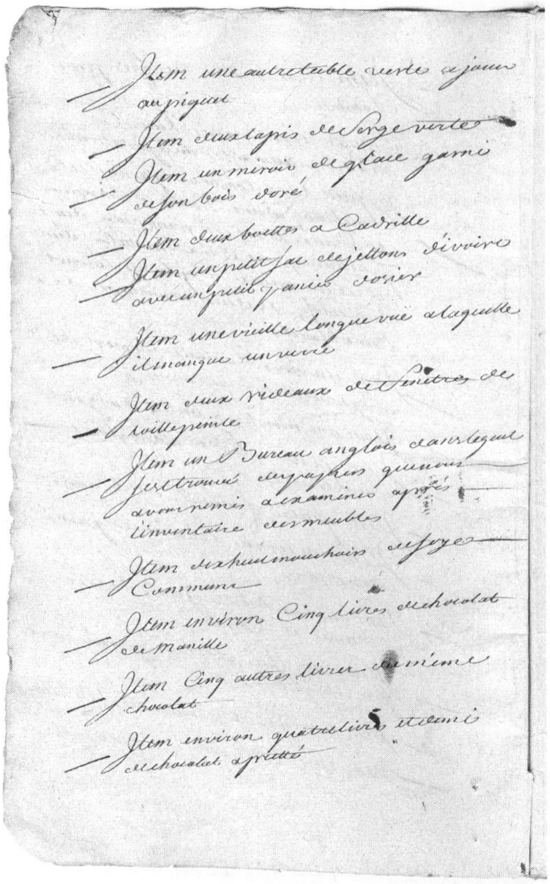

FIGURE 25.17. Dossier relating to the estate of Jean-Baptiste-Louis LePrévost Duquesnel, dated 1744. Page 47 of the inventory itemizes objects in his *cabinet*, including a gold-framed mirror, an English desk, and chocolate from Manila. *Source*: CAOM DPPC GR 2124. Courtesy of Archives nationales d'outre-mer (Aix-en-Provence), France. (Used with permission.)

FIGURE 25.18. Private cabinet of Governor Duquesnel's reconstructed apartments at the Fortress of Louisbourg. Photograph highlights the antique gold-framed mirror, antique English desk, reproduction porcelain, and antique chocolate/coffeepot, reflecting items listed in his inventory (Fig. 25.17). *Source*: Photograph by Ruby Fougère. Image Accession Number 2007-21S-1. Courtesy of the Fortress of Louisbourg National Historic Site of Canada, Parks Canada. (Used with permission.) (See color insert.)

wig powder; an accessory reserved for the upper classes of French society, not laborers [33, 34]. Whatever Valerien's motivation, chocolate was among the luxuries that he stole from his employer and kept, rather than sold. Eighteenth century French society guarded its private property fiercely. For his crime, Valerien was tortured and then condemned to life service on the Mediterranean galley ships, a protracted death sentence. The Louisbourg authorities, however, did not anticipate that a stonecutter might be a hard person to keep in prison. One night before he was to be sent to the galleys, his guard found a hole cut in the cell wall, enlarging the window, and Valerien was gone [35].

The consumption of luxury goods demonstrated wealth, promoting the status of their consumers in this hierarchical society. The hazard of this is exposed in the complaint of a bankrupt young merchant, whose new enterprise was ruined by either bad luck, his lack of skill, or, as he claimed in the ensuing court case, the extravagance of his wife and her mother. In the legal proceedings following the bankruptcy of Jacques Rolland, he claimed that his wife and mother-in-law had squandered their income on "*des parties de caffée et d'autres friandises*" [36]. Social gatherings such as these at Louisbourg provided the opportunity for hosts to impress their guests with the sophistication of their offerings by including chocolate as a beverage or as an ingredient in the *friandises*, sweet desserts. These fashionable food products appealed to new colonials wishing to emphasize their status, and ironically, their European-ness through their awareness of current European taste fashions [37–39]. Thus, colonial products such as chocolate, coffee, and sugar became symbols of European sophistication in other colonies.

French officials, merchants, and seafarers from throughout France who based their operations in Louisbourg shared their tastes and ideas about foods with their neighbors. A significant number of Louisbourg's merchants, 26 out of 147, were from the Basque region of southwestern France [40]. The Basques had been in the chocolate trade well ahead of the northern French because of their close ties with Spain. Spanish Jews, many involved in chocolate trade in Spain, had fled religious repression across the border into the Basque region of France. Although chocolate was elite and exotic for most French, the Basques in Louisbourg were routinely consuming chocolate, as well as trading it (see Chapter 42).

The settlements of the estates of three Basques who died at Louisbourg show their role in the trade and consumption of chocolate at Louisbourg. Bernard d'Etcheverry, a Basque-born merchant, was one of many Louisbourgeois who rented a room to a visiting captain during the summer months. He provided his guest and fellow countryman, Bernard Dargaignarats, with chocolate on a regular basis, indicating that this Basque sea captain viewed chocolate more as a staple than a luxury. Another Basque merchant and captain, Bertrand Larreguy, who regularly came to Louisbourg in the summer to trade, also had chocolate among his possessions as did the Louisbourg-based Basque merchant Jean-Baptiste Lascorret. These men were consuming chocolate and importing it to Louisbourg for sale [41, 42].

The majority of those in Louisbourg with a taste for chocolate were recent immigrants from Europe. In 1713, Île Royale was a new colony and most of the population was European-born, but by the 1740s many officers and merchants had been in the colony for their entire careers and many Louisbourg residents were born in the colony. Among the 16 identified chocolate owners and traders in Île Royale, 12 had come to the colony as adults or were based in Europe.[1] Of the four colonial-born owners of chocolate, two were innkeepers.[2] Although chocolate was a colonial commodity, the taste for it reached Louisbourg via France.

Louis Franquet, a royal engineer in Louisbourg in the 1750s in charge of the repair and improvement of the fortifications after the conclusion of the War of the Austrian Succession, traveled to Québec and Montréal during his stay in the colonies and wrote a detailed journal of his travels. He recorded that during his stay with the *Intendant* of New France, François Bigot, who had been the civil administrator of Louisbourg in the 1740s, coffee, tea, and chocolate were served for breakfast. In February of 1753, Franquet accompanied the new Governor and *Intendant* of New France as they journeyed north of Montréal to meet with the Native allies of France. Franquet complained of the harshness of this journey during the winter but praised the governor's kitchen staff who traveled ahead to prepare a meal for the travelers at the Sulpicien seminary where the party would stop. Franquet wrote that they experienced an excellent meal, after which "*on y prit du thé, du caffé, du chocolat, en un mot tout ce qu'on voulait* [we had tea, coffee and chocolate, in a word, all that one could have wanted]" [43].

To Franquet the offer of these fashionable beverages was a sign of sophistication and luxury, made more remarkable because of the rough environment in which they were served.

A Swedish botanist, Pehr Kalm, traveled in New France in the 1740s and carefully recorded his observations of plants, the landscape, and the people. He observed that for the upper classes chocolate was a common breakfast. Both Franquet and Kalm saw chocolate as luxury for the elite, but in a letter written in 1745, the author upon hearing that a female relative was ill, resolved to send her the best chocolate money could buy [44]. For that author, a member of the elite of New France, chocolate had a therapeutic value. Kalm remarked, as is clear from the import records of Île Royale, that coffee was the most common drink, chocolate less so, but quite popular, and tea was extremely rare in New France. He attributed this to the trade connections among the French colonies [45–47].[3]

Chocolate began to appear in French records more frequently toward the mid-century. It was not mentioned in Île Royale until 1726, and then not again until 1737; there were seven records that referred to it in the 1740s and three in the 1750s. Christopher Moore wrote a detailed study of imports to Île Royale focused on six sample years, chosen because of the distribution and the quality of the available data. Chocolate and cocoa only appeared once in the first three sample years, 1737, and in each of the last three years, 1743, 1752, and 1753 [48]. Its first uses were medicinal but by the mid-century it had become a symbol of the wealth and worldliness of those who consumed or offered it.

From Martinique to Louisbourg: The Colonial Trade of Chocolate

Although chocolate, or cocoa beans, did not appear in large volumes among West Indies cargoes coming to New France, these products were named in discussions of the trade and its value throughout the period [49–51]. Jean Pierre Roma, a Parisian merchant granted a company charter on Île St. Jean (Prince Edward Island), part of the colony of Île Royale, expressed optimism about the potential value of trade between the French West Indies and New France in his letters to the Minister of the Marine. Roma's Company of the East of Île St. Jean was his second attempt to make a fortune in colonial trade; his first such venture had been based in Saint Domingue (Haiti). The elaborate base he built for his fishing and trading company on the eastern end of Île St. Jean was destroyed after the siege of Louisbourg in 1745. Among the ruins of Roma's Canadian estate, archaeologists have found shards of tall, narrow porcelain cups, a style preferred for drinking chocolate, indicating that Roma was a consumer of chocolate [52–54]. Roma had planned to sell cod from the North Atlantic to the West Indian slave plantations and return to Île Royale with West Indian products. After the loss of their establishment in Île St. Jean, the Romas fled to Québec; there Roma's son sold 130 *livres* of chocolate to the military in 1747 [55]. At the conclusion of the War of the Austrian Succession, after the loss of his Île St. Jean establishment, Roma moved his operations to Martinique.

Although Roma tried to develop trade based on the charter company model, as had been very successful in the East Indies, large companies imposing monopolies did not succeed in the North Atlantic, where, by the 18th century, trade was shared among a large number of private merchants and small companies. Louisbourg merchants frequently traveled to the West Indies, whose slave plantations were important consumers of Louisbourg's dried salt cod. They returned with rum and sugar, as well as smaller quantities of coffee and chocolate. The production of chocolate in Martinique and Saint Domingue was greatly reduced after crop disasters of the 1720s [56]; they were, however, still an important source for Louisbourg's chocolate, whether imported directly or processed in France first and then reexported via the West Indies. The mercantile model of trade governed colonial commerce in the 18th century, and so to encourage trade among the developing colonies, the French crown exempted Canada and Île Royale from having to pay taxes on raw goods received from the French West Indies [57]. Louisbourg, nevertheless, also had permission to trade with British colonies when judged necessary, in recognition of the reality of Louisbourg's geographic position and available sources of supply [58]. In 1743, a vessel from New York, the *Midnight*, was allowed to unload its chocolate along with other food items because Louisbourg's port officials judged them to be needed commodities [59–60]. Louisbourg received cargoes of chocolate from the British colonies, France, and the West Indies.

As well as the large cargoes recorded by the authorities, there were smaller quantities of chocolate that regularly arrived in Louisbourg. Ships' officers and merchants who traveled between the colonies had the right to bring back private cargoes, *pacotilles*, that they were not obliged to report in detail [61]. These cargoes were a supplement to officers' earnings and another route by which chocolate arrived in Louisbourg. One such *pacotille* was inventoried after the death at sea of Jean-Baptiste Lascorret, a young Basque merchant. When Lascorret left his post as a clerk in San Sebastian, near Bayonne, to work for a merchant in Louisbourg, he was lured by the promise that he could make a fortune in this lucrative new Atlantic

trading entrepôt. His first experiences were not good, as he ended up disputing his wages with his new employer in civil court [62]. His prospects must have seemed brighter when he found a position as clerk for François Dupont Duvivier, a captain with the *compagnies franches* in Louisbourg, who had a profitable sideline of trade (much of it illegal) along the North Atlantic coast. In 1737, Lascorret owned his own vessel, *La Reine du Nord* [63]; and in 1739, Duvivier brought him a *habit*, a gentleman's suit of clothes, worth over 100 *livres* from Paris. The suit's cost equaled almost two and a half months' wages at Lascorret's previous position. Lascorret died in 1741 returning from a commercial voyage to Martinique. In the cases of goods, the *pacotille*, that he was bringing back to Louisbourg there were 12 half-*livre* (weight) balls of chocolate worth 18 *livres* [64].

The scattered bills and accounts left in the estate papers and after-death inventories of merchants who died at Louisbourg complement the less-detailed official import lists with examples of how chocolate was traded and used. The chocolate from estates was sold at public auction in Louisbourg. Otherwise, it is not clear how chocolate made it from the hold of a ship to the individual consumer since chocolate was not found in any of the *boutique* inventories for Louisbourg. The court usher, the *huissier*, publicly announced information about auctions and official sales, but New France did not have any newspapers, so there were no printed advertisements to create an historical record of retail trade. Consumers may have asked a traveling acquaintance to seek out some chocolate for them, as did the author of a letter carried on board *La Surprise*, a small coastal trader out of La Rochelle [65]. Louisbourg consumers purchased chocolate from some of the many inns and taverns that served the port's itinerant population. One *aubergiste* (innkeeper) of Louisbourg had several *livres* of chocolate in her inventory; another had a chocolate grater in her kitchen, to prepare the chocolate for frothing into warm milk or water [66, 67].

Although chocolate was growing in importance as a commodity of colonial trade, it did not become an object of exchange in the well-established relationship between the French and Native North Americans, as it did for the British and Natives in the Northeast. The officials at Louisbourg traveled annually to the seasonal meeting place of the Mi'kmaq, the indigenous people of the Maritime region, and their Catholic missionaries, but chocolate did not figure among their exchanges. In a discussion of the exchanges taking place between Native peoples of the *pays d'en haut* and the administration of New France in the late 17th century, Baron de Lahontan recorded an extensive list of items for exchange, but did not include chocolate [68]. In the 1740s, when French officers regularly received chocolate, it still never appeared as a gift or provision for the Natives [69].

Maritime War: Chocolate Prizes

In the wars between European empires, seaborne trade was an important target. In wartime the European sea powers licensed private ships, *corsairs* or privateers, to take part in the commercial war. Privateers and navies alike profited from the sale of seized cargoes of their enemy's commercial shipping. Louisbourg was a valuable base for French privateers targeting New England's shipping and fisheries, but their West Indies trade was also vulnerable to British and New England privateers. A valuable cargo like chocolate was a lure for these vessels. France's navy was considerably smaller than that of Britain, so the French organized their large trading vessels into convoys to be protected by naval escorts, rather than positioning squadrons near key ports or shipping lanes [70]. Louisbourg did not normally see the large convoys organized to protect the East Indies vessels on their voyages to and from France; however, they sought shelter there from British privateers during wartime.

At the beginning of the War of the Austrian Succession in 1744, the French Admiralty Court auctioned cocoa from a British prize captured by Louisbourg privateers. The captain and the crew of the *corsair* shared the proceeds of the sale, after paying dues to the court. The purchaser of the cocoa promptly shipped it to the Basque port of St. Jean de Luz [71]. In early 1745, as the New England privateers gained the upper hand, a New England privateer captured *La Victoire*, a French trader headed to Louisbourg with rum, coffee, molasses, and cocoa from Martinique. The ship was originally the *Victory*, from New England, captured by the French the year before [72].

Cocoa as a prize cargo was an important part of the story of Louisbourg's occupation in 1745 by a combined New England land force and British naval squadron. The British naval squadron, commanded by Commodore Peter Warren, was key to the success of the siege of Louisbourg. Warren also profited enormously by his involvement in the campaign. After the fall of the fortress, New England militia troops unwillingly stayed in garrison over the winter in the destroyed fortress town to hold the capture until British regular troops arrived the following year. After the fall of the fortress, Warren and his squadron, flying French colors, lay in wait in Louisbourg harbor for French East Indies vessels to come to shelter there, unaware of the danger. Three cargo-laden French trade vessels were taken in this way, the richest of which was from Peru: *Notre Dame de la Deliverance*, carrying bullion and cocoa. This prize was the basis of Warren's fortune, as he took the largest share, and the source of deep resentment from the New England garrison in town, who had captured the fortress but got no part of the prize [73].

The soldiers of the garrison obtained some of the cocoa from the prize. A part of the cargo was shared among the ships that had participated in the capture. One of them, the *Vigilant*, received 49 bags of cocoa but remained in Louisbourg over the winter. That winter dysentery swept through the garrison troops, killing more than half of the garrison, over 1000 men. The lieutenant of the *Vigilant*, Ben Gardner, recorded in his log that on seven different occasions in January 1746 his shallop carried cocoa to the town [74]. The cocoa, after being processed into chocolate with mortar and pestle, as there were no *matate* (grinding stones) in Louisbourg, may have nourished or soothed the suffering soldiers. The archaeological remains from the New England occupation indicate that they consumed chocolate as a hot drink, as the French had.

Conclusion

Discovering the patterns of consumption and trade of chocolate in 18th century New France—who had it, how they got it, and how they used it—provides insight into the function of that society and its economy. In civilian society and in the *compagnies franches* in Canada, the ability to own and consume chocolate depended on class and economic status. The desire to consume chocolate was the result of the exchange of ideas between merchants and metropolitan officials in New France, and the Basque community in Louisbourg, who brought and shared their tastes and beliefs about food with their new neighbors. Thus, Louisbourg, a port far from the chocolate-producing regions, became familiar with this new product and incorporated it into their ideas of healthfulness and food luxury. Whereas New France's civilian society extended Basque ideas of chocolate as a staple food and northern French aristocratic ideas of chocolate as luxurious and for the elite, Canada, and the French colonial troops, adapted the concept of chocolate's healthfulness to their need for portable nourishing food during the wars of the mid-century. Even in the military context, the French remained true to their *ancien régime* concepts of hierarchy and reserved chocolate for the privileged, rather than distributing it freely. Elite and sophisticated, although chocolate was adapted to the colonial context, it represented European-ness in New France.

Chocolate also illustrates the function and limitations of mercantile trade in the North Atlantic. The means of acquiring and distributing chocolate illustrate the dynamism of the small colonial merchants in the world of highly regulated trade. Chocolate as a war prize also reminds us of the commercial aspect of war in this period. Its trade illustrates the evolving regional economy of the North Atlantic, with growing links between French colonies, as well as the inevitable exchanges between the British and French colonies.

The process by which the knowledge of chocolate, the desire for it, and the means to acquire it spread in the colonial world illustrates the evolution of colonial society in the North Atlantic. The trade and consumption of chocolate increased as the century progressed; its use was becoming common for the European colonists of North America as the French period closed. However, the cosmopolitan society of the port of Louisbourg disappeared after 1758, and at the end of the Seven Years' War the culture that introduced chocolate to this region was either deported or adapted to North America under the British regime.

Acknowledgments

The authors would like to thank their colleagues at the Fortress of Louisbourg National Historic Site of Canada, and Parks Canada, Atlantic Service Centre for their helpful suggestions in the preparation of this chapter. In particular, we would like to thank Sandy Balcom, Ken Donovan, John Johnston, Heather Gillis, and Rob Ferguson.

Endnotes

1. Identified as: Muiron, Valerien, Hegey, the Captain of the *Hermione*, Larreguy, de Mezy fils, Lascorret, Dargaiganarats, Duquesnel, Schonherr, Prevost, and Roma.

2. Terriau, Petit, the purchaser of her chocolate, Menard, Marin de la Malgue fils.

3. Moore's 1975 study of six years of Louisbourg trade found a total importation of 407 *livres* of chocolate plus two references without a specified amount (p. 37) and one reference to cacao without a specified amount, 13,035 *livres* of coffee plus three references without a specified amount (p. 46), and 12 *livres* of tea from New England (Christopher Moore, *Commodity Imports at Louisbourg, 1713–1758*, MRS 317, Ottawa: Parks heather Gillis Canada, 1975; pp. 37, 46, and 47).

References

1. Braudel, F. *The Structures of Everyday Life: The Limits of the Possible. Civilization and Capitalism 15th–18th Century.* New York: Harper & Row, 1981. Originally published as *Les Structures du quotidien: Le Possible et l'Impossible.* Paris, France: Librairie Armand Colin, 1979; Volume 1, p. 249.

2. Dull, J. R. *The French Navy and the Seven Years' War,* Lincoln: University of Nebraska Press, 2005.

3. McLennan, J. S. *Louisbourg from Its Foundation to Its Fall.* Sydney, Nova Scotia: Fortress Press, 1969. (Originally published in 1918.)

4. Moore, C. *Merchant Trade in Louisbourg, Île Royale.* Ottawa, Canada: University of Ottawa, unpublished M.A. thesis, 1977.

5. Pritchard, J. *Anatomy of a Naval Disaster: The 1746 French Expedition to North America.* Montréal, Kingston, Canada: McGill–Queen's University Press, 1995.

6. Johnston, A. J. B. *Control and Order in French Colonial Louisbourg, 1713–1758.* East Lansing: Michigan State University Press, 2001.

7. Moogk, P. N. *La Nouvelle France: The Making of French Canada—A Cultural History.* East Lansing: Michigan State University Press, 2000.

8. Dull, J. R. *The French Navy and the Seven Years' War.* Lincoln: University of Nebraska Press, 2005; pp. 4–5.

9. Harris, R. C., and Matthews, G. J. *The Historical Atlas of Canada, Volume I: From the Beginning to 1800.* Toronto, Canada: University of Toronto Press, 1987; Plate 24.

10. Dull, J. R. *The French Navy and the Seven Years' War.* Lincoln: University of Nebraska Press, 2005; pp. 107–108.

11. McLennan, J. S. *Louisbourg from Its Foundation to Its Fall.* Sydney, Nova Scotia: Fortress Press, 1969; pp. 161–163.

12. Savary, J. *Dictionnaire Universel De Commerce, D'Histoire Naturelle, Et Des Arts Et Métiers.* Paris, France: Chez la Veuve Estienne, 1742; Tome Première, A–C, column 885.

13. Centre des archives d'outre-mer, Colonies (CAOM COL), France, C11B, Volume 8, folios 181–183, 184, and 189, letters 11, 18, 20 September 1726.

14. Storm, A. Seaweed and gold: the discovery of the ill-fated *Chameau* 1961–1971. In: Donovan, K., editor. *The Island: New Perspectives on Cape Breton's History 1713–1990.* Sydney, Nova Scotia: Acadiensis and University College of Cape Breton Press, 1990; pp. 221–242.

15. CAOM COL. C11A, Volume 84, folios 147–150v.

16. Chaput, D. Joseph Marin de la Malgue. In: *Dictionary of Canadian Biography,* Volume IV. Available at http://www.biographi.ca/EN/. (Accessed January 10, 2007.)

17. CAOM COL. C11A, Volume 115, folios 254–276.

18. CAOM COL. C11A, Volume 86, folios 178–236.

19. CAOM COL. C11A, Volume 117, folios 168–302v, 1 September 1747.

20. CAOM COL. C11A, Volume 117, folios 176–177.

21. Russ, C. J. Louis La Corne. In: *Dictionary of Canadian Biography,* Volume III.

22. CAOM COL. C11A, Volume 117, folio 309.

23. Gadoury, L. *La Noblesse de Nouvelle-France: Familles et Alliances,* Québec Canada: Éditions Hurtubise, 1991.

24. CAOM COL. C11A, Volume 117, folios 168–320v, 1 September 1747, f. 254.

25. CAOM COL. C11A, Volume 88, folios 184–198, 31 October 1747.

26. CAOM COL. C11A, Volume 88, folios 248–254v, 15 October 1747.

27. CAOM COL. C11A, Volume 104, folios 150–154, 1759, Bigot au ministre.

28. Franquet, L. *Voyages et Mémoires sur le Canada en 1752–1753*. Toronto, Canada: Canadiana House, 1968; p. 132.

29. J.-C. B. *Voyage au Canada, fait depuis l'an 1751 à 1761*. Paris, France: Éditions Aubier Montaigne, 1968; p. 124.

30. Public Record Office (PRO), High Court Admiralty (HCA) Kew, England, Volume 32, document 198, 1758, *Hermione*.

31. CAOM COL. Depôt des papiers publics des colonies (DPPC) G2, Volume 199, document 189.

32. CAOM DPPC. G2, Volume 199, document 189, folios 124–126.

33. Jonah, A. M. L., and Tait, E. Filles d'Acadie, femmes de Louisbourg: Acadian women and French colonial society in 18th century Louisbourg. In: *French Colonial History*, Volume 8. East Lansing: Michigan State University Press, 2007.

34. Donovan, K. Tattered clothes and powdered wigs: case studies of the poor and well-to-do in 18th century Louisbourg. In: Donovan, K., editor. *Cape Breton at 2000*. Sydney, Nova Scotia: University College of Cape Breton Press, 1985; pp. 1–20.

35. CAOM DPPC. G2, Volume 187, folios 12–334, 1743. Valerien Louis dit Bourguignon accusé de vol.

36. CAOM DPPC. G2, Volume 198, file 180, document 14, succession de Jacques Rolland, marchand.

37. White, S. A baser commerce: retailing class and gender in French Colonial New Orleans. *William and Mary Quarterly* (3rd series) 2006;LXIII(3):517–550.

38. White, S. This gown was much admired and made many ladies jealous: fashion and the forging of elite identities in French Colonial New Orleans. In: Harvey, T., and O'Brien, G., editor. *George Washington's South*. Gainesville: University Press of Florida, 2004; pp. 86–118.

39. Ketcham Wheaton, B. *Savoring the Past: The French Kitchen and Table from 1300 to 1789*. New York: Scribner, 1983; pp. 87–94.

40. Moore, C. *Merchant Trade in Louisbourg, Île Royale*, Ottawa, Canada: University of Ottawa, unpublished M.A. thesis, 1977; p. 43.

41. CAOM DPPC. G2, Volume 199, document 18, 1744 succession Dargaignarats.

42. CAOM DPPC. G2, Volume 185, folios 52–64, succession de Bertrand Larreguy, 1 mars 1738.

43. Franquet, L. *Voyages et Mémoires sur le Canada en 1752–1753*. Toronto, Canada: Canadiana House, 1968; p. 148.

44. National Archives of Canada. Baby Collection: Correspondence, 2, Havy and Lefebvre to Pierre Guy, Québec, 29 October 1745, 742. Cited in: Katherine Young, Sauf les périls et fortunes de la mer: merchant women in New France and the French transatlantic trade, 1713–46. *The Canadian Historical Review* 1996;77(3):388–407.

45. Rousseau, J., and Béthune, G. *Voyage de Pehr Kalm au Canada en 1749: traduction annotée du journal de route*. Montréal, Canada: Pierre Tisseyre, 1977; p. 297.

46. Moore, C. *Merchant Trade in Louisbourg, Île Royale*, Ottawa, Canada: University of Ottawa, unpublished M.A. thesis, 1977; pp. 16–23.

47. Moore, C. *Commodity Imports At Louisbourg, 1713–1758*, MRS 317. Ottawa, Canada: Parks Canada, 1975; pp. 37, 46, and 47.

48. Moore, C. *Commodity Imports At Louisbourg, 1713–1758*, MRS 317. Ottawa, Canada: Parks Canada, 1975; p. 37.

49. Harms, R. *The Diligent: A Voyage Through the Worlds of the Slave Trade*. New York: Basic Books, 2002; pp. 342–346.

50. Pritchard, J. *The French in the Americas, 1670–1730*. Cambridge, England: Cambridge University Press, 2004; pp. 125–129.

51. CAOM COL. C11A, Volume 105, folios 322–325, 4 November 1761, Duvernet au ministre.

52. Blanchette, J.-F. *The Role of Artifacts in the Study of Foodways in New France, 1720–1760*. History and Archaeology Series, 52. Ottawa, Canada: Parks Canada, 1981; pp. 122, 130.

53. McLean, J. *Jean Pierre Roma and the Company of the East of Isle St. Jean*. Charlottetown, Prince Edward Island, Canada: Prince Edward Island Heritage Foundation, 1977.

54. Coleman, M. Jean-Pierre Roma. In: *Dictionary of Canadian Biography*, Volume III.

55. CAOM COL. C11A, Volume 88, folios 248–254v, 15 October 1747.

56. Pritchard, J. *In Search of Empire: The French in the Americas, 1670–1730*. Cambridge, England: Cambridge University Press, 2004; pp. 125–126.

57. CAOM COL. C11B, Volume 19, folio 28v, 26 October 1737.

58. McNeill, J. R. *Atlantic Empires of France and Spain: Louisbourg and Havana, 1700–1763*. Chapel Hill: University of North Carolina Press, 1985; pp. 46–52.

59. Archives départementales de la Charente-Maritime (ADCM). Amirauté, B, 272,204, August 11, 1743, the *Midnight*.

60. Moore, C. *Merchant Trade in Louisbourg, Île Royale*, Ottawa, Canada: University of Ottawa, unpublished M.A. thesis, 1977; pp. 97–100.

61. Moore, C. *Merchant Trade in Louisbourg, Île Royale*, Ottawa, Canada: University of Ottawa, unpublished M.A. thesis, 1977; p. 26.

62. CAOM DPPC. G2, Volume 183, folios 434–470, September 1736.

63. CAOM DPPC. G3, Volume 2046-1, folio 47, 9 December 1737, vente du brigantin neuf nommé La Reine du Nord.

64. ADCM B. Volume 6113, document 26, September 1741.

65. PRO, HCA. Volume 32, document 242, 1757, *La Surprise*.

66. CAOM DPPC. G3, Volume 2044, document 19, 1756 Succession de Marguerite Terriau.

67. CAOM DPPC. G2, Volume 202, document 286, 1753 Succession de Thérèse Petit.

68. Baron de Lahontan. *New Voyages to North America*, 1703. Translated by Reuben Gold Thwaites. New York: Burt Franklin, 1905; reprinted 1970, Volume 1, pp. 373–378.

69. CAOM COL. C11A, Volume 86, folios 178–236.

70. Dull, J. R. *The French Navy and the Seven Years' War*. Lincoln: University of Nebraska Press, 2005; pp. 9–11.

71. PRO HCA. Volume 32, documents 130–1, 1744–1745, *The Marie*.

72. Effingham de Forest, L. E., editor. *The Journals and Papers of Seth Pomeroy*, The Tuttle, Morehouse and Taylor Company, Publication #38. New Haven, CT: The Society of Colonial Wars, 1926; p. 18.

73. Gwyn, J. *An Admiral for America Sir Peter Warren, Vice Admiral of the Red, 1703–1752*. Gainesville: University Press of Florida, 2004; pp. 102–103.

74. PRO, HCA/L. Volume 67, document 33, Journal of Lieutenant Ben Gardner, 6th July 1745–7th July 1746.

26

Chocolate Manufacturing and Marketing in Massachusetts, 1700–1920

Anne Blaschke

Introduction

Since the initial mercantile capitalist days of the Massachusetts Bay Company in the mid-17th century, New England has produced commodities of interest to epicures both on the continent and abroad, among them fruits, grains, legumes, and shellfish. Less well-known than the lobsters and cranberries of New England fame, however, is the region's historical production and consumption of chocolate. Indeed, though San Francisco, California; Hershey, Pennsylvania; and McLean, Virginia, are headquarters of the lauded American mega-producers today, chocolate and cocoa production, marketing, and broadly distributed consumption began in the New England colonies.

Unlike the more common crops noted above, chocolate has been considered a delicacy and health aid, rather than a food staple, since its introduction to North America by the Spanish and English in the early 17th century (see Chapter 6). In the Colonial period, prior to the establishment of chocolate works in New England, local merchants seem to have imported cocoa beans and traded chocolate in relatively small quantities, regarding it as a luxury item, the value of which lay in its rich taste and specialized healing properties [1]. In the late 18th century, however, the founding of the Walter Baker & Co. chocolate manufactory dramatically increased New Englanders' exposure to and enthusiasm for chocolate by producing it in relative mass quantities with innovative technology and labor practices, and deliberately marketing it as an affordable, healthful indulgence. In the 19th and early 20th centuries, the company maintained this strategy of mass production, industrial innovation, and strategic advertising, and as a result it revolutionized the chocolate industry in New England. Because of this distinct regional marketing and consumption pattern, tracing cocoa and chocolate in New England reveals an unusual local history, one that elucidates the development of key political, economic, and sociocultural movements important in North America and refracted on a microcosmic level to the regional chocolate industry as well. Thus, the purpose of this chapter is to examine cocoa and chocolate manufacturing in New England—with a focus on Massachusetts, the locus of this industry—to understand the role that a gourmet foodstuff can play in shedding new light on local and national history from 1700 to 1920.

Chocolate: History, Culture, and Heritage. Edited by Grivetti and Shapiro
Copyright © 2009 John Wiley & Sons, Inc.

Small-Scale, Individualized Colonial Chocolate Use

Like coffee, tea, spices, and other delicacies imported from across the Atlantic and South America to this day, cacao trees thrive in warm, humid climates [2]. For this reason, cocoa and chocolate manufacturing industries began in the New England colonies only when the importation of cocoa beans to the area began in earnest around 1755 [3]. Prior to this point, the sale and consumption of cocoa and chocolate was isolated and of small quantity, although those colonists who were literate in English, Spanish, or Latin may have read a growing literature on the history, production, and health benefits of chocolate crossing the Atlantic from Europe in the mid-17th century. Indeed, doctors, merchants, ship captains, and others published works on chocolate ranging from instructions in chocolate preparation to accounts of its medicinal use [4–6]. In 1652, for example, New England residents who imported books along with manufactured goods from the mother country could have read Captain James Wadsworth's translation of a recent Spanish publication on the health benefits of chocolate, which lauded the beverage in its title: *Chocolate, or, An Indian drinke: by the wise and moderate use whereof, health is preserved, sickness diverted and cured, especially in the plague of the guts* . . . [7]. Since Puritans saw the Bible as the ultimate source of revelation, literacy was central to their society, and it is likely that this relatively educated group with significant economic ties to England would have encountered publications referencing chocolate as a delicacy or medicine during this period [8].

Nevertheless, as historian W. G. Clarence-Smith has noted, establishments that served drinking chocolate (nearly all chocolate was consumed as a beverage until the mid-19th century) were rare, and only the well-to-do could afford to consume it at home [9]. The account books of Boston merchants in the early 18th century support this conclusion—the vast majority recording no trade in chocolate or cocoa, and those few that do revealing only sporadic sales of chocolate at comparatively high prices. Boston merchant Ann Greene, for example, traded only four bags of cocoa during 1725–1727, in comparison to much larger quantities of flour, textiles, salt, and other sundries [10]. The relatively high price of "twenty four pounds one shilling as in full for 2 bags cocoa sold her" on April 11, 1726—her sole chocolate or cocoa purchase recorded in that year—strongly suggests that Greene considered cocoa a rare commodity and sold it only sporadically in her store, and at expensive rates. This early pattern of sparse, scattered chocolate consumption is further supported by the account book of Boston distiller and merchant Thomas Amory, who, as a member of one of Boston's more prosperous families, operated on a much larger scale than did Greene. Despite the comparatively large, diversified nature of Amory's inventory, he too recorded only 11 commercial chocolate shipments of 20 or fewer hundredweight each, to clients in North Carolina, in a ledger maintained during 1720–1728 [11]. The dearth of chocolate sales in Amory's record book speaks to the irregularity with which his clients sought it, in comparison to other such sundries as textiles, sugar, and rum, which sold in much higher quantities. Furthermore, Amory's purchase of seven hundredweight of chocolate "for the house," presumably his family, speaks to the specialized, inaccessible way chocolate and cocoa were understood to be gourmet foodstuffs by contemporary New Englanders. As one of the most successful merchants in Boston, Amory could afford to buy chocolate for his family to enjoy, while many local consumers likely could not pay for such an indulgence at home.

Although Amory's large business operation and Greene's middling enterprise speak to the variety within Boston's early 18th century merchant class, the ledgers of both point to the limited consumption and high price of chocolate during this period in New England. At mid-century, this trend continued apace, as demonstrated in the accounts of Salem, Massachusetts, merchant John Higginson. In 1756, Higginson noted one sale of a small quantity of chocolate, for the price of £11.6 shillings—a high price compared to that of the other goods he sold. This transaction was his sole sale of chocolate or cocoa in three years of record-keeping [12]. Although scores of merchants assiduously maintained ledgers in this period, the paucity of businesses recording trade in chocolate or cocoa speaks to the irregularity with which it was bought or consumed by New England residents.

Increased Trade Prior to Revolution

In the decade or so prior to the American Revolution, however, New England merchants' trade in chocolate expanded markedly. Increased population, through immigration and natural growth, and corresponding increasing mercantile activity (both intercolonial and Atlantic) likely explain this jump. Heightened interest in chocolate consumption is thus reflected in increased importation and trade in the 1760s. Demonstrating this trend are docking records of Boston shipper George Minot, who in 1765 oversaw the importation of 48 bags of cocoa powder and 69 bags of "Cocoa & Coffee" for local retailers [13]. Striking in Minot's ledger are the prices for cocoa in comparison to other commodities shipped; rather than being significantly higher, as in previous decades, rates parallel those of the other sundry items—such as flour, tar, lime, and vegetables—passing through this port [14]. Here we see that cocoa was now imported in larger quantities, with lower rates, likely indicating its increasingly widespread marketability and consumption. Further supporting the assertion that chocolate use became increasingly popular is its reference in late 1760s correspondence of Bostonian politicians and merchants Samuel and Joseph Otis. In recounting his sundries order, which included 190 hundredweight of chocolate at £11, to his brother Joseph, Samuel Otis noted that presently he had run out of "chocolate rum pork or molasses," and that he "shall purchase them upon my own [*sic*] but then you must punctually support [*sic*] one or I shall suffer for nothing" [15]. In this correspondence, Otis not only notes significant amounts of chocolate among such dietary staples as corn, sugar, flour, rum, and pork, but further notes the necessity of replacing chocolate immediately—demonstrating its importance in their inventory. Moreover, Otis needed chocolate to the extent that he purchased it "on his own," rather than waiting for alternative funding. Whereas small quantities of chocolate and cocoa were purchased but once a year, for high prices, just decades before, in the pre-Revolutionary period, major merchants displayed determination to procure it in large quantities, keep it in stock, and pay less for it.

Although dry goods merchants and grocers like Ann Greene or the Otis brothers were the main purveyors of chocolate before the Revolution, apothecaries and druggists sold chocolate as well, and continued the early 16th century Native American and Spanish uses of chocolate as a nutritious food with healing properties [16]. As such, these colonial pharmacists simultaneously acknowledged the health benefits of chocolate consumption touted by their predecessors and presaged the more concerted effort to link chocolate with good health that would mark 19th century sales and marketing so prominently. Indeed, Clarence-Smith points out that some Boston pharmacists advertised chocolate for medicinal use as early as 1712 (see Chapter 27) [17]. One such mid-century vendor was Benedict Arnold, who prior to his military career ran a successful drug shop in New Haven, Connecticut. In a broadside tentatively dated 1765, Arnold advertised all manner of sundries for the maintenance of one's physical, mental, and spiritual health, including "choc [*sic*]" among these:

> *Benedict Arnold, Has just imported (via New-York) and sells at his Store in New-Haven, A very large and fresh Assortment Of Drugs and Chymical [sic] Preparations: Essesnce [sic] Water, Dock, Essence Balm Gilded, Birgamot . . . TEA, Rum, Sugar, m [sic], Fine Durham Flower, Mustard, &_choc . . . many other Articles, very cheap, for Cash or Short Credit.* [18]

In addition to grouping the chocolate with herbs, spices, and tonics promoting good health, Arnold also emphasized the economic benefits of

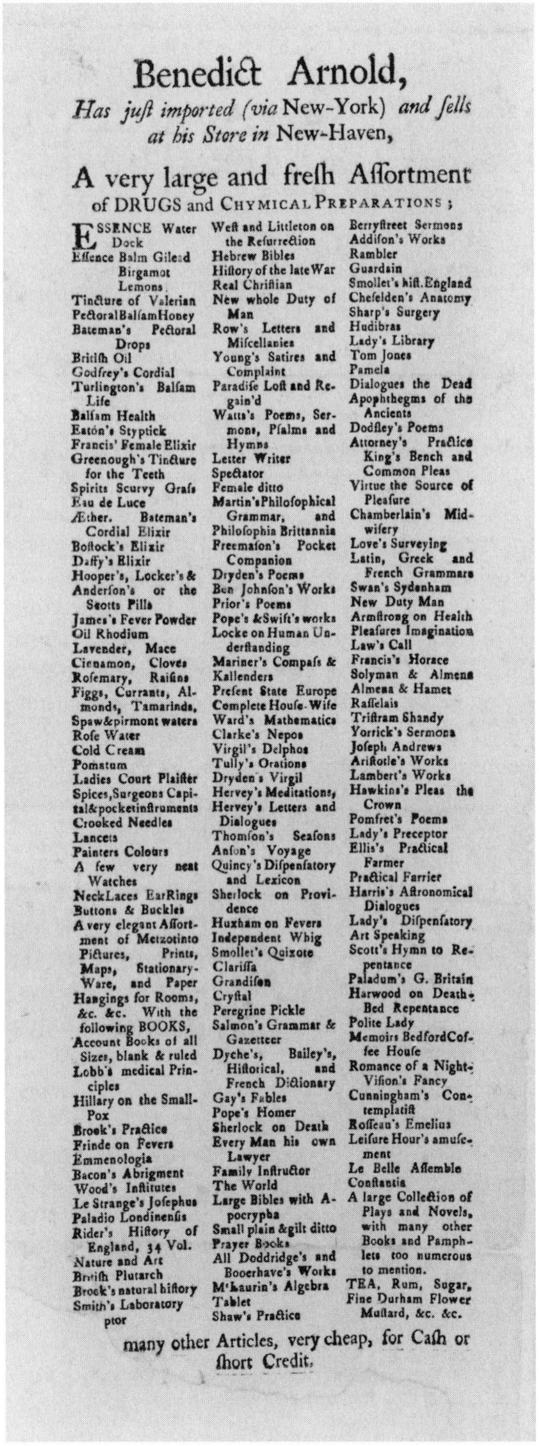
Benedict Arnold,

Has juſt imported (via New-York) *and ſells at his Store in* New-Haven,

A very large and freſh Aſſortment of DRUGS and CHYMICAL PREPARATIONS;

ESSENCE Water Dock
Eſſence Balm Gilead Birgamot Lemons
Tincture of Valerian
PectoralBalſamHoney
Bateman's Pectoral Drops
Britiſh Oil
Godfrey's Cordial
Turlington's Balſam Life
Balſam Health
Eatón's Styptick
Francis' Female Elixir
Greenough's Tincture for the Teeth
Spirits Scurvy Graſs
Eau de Luce
Æther. Bateman's Cordial Elixir
Boſtock's Elixir
Daffy's Elixir
Hooper's, Locker's & Anderſon's or the Scotts Pills
James's Fever Powder
Oil Rhodium
Lavender, Mace
Cinnamon, Cloves
Roſemary, Raiſins
Figgs, Currants, Almonds, Tamarinds,
Spaw&pirmont waters
Roſe Water
Cold Cream
Pomatum
Ladies Court Plaiſter
Spices,Surgeons Capital&pocketinſtruments
Crooked Needles
Lancets
Painters Colours
A few very neat Watches
NeckLaces EarRings
Buttons & Buckles
A very elegant Aſſortment of Metzotinto Pictures, Prints, Maps, Stationary-Ware, and Paper Hangings for Rooms, &c. &c. With the following BOOKS,
Account Books of all Sizes, blank & ruled
Lobb's medical Principles
Hillary on the Small-Pox
Brook's Practice
Frinde on Fevers
Emmenologia
Bacon's Abrigment
Wood's Inſtitutes
Le Strange's Joſephus
Paladio Londinenſis
Rider's Hiſtory of England, 34 Vol.
Nature and Art
Britiſh Plutarch
Brook's natural hiſtory
Smith's Laboratory
ptor

Weſt and Littleton on the Reſurrection
Hebrew Bibles
Hiſtory of the late War
Real Chriſtian
New whole Duty of Man
Row's Letters and Miſcellanies
Young's Satires and Complaint
Paradiſe Loſt and Regain'd
Watts's Poems, Sermons, Pſalms and Hymns
Letter Writer
Spectator
Female ditto
Martin'sPhiloſophical Grammar, and Philoſophia Brittannia
Freemaſon's Pocket Companion
Dryden's Poems
Ben Johnſon's Works
Prior's Poems
Pope's &Swift's works
Locke on Human Underſtanding
Mariner's Compaſs & Kallenders
Preſent State Europe
Complete Houſe-Wife
Ward's Mathematics
Clarke's Nepos
Virgil's Delphos
Tully's Orations
Dryden's Virgil
Hervey's Meditations,
Hervey's Letters and Dialogues
Thomſon's Seaſons
Anſon's Voyage
Quincy's Diſpenſatory and Lexicon
Sherlock on Providence
Huxham on Fevers
Independent Whig
Smollet's Quixote
Clariſſa
Grandiſon
Cryſtal
Peregrine Pickle
Salmon's Grammar & Gazetteer
Dyche's, Bailey's, Hiſtorical, and French Dictionary
Gay's Fables
Pope's Homer
Sherlock on Death
Every Man his own Lawyer
Family Inſtructor
The World
Large Bibles with Apocrypha
Small plain &gilt ditto
Prayer Books
All Doddridge's and Booerhave's Works
M'Laurin's Algebra
Tablet
Shaw's Practice

Berryſtreet Sermons
Addiſon's Works
Rambler
Guardain
Smollet's hiſt.England
Cheſelden's Anatomy
Sharp's Surgery
Hudibras
Lady's Library
Tom Jones
Pamela
Dialogues the Dead
Apophthegms of the Ancients
Dodſley's Poems
Attorney's Practice King's Bench and Common Pleas
Virtue the Source of Pleaſure
Chamberlain's Midwifery
Love's Surveying
Latin, Greek and French Grammars
Swan's Sydenham
New Duty Man
Armſtrong on Health
Pleaſures Imagination
Law's Call
Francis's Horace
Solyman & Almena
Almena & Hamet
Raſſelais
Triſtram Shandy
Yorrick's Sermons
Joſeph Andrews
Ariſtotle's Works
Lambert's Works
Hawkins's Pleas the Crown
Pomfret's Poems
Lady's Preceptor
Ellis's Practical Farmer
Practical Farrier
Harris's Aſtronomical Dialogues
Lady's Diſpenſatory
Art Speaking
Scott's Hymn to Repentance
Paladum's G. Britain
Harwood on Death-Bed Repentance
Polite Lady
Memoirs BedfordCoffee Houſe
Romance of a Night-Viſion's Fancy
Cunningham's Contemplatiſt
Roſſeau's Emelius
Leiſure Hour's amuſement
Le Belle Aſſemble
Conſtantia
A large Collection of Plays and Novels, with many other Books and Pamphlets too numerous to mention.
TEA, Rum, Sugar, Fine Durham Flower Muſtard, &c. &c.

many other Articles, very cheap, for Caſh or ſhort Credit.

FIGURE 26.1. Benedict Arnold advertisement: sale of chocolate, mustard, rum, sugar, and tea—dated 1765. *Source*: Call Number 41515. Courtesy of Massachusetts Historical Society, Boston, MA. (Used with permission.)

patronizing his store—it was affordable—and he accepted payment in either cash or credit (Fig. 26.1). Here Arnold demonstrated his desire to appeal to shoppers in a wide income bracket, most of whom would have read his advertisement and understood chocolate to be a commodity they could afford. Furthermore, by including chocolate in his list of health-related wares, Arnold identified chocolate as a nutritive aid—but one of many, rather than a delicacy above the sophistication level of the average consumer. Although his inclusion of chocolate on an apothecary's inventory was a far cry from an overt, aggressive marketing attempt to link cocoa and nutrition, Arnold's appeal to New Haven residents' pocketbooks and physical constitutions subtly speaks to chocolate's increasing accessibility as a food in the decade before the war.

Revolution

Although the Second Continental Congress closed trade with Britain during the American Revolution, New England merchants continued to conduct a markedly higher trade in chocolate than earlier in the century during these years of conflict and nation-building. The war profoundly affected international commerce and necessitated, at its end, the creation of a new domestic financial structure, but evidence indicates that even in this time of uncertainty New Englanders continued to see chocolate as an increasingly commonplace food, rather than exclusively a luxury item or medicine. One crucial reason Americans gravitated toward chocolate as a beverage during these turbulent years was the Patriots' boycott of British tea, a commodity that "became a symbol of the iniquities of colonial taxation and regulation" in the early 1770s [19]. However, such was its increased integration into the regional market, lexicon, and diet, that British soldiers stationed in New England prior to the Revolution also purchased chocolate, on their own time and wages. Hence, the growing popularity of chocolate as an alternative beverage superseded political connotations associated with tea. The account book of Boston merchant George Deblois notes that in March 1775, Ensign Samuel Murray "of [the] 43 Reg[iment]" bought a shilling's worth of chocolate, along with a shoe brush and several yards of ribbon [20]. Murray's purchase of chocolate while in camp, or on the march, supports the argument that chocolate became an increasingly well-liked and familiar food in the late 18th century, as its refreshing taste and energizing faculties became more widely known. Although his canteen rations likely comprised most of his food intake, Murray supplemented his diet with chocolate, the cheapest of his purchases that day, indicating that he understood it to be an affordable, endurance-boosting addition.

The First American Large-Scale Chocolate Works

Although Murray purchased but one small piece of chocolate in 1775, the founding of the first significant American chocolate works just prior to and following the Revolution would enable him, and myriad other New Englanders, to consume cocoa and chocolate with increasingly more ease and less money. In 1765, chocolate maker John Hannon, financed by John Baker, set up shop in a newly constructed mill in Dorchester, Massachusetts, and began grinding imported cocoa beans. Over the following decade, Baker assumed increasing responsibility for the culinary craftsmanship of chocolate from Hannon, a "penniless Irish immigrant" [21].

At this seminal stage, Baker outfitted his factory with culinary implements necessary for commercial manufacture of chocolate. To convert a paper mill, which he rented in 1772, into a chocolate mill, for example, Baker installed "a run of stones and a set of kettles" [22]. He also bought iron pestles and thimbles, likely to refine the process of grinding cocoa beans, and an iron roaster that same year. For his recipes, Baker purchased such key ingredients as rum, salt, and molasses, and also commissioned a sign for his property—the first recorded advertisement of sorts for a chocolate manufactory in New England. Baker faced competition in the chocolate industry from the outset, however. Although a temporary adversary at best, Rhode Island businessman John Waterman also sought to establish himself at this time as a regional chocolate maker and purveyor. His intention, revealed in the increasingly aggressive advertising that would come to characterize the New England chocolate industry in the 19th century, is manifest in the following notice published in the *Newport Mercury* and as a Southwick and Newport broadside in 1774:

> John Waterman hath lately built the completest [sic] Works in the colony, for making CHOCOLATE, Near the Paper-Mills, in Providence. He wants to buy Cocoa-Nuts, or will take them to make up, at Two Dollars per Hundred Weight, with a quick return and will at all Times be ready to furnish his customers with CHOCOLATE, Made in the best and purest Manner, as cheap as it can be had in this colony, or Boston. [23]

Commissioned less than a year prior to the first battles of the Revolution at Lexington and Concord in April 1775, this advertisement ran alongside Great Britain's recently issued *Act For Blocking Up the Harbour of Boston*, one of the mother country's punitive measures issued in 1774 in response to the Boston Tea Party. Because Boston merchants faced much consternation over whether or not to cease trade with Britain over the harbor blockage and other "Intolerable Acts" of the crown, Waterman's assertion that he will "at all Times be ready to furnish his customers with CHOCOLATE," was well-timed; regardless of whether local patriots refused to import British-controlled cocoa beans or chocolate, Waterman promised he would be available as a domestic, continental producer [24]. Furthermore, in offering his clientele an alternative to tea, that hated symbol of Parliament's tyranny, Waterman appealed on an ideological level to colonists who saw themselves as oppressed by the metropole. In replacing British imperial tea with chocolate, Waterman's customers could exercise political agency and continue to enjoy the habit, originally carried to the colonies from Europe, of drinking a hot beverage during the day and evening.

Yet another notable factor in Waterman's advertisement was his claim to sell chocolate as cheaply as any other New England producer. As in Benedict Arnold's apothecary advertisement of the previous decade, Waterman attempted to resonate with the myriad small farmers, merchants, journeymen, servants, and other regional residents who had little cash to spend, bought on credit, or still relied on the traditional bartering system that survived from the early 17th century. Although Waterman marketed only chocolate, as opposed to the myriad health products of Arnold's business, he nevertheless in this advertisement took a major step in making a specialized commodity available to an increasingly widening group of Americans who dealt in cash. Thus, although Waterman's business disappeared from the historical record with this advertisement, he nonetheless opened a door for larger producers like Baker to access the New England masses, and their burgeoning economic power, in the years of the early Republic.

Chocolate: Beginning of Mass Production and Consumption in New England

Although regional interest in chocolate increased markedly during the Revolutionary era, the chocolate industry experienced a boom that trumped these earlier gains during the formation of the Republic and would continue to burgeon into the 19th century. Heading the new wave of production, innovation, and advertising of cocoa and chocolate products was the Baker family, which officially dates its professional entry into the business in 1780, when James Baker abandoned all other professional ventures to buy out the Hannon family, and first produced chocolate stamped with the Baker name [25]. Although the company began without ownership of a single mill and struggled against rampant inflation of Continental paper currency owing to the high costs of fighting the Revolution, Baker turned a profit rapidly. Although no record remains indicating that Baker sought an arrangement with the Second Continental Congress to supply chocolate to American troops, nevertheless, he expanded his business during the war by renting new facilities and hiring innovative chefs. Equipment and staff expansions, as well as lucrative contracts with wholesalers and shippers, enabled Baker to produce sufficient quantities of chocolate his first year in business to account for such impressive

sales as that of March 19, 1781, when Boston retailers Fox & Gardner bought 2210 pounds of chocolate [26].

Such striking jumps in production and profit were no accident. Baker's early success was contingent upon his ability to exploit new technology and to market chocolate with an approach to expansion and mass production that was entirely new to the American industry. In 1771, Hannon and Baker published an advertisement for Hannon's chocolate affirming the product to be:

Warranted pure, and ground exceeding fine. Where may be had any Quantity, from 50wt. to a ton, for Cash or Cocoa, at his Mills in Milton . . . If the Chocolate does not prove good, the Money will be returned. [27]

In this promotion, the manufacturers established several founding tenets of the company—high quality, unadulterated cocoa and chocolate; the ability to mass-produce; and guaranteed merchandise. These principles, along with affordable price ranges, would become the core Baker philosophy in the 19th century.

Foremost among these company pillars was high-quality chocolate and cocoa. Although James Baker patronized domestic wholesalers of cocoa beans at the outset of his business, during the early 19th century (following the cessation of Jefferson's embargo in 1809), the company consistently bought cacao from Caribbean and Central and South American producers [28]. Dealing directly with representatives of cacao plantations from various producing countries allowed the Bakers to experiment with regional characteristics, adapt their chocolate blends accordingly, and buy from any given raw producer when the market favored them [29].

This culinary amalgamation was made possible by a major expansion of the Bakers' facilities. During his ownership of the business, Dr. James Baker increased production capability by not only renting increased mill space, but by investing in increasingly specialized equipment and hiring creative staff who perfected a number of techniques particular to the company in the 19th century [30–32]. Upon Baker's retirement and bequest of the business to his son, Edmund Baker, in 1804, the latter increased the company's rented mill capacity further still, bought its first mill, and built yet another one equipped with new technology, enabling the family to vastly increase its output as sales climbed [33]. To that end, although the embargo of 1807 hit New England merchants hard, the Baker chocolate manufactory consistently turned a profit during the decade, in part through its own well-managed advertising. In addition, as the brand gained notoriety near Boston, the company reaped the benefit of local grocers' advertising. One such ad, written and published by Dorchester schoolteacher Samuel Temple, advised *Norfolk Repository* readers:

*To be sold at the Store opposite the Arch Over Milton Bridge,
the following articles . . . Shells, Chocolate* & Stetson's
Hoes As good as can be (I suppose)
BAKER'S, of course! . . .

*I've many things I shall not mention
Too sell them cheap is my intention
Lay out a dollar when you come
And you shall have a glass of RUM . . .
Since mane [sic] to man is so unjust
Tis hard to say whom I can trust
I've trusted many to my sorrow
Pay me to-day. I'll trust to morrow [sic].*

Dorchester, June, 1805. [34]

This advertisement articulates precisely the philosophy that the Baker Company sought to express in its own promotions: a quality name, which was to be emphasized as a mark of excellence, and a bargain price. Rather than paying dearly for name-brand superiority or saving money on a mediocre article, Temple confided to his potential consumers that one could now have both quality and economy. Although the Baker family dealt in monthly payments and credit with its clients, unlike Temple, who had "trusted many to my sorrow," the philosophies of manufacturer and retailer were otherwise quite similar.

Edmund Baker continued to expand the company's mill ownership during the War of 1812, and in the following decade introduced his son, Walter Baker, to the family business. Despite increasing tariffs on imported cocoa beans following the war, Edmund Baker rebuilt and expanded a main mill and increased the company's output [35]. Furthermore, he regularly bought chocolate for his own family, recorded as a personal expense in his ledgers, seemingly considering it an everyday foodstuff rather than a specialty item—precisely the attitude toward the commodity the Bakers sought to convey to New England consumers [36]. When Edmund Baker retired in 1824, leaving the company to Walter, the family business was ambitiously pursuing the redefinition of the New England chocolate industry it had begun prior to the Revolutionary War.

Walter Baker and the Marketing Revolution of the Mid-19th Century

If the Baker family had eagerly introduced a new philosophy of quality chocolate and cocoa at affordable prices to the New England populace in the late 18th and early 19th centuries, Walter Baker aggressively pushed this creed to new heights in the following 20 years. His ambition made Walter Baker & Co. a household name regionally and nationally, substantially increased the company's profits, and furthered the New England chocolate industry's industrial revolu-

tion—from small, independent artisan chocolate makers to mass-producing, highly diversified, technologically innovative companies. Although Baker assiduously pursued the basic tenets of quality at a bargain and mass production that his father and grandfather had instilled, he proved particularly adept at two areas of expansion: insistent correspondence with wholesalers and retailers, and innovative diversification of chocolate and cocoa products.

Perhaps remembering fondly the benefits of free incidental advertising in such grocers' advertisements as Samuel Temple's from his boyhood in Dorchester, Walter Baker relentlessly inquired after the sales and marketing practices of his retailers, sometimes through his agents, and sometimes in blunt queries to the grocers themselves. Indeed, his letter book from the early 1820s reveals letters to local associates, as well as his agents, Grant & Stone in New York, regarding such matters as bad debt and collection, sales, advertising, and shelf-life of chocolate over time. The latter concern, in particular, occupied much of his correspondence. Because the Bakers refused to adulterate their chocolate, and preservatives had not yet been invented, chocolate and cocoa products that had traveled by ship through inclement weather or had been in shipwrecks often arrived at retailers unfit for consumption. These Baker insisted be returned to Dorchester or disposed of. In other cases, warm weather or damp storage caused the chocolate to change appearance, becoming "old & rusty," but still suitable for eating [37]. In those cases in which the color but not the taste or smell of the product was altered, Baker requested that the retailer sell it at reduced rates. In still other cases where chocolate had deteriorated beyond an edible state, however, Baker sometimes insisted either that the retailer return the affected chocolate to Dorchester to be "run" again, or that the merchant sell it at a reduced price. Very seldom did Baker offer to refund the value of the chocolate.

In addition to following the condition of his chocolate closely, Baker insisted on his clients apprising him of their particular marketing, wholesale, and retail practices. Discovering an agency that sold individual boxes of Baker's No. 1 (the company's premium brand) at the wholesale price in order to introduce it to more retailers, Baker sharply urged:

*I would not wish you to continue the practice . . . and never
sell a single 13 oz. Box twice to the same grocer at the
wholesale price.* [38]

This insistence on controlling every aspect of the commercial process, from factory to sales floor, garnered prompt responses from Baker's associates, most of whom seem to have eagerly responded to his requests for more information on the fate of his chocolate. The short shelf-life of chocolate in this period before refrigeration, fast overland shipping, and preservatives likely explains Baker's clients' strong desire to maintain frequent correspondence, for a shipment of chocolate that turned "rusty" in a retailer's store was a lost profit for all involved. Consistent mailing of letters, orders, and bills connected Baker and his wholesalers for their mutual benefit, enabling them to increase profits by micromanaging their output and marketing, despite Baker's admission in 1823 that:

Chocolate is not a very saleable article, and many grocers will [not] purchase it after it loses its bright colour, & we do not desire it returned. [39]

Because chocolate was not, in the 1820s, as popular a foodstuff as many other comestibles, Baker and his clients monitored their markets assiduously and adjusted the size and frequency of shipments accordingly.

Perhaps Baker's clients also tolerated his insistence on his standards in their own sales practices because his diverse range of products sold increasingly well in the early 19th century. In addition to his tenacious desire for control, Baker also ambitiously strove to diversify the company's confections so that shoppers of varied income brackets and purposes could satisfy their cravings for chocolate. Under Baker's tenure, Walter Baker & Co. developed a wide range of innovative confections specifically marketed for particular uses and customers. While the company continued to sell Bakers "No. 1" as their highest-quality and most expensive chocolate, during the 1820s they offered more frugal customers the options of cracked cocoa, cocoa shells, "Mass[achusetts] Chocolate," and the "No. 2" blend [40]. Walter Baker & Co. marketed these products, which were made of lesser-quality raw materials than the "No. 1," as low-priced articles that could nevertheless produce a satisfying cocoa drink. Baker included instructions for making cocoa with these items. Further diversifying the company's product lines in the 1830s, Baker created his "Lapham" brand of chocolate, named for a longtime employee, and aggressively touted it as an economical alternative to his "No. 1" [41]. By stressing the fine quality and healthful results of these lower-cost options, Baker sought to retain the company's reputation for pure, gourmet-quality chocolate at a price that made this former luxury available to all New Englanders, every day. He also sought to diversify by creating indulgent confections that could be sold individually at low single-serving prices. The company's "chocolate sticks" and "spice cocoa sticks," created around 1840, speak to Baker's intention to offer consumers luxury in attainable amounts reasonable for daily indulgence. His promotion of these "cocoa sticks" in an 1845 letter to an agent speaks to this desire to convey both sophistication and affordability:

There is quite a sale in Boston—by all retail grocers—to children & others—at a cent a stick for small & two cents for large sticks. I learned to make them in London, and there as in Boston the sticks are piled in the retail shop window, cross each other, & rather ornament the store. If it be no trouble, you could hand to some of your best retailers some of the sticks—always to be eaten raw or melted on the tongue to taste; & children once getting the taste would purchase them for enough; being much more healthy & suitable for them than Candy or Sugar Plums etc. [42]

While years earlier Baker had berated an agent for selling boxes of Baker's No. 1 individually, he now embraced that concept in a new product line. As such, he critically redefined the parameters within which the New England chocolate industry worked by simultaneously expanding and diversifying its options for consumers. Baker continued to push aggressively for more retailers and specific new consumers, despite the company's economic struggles resulting from devalued cash during Andrew Jackson's "Bank War" and the corresponding specie crisis of the 1830s and 1840s. Indeed, his rant to New York wholesalers Grant & Stone, that "I should like to place your United States Bank Directors—for their bolstering up Southern rottenness—& Cotton Speculations—between my two Chocolate Mill Stones," throws into sharp relief the challenge of maintaining a secure profit line during a period of wildcat banking and national economic depression [43]. Nevertheless, Baker courted more retailers and marketed his products through increasingly narrow consumer windows in the 1840s and 1850s.

Reform, Health, and the Civil War

Baker's concerted efforts to attract new customers and make chocolate an indispensable foodstuff, rather than a luxury, on the American palate bore fruit during the antebellum period. Reform movements, a consequence of the impact of the Second Great Awakening on Americans' consciences, were likely an impetus behind his new customers' demands. Abolition and antislavery exemplify this new opportunity for Baker to profit by tailoring specific products to fit cultural moral standards.

As sectional tensions between North and South grew increasingly strained in the 1840s, more New Englanders opposed slavery on ideological grounds, arguing that "free labor" was inherently more moral and promoted more fair competition among white men than did slave labor [44]. Additionally, abolitionists, previously seen as irrational extremists in the public mind, began to garner more sympathy from the northern populous in the late 1830s following such events as the petition scandal and imposition of the Gag Rule in Congress, and the murder of abolitionist Elijah P. Lovejoy in Alton, Illinois [45–47]. New Englanders' ideological and moral opposition to slavery, therefore, became increasingly pronounced during this period.

Baker reaped the bounty of this intellectual shift in Boston by producing "Free Labor" chocolate in the late 1840s for a Dorchester buyer, Mr. Rob. L. Murray, with whom he was connected through his New York agents. In producing "Free Labor" chocolate, Baker predated Great Britain's early 20th century boycott of Portuguese cocoa from São Tomé, which was grown by slaves at that time. However, Baker also sold chocolate to southern slave owners in the antebellum period. Thus, it is difficult to argue that his level of commitment to antislavery transcended the financial benefit of his connection to his abolitionist client, Murray.

Although it appears that Murray was Baker's sole "free labor" customer, the company seems to have profited from the connection financially, because Baker eagerly sought these sales, writing seven letters to his wholesalers seeking new transactions with Murray. In 1848, Baker responded to his agents' inquiry into "Chocolate and Cocoa made up from free products entirely," by affirming his belief that "my Baker's . . . Cocoa is entirely of that character; the Cocoa coming from countries where slavery does not exist" [48].

Although Baker was a dedicated Whig and seems to personally have advocated free labor ideology as a political philosophy, he appears here to have no second thoughts about involving his business in a hotly debated sectional issue; by making a product called "Free Soil" chocolate, he in essence affirmed that the need for such a product was legitimate [49]. Baker continued production of high-grade, free-soil chocolate for Murray prior to the Civil War, using at times both free labor sugar and cocoa beans. However, it is difficult to ascertain whether he stepped far out of his way, at the expense of profit, to draw this moral line. He affirms, for instance, in later 1848 correspondence:

> In regard to free Cocoa Most of the Pd Cocoa is made from Cocoa from the British Island of Trinidad, or Guayaquil in South America where slavery has I believe been abolished. Also from Cayenne, one of the French Provinces where the Provisional Government of France decreed a short time since that the slaves should be emancipated. . . . Our WB [i.e., Walter Baker] Mass, & Norfolk Co. chocolate is made from Cocoa from the free Island of Santo Domingo or Hayti [sic]. [50]

Here Baker does not reveal whether these items are incidentally produced from free chocolate, or whether Walter Baker & Co. has made a concerted effort to assure the ideological purity of its product. Because the "slavery question" became so sectionally and politically divisive at this moment, however, Baker did not likely succeed in his consumers' eyes in remaining neutral. Furthermore, he simultaneously made "free" chocolate for antislavery advocates and sold "a poorer cocoa," made with "ground rice enough to

make it about as stiff as thick mud . . . mostly . . . to go South for the slaves and to the West Indies," during the antebellum period, clearly willing to jettison his high-quality aspirations to sell to large plantations [51]. Also on the issue of low-quality chocolate mixed with rice for slaves is the following comment from Baker to Messrs. Leland & Brothers, Boston, dated December 28, 1847:

> I use large quantities of Paraguay [sic] Chocolate & when it is very high in price, I buy the broken Para; and in former years frequently received the very broken from Charleston, Which I believe is reserved for the Colored population. [52]

By dealing financially with both slaveholders and abolitionists, Baker demonstrated either moral ambivalence or his more powerful desire for profit in the context of the slavery issue. Additionally, Baker's use of imported chocolate from Brazil, a nation that produced cocoa through slave labor until abolition became law in 1888, precludes the possibility of his own extreme moral opposition to slavery [53]. It also may be noted that Brazilians continued to import African slaves and coerced Amerindians to plant and harvest cocoa beans following independence from Portugal, and that "slaves were still being purchased shortly before 1888 by planters in southern Bahia" following abolition of slavery in Brazil [54]. Thus, he effectively exploited divisions between proslavery and antislavery advocates, successfully diversifying the company—and industry—to include grades of chocolate ranging from a low-quality article that was, in his words, "reserved for the Colored population," to more expensive "free labor" varieties produced at the insistence of abolitionist reformers [55].

As the country dissolved into civil war, the reformist impulse continued to create business opportunities for Walter Baker & Co., now under the leadership of Henry L. Pierce following Baker's death in 1852. Because disease in camp claimed more lives than battlefield casualties, health care became a key issue during the war. The central northern agency to address the issue of soldiers' health was the United States Sanitary Commission (USSC), a federal organization sanctioned by President Abraham Lincoln and begun by northern wives, mothers, daughters, and sisters who longed to contribute to the war effort (see Chapter 56). Although a Democrat prior to the conflict, Pierce demonstrated his financial (if not ideological) support for Lincoln's war effort by selling the USSC more than 20,000 lb of chocolate during 1862–1864 [56, 57]. It also is worth noting that Deborah Smith Mott Baker, Walter Baker's widow, shared her husband's Whiggish political views and free labor-orientated support for the Republican Party and Union war effort. Upon her death in 1891, she was lauded locally for her patriotism in the face of secession:

> When the War of the Rebellion broke out, Mrs. Baker opened her house and furnished quantities of material for the women

of Dorchester to make into clothing for the volunteers. She visited Army hospitals, and gathered a large number of books for the Soldier's Free Library at Washington. [58]

In addition to this economic sanction of Republican policy, Pierce's support of the USSC also speaks to the federal government's efforts to stem the tide of disease-related deaths of soldiers made vulnerable to illness by malnutrition. By buying mass amounts of chocolate to feed to tens of thousands of ailing men, USSC nursing officials indicated their belief that it had nutritional and wellness properties lacking in Union Army rations. Northern physician E. Donnelly, M.D., confirmed the desirability of chocolate in troops' diets to prevent disease in 1864. In a letter to the company, he suggested the idea of introducing Baker chocolate into field rations, touting its purity and strengthening effects on the body:

I have just returned from serving my country as Surgeon for over 3 years, and expect to return to the field again. . . . Why do you not make an effort to introduce some of your preparations in to the army, it would be much better than the adulterated coffee the soldiers now drink: a chocolate should be made to keep in a powdered condition, not too sweet, and free from all husks or other irritating substances. Chocolate . . . would be much more nutritious than coffee, not so irritating to the bowels, and it is an excellent antiscorbittic [sic]. Diarrhea in the army is often the result of continually drinking impure and imperfectly made coffee: it would not be so with chocolate. [59]

Rather than waiting until such ailments as dysentery and scurvy killed soldiers or sent them to hospitals, in which they might be treated with chocolate, Donnelly wished to see chocolate boost soldiers' constitutions in the field, enabling them to remain healthy during active duty. While Donnelly was incorrect

about the "antiscorbittic" properties of chocolate, his assertion that it would prove more beneficial to men than adulterated, cheap coffee was well founded. No record remains regarding further sales of Baker chocolate to the Union forces, but both physicians and women continued to play pivotal roles in U.S. health reform following the war. Henry L. Pierce, in turn, aggressively targeted these two groups in late 19th and early 20th century marketing campaigns, as Walter Baker & Co. sought to engage an increasingly diverse mass market by appealing to the contemporary sociocultural values of Reconstruction, the Gilded Age, and the Progressive Era—particularly those mores of gender and patriotism.

Turn-of-the-Century Mass Marketing to "Proper" American Women and Men

Although the Civil War lasted but four years, its effects on the Northern economy reverberated for decades following the conflict. Unlike the South, which was left with worthless currency, a ravaged countryside, no infrastructure, and the illegitimacy of defeat, the North experienced an economic boom and surge in manufacturing advancements during Reconstruction [60]. As industrial capitalism and technical innovation surged ahead, Walter Baker & Co. rode this wave of prosperity to unprecedented expansion and profit. Despite depressions in the early 1870s and the 1890s, Henry Pierce continued to expand the company's product lines, technologies, mill and factory space, and staff (Fig. 26.2). Of the company's more

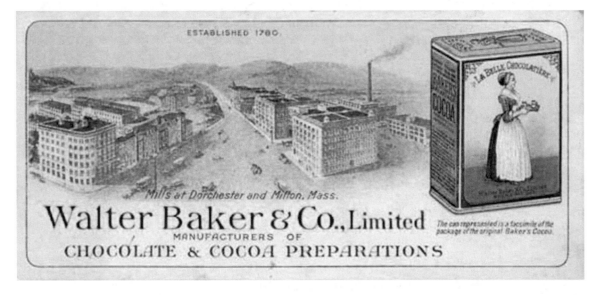

FIGURE 26.2. Late 19th century image of Walter Baker & Co., Limited, factory complex. *Source*: Dorchester Atheneum Website Image 186. Courtesy of Dorchester Atheneum. (Used with permission.) (See color insert.)

than 400 employees in 1897, most were factory laborers, including "girls" whose job consisted of tying ribbon around stacks of chocolate bars wrapped by machines installed earlier in the decade [61]. Thus, in the Gilded Age, Walter Baker & Co. upgraded to mechanized technology when expedient, but also continued to employ laborers, including children, to handle a variety of repetitive tasks. By expanding technologically and hiring more wage laborers, Pierce enabled the company to aggressively pursue a reputation of respect and dominance outside New England. Indeed, the company's awards for excellence at such prestigious events as the Vienna Exposition and the Philadelphia Centennial in the 1870s speak to its heightened prominence both at home and abroad (Fig. 26.3) [62].

The company's adoption of the "Chocolate Girl," as its trademark in 1883 reflected Pierce's twofold desire for mass consumption and cosmopolitanism. The highly romanticized legend behind *La Belle Chocolatière* held that as the waitress at a chocolate shop in mid-18th century Vienna, and despite her modest background, she captured the heart of the region's prince with her beauty and her delicious daily servings of chocolate [63]. This tale embodied the Baker philosophy of the previous century in several key ways: although the chocolate girl was poor, she moved quickly to the top of society, and the prince's strong desire for the chocolate she served demonstrated how the beverage appealed to all, across class lines. The most spare and most sophisticated palates could appreciate chocolate on equal terms; for the waitress and her prince, it proved the socioeconomic leveler enabling them to access each other's worlds to discover their love. Significantly, these values—making one's own

destiny, an equal-opportunity society—would have resonated strongly with Walter Baker & Co.'s late 19th century customers as well, making this choice of trademark a canny connection between tradition and modernization in a country struggling to define its identity within and outside its borders. Thus, *La Belle Chocolatière* represents a key example of the Bakers' using a highly gendered symbol as a marketing tool to link the company in consumers' minds with distinctly "American" roles for men, women, and children, and with national patriotism at the turn of the 20th century (Fig. 26.4).

As other Western powers honed imperial aspirations and asserted forceful national identities abroad, some historians have recently argued, the United States experienced a "crisis of gender" in the Gilded Age and early Progressive Era [64–66]. Marketing from this

FIGURE 26.4. Walter Baker & Co.'s Caracas Sweet Chocolate. The Chocolate Girl is the subject of the advertisement, framed by an announcement of international expositions at which the company has been honored. *Source:* Courtesy of Dorchester Atheneum. (Used with permission.) (See color insert.)

FIGURE 26.3. Baker's Breakfast Cocoa advertisement, 1878. *Source:* Dorchester Atheneum Website Image 2511. Courtesy of Dorchester Atheneum. (Used with permission.)

period at Walter Baker strongly supports this conclusion; the company under Pierce's management was striving to capitalize on Americans' search to define themselves nationally and socially. Following Pierce's death in 1896, an associate, H. Clifford Gallagher, ascended to the helm, furthering this advertising strategy as the company continued to try to attract new consumers by appealing to specific segments of society and of the family.

While some advertisements in this period are marketed toward young boys working at soda fountains who notice that other boys order chocolate, others are aimed at young girls in the kitchen with their mothers [67]. More numerous still are advertisements directed at middle-class women, who despite the burgeoning undercurrent of first-wave feminism, continued to be relegated to the kitchen and home as men engaged deeply in an emergent world of stocks, bonds, and capital. Gallagher and the Baker Co. specifically appealed to mothers' twofold desire to raise a happy, prosperous family that could afford to buy confections for its children, and to provide nutritious food. Chocolate, early 20th century Baker pamphlets assured these women, served both purposes, as children loved its rich, sweet flavor and grew healthy and strong while consuming it:

What nourishment more delicious than a hot drink, with the refreshing qualities that a cup of full-flavored, well-made cocoa provides? What simpler or more satisfying way to find a family beverage that everyone can enjoy without a qualm, safe in the knowledge that many leading child specialists recommend it—that children like it and thrive on it? What better way to take milk? Many a mother knows how Baker's Cocoa has helped her get more milk into the children's diet. [68]

Additionally, such marketing campaigns appealed to mothers' homemaking skills, emphasizing women's dominion over their kitchens and parlors and, consequently, their circumscribed roles within the home. Not only did the accomplished hostess serve Baker's cocoa, but she prepared it in the most nutritious, flavorful, attractive manner. Indeed, an 1891 New York promotion of Baker's cocoa speaks to this pressure on women to present chocolate in a way befitting the consummate mistress of the house. Observed a reporter from *Frank Leslie's Illustrated Newspaper*:

In the Fifty-ninth Street Annex of the Lyceum the ladies' parlors were occupied by the famous firm of Walter Baker & Co. . . . The taste displayed in the decoration of the parlors . . . excited the admiration of every visitor. The handsome young ladies, in becoming and attractive costumes, who served refreshments so generously and so kindly, made Baker's Chocolate and Baker's Cocoa taste all the sweeter, richer, and more delicious because of their grace and beauty. [69]

Here, the physical attractiveness and accommodating natures of the staff women actively enhance

the reviewer's opinion of the company, and by extension the chocolate. Inherent in this article is, furthermore, the possibility for the average American housewife that, should she serve Baker's cocoa in a similarly beguiling and attractive way, she herself can attain the success of the perfect hostesses in the article, becoming as beautiful, desired, and sweet as they are [70]. In an era when men struggled to carve out niches in the cutthroat business landscape of speculation, risk, and big corporations—of which Walter Baker & Co. was one—the company appealed to female consumers with the promise of social rewards for those who remained unthreateningly within their proscribed gender roles, as hostesses and mothers. The following piece published in the *Brooklyn Eagle*, October 19, 1889, illustrates this point:

How to Make Chocolate. First Lesson to the Class at Maillard's School. . . . Thirty-one ladies gathered yesterday afternoon in the pretty little semi parlor attached to Mr. Henry Maillard's chocolate establishment at 114 West Twenty-fifth street, New York, to receive instruction in the art of properly making a cup of chocolate or cocoa. [71]

In addition to appealing to families within circumscribed gender roles, Walter Baker & Co. also sought to exploit national questions of identity. As Americans struggled to find a national character in a global climate of international chauvinism and European dominance over the colonized world, the company increasingly sought to define itself as an "American" business. Following the Spanish–American War in 1898 and the United States annexation of the Philippines, Hawaii, Puerto Rico, Guam, and Cuba, Americans increasingly understood their nationality in opposition to the residents of these satellites, and adopted a twofold justification for interference: "civilizing" the natives and profiting from their natural resources. Walter Baker & Co., under Gallagher, sought to capitalize from this national identity definition and patriotism in opposition to the foreign "others" newly subjected to United States authority. A 1905 Baker advertisement speaks to this nascent national pride in strength, moral rectitude, and commercial hegemony of the United States. The advertisement is a fictional account of an American colonel enjoying an afternoon cup of cocoa with close friends, and heavily emphasizes the connection between Baker chocolate products and American nationalism.

Traveling in an "out-of-the-[way]-place [*sic*] in an island of Jamaica" [72], the colonel stopped at a small hut staffed by a native resident to eat, and was served biscuits, eggs, and cocoa. Of the last, the waiter pointed out, "Eet ees the chocolate-cocoa Bakaire," and produced the tin with *La Belle Chocolatiere* on it for the colonel's inspection. Upon this contact with Baker chocolate, the officer instantly felt at home, connected by the taste of the cocoa and the sight of the chocolate girl to American national identity, despite his solitary adventure in a country so far from his own. He affirmed:

> When I saw that familiar trade-mark, I felt very much as one feels after long traveling in the interior of Africa or China and comes to a port where he sees the American flag floating from a topmast. I took off my hat to that can and said, "Good old Baker—three cheers for La Belle Chocolatiere." [73]

Here, the American flag and Baker chocolate tin are fused as symbols of national safety and prosperity. Furthermore, the colonel's confidante, a physician, affirmed the connection of Walter Baker & Co. with a distinctly American identity, and lauded the company's longevity and innovative business practices as uniquely national characteristics:

> "I can understand your feelings," said the Doctor. "When you think of the business as having been established before we secured our independence from Great Britain, and remember that it has gone on for a century and a quarter without a break, without a failure or strike, keeping well up to the times in its equipment and methods of manufacture, and that it is now the largest establishment of its kind in the world, you cannot help taking pride in it as something more than a merely manufacturing concern—as, in fact, an educational and health-improving American institution." [74]

The doctor's reference to the American Revolution and the company's subsequent progressive labor and business strategy, as well as its industrial growth and international notoriety, reflect the efforts of Walter Baker & Co. to link its image as a high-quality, mass-production enterprise to that of the country, a democracy founded on revolutionary principles in which small businessmen built great companies with their own hands and deservedly attained domestic consumers' loyalty and international acclaim. In this advertisement, then, the company sought to appeal to millions of domestic consumers who appreciated the perception of uniquely American values—capitalism, democracy, militarism, civilization, prosperity—circulating abroad. By buying Baker chocolate, Americans evinced an increasingly coherent national identity and patriotism, as well as discerning palates and strong bodies.

In addition to connecting American patriotism with Walter Baker & Co. in consumers' minds, Clifford Gallagher sought to fuse the association of country and company at work as well. Implicitly, he accomplished this goal by adopting such progressive labor practices as pensions, reduced work weeks, and life insurance plans before they became national law in the early 20th century [75]. More overt efforts, heartily supported by company employees, included collecting money from workers "for the purpose of hanging a flag from the plant" during the Spanish–American War, and hanging in effigy McKinley's assassin, Leon Czolgosz, "on wire stretched between the Baker and Webb mills" [76]. These efforts to inculcate national pride at work bore further fruit during World War I, when Baker employees produced mass amounts of chocolate for Allied forces, stamping it "W.T.W" (Win The War), and many young men from Baker's factories joined up to fight against European aggressors [77]. In their honor, Gallagher installed a plaque at the Baker mills commemorating effort and valor—in so doing, permanently inscribing American martial and democratic prowess within the company's walls.

Conclusion

Following World War I, Walter Baker & Co. was acquired by General Foods and grew exponentially in the next decades as a satellite of that American mega-corporation. Its absorption into a billion-dollar, broad-spectrum conglomerate, in effect, represents the logical conclusion of a company whose initial goals were to mass-produce a specialized product at affordable prices. Despite its major growth and its loss of individual ownership in 1927, Walter Baker & Co. continued to emphasize its seminal New England origins in the following decade. Symbolically, the company reinforced its connection to its earliest production by purchasing the late 18th century desk of the first Baker chocolate maker, John Hannon. Although Walter Baker & Co. revolutionized the New England chocolate industry following the American Revolution by relying on the strategy of mass production, high-quality products, and aggressive advertising to attract diverse consumers, its ability to succeed was based on Americans' earliest interest in chocolate as a delicacy and health aid. Hannon and John Baker's original chocolate works built on the foundation of isolated chocolate sales by apothecaries and grocers, who promoted its purchase for specific, if rare, uses. The remarkable success of Walter Baker & Co. in the following 150 years speaks to its leaders' ability to capitalize on the pleasant taste and perceived health benefits of chocolate to promote it by exploiting popular American conceptions of family, society, and nation.

Acknowledgments

I wish to acknowledge the following for research assistance during the course of this project: Captain Richard Bailey, Ocean Classroom Foundation; Esther Friedman, freelance journalist; and E. Jeanne Harnois, freelance journalist. In addition, the author would like to express her gratitude to the staff at The Massachusetts Historical Society, particularly Librarian Peter Drummey and Kathy Griffin; and the de Gaspé Beaubien Reading Room at the Baker Library at Harvard Business School. Finally, I wish to thank Dr. Timothy D. Walker, Assistant Professor of History at the University of Massachusetts Dartmouth, who provided assistance and direction to my research.

1. Clarence-Smith, W. G. *Cocoa and Chocolate, 1765–1914*. London, England: Routledge, 2000; p. 16.

2. Clarence-Smith, W. G. *Cocoa and Chocolate, 1765–1914*. London, England: Routledge, 2000; p. 163.

3. Baker Business Library (BBL), Harvard Business School, K-1. *A Calendar of Walter Baker & Company, Inc. and its Times, 1765–1914*, 1775.

4. Stubbe, H. *The Indian nectar, or A discourse concerning chocolate*. London, England: James Cottrell, 1662.

5. Lightbody, J. *Every man his own gauger: wherein not only the artist. . . .* London, England: Printed for G. C., 1695.

6. Simon Paulli, S. *Commentarius de abusu tabaci Americanorum veteri, et herbae thee Asiaticorum in Europa novo. . . .* Argentorati, Sumptibus B. authoris filii Simonis Paulli, 1681.

7. Colmenero de Ledesma, A. *Curioso de la naturaleza y calidad del chocolate*, Translated by Capt. James Wadsworth. London, England: Printed by J.G. for John Dakins, 1652.

8. Taylor, A. *American Colonies*. New York: The Penguin Group, 2001; pp. 178–179.

9. Clarence-Smith, W. G. *Cocoa and Chocolate, 1765–1914*. London, England: Routledge, 2000; p. 16.

10. Massachusetts Historical Society (MHS). Ms.SBd-222, Ann Greene Account Book, 1725–1890, bulk: 1725–1727.

11. MHS. Ms. N-2022, Thomas Amory Account Book, 1720–1728.

12. MHS. Ms.SBd-102, John Higginson Account Book, 1755–1758.

13. MHS. Ms. N-107, George Minot Account Book, 1765–1767.

14. MHS. Boston Pier Accounts, 1765.

15. MHS. Ms.N-621, Otis Family Papers, 1701–1800.

16. Clarence-Smith, W. G. *Cocoa and Chocolate, 1765–1914*. London, England: Routledge, 2000; pp. 11–12.

17. Clarence-Smith, W. G. *Cocoa and Chocolate, 1765–1914*. London, England: Routledge, 2000; p. 16.

18. MHS. Amory Fiche, 41515, Broadside. Benedict Arnold, 1741–1801 (New Haven, 1765).

19. Clarence-Smith, W. G. *Cocoa and Chocolate, 1765–1914*. London, England: Routledge, 2000; p. 16.

20. MHS. Ms.N-1106, George Deblois Account Book, 1769–1792.

21. BBL. K-1, *A Calendar of Walter Baker & Company, Inc. and its Times, 1765–1914*, May 17, 1771.

22. BBL. K-1, *A Calendar of Walter Baker & Company, Inc. and its Times, 1765–1914*, 1772.

23. MHS. Evans Fiche, 42610, *Act For Blocking Up the Harbour of Boston*, 1774.

24. MHS. Evans Fiche, 42610, *Act For Blocking Up the Harbour of Boston*, 1774.

25. BBL. K-1, *A Calendar of Walter Baker & Company, Inc. and its Times, 1765–1914*, 1780.

26. BBL. K-1, *A Calendar of Walter Baker & Company, Inc. and its Times, 1765–1914*, March 19, 1781.

27. BBL. K-1, *A Calendar of Walter Baker & Company, Inc. and its Times, 1765–1914*, May 17, 1771.

28. BBL. B-1, June 22, 1822; July 19, 1822.

29. BBL. B-1, 1823–1824, pp. 130–133.

30. BBL. L-1, December 25, 1822, p. 63.

31. BBL. A-1, December 4, 1840.

32. BBL. LA-1, Dorchester, May 4, 1843.

33. BBL. K-1, *A Calendar of Walter Baker & Company, Inc. and its Times, 1765–1914*, October 17, 1805.

34. BBL. K-1, *A Calendar of Walter Baker & Company, Inc. and its Times, 1765–1914*, June 1805.

35. BBL. K-1, *A Calendar of Walter Baker & Company, Inc. and its Times, 1765–1914*, 1816.

36. BBL. B-1, 1815.

37. BBL. L-1, To Grant & Stone, February 5, 1823.

38. BBL. L-1, To Hussey & Mackay, December 25, 1822.

39. BBL. L-1, To S & BJ Sweetser, January 22, 1823.

40. BBL. B-1, June 2–26, 1821 and October 19 to November 7, 1822.

41. BBL. L-1, To King & Richardson, December 10, 1820.

42. BBL. LA-1, To Messrs Smith & Wood, Dorchester, Massachusetts, March 18, 1845.

43. BBL. LA-1, To Messrs Grant & Stone, Dorchester, Massachusetts, October 17, 1839.

44. Foner, E. *Free Soil, Free Labor, Free Men: The Ideology of the Republican Party Before the Civil War*. Oxford, England: Oxford University Press, 1970.

45. Faust, D. *James Henry Hammond and the Old South: A Design For Mastery*. Baton Rouge: Louisiana State University Press, 1982.

46. Mayer, H. *All on Fire: William Lloyd Garrison and the Abolition of Slavery*. New York: St. Martin's Press, 1998.

47. Wilentz, S. *The Rise of American Democracy: Jefferson to Lincoln*. New York: W. W. Norton & Company, 2005.

48. BBL. LA-2, To Messrs Hussey & Murray, Dorchester, Massachusetts, May 23, 1848.

49. BBL. LA-2, To Messrs Hussey & Murray. Dorchester, Massachusetts, October 24, 1848.

50. BBL. LA-2, To Messrs Hussey & Murray, Dorchester, Massachusetts, June 3, 1848.

51. BBL. K-1, *A Calendar of Walter Baker & Company, Inc. and its Times, 1765–1914*, January 3, 1842.

52. BBL. LA-2, To Messrs Leland & Brothers, Boston. December 28th, 1847.

53. Clarence-Smith, W. G. *Cocoa and Chocolate, 1765–1914*. London, England: Routledge, 2000; pp. 204–212.

54. Clarence-Smith, W. G. *Cocoa and Chocolate, 1765–1914.* London, England: Routledge, 2000; p. 212.

55. BBL. LA-2, To Messrs Leland & Brothers, Boston. December 28, 1847.

56. BBL. A-4, 1863–1868, p. 443.

57. BBL. A-3, 1857–1863, p. 569.

58. BBL. K-1, *A Calendar of Walter Baker & Company, Inc. and its Times, 1765–1914,* 1891.

59. BBL. K-1, Folder Three: Incoming Correspondence, 1863–1927. Philadelphia, Pennsylvania, letter dated July 19, 1864.

60. Woodward, V. *Origins of the New South, 1877–1913.* Baton Rouge: Louisiana State University Press, 1951.

61. BBL. K-1, *A Calendar of Walter Baker & Company, Inc. and its Times, 1765–1914,* 1894 and 1897.

62. BBL. K-1, *A Calendar of Walter Baker & Company, Inc. and its Times, 1765–1914,* 1873–1877.

63. BBL. K-1, 1928, *Famous Recipes for Baker's Chocolate and Breakfast Cocoa,* established 1780, Walter Baker & Co., Inc. of Dorchester, Massachusetts.

64. Bederman, G. *Manliness & Civilization: A Cultural History of Gender and Race in the United States, 1880–1917.* Chicago, IL: University of Chicago Press, 1995.

65. Hoganson, K. L. *Fighting for American Manhood: How Gender Politics Provoked the Spanish–American and Philippine–American Wars.* New Haven, CT: Yale University Press, 1998.

66. Rotundo, A. *American Manhood: Transformations in Masculinity from the Revolution to the Modern Era.* New York: Basic Books, 1993.

67. BBL. K-1, Walter Baker & Co., Inc., *Famous Recipes for Baker's Chocolate and Breakfast Cocoa.* Dorchester, MA: Walter Baker & Co., Inc., 1928.

68. BBL. K-1, Walter Baker & Co., Inc., *Famous Recipes for Baker's Chocolate and Breakfast Cocoa.* Dorchester, MA: Walter Baker & Co., Inc., 1928.

69. BBL. H-3, Walter Baker & Company's Preparations. *Frank Leslie's Illustrated Newspaper,* March 28, 1891.

70. BBL. H-3, on the "proper" training of women as hostesses under the tutelage of Walter Baker & Co.

71. *Brooklyn Eagle,* October 19, 1889.

72. BBL. K-1, Walter Baker & Co., Ltd., *Over the Cocoa Cups.* Dorchester, MA: Walter Baker & Co., Limited, 1905.

73. BBL. K-1, Walter Baker & Co., Ltd., *Over the Cocoa Cups.* Dorchester, MA: Walter Baker & Co., Limited, 1905.

74. BBL. K-1, Walter Baker & Co., Ltd., *Over the Cocoa Cups.* Dorchester, Massachusetts: Walter Baker & Co., Limited, 1905.

75. BBL. K-1, *A Calendar of Walter Baker & Company, Inc. and its Times, 1765–1914;* on late 19th and early 20th century labor practices benefiting employees.

76. BBL. K-1, *A Calendar of Walter Baker & Company, Inc. and its Times, 1765–1914;* on late 19th and early 20th century labor practices benefiting employees. June 7, 1898 and September 6, 1901.

77. BBL. K-1, *A Calendar of Walter Baker & Company, Inc. and its Times, 1765–1914;* on late 19th and early 20th century labor practices benefiting employees. 1917.

27

Boston Chocolate

Newspaper Articles and Advertisements, 1705–1825

Louis Evan Grivetti

Introduction

Boston's unique geographical location and long history present scholars with a rich source of information related to everyday life in the Colonial and early Federal eras of American history. Many of these activities were dominated by merchants who sold chocolate. Colonial era newspaper advertisements and articles provide researchers with tens of thousands of texts that contain unusual insights regarding the development and expansion of chocolate-related enterprises in New England. Several hundred newspapers were published in North America in the period 1700–1825; of these, more than 60 were published in Boston [1, 2].

University of California digital collections available through Shields Library on the Davis Campus were used to conduct a systematic search of Boston Colonial era newspapers for chocolate-associated information (Table 27.1). The primary database used was *Readex* [3]; their digitized collections of North American documents allowed searches by date, document type (e.g., advertisements, articles, letters to editors, obituaries), by keyword (cacao, cocoa, chocolate), and by specific newspaper. Because of the volume of information available, we established our time period search limit at 1700–1825. Early experience with the Readex data set revealed the expected finding that chocolate merchants submitted the same advertising copy to different newspapers, and identical advertise-

ments commonly repeated for weeks or months. We decided, therefore, to record only the merchant's first advertisement, although subsequent advertisements from the same merchant that reported any change in address or content also were captured on our spreadsheets.

Each "hit" retrieved was opened, scanned for content, then categorized into one of 18 chocolate-associated themes: (1) accidents (employee injury, fire), (2) crime (fraud, theft), (3) culinary (ingredients, serving equipment), (4) disease (smallpox or other epidemics or health considerations), (5) economics (to document prices and price changes through time), (6) ethics (issues associated with child labor or slavery), (7) festivals/holidays (national or religious), (8) gender (indentured female servants, widows in the chocolate trade), (9) health/medicine (chocolate sold at apothecary shops, health claims, chocolate in curative or preventive medicine reports), (10) legal (bankruptcy,

Chocolate: History, Culture, and Heritage. Edited by Grivetti and Shapiro
Copyright © 2009 John Wiley & Sons, Inc.

Table 27.1 Boston Newspapers, 1690–1826

Newspaper	First Issue	Last Issue	Newspaper	First Issue	Last Issue
Agricultural Intelligencer and Mechanic Register	1820	1820	*Gazetteer*	1803	1803
			Herald of Freedom	1788	1791
American Apollo	1792	1794	*Idiot*	1817	1819
American Herald	1784	1788	*Independent Chronicle*	1783	1820
Argus	1791	1793	*Independent Ledger*	1778	1786
Boston Chronicle	1767	1770	*Kaleidoscope*	1818	1819
Boston Commercial Gazette	1821	1826	*Ladies' Port Folio*	1820	1820
Boston Daily Advertiser	1813	1821	*Massachusetts Centinel*	1784	1788
Boston Evening-Post	1735	1784	*Massachusetts Gazette*	1785	1788
Boston Gazette	1719	1798	*Massachusetts Mercury*	1793	1803
Boston Gazette Commercial & Political	1800	1820	*Massachusetts Spy*	1770	1775
			New-England Chronicle	1776	1782
Boston Intelligencer	1816	1820	*New-England Courant*	1721	1726
Boston Mirror	1808	1810	*New-England Galaxy*	1817	1820
Boston News-Letter	1704	1776	*New-England Palladium*	1803	1820
Boston Patriot	1809	1820	*New-England Weekly Journal*	1727	1741
Boston Post-Boy	1735	1775	*Pilot*	1812	1813
Boston Price-Current	1795	1798	*Polar-Star*	1796	1797
Boston Recorder	1816	1820	*Publick (sic) Occurrences*	1690	1690
Boston Spectator	1814	1815	*Repertory*	1804	1820
Censor	1771	1772	*Republican Gazetteer*	1802	1803
Christian Watchman	1819	1820	*Russell's Gazette*	1798	1800
Columbian Centinel (sic)	1790	1823	*Satirist*	1812	1812
Columbian Detector	1808	1809	*Saturday Evening Herald*	1790	1790
Constitutional Telegraph	1799	1802	*Scourge*	1811	1811
Continental Journal	1776	1787	*Times*	1794	1794
Courier	1795	1795	*Times*	1807	1808
Courier de Boston	1798	1798	*Weekly Messanger (sic)*	1811	1820
Daily Advertiser	1809	1809	*Weekly Rehersal (sic)*	1731	1735
Democrat	1804	1809	*Weekly Report*	1819	1820
Federal Gazette	1798	1798	*Yankee*	1812	1820
Federal Orrery	1794	1796			
Fredonian	1810	1810			

business relations, formation and dissolution of partnerships), (11) location (general or specific address information), (12) manufacturing (types of production equipment, different types of chocolate mills), (13) military (rations, sales to officers and soldiers, Revolutionary War-related chocolate information), (14) obituaries (to identify chocolate-associated families), (15) politics (chocolate advertisements that differentiated royalist vs. rebel affiliations), (16) product quality (issues of adulteration, branding, certification, purity and guarantees), (17) shipping information (cacao bean country of origin, ship types, associated cargo), and (18) general/miscellaneous accounts.

Data from keyword "hits" were entered into an Excel spreadsheet under the following columns: publication date, name of individual or company, profession (auctioneer, broker, importer, manufacturer, or merchant/vendor), address (street number or general

geographical location), document theme, general comments, and URL links to the specific newspaper image.

Boston Chocolate

The names and addresses of more than 900 Bostonians were associated with the chocolate trade between 1700 and 1825 (see Appendix 6). The first Boston advertisement with cacao- or chocolate-associated information identified was published in *The Boston News-Letter* on December 3, 1705 (Fig. 27.1). This early advertisement for cacao identified the warehouse of James Leblond, its general location on Long Wharf near the Swing-Bridge, and that he offered his customers retail or wholesale options [4].

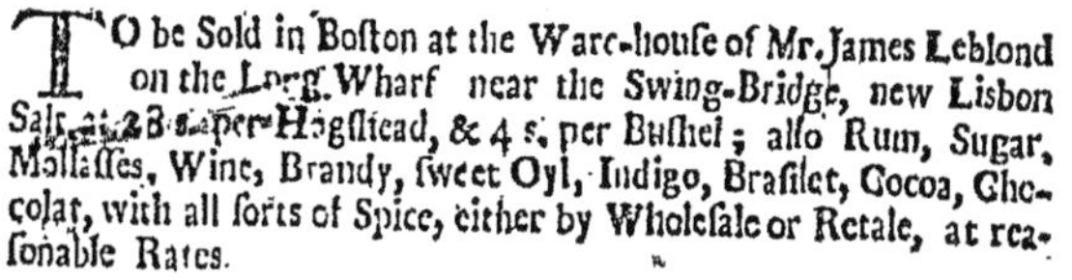

To be Sold in Boston at the Ware-house of Mr. James Leblond on the Long Wharf near the Swing-Bridge, new Lisbon Salt at 2 s. per Hogshead, & 4 s. per Bushel; also Rum, Sugar, Mollasses, Wine, Brandy, sweet Oyl, Indigo, Brasilet, Cocoa, Chocolat, with all sorts of Spice, either by Wholesale or Retale, at reasonable Rates.

FIGURE 27.1. First North American newspaper advertisement for chocolate. *Source: The Boston News-Letter*, December 3, 1705, p. 4. Courtesy of Readex/Newsbank, Naples, FL. (Used with permission.)

ACCIDENTS

A significant number of advertisements and announcements published in the Boston newspapers reported on fires at chocolate mills, whether large or small establishments. An article published the *Boston Post Boy*, March 30, 1772, reported that the house of Daniel Jacobs, a resident of nearby Danvers north of Boston, burned with substantial loss of personal possessions:

The building was large, and contained, besides the furniture, provisions, &c 3,000 lb. of cocoa, and several 100 lb [sic] of chocolate, which were nearly or quite all destroyed. The loss, at a moderate computation is said to amount to Five Hundred Pounds Money. [5]

The Welsh chocolate company of Boston suffered two serious fires at their North Mills, the first in 1783 followed by another in 1785:

1783. *Yesterday morning [January 21] about two o'clock, a fire broke out at the North Mills in this Town, improv'd by Mr. Welsh, which entirely consumed the same, with a large quantity of grain, cocoa, chocolate, &c. Also a large barn with five horses therein, and a store containing a number of valuable articles. The Flakes [i.e., embers/ashes] communicated to a number of buildings at the South part of the town, but being timely discovered were soon extinguished. [6]*

1785. *Friday last [prior to April 28] about three o'clock a fire broke out in the Chocolate Mills occupied by Mr. Welch, at the North-End, which, notwithstanding the exertions of the inhabitants, ever conspicuous on such occasions, was entirely consumed. A large quantity of cocoa, &c. was destroyed by the flames. The case of Mr. Welch, the principal sufferer, is peculiarly disastrous, as by the destruction of the mills by fire, about two years since, he lost the greater part of his property. [7]*

Fires associated with chocolate mills eventually led to local and regional legislation. On Wednesday, December 28, 1785, a Boston town-hall meeting was held at Faneuil Hall to consider and revise the city Town-Orders and By-Laws. The proposition to regulate the location of chocolate mills within town limits passed. The legislation was called *An Act to prevent damage by fire from Chocolate-Mills and Machines for roast-*ing Cocoa in the town of Boston and appeared in the *Mass [Massachusetts] Gazette*, January 9, 1786—"for the future good government of the town" [8].

The Chocolate-Mill Act and its adoption only restricted the location of new and future mills and did nothing to prevent fires in already established locations. Not unexpectedly, fires continued. One of Welsh's competitors, Boies and M'Lean of Milton, south of Boston, suffered a serious fire during late December 1788 at their mill as reported in the *Vermont Gazette or Freemen's Depository*:

Friday morning last [prior to January 5], about five o'clock, the Grist and Chocolate Mills, belonging to Messieurs. Boies and M'Lean, of Milton, were entirely destroyed by fire, together with all the property therein; estimated at eight hundred pounds. The latter was the earnings of many years industry of Mr. J. Boies, of that town. [9]

Fires continued to consume chocolate mills in Massachusetts throughout the 18th century. Edward Rumney's mill located in Malden, north of Boston, burned as reported in the *Columbian Centinel* (sic), October 23, 1793:

On Wednesday night late the Chocolate Mills at Malden, belonging to Mr. Edward Rumney, of this town, were with their contents, entirely [sic] destroyed by fire. By this event an industrious and respectable citizen is not only deprived of considerable property, but is deprived of the means of maintaining a large and amiable family—to relieve whom, a number of benevolent citizens have commenced a generous subscription; to which we doubt not will be added by those, who with the ability to give, enjoy the pleasure of feeling for the woes of others. [10]

Accidental deaths sometimes were associated with chocolate. The following obituary (name not identified) ran in the *Boston Evening Post*, December 2, 1765, and reported a terrible accident:

One morning last week [last week of November 1765] a child about 3 years old accidentally overset a large sauce-pan of hot chocolate into its bosom, whereby it was so terribly scalded that it died soon after. [11]

Accident and negligence are companions not easily differentiated. The *Middlesex Gazette* reported the tragic death of a child, Robert Clough, in their September 3, 1787 issue:

On Saturday last, master Robert Clough, son of Mr. Robert Clough, of this town, being at play with another child, in Mr. Makepeace's chocolate mills, north-end [of Boston], by some accident the former got entangled in the machinery of the works, while they horses were on the go, and was so shockingly mangled in his bowels, &c. that he died at two o'clock on Sunday morning. [12]

The death of one's child is always a sad event, made even more poignant when negligence has been involved. While not associated with a specific mill or shop, the following chocolate-associated tragedy is, indeed, linked to chocolate through John Wait, the well-known Boston chocolate maker. His two sons were scalded to death when a steam ship boiler exploded when the boys were returning to visit their family—most likely at Christmas—since the publication date of the tragedy was December 21, 1816:

In New-York, two sons of Mr. John Wait, jun. Chocolate-maker, late of Boston, one aged 8, the other 6 years. They had been at home on a visit from school, at Elizabeth-town and had taken their passage on their return in a new Steam-boat, when the boiler burst, and so scalded them as to occasion their death in a short time. Several others were injured. Those interested in steam-boats say, the boat was not finished; and that had there been a partition between the fire room and cabin, no great injury would have occurred. But why did the owners take passengers before the boat was finished? [13]

CRIME: FRAUD AND THEFT

Two essays on chocolate-associated crimes are presented in the present book (Chapters 20 and 21). It is appropriate in this chapter, however, to provide several examples that appeared in Boston newspaper advertisements. An unusual example of a "good Samaritan" who attempted to locate the owners of goods stolen was reported in *The Boston News-Letter and New-England Chronicle* in 1766. Timothy Paine, occupation and address not provided, paid for the following advertisement to locate the owners of property previously stolen:

Now in the Hands of Timothy Paine, Esq. [Esquire]; of Worcester, sundry articles viz. five black Barcelona Handkerchiefs; one striped ditto; about seven Yards Garlix [?]; about 6 lb. Tea; two Cakes Chocolate; one Cake Soap, &c. supposed to be stolen by one Timothy Williams, now confined in Goal [Jail] there [in Worcester] for breaking up a Shop as they were found nigh the said Shop in Worcester. Any Person or Persons claiming all or any of the aforesaid Articles may apply for them to the said. [14]

An advertisement placed by an unidentified newspaper subscriber appeared in *The Boston News-Letter and New-England Chronicle* on October 27, 1768, and announced a household burglary:

Taken out of a House at the South End, some Time last Week, one Two-Quart Copper Tannel [?] also one Silver Chocolate Ladle, that weighs about 3 ounces and a Half, marked' M.S. on the bottom, it's desired if either of them [the items] are offered for sale the Printer hereof may be informed; if the present Possessor will Return them, or either of them, they shall receive One DOLLAR Reward for each and no Questions asked. [15]

A general store managed by Bossenger & William Foster located at Spear's Wharf on Boston Harbor was robbed on February 8, 1769. The brothers published an announcement of the theft in *The Boston News-Letter and New-England Chronicle* and mentioned a reward (amount not identified):

Whereas some evil-minded person or persons have again broke open the store of the subscribers, on Spear's Wharf, and last night took from thence a number of articles amongst which was two firkins [of] butter, about 50 or 60 lb. of chocolate mark'd W. Call, and S. Snow, half a barrel of coffee, nine or ten Pair[s'] [of] Lynn shoes, some Cocoa and Sugar. Any Person that will discover the Thief or Thieves, so that he or they may be brought to Justice, shall be handsomly [sic] rewarded. [16]

The store of John Winslow, Junior, Boston merchant, was robbed July 25, 1774. Winslow published an advertisement identifying a 20 dollar reward on August 8, 1774 for "intelligence" regarding return of the goods:

20 Dollars Reward. STOLEN out of the Store of the Subscriber, the Night of the 25th Instant [i.e., the previous July], the following Goods, viz. A half Piece superfine blue Cloth, half Piece superfine garnet Cloth, two Bags Buttons, three Dozen Women's fine cotton Stockings, ten Dozen Spitalfield Handkerchiefs, one Piece Bandanno Handkerchief, three Bandanno Handkerchiefs, some Chocolate, Loaf-Sugar, a parcel [of] [illegible], some Metal Buttons. Whoever will give Intelligence of the above Goods, so that the Owner may have them again shall be entitled to the above Reward. [17]

An interesting advertisement by an unidentified Boston merchant identified a theft of chocolate in late December 1798. The text also revealed the relative "price" of the 50 pounds of chocolate stolen, since the shop-owner considered the $10.00 reward and the stolen goods equal in value. A small notation at the bottom of the text revealed that the advertisement was given to the printers on December 22:

TEN DOLLARS REWARD. STOLEN from No. 38, Long-Wharf, on Monday Morning, one box Chocolate, containing 50 wt. [i.e., 50 pounds]. The box [was] branded Watertown. Any person who shall bring the Thief to justice shall be entitled to the above reward—or the Chocolate, for their trouble. [18]

Nine years later a different shop on Boston's Long Wharf, one owned by Edward Holden, was burgled on February 16, 1808. While a near equivalent quantity of chocolate was stolen from Holden (47 pounds) compared to the 50-pound theft identified in the 1798 advertisement cited above, Holden offered just half the previous reward:

STOLEN yesterday, from Store No. 22, Long Wharf, 2 boxes of Chocolate, one of them Branded E. Baker, of Dorchester, [illegible] No 1—the other not branded—one contained 24 pounds, the other 23. Whoever will give information, so that the Goods may be recovered, and the Thief or Thieves brought to punishment, shall receive FIVE DOLLARS Reward. [19]

CULINARY: CHOCOLATE-RELATED INGREDIENTS AND SERVING EQUIPMENT

Boston newspaper articles/advertisements we identified did not provide recipes or list specific ingredients to be mixed with chocolate. At the same time, the merchants selling chocolate commonly advertised a range of items for sale that could have been purchased and used to prepare chocolate beverages, for example, various spices and sugar. Two examples of such advertisements are presented here.

> Choice Rhode-Island & Cheshire CHEESE, good Chocolate, Raisins, Corants [sic]: All Sorts of Sugars, Spices, &c. Sold by Thomas Procter, in Queen-Street, next Door to [the] Convert-Hall. [20]

The following advertisement published on March 5, 1770—date of the Boston Massacre[1]—provided information regarding the general types of goods sold at Colonial era grocery shops:

> Choice Capers and Anchovys [sic], by the Keg or single Pound TO BE SOLD BY Archibald Cunningham, At his Shop near the Draw-Bridge, Fore-Street, Boston. ALSO, New Raisins and Turkey Figgs [sic] by the Cask. CURRANTS, Citron, Mace, Cloves, Cinnamon, Rice, Oatmeal, Piemento [sic] per Hundred weight, Pepper, Almonds, Salt-Petre [sic], Indigo, Starch, Pipes, Lyn [sic, Lynn meant] Shoes, Wine Glasses by the Groce [sic], Allum [sic], Ginger . . . , Chocolate, Loaf and brown Sugar, Crown Soap, Mustard per pound or in Bottles, long-handled Hearth Brushes . . . , N.B. A Quantity of NUTMEGS to be sold cheap at said Shop. [21]

Chocolate pots, whether manufactured of silver or imported chinaware were part of serving equipment of wealthy Bostonians (see Chapters 10 and 11). Less wealthy residents used chocolate pots made of pewter, copper, or base metal (Fig. 27.2). Boston advertisements provide only limited information on the type/kind of chocolate pots sold within the city, as these selections document:

- 1768. John Mazro, merchant; shop located at the establishment lately improved by Dr. William Greenleaf in Cornhill, just below the Court House, announced the sale of chocolate, cocoa, and general items; also sold chocolate pots and bone paddle sticks (to stir and froth the chocolate). [22]

- 1768. J. Russell, auctioneer, auction house located on Queen Street, announced the sale of chocolate pots. [23]

- 1809. John Ballard Junior and Company, auctioneers, auction house located at Number 32, Marlboro' Street, announced the sale of chocolate pots among the items to be auctioned. [24]

- 1817. Francis Wilby and Company, auctioneers, auction venue at their Assembly Room at the

FIGURE 27.2. Chocolate/coffeepot: 19th century, Salem, Massachusetts. Private Collection. *Source:* Courtesy of the Grivetti Family Trust. (Used with permission.)

Coffee Exchange House, announced the sale of copper and burnished tin dish covers of various sizes, stew and saucepans, tea kettles, coffee and chocolate pots, powder flasks, snuff boxes, &c. [25]

Other newspaper advertisements announced the sale or importation of chocolate serving cups, and sometimes provided the names of English ships from London that brought the goods to Boston. The color of chocolate cups was identified in only one advertisement we located:

- 1748. Elizabeth Parker, merchant; shop located at the corner house next to Dr. Cutler's church, sold chocolate cups and saucers. [26]

- 1767. Daniel M'Carthy, merchant; shop located in the street leading from the Cornfields to the Mill Bridge, sold chocolate cups imported from the ship *Amazon*, Captain Hall, from London. [27]

- 1768. Joseph Barrell, merchant; shop located at Number 3, on the south side of the Town Dock, sold chocolate, cocoa, and chocolate cups and general items. [28]

1773. John Head, merchant; shop located at Number 5, south side of the Town Dock, sold chocolate and general items, imported from the ship *Haley* from London also a fresh assortment (of goods) and white chocolate cups with handles. [29]

1783. William Greenleaf, merchant; shop located in Cornhill, sold chocolate cups and general items. [30]

A chocolate culinary-related advertisement published in 1807 by Peter Pailhes, merchant, whose shop was located at the store opposite William Smith, Esquire, near the Concert Hall on Court Street, informed customers that he sold ice cream in a variety of flavors:

> *He intends to keep a constant supply of ICE CREAM, such as Vanilla, Rose, Lemon, Orange, Pine-Apple, Strawberry, Raspberry, Currant, Coffee, Chocolate, and other kinds, too many to describe!* [31]

DISEASE

One would not normally expect to see references to disease or illness in food-related advertisements. But events in Boston during 1764 changed that perception. Rebecca Walker managed a small business in Boston located opposite the Blue-Ball pub near Mill Bridge. During the Boston smallpox outbreak of 1764, many merchants relocated out of fear. Her advertisement stated that she had moved to Roxbury, a southern suburb of Boston, into the home of Nathaniel Fetton, Scythe-Maker. Walker's advertisement that ran in the *Boston Post-Boy* on April 2, 1764, reported the sale of chocolate and general items imported from London, along with garden seeds, and contained the following information:

> *ALL Sorts of Garden Seeds imported in the last Ship from London, and to be Sold at the House of Mr. Nathaniel Felton, Scythe-Maker in Roxbury; likewise early Peas and Beans; red and white Clover and other Grass Seeds; Hemp Seed; Cheshire Cheese; Flour of Mustard; Jordan Almonds; Florence Oyl [sic]; split and boiling Peas; Stone and Glass Ware; Chimney Tiles; Kippen's Snuff; Pipes, Spices, Sugar, Chocolate; English and Scotch GOODS, &c. N.B. The above Seeds, &c were imported to be Sold by Rebecca Walker, who lately kept the Shop opposite the Blue Ball, near the Mill-Bridge, but were removed out of Boston, on Account of the Small-Pox.* [32]

Other advertisements and letters to the editor regarding the 1764 smallpox outbreak are reviewed and discussed elsewhere (see Chapter 7).

ECONOMICS

A significant number of Boston advertisements listed prices for various grades and quantities of chocolate. A

brief example to illustrate initial price stability then gradual rise in chocolate prices can be seen through examination of Boston advertisements through the decade 1730–1740. One issue that needed to be explored and resolved, however, was value. Was chocolate expensive when compared to worker's daily wages during the 1730s? This question can be answered relatively easily as chocolate *became* expensive during this decade! Daily wage for a master craftsman in Colonial America between the years 1730 and 1740 was 12 shillings/day; the wage for an average person with a trade was 7 shillings, 9 pence; while the daily wage for a common laborer was 7 shillings, 6 pence [33].

The first Boston advertisement identified for the 1730–1740 decade that contained information on chocolate prices was published in *The Boston Gazette*, February 16, 1730. John Brewster, chocolate-maker, shop located in the North End of Boston at the Boot (i.e., sign over the entrance to his shop), announced the sale of chocolate and general items, and posted his price of chocolate at 6 pence (d) per pound [34]. Three years later in 1733 the price of chocolate in Brewster's advertisements had risen to 12 shillings/pound, and offered his customers an unspecified lower price if they purchased chocolate by the "dozen," that is, a dozen bars/tablets [35]. John Pinkney (sometimes spelled Pinkny or Pinckny in subsequent advertisements), merchant, shop located near the Red Lyon at the North End of Boston, undercut Brewster's price in 1734 when he listed chocolate at 11 shillings/pound [36]. The same year John Merrett, grocer, shop located at the sign of the Three Sugar Loaves and Canister on King Street near the Town House, advertised his goods for the first time; while chocolate was among the items presented, he did not provide pricing information. Merrett's advertisement, however, is an outstanding example of Colonial era graphics (Fig. 27.3) [37]. Two years later in 1736 Merrett provided his customers with information on price and cacao bean source and reported that his chocolate was "made of the best Maraciabo [sic] and Carraca [sic] cacao, and sold for 14 Shillings and 6 Pence per single Pound [weight], or 14 shillings per dozen or larger quantities" [38]. Pinkey subsequently raised his chocolate price the same year to 14 shillings and 6 pence [39]. Also the same year (1736), John Tinkum, chocolate maker, whose shop was located "opposite to the house where the late Mrs. Smith dwelt, in Milk Street, below the South Meeting House in Boston," developed his chocolate trade and established a base price at 14 shillings/pound [40]. John Brewster subsequently adjusted his chocolate price in 1738 and published an announcement that he had dropped the price to 13 shillings and 12 shillings and 6 pence for the dozen [41]. The most important decision that Brewster made during 1738, however, was to initiate the process of branding his initials, I. B. (for John Brewster), into his bars of manufactured

John Merrett
GROCER,
At the Three Sugar-Loaves and
Canister in King-street, near
the Town-house BOSTON

SELLS Cocao, Chocolate, Tea Bohea and Green, Coffee raw and roasted All sorts of Loaf, Powder and Muscovado Sugar, Sugar Candy brown and white, Candid Cittron, Pepper, Pimienta or all Spice, white Pepper, red Pepper, Cinamon, Cloves, Mace, Nutmegs, Ginger Race and Powder, Raisins, Currants, Almonds sweet and bitter, Prunes, Figs, Rice, ground Rice, Pearl and French Barley, Sago, Starch, Hair-powder, Powder-blew, Indigo, Annis Corriander and Carraway Seeds, Saltpetre, Brimston, Flower of Brimston. All sorts of Tobacco and Snuff, Rozin, Bee's-Wax, Tarmarins, Castile Soap, Olive Oyl, Fine Florence Oyl, Capers, Anchoves, &c

Also Painters Colours, Oyls and Tools, Dye woods, Allum, Copperas and other Dyers Wares, &c. by Wholesale and Retail at reasonable Rates.

FIGURE 27.3. Merritt's grocery store advertisement at the Three Sugar-Loaves. *Source: The Boston Gazette*, November 11, 1734, p. 4. Courtesy of Readex/Newsbank, Naples, FL. (Used with permission.)

chocolate. He announced he did so in *The Boston Gazette* for Monday, January 2, 1738:

> This is to inform my Customers and Others [presumably his competitors] that all my Chocolate is Marked I. B. to prevent them being imposed upon, as several have informed me. [42]

This reconfirmation that he marked his chocolate with his initials is significant since his actions represent one of the earliest attempts in North America to "brand" or certify that chocolate came from a specific source.

John Merrett reduced his chocolate price in March 1738 to 12 shillings/pound, 11 shillings 6 pence per dozen [43]. On January 8, 1740, Merrett announced to his customers that he was "designing to leave the province" and wished to settle all accounts [44]. Merritt remained in business, however, through at least November 3, 1740, and still had not liquidated his stock of items, which included chocolate [45]. The assumption is, however, that Merrett ultimately sold his store and left Boston. Was Merrett a "victim" of a Boston "chocolate price war" whereby the stronger, more influential merchants flourished, while others like Merrett did not? Alternatively, might Merrett have accumulated enough wealth through shrewd pricing decisions that he was able leave the Colonies well in advance of the Revolution and return to mother England? What other possibilities come to mind? Perhaps it should be noted here that John Brewster, among the earliest to "brand" chocolate with his personal initials, died in 1766 [46].

Perhaps the most interesting North American chocolate-associated economic documents encountered during research on this project, however, were

two letters sent to Boston newspaper editors by patrons who complained how much prices had risen in succeeding decades. These documents are especially noteworthy for their candor, and for identification of Colonial era household budgets. The letters may be appreciated further given knowledge of currency equivalents and that a daily working wage in early 18th century Boston would have been less than 7 shillings/day for a common laborer while a master craftsman would have earned only 12 shillings/day [47].

The first letter of this type was anonymous and the author's profession cannot be determined from the text. The document is dated Boston, November 25, 1728, and presented the argument that persons and families "today" were experiencing economic difficulties because they "live at a higher Rate than their Incomes bear, & so become involved in surprising Straits & Difficulties." The author then presented an argument for what he considered would be "The Necessary Expenses in a Family of but middling Figure, and no more than Eight Persons," that he claimed amounted to 10 shillings and 6 pence/week for the most basic items:

> Breakfast [of] Bread & 1 pint of Milk; 3 pence
> Dinner [of] Meat, Roots, Salt, and Vinegar [this is for the noon meal]; 10 pence
> Supper as the Breakfast [and]; 3 pence
> Small Beer for the whole Day Winter & Summer; 2 pence
>
> ───────────────────────────────
>
> [Total] one Person, one Week [i.e., 7 days]: 10 shillings 6 pence
> [Total] For Eight Persons for one Day: 12 shillings

The author then calculated these costs for eight persons for one year to be 219£, then expanded his calculations for three additional items expressed on a yearly basis:

> Candles, 3 a night Summer and Winter; 7£ 12 shillings and 1 pence
> Sand [to scrub clothes] Soap, and Washing; 5£
> One Maid's Wages; 10£.

He then tallied the amount if the cost of dinner was set at 1 shilling instead of 10 pence (i.e., twice the amount). At the higher value the author then claimed the families' annual budget would be 265£ 18 shillings and 9 pence, or 5£ and 2 shillings/week.

The author then identified other household expenses not covered in the above basic household budget:

1. *No House Rent . . . , nor Buying, Carting or Sawing fire-wood;*
2. *No Butter, Cheese, Sugar, Coffee, Tea, nor Chocolate.*
3. *No Wine, nor Cyder [sic] not any other Spiritous [sic] Liquors;*
4. *No Tobacco, Fruit, Spices, nor Sweet-meats;*
5. *No Hospitality, or Occasional Entertaining;*
6. *No Acts of Charity, nor Contribution for Pious Uses;*

7. *No Pocket Expences [sic], either for Horse-hire, Travelling [sic], Convenient Recreation;*
8. *No Charge of Nursing;*
9. *No Schooling for Children;*
10. *No Buying of Books of any sort, or Pens, Ink, & Paper;*
11. *No Lyings in [i.e., costs associated with illness];*
12. *No Sickness—nothing to the Apothecaries or Doctors;*
13. *No Buying, mending or repairing Household Stuff, or Utensils;*
14. *Nothing to the Simstress [sic], nor to the Taylor, nor to the Barber, nor to the Hatter, nor to the Shoemaker, nor to the Shopkeeper—and therefore NO CLOATHS [sic].* [48]

Even more interesting, however, is a second letter written by an ordained pastor identified only as "T." This remarkable document presented two expense ledgers, the first identified as current expenses for 1747; the second list is what the same items cost 40 years previously (ca. 1707) when "T." began his service as pastor. He stated, initially, that when he was first hired in 1707(?), his salary was 100£ per annum, or approximately 40 shillings/week, that is, less than 6 shillings/day. With this income he was able to purchase the following items weekly (except shoes, as this expense was annual):

1 Dunghill Fowl [cost illegible], [cost illegible] Goose [cost illegible], and [cost illegible] Turkeys [cost illegible];

1 pound [each] Butter (6 pence), Cheese (2 pence), 1 pound Candles (5 pence), and 1 dozen Eggs (2 pence);

1 pound [each] Beef (2$\frac{1}{2}$ pence), Mutton (2$\frac{1}{2}$ pence), Pork (3 pence), and Veal (3 pence);

1 pound Sugar (6 pence), 1 gallon Molasses (1 shilling 10 pence), 1 pound Chocolate (2 shillings; 6 pence);

1 Bushel [each] Indian Corn (3 shillings), Rye (4 shillings) Wheat (5 shillings);

1 pair Shoes for my self (5 shillings), for my Wife (3 shillings; 6 pence);

1 gallon milk (8 pence); half a Barrel beer (4 shillings).

The author then presented prices for the same items and what they cost him during 1747. In his letter to the editor he lamented the extraordinary rise in prices:

1 Dunghill Fowl (3 shillings 6 pence), 1 Goose (10 shillings), 1 Turkey [at a weight of 8 pounds] [illegible];

1 pound [each] Butter (5 shillings), Cheese (2 shillings), 1 pound Candles (5 shillings), and 1 Doz. Eggs (3 shillings);

1 pound [each] Beef (1 shilling), Mutton (1 shilling; 6 pence), Pork (2 shillings), Veal (2 shillings);

1 pound Sugar (5 shillings), 1 Gallon Molasses (20 shillings), 1 pound chocolate (14 shillings);

1 Bushel [each] Indian Corn (20 shillings), Rye (22 shillings), Wheat (25 shillings);

1 pair Shoes for my self (60 shillings), for my Wife (35 shillings);

1 Gallon Milk (4 shillings), half a Barrel Beer (18 shillings). [49]

Another economic-related theme extracted from the Boston chocolate advertisements was appreciation of the high quantities of cacao beans and/or ground cocoa and chocolate maintained by merchants at their stores or warehouses. The assumption can be made that certain merchants, when able, stockpiled large quantities of cacao beans to offset vagaries of international politics, piracy, and weather conditions that could have interdicted regular shipments of beans from South America and the Caribbean to East Coast American ports. Some of these quantities are quite staggering:

1812. F. Beck, merchant, located at Number 11 India Street—"20,000 pounds of first quality Cocoa" [50].

1814. Munson & Barnard, merchants, located at Number 27, India-Street—"120,000 wt. Caraccas [sic] Cocoa, 26 hhds [hundredweight] and 90 bags Para Cocoa . . . 100 boxes of Number 2 Chocolate" [51].

1815. Moseley & Babbidge, Merchants, located at Number 39 India Street—"have for sale 10,000 lb. COCOA, from the Portugese [sic] Main, intitled [sic] to debenture, and a superior quality for manufacturing" [52].

ETHICS

Several merchants during the survey period printed apologies to their customers for supplying improper weights for goods purchased and argued that their business scales were set incorrectly or damaged without their knowledge. Sometimes these advertisements contained convoluted explanations; other times the shopkeeper clearly wished to maintain customer loyalty and not be accused of fraud. Luke Baldwin and Company, merchants, shop located at Number 44 Long Wharf, Boston, published an announcement in the *Columbian Centinel* (sic) on March 15, 1806, where they apologized to their customers. Baldwin and Company had received their chocolate from Gideon Foster, Boston chocolate maker, whose scale was "light." They informed their customers that they should return the packages of chocolate and that the company would "make good to them any less they may have suffered, in consequence of such deficiency in weight" [53]. But would it be logical that customers would "heft" their paper-wrapped package of chocolate, sense a weight deficiency, and return the package? Would it not be more logical that much of the "evidence" already would have been prepared as beverages? A follow-up notice published in *The Boston Gazette*, March 17, 1806, added more details to the problem with Gideon Foster's scales. The chocolate maker, himself, testified that because of an "accidental loss of a wire from the scale beam in his manufactory," the chocolate that he had consigned to Luke Baldwin and Company was of an inappropriate weight [54].

Beyond the issue of incorrect weights as an ethical concern, Colonial and Federal era merchants commonly used slave labor in the production of chocolate products. Newspaper advertisements confirmed this practice and announced rewards for runaway slaves who had specific chocolate-associated training. Sometimes the advertisements merely identified the slaves as being owned by chocolate makers. James Lubbuck, chocolate maker (no address provided), published the following advertisement in the *New England Weekly Journal* on September 4, 1727:

> Ran-away from his Master, Mr. J. Lubbuck of Boston, Chocolate-Grinder, on the 28th of last. A Young Negro Man-Servant, about 20 years of age. Fellow speaks pretty good English. . . . Whoever shall take up the above said Runaway, safely convey to his above said master, living near Mr. Colman's house, shall have three pounds reward, and all necessary charges paid. [55]

The obituary notice for John Maxwell, merchant, shop located in his house in Long Lane, Boston, was published in *The Boston News-Letter* on May 31, 1733, and noted that as part of Maxwell's estate to be sold was:

> A Negro Woman, about Twenty Years of age, can do household work and grind chocolate very well; at the above said house. [56]

John Brewster, well-known Boston chocolate merchant, issued a slave advertisement that was published in the *Boston Evening Post*, December 27, 1762: "To be sold by John Brewster, a lusty NEGRO MAN, used to Grinding Chocolate" [57]. Brewster's advertisement did not yield the expected results, since the same advertisement repeated a year later [58].

Daniel Sharley, Boston merchant (no address provided), published the following announcement in the *Boston Evening Post*, January 24, 1763:

> Ran away from his Master . . . A Negro fellow about 35 years of age, named Sam, has a large bump over his right eye . . . He has been used to grind Chocolate and carry it about for Sale. Whoever takes up said Negro, and will bring him to his Master, shall have TWO DOLLARS Reward, and all necessary Charges paid by Daniel Sharley. [59]

FESTIVALS/HOLIDAYS

The American celebration of Thanksgiving is deeply rooted in myth and tradition. While much has been written, little actually is known regarding the event that many call the "First Thanksgiving." Tradition holds that the Plymouth colonists invited 90 Native Americans from the local Wampanoag Nation to a celebration that lasted for three days. The date of this event is not known with certainty, but logically would have been in the fall season sometime between September 21 and November 11, 1621 [60]. Documented evidence exists that relatively few foods were shared that day: venison, wild turkey, unidentified wild fowl, bass, codfish, and Indian corn or maize [61].

Almost 200 years passed from the time of that first Thanksgiving to the first publication of a Boston merchant advertising Thanksgiving-related items, with chocolate identified among the "Good Things for Thanksgiving." A Mr. Beck (first initial F), merchant, shop located at Number 11, India Street, published the following advertisement in *The Boston Gazette* on November 1, 1813:

> A Small Assortment of Good Things for Thanksgiving . . . For Sale by F. Beck, at No. 11, India-street, 20 casks Malaga Raisins, 100 boxes best Muscatel and bloom Raisins, A few boxes Lemons in prime order, A few pipes [barrels] old Cognac Brandy, from the house of Otard, Duprey & Co (acknowledged by Connoiseurs [sic] to be equal to any ever imported), A few bbls [barrels] Muscovado Sugar, 100 bales prime Cotton, 450 bags Laguira Coffee, entitled to debenture, Dan Fish, Chocolate, Cider Brandy &c. [62]

While "good things" were offered to customers for Thanksgiving in Boston, our search through Boston newspapers did not reveal the association of chocolate with Christmas food or gift sales. We identified, however, an advertisement from New London, Connecticut, dated December 13, 1792, where the merchant Samuel Woodbridge offered:

> West-India and New-England Rum, Geneva, Madeira and Malaga Wines, Loaf and Brown Sugar, Bohea and Green Teas, Coffee, Chocolate, Pepper, Spices, Codfish, Molasses; and many other good things for CHRISTMAS. [63]

GENDER ISSUES

Advertisements of several types document the important role played by women as merchants in 18th century Boston. Some advertisements reveal that widows had taken over the day-to-day management of shops after the death of their husbands, and requested that their "loyal patrons" resume or maintain their shopping custom by having their cacao ground or purchasing ready-made chocolate under her new management. Other advertisements, however, are less clear and suggest that male–female business partnerships were established by relatives (i.e., brother–sister business partnerships). Of the more than 900 chocolate-related auctioneers, brokers, manufacturers, merchants, and vendors identified from Boston (see Appendix 6) only 2–3% of these were women.

A Mrs. Read (no first name provided in the advertisement), identified as a manager, opened her chocolate house on King Street, on the north side of the Town House, Boston, in 1731. Her advertisement in *The Boston Gazette* noted that patrons may:

> Read the news, and have chocolate, coffee, or tea ready made any time of the day. [64]

The same year, 1731, Mrs. Hannah Boydell, widow of John Boydell, announced that she continued to sell:

Sugar, Coffee, Chocolate, Starch, Indigo, Spices, and Grocery Ware, reasonably, in a Shop adjoyning [sic] to the Office [County Probate Office], over against the Bunch of Grapes Tavern. [65]

Mary Watts, related in an undefined way to chocolate manufacturer Samuel Watts, operated a shop on Middle Street, Boston, and advertised her chocolate at prices in 1751 as being similar to Samuel's [66].

Mary Minot, chocolate merchant, shop located in Cole Lane, advertised her goods in 1754 and the text read: "GOOD Chocolate by the Quantity or single Pound, very cheap" [67].

Mary Ballard, coffee house manager of her shop located on King Street, published an advertisement in the *Boston Evening Post*, December 22, 1755, and stated that her coffee house was opened:

For the Entertainment of Gentlemen, Benefit of Commerce and Dispatch of Business, a Coffee-house is this Day opened in King Street. All the News-Papers upon the Continent are regularly taken in and several English Prints, and the Magazines are ordered. Gentlemen who are pleased to use the House, may at any Time of the Day, after the Manner of those in London, have Tea, Coffee, or Chocolate, and constant Attendance given by their humble Servant, Mary Ballard. [68]

Hepzibah Goldsmith, merchant, no address provided, issued the following statement that she was the:

Sole Executrix to the estate of the late Mr. John Goldsmith, late of Boston, Chocolate-grinder, deceased, that she carries on the business of chocolate grinding, where the former customers and others may be supply'd at the cheapest rate. [69]

HEALTH AND MEDICINE

Early Boston newspaper advertisements reveal that commercial chocolate sometimes was sold in a medicinal context, as an item enumerated within lists of drugs and medicines available at apothecary shops. Zabeiel Boylston, physician, for example, managed an apothecary shop located on Dock Square, Boston. He was among the more prominent Bostonians, a friend and colleague of the noted jurist Cotton Mather, and was an important figure in the debate whether or not to inoculate patients against smallpox (see Chapter 7). His apothecary shop advertisement, dated March 17, 1711, is among the earliest to identify chocolate in Colonial era newspapers and clearly documents the important place held by chocolate as a medicine during this early period of American history [70].

Benjamin Andrews, Junior, was a merchant was proprietor of an apothecary shop located opposite the Swing Bridge, Boston. He sold chocolate and general household items but also advertised in 1769 that he sold a large assortment of patent medicines such as Jesuits' Drops, an infallible cure for "the veneral [sic] disease" [71].

Penuel Bowen, merchant, managed a drug and medicine store on Dock Square in 1784, opposite Deacon Timothy Newell's (residence?), where he imported drugs and medicines from London and supplied patrons with "preparations and compositions of Apothecaries." He also sold coffee, chocolate, cask and jar raisins, tamarinds, prunes, currants, nutmegs, mace, cinnamon, cloves, loaf and powdered sugar [72].

One of the more curious chocolate-related medical announcements to appear in Boston newspapers was publication of an anonymous account in *The Boston Gazette, or Weekly Journal* on March 5, 1751, with the title "A Friendly Caution against drinking Tea, Coffee, Chocolate, &c. very hot." The argument advanced by the author was that hot beverages posed essential health risks to consumers when beverage temperatures exceeded 100 "degrees of warmth," a point which the author claimed could coagulate blood serum and endanger life. The author argued that since tea, coffee, and chocolate regularly were served 30 degrees hotter than blood, the practice would:

Thicken the Blood [and] relax and weaken the Nerves and Stomach, and thereby hurt the Digestion, & produce Colics, &c. [73]

LEGAL ISSUES

Advertisements commonly listed the names of business owners. When such documents are arranged in chronological sequence, they sometimes reveal business expansion, formation of partnerships, and evidence of long-term associations. Other advertisements revealed dissolution of partnerships, the adoption of and naming of new associates, going-out-of-business sales, and bankruptcies.

As mentioned earlier, John Merritt announced in 1740 that he was "designing to leave the Province [of Massachusetts]." Merritt's shop was located at the Three Sugar-Loaves and Cannister on King Street, Boston, near the Town House. He ultimately reduced his inventory, sold his stock of goods that included chocolate, and presumably left Boston [74].

An announcement published in the *Columbian Centinel* (sic), on March 1, 1806, identified a new business partnership that linked Luke Baldwin, Theodore Mansfield, and John A. Parkman. Their business sold chocolate and general items at their shop located at Number 44, Long Wharf [75].

An advertisement dated April 24, 1802, by S. Bradford (auctioneer) announced a sale to be held opposite Shop Number 10 on Long Wharf in Boston, where a portion of the effects of a Mr. John M'Lean, who had declared bankruptcy, were to be auctioned. Included in the auction were 100 boxes of chocolate that belonged to M'Lean [76].

MANUFACTURING EQUIPMENT AND TYPE OF MILLS

Other advertisements in Boston newspapers identified importation of chocolate-associated manufacturing equipment, whether grinding stones and rollers, mills, or other equipment. Some advertisements listed complete chocolate manufacturing mills with full equipment as components of land or estate sales. Still others, especially auction announcements, provided information on chocolate manufacturing items.

George Fechem, chocolate maker, owned a mill located near the Watertown Bridge in Boston. He elected to sell his mill, perhaps due to impending hostilities between the Americans and British, and published the following advertisement in *The Boston Gazette, and Country Journal* on September 9, 1776, that described the property:

A good chocolate mill, which will go with a horse and grind 120 wt of chocolate in a day. Said mill consists of three good kettles, with twelve pestles in a kettle well leaded, nine dozen pans, one good nut cracker, a good musline [cloth] to clean the shells from nuts. Any Person wanting the same may apply to GEORGE FECHEM, near Watertown Bridge. [77]

One of the more detailed chocolate manufacturing announcements that appeared in Boston newspapers, however, was an account from Salem, Massachusetts, dated September 13, 1737, that appeared in the *New England Weekly Journal*:

By a Gentleman of this Town [Salem] is this Day bro't [sic] to Perfection, an Engine to Grind Cocoa; it is a Contrivance that cost much less than any commonly used; and will effect all that which the Chocolate Grinders do with their Mills and Stones without any or with very Inconsiderable Labour; and it may be depended on for Truth, that will in less than Six Hours bring one Hundred weight of Nuts to a consistence fit for the Mold. And the Chocolate made by it, is finer and better, the Oyly [sic] Spirit of the Nut being almost altogether preserved. And there is little or no need of Fire in the making. [78]

MILITARY ASSOCIATIONS

Use of chocolate as a military ration, while well documented (see Chapter 31), would not appear commonly in newspaper advertisements. Still, other military associations with chocolate can be gleaned from examination of Boston newspaper advertisements. Daniel Jones, merchant, for example, managed a shop located adjacent to the Hat and Helmet pub on Newbury Street. On December 15, 1760, he published the following advertisement in the *Boston Post-Boy* where he identified a range of goods sold, that included chocolate:

Officers and Soldiers who have been in this Province Service may be supplied with any of the above articles on short credit till their muster rolls are made up, as usual, the soldiers being well recommended. [79]

Approximately six weeks later, on January 26, 1761, the same merchant published another advertisement in the *Boston Evening Post*, where he listed his intent to cater to military officers (British). The goods advertised, which included chocolate, had just been imported from London on ships commanded by Captains Aitken and Bull. The bottom of the advertisement stated:

Officers and Soldiers who have been in this Province Service, may be supply'd with any of the above articles on credit. [80]

OBITUARIES

Obituary announcements in historical newspapers provided hints or clues regarding occupations that sometimes revealed that chocolate-related occupations rested within family lines, from father to son. The documents sometimes provided insights on the character of the deceased and revealed customs and traditions associated with Colonial and early Federal era North America as noted in the following obituary notices:

1733. John Maxwell, chocolate merchant. To be sold as part of Maxwell's estate was a "Negro Woman, about Twenty Years of age, can do household work and grind chocolate very well; at the above said house" [81].

1771. Miss Eunice Feacham. Ms. Feacham died at the age of 26, of undefined causes; identified as the daughter of Mr. Feacham, chocolate grinder [82].

1772. Bethiah Oliver, chocolate merchant. Ebenezer Oliver, her son, announced the death of his mother to her customers, and alerted customers that he had to sell his mother's shop [83]. The assumption can be made that Ebenezer continued to work at the shop since he subsequently produced his own chocolate sales advertisements [84].

1773. John Goldsmith, chocolate-grinder. Only information provided is the date of his death, given as September 27 [85].

1773. John Ruddock, chocolate merchant. The obituary notice alerted customers that "all persons indebted to the Estate of John Ruddock, late of Boston, Esq., deceased, are desired to settle the same with Abiel Ruddock, Administrator, or they will be Sued at next Term" [86].

1819. Ezekiel Merrill, chocolate merchant. Advertisement announced a liquidation or administrator's sale. The document was published on October 6th and read:

Tomorrow, at 9 o'clock, at No. 35, Long Wharf, By order of the Administratrix [sic] on the estate of Mr. Ezekiel Merrill, Deceased, His whole stock of groceries, consisting of rum, sugar, molasses, cordials, tea, gin, coffee, chocolate, wine, scales and weights desk, empty barrels and kegs, &c. [87]

POLITICS

Boston on the eve of the American Revolution was a city of conflicting loyalties. Some chocolate-associated advertisements reflected this conflict. Daniel Jones, for example, was typical of some Boston merchants who extended credit to British officers and common soldiers [88]. In other instances, advertising copy identified the address of the chocolate shops as proximal to well-identified British establishment. Elizabeth Perkins, for example, advertised the location of her coffee house, on August 30, 1773, as "two doors below the British Coffee-House, North Side of King Street," where she also sold chocolate and general food items [89]. By publishing such advertising copy, Ms. Perkins made a supportive, political statement by associating herself and her establishment with the British. Had she been a supporter of the American Revolution, she could have written her advertising copy in a manner that provided information on location, but without mentioning the British. The last newspaper advertisement for the Perkins Coffee-House was dated April 17, 1775 [90]. Interesting questions remain. Did Ms. Perkins remain in Boston during the Revolution? Was her shop damaged by rebels? At the conclusion of the Revolution did she return to England or move to Canada?

The event known historically as the Boston Massacre occurred on March 5, 1770. On this date, Archibald Cunningham, merchant, shop near the drawbridge on Fore Street, advertised "choice capers and anchovys [sic] by the keg of pound" and also supplied a broad range of local and imported foods and general items to his customers, among them:

New Raisins and Turkey Figgs [sic] by the Cask. CURRANTS, Citron, Mace, Cloves, Cinnamon, Rice, Oatmeal, Pimento per Hundred weight, Pepper, Almonds, Salt-Petre [sic], Indigo, Starch . . . Wine Glasses by the Groce [sic], Allum [sic], Ginger . . . Kippen's Snuff per Dozen, Kippen's Tobacco per Dozen or les Quantity, Chocolate, Loaf and brown Sugar . . . Mustard per pound or in Bottles, long handled Hearth Brushes . . . [also] A Quantity of NUTMEGS to be sold cheap at said shop. [91]

As hostilities progressed, the Massachusetts Bay Colony passed legislation that regulated prices to prevent "Monopoly and Oppression." These notices were published in *The Boston Gazette and Country Journal*, February 3, 1777, and established the prices for cocoa and chocolate:

Best cocoa at 6 pounds, 10 shillings a cwt. American manufactured chocolate, 1 shilling and 8 pence a lb. [92]

As the Revolutionary War continued, the Massachusetts Bay Colony passed further legislation to restrict the export of basic commodities from their territory to adjacent states. The legislation was published in the *Continental Journal* on October 7, 1779:

Be it therefore enacted by Council and House of Representatives in General Court assembled, and by the authority of the same, that no exportation be permitted of rum, wine, or any kind of spirits, molasses, sugar, cotton wool, sheep's wool, wool cards, flax, salt, coffee, cocoa, chocolate, linen, cotton, and linen, woolen and cotton goods of all kinds; provisions of all and every sort; live-stock, shoes, skins, and leather of all kinds; either by land by or water from any part of this State, after the twenty-third day of September instant, to be carried to any place not within this State, and except reasonable ship stores. [93]

PRODUCT QUALITY

A significant number of newspaper advertisements considered issues such as adulteration, branding, certification, and purity guarantees. Human nature being what it is, unscrupulous chocolate manufacturers and merchants in 18th century Boston sometimes looked for ways to defraud customers, whether through "light," incorrect product weights or through the addition of adulterants during the manufacturing process (see Chapter 47). A sampling of such advertisements reveals merchant concerns and attempts to set themselves apart from the crooks.

1769. Joseph Mann, merchant, at the sign of the Wheat-Sheaf in Water-Street near Oliver's Dock, Boston:

MAKES and Sells Chocolate, which will be warranted free of any Adulteration, likewise New-England Mustard, manufactured by said Mann, who will be glad of the Continuance of his former Customers, and thankfully receive the Favour [sic] others. [94]

1770. John Goldsmith, merchant, at the corner shop leading down to John Hancock, Esquire's Wharf:

CHOICE CHOCOLATE . . . by the large or small Quantities. Also all Sorts of Groceries. The Chocolate will be warranted good, and sold at the cheapest Rates—Cash given for Cocoa. Cocoa manufactured for Gentlemen in the best Manner. ALSO, Choice Cocoa and Cocoa Nut Shells. [95]

1771. Joseph Mann, merchant, shop located at his house in Water-Street near Oliver's Dock:

CHOCOLATE *Choice Chocolate Ground and Sold
. . . Warranted pure from any Adulteration . . .
Cocoa taken in to Grind for Gentlemen and done
with Fidelity and Dispatch. Cocoa Shells to Sell.*
[96]

1771. J. Cooke, merchant, shop located at the Slitting Mills in Milton—and also at his store joining to Mr. Sweetfer's (spelling uncertain) on the Town Dock, Boston:

> *In large or small quantities, at the lowest Prices for Cash, or exchang'd for good Cocoa at the Market Price. Cocoa taken in to manufacture at the above Places and warranted to be done in the best and neatest Manner, and with the greatest Dispatch, at Ten Shillings Lawful Money per C.Weight.* [97]

SHIPPING INFORMATION

Shipping News (see Chapter 17) and port documents provide the primary source of information on imported goods. Sometimes, however, Boston newspaper advertisements identified specific shops and ship captains/ masters with their associated cargo of cacao and chocolate. Some advertisements also listed the geographical origin of cacao beans as a means of providing source-related information to customers. The Boston advertisements searched during the survey period revealed that chocolate was imported from a range of South American and Caribbean ports of origin. John Merrett, merchant, shop located in King Street, for example, published the following advertisement in 1736 that touted chocolate from South America and provided price information:

> *[Chocolate] made of the best Maraciabo [sic] and Carraca [sic] cacao, at 14 Shillings and 6 Pence per single Pound and 14 Shillings per the Dozen or larger Quantity. N.B. [i.e., note] The nuts are roasted, well clean'd, and ground after the best method.* [98]

GENERAL INFORMATION

An unusual advertisement published on October 8, 1751, in *The Boston Gazette, or Weekly Journal*, by Samuel Watts, chocolate maker, shop located up by the Boston Prison Yard, reported that his chocolate sales were down due to summer heat:

> *By reason of the heat of the weather, [he] has been unable to supply his customers with chocolate for these few weeks past, [and] fears to the loss of many of them.* [99]

Hot weather, however, was good for other chocolate-associated merchants. Peter Pailhes published an advertisement in the *Columbian Centinel* (*sic*) on May 15, 1807, where he informed his customers he had opened a new store opposite William Smith, Esquire, near the Concert Hall on Court Street where he:

Intends to keep a constant supply of ICE CREAM, such as Vanilla, Rose, Lemon, Orange, Pine-Apple, Strawberry, Raspberry, Currant, Coffee, Chocolate, and other kinds, too many to describe—which any person or family may be supplied with, in any quantity, shape or color, made with taste and neatness. Families wishing for a quantity will please to deliver their orders the day before at the above place. He has a convenient room to receive company, where they may be accommodated with ice Cream, and other refreshments, Pastry and Confectionary. Families may be supplied with Ice at any hour of the day. [100]

Conclusion

Advertisements in the Boston newspapers provide a rich mine of information as represented by the themes reported here. Sorting the Boston chocolate-related merchant data by date reveals price changes, address changes, bankruptcies, the impact of death upon widows, association of chocolate with holidays, sales of medicinal chocolate, and a host of other relationships.

Some Boston merchants began small and conducted their chocolate-related business selling on the street; others sold chocolate out of their houses; still others managed small shops, grocery stores, or warehouses where they conducted their business. Merchants were known both by the product quality of the chocolate they sold and by their address. Address changes necessitated by economic, military, or personal needs required that customers be alerted through newspaper advertisements. Sometimes merchants moved their places of business voluntarily; at other times they vacated out of fear, as during the 1764 Boston smallpox epidemic or because of impending hostilities with England. Some advertisements provided precise, numerical addresses, while others listed only general information—difficult to interpret today. Such general locations, however, would have been familiar to customers of the 17th and 18th centuries (i.e., near a specific bridge, church, meeting hall, doctor's house). Merchants sold chocolate from a variety of locations that might better be described as apothecary shops, bookstores, even clothing and garment stores. Given that detailed street maps of Boston are widely available through the Library of Congress, perhaps interested readers would like to revisit the Boston chocolate-associated addresses to determine whether or not chocolate-associated commerce has been maintained at these 18th and early 19th century addresses.

Other questions remain: What happened to Loyalist chocolate merchants during the Revolutionary War era? It would be interesting to pursue whether or not some moved to Canada and reopened chocolate-associated establishments there. Although members of our team—Canadian and American—have shared the names of presumed Loyalist merchants, at present we have no evidence for this presumption.

Interesting advertisements published in the Boston newspapers were related to economic complaints about current prices for local goods, when compared to prices decades earlier. Others were related to the Boston smallpox epidemic of 1764 and merchant decisions whether or not to keep their stores open, or to relocate to presumed disease-free localities.

But perhaps the most interesting finding of all was how these advertisements related to human nature and how the lives and emotions of chocolate-associated merchants were expressed through the context and wording of their advertisements: anger as when merchants chided and harangued other merchants in print over presumed slights or misrepresentation; fear of smallpox; joy in alerting customers to newly imported items; pride in shop ownership and in the diversification of goods sold; sadness of going-out-of-business sales; and uncertainty regarding potential hostilities that eventually would culminate in the American Revolution. Are these advertisements not snapshots of human nature, and do we not feel connected to these people more than 200 years ago?

Acknowledgments

I wish to thank the Readex/Newsbank Company for permission to use selected advertisements to illustrate this chapter.

Endnote

1. A provocative account of the events called the Boston Massacre appeared in the *Boston Evening Post*, March 12, 1770. The article identified events leading up to the deaths, provided names of the dead, and described in specific detail the wounds suffered.

References

1. http://www.readex.com/readex/. (Accessed January 10, 2007.)

2. http://firestone.princeton.edu/microforms/. (Accessed January 10, 2007.)

3. http://www.readex.com/readex/. (Accessed January 10, 2007.)

4. *The Boston News-Letter*. December 3, 1705, p. 4.

5. *Boston Post-Boy*. March, 30, 1772, p. 3.

6. *The New Jersey Gazette*. January 22, 1783, p. 2.

7. *The Boston Post*. April 28, 1785, p. 3.

8. *Mass [Massachusetts] Gazette*. January 9, 1786, p. 4.

9. *The Vermont Gazette or Freemen's Depository*. January 5, 1789, p. 3.

10. *Columbian Centinel (sic)*. October 23, 1793, p. 2.

11. *Boston Evening Post*. December 2, 1765, p. 3.

12. *Middlesex Gazette*. September 3, 1787, p. 2.

13. *Columbian Centinel (sic)*. December 21, 1816, p. 2.

14. *The Boston News-Letter and New-England Chronicle*. June 4, 1766, p. 2.

15. *The Boston News-Letter and New-England Chronicle*. October 27, 1768, p. 2.

16. *The Boston News-Letter and New-England Chronicle*. February 9, 1769, p. 2.

17. *The Boston Gazette, and Country Journal*. August 8, 1774, p. 4.

18. *Columbian Centinal (sic)*. January 5, 1799, p. 4.

19. *Columbian Centinel (sic)*. March 2, 1808, p. 4.

20. *The Boston Weekly News-Letter*. August 2, 1759, p. 4.

21. *The Boston Gazette, and Country Journal*. March 5, 1770, p. 4.

22. *Boston Post-Boy*. September 19, 1768, p. 4.

23. *Boston Evening Post*. October 24, 1768, p. 3.

24. *The Repertory*. March 28, 1809, p. 3.

25. *The Boston Gazette*. January 20, 1817, p. 3.

26. *Boston Evening Post*. October 24, 1748, p. 2.

27. *Boston Evening Post*. July 6, 1767, p. 4.

28. *The Boston Gazette and Country Journal*. December 12, 1768, p. 4.

29. *Boston Evening Post*. November 22, 1773, p. 3.

30. *Continental Journal*. April 10, 1783, p. 1.

31. *Columbiana Centinel (sic)*. April 15, 1807, p. 2.

32. *Boston Post-Boy*. April 2, 1764, p. 4.

33. Steward, E. M., and Bowen, J. C. *History of Wages in the United States from Colonial Times to 1928*. Washington, DC: Bureau of Labor Statistics and United States Government Printing Office, 1929.

34. *The Boston Gazette*. February 16, 1730, p. 2.

35. *The Boston Gazette*. December 17, 1733, p. 4.

36. *New England Weekly Journal*. March 20, 1734, p. 2.

37. *The Boston Gazette*. November 11, 1734, p. 4.

38. *Boston Evening Post*. February 9, 1736, p. 2.

39. *The Boston News-Letter*. January 29, 1736, p. 2.

40. *The Boston News-Letter*. March 18, 1736, p. 2.

41. *The Boston Weekly News-Letter*. February 16, 1738, p. 2.

42. *The Boston Gazette*. January 2, 1738, p. 4.

43. *New England Weekly Journal*. March 28, 1738, p. 2.

44. *New England Weekly Journal*. January 8, 1740, p. 2.

45. *Boston Post-Boy*. November 3, 1740, p. 4.

46. *The Boston Gazette and Country Journal*. March 5, 1766, p. 2.

47. Steward, E. M., and Bowen, J. C. *History of Wages in the United States from Colonial Times to 1928*. Washington, DC: Bureau of Labor Statistics and United States Government Printing Office, 1929.

48. *New England Weekly Journal*. November 25, 1728, p. 2.

49. *Boston Evening Post*. December 14, 1747, p. 2.

50. *Columbian Centinel (sic)*. October 10, 1812, p. 1.

51. *The Boston Gazette*. February 3, 1814, p. 3.

52. *The Boston Gazette*. December 18, 1815, Supplement, p. 2.

53. *Columbian Centinel (sic)*. March 15, 1806, p. 4.

54. *The Boston Gazette*. March 17, 1806, p. 4.

55. *New England Weekly Journal*. September 4, 1727, p. 2.

56. *The Boston News-Letter*. May 31, 1733, p. 2.

57. *Boston Evening Post*. December 27, 1762, p. 3.

58. *Boston Evening Post*. January 24, 1763, p. 4.

59. *The Boston News-Letter and New-England Chronicle*. June 2, 1763, p. 3.

60. Grivetti, L. E., Corlett, J. L., and Lockett, C. T. Food in American history. Part 2. Turkey. Birth of a nation: colonialization to the Revolution (1565–1776). *Nutrition Today* 2001; 36(2): 88–96.

61. *Mourt's Relation. A journal of the Pilgrims at Plymouth; Mourt's relation, a relation or journal of the English plantation settled at Plymouth in New England, by certain English adventurers both merchants and others*. New York: Corinth Books, 1963.

62. *The Boston Gazette*. November 1, 1813, p. 3.

63. *Connecticut Gazette*. December 13, 1792, p. 4.

64. *The Boston Gazette*. September 13, 1731, p. 2.

65. *New England Weekly Journal*. May 31, 1731, p. 2.

66. *The Boston Gazette, or Weekly Journal*. October 8, 1751, p. 2.

67. *Boston Evening Post*. December 16, 1754, p. 2.

68. *Boston Evening Post*. December 22, 1755, p. 2.

69. *The Boston Gazette and Country Journal*. October 25, 1773, p. 4.

70. *The Boston News-Letter*. March 17, 1711, p. 2.

71. *The Boston Gazette and Country Journal*. February 20, 1769, p. 4.

72. *The Boston Gazette, and the Country Journal*. January 12, 1784, p. 4.

73. *The Boston Gazette, or Weekly Journal*. March 5, 1751, p. 1.

74. *Boston Post-Boy*. November 3, 1740, p. 4.

75. *Columbian Centinel (sic)*. March 1, 1806, p. 3

76. *Columbian Centinel (sic)*. April 24, 1802, p. 3.

77. *The Boston Gazette and Country Journal*. September 9, 1776, p. 4.

78. *New England Weekly Journal*. September 13, 1737, p. 1.

79. *Boston Post-Boy*. December 15, 1760, p. 4.

80. *Boston Evening Post*. January 26, 1761, p. 4.

81. *The Boston News-Letter*. May 31, 1733, p. 2.

82. *Boston Evening Post*. March 25, 1771, p. 3.

83. *The Boston Gazette and Country Journal*. July 13, 1772, p. 3.

84. *Boston Evening Post*. July 12, 1773, p. 4.

85. *Essex Gazette*. September 21, 1773, p. 35.

86. *The Boston Gazette and Country Journal*. August 23, 1773, Supplement, p. 2.

87. *Boston Patriot & Daily Mercantile*. October 6, 1819, p. 3.

88. *Boston Post-Boy*. December 15, 1760, p. 4.

89. *Boston Evening Post*. August 30, 1773, p. 4.

90. *Boston Evening Post*. April 17, 1775, p. 3.

91. *The Boston Gazette and Country Journal*. March 5, 1770, p. 4.

92. *The Boston Gazette and Country Journal*. February 3, 1777, p. 1.

93. *Continental Journal*. October 7, 1779, p. 4.

94. *Boston Post-Boy*. March 6, 1769, p. 2.

95. *The Boston News-Letter*. March 8, 1770, p. 2.

96. *Boston Post-Boy*. May 27, 1771, p. 3.

97. *The Boston Gazette, and Country Journal*. June 24, 1771, Supplement, p. 1.

98. *Boston Evening Post*. February 2, 1736, p. 2.

99. *The Boston Gazette, or Weekly Journal*. October 8, 1751, p. 2.

100. *Columbian Centinel (sic)*. May 15, 1807, p. 2.

VI

Colonial and Federal Eras (Part 2)

Chapter 28 (Rose) considers the Dutch cacao/chocolate trade in the Lower Hudson Valley of New York State during the 17th and 18th centuries, and traces purchases of raw commodities in the Caribbean to chocolate processing in Holland, with subsequent export to North America during the Colonial era. Chapter 29 (McCombs) builds on the Dutch experience and further examines chocolate manufacturing in the Upper Hudson Valley, especially the development and expansion of cocoa processing in Albany, New York. Chapter 30 (Gay) considers the rise and development of chocolate manufacturing in Philadelphia, Pennsylvania, and explores business relationships that evolved among chocolate makers, the impacts of war and peace during the Revolutionary era on sales, and the changing complexity of chocolate manufacturing in this important Colonial North American city. Chapter 31 (Westbrook, Fox, and McCarty) considers military aspects of chocolate use during the Colonial era and Revolutionary War within the Northern Frontier, especially its role as a dietary component and medical product used at Fort Ticonderoga, New York. Chapter 32 (Kelly) considers chocolate as a common provision taken aboard 19th century whaling ships, and how chandlers and grocers in New Bedford, Massachusetts, obtained and supplied chocolate for long-term whaling voyages.

Dutch Cacao Trade in New Netherland During the 17th and 18th Centuries

Peter G. Rose

Introduction

In 2009, it will be 400 years since Henry Hudson explored the river that would later receive his name. He was involved in this exploration on behalf of the Dutch East India Company with the goal of finding a northerly passage to the Orient. While he did not succeed, he found a promising area for extensive fur trade that would become the Dutch colony of New Netherland. In 1621, the West India Company was founded, modeled on the lucrative Dutch East India Company, and was given a monopoly of trading rights in the Western Hemisphere.

The company brought settlers to the new "province" of New Netherland, a vast area wedged between New England and Virginia that comprised the present day American states of New York, New Jersey, Delaware, and portions of Pennsylvania and Connecticut. The settlers in this "new" land contributed the foodstuffs commonly associated today with the Hudson River Valley: vegetables such as cabbages, carrots, onions, peas, parsnips, and turnips; herbs such as chives, parsley, and rosemary; and fruits such as apples, peaches, and pears. In *A Description of the New Netherlands*, written in 1655, Adriaen van der Donck reported that these foods "thrive well" [1].

Trade

In New Netherland, the emphasis was on the fur trade, but settlers also raised crops to supply foodstuffs required to outfit ships that were part of the vast trade network of the Atlantic world. In the 16th century the Dutch had obtained sugar from the Iberian Peninsula, but by the turn of the century and during the truce with Spain in 1609, they were allowed to obtain sugar directly from Brazil [2]. To own and monopolize that oceanic trade, as well as to control the production of sugar, was an early goal of the West India Company [3]. By 1630 they achieved that goal when they conquered Recife and gradually expanded their Brazilian territory. The Portuguese recaptured Recife in 1654 and historian Wim Klooster calls the loss of Brazil "the kiss of death for the West India Company." He argued that the West India Company, which would continue

Chocolate: History, Culture, and Heritage. Edited by Grivetti and Shapiro
Copyright © 2009 John Wiley & Sons, Inc.

to exist for another 20 years (or its successor), would never equal in importance the role of the Dutch East India Company in Asia [4].

Dutch traders along with the French and English, however, remained active in the Atlantic world and retained six important islands in the Caribbean: Saba, St. Maarten, St. Eustatius, and what commonly were called the "ABC Islands," or, Aruba, Bonaire, and Curaçao. The Dutch subsequently acquired Surinam in 1667. When Curaçao was granted free port status by the second West India Company in 1675, it became a base of operation for many merchants, with bonded warehouses where goods could be stored for the inter-Caribbean trade. Some even described Curaçao as "the local counterpart of Amsterdam" [5]. Later, St. Eustatius also would become an important staples market.

By the end of the 1670s, Frederick Philipse, who owned Philipsburg Manor, now a living history museum administered by Historic Hudson Valley in Sleepy Hollow, New York, was an important Caribbean trade merchant, and his activities included shipping cacao to Europe. This is confirmed in a journal written by Jasper Danckaerts in 1679/1680, where he wrote: "and here it is to be remarked that a Frederick Filipsen has the most shipping and does the most trade" [6]. Shipping records available for research inspection at the London Public Record Office also reinforce the importance of Frederick Filipsen/Frederick Philipse: on December 4, 1679, *The Charles* out of New York, a vessel owned by Frederick Philipse, paid taxes on "hhds [hogsheads of] Coco valued at 114:10 English pounds." By about this time the Dutch colony of New Amsterdam had fallen into British hands (as of 1674) and ships departing New York for England or other parts of Europe were required to pay taxes to the British Crown. On March 11, 1681, a ship owned by Frederick Philipse arrived at Cowes, England, on the isle of Wright, bound for Amsterdam, paid taxes on the cargo of various skins and "cocoanuts [cacao beans] damnified val. 5 English pounds." The records continue in this manner. October 3, 1686: in *the Beaver* bound for Amsterdam, quantities shipped were "50 pounds cocow nutts [*sic*] v. [value] 30 shillings" [7].

The various ways one can spell cacao is amusing, but the other contents of the ships make for entertaining reading as well. They create eye-opening insights into what was traded and transported across the ocean at the time. A ship like the *Charles* carried "55 packs, 10 bundles, 17 cases, one chest, 4 barrels and 2 hogsheads." It not only contained the 2 hogsheads of "Coco," but also "1201 buckskins in the haire; 104 black bear skins in the haire; 438 otter skins; 1355 fox skins; 2345 beaver skins; 2694 small furs valued at 45 pounds; 14 cat skins; 1382 raccoon skins; 1 barrel unwrought copper; 512 ox and cow hides in the haire; 287 elk skins or buffalo hides in the haire; Sassafras; 181 hogshead, 3 barrels and 95 rolls of tobacco; and 1661 sticks of logwood;" among other items. Subsequent ship-

ments give an equally clear and interesting insight into the commerce of the day. Skins were the main trading items, as well as tobacco, but cacao in its various spellings appears frequently as well.

Adolph Philipse (1665–1750) inherited Philipsburg Manor and continued his father's business as an international merchant. He was one of the wealthiest men in New York City. He did not live in Philipsburg Manor, but the estate served as headquarters for his business [8]. His lands and businesses produced foodstuffs for New York's urban population and for the plantations in the West Indies, where such items as preserved meats, dairy products, fish, and seafood such as oysters, and wheat products such as flour and ship's biscuits, found a ready market. It could be said that, "in essence, the plantations of the Caribbean gave rise to the plantations of the Hudson Valley, like Philipsburg Manor" [9]. Jacobus Van Cortlandt's *Shipping Book* entries for 1702/1703 reveal ship's cargoes bound predominantly for Curaçao, but also Jamaica and Barbados. Common entries include bread, flower (*sic*), butter, yew beef, pease (*sic*), pork, bacon, and meal. Entries also list *cocoa* and beef hides for shipment to London in 1703, but quantities are not noted [10].[1]

Curaçao with its bonded warehouses and free port was ideally located close to the coast of Venezuela, and the Dutch were deeply involved in the Venezuelan cacao trade, so much so that in 1701 the governor informed the Spanish king that "the Dutch were familiar with all the roads leading to the major cacao plantations" [11]. Klooster's book, *Illicit Riches*, carries a lengthy appendix of ships carrying cacao from Curaçao to the Netherlands between 1701 and 1755; cacao cargoes ranged from 5000 to 485,000 pounds with destinations identified as Amsterdam, Flushing (Vlissingen), Middelburg, and Zeeland [12].

Since the fourth quarter of the 17th century there had also been an increasingly important trade between New York and Curaçao, bringing Venezuelan cacao to New York City and its hinterlands. Younger sons involved in Dutch firms such as Van Ranst, Depeyster & Duyckinck, or Cuyler & Lansingh were sent to Curaçao to be involved in mercantile activities. There were also Jewish connections between New York and the island; for instance, such well-known traders as Nathan Simson and Daniel Gomez were heavily involved during the 18th century (see Chapter 5) [13]. The *Books of Entry* in the New York State Library show that Daniel Gomez imported two casks of cacao and paid duty on October 18, 1742 and on November 26, 1742, he imported 46 casks of the luxury item [14].

Peter Kalm, assistant to Linnaeus, who came to America in 1749–1750, reported on life in the former New Netherland after the English takeover. In this mid-18th century account he mentioned that, in Albany, the Dutch sometimes took chocolate at supper:

Their supper consists generally of bread and butter and milk with small pieces of bread in it. The butter is very salt. Sometimes too they have chocolate. [15]

Merchants in New York City and the Hudson Valley made certain that supplies of this nutritious beverage were available to their customers, as we can surmise from the account book of Joshua Delaplaine, a carpenter, joiner, and New York City merchant. He married Esther Lane of Hempstead in 1716. Three of his sons also became merchants. His *day book* was among the papers of Abraham De Peyster (a well-known New Netherland name with various spellings). It spans the years 1752–1756 and is filled with chocolate sales, including an entry in 1754 where he sold four 25-pound boxes of chocolate for 1 pound 8 shillings each to Abram Depeister. Another notation related that Jacobus Rosevelt purchased two 25-pound boxes of chocolate at 1 pound 9 shillings, and shortly thereafter is recorded another sale of three more boxes. During the same time period another customer, William Palmer, purchased 3 pounds of sugar and $\frac{1}{2}$ pound of chocolate on a regular basis, but sometimes varied his purchase and bought 3 pounds sugar and $\frac{1}{2}$ pound tea. In another instance he bought 3 pounds of sugar and $\frac{1}{2}$ pound each of coffee, tea, and chocolate [16].

In the Hudson Valley, the Huguenot Historical Society of New Paltz has several account books in its library that list inventories and chocolate sales or purchases. Account books numbers 4, 5, and 6 of Josiah Hasbrouck span the period 1793–1797 and provide notations for many chocolate sales to persons with Huguenot names, among them Bevier, Dubois, and Deyo. Hasbrouck's account books numbers 8, 9, and 10 provide additional information for the period 1797–1801, with many ledger entries for chocolate sales, most of either $\frac{1}{2}$ pound or 1 pound each and on occasion purchases of 2 pounds. An interesting entry in account book number 17, Daybook "A," of Josiah Hasbrouck (1800–1802) records that on August 22, 1802, he sold 1 pound of chocolate to the "heirs of Poulis Freer, Deceased." Genealogical records show that Poulis Freer died in 1802 and apparently the chocolate sold to the heirs was prepared as a beverage used at the time of Freer's funeral. Another interesting entry in the Daybook of Josiah DuBois (1807–1820) recorded that $\frac{1}{2}$ pound of chocolate was dispensed on September 15, 1819, and again on October 19, 1819, and paid for by the "Town of New Paltz for relief of John A. Freer and family—A family of color" [17].

Artifacts

Roelof J. Elting's store inventory of 1768–1769, also in the Huguenot Historical Society's archives, includes the notation: "1 doz. Black chockcolat [*sic*] cups and 4 pounds and 3 pounds of chockcolate [*sic*]." Curator Leslie Lefevre informed the author that black chocolate

cups may be basaltware of the type produced by Wedgwood and others, or Jackfield-ware, a type of shiny black-glazed earthenware. In the times when drinking hot chocolate was a fashionable and social affair, special tall, columnar cups were used for the beverage. The drink would be served from a special chocolate pot commonly with its handle at a right angle to the spout and with a lid with a removable filial so that a stick with flanged end (Spanish: *molinillo*) could be inserted (see Chapters 10–13). The stick would be rubbed between the hands—a movement like rubbing one's hands to warm them—to agitate the cocoa and create a delicious foam, which was especially desired at the time. Pouring the liquid chocolate from one pot into another was another way of creating the highly prized froth. Some of these old-style chocolate pots remain in museums in New York City and the Hudson Valley.

The most interesting chocolate-related artifact identified during research for this chapter, however, was a double-handled blue and white earthenware chocolate cup from Delft, the Netherlands, manufactured ca. 1655–1675. This artifact was found in the *Dann Site*, a Seneca-Iroquois archeological location, in New York State (Fig. 28.1). Rather than using the cup for its intended purpose, Native Americans might have used it as an item of personal adornment. In their article entitled "Beavers for Drink, Land for Arms," Jacobs and Shattuck reported that "ceramics, such as pieces of majolica or Delft ware, were used by the Indians primarily as ornamentation" [18].

The Museum of the City of New York owns a copper pot, and in the American Wing of the Metropolitan Museum of Art, three silver chocolate pots are on display. The oldest, manufactured by Edward Winslow, dates to ca. 1710, and belonged to Thomas Hutchinson of Boston. The museum also has in its collection two early 19th century chocolate pots. One of

FIGURE 28.1. Chocolate cup used by Native Americans: blue and white earthenware, manufactured in Delft, the Netherlands, dated ca. 1655–1675. *Source:* Excavated at the Dann Site, Seneca-Iroquois, New York State. Courtesy of Rochester Museum and Science Center, Rochester, NY. (Used with permission.) (See color insert.)

these, dated 1827, was made by Frederick Marquand and belonged to a prominent New York surgeon with the enticing name of Dr. Valentine Mott.

The New York Historical Society possesses a Chinese export porcelain chocolate pot dated to ca. 1730 that later belonged to the Society of the Cincinnati, the oldest military hereditary society in the United States, founded by Major General Henry Knox. The birthplace of this society, our country's first veterans' organization, is Mount Gulian in Fishkill, now a State Historic Site, where the New York Chapter presently is headquartered. The New York Historical Society also owns chocolate cups dated to 1805, previously owned by Isaac Beekman-Cox and Cornelia Beekman, a well-known Hudson Valley Dutch name.

The former New Netherland area still abounds with Dutch family and place names, and Dutch house and barns dot the landscape. Collections in museums in New York City along the Hudson River and other parts of the former colony are filled with fine samples of other crafts and artistry, from beautiful portraits to painted furniture, colorful earthenware, and gleaming silver that remain as vestiges of its colonial beginnings [19]. We can now add cacao to the long list of other foodstuffs introduced here by the Dutch: coleslaw, donuts, pancakes, pretzels, wafers, and waffles, and above all cookies—that have become part of America's culinary heritage brought to New Netherland by the Dutch in the 17th century. Recipes for these and many other Dutch foods can be found in hand-written manuscript cookbooks, spanning more than three centuries, belonging to the descendants of these settlers. The Dutch touch left a lasting mark on America's kitchens [20].

Acknowledgments

I would like to thank the following for their help with this chapter: Doug McCombs, Curator of History, Albany Institute of History & Art; Charles T. Gehring, New Netherland Project, Albany, New York; Noah L. Gelfand; Kathleen Eagen Johnson, Curator, Historic Hudson Valley; Catalina Hannan, Librarian, Historic Hudson Valley; Leslie Lefevre-Stratton, Curator, Huguenot Historical Society; Eric Roth, Director, Huguenot Historical Society; and David W. Voorhees, Editor *De Halve Maen*.

Endnote

1. Cacao commonly was brought from the West Indies for processing in Europe, an economic pattern still practiced in the 21st century. In Ghana and Ivory Coast, cacao represents 80 percent of the national product, but only 10 percent is processed locally, the remainder primarily in the United States or Europe.

References

1. Van der Donck, A. A *Description of the New Netherlands*. Edited with an introduction by T. F. O'Donnell. Syracuse, NY: Syracuse University Press, 1968; p. 24.

2. Klooster, W. *Illicit Riches*. Leiden, the Netherlands: KITLV Press, 1998; p. 35.

3. Klooster, W. *Illicit Riches*. Leiden, the Netherlands: KITLV Press, 1998; p. 37.

4. Klooster, W. *Illicit Riches*. Leiden, the Netherlands: KITLV Press, 1998; p. 37.

5. Klooster, W. *Illicit Riches*. Leiden, the Netherlands: KITLV Press, 1998; p. 59.

6. Danckaerts, J. In: James, B. B., and Jameson, J. F., editor. *Journal of Jasper Danckaerts, 1679–1680*. New York: Charles Scribner's Sons, 1913; p. 69.

7. London Public Record Office, Chancery Lane, London. Archives, Historic Hudson Valley.

8. Vetare, M. L. *Philipsburg Manor Upper Mills*. Tarrytown, NY: Historic Hudson Valley Press, 2004; p. 7.

9. Vetare, M. L. *Philipsburg Manor Upper Mills*. Tarrytown, NY: Historic Hudson Valley Press, 2004; p. 14.

10. Jacobus Van Cortlandt, J. Shipping Book. Archives, Historic Hudson Valley.

11. Klooster, W. *Illicit Riches*. Leiden, the Netherlands: KITLV Press, 1998; p. 126.

12. Klooster, W. *Illicit Riches*. Leiden, the Netherlands: KITLV Press, 1998; Appendix 2, pp. 207–223.

13. Klooster, W. *Illicit Riches*. Leiden, the Netherlands: KITLV Press, 1998; p. 100.

14. *Books of Entry*. New York State Library.

15. Kalm, P. *Travels in North America: The English Version of 1770*. Mineola, NY: Dover, 1966; p. 347.

16. Joshua Delaplaine Day Book, 1752–56. New York Historical Society Archives, p. 22.

17. Account and Day Books. Huguenot Historical Society Archives, New Paltz, NY.

18. Jacobs, J., and Shattuck, M. D. Beavers for drink, land for arms: some aspects of the Dutch–Indian trade in New Netherland. In: *One Man's Trash is Another Man's Treasure*. Rotterdam, the Netherlands: Museum Boijmans-van Beuningen, 1995; p. 109.

19. Rose, P. G., translator and editor. *The Sensible Cook: Dutch Foodways in the Old and the New World*. Syracuse, NY: Syracuse University Press, 1998; p. 20.

20. Rose, P. G., translator and editor. *The Sensible Cook: Dutch Foodways in the Old and the New World*. Syracuse, NY: Syracuse University Press, 1998; p. 35.

Chocolate Consumption and Production in New York's Upper Hudson River Valley, 1730–1830[1]

W. Douglas McCombs

Introduction

Stretching for more than 150 miles from Manhattan to the falls at Troy, New York, the navigable section of the Hudson River has served as a conduit for trade, commerce, and cultural exchange for centuries. With the arrival of the Dutch in 1609, and later the English, Palatine Germans, French Huguenots, and African slaves, the Hudson River Valley became increasingly connected to the world beyond America's shores. The people who lived and worked throughout the valley brought various cultural beliefs and practices to the region, while merchants and traders in New York, Kingston, Albany, and other communities along the banks of the Hudson River brought consumer goods from around the world. All contributed to create a cosmopolitan atmosphere that has characterized the Hudson River Valley for 400 years.

As Peter Rose has discussed in Chapter 28, chocolate from the Caribbean Islands and South American mainland was traded, sold, consumed, and enjoyed by regional inhabitants. Chocolate was available in the port city of New York, at the mouth of the Hudson River, by at least the early 18th century, as demonstrated in the numerous account books showing the sale of chocolate by city merchants. Farther north along the upper stretches of the Hudson River at Albany and along the Mohawk River in Schenectady, surviving documents reveal a similar availability of chocolate by the 1730s. In Albany, chocolate was not only sold by storeowners, it was ground and processed at mill sites near town beginning around 1750 and continuing well into the 19th century. An examination of the sale and production of chocolate in the Upper Hudson Valley offers a window onto a burgeoning consumer economy that had links to both trans-Atlantic trade as well as emerging markets within the newly established United States [1].[1]

Chocolate: History, Culture, and Heritage. Edited by Grivetti and Shapiro
Copyright © 2009 John Wiley & Sons, Inc.

Early Consumption of Chocolate in the Upper Hudson Valley

Several documents and artifacts in the collection of the Albany Institute of History & Art, Albany, New York, provide valuable information about the distribution and consumption of chocolate in the Upper Hudson River Valley throughout much of the 18th and early 19th centuries. What these historical materials detail is a consumer market rich in imported products as well as locally produced goods. Chocolate played an increasingly visible role in that market.

A surviving account book for Henrick Van Rensselaer IV (1712–1793), who operated a store in Schenectady and later in the ancestral landholdings of Claverack, New York, lists a few sales of chocolate [2]. On April 28, 1735, one Joseph Yeats, Jr., made a purchase of one pound of "cocolath" together with twelve pounds of sugar and one gallon of rum. The chocolate that he purchased cost 5/0 per pound (5 shillings/no pence per pound) in comparison to the sugar that was 0/7 per pound and the rum that brought 4/0 per gallon [3].

During the same period, the widow Jannitie Van Slych purchased half a pound of chocolate on May 3, 1734, also at the Schenectady store. Over the course of the next three years, her account listings show that she bought numerous articles from Van Rensselaer, including nails, sugar, tobacco, knee buckles, various types of fabrics, and on June 8, 1735, she paid 12/0 for the writing of a lease. Yet, throughout those years she made no additional purchase of chocolate. What is even more surprising to the modern researcher is that within the large account volume of Henrick Van Rensselaer, which spans more than 400 pages and includes thousands of entries, only two other sales of chocolate appear. In May 1735, Johannis Barhuyt bought half a pound of chocolate in Schenectady, and several years later, after Van Rensselaer relocated to Claverack, Samuel Ten Broeck made a purchase of one pound of chocolate on October 11, 1743 for 3/9. The price charged to Ten Broeck, resident of Claverack, represents a decrease in price of 1/3 from what Van Rensselaer was charging eight years earlier in Schenectady. The price reduction may reflect the greater availability of chocolate by the 1740s or the greater availability at Van Rensselaer's new location in Claverack [4].

One other document in the collection of the Albany Institute confirms the early availability of chocolate in the Upper Hudson Valley. A single receipt for Captain Peter Winne[2] (1690–1759) lists two separate purchases of chocolate, both of considerable size, indicating they were probably intended for wholesale to an Albany merchant or merchants. Winne was the skipper of his own boat, which carried cargo between Albany and New York. He also made stops at various ports along the length of the Hudson River. On September 4, 1735, Winne acquired one 60 pound box of chocolate at 4/0 per pound for a total of £12/0/0. A few months later on May 15, 1736, he purchased an 84 pound box of chocolate, again for 4/0 per pound or a total of £16/16/0. The receipt also lists "3 boxes not returned 4/6." Apparently Winne had to pay for the boxes used to store the chocolate. In a separate column on the same receipt, a list of credits shows how Winne eventually paid for the large amounts of chocolate. Boards, barrels of tar, cash payments, and a balance transferal to one Peter Low took care of Winne's debt of £29/0/6 [5].

By the 1750s, chocolate seems to have been readily obtainable in Albany. An account book for Albany merchant and trader Harmen Van Heusen abounds with sales of chocolate between 1758 and 1760. Van Heusen, who operated his shop and storeroom on Broadway between State and Beaver Streets in Albany, was dealing with a variety of clients, from individual consumers wanting basic goods to frontier traders, city merchants, and the British military. According to the account book, Van Heusen's storeroom held a vast assortment of consumer goods. Most accounts consisted of basic supplies such as rum, sugar, and tobacco, but other articles including fabrics, buttons, china sets, punch bowls, and decanters appear. In terms of cost, chocolate was relatively inexpensive, ranging in price between 2/0 and 2/8 per pound. By comparison, rum fluctuated between 4/6 and 7/0 per gallon, a heavy blanket termed a "rugg" cost 16/0, and tea averaged around 9/0 per pound [6].

Most purchases of chocolate from Van Heusen's storeroom were small, usually from half a pound to four pounds. Occasionally an order from a military regiment included chocolate purchases in slightly larger quantities. On November 25, 1759, Captain Maxfield of the 46th Regiment bought six pounds of chocolate at 2/6 per pound. Other items purchased at the same time include 14 pairs of shoes at 9/6 per pair, 36 pairs of shoes at 8/0 per pair, 50 pairs of stockings, and ten gallons of spirits. On March 29, 1760, Mr. Gaskall placed an order for Captain William Howard of the 17th Regiment. In that order, he bought six pounds of chocolate, as well as two pairs of stockings, one silk purse, and a few additional articles in small quantities. Other military officers appear throughout the account book, some also making small purchases of chocolate.

The Van Heusen account book contains several instances where chocolate was purchased in large quantities either for resale or for use in public taverns. Isaac Sawyer bought 44 pounds of chocolate on June 9, 1758, for the very low price of 1/10 per pound. At the same time he bought a barrel of sugar. Later in the account book, Sawyer is identified as a tavern keeper. Probably the large quantities of chocolate and sugar he purchased were served at his tavern. Almost a year later, on May 24, 1759, an account for "Kidd & Fergison, treaders [sic]," shows a purchase of 50 pounds of chocolate, and on the same day, "Mathew Noalls, Suttlar," bought 54 pounds. Both Kidd & Fergison and Noalls paid 2/6 per pound for their chocolate, an increase of 0/8 per pound over what Sawyer paid the preceding year. Throughout the late spring and summer of 1759, Van Heusen sold other large quantities of chocolate to merchants and traders: "Duglas, Fillas, & Morres" bought 57 pounds of chocolate on May 29th; Henry Morris (perhaps a partner in the business of Duglas, Fillas, & Morres) purchased one box of chocolate containing 120 pounds on August 7th; Captain Jonethan Ogdan bought one box of chocolate containing 50 pounds on August 15th; and, on September 19th, Timothy Northam bought one box of chocolate consisting of 110 pounds. Apparently, Van Heusen received a large shipment of chocolate some time during the spring of 1759, and traders and merchants were placing sizeable orders for resale throughout the summer and autumn of that year.

One individual chocolate consumer in the Van Heusen account book stands out. A man named Frederick Garison appears regularly during the first few months of 1758, making singular purchases of half a pound of chocolate. These purchases continue through June 13th of that year. The Van Heusen account book shows a hiatus of chocolate purchases from June until December 7, 1758, when John Van Nalan's account lists two pounds of chocolate. For several months, Van Heusen must not have been able to obtain chocolate for resale, and throughout that period, Garison had to content himself with substitutes. On June 18, 1758, Garison bought half a pound of tea, and again, on September 9, 1758, another half a pound of tea. The following year, on January 8, 1759, Garison bought one pound of coffee and two pounds of sugar. He does not purchase chocolate again until February 22, 1759. The absence of chocolate forced Garison to drink tea and coffee, but not long after chocolate reappeared in Van Heusen's account book, Garison reverted back to what seems to have been his drink of choice.

Another account book, kept by merchant Robert Montgomery, provides additional details about the sale and consumption of chocolate in the upper portions of the Hudson River Valley during the 18th century. Montgomery operated a business in Kingston, New York, between 1773 and 1782. Following the American Revolution, he moved to New York City and then eventually relocated to New City, a small community in Rockland County, north of New York City on the western side of the Hudson River. The account for John Slight, Esq., lists numerous articles purchased during November and December 1773. Rum, cloth, salt, and a cape were among some of the items he bought. On December 20, 1773, he purchased "1/2 Dozen of Coholate Boels" for 2/9. One month earlier, on November 20, 1773, William Euistin

[?] Cox likewise made a purchase of "1/2 Dozen of Cholate Boals" for 2/6, a little less than what Slight paid. Unfortunately, the account book offers no further description of what is meant by these terms, and there is no further listing in Montgomery's account book for such items. Possibly, the chocolate "boals" Montgomery sold were ceramic cups used for drinking a hot chocolate beverage. Since they were listed as boals (bowls), they may have been handleless cups sold without matching saucers. Their origin also remains a mystery. Were they Chinese export porcelain or English earthernware? Regardless, the bowls appear to have been a one-time purchase for Montgomery since no others are mentioned in his account book, which ends several years later in 1789 [7].

In addition to the chocolate bowls, Montgomery's patrons purchased chocolate, usually in small quantities enough for domestic use, but a few listings for sizeable quantities suggest that some chocolate was sold to merchants and traders for resale. One noteworthy account for Dinnis McReady, dated May 14, 1777, reads: "265 lb Cholate [sic] which he sold in Philadelphia for 3/6 pr lb." Two days prior to McReady's purchase, Thomas Leland bought 254 pounds of chocolate for 4/6 per pound. Despite the inconveniences and hardships caused by the American Revolution, chocolate was sold and traded during the early years of conflict, but following these two transactions, chocolate is conspicuously absent from Montgomery's account book until the autumn of 1781, near the end of the war.

Throughout the last quarter of the 18th century, chocolate seems to have been a common consumer good available at stores in Albany and surrounding communities. Merchant Thomas Barry, for example, identified chocolate in a long list of commodities available at his store "near the *Dutch-Church*, Albany," in a December 1771 Albany newspaper. In June 1772, John Heughan sold chocolate at his store "next Door to 'Squire CAMPBELL's in *Schenectady*," and merchant Edward Cumpston advertised in January 1788 that he had chocolate at his store "opposite the Northeast Corner of the Low Dutch Church, Albany." Numerous other newspaper advertisements list chocolate during the waning years of the 18th century, and the occurrences of chocolate only increased throughout the 19th century [8–10].

Chocolate Manufacturing in Albany

Consumers throughout the Hudson River Valley had easy access to chocolate through shops and merchant storerooms from the early 1700s onward, at prices that were not prohibitive. By the 19th century, chocolate was a commonly stocked item found in the inventories of most stores. Albany residents also had the opportunity to buy freshly ground chocolate from about 1790 to about 1830, and a source of locally ground chocolate may possibly have existed as early as 1750.

In the 18th century, the Wendell family of Albany owned land just to the south of town (in an area now called Lincoln Park). A deep ravine with a swift-flowing stream named the Beaver Kill divided their landholdings, making it an ideal site for milling operations. A portrait of Abraham Wendell (b. 1715), probably painted by John Heaten around 1737 (Fig. 29.1), depicts a rare view with one of the family mills in the background [11]. In a will dated July 29, 1749, Abraham's father, Evert Wendell (1681–1750), left the mills, the dwelling house, and all the land on both sides of the stream to Abraham. Evert also left specific instructions in his will that his sons Abraham and Philip "shall build a Chocolate mill for the use of my son Philip" [12]. The chocolate mill was probably constructed near the other mills, or chocolate milling equipment may have been retrofitted to one of the existing buildings. As late as 1794, the Wendell mills still appeared on the Simeon DeWitt map of Albany, although identification for each of the buildings was not given. We cannot be certain whether the chocolate mill was actually built or how long it operated, but if Abraham was as "trusty, faithful, beloved, honest" as Evert's will states, he would not have wanted to disobey his father's last wishes.

About 20 years after Abraham was instructed to build a chocolate mill, a Scots-Irish immigrant named James Caldwell (1747–1829) arrived in Albany and soon commenced the production of chocolate at a mill that was touted to be a technological marvel. James was an ambitious and enterprising Presbyterian who learned quickly that commercial opportunities abounded in Albany and the regions to the west and north [13, 14]. In the early 1770s he began his professional career in Albany as a grocer, in partnership with his brother Joseph. An advertisement placed in 1772 in the *Albany Gazette* offers a tantalizing look at the specialty foods and dry goods the two carried at their grocery store on Market Street: "best-cured gammon, Gloucester, Cheshire and Jersey Cheese, Scotch Barley, All-spice, Best French Indigo, Kirby and common Fish Hooks, Seal-skin Trunks." They also stocked imported liquors and made special note that they "embrace every Opportunity to purchase Lemons, Oranges, Limes, and all sorts of exotic Fruits and Spices." In addition, James and Joseph Caldwell sold chocolate, probably purchased from New York or Philadelphia, where the Caldwell's first lived when they arrived in America and where they maintained business contacts. The Caldwells accepted cash or country produce for payment [15].

James enjoyed a lucrative business supplying traders and merchants in the Mohawk River Valley, northern New York, and Canada. He eventually established stores in Montreal and Bennington, Vermont. With the profits of his business, Caldwell invested in land, especially around Lake George, and in 1790, he opened one of the most technologically advanced mills in the country for the processing of tobacco, chocolate, hair powder, mustard, and the grinding of barley, peas, and wheat. Caldwell's Manufactory, also called the Rensselaerville Manufactory, used both water and steam to power grinding stones, cutting machines, and other equipment (Fig. 29.2) and employed around 100 men and boys seasonally, making it one of the first factories in the nation. The mills were located in a valley about one mile north of Albany and about 400 yards west of Stephen Van Rensselaer's mansion [16].

Snuff and cut tobacco were the primary products processed at Caldwell's Manufactory, but his chocolate must have received recognition, since one of his advertisements announced that, "he has also the satisfaction to find his Chocolate and Mustard in the highest repute." Caldwell guaranteed the quality of his products by placing stamps and warranties on the packaging and promised dissatisfied customers that he would exchange their goods if "returned within thirty days from the purchase" [17]. In addition to guaranteeing quality, Caldwell's packages offered an appealing way of advertising and promoting his products. A copper printing plate engraved "Caldwell's Super Fine Chocolate Albany" set within an elegant octagonal border and decorated with roses and ribbon was used for printing the labels that marked Caldwell's chocolate

FIGURE 29.1. Portrait of Abraham Wendell (b. 1715), attributed to John Heaten; dated ca. 1737. *Source*: Accession Number 1962.47. Courtesy of the Albany Institute of History and Art, Albany, NY. (Used with permission.) (See color insert.)

FIGURE 29.2. View of Rensselaerville Manufactory, artist unidentified; dated 1792. *Source*: Caldwell Family Papers, GQ 78-14. Box 1/F-2. Courtesy of the Albany Institute of History and Art, Albany, NY. (Used with permission.)

FIGURE 29.3. Copper engraving plate inscribed Caldwell's Super Fine Chocolate, Albany; dated ca. 1790–1820. *Source*: Accession Number 1959.123.20. Courtesy of the Albany Institute of History and Art, Albany, NY. (Used with permission.)

(Fig. 29.3). Caldwell was not the first in this region, however, to mark his chocolate. A receipt dated 1759 from Captain Marte G. Van Bergon to William DePeyster listed two separate purchases of chocolate. One was marked with the letters "FYP" and the other "MG," probably for the initials of the manufacturers or importers (Fig. 29.4) [18]. Caldwell's printing plate, nonetheless, demonstrates that he was marking his chocolate more attractively and with clearer information than his predecessors.

FIGURE 29.4. Chocolate receipt prepared by Captain Marte G. Van Bergon for William DePeyster; dated 1759. *Source*: Manuscript Number 2229. Courtesy of the Albany Institute of History and Art, Albany, NY. (Used with permission.)

On July 12, 1794, Caldwell's extensive mill complex burned to the ground, resulting in damages estimated around £13,000. The fire was reported to have started in the chocolate mill but spread to nearby buildings (Fig. 29.5). Only a single kitchen building survived. By the following year, Caldwell had rebuilt his mills incorporating the latest technology. A notice in the *Albany Gazette*, for December 28, 1795, included a woodcut of the manufactory and Caldwell's announcement that he had reopened for business: "James Caldwell with pleasure informs his friends and the public, that he has now got his manufactory rebuilt and fully completed." Chocolate, of course, was among the list of products he manufactured [19].

According to newspaper advertisements, Caldwell continued to process chocolate at his mill well into the 19th century, although with various business partners. In 1802, an advertisement announced "Fresh Chocolate. From *Caldwell, Fraser, & Co's.* new Manufactory, a constant supply of the first quality *fresh chocolate.*" Fifteen years later, chocolate was still one of the products manufactured by Caldwell & Solomons and sold at their store at "No. 346, North Market Street, *Opposite the Store of Dudley Walsh & Co. at the sign of the Indian Chief and hand of Tobacco*" [20]. Levy Solomons eventually took control of the store at 346 North Market Street and may have continued operations at Caldwell's mill since an 1828 notice from Levy Solomons in the *Albany Daily Advertiser* informed customers that he was manufacturing "Fresh Chocolate. Never Better," which was available at his store, No. 346 North Market Street [21].

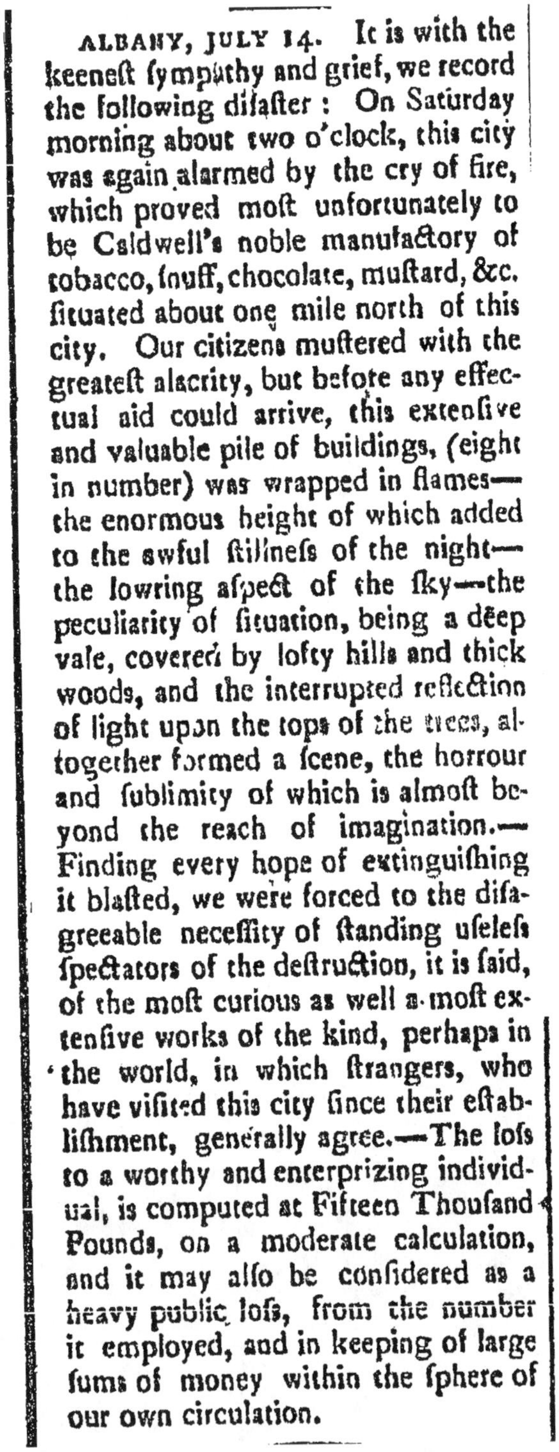

Although the Upper Hudson Valley rested on the edge of a vast frontier throughout most of the 18th century, residents from its flourishing communities did not need to abstain from tasting the delightful, bittersweet flavor of chocolate. Merchants and traders supplied the delicious commodity on a fairly regular basis, although periodically shortages may have limited con-

ALBANY, JULY 14. It is with the keenest sympathy and grief, we record the following disaster: On Saturday morning about two o'clock, this city was again alarmed by the cry of fire, which proved most unfortunately to be Caldwell's noble manufactory of tobacco, snuff, chocolate, mustard, &c. situated about one mile north of this city. Our citizens mustered with the greatest alacrity, but before any effectual aid could arrive, this extensive and valuable pile of buildings, (eight in number) was wrapped in flames—the enormous height of which added to the awful stillness of the night—the lowring aspect of the sky—the peculiarity of situation, being a deep vale, covered by lofty hills and thick woods, and the interrupted reflection of light upon the tops of the trees, altogether formed a scene, the horrour and sublimity of which is almost beyond the reach of imagination.— Finding every hope of extinguishing it blasted, we were forced to the disagreeable necessity of standing useless spectators of the destruction, it is said, of the most curious as well as most extensive works of the kind, perhaps in the world, in which strangers, who have visited this city since their establishment, generally agree.—The loss to a worthy and enterprizing individual, is computed at Fifteen Thousand Pounds, on a moderate calculation, and it may also be considered as a heavy public loss, from the number it employed, and in keeping of large sums of money within the sphere of our own circulation.

FIGURE 29.5. James Caldwell: chocolate advertisement. *Source: Albany Gazette*, December 28, 1795. Courtesy of Readex/Newsbank, Naples, Fl. (Used with permission.)

sumers' access to it. By the end of the 18th century, Albany could boast having one of the largest, best-equipped chocolate mills in the nation, which provided freshly ground chocolate for both local consumers and more distant markets in Canada and elsewhere. The availability of chocolate in Albany and other communities in the Upper Hudson Valley allowed regional consumers to participate in a broader national and international market, while reaffirming connections to European social practices and cultural customs.

Acknowledgments

I wish to thank the following for their assistance, advice, and expertise: Peter Rose, Food Historian; Nicholas Westbrook, Fort Ticonderoga National Historic Landmark; Norman S. Rice, Tammis Groft, Rebecca Rich-Wulfmeyer, Megan Gillespie, Ruth Greene-McNally, Barbara Bertucio, Albany Institute of History & Art; Tricia Barbagallo, Colonial Albany Social History Project; and Jennifer Lemak, New York State Museum.

Endnotes

1. Matson's text, [1] provides a thorough history of New York's merchants and their role in colonial trade and an emerging capitalism.

2. Information about Peter Winne can be found at http://www.nysm.nysed.gov/albany/bios/w/piwinne3032.html. (Accessed January 10, 2007.)

References

1. Matson, C. *Merchants & Empire: Trading in Colonial New York*. Baltimore, MD: Johns Hopkins University Press, 1998.

2. Van Rensselaer, K. *The Van Rensselaer Manor*. Baltimore, MD: Order of Colonial Lords of Manors in America Publications, 1929.

3. Account book for Hendrick Van Rensselaer, Albany Institute of History & Art Library, Albany, NY, MS-697, p. 22.

4. Account book for Hendrich Van Rensselaer, Albany Institute of History & Art Library, Albany, NY, MS-697. Barhuyt account listed on p. 61 and Ten Broeck account listed on p. 151.

5. Captain Peter Winne receipt, Albany Institute of History & Art Library, Albany, NY, MS-2108.

6. Account book for Harmen Van Heusen, Albany Institute of History & Art Library, Albany, NY, MS-111.

7. Account book for Robert Montgomery, Albany Institute of History & Art Library, Albany, NY, MS-96.

8. Barry advertisement in *Albany Gazette*, December 16, 1771.

9. Heughan advertisement in *Albany Gazette*, June 22, 1772.

10. Cumpston advertisement in *Albany Gazette*, January 24, 1788.

11. Mackay, M. A. In: Groft, T., and Mackay, M. A., editors. *Albany Institute of History & Art: 200 Years of Collecting*. New York: Hudson Hills Press, 1998; pp. 51–53.

12. http://www.nysm.nysed.gov/albany/wills/willevwendell2657.html. (Accessed January 10, 2007.)

13. Barbagallo, T. A. James Caldwell, immigrant entrepreneur. *The Hudson Valley Regional Review* 2000; 17(2):55–68.

14. Bielinski, S. Episodes in the coming of age of an Early American community: Albany, N.Y., 1780–1793. In: Schechter, S. L., and Tripp, W., editors. *World of the Founders: New York Communities in the Federal Period*. Albany, NY: New York State Commission on the Bicentennial of the United States Constitution, 1990; pp. 109–137.

15. *Albany Gazette*. July 27, 1772.

16. Munsell, J. A tobacco establishment of 1790. *Annals of Albany*, 1850;1:339–340.

17. *Albany Register*. July 11, 1791.

18. Receipt from Captain Marte G. Van Bergon to William DePeyster, 1759, Albany Institute of History & Art Library, Albany, New York, MS-2229.

19. *Albany Gazette*. December 28, 1795.

20. *Albany Register*. June 29, 1802. Albany Gazette. June 16, 1817.

21. *Albany Daily Advertiser*. February 13, 1828.

Chocolate Makers in 18th Century Pennsylvania

James F. Gay

Introduction

An enduring myth of American chocolate manufacturing is that it all began in 1765 in Dorchester, Massachusetts. The story goes that Dr. James Baker met an itinerant chocolate maker named John Hannon who established the first chocolate mill in this country. The legend is sustained by its retelling again and again by historians, marketing departments, and food writers. The more extreme telling of the tale has Americans completely ignorant of chocolate until 1765. Actually, the truth is much more interesting. While Americans have been eating a brand of chocolate called "Baker's" since 1780, the company was neither the first nor even particularly remarkable in the story of chocolate in the 18th century. Chocolate had been machine-made in America for many decades and in several places prior to Hannon meeting Baker in 1765. While only Baker's name has survived the ages, he had many predecessors and contemporary competitors. Of the approximately 70 commercial chocolate makers in North America during the 18th century, one-third were living and working in Philadelphia.

Founded in 1682, Philadelphia had a population of 2500 with 350 houses within two years [1]. Almost immediately, Pennsylvania began exporting foodstuffs to the West Indies. Philadelphia became the financial center of North America. When John Welsh, master of the 35 ton ship *Adventure*, entered Boston Harbor from the island of Nevis on August 23, 1686 carrying 100 cwt of cocoa, his bond had been given in Pennsylvania. This was the earliest arrival of cocoa beans on record for North America [2].

On November 3, 1718, Samuel Payton from New York guided the *Royal Prince* into Boston Harbor carrying 18 bags of cocoa and chocolate, one bag of cotton, 20 small casks of molasses, along with four casks of pork, salt, and skins [3]. The Boston customs official who recorded this routine entry wrote the oldest surviving record of *chocolate* exchanged between the North American colonies. Colonial shipping records were not contiguous and have gapping holes in coverage. Boston coffeehouse owners had been banking their survival on cocoa imports and satisfying chocolate consumers for half a century before Payton's arrival. But only one name survived the ages . . . and that name was Baker. And only one place: Dorchester, Massachusetts; and only one date: 1765. Now it is time to tell the rest of the story.

Quakers

North America is blessed with rivers with waterfalls close to the coast, necessary for the construction of watermills. An ideal setting for a water mill included "a steady stream of water with a good dam ready built, sufficient for a Blacksmith's works, or Mill for grinding Scyths, Chocolate Mill, or Fulling Mill" [4]. Flowing

Chocolate: History, Culture, and Heritage. Edited by Grivetti and Shapiro
Copyright © 2009 John Wiley & Sons, Inc.

water turning a wheel to grind grain into flour had been known to Europeans since the time of the Romans. Mills used exclusively to grind cocoa into chocolate, however, were not part of the American equation until the 19th century. Eighteenth century watermills, on the other hand, could produce chocolate, mustard, tobacco, and oils all at the same time. Chocolate maker Cornelius Roosevelt of New York advertised "chocolate, Lynseed oil and painting-colours &c." [5]. In 1773, a Dorchester, Massachusetts, wool fulling mill and a paper mill[1] were both set up to simultaneously grind chocolate [6]. Pennsylvania surveyor Samuel Preston constructed a chocolate mill also capable of producing beechnut oil. Eighteenth century water mills had a production capacity of about five to six hundredweight of chocolate a day, which could easily account for a large quantity of chocolate for export [7]. The vast majority of chocolate exported coastwise was manufactured in New England. But American consumers could also obtain chocolate from New York and Pennsylvania. Practically none came from Britain.

A common characteristic of American Quakers of the period, that applied to the original Quaker settlers of the Delaware Valley as well, was their industriousness. They were mainly artisans and yeomen from the middling ranks. David Hackett Fischer has written of the Quakers that they created an extraordinary complex industrial economy within a few years of their arrival. One observer reported in 1681 that they:

> have coopers, smiths, bricklayers, wheelwrights, plowrights, and millwrights, ship carpenters and other trades, which work upon what the country produces and for manufacturories.
> . . . There are iron-houses, and a Furnace and Forging Mill already set up in East Jersey, where they make iron. [8]

Quaker religious tenets of toleration for all people resulted in a wide diversity of religions and nationalities immigrating to the Delaware Valley. Their pacifist beliefs ensured good relations with the Native Americans. Sandwiched between the cold New England colonies and the hotter Chesapeake with malarial summers, Pennsylvania and New Jersey attracted large numbers of immigrants, from abroad and from other colonies. The long growing season resulted in Pennsylvania becoming the breadbasket of North America. Wheat was as synonymous with Philadelphia as tobacco with Virginia, or cod and New England. Population growth was explosive. Today, eastern Pennsylvania still hosts a disproportionate number of chocolate manufacturers, some of which are the world's largest (Fig. 30.1).

One might make a general assumption that chocolate making was a trade learned in England or the West Indies and that the technology was imported. Britain did not have a huge chocolate industry so there would not have been many journeymen. England recorded only one chocolate maker in 1776, Joseph

FIGURE 30.1. Wythe kitchen, Colonial Williamsburg (contemporary photograph). *Source:* Courtesy of The Colonial Williamsburg Foundation, Williamsburg, VA. (Used with permission.) (See color insert.)

Fry. Since there were more chocolate manufacturers in North America, the innovations that were here probably were homegrown. Franklin's *The Pennsylvania Gazette* has numerous advertisements describing manufacturing equipment for sale, some of it locally made:

> *May 23, 1765, The Pennsylvania Gazette*
> To be SOLD, and may be entered on immediately, THE Vendue House, belonging to Bowman and Smith, the Corner of Second and South streets, on Society hill; the House if very commodious, and well calculated for the Vendue Business. For Particulars, apply to WILLIAM SMITH, who has likewise to dispose of, an exceeding good Chocolate Mill, double geared, Paschallmake [sic], and almost new, with all Appurtenances necessary for carrying on the Chocolate Business. The Mill may either be worked by a tread Wheel, or by a Horse, and is esteemed as compleat [sic] a Machine for the Purpose, as any in the City. [9]

The Paschall mentioned in the advertisement was Thomas Paschall, a Quaker who advertised himself as a hatter. However, the sign outside his door was a "crosscut saw" [10]. He sold numerous objects made of metal. Whether he was the manufacturer or not cannot be confirmed; however, his name was synonymous with quality metal goods. Smith also described his "Paschallmake" mill turned by a balance weight capable of producing chocolate by the hundredweight. He also offered other chocolate making equipment for sale including a "Roaster, Stand, Fan, Receivers, Kettles, and near a Groce [sic] of Pans, &c. &c." [11]. Other equipment sold in Philadelphia to manufacture chocolate included "a steel chocolate mill" [12], and "a very Neat pair of Steel Rollers, faced with Brass half inch Thick, with Steel Coggs, and Brass cudgeons, the rollers 7 Inches in Length, for the use of grinding Cocoa Nuts" [13]. Rollers of this type were probably fitted to a frame like the spokes of a wheel, each turning on its own axis. These were placed between two millstones, whose motion was imparted by a cogwheel and attached to the horse or treadmill [14].

Philadelphia chocolate makers did not confine themselves to just grinding chocolate. John Browne advertised a mill that could grind chocolate and tobacco at the same time:

> March 8, 1764, The Pennsylvania Gazette
> To be SOLD by the Subscriber, A Chocolate and Tobacco Manufactory, situate in Sixth street between Arch and Market streets, all in good Order for carrying on both Businesses in a very easy Way; the Chocolate Mill and Tobacco Engine both turned by one Wheel, and may grind Chocolate and cut Tobacco at once, likewise all other the Implements for carrying on both Businesses; the Wheel is turned by a Horse at a very easy Draugh [sic]. . . . The Subscriber has likewise a Number of Lots yet remaining on Hand, in the Northern Liberties, situate [sic] on Front, Second, Third, and other adjacent Streets, which he will either sell or lett [sic] on a reasonable Groundrent [sic] for ever. A Plan of the said Lots may be seen at the Subscriber. JOHN BROWNE. [15]

One of the more curious commodities that Philadelphia chocolate makers manufactured besides chocolate was flour of mustard (Fig. 30.2):

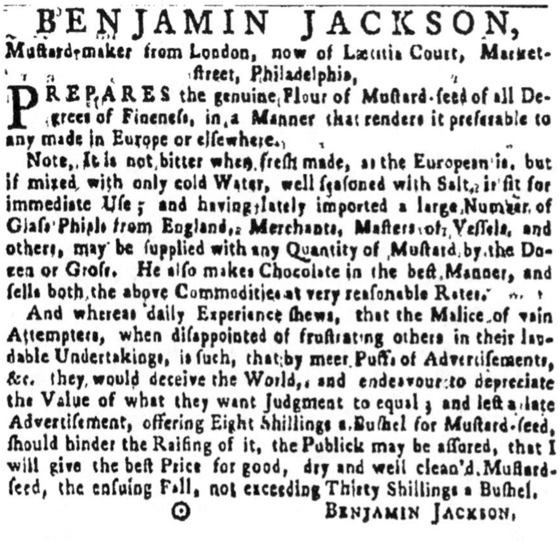

> December 23, 1756.
> Whereas BENJAMIN JACKSON, Mustard Maker from London, now at Laetitia Court, Market Street, Philadelphia, PREPARES the genuine Flower of Mustard seed of all degrees of fineness, in a manner that renders it preferable to any from Europe. Note, it is not bitter when fresh made (as the European is) but if mixed with only cold water, well seasoned with salt, is fit for immediate use. P.S. Said Jackson also makes chocolate in the best manner, and grinds cocoa by the hundred; where merchants, masters of vessels, and others, may depend on being supplied with the above commodities wholesale, or retail, at very reasonable rates. [16]

Jackson was the first Philadelphia mustard and chocolate maker to advertise that his mill was powered by water; however, there may have been others. In December 1759, he advertised "at a very considerable expence [sic], [he has] erected machines, proper for those businesses, at the mill, in the Northern Liberties of this city, formerly known by the name of Governor, alias Globe Mill, where they all go by water" [17]. Over the years, Jackson had many partners, disputes, and breakups . . . all played out in Franklin's The Pennsylvania Gazette. When he and his partners split up, the former partner would offer flour of mustard and chocolate in competition with Jackson. Even Franklin sold chocolate in his print shop. Being the politician he was, he never identified the maker (Fig. 30.3):

> May 22, 1735, The Pennsylvania Gazette
> To be SOLD, BY the Printer hereof, very good Chocolate at 4 s. per Pound by the half Dozen, and 4 s. 6 d. by the single Pound. [18]

Mary Keen Crathorne Roker of Philadelphia

The modern 21st century conception of a chocolate manufacturer is a large-scale candy producer or

BENJAMIN JACKSON,
Muſtard-maker from London, now of Lætitia Court, Market-street, Philadelphia,
PREPARES the genuine Flour of Muſtard-feed of all Degrees of Fineneſs, in a Manner that renders it preferable to any made in Europe or elſewhere.
Note, It is not bitter when freſh made, as the European is, but if mixed with only cold Water, well ſeaſoned with Salt, is fit for immediate Uſe; and having lately imported a large Number of Glaſs Phials from England, Merchants, Maſters of Veſſels, and others, may be ſupplied with any Quantity of Muſtard by the Dozen or Groſs. He alſo makes Chocolate in the beſt Manner, and ſells both the above Commodities at very reaſonable Rates.
And whereas daily Experience ſhews, that the Malice of vain Attempters, when diſappointed of fruſtrating others in their laudable Undertakings, is ſuch, that by meer Puffs of Advertiſements, &c. they would deceive the World, and endeavour to depreciate the Value of what they want Judgment to equal; and left a late Advertiſement, offering Eight Shillings a Buſhel for Muſtard-ſeed, ſhould hinder the Raiſing of it, the Publick may be aſſured, that I will give the beſt Price for good, dry and well clean'd Muſtard-ſeed, the enſuing Fall, not exceeding Thirty Shillings a Buſhel.
BENJAMIN JACKSON.

FIGURE 30.2. Benjamin Jackson: advertisement for chocolate and mustard. *Source: The Pennsylvania Gazette*, April 14, 1757. Courtesy of Readex/Newsbank, Naples, FL. (Used with permission.)

FIGURE 30.3. Handmade chocolate (contemporary photograph). *Source*: Courtesy of The Colonial Williamsburg Foundation, Williamsburg, VA. (Used with permission.) (See color insert.)

confectioner. By contrast, the 18th century conception was more of a miller. By focusing on the milling trade one would be closer to describing the chocolate experience. And while mills were generally run by men, there were exceptions.

Mary Keen was married in 1760 to a retired sea captain who had formed a partnership with Benjamin Jackson three years earlier [19, 20]. Mary Keen's family history can be traced back to the small 17th century colony of New Sweden in the Delaware Valley. In the 1750s, her husband, Captain Jonathan Crathorne, had been involved primarily with the wine trade to southern Europe and the so-called Wine Islands, sailing to and from Lisbon, the Azores, and Madeira with occasional runs to Halifax and Britain [21–24].

Benjamin Jackson was an established mustard and chocolate maker in Philadelphia and began advertising in *The Pennsylvania Gazette* in 1756 [25]. Crathorne's last voyage was reported in 1758 [26]. Jonathan Crathorne had been involved with chocolate and mustard making for two years before he and Mary married. Crathorne and Jackson manufactured their products with "machines . . . at the Globe Mill where they all go by water." Initially, Crathorne remained a silent partner behind the more outspoken Benjamin Jackson:

December 27, 1759, The Pennsylvania Gazette
Whereas BENJAMIN JACKSON, Mustard and Chocolate maker, late from London, now of Laetitia Court, near the lower end of the Jersey Market, in Philadelphia, finding that, by his former method of working those articles, he was unable to supply all his customers, he therefore takes the liberty of thus informing them, and the publick [sic], that he has now, at a very considerable expence [sic], erected machines, proper for those businesses, at the mill, in the Northern Liberties of this city, formerly known by the name of Governor, alias Globe Mill, where they all go by water, altho' he sells them only at his Mustard and Chocolate store, in Laetitia Court, as usual. And as the machines, and method of working both the above commodities, are, great part of them, from an entire new invention (that of mustard particularly) his Flour of Mustard is rendered preferable to the English, Durham, or any other yet made. He manufactures it in different degrees of fineness as in England, and it excels all other for exportation, as it will be warranted to keep perfectly good any reasonable time, even in the hottest climates; and altho' mixing [sic] it with hot water, agreeable to the directions given with each bottle, is the best way, yet in case of neglect, or forgetfulness, if it is mixed with only cold water, well seasoned with salt, is not bitter when fresh made as other Mustard is, but is fit for immediate use so that I believe it may, without vanity, be said to merit the Character it has universally gained, viz. Of being the best Flour of Mustard Seed yet made; and as I am at present the only proper mustard manufacturer in this City, or on this Continent, others being only imperfect imitators of my method; therefore, to prevent deception, all my bottles are

sealed with this City Arms, and this inscription around, B. Jackson Philadelphia Flour of Mustard, as at the Top of this Advertisement. He also prepares Chocolate in the very best manner as aforesaid for the perfecting which no cost nor pains is spared: And as his machines for this Business are also very compleat [sic] no one else in this city has conveniencies [sic] for manufacturing it with equal dispatch; therefore merchants, masters of Vessels, and others, may depend on being, at all times, supplied with any Quantities of that, or flour of mustard at very short warning, and the most reasonable Rates. N.B. Said Jackson likewise sells the best Sallad or eating Oil, Pickles of various sorts, and White Wine and other Vinegars. * * * He has also a small parcel of Books to dispose of, chiefly the writings of some of the first people called Quakers; the whole are in good case, and may be seen at Laetitia Court as aforesaid. [27]

In 1762, Mary gave birth to a son while her husband and his partner continued their partnership [28]. Jackson called the company "Benjamin Jackson and Company." Crathorne and Jackson's partnership continued until 1765. In that year, Mary gave birth to a daughter and Jonathan Crathorne bought out Jackson's interest in the Globe Mill. Jackson went into business for himself and then eventually formed another partnership with John Gibbons [29]. Captain Crathorne became "Jonathan Crathorne, Mustard and Chocolate Maker."

October 31, 1765, The Pennsylvania Gazette
This is to inform the PUBLIC, THAT the Copartnership [sic] of JONATHAN CRATHORNE and BENJAMIN JACKSON is now dissolved, and that Jonathan Crathorne has bought of said Jackson all the Works late belonging to the Company, viz. the Mustard and Chocolate Mills, and all Materials thereunto belonging, in the Northern Liberties, a little Way from Town, on the Germantown Road, and commonly called the Globe Mill, where he continues to manufacture Mustard, Chocolate, Vinegar, &c. and grinds Ginger. Any Person having a Quantity to grind, he will send his Cart for it, and when ground will return it to any Part of the Town. Said Crathrone continues to live at the Old Mustard and Chocolate Store, in Letitia Court, nearly opposite the lower End of the Jersey market, Philadelphia, where he has lived ever since the Year 1759 and acquaints the Public, that the Company Sign is taken from over the old Store Door, in Letitia Court, and another Placed in its Room, with JONATHAN CRATHORNE, Mustard and Chocolate maker, in large Letters, where he continues to sell the genuine Flour of Mustard, of all Degrees of Fineness, in such a Manner that he will warrant it to exceed any from England or elsewhere, and keep good even in the hottest Climates; he sells it in proper Glass Bottles, or any other Package which may best suit the buyer, and at very reasonable Rates, particularly to those who buy a Quantity for Exportation. His Chocolate is manufactured in the best Manner, which he will sell at the most reasonable Rates; likewise good Vinegar, very cheap by the Hogshead, Barrel, and by Retail, having a large Quantity by him; raw West India Coffee, Ditto roasted and ground, to as great Perfection

as in England, Oat Grots [sic], Oatmeal, English and split Pease, mace and ground Ginger, Spices in general, Pickles of sundry Sorts, both foreign and home made, Madeira and Fyal Wines, in Pipes, Hogsheads and Quarter Casks, Lisbon by the Gallon, and very good Claret in bottles, by the Dozen or smaller Quantity. He also sells Mustard, Chocolate, and West India Coffee, ready roasted and ground, at his Mill on Germantown Road, with a Sign over the Door, Jonathan Crathorne Mustard and Chocolate maker. [30]

In April 1767, Mary gave birth to another daughter, and four months later, Jonathan Crathorne died [31]. Mary's oldest child by that time was five years old. She was named the "Administratrix" of her husband's estate. Besides dealing with the emotional loss to herself and her children, she was also expected to deal with creditors and debtors and run the manufacturing and retail business, literally with a newborn on her hip. While it is possible that she might have had a "hidden hand" in her husband's business activities while he was alive, now her activities became public. In English common law, a wife's legal identity, *femme covert*, was submerged in the marriage and the husband represented her in a court of law. A married woman could not make contracts, be sued, pay taxes, or vote. Her property rights were also submerged and were controlled by her husband. She could not buy or sell property. As a widow, however, Mary became a *femme sole* with the right to make contracts, run a business, buy and sell property, sue and be sued, and pay taxes, though she still could not vote. As a married woman, her husband made all the decisions. As a widow, she could make her own.

September 3, 1767, The Pennsylvania Gazette
ALL persons that are any ways indebted to the estate of JONATHAN CRATHORNE, late of the city of Philadelphia deceased, are desired to make immediate payments to the subscriber; and all those who have any demands against said estate, are desired to bring in their accounts, properly attested, that they may be settled and paid, by MARY CRATHORNE, Administratrix. N.B. The mustard and chocolate business is carried on as usual, and the highest price for mustard seed is given being 32 shillings per bushel. [32]

In the next year, she placed the now familiar sounding advertisement extolling the Globe Mill's "incomparable mustard and chocolate works:"

February 11, 1768, The Pennsylvania Gazette
MARY CRATHORN, Begs leave to inform the public (and particularly those that were her late husband customers) that she has removed from the house she lately occupied in Laetitia Court, to the house lately occupied by Mrs. Aris, at the corner of the said court, in Market street, where she continues to sell by wholesale and retail, THE genuine FLOUR of MUSTARD, of different degrees of fineness; chocolate, well manufactured, and genuine raw and ground coffee, tea, mace and ground ginger, whole and ground pepper, allspice [sic], London fig blue, oat groats, oatmeal, barley, rice, corks; a fresh

assortment of spices, domestic pickles, London loaf sugar, by the loaf or hundred weight, Muscovado sugars, choice raisins by the keg or less quantity, best thin shell almonds, olives and capers, with sundry other articles in the grocery way; likewise Madeira, Lisbon and Fyall wines in half pipes and quarter casks, and claret in bottles. As the articles of mustard and chocolate are manufactured by her, at those incomparable mustard and chocolate works at the Globe mill, or Germantown road, which her late husband went to a considerable expence [sic] in the erecting, and purchasing out Benjamin Jackson['s] part; and as she has a large quantity of choice clean mustard seed by her, and the singular advantage of being constantly supplied with that article, she flatters herself, that upon timely notice she can supply any person with large quantities of the said articles of mustard and chocolate, either for exportation, or for retailing again, when a good allowance will be made, and the same put up in any kind of package as may best suit the buyer. N.B. All the mustard put up in bottles, has the above stamp pasted on the bottles, and also the paper round each pound of chocolate has the said stamp thereon; and least any person may be discouraged from bringing small quantities of mustard seed to her, from the singular advantages already mentioned, she therefore informs those persons, that may either have great or small quantities to dispose of, that she will always be ready to purchase of them, and give the highest price. She also has two genteel eight day clocks, London make, with mahogony [sic] cases, which will be disposed of at a reasonable price. [33]

An 18th century mill involved heavy labor and possibly, for a time, Mary Crathorne was assisted by her family, friends, and, in Quaker Philadelphia, probably her neighbors. Her husband's customers might have carried their loyalty over to her, also. But as much as Mary might have "stuck it out" against the odds, she probably did not do the actual millwork singlehanded. Chocolate makers in her situation relied on their foremen as the experienced hands in the actual mechanics of running mills. Philadelphia's William Norton bought out the partnership that the (late) Benjamin Jackson and John Gibbons had established after the break up of Jackson and Crathorne (Fig. 30.4). Norton "purchased the Mills" and intended to "carry on the Business with the same head Workman as manufactured for Jackson and Gibbons" [34]. Perhaps faced with the loss of her experienced foreman, Mary Crathorne searched for a replacement:

Philadelphia, November 30, 1769. A MUSTARD and CHOCOLATE MAKER is wanted, ANY Person, capable of undertaking the above Branches, may hear of Employ, by enquiring of MARY CRATHORN, 4 Doors above the Coffee House, in Market street. [35]

Mary Crathorn disappeared from the pages of *The Pennsylvania Gazette* after that and eventually married her neighbor, merchant Thomas Roker, in October 1771 [36]. Thomas Roker, however, never advertised himself as a chocolate or mustard maker and no

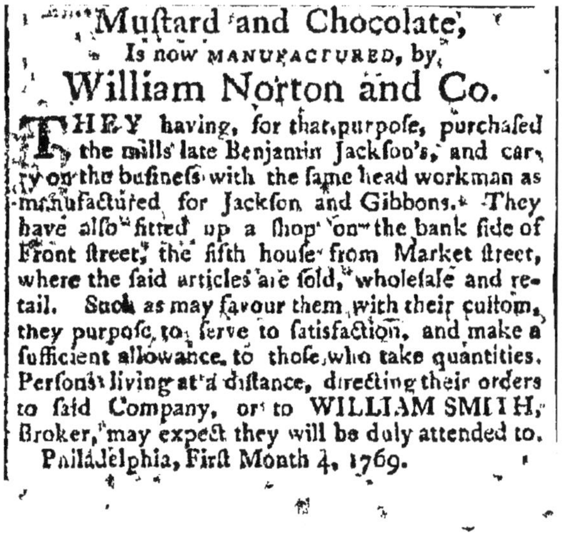

Muſtard and Chocolate,
Is now MANUFACTURED, by,
William Norton and Co.
THEY having, for that purpoſe, purchaſed the mills late Benjamin Jackſon's, and carry on the buſineſs with the ſame head workman as manufactured for Jackſon and Gibbons. They have alſo fitted up a ſhop on the bank ſide of Front ſtreet, the fifth houſe from Market ſtreet, where the ſaid articles are ſold, wholeſale and retail. Such as may favour them with their cuſtom, they purpoſe to ſerve to ſatisfaction, and make a ſufficient allowance to thoſe who take quantities. Perſons living at a diſtance, directing their orders to ſaid Company, or to WILLIAM SMITH, Broker, may expect they will be duly attended to.
Philadelphia, Firſt Month 4, 1769.

FIGURE 30.4. William Norton: chocolate advertisement. *Source: Pennsylvania Chronicle and Universal Advertiser*, January 2, 1769. Courtesy of Readex/Newsbank, Naples, FL. (Used with permission.)

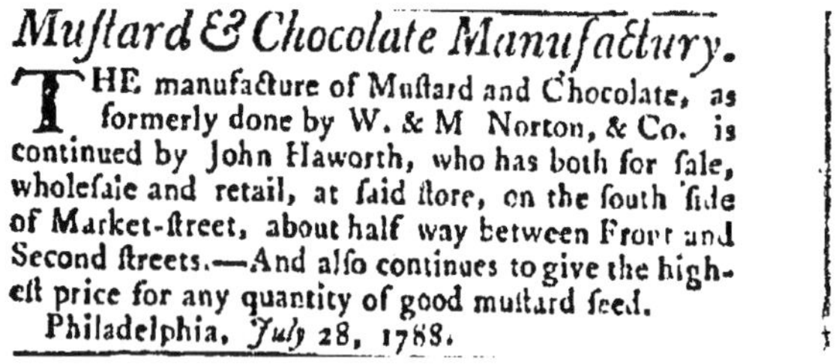

Muſtard & Chocolate Manufactury.
THE manufacture of Muſtard and Chocolate, as formerly done by W. & M Norton, & Co. is continued by John Haworth, who has both for ſale, wholeſale and retail, at ſaid ſtore, on the ſouth ſide of Market-ſtreet, about half way between Front and Second ſtreets.—And alſo continues to give the higheſt price for any quantity of good muſtard ſeed.
Philadelphia, *July* 28, 1788.

FIGURE 30.5. John Hawthorn: chocolate advertisement. *Source: Freeman's Journal or The North-American Intelligencer*, July 30, 1788. Courtesy of Readex/Newsbank, Naples, FL. (Used with permission.)

Philadelphia chocolate maker makes further mention of the "incomparable Globe Mill." Not much is known about Thomas Roker except that he must have been a controversial figure in revolutionary wartime Philadelphia since he was a Tory. In 1778, Roker's estate, including what Mary brought to the marriage, was confiscated by the State of Pennsylvania. Records reveal that Mary died two years later in May 1780, at the age of 52 [37]. The chocolate mill that Benjamin Jackson, Jonathan Crathorne, and Mary Keen Crathorne Roker ran and maintained was not particularly unique in the American chocolate experience. And while we may never know all of the particulars, we can surmise the processes required to make chocolate by looking at a similar circumstance from Rhode Island (Fig. 30.5).

A Look Inside a Mill

In November 1824, auctioneer Edward Shelton placed the following advertisement in the *Providence Gazette*:

> *Nov. 6, 1824*
> *Public Auction*
> *To Be Sold at Public Auction, at Olneyville, in the Brown George Mill, On Monday, the 15th instant, at 11 o'clock, A.M. All the Machinery, Apparatus, Utensals [sic] and Implements generally used for manufacturing Chocolate, consisting of the following articles, which were mortgaged as security for the payment of the rents of the said Brown George Mill, and the mansion house of the late Colonel Christopher Olney, viz: 1 pair Millstones for grinding Chocolate, with the whole apparatus thereto belonging, Frame Gearing, &c., attached to the same; 3 Kettles, with Stirrers, Shafts, and Cogs; 1 other Kettle, for melting Chocolate; 1 cracking machine and Apparatus; 1 Fanner and Apparatus; 1 Roaster with Crane and Apparatus; 6000 tin Chocolate Pans; 2 small Scale Beams and Weights; Set of Press Screws with Bolts; Box, containing stamps and four Plates; Sundry other small articles; 1 large Water Wheel. The whole of the above articles will be positively sold to the highest bidder, without reserve— or so much of them as shall be sufficient to pay the whole rent that is due on both estates. Conditions made known at the time and place of sale. Edward S. Sheldon, Auct'r.* [38]

Auctioneer Sheldon's listing provides valuable insights into the technology of chocolate manufacturing available in the period. Some of the equipment, such as the waterwheel, would have been common to any water mill. However, it is clear that there were specific machines related to various steps in the chocolate making process. Taken in sequence from roasting cocoa to molding chocolate cakes, here is a synopsis of how chocolate was made using the listed machines.

STEP ONE: ROASTING

The "1 Roaster with Crane and Apparatus" in the above advertisement refers to the first step in chocolate making, whether by an Aztec in pre-Columbian Mexico or a 21st century chocolate manufacturer: roasting the cocoa. Roasting can be done in a pan for small amounts. For large amounts, however, roasting was accomplished by turning a cylinder filled with cocoa beans before a fire. These cylinders were probably similar to coffee roasters of the period. Coffee roasters were metal cylinders punctured with holes and were able to hold a pound or two of coffee beans. The cocoa roaster, on the other hand, would have been larger and without holes. Whereas a coffee roaster could be used over a fire, one for roasting cocoa could not. Cocoa is very absorbent of odors of whatever it is near. Roasting cocoa over a hickory fire would result in chocolate that tasted like Virginia ham. This cylinder would have had to hold several pounds of cocoa beans at a time. The roaster would have been attached horizontally to a "crane," which probably refers to a swinging iron king-post bracket common in hearths of the day. This *crane* allowed the roaster to be brought away from the heat for loading and unloading. It is not clear what the "Apparatus" was but it was possibly some type of device to assist in turning the roaster. Conceivably, this *apparatus* could have connected the roaster to the waterwheel or perhaps was a stand-alone spit jack. A description of a Massachusetts chocolate mill said "the machinery for roasting, cracking and fanning the cocoa was run by chains from horizontal shafts. The noise and din of such machinery was indescribable" [39, 40]. However, it did not have to be that elaborate. Other period sales advertisements for cocoa roasters and business records simply referred to "stands and spits" [41, 42].

STEP TWO: CRACKING AND WINNOWING

The next process was shelling the cocoa. Cacao seeds when fresh have a white outer membrane that turns brown during fermentation. When dry, this membrane is referred to as the shell. After roasting, this shell becomes brittle and is removed. The process resembles removing the skin from a peanut. When the shells are removed, the kernels called "nibs" are ground. The "cracking machine and Apparatus" referred to in the advertisement accomplished the first step of the process, which was to crack the shell. The "Fanner and Apparatus" then winnowed the shell with moving air. Cocoa shells could have been collected and sold separately. These machines were very likely invented and made in America. Six decades earlier in 1767, Boston chocolate maker Henry Snow advertised "A Machine, the newest that has been made in Boston . . . a Cleaner of Cocoa for Chocolate, fit to grind 500 wt. in Ten Hours. This Mill is warranted to 14 wt. in Two Hours" [43].

STEP THREE: GRINDING

The cocoa nibs would then be ground into cocoa liquor. Cacao seeds contain between 40 and 60 percent fat, called "cocoa butter." When the nibs are ground, they turn to a liquid. This was accomplished with "a pair [of] Millstones for grinding Chocolate. . . ." Since this was a water mill, and attached to "one Large Waterwheel," the work could have gone on for several hours, with the chocolate maker finishing one batch, then another, and another . . . until done. Probably the "Frame Gearing . . . [was] attached . . . to" an opposing "pair [of] Millstones" and was set inside a circular tub that held a measured quantity of cocoa nibs. The chocolate maker would weigh the nibs using "2 small Scale Beams and Weights." The longer the grinding went on, the smoother and "finer" the resulting chocolate. After the chocolate was ground, the batch would be put into one of "3 Kettles." The chocolate in these kettles was maintained in the liquid state by "Stirrers" connected to "Shafts, and Cogs." This allowed the more volatile flavors in the chocolate mass

to evaporate, making the chocolate flavors mellower. Additionally, this kept the chocolate in a liquid state for molding. An alternative usage for these kettles would be to keep the chocolate nibs warm when they came out of the roaster, which would intensify their flavor. Whatever the process used, it could not be rushed. If the process continued to the next day and the chocolate solidified, then the "Kettle, for melting Chocolate" would be used to bring it back to liquid.

STEP FOUR: MOLDING

When the kettles were filled, the chocolate liquor was molded in some of the "6000 tin Chocolate Pans." While still wet, the chocolate probably was "stamped" or embossed with the chocolate maker's name or emblem. The shape of these tins could have been arbitrary. However, there were many references to a chocolate "cake" being one pound. Additionally, advertisements for chocolate makers Benjamin Jackson, and John and Mary Crathorne depicted rectangular shaped chocolate cakes, wrapped in paper and stamped with their name on it [44].

STEP FIVE: WRAPPING

The wrapping paper might have been included in the "Sundry other small articles" notation. Little is known about the paper used to wrap the product, but one can expect it to be rather plain in appearance compared to modern packaging. While wrapping obviously was done by hand, the fact that there were no sanitary wrapping systems and no concept of bacterial contamination, means chances for off flavors were high. Additionally, the wooden boxes used to ship the chocolate might have added to the off flavors if made of fresh cut pine or cedar.

Only limited examples of such equipment have survived to be examined by modern researchers. Chocolate mills, unless made out of brick or stone, were prone to destruction by fire or lightening. While place names like Valley Forge or Owings Mills give hints to our nations preindustrial past, few such sites remain undisturbed. Mills made of wood were notorious for being struck by lightning and burning, as generally they were the largest and highest structures in a given area and therefore prone to receiving lightening bolts. Fires also could be started by equipment malfunction; large brakes overheating or the friction of stones rubbing together sometimes generated enough heat for combustion [45]. Eighteenth century mills still standing today most probably were constructed of brick or stone. Additionally, cocoa required roasting, and roasting required fuel. Furthermore, stoves were needed to heat the workers during Pennsylvania winters. The fuel used was coal or possibly charcoal [46]. Highly combustible cocoa shells accumulated more rapidly than they could be removed or sold and would have served as a ready source of fuel during an accidental fire. A chocolate mill in Dorchester, Massachusetts, burned in 1775 [47], and another in 1782 [48]. In 1786, the Massachusetts Legislature passed a law entitled "Chocolate Mills, To prevent Danger Therefrom by Fire&c. . . . An Act to prevent Damage from Fire being communicated from Chocolate Mills and Machines for roasting Cocoa in the Town of Boston" [49]. In 1787, Boston chocolate maker/ merchant/miller Jonas Welsh gave a mortgage to another miller for two remaining grist mills after five other structures had been consumed by fire [50]. Finding the site of an 18th century chocolate mill left undisturbed by urbanization, therefore, would not be likely—but potentially possible in a rural setting. Even then, the ravages of time would not be kind to whatever remained to be discovered.

Acknowledgments

I would like to acknowledge the following institutions and individuals whose cumulative guidance and support have been instrumental in writing this chapter. Associated with Colonial Williamsburg Foundation: Barbera Scherer, Robert Brantley, Dennis Cotner, Susan Holler, and Frank Clark, Historic Foodways; Jay Gaynor, Director of Trades; Gayle Greve, Curator of Special Collections Library; Juleigh Clarke. Colonial Williamsburg Foundation Rockefeller Library: Pat Gibbs and Martha Katz-Hyman, Research; Donna Seale, Nicole Gerding, and Kristi Engle, Historic Foodways. Fort Ticonderoga, New York: Nick Westbrook and Virginia Westbrook. Masterfoods USA/Mars, Incorporated: Rodney Synder, Doug Valkenburg, and Mary Myers. Oxford, England: Pamela Richardson. Family members: Bill Gay, Andrea West, and Judy Whitacre.

Endnote

1. The fulling mill was owned by Edward Preston and the paper mill by Jeremiah Smith.

References

1. Taylor, A. *American Colonies*. New York: Penguin Books, 2001; p. 267.
2. Massachusetts Naval Office Shipping Lists, 1686–1765, PRO, Reel 1, 1979.
3. Massachusetts Naval Office Shipping Lists, 1686–1765, PRO, Reel 1, 1979.
4. *The Boston News-Letter and New-England Chronicle*. March 17, 1768, p. 4.
5. *New York Gazette*, or *Weekly Post Boy*, published as *New York Gazette Revived in the Weekly Post-Boy*. July 7, 1752, p. 3.
6. James Baker Records, Box 1, Folder 2, held at the American Antiquarian Society, Worcester, MA.
7. James Baker Records, Box 1, Folder 19, held at the American Antiquarian Society, Worcester, MA.
8. Fischer, D. H. *Albion's Seed*. New York: Oxford University Press, 1989; p. 560.
9. The Accessible Archives CD-ROM *The Pennsylvania Gazette* Folio II 1751–1765. Malvern, PA: Accessible Archives, Inc., 1993; #35883.
10. Accessible Archives CD-ROM *The Pennsylvania Gazette* Folio II 1751–1765. Malvern, PA: Accessible Archives, Inc., 1993; #20389.
11. The Accessible Archives CD-ROM *The Pennsylvania Gazette* Folio III 1766–1783. Malvern, PA: Accessible Archives, Inc., 1993; #37935.
12. The Accessible Archives CD-ROM *The Pennsylvania Gazette* Folio III 1766–1783. Malvern, PA: Accessible Archives, Inc., 1993; #62632.
13. The Accessible Archives CD-ROM *The Pennsylvania Gazette* Folio II 1751–1765. Malvern, PA: Accessible Archives, Inc., 1993; #27929.
14. Coe, S. D., and Coe, M. D. *The True History of Chocolate*. New York: Thames and Hudson, 1996; p. 231.
15. The Accessible Archives CD-ROM *The Pennsylvania Gazette* Folio II 1751–1765. Malvern, PA: Accessible Archives, Inc., 1993; #32819.
16. The Accessible Archives CD-ROM *The Pennsylvania Gazette* Folio II 1751–1765. Malvern, PA: Accessible Archives, Inc., 1993; #34034.
17. The Accessible Archives CD-ROM *The Pennsylvania Gazette* Folio II 1751–1765. Malvern, PA: Accessible Archives, Inc., 1993; #24206.
18. The Accessible Archives CD-ROM *The Pennsylvania Gazette* Folio I 1728–1750. Malvern, PA: Accessible Archives, Inc., 1993; #1881.
19. U.S. and International Marriage Records, 1560–1900 (database on-line). Provo, Utah: The Generations Network, Inc., 2004. Accessed through Ancestry.com. http://search.ancestry.com/cgi-bin/sse.dll?indiv=1&db=WorldMarr_ga%2c&rank=0&gsfn=Keen&gsln=Mary&sx=&gs1co=1%2cAll+Countries&gs1pl=1%2c+&year=1728&yearend=1780&sbo=0&srchb=r&prox=1&ti=0&ti.si=0&gss=angs-d&o_iid=21416&o_lid=21416&o_it=21416&fh=1&recid=291165&recoff=14+15+16+27. (Accessed February 13, 2007.)
20. *Pennsylvania Magazine of History and Biography*, Volume 4, p. 493.
21. Keen, G. B. *The Descendants of Jöran Kyn of New Sweden*. Philadelphia: Swedish Colonial Society, 1913; p. 95.
22. The Accessible Archives CD-ROM *The Pennsylvania Gazette* Folio II 1751–1765. Malvern, PA: Accessible Archives, Inc., 1993; #1765.
23. The Accessible Archives CD-ROM *The Pennsylvania Gazette* Folio II 1751–1765. Malvern, PA: Accessible Archives, Inc., 1993; #7233.

24. The Accessible Archives CD-ROM *The Pennsylvania Gazette* Folio II 1751–1765. Malvern, PA: Accessible Archives, Inc., 1993; #8492.

25. The Accessible Archives CD-ROM *The Pennsylvania Gazette* Folio II 1751–1765. Malvern, PA: Accessible Archives, Inc., 1993; #34034.

26. The Accessible Archives CD-ROM *The Pennsylvania Gazette* Folio II 1751–1765. Malvern, PA: Accessible Archives, Inc., 1993; #9963.

27. The Accessible Archives CD-ROM *The Pennsylvania Gazette* Folio II 1751–1765. Malvern, PA: Accessible Archives, Inc., 1993; #24206.

28. West, E., compiler. *Family Data Collection-Individual* Records (database on-line). Provo, UT: The Generations Network, Inc., 2000; accessed through Ancestry.com.

29. The Accessible Archives CD-ROM *The Pennsylvania Gazette* Folio III 1766–1783. Malvern, PA: Accessible Archives, Inc., 1993; #40685.

30. The Accessible Archives CD-ROM *The Pennsylvania Gazette* Folio II 1751–1765. Malvern, PA: Accessible Archives, Inc., 1993; #37058.

31. West, E., compiler. *Family Data Collection-Individual* Records (database on-line). Provo, UT: The Generations Network, Inc., 2000; accessed through Ancestry.com.

32. The Accessible Archives CD-ROM *The Pennsylvania Gazette* Folio III 1766–1783. Malvern, PA: Accessible Archives, Inc., 1993; #40956.

33. The Accessible Archives CD-ROM *The Pennsylvania Gazette* Folio III 1766–1783. Malvern, PA: Accessible Archives, Inc., 1993; #41937.

34. The Accessible Archives CD-ROM *The Pennsylvania Gazette* Folio III 1766–1783. Malvern, PA: Accessible Archives, Inc., 1993; #43887.

35. The Accessible Archives CD-ROM *The Pennsylvania Gazette* Folio III 1766–1783. Malvern, PA: Accessible Archives, Inc., 1993; #45829.

36. West, E., compiler. *Family Data Collection-Individual* Records (database on-line). Provo, UT: The Generations Network, Inc., 2000; accessed through Ancestry.com.

37. *Pennsylvania Magazine of History and Biography*. Volume 4, p. 493.

38. *Providence Gazette*. November 6, 1824, Vol. LXIII, Issue 3344, p. 3.

39. *History of Essex County, Massachusetts: With Biographical Sketches of Many of Its Pioneers and Prominent Men, Volume 1, Cities and Towns*. Chapter XXVII, Saugus, Massachusetts, p. 409.

40. Ancestry.com database on-line. Provo, UT: The Generations Network, Inc., 2005.

41. The Accessible Archives CD-ROM *The Pennsylvania Gazette* Folio III 1766–1783. Malvern, PA: Accessible Archives, Inc., 1993; #37935.

42. James Baker Records, Box 1, Folder 15, held at the American Antiquarian Society, Worcester, MA.

43. *The Boston Gazette and Country Journal*. November 30, 1767, p. 1.

44. The Accessible Archives CD-ROM *The Pennsylvania Gazette* Folio III 1766–1783. Malvern, PA: Accessible Archives, Inc., 1993; #41937.

45. Williamsburg Craft Series. *The Miller in Eighteenth Century Virginia*. Williamsburg, VA, Colonial Williamsburg Foundation, 1958; p. 26.

46. James Baker Records, Box 1, Folder 5, held at the American Antiquarian Society, Worcester, MA.

47. James Baker Records, Box 1, Folder 1, held at the American Antiquarian Society, Worcester, MA.

48. Dorchester Antiquarian and Historical Society, History of the Town of Dorchester, Massachusetts, Ebenezer Clapp, 1859, p. 638.

49. The by-laws and town-orders of the town of Boston, made and passed at several meetings in 1785 and 1786. And duly approved by the Court of Sessions. 1786. Early American Imprints, Series I. Evans (1639–1800), No. 19515(filmed).

50. *Twing Index of Early Boston Inhabitants*. Boston: Massachusetts Historical Society, 2005.

Breakfasting on Chocolate

Chocolate in Military Life on the Northern Frontier, 1750–1780

Nicholas Westbrook, Christopher D. Fox, and Anne McCarty

Introduction

During the spring of 1777, on a cloudy, windy, raw day in April, 22-year-old Captain Moses Greenleaf began his second full day at Fort Ticonderoga by brewing himself a breakfast cup of hot chocolate. He needed the energy. Greenleaf was defending Ticonderoga as part of a minuscule army of fewer than 3000 men, half of them sick, protecting the fragile attempt at American democracy, then less than a year old, against an invading British and German army of more than 9000 men. At that time, Greenleaf had been married less than nine months and kept a diary[1] of his wartime experiences to share with his new family [1, 2].

During the previous 110 years since the introduction of chocolate to the English colonies in North America, chocolate, too, had become part of the democratizing consumer revolution of the 18th century [3]. For Greenleaf's generation, the once-exotic had become commonplace. In this chapter we examine the broad penetration of chocolate into the many 18th century worlds spanning the northeastern portion of the United States and Canada, with particular emphasis on military and Native American uses.

In 1697 Dorchester, Massachusetts, the highest-ranking Crown representative in the colony, Lieutenant-Governor William Stoughton, and a leading Congregational cleric, Rev. Samuel Sewell, made "breakfast together on Venison and Chockalatte." Sewell recorded his reaction in his diary: "I said Massachusetts and Mexico met at his Honour's table" [4]. Here were men at the pinnacle of New England society "breakfast-ing on chocolate"—at that time, an exotic product enjoyed largely by the upper strata of society.

Reverend Sewell's witticism also reminds us that chocolate had been a familiar drink among indigenous peoples in Central and South America for centuries before Spanish explorers brought it to Spain in the 16th century. Slowly, chocolate migrated across Europe and then back across the Atlantic to the British colonies more than a century later. By the 18th century, documents record the increasingly widespread use of chocolate among both Euro-Americans and Native Americans in matters of diplomacy and personal consumption.[2]

Half a century later, chocolate was an essential part of an army officer's life during the Seven Years' War in North America (also known as the French & Indian War). In the spring of 1755, as General Braddock's British army began its disastrous march westward toward the Forks of the Ohio, Benjamin Franklin provided a special supplement of sugar, tea, coffee, vinegar, cheese, Madiera, Jamaican spirits, mustard, and "6 lbs. of chocolate" for each subaltern officer [5]. In March 1757, as a French and Indian

Chocolate: History, Culture, and Heritage. Edited by Grivetti and Shapiro
Copyright © 2009 John Wiley & Sons, Inc.

force of 1200 men prepared to move out from Fort Carillon (Ticonderoga) to attack the British Fort William Henry, the raiders were issued rations for 12 days. The French and Canadian officers were issued an additional "three pints of brandy and two pounds of chocolate," the former to fuel their spirits, the latter for energy [6].

In late July 1758, a 400-man French and Indian war party ambushed a British supply train of 60 ox-carts loaded with supplies, destined for General James Abercromby's army encamped at the head of Lake George, and accompanied by an escort of 150 troops. Thirty teamsters and soldiers were massacred, 250 oxen killed, three women slaughtered and 20 more missing. The provisions were plundered. Supplies that could not be carried away by the raiding party were smashed to the ground. Decades later, New England soldiers remembered the sight of the destroyed crates and barrels, the mutilated corpses, and the rivulets of melting chocolate mingling with rivers of blood. In the annals of this final North American war between Britain and France, this was remembered as the "Chocolate Massacre" [7, 8].

Those Indian warriors sought chocolate as an energy food as well. In late August 1757, just several weeks after the successful French siege of Fort William Henry (and subsequent "massacre" of some of the British garrison) [9], British provincial ranger Captain Israel Putnam led a 12-day scouting party to examine the strength of French forces at Carillon. When the raiding party returned, Putnam was forced to leave behind "a negro sick in the woods and two Indians to look after him." A rescue party of 40 men was sent out to recover the abandoned three. The rescue team "met the two Indians that were left with the negro. . . . [but] As one of them was some way off boiling some cocolatt [*sic*]," they were discovered by the enemy; the two Indians made their escape "and left the negro." The rescuers apparently made no further effort to locate the sick black man [10].

A generation later, during the American Revolution, chocolate was more commonly available. Soldiers like Moses Greenleaf "breakfasted on chocolate." On his way to Fort Ticonderoga in August 1776 to help stop the impending British invasion from the north, Chaplain William Emerson[3] of Concord, Massachusetts, wrote plaintively to his wife, thinking about supplies he forgot to pack. "I hope my dear you will not forget to send me a little Skillet and 2 or 3 lb. of Chocolat,—and my baiz Waisscoat by the first Opportunity." A week later on his arrival at "the Fort at Tyconderoga," Rev. Emerson reported the simple joys of his new military life:

Dear Phebe . . .
This morning I breakfasted just as I would at Home, my Porringer of Chocolate was brought in, in as much Order as

need be. We live in as comfortable Apartment of the barracks as any Room in our House (at present). We have another Apartment where the Servants keep and dress our Provision and bring it in à la mode Pierre.

Reverend Emerson's homey pleasures did not last long. Three weeks later, he had succumbed to "nothing more than a Sign of mongrell Feaver & Ague . . . The Nature of all these Camp Diseases strikes at the very Life and Spirit of a Soldier" and was buried in West Rutland, Vermont, on his journey home to family [11].

Chocolate in Hospitality and Diplomacy

Gift-giving to create a reciprocal obligation was an essential strategy in hospitality and diplomacy. Chocolate soon became an effective tool in New World diplomacy. In May 1698, the War of the League of Augsburg (called King William's War in the British colonies) had been over for a year. The Dutch *domine*, Reverend Godfrey Dellius, and Col. Peter Schuyler were sent by the Earl of Bellomont, Royal Governor of New York province, to Canada to deliver news of the peace to the French governor and to exchange prisoners held by both sides [12]. Their Canadian hostess served them fresh salmon; Dellius and Schuyler boasted of New York oysters, each bragging about resources not available in the other place. Dellius and Schuyler went on to praise "the lemons, oranges, cacao nuts, and such like fruits, which [were] sent thither from Barbados, Curaçao, and other places."[4] New York's ambassadors had taken a few specimens with them as gifts, but had already dispersed them; they promised to send some "at the first opportunity." A year later, their hostess in Québec wrote her erstwhile guests an exquisitely sarcastic letter: "I greatly doubt whether you have thought of it, for I believe that Mr. Dellius is too gallant not to have done for me what I did for him that day, with so much pleasure." Such neglect germinates the seeds of war. Three years later, war had broken out again between France and England, and across the frontiers of New France and New England [13].

The failure to respect the *necessity* of reciprocity in gift exchange always caused problems. In the 1770s and 1780s, as the Moravian United Brethren moved westward into the Ohio country, missionary John Heckewelder noted their sequential gift-exchanges with the Delaware (Lenapi) in the valley of the Tuscarawas river in east-central Ohio. "We presented them with some flour & as they sat down quietly beside us & smoked their Gilickinik, we regarded each other as silent guests. . . . In the afternoon the Indians left us. To [their leader] we presented some chocolate, sugar &c." Later, during a time of great famine in 1782, several Wyandot (or "Huron") Indians, including the "notorious" Simon Girty,[5] entered the house of one of the Moravian missionaries and began helping themselves to the available food. The next time that happened, the hungry Wyandot were emphatically reminded by the also-hungry converted Indian Brethren, "When you came ... with Your Men, we gave You not only enough to eat of such Provisions, which the Indians generally make use of; but we supplied You with anything You wished for that we had; Bread, Pork, Butter, Milch [sic], Sugar; Tea, Chocolate &." [14].

Greater diplomatic *savoir faire* led to great success in New York's Mohawk Valley on the New York frontier. In 1738, a young Irishman named William Johnson established himself as a merchant and fur trader dealing primarily with the *Haudenosaunee* (Iroquois) of the Six Nations. Johnson's honesty and compassion toward them gained him their trust. As a result he formed a special relationship with Indians, particularly the Mohawks, which would eventually lead to his involvement in politics and military affairs. In 1755, William Johnson acquired the title of Baronet and the following year he was appointed Superintendent of Indian Affairs, a position he held until his death in 1774 [15–17].

A compilation of invoices, accounts, and letters indicate that chocolate was a regular staple at Fort Johnson and then at Johnson Hall (Fig. 31.1), Sir William Johnson's baronial estate, where hundreds of Indians gathered to attend the Indian councils frequently held there. "I have ... every Room & Corner in my House Constantly full with Indians" [18]. The Indians were fed, sheltered, and provided with presents. Typical supplies ordered by Johnson included rum, wine, tea, loaf sugar, muscavado sugar, candles, salt, allspice, cinnamon, nutmegs, cloves, mace, 'Raisins of the Sun,' vinegar, cider and beer ... 28 # Coffe,— & 50 # Chacolite ..." [19]. Sir William Johnson's Mohawk wife, Molly Brant, also played a diplomatic role managing relationships between the British Crown and the Indians. Together they entertained numerous visitors to Johnson Hall on a regular basis [20]. Judge Thomas Jones, Justice of the Supreme Court of New York province, described a visit to Johnson Hall:

> The gentlemen and ladies breakfasted in their respective rooms, and, at their option, had either tea, coffee, or chocolate. . . . After breakfast, while Sir William was about his business, his guests entertained themselves as they please . . . until the hour of four, when the bell punctually rang for dinner, and all assembled. . . . Sometimes seven, eight, or ten, of the Indian Sachems joined the festive board. [21]

Even those who could not join "the festive board" asked for chocolate. In a letter to Sir William, "Isaac the Indian" wrote: "Sarah the wife of Isaac Gives her kind

FIGURE 31.1. *The Council Fire at Johnson Hall*, by Edward Lamson Henry; dated 1903. In this painting, Sir William Johnson convenes an Indian council at Johnson Hall, Britain's foremost site in North America for negotiating with the Indians. *Source*: Courtesy of the Albany Institute of History and Art, Albany, NY. (Used with permission.) (See color insert.)

love to your honour and Desires the favour of a little Chocolate if you please" [22]. An inventory of Johnson Hall taken in 1774 shortly after the death of Sir William Johnson included "Articles in Store under the old Study 1 Box almost whole of Chocolate [valued at] 1[£] 10 ..." [23].

Oh those ever-hospitable Irishmen! On May 10, 1775, Ethan Allen, a Green Mountain Boy from Vermont, became one of the first certifiable rebels against Britain when he and Benedict Arnold and a modest force of 83 Green Mountain Boys captured Fort Ticonderoga in America's first victory of the Revolution. By September 1775, however, Allen was a prisoner of war; by November he was on his way to Britain with 33 other rebels taken with him at Montréal. Allen was imprisoned at Pendennis Castle in Cornwall, where he became a celebrity because of his size, exotic "Canadian" clothing, and "harangues on the impracticability of Great Britain's conquering the then-colonies of America." People came from 50 miles around to see Allen and to quiz him endlessly: "One of them asked me what my occupation in life had been? I answered him, that in my younger days I had studied divinity, but was a conjurer by passion. He replied, that I had conjured wrong at the time I was taken; and I was obliged to own ... that I had conjured them out of Ticonderoga." The British authorities quickly realized that Allen was too hot to handle at home. So, just weeks after he had been landed at Falmouth, he was placed onboard ship again in January 1776 to be sent back to the American colonies. HMS *Solebay* sailed for a fleet rendezvous at Cork, Ireland. There Allen was again greeted as a celebrity. Anti-British sympathizers immediately sent him baskets of clothing and supplies, shirts,

ALAMEDA FREE LIBRARY

shoes, beaver hats, "wines of the best sort, old spirits, Geneva, loaf and brown sugar, coffee, tea and chocolate . . . a number of fat turkies [sic], with many other articles . . . too tedious to mention here" [24–26]. Allen, ever the egomaniac, accepted their tribute as only his due. He missed the implicit message that this welcoming gift of chocolate and sundries came spiced with the equally fierce anti-British politics of others!

Supplying the Armies with Chocolate

In the earliest years, chocolate moved from importers and colonial manufacturers on the Atlantic coast out through a network of petty merchants to frontier traders. As warfare intensified during the latter half of the 18th century, and temporary frontier (military) populations became as great or greater than those located in the largest urban centers, new, complex, and temporary supply chains had to be created. During half-a-century of warfare, French, British, and American army establishments struggled to keep supplies moving forward. Sutlers supplemented the official commissary departments to meet the broad needs of troops in the field. Then, as shortages occurred, particularly during the Revolution, state governments worked to control spiraling costs and to impose embargoes against export. In the meantime, privateers bringing in prizes laden with cacao or chocolate were welcomed.

Numerous newspaper advertisements testify to the increasing availability of chocolate in colonial cities, especially such major ports as Boston, New York, and Philadelphia [27–30]. The most important cities for supplying the northern frontier were New York and Albany—a deep-water port located on a fjord penetrating into the heart of the contested frontier between British and French territories in North America. Significantly, Albany was located just a couple of hundred miles by water (only three major portages away) from the second major French city, Montréal, to the north; just under 300 miles away to the west, mostly by water, lay the major French supply base for the Great Lakes fur trade, Niagara. As modern readers, we are perhaps not alarmed to discover a globalizing economy in furs, chocolate, and other commodities breaking down protonational barriers. An active, semilegitimate smuggling trade moved commodities across contested borders in northeastern North America throughout the 18th century, and beyond [31–33].

Thanks to competitive pricing, quality of goods, and aggressive entrepreneurs (particularly William Johnson), during the 18th century an increasing amount of the northeastern exchange of furs and manufactured goods funneled toward Albany (Fig. 31.2) [33, 34]. Albany merchant and sometime mayor Robert Sanders recorded in "execrable French" in his ledgers the trade goods (including chocolate) he exchanged for smuggled furs arriving from Montréal [35].

William Johnson held the British government contracts to supply the modest trading posts and garrisons at Oswego (April 1746, in the aftermath of King George's War) and Fort Bull (October 1755, at the outset of the Seven Years' War). All of that investment in supply inventory and an enormous volume of trade was lost in 1756 with the rapid fall of Fort Bull

FIGURE 31.2. Cartouche inset from *A Map of the Inhabited Part of Canada from the French Surveys; With the Frontiers of New York and New England*, by C. J. Sauthier. Published by William Faden, London; dated 1777. A symbolic representation of trade at one of the strategic portages beyond the frontier. *Source*: Courtesy of Fort Ticonderoga, Ticonderoga, NY. (Used with permission.)

(March 27) and Oswego (August 10). The French accounts of the destruction of the new but undermanned Fort Bull are dramatic: After looting everything portable from the magazines at the little fort, French commander de Lery ordered the utter destruction of the remainder. Splintered barrels and crates spilled flour, sugar, and brine from pork barrels; even cold chunks of chocolate and butter were thrown into the river [36–38].

By the mid-18th century, chocolate was an affordable staple for nearly all classes of American society. Advertisements in early American newspapers regularly include chocolate as one of the many goods "to be sold at the most affordable prices." So inexpensive and popular was chocolate that, even in the military, it was regarded as a regular nourishing staple among all ranks.

Soldiers in the British or American provincial armies in the 18th century had access to a variety of goods from regimental sutlers or merchants licensed to sell alcohol and other provisions to the army. Each sutler was allowed to sell his goods to only one regiment. Like any business of the period, sutlers kept regular account books to keep tabs on individual soldiers' accounts. Few sutler account books have survived into the 21st century. Those that have are remarkable documents and provide scholars with a unique view of the wide variety of goods available to the common soldier stationed at even the most remote frontier outposts. Sutlers' account books record that alcohol was one of the most commonly purchased commodities by soldiers, but at times, chocolate is a clear close second.

One of the rare sutler account books known to have survived documents the purchases of soldiers in Captain Edward Blake's Company of Colonel Joseph Ingersoll's Massachusetts provincial regiment. This account book was kept at Crown Point, New York, by the Boston-based partnership of William Tailer and Samuel Blodgett during the period of November 20, 1762 to June 1763 (Fig. 31.3) [39]. Records documenting Tailer and Blodgett's sutlery are very scant. Apart from this single account book, their work as military sutlers is documented by only a few Boston newspaper advertisements. It appears from newspaper accounts[6] that Tailer and Blodgett served Colonel Ingersoll's regiment as early as 1761 [40–42]. The latest accounts recorded in the Tailer and Blodgett account book date to June 1763. It is likely that when the Massachusetts provincial regiments were disbanded at the conclusion of the Seven Years' War in 1763, so ended the need for the services of Tailer and Blodgett as sutlers. In November 1763, their partnership was dissolved[7] and the company's accounts were left in the care of Samuel Blodgett [42].

C-013314

FIGURE 31.3. *A South View of Crown Point*, by T. Davies; dated 1759. *Source*: Courtesy of the Library and Archives of Canada, Ottawa, Ontario. (Used with permission.)

FIGURE 31.4. Chocolate purchases by George Prince Eugane, dated 1762–1763. *Source:* Tailer & Blodgett Account Book, Manuscript Number M-2178. Courtesy of Fort Ticonderoga, Ticonderoga, NY. (Used with permission.)

FIGURE 31.5. Chocolate pot (copper), dated ca. 1750–1780. *Source:* Courtesy of Fort Ticonderoga, Ticonderoga, NY. (Used with permission.) (See color insert.)

Tailer and Blodgett's account book itemizes the sale of goods to 60 soldiers and officers at Crown Point (Fig. 31.4). Over half of the entries include the sale of chocolate. The quantities of chocolate purchased by the men on any given date vary widely. Units of weight are expressed in relative terms. Chocolate quantity is usually described in terms of the amount of chocolate being purchased and is expressed as "weight." While the term "weight" is not by itself quantifiable, it indicates that a balance scale was used to measure a quantity of a product.[8] In this case, one "weight" most likely indicates a mass of one pound. One "weight" or pound of chocolate was priced by Tailer and Blodgett at three shillings.[9] The smallest quantity documented in the account book is noted as having a value of nine pence— this equates to about four ounces. The largest single purchase, four and a half "weight" (pounds) is valued at thirteen shillings, six pence. Often, chocolate is listed as being sold by the cake. In the English colonies of North America, chocolate was generally available in block or "cake" form. Occasionally, advertisements describe chocolate in "ball" form, but this seems to have been much more common in New France.[10] A rare illustration of chocolate in cake form appears in an advertisement for Benjamin Jackson, mustard and chocolate maker, in the August 10, 1758 edition of *The Pennsylvania Gazette.* It illustrates a rectangular block with molded seams similar to modern, commercially produced, chocolate bars. The seams are presumably intended to facilitate easy breakage of the cake into eight smaller, more manageable squares. Chocolate described in military sutler account books of the period of the Seven Years' War is almost always noted as being sold in cake form. Cakes of chocolate sold by Tailer and Blodgett are always priced at one shilling, six pence each—exactly half the cost of one pound of chocolate. Therefore, the weight of a single cake of chocolate can be documented as one-half pound.

FIGURE 31.6. How to "breakfast on chocolate." Phil Dunning, Parks Canada, shaving chocolate off the block to begin the process (contemporary photograph, June 2004). *Source:* Courtesy of Fort Ticonderoga, Ticonderoga, NY. (Used with permission.) (See color insert.)

In domestic settings chocolate was generally prepared in specifically designed "chocolate pots." A 1675 recipe for making chocolate describes a typical chocolate pot as "a deep pot of Tin, copper or stone[ware pottery] with a cover with a hole in the middle of it, for the handle of the mill to come out at" (Fig. 31.5) [44]. Published recipes for the preparation of chocolate describe a relatively simple process. Typically, chocolate would be shaved or grated into a pot into which was then poured hot water (Fig. 31.6). Using a wooden mill inserted through the hole in the cover of the pot, the chocolate and water mixture was whisked to melt the chocolate into liquid suspension (Fig. 31.7). Sugar, spices, and a variety of other flavor-enhancing materials were added to taste and the chocolate was ready for consumption. In the military, however, these specialized pots must have been a luxury item. Unfortunately, soldiers' journals are

FIGURE 31.7. How to "breakfast on chocolate." Phil Dunning, Parks Canada, and Christopher D. Fox, Fort Ticonderoga, whisking a froth on breakfast chocolate and warming over a brazier (contemporary photograph, June 2004). *Source*: Courtesy of Fort Ticonderoga, Ticonderoga, NY. (Used with permission.) (See color insert.)

Very good syrup is made from ferns, although it has the taste of burnt paper. It is also good for all sorts of jams, gives an added flavor to chocolate & blends well with either milk or coffee which, however, it causes to taste like unpleasant medicine. [46]

Exotic flavorings were fashionable and relatively available at home. But in an 18th century military camp on the frontier, most of these luxuries were virtually impossible to obtain at any cost. Tailer and Blodgett's account book, however, documents that not all chocolate-related ingredients were inaccessible when soldiers went off to war. A survey of a typical soldier's accounts reveals that in nearly all cases when chocolate was purchased, the soldier also purchased a quantity of wine and sugar. It is perhaps no coincidence that sugar and wine are the two principal ingredients of the simplest 18th century chocolate recipes. One such recipe is John Nott's recipe for wine chocolate published in his popular 1726 cookbook, *The Cooks and Confectioners Dictionary*. Article 130, "To make Wine Chocolate," of Nott's *Dictionary* instructs mixing into a pint of sherry or a pint-and-a-half of red port, four-and-a-half ounces of chocolate, six ounces of sugar, and a half-ounce of white starch or fine flour and bringing it all to a boil before serving [47]. The frequency of the simultaneous purchase of wine and sugar with chocolate as documented in the Tailer and Blodgett account book suggests that simple wine chocolate drinks as described by John Nott were very popular with soldiers at Crown Point in 1762–1763.

Unfortunately, the Tailer and Blodgett account book only records sales to soldiers over a six-month period. Even though the time covered by the account book is brief, its individual accounts do comprehensively cover the purchases of one complete company of soldiers. Furthermore, it documents chocolate consumption during a period when the entire company was camped for an extended period of time. Because of this stable camp environment, many soldiers maintained accounts with Tailer and Blodgett for the entire six-month period that encompassed the winter months of 1762–1763. During November, the first month covered in the account book, very little chocolate (only $1\frac{1}{2}$ pounds) was purchased by soldiers in Blake's Company. In December, total purchases increased to $6\frac{1}{2}$ pounds. As winter set in and temperatures plummeted to midwinter lows, chocolate consumption rose significantly, to 19 pounds in January and nearly 29 pounds at the heart of winter in February. As the days grew longer and temperatures began to rise with the approach of spring, chocolate consumption began to fall. In March, purchases of chocolate fell to 16 pounds. With the arrival of spring only one pound of chocolate was sold in April and finally, only $\frac{1}{2}$ pound of chocolate was listed in Tailer and Blodgett's sales for May (Fig. 31.8). In all, Tailer and Blodgett sold nearly 73 pounds of chocolate, to 32 soldiers in Captain Edward Blake's

silent regarding the details of how chocolate was prepared in military camps. Chocolate pots, either complete or fragmentary, rarely appear in archeological contexts from 18th century military sites. Soldiers generally carried few personal luxuries from home. A chocolate pot, even in its smallest form, would likely be considered unnecessarily cumbersome in a soldier's knapsack. Soldiers did, however, commonly carry basic cooking equipment including tinned-iron camp kettles, spoons, and tinned iron or sometimes horn drinking cups. The tinned-iron camp kettle could be used to heat a quantity of water into which chocolate could be shaved with a pocketknife. If available, other flavor additives could be mixed to satisfaction with a simple spoon. These simple utensils were more than adequate for the production of perfectly palatable chocolate.

When chocolate was prepared in domestic settings, it was commonly mixed with a variety of sweeteners and spices to enhance its flavor. Chocolate recipes dating to the 17th and 18th centuries describe a seemingly unending combination of flavor-enhancing ingredients for chocolate, including salt, cinnamon, aniseed, chilies, pepper, flour, vanilla, orange-flower or rose water, powdered cloves, ground almonds or pistachios, musk, and ambergris, to name just a few (see Chapter 8). Nor was the habit of spicing chocolate limited to Euro-Americans. Native Americans did so as well. From time immemorial, in the origin lands of chocolate, Aztec and Mayan Indians had spiced chocolate with chilies, annatto, and maize [45]. By the mid-18th century, Indians at Niagara and along the Saint Lawrence River had incorporated into their lifeways chocolate acquired from Euro-American traders. They, too, spiced up the chocolate, using an infusion made from maidenhair fern:

Purchases of Chocolate at Crown Point During the Winter of 1762-1763

Source: Account book of Tailer & Blodgett, sutlers to Capt. Blake's Co. Mass. Militia at Crown Point, Nov. 1762 - June 1763.
Coll. of the Fort Ticonderoga Museum, compiled. by C. Fox.

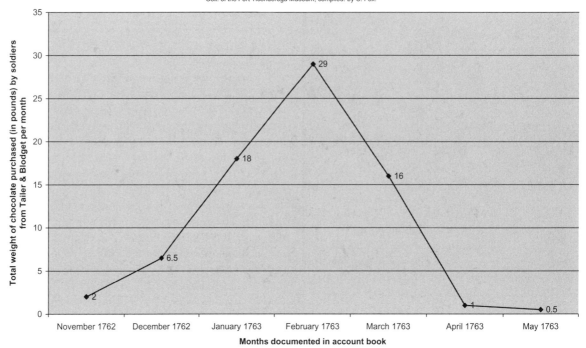

FIGURE 31.8. Graph: miscellaneous chocolate purchases at Crown Point, New York, winter of 1762–1763. *Source*: Tailer & Blodgett Account Book, Manuscript Number M-2178. Courtesy of Fort Ticonderoga, Ticonderoga, NY. (Used with permission.)

Company of Colonel Joseph Ingersoll's Massachusetts provincial regiment at Crown Point during the winter of 1762–1763. The dramatic rise in consumption of chocolate during the winter months suggests that chocolate was viewed as a drink that had warming as well as nourishing and invigorating effects on the body.

Tailer and Blodgett's account book also allows a detailed price comparison of chocolate with other common commodities. Chocolate was most commonly purchased in one-pound quantities at a cost of three shillings. The prices of dozens of other goods are also documented in the account book. For the same price as a pound of chocolate, a soldier could also purchase two quarts of wine, two quarts of molasses, or two ounces of green tea. Three shillings would also buy a soldier two-and-a-half pounds of sugar, two pounds of raisins, two pounds of cheese, four pounds of onions, or six ounces of pepper. For the same price, a number of articles of clothing or personal items could be acquired. These items include three pounds of tobacco, nearly 50 clay smoking pipes, two pounds of candles, four "regular quality" or two "best quality" combs, three pounds of soap, two pairs of garters, a pair of mittens, or a brass ink pot. The wide variety of competitively priced everyday foods and goods in comparison to chocolate indicates that chocolate was easily within the economic reach of and enjoyed by the common American soldier of the mid-18th century.

Moving supplies to the front, whether propelled by sutlers or the official army commissaries, required a staggering amount of labor to navigate around obstacles in the water highways and over portages. To move supplies forward just 100 miles from the supply base in Albany to troops on the northern frontier during either the Seven Years' War or the American Revolution required that each barrel of flour or salt pork, each box of chocolate, had to be handled ten times: into a *bateau*, out of the *bateau* and into an ox-cart, out of the ox-cart, and into another *bateau*, and so on. In 1758, commissaries estimated that 1000 bateaux, 800 wagons, and 1000 ox-carts would be required to haul the supplies (5760 barrels of just the most basic commodities: pork, beef, bread, and flour, the standard measure by which an army calculated provisions) necessary to sustain 20,000 men during their first month in the field [48]. Then the men loaded and unloaded just as many times the essential tents, cannon, powder, and shot for the impending assault. And this was the easiest, quickest path for delivering men and matériel into the cockpit of war! We have seen above the 1758 French and Indian ambush of a weary supply train of ox-carts headed for the British army passing over one of these portages.

Every time goods have to be handled, one runs the risk of pilferage. Armies try to control those losses with detailed recordkeeping. In 1759, a young

Connecticut soldier, Daniel Sizer, was stationed at Half-Way Brook, halfway along the portage route that cost General Abercromby his chocolate the previous year. Sizer's task was to reckon the passing ox-carts and their contents. But he brightened his days by purchasing chocolate from passing sutlers: one-and-a-half pounds of "Chalkolat" one day for £3/3 and three weeks later buying another half pound for £1/1 [49].

In the chaos of the failed American invasion of Canada in March 1776, Montréal merchants Wooden & Bernard attempted to send 29 sleighs of goods into "Indian Country" to prevent those supplies from falling into the hands of the rebel Americans. The American commander ordered a pursuit; seven sleighs were recovered, and their contents (mostly food and "5 bales of goods as are usually sent in the Indian Trade") were "delivered to the keepers of stores for the use of the Army." But when accounts were reconciled, it appeared that two of the American soldiers had pilfered for their own use a keg of red wine, a keg of New England rum, 4 pounds of chocolate, and 3 pounds of sugar. It certainly sounds like John Nott's recipe for wine chocolate [50].

The new rebel American governments had to organize *everything* from scratch: establishing a hierarchy of officers and appropriate rations for officers and troops, locating and controlling supplies, organizing commissary departments for safeguarding the distribution of the Public Stores—and *everything else*. In what was still the "Colony of Connecticut," in May 1775, just days after Ethan Allen had seized Fort Ticonderoga in Connecticut's name (and, he boasted, "in the name of the Great Jehovah and the Continental Congress"), the Assembly began to define the rations to be issued "to the Troops by them raised for the defence [sic] of their rights and privileges." In addition to a pound of beef per man per day, fish three times a week, three pints of beer a day, and so on, a company of about 75 men was allowed three pounds of candles a week, two gallons of vinegar, and six pounds of chocolate [51–53]. That launched a steady litany of supplies received in Albany, supplies then forwarded, and ultimately "Provisions on hand in the Magazine at Ticonderoga." In mid-July 1775, Commissary Elisha Phelps at Albany "Received 32 barrels of pork; forwarded 24. Received 507 barrels of flour; forwarded 371." But he forwarded every bit of chocolate ("150 weight"), all of the rum ("260 gallons"), molasses ("56-1/2 barrels"), and paper ("14 quires") that came to hand [54]. Nonetheless, on the receiving end at Ticonderoga ten days later, a Connecticut officer complained to his governor that the supplies were totally inadequate:

At the present we have some beer, but it wont [sic] last long . . . Rum and molasses are wanted. The rum that comes, as far as I have seen, is worse than none. . . . I think there has not been one pound of soap bought for the Army. A small matter of coffee and chocolate was bought about two weeks

ago, so the sick have a small matter, but none for them that can keep about. [55]

Upon that abject want of supplies was launched the disastrous invasion of Canada, and the situation did not improve [56, 57].

After the retreat of some 3000 healthy and 3000 sick from Canada to Ticonderoga in the summer of 1776, the American army made its strongest defense at Ticonderoga. Sutlers arrived before the commissary could organize the flow of supplies. The sick men were recovering, "having good ground to camp upon, and good water to drink, and some fresh provisions; but have not yet had any vegetables of any kind. Here are likewise sutlers who have spirits of all kinds—wines, sugar, chocolate, &c., to sell, though at a very dear rate—sugar three shillings, lawful money, per pound, &c." [58]. General Schuyler estimated that 120 barrels of pork and flour, loaded on a dozen or more *bateaux*, would need to head north from Albany every single day, rain or shine, to keep the troops fed [59]. On the receiving end, Deputy Commissary Elisha Avery recorded the ebb and flow of supplies: "565 barrels of pork . . . 300 pounds Coffee . . . 800 pounds Chocolate." "Returns," or formal reports, record the northward movement of supplies to support the army retreating toward Ticonderoga. A return drawn in late July 1776, a couple of weeks after the troops arrived at Ticonderoga, records "4 boxes of chocolate" at Fort George, just 35 miles away. Two weeks later in mid-August, Deputy Commissary recorded all manner of supplies, including 400 pounds of coffee, but no chocolate. By late August, the troops had consumed 100 pounds of coffee, but 800 pounds of chocolate had finally arrived [60, 61].

Given the difficulty of securing supplies of chocolate, governments resorted to a variety of embargoes and prohibitions. Embargoes shut down ports immediately. Just a month after fighting began at Lexington and Concord, in May 1775, Gen. Thomas Gage, British military governor of Massachusetts, imposed an escalating series of transit restrictions, ultimately examining all outgoing "trunks, boxes, beds, and all else to be carried out." These restrictions now "take from the poor people a single loaf of bread, and half a pound of chocolate, so that no one is allowed to carry out a mouthful of provisions" [62]. Colony and then state governments also imposed embargoes on the shipment of cacao and chocolate, to ensure that supplies would be available for the troops on the frontier, and those on the home front [63].

At the same time, the extraordinary supply/demand pressures of a wartime economy forced local Committees of Safety and state governments to set price controls on various commodities, including chocolate. As early as March 1776, the Committees of Safety in Philadelphia, Pennsylvania, and Newark, New Jersey, set prices for chocolate to control profiteering [64]. At the same time, troops in the field complained about escalating prices:

> Everything that is to be had is very exorbitant—spirits 20/p Gal. wine 30/ chocolate 4/ loaf sugar 3/6 & every other matter equally dear though the expense is of bringing articles is trifling. A person properly qualified as a sutler might make his fortune in 2 or 3 months. [65]

As the war ground on for years to come, the Continental Congress defined prices and rations. In 1779, officers should receive chocolate valued "at half a dollar per pound" and apportioned rations to Continental officers: Colonels and chaplains would receive four pounds of chocolate a month, majors and captains three pounds, lieutenants two pounds, and so on [66].

Much of the chocolate coming to Ticonderoga was intended for the hospitals. Chocolate was an energy food, readily digestible, even by men struck down with ague, smallpox, or camp fever. Doctor Morgan laid out the requirements for a military hospital just months after independence was declared:

> Wherever a General Hospital is established, it is necessary that the Commissary in that department furnish the necessary hospital stores, in sufficient abundance, so that the Surgeon on drawing for them be immediately provided with flour or bread, fresh-meat, salt, vinegar, rum, soap, candles, vegetables, Indian-meal, oatmeal, barley, rice, chocolate, coffee, tea, sugar, wine, butter &c. [67]

The practical realities on the ground at a frontier post like Ticonderoga forced commanding officers to juggle the foods issued to sick soldiers:

> Doctor Johnson will order Chocolate and Sugar for the Sick in the Hospital and such other sutible [sic] Refreshments as may be had on the Ground, in place of one half of the Beef they useless draw. [68]

Sometimes chocolate was used by shirkers to try to beat the army's health system. General Philip Schuyler lamented the number of men from Connecticut who seemed to be ill and were discharged and sent home as "unfit for service." Schuyler darkly hinted to Connecticut Governor Trumbull that some men "had procured their discharges by swallowing tobacco juice to make them sick. Others had scorched their tongues with hot chocolate to induce a belief that they had a fever, &c." [69].

Then in early July 1777, a nightmare, a disaster, happened. Congress, General Washington, and other American commanders had anticipated that the major British summer offensive would be toward New York or Philadelphia, and so had concentrated troops there. Instead, the British launched an attack of some 9000 men—British, Germans, and Indians, led by Gen. John Burgoyne—against an American garrison at Ticonderoga of fewer than 3000 marginally healthy men. Ticonderoga was abandoned by its defenders, including Captain Moses Greenleaf, on the night of

July 5, 1777. He noted that the day had been "pleasant," unlike his cold, windy arrival three months earlier in April. He did not note in his diary whether he had had a bracing "breakfast on chocolate" on that July day. Around noon they "spy'd a part of the British Troops on the mountains which overlook Ti." [11] The American troops were immediately issued provisions for four days and "48 rounds of Cartridges per man." Then at 9 pm, they received orders to evacuate: "the disagreeable News of Leaving the Ground . . . This night was employ'd in packing up stores & preparing for a retreat" [70].

Abandoned at Ticonderoga were huge quantities of munitions and supplies, despite the American garrison's best efforts to remove or destroy them. Supplies and artillery were loaded into *bateaux* and sent off toward Skenesborough (now Whitehall, New York). Cannons were spiked, trunnions knocked off, and dumped into Lake Champlain. Food supplies were divvied up into portions each man could carry, and the rest supposedly destroyed. In fact, the attackers were astounded at the quantities of artillery, powder, and provisions that had been abandoned during the hasty retreat:

> The foe had abandoned over a hundred [cannon]. . . . The magazines—crammed with flour, meat, coffee, wine, porter beer, sugar, medicines, etc.—held stock in super-abundance. Since we were the first regiment of Germans to reach the site, certain of our men were able to grab their share of the loot. . . . The quantities of supplies were simply amazing. [71]

Then came the recriminations. The accusations began in Congressional hearings in 1778. In May 1777, Deputy Commissary Avery had reported that there were at Ticonderoga sufficient provisions for 4000 men for 65–75 days. A month later, Deputy Commissary Jancey reported that supplies at Ticonderoga, on June 20, 1777, included all the basics: adequate beef, pork, and flour, as well as 12 boxes of chocolate, and other sundries. Jancey reported that sufficient supplies had existed at Ticonderoga on July 5, 1777, the day of the evacuation, to support a force of 5639 [72, 73].

Two valiant American commanders, Schuyler and St. Clair, were court-martialed for their failure to hold Ticonderoga, the "Gibraltar of North America." At his court-martial, Gen. St. Clair testified: "If I evacuate the place, my character will be ruined; if I remain here, the army will be lost. But I am determined to evacuate it, altho' it will give such an alarm that has not happened in the country since the war has commenced" [74]. But Schuyler and St. Clair were acquitted; Congress had failed to support the troops in the Northern Department.

On that fateful night of 5/6 July 1777, all appeared lost in the evacuation of Ticonderoga: American munitions, American food supplies, even American chocolate. The next day, in desperate retreat, Moses Greenleaf sought chocolate energy after total exhaustion: "day after [6 July] as fatigueing [*sic*] a March as ever known. We arriv'd at a Town Call'd Hubbardton. . . . Supp'd this night with Colo Francis on Chocolate. Encampt [*sic*] In the woods this night." The next morning, he breakfasted again on chocolate. Then "at $\frac{1}{4}$ past 7 [Col. Francis] came In haste to me told me an Express had arrived from Genl [*sic*] St. Clair Informing that we must march with the greatest Expedition or the Enemy would be upon us" [75].

Three months later, the tides of fate turned toward Greenleaf and his comrades in arms. British General Burgoyne surrendered his army at Saratoga. Was the success due to American arms? To American manpower? To American stamina? To American chocolate?

Acknowledgments

We would like to acknowledge the following institutions where we worked and individuals who assisted us at different stages of our research: Thompson-Pell Research Center, Fort Ticonderoga, Ticonderoga, New York; David Library of the American Revolution, Washington's Crossing, Pennsylvania; New York State Library and Archives, Albany, New York; National Archives, Washington, DC; Dean Barnes, Living Historian; Wanda Burch, Site Manager, Johnson Hall State Historic Site, New York City, Office of Parks, Recreation and Historic Preservation; Karl Crannell, Public Programs Coordinator, Fort Ticonderoga; John Kukla, Director, Patrick Henry Memorial Foundation; Jack Lynch, Associate Professor, Rutgers University; Kenneth Minkema, Yale University; William Sawyer, Ranger, Fort Stanwix National Monument, National Park Service; Eric Schnitzer, Ranger, Saratoga National Historic Site, National Park Service; Richard M. Strum, Director of Interpretation & Education, Fort Ticonderoga; LeGrand J. Weller, Jr., whose digital editions of *American Archives* (1837–1853), *Documentary History of the State of New York* (1849–1851), and *Documents Relative to the Colonial History of the State of New York* (1853–1887) remain invaluable resources for colonial historians.

Endnotes

1. Greenleaf's diary is distinctive for his consistent attention to his daily meals; that is, "Breakfast on Chocolate . . . Dined on Beef & Greens . . . supp'd on fry'd Pidgeons" (entry for May 4, 1777), and for perhaps the earliest mention of baseball (entry for June 2, 1777).

2. Spanish missionaries extended the world of chocolate consumption northward from Mexico across the American Southwest beginning in the late 17th century, and with particular alacrity during the next 75 years. (See Appendix 8.)

3. Reverend William Emerson was the grandfather of Ralph Waldo Emerson.

4. This was just one year after comparable clerical and political figures in the Massachusetts Bay Colony were breakfasting on chocolate.

5. Simon Girty was of Scots-Irish origin. He had been captured as a boy by the Seneca during the Seven Years' War and kept prisoner for seven years. During his captivity, he adopted Seneca ways. Later, during the American Revolution, he tried to occupy a middle ground and distanced himself from British, American, and Indian ways.

6. Advertisements for Samuel Blodgett as sutler to Col. Joseph Ingersoll's Massachusetts regiment appeared simultaneously and identically in both Boston newspapers. These advertisements appeared for several weeks following the original insertion. A footnote at the end of each advertisement mentions Blodgett's work as a sutler in 1761. Curiously, these advertisements do not mention William Tailer as a partner [40–42].

7. The advertisements documenting the dissolution of their partnership reads: "The Co-partnership between *Samuel Blodgett* and *William Tailer*, in supplying the Officers and Soldiers at *Crown-Point*, being dissolved, and the Company's Accounts in the hands of *Samuel Blodget*: . . . These are to desire all Officers and others who are still Indebted to them to make speedy Payment to said *Blodgett*."

8. "Weight" is defined in Samuel Johnson's 1768 *A Dictionary of the English Language* as: "1. Quantity measured by the balance. 2. A mass by which, as the standard, other bodies are examined."

9. One shilling is the equivalent of twelve pence.

10. Chocolate produced in St. Malô and Nantes was sold in balls weighing between $\frac{1}{2}$ and 2 *livres* [42].

11. Mount Defiance.

References

1. Wickman, D. H. "Breakfast on Chocolate": The Diary of Moses Greenleaf, 1777. *Bulletin of the Fort Ticonderoga Museum* 1997;**15**(6):483–506.

2. Return of Provisions at Sundry Posts in the Northern Department, Albany, March 13, 1777. New York State Library, Manuscripts and Special Collections, Jacob Abbott Papers [BA9691:26].

3. McKendrick, N., Brewer, J., and Plumb, J. H. *The Birth of a Consumer Society: The Commercialization of Eighteenth-Century England*. Bloomington: Indiana University Press, 1982.

4. Sewell, S. *Diaries*, Volume 1. Thomas, M. H., editor. New York: Farrar, Strauss, Giroux, 1973; p. 380.

5. Franklin, B. *Autobiography*. Eliot, C. W., editor. New York: PF Collier & Son, 1909; Chapter 9, p. 139.

6. Bougainville, L. A. *Adventure in the Wilderness: The American Journals of Louis Antoine de Bougainville,*

1756–1760. Hamilton, E. P., translator and editor. Norman: University of Oklahoma Press, 1964; p. 87.

7. Holden, A. W. *A History of the Town of Queensbury.* Albany, NY: Joel Munsell, 1874; pp. 321–324.

8. Parsons, J. *A Journal of an Expedition Design'd Against the French Possessions in Canada 1758 and 1759.* Manuscript. Collection of Dr. Gary M. Milan. Westbrook, N., transcriber and editor. pp. 38–40.

9. Steele, I. K. *Betrayals: Fort William Henry and the "Massacre."* New York: Oxford University Pres, 1990.

10. Dawes, E. C., editor. *Journal of Gen. Rufus Putnam . . . 1757–1760.* Albany, NY: Joel Munsell's Sons, 1886; p. 44.

11. Emerson, W. *The Diaries and Letters of William Emerson, 1743–1776.* Emerson A. F., editor. Boston: Thomas Todd, 1972; pp. 108–118.

12. Pell, J. H. G. Schuyler Peter (1657–1724). In: *Dictionary of Canadian Biography.* Toronto, Canada: University of Toronto Press, 1969; Volume II, pp. 602–604. Also available on-line at http://www.biographi.ca.

13. *The Ecclesiastical Records of the State of New York.* Volume 2, 1901–1915. Corwin, E. T., editor. Albany: State of New York, pp. 1409–1411.

14. Heckewelde, J. G. E. *A Narrative of the Mission of the United Brethren among the Delaware and Mohegan Indians from Its Commencement in the Year 1740 to the Close of the Year 1808.* Cleveland, OH: Burrows Brothers Co., 1907; pp. 24, 42, 52–53, 406–407, 422–423. Also available at http:www.americanjourneys.org/aj-120.

15. Gwyn, J. Johnson Sir William (c1715–1774). In: *Dictionary of Canadian Biography.* Toronto, Canada: University of Toronto Press, 1979; Volume IV, pp. 394–398. Also available on-line at http://www.biographi.ca.

16. Hamilton, M. W. *Sir William Johnson: Colonial American, 1715–1763.* Port Washington, NY: Kennikat Press, 1976.

17. O'Toole, F. *White Savage: William Johnson and the Invention of America.* New York: Farrar, Straus, Giroux, 2005.

18. Sir William Johnson to Thomas Gage, April 8, 1768. In Flick, A., Sullivan, J., et al., editors. *Papers of Sir William Johnson,* Volume 13. Albany: The University of the State of New York, 1927; p. 650.

19. Johnson's Order to messrs Wallace, Johnson Hall, Septr 11th 1772. In: Flick, A., Sullivan, J., et al., editors. *Papers of Sir William Johnson.* Volume 8. Albany: The University of the State of New York, 1927; p. 595.

20. Feister, L. M., and Pulis, B. Molly Brant: her domestic and political roles in eighteenth-century New York. In: Grumet, R., editor. *Northeastern Indian Lives: 1632–1816.* Amherst: University of Massachusetts Press, 1996; pp. 295–320.

21. Delancey, E. F., editor. *History of New York During the Revolutionary War, and of the Leading Events in the Other Colonies at that Period, by Thomas Jones,* Volume 2. New York: New-York Historical Society, 1879; pp. 373–374. Quoted in: Huey, L. M., and Pullis, B. *Molly Brant: A Legacy of Her Own.* Youngstown, NY: Old Fort Niagara Association, Inc., 1997; pp. 25–26.

22. Letter from Isaac the Indian to Sir Wm Johnson, April 10, 1762. In: O'Callghan, E. B., editor. *Documentary History of the State of New York,* Volume 4. Albany, NY: Weed, Parsons & Co., 1849–1851; p. 312.

23. An Inventory [of Johnson Hall], August 2d. 1774. In: Flick, A., Sullivan, J., et al., editors. *Papers of Sir William Johnson Papers,* Volume 13. Albany: The University of the State of New York, 1927; p. 650.

24. Allen, E. *A Narrative of Colonel Ethan Allen's Captivity, Containing His Voyages and Travels.* Ticonderoga, NY: Fort Ticonderoga Museum, 1930; pp. 21–50.

25. Extract of a Letter from Cork (Ireland) to a Gentleman in Philadelphia, dated January 1776. In: Force, P., editor. *American Archives,* Series 4, Volume 4, p. 836.

26. Pell, J. H. G. *Ethan Allen.* Boston: Houghton Mifflin Company, 1929; pp. 122–130.

27. Swigard, P. Chocolate maker. *New-York Gazette or the Weekly Post-Boy,* 18 September 1758. In: Gottesman, R. S., compiler. *The Arts and Crafts in New York: 1726–1776.* New York: New-York Historical Society, 1938; p. 295.

28. Gilliland, J. Stone, delft and glass ware. *New-York Gazette,* 4 April 1763. In: Gottesman, R. S., complier. *The Arts and Crafts in New York: 1726–1776.* New York: New-York Historical Society, 1938; p. 89.

19. Low, P. Chocolate maker. *New-York Gazette or the Weekly Post-Boy,* 23 October 1769. In: Gottesman, R. S., complier. *The Arts and Crafts in New York: 1726–1776.* New York: New-York Historical Society, 1938; p. 296.

30. Arthur, J. Split peas manufactured. *New-York Gazette or the Weekly Post-Boy,* 22 November 1773. In: Gottesman, R. S., complier. *The Arts and Crafts in New York: 1726–1776.* New York: New-York Historical Society, 1938; p. 316.

31. Armour, D. A. *The Merchants of Albany, New York, 1686–1760.* New York: Garland, 1986; 275 pp. Based on 1965 Ph.D. dissertation at Northwestern University.

32. Trelease, A. W. *Indian Affairs In Colonial New York: The Seventeenth Century,* revised edition. Lincoln: University of Nebraska Press, 1997.

33. Lunn, J. The illegal fur trade out of New France, 1713–1760. *Annual Report of the Canadian Historian Association* 1939:61–76.

34. An Inventory of Goods left at Stone Arabia with Thomas Smith Diamond which he is to sell for William Spotten 1770–71. Files of the Herkimer Home NY State Historic Site.

35. Sanders, R. Letterbook. 1754–1755. National Archives of Canada.

36. DeLery, C. In: O'Callaghan, E. B., editor. *Documentary History of the State of New York* (commonly DHNY), Volume 1. Albany, NY: Weed, Parsons & Co., 1849; p. 513.

37. Hagerty, G. *Massacre At Fort Bull: The De Lery Expedition Against Oneida Carry, 1756*. Providence, RI: Mowbray Company, 1971; pp. 60–61.

38. MacLeod, D. P. The Franco-Amerindian Expedition to the Great Carrying Place in 1756. Paper read at the 18th Annual Meeting of the French Colonial Historical Society, McGill University, Montréal, May 23, 1992. Available at http://www3.sympatico.ca/donald.macleod2/struggle.html.

39. *Tailer and Blodgett ms. account book, 1762–1763*. Fort Ticonderoga Museum collection FTA #M-2178. Acquired in 1943.

40. Tailer and Blodgett advertisement. *Boston Post-Boy*. December 20, 1762.

41. Tailer and Blodgett advertisement. *Boston Evening-Post*. December 20, 1762.

42. Tailer and Blodgett advertisement. *Boston Post-Boy*. November 28, 1763.

43. Savary, J. *Dictionnaire universel de commerce*. Genève, Switzerland: Chez les Frères Cramer et Claude Philibert, 1750; p. 887.

44. John Worlidge (1675). Quoted in: Davis, J. D. *English Silver at Williamsburg*. Williamsburg, VA: Colonial Williamsburg Foundation, 1976; p. 85.

45. Dalby, A. *Dangerous Tastes: The Story of Spices*. Berkeley: University of California Press, 2000; pp. 144–147.

46. Pouchot, P. *Memoirs on the Late War in North America between France and England*, revised edition. Cardy, M., translator; Dunnigan, B. L., editor. Youngstown, NY: Old Fort Niagara Association, 2004; p. 477.

47. Nott, J. *The Cooks and Confectioners Dictionary, or The Accomplish'd Housewives Companion: containing . . . revised and recommended by John Nott, late cook to the Dukes of Somerset, Ormond, and Bolton*, 3rd edition revised. London, England: Printed by HP for Charles Rivington, 1726.

48. Hamilton, E. P. *The French and Indian Wars*. Garden City, NY: Doubleday, 1962; p. 16.

49. Sizer, D. His Book Bought at New York May ye 23, 1759 (manuscript diary for 1759–1760). Connecticut Historical Society (#81911). Entries for 4 July and 25 July 1759.

50. Ford, W. C., et al., editors. *Journals of the Continental Congress, 1774–1789*. Washington, DC: Government Printing Office, 1904–1937. October 7, 1776; pp. 852–853.

51. Allen, E. *A Narrative of Colonel Ethan Allen's Captivity, Containing His Voyages and Travels*. Ticonderoga, NY: Fort Ticonderoga Museum, 1930; p. 8.

52. The Order and Direction from the General Assembly of the Colony of Connecticut . . . for issuing Provisions. In: Force, P., editor. *American Archives*, Series 4, Volume 3, p. 31.

53. Acts extending the Boundaries of the Town of Westmoreland, and making it one of the Regiment. In: Force, P., editor. *American Archives*, Series 4, Volume 2, p. 562.

54. Return of Provisions, &c., forwarded to the Northern Army by Elisha Phelps, Commissary from the 3d day of July 1775 to the 20th of July 1775 inclusive. In: Force, P., editor. *American Archives*, Series 4, Volume 2, p. 1701.

55. David Welsh to Governour Trumbull, Ticonderoga, August 5, 1775. In: Force, P., editor. *American Archives*, Series 4, Volume 3, pp. 46–47.

56. Return of Men sent northward since the one hundred and forty of Easton's, mentioned in my Letter of the 19th inst. In: Force, P., editor. *American Archives*, Series 4, Volume 3, p. 955.

57. Hatch, R. McC. *Thrust for Canada: The American Attempt on Québec in 1775–1776*. Boston: Houghton Mifflin Co., 1979.

58. Caleb Cushing to his brother, Camp at Crown-Point, July 8, 1776. In: Force, P., editor. *American Archives*, Series 5, Volume 1, pp. 128–132.

59. Hamilton, E. P. *Fort Ticonderoga: Key to a Continent*. With an introduction by N. Westbrook. Ticonderoga, NY: Fort Ticonderoga Museum, 1995; p. 138.

60. Return of Stores & Provisions In Store at the Undermentioned places Vizt: . . . Half Moon, Stillwater, Fort Edward, Fort George. Return made on 27–29 July 1776. New York State Library, Manuscripts and Special Collections, John N. Bleecker Papers [DA 10431:42]. In a Memorandum Book of Stores and Provisions at Different Places &ca.

61. Provisions on hand in the Magazine at Ticonderoga, August 13 and 23, 1776. In: Force, P., editor. *American Archives*, Series 5, Volume 1, pp. 1201–1202.

62. Extract of a letter to a Gentleman in Philadelphia, dated Boston, May 21, 1775. In: Force, P., editor. *American Archives*, Series 4, Volume 2, p. 666.

63. State of Massachusetts-Bay. In the House of Representatives, 5 February 1777, Boston: Benjamin Edes; 1777. A Resolution: "Whereas the rum, molasses, and sundry other articles herein [incl. chocolate] needed for the supply of the army and the inhabitants of this state: It is therefore resolved . . . [to prohibit export]. Copy in American Memory, Library of Congress. Available at http://memory.loc.gov/cgi-bin/query/r?ammem/rbpe:@field(DOCID+@lit(rbpe04000800)).

64. Prices of Sundry Articles. . . . In: Force, P., editor. *American Archives*, Series 4, Volume 5, pp. 86, 236.

65. Persifer Fraser to "My dearest Polly.". Ticonderoga, 6 August 1776. *Bulletin of the Fort Ticonderoga Museum* 1961;10(5):393.

66. Ford, W. C., et al., editors. *Journals of the Continental Congress, 1774–1789*. Washington, DC: Government Printing Office, 1904–1937. April 14, 1779; pp. 450–451.

67. Dr. Morgan to General Heath, 20 November 1776. In: Force, P., editor. *American Archives*, Series 5, Volume 3, pp. 781–782.

68. Entry for Headquarters (Ticonderoga), 19 December 1776. Wayne A. Orderly Book. *Bulletin of the Fort Ticonderoga Museum* 1935; 3(6):249.

69. General Schuyler to Governour Trumbull, Ticonderoga, 12 October 1775. In: Force, P., editor. *American Archives*, Series 4, Volume 3, p. 1035.

70. Wickman, D. H. "Breakfast on Chocolate": The Diary of Moses Greenleaf, 1777. *Bulletin of the Fort Ticonderoga Museum* 1997;15(6):497.

71. Barker, T. M. The Battles of Saratoga and the Kinderhook Tea Party: The Campaign Diary of a Junior Officer of Baron Riedesel's Musketeer Regiment in the 1777 British Invasion of New York. *The Hessians: Journal of the Johannes Schwalm Historical Association* 2006;9: 31.

72. Ford, W. C., et al., editors. *Journals of the Continental Congress, 1774–1789*. Washington, DC: Government Printing Office, 1904–1937. June 12, 1778; pp. 593–594.

73. Jancey, J. A Return of Provisions and Stores, &c. remaining on hand at Ticonderoga and Mount Independence, 20 June 1777. *Trial of Major General St. Clair. Collections of the New-York Historical Society for the Year 1880*. New York: New-York Historical Society, 1881; pp. 28, 125. 12 boxes of chocolate.

74. *Trial of Major General St. Clair. Collections of the New-York Historical Society for the Year 1880*. New York: New-York Historical Society, 1881; pp. 54, 83.

75. Wickman, D. H. "Breakfast on Chocolate": The Diary of Moses Greenleaf, 1777. *Bulletin of the Fort Ticonderoga Museum* 1997;15(6):497.

Chocolate and North American Whaling Voyages

Christopher Kelly

Introduction

Before the discovery of petroleum in 1859 and its subsequent use as a lubricant and fuel, the machines that were the driving force for North American industry ran smoothly with the oil obtained by hunting whales. Thus, the whaling industry, which had existed for centuries, exploded into an extremely lucrative business. After 1820, this burgeoning industry was primarily centered in the port of New Bedford, in southeastern Massachusetts, which by 1846 was sending 735 whaling vessels to sea each year—more than twice that of any other port in the world [1]. The rise of chocolate as a popular comestible in the 19th century and the burgeoning business of whaling were concurrent; not surprisingly, these industries converged in several ways.

A number of technological advances during the 19th century created an environment in which chocolate began to become a product of mass consumption. The process of extracting cocoa butter and the advent of steam power made chocolate candy more cost-effective to produce; at the same time industrial growth created a larger class of consumers who now had greater purchasing power to obtain cocoa-based luxuries. The industrial revolution that pushed forth these unprecedented developments was driven in large part with the aid of whale oil—not only to lubricate the new machinery, but to light America's homes, factories, and coastal lighthouses.

Whaling and Chocolate

Although whale hunting was practiced for centuries by both Europeans and Native Americans, the origins of New England sperm whaling, which would profoundly change the nature of New England maritime history, have their roots in a semiapocryphal tale about ship captain Christopher Hussey's voyage from Nantucket in 1712. It is written that, after being blown off course by a storm, Hussey accidentally ran into a school of sperm whales, caught and killed several of them, and returned home with a valuable cargo of ivory teeth and spermaceti (a waxy and highly valuable amber oil contained in the head of the sperm whale). Whatever the truthfulness of the tale, whale hunting soon became a lucrative trade that brought vast wealth to New England's maritime industry [2].

Whaling and chocolate, in fact, have common links that long predate the North American whaling industry. A 17th century Tuscan nobleman, Duke Cosimo de Medici, enjoyed his drinking chocolate flavored with ambergris, a rare and valuable substance formed in the stomachs of sperm whales. Fortunate sailors occasionally encountered this highly aromatic material floating in the sea; whaling captains learned to seek ambergris to augment their earnings in the already

Chocolate: History, Culture, and Heritage. Edited by Grivetti and Shapiro
Copyright © 2009 John Wiley & Sons, Inc.

lucrative whale hunt [3]. Prized by perfumers, apothecaries, and chefs in early modern Europe, ambergris became an accepted ingredient in many of the finest chocolate recipes of colonial times.

Throughout the 18th century, coastal whaling or very short off-coastal trips were the norm for the new industry. But by mid-century the coasts had been overfished and new methods were needed. Around 1750, whaling vessels began employing "try works," large iron cauldrons that enabled crewmembers to process whale blubber into oil while at sea, a process that was formerly accomplished on land [4]. Due to this innovation, journeys that were once shortened because of whale blubber's quick deterioration were now able to stretch into far longer times at sea [5]. After the Seven Years' War ended in 1763, new whaling grounds were discovered in West Africa and the South Atlantic [6]. Soon American whaling ships and barks were departing for voyages that lasted years rather than months, and new industries were needed to supply them for their long trips. Longer voyages required greater attention to ships' stores. In this context, chocolate emerged as one of several sustaining victuals that whaling captains relied on and supplied to their crews to safeguard their health under grueling working conditions.

Chocolate consumption in the British colonies of North America had a particularly early start in New England. It is apparent that a chocolate drinking public existed in Massachusetts Colony in the early part of the 18th century, as apothecaries in Boston sold chocolate as far back as 1712 [7]. In 1765, Dr. James Baker of Dorchester pioneered the use of power machinery in chocolate manufacturing in the colonies when he joined with an Irish chocolate maker named John Hannon to start the Hannon's Best Chocolate Company, which by 1820 would become The Walter Baker Company, one of the world's best-known chocolate makers [8].

Chocolate and whaling seem to have been conjoined almost from the beginning. In June 1753, the Quaker-owned sloop *Greyhound*, based in Nantucket, was cruising for Right whales at the mouth of the Davis Strait off Labrador. To provide some variety to their routine provisions, the crew killed a brace of seabirds and baked them into meat pies. These the sailors ate along with copious hot chocolate "poured down into our bellies," as recorded by the *Greyhound*'s captain in the vessel's log [9].

Because of a strong Quaker influence in Nantucket and, later, New Bedford, whaling ships were usually "dry" by order of the owners—that is, they carried none but medicinal alcoholic beverages and allowed no recreational consumption on board by the crew. Contemporary warships in the British and American navies, by contrast, issued a daily ration of "grog" (a measure of strong rum mixed with water) to their sailors. To some extent, chocolate on whaling ships would serve as a palliative substitute for grog, though cocoa was not a regular daily indulgence.

Whaling and chocolate made an early commercial alliance in northern Massachusetts as well. In 1819, Hezekia Chase bought a small grinding mill at the head of the Saugus River in Lynn, Massachusetts. Here he manufactured chocolate, milled corn, and traded in ground spices, using the sloop *Patriot* to send his goods to Boston and other cities along the coast. Some of these products were likely shipped to New Bedford for use on board ships of the whaling fleet. Perhaps because of this connection, Chase would later use his chocolate profits to build a wharf in Lynn and outfit a pair of ships, the *Atlas* and the *Louisa*, for whaling, entering this lucrative industry in the late 1820s [10].

Since Massachusetts was already a center for chocolate production and consumption, and its economy was based primarily on maritime trade, it is not surprising that chocolate became popular among sailors as well. By the middle part of the 18th century, sea merchants from Massachusetts were regularly returning from the tropics with large consignments of cacao beans among their cargoes [11, 12].

In 1780, the Royal Navy began issuing slabs of chocolate to its seamen for a dietary supplement [13]. Believing chocolate to be a highly nutritious and sustaining foodstuff, the British Admiralty hoped thereby to improve the health of sailors working under arduous shipboard conditions. After the American Revolution, the new United States Congress passed legislation that stipulated minimum daily rations of meat, rice, bread, and sugar for its sailors. These dietary allotments also included "one ounce of coffee or cocoa" per day for each common mariner [14]. Such regulations no doubt served as a precedent that whalemen looked to when fitting out and provisioning their vessels.

One important early voyage from New England provides a telling insight into the significance of chocolate carried as ships' provisions by sailors of the young American Republic. From 1787 to 1793, the ship *Columbia*, a Boston-based merchant vessel, made two pioneering trading voyages around Cape Horn, bound for the Pacific Northwest and China. The *Columbia* became the first American-flagged vessel to circumnavigate the globe when she returned home to Boston via the Cape of Good Hope [15]. On the *Columbia*'s second voyage in 1790, the ship left port carrying more than 500 pounds of chocolate packed into 50-pound crates, a massive amount for the time period [16]. It is unclear, however, whether the chocolate on this voyage was intended for use as crew provisions or for trade, but it was probably used for both.

As the New Bedford-based whaling fleet began to rise to the forefront of the industry in the 1830s, the city's ancillary industries grew rapidly. A number of "chandlers," portside merchants who supplied ships with basic nautical tools, hardware, and provisions, expanded their inventories to include wholesale food

ARTICLES

Of Agreement relative to the Whale-Fishery.

STORES agreed upon as neceſſary for a Whaling Voyage, viz.

PORK, Beef, Bread, Corn and Meal, as may be thought ſufficient,
    1 Barrel of Beans,
    1 Barrel of Salt,
    2 Barrels of Flour,
    1 Barrel of Rum,
    2 Barrels of Molaſſes,
    12 Buſhels of Potatoes,
    14 lb. either Coffee or Chocolate,
    1 lb. Pepper,
    56 lb. Sugar,
    50 lb. either Butter or Cheeſe,
    75 lb. Fat,
    112 lb. Rice,
    Dry Fiſh if they go to the Weſtern Iſlands.

THE following Charges to be deducted from the whole Stock, viz.
The Uſe of the Try Kettles at 12ſ. each, if broke what they Coſt,
All the Barrels fill'd, ~~a~~ at 4ſ. each;
The Uſe of Blubber Hogſheads and Truckage, a~~~~
The Wood, Candles, Nails, Chalk and Hoops, at what they coſt,
Clearing and entering the Veſſel, and Mediterranean Paſs, if any,

After which the following Diviſion of the Neat Stock is to be made, viz.
    To the Veſſel, ——
    To the Supplies, ——
    To the Maſter & Crew

INASMUCH as it is found from ſeveral Years Experience in the Whale-
Fiſhery, that the Method of fixing out for that Buſineſs has been unequal,
and much to the Prejudice of the Fixers: Wherefore, in Order to put the ſame
upon a more juſt and equal Footing for the future, We the Subſcribers do agree
upon the foregoing Method reſpecting the Diviſion of the Stock of Oil, Bone, &c.
taken by each Veſſel, and alſo the Stores neceſſary to be furniſhed for the Crew of
each Veſſel employed in the Whaling Buſineſs——And we agree to grant ſuch a
Diviſion of the Voyage, and ſuch Stores to each Crew by us fitted out reſpectively,
as is ſpecefied above, and no more.

To the Performance of which, we bind ourſelves each to the other, in the penal
Sum of Two Hundred Pounds. Witneſs our Hands at Boſton, New-England,
the Tenth Day of February, 1769.

FIGURE 32.1. Chocolate as a provision on whaling ships. *Source*: Articles of Agreement Relative to the Whale-Fishery, Boston; dated February 10, 1769 (broadside). Courtesy of Readex/Newsbank, Naples, FL. (Used with permission.)

supplies. Chandlers began to issue convenient catalogs of goods that enabled vessels to equip for much longer journeys (Fig. 32.1) [17]. New Bedford town and city directories listed a large increase in maritime grocers and suppliers from the late 1700s through the peak years of New Bedford whaling in 1839–1859 [18].

Small New Bedford merchants like Butler & Allen, Samuel Rodman, James H. Howland, and Charles Morgan became wealthy by supplying whaling vessels with a long list of provisions for their very lucrative voyages. Generally, the ships' chandlers published their inventories in small "outfitting books,"

which were given to captains or shipping agents to assist them in ordering a list of necessary supplies for their expeditions. Most outfitting books were divided into sections and subsections according to the types of provisions; these included everything from groceries to various grades of cordage to shackles, anchor chain, and rifles.

The chandlery catalog grocery section in most cases listed both bar chocolate and cocoa powder as purchasable items. Typically, bar chocolate was sold by whole crates of 50 pounds each; ships' agents could also buy half boxes, or 25 pounds of bar chocolate.

Cocoa powder, purchased less frequently due to its higher cost, was sold in sealed tins of varying sizes ranging from 1 to 10 pounds [19, 20]. In 1844, Fredrick S. Allen and his brother-in-law, Jeremiah Swift, became the primary supplier of grocery provisions to the New Bedford whaling industry. Most of the chocolate that went to sea through New Bedford port, from the mid-19th century to the 1880s, was supplied by Swift & Allen [21].

Loading chocolate and other goods onto whaling vessels entailed the payment of a fee. In 19th century New Bedford, as in most any port, it was common for ship owners to pay wharfage costs at the town's privately owned docks. Wharfage fees included a daily charge for docking a vessel, as well as handling fees for any cargo or goods loaded or disembarked during a ship's time at the dock. In February 1856, local New Bedford dock owners posted a "broadside," or public notice, that explained the various fees associated with doing business on the wharves of New Bedford. When calculating the costs of fitting out a voyage, whaling ship owners and captains had to take these costs into account [22].

According to the 1856 "Regulations at the Wharves of the City of New Bedford," bulk chocolate was charged at the rate of two cents per 50-pound box; logically, half-boxes were charged at half the rate, or one penny each. These rates were the same for other similarly boxed luxury goods like soap, candles, and tobacco. Such modest fees paid for what we may term "indulgence" commodities compared well with those charged for common staple goods. For example, foreign cheese paid four cents per box at this time, while "American" cheese paid wharfage fees of only one penny per box. Each 50-pound box of "Havana sugar" that passed through the hands of New Bedford's longshoremen paid seven cents before it could be used to sweeten a whaleman's chocolate or tea, while entire crates of opium, used medicinally as an indispensable painkiller aboard the whalers and other offshore ocean-going vessels, paid only six-and-a-quarter cents each [23].

Speaking of medicines, it is important to note that, in the 18th and 19th centuries, contemporaries recognized the healing properties of cocoa as a matter of course. In addition to its uses as a delectable snack, chocolate was certainly used for medicinal purposes aboard New England whaling ships. While their English whaling counterparts often sailed with at least one medical expert on board (usually a surgeon), American whalers were almost never afforded such a luxury. Instead, the captain, armed only with a medical chest filled with enumerated medicinal phials, a few instruments, and a rudimentary guidebook, served as the vessel's "doctor" [24].

Chocolate did not seem to be a part of the standard maritime medical supply kit, at least insofar as the substance counted as a conventional drug. A list of drugs and medical supplies used aboard the early 19th century American Navy frigate *New York,* by the onboard doctor, Peter St. Medard, listed no chocolate [25]. Still, until around the 1850s, chocolate was widely believed to provide important health benefits [26]. According to varying contemporary sources, chocolate was used as a treatment for indigestion and was believed to help recover strength. In addition, it was believed to reduce the effects of scurvy, one of the terrible scourges that faced sailors at the time [27]. It is likely that ship captains would have bought in to the conventional wisdom of the time and, seeing that chocolate was relatively inexpensive and readily available, it can be assumed that it was regularly used for medicinal purposes aboard whaling vessels.

That said, it does not seem that chocolate was consumed widely on shore in the quintessential whaling city during the early 19th century. Mary Russell, a prominent figure in New Bedford at the time, regularly published a household cookbook, which featured a lengthy dessert section. An edition of her "recipe book" published in 1810 featured a cornucopia of cakes, puddings, and breads flavored with all manner of agents, such as molasses, wine, brandy, and fruit, but no recipe contained chocolate [28].

By the end of the 19th century, however, chocolate had become a regular part of the New Bedford diet. The city's foremost apothecary, Lawton's Drug and Seed store, sold pamphlets and small books with drink and medicinal recipes, which contained instructions for making at least four chocolate beverages and both hot and cold chocolate syrups, and also sold imported gourmet chocolate pastes and powders [29]. Given New Bedford's small commercial circles, it is highly likely that Lawton's medicinal chocolate recipes made their way onto New Bedford whaling ships, where they were used to revive exhausted crewmen.

It would seem clear, however, that whaling voyages carried chocolate among their provisions from an early date. Whalemen's outfitting books began listing chocolate in their grocery sections as early as 1839 [30]. There is also earlier evidence of merchants and shipmasters purchasing chocolate. Financial accounts of Sandwich, Massachusetts, food merchant William Bodfish, whose waterfront shop catered to mariners, show sales of chocolate by the pound dating to December of 1792. Records of these transactions indicate that his primary chocolate consumers were prominent whaling vessel captains and ships' agents. These included Jonathan Bourne (father of the prominent New Bedford whaling mogul Jonathan Bourne, Jr.) and Captain Elisha Bourne, who earned his living as the master of a whaling ship [31].

By the time the New Bedford whaling industry reached its peak, generally considered to be during the later 1840s and early 1850s, chocolate was a standard provision on local whaling vessels. During this period, most outfitting books show between one and two boxes of chocolate being purchased per voyage [32]. Thus, the average whaling crew of 30 men generally left port with between 25 and 50 pounds of chocolate on board. While this may seem like a small amount for a long voyage (whaling trips averaged about three years in duration throughout this period), whaling crews could replenish their stocks of chocolate when touching at ports in the Portuguese Atlantic Islands or South America.

Nautical grocer and outfitter Swift & Allen was the largest provider of chocolate and other comestible provisions to New Bedford whaling vessels. However, records for a number of voyages indicate that, while most of a vessel's needs for rope, nautical equipment, and other tools were supplied by Swift & Allen, the outfitting books specifically listed no chocolate among the other food items purchased. It is probable that, in these cases, other local grocers or smaller provisioning companies, possibly even specialty shops like Lawton's Drug and Seed store, provided chocolate, a key foodstuff for any whaling journey.

It is difficult to discern how much chocolate was consumed on whaling voyages during the peak years of the industry. Whether or not the initial 25–50 pounds of chocolate that left New Bedford on each whaling ship (thus averaging just one to five pounds per crewman) was enough to maintain the onboard consumers for an entire journey, many of which lasted as long as four or even five years, is difficult to say. This is certainly possible, as there is evidence that chocolate was only issued to all hands on certain special occasions, and was used regularly by only a select few members of the crew. It is also possible that chocolate was consumed often and was simply replenished at one of several ports where ships regularly reprovisioned [33]. It is likely that cacao-producing centers such as Brazil, the Azores, or Valparaiso, ports that whalers often visited, were places where ship stores were resupplied, possibly with exotic forms of chocolate to which New Englanders were unaccustomed. Although early purchasing logbooks of vessels at different ports almost never mention chocolate, they often logged purchases under the generic title of "provisions," which could easily have included chocolate. Since chocolate prices decreased considerably during this period, chocolate purchases may have simply been factored in with other low cost food staples such as butter, salt, and coffee [34].

Chocolate, for example, was widely available in the Caribbean during the late 19th century, where it had become a staple foodstuff for island mariners. Indeed, Caribbean native near-coastal whale hunters relied on chocolate to sustain them in their small open boats; cocoa became a central dietary component on land and at sea. Early 20th century explorer Frederic Fenger, who in 1911 tried to recreate the experience of the Caribbean whale hunters by building a native

canoe and sailing among the islands, proclaimed that "chocolate would make an excellent stop-gap till I could reach shore and cook a substantial evening meal" [35]. In his book about these experiences, Fenger often praised the nourishing power of cocoa:

We drank the coarse native chocolate sweetened with the brown syrupy [muscovado] sugar of the islands. I did not like it at first; there was a by-taste that was new to me. But I soon grew fond of it and found that it gave me a wonderful strength for rowing in the heavy whaleboats, cutting blubber and the terrific sweating in the tropical heat. [36]

———————

[When] whaling was over for that day and we sailed back to the cove to climb the rocks . . . where we filled our complaining stomachs with manicou and chocolate. [37]

In extolling the virtues of chocolate, the obviously well-read Fenger even invoked the glowing appraisal made by a 17th century French physician, Père Labat, who had also traveled among the islands of the West Indies:

As early as 1695 Père Labat in his enthusiasm truly said, "As for me, I stand by the advice of the Spanish doctors who agree that there is more nourishment in one ounce of chocolate than in half a pound of beef." [38]

But who was using chocolate aboard New England whaling vessels during the 19th century, and how was that chocolate being used? There is little direct documentation discussing chocolate use by the American whalers' crewmembers. In Sandra L. Oliver's *Saltwater Foodways: New Englanders and Their Food at Sea and Ashore on the Nineteenth Century*, the author mentions an 1853 account by Captain John Whidden of the fishing vessel *Ceres*, wherein the regular Saturday night treat served aboard was "chocolate rice," a simple dish prepared in the galley by boiling rice with chocolate and molasses to form a "pudding-like" dessert [39]. It is important to remember that the vastly longer voyages that whaling vessels undertook (as compared to those of fishing vessels) would have necessitated a stricter rationing of food. Still, all of the aforementioned ingredients were generally stocked on most whaling voyages, and both came from the same maritime cookery tradition. It is therefore extremely likely that whalers enjoyed similar treats, just like their fishing counterparts.

Without clear accounts of chocolate usage aboard whaling vessels, it becomes important to discuss the nature of food preparation and consumption aboard such working vessels, and the onboard issues of class and hierarchy that such food consumption raised. Based on known evidence of chocolate use, and eating habits and customs aboard 19th century whaling vessels, it would seem that chocolate played a dual role. On one hand, it was a specialty item used mainly by the ships' privileged commanding officers; and on the other hand, cocoa was either a reward for the hard work of an ordinary sailor or employed medicinally to maintain his health.

Life on a whaling ship operated according to a rigid hierarchical structure. At the top of the ship's pyramid was the captain, followed closely by his first and second mates. The officers slept in relatively spacious quarters compared to the foremast hands in the crowded forecastle (known to seamen as the "fo'c'sle"), ate higher quality food, and generally enjoyed a much better standard of living than the rest of the crew. Just below the officers in the ship's hierarchy were a few skilled specialists, such as carpenters, steersmen, and harpooners, who did not share the same luxuries as the officers but nonetheless led a relatively comfortable existence that included small but private quarters and decent earnings. At the bottom of the ladder were the ordinary sailors, often known as "greenhands." These men served as a sort of maritime proletariat who led a difficult life that deprived them of most amenities. For the greenhand, life was an endless cycle of poor sleep, monotonous and low quality food, and often back-breaking work [40].

The rationing of food corresponded closely with this strict hierarchy. Ship officers enjoyed their meals in separate eating quarters. Most vessels left port with some livestock that, while they lasted, provided the captain and his mates with regular servings of fresh meat [41]. The captain's quarters were also privy to certain treats. Apart from the vessel's normal stores of staple foods, the captain often had access to such luxuries as olive oil, butter, cinnamon, canned hams, cloves, cheese, and pickles [42]. It may be assumed that the regular stores of chocolate aboard the ship were kept with these other culinary delights; chocolate could have been stored with the more universal provisions as well. More likely, chocolate may have been part of the captain's "slop chest," which consisted of nonessential items (including tobacco and ink) that sailors could purchase from the captain on credit. Moreover, since chocolate was considered semimedicinal and classed as a recuperative foodstuff, it may have been kept in reserve on whaling voyages, to be doled out only to the sick, or generally to the crew as a bonus after some particularly arduous labor such as rounding Cape Horn or the travail of catching, cutting in, and trying out a whale after a long chase, a task that had to be performed in one unbroken shift that could occupy the entire crew for up to twenty hours [43].

For the greenhand, food was one of the few pleasures available aboard the whaling vessel [44]. Yet eating was a monotonous exercise. Crew meals were shared communally in the ship's galley, located in the forecastle, the forward part of the ship where the sailors slept. The fare was, for the most part, limited to a number of nonperishable staples such as salted beef, pork, or cod, hardtack (a hard biscuit made with flour, salt, and water), beans, and rice.

But at times there were other foods available. During the first few weeks after leaving port, crewmembers often enjoyed fresh potatoes and onions. After a resupplying stop in a tropical port, whalemen might have procured fresh coconuts, yams, and oranges to eat, as well as such rare delicacies as Galapagos turtle meat [45].

The possibility that sailors also tried new forms of chocolate in foreign ports is reasonable as well. One can only imagine chocolate being drunk with exotic tropical spices or even eaten in ways not well documented in the United States. Fresh fruits and meats spoiled quickly, however, so a vast majority of the time, the crew was treated to the same unchanging fare on a weekly basis. The food was monotonous and at times the only way to tell what day it was, was by noting which foods were being served, since menus rotated according to a set schedule [46].

It is difficult to imagine that ordinary sailors on whaling ships, unlike their counterparts on naval vessels, had access to chocolate on a daily basis. Chocolate was certainly a scarce, rare, and precious item at sea. As mentioned earlier, a whaling ship usually stocked up to 50 pounds of chocolate at the outset of a voyage. By contrast, coffee and tea, regular staples of a sailor's diet, were carried in much greater abundance. The famous New Bedford whaling ship *Charles W. Morgan*, for example, left port for a four-year voyage in 1841 and carried 1600 pounds of coffee and 200 pounds of Souchong tea [47]. Since there is ample explicit documentation of sailor's coffee and tea consumption, and very little regarding chocolate, it is likely that chocolate was not drunk regularly by common sailors, but rather was served as a treat for special occasions or to address exceptional fatigue or illness.

An important question arises as to how greenhands would have drunk their chocolate. While officers used refined sugar for their coffee and tea, and skilled crewmen were granted a ration of molasses, greenhands usually drank their beverages black without addition [48]. It is doubtful that sailors who were deprived of sugar would have access to cinnamon, pepper, mustard, or any other popular chocolate flavoring agents, although it is certainly possible that, on occasions when sailors had access to chocolate, they also were able to obtain seasonings, even if just a bit of molasses. Considering the hardiness of the common sailor with his regular diet of old salted meat and stale bread, the idea of seaman eagerly gulping down cups of bitter unflavored chocolate is not difficult to imagine.

If these assumptions that greenhands did not have daily access to chocolate are correct, it is then left to the officers to be the primary, regular users of chocolate. The presence of 50 pounds of chocolate for fewer than five people (the captain and his mates, although sometimes a captain might bring his family on a voyage) would suggest that officers may have consumed chocolate on a regular basis. Without knowing if, and if so how often, chocolate was restocked, it can be assumed that each officer might have been allotted approximately 10 pounds of chocolate or more for the entire journey. If one uses the aforementioned Royal Navy allotment of one ounce of chocolate per day as a rough guide for daily usage, it is likely that officers would have had enough chocolate aboard for drinking at least once a week on a four-year journey.

Despite the strict rationing hierarchy that existed, naturally there was a wide variation of experiences on different ships. The period that saw the rise of the whaling industry in New Bedford also coincided with a rise in chocolate production. The worldwide chocolate industry, which suffered a slump in production after successive European wars, began to recover by the 1840s [49]. In addition, advances in manufacturing technology, most notably Van Houten's hydraulic press for extracting cocoa butter, allowed for the cheaper manufacture of chocolate and thus greater mass consumption of the product [50]. These events would be reflected in the prices shipping agents in New Bedford paid to purchase chocolate for whaling voyages. William Bodfish's financial records from 1793 to 1798 show that the going price for chocolate—or at least that paid by well established merchants—was roughly 13 dollars per pound [51]. By the 1840s and 1850s, when whaling was at its peak, Swift & Allen's outfitting books regularly listed chocolate at between 9 and 13 cents per pound, roughly the same price as such everyday staples as sugar and black pepper [52]. Such great price variations suggest that costs had dropped due to increasing ease of manufacture, but also indicate that the chocolate purchased for whaling voyages was probably not of the highest quality. Although little documented evidence exists as to how these price changes affected onboard access to chocolate, they allow for the possibility that greenhands would have been able to enjoy chocolate more frequently in the latter years of whaling's ascendancy.

Consumption of chocolate and other treats, such as butter and refined sugar, among the common whaling crewmen would have also depended on the ship's captain and the shipmaster. The initial provisioning of vessels was dependent on the budget set forth by whaling agents who owned the vessels and commissioned the voyages. Profits on whaling voyages were great, and reducing provisions was not an economic necessity, but rather grew from a desire to increase profits by limiting costs [53]. It was, therefore, at the ship owner's discretion how much variety of food would be allowed aboard. The generosity of the captain was also important, and some whaling ship commanders attracted crews through their reputations of being liberal with victuals.

Despite the rigid hierarchy that was the norm on whaling ships, it is difficult to believe that 20–30 men sailing together for up to four years did not form bonds that transcended this traditional social structure. Journals indicate that on sailors' birthdays, or occasions such as the ship's thousandth barrel of whale oil logged, the captain would have allowed the preparation of delicacies like fruitcake or macaroni and cheese [54]. On holidays it was customary for officers to share their regular allotment of fresh meat, provided by livestock on board, with the rest of the crew. This was usually served as a soup or a stew [55]. It is possible that these holidays and other special events would have occasioned the issue of chocolate and any number of flavoring agents to regular crewmembers. Perhaps some of the aforementioned hearty chocolate rice treat would have been served after a particularly trying day of cutting, hauling, and boiling down blubber. Chocolate, with its caffeine and relatively high protein and caloric content, would have been a more desirable alternative to whalemen than coffee or tea after an around-the-clock shift of hard labor.

The captain, himself, did not directly dole out rations; this duty fell to the second mate [56]. It is plausible that the second mate would have allowed crewmembers chocolate rations as a substitute for coffee or tea, or as a special indulgence after a particularly exhausting day. Another factor that determined the use of chocolate on whaling voyages was that of the disposition of the ship's cook. Cooks held unique position in the crew structure or hierarchy. They were neither officer nor sailor. They were both looked down upon for doing "women's work" but ultimately respected as persons who prepared everyone's food [57]. Cooks served six meals a day, three each for officers and crew. As those responsible for the preparation of higher quality meals to be served to the crew's officers, cooks had access to the vessel's premium grade items, including chocolate. It is difficult to imagine, therefore, that a certain amount of pilfering did not occur. Although fairly strict inventories were kept of provisions like chocolate, items sometimes would simply "vanish" from the ship stores. In his journal, Captain Baker of the whaling ship *Ohio* recorded an instance of eggs disappearing from the ship's store. Although he was suspicious that the cook had stolen them, Captain Baker could not prove the allegation [58]. Failure to notice missing chocolate bars or cocoa powder would not have been difficult.

Despite the high degree of standardization of food that seems to have existed on whaling vessels, some innovation and creativity transpired. Cooks, of necessity, developed varied recipes that produced relatively "exotic" dishes even with a limited number of ingredients. Dishes such as lobscouse (stew of salted meat, potatoes, and vegetables), sea pie (a fish or meat flour dumpling), and duff (a molasses-sweetened fruitcake) showed innovative variation achieved by maritime cooks [59]. It is likely, therefore, that some of the cooks would have used chocolate in new or interesting ways and that chocolate was occasionally part of inventive galley dishes.

Besides cooks, some voyages carried an additional resource of culinary inspiration: captains' wives.[1] Long voyages with little else to do allowed these "sister sailors" the opportunity to provide some extra food magic, but it was usually just the captain who benefited. Dorothea Moulton, a mariner's wife, recorded several instances of spoiling her husband with her maritime cookery during a North Atlantic voyage in 1911: "Made chocolate [today] and did it ever taste good," she proclaimed in a journal entry dated June 30th [60]. But because of the rudimentary cooking facilities available to her aboard ship, Mrs. Moulton could never make her chocolate fudge quite to his liking [61].

As would be expected, tension often arose between cooks and captains' wives. Women accustomed to their own home-cooking often found it difficult to stomach the monotonous fare produced by ships' cooks. Furthermore, the galley—located in the crew's sleeping quarters—was strictly off-limits to women [62]. Still, the captain's wife often played a culinary role that served all of the crew. If a cook fell ill, for example, or was simply not very skilled at his job, the lady of the ship could take it upon herself to "train" him in the right way of food preparation [63].

According to the prominent whaling agent Charles W. Morgan, "there is more decency on board [whaling vessels] when there is a woman" [64]. It was often noted that fewer instances of corporal punishment occurred, and that the captain generally had a better temperament when his wife was aboard. It is likely that a number of kind-hearted captains' wives, unaccustomed to the rigid social order of the maritime world, would have shared their cooking with the entire crew. For instance, Mary Rowland, the captain's wife, recorded during a journey that she had baked apple dumplings in "ample supply to both forward and aft," suggesting that her special desserts were not reserved solely for officers and skilled crewmen [65]. This was not likely a random occurrence but a regular event when a woman's touch was present. Similarly, whaling crews most likely enjoyed delicacies prepared with chocolate more frequently when an officer's wife sailed aboard, especially later in the century, when this became a more regular circumstance, and chocolate grew in popularity.

The discovery of abundant supplies of petroleum in 1859 and the shift to other lower cost fuels caused the steady decline of the North American whaling industry in the late 19th century. On August 25, 1924, the ship *Wanderer* became the final whaling vessel to depart from New Bedford port, but the next day she was wrecked on the rocks off Cuttyhunk Island in Buzzard's Bay [66].

Conclusion

Chocolate in a small but crucial way was one of the sustaining forces of whaling. Although it was not often a regular foodstuff for most members of the average whaling ship's crew, nevertheless, it served an important function as an incentive, reward, and health-sustaining product. A lengthy whaling voyage, without chocolate as a means to perk up tired sailors or as a special treat for officers and their families, would have been a much drearier endeavor. The concurrent expansion of these affiliated American industries—chocolate and whaling—is an important, underappreciated theme in New England history.

Acknowledgments

I wish to thank Amanda Zompetti, Department of History, University of Massachusetts, Dartmouth, for her assistance with research on this chapter. Dr. Timothy D. Walker, Assistant Professor of History at the University of Massachusetts Dartmouth, advised and coordinated the writing of this chapter. I also wish to thank Mars, Incorporated for their financial support through the Colonial Chocolate Society Project that made possible the research used in this chapter. In addition, the author expresses his gratitude to Michael Dyer, Michael Lapides, and Laura Pereira, staff members of the Kendall Institute Archive of the New Bedford Whaling Museum, New Bedford, Massachusetts, for valuable research assistance in location key documents used in the preparation of this chapter.

Endnote

1. This seems to run counter to the oft-told tradition that women about sailing shiops foretold bad luck.

References

1. Ellis, R. *Men and Whales*. New York: Alfred A. Knopf, 1991; p. 160.

2. Ellis, R. *Men and Whales*. New York: Alfred A. Knopf, 1991; p. 141.

3. Coe, S. D., and Coe, M. D. *The True History of Chocolate*. London, England: Thames and Hudson, 1996; pp. 148–149.

4. Stackpole, E. A. *The Sea Hunters: The New England Whalemen during Two Centuries, 1635–1835*. Philadelphia: J. B. Lippincott Company, 1953; p. 32.

5. Ellis, R. *Men and Whales*. New York: Alfred A. Knopf, 1991; p. 143.

6. Ellis, R. *Men and Whales*. New York: Alfred A. Knopf, 1991; p. 145.

7. Clarence-Smith, W. G. *Cocoa and Chocolate: 1765–1914.* London, England: Routledge, 2000; p. 16.

8. Coe, S. D., and Coe, M. D. *The True History of Chocolate.* London, England: Thames and Hudson, 1996; p. 230.

9. Stackpole, E. A. *The Sea Hunters: The New England Whalemen during Two Centuries, 1635–1835.* Philadelphia: J. B. Lippincott Company, 1953; pp. 42–44.

10. Wrynn, K. *The Saugus River and the Whaling Industry.* Lynn, MA: The Saugus River Watershed Council, 2001.

11. Clarence-Smith, W. G. *Cocoa and Chocolate: 1765–1914.* London, England: Routledge, 2000; pp. 66–68.

12. Coe, S. D., and Coe, M. D. *The True History of Chocolate.* London, England: Thames and Hudson, 1996; p. 230.

13. Clarence-Smith, W. G. *Cocoa and Chocolate: 1765–1914.* London, England: Routledge, 2000; pp. 15–16.

14. Oliver, S. L. *Saltwater Foodways: New Englanders and Their Food at Sea and Ashore in the Nineteenth Century.* Mystic, CT: Mystic Seaport, Inc., 1995; p. 93.

15. Howay, F. W., editor. *Voyages of the Columbia to the Northwest Coast: 1787–1790 and 1790–1793.* Portland Oregon: Oregon Historical Society Press, 1990; p. viii.

16. Howay, F. W., editor. *Voyages of the Columbia to the Northwest Coast: 1787–1790 and 1790–1793.* Portland Oregon: Oregon Historical Society Press, 1990; pp. 453–455.

17. Oliver, S. L. *Saltwater Foodways: New Englanders and Their Food at Sea and Ashore in the Nineteenth Century.* Mystic, CT: Mystic Seaport, Inc., 1995; p. 85.

18. Ellis, R. *Men and Whales.* New York: Alfred A. Knopf, 1991; p. 160.

19. NBWM, Kendall Institute Mss 56: Series B: Bailey-Butler and Allen Folder 4: Subseries 16.

20. NBWM, Kendall Institute Mss 7: Subgroup 1: Series B: Subseries 3: Volume 1.

21. NBWM, Kendall Institute, Mss 5, biographical notes.

22. NBWM. Regulations at the Wharves of the City of New Bedford. Broadside poster dated February 26, 1856.

23. NBWM. Regulations at the Wharves of the City of New Bedford, Broadside Poster dated February 26, 1856.

24. Druett, J. *Petticoat Whalers: Whaling Wives at Sea, 1820–1920.* Aukland, New Zealand: Collins Publishers, 1991; p. 45.

25. Estes, J. W. *Naval Surgeon: Life and Death at Sea in the Age of Sail.* Boston: Science History Publications, 1997; p. 216.

26. Keen, C. L. Chocolate: food as medicine/medicine as food. *Journal of the American College of Nutrition* 2001;20: 436S.

27. Dillinger, T. L., Barriga, P., Escarcega, S., Jimenez, M., Salazar Lowe, D., and Grivetti, L. E. Food of the gods: cure for humanity? A cultural history of the medicinal and ritual use of chocolate. *Journal of Nutrition* 2000;130(Supplement):2057S–2072S.

28. NBWM. Kendall Institute Mss 64: Series G: Subseries 49.

29. NBWM. Kendall Institute, Mss 56: Series L: Subseries 2: Volume 4 and Folder 5.

30. NBWM. Kendall Institute Mss 56: Series B: Bailey-Butler and Allen Folder 4: Subseries 16.

31. NBWM. Kendall Institute, Mss 56: Series B: Folder 26.

32. NBWM. Kendall Library, Mss. 7: Subgroup 1: Series B: Subseries 3: Volume 1.

33. Draper, C. L. *Cooking on Nineteenth Century Whaling Ships.* Mankato, MN: Blue Earth Books, 2001; p. 23.

34. Coe, S. D., and Coe, M. D. *The True History of Chocolate,* London, England: Thames and Hudson, 1996; p. 243.

35. Fenger, F. A. *Alone in the Caribbean: Being the Yarn of a Cruise in the Lesser Antilles in the Sailing Canoe "Yakaboo."* Whitefish New York: Kessinger Publishing, 1917; Chapter 7.

36. Fenger, F. A. *Alone in the Caribbean: Being the Yarn of a Cruise in the Lesser Antilles in the Sailing Canoe "Yakaboo."* Whitefish New York: Kessinger Publishing, 1917; Chapter 2.

37. Fenger, F. A. *Alone in the Caribbean: Being the Yarn of a Cruise in the Lesser Antilles in the Sailing Canoe "Yakaboo."* Whitefish New York: Kessinger Publishing, 1917; Chapter 8.

38. Fenger, F. A. *Alone in the Caribbean: Being the Yarn of a Cruise in the Lesser Antilles in the Sailing Canoe "Yakaboo."* Whitefish New York: Kessinger Publishing, 1917; Chapter 2.

39. Oliver, S. L. *Saltwater Foodways: New Englanders and Their Food at Sea and Ashore in the Nineteenth Century.* Mystic, CT: Mystic Seaport, Inc., 1995; p. 138.

40. Stackpole, E. A. *The Sea Hunters: The New England Whalemen during Two Centuries, 1635–1835.* Philadelphia: J. B. Lippincott Company, 1953; pp. 434–442.

41. Draper, C. L. *Cooking on Nineteenth Century Whaling Ships.* Mankato, MN: Blue Earth Books, 2001; p. 8.

42. Oliver, S. L. *Saltwater Foodways: New Englanders and Their Food at Sea and Ashore in the Nineteenth Century.* Mystic, CT: Mystic Seaport, Inc., 1995; p. 101.

43. Stackpole, E. A. *The Sea Hunters: The New England Whalemen during Two Centuries, 1635–1835.* Philadelphia: J. B. Lippincott Company, 1953; pp. 37–39 and 44–46.

44. Oliver, S. L. *Saltwater Foodways: New Englanders and Their Food at Sea and Ashore in the Nineteenth Century.* Mystic, CT: Mystic Seaport, Inc., 1995; p. 84.

45. Busch, L., and Busch, B. C. The seaman's diet revisited. *International Journal of Maritime History* 1995;2:164.

46. Draper, C. L. *Cooking on Nineteenth Century Whaling Ships.* Mankato, MN: Blue Earth Books, 2001; p. 15.

47. Oliver, S. L. *Saltwater Foodways: New Englanders and Their Food at Sea and Ashore in the Nineteenth Century.* Mystic, CT: Mystic Seaport, Inc., 1995; p. 107.

48. Draper, C. L. *Cooking on Nineteenth Century Whaling Ships.* Mankato, MN: Blue Earth Books, 2001; p. 15.

49. Clarence-Smith, W. G. *Cocoa and Chocolate: 1765–1914.* London, England: Routledge, 2000; p. 69.

50. Coe, S. D., and Coe, M. D. *The True History of Chocolate*. London, England: Thames and Hudson, 1996; p. 242.

51. NBWM. Kendall Library, Mss 56: Folder 26.

52. NBWM, Kendall Library, Mss 7: Subgroup1: Series B: Subseries 13: Volume 2.

53. Oliver, S. L. *Saltwater Foodways: New Englanders and Their Food at Sea and Ashore in the Nineteenth Century*. Mystic, CT: Mystic Seaport, Inc., 1995; p. 113.

54. Oliver, S. L. *Saltwater Foodways: New Englanders and Their Food at Sea and Ashore in the Nineteenth Century*. Mystic, CT: Mystic Seaport, Inc., 1995; p. 103.

55. Draper, C. L. *Cooking on Nineteenth Century Whaling Ships*. Mankato, MN: Blue Earth Books, 2001; p. 26.

56. Oliver, S. L. *Saltwater Foodways: New Englanders and Their Food at Sea and Ashore in the Nineteenth Century*. Mystic, CT: Mystic Seaport, Inc., 1995; p. 89.

57. Oliver, S. L. *Saltwater Foodways: New Englanders and Their Food at Sea and Ashore in the Nineteenth Century*. Mystic, CT: Mystic Seaport, Inc., 1995; p. 87.

58. Druett, J. *Petticoat Whalers: Whaling Wives at Sea, 1820–1920*. Aukland, New Zealand: Collins Publishers, 1991; p. 44.

59. Draper, C. L. *Cooking on Nineteenth Century Whaling Ships*. Mankato, MN: Blue Earth Books, 2001; p. 15.

60. Balano, J. W. *The Log of a Skipper's Wife*. Camden, ME: Down East Books, 1979; p. 25.

61. Balano, J. W. *The Log of a Skipper's Wife*. Camden, ME: Down East Books, 1979; p. 19.

62. Druett, J. *Petticoat Whalers: Whaling Wives at Sea, 1820–1920*. Aukland, New Zealand: Collins Publishers, 1991; p. 44.

63. Druett, J. *Hen Frigates*. New York: Touchstone, 1998; p. 135.

64. Druett, J. *Petticoat Whalers: Whaling Wives at Sea, 1820–1920*. Aukland, New Zealand: Collins Publishers, 1991; p. 169.

65. Druett, J. *Hen Frigates*. New York: Touchstone, 1998; p. 136.

66. Ellis, R. *Men and Whales*, New York: Alfred A. Knopf, 1991; p. 167.

PART

VII

Southeast/Southwest Borderlands and California

Chapter 33 (Cabezon, Barriga, Grivetti) describes the earliest documented uses of chocolate in North America, associated with Spanish Florida (St. Augustine). Chocolate-related documents reveal other introductions through Texas, New Mexico, and into Arizona and show both religious and military uses of chocolate as desired energy and pleasurable products within this geographical region of the continent. The authors also identify how chocolate was used as a reward and was provided to friendly Native American populations during the early years of Spanish colonization. Chapter 34 (Grivetti, Barriga, Cabezon) considers the introduction and spread of cocoa and chocolate in California, from the time of earliest European exploration, through the Mexican period, to the eve of the California Gold Rush. Their chapter draws heavily on chocolate-content in original letters penned by Franciscan priests who were responsible for much of the distribution of chocolate and chocolate products in California. Chapter 35 (Gordon) continues the California chocolate story and explores its use during the Gold Rush and post-Gold Rush eras when chocolate played important roles in miners' food patterns, in the social life of mid-19th century San Francisco. His chapter concludes with an examination of California chocolate manufacturing from its beginnings to a place of prominence in the 21st century.

Blood, Conflict, and Faith

Chocolate in the Southeast and Southwest Borderlands, 1641–1833

Beatriz Cabezon, Patricia Barriga, and Louis Evan Grivetti

Introduction

Chocolate was introduced to North America initially by the Spanish. How and when the first chocolate/cacao arrived, however, is a convoluted story revealed through documented evidence, interesting suppositions, and unsubstantiated possibilities. Part of the story is an oft-told one, whereby the Spanish gained a foothold in North America with their 1565 establishment of a port in northeastern Florida named after Saint Augustine. Documents located in the Archive of the Indies, Seville, Spain, reveal an interesting background relative to Saint Augustine's foundation and offer tantalizing hints about the possibilities of when and how chocolate may have been introduced to North America.

Spanish Florida

Pedro Menéndez de Avilés was commissioned by King Philip II of Spain to establish a Spanish presence in Florida. In July 1565 he sailed from Spain with a modest armada and soldiers. After he and his men reached Puerto Rico, Menéndez de Avilés wrote the Spanish king on August 13, 1565 and related his detailed plans for the military expedition to Florida [1]. The document stated that the Spaniards would spend the month of August in Florida and remain there through September 10th. The text also confirmed that the governor of Puerto Rico supported the venture and that the governor had provided all required additional boats, soldiers, and supplies [2]. The military expedition left Puerto Rico on an unidentified date after September 10, 1565 and landed on the Florida coast August 28, 1565 (Chris-

tian feast day of Saint Augustine), where a rudimentary fort and settlement named after the saint was established [3].

A second letter from Menéndez de Avilés to King Philip II of Spain, dated December 5, 1565, written in Matanzas, Cuba, described his victory over the French at Fort Caroline, north of Saint Augustine [4]. The document was brutal in details of the battle and put forth the case that French corsairs (pirates) off the coast of Florida should be attacked as well [5]. In this second letter, Menéndez de Avilés reported the hardships of the soldiers remaining at the St. Augustine fort: food was so scarce—there was nothing to eat or drink except palm hearts, wild herbs, and fresh water [6].

A third letter from Menéndez de Avilés to King Philip II, dated January 30, 1566, written in Havana, Cuba, resummarized his victory over the French and emphasized the need to explore further coastal North America in order to establish settlements at ports that would allow easy provisioning and access to Cuba [7]. Menéndez de Avilés announced

Chocolate: History, Culture, and Heritage. Edited by Grivetti and Shapiro
Copyright © 2009 John Wiley & Sons, Inc.

that he would bring additional soldiers to the forts of San Agustin (sic) and San Mateo, and that he would leave one captain and 150 soldiers at each fort [8].

While no mention of chocolate appears in these three documents, Menéndez de Avilés's discussion of resupplying troops in St. Augustine remains tantalizing since he provided the following information. In his August 13th letter, neither ship nor captain is identified by name but in his December 5th letter, he identified the captain as Diego Anaya, and a subsequent January 30th letter reported that:

Captain Diego Anaya came in a 70 ton ship loaded with provisions for the forts of San Agustin and San Mateo. He unloaded in San Agustin whatever provisions were destined and pregnant pigs and after three days left with the provisions for the fort of San Mateo. [9]

Menéndez de Avilés then ordered Captain Anaya to return to Puerto Rico in search of a galleon named the *San Pelayo*. With these two additional clues (name of Captain Anaya and the *San Pelayo* galleon) it may be possible for scholars at a later date to pursue this lead, locate the supply ship's manifest, and determine whether or not chocolate was listed among the supplies that later were sent to St. Augustine.

From 1565 to the mid-17th century, the record regarding chocolate in the North American Southeast is basically blank. The earliest chocolate-associated document located by members of our research team associated with this geographical region was dated to 1641 and mentioned chocolate in Saint Augustine. The paleography and content of this unusual document—written using a mixture of French and Spanish syntax and spellings—provides the names and occupations of 222 of the residents (221 males; 1 female) living at Saint Augustine, an interesting mixture of administrators, soldiers, and sailors. The document is unique given that it lists specific weights/quantities of chocolate allocated to each person who is identified by name (see Appendix 5). This distribution resulted from a legal settlement over who owned the chocolate carried aboard the crippled, storm-damaged ship, the *Nuestra Señora del Rosario y el Carmen*, that put into Saint Augustine for repairs [10–14]. These important documents are reviewed in depth and analyzed elsewhere in this book (see Chapter 51).[1]

In 1686, Pensacola Bay was surveyed by Andres de Pez, who recommended development of a Spanish settlement. Initial attempts to settle this region of Spanish Florida, however, were unsuccessful and it was not until 1698 that a fortified presidio, Santa Maria de Galve, was established [15]. A document dated to 1712 identified chocolate given as medicine to the sick housed in the presidio:

Account of the provision of medicines that are requested for the sick people of the Presidio Santa Maria de Galve . . . 30 pounds of good chocolate for the sick . . . and enough sugar for the chocolate. [16]

This Spanish outpost on Pensacola Bay experienced a wave of military and political activity. The settlement was captured by the French in 1719 but returned to Spain in 1722. The British held Pensacola from 1763 until forced to leave in 1781. Ultimately, Spain ceded Pensacola and Florida in 1821 to the United States [17].

Southwest Borderlands (Arizona, New Mexico, Texas)

To study the history of the American Southwest, from modern Texas westward through New Mexico and Arizona into California, is to study the lives and times of Spanish explorers and priests who opened new lands and founded Christian missions. Diaries and letters produced by the earliest explorers to the region, whether Francisco Vazquez de Coronado, Pedro de Castañeda, Juan de Oñate, or others, contain abundant information regarding hunting, transportation of food, and foods procured/stolen from Native American settlements encountered along the various routes [18–20]. Although the accounts of Coronado and Oñate expeditions were written years after Hernán Cortés and his literate officers had drunk chocolate with King Montezuma in the Mexica/Aztec capital of Tenotichlan [21], none of these earliest exploration accounts through the North American Southwest mentioned chocolate as a food or medicinal item. This lack, however, would be remedied in subsequent decades as missionaries to the North American Southwest were escorted through these territories and missions and presidios were established. With churches and protective administrative centers in place, settlers arrived, villages were founded, and the need for food supplies and medicines increased.

Despite the widespread use of chocolate in the heartland of New Spain (Fig. 33.1), documented to the third decade of the 16th century [22, 23], the earliest chocolate-related document from the Southwest borderlands that we located is dated only to 1690, nearly 50 years after the Saint Augustine document identified earlier. The reference appeared in a letter written to Don Carlos de Sigüenza by Father Damián Massanet, founder of the first Spanish mission in east Texas. In 1689, Governor Alonso de León organized a search for the fort constructed by La Salle. Father Massanet accompanied this expedition

MEXICO

SONORA
1. Caborca Town
2. Altar Mission and Presidio
3. Tubutama Mission
4. San Ignacio Mission and Settlement
5. Mission Nuestra Señora de Los Dolores del Corsari (Mission Dolores)
6. Mission Populo
7. Mission San Luis Gonzaga de Baca de Huachi (Mission Baca de Huachi)
8. San Antonio de Las Huertas Town

CHIHUAHUA
9. Chihuahua Province
10. Chihuahua Settlement: Named Villa de San Felipe El Real Town
11. Coahuila Province

NUEVO LEON
12. Nuevo Leon Province
13. Guadalupe Town

Sinaloa
14. Sinaloa Province
15. Mocorito Mission

DURANGO
16. Durango Settlement and Santo Domingo Mission
17. Mission/Town San Joseph del Tizonaso
18. Zacatecas Town
19. San Luis Potosi Town and Mission

TAMAULIPAS
20. Tamaulipas Province
21. Tampico Town

NAYARIT
22. Nayarit Province
23. San Blas Port, Presidio and Hospital
24. Matanchel Town

JALISCO
25. Jalisco Province
26. Guadalajara Town
27. Colima Province

MOCHOACAN
28. Morelia and College/Mission Valladolid

Mexico City and surroundings
29. Mexico City- Ciudad Real- Tenochtitlan
30. Texcoco Lake
31. Xochimilco District

TLAXCALA
32. Tlaxcala Province/Town- and San Lazaro Hospital
33. Cacaxtla Archeological Site

PUEBLA
34. Puebla Province
35. Tepeaca (Tepeca)
36. Puebla de Los Angeles Town and Hospital Sancti-Spirit

QUERETARO
37. Queretaro Province
38. Santa Cruz de Queretaro Mission College
39. Xalpam Mission- Sierra Gorda-

GUERREO
40. Guerrero Province and Acapulco Port

VERACRUZ
41. Vera Cruz Town (port)
42. Xalapa Town (Jalapa)

OAXACA
43. Oaxaca Province and town
44. A Quatulco Town-(In today Huatulco)
45. Copalita River (In Huatulco Bay)
46. Teotitlan Town
47. Mixteca Alta

TABASCO
48. Tabasco Province

CHIAPAS
49. Chiapas Province
50. Soconusco (County in Chiapas)

CAMPECHE
51. Campeche Town and San Roque Hospital

YUCATAN
52. Yucatan Penninsula and Province
53. Chichen Itza Archeological Site

54. Aztapalapa (Iztapalapa) Town (Mexico City DF)
55. Coatepeque (In Queztaltenango, Guatemala)

FIGURE 33.1. Map of New Spain: Mexico.

[24]. In his letter to Sigüenza, Father Massanet wrote—cynically:

Captain Leon had a compadre along, Captain [?], so honorable that he never failed to play the tale-bearer and excite quarrels; so kind-hearted that only his friend Leon drank chocolate, and the others the lukewarm water; so considerate of others that he got up early in the morning to drink chocolate, and would afterward drink again with the rest; so vigilant that he would keep awake and go at midnight to steal the chocolate out of the boxes; perhaps this vigilance was the reason why, while, by order of His Excellency, Captain Leon should have left for the priests three hundredweight of chocolate and the same quantity of sugar, but he left only one and one-half hundredweight of each.

[25]

The next earliest chocolate-associated document we located was dated five years later. This record is interesting given the date, since the Spanish in New Mexico at this time were under attack by both Apache and Pueblo Native Americans. Part of the Southwestern chocolate story—and the Spanish Apache/Pueblo wars—begins more than a century earlier, with the founding of Santa Fe in 1598. The initial settlement that became Santa Fe was established by Pedro de Peralta and named La Villa Real de la Santa Fe de San Francisco de Asís (after Saint Francis of Assisi) [26]. Santa Fe was named the capital of Nuevo México, New Spain, in 1607. The period between the years 1680 and 1692 was one of unrest and hostilities between the Spanish and the Apaches and Pueblo Indians. In 1680, Diego de Vargas Zapata y Luján Ponce de León was appointed Governor of New Mexico with orders to pacify and reestablish administrative and political control. By 1692, Diego de Vargas reported that the territory still was not pacified and settled, and he was recalled to Mexico City in 1693 for "consultations." Upon his return to New Mexico, he discovered that 70 Pueblo families had retaken the settlement at Santa Fe. The decision was made to storm the city by force, a military action that resulted in the deaths (and subsequent executions) of many hundreds of Pueblo warriors [27–29].

After order had been established in Santa Fe, the soldiers led by Antonio Ziena were ordered to the town of Santo Domingo to recover horses and stores of corn and chocolate that had been held by the Indian rebels. As the military unit approached Los Alamos, two soldiers named Miguel Maese and Juan Rogue Guetierrez entered the presidio, found three Indians, and took them prisoner. Two tried to escape and were killed. The officers interrogated the surviving captive, Sebastián Pascure, and learned that the corn had been hidden in a nearby ravine adjacent to the old Presidio of Cochiti. When asked about horses, Pascure replied that they were also near the old presidio. When Pascure was asked where the Pueblo Indians had buried the chocolate, he replied that he was not certain but that someone had told him that the chocolate was buried on the other side of the river (*alguien le dijo que el chocolate estaba enterrado del otro lado del rio*) [30].

After Santa Fe had been retaken in 1695, hostilities continued elsewhere against the Apaches in New Mexico. In a document dated July 6, 1695, Father Francisco Vargas asked a Sergeant Major Roque de Madrid, commander of the presidio located at San Joseph, to supply food for the troops [31]. Father Vargas then wrote a second letter to the district governor, not named but certainly Diego de Vargas (no relation), who complied and sent a limited quantity of supplies to the garrison at the San Joseph Presidio. This document continued with Father Vargas's notation that the provisions were received and he provided a passing comment that the Spanish soldiers were fighting local Apache Indians and during rest periods

between raids the soldiers were supplied with cold chocolate to drink:

> . . . *le llebaron refresco de chocolate, tabaco y papel.* [Chocolate refreshments [cold drinks], tobacco, and paper [probably to roll cigarettes]]. [32]

As a result of Governor Diego de Vargas's brutality in the retaking of Santa Fe, the Pueblos renewed hostilities again in 1696, an event known as the Second Pueblo Revolt [33]. A 300-page file we reviewed provided details of this revolt and described Pueblo Indian attacks on Spanish missions and presidios at San Lázaro [34], San Ildefonso [35], Pecos [36], and Santa Ana [37]. Manuscripts in the file also described Indian thefts of cattle, horses, and grain, even religious objects such as statues of the Virgin Mary [38, 39]. The file also contained a letter by Diego de Vargas to the King of Spain that reported he had suppressed the Second Pueblo Revolt [40].

The same year, 1696, Father Francisco Vargas, a Franciscan priest (again no relation to the governor) was reappointed as "Guardian of all Missions" in New Mexico. He wrote to Governor Diego de Vargas on July 6, 1696, and requested an escort to recover provisions left behind at the Presidio of Pecos during the initial Pueblo Revolt:

> *I am here in this town of Santa Fe with the nine Fathers that left their missions because of the treason of the Indians that killed five priests and many Spaniards, profaned the sacred vases and stole cattle. I am pleading for you to provide an escort so that we can recover the provisions that are in the Presidio de Pecos, that amount [estimated] to be 66 corn sacks [costales], 8 fanegas of wheat, and 64 sheep, and also to recover the sacred vases, ornaments and other things related with the religious services.* [41]

While Father Francisco Vargas's letter contained no mention of chocolate, the governor wrote from Santa Fe to Father Vargas on July 6, 1686,[2] and invited the priests to drink chocolate with him:

> *I have decided that I am determined to provide the necessary escort [you requested] with eight soldiers and the necessary muleteers for the transport of the provisions. . . . I would like to have the great pleasure to have them [the Fathers] at my table. I request that you honor me by accepting my sincere affection and coming to my house. Also, in order that you consider the strength of my ardent affection I made public the remission of the debt of 600 pesos with which I supplied the necessities of the four reverend Fathers that died. I will make my table available [mesa franca] for the nine fathers that our Reverence tells me are in this town, if they would agree to honor my house. And I will provide to all of them and their servants if they have any with more chocolate in case they were lacking [chocolate], with wine and waxes [i.e., candles] for the celebration of the mass. I will provide all this with my salary and whatever I have I will spend it on your service. Signed: Don Diego de Vargas and Domingo de la Barreda.* [42]

Documents dated 1712–1713 represented separate requisitions provided by four Jesuit missionaries stationed in Sonora and Sinaloa (today Northern Mexico). One was written by an Italian Father Miguel Xavier de Almanza who was assigned to the Mission Populo, Sonora, ca. 1703–1704.[3] Father Almanza's account, dated March 6, 1712, specifically identified quantities of ordinary and fine chocolate, sugar, and the spices cinnamon, cloves, and nutmeg used to prepare chocolate beverages [43]. His requisition was basically similar in content and quantities to those filed by Father Nicolas de Oro, missionary at Bacadeguachi in 1712 [44]; by Father Bello Pereira, missionary in the county of Tizonaso in 1712 [45]; and by Father Joseph Xavier Cavallero, missionary in the county of Mocorito (Sinaloa), in 1713 [46].

During the next three decades the Southwestern borderlands remained in turmoil as the Apache Indians resisted pacification and continued raiding. Military documents from this era provided supply and food-related information associated with these Spanish/Apache wars. A letter from Jesuit Father Tomás de Nolchaga from the College of Durango to Father Francisco Echeverría, dated 1715, related that chocolate beverages were offered to Apache Indians and that they:

> drank with pleasure—after offering the first morsel to the sun according to their custom [y chocolate que bebieron gustosos, haciendo antes sus ceremonias gentilicas de ofrecer al sol el primer bocado]. [47]

A subsequent document, dated 1731, written by the governor of Texas, Don Juan Antonio de Bustillo y Lerna, requested food supplies for his Spanish soldiers and for the local Comanche Indians who were identified as "friendly:"

> I need to provide for the sustenance of more than 90 men, that as helpers, I invited from the Kingdom of Leon [located in Mexico; not Spain] with some neighbors of this presidio in the town of San Antonio de Valero—60 friendly Indians of this town and also from the nations Pampopas and Sanes: 44 loads of flour; 50 fanegas of corn; 2 ground loads of corn for [making] pinole for the Indians; [and] 250 pounds of "chocolate de regalo" [gift chocolate]. [48]

In the early 18th century, the Southwest borderlands remained unsettled in terms of Spanish–Indian relationships. Mission Nuestra Señora de los Dolores del Cosari was established in Pimería Alta (Arizona) by Father Eusebio Francisco Kino[4] on March 13, 1687. In a letter signed at Mission Nuestra Señora de los Dolores, written after the victory of the Pimas who allied with the Jesuit missionaries against their common enemies the Jocomes, Father Kino described the celebrations that followed that victory and that the missionaries offered the Pima gifts of chocolate as tokens of appreciation. He then departed the mission for a land expedition to determine a connection by land to the "Sea of California"—because California at that time was thought to be an island [49]. In his letter he wrote:

> On September 15, of the year 1697 [when] the Pimas and Pima-Sobaipurís struck the first blow against the enemy Jocomes, we had a fiesta for this Great Lady and Sovereign Queen of the Angels. . . . There was a procession in which the Virgin was carried under a canopy. . . . We killed some cattle, for our native Pima, and we gave chocolate to some of them and told them goodbye-and in saying goodbye to these Pimas of the North and Northwest I told them how, within ten days I had to go to the Rio Grande of the Sea of California. [50]

As the years passed by, Father Kino's zeal for exploration became more and more apparent. In 1706, he would explore the Tepoca coast located along the eastern shore of the Gulf of the California north of the location now called after him—Kino Bay. After preparing the necessary supplies for his long journeys, he departed from Mission Dolores. Before departure, however, Father Kino also received an unexpected contribution for his journey from Father Jerónimo Minutuli, a Jesuit priest from Mission Tubutama:

> His Reverence with great love and generosity supplied us with wine for masses, with wax candles, chocolate, bread, and biscuit, pinole, mutton, beef, and even with his own saddle mule. [51]

Father Kino ultimately discovered that Baja California was not an island, but a peninsula—characterized by blue seashells[5] [52].

Father Luis Xavier Velarde succeeded Father Kino at Mission Dolores two years after Kino's death in 1711 [53]. Between 1716 and 1720, Velarde's work focused on children and the sick.[6] On June 17, 1716, Father Velarde prepared a requisition for items to be sent to Mission Dolores that included the following items:

> 3 Arrobas chocolate; 3 arrobas sugar; ½ pound saffron; ½ pound tobacco powder [i.e., snuff]; ½ pound cloves; 1 pound anise; 1 pound cilantro; 1 cassock [sotana] made of Cholula cloth [i.e., cotton]; 1 missal (religious book used in Mass celebrations); 3 pairs of shoes (size 11). [54]

Several accounts provided to us by Bernard Fontana, Historian at the University of Arizona Museum, Tucson, revealed the difficult interplay between desire for chocolate and desert heat. One document, written by the Swiss Jesuit Phelipe Segesser to his superiors in Mexico City identified a list of needed supplies; chocolate was among these items. Father Segesser urged that the mule train carrying the supplies be sent as early as possible because "here, the chocolate melts in June" [55].

A second document from Father Gaspar Stiger, S.J., to Father Procurator Joseph Herrera, written at San Ignacio, Sonora, April 25, 1737, also commented on the problem of shipping chocolate during the summer due to the desert heat:

If only Your Reverence could send the mule train driver earlier every year because the chocolate melts when the alms come in June. But I see that Your Reverence is not at fault for the late arrival of the mule trains in Mexico. [56]

Relationships between the Spanish and Native Americans in the Southwest continued to deteriorate throughout the mid-19th century, either through overt warfare or through mistreatment of Indians. Joseph de Goxxaes (Gonzales) of the Mission La Punta de Lamparo denounced the ill treatment of the Chichimecos Indians by missionaries and soldiers in a document dated to 1738.[7] He identified the misconduct and concluded with a list of provisions requested to be sent by Father Gabriel de Vergara to the Santa Maria de los Dolores mission located at Saltillo-Cohauila (located today in southern Texas).

3 ½ arrobas of ordinary chocolate cut in 12 parts, and 1 arroba of ordinary chocolate [no further designation]. [57]

In a separate document written the following year in 1739, Goxxaes (Gonzales) reported that a Father Guardian, Pedro del Barco had complied with a request for supplies to be sent to the Mission de Lamparo. Included among the items were boxes of chocolate:

4 arrobas of good chocolate, at 11 pesos 2 reales [for a total cost of] 45 pesos. [58]

Twenty-five years later, the Southwest borderlands still remained unsettled. A document written in Sonora dated to 1764 presented a list of foods consigned to the presidios located at Altar, Tubac, and Xavier del Bac. Apache raids were mentioned along with notations that the Indians had stolen Spanish cattle. Chocolate was mentioned on this list of items supplied, but only in small quantities. The document revealed that the items supplied, including the chocolate, were not specified as military rations, but items to be sold to the soldiers [59].

The chocolate-associated history of the North American Southwest borderlands also is linked with Don Juan Bautista de Anza, who in 1774–1776 explored and established a permanent overland route that linked Sonora, Mexico, and modern Arizona with coastal *Alta* California (see Chapter 34). During his early military career, Anza served in northern Sonora and would have been familiar with the geography of this region. After he completed his two famous overland trips to California and returned to Tubac (south of Tucson, Arizona), Anza was appointed governor of the region that is now New Mexico in 1777 [60]. During his administration he continued to explore the northern portions of New Mexico and in 1779 led an army of 800 soldiers against the Comanches but only was able to pacify them in 1786 when he signed a treaty with the Comanche chief, Ecueracapa [61].

The Mission Puerto de la Purísima Concepción located in the vicinity of the Gila and Colorado River junction was founded by Father Francisco Garcés in October 1780. Father Garcés, who had accompanied Anza during his expeditions, was himself an explorer and previously had founded the Tucson Presidio in 1775, to protect the Mission San Xavier del Bac. According to historians, the Concepción mission site was poorly administered and rights of the local Yuman Indians were ignored. Teodoro de Croix, the viceroy of New Spain, ignored Father Garcés's advice and authorized a Spanish settlement adjacent to the newly founded mission. Not unexpectedly, the Spanish settlers took the best lands, destroyed Yuman crops, and set the stage for a terrible regional Spanish–Indian war [62]. The Yumans honored and respected Father Garcés and had accompanied him during his regional explorations. Nevertheless, tensions continued to mount and exploded July 17–19, 1781, when local Yumans attacked the mission and precipitated what became known as the Yuma Massacre [63].

A letter written by Maria Ana Montielo, wife of Ensign Santiago Islas, commander of the Colorado River settlement, described the events that took place during a Catholic service when the Yuma attacked [64]. She wrote that the parishioners attempted to flee, and while some escaped others remained behind and were protected by a "friendly" Yuman who hid Father Garcés and the others in his house. Montielo wrote that while they were under protection, chocolate was served. Then the leader of the Yumans shouted from outside the house—"Stop drinking that and come outside. We are going to kill you." Montielo recorded later that Father Garcés replied: "We would like to finish our chocolate first" [65, 66]. Montielo's account continued and documented that Father Garcés and Father Juan Barreneche ultimately left the house and both were killed[8] [67]. Maria Ana Montielo survived the massacre through the advice of Father Garcés, who had told her during the events: "Stay together with the other women; do not resist capture and the Yumans will not harm you" [68].

After the Yuma Massacre the borderlands still remained in turmoil. Two years later, Teodoro de Croix, nephew of the viceroy of New Spain, served as Commander General of the Interior Provinces (*Provincias Internas*), of New Spain (a region that extended from what is now Texas westward through New Mexico and Arizona, and included *Alta* and *Baja* California, as well as the northern Mexican states of Coahuila, Neuva Vizcaya, Sinaloa, and Sonora). In 1783, however, Teodoro de Croix was reposted and left the Southwest region to become viceroy of Peru. He was succeeded in the Southwest borderlands by Brigadier Don Phelipe (*sic* Felipe) Neve de Arispe[9] [69].

As Brigadier Neve de Arispe assumed his new duties in 1783, he needed additional supplies. We identified an account and packing list of the goods sent to him by Don Joaquin Dongo via a muleteer named Don Rafael Villagran, who lived at San Antonio de las Huertas in Sonora. The following items appeared on the requisition:

4 Boxes of 30 arrobas of chocolate [750 pounds] [of this]
10 arrobas of the very dark chocolate [prieto];
20 arrobas of the white chocolate [blanco];
[Total] price 411 pesos and 3 3/8 reals;
2 Boxes with 63 hams well packed—of a weight of 14 arrobas and 8 pounds
Price 68 pesos
2 Boxes with 12 arrobas and 22 ½ pounds of white sugar
Price 31 pesos and 3 reals. [70]

The identification of "white chocolate" in this document represents an unusual curiosity.[10] This term cannot mean cocoa butter given the position and context of the words within the document. Since the initial chocolate on the list was described using the word *prieto* (i.e., very dark), the interpretation for "white chocolate" could be merely chocolate of a lighter color. However, two additional explanations also are possible. First, there is a variety of *Theobroma* called *Theobroma alba*, or white cacao, in the early Spanish literature. In such documents the beans are described as white. This variety of cacao, *criollo*, originally came from Venezuela. It is problematical but not unreasonable that such beans could have been imported to the Southwest borderlands. On the other hand, the designation also could be processed "white chocolate" beans. Whatever the origin of the product it may be surmised that the "white chocolate" was for the very wealthy.

As governor of New Mexico, Juan Bautista de Anza also was active in campaigns fighting the Apaches. During his administrative period that spanned 1776 to 1788, he signed pacts with the Comanches, Navajos, and Yutas (Utes) and the men were used as military scouts and as "irregulars" to attack the Apaches. A document signed April 5, 1786, by both the paymaster of the Presidio de Santa Fe (New Mexico) and Governor Juan Bautista de Anza, identified chocolate as an important military supply:

Account of what was received and of the expenditures, for the extraordinary expenditures of the Peace and War of this Province of New Mexico, according to the orders that were given to me [José Maldonado] the Lieutenant and Paymaster of the Presidio Santa Fe in the above mentioned Province in the year 1787: 3 arrobas [= 75 pounds] of ordinary chocolate at 3 reales and 6 grains [?] 32 pesos 6 reales 6 grains. [71]

A companion document issued a year later, on June 10, 1787, reported that chocolate was used to pay a grinding fee to prepare processed flour given to selected Comanches as partial payment for their loyalty and assistance in the Spanish/Apache wars:

I received from Lieutenant José Maldonado, 5 ½ pounds of chocolate and 15 bars of soap as fee for the grinding of flour that was ground for the Comanches. [72]

A document dated to April 5, 1786 also revealed the use of chocolate during the Spanish/Apache wars and provided the following information:

3 pesos 52 reales for indigo blue and chocolate; Chocolate [box] 2 vs de mta de 7/8 [undecipherable notations for quantity]; [73]

---

3 arrobas [75 pounds] of ordinary chocolate at 3 reales, 6 grains. [74]

Governor Anza also authorized a Spanish expedition to explore lands to the east of Texas. In a document dated and signed June 16, 1788, Anza approved chocolate and sugar for "explorers on the road" and identified a Pedro Viale as the leader of the expedition to Louisiana:

Exchequer bills for the extraordinary expenditures to equip the individuals that will march with Pedro Vial, with the purpose to find a direct road to Nachitoches [el camino derecho para Nachitoches]: 6 lbs. brown solid sugar cone [piloncillo]; 6 lbs of chocolate; 10 lbs of sugar. [75]

A letter written by a Spanish military inspector, Diego de Borica, on October 31, 1788, described how troops were supplied with commodities purchased at the Villa de San Felipe el Real, Chihuahua, Mexico. The document reported a quarrel between Borica and the military supplier, Francisco Guizarnotegui, and the text identified the supplies and agreed prices to be charged to the military. A significant quantity of chocolate was among the items to be supplied by Guizarnotegui:

8 boxes of gift chocolate [chocolate de regalo], containing 12 quantities of 7 arrobas each [175 pounds] at 2 ¼ reales [a pound]; [76]

---

3 boxes of ordinary chocolate [chocolate ordinario], containing 12 quantities of 7 arrobas each [175 pounds] at 1 3/8 reales [a pound]. [77]

Another document that revealed the importance of chocolate as a military food was dated to 1789 and signed in Chihuahua. The text identified the same supply agent (Don Francisco Guisarnotegui) but used a different spelling for his name (Francisco Guizarnotegui). The manuscript stated that the military supplies were to be sent to presidios located in New Mexico and Chihuahua, and specifically for the "road troops" (*compañia volante*), and that the chocolate was to be sent via the muleteer Don Rafael Villagran [78]. This chocolate was supplied from Mexico City, as apparently local supplies were exhausted. The repetitive use of similar box sizes, references to specific cuts/shapes of the packed chocolate, and specific weight limits per box permit the calculation that at least 8 mules would have been required to carry only the chocolate. Presented here are two examples from the document:

> *8 boxes of gift chocolate cut in 12 [pieces] each box with 7 arrobas—a total of 1400 pounds; [79]*

> *3 boxes of ordinary chocolate cut in 20 [pieces] each box of 7 arrobas—a total 525 pounds. [80]*

Governor Anza died December 19, 1788. Three years after his death, a document commemorating the governor was prepared in Durango and signed by Philipe de Hortega on September 27, 1791. This written tribute to Anza contained the notation that three years earlier in 1788 the governor had used chocolate to pay friendly Comanches for their assistance as scouts and for fighting with the Spanish against the Apaches. This invoice signed by Anza authorized the chocolate payment to be delivered at the presidio in Santa Fe (New Mexico):

> *Merchandise that has been delivered to the Lieutenant Jose Maldonado, Pay Master of the Presidio of Santa Fe for the reward of the Comanches: a large box of good chocolate [un cajon de buen chocolate].*
> *Jose Maldonado, Pay Master*
> *Juan Batista de Anza, Governor Colonel. [81]*

A coda to the Spanish/Apache wars also is appropriate here. When chocolate availability in the Southwest borderlands was first noted in 1695, it was within the context of the Spanish/Apache wars, and this linkage between conflict and chocolate continued until Apache pacification in 1791. Several documents from the posthostility period itemized expenditures at the Presidio el Carrizal, where an undefined number of Apaches had been imprisoned. One of these documents, entitled *An account of the expenditures made in this Province and in the Pueblo del Paso [Texas] with the Apaches in Peace and War for prisoners of the year 1791*, is unusual in that it listed more than 120 items (many repeated) provided to both "friendly" and "captive" Apaches [82]. Of special interest here, however, is a specific notation in a portion of the document that the chocolate was ordered to be distributed specifically to wounded Apache prisoners:

> *6 tablets of chocolate for the wounded [erido; in original manuscript]. [83]*

Era of Political Transition and Independent Mexico

The late 18th through early 19th century was a period of military and political transition in Mexico and along the Southwest borderlands. The French Revolution of 1789 provided the textual doctrines for freedom and liberty that began to spread through New Spain. The French established themselves in Louisiana in 1803, and then during the administration of Thomas Jefferson came the Louisiana Purchase and the exploratory journeys of Lewis and Clark.

We identified an unusual chocolate-associated document dated June 8, 1814, that related how two women, Antonia Villavicencio and Dolores Blanquer, along with three unnamed male companions, purchased 19 *tercios* (approximately 4180 pounds) of cacao from Acapulco to be delivered to insurgents fighting for Mexican independence [84].

Don Jose Maria Arze was a cadet at the military school in San Antonio, Texas, in 1803. Fifteen years later, in 1818, after he had established his military credentials, Arze submitted an invoice to Hipolito Acosta, an army supplier based in Chiguagua (i.e., Chihuahua, Mexico) for goods to be sent from Chiguagua to San Antonio, Texas [85, 86]. Because of the quantities and weights listed in the requisition document, the invoice could not represent goods for personal use. Although the requisition document does not list the ultimate receiver of the goods—we suggest that it could have been ordered on behalf of the presidio at San Antonio. In addition, the document is important because of the considerable quantities of items requested, and the division of the chocolate into two distinctive categories, "fine" (*fino*) and "ordinary" (*ordinario*)—and clearly differentiated by social status: *fino* for the officers and *ordinario* for the foot-soldiers:

> *A box of fine chocolate weighing 6 arrobas 10 pounds at 592 pesos;*
> *A box of ordinary chocolate weighing 9 arrobas 24 pounds at 182 pesos;*
> *A box of ordinary chocolate weighing 7 arrobas 5 pounds at 140 pesos;*
> *A box of ordinary chocolate weighing 8 arrobas 5 pounds at 154 pesos;*
> *A box of ordinary chocolate weighing 7 arrobas 20 pounds at 153 pesos.*
> [87]

The invoice also identified the prices paid for the boxes used to transport the chocolate (10.0 pesos), cost for the canvas coverings to protect the goods (6.2 pesos), and sack-cloth for use under the canvas (5.0 pesos), and 3 wooden boards (11.2 pesos) [88].

We identified a letter from a resident at Santa Fe, New Mexico, dated to May 11, 1820, that revealed issues of social power and class structure, set within the way that subordinates requested chocolate from their superiors. The letter clearly differentiated soldiers and priests from "commoners," as the former would never use the language and style expressed in the following document:

Santa Fe May 11ᵗʰ, 1820

My distinguished Lieutenant-General, I humbly beg you to have the kindness to provide me with one pound (16 ounces) of chocolate and two piloncillos [cones of solid, brown sugar]. Wishing you—my Lieutenant-General—a thousand [sic] of happiness.

The lesser of your subjects [súbdito],

Joaquin [last name illegible]. [89]

Mexico received independence from Spain in 1821. A requisition dated three years later in 1824 (month and day not recorded) provided a list of foods that were forbidden to be imported to Mexico: chocolate was identified as one of the items [90]. The context for this document can be seen in how the Mexican revolution had disrupted the flow of chocolate into Mexico City, and many of the plantations had been destroyed during hostilities. To counter this problem, the Mexican Ministry of the Interior published an announcement in 1826 that stated priority would be given to settlers and investors who would develop (and restore) lands within Mexico in geographical areas suitable for cacao production [91].

Given the important role played by chocolate in the Southwest borderlands—whether Arizona, New Mexico, or Texas—we placed one of our highest research priorities to answering the following question: Was chocolate associated with important revolutionary events in Texas during February 1836 and specifically events during February 23 to March 6, 1836 during the Mexican siege and Texan defense at the Mission San Antonio Béxar (The Alamo)?

It would be logical that the military supply train that accompanied General Antonio López de Santa Anna north to San Antonio, Texas, would have included chocolate among the necessary supplies carried for the soldiers. We searched for but did not find textual evidence that chocolate was among the items used by Mexican soldiers en route to or during The Alamo siege and our research on this topic continues. We were successful, however, in locating a document dated November 18, 1833, countersigned by General Santa Anna, requiring chocolate be served at the officer training school mess located in Mexico City. This interesting and important document was countersigned as well by General José Maria Tornel.[11] The Santa Anna/Tornel document provided a summary of the provisions and general expenditures for the month of December 1830. For the officers' mess the following items were to be procured:

For the breadmaker—50.00 pesos

Lard [manteca]—	*10.47*
Chocolate—	*43.00*
Beans—	*17.00*
For the butcher—	*240.00*
Chickpeas—	*27.00*
Coal [for cooking]—	*30.00*

The list was followed by four unidentified items given merely as &c. for a total cost of 1300.50 pesos [92].

Given the prominent role of chocolate as an exploration, mission, and military beverage, it would be curious, indeed, if chocolate did not play a role in the food patterns of the Mexican soldiers at The Alamo battle. Perhaps evidence will be forthcoming.

Conclusion

Spanish interests in North America stemmed from the foundation and development of their settlement at Saint Augustine, Florida, in 1565. It would be expected that chocolate would play a role in St. Augustine daily life as the settlement developed, but the earliest record of chocolate use at the fort we located was dated only to 1642 (see Chapter 53 and Appendix 5). Close inspection of the document, however, suggests prior knowledge of chocolate by the Spanish "Flordinos," since members of the community commented on the arrival of chocolate grinding stones (*piedras de moler*) aboard ship, which would imply, logically, that the residents knew chocolate prior to 1642 [93].

Documents reviewed that spanned the period between 1565 and the late 17th century basically were devoid of chocolate-associated information. This lack, however, does not imply that future work in additional archives and libraries would not be worthwhile. Chocolate clearly was part of daily life in the Spanish Caribbean and mainland New Spain, so it would be logical, therefore, that as the Spanish moved northward within and along the Florida peninsula, and westward through what is now Louisiana and into the borderlands between what is now Mexico and the United States, that early chocolate documents could be found.

The Southwest borderland documents also reveal a long history of chocolate that intersected with religious and military expansion into North America, as evidenced by the considerable number of indigenous uprisings and conflicts with Native Americans. That Comanche scouts for the Spanish were paid partially with chocolate was an unexpected finding [94]. Similarly, that chocolate was served to captured, injured Apache Indians recuperating in Spanish hospitals likewise drew our attention [95]. That we could not document Spanish military requisitions for chocolate supplies, and their transportation by mule-train northward into Texas prior to the battle of The Alamo, was a disappointment. But this disappointment paled when set against the extraordinary chocolate documents associated with St. Augustine in 1642.

Captain Ermenexildo Lopez of the *Nuestra Señora del Rosario* never could have guessed how his decision to guide his damaged ship into St. Augustine harbor would ring out across time and space. These documents, lying basically unnoticed in the Archives of the Indies, Seville—and probably unread for many centuries—tell an interesting tale of chocolate, its importance and value, human nature and criminal behavior, and ultimate justice, when the boxes that contained the chocolate were distributed to the persons residing in St. Augustine in what was probably mid-March or April of 1642. The names of the administrators, religious leaders, soldiers, and tradesmen represent links in a human chain that stretch from the mid-17th to 21st century Florida. And what a thrill to see the handwritten name of Nicolás Ponce de León jump off the page—a presumed relative of the Florida explorer Ponce de Leon!

We expect that more chocolate-associated documents related to St. Augustine can be identified in the Archive of the Indies in Seville, Spain, perhaps some earlier and certainly later documents. And when found, what stories will they tell?

Acknowledgments

We wish to thank the following individuals who assisted during the archive research component of this chapter: Roberto Beristain Rocha, Director, Reference Center, Archivo General de la Nación (General Archive of the Nation), México City; Enrique Melgarejo Amezcua, Public Service Officer, Archivo General de la Nación (General Archive of the Nation); and Liborio Villagomez, Director, Reverence Center, Biblioteca Nacional, Universidad Nacional Auténoma de México (National Library), México City.

Endnotes

1. We identified another letter from Menéndez Avilés written to the Spanish king, on December 16, 1565, that provided early evidence for the *Manila* galleon. The document reported that four ships had returned from China, two of them to the port of Navidad (modern Acapulco), after 60 days at sea (AGI. SantoDomingo.224/Santo_Domingo.224.R.1,N.3/1/1 Recto).

2. Letter was recorded and signed by Governor Diego de Vargas's personal secretary, Domingo de la Barreda.

3. Father Almanza was affiliated with Father Juan María Salvatierra, who founded the Jesuit mission Loreto, in Baja California (see Chapter 34).

4. Religious efforts in the Southwest borderlands and California were conducted by many Jesuit and Franciscan priests. Of these, however, Fathers Junípero Serra and Eusebio Francisco Kino stand out. In contrast with Father Serra, Father Kino did not experience a food scarcity problem and difficulties of supply. Kino regularly is described as riding horses and wearing good clothing. Kino also was an accomplished administrator, astronomer, cartographer, and rancher. In personality, Father Kino commonly is described as "extremely austere."

5. Possibly the iridescent shells of abalone (*Haliotis* spp.)?

6. Especially notable were the efforts of Father Luis Xavier Velarde during epidemics of measles. Between November 1, 1725 and April 20, 1729, Father Velarde reported that "57 marriages have been celebrated in facie ecclesiae for the natives from the villages of Santa María and Guevavi and their adjacent rancherías. From the same villages and rancherías and others farther into the interior whose people frequent our villages from the end of November until the end of January, 276 children have been baptized, most of whom have died of measles, but with the grace of baptism." Some scholars have reported that more people died from smallpox and measles than during the Apache War hostilities. Available at http://www.nps.gov/tuma/historyculture/luis-xavier-velarde.htm. (Accessed January 10, 2007.)

7. Although this letter was written and signed by Joseph de Goxxaes (Gonzales) in 1738, the document was located in the General Archives of the Nation (AGN), Mexico City, in a file marked 1741. This is a recurring problem in many archives where documents are clustered by theme, not specifically by date—or misfiled.

8. After these tragic events the Yuman leaders informed the Spanish authorities that they did not mean to kill Father Garcés, but that he died during an unfortunate outbreak of violence; the intent had been to kill the settlers, not the priests, since the religious fathers had sympathized with the Yumans after the Spanish settlers took the Indian lands.

9. Felipe Neve de Arispe ultimately became governor of California.

10. These are two logical explanations for white chocolate. (1) Botanist to the Spanish king, Juan Pulgar, sent specimens of *Theobroma* to the Royal Botanical Gardens in Madrid in 1791 that he called *Theobroma alba* (white cacao). (2) An ancient method of preparing drinking chocolate still practiced in the 21st century also was named "white cacao." To prepare this beverage beans are placed inside a water-filled bag and buried in the ground for up to six months. This produces a defatted bean. A small feather-like tool is used to remove the cocoa bean shell and to retrieve the small white bead that remains. These white beans are frothed and poured over regular chocolate to form a "head"—much like the froth when cappuccino coffee is served. White chocolate of this type is a highest status food served to guests (Dr. Howard-Yana Shapiro, personal communication, 2007).

11. General José María Tornel served General Antonio López de Santa Anna as Minister of War and helped plan the campaign that led to The Battle of The Alamo.

References

1. Archivo General de Indias (Archive of the Indies, Seville, Spain). Menéndez de Avilés letter to the King

of Spain, Philip II, dated August 13, 1565. Santo Domingo, 224/Santo Domingo,224,R.1,N.1/21/recto. (Note: The repetition of Santo Domingo is not an error; this is the way the document is written by the archive office along the bottom of this page and subsequent documents in this file.)

2. AGI. Menéndez de Avilés letter to the King of Spain, Philip II, dated August 13, 1565, Santo Domingo, 224/Santo Domingo,224,R.1,N. p. 1/2/1 Verso.

3. AGI. Menéndez de Avilés letter to the King of Spain, Philip II, dated December 5, 1565, Santo Domingo, 224/Santo Domingo,224,R.1,N.2/1/1Recto-2/1/1 Verso.

4. AGI. Menéndez de Avilés letter to the King of Spain, Philip II, dated December 5, 1565, Santo Domingo, 224/Santo Domingo,224,R.1,N.2/1/1Recto.

5. AGI. Menéndez de Avilés letter to the King of Spain, Philip II, dated December 5, 1565, Santo Domingo, 224/Santo Domingo,224,R.1,N.2/1/1Recto.

6. AGI. Menéndez de Avilés letter to the King of Spain, Philip II, dated December 5, 1565, Santo Domingo, 224/Santo Domingo,224,R.1,N.2/1/3 Recto.

7. AGI. Menéndez de Avilés letter to the King of Spain, Philip II, dated January 30, 1566. Santo Domingo, 224/Santo Domingo,224,R.1,N. 3 BIS/1/1 Recto.

8. AGI. Menéndez de Avilés letter to the King of Spain, Philip II, dated January 30, 1566, Santo Domingo, 224/Santo Domingo,224,R.1,N. 3 BIS.1/2 Recto.

9. AGI. Menéndez de Avilés letter to the King of Spain, Philip II, dated January 30, 1566, Santo Domingo, 224/Santo Domingo,224,R.1,N. 3 BIS.1/1 Recto.

10. AGI. Escribanía 155 A. Dated September 1641, fs. 289–285.

11. AGI. Escribanía 155 A. Dated March 6, 1642, fs. 163–168.

12. AGI. Escribanía 155 A. Dated March 12, 1642, fs. 184–188.

13. AGI. Escribanía 155 A. Dated May 2, 1642, fs. 226–227.

14. AGI. Escribanía 155 A. (Undetermined date between March 6 and March 20, 1642. The cited document is attached to the back of a another dated March 6, 1642. The cited document consists of nine pages with all corners damaged and dates missing.)

15. Bense, J. A. *Presidio Santa Maria de Galve. A Struggle for Survival in Colonial Spanish Pensacola.* Gainesville: University Press of Florida, 2003.

16. AGI. Contaduria # 803.

17. http://en.wikipedia.org/wiki/Pensacola,_Florida. (Accessed January 10, 2007.)

18. Coronado, F. V. de. Letter to [Antonio] Mendoza. Dated August 3, 1540. Report Given by Francisco Vázquez de Coronado, Captain General of the Force That Was Sent in the Name of His Majesty to the Newly Discovered Country, of What Happened on the Expedi-

tion After April 22nd of the Present Year, 1540, When He Started From Culiacan, and of What He Found in the Country Through Which He Passed. In: Hammond, G. P., and Rey, A., editors. *Narratives of the Coronado Expedition 1540–1542.* Albuquerque: University of New Mexico Press, 1940; pp. 162–178.

19. Pedro de Castañeda, P. de. Account of Pedro de Castañeda, Dated Saturday, October 26, 1596, written in Seville. In: *The Journey of Francisco Vázquez de Coronado 1540–1542. As Told by Pedro de Castañeda, Francisco Vazquez de Coronado, and others.* Translated and edited by G. P. Winship. San Francisco, CA: The Grabhorn Press, 1933; pp. 1–78.

20. Simmons, M. *The Last Conquistador: Juan de Oñate and the Settling of the Far Southwest.* Norman: University of Oklahoma Press, 1991.

21. Díaz del Castillo, B. 1560–1568. *The Conquest of New Spain.* New York: Penguin Books, 1983; pp. 226–227.

22. Sahagun, B. 1590. (Florentine Codex). *General History of the Things of New Spain.* Santa Fe, NM: The School of American Research and the University of Utah Monographs of The School of American Research and The Museum of New Mexico, 1981; Part 12: pp. 12, 119–120, 176–178, 189.

23. Hernández, F. *Historia de las Plantas de la Nueva España.* México City, Mexico: University Press, 1577.

24. http://www.tsha.utexas.edu/handbook/online/articles/MM/fma71.html. (Accessed January 10, 2007.)

25. Massanet, D. Letter of Fray Damián Massanet to Don Carlos de Sigüenza. In: *Spanish Exploration in the Southwest, 1542–1706.* New York: Charles Scribner's Sons, 1916; p. 386. Available at http://www.americanjourneys.org/aj-018/index.asp. (Accessed January 10, 2007.)

26. http://www.thesantafesite.com/history.html. (Accessed January 10, 2007.)

27. http://www.newmexico.org/go/loc/about/page/about-history.html. (Accessed January 10, 2007.)

28. http://www.ppsa.com/magazine/NMtimeline.html. (Accessed January 10, 2007.)

29. AGI. Provincias Internas, pp. 34–42.

30. AGI. Provincia Internas, pp. 58–62.

31. Archivo General de la Nación (AGN), Mexico. Provincias Internas, Volume 37, exp.6, fs. 209–289, 1685—p. 240 verso.

32. AGN. Provincias Internas, Volume 37, exp.6, fs. 209–289, 1685—p. 241 verso.

33. Espinosa, J. M. *The Pueblo Indian Revolt of 1696 and the Franciscan Missions in New Mexico. Letters of the Missionaries and Related Documents.* Norman: University of Oklahoma Press, 1991.

34. AGI. Provincias Internas, pp. 122f.–123v.

35. AGI. Provincias Internas, pp. 64v.–67f.

36. AGI. Provincias Internas, pp. 83f.–84v.

37. AGI. Provincias Internas, pp. 99v.–67f.

38. AGI. Provincias Internas, pp. 57f.–61f.; 124v.–125v.

39. AGI. Provincias Internas, pp. 58f.–59v.

40. AGI. Provincias Internas, pp. 96f.–98v.

41. AGI. Provincias Internas, dated July 6, 1696, pp. 107–108.

42. AGI. Provincias Internas, dated July 6, 1696, pp. 110v.–113v.

43. AGI. Provincias Internas, dated March 6, 1712, Volume 230, no pagination.

44. AGI. Provincias Internas, dated March 14, 1712, Volume 230, no pagination.

45. AGI. Provincias Internas, dated 1712, Volume 230 no pagination.

46. AGI. Provincias Internas, dated 1713, Volume 230 no pagination.

47. Ortega, J. *Historia del Nayarit, Sonora, Sinaloa y ambas Californias con el título de "Apostólicos Afanes de la Compañía de Jesús en la América Septentrional" escrito por el Padre José Ortega*. Barcelona, Spain: 1754. Chapter VII. Expedición al Gran Nayarit, p. 83. (Version in Mission Archives, Santa Barbara was printed in Mexico, 1883.)

48. AGN. Provincias Internas, Volume 32, 2nd parte, exp. 15, fs. 410–412v., 1731.

49. Kino, E. F. *Kino's Historical Memoir of Pimeria Alta. A Contemporary Account of the Beginnings of California, Sonora, and Arizona, by Father Eusebio Francisco Kino, S.J., Pioneer Missionary Explorer, Cartographer and Ranchman. 1683–1711*, 2 Volumes. Translated and edited by H. E. Bolton. Cleveland, OH: The Arthur H. Clark Company, 1919; Volume 1, pp. 229–234.

50. Bolton, H. E. *Rim of Christendom. A Biography of Eusebio Francisco Kino. Pacific Coast Pioneer*. Tucson: The University of Arizona Press, 1936; p. 392.

51. Bolton, H. E. *Rim of Christendom. A Biography of Eusebio Francisco Kino. Pacific Coast Pioneer*. Tucson: The University of Arizona Press, 1936; p. 553.

52. Bolton H. E. *Rim of Christendom. A Biography of Eusebio Francisco Kino. Pacific Coast Pioneer*. Tucson: The University of Arizona Press, 1936; p. 430–433.

53. Bolton, H. E. *Rim of Christendom. A Biography of Eusebio Francisco Kino. Pacific Coast Pioneer*. Tucson: The University of Arizona Press, 1936; pp. 249, 252–253, 584–585, and 593.

54. AGI. Provincias Internas, dated June 17, 1716, Volume 230, no pagination.

55. Personal communication from Professor Bernard Fontana, Historian, University of Arizona Museum, Tucson, June 2005. Manuscript deposited in the archives, Documentary Research of the Southwest (DRSW), at the Arizona State Museum, University of Arizona, Tucson.

56. Personal communication from Professor Bernard Fontana, Historian, University of Arizona Museum, Tucson, June 2005. Manuscript deposited in the archives, Documentary Research of the Southwest (DRSW), at the Arizona State Museum, University of Arizona, Tucson.

57. AGN. Provincias Internas, Volume 32, 2nd parte, exp. 1 fs. 228–265, p. 233.

58. AGN. Provincias Internas, Volume 32, 2nd parte, exp. 1 fs. 228–265, p. 235.

59. AGN. Provincias Internas, Volume 86, fs. 362–397v.

60. Bolton, H. E. Preface. In: *Anza's California Expeditions*. Berkeley: University of California Press, 1930; Volume 1, p. xiv.

61. http://anza.uoregon.edu/people/anzabio.html. (Accessed January 10, 2007.)

62. www.library.arizona.edu/exhibits/desertdoc/massacre.htm. (Accessed January 10, 2007.)

63. www.library.arizona.edu/exhibits/desertdoc/massacre.htm. (Accessed January 10, 2007.)

64. www.library.arizona.edu/exhibits/desertdoc/massacre.htm. (Accessed January 10, 2007.)

65. Diehl, M. W., and Sugnet, C. Chocolateros. *Archaeology Southwest* Spring 2001:13.

66. Letter dated December 21, 1786 from María Ana Montielo to Father Francisco Antonio Barbastro. In: McCarty, K. *Desert Documentary. The Spanish Years, 1767–1821. The Yuma Massacre of 1781*. Tucson: Arizona Historical Society, 1976; pp. 35–40. Historical Monograph No. 4. Translated from the original on two folios in folder 40 of box 202, Civezza Collection, Antonianum Library, Rome Italy.

67. Letter dated December 21, 1786 from María Ana Montielo to Father Francisco Antonio Barbastro. In: McCarty, K. *Desert Documentary. The Spanish Years, 1767–1821. The Yuma Massacre of 1781*. Tucson: Arizona Historical Society, 1976; pp. 35–40. Historical Monograph No. 4. Translated from the original on two folios in folder 40 of box 202, Civezza Collection, Antonianum Library, Rome Italy.

68. Letter dated December 21, 1786 from María Ana Montielo to Father Francisco Antonio Barbastro. In: McCarty, K. *Desert Documentary. The Spanish Years, 1767–1821. The Yuma Massacre of 1781*. Tucson: Arizona Historical Society, 1976; pp. 35–40. Historical Monograph No. 4. Translated from the original on two folios in folder 40 of box 202, Civezza Collection, Antonianum Library, Rome Italy.

69. AGI. Provincias Internas, Volume 230, p. 113v.

70. AGI. Provincias Internas, Volume 230, pp. 112f.–113f.

71. AGN. Provincias Internas, Volume 67, exp. 1, f.1, 1786, p. 114v.

72. AGN. Provincias Internas, Volume 67, exp. 1, f.1, 1787, p. 259.

73. AGN. Provincias Internas, Volume 67, exp. 1, f.1, 1787, p. 270f.

74. AGN. Provincias Internas, Volume 67, exp. 1, f.1, 1787, p. 270v.

75. AGN. Provincias Internas, Volume 67, exp. 1, f.1, 1788, p. 329.

76. AGN. Provincias Internas, Volume 13, exp. 11, fs. 210–386, 1789, p. 272.

77. AGN. Provincias Internas, Volume 13, exp. 11, fs. 210–386, 1789, p. 272.

78. AGN. Provincias Internas, Volume 13, exp. 11, pp. 245–308v., 1790, pp. 210–211.

79. AGN. Provincias Internas, Volume 13, exp. 11, pp. 245–308v., 1790, p. 272.

80. AGN. Provincias Internas, Volume 13, exp. 11, pp. 245–308v., 1790, p. 275.

81. AGN. Provincias Internas, Volume 67, exp. 1, fs. 16–19v.

82. AGN. Provincias Internas, Volume 66, exp. 1, p. 122f–125v.

83. AGN. Provincias Internas, Volume 66, exp. 1, p. 125.

84. AGN. Infidencias, Volume 157, exp. 64, s/fs.

85. Bexar Archives on Microfilm: Roll 031, Frame 024. Transcription by Robert L. Tarín. Available at http://www.tamu.edu/ccbn/dewitt/adp/archives/translations/trans004.html? (Accessed January 10, 2007.)

86. http://www.tsha.utexas.edu/handbook/online/articles/SS/uqs8.html. (Accessed January 17, 2007.)

87. AGI. Provincias Internas, Volume 230, p. 6.

88. AGI. Provincias Internas, Volume 230, p. 6.

89. AGI. Provincias Internas, Volume 230, fs. 38–50, p. 38.

90. AGN. Folleteria, Caja 1, Folleto No. 31.

91. AGN. Folletería, Caja 2, Folleto No. 54. Memoria de los Ramos del Ministerio de Relaciones Interiores y Esteriores de la República, leída en las Cámaras del Soberano Congreso en los Días 9 y 14 de Enero del Año 1826. Mexico: 1826. Imprenta del Supremo Gobierno; pp. 15–16.

92. Centro de Estudios de Historia de México. CONDUMEX. 355.07 MEX. Núm.33825. Secretaria de Guerra y Marina, Secion 7a. Reglamento para el Colegio Militar.

93. AGI. Escribanía 155 A, dated September 1641, fs. 289–285.

94. AGN. Provincias Internas, Volume 32, 2nd parte, exp. 15, fs. 410–412v., 1731.

95. AGN. Provincias Internas, Volume 66, exp. 1, p. 125.

CHAPTER

Sailors, Soldiers, and Padres

California Chocolate, 1542?–1840

Louis Evan Grivetti, Patricia Barriga, and Beatriz Cabezon

Introduction

Documenting the first arrival of chocolate as a food or medicine in the geographical region now known as California has remained elusive. Spanish administrators, doctors, and priests in mid-16th century Mexico knew the importance of chocolate as both a food and medicine and by 1521 Hernán Cortés (commonly spelled Hernando Cortés or Cortez in English) had defeated the Mexica/Aztecs and already had drunk chocolate with King Montezuma (See Chapter 8). As the Spanish administration became established in New Spain in what is now Mexico City, they continued their search for gold. This quest, in part, was responsible for the initial Spanish westward exploration toward the Pacific Coast of New Spain, and administrators and explorers ultimately turned their attention northward toward lands called California.

It may be, however, that the idea of a California rich in gold existed well before Spanish contact with the New World [1]. In 1510, the Spanish author García Rodríguez de Montalvo published a romance novel in Seville entitled *Las Sergas de Esplandián*, where he described an island in the Indies where Amazon-like women were ruled by a black queen named Califia[1] (California = the land of Califia); the weapons of these women were said to be manufactured from gold [2].

Despite Montalvo's fictional account of a golden California—an idea that would prove correct in 1848 with the discovery of gold—the history of California was intimately associated with both military and religious opportunities. Missionaries and soldiers of the Spanish Crown traveled from the Mexican mainland to *Baja* (Lower) California, established settlements there, and ultimately reached uncharted lands to the north that they called *Alta* (Upper) California. These exploratory journeys to *Alta* California with their joys and sorrows have been well documented by diaries and correspondence. Through these manuscripts, available through archives and libraries in Mexico, North America, and Spain, we can see with extraordinary detail how life and times in early California were influenced by chocolate.

In this chapter we focus on chocolate-related information associated with the Spanish and Mexican periods of California history. In our efforts we identified and summarized documents of several types, especially the correspondence of Father Junípero Serra, the Franciscan missionary who left Spain for Mexico and who would become president of the missions

Chocolate: History, Culture, and Heritage. Edited by Grivetti and Shapiro
Copyright © 2009 John Wiley & Sons, Inc.

established in California. Through his letters and those of others, we learn that the missionaries responded on one hand to the Pope in Rome, who gave them the task to convert *naturals* (i.e., Indians or Native Americans) to the Christian faith, while other responsibilities included a strong role in the pacification and security of lands claimed as Spain's possessions by the military. These Californian Franciscan fathers—few in numbers but strong in faith and administrative qualities—ultimately accomplished both political and religious objectives and established a mission system alongside military forts and presidios that remain part of the California landscape in the 21st century.

The Father Junípero Serra collection, housed at the Santa Barbara Mission, Santa Barbara, California, represents an archive of 1075 letters that document the foundation of the California missions. Included, too, is correspondence between Father Serra and other missionaries, governors, presidio commanders, viceroys, and expedition leaders such as Gaspar de Portolá and Juan Bautista de Anza. While the Santa Barbara Mission archive is the primary repository of 18th and 19th century documents associated with the 21 California missions, other California archives provided additional documents. Especially noteworthy in our search were the Presidio Archive Library and Historical Society, located in Santa Barbara, and the Public Library and Historical Society Archives in Monterey, California. Augmenting these important California documents were additional manuscripts located in the *Archivo General de la Nación* (General Archive of the Nation) located in Mexico City.

It is the Father Serra archive, however, that contained the most telling documents. The initial letters in his diary were written in Catalan since his homeland was Mallorca, Spain, but he subsequently wrote in Castilian Spanish. His handwriting is distinctive and easily differentiated from that of other California Franciscan friars of his time. Of the 1075 letters examined, 44 contained references to chocolate and revealed the importance of chocolate in everyday mission life, as well as the importance of this food to both land and sea journeys during the Spanish and Mexican eras of California history. In total, the California chocolate-associated documents revealed a story not heretofore told, one that stretched across a vast geographical region from central Mexico to the Northwest lands in what is now British Columbia. The documents told stories of brave men, some driven by the search for gold, others driven by their religious calling to bring Christianity to a new land. It is to this story we now turn.

Cacao and Chocolate in California

It may be that the first Spaniard to enter California by land was Alvar Núñez Cabeza de Vaca,[2] who reached what is now called the Gulf of California in 1536 after an arduous overland trek through Texas, New Mexico, Arizona, and ultimately Mexico [3]. Given his destitution and the trials and tribulations associated with his trek, however, chocolate use by Cabeza de Vaca would not be expected. When Vaca ultimately reached Mexico he met with Viceroy Antonio de Mendoza and reported tales of the seven cities called Cibola—wealthy in gold. Mendoza subsequently sent Father Marcos de Niza (alternate spelling Marco da Nizza) and a small expeditionary group to search for Cibola and confirm Vaca's report. Niza returned to Mexico and provided (what would turn out to be) an exaggerated account to Mendoza, whereupon the viceroy ordered a military expedition be assembled to verify Niza's report—an expedition commanded by Francisco Vázquez de Coronado [4, 5].

While Mendoza sent Coronado northward into what is now the American Southwest in search of the seven cities of Cibola, the viceroy also planned and organized a second expedition to map and explore coastal California. This effort involved three ships—*San Diego, San Salvador*, and *Vittoria*—outfitted and captained by João Rodrigues Cabrilho (commonly spelled Juan Rodriguez Cabrillo in English). The expedition departed New Spain from the west coast port of Navidad (now Acapulco) on June 27, 1542. Expedition members landed several times along what is now the California coast and documented in the ship's log sites where the settlements of San Miguel (later called San Diego) and Santa Barbara ultimately would be developed [6].

The story of Cabrillo's injury, death, and presumed burial on one of the Channel Islands off the California coast are well known. Because gold was not discovered during the course of his expedition, Cabrillo's journey was not considered an influential endeavor and California basically remained a Spanish "backwater" until political pressures to establish and maintain a physical presence in this region emerged in the late 18th century [7].

Regarding the quest to determine the first use of chocolate in California, a basic question may be posed: Was chocolate among the supplies aboard Cabrillo's ships during his 1542 exploration of coastal California? The answer to this question, however, is not easy to verify. The first problem is that the original ship's log kept by Cabrillo has been lost. A secondary trip report exists, however, authored by either Bartolomé Ferrelo, the expedition's senior pilot who assumed command after Cabrillo's death, or by another mariner named Juan Páez, but chocolate is not mentioned in this document. The Ferrelo/Páez log, however, contains several tantalizing passages that pose even further questions regarding chocolate possibilities as illustrated by this example:

Latitude 30 degrees 24 minutes North; 27 miles northwest of Punta Baja, and five or six miles southwest of the village of San Quentín. . . . On Thursday they saw some smokes and, going to them with the boat, they found some thirty Indian fishermen, who remained where they were. They brought to the ship a boy and two women, gave them clothing and presents, and let them go. [8]

What were these presents offered to the California Indians? Katherine Abbott Sanborn, author of *A Truthful Woman in Southern California*, wrote on the Cabrillo expedition during her visit to Santa Barbara, California, in the late 19th century and published the following statement:

These Channel [Island] Indians let their hair grow so long that they could make braids and fasten them round the face with stone rings. The visitors [Spanish] spoke of the "Island of the Bearded People". . . . For gifts, they [the Indians] most desired red calico and chocolate. [9]

What is to be made of Sanborn's account since it is at odds with the Ferrelo/Páez log? Is it reasonable to suggest that, during her visit to Santa Barbara, Sandborn had access to the original Cabrillo log? We would suggest no. But might there have been other expedition documents available in the Santa Barbara Presidio or Mission that she could have examined? This could be a possibility. It also seems to us that given the widespread use of chocolate in late 19th century Santa Barbara, the calico–chocolate tale told to Sandborn more likely could have been part of local lore and oral tradition. Alternatively, her "red calico and chocolate" reference might be an artistic "flight," since she could have perceived it to be logical that chocolate could have been given as gifts to the Indians. Because the original Cabrillo log is lost, it is unlikely that a positive chocolate connection with this early expedition along the California coast can be determined, and the question will remain in the realm of speculation. While it could have been possible for Cabrillo and subsequent explorers of Spanish California, whether Sebástian Cermeño[3] in 1595 or Sebastian Vizcaíno[4] in 1602, to have carried chocolate on their sea or land journeys, textual documentation for such assertions also are problematical at present.

EARLY *BAJA* CALIFORNIA

After the second voyage of Vizcaíno to California, the Spanish Crown established the first settlements in *Baja* California in order to exploit the collection and commerce of natural pearls found in coastal waters (Fig. 34.1). Early 16th century attempts to establish forts or colonies, however, failed due to the opposition of the natives who rebelled after brutal treatment by the Spanish military [10]. In 1632, Francisco Ortega led an expedition and constructed a fort in La Paz on the southeastern side of the *Baja* Peninsula, but this

effort also was abandoned. The next 42 years saw additional expeditions to the region but each met with little success due to sustained hostilities between the Spanish and Native Americans. In 1679, however, Admiral Otondo y Antillón and the Jesuit missionary and explorer Father Eusebio Kino were successful and established a Spanish colony at La Paz (see Chapter 33). Four years later, in 1683, a mission was added to the colony—but like earlier attempts, this effort also was abandoned due to resumed Indian attacks caused by unjustified and senseless killings ordered by Otondo y Antillón [11].

The Italian-born Jesuit missionary Father Juan María de Salvatierra was appointed inspector of the Jesuit missions of the northwestern region of Mexico in 1690. After meetings and conversations with Father Kino, he conceived a plan for the colonization and evangelization of *Baja* California. Permission was granted and, on October 25, 1697, Salvatierra founded the first California mission, Our Lady of Loreto (*Nuestra Señora de Loreto*), at Concepcion Bay on the east coast of *Baja* California to honor the Loreto Sanctuary located in his Italian homeland. Soon after foundation of the mission, the Spanish started construction of a garrison that would become the Real Presidio of Loreto (Royal Presidio of Loreto),[5] the first presidio in California, sometimes called *Real Presidio de Las Californias* [12, 13].

The first mention of chocolate in California, whether *Alta* or *Baja* California, that we located was a document dated to March 18, 1726. While the document clearly is dated 1726, the date on the archive file is written 1739, which we interpret to mean that the document was misfiled or that the file at one time held a collection of earlier and later documents. The manuscript in question reported that goods were sent to the Spanish presidio at Loreto, *Baja* California. The author was Father Visitor Joseph de Echeverría, who listed the title as: *Invoice and Inventory of all the necessary goods that are brought to the Real Presidio de Las Californias [Loreto] from the Port of Matanchel [near modern San Blas].* The contents related that a merchant, Miguel Rodriguez Vaca, owner of a pack-train (*requa*), was responsible for transportation of goods to the presidio. Included among the items to be transported by Vaca were chocolate and chocolate preparation and serving equipment:

10 dozen hand mills [120 molinillos]; 1 box of chocolate cup gourds [jícaras[6]] 16 arrobas of superior chocolate [i.e., 400 pounds] at 10.2 reales per pound; 36 arrobas of ordinary chocolate [i.e., 900 pounds] at 7 reales per pound; 6 arrobas and 18 pounds of white sugar, at 18 reales per pound. [14]

Such a requisition was not an isolated event. Documents we examined from the period 1726–1739 revealed that chocolate and utensils used in chocolate preparation and serving were present and in similar amounts during 1774, 1777, and 1780:

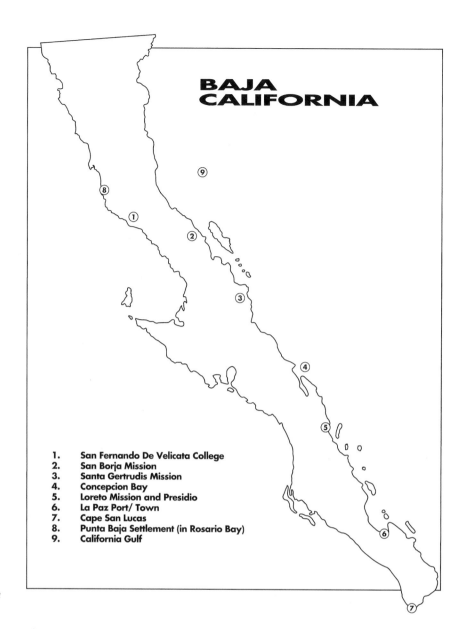

BAJA CALIFORNIA

1. **San Fernando De Velicata College**
2. **San Borja Mission**
3. **Santa Gertrudis Mission**
4. **Concepcion Bay**
5. **Loreto Mission and Presidio**
6. **La Paz Port/ Town**
7. **Cape San Lucas**
8. **Punta Baja Settlement (in Rosario Bay)**
9. **California Gulf**

FIGURE 34.1. Map of *Baja* California.

Year 1774 One copper chocolate pot; 17 medium size copper chocolate mugs; 3 sealed boxes of ordinary chocolate. [15]

Year 1777 12 solid boxes of chocolate with 7 arrobas each box [i.e., 2100 pounds]; 1 crater with 50 chocolate mugs. [16]

Year 1780 100 copper chocolate mugs; 144 molinillos; 25 boxes of fine chocolate, not loaded with too much sugar, and well dried at the time of packing; 1 crater of 30 pantles[7] of thick jícaras [i.e., 600 jícaras]. [17]

The 1780 document also contained the interesting notation that the requisition was an emergency request, where the goods were needed urgently, and concluded with the statement that the presidio at Loreto had been "entirely deprived [of chocolate]."

FRANCISCAN ARRIVAL IN MEXICO AND *BAJA* CALIFORNIA

Franciscan missionaries initially arrived in New Spain in 1524, just at the time of the Spanish conquest. But it would not be until more than two centuries later in 1749 that the Franciscan missionary Father Junípero Serra would leave Europe for religious work in the New World. When Serra left the Spanish port of Cadiz on August 28, 1749, aboard the ship *Villasota* (also called *Nuestra Señora de Guadalupe*), he boarded with his personal belongings and a quantity of chocolate. Accompanying Father Serra aboard were another 20 Franciscan missionaries. Also among these passengers were his student and disciple, Father Juan Crespi, who would become Serra's diarist on the subsequent expedition through *Alta* (Upper) California, and Father Francisco Palou, who later would become Serra's biographer.[8]

The journey to New Spain was especially long and arduous, a trial that lasted 99 days with two stops at Puerto Rico (the second necessitated by a storm) before the fathers arrived at the port of Vera Cruz (modern Veracruz) on December 6, 1749. Father Serra wrote a long account of the voyage in a letter dated December 14, 1749, addressed to his dear friend Fray (i.e., Father) Franchesco Serra (no relation), at the San Bernardino convent in Petra, Mallorca. Serra's letter contained three references to chocolate:

> [While aboard ship] *on the feast of "Our Lady of the Holy Rosary" in the fear that there was not enough water, rationing was introduced for all and the ration was so small that we could not drink the chocolate we brought with us because there was not water to boil it with; and this [rationing] lasted for all the 15 days that took us to reach the port of Puerto Rico that is 1200 to 1300 leagues from Cádiz.* [18]

Father Francisco Palou, Father Serra's traveling companion aboard ship, also mentioned this specific event in his biography of Serra written in 1787 three years after Serra's death:

> *On August 28, 1749 he (Father Serra) embarked in Cadiz; with him were 20 other religious [persons]. During the very long voyage—it took us 99 days to reach Veracruz we had many inconveniences and scares, due in part to the reduced space of the boat since they have also to accommodate besides our mission the Dominican's missions and many other passengers. Also 15 days before our arrival to Puerto Rico the scarcity of water was such that our ration of water diminished sharply, to the point that the amount of water that was given to us for 24 hours was just a "quartillo" [half a liter] and we could not even make chocolate.* [19]

Father Serra's account of his journey to New Spain mentioned that the ship put into Puerto Rico where the captain deceived the passengers regarding provisions. Still, Father Serra and his companions reported that they did not want for food because Christians associated with the local mission cared for them:

> *I will say it in a few words; we have arranged with the Captain of the boat that he would provide food for us while we were in the land, but once there he refused . . . so it was that we disembarked 20 religious and 3 servants without a penny to eat or drink. And yet for 18 days we ate better that in any convent, all drinking chocolate everyday and having tobacco to smoke and snuff, lemonade and refreshments in the afternoon and everything we could wish to have, and still we had more to bring aboard for the rest of our voyage.* [20]

The voyage continued, but after encountering a severe storm the ship was forced to return to Puerto Rico:

> *We disembarked, although not all of us, and soon [the plaza] was crowded; some [people] came with dishes [of food], other with chocolates, and that night we had everything in*

> *abundance, and chocolate in the morning and we still had a supply [of chocolate] to last for the rest of our journey.* [21]

Finally reaching the Mexican coast, Father Serra along with an unnamed colleague from Andalusia (probably either Father Francisco Patiño or Father Pedro Pérez) began their long walk of 250 miles to Mexico City and arrived 24 days later at the San Fernando College, Mexico City, on January 1, 1750 [22]. While working in the College of San Fernando—founded by the Franciscan Friars in 1731—Father Serra reviewed and revised a set of behavioral rules written for this college, a document that he would carry to the missions where he subsequently was posted, specifically, the Sierra Gorda Mission north of Mexico·City and later the College-Mission of San Fernando de Velicatá in *Baja* California that he founded prior to his departure north to *Alta* California. When Father Serra left Mexico City in 1752, he brought with him a copy of these revised rules to *Baja* California that defined administrative responsibilities. This document, entitled *Proposed Constitution for the San Fernando College 1751–1752 [México]*, described and defined in detail priestly activities, behaviors, and responsibilities and contained chocolate-associated passages. Section Seven of this document considered missionary responsibilities regarding chocolate and conversations with women:

> *The conversations with women no matter how brief and light they are always distracting and so it is commanded that the Missionaries should avoid them. As a consequence chocolate should not be administered or given to any women even if she is a Syndic,[9] during breakfast, lunch, meal or afternoon merienda [snack] at the porteria[10] or talk to her in the said place, in the Church, or in the Sacristy, or in any other part that belongs to the convent.* [23]

Section Sixteen of the document considered what was to be done at the *chocolatería* (special room for making chocolate)[11] and the traveler's lodge (*hospedería*) at the San Fernando College:

> *The Chocolate Shop [chocolatería] and the Traveler's Lodge [hospedería] have been under the care of a religious lay person; and because he must assist in the chocolate shop from 4:30 in the morning until 3:30 in the afternoon, when there is a conference, and when there is no conference until 3:00 in the afternoon, he is to be relieved to attend the Chorus hours.* [24]

THE *ALTA* CALIFORNIA EXPEDITIONS

Faced with the threat of Russian colonization from the north and Dutch incursions along the California coast, the King of Spain ordered a military expedition into upper or *Alta* California (Fig. 34.2). The objectives were to consolidate Spanish claims to the land then differentiated as *Baja* (lower) and *Alta* (upper) California. José de Gálvez, Spanish Inspector General, planned the California expedition that consisted of both land

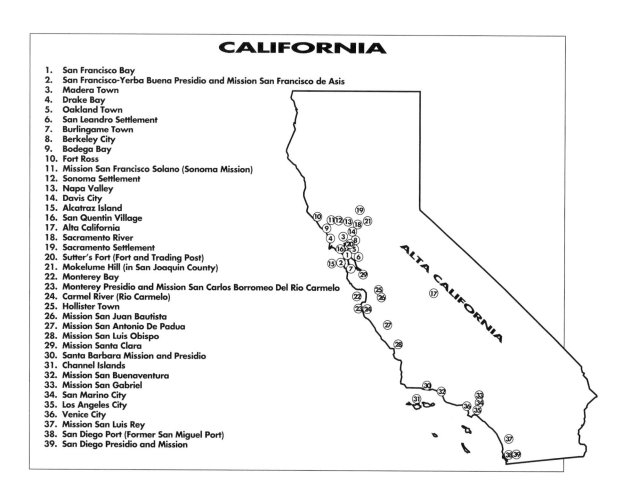

FIGURE 34.2. Map of *Alta* California.

and sea ventures. Gaspar de Portolá was identified to lead the land expedition that included the missionaries Junípero Serra, Miguel Campa, and Juan Crespi; their initial destination was the Spanish settlement at San Diego [25].

The expedition planned by Gálvez and his staff consisted of four enterprises to minimize risk of failure: two by land and two by sea. The two land expeditions to *Alta* California left from Loreto but on different dates. The first land expedition was led by Captain Fernando de Rivera y Moncada, commandant of the Loreto Presidio; he and his party left Loreto on September 30, 1768. On their way north they visited the older *Baja* missions to gather livestock and supplies. Father Juan Crespi, a personal friend and subordinate of Father Serra, the engineer José de Cañizares, along with additional soldiers and Indian archers accompanied the Rivera expedition. They arrived at the settlement of San Miguel (modern San Diego) on May 15, 1769 [26].

The second land expedition was commanded by Gaspar de Portolá, the civil and military governor of *Baja* Province. This expedition included Father Serra, Father Miguel de la Campa, 25 *soldados de cueras* (leather-jacket soldiers), and 30 Indian archers. Father Miguel de la Campa and Father Serra went as chaplain and diarist of this group. The Portolá expedition left Loreto

six months after the initial group on March 9, 1769. On May 13, 1769, this arm of the expedition had reached a small settlement just south of modern San Diego. Here, Father Serra and Father Campa founded the College/Mission of San Fernando de Velicatá that in subsequent years would provide supplies and provisions for the missions that ultimately would be founded in *Alta* California. Father Campa remained in Velicatá to serve the new mission. Moving forward, the expedition reached San Diego on July 1, 1769 [27].

A letter written to Father Campa by Father Serra and Father Crespi, dated May 14, 1769, identified the supplies left behind at Velicatá for the new mission, and mentioned that the chocolate had been provided to the priests by Governor Gálvez:

By order of His Excellency Visitador General it was given to Father Miguel de la Campa a fifth of the cattle that was given for the expedition; I gave him one of the four loads of biscuits, one third of the flour, and some soap we have brought with us. The Governor gave them some chocolate, raisins and figs; more than 40 fanegas of corn, so he [Father Campa] would remain with provisions and also with something to offer to the gentiles [Indians] until the arrival of new supplies. [28]

The first of the two sea expeditions to reach *Alta* California was commanded by an experienced mariner, Juan Pérez. His ship, the *San Antonio* (also known as *El Príncipe*), sailed from Cabo (Cape) San Lucas, *Baja* California, on February 15, 1769 and arrived in what is now San Diego on April 11, 1769. Upon arrival they found no other Spaniards whether from the land or sea expeditions. The second ship, the *San Carlos*, commanded by Vicente Vila, had left the Port of La Paz, *Baja* California, a month earlier on January 10, 1769 and after an extensive delay arrived in San Diego on April 29, 1769, after an arduous 110 days at sea. Aboard were several personages who would play important and interesting roles in the chocolate history of California, among these Lieutenant Pedro Fages (along with 25 Catalonian volunteers), Miguel Angel Constansó (engineer and acting cosmographer), and Don Pedro Prat (Frenchman and surgeon of the Royal Spanish Army) [29].

While the *San Antonio* commanded by Pérez had arrived in *Alta* California with its crew basically intact and only a few suffering from illness, the crew of the *San Carlos* was less fortunate. Upon reaching San Diego, sailors on the *San Carlos* were so weakened with illness that they were unable even to launch boats to reach shore and many had to be rescued by fellow crew members from the *San Antonio*. Curiously, the officers, chaplains, and the cook of the *San Antonio* had not become ill on the voyage.[12] The sick, rescued sailors were cared for by the Franciscan fathers and Dr. Pedro Prat, himself a convalescent. While it would be logical to suggest the possibility that the illness mentioned was scurvy, since it was identified variously as "pestilence," "plague," "sickness," or *mal de Loanda* [30–32], the problem is confounded and not easy to resolve since members of the *San Antonio* crew subsequently became "infected" and many died. Scurvy, of course, is not contagious.

On May 15, 1769, after 51 days on the road, the first land expedition led by Rivera y Moncada, along with the 25 *soldados de cuera* and three muleteers, arrived in San Diego. It was not until July 1, 1769 that the second land expedition led by Portolá—with Father Serra—arrived and found a host of sick men. Documents written at the time reveal an interesting clue as to the disease origin, for it was noted that the disease also had killed most of the live pigs transported aboard ship [33]. Writing from the perspective of the 21st century, might this event in early California history be a form of swine flu?

Still, those treating the sick considered it to be *mal de Loanda* (scurvy), which at the time was believed to be infectious. Father Serra described the tents used to house and to isolate the sick [34]. By one estimate only one-third of the original crews from both sea expeditions survived—and tragically all the sailors aboard the *San Carlos* with the exception of one (the cook) died[13] [35].

With almost no able crew, it was not possible for the two ships to continue northward toward Mon-terey Bay as previously planned. The *San Antonio* was ordered back to the port of San Blas to procure additional supplies and to obtain a new crew. With these difficulties at hand it was not until July 16, 1769 that the first European settlement was formally established in *Alta* California, with the construction of the mission and presidio at San Diego [36].

After the *San Antonio* returned to San Diego from San Blas with its new crew, Father Serra went on board and continued his journey northward to the Monterey region. Father Crespi, however, remained behind at San Diego. Serra arrived in Monterey on May 31, 1770. Shortly after, on June 3, he consecrated and founded the second mission in California, San Carlos Borromeo del Carmelo at the Monterey Presidio. On June 12, 1770, Father Serra wrote to Father Guardian Juán Andrés and described an unusual cold spell in early summer, and how important chocolate was in the daily life of the missionaries:

What we really need is wax for the masses, underwear robes, because, as I said it in other occasion is very cold here, also two good blankets for each of the new missions [San Diego and Monterey] and for the next one I hope soon will be established [Buenaventura] with a label each of them [2 for San Diego, 2 for Monterey and 2 for Buenaventura] so they will remain where they belong; we also need some chocolate, that Thanks God, we were not lacking up to now, due to the care of the Illustrious General Surveyor; also many heavy wool clothing and flannels to cover so many naked poor people that here are peaceful and docile. [37]

Food scarcity was common on long hard journeys to the distant missions, and on land and sea expeditions. Documents that describe these efforts commonly reveal that the missionaries, soldiers, and others accompanying on the journey existed only on tortillas and limited quantities of chocolate. Sometimes the documents are stark and describe the scarcity of food as very painful, so much so that members of the various parties complained that they did not have even a single *tablilla de chocolate* (*tablilla* = small board, in this sense a small, rectangular flat piece of chocolate) and as noted above, they regularly offered thanks to God for the chocolate supply.

Supplying missions and presidios became a difficult task in *Alta* California. Ships were delayed by storms or simply unable to reach their destination because the ravages of scurvy could leave them without able crews. In the almost nonexistent land routes, adverse weather conditions commonly determined whether or not muleteers with their trains loaded with provisions would arrive on time as requested. But even if ships and mules arrived on time, the supplies commonly were spoiled; missionaries' letters reveal that precious loads of beans (*frijoles*), chickpeas, or lentils arrived half eaten by insects or that jerked beef already had decomposed. Chocolate, on the other hand, did not suffer these problems. Chocolate did not spoil and was easy to carry even in the missionaries' or soldiers' pockets when on the road, and even small amounts provided much-needed energy. Chocolate often was distributed in the form of *tablillas*. Chocolate, too, was a substantial part of breakfast—either at the missions or for the road—and provided a good start for arduous journeys ahead. As a type of "road food" or travel provision, chocolate was highly appreciated [38, 39].

From the time of his arrival in Mexico until the year if his death in 1784, Father Serra made six voyages by sea back and forth between *Alta* and *Baja* California or overland to mainland Mexico City, commonly a round-trip land journey of more than 5500 miles. Serra knew well the hardships of the road and what it meant to run out of provisions. He also recognized that supply ships were irregular and might not arrive on time when anticipated. These worries might have contributed to his frequent mention of chocolate in his letters, since this commodity was food and energy for the road, and its use commonly was described as comforting [40].

Father Serra sometimes distributed chocolate to the sick, and also as an incentive for hard work. A letter written at the Presidio San Blas in Mexico by him, dated July 1, 1770, recorded the following:

The method I follow to ration the people that must remain with me in this new Presidio is to distribute daily to each one: a pint [the fourth part of a peck] of corn, half a pound of meat, one ounce of panocha [solid brown sugar], two ounces of pinole and 4 miniestora [ration of cooked legumes]; sometimes I stretched myself in giving a chocolate tablet to those that work harder; to the sick I always give [the chocolate] they need. [41]

Chocolate also is mentioned in documents where mule trains were scheduled to supply provisions to the respective missions in California. On October 12, 1772, Father Serra wrote from San Diego to Captain Pedro Fages, commandant of the presidio at Monterey:

The corn, flour and beans are the only foods that remain there because they ran out of meat. At the most, the mules will be able to make two more trips . . . and I doubt it. And what they carry cannot be just corn and flour, since it is necessary to carry some crates for the Fathers, like crates with chocolate, lard, [and] wine for the Masses . . . and therefore, it is clear the necessity that the mules of the King [are needed/are required] to help in the transport of the provisions for the Mission San Gabriel, that is the most important. [42]

One of Father Serra's letters, written April 22, 1773, to Viceroy Antonio Maria Bucarelli y Úrsua related the need to maintain the port of San Blas and to keep it open for commerce and supplies:

The notice of the intent of some persons that are promoting to close the Port of San Blas and the routes of the ships from that port to the ports of San Diego and Monterey, that had been supplying the new missions, has reached me . . . Last

year, both ships brought 800 fanegas of corn, that if this had to be carried by land it would have been necessary to do so with 400 mules . . . and not only for the corn but for the chocolate loads, clothing and other effects for the 13 religious [priests/companions]. [43]

The themes of chocolate, illness, and human greed appear in an unusual letter found in the Serra correspondence. Corporal Miguel Periquez wrote a letter in his own name and speaking for other soldiers; he denounced Pedro Fages, commandant of the presidio at Monterey.[14] The letter was generated because of ill treatment and abuse of soldiers identified as being sick with scurvy (*mal de Loanda*). Periquez sent his letter in either May or June of 1773 (exact month is unclear) to his captain at the Monterey Presidio, Augustin Callis. The letter then was forwarded to Father Serra who at that time was in Mexico City. The letter from Periquez was poorly written, as few ordinary soldiers were able to read or write in Spanish California. Father Serra rewrote the document, smoothing the spelling and context, then forwarded the revised Periquez letter to Viceroy Bucarelli y Úrsua under the title "Charity for the sick". This heading was followed by an explanation how this letter, written by Sergeant Periquez, came to his hands. Such a letter was highly unusual and the chocolate-associated content most interesting (Fig. 34.3):

When the ship San Carlos arrived to the Port of San Diego, being the troops were quite sick, the surgeon asked me to go to the boat and ask the Captain of the ship Don Vicente for the ration that belonged to the healthy and the sick; and chocolate, brandy [aguardiente], and vinegar to feed and to cure the sick. All this Don Vicente gave it to me generously, but when I was coming back (midway back) Don Pedro Fages intercepted me and took all the chocolate, brandy, and vinegar I was bringing for the sick; he did so saying that he was the Commandant and that he would give them [the patients] what he wanted. He did so, given them a very small amount and not the amount they needed, and this caused a great deal of pain to the surgeon Don Pedro Prat; the rest he [Fages] kept it for himself, and filled a whole chest with the chocolate that was destined for the sick. When we left San Diego to reach Monterey, he [Fages] left 14 sick soldiers in San Diego with no more food than some corn; the Governor Portolá gave them a little bit of chocolate, but he [Fages] left for them nothing, even though he had taken a box from the boat that belonged to the expedition. Along the entire road he [Fages] did not give a single chocolate tablet [to the soldiers] but when they would wash or mend his clothes. [44]

The outcome of Corporal Periquez's complaint, as transmitted to the viceroy for New Spain, Antonio María de Bucarelli y Úrsua via Father Serra's letter, was that Commandant Fages ultimately was dismissed. Tragically, the surgeon associated with the letter, Don Pedro Prat, later became ill and died [45]. The importance of this event also can be set within global maritime issues regarding shipboard illnesses, especially

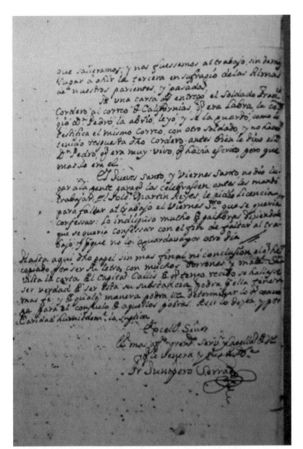

FIGURE 34.3. Father Junipero Serra correspondence: Serra's transcription of the Captain Pedro Fages misconduct letter (theft of chocolate meant for sick soldiers); dated May (no day) 1769. *Source*: Father Serra Collection, Document 332. Courtesy of the Santa Barbara Mission Archive-Library, Santa Barbara, CA. (Used with permission.) (See color insert.)

scurvy, and the important contribution by Surgeon James Lind, who published his *Treatise on Scurvy* in 1753 that documented the important role of citrus (limes) in offsetting this scourge [46–49].

CHOCOLATE IN TIMES OF SCARCITY AND HARDSHIP

Travelers to these unexplored lands of *Baja* and *Alta* California regularly experienced hardships. Chocolate was considered a valuable commodity not only for the missionaries and travelers but also for members of the military of the various sea and land expeditions that traveled up and down coastal California. Numerous archive documents identify what might be called today "times of scarcity" and related phrases such as: "we didn't even have a chocolate tablet" or "for breakfast we had coffee instead of chocolate." Such comments implied that chocolate was part of the daily life in early California and whenever missionaries, soldiers, or travelers asked for "emergency supplies" chocolate was on the list along with wine so that mass could be cele-

brated. In addition, chocolate also was viewed as a treat—as a comfort food—something to be drunk in the midst of hardships. The value placed on chocolate also is revealed when this commodity was provided as a reward for a job well done or offered to the sick.

A number of documents and letters associated with Father Serra reveal difficulties at his early mission post in northern Mexico and subsequently in *Alta* California. An early document of this type was written by Serra while at Mission Xalpam (in the Sierra Gorda Mountains), located in the state of Querétaro, Mexico, where the Franciscans had established five missions in the 18th century:

> *January 11th, 1762, Sierra Gorda Missions: The hardships that the Missionaries and the Indians endured were indescribable as they sustained themselves for many days with a little bit of chocolate and sometimes atole.* [50]

Elsewhere in the same 1762 document Father Serra reported that both priests and Indians suffered from local epidemics and subsequent lack of provisions:

> *Many Indians died because the epidemics we were suffering and those that survive fled because the hunger and necessities that we went through in those years because there was not any contribution from the Royal Treasury, nor with the necessary animals to till the soil, nor with cattle, nor with the necessary corn that is customary to give to the new reducciones [converted Indian people who lived on Mission land]; the hardships they went through, the Indians and their ministers, were many and for many days the ministers [i.e., the friars] sustained themselves with a little bit of chocolate and sometimes atole [corn-based gruel], and some comistrajo [a medley of foods: i.e., food of low quality/leftovers] that the charity of some persons would provide for them, and the Indians with some clothing, like chamal [cotton or woolen blankets that Indian women used to wear as tunic and men around the waist] and guapilla [a fruit]and other wild fruits, with a little bit of corn and beans the Fathers gave them.* [51]

Father Francisco Palou described the founding of the Mission San Fernando de Vellicata in northern *Baja* California in 1769 and identified Father Miguel de Campa as being in charge. Palou repeated Serra's request to the governor of Mexico (identified earlier) that Father Campa be supplied with needed food supplies:

> *The governor gave him a fifth of the cattle from Mission Borja. i.e. 46 animals. . . . Also they left with him 40 bushels of corn, half a load of flour, a load of biscuits, some chocolate, some meat, as well as a portion of figs, raisins and grapes to win over the native people; so now he has these assistance while waiting for more provisions sent from Loreto.* [52]

After Father Serra's departure to *Alta* California, Father Palou was appointed as the president of the *Baja* California missions that previously had been administered by the Jesuits before their expulsion from

Spanish territories for political intrigue. By the end of November 1769, Palou arrived at Loreto to assume administrative responsibilities over all church-related activities. During this time the governor of Loreto, Felipe Barri, had demonstrated his open dislike for the Franciscan friars, particularly for Father Palou, who then was in charge of the missions in his territory. In a letter written in Loreto–*Baja* California on November 24, 1769, Father Palou described the poor relations between the Franciscan fathers and Governor Barri. This deterioration seemed to be based on the governor's accusation that the Franciscans had plundered the Jesuit missions, taking valuable ornaments, sacred vases, and church jewels, and had stolen the Jesuits' store of chocolate [53]. In order to clarify and defend the Franciscan position, Father Palou wrote to Father Guardian Juan Andrés:

At the Presidio of Loreto the principal mission we were only given the church and the things concerning to the church, since the governor refused to give us all the rest that was destined to the Fathers; [he gave us] two rooms without any other furniture than a table, a chair, a leather cot, a candleholder and a bookshelf; all the other things remained in the hands of this [governor] and [in the hands] of his family; although he offered his table [to eat] to the Fathers, he would do it at the cost of the Mission, and did not even give them a tablilla [small tablet] of chocolate; and the same occurred-by order of the above mentioned Sir in all the other missions, where the soldiers were under orders of not to give us the chocolate that had been left by the Jesuits Fathers. [54]

Father Palou recalled another similar episode in his *Noticias de la Nueva California*—the first historical account of California:

We left [Mission] San Fernando de Velicatá [at] the frontier [with California] without any other load [except] some food and only some clothing for the six Fathers myself included that were going up north to San Diego. . . . Father Pedro Cambón wrote a letter begging him [the Governor] to allow him [while he was at Mission San Fernando de Velicatá] to send a box of chocolate, a set of clothes for the six fathers that just have reached San Diego, because thinking that soon all the other provisions will arrive have brought with them only the strictly necessary for the road; and he was sure that everybody was without a chocolate tablet, but the petition was denied. . . . [But] nothing of what we could say could stop his actions [the governors' actions]; he stopped all the provisions at the frontier. We were lacking of all and suffered until the frigate arrived with the provisions; for many months we were without even a chocolate tablet, without another set of clothes to wear, and other necessities that I leave out. [55]

During the first months of 1772, the decision was made at the highest levels of the Church in Rome that the *Baja* California missions were to be turned over to and administered by the Dominican order, a decision approved by the viceroy of New Spain. The Franciscan fathers, therefore, were forced to leave *Baja*

California, and after Mission Loreto buildings and their contents were turned over to the Dominicans, Father Palou departed for *Alta* California, leaving behind the mission he had served so well [56].

Palou left Loreto, walked overland to Mission Santa Maria (just south of modern San Diego), and was conducted safely to the San Diego Presidio. He then organized a mule-train to transport goods northward along the mission chain. He reached the newly founded mission at San Luis Obispo and wrote:

For the maintenance of the fathers Missionaries, the five soldiers and the two Indians, the Captain [perhaps Fages?] left 50 lb. of flour, half a half a load of wheat, a box of "panocha" [brown solid sugar] and another box with 112.5 lb of ordinary chocolate; he said that he will send more assistance from San Diego. [57]

Life at the missions was not easy. Father Palou wrote to Serra on January 14, 1774, regarding food needs at the Mission San Carlos de Monterey:

The necessities are reaching a high point and more than last year. . . . Here we are with only half a load of flour, a little bit of beans and some garbanzos for the Fathers and for the Indians; chocolate we have none and we do not have the wine we need to celebrate the mass, so we do not celebrate it everyday. [58]

That chocolate was an essential part of mission hospitality commonly was reflected in Father Serra's correspondence. On March 31, 1774, Father Serra wrote to Father Guardian Francisco Pangua in charge of provisions for the California missions and discussed the impending arrival of the Juan Bautista de Anza expedition from Sonora:

The Captain and the Fathers stated that they did not have even a chocolate tablet. . . . I will go to [Mission] San Gabriel to see how we can get them food supplies since they have not heard about the arrival of the boat and they do not have even a chocolate tablet or a tortilla to offer to the Fathers and to the Captain when they arrive. [59]

Several documents in the California Mission Archives reveal that food rationing commonly was required by the middle of each month. In a document written in San Carlos de Monterey dated January 8, 1775, Father Serra wrote again to Father Guardian Pangua to clarify issues regarding rations sent to the various California missions. In his letter Father Serra developed two lists of items needed to be sent on a regular basis. The letter also clarified a specific point that the chocolate supplies—when sent—were assigned as personal possessions for each father and therefore could be carried from one mission to another:

The charge for the 5 years ration that correspond to the Fathers can be applied to the Missions where they reside, discounting wherever they need for their clothes and other provisions; their chocolate although can be set apart in another list in case that one of the Fathers has to move to another mission. [60]

Another example of times of scarcity and hardships on the road and the importance of chocolate is a letter written on March 11, 1775, signed by two soldiers, a Sergeant Juan Puig and a Sergeant Miguel Periquez,[15] members of the Compañia Franca de los Voluntarios de Cataluña serving under Captain Callis in Monterey. In the letter they described in detail the small amounts of food that were given to them:

In Monterey since March 8, 1772 until June 15 of the same year it was given to us 8 ounces of meat and half a pound of flour daily; from the 16 of June until, September 8 we were not given more than five pounds of flour; and for orders of the Commandant, the sergeant that departed on August 3 in search of the deserters was given a ration for 25 days with no more than twelve ounces of bull meat, seven jícaras [small cups] of wheat, two pounds of flour and a chocolate tablet daily. [61]

A letter written by Captain Juan Bautista de Anza to the viceroy of New Spain, Don Antonio Bucarelli y Urúsa, dated December 8, 1775, provided a list of foods sent to the Río Carmelo Mission (Carmel) near Monterey, consigned to Father Francisco Garcés and unnamed companions (another priest, three interpreters, and a servant):

Following the orders of His Excellence I have mandated that the following list of supplies remain here [San Carlos Borromeo del Río Carmelo at Monterey] for Father Garcés and the other Father that accompany him, for their subsistence; also will remain here three interpreters and a servant. A list of supplies (for two persons): 2 boxes of beads (to trade with the Indians); 1 box of tobacco; 1 load of flour; 1 load of beans; 1 load of pinole; 450 lb of jerked beef ; 100 lb of biscuits; 25 lb. of chocolate; 1 lb. of sugar; 5 live cows; 6 medium size cheeses; 3 jams; 12 bars of soap; 1 jug of wine; 1 jug of aguardiente [Brandy]; 1 ax; 1 comal [a flat earthenware griddle used in Mexico to cook maize cake/tortillas]. [62]

When Father Francisco Palou completed his book, *Noticias de la Nueva California*, in 1780, he identified the standard provisions that the Reverend Father Junipero Serra had obtained for the California missions:

Of other provisions that the reverend Father Serra obtained for the Missions: 10 boxes of ham; 6 craters [large crates] of ordinary chocolate with a weight of 155 pounds each [a total of 930 pounds]; 4 barrels of wine from Castile; 960 lb of good flour; 900 bushels of corn; 250 bushels of beans; 16 boxes of sugar. [63]

On December 8, 1781, while at the Monterey Mission, Father Serra wrote to Father Lausén at the Santa Clara and San Antonio (California) Missions. The letter discussed crops growing at the missions and also reported that he had been injured after being kicked in the ribs by a mule, but that he was recovering:

I also wish that the corn has been growing well and that the grape vines are alive and produce fruits; the lack of wine for the masses it is already unbearable, and that we have ran out of chocolate and snuff is a sensible lost. [64]

A document dated to May 28, 1791, written by Father Manuel Rodríguez at the Mission Santa Gertrudis (in *Baja* California), complained that after six years of waiting for an answer (the original request for supplies had been written in 1785) mission supplies had still not arrived from Mexico City. The document is enlightening as it provides insights on personal and religious preferences, mentions the *Manila Galleon* trade with Asia, mentions the need for firearms and ammunition, and reveals the greater importance of chocolate over coffee in terms of quantities requested (2 pounds coffee vs. 225 pounds of chocolate). The manuscript repeated the original request for supplies:

5 arrobas [125 pounds] of carved, good chocolate [chocolate labrado] to be sent at all cost [a toda costa]; 4 arrobas [100 pounds] of the more appropriate cacao [beans meant] used to make chocolate; 8 arrobas [200 pounds] of sugar; 1.5 pounds of cinnamon; 2 pounds of candy sugar; 1.5 pounds of nutmeg; 2 pounds of coffee; 6 pounds of fine gunpowder; 1/5 arroba [12.5 pounds] of bullets and ammunition; 10 pairs of white cotton socks; 10 pairs of shoes; 20 blankets; 1 dozen of scarves [pañuelos] from China [via the Manila Galleon]; 4 dozen large heavy knives; and 20 bundles of tobacco. [65]

The documents cited here revealed that difficulties and hardships characterized mission life in Spanish California. The role of chocolate was perceived by the expedition leaders and church fathers as a necessity. The documents also revealed ethical considerations, as in the important letter written by Miguel Periquez to Father Serra, where chocolate assigned to ill soldiers was taken for personal use by Commandant Fages. One must marvel, too, over the patience and perseverance of Father Manuel Rodríguez, who dedicated his life to assisting others, who was seemingly abandoned/ignored and waited six years for promised supplies to arrive at his mission station in Santa Gertrudis, *Baja* California. After we located Father Rodríguez's second request for supplies, we searched further but were unable to locate subsequent correspondence indicating whether or not his request for supplies (and chocolate) was honored.

THE ANZA EXPEDITION

One of the key events in California history was the expedition by Juan Bautista de Anza overland from Sonora, Mexico, to *Alta* California. The objective was to open a land route, making it possible for year-around supplies to be sent to the California presidios and to the mission chain being founded by Fathers Crespi, Palou, and Serra.

Juan Bautista de Anza was an important figure in two geographical regions of New Spain, the

Southwest borderlands (see Chapter 33) and *Alta California*. Anza led an overland expedition to *Alta California* in 1774, leaving Tubac, south of modern Tucson, Arizona, on January 8, with a retinue that included priests, servants, and soldiers. After a trek of nearly three months he arrived at Mission San Gabriel in California on March 22, 1774. Having achieved success with his first overland venture, Anza led a second expedition to California in October 1775 that included nearly 300 potential civilian colonists and arrived in San Gabriel in January 1776 [66, 67].

Herbert Bolton summarized a letter that identified the provisions Anza obtained at Tubac in advance of the first successful overland journey to California:

For provision on the march from Tubac, estimated at a hundred days, Anza asked for a hundred beef cattle [they drove cattle along the way]—a barbecue every day; thirty pack loads of flour for tortillas; sixty bushels of pinole [corn meal]; sixty bushels of beans; six boxes of chocolate; and sixteen arrobas [400 pounds] of sugar. [68]

Elsewhere, Bolton identified the foods that would have been available to the Spanish officers and priests during the arduous journey:

Echeveste [government official] recommended for the table of the commander and the chaplain: a box of hams [175 pounds]; twenty-five pounds of sausage; six boxes of biscuits; a hundred and seventy-five pounds of fine chocolate [weighing 7 arrobas at 3.75 reales per arroba]; a barrel of wine [worth sixty-five pesos]; six arrobas of cheese [150 pounds]; four pounds of pepper; half a pound saffron; four ounces of cloves; four ounces of cinnamon; a jar of olive oil; and a jar of vinegar. [69]

It also may be noted that the Anza expedition carried 12 large chocolate pots [70].

Father Pedro Font accompanied Anza to California on the second expedition and kept a daily diary. His entry for November 16, 1775 recorded the use of chocolate as a "comfort food" provided to weary travelers:

The commander, seeing that we had gone ahead, thought we had found a better camp site than the one which he had told us of, and so he continued with all the train. But seeing that night was coming on and that he did not find us, he halted at half past four in the afternoon in the alameda of the river not far from the water, at once sending the sergeant to look for us. Having found us about a league below the place where the camp halted, the sergeant returned to report, and the commander sent two soldiers to stay with us during the night, bringing some cakes of chocolate, some dried fruit and a little biscuit. Here we passed the night, Father Thomás and I, both suffering from fever. [71]

Father Font's entry for December 3, 1775 identified the supplies necessary for seven civilians[16] who accompanied the expedition:

The rest of the day was spent trying to finish the cabin which, although it was not completed, was well along, and also the fathers were well satisfied with what was being left for the two of them and the seven persons who remained with them. It was as follows: A tierce of tobacco; two boxes of glass beads; an arroba of chocolate; an arroba of sugar; an arroba of tallow; five beeves [sic]; three tierces of dried meat; a pack load of beans; a pack load of fine flour; a little superfine flour; an almud of chickpeas; a box of biscuits; three hams; six cheeses; one frying pan; one griddle; one ax; two cakes of soap; twelve wax candles; a bottle of wine, with which it was not possible to say Mass because it was so bad that it did not resemble wine either in color or in taste so that it was necessary to go to Caborca to get some. All this was something, but not much in view of the nine mouths to feed and the time they were to remain and during which the provisions had to last them, which was until our return. [72]

In an unusual diary entry dated December 5, 1775, Father Font described drinking chocolate with Anza and local Native Americans in relatively strong language:

We halted in a plain with plenty of pasturage near one of the many lagoons in the bottom lands which are left full by the river when it goes down. Many Indians came to the camp, bringing calabashes, beans, and other crops of the kinds which they raise, and making their trades with the soldiers for beads, which the commander gave to the people for that purpose. Near the tent of the commander a beef was killed for today, to give rations to the people, as was done every six days. I was seated with the commander near the beef, taking chocolate. The Indians became such a mob and were so filthy, because of their vile habits, that we could not breathe, and there was no way by which to get away from them. So I stood up and, asking an Indian for a long stick, some ten palms long, with which they are accustomed to go about playing the wheel, I took hold of it at the bottom and with it gently and in good nature, as if I were laughing, made them get away from me and behind me. Thereupon an Indian immediately appeared offended, and, taking hold of the stick at the top, he again pushed in and others followed his example. Then the owner of the stick took it from my hands, and the one who was offended assumed a haughty air, and kept his eyes on me until I went into my tent. From this I inferred that all their affability, which is more due to the gifts of beads than to their gentleness, might easily be converted to arrogance whenever an attempt is made to reduce them to the catechism and to obedience, especially if we take into account their mode of living, of which I shall speak later on. [73]

On December 25, 1775, Christmas Day, Father Font wrote in his diary that he was ill and unable to drink chocolate. The letter is important as it also identified that drinking chocolate was the standard daily regimen along the march as directed by Anza. Curiously, Font made no mention of the significance of the date in his diary entry:

Since I was so ill, I was not able to conform to the regimen which the commander followed in the meals, which was chocolate in the morning, and then during the whole day nothing to eat until the day's march was ended, and at times not until night. I therefore many times asked for something to carry with me to eat during the day, although I might be traveling, and ordinarily I obtained it with a great deal of difficulty. Many times I went without supper because it consisted only of Chile and beans and I was better off without it, going to bed early without waiting for such supper and so late. [74]

Father Font made no further mention of chocolate in his diary until the expedition arrived in San Diego. Juan Bautista de Anza, however, filled the chronological "chocolate gap" with two entries that concerned chocolate stolen from the expedition stores by deserters. Both entries were approximately one month apart, the first dated February 12, 1776, followed by notations for March 7, 1776:

The first that was known of this matter was when one of muleteers who have remained returned at midnight from the ranch of the soldiers to sleep at his camp and noted some fragments scattered about. Inferring from this that there had been a robbery, he immediately reported to the officer and commissary, and they went at once to examine the site and the pieces, inquiring for the guard in whose charge they had been. He was not found, but they did find that they had stolen some glass beads, tobacco, and chocolate, which indicated a desertion. A review of all the men being held, it was found that the persons mentioned were lacking, besides two muskets, a saddle, and other things of less importance which they were able to lay their hands on. [75]

In various depositions which an officer has taken from the deserters for the purpose of learning whether the soldiers or any of the other members of the expedition wished to accompany them, and what may have been their motives for committing the treason of which they are guilty, they reveal nothing except what caused it. This was that a muleteer, having stolen two arrobas of chocolate, the sealed box of which was in his charge, as well as some brandy [aguardiente] taken from a sealed barrel, he became frightened lest as soon as I should notice this shortage and others of similar import I should punish him as he merited, because these things were property of the expedition and entrusted to his care. Not having any way of exculpating himself from everything, he now invited the five persons mentioned to flee with him, the plan being to go out to the province of Sonora to hide themselves there in order not to be discovered. [76]

Of the chocolate-associated entries in the Anza and Font diaries, the most interesting (and touching) were comments noted by Father Font on March 7, 1776, regarding how chocolate was used by a Spanish soldier as an engagement gift to his girlfriend:

While we were in San Diego the corporal of the guard of San Gabriel fell in love with a girl of our expedition; and since he had nothing to give her as a means of getting into her good graces, he urged the muleteers to give him something of what

came in their charge, and, condescending, they gave him chocolate and other things. [77]

After he described the chocolate "love gift," Font continued on the same entry date that a different corporal had plotted with the muleteers, raising the serious crime of desertion. Font's diary mentioned that the men were caught and interrogated and the plan to lead an Indian uprising in San Diego had been thwarted [78].

Father Pedro Font's last reference to chocolate was dated April 2, 1776, in the context of negative views of Native American behavior. He wrote that the items most coveted by the Indians were clothing, and while the Indians were in camp, they had stolen a "chocolate beater" (i.e., molinillo) [79].

CHOCOLATE IN CALIFORNIA PRESIDIO DOCUMENTS

As the Spanish pushed northward into *Alta* California, they established forts and military garrisons at strategic locations. Presidios were founded at San Diego (1769), Monterey (1770), San Francisco (1776), and Santa Barbara (1782); while the last was established in Sonoma, north of San Francisco (1836) to counter Russian colonial advances southward along the coast of north-central California. The Spanish colonial pattern in California was to establish centers of administrative control, with associated military garrisons and offices. Missions were established at each of the presidios and subsequently constructed at reasonable horseback travel distances between the presidios, along the route that became known as the "King's Highway" or *el Camino Real*, for a total of 21 missions established between 1769 and 1834 [80].

The sea and land expeditions to *Alta* California—planned by Joseph Gálvez and led by Portolá—needed provisions, goods, and arms. While the land expedition could carry only what they might need on the road, the sea expedition could carry all the supplies needed aboard ship as well as items necessary to supply the land expedition once they reached San Diego. These supplies—as planned—were enough to establish the three missions initially in *Alta* California: San Diego, San Buena Ventura, and Monterey [81]. A report signed by Gálvez in Cabo San Lucas, dated February 16, 1769, provided a detailed account of provisions, goods, and armaments for an estimated stay of eight months to be shipped aboard the *San Antonio* (sometimes called *El Príncipe*) and included the following:

8182 pounds of bread, 13525 pounds of flour, 19690 pounds of new corn, 2928 pounds of salted meat, 12 alive hens, 611 pounds of fish, 490 pounds of lard [manteca], 11 oil jars, 488 pounds of rice, 2295 pounds of chickpeas, 985 pounds of lentils, 1100 pounds of beans [frijoles], 490 pounds of cheese, 2 pounds of fine spices, 500 pounds of chili,

250 pounds of garlic, 240 pounds of figs, 660 pounds of salt, 2 barrels of rum [aguardiente], 2 large earthen jars of wine, 4 large earthen jars of vinegar, 839 pounds of sugar and 750 pounds of chocolate. [82]

A separate list (in the same document) identified kitchen utensils and among them a number of items used for the preparation of chocolate:

2 large copper saucepans with iron handles to prepare pozol [i.e., corn flour and ground cacao mixed with water], 2 spoons to prepare pozol, 2 copper kettles, 4 round copper pots, 4 frying-pans, 22 earthenware dishes, 2 metates with their manos [i.e., grinding stones and rollers], 2 copper pitchers, 17 small chocolate cups [pocillos] 3 pantles [bundle] of jícaras [used to serve chocolate] and two chocolate pots [to pour chocolate] and 4 molinillos [to froth the chocolate]. [83]

A second list identified the number of passengers aboard the companion vessel, the *San Carlos* (sometimes called *El Toisón de Oro*), and their occupations (i.e., captain, pilot, engineer, missionary, and surgeon) along with 56 soldiers and crew. The document also provided a list of provisions and goods that included:

275 pounds of chocolate, 32 small chocolate cups from China and earthenware [loza], 10 metates with their manos, 45 copper chocolate mugs, and 36 molinillos. [84]

After this initial supply of chocolate for the permanent settlement of California, the California presidios continued to enjoy an almost routine supply of chocolate. A document dated March 3, 1778 detailed quantities of an eight-month food supply shipped on the frigate *Santiago* as rations for the captain, pilot, chaplain, surgeon, two cooks, and 64 sailors/crew and garrison soldiers (a total of 70 persons who departed San Blas for new destinations of Monterey or San Francisco). The long list of provisions included jerked beef, fish, bread, beans, rice, lentils, chickpeas, lard, cheese, vinegar, salt, chili, water, firewood, and three *arrobas* (75 pounds) of chocolate [85].

An inventory of supplies and provisions for troops at the Presidio San Carlos de Monterey dated to 1781 documented the arrival of 12 boxes that contained 2100 pounds of chocolate [86]. Another interesting document also related to the Monterey Presidio was a requisition for chocolate prepared by the presidio surgeon dated February 25, 1794 and entitled *Requisition of the goods Don Pablo Soler, Surgeon of the Royal Army and a Resident of This Presidio [Monterey] is Asking to be Brought from Mexico, on Account of His Salary for the Year 1795*. Included in Don Soler's request were the items 150 pounds of ordinary chocolate and another 150 pounds of fine chocolate—to be supplied in a trunk [87].

Less than a month later, on March 12, 1794, Lieutenant Joseph Arguello, commander of the Monterey Presidio, requested one medium-size trunk with

five *arrobas* (125 pounds) of fine chocolate (perhaps for personal use?) [88]. Arguello's requisitions for chocolate continued and three weeks later, on March 30, 1794, he ordered more, this time however, not for personal use but for the presidio troops:

Requisitions of the goods and effects that the Lieutenant Joseph Arguello of the Presido of Monterey consider necessary for the provision of this regiment, for the invalids and for other people, craftsmen (artisans) and those that are dependents of this presidio.

For the troops and dependents
1400 pounds of ordinary chocolate in common trunks with a lock; 500 pounds of fine chocolate in three good trunks with a lock; 100 or 125 pounds of fine burned [dark roasted] chocolate in one good trunk with a lock. [89]

Chocolate was also considered part of the normal provisions for the San Diego Presidio troops as reported in a document from 1788. Lieutenant Zuñiga sent a requisition for goods on behalf of troops stationed at San Diego that included:

60 blankets, 60 jackets, 8 dozen beige trousers, 10 dozen wool socks, 24 dozen buttons, one barrel Catalonian brandy, 1400 pounds ordinary chocolate, 175 pounds of superior gift chocolate, 70 chocolate mugs, 16 boxes of small jícaras and 4 boxes of very fine chocolate cups. [90]

Six years later a similar document from San Diego, dated March 20, 1794, indicated not only the amount of chocolate ordered by box[17] but also the number of chocolate drinking cups requested, and continued to reveal that chocolate was part of the daily life of troops at the San Diego Presidio. In his request, Lieutenant Antonio Grajera itemized customary goods and effects "for the provision of the regiment of the Real Presidio de San Diego" and asked for not only three different qualities of chocolate but for 20 dozen chocolate drinking cups of various sizes:

six boxes of ordinary chocolate; one box of fine chocolate; one box of superior gift chocolate and 240 of small and large jícaras. [91]

Part of the California presidio chocolate story is associated with exploration of the north-central coast. Sir Francis Drake landed on the California coast north of San Francisco Bay in 1579 but missed the entrance to the bay [92]. Sebastián Cermeño explored the north-central coast as well in 1595 and also missed fog-shrouded San Francisco Bay [93]. Sebastián Vizcaíno explored the California coast and Monterey Bay region but made no report of a large bay [94]. It was not until 1769 that Sergeant José Ortega discovered the entrance, *La Boca del Puerto* (The Mouth of the Port), to what later would be called San Francisco Bay [95].

Further exploration of the entrance was delayed due to lack of Spanish administrative interest in *Alta California* at this time. It was not until 1775 that an expedition was organized to determine whether or not navigation was possible. The ship, the *San Carlos*, was assigned to this venture and this time was commanded by Captain Juan Manuel Ayala. Aboard were José de Cañizares, the first sailing master (who confirmed the navigational information), and Father Vicente Santa María, chaplain and diarist, who provided the following account:

> As we came to the shore, we wondered much [sic] to see Indians, lords of these coasts, quite weaponless and obedient to our least sign to them to sit down, doing just as they were bid. . . . The Indians made various signs of good will by way of salutations. . . . One of the sailors had brought some chocolate. He gave it to an Indian who finding it sweet, made signs that he would get something of similar flavor. He did so, bringing back to him a small tamale that had a fairly sweet taste and is made from a seed resembling polilla [Caesalpinia palmeris]. [96]

Is it not interesting that the first piece of chocolate offered to the indigenous peoples of the San Francisco Bay region was given as a token of friendship?

A presidio and mission would be established seven years later in 1776 in what is now San Francisco. At the San Francisco Presidio, chocolate was also considered a necessity for daily life. The annual request for provisions to be sent to the San Francisco Presidio signed by Joseph Arguello, dated May 21, 1787, itemized the following:

> Requisition of the goods, effects, and reales [currency] that are considered necessary for the provision and sustenance of the officials and troops that guard the Royal Presidio of San Francisco: Supplies for the troops [among them] . . . 4 dozens black hats; 6 dozens of white wool socks; 72 dozens of yellow metal buttons; 100 blankets; 525 pounds of ordinary chocolate in a trunk with iron fittings; 350 pounds of gift chocolate in a trunk with iron fittings; 30 strong chocolate mugs, big and small. [97]

We confirmed that chocolate remained a necessary supply at the San Francisco Presidio into the next decade. A document dated April 1, 1794, to the governor, Juan Joaquín Arrillaga, signed by Lieutenant Hermenegildo Sal, requested "all the things he [the governor] considered necessary to be sent from Mexico for the provision of the Presidio of San Francisco." In one section of the document he itemized provisions for the troops that included:

> 3 trunks of 525 pounds of fine chocolate with not too much sugar; 2 trunks with 350 pounds of ordinary chocolate. [98]

The Santa Barbara Presidio founded in 1782 contained invoices and ledger accounts that detailed the flow of goods, foods, and food preparation equipment. Provisions and supplies for the soldiers and their families shipped from the port of San Blas (Mexico) had to be requested months in advance. This pattern is documented through a remarkable 30-year list of requested goods and effects sent to Santa Barbara

from San Blas between the years 1779 and 1810. These requests, authorized and signed by Governor Neve, identified the items as "necessary for the provision and subsistence of the troops and relatives, and for the invalids [the sick or wounded soldiers that could no longer serve] living at the Real Presidio of Santa Barbara." Chocolate regularly was identified throughout the 30-year requisition sequence among the necessary products [99].

Year after year, through the last two decades of the 18th century and the first decade of the 19th century, cacao, chocolate, and the utensils used to prepare and serve chocolate (pots, mugs, *molinillos*, and *jícaras*) all were imported to the Santa Barbara Presidio. These chocolate-associated requisitions, in sharp contrast to documents usually associated with the missionary fathers, were specific in their descriptions. While missionaries commonly wrote in their letters and requisitions "please send some chocolate," military requests were precise and identified specific types of chocolate, whether so-called common or ordinary chocolate given to the foot soldiers and servants or fine chocolate reserved for the high ranking officials. Furthermore, a third type of chocolate also appeared in the Santa Barbara Presidio documents—identified as superior gift chocolate—a product reserved for special visitors and occasions (Table 34.1).

The documents between 1779 and 1810 do not identify the numbers of personnel associated with the presidio, guests/visitors, or ill persons—all of whom would have been served chocolate—but given the quantities identified, more than enough would have been available to meet presidio needs (Table 34.1). In most years, 1300 or more pounds of chocolate were available and in 1779 the quantity of imported chocolate exceeded 3000 pounds! The arrival of copper chocolate serving jars in 1779 also was augmented by many "bundles" of *jicaras* or ornamented gourd cups (each bundle was composed of approximately 20 cups). Beginning in 1804, the requests state that chocolate imported "should not be too sweet." While gift chocolate contained cinnamon; ordinary chocolate had no such designation.

EMERGENCE OF SPANISH–ENGLISH CONFLICT IN THE PACIFIC NORTHWEST

The Nuu-chah-nulth, also referred to as Aht, Nootka, Nutka, or Nuuchahnulth, are an indigenous First Nation society of Canada whose traditional home was in the Pacific Northwest along the west coast of Vancouver Island [100]. This geographical location in an isolated corner of North America became a flash point in a bitter international dispute between Spain and England. In 1761, King Carlos III of Spain appointed José Gálvez as Minister of the Indies to develop a plan for the exploration of the Pacific Northwest in response to an expanding Russian presence. As noted earlier in

Table 34.1 **Chocolate Requisitions: Santa Barbara Presidio, 1779–1810**

Year	Quantity	Weight	Jars or Pots	Jícaras or Cups
1779	18 boxes of fine chocolate	3150 lb	60 copper chocolate mugs	6 pantles (= 120)
1782	6 boxes of fine chocolate with metal plates and hinges	1050 lb	60 chocolate mugs	x
1784	6 boxes of fine chocolate	1050 lb	36 tinned copper chocolate (cobre estañado) mugs	6 pantles (= 120)
1785	1 box good gift chocolate; 3 boxes fine chocolate	Total weight 700 lb	x	6 pantles (= 120)
1787	6 boxes ordinary chocolate	1050 lb	x	x
1788	6 boxes: ordinary chocolate (1 box) gift chocolate	1050 lb 175 lb	x	2 pantles (= 40)
1789	2 boxes superior gift chocolate 6 boxes ordinary chocolate	350 lb 1050 lb	x	6 pantles (= 120)
January 1790	6 boxes ordinary chocolate 2 boxes superior gift chocolate	1050 lb 350 lb	x	10 pantles of fine medium size (= 200)
December 1791	2 boxes superior gift chocolate	350 lb	50 large copper chocolate mugs	x
1792	2 boxes superior gift chocolate 6 boxes ordinary chocolate Cacao Colorado Cacao from Caracas	350 lb 1050 lb 100 lb 100 lb	x	18 pantles of fine large jícaras (360)
1793	6 boxes ordinary chocolate 1 box very fine, superior chocolate	1050 lb 175 lb	x	24 pantles fine large jícaras (= 480)
1794	6 boxes ordinary chocolate Cacao Caracas and Colorado, fresh and clean	1050 lb 200 lb	2 large chocolate pots with lids, and 4 molinillos	1 box of fine jícaras of similar size (200 ?)
January 1795	6 boxes ordinary chocolate 2 boxes gift chocolate Cacao from Caracas Cacao from Tabasco	1050 lb 600 lb 100 lb 100 lb	2 chocolate pots with lids	18 pantles (= 360)
March 1795	6 boxes ordinary chocolate 1 box of gift chocolate for officials Fine chocolate for the troops Cacao Colorado and cacao from Caracas, fresh and good quality for the officials	1050 lb 175 lb 300 lb 150 lb	x	x
December 1795	6 boxes ordinary chocolate 2 boxes gift chocolate Cacao from Caracas Cacao Colorado	1050 lb 300 lb 75 lb 75 lb	72 chocolate copper mugs 72 molinillos	x
1796	6 boxes ordinary chocolate 2 boxes good gift chocolate Superior gift chocolate for officials Cacao from Caracas, fresh and clean Cacao Colorado, fresh and clean	1050 lb 300 lb 175 lb 75 lb 75 lb	72 large tinned copper mugs 72 molinillos	x
1797	6 boxes ordinary chocolate 1 box good, fine chocolate 1 box of the best fine chocolate that can be found; if this cannot be found please send the most fresh cacao—half of Caracas and half Colorado	1050 lb 175 lb (175 lb)	x	x
1798	6 boxes ordinary chocolate 6 boxes fine chocolate	1050 lb 1050 lb	80 chocolate mugs 80 molinillos	12 pantles (= 140) large, fine jícaras

Table 34.1 Continued

Year	Quantity	Weight	Jars or Pots	Jícaras or Cups
1801	6 boxes of ordinary chocolate 1 box fine chocolate 1 box of fine, dark roast chocolate [chocolate quemado].	1050 lb 175 lb 175 lb	x	x
1802/1803	6 boxes of ordinary chocolate 2 boxes of fine chocolate	1050 lb 350 lb	72 tinned copper mugs 72 molinillos	x
1804	6 boxes of fine chocolate 2 boxes gift chocolate with some cinnamon and not too much sugar	1050 lb 350 lb	x	x
1806	6 boxes ordinary chocolate 2 boxes very fine chocolate	1050 lb 350 lb	x	x
1807	6 boxes ordinary chocolate 2 boxes fine chocolate	1050 lb 350 lb	30 chocolate mugs	x
1808	Ordinary chocolate in boxes Gift chocolate not too sweet in boxes	700 lb 250 lb	x	8 pantles (= 160) fine jícaras
1809	Ordinary chocolate Gift chocolate not too sweet in boxes	700 250		5 pantles of fine jícaras (= 100)
1810	Gift chocolate in boxes Cacao, half Caracas and half Colorado	300 lb 200 lb	2 chocolate pots with handles 18 copper chocolate jars	x
Totals amounts From 1779 to 1810	Chocolate Cacao	33800 lb 1250 lb	6 chocolate pots 18 chocolate jars 550 chocolate mugs 300 molinillos	2520 jícaras

this chapter, Gálvez established the Naval Department at the strategic port of San Blas in 1767 and used the facilities to support the *Alta* California expeditions and as a base to provide the missions with supplies. He also developed a plan to explore the Northwest Pacific coast region [101].

Gálvez's plan, however, was slow to develop. The initial expedition to Nootka Sound took seven years to organize: boats had to be built; crews and military escorts assembled; and supplies obtained. By 1774, however, Gálvez's plan was ready for implementation and he identified Captain Juan Pérez to lead the expedition to the northwest coast of North America. Captain Pérez left San Blas, Mexico, on January 25, 1774, in the 82-foot frigate *Santiago*. His second-in-command was Esteban José Martínez. Their specific objective was to explore and claim the Northwest Pacific coastal territories for the Spanish Crown. Pérez reached latitude 55° North, sailed around Nookta Sound in what is now British Columbia and named it San Lorenzo. Neither he nor the crews of the two ships disembarked and after brief observations all returned to San Blas [102]. A second expedition to the Northwest coast the following year in 1775, commanded by Bruno de Hezeta, included a second ship captained by Juan Francisco de la Bodega y Quadra. On July 14,

1775, near Point Grenville (Washington State), some members of Hezeta's crew went ashore for water and were killed by local Native Americans. After this disaster Captain Hezeta returned to Mexico, but Bodega y Quadra[18]—true to his orders—continued northward and ultimately reached Sitka, in what is now Alaska [103].

A decade after the Hezeta/Bodega expedition, Esteban José Martínez commanded a routine supply voyage in 1786 and landed in Monterey, California. While awaiting clearance to return to San Blas, he met and conversed with the Comte de la Pérouse, captain of a French vessel that had anchored in the port of Monterey for repairs and supplies. Captain la Pérouse informed Martínez that the Russians were active in the Pacific Northwest and pushing southward. Upon his return to San Blas, Martínez reported this conversation to the Minister of Indies, José Gálvez [104].

Gálvez, as before, did not act immediately. It took almost three years after initial knowledge of the Russian incursions to implement his plan for a military expedition, a plan that included Spanish settlement of the Nootka region. When ready, this expedition consisted of three ships: the *Princesa* (commanded by Martínez), the *San Carlos*[19] (captained by López de Haro), and a frigate, *La Concepción* [105].

Captain Esteban José Martínez prepared final documents for his Nootka expedition and wrote Gálvez at San Blas on January 22, 1789. The document requested provisions and supplies needed to establish a Spanish settlement in Nookta. All items shipped aboard *La Concepción* were considered—in his words—"absolutely essential." His provision list was divided into two parts: food-related items and other goods differentiated from foods. An undefined quantity of chocolate appeared on the list—enough to provide for the garrison (*guaxnicion*), and the crews of the expedition ships (perhaps as rewards for their participation?). The document also mentioned seeds for planting, further indication of the Spanish intent to settle the Pacific Northwest region:

> Rice; garbanzos [chickpeas], lentils, beans, salted meat; ham; lard; oil; vinegar; chili; dry bread [hardtack-like] for 6 months; wine; brandy; fish [dried]; tobacco; seeds of cabbage, radish, onions, and some other vegetables; 4–6 fanegas of wheat; some metates [sic] with their hand-stones for grinding. [106]

The second list of goods in the document contained reference to equipment used to prepare chocolate. Neither list contained any items that could have been associated with women; hence, this initial expedition was purely military and to establish a basic settlement:

> 2 blankets; wool socks; buttons of various sizes; silk; soap; needles; shoes; pots and pans; chocolate pots [chocolattexos = chocolateros]. [107]

Since maize or wheat did not grow in the Nootka Sound region of North America at this time,[20] the grinding stones most probably were to be used to grind the cacao beans.

ERA OF MEXICAN INDEPENDENCE FROM SPAIN

The initial decades of the early 19th century were a transition period for California. Dissatisfaction and unrest associated with French occupation ultimately led to insurrection, and by 1821 Mexico had become an independent country. California was a province of Mexico from 1821/1822 through 1848. The United States declared war on Mexico on May 11, 1846, invaded the Mexican state of California, and occupied San Diego on July 29, 1846. California subsequently became a United States "holding" after signing of the Treaty of Guadalupe on February 2, 1848, which ended hostilities.

The Mexican period of California history is characterized by numerous chocolate-associated accounts. We identified an original document produced by the new government of Mexico entitled *El Territorio de la Federación Mexicana*, dated May 20, 1824, that provided the names of commercial items forbidden to be imported to Mexico, chocolate being among the prohibited food articles [108].

The new nation of Mexico also had a navy. A document entitled, *Budget for the Provisions for Three Months for the Crew of the Brig Morelos, Anchored in Acapulco*, provided a list of rations for this Mexican warship that had a standard crew of 105. The supply manifest listed a broad range of foods, among them salted meat, rice, chickens, vinegar, garlic (300 bunches), cherry wine, and 6 pounds of chocolate [109].

Chocolate also was associated with early 19th century festivities in Mexican California. The 1825 marriage of William Hartnell in Santa Barbara was considered by some to be the highlight of the social season. Hartnell was born in England and through a business partnership arrived in California in 1822. On April 30, 1825, he married Maria Teresa, daughter of Don José de la Guerra y Noriega, the fifth commandant of the presidio at Santa Barbara [110]:

> The marriage of a most prominent daughter of the country [hija del pais] to the most prominent foreigner in California. Such meals! At daybreak came desayuno [breakfast] hot chocolate and tortillas made of flour or corn followed by almuerzo [i.e., brunch?] after morning mass, between 8:30 and 9:00. . . . Tables were laden with produce from private gardens and mission lands. A family breakfast would include eggs or frijoles [beans] prepared in delicious Spanish style by Indian cooks, coffee with rich cream or chocolate, or tea, honey and tortillas. [111]

The war for independence in Mexico had only modest disruptive effects in California. This was not the case, however, elsewhere in Mexico, as in the southern portions of the country where the cacao plantations were concentrated. In fact, many of the cacao plantations in southern Mexico were destroyed during hostilities. In 1826, the government of Mexico issued a proclamation encouraging the settlement of lands especially suitable for cocoa production, in order to recover and redevelop the fields damaged during the war [112].

And following the reinstitution of cacao production, chocolate resumed its "flow" into California. An account of food customs of the *Californianos* (initial Spanish settlers of California) written in 1827 identified chocolate used on fast days:

> Fast days are distinguished from others only because one must eat either fish or flesh, without a mixture of both; in these days one sees even the table of the devotees as usual with meat, fish and vegetables and everyone according to his taste, devotes himself to one or the other of these dishes. The missionary fast is limited to not eating, ragouts at morning and evening, but with a cup of chocolate and a tart they wait patiently for dinner. [113]

In 1829, Alfred Robinson traveled to California and described the country and mission system. When visiting Mission San Luis Rey de Francia, located at modern Oceanside near San Diego, he wrote:

> The Reverend Father was at prayers and sometime elapsed ere he came, giving us a most cordial reception. Chocolate and refreshments were at once ordered for us, and rooms where we might arrange our dress, which had become somewhat soiled. [114]

Traveling north, Robinson and his party reached the San Juan Bautista Mission near modern-day Hollister, where they were assigned rooms with flea-infested beds. Robinson described the terrible night in his diary and how relief came only with chocolate served the following morning:

> Thus the whole night was passed in scratching and complaining till morning broke, when worn out with fatigue and loss of sleep, we finally closed our eyes and slept till roused [by the Fathers] to chocolate. [115]

Long after the Americans declared war on Mexico and invaded California, subsequently making California the 31st state of the Union, older *Californianos* sometimes sat for interviews recalling "earlier days." One such interview conducted with Guadalupe Vallejo in December 1890 described how the missions and presidios were supplied on a regular basis from Mexico:

> In 1776 the regular five year' supplies sent from Mexico to the Missions were as follows: 107 blankets, 480 yards striped sackcloth, 389 yards blue baize, 10 pounds blue maguey cloth, 4 reams paper, 5 bales red pepper, 10 arrobas of tasajo [dried beef], beads, chocolate, lard, lentils, rice, flour, and four barrels of Castilian wine. [116]

CALIFORNIA–RUSSIAN CONNECTIONS

As seen earlier, Spanish interests in California and the Pacific Northwest conflicted with Russian territorial expansion in North America. The Russians settled in Kodiak Island in 1784 and Russian fur trappers/traders established a trading post at Sitka (Alaska) in 1799, while the Russian explorer Ivan Kuskov explored the California coast in 1804 [117]. George Heinrich von Langsdorff (Grigory Ivanovitch), a Prussian physician and expatriate who lived in Russia, participated in another Russian voyage in 1806, visited the Aleutians and Sitka, and explored the coast of northern California [118]. Langsdorff's diary identified the potential for Russian trade with California and he itemized the goods needed in California that could be supplied by Russia:

> Goods required in Neuva [sic] California are, cloths, manufactured articles, sugar, chocolate, wine and brandy, tobacco, iron and iron tools, etc. Of these the Russian establishments, from Okhotsk to the northwest coast of America, are no less in want—perhaps even more in want—than the Spanish in Neuva California. If this plan could be carried out, and ships were sent from Europe to Neuva California to purchase, either with money or by barter, the breadstuffs and salted meat needed by the Russian possessions, and above all to collect sea-otter skins, this would, in my estimation, be to procure them at a much higher cost than if the Russian American Company should draw the supplies directly from Kronstadt [west of Saint Petersburg]. [119]

Kuskov continued his exploration of the North American west coast and in 1812 returned to California and established a settlement along the north-central coast of California known today as Fort Ross.[21] Fort "Russia" became the southernmost Russian colony in North America, and because of its fair climate and rich soil supplied food to the Russian settlements in Alaska [120]. Ultimately, Russian interests in Alaska and California flagged and the Fort Ross property and goods were sold to John Sutter in December 1841. Sutter sent his friend, John Bidwell, to the fort to take charge of the "arms, ammunition, hardware, and other valuables, including herds of cattle, sheep, and other animals." These were transferred to Sutter's Fort in Sacramento [121, 122].

Excavations at Fort Ross through the years have unearthed numerous cultural and historical items from bone harpoons; flint and glass arrowheads; English, Chinese, and Russian ceramics; cannon balls; gun flints; and glass beads [123]—but no evidence of importation or local manufacture of chocolate, or chocolate-associated tools, serving equipment, or utensils. This lack of chocolate history at Fort Ross represents a curious enigma, given the strong desire for and extensive use of chocolate in Tsarist Russia through centuries.[22] And while the Russians and Spanish/Mexicans of California tolerated each other—and only a relatively short distance separated Fort Ross from the Mexican presidio at what is now San Francisco—it is curious indeed that while the San Francisco Presidio was rich in chocolate drink, it cannot be as yet documented at Fort Ross. Further efforts along this topic could be rewarding.

Conclusion

The story of chocolate in Spanish and Mexican California is one of military and religious events. Associated with this rich history are some of the most prominent and influential persons in California history, among them Juan Bautista de Anza, Juan Francisco de la Bodega y Quadra, João Rodrigues Cabrilho, Juan Crespi, Eusebio Kino, Francisco Palou, Gaspar de Portolá, Juan María de Salvatierra , Junípero Serra, and Antonio Maria Bucarelli y Ursúa.

The first conclusive evidence we found for the introduction of chocolate into either *Alta* or *Baja* California was the March 18, 1726 document discussed earlier. Evidence at this time does not permit identification of an earlier date for chocolate use by Cabrillo or other Spanish explorers of California during the 16th and 17th centuries. We found no evidence that chocolate accompanied the Ortega expedition to *Baja* California or was part of the settlement stores after initial construction of the fort at La Paz. It could be, however, that documents in other archives may one day identify such use. It also would be logical that chocolate would have been associated with the earliest Jesuit missions in *Baja* California, potentially with Father Eusebio Kino's work there. The document on which to build—to seek potentially earlier records—is the March 18, 1726 letter, where Father Joseph de Echeverría requested that chocolate be transported to the Loreto Presidio from the port of Matanchal (near modern San Blas), and that the chocolate was shipped to Loreto by the owner of a mule-train, Rodríguez Vaca.

The documents examined and summarized in the present chapter reveal a long, sustained use of chocolate by missionaries, soldiers, and settlers. In order to develop and maintain a stable supply of chocolate to the California provinces, cacao beans had to be harvested in southern Mexico or elsewhere within the Central American provinces of New Spain, then prepared as chocolate. Once chocolate had been requisitioned by a mission or presidio in California, a year's supply commonly was ordered in advance. The chocolate then had to be transported whether by land (mule-train) or by sea aboard the various vessels sent by the government of New Spain. The documents we reviewed suggest that these supply ships and overland mule-trains, while dispatched at different times during the year, were unreliable in terms of meeting general target dates for receipt of the goods. And, of course, not all requisitions were handled on a timely basis, and some clearly were lost, as evidenced by the poignant message from Father Manuel Rodríguez at the Mission Santa Gertrudis in Baja California, who waited more than six years for his supplies to arrive from Mexico City!

Two events in the chocolate history of California captivated us during our research. Both selections have nothing to do with the enormous amounts of chocolate sent northward to California from Mexico, in some instances quantities of 3000 pounds or more being shipped. Furthermore, the two events that caught our attention would not have been especially notable at the time they occurred. Still, they are important and interesting due to their connection with human nature.

The first was the May/June 1773 letter written by Corporal Migual Periquez to Captain Augustin Callis, a document that revealed to us the strength and character of a soldier who risked his military career when he denounced Pedro Fages, the arrogant commandant of the Monterey Presidio, who had stolen the chocolate that was to have been given to Dr. Don Pedro Prat and used to treat sick soldiers. The ethical issues in Periquez's letter and how he drew the sharp line between what was right and what was wrong, still echo across the centuries.

The second was the diary entry by Father Pedro Font, who, on March 7, 1776, identified an unnamed Spanish soldier garrisoned in San Diego, who fell in love with a woman from the second Anza expedition. Having essentially little wealth and nothing to give the woman, muleteers from the Anza expedition presented the soldier with some of the expedition's chocolate stores so he could offer an item of quality to the woman and seal their engagement. Font's simple diary text does not convey the importance of the act, and seen through the perspective of 2007, two facts can be appreciated: women and men were arriving in California to settle a new land, and this unnamed woman was strong indeed, for she had walked from Arizona to California to seek a new life. Let us assume (although Father Font's diary is silent on this issue) that the unnamed couple married and had children. How did this family contribute to history of California? Perhaps one day, archivists working with genealogists and California military history experts might determine who they were.

Acknowledgments

We wish to thank the following individuals who assisted during the archive research component of this chapter. In México we thank: Roberto Beristain Rocha, Director, Reference Center, Archivo General de la Nación (General Archive of the Nation), México City; Enrique Melgarejo Amezcua, Public Service Officer, Archivo General de la Nación (General Archive of the Nation); and Liborio Villagomez, Director, Reverence Center, Biblioteca Nacional, Universidad Nacional Autónoma de México (National Library), México City. In California we thank: Dennis Copeland, Archivist, Monterey Public Library History Archive, Monterey; Patricia Livingstone, Former Director, Archive Library, Santa Barbara Mission, Santa Barbara; Brother Timothy Arthur, Director, Friar's Archive, Santa Barbara Mission, Santa Barbara (especially for his assistance in granting access to the Father Junípero Serra Collection); Lynn Bremer, Director, Archive Library, Santa Barbara Mission, Santa Barbara; Brother Joachim Grant, Santa Barbara Mission, Santa Barbara; Rafaela Acevedo, Presidio Trust of Santa Barbara Research Center, Santa Barbara; Kathleen Brewster, Founder of the Santa Barbara Presidio Trust, and Member, Board of Directors, Santa Barbara Mission Archive Library, Santa Barbara; Teresa Carey, Curator, San Antonio de Padua Mission Museum, Fort Hunter Liggett Military Reservation; and Louis Sanna, Curator, Carmel Mission Basilica Museum, Mission San Carlos Borromeo del Carmel, Carmel.

Endnotes

1. The 11th century epic *Chanson de Roland (Song of Roland)* identified a geographical location in Africa (Affrike) named Califerne (see *Chanson de Roland*. Boston, MA: Houghton, Mifflin, and Company, 1904; verse 209, lines 2920–2924).

2. Alvar Núñez Cabeza de Vaca was shipwrecked off the coast of Texas and for eight years lived among Native Americans and ultimately explored the Southwest regions of North America.

3. Sebastián Cermeño sailed from Manila aboard the *San Agustín* and reached the coast of California (in and around Drake's Bay), where his ship was wrecked. There is no evidence that he carried chocolate aboard the *San Agustín*.

4. Sebastián Vizcaíno explored the coast of California in 1602 but there is no evidence that he carried chocolate aboard his ship, the *Santo Tomás*.

5. Loreto served as the capital of *Alta* and *Baja* California for more than 130 years.

6. The word *jicara* comes from the name of the Jícaro tree (*Crescentia alata* and *C. cujete*). *Jícaras* are drinking cups made from the dried fruit/gourds of this tree.

7. Packing-related term: 1 *pantle* = 20 sets of matched *jicaras*.

8. Father Francisco Palou was a lifelong friend of Father Serra. In 1774, he joined the Rivera y Moncada expedition to the San Francisco Bay region and in 1776 founded Mission Dolores in San Francisco. While serving in San Francisco he compiled his book *Noticias*, a four-volume work considered to be the standard history of the California missions during the period 1767–1784. In 1784, Father Palou was called to Mission San Carlos del Borromeo del Río Carmelo (Carmel Mission) to assist the elderly Father Serra who lay dying. After Father Serra's death, Father Palou was appointed Acting President of California Missions. While serving at the Carmel Mission, he wrote his definitive *Life of Father Serra*.

9. Syndic: representative of the local religious community. Commonly, treasurer of alms/charity money used to support religious establishments.

10. Porter's lodge at the main entrance of a monastery or convent.

11. The lower apartments (cellars) in houses or convents, where foods or beverages were prepared; in the sense used here a specific location for the preparation of chocolate.

12. There is a suggestion that the officers and missionaries aboard the *San Carlos* did not sicken as did the crew, which possibly could imply two different diets based on work/social status. Given that the *San Carlos* voyage took approximately seven weeks, this would not be sufficient time for scurvy to erupt, unless the sailors already were malnourished. Might it also be that because the officers and friars drank chocolate—and because the water used to prepare the chocolate was heated, perhaps boiled—that it was this action that protected them?

13. Dr. Prat carried another manifest for the supplies to be shipped aboard the *San Carlos*. However, it would appear that these items were placed instead on the third ship, the *San Jose*, that wrecked and never appeared in San Diego. This document, cited by Graves, is in the archives of Guadalajara, Mexico, Document 1440, dated January 7, 1770. It listed 10 barrels of meat, 33 sacks of flour, 6 sacks of rice, 10 sacks of beans, 29 bushels of corn, 2 sacks of barley seed, 10 bushels of brown sugar, 1 box of coarse chocolate, 18 pounds of melon seeds, 10 pounds of watermelon seeds, 10 pounds of pumpkin seeds, an unspecified quantity of crackers, 1 box of fine chocolate, 1 barrel of California vinegar, and an unspecified quantity of lemon juice. This last entry is of great interest and suggests that Dr. Prat knew the medicinal value of lemon juice, 16 years after Lind's publication on scurvy.

14. Pedro Fages was appointed by Gálvez to accompany the sea component of the Portolá expedition in 1769. He was part of the first land expedition from San Diego to Monterey, then was appointed commander of the Monterey Presidio, a position he held from 1770 until 1774. According to the California State Military Museum, Fages was identified in accounts as "hot-tempered and inclined to storm over trifles, always ready to quarrel with anybody from his wife to the Padre Presidente" (i.e., Father Serra).

15. This is the same Miguel Periquez who wrote to Father Serra and condemned Captain Pedro Fages for misbehavior when the captain took chocolate that was to be given to ill soldiers at the Monterey Presidio.

16. Father Pedro Font did not name or describe these persons further. They may have been civilian settlers who accompanied the expedition.

17. Based on information we identified at the presidio at Santa Barbara, a normal box of chocolate supplied to the soldiers was 175 pounds; therefore, the six boxes identified in Grajera's request would equal 1050 pounds of ordinary chocolate. Ordinary chocolate was provided in units of 6 boxes (each box being as noted 175 pounds), while fine or superior gift chocolate was supplied in only one unit, an individual box that weighed 175 pounds.

18. Subsequently, Bodega y Quadra and British captain George Vancouver negotiated the Nootka Sound Convention. The land directly opposite Nootka Sound, in fact, was an island and Vancouver suggested that it be called Quadra and Vancouver's Island . . . but sadly for history Bodega y Quadra's name was removed by subsequent English cartographers, and only Vancouver's name remains today! Still, Bodega's name is not forgotten, as he was honored and associated with the beautiful, narrow bay just north of San Francisco, California.

19. This was the same *San Carlos* that was part of the Portolá sea expedition to San Diego.

20. Personal communication, Howard-Yana Shapiro, January 10, 2007.

21. Fort Ross is the English corruption of *Fort Russia*.

22. As evidenced by numerous examples of Russian chocolate pots, serving dishes, and chocolate-related advertising posters of past centuries.

References

1. http://en.wikipedia.org/wiki/Origin_of_the_name_California. (Accessed January 10, 2007.)

2. Rodriguez de Montalvo, G. *The Labors of the Very Brave Knight Esplandián*. Translated by W. T. Little. Binghamton, NY: Center for Medieval and Early Renaissance Studies, State University of New York at Binghamton, 1992.

3. Cabeza de Vaca, A. N. *Relación de los Naufragios y Comentarios de Alvar Núñez Cabeza de Vaca, Adelantado y Gobernador del Río de la Plata*, 2 Volumes. Madrid, Spain: Victoriano Suárez, 1906.

4. Coronado, F. V. de. Account by Francisco Vázquez de Coronado, Captain-General, written August 3, 1540. In: *The Journey of Francisco Vázquez de Coronado 1540–1542. As Told by Pedro de Castañeda, Francisco Vázquez de Coronado, and Others*. Translated and edited by G. P. Winship. San Francisco: The Grabhorn Press, 1933; pp. 86–99.

5. Bancroft, H. H. *The Works of Hubert Howe Brancroft. Volume XVIII, History of California, 1542–1800*. San Francisco: The History Company, 1886; pp. 6–10.

6. Bancroft, H. H. *The Works of Hubert Howe Brancroft. Volume XVIII, History of California, 1542–1800*. San Francisco: The History Company, 1886; pp. 64–72.

7. Bancroft, H. H. *The Works of Hubert Howe Brancroft. Volume XVIII, History of California, 1542–1800*. San Francisco: The History Company, 1886; pp. 69–73.

8. *Relation of the Voyage of Juan Rodriguez Cabrillo, 1542–1543. American Journeys Collection. Document Number AJ-001*. Wisconsin Historical Society Digital Library and Archives, Wisconsin Historical Society, 2003. Available at www.americanjourneys.org and www.wisconsinhistory.org. (Accessed January 10, 2007.)

9. Sanborn, K. A. *A Truthful Woman in Southern California*. New York: Appleton, 1893; pp. 140–141.

10. Bolton, H. E. *Rim of Christendom. A Biography of Eusebio Francisco Kino Pacific Coast Pioneer*. Tucson: The University of Arizona Press, 1936; pp. 100–120.

11. Bolton, H. E. *Rim of Christendom. A Biography of Eusebio Francisco Kino Pacific Coast Pioneer*. Tucson: The University of Arizona Press, 1936; pp. 105–124.

12. Bolton, H. E. *Rim of Christendom. A Biography of Eusebio Francisco Kino Pacific Coast Pioneer*. Tucson: The University of Arizona Press, 1936; pp. 22–23.

13. Bancroft, H. H. *The Works of Hubert Howe Brancroft. Volume XVIII, History of California, 1542–1800*. San Francisco: The History Company, 1886; pp. 69–73.

14. Archivo General de la Nación, México (AGN). Cárceles y Presidios, Volume 5, exp. 2, fs. 37f–81f.

15. AGN. Cárceles y Presidios, Volume 1, exp. L, fs. 1f.–14v.

16. AGN. Cárceles y Presidios, Volume 2, exp. l, fs. 1f.–7f.

17. AGN. Cárceles y Presidios, Volume 2, exp. 5, fs. 23f–29f.

18. Santa Barbara Mission Archives (SBMA). Father Serra Collection, Document 37, dated May 14, 1769.

19. Palou, F. *Relación Histórica de la Vida y Apostólicas Tareas del Venerable Padre Fray Junípero Serra, y de las Misiones que Fundó en la California Septentrional, y Nuevos Establecimientos de Monterey*. Mexico City, Mexico: Imprenta de Don Felipe de Zuñiga y Ontiveros, 1787; Chapter 13, p. 14.

20. SBMA. Father Serra Collection, Document 37, dated December 14, 1749.

21. SBMA. Father Serra Collection, Document 37, dated December 14, 1749.

22. Geiger, M. J. *The Life and Times of Fray Junípero Serra, O.F.M. or The Man Who Never Turned Back (1713–1784)*, 2 Volumes. Washington, DC: Academy of American Franciscan History, 1959; Volume 1, p. 83.

23. SBMA. Father Serra Collection, Document 53, dated 1751–1752.

24. SBMA. Father Serra Collection, Document 53, dated 1751–1752.

25. Geiger, M. J. *The Life and Times of Fray Junípero Serra, O.F.M. or The Man Who Never Turned Back (1713–1784)*, 2 Volumes. Washington, DC: Academy of American Franciscan History, 1959; Volume 1, pp. 208–217.

26. Geiger, M. J. *The Life and Times of Fray Junípero Serra, O.F.M. or The Man Who Never Turned Back (1713–1784)*, 2 Volumes. Washington, DC: Academy of American Franciscan History, 1959; Volume 1, p. 210.

27. Englebert, O. *The Last of the Conquistadores. Junípero Serra (1713–1784)*. Translated from the French by K. Woods. New York: Harcourt, Brace and Company, 1956; pp. 68–78.

28. SBMA. Father Serra Collection, Document 184, dated May 14, 1769.

29. Englebert, O. *The Last of the Conquistadores. Junípero Serra (1713–1784)*. Translated from the French by K. Woods. New York: Harcourt, Brace and Company, 1956; pp. 68–78.

30. Englebert, O. *The Last of the Conquistadores. Junípero Serra (1713–1784)*. Translated from the French by K. Woods. New York: Harcourt, Brace and Company, 1956; p. 80.

31. Geiger, M. J. *The Life and Times of Fray Junípero Serra, O.F.M. or The Man Who Never Turned Back (1713–1784)*, 2

Volumes. Washington, DC: Academy of American Franciscan History, 1959; Volume 1, p. 231.

32. Graves, C. Don Pedro Prat. A great and ineffable tragedy. *The Journal of San Diego History* Spring 1976;22(2):1–8.

33. Graves, C. Don Pedro Prat. A great and ineffable tragedy. *The Journal of San Diego History* Spring 1976;22(2):1–8.

34. Englebert, O. *The Last of the Conquistadores. Junípero Serra (1713–1784)*. Translated from the French by K. Woods. New York: Harcourt, Brace and Company, 1956; p. 80.

35. Geiger, M. J. *The Life and Times of Fray Junípero Serra, O.F.M. or The Man Who Never Turned Back (1713–1784)*, 2 Volumes. Washington, DC: Academy of American Franciscan History, 1959; Volume 1, pp. 208–217.

36. Bancroft, H. H. *The Works of Hubert Howe Brancroft. Volume XVIII, History of California, 1542–1800*. San Francisco: The History Company, 1886; pp. 115–130.

37. SBMA. Father Serra Collection, Document 217. Letter of Father Serra to Father Guardian Juán Andrés, dated June 12, 1770.

38. SBMA. Father Serra Collection, Document 217, dated June 12, 1770.

39. SBMA. Father Serra Collection, Document 332, dated May 1773.

40. Geiger, M. J. *The Life and Times of Fray Junípero Serra, O.F.M. or The Man Who Never Turned Back (1713–1784)*, 2 Volumes. Washington, DC: Academy of American Franciscan History, 1959; Volume 1, p. 416.

41. SBMA. Father Serra Collection, Document 227-A dated December 1, 1770.

42. SBMA. Father Serra Collection, Document 302, dated October 12, 1772.

43. SBMA. Father Serra Collection, Document 331, dated April 2, 1773.

44. SBMA. Father Serra Collection, Document 332, dated May (no day) 1773.

45. SBMA. Father Serra Collection, Document 333, dated March 23, 1773.

46. Geiger, M. J. *The Life and Times of Fray Junípero Serra, O.F.M. or The Man Who Never Turned Back (1713–1784)*, 2 Volumes. Washington, DC: Academy of American Franciscan History, 1959; Volume 1, p. 377.

47. Lind, J. *A Treatise of the Scurvy. In Three Parts. Containing an Inquiry into the Nature, Causes and Cure, of That Disease. Together with a Critical and Chronological View of What Has Been Published on the Subject*. Edinburgh, Scotland: Printed by Sands, Murray and Cochran for A. Kincaid and A. Donaldson, 1753.

48. Carpenter K. J. *The History of Scurvy and Vitamin C*. Cambridge, England: University Press, 1986.

49. Graves, C. Don Pedro Prat. A great and ineffable tragedy. *The Journal of San Diego History* Spring 1976;22(2):1–8.

50. SBMA. Father Serra Collection, Document 74, dated January 11, 1762.

51. SBMA. Father Serra Collection, Document 74, dated January 11, 1762.

52. Palou, F. *Noticias de la Nueva California*. San Francisco: no publisher, 1780; Volume 2, Chapter 6, p. 28.

53. Geiger, M. J. *The Life and Times of Fray Junípero Serra, O.F.M. or The Man Who Never Turned Back (1713–1784)*, 2 Volumes. Washington, DC: Academy of American Franciscan History, 1959; Volume 2, p. 13.

54. SBMA. Father Serra Collection, Document 192, dated November 24, 1769.

55. Palou, F. *Noticias de la Nueva California*. San Francisco: no publisher, 1780; Volume IV, Chapter 42, pp. 27–30.

56. Geiger, M. J. *The Life and Times of Fray Junípero Serra, O.F.M. or The Man Who Never Turned Back (1713–1784)*, 2 Volumes. Washington, DC: Academy of American Franciscan History, 1959; Volume 1, pp. 13–15.

57. Palou, F. *Noticias de la Nueva California*. San Francisco: no publisher, 1780; Volume III, Chapter 35, pp. 34–35.

58. SBMA. Father Serra Collection, Letter from Palou to Serra, Document 387, dated January 14, 1774.

59. SBMA. Father Serra Collection, Letter from Serra to Father Guardián, Document 410, dated March 31, 1774.

60. SBMA. Father Serra Collection, Document 501, dated January 8, 1775.

61. SBMA. Father Serra Collection, Document 512, dated March 11, 1775.

62. SBMA. Father Serra Collection, Document 587, dated December 8, 1775.

63. Palou, F. *Noticias de la Nueva California*. San Francisco: no publisher, 1780; Volume 3, Chapter 41, p. 128.

64. SBMA. Father Serra Collection, Document 879, dated December 8, 1781.

65. AGN. Misiones, Volume 5, exp. 4, fs. 63f–65v, 1791.

66. Bowman, J. N., and Heizer, R. F. *Anza and the Northwest Frontier of New Spain*. Southwest Museum Papers Number 20. Los Angeles: Southwest Museum, 1967.

67. Bolton, H. E. *Anza's Expeditions de Anza*, 5 Volumes. New York: Russell and Russell, 1966; Volume 2, passim.

68. Bolton, H. E. *Anza's Expeditions de Anza*, 5 Volumes. New York: Russell and Russell, 1966; Volume 1, p. 223.

69. Bolton, H. E. *Anza's Expeditions de Anza*, 5 Volumes. New York: Russell and Russell, 1966; Volume 4, pp. 223–224, 231.

70. Bolton, H. E. *Anza's Expeditions de Anza*, 5 Volumes. New York: Russell and Russell, 1966; Volume 1, p. 224.

71. Font, P. Diary. Entry for November 16, 1775. Available at http://anza.uoregon.edu/. (Accessed January 10, 2007.)

72. Font, P. Diary. Entry for December 3, 1775. Available at http://anza.uoregon.edu/. (Accessed January 10, 2007.)

73. Font, P. Diary. Entry for December 5, 1775. Available at http://anza.uoregon.edu/. (Accessed January 10, 2007.)

74. Font, P. Diary. Entry for December 25, 1775. Available at http://anza.uoregon.edu/. (Accessed January 10, 2007.)

75. Anza Diary. Entry for February 12, 1776. Available at http://anza.uoregon.edu/. (Accessed January 10, 2007.)

76. Anza Diary. Entry for March 7, 1776. Available at http://anza.uoregon.edu/. (Accessed January 10, 2007.)

77. Font, P. Diary. Entry for March 7, 1776. Available at http://anza.uoregon.edu/. (Accessed January 10, 2007.)

78. Font, P. Diary. Entry for March 7, 1776. Available at http://anza.uoregon.edu/. (Accessed January 10, 2007.)

79. Font, P. Diary. Entry for April 2, 1776. Available at http://anza.uoregon.edu/. (Accessed January 10, 2007.)

80. Geiger, M. J. *The Life and Times of Fray Junípero Serra, O.F.M. or The Man Who Never Turned Back (1713–1784)*, 2 Volumes. Washington, DC: Academy of American Franciscan History, 1959; Volume 1, passim.

81. Geiger, M. J. *The Life and Times of Fray Junípero Serra, O.F.M. or The Man Who Never Turned Back (1713–1784)*, 2 Volumes. Washington, DC: Academy of American Franciscan History, 1959; Volume 1, passim.

82. SBMA. Father Serra Collection, Document 161A, dated February 16, 1769.

83. SBMA. Father Serra Collection, Document 161A, dated February 16, 1769.

84. SBMA. Father Serra Collection, Document 161B, dated February 16, 1769.

85. AGN. Provincias Internas, Volume 33, SIN exp. fs. 119–121.

86. AGN. Cárceles y Presidios, Volume 3, exp. 4, fs. 129f.–132v.

87. AGN. Provincias Internas, Volume 8, exp. 19, fs. 115–116.

88. AGN. Provincias Internas, Volume 8, exp. 19, fs. 115–116. (*Note*: This is the second document in the file with the same number.)

89. AGN. Provincias Internas, Volume 8, exp. 17, fs. 109–113v.

90. AGN. Cárceles y Presidios, Volume 3, exp. 10, fs. 329f.–338v.

91. AGN. Provincias Internas, Volume 8, exp. 14, fs. 103–106v.

92. Drake, F. *Account by Sir Francis Drake held from the haven of Guatulco, in the South Sea, on the east side of Nueva Espanna, to the Northwest of California, as far as fourtie three degrees: and his returne back along the said Coast to thirty eight degrees: where, finding a faire and goodly hauen, he landed, and staying there many weekes, and discouering many excellent things in the countrey, and great shewe of rich minerall matter, and being offered the dominion of the countrey by the Lord of the same, hee tooke possession thereof in the behalfe of her Maiestie, and named it Noua Albion*. In: *Sir Francis Drake. 1628. The World Encompassed. Being His Next Voyage to That to Nombre de Dios Formerly Imprinted; Carefully Collected Out of the Notes of Master Francis Fletcher, Preacher in This Imployment, and Diuers Others His Followers in the Same; Offered Now at Last to Publique View, Both for the Honour of the Sactor, but Especially for the Stirring up of Heroick Spirits, to Benefit Their Countrie, and Eternize Their Names by like Noble Attempts*. London, England: Nicholas Bourne, 1628; pp. 221–226.

93. http://www.californiahistory.net/3_PAGES/manilla_cermeno.htm. (Accessed January 10, 2007.)

94. http://www.books-about-california.com/Pages/History_of_California_HEB/History_of_California_Ch03.html#anchor-Vizcaino. (Accessed January 10, 2007.)

95. http://www.books-about-california.com/Pages/History_of_California_HEB/History_of_California_Ch04.html#anchor-ortega. (Accessed January 10, 2007.)

96. Santa María, V. *The First Spanish Entry into San Francisco Bay, 1775. The Original Narrative, Hitherto Unpublished by Fr. Vicente Santa María, and Further Details by Participants in the First Explorations of the Bay's Waters*. Edited by J. Galvin. San Francisco: John Howell-Books, 1971; p. 45.

97. AGN. Cárceles y Presidios, Volume 3, exp. 16, fs. 329f.–338b.

98. AGN. Provincias Internas, Volume 8, exp. 22, fs. 119–120v.

99. Santa Barbara Presidio Archive. File entitled: *Memorias y Facturas, dated 1779–1810*. Document 1 dated July 19, 1779; Document 2 dated October 29, 1782; Document 6 dated October 14, 1783; Document 7 dated March 31, 1785; Document 10 dated August 25, 1785; Document 11 dated March 21, 1787; Document 13 dated February 24, 1788; Document 14 dated January 25, 1789; Document 15 dated December 29, 1790 (whoever filed Documents 15 and 16 reversed the numbers since #15 has a later date than #16); Document 16 dated January 16, 1790; Document 20 dated December 9, 1791; Document 23 dated May 10, 1792; Document 26 dated March 31, 1793; Document 27 dated March 31, 1794;

Document 28 dated January 31, 1795; Document 29 dated March 7, 1795; Document 30 dated December 29, 1795; Document 32 dated June 8, 1796; Document 33 dated February 28, 1797; Document 36 dated February 7, 1798; Document 37 dated January 22, 1799; Document 40 dated January 3, 1802; Document 42 dated June 2, 1802; Document 43 dated June 31, 1804; Document 45 dated January 31, 1806; Document 47 dated January 31, 1806; Document 49 dated January 14, 1808; Document 51 dated February 15, 1810. See also: Perissinotto, G. *Documenting Everyday Life in Early Spanish California: The Santa Barbara Presidio Memorias y Facturas, 1779–1810.* Santa Barbara, CA: Santa Barbara Trust for Historic Preservation, 1998.

100. http://en.wikipedia.org/wiki/Nuu-chah-nulth. (Accessed January 10, 2007.)

101. Bancroft, H. H. *The Works of Hubert Howe Brancroft. Volume XVIII, History of California, 1542–1800.* San Francisco: The History Company, 1886; pp. 110–113.

102. Whitehead, R. S. Introduction. In: Whitehead, R. S., editor. *The Voyage of the Princesa as Recorded in the Logs of Juan Pantoja y Arriaga and Esteban José Martínez. Special Publication Number 1. Part One: The Log of Juan Pantoja y Arriaga.* Translated by G. V. Sahyun. Glendale, CA: Arthur H. Clark, 1982; pp. 3–6.

103. Bancroft, H. H. *The Works of Hubert Howe Brancroft. Volume XVIII, History of California, 1542–1800.* San Francisco: The History Company, 1886; pp. 240–249.

104. Whitehead, R. S. Introduction. In: Whitehead, R. S., editor. *The Voyage of the Princesa as Recorded in the Logs of Juan Pantoja y Arriaga and Esteban José Martínez. Special Publication Number 1. Part Two: The log of Estaban José Martínez.* Translated by G. V. Sahyun. Glendale, CA: Arthur H. Clark, 1982; pp. 93–95.

105. Bancroft, H. H. *The Works of Hubert Howe Brancroft. Volume XVIII, History of California, 1542–1800.* San Francisco: The History Company, 1886; pp. 240–249.

106. AGN. Historia, Volume 65, exp. 1, expedition 2nd volume, no page number.

107. AGN. Historia, Volume 65, exp. 1, expedition 2nd volume, no page number.

108. AGN. Folleteria, Caja 1, Folleto No. 31.

109. AGN. Intendencias, Volume 57, exp. 19, sin fols., 1824.

110. Dakin, S. B. *The Lives of William Hartnell.* Stanford, CA: Stanford University Press, 1949; passim.

111. Dakin, S. B. *The Lives of William Hartnell.* Stanford, CA: Stanford University Press, 1949; p. 74.

112. AGN. Folleteria, Caja 2, Folleto No. 54.

113. Duhaut-Cilly, A. Duhaut-Cilly's Account of California in the Years 1827–1828. *Quarterly of the California Historical Society* 1929;8(4):306–313.

114. Robinson, A. *Life in California during a Residence of Several Years in That Territory Comprising a Description of the Country and the Missionary Establishments, with Incidents, Observations, etc. ...* San Francisco: William Doxey Publisher, 1891; p. 37.

115. Robinson, A. *Life in California during a Residence of Several Years in That Territory Comprising a Description of the Country and the Missionary Establishments, with Incidents, Observations, etc. ...* San Francisco: William Doxey Publisher, 1891; p. 119.

116. Vallejo, Guadalupe. Ranch and mission days in Alta California. *The Century Magazine* 1890;41(2):113.

117. http://www.parks.sonoma.net/rosshist.html. (Accessed January 10, 2007.)

118. Langsdorff, G. H. von. *Voyages and Travels in Various Parts of the World, during the Years 1803, 1804, 1805, 1806, and 1807. Illustrated by Engravings from Original Drawings.* London, England: Printed for Henry Colburn, 1813.

119. Langsdorff, F. von. *G. H. Langsdorff's Narrative of the Rezanov Voyage to Neuva California in 1806.* San Francisco: Thomas C. Russell, 1927; p. 88.

120. http://www.fortrossstatepark.org/. (Accessed January 10, 2007.)

121. http://www.fortrossstatepark.org/chronology.htm. (Accessed January 10, 2007.)

122. http://www.fortrossinterpretive.org/Frames/fr_cultural_frameset.html. (Accessed January 10, 2007.)

123. http://www.fortrossstatepark.org/archaeol.htm. (Accessed January 10, 2007.)

CHAPTER

From Gold Bar to Chocolate Bar

California's Chocolate History

Bertram M. Gordon

Introduction

Chocolate became highly fashionable in California at the beginning of the 21st century to the point that some write of a "chocolate revolution" in the Golden State, while others see San Francisco as a world chocolate capital. The state boasts some 218 chocolate makers and wholesalers, with approximately 20 in the San Francisco Bay area [1]. The standard histories of chocolate, however, say little about chocolate in California. Aside from a brief reference to the Guittard and Ghirardelli chocolate companies, Nikita Harwich's *Histoire du Chocolat* is virtually mute on the subject [2]. *The True History of Chocolate* by Sophie D. and Michael D. Coe, generally considered a standard history of chocolate in English, makes only two passing references to California [3]. Looking back in California history, coffee was more common than chocolate (which was invariably a beverage prior to the 20th century). Mark Twain's *Roughing It*, an account of his activities in Gold Rush California, mentions coffee repeatedly but never chocolate [4]. Chocolate generally was not featured in the rise of California cuisine in the 1970s and 1980s [5].

Chocolate, however, has played a significant, if not dominant, role in California history (see Chapter 34). Its marketing and consumption is tied into the vortex of the different national groups that have comprised California's population. Introduced by the Franciscan missionaries while California was part of New Spain, it grew with the territory and was popularized by Anglo-American and German immigrants in the 19th century. Its history in California reflects the general history of chocolate styles, which went through, schematically speaking, three phases from the

European discovery of cacao in Mexico to the present. The first phase was the introduction of boiled and spiced beverages as well as thick and *molé*-type chocolate sauces used by Mayans and Aztecs in Mexico. Spanish and Spanish-style breakfast marked the second phase, with chocolate as a hot beverage drunk with *churros*. The beverage was seasoned with spices, including salt and peppers or cinnamon and sugar. Milk might be added but as it was often difficult to keep fresh and was expensive, the resulting chocolate was often a thin, watery drink. The third, and most recent, phase was the popularization of chocolate confectionery, often in the form of candy bars, facilitated by Coenraad Johannes van Houten's invention in 1828 of a process to more efficiently separate cocoa powder

Chocolate: History, Culture, and Heritage. Edited by Grivetti and Shapiro
Copyright © 2009 John Wiley & Sons, Inc.

from cocoa butter. In Switzerland, Daniel Peter developed milk chocolate in 1875 and four years later the first milk chocolate bar was produced (see Chapter 46) [6]. The result of the van Houten and Peter inventions was the spread of milk chocolate bars, highlighted in the early 20th century in the United States with the coming of Hershey's chocolate kisses in 1907 and the Mars Milky Way in 1923 [7].

Before the Gold Rush

Spanish missionaries, notably the Franciscan Junípero Serra, introduced chocolate into California in the middle of the 18th century. Explorers and settlers from Spain had learned to consume chocolate in Mexico and Central America. To make their cacao more palatable, the Spanish prepared it with vanilla and added sugar [8]. Marcy Norton points out that the Spanish fashion for chocolate was the result of a complex process of appropriating the Indian beverage, often through the intermediaries of colonists who returned from New Spain to the Continent [9].

The missionaries, in effect, reexported a taste for chocolate to the Spanish colonies in Mesoamerica, including *Alta* California. In contrast to the Jesuits, who preceded them in California, there is ample evidence for the Franciscans' use of chocolate, largely as a beverage. Father Serra, especially, appears to have been an aficionado. In 1769, Serra and several companion Franciscans were part of a Spanish expedition that established the mission in San Diego. The Franciscans brought chocolate and coffee, both of which they used on social occasions, into the Salinas missions of New Mexico [10], and substantial evidence shows they did the same in *Alta* California (see Chapter 34) [11]. According to Irving Berdine Richman, the "better-class ranchero" usually arose early in the morning to a breakfast of chocolate, then mounted his horse to survey his lands, and only between 8:00 and 9:00 a.m. returned home for a breakfast of *carne asada* (meat broiled on a spit), eggs, beans, tortillas, and, occasionally, coffee [12].

It is probable that the California Indians who came into contact with the missions tasted chocolate at these localities but there is no evidence to indicate that chocolate use spread among them during the period of Spanish rule. This was despite the story told by Katherine Abbott Sanborn in 1893 to the effect that the Channel Indians, off the Santa Barbara coast, prized red calico and chocolate highly as gifts from Juan Rodríguez Cabrillo during his expedition of 1542 [13]. Sanborn, who had formerly taught English literature at Smith College and traveled and lectured widely as a writer, was said to present her talks in a "humorous, entertaining manner" [14]. Her story is improbable in the absence of confirming evidence but it highlights the research that remains to be done on the use of

chocolate among the Indian population of California [15].

Non-Hispanic travelers and immigrants also brought chocolate into California during the first half of the 19th century. Auguste Bernard Duhaut-Cilly, a French sea captain seeking to trade for furs on the Pacific coast of North America, was served chocolate at the San Luis Rey mission in 1827 [16]. Richard Henry Dana refers to the use of chocolate on board the *Pilgrim*, which sailed around Cape Horn to California in 1834–1835 [17]. Referring to an execution of Indians in California without the benefit of legality in 1836, Don Agustin Janssens, a Belgian who had joined the Hijar-Padres expedition to California, noted that Sergeant Macedonio Gonzalez of San Diego said: "shooting an Indian was as easy as taking a cup of chocolate" [18]. In 1841, an Ecuadorian ship, the *Joven Carolina*, arrived at Yerba Buena (San Francisco) from Guayaquil, bringing a cargo of cocoa from South America. This cocoa was sold locally as well as to other ships plying the West Coast. Using hand-mills (*metates*), local women made the cocoa into chocolate, which was said to be popular among the "Californians," although who the Californians were is not made clear [19]. Alfred Robinson, writing in 1846, recalls having been given "chocolate [almost certainly as a beverage] and refreshments" by the missionaries at San Luis Rey [20].

The iconicity of chocolate was such that by the time of the Mexican–American War in 1846, during an American siege of the San Diego garrison still held by Mexicans, one of the defenders, Manuel Rocha, is said to have called out to his aunt, Doña Victoria, mischievously asking her to send him some clothing and chocolate [21]. Although it appears that coffee was more widely used than chocolate, the latter appears occasionally in the period of the Mexican War. Military rations at the time California was ceded by Mexico included cacao, as evidenced in the Bidwell ledger for December 1846. Ledger entries include purchases of cocoa on December 18, 1846 [22]. Instrumental in organizing a rescue mission for the Donner party trapped in the high Sierras during the 1846–1847 winter, Bidwell purchased 100 pounds of cocoa from a naval vessel commanded by Captain Hull and probably the *U.S.S. Warren* (Fig. 35.1) [23].

Chocolate and the Gold Rush

Not surprisingly, the 1849 Gold Rush increased the flow of chocolate to California. Forty-niners who drank chocolate included Enos Christman, who sailed around Cape Horn and en route collected rainwater, which, free of salt, was good for making cups of chocolate [24]. Ernest de Massey, a Frenchman prospecting for gold in 1849, complained of a colleague, who, charged

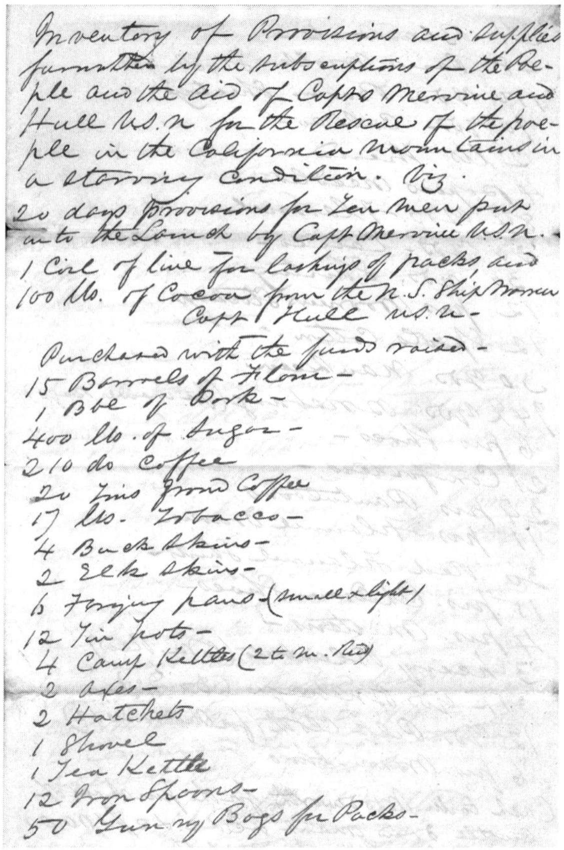

FIGURE 35.1. John Bidwell Donner party rescue document: cacao identified as a supply food to be sent to those stranded in the Western mountains; dated December 1846. *Source*: Handwritten Note. Courtesy of the California State Parks, Sutter's Fort Archives, Patty Reed Lewis Collection. (Used with permission.)

with buying provisions, bought sugar, chocolate, tea, tools, ammunition, and a gun "but he forgot to put in any real provisions." Consequently, "we had nothing to eat but tea, sugar, and chocolate" [25]. Although available at the time of the Gold Rush, chocolate was expensive. Felix Paul Wierzbicki, an immigrant from Poland, obtained his medical degree, then enlisted in the American army at the time of the Mexican War and was sent to California. Once demobilized, he practiced medicine in California until lured by the prospect of gold to Mokelumne Hill. He returned to San Francisco in 1849, where he wrote his memoirs [26]. He wrote that sugar sold for $15 per pound in San Francisco in 1849, coffee cost $12.50, and chocolate was $40 per pound. Tea, in contrast, sold for $1 per pound. Prices were higher the further away from San Francisco one traveled [27].

As a physician, Wierzbicki also commented on the medicinal qualities of chocolate, long in dispute in America and Europe. He saw chocolate as a harmless substitute for tea and coffee, both of which, he argued, caused dyspepsia and tooth decay. Tea, he argued, was more insidious and permanent in its adverse effects because the results were less immediately perceptible to the person drinking it [28]. Wierzbicki wrote:

> As a substitute for tea or coffee chocolate may be used advantageously; it is not a drug; it possesses no remedial powers, in the proper sense of the word; it is only alimentative, nourishing by its natural oil and substance. And as an aliment, an excessive quantity must be guarded against, as it is the case with any other article of food. If it disagrees with people generally, as it does with some persons, the cause is within them, their digestive powers being weak, its oily particles prove too much for their stomachs. The remedy for such cases may be to take less of it, or take it alone without any other substantial food. The best way of preparing and using it is unquestionably the Spanish way, as it is thus made more palatable and less greasy. [29]

Wierzbicki's mention of Spanish-style chocolate is a clear, if rare, reference to the thin beverage drunk in Spain prior to the coming of milk chocolate. By early September 1849, according to Bayard Taylor, who sailed to California to report on the Gold Rush for his friend Horace Greeley's *New York Tribune*, one could find a good meal of beefsteak, potatoes, and a cup of good coffee or chocolate, in San Francisco for one dollar [30]. In 1850, Friedrich Gerstacker found in a San Francisco gambling house:

> . . . a young and very beautiful girl coquettishly dressed in a gown of black silk; with her slender white hands, fitted with rings, she serves to some tea, to others coffee, and chocolate with cake or jam, while in the opposite corner of the room a man sells wine and liquors. [31]

Ships brought chocolate from Mexico to the port of San Francisco and up the Sacramento River to the city of Sacramento, where it was off-loaded and sold to wholesalers such as J. B. Starr and Company and the Lady Adams Company, two local merchants in the early 1850s. A ledger of 1850–1851 kept by J. B. Starr recorded more than 25 chocolate or cocoa transactions beginning October 18, 1850 and extending through August 13, 1851. The sales included tins and boxes (which appear to have been identical) of chocolate and cocoa, cases of chocolate, and boxes of cocoa paste [32]. Starr's customers, including the Lady Adams Company, most likely resold their wares, and prices varied depending on availability [33]. Chocolate was also advertised as available in the Miners' Variety Store in Sacramento in 1851 [34].

Chocolate drunk in the Gold Rush era often lacked milk, as Elizabeth Le Breton Gunn wrote to her family in 1851 and 1852. Her husband, Lewis Carstairs Gunn, had traveled from Philadelphia to California in 1849, where he established himself in medicine, journalism, and government service, first in Sonora and later in San Francisco. In January 1851, Elizabeth left Philadelphia with her four young children to join him.

Sailing around Cape Horn, she noted, on March 9:

> We have no milk, but something that they mix with water and use in chocolate. It looks like milk and tastes like it in the chocolate. [35]

Ambergris, sometimes used as a thickener for chocolate prior to the availability of inexpensive milk, may have been the substitute. Once in Sonora, Elizabeth emphasized the high cost of milk again when she wrote of the inflated prices in the gold country:

> As milk is now fifty cents a pint, I often give the children chocolate to drink. [36]

Louise Clappe, who in 1852 lived with miners in the gold country, offered another example of the watery chocolate drunk in Gold Rush days. She found spring water there:

> . . . very useful to make coffee, tea, chocolate, and other good drinks. [37]

The Lady Adams Company, founded by German immigrants [38], carried on a brisk trade in cocoa and chocolate. Their *Day Book* of 1852–1853, now in the Sutter's Fort Archive in Sacramento, listed repeated cocoa and chocolate transactions in tins and boxes during those years, including a sale of one box of chocolate for $3.00 to the Antelope Restaurant in Sacramento [39].

By the 1850s, the United States was exporting processed cocoa to various parts of the world and California seems to have played a minor role in this trade. In 1856, 6000 pounds of cocoa were exported through the port of San Francisco, with an *ad valorum* duty paid of $447. In contrast, the export of cocoa passing through New York was 704,456 pounds with $41,560 paid in duty. The respective figures for brown sugar exports for the same year from San Francisco were 14,112,448 pounds with $604,417 paid in duty and for New York, 337,370,868 pounds and $13,744,668 paid in duty [40].

Ghirardelli and Guittard Put California Chocolate on the Map

Domingo Ghirardelli in 1852 and Étienne Guittard in 1868 began making and selling chocolate in San Francisco. Originally from Rapallo (present day Italy), Ghirardelli entered the confectionary business first in Montevideo, Uruguay, then in Lima, Peru. In Lima, he set up shop next to James Lick, a cabinetmaker, who left for California in 1847, taking with him 600 pounds of Ghirardelli's chocolate [41]. Following the news of the discovery of gold, Ghirardelli left Lima for California, arriving in early 1849. After launching several businesses, some of which were destroyed by fires, he established *D. Ghirardely and Girard* in 1852 in central San Francisco [42]. In 1856, the company was advertised in the newspaper *Alta California* as making chocolate, as well as syrups, liquors, and ground coffee [43].

Ghirardelli's personnel discovered in 1867 that hanging a sack of chocolate in a warm room would allow the cocoa butter to drip out, leaving a residue that could readily be made into ground chocolate, which they named *broma* and began to sell [44]. Ghirardelli marketed its products under the names *Broma* and *Eagle*, with the former abandoned by 1885 and the latter continuing to the present [45]. When Domingo died in 1895, the firm passed into the control of his son, Domingo, Jr. It had become one of California's largest businesses [46]. After several relocations, the company's headquarters were moved to North Point Street, now Ghirardelli Square, in San Francisco in 1897 and the entire operation migrated there after the 1906 earthquake (Figs. 35.2 and 35.3) [47]. In 1910, the company adopted its long-time mascot, a parrot, for an advertising campaign to teach the public to "Say Gear-Ar-Delly" [48]. Following the close of the 1915 Panama–Pacific International Exposition, Ghirardelli authorized the addition to its factory of a clock tower in the style of the château of Blois in France; a 125-foot-long electrified sign with the name "Ghirardelli," added in 1923, has since become a San Francisco landmark. Ghirardelli Square became a marketplace in 1964 and the company moved its operations to San Leandro three years thereafter; in 1982, the Square was named a historic site by the National Park Service (Fig. 35.4) [49].

Étienne Guittard arrived in California from Lyon shortly after the Gold Rush but returned to France to learn the confectionary trade from his father, also a chocolatier. Back in San Francisco in 1868, he opened a chocolate shop. Unlike Ghirardelli, Guittard focused on the wholesale rather than the retail trade [50]. Currently located in Burlingame, Guittard remains one of the oldest chocolate producers still owned by the original family and continues to produce chocolate for the wholesale market.

Diffusion and Diversity in California Chocolate in the Second Half of the 19th Century

Ghirardelli and Guittard, better known now than their competitors because of their operational longevity, were hardly the only entrants in the California chocolate business. A ledger from the H. Winters Company, a Sacramento wholesaler with German origins, covers the period from 1863 through 1867, with many entries for sales of chocolate. To put the chocolate volume into context, the first Winters entry for a chocolate sale,

FIGURE 35.2. Ghirardelli wagon used to transport chocolate (ca. 1919). *Source*: Ghirardelli Chocolate Company, San Leandro, CA. (Used with permission.)

FIGURE 35.3. Chocolate trade card (Ghirardelli), early 20th century. *Source*: Courtesy of the Grivetti Family Trust. (Used with permission.)

January 28, 1863, appears on a page that lists one transaction for chocolate, as well as three for sugar and 13 for coffee [51]. Chocolate was served as a drink in the Olympia Coffee Parlor, whose name symbolizes the relationship between the two beverages, according to an undated menu from early Sacramento [52]. Californians also used chocolate for pharmaceutical purposes, as did people elsewhere. A patent medicine list of early 1864, published by Redington and Company, wholesale druggists in San Francisco, mentioned Barney's Cocoa Castorine, a product for the hair, at $3.75 per dozen [53]. Made in South Malden, Massachusetts, Barney's Cocoa Castorine was advertised in the *New York Times* in 1860 as "one of the best compounds for beautifying the Hair," and was sold by druggists throughout the United States [54]. Pharmacies also used cocoa butter to flavor or coat medications, and for suppositories [55].

FIGURE 35.4. Ghirardelli Square, San Francisco, 20th century view. *Source*: Ghirardelli Chocolate Company, San Leandro, CA. (Used with permission.) (See color insert.)

Most references to chocolate, however, were as a beverage and it was in this form that it entered the California restaurant world. The Pullman Palace Car on the Union Pacific trains from Omaha to California in 1872 served chocolate along with French coffee and English breakfast tea. Each cost 15 cents if accompanying a meal, which featured items such as buffalo, elk, antelope, beef steak, muttonchops, and grouse. If ordered without food, the beverages were 25 cents [56]. The 25th anniversary of the 1849 arrival of the steamship *California* in San Francisco was celebrated in 1874 with coffee, tea, chocolate, lemonade, and wine, although it is not clear where this meal was served, and the menu itself is a later reproduction [57].

As hotels, restaurants, and banquets expanded during the second half of the 19th century, chocolate appeared on their menus, usually in the form of cocoa or a hot chocolate beverage for breakfast, but its proliferation into confectionary, pies, and ice creams was evident. Falling sugar prices and the industrialization of chocolate in the late 19th century made it more affordable to the middle and lower classes in California as it did across the United States and around the industrializing world. By 1881, the Pico House in Los Angeles was serving chocolate cream cake as a dessert and chocolate as a beverage (Fig. 35.5) [58]. The following year, Lloyd Vernon Briggs reported arriving in Los Angeles at 8:00 a.m. on January 29 and having a breakfast imme-

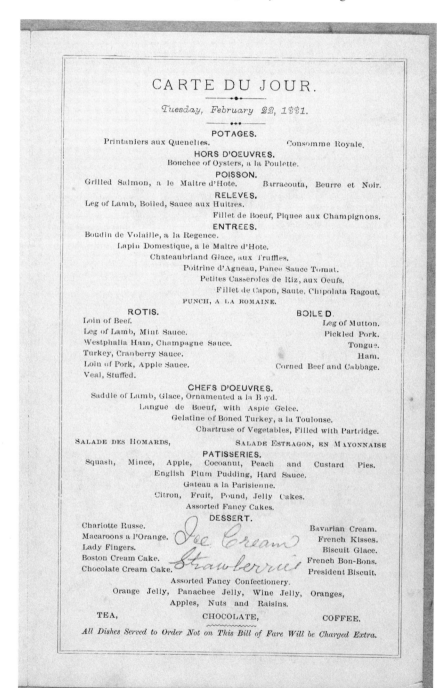

FIGURE 35.5. Menu (carte du jour) offering chocolate cream cake, Pico House, Los Angeles; dated Tuesday, February 22, 1881. *Source:* Ephemera, E38-136. Courtesy of the Huntington Library, San Marino, CA. (Used with permission.)

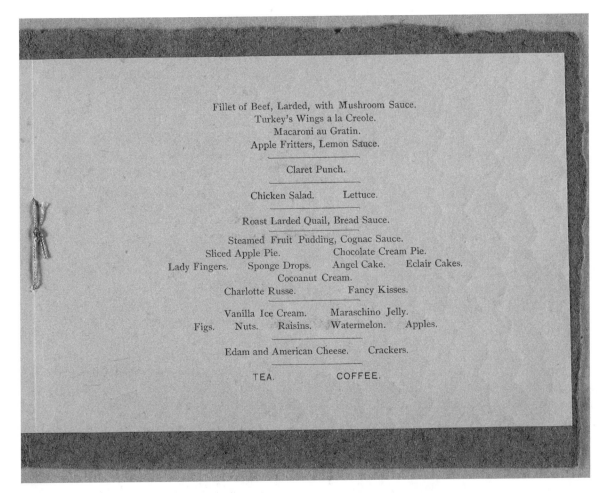

Fillet of Beef, Larded, with Mushroom Sauce.
Turkey's Wings a la Creole.
Macaroni au Gratin.
Apple Fritters, Lemon Sauce.

Claret Punch.

Chicken Salad.   Lettuce.

Roast Larded Quail, Bread Sauce.

Steamed Fruit Pudding, Cognac Sauce.
Sliced Apple Pie.   Chocolate Cream Pie.
Lady Fingers.   Sponge Drops.   Angel Cake.   Eclair Cakes.
Cocoanut Cream.
Charlotte Russe.   Fancy Kisses.

Vanilla Ice Cream.   Maraschino Jelly.
Figs.   Nuts.   Raisins.   Watermelon.   Apples.

Edam and American Cheese.   Crackers.

TEA.   COFFEE.

FIGURE 35.6. Menu offering chocolate cream pie, The Raymond, South Pasadena; dated November 13, 1886. *Source*: Ephemera, E38-187. Courtesy of the Huntington Library, San Marino, CA. (Used with permission.)

diately thereafter at the Pico House on the plaza, or town square, in the lower part of the town:

We were hungry and enjoyed a delicious breakfast of fried chicken, eggs, chocolate, strawberries, oranges, and griddle cakes. [59]

Chocolate cream pie and chocolate cake appeared in banquets at The Raymond, then a grand hotel and now a restaurant, in South Pasadena, within a week in November 1886 (Figs. 35.6 and 35.7) [60]. "Chocolate Cream Cake" appeared on a dessert list for Christmas dinner at The Arlington in Santa Barbara, also in 1886 (Fig. 35.8) [61]. The Santa Fe Railroad, serving California and the West, offered an unusual entrée listed as "Sweetbreads, Saute [*sic*] with Mushrooms, Spanish Puffs, Chocolate Glace," in 1888, most likely under Hispano-Mexican culinary influence (Fig. 35.9) [62].

Chocolate also appears in the menus of steamships serving California in the early 20th century, if not before. The Pacific Coast Steamship Company [63], established in 1875, plied routes from Panama to San Francisco and from California to Japan and China. One of their menus, of December 1906, shows coffee, tea, and cocoa served as beverages on the *S.S. Spokane*. In the following years, PCSSCO ships based in San Francisco, carrying the tourist trade from Seattle to Alaska and back, occasionally listed chocolate and cocoa, as beverages, at the ends of their menus [64]. The proliferation of chocolate products in the late 19th century is reflected in a San Francisco cookbook of 1879 that included recipes for chocolate pudding, cake, confectionary, and beverages [65]. The confectionary, chocolate "caromels" (*sic*), and the pudding both made use of milk mixed in with chocolate, as familiarity with Daniel Peter's process had spread. An 1886 menu from St. Elmo, a boarding house that appears to have been in Los Angeles, listed "chocolate eclairs" under desserts, and chocolate as a beverage [66].

The Hispanic influence on chocolate added to its growing popularity. Describing a gastronomic tour in Los Angeles in 1888, Harriet Harper observed that a traveler no longer had to dine in the hotels or boarding houses where he slept because there were so many different dining options in the city:

If he discovers the aching void while he is in the adobe town, he asks the handsome, dark-eyed senorita for hot tamales and chocolate. [67]

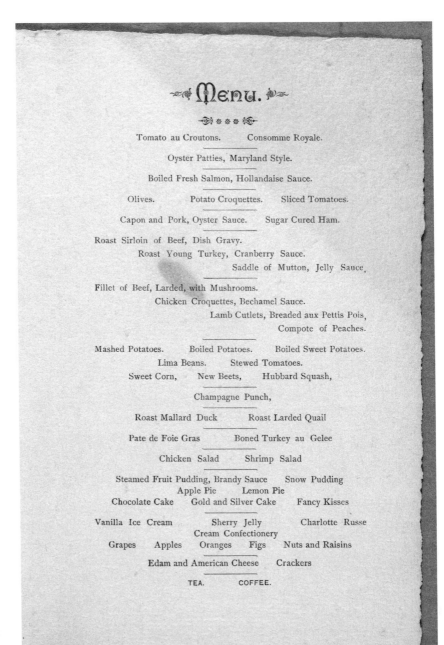

FIGURE 35.7. Menu offering chocolate cake, The Raymond, South Pasadena; dated November 20, 1886. *Source*: Ephemera, E38-188. Courtesy of the Huntington Library, San Marino, CA. (Used with permission.)

By 1891, The Weinstock and Lubin mail order house sold chocolate in several variations that included chocolate cream candies, along with chocolate bars wrapped in paper by Ghirardelli, Baker, and other companies. Also available were cocoa in cans by Ghirardelli, breakfast cocoa by Baker, half-pound cans of Epp's homeopathic cocoa, and cocoa shells, the husks removed from roasted cacao beans, discarded, and used as mulch [68].

Chocolate use before refrigeration was limited, especially for home consumption, as it had to be purchased from restaurants or vendors such as Weinstock's and was invariably in the form of powder listed as ground chocolate or cocoa. By 1894, chocolate had entered cakes, cake frostings, puddings, and pies, in addition to continuing as a beverage and confectionary. The *Ladies Social Circle* cookbook, collected from local women in Los Angeles, offers six different recipes for chocolate used in various ways and also lists complete meals. For example, chocolate appears as an item, presumably a beverage, served with a hot biscuit toward the end of a luncheon menu [69]. Separate French, German, Spanish, and Russian "departments" are included and it is noteworthy that chocolate appears frequently in the German department and in none of the others. One German menu lists chocolate as an option to accompany afternoon coffee [70]. Chocolate cookies and chocolate cake

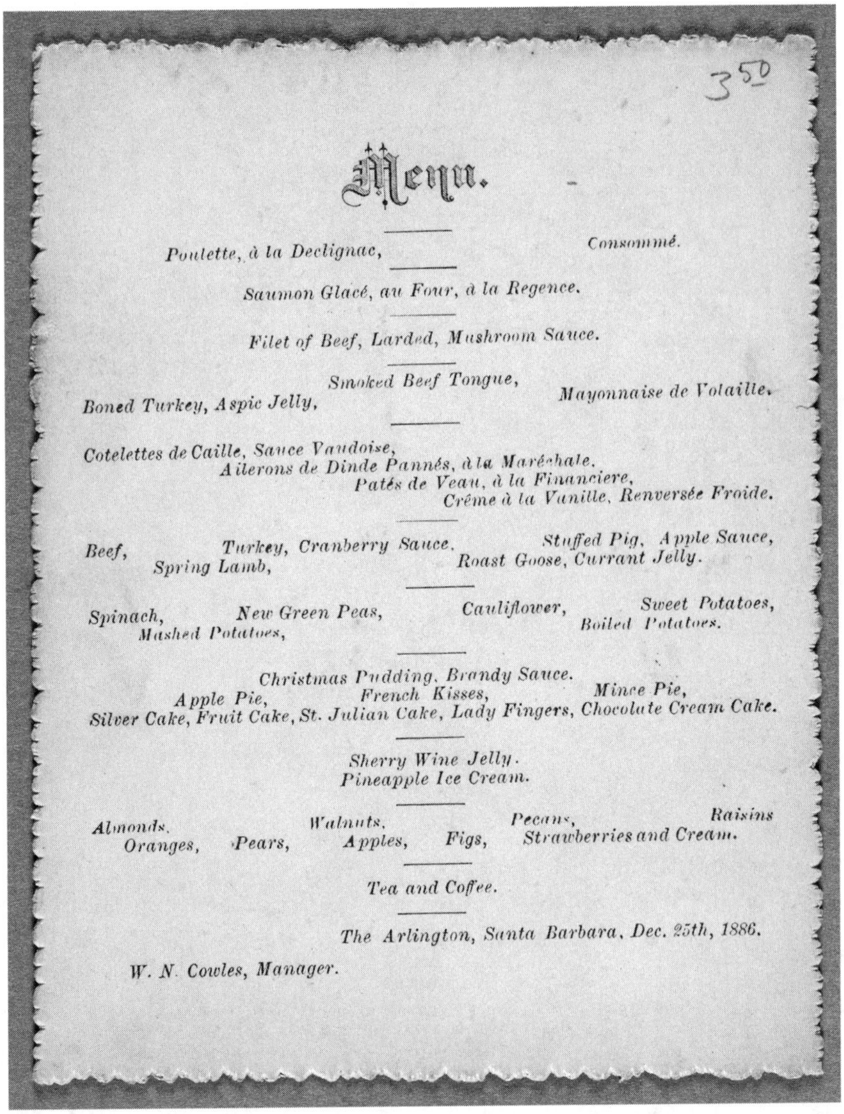

350

Menu.

Poulette, à la Declignac,   Consommé.

Saumon Glacé, au Four, à la Regence.

Filet of Beef, Larded, Mushroom Sauce.

Smoked Beef Tongue,

Boned Turkey, Aspic Jelly,   Mayonnaise de Volaille.

Cotelettes de Caille, Sauce Vaudoise,
     Ailerons de Dinde Pannés, à la Maréchale.
         Patés de Veau, à la Financiere,
           Crème à la Vanille, Renversée Froide.

Beef,   Turkey, Cranberry Sauce.   Stuffed Pig, Apple Sauce,
    Spring Lamb,   Roast Goose, Currant Jelly.

Spinach,   New Green Peas,   Cauliflower,   Sweet Potatoes,
    Mashed Potatoes,   Boiled Potatoes.

Christmas Pudding, Brandy Sauce.
   Apple Pie,   French Kisses,   Mince Pie,
Silver Cake, Fruit Cake, St. Julian Cake, Lady Fingers, Chocolate Cream Cake.

Sherry Wine Jelly.
Pineapple Ice Cream.

Almonds,   Walnuts,   Pecans,   Raisins
   Oranges,   Pears,   Apples,   Figs,   Strawberries and Cream.

Tea and Coffee.

The Arlington, Santa Barbara, Dec. 25th, 1886.

W. N. Cowles, Manager.

FIGURE 35.8. Menu offering chocolate cream cake, The Arlington, Santa Barbara; dated December 25, 1886. *Source:* Ephemera, E38-212. Courtesy of the Huntington Library, San Marino, CA. (Used with permission.)

also appear in the German department, and another German daily menu starts with a breakfast of coffee or chocolate, followed by rolls and Kuchen (*sic*) [71]. The chocolate cookies were prepared with Ghirardelli's chocolate, which was grated with brown and white sugar, butter, and eggs [72]. There is no reference to chocolate in the Spanish department of the *Ladies' Social Circle* cookbook, again suggesting that the popularization of chocolate in late 19th and early 20th century California had less to do with the original Spanish settlers than with their German and English successors.

The Maison Faure ledger book in Sacramento offers a rare example of chocolate consumption. A boarding house, the Maison Faure recorded its sales of meals, to locals as well as guests. A rare example of a specific consumer ordering chocolate in a shop, restaurant, or hotel is the record of Mr. Joseph Thieben, the manager of a crockery establishment in Sacramento, ordering "chocolate and lunch" on three successive days

from December 20 through 22, 1894. On the 24th, he ordered chocolate and six lunches [73]. Undoubtedly, the chocolate in question was a beverage. By the beginning of the 20th century, the popularity of cocoa as a breakfast beverage in California was such that students at Mills College in Oakland recommended it in 1900 as part of a facetious weight-gain diet. A "Splendid Flesh-Producing Diet," to help "thin women gain flesh," included a breakfast of porridge and milk, followed by cocoa and Fritzie (a shortbread cookie with nutmeg, popular in the early 20th century), also, "weak tea and coffee or milk, with fat bacon, fish, and jam." Afternoon tea was to be replaced by cocoa [74]. Chocolate proliferation in 1908 included the Hotel Alexandria in downtown Los Angeles, which listed chocolate ice cream and chocolate éclairs, as well as both chocolate and cocoa as beverages [75]; and the Bismarck Café in downtown San Francisco with chocolate ice cream for dessert and chocolate with whipped cream by the pot as a beverage [76].

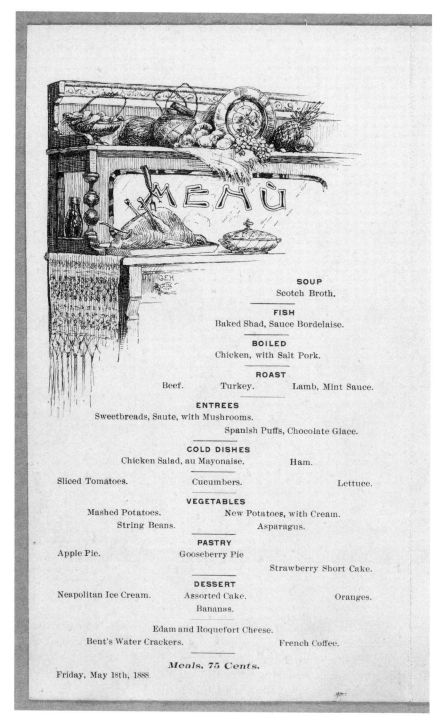

FIGURE 35.9. Menu offering chocolate glace, Santa Fe Railroad; dated May 18, 1888. *Source*: Ephemera, E38-196. Courtesy of the Huntington Library, San Marino, CA. (Used with permission.)

Milk Chocolate, Refrigeration, and the Bar

The coming of refrigeration by the early 20th century changed the world of chocolate in California, as elsewhere. Electric refrigerators offered stabilization of temperatures for chocolate in a way that the ice blocks could not. Together with falling sugar prices and abundant milk supplies by the 1920s, electric refrigeration facilitated the development of ice cream, Hershey's bars, Baby Ruth, and Mars bars. A dynamic "greater

California," in a culinary sense, came to extend from the elite restaurants of *Baja* California, with the creation of Caesar salad and the other Hollywood dishes, to Seattle, Vancouver, and Hawaii, and points beyond, later to include the luxury food business in Reno and Las Vegas. In this greater California region, the topologies of chocolate are tested and retested. The menus in restaurants generally define fashionable chocolates in drinks, ice creams, and cakes.

The coming of a new chocolate era also was symbolized in the name of the Chocolate Garden, a restaurant in the Dance Pavilion of Venice, California, in the 1920s. The Chocolate Garden may have been the first such establishment bearing the name "chocolate" in California and was clearly a resort-style ice cream parlor rather than a hotel or full-service restaurant. One of its menus offers a variety of chocolate treats, all priced between 15 and 25 cents. A list of "fancy sundaes" included a Chocolate Garden Special made with "chocolate ice cream, 1 banana, bittersweet chocolate, nuts, and ice cream" and another sundae of "milk chocolate sunday [sic] with whipped cream." Additional dishes included "French ice cream," sold in vanilla, chocolate, and strawberry flavors, and a "special mixed drink" of "creme de chocolate" [77]. A transitional menu of December 6, 1924 from the Hotel Mills in Oakland offered "vanilla, chocolate or strawberry ice cream," and "coffee, tea or chocolate, single pot" [78]. In contrast, another menu, from the Los Angeles Biltmore Hotel in 1933, offered chocolate custard and chocolate ice cream for dessert but no cocoa [79].

It is in the middle level of usage, however, that chocolates stand out, for gifts on the major holidays and special occasions, such as birthdays or farewells. The major California purveyor in the early 20th century was See's Candies, established in November 1921 in Los Angeles by Charles See and his mother, Mary See.

Chocolate cakes, icings, and puddings, with a few days' shelf life, could all be prepared easily from powder or bar chocolate, but the real takeoff in the 1920s was in commercial ice cream, which most effectively brought chocolate to the mid-level market. Dreyer's ice cream was established in 1928 in Oakland and the next year began producing Rocky Road, with walnuts, later almonds, and cut marshmallows added to their chocolate ice cream [80]. Thus it was that chocolate had evolved from its Gold Rush days as a beverage and the era of Hershey's and Mars was about to begin. . . .

Acknowledgments

I wish to thank the following persons who assisted during the preparation of this chapter: John S. Aubrey, Ayer Librarian, Newberry Library, Chicago, Illinois; Steve Beck, Archivist, Sutter's Fort State Historic Park, Sacramento, California; Anthony S. Bliss, Curator, Rare Books and Manuscripts, The Bancroft Library, University of California, Berkeley; Janice Braun, Curator of Special Collections and Director of the Mills Center for the Book, Oakland, California; Cathy Cherbosque, Curator of Historical Prints and Ephemera, Rare Book Department, The Huntington Library, San Marino, California; Kathleen Correia, California History Section, California State Library, Sacramento, California; Steven Davenport, San Francisco Maritime National Historic Park, J. Porter Shaw Library, San Francisco, California; Mario Einaudi, Archivist, Kemble Maritime Ephemera, John Haskell Kemble Collection, Huntington Library, San Marino, California; Cristina Favretto, Rare Books Librarian, UCLA Library, Los Angeles, California; James Henley, Manager of the Sacramento History and Science Division, Sacramento Archives and Museum Collection Center, Sacramento, California; Patricia Johnson, Archivist, Sacramento Archives and Museum Collection Center, Sacramento, California; Alan Jutzi, Avery Chief Curator of Rare Books, The Huntington Library, San Marino, California; Alison Moore, Readers Services Librarian, California Historical Society, San Francisco, California; Laura Stalker, Avery Head of Collection Development and Reader Services, The Huntington Library, San Marino, California; and Andrew Workman, Department of History, Mills College, Oakland, California.

References

1. Gordon, B. M. Les circuits touristiques du chocolat en Californie. In: *Colloque Gastronomie et rayonnement touristique: contribution à l'étude des hautes lieux et capitales gastronomiques.* Presentation (with abstract) delivered in Lyon, France, December 6, 2006.

2. Harwich, N. *Histoire du Chocolat.* Paris, France: Éditions Desjonquères, 1992; p. 175.

3. Coe, S. D., and Coe, M. D. *The True History of Chocolate.* London, England: Thames and Hudson, 1996; pp. 235 and 265.

4. Twain, M. (Samuel Langhorne Clemens). *Roughing It.* Hartford, CT: American Publishing Company, 1886; pp. 45, 88, 105, 170, 201, 202, 261, 348, 432, and 434.

5. Gordon, B. M. Shifting tastes and terms: the rise of California cuisine. *Revue Française des Études Américaines* 1986;11(27/28 Feb):109–126.

6. Coe, S. D., and Coe, M. D. *The True History of Chocolate.* London, England: Thomas and Hudson, 1996; pp. 242 and 250.

7. Harwich, N. *Histoire du Chocolat.* Paris, France: Éditions Desjonquères, 1992; p. 195.

8. Coe, M. D. KAKAW El chocolate en la Cultura de Guatemala/Chocolate in Guatemalan Culture. Available at http://www.popolvuh.ufm.edu/eng/Kakaw01. htm. (Accessed May 28, 2006.)

9. Norton, M. Tasting empire: chocolate and the European internalization of Mesoamerican aesthetics. *American Historical Review* 2006;111(3):660–661.

10. Ivey, J. E. Daily life in the Salinas Missions—life and times in the convento. In: In the Midst of a Loneliness, Chapter 7. The Architectural History of the Salinas Missions, Salinas Pueblo Missions National Monument Historic Structures Report, Southwest Cultural Resources Center Professional Papers No. 15, Southwest Regional Office, National Park Service, Santa Fe, NM, 1988. Available at http://www.nps.gov/archive/sapu/hsr/hsr7e.htm. (Accessed June 11, 2006.)

11. Santa Barbara Mission Archive, Junipero Serra Collection, May 14, 1769. In: Cabezon, B. Chocolate and cacao in California: the missions and the presidios. Chocolate Research Project, p. 2. Available at https://cocoaknow.ucdavis.edu/ChocolateResearch/. (Accessed July 14, 2006.)

12. Richman, I. B. *California under Spain and Mexico 1535–1847*. Boston: Houghton Mifflin Company, 1911; p. 350.

13. Sanborn, K. A. *A Truthful Woman in Southern California*. New York: D. Appleton and Company, 1893; pp. 140–141.

14. Five College Archives and Manuscripts Collections. Smith College Archives. Kate Sanborn Papers, 1878–1996. Available at http://asteria.fivecolleges.edu/find-aids/smitharchives/manosca113bioghist.html. (Accessed December 20, 2006.)

15. Grivetti, L. Personal correspondence. E-mail dated November 19, 2006.

16. Pourade, R. History of San Diego, San Diego Historical Society. Available at http://www.sandiegohistory.org/books/pourade/silver/silverchapter5.htm; http://sandiegohistory.org/books/pourade/explorers/explorerschapter7.htm; and http://sandiegohistory.org/books/pourade/time/timechapter10.htm. (Accessed June 11, 2006.)

17. Dana, R. H., Jr. Two Years Before the Mast; A Personal Narrative, by Richard Henry Dana, jr.; with a Supplement by the Author and Introduction and Additional Chapter by His Son . . . with illustrations by E. Boyd Smith, p. 386, in Library of Congress. American Memory, California As I Saw It: First-Person Narratives of California's Early Years, 1849–1900. Available at http://memory.loc.gov/cgi-bin/query/r?ammem/calbk:@field(DOCID+@lit(calbk139div34). (Accessed December 5, 2006.)

18. Janssens, V. E. A. In: Ellison, W. H., and Price, F., editors. *The Life and Adventures in California of Don Agustin Janssens, 1834–1856*. Translated by F. Price. San Marino, CA: Huntington Library Press, 1953; pp. 102–103.

19. Davis, W. H. In: Watson, D. S., editor. *Seventy-five Years in California, A History of Events and Life in California: Personal, Political and Military*. San Francisco: J. Howell, 1929.

20. Robinson, A. *Life In California: During a Residence of Several Years in That Territory . . . by an American*. New York: Wiley and Putnam, 1846; p. 24.

21. Pourade, R. History of San Diego, San Diego Historical Society. Available at http://www.sandiegohistory.org/books/pourade/silver/silverchapter5.htm; http://sandiegohistory.org/books/pourade/explorers/explorerschapter7.htm; and http://sandiegohistory.org/books/pourade/time/timechapter10.htm. (Accessed June 11, 2006.)

22. Bidwell, J. Ledger. Handwritten manuscript, Sutter's Fort Archives, Sacramento; pp. 18, 20, and 143.

23. Colton, W. *Three Years in California*. Cincinnati, OH: H. W. Derby and Company, 1850; pp. 187 and 213. Available at http://delta.ulib.org/ulib/data/moa/0ad/640/d5a/e86/722/e/data.txt. (Accessed December 16, 2006.)

24. Christman, E. *One Man's Gold. The Letters and Journal of a Forty-Niner*. Compiled and edited by Florence Morrow Christman. New York: Whittlesey House McGraw-Hill, 1930; p. 41.

25. de Massey, E. *A Frenchman in the Gold Rush. The Journal of Ernest de Massey, Argonaut of 1849*. Translated by Marguerite Eyer Wilber. San Francisco: California Historical Society, 1927; p. 97.

26. California As I Saw It: First-Person Narratives of California's Early Years, 1849–1900. Library of Congress, American Memory. Available at http://lcweb2.loc.gov/cgi-bin/query/r?ammem/calbkbib:@field(AUTHOR+@band(Wierzbicki,+Felix+Paul,1815-1860.+). (Accessed December 20, 2006.)

27. Wierzbicki, F. P. *California As It Is and As It May Be; Or, A Guide to the Gold Region*, 2nd edition. San Francisco: The Grabhorn Press, 1933; pp. 47–48.

28. Wierzbicki, F. P. *California As It Is and As It May Be; Or, A Guide to the Gold Region*, 2nd edition. San Francisco: The Grabhorn Press, 1933; pp. 95–96.

29. Wierzbicki, F. P. *California As It Is and As It May Be; Or, A Guide to the Gold Region*, 2nd edition. San Francisco: The Grabhorn Press, 1933; pp. 98–99.

30. Taylor, B. *EldOrado, Or, Adventures in the Path of Empire: Comprising a Voyage to California, via Panama; Life in San Francisco and Monterey; Pictures of the Gold Region, and Experiences of Mexican Travel*, 2nd edition. New York: G. P. Putnam, 1850; p. 113. Reproduced in "California As I Saw It: First-Person Narratives of California's Early Years, 1849–1900," Library of Congress, American Memory. Available at http://lcweb2.loc.gov/cgibin/query/r?ammem/calbk:@field(DOCID+@lit(calbk122div16). (Accessed December 20, 2006.)

31. Gerstacker, F. W. C. 1942. *Scenes of Life in Californiaz by Friedrich Gerstacker*. Translated from the French by George Cosgrave. San Francisco: J. Howell, 1942; pp. 70–72.

32. Starr, J. B. Ledger Book, 1850–1851, #1978/93, City of Sacramento, History and Science Division, Sacramento Archives and Museum Collection Center (SAMCC), p. 20.

33. Starr, J. B. Ledger Book, 1850–1851, #1978/93, City of Sacramento, History and Science Division, Sacramento Archives and Museum Collection Center (SAMCC), p. 194.

34. Miners. "Miners' Variety Store," advertisement, in *Placer Times*, June 15, 1851, p. 45. Reproduced in: Neasham, A. A., and Henley, J. E. *The City of the Plain: Sacramento in the Nineteenth Century*. Sacramento: Sacramento Pioneer Foundation and Sacramento Historic Landmarks Commission, 1969; p. 142.

35. Gunn, L. C. In: Marston, A. L., editor. *Records of a California Family. Journals and Letters of Lewis C. Gunn and Elizabeth Le Breton Gunn.* San Diego, CA: Johnek and Seeger, 1928; p. 102.

36. Gunn, L. C. In: Marston, A. L., editor. *Records of a California Family. Journals and Letters of Lewis C. Gunn and Elizabeth Le Breton Gunn.* San Diego, CA: Johnek and Seeger, 1928; p. 165.

37. Clappe, L. A. K. S. In: Russell, T. C., editor. *The Shirley Letters from California Mines in 1851–1852; Being a Series of Twenty-three Letters from Dame Shirley (Mrs. Louise Amelia Knapp Smith Clappe) to Her Sister in Massachusetts, and Now Reprinted from the Pioneer Magazine of 1854–1855.* San Francisco: T. C. Russell, 1922; p. 290.

38. Terry, C. C. Germans in Sacramento, 1850–1859. Psi Sigma Historical Journal (University of Nevada Las Vegas) Summer 2005;3:8. Available at http://www.unlv.edu/student_orgs/psisigma/PAT%20Articles/CTerry.pdf. (Accessed December 16, 2006.)

39. Lady Adams. Account Book: Day Book of Lady Adams Co. Wholesale Grocers, Sacramento, 1852–1853. Sutter's Fort Archives, Sacramento. Catalogue No. 308-408-1, p. 205.

40. United States Treasury Department. *Report of the Secretary of the Treasury, Transmitting a Report from the Register of the Treasury, of the Commerce and Navigation of the United States for the Year Ending June 30, 1856.* Washington, DC: A. O. P. Nicholson, 1856; p. 477.

41. Teiser, R. *An Account of Domingo Ghirardelli and the Early Years of the D. Ghirardelli Company.* San Francisco: D. Ghirardelli Co., 1945; p. 4.

42. Teiser, R. *An Account of Domingo Ghirardelli and the Early years of the D. Ghirardelli Company.* San Francisco: D. Ghirardelli Co., 1945; pp. 11–13.

43. Teiser, R. *An Account of Domingo Ghirardelli and the Early Years of the D. Ghirardelli Company.* San Francisco: D. Ghirardelli Co., 1945; p. 15.

44. Teiser, R. *An Account of Domingo Ghirardelli and the Early years of the D. Ghirardelli Company.* San Francisco: D. Ghirardelli Co., 1945; p. 22.

45. Teiser, R. *An Account of Domingo Ghirardelli and the Early years of the D. Ghirardelli Company.* San Francisco: D. Ghirardelli Co., 1945; p. 27.

46. Lawrence, S. The Ghirardelli story. *The Magazine of the California Historical Society* 2002;81(2):96.

47. Teiser, R. *An Account of Domingo Ghirardelli and the Early years of the D. Ghirardelli Company.* San Francisco: D. Ghirardelli Co., 1945; p. 29.

48. Lawrence, S. The Ghirardelli Story. *The Magazine of the California Historical Society* 2002;81(2):105.

49. Lawrence, S. The Ghirardelli Story. *The Magazine of the California Historical Society* 2002;81(2):102.

50. Lawrence, S. The Ghirardelli Story. *The Magazine of the California Historical Society* 2002;81(2):96.

51. Winters, H. (Hermance) and Co, Ledger, Book 7 (1863–1867), 98/721/033, City of Sacramento, History and Science Division, Sacramento Archives and Museum Collection Center (SAMCC), p. 60.

52. Olympia. "Bill of Fare, Olympia Coffee Parlor," undated, Thomas Prittie Collection, 73/10/03b, City of Sacramento, History and Science Division, Sacramento Archives and Museum Collection Center (SAMCC).

53. Redington and Company, Wholesale Druggists, 416 and 418 Front Street, San Francisco, Patent Medicine List, January 22, 1864, Ephemera F16-H20. Huntington Library, San Marino, CA.

54. *New York Times.* September 15, 1860, p. 5.

55. Brooks, D., and Donald, F. Salvatori California Pharmacy Museum, Sacramento. Telephone conversation, July 25, 2006.

56. Nordhoff, C. California. *Harper's New Monthly Magazine* 1872;44(264):872.

57. Clipper Restaurant Menu, 1871–1884, Ephemera 38-47. Huntington Library, San Marino, CA.

58. Pico House Carte du Jour, February 22, 1881, Ephemera 38-136. Huntington Library, San Marino, CA.

59. Briggs, L. V. *California and the West, 1881 and Later.* Boston: Wright and Potter, 1931; p. 113.

60. The Raymond. "Complimentary to the First Raymond Party Saturday, Nov 13th," Menu, November 13, 1886; and The Raymond Complimentary Banquet by the Citizens of Pasadena to Messers Emmons Raymond and Walter Raymond, Menu, November 20, 1886. Ephemera 38-187 and 38-188, Huntington Library, San Marino, CA.

61. The Arlington. Menu, Santa Barbara, December 25, 1886. Ephemera 38-212, Huntington Library, San Marino, CA.

62. Santa Fe Route, Menu, May 18, 1888. Ephemera 38-196, Huntington Library, San Marino, CA.

63. P.C.S.S.CO. The Pacific Coast Steamship Company, Kemble Maritime Ephemera. John Haskell Kemble Collection, Huntington Library, San Marino, CA.

64. P.C.S.S.CO. The Pacific Coast Steamship Company, Kemble Maritime Ephemera. John Haskell Kemble Collection, Huntington Library, San Marino, CA.

65. California Recipe Book by Ladies of California, 4th edition. San Francisco: Pacific Press Print and Publishing House, 1879; pp. 24, 41, 53, and 65.

66. St. Elmo, Menu, August 29, 1886. UCLA Special Collections 1306, Box 10.

67. Harper, H. *Letters from California.* Portland, ME: Press of B. Thurston & Company, 1888; pp. 13–14.

68. Weinstock Lubin and Company. Catalogue, Spring and Summer 1891. Reprinted, Sacramento, CA: Sacramento American Revolution Bicentennial Committee, 1975; p. 110.

69. Ladies' Social Circle, Simpson Methodist Episcopal Church, Los Angeles. *How We Cook in Los Angeles. A*

Practical Cook-Book Containing Six Hundred or More Recipes Selected and Tested by over Two Hundred Well Known Hostesses, Including a French, German and Spanish Department, with Menus, Suggestions for Artistic Table Decorations, and Souvenirs. Los Angeles: Commercial Printing House, 1894; p. 52.

70. Ladies' Social Circle, Simpson Methodist Episcopal Church, Los Angeles. *How We Cook in Los Angeles. A Practical Cook-Book Containing Six Hundred or More Recipes Selected and Tested by over Two Hundred Well Known Hostesses, Including a French, German and Spanish Department, with Menus, Suggestions for Artistic Table Decorations, and Souvenirs.* Los Angeles: Commercial Printing House, 1894; p. 282.

71. Ladies' Social Circle, Simpson Methodist Episcopal Church, Los Angeles. *How We Cook in Los Angeles. A Practical Cook-Book Containing Six Hundred or More Recipes Selected and Tested by over Two Hundred Well Known Hostesses, Including a French, German and Spanish Department, with Menus, Suggestions for Artistic Table Decorations, and Souvenirs.* Los Angeles: Commercial Printing House, 1894; pp. 295 and 296

72. Ladies' Social Circle, Simpson Methodist Episcopal Church, Los Angeles. *How We Cook in Los Angeles. A Practical Cook-Book Containing Six Hundred or More Recipes Selected and Tested by over Two Hundred Well Known Hostesses, Including a French, German and Spanish Department, with Menus, Suggestions for Artistic Table Decorations, and Souvenirs.* Los Angeles: Commercial Printing House, 1894; p. 295.

73. Maison Faure. Ledger Book, 1892–1896, Michael Foley Collection, #2002/064/02, City of Sacramento, History and Science Division, Sacramento Archives and Museum Collection Center (SAMCC), p. 47.

74. "Greetings: To the Senior Class," *White and Gold* (Mills College), 6:4 (May 1900), p. 134.

75. Hotel Alexandria, Menu, 1908, UCLA Special Collections 1306, Box 8.

76. Bismarck Café, Menu, Market and Fourth Streets, San Francisco, 1908, pp. 13 and 14. California Historical Society, North Baker Research Library, San Francisco.

77. Chocolate Garden Restaurant, Dance Pavilion, Venice, california, Ca. 1920s, Los Angeles Public Library. Available at http://www.lapl.org/resources/guides/food_drink.html. (Accessed January 28, 2006.)

78. Hotel Mills, Menu (pro tem), Hotel Oakland, Supper Menu, December 6, 1924, in HMDS (Housing, Management, and Dining Services) folder, Mills College Archive.

79. Biltmore Hotel, Restaurant menu, April 30, 1933, Los Angeles Public Library, Regional History, Menu Collection. Available at http://www.lapl.org/resources/guides/food_drink.html. (Accessed January 3, 2007.)

80. Dreyer's Ice Cream. About Dreyer's. Available at http://www.dreyersinc.com/about/index.asp. (Accessed January 3, 2007.)

PART

VIII

Caribbean and South America

Chapter 36 (Momsen and Richardson) offers a broad historical overview of the place of cocoa in the Caribbean, from the 16th century until shortly after World War I. Their chapter focuses on the development of cocoa production especially in Jamaica, Trinidad, and the Windward Islands, and describes 19th century cacao farming and its impact on Caribbean society. Chapter 37 (Momsen and Richardson) reviews 17th and 18th century methods of chocolate preparation in the Caribbean, and the colonial trade patterns between Caribbean cocoa producers and consumers in Europe and North America. Chapter 38 (González and Noriega) traces the history of cacao introduction to Cuba and explores the economic history of this important crop, using evidence based on travel accounts and archive documents. Chapter 39 (Noriega and González) presents a historical survey of cocoa and chocolate in Cuban literature and cultural life as evidenced in short story and novel genres, oral and folkloric traditions, musical texts, and traditional cookbooks that reflect the role of chocolate in Cuban cultural history. Chapter 40 (Walker) identifies the origins of commercial cacao production in the Portuguese Atlantic colonies, explores the connections between cocoa and forced labor, and how Brazil and West Africa evolved into the world's leading cacao production zone during the 19th and subsequent 20th and 21st centuries.

Caribbean Cocoa

Planting and Production

Janet Henshall Momsen and Pamela Richardson

Introduction

Cacao, a profitable hill-grown species that did not compete spatially with sugar cane, has been periodically significant both as an estate and a peasant commodity throughout the history of the Caribbean [1]. Even after sugar cane was introduced to the Caribbean in the mid-17th century, cacao continued to be grown in the Spanish and French islands. In Puerto Rico, ginger and cacao had become the most important export crops by 1665. Trinidad, nominally under the authority of Spain, also grew crops of tobacco and cacao. Cacao was grown on small plantations in Jamaica and as early as the 1680s it was said that a return of £480 per year was to be expected from an area of 8.5 hectares of cacao trees [2]. However, by 1800 the only remaining exporters of any significance in the region were Grenada and Trinidad, the former sending most of its crop to the United States and the latter sending it to Spain by way of Cuba and Puerto Rico.

Cacao is indigenous to Central America. In Belize, cacao has been grown since precontact times, possibly more than 3000 years ago with more intensive cacao farming beginning by 250 BCE [3]. Cacao is still grown by Mayan Indians in the Toledo district in the south of the country and various unsuccessful efforts have been made to encourage commercial production by the Hershey Corporation with assistance from the United States Agency for International Development (USAID) and others [4]. The export of organic cocoa began in 1992 when the British firm Green and Black introduced it as the first Free Trade commodity in the United Kingdom, marketed as Maya Gold [5]. Belize is a relatively high-cost producer and with the growth of off-farm employment following the building of new roads in the region, government subsidies of other crops, and the damage caused by Hurricane Iris in

2001, production has not expanded as expected despite a secure market and Fair Trade prices paid to the farmer [6].

In Costa Rica, cacao growing was introduced by Jamaicans brought in to the Caribbean coast to work on the railway and in the banana industry in the 19th century. Cacao was grown by small farmers under the shade of forest trees using traditional methods of processing. Fungal damage was first noted in the 1960s. This pod rot, known as *monilia roreri*, was finally identified in 1978, by which time it had destroyed 95 percent of cocoa production in the country [7].

This chapter offers a broad historical overview of the place of cocoa in the Caribbean, from the 16th century until shortly after the First World War, by which time West Africa had become the dominant world producer of cocoa [8]. We start with a historical review of the development of cocoa production in the Caribbean, focusing mainly on Jamaica, Trinidad, and the Windward Islands. We then consider in more detail

Chocolate: History, Culture, and Heritage. Edited by Grivetti and Shapiro
Copyright © 2009 John Wiley & Sons, Inc.

19th century cocoa farming and its impact on land and society in the Caribbean.

Sixteenth and Seventeenth Century Cocoa Planting in the Caribbean

Chocolate, the "Food of the Gods" [9], was introduced as a crop to the Caribbean by the Spanish who took it from Venezuela and planted it in Trinidad probably in 1525 [10] and to Jamaica from "Caracus [sic] and Guatemala" [11]. In 1579, the British, we are told, "contemptuously burned a shipload of the stuff, thinking the beans were but sheep droppings" [12]. By the early 1600s, through the depredations of the English buccaneers in the Caribbean and knowledge gained from contact with Spanish physicians [13], chocolate became known in England for its (heavily disputed) medicinal qualities.

The transformation of chocolate from a product for the elite to the mass-consumed item of the 21st century is imbricated in the complex colonial history of the Caribbean region. At the beginning of the 18th century Venezuela was the leading producer, although cocoa cultivation had already spread to the Caribbean islands. Spain tried to maintain a monopoly by prohibiting the export of raw beans from Venezuela. From 1728 to 1780, the cacao trade of Venezuela was controlled by a group of noblemen who obtained an exclusive license from Philip V of Spain [14]. Yet much cacao entered international trade either illegally or from other countries' production.

The early tobacco industry in Trinidad, for example, was destroyed by "competition from the English colonies in North America and by the efforts of Spain to end the contraband trade with foreign ships" [15], leading to the planting of cocoa, a crop that was not grown in the colonies of other European powers. Caribbean privateers and pirates were an indisputable nuisance to the early efforts of the colonial planters. There were several pirate raids on Trinidad including five serious ones between 1594 and 1674 [16] and as late as 1716 the pirate Captain Eduard Tench (Teach), known as Blackbeard, plundered a brig loaded with cocoa bound for Cadiz, and then set fire to her in sight of the town or rather village of Port of Spain [17].

The privateers ruled the Caribbean seas, and indeed the burgeoning cocoa trade, well into the 18th century despite the steady settlement of the islands and the development of large plantations. The letter of October 1708 from Governor Handasyd of Jamaica to the Council of Trade and Plantations in London provides evidence of the complex alliances between buccaneering privateers and nationalist interests:

Two of our privateers have sent in two Spanish vessels loaden [sic] with cocoa and other goods, they took severall [sic] more off Campeachy, but were not able to bring them up to windward, and have either burnt or sold them. It is reported that these they have brought in are very valuable, but the truth of that I am not certain. [18]

COCOA PLANTING IN JAMAICA

In 1655, the British captured Jamaica from the Spanish and took over the cocoa plantations, which were known as cocoa "walks." The early success of the British planters in Jamaica was short-lived, however, and by 1673, the island was reported to be in a terrible state afflicted by drought and so-called "blasts" [19]. In 1672, Sir Thomas Lynch discusses the possible causes of the cocoa blasts in a letter to H. Slingesby, Secretary to the Council for Plantations. The compiled Calendar of State Papers transcribes the letter as follows:

None here, nor any Spaniards he has met, can guess at the cause of these blasts of the cocoa trees, or what remedy to apply; on Hispaniola and Cuba the cocoa is also gone; the young walks to the eastward and some few on the north side are in good condition; but all that are old and some of the young are gone or going, though planted in different places and soils; believes the three former dry years occasioned it, by exsiccating [sic] the radical moisture, for after spring is past the leaves fall, and little by little they die, and some breed worms; but the certain sign of mortality is a dryness of the bark and dustiness about the root. [20]

By the 1690s, the cocoa crop in Jamaica seems to have recovered from the "blasts" perhaps because the failing crop, as reported by Lynch in 1672, "has not discouraged them from planting, and in some years they are likely to have much better and greater walks than ever" [21].

As settlement of Jamaica became established, rich planters with experience in sugar growing in Barbados obtained land, and the production of sugar cane and cotton expanded while cacao production declined [22]. The remaining cacao planters suffered from competition for labor, from the Port Royal earthquake of 1692, which particularly affected the cacao area around Spanish Town, and from the disturbances caused by the French invasion in 1688 and the Maroon Wars of the 1730s. New cocoa walks were laid out after the Treaty of Paris in 1763 but by the end of the century supply exceeded demand and prices collapsed. Labor shortages following slave emancipation meant that cocoa did not recover until the end of the 19th century when market prices improved (Fig. 36.1). Cocoa became a postemancipation crop for small peasant producers in many islands but in Jamaica, as in Dominica, coffee filled this role although a few cocoa plantations survived. In 1876, there were 45 acres of cocoa in Jamaica with holdings falling to 26 acres in 1881 [23]. By 1892, the cocoa acreage had

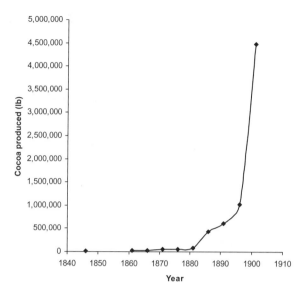

FIGURE 36.1. Jamaica: cocoa production 19th century. *Source*: Jamaica Blue Books for various years.

risen to 1182 and ten years later reached 3548 [24]. Reaction to price rises that began in 1870 were delayed for a tree crop such as cocoa, where young trees take several years to come into bearing, but some of the increase in acreage may reflect abandoned trees brought back into production.

COCOA IN THE WINDWARD ISLANDS

In the Windward Islands of Dominica, St. Lucia, St. Vincent, and Grenada, the former slaves who were able to obtain farmland became the innovators in the face of the decline of plantation agriculture [25]. In the second half of the 19th century, there was a dramatic shift in peasant agriculture from subsistence production to export crops [26]. In St. Lucia sharecropping of cocoa and sugar and in Grenada cocoa and spices came to be regarded as peasant crops, while in Dominica coffee and in St. Vincent arrowroot and cotton as well as cocoa were grown by peasants. By the 1890s, St. Vincent's exports of cocoa rose from 11,996 lb in 1878 to 73,862 lb in 1888, 193,073 lb in 1906–1907, 241,294 lb in 1909–1910, and 285,778 lb in 1913, with 20,000 acres planted in cocoa [27]. At first, cocoa was exported to the United Kingdom and to other British West Indian possessions. In 1886, one bag of St. Vincent cocoa was shipped to the United States [28], and in 1893, 672 lb were shipped to the United States increasing annually after that [29].

Dominica is a mountainous, densely forested island with low population density. In the 19th century there was little agriculture, as land suitable for sugar plantations was scarce and the coffee grown by the French had largely been abandoned when the British took over. The former slaves, however, sharecropped the remaining coffee and cocoa estates. Dominica

exported 820,280 lb of cocoa in 1891, but production fluctuated widely. Barber explained these fluctuations in terms of hurricanes in 1891 and 1893, drought interspersed with heavy rainfall, landslips, and lack of manure and proper care [30].

In Dominica, despite the lack of suitable flat land, English settlers were encouraged to produce sugar cane, while both coffee and cocoa were grown by French settlers:

> The production of cocoa is not much attended by the English planters; and the small quantity exported is chiefly raised on the plantations of the French inhabitants. [31]

This link between French settlers and cocoa was also seen in Trinidad, and in St. Vincent, St. Lucia, and Grenada. These cultural–national links influenced the history of cocoa production in the region.

COCOA IN ST. LUCIA

The search for new locations for cocoa plantations continued in the Caribbean. With reference to cocoa production in St. Lucia, a Reverend W. Gordon proposed to the Council of Trade and Plantations in 1720 that:

> Settlers [in St. Lucia] should be confined to planting cocoa trees, indigo, cotton and ginger, but especially cocoa trees; for the soil is of the same nature with Martinique, and in a few years be brought to produce cocoa enough for all H.M. Dominions, which we are now obliged to have from foreign nations. [32]

Saint Lucia, which was intermittently fought over by the English and the French and changed hands some 13 times, was reported to have 1,321,600 cocoa trees under French occupation in 1772 [33]. In 1787, there were 30 holdings growing only cocoa in St. Lucia, 82 growing mixed coffee and cocoa, 7 growing cocoa and cotton, and 53 growing cocoa, coffee, and cotton out of a total of 556 nonsugar holdings—with most of the cocoa grown in the southwestern district of Soufrière. The concentration of cocoa production in Soufrière, which Lefort de Latour considered as the best in the island despite severe damage caused by the hurricane of 1780, may be explained by the relative shelter from winds on the leeward coast and the excellent harbor in Soufrière town that would have facilitated exportation [34]. The treaties of 1814 and 1815 (Treaty of Paris, Congress of Vienna) allocated St. Lucia and Tobago to the British on a permanent basis.

Agriculture had suffered severely from the vagaries of war and much formerly cultivated land was abandoned. The new colonial occupier had less interest in cotton and tree crops than the French and instead encouraged the monoculture of sugar cane. The number of slaves in St. Lucia declined from about 16,000 in 1787 to 9748 in 1835, probably because of the migration of French settlers with their slaves to Martinique

Year	Acres	Produced (lb)	Exported (lb)
1831	316.5	33,515	
1835	200		
1836	205	163,486	
1841	103	68,400	
1846	71	18,524	
1851	134	15,143	
1856	59.5	42,106	209,720
1861	489.5	76,840	127,610
1866	590	168,116	192,885
1876			399,030
1881			524,,612
1886			546,768
1891			988,200
1996			1,136,843

FIGURE 36.2. St. Lucia: cocoa acreage, production, and exports—1831–1910. *Source*: St. Lucia Blue Books for various years.

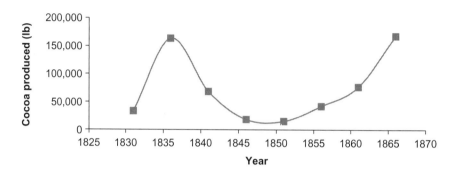

FIGURE 36.3. St. Lucia: cocoa production, 1831–1866. *Source*: St. Lucia Blue Books for various years.

and Guadeloupe. There was a large increase of cocoa production in the 1830s as the trees slowly recovered from the hurricane of 1830. But sugar production also increased from 5,861,379 lb in 1835 to 7,786,270 lb in 1845, largely through the efforts of former slaves who grew sugar cane on the sharecropping system [35]. Through to the 1850s, as sugar production increased, cocoa production fell dramatically (15,143 lb produced in 1851), but then increased again in the 1860s with exports increasing until the end of the 19th century (Figs. 36.2 and 36.3).

After slave emancipation, cocoa in St. Lucia was produced by former slaves working as sharecroppers or *metayers*. This system worked to the advantage of the estates, as the *metayers* were easily evicted from the estates after the trees matured. "Most of the *metayers* who cultivated cocoa were exploited and pauperized" even more than those sharecropping sugar cane [36]. The cultivators were prohibited by law from selling cocoa without the permission of the proprietor of the land. As sugar prices fell, cocoa became the major source of foreign exchange, especially after the 1880s when cocoa prices rose rapidly (Fig. 36.4). A few *metayers* were able to become propertied freeholders after 1897 and in this way cocoa was instrumental in changing the island's social system [37]. By the late 1930s, cocoa production had declined because of disease, weather extremes, and price fluctuations, and bananas began to be planted, although in 1938 St. Lucia still produced 694,176 lb of cocoa with a value of £8332

Year	Shillings per cwt.
1831	16/-
1846	14/ to 20/
1851	30/ to 36/
1856	30/- to 35/-
1861	28/- 33/-
1866	30/ to 36/
1871	24/ to 32/
1881	70/

FIGURE 36.4. St. Lucia: cocoa prices, 1831–1881. *Source*: St. Lucia Blue Books for various years.

(at 27 shillings per cwt) from 4500 acres, while sugar only occupied 3000 acres, and bananas 450 acres [38].

COCOA PRODUCTION IN GRENADA

In Grenada, the story of cocoa was quite different. Although cocoa seems to have had a slower start here than in Trinidad, St. Lucia, or Jamaica, throughout the late 18th and 19th centuries, the island became relatively important for British cocoa production. In 1753, there were only 150,300 cocoa trees in Grenada [39]. When the island was ceded to Britain in 1763, Grenada had 77 sugar plantations, 151 coffee plantations, 4 cocoa estates, and 42 estates with a mixture of coffee, cocoa, and pasture [40]. At this time, sugar occupied 20,384 acres, while estates growing cocoa (usually with other crops) occupied 15,558 acres [41]. By 1776, sugar estates employed the vast number of slaves (18,293), with estates growing other crops such as cocoa and cotton employing 8858 slaves [42]. However, the disturbances of the latter part of the century including slave rebellions, a plague of red ants, a hurricane, and the opening up of Trinidad to foreign settlers, led many people to leave the island.

Following slave emancipation and the shortage of labor that ensued, cocoa production declined to half that of the 1770s, falling to 222,480 lb by 1841 (Fig. 36.5). In addition to the shortage of labor, a worsening of the terms of trade for the island's products hindered the return to prosperity. However, ex-slaves gradually began to grow cocoa as a smallholder crop and there was a large increase in acreage and production of cocoa in the 1860s. There was a tenfold increase of cocoa production in Grenada between 1856 and 1886, which was linked to the increase in land ownership among the former slaves.

They [the laborers] appear more inclined to lay out their savings in purchasing a portion of land which they convert into Gardens and Cocoa plantations, generally erecting thereon a Hut for themselves and family, they seem very wishful of leaving Estates' residences; this arises from Managers giving them notice to quit possession upon the most trifling misunderstanding, consequently freeholders are on the increase. [43]

Year	Acres	Cocoa produced (lb)
1831		337,903
1836		301,172
1841		222,480
1846		226,564
1850		438,637
1851	1130	416,481
1856	1538	470,252
1861	2117.5	618,000
1866		820,643
1870	20,168.50	774,858
1871	15,647.50	1,036,384
1876	5,038	1,223,557
1881	8364.75	2,502,440
1886	10,367	4,763,696
1891	12,606.75	

FIGURE 36.5. Grenada: cocoa production, 19th century. *Source:* Grenada Blue Books for various years.

Most of the cocoa in Grenada was grown at altitudes above 1000 feet and 79 percent of the production came from the two parishes of St. Mark and St. John [44]. Paterson described these parishes as being thinly settled because the people were poor, the terrain mountainous, and "the land is very cold, and requires a great deal of labour to keep it in order" [45]. It would seem that cocoa was one of the more successful crops in these conditions, and cocoa production steadily gained momentum in Grenada throughout the 19th century (Figs. 36.5 and 36.6).

COCOA IN TRINIDAD

Although cocoa was planted with varying success throughout the 18th and 19th centuries in Jamaica and the Windwards, the scale of production was never to match that of the larger island of Trinidad, which became a British colony in 1803. In 1668, a British officer reported that the cocoa grown in Trinidad was the best in the Indies [46]. The Spanish had first planted the *criollo* variety of cocoa (*Theobroma cacao* L.) in Trinidad in 1525, establishing permanent settlement in 1591 [47]. Linda Newson has written that cocoa had been growing wild in Trinidad from 1616 and the Spanish "reported" its discovery there in 1617, and it had come to rival tobacco as one of the two principal crops of the island as early as 1645 [48, 49]. On the other hand, Borde [50] reported that cacao was not introduced until the 1680s because there was not enough labor before then. Kathleen Phillips-Lewis [51] has suggested that cacao may have been brought to Trinidad by the Dutch, from the coast of Guiana or from Caracas via Curaçao, but others have reported that cacao was imported to Trinidad from Venezuela in 1678 [52].

Cocoa became Trinidad's staple export by the beginning of the 18th century [53] and was thought to be superior to that of Venezuela (Fig. 36.7) [54]. It was cultivated by Amerindian laborers on Spanish-owned plantations [55]. The Capuchin friars came to Trinidad in 1687 and established mission towns in an attempt to convert or "civilize" the Indians [56]. The townships cultivated root crops and cocoa, in which the friars had a vested interest, as Borde has argued:

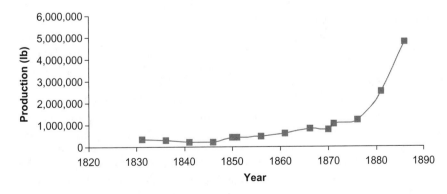

FIGURE 36.6. Grenada: cocoa production (pounds produced), 19th century. *Source:* Grenada Blue Books for various years.

Year	Acres	Produced (lb)	Exported (lb)
1826			3,011,091
1831			19,521
1835	10,468	2,315,937	
1838	10,307	2,312,197	2,466,059
1841	6,910	1,257,616	2,964,983
1847	8,767	1,977,188	3,494,368
1851			5,552,437
1859	7,000		
1861		6,530,906	6,530,906
1876		9,710,020	1,265,000
1881			10,809,796
1886		17,913,037	18,000,000
1891	94,500	16,188,493	16,188,493
1894	94,500	21,608,384	

FIGURE 36.7. Trinidad: cocoa acreage, production, and exports—19th century. *Source:* Grenada Blue Books for various years.

The cocoa house was the community home of the mission. On the products of cocoa depended the ornamentation of the church, the furniture and the utensils of the presbytery, the agricultural instruments, the rations for the missionaries and the clothing for the neophytes. What was left was put in the community chest of the Catalan Capuchin priests. [57]

The planters misused the Amerindians, and the missionaries were forced to leave by Royal decree in 1708, whereupon the Indians fled into the mountains [58]. Planters then became dependent on African slave labor. By 1704, about 250,000 lb of cocoa were being exported from Trinidad to Spain and to the Spanish colonies annually.

By 1713, however, Mexico had entered the European market for cocoa and prices subsequently fell, though Trinidad cocoa continued to fetch high prices because of its superior quality [59]. By 1725, the settlers were almost entirely dependent on the export of cocoa, and in 1719, they could afford to equip a warship to protect the island's cocoa trade [60]. But in 1727, the cocoa industry in Trinidad was almost destroyed by a "blast" and many settlers emigrated. In 1739, the War of Jenkin's Ear between Spain and England, together with an outbreak of smallpox in Trinidad, led to abandonment of most of the remaining cocoa plantations [61]. The industry was revived in 1756 after more resistant stock of the *forestero* variety had been imported from Venezuela, but the cacao trade never regained its earlier preeminence [62]. The *forestero* variety of cacao eventually interbred with the remnant *criollo* type to produce the esteemed hybrid cacao referred to as *trinitario*; a variety that combines the hardiness of *forestero* with the aromatic properties of *criollo* [63]. This signaled the beginning of a period marked by a large increase in sugar estates and a marginalization of cocoa production.

The latter part of the 18th century saw a series of efforts to populate the island and increase agricultural production. In 1777, a Frenchman from Grenada, M. Roume de St. Laurent, who was impressed by Trinidad, persuaded the *Intendant* of Caracas to offer special favors to French and Irish colonists to settle in Trinidad [64]. Furthering these attempts to lure wealthy planters to the island, under the Spanish *Cedula* of Population in 1783, settlers were granted 32 acres of land for each member of either sex of a white family plus 16 acres for each slave. Free Negroes and people of color were allowed 16 acres for each free person and eight acres for each slave. They could bring with them their belongings free of tax and were allowed to trade with the French West Indies but had to profess that they were Roman Catholic and declare allegiance to the King of Spain. At the time of the *Cedula*, the population of Trinidad numbered only 2813 persons and of these 2082 were Indians [65]. After the *Cedula*, however, the population of Trinidad grew to 18,918 by 1789, consisting of 10,000 slaves, 2200 Indians, 4467 free colored people, and 2151 whites. Most of these immigrants came from Grenada, which had been ceded to Britain in 1783.

The location of sugar cane was constrained by the need for easy transport routes for a bulky product, so it was mainly grown on the flat land along the Gulf of Paria [66]. Cocoa, on the other hand, was less bulky and so could be grown in more isolated areas, mainly in the foothills and valleys of the Northern Range and along the Caroni River according to the map prepared by Joseph Basanta in 1846 [67]. Land holding patterns had important implications for cocoa production in British Trinidad. In 1797, when Trinidad was unofficially captured by the British, it was found that there was a problem with insecure land titles because land had been granted by different colonial powers and there had been private sales of land without official title endowments. All these problems led to underutilization of land.

Cocoa Producers

Labor has been a central factor in the history of cocoa production in the British Caribbean. At the end of the 18th century, much of the land owned by free blacks and free coloreds had come into their hands as a result of the Spanish *Cedula* of 1783 [68] and they became the majority of the small farmers growing cocoa. After the abolition of the slave trade in 1807, Crown Lands in Trinidad were distributed at low cost. Most of this land was cultivated in cocoa by a new class of small farmers who occupied these lands under squatters' rights [69]. These preemancipation farmers founded villages and communities away from the plantations.

Further efforts to increase the population of the island by encouraging immigration led to Chinese laborers arriving in 1806, 4000 Africans and a few workers from Madeira and Brazil in 1835, followed by East Indians in 1838 [70]. Other migrant laborers came from the British West Indies, the United States, Malta, and Germany, attracted by the relatively high wages in Trinidad [71]. The expenses of the introduction of such labor were borne by the Colonial Treasury. However, cocoa did not benefit from this new source of labor, as most estates were so scattered, small, and isolated that they could not comply with the hospital requirements of the immigration laws and so did not qualify to employ indentured immigrant workers [72]. But from the 1880s as cocoa prices rose, the larger cocoa estates managed to meet the requirements by collaborating to build a common hospital and so were able to obtain indentured East Indian workers, which resulted in forcing down wages for free workers [73].

Trinidadian cocoa production also benefited from the arrival of Venezuelan peons, known as *panyols*, who immigrated to Trinidad early in the 19th century [74]. This was a time of great instability in Venezuela, as the country fought for independence from Spain. The Venezuelans became small-scale cultivators squatting in the forests. As a peasant class, familiar with cocoa farming in Venezuela, they came to form the backbone of the then-minor Trinidadian cocoa industry [75]. With new processing techniques, exports increased. According to the 1826 *Blue Book of Trinidad*, cocoa production in Trinidad more than doubled between 1816 and 1826. By 1830, Trinidad was the third-largest producer of cocoa in the world after Venezuela and Ecuador [76].

Phillips-Lewis [77] has held that cocoa maintained a much larger number of peasant proprietors than any other crop, including sugar. Cocoa growing, therefore, became the most popular agricultural activity of the ex-slave class in Trinidad because it was more conducive to small-scale cultivation than sugar. By 1866, almost half of all cocoa in Trinidad was grown on squatter holdings [78]. Squatting was undesirable from the colonial planters' perspective, but impossible to prevent on a large, unevenly populated and forested island. From the mid-19th century, some efforts were made to persuade the squatters to become legitimate landowners. However, as Phillips-Lewis argues: "[T]he cocoa peasantry . . . by 1858 was well entrenched and its resilient nature frustrated all attempts to annihilate it completely" [79]. Cocoa led to the development of previously unpopulated areas such as the valleys of the Northern Range, and parts of central and south Trinidad, as well as some of the hillsides of Tobago [80]. As Brereton has pointed out for Trinidad: "The peasant was the real pioneer and the planter simply followed" [81].

The second method linking the peasantry to the plantation was the contract system. Under this system the landowner would clear the forest and then arrange with contractors to plant cocoa on a *quarrée* of land (3.2 acres). When the cocoa trees came into bearing the owner took over the land, paying agreed sums for each tree according to the stage of growth it had reached.

The contractor was free to plant food crops beneath the young trees and often provided wage labor to nearby cocoa estates. The contracts usually lasted for five years and contractors generally were able to get a reasonable return on their labor [82]. This arrangement also suited the planter as his start-up expenses were low and the growing trees constituted security on which he could raise loans when the five years were up. Many of the freeholders and contractors were Venezuelan *panyols*, but this group also included Creoles and Africans and, from the 1870s, East Indians who had completed their period of indenture. All members of the family took part in the labor of cocoa production, with women being involved especially in the picking and breaking of the cocoa pods [83]. Harwood also noted that:

> On peasant cocoa holdings, women can attend to their domestic duties and at the same time supervise the fermentation and drying of the cocoa. Under these circumstances, the women's contribution to family agricultural labour may be more than is apparent at first sight. [84]

The larger cocoa plantation owners were often French Creoles, as were the cocoa dealers in Port of Spain. Many of these people, forced to sell out of sugar production because they could not afford to construct large modern sugar factories, used the proceeds of the sale of their sugar estates to build up large cocoa estates. Some British companies, such as Cadburys, also bought cocoa plantations, so starting the vertical integration of the industry.

To summarize, in the last decades of the 19th century and the first two of the 20th century, it could be said that cocoa provided a relatively comfortable livelihood for many peons, Creoles, and East Indians and made fortunes for the larger French Creole planters [85]. Despite the difficulties and fluctuations discussed here, Trinidad was the undisputed leader in Caribbean cocoa production throughout the 19th century, with acreage increasing in the last third of the century from 19,000 acres in 1871, to 94,500 acres in 1890, and 200,252 in 1920, which constituted about a quarter of the cultivable land [86]. In 1921, Trinidad reached its peak production of 75 million pounds and was the fourth- or fifth-largest producer of cacao in the world [87]. Sylvia Moodie-Kublalsingh has pointed out that: "[I]n 1898, for the first time in the history of British Trinidad, cacao exports exceeded those of the sugar industry in value," but it is evident from the Blue Book of 1886 that even by the mid-1880s, acres in cocoa exceeded those in sugar cane by about a third [88, 89]. In 1894, Trinidad produced 21,608,384 lb of cocoa worth £509,808 [90]. This contrasts with Jamaica, and the small islands of Grenada and St. Lucia, which were producing about five, four, and one million pounds of cocoa, respectively, at this time. By 1912, there were about 700 cocoa plantations in Trinidad covering over 300,000 acres [91].

Technological Changes

In addition to land, labor, and cultural issues, as well as the development of new cocoa varieties, cocoa production underwent several technological developments throughout the late colonial period. The Grenada Blue Book of 1896 reveals that: "The application of manure to cocoa cultivation is on the increase throughout the island, and the drying of that product by artificial means, instead of relying upon sunlight is being gradually adopted" [92]. A cocoa tree of the *forestero* variety was planted in the St. Vincent Botanical Garden in 1890 and produced pods by 1894 [93]. Cacao yields in St. Vincent, at 7 to 11 bags per acre per year, were higher than in Grenada because the Vincentian farmers used manure on their fields [94]. In 1892, the Trinidad Blue Book noted that a prize of £100 offered by the Agricultural Board for the best cocoa-drying machine was awarded to a Mr. Scheulte [95]. In 1894, the Trinidad Blue Book recorded that glass-covered houses had been introduced for drying the cocoa in 1893 and artificial drying using hot air was making steady progress [96]. Hart described new machines for polishing cacao beans, a process previously done by hand, and for grading cacao for size as practiced on large estates and on cocoa merchants' drying floors, producing about one-third of cacao exports [97]. These technological developments would have influenced the yields, quality, and labor requirements of cocoa estates, thus having some effect on the production of cocoa over time.

Conclusion

Cocoa played a major role in the settlement of former slaves as independent farmers and throughout its history has influenced the social and cultural development of the Caribbean region. Trinidad led in technological development of the cocoa industry through early research at the Imperial College of Tropical Agriculture and today at the Cocoa Research Unit, both based at the Trinidadian campus of the University of the West Indies. The British Caribbean had become a major world producer of cocoa by the end of the 19th century but cocoa has repeatedly suffered from hurricanes and diseases, contributing to the constantly fluctuating fortunes of the region's producers. After 1920, Caribbean cocoa became increasingly marginalized by the expansion of cocoa production in Ceylon, Malaysia, Philippines, and ultimately West Africa.

Acknowledgments

We would like to acknowledge the help and assistance of Dr. Glenroy Taitt, Senior Librarian in the West India Collection of the University of the West Indies, St. Augustine Library, who aided in the identification of numerous chocolate-related documents within the collection. We also thank Dr. J. A. Spence, Head of the Cocoa Research Unit at the University of the West Indies, Trinidad, for giving us access to the CRU library. Also thanks to Professor Bridget Brereton of the History Department at the University of the West Indies, Trinidad, for guidance to historical sources on cocoa in Trinidad.

References

1. Watts, D. *The West Indies: Patterns of Development, Culture and Environmental Change since 1492.* Cambridge, England: Cambridge University Press, 1987.
2. Watts, D. *The West Indies: Patterns of Development, Culture and Environmental Change since 1942.* Cambridge, England: Cambridge University Press, 1987; p. 551.
3. Young, A. *The Chocolate Tree: A Natural History of Cacao.* Washington, DC: Smithsonian Institution Press, 1994.
4. Marcotte, T. P. *Competing land use systems in the Toledo Uplands of Belize: An analysis of limited supply response and adoption of organic shade-grown cacao.* Unpublished M.S. thesis in International Agricultural Development, University of California, Davis, 2003.
5. Sutcliffe, W. Counting beans. *The Guardian Weekend* August 7, 2004: 18–23.
6. Marcotte, T. P. *Competing land use systems in the Toledo Uplands of Belize: An analysis of limited supply response and adoption of organic shade-grown cacao.* Unpublished M.S. thesis in International Agricultural Development, University of California, Davis, 2003.
7. Martin, G. Conservation and recreation planning on the Caribbean coast: Cahuita, Costa Rica. In: Pugh, J., and Momsen, J. H., editors. *Environmental Planning in the Caribbean.* Aldershot, England: Ashgate, 2006; p. 133.
8. Coe, S. D., and Coe, M. D. *The True History of Chocolate.* New York: Thames and Hudson, 1996.
9. Head, B. *The Food of the Gods. A Popular Account of Cocoa.* London, England: George Routledge and Sons, 1903.
10. Bekele, F. L. The history of cocoa production in Trinidad and Tobago. In: *Proceedings of a Seminar/Exhibition on Revitalization of the Trinidad and Tobago Cocoa Industry, 20th September 2003.* University of the West Indies, St. Augustine, Trinidad and Tobago: The Association of Professional Agricultural Scientists of Trinidad and Tobago, The Cocoa Research Unit, 2003; p. 4.
11. Lucas, N. The Lucas Manuscript compiled from "An Historical Account of the Rise and Growth of the West Indian Colonies and of the Great Advantages they are to England in respect to Trade." London, England: Printed 1690. Reprinted in the *Journal of the Barbados Museum and Historical Society* Nov. 1949;17(1).
12. Coe, S. D., and Coe, M. D. *The True History of Chocolate.* New York: Thames and Hudson, 1996; p. 165.

13. de Ledesma Colmenero, A. *A Curious Treatise of the Nature and Quality of Chocolate*. Madrid, 1631. Translated from the original Spanish by Don Diego de Vade-Forte for Edward Lord Viscount of Conway and Killultah, Baron of Ragley, Lord Marshal of Ireland and one of the Councell of Warre to His Majesty of Great Brittaine. London, England: J. Oakes dwelling in Little St. Bartholmewes, 1640.

14. Shephard, C. Y. *Cacao—Part II: General History of the Production and Consumption of Cacao*. Port of Spain, Trinidad and Tobago: The Government Printing Office, 1932; p. 1.

15. Brereton, B. *A History of Modern Trinidad 1783–1962*. Kingston, Port of Spain and London: Heinemann Educational Books Inc., 1981; p. 2.

16. Shephard, C. Y. *Cacao—Part III: History of the Industry up to 1870*. Port of Spain, Trinidad and Tobago: The Government Printing Office, 1932; p. 1.

17. Clark, H. J. *The Land of the Humming Bird, being a Sketch of the Island of Trinidad* (Specially written for the Trinidad exhibition at the Chicago World's Fair). Port of Spain, Trinidad and Tobago: Government Printing Office, 1893; p. 76.

18. Colonial Office "Letter from Governor Handasyd of Jamaica to the Council of Trade and Plantations in London." 1708, C.O. 137/8, Nos. 27i, ii.

19. Lynch, Lieut-Governor Sir Thomas to Dr Benj. Worsleu, Secretary to the Council for Plantations, Jamaica, 1673, Public Record Office, CO 1/30, No. 21.

20. Sir Thos. Lynch to H. Slingesby, Secretary to the Council for Plantations, Jamaica, 1672, Public Record Office, CO 1/29, Nos. 43, 43i., ii.

21. Sir Thos. Lynch, to H. Slingesby, Secretary to the Council for Plantations, Jamaica, 1672, Public Record Office, Co 1/29, Nos. 43, 43i., ii.

22. Watts, D. *The West Indies: Patterns of Development, Culture and Environmental Change since 1942*. Cambridge, England: Cambridge University Press, 1987; p. 341.

23. Jamaica Blue Books, 1876 and 1881, Kingston 1877 and 1882.

24. Jamaica Blue Books, 1892 and 1902, Kingston 1893, and 1903.

25. Momsen, J. D. *The Geography of Land Use and Population in the Caribbean (with Special Reference to Barbados and the Windward Islands)*. Ph.D. thesis, University of London, King's College, 1969.

26. Momsen, J. D. *The Geography of Land Use and Population in the Caribbean (with Special Reference to Barbados and the Windward Islands)*. Ph.D. thesis, University of London, King's College, 1969.

27. St. Vincent Blue Books for 1878, 1888, 1906–1907, 1909–1910, and 1913, Kingstown, 1879, 1889, 1911, and 1914.

28. St. Vincent Blue Book, 1887, Kingstown, 1888.

29. St. Vincent Blue Books, 1894, Kingstown, 1895.

30. Barber, C. A. Report on the failure of the Dominica cacao crop 1892-3. Supplement to the Leeward Islands Gazette, Thursday, 27 April 1893:44.

31. Atwood, T. *The History of the Island of Dominica*. London, England: J. Johnson, 1791; p. 82.

32. Gordon, Rev. W. *Reverend Gordon to the Council of Trade and Plantations*, March 3, 1720. Public Record Office, CO 28/15.

33. Haye, La. Present State of French Settlements in the West Indies. *Martinico* 1774;5:28. SP78/295, 1775.

34. Lefort de Latour, M. *A General Description of the Island of St Lucia, 1787*. London, England: Colonial Office, West Indies, 1883; No. 44.

35. St. Lucia Blue Books, 1836 and 1846, CO258/32 and CO258/42.

36. Adrien, P. *Metayage, Capitalism and Peasant Development in St. Lucia 1840–1957*. New Generation Series, Consortium Graduate School of Social Sciences. Mona, Jamaica: University of the West Indies, 1996; p. 47.

37. Adrien, P. *Metayage, Capitalism and Peasant Development in St. Lucia 1840–1957*. New Generation Series, Consortium Graduate School of Social Sciences. Mona, Jamaica: University of the West Indies, 1996; p. 47.

38. Adrien, P. *Metayage, Capitalism and Peasant Development in St. Lucia 1840–1957*. New Generation Series, Consortium Graduate School of Social Sciences. Mona, Jamaica: University of the West Indies, 1996; p. 57.

39. Gittens-Knight, E. *The Grenada Handbook and Directory*. Bridgetown, Barbados, Government Printer, 1946; p. 22.

40. Paterson, D. *A Topographical Description of the Island of Grenada. Surveyed by M. Pinel in 1763, with the addition of English names, alterations of Property, and other Improvements to the present time by Lieut. Daniel Paterson, Assistant to the Quarter-Master-General*. London, England: n.p., 1780.

41. Paterson, D. *A Topographical Description of the Island of Grenada. Surveyed by M. Pinel in 1763, with the addition of English names, alterations of Property, and other Improvements to the present time by Lieut. Daniel Paterson, Assistant to the Quarter-Master-General*. London, England: n.p., 1780.

42. Paterson, D. *A Topographical Description of the Island of Grenada. Surveyed by M. Pinel in 1763, with the addition of English names, alterations of Property, and other Improvements to the present time by Lieut. Daniel Paterson, Assistant to the Quarter-Master-General*. London, England: n.p., 1780; p. 13.

43. Colonial Office *Stipendiary Magistrates' Reports, Grenada*. December, 1845, CO 106–113.

44. Paterson, D. *A Topographical Description of the Island of Grenada. Surveyed by M. Pinel in 1763, with the addition of English names, alterations of Property, and other Improvements to the present time by Lieut. Daniel Paterson, Assistant to the Quarter-Master-General*. London, England: n.p., 1780; p. 2.

45. Paterson, D. *A Topographical Description of the Island of Grenada. Surveyed by M. Pinel in 1763, with the addition of English names, alterations of Property, and other Improvements to the present time by Lieut. Daniel Paterson, Assistant to the Quarter-Master-General.* London, England: n.p., 1780; p. 2.

46. Brereton, B. *A History of Modern Trinidad 1783–1962.* London, England: Heinemann Educational Books, Inc., 1981; p. 2.

47. Bekele, F. L. The history of cocoa production in Trinidad and Tobago. In: *Proceedings of a Seminar/Exhibition on Revitalization of the Trinidad and Tobago Cocoa Industry, 20th September 2003.* University of the West Indies, St. Augustine, Trinidad and Tobago: The Association of Professional Agricultural Scientists of Trinidad and Tobago, The Cocoa Research Unit, 2003; p. 4.

48. Shepard, C. Y. *Cacao—Part II: General History of the Production and Consumption of Cacoa.* Port of Spain, Trinidad and Tobago: The Government Printing Office, 1932; p. 1.

49. Newson, L. A. *Aboriginal and Spanish Colonial Trinidad: A Study in Culture Contact.* London, England: Academic Press, 1976; p. 133.

50. Borde, P. G. L. *The History of the Island of Trinidad Under the Spanish Government.* Volume 1: 1498–1622; Volume II: 1622–1797. First published Paris, France: Maisonneuve et Cie, 1876 and 1883. Translated from the French—Volume 1 by James A. Bain (1932) and Volume II Brigadier-General A. S. Mavrogordato (1961). Port of Spain, Trinidad and Tobago: Paria Publishing Company Ltd., 1982.

51. Phillips-Lewis, K. E. *British Imperial Policy and Colonial Economic Development: The Cocoa Industry in Trinidad 1838–1939.* Unpublished Ph.D. thesis, University of Manitoba, Winnipeg, Canada, 1994.

52. Shephard, C. Y. *Cacao—Part III: History of the Industry up to 1870.* Port of Spain, Trinidad and Tobago: The Government Printing Office, 1932; p. 1.

53. Shephard, C. Y. *Cacao—Part III: History of the Industry up to 1870.* Port of Spain, Trinidad and Tobago: The Government Printing Office, 1932.

54. Shephard, C. Y. *Cacao—Part III: History of the Industry up to 1870.* Port of Spain, Trinidad and Tobago: The Government Printing Office, 1932.

55. Brereton, B. *A History of Modern Trinidad 1783–1962.* London, England: Heinemann Educational Books, Inc., 1981; p. 2.

56. Brereton, B. *A History of Modern Trinidad 1783–1962.* London, England: Heinemann Educational Books, Inc., 1981; p. 2.

57. Borde, P. G. L. *The History of the Island of Trinidad Under the Spanish Government.* Volume II: 1622–1797. Paris, France: Maisonneuve et Cie, 1876. Translated from the French by Brigadier-General A. S. Mavrogordato (1961). Port of Spain, Trinidad and Tobago: Paria Publishing Company Ltd., 1982; p. 45.

58. Shephard, C. Y. *Cacao—Part III: History of the Industry up to 1870.* Port of Spain, Trinidad and Tobago: The Government Printing Office, 1932; p. 1.

59. Brereton, B. *A History of Modern Trinidad 1783–1962.* London, England: Heinemann Educational Books, Inc., 1981; p. 2.

60. Brereton, B. *A History of Modern Trinidad 1783–1962.* London, England: Heinemann Educational Books, Inc., 1981; p. 2.

61. Brereton, B. *A History of Modern Trinidad 1783–1962.* London, England: Heinemann Educational Books, Inc., 1981; p. 2.

62. Bekele, F. L. The history of cocoa production in Trinidad and Tobago. In: *Proceedings of a Seminar/Exhibition on Revitalization of the Trinidad and Tobago Cocoa Industry, 20th September 2003.* University of the West Indies, St. Augustine, Trinidad and Tobago: The Association of Professional Agricultural Scientists of Trinidad and Tobago, The Cocoa Research Unit, 2003; p. 4.

63. Phillips-Lewis, K. E. *British Imperial Policy and Colonial Economic Development: The Cocoa Industry in Trinidad 1838–1939.* Unpublished Ph.D. thesis, University of Manitoba, Winnipeg, Canada, 1994; p. 72.

64. Shephard, C. Y. *Cacao—Part III: History of the Industry up to 1870.* Port of Spain, Trinidad and Tobago: The Government Printing Office, 1932; p. 2.

65. Shephard, C. Y. *Cacao—Part III: History of the Industry up to 1870.* Port of Spain, Trinidad and Tobago: The Government Printing Office, 1932; p. 2.

66. Shephard, C. Y. *Cacao—Part III: History of the Industry up to 1870.* Port of Spain, Trinidad and Tobago: The Government Printing Office, 1932; p. 3.

67. Shephard, C. Y. *Cacao—Part III: History of the Industry up to 1870.* Port of Spain, Trinidad and Tobago: The Government Printing Office, 1932; Figure V.

68. Campbell, C. Black testators: fragments of the lives of free Africans and free Creole Blacks in Trinidad 1813–1877. In: Brereton, B., and Yelvington, K. A., editors. *The Colonial Caribbean in Transition: Essays on Postemancipation Social and Cultural History.* Mona, Jamaica: The Press University of the West Indies and Gainesville: University Press of Florida, 1999; footnote 8.

69. Bekele, F. L. The history of cocoa production in Trinidad and Tobago. In: *Proceedings of a Seminar/Exhibition on Revitalization of the Trinidad and Tobago Cocoa Industry, 20th September 2003.* University of the West Indies, St. Augustine, Trinidad and Tobago: The Association of Professional Agricultural Scientists of Trinidad and Tobago, The Cocoa Research Unit, 2003; p. 5.

70. Shephard, C. Y. *Cacao—Part III: History of the Industry up to 1870.* Port of Spain, Trinidad and Tobago: The Government Printing Office, 1932; p. 6.

71. Shephard, C. Y. *Cacao—Part III: History of the Industry up to 1870.* Port of Spain, Trinidad and Tobago: The Government Printing Office, 1932; p. 6.

72. Shephard, C. Y. *Cacao—Part III: History of the Industry up to 1870.* Port of Spain, Trinidad and Tobago: The Government Printing Office, 1932; p. 6.

73. Millette, J. The wage problem in Trinidad and Tobago, 1838–1938. In: Brereton, B., and Yelvington, K. A., editors. *The Colonial Caribbean in Transition: Essays on Postemancipation Social and Cultural History*. Mona, Jamaica: The Press University of the West Indies/and Gainesville: University Press of Florida, 1999.

74. Moodie-Kublalsingh, S. *The Cocoa Panyols of Trinidad: An Oral Record*. London, England: British Academic Press, 1994.

75. Phillips-Lewis, K. E. *British Imperial Policy and Colonial Economic Development: The Cocoa Industry in Trinidad 1838–1939*. Unpublished Ph.D. thesis, University of Manitoba, Winnipeg, Canada, 1994; p. 74.

76. Bekele, F. L. The history of cocoa production in Trinidad and Tobago. In: *Proceedings of a Seminar/Exhibition on Revitalization of the Trinidad and Tobago Cocoa Industry, 20th September 2003*. University of the West Indies, St. Augustine, Trinidad and Tobago: The Association of Professional Agricultural Scientists of Trinidad and Tobago, The Cocoa Research Unit, 2003; p. 5.

77. Phillips-Lewis, K. E. *British Imperial Policy and Colonial Economic Development: The Cocoa Industry in Trinidad 1838–1939*. Unpublished Ph.D. thesis, University of Manitoba, Winnipeg, Canada, 1994; p. 237.

78. Phillips-Lewis, K. E. *British Imperial Policy and Colonial Economic Development: The Cocoa Industry in Trinidad 1838–1939*. Unpublished Ph.D. thesis, University of Manitoba, Winnipeg, Canada, 1994; p. 139.

79. Phillips-Lewis, K. E. *British Imperial Policy and Colonial Economic Development: The Cocoa Industry in Trinidad 1838–1939*. Unpublished Ph.D. thesis, University of Manitoba, Winnipeg, Canada, 1994; p. 181.

80. Bekele, F. L. The history of cocoa production in Trinidad and Tobago. In: *Proceedings of a Seminar/Exhibition on Revitalization of the Trinidad and Tobago Cocoa Industry, 20th September 2003*. University of the West Indies, St. Augustine, Trinidad and Tobago: The Association of Professional Agricultural Scientists of Trinidad and Tobago, The Cocoa Research Unit, 2003; p. 6.

81. Brereton, B. *A History of Modern Trinidad 1783–1962*. London, England: Heinemann Educational Books, Inc., 1981; p. 92.

82. Shephard, C. Y. *Cacao—Part IV: Historical 1870–1920*. Port of Spain, Trinidad and Tobago: The Government Printing Office, 1932; p. 1.

83. Hart, J. H. The consumption of cacao. *The West India Committee Circular* 21 June 1910: 2914.

84. Harwood, L. W. *Peasant Cocoa Farming in the Maracas Valley*, Report for Diploma in Tropical Agriculture, 1953, p. 64.

85. Brereton, B. *A History of Modern Trinidad 1783–1962*. London, England: Heinemann Educational Books, Inc., 1981; p. 93.

86. Trinidad Blue Books, 1871, 1890, 1920, Port of Spain, 1872, 1891, 1921.

87. Bekele, F. L. The history of cocoa production in Trinidad and Tobago. In: *Proceedings of a Seminar/Exhibition on Revitalization of the Trinidad and Tobago Cocoa Industry, 20th September 2003*. University of the West Indies, St. Augustine, Trinidad and Tobago: The Association of Professional Agricultural Scientists of Trinidad and Tobago, The Cocoa Research Unit, 2003; p. 6.

88. Moodie-Kublalsingh, S. *The Cocoa Panyols of Trinidad: An Oral Record*. London, England: British Academic Press, 1994; p. 1.

89. Trinidad Blue Book, 1886, CO 300/97.

90. Trinidad Blue Book, 1894, CO 300/105.

91. Collens, J. H., editor. *Handbook of Trinidad and Tobago for the Use of Settlers Prepared by a Committee of the Board of Agriculture*. Port of Spain, Trinidad and Tobago: Government Printing Office, 1912; p. 58.

92. Grenada Blue Book, 1896, CO 106/90.

93. Powell, H. *Report on the St. Vincent Botanical Garden*, March 26, 1894, St. Vincent Blue Book, 1893, Kingstown, 1894.

94. Powell, H. *Report on the St. Vincent Botanical Garden*, March 26, 1894, St. Vincent Blue Book, 1893, Kingstown, 1894.

95. Trinidad Blue Book, 1892, Port of Spain, 1893.

96. Trinidad Blue Book, 1894, Port of Spain, 1895.

97. Hart, J. H. The consumption of cacao. *The West India Committee Circular* 21 June 1910: 80–81.

37

Caribbean Chocolate

Preparation, Consumption, and Trade

Janet Henshall Momsen and Pamela Richardson

Introduction

This chapter begins with a discussion of 17th and 18th century methods of chocolate preparation in the Caribbean and the exchange of this knowledge through the metropolitan countries. We next consider beliefs as to the medicinal and nutritional uses of chocolate and how consumption patterns changed over time. Consumption in Europe was at first only for the elite and places where chocolate was drunk became important social centers for the intelligentsia. By the 19th century, chocolate had become available to the masses and we examine the role of chocolate in Britain and the Caribbean as a source of nutrition in the institutional context of prisons and the military. The chapter concludes with a discussion of the changing colonial trade patterns between Caribbean cocoa producers and consumers in Europe and North America.

The consumption of chocolate quickly became popular among the early Spanish settlers in the Americas but was kept secret from other nations. For many years drinking chocolate was only exported to Spain in the form of concentrated tablets of chocolate, not raw beans [1]. However, Colmenero's treatise on chocolate, published in Madrid in 1631, was soon translated from Spanish into several European languages. After it was translated into English and published in London in 1640, chocolate quickly became a fashionable drink in England, in many cases introduced by foreigners. It was introduced early and soon became popular in the university city of Oxford: in 1648, Nathaniel Copius, from Crete, made cocoa while at Balliol College; in 1650, coffee and chocolate were drunk at *The Angle*, prepared by Jacob, a Jew; and in 1654, Cirques Jobson, a Jew, and another person, a Jacobite (from Lebanon), sold coffee and chocolate "at or neare *The Angle*, within the East Gate of Oxon" [2]. Chocolate houses subsequently opened in London and in Amsterdam and began to replace coffeehouses as the resort of the "elegant and refined" [3]. The first London chocolate house opened in 1657 in Bishopgate Street, Queen's Head Alley [4]. According to Head [5], quoting a notice in the *Public Advertiser* of 1657, this was at a Frenchman's house where "is an excellent West India drink, called chocolate, to be sold, where you may have it ready at any time and also made at reasonable rates." The Cocoa Tree, early the headquarters of the Jacobite party, became subsequently recognized as the club of the literati, including Garrick and Byron [6]. Others such as "White's Chocolate House" developed into a famous and exclusive club, until recently only open to men. By the end of the 17th century, chocolate was widely appreciated as a beverage, but its high cost of 10–15 shillings per pound weight limited its spread [7]. In fact, it was said that "the price of chocolate made it rather than coffee the beverage of the aristocracy" [8]. In the 1640 edition of *A Curious*

Chocolate: History, Culture, and Heritage. Edited by Grivetti and Shapiro
Copyright © 2009 John Wiley & Sons, Inc.

Treatise on the Nature and Quality of Chocolate, originally written by the Spanish physician Antonio Colmenero de Ledesma and published in Madrid in 1631, it is claimed that:

> So great is the number of those persons, who at present do drink of Chocolate, that not only in the West Indies, whence this Drink has its Original and beginning, but also in Spain, Italy, Flanders, &c. it is very much used, and especially in the Court of the King of Spain; where the great Ladies drink it in a morning before they rise out of their beds, and lately much used in England, as Diet and Physick with the gentry. [9]

We hear that in the Spanish West India colonies:

> It alone makes up both the necessary provision for their sustenances, and their delicacies for extraordinary entertainments for pleasure. [10]

The popularity of chocolate drinking in 17th century Spain and her colonies is conveyed below:

> In Spain, it is drunk all Summer, once, or twice a day, or indeed at any time, by way of entertainment: for however Physicians there endeavour [sic] to confine the people to Rules, yet it is generally drunk without regard to any: and it's there, as well as in the Indies, all the year long. [11]

Cocoa and chocolate became known to the English as a result of colonization and confrontation in the Spanish territories of the Caribbean. In Jamaica, taken from the Spanish by the English in 1655, Spanish Creole methods of chocolate preparation were assimilated, practiced, and enjoyed by the English and transported back to the metropole:

> In Jamaica, there is a sort of Chocolate made up of only the Paste of the Cacao itself . . . and this is one of the best masses of Chocolate-Paste that is, and it may be had often here in England, neat and good, of Merchants and Sea-men that travel to those parts. [12]

Chocolate and Its Medicinal Properties

In the British West Indies, England, and Europe at large, this new drink, chocolate, was first considered as a medicine of sorts and was classified in terms of the Galenic humoral system (see Chapter 6) [13]. By the time the monarchy was restored, the Royal Physician, Dr. Henry Stubbe, called it the "The Indian Nectar," making it clear that chocolate was highly prized by the elite for its apparently healthful properties [14]. Humphrey Broadbent of London, a "Coffee-Man," wrote in 1722 that chocolate:

> Is recommended to be a Nourisher [sic] and Restorer of the Body on Consumption, causes a Cheerful and Healthful

Countenance, it comforts and Strengthens Weak and Infirm People. [15]

Opinions about chocolate, however, were not everywhere as positive as Colmenero professed. There was fierce opposition to the introduction of chocolate, together with tea and coffee, to English society and culture. Warnings against the consumption of chocolate were often made in terms of medical and bodily effects. For example, in Daniel Duncan's 1706 book *Wholesome Advice Against the Abuse of Hot Liquors, particularly of Coffee, Chocolate, Tea, Brandy, and Strong-Waters*, he suggested that people should:

> Abstain therefore from Hot Things and Coffee, or let the Use of them be with Discretion, on pain of making your Humours sower [sic] enough to gnaw your Intrails [sic]. [16]

In the 1746 edition of *A Treatise of Tobacco, Tea, Coffee and Chocolate* written originally by Simon Pauli and translated by Dr. James of London, it was argued that there was no need to drink the new nonalcoholic drinks as a guard against drunkenness. This treatise was especially critical of the alleged sexual and aphrodisiac properties of chocolate, arguing instead that chocolate caused sterility:

> I do not deny, but Coffee, Chocolate, and Tobacco, have a Power of stimulation to Venery, but may yet induce Sterility, because they consist of heterogeneous Parts, or rather act by their whole Substances . . . all these Things are unfit for fecundating the semen. . . . [W]e may reasonably infer, that as Chocolate agrees with Coffee and Tea, in one Third of its Qualities, so all these three exactly agree with each other, in producing Effeminacy and Impotence. [17]

Chocolate was often considered "cold" and "dry," but this was never undisputed and was endlessly complicated by consideration of the many constituent parts of the chocolate drink, and the nature of cocoa itself, regardless of Colmenero's claim that "I hold therefore, with the common opinion of all the world, that the *Cacao* is cold and dry" [18]. In his 1640 treatise, Colmenero suggested that:

> We must acknowledge that the Chocolate is not so cold as the Cacao, nor is it so hot as the other ingredients, but from the action and reaction thereof, there proceeds a moderate complexion, or temper which may agree with, and serve as well for the stomachs that be cold, as those that are hot. [19]

Demonstrating the various exchanges and transformations of chocolate-related knowledge, Dr. Stubbe, argued that:

> The Spaniards coming amongst them [the natives of Mexico], made several mixtures and Compounds; which instead of making the former better (as they supposed) have made it much worse: And many of the English do now imitate them; for in Jamaica, as well as other places, in making it into Lumps, Balls, Cakes, etc. they add to it Cacao-Paste, Chille,

[sic] or red pepper; *Achiote, Sweet Pepper*, commonly known by the name of *Jamaica Pepper*. [20]

Dr. Stubbe was also physician to "The Right Honourable Thomas Lord Windsor in the Island of Jamaica in the West Indies," which suggests a good deal of contact with and knowledge of the uses of cocoa and chocolate in the colonies. The various spices that were added to chocolate in the 17th century were linked to medicinal beliefs of the period. For example, black pepper was considered to be "very hot and dry [and did] not agree with those whose Liver is very cold" [21]. Chilli was recommended as "good, not too hot or drying." Dr. Stubbe offers some chocolate medicinal remedies for various ailments. For example, to strengthen the stomach, achiote or saffron should be added; for coughs, almonds or almond oil and sugar or candied sugar are recommended [22]. Cocoa butter was also recognized for its remedial properties:

Being internally administered it is good against all Coughs, shortness of breath, opening, and making the roughness of the Artery smooth, palliating all sharp Rheums, and contributing very much to the Radical Moisture, being very nourishing, and excellent against Consumptions. [23]

The oil was also applied externally on inflammations:

The oil is a cooler therefore assuages all pains proceeding from heat. . . . [The cocoa butter could be applied to]

Crustiness or Scars on Sores, Pimples, chapped Lips and Hands; it is an Anodyne, and exceeding good to mitigate the pain of Gout, as also Itches by reason of old Age: it wonderfully refresheth [sic] wearied limbs, being anointed therewith; and it maketh the skin smooth. [24]

Colmenero was a steadfast supporter of chocolate and its apparently multifarious medicinal properties. He concluded that chocolate was:

So good and wholesome as a Liquor [and that in] all the world generally using it, there is scarce any one, that does not highly approve of it through all Europe, as well as in the Indies. [25]

Head wrote that "Colonel Montague James, the first Englishman born in Jamaica, who lived to the age of 104, took scarcely any food but cocoa and chocolate for the last thirty years of his life" [26]. Perhaps this is an early dietary recommendation! Today it is reported that the Cuna Indians of Panama drink large quantities of bitter chocolate and have low blood pressure, and low rates of cancer, cardiovascular disease, and diabetes [27].

Head also pointed out in 1903 that the introduction of cocoa into Europe and its cultivation for the European market was due to Jesuit (and Capuchin) missionaries. It was also monks who about 1661 made it known in France: "It is curious therefore, to notice the contest that at one time raged amongst ecclesiastics as to whether it was lawful to make use of chocolate in Lent" [28].

Methods of Preparation

The Mexicans made a drink by boiling ground corn and cacao and flavoring it with hot pepper (see Chapter 8) [29]. According to Shepherd, the nuns of Guanaco began to mix sugar, vanilla, and cinnamon with cacao beans and the Spanish came to like this concoction [30]. Sugar had not been known in Mexico before European contact. Shephard gives the following version of Colmeneros's 1631 recipe:

*Take a hundred cacao kernels (that is about a quarter of a pound), two hands of chilli or long peppers, a handful of anise of orjevala, and two of mesachusil or vanilla—or instead, six Alexandria roses, powdered—two drachmas of cinnamon, a dozen almonds and as many hazelnuts, a half-pound of white sugar, and annotto enough to colour it,
and you can have the King of chocolates.* [31, 32]

In the 1695 edition of *Every Man his own Gauger* by James Lightbody in London, we learn how to make chocolate "Cakes and *Rowles*:"

Take Cocoa Nutts, and dry them gently in an Iron pan, or pot, and peel off the husks, then powder them very small, so that they may be fitted through a fine searce [sieve], then to every pound of the said Powder add seven Ounces of fine white Sugar, half an ounce of Nutmegs, one Ounce of Cinnamon, Ambergreace and Musk each four Grains, but these two may be omitted unless it be for extraordinary use. [33]

And so we see the profuse and exotic mixtures of spices that accompanied chocolate. The 1685 edition of Colmenero's treatise would add:

Seven hundred Cacao Nuts, a pound and a half of White Sugar, two ounces of Cinâmon, fourteen grains of Mexico Pepper, call'd Chile or Pimiento, half an ounce of Cloves, three little Straws or Vanilla's de Campeche, or for want thereof, as much Annis-seed as will equal the weight of a shilling of Achiot, a small quality as big as a Filbeard, which may be sufficient only to give it a colour; some ad thereto Almonds, Filbeards, and the Water of Orange Flowers. [34]

From the dried preparations of chocolate cakes or *rowles*, the chocolate drink would be concocted. James Lightbody gives us a good description of this process in the 17th century:

To make Chocolate: Take of Milk one Pint, and of Water half as much, and Boil it a while over a gentle Fire: then grate the quantity of one Ounce of the best Chocolate, and put therein, then take a small quantity of the Liquor out, and beat with six Eggs, and when it is well beat, pour it into the whole quantity of Liquor, and let it boil half an Hour gently, stirring it often with your Mollinet; then take it off the Fire, and set it by the Fire to keep it hot; and when you serve it, stir it well with your Mollinet. If you Toast a thin slice of white Bread, and put therein, it will eat extraordinarily well. [35]

Although most of the recipes and publications regarding the early use of chocolate come from the metropole, where chocolate and coffeehouses were becoming well-established places of business, there is ample reference to chocolate drinking taking place in the Caribbean island colonies. For example, in a letter dated July 14, 1720, Mr. Gordon wrote to the Council of Trade and Plantations in London regarding "the advantages and disadvantages that would arise from the settlement of St. Lucia." In this he states that:

The soil is of the same nature as Martinique, and very proper for producing cocoa; a species of merchandize which the English have never yet produced, and where they consume great quantitys [sic] of, especially in our Colonies, where every mechanic drinks chocolate for breakfast and supper. [36]

Indeed, many of the recipes quoted in London publications refer to Caribbean sources for their recipes. In the 1672 edition of *The American Physitian: or, a treatise of the roots, plants, trees, shrubs, fruit, herbs, etc. growing in the English Plantations in America*, Hughes describes the way that he saw chocolate made during his travels in the West Indies. He described how chocolate was made in the way "which is most generally known, and that which all common people, and servants also, before they go forth to work in the Plantations in a morning, take a draught or two of, for the better support of Nature in their hard labour" [37]. He gave details of the making of chocolate in Jamaica and the use of West Indian cassava bread as an accompaniment:

For this, they take of the Balls or Lumps made up only of the Cacao, when they are thorough dry and hard (for they ought to be at least nine days or a fortnight old fermenting, before they be made use of) and grate it very small in a Tin-Grated, such as Cooks use to grate their bread on, or the like . . . holding a piece of Chocolate lightly on it, that it may be grated very fine into some dish, or the like convenient thing, as may be fit to receive it. The quantity to be grated is as much as shall be thought fit . . . then take as much fair water out of the Spring or Jars . . . as they think will be sufficiently answerable to their ingredients; and in that water they put as much, or else somewhat less Casava-bread [sic], as the quantity of the grated Chocolate. This bread being a while in the cold water, although it not be grated, but only broken into bits or small pieces, and put therein, will dissolve in a little time; which in to water it is not and apt to so . . . when it is dissolved, they set that water in the fire in the Chocolate-Pot, kettle, or what other vessel they see good, and when it boils, they put in the grated Chocolate, and make it boil again a quarter of an hour, or less, stirring it a little in the interim: and then taking it off the fire, they pour it out of the Pot, or what else it was boiled in, into some handsome

large dish or Bason [sic]: and after they have sweetened it with a little Sugar, being all together, and sitting down round about it like good Fellows, everyone dips in his Calabash, or some other Dish, supping it off very hot. [38]

Hughes also refers to a Mr. A. Mevis, "a planter-friend," and describes how he made his chocolate:

First, he took some water and made it ready to boil; and in the interim, he grated as much Chocolate as he thought fit, according to the company, to drink thereof at that time. Also he beat 3 Eggs very well; then he put in the Chocolate, and a little cassava beaten small, and some Maiz-Flower [sic]: All which made it indifferently thick, and when it was just boiled up, he put in the Eggs, and sweetened it with a little sugar, and when it had boiled less than a quarter of an hour after, he frothed it, and poured it forth into a Dish. This seemed very and some to the eye, and I am sure it tasted much better, being drunk not only very hot, but also very merrily. [39]

Hughes claimed that chocolate was a remedy for most ailments, that it was a restorative and curative food, and especially so in the West Indies. However, he claimed that chocolate should not be drunk regularly by healthy persons in England.

Nineteenth century visitors to Trinidad came to appreciate other parts of the cocoa pod in addition to the beans. Lady Brassey stated that:

When cut open, the interior of the pod is found to be filled with . . . what looks like custard, which when quite fresh, tastes like the most delicious lemon ice-cream, with a delicate soupçon of vanilla chocolate. I know nothing more agreeable in the way of refreshment than to have two or three large cacao-pods sat before you in some cool shady spot, where the cream-like contents can be quietly discussed and enjoyed. [40]

Humboldt estimated in 1806 that the entire European consumption of cacao was only 20 million pounds annually, of which half was consumed by Spain [41]. Zipperer noted that the consumption of cacao products increased in the last part of the 19th century:

In 1874 the consumption in the United Kingdom amounted to 8,300,000 pounds, and in 1894 it was estimated at more than 22 million pounds. In 1895 the English cacao factories sold 24,484,502 pounds of cacao preparations and 20/24 percentage of the cacao they consumed was grown in British Colonies. [42]

Between 1896 and 1909 the consumption of chocolate in the United States increased 414 percent from 22.9 million pounds to 117.7 million pounds [43]. No other country increased its consumption as much over this period despite price increases. Hart noted that:

It is only during the last fifteen years [prior to 1909] that the world has recognised the importance of cacao and its products as a food. The people of the United States were the last to do so. [44]

Institutional Use of Cocoa

Cocoa became an important food for the military and for those in institutions in the 19th century (see Chapters 31 and 55). In November 1893, there was a request from the Chief Medical Officer of *H.M.S. Serapis* that sailors changing watch at 4.00 a.m. should be allowed a drink of hot chocolate [45]. At first the director of transports refused the request, but by August 1894 he wrote:

I have to acquaint you that it has been decided to sanction the issue on board H.M. Troop Ships a ration (cup) of Cocoa or soluble chocolate to soldiers of the guard (including sentries), about 4.am daily—such issue to be made at the discretion of the Office commanding troops on board. [46]

In 1844, the Admiralty requested that cocoa was to be supplied for prisoners at London's Millbank Prison. In reply, the Comptroller of Victualling informed the Secretary of State that there would be:

No difficulty in providing Cocoa of the same quality as that purchased for the Navy on timely notice given of the quantity which will from time to time be required from said prison. It is generally the produce of Guayaquil and undergoes at Deptford before it is manufactured into Chocolate, the process of screening, being for the most part in a dirty state. [47]

In St. Vincent, male prisoners were employed in making cocoa for public institutions and made a profit for the prison [48]. In Trinidad, the patients in the lunatic asylum were fed coffee, tea, or chocolate on alternate days for breakfast [49]. During the First World War, Grenada and Trinidad provided cocoa free for the use of the British troops as their contribution to the war effort, and during the Second World War, the British Ministry of Food purchased Trinidad cocoa at a fixed price to provide chocolate as iron rations for the troops [50].

Manufacturing

Until the end of the 18th century, the manufacture of chocolate was a small-scale and labor-intensive operation. However, as operations and industrialization grew, and as increased cocoa production led to a decrease in price and thus an increase in consumption and demand, chocolate became a more and more commonplace commodity, no longer reserved for the upper echelons of society. In 1776, Joseph Fry, the chocolate manufacturer of Bristol, sang praises for English (i.e., Fry's) chocolate and predicted a vast export market for chocolate,[1] were trade duties to be more favorably aligned to these interests:

As it is certain that good and beautiful Chocolate cannot be made in warm Climates, and that abroad that English Chocolate is prefferr'd [sic] to most if not any other sort, doubtless very large Quantities would be exported to the East

and West Indies and many other Places, if a suitable Drawback were to be allowed in Exportation. [51]

A century later, Lady Brassey bemoaned the failure to manufacture chocolate in Trinidad:

It is a curious fact and one showing a certain want of enterprise, and also a decided amount of prejudice, on the part of the Trinidadians, that although sugar, cacao and vanilla are cultivated one may say, side by side, on many estates, not a single ounce of chocolate is manufactured in the island. The raw materials are all sent over to France, whence all the manufactured chocolate consumed in Trinidad is imported, though both the inward and outward duties are high. [52]

She further notes that local chocolate manufacturing was once started but the islanders would not buy the local product because it was not in the familiar wrapper of the French import they were used to. However, the existence of a chocolate factory is noted in Trinidad Blue Books from 1900 but disappeared from the record within a few years [53]. Paul Zipperer reported that in Britain "hand labour has now been replaced . . . in the manufacture of chocolate, by manufacturing" and that chocolate was a commonplace luxury [54].

The Caribbean Cocoa Trade

The first cocoa transactions of the British colonial forces involved piracy, looting, and illicit dealings. Trade was precarious and involved considerable risk.[2] In addition to the battles and bandits faced at sea, early traders also faced challenges in terms of navigation.[3] Weather continued to limit cocoa exports until the late 19th century when the establishment of steamship connections made trade more reliable for the islands, but as late as 1897 bad weather was blamed for a fall in exports of cocoa from Trinidad [55].

Documents from the late 17th and early 18th centuries reveal that colonial British North America was involved in this early trading. For example on May 11, 1711, Governor Hunter reported to the Commissioners of Customs in New York that:

In Sept. last the captains of two privateers brought in hither a large ship loaden [sic] with cacoa [sic] upon my promise that no injury or hardship should be offered them; the collector agreed to the unloading of the vessell [sic] after condemnation putting the effects into safe store-houses under lock and key in his possession for securing Queen's dutys [sic], condescending to let them sell from time to time what they could paying the dutys [sic] as they sold. [56]

This report, together with other correspondence of the period, shows that cocoa was being transported directly from British and foreign Caribbean colonies to British colonies in North America. For example, on May 15, 1712, Lt. Governor Spotswood of Virginia reported to the Council of Trade and Plantations that:

The Bedford galley arrived here the other day, and brought in a French merchant ship loaded with sugar, indico [sic] and cocoa, and I hear Capt. Pudner in the Severn, one of the convoys to the Virginia Fleet, has taken and carryed [sic] into New York a French privateer of 180 men, wch. Very much infested this coast. [57]

In 1717, French, Dutch, Danish, and Spanish ships are recorded landing "Cocoa, from St. Thomas, 144 barrls.; from Martinico, 133" [58]. Such reports contribute to some of the confusion over the directions of early cocoa routes into North America.

It was probably not long after Pepys first tasted it [chocolate] in London that it went back across the Atlantic to England's North American colonies (although it is possible that it traveled there directly from Jamaica after that island had been wrestled from Spanish control). [59]

Thus, it can be seen that it is not just possible, but absolutely certain that chocolate reached North America directly from the European Caribbean colonies. Cocoa was both imported and exported from many ports in the Caribbean, as well as from ports in colonial North America and the European metropolises.

It is difficult to obtain a clear idea of the volumes of cocoa that were being exported early on, due to variable measures and scanty records, but a report of exports from Jamaica over the period dated March 25, 1709 to September 29, 1711 tells us that "41 pipes" and "189 hhds" of cocoa were shipped to England, and 1 barrel to the "Plantations," which likely refers to North America [60]. Later on, the Jamaica Shipping Returns of 1763 (CO 142/18) show that 66 bags of cocoa left Jamaica on May 2 bound for Amsterdam, and that about 25–30 percent of departing ships were bound for North American ports, including Philadelphia, New London, Boston, New Haven, Rhode Island, and New York [61]. An "Abstract of the Returns to the Boards Order of the 25th of October, 1775" gives an account of vessels arriving in Bristol from North America; showing, for example, that cocoa was brought in from South Carolina, on board James Ramsay's ship, *The Providence*, on October 24, 1775 [62]. Thus, a triangular trade in cocoa between the Caribbean, North America, and England was established by the 18th century.

In 1778, the Commissioners of Excise for England and Wales recommended a duty to be placed on cocoa beans as well as on chocolate, to stop duty evasion by unlicensed chocolate traders and makers in England. The 1778 Commissioners' report tells us that these illicit chocolate makers would avoid paying the duty that had been placed on chocolate by buying cocoa beans under pseudonyms and fleeing, so that bills could not be issued by the taxmen [63].

There was some direct trade between Trinidad and the United States at the beginning of the 19th century.[4] The Trinidad Blue Books show that cocoa exports to foreign destinations were high at the beginning of the 19th century, with 92 percent, or £63,212 of export value, going to unspecified foreign states in 1826 [64]. By 1831, this proportion had fallen significantly, along with all cocoa exports generally (Table 37.1); the United States and other foreign states accounted for 43 percent of the export value of cocoa [65]. In 1847, by which time Trinidad cocoa production had recovered to pre-1830s levels, the United States and other foreign states accounted for 27 percent of total cocoa export value, with countries other than the United States accounting for most of this [66]. By the 1850s, this proportion had fallen further, to 14 percent, but North America now accounted for the majority of cocoa exports to foreign states [67]. In the 1870s, the trend reversed again, with more cocoa, constituting 30 percent of export value, going to foreign states. By then, France had become the dominant foreign market, accounting for £34,333 of export value as compared to the United States, which imported only £15,759 of the £172,873 total value of cocoa exports [68]. The pattern of export trade is similar a decade later in 1881, with foreign states accounting for 35 percent of the weight of cocoa exports and France accounting for about two-thirds, or 2,703,487 lb, of this [69]. In 1886 the Trinidad Blue Book noted that the island was looking forward to having direct steamship communication with the United States. In that year, 51 percent of Trinidad's exports of cocoa went to the United Kingdom, 32 percent to France, and 16 percent to the United States. By 1889, when direct shipping links were established, 44 percent of cocoa exports went to the United Kingdom, 25 percent to France, and 21 percent to the United States [70]. In 1903–1904, for the first time, the proportion of Trinidadian cocoa exported to the United States was second only to that going to the United Kingdom [71].

Trinidad was also a transshipment point for cocoa from Venezuela and from other parts of the British West Indies. The pattern of trade is similar for St. Lucian cocoa, with foreign markets becoming increasingly important throughout the 19th century. For the first half of the century, all of St. Lucian cocoa is recorded as exported to Great Britain, the British West Indies, or British North America. By the 1880s, the French market accounted for between 20 and 35 percent of total cocoa exports and by 1891, France was the largest market, accounting for 66 percent of cocoa exports: 661,600 lb or £19,186–8–0 [72, 73].

The Jamaica Blue Books only become useful for gauging this information after the 1860s. These reports show that, in 1861, foreign markets accounted for only 9 percent of the value of cocoa exports. The United States was the most important market at this time after the United Kingdom, buying £91,676 worth [74]. By

Table 37.1 Nineteenth Century Cocoa Exports from Grenada and Jamaica

Year	Total Amount of Cocoa Exported (lb unless otherwise stated)	Cocoa Exported to Great Britain (lb unless stated otherwise)	Cocoa Exported to United States (lb unless stated otherwise)
Grenada			
1821	1641		
1826	788 Casks		
1831	317,717		
1836	195 barrels and 1864 bags		
1841		260 barrels, 1837 bags	
1846	8 barrels and 6399 bags		
1851	71 barrels and 3877 bags	71 barrels, 3877 bags	
1856	837,760	7268 cwt	
1861		591 tons. 15 cwt, 1 qu. 15 lb	37 cwt 1 qu. 13 lb
1866	631 tons	611 tons	
1870	1050 tons	18 cwt	9 lbs
1871	1050 tons		
1876		1676 tons	
1881	2619 tons, 13 cwt, 2 qu. 22 lbs	2617 tons, 2 cwt, 0 qu 26 lb	
1886	5,717,745	49,573$\frac{1}{2}$ cwt	3125 cwt
1891		64,994$\frac{3}{4}$ cwt	823$\frac{1}{2}$ cwt
1896	90,193$\frac{1}{2}$ cwt and 57,691 bags	86,035 cwt	2806$\frac{1}{4}$ cwt
1901	93,856 cwt and 58,312$\frac{1}{2}$ bags	72,911 cwt	8193 cwt
1906	111,000 cwt		
Jamaica			
1836		150 lbs	
1840	222 lb	222 lbs	
1846	154. 3. 21 (and additional 96 lb)	96 lbs	
1851	6. 6. 23		
1856	2 qu. 19 lb		
1861	235.1.13	221.3.2	13.2.11
1866	24,446	207 .2 .0	4. 3. 17
1871	51,597	330. 2. 13	34. 2. 14
1876	459.1.27	280. 0. 23	0. 2. 4
1881	698. 3. 2	487. 1. 9	177. 2. 21
1886	3855. 0. 21	2528. 2. 1	435. 0. 18
1891	5485. 0. 15	2795. 1. 9	193. 1. 8
1896	9178. 9. 2	3683. 0. 15	2786. 0. 11
1901	39,953. 3. 15	12,978. 3 . 13	10,660. 1. 20

the 1880s, this value had significantly increased, with foreign states accounting for 30–40 percent of cocoa exports. Although the United States was the largest foreign market in 1881, importing over 177 cwt of cocoa [75], it was commonly second to France as the largest foreign market, as it was in 1886 [76]. By the 1890s, reliance on foreign markets had increased, but not to the extent that had occurred in Trinidad and St. Lucia. For Jamaican cocoa, foreign markets accounted for about a third of total exports, with France taking the majority—about 534 cwt (hundred-weight) of the quantity [77]. For Grenada and St. Vincent, Great Britain remained by far the major market for cocoa throughout the 19th century, with minimal cocoa

exported to foreign countries, although some of the cocoa from these islands may have been exported via Trinidad. Indeed, foreign markets accounted for about 2 percent of total cocoa exports from these islands throughout the period. In the year 1892–1893, the value of cocoa imports into the United States amounted to $3,195,811, decreasing to $2,402,382 in 1893–1894 and then rising to $4,017,801 in 1894–1895 [78]. A significant proportion of this cocoa originated from the British West Indies, especially Trinidad.

At the end of the 19th century, there were transformations in agricultural markets and production throughout the British West Indies due to the Spanish–American War and the changes in United States territory that followed. By the end of the war, the United States had claimed Puerto Rico, Cuba, the Philippines, and Hawaii and in doing so had acquired vast plantations. This impacted the British West Indian islands, which had formerly been a major supplier of tropical products such as sugar, cocoa, and coffee to the United States. The impact of the diminished American market for West Indian sugar at the end of the 19th century goes some way to explaining the replacement of sugar by cocoa as the principal export from the British West Indies to the United States. Clark noted that a "marked feature of the cocoa industry in recent years is the large increase that has taken place in the exports to the United States which have risen from 11,765 cwts in 1883 to 64,213 cwts in 1892" [79]. Clark, writing in support of Trinidad's presence at the Chicago World's Fair, also suggests that the demand for Trinidadian cocoa will increase after its quality has become better known at the World's Fair and pointed out that the United States is "not only our nearest market for it, but one large enough to absorb all that the colony can produce for years to come" [80]. By 1909, the United States imported more cocoa than any other country [81].

Between 1894 and 1903 world cocoa production doubled and by 1909 it had almost doubled again to 452 million pounds [82]. In 1909, the largest amount was produced by Brazil (74.7 millions pounds), Ecuador (68.3 million pounds), San Tomé (61.2 million pounds), British West Africa (52.9 million pounds), and Trinidad (51.1 million pounds) [83]. At this time, Venezuela, Grenada, and San Domingo followed with production of 38.5, 32.6, and 13.8 million pounds, respectively [84]. Jamaica, Cuba, Surinam, and Haiti also produced small amounts so that the total production of cocoa from the Caribbean region, including Venezuela, was 158.7 million pounds in 1909 or 35 percent of total world production [85]. The cocoa trade was disrupted by World War I and in Trinidad there was prolonged drought from 1911 to 1916 [86]. By 1920, the Trinidadian cocoa industry had recovered and the island produced 20 percent of the world's supply [87]. In 1921, Trinidad reached its peak production, largely in response to the favorable market price, and was the fourth- or fifth-largest producer in the world [88].

Conclusion

After 1921, Trinidadian cocoa production fell in response to a decline in world cocoa prices caused by a glut on the market as production increased in Africa, and by the depression of the 1920s and the appearance of Witches Broom disease in Trinidad in May 1928 [89]. At the same time the oil industry began to develop and the price of sugar improved, offering alternative opportunities for the Trinidadian economy. By 1945,

the cocoa industry was faced with extinction but government rehabilitation and subsidy schemes were introduced and continue to this day. Trinidad produces a superior brand of cocoa known as fine or flavor cocoa from *trinitario* beans and has a reputation for purity with low levels of contaminants [90]. There is a niche market for this cocoa today but there are higher production, processing, and material costs associated with the production of such cocoas [91]. In 2003, in the two-island state of Trinidad and Tobago, there were only 3500 cocoa farmers (45 in Tobago) compared to 10,000 in 1966, and yields were low [92]. Production of cocoa is declining throughout the region as farmers grow tired of coping with damage to their trees from hurricanes, drought, and disease while fluctuating world prices add to the risk. On the other hand, Trinidad and Tobago has a ready market for all the cocoa it can produce because of its high quality and lack of restrictive quotas. Production has more or less stabilized since 1985 at about one million kilograms compared to almost 35 million kilograms in 1921 [93]. The government of Trinidad and Tobago has introduced central processing to ensure consistently good quality and is developing value-added products, but all the efforts for resuscitation of the industry over the last 60 years have failed to reverse the decline in production in a country that has been growing cocoa for over 500 years [94].

Acknowledgments

We would like to acknowledge the help and assistance of Dr. Glenroy Taitt, Senior Librarian in the West India Collection of the University of the West Indies, St. Augustine Library, who aided in the identification of numerous chocolate-related documents within the collection. We also thank Dr. J. A. Spence, Head of the Cocoa Research Unit at the University of the West Indies, Trinidad, for giving us access to the CRU library. Also thanks to Professor Bridget Brereton of the History Department at the University of the West Indies, Trinidad, for guidance to historical sources on cocoa in Trinidad.

Endnotes

1. Imports of manufactured chocolate from the United Kingdom appear in the Trinidadian trade statistics after 1897.

2. For example, on January 9, 1714, Lt. Governor Pulleine to the Council of Trade and Plantations reported from Bermuda:

   The Spaniards, from several of [the] ports, here in [the] North Seas, arm out sloops with commissions to seize all English vessels in which they find any Spanish money (even to [the] value of but ten pieces of eight), any salt, cacao, or hides, for wch. [which] reasons any vessels that trade in these parts, from port to port, are certainly prizes, if they can overpower them, having one or other of these commoditys [sic] always aboard 'em' [sic]. (CO 37/9, Nos. 27, 27i, ii)

Similarly, the deposition of John Williams, a mariner of Bermuda, reported on January 19, 1713(/14) that:

The sloop Swan whilst riding at anchor in the Rode of Bonaire . . . where she intended to take in salt, was seized by a sloop under Spanish colours, the masters and mariners whereof pretended that they had a commission from the Govermt. of Porto Rico [sic], but that they had lost the same. The Swan was carried to Porto Rico [sic] and condemned as prize. The master and mariners pretended to justify their seizure for this cause only, that she had on board 7 bags of cocoa nuts [taken on board at Curacao]. (CO 37/9, Nos. 27, 27i, ii)

3. Such problems are illustrated by Samuel Beeckman's letter to the Directors of the Dutch West India Company dated August 1700:

On his [Capt. Evertsen] return thence [from Barbados] resolved again to cause a little journey to be undertaken to Rio Orenocque [sic] and Triniedados to trade there the aforesaid goods [salt fish] for cocoa; but this too has not turned out according to desire, but fruitless; coming after the current and contrary winds, got too low in returning hither against their will to Martinique, whence, having been out for more than four months and having endured straights, consequently [they] arrived here again. (CO 116/19, Nos. 4, 4i–xii)

4. From April 7 to July 4, 1804, destination of ships leaving the Port of Spain included Philadelphia, Savannah, Washington, Norfolk, Baltimore, Boston, and New York: exporting a total of 292 bags and 67 casks and cocoa. For July 13 to October 6, 1804, United States destinations included New York, Boston, New London, and Philadelphia. A total of 122,211 lb of cocoa were exported to the United States during this period. On July 13, 1804, 3500 lb of cocoa departed from Trinidad bound for New York.

References

1. Shephard, C. Y. The cacao industry of Trinidad—some economic aspects. In: *Part II. General History of the Production and Consumption of Cacao.* Port of Spain, Trinidad and Tobago: The Government Printing Office, 1932; p. 1.

2. Robinson, E. F. *The Early History of Coffee Houses in England.* London, England: Kegan Paul, Trench and Trubner and Co., 1893.

3. Shephard, C. Y. The cacao industry of Trinidad-some economic aspects. In: *Part II. General History of the Production and Consumption of Cacao.* Port of Spain, Trinidad and Tobago: The Government Printing Office, 1932.

4. Shephard, C. Y. The cacao industry of Trinidad-some economic aspects. In: *Part II. General History of the Production and Consumption of Cacao.* Port of Spain, Trinidad and Tobago: The Government Printing Office, 1932.

5. Head, B. *The Food of the Gods. A Popular Account of Cocoa.* London, England: George Routledge and Sons, 1903; p. 86.

6. Head, B. *The Food of the Gods. A Popular Account Of Cocoa.* London, England: George Routledge and Sons, 1903; p. 86.

7. Shephard, C. Y. The cacao industry of Trinidad-some economic aspects. In: *Part II. General History of the Production and Consumption of Cacao.* Port of Spain, Trinidad and Tobago: The Government Printing Office, 1932; p. 1.

8. Head, B. *The Food of the Gods. A Popular Account of Cocoa.* London, England: George Routledge and Sons, 1903; p. 86.

9. Colmenero, A. de Ledesma. *A Curious Treatise of the Nature and Quality of Chocolate. Madrid, 1631. Translated from the original Spanish by Don Diego de Vade-Forte for Edward Lord Viscount of Conway and Killultah, Baron of Ragley, Lord Marshal of Ireland and one of the Councell of Warre to His Majesty of Great Brittaine.* London, England: J. Oakes dwelling in Little St. Bartholmewes, 1640; p. D4.

10. Colmenero, A. de Ledesma. *A Curious Treatise of the Nature and Quality of Chocolate. Madrid, 1631. Translated from the original Spanish by Don Diego de Vade-Forte for Edward Lord Viscount of Conway and Killultah, Baron of Ragley, Lord Marshal of Ireland and one of the Councell of Warre to His Majesty of Great Brittaine.* London, England: J. Oakes dwelling in Little St. Bartholmewes, 1640; p. 3.

11. Colmenero, A. de Ledesma. *A Curious Treatise of the Nature and Quality of Chocolate. Madrid, 1631. Translated from The original Spanish by Don Diego De Vade-forte For Edward Lord Viscount of Conway and Killultah, Baron of Ragley, Lord Marshal of IrelAnd And one of The Councell of Warre to His Majesty of Great Brittalne.* London, England: J. Oakes DwellIng In Little St. Bartholmewes, 1640; p. 111.

12. Colmenero, A. de Ledesma. *A Curious Treatise of the Nature and Quality of Chocolate. Madrid, 1631. Translated from the original Spanish by Don Diego de Vade-Forte for Edward Lord Viscount of Conway and Killultah, Baron of Ragley, Lord Marshal of Ireland and one of the Councell of Warre to His Majesty of Great Brittaine.* London, England: J. Oakes dwelling in Little St. Bartholmewes, 1640; p. 111.

13. Coe, S. D., and Coe, M. D. *The True History of Chocolate.* New York: Thames and Hudson, 1996.

14. Stubbe, H. *The Indian Nectar. Or a Discourse Concerning chocolate: wherein The Nature of the cocoa-nut, and the other Ingredients of that Composita, is examined, and stated according to the judgment and experience of the Indians and Spanish*

writers, 1662. University of Cambridge Library, B125:2.9 Reel 15:37:23.

15. Broadbent, H. *The domestick Coffee-Man, shewing the True Way of Preparing a Making of Chocolate, Coffee and Tea,* London, 1722 British Library T357 (No. 2).

16. Duncan, Dr. D. *Wholesome Advice Against the Abuse of Hot Liquors, particularly of Coffee, Chocolate, Tea, Brandy, and Strong-Waters,* 1706. University of Cambridge Library, L5.59.

17. Pauli, S. A *Treatise of Tobacco, Tea, Coffee and Chocolate.* Translated by Dr. James of London, 1746, p. 166. Natural History Museum London 633.7 Pau.

18. Colmenero, A. de Ledesma. *A Curious Treatise of the Nature and Quality of Chocolate. Madrid, 1631. Translated from the original Spanish by Don Diego de Vade-Forte for Edward Lord Viscount of Conway and Killultah, Baron of Ragley, Lord Marshal of Ireland and one of the Councell of Warre to His Majesty of Great Brittaine.* London, England: J. Oakes dwelling in Little St. Bartholmewes, 1640; p. 58.

19. Colmenero, A. de Ledesma. *A Curious Treatise of the Nature and Quality of Chocolate. Madrid, 1631. Translated from the original Spanish by Don Diego de Vade-Forte for Edward Lord Viscount of Conway and Killultah, Baron of Ragley, Lord Marshal of Ireland and one of the Councell of Warre to His Majesty of Great Brittaine.* London, England: J. Oakes dwelling in Little St. Bartholmewes, 1640; p. 98–99.

20. Stubbe, H. *The Indian Nectar. Or a Discourse Concerning chocolate: wherein The Nature of the cocoa-nut, and the other Ingredients of that Composita, is examined, and stated according to the judgment and experience of the Indians and Spanish writers,* 1662. University of Cambridge Library, B125:2.9 Reel 15:37:23; pp. 119–120.

21. Stubbe, H. *The Indian Nectar. Or a Discourse Concerning chocolate: wherein The Nature of the cocoa-nut, and the other Ingredients of that Composita, is examined, and stated according to the judgment and experience of the Indians and Spanish writers,* 1662. University of Cambridge Library, B125:2.9 Reel 15:37:23; p. 71.

22. Colmenero, A. de Ledesma. *A Curious Treatise of the Nature and Quality of Chocolate. Madrid, 1631. Translated from the original Spanish by Don Diego de Vade-Forte for Edward Lord Viscount of Conway and Killultah, Baron of Ragley, Lord Marshal of Ireland and one of the Councell of Warre to His Majesty of Great Brittaine.* London, England: J. Oakes dwelling in Little St. Bartholmewes, 1640; p. 124.

23. Colmenero, A. de Ledesma. *A Curious Treatise of the Nature and Quality of Chocolate. Madrid, 1631. Translated from the original Spanish by Don Diego de Vade-Forte for Edward Lord Viscount of Conway and Killultah, Baron of Ragley, Lord Marshal of Ireland and one of the Councell of Warre to His Majesty of Great Brittaine.* London, England: J. Oakes dwelling in Little St. Bartholmewes, 1640; p. 125.

24. Colmenero, A. de Ledesma. *A Curious Treatise of the Nature and Quality of Chocolate. Madrid, 1631. Translated from the original Spanish by Don Diego de Vade-Forte for Edward Lord Viscount of Conway and Killultah, Baron of Ragley, Lord Marshal of Ireland and one of the Councell of Warre to His Majesty of Great Brittaine.* London, England: J. Oakes dwelling in Little St. Bartholmewes, 1640; p. 126.

25. Colmenero, A. de Ledesma. *A Curious Treatise of the Nature and Quality of Chocolate. Madrid, 1631. Translated from the original Spanish by Don Diego de Vade-Forte for Edward Lord Viscount of Conway and Killultah, Baron of Ragley, Lord Marshal of Ireland and one of the Councell of Warre to His Majesty of Great Brittaine.* London, England: J. Oakes dwelling in Little St. Bartholmewes, 1640; pp. 56 and 99.

26. Head, B. *The Food of the Gods. A Popular Account of Cocoa.* London, England: George Routledge and Sons, 1903; p. 23.

27. Dahlberg, C. P. Candy maker's a sugar daddy for UCD cocoa health studies. *Sacramento Bee* 19 February 2007:A10.

28. Head, B. *The Food of the Gods. A Popular Account of Cocoa.* London, England: George Routledge and Sons, 1903; p. 84.

29. Shepherd, C. Y. The cacao industry of Trinidad—some economic aspects. In: *Part II. General History of the Production and Consumption of Cacao.* Port of Spain, Trinidad and Tobago: The Government Printing Office, 1932; p. 1.

30. Shephard, C. Y. The cacao industry of Trinidad—some economic aspects. In: *Part II. General History of the Production and Consumption of Cacao.* Port of Spain, Trinidad and Tobago: The Government Printing Office, 1932; p. 1.

31. Shephard, C. Y. The cacao industry of Trinidad—some economic aspects. In: *Part II. General History of the Production and Consumption of Cacao.* Port of Spain, Trinidad and Tobago: The Government Printing Office, 1932; p. 1.

32. Colmenero, A. de Ledesma. *A Curious Treatise of the Nature and Quality of Chocolate. Madrid, 1631. Translated from the original Spanish by Don Diego de Vade-Forte for Edward Lord Viscount of Conway and Killultah, Baron of Ragley, Lord Marshal of Ireland and one of the Councell of Warre to His Majesty of Great Brittaine.* London, England: J. Oakes dwelling in Little St. Bartholmewes, 1685 edition, p. 72.

33. Colmenero, A. de Ledesma. *A Curious Treatise of the Nature and Quality of Chocolate. Madrid, 1631. Translated from the original Spanish by Don Diego de Vade-Forte for Edward Lord Viscount of Conway and Killultah, Baron of Ragley, Lord Marshal of Ireland and one of the Councell of Warre to His Majesty of Great Brittaine.* London, England: J. Oakes dwelling in Little St. Bartholmewes, 1685.

34. Colmenero, A. de Ledesma. *A Curious Treatise of the Nature and Quality of Chocolate. Madrid, 1631. Translated from the original Spanish by Don Diego de Vade-Forte for*

Edward Lord Viscount of Conway and Killultah, Baron of Ragley, Lord Marshal of Ireland and one of the Councell of Warre to His Majesty of Great Brittaine. London, England: J. Oakes dwelling in Little St. Bartholmewes, 1685.

35. Lightbody, J. *Every Man his own Gauger*, 1695. University of Cambridge Library B125:2.9 Reel 766:31.

36. Gordon, Rev. W. *Reverend Gordon to the Council of Trade and Plantations*, 1720. Public Record Office, CO 28/17, ff. 97–99.

37. Hughes, W. *The American Physitian: or, a treatise of the roots, plants, trees, shrubs, fruit, herbs, etc. growing in the English Plantations in America. Describing the Place, Time, Names, Kindes, Temperature, Vertues and Uses of them, either for Diet, Physick, etc. Whereunto is added a Discourse of the cacao-nut tree, And the use of its Fruit; with all the ways of making of CHOCOLATE, the like never extant before.* London, 1672. University of Cambridge Library 449.a.24; pp. 127–129.

38. Hughes, W. *The American Physitian: or, a treatise of the roots, plants, trees, shrubs, fruit, herbs, etc. growing in the English Plantations in America. Describing the Place, Time, Names, Kindes, Temperature, Vertues and Uses of them, either for Diet, Physick, etc. Whereunto is added a Discourse of the cacao-nut tree, And the use of its Fruit; with all the ways of making of CHOCOLATE, the like never extant before.* London, 1672. University of Cambridge Library 449.a.24; pp. 127–129.

39. Hughes, W. *The American Physitian: or, a treatise of the roots, plants, trees, shrubs, fruit, herbs, etc. growing in the English Plantations in America. Describing the Place, Time, Names, Kindes, Temperature, Vertues and Uses of them, either for Diet, Physick, etc. Whereunto is added a Discourse of the cacao-nut tree, And the use of its Fruit; with all the ways of making of CHOCOLATE, the like never extant before.* London, 1672. University of Cambridge Library 449.a.24; p. 137.

40. Brassey, Lady. *In the Trades, the Tropics and the "Roaring Forties".* London, England: Longmans, Green and Co., 1895; p. 19.

41. Hart, J. H. The consumption of cacao. *The West India Committee Circular* 21 June 1910:291.

42. Zipperer, P. *The Manufacture of Chocolate and Other Cacao Preparations*, 2nd edition, Berlin and New York: M. Krayn, Spon, and Chamberlain, 1902; p. 29.

43. Hart, J. H. The consumption of cacao. *The West India Committee Circular* 21 June 1910:292.

44. Hart, J. H. The consumption of cacao. *The West India Committee Circular* 21 June 1910:291.

45. Forsyth, W. N. Memorandum from Chief Medical Officer, Major Commander and Captain requesting issue of chocolate to Troops on board Transports, January 11, 1893. Public Record Office T.2800.

46. Forsyth, W. N. Memorandum from Chief Medical Officer, Major Commander and Captain requesting issue of chocolate to Troops on board Transports, January 11, 1893. Public Record Office T.2800.

47. Request for Deptford Victualling Yard to supply cocoa for prisoners at Millbank Prison, Admiralty, February 5, 1844 (HO 45/788).

48. St. Vincent Blue Books, 1916 to 1920, Kingstown, St. Vincent, 1917–1921.

49. Trinidad Blue Books, 1886 and 1892, Port of Spain, Trinidad, 1887 and 1893.

50. Brereton, B. *A History of Modern Trinidad 1783–1962*, Kingston, Port of Spain and London: Heinemann Educational Books (Caribbean) Ltd., 1981; p. 213.

51. Fry, J. *Correspondence from the Commissioners of H.M. Excise at the Excise Office, London: Report of enclosed memorial of Joseph Fry, chocolate manufacturer of Bristol: his loss of trade due to illegal hawkers*, June 27, 1776. Public Record Office TI 523/332–335.

52. Brassey, Lady. *In the Trades, the Tropics and the "Roaring Forties".* London, England: Longmans, Green and Co., 1895; p. 19.

53. *Trinidad Blue Books*, 1900 to 1906, Port of Spain, 1901–1907.

54. Zipperer, P. *The Manufacture of Chocolate and Other Cacao Preparations*, 2nd edition, Berlin and New York: M. Krayn, Spon, and Chamberlain, 1902; p. 98.

55. *Trinidad Blue Book*, 1896–1897, Port of Spain, 1897.

56. Hunter, Governor. *Report to the Commissioners of Customs, New York*, May 7, 1711. CO 5/1050.

57. Spotswood, Lt. Governor. *Lt. Governor Spotswood to the Council of Trade and Plantations, Virginia*, 1712. Public Record Office CO 5/1316, No. 89.

58. Cumings, Archdeacon. *An Account of Foreign Enumerated Commodities Imported in the Port of Boston*, June 24, 1717. CO 5/867, Nos. 10, 10i.

59. Coe, S. D. and Coe, M. D. *The True History of Chocolate.* New York: Thames and Hudson, 1996; p. 176.

60. *Jamaica exports*, 1709–1711. CO 137/10, Nos. 6, 6i, ii.

61. Commissioners of Excise for England and Wales, *Report*, 1778. T1 542/222–223.

62. Jamaica Shipping Returns of 1763. CO 142/18.

63. Abstract of the Returns to the Boards Order of the 25th of October 1775. T1 523/89.

64. *Trinidad Blue Book*, 1826. CO 300/40.

65. *Trinidad Blue Book*, 1831. CO 300/45.

66. *Trinidad Blue Book*, 1847. CO 300/58.

67. *Trinidad Blue Book*, 1851. CO 300/62.

68. *Trinidad Blue Book*, 1871. CO 300/82.

69. *Trinidad Blue Book*, 1881. Port of Spain, 1882.

70. *Trinidad Blue Book*, 1894. Port of Spain, 1895.

71. *Trinidad Blue Book*, 1904. Port of Spain, 1905.

72. *St. Lucia Blue Book*, 1871. CO 258/77.

73. *St. Lucia Blue Book*, 1881. CO 258/87.

74. *Jamaica Blue Book*, 1861. CO 142/75.

75. *Jamaica Blue Book*, 1881. CO 142/95.

76. *Jamaica Blue Book*, 1886. CO 142/100.

77. *Jamaica Blue Book*, 1891. CO 142/ 105.

78. Zipperer, P. *The Manufacture of Chocolate and Other Cacao Preparations*, 2nd edition, Berlin and New York: M. Krayn, Spon, and Chamberlain, 1902; p. 30.

79. Clark, H. J. *The Land of the Humming Bird, being a Sketch of the Island of Trinidad* (Specially written for the Trinidad exhibition at the Chicago World's Fair). Port of Spain, Trinidad and Tobago: Government Printing Office, 1893; p. 78.

80. Clark, H. J. *The Land of the Humming Bird, being a Sketch of the Island of Trinidad* (Specially written for the Trinidad exhibition at the Chicago World's Fair). Port of Spain, Trinidad and Tobago: Government Printing Office, 1893; p. 79.

81. Hart, J. H. The consumption of cacao. *The West India Committee Circular* 21 June 1910:292.

82. Hart, J. H. The consumption of cacao. *The West India Committee Circular* 21 June 1910:291.

83. Hart, J. H. The consumption of cacao. *The West India Committee Circular* 21 June 1910:291.

84. Hart, J. H. The consumption of cacao. *The West India Committee Circular* 21 June 1910:291.

85. Hart, J. H. The consumption of cacao. *The West India Committee Circular* 21 June 1910:291.

86. Bekele, F. L. The History of Cocoa Production in Trinidad and Tobago. In: *Proceedings of a Seminar / Exhibition on Revitalization of the Trinidad and Tobago Cocoa Industry, 20th September, 2003.* University of the West Indies, St. Augustine, Trinidad and Tobago: The Association of Professional Agricultural Scientists of Trinidad and Tobago, The Cocoa Research Unit, 2003; p. 6.

87. Bekele, F. L. The History of Cocoa Production in Trinidad and Tobago. In: *Proceedings of a Seminar / Exhibition on Revitalization of the Trinidad and Tobago Cocoa Industry, 20th September, 2003.* University of the West Indies, St. Augustine, Trinidad and Tobago: The Association of Professional Agricultural Scientists of Trinidad and Tobago, The Cocoa Research Unit, 2003; p. 6.

88. Bekele, F. L. The History of Cocoa Production in Trinidad and Tobago. In: *Proceedings of a Seminar / Exhibition on Revitalization of the Trinidad and Tobago Cocoa Industry, 20th September, 2003.* University of the West Indies, St. Augustine, Trinidad and Tobago: The Association of Professional Agricultural Scientists of Trinidad and Tobago, The Cocoa Research Unit, 2003; p. 6.

89. Bekele, F. L. The History of Cocoa Production in Trinidad and Tobago. In: *Proceedings of a Seminar / Exhibition on Revitalization of the Trinidad and Tobago Cocoa Industry, 20th September, 2003.* University of the West Indies, St. Augustine, Trinidad and Tobago: The Association of Professional Agricultural Scientists of Trinidad and Tobago, The Cocoa Research Unit, 2003; p. 6.

90. Bekele, F. L. The History of Cocoa Production in Trinidad and Tobago. In: *Proceedings of a Seminar / Exhibition on Revitalization of the Trinidad and Tobago Cocoa Industry, 20th September, 2003.* University of the West Indies, St. Augustine, Trinidad and Tobago: The Association of Professional Agricultural Scientists of Trinidad and Tobago, The Cocoa Research Unit, 2003; p. 9.

91. Bekele, F. L. The History of Cocoa Production in Trinidad and Tobago. In: *Proceedings of a Seminar / Exhibition on Revitalization of the Trinidad and Tobago Cocoa Industry, 20th September, 2003.* University of the West Indies, St. Augustine, Trinidad and Tobago: The Association of Professional Agricultural Scientists of Trinidad and Tobago, The Cocoa Research Unit, 2003; p. 9.

92. Bekele, F. L. The History of Cocoa Production in Trinidad and Tobago. In: *Proceedings of a Seminar / Exhibition on Revitalization of the Trinidad and Tobago Cocoa Industry, 20th September, 2003.* University of the West Indies, St. Augustine, Trinidad and Tobago: The Association of Professional Agricultural Scientists of Trinidad and Tobago, The Cocoa Research Unit, 2003; p. 8.

93. Bekele, F. L. The History of Cocoa Production in Trinidad and Tobago. In: *Proceedings of a Seminar / Exhibition on Revitalization of the Trinidad and Tobago Cocoa Industry, 20th September, 2003.* University of the West Indies, St. Augustine, Trinidad and Tobago: The Association of Professional Agricultural Scientists of Trinidad and Tobago, The Cocoa Research Unit, 2003; p. 3.

94. Bekele, F. L. The History of Cocoa Production in Trinidad and Tobago. In: *Proceedings of a Seminar / Exhibition on Revitalization of the Trinidad and Tobago Cocoa Industry, 20th September, 2003.* University of the West Indies, St. Augustine, Trinidad and Tobago: The Association of Professional Agricultural Scientists of Trinidad and Tobago, The Cocoa Research Unit, 2003; p. 10.

FRONTISPIECE AND FIGURE 6.1. Image of
Cacao Tree from the Badianus Codex.

FIGURE 1.1 Vessel from Tomb 19, Rio Azul, Guatemala.
Hieroglyphic caption includes the glyph for cacao (at left on
vessel lid).

FIGURE 1.6. God "L" and cacao tree from the Red Temple
at Cacaxtla, Tlaxcala, Mexico.

FIGURE 3.1. Religious procession in Oaxaca, Mexico with participants carrying cacao pods.

FIGURE 3.2. Image of Saint with cacao bean necklaces.

FIGURE 3.3. Altar constructed in celebration of Day of the Dead in Mexico.

FIGURE 3.4. Cacao pods on a Cuna textile from Panama.

FIGURE 5.1. *Auto de Fe* of June 30th, 1680, Plaza Mayor, Madrid, by Francisco Ricci, dated 1683.

FIGURE 6.2. Native American presenting gift of cacao to the god, Europe.

FIGURE 9.2. Nurse serving cocoa to the infirmed.

FIGURE 9.3. Chocolate and Seduction: woman stirring cocoa.

FIGURE 9.4. Chocolate and righteous living: post Civil War cookbook.

FIGURE 9.5. Chocolate and health from *Food and Cookery for the Sick and Convalescent* by F. M. Farmer, 1869.

FIGURE 10.2. Chocolate pot manufactured by Zachariah Brigden (1734–1787), Boston, Massachusetts, circa 1760.

FIGURE 10.4. Chocolate pot in the form of a Roman askos, manufactured by Anthony Simmons and Samuel Alexander, Philadelphia, dated 1801.

FIGURE 10.6. Chocolate pot, manufacturer unknown, either The Netherlands or possibly New York, dated 1703.

FIGURE 10.8. Chocolate cup and saucer from Staffordshire, or possibly Yorkshire, England, circa 1770.

FIGURE 10.10. Chocolate cup with cover and saucer, manufacturer unknown, from Jingdezhen, China, circa 1795.

FIGURE 10.11. Chocolate stand (trembluese), manufacturer unknown, made of hard-paste porcelain with overglaze enamels and gilding, Jingdezhen, China, circa 1740–1750.

FIGURE 11.1. Portrait of Judge Samuel Sewall, by John Smibert (1688–1751), dated 1729.

FIGURE 11.2. Still-life painting of chocolate service by Luis Melendez (1716–1780), circa 1760.

FIGURE 11.3. Chocolate pot made by John Coney (1665–1722), dated 1701.

FIGURE 11.4. Chocolate pot made by
Isaac Dighton, London, England, dated 1607.

FIGURE 11.5. Chocolate pot made
by John Coney, circa 1710–1722.

FIGURE 11.6. Chocolate pot made by Edward Winslow (1669–1753), Boston, Massachusetts, circa 1705.

FIGURE 11.7. Chocolate pot made by Peter Oliver (ca. 1682–1712), Boston, Massachusetts, circa 1705.

FIGURE 11.8. Chocolate pot made
by Edward Webb (1666–1718),
Boston, Massachusetts, circa 1710.

FIGURE 11.9. Monteith (large
silver punch bowl) made by
William Denny, London, England,
dated 1702.

FIGURE 11.10. Chocolate pot made by Zachariah
Brigden (1734–1787), Boston, Massachusetts, circa
1755–1760.

FIGURE 11.13. Silver sugar box made by John Coney, Boston, Massachusetts, circa 1680–1690.

FIGURE 11.14. Silver box made by Edward Winslow (1669–1753), Boston, Massachusetts, circa 1700.

FIGURE 11.15. *Un Cavalier et une Dame Beuvant du Chocolate* (*Cavalier and Lady Drinking Chocolate*), engraving on paper, by Robert Bonnart, circa 1690–1710.

FIGURE 11.16. *La Jolie Visiteuse* (*The Pleasant Houseguest*) by Jean-Baptiste Mallet (1759-1835), circa 1750.

FIGURE 11.17. *La Crainte* (*The Fear*) by Jean-Baptiste Leprince (1734–1781), oil on canvas dated 1769.

FIGURE 12.2. *Chocolatière* in the Louis XV silver-gilt rococo style, manufactured by Henry-Nicolas Cousinet, circa. 1729–1730.

FIGURE 12.3. *Chocolatière* and *cafetière* made of silver-gilt with an asymmetrical form, manufactured by Jean-Jacques Ehrlen, Strasbourg, France, circa 1736–1750.

FIGURE 12.5. Silver *chocolatière* commissioned by Catherine the Great of Russia as a gift for Count Gregorii Orlofv, manufactured by Jacques-Nicolas Roettiers, Paris, France, dated early 1770s.

FIGURE 12.6. Earthenware *chocolatière* of the faïence fine style, manufactured by the Rue de Charenton factory, Paris, circa 1750s–1760s.

FIGURE 12.7. Sevres porcelain *chocolatière* owned by Prince Yusupov. Chinoiserie decor (black background to imitate lacquer work) inspired by Oriental sources, dated 1781.

FIGURE 12.8. Sevres porcelain *chocolatière* with white background, part of a breakfast service (déjeuner) belonging to Count Shuvalov, dated to the 1780s.

FIGURE 12.9. Detail of the lid top of *chocolatière* in FIGURE 12.7.

FIGURE 12.10. Detail of the lid bottom of *chocolatière* in FIGURE 12.7

FIGURE 12. 11. Detail of the bottom of *chocolatière* in FIGURE 12.7.

FIGURE 12.12. Sèvres porcelain *chocolatiere*, off-white with gold decoration, manufactured for Madame Victoire (daughter of King Louis XV) in 1786.

FIGURE 12.13. *Déjeuner (Breakfast)* by François Boucher. Informal scene of drinking chocolate and children, dated 1739.

FIGURE 12.14. *La Tasse de Chocolate (A Cup of Chocolate)* by by Jean-Baptiste Charpentier. An interior scene of men and women drinking chocolate, dated 1768.

FIGURE 12.15. *Portrait de Mme. du Barry à sa Toilette* (*Madame du Barry at her Dressing Table*), by Jean-Baptiste André Gautier d'Agoty. Madame du Barry being served chocolate by her servant, Zamore, dated between 1769–1786.

FIGURE 12.16. *La Tasse de Thé* (*The Cup of Tea*), by Nicholas Lancret. A detail showing a family taking tea and chocolate outdoors, circa 1740.

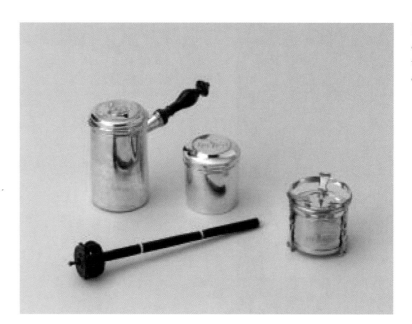

FIGURE 12.18. Silver *chocolatière* owned by Marie Antoinette, manufactured by Jean-Pierre Charpenat, Paris, dated 1787–1788.

FIGURE 12.19. Silver *verseuse* (pitcher) owned by Napoleon, converted into a *chocolatière,* manufactured by Martin-Guillaume Biennais, dated 1798–1809.

FIGURE 12.20. Close-up showing of FIG-URE 12.19 showing the Imperial Arms of Napoleon as King of Italy.

FIGURE12.21. Painting on a tray showing cocoa production and preparation of hot chocolate.

FIGURE 12.22. Detail of FIGURE 12.21, painting showing three seated women separating beans from cacao pods and another woman preparing hot chocolate

FIGURE 12.23. Coffeepot, one of four pieces, from a coffee set made at the Sevres factory in 1836.

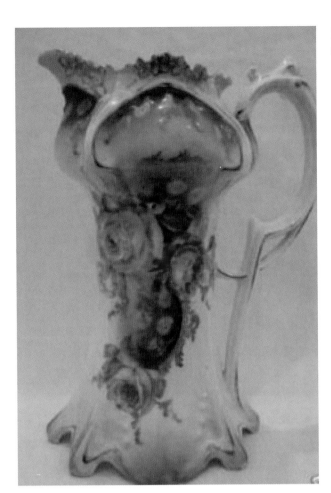

FIGURE 13.4. Twentieth century German chocolate pot.

FIGURE 13.5. Twentieth century, English Wedgewood chocolate pot.

FIGURE 14.1. Runkel chocolate trade card.

FIGURE 14.2.
Walter H. Baker chocolate trade card.

FIGURE 14.3. Millard chocolate trade card.

FIGURE 14.4. Van Houten chocolate
trade card.

FIGURE 14.5. Huyler chocolate trade card.

FIGURE 14.7. Bendorp chocolate postcard.

FIGURE 14.8. Phillips chocolate trade card.

FIGURE 14.9. Cherub chocolate trade card.

FIGURE 14.10. Van Houten chocolate trade card.

FIGURE 15.1. *Compagnie Francaise des Chocolats et des Thes* chocolate poster, color lithograph by Theophile Alexandre Steinlen, dated 1898.

FIGURE 15.2. Gala-Peter chocolate poster, color lithograph from the French school.

FIGURE 15.3. Suchard chocolate poster, color lithograph from the Swiss school, dated late 19th century.

FIGURE 15.4. Kasseler chocolate poster, color lithograph from the German school, dated early 20th century.

FIGURE 15.5. Vinay chocolate poster, color lithograph from the French school.

FIGURE 15.7. Bunte chocolate poster for a Christmas advertisement, dated 1930.

FIGURE 16.2. Chocolate exhibition for Stephen F. Whitman and Son of Philadelphia, Machinery Hall,
Centennial Exposition, Philadelphia, 1876.

FIGURE 16.3. Chocolate exhibition for Henry Maillard of New York, Machinery Hall, Centennial Exposition, Philadelphia, 1876.

FIGURE 16.5. Walter Baker & Company chocolate trade card depicting the Exhibition Building at the Columbian Exposition, Chicago, 1893.

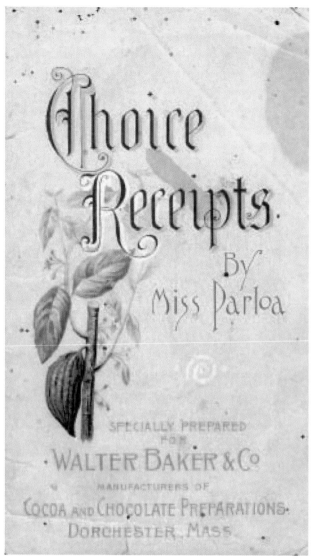

FIGURE 16.6. Walter Baker & Company's *Choice Receipts by Miss Parloa* chocolate trade card from the Columbian Exposition, Chicago, 1893.

FIGURE 16.7. Stollwerck chocolate trade card (obverse) depicting the Columbian Exposition.

FIGURE 16.10 Blooker chocolate trade card (obverse), from the Columbian Exposition, Chicago, 1893.

FIGURE 20.1. The Old Bailey Criminal Court, by Thomas Rowlandson and Augustus Pugin, dated 1808.

FIGURE 22.1. Toby Philpot.

FIGURE 22.8. *The Coffee Seller* (*Un Caffetier*) by Martin Englebrecht, Germany, circa 1720.

FIGURE 24.1. Metal chocolate pot (reproduction) used for the interpretative program at the Fortress of Louisbourg.

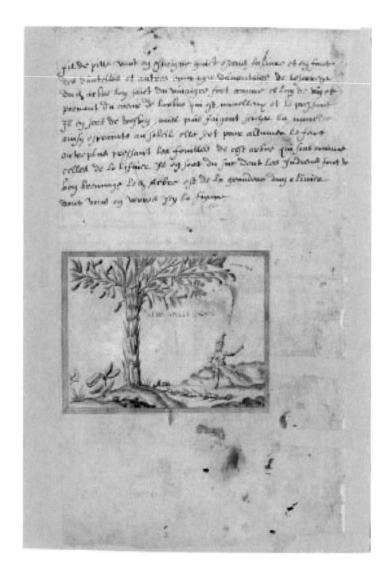

FIGURE 24.2 *L'arbe quis'appelle cacao.* (*The Tree That is Called Cacao*), manuscript version.

FIGURE **24.6.** Cover of *Cowan's Dainty Recipes*, circa 1915.

FIGURE **25.1.** Modern depiction of 1744 Louisbourg showing commercial activity on the busy quay.

FIGURE 25.2. The modern reconstruction of the Fortress of Louisbourg, operated by Parks Canada, representing approximately one quarter of the original fortifications and one fifth of the 18th-century town of Louisbourg.

FIGURE 25.7. Chinese export porcelain, handle-less narrow cups preferred for the consumption of chocolate, excavated at Fortress of Louisbourg.

FIGURE 25.8 Chocolate cups in the French faience style excavated at Fortress of Louisbourg.

FIGURE 25.9. Chinese export porcelain, handle-less chocolate cup, 18th-century Batavian style with brown outer glaze with blue under glazing in interior, excavated at Fortress Louisbourg.

FIGURE 25.10. Chinese export porcelain, handle-less chocolate cup, 18th-century, Chinese Imari-style with blue under glaze with over painting of gold and red, excavated at Fortress of Louisbourg.

FIGURE 25.11. English-manufactured agateware, handled chocolate cup, 18th-century, named for use of multicolored clay that mimics agate. John Astbury, Staffordshire Potter, is credited with inventing agateware, and is also produced as white salt glaze stoneware. This cup was excavated at Fortress of Louisbourg.

FIGURE 25.15. Silver, 18th-century chocolate/coffee pot.

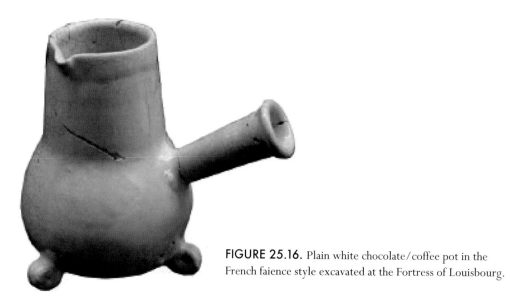

FIGURE 25.16. Plain white chocolate/coffee pot in the French faience style excavated at the Fortress of Louisbourg.

FIGURE 25.18. The private cabinet of Governor Duquesnel's reconstructed apartments at the Fortress of Louisbourg.

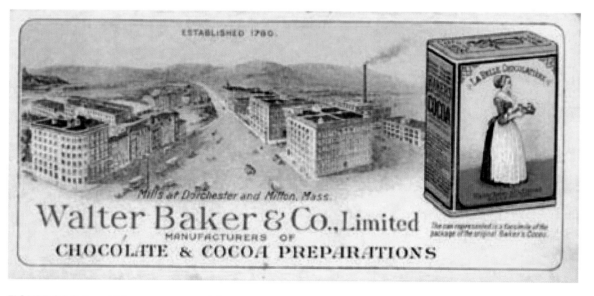

FIGURE 26.2. Late 19th-century image of the Walter Baker and Company factory complex.

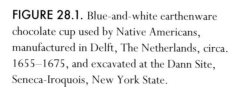

FIGURE 26.4. Advertisement for Walter Baker & Co.'s Caracas Sweet Chocolate with a chocolate girl is the subject of the advertisement and framed by an announcement of International Expositions at which the company has been honored.

FIGURE 28.1. Blue-and-white earthenware chocolate cup used by Native Americans, manufactured in Delft, The Netherlands, circa. 1655–1675, and excavated at the Dann Site, Seneca-Iroquois, New York State.

FIGURE 29.1. Portrait of Abraham Wendell (b. 1715), attributed to John Heaten, circa 1737.

FIGURE 30.1. The Wythe Kitchen, Colonial Williamsburg, Virginia.

FIGURE 30.3. Handmade chocolate, Colonial Williamsburg, Virginia.

FIGURE 31.1. Painting of Sir William Johnson convening an Indian council at Johnson Hall, Britain's foremost site in North America for negotiating with the Indians, by Edward Lamson Henry, dated 1903.

FIGURE 31.5. Copper chocolate pot, dated circa 1750–1780.

FIGURE 31.6. Shaving chocolate off the block to begin the process of drinking chocolate for breakfast, Fort Ticonderoga, New York.

FIGURE 31.7. Whisking a froth on breakfast chocolate and warming over a brazier at Fort Ticonderoga, New York.

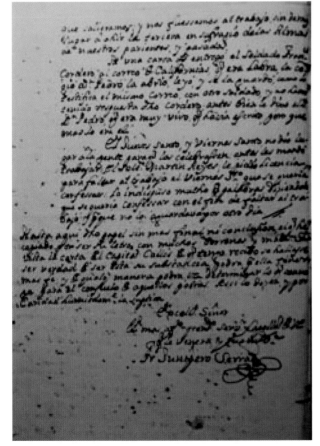

FIGURE 34.3. Father Junipero Serra correspondence of Captain Pedro Fage's misconduct letter (theft of chocolate meant for sick soldiers, dated May 1769

FIGURE 35.4. A 20th-century view: Ghirardelli Square, San Francisco.

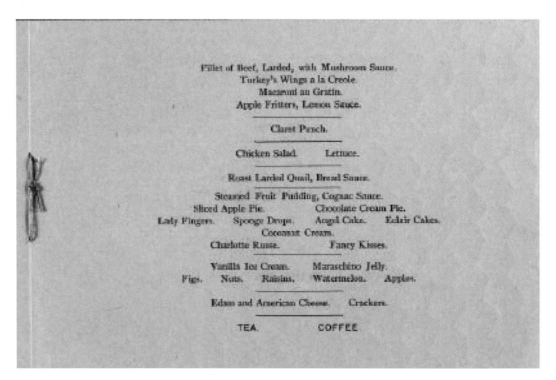

FIGURE 35.6. Menu offering chocolate cream pie at The Raymond, South Pasadena, California, dated November 13, 1886.

FIGURE 41.1 Chocolate pot of Portuguese King João V, dated 1720–1722.

FIGURE 41.2 Painting of Portuguese King João V with chocolate, dated 1720.

FIGURE 41.3. Chocolate tile panel from the factory of Vicente Navarro, Valencia, Spain, circa 1775.

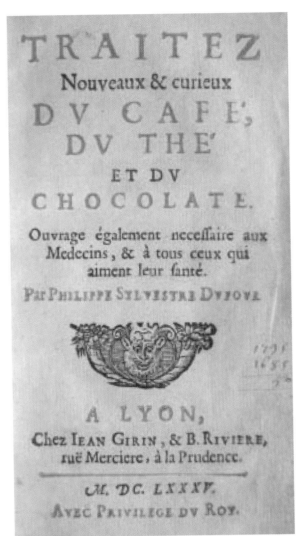

FIGURE 42.3. Title page of the *Traitéz Nouveaux & Curieux du café, du thé et du chocolat*, by Sylvestre Dufour, dated 1685.

FIGURE 43.2. Chocolate pot, manufactured by George Garthorne. Arms of original owner [Moore] are emblazoned on surface and the shape is inspired by a Chinese ginger jar, dated 1686.

FIGURE 43.3. Gold chocolate cup manufactured by Ralph Leake, circa 1690.

FIGURE 43.4. Tin-glazed earthenware chocolate cups collected by Sir Hans Sloane.

FIGURE 44.1. Chinese chocolate cup and saucer, dated 1729-1735.

FIGURE 48.1. Chocolate cream and pot.

FIGURE **48.2.** Roasting cacao beans.

FIGURE **48.3.** Cracking roasted
cacao beans to remove nibs.

FIGURE **48.4**. Winnowing
roasted cacao beans.

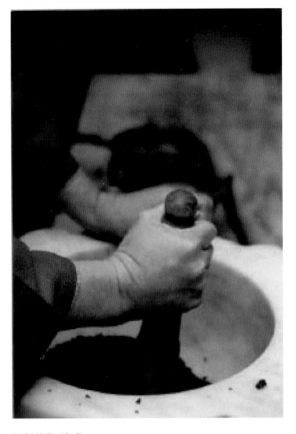

FIGURE 48.5. Crushing nibs.

FIGURE 48.6. Grinding cacao beans.

FIGURE 48.7.
Pouring chocolate.

FIGURE 54.3. Chocolate pot with a smooth, rounded-shoulder on baluster form.

FIGURE 54.6. European chocolate pot from the 19th–20th centuries.

FIGURE 54.7. Chocolate pot with unusual baluster form lacking distinctive flared footing common for this style.

FIGURE 54.8. Chocolate pot with footed-baluster form with spout.

FIGURE 54.9. Chocolate pot with baluster form with legs.

FIGURE 54.10. Chocolate pot with tapered-cylinder form, perhaps from the early 18th century.

FIGURE 56.1. Suchard chocolate trade card, date unknown.

FIGURE 56.2. Ghirardelli Birds of North America chocolate trade card, date unknown.

FIGURE 56.3. Poulain chocolate trade card, date unknown.

FIGURE 56.4. Ghirardelli general advertising trade card, date unknown.

FIGURE 56.5. Bensdorp chocolate card, date unknown.

FIGURE 56.6. Exterior view of Shackleton's hut in Antarctica.

FIGURE 56.7. Rowntree's cocoa canister inside Shackleton's hut in Antarctica.

FIGURE 56.8. Exterior view of Scott's hut, Cape Evans, Ross Island, Antarctica.

FIGURE 56.9. Scott's hut at Cape Evans, Ross Island, Antarctica, showing Fry's chocolate box and canister.

FIGURE 56.14. Chocolate service from the *H.M.S. Titanic,* dated 1912.

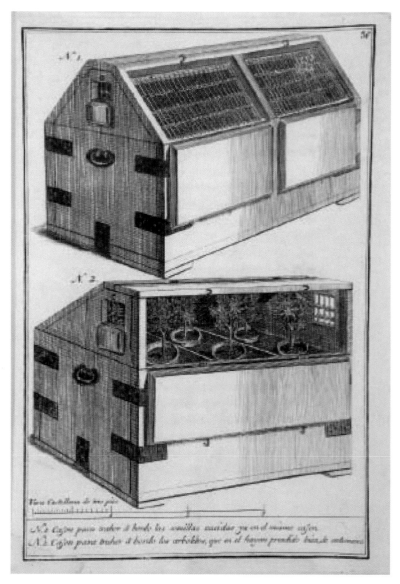

FIGURE 56.16. Transportation of cacao tree seedlings aboard ship, dated 1779.

History of Cacao Cultivation and Chocolate Consumption in Cuba

Niurka Núñez González and Estrella González Noriega

Introduction

To conduct research on cacao and chocolate in Cuba has been a wonderful and passionate adventure. Our two chapters—better seen as companion pieces—provide a panoramic view of the historic evolution of Cuban chocolate from earliest times to the present. We have tracked cacao and chocolate from the perspectives of economic history using travel accounts, even census data, and have searched for information in places sometimes obscured by the economic transcendence of sugar cane, tobacco, and, at certain historical periods, coffee. Our second chapter is dedicated to understanding cocoa and chocolate in Cuban literature and cultural life. In addition to studying these themes as evidenced in short story and novel genres, we investigated works dedicated to oral and folkloric traditions, musical texts, and linguistic themes, and especially sought traditional cookbooks, from which we have extracted recipes typical of Cuban cultural history (see Chapter 39).

Cuban Cacao and Chocolate: The Beginnings

Tracking the history of cacao in Cuba is challenging. Cacao-related evidence during the first two centuries after the Spanish conquest and colonization is sporadic due to the low importance of cacao at that time in contrast with agricultural and economic development based on coffee, sugar, and other activities. During the first half of the 16th century, the Spanish introduced into Cuba many different crops originally from diverse parts of the world, along with a range of domesticated animals not previously present in the Indies [1–3].[1] Nevertheless, during the early colonial years, Cuba experienced a brief flourishing period that, in addition to the search for gold and extensive cattle ranching, included the development of a diversified subsistence agriculture [4].

After 1555, the island suffered an almost total abandonment, with the enlistment of the Spanish military for further conquest on the American continent. Furthermore, the accidental introduction of diseases into the Indies from Europe caused the near extinction of the original aboriginal population and led to the depopulation of the island. Cattle-raising, however, remained a prominent economic activity until approximately 1762. Beef produced in Cuba during this time served as one of the main provisions for Spanish military troops on the American mainland and to supply

Chocolate: History, Culture, and Heritage. Edited by Grivetti and Shapiro
Copyright © 2009 John Wiley & Sons, Inc.

the Spanish fleet. Nevertheless, agriculture continued to expand and developed into several commercial forms, especially sugar and tobacco—crops raised primarily for export and controlled via a monopoly by the Spanish government. As a result, almost all other forms of agriculture on Cuba at this time were subsistence based, primarily for individual and home consumption [5].

Regarding subsistence agriculture, the people of Cuba raised numerous fruits already present since pre-Columbian times. The first mention of cacao in Cuba, however, stems from documents related to cocoa cultivation that appear toward the end of the 16th century through the mid-17th century [6, 7]. From the beginning to the middle of the 17th century, there are reports of numerous cocoa plantings in the geographical regions of Bayamo[2,3] and Santiago de Cuba[4] [8]. But even at this time, it would appear that cacao did not reach the level of an export crop compared to other locations such as Caracas,[5] Cartagena[6], Cumaná[7], Honduras, or Portobelo,[8] although it regularly was used by Cubans as a breakfast beverage [9, 10]. The situation just described continued until the beginning of the second half of the 18th century. Between the years 1754 and 1757, the Bishop of Havana, Pedro Agustín Morell de Santa Cruz, traveled the island and produced a report; especially enlightening was his description regarding chocolate in the region of Baracoa:

> Everybody has here his ranch, a hand mill to grind Guarapo[9] and to make furrio[10] which is equivalent to chocolate. There are also some cattle so that each ranch holds the four[11] [sic] items [earlier in his manuscript he had mentioned pigs] that are necessary for their sustenance. [11]

Such assertions in the literature allow scholars to conjecture about how cacao was introduced and spread throughout Cuba, and the location of various sites and ranches where cocoa was cultivated for local consumption.

During the same relative time period (1757), Nicolas Joseph de Ribera wrote in his *Description of the Island of Cuba* that "cacao grows with quite promptness" and he lamented that:

> The agriculture of the Island is reduced to sugar and tobacco. All the rest is sowed just to feed its towns, no more . . . other genres that produce well . . . they sow to consume them each in his own house or for recreation, and they are not cultivated formally. . . . But the more strange is that they follow the same rule with cocoa, and wheat, genres that there are highly estimated and highly consumed. Very few sow them, they come in great abundance, and those that do not sow them purchase them with money. The cacao comes from Caracas, Cartagena, Portobelo, and Campeche. [12]

He continued and wrote:

> Few sow cocoa because the cocoa tree needs a continuous care, and it takes time before they are ready to produce, and principally because the canes [sugar cane] and tobacco are the most important crops, and the farmers are more interested on them. [13]

In relation to the internal commerce of the island, he also mentioned that Bayamo received cacao from Santiago de Cuba, but he also indicated that this cacao might have been shipped from an overseas location to the port of Santiago de Cuba [14].

During the mid-18th century, around 1760, some effort was made to diversify the commercial Cuban agriculture that produced a boom in cocoa and coffee. Coffee production in Cuba generally was linked with cacao and in 1748[12] the first plantation of both crops was promoted at Wajay, a locality near Havana [15]. Cacao, along with other crops, received tax exemptions from the government. This boom, however, was not sustained and agriculture lagged until the relative liberalization of Cuban commerce that began in the second half of the 18th century[13] and peaked during the 1930s [16, 17]. Agricultural diversification, however, found an almost insuperable obstacle in the widespread domination of sugar cane, whose supremacy was defended by the majority of the landowners.

Nineteenth Century Developments

The commercial agriculture offensive did not exclude the development of other types of crops, in particular, those of subsistence agriculture. During most of the 19th century, commercial agriculture expansion was limited to the region of western Cuba. It was not until 1868 that portions of the central and eastern regions also would be developed as commercial agriculture areas. *Remedios*, one of the oldest Spanish settlements in the central area, traditionally was the main cocoa producer. In the eastern region cultivation of cocoa has prospered since the 18th century,[14] as promoted by the French immigrants from Haiti [18–20].

During the 18th century, chocolate was the most consumed beverage in Cuba. A 1791 reference cited by Eguren points to its supremacy and even notes that coffee was "prepared . . . in a kind of chocolate pot" [21]. By the first half of the 19th century, cocoa and coffee had the same level of consumption. Then coffee finally prevailed and was considered the national beverage. Still, throughout the 19th century, references to the consumption of cacao and coffee without distinction are found. To illustrate the widespread consumption of chocolate at this time, we can quote diverse travelers who visited Havana in different times and left records of the "more popular" cafés, where chocolate or coffee was served. In 1819, a traveler visiting Havana discussed chocolate served at the Paloma (café) and at the Comercio (café) [22]; in 1838, another traveler mentioned the café La Lonja [23]; and during the 1860s, another visitor indicated chocolate at the café La Dominica [24].

Through the 19th century there are abundant accounts and anecdotes regarding the manufacture of chocolate and its consumption in public places. Baron Alexander von Humboldt[15] wrote an essay published in 1826 entitled "Political Essay About the Island of Cuba," where he described the manufacture of chocolate [25]. The Countess of Merlín[16] returned to Havana in 1840 and wrote: "there are signs everywhere that say coffee, sugar, [and] cocoa." She also described the consumption of coffee and cocoa at funerals, and the existence of cocoa groves at Artemisa in the San Marcos Valley, Province of Havana [26].

Among the many references to domestic consumption of cocoa are those provided by the North American physician John G. Wurdemann, who visited Cuba several times during the 19th century. Included among his more revealing statements are: "Coffee and chocolate, [are] here are almost universally consumed" [27]; and during his visit in 1844, he observed that "in Havana . . . the ladies remain in their homes during the heat of the day, reclined, with their chocolates and tobaccos" [28]. Finally, toward the end of the century in 1883, the following also is attributed to Wurdemann, in a passage where he is being urged on by a servant (?): "Hurry up Sir, they are having the chocolate" [29].

The attempt to diversify agriculture as an alternative to the prevalent monoculture (sugar cane) was reflected in debates that took place during the early 19th century, where the development and expansion of cocoa production received particular attention. Key players in these debates were the Sociedad Económica de Amigos del País (Economic Society of the Friends of the Country), founded in 1793, and the Real Consulado de Agricultura y Comercio (Royal Consulate of Agriculture and Commerce), founded in 1794. Several documents from the early and mid-19th century reveal some of the debates' outcomes; some, from 1818, suggested the improvement of different crops, instead of continuing with the prevailing sugar cane monoculture; another, from 1849, includes in a list of tasks the promotion of the cocoa crop by bringing seeds, plants, and workers from the Costa Firme;[17] and a document from 1850 was specifically dedicated to cocoa cultivation [30].

In relation to this controversy, it is important to mention here the well-known discourse by von Humboldt against the monoculture in his aforementioned essay and his criticism about the lack of interest in production of cacao, cotton, and indigo in Cuba: "the cultivation of cacao, cotton and indigo have been almost null" [31]. It is also worthy to emphasize the efforts of Ramón de la Sagra, who in 1863 tried to systematize the problems of diversification by grouping the crops; cacao was among the crops that he indicated have sub-

sisted in a precarious state but showed a great probability of success [32]. Although cacao did not reach the status of an important commercial crop, it did remain widely cultivated in small land areas and in the hands of the small farmers in subsistence agriculture settings [33, 34].

Cacao Production

The 1827 census reported the existence of 60 cocoa plantations (cacaotales or cacahuales) distributed along the three Cuban regions identified as departments: two in the western region, 54 in the central area, and four in the eastern region [35, 36]. The total harvest in 1827 was 23,806 arrobas of cacao, that is, 595,150 pounds [37].

In 1828, the North American Reverend Abiel Abbot[18] visited the Occidental (Western) region. In his "Letters" he wrote that he saw for the first time cocoa tree production at a ranch (hacienda) in Camarioca.[19] The fruit of the cacao tree astonished him and he described it by saying: "the fruit has all the size [development stages] gradations, since the beautiful flowers in the spike . . . until the berry already mature" [38]. He also observed that cacao trees were cultivated in other areas: in the mountainous region near Guanajay, in the Province of Havana; in the ranch (hacienda) of the Marquis de Beytia (sic), near the capital (it was the marquis who was the first to establish a cocoa plantation); and in the Sierra del Rosario (Rosario Sierra), in the proximity of Soroa, in the Province of Pinar del Río in northwestern Cuba. In this last area he described a new cocoa plantation belonging to an Italian rancher and wrote: "that is the first one that had begun this interesting crop in the island at a considerable scale" [39].

Abbot's observations not only provide detailed descriptions of the cacao plantations but also insights on production and manufacturing strategies:

It has 20,000 trees in flowering stage, 2 to 8 years old. They are intercalated among the coffee bushes and can serve to give these delicate bushes the needed shade and at the same time to proportion themselves a valuable harvest. . . . This gentleman has proceed with caution while fomenting its cocoa plantation, beginning with a few trees. He presented a sample of his cocoa, manufactured into the form of chocolate, to the Consulate [the Royal Consulate and the Board of Commerce—Real Consulado and Junta de Comercio], and it was judged as of an equal quality to the best of those that are being offered from outside [of Cuba], and they offered him two dollars more per quintal because it was more fresh and more succulent [than the others], what makes it very appropriated to mix it with the more drier foreign cacao. [40]

Abbot continued with his description of the cacao habitat, cultivation, and harvesting and provided information on the cacao markets and prices:

It grows equally well at the top of the mountains and in the valleys. It likes a northern orientation and a cool place; does not need to be fertilized and although does not prosper in an exhausted soil, for the most part it has been sowed among coffee bushes that are 20 years old. . . . Large amounts of cacao are harvested for the market of San Juan de los Remedios. [40]

It is sowed, pruned and weeded; it is harvested and the berry is broken; the seeds can be separated [from the fruit] very easily and they are brought to drying sheds that have been already prepared for the coffee. The seeds must remain two days in a pile and then they should be exposed gradually to the heat of the sun until the product is ready for the market. [41]

During 40 years of field work, Esteban Pichardo,[20] the first Cuban geographer, registered many traditional names inherited through oral tradition by the Cuban inhabitants of his time. In his work *Diccionario provincial casi razonado de vozes y frases cubanas* (*Provincial Dictionary with Explanatory Notes of Cuban Voices and Phrases*), published in 1836, he wrote that cacao was rare in many regions but also grew wild. He affirmed that:

It [cacao] is sown in a reduced scale . . . the ones [cacao trees groves] from Vueltarriba and from Remedios are better than the ones from Vueltabajo. [42]

Pichardo also mentioned a *Catecismo de la Agricultura Cubana* (*Catechism of the Cuban Agriculture*) that reportedly mentioned the cultivation of cocoa in the country, but we could not find any other information about this text.

According to the North American physician Wurdemann, there were reports of high cocoa yields in the early 1840s, especially in the area between the Western region and Sagua la Grande [43]. Still, cocoa production could not be maintained, as revealed in the 1846 census, where a severe decline in cocoa production was reported. The data provided by this census showed that although there was an increase in the number of cacao plantations in Cuba, the cacao production/yield was only a quarter of that reported in the 1827 census. The census 1846 registered a total of 69 cocoa plantations with a production of approximately 4000 *arrobas* (i.e., 100,000 pounds) [44, 45].

The *Diccionario geográfico, estadístico , histórico de la Isla de Cuba* (*Geographic, Statistical and Historic Dictionary of the Island of Cuba*) is a four-volume work by Jacobo de la Pezuela, published in 1863, that summarized information on Cuban economics and history. Pezuela provided information about cacao production areas and yields, chocolate manufacture, cacao trade and prices, and the influence of the French immigration on the Cuban economy. According to Pezuela:

After covering the internal consumption of a country that manufacture little amount of chocolate,[21] the general exportation of cacao was under 9,165 arrobas [i.e., 229,125

pounds or 115 tons]. In 1848 went down to 3,485 arrobas [i.e., 87,125 pounds or 44 tons], sold in the amount of 8,705 strong pesos [written as ps. = fs. or pesos fuertes[22]]. In the following years, cacao, experienced some progressive increases, and in 1861 its production rose to 96,101 arrobas [i.e., 2,402,525 pounds or approximately 75,079 tons] sold at an average price of 4 strong pesos per arroba . . . paid at the rate of 16 strong pesos per quintal.[23] . . . In 1862, 43,553 arrobas [i.e., 1,088,825 pounds or 34,026 tons] of cacao was harvested with a value of 10,888.25 pesos fuertes. [46, 47]

Pezuela's *Diccionario geográfico* provided detailed accounts of cacao plantations[24] (*cacahuales*), their production, and chocolate factories within Cuba. Most production was localized in the central region of the island within the jurisdiction of San Juan de los Remedios and Sancti Spiritus, and in the eastern part of the island, specifically Baracoa Guantánamo, Holguín, and Santiago de Cuba.[25] At the beginning of Spanish settlement of Cuba, the primary economic activity at San Juan de los Remedios had been cattle production, and the region supplied meat to the Spanish armada via the port of Havana.[26] These cattle-raising activities changed, however, at the beginning of the 18th century:

The [jurisdiction of San Juan de los Remedios] began to cultivate cacao and cotton, with not too much success. . . . Nevertheless, still today [i.e., in 1863] the only activity [here] with some success is the dark almond [cacao] with which the chocolate is manufactured . . . a little bit more than 1,000 arrobas [i.e., 25,000 pounds]. [48]

With respect to the district of Yaguas, within the jurisdiction of Santiago de Cuba, Pezuela affirmed that:

The nature of the territory in many locations is suitable for the cultivation of coffee and even cacao. . . . [Cacao production] in 1857 reached about 11,000 arrobas [i.e., 275,000 pounds]. [49]

It is important to mention that in Pezuela's monumental work, cacao for local consumption was not emphasized. In fact, he described the absence of cacao cultivation in some areas[27] and mentioned the presence of cacao only as a secondary crop in predominantly cattle regions.[28] Nevertheless, he affirmed that cacao constituted one of the "most common crops in the island."[29]

Even when cacao is not mentioned, this would not exclude the possibility that chocolate could have been prepared and used in local ranches and other working places. We also note here that in most of the articles written during the 19th century regarding different towns or other settlements in Cuba, chocolate factories are mentioned and the various authors also identify many other places where chocolate could have been offered to patrons, such as billiard parlors, cafés, candy stores, confectioner's shops, general stores, food stores, hotels, inns, taverns, and wine cellars.

The importance of French immigration to the promotion of cacao cultivation in Cuba during late 18th and early 19th centuries did not escape Pezuela's attention.[30] In his description of cacao production patterns in the jurisdiction of Holguín, he wrote that while cacao never reached the importance of tobacco "the industry of the emigrated from Santo Domingo included in some of the regions the cultivation of coffee and cacao" [50].

Pezuela also identified commerce and trade in cacao between Cuba and Caracas during the 17th century. He cited a document dated to 1635 that mentioned cacao and payment of taxes:

From the Firm Coast [Costa Firme], per each fanega of cacao from Caracas of 110 pounds of weight, 16 silver reales.[31] [51]

Two centuries later, data on exportation and importation tariffs presented by Pezuela reveal a more diversified cacao trade. Imported cacao was classified into two categories. The first included that arriving in Cuba from Caracas, Soconusco,[32] Maracaibo,[33] Costeño,[34] and Trinidad[35] "and others of equal quality." The second category was identified as cacao that originated from Guayaquil,[36] Marañón,[37] and ports along the Gulf of Para,[38] "and others [ports?] of equal quality." Pezuela also documented additional cacao/chocolate-related imports, among them cacao butter, chocolate, and brass, copper, iron, or tin chocolate pots (*chocolateras*). In turn, he noted that Cuba also exported some cacao [52].

Chocolate and Chocolate Making in Travelers' Accounts

The West Indies of the 19th century had a strong charm for many Europeans and North Americans. Some not only visited Cuba but made the island their home for several years. A wealth of information on the social customs in Cuba during the 19th century can be gathered by examining accounts produced by these expatriates, commonly painters and writers. The British artist and journalist Walter Goodman[39] lived in Cuba between 1864 and 1869, with most of his time spent in Havana and Santiago de Cuba, where he worked. In one of his accounts he described beverages served at breakfast:

[For breakfast I was served] a large bowl of milk with coffee, or, if I prefer it, a large bowl of extraordinary thick chocolate. [53]

As a long-term resident of Cuba, Goodman was invited to important social and family events. He wrote the following lines that described foods served at a funeral:

A tray filled with sweets, biscuits, coffee, chocolate and a good tobacco, was offered from time to time. [54]

Cuba and, in particular, the city of Santiago de Cuba during the 19th century were famous for carnivals.[40] Describing one of those large public festivals where music, dance, and food was abundant, Goodman wrote:

[During the carnivals in a large plaza] . . . whoever wanted to drink coffee or a hot chocolate, there they have little stoves ready to use. [55]

Besides his detailed observations of city life, Goodman also recorded the presence of cacao groves among the coffee plantations along the island [56].

Another chronicler of 19th century Cuba was the American writer and illustrator Samuel Hazard,[41] who visited the island just before the beginning of the Ten Years' War.[42] He left numerous descriptions regarding the consumption, cultivation, and benefits of cacao. From his book, *Cuba a pluma y lápiz (Cuba with Pen and Pencil)*, considered one of the best documents about Cuban customs and traditions during the 19th century, Hazard provided a direct glimpse into the Havana cafés and hotels, and especially the preparation of chocolate that he identified as the drink *par excellence*. His travels through the Cuban landscape made him familiar with the various aspects of cacao production: cacao plantation activities, descriptions of cacao trees and fruits, the cacao bean drying process, and market prices. His frequent visits to cafés and food shops in Havana allowed him to observe first-hand the process of chocolate making in the 19th century, knowledge that he recorded in this evocative book that was enhanced with his own illustrations. Early on, Hazard described two of the most important hotels in 19th century Cuba, the Santa Isabel[43] and El Telégrafo,[44] and wrote:

The facilities and comforts of the Telégrafo Hotel or those of Santa Isabel, become more acceptable with their coffee with abundant milk [café con leche] or chocolate each early morning. [57]

Further on, Hazard described the Cuban custom of beginning the day with coffee or chocolate:

Only a cup of coffee or chocolate with a toast or a biscuit that satisfied the appetite until lunch time . . . in Havana, we can find in each park, restaurants, coffee houses, candy shops and other stores known as dairies [lecherías], where all sorts of beverages made of milk[45] *are prepared and sold.* [58]

Among the Havana coffeehouses Hazard identified "La Domínica" as not yet a *place a la mode* (i.e., a fashionable place), famous for its confectionery and above all for its bonbons [59].

In his second volume, Hazard registered with utmost detail and care the art of making a good chocolate and provided thorough description of the technicalities of the proper use of the *molinillo*[46] to froth a fragrant cup of chocolate before serving. In his description, he emphasized the nutritional properties of the chocolate, particularly if one were about to debark on a long journey:

But the beverage "par excellence" that can be taken in Havana is a cup of chocolate, whose confection has made famous all the Hispanic countries. It is served in delicate small cups (jícaras), [and the chocolate] is very rich and has the consistency of a thick atole, and because is highly nutritive is the most excellent beverage that one can drink in early morning before starting a long journey. A Spanish lady, friend of mine, gave me this recipe for its confection: In a small tea cup (or demitasse) of milk, put an ounce of chocolate cut in small pieces; bring it to a boil, and while is boiling, beat it with an egg beater until becomes completely thick and foamy; then it is ready to serve it. Per each cup of chocolate desired, you must use one cup of milk and one ounce of chocolate; so that, for six cups of chocolate you will need six cups of milk and six ounces of chocolate. If this is not rich enough for you, add a half an ounce more of chocolate; if, to the contrary, you do not like it so rich, make it using half the amount of milk and half the amount of water, or all water. To make chocolate, the Spanish use what they called molinillo, that is nothing but a metal chocolate pot, with an inside beater,[47] that it is operated by means of a rod that goes through the chocolate pot's lid hole; the rod is held with the palm of both hands, given to it a rotational movement while the chocolate is still boiling. With one of these devices and a small alcohol lamp underneath [the chocolate pot], anybody can make a delicious chocolate on his table or in his own bedroom. It is customary in Cuba, to take chocolate in the morning with a light biscuit, or even better, with a piece of well toasted bread; after this they drink a glass of water. [60]

In his third volume and regarding the cultivation of cacao, Hazard confirmed the close association between coffee and cacao, when he described the process of inter-cropping:

As we have said before, the coffee plant it is not the only crop in the coffee plantations. There are several other crops, that together with the coffee result in a very profitable venture ... the most important of those crops is cacao, from which they make chocolate. There are three types and colors: the cacao Caracas that is red; the cacao from Guayaquil is purple, and the cacao Criollo[48] or Cuban cacao has a fruit with a yellow skin. [61]

The cacao tree and the appearance of its fruits also stirred Hazard's curiosity:

It is very singular because instead of forming racemes hanging from the branches, they grow from the same trunk, adhered to it by a single stem without leaves, and even sometimes resting on the ground, close to the roots. [62]

Details of cacao bean processing as well as the profitability of cacao growing in good soil also did not escape his observation:

Between the months of June and December the cacao beans are placed in piles to dry. They allow them to ferment for 3 or 4 days, and during this period some cacao hacienda owners add red ochre to each pile, to give the seeds a strong red color, changing in this way their natural color, what is a very dark brown, almost black. Then they are passed through sieves. In this stage, the grains, well dry, have an almost bitter and bad taste and they look like almonds, though of a darker color. ... The cultivation of this crop provides a high return with the excess work that requires if it is done in a good soil. The cacao of the best quality has a high price: 17 pesos for 100 pounds, and the one of an inferior quality, 12 or 13 pesos [for 100 pounds]. The work that it demands and the time that it takes to cultivate it, is relatively small. ... The most simple and better way to use the cacao beans is to take the beans, barely crushed and boil them for two hours ... a great quantity of chocolate is manufactured in Cuba in the form of bonbons, and the ladies are very fond of them. ... The liquid chocolate that is made in Cuba is very delicious and nutritive; a demitasse is equivalent to a meal and is capable of sustaining us during several hours of exercise. [63]

With respect to the manufacture of chocolate itself, he remarked:

After being carefully selected and after any foreign matter has been removed, the grains are toasted, and then placed inside of an iron cylinder with holes in one side to allow the vapor to escape. When the aroma becomes perceptible, the operation is finished. Once cooled down, the shells are removed and the beans are placed into a grinding mill that consists of an iron cylinder that rotates around a copper channel and in this way the beans are ground. Then, this ground cacao is transformed into a paste and packaged, or it is aromatized to make different types of candies with it. This operation can be seen in any confectioner's shop in Havana. [64]

Hazard also listed articles for export and indicated that cacao—along with other items (not identified)—represented only 7 percent of total Cuban exports. This statistic revealed that during Hazard's time cacao had only a minor role in the Cuban national economy [65].

Cacao and Chocolate During the Mid-19th Century

Wars bring destruction, but also foreign war correspondents—journalists who, in addition to covering the hostilities, often describe the country, the people, and their traditions. Cuban patriots revolted in 1868 due to discontent caused by excessive Spanish taxes, commerce restrictions, and exclusion of native Cubans from governmental posts. This war lingered for many years (1868–1878) and is known as the Ten Years' War. In 1874, James J. O'Kelly, an Irish journalist, arrived in Cuba as a war correspondent for the New York Herald.[49] He documented his experiences in a book titled The Land of the Mambi[50] (La tierra del Mambi) and described the daily routine of soldiers and what they ate. Through O'Kelly's complaint about not having coffee, we learn that the mambises (revolutionaries) drank chocolate:

What caused me the greatest pain was the lack of coffee . . .
we have to substitute it with chocolate, that in no way could
replace my favorite campaign drink. [66]

While we may sympathize with O'Kelly for the lack of caffeine that a war correspondent doing field work needed to boost his stamina, coffee might not have been the best beverage to sustain his endurance (or that of the soldiers) when provisions were scarce. Wisely, when other foods were not available, the soldiers were provided chocolate for lunch instead of coffee, as noted by O'Kelly:

[On another occasion] our lunch meal was reduced to a small
cup of not too sweet chocolate. [67]

At other times both beverages were available and were sweetened with honey instead of sugar:

Sometimes we had coffee and chocolate, and in the
mountainous district the sugar was replaced by bee honey.
The majority of these elements of subsistence can be found
in small quantities in all free Cuba. Due to the marvelous
soil fertility, the abandoned farms do not stop producing,
regardless that their fields were not longer cultivated by
man. [68]

On rare occasions and when circumstances allowed it, the soldiers also received a well-deserved treat—solid chocolate—that had been imported by Spanish soldiers (?) from France:

Coronel Noguera captured [an enemy military] convoy. And
thanks to this he could supply [with the provisions of the
enemy] the camp with some luxury articles, such as chocolate
Menier.[51] *[69]*

This brief observation revealed that the army loyal to Spain was better provisioned than the rebels and was supplied with imported solid chocolate. It is important to note, too, that O'Kelly did not describe the chocolate as a beverage, but as a luxury item that implied solid chocolate.

From the conclusion of the Ten Years' War in 1878 until the Spanish–American War[52] in 1898, we located only fragmentary information regarding cacao or chocolate use in Cuba. The most valuable source of information from this time, however, was the *Revista de Agricultura* (*Journal of* [*Cuban*] *Agriculture*) published between 1879 and 1894. We were only able to consult volumes for the years 1883, 1889, 1891, and 1893 due to incomplete archive/library holdings.[53] Through this journal, however, we learned that the chocolate factory "Mestre y Martinica" was inaugurated in 1883 and said to be:

The best equipped in Havana . . . [in which] . . . almost all
operations that cacao demands in order to be transformed into
chocolate are accomplished by mechanical devices, unlike the
other factories [where] much human labor is required. [70]

From such assertions as these, we can deduce that other unnamed chocolate factories also existed in Havana at this time.

The year 1883 brought a sudden increase in cacao production, a trend that would continue through the next decade. According to the *Journal*:

[The increase in cacao production] was due to the sudden rise
in the price of cocoa in foreign markets produced by the
blockade of some ports of the Pacific, due to the wars between
those republics.[54] *[71]*

Friedlaender, in his book *Historia económica de Cuba* (*Economic History of Cuba*), also documented a sharp increase in cacao production during the "inter-war era" and mentioned that in 1891 Cuban exports of cacao represented a value of 314,000 pesos [72].

The "inter-war era" was not an impediment for Spain to recognize the value of Cuban cacao. In January 1889, Cuban cacao was recognized as being of excellent quality in Barcelona, Spain, and a gold medal was awarded to the coffee and cacao plantation owner Cástulo Ferrer:

Because his coffee samples and his cacao beans samples of
cacao Criollo and Guayaquil from his hacienda "María de
Loreto". . . . Mister Ferrer has shown his efforts to raise again
in the Oriental region [of the island] old and excellent coffee
and cacao plantations. [73]

The (Cuban) *Journal of Agriculture* also reported on the quality of cacao in a publication dated to 1889:

After the peace of 1878 coffee and even cacao have been
progressing on the hillocks of the central region of the country
(lomas trinitarias) with a great impulse, and so [great is this
increase] that they export to Cienfuegos[55] *and even to*
Trinidad.[56] *[74]*

The journal also reported that the coffee and cacao plantations in western Cuba previously displaced by sugar cane production were recovering:

Each day they make efforts to install again the coffee and
cacao plantations in the occidental region of the Island, that
were displaced by sugar cane [when the coffee prices fell],
imitating the farmers from the oriental region, who each day
are making efforts to make a good use of the richness of the
soil. [75]

The (Cuban) *Journal of Agriculture* also reported the same year (1889) that cacao cultivation was being promoted in Samá, an area known in the 21st century as Holguín Province in the eastern region:

A ship will bring them with 3,000 cacao spikes [fruits] from
Baracoa[57] *for the sowing. [76]*

It is important to mention here that the *Journal of Agriculture* consistently supported the struggle against the monoculture and published an article/essay by Juan Bautista Jiménez entitled "Aventuras de un Mayoral"[58] ("Adventures of a Farm Foreman"), wherein the author warned against the dangers of a commercial extensive agriculture (monoculture) and defended agricultural diversity [77].

In 1891, the *Journal* published a brief commentary offering technical agricultural information about all the cultivation phases of cacao. It was written in a language accessible for the farmers and used a highly readable, simple style:

> The cacao is wild in some areas of Cuba. In Remedios . . . it is very abundant. . . . Certainly, in those places they barely care about exploiting them . . . here the cacao, as the rest of the fruit trees, are not cultivated at all. They themselves (the trees), in their wild stage, produce fruits without the intervention of the intelligent care of mankind. [78]

Also in the same year (1891), in the miscellanea section of the *Journal*, is a short piece on methods for cacao butter extraction using pressure and by means of water [79].

Toward the end of the 19th century, a variety of sources document the presence of chocolate factories in Cuba. One factory was located in the city of Santa Clara and another at Baracoa [80, 81].[59] Other authors describe drinking chocolate during this "interwar era," among them Eguren's observations, dated to1884:

> As a general rule, businessmen take a cup of chocolate or coffee with milk when they get up from bed. [82]

We also have identified chocolate drinking during times of the *altares de la cruz* (cross altars) celebrations,[60] truly social gatherings of the times, where courtship played an important role:

> [The celebration around the altars] . . . will last several days, and they [the people] will narrate all sorts of happenings and stories, they will ask riddles and play different entertainment games; the young men [at this time] would court young women, they drank coffee and chocolate and they ate cracklings and other small things, molasses candy and other sweets. [83]

Chocolate consumption during the Second War of Cuban Independence was documented by Grover Flint,[61] an American lieutenant who fought alongside the Cubans. In his book *Marching with Gomez (Marchando con Gómez)*, published in 1898, he provided testimony of events from the battlefield, where he offered an oblique reference to the availability of chocolate tablets. While in Matanzas,[62] he described captured military documents:

> [These] were documents of a strange appearance. They were written on anything; sometimes on the blue or orange wrappings of a chocolate tablet. [84]

Later on, he described a speech delivered by General Gómez, a high-ranking military officer of the Cuban Liberation Army. Speaking to his soldiers and showing his profound irritation, General Gómez scolded them and reported that:

> It is possible to buy the highest chief of Camagüey[63] with a chocolate tablet. [85]

As would be expected in times of war, Flint spoke frequently about food, particularly mentioning the consumption of coffee and *guarapo*, a freshly squeezed juice of the sugar cane, but he also reported on the procurement and identification of supplies. When he was in San Andrés, about 6–8 miles from Nasaja, he described the following:

> [During the war] a dispatch carrier left for one of the smallest villages, where he had the ability to enter and purchase things . . . and the young officials of army wrote what they wanted on small discarded pieces of paper. The following was typical: 4 pounds of coffee; 2 pounds of sugar; 4 bundles of tobacco; 2 liters of rum; and 1 pound of chocolate. [86]

On another occasion, while stationed in the area of Yatal, close to Nuevitas, the military purchased, through a third party, "a provision of coffee, sugar, rum, tobacco, crackers and chocolate" [87].

Consequences of the Independence Wars and the First Cuban Republic

After the prolonged Independence Wars, Cuban agriculture was devastated. A report dated to 1903 affirmed that the majority of the cacao plantations in the Oriental Province (eastern) "were destroyed by insurgents and had not been re-established" [88]. When the Cuban Republic was proclaimed (1902), 50 percent of farmland was cultivated with sugar cane, 11 percent in sweet potatoes, 9 percent in tobacco, 8 percent in bananas, and 7 percent in corn/maize. The remainder of Cuba's agricultural land was dedicated to the cultivation of cacao, coconuts, coffee, *malanga*,[64] onions, oranges, pineapples, potatoes, rice, yams (*ñames*), and yucca (manioc) [89]. This year marked the beginning of the large extensions of lands (*latifundios*) dedicated to sugar cane plantations. These large sugar cane plantations increased constantly in number and displaced cacao plantations.[65] Nevertheless, cacao as *a complementary crop with coffee*, as grown during the first half of the 19th century, continued; even today in the early 21st century in some areas of the Oriental region where agriculture is more diversified, cacao as a crop continues to expand [90, 91]. Those times also were characterized by the strong economic interests of the United States that extended to all sectors of the Cuban national economy. The *Report on Cuba*, published in 1903, included mention of both commercial and economic perspectives on Cuban agriculture and reported:

> In the district of Guantánamo, cacao is on the rise and it would be a highly promising business if [it] could be developed as an industry. [92]

The cacao industry in Cuba was an appealing business to American investors, as reported by Riccardi [93]. In 1902, the Reciprocity Treaty was signed with the United States. The first article of the treaty identified all the products of both nations that were exempt from export duties, among them cacao beans from Cuba. The third article of the treaty established a reduction of 20 percent for the importation of American goods; among the goods covered were processed chocolate and bonbons [94].

The 1919 census of the Cuba Republic revealed how coffee and cacao were intertwined in Cuban agriculture. Based on an estimated yield of 1500 pounds per acre for cacao and a market price of $10.00 per quintal that would produce $150,000 per year, the census analysis concluded with mixed message:

> The cultivation of cacao, especially when planted with coffee, is one of those crops that promises satisfactory results, particularly taking into account that the demand of both products increases constantly through all the world. . . . Because sugar and tobacco have absorbed, in the last years, a great part of capital in Cuba, the cultivation of cacao, as well as coffee, and cacao even more than coffee, has declined greatly in Cuba, after having been an important source of its wealth. [95]

In the census chapter dedicated to commerce, cacao can be seen among the exportation products of the years 1908–1919, but classified together with fruits and legumes and not a separate item. Through the census appendix, dedicated to the reports of regional inspectors, we know about the establishment of a chocolate factory in the city of Santa Clara [96]. With respect to the Oriental region, the census report complained that:

> The enormous increment that the sugar cane industry has had in the last years . . . has affected the coffee production; on one hand by reducing the available hand labor, and on the other hand by taking land. [97]

The 20th Century

The *Geography of Cuba*, written by Alfredo M. Aguayo and Carlos M. de la Torre y Huerta, was published in 1928. They mentioned cacao among the main crops in Cuba with 9,800,000 pounds (i.e., 4900 tons of cacao) produced in that year [98]. Primary cacao production was associated with Oriental Province, particularly the towns of Alto Songo, Baracoa, Caney, Guantánamo, El Cobre, Palma de Soriano, San Luis, and Yateras. Additional cultivation was noted for Bayamo, Jiguaní, Morón, and Santa Clara in Camagüey Province [99–103]. The *Geography of Cuba* also mentioned the existence of an unnamed chocolate factory in Havana [104].

A book dedicated to the diverse spheres of Cuban national life was published in 1940 entitled *Cuba en la mano (Cuba in the Hand)*. It included a business summary, and listed Cuban chocolate factories as well as chocolate importers and exporters.[66] Companies

and establishments in Havana included: La Ambrosía Industrial, S.A.; Cuba Industrial y Comercial, S.A.; La Española; Rubine e Hijos; Solo y Cía; Suárez Ramos y Cía; The Sondy Corporation, S.A.; and shops managed by Alberto Bobak, Juan S. Eliskin, and Gumersindo Rey Suárez [105]. Havana importers of cacao included: Candied Fruit Packaging Co., S.A.; Solo y Cía S. en C.; Rubine S.A.; La Ambrosía Industrial, S.A.; and The Sondy Corporation, S.A. Cacao exporters in Havana were identified as: Alvarez Parreño; La Flor de Tibes S.A.; López Serrano & Co., Ltd.; Llopart y Cía; J. M. Rodríguez y Cía; and Trueba Hno y Cía; among the exporters of chocolates, again they mention La Ambrosía Industrial, S.A.; Cuba Industrial y Comercial, S.A.; and Solo y Cía S. en C.; and they add Urbano Soler [106].

Elsewhere in Cuba, in Camagüey, G. Cases y Cía were listed as chocolate importers [107], as were a chocolate factory in Guantánamo, Bueso y Cía [108]; and another chocolate factory located in Santiago de Cuba, Colomé, S. en C. [109].

During the Second World War, cacao experienced a temporary resurgence but prices fell again shortly after the conclusion of the war. Toward, 1952, once again, there was a slight recuperation of cacao production [110]. This rise in production, according to Riccardi, was due to increased internal consumption within Cuba, where cacao supplied local chocolate, cookie, and ice cream production [111]. Riccardi also reported the Cuban industries in the mid-1950s had a total of 532 work centers (*centros de trabajo*) associated with the processing and/or manufacturing of cacao and coffee and their products, which employed 6550 workers [112].

Chocolate in Cuba After 1959

With the triumph of the Cuban Revolution in 1959, the agricultural diversification that had been debated during past centuries was given a notable impulse and cacao production was included in the new agriculture development plans. Institutions from the Revolutionary Government devoted themselves to the rescue of the cacao plantations in the eastern region and provided fertilizers and diverse products for the treatment of plant diseases. An important aspect of this governmental plan was the implementation of a training program for workers associated with cacao production, since cacao bean processing had resulted in low yields [113]. This program included publication of technical books and manuals to train personnel in the technical and social transformation of lands dedicated to agricultural production. Among those books there were manuals exclusively dedicated to the training of competent personnel and qualified technicians in cacao production.[67]

Among the private enterprise that became state property on October 13, 1960 were two chocolate factories [114]. In 1963, a new chocolate factory was inaugurated in Baracoa that would receive all the cacao produced in the Eastern region of Cuba,[68] after being processed in Jamal (Baracoa, Guantánamo) and Contramaestre (Santiago de Cuba). Today in the first decade of the 21st century the Baracoa chocolate factory produces a broad range of items, among them pure dark chocolate, milk chocolate, chocolate syrup, chocolate cream, ice cream and ice cream cakes, chocolate fondues, cacao paste, cacao powder, and cacao butter.[69] These items are utilized or produced at the Havana chocolate factory La Estrella (we believe that the factory was owned by Gerardo Abreu Fontán prior to the Revolution).

The majority of cacao farms in Cuba today are in the hands of small-scale farmers.[70] These cacao growers are associated in Credit and Services Cooperatives (*Cooperativas de Crédito y Servicios*) (CCS) or Agricultural Production Cooperatives (*Cooperativas de Producción Agropecuaria*) (CPA), and also in the Basic Units of Cooperative Production (Unidades Básicas de Producción Cooperativa) (UBPC) of a more recent creation; some farms that belong to the Working Youth Army (*Ejército Juvenil del Trabajo*) and some others from national organizations are included in the national sector.[71,72]

Conclusion

The revitalization of chocolate consumption as food—chocolate in a cup or solid chocolate in tablets as part of the common diet and not as a luxury candy—has been stressed by the Latin American and Caribbean School of Chocolate Confectionery and Pastry (*Escuela Latinoamericana y del Caribe de Chocolatería, Repostería y Pastelería*) and the Research Institute of the Food Industry (*Instituto de Investigación de la Industria Alimenticia*), on the occasion of the First Latin American Encounter of Cacao and Chocolate (*Primer Encuentro Latinoamericano de Cacao y Chocolate*),[73] celebrated February 27 to March 2, 2001, in Havana. These organizations date to the 1980s when they were formed to conduct research services to the cacao industry. From its beginning, the Confectioner Group (*Grupo de Confitería*) developed a pilot plant for chocolate research with the objective of helping the national chocolate industry. The existence of this pilot plant, coupled with the high-quality professionals associated with cacao research in Cuba, has led to the development and expansion of an international training center for cacao production and manufacturing. The training center already has developed a wide program of consultancy, seminars, and theoretical–practical courses covering a wide range of topics, from microbiology to the diverse technological process in the manufacture of chocolate, including a course in fine chocolate crafting.[74,75] A small amount of select bonbons are made at the school facilities. Because of the almost total disappearance of the bonbon industry in Cuba, the development of this school is a basic objective at a national level. The school's efforts are directed toward promoting the advancement of the regional chocolate industry and reclaiming chocolate's role as a food recognized for its highly concentrated nutritive value relative to its volume and weight.[76]

Acknowledgments

We would like to thank James I. Grieshop and Rami Ahmadi for their strong support and encouragement of our investigation of chocolate in Cuba. The Spanish language text of this chapter prepared by Niurka Nuñez González and Estrella González Noriega was translated into English by Beatriz Cabezón. Minor editorial revisions were made by Louis Evan Grivetti.

Endnotes

1. Hernández proposed in 1987 that the Spanish introduced cocoa to Cuba in 1540 from trees that originated in Mexico, which were sown for the first time at the location called *Mi Cuba* (My Cuba) located in Cabaiguán province in the central region of the country. This version has been favored by cocoa specialists from the Ministry of Agriculture (Ministerio de Agricultura). Others (unidentified) attribute the introduction of cocoa to Cuba to the French, who established the first cocoa plantations in the area of Ti Arriba in Oriente Province. We searched but could not find evidence to support the proposition that cacao was introduced to Cuba by the Spanish. It should be noted, though, that there are abundant references in the literature against the French hypothesis as well.

2. Bayamo is one of the largest cities in Oriente Province. It lies on a plain dissected by the Bayamo River. The settlement was founded by Diego Velázquez de Cuéllar in 1513. During the 16th century, Bayamo was one of the more important agricultural and commercial settlements on Cuba. Its inland location provided relative security against the pirates who then infested the West Indian seas. At the beginning of the 17th century, Bayamo became one of the leading towns of Cuba, partially due to a thriving contraband trade. As the result of a disastrous flood in 1616, a sandbank was formed at the river mouth, blocking access to the sea for ships that previously had sailed upriver to Bayamo from the coast, and the importance of the town declined [Ref. 8, p. 340].

3. Marrero [8] pointed out that the importance of cacao was because it was "the main beverage for breakfast." Use of coffee did not become common in Cuba until the beginning of the 19th century.

4. Santiago de Cuba, capital of Santiago de Cuba Province, is located in the southeastern part of the island. It lies on a bay connected to the Caribbean Sea and has remained an important seaport since the early 16th century. Santiago de Cuba was founded by Spanish conquistador Diego Velázquez de Cuéllar, June 28, 1514. This port became the starting point for the 1518 expeditions to the coast of Mexico led by Juan de Grijalba and Hernán Cortés. Hernando de Soto's expedition to Florida in 1538 also started from Santiago de Cuba. Historically, Santiago de Cuba has been the second most important city on the island after Havana. The settlement was attacked and plundered by the French in 1553 and by the British in 1662.

5. Taxes were initiated in 1688 on imports and exports to sustain the Spanish fleet anchored at the port of Havana. Although *fruttos desta Ysla* (fruits of this Island) are mentioned among the exports, cacao is not named. In the section entitled "Imported Goods," cacao is among the items identified, specifically as *merchandise from Caracas*. Appended to the notation is the statement: "from each fanega of cacao, that used to cost four, it will be charged [from now on] sixteen."

6. Cartagena, formally known as *Cartagena de Indias* (Cartagena of the Indies), is a large seaport on the northern coast of Colombia. Cartagena was founded in 1533 by Pedro de Heredia and was a major center of early Spanish settlement in the Americas. The settlement was protected from plundering by Dutch, English, and French pirates by a strong, walled military fortress. Despite these military and defensive precautions, the city was attacked many times. Cartagena and Veracruz (México) were the only cities authorized to trade in human slaves.

7. Cumaná was founded by Franciscan monks in 1515. The site is located east of Caracas, Venezuela, and was the first European settlement on the South American mainland. Due to repeated attacks by indigenous peoples, the site had to be rebuilt several times.

8. Portobelo, known formerly as Puerto Bello (Beautiful Port), is located in Colon Province, Panamá. Columbus mentioned the site during his fourth voyage in 1502. A settlement was established there in 1597 that became one of the most important ports for the silver trade from the 16th to the 18th centuries and one of the departure points for the Spanish fleet (*Flota de Indias*).

9. In Cuban Spanish, the word *guarapo* is used to describe a beverage made of fresh sugar cane juice. In Colombian Spanish, the word refers to a fermented beverage made from a solidified form of sugar cane juice called *panela* (sugar cane honey).

10. *Furrio* is the local word for chocolate in Baracoca, Cuba.

11. Sometimes basic transcriptions and translations from Old Spanish do not provide clear understanding of the original text. The four items mentioned were the ranch, the hand mill to grind *guarapo* (sugar cane) and to grind chocolate (*furrio*), cattle, and swine. The passage also could be interpreted as a mill to grind *guarapo*, a mill to grind chocolate, some cattle, and some pigs.

12. This date, also cited by Hernández, marks the beginning of the commercial exploitation of cocoa in Cuba.

13. A document dated to 1771 recorded the departure of ships from the port of Havana loaded with merchandise for Spain. Among the items identified were two large bags of cocoa from Guayaquil, and other 31 bags (?) and 2 *small barrels of cocoa paste,* presumably from Cuba. Toward the end of the 18th century, cocoa also was among the products purchased from Cuba by the United States.

14. In 1863, the natural grasslands in the eastern region of Cuba had a value of approximately 100 pesos per *caballería* (a measurement usually defined as 42 hectares or approximately 84 acres). In contrast, agricultural land for basic crops like coffee and cocoa, reached a value of at least 2500 pesos per *caballería* [19].

15. Alexander von Humboldt (1769–1859) was a Berlin-born Prussian naturalist and explorer. Humboldt's quantitative work on botanical geography established the foundation for the field of biogeography. He traveled within Central and South America between 1799 and 1804, where he explored and described the land and peoples.

16. The married name of the Countess of Merlín was Maria de las Mercedes Santa Cruz y Montalvo. She was born in colonial Havana in 1798 to an aristocratic Creole family. She left Cuba for Spain at an early age and later married the French Count Merlín. The invasion of Spain by French Napoleonic troops precipitated another move to France, where she began her literary career. After the death of her husband, she returned to Cuba in 1840. She is among Cuba's most engaging authors and while she published novels, memoirs, and travel writings, her works largely have been ignored. As Cuba's first female historian, she articulated a sense of national identity through her three-volume account of the political, social, and economic organization of the island [26].

17. Costa Firme (Firm Coast) is the name given during colonial times to the shores of the American continent (in contrast to the shores of the Caribbean islands).

18. The Reverend Abiel Abbot established the Peterborough Lyceum in 1829, the forerunner of the Monadnock Summer Lyceum that supported free speech and open debate. Despite his traditional Andover/Harvard education, Abiel Abbot refused to condemn unorthodox opinion and preached a message of tolerance. For this he was accused of heresy.

19. Camarioca is located on the northwest of Cuba between the cities of Matanza and Cárdenas.

20. Esteban Pichardo is known as the Father of Cuban geographers. Born in the Dominican Republic, he spent most of his life traveling and conducting field work through the different regions of Cuba. His detailed observations

of Cuban geography are revealed in an important map he published in 1850. He also wrote a general geography of Cuba.

21. This statement does not take into account the cacao imported to Cuba.

22. *Pesos fuertes* is translated literally as "strong pesos." These were old silver Spanish coins with a weight of one ounce, equivalent to eight *reales*.

23. Quintal is a variable unit of weight. In the United States, it was equivalent to 100 pounds; in Spain and in 19th century Cuba, a quintal had the value of 100 kilograms.

24. According to Friedlander there were only 18 cacao plantations on the island in 1862 compared to 69 plantations in 1846.

25. We could find no explanation for the 36 cacao plantations reported in Matanzas with no production data.

26. The region of modern Havana and its natural bay were first visited by Europeans in 1509 when Sebastián de Ocampo circumnavigated the island. Shortly thereafter, in 1510, the first Spanish colonists arrived from La Hispaniola, initiating the conquest of Cuba. Conquistador Diego Velázquez de Cuéllar founded Havana on August 25, 1515. Shortly after the founding of Cuba's first settlements, the island served as little more than a base for the *conquista* of other lands. Hernán Cortés organized his expedition to Mexico from there. Cuba, during the first years of the discovery, provided no immediate wealth to the conquistadores, as it was poor in gold, silver, and precious stones, and many of its settlers moved to the more promising lands of Mexico and South America that were being discovered and colonized at the time. To counteract pirate attacks on galleon convoys headed for Spain loaded with prized New World treasures, the Spanish crown decided to protect its ships by concentrating them in one large fleet that traversed the Atlantic Ocean as a group. Following the decree signed in 1561 by King Philippe II of Spain, all ships headed for Spain were required to join this fleet in the Havana Bay. Ships arrived from May to August, waiting for the best weather conditions, and after being loaded with supplies, departed Havana for Spain by September. These activities boosted commerce and development of the adjacent settlement of Havana.

27. Describing events of the mid-19th century, Pezuela wrote that cocoa was not cultivated in Guabasiabo or in San Andrés, a district of the Holguín jurisdiction, and was nearly nonexistent in the Nuevitas jurisdiction; while in Villa Clara "coffee and cocoa are not harvested."

28. Pezuela identified only limited cacao production in Jatibonico, located in the Sancti Spiritus jurisdiction, where the fundamental economic activities were cattle ranching. He also noted that in the San Juan de los Remedios jurisdiction, cacao was harvested only for local consumption.

29. Pezuela wrote of Porcallo, a cattle district of the Puerto Príncipe jurisdiction, that cacao cultivation was not known, and at Santa Cruz, within the same jurisdiction, the land was not suitable for cacao.

30. The French immigration during the 18th and 19th centuries—initially to Puerto Rico and from there to Cuba—resulted from economic and political situations that occurred in Louisiana, Saint Domingue (i.e., Haiti), and Europe. French settlements in the New World during the 17th century, collectively were known as New France, and Louisiana was an administrative district. Upon the outbreak of the French and Indian War, also known as the Seven Years' War (1754–1763), many French settlers fled to the Caribbean islands of Cuba, Hispaniola, and Puerto Rico. In 1697, the Spanish crown ceded the western half of the island of Hispaniola to the French. The Spanish part of the island was named Santo Domingo (now the Dominican Republic) and the French named their portion of the island Saint Domingue (later renamed Haiti). These Caribbean French settlers dedicated themselves to the cultivation of sugar cane and owned plantations, which required a huge amount of manpower. They imported slaves from Africa to work in the fields. However, soon the population of the slaves outnumbered the French. In 1791, the slaves were organized into an army led by the self-appointed general Toussaint Louverture and rebelled against the French. The ultimate victory of the slaves over their French masters came about after the Battle of Vertières in 1803. The French fled to Santo Domingo and ultimately made their way to Puerto Rico.

31. Silver *reales (reales de plata)*. An old Spanish silver coin equivalent in Spain to a quarter of a peseta. In Colonial America this coin had different values.

32. Soconusco originates from the Nahuatl word *xoconostle*, meaning fruit of the prickly pear cactus. Under the Mexica/Aztec culture, it was the furthest region of trade, providing jaguar pelts, cacao, and quetzal feathers for the ruling classes in the Mexica/Aztec capital of Tenochtitlan. Today, it is the region of the Mexican state of Chiapas, located in the extreme south of the state and bounded by Guatemala on the southeast and the shore of the Pacific Ocean on the southwest. It is a region of rich lowlands and foothills.

33. Maracaibo is today (2009) the second largest city in Venezuela. It was founded in 1529 on the western side of Lake Maracaibo. A combination of Maracaibo alluvial soils, high humidity, and high temperatures provided suitable conditions for cacao production.

34. In the sense used here *costeño* is a geo-sociological definition for the inhabitants of the regions near or on the Caribbean coast in the north of Colombia. Cartagena, Barranquilla, and Santa Marta are the more representative cities of this area. As a consequence, we can deduce that *cacao costeño* was the name given to the cacao growing in that area.

35. Trinidad (from the Spanish for Trinity) is the largest and most populous of the 23 islands that today form the country of Trinidad and Tobago, located off the northeastern coast of Venezuela. The history of Trinidad and Tobago begins with the settlements of the islands by Amerindians of South American origins. Both islands were encountered by Columbus on his third voyage in 1498. Tobago changed hands between the Courlanders (the Duchy of Courland was a fief of the United Kingdom of Poland and Lithuania), Dutch, English, and French but eventually became English. Cacao from Trinidad, known as *cacao trinitario* or Trinity cacao, was a hybrid of two cacao varieties: *criollo* and *forastero* types. Forastero cacaos originated in the Amazon Basin and are characterized by a thick, smooth, usually yellow fruit with flattened, purple seeds. Cacao trinitario is highly variable and considered high quality for chocolate production.

36. The settlement of Guayaquil was founded by the Spanish in 1538. Guayaquil is today (2009) the largest and most populous city in Ecuador and the nation's main port.

37. Marañón is one of 11 provinces of the Huánuco region in central Peru. Huánuco is characterized by mountainous topography comprising parts of the Sierra and the high jungle (mountain rim) regions.

38. Para is a state in northeastern Brazil (Estado do Para) with Belem at the mouth of the Amazon River as its capital. Belem was founded in 1616 by the Portuguese to defend the region against British, Dutch, and French attempts at colonization.

39. Walter Goodman was a British painter, illustrator, and author. He was born in London in 1838 and was accepted into the Royal Academy in 1857. In 1864, he accompanied the Spanish artist Joaquin Cuadras on a visit to the West Indies and lived there for five years, most of the time in Cuba. While there he painted and wrote newspaper articles for the New York press. He contributed to local Cuban newspapers as well and published a series of humorous sketches entitled *Un Viaje al Estranjero* (*A Trip Abroad*). While in Cuba, Goodman contributed articles and letters to the *New York Herald* under the nom de plume *el Caballero Inglés* (The English Gentleman). Civil unrest in Cuba forced him to flee to New York in 1870.

40. Carnival (Spanish *carnaval*) is a pre-Lenten festival commonly held in Roman Catholic countries that became popular in Spain from the middle of the 16th century and was introduced to Cuba by Hispanic colonists. What is today called the *Carnaval* of Santiago de Cuba is not a manifestation or reflection of the typical pre-Lenten carnival that would be celebrated in February or March, but a different tradition that evolved from summer festivals called *Mamarrachos* (buffoons). A typical 19th century *Mamarrachos* was the occasion for a number of customs and activities that involved the preparation and consumption of traditional foods, especially *ajiaco* (a hot stew of various meats), empanadas, omelets, fritters,

fruit of all types, and fried or roast pork with boiled plantains. Also associated with the *Mamarrachos* is the erection of *mesitas*—tables covered with canopies where traditional beverages and refreshments such as rum (*aguardiente*), beer, chocolate, coffee, and natural fruit juices were sold.

41. Samuel Hazard, 1834–1876, a Philadelphia-born American writer, arrived in Cuba after the American Civil War to begin work on a guidebook to the island. His arrival coincided with the decade when the first concerted struggle for independence was underway in Cuba. While Hazard's sympathies clearly were with the pro-independence patriots, his main objective was to produce a guidebook to the island aimed at visitors. His efforts are informative regarding local establishments and the idiosyncrasies of Cuban social life.

42. The Ten Years' War, 1868–1878, refers to the period of struggle for Cuban independence from Spain. Discontent was caused in Cuba by excessive taxation, trade restrictions, and virtual exclusion of native Cubans from governmental posts. In 1868, Carlos Manuel de Céspedes and other patriots raised the standard of revolt in 1868 and on April 20, 1869, the revolutionary republic of Cuba was established with Bayamo as the provisional capital. The war dragged on without decisive incident. The costly and bitter war—seemingly without result—actually foreshadowed the Cuban war of independence that broke out in 1895 and the subsequent Spanish–American War.

43. The Santa Isabel is today (2009) one of the grandest of Old Havana's hotels and represents the colonial-style architecture of Cuba. The building is an 18th century mansion constructed for the Countess of San Juan de Jaruco and later purchased by the Count of Santovenia. In 1867, artists, businessmen, celebrated travelers, and ship owners chose the magnificent mansion as their favorite hotel when staying in the capital city because of its comfort and elegance.

44. The Hotel Telégrafo, located in Old Havana, was the first building specifically designed as a hotel in Cuba. It was constructed during the 1860s. According to Samuel Hazard, it "is located on the Amistad street across Campo del Mar and near the cafés and the Tacón Theater. All languages can be heard there and it has the advantage to have bathrooms."

45. A beverage of chocolate made with milk, cold or hot, was most likely among them.

46. *Molinillo* is a Spanish word. The diminutive *molino* (mill) means "little mill" but when used in the context of making chocolate it is a wooden beater or a whisk used to froth the liquid.

47. Hazard's description of a *molinillo* is confusing. A *molinillo* is not actually the chocolate pot with the beater, but the beater itself.

48. Criollo cacao developed in Central America and northern portions of South America is characterized by thin-

walled, red or yellow fruits. The seeds are large, round, white or pale purple, not astringent, and produce the highest quality chocolate. Unfortunately, criollo types are low yielding and susceptible to many diseases.

49. The *New York Herald* was launched in 1835 by James Gordon Bennett, a Scottish immigrant. Bennett originally charged readers two cents for the paper, but within a year dropped the price to one cent—making the *Herald* one of the notorious *penny-papers* that became popular in American cities in the mid-1800s.

50. According to the Cuban historian Carlos Márques Sterling, the word *Mambí* is of Afro-Antillean origin and was applied to revolutionaries from Cuba and Santo Domingo (now Dominican Republic), who fought against Spain in the War of Independence (Ten Years' War). According to the fiction writer Elmore Leonard in his adventure novel *Cuba Libre*, the word Mambí comes from Eutimio Mambí, a leader who fought the Spaniards in Santo Domingo 50 years previously. The Spanish soldiers, noting the similar machete-wielding tactics of the Cuban revolutionaries, started referring to them as the *men of Mambí*, which later was shortened to *Mambís* or *Mambíses*. There also were women among the *men of Mambí*. Some of these women were part of the medical corps but others who achieved officer ranks formed part of regular army units. Recent estimates of the participation of Cubans of African descent in the Mambí Army are as high as 92 percent.

51. Pharmacist Jean-Antoine-Brutus Menier created a chocolate factory in France in 1825, where he coated medicines with chocolate. His sons ultimately transformed the Menier business into a candy/chocolate factory, which became more lucrative.

52. Spanish–American War (1898) is the name given to a conflict between Spain and the United States of America. Its roots had two origins: the struggle for independence (Cuban objectives) and economic ambitions (American objectives). The United States used the mysterious blowing up of the battleship *Maine* in Havana harbor as a pretext for declaring war. The Spanish navy suffered serious defeats in Cuba and the Philippines. A United States expeditionary force defeated Spanish ground forces in Cuba and in Puerto Rico. Spain surrendered at the end of 1898; Puerto Rico was ceded to the United States and Cuba was placed under U.S. protection. The Pacific island of Guam was also ceded to the United States. The Philippines, a former Spanish colony, was purchased by the United States for $20 million.

53. In 1883, the *Revista de Agricultura* (*Journal of Agriculture*) was a monthly publication but became a weekly journal after 1889.

54. The United States had blockaded Philippine ports during the Spanish–American War.

55. Cienfuegos is one of the chief seaports along the southern coast of Cuba. This geographical area was called Jagua when the Spanish arrived and was settled by indigenous people. In 1819, the location was settled by French immigrants from Bordeaux and Louisiana. The name *Cienfuegos,* literally meaning *one hundred fires,* was applied to honor a Spanish governor of the times.

56. Trinidad, named after the Caribbean country of Trinidad and Tobago, is a town in the province of Sancti Spíritus located in central Cuba. Together with the nearby Valle de los Ingenios, it was identified in 1988 as a UNESCO World Heritage site. Trinidad was founded in 1514 by Diego Velázquez de Cuéllar and its original name was Villa De la Santísima Trinidad. It is one of the best-preserved cities in the Caribbean from the time when the sugar trade was the main industry in the region. In 1790, there were 56 sugar mills operating here and the city had a population of over 28,000. Of these, 12,000 were slaves who worked the sugar cane fields. In the early 19th century there was a slave revolt in Haiti that caused French planters to flee to Trinidad. Once emancipated, many former slaves immigrated to Cuba and contributed to the formation of Cuban nationality, as they brought along their African customs, traditions, songs, and dances, as well as their African religions.

57. Baracoa is located along the northern coast of Guantánamo Province, over 620 miles east of Havana. The original habitants of Cuba, the Aruacos or Arawacos (Arawaks), called this area Baracoa, which meant the *presence of the sea.* Baracoa was the first Spanish settlement in Cuba, founded in 1512. It is a land of great rainfall and many rivers. The region is characterized by lush vegetation and has been called the "Cuban Capital of Cacao."

58. From the same time, and with an identical style, is the book *El Tesoro del Agricultor Cubano* (*The Treasure of the Cuban Farmer*), a three-volume work written by Francisco Balmaseda (1890–1896).

59. Due to the geographical isolation of Baracoa and lack of reliable roads, the chocolate factory then named Vilaplanas y Co. was transferred to Havana (J. Martínez, personal communication).

60. *Altares de la cruz*: these altars were constructed in stone by pre-Columbian Amerindians and used for religious offerings, rituals, and sacrifices. After arrival of the Spanish and after Christianity was adopted by the native population, a cross was added. Later on, they were built for special celebrations and became the epicenter of social gatherings.

61. Grover Flint was an American lieutenant and an active war correspondent who fought alongside the Cubans, during the Second Cuban War of Independence (1895–1898). During the four months he fought alongside the Cuban army, he kept a field notebook with his observations. Later on, he published these observations in a book entitled *Marching with Gómez: A War Correspondent's Field Notebook, Kept During Four Months with the Cuban Army.*

62. Matanzas is the capital of the Cuban province of Matanzas. It is famed for its Afro-Cuban folklore. The town lies on the northern shore on the Bay of Matanzas

(Bahía de Matanzas), 90 km east of the capital Havana. The name Matanzas means "slaughterings" and refers to a putative slaughter at the port of the same name in which 30 Spanish soldiers were crossing one of the rivers to attack an aboriginal camp on the far shore. The Spanish soldiers, however, did not have boats and enlisted the help of native fishermen. Once the soldiers reached the middle of the river, the fishermen flipped the boats, and due to the Spanish soldiers' heavy metal armor, most of them drowned.

63. Camagüey today is a city in central Cuba. It was founded under the name Santa María del Puerto del Príncipe ca. 1515. It is the capital of the Camagüey Province and played an important role during the Ten Years' War against Spain in 1868–1878.

64. *Malanga* or *yautia*, also known as *tannia* or *tannier* and in English cocoyam, is a species of the genus *Xanthosoma*. It is a type of tuberous vegetable of more than 40 species similar to the related taro or dasheen (*Colocasia esculenta*).

65. The cultivation of cocoa as well as the cultivation of coffee had been considered characteristic of mountainous areas.

66. We were not able to confirm the chocolate-related industries mentioned in the text. We only found a brief mention of *La Ambrosía* and *Chocolate La Estrella*, listed among the companies that "contributed on the Hospital Day to the Costales and San Martin Wards of the Calixto García Hospital." Our personal interviews with J. Martínez also revealed the names of other chocolate factories in Havana, among them *Armada*, *Cacusa*, and *Siré*.

67. According to Cacao and Coffee National Direction (*Dirección Nacional de Cacao y Café*): 1972, 1981, 1987; Tito: 1978; Hernández: 1978 and 1987.

68. According to *Granma* (January 1, 2000, and November 29, 2001) some cacao is exported today as cacao beans. It is widely hoped, however, that in the future, all cacao produced in Cuba could be utilized for the national industry and the final manufactured products then would be exported.

69. Although cacao butter is utilized by the Cuban pharmaceutical industry, considerable quantities are exported.

70. Although cocoa is included among the most important perennial crops for Cuban agriculture, cacao occupies the smallest area dedicated to agricultural production.

71. As for the industrial production, the general manufacturing industries that utilize cocoa (i.e., ice cream, candy, and the confection industries) are registered in these annals but data are not provided about how cocoa is used.

72. According to *Granma* (May 17, 2001, and August 22, 2001), during recent years cocoa is being produced on state lands given to peasants.

73. The idea to celebrate these Encounters emerged with the need to propitiate the relationships between the cocoa growers and the chocolate makers. The first activity was the creation of the Latin American and Caribbean Association of Cacao Growers and Chocolate Makers. The bylaws of this association were presented at the second meeting held in Havana in April 2002.

74. Financial constraints were also a consideration for the creation of this Cuban Chocolate School given that the costs for training Cuban food industry professionals and technicians in schools already extant in Germany and the United States were beyond financial means.

75. The creation of the School of Chocolate became a reality thanks to the Catalonian Confectioner Master Quim Capdevila. After retirement, he volunteered his professional experience as well as a portion of his personal financial resources to assist at the school. The school also has been supported by the Catalonian Council, the University of Vic (near Barcelona), and the ChocoVic School (*Escuela Chocovic*) managed by the largest fine chocolate manufacturer in Spain (founded in 1872).

76. We would like to thank Dr. María Cristina Jorge Cabrera, founder of the school, for all the information and documentation she provided during our interview.

References

1. Liogier, H. A. *Las plantas introducidas en las Antillas después del Descubrimiento y su impacto en la ecología*. Havana, Cuba: Jardín Botánico Nacional and Universidad Pinar del Río, 1992; p. 78.

2. Nosti Nava, J. *Cacao y café*. Havana, Cuba: Edición Revolucionaria, 1970; p. 5.

3. Hernández, J. *Fitotecnia del cacao*. Havana, Cuba: Editorial Pueblo y Educación, 1987; pp. 7–9.

4. Roig de Leuchsenring, E. *Curso de Introducción a la Historia de Cuba*. Havana, Cuba: Municipality of Havana, 1937; Volume 1 (no pagination).

5. Pezuela J. de la. *Diccionario geográfico, estadístico, histórico de la Isla de Cuba*, 4 Volumes. Madrid, Spain: Imprenta del Establecimiento de Mellado, 1863; Volume 1, p. 31.

6. Friedlaender, H. E. *Historia económica de Cuba*. Havana, Cuba: Editorial de Ciencias Sociales, 1978; Volume 1, p.12.

7. Le Riverend, J. *Historia económica de Cuba*. Havana, Cuba: Instituto del Libro, 1967; pp. 88–90.

8. Marrero, L. *Historia económica de Cuba*. Havana, Cuba: Universidad de la Habana, 1956; Volume 1, p. 268.

9. Marrero, L. *Historia económica de Cuba*. Havana, Cuba: Universidad de la Habana, 1956; Volume 1, p. 268.

10. Le Riverend, J. *Historia económica de Cuba*. Havana, Cuba: Editorial Ciencias Sociales, 1985; p.111.

11. Morell de Santa Cruz, P. A. *La visita eclesiástica*. Havana, Cuba: Editorial Ciencias Sociales, 1985; p.120.

12. Ribera N. J. de. *Descripción de la Isla de Cuba*. Havana, Cuba: Editorial Ciencias Sociales, 1975; p. 96.

13. Ribera N. J. de. *Descripción de la Isla de Cuba*. Havana, Cuba: Editorial Ciencias Sociales, 1975; pp. 109–130.

14. Ribera N. J. de. *Descripción de la Isla de Cuba*. Havana, Cuba: Editorial Ciencias Sociales, 1975; p. 108.

15. Eguren, G. *La Fidelísima Habana*. Havana, Cuba: Editorial Letras Cubanas, 1986; p. 152.

16. Le Riverend, J. *Historia económica de Cuba*. Havana, Cuba: Editorial Ciencias Sociales, 1985; pp. 104–107.

17. Hernández, J. *Fitotecnia del cacao*. Havana, Cuba: Editorial Pueblo y Educación, 1987; pp. 12–13.

18. Acuña, J., and Díaz Barreto, R. *Baracoa (Oriente)*. Havana, Cuba: Estudios Económicos-Sociales, Publicaciones del Banco de Fomento Agrícola e Industrial de Cuba, 1952; p. 25.

19. Le Riverend, J. *Historia económica de Cuba*. Havana, Cuba: Editorial Ciencias Sociales, 1985; pp. 172, 315.

20. Friedlaender, H. E. *Historia económica de Cuba*. Havana, Cuba: Editorial de Ciencias Sociales, 1978; Volume 1, p. 385.

21. Eguren, G. *La Fidelísima Habana*. Havana, Cuba: Editorial Letras Cubanas, 1986; p. 194.

22. Eguren, G. *La Fidelísima Habana*. Havana, Cuba: Editorial Letras Cubanas, 1986; p. 123.

23. Eguren, G. *La Fidelísima Habana*. Havana, Cuba: Editorial Letras Cubanas, 1986; pp. 247–248.

24. Eguren, G. *La Fidelísima Habana*. Havana, Cuba: Editorial Letras Cubanas, 1986; pp. 322, 335–336.

25. Humboldt, A. de. *Ensayo Político sobre la Isla de Cuba*. Havana, Cuba: Fundación Fernando Ortiz, 1998; p. 120.

26. Merlín, Condesa de. *Viaje a La Habana*. Havana, Cuba: Editorial Arte y Literatura, 1974; pp. 89, 153, and 172.

27. Wurdemann, J. G. *Notas sobre Cuba*. Havana, Cuba: Editorial Ciencias Sociales, 1989; p. 130.

28. Eguren, G. *La Fidelísima Habana*. Havana, Cuba: Editorial Letras Cubanas, 1986; p. 272.

29. Eguren, G. *La Fidelísima Habana*. Havana, Cuba: Editorial Letras Cubanas, 1986; p. 332.

30. Friedlaender, H. E. *Historia económica de Cuba*. Havana, Cuba: Editorial de Ciencias Sociales, 1978; Volume 1, pp. 217–218, 326.

31. Humboldt, A de. *Cuadro Estadístico de la Isla de Cuba. 1825–1829*. Havana, Cuba: Bayo Libros, 1965; p. 77.

32. Friedlaender, H. E. *Historia económica de Cuba*. Havana, Cuba: Editorial de Ciencias Sociales, 1978; Volume 1, pp. 383–385.

33. Friedlaender, H. E. *Historia económica de Cuba*. Havana, Cuba: Jesús Montero Editor, 1944; p. 175.

34. Le Riverend, J. *Historia económica de Cuba*. Havana, Cuba: Editorial Ciencias Sociales, 1985; p. 310.

35. Guerra, R. *Azúcar y población en las Antillas*. Havana, Cuba: Imprenta Nacional de Cuba, 1961; p. 66.

36. Le Riverend, J. *Historia económica de Cuba*. Havana, Cuba: Editorial Ciencias Sociales, 1985; pp. 171–172.

37. Pezuela, J. de la. *Diccionario geográfico, estadístico, histórico de la Isla de Cuba*. Madrid, Spain: Imprenta del Establecimiento de Mellado, 1863; Volume 1, pp. 60–61.

38. Abbot, A. *Cartas*. Havana, Cuba: Editora del Consejo Nacional de Cultura, 1965; p. 82.

39. Abbot, A. *Cartas*. Havana, Cuba: Editora del Consejo Nacional de Cultura, 1965; p. 203.

40. Abbot, A. *Cartas*. Havana, Cuba: Editora del Consejo Nacional de Cultura, 1965; pp. 278–279.

41. Abbot, A. *Cartas*. Havana, Cuba: Editora del Consejo Nacional de Cultura, 1965; pp. 355–356.

42. Pichardo, E. *Diccionario provincial casi razonado de vozes y frases cubanas*. Havana, Cuba: Editorial Ciencias Sociales, 1976; p. 116.

43. Wurdemann, J. G. *Notas sobre Cuba*. Havana, Cuba: Editorial Ciencias Sociales, 1989; p. 130.

44. Friedlaender, H. E. *Historia económica de Cuba*. Havana, Cuba: Jesús Montero Editor, 1944; p. 544.

45. Pezuela, J. de la. *Diccionario geográfico, estadístico, histórico de la Isla de Cuba*. Madrid, Spain: Imprenta del Establecimiento de Mellado, 1863; Volume 1, pp. 60–61.

46. Pezuela, J. de la. *Diccionario geográfico, estadístico, histórico de la Isla de Cuba*. Madrid, Spain: Imprenta del Establecimiento de Mellado, 1863; Volume 1, pp. 38–39.

47. Pezuela, J. de la. *Diccionario geográfico, estadístico, histórico de la Isla de Cuba*. Madrid, Spain: Imprenta del Establecimiento de Mellado, 1863; Volume 3, p. 390.

48. Pezuela, J. de la. *Diccionario geográfico, estadístico, histórico de la Isla de Cuba*. Madrid, Spain: Imprenta del Establecimiento de Mellado, 1863; Volume 4, pp. 461–462.

49. Pezuela, J. de la. *Diccionario geográfico, estadístico, histórico de la Isla de Cuba*. Madrid, Spain: Imprenta del Establecimiento de Mellado, 1863; Volume 4, pp. 677–678.

50. Pezuela, J. de la. *Diccionario geográfico, estadístico, histórico de la Isla de Cuba*. Madrid, Spain: Imprenta del Establecimiento de Mellado, 1863; Volume 3, pp. 403, 678.

51. Pezuela, J. de la. *Diccionario geográfico, estadístico, histórico de la Isla de Cuba*. Madrid, Spain: Imprenta del Establecimiento de Mellado, 1863; Volume 3, p. 376.

52. Pezuela, J. de la. *Diccionario geográfico, estadístico, histórico de la Isla de Cuba*. Madrid, Spain: Imprenta del Establecimiento de Mellado, 1863; Volume 2, pp. 65, 77–78, 97, and 123–124.

53. Goodman, W. *Un artista en Cuba*. Havana, Cuba: Editorial Arte y Literatura, 1986; p. 24.

54. Goodman, W. *Un artista en Cuba*. Havana, Cuba: Editorial Arte y Literatura, 1986; p. 40.

55. Goodman, W. *Un artista en Cuba*. Havana, Cuba: Editorial Arte y Literatura, 1986; p. 142.

56. Goodman, W. *Un artista en Cuba*. Havana, Cuba: Editorial Arte y Literatura, 1986; p. 197.

57. Hazard, S. W. *Cuba a pluma y lápiz*, 3 Volumes. Havana, Cuba: Editorial Cultural S.A., 1928; Volume 1, p. 27.

58. Hazard, S. W. *Cuba a pluma y lápiz*. Havana, Cuba: Editorial Cultural S.A., 1928; Volume 1, p. 30; Volume 2, pp. 28–29.

59. Hazard, S. W. *Cuba a pluma y lápiz*. Havana, Cuba: Editorial Cultural S.A., 1928; Volume 1, p. 62.

60. Hazard, S. W. *Cuba a pluma y lápiz*. Havana, Cuba: Editorial Cultural S.A., 1928; Volume 2, pp. 24–25.

61. Hazard, S. W. *Cuba a pluma y lápiz*. Havana, Cuba: Editorial Cultural S.A., 1928; Volume 3, p. 52.

62. Hazard, S. W. *Cuba a pluma y lápiz*. Havana, Cuba: Editorial Cultural S.A., 1928; Volume 2, pp. 85–87.

63. Hazard, S. W. *Cuba a pluma y lápiz*. Havana, Cuba: Editorial Cultural S.A., 1928; Volume 3, p. 87.

64. Hazard, S. W. *Cuba a pluma y lápiz*. Havana, Cuba: Editorial Cultural S.A., 1928; Volume 3, pp. 87–88.

65. Hazard, S.W. *Cuba a pluma y lápiz*. Havana, Cuba: Editorial Cultural S.A., 1928; Volume 3, pp. 187–188.

66. O'Kelly, J. *La tierra del Mambí*. Havana, Cuba: Ediciones Huracán, 1968; p. 318.

67. O'Kelly, J. *La tierra del Mambí*. Havana, Cuba: Ediciones Huracán, 1968; p. 320.

68. O'Kelly, J. *La tierra del Mambí*. Havana, Cuba: Ediciones Huracán, 1968; p. 376.

69. O'Kelly, J. *La tierra del Mambí*. Havana, Cuba: Ediciones Huracán, 1968; p. 387.

70. *Revista de Agricultura* Havana, Cuba, 1883. Volume 4, pp.170–171.

71. *Revista de Agricultura* Havana, Cuba, 1883. Volume 4, pp. 173–174.

72. Friedlaender, H. E. *Historia económica de Cuba*. Havana, Cuba: Jesús Montero Editor, 1944; p. 446.

73. *Revista de Agricultura* Havana, Cuba, 1889. No volume number, p. 33.

74. *Revista de Agricultura* Havana, Cuba, 1889. No volume number, pp. 96 and 583.

75. *Revista de Agricultura* Havana, Cuba, 1889. No volume number, p. 129.

76. *Revista de Agricultura* Havana, Cuba, 1889. No volume number, p. 563.

77. *Revista de Agricultura* Havana, Cuba, 1889. No volume number, p. 130.

78. *Revista de Agricultura* Havana, Cuba, 1891. No volume number, pp. 416–418.

79. *Revista de Agricultura* Havana, Cuba, 1891. No volume number, p. 504.

80. Martínez, F. *Ayer en Santa Clara*. Havana, Cuba: Universidad Central de Las Villas, 1959; p.16.

81. Acuña, J., and Díaz Barreto, R. *Baracoa (Oriente)*. Havana, Cuba: Publicaciones del Banco de Fomento Agrícola e Industrial de Cuba, Estudios Económico-Sociales, 1952; p. 119.

82. Eguren, G. *La Fidelísima Habana*. Havana, Cuba: Editorial Letras Cubanas, 1986; p. 390.

83. Guerra, R. *Mudos Testigos*. Havana, Cuba: Editorial Ciencias Sociales, 1974; p. 147.

84. Flint, G. *Marchando con Gómez*. Havana, Cuba: Editorial Ciencias Sociales, 1983; p. 62.

85. Flint, G. *Marchando con Gómez*. Havana, Cuba: Editorial Ciencias Sociales, 1983; p. 183.

86. Flint, G. *Marchando con Gómez*. Havana, Cuba: Editorial Ciencias Sociales, 1983; p. 194.

87. Flint, G. *Marchando con Gómez*. Havana, Cuba: Editorial Ciencias Sociales, 1983; p. 222.

88. Dumont, H. D. *Report On Cuba*. New York: The Merchants Association of New York, 1903; p. 29.

89. Riccardi, A. Visión económica de Cuba. In: Jenks, Leland H., editor. *Nuestra Colonia de Cuba*. Havana, Cuba: Edición Revolucionaria, 1966; p. 288.

90. Hernández, J. *Fitotecnia del cacao*. Havana, Cuba: Editorial Pueblo y Educación, 1987; pp. 14–15.

91. Riccardi, A. Visión económica de Cuba. In: Jenks, Leland H., editor. *Nuestra Colonia de Cuba*. Havana, Cuba: Edición Revolucionaria, 1966; p. 288.

92. Dumont, H. D. *Report On Cuba*. New York: The Merchants Association of New York, 1903; p. 28.

93. Riccardi, A. Visión económica de Cuba. In: Jenks, Leland H., editor. *Nuestra Colonia de Cuba*. Havana, Cuba: Edición Revolucionaria, 1966; p. 211.

94. Le Riverend, J. *Historia económica de Cuba*. Havana, Cuba: Editorial Ciencias Sociales, 1985; pp. 594–595.

95. *Censo de la República de Cuba*. Havana, Cuba: República de Cuba, 1919; pp. 56–57

96. *Censo de la República de Cuba*. Havana, Cuba: República de Cuba, 1919; p. 230

97. *Censo de la República de Cuba*. Havana, Cuba: República de Cuba, 1919; p. 938.

98. Aguayo, A. M., and de la Torre y Huerta, C. Ma. *Geografía de Cuba*. Havana, Cuba: Editorial Cultural S.A., 1952; pp. 117–121.

99. Aguayo, A. M., and de la Torre y Huerta, C. Ma. *Geografía de Cuba*. Havana, Cuba: Editorial Cultural S.A., 1952; pp. 121–243.

100. Aguayo, A. M., and de la Torre y Huerta, C. Ma. *Geografía de Cuba*. Havana, Cuba: Editorial Cultural S.A., 1952; pp. 254–255.

101. Aguayo, A. M., and de la Torre y Huerta, C. Ma. *Geografía de Cuba*. Havana, Cuba: Editorial Cultural S.A., 1952; p. 108.

102. Aguayo, A. M., and de la Torre y Huerta, C. Ma. *Geografía de Cuba*. Havana, Cuba: Editorial Cultural S.A., 1952; p. 235.

103. Aguayo, A. M., and de la Torre y Huerta, C. Ma. *Geografía de Cuba*. Havana, Cuba: Editorial Cultural S.A., 1952; p. 165.

104. Aguayo, A. M., and de la Torre y Huerta, C. Ma. *Geografía de Cuba*. Havana, Cuba: Editorial Cultural S.A., 1952; p. 263.

105. García y Cia. *Cuba en la mano*. Havana, Cuba: Imprenta Ucar, 1940; pp. 1242 and 1271.

106. García y Cia. *Cuba en la mano*. Havana, Cuba: Imprenta Ucar, 1940; pp. 1251 and 1272.

107. García y Cia. *Cuba en la mano*. Havana, Cuba: Imprenta Ucar, 1940; p. 1289.

108. García y Cia. *Cuba en la mano*. Havana, Cuba: Imprenta Ucar, 1940; p. 1294.

109. García y Cia. *Cuba en la mano*. Havana, Cuba: Imprenta Ucar, 1940; p. 1298.

110. Acuña, J. and Díaz Barreto. *Estudios Económico-Sociales*. Baracoa (Oriente), Cuba: Publicaciones del Banco de Fomento Agrícola e Industrial de Cuba, 1952; pp. 84–85, 121.

111. Riccardi, A. Visión económica de Cuba. In Jenks, Leland H., editor. *Nuestra colonia de Cuba*. Havana, Cuba: Edición Revolucionaria, 1996; p. 288.

112. Riccardi, A. Visión económica de Cuba. In Jenks, Leland H., editor. *Nuestra colonia de Cuba*. Havana, Cuba: Edición Revolucionaria, 1996; p. 293.

113. Acuña, J. and Díaz Barreto. *Estudios Económico-Sociales*. Baracoa (Oriente), Cuba: Publicaciones del Banco de Fomento Agrícola e Industrial de Cuba, 1952; p. 86.

114. Ediciones políticas. *Amèrica Latina: economía e intervención*. Havana, Cuba: Instituto del Libro, 1968; Volume III, p. 162.

39

History of Cacao and Chocolate in Cuban Literature, Games, Music, and Culinary Arts

Estrella González Noriega and Niurka Núñez González

Introduction

Cuba became the epicenter of 16th century Spanish activities in the Caribbean and New Spain. The island was a strategic location for three reasons: it was a center for Spanish fleet operations, a reunion point for Spanish galleons returning to Spain loaded with treasures from the New World, and a debarkation point for conquest expeditions of the American continent. Providing food and supplies for the Spanish fleet and the conquistadores required considerable hand labor. By the beginning of the 18th century, however, most of the indigenous peoples of Cuba, the Taínos[1] and the Siboneys[2], were decimated due to diseases introduced from the Old World by Spanish colonists and because of inhumane conditions of forced labor. The disappearance of local hand labor prompted the beginning of the slave trade—first Amerindians from Venezuela, and then Africans from West Africa—to supply the hand labor required. The arrival of different ethnic African groups to Cuban shores marked the birth of the Afro-Cuban traditions and their influence on different aspects of Cuban life, especially literature, music, and culinary practices. As we will see in this chapter, literature, music, and food are powerful reflections of Cuban history and each provides images and feelings of past times that otherwise would have remained veiled.

Early 19th century Cuban writers devoted their works in one way or another to the proclamation of the idea of freedom from the grip of Spain. Black slaves played crucial roles in the struggles for independence. During the 20th century the growing importance of the African inheritance, already an integral part of Cuba, clearly was recognized when Nicolás Guillén,[3] one of the glories of Cuban poetry, founded the Afro-Cuban School of Literature. But it is perhaps most with the music of Cuba where the rhythms and percussions of Africa can be found. Good food and music also are tightly woven in Cuba and the lyrics of popular songs are impregnated with the fragrances and tastes of local foods. As with Cuban music, island culinary traditions are an amalgam of predominantly Spanish and African cuisines enriched with local indigenous foods. Cacao

Chocolate: History, Culture, and Heritage. Edited by Grivetti and Shapiro
Copyright © 2009 John Wiley & Sons, Inc.

and chocolate, introduced to the island in earlier centuries, found fertile ground and became important components of Cuban traditions.

Cocoa and Chocolate in Cuban Literature and Proverbs

Cuban literature has enriched the world with the work of many prominent writers and poets. It was not until the 19th century that Cuban authors gained worldwide recognition with their ideals of freedom in their struggle for independence from Spain. Among these literary revolutionaries we can cite José María Heredia (1803–1839); Ramón de Palma y Romay (1812–1860); Gertrudis Gómez Avellaneda (1814–1873), novelist and playwright with her novel *Dos mugeres* (*Two Women*) and her play *Baltasar*, both literary icons against slavery; José Martí (1853–1895), a leader of the Cuban independence movement and renowned poet and writer; Cirilo Villaverde (1812–1894), journalist and novelist credited with writing the first book published in Cuba; Dolores María Ximeno y Cruz (1866–1934); and Miguel de Carrión (1875–1929).

The 20th century brought the works of the poet and journalist Nicolás Guillén (1902–1989), one of Cuba's foremost poets who used his efforts as a form of social protest. He was also responsible for the rescue of the cultural heritage of the descendants of African slaves and for valuing Afro-Cuban traditions. Others included Dulce María Loynaz (1902–1997), one of the greatest Cuban poets, member of the Royal Spanish Academy of Language and recipient of the Cervantes Prize in Literature; Alejo Carpentier (1904–1980), novelist, essay writer, and musicologist, who greatly influenced Latin American literature; Jose Lezama Lima, Virgilio Piñera, Dora Alonso, Onelio Jorge Cardoso, Concepción T. Alzola, Samuel Feijóo, and Antonio Núñez Jiménez. These writers and poets each produced works where chocolate or cacao played important roles. These stories, either in the form of essays, novels, or poems, lie at the heart of this chapter.

CHOCOLATE IN CUBAN LITERATURE

We begin with Cirilo Villaverde,[4] one of the pioneers of the Cuban novel and considered by many the Cuban novelist par excellence. The historic details of his novels, as well as the strong regional flavor of his vivid descriptions of the social customs of the times, have made his work an important source for research. In his most acclaimed novel, *Cecilia Valdés*, considered the most important work of 19th century Cuban literature, he exposed the corruption of Spanish-ruled Cuba and also revealed the traditions of mistresses during the

Spanish colonial period. The 1882 expanded version of his novel constituted an attack on the evils of slavery on sugar plantations. In the first volume he described a dinner at the elegant Havana Philharmonic Society in 1830, attended by the wealthy:

The dinner began around midnight [between 12:00 and 1:00 a.m.] and included cold turkey, Westfalia ham, cheese, excellent lamb, shredded beef [ropa-vieja[5]], dry sweet biscotti, preserves, sweet dessert wines from Spain and from another countries, a succulent chocolate, coffee and fruits from all the countries that were trading with Cuba. [1]

In the second volume he narrated:

The young ladies, the aunt and Mr. Ilincheta that obligingly had sat together on one side of the table, participated (in the dinner) only at the time that chocolate or the coffee with milk [café con leche[6]] was served. [2]

As the novel progresses, Villaverde describes a quarrel between a man and a woman: Nemesia, one of the female characters of the novel, abruptly shouts the well-known proverb to Leonardo (the main character):

Quien tiene la sangre como agua para chocolate no puede burlarse.[7] [Those who had the blood [as hot as] as water [used] to make chocolate cannot joke about it.] [3]

Villaverde also wrote a series of short novels that were compiled as an anthology with the title *La joven de la flecha de oro y otros relatos* (*The Young Woman with the Golden Arrow and Other Stories*). In one of the contributions, entitled *El ave muerta* (*The Dead Birth*), he uses chocolate as a color to emphasize the description of clothing:

His grey head [i.e., his grey hair] was covered by a black cap made of silk; his untidy pigtail, his chocolate [color] waistcoat, with beautiful embroidered flowers, [now] wet, with mud stains, gave him a very sad appearance. [4]

But it is in his work *El penitente* (*The Penitent*), written in 1844 and included in the previously mentioned compilation, that we find a scene where the preparation and consumption of chocolate are described in great detail. The action takes place during the month of November 1780. It is evening time and guests are arriving at the house of a traditional family of Havana. The situation is tense: Two ladies (one very old) and a gentleman are playing cards on a mahogany table. The salon is spacious, and the owners of the house have domestic slaves. The voice of the authoritarian old lady rises as she orders the slaves to bring food to her guests. Silently both slaves began to bring food. What are they going to eat? The answer is simple, the habitual cup of chocolate with biscuits traditionally offered to guests or travelers immediately upon their arrival:

Tomorrow, Recio continued saying, we will fix that. For the time being, it would be convenient that they serve us a cup of chocolate, because the joy to see and hug our friends, the bay breeze and the movement had awakened, in me at last, a very

strong appetite. ... Soon a female and a male servant appeared in the salon; he was carrying a tray made of willow twigs filled with small China cups painted with very vivid colors, and she with an enormous brass saucepan and a small mill made of wood. While the female slave was whipping and dissolving skillfully the chocolate, rubbing for this purpose the small mill between both hands, the male slave was collecting [into the China cups] that frothy and fragrant beverage, and serving it to the people gathered around the table. Giaraco received it from the hands of doña Margarita, who immediately brought it to Rosalinda, that impatient and agitated was waiting for him in her bedroom. Eguiluz, Recio y doña Elvira, his wife, accompanied the [chocolate] beverage with some very delicate biscuits from Spain that have been served in a plate in the middle of the tray. [5]

Villaverde, with his love for details, continues his narrative ... a young lady has been murdered: chocolate was present at the crime scene—perhaps the perfect vehicle to administer poison:

Taking in account the time she began to feel sick and the horrible aspect that her face, whose habitual expression was that one of the candor of an angel, showed just a few hours after her death would make you think that it was caused by poison. How did they give it to her? Mixed with what? With the chocolate that was her dinner? If everybody were accustomed to drink the same [chocolate] beverage, why is it that nobody else in the family died from the same cause? [6]

Villaverde also excelled in the description of places and people. A good example of his craft can be found in his *Excursión a Vuelta Abajo*[8] (*Excursion to Vuelta Abajo*). The first part of this book was published in 1838 and the second portion in 1842. When he described the preparations required to complete the trek to the Guajaibón,[9] he does not forget to include chocolate as an important hikers' food:

We will bring biscuits, crackers, chocolate, cheese and sausage. [7]

And when the hikers stop at the town of Los Palacios, where they were served chocolate that did not fulfill their expectations, the author does not hesitate to let the reader know of their disappointment:

[In the town of Los Palacios] we had a cup of chocolate too watery and bad. [8]

From the mid-19th to early 20th century the works of Álvaro de la Iglesia (1859–1940), especially his *Relatos y retratos históricos* (*Historic Narratives and Portraits*) and *Cuadros Viejos* (*Old Pictures*), both published during the early 20th century, deserve special attention. In 1969, these two books were compiled into a single volume and published with the title *Tradiciones Cubanas* (*Cuban Traditions*), which contains the essay *Viaje alrededor de un mango* (*Travel Around a Mango*). Iglesia wrote:

[It was not] ... until more or less than one and a half century ago, towards the end of the 18th century that people

began to drink coffee; before people used to drink chocolate, and the cacao from which chocolate is made, although still growing wild in Santo Domingo and in Jamaica, was unknown by ours Siboneys. The Spanish brought it. [9]

One of these historic accounts by Iglesias, *Secreta venganza* (*Secret Vengeance*), offers a glimpse not only into Cuban traditions but into the persistent use of chocolate as a vehicle for poisoning during the early 19th century—this time in Madrid, Spain. The scene is Madrid in 1813, where Iglesias describes the murder of Don Salvador de Muro y Salazar, Marquis of Someruelos,[10] after poison had been dissolved in his chocolate [10]. The murder was accomplished because in both Cuban and Spanish society the offer of an aromatic cup of chocolate would be very difficult to decline.

In Iglesias's *La expulsión de los jesuitas* (*The Expulsion of the Jesuits*),[11] he writes:

It will only be allowed that each priest keep his prayer book, his tobacco and his chocolate. [11]

From the early 20th century, we selected several chocolate-related passages from Miguel de Carrión (1875–1929), the well-respected journalist, physician, and writer. Two of his most famous novels include *Las [mujeres] honradas* (*The Honest Women*), written in 1917, and *Las [mujeres] impuras* (*The Impure Women*), a companion volume written in 1918. In one of the scenes of *The Impure Women* neighbors visit a friend in distress:

As in any humble home where a misfortune strikes, women friends and neighbors rivaled in diligence and did—as if it were the most natural thing in the world—all the house chores. Anita, herself, risking to ruin her pretty dress, had gone to the kitchen to prepare, herself alone, the chocolate. [12]

Women writers were not absent from the late 19th and early 20th century Cuban literary scene. One of them, Dolores María Ximeno y Cruz,[12] novelist, journalist, and Havana social figure, habitually hosted literary gatherings. In her novel *Memorias de Lola María* (*Memoirs of Lola María*), she uses the word *chocolate* to describe the color of a female dog named Marquesa: *white with chocolate spots* [13].

Later in these memoirs, however, the word *chocolate* appears in an unexpected way, in a context closely linked with the dishonesty and greed of certain government officials. The word *chocolate* in this instance referred to the acceptance of bribes by high officials. Ximeno y Cruz described the way in which Spanish political policies were destroying the Cuban economy:

Those insatiable desires [of wealth] beyond measure and the daily small thievery; the false reports and the chocolates [i.e., bribes], the big frauds carried out by the majority of officers that once they were discovered they were severely punished. [14]

Among 20th century Cuban authors, Alejo Carpentier,[13] novelist, writer of essays, and musicologist, greatly influenced Latin America and was a prominent political figure as well. Some of his writings vividly describe small Cuban villages and foods that people ate, as well as the social environment. His last novel, *La Consagración de la primavera* (*The Rite of Spring*), written in 1978, is of special interest to the theme of chocolate and literature because part of the story is set in Baracoa, where his description of the eclectic local food and village homes surrounded by cacao trees and their fruits are evocative and full of life:

A stew with all sorts of ingredients, chorotes[14] of cocoa and dried shark. . . . There the vegetation is so thick that certain peasants' huts seem to have been inlayed into it, particularly when they are surrounded by cacao groves, shady as the nave of the Roman churches, [and the trees] with their udders sprouting from their trunks. [15, 16]

Carpentier had a deep knowledge of the Spanish crown, colonization enterprises, and the Inquisition (see Chapter 4), policies that did not escape his sharp pen. In his *El camino de Santiago* (*The Road to Santiago*), one of the short stories of his book *Novelas y Relatos* (*Novels and Stories*), he summarized with piercing irony the activities of the Inquisition in Spanish America:[15]

And because the attenuating circumstances of sudden impulse was much more valid in hot lands the Holy Office [of the Inquisition] in the Americas had opted, from the beginning, to heat up cups of chocolate in their burner, avoiding the difficult task to establish the differences between obstinate, negative, minute, impenitent, perjure or possessed heresies. [17]

And further in the same work, he wrote:

What was now rewarding in the Indies was a keen sense of opportunity, to pry in other people's affairs, to abuse people, without paying too much attention neither to the ordinances of Royal Letters, recriminations of academics, nor to the screaming of Bishops. There [in the Americas] where the Inquisition had a soft hand, the fires would heat more cups with chocolate than flesh of heretics. [18]

In the same collection may also be found the story *Viaje a la semilla* (*A Voyage to the Seed*), where Carpentier provided a glimpse into the highlights of social gatherings in an elegant Cuban theater:

And in all the galleries, people dressed in black murmured at the rhythm [beat] of the spoons stirring their chocolate. [19]

In 1956, Carpentier wrote *El siglo de las luces* (*The Age of Enlightenment*), a historical novel with philosophical implications. Much of the action takes place in Havana at the end of the 18th and beginning of the 19th centuries, when Cuban commerce and consumption of cocoa and chocolate were still at their apogee. The following three excerpts transport us back in time to the bustling port of Havana:

As soon as the coach reached the first street, hurling mud right and left, the sea smells were left behind, swept away by the smells of enormous buildings filled with hides, meat salting, wax and hard sugar cakes, with the long time ago stored onions that were sprouting in the dark places, together with the green coffee and the cocoa spilled from the scales. [20]

Cheeses from La Mancha[16] [manchego cheeses] were aligned over the parallel wooden boards, leading to the Patio of Oils and Vinegars, where at the back and under vaults a most diverse assortment of merchandise was stored: packs of cards, barber cases, clusters of lockers, green and red parasols, cocoa hand-mills. [21]

The fires were lighted in the kitchens at all times, and the days passed by between breakfasts, table services of inexhaustible delicious foods, teas and collations between lunch and dinner [meriendas] and always at hand some cup of chocolate or a glass of Sherry [Jerez[17]]. [22]

José Lezama Lima,[18] a contemporary of Carpentier and regarded as the true father of the current Renaissance in Latin American poetry, in 1949 wrote the novel *Paradiso*,[19] where chocolate plays an interesting symbolic role. This novel, considered one of the most accomplished in Cuban history, describes the physiological collapse of one of the main characters when he learns that his mother has a terminal disease. The text contains a passage where chocolate symbolism depicts the strong affection between mother and son as during special childhood days when the mother prepared a mouthwatering cup of chocolate instead of the ordinary beverage of coffee with milk:[20]

Alberto knew that Doña Augusta was the one that sustained his life; she gave him the joy to feel safe and still young, since in reality the old age of a man begins the day of his mother's death. There are men in their old age that when they arrive to their mother's house, their mother gives them—as a gift—a piece of chocolate, perhaps a gift given to her by one of her grandchildren. Here then, it is established a kind of homologous youth relationship between that little chocolate bar, a gift from a grandson to his grandmother, and the relationship between a mother and her son. The son that arrives to visit his mother is a bachelor in his fifties, with a moustache covered by an autumnal frost, but his mother has saved that magical little bar, for the only day on the week in which her son will visit her, and with the same juvenile act that her grandson had given the chocolate bar to her, she gives it to her son. At that very moment [when the mother gives the chocolate bar to her son] her son begins to evoke María's cookies dipped in the chocolate [chocolate prepared with milk] that his mother would prepare in the school-free days to set them apart from the other days of the week where it was the coffee with milk the one that received the enthusiastic absorptions of the still crunchy bread with its copper crust. And while this old man was savoring this chocolate, a gift from his mother already into a venerable old age, he felt transported to the morning of the world, like a deer discovering the moment in which a secret river emerges to the surface and goes into his mouth, a mouth that was chewing red currants [grosellas]. [23]

The symbolic meaning of chocolate as a good companion for young students is expressed in another passage where mathematics, music, and chocolate become deities and chocolate, in particular, is seen as a reward:

"Come to study with Albertico tonight," said Andrés Olalla to José Eugenio. "Afterwards, Rialta will play piano and we'll have chocolate. Mathematics, music and chocolate, are excellent deities for your age." [24]

There is another scene about student life where a professorial tribunal attempts to shock a student with absurd questions:

One of the professors who was . . . eating a piece of chocolate with almonds and cherry liquor, of ignoble manufacture from Florida asked the following question: Where can we find the best chocolate in the world? The straight answer was: In the number 17 of Rue de Rívoli,[21,22] exposition salon, first floor. [25]

In Paradiso we also find mention regarding the consumption of chocolate at breakfast, as a collation between lunch and dinner (meriendas[23]), or late at night before retiring:[24]

He/she brought him/her to bed a large cup of chocolate with a teaspoon of anise liquor.[25] [26]

And it came as no surprise to read that an individual had taken a midnight chocolate cup. Did he have an upset stomach and did he drink chocolate to find relief? The result, however, perhaps was unexpected:

One of the rowers, urged by the midnight chocolate [he had drunk], rushed to evacuate the intestinal coil. [27]

Lezama Lima, as with Villaverde in his novel Cecilia Valdés, used chocolate as a proverb to indicate clear and direct speech. In this passage he describes making light chocolate, where water was used instead of milk, and prepared in front of everybody:

Quería que le hicieran el chocolate en reverbero y con agua sola. [He wanted his chocolate to be made with (only) water and in full light.] [28]

Virgilio Piñera,[26] a contemporary of Lezama Lima, wrote The Flesh of René (La carne de René),[27] an unusual novel with chocolate at its very center. A ferocious dictator has prohibited the consumption of chocolate under the penalty of death, because chocolate was considered a very powerful food, and so that his subjects (the common people) should be maintained "in a state of semi-hunger." The Flesh of René contains a disturbing plot, a "flesh cult," a dichotomy of pain and pleasure, where conflict that moves the main characters is caused by chocolate. When René asks his father why they are fighting the answer is: for a piece of chocolate! Indeed, it was a chocolate revolution: a revolution for the right to eat and drink chocolate whenever they

wanted. It was the chocolate cause: It was an ode to the goodness of chocolate and the passions of the flesh [29, 30].

The emblematic importance of chocolate is shown in the following dialog between father (Ramón) and son (René) that appears in The Flesh of René. We considered it worthy to reproduce here, given its importance for a better understanding of the symbolic influence that chocolate has in the novel. This surrealistic story by Piñera contains elements of freedom and passion, as well as the unavoidable revolution to obtain freedom accompanied with the sacrifice of life to obtain it. Corruption of power and political games are also clearly symbolized:

Father speaking: Many years ago, after a bloody battle, the chief that now haunts me, managed to defeat the powerful and ferocious ruler that had forbidden in his states, under death penalty, the use of chocolate. This ruler had maintained rigorously such prohibition that had started many centuries ago. His ancestors, the founders of the monarchy, had prohibited the use of chocolate in their kingdoms. They affirmed that chocolate could undermine the security of the throne. Imagine the struggles that took place during centuries to impede the use of such food. Millions of people died, others were deported. Finally, the chief that now pursues me, obtained an overwhelming victory over the last sovereign and we had the happiness, very short although, to fill our territories with chocolate.

Son speaking: Tell me father, in which way the chocolate would undermine the security of the throne?

Father speaking: Very simple: the founder of the dynasty declared that chocolate is a powerful food and that the common people should be maintained in a state of semi-hunger. It was the best measure to ensure the perdurability of the throne. Imagine then our joy when, after centuries of horrendous battles, we could inundate the country with chocolate. The masses, that have inherited this pathetic predisposition to drink it, devoted themselves to consume it wildly. At the beginning everything went smoothly. One dark day, the chief began to restrict its use. Your grandfather, that has seen his father and his grandfather die because of the chocolate's control measures, resolutely opposed such restriction. And so it was that the first friction with the chief took place. As in any battle where life and death are at stake, there were very careful considerations and deceptive settlements. One day we woke up with an overflow of hope: the chief gave us carte blanche for the use of chocolate; another day he will limit its use to three days per week. Meanwhile the altercations grew stronger. Your grandfather, the most influential personage close to the chief, reproached him for such a disastrous politics, to the point of calling him "a reactionary." A bitter dispute then took place which resulted next in the murder of the secretary of my father, the Secretary of War. He was found agonizing in his house: someone had forced him to drink a gallon of hot chocolate. This was the last straw. My father opposed openly the government and the group of chocolatophilos [i.e., chocolate lovers] was formed. At that time I was very young, but I remember sharply a parade under the balconies of the Government House eating small bars of chocolate. As a reprisal, the chief confiscated all the existent chocolate in the country. We did not give up it and we all dressed in chocolate color. The chief, considering that this could rise up the people against him, declared us criminals and accused us of acts of treachery to our country (*reos de lesa patria*) and gave the order to prosecute us. My father barely could cross the frontier and ask for asylum in a neighboring country. The result of this prosecution was the death of thousands of our people.

Son speaking: *If they were not guilty, why did they execute them? screamed René horror-struck.*

Father speaking: *Why . . . ? Ask the Chief—and Ramón burst into a loud laugh. Meanwhile my father and his adepts maintained the holy cause of the chocolate from the neighbor country. The chief had betrayed the most holy principles of the chocolate revolution; as a consequence he should die. He knew it and he also knew that the people, who would not have had openly protested against the prohibition [to use chocolate] if it were not for the campaign carried on by the opposition, now they would start a battle to death for the right to eat and drink chocolate where and when they would want it. Soon, the events confirmed the chief's anxieties. The peasants revolted. The result was the death of thousands of them and the deportation of many thousands to the frozen regions of the country. Almost everybody died. Then, your grandfather launched the first attack. The principal secretary of the chief was one of ours. I won't linger here with the details of this laborious commission. I will limit myself to say to you that the cup of chocolate that the chief would drink next morning was poisoned.*

Son speaking: *Did he drink chocolate? exclaimed René.*

Father speaking: *How can you be so naive! Of course he did use chocolate and in large quantities. He and his underlings knew that this beverage was highly stimulant, and that the reason why the enjoyment of chocolate was forbidden to the common people was precisely because if both—the people and the government—would drink it, it would debilitate the rulers politically. Do not forget that the chief and his underlings aspired, through the secret goodness of the chocolate, to dominate the world.*

Son speaking: *Father, interrupted René, now I realized I never saw you drinking chocolate. And as for me, I do not know the taste of it.*

Father speaking: *You continue to be a naive. So, do you think that we will lessen ourselves to the point to drink chocolate? Do you think we are so simple that you would see us with a cup of chocolate in our hands? What we were defending was the cause of chocolate. It wouldn't have had any sense that being, as we were, at a distance of thousands of leagues from the chocolate battle ground we would set ourselves to drink it as desperate people . . . And to finish and to illustrate for you: I confess to you that we are tired of chocolate. Your own grandfather used to smile with sarcasm when they talked about chocolate, which did not prevent*

him to have spent his life defending it in hand-to-hand combat. [30]

And elsewhere in *The Flesh of René* . . .

[René approaches the picture and lays a finger on the hand that held the arrow piercing the forehead of Saint Sebastian,[28] *and candidly asks his father]:*

Son speaking: *Why didn't you order the painter to put in my hand a cup of chocolate instead of an arrow?*

[Ramón's [the father] countenance was altered.]

Father speaking: *It was for propaganda purposes.*

[The father walks toward the table, searches in a drawer, removes a photograph, and shows it to his son.]

Father speaking: *Look. Here we are in the chocolate prohibition's anniversary banquet. Don't you see that we all are holding a cup? Nevertheless, we never gave such a banquet and much less drank chocolate, but this did not prevent us from making a million copies of photographs and circulating them throughout the world. But, let's leave aside your naiveté and let's go back to the chief. He had discovered the conspiracy. The morning that I was talking about, he went to the restaurant of the Chancellery, holding in his hand a cup of steaming chocolate. In seeing it, the secretary froze with fear. Without more ado, the chief told him that he had to drink it. . . . You will probably imagine the rest of the scene: the secretary was compelled to consume his chocolate with hemlock. In a few moments he became a cadaver. This same day, the government of the country in which we were given asylum, declared that we were pernicious foreigners. Many things happened since then. Your grandfather was murdered, I am about to die. The circle keeps narrowing. This is why, and on the occasion of your birthday, I have called you here to tell you about the will of the party [the political party] and my own.*

Son speaking: *The will of the party . . . ? René barely could stammer.*

Father speaking: *It is the will of the party that you be my successor.* [30]

Throughout the novel, René struggles with his destiny: Should he take up "the torch of the Chocolate Holy Cause" *to whom he must sacrifice his own flesh* [31]? Initially, he refuses and is unwilling to "sacrifice his life for a cause like the one of chocolate" [32]. But upon his father's violent death, the followers of the *Chocolate Cause* force him to accept his fate. There follows a short dialog with one of the followers that clearly elucidates the meaning of chocolate in this book, when one of the followers explains to René:

The Follower speaking: *Those who have a simple spirit think that we defend chocolate. Let them believe it. To you—a chief—I will tell the truth. The issue is not the chocolate. The very flesh is the one who is the matter at play. Now then, why lose it? Here they are the precious services of chocolate. Nobody ever will praise enough this infusion. The chocolate presents itself to the flesh with tears in his eyes and tells her* *[the flesh]: I am in mortal peril, save me, my enemies are pursuing me. Do not forget that in saving me you give yourself what you more desire in this life, your own ruin. I know that you desire to be the target of bullets and knives. And with a chocolate smile he [the chocolate] jumps and shields himself behind the flesh, that at this very moment [the flesh] falls—cut down by the persecutors in the spring of her life.*

Son speaking: *It is amazing—muttered René, looking at the walls—[he sees them as if they were made of chocolate and it is as if the walls begin to speak and implore him to protect them].*

The Follower speaking: *Amazing—emphasized the little old man, in the middle of his eternal reverences. And there is still something more amazing: each time that the chocolate comes to shield himself behind the flesh of the hunted, the hunters wait tremulous that she [the flesh] rejects them [the hunted]. If this would happen the chocolate would not have any other option other than seek the protection in their flesh [in the flesh of the hunters]; and in doing so, ipso facto, they, the hunters, will become the hunted, and would be killed by the thousands.*

Son speaking: *And if the flesh . . . ? asked René without daring to complete his thought . . .*

The Follower speaking: *I know what you want to ask me. If the flesh refuses to run, then, there is the chocolate to give her a push. Can't you see that it is he and only he the one that keeps her on the track? It looks absurd at first sight that something that you do not care about would make us run through mountains and valleys until we fall dead by bullets or knives. Nevertheless, think that it would be utterly absurd that the flesh, without any pretext, would get herself to run through mountains and valleys. Oh! he [the follower] screamed leaping with exaltation: The chocolate is a powerful stimulant.* [33]

And in spite of René's desire to die "in his bed far from bullets and knives, far from the besieged chocolate, absolutely useless as shield" [34], finally he has no other choice than to enter the game. And with this decision Virgilio Piñera challenges his readers to accept "the first truth: we are made of flesh" [35].

Dora Alonso[29] described the life of Cuban farmers. Most of her characters and stories were set within rural environments. Her works include the anthology *Letters*. Among these letters is one entitled *Ana and the Love Affairs (Ana y los amores)*, where the word chocolate is used in an unexpected way to describe the sound coming from a tight taffeta dress stretched by body motions. Only someone growing and living in a society where chocolate was part of daily life could have employed this metaphor:

You used to wear tight dresses made of taffeta creaking as chocolate paper. [36]

Alonso also wrote a collection of poems for children entitled *La flauta de chocolate (The Chocolate Flute)*, although it contains no specific mention of chocolate.

Yet, it is interesting to note here that the words she chose for the title—flute (music) and chocolate (a traditional food item associated in Cuban traditions with celebrations and friendship)—combined to produce a lively image: "The well chosen words [of the title] represent the vivacity of movement and rhythm and the exultance of the vital environment of the children's soul" [37].

Onelio Jorge Cardoso,[30] considered the national short story writer of Cuba, gained a profound knowledge of fishermen, peasant life, and the rural environment during his travels across Cuba. Three of the short stories included in his anthology called *Cuentos (Stories)* provide references to cacao or chocolate. In the narrative entitled *Moñigüeso* the author describes cacao harvesting:

Among his/her obligations he/she has to sweep the floor of the tree groves and take outside the dirt. He/she harvested the cocoa fruits and held firmly the rope through which the clusters of Royal Palm [palmiche[31]] were hanging. [38]

In another of his works, *El cuentero (The Storyteller)*, published in the same compilation, Cardoso used the word chocolate to describe the color and texture of agitated river waters:

I had already crossed rivers larger than that one, and I knew the area very well so I could go around with only the mule's eyes. So, I started the journey looking for the river, and by the afternoon I was already wading through its waters. The river still was carrying waters of stirred chocolate, but it hardly bounced more than half a meter over the riverbank. [39]

In Cardosos's *Los sinsontes*[32] *(The Mockingbirds)*, he uses the word chocolate in a poetic context to describe the color of a young tree trunk:

Then, he would start telling a story about anything [for instance] the case of the buttonwood [patabán[33]] that refused to be cut down: It was a new, small buttonwood with a trunk of the color of chocolate, standing very straight pointing to a star in the sky. [40]

Samuel Feijóo[34] dedicated the greater part of his life to compiling oral traditions of the Cuban people. Although he also was an accomplished novelist, Feijóo excelled in the narrative of Cuban folkloric traditions. His rich anthology *Cuentos populares cubanos de humor (Cuban Popular Humor Stories)* contains the short story *El isleño que le mandó un regalo a su mujer (The Islander That Sent a Gift to His Wife)*:

An islander came from Canary Islands to Cuba to work and left his wife there [in the Canary Islands]. After a year he sent her as gifts a chocolate bar and a small parrot. And the woman sent him a letter that said: "I would like to tell you that I put the chocolate bar you sent me on the stove to melt, and it was ruined. And that ugly bird you sent me began to speak as people do, and so I twisted its neck". [41]

In another story by Feijóo, *Cuartetas en la parranda (Party Verses)*, he incorporated a very popular saying, one commonly used by other authors:

> *Porque esta noche estoy yo como agua pa' chocolate. [Because this night I am like water for chocolate.*[35]*]* [42]

Afro-Cuban[36] culture has remained a vital and deeply rooted element in Cuban traditions and folklore and is reflected in Cuban literature and music. This theme attracted Feijóo in his *El negro en la literatura folklórica cubana (The Black in Cuban Folkloric Literature)*, a superb anthology that portrays the influence of black people in Cuban literature. Among the collection of verses considered by Feijóo as truly black in origin is the following delightful rhyme:

> *Si quieres saber mi nombre*
> *en mi canto te diré*
> *que me llamo chocolate,*
> *primo hermano del café.*
> *If you want to know my name*
> *I can tell you with my song*
> *That chocolate is my name*
> *and coffee my first cousin is.* [43]

In black Cuban literature, chocolate sometimes reaches the category of a heavenly gift that is brought to earth in the company of a specialty cake/biscuit! This is the case in Feijóo's *Las dos hijas (The Two Daughters)*, where one of the main characters says:

> *[She] brought chocolate and bizcochuelos*[37] *from heaven.* [44]

As a novelist, Feijóo also brought to the 20th century older Afro-Cuban religious traditions. Black slaves combined the Christian Catholic religion brought by the Spanish with their African religious traditions, creating a new religion known as *Santería,*[38] a blend of Catholicism and traditional West African religions. This *Santería* tradition is revealed in his most famous novel, *Juan Quinquín en Pueblo Mocho (Juan Quinquín in Pueblo Mocho)*, published in 1963. In this work Feijóo narrates the life and miracles of Juan Quinquín, a man of incorruptible ethics and loyalty, and describes in detail the festivities surrounding the celebration of a local saint. This religious event, commonly on the saint's name day (*Velorio de Santo*[39]), is an integral part of the Cuban folkloric traditions where chocolate is habitually consumed. Feijóo describes a party that concluded with a colossal fight and, regrettably, the precious chocolate was spilled on the floor when one of the characters fell and hit the stove:

> *... and he fell where the chocolate was boiling and knock[ed] down the earthen pan spilling the aromatic liquid on the kitchen's floor.* [45]

Elsewhere in the same text he wrote that chocolate was one of the beverages commonly served at real funerals [46].

Feijóo also wrote the novel *Vida completa del poeta Wampampiro Timbereta (The Complete Life of the Poet Wampampiro Timbereta)*. Here may be found a paragraph describing the tenderness of a son toward his blind and ailing father and the comforting presence of chocolate:

> *Sebastián, the blind man, remained silent. . . . Alejo stood up, wrapped his arm around his father's shoulders and told him: "Let's go my old man, let's go . . . There is a very tasty chocolate at home . . . Come . . . Tomorrow you will feel better. . . ." Alejo held tight his father's arm and disappeared with him into the night.* [47]

Antonio Núñez Jiménez,[40] a renowned national academic figure, described poetically the geography of the island in his work *Cuba: La naturaleza y el hombre (Cuba: Nature and Man)*. In Volume 2, he referred to the banks of the river Toa in Baracoa located on the north coast of Guantánamo Province:

> *The banks are covered . . . by cocoa plantations [cacahuales], banana plantations and coconut plantations [cocales].* [48]

As the title of his geography indicates, his observations also focused on the people who inhabited the land and their labors. In one passage he described in detail the harvesting of cacao:

> *[They use a] long pole with a small scythe at the end, used to cut the cocoa fruits and coconuts in the area.* [49]

The hospitality of the people Jiménez encountered in his journeys is also registered in his writing:

> *[A peasant] after the irreplaceable coffee sip [buchito de café] . . . asked his wife to give us milk with chocolate; [chocolate that has been made] with the cocoa cultivated by his own hands; and she always added cinnamon.* [50]

In his third volume, Jiménez included a passage that he designated as *Autorretrato de Cuba (Self Portrait of Cuba)*, where he wrote:

> *The valley of the river Toa is one of the richest regions [of Cuba]. . . . Cocoa, coconuts and bananas are harvested near the mouth of the river.* [51]

In another chapter of the third volume of his *Geography*, Jiménez included the lyrics of a song (entitled *Alma guajira) (The Soul of Guajira)*, written in 1925 by Ignacio Piñeiro, that includes references to cacao (see below).

CHOCOLATE IN CUBAN PROVERBS

Proverbs are popular sayings that contain advice or state generally accepted truths. Because most proverbs have their origins in oral tradition, they generally are worded in such ways as to be remembered easily, tend to change little from generation to generation, and commonly reflect the way of life and environment of a region during past centuries. In our search of oral traditions, we interviewed farmers and local residents for proverbs where chocolate was mentioned. These are reported here along with the names of the respondents who provided the information.

It has been said that the first cacao tree was sown in Mosquitero,[41] and this is what the following proverb reveals:

Es más Viejo que el cacao de Mosquitero. [He is older than the cacao from Mosquitero.] [52]

But perhaps the most famous of all chocolate proverbs is the following:

Está como agua para chocolate. [He or she is like water for chocolate.] [53]

We have already seen this popular proverb earlier in the chapter, but we learned during our research that the same words can have different meanings. For Arístides Smith, Baracoa, for example, the proverb means: "hot," "boiling," "very brave." However, in other interviews with Jesús M. Quintero from San Luis, and Dionisio José (Pepe) Reyes from Jobo Dulce, they affirmed that this phrase meant that something has a very good taste (i.e., really tasty), or as North Americans might say—"finger-licking good."

Another interesting proverb forms part of the refrain in the song The Grocer (see below):

Toma chocolate, paga lo que debes. [Drink your chocolate, and pay what you owe.] [54]

This Havana proverb has an identical meaning to another one commonly used in Spain:

Las cuentas claras, y el chocolate espeso. [The accounts in order and the chocolate well done (thick).]

Both proverbs consist of two components: the value of friendship and issues that arise with debt repayment. The meaning is clear: Friends share thick, good chocolate, but if one borrows money from a friend—always repay it!

Chocolate in Cuban Games and Music

Children's games and music commonly go hand-in-hand and complement each other as is seen throughout Cuba.

GAMES

Concepción T. Alzola[42] published her Folklore del niño cubano (Folklore of the Cuban Child) in 1962. This compilation included descriptions of children's traditional games. In some instances, references to chocolate appear in the songs that accompany the games. One such hand-game called pun–puñete (fist-little-fist) starts by children grouping together and layering their fists one atop another. Once the tower of fists has been formed, the children begin to rotate their fists as if they were pounding with a mortar, and they recite and chant at the same time:

¿Qué hay aquí?	What (who) is here?
Un negrito	A little black child
¿Qué está haciendo?	What is he doing?
Chocolate	Chocolate
Pues bate que bate el chocolate	Well, beating and beating the chocolate
Bate que bate el chocolate.	Beating and beating the chocolate. [55]

Among the hand games commonly played by Cuban children is one where two children sit facing each other. They start the game by slapping their palms on their thighs, then raise their hands and clap their palms against the hands of the other player sitting opposite. The actions are rhythmic, slow at first, and different recitations accompany the movements. The actions are repeated several times, and with increased speed, until one of the participants misses (loses). One such recitation goes like this:

Té, chocolate y café,	Tea, chocolate and coffee,
Pluma, tintero y papel	Paper, ink bottle and pen
Para escribirle una carta	To write a letter
A mi querido Manuel.	To my beloved Manuel. [56]

Sometimes children take part in another game where participants must recite words in correct order as part of rhythmic chanting. The word list grows longer with each round. One of these, known as El Velorio (The Funeral), starts where the first participant identifies a common food or beverage customarily served at funerals, like milk or chocolate; the next child repeats that word and adds another, the third adds another word and so on until one of the participants makes a mistake and drops out while the others continue. The "winner" is the child with the best reaction time and accurate memory. The game begins with a statement said by all—and subsequent single word additions follow:

In that funeral there wasn't any milk.

MILK (shouted) . . .

Milk there was; what wasn't there was chocolate.

CHOCOLATE (shouted) . . .

Milk there was; Chocolate there was; what wasn't there was guava jam

GUAVA (shouted) . . .

Milk there was; Chocolate there was; Guava there was; what wasn't there was etc. [57]

In one children's game, mud is a central theme and called chocolate. The children prepare chocolate in a bucket by mixing dirt and water, and then invite their little friends to drink it: the chocolate is "served;" the children act as if they savor it and make comments like "this chocolate is very thick" or "this chocolate is light," reflecting what they heard at home [58].

MUSIC

Alejo Carpentier's *Música en Cuba* (*Music in Cuba*) describes the history of Cuban music with its principal roots in Spain and West Africa, later influenced by France and the nearby islands of the Caribbean [59]. The musical expressions of the original inhabitants of Cuba, the Taínos and Siboneys, were basically ignored by the Spanish conquerors. While little is known regarding the music of the Siboneys, the Taínos celebrated their ancestors with a combination of music and dance called the *areito*.[43] According to Carpentier, the only *areito* that had reached the 20th century was the *Areito de Anacaona*. Although most of the areitos have been lost, we still can find them metaphorically in French 18th century lullabies sung throughout the Caribbean [60]. It is also noteworthy to mention that Cuba had a composer[44] even before the first theaters or the first newspapers were developed on the island [61].

During the 19th century black Cubans began to embrace music as a profession. In a colonial society, although legally no longer slaves,[45] black Cubans still were barred from the classical professions of law and medicine and from important administrative positions. The musical profession, however, was an open field and offered the opportunity to reach a higher social status as Afro-Cuban music developed new, well-defined genres with universal influence [62].

It is not difficult to find chocolate mentioned in Cuban music, particularly in popular and folk genres. We conducted interviews in the city of Baracoa among local farmers at the towns of Jamal, La Alegría, Mosquitero, and San Luis. Some respondents reported song lyrics that mentioned chocolate in a religious context (altar songs); others provided lyrics that were more romantic.

The intent of the following altar song is to attract the attention of the godmother, the woman responsible for assembling the structure, so that she will bring chocolate to her guests:

> *Pedro go to the kitchen*
> *and bring me chocolate*
> *and if your wife gives you trouble*
> *you tell her that you bought it.* [63]

The following song reveals that some altar godmothers may not be good hostesses. The lyrics show that she has built the altar and invited her friends to fulfill the promise she has made to the cross, but that she has forgotten to offer the chocolate—something that no hostess should ever forget to serve to guests:

> *The godmother of the altar*
> *has walked over gold*
> *she has paid her promise*
> *and she has much gained*

> *but I don't see she is bringing*
> *for us the hot chocolate.* [64]

The two following chocolate songs have nothing to do with religion. Instead, they reveal a sense of dance and seduction. We leave interpretation of the passages to the reader:

> *I feel a turuntuntún*[46]
> *and this is what makes me flutter:*
> *because there in the kitchen*
> *they drank the chocolate.* [65]

The second example reveals a similar, although more direct and unmistakable, intention:

> *I am leaving, I am going,*
> *I am feeling that all my body flutters*
> *the women are in the kitchen*
> *and have drunk the chocolate.* [66]

Cuban music lyrics, under a strong Afro-American influence and impregnated with the fragrances and colors of Caribbean life, could not escape the aromas and hues of cacao and chocolate: The Cuban musical texts associated with this music are abundant. Among the several 20th century commercial songs with lyrics that include chocolate themes may be found a popular carnival song, another of international fame known as *Alma Guajira*, one of the most famous cha-cha-cha from the 1950s, and others written in the last decade and performed by Cuban popular music groups.

Carnival[47] in Cuba is a time for merriment, music, and dance. Goodman, in his *Un artista en Cuba* (*An Artist in Cuba*), reproduced a carnival folk song from the decade of the 1960s, where the theme reveals the advantages of being the wife of a baker—since she will always have bread and chocolate.

> *The tobacco merchant's wife*
> *has nothing that makes her feel safe*
> *while for the bread maker's wife*
> *everything is certain,*
> *at five in the morning*
> *she has bread and chocolate*
> *and some money on her pocket.* [67]

The song *Alma Guajira*[48] (*Guajira Soul*) was written by the composer Ignacio Piñeiro,[49] one of the most important musicians in the history of Cuban music. The geographer Núñez Jiménez, as noted earlier, incorporated lines from Piñeiro's song in one of his volumes on Cuban geography:

> *Guajira soul*
>
> *From my Cuba, full of pride*
> *About her fruits I will tell:*
> *Produces sugar and coffee*
> *And cacao in abundance,*
> *The fragrant pineapple,*
> *And humble abundant fruit,*
> *And in my fruitful yard,*
> *Magnificent coconuts, macaco*

And above all tobacco
The best of all the world.
Listen to my genuine song,
That always the "song" inspires me
Listen to the guajira soul
And dance my song. [68]

Lyrics from a very famous cha-cha-cha from 1956 called *El Bodeguero* (*The Grocer*) were written by the flute player and composer Richard Egües. The North American artist/singer Nat King Cole popularized a version of this song that included the following lines:

The Grocer

The cha-cha-cha and the grocer
Always at home they are.
Go to the corner and you will see him,
And always polite he will serve you.
Go quick now, run there,
That on the counter
And with the money,
Pleasing and helpful.
You will find him . . .
Chorus:
Drink chocolate, pay what you owe
Drink chocolate, pay what you owe. [69]

From recent years we have identified two songs performed by the more important popular Cuban music groups. The first is called *El pregón del chocolate* (*The Chocolate Proclamation*);

Come, I have chocolate here!
On the street, the vendors
Are announcing their products
To let people know which is the best.
But I still didn't meet—around my house—
The more important one
That will bring the chocolate to your door . . .

I am bringing chocolate for you
Because it is the sweet you like. [70]

Another recent song is associated with the Cuban musical group Los Van Van. These lyrics reveal how chocolate takes the shape of the body of a mulatto woman (i.e., mulata),[50] whose desperate jealous husband tries to hide her from the sight of other men:

My chocolate

What a mulata! Pure chocolate!
Tell me where is my chocolate,
Tell me where it is, where it is hiding?
My chocolate is a mulata,
That has light eyes and black hair
And her mouth is like a ripe mango
And front and behind
She has too much . . .
For the love of that woman
I am going to become crazy and die.
Tell me where has she gone,
Where is she hiding. [71]

Like Cuban music, the Cuban culinary approach to food is the result of integration of many cultures and has produced a distinct and attractive variety of appetizing fare. Traditional Cuban food is the result of an amalgamation of Amerindian, African, Caribbean, and Spanish foods, with a dash of French influence brought to the island by the French settlers escaping from Haiti during the War of Haitian Independence at the beginning of the 19th century.

Among the ingredient contributions from the native cultures of Cuba, the Taínos and Syboneys, were cassava (*Manihot esculenta*), corn (*Zea mays*), peanuts (*Arachis hypogea*), and green/red peppers (*Capsicum* sp.) The Spanish brought with them their homeland foods that also included Moorish recipes that for eight centuries had influenced the Iberian Peninsula. The Spanish introduced to Cuba an astonishing array of Old World foods, among them citrus, garlic, olive oil, onions, saffron, and tomatoes. They also introduced livestock (i.e., cattle, goats, and sheep) and various dairy products. While Spanish influences were stronger in the western portions of the island and especially in the region around Havana, the Afro-Caribbean influences were particularly felt in the eastern region of Santiago de Cuba. Some *criollo* (Creole/ traditional) Cuban dishes, furthermore, have linguistically picturesque names, as evidenced by *Moros y Cristianos* (*Moors and Christians*) prepared using white rice and black beans; *Negro en Camisa* (*Black Man in a Shirt*); and the intriguing chocolate dish *Brazo Gitano de Chocolate* (*Chocolate Arm of the Gypsy*).

Cacao and chocolate were readily incorporated into Cuban culinary traditions, so much so that they influenced names on the landscape. Toponyms, or place names, reveal a number of locations influenced by cacao and chocolate. The *Atlas de Cuba* (*Atlas of Cuba*) lists two small islands (keys): *Cayo Chocolate* (Chocolate Key) located in the archipelago known as *Jardines de la Reina* (The Queen's Gardens) and *Cayo Cacao* (Cacao Key) in the group of islets south of *Ciénaga de Zapata* (Zapata's Marsh). There is a *Presa El Cacao* (Cacao Dam) constructed in the vicinity of Santa María del Rosario, in the province Havana. Rural settlements also have received such names: *Cacaito* (Little Cacao) is situated north of Sierra Maestra in the province of Granma; *Cacaotal* (Cacao Grove) is a location identified in the Sierra de Najasa, in the province of Camagüey; and the atlas lists three village settlements with the name *Cacahual* (Nahuatl for Cacao Grove), located in *Ciego de Ávila*, *Sancti Spíritus*, and *Villa Clara* [72].

TRADITIONAL CULINARY USES OF CACAO

Baracoa the oldest Spanish settlement in Cuba, founded in 1512, has a strong family tradition regarding the manufacture of *bolas de cacao* (cacao balls) used to prepare *chorote*, the local word for chocolate. Both cacao and the cacao beverage named *chorote* are distinctive elements of this region; visiting travelers return home after purchasing some of these cacao balls and use the chocolate to entertain family or friends as a token of friendship. In this section we present our results of cacao-related interviews with local residents in Baracoa.

The Baracoa peasant families we interviewed considered cacao a food, and to prepare it they used procedures similar to those used in pre-Columbian times. Once the cacao beans are dried, they are roasted to remove the shells.[51] After winnowing the shells, the clean beans are pounded or ground. During the final preparation procedures, various spices like nutmeg (the favorite Cuban spice mixed with cacao), cloves (*Clavo de Castilla*), anise, or vanilla may be added.

The cacao balls are prepared using pure cacao. Some people dust the balls with wheat flour to facilitate the process of shaping them, while others actually incorporate flour into the dough with the same purpose. Wheat flour is one of the ingredients used to give the necessary thickness to *chorote*. Some respondents told us that they boiled the cacao balls to extract some of the butter, and that boiled cacao balls could be preserved for a long period of time without deterioration or spoilage.

Other respondents mentioned that they preferred to prepare cacao from powder, and that they stored the cacao in lid-covered cans.[52] Cacao balls, in contrast, were seen by some to be inconvenient, in that they had to be grated when cacao beverages were prepared and sometimes this was a little difficult. We found that cacao balls were made primarily to offer as gifts, or for special orders, and that most chocolate was prepared using powder already ground.

To prepare *chorote* (a thick chocolate beverage), cacao powder is mixed with milk or water at room temperature, since hot water would make the powder difficult to dissolve. The mixture is then heated and stirred until the desired thickness is obtained, and more flour is added if a thicker *chorote* is desired. In Baracoa, the people compare *chorote* to a very thick homemade chocolate—*a la española* (made in the Spanish fashion). To obtain the required thickness, we were informed that it was better to use corn starch or corn flour, although flour from bananas (*bananina*), cassava (i.e., manioc or yucca), sago[53] (*sagú*), or wheat commonly could be used.

Chorote is considered an excellent food for children and adults, very appropriate for breakfast and for an afternoon snack, and particularly good for supper when the weather is cold. It is considered an exceptionally good food to administer to women after childbirth and during nursing. Our respondents stated that such thick chocolate beverages were considered vital foods, since cacao reportedly stimulated the mammary glands to produce abundant milk. One male respondent told us that when the day of childbirth approached, cacao balls were prepared to have them ready for the moment of delivery. If the family did not have cacao at this time, they would ask someone else for it. Cacao balls could not be absent at childbirth time. Another of our interviewees affirmed that some people cooked (i.e., fried) using cacao butter, and several others mentioned the production of homemade cacao wines. One peasant lady kindly offered the following two recipes:

> *Cream of Cacao Liquor*
>
> *Put ten dry, roasted, hulled cacao beans into a bottle of alcohol or rum. Leave it alone for a period of 30–40 days. Afterwards, filter it and add sugar syrup to thicken. Then add ground or finely chopped peanuts.*

> *Bonbons*
>
> *Mix one cup of wheat flour and one cup of milk; add half a cup of cacao paste. If the cacao paste has already flour in it, mix it directly with the milk. Stir well, then add syrup and make small balls.*

Cacao and Chocolate in Contemporary Cuban Cookbooks

From the abundant collection of contemporary Cuban cooking books, we have identified for consideration three works to illustrate the diversity of chocolate preparations: *¿Gusta Usted? Prontuario culinario y necesario* (*Would you like? Necessary Culinary Compendium*), a compendium of recipes by different authors published in Havana by Ucar García and Company; *Cocina al minuto* (*Cooking in a Minute*), written by the Cuban culinary guru Nitza Villapol[54] and M. Martínez, and *Cocina ecológica en Cuba* (*Ecological Cooking in Cuba*), written by Madelaine Vázquez Gálvez,[55] a nutritional and agricultural sustainability specialist.

From *¿Gusta Usted? Prontuario culinario y necesario*, we summarize here two recipes, interesting because of their unusual names: *Negro en camisa* (*Black Man in a Shirt*) and *Brazo gitano de chocolate* (*Chocolate Arm of the Gypsy*).

Black Man in a Shirt: This is a two-step recipe. The first step is to prepare dough consisting of almonds (chopped and finely ground), "French-style" chocolate (grated), egg yolks and whites, butter, and sugar. The almonds, butter, and sugar are mixed with the egg yolks; the egg whites are beaten until stiff, and then folded into the chocolate mixture. The dough then is baked. The second step is to prepare a sauce from cornstarch, egg yolks, grated lemon peel, milk, sugar, and vanilla. The sauce is cooked, cooled, and then poured over the dough. [73, 74]

Chocolate Arm of the Gypsy: This also is a two-step recipe. First, a paste is prepared from cacao, eggs, flour, sugar, and a small quantity of baking powder. The eggs are separated and the whites are beaten until stiff, then sugar is added. The cacao and baking powder are blended together, added to the mixture, and the egg yolks added last. The paste is then molded into the shape of an arm, then baked. A second paste is prepared using butter, milk, salt, sugar, and walnuts. After blending/beating this will take on a creamy white appearance. At this point the rum is added. This second paste is used to "fill" the gypsy arm, and what remains is then spread over the outside. [75]

Included in the Ucar Garcia recipe compendium are a number of references to alcoholic beverages where cacao/chocolate served as ingredients. Several of these have interesting names: *Doncellita* (*Young Maid*), prepared with cream of cacao [76]; *Costa Bárbara* (*Savage Coast*), a mixture of heavy cream, cream of cacao, gin, and Scotch whisky served over crushed ice [77]; and *Magia Negra* (*Black Magic*), a blend of sugar, grapefruit juice, cream of cacao, and rum [78].

The book *Cocina al minuto* (*Cooking in a Minute*) by Villapol and Martínez, originally published in 1956 (reprinted in 1980), included a section on ice cream and liquors used to prepare characteristic desserts:

> To make a very simple but special dessert we only need a commercial ice-cream and some liquor or sweet wine. Serve the ice-cream in small glasses or cups and pour over them a small glass of sweet wine or liquor, that match well with the ice-cream flavor. We suggest: Vanilla ice-cream with cream of pineapple, banana, coffee or cacao; Chocolate ice-cream with coffee or cacao cream; or Strawberry ice-cream with cream of cacao. [79]

In the book *Cocina ecológica en Cuba* (*Ecological Cooking in Cuba*), we found several recipes for ice cream prepared using cacao and cassava, and we summarize one here:

Helado de cacao, yuca y ajonjolí (*Cacao, Cassava, and Sesame Ice Cream*): The ingredient list consists of cacao, cassava, salt, sesame seeds, sugar, and water. The cassava root is cooked until it becomes soft and the pulp is removed. Then the pulp is blended with half of the water, and the cacao, roasted ground sesame seeds, sugar, and salt are added. The mixture is then beaten. After beating, the mixture is filtered and the remaining portion of water is added. The mixture is poured into a container and cooled until half frozen. The mixture is beaten again and the process repeated 2–3 times until it solidifies and reaches the desired consistency. [80]

Traditional Medicinal Uses of Cacao

The men and women we interviewed regularly told us that cacao and chocolate were both foods and medicines. Almost all of the interviewees agreed that cacao or some of its derivatives had healing properties, in particular, cacao butter. Cacao butter has always been considered a refreshing salve: It alleviates children's chapped skin (diaper rashes), sunburn psoriasis, and skin eruptions caused by the sun; and it is used to eliminate dark skin spots. The use of cacao butter in the preparation of unguents and creams to protect and heal cracked lips caused by cold weather and to heal the nipples of nursing women was also reported.

Cacao butter it is also considered by the persons we interviewed to be an antiseptic and revitalizing product, good to fight wrinkles and cellulitis, for use during massage, and as an emulsion to treat bruises and dislocations. Some also claimed that cacao butter had anti-inflammatory and wound-healing properties, and as a consequence it was recommended to alleviate inflammation of glands caused by throat infections and mumps. It is also taken by some respondents to alleviate colds and asthma.

Some adult women we interviewed used cacao butter as a method to prevent pregnancy. To achieve this aim, cacao butter was shaped into the form of little balls and inserted vaginally about half an hour before sexual contact. One of the women who described the procedure to us said that she had used this method and never became pregnant. Using cacao butter in this way also was said to be useful to relieve pelvic inflammation.

Another medicinal use of cacao was reported to relieve arthritis pain: the green (untoasted) cacao beans of the *morado* (purple) variety are mashed and placed in either alcohol or gasoline (!); then ginger and hot pepper are added. The liquid is applied to the arthritic joints during massage and this is said to reduce pain.

Still another interesting traditional medicinal cure was to take portions of cacao pods (the shell/peel), add indigo dye, and boil. This liquid was said to help eliminate body lice! Other medicinal uses of cacao were identified during interviews, but their efficacy could not be confirmed: boiled cacao tree root used to treat leprosy, syphilis, and urinary track infections, and cacao tree leaves and pod shell/peel used to treat tuberculosis.

Conclusion

The importance of chocolate to Cuban history and culture is clearly demonstrated through the abundant references in Cuban literature, music, and culinary arts. Not only have we seen chocolate in the description of scenes with regional flavor, where the consumption of chocolate is often present, but also as an integral role in daily meals throughout the year, especially as a warming food during winter. We have seen, too, that it is difficult to separate chocolate as food from chocolate as medicine.

Chocolate appears in Cuban life as a companion during times of joy and times of sorrow, during religious festivities and funerals. As part of proverbs and popular sayings, chocolate became an important element of popular wisdom. Chocolate also is used to describe a color: the *chocolate* color or *achocolatado* (similar to chocolate color); it is also used to describe mud, turbid liquids, and the color of landscape features. Chocolate is a powerful symbol throughout Cuba: Chocolate is the black or mulatto Cuban; chocolate is the incarnation of childhood and youth; chocolate is the image whereby mothers show the intensity of their love for their children. Chocolate, too, is allegorical, sometimes so desired that its presence or absence can provoke hostilities.

What we also found during the course of our research on the history of chocolate in Cuba and the various roles it has played in Cuban culture is that there remains much more to find and investigate. What we learned, too, is that this chocolate-related work is not exhausted, as there is still much to be done in tracing even more literature, proverbs, regional recipes, and traditional uses in medicine. What we accomplished in our work left us—as the proverb says, "like water for chocolate"—anxious to continue and further our efforts in the near future. It is our hope that through our chapters and contributions to the present work that we might stimulate the desires and efforts of others to become involved further in this almost unexplored theme of Cuban history—and of the history of all nations.

Acknowledgments

We would like to thank James I. Grieshop and Rami Ahmadi for their strong support and encouragement of our investigation of chocolate in Cuba. The Spanish language text of this chapter prepared by Estrella González Noriega and Niurka Nuñez González was translated into English by Beatriz Cabezon. Minor editorial revisions were made by Louis Evan Grivetti.

Endnotes

1. The Taíno were pre-Columbian indigenous people of the Bahamas, Greater Antilles, and some of islands of the Lesser Antilles. The word Taíno means "good" or "noble," and the Taíno people used it to differentiate themselves from the island's fierce Caribs, their enemies. Although their place of origin is uncertain, it is believed that they were relatives of the Arawak people of the South American Amazon Basin, who migrated to the West Indies by way of Guyana and Venezuela, then into Trinidad, and from there to the Lesser Antilles and Cuba. By the 18th century the Taíno had been decimated by Old World diseases carried by the European colonists and by hardships of the forced labor and brutalities of the plantation system imposed on them.

2. The Siboneys were a pre-Columbian people also called "rock-dwellers." They were the earliest inhabitants of the Caribbean and settled in Haiti, Cuba, and possibly Jamaica. It is believed that the Siboneys migrated to the Caribbean from Venezuela.

3. Nicolás Guillén, writer, poet, and journalist, was active in social protest and a leader of the Afro-Cuban movement in the late 1920s and 1930s. He is considered one of the remarkable voices of Ibero American poetry of the 20th century. He was born in Camagüey, in 1902, and died in Havana in 1989. His work integrates the poetic tradition of Castilian language with Afro-Antilles roots, a blending that results in a deep humanism and beauty that represents the kaleidoscopic culture of Cuba. He was president and founder of the National Writers and Artists Union of Cuba, a responsibility he kept until his death.

4. Cirilo Villaverde was born in 1812. His father was a physician employed by the owners of a sugar plantation who also owned more than three hundred slaves. Villaverde's early exposure to the evils of slavery provided inspiration for his antislavery novel. He studied Latin with his grandfather and later on attended the *Seminario de San Carlos* (San Carlos Seminary/School), where he studied philosophy and earned a law degree in 1834. In 1837, he produced four short novels and the following year wrote what is widely considered the first book published in Cuba, *El espetón de oro* (*The Golden Spear*). In 1839, he completed the first part of his most important work, *Cecilia Valdés*. In 1848, Villaverde was arrested and fled to the United States the following year. While in New York, he completed the second part of *Cecilia Valdés*, producing the definitive version of the work. He died in New York on October 20, 1894 and is buried in Cuba.

5. *Ropa vieja* is the term for shredded beef. The Spanish literally means *old clothes*, not the best name for a dish served at the Philharmonic Society reception where Cuban notables would have been present. Ropa vieja commonly is prepared as a beef stew and includes bay leaves, garlic, onions, green bell peppers, and dry white wine.

6. *Café con leche* literally means coffee with milk but the translation "milk with coffee" would be closer to reality. It is one of several traditional breakfast beverages throughout Cuba and Latin America and commonly is served in the mid-afternoon and evenings as well.

7. This proverb is sometimes given as a synonym for "to be like water for chocolate, that he/she woke up with

the ribbon twisted" means that he/she woke up in a very bad mood.

8. In his book *Cuba with Pen and Pencil*, Samuel Hazard wrote:

   Cuba is divided rather indefinitely, into two unequal portions, the "Vuelta Arriba" or higher valley, and the "Vuelta Abajo" or lower valley. General usage seems to settle the point that the Vuelta Abajo is all that low country lying to the West of Havana.

9. Guajaibón is the highest mountain of the Sierra del Rosario region of western Cuba (2270 feet or 692 meters) and is located approximately 35 miles northeast of Pinar del Río.

10. Don Salvador de Muro y Salazar was the General Captain of Cuba between 1799 and 1812. He faced the first conspiracies against Spanish rule and arrival of the first refugees from Santo Domingo. He defended Cuban interests and championed free trade (without taxation) and promoted a more humane treatment of slaves. He was murdered in Madrid in 1823.

11. The religious order of the Society of Jesus was founded in 1540 by Ignatius of Loyola. It grew rapidly and assumed an important role in the renewal of the Catholic Church. Jesuits were educators, scholars, and missionaries throughout the world. By the 18th century the intellectual standards, economic power, and social influence of the Jesuits were unmatched. Their work in the missions on behalf of the indigenous peoples and their apparent power at royal courts and in the Church aroused the hostility of many lay and clerical adversaries. Finally in 1776, King Carlos III of Spain decreed the expulsion of Jesuits from Spain and from all Spanish possessions.

12. Dolores María Ximeno y Cruz (1866–1934) was born to a wealthy family from Matanzas. Her home was the center of literary gatherings. She collaborated with the Cuban publications *El Fígaro* and *Archivos del Folklore Cubano*. She published her *Memorias de Lola María* (1928–1929) in *Revista Bimestre Cubana*.

13. Alejo Carpentier (1904–1980) is considered Cuba's most important 20th century intellectual. Born in Switzerland (Russian mother, French father), the family moved to Havana when he was an infant. Despite his ethnic heritage he always considered himself Cuban. Carpentier was exiled to France several times for political reasons. In 1966, he became the Cuban Ambassador to France. Carpentier greatly influenced the development of Latin American literature and served as a cultural bridge between Europe and the Americas. He also was a classically trained pianist and in 1946 published *Música en Cuba* (*Music in Cuba*), considered the most comprehensive study of Cuban musical history.

14. *Chorote* is a Caribbean Spanish word equivalent to the Mexican Spanish word *atole*. *Chorote* is a traditional thick beverage made from the flour of roasted corn kernels, with the addition of sugar/honey and cinna-

mon. A *chorote* of cocoa (*chorote* + cocoa) basically is equivalent to the Mesoamerican *champurrado* (see Appendix 1).

15. The Inquisition was a religious tribunal created by the Spanish crown and Christian Catholic Church during the 13th century to combat heresy (see Chapter 4). Inquisition tribunals were widespread not only in Spain and its possessions but throughout medieval Europe, and appeared in the Americas in the 16th century. Carpentier's comments regarding Inquisition policies reveal a "more tolerant" Inquisition regimen in the colonies.

16. The term *La Mancha*, according to some reports, stems from the classical Arabic word *al-Mancha* (wilderness, dry land). La Mancha is the largest plain in the Iberian Peninsula. Cervantes, one of the most influential writers of 16th century Spain, during the so-called *Siglo de Oro*, made La Mancha famous with his classic novel *Don Quixote de La Mancha*. The *manchego* cheese is prepared from sheep's milk and made in this region.

17. Jerez de la Frontera is a municipality in the province of Cádiz in the Andalusia in southwestern Spain. Jerez is widely known for its sherry and brandy production.

18. José Lezama Lima, born in Havana in 1910, was a major Latin American literary figure. He is highly regarded as the father of the current Renaissance in Latin American poetry and recognized as one of the most influential Latin American writers of the 20th century. Lezama Lima's vision of America—in a continental sense— reflects the influences of Amerindians, Africans, and Europeans. He is perhaps best known in English through his novel *Paradiso*, since little of his poetry has been translated.

19. The novel *Paradiso* is a narrative of the childhood and youth of José Cemí, the main character, told in a highly baroque experimental style, that depicts many scenes and experiences consistent with Lezama Lima's own life as a young poet in Havana. The author's choice of *Cemí*, as the family name of the main character, is connected to Taíno religious beliefs. The Taíno religion is centered on the worship of *zemís* or *cemís*: gods, spirits, or ancestors. Some of the gods were agricultural; others were associated with life processes, creation, and death.

20. In many countries coffee is considered an adult beverage inappropriate for children. In Latin America, however, coffee is given to children mixed with milk.

21. Before 1959, Cuba imported chocolate products as well as many others goods from the United States.

22. Rue de Rivoli is one of the most famous avenues of Paris, a commercial street whose shops include the most fashionable names in the world. The name Rivoli commemorates Napoleon's early victory against the Austrian army at the battle of Rivoli, Italy, fought in 1797.

23. *Merienda* is to Spain and Latin America what 5 o' clock tea is to Anglo-Saxon traditions. The verb meaning "to take a *merienda*" is *merendar*. It is a nutritious snack,

particularly for children, one that often includes coffee and milk, or chocolate (as beverage) with bread and butter and sometimes cheese and cold meats.

24. The custom to drink a cup of hot chocolate late at night particularly in cold weather is an old Spanish and French tradition, especially during special celebrations like returning home from midnight mass on Christmas Eve.

25. Translator's Note: Because we did not have available for inspection the whole text, we could not determine if the person who brought the chocolate—as well as the person to whom the chocolate was destined—was male or female. In the Spanish language, subject pronouns, in this case "he" and "she," are omitted, although gender (feminine/masculine) and noun match (singular/plural) is a characteristic of the Spanish language. But with indirect object pronouns such as "him" or "her," there are no differences and both are translated as *le*. Still, it is widely known that anise liquor is favored by Spanish women, whereas Spanish men commonly favored stronger beverages like cognac or rum. The presence of anise in the chocolate cup, therefore, might indicate that in this instance it was chocolate being provided to a woman.

26. Virgilio Piñera (1912–1979) was a Cuban essayist, playwright, poet, and author of short stories. He was well known not only for his artistic works, but for his Bohemian lifestyle. Following a long period of exile in Buenos Aires, Argentina, Piñera returned to Cuba in 1958 several months before Fidel Castro assumed governmental responsibilities. His work includes essays on literature and literary criticism, several collections of short stories, a great number of dramatic works, and three novels: *La carne de René* (*The Flesh of René*), *Presiones y Diamantes* (*Pressures and Diamonds*), and *Las pequeñas maniobras* (*Small Maneuvers*). His work is seen today as a model by new generations of Cuban and Latin American writers.

27. Mercedes Melo Pereira, in her literary analysis of Piñera's novel, *La carne de René* (*The Flesh of René*) wrote: "Regardless [of] the bizarre circumstances and characters, the themes of this novel are based on the transgression of the established norms of the state, the society and the family, i.e. the transgression of all the cultural scaffolding of Western civilization. . . . From this story of conspiracies, assaults and power struggles, Piñera develops a profound reflection about the human being destiny."

28. Saint Sebastian was martyred in the year 287 during persecution of Christians by the Roman emperor Diocletian. He is commonly depicted in art as tied to a post and being shot with arrows. He is the patron saint of athletes and soldiers. In the novel *La carne de René* (*The Flesh of René*), Ramón, the father of René, is a fervent militant for the cause of the flesh and pain. Consequently, he has his son, René, painted in the style of Saint Sebastian, but the painting exhibits a "twist," as Saint Sebastian (i.e., René) is depicted holding a cup of chocolate in his hand—instead of an arrow.

29. Dora Alonso (1910–2001) was a narrator, a journalist, a poet, and a war correspondent. Her writings reveal a deep appreciation of rural Cuban life and human values inspired by a love of nature. Alonso was awarded the Cuban National Prize of Literature in 1988. Her novels *Tierra inerme* (*Defenseless Land*), *Tierra Brava* (*Brave Land*), and *Sol de Batey* (*Sun of Batey*) gained international recognition. She began her career writing poems and stories for children and her publications include *La flauta de chocolate* (*The Chocolate Flute*), *Teatro para niños* (*Theater for Children*), *Tres lechuzas en un cuento* (*Three Owls in a Story*), and *El cochero azul* (*The Blue Coachman*).

30. Onelio Jorge Cardoso (1914–1986) was master of the narrative and short story. He was born in a village in central Cuba and due to poverty could not continue his studies. Nevertheless, Cardoso traveled through Cuba as a merchant, and his experiences allowed him to accumulate a deep knowledge of the Cuban rural environment. Through his journeys he also became acquainted with many persons and later would base many of his characters on the people he had met. His works portray the hardships of fishermen and peasants, while reflecting the joy of life. In 1952, Cardoso was awarded the Cuban National Peace Prize (Premio Nacional de la Paz) for his book *Hierro Viejo* (*Old Iron*). The first edition of his *Cuentos* (*Stories*) was published in 1962 and in the same year he also published *Gente de pueblo* (*Rural Town People*), a collection of interviews with a preface written by Samuel Feijóo. His works have been translated into many languages and have been adapted for ballet, film, and theater.

31. *Palmiche* or *Palma Real* (Royal Palm: *Roystonea regia*) is a native of Cuba.

32. *Sinsonte* is the Caribbean word for mockingbird. Also known in Mexico as *cenzotle*, the name is derived from the Nahuatl word *centzuntli*, meaning "the one who has 400 voices" because it is said to be able to imitate the songs of other birds as well.

33. Patabán, or buttonwood, is the genus *Conocarpus*, native to North America and the shores of tropical America and Africa.

34. Samuel Feijóo (1914–1992) was an eminent Cuban folklorist, novelist, painter, and poet. His extensive works in poetry as well as in the narrative have been compiled in many collections and anthologies. His *Cuentos populares cubanos de humor* (*Popular Humorous Cuban Tales*) (1960–1962) is one of the largest anthologies of Cuban folklore and traditions. His novel *Juan Quinquín en Pueblo Mocho* is probably one of his best known novels. After a lifetime of research and collection of the oral Cuban folkloric traditions he founded, at the Universidad Central Marta Abreu de Las Villas (Central University Marta Abreu de Las Villas), the magazine *Islas* (Islands), a literature and art magazine

that reflects the folklore and linguistic traditions of Cuba.

35. The translation of verses or poetry often robs part of the beauty of it, since a poem, or the lyrics of a song, are not just words but words with a particular rhythm. Nevertheless, we have attempted an English translation of the original Spanish poems or lyrics.

36. Until the last decades of the 18th century, Cuba was a relatively underdeveloped island with an economy based mainly on cattle raising and tobacco farms. The intensive cultivation of sugar that began at the turn of the 19th century transformed Cuba into a plantation society, and the demand for African slaves, who had been introduced into Cuba under the Spanish rule at the beginning of the 16th century, increased dramatically. The slave trade with the West African coast exploded, and it is estimated that almost 400,000 Africans were brought to Cuba during the years 1835–1864. As early as 1532, the blacks formed 62.5 percent of the population. In 1841, African slaves made up over 40 percent of the total population. The flourishing of the Cuban sugar industry and the persistence of the slave trade into the 1860s are two important reasons for the remarkable density and variety of African cultural elements in Cuba. Fernando Ortiz counted the presence of over one hundred different African ethnic groups in 19th century Cuba and estimated that by the end of that century fourteen distinct African ethnicities had preserved their identity in the mutual aid associations and social clubs known as *cabildos*, societies of free and enslaved blacks from the same African group which later included their Cuban-born descendants. Forged in the *cabildos* and amidst the grueling labor at the sugar mills, four major Afro-Cuban groups (Arará, Abakuá, Kongo, Lucumí) are represented in Cuba. According to official statistics, 30 percent of Cuba's population is now black, with the rest being mulatto and white.

37. The Spanish word *bizcochuelo* refers to a traditional cake made with flour, milk, sugar, and eggs. However, in Cuba there is also a variety of mango called *bizcochuelo mango*. This mango variety grows in El Caney (Camagüey), and it is considered unique by its flavor and pulp, due to the climate conditions of the territory. This variety of mango was used by the Cuban pharmaceutical industry to produce an efficient immune system modulator and analgesic, which is applied to patients with HIV/AIDS, and it works as a complement in the antiretroviral therapy. So either way, a tasty cake or a flavorful mango, the chocolate brought from heaven in the story was in very good company.

38. In her book *Afro-Cuba Religions Diaspora: A Comparative Analysis of the Literature*, Sara M. Sánchez explains: "The expression *Santería*, literally, 'the way of the saints,' is often used as an umbrella term for various forms of Afro-Cuban religious traditions, originated from the influences of all ethnic groups that reached the island. The term *santería*, however, is perceived by devotees as a pejorative term. The Afro-Cubans of Yoruba ancestry preferred to call it *Religión Lucumí* (Lucumí Religion). The Lucumí/Yoruba pantheon of gods, reminiscent of Greek mythology, was juxtaposed with those Catholic saints widely venerated in Spanish popular religiosity. The Yoruba are a large ethnic group in Africa, descendants from the Nok Civilization (900 BCE–200 CE) of Central Nigeria and today constitutes 21% of Nigeria's total population with approximately 30 million individuals throughout the region of West Africa."

39. The Saint Funeral (*Velorio de santo*) is a religious festivity, a funeral and a vigil, celebrated in the Caribbean including the Caribbean regions of Venezuela and Colombia. It takes place on the birthday of the saint protector of the town, or on any other day, to fulfill a promise that has been made to the saint. In this case, the person who organizes it invites his friends and pays for all the candles as well as for the beverages and food for his guests. Music and dancing are an important part of these celebrations. At night time the candles are lighted and placed around the figure of the saint, and the celebration with food and music begins.

40. Antonio Núñez Jiménez (1923–1998) is internationally known through his scientific work in geography. His spirit of exploration and taste for adventure brought him to all continents. He was particularly interested in the relation between human activities and nature. After his research on the Cuban archipelago and its native population, he continued with the study of the Mayan civilization of Mesoamerica and the Incas in Peru.

41. Mosquitero is a municipality of Baracoa in the province of Guantánamo.

42. Concepción T. Azola, writer of children's literature, was born in Cuba in 1930. Her work, dedicated to children's stories, games, and poems, has become popular among Latin American children playgrounds.

43. The Spanish chronicler Gonzalo Fernández de Oviedo describes the Taíno celebrations saying: "They [the Taínos] remember past and ancient things, and this was in their songs and dances, and they called them areito, which is what we call dancing as we sing. And to greater propagate their joy and mirth they would take each other by the hand . . . and form a circle of dancers all singing . . . these dances can last three or four hours."

44. Esteban Salas, the first Cuban composer, was born in 1725 in the city of Havana and died in 1803. He studied philosophy and theology at the *Seminario de San Carlos*. In 1764, he settled in Santiago de Cuba, capital of the Oriente region and was named Cathedral Master of Music.

45. In 1880, four years after the end of the Ten Years' War, the Spanish courts approved the abolition law, which provided for an eight-year period of *patronato* (tutelage) for all slaves liberated according to the law. This act still amounted to indentured servitude as slaves were required to spend those eight years working for their masters at no charge. On October 7, 1886, slavery was

finally abolished in Cuba by a royal decree that also made the *patronato* illegal.

46. Turuntutún, in this context, reflects onomatopoeia as used in the Caribbean to describe strong pleasurable emotions. It is also a joyful song.

47. Carnival is celebrated in all Cuba, but it is in the city of Santiago de Cuba and Havana where carnival activities reach their peak. Since colonial times in Santiago de Cuba, neighborhood groups have taken to the streets in June and July in masked celebrations known as *fiestas de mamarrachos*, which extend from St. John's Day (June 24) to St. Ann's Day (July 26). In Havana, the *cabildos*, or associations of black ethnic groups, hold public celebrations on the *Día de los Reyes*, or Epiphany (January 6). In both cities, these Catholic holidays have been opportunities for public display of African dress, dance, and musical instruments.

48. In this context the word *guajira* is a musical form that evolved from a style called *Punto Guajiro* (also called *Punto Cubano*). The origins are Andalusian. It is the predominant local music from the western and central provinces of Cuba. This style began to become popular around the end of the 18th century. The performers are known more as poets, not singers, and a distinguishing feature is that the lyrics often are improvised.

49. Ignacio Piñeiro (1888–1969), born in Havana, was a composer and singer. He gained great fame for the acceptance and recognition of the *son* as a traditional Cuban style rhythm. Cuban *son* is both music and dance. The word *son* means "sound" but also "way" (way of the sound), and it is used today in Cuba to identify a genre of music that began in the 20th century in the eastern mountainous area of Cuba, the land of the *son*.

50. Cubans often use the word *chocolate* to characterize black, mulatto, and mestizo peoples, as of "chocolate color." In the past, such designations could have been derogatory. Today, in Cuba, we did not find any derogatory meaning or racist connotations; it is simply used to indicate a skin color.

51. After 1959, the cacao bean shells were used to feed animals.

52. One peasant family we interviewed reported that long ago people used to keep cacao balls in large containers called *güiros* or galleons, covered with a lid attached to the gourd with hinges. Today, the Afro-Cuba *güiro* is a fragile musical instrument, a hand drum, traditionally made from a smaller gourd. This *güiro* is used in traditional Cuban music such as *danzón, cha-cha-cha, guajira, charanga*, and others.

53. Sago is a powdery starch made from the processed pith found inside the trunks of the sago palm (*Metroxylon sagu*).

54. Nitza Villapol is a well-known person who taught generations of Cubans how to cook. A home economist, Villapol became well versed in making do with little while studying in England during the Second World War. She put these skills to work decades later in what became her greatest challenge of the last decades: how to cope with food shortages. Although Villapol has been known for her cookbook *Cocina al minuto* (*Cooking to Order*), she has written two other important cookbooks: *Sabor a Cuba* (*The Flavor of Cuba*) and *El arte de la cocina cubana* (*The Art of Cuban Cuisine*).

55. Madelaine Vázquez Gálvez, born in Havana in 1959, is a food technology specialist and consultant for the Eco-Restaurant, *El Bambú*, located in the Havana Botanical Garden. She published *Cocina ecológica en Cuba* (*Ecological Cooking in Cuba*) in 2001 and *Educación alimentaria para la sustentabilidad* (*Nutrition and Sustainability Education*) in 2003. Today she lives in Havana.

References

1. Villaverde, C. *Cecilia Valdés*. Havana, Cuba: Editorial Letras Cubanas, 1980; Volume 1, p. 233.

2. Villaverde, C. *Cecilia Valdés*. Havana, Cuba: Editorial Letras Cubanas, 1980; Volume 1, pp. 19–20.

3. Villaverde, C. *Cecilia Valdés*. Havana, Cuba: Editorial Letras Cubanas, 1980; Volume 1, p. 244.

4. Villaverde, C. *La joven de la flecha de oro y otros relatos: EL ave muerta*. Havana, Cuba: Editorial Letras Cubanas, 1984; p. 42.

5. Villaverde, C. *La joven de la flecha de oro y otros relatos: El Penitente*. Havana, Cuba: Editorial Letras Cubanas, 1984; pp. 601–604.

6. Villaverde, C. *La joven de la flecha de oro y otros relatos: El Penitente*. Havana, Cuba: Editorial Letras Cubanas, 1984; p. 664.

7. Villaverde, C. *Excursión a Vuelta Abajo*. Havana, Cuba: Editorial Letras Cubanas, 1981; p. 181.

8. Villaverde, C. *Excursión a Vuelta Abajo*. Havana, Cuba: Editorial Letras Cubanas, 1981; p. 237.

9. Iglesia, A. de la. *Tradiciones Cubanas. Viaje alrededor de un mango*. Havana, Cuba: Instituto del Libro, Ediciones Huracán, 1969; p. 243.

10. Iglesia, A. de la. *Tradiciones Cubanas. Secreta Venganza*. Havana, Cuba: Instituto del Libro, Ediciones Huracán, 1969; p. 278.

11. Iglesia, A. de la. *Tradiciones Cubanas. La expulsión de los Jesuitas*. Havana, Cuba: Instituto del Libro, Ediciones Huracán, 1969; p. 341.

12. Carrión, M. de. *Las impuras*. Havana, Cuba: Instituto del Libro, Ediciones Huracán, 1972; p. 350.

13. Ximeno y Cruz, D. M. *Memorias de Lola María*. Havana, Cuba: Editorial Letras Cubanas, 1983; p. 107.

14. Ximeno y Cruz, D. M. *Memorias de Lola María*. Havana, Cuba: Editorial Letras Cubanas, 1983; p. 150.

15. Carpentier, A. *La consagración de la primavera*. Havana, Cuba: Editorial de Arte y Literatura, Letras Cubanas, 1979; p. 382.

16. Carpentier, A. *La consagración de la primavera*. Havana, Cuba: Editorial de Arte y Literatura, Letras Cubanas, 1979; p. 389.

17. Carpentier, A. *Novelas y Relatos, El Camino de Santiago*. Havana, Cuba: Editorial de Arte y Literatura, Letras Cubanas, 1981; pp. 306–307.

18. Carpentier, A. *Novelas y Relatos, El Camino de Santiago*. Havana, Cuba: Editorial de Arte y Literatura, Letras Cubanas, 1981; pp. 333–334.

19. Carpentier, A. *Novelas y Relatos, Viaje a la Semilla*. Havana, Cuba: Editorial de Arte y Literatura, Letras Cubanas, 1981; pp. 340–341.

20. Carpentier, A. *El Siglo de las Luces*. Havana, Cuba: Editorial de Arte y Literatura, Letras Cubanas, 1974; p. 15.

21. Carpentier, A. *El Siglo de las Luces*. Havana, Cuba: Editorial de Arte y Literatura, Letras Cubanas, 1974; p. 24.

22. Carpentier, A. *El Siglo de las Luces*. Havana, Cuba: Editorial de Arte y Literatura, Letras Cubanas, 1974; pp. 294–295.

23. Lezama Lima, J. *Paradiso*. Havana, Cuba: Ediciones Unión, 1966; p. 248.

24. Lezama Lima, J. *Paradiso*. Havana, Cuba: Ediciones Unión, 1966; p. 148.

25. Lezama Lima, J. *Paradiso*. Havana, Cuba: Ediciones Unión, 1966; pp. 573–574.

26. Lezama Lima, J. *Paradiso*. Havana, Cuba: Ediciones Unión, 1966; p. 312.

27. Lezama Lima, J. *Paradiso*. Havana, Cuba: Ediciones Unión, 1966; p. 327.

28. Lezama Lima, J. *Paradiso*. Havana, Cuba: Ediciones Unión, 1966; p. 145.

29. Lezama Lima, J. *Paradiso*. Havana, Cuba: Ediciones Unión, 1966; p. 574.

30. Piñera, V. *La carne de René*. Havana, Cuba: Ediciones Unión, 1995; pp. 34–37.

31. Piñera, V. *La carne de René*. Havana, Cuba: Ediciones Unión, 1995; p. 82.

32. Piñera, V. *La carne de René*. Havana, Cuba: Ediciones Unión, 1995; p. 144.

33. Piñera, V. *La carne de René*. Havana, Cuba: Ediciones Unión, 1995; pp. 176–177.

34. Piñera, V. *La carne de René*. Havana, Cuba: Ediciones Unión, 1995; pp. 187–188.

35. Piñera, V. *La carne de René*. Preface by Antón Arrufat. Havana, Cuba: Ediciones Unión, 1995; p. 16.

36. Alonso, D. *Ana y los amores*. Havana, Cuba: Editorial Letras Cubanas, 1980; p. 168.

37. Alonso, D. *La flauta de chocolate*. Havana, Cuba: Editorial Gente Nueva, 1980.

38. Cardoso, O. J. *Cuentos. Moñigüeso*. Havana, Cuba: Editorial Arte y Literatura, 1975; p. 50.

39. Cardoso, O. J. *Cuentos. El Cuentero*. Havana, Cuba: Editorial Arte y Literatura, 1975; p. 59.

40. Cardoso, O. J. *Cuentos. Los Sinsontes*. Havana, Cuba: Editorial Arte y Literatura, 1975; p. 263.

41. Feijóo, S. *Cuentos populares cubanos de humor. El isleño que le mandó un regalo a su mujer*. Havana, Cuba: Editorial Letras Cubanas, 1982; pp. 60–61.

42. Feijóo, S. *Cuentos populares cubanos de humor. Cuartetas en la parranda*. Havana, Cuba: Editorial Letras Cubanas, 1982; pp. 309–310.

43. Feijóo, S. *El negro en la literatura folklórica cubana*. Havana, Cuba: Editorial Letras Cubanas, 1980; p. 46.

44. Feijóo, S. *El negro en la literatura folklórica cubana: Las dos hijas*. Havana, Cuba: Editorial Letras Cubanas, 1980; p. 166.

45. Feijóo, S. *Juan Quinquín en Pueblo Mocho*. Havana, Cuba: Editorial de Arte y Literatura, 1976; p. 119.

46. Feijóo, S. *Juan Quinquín en Pueblo Mocho*. Havana, Cuba: Editorial de Arte y Literatura, 1976; p. 205.

47. Feijóo, S. *Vida completa del poeta Wampampiro Timbereta*. Havana, Cuba: Editorial Letras Cubanas, 1981; p. 192.

48. Núñez Jiménez, A. *Cuba: la naturaleza y el hombre*. Havana, Cuba: Editorial Letras Cubanas, 1984; *Volume 2, Bojeo*, p. 471.

49. Núñez Jiménez, A. *Cuba: la naturaleza y el hombre*. Havana, Cuba: Editorial Letras Cubanas, 1984; *Volume 2, Bojeo*, p. 485.

50. Núñez Jiménez, A. *Cuba: la naturaleza y el hombre*. Havana, Cuba: Editorial Letras Cubanas, 1984; *Volume 2, Bojeo*, p. 487.

51. Núñez Jiménez, A. *Cuba: la naturaleza y el hombre*. Havana, Cuba: Editorial Letras Cubanas, 1983; *Volume 3, Geopoética*, p. 18.

52. Personal communication provided by Angel Centeno, La Alegría. Baracoa.

53. Personal communication provided by Aristides Smith, Baracoa.

54. Personal communication provided by Armando Romero, La Alegría, Baracoa.

55. Alzola, C. T. *Folklore del niño cubano*. Havana, Cuba: Universidad Central de Las Villas, Santa Clara, 1962; pp. 49–50.

56. Alzola, C. T. *Folklore del niño cubano*. Havana, Cuba: Universidad Central de Las Villas, Santa Clara, 1962; pp. 50–51.

57. Alzola, C. T. *Folklore del niño cubano*. Havana, Cuba: Universidad Central de Las Villas, Santa Clara, 1962; p. 60.

58. Alzola, C. T. *Folklore del niño cubano*. Havana, Cuba: Universidad Central de Las Villas, Santa Clara, 1962; p. 75.

59. Carpentier, A. *Music in Cuba*. Translated by Alan West-Durán. Minneapolis: University of Minnesota Press, 2001.

60. Carpentier, A. *Music in Cuba*. Translated by Alan West-Durán. Minneapolis: University of Minnesota Press, 2001; pp. 72–74.

61. Carpentier, A. *Music in Cuba*. Translated by Alan West-Durán. Minneapolis: University of Minnesota Press, 2001; pp. 106–110.

62. Carpentier, A. *Music in Cuba*. Translated by Alan West-Durán. Minneapolis: University of Minnesota Press, 2001; pp. 59–60.

63. Personal Communication provided by Manuel (Tano) Durán from Mosquitero, Baracoa.

64. Personal communication provided by José U. Toirac, San Luis, Baracoa.

65. Personal communication provided by Manuel (Tano) Durán from Mosquitero and Matías Suárez from Jamal. Both from the area of Baracoa.

66. Personal communication provided by Angel Centeno from La Alegría, Baracoa.

67. Goodman, W. *Un artista en Cuba*. Havana, Cuba: Editorial Arte y Literature, 1965; p. 130.

68. Núñez Jiménez, A. *Cuba: la naturaleza y el hombre*. Havana, Cuba: Editorial Letras Cubanas, 1984.

69. The Grocer (El Bodeguero). Released in a 45 RPM as El Bodeguero/Quizas, Quizas, Quizas. Lyrics by Richard Egues. Performer: Nat King Cole. Credits: Conductor, Armando Romeu, Jr. Label: Capitol Records. Catalog #: F 17869, F 17874. Recording first published 1958: Electric and Musical Industries, Limited, Great Britain.

70. El pregon del chocolate (The Chocolate Proclamation). Lyrics: Isaac Delgado. From the Album, La Formula: El Pregon del chocolate. Original release date: December 19th, 2000. Label: Ahi Name. Copyright 2007 Ahi Nama. ASIN: B000WP7Z9C.

71. Mi chocolate (My Chocolate). Performers: Van Van. From the album: Llego Van Van (Van VAn Is Here). Comjposer: Juan Formell; Arranger: Juan Formell; Lead Singer: Pedro Calvo. Label: Havana Caliente. Catalog Number 83227.2. Released November 1999. CD track number 9.

72. Núñez González, N., and González Noriega, E. *El cacao y el chocolate en Cuba*. Havana, Cuba: Mar y Pesca, 2005; p. 96.

73. Ucar García, compiler. *¿Gusta Usted? Prontuario culinario y necesario*. Havana, Cuba: Ucar García, 1956; p. 187.

74. Ucar García, compiler. *¿Gusta Usted? Prontuario culinario y necesario*. Havana, Cuba: Ucar García, 1956; pp. 571–572.

75. Ucar García, compiler. *¿Gusta Usted? Prontuario culinario y necesario*. Havana, Cuba: Ucar García, 1956; p. 214.

76. Ucar García, compiler. *¿Gusta Usted? Prontuario culinario y necesario*. Havana, Cuba: Ucar García, 1956; p. 286.

77. Ucar García, compiler. *¿Gusta Usted? Prontuario culinario y necesario*. Havana, Cuba: Ucar García, 1956; p. 289.

78. Ucar García, compiler. *¿Gusta Usted? Prontuario culinario y necesario*. Havana, Cuba: Ucar García, 1956; p. 292.

79. Villapol, N., and Martínez, M. *Cocina al minuto*. Havana, Cuba: Artes Gráficas Roger A. Queralt, 1956/1980; pp. 250–251.

80. Vázquez Gálvez, M. *Cocina ecológica en Cuba*. Havana, Cuba: Editorial José Martí, 2001; p. 200.

CHAPTER

Establishing Cacao Plantation Culture in the Atlantic World

Portuguese Cacao Cultivation in Brazil and West Africa, Circa 1580–1912

Timothy Walker

Introduction

Theobroma cacao, or cocoa, originated near the headwaters of the Amazon River of South America and grew wild in the rain forests of the Amazon basin. Although they did not popularize chocolate consumption among Europeans, Portuguese colonists in Brazil were pioneers in the commercial production of cacao.[1] They used forced labor (Native Americans and Africans) to collect or cultivate and export a highly lucrative cacao crop from the 16th to the 19th centuries. When Brazilian independence loomed, Portuguese authorities became the first to transplant not only the cacao plant to their West African colonies, but also an entire plantation system for cacao production based on forced labor that endures to the present day [1]. The destinies of millions of African slaves on both sides of the Atlantic Ocean were shaped by the transatlantic dissemination—and vast expansion in production—of cacao under Portuguese colonial rule.

This chapter describes the origins of commercial cacao production in the Portuguese Atlantic colonies. Further, I explore the connection between cocoa and forced labor across the tropical Atlantic world and discuss the nexus within the Portuguese maritime colonial network of two enormously important commodities: slaves and chocolate. In the 18th and 19th centuries, Portuguese planters and colonial officials expanded the cocoa trade in the Atlantic world and helped to establish the parameters of the modern industry's growing regions. In the 20th century, Brazil and West Africa evolved into the world's leading cacao production zones.[2]

Furthermore, this chapter focuses on the development of cocoa plantation systems in Brazil and West Africa, and on consignments of young cacao trees sent across the Atlantic to São Tomé and Príncipe, destined for the large-scale production of chocolate, which resulted in great profits that accrued

Chocolate: History, Culture, and Heritage. Edited by Grivetti and Shapiro
Copyright © 2009 John Wiley & Sons, Inc.

in Lisbon. The cacao seedlings sent to West Africa followed centuries-old slave routes to Portuguese colonies that supplied laborers who grew cacao in Brazil. The Portuguese carried out this Afro-Brazilian exchange of slaves and cacao during the 18th and early 19th centuries, at the exact moment when abolition movements began to gain traction in Europe and North America. Although chocolate consumption burgeoned while demand for slaves diminished due to the 19th century British-led abolition movement among colonizing nations, throughout most of the Atlantic world, cocoa grown in Portuguese-speaking Atlantic enclaves actually helped to perpetuate plantation agriculture—with its attendant forced labor systems—even into the 20th century. This chapter closes with comments about the roles of British abolitionism and of British chocolate confectioners in curbing labor abuses in the Portuguese African colonies.[3]

Cultivation in Brazil

Cacao is indigenous only to the tropics of the Americas; it was not known in Europe prior to the Columbian Exchange [2]. In northern Brazil, cacao grew in abundance along the waterways of *Amazônia* and in the dense forests of the Grão Pará and Maranhão territories; recent genetic research has established the origins of all cacao in the headwaters of the Amazon River [3]. Since time immemorial, indigenous South American peoples had gathered this succulent fruit in the wild, never needing to cultivate the plant to satisfy the requirements of their subsistence societies. *Túpi* natives opened cacao pods and sucked the nutritious and tasty pulp from the beans, which they fermented in heaps covered with large leaves and then spread out on woven mats to dry in the sun [4]. The *índios* in Brazil believed that cacao had medicinal qualities, as well; they recognized chocolate as a mild stimulant that could provide sustaining energy to combat hunger and fatigue [5–8].

By the middle of the 17th century, Spanish *hidalgos* and their ladies, both in the American colonies and in the Spanish metropôle, fancied drinking petite cups of chocolate morning, noon, and night. They had changed the native recipe, however, mixing cocoa with American-grown vanilla or sugar and imported seasonings like anise seed, pepper, and cinnamon, all brought from outside the Americas, and whipping them into a hot frothy beverage [9]. Contemporary belief among European colonists held that drinking chocolate sustained the notoriously delicate health of the Spanish elites in *América Latina*, as well as in the Philippines. But consumption of cacao in Spanish-controlled areas extended down through colonial society to the lowest levels: *mestizo* laborers in Central America were also

fond of their cocoa breaks and would down tools sometimes twice a day to imbibe a reviving cup of the flavorful concoction [10, 11].

The Spanish craving for chocolate created an auspicious market opportunity for exporting cacao from Brazil. Because of its value and nutritional utility, the Portuguese crown and colonial administrators wished to encourage Brazilian cacao production. Between 1580 and 1640, after the young Portuguese King Sebastião died without an heir, Portugal and its colonies were subsumed under the control of the Spanish crown. Earlier trade barriers fell and New Spain became one of the main regions to import and consume Brazilian cacao. During this early period, native *Túpi* labor gangs controlled by Jesuit missionaries gathered most of the wild, forest-grown cacao exported from northeastern Brazil [12].

Cacao has been an important economic commodity in Brazil for over 400 years, both as a culinary and medical product. Cacao came to be exported from two Brazilian regions: the Amazon, where it grew in abundance naturally, and Bahia, where it had to be transplanted and cultivated on plantations. Although Bahia state would not develop into the primary location of cacao production in Brazil until the late 1800s, the hinterland of the original colonial capital city, Salvador de Bahia, was the first place in Portuguese America where European settlers systematically cultivated cacao trees. In Bahia, Brazilian cacao was first planted and tended as a crop, as opposed to being gathered from indigenous trees growing in the wild. Of necessity, compelled Native Americans and Africans provided much of the labor for this initiative.

The very earliest cacao grown systematically as a commodity in Bahia was cultivated by the Jesuits in their missionary gardens at Olivença, Camamu, and Canavieiras, all coastal communities south of Salvador (Fig. 40.1). Olivença, approximately 10 kilometers south of Ilhéus, was probably the first site, given that the Society of Jesus had established a mission and sugar estate there in 1563 [13]. It is probable that they transplanted cacao to this site from the Amazon jungle, more than 2000 kilometers to the north, during the second half of the 17th century [14]. From there, cacao cultivation spread to the large Jesuit mission at Camamu, situated on a bay at the mouth of the Trindade River, 80 kilometers north of Ilhéus [15]. Jesuits had recognized the nutritional and medicinal properties of cacao since at least the mid-16th century [16]. It is probable, therefore, that medicinal uses were their motivation for cultivating the plant, which was accomplished initially on a modest scale.

In 1664, the viceroy of Brazil, Dom Vasco de Mascarenhas, wrote from Salvador da Bahia to a Jesuit missionary in the *capitanéa* (colonial administrative

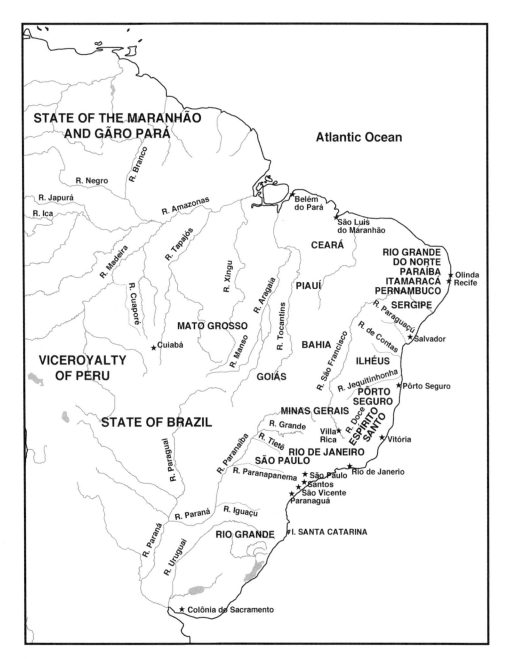

FIGURE 40.1. Portuguese America in 1750.

zone) of Ceará, inquiring about introducing cacao trees into Bahia *en masse* from the Amazon rain forest [17]. The following year, the governor general of Brazil wrote to his subordinate governor in Belém, capital of Grão Pará, to formally request that a quantity of cacao plant cuttings, along with directions for their proper cultivation, be carefully packed and sent for transplant in the fertile hinterlands of Bahia [18].

The Portuguese Crown and Overseas Colonial Council explicitly backed this initiative in 1679, when King Pedro II issued a royal directive (*carta régia*) that not only authorized but encouraged all Brazilian landowners to plant cacao trees on their property [19].

Through his letter, the Portuguese monarch sought to expand production of a lucrative crop among agricultural producers whom colonial agents in Brazil had authority to tax. The number of colonial landholders who attempted to take advantage of the king's ruling can never be known, but certainly some farmers in *América Portuguesa* began to experiment with cocoa growing around Bahia; systematic plantation-based cultivation of cacao in Brazil dates from this era, having been stimulated by royal permission [20].

The small plantations experimenting with cacao that grew up initially around the "Bay of All Saints" and on the southern coast of Bahia in the 17th century would serve as examples for later large-scale production facilities. In the 19th century, cacao plantations would expand in force to southern Bahia, making the area around Ilhéus one of the top cacao-producing regions in the world by the early 20th century [21].

Prior to the 1880s, the cacao trade in Brazil relied universally (but not exclusively) on compelled, exploited, or enslaved labor. Whether collected from wild-growing *cacaueiros* (cocoa trees) in the bush or from trees cultivated in neat plantation rows, at every stage of Brazilian cacao production, coerced human labor played a central role in the gathering and processing of this exceptionally valuable commodity [22].

To focus mainly on Bahia, however, risks providing a distorted picture of the region's importance for chocolate production, especially for the period prior to the 19th century. In fact, the majority of cacao production in Brazil until the latter quarter of the 1800s remained centered in the northern *capitanéa* of Pará, in the environs of the regional capital, Belém. Organized collecting and husbanding of *Theobroma cacao* in Pará began after 1678, following the arrival of a royal order from Lisbon, dated November 1, 1677, that mandated systematic cultivation [23]. By 1749, about 7000 *cacaueiros* had been planted along rivers in Amazonia, generating admirable profits for merchants in Belém de Pará and Lisbon.

Cacao as a Medicinal Plant

There was, in fact, an important concurrent medical dimension that shaped imperial and ecclesiastical interests regarding cacao. Contempory Europeans during the early modern period—missionaries and colonial authorities alike—also considered cacao a medicinal plant. Jesuit *padres* in South America, as elsewhere in the Iberian colonial world, excelled in discovering and experimenting with indigenous medicinal substances, knowledge that they carefully recorded and disseminated among their brethren. Chocolate had long been known to the Society of Jesus as a medicinal substance. About 1580, Jesuits working in Paraguay compiled a detailed manuscript volume of native remedies found in that land, as well as in Chile and Brazil. The weighty tome's very first entry describes the "Virtues of Cacao." According to the text, prepared chocolate had a number of salubrious qualities and the power to "open the [body's] passages . . . comfort the mind, the stomach, and the liver, aid asthmatics . . . and those with cataracts" [24]. Such attributes gave an added incentive to cultivate cacao trees at Jesuit communities, even if on a limited scale.

Most permanent Jesuit colonial missions around the world operated medical facilities—typically an infirmary and a pharmacy—from which they dispensed medical compositions for a profit [25]. In many regions of colonial Brazil, Jesuit pharmacies were the only source of prepared medicines. One standard piece of pharmacy equipment was a grinding stone, necessary to prepare many medicines, including chocolate. For example, in 1757, an inventory of the Jesuit pharmacy of Belém de Pará included some objects called "*chocolateiros*"—stone or ceremic grinding devices (similar to the Aztec *metate*) used for the preparation of roasted cacao beans, as well as other medicinal substances that had to be pulverized [26].

The missionaries also learned how to extract cocoa butter from cacao and sold it as a remedy for skin maladies. By the late 18th century, cocoa butter (*manteiga de cacão*) was being used therapeutically in colonial military hospitals and infirmaries throughout the Portuguese empire. Colonial soldiers and officials in the tropics soothed chafed, dry, or abraded skin with cocoa butter; they employed it as a standard regular treatment for heat rashes, or for more serious skin disorders like shingles. In Portuguese India, missionary pharmacies stocked quantities of Brazilian cocoa butter for daily application to infirmary patients, and for retail distribution to the general population; the *botica* (pharmacy) of the Convent of Santo Agostinho in Goa listed regular monthly purchases of *manteiga de cacão* between 1807 and 1835 [27]. In fact, cocoa butter from Brazil could be found in the stocks of colonial pharmacies from Macau to Timor, Mozambique, and São Tomé, and in the medical chests of most ships of the Portuguese India fleet during the late 18th and 19th centuries [28].

Cacao, then, assumed a key role as a cash crop for Jesuit missions in Brazil. The cacao plant provided several ways to earn revenue that in turn helped support Jesuit missionary activities. The Jesuits expanded their endeavors to collect, process, and export cacao during the 17th and 18th centuries because they depended on the income that cacao earned. Production during this period was centered in the Amazon rain forest; Native American workers performed most of the labor, either by compulsion or in exchange for meager wages.

Missionaries, Native Americans, and Wild Cacao Collection in the Amazon

Coerced labor and cacao have a long mingled history in Brazil. Jesuit missionaries in the 17th century had attempted, with notable success, to monopolize the cacao trade in Brazil, but not through systematic cultivation. Cacao's abundance in the Amazon rain forests was so great that, instead of founding cacao plantations, the Jesuits simply organized expeditions of knowledgeable native workers to scour the interior jungle and riverbanks, gathering the wild-growing *foresteiro* variety of cacao pods [29]. However, these indigenous laborers required a great deal of autonomy as they spread out through the forest to work. Because they were being forced to work in lands that they knew intimately, those who became dissatisfied with the arduous travail of cacao harvesting could simply flee.

Hence, the Jesuits' method of forcing natives to collect cacao in the wild with little supervision was fraught with difficulties, since the opportunities for escape were countless [30]. Harvesting wild cacao in the Amazon thus did not lend itself to the exclusive employment of compulsory labor. Although enslaved natives participated as porters in rain forest cacao expeditions, the Jesuits had to find other incentives to convince skilled *índios* to collect cacao pods without absconding into the bush.

Chocolate's value was sufficient for the Jesuits to experiment with hiring free native laborers to gather cacao. Free indigenous laborers, however, inevitably demanded relatively high compensation from the Jesuit brothers for cacao pod collection. Knowledgeable rain forest natives understood that the Jesuits prized them for their skill in consistently finding the most valuable *cacaueiro* trees. The Jesuits also recognized that *Túpi* Indians displayed a conscientious habit in stripping only the ripest pods, thus leaving green pods to mature for future collection [31]. In practice, therefore, cacao expeditions into the Amazon jungle frequently employed a mixture of paid indigenous workers—"free" native laborers familiar with cacao who received wages (based on volume gathered and time worked) to locate and harvest the ripe pods—and compelled laborers or chattel slaves to do most of the heavy common jobs: carrying cacao out of the bush, paddling dugout canoes, and making camp.

Nevertheless, despite their chronic labor problems, Jesuit attempts to make Amazonian forest-grown cacao profitable persisted into the 18th century. In lower *Amazônia*, the Jesuits planted a cacao zone "consisting of 40,000 trees on the right [south] bank of the river, a little below Óbidos" in the middle of the 18th century [32]. Poorly compensated native laborers tended these *cacaueiros*, spread broadly and unmethodically along the river shoreline. Collection and transport of the ripened pods using large dugout canoes (*pirogues*) was thus simplified; merchant houses in the regional capital, Belém de Pará, coordinated export of the fermented and dried beans.

By 1678, expanding Jesuit missionaries had begun to send canoes laden with enslaved native *índio* work gangs far into the *Sertão* (the interior of Maranhão and Pará) to collect wild cacao pods. Such expeditions often met with grave difficulties when confronted by roaming slave hunters and crown revenue collectors [33]. In a rare petition from Jesuit Padre Filipe de Borja to the governor of Pará, dated October 31, 1731, the missionary leader asked that the *pirogues* of the Jesuit

College of Maranhão, while engaged in collecting cacao deep in the interior along the river Solimões, may not be "molested" by government agents, nor that their native Indian "servants" (compelled laborers) be taken from them by meddling parties of *bandeirantes* (roving slave raiders) [34].

An Early Attempt to Transplant Cacao to Portuguese Colonies Outside Brazil

The promise of rich chocolate profits prompted Italian architect and Portuguese royal appointee António Giuseppe Landi to ship cacao seeds from Grão Pará to lands he held in the Portuguese Madeira Islands off the coast of Morocco. Portuguese King José I had dispatched Landi to Brazil to collaborate on a cartographic survey of the colony's borders. In a letter to Brunelle written from Grão Pará on August 23, 1762, Landi informed his correspondent about his impending voyage to Funchal, Madeira, about the progress of his "experiments with the cultivation of seeds of diverse plants," and that he was sending cacao seeds to Madeira "for the production of chocolate" [35]. Landi's hopes for founding an estate closer to Europe with which to make a chocolate fortune would come to naught, however, since Madeira's climate and northerly latitude would not support cacao cultivation.

Landi's idea for transplanting cacao to islands off Africa was both prescient and sound, although his Madeira folly had little impact on the expanding Portuguese colonial cacao industry in the Atlantic. However, cacao production in Brazil would expand enormously at that time, during the reign of King José I (1750–1777), due to the intervention of the Portuguese prime minister, the Marquês de Pombal. In 1755, Pombal, his attention drawn to the potential for profit through expanded agricultural commerce in northern Brazil, initiated a new monopolistic state-regulated trading company, the *Companhia do Grão-Pará e Maranhão*. After ten years of intensive state-sponsored development, Brazilian cacao exports rose to constitute, on average, more than 10 percent of total world production, a level sustained for three decades [36]. By the late 18th century, plantations in the environs of Belém de Pará were exporting about 1000 tons of cacao annually, drawn from an estimated 700,000 cultivated cacao trees [37]. Once again, most of the labor came from imported African slaves or compelled native *índios*.

Cacao exports from Brazil rose markedly during the Napoleonic Wars (1790s to 1815), when production and shipment elsewhere were disrupted by hostilities among Europe's great colonial powers [38, 39]. Brazil's isolation from the fighting allowed for expanded production and the capture of a larger percentage of world market share. Thus, from 1794 to 1822, Brazilian cacao production increased to account for between 12 and 27 percent of average annual global production, depending on the year [40]. Almost all of this production, most of it still collected in the wild, was centered in the Grão Pará and Maranhão districts—in the Amazon rain forest of northern Brazil.

By 1800, cacao had become the principal and most lucrative crop exported from Belém de Pará by the *Companhia do Grão-Pará e Maranhão* [41]. During the six decades prior to Brazilian independence (1823), cacao accounted for over half of Grão Pará's total export earnings [42]. The extent or volume of cacao production further south in Bahia did not approach these levels until the end of the 19th century, just at the end of the legal slavery era.

Cacao Plantations and Production in Bahia

Despite cacao's medicinal utility and expanding value in the global market, large-scale commercial expansion of production in southern Bahia (about 2200 kilometers south of Grão Pará and Maranhão) did not begin until 1746. The earliest site was on a plantation called the *Cubículo*, on the north bank of the Pardo River in the municipality of Cannavieiras, approximately 300 kilometers south of Salvador by sea. The Portuguese plantation owner, António Dias Ribeiro, used seedlings brought from Grão Pará by an associate, the French immigrant Colonel Louis Frederic Warnaux [43]. Ribeiro employed African slave laborers purchased in Salvador to cultivate his large holdings with cacao trees that thrived and, in an astonishingly short time (as little as three to four years), proved profitable. From there, larger-scale cacao cultivation spread into the coastal lowlands of the Ilhéus region, north of Cannavieiras, beginning in 1752 [44].

By the 1770s, most cacao production within Bahia state had become concentrated in the region around Ilhéus (prior to 1759 an autonomous *capitânia*) and expanded tremendously. The Ilhéus district benefited from good navigable waterways (like the Pardo, de Contas, and Jequitinhonha rivers) that gave access for cargo boats to plantation fields in the interior [45], and a small port, Ilhéus town, that could accommodate coastal transport vessels for the lading of cacao. The first recorded cargo of cacao beans shipped from Ilhéus, a humble consignment weighing just under one ton, left the harbor for export from Salvador in 1778 [46]. During the 1780s, the Marquês de Valença, then the governor general of Bahia, encouraged further cultivation, introducing the culture of cacao at different points along the coastal waterways that led into the interior of the former *capitânia* of Ilhéus [47]. On August 5, 1783, the *corregedor* (customs official) of the

district of Ilhéus wrote to the governor general of Bahia to inform him that "the planting of coffee and cacao trees, formerly unknown [in Ilhéus], has reached an excellent beginning, with four hundred thousand plants" under cultivation [48].

In the 19th century, Ilhéus would grow to become the new center of cacao production in Brazil. The region lies only about 160 kilometers coastwise south of Salvador de Bahia. For small steamers or cargo schooners under sail, it was an easy one- to two-day journey to take sacks of dried fermented cacao beans up the coast to Salvador, the large international commercial port on the Bay of All Saints. There the cacao was loaded onto ocean-going ships for export to processing centers and markets in Europe and North America [49]. Prior to the 20th century, all Bahian-grown cacao passed through the port of Salvador for export, since the large merchant houses with links to international markets operated from there [50].

Postindependence internal political turmoil hurt cacao production in Brazil; annual production fell to just 9 to 11 percent of total global production, on average, between 1823 and 1837. Thereafter, Brazilian cacao production recovered steadily for nearly half a century: from 1838 to 1884, production increased to account for, on average, 16 to 21 percent of annual global output [51]. In the mid-19th century, Brazilian cacao, liberated from the restrictions of the Portuguese colonial system, found direct markets in Britain, the United States, Germany, and France [52].

Although the initial phase of cacao production in Ilhéus would taper off by the 1820s, it revived strongly between 1836 and 1852, cultivated primarily by new German, French, and Portuguese immigrants (and their modest numbers of slaves) who were dissatisfied with the results of coffee growing in the region [53–55]. Thereafter, cacao cultivation expanded steadily in southern Bahia for the next 70 years; in 1890, two years after the emancipation of Brazil's remaining slaves, Ilhéus surpassed Amazonia as Brazil's top cacao-producing region. Properly cultivated and managed, financial returns per acre for cacao in southern Bahia averaged three to seven times that of other cacao-producing regions in Brazil [56]. By 1900, Bahia ranked among the world's top cacao-producing areas. Contemporary Brazilian cacao expert Gregório Bondar, an agronomist who worked for the Bahian state government to promote cacao cultivation, asserted that Bahia had 400,000 cacao trees by the turn of the 20th century [57]. One contemporary Brazilian agricultural economist noted that:

> The farming of cacao is the principal factor of the economic wealth of the State of Bahia, and occupies the fourth place among the principle export products of [Brazil]. [58]

News of the Bahian cacao growers' successes also eventually reached the national capital in Rio de Janeiro, prompting landowners there to seek Bahian cacao seedlings to be planted on their agricultural properties. For example, on May 31, 1880, the aristocrat José Botelho de Araújo sent an official communiqué to the government of Bahia "requesting that he [Araújo] be furnished with cacao seeds for cultivation on his *fazenda* [estate]" [59]. Cacao production never thrived in the environs of Rio de Janeiro, however; the city lies too far south, near the Tropic of Capricorn, outside the tropical belt (between 20°N and 20°S latitude) wherein cacao grows best.

Expanding cacao plantations in coastal West Africa after 1885 and problems with agricultural diseases gradually cut into Brazil's share of annual production. Brazil's percentage of global cacao exports declined and did not recover until the mid-20th century, when trade disruptions and demand caused by World War II benefited Brazilian producers [60].

Cacao cultivation in equatorial Brazil was of foremost economic importance to Amazonia and Bahia during different periods, both under Portuguese colonial and, after 1823, Brazilian independent rule. But all cacao production in the peak growing areas prior to 1888 depended largely on coerced or enslaved Native American and African workers [61]. Even after chattel slavery's *de jure* demise, exploited and coercive labor practices remained common on Brazil's cacao plantations. Through the exploitation of slaves and coerced laborers, the early Portuguese colonial cacao industry in Brazil became successful and profitable. The legacy of these labor practices is evident across Brazil's "cacao lands"—Bahia, Ceará, Grão Pará, and Maranhão—where indigenous peoples and innumerable descendants of African slaves still make their homes. However, because the Portuguese transplanted cacao trees—and then replicated the entire plantation system that cultivated them most efficiently—across the Atlantic Ocean in West Africa, the production methods peculiar to Portuguese plantation operations were perpetuated into the 20th century. Indeed, on cacao plantations in present-day West Africa, methods pioneered by the Portuguese in Brazil continue as current standard operating procedure [62, 63].

The Portuguese and Cacao in São Tomé: Early Introduction of Cacao from Brazil

Although the opinion is widespread in Portuguese popular historic literature that cacao was introduced into the tiny West African colonial islands of São Tomé and Príncipe—and subsequently the rest of Africa—by happenstance, as an ornamental plant and not as part of a deliberate commercial endeavor, the truth is made plain in Portuguese colonial correspondence of the early 19th century[4] [64]. In fact, just a few years before

Brazil won its independence (1823), in a remarkable act of prescient imperial pragmatism, the Portuguese monarchy ordered the transplant of cacao seedlings from Brazil to São Tomé and Príncipe.

João Baptista da Silva de Lagos, governor of São Tomé and Príncipe, at the behest of a royal order from King João VI dated October 30, 1819, soon accomplished the transplant of "boxes of small cacao trees" from Bahia [65]. This deliberate, premeditated commercial coup was completed by 1820, two years before Brazil severed sovereign ties with Portugal. In his report, Governor Silva de Lagos claimed that the marvelously fecund equatorial volcanic soil of São Tomé "embraced" the Bahian plants; he predicted that, in time, given the "foreign demand" for cacao, its large-scale cultivation in São Tomé and Príncipe would provide "great lucre" to the royal treasury and the inhabitants of those islands [66, 67].

But the Portuguese transplanted more than just Brazilian cacao plants to São Tomé: Portuguese landowners replicated Brazil's entire slave-driven cacao plantation system and perpetuated it for more than 100 years [68]. Thus, for three generations, São Tomé and Príncipe became a major global cacao production zone. During the first decade of the 20th century, tiny São Tomé was often the top annual cacao-producing region in the world[5] [69]. By the late 19th century, however, cacao production had begun to move definitively from plants grown on the Portuguese islands to new plantations organized by citizens of other colonial nations on the African mainland. Widespread cacao cultivation soon extended across vast planned plantation facilities in coastal Ghana, Guinea, Cameroon, Nigeria, Ivory Coast, and Sierra Leone, where the global production of cacao remains centered today [70, 71].

Transplanting the Cacao Plantation System from Brazil to West Africa

By the time of cacao's arrival from Brazil, São Tomé and Príncipe had already been occupied by Portuguese colonizers for over 350 years. These diminutive islands, located at the equator approximately 300 kilometers off the northwestern coast of Gabon, were uninhabited in 1470 when discovered by Portuguese navigators seeking a sea route to India. Slave laborers brought from the African mainland quickly planted São Tomé's rich volcanic soil with highly profitable sugarcane fields. Within 60 years, however, Brazil's burgeoning sugar industry in Bahia began to eclipse that of São Tomé, both in size and quality of the product [72].

Beginning in the late 16th century, then, to meet the ever-expanding demand for African slaves on plantations in the New World, the economy of São Tomé and Príncipe gradually became centered on slave trading [73]. The islands filled an imperative commercial function as a convenient slaving entrepôt, an inescapable concentration and holding area for Africans captured on the mainland. Compact packet vessels would ferry small consignments of new slaves to São Tomé from various points along the African coast; there they would be detained until large ocean-going ships came to fill their holds with human cargoes for transport to Brazil, New Spain, and the Caribbean.

São Tomé and Príncipe, therefore, had a long-established supply system for slave labor; this facilitated the creation of an intensive plantation culture on the islands. Even though coffee and cacao were introduced at approximately the same time from Brazil,[6] planters in São Tomé and Príncipe initially focused their efforts on coffee cultivation simply because it grew faster and allowed them to realize financial returns sooner [74]. But coffee is a high-altitude crop that left much of São Tomé's fertile lowlands underutilized [75]. A generation of slave-based coffee crop development in the early 19th century helped lay the foundations for a widespread cacao plantation system that would grow and flourish after the mid-1800s.

Clearly, the colonial government expected that cacao would eventually become economically important to São Tomé, and in fact the plant's commercial exploitation started earlier than is commonly supposed. An administrative structure for raising revenue by exporting cacao from São Tomé and Príncipe was in place as early as 1854. Colonial customs documents dated September 2 of that year set export duties on cacao at half the rate of those for coffee[7] [76]. No coinciding documents reveal the actual quantities of cacao exported from the islands at that time, but the codification of regulations levying duties on this commodity indicates strongly that cacao exports had already begun or were imminent.

By 1860, however, archival evidence makes clear that São Tomé and Príncipe's incipient cacao plantations (locally referred to as roças—a term itself borrowed from Brazilian usage) had begun to turn out a sufficient crop for significant dutiable exports. Customs service documents for São Tomé port record that, in 1860, cacao was exported on three categories of vessels. Portuguese sailing vessels carried away the largest quantity: 428 arrobas and 28 arrateis of cacao.[8] Steam vessels of the national Portuguese Merchant's Union Company laded 388 arrobas and nine arrateis of cocoa beans, while foreign ships exported a mere 27 arrobas and 16 arrateis of cacao [77]. Clearly, these quantities are minor, but production capacity was growing steadily as newly planted cacao trees matured. The decisive shift from coffee to cacao on São Tomé's roças was already well underway.

Then, on March 8, 1861, the São Tomé customs house dispatched one small sack of cocoa beans to a Mister Robert Young of the English schooner *Jason*, bound for Liverpool [78]. By all appearances, this consignment represented a sample that would eventually come to the attention of British chocolate manufacturers. Within five years, São Tomé and Príncipe would be exporting hundreds of tons of cacao to brokers in Portugal, who sold them to northern European confectioners. Two decades later, English purchasing agents would be bargaining for the expanding output of the islands' large cacao estates [79].

After 1880, cacao exports from São Tomé and Príncipe began to grow at a tremendous rate. By the end of the 19th century, the very best agricultural land in São Tomé had been divided into 28 main plantation estates (roças) and more than 200 smaller landholdings, almost all of which had turned to cacao as their main cash crop. Ninety percent of the arable land was in the hands of Portuguese landholders, many of whom resided in Lisbon and rarely if ever saw their estates [80]. Planters and Lisbon-based agricultural corporations rapidly built up a complex, highly efficient plantation infrastructure for the exploitation of São Tomé and Príncipe's geographic location, extraordinarily productive soil, and high annual rainfall, which, taken together, created perfect growing conditions for cacao (Table 40.1) [81].

Essentially, Portuguese colonizers took the principles and methods perfected on the cacao plantations of Bahia, Brazil, and applied them in a more concentrated, intensive way in São Tomé. Because of São Tomé's uniquely fecund volcanic equatorial agricultural environment, Portuguese plantations, after replicating Brazilian cacao growing techniques, became even more productive, turning out more cocoa pods per hectare than Bahian plantations [82]. The only element in short supply for profitable production was labor. For that, the Portuguese had to look abroad to their mainland African colonies.

But cacao planters in the late 19th century faced legal impediments to large-scale labor procure-

Table 40.1 **Comparative Cacao Exports: Brazil and São Tomé/Príncipe (1860–1914)**

Year	Brazil	São Tomé/Príncipe
1860	3,181 tons	176 tons
1870	4,578 tons	361 tons
1880	4,972 tons	702 tons
1885	6,214 tons	1,200 tons
1890	6,815 tons	3,208 tons
1895	10,846 tons	7,023 tons
1900	16,916 tons	13,935 tons
1905	21,090 tons	25,669 tons
1910	29,158 tons	37,810 tons
1914	40,767 tons	32,064 tons

ment that earlier generations of colonial plantation owners had not. The vast expansion of cacao cultivation in São Tomé and Príncipe coincided with the final legal abolition of slave trading within the Portuguese empire. On February 23, 1869, under intense international pressure, the Portuguese government in Lisbon finally outlawed slavery throughout the nation's overseas possessions. However, according to this legislation, liberated slaves could be obliged to work for their former masters until April 1878 [83, 84]. Although legal slavery had ostensibly been abolished, the islands' contemporary burgeoning cacao plantation system still had an enormous and growing annual need for new laborers, almost none of whom could be acquired in São Tomé and Príncipe. Economic exigency combined with greed and an unscrupulous willingness to circumvent the spirit if not the letter of abolitionist laws led Portuguese planters and administrators in the colonies to devise a system of coercive labor that fed contracted or convict workers from Angola, Cape Verde, or Mozambique into the *roças* of São Tomé and Príncipe.

Finding Laborers: Contracts and Convicts from Cape Verde, Angola, and Mozambique

Filling the islands' plantation labor quota was achieved through three primary methods of legal trickery: inducing illiterate Africans to "sign" long-term indentured labor contracts, manipulating the colonial penal system to allow for the deportation of petty criminals to São Tomé and Príncipe as convict laborers, or by outright slave purchases in Portuguese Africa masked by corrupt bureaucrats who turned a blind eye in return for financial gain (instead of bribes, they collected processing fees that bore a veneer of legitimacy). The result was the creation of a captive workforce on São Tomé and Príncipe that differed from chattel slavery only in name, not in effect. Moreover, throughout the Portuguese colonial system, there existed neither the administrative capacity nor the moral will to stop these practices [85]. As British antislavery investigator Joseph Burtt would observe in 1907:

> The root of the whole matter lies in S. Thomé and Principe. The Islands want labour, and are ready at any time to pay down half a million [pounds] sterling for it, without asking questions. . . . What are the barriers between supply and demand? Two hundred miles of sea, a weak and corrupt Government [in Angola], with a few forts spread over half a million square miles of territory. [86]

Typically, Africans captured deep in the interior of Angola or Mozambique, where Portuguese administrative control barely penetrated, would be marched to an embarkation point on the coast (suffering brutal conditions and high mortality along the way). At dockside, agents of São Tomé's planters paid the slave drovers a fee per laborer; a colonial government clerk would then witness and register each worker's five-year contract. Upon expiration of the term of service a homeward-bound passage, paid by his employer, was supposed to be provided each worker [87].

Between 1875 and 1900, approximately 56,000 contract laborers were supplied to São Tomé and Príncipe by these methods; almost all were shipped from Angola, Mozambique, and Cape Verde, but some had originally been taken from neighboring areas, like the Belgian Congo. Mortality rates were staggering—a damp climate, hard labor, and despair carried off 20 percent of new laborers (called *serviçais*, or "servants" in Portuguese) in the first year. On average, only 50 percent lived out the terms of their contracts. Some European observers who visited São Tomé and Príncipe during this period noted that living conditions on the *roças* were, compared to circumstances prevailing in their home villages on the African mainland, of quite good quality, modern and well-organized. While the *serviçais* in São Tomé and Príncipe were generally well treated—it was, after all, in the plantation managers' best interest to do so—prior to reforms in the early 20th century, contract laborers were universally prohibited from repatriation and instead obliged (often due to debts accrued during their service) to be recontracted [88].

Scores of bound volumes of workers' contracts held in the Historical Archive of São Tomé and Príncipe testify to the brutality of the colonial labor regime in the islands—a system that lasted almost until independence in 1975. Each individual contracted laborer from Cape Verde or the African mainland colonies "signed" a complex legal document binding him or her to at least three and up to five years of labor for scandalously low wages. The employee's "signature" was typically a thumbprint, providing mute evidence that the individual could not read the document that amounted to a legal indenture [89, 90]. In theory, the employer was supposed to deduct a part of the workers' wages for contribution to a repatriation fund to be paid out at the end of the term of service, but in practice very few *serviçais* ever applied successfully for return passage to their homelands. One extraordinary archived volume of labor documents, entitled "Contracts for agreed terms of service sworn in Mozambique and Cape Verde; Certificates of Repatriation" covers the period 1903 to 1976 [91]. Even after independence, contracted workers who had been in the islands for years were still applying for their promised funds that would return them to their homes in the Lusophone African mainland.

Rhythms of Work on Portuguese Colonial Cacao Plantations

Large plantations, whether in Brazil or São Tomé and Príncipe, were organized in approximately the same way. They operated most efficiently when the labor was carefully managed and divided into specialized gangs. Generally, the lands of the largest plantations were subdivided into working units based on the amount of land each labor gang could manage: checking the trees, collecting pods, stripping out the pulp-covered cacao seeds, repairing irrigation canals, and planting new trees. Each subdivision typically had its own laborers' quarters and overseer's house; these tended to be located near the center of the parcel that the gang oversaw. During harvest times, workers transported wet cacao beans from the subdivisions to a central processing area that had facilities for fermentation, drying, sacking, and shipment. In addition, specialized gangs of laborers performed all of the ancillary support roles necessary for the functioning of the plantation as a whole: Some cleared land for new cacao groves; others built and maintained buildings and transportation ways (wagon roads or light railways). Some worked as carpenters; others as teamsters or watermen, transporting cacao to export warehouses along the harbor front of São Tomé port. Still others maintained and operated the equipment for drying raw cacao beans, readying them for shipment[9] [92, 93].

On large cacao estates, workdays were highly regimented. Overseers mustered their laborers at least once a day to issue orders and check for runaways or malingerers. Hours were long, particularly during peak seasonal harvest periods, when planters monopolized their workers' time from dawn to dusk. Respites were brief; the workers' midday meal was generally taken communally in the fields.[10] Working conditions of forced laborers on São Tomé's roças varied greatly, depending in part on the size and wealth of the operation, and also on the character and disposition of the slave owners or their hired overseers [94].

Working rhythms on cacao plantations on São Tomé and Príncipe proceeded steadily year-round. Three major pod collection periods annually followed seasonal rains. Cacao plants might produce up to three crops per year, but pods ripened at different rates, so plants had to be monitored regularly to collect the individual pods as they became ready. Workers also had to remain vigilant in their supervision of the cacao trees to ensure that troublesome animals (monkeys, pigs, and birds) did not eat the pods and damage a lucrative harvest [95].

Harvesting was a delicate process involving sure-handed machete work; pods had to be cut from the trees without harming the integrity of the plant. Workers then had to split the pods open, usually near the groves where they had been picked, and strip the pulp and seeds out of the pod shell [96]. Piles of empty pod husks were typically left in the fields as compost, or heaped along carriageways as fodder for cattle and pigs. Next, the wet, pulp-covered cacao seeds were transported to processing areas—usually in baskets on workers' backs or, on larger roças, in special narrow-gauge railway wagons drawn by hand or by draught animals.[11]

The wet seeds next had to be left for several days in wooden bins or heaped in palm-leaf covered mounds to ferment. Fermentation periods depended on the type of cacao tree being harvested. The wine-colored beans of foresteiro cacao, the most common variety in São Tomé and Príncipe, required a fermentation period of five to six days [97].

After fermentation, the cacao seeds needed to be thoroughly dried to protect them from rot. Typically, beans were air dried in the sun, either in forest clearings on woven mats on the ground, on paved drying "balconies" or "patios," or atop raised wooden platforms constructed specifically for this purpose. Some raised platforms were built on rollers so that they could be retracted under a roof in the event of rain. No matter what the drying technique, plantation workers spread out the cacao beans, turning and stirring them with wooden rakes until they were fully desiccated.[12]

Sun drying was the simplest and cheapest method, but also the least certain to completely dry the seeds. Rushing this process could spoil the product; premature bagging of cacao seeds that had not been thoroughly dried after fermentation contributed to the rot that ruined a high percentage of cacao in the Portuguese colonies, both in Brazil and West Africa [98].

In damp weather, wet fermented cacao beans could also be dried indoors on a long rectangular platform—a metal grid or broad ceramic tiles built above a fire pit or furnace. Typically, the platform would cover a broad earthen pit through which ran metal or ceramic pipes to convey the heat of a furnace; the fire would be stoked by workers in a deeper pit on one end of the drying platform, and smoke from the fire exited through a chimney on the other end. Heat from the fire drawn through the pipes warmed the tiles on which the cacao beans rested. Thus, the beans were desiccated with a dry, even heat, with very little contact with the firewood smoke that could penetrate the beans and contaminate their taste.[13]

In the later 19th century, workers began to spread cacao beans on porous metal drying racks (some purely solar-warmed, others heated by purpose-built wood-fired ventilating engines). Over time, as plantation-based cacao production grew in economic importance, the methods of preparing cacao for export became increasingly efficient. Still, artificial drying of cacao was expensive, requiring the burning of enormous quantities of fuel—wood, imported coal, and, later, oil.

Controversy Over Labor and The British Boycott

Beginning in 1901, after hearing rumors of corrupt, illegal recruiting and labor practices in the Portuguese cacao industry that seemed to approximate enslavement, William Cadbury, head of the British chocolate firm Cadbury's, began an inquiry into the roças of São Tomé and Príncipe. In 1903, he visited Lisbon, personally interviewing members of the Portuguese government as well as local businessmen with interests in the Portuguese colonial cacao plantations. Cadbury pressed the Portuguese to reform their labor practices and left with assurances that abuses would be corrected. Rumors of malfeasance persisted, however, so in 1905 Cadbury's sent an agent, Joseph Burtt, to investigate the São Tomé and Príncipe roças and their labor sources in Angola. His report, published in October 1908, thoroughly damned the Portuguese system as perpetuating de facto slavery [99].

The following year Cadbury's, along with fellow British chocolate manufacturers Fry's and Rowntree, announced a purchasing boycott of all cocoa beans grown in the Portuguese territories, an embargo they vowed to maintain until the roça managers undertook meaningful reforms of their labor procurement and repatriation system [100]. In 1910, William Cadbury published a book, Labour in Portuguese West Africa, in which he exposed and excoriated the Portuguese serviçais contract system. Furthermore, he laid out a specific set of conditions that he asserted must be met before the boycott would be lifted. Among them were guarantees to enable laborers to contract freely for terms of service on the islands, effective Portuguese government oversight of recruitment, and assurances of repatriation following terms of service, if desired [101].

For their part, the Portuguese responded with indignation, publishing numerous works that attempted to counter the mounting British criticism. The Portuguese argued that radical changes to their labor system could not be imposed in so few years and that, insofar as they were able, they had already begun to live up to the expectations set out by Cadbury's call for reforms, particularly in providing repatriation for serviçais upon the expiration of their contracts [102]. Portuguese planters also protested that they provided "civilized" amenities—religious instruction, housing, and hospital facilities—that created far better living conditions than their workers would ever have known had they stayed in their home territories [103, 104]. Furthermore, they pointed out that their workers had all been brought to São Tomé and Príncipe by legitimate legal means, having signed and registered labor contracts or, in some cases, been sent to the cacao islands as convicts, a reasonable punishment for crimes according to several hundred years of tradition within the Portuguese empire [105].

Following the British cacao boycott, Portuguese authorities in São Tomé did take pains to demonstrate their compliance with labor reforms. They kept meticulous records of the few workers who were returned from São Tomé's primary roças to their home territories each year. In 1909, for example, colonial records show that the 28 main plantations collectively repatriated 275 workers to Cape Verde, Angola, Mozambique, and other destinations in mainland Africa [106]. Still, this effort was largely a façade—the colonial government's posture was só para o inglês ver (literally, "just for the Englishman to see"), an old Portuguese idiom that means, figuratively, "merely for show" [107]. In actuality, the great majority of imported workers resident in São Tomé and Príncipe would be bullied or tricked into renewing their terms of service, eventually ending their lives on the cacao estates [108].

Even so, for all its shortcomings, the British boycott of the early 20th century had brought labor practices in Portuguese Africa generally, and the plight of workers in the cacao industry specifically, to the world's attention. Cadbury, Fry's, and Rowntree had used their consumer power to force the Portuguese cocoa planters to undertake substantive reforms that, over time, genuinely improved working conditions for the serviçais in São Tomé and Príncipe [109].

Among African subjects of the state throughout the Portuguese colonies, however, São Tomé and Príncipe continued to be thought of during the 20th century with horror as a terminal destination, an isolated place of brutal toil from which very few convicts or contract laborers, if any, ever returned. To be sent to São Tomé as a laborer was, in the popular vision, effectively the end of life in one's home territory [110].

The line separating the practical circumstances of life for an African convict or contract laborer from that of a chattel slave was virtually a nonexistent one—particularly for untrained workers from Cape Verde, Angola, or Mozambique being conveyed unwillingly along the African coast to their service destinations in São Tomé or Príncipe. In experiential terms, the difference between these two statuses, slave or legitimate contract laborer, was mostly a matter of semantics—the thin veil of a legal definition. However, in the political climate extant in Portuguese African colonies during the late 19th and early 20th centuries, the Portuguese recognized that this key legal distinction could be used to admit illicit, forced African laborers into São Tomé and Príncipe and mask the underlying nature of their compulsory employment. Moreover, given the contemporary social climate among Portuguese elites in São Tomé and Príncipe who depended financially on exploited labor and had demonstrated that they resented international policies designed to curtail *de facto* slavery, Portuguese authorities in Lisbon could expect little cooperation from local colonists to fully eliminate the system.

Conclusion

From the 16th to the 19th centuries, Portuguese colonists in Brazil pioneered the commercial plantation-based production of cacao in the Atlantic world. When imminent Brazilian independence threatened the loss of this valuable crop to the empire, Portuguese authorities became the first to transplant not only the cacao plant to their West African colonies, São Tomé and Príncipe, but also the entire plantation system for cacao production. Based mainly on forced labor, this system proved durable and economically efficacious; it became the basic model for virtually all West African cacao plantations well into the 20th century. Thus, during the 18th and 19th centuries, Portuguese planters and colonial officials vastly expanded cocoa cultivation in the Atlantic world and helped to establish the modern industry's growing regions: In the 20th century, Brazil and West Africa became the world's leading cacao production zones. This transatlantic dissemination of cacao concomitantly shaped the destinies of millions of African slaves or forced laborers on both sides of the Atlantic Ocean.

By describing the origins of commercial cacao production in the Portuguese Atlantic colonies and exploring the connection between cocoa and forced labor across the tropical Atlantic world, some points of departure for further investigation into these fields of research may be suggested. To fully understand the role of forced labor in the production of cacao and chocolate in world history (indeed, to examine exploited labor as a fundamental element in supplying world markets with many crops, both historically and currently), much more detailed research remains to be done. The archive investigation presented here provides insights regarding the transfer of plantation-based cacao production from Brazil to West Africa and offers a glimpse into the historical working conditions of slaves and exploited laborers on cacao plantations in Brazil and São Tomé. Such topics are of elemental importance to Atlantic world foodways and folkways, but have heretofore received only meager scholarly attention. We would encourage future scholars to continue this line of inquiry.

Acknowledgments

I would like to thank the organizers of the Boston University Symposium *Chocolate Culture* (2004) for the opportunity to develop the present chapter theme. Special thanks to Professor Mary Ann Mahony of Central Connecticut State University for her pioneering work on Bahia and for her kind assistance. The author would like to extend his deep gratitude to the staffs of the following libraries and archives, where primary source research for this chapter was undertaken: the Biblioteca Nacional in Rio de Janeiro (BNRJ), the Public Archive of Bahia in Salvador (APS), the Public Archive of Ilhéus (API), the Portuguese Historical Overseas Colonial Archive in Lisbon (AHU), and the Historical Archive of São Tomé and Príncipe (AHSTP).

Endnotes

1. Primary archives used during the course of research for this chapter were the Biblioteca Nacional in Rio de Janeiro, Brazil (BNRJ); the Public Archive of Bahia in Salvador, Brazil (APS); the Public Archive of Ilhéus (API); the Portuguese Historical Overseas Colonial Archive in Lisbon, Portugal (AHU); and the Historical Archive of São Tomé and Príncipe (AHSTP).

2. Former Portuguese Africa is defined as the 21st century states of Angola, Cape Verde, Guiné-Bissău, Mozambique, and São Tomé and Príncipe.

3. To obtain a firsthand understanding of cacao labor and production exigencies, the author visited working cacao plantations in Bahia, Brazil (2003–2004 and 2006) and in São Tomé (2005), where basic traditional cultivation, collection, and processing methods remain essentially unchanged since the 19th century. Original 19th century laborers' houses, drying machinery, and narrow-gauge transport railways, in fact, were frequently found in situ, still in use, in both Bahia and São Tomé.

4. As late as 2002, official cultural publications in São Tomé stated erroneously that cacao was first transplanted to the islands in the late 19th century, and then only as a decorative plant.

5. The diminutive Portuguese colonial West African island of Fernando Pô was important and in 1910 ranked 14th among world cacao-producing regions.

6. Accounts vary; most, including early 19th century governor of São Tomé and Príncipe João Baptista da Silva de Lagos, say coffee and cacao were contemporary transplants in 1819–1820, while others fix the date of coffee's arrival as early as 1787.

7. Exported coffee was taxed at 800 *reis* per 100 *arrateis* laded; cacao was taxed at 400 *reis* per 100 *arrateis* laded. For comparison, manioc flour was taxed at 100 *reis* per 100 *arrateis* laded.

8. One *arroba* is equivalent to approximately 32 English pounds; one *arrátel* is equivalent to approximately one English pound (see Appendix 1).

9. Analysis based on observations of 19th century cacao plantation physical plant ruins and interviews with descendants of enslaved cacao plantation workers made in situ in Bahia, 2003–2004, and in São Tomé, 2005.

10. Analysis based on interviews with descendants of enslaved cacao plantation workers in São Tomé, 2005.

11. Analysis based on observation of cacao plantation facilities in situ and interviews with descendants of compelled cacao plantation workers in São Tomé, 2005.

12. Analysis based on observation of cacao plantation facilities in situ in São Tomé, 2005.

13. Analysis based on observations of 19th century cacao plantation physical plant ruins and interviews with descendants of enslaved cacao plantation workers made in situ in Bahia, 2003–2004, and in São Tomé, 2005.

References

1. Bales, K. *Understanding Global Slavery*. Berkeley: University of California Press, 2005; p. 21.

2. Schlesinger, R. *In the Wake of Columbus: The Impact of the New World on Europe, 1492–1650*. Wheeling, IL: Harlan Davidson, 1996; pp. 88–89.

3. Motamayor, J. C., Risterucci, A. M., Lopez, P. A., Ortiz, C. F., Moreno, A., and Lanaud, C. Cacao domestication I: the origins of the cacao cultivated by the Mayas. *Heredity* 2002;89:380–386.

4. Clarence-Smith, W. G. *Cocoa and Chocolate, 1765–1914*. New York: Routledge, 2002; p. 18.

5. Maxwell, K. *Naked Tropics; Essays on Empire and Other Rogues*. New York: Routledge, 2003; pp. 39–40.

6. Maxwell, K. *Naked Tropics; Essays on Empire and Other Rogues*. New York: Routledge, 2003; pp. 39–40.

7. Clarence-Smith, W. G. *Cocoa and Chocolate, 1765–1914*. New York: Routledge, 2002; pp. 10–11.

8. Dillinger, T. L., Barriga, P., Escárcega, S., Jimenez, M., Lowe, D. S., and Grivetti, L. E. Food of the gods: cure for humanity? A cultural history of the medicinal and ritual uses of chocolate. *The Journal of Nutrition* 2000;130:2057S–2072S.

9. Coe, S. D., and Coe, M. D. *The True History of Chocolate*. New York: Thames and Hudson, 1996; pp. 115, 133–176.

10. Coe, S.D., and Coe, M. D. *The True History of Chocolate*. New York: Thames and Hudson, 1996; pp. 115, 133–176.

11. Clarence-Smith, W. G. *Cocoa and Chocolate, 1765–1914*. New York: Routledge, 2002; pp. 14, 16.

12. Leite, S., Society of Jesus. *História da Companhia de Jesus no Brasil*. São Paulo, Brazil: Edicões Loyala, 2004; Volume V, p. 262 and Volume VIII, pp. 105, 238, 243.

13. da Silva Campos, J. *Crônica da Capitania de São Jorge dos Ilhéos*. Rio de Janeiro, Brazil: Ministério da Educação e Cultura, 1981; pp. 103–105.

14. Leite, S., Society of Jesus. *História da Companhia de Jesus no Brasil*. São Paulo, Brazil: Edicões Loyala, 2004; Volume V, p. 262.

15. Leite, S., Society of Jesus. *História da Companhia de Jesus no Brasil*. São Paulo, Brazil: Edicões Loyala, 2004; Volume V, p. 252.

16. Biblioteca Nacional do Rio de Janeiro (BNRJ). Manuscripts Division, Reference No. I-15, 02, 026, Capítulo I, p. 1.

17. Russell-Wood, A. J. R. *A World on the Move: The Portuguese in Africa, Asia and America, 1415–1808*. Manchester, UK: Carcanet Press, 1992; p. 155.

18. da Silva Campos, J. *Crônica da Capitania de São Jorge dos Ilhéos*. Rio de Janeiro, Brazil: Ministério da Educação e Cultura, 1981; p. 126.

19. Arquivo Público do Salvador, Bahia (APSB). Seção Colonial e Provincial, Volume 1 (*Cartas Régias, 1648–1690*), ff. 22–23.

20. Alvim, P. de T., and Rosário, M. *Cacau Ontem e Hoje*. Itabuna, Bahia, Brazil: CEPLAC (Commissão Executiva do Plano da Lavoura Cacaueira), 1972; p. 14.

21. Ministério da Agricultura, Industria e Commércio; Serviço de Informações. *Producção, Commércio e Consumo de Cacão*. Rio de Janeiro, Brazil: Imprensa Nacional, 1924; pp. 3, 6–7, 14.

22. Mahony, M. A. In the footsteps of their fathers? Family labor, enslaved and free, in Brazil's cacao area, 1870–1920. Presented at the Boston Area Latin American History Workshop; Harvard University, November 16, 2005.

23. Serrão, J., editor. *Dicionário de História de Portugal*. Porto, Portugal: Livraria Figuerinhas, 1979; Volume I, p. 420.

24. BNRJ. Manuscripts Division, Reference No. I-15, 02, 026, Capítulo I, p. 1.

25. Licurgo de Castro Santos Filho. *História de Medicina no Brazil, do Século XVI ao Século XIX*. São Paulo, Brazil: Editora Brasiliense Ltda., 1947; Volume I, p. 112.

26. Licurgo de Castro Santos Filho. *História de Medicina no Brazil, do Século XVI ao Século XIX*. São Paulo, Brazil: Editora Brasiliense Ltda., 1947; Volume II, p. 32.

27. Historical Archives of Goa (HAG). Manuscripts, Volume 8030.

28. Arquivo Histórico Ultramarino (AHU). São Tomé and Príncipe Collection; caixa (box) 55, doc. 75.

29. Coe, S.D., and Coe, M. D. *The True History of Chocolate*. New York: Thames and Hudson, 1996; pp. 193–194.

30. Clarence-Smith, W. G. *Cocoa and Chocolate, 1765–1914*. New York: Routledge, 2002; pp. 195–196.

31. Clarence-Smith, W. G. *Cocoa and Chocolate, 1765–1914*. New York: Routledge, 2002; pp. 204–207.

32. Ministério da Agricultura, Industria e Commércio; Serviço de Informações. *Producção, Commércio e Consumo de Cacão*. Rio de Janeiro, Brazil: Imprensa Nacional, 1924; p. 11.

33. Leite, S., Society of Jesus. *História da Companhia de Jesus no Brasil*. São Paulo, Brazil: Edicões Loyala, 2004; Volume VII, p. 105.

34. Cited in: Leite, S., Society of Jesus. *História da Companhia de Jesus no Brasil*. São Paulo, Brazil: Edicões Loyala, 2004; Volume VIII, p. 243. Original manuscript held in the Arquivo Provincial do Portalegre (Brazil), pasta 176, doc. no. 38.

35. BNRJ. Manuscripts Division, Reference No. I-04, 25, 078.

36. Clarence-Smith, W. G. *Cocoa and Chocolate, 1765–1914*. New York: Routledge, 2002; tables, pp. 234–235.

37. Serrão, J., editor. *Dicionário de História de Portugal*. Porto, Portugal: Livraria Figuerinhas, 1979; p. 420.

38. Coe, S.D., and Coe, M. D. *The True History of Chocolate*. New York: Thames and Hudson, 1996; p. 196.

39. Clarence-Smith, W. G. *Cocoa and Chocolate, 1765–1914*. New York: Routledge, 2002; pp. 234–235.

40. Clarence-Smith, W. G. *Cocoa and Chocolate, 1765–1914*. New York: Routledge, 2002; p. 45, and tables, pp. 234–235.

41. Schwartz, S. B. *Sugar Plantations in the Formation of Brazilian Society: Bahia, 1550–1835*. Cambridge, England: Cambridge University Press, 1985; p. 417.

42. Russell-Wood, A. J. R. *A World on the Move: The Portuguese in Africa, Asia and America, 1415–1808*. Manchester, UK: Carcanet Press, 1992; p. 171.

43. Ministério da Agricultura, Industria e Commércio; Serviço de Informações. *Producção, Commércio e Consumo de Cacão*. Rio de Janeiro, Brazil: Imprensa Nacional, 1924; p. 13.

44. de Castro, R. B. *O Cacao na Bahia*. Rio de Janeiro, Brazil: Pimenta de Mello, 1925; p. 10.

45. Bondar, G. *Terras de Cacau no Estado da Bahia*. Bahia, Brazil: Typographia de São Joaquim, 1923; pp. 20–21.

46. Ministério da Agricultura, Industria e Commércio, Serviço de Informações. *Producção, Commércio e Consumo de Cacão*. Rio de Janeiro, Brazil: Imprensa Nacional, 1924; p. 14.

47. de Castro, R. B. *O Cacao na Bahia*. Rio de Janeiro, Brazil: Pimenta de Mello, 1925; p. 11.

48. de Castro, R. B. *O Cacao na Bahia*. Rio de Janeiro, Brazil: Pimenta de Mello, 1925; p. 11.

49. Clarence-Smith, W. G. *Cocoa and Chocolate, 1765–1914*. New York: Routledge, 2002; pp. 106 and 110.

50. Mahony, M. A. In the footsteps of their fathers? Family Labor, enslaved and free, in Brazil's cacao area, 1870–1920. Presented at the Boston Area Latin American History Workshop; Harvard University, November 16, 2005.

51. Clarence-Smith, W. G. *Cocoa and Chocolate, 1765–1914*. New York: Routledge, 2002; tables, pp. 236–237.

52. Clarence-Smith, W. G. *Cocoa and Chocolate, 1765–1914*. New York: Routledge, 2002; pp. 109–111.

53. de Castro, R. B. *O Cacao na Bahia*. Rio de Janeiro, Brazil: Pimenta de Mello, 1925; p. 12.

54. Clarence-Smith, W. G. *Cocoa and Chocolate, 1765–1914*. New York: Routledge, 2002; p. 110.

55. Arquivo Público da Bahia (Salvador). Seção Colonial e Provincial, No. 173, ff. 57v and 130v; and No. 423, Pasta 8, Doc. 2.

56. Bondar, G. *Terras de Cacau no Estado da Bahia*. Bahia, Brazil: Typographia de São Joaquim, 1923; p. 21.

57. Bondar, G. A Cultura do Cacau na Bahia. In: *Boletim Técnico*, No. 1. São Paulo, Brazil: Instituto de Cacau da Bahia, 1938.

58. de Castro, R. B. *O Cacao na Bahia*. Rio de Janeiro, Brazil: Pimenta de Mello, 1925; p. 7.

59. BNRJ. Manuscripts Division, Reference No. I-47, 34, 8.

60. Clarence-Smith, W. G. *Cocoa and Chocolate, 1765–1914*. New York: Routledge, 2002; pp. 234–239.

61. Mahony, M. A. In the footsteps of their fathers? Family Labor, enslaved and free, in Brazil's cacao area, 1870–1920. Presented at the Boston Area Latin American History Workshop; Harvard University, November 16, 2005; pp. 18–23.

62. Bales, K. *Understanding Global Slavery*. Berkeley: University of California Press, 2005; p. 21.

63. *Child Labor in the Cocoa Sector of West Africa*. International Institute of Tropical Agriculture, August 2002; pp. 12–22.

64. Dória, A. *Cultura do Cacau*, 2nd edition. São Tomé: Instituto Camões, Centro Cultural em S. Tomé e Príncipe, 2002; p. 15.

65. AHU. São Tomé Collection, *caixa* (box) 54; doc. 15.

66. AHU. São Tomé Collection, *caixa* (box) 54; doc. 15.

67. AHU. São Tomé Collection, *caixa* 54; doc. 31.

68. Clarence-Smith, W. G. *Cocoa and Chocolate, 1765–1914*. New York: Routledge, 2002; p. 208.

69. Ministério da Agricultura, Industria e Commércio; Serviço de Informações. *Producção, Commércio e Consumo de Cacão*. Rio de Janeiro, Brazil: Imprensa Nacional, 1924; pp. 6–7.

70. Lopez, R. *Chocolate: The Nature of Indulgence*. New York: Harry N. Abrams, Inc., 2002; pp. 78–79.

71. Clarence-Smith, W. G. *Cocoa and Chocolate, 1765–1914*. New York: Routledge, 2002; pp. 238–239.

72. Boxer, C. R. *The Portuguese Seaborne Empire: 1415–1825*. New York: Alfred A. Knopf, 1969; pp. 88–89, 96–97.

73. Seuanes Serafim, C. M. *As Ilhas de São Tomé no século XVII*. Lisbon, Portugal: Centro de História de Além-mar, Faculdade de Ciências Sociais e Humanas, Universidade Nova de Lisboa, 2000; pp. 215–221.

74. Satre, L. *Chocolate on Trial: Slavery, Politics, and the Ethics of business*. Athens, Ohio: Ohio University Press, 2005; p. 42.

75. César, A. *O 1° Barão d' Água-Izé*. Lisbon, Protugal: Agência Geral do Ultramar, 1969; pp. 79–86.

76. Arquivo Histórico do São Tomé e Príncipe (AHSTP). Secretária Geral do Governo, Series "A," *caixa* (box) 5, *pasta* (folder) 1, doc. 42.

77. AHSTP. Secretária Geral do Governo, Series "A," *caixa* 9, *pasta* 4, doc. 59.

78. AHSTP. Secretária Geral do Governo, Series "A," *caixa* 9, doc. 110: "Mappa do Rendamento da Alfândega da Ilha de São Thomé na Mez de Março de 1861."

79. Clarence-Smith, W. G. *Cocoa and Chocolate, 1765–1914*. New York: Routledge, 2002; pp. 55, 236.

80. Satre, L. *Chocolate on Trial: Slavery, Politics, and the Ethics of business*. Athens, Ohio: Ohio University Press, 2005; pp. 10–11, 42.

81. Clarence-Smith, W. G. *Cocoa and Chocolate, 1765–1914*. New York: Routledge, 2002; pp. 236–239.

82. Conselho Fiscal do São Tomé. *Sociedade De Emigração para S. Thomé e Principe: Relatorio da Direcção, Parecer do Conselho Fiscal, Lista dos Accionistas, 2a. Anno—1914*. Lisbon, Portugal: Centro Typográfia Colonial, 1914; pp. 13–15.

83. Clarence-Smith, W. G. *Cocoa and Chocolate, 1765–1914*. New York: Routledge, 2002; pp. 221–222.

84. *Os Negros em Portugal, séculos XV a XIX*. Lisbon, Portugal: Comissão Nacional para as Comemorações dos Descobrimentos Portugueses, 1999; p. 98.

85. Satre, L. *Chocolate on Trial: Slavery, Politics, and the Ethics of business*. Athens, Ohio: Ohio University Press, 2005; pp. 47–48.

86. British Cadbury Company envoy Joseph Burtt's report, 1907. Cited in: Satre, L. *Chocolate on Trial: Slavery, Politics, and the Ethics of business*, Athens Ohio: Ohio University Press, 2005; p. 93.

87. Satre, L. *Chocolate on Trial: Slavery, Politics, and the Ethics of business*. Athens, Ohio: Ohio University Press, 2005; pp. 7–9, 20, 45–50.

88. Satre, L. *Chocolate on Trial: Slavery, Politics, and the Ethics of business*. Athens, Ohio: Ohio University Press, 2005.

89. AHSTP. Cota 3.22.2.10, doc. 1 (three-year renewable contract for a Mozambican worker).

90. AHSTP. Cota 3.3.2.3 (Contracts and Re-contracts, 1917–1923), unnumbered bound folios.

91. AHSTP. Cota 3.22.2.10 (Contratos para prestação de serviço celebrados em Moçambique e Cabo Verde; Boletims de Repatriação, 1903–1976), unnumbered bound folios.

92. See also Lopez, R. *Chocolate: The Nature of Indulgence*. New York: Harry N. Abrams, Inc., 2002; pp. 19–24;

93. Coe, S.D., and Coe, M. D. *The True History of Chocolate*. New York: Thames and Hudson, 1996; pp. 22–25.

94. Nascimento, A. *Poderes e Quotidiano Nas Roças De São Tomé e Príncipe de finais de Oitocentos a meados de novecentos*. Lousã: Tipografia Lousanense, Ltda., 2002; pp. 82–86, 131–138, 363–373.

95. Dória, A. *Cultura do Cacau*, 2nd edition. São Tomé: Instituto Camões, Centro Cultural em S. Tomé e Príncipe, 2002; pp. 41–44.

96. Lopez, R. *Chocolate: The Nature of Indulgence*. New York: Harry N. Abrams, Inc., 2002; pp. 19–20.

97. Dória, A. *Cultura do Cacau*, 2nd edition. São Tomé: Instituto Camões, Centro Cultural em S. Tomé e Príncipe, 2002; pp. 55–56.

98. Dória, A. *Cultura do Cacau*, 2nd edition. São Tomé: Instituto Camões, Centro Cultural em S. Tomé e Príncipe, 2002; pp. 55–56.

99. Satre, L. *Chocolate on Trial: Slavery, Politics, and the Ethics of business*, Athens, Ohio: Ohio University Press, 2005; pp. 30–32.

100. Conselho Fiscal do São Tomé. *Sociedade De Emigração para S. Tomé e Príncipe: Relatorio da Direcção, Parecer do Conselho Fiscal, Lista dos Accionistas, 2a. Anno—1914*. Lisbon, Portugal: Centro Typográfia Colonial, 1914; p. 58.

101. Cadbury, W. A. *Labour in Portuguese West Africa*, 2nd edition. London, England: George Routledge & Sons, 1910; p. 99.

102. Conselho Fiscal do São Tomé. *Sociedade De Emigração para S. Thomé e Príncipe: Relatorio da Direcção, Parecer do Conselho Fiscal, Lista dos Accionistas, 2a. Anno—1914*. Lisbon, Portugal: Centro Typográfia Colonial, 1914; pp. 31–33.

103. Conselho Fiscal do São Tomé. *Sociedade de Emigração para S. Thomé e Principe: Relatorio da Direcção, Paracer do Conselho Fiscal, Lista dos Accionistas, 2a. Anno—1914*. Lisbon, Portugal: Centro Typográfia Colonial, 1914; pp. 55–59, 104–114.

104. See also: Anonymous. *O cacau de S. Tomé: Resposta dos Agricultores da Provincia de S. Thomé e Principe ao Relatorio do inquerito mandado fazer pelos industriaes inglezes MM. Cadbury, Fry, Rowntree e Stollwerck às condições do trabalho indigena nas colonias portuguezas*. Lisbon, Portugal: Typographia "A Editor," 1907; pp. 4–12.

105. Coates, T. J. *Convicts and Orphans: Forced and State-Sponsored Colonization in the Portuguese Empire, 1550–1755*. Stanford, CA: Stanford University Press, 2002; pp. 78–85.

106. AHSTP. Cota 3.60.3.5, doc. 33: "Repatriation notice for workers" (1909).

107. Wheeler, D. L. *Historical Dictionary of Portugal*. Metuchen, NJ: The Scarecrow Press, Inc., 1993; p. 38.

108. Nascimento, A. *Poderes e Quotidiano nas Roças de São Tomé e Príncipe de finais de Oitocentos a meados de novecentos.* Lousã: Tipografia Lousanense, Ltda., 2002; pp. 131–138, 363–373.

109. Conselho Fiscal do São Tomé. *Sociedade de Emigração para S. Tomé e Príncipe: Relatorio da Direcção, Paracer do Conselho Fiscal, Lista dos Accionistas, 2a. Anno—1914.* Lisbon, Portugal: Centro Typográfia Colonial, 1914; pp. 55–59, 104–114.

110. Satre, L. *Chocolate on Trial: Slavery, Politics, and the Ethics of business.* Athens, Ohio: Ohio University Press, 2005; pp. 47–48.

PART

IX

Europe and Asia

Chapter 41 (Walker) considers cacao both as a medicine within the Portuguese colonial empire and also as a confection enjoyed by Portuguese elites at home or at court. His chapter explores uses of cacao and its by-products as a medical substance throughout the Portuguese empire from 1580 to 1830 and examines chocolate-associated themes in 18th to 19th century Portuguese art. Chapter 42 (Gordon) traces the history and evolution of French chocolate from its possible introduction by Jews fleeing the Spanish Inquisition, through its adoption as an aristocratic beverage in the 17th and 18th centuries and popularization in the second half of the 19th century. Chapter 43 (Gordon) reveals how the history of English chocolate was conected to 17th and 18th century popularization of hot beverages and the industrialization that followed. His chapter shows how the development of milk chocolate, that followed Daniel Peter's technical innovations in Switzerland in 1875, ushered in the modern chocolate era in England and elsewhere. Chapter 44 (Gordon) challenges the widely held view that chocolate was unknown until recently in China and presents a solid case for its introduction as early as the 17th century.

Cure or Confection?

Chocolate in the Portuguese Royal Court and Colonial Hospitals, 1580–1830

Timothy Walker

Introduction

Chocolate played a dual role in the story of Portuguese colonial expansion. Initially valued for its acclaimed medicinal qualities, the confection did not gain acceptance as a sweet delicacy among Portuguese consumers until the 18th century. Even then, outside of hospitals and infirmaries, chocolate was served only in elite social circles—at aristocratic households and to nobles of the royal court. The Portuguese king and his advisors—architects of a vast overseas empire—thus saw cocoa in a curious double light: as a healing plant, fundamental to the health of colonial garrison soldiers (usually conscripts drawn from the dregs of society) who protected the empire, and simultaneously as a stimulating indulgence fit for use by those of the highest class, lineage, and privilege.

In the rarified environment of the monarch's palace in 18th century Lisbon, the court actually employed a "royal chocolatier" whose twofold job, being both culinary and medical, neatly encapsulates the dual role of cocoa in the Portuguese imperial realm. This special functionary, called the *Chocolateiro da Casa Real* (Chocolatier of the Royal Household), was responsible for procuring, managing, handling, and preparing the monarch's stock of cocoa. The *chocolateiro* created sumptuous chocolate confections for the royal family and nobility, of course, but he also was attached to the

staff of the Royal Military Hospital, supplying the rich cocoa beverage prescribed for its supposed recuperative powers to invalid troops. His work also entailed extracting the valuable medicinal cocoa butter provided to interned patients for their skin diseases, or sold for profit in the hospital pharmacy [1].

This chapter considers cacao both as a medicinal substance within the Portuguese colonial empire and as a confection enjoyed by Portuguese elites at home or at court. The person of the *chocolatier* to the Portuguese royal household will be used as a lens through which to view these two issues. Further, this chapter explores the uses of cacao and its by-products as medical substances throughout the Portuguese empire from approximately 1580 to 1830 and

Chocolate: History, Culture, and Heritage. Edited by Grivetti and Shapiro
Copyright © 2009 John Wiley & Sons, Inc.

considers contemporary court rituals surrounding chocolate in continental Portugal and Brazil. Finally, the chapter examines chocolate consumption as represented in Portuguese art during the 18th and 19th centuries. Thus, this work places into context the role Portuguese-speaking peoples have played in the global history of chocolate, as major disseminators of medical knowledge about cacao from Brazil to Europe, Africa, and Asia.

Royal Chocolatier: Confectioner and Pharmacy Technician

During the opening years of the 19th century, when Napoleon's expansionist wars convulsed Europe, the Portuguese "royal chocolatier" was a man named Vicente Ferreira. When an imminent French conquest of Lisbon forced the Portuguese royal family to flee to Rio de Janeiro in 1807, Ferreira traveled as part of the official court entourage. In Brazil, Ferreira continued his dual duties as the royal palace *chocolateiro* and *manipulador* (handler) of the Military Hospital of Rio de Janeiro's supply of medicinal cocoa butter. He also sought royal recognition—and additional compensation—for his work. In 1811, *senhor* Ferreira petitioned the prince regent, Dom João, and the Ministry of the Empire, asking to be nominated as an alderman (*vereador*) of the Rio de Janeiro town council. He hoped through the ruler's largesse to be given an additional office that would allow him to increase both his income and influence [2].

In support of his petition, Ferreira obtained a letter of recommendation from Frei Custódio de Campos e Oliveira, the chief surgeon of the Portuguese army and navy and inspector of the Military Hospital of Rio de Janeiro. This letter, dated October 21, 1811, strongly attested to Vicente Ferreira's skill as a *chocolateiro*. The medical officer commented that:

> He [Ferreira] has served since Lisbon as the Chocolateiro of the Royal Household, and fulfilled his duties with promptness and zeal . . . as with his service to Your Royal Highness in . . . the Military Hospital . . . finding choice chocolate when necessary to nourish the sick, and also extracting the cocoa butter so indispensable to the curing of certain maladies. [3]

The letter continued, saying explicitly that Ferreira's ability to provide medicinal cocoa butter and chocolate had "been of great utility to the overseas States and Dominions [colonies] of Your Royal Highness" [4].

Clearly, the perception of chocolate as an exceptionally beneficent foodstuff and useful medicine that helped to sustain the Portuguese empire was not lost on the man who was simultaneously the inspector of the Royal Military Hospital and chief surgeon of all Portuguese military forces. In an age when the Portuguese grip on many of its overseas enclaves was precarious and the efficacy of few drugs was certain, top administrative officials within the Portuguese colonial apparatus were willing to exploit any substances they felt would help reduce mortality among their soldiers and other essential personnel. Holding the global empire meant keeping garrison troops alive and treating their chronic vexatious maladies. In recognition of Vicente Ferreira's contribution to the health of the empire, the prince regent granted his request for a public office [5].

JESUIT MISSIONARIES AND EARLY MEDICINAL CACAO IN SOUTH AMERICA

Portuguese beliefs regarding the medical benefits of cacao date nearly to the founding of Brazil as a royal colony in 1549. Contempory Europeans of the mid-16th century—missionaries and colonial authorities alike—already considered cacao one of South America's most important medicinal plants. Such conceits in turn shaped and drove colonial and missionary administrative policies designed to best advance Portuguese imperial interests. In the case of cacao, interest in cultivating the indigenous plant in Portuguese South America was first and foremost a pragmatic attempt at increasing resources for health preservation, long before exploiting the commercial potential of Brazilian chocolate became a colonial economic priority. And, as was often the case throughout the worldwide empire, proselytizing missionary *padres*—learning native folkways—pioneered this effort.

Jesuit missionaries in South America, as elsewhere in the Iberian colonial world, excelled in discovering and experimenting with indigenous medicinal substances, knowledge that they carefully recorded and disseminated. Chocolate had become known to the Society of Jesus as a medicinal substance during the earliest days of colonization in Brazil. The European missionaries had learned this lore directly from Native American practices—the *índios* in Brazil also believed that cacao had healing qualities. Since time immemorial, *Túpi* and *Guaraní* peoples had recognized chocolate as a mild stimulant that could provide sustaining energy to combat hunger and fatigue [6, 7].

In approximately 1580, Jesuits working in Paraguay compiled a detailed manuscript volume of native remedies found in that land, as well as in their other mission fields in Chile and Brazil. The tome runs to over 230 manuscript pages and contains listings for scores of South American medicinal plants, but its very first entry describes the "Virtues of Cacao" [8]. According to the text, an ounce and a half of prepared chocolate had the power to:

open the [body's] passages . . . comfort the mind, the
stomach, and the liver, provide aid to asthmatics . . . and
those with cataracts . . . and maintain the body in case of
cold. [9]

Such native medicinal understanding of chocolate gave the Jesuits an added incentive to cultivate cacao trees for health purposes in their mission communities, even if on a limited scale.

It is important to note, however, that most of the maladies described as treatable with chocolate in the above passage were chronic ailments suffered primarily by the colonizers, not the native *Túpi* and *Guaraní* peoples who initially led the Portuguese to cacao as a healing plant. After initial discovery, the Jesuits tended to experiment subjectively with medicinal plants, assigning them efficacies according to their own requirements and exigencies. Thus, the 1580 medical compendium should be seen as a singularly European document that reflected the needs and medical interpretation of the Old World, not necessarily the medical qualities and "virtues" accorded to cacao solely by the native peoples of Brazil.

CACAO CULTIVATION ON JESUIT MISSION LANDS: A MEDICINE AND EXPORT COMMODITY

Jesuit priests first started to harvest cacao regularly near their missions along the Amazon River in the mid-17th century [10]. Because the Jesuits had recognized the healthful, stimulating, and nourishing properties of cacao through their interaction with native peoples, this medical motivation was their most probable reason for culling the plant. From *Amazônia* the Jesuits moved cacao south to the province of Bahia, the colonial capital of Brazil, site of the earliest "industrial cultivation" of cacao—where the plant was first grown systematically as a medicinal commodity in *América Portuguesa* [11].

In southern Bahia, known until 1754 as São Jorge dos Ilhéus [12], the Jesuits cultivated cacao in their gardens at Olivença, Camamu, and Canavieiras, all coastal missionary communities south of the main port and capital city, Salvador de Bahia. Olivença, located approximately 10 kilometers south of the municipality of Ilhéus, Bahia (destined to become the epicenter of Brazilian cacao production in the late 19th century), was probably the first site; the Society of Jesus had established a mission and sugar estate there in 1563 [13], to which they transplanted cacao trees from *Amazônia* during the 1670s or 1680s [14]. From there, cacao cultivation spread to the large Jesuit mission at Camamu, situated on a bay at the mouth of the Trindade River, 80 kilometers north of Ilhéus [15].

Most permanent Jesuit colonial missions around the world operated medical facilities—typicallly an infirmary and a pharmacy—from which they dispensed medical compositions for a profit [16]. Sales of remedies and drugs helped fund their further missionary work. In many regions of colonial Brazil, Jesuit pharmacies were the only source of prepared medicines. One standard piece of pharmacy equipment was a grinding stone necessary to prepare many remedies, including chocolate. For example, in 1757, an inventory of the Jesuit pharmacy of Belém do Pará included some objects referred to as *chocolateiros*—stone and ceramic grinding devices (similar in function and form to *metates*) used for the preparation of roasted cacao beans, as well as other medicinal substances that had to be pulverized [17].

CACAO BUTTER AS MEDICINE THROUGHOUT THE PORTUGUESE EMPIRE

The missionaries also learned how to extract cocoa butter from cacao and sold it as a remedy for skin maladies. This Brazilian cacao derivative product found a steady global market as a medicinal substance. By the late 18th century, cocoa butter (*manteiga de cacão*) was being used therapeutically in colonial military hospitals and infirmaries throughout the Portuguese empire—in South America, Africa, India, and the East Indies [18–20]. Colonial soldiers and officials in the tropics soothed chafed, dry, or abraded skin with cocoa butter; they employed it as a standard regular treatment for heat rashes, or more serious skin disorders like shingles.

In Portuguese India, missionary pharmacies stocked quantities of Brazilian cocoa butter for daily application to infirmary patients and for retail distribution to the general population; the *botica* (pharmacy) of the Convent of Santo Agostinho in Goa listed regular monthly purchases of *manteiga de cacão* between 1807 and 1835 [21]. In fact, cocoa butter from Brazil could be found in the stocks of colonial pharmacies from Macau to Timor, Mozambique, and São Tomé, and in the medical chests of most ships of the Portuguese India fleet, as well, be they merchantmen or men-o-war [22].

Expansion of Brazilian Cacao Production and Consumption Within the Empire

Despite cacao's medicinal utility and growing value in the global market, large-scale commercial expansion of production in southern Bahia did not begin until 1746. Shortly thereafter, during the reign of King José I (1750–1777), cacao production in the lower Amazon region—the Grão Pará and Maranhão provinces—would expand enormously due to the intervention of the autocratic Portuguese prime minister, the Marquês de Pombal. In 1755, Pombal, his attention drawn by the profit potential of expanded agricultural commerce in northeastern Brazil, initiated a new state-sponsored trading company, the *Companhia do Grão-Pará e Maranhão*. Pombal's plan gave the *Companhia* a complete monopoly on the export of cacao from *Amazônia* to Lisbon, from whence about two-thirds was reexported (mainly to Italy) [23]. After ten years of development, Brazilian cacao exports rose to constitute, on average, more than 10 percent of total world production, a level sustained for thirty years [24]. By 1800, cocoa beans were the *Companhia do Grão-Pará e Maranhão*'s primary and most lucrative export crop [25].

Ironically, despite Brazil being one of the world's most important cacao production zones, cocoa consumption was not widely popular there, nor in the continental Portuguese metrópole, nor for that matter anywhere else in the Lusophone maritime empire. During the late 18th century, Brazilian per capita use of chocolate was the lowest in Latin America, with Portuguese colonists' total consumption averaging only 50 tons of cacao per year [26]. Regardless of the considerable import costs, Portuguese colonists much preferred drinking tea and coffee, a habit they had learned in their Asian colonies [27]. Brazilian native groups, on the other hand, were more fond of drinking an infusion produced with *guaraná*, a common plant found in Amazonia. *Guaraná* is a strong natural stimulant, containing a substance similar in effect to caffeine. Although native peoples sometimes imbibed *guaraná* mixed with ground cacao powder that they produced themselves, their demand for chocolate was never significant commercially [28].

Because of low demand locally among European colonists in Brazil, during the 17th, 18th, and 19th centuries, the vast majority of Brazilian cacao was produced for export. Dried cacao beans were bagged and shipped to Portugal for immediate reexport or domestic processing into products destined primarily for foreign markets. Besides edible products, items manufactured in Brazil and Portugal from raw cacao included cocoa butter, soap, oil, and a chocolate-flavored liqueur [29]. These luxury items, never produced in great quantities, found consumers among elites in Portugal, the rest of Europe, and throughout the Portuguese maritime colonial network.

Chocolate as a Recuperative Foodstuff in Portuguese Colonial Hospitals

Since at least the first quarter of the 19th century (and probably by the late 18th century), grated powder or pressed tablets of cocoa from Brazil were being used medicinally to make a sustaining beverage administered to sick soldiers in the garrison hospitals of far-flung

Portuguese colonies, from West Africa to Goa and Macau [30]. News of chocolate's utility as an agent to restore corporeal vitality had spread through colonial channels to be applied to good effect in state medical facilities, far from its origins in South America. In the tropics, where wasting diseases caused tremendous mortality and sapped Portuguese colonial manpower, medical authorities were keen to adopt and apply any substance that promised to preserve the lives of troops and European settlers [31–34].

In Brazil, at the Royal Military Hospital of Rio de Janeiro, just prior to the imperial schism that led to independence from Portugal, new regulations for regimental hospitals situated across the colony became effective in 1820. These rules stated explicitly that chocolate was to be reserved for wounded or sick soldiers, to be fed to them on a regular schedule as part of an invalid diet meant to provide extra energy for their recuperation [35]. The reform regulations also provided a graphic dietary chart, called the "Table of Diets for the Troops of Portugal," prescribing the precise days and times that chocolate, as well as other special foods, was to be distributed to patients admitted to military hospitals. Such dietary supplements were to be dispensed at the discretion of the chief surgeon, "as necessary, for the improvement of soldiers' lunches, ... as the health and security of his patients requires" [36].

Incidentally, well into the late 19th century, in Portugal as well as Brazil, patent medicines traded on the medicinal efficacy of chocolate. In Lisbon, Salvador de Bahia, and Rio de Janeiro, pharmacy newspaper advertisements in the 1880s and 1890s carried notices for preparations of "chocolate [blended] with vanilla, musk and iron," designed to promote corporeal force and vigor [37].

Chocolate and the Portuguese Royal Court in the 18th and 19th Centuries

Eighteenth century Portugal was flush with cash; gold, silver, and diamond mines in colonial Brazil provided the European kingdom with a reliable fountain of wealth. Dom João V (reign 1706–1750) and his son, Dom José I (reign 1750–1777), could thus afford the best of everything for the royal table, so they turned to master artisans in France who, in the baroque age, established the modes of fashion for all of Europe (Fig. 41.1). The French court at Versailles became the standard for aristocratic tastes and elegance, to be emulated by ruling households across the continent.

The Portuguese royal household's premier chocolate service was crafted by the master silver-

FIGURE 41.1. Chocolate pot of Portuguese King João V, dated ca. 1720–1722. *Source*: Museu do Arte Antiga Inventory Number 1872. Courtesy of the Instituto Português dos Museus, Lisbon, Portugal. (Used with permission.) (See color insert.)

smiths to French king Louis XV's court as part of a massive commission, weighing more than $1\frac{1}{2}$ tons in silver, ordered in the 1720s [38]. The most renowned gold- and silversmiths of the era, Thomas Germain and his son, François-Thomas Germain, executed the Portuguese king's chocolate pots and accoutrements at their manufactory at the Louvre palace, Paris. Portuguese patronage of these workshops lasted over half a century, providing the Germain artisans an opportunity to "create some of the most accomplished French works in silver of the 18th century that have survived to the present day" [39]. The elder silversmith died in 1748, and much of the original service was subsequently lost in the 1755 Lisbon earthquake. Dom José ordered a replacement set on June 24, 1756, of which an elegant silver chocolate pot held in the collections of the Museu de Arte Antiga (Museum of Antique Art) in Lisbon is a part.

This richly decorated *chocolateira*, made in Paris circa 1760–1761, is a fine example of the silver craftsman's peerless skill. The cylindrical body of the pot, approximately nine inches tall, tapers gracefully upward; its sides are decorated with parallel striations in an undulating wave pattern that emerge from a band of shell adornments at the base of the pot. The smooth handle is of turned ebony, which meets the body of the pot in a chalice-shaped silver structure decorated with the leaves of a cocoa tree (or *cacaueiro* in Portuguese) [40]. The lid of the pot is pierced to admit a blending stick (a *molinillo*; in Portuguese, *batedor*); the aperture is fitted with a removable turned ebony plug, which

also serves as a knob to lift the lid [41]. The royal arms of Portugal are engraved on the interior of the lid, while makers' and proof marks are stamped on the base of the piece.

Another French-made chocolate pot, this one manufactured in the late 18th century for the table of Portuguese Queen Maria I (reigned 1777–1816), can also be found in the Museu de Arte Antiga collections. This pear-shaped pot stands approximately 11 inches tall with a turned ebony handle that projects directly from the pot body. The pot is pure cast silver, with an absolutely smooth polished finish both on the metal and the wooden handle. A Portuguese royal coat of arms is engraved on the base [42]. The hinged pot lid, adorned with an engraved letter "P" topped by a crown, has an aperture through which a finely carved heartwood *batedor* could pass. With this implement an attending servant—a specialist confectioner, or *chocolateiro*—could whip and froth the spiced cocoa drink by rotating the stick rapidly between the palms of his hands. The threaded *batedor* handle can be taken apart for storage and transport. A second solid silver vessel in Queen Maria's service, a milk-warming urn, sits on elegant bowed tripod legs above its matching alcohol burner [43]. Servants would pour hot milk from this vessel into the mixing pot, there to be blended, using the *batedor*, with a prepared chocolate tablet and spices [44].

Images of Portuguese Elites' Chocolate Consumption in the 18th and 19th Centuries

To serve hot chocolate to the king and other members of the royal court was either a jealously guarded right of a prized servant (the court *chocolateiro*) or a special privilege for which favored courtiers competed and intrigued. Laden with significance, this ceremonial act—a public display of subservience and loyalty from a dependent vassal—is the subject of a miniature painting held in the collections of the Museu de Arte Antiga in Lisbon (Fig. 41.2) [45]. A visiting Italian miniaturist, the renowned Alessandro Castriotto of Naples, executed this delightful peek into the world of the Portuguese royal court in 1720. In this small scene, we see the sovereign, Dom João V, drinking chocolate served from a delicate gold-inlaid pot in the company of several courtiers, including the *Infante* Dom Miguel. Significantly, the cocoa has been prepared by one of the king's devoted subjects, an anonymous court favorite who, in the background, solemnly wields the *batedor* to blend the chocolate that will be offered to the monarch [46].

FIGURE 41.2. Portuguese King João V with Chocolate, dated 1720. *Source*: Museu do Arte Antiga Inventory Number, 58 Min. Courtesy of the Instituto Português dos Museus, Lisbon, Portugal. (Used with permission.) (See color insert.)

FIGURE 41.3. Chocolate tile panel from the factory of Vicente Navarro, Valencia, Spain; dated ca. 1775. *Source*: Inventory Number 1/803-804. Courtesy of the Museu Nacional de Cerâmica de Valência, Spain. (Used with permission.) (See color insert.)

Castriotto's miniature captured a poignant moment of contemporary court diplomatic machinations; he intended the image as a gentle gibe at Dom João V and his envoy to the Vatican. Pictured with the king are the portrait artist Giorgio Domenico Duprà, commissioned from Rome at great cost to paint a likeness of the king, and the papal nunzio Vincenzo Bichi, whose scandalous dissolute behavior at court prevented his being accepted as a candidate for cardinal by the Vatican and eventually led to a diplomatic rift between Portugal and the papacy [47]. Within this scene, chocolate symbolizes the indulgence and decadence of aristocratic courtiers—the smirking papal nuncio, wine flask in hand, certainly appears eager to get his share of the proffered sweets.

Images of elite consumption of chocolate in early modern Portugal are few, but we can get a hint of what domestic rituals were like by looking across the Iberian Peninsula into contemporary Spanish Valencia. Chocolate being served in an aristocratic home is the subject of a large panel of Spanish traditional glazed tiles (*azulejos*) recently displayed in a visiting exhibition at the Museu Nacional do Azulejo (National Tile Museum) in Lisbon. In the panel we see a pair of elite domestic servants, a young woman and man, dressed in matching livery from an aristocratic household of the late 18th century. These polychromatic tiles are in fact

a section from a larger wall panel used to decorate the kitchens of a great estate home in Valencia. The woman holds a tray bearing toasted sweet cakes or confections, while the male servant carries a tray of porcelain beakers filled to the brim with hot cocoa. In a playful twist, the young man, obviously distracted by his female colleague's compelling charms, has allowed one of the cups to tip, spilling its tasty contents to the floor (Fig. 41.3) [48].

Tile panels like these were a typical feature of many Portuguese aristocratic households in the 18th century, too, where they can be found as wall decorations in a parlor, main hallway, or kitchen. Artisans produced such *azulejo* panels on commission, so a themed tile scene focused on the consumption of chocolate is a telling artifact. This particular panel, from a Spanish urban residence, shows how conspicuous indulgence in chocolate had entered elite habits in Iberia. Similarly, chocolate had emerged as a luxury product in contemporary Portugal, to be served proudly and for prestige while entertaining guests. However, if this practice was ever commemorated in decorative art within a Portuguese home, as it was in Spain, representative examples have either been lost or forgotten.

Viewed through the prism of the Portuguese royal chocolatier, it is easy to see how cocoa assumed a dual role in the early modern Portuguese empire.

Within hospitals and infirmaries, chocolate was served mainly to the conscript garrison soldiers, who were of the lowest social classes. Still, maintaining their health was deemed vital to perpetuating the integrity of the empire. Conversely, within Portuguese elite social circles, in the 18th century, chocolate emerged as a commodity of priviledged indulgence and a tool for conveying signs of royal favor. In contemporary Portuguese art, chocolate became a motif—a mark of conspicuous consumption, or one portraying episodes of court intrigue.

Acknowledgments

The author would like to thank Mars, Incorporated for support through the Colonial Chocolate Society Project that made possible most of the research used in this chapter. In addition, the author is grateful to the organizers of the Boston University Symposium *Chocolate Culture* (2004) for the opportunity to develop this theme, and to Professor Mary Ann Mahony of Central Connecticut State University for her pioneering work and kind assistance. The author would like to extend his deep gratitude to the staffs of the following libraries and archives, where primary source research for this chapter was undertaken: the Biblioteca Nacional in Rio de Janeiro (BNRJ), the Public Archive of Bahia in Salvador (APS), and the Portuguese Historical Overseas Colonial Archive in Lisbon (AHU).

References

1. Biblioteca Nacional do Rio de Janeiro (BNRJ). Manuscripts Division, Reference No. C-1052, 102, docs. 1 and 2.

2. BNRJ. Manuscripts Division, Reference No. C-1052, 102, docs. 1 and 2.

3. BNRJ. Manuscripts Division, Reference No. C-1052, 102, doc. 1.

4. BNRJ. Manuscripts Division, Reference No. C-1052, 102, doc. 2.

5. BNRJ. Manuscripts Division, Reference No. C-1052, 102, doc. 1.

6. Clarence-Smith, W. G. *Cocoa and Chocolate, 1765–1914*. London, England: Routledge, 2002; pp. 10–11.

7. Dillinger, T. L., Barriga, P., Escarcega, S., Jimenez, M., Salazar Lowe, D., and Grivetti, L. E. Food of the gods: cure for humanity? A cultural history of the medicinal and ritual use of chocolate. *Journal of Nutrition* 2000; 130(Supplement):2057S–2072S.

8. BNRJ. Manuscripts Division, Reference No. I-15, 02, 026, Capítulo I, p. 1.

9. BNRJ. Manuscripts Division, Reference No. I-15, 02, 026, Capítulo I, p. 1.

10. Leite, S., Society of Jesus. *História da Companhia de Jesus no Brasil*. São Paulo, Brazil: Edições Loyala, 2004; Volume V, p. 252.

11. Leite, S., Society of Jesus. *História da Companhia de Jesus no Brasil*. São Paulo, Brazil: Edições Loyala, 2004; Volume V, p. 262.

12. da Silva Campos, J. *Crônica da Capitania de São Jorge dos Ilhéos*. Rio de Janeiro, Brazil: Ministério da Educação e Cultura, 1981; pp. 233–234.

13. da Silva Campos, J. *Crônica da Capitania de São Jorge dos Ilhéos*. Rio de Janeiro, Brazil: Ministério da Educação e Cultura, 1981; pp. 103–105.

14. Leite, S., Society of Jesus. *História da Companhia de Jesus no Brasil*. São Paulo, Brazil: Edições Loyala, 2004; Volume V, p. 262.

15. Leite, S., Society of Jesus. *História da Companhia de Jesus no Brasil*. São Paulo, Brazil: Edições Loyala, 2004; Volume V, p. 252.

16. Licurgo de Castro Santos Filho. *História de Medicina no Brazil, do Século XVI ao Século XIX*. São Paulo, Brazil: Editora Brasiliense Ltda., 1947; Volume I, p. 112.

17. Licurgo de Castro Santos Filho. *História de Medicina no Brazil, do Século XVI ao Século XIX*. São Paulo, Brazil: Editora Brasiliense Ltda., 1947; Volume II, p. 32.

18. BNRJ. Manuscripts Division, Reference No. C-1052, 102, docs. 1 and 2.

19. Historical Archives of Goa (HAG). Manuscripts, Volume 8030.

20. Arquivo Histórico Ultramarino, Lisbon (AHU). São Tomé and Príncipe Collection, *caixa* (box) 55, doc. 75.

21. HAG. Manuscripts, Volume 8030.

22. AHU. São Tomé and Príncipe Collection, *caixa* (box) 55, doc. 75.

23. Alden, D. The significance of cacao production in the Amazon region during the late colonial period: an essay in comparative history. *Proceedings of the American Philosophical Society* 1976;120:124, 130.

24. Clarence-Smith, W. G. *Cocoa and Chocolate, 1765–1914*. London, England: Routledge, 2002; pp. 234–235.

25. Schwartz, S. B. *Sugar Plantations in the Formation of Brazilian Society: Bahia, 1550–1835*. Cambridge, England: Cambridge University Press, 1985; p. 417.

26. Clarence-Smith, W. G. *Cocoa and Chocolate, 1765–1914*. London, England: Routledge, 2002; p. 35.

27. Clarence-Smith, W. G. *Cocoa and Chocolate, 1765–1914*. London, England: Routledge, 2002; pp. 18, 21, 23, 30.

28. Clarence-Smith, W. G. *Cocoa and Chocolate, 1765–1914*. London, England: Routledge, 2002; p. 26.

29. Serrão, J., editor. *Dicionárin de História de Portugal*. Porto, Portugal: Livraria Figuerinhas, 1979; p. 420.

30. AHU. São Tomé and Príncipe Collection, *caixa* (box) 55, doc. 29.

31. da Silva Correia, A. C. G., La Vieille-Goa, Bastorá. 1931; pp. 274–275.

32. da Luz, M. Livro das Cidades. *Studia* 1960; 6:8.

33. HAG. Manuscripts, No. 181A, ff. 65, 194–201.

34. HAG. Manuscripts, No. 212A, f. 200v.

35. Arquivo Público de Salvador da Bahia (APS). Volume 454–2, ff. 347–360.

36. APS. Volume 454–2, f. 361.

37. APS. *Bahia Ilustrada*, dated September 23, 1899; p. 4.

38. d'Orey, L. The silver table service of Dom José I of Portugal. In: Levenson, J. A., editor. *The Age of the Baroque in Portugal*. New Haven, CT: Yale University Press, 1993; pp. 167–176.

39. d'Orey, L. The silver table service of Dom José I of Portugal. In: Levenson, J. A., editor. *The Age of the Baroque in Portugal*. New Haven, CT: Yale University Press, 1993; p. 168.

40. Museu Nacional de Arte Antiga, Lisbon. Inventory No. 1872 (from the table service of King Dom José I).

41. Coe, S. D. and Coe, M. D. *The True History of Chocolate*. New York: Thames and Hudson, 1996; pp. 116–117.

42. Museu Nacional de Arte Antiga, Lisbon. Inventory No. 1089 *Our* (from the table service of Queen Dona María I).

43. Museu Nacional de Arte Antiga, Lisbon. Inventory No. 1090 *Our* (from the table service of Queen Dona María I).

44. Coe, S. D., and Coe, M. D. *The True History of Chocolate*. New York: Thames and Hudson, 1996; pp. 135–164.

45. *Dom João V Drinking Chocolate with the Infante Dom Miguel*. Artist: Alessandro Castriotto, 1720; oil on ivory, silver frame. Museu Nacional de Arte Antiga, Lisbon, Portugal; Inventory No. 58 Min (58, 1717).

46. de Carvalho, A. A. Dom João V and the artists of papal Rome. In: Levenson, J. A., editor. *The Age of the Baroque in Portugal*. New Haven, CT: Yale University Press, 1993; pp. 38–39.

47. de Carvalho, A. A. Dom João V and the artists of papal Rome. In: Levenson, J. A., editor. *The Age of the Baroque in Portugal*. New Haven, CT: Yale University Press, 1993; pp. 38–39.

48. Museu Nacional de Cerâmica De Valência, Spain. Inventory 1/803-804 (From the factory of Vicente Navarro, Valencia, ca. 1775).

Chocolate in France

Evolution of a Luxury Product

Bertram M. Gordon

Introduction: Chocolate Comes to France

The history of French chocolate reflects the general history of chocolate styles, extending from the European discovery of cacao in Mexico to the present. From the boiled and spiced beverages and thick, *molé*-type chocolate sauces used by Mayans and Aztecs in Mexico to the Spanish hot breakfast beverage seasoned with salt and peppers or cinnamon and sugar, drunk with *churros* [1–4], chocolate evolved into the milky candy confections of the 20th century. In 20th century France, chocolate use accelerated in a more socially upscale direction.

Chocolate was introduced into France from Spain. Cacao was known, or known of, in France by the middle of the 16th century with the publication in French of Spanish accounts of the West Indies, although the first reference to it appears to have been a 1532 translation from the Latin account of the Milanese writer Pietro Martire d'Anghiera, whose *De Insulis* addressed the Spanish explorations in the Western Hemisphere. The French translation contains a reference to *cacap*, as a fruit similar to almonds and used as money in Mexico [5]. There are several other early French references to *cacap*, which may have resulted form a typographical error substituting "p" for "o" [6]. Francisco López de Gómara's *Historia general de las Indias*, published in Spanish in 1552, appeared in French translations by Martin Fumée, in 1568 and 1584, and Guillaume Le Breton in 1588 [7, 8]. Editions of Fumée's translation in 1584 and 1605, following the description by Martire d'Anghiera, also refer to cacao used in Guatemala as a form of money

[9]. The earliest reference in French to chocolate may have been in José de Acosta's *Histoire naturelle et morale des Indes tant occidentales qu'orientales*, also translated from the Spanish and published in French in 1598. It describes the brewing of a beverage called "Chocholaté," prepared both hot and cold and strongly spiced with chilies, said to be good for the stomach and to counteract catarrh [10].

In 1602, French King Henry IV commissioned the Gramont family, who governed the city of Bayonne in the French Basque region, to offer protection to Jews fleeing the Inquisition in Spain. Jewish refugees from Spain and Portugal were granted asylum in the Basque village of Saint-Esprit, then a suburb and now a part of Bayonne. Among these refugees were chocolate makers but it is not clear when they first arrived in the Bayonne region or when they began to sell chocolate (see Chapter 5) [11].

A frequently told story holds that Anne of Austria, daughter of Spanish King Philip III, introduced chocolate into France when she married Louis XIII in 1615. Told by Jean Anthelme Brillat-Savarin in his *Physiology of Taste* in 1825 and often repeated since, this account lacks contemporary documenta-

Chocolate: History, Culture, and Heritage. Edited by Grivetti and Shapiro
Copyright © 2009 John Wiley & Sons, Inc.

tion [12]. Where Brillat-Savarin heard this story is uncertain; he might have confused Anne of Austria with Marie Thérèse of Austria, who later married Louis XIV, and did play a role in popularizing chocolate in France. As the story goes, Anne of Austria, age 14 when she married Louis XIII, brought a coterie of ladies-in-waiting and was attached to her Spanish ways, including, it is said, the drinking of chocolate as a beverage. Her taste for chocolate supposedly was passed to the king and his court, from where it spread to the rest of France. In reality, however, the marriage had been arranged to end a long period of Franco-Spanish enmity and there was considerable mistrust in high French court circles of Anne and her Spanish retinue, who were sent back to Spain in 1618. Louis Batiffol, whose biography of the young king was based heavily on accounts of the day, notes that Louis, also 14 at the time of the wedding, was not attracted to Anne and that they lived in separate apartments in the Louvre, then the royal palace. They saw one another only briefly, at set intervals during the day, with fixed rituals prescribed by ceremonial rules, and they said little to one another. Nor did they dine together [13]. The Spanish courtiers who came to France with Anne were mistrusted as agents of a hostile power and their presence aggravated the situation [14]. Until 1616, Anne was served her meals "in the Spanish fashion," according to reports from the Spanish embassy at the French court [15]. Batiffol's study supports the likelihood that Anne may have been served chocolate in France but makes it highly improbable that she inspired any fashions there. Madame Françoise de Motteville, later a courtier for Anne, whose memoirs describe the Queen's wedding and subsequent court life in great detail, does not mention chocolate. Lastly, Dr. Jean Héroard, Louis XIII's personal physician, kept a detailed record of his dining habits, listing everything he ate and drank during the early years of his marriage to Anne, and does not mention chocolate.

The demonstrable onset of chocolate in France occurred in the middle of the 17th century. Jean de Laet's *Histoire du Nouveau Monde* mentions chocolate, in reference to its use in Mexico, in 1640. In 1643, René Moreau published his *Du chocolate: Discours curieux divisé en quatre parties*, an annotated translation of another Spanish text (Fig. 42.1). Moreau dedicated his book to Cardinal Alphonse de Richelieu of Lyon (brother of the more famous Richelieu), who had consulted him regarding the "virtues" of chocolate, a "foreign drug" [16]. Cardinal Richelieu's interest in chocolate was also addressed in 1725 by Bonaventure d'Argonne, under the nom-de-plume Vigneul-Marville. By this time, chocolate was of sufficient interest that Argonne speculated about its first use in France. He wrote that Cardinal Richelieu had been the first in France to use "this drug" (chocolate). The cardinal, who had learned of this remedy from Spanish

DV CHOCOLATE DISCOVRS CVRIEVX, DIVISÉ EN QVATRE PARTIES.

Par *Antoine Colmenero de Ledesma Medecin & Chirurgien de la ville de Ecÿa de l'Andalouzie.*

Traduit d'Espagnol en François sur l'impression faite à Madrid l'an 1631. & esclaircy de quelques Annotations.

Par *RENÉ MOREAV Professeur du Roy en Medecine à Paris.*

Plus est adjousté vn Dialogue touchant le mesme Chocolate.

Dedié à Monseigneur l'Eminentissime Cardinal de Lyon, grand Aumosnier de France.

A PARIS.
Chez SÉBASTIEN CRAMOISY, Imprimeur ordinaire du Roy, ruë S. Iacques, aux Cicognes.

M. DC. XLIII.

FIGURE 42.1. *Du chocolate: Discours curieux divisé en quatre parties*, by René Moreau (title page); dated 1643. *Source:* Courtesy of the Peter J. Shields Library, University of California, Davis. (Used with permission.)

priests, used it to ease "the vapors of his spleen" (*vapeurs de sa rate*). The source for this information, according to Argonne, was one of Richelieu's servants [17]. Another example of the knowledge of chocolate in mid-17th century France oddly comes from London (see Chapter 43), where, in 1657, an unnamed entrepreneur placed an advertisement in the *Public Advertiser* announcing that "in Bishopsgate Street, in Queen's Head Alley, at a Frenchman's house, is an excellent West India drink, called chocolate, to be sold, where you may have it ready at any time, and also unmade, at reasonable rates" [18].

A second, verifiable origin story concerns another Spanish Infanta, Marie Thérèse, who was said to have introduced chocolate into France and spread its popularity when she married Louis XIV in 1660. This account is based on the *Memoirs* published twenty years later by the Duchess de Montpensier, a courtier at the time of the wedding. Montpensier wrote of the queen having two ladies-in-waiting brought from Spain, La Molina and, later, La Philippa, who prepared the chocolate for her in the Spanish style, which she took in secret, although, according to Montpensier, everyone knew about it [19, 20]. A century later, in a history of French social customs, published in 1782, Pierre Jean Baptiste Le Grand d'Aussy argued that Marie Thérèse could not have been responsible for popularizing chocolate in France because she drank her chocolate in secret. Le Grand d'Aussy, who is said to have written the first "bona fide French culinary/food history" [21] and was also involved in the production of chocolate in the late 18th century [22, 23], found the Richelieu story more credible [24].

The role of Marie Thérèse in popularizing chocolate among the French aristocracy, however, is also supported by another female courtier of the era, the Marquise de Montespan, maid of honor to Marie Thérèse and subsequently mistress of Louis XIV. In her memoirs, Montespan wrote of the young queen:

The Senora Molina, well furnished with silver kitchen utensils, has a sort of private kitchen or scullery reserved for her own use, and there it is that the manufacture takes place of clove-scented chocolate, brown soups and gravies, stews redolent with garlic, capsicums, and nutmeg, and all that nauseous pastry in which the young Infanta revels. [25]

Elizabeth-Charlotte, Duchesse d'Orléans, married to Louis XIV's younger brother, wrote of Marie Thérèse: "Her teeth were very ugly, being black and broken. It was said that this proceeded from her being in the constant habit of taking chocolate; she also frequently ate garlic" [26]. The French archives show a Jean de Herrera appointed chocolate maker to the queen in 1676 [27]. Another possible connection was through the Duke Antoine III de Gramont, the French ambassador to Venice, and Cardinal Jules Mazarin, who helped negotiate the marriage of Marie Thérèse and Louis XIV. Gramont and Mazarin are said to have had cooks brought from Italy to prepare chocolate for them [28, 29].

Chocolate was gaining popularity, at least among prominent social circles, in France in the 1650s and 1660s, but it was difficult to prepare and expensive. Bayonne and the surrounding French Basque country may have been exceptions where Iberian refugee Jews, called "Portuguese Jews," had been admitted since the beginning of the 17th century. Jews imported cacao from Cadiz and elsewhere in Spain and transformed the cacao into a chocolate paste, which they then sold to the Bayonne customers, but it is difficult to say when this began. The earliest archival record of chocolate in Bayonne is a request on April 6, 1670, to the city authorities to purchase chocolate brought from Spain, to be given to "persons of consideration" (Fig. 42.2) [30]. Only in 1687 is there mention of a chocolate maker in Bayonne.

In 1691, local non-Jewish merchants gained passage of a municipal ordinance prohibiting Jews from establishing workshops or boutiques in Bayonne and, although the document did not refer specifically to chocolate vendors, they were henceforth restricted to wholesale trade in town or retail trade outside the city gates [31]. Documents in the Bayonne archives, such as a police ordinance of 1735, indicate that the ordinance was not enforced [32]. In 1761, the non-Jewish chocolate makers organized a guild, which attempted to enforce the prohibition against the Jews selling chocolate in the town. The guild statutes excluded Jews and stipulated that only guild masters could establish shops or sell chocolate within the town of Bayonne [33]. A

FIGURE 42.2. Request of Bayonne City authorities to purchase chocolate from Spain, dated April 6, 1670. *Source:* CC 314, #105 (entitled) "Pour le Sieur Martin de Lafourcade bourgeois et trésorier de la ville de cette présente année 1670," 6 April 1670. Courtesy of the Archives Communales de Bayonne, France. (Used with permission.)

codicil added that an "infinite number" of foreigners had flooded the city selling poor quality chocolate [34]. By the second half of the 18th century, guilds in general were increasingly seen as inhibiting trade and economic development. Jewish leaders appealed to regional officials and in 1767 the Bordeaux *parlement* dissolved the guild, handing the Jews a victory in the Bayonne chocolate war [35, 36]. The 1789 Revolution gave Jews legal citizenship in France and in 1790 all guilds were dissolved. The Bayonne area may have produced France's first female chocolate maker, Jeanne Dunatte, who in 1801 ran an atelier in Ustaritz, a neighboring town [37].

Chocolate's introduction and spread may be seen as a microcosmic way of viewing much of French political, social, and cultural life in the early modern period. The fashion for chocolate, which expanded during the second half of the 17th century, played a role in the modernization of French dining styles and appeared in economic, medical, and theological discourse, as well as in popular plays and songs. France had a powerful monarch who headed a centralized interventionist state in a society more dominated by the aristocracy and the merchant guilds than was true in the English-language countries. The mercantilist state-directed economy of Jean-Baptiste Colbert, Louis XIV's finance minister, gave rise to the term *Colbertisme*, still used for the French interventionist state. Louis XIV and Colbert established state-supported monopolies to increase production and trade, and chocolate was no exception. In 1659, Louis XIV granted David Chaliou, a member of the queen's retinue, the privilege to "manufacture and sell for 29 years a composition called chocolate." The royal patent gave Chaliou a monopoly, stipulating that he could make chocolate in liquid, pastille (lozenge or chocolate drop), or box form [38, 39]. Chaliou was a "protégé" of Olympe Mancini, former mistress of the king and future superintendent of the queen's household [40]. After the expiration of Chaliou's monopoly, royal ordinances in 1693 gave all grocers and merchants in France the freedom to buy and sell chocolate, cacao, and tea, and allowed anyone to open a café [41]. Cafés such as Le Procope, which opened in 1686 and still exists in Paris, are said to have offered chocolate ice cream, which may have helped spread chocolate use into the lower classes, though it is not clear when.

French mercantilism included the acquisition of colonies, the importation of slaves from Africa, and the creation of trading monopolies to protect and promote national commerce. The French arrived in Martinique in 1635 and by the end of the century that island had become a major exporter of cacao to France. Describing cacao in the West Indies in 1654, Jean-Baptiste du Tertre predicted that the local population would make a good profit if they chose to cultivate it [42]. The first cacao trees appear to have been cultivated in Martinique by Benjamin Andreade D'Acosta, a Marrano Jew,

fleeing possible Portuguese encroachment upon Curaçao, in or around 1660 [43, 44]. In 1664, however, Colbert created the West Indies Company and gave it monopolistic trading privileges in the French Antilles, and D'Acosta and other Jews in Martinique lost the right to trade in cacao. In 1679, a ship, the *Triomphant*, arrived in Brest from Martinique with a cargo of cacao, opening the official trade [45]. Six years later, in 1685, the Jesuits persuaded Louis XIV to expel the Jews from the French islands.

Problems with pirates from England and elsewhere interfered with French trade with Martinique, and economic issues of free trade versus state regulations were played out in the cacao discussions that took place between island governors and state officials back in France [46]. Gabriel Dumaitz de Goimpy, the governor of the French Caribbean islands, wrote to the king in 1694 that excessive entry taxes in the port of Bayonne threatened to drive the Martinique cacao producers out of business [47]. Pressure from officials in Martinique and, later, local merchants and manufacturers of chocolate in France to lower tariffs and other taxes on their trade would be a recurring theme in French chocolate history.

French shipping records, partial at best, give only a general idea of the smuggling of cacao to avoid taxes or the booty taken by pirate and corsair ships. In 1711, for example, a dispute over cacao booty taken in Nantes from three captured enemy ships led to a decree by the government, ever desperate for funds, that cacao taken from such ships was subject to special taxes [48]. Against pressure from Paris to maintain high taxes and tariffs for France's wars in the 18th century, Martinique officials struggled to persuade the royal government to allow freer trade in cacao to stimulate the island's economy. Earthquakes in 1727 and 1728 destroyed Martinique's cacao plantations, requiring substantial replanting [49–53].

Chocolate Enters French Culture

By the end of the 18th century, chocolate had crossed into French culture in a variety of ways, from food to medicine to theater and song. Its use, largely among the affluent, reflected the coming of an increasingly refined dining style characterized also by the introduction of coffee and tea, and increased use of tablecloths and forks (Louis XIV was the last major figure to resist the fork) (see Chapter 12). Prices were kept high by the French preference for Spanish and Italian chocolates because, until the revolutionary period, the local chocolate was made from inferior French colonial cacao, at least according to Antoine Gallais, a 19th century producer in Paris [54].

The spread of chocolate in France and other Catholic countries in Europe raised the issue of its suitability for fasting days. In 1662, Cardinal Francesco Maria Brancaccio ruled for the Catholic Church that chocolate was a beverage rather than a food and therefore could be consumed on religious fast days. For Catholic France, this ruling was important in opening the door to chocolate, although it did not fully quiet a theological debate on the nature of chocolate that extended into the 18th century. The medicinal value of chocolate, together with its role in aristocratic fashion, is seen in the Marquise de Sévigné's correspondence with her adult daughter, Madame de Grignan, in which she first advised her daughter to take chocolate as a restorative in 1671, then changed her mind, and told the story of a pregnant acquaintance who ate too much chocolate and gave birth to a black baby, who died shortly thereafter [55]. In the last letter, Madame de Sévigné, in line with the humorial argument that chocolate was hot, asked her daughter if she were not "afraid of burning your blood." Madame de Sévigné, who continued to mention the heating effects of chocolate in letters through 1689, became so closely identified with chocolate that a French chocolate maker, who has also created a museum of chocolate in Alsace, now uses her name.

Literary references to chocolate increased in the 1680s. The first appearance of the word "chocolate" in a French dictionary came with the publication of Pierre Richelet's dictionary of 1680 [56]. Two years later, on October 4, 1682, a *limonadier* (a member of the café owners' guild) was depicted in a play, *Arlequin, Lingère du Palais*, performed in Paris by the *Comediens Italiens du Roy*. Arlequin appears in a café, where he calls out: "Biscuits, macaroons, coffee, iced chocolate, sirs" ("*chocolat à la glace*," which could have been a beverage, a sorbet, or an ice cream). In a dissertation, *An Chocolatae usus salubris?*, published in 1684, François Foucault, under the direction of Stéphane Bachot, argued for the medicinal benefits of chocolate. Sylvestre Dufour, *Traitéz Nouveaux & curieux du café, du thé et du chocolate*, first published in 1671, and Nicolas de Blégny, *Le Bon Usage du thé, du Caffé, et du Chocolat pour la Preservation & pour la Guerison des Maladies*, published in 1687, also addressed chocolate as medicine (Figs. 42.3 and 42.4). Noting that chocolate had become widely used in France and often simply for pleasure, Dufour warned that it was fattening [57]. Blégny suggested taking "*une tablette du Chocolat anti-venerien*" as a remedy for venereal or "gallant" diseases [58].

A developing connoisseurship of chocolate was evident by the end of the 17th century, most frequently as a beverage for the upper classes. In 1691, François Massaliot's *Le Cuisinier Roial et Bourgeois*, a guide to organizing meals for the French upper classes, offered recipes for *Macreuse en ragoût au Chocolat*, a savory poultry dish with a chocolate sauce and *Crême de Chocolat*, using milk, sugar, and egg yolk, in addition to chocolate (Figs. 42.5 and 42.6) [59, 60]. At least one historian argues that Massaliot's recipes mark the intro-

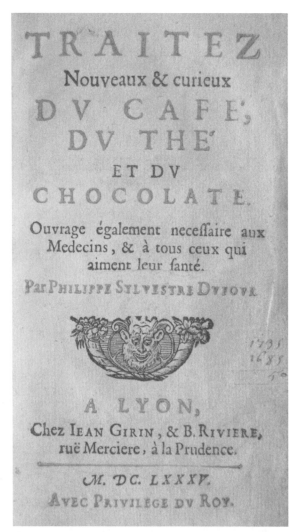

FIGURE 42.3. *Traitéz Nouveaux & curieux du café, du thé et du chocolate*, by Sylvestre Dufour (title page), dated 1685. *Source*: Courtesy of the Peter J. Shields Library, University of California, Davis. (Used with permission.) (See color insert.)

duction of chocolate into French cuisine, meaning that it was no longer solely a medicine or exotic specialty [61]. A catalog of cacao and chocolate varieties on sale in the shop (*cabinet*, or office) of Pierre Pomet in Paris in 1695 included six varieties of cacao from Venezuela and the West Indies, and chocolate from Spain, Saint-Malo, Paris, and "America" [62].

The statutes of the *limonadiers'* guild, established in 1676, were broadened in 1705, specifically allowing them to sell chocolate in pastry, chocolate after the butterfat had been removed, and chocolate *dragées* (bonbons), as well as cacao and vanilla, although, as Alfred Franklin has noted, they had been selling chocolate with or without authorization at least since the performance of *Arlequin* in 1682 [63]. Addressing the popularity of chocolate in France, Philippe Hecquet in 1710 referred to it as the "queen of beverages" and the "beverage of the gods," and added that old vintage wines were now obliged to cede their places to

FIGURE 42.4. Sylvestre Dufour, coffee, tea, and chocolate consumption. *Source*: Sylvestre Dufour, *Traitéz Nouveaux & curieux du café, du thé et du chocolate*, 1685. Courtesy of the Peter J. Shields Library, University of California, Davis. (Used with permission.)

LE
CUISINIER
ROIAL
ET BOURGEOIS,

Qui apprend à ordonner toute ſorte de Repas, & la meilleure maniere des Ragoûts les plus à la mode & les plus exquis.

Ouvrage tres-utile dans les Familles, & ſin_gulierement neceſſaire à tous Maîtres d'Hôtels, & Ecuiers de Cuiſine.

A PARIS,

Chez CHARLES DE SERCY, au Palais au ſixiéme Pilier de la Grand'Salle, vis-à-vis la Montée de la Cour des Aides, à la Bonne-Foi couronnée.

M. D C. X C I.
AVEC PRIVILEGE DU ROI.

FIGURE 42.5. *Le Cuisinier Roial et Bourgeois*, by François Massaliot (title page), dated 1691. *Source*: Private Collection of Mary and Philip Hyman. (Used with permission.)

chocolate [64]. A diet of chocolate, he argued, had cured a consumptive couple and the wife had recovered to the point that she was able to bear children [65]. Because chocolate was a nutritive food, according to Hecquet, it should be given during fasting days only to pregnant women, the elderly, or those exempt from fasting because of health issues. Those whose health allowed them to eat only small meals could be given chocolate as a substitute on fasting days, and then only in the morning. Their limited ability to digest food would render them continually hungry and, Hecquet reasoned, keep them in a sort of fast anyway [66].

Following the death of Louis XIV, Philippe d'Orléans, regent from 1715 to 1723 for the young Louis XV, held court over chocolate at breakfast. Those fortunate enough to be seen were "admitted to the

Prince's chocolate" [67]. In 1732, a Monsieur de Buisson devised a table for the "chocolate-worker," who ground the beans, allowing him to stand upright rather than having to bend over the grinding stone [68].

The developing uses of chocolate are evidenced in the reissues of Massaliot's *Cuisinier Roial et Bourgeois*, republished and expanded during the 1730s. The two chocolate recipes of the 1691 edition were repeated, along with a *Crême de Chocolat au bain-marie*, a *Tourte de Chocolat*, with rice flour, egg yolk, and cream, and a *rissole de chocolat*, with orange flower water added [69–71]. The new edition included sample menus, which in the 18th century were structured as *Service Française*, meaning relays of courses in groaning board style rather than today's item-by-item sequence, known as *Service Russe*. A fasting menu, with fish highlighted, for 30

bien on l'accommode de l'une des manieres qui fuivent.

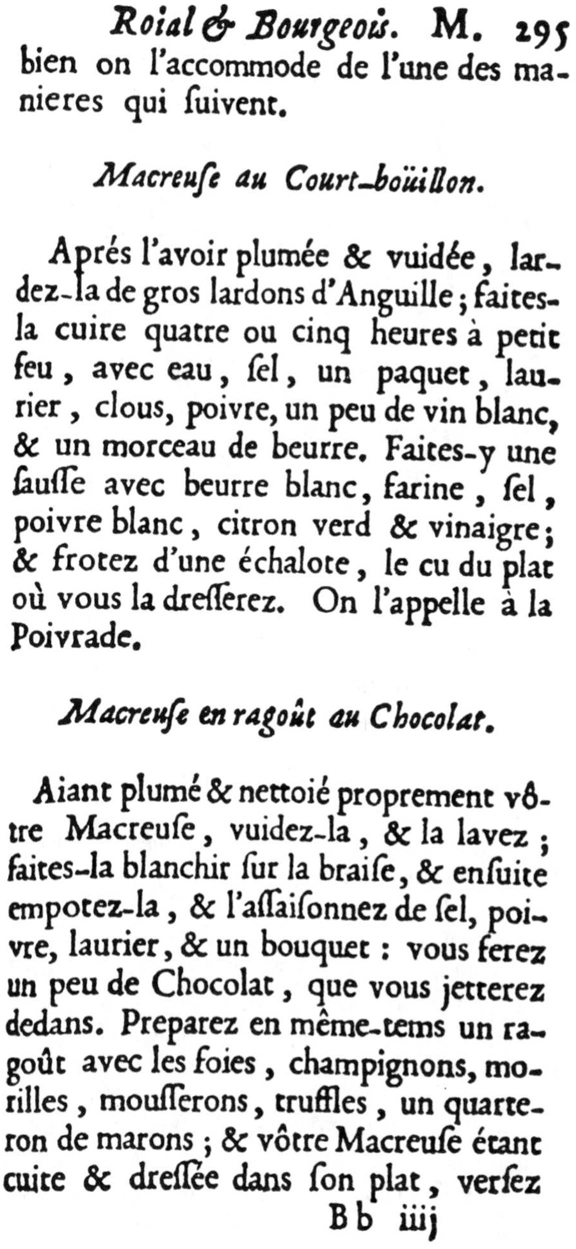

Macreuſe au Court-boüillon.

Aprés l'avoir plumée & vuidée, lardez-la de gros lardons d'Anguille ; faites-la cuire quatre ou cinq heures à petit feu, avec eau, ſel, un paquet, laurier, clous, poivre, un peu de vin blanc, & un morceau de beurre. Faites-y une ſauſſe avec beurre blanc, farine, ſel, poivre blanc, citron verd & vinaigre ; & frotez d'une échalote, le cu du plat où vous la dreſſerez. On l'appelle à la Poivrade.

Macreuſe en ragoût au Chocolat.

Aiant plumé & nettoié proprement vôtre Macreuſe, vuidez-la, & la lavez ; faites-la blanchir ſur la braiſe, & enſuite empotez-la, & l'aſſaiſonnez de ſel, poivre, laurier, & un bouquet : vous ferez un peu de Chocolat, que vous jetterez dedans. Preparez en même-tems un ragoût avec les foies, champignons, morilles, mouſſerons, truffles, un quarteron de marons ; & vôtre Macreuſe étant cuite & dreſſée dans ſon plat, verſez

B b iiij

FIGURE 42.6. Recipe for *Macreuse en ragoût au Chocolat*: Part 1; dated 1691. *Source:* François Massaliot, *Le Cuisinier Roial et Bourgeois.* Paris, France: Charles de Cercy, 1691; pp. 295–296. Private Collection of Mary and Philip Hyman. (Used with permission.)

place settings included two *Crèmes de Chocolat*, among 12 hors d'oeuvres and entremets featured between fish and salad courses [72]. Subsequent cookbooks sometimes included diagrams of table settings for the reader to see where the various dishes were to be placed. Massaliot also listed foods for feast and fast days, with *Crèmes de Chocolat* and *Tourtes de Chocolat* included as acceptable entremets for both [73]. A companion *Nouvelle Instruction pour les confitures, les liqueurs, et les fruits,* also published by Massaliot in 1734, noted that chocolate could be made as a beverage or in solid form for candies and biscuits [74]. In an expanded version of his earlier catalog, Pierre Pomet ranked the cacao beans with the *Gros Caraque*, shipped from Nicaragua, the best for making chocolate. Cocoa oil (or butter) separated from the cacao solids, via processes known prior to the Spanish conquest, could be used as a conditioner for the skin [75]. Chocolate had filtered sufficiently into popular culture by 1738 that a song entitled *Cresme au Chocolat*, sung to the tune of another popular song of political protest, was circulating in France and published in a book of food-related songs [76]. Several years later, in 1744–1745, Jean-Etienne Liotard, a French Huguenot artist living in Geneva, drew the often-reproduced pastel on parchment *La Belle Chocolatière*. A portrait of a young woman serving a chocolate beverage, it is now in the Dresden Gemäldegalerie. A guidebook for *maîtres d'hôtel* in 1750 offered recipes for chocolate ice cream [77], chocolate mousse [78], a *fromage glacé de chocolat*, which was made in a cheese mold [79], *chocolat en olives*, little pieces of hard chocolate shaped as olives [80], and chocolate *dragées* [81]. In 1755, Louis Lémery wrote that "chocolate is the best thing that came to us from America, after gold and silver" [82].

Chocolate also appears in the literature of the 18th century Enlightenment. Voltaire refers to it several times in *Candide* in 1759, once in a motif of murder by poison-laced chocolate, a story told in Chapter 11 by the "daughter of Pope Urban X." A former mistress who secretly placed poison in his chocolate beverage killed her husband, in the story. This motif would recur in French history. *L'Encyclopédie* of Diderot et d'Alembert included an illustration of a worker grinding chocolate, in a manner that had not changed since its importation into Europe (Figs. 42.7 and 42.8) [83]. Mail order advertisements in the *Gazetin du comestible* for chocolate for Morlaix in Brittany appear in 1767 (Fig. 42.9) [84]. In 1778, a Monsieur Doret was cited by the government for developing a hydraulic machine to mix ground chocolate [85].

Chocolate's aphrodisiac qualities in late 18th century France were extolled by Madame du Barry and Madame de Pompadour, both of whom were said to have used it to excite the ardors of their lovers [86]. Describing an orgiastic event in *The 120 Days of Sodom and other Writings*, in 1785, the Marquis de Sade wrote:

> The company rose the 1st of November at ten o'clock in the morning, as was specified in the statutes which Messieurs had mutually sworn faithfully to observe in every particular. . . . At eleven o'clock they passed into the women's quarters where the eight young sultanas appeared naked, and in this state served chocolate, aided and directed by Marie and Louison, who presided over this seraglio. [87]

Shortly before the French Revolution, Pierre-Joseph Buc'hoz, a leading naturalist and private physician for the king's brother, after describing the cacao

FIGURE 42.7. Artisans making chocolate, dated 1751. *Source*: "Confiseurs." In: *Artisanats au 18ème Siècle, l'Encyclopédie Diderot et d'Alembert.* Courtesy of the Peter J. Shields Library, University of California, Davis. (Used with permission.)

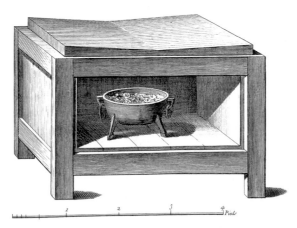

FIGURE 42.8. Detail of chocolate grinding stone with fire box below, dated 1751. *Source*: "Confiseurs." In: *Artisanats au 18ème Siècle, l'Encyclopédie Diderot et d'Alembert.* Courtesy of the Peter J. Shields Library, University of California, Davis. (Used with permission.)

plant, noted that "twenty negroes can maintain fifty thousand cacao trees" [88]. Cacao oil, he wrote, could be used as a substitute for olive oil and was especially good for hemorrhoids. Following Hecquet's arguments regarding the health benefits of chocolate, Buc'hoz praised chocolate's restorative qualities for consumptives. He also repeated in detail Hecquet's story of the woman who had recovered from consumption with a diet of chocolate, after which she was able to bear children [89]. It may have been Buc'hoz who influenced Brillat-Savarin's praise of chocolate as a cure for exhaustion [90]. At the end of his discussion of chocolate, Buc'hoz listed 17 different uses, including hot beverages with water and with milk, chocolate biscuits, iced chocolate in molds, chocolate mousse, chocolate conserves, chocolate cream, chocolate formed in the shape of olives, *dragées* (candies), and *eau de chocolat* (chocolate mixed with eau de vie) [91].

Chocolate's Democratization

Chocolate's evolution into a more broadly popular product would await the technological changes of the 19th century. The wars of the French Revolution and Napoleon, together with the British blockade, cut imports and raised the price of chocolate [92]. Pierre Clement de Laussat, the French colonial prefect in Martinique, reported to Paris in 1807 that revenues from cacao, as well as other products from the islands, had fallen because of reduced commerce in wartime and would need to be raised [93]. Chocolate, however, continued to be sold for gastronomic and medicinal use. In 1800, Sulpice Debauve, a pharmacist, established a chocolate shop in Paris, which was recommended in Grimod de la Reynière's *Almanach des Gourmands* in 1805 [94]. Debauve's nephew, Antoine Gallais, also a pharmacist, published a *Monographie du cacao* and joined him in the business in 1827. According to Gallais, Debauve was the first to blend chocolate with the aroma of coffee [95]. Their shop, now known as Debauve and Gallais, still exists as a center for fine chocolate in Paris. Grimod de la Reynière also remarked that chocolate, used twenty years earlier primarily by the elderly at breakfast, had become by 1805 the breakfast of all those who wished to maintain the "freshness of their imagination" or "whose abilities were not raised above the level of a chicken." Formerly, he added, only apothecaries prepared chocolate and its quality could be trusted, but now, ignorance and fraud were common and the consumer had to be wary [96]. In addition to the fears of fraudulent chocolate raised by Grimod and others, the theme of assassination by poisoning with chocolate reappeared in 1805. Writing under the nom-de-plume of Stewarton, Lewis Goldsmith, an English writer, who had visited France in the early 1800s while France and England were at war, told the story of an attempted assassination of Napoleon by poison in his chocolate drink (see Chapter 21). In the story, Pauline

(2)

NOMS DES DENRÉÈS.	VILLES QUI LES PRODUISENT	PRIX.	OBSERVATIONS.
Bajoues de porc ..	de Châlons	7 fols piecé , fur le lieu.	
Bœuf à l'écarlatte.	de Saint-Malo	18 à 20 f. la livre au Bureau.	Il vient en baril & fe conferve.
Becaffes	de Picardie , de Pithiviers & du Maine...	de 40 à 45 f. chez foi.	
Becaffines	Idem	12 à 15 f. piece, chez foi.	
Beurre	de Gournay ,........	de 20 à 21 f. chez foi.	Les beurres font fort augmentés dans la Province ; on mande que ceux qui valloient 6 à 7 fols , fe vendent aujourd'hui 12 à 15 fols.
Cannetons	de Rouen		
Canards fauvages .	de différens endroits..	de 40 à 55 f. chez foi.	Suivant la beauté ; inceffamment ils feront a meilleur compte.
Chapons	de différens endroits..	de 45 f. à 3 l. piéce,chez foi.	Il y en a du Mans à 3 l. 10 f. piéce au Bureau. On ne peut rien voir de plus beau.
Idem.	de Breffe...........	5 liv. piece , chez foi.	
Cervelats	de Troyes...........	depuis 20 f. jufqu'à 3 l. 10 f. piece, chez foi	Ils reviennent à 15 ou 18 fols la livre.
Cercelles........	de différens endroits..	de 12 à 18 f. piece, chez foi.	
Cochons de lait ..	d'Arras & d'autres endroits	depuis 6 l. jufqu'à 10 l. chez foi.	Ceux d'Arras font fort eftimés.
Cocqs-vierges....	de Normandie	5 & 6 livre chez foi.	
Id. de Bruyere	d'Alface	depuis 5 l. jufqu'à 7 l. 10 f. piece, chez foi.	
Chocolat	de Morlaix..........	32 f. la livre, chez foi.	Il y en a avec apprêt & fans apprêt, les deux façons font tres-eftimées.

FIGURE 42.9. Chocolate available via mail order. Advertisement from Morlaix in Brittany, France; dated December 1767. *Source: Gazetin du comestible*, No. 12. December 1767. Courtesy of Bibliothèque Nationale de France; online Gallica site. (Used with permission.)

Riotti, a Corsican woman whom Napoleon had seduced and jilted, subsequently became a kitchen assistant in Lyon, where Napoleon stayed while en route to Italy in August 1805. Riotti supposedly planted poison in his chocolate. The plot was discovered, Riotti was forced to drink the chocolate herself and shortly thereafter died. Goldsmith's account, given his known anti-Napoleon and propagandistic proclivities at the time, is thought to be apocryphal [97].

The end of the Napoleonic wars brought an increase in the cacao trade. An increase in restaurants that had begun in Paris before the Revolution and intensified during the revolutionary and Napoleonic years was accompanied by the increased use of chocolate as well. By 1815, Parisian restaurants, such as Les Frères Provençaux and Grignan, served *pots de chocolat* as *entremets de douceur* [98] and *entremets sucrés* [99], respectively, and Lambert offered a *Crème au chocolat* as an hors d'oeuvre [100]. Legacque served a *crème à la fleur d'orange ou au chocolat* as an entremet [101], and Quénard a *pot de crème au chocolat* as an entremet [102]. Corazza, a *limonadier*, located in the Palais Royal, was praised for his "excellent chocolate" [103]. A report comparing the consumption of cacao in Paris between 1789 and 1817 showed that it had grown from a value of 500,000 livres for a city of 600,000 in 1789 to a value of 750,000 francs for a city of 714,000 in 1817 [104]. Given the livre to franc exchange ratio (81 livres equaled 80 francs) and the inflation that accompanied the revolutionary period, the growth was modest at best. Change, however, was on the way. In 1819, François Pelletier, known as "the American," because he had participated with Lafayette in America's War of Independence, established a chocolate factory in Paris, which he called "The American's." In it he installed a four-horsepower steam engine that drove the machines he used to produce chocolate [105].

By the 1820s, the world of chocolate was changing. François Cailler opened a chocolate factory near Vevey in Switzerland in 1819, and five years later, in 1824, Portuguese *forestero* cuttings were brought from Brazil to São Tomé in the Gulf of Guinea, off the coast of Gabon in Africa, sparking the spread of cacao cultivation to Africa (see Chapter 40). In 1825, when

Brillat-Savarin published *The Physiology of Taste*, Jean-Antoine Brutus Menier started producing chocolate at Noisiel near Paris, and a year later, Philippe Suchard began production at Serrières near Neufchatel, Switzerland. When the Menier plant began production, chocolate consumption in France was barely 300,000 kilograms, according to tariff statistics. Good chocolate was costly, and inexpensive chocolate was of poor quality. In 1853, consumption was 3,107,523 kilograms and ten years later, according to the tariff statistics, it was 5,513,107 kilograms [106].

An important shift occurred in 1828 with the development of a new process to separate cocoa powder from the cocoa butter by Coenraad Johannes van Houten. Van Houten's process contributed significantly to the subsequent mechanization of chocolate production. The Poulain Chocolate Company opened in Blois in 1848. Elaborate chocolate statuettes made with molds, used to shape *chocolat ouvragé* (worked or sculpted hard chocolate), were displayed at the 1855 Paris Exposition [107]. Arthur Mangin, a zoologist, wrote a study of chocolate in 1860, in which he argued that it was healthful and tasty, provided employment to people in France, and was the only food that could be called moralizing (*aliment moralisateur*). He wanted the government, then under Napoleon III, to lift all taxes on its sale [108].

In 1867, Henri Nestlé developed powdered milk, which in 1875 was used by Daniel Peter of Vevey, Switzerland, to make milk chocolate [109]. Milk was often added to chocolate prior to 1875 but it was difficult to keep fresh and was expensive. Rodolphe Lindt, also Swiss, developed "conching," a process of agitating the chocolate to make it more suave and mellow, leading to the production of what he called *fondant* in 1879 (see Chapter 46). By the end of the 19th century, cacao, once limited to the wealthy, had become available to "all classes" [110]. France was importing nearly 10 million kilograms of chocolate and had become the largest consumer of chocolate after Spain [111]. Colorful advertisements for companies such as Menier and Poulain proliferated and subsequently became collectors' items (see Chapter 15). Reproductions of late 19th and early 20th century chocolate advertising posters may still be found for sale in Paris today (2009) along the banks of the Seine. In 1905, cacao was planted in the Côte d'Ivoire in French Africa.

The democratization of chocolate in France that accompanied its industrialization paralleled developments in other countries but was followed toward the end of the 19th century by a renewed socially upscale direction in the production and distribution of fine chocolates that would continue to the present. France, a long-time producer of luxury goods, carried this tradition into the 20th century. In 1886, Fauchon's fine foods shop was established in Paris [112]. Auguste and Clementine Rouzaud created the Marquise de Sévigné brand of chocolate in Vichy in 1898. The Sévigné shops proliferated in France and, in 1917, in the middle of World War I, Premier Georges Clemenceau visited one in Paris. "The Tiger at the Marquise's," ran the headlines of French newspapers reporting the event [113]. The story of Clemenceau's visit to the Marquise de Sévigné's chocolate shop in wartime Paris appears in a book by Robert Linxe, who, in 1977, began a new page in the history of French luxury chocolates when he opened his first Maison du Chocolate shop in Paris [114].

Acknowledgments

I would like to acknowledge the assistance of the following persons during the course of research on the present chapter: Evelyne Bacardatz, Archivist, Archive Communale de Bayonne, France; Michèle Bimbenet-Privat, Conservateur en chef du patrimoine, Archives Nationales, Paris; Rachel Brishoual, Responsable de la Photothèque des Musées des Arts décoratifs, Paris; Helen Davies, Research Fellow, Department of Historical Studies, University of Melbourne; Philip and Mary Hyman, Historians, Paris; Karen Offen, Senior Scholar, Institute for Research on Women & Gender, Stanford University; Dany Pelong, Musée du Chocolat, Biarritz, France; Suzanne Perkins, Art Historian, Berkeley, California; Joanne Ryckelynck, Salon du Chocolat, Paris; Barbara Santich, Program Manager, Graduate Program in Gastronomy, The University of Adelaide, South Australia; Stuart Semmel, Associate Professor, Department of History, University of Delaware; and Stephen Shapiro, Assistant Professor of French, Department of Modern Languages and Literatures, College of the Holy Cross.

References

1. Franklin, A. *La Vie Privée d'autrefois. Le Café, le Thé et le Chocolat*. Paris, France: Plon, 1893; pp. 157–158.

2. Labat, J.-B. Cited in: Le Grand d'Aussy, P. J. B. *Histoire de la vie privée des Français*, 3 Volumes. Paris, France: Ph.-D. Pierres, 1782; Volume III, p. 108.

3. Mangin, A. *Le cacao et le chocolat suivi de la légende de Cacahuat par Ferdinand Denis*. Paris, France: Guillaumin, 1860; p. 161.

4. Franklin, A. *La Vie Privée d'autrefois. Le Café, le Thé et le Chocolat*. Paris, France: Plon, 1893; pp. 175–176.

5. Arveiller, R. *Contribution à l'Étude des Termes de Voyage en Français (1505–1722)*. Paris, France: La Rochefoucauld, 1963; p. 105.

6. "Cacao." In Trésor de la Langue Française Informatisée. Available at http://atilf.atilf.fr/dendien/scripts/tlfiv5/advanced.exe?8;s=3826981095;. (Accessed February 10, 2007.)

7. Arveiller, R. *Contribution à l'Étude des Termes de Voyage en Français (1505–1722)*. Paris, France: La Rochefoucauld, 1963; p. 105.

8. Gerbault, M. *López de Gómara dans les controverses sur le Nouveau Monde: les traductions françaises de la Historia general de las Indias y conquista de Mexico. Édition critique et commentaire comparé*. Paris, France: Thèse de l'École des Chartes, 2003. Available at http://theses.enc.sorbonne.fr/document79.html. (Accessed February 10, 2007.)

9. López de Gómara, F. *Histoire génerálle des Indes occidentales et terres neuves qui jusques à présent ont esté descouvertes*. Paris: Michel Sonnius, 1584 and 1605; p. 461.

10. de Acosta, J. *Histoire naturelle et morale des Indes tant occidentales qu'orientales: où il est traicté des choses remarquables du ciel, des élémens, métaux, plantes & animaux qui sont propres de ces païs . . . composée en castillan par Joseph Acosta; traduite en français par Robert Regnault Cauxois*. Paris, France: Marc Orry, 1598; pp. 171–172.

11. Terrio, S. J. *Crafting the Culture and History of French Chocolate*. Berkeley: University of California Press, 2000; pp. 74–75.

12. Brillat-Savarin, J.-A. *The Physiology of Taste (1825)*. Translated and annotated by M. F. K. Fisher. San Francisco, CA: North Point Press, 1986 (1949); p. 110.

13. Batiffol, L. *Le Roi Louis XIII à Vingt Ans*, 2nd edition. Paris, France: Calmann-Lévy, 1910; p. 389.

14. Batiffol, L. *Le Roi Louis XIII à Vingt Ans*, 2nd edition. Paris, France: Calmann-Lévy, 1910; p. 391.

15. Batiffol, L. *Le Roi Louis XIII à Vingt Ans*, 2nd edition. Paris, France: Calmann-Lévy, 1910; p. 391, note 1.

16. Moreau, R. *Du chocolate: Discours curieux divisé en quatre parties. Par Antoine Colmenero de Ledesma Medecin & Chirurgien de la ville de Ecija de l'Andalouzie. Traduit d'Espagnol en François sur l'impression faite à Madrid l'an 1631, & éclaircy de quelques annotations. Par René Moreau Professeur du Roy en Medicin à Paris. Plus est ajouté un Dialogue touchant le même chocolate*. Paris, France: Sébastien Cramoisy, 1643; p. ii.

17. d'Argonne, B. (Vigneul-Marville). *Mélanges d'Histoire et de Littérature Quatrième Édition, reviïe, corrigée, et augmentée*. Paris, France: Claude Prudhomme, 1725; I, pp. 4–5.

18. Head, B. *A Popular Account of Cocoa*. London, England: R. Brimley Johnson, 1903. Electronic text: www.gutenberg.net. (Accessed June 10, 2005.)

19. Montpensier, Anne Marie Louise d'Orléans, duchesse de Montpensier. *Mémoires de Mlle de Montpensier, petite-fille de Henri IV, collationnés sur le manuscrit autographe avec notes biographiques et historiques, par A. Chéruel*. Paris, France: Charpentier, 1858–1859; IV, p. 414.

20. Havard, H. *Dictionnaire de l'ameublement et de la décoration depuis le XIIIe siècle jusqu'à nos jours*. Paris, France: Quantin, 1894; I, p. 816.

21. Fink, B. When readables become edibles. *Gastronomica* Summer 2002;2(3):93.

22. Harwich, N. *Histoire du chocolat*. Paris, France: Éditions Desjonquères, 1992; pp. 78–79.

23. Huetz de Lemps, A. Boissons coloniales et essor de sucre. In Flandrin, J.-L., and Montanari, M., editors. *Histoire de l'Alimentation*. Paris, France: Fayard, 1996; p. 632.

24. Le Grand d'Aussy, P. J. B. *Histoire de la vie privée des Français*, 3 Volumes. Paris, France: Ph.-D. Pierres, 1782; Volume III, pp. 102–103.

25. Montespan, F. A. M. *The Memoirs of Madame de Montespan, Being the Historic Memoirs of the Court of Louis XIV*. Book I, Chapter XI. Available at http://www.knowledgerush.com/gutenberg/3/8/5/3854/3854.txt. (Accessed December 29, 2006.)

26. Elizabeth-Charlotte, Duchesse d'Orléans. "The Queen—Consort of Louis XIV," No. 7. In: *The Memoirs of the [sic] Louis XIV*. Available at http://www.public-domain-content.com/books/Louis14/C7P1.shtml. (Accessed January 10, 2007.)

27. Archives Nationales, Paris. O¹ 3713 fol. 9. "Maison de la Reine Marie Thérèse."

28. Cuzacq, R. *Triptyque Bayonnais: Jambon, Baionette, Chocolat de Bayonne*. Mont-de-Marsan, France: J. Glize, 1949; p. 65.

29. Franklin, A. *La Vie Privée d'autrefois. Le Café, le Thé et le Chocolat*. Paris, France: Plon, 1893; p. 164.

30. Archives Communales de Bayonne. CC 314, #105, "Pour le Sieur Martin de Lafourcade bourgeois et trésorier de la ville de cette présente année 1670," 6 April 1670.

31. Archives Communales de Bayonne. GG 229, #4, 23 August 1691; and Léon, H. *Histoire des Juifs de Bayonne*. Paris: Armand Durlacher, 1893; p. 114.

32. Archives Communales de Bayonne. GG 229, #21, 3 January 1735. "3 Jan 1735, "Ordonnance de police portant defenser aux Juifs de tenir aucune boutique ou chays pour vendre des marchandises en detail."

33. Archives Communales de Bayonne. HH 139, pp. 435–446. "Statutes et reglemens pour la Communauté des marchands chocolatiers," 1761, Articles 17 and 9, respectively, pp. 438 and 436.

34. Archives Communales de Bayonne. HH 139, pp. 435–446. "Statutes et reglemens pour la Communauté des marchands chocolatiers," 1761; p. 443.

35. Archives Communales de Bayonne. GG 229, #54, "Signification d'un arrêt obtenu du Parlement de Bordeaux par le marchands Portuguais et le sindic des marchands epiciers de cette ville—Contre les maîtres chocolatiers," 4 September 1767.

36. Léon, H. *Histoire des Juifs de Bayonne*. Paris, France: Armand Durlacher, 1893; pp. 73–74.

37. Curutcharry, M., and Douyrou, M. *Le Chocolat en Pays Basque. Dossier Educatif de la Chambre de Commerce de*

Bayonne. Bayonne, France: Chambre de Commerce, 1994; n.p.

38. Archives Nationales, Paris. Registre X^{1a} 8665 f° 68. "Commission pour vingt-neuf ans à David Chaliou de composer et vendre le chocolat en liqueur, pastilles, boîte, ou autres manières qu'il luy plair," 9 February 1666.

39. Franklin, A. *La Vie Privée d'autrefois. Le Café, le Thé et le Chocolat.* Paris, France: Plon, 1893; pp. 164–166.

40. Douyrou, M. "Les Fagalde," Chocolatiers du Pays Basque Cambo—Bayonne. No. 20 spécial. *Cercle Généalogique du Pays Basque et Bas Adour.* Bayonne, France, 1996; p. 6.

41. Mangin, A. *Le cacao et le chocolat suivi de la légende de Cacahuat par Ferdinand Denis.* Paris, France: Guillaumin, 1860; pp. 180–181.

42. Du Tertre, J.-B. *Histoire générale des isles de S. Christophe, de la Guadeloupe, de la Martinique et autres dans l'Amérique.* Paris, France: E. Kolodziej, 1978 (1654); p. 199.

43. de Quélus, D. *Histoire naturelle du cacao, et du sucre, divisée en deux traités.* Paris, France: L d'Houry, 1719; p. 18.

44. Labat, J.-B. *Nouveau Voyage aux Isles de l'Amérique.* Paris, France: Ch. J. B. Delespine, 1742; I, p. 353.

45. Constantin, A. A propos du chocolat de Bayonne. *Bulletin du Musée Basque* 1932;2(3):403.

46. Archives Nationales, Paris. Gabriel Dumaitz de Goimpy, Intendant aux Isles (Governor of the Islands), Martinique, Letter to the King, 2 July 1692, C^{8A} 7-1692, folios 134 and 137, respectively.

47. Archives Nationales, Paris. Gabriel Dumaitz de Goimpy, Intendant aux Isles (Governor of the Islands), Martinique, Letter to the King, 25 June 1694, C^{8A} 8-1694. folio 191.

48. Mangin, A. *Le cacao et le chocolat suivi de la légende de Cacahuat par Ferdinand Denis.* Paris, France: Guillaumin, 1860; pp. 182–183.

49. Archives Nationales, Paris. Blondel de Jouvencourt, Charles François, Intendant, Letter to the King, 1 December 1727, C^{8A} 38-1727, folios 403–404.

50. Jean-Charles de Bochart de Champigny, Governor General of Martinique, and Blondel, Letter to the King, 13 April 1728, C^{8A} 39-1728, folios 52–54.

51. Blondel, Letters, 13 January 1728, C^{8A} 39-1728, folio 195; and 13 May 1728, C^{8A} 39-1728, folios 287–289.

52. Labat, J.-B. *Nouveau Voyage aux Isles de l'Amérique.* Paris, France: Ch. J. B. Delespine, 1742; I, pp. 342–343.

53. Franklin, A. *La Vie Privée d'autrefois. Le Café, le Thé et le Chocolat.* Paris, France: Plon, 1893; p. 189.

54. Gallais, A. *Monographie du Cacao, ou Manuel de l'Amateur de Chocolat.* Paris, France: Chez Debauve et Gallais, 1827. Facsimile reprint—Paris, France: Phénix, 1999; pp. 143–144.

55. Sévigné, M. R.-C. *Lettres de Madame de Sévigné, de sa famille et de ses amis recueillies et annotées par M. Monmer-*

qué. Paris, France: L. Hachette, 1862–1868; Letters to Madame de Grignan, 11 February 1671, No. 133, II, p. 60; 15 April 1671, No. 157, II, pp. 164–165; and 25 October 1671, No. 214, II, p. 399; respectively.

56. Richelet, P. *Dictionnaire françois: contenant les mots et le choses, plusieurs nouvelles remarques sur la langue françoise.* Geneva, Switzerland: Slatkine Reprints, 1994 (original, Geneva, Switzerland: J. H. Widerhold, 1680; I, p. 136).

57. Dufour, S. *Traitéz Nouveaux & Curieux du café, du thé et du chocolate.* Lyon, France: Jean Girin and B. Rivière, 1685; p. 417.

58. Blégny, N. de. *Le bon usage du thé, du caffé et du chocolat pour la préservation et pour la guérison des maladies.* Paris, France: E. Michallet, 1687; p. 297.

59. Massaliot, F. *Le Cuisinier Roial et Bourgeois.* Paris, France: Charles de Cercy, 1691; pp. 295–296.

60. Massaliot, F. *Le Cuisinier Roial et Bourgeois.* Paris, France: Charles de Cercy, 1691; pp. 223–224.

61. Michel, D. *Vatel et la naissance de la gastronomie.* Paris, France: Fayard, 1999; p. 189.

62. Pomet, P. *Catalogue des drogues simples et composées.* Paris, France: The Author, 1695; pp. 11–13.

63. Franklin, A. *La Vie Privée d'autrefois. Le Café, le Thé et le Chocolat.* Paris, France: Plon, 1893; pp. 203–204.

64. Hecquet, P. *Traité des dispenses du carême,* 2nd edition. Paris, France: F. Fournier, 1710; Volume II, pp. 345–346.

65. Hecquet, P. *Traité des dispenses du carême,* 2nd edition. Paris, France: F. Fournier, 1710; Volume II, p. 420.

66. Hecquet, P. *Traité des dispenses du carême,* 2nd edition. Paris, France: F. Fournier, 1710; Volume II, pp. 475–477.

67. Harwich, N. *Histoire du chocolat.* Paris, France: Éditions Desjonquères, 1992; p. 77.

68. Le Grand d'Aussy, P. J. B. *Histoire de la vie privée des Français,* 3 Volumes. Paris, France: Ph.-D. Pierres, 1782; Volume III, p. 107.

69. Massaliot, F. *Le Cuisinier Roial et Bourgeois.* Paris, France: Charles de Cercy, 1734; Volume I, pp. 197–198, 283.

70. Massaliot, F. *Le Cuisinier Roial et Bourgeois.* Paris, France: Charles de Cercy, 1734; Volume II, p. 171.

71. Massaliot, F. *Le Cuisinier Roial et Bourgeois.* Paris, France: Charles de Cercy, 1734; Volume II, pp. 384–387.

72. Massaliot, F. *Le Cuisinier Roial et Bourgeois.* Paris, France: Charles de Cercy, 1734; Volume I, p. 32.

73. Massaliot, F. *Le Cuisinier Roial et Bourgeois.* Paris, France: Charles de Cercy, 1734; Volume I, pp. 56–61.

74. Massaliot, F. *Nouvelle instruction pour les confitures les Liqueurs, et les Fruits.* Amsterdam: Aux Depens de la compagnie, 1734; p. 225.

75. Pomet, P. *Histoire générale des drogues simples et composées.* Paris, France: Étienne Ganneau and Louis-Étienne Ganneau Fils, 1735; pp. 237–238.

76. *Festin Joyeux, ou, La Cuisine en Musique, en vers libres*, Parts 1 and 2. Paris, France: Lasclapart Père et Fils, 1738; Part 1, p. 185.

77. Menon. *La Science du Maître d'Hôtel Confiseur à l'usage des officiers, avec des observations sur la connaissances et les propriétés des fruits. Enrichie de dessins en décorations et parterres pour les desserts.* Paris, France: Paulus Du-Mesnil, 1750; p. 171.

78. Menon. *La Science du Maître d'Hôtel Confiseur à l'usage des officiers, avec des observations sur la connaissance et les propriétés des fruits. Enrichie de dessins en décorations et parterres pour les desserts.* Paris, France: Paulus Du-Mesnil, 1750; p. 175.

79. Menon. *La Science du Maître d'Hôtel Confiseur à l'usage des officiers, avec des observations sur la connaissances et les propriétés des fruits. Enrichie de dessins en décorations et parterres pour les desserts.* Paris, France: Paulus Du-Mesnil, 1750; p. 185.

80. Menon. *La Science du Maître d'Hôtel Confiseur à l'usage des officiers, avec des observations sur la connaissances et les propriétés des fruits. Enrichie de dessins en décorations et parterres pour les desserts.* Paris, France: Paulus Du-Mesnil, 1750; pp. 408–409.

81. Menon. *La Science du Maître d'Hôtel Confiseur à l'usage des officiers, avec des observations sur la connaissances et les propriétés des fruits. Enrichie de dessins en décorations et parterres pour les desserts.* Paris, France: Paulus Du-Mesnil, 1750; pp. 411–412.

82. Lémery, L. *Traité des Alimens*, 3rd edition, Volume I. Paris, France: Durand, 1755; p. 94.

83. "Confiseurs." In Artisanats au 18ème Siècle, *L'Encyclopédie Diderot et d'Alembert*. Paris, France: Inter-Livres, 1986; pp. 3–4.

84. *Gazetin du comestible*, No. 12. December 1767; p. 2.

85. Le Grand d'Aussy, P. J. B. *Histoire de la vie privée des Français*, 3 Volumes. Paris, France: Ph.-D. Pierres, 1782; Volume III, p. 107.

86. Harwich, N. *Histoire du chocolat*. Paris, France: Éditions Desjonquères, 1992; pp. 98–99.

87. de Sade, D. A. F. (Marquis). *The 120 Days of Sodom and other Writings*. (Originally published 1785.) Compiled and translated by Richard Seaver and Austryn Wainhouse. New York: Grove Press, 1966; p. 263.

88. Buc'hoz, P.-J. *Dissertations sur l'utilité, et les bons et mauvais effets du Tabac, du Café, du Cacao et du Thé, ornées de Quatre planches en taille-douce*, 2nd edition. Paris, France: Chez l'Auteur, De Burc l'aîné, et la Veuve Tilliard et Fils, 1788; p. 112.

89. Buc'hoz, P.-J. *Dissertations sur l'utilité, et les bons et mauvais effets du Tabac, du Café, du Cacao et du Thé, ornées de Quatre planches en taille-douce*, 2nd edition. Paris, France: Chez l'Auteur, De Burc l'aîné, et la Veuve Tilliard et Fils, 1788; pp. 121–122.

90. Brillat-Savarin, J.-A. *The Physiology of Taste (1825)*. Translated and annotated by M. F. K. Fisher. San Francisco, CA: North Point Press, 1986 (1949); pp. 112, 371.

91. Buc'hoz, P.-J. *Dissertations sur l'utilité, et les bons et mauvais effets du Tabac, du Café, du Cacao et du Thé, ornées de Quatre planches en taille-douce*, 2nd edition. Paris, France: Chez l'Auteur, De Burc l'aîné, et la Veuve Tilliard et Fils, 1788; pp. 124–131.

92. Brillat-Savarin, J.-A. *The Physiology of Taste (1825)*. Translated and annotated by M. F. K. Fisher. San Francisco, CA: North Point Press, 1986 (1949); p. 111.

93. Archives Nationales, Paris. de Laussat, Pierre Clement, Colonial Prefect in Martinique, Letter to the Emperor, 10 March 1807, C^{8A} 115-1807, folios 96–98.

94. Grimod de la Reynière, A. B. *Almanach des Gourmands*. Paris: Maradan, 1805; p. 47.

95. Gallais, A. *Monographie du Cacao, ou Manuel de l'Amateur de Chocolat*. Paris, France: Chez Debauve et Gallais, 1827. Facsimile reprint—Paris, France: Phénix, 1999; p. 167.

96. Grimod de la Reynière, A. B. *Almanach des Gourmands*. Paris: Maradan, 1805; p. 45.

97. Stewarton (Goldsmith, L.). *Memoirs of the Court of St. Cloud, Being Secret Letters from a Gentleman at Paris to a Nobleman in London*, Volume 4. Available at http://www.gutenberg.org/files/3899/3899.txt. (Accessed February 1, 2007.) (Original—Philadelphia: John Watts; 1806, Letter 34.)

98. Blanc, H. *Le Guide des Dineurs, ou Statistique des principaux restaurans de Paris*. Paris, France: L'Étincelle, 1985 (original edition—Paris: Marchands de Nouveautés, 1815), p. 100.

99. Blanc, H. *Le Guide des Dineurs, ou Statistique des principaux restaurans de Paris*. Paris, France: L'Étincelle, 1985; p. 111.

100. Blanc, H. *Le Guide des Dineurs, ou Statistique des principaux restaurans de Paris*. Paris, France: L'Étincelle, 1985; p. 138.

101. Blanc, H. *Le Guide des Dineurs, ou Statistique des principaux restaurans de Paris*. Paris, France: L'Étincelle, 1985; p. 163.

102. Blanc, H. *Le Guide des Dineurs, ou Statistique des principaux restaurans de Paris*. Paris, France: L'Étincelle, 1985; p. 183.

103. Blanc, H. *Le Guide des Dineurs, ou Statistique des principaux restaurans de Paris*. Paris, France: L'Étincelle, 1985; p. 213.

104. Benoiston de Chateauneuf, L. F. *Recherches sur Les Consommations de tout genre de la ville de Paris en 1817; comparées à celles qu'elles étaient en 1789*, 2nd edition. Paris, France: Chez l'Auteur and Martinet, 1821; pp. 156–157.

105. Roret. *Nouveau Manuel Complet du Confiseur et Chocolatier*. Paris, France: Charles Moreau, René Baudouin Successeur, n.d.; p. 385.

106. Turgan, J. Chocolat Usine de Mr Menier à Noisiel. In Turgan, *Les Grandes Usines de France: Tableau de l'industrie française au XIXe siècle*. Paris, France: Michel Lévy Frères, 1870; Volume 7, p. 111.

107. Dorchy, H. *Le Moule à chocolat. Un nouvel objet de collection*. Paris, France: Éditions de l'Amateur, 1987; p. 12.

108. Mangin, A. *Le cacao et le chocolat suivi de la légende de Cacahuat par Ferdinand Denis*. Paris, France: Guillaumin, 1860; p. 211.

109. Coe, S. D., and Coe, M. D. *The True History of Chocolate*. London, England: Thames and Hudson, 1996; pp. 250–251.

110. Colombie, A. *Manuel des Éléments Culinaires à l'usage des jeunes filles*. Paris, France: L. Mulo, 1897; p. 30.

111. Roret. *Nouveau Manuel Complet du Confiseur et Chocolatier*. Paris, France: Charles Moreau, René Baudouin Successeur, n.d.; p. 385.

112. Fauchon. La Maison FAUCHON/FAUCHON au fil du temps. Available at http://www.fauchon.fr/cadeauxprives/info.asp?id=1403&codeM=SU31 Active. (Accessed February 2, 2007.)

113. Linxe, R. *La maison du Chocolat*. Paris, France: Robert Laffont, 1992; p. 34.

114. Linxe, R. *La maison du Chocolat*. Paris, France: Robert Laffont, 1992; p. 52.

CHAPTER

Commerce, Colonies, and Cacao

Chocolate in England from Introduction to Industrialization

Bertram M. Gordon

Introduction: The English Adopt Chocolate

The history of English chocolate is tied to the 17th and 18th century popularization of the hot beverage, the industrialization of the 18th and 19th centuries, and the coming of milk chocolate after 1875. Chocolate entered England primarily as a drink consumed by the affluent social classes. Industrialization led to an increased production of hard chocolate, highlighted by Coenraad van Houten's mechanized process to separate cocoa powder from cocoa butter in 1828, although the industrialization of chocolate production began earlier. The coming of milk chocolate that followed Daniel Peters's technical innovations in Switzerland in 1875 ushered in the modern chocolate era. Woven through this history is the British involvement in commerce and colonization through the course of the early modern period and the 19th century. This chapter explores the introduction of chocolate in 17th century England through its popularization by firms such as Fry's and Cadbury's in the late 19th and early 20th centuries and is based in substantial measure on the archival research in England by Pamela Richardson, a member of the chocolate history research team.

As with many other Western and Central European peoples, the English adopted chocolate in the middle of the 17th century. The earliest English encounters with cacao appear to have been through pirate raids on Spanish ships. In *The True History of Chocolate*, Sophie and Michael Coe discussing early English accounts of chocolate, cite Thomas Gage's *The English American: His Travail by Sea and Land*, published in London in 1648, to the effect that in 1579 English pirates raided a Spanish ship loaded with cacao beans. Believing the beans to be sheep dung, the English burned the ship.

The Coes also mention José de Acosta's account of the destruction in 1587 of another cacao-laden Spanish ship by English corsairs who deemed the cargo useless. De Acosta was a Spanish Jesuit missionary in Latin America and his *Historia natural y moral de las Indias*, published in 1590, became one of the first works to address cacao in English when it was translated as *The Natural and Moral History of the Indies* in 1604. Cacao also appears in a 1633 edition of John Gerard's *Herbal or General History of Plants*, which refers to its use as money in the Western Hemisphere. Girard refers to cacao only in general terms, indicating that the local populations made a bitter drink from it, leading the Coes to conclude that he was acquainted with it only by reputation [1].

Chocolate: History, Culture, and Heritage. Edited by Grivetti and Shapiro
Copyright © 2009 John Wiley & Sons, Inc.

In addition to mentioning chocolate as a means to increase urine flow and a remedy for kidney ailments [2], Thomas Gage tells a much repeated story of Creole women who were so fond of their chocolate that they insisted on drinking it during church services in Chiapas (present town of San Cristóbal de las Casas). Attempting to stop the practice, the bishop excommunicated those who continued to have their chocolate brought into the church. A row ensued and Gage, there as a Dominican missionary, tried to mediate. He failed and the bishop was murdered by chocolate laced with poison, presumably by one of the unhappy local women. The story gave rise to a ditty—"Beware the chocolate of Chiapas"—and became the motif for subsequent murder-by-chocolate stories [3, 4]. Henry Stubbe retold Gage's story in 1662 [5].

Cacao, Colonization, and Clubs

Gage was in Jamaica when the English seized the island in 1655, creating a turning point in English chocolate history. The English would establish cacao plantations on the island but only after they had acquired a taste for hot chocolate, a process begun during the 1650s and 1660s. When chocolate use spread into England in the 1650s, high import duties on cacao beans restricted its use to the affluent. Chocolate cost the equivalent of 50–75 pence a pound, a considerable sum in the middle of the 17th century [6].

One of the earliest known advertisements for chocolate in London appeared in a 1657 issue of the *Public Advertiser*, which contained a notice that "in Bishopsgate Street, in Queen's Head Alley, at a Frenchman's house, is an excellent West India drink, called chocolate, to be sold, where you may have it ready at any time, and also unmade, at reasonable rates." These rates appear to have been from 10 to 15 shillings per pound, a price that made chocolate, rather than coffee, the beverage of the aristocracy [7]. The anonymous Frenchman has yet to be identified and little else is known about this chocolate house, which opened while the theocratic Puritan Commonwealth of Oliver Cromwell governed England.

Cromwell's death in 1658 led to the restoration of the monarchy under Charles II, under whom moral and social behavior codes were relaxed in reaction to Cromwell's rule. Coffeehouses, which sometimes served chocolate, proliferated, as did chocolate houses. In 1674, a London coffeehouse, "At the Coffee Mill and Tobacco Roll," was said to serve chocolate in cakes and rolls, "in the Spanish style" [8]. Nonetheless, as coffeehouses and chocolate houses grew in popularity, often becoming centers of political discussion and intrigue, in 1675 Charles II attempted—unsuccessfully as it turned out—to suppress them [9]. Chocolate houses developed as centers for political factions and became the forerunners of modern British clubs. During the reign of Queen Anne (1702–1714), The Cocoa Tree became the headquarters of the Jacobite party, which supported the restoration of the Stuarts after their overthrow in the Glorious Revolution of 1689. Following an unsuccessful attempt by the Stuarts to regain power, the name of The Cocoa Tree was changed in 1746 and became the Cocoa-tree Club for "Tories of the strictest school," according to Brandon Head [10]. Members included Jonathan Swift and Edward Gibbon [11, 12]. Brandon Head recorded that "De Foe tells us in his 'Journey through England,' that a Whig will no more go to the 'Cocoa Tree' . . . than a Tory will be seen at the coffee-house of St. James's." The Cocoa Tree also became a center for the London literati, including the Shakespearean actor David Garrick and the poet Lord Byron. The Cocoa Tree eventually developed into a gambling house and a club, as did the even better known White's Cocoa House, adjoining St. James' Palace [13].

White's was founded in 1693 by Francesco Bianchi, an Italian immigrant, who conducted business under the name of Francis White. White's served chocolate drinks as well as ale, beer, coffee, and snacks. White's chocolate was made from blocks of solid cocoa, probably imported from Spain. A pressed cake from which the drink could be made at home was also sold [14]. Another source suggests, however, that it was unlikely that chocolate was served at White's, despite its name. Because of its reputation as a gambling house, Jonathan Swift referred to White's as the "bane of half the English nobility," and those who frequented the establishment were known as "the gamesters of White's." The sixth print in William Hogarth's "Rake's Progress" series shows a distraught man having just lost his fortune in the gambling room at White's [15]. By the end of the 18th century, London's chocolate houses began to disappear, with many of the more fashionable ones becoming gentlemen's clubs. White's Chocolate House remains an exclusive London gentlemen's club.

Chocolate Penetrates English Culture

The growth of chocolate's popularity in mid-17th century England was reflected in the increasing commentaries about it in literary circles. Samuel Pepys, the noted diarist, evidently enjoyed chocolate, which he mentions several times. The Coes excerpt several of his references to chocolate, starting in 1660, when his diary began and when Charles II returned to England (the Stuart Restoration of 1660). Pepys first mentions some chocolate having been left for him at his home.

Three of his entries refer to drinking chocolate in the morning and in one he drank it at noon. In an entry of 1664, he refers to drinking chocolate in a coffeehouse [16].

As elsewhere in Europe, the acceptance of chocolate in England was due in part to various medical benefits claimed for it. Henry Stubbe, who had visited Jamaica, became the Royal Physician following the Stuart Restoration. As such, he prepared chocolate for King Charles II and, in 1662, published his *Indian Nectar* (Fig. 43.1). A compendium of much of the literature on chocolate of his time, Stubbe's book recommends taking chocolate twice a day as quick refreshment for someone "tyred [*sic*] through business" [17]. Chocolate, together with spices from the West Indies, Stubbe added, strengthened the heart and the stomach, helped women in menstruation, and reduced gas [18]. By the time Stubbe wrote his book, a connoisseurship of cacao beans had developed in England and Stubbe was well aware of it. In serving the king, he preferred beans from Caracas and Nicaragua to those of Jamaica, which he deemed less well cured [19]. Stubbe's overall view of chocolate was positive. "I have heard and read Discourses of Panaceas and Universal Medicines: and truly I think Chocolata [*sic*] may as justly at least pretend to that Title, as any" [20]. Chocolate was also an aphrodisiac in Stubbe's view, although it was to be taken mildly, and not heavily spiced with pepper, to avoid too great a stimulation [21]. His discussion of the aphrodisiacal properties of chocolate, together with his Biblical references, was rewritten in a subsequent edition of his book in 1682 and is quoted at length by the Coes [22]. Ten years after Stubbe, in 1672, William Hughes published a treatise on chocolate in which he contended that chocolate drunk with cinnamon or nutmeg would cure coughs. His contention that chocolate was also effective against the symptoms associated with scurvy may have contributed to the British navy's eventual decision to include it in the daily rations of its seamen [23, 24].

In England, the earliest known silver chocolate pot was made in 1685 by the silversmith George Garthorne (Fig. 43.2) [25, 26]. Accoutrements for chocolate service in the late 17th and into the 18th century

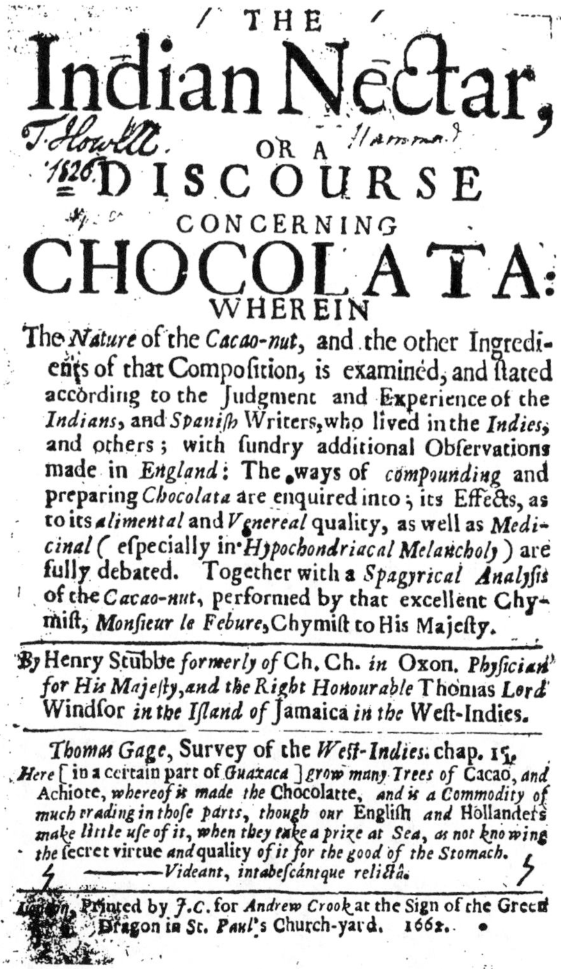
THE
Indian Nectar,
OR A
DISCOURSE
CONCERNING
CHOCOLATA:
WHEREIN
The *Nature* of the *Cacao-nut*, and the other Ingredients of that Compofition, is examined, and ftated according to the Judgment and Experience of the *Indians*, and *Spanifh* Writers, who lived in the *Indies*, and others; with fundry additional Obfervations made in *England*: The ways of *compounding* and preparing *Chocolata* are enquired into; its Effects, as to its *alimental* and *Venereal* quality, as well as *Medicinal* (efpecially in *Hypochondriacal Melancholy*) are fully debated. Together with a *Spagyrical Analyfis* of the *Cacao-nut*, performed by that excellent Chymift, *Monfieur le Febure*, Chymift to His Majefty.

By Henry Stubbe *formerly of* Ch. Ch. *in* Oxon. *Phyfician for His Majefty, and the Right Honourable* Thomas *Lord* Windfor *in the Ifland of* Jamaica *in the* Weft-Indies.

Thomas Gage, Survey of the *Weft-Indies*. chap. 15.
Here [in a certain part of *Guaxaca*] *grow many Trees of* Cacao, *and* Achiote, *whereof is made the* Chocolatte, *and is a Commodity of much trading in thofe parts, though our* Englifh *and* Hollanders *make little ufe of it, when they take a prize at Sea, as not knowing the* fecret virtue *and* quality *of it for the good of the Stomach.*
——*Videant, intabefcántque relictâ.*

London, Printed by *J.C.* for *Andrew Crook* at the Sign of the Green Dragon in St. *Paul's* Church-yard. 1662.

FIGURE 43.1. *The Indian Nectar, or A Discourse Concerning Chocolata*, by Henry Stubble. London, England: J. C. for Andrew Crook (title page); dated 1662. *Source*: Courtesy of the Peter J. Shields Library, University of California, Davis. (Used with permission.)

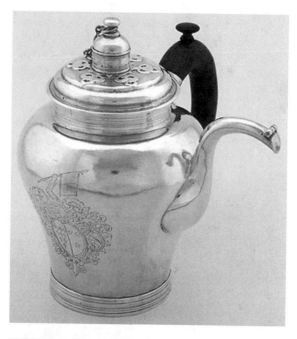

FIGURE 43.2. Chocolate pot, manufactured by George Garthorne. Arms of the original owner (Moore) are emblazoned on the surface; the shape was inspired by a Chinese ginger jar: dated 1686. *Source*: Courtesy of the Minneapolis Institute of Arts, Minneapolis, MN. (Used with permission.) (See color insert.)

included cups and *chocolatières*, or chocolate pots, as England's aristocracy and middle classes embraced the new beverage [27]. An especially attractive piece is a gold chocolate cup manufactured by Ralph Leake, dated ca. 1690, now housed at Temple Newsam House in Leeds, England (Fig. 43.3). The Victoria and Albert Museum in London, in its Silver Galleries, currently displays a chocolate pot made by the silversmith John Fawdery and dated to 1704–1705. Everard Fawkener, a diplomat, spent nearly £2000 on a dinner service in 1735, including a chocolate pot for £20 and 18 shillings [28].

While visiting Jamaica from 1687 to 1689, Hans Sloane, a physician and botanist, who found the chocolate drunk by the locals to be "nauseous," introduced the addition of milk to it [29]. Alternatively, the Coes suggest that the first person known to have com-

bined milk with chocolate was Nicholas Sanders, an Englishman, who in 1727 produced a milk and chocolate drink for Sloane, who by that year had returned from Jamaica and had become first surgeon to King George II. The Coes point out, however, that this was not modern milk chocolate but simply milk mixed with chocolate liquor [30]. Nikita Harwich, the author of *Histoire du Chocolat*, argues the point even more strongly, writing that Sloane's decoction was merely a replacement of water by milk in a cup of hot chocolate and was not "especially original" [31]. Upon Sloane's return to England, his milk and chocolate recipe was copied by pharmacists and sold as medicine. The Cadbury Company later adopted it. Sloane, who amassed a large collection of flora, fauna, and coins, later became a founder of the British Museum. Upon his death in 1753, the British government bought his collection, which includes Italian-made ceramic chocolate cups now on display at the British Museum (Fig. 43.4) [32].

Popularized by Stubbe, Hughes, Sloane, and others, as well as in chocolate houses with the relevant accoutrements, chocolate also appeared in cookbooks. A handwritten recipe book compiled during the period from 1701 through 1706 by Diana Astry contains recipes for chocolate cakes, chocolate cream, and plain chocolate. Little appears to be known about the compiler, and the recipes are attributed to different people. They are similar to those of landed gentry households at the beginning of the 18th century [33]. The recipe for chocolate cakes called for the addition of fine chocolate shavings, sugar, and orange flower or plain water boiled together to the consistency of candy. These chocolate cakes were said to be the size of a shilling coin [34]. Chocolate cream called for blanched almonds and chocolate in cream, to which egg yolks were added. The chocolate was milled (churned) until thick and the froth removed [35]. Astry's book confirms the time-consuming and difficult process required to make chocolate. The reader wishing to make chocolate began with "cocoa nuts," whose husks were to be removed. The

FIGURE 43.3. Gold chocolate cup, manufactured by Ralph Leake; dated ca. 1690. *Source*: Courtesy of Temple Newsam House, Leeds, England. (Used with permission.) (See color insert.)

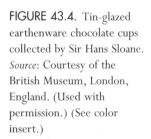

FIGURE 43.4. Tin-glazed earthenware chocolate cups collected by Sir Hans Sloane. *Source*: Courtesy of the British Museum, London, England. (Used with permission.) (See color insert.)

nuts then were beaten in a hot iron mortar and made into a paste. This paste was then worked into cakes that were then finely ground into a powder. Refined sugar and cinnamon were added, and heated and beaten again until forming a second paste. This new paste was to be put into tin pans and allowed to cool [36].

The medical benefits of chocolate continued to be extolled by physicians, such as Daniel Duncan, a French Huguenot transplant in London, whose widely read treatise promoted the moderate use of chocolate, coffee, and tea. First published in 1703 in France, Duncan's book was brought out in an English edition in 1706. Arguing in the humoral medical theory of the day, he counseled that taken in excess, chocolate made the blood too sharp, hot, and thin. Summarizing Duncan's position, the Coes note that he feared that too much chocolate would be harmful for the "sanguine," making their blood "over-inflammable," and for the choleric [37, 38]. One of the more intriguing findings in the British archives is a proposal made by Marcus Meibomius in 1711 for making tea, coffee, and chocolate out of indigenous herbs and fruits of England, and for placing a tax on beer to fund the schooling of children [39]. An antiquarian and librarian, and also a philologist and mathematician, Meibomius, who was from Holstein and may have been Danish or German, was best known as a historian of music. Given his interest in taxing beer and encouraging the production of chocolate and coffee, he might have been a forerunner of 19th century temperance activists who saw chocolate as a morally acceptable alternative to alcoholic beverages.

Cacao was also involved in the mercantile struggles for empire in the early 18th century. A letter of November 30, 1717, from H. S. Worsley in Lisbon to Joseph Addison, the Secretary of State for the Southern Department, the English office that was responsible for overseeing the American colonies, reported that Spain had blocked the import of sugar, sweetmeats, and cacao from Britain, which meant the entire British Empire. This action had been taken in reprisal for a British ban on imported Spanish wines and brandies [40]. The cacao ban followed in the wake of the War of the Spanish Succession, which had ended in 1714 with Britain acquiring Gibraltar and Minorca from Spain, together with the Asiento, a British monopoly over the import of African slaves into Spanish colonies.

Joseph Fry and Fraud in Chocolate

The earliest records for Fry's chocolates date to 1728 when Walter Churchman, an apothecary, began producing chocolate in Bristol, then one of England's major ports handling commerce with the Americas [41]. In the following year, Churchman applied for a patent for an "invention and new method for the better

making of chocolate by an engine set forth in his annexed Petition." Churchman declined to specify the details of his engine, presumably afraid that his device would be copied. The Attorney General handling his petition supported it for a term of 14 years [42]. Following Churchman's death, his son took over the business until his own death in 1761, when it was acquired by Joseph Fry (1728–1787), a Quaker and apothecary who had settled in Bristol [43]. By the late 18th century—1789 according to the Coes and 1795 according to Harwich—the Fry Company had purchased a steam engine to grind chocolate [44, 45].

Joseph Fry and his chocolate manufacturing company in Bristol appear in the public record in a complaint he registered with the excise authorities in which he complained of illegal selling of chocolate that was cutting into his trade. Chocolate imported into Bristol, a major port as noted earlier, was often resold elsewhere in England. Fry noted that illegal chocolate shipments had been seized, one weighing 500 pounds, but that the Treasury officials had not been compensated for their efforts and had to pay the costs of seizing the contraband. The solution, according to Fry, was to sell the seized chocolate for whatever price it would command to support the enforcement of the duties on subsequent imports. He also asked for a reduction of the excise duty on "cocoa nuts" and chocolate to lower the retail price of chocolate and thereby reduce the incentives for smugglers [46]. Fry appears to have been unsuccessful in his lobbying. A Treasury official refuted all his arguments, holding that the illegal sale of chocolate had been greatly exaggerated, that Treasury officers did receive payment from duties collected, and that lowering the duties on cocoa nuts and chocolate would be detrimental to Treasury revenues [47].

Fraud and smuggling were continual problems as the British, along with other European countries, taxed colonial products for the enrichment of the homeland. A report in 1778 complained that tea import records of the East Indies Company had not kept pace with a rise in consumption of tea among the middle classes in England and called for remedies. The government would now require sworn affidavits in excise books pertaining to tea, coffee, "cocoa nuts," and chocolate brought into the country [48]. Also in 1778, several people, presumably in the chocolate trade, signed a petition to the Treasury with the request that the duty on imported chocolate be replaced with a lower one on "cocoa nuts." The signers claimed that some people were fraudulently buying large quantities of cocoa nuts, getting approval for them to be imported under fictitious names, and selling it at inflated prices as chocolate. Taxing raw cocoa (or cacao) at a lower rate would yield more revenue, the petitioners argued, because the income would be more certain [49]. A Treasury report of 1779 proposed undetermined fines for smugglers of chocolate, with sellers of the contraband material to pay twice that amount. Contraband material would be seized. Nonpayment of the fines would subject the offender to a two year prison sentence. A second offence brought a term of five years labor on the Thames or five years aboard a warship. A third conviction could lead to banishment for life to the East Indies or the coast of Africa. Informers were to be rewarded [50].

Chocolate and the Struggle for Empire: Fighting the Americans and the French

The world of chocolate in England shifted again in the late 18th century as the English first lost the North American colonies that would become the United States, and then fought a 20 year battle for empire with the French. In 1780, a flagon of chocolate was included in the daily rations of English sailors serving in the Caribbean and the practice was generalized for all English naval vessels in 1824. In addition to Bristol's being an important port, it was also a leading British naval base. Fry's, located there, became an important chocolate supplier to the British navy [51].

The loss of their American colonies in 1783 led the British to reassess trade policies. In 1784, as Britain was reviewing its trade patterns with its colonies, Colonel Robert Pringle, assessing Newfoundland's trade with the West Indies, noted that the province's imports from the West Indies were mainly rum, sugar, molasses, coffee, and chocolate [52]. Addressing whether the newly independent United States should be allowed to renew its trade with the British West Indies colonies, Thomas Irving, formerly Inspector General of the Imports and Exports of North America and Register General of Shipping, concluded that the high demand for imports of lumber to the West Indies and exports of sugar to the United States and the rest of North America would also help support the development of the cocoa trade between the West Indies and North America [53].

Eventually, British policy-makers reconsidered mercantilist colonial expansion and a growing industrial sector, and progressively adopted the free trade principles advocated by Adam Smith, among others. As noted previously, during the last decade of the 18th century, the Joseph Fry Company purchased one of James Watt's steam engines, which was used for grinding cocoa beans, leading to the manufacture of chocolate on a scale large enough to become accessible to the middle classes [54]. Fry's chocolates would become a leader in the industrialization of chocolate processing.

The loss of their American colonies did not deter British imperial policy, rather the opposite, as the British seized substantial territory and blockaded the Continent during their wars against revolutionary and Napoleonic France between 1792 and 1815. In 1797, the British took control of Trinidad and several years later, Governor Sir Thomas Picton reported on conditions there to Lord Melville, First Lord of the Admiralty. In a letter of 1804 that highlights the social and racial inequalities in the cultivation of cacao, Picton, who had been governor since the British seized the island, addressed the difficulties of starting a plantation there:

> It is, I know, held by persons who have their reasons for recommending this ruinous speculation, that there are Species of Cultivation (requiring little labour and less exposure to the Sun) which are peculiarly adapted to this class of people [European colonists], and they cite that of Cacao: They had read in Raynal [whose Histoire philosophique et politique des établissements et du commerce des Européens dans les deux Indes, published in 1770, described the profit to be made in raising cacao] and in other picturesque Historians, that the lazy Spaniard strung his Hammock up in the delightful Walks of Cacao, and dozed away a Life of Indolence: but they do not tell you, how the Spaniard became possessed of these imaginary Elysiums; they do not tell you, that they were purchased by the best years of his Life, employed in hard and incessant Labour and privation; nor that the Labour was performed, not by the European Spaniard, but by the Native Creole of the Country.—Their only object is to mislead the Country by delusory prospects, and they cautiously conceal every kind of Information, which is not favourable to their Projects.
>
> The fact is, that the first five or six years of a Cacao Plantation, is attended with as much Labour, and exposure to the Sun, as any other Cultivation in the Tropical Climates— The Timber must be cut down,—the Land must be well cleared of Weeds, The plants must be continually attended to, for five or six years, before the Holocaust affords any Shade, or protection, from the Sun—The Spaniard calculates that he must have at least one working Negroe for every quadrado (or 3 and 1/5 acres of Cacao) even after it is brought to maturity. It is a cultivation which may well suit Whites, who have five or six Negroes; but the single European will always fail in the attempt. [55, 56]

Cacao had been introduced in Trinidad under Spanish rule in approximately 1725 but, as Harwich notes, had suffered from an unknown "disaster" in 1727, the year that an earthquake destroyed much of the cacao plantings in Martinique. Toward the middle of the century, Catalan and Aragonese Capuchin friars came to Trinidad, where they planted forestero (foreign) seedlings from the middle Orinoco area, the present-day Venezuela–Colombia region. Harwich writes that the foresteros (another cacao variety) were later hybridized with criollo (cacao variety) plants remaining from the Capuchins, and the trinitario variety was born. The British takeover of the island led to a modest increase in cacao production [57].

Grenada was another Antilles island where the British became active in cacao planting. Britain had taken Grenada from France at the end of the Seven Years' War in 1763 and had lost it again at the end of the American Revolutionary War in 1783. The British intensified the cultivation of cacao, cotton, and nutmeg in Grenada. By the time of emancipation in 1833, the island's slave population had reached 24,000 [58]. England used the blockade against Napoleon and French shipping as an opportunity to increase trade with the newly liberated colonies of the Western Hemisphere. In 1805, for example, William Davis Robinson, an American then living in Caracas, requested a license to import products from Spanish South America into Grenada. The list of products included fifty to sixty thousand *fanegas* (bushels) of cacao, mentioned as "cocoa," weighing 110 pounds each and estimated to have a value of $1 million, a considerable quantity and more than the value of the coffee and tobacco to be included in the shipments. Clearly, the British intended the cacao to be transshipped elsewhere in the British Empire [59]. Governor Maitland granted the requested permission [60].

Beginnings of the Industrialization of Chocolate Production

The British victory over Napoleon sealed England's gains in the Caribbean and ushered in a period of sustained industrial expansion, which carried chocolate with it in respect to increased demand and enhanced technology. Fry's was joined on March 1, 1824, when John Cadbury, another young Quaker, as was Joseph Fry, announced the opening of a shop in Bull Street in Birmingham. The announcement, in the *Birmingham Gazette*, read as follows: "John Cadbury is desirous of introducing to particular notice 'Cocoa Nibs,' prepared by himself, an article affording a most nutritious beverage for breakfast" [61].

A government report in 1826 indicated that cocoa production in Trinidad had more than doubled in 10 years and because many of the plantings were new, the yield could be expected to increase. Many of the farms, however, "belong to small settlers who being needy Persons are often in haste to convey their produce to the market." The result, according to the report, was that the quality of Trinidad cocoa, once nearly the equal to that of Caracas, had fallen. More government inspections were required [62].

Development of the van Houten process for separating cocoa powder from cocoa butter in 1828 intensified an industrial process already underway in Britain. The 1831 Blue Book of Grenada showed nearly 338,000 pounds of cocoa produced on the island, more than coffee, but less than a tenth the quantity of sugar. Most of the cocoa was exported to Great Britain (see Chapters 26 and 27) [63].

In 1831, 1,943,953 pounds of cocoa were exported from Trinidad with a value of £12,635. About half went to Britain, with the rest divided among "foreign states," the United States, and other countries [64]. Some 21,608,384 pounds of cocoa were produced by Trinidad in 1894, marking more than a tenfold increase since 1831 [65]. In contrast, Jamaica, which had seen cocoa planting since the time of Thomas Gage, witnessed a drop in production such that, in 1881, only 26 acres were planted, in contrast to 18,456 for coffee, and 39,712 for cane sugar [66]. By 1901, Jamaica's acreage of cocoa production had increased to 3548 but it was still far behind the 31,265 acres devoted to coffee and 27,342 to cane sugar [67].

Chocolate remained a beverage primarily for those with means in England during the first half of the 19th century. The 1837 diary of William Taylor, a servant in a house in Mayfair, London, describes a breakfast chocolate served in much the same manner as the foamy spicy brew from pre-Columbian Mexico and later Spain, as "something like coffee but of a greasey [*sic*] and much richer nature" [68]. By mid-19th century, medical arguments about chocolate appear to have been favorably settled in England. A growing temperance movement also saw chocolate, tea, and coffee as morally acceptable alternatives to alcoholic beverages. In 1844, a request was made to the Deptford Victualling Yard to supply cocoa for prisoners at Millbank Prison. The reasons given were the health of the prisoners, although it is also possible that unstated temperance issues were behind the request [69].

Increased chocolate consumption heightened the possibilities for fraudulent product, especially as the methods of chocolate preparation leant themselves to adulteration. The chemist Frederick Accum alerted the English public to the dangers of food adulteration with his publication of *A Treatise on Adulterations of Food and Culinary Poisons* in London in 1820. In the years from 1851 through 1854, the medical journal *The Lancet* published a series of articles by Arthur Hill Hassall, a London physician, who showed that cacao and chocolate were frequently adulterated with Venetian red (a reddish brown pigment), red ochre, and iron compounds to improve their color, and with arrowroot, wheat, Indian corn, sago, potato, tapioca flour, and chicory to add bulk and weight [70]. *The Lancet's* publication of Hassall's and other studies showing adulteration across a wide swath of English foods led to the Food and Drug Act of 1860, considered a milestone in the struggle for pure foods (see Chapter 47) [71].

The Modern Era: Fry, Cadbury, and the Chocolate Bar

If there is any one date that could be said to usher in the modern age of chocolate in England, it would be 1847, the year that Fry and Sons began sale of a "Chocolat Delicieux à Manger" (chocolate delicious to eat), commonly considered to be the first chocolate bar [72]. In 1849, Fry and Cadbury chocolate bars were displayed publicly at an exhibition in Bingley Hall, Birmingham, and two years later, bonbons, chocolate creams, hard candies (called "boiled sweets"), and caramels were shown at the Crystal Palace Exposition, which showcased British industry, in London [73]. Cadbury, however, was tainted by the mid-19th century food adulteration scandals, as their chocolate was found to have been diluted with starch and flour. George Cadbury, forced to admit that his chocolate was adulterated, went on an advertising blitz with posters, showing among others things, a solemn looking scientist attesting that his chocolate was "absolutely pure," and therefore the best. Fry's responded that the addition of other products to chocolate did not necessarily cheapen its quality nor threaten the health of the consumer. The battle between the two companies went on for some 30 years but British public opinion had turned in favor of "purity" and Cadbury supplanted Fry's as England's number one chocolatier [74].

By the middle of the 19th century, Britain had become the "workshop of the world." Committed to an increasingly free trade economic policy, the British reduced import duties, which, combined with falling sugar prices, led to lower prices and the popularization of chocolate among the less affluent social strata in England and around the world [75]. Tariff reductions on cacao in 1853 by a Liberal government in which William E. Gladstone was Chancellor of the Exchequer opened the way for additional companies to enter the chocolate business [76]. Export figures from the British colony of Saint Lucia illustrate the increase in cacao and chocolate commerce. In 1851, Saint Lucia exported 122,875 pounds of cocoa, a figure that increased nearly tenfold to 1,136,843 pounds in 1896. Saint Lucia sent a value of £143/10/0 (measured in pounds, shilling, and pence) in cocoa to Britain and £1,009/5/9 to the West Indies, for a total cocoa export value of £1,152/15/9 in 1851 [77]. Of the 1,136,843 pounds of cocoa exported from Saint Lucia in 1896, 261,300 went to Britain and 795,200 to France. The total cocoa export value in sterling for that year was £15,402/13/7 (see Chapter 42) [78].

In 1867, Henri Nestlé developed powdered milk in Switzerland and eight years later, Daniel Peter, of Vevey, used Nestlé's powdered milk to make chocolate, launching a new era of milk chocolate and increasing demand still further. Britain, as did other European countries, contributed to introducing cacao implantations to West Africa in the late 19th century. In 1874, cacao was introduced into Nigeria. More importantly for the world chocolate trade, it was brought into the Gold Coast, now Ghana, in 1879 [79]. Forty kilograms of cocoa exported from the Gold Coast in 1891 grew to 40 million in 1911. In 1834–1835, Sir Robert Horton, then governor of Ceylon (present-day Sri

Lanka), introduced cacao plants from Trinidad. A modest cacao production resulted in 508 kilograms exported from Ceylon in 1875. By the eve of World War I, it had reached 3.5 million kilograms [80].

Chocolate consumption quadrupled in England between 1880 and 1902 [81]. In 1879, reflecting the growth in demand and of general interest in chocolate, Cadbury moved its plant to Bournville on the south side of Birmingham. There they created a model town, which by the end of the century had become a tourist site, similar to the development of Menier, near Paris, and Hershey, in Pennsylvania, which also evolved from religiously based paternalistic model factory towns into tourist attractions. Describing the Cadbury Cocoa Offices in Bournville at the beginning of the 20th century, Brandon Head wrote:

> Taking [the] train from the city, glimpses can be caught, as we near our destination, of the pretty houses and gardens of the village, forming a great contrast to the densely populated district of Stirchley on the other side of the line. Stepping on to the station, we are greeted by a whiff of the most delicious fragrance, which is quite enough of itself to betray the whereabouts of the great factory lying beneath us, of which from this point we have a fairly good bird's-eye view. [82]

The increased popularity of chocolate in England led to renewed calls to extend chocolate rations to seamen, beyond those serving in the British navy. A request for extra chocolate for the men on British warships in 1893 led to a directive the next year to extend chocolate rations to those serving night watch on all British ships sailing to India. In 1894, Henry Campbell Bannerman, the Secretary of State for War, and later prime minister, authorized rations of cocoa or chocolate to be given at 4:00 a.m. to guards and sentries on ships at the discretion of their commanding officers. The Queen's Regulations and Admiralty Instructions of 1893 provided these rations to men in the Royal Navy and a letter from Campbell Bannerman extended them to those serving in troop ships, transports, and freight ships, adding:

> Mr. Campbell Bannerman considers that much advantage would accrue to the health of the troops especially in view of the large number of cases of diseases of the respiratory organs which have occurred on some homeward voyages from India in recent seasons, and that the issue would afford a break in the long interval which elapses between the evening and morning meals, which is specially felt by men on duty. [83]

Ultimately, the 1893 Instructions mentioned earlier were amended to provide half a pint of "chocolate with sugar" to all soldiers on night duty, including sentries at about 4:00 a.m. [84].

By the end of the century, Cadbury, with an intense advertising campaign featuring colorful posters with cherubic children enjoying their chocolate, had effectively won the battle against Fry's. In 1875,

Cadbury premiered the first chocolate Easter eggs and, although fragile at first, they were improved as the years went by [85]. In 1905, Cadbury launched what became its top selling brand: Dairy Maid, which later became Dairy Milk, with a unique flavor and creamy texture that was positioned to challenge Swiss domination of the milk chocolate market [86]. In 1918, Cadbury took control of its old rival, Fry's, although the two retained separate brand identities. They also tried to gain control of Rowntree, a third key British chocolate maker, established by a Quaker family in 1862 [87]. Rowntree, however, managed to remain independent. In the post-World War I years, both Rowntree and Cadbury, in addition to struggling against the Swiss milk chocolate giants, would face new competition from the American companies, Hershey and Mars.

Acknowledgments

I wish to thank the following persons who assisted in the research for this chapter: Deborah Gage (Works of Art) Ltd., London (descendent of Thomas Gage); Stephen Mennell, Professor, School of Sociology, University College, Dublin, Ireland; Suzanne Perkins, Art Historian, Berkeley, California; Pamela Richardson, Archivist and Cultural Geographer, Independent Scholar, Oxford, England; John Walton, Professor of Social History, University of Central Lancashire, Preston, England; and Richard White, Senior Lecturer, Department of History, University of Sydney, Australia.

References

1. Coe, S. D., and Coe, M. D. *The True History of Chocolate.* London, England: Thames and Hudson, 1996; p. 165.

2. Grivetti, L. E. From aphrodisiac to health food: a cultural history of chocolate. *Karger Gazette* No. 68, Chocolate. Available at http://www.karger.com/gazette/68/grivetti/art_1.htm. (Accessed February 20, 2007.)

3. Sebens, H. "Gage, Thomas," Historical Text Archive. Available at http://historicaltextarchive.com/sections.php?op=viewarticle&artid=451. (Accessed February 24, 2007.)

4. Coe, S. D., and Coe, M. D. *The True History of Chocolate.* London, England: Thames and Hudson, 1996; pp. 185–186.

5. Stubbe, H. *The Indian Nectar, or A Discourse Concerning Chocolata.* London, England: J.C. for Andrew Crook, 1662; pp. 93–97.

6. "Chocolate Houses," in Cadbury. Available at http://www.cadbury.co.uk/EN/CTB2003/about_chocolate/history_chocolate/chocolate_houses.htm. (Accessed May 30, 2006.)

7. Head, B. *A Popular Account of Cocoa.* London, England: Brimley Johnson, 1903. Electronic text: www.gutenberg.net. (Accessed June 10, 2005.)

8. The Gourmet Chocolate of the Month Club. Available at http://www.chocolatemonthclub.com/chocolatehistory.htm. (Accessed May 28, 2006.)

9. Coe, S. D., and Coe, M. D. *The True History of Chocolate.* London, England: Thames and Hudson, 1996; p. 172.

10. Head, B. *A Popular Account of Cocoa.* London, England: Brimley Johnson, 1903. Electronic text: www.gutenberg.net. (Accessed June 10, 2005.)

11. Chocolaterie Cantalou-Cemoi. Electronic text: www.ceomi.fr. (Accessed December 11, 2005.)

12. Coe, S. D., and Coe, M. D. *The True History of Chocolate.* London, England: Thames and Hudson, 1996; p. 227.

13. Appendix III, "The Early Cocoa Houses." In Head, B. *A Popular Account of Cocoa.* London, England: R. Brimley Johnson, 1903. Electronic text: www.gutenberg.net. (Accessed June 10, 2005.)

14. "Chocolate Houses," in Cadbury. Available at http://www.cadbury.co.uk/EN/CTB2003/about_chocolate/history_chocolate/chocolate_houses.htm. (Accessed May 30, 2006.)

15. "The English Coffee Houses," in Waes Hael Poetry and Tobacco Club. Available at http://waeshael.home.att.net/coffee.htm. (Accessed June 1, 2006.)

16. Coe, S. D., and Coe, M. D. *The True History of Chocolate.* London, England: Thames and Hudson, 1996; p. 170.

17. Stubbe, H. *The Indian Nectar, or A Discourse Concerning Chocolata.* London, England: J.C. for Andrew Crook, 1662; p. 3.

18. Stubbe, H. *The Indian Nectar, or A Discourse Concerning Chocolate.* London, England: J.C. for Andrew Crook, 1662; pp. 50, 53.

19. Stubbe, H. *The Indian Nectar, or A Discourse Concerning Chocolate.* London, England: J.C. for Andrew Crook, 1662; p. 70.

20. Stubbe, H. *The Indian Nectar, or A Discourse Concerning Chocolate.* London, England: J.C. for Andrew Crook, 1662; p. 125.

21. Stubbe, H. *The Indian Nectar, or A Discourse Concerning Chocolate.* London, England: J.C. for Andrew Crook, 1662; pp. 132, 152.

22. Coe, S. D., and Coe, M. D. *The True History of Chocolate.* London, England: Thames and Hudson, 1996; pp. 175–176.

23. Grivetti, L. E. From aphrodisiac to health food: a cultural history of chocolate. *Karger Gazette* No. 68, Chocolate. Available at http://www.karger.com/gazette/68/grivetti/art_1.htm. (Accessed February 20, 2007.)

24. Hughes, W. *The American Physitian.* London, England: British Library. University of Cambridge, 1672; 449.a.24.

25. "Chocolate Pot." *Encyclopædia Britannica*. 2007. Encyclopædia Britannica Online. Available at http://www.britannica.com/eb/article-9082279. (Accessed February 25, 2007.)

26. Deitz, P. Chocolate pots brewed ingenuity. *New York Times*, February 19, 1989, Section 2, Arts and Leisure, p. 38.

27. Perkins, S. Is it a chocolate pot? Chocolate and its accoutrements in France from cookbook to collectible. In Grivetti, L.E., and Shapiro, H.-Y., editors. *Chocolate: History, Culture, and Heritage*. Hoboken, NJ: Wiley, 2009; p. 160.

28. Fawdery, J. *Chocolate Pot*. London, England, early 18th century. Victoria and Albert Museum, London, England, Case 12, Number 10, Museum Number: M.1819–1944.

29. The Sloane Herbarium, Natural History Museum, London, England. Available at http://www.nhm.ac.uk/research-curation/projects/sloane-herbarium/hanssloane.htm. (Accessed February 25, 2007.)

30. Coe, S. D., and Coe, M. D. *The True History of Chocolate*. London, England: Thames and Hudson, 1996; p. 249.

31. Harwich, N. *Histoire du chocolat*. Paris, France: Éditions Desjonquères, 1992; p. 173, note 1.

32. The Sloane Herbarium, Natural History Museum, London, England. Available at http://www.nhm.ac.uk/research-curation/projects/sloane-herbarium/hanssloane.htm. (Accessed February 25, 2007.)

33. Stitt, B. *Diana Astry's Recipe Book, c. 1700*. The Publications of the Bedfordshire Historical Record Society, 37 (1957), Public Record Office, 942 BEDS, Volume 37, p. 87.

34. Stitt, B. *Diana Astry's Recipe Book, c. 1700*. The Publications of the Bedfordshire Historical Record Society, 37 (1957), Public Record Office, 942 BEDS, Volume 37, p. 104.

35. Stitt, B. *Diana Astry's Recipe Book, c. 1700*. The Publications of the Bedfordshire Historical Record Society, 37 (1957), Public Record Office, 942 BEDS, Volume 37, p. 105.

36. Stitt, B. *Diana Astry's Recipe Book, c. 1700*. The Publications of the Bedfordshire Historical Record Society, 37 (1957), Public Record Office, 942 BEDS, Volume 37, p. 134.

37. Coe, S. D., and Coe, M. D. *The True History of Chocolate*. London, England: Thames and Hudson, 1996; p. 208.

38. Duncan, D. *Wholesome Advice against the Abuse of Hot Liquors, particularly Coffee, Chocolate, Tea, Brandy, and Strong-Waters*. London, England: H. Rhodes, 1706; University of Cambridge, UL, L5.59.

39. Public Record Office, SP 34/24/37.

40. Worsley, H. S. Letter to Joseph Addison, 30 November 1717, Public Record Office, SP 89/25.

41. "Cadbury & JS Fry & Sons," in Cadbury. Available at http://www.cadbury.co.uk/EN/CTB2003/about_chocolate/history_cadbury/cabury_fry/. (Accessed March 3, 2007.)

42. Churchman, W. Petition, Bristol, England, July 1729 and P. Yorke, Report, 16 December 1729, Public Record Office, SP 36/15, Part 2, folios 140–143.

43. "Cadbury & JS Fry & Sons," in Cadbury. Available at http://www.cadbury.co.uk/EN/CTB2003/about_chocolate/history_cadbury/cabury_fry/. (Accessed March 3, 2007.)

44. Coe, S. D., and Coe, M. D. *The True History of Chocolate*. London, England: Thames and Hudson, 1996; p. 242.

45. Harwich, N. *Histoire du chocolat*. Paris, France: Éditions Desjonquères, 1992; p. 79.

46. Fry, J. Memorial to the Right Honourable The Lords Commissioners of the Treasury, 2 April 1776, Public Record Office, T1 523/333.

47. "To the Right Honourable The Lords Commissioners of His Majesty's Treasury," 9 April 1776, Public Record Office, T1 523/332-336.

48. Excises; Rules for Traders in Tea, Coffee, and Chocolate, undated (1778), Public Record Office, TI 542/229-230.

49. "To the Right Honourable The Lords Commissioners of his Majesty's Treasury," 18 March 1778, Public Record Office, TI 542/222-223, and 30 April 1779, Public Record Office, TI 552/380-381.

50. Treasury Board Papers, 1779, Public Record Office, TI 552/308, pp. 308–312.

51. Harwich, N. *Histoire du chocolat*. Paris, France: Éditions Desjonquères, 1992; pp. 162–163.

52. Pringle, R. "Memorandum," 20 March 1784, Public Record Office, 30/29/3/10/11.

53. Irving. "Memorandum," 1784, Public Record Office, 30/29/3/10/12.

54. "A Brief History of Chocolate: A chronicle of when and where it was discovered and how it evolved to the confection we enjoy today." Available at http://www.waynesthisandthat.com/chocolatehistory2.htm. (Accessed March 3, 2006.)

55. Picton, Sir Thomas To Lord Melville; Conditions and Affairs of Trinidad, 25th May 1804, University of Oxford, Rhodes House Library, MSS. W. Ind s8, f.109a,b, 1804, pp. 13–14.

56. Raynal, G. T. F. Abbé. *Épices & produits coloniaux*. Paris, France: La Bibliothèque, 2004 (1992); pp. 13–16.

57. Harwich, N. *Histoire du chocolat*. Paris, France: Éditions Desjonquères, 1992; pp. 61–62.

58. Commonwealth Secretariat. Available at http://www.thecommonwealth.org/YearbookInternal/145158/history/. (Accessed February 25, 2007.)

59. Maitland, F. Governor, Grenada, 26 July 1805, Public Record Office, CO 5/2, pp. 89–92.

60. Robinson, W. D. Petition to Brigadier General Frederick Maitland, Governor of Grenada, 1805, Public Record Office, CO 5/2, pp. 95–100.

61. CHOCOLATE. The history of Talleyrand's "Culinary Fare" macht alles mit liebe. Available at http://www.geocities.com/napavalley/6454/chocolate.html. (Accessed May 30, 2006.)

62. Blue Book of Trinidad, Supplementary Report, 1826, Public Record Office, CO 300/40, pp. 7–8.

63. Blue Book of Grenada, 1831, Public Record Office, CO 106/25, pp. 144, 154–155.

64. Blue Book of Trinidad, 1831, Public Record Office, CO 300/45, pp. 144–145.

65. Blue Book of Trinidad, 1894, Public Record Office, CO 300/105, p. Y4.

66. Blue Book of Jamaica, 1881, Public Record Office, CO 142/95, p. 4X.

67. Blue Book of Jamaica, 1901, Public Record Office, CO 142/115, p. 2Z.

68. Taylor, W. Diary of William Taylor, Footman, 1837. Cited in: Allen, B., editor. *Food, An Oxford Anthology*. Oxford, England: Oxford University Press, 1994.

69. Correspondence, 1844, Public Record Office, HO 45/788.

70. "The Fight Against Food Adulteration," in RSC/Advancing the Chemical Sciences. Available at http://www.rsc.org/Education/EiC/issues/2005Mar/Thefightagainstfoodadulteration.asp. (Accessed March 4, 2007.)

71. Coe, S. D., and Coe, M. D. *The True History of Chocolate*. London, England: Thames and Hudson, 1996; p. 247.

72. "Chocolate Corner—Timeline of Chocolate History." Available at http://www.geocities.com/chocolate-corner/historyp3.html. (Accessed June 8, 2005.)

73. Bellis, M. "The Culture of the Cocoa Bean, Timeline of Chocolate," Inventors. Available at http://inventors.about.com/library/inventors/blchocolate.htm. (Accessed March 3, 2007.)

74. Harwich, N. *Histoire du chocolat*. Paris, France: Éditions Desjonquères, 1992; pp. 166–167.

75. Mintz, S. W. *Sweetness and Power, The Place of Sugar in Modern History*. New York: Elizabeth Sifton Books/Viking, 1985; pp. 149, 162.

76. "The Gourmet Chocolate of the Month Club." Available at http://www.chocolatemonthclub.com/chocolate-history.htm. (Accessed May 28, 2006.)

77. Blue Book of Saint Lucia, 1851, Public Record Office, CO 258/47, pp. 174–175, 184–185, 194–195.

78. "General Exports from the Colony of Saint Lucia in the Year 1896," Chart 3, in Blue Book of Saint Lucia, 1896, Public Record Office, CO 258/92, pp. W-33–34.

79. Young, A. M. *The Chocolate Tree, A Natural History of Cacao*. Washington DC: Smithsonian Institution Press, 1994; p. 40.

80. Van Hall, C. J. J. *Cocoa*. London, England: Macmillan and Company, 1914; pp. 8–9.

81. Head, B. *A Popular Account of Cocoa*. London, England: Brimley Johnson, 1903. Electronic text: www.gutenberg.net. (Accessed June 10, 2005.)

82. Head, B. *A Popular Account of Cocoa*. London: Brimley Johnson, 1903. Electronic text: www.gutenberg.net. (Accessed June 10, 2005.)

83. Thompson, R. War Office, 21 June 1894, Public Record Office, MT 23/97, T 1949-94, in "Issue of Chocolate to Troops on Board Transports &c," T 2800, 1896.

84. Chapter XLI, "Victualling Instructions," Public Record Office, MT 23/97, T 2800/1896.

85. Harwich, N. *Histoire du chocolat*. Paris, France: Éditions Desjonquères, 1992; p. 176.

86. "Chocolate Corner—Timeline of Chocolate History". Available at http://www.geocities.com/chocolate-corner/historyp3.html. (Accessed June 8, 2005.)

87. Harwich, N. *Histoire du chocolat*. Paris, France: Éditions Desjonquères, 1992; p. 194.

Chinese Chocolate

Ambergris, Emperors, and Export Ware

Bertram M. Gordon

Introduction: Chocolate in China ... An Untold Story

Historical accounts often suggest that chocolate was unknown until recently in China but in reality it was present there as early as the 17th century. Although little used until recent times, and even now not a major constituent in Chinese gastronomy, chocolate appears peripherally on occasion from the time of its introduction by Catholic missionaries, and its story in China has not been examined previously. Most intriguing is the mention of Chinese chocolate in at least three French sources, discussed here, that were published during the 19th century. It was the discovery of these references that resulted in the decision to write this chapter on Chinese chocolate: it is likely that this chapter will raise as many questions as it resolves.

Specialists in the history of Chinese gastronomy rarely mention chocolate or address it only in recent history. Eugene Anderson writes that chocolate, just as coffee and opium, reached China in the early modern period, but he then passes over it [1]. Françoise Sabban finds no "old" mentions of chocolate in China [2]. However, there is a chocolate history to uncover in China, even if sketchy, and even if much remains to be filled in.

Jesuits and Franciscans

The Jesuits preceded the Franciscans in China but the available documentation shows more about Franciscan use of chocolate. Dr. Noel Golvers, a specialist on the early history of the Jesuits in China, does not record any mention of chocolate in the account books of Fran-

çois de Rougemont, a Jesuit missionary in Ch'ang-Shu (Chiang-Nan) from 1674 through 1676, nor in any of the other Jesuit sources there [3]. On the other hand, there is substantial evidence for Franciscan use of chocolate in early modern China. Writing from Macao on March 27, 1674, Fr. Buenaventura Ibañez referred to himself as an old man aged sixty-five, with difficulty consuming anything other than rice and liquids, and asked a visiting colleague to bring cacao with vanilla that he could then eat in the mornings from a chocolate bowl (*una escudilla de chocolate*), which could be prepared locally [4].

Writing from Macao to his Minister Provincial on March 5, 1678 regarding Franciscan efforts to build a church in Canton, Ibañez noted that he had requested a cargo of cacao to be sent from New Spain with all the accoutrements necessary to make chocolate as a present for the "king of Canton" [5]. In another letter, dated March 4, 1681, Ibañez mentioned chocolate received as annual alms to the Franciscan mission in China [6], and a letter of February 15, 1683 contained a reference to the Franciscans

Chocolate: History, Culture, and Heritage. Edited by Grivetti and Shapiro
Copyright © 2009 John Wiley & Sons, Inc.

using it during Christian fast days [7]. In still another letter, dated February 24, 1686, Ibáñez referred to the use of chocolate as a pharmaceutical and complained that he could eat neither meat nor chicken and only a little fish, but that a good bowl of chocolate with vanilla, pimento, and cinnamon in the mornings was helpful [8].

By the early 18th century, chocolate as a pharmaceutical appeared frequently in the requests of the Franciscans' medical surgeon, Fr. Antonio Concepción, who worked in southern China from 1705 to 1743. In a collection of 48 of his letters, published in *Sinica Franciscana* in 1995, Fr. Concepción referred to chocolate on ten occasions. These references are nearly all in letters to his superiors in the Order, in which he either thanked them for chocolate previously sent or requested it for his mission. Writing to the provincial ministry in Canton on November 17, 1707, for example, he noted that chocolate, presumably requested earlier, had not arrived [9]. Another letter from Fr. Concepción, dated February 22, 1711, acknowledged receipt of chocolate sent as alms to the missionaries. It seems that this chocolate had already been shared among the missionaries present and Fr. Concepción hoped to have occasion to distribute some to those who had been away from the mission [10]. Several months later, on May 13, 1711, Fr. Concepción again thanked the Provincial Ministry for the charitable gifts of "medicine and chocolate" [11]. During the next year, on April 16, 1712, he thanked a superior in Canton for the receipt of eight *gantas* (one ganta = three liters) of chocolate that had been distributed to his fellow missionaries as instructed [12].

Once again, on February 16, 1713, Fr. Concepción expressed thanks for a shipment of chocolate that had arrived on a boat from Macao and had been distributed as per instructions to the other brothers in the Order [13]. Additional requests on April 30, 1714 included mention of two Chinese earthenware chocolate jars (*tibores de chocolate*) [14]. Chocolate received four years later, as noted in a letter dated April 23, 1718, was duly distributed to the brothers [15], and again noted in a letter of April 16, 1719 [6]. In a letter written April 20, 1729, Fr. Concepción asked approval to allow one of the Franciscan brothers to make 130 pesos of chocolate to give to the Jesuit fathers at the Canton court, who, he wrote, had already received the larger share [17]. This approval may not have been forthcoming because it was requested again a year later as noted in his letter of April 18, 1730, and Fr. Concepción also added that he wished to be able to have the value of 130 pesos of chocolate made to give to the Jesuits at court, "from whom we receive many favors" [18].

The chocolate sent to Fr. Concepción appears to have been distributed exclusively as a pharmaceutical to his fellow Franciscans. It is unclear whether he ever received permission to have chocolate made for the Jesuits and, unlike the case of Ibáñez, there is no evidence in the letters that any of it was given to local Chinese, except possibly for Chinese who were accepted into the mission. Further research might change this picture. To the degree that chocolate was known in China, it must undoubtedly have been restricted to the upper levels of society, perhaps only the court, as implied in Ibáñez's letter of 1678.

How Chocolate Came to China

There are several routes that cacao could have taken to China. Frederick W. Mote has written that new foods from the Americas began to enter China in the 16th and 17th centuries via three trade routes: southeastern Chinese ports, via Chinese or foreign ships; from land routes through southeast Asia into Yunnan; and via the silk route from Persia and Turkey [19]. Golvers has suggested that in the absence of Jesuit references to chocolate, but with extensive Franciscan documentation, the entry route to China was via Spain rather than Portugal. If correct, the most likely trade route went from Mexico either through Spain or via ocean traffic to the Philippines. The Franciscans, who were also active as missionaries in the Philippines, introduced new crops there such as avocado, cacao, maize, and tomato. Several accounts cite Fr. Gaspar de San Augustine's *Historia de Filipinas* to the effect that cacao (trees?) were first brought there in 1670 from Acapulco by a sea pilot named Pedro Brabo (sometimes written as Bravo) of Laguna Province, who gave them to a priest named Bartolome Brabo in the Camarines, a southeastern region of Luzon [20, 21]. According to another account, the Jesuits introduced chocolate to the Philippines during the Spanish conquest by Miguel López de Legazpi and Juan de Salcedo, between 1663 and 1668 [22]. Simon de La Loubere, sent by Louis XIV to Siam in 1687 and 1688, wrote that the Portuguese there drank chocolate, which had come from the Spanish West Indies via Manila [23].

Whatever the manner of chocolate's introduction into the Philippines, there was a substantial colony of Chinese living there, largely from the coastal areas of Canton and Fokien (Fukien, Fujien), estimated by Ferdinand Blumentritt, who also relied heavily on San Augustine's *Historia de Filipinas*, at some 30,000 in Manila, where many traded for local products and Mexican silver [24]. By the late 17th century, the Spanish monarchy was in decline and faced growing economic competition from the Dutch, who had also been trading actively in Taiwan. Blumentritt described a continual flow of Chinese to and from the Philippines. Under Spanish rule, Chinese immigration changed the Filipino diet. Noodles that arrived from China were called *pancit* by the Filipinos. Sweet potatoes, brought from Spain to the Philippines, made their way to China, where they were often substituted for rice among the poor. Chinese immigrants later brought sweet potato recipes back to the Philippines, making sweet potatoes into a Filipino staple food [25]. Whether cacao and chocolate followed this route remains to be explored.

There is also a possibility that chocolate made its way across a land route to East Asia. In his *Indian Nectar*, published in 1662, Henry Stubbe wrote that the use of chocolate had spread to Turkey and Persia [26]. Both Fernand Braudel and Michael and Sophie Coe tell of the Italian world traveler Giovanni Francesco Gemelli Careri, and how he offered chocolate to a Turkish Aga at Smyrna (Izmir) in 1693, but with unfortunate results. The Aga reportedly flew into a rage and accused Gemelli Careri of trying to intoxicate him in order to take away his powers of judgment [27, 28]. Another hint of a possible land route for chocolate is the Italian recipe for *Cuscusu Dolce*, a Sicilian dish, which, according to the San Francisco restaurateur and cookbook author Carlo Middione, may have North African roots. The dish is a sweet dessert made with couscous and a custard sauce to which is added sugar, vanilla extract, blanched almonds, citron, orange peel, cinnamon, and bittersweet chocolate [29]. On the other hand, a Muslim website gives a recipe for *Cuscusu Dolce* without chocolate, leading to the possibility that chocolate was added only in Italy [30].

Another hint of an overland route related to chocolate and its use, if not actually to the substance itself, is the product known as *chocolat analeptique au salep de Perse* (analeptic chocolate of Persian *salep*). This is a floury mix of chocolate with dried, ground tubers of different species of orchids from Persia and the Ottoman Empire [31]. This version of chocolate, highly praised as a restorative, and sometimes as an aphrodisiac, was produced in the early 19th century by Sulpice Debauve, a pharmacist who established a chocolate shop in Paris in 1800. His nephew, Antoine Gallais, also a pharmacist, who joined the firm in 1827, described it [32]. Debauve and Gallais subsequently shifted from a focus on chocolate as a pharmaceutical to gourmet confectionary; their shop still exists in Paris [33]. Hints such as a product used with chocolate coming from Persia and couscous recipes with chocolate, even if added only after their arrival in Italy, offering the possibility that chocolate was brought from Italy to the Ottoman and Mogul empires, and then overland to Eastern Asia, cannot be fully excluded.

Future research should elaborate the patterns of cacao and chocolate diffusion, as far east as China, following the model of maize in the Balkans by Traian Stoianovich, who used the many different local linguistic equivalents for maize to map its diffusion into the Balkans [34]. The diffusion pattern for chocolate

from Latin America to China—most likely to the Chinese court—now seems at this stage of research to be along the routes traveled by the Franciscans, probably with the Philippines as an intermediate point.

Chinese Chocolate Export Ware

Another indication of Chinese awareness of chocolate and a connection to Spain's Latin American colonies may be found in ceramic ware made in China for the export trade. Because the earliest non-Native American chocolate drinking occurred in New Spain, Jean McClure Mudge has suggested that the earliest references to Chinese chocolate pots should be available and, in fact, reference to a *jocolato pot* appears in an inventory of Chinese export ware belonging to William Pleay in New York in 1690 [35]. In addition to the 1690 Chinese chocolate pot in New York, Mudge found another that belonged to a Captain Nicholas Dumaresq (died 1701), who had been active in the Spanish trade network. Mudge adds that Japanese porcelain "coffee/chocolate" pots appeared in the 1680s and this is consistent with the presence of Japanese-made chocolate pots given by King Narai of Siam to King Louis XIV of France in 1686 (see Chapter 42) [36]. However, by the 1690s, newly revived kilns at Jingdezhen (Ching-tê-chên), the center of the porcelain industry in China, were producing lines of porcelain that may have included chocolate pots for sale to the West [37]. A cup dated ca. 1710 to the Kangxi period (Emperor Kangxi: 1661–1722) has been excavated. Although this cup might have been used for a variety of beverages, it is described as a chocolate cup in the period records of the Dutch East India Company, which had come to dominate the trade route between China and Latin America [38].

A letter written in 1722 by the Jesuit Father François Xaver d'Entrecolles indicated that Chinese chocolate cups were being made in Jingdezhen [39]. Additional porcelain fragments of similar Chinese-made cups from the same era were recovered from the Spanish colonial archaeological site at the Templo Mayor in central Mexico City [40]. Chocolate pots of the 18th century with Asian motifs also included jars from Puebla mounted with metal covers, "in which chocolate and vanilla could be securely locked" [41]. It has been suggested that by 1839 Mexico had more trade with China than with neighboring California [42].

Ship registers recording the import of Chinese chocolate cups and pots included those of the *Coxhoorn*, which brought 9457 cups and pots in 1730 from Canton to Amsterdam for the Dutch East India Company (Fig. 44.1) [43]. By the mid-18th century,

FIGURE 44.1. Chinese chocolate cup and saucer, dated 1729–1735. *Source*: William R. Sargent, a Preissler Decorated Chinese Chocolate Cup and Saucer, 1729–35. In: *Mélanges en souvenir d'Elisalex d'Albis, 1939–1998*. Sèvres, France: Société des Amis du Musée National de Céramique, 1999; p. 49. Courtesy of the Peabody Essex Museum, Salem, MA. (Used with permission.) (See color insert.)

FIGURE 44.2. Chinese chocolate pots, Imari style, 18th century. *Source*: Luisa Ambrosia, Caffè, tè e cioccolate mei servizi di porcellana de Museo. In: *Un mondo in tazza, porcellane per caffè, tè e cioccolata*. Naples, Italy: Electa Napoli, 2003; p. 13. Courtesy of the Museo Nazionale della Ceramica Duca di Martina, Naples, Italy. (Used with permission.) (See color insert.)

the volume of Chinese exports ware had increased significantly. The Dutch East India Company's ship *Geldermalsen*, wrecked during its return to Amsterdam from Canton in January 1752, contained a cargo of some 150,000 porcelain items; among the items listed were 9735 chocolate cups and saucers [44]. Supporting evidence for the listing of chocolate cups among the *Geldermalsen's* cargo includes a drawing of a chocolate cup from 1758, an exact replica of those found in the wreck [45]. The Chinese produced more chocolate cups than pots, but an example of the latter, dated to 1767, is on display at the museum in Kampen, in the Netherlands [46]. Mid-18th century Chinese chocolate pots are also to be found in the Museo Duca di Martina in Naples (Fig. 44.2) [47]. A set of ten silver chocolate pots with covers, from approximately 1810 to 1820, similar to those produced in porcelain, indicates that chocolate-related technological knowledge in China was not limited to the porcelain factories [48].

Chinese Chocolate in 19th Century French Texts

At least three French texts refer to "Chinese chocolate" in the 19th century. The earliest, in chronological sequence, is a confectionary encyclopedia by J.-J. Machet, entitled *Le Confiseur Moderne, ou l'Art du Confiseur et du Distillateur*, and first published in 1803. *Le Confiseur Moderne* enjoyed a measure of popularity as it went through nine editions by 1852. In the section on chocolate, in the third edition published in 1817, the earliest available for inspection, Machet describes

vacaca chinorum as composed of four ounces of cacao almonds (*amande de cacao*), one ounce of vanilla, one ounce of fine cinnamon (*cannelle*), 48 grains of ambergris, and three ounces of sugar. The powdered spices are added to the cacao and sugar to form a paste (*pâte*). The *pâte* is then placed inside a tin box and stored, to be used as desired. Consumers could then put ten to twelve grains of the *pâte* into a chocolate cup, which, accordingly, was "agreeably spiced." This drink, Machet wrote, was excellent for giving tone to the stomach and for repairing one's strength lost by exhaustion. Machet wrote: "*Les Chinois font grand usage de cette pâte dans leur chocolat* [The Chinese make great use of this paste in their chocolate]" [49]. This same *vacaca chinorum* recipe appeared on nearly the same page and with the exact same text, through the ninth edition—although with a different publisher in 1852 [50].

The second French reference to Chinese chocolate is a manual about cacao, written by Antoine Gallais, of Debauve and Gallais, who described the *chocolat analeptique au salep de Perse* discussed earlier. In a brief geographical survey of chocolate styles of the time, Gallais noted that:

Les Chinois prennent beaucoup de chocolat; mais comme leur cacao arrive en pâte, sans arromate, ils ont, dans des boîtes de fer-blanc ou de porcelaine, une poudre composée de vanille, de cannelle et d'ambre gris. [The Chinese take much chocolate but as their cacao arrives in paste (form), without spices, they have, in tin or porcelain boxes, a powder composed of vanilla, cinnamon, and ambergris.] [51]

Vacaca chinorum appears again in Alexandre Dumas's *Grand dictionnaire de cuisine*, published in 1873, in language very close to that of Machet, except that Dumas transformed the measurements into metric terms and reduced the quantity of sugar. He also omitted Machet's injunction to the effect that this chocolate beverage gave good tone to the stomach, but he followed the latter in suggesting that it was a good restorative. According to Dumas's last line:

Les Chinois font un très grand usage de cette pâte et s'en trouvent très bien. [The Chinese make great use of this paste and find it very good.]

This text, of course, is but a slight variation of Machet's [52].

The context for the early 19th century French references to *vacaca chinorum* or *chocolat Chinois* has yet to be determined. *Vacaca*, a term that appears in virtually no dictionaries, might have been a transformation of the Mixe-Zoquean word *kakawa*, the Mesoamerican linguistic root of the word cacao (see Chapter 2) [53]. It may also have been derived from the terms for the cacao tree that, in 1687, Nicolas de Blégny referenced as Indian (*Cacavaqua huit*) and European (*Cacaotal* or *Cacavifere*) [54]. Describing how to make chocolate as a beverage in 1755, Louis Lémery listed a series of hot spices including amber, musk, vanilla, and *poudre d'Ouacaca* [55]. *Houacaca* powder (Oaxaca powder, from the state of Oaxaca in southern Mexico) was described in 1768 as brown or cinnamon in color, "a little like Spanish tobacco," made of cinnamon and amber and imported into Paris from Portugal [56]. Despite the statements about Chinese use of chocolate in Machet, Gallais, and Dumas, the available evidence does not support the argument for widespread use of chocolate in China at the time. Nineteenth century French cultural associations, however, with *salep de Perse* and cinnamon, the latter an ingredient in "Chinese chocolate," linked both of these to the East and sometimes to China. In his *Traité usuel du chocolat* (*Common Treatise on Chocolate*), published in 1812, Pierre Joseph Buc'hoz, a botanist and former personal physician to King Stanislaus Leszczynski of Poland, mentioned that the Chinese and Persians used *salep* as an aphrodisiac [57]. Discussing different ways of making chocolate, Machet himself referred to cinnamon (*cannelle*) from Ceylon (modern Sri Lanka) and China, arguing that the latter was inferior to the former [58]. The use of cinnamon in making chocolate had been known in France at least since René Moreau's 1643 translation of Antonio Colmenero de Ledesma's *Curiosa tratado de la naturaleza y chocolate*, published in Madrid in 1631 [59]. Chocolate-related associations, such as *salep* and cinnamon, with China may have given rise to the concept of Chinese chocolate. After all, not all porcelain called "china" is made in China, but the term has a historic connection

with the place and this may also be the case for *vacaca chinorum*, especially given the ingredients and the phrase about the Chinese using this paste in their chocolate. Possibly *vacaca chinorum* was an association with Asians other than Chinese, although it is highly probable that in the 19th century the French would have differentiated between Chinese and Filipinos or other Asians. The possibility that the reference is to Chinese living in France is equally remote, as the size of the Chinese community in France was negligible before World War I [60].

The *vacaca chinorum* recipes describe a steeped beverage with spices added, together with the use of ambergris, an excretion from the sperm whale. As an oily sweet perfume or flavoring, much as rose water and butter, a kind of waxy binder, ambergris, similarly to the *salep* described by Gallais, would be used typically in a sweet dish, such as chocolate, and it seems to have played the role that milk later assumed in milk chocolate. Familiarity with ambergris and its use in making chocolate goes back in France at least to Moreau's 1643 treatise [61]. The name seems to be derived from the French *ambre gris* (gray amber, ambergris). Ambergris became a significant culinary ingredient with the building of larger ships in the 17th and 18th centuries and with the development of the whaling industry, and is the subject of a chapter in Herman Melville's *Moby Dick*. In his *Dissertations sur l'utilité, et les bons et mauvais effets du Tabac, du Café, du Cacao et du Thé* (*Dissertations on the Utility, and the Good and Bad Effects of Tobacco, Coffee, Cacao, and Tea*), published in 1788, Buc'hoz gave instructions for a chocolate paste (*pâte*) that included amber, cinnamon, sugar, and vanilla. It was close to Machet's *vacaca chinorum* recipe as presented earlier [62].

Ambergris was widely traded and was known in ancient China as *lung sien hiang* (dragon's spittle perfume). François Foucault praised the healthful benefits of chocolate with ambergris, cinnamon, sugar, and vanilla in a doctoral dissertation published in Paris in 1684 [63]. Appearing in the Chinese, as well as the French, cuisine of the 1820s, ambergris was probably used elsewhere as well. In his *Physiology of Taste*, Brillat-Savarin in 1825 praised the restorative virtues of *chocolat amber* [64]. The style of *vacaca chinorum* and the chocolate drunk by Brillat-Savarin, most likely disseminated from Spain, followed the pre-Columbian Mexican practice of drinking a foamy spicy brew. Such a beverage was described in the 1837 diary of William Taylor, a servant in a house in Mayfair, London, as a breakfast chocolate, "something like coffee but of a greasey [*sic*] and much richer nature" [65].

Sperm whales were hunted until World War I, after which the development of the petroleum industry rendered the whales' oil function obsolete. The culinary niche for ambergris also was ended by industrialization. China traditionally lacked a significant dairy industry [66] and, as the worldwide popularity of chocolate has been based on milk chocolate since 1875, the absence of a local dairy industry in China limited chocolate's culinary role. Not only ambergris, but the addition of vanilla also characterized the Chinese chocolate recipes of the early 19th century. Vanilla often accompanied chocolate in the 16th century and, later, in a diffusion pattern transmitted through the Spanish and possibly the Portuguese. Chinese chocolate, in other words, was part of the diffusion process of a chocolate beverage usually without milk but with the addition of sugar, cinnamon, spices, and vanilla, much as that described by William Taylor in 1837.

Chocolate and China to the Coming of the Bar

Three technical innovations in the 19th century rendered obsolete the ambergris and spiced chocolate model of the kind described as Chinese. In 1828, Coenraad Johannes van Houten developed a hydraulic press that more efficiently separated the cacao butter from the chocolate, effectively defatting the chocolate and facilitating the large-scale manufacture of inexpensive chocolate in both powdered and hard form. A second breakthrough was the invention of the crank ice cream maker by Nancy Johnson in 1846. These two developments, together with Daniel Peter's development of milk chocolate in 1875, shifted the locus of chocolate production to dairy areas such as Switzerland. As noted, China and its gastronomy did not focus on dairy foods and this may have been the single most important reason that China did not experience the "takeoff" in chocolate use that characterized so much of the West in the late 19th and early 20th centuries.

The mid to late 19th century Industrial Revolution also transformed food service sequences, especially in the case of desserts in the West. Improved and cheaper ways to freeze foods contributed to the increased popularity of ice cream; as it gained in market share, many desserts with older style flavorings, including ambergris, which has no functional role in ice cream, dropped out. Milk and cream replaced ambergris as thickeners for chocolate. *Service à la Russe*, courses brought sequentially to the table, replaced *Service à la Française*, similar to the older buffet style with clusters of dishes brought in relays to the table, in public dining. *Entremets* (side dishes) and *entremets douces* (sweet side dishes) gave way to desserts. The new dessert order was codified in late 19th century cookbooks, often French, exemplified in *Le Grand livre des pâtissiers et des confiseurs* by Urbain-Dubois in 1883 (see Chapter 42).

The increase in ice cream production was also linked to the growing industry of milk chocolate. Ice cream requires the availability of milk, and the dairy industry must be developed to produce milk chocolate. In fact, European cattle herds increased and the spread of railway transport, together with pasteurization, led to a surge in ice cream and milk chocolate at the same time, transforming the chocolate industry in Switzerland and the Netherlands. Cow's milk was either unavailable or limited in much of the world, including China and Japan [67].

By the late 19th and early 20th centuries, chocolate occasionally appeared at the intersections of Chinese and Western cultures in both China and among overseas Chinese. The Debauve et Gallais website tells of the company's having arranged to sell chocolates through "Marcel," a local reseller of French patisserie and other food on the Nanjing Lu (Road) in the French concession in Shanghai from 1886 to 1888. The shop in China sold mostly *pistoles* (chocolate drops or large chocolate chips) and chocolate bars. The author of the website hypothesizes, probably correctly, that the clientele were more Europeans living in Shanghai than local Chinese. Marcel is said to have introduced a friend, presumably European, who supplied him with ginger-flavored chocolate made in Shanghai. From their experience in China, Debauve et Gallais eventually created two chocolates inspired by China: a ginger *Aiguillette*, a pepper-flavored candied ginger imported from Canton called "Le Cantonais" (the Cantonese); and a *ganache* made with a peppered ginger paste called the "Marco Polo" [68].

In China, an English-language menu from the Canton Hotel, for January 9, 1910, listed cocoa as an item along with coffee and several varieties of tea on its dinner bill of fare (Fig. 44.3) [69]. The next day's tiffin bill of fare—a light lunch or other meal in overseas English parlance—also listed cocoa, along with coffee and tea, at the bottom of the offerings (Fig. 44.4) [70]. A Hong Kong advertising billboard for California's Ghirardelli Chocolate Company showed a can of the company's ground chocolate with the slogan: "'One Minute' and a delicious cup of chocolate. From D. Ghirardelli's Strictly Pure." The sign was in English and, most likely, appealed primarily to the British and American communities in the city (Fig. 44.5) [71].

Among overseas Chinese, examples of familiarization with chocolate are the references to it in the *Chinese and English Cook Book*, published in San Francisco in 1910. In this case non-Chinese dishes are introduced to teach Chinese cooks in America how to prepare American dishes with which they were unfamiliar [72]. Included are recipes for a hot chocolate beverage and chocolate cream candy [73]. Even at the time of the "opening" of China to the West at the beginning of the 20th century, Chinese acceptance of Western culinary tastes was slow and grudging [74].

Chocolate came into widespread use in China only recently. In June 2004, China's first Salon du Chocolat, an international exposition of fine chocolates, was held in Beijing [75]. Chocolate, however, has met significant resistance, especially outside large cities. World Trade Organization (WTO) figures show

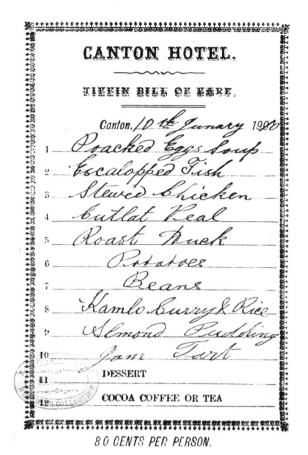

CANTON HOTEL.

DINNER BILL OF FARE

Canton, 9th Junary 1910.

1 Mushroom Soup
2 Boiled Fish
3 Duck Liver on Toast
4 Ox-Tongue Petits Pois
5 Roast Chicken
6 Cold York Ham
7 Potatoes
8 Beans
9 Kamlo curry & Rice
10 Tasfy Cake
11 Fried Pudding
12 CHEESE
13 DESSERT
14 COFFEE
15 COCOA
16 OO LUNG TEA
   HUNG MULY TEA
17 MILK & SUGAR
18 BUTTER
19 BREAD
20 JAM
21
22 PICKLES
23 GOLDEN SYRUP

N.Y. PUBLIC LIBRARY
1900-1910
BUTTOLPH COLLECTION

$ 1.00 PER PERSON.

FIGURE 44.3. Menu from the Canton Hotel, dinner bill of fare; dated January 9, 1910. *Source*: New York Public Library Menus Collection, Buttolph Collection, New York. Courtesy of the New York Public Library. (Used with permission.)

CANTON HOTEL.

TIFFIN BILL OF FARE

Canton, 10th Junary 1910

1 Poached Eggs Soup
2 Escalopped Fish
3 Stewed Chicken
4 Cutlat Veal
5 Roast Duck
6 Potatoes
7 Beans
8 Kamlo Curry & Rice
9 Almond Pudding
10 Jam Tart
11 DESSERT
12 COCOA COFFEE OR TEA

80 CENTS PER PERSON.

FIGURE 44.4. Menu from the Canton Hotel, tiffin bill of fare; dated January 10, 1910. *Source*: New York Public Library Menus Collection, Buttolph Collection, New York. Courtesy of the New York Public Library. (Used with permission.)

Chinese chocolate imports growing from $17.7 million in 1999 to $49.2 million in 2003. The average annual per capita consumption of chocolate in China, however, is estimated at 0.7 kilogram, below the average of 5.7 kilograms for consumers in the European Union countries [76]. Clearly, chocolate's penetration into Chinese cuisine has historically been peripheral at best and remains problematic even today. Even at the margins, however, there is a history of Chinese chocolate from the 17th century to the present that appears in the Catholic missionaries' use of it, the chocolate-related export ware of the 18th century, and the curious French references of the 19th century. Hopefully, this brief study will encourage future researchers to help clarify the Chinese chocolate story.

FIGURE 44.5. Advertising billboard: Ghirardelli Chocolate Company, Hong Kong, China; ca. 1900. *Source*: Ghirardelli Chocolate Company, San Leandro, CA. (Used with permission.)

Acknowledgments

I wish to thank the following for assistance during the preparation of the present chapter: Carol Benedict, Department of History, Georgetown University; Wah Cheng, Department of History, Mills College, Oakland, California; Noël Golvers, F. Verbiest Instituut — Katholieke Universiteit Leuven, Belgium; Philip and Mary Hyman, Historians, Paris; Lupe Martin, Alameda, California; Anna Naruta, Second Vice President, Chinese Historical Society of America, San Francisco, California; Suzanne Perkins, Art Historian, Berkeley, California; Erica J. Peters, Culinary Historians of Northern California; Françoise Sabban, Directrice d'études École des Hautes Études en Sciences Sociales, Paris; Directrice française à la Maison Franco-Japonaise, Ebisu, Shibuya-ku, Tokyo, Japan.

References

1. Anderson, E. N. *The Food of China*. New Haven, CT: Yale University Press, 1998; p. 169.

2. Sabban, F. Email to the author, dated July 25, 2006.

3. Golvers, N. *François de Rougemont, S. J., Missionary in Ch'ang-Shu (Chiang-Nan) A Study of the Account Book (1674–1676) and the Elogium*. Louvain, Belgium: Louvain University Press/Ferdinand Verbiest Foundation, 1999; p. 511. Golvers, N. Email, dated October 14, 2006.

4. Alcobendas, R. P. *Fr. Severiano. Las Misiones franciscanas en China. Cartas, Informes y Relaciones del Padre Buenaventura Ibañez (1650–1690). Con Introduccion, Notas y Apéndices. Bibliotheca Hispana Missionum, V*. Madrid, Spain: Estanislao Maestre, 1933; p. 61.

5. Alcobendas, R. P. *Fr. Severiano. Las Misiones franciscanas en China. Cartas, Informes y Relaciones del Padre Buenaventura Ibañez (1650–1690). Con Introduccion, Notas y Apéndices. Bibliotheca Hispana Missionum, V*. Madrid, Spain: Estanislao Maestre, 1933; p. 111.

6. Alcobendas, R. P. *Fr. Severiano. Las Misiones franciscanas en China. Cartas, Informes y Relaciones del Padre Buenaventura Ibañez (1650–1690). Con Introduccion, Notas y Apéndices. Bibliotheca Hispana Missionum, V*. Madrid, Spain: Estanislao Maestre, 1933; p. 148.

7. Alcobendas, R. P. *Fr. Severiano. Las Misiones franciscanas en China. Cartas, Informes y Relaciones del Padre Buenaventura Ibañez (1650–1690). Con Introduccion, Notas y Apéndices. Bibliotheca Hispana Missionum, V*. Madrid, Spain: Estanislao Maestre, 1933; p. 167.

8. Alcobendas, R. P. *Fr. Severiano. Las Misiones franciscanas en China. Cartas, Informes y Relaciones del Padre Buenaventura Ibañez (1650–1690). Con Introduccion, Notas y Apéndices. Bibliotheca Hispana Missionum, V*. Madrid, Spain: Estanislao Maestre, 1933; p. 200.

9. *Sinica Franciscana*. Relationes et Epistolas Fratrum minorum hispanorum in Sinis qui annis 1697–98 missionem ingressi sunt. Madrid, Spain: Volume 9 (1995), p. 638.

10. *Sinica Franciscana*. Relationes et Epistolas Fratrum minorum hispanorum in Sinis qui annis 1697–98 missionem ingressi sunt. Madrid, Spain: Volume 9 (1995); p. 650.

11. *Sinica Franciscana*. Relationes et Epistolas Fratrum minorum hispanorum in Sinis qui annis 1697–98 missionem ingressi sunt. Madrid, Spain: Volume 9 (1995); p. 657.

12. *Sinica Franciscana*. Relationes et Epistolas Fratrum minorum hispanorum in Sinis qui annis 1697–98 missionem ingressi sunt. Madrid, Spain: Volume 9 (1995); pp. 667–668, 669.

13. *Sinica Franciscana*. Relationes et Epistolas Fratrum minorum hispanorum in Sinis qui annis 1697–98 missionem ingressi sunt. Madrid, Spain: Volume 9 (1995); p. 683.

14. *Sinica Franciscana*. Relationes et Epistolas Fratrum minorum hispanorum in Sinis qui annis 1697–98 missionem ingressi sunt. Madrid, Spain: Volume 9 (1995); p. 688.

15. *Sinica Franciscana*. Relationes et Epistolas Fratrum minorum hispanorum in Sinis qui annis 1697–98 missionem ingressi sunt. Madrid, Spain: Volume 9 (1995); p. 703.

16. *Sinica Franciscana*. Relationes et Epistolas Fratrum minorum hispanorum in Sinis qui annis 1697–98 missionem ingressi sunt. Madrid, Spain: Volume 9 (1995); p. 705.

17. *Sinica Franciscana*. Relationes et Epistolas Fratrum minorum hispanorum in Sinis qui annis 1697–98 missionem ingressi sunt. Madrid, Spain: Volume 9 (1995); p. 719.

18. *Sinica Franciscana*. Relationes et Epistolas Fratrum minorum hispanorum in Sinis qui annis 1697–98 missionem ingressi sunt. Madrid, Spain: Volume 9 (1995); p. 724.

19. Mote, F. W. Yüan and Ming. In: Chang, K. C., editor. *Food in Chinese Culture: Anthropological and Historical Perspectives*. New Haven, CT: Yale University Press, 1977; p. 198.

20. Lyon, W. S. *Cacao Culture in the Philippines*. Philippines Bureau of Agriculture, Farmers' Bulletin No. 2. Manila, Philippines: Bureau of Public Printing, 1902; p. 11, Note 1.

21. Wright, H. *Theobroma Cacao or Cocoa, Its Botany, Cultivation, Chemistry and Diseases*. Colombo, Sri Lanka: A. M. and J. Ferguson, 1907; p. 11.

22. Jagor, F., de Comyn, Wilkes, C., and Virchow, R. *The Former Philippines thru Foreign Eyes*, 1916; Chapter X. Presented by Authorama Public Domain Books. Available at http://www.authorama.com/former-philippines-11.html. (Accessed December 24, 2006.)

23. La Loubere, S. de. *A New Historical Relation of the Kingdom of Siam*. London, England: Printed by F. L. for Tho.

Horne at the Royal Exchange, Francis Saunders at the New Exchange, and Tho. Bennet at the Half-Moon in St. Pauls Church-yard, 1693; p. 23.

24. Blumentritt, F. *Die Chinesen auf den Philippinen, Historische Skizze, Dreizehnter Jahres-Bericht der Communal-Ober-Realschule in Leitmeritz, 1879*. Leitmeritz, Austria (presently Litomerice, Czech Republic): Verlag der Communal-Ober-Realschule, 1879; p. 23.

25. Neumann, C. Asian Vegetarian Cooking, Lesson 1: The Philippines, Suite 101. Available at http://www.suite101.com/lesson.cfm/18242/1536/2. (Accessed August 14, 2005.)

26. Stubbe, H. *The Indian Nectar, or A Discourse Concerning Chocolata*. London, England: J.C. for Andrew Crook, 1662; Chapter 1, p. 2.

27. Braudel, F. *Capitalism and Material Life 1400–1800*. Translated by M. Kochan. New York: Harper Colophon, 1973; p. 179. (Original French publication, dated 1967.)

28. Coe, S. D., and Coe, M. D. *The True History of Chocolate*. London, England: Thames and Hudson, 1996; p. 177.

29. Middione, C. Email, dated June 9, 2006.

30. MuslimRichtey.com no. 1 Muslim matrimonial website, "Cuscusu Dolce." Available at http://www.muslimrishtey.com/africa_food_recipes/african_cuscusu_dolce.php. (Accessed December 26, 2006.)

31. Almeida-Topor, H. d'. *Le goût de l'étranger, Les saveurs venues d'ailleurs depuis la fin du XVIIIe siècle*. Paris, France: Armand Colin, 2006; p. 75.

32. Gallais, A. *Monographie du Cacao, ou Manuel de l'Amateur de Chocolat*. Paris, France: Chez Debauve et Gallais, 1827; pp. 167, 195. (Facsimile reprint: Paris: Phénix, 1999.)

33. Harwich, N. *Histoire du chocolat*. Paris, France: Éditions Desjonquères, 1992; pp. 161–162.

34. Stoianovich, T. Le maïs dans les Balkans. In: Hémardinquer, J.-J., editor. *Pour une histoire de l'alimentation*. Paris, France: Armand Colin, 1970; pp. 282–284.

35. Mudge, J. Mc. *Chinese Export Porcelain in North America*. New York: Clarkson N. Potter, 1986; p. 102.

36. Coe, S. D., and Coe, M. D. *The True History of Chocolate*. London, England: Thames and Hudson, 1996; pp. 161–162.

37. Mudge, J. Mc. *Chinese Export Porcelain in North America*. New York: Clarkson N. Potter, 1986; p. 102.

38. Sargent, W. R. A Preissler decorated Chinese cup and saucer, 1729–35. In: *Mélanges en souvenir d'Elisalex d'Albis, 1939–1998*. Sèvres. France: Société des Amis du Musée National de Céramique, 1999; p. 45.

39. Lunsingh Scheurleer, D. F. *Chinese Export Porcelain. Chine de Commande*. New York: Pitman Publishing Corporation, 1974; p. 108.

40. Sargent, W. R. A Preissler decorated Chinese cup and saucer, 1729–35. In: *Mélanges en souvenir d'Elisalex d'Albis,*

1939–1998. Sèvres, France: Société des Amis du Musée National de Céramique, 1999; p. 46.

41. Barber, E. A. *Pottery and Porcelain of the United States: An Historical Review of American Ceramic Art from the Earliest Times to the Present Day: To Which Is Appended a Chapter on the Pottery of Mexico*. New York: G. P. Putnam's Sons, 1909; p. 585.

42. Forbes, A. *California: A History of Upper and Lower California*. London, England: Smith, Elder and Company, 1839; p. 298.

43. Lunsingh Scheurleer, D. F. *Chinese Export Porcelain, Chine de Commande*. New York: Pitman Publishing Corporation, 1974; p. 108.

44. Jörg, C. J. A. *The Geldermalsen: History and Porcelain*. Translated from the Dutch by M. Steenge-de Waard and P. de Waard-Dekking. Groningen, Holland: Kemper and London, 1986; pp. 57, 59.

45. Jörg, C. J. A. *The Geldermalsen: History and Porcelain*. Translated from the Dutch by M. Steenge-de Waard and P. de Waard-Dekking. Groningen, Holland: Kemper and London, 1986; p. 69.

46. Lunsingh Scheurleer, D. F. *Chinese Export Porcelain, Chine de Commande*. New York: Pitman Publishing Corporation, 1974; p. 108.

47. Ambrosia, L. Caffè, tè e cioccolate mei servizi di porcellana de Museo. In: *Un mondo in tazza, porcellane per caffè, tè e cioccolata*. Naples, Italy: Electa Napoli, 2003; pp. 12–13.

48. Forbes, H. A. C., Kernan, J. D., and Wilkins, R. S. *Chinese Export Silver 1785 to 1885*. Milton, MA: Museum of the American China Trade, 1975; pp. 62, 227.

49. Machet, J.-J. *Le Confiseur Moderne, ou l'Art du Confiseur et du Distillateur, contenant toutes les opérations du confiseur et du distillateur, et, en outre, les procédés généraux de quelques arts qui s'y rapportent, particulièrement ceux du parfumeur et du limonadier*, 3rd edition. Paris, France: Maradan, 1817; p. 90.

50. Machet, J.-J. *Le Confiseur Moderne, ou l'Art du Confiseur et du Distillateur, contenant toutes les opérations du confiseur et du distillateur, et, en outre, les procédés généraux de quelques arts qui s'y rapportent, particulièrement ceux du parfumeur et du limonadier*, 9th edition. Paris, France: Corbet, 1852; p. 90.

51. Gallais, A. *Monographie du Cacao, ou Manuel de L'Amateur de Chocolat*. Paris, France: Chez Debauve et Gallais, 1827; p. 157.

52. Dumas, A. *Le Grand dictionnaire de cuisine*. Paris, France: Alphonse Lemerre, 1873; p. 330.

53. Coe, S. D., and Coe, M. D. *The True History of Chocolate*. London, England: Thames and Hudson, 1996; p. 39.

54. Blégny, N. de. *Le bon usage du thé, du caffé et du chocolat pour la préservation et pour la guérison des maladies*. Paris, France: E. Michallet, 1687; pp. 200–202.

55. Lémery, L. *Traité des Alimens*, 3rd edition. Paris, France: Durand, 1755; Volume I, p. 97.

56. Emy, M. *L'Art de bien faire Les Glaces d'Office; [or] Les Vrais Principes Pour congeler tous les raffraichissemens.* Paris, France: Le Clerc, 1768; pp. 168–169.

57. Robert, H. *Les Vertus thérapeutiques du chocolat, Ou comment en finir avec les idées reçues.* Paris, France: Artulen, 1990; pp. 113–114.

58. Machet, J.-J. *Le Confiseur Moderne, ou l'Art du Confiseur et du Distillateur, contenant toutes les opérations du confiseur et du distillateur, et, en outre, les procédés généraux de quelques arts qui s'y rapportent, particulièrement ceux du parfumeur et du limonadier,* 3rd edition. Paris, France: Maradan, 1817; p. 93.

59. Moreau, R. *Du chocolate: Discours Curieux divisé en quatre parties. Par Antoine Colmenero de Ledesma Medecin & Chirurgien de la ville de Ecija de l'Andalouzie. Traduit d'Espagnol en François sur l'impression faite à Madrid l'an 1631, & éclaircy de quelques annotations. Par René Moreau Professeur du Roy en Medicin à Paris. Plus est ajouté un Dialogue touchant le même chocolate.* Paris, France: Sébastien Cramoisy, 1643; p. 54.

60. Ma, L. Northern Chinese migrants in France: insertion problems and the economic reforms in China, East Asia Institute, Lyon. Available at http://192.38.121.218/issco5/documents/MaLipaper.doc. (Accessed January 1, 2007.)

61. Moreau, R. *Du chocolate: Discours Curieux divisé en quatre parties. Par Antoine Colmenero de Ledesma Medecin & Chirurgien de la ville de Ecija de l'Andalouzie. Traduit d'Espagnol en François sur l'impression faite à Madrid l'an 1631, & éclaircy de quelques annotations. Par René Moreau Professeur du Roy en Medicin à Paris. Plus est ajouté un Dialogue touchant le même chocolate.* Paris, France: Sébastien Cramoisy, 1643; p. 54.

62. Buc'hoz, P.-J. *Dissertations sur l'utilité, et les bons et mauvais effets du Tabac, du Café, du Cacao et du Thé, ornées de Quatre planches en taille-douce,* 2nd edition. Paris, France: Chez l'Auteur, De Burc l'aîné, et la Veuve Tilliard et Fils, 1788; p. 125.

63. Foucault, F. *An Chocolatae usus salubris?* Paris, France: Parisiis, 1684; p. 3.

64. Brillat-Savarin, J.-A. *The Physiology of Taste (1825).* Translated and annotated by M. F. K. Fisher. San Francisco, CA: North Point Press, 1986; pp. 112, 371.

65. Taylor, W. Diary of William Taylor, Footman, 1837. Cited in: Allen, B., editor. *Food, An Oxford Anthology.* Oxford, England: Oxford University Press, 1994; p. 166.

66. Mote, F. W. Yüan and Ming. In: Chang, K. C., editor. *Food in Chinese Culture: Anthropological and Historical Perspectives.* New Haven, CT: Yale University Press, 1977; pp. 200, 206.

67. Anderson, E. N. *The Food of China.* New Haven, CT: Yale University Press, 1998; p. 66.

68. Debauve et Gallais, Canton / Marco Polo. Available at http://www.debauveandgallais.com/main/history_canton.asp. (Accessed December 30, 2006.)

69. Canton Hotel, Dinner Bill of Fare, 9 January 1910, in Buttolph Collection, New York Public Library.

70. Canton Hotel, Tiffin Bill of Fare, 10 January 1910, in Buttolph Collection, New York Public Library.

71. Lawrence, S. The Ghirardelli story. *The Magazine of the California Historical Society* 2002; 81(2):95.

72. Longone, J. What is your name? My name is Ah Quong. Well, I will call you Charlie. *Gastronomica* May 2004;4(2):89.

73. Fat, M. *Chinese and English Cook Book.* San Francisco, CA: Fat Ming Company, 1910; pp. 4, 926.

74. Spence, J. Ch'ing. In: Chang, K. C., editor. *Food in Chinese Culture: Anthropological and Historical Perspectives.* New Haven, CT: Yale University Press, 1977; p. 286.

75. "Chocolat: Le salon 2006 et ses tendences," Le Journal des Femmes/Cuisinier. Available at http://www.linternaute.com/femmes/cuisine/magazine/reportage/chocolat2006/infos.shtml. (Accessed December 31, 2006.)

76. Freeman, D. EU chocolatiers chase Chinese market. *Asia Time Online—Daily News, Greater China,* July 28, 2005. Available at http://www.atimes.com/atimes/China/GG28Ad02.html. (Accessed December 31, 2006.)

PART

X

Production, Manufacturing, and Contemporary Activities

Chapter 45 (Cabezon) explores the content of selected Jesuit letters (1693–1751) written in New Spain. The four documents selected provide unusual insights regarding cacao production under Jesuit supervision during the 17th and early 18th centuries and reveal how income provided from cacao was used to offset annual expenses/debts and to pay *censos* (annual alms) to the Catholic Church. Chapter 46 (Snyder, Olsen, and Brindle) presents an overview of technological developments associated with chocolate manufacturing, from stone grinding stones to stainless-steel equipment developed during the late 19th and early 20th centuries. The authors explore the evolution of cocoa processing that began using equipment commonly used to process other foods, to the development of specialized equipment, and how changes in production methods affected the flavor, texture, and form of finished chocolate. Chapter 47 (Brindle and Olsen) considers the long history of unscrupulous merchants who adulterated chocolate by mixing and blending unnatural ingredients with the intent of defrauding customers. This dark world of adulterated chocolate included additions of animal, vegetal, and mineral products, commonly used to add color or serve as extenders and replacements for fat. The analysis reveals that while many adulterants were harmless, some were potentially toxic. Chapter 48 (Gay and Clark) describes the creation and development of the historic chocolate program implemented at the Colonial Williamsburg Foundation. This chapter identifies the stages taken in an effort to produce authentic, historic chocolate and offers advice on how other historical institutions and museums might establish their own chocolate programs. Chapter 49 (Whitacre, Bellody, and Snyder) relates how historical research and technical knowledge can be blended to produce a product that imitates the taste, texture, and quality of Colonial era chocolate within reasonable parameters of historical accuracy. Chapter 50 (Pucciarelli and Barrett) relates that since earliest times chocolate has been viewed as both a food and a medicine and has been used to treat disease and promote health. The authors report data that measured attitudes, behavior, and knowledge of chocolate's role as food and medicine among college students and members of the general population.

Cacao, Haciendas, and the Jesuits

Letters from New Spain, 1693–1751

Beatriz Cabezon

Introduction

Four documents dated between 1693 and 1751 provide unusual insights regarding cacao production in 17th and early 18th century Mexican history. The documents, treated here in chronological sequence, reveal how Jesuit priests produced cacao, cared for the plantations (*haciendas*), and used income from the cacao to sustain their Order and to pay obligatory fees (*censos*) to the Catholic Church. The documents also mention Jesuits buying and selling slaves. Whether the slaves were Indian or African in origin, however, is not detailed in the letters. The records also reveal that under normal conditions production of cacao increased annually under Jesuit supervision, but that in certain years yield was so low that the priests could barely sustain themselves.

Presentation of the Documents

Document 1: This letter was written in 1693 by the Jesuit priest Father Alfonso Avirrillaga to the Jesuit Provincial Father Diego de Almonacir,[1] and signed in the College of Valladolid[2] (today Morelia[3]), Mexico.

Obeying Your Reverence I reply.[4] I handed the College of Chiapas to Father Ignacio Guerrero, because father Nicolás de Veza had not yet arrived. [I have built this College] from the basement to the cloister; the cloister-built in the form of an arc using bricks has been finished; upstairs it has ten rooms, a chapel and other places, and downstairs it has a classroom, a school, a refectory, a kitchen, and several pantries. Outside

there is a secular patio surrounded by a low wall, a room for kitchenware, the secular porter's lodge, the stable for the horses, and the hen house. The church is a small one in relation to what we need; it has three altars, an organ and a pulpit. The sacristy is decorated with ornaments of all colors, some [made] of cotton, others [made] of mountain wool, with silver jewels, several paintings and statues [hechuras], all [with a value of] more than 6000 pesos. During my time here, two cacaotales [cacao haciendas], the cacaotal of San Antonio and the cacaotal La Concepción became part of this College; there will probably be more than 75,000 trees, plus 2000 that were added. Both cacaotales became a single one; 600 pesos for censo[5] were taken from the San Antonio cacaotal and a stable for horses with a dirt floor was added to La Concepción. In the hacienda del Rosario, we have added 20,000 [cacao] trees, and in both haciendas we built up a new group of dwellings [caserío]. Now we have in both haciendas 190,000 [cacao] trees. A mill [molino] also became part [of this College] from the estate of Father Manuel de Valtierra, one

Chocolate: History, Culture, and Heritage. Edited by Grivetti and Shapiro
Copyright © 2009 John Wiley & Sons, Inc.

stone, several harnesses and saddles, and I took 800 pesos for censo. I bought one labor for 3000 pesos.[6] There is no debt remaining against this College, and we still have 2300 pesos to be paid [to us]. I left 300 pesos in reales for the haciendas and more than 500 pesos [worth] in corn, wheat, oil, wax, and other provisions. These provisions remained in the pantries. In the six years that I governed this College there was an increase of 40,000 pesos. There are four priests living in this College without counting the one that lives in the Hacienda. These priests are the ones that were requested by the founder [of this College]: the Reverend Father [the Prior], another for teaching moral precepts, a Preacher, and a Grammar teacher. There are also two brothers, one at the School and the other at the office. This is the actual condition today in those [haciendas] and in the College. We did not grow less [cacao] on the temporal than on the spiritual [aspect], because sometimes we needed three priests to be able to provide confession [to all that wanted]. [1]

Document 2: This letter has its front page missing, so the date and names of the correspondents could not be determined. The document was attached to the back of a third document (below) that described a list of alms received from the years 1704 to 1707. Logically, the letter is from 1707, the same year as the list. Document 2 begins with an inventory and then turns to a description of the cacao haciendas.

Two small altar ornaments [retablos], one is being gilded, the other we want to gild, that two benefactors gave us. Three slaves I bought for the cacao haciendas and finally a large collection of books that Our Lordship the Bishop left; [all] that have a value of 3000 pesos.

Haciendas

All the cacao haciendas have been supplied with goods and provisions and even though one slave was sold, three were bought. The cacao haciendas have provided—as appears in the books during the 26 years we have had them with 319 cargas[7] of cacao [i.e., 140,360 pounds per year]. After the cacao [beans] have been cleaned [the volume is reduced] and we deduce the cost of the shipment—[having to transport the beans] more than 40 leagues—and we profit around 4500 pesos. If the cacao haciendas would produce this amount every year, the Church expenses and the sustenance of this College could be covered decently; this [would be the situation] if there were no accumulated debts. The problem arises if one year the haciendas do not produce this amount; this would cause [us] to remain in debt, unless God would remedy it by giving us another hacienda. And these last years we were in such a desperate situation that we were about to beg for alms for our sustenance. This year the haciendas produced 300 cargas of cacao [i.e., 132,000 pounds] and with this we were able to sustain ourselves and pay the debts we had. This College of labor[8] has another hacienda that, when the year is good, this allows us to harvest the wheat we grow to make bread; it also has some mares and other animals that we eventually sell. This is the actual situation of this College, so now Your Lordship

knows about it and can have some idea of what it is and what you have, and not [believe] what they are saying around there, because they think that because we have cacao we have a Potosí [great wealth]. [2]

Document 3: The top of the document is torn away; date and names of the correspondents are missing. The letter suggests that the steward of the Jesuit College of Ciudad Real (i.e., Mexico City)—Joseph Lorenzo—has died. A Juan de Murguía, upon his arrival at the college, wrote a detailed report to the Visitador General (i.e., Inspector General). The document consists of an inventory of the college and objects are itemized that had been registered in college account books between 1704 and 1751. The report specified expenditures, debts, inventory of goods, and description of three cacao haciendas (*cacaotales*) and cattle haciendas. The document also contained paragraphs that describe improvements that had been made previously to the college and offered additional suggestions. The account is countersigned by both Juan de Murguía and Francisco de la Cavada (administrative positions unknown).

Two cattle haciendas in the Cutepeque [Coatepeque] valley [i.e., located in modern Guatemala near the border with Mexico].

In [i.e., on the] books: 3331 cows—Lost: 1787

In [i.e., on the] books: 920 horses—Lost: 157

Miscellaneous Inventory:

74.5 yards of linen manufactured at Rouen;

48.5 yards of Spanish linen;

21 yards of cloth made with that linen;

13 ounces of thread;

20 pounds of incense;

70 pounds of wax from Castile;

3096 pounds of cacao;

400 pounds of rice;

425 pounds of fish [dried?];

4.5 almuds of garbanzos [chickpeas];

2 fanegas of salt;

4 fanegas of chile;

15 fanegas of corn;

3 earthen jugs of oil;

4 casks of wine;

1 box of glazed earthenware;

1 ream [resma] of paper;

2 bundles of canons [i.e., religious texts];

1 arroba and 17 pounds of wire; and

3 arrobas and 2 pounds of lead

Warnings. First, I would like to warn you that in the cacao haciendas and in the cattle haciendas there remain only 2200 pesos.

Cacaguatales [Cacao haciendas]

In the three haciendas of cacao, named San Antonio, Nuestra Señora del Rosario y Nuestra Señora del Carmen in the past there were 110,000 cacao trees, 11,000 mother of cacao trees [madrerías] and from them, 84,000 cacao trees already had produced fruits, while the rest were young trees. Now, there are 140,000 cacao trees and 5000 mother of cacao trees; 100,000 of those cacao trees are producing fruits and the rest are new plants. So it seems that there has been an important improvement [in these haciendas] increasing the number of cacao trees; of those [cacao trees] that have been added, 6000 are yielding fruits and 34,000 are still young trees. This is due to the great dedication of the Father Administrator Antonio Joseph Cárdenas, who has been working gloriously in the administration and cultivation of these haciendas. In the hacienda Nuestra Señora del Rosario he has opened and drained new and better land, free of the inundation from the river; at the hacienda San Antonio he also rebuilt the hermitage [hermita]. An altar was donated for the hacienda Nuestra Señora del Rosario, and a statue of Our Lady of Rosario and another of Saint Rosalía was placed in the altar. In the three haciendas 5000 trees of achiote have been planted; this improvement was not made before because there was not much knowledge about the product and about the benefits that may yield [from] this [i.e., from these new trees]. [3]

Document 4: This report, dated May 12, 1751, was prepared by the Visitador General (General Inspector) Father Nicolás Prieto who described the *censos* (annual alms paid to the Church), debts, the cattle hacienda, along with a cacao hacienda inventory.

Cacao Hacienda

There are three haciendas: Nuestra Señora del Rosario, Nuestra Señora del Carmen, y San Antonio. They have 200,000 cacao trees that are producing fruits, and 30,000 new young trees with their "mother of cacao" companion trees.

College

This College has 148 cargas of cacao [62,120 pounds]; 90 cargas [39,600 pounds] in Oaxaca and the rest in another College. This cacao if sold at 20 pesos, would amount to 2960 pesos [i.e., 20 pesos per carga: 440 pounds]. [4]

Conclusion

These four documents, taken as a unit, provide unusual insights regarding the interplay between the Catholic Church, cacao-growing plantations (*haciendas*), and maintenance/production of the cacao trees. They relate that the Jesuits used income from cacao plantations to pay their annual religious "obligations/rent" to the Mother Church and to sustain themselves. During years of good production, the Jesuits flourished; when cacao yields declined, however, they were hard-pressed to meet the financial obligations of their Order.

One clue from the documents (letter 4) is that some of the Jesuit haciendas were located in the modern state of Oaxaca. Another insight from the documents was that by at least 1751 the priests had started their own plantings of *achiote* (*Bixa orellana*) since the letter stated that 5000 *achiote* trees were flourishing. Curiously, the same letter related that the priests—before this time—had not planted *achiote* because they did not recognize potential benefits from the product. What benefits there were and how *achiote* was used by the Jesuits were not identified in the document.

Production figures supplied by the documents revealed considerable growth of the cacao *haciendas*. The first letter dated 1693 identified a total of 77,000 cacao trees under Jesuit control. Letter 3 reported an expansion from 110,000 trees (84,000 producing; 36,000 new plantings) to 140,000 (100,000 producing; 40,000 new plantings), significant growth by any measure. Accompanying the multiplication of cacao trees was a corresponding increase in cacao production. In 1707, the *haciendas* produced 319 *cargas* of cacao, an amount equal to 140,360 pounds or approximately 63,666 kilograms/per year. When expressed as tons, the quantity seems more impressive: 70 tons!

The documents also cast light on slavery as practiced by the Jesuits. In letter 2, dated 1707, they report the purchase of three slaves … something commonly omitted when describing religious "good works," but a working reality of life on these religious-operated haciendas during the 17th and 18th centuries.

Acknowledgments

I would like to acknowledge Ms. Patricia Barriga for her efforts working in the Archive General of the Nation (Mexico City). It is she who identified these early 17th and 18th century documents that shed light on the role of Jesuit priests in managing cacao plantations in New Spain.

Endnotes

1. Father Diego de Almonacir was born in Puebla in 1642 and entered the Society of Jesus in 1658. He served as *Visitador General* of the missions of Sonora and Sinaloa between 1677 and 1685. In 1693, he was named Provincial of the Compañía de Jesús; he held that position until 1696. He died in Mexico in January 1706.

2. The College of Valladolid, built in the mid-17th century, is known today as the Palacio de Clavijero (named after a Jesuit priest). The structure also is called the Temple and College of the Jesuits. Miguel Gregorio Antonio Hidalgo y Costilla Gallaga, hero of

the Mexican Revolution, graduated from this Jesuit college.

3. Located today in the modern state of Morelia, birthplace of both José María Morelos y Pavon and Miguel Gregorio Antonio Hidalgo y Costilla Gallaga, both revolutionary heroes.

4. The original request for information was not included in the archive file. The letter cited was directed to a superior (unidentified).

5. Pension annually paid by the Jesuits to their prelate.

6. In the context of this passage the term *labor* means building supplies, that is, bricks and tiles, where 1 *labor* = 1000 bricks/tiles.

7. One *carga* of cacao was equivalent to 1 *load* or two *tercios*, or 440 pounds.

8. In the context of this passage the term labor means that all persons at the college worked—but, of course, some more than others.

References

1. AGN. Jesuitas, leg. 1–9, exp.1, fs. 1–13. Año 1693, pp. 12r and 12v.

2. AGN. Jesuitas, leg. 1–9, exp.1, fs. 1–13. Año 1707, p. 10r.

3. AGN. Jesuitas, leg. 1–9, exp.1, fs. 1–13. Año 1725–1751 presumed pages 1–6 (all original page numbers torn away).

4. AGN. Jesuitas, leg. 1–9, exp.1, fs. 1–13. Año 1751, pp. 6r–7r.

From Stone Metates to Steel Mills

The Evolution of Chocolate Manufacturing

Rodney Snyder, Bradley Foliart Olsen, and Laura Pallas Brindle

Introduction: The Early Evolution of Chocolate Manufacturing

Although there is no recorded history regarding the first manufacture of chocolate, it is known that natives along the Orinoco River in South America made use of the cocoa tree. Because of the strong bitter flavor of beans straight from the pod, the natives only sucked the sweet pulp surrounding the beans [1]. At some point, the beans may have been thrown in or near a fire, causing the aroma from the roasting beans to fill the air. The seductive aroma and flavor of the roasted cocoa beans would have been just as enticing then as it is now. It might also be that the early use of the cacao bean was associated with medicine, for its bitter and stimulative properties (see Chapter 6).

Whatever the origin of chocolate manufacturing in the Americas, there would have come a time when the roasted beans were ground on stone slabs or *metates* (Fig. 46.1). The resulting chocolate would have hardened into a solid mass or formed into small cakes or tablets. The earliest manufacturing processes would have resulted in chocolate that had quite variable flavor and quality characteristics. But once techniques were developed—whether through planning or accident—on how to avoid burning the beans during roasting and how to grind and mix products with the beans, chocolate manufacturing was born.

The Mexica/Aztec method of chocolate manufacture prior to European contact was simple and basic:

The Indians, from whom we borrow it, are not very nice in doing it; they roast the kernels in earthen pots, then free them from their skins, and afterwards crush and grind them between two stones, and so form cakes of it with their hands. [2]

While basic, this short process summary outlines some of the key process steps still followed today: roasting, winnowing, milling, and molding.

Once the Spaniards became aware of chocolate, they modified the Mexica/Aztec procedure. Early accounts relate that the beans were first cleaned with a sieve to remove dirt, stones, and broken/moldy beans. About two pounds of cacao beans were poured onto the blade of a shovel, and then held over a smokeless fire. The beans on the shovel blade were stirred as

Chocolate: History, Culture, and Heritage. Edited by Grivetti and Shapiro
Copyright © 2009 John Wiley & Sons, Inc.

FIGURE 46.1. Stone metate used to grind cacao beans. *Source*: Brandon Head. *A Popular Account of Cocoa*. New York; George Routledge & Sons, 1903; p. 104. Courtesy of Rodney Snyder. (Used with permission.)

FIGURE 46.2. Hand-turned cylindrical bean roaster. *Source*: P. F. Campbell, 1900. Courtesy of Rodney Snyder. (Used with permission.)

the heat from the fire dried the beans and loosened the shells. After drying, the warm beans were spread on a tabletop, cracked with an iron roller to loosen the brittle shells, and then winnowed with a basket/sieve to separate the cocoa nibs from the shells. The nibs were then roasted to remove the remaining moisture and to develop the flavor. Because the shells were also used to produce a beverage, their removal prior to roasting prevented the shells from burning during the nib roasting process. Underroasted beans were described as having harsh bitter flavors and overroasted beans as having a burnt flavor [3].

The roasted nibs were placed into a mortar and hand-ground into coarse liquor. The liquor was transferred to a heated stone for further grinding with an iron roller with wooden handles to achieve the desired fineness. Stone mills were developed to increase production, although the source of power for such mills is not recorded, that is, whether turned by human, animal, or water power [4]. After grinding, the hot liquor (actually more like a paste) was poured into molds, cooled (sometimes wrapped in paper or vegetable fiber), and stored in a dry place. To produce chocolate, ingredients such as sugar, cinnamon, and vanilla were added during the stone milling step to allow for good mixing with the liquor [5]. The primary use for the chocolate was to produce a beverage. But while it is commonly suggested that "eating chocolate" is a product of the mid-19th century, Richard Brooks commented in 1730 that chocolate also was eaten in solid form:

> When a person is obliged to go from home, and cannot stay to have it made into drink, he may eat an ounce of it, and drinking after it, leave the stomach to dissolve it. [6]

Chocolate Manufacturing, 1700–1850

As the basic processing steps of cleaning and roasting the beans, removing the shells, milling the nibs into liquor, and molding the chocolate were adapted to

European technologies, they continued to be improved and expanded. Typical early bean roasters were cylinders that were turned by hand to mix the beans while roasting (Fig. 46.2), while later versions were powered by water. Spain and Portugal were the first countries to introduce the process of chocolate manufacture into Europe, but the result was a very coarse chocolate [7]. The Spanish process was described as barely drying the cocoa (powder), producing a less bitter, greasy chocolate, whereas cacao (beans) imported to Italy were roasted at a higher temperature that resulted in a more bitter aromatic chocolate [8]. In France, in contrast to Italy and Spain, the beans were roasted at higher temperatures and for longer periods of time and were described as producing chocolate that tasted like burnt charcoal [9].

Milling of the roasted nibs into liquor was the most labor-intensive activity, and from the beginning efforts most likely were concentrated on improving the process. Early chocolate manufacture often was associated with chemists and doctors because the apothecaries had the necessary mortar and pestle equipment to grind roasted nibs. This method was effective in producing chocolate for personal consumption but could not supply the retail trade with sufficient quantities.

Expanding chocolate production for commercial sale required milling equipment on a large scale. Walter Churchman started manufacturing chocolate in Bristol, England, in 1728. He developed a water engine to power his stone mills and was granted a Letters Patent by King George II in 1729, commending him on his advanced method of chocolate making. Dr. Joseph Fry began making chocolate by hand in Bristol by the 1750s. When Walter Churchman's son, Charles, died in 1761, the business, water engine, patent, and recipes were purchased by Fry [10]. By 1770, James Watt had perfected the steam engine, which decoupled large-scale chocolate production from the water-driven mill. Fry installed a steam engine in Bristol in 1795 [11].

Although a Boston newspaper carried an advertisement in 1737 for a hand-operated machine for making chocolate, most of the chocolate available in North America was imported until the mid-18th century [12]. James Baker graduated from Harvard in 1760 and, after brief stints in the ministry and medicine, became a storekeeper in Dorchester, Massachusetts. In 1765, he formed a partnership with John Hannon to produce chocolate in a small building attached to a sawmill on the Neponset River. Records showed that they utilized a kettle and a grinding mill [13]. By 1775, importing cocoa beans into North America had become very difficult and John Hannon sometimes traveled abroad to secure supplies. He was reported lost at sea in 1779 while traveling to the Caribbean to obtain sources for importing cocoa beans, although other reports had Hannon returning to Ireland to escape from his overbearing wife [14]. By 1780, Baker had taken over the chocolate business, and while production of chocolate moved from mill to mill over the next 100 years, Baker and the company he expanded developed a sound reputation for high-quality chocolate (see Chapter 26).

Chocolate production started in Switzerland when François-Louis Cailler established a mechanized factory in Corsier on Lake Geneva in 1819 [15]. Cailler learned the art of making chocolate by hand in northern Italy. He envisioned making chocolate with machines to improve the quality and reduce the cost, and designed his own equipment to produce it in Corsier. He was a pioneer in mechanizing individual steps in the chocolate process such as using a stone roller driven by water power [16]. Philippe Suchard started a factory in Serrieres in 1826 with just one worker who could produce about 30 kg of chocolate per day [17]. Suchard also designed his own chocolate-making equipment and harnessed water power to drive the machines [18]. Rudolf Sprüngli followed his father into the Switzerland chocolate business around 1845 using mechanized chocolate-making methods. In 1853, following local wedding traditions, a bride saved some Sprüngli chocolate in its wedding wrapping as a memento. The chocolate was passed down through subsequent generations and finally returned to the company in 1964. The flavor of this chocolate—when tasted in 1964—was more bitter than sweet, with a coarse, sandy texture. Although the flavor and texture failed to meet current-day standards, the 100-year-old chocolate was still edible [19].

Chocolate Manufacturing, 1850–1900

By 1850, the chocolate process was relatively standardized as manufacturers followed the same basic steps. Starting with liquor and sugar, the process began with the mélangeur, which consisted of two heavy, round millstones supported on a granite floor (Fig. 46.3). The millstones remained stationary while the floor rotated to mix the sugar and liquor. When the mixture reached the consistency of dough, it was removed with a shovel and fed into either granite or cast iron roller refiners (Fig. 46.4). The refiners transformed the dough into a

FIGURE 46.3. Mélangeur. *Source:* J. M. Lehmann, undated. Courtesy of Rodney Snyder. (Used with permission.)

FIGURE 46.4. Roller refiner. *Source:* J. M. Lehmann, undated. Courtesy of Rodney Snyder. (Used with permission.)

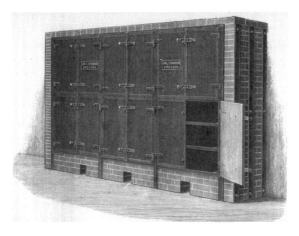

FIGURE 46.5. Holding room for chocolate liquor. *Source*: Paul Zipperer, *The Manufacture of Chocolate*, 2nd edition, 1902; p. 143. Courtesy of Rodney Snyder. (Used with permission.)

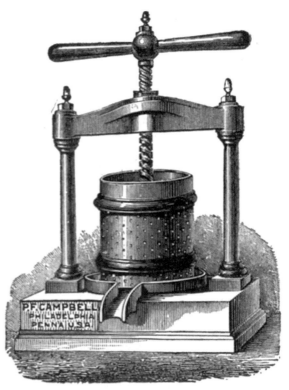

FIGURE 46.6. Hand-operated cocoa butter press. *Source*: P. F. Campbell, ca. 1900. Courtesy of Rodney Snyder. (Used with permission.)

dry powder, as the milled sugar produced more surface area for the cocoa butter to coat. The dry powder was returned to the mélangeur, where the powder was again turned into dough. Grinding alternated between the roller refiner and the mélangeur until the chocolate reached a desired degree of fineness [20]. Manufacturers of the era believed that the thorough mixing of the sugar and liquor was critical to the manufacture of fine chocolate [21]. Since the mixing had to occur at temperatures above the melting point of the cocoa butter (about 35 °C), steam jacketing was added to the holding tanks and the mélangeur. After mixing, the chocolate mass was held at a temperature of about 25 °C for several days to allow a thorough penetration of the flavors into the sugar (Fig. 46.5).

Next, the thick dough traveled through a chocolate kneader, a single granite roller that conditioned the dough so it could be spread on a hard wooden table for placement into molds. Operators filled the molds, pressing the chocolate to remove air bubbles, and transported the molds to a cooling room. As the chocolate cooled, it contracted, which allowed for easier removal. The chocolate then was wrapped in foil and covered with a paper bearing the name and mark of the manufacturer [22].

Although drinking chocolate remained popular into the early 1900s, attempts to resolve some of the inherent difficulties with drinking chocolate identified other chocolate product opportunities. Drinking chocolate was hindered by the high cocoa butter content of cocoa beans. Small-scale methods of extracting butter from liquor, such as the hand-operated butter press (Fig. 46.6), were used throughout the years. Other creative methods also were employed. In 1864, the Ghirardelli chocolate factory in San Francisco solved the problem in a unique way: liquor was poured into a cloth bag and hung in a hot room so the fat dripped through the cloth while retaining the cocoa solids in

the bag [23]. This method reduced the fat content of the liquor to produce lower-fat cocoa solids for improved drinking chocolate. The butter was a by-product that found utility elsewhere as a fat or in cosmetics. The reduced-fat cocoa solids were then milled into cocoa powder. Ghirardelli's survival as a company after its bankruptcy of 1870 was attributed to the sales of its Broma cocoa powder [24].

Caspar van Houten built a chocolate mill in Amsterdam in 1815 and started with two pairs of stones driven by manpower. Coenraad van Houten, Caspar's son, developed a mechanical method to press cocoa butter from liquor. Previous methods had mixed water into the liquor to form a stiff paste, which was packed into cloth bags. As the bags were pressed, the cloth retained the cocoa solids while the fat pressed through the cloth. The cocoa solids were compressed into hard cocoa cakes, which turned gray in color and moldy because of the added water. Van Houten developed a process without the added water, producing cocoa cakes that could be milled into reduced-fat cocoa powder. He filed a patent application for a "Method for Pressing the Fat from Cocoa Beans." The patent examiner recommended the application be rejected because he "did not see an invention in the method and . . . even less a useful invention, and . . . because pressing is simple and public knowledge, and the oil from cocoa far from being harmful to health is beneficial; it is

FIGURE 46.7. Four-pot vertical cacao butter press. *Source*: Paul Zipperer, *The Manufacture of Chocolate*, 2nd edition, 1902; p. 186. Courtesy of Rodney Snyder. (Used with permission.)

therefore useful to retain it in the chocolate." Ignoring the examiner's recommendation, the Patent Office awarded the patent on April 4, 1828 to C. van Houten for the duration of ten years [25].

It took many years for the butter press (Fig. 46.7) to become a standard process in the chocolate industry. By 1861, there were about 30 manufacturers of chocolate in England. At the time, Cadbury had a reputation for producing inferior chocolates. To save money, Cadbury reduced the actual cocoa content of their chocolate to about 20 percent, with the remaining 80 percent consisting of potato starch, sago, flour, and treacle. However, Cadbury was not content in producing common chocolate and looked for a way to differentiate its chocolate from competitors. George Cadbury later related the story of how he "heard of a machine in Holland that was necessary for the manufacture of the finer cocoas. I went off to Holland without knowing a word of Dutch, saw the manufacturer, with whom I had to talk entirely by signs and a dictionary, and bought the machine. It was by prompt action such as this that my brother and I made our business" [26]. The addition of cocoa presses at Cadbury produced a reduced-fat cocoa powder that the company named *Cocoa Essence*, which became an immediate success. By the end of 1866, Cadbury had dropped their entire line of chocolates and *Cocoa Essence* became their only chocolate product. Their early slogan—*Absolutely Pure: Therefore Best*—helped to provide the company with a new reputation as *Makers of Quality Goods*. Cadbury *Cocoa Essence* was the first reduced-fat cocoa powder available in England at the time and it launched the long-term success of the Cadbury business [27]. In 1896, Richard Cadbury described *Cocoa Essence* as the most important process since it was a specialty of the Cadbury Company [28].

Because only a few manufacturers had adopted pressing methods to reduce the fat of cocoa powder, the process was considered a trade secret. The pressing process was described as pouring liquor "into round metal pots, the top and bottom of which are lined with pads of felt, and these are, when filled, put under a powerful hydraulic press, which extracts a large percentage of the natural oil or butter" [29]. By the end of the 1800s, presses had increased in size to four pots that held about 88 pounds, and the pressing cycle lasted about an hour. The cocoa solids from the press pots were compressed into hard disks called cocoa cakes. The cakes then were broken into smaller pieces in a cake crusher and ground in a cocoa cake mélangeur to produce cocoa powder [30]. In the United States, the Walter Baker Chocolate Company had presses installed by the 1890s and marketed their cocoa powder as being superior to the dull, brown-colored powder from the competition [31]. By the end of the 1800s, the pressing process was commonplace throughout the industry [32].

Throughout the years, the natural acidity of cocoa beans has also been a manufacturing problem. Even the Aztecs attempted to reduce the acidity by mixing wood ashes into the chocolate [33]. Because of the lack of sophistication of the Aztec method, European chocolate manufacturers did not appreciate the potential of alkalization [34]. In the 1860s, the Dutch company van Houten was the first to capitalize on the alkalization process. Manufacturers in other countries had trouble matching van Houten's process, so initially most alkalized cocoa powder was sourced from the Netherlands [35]. Even today, the alkalization process is referred to as "dutching," based on its origins.

Alkalization produces various flavors and colors ranging from brown to red to black. Many products such as chocolate cakes, puddings, and drinks use alkalized powder to enhance the chocolate color. For example, the distinctive color and flavor of Oreo® cookies is derived from a heavily alkalized black cocoa powder. By the 1900s, a large proportion of the public preferred alkalized cocoa powder. Cadbury introduced Bourneville alkalized cocoa powder in 1906, and it soon outsold *Cocoa Essence*, Cadbury's flagship brand, by a 50 to 1 ratio [36].

First references to milk chocolate in Europe have been attributed to Sir Hans Sloane and his recipe for a chocolate milk drink. By the 1600s in Jamaica, chocolate commonly was prepared by being grated into a pot, mixed with an egg, sugar, and cinnamon; then either boiling water or milk was added to prepare a chocolate drink [37]. In 1687, while traveling in Jamaica, Sloane complained that the flavor of chocolate was too intense for his taste, but found a local native recipe to be a very pleasant drink. He brought this chocolate recipe back to England, where it was manufactured and sold by apothecaries as a medicine. Cadbury later produced a "Milk Chocolate Prepared after the Sloane Recipe" from 1849 to 1885 [38].

The first milk chocolate designed for eating was invented by Daniel Peter in Vevey, Switzerland, in 1876. Peter believed manufacturing food products gave the best prospects for business success, and chocolate was becoming more and more popular. His job as an ordinary worker in a chocolate factory in Lyons, France, helped him to learn the art of chocolate making. After returning from Switzerland, he intended to continue the operation of the family candle business while diversifying into chocolate manufacture. The presence nearby of his friend Henri Nestlé gave Peter the idea of producing a milk-based form of chocolate. While working on a method of preserving fresh milk, Nestlé had developed powdered milk. Since milk was plentiful in Switzerland, adding milk to the chocolate recipe also added volume and replaced some of the expensive, imported cocoa beans. This process both supported the local milk industry and lowered the ingredient costs for the chocolate [39].

Peter ran his candle/chocolate factory with one employee and his wife while experimenting at night to discover a way to combine milk with chocolate [40]. Because of the high water content of milk, he found that milk did not form a stable emulsion with cocoa butter and resulted in rancid chocolate. He also found that using the coarse milk powder from Nestlé resulted in gritty chocolate. The Anglo-Swiss Condensed Milk Company—Nestlé's competitor—had developed condensed milk in 1867, and after experimentation Peter concluded that a recipe that incorporated condensed milk, cocoa powder, sugar, and cocoa butter produced a very smooth chocolate. A key step to his recipe was the precise regulation of temperature. Peter's *chocolat au lait* was an immediate success, and soon the chocolate manufacturing industry adopted the new process [41].

Much as Starbucks redefined the public's expectations of quality coffee starting in the 1990s, a similar opportunity in the late 1800s occurred with chocolates suitable for eating. As eating chocolates appeared in the marketplace, they typically were characterized as having a coarse and gritty texture. While achieving a measure of popularity, eating chocolate was not perceived as a viable alternative to chocolate beverages. Development of the manufacturing process called conching and the subsequent introduction of *chocolat fondant*, or melting chocolate, would lead to a redefinition of the word "chocolate."

Conching started in 1879 in Berne, Switzerland, originating with Rodolphe Lindt, son of a local pharmacist. Lindt had trained as a confectioner apprentice and bought two fire-damaged factory buildings and some roasting machinery from a bankrupt mill to manufacture chocolate. In the beginning, the roasters were unable to sufficiently dry and roast the cocoa, and grinding the damp nibs produced a very coarse chocolate. When put into molds, his chocolate developed a whitish coating that was unappealing to consumers. He enlisted the assistance of his brother, August, also a pharmacist, to help investigate the source of the white coating. August determined that the cause was too much water in the chocolate, which allowed the migration of fat to the surface of the product. He advised Rodolphe to heat his roller grinder and let the chocolate mix longer to drive the excess water from the chocolate. Rodolphe modified an old water-powered grinding machine developed by an Italian named Bozelli, by embedding iron troughs in granite with the upper edges curved inward. A vertical profile of the trough resembled a shell, and Lindt called his invention a *conche* from the Spanish word for shell, *concha*.

The curved edges allowed for more chocolate mass to be added to the trough without splashing out. At the end of each stroke of the roller, the chocolate broke like a wave, incorporating air into the mass. Rodolphe added some cocoa butter to reduce the viscosity of the chocolate so it flowed more efficiently over the rollers in the trough. After three days of uninterrupted rolling, the chocolate did not resemble regular chocolate. The aeration reduced the bitter and sour flavors and helped to develop the chocolate aroma. Instead of pressing chocolate paste into the molds, the new chocolate could be poured into molds. When eaten, this new chocolate melted on the tongue and possessed a very appealing aroma. In this way began the production of *chocolat fondant* [42].

History does not record why Lindt let the chocolate mix for three days. Perhaps it was part of an experimental plan developed with his brother. One anecdote related to the process involved Lindt leaving for a long weekend and forgetting to turn off the machine, which was powered by water from the Aare River [43]. Regardless of the reason for the discovery, Lindt realized this new process had to be maintained as a trade secret. A separate conching building was built, with access limited only to authorized personnel. In 1899, the German magazine *Gordian* published a discussion entitled: "Why does this chocolate taste so different from all the others?" The magazine received many ideas from readers speculating on the Lindt process, ranging from using a new kind of grinding machine, to adding peppermint oil, even to the addition of more cocoa butter. Their conclusion was Lindt's secret could not be cracked. Lindt was able to maintain the secret of his conching process for more than 20 years and eventually sold his company (and the conching secret) to Rudolf Sprüngli [44].

One of the ingredients needed to produce *chocolat fondant* was cocoa butter. About 50 percent of a cocoa bean is cocoa butter, which has been extracted from cocoa beans since they were first discovered for food and cosmetic applications. The Aztecs put liquor into a pan of boiling water until almost all of the water evaporated. They then re-filled the pan with water and the butter floated to the surface of the water. Cocoa butter was stable, and even 15-year-old butter could be substituted for olive oil [45].

The original purpose for the van Houten butter press was to remove the excess cocoa butter from the liquor to produce reduced-fat cocoa powder. While considered a by-product at the time, it was still regarded as an expensive fat as compared to other available alternatives. However, it possessed a unique characteristic of having a melting point very similar to body temperature. Because many other fats did not mix well into chocolate, cocoa butter became a key ingredient for eating chocolate and milk chocolate.

Chocolate Manufacturing in the 1900s

By the beginning of the 1900s, machines developed for each step in the process prompted the Walter Baker Chocolate Company to describe the chocolate-making process as "modified and greatly improved by modern science" [46]. Richard Cadbury regarded the development of chocolate processing machines as "an interesting study, for we are apt to forget, in this age of luxuries, the years of toil and thought that it has taken to perfect them." Cadbury also believed that the production of chocolate required great experience, skill, and special knowledge [47].

As demand for chocolate increased, companies were consolidated to gain market share and production efficiencies. In 1902, it was observed that "the art of chocolate making is no longer the same as it was some twenty or thirty years ago; but it has for the most part passed from small operators into the hands of large manufacturers" [48]. As economies of scale required larger investments in equipment, specialized machinery was developed to optimize the process.

France was credited with developing a better system of roasting, grinding, and mixing chocolate and Germany was noted for producing the best chocolate-

making equipment [49]. Companies investing in improving the efficiency of their processes to meet consumer expectations had lower costs and higher profit margins than companies maintaining the status quo. Companies that failed to modernize were purchased by larger and more efficient companies. As companies increased the scale of their operations, they became more concerned with keeping their process advancements secret. Although many companies such as the Walter Baker Chocolate Company continued to offer factory tours to visitors, areas with proprietary processes were off-limits [50].

As eating chocolate increased in popularity, the flavor of the chocolate became a major selling point. Most cocoa bean shipments arriving in Europe were damaged by seawater, mold, or infestation; the use of moldy and inferior beans had to be minimized to avoid off-flavors in the chocolate [51].

Some manufacturers roasted their beans without removing the trash and extraneous materials. The recommended practice was to store the beans in a dry warehouse and turn them at regular intervals to dry them in order to prevent damp and mold [52]. For cleaning the beans, the method of hand-picking foreign materials from rotating casks was soon replaced by rotary cylindrical screens driven by either steam or water power (Fig. 46.8) [53]. By the 1930s, the beans were automatically metered into the cleaning machines, automatic brush sieve cleaners were added to keep the screens open, and variable air-blowing capabilities were added to remove the dust. A machine to separate stones was developed that used differences in density to separate the lighter beans from the heavier stones [54]. As late as the 1950s, manual picking of trash after cleaning was still required [55]. Modern cleaning

FIGURE 46.8. Cacao bean cleaner. *Source*: Paul Zipperer, *The Manufacture of Chocolate*, 2nd edition, 1902; p. 113. Courtesy of Rodney Snyder. (Used with permission.)

processes use screens with air aspiration capabilities to separate the light trash and vibrating density tables to remove heavy impurities such as stones. Powerful electromagnets ensure that metallic objects are removed prior to roasting and milling.

Nib roasting required a separate step of drying the beans prior to winnowing and roasting; to improve the process, manufacturers invented a bean roasting method that dried and roasted in one step. To develop the desired color and taste, the roasting process was viewed as "one of the most delicate processes from beginning to end" [56]. Roasters increased in capacity to one-ton rotating drums heated by high-pressure steam. However, the roast level depended on the roasting operators. The operators monitored the sound of the falling beans and the smell of the beans to judge the degree of roast. Experience in the art of roasting was critical: "[It is] . . . impossible to give any technical explanation of this; only practice can give this knowledge, which is the first condition to manufacture of the highest class" [57]. Drinking chocolate and eating chocolate required different roast levels. Some beans were partially roasted over a low fire to about 85 °C to dry and loosen the shells. Others were completely roasted to about 130 °C until almost all of the moisture was driven off through evaporation. Great care was taken to prevent overroasting due to the flavor impact of burnt and ashy flavors in the finished products [58].

By 1920, it was observed that "although attempts were being made to replace the æsthetic judgment of the operator, the complexity of variables associated with roasting cocoa made the application of science to the roasting process difficult" [59]. By the 1950s, better instruments were available to monitor the temperature, moisture, and color of the roasted beans [60]. By the 1970s, roasting was controlled more by science than art: "Science has given us close automatic control of time, temperature, motion, and air flow in roasting; but skill is still needed to bring cocoa beans from the roaster with just the flavor needed for any given purpose" [61]. By the beginning of the 21st century, manufacturers had increased the capacity of a roaster to five tons per batch, they had the ability to roast either beans or nibs, and they had the ability to automate the process to repeatedly produce the same roast level. The art of roasting had transitioned from the operator to the research scientists who developed the optimal roasting recipes for beans from each country of origin.

Even from the beginnings of cocoa processing, it always has been difficult to effectively separate shells from nibs. In the late 1800s, the association of German Chocolate Manufacturers offered a prize to anyone who could develop an effective and economical machine to separate shells from nibs while minimizing the amount of nibs in the discarded shells. Equipment manufacturer J. M. Lehmann subsequently developed

a winnower to improve separation and nib yield [62]. These winnowing machines were described as having an "ingenious arrangement and fine divisions of the cylindrical sieves" [63]. Warm nibs were fed into crushing cylinders to crack the shells and then fed into the winnower. Operators adjusted the airflow over the screens to minimize the loss of small nibs carried into the shell stream. Winnowing machines evolved into a rectangular shape that vibrated a set of stacked screens to separate the nibs and shells by size. Since shells were much lighter, air aspiration was utilized to blow the shells away from the nibs. Because of the constant shaking of the screens, the body of the winnower was changed from wood to steel to improve stability and durability. Despite all of the advances in winnowing technology, about one percent of the shell still remains with the nibs regardless of the winnowing equipment utilized.

The nibs were ground into liquor using a variety of stone mill designs by the 1900s (Fig. 46.9). Cadbury used three lines of millstones that fed granite millstones to further reduce the particle size [64]. The heat generated during the milling process liquefied the butter in the nibs, reducing the viscosity of the liquor. Double or triple stone mills, able to grind up to 140 pounds per hour, were introduced to increase milling capacity [65]. Cadbury employed hundreds of circular stone mills with one stationary stone and one revolving stone. The nibs were fed through a hole in the center of the top stone, and grooves cut into the stones allowed the nibs to gradually be reduced in particle size. The liquefied mass was used for either cocoa powder or chocolate production [66]. The standard for stone mills gradually evolved into the triple stone mill. By the 1950s, impact mills such as disk mills, hammer mills, and pin mills were introduced. These mills relied on impacting the nibs with the

FIGURE 46.9. Set of four stone mills for grinding nibs. *Source*: J. M. Lehmann, undated. Courtesy of Rodney Snyder. (Used with permission.)

FIGURE 46.10. Conching room. *Source*: J. M. Lehmann, undated. Courtesy of Rodney Snyder. (Used with permission.)

disks, hammers, or pins until the particles were small enough to pass through a screen or grate. For fine grinding of the liquor, agitated ball mills were introduced that pumped the liquor through a cylinder filled with steel balls. The impact of the liquor particles between the agitated balls reduced the size of the particles.

Due to the mystery surrounding the production of melting chocolates and the additional equipment needed to produce it, drinking chocolate continued to form the backbone of the chocolate trade into the beginning of the 1900s. Melting chocolates were considered a high-class chocolate only preferred by a certain class of consumers, and could "only be produced in factories which are fitted up with special machinery of modern construction, and their manufacture not only consists of a series of operations of extreme delicacy, but also requires the attendance of an experienced person who has made a special study of such preparations" [67]. By 1900, it was predicted that "chocolate creams are, from the gastronomic point of view, the finest pure chocolate preparations that are made, and there is reason for believing that this branch of chocolate manufacture will soon be very much extended" [68].

The Swiss were renowned for their high degree of perfection in the quality and manufacture of *chocolats fondant*. The Swiss production methods were imitated by most manufacturers, but lack of understanding and failure to pay attention to the details of the Swiss methods often did not lead to success [69]. To make Swiss chocolates, the chocolate had to be ground very fine and additional cocoa butter needed to be added to allow the chocolate to be in a liquid form when warm. Conching lasted up to 48 hours at a temperature of around 55 °C. But for many imitators, by the beginning of the 1900s, it was still unclear how Swiss chocolates

were produced. It was first believed that additional cocoa butter had to be added to the chocolate. However, high levels of fat in the chocolate were objectionable in flavor and texture, so the melting characteristics of the chocolate were then attributed to the chocolate process itself [70].

To replicate the caramel-like taste of *chocolat fondant*, experiments over many years were conducted to discover the method of imparting this flavor to the chocolate. Some manufacturers poured boiling sugar into the cacao mass and others added caramelized sugar. None of these experiments was successful in replicating the flavor of Lindt chocolate. Unknown to others, the secret was not in the form of the sugar added, but the subsequent processing of the chocolate.

Conching (Fig. 46.10) was described as an extraordinary process of which the science behind the effects was unknown: "There is no general agreement as to exactly how the conche produces its effects— from the scientific point of view the changes are complex and elusive" [71].

By 1923, it was recorded that the "crunchy chocolates which sold in quantity only five to ten years ago have gone, the public now demanding a chocolate with a smooth velvet feel. This is most effectively obtained by means of a machine peculiar to chocolate, called a conche, in which the chocolate is worked continuously for three of four days. Along the concave bottom of the conch a heavy roller is thrust backwards and forwards under the chocolate. Many clever attempts have been made to supersede this tedious and costly process, but, in spite of local secessions, it still holds the field" [72]. Although mélangeurs were still in common use in the 1950s, further refining and conching were the standard accepted practice [73]. Highly polished five-roll steel refiners were used to reduce the

FIGURE 46.11. Five-roll steel refiner. *Source*: J. M. Lehmann, undated. Courtesy of Rodney Snyder. (Used with permission.)

particle size of the chocolate, and many new conche designs were being developed (Fig. 46.11).

Although the concepts of chocolate tempering and fat bloom were still unknown in 1852, their effects on the appearance of chocolate were noted: "A little while after it is made the surface grows dull, and is covered with an efflorescence of cacao-butter; but this slight alteration should not cause it to be rejected" [74]. The introduction of melting chocolate required a process to cool the chocolate to ensure that the finished chocolate had a glossy appearance and crisp snap when broken. In 1902, it was believed that the "more rapidly the molded chocolate is cooled the finer is its texture and the more uniform is the appearance of the fractured surface" [75]. By 1931, the tempering process was developed to control bloom in chocolate molding but the underlying principles were not fully understood [76]. By the 1950s, to produce proper temper to avoid fat bloom, the chocolate was cooled to about 30 °C until "mushy," then raised to 33 °C prior to molding [77]. The effects of tempering on the crystal forms of the cocoa butter in the chocolate were better understood by the 1970s. Raising the temperature slightly during tempering to melt out the unstable crystal forms produced a chocolate with a glossy appearance and good snap.

Although Cadbury's Milk Chocolate dates from 1897, their first milk chocolate was not as sweet and milky as consumers preferred. It took the company until 1902 to have operational equipment for condensing the milk, refining the chocolate, and conching it. By 1911, Cadbury opened its first subsidiary factory to condense the milk. The liquor and sugar were shipped to the condensing factory to be mixed and dried in large ovens to produce a product called *crumb*. Milk in *crumb* form could be stored for long periods of time without turning rancid, allowing surplus milk produced during the summer to be stored and manufactured into chocolate in the winter months.

Milton Hershey was manufacturing caramels in Lancaster, Pennsylvania, when he visited the World Colombian Exposition in Chicago in 1893 (see Chapter 16). He saw the new mechanized chocolate-making equipment displayed by the Lehmann Company of Germany. Lehmann had set up a display showing a working chocolate-making process, including roasting, winnowing, milling, mixing, and molding. After tasting the chocolate, Hershey bought the entire Lehmann display and had it shipped to Lancaster. He then hired two chocolate makers from the Walter Baker Chocolate Company to learn how to make melting chocolate. In 1900, he made headlines across the state of Pennsylvania when he sold the Lancaster Caramel Company for one million dollars [78].

After tasting many European milk chocolates and touring some of their factories, Hershey decided to manufacture his own milk chocolate. Although he had seen the equipment used to produce milk chocolate, he did not understand how to make it. He sent William Klein, one of his most trusted workers, to work at the Walter Baker Chocolate Company to discover the methods used by the Swiss. However, Baker had not yet learned how to make milk chocolate either, and Hershey struggled to develop a milk condensing method. According to Anne Klein, daughter-in-law of William Klein, Hershey hired a German chemist who claimed to know how to make milk chocolate and assigned Klein to assist him since he also spoke German. After this attempt failed, Hershey finally called on a worker from the Lancaster Caramel Company, who demonstrated in one day how to add the sugar to the milk, then boiled the milk at low heat under vacuum to remove the moisture. The finished sweetened and condensed milk was smooth like taffy and blended easily with the other chocolate ingredients [79]. After a falling-out with Hershey in 1913, Klein moved 10 miles south to Elizabethtown and started the Klein Chocolate Company, which produced chocolate until 1970.

By 1912, the process of making milk chocolate was known to the entire industry and a turnkey milk condensing plant could be purchased for £275 [80]. Milk chocolate began to be conched to help develop its special flavors, and by the 1930s, milk chocolate had become the most popular chocolate type [81].

The basic concept of pressing has remained constant since it was patented in 1828. The equipment to pulverize the cocoa cakes into cocoa powder transitioned from stone rollers on wooden frames to steel hammer mills. The presses were manufactured in a vertical configuration through the 1950s. By the 1960s, the presses were converted to a horizontal configuration to allow for more capacity and higher pressures. Current presses can have up to 18 horizontal pots with a total pressing capacity of five tons of liquor per hour. Cocoa powder has between 10 and 12 percent fat remaining from a starting fat content of about 54 percent in the liquor. The cocoa butter is filtered from the presses to remove any entrained cocoa solids. While in early presses the cocoa butter was filtered through horsehair or camelhair bags, today continuous centrifuges and paper filters produce the clear yellow cocoa butter [82].

Conclusion

After almost 400 years of innovations in chocolate manufacturing, the captivating aroma and seductive flavor of chocolate still entice scientists to discover new advances in both technology and product development. The evolution from stone metates to automated steel mills has resulted in a variety of new products. Today, the health benefits of dark chocolate are well known, and a trip through local grocery stores provides evidence that chocolate has infiltrated a diverse selection of edible products. Some new trends can be connected to the past. The recent popularity of chocolates made from a single origin is reminiscent of the early years of chocolate production, and the *American Heritage Chocolate* from The Historical Division of Mars, Incorporated uses authentic recipes from the Colonial era of America (see Chapter 49). The blending of old and new will continue, but, unlike the Aztec's methods that were described as "not very nice" [83], modern techniques are sleek and efficient. In other words, very nice indeed.

Acknowledgments

I would like to thank the following individuals who have contributed invaluable support in the writing of this chapter: Doug Valkenburg, Mars Snackfood Master Chocolatier from Elizabethtown, Pennsylvania, who started my cocoa education journey many years ago; Bill Spichiger, cocoa buyer for M&M/Mars from Hackettstown, New Jersey, who first exposed me to the vast and incredibly diverse world of cocoa; and my wife, Maria Snyder, professional author, whose extensive writing knowledge provided the final polish to the chapter.

References

1. Knapp, A. *The Cocoa and Chocolate Industry*. Bath, England: Sir Humboldt, 1923. Travels in America, 1799–1804; p. 12.

2. Brooks, R. *The Natural History of Chocolate*, 2nd edition. London, England: J. Roberts, 1730; p. 33.

3. Brooks, R. *The Natural History of Chocolate*, 2nd edition. London, England: J. Roberts, 1730; pp. 36–37.

4. Hughes, W. *The American Physitian and A Discourse of the Cacao-Nut Tree*. London, England: J.C. for William Crook, 1672; pp. 121–122.

5. Brooks, R. *The Natural History of Chocolate*, 2nd edition. London, England: J. Roberts, 1730; p. 37.

6. Brooks, R. *The Natural History of Chocolate*, 2nd edition. London, England: J. Roberts, 1730; p. 67.

7. Cadbury, R. *Cocoa: All About It*, 2nd edition. London, England: Sampson Low, Marston and Co., 1896; p. 57.

8. Saint-Arroman, A. *Coffee, Tea and Chocolate*. Philadelphia: Crisy & Markley, 1852; p. 83.

9. Brooks, R. *The Natural History of Chocolate*, 2nd edition. London, England: J. Roberts, 1730; p. 35.

10. Wagner, G. *The Chocolate Conscience*. London, England: Chatto & Windus, 1987; p. 13.

11. Knapp, A. *The Cocoa and Chocolate Industry*. Bath, England: Sir Humboldt, 1923; p. 22.

12. Hamilton, E. P. *Chocolate Village—The History of the Walter Baker Chocolate Factory*. 1966; p. 2.

13. Hamilton, E. P. *Chocolate Village—The History of the Walter Baker Chocolate Factory*. 1966; p. 4.

14. Hamilton, E. P. *Chocolate Village—The History of the Walter Baker Chocolate Factory*. 1966; p. 7.

15. *Chocology: The Swiss Chocolate Industry, Past and Present*. Langnau, Switzerland. Chocosuisse, Association of Swiss Chocolate Manufacturers, 2001; p. 8.

16. *150 Years of Delight*. Kilchberg, Switzerland: Chocoladefabriken Lindt & Sprüngli, 1995; p. 34.

17. *Chocology: The Swiss Chocolate Industry, Past and Present*. Langnau, Switzerland: Chocosuisse, Association of Swiss Chocolate Manufacturers, 2001; p. 8.

18. *150 Years of Delight*. Kilchberg, Switzerland: Chocoladefabriken Lindt & Sprüngli, 1995; p. 35.

19. *150 Years of Delight*. Kilchberg, Switzerland: Chocoladefabriken Lindt & Sprüngli, 1995; p. 18.

20. Knapp, A. *Cocoa and Chocolate: Their History from Plantation to Consumer*. London, England: Chapman and Hall Ltd., 1920; p. 144.

21. Zipperer, P. *The Manufacture of Chocolate*, 2nd edition. Berlin, Germany: Verlag von M. Krayn, 1902; p. 125.

22. Jacoutot, A. *Chocolate and Confectionery Manufacture*. London, England: MacLaren & Sons, 1900; p. 16.

23. Beach, N. *The Ghirardelli Chocolate Cookbook*. Berkeley, CA: Ten Speed Press, 1995; p. 6.

24. Gervase-Smith, W. *Cocoa and Chocolate, 1765–1914*. London, England: Routledge, 2000; p. 75.

25. *Cocoa—A Legacy from the Golden Age*. Amsterdam, the Netherlands: Netherlands Cocoa and Cocoa Products Association, 1988; p. 9.

26. Gardiner, A. G. *Life of George Cadbury*. London, England: Cassell and Company, 1923; p. 29.

27. Rogers, T. B. *Cadbury Bournville—A Century of Progress 1831–1931*. Bournville, England: Cadbury Bournville, 1931; p. 24.

28. Cadbury, R. *Cocoa: All About It*, 2nd edition. London, England: Sampson Low, Marston and Co., 1896; p. 60.

29. Head, B. *The Food of the Gods: A Popular Account of Cocoa*. London, England: George Routledge & Sons, 1903; p. 56.

30. Jacoutot, A. *Chocolate and Confectionery Manufacture*. London, England: MacLaren & Sons, 1900; p. 52.

31. Walter Baker Chocolate Company. *The Chocolate-Plant and Its Products*. Dorchester, MA: Walter Baker & Co., 1891; p. 31.

32. Zipperer, P. *The Manufacture of Chocolate*, 2nd edition. Berlin, Germany: Verlag von M. Krayn, 1902; p. 179.

33. Dand, R. *The International Cocoa Trade*, 2nd edition. Cambridge, England: Woodhead Publishing Limited, 1999; p. 10.

34. Lees, R. *A History of Sweet and Chocolate Manufacture*. Surrey, England: Specialised Publications Ltd., 1988; p. 133.

35. Gervase-Smith, W. *Cocoa and Chocolate, 1765–1914*. London, England: Routledge, 2000; p. 74.

36. Williams, I. *The Firm of Cadbury*. London, England: Constable and Company, 1931; p. 83.

37. Brooks, R. *The Natural History of Chocolate*, 2nd edition. London, England: J. Roberts, 1730; pp. 67–68.

38. *Cocoa and Chocolate from Grower to Consumer*. Bournville, England: Cadbury Publication Department, 1959; p. 22.

39. Heer, J. *Nestlé 125 years, 1866–1991*, 2nd edition. Vevey, Switzerland: Nestlé S.A., 1992; p. 138.

40. Heer, J. *Nestlé 125 years, 1866–1991*, 2nd edition. Vevey, Switzerland: Nestlé S.A., 1992; p. 138.

41. *150 Years of Delight*. Kilchberg, Switzerland: Chocoladefabriken Lindt & Sprüngli, 1995; p. 36.

42. *150 Years of Delight*. Kilchberg, Switzerland: Chocoladefabriken Lindt & Sprüngli, 1995; p. 47.

43. *150 Years of Delight*. Kilchberg, Switzerland: Chocoladefabriken Lindt & Sprüngli, 1995; p. 47.

44. *150 Years of Delight*. Kilchberg, Switzerland: Chocoladefabriken Lindt & Sprüngli, 1995; p. 53.

45. Brooks, R. *The Natural History of Chocolate*, 2nd edition. London, England: J. Roberts, 1730; p. 76.

46. Walter Baker Chocolate Company. *The Chocolate Plant and Its Products*. Dorchester, MA: Walter Baker & Co., 1891; p. 26.

47. Cadbury, R. *Cocoa: All About It*, 2nd edition. London, England: Sampson Low, Marston and Co., 1896; p. 60.

48. Zipperer, P. *The Manufacture of Chocolate*, 2nd edition. Berlin, Germany: Verlag von M. Krayn, 1902; p. III.

49. Cadbury, R. *Cocoa: All About It*, 2nd edition. London, England: Sampson Low, Marston and Co., 1896; p. 57.

50. Walter Baker Chocolate Company. *The Chocolate Plant and Its Products*. Dorchester, MA: Walter Baker & Co., 1891; p. 28.

51. Jacoutot, A. *Chocolate and Confectionery Manufacture*. London, England: MacLaren & Sons, 1900; p. 1.

52. Philemon, A. *The Manufacture of Chocolate*. J. P. D., 1900; p. 8.

53. Zipperer, P. *The Manufacture of Chocolate*, 2nd edition. Berlin, Germany: Verlag von M. Krayn, 1902; p. 101.

54. Bywaters, B. W. *Modern Methods of Cocoa and Chocolate Manufacture*. London, England: J. and A. Churchill, 1930; p. 74.

55. Chatt, E. *Cocoa*. New York: Interscience Publishers, 1953; p. 134.

56. Walter Baker Chocolate Company. *The Chocolate Plant and Its Products*. Dorchester, MA: Walter Baker & Co., 1891; p. 26.

57. Jacoutot, A. *Chocolate and Confectionery Manufacture*. London, England: MacLaren & Sons, 1900; p. 8.

58. Philemon, A. *The Manufacture of Chocolate*. J. P. D., 1900; p. 19.

59. Knapp, A. *Cocoa and Chocolate: Their History from Plantation to Consumer*. London, England: Chapman and Hall Ltd., 1920; p. 128.

60. Chatt, E. *Cocoa*. New York: Interscience Publishers, 1953; p. 134.

61. Cook, L. R. *Chocolate Production and Use*, 2nd edition. New York: Books for Industry, 1972; p. 136.

62. Zipperer, P. *The Manufacture of Chocolate*, 2nd edition. Berlin, Germany: Verlag von M. Krayn, 1902; p. 110.

63. Philemon, A. *The Manufacture of Chocolate*. J. P. D., 1900; p. 12.

64. Cadbury, R. *Cocoa: All About It*, 2nd edition. London, England: Sampson Low, Marston and Co., 1896; p. 62.

65. Jacoutot, A. *Chocolate and Confectionery Manufacture*. London, England: MacLaren & Sons, 1900; p. 10.

66. Knapp, A. *Cocoa and Chocolate: Their History from Plantation to Consumer*. London, England: Chapman and Hall Ltd., 1920; pp. 134–135.

67. Philemon, A. *The Manufacture of Chocolate*. J. P. D., 1900; p. 17.

68. Zipperer, P. *The Manufacture of Chocolate*, 2nd edition. Berlin, Germany: Verlag von M. Krayn, 1902; p. 142.

69. Philemon, A. *The Manufacture of Chocolate*. J. P. D., 1900; p. 29.

70. Zipperer, P. *The Manufacture of Chocolate*, 2nd edition. Berlin, Germany: Verlag von M. Krayn, 1902; p. 142.

71. Knapp, A. *Cocoa and Chocolate: Their History from Plantation to Consumer*. London, England: Chapman and Hall Ltd., 1920; pp. 145–146.

72. Knapp, A. *The Cocoa and Chocolate Industry*. Bath, England: Sir Humboldt, 1923; p. 128.

73. Chatt, E. *Cocoa*. New York: Interscience Publishers, 1953; p. 178.

74. Saint-Arroman, A. *Coffee, Tea and Chocolate*. Philadelphia: Crisey &Markley, 1852; p. 83.

75. Zipperer, P. *The Manufacture of Chocolate*, 2nd edition. Berlin, Germany: Verlag von M. Krayn, 1902; p. 156.

76. Jensen, H. R. *The Chemistry Flavouring and Manufacture of Chocolate Confectionery and Cocoa*. Philadelphia: Blakiston's Son & Co., 1931; p. 127.

77. Chatt, E. *Cocoa*. New York: Interscience Publishers, 1953; p. 192.

78. Brenner, J. G. *The Emperors of Chocolate*. New York: Random House, 1999; pp. 85–89.

79. D'Antonio, M. *Hershey: Milton S. Hershey's Extraordinary Life*. New York: Simon & Schuster, 2006; pp. 107–108.

80. Whymper, R. *Cocoa and Chocolate—Their Chemistry and Manufacture*. London, England: J. & A. Churchill, 1912; p. 170.

81. Bywaters, B. W. *Modern Methods of Cocoa and Chocolate Manufacture*. London, England: J. and A. Churchill, 1930; p. 227.

82. Zipperer, P. *The Manufacture of Chocolate*, 2nd edition. Berlin, Germany: Verlag von M. Krayn, 1902; pp. 184–185.

83. Brooks, R. *The Natural History of Chocolate*, 2nd edition. London, England: J. Roberts, 1730; p. 33.

Adulteration

The Dark World of "Dirty" Chocolate

Laura Pallas Brindle and Bradley Foliart Olson

Introduction

Mixing unnatural ingredients with chocolate—with the intent of extending the quantity of raw product and defrauding customers—has a long history. During the pre-Columbian era, "bones of giants" (possibly vertebrate fossils) sometimes were ground and mixed with chocolate [1]. It is difficult, sometimes, to differentiate between adulteration in the sense of adding external products during the preparation of medicinal or culinary chocolates and adulteration in the sense of "cheating" the consumer. In other instances, however, the intent is clear. Chocolate, being an expensive item, commonly was mixed with a variety of "extenders" to reduce cost and to increase the profits of unscrupulous manufacturers/merchants.

This problem of "impure" chocolate increased during the Colonial era and continued into the early modern period of North American history, so much so that chocolate merchants paid for newspaper advertisements that guaranteed product purity in order to maintain consumer loyalty [2–4]. Even into the 19th century, unscrupulous merchants removed the natural, fatty constituents of chocolate, sold cocoa butter for a high profit, and then substituted paraffin during the final chocolate manufacturing process [5]. George T. Angell sounded a warning to customers in 1879 regarding continued adulteration of chocolate throughout North America:

Cocoa and chocolate are adulterated with various mineral substances. Several mills in New England are now engaged in grinding white stone into a fine powder for purpose of adulteration. [6]

Still the problem continued in succeeding decades as unscrupulous manufacturers used unethical practices to deceive consumers. By the early 20th century, the United States government issued a landmark report on food purity in 1906:

Confectionary shall be deemed to be adulterated if they contain terra alba, barites, talc, chrome yellow, or other mineral substances, or poisonous colors or flavors, or other ingredients deleterious or detrimental to health. [7]

What products did merchants add to chocolate? Adulterants could be of animal, vegetal, or mineral origin. They were always cheaper than cocoa powder or cocoa butter and had inconspicuous flavors that were hard to identify when mixed into chocolate. Some were colorants (occasionally poisonous) while others served as extenders or fat replacements. Most chocolate adulterations were contrived to increase profits, but some adulterants, implemented without intention of monetary gain, were added for purported medicinal purposes. Examining the various types of products added to chocolate shows they differ primarily in how they accomplished monetary savings. An article in

Chocolate: History, Culture, and Heritage. Edited by Grivetti and Shapiro
Copyright © 2009 John Wiley & Sons, Inc.

Peterson's Magazine dated to 1891 reported that adulterations to chocolate could be divided into two primary categories:

> 1. *Those which are simply fraudulent, but not necessarily injurious to health—the use of some cheap but wholesome ingredient with the pure article for the purpose of underselling and increasing profits.*
>
> 2. *Those which are injurious to health—the use of drugs or chemicals for the purpose of changing the appearance or character of the pure article, as for instance, the admixture of potash, ammonia, and acids with cocoa to give the apparent smoothness and strength to imperfect and inferior preparations.* [8]

Of the two categories identified by the anonymous author, animal and vegetal adulterants generally fit into the first of these two categories, while most toxic additions were mineral products.

Adulterations have occurred in chocolate since remote antiquity and for as long as it has been produced for consumption. The initial articles that mention chocolate adulteration were published in the 1700s; by the mid-1800s, most of the chocolate being produced, especially in urban areas, was adulterated in some fashion. Investigations in London traced adulteration of chocolate throughout the production process until the final product reached retail stores, and reported that unadulterated products were the minority:

> *The result of these inquiries has exposed deep and wide-spread systems of adulterations, commencing often with the manufacturer and terminating only with the retail dealer. It has shown that in purchasing any article of food or drink in that metropolis the rule is that one obtains an adulterated product—the genuine commodity being the exception.* [9]

Between 1998 and 2002, members of the history of chocolate research team at the University of California, Davis, amassed a substantial amount of information regarding chocolate-related technology using the University of California library system, interlibrary loan services, and digitized databases including Evans 1 [10], Early English Books [11], Chadwyck [12], and Proquest [13]. Materials gathered were surveyed and key documents related to chocolate adulteration filed for further analysis. We joined the history of chocolate research team in 2005 and were drawn to this theme. Subsequently, we sought out and identified additional information specific to adulteration of chocolate using the previously collected resources as a starting platform. We then conducted our search using Chadwyck, ProQuest, and associated digitized databases [14]. Keywords used included adulterant, adulteration, deception, and fraud, cross-referenced with cacao, chocolate, and cocoa. Resources retrieved were entered into an Excel spreadsheet and sorted by theme and subtopic, type of adulterant (animal, vegetal, mineral), and use in chocolate production.

Animal Adulterants

In almost all cases of adulteration using animal products, the goal was to replace cocoa butter with a lesser-valued fat in an effort to cut production costs (Table 47.1). This practice also allowed the manufacturers to sell the separated and purified cocoa butter for a side profit:

> *The noxious and blamable adulterations are the following: to express the cocoa oil, in order to sell its butter to the apothecaries and surgeons; then to substitute the grease of animals.* [15]

Products with textures similar to cocoa butter often were called "replacements"—a softer less "fraudulent" term that shaded the true meaning of adulterant. Such replacements included inferior kinds of butter, lard, mutton suet, and tallow [16]. In addition to agricultural notices on the problem with chocolate, religious groups also commented:

> *Manufacturers of chocolate in the shape of cakes often extract the plenteous oil naturally contained in the seeds of* Theobroma cacao *and sell it to druggists and perfumers at a high price. Then they use some other and cheaper fat, such as suet, lard, or poor butter, to mix in their chocolate paste.* [17]

Although suet, lard, and tallow had compositions that made them reasonable replacements for cocoa butter, their high melting points prevented the adulter-

Table 47.1 Animal Adulterants

Substance	Use as an Adulterant	Description
Butter	Cocoa butter replacement	The result of churning cream or milk, butter exhibits a pleasant taste and lower melting point (closer to that of cocoa butter) than other fat substitutes. Butter's distinctive taste aids in detection as a chocolate adulterant, especially when inferior or spoiled butter is used.
Lard	Cocoa butter replacement	Rendered from the fatty tissue of swine; a cheap by-product from pig slaughter.
Suet	Cocoa butter replacement	Defined as raw, unrendered tallow. Unlike rendered tallow or lard, suet must be refrigerated to maintain freshness.
Tallow	Cocoa butter replacement	Commonly rendered from the fat of cattle or sheep; may contain lard.

ated chocolate from melting in the consumer's mouth. This physical trait of chocolate, now used in the 21st century as a slogan for many chocolate products, results because cocoa butter has a melting point just below that of human body temperature (98.6°F). Adulterated chocolate that did not melt in the mouth commonly had a very unpleasant, slimy mouth-feel. Consumers and inspectors, therefore, were able to detect chocolate adulteration due to the adverse mouth-feel. The flavor of certain animal fats also made their use easy to detect:

> When chocolate has a kind of cheesy taste, animal fat has been added. [18]

Portions of insects, reptiles, and other vertebrates purposely were added to some chocolates. Cited examples included millipedes, earthworms, vipers, and the liver of eels. In these cases, however, the "additions" theoretically were added for medicinal purposes, not as replacements:

> After this manner may you mix with the Chocolate the Powders of Millepedes [sic], Vipers, Earthworms, the Liver and Gall of Eels, to take away the distasteful Ideas that the Sick entertain against these Remedies. [19]

Vegetal Adulterants

Numerous vegetal products were used as adulterants (Table 47.2). They commonly were employed as extenders, cocoa butter replacements, and sweeteners to mask off-flavors produced by other adulterants. Extending the bulk of produced chocolate, perhaps, was the most common reason for adulteration using vegetal ingredients. As with most forms of adulteration, however, financial gain was the motivation: With more total chocolate product to sell, more money would be received and more profit gained.

STARCHES

Starches, with their functional roles providing energy storage and structural integrity, were most often used to increase the mass of chocolate produced. Starch powders and/or flours were produced from various plants, among them arrowroot, potato, rice, sago, and wheat:

> The late Dr. Edmund Parkes, professor of military hygiene, and one of the highest English authorities on the subject of the adulteration of food, stated that he found the cocoa sold in England very commonly mixed with cereal grain, starches, arrowroot, sago, or potato starch. [20]

Sometimes more mealy or finely ground starch derivatives were obtained from a wide variety of sources and used to adulterate chocolate: amaranth, beans, farina (finely ground meal from different cereals), fecula (potato starch), potatoes, rice, and wheat. When mixed with hot water, starch proteins

Table 47.2 **Vegetal Adulterants**

Substance	Use as an Adulterant	Description
Almonds	Extender	The seed of a drupe related to a peach.
Amaranth	Extender	New World domesticate; a grain crop of moderate importance in Africa and Asia. Amaranth was one of the staple food commodities of the Inca.
Arrowroot	Extender	An easy to digest starch from the roots of the West Indian arrowroot. Acting as a thickener, it is used in much the same way as cornstarch.
"Bone" of avocado	Coloring	The seed of the avocado fruit. Usually disposed of when opening the fruit for consumption. Used sometimes in Mexican cooking. Can make the food taste overly bitter if too much is added.
Cocoa-nut shells	Extender	The outer hull of the cocoa seeds. Usually removed, they sometimes are ground into a powder then added in the mixing process to increase bulk.
Farina	Extender	Any meal made from grains, nuts, or starchy roots. Most often from the germ and endosperm of wheat after the hull and bran have been removed.
Fecula	Extender	Any plant matter obtained by breaking down the texture and washing with water. Often made from potatoes.
Grape sugar	Flavor masking	Basically glucose and dextrose. Commonly termed grape sugar because of its high concentration in this fruit.
Honey	Flavor masking	Product of the inversion of sucrose by honeybees into glucose and fructose. Significantly sweeter than table sugar by weight.
Paraffin	Extender	A white, odorless, and tasteless waxy substance obtained from crude petroleum.
Potato	Extender	A perennial tuber very high in starch.
Rice	Extender	A cultivated grass and the most consumed grain worldwide. As an adulterant, rice was used whole and ground, then used as flour.
Sago	Extender	A powder made from the pith of the sago palm.
Syrup	Flavor masking	Any viscous liquid containing a high level of sugars that do not crystallize out of solution. May be made from many different sources.
Wheat	Extender	A grass cultivated worldwide. When finely ground, it is called flour.

would form a sticky substance due to the high gluten content. While writers in the mid-19th century would not have known the reason why such "flours" thickened (due to gelatinization of starch proteins upon contact with hot water), this change was a useful adulteration detection method:

Some greedy merchants (say Messrs, Bussy and Bourtron-Charlard) add to chocolate a greater or less quantity of rice flour, or the fecula of potatoes. Chocolates that have undergone this fraud, have the peculiarity of thickening water to such a degree, that on growing cold, the liquid finally turns into a jelly. [21]

Potato starch commonly was present in adulterated chocolates and detected by mouth-feel:

If, in breaking chocolate it is gravelly—if it melts in the mouth without leaving a cool, refreshing taste—if, on the addition of hot water it becomes thick and pasty—and lastly, if it forms a gelatinous mass on cooling—it is adulterated with flour, or potato starch. [22]

Potatoes also were the primary source of starch used in the production of fecula, a product mentioned numerous times as an adulterant in chocolates:

An analogous preparation is sold at Paris under the name of analeptic chocolate, or sago; it is the ordinary chocolate; into which has been put, not sago, which would make it too dear, but the fecula of potatoes. [23]

SEEDS/OILS/OTHER

Other vegetal products, whether whole nuts, oils, seeds, or miscellaneous plant parts, sometimes were used to reduce the cost of chocolate production. Such ingredients included almonds, cacao shells, various seeds, vegetable oils, and even paraffin:

When very rancid, when it has been exposed for some time to the action of the air, in a tolerably warm place, bad butter, and either vegetable oil, or even the seeds themselves from which the oils were extracted, have been made use of in the sophistication. [24]

Almonds, because of their texture, often were finely ground and used as the basis for producing "fake chocolate." Typically, this process was performed after the almonds were roasted, as noted in this passage, dated to 1848:

Quackery, which corrupts our food as well as our medicines, has also contrived to fabricate indigenous chocolate, without cacao or sugar, from the sugar cane. This composition, which is kept secret, seems to have for its basis roasted almonds, rendered more greasy by butter, and sweetened with grape sugar. [25]

Other plant products documented as chocolate extenders included the "bones" of avocadoes (actually the avocado pit or stone) and the powder from "cocoa-nut" shells (i.e., shells of cacao beans):

The chocolate sold in Paris is often mixed with fecula, farinas and sometimes with an inert powder prepared by bruising cocoa-nut shells. [26]

Such powders, when finely ground, increased the overall mass of the final chocolate product:

The shell or portions of it is sometimes ground up with the kernel for the purpose of increasing the bulk. [27]

Avocado "bones," as well as amaranth grains, were broken into fragments and sometimes sold fraudulently as cacao beans. The length to which unscrupulous traders would go to maximize their profits is illustrated by this passage:

The bad cacao trader is a trickster, he arranges the cacao, he sells the cacao, green ash-colored cacao [baked], toasted, swollen in the fire, inflated with water, mature. He recovers the whiteness of the beans which are still green, the green ones are covered with ashes, are worked with ashes, they are covered with magnesium dirt, they are covered with dirt, they are treated with dirt, [with] grains of amaranth, [with] wax, [with] bones of avocado, they pretend to be beans of cacao. In the shells of cacao they wrap this; in the shells of cacao they put this. The white ones, the green ones, the wrinkled ones, those that are like pepper grains, the broken ones, the sticky ones, those that look like the ear of a mouse, they are all mixed. He turns them around, he substitutes some for others, [making this] he deviates them all, the grains of quappatlachtli with caro, among the others he puts them, because he makes fun of other people. [28]

Paraffin and other waxes were also added as bulking agents or extenders. Because of their natural texture, the addition of these products could be better hidden. Still, their waxiness and different melting points would have alerted consumers:

[Paraffin] has also been employed for the adulteration of chocolate and candies. [29]

SWEETENERS/FLAVORINGS

When adulterations were made to chocolate or cocoa, there developed a parallel need to flavor the final mixture so the products would go unnoticed. Sweeteners were most often used for this purpose. Saint-Arroman, writing in 1846, cited the use of grape sugar for this purpose [30], but the use of honey and sweet fruit syrups also has been identified:

Roast the cocoa to excess in order to destroy this foreign taste, to mix it with rice, meal, potatoes, honey, syrup, etc. [31]

Mineral Adulterants

Of the three chocolate adulterant categories—animal, vegetable, and mineral—those of mineral origin exhibited the widest range of uses (Table 47.3). While many mineral adulterants were used to color chocolate or for medicinal purposes, other compounds were employed as general extenders and/or processing additives. In some instances, the mineral substances were potentially toxic and could have caused serious injury or health complications to unsuspecting chocolate consumers.

Table 47.3 Mineral Adulterants

Substance	Use as an Adulterant	Description
Alkali	Coloring	Basic compounds (pH > 7) soluble in water. References commonly state alkali or alkaline salts, or soda.
Ammonia	Coloring	NH3, a toxin with a pungent odor.
Ashes	Coloring	Residual mineral salts after organic substances have been burned.
Brick dust	Coloring	Bricks, commonly with a clay base, could be ground into a fine dust.
Chalk	Coloring	Soft, white, porous limestone.
Earthy matter	Extender	Term generally applied to dirt or composted material.
Laudanum	Medicinal	Narcotic consisting of an alcohol solution of opium or any preparation in which opium is the main ingredient.
Lead	Medicinal	A soft, malleable, heavy metal toxic to humans.
Magnesium dirt	Coloring	The earth's eighth most abundant element, magnesium dirt being its powdered form.
Peroxide of iron	Coloring	Iron oxide.
Pompholix	Medicinal	Unrefined zinc oxide.
Potash	Coloring	Any of several compounds containing potassium, especially soluble compounds such as potassium oxide, potassium chloride, and various potassium sulfates.
Red lead	Coloring	A poisonous, lead oxide; bright red powder, used in the manufacturing of glass, paint, and pottery.
Reddle	Coloring	A variety of ocher used in coloring and to mark sheep.
Seawater	Extender	Used to make cocoa beans swell, so to make them appear larger.
Steel filings	Medicinal	Small pieces of steel (filings), although more likely iron (filings).
Sugar of lead	Medicinal	A poisonous form of lead, a white solid used to dye cotton and in the manufacture of enamel and varnish.
Sulfate of lime (plaster)	Coloring	Calcium sulfate, commonly known as gypsum.
Umber	Coloring	A natural, brown, earth pigment containing iron (ferric) oxide and manganese oxide.
Vanilla substitute	Replacement	Imitation vanilla: any synthetic compound manufactured to simulate vanillin, the flavor compound present in vanilla beans.
Venetian red	Coloring	A light, warm, unsaturated scarlet pigment derived from nearly pure ferric oxide.
Vermillion	Coloring	A bright red pigment prepared from mercuric sulfide.

COLORING AGENTS

Minerals cited as colorants in chocolate included alkali and/or their salts, brick dust, iron peroxide, magnesium dirt, red lead, reddle (red chalk), umber, Venetian red, and vermillion, along with ashes from undefined burned plants. Reddle, umber, and Venetian red would have been added during the mixing process to color the chocolate and perhaps mask an undesired or off-color:

In Dr. Hassall's well-known work on "Food and its Adulteration," it is stated that out of sixty-eight samples of cocoa examined thirty-nine contained earthy coloring matter, such as reddle, Venetian red, and umber. [32]

Ash and magnesium dirt were not used in the processing of chocolate; instead, they were added to replicate the white coating on fresh cacao beans and mask the use of green beans (unripe), or those fabricated from avocado pits or amaranth grains [33]. Red lead and vermillion were toxic compounds, reddish in color, used to modify the chocolate. These red ochres (colorants) were mixed with yellow compounds including sulfate of lime (plaster) and chalk to alter the appearance of chocolate:

The mineral substances employed in making up chocolate, are some of the ochres, both red and yellow, together with red lead, vermillion, sulphate of lime (plaster), chalk, etc. [34]

In some cases, 18th and 19th century manufacturers may not have been aware of the toxicity of the adulterant, but in the case of lead, its toxicity certainly was known by the medical community:

Chocolates so adulterated, more especially with the preparations of lead, are highly injurious to health. [35]

As the concern about toxic adulterants spread, companies began using the purity of their product as a marketing tool. This marketing strategy is illustrated in advertisements from chocolate manufacturers Walter Baker, Lowney, and Van Houten.

Alkali and soda were used in chocolate production as processing additives, performing a function now known as dutching (see Chapter 46). Dutching darkens the appearance of chocolate; as the popularity of cocoa beverages grew, use of the alkaline treatment also increased:

The attempt to prepare cocoa in a soluble form has tempted some foreign firms to add alkaline salts freely. [36]

Many individuals in the late 19th century, however, thought the dutching process made chocolate products unnatural and no longer healthy to consume:

The Birmingham [England] Medical Review for October, 1890, contains an article on "Food and its Adulterations," in which it is stated that quite apart from any question as to the injury resulting to the human system from taking these salts it would be only right that the medical profession should resolutely discountenance the use of any and all secret preparations confessedly adulterations, and adulterations, too, of a sort not justified by any of the exigencies of the circumstances. . . . Cocoa is only to be recommended as a beverage when it is as pure as possible. [37]

German scientists cautioned, too, that alkali treatment not only affected the healthfulness of chocolate, but the aroma as well:

The cocoa of those manufacturers who employ the alkaline method is sometimes subjected to a perfectly barbarous treatment in order to secure solution by means of the alkali. For instance the roasted cocoa-beans are boiled with an aqueous alkaline solution; the product is then dried, deprived of its oil, afterwards ground. Or the crusted cocoa is roasted, deprived of its oil, powdered, and boiled with water containing an alkali. Both methods of treatment are in the highest degree destructive to those bodies, which are essential constituents of cocoa. It is especially the cacao-red which is attacked, and with it disappear also the aroma. [38]

Researchers in the late 19th century also published on the negative health effects of related compounds like ammonia, potash, and soda used during the chocolate manufacturing processes:

They possess a strong alkaline reaction, are freely soluble in water, have a high diffusion power, and dissolve the animal textures. . . . If administered too long, they excite catarrh of the stomach and intestines. [39]

In response to such negative publicity, chocolate manufacturers like Walter Baker condemned the addition of alkali, or similar processing additives, and used the lack of these substances in his chocolate as an advertisement of purity:

All of Walter Baker & Co.'s cocoa preparations are guaranteed absolutely free from all chemicals. These preparations have stood the test of public approval for more than one hundred years, and are the acknowledged standard of purity and excellence. The house of Walter Baker & Co. has always taken a decided stand against any and all chemically-treated cocoas, and they believe that the large and increasing demand for their goods has proved that the consumer appreciates this decision. [40]

MEDICINAL ADULTERATION OF CHOCOLATE

Medicinals added to chocolate sometimes included laudanum (an opiate derivative), lead, sugar of lead, pompholix (slag on the surface of smelter ore, sometimes

impure zinc oxide), and even steel filings. In one medicinal recipe, laudanum, lead, pompholix, and sugar of lead were combined with crushed millipedes, and the chocolate so concocted—used externally—was deemed a cure for hemorrhoids:

In the American Islands they make use of this Oil to cure the Piles [hemorrhoids]; some use it without Mixture, others melt two or three Pounds of Lead, and gathering the Dross, reduce it into fine Powder, and after it is finely searced [sifted] incorporated it with this Oil, and make a Liniment of it very efficacious for this Disease. Others for the same Intention mix with this Oil the Powder of Millepedes [sic], Sugar of Lead, Pompholix, and a little Laudanum. [41]

According to physicians, when steel fillings were added to chocolate as a remedy, they had to be fine enough to prevent separation and caking in the stomach or intestines. To obtain fine enough steel filings, they were mixed with the spices and ground up in the spice mills:

Because all the Particles of the Steel uniting together, by their Weight, at the bottom of the Stomach, for a kind of a Cake, which fatigues it, and makes it very uneasy. To remedy this, after the Filings have been ground into a very fine Powder upon a Porphyry [i.e., grinding stone]; you must mix it with the Cinnamon, when you make your Chocolate, and it is certain that the Particles of the Steel will be so divided and separated by the Agitation of the Mill, and so entangled in the Chocolate, that there will be no danger of a future Separation. [42]

OTHER

Other mineral adulterants included seawater and "earthy matter." A technique used by some fraudulent manufacturers of chocolate to make the "beans" swell, and therefore increase the total mass of chocolate produced, was to soak cacao beans in seawater. The use of seawater was discernable to consumers, however, by an extreme bitter taste in the final product:

If the chocolate be very bitter, the bean has either been burnt in the roasting, or it has been impregnated with sea water. [43]

Furthermore, seawater itself was thought by some commentators to be a very dangerous additive when used to soak cacao beans:

Sometimes, however, the falsification is dangerous, particularly when it is adulterated with the cocoa bean that has been spoilt by sea water. [44]

"Earthy matter," most likely merely dirt or composted materials, was blended directly into chocolate to increase product bulk. To detect earthy matter adulterants in chocolate, a process similar to that of starch detection was developed:

In order to detect earthy matter in chocolate, a considerable quantity of the suspected article must be finely scraped and steeped in hot water for some minutes, stirring it well during the time. After about a quarter of an hour, the supernatant liquid may be poured off, and the residual matter again treated with hot water until nearly tasteless. The liquid part must then be poured off, and the remainder, or mineral portions, collected and dried. [45]

One mineral-based product cited that did not fit into the three broad adulterant categories was synthetic vanilla substitute. This product would be familiar to 21st century consumers, who can see the supermarket price difference between "pure vanilla" and various "vanilla-flavored" extracts. To save money in the production of chocolate, artificial flavorings such as "vanilla substitute" would have been used as a replacement for the more expensive vanilla beans:

It is generally known that certain substitutes for the flavor of vanilla are widely employed, on account of cheapness, in the manufacture of chocolates; but the firm of Walter Baker & Co. has held aloof from all of these . . . and confines itself to-day . . . to the pure flavor of the choicest vanilla-beans. [46]

Perspectives on Chocolate Adulteration

The use of adulterants has been of great concern ever since chocolate became a common commodity. Unknowing consumers would have been shocked to learn that they could be eating toxic substances:

It is edifying to think that we cannot take a spoonful of sugar, drink a cup of coffee or chocolate, season our edibles with a little mustard, pepper of vinegar, take a mouthful of bread, eat a bit of pickle, or enjoy our pastry, but that the chances are we are swallowing some vile poison. [47]

By the mid-19th century, a definition of manufactured chocolate was established in an effort to discourage adulteration and improve detection methods:

As pure cocoa contains, on an average, fifty per cent of fat, the quantity of fat found in manufactured chocolate ought to be about equal to one-half the weight of the chocolate after deducting the sugar. For example, chocolate made from 63 parts sugar and 37 of cocoa mass would normally give about the following results:

Moisture	*2 per cent*
Fat	*17″*
Residue	*18″*
Sugar	*63″*
	100

If this chocolate contained much more or less than 100 − 63 + 3 percent of fat, we should suspect adulteration, and carefully test the extracted fat. [48]

In some cases, however, deciding whether or not an ingredient or additive was intended as a true adulteration or was just contributing to the ever-evolving process of product development can be very difficult. During the mid-19th century, Arthur Hill Hassall described chocolate in the following way:

Genuine cocoa contains 53 per cent of fatty matter, 16 per cent of aromatic albuminous matter, 10 per cent of starch, and nearly 8 of gum. It will thus be seen, that in its pure state it is a very concentrated and nutritious, as well as agreeable food. [49]

In the same article, Hassall stated that arrowroot was the best starch to use when making chocolate [50]. Based on Hassall's report, it would appear that the addition of starches, especially arrowroot, and gum were perceived by some manufacturers as reasonable and necessary steps in producing quality chocolate. Nevertheless, Hassall's comment seems surprising since other commentators listed the use of starches like arrowroot as adulterations [51].

What compounds, therefore, can be considered adulterants? In many cases these included animal fats, "fake beans," even toxins. But applying the term "adulteration" to other products such as starches and medicinal compounds may be incorrect. Consider the starch amaranth: Although listed as an "adulterant" in the production of fake cocoa beans, amaranth seed (*huautli*) was used by the ancient Aztecs to prepare ritual chocolate beverages. Is it correct, therefore, to call amaranth seed an adulterant?

While the Mayans consumed cacao more than 2000 years ago, Europe and the rest of the world did not encounter chocolate until Cortes and subsequent Spaniards introduced it to the Old World. *Xocolatl* (Mexica/Aztec word for bitter water) was not sweetened and was served to consumers with spices added. Its composition would have been almost 50 percent fat (see Appendix 10), making chocolate a very thick drink that also contained quantities of cocoa solids. If this original Mexica/Aztec concoction stands as the original model for chocolate, then all changes/additions thereafter through the centuries could be considered adulterations. Indeed, one of the earliest such "adulterations" would be the addition of sugar by the Spanish—a change from the original that made the beverage more acceptable to European tastes.

Defining Chocolate

Considering chocolate that consumers know and enjoy in the 21st century, the composition has changed through the years. Who knows what ancient ancestors would have thought of white chocolate, a "chocolate" that contains no cocoa solids? The evolution of chocolate also may be the result of public demand, or from adulterations made by the manufacturers that resulted in producing pleasant tastes to consumers. Definitions have been established for the three most popular forms of chocolate in the 21st century—milk chocolate, dark/bittersweet chocolate, and white chocolate—established in the form of "standard of identities," known as SOIs in the food industry. Reproduced here is a sample SOI written for milk chocolate:

§ 163.130 Milk chocolate.

(a) Description. (1) Milk chocolate is the solid or semiplastic food prepared by intimately mixing and grinding chocolate liquor with one or more of the optional dairy ingredients and one or more optional nutritive carbohydrate sweeteners, and may contain one or more of the other optional ingredients specified in paragraph (b) of this section. (2) Milk chocolate contains not less than 10 percent by weight of chocolate liquor complying with the requirements of § 163.111 as calculated by subtracting from the weight of the chocolate liquor used the weight of cacao fat therein and the weights of alkali, neutralizing and seasoning ingredients, multiplying the remainder by 2.2, dividing the result by the weight of the finished milk chocolate, and multiplying the quotient by 100. The finished milk chocolate contains not less than 3.39 percent by weight of milk fat and not less than 12 percent by weight of total milk solids based on those dairy ingredients specified in paragraph (b)(4) of this section, exclusive of any added sweetener or other dairy-derived ingredient that is added beyond that amount that is normally present in the specified dairy ingredient. [52]

There are, however, other adulterants—not purposely added—that find their way into chocolate manufactured products. Insect parts and rodent hairs, for example, both defined as "filth," are contaminants because it is impossible to manufacture chocolate at a level of quality where all such items would be excluded. Knowing this, the United States Code of Federal Regulations (CFR) permits specific acceptable levels of "filth," specifically:

Chocolate and Chocolate Liquor

· Insect filth: Average is 60 or more insect fragments per 100 grams when 6 100-gram sub-samples are examined OR any 1 sub-sample contains 90 or more insect fragments

· Rodent filth: Average is 1 or more rodent hairs per 100 grams in 6 100-gram sub-samples examined OR any 1 sub-sample contains 3 rodent hairs. [53]

While accidental contamination of chocolate with insect and rodent parts may be unaesthetic, their inclusion is not dangerous, as noted by University of Illinois entomologist Phillip Nixon:

If [consumers] were more willing to accept certain defect levels such as insects and insect parts, growers could reduce pesticide usage. Some of the spraying that goes on is directly related to the aesthetics of our food. [54]

Contamination from insects, rodents, or other forms of filth accounts for most of the chocolate adulteration that occurs today. Such contamination was the subject of a recent article published in the Times of India in regard to certain chocolate samples identified in Mumbai (Bombay), after a previous complaint from consumers in the city of Nagpur:

After Nagpur, it was once again Mumbai's turn to complain about worms in some bars of Cadbury chocolate, sold in Kurla in central Mumbai, Food and Drugs Commissioner Uttam Khobragade told [the Times]. Ten chocolates were found by a consumer, which have been now sent for analysis. There has been a flood of complaints from all over the state after the [Indian] Food and Drugs Administration proposed to take action against the multinational company, which is number one in the business, after finding worms in its milk chocolate recently. [55]

Another form of "adulteration" has begun to arouse consumer concern and this is the use of cocoa butter replacements. Such substitutions have increased given the reality that it is difficult to analyze cocoa butter content and even more difficult for consumers to detect a change in content or percentage [56]. Systematic monitoring of any changes in the fat content of chocolate is difficult—and the chances of being apprehended are slim. Indeed, food technologists in the Asian-Pacific region have been warned that regional chocolate manufacturers may be using artificial cocoa butter to save on economics, as noted in the following statement:

Others estimate that 90 per cent of Chinese chocolate makers rely on cocoa butter substitutes. This is a result of the significant price differential—a ton of cocoa costs around [US $4000] while cocoa substitutes range from [US $800–2000] depending on quality and type. [57]

Use of artificial cocoa butter substitutes, however, may cause problems. Depending on what substitute is used, proper and thorough ingredient labeling should be required to prevent allergic and other adverse reactions. Consumer acceptance of the artificial cocoa butter also is a serious issue for consideration as chocolate purists of the 21st century despise the use of these artificial products just as the purists of centuries ago decried the use of animal fats:

"Chinese customers are more and more concerned about quality as well as price. As a result, in order to better satisfy people's taste, more producers will use pure cocoa butter instead of substitute which has been adopted by most local companies," Chen from the food industry association said. [58]

Conclusion

The use of adulterants in chocolate has a long history, dating before written documentation. When reviewing additions to chocolate, it is difficult to differentiate between medicinal ingredients and "true" adulterants. Most adulterants identified in this chapter were employed and found their way into chocolate production to increase the profits of unscrupulous merchants, with only a few minerals and certain ground insect parts purposely added for medicinal purposes. Adulterations were mentioned in the early 18th century, and concern with such practices continued to rise through the mid to late 19th century. As concern with adulteration grew, new methods of detection were developed. When adulterations were documented, their reporting understandably resulted in great public concern. When the public discovered adulterations, they felt cheated by unscrupulous manufacturers. Consumer disgust, in turn, also served as a corrective device as people "voted" with their pocketbooks: Once chocolate companies were found cheating, trust was lost and business went elsewhere. Entrepreneurs such as Walter Baker and Van Houten recognized consumer purchasing trends and they advertised their products as "pure;" they delivered quality products and as a result had great success in the marketplace.

Most adulterants identified in this chapter were fairly harmless (except for the lead-based products); at worst many left only bad aftertastes. Many of the mineral adulterants, however, potentially were toxic. Addition of these compounds raised concern in the scientific community and subsequent publications in the popular press, as through outlets such as the *American Agriculturist, American Register, Colman's RuralWorld, Godey's Lady's Book*, and *Peterson's Magazine*, shocked the public. In attempts to reduce the incidence of chocolate adulteration, definitions as to what constituted chocolate were published in the 19th century. About a century ago amendments were made to the United States Pure Food and Drug Act and more recently chocolate definitions for use by federal inspectors appear in the United States Code of Federal Regulations (CFR). To accompany the more sophisticated detection process, there are now are in place what are called *defect action levels* that differentiate between accidental contamination (e.g., insect fragments) and malicious examples of adulteration performed by manufacturers or merchants to increase bulk and cheat consumers. All this said, as long as adulterations have occurred, and assuming the adulterants to be nontoxic, modified chocolate products have been consumed and sometimes . . . even enjoyed!

Acknowledgments

We would also like to thank our chocolate-researching predecessors who dedicated countless hours collecting information that we used as a starting block for the current chapter. At the University of California, Davis, we wish to thank Deanna Pucciarelli, Department of Nutrition, our guide early on in the chocolate research project, who helped us locate adulteration data already amassed from the project and who provided insights on how and where to search next; Juri Stratford and Marcia Meister in the Government Information and Maps Department, and Axel Borg in the Biological and Agricultural Sciences Department of the Shields Library, University of California, Davis, for their hours of assistance that helped us gather information on the United States Food and Drug Administration codes and infringements during the past century.

References

1. Sahagun, B. 1590. (Florentine Codex). *General History of the Things of New Spain*. Santa Fe, New Mexico: The School of American Research and the University of Utah Monographs of The School of American Research and The Museum of New Mexico, 1981; Part 12, p. 189.

2. *Boston Post Boy*. March 6, 1769, p. 2.

3. *The Boston News-Letter*. March 8, 1770, p. 2.

4. *Boston Post Boy*. May 27, 1771, p. 3.

5. Anonymous. Paraffine [*sic*] industry. *Scientific American* 1870; XXIII: 10.

6. Angell, G. T. What people eat. *The Farmer's Cabinet*, February 25, 1879, p. 1.

7. United States Pure Food an Drug Act. Document 3915, 34. Statute p. 768, Dated 1906.

8. Anonymous. Some notes on the adulteration of food. *Peterson's Magazine* 1891;XCIX:269.

9. Anonymous. Adulteration of articles of food. *New York Daily Times*, May 6, 1852, p. 1.

10. HTTP://infoweb.newsbank.com [Evans 1]. (Accessed 2005–2007).

11. HTTP://infoweb.newsbank.com [shaw-shoemaker] (Accessed 2005–2007).

12. HTTP://eebo.chadwyck.com [Early English Books] (Accessed 2005–2007).

13. HTTP://eebo.chadwyck.com [Americam periodicals Series Online 1740–1900] (Accessed 2005–2007).

14. Available through Shields Library, University of California, Davis.

15. Anonymous. Selections. Chocolate. *The Literary Magazine, and American Register* 1804;1:427.

16. Anonymous. Adulteration of food—No. 5. *American Agriculturist* 1848;7:11.

17. Bennett, E. T. The cups that cheer. *Christian Union* 1887;36: p. 467.

18. Anonymous. Adulteration of food—No. 5. *American Agriculturist* 1848;7:11.

19. Anonymous. Adulteration of food—No. 5. *American Agriculturist* 1848;7:11.

20. Anonymous. Some notes on the adulteration of food. *Godey's Lady's Book* 1891;122(729):272.

21. Saint-Arroman, A. *Coffee, Tea and Chocolate. Their Influence upon The Heath*. Philadelphia: T. Ward, 1846. (Accessed 2005–2008).

22. Anonymous. Adulteration of food—No. 5. *American Agriculturist* 1848;7:11.

23. Saint-Arroman, A. *Coffee, Tea and Chocolate. their Influence upon The Health*. Philadelphia: T. Ward, 1846.

24. Anonymous. Adulteration of food—No. 5. *American Agriculturist* 1848;7:11.

25. Saint-Arroman, A. *Coffee, Tea and Chocolate. Their Influence upon the Health*. Philadelphia: T. Ward, 1846.

26. Chevallier, A. On the alterations and adulterations of alimentary substances. *American Journal of Pharmacy*. 1845;10.

27. Shearer, H. Chocolate. *Colman's Rural World* 1901;54(34):6.

28. Durand-Fovest, J. de. El Cacao entre los Aztecas, *Estudios de Cultura Náhuatl*. Mexico, D.F.: Universidad Nacional Aotónoma de México, 1968; pp. 155–181.

29. Anonymous. Paraffine [*sic*] industry. *Scientific American* 1870; XXIII:10.

30. Saint-Arroman, A. *Coffee, Tea and Chocolate. Their Influence upon the Health*. Philadelphia: T. Ward, 1846.

31. Anonymous. Selections. Chocolate. *The Literary Magazine, and American Register* 1804;1:427.

32. Anonymous. Some notes on the adulteration of food. *Godey's Lady's Book* 1891;122(729):272.

33. Durand-Fovest, J. de. El Cacao entre los Aztecas, *Estudios de Cultura Náhuatl*. Mexico, D.F.: Universidad Nacional Aotónoma de México, 1968; pp. 155–181.

34. Anonymous. Adulteration of food—No. 5. *American Agriculturist* 1848;7:11.

35. Anonymous. Adulteration of food—No. 5. *American Agriculturist* 1848;7:11.

36. Anonymous. Some notes on the adulteration of food. *Peterson's Magazine* 1891;XCIX:269.

37. Anonymous. Some notes on the adulteration of food. *Godey's Lady's Book* 1891;122(729):272.

38. Anonymous. Some notes on the adulteration of food. *Godey's Lady's Book* 1891;122(729):272.

39. Anonymous. Some notes on the adulteration of food. *Peterson's Magazine* 1891;XCIX:269.

40. Anonymous. Some notes on the adulteration of food. *Peterson's Magazine* 1891;XCIX:269.

41. Quélus, de. *The natural history of chocolate; a distinct and particular account of the cocoa-tree; its growth and culture, and the preparation, excellent properties, and medicinal vertues of its fruit. Wherein the errors of those who have wrote upon this subject are discover'd; the best way of making chocolate is explain'd; and several uncommon medicine drawn from it, are communicated*. Translated by R. Brookes. London: J. Roberts, 1730.

42. Quélus, de. *The natural history of chocolate; a distinct and particular account of the cocoa-tree; its growth and culture, and the preparation, excellent properties, and medicinal vertues of its fruit. Wherein the errors Of those who have wrote upon this subject are discover'd; the best way of making chocolate is explain'd; and several uncommon medicine drawn from it, are communicated*. Translated by R. Brookes. London: J. Roberts, 1730.

43. Anonymous. Adulteration of food—No. 5. *American Agriculturist* 1848;7:11.

44. Anonymous. Adulteration of food—No. 5. *American Agriculturist* 1848;7:11.

45. Anonymous. Adulteration of food—No. 5. *American Agriculturist* 1848;7:11.

46. Anonymous. The cacao-plant. *Peterson's Magazine* 1890;XCVII: p. 489.

47. Anonymous. Adulteration of articles of food. *New York Daily Times*, May 6, 1852, p. 1.

48. Hassall, A. H. Food and its adulterations; comprising the Reports of the Analytical Sanitary Commission of *The Lancet*, for the Years 1851–1854 inclusive. Available at http://memory.loc.gov/cgi-bin/query/D?ncps:22:./temp/~ammem_hf3U::. (Accessed January 27, 2007.)

49. Hassall, A. H. Food and its adulterations, comprising the Reports of the Analytical Sanitary Commission of *The Lancet*, for the Years 1851–1854 inclusive. Available at http://memory.loc.gov/cgi-bin/query/D?ncps:22:./temp/~ammem_hf3U::. (Accessed January 27, 2007.)

50. Hassall, A. H. Food and its adulterations, comprising the Reports of the Analytical Sanitary Commission of *The Lancet*, for the Years 1851–1854 inclusive. Available at http://memory.loc.gov/cgi-bin/query/D?ncps:22:./temp/~ammem_hf3U::. (Accessed January 27, 2005.)

51. Anonymous. Some notes on the adulteration of food. *Godey's Lady's Book* 1891;122(729):272.

52. Milk Chocolate. *Code of Federal Regulations*. 21 CFR Ch 1(4-1-05 Edition) p. 522.

53. http://www.sixwise.com/newsletters/05/06/29/how_many_insect_parts_and_rodent_hairs_are_allowed_in_your_food.htm. (Accessed February 14, 2007.)

54. http://www.sixwise.com/newsletters/05/06/29/how_many_insect_parts_and_rodent_hairs_are_allowed_in_your_food.htm. (Accessed February 14, 2007.)

55. Kamdar, S. Worms in Cadbury chocolate. *Times of India*. October 14, 2003. Available at http://timesofindia.indiatimes.com/articleshow/232235.cms. (Accessed February 14, 2007.)

56. Breaking News on Food and Beverage in Asia Pacific. AP—food technology. Available at http://www.ap-foodtechnology.com/news-by-product/news.asp?id=71928&idCat=0&k=new-regulations-back. (Accessed February 12, 2007.)

57. Breaking News on Food and Beverage in Asia Pacific. AP—food technology. Available at http://www.ap-foodtechnology.com/news-by-product/news.asp?id=71928&idCat=0&k=new-regulations-back. (Accessed February 12, 2007.)

58. Breaking News on Food and Beverage in Asia Pacific. AP—food technology. Available at http://www.ap-foodtechnology.com/news-by-product/news.asp?id=71928&idCat=0&k=new-regulations-back. (Accessed February 12, 2007.)

Making Colonial Era Chocolate

The Colonial Williamsburg Experience

James F. Gay and Frank Clark

Introduction

This chapter examines the creation of the historic chocolate program at The Colonial Williamsburg Foundation. It describes the experience of the museum staff in making historic chocolate and offers advice as to how other institutions might establish their own programs. The first part discusses the historic research that led to the creation of Colonial Williamsburg's "Secrets of the Chocolate Maker" program. The second section addresses the cost and practicality of establishing a historic chocolate-making program at other museums and considers their potential educational and financial benefits.

Colonial Williamsburg in Williamsburg, Virginia, is the world's oldest and largest living history museum. It is the restored and reconstructed site of colonial Virginia's second capital with 500 structures on 301 acres. Each year 700,000 or more visitors stroll around the 18th century buildings and gardens, where visitors encounter costumed interpreters depicting various aspects of 18th century life. The stories told are of events, big and small, about people famous and anonymous. On an average visit, one might encounter Thomas Jefferson or Patrick Henry on a podium, or see a street theater called "Revolutionary City." Within the context of this setting, there are approximately 90 skilled craftsmen and women in 22 trade shops demonstrating the hand skills of the pre industrial world [1]. Unlike the actors who populate the street scenes

speaking words of the 18th century, tradesmen use their hands to reproduce an authentic rendering of a period silver cup or wooden bucket, all the while explaining the process or other aspects of the period. Included in this milieu, food-related activities include cooking, brewing, baking, butchering, confectionary, and chocolate making (Fig. 48.1). All of these were distinct trades in the 18th century and all of them fit under the umbrella title of Historic Foodways as practiced at historic Colonial Williamsburg.

Why Make Chocolate?

Reay Tannahill's prologue to her classic *Food in History* begins: "It is obvious truth, all too often forgotten, that food is not only inseparable from the history of the human race, but basic to it. Without food there would be no human race, and no history" [2]. Within Tannahill's context, it is obvious that some foods receive

Chocolate: History, Culture, and Heritage. Edited by Grivetti and Shapiro
Copyright © 2009 John Wiley & Sons, Inc.

FIGURE 48.1. Chocolate cream and pot. *Source*: Courtesy of The Colonial Williamsburg Foundation, Williamsburg, VA. (Used with permission.) (See color insert.)

more attention than others. Educators need the attention of their students before "teaching" happens. Chocolate making is an ideal way to get noticed and, not surprisingly, historic chocolate can lead to discussions of politics, family, labor, religion, revolution, economics, and just about anything else. It is not likely you would achieve the same results with broccoli!

Chocolate's universal popularity attracts so much interest that researchers do not have to wade too deep before they encounter an array of chocolate histories. However, if one is a museum tradesman, intent on reproducing 18th century chocolate by hand, modern sources are poor guides through these uncharted waters. Most serious scholarship on the history of chocolate has focused on Latin America and Europe. The North American chocolate experience for the 17th and 18th centuries basically has been ignored by researchers. This omission has left the story to be told by food writers and marketing departments, commonly repeating each other's "facts" without much attribution and without concern for accuracy. In addition, no other museum in North America had ever attempted to reproduce 18th century chocolate, so the entire Colonial Williamsburg effort was homegrown. This would not be the case if the trade were historic blacksmithing or woodworking. There are a relatively large number of artisans who work in museums or as independent craftsmen. But there were no master chocolate makers in any museum specializing in the 18th century—anywhere in North America—with whom we could consult.

In order to establish a historic chocolate-making program at Colonial Williamsburg or any museum, the guiding ethic is accuracy and authenticity. Additionally, as far as craftwork goes, functionality also is required. In this case, the chocolate produced must be manufactured from period recipes and to the same degree of acceptability as in the 18th century. Some trades have

access to original articles, whether chairs or rugs, which can be measured with a micrometer and inspected using a magnifying glass. The results of the labor needed to reproduce such objects can be compared against the originals and the degree of accuracy measured. The final product can be quite exact—in fact, when manufacturing a teaspoon reproduction, for example, the item should carry the tradesman's specific hallmark so the object is not confused with an antique. But unlike other period tradesmen, the food-related trades produced something that was, by definition, perishable. Archaeologists have discovered traces of chocolate in Mesoamerican sites [3] but none in North America. While archaeologists may excavate rabbit bones or fish scales, they are not likely to excavate chunks of 18th century chocolate. The historic foodways process, therefore, is to consult period cookbooks and other sources and replicate their techniques using period equipment but modern ingredients.

The authenticity of the process also involves use of specific tools and techniques. What makes historic craftwork challenging is mastering a process while keeping in mind that historical accuracy is the goal, not the production of just another teaspoon. When such spoons are sold, the intrinsic value comes from the fact that the historical technique was followed by a costumed tradesman using tools from the period, and nothing else. But, unlike an 18th century tradesman, whose sole job was production, the museum interpreter's role is education. With this in mind, the Colonial Williamsburg chocolate program was established to educate the public on historic chocolate.

Importance of Background Research

Research on historical chocolate at Colonial Williamsburg began in spring 2001, with initial investigations on the feasibility of interpreting chocolate making. At first, the initial goal was merely to show guests some cocoa beans. This activity in itself was a challenge because, in 2001, no merchants known to us other than chocolate manufacturers had cocoa beans for sale. Eventually, however, we were able to purchase 50 pounds of cocoa beans from Scharffenberger, a chocolate company based in Berkeley, California. While on vacation, a staff member purchased and shipped an authentic metate (grinding stone) from Mexico. Additionally, Colonial Williamsburg blacksmiths fashioned a metal roller, and the basket makers fashioned a winnowing basket. By summer 2001, all the tools and ingredients were in place.

When trying to rediscover historic trade processes, the first attempts commonly turn out to be "learning experiences." As time and experimentation progressed, members of the historic foodways staff made discoveries about the various processes and contributed their insights, yielding better and better tasting results. Also, with these new insights, the few historical references that described the process of chocolate making began to make more sense. During this period, historic craftwork often required "reading between the lines" of the written record, a process required when making chocolate by hand. These initial experiments were highly consumptive of the cocoa beans on hand because so many processes had to be mastered. Because of the high degree of learning required, no formal program in chocolate making was initiated at Colonial Williamsburg until winter 2002.

Besides the actual craftwork, which was largely experimental, the historical questions that required answers were obvious: the "who, what, where, when, and how" of the 18th century chocolate-making process. These same questions were asked specifically:

Who: Governor, gentry, middling sort, and slaves;

What: Identify the unique resources (cocoa and ingredients) and specific tools available;

Where: Colonial Williamsburg;

When: 1774–1783; and

How: Could we master the specific skills necessary to make chocolate.

The results of the research indicated sufficient historical support for a colonial chocolate demonstration at Colonial Williamsburg.

Eighteenth century chocolate making in eastern North America was a trade primarily located in four major urban areas associated with Boston, Philadelphia, New York City, and Newport, Rhode Island [4]. Nearly all of the chocolate imported to Virginia during this period arrived by ships from these areas. While chocolate making was a trade in these cities, there were no advertisements from chocolate makers in the *Virginia Gazette* [5] or other Virginia newspapers during the 18th century. Yet the records show that cocoa, called "chocolate nuts" and the main ingredient necessary to make chocolate, was imported into Virginia. Also, chocolate stones were listed in a few inventories, but on plantations away from Williamsburg. While this might have been a "showstopper" as far as a Williamsburg-specific foodways program goes, still, there were tantalizing clues that chocolate making occurred in 18th century Williamsburg.

Records revealed that Williamsburg carpenter James Wray had 50 pounds of "chocolate nuts" in his inventory [6], and tavern owner James Shields owned a chocolate stone [7]. Wray was actually more than a carpenter. He had trade dealings with the West Indies so chocolate nuts in his inventory could be explained in a number of ways: James Wray, the chocoholic; James Wray, the cocoa trader; perhaps James Wray, the food dealer. That James Shields had a chocolate stone made perfect sense if he catered to an "upscale" clien-

tele, as was the case for several taverns in wealthy Williamsburg. So, with circumstantial evidence pointing to chocolate making in Williamsburg, the next issue was to decide on the location where Historic Foodways would present their chocolate education program.

Due to practicality, the detached kitchen complex associated with the Governor's Palace was chosen as the location for chocolate-making demonstrations. Besides having the most space for making chocolate, the kitchen area also could accommodate a large volume of visitors. Although this was a logical location due to the prominence of the occupants, other possibilities might have been in or around Shields Tavern, a coffeehouse that serves snacks and beverages to 21st century visitors. However, if there was one place that covered the life of Williamsburg as a capital—with chocolate being the hook to tell the story—then the Governor's Palace kitchen area best served our needs.

Spanning nearly 60 years of occupancy, the original palace was the residence of seven colonial royal governors and two state governors. The first and last occupants, Governor Alexander Spotswood and Governor Thomas Jefferson, had chocolate paraphernalia listed in their records. Spotswood's probate inventory listed a chocolate stone [8]. Jefferson's account book shows that he purchased "chocolate nuts" in April 1780, the month that the Virginia capital moved to Richmond [9]. From the viewpoint of a Colonial Williamsburg interpreter, these facts present the opportunity to turn the straightforward discussion of chocolate production into broader issues dealing with the rise and fall of Williamsburg as a capital city.

Having established a historical basis for a chocolate-making program in Colonial Williamsburg, we then needed to look at period sources to determine the types of raw materials, equipment, and processes we would need in order to create the program. The first issue involved the cocoa. While modern cocoa grows worldwide, 18th century cocoa was exclusively grown in the New World [10]. For the sake of authenticity, therefore, the cocoa sought for the Colonial Williamsburg program had to have a New World origin. Import and export records indicated that the cocoa coming to Virginia originated in the British West Indies [11]. These records, however, more accurately reflect the ship's debarkation point, rather than where the cocoa actually was grown. In fact, 18th century preferences seem to have been for cocoa grown in Venezuela, the major producer at that time. Which specific variety of Venezuelan criollo cacao beans were used by our colonial ancestors, however, was unknown, as cocoa was linked in most documents to the port of origin, for example, Caracas or Maracaibo [12]. The Dutch island of Curaçao was particularly well suited for exploitation of cacao within the Spanish Main. Curaçao lies only 38 miles off the coast of Venezuela. By the 1730s, Spanish traders went to great lengths to avoid strictures against contraband trading. Isaac and Suzanne Emmanuel write:

Spain prohibited her subjects from exporting cacao, Mexican silver, gold dust, and so forth, to other lands since she wanted a monopoly on the sale of all of her colonial products in Europe. But rather than wait for months and months for a sailing from Spain, her colonists chanced selling their products to the Dutch and English of the Caribbean area. So, in spite of the ban, the Spanish Creoles conducted a lively "business as usual" with Curaçoans in cacao. As a precaution, however, they would engage in sham battles in Venezuelan waters. The Curaçoan "pirates" would "seize" the cacao from the Venezuelans after paying them in gold doubloons or in finished goods. [13]

Called *criollo* today, this cocoa is still highly prized by 21st century chocolate makers. For purposes of the modern chocolate-making program at Colonial Williamsburg, choosing Venezuelan criollo as the bean source provided several educational benefits. Besides lending authenticity, it also yielded interpretive opportunities to discuss the West Indian trade with North America, the trade between Atlantic ports, specific types of cocoa, smuggling, piracy, and relations with the Dutch and Spanish.

We then needed to determine what processes were to be used to make the chocolate. Most chocolate consumed in North America in the 18th century was machine-made. In Virginia, chocolate was sold in the stores and listed in inventories throughout the period. On the other hand, chocolate stones in the inventories of Robert "King" Carter [14] and Phillip Ludwell Lee [15], two members of the Virginia planter aristocracy, point to a small exclusive club of wealthy men who had their chocolate custom-made. While there were other Colonial era families in Virginia with records that documented either chocolate nuts or stones, the Carter household was the only one that recorded both. With no chocolate maker tradesman in Virginia using a machine in the 18th century, the only possible alternative was making chocolate by hand. However, there remained the ambiguity as to the style of chocolate stone actually used in Virginia: Was it in the style of a Mexican *metate* with three legs, or was it similar to a European saddle quern with four legs?

Hand grinding of cacao beans was done on a *metate*, the term still used today in Mexico and Central America, or on a *saddle quern*, the term generally used by Europeans. Americans, however, called such objects *chocolate stones*. Whatever the name, it was a heated stone used for grinding, worked with a back and forth motion using what the Spanish called a *mano*, and what the English called a *roller*. Such *metates* generally rest on three legs with two of the legs supporting the lower end of the sloping concave stone, which is supported on the higher end by a longer third leg. A saddle quern, on the other hand, specifically converted to a chocolate stone, was a concave stone raised above and set on a wooden or metal frame supported by four legs. The Roman word for a saddle quern was *mola trusatilis* (from *mola*, or mill, and *trudo*, to push) [16]. The converted chocolate stone had to be high enough to allow a pan of burning charcoal or coal to be fitted underneath it. In 1730, Brookes described one and its dimensions:

They make choice of a Stone which naturally resists the Fire, not so soft as to rub away easily, no so hard as to endure polishing. They cut it from 16 to 18 Inches broad, and about 27 to 30 long, and 3 in thickness, and hollowed in the middle about an inch and a half deep. This Stone should be fix'd upon a Frame of Wood or Iron, a little higher on one side than the other: Under, they place a Pan of Coals to heat the Stone, so that the Heat melting the oily Parts of the Kernals, and reducing it to the Consistence of Honey, makes it easy for the Iron Roller, which they make use of for the sake of its Strength, to make it so fine as to leave neither Lump, nor the least Hardness. This Roller is a Cylinder of Polished Iron, two inches in diameter, and about eighteen inches long, having at each End, a wooden handle of the same Thickness, and six Inches long, for the Workman to hold by. [17]

Since ambiguity still remained, the solution to the stone problem was to have both on hand and make the visitor aware of the differences. The least expensive route was to only use the metate. These can readily be purchased in Mexico (or at many locations in the American Southwest and California) and are relatively inexpensive. Saddle querns of the type described by Brookes, however, are rare and their manufacture today would require the services of a stonemason. Additionally, the metal frame and iron roller also requires the services of a blacksmith. And since there were no examples of chocolate stones in North American museums to inspect, North Carolina granite was chosen as the stone material for manufacture of a saddle quern. In 1736, there is a record of a chocolate stone imported into Virginia from North Carolina [18]. The assumption was made, therefore, that this chocolate stone was manufactured in North America from local North Carolina stone. The style of stone gives interpreters the opportunity to discuss these differences, the role of the Mesoamerica versus European/North American grinding stones, as well as the inherent uncertainties in modern historical research. The differences also put chocolate into historical perspective for the visitor.

The Hand-Grinding Process

The London Tradesman was written by Campbell in 1747 for the "Information of Parents and Instruction of Youth in the Choice of Business." Much like a modern college guide, Campbell described for parents what was in store for their child in various trades. Filled with rather pithy comments, Campbell probably reflected the condescending attitude of educated British and Americans toward the tradesmen of the day and their attitudes toward hand labor. Regarding the occupation of chocolate maker he wrote:

> Chocolate is made of Cocoa, the Product of the West-Indies. It is stripped of its Shell, or rather Husk, and wrought upon a Stone over Charcoal Fire till it is equally mellow put into Moulds, which shapes it into Cakes. To perfume it they mix it Venello [sic]. It is a hot laborious Business, but does not require much Ingenuity. Journeyman's Wages is [sic] from Twelve to Fifteen Shillings a Week, are not employed in much in the Summer. They require Heat to work with, but cold Weather is necessary to dry it. [19]

Examination of other period resources informs 21st century readers that Campbell omitted several steps in the process. First, the cacao beans were removed from the pods and fermented on the forest floor for several days. This fermentation process is the necessary step to begin the enzymatic action that ultimately produces natural chocolate flavors. Cocoa beans, having been fermented on the ground and then dried, were bagged or put into wooden containers. When the cocoa bags or boxes arrived at the port of destination and were unpacked, they were inspected for insects, pieces of lint, rocks, and other extraneous matter. Some of the cocoa beans might have been partially eaten by insects; others may have molded or spoiled. Furthermore, there would have been pieces of broken shell, dust, and dirt. Most such items—but not all— were removed during inspection. The raw cocoa would have no chocolate flavor at this point.

Once inspected, and most of the detritus removed, the nuts were pan-roasted over low heat. Next to the quality of the cocoa itself, roasting was (and remains today) the most critical step. At this processing stage, cocoa shares a commonality with coffee beans. The enzymes created in the fermenting process are converted by heat, and the aroma of chocolate is released. The roasting is what consumers "taste" as much as anything (Fig. 48.2). Chocolate makers of the era had to find the "sweet spot" of each batch roasted: This would be the perfect combination of natural cocoa flavor combined with the flavor produced by roasting.

Cocoa can have many different flavors. Some varieties taste fruity, some nutty, and others woody. If underroasted, cocoa can taste too fruity and the chocolate flavor is less pronounced. Overroasted cocoa has more pronounced chocolate flavors but can also express bitter or burnt flavors. Such "off" flavors also could mask worse possibilities, for example, flavors that might have been picked up in shipment or during storage (i.e., rotting fish). The ideal would have been what chocolate makers in the 21st century call "medium roast." This would be the case where the chocolate flavor is increased (over underroasted beans) and also

FIGURE 48.2. Roasting cacao beans. *Source*: Courtesy of The Colonial Williamsburg Foundation, Williamsburg, VA. (Used with permission.) (See color insert.)

has well-developed nutty flavors—especially characteristic of criollo cacao beans. The physician Brookes, writing in 1730, perhaps said it best:

> The whole Art consists in avoiding the two Extremes, of not roasting them enough and roasting them too much; that is to say, till they are burnt. If they are not roasted enough, they retain a disagreeable Harshness of Taste; and if they are roasted so much as to burn them, besides the Bitterness and ill Taste that they contract, they lose their Oilyness [sic] entirely, and the best part of their good Qualities. [20]

Our chocolate-making ancestors would have had to taste the cocoa beans being roasted to determine the proper degree of heat (called "roast"). Those more experienced might have been able to tell when to taste by the aroma rising from the pan and by listening to the shells pop. Our experience at Colonial Williamsburg, however, has shown that it is difficult to achieve consistency each time chocolate is made. Even within the same batch, some cocoa beans will be underroasted, while others will be overroasted, and others just right. The process is slow and it cannot be rushed by applying more heat. Roasting is the only true "cooking" in the entire process of making chocolate.

Next comes shelling of the beans. Cacao seeds when fresh have a white outer membrane, which turns brown during fermentation. When dry, this

FIGURE 48.3. Cracking roasted cacao beans to remove nibs. *Source*: Courtesy of The Colonial Williamsburg Foundation, Williamsburg, VA. (Used with permission.) (See color insert.)

FIGURE 48.4. Winnowing roasted cacao beans. *Source*: Courtesy of The Colonial Williamsburg Foundation, Williamsburg, VA. (Used with permission.) (See color insert.)

membrane is referred to as the shell. When the shells are removed, the kernels, called "nibs," are ground. Shelling is tedious if done by hand, but at this stage visitors can become involved and join in the process. Young and old alike are often drafted at Colonial Williamsburg to shell the nibs as they are removed from the pan. The trick to shelling is to get the largest nib possible from every "chocolate nut." Nibs become brittle from the roasting process, however, and shatter easily. Inexperienced persons shelling cocoa commonly produce a pile of "trash." Imagine taking the outer skin off every peanut in the can with the peanut being as brittle as an eggshell. A rolling pin with light pressure will break the shell but can also shatter the nib if too much pressure is used (Fig. 48.3). Most of the shells then are separated from the nibs by winnowing, and any that remain are removed by hand. Small bits of chaff and ash can be removed by the wind during this process (Fig. 48.4). Even so, some bits of extraneous materials may remain; it should be noted that even 21st century chocolate makers are unable to remove all the ash, although the amount that remains usually is very small. Brookes, writing in 1730, informed his readers:

> There [sic] Skins come off easily, which should be done one by one, laying them apart, and taking great heed that the rotten and mouldy Kernals [sic] be thrown away, and all that comes off the good ones: for these Skins being left among the Chocolate, will not desolve [sic] in any Liquor, nor even in the Stomach, and fall to the bottom of Chocolate-Cups, as if the Kernels had not been cleaned. [21]

While Brookes's method describes the process probably preferred by most consumers during his era, others had different ideas about the hulls. A 1743 Virginia recipe for making chocolate from Jane Randolph recommended inclusion of some hulls in the grind, as reported here in her exact spelling:

> Chocolate to make up You must wash the Nuts very clean, the rost then to get the huls off, then roul them on the stone till

tis a fine past. Yu must mix a few of the best huls with it, to make it froth. [22]

The Randolph recipe provides further confirmation that the hand-grinding process occurred in Virginia households. It also provides insight and recognition regarding the use of the cacao bean shells. While modern manufacturers turn such shells into mulch or discard them directly, 18th century chocolate makers used them to make a type of tea (i.e., shell tea).

Once the nibs are winnowed, and extraneous materials removed, they are crushed using a mortar and pestle. Again, young children visiting Colonial Williamsburg can be drafted to pound away (Fig. 48.5). Next comes the grinding. Most today would be familiar with what happens to coffee beans when ground: They turn to powder. Cocoa nibs, in contrast, liquefy. This difference is because the fat content in the cocoa nibs (the cocoa butter) approaches 50 percent or more. Cocoa butter melts at human body temperature, which is why chocolate chunks disappear and melt in one's mouth. Essentially, cocoa butter has no chocolate flavor and becomes clear oil when melted. The white "chocolate" mochas of 21st century upscale coffee bars are made from cocoa butter, milk, and sugar—not with chocolate.

The flavor of chocolate comes from the dark cocoa solids. The essence of grinding is to make the particles of cocoa solids as small as possible while suspended in the liquid cocoa butter: the smaller the particle size, the more chocolate flavor is released and the smoother the sensation on consumers' tongues

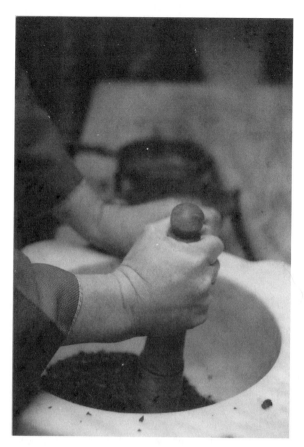

FIGURE 48.5. Crushing nibs. *Source*: Courtesy of The Colonial Williamsburg Foundation, Williamsburg, VA. (Used with permission.) (See color insert.)

FIGURE 48.6. Grinding cacao beans. *Source*: Courtesy of The Colonial Williamsburg Foundation, Williamsburg, VA. (Used with permission.) (See color insert.)

(Fig. 48.6). The more finely the cocoa is ground, the more liquid the chocolate becomes. Chocolate is naturally gritty, but prolonged grinding can reduce particle size to where they are imperceptible by taste. Modern chocolate makers usually grind nibs to a size somewhere between 12 and 40 microns with very little variation throughout the mass. Our ancestors' chocolate, in contrast, would have been grittier than what consumers would be used to today. The degree of polish of the grinding surfaces was a major reason, along with the difference between human muscle power versus horsepower (four-legged or otherwise). Hand-ground chocolate sometimes can achieve the 12–40 micron range, but commonly it will have noticeable "chunks" in various sizes.[1]

As the dry nibs are ground on the stone, the mass starts to thicken and glossy specks of cocoa butter become evident. As the grinding continues, the cocoa butter becomes smoother and the mass wetter. With more grinding, the mass eventually liquefies. The chocolate maker knows that "enough is enough" by the viscosity of the liquid and by taste. The semiliquid chocolate then is scraped off the surface of the grinding stone and put aside, more dry nibs are added to the stone surface, and the process is repeated. The stone's surface area is only large enough to make a few

ounces of chocolate at a time. When the small batches are all ground and liquefied, the whole mass is set on the stone at once and sugar and spices are added. A chocolate stone the size that Brookes described could accommodate about three pounds of chocolate at a time.

The back and forth grinding continues but with the added purpose of mixing the other ingredients into the chocolate mass. The amount of sugar added is by taste but is generally about the same as the starting weight of the raw cocoa. Cocoa loses moisture in the roasting. The shells account for 10–15 percent of the original weight. Add to that spillage during winnowing and grinding, and the amount lost represents approximately 20 to 50 percent of the starting weight. If the process is begun using 3 pounds of raw cocoa then the yield will be $1\frac{1}{2}$ pounds of unsweetened chocolate. The complete process, of course, is labor intensive. Depending on the number of people involved in the shelling, a pound of chocolate could take an hour to make. Working from sunup to sundown, about 10–15 pounds a day might have been possible in the 18th century.

The one main difference between European and American chocolate making and the ancient Maya was the use of iron rollers. A Colonial era polished-iron roller weighed about 8–10 pounds. The Maya used *manos* or lighter rollers made of stone that could be broken easily if dropped. Another difference was that, in Latin America, *metates* and chocolate making were the primary domain of women—who completed the grinding work while on their knees. In contrast, European chocolate grinders were usually men who commonly worked while standing. In Williamsburg, we can presume that the work was done by anonymous enslaved male or female adults or children, probably overseen by the mistress of the house. There are no records to document whether or not the Colonial Williamsburg chocolate grinders stood or kneeled.

Is Making Chocolate Worth the Effort?

The present discussion is intended to help the staff of other museums to decide if it is worth the effort and expense to add a chocolate-making program. There are a number of factors that must be considered including staffing, material costs, administration, site appropriateness, and interpretive goals. The expenses of adding chocolate making will vary depending on the individual museum along with frequency of the program. While research and staff resources are obviously the highest recurring expenses, they are budgeted annually and beyond the scope of this chapter. But it should be noted that, in the 21st century museum setting, to make 3 pounds of chocolate generally takes 5–6 hours using the labor of three skilled interpreters.

A chocolate program also has special requirements relating to equipment and cocoa beans. The *metate* or saddle quern will be one of the largest investments. From experience, these stones are heavy but quite fragile. If only one stone is purchased or manufactured and it is dropped, then you are out of business. A superior plan, therefore, would be to have two. *Metates* can be found in Mexico and Central America, elsewhere in the North American Southwest and California, or on the Internet, and most are quite inexpensive. Even with transportation costs, it would be worthwhile to obtain two because of breakage possibilities. The total expense of two *metates* should not be more than a few hundred dollars including shipping. Saddle querns, on the other hand, are not available in Mexico and even antique examples in museums are rare. The option here would be to have a stonemason fashion one out of granite.

While stonemasons can be found on the Internet, an order would have to specify the type and origin of the stone. The specifications provided by Brookes in 1730 are adequate for a modern stonemason to envision and fashion a saddle quern. Primary sources revealed that a North Carolina chocolate stone had been imported to Virginia: This is why Colonial Williamsburg specified that its chocolate stones be made of granite from that state. Two were produced to 18th century specifications for $1700, excluding shipping costs. Such a saddle quern also needs a metal or wooden frame constructed for support. Since no period example exists, the blacksmith's frame interpretation has to stay within the historical style. A metal brazing pan also needs to be fashioned or procured. Lastly, a saddle quern, if made as described by Brookes, will weigh close to 100 pounds not including the frame. *Metates* weigh half that. The saddle stone represents the single largest cost in creating a chocolate-making program, while the *metate* is significantly less expensive. Future research may eventually solve the saddle quern verses *metate* dilemma. As previously stated, Colonial Williamsburg's solution was to have both styles and explain the reason to the visitor.

The metal roller is a fairly easy item for an artisan blacksmith to produce and would be relatively inexpensive. If not, it is possible that some round metal bar stock might work if it is polished.

The next expense is the cocoa beans. Eighteenth century sites should be using either Ecuadorian or Venezuelan beans while 19th century sites could include varietals from Trinidad. A few phone calls to a favorite chocolate manufacturer might yield a free donation of cocoa beans, but do not count on it. The major expense here would be the shipping costs, which are significantly higher than the cost of the beans. The quantity of beans needed, of course, will be determined by the frequency of the program. At Colonial Williamsburg, 2–3 pounds of raw cocoa beans are used for the program (once a month/ ten times a year). Special programs, on the other hand, increase the bean requirement substantially and exceed the cost of regular program needs.

The final cost that should be considered would be a chocolate pot with a chocolate mill or *molinillo* for demonstrating the making of the beverage. For some demonstration programs this might be the *only* expense if the program is about chocolate consumption instead of chocolate making (Fig. 48.7). However, a chocolate pot is optional since chocolate can be made in a pan. Birch whisks can be used to produce the foam if a chocolate mill is not available.

In total, the cost of establishing a historic chocolate-making program should be no more than a few thousand dollars; whereas a chocolate consumption program would require much less funding. In today's museum world of ever-shrinking budgets, two or three thousand dollars can be a considerable expense. However, the popularity of chocolate may provide some relief in this area.

Our program at Colonial Williamsburg, entitled "Secrets of the Chocolate Maker," was the first regularly scheduled historic chocolate-making program in the United States. Accordingly, it attracted media

FIGURE 48.7. Pouring chocolate. *Source*: Courtesy of The Colonial Williamsburg Foundation, Williamsburg, VA. (Used with permission.) (See color insert.)

attention and has been featured in newspapers, magazines, and on the Internet and television. The value of this free advertising alone has more than paid for the program several times over. Additionally, this exposure resulted in receiving support from Masterfoods/Mars Corporation, one of the world's largest and most successful chocolate makers. Members of the Mars Corporation became interested in the program after reading a description of it in a newspaper article. In 2004, Masterfoods/Mars, Incorporated, Colonial Williamsburg, Fort Ticonderoga, and the University of California, Davis, Department of Nutrition, formed a partnership informally known as the "Colonial Chocolate Society," with the objective to further the research of Colonial era chocolate use in America. In 2006, the Mars Foundation also generously endowed Colonial Williamsburg's chocolate education efforts with financial resources to support the program.

Unfortunately, most museums cannot count on a major company to fund their chocolate-making program. However, there is another alternative. One of the offshoots of the Colonial Chocolate Society was the creation of the Historic Division of Mars, organized to create an authentic historic chocolate product. In February 2006, Mars launched their product line called American Heritage Chocolate (see Chapter 49). Cocoa scientists and food technicians consulted with the various members of the Colonial Chocolate Society including Colonial Williamsburg foodways staff and used a wide variety of period recipes to develop and manufacture the product. American Heritage Chocolate is made from cocoa beans that were available in the period and spiced according to period recipes. Most importantly, it retains the slightly gritty texture of hand-ground chocolate. Its appearance is authentic and virtually identical to untempered hand-ground chocolate. Sales of this product might be used to offset the cost of the historic chocolate program, as American

Heritage Chocolate has achieved significant sales in historic era stores. For Colonial Williamsburg, the sales figures in one year have exceeded the total costs of creating and running the program for the last six years.

One of the problems facing historic foodways programs in the United States is that state and federal health codes prevent the sale of food products produced in a historic museum setting. There is little chance that the chocolate being produced by a museum would be approved for sale to the general public. Technically, these same health laws prevent anyone other than the immediate staff from even tasting it. The American Heritage Chocolate product—in contrast—provides an authentic alternative that can be sold in museum gift shops, thereby offsetting the costs of the program and potentially providing a profit to the museum.

Cost alone, of course, is not the only consideration in determining whether a chocolate program is right for a historic site. There also needs to be both a historical and educational justification for the program. Obviously, research is required at each museum to determine the validity of chocolate making and consumption on that site. Colonial era museums that reflect city life, or are near or on the coast, have the most flexibility in determining whether to make chocolate on site or to merely consume it. In these urban or coastal areas, Colonial era consumers had access to raw cocoa beans as well as manufactured chocolate. While research has established that Boston, Philadelphia, New York, and Newport, Rhode Island, were the principal colonial manufacturing areas, Boston emerged from the Revolution with overwhelming dominance on the East Coast. Chocolate, however, also was made in New Orleans [23], in California missions [24], and in San Francisco during the Gold Rush [25]. Lewis and Clark consumed chocolate in Missouri in 1806 [26]. Therefore, the question each site must determine is whether or not the chocolate was handmade on site, or machine-made and purchased from a merchant. If it was from a merchant, where was the chocolate made and by whom? In the 19th century, North American residents relied on manufactured chocolate rather than grinding their own. Therefore, a small farm museum set in the 1840s Ohio Valley would have a much more difficult time finding historical precedents for hand-grinding chocolate.

On the other hand, there may be unique opportunities to create chocolate-making demonstrations at historic mill sites, whether 18th or 19th century. Working mill sites might be able to follow the example of period chocolate makers and add chocolate to the list of products produced at their mill. There are numerous references to mills producing mustard, chocolate, oils, or snuff tobacco all at the same mill. Obviously, research is needed on what types of stones were used along with other machinery such as roasters, kettles, crackers, and fanners. Additionally, molds would have to be researched and procured. While the costs of this would seem excessive, the potential for sale of this chocolate might justify the expense if modern health codes are followed.

Conclusion

When considering whether or not to implement a chocolate-related program, the most important factor is its educational value. The program developed at Colonial Williamsburg is "family friendly" with numerous opportunities to involve guests of all ages in the various processes. Interpreters can discuss family issues and children's roles in the period, and provide young people opportunities for hands-on learning while at the same time participating in a task necessary for chocolate production. Some visitors are entranced by watching the fire and smelling the chocolate aroma. This aroma is not confined to the kitchen and serves as "advertising" to those passing by.

The educational value of making chocolate is not exclusive to the guests, but to the staff as well. Chocolate is one of the most complex foods anyone can make, so achieving a consistent result from month to month is a challenge. Holding an audience, on the other hand, has never been a challenge (although finding space for all the guests who wish to view the process has been!). Creating a historic chocolate demonstration at historical or museum sites initially requires a moderate monetary investment and time for research, but the efforts can be rewarding both interpretively and financially. Plus—the process is extremely enjoyable!

Acknowledgments

We would like to acknowledge the following people whose cumulative advice, guidance, and support were instrumental in the writing of this chapter. The Colonial Williamsburg Foundation—Barbara Scherer, Historic Foodways Division; Robert Brantley, Historic Foodways Division; Dennis Cotner, Historic Foodways Division; Susan Holler, Historic Foodways Division; Jay Gaynor, Director of Trades; Gayle Greve, Curator of Special Collections Library; Juleigh Clarke, Rockefeller Library; Pat Gibbs, Research; Martha Katz-Hyman, Research; Donna Seale, Historic Foodways Division; Nicole Nelson, Historic Foodways Division; and Kristi Engle, Historic Foodways Division; University of California, Davis—Deanna Pucciarelli, Department of Nutrition; Fort Ticonderoga—Nick Westbrook and Virginia Westbrook; Masterfoods USA/Mars, Incorporated—Deborah Mars, Forrest Mars, Mary Myers, Harold Schmitz, Rodney Snyder, Eric Whitacre, Judy

Whitacre, and Doug Valkenburg; Oxford, England—Pamela Richardson; and family members—Bill Gay, Jan Gay, and Andrea West.

Endnote

1. An unsweetened hand-crafted chocolate sample measured by Masterfoods cocoa scientists yielded Coulter results with a mean of 19.43 microns on particle size distribution and micrometer readings of 86, 83, 78, 96, and 88 microns. A sweetened chocolate sample had Coulter results with a mean of 28.92 microns and micrometer readings of 107, 110, 112, 100, 103 microns.

References

1. Colonial Williamsburg Trades. Available at http://www.history.org/Almanack/life/trades/tradehdr.cfm. (Accessed January 15, 2007.)

2. Tannahill, R. *Food in History*. New York: Crown Trade Paperbacks, 1988; p. xv.

3. Hall, G. D., Tarka, Jr., S. M., Hurst, W. J., Stuart, D., and Adams, R. E. W. Cacao residues in ancient Maya vessels from Rio Azul, Guatemala. *American Antiquity* 1990;55(1):138.

4. Ledger of Imports and Exports (American); January 5, 1768–January 5, 1773, PRO/Customs 16/1, M-532 at the Rockefeller Library, The Colonial Williamsburg Foundation.

5. Tradesmen in the Virginia Gazette, The Colonial Williamsburg Foundation. Available at http://research.history.org/JDRLibrary/Manville/TradesProject.cfm. (Accessed January 16, 2007.)

6. *York County Wills & Inventories, Book 20*, 1745–1759, pp. 204–208.

7. *York County Records, Orders, Wills, & Etc., Book 16*, 1730–1729, p. 509.

8. Orange County Will Book No. 1, 1735–1743, The Governor's Palace Historical Notes, The Colonial Williamsburg Foundation. Available at http://pastportal.org/cwdl_new/archive/research%20reports/html/rr0219.htm. (Accessed January 16, 2007.)

9. Bear, Jr., J. A., and Stanton, L. C., editors. *Jefferson's Memorandum Books: Accounts, with Legal Records and Miscellany, 1767–1826,* Volume I. Princeton, NJ: Princeton University Press, 1997; p. 495.

10. Coe, S. D., and Coe, M. D. *The True History of Chocolate*. New York: Thames and Hudson, 1996; p. 187.

11. Colonial Port Commodity Import and Export Records for Virginia (Accomac, Rappahannock, South Potomac), Jamaica, New York, and South Carolina (1680–1769), M-1892 1–9 at the Rockefeller Library, *The Colonial Williamsburg Foundation.*

12. Presilla, M. *The New Taste of Chocolate: A Cultural and Natural History of Cacao with Recipes*. Berkeley, CA: Ten Speed Press, 2001; p. 27.

13. Emmanuel, I. S., and Emmanuel, S. A. *History of the Jews of the Netherlands Antilles*, First Volume, *Royal Vangorcum, LTD*. Cincinnati, OH: American Jewish Archives, 1970, p. 216.

14. For Carter, see *The Virginia Magazine of History and Biography, Volume VI*. Richmond, VA: The Virginia Historical Society, 1899; pp. 395–416.

15. For Lee, see Gunston Hall Plantation Probate Inventory Database. Available at http://www.gunstonhall.org/probate/LEE76.PDF. (Accessed January 16, 2007.)

16. Thayer, W. University of Chicago. Available at http://penelope.uchicago.edu/Thayer/E/Roman/Texts/secondary/SMIGRA*/Mola.html. (Accessed January 17, 2007.)

17. Brookes, R. *The Natural History of Chocolate*. London, England: R. Roberts, printer 1730; p. 36.

18. Colonial Port Commodity Import and Export Records for Virginia (Accomac, Rappahannock, South Potomac), Jamaica, New York, and South Carolina (1680–1769), M-1892 1–9 at the Rockefeller Library, The Colonial Williamsburg Foundation.

19. Campbell, R. *The London Tradesman*. London, England: T. Gardner, printer, 1747, page number unreadable.

20. Brookes, R. *The Natural History of Chocolate*. London, England: R. Roberts (printer), 1730; p. 35.

21. Brookes, R. *The Natural History of Chocolate*. London, England: Roberts (printer), 1730; p. 34.

22. Harbury, K. E. *Colonial Virginians Cooking Dynasty*. Columbia: University of South Carolina Press, 2004; p. 402.

23. Hall, G. M., compiler. *Afro-Louisiana History and Genealogy, 1718–1820 (Slave)* [database on-line]. Provo, UT: The Generations Network, Inc., 2003. Accessed at Ancestry.com at http://search.ancestry.com/cgi-bin/sse.dll?db=AfroLASlave&so=2&rank=1&gskw=chocolate+maker&gs2co=1%2cAll+Countries&gs2pl=1&sbo=0&srchb=p&prox=1&ti=0&ti.si=0&gss=angs-b&o_iid=21416&o_lid=21416&o_it=21416. (Accessed February 21, 2007.)

24. Linse, B. Book review of *Live Again Our Mission Past*. *California Mission Studies Association*, August 13, 2003. Available at http://www.ca-missions.org/linse.html. (Accessed February 21, 2007.)

25. Ghirardelli History Page. Available at http://www.ghirardelli.com/chocopedia/history.aspx. (Accessed February 21, 2007.)

26. University of Nebraska Press / University Of Nebraska–Lincoln Libraries-Electronic Text Center (March 2005). "The Journals of the Lewis and Clark Expedition." Available at http://lewisandclarkjournals.unl.edu/hilight.php?id=1082&keyword=chocolate&keyword2=&keyword3=. (Accessed February 21, 2007.)

49

American Heritage Chocolate

Eric Whitacre, William Bellody, and Rodney Snyder

Introduction

Foodways is an anthropological term describing the eating habits and culinary practices of a people, region, or historic period. When researching human culture and history, few things convey more information about life and times than foods consumed. Food patterns reveal the consumer's origin, ethnicity, religion, affluence or poverty, political orientation, and general lifestyle. The foodways associated with a region or people also influenced the type of businesses and support structures that developed. The need for grains and meat gave birth to millers and butchers, blacksmiths and coopers, masons and lumbermen, all progressing forward as societies moved from primitive hunter/gatherers ultimately to businesspeople involved with trade and commerce. The study of 18th century chocolate makers and their goods, recipes, and processes supports this notion of anthropological progression, as time-honored traditions, values, and craftsmanship were challenged by slowly emerging technologies, religious temperaments, and survival at the interface with aboriginal peoples during early contact and the Colonial era. Certainly, this evolution and progression can be seen when studying the works of the 18th century author Diderot, especially the progression from simple to complex manufacturing techniques from earliest times into the 21st century [1, 2]. It is within this context that we started our research that led to the production of *American Heritage Chocolate*®, using knowledge of modern manufacturing techniques but with an understanding of 18th century processes and limitations.

Background Research

The first portion of our work was to investigate and research 17th and 18th century food processing techniques. To this end, Denis Diderot proved invaluable as a resource. Many of Diderot's descriptions in his *Confiseur* were related to medium-sized or larger-scale chocolate production houses of the 18th century (Fig. 49.1). We also needed to understand aspects of 18th century non commercial production, as these may have been more widely practiced by smaller "home-style" businesses, such as those that began early production of chocolate in the American colonies.

The development phase of American Heritage Chocolate, therefore, stemmed from two basic questions:

1. How was chocolate manufactured or processed in the 18th century?

2. What formulas, ingredients, and recipes were used during the 18th century?

The manufacturing question could be answered through descriptions typical of the era, as revealed in

Chocolate: History, Culture, and Heritage. Edited by Grivetti and Shapiro
Copyright © 2009 John Wiley & Sons, Inc.

FIGURE 49.1. Artisans making chocolate, dated 1751. *Source*: "Confiseurs." In: Artisanats au 18ème Siècle, *l'Encyclopédie* Diderot et d'Alembert. Courtesy of the Peter J. Shields Library, University of California, Davis. (Used with permission.)

the following three citations. Diderot's *Confiseur*, Plate V, 1762, provides the following description:

The chocolate, which has a reddish brown colour, is a hard dry paste, quite dense, made into the form of a loaf or roll. To make it, one must have the best cacao, a roaster, and to separate from the almonds (cocoa-nuts) all that is not meat [i. e., shells and detritus] to scrutinize them carefully to remove all those that are damaged or rancid, to roast them over and over again, to put them in a hot mortar, to break them and crush them on a hard stone, to work them into a sweet paste, then throw in some sugar; this will be called superb chocolate. The other chocolate is vanilla. For vanilla chocolate, one takes 4 pounds of superb chocolate which one puts on a hard slab. Three pounds of sugar are incorporated by means of a rolling pin, eighteen dried vanilla pods are ground up with a dram and a half of cinnamon, eight cloves, and two grains of ambergris if desired. All of these are well mixed together, and made into tablets or cakes, and dried in a drying stove on a piece of white paper. The chocolate must be very soft before adding the other ingredients, and the vanilla mixture should not touch the hot slab. [3]

Accompanying Diderot's text is an engraving with four separate workmen that illustrate the chocolate-making process:

1: Workman who tans [*sic*] or roasts the cocoa in an iron kettle, over a fire;

2: Workman who winnows [*sic*] the almonds (cocoa-nuts);

3: Workman who piles [*sic*] them into an iron mortar, which has already been heated and under which is held a fire; and

4: Workman who grinds [*sic*] the chocolate on a hard, heated stone, with an iron rolling pin. [4]

Another historical perspective on chocolate production from the late 18th century appeared in the 1791 edition of the *Encyclopaedia Britannica*:

The Indians, in their first making of chocolate, used to roast the cacao in earthen pots; and having afterwards cleared it of the husks, and bruised it between two stones, they made it into cakes with their hands. The Spanish improved this method; when the cacao is properly roasted, and well cleaned, they pound it in a mortar, to reduce it into a coarse mass which they afterwards grind on a stone, till it be of the most fineness; the paste being sufficiently ground, it was put quite hot into tin moulds, in which it congeals in a very little time. The form of these moulds is arbitrary; the cylindrical ones, holding two or three pounds, are the most proper, because the bigger the cakes are, the longer they keep. Observe that these cakes are very liable to take any good or bad scent, and therefore they must be carefully wrapt [sic] up in paper, and kept in a dry place. Complaints are made that the Spanish mix with the cacao nuts with too great a quantity of cloves and cinnamon, besides other drugs without number, as deer musk ambergris, and etc. The grocers of Paris use few or none of these ingredients; they only chuse [sic] the best nuts, which they are called Caracca, from the place from whence they are bought, and with these they mix a very small quantity of cinnamon, the freshest vanilla and the finest sugar, but very seldom any cloves. Among us in England, the chocolate is made of the simple cacao, excepting that sometimes sugar, and sometimes vanilla is added. [5]

A third useful document, this one dated to the early 19th century, offered still another explanation regarding the production of chocolate.

The manner of preparing chocolate for use is regularly thus— Gently parch the cacao-nuts in an iron vessel over a slow fire to facilitate the taking off of their exterior shells, the future coco, which would be injured or destroyed by too much heat; then bruise and work the kernels into a paste on a smooth concave stone, with a moderate charcoal fire beneath; occasionally adding a little water, and a small quantity of water, and a small quantity of sugar, vanilla, and Spanish annatto. As soon as the paste is sufficiently fine and smooth, put it quite hot into tin moulds where it will speedily congeal and become hard cakes similar to those usually sold. . . . Genuine chocolate, when made in the usual manner, by slicing it small with a knife, boiling it in a proper chocolate pot, well milling and frothing it as poured into the cups, sweetening it and softening it with cream forms a most nourishing and agreeable food for valetudinarians.[1] [6]

Regarding the second question, what formulas, ingredients, and recipes were used during the 18th century, the information available to answer this question is abundant (see Chapter 8). Many of the recipes and ingredients used during the development stages of making American Heritage Chocolate came from 17th and 18th century references and cookery books. In his *Arte de repostería* (*Art of Confectionery*), published in 1747, Juan de la Mata offered a relatively simple recipe for drinking chocolate in which the quantity of sugar should equal the quantity of cacao, with the only additional flavors being cinnamon, orange water, and vanilla [7].

Hannah Glass, in *The Art of Cookery Made Plain and Easy*, published in 1796, offered two clear, non technical descriptions of how to make chocolate:

Recipe 1:

Six Pounds of cocoa-nuts, one of anise-seeds, four ounces of long pepper, one of cinnamon, a quarter of a pound of almonds, one ounce of pistachios, as much achiote as will make it the colour of brick, three grains of musk, and as much ambergris, six pounds of loaf-sugar, one ounce of nutmegs, dry and beat them, and searce [i.e., sift] them through a fine sieve; your almonds must be beat to a paste and mixed with the other ingredients; then dip your sugar in orange-flower or rose water, and put it in a skillet on a very gentle charcoal fire; then put in the spice and stew it well together, then the musk and ambergris, then put in the cocoa-nuts last of all, then achiote, wetting it with the water was dipt in [sic]; stew all these very well together over a hotter fire than before; then take it up and put it into boxes, or what form you like, and set it to dry in a warm place: the pistachios and almonds must be a little beat in a mortar, then ground on a stone. [8]

Recipe 2:

Take six pounds of the best Spanish nuts, when parched and cleaned from the hulls, take three pounds of sugar, two ounces of the best beaten and sifted very fine; to every two pounds of nuts put in three good vanillas or more or less as you please; to every pound of nuts half a drachm[2] *of cardamom-seeds, very finely beaten and searced [sifted].* [9]

It was from these and other resources that we started preparing lists of ingredients, spices, and processing equipment common to 18th century chocolate making. We then turned to the consideration of process, and how different tools and their composition might produce large variations in flavor and consistency that in turn would lead to varying degrees of quality in the final product. Having reviewed a very broad range of resource materials (see Chapters 46 and 48), we sum-

marized the information and drew two basic conclusions.

First, chocolate was stored as two- to three-pound cakes or rolls for later use by consumers and was never intentionally tempered.[3] A crude form of tempering may have occurred inadvertently during the process of removing the mass of heated chocolate paste from grinding stones and placing it into cooler bowls or dishes, but tempering was not a consistent chocolate manufacturing process until the 19th century. Second, we concluded that many of the adaptations to 16th–19th century English, European, and North American recipes were made to accommodate local tastes and the types of flavoring agents and spices available to 18th century North American chocolate makers. Spices, both Old and New World (see Chapter 8, Tables 8.1 and 8.2) were commonly available, especially in coastal port cities. Common spices and ingredients mixed with chocolate in early North America included aniseed, cassonade,[4] cinnamon, cloves, nutmeg, orange or rose water, and salt. Less common in the Colonies were ambergris, chilies, deer musk, pepper (long pepper), and vanilla. Because many merchants who made chocolate also used their equipment to grind other grains and spices, whether barley, ginger, mustard, wheat, or even tobacco, incident or trace levels of these adulterations commonly found their way into the chocolate paste (see Chapter 47).

The next step in the process was the translation of 18th century recipes and processes into 21st century production. Our research up to this point had identified a range of methods and ingredients used in the manufacturing of chocolate. This information provided insights regarding how 18th century chocolate tasted, and its texture and "look" at the time of purchase and during drinking.

Our initial goal, therefore, was to develop a recipe that would be an integrated form or culmination of the many, more prevalent recipes researched and identified from the 18th century. This process involved not taking a single recipe, but rather testing and incorporating several together, using the more commonly available spices and ingredients in order to fully capture the diversity of the many geographic regions and cultural temperaments of the original American Colonies. An emphasis on recipe accuracy was critical during this formulation testing stage. And while recipe accuracy was critical, the "end-product" chocolate also had to taste good in order to be commercially viable in the marketplace. Only after a final recipe was developed, could we concentrate on the type of equipment needed to scale up the process for commercial production.

Findings from the recipe and ingredient research provided us with mental images of the product. The chocolate should be low in relative sweetness (similar to 21st century bittersweet or semi sweet chocolates), should contain several common spices (selections from aniseed, cinnamon, nutmeg, orange powder/orange water, and perhaps traces of chili pepper), and just a hint of vanilla. Furthermore, the finished product should be rough and crude in texture and appearance in order to match that commonly available during early North American history.

In the modern chocolate production environment, the product is milled or refined to very small tolerance range: Most chocolate today is milled to the 15–35 micron range for the largest particles, while most of the particle distribution would be within the single digit or submicron range. When milling chocolate on an 18th century chocolate stone, however, the smallest particles achievable typically would be in the 20–50 micron range for the smallest particles, with most particle sizes being within the 100–200 micron range. Such crude refining or milling would give 18th century chocolate a grittiness or "rough mouth feel." Additionally, many of the spices available and used during the 18th century would exhibit higher moisture content due to primitive drying practices, with the net result being they could not be milled as finely as spices in the 21st century.

With these facts in consideration, a process would have to be developed that would mill the chocolate paste and spices to higher particle size values and limit the amount of mixing or conching.[5] Finally, a new technique would have to be developed to simulate the incidental tempering that may have occurred during the folding and mixing of the chocolate paste when passing from a hot stone to a cold plate/dish, and subsequent mixing when the cakes of chocolate were used for other applications, such as baking.

Modern chocolate industry processes for tempering involve several steps that would not have been followed when making early American chocolate:

1. Heating the chocolate beyond its melting point of ca. 100 °F.

2. Cooling the chocolate slowly to a point where seed crystals formed (usually ca. 80 °F).[6]

3. Reheating the chocolate slowly to ca. 90 °F.[7]

4. When proper tempering has been completed, the chocolate should exhibit high gloss, a nice, even snap (when broken), and smooth texture.

For an 18th century inferred method of tempering, none of the above attributes or conditions would have been met, but a commercially stable product would have been desired by manufacturers. Developing a tempering process for a modern product—while still maintaining historical accuracy—required a careful balance. Furthermore, the resulting product needed to have a stable shelf life—but also should appear crude or coarse like 18th century chocolate.

The tempering process that was developed for American Heritage Chocolate involved only the first step of a normal tempering regime, which was to fully melt the chocolate. The other process steps developed were unique and resulted in a rapidly cooled chocolate paste under high shear (or tearing/mixing), creating nonuniform seed crystals. While these crystals were relatively stable, they did not exhibit uniform crystal patterns. This created a stable product that exhibited poor gloss and "snap" and a rough texture. At the same time the product had a sufficiently high melting point as to be stable in the humidity and heat of 21st century East Coast summers—while not exhibiting any waxy or disagreeable "mouth feel" during consumption.

To further create the look of 18th century texture on the finished chocolate, a spicy dusting powder was hand applied, then brushed off. This final process complemented the overall appearance, while maintaining the desired look and feel of the historical time period. In addition to this process, a unique extrusion type process was developed to produce crude or irregularly shaped rolls and blocks.

Conclusion

The processes used to develop and manufacture American Heritage Chocolate involved identification and evaluation of authentic 18th century recipes, considerable testing and hand batching, development of rough or incomplete milling processes, creation of a unique tempering and crude extrusion process, product hand dusting, and, eventually, the development of authentic looking 18th century packaging that met 21st century U.S. federal guidelines for product safety. Other team members identified and provided accurate historical texts and background information ultimately incorporated into product inserts and advertising copy.

Acknowledgments

We would like to acknowledge the following individuals whose cumulative advice, guidance, and support have been instrumental in the writing of this chapter. From Colonial Williamsburg Foundation: Frank Clark, Historic Foodways Division; and Jim Gay, Historic Foodways Division; from Fort Ticonderoga: Nick Westbrook and Virginia Westbrook: and from Masterfoods USA/Mars, Incorporated: Deborah Mars and Mary Myers.

Endnotes

1. Invalids or persons excessively concerned with their health.

2. Apothecary weight = 1 dram or 1/8th of an ounce = 60 grains.

3. Tempering is a thermal treatment used on finished or semi finished chocolate to improve its texture and appearance.

4. Raw, unrefined sugar.

5. Conching is a grinding/heating process developed in the 19th century and used to enhance the flavor and "roundness" or balance of the finished chocolate.

6. Cocoa butter, the primary fat found in the cacao bean, is polymorphic in nature, which means it can form numerous crystal growth patterns, based on thermal conditioning. Each crystal pattern has unique kinetics relative to heat of enthalpy, latent heat, and melting point.

7. The slow warming process during tempering eliminates (i.e., melts out) some of the unstable crystal patterns that cause poor stability in the final product.

References

1. Diderot, D. *Diderot/ D'Alembert L'Encyclopedie, Confiseur*, Volume I, Plate V, Paris: Andre le Breton, 1762.

2. Minifie, B. W. *Chocolate, Cocoa, and Confectionery*. New York: Van Nostrand Reinhold, 1989.

3. Diderot, D.*Confiseur*. Reproduced from an original folio by Diderot and printed as a facsimile. London, England: J.G. Kennedy and Company, 1975; No. 10, Plate V, p. 124.

4. Diderot, D. *Confiseur*. London, England: J.G. Kennedy and Company, 1975; No. 10, Plate V, p. 124.

5. *Encyclopaedia Britannica*, Volume 2. Edinburgh, Scotland: Colin MacFarquhar, 1771; p. 193.

6. Anonymous. *The Family Receipt-Book or useful Repository of Useful Knowledge and Experiences in all the various Branches of DOMISTIC ECONOMY, various editors*. London, England: Oddy and Company, 1808; pp. 99–104.

7. de la Mata, J. *Arte de repostería (Art of Confectionery)*. Madrid: Antonio Martin, 1747.

8. Glass, H. *The Art of Cookery Made Plain and Easy*, revised edition of 1796. Schenectady, NY: United States Historical Research Service, 1994 (reprint); p. 341.

9. Glass, H. *The Art of Cookery Made Plain and Easy*, revised edition of 1796. Schenectady, NY: United States Historical Research Service, 1994 (reprint); p. 342.

CHAPTER

Twenty-First Century Attitudes and Behaviors Regarding the Medicinal Use of Chocolate

Deanna Pucciarelli and James Barrett

Introduction

The thought of using chocolate, or more specifically flavanols, an inherent nutrient of cacao, to decrease risk of certain chronic diseases in the 21st century recently has gained credibility among nutritionists [1–4]. Certainly the contemporary discovery of the physiology and health implications of this phytochemical has renewed interest in using chocolate as a medicine [5–7]. The medicinal use of chocolate has a long history, however. The origin of using chocolate as a medicine has been traced to the second half of the 16th century [8]. Researchers have documented the medicinal use of chocolate in Central America during the 16th–19th centuries and its trajectory to Western Europe primarily through manuscripts, but also through societal artifacts such as diaries, paintings, and pottery.

With the initiation of germ theory and modern drug therapy during the late 19th century, the concept of food as medicine fell out of favor in the United States and would not become popular again until the discovery, in the first 50 years of the 20th century, of vitamin essentiality in curing specific nutrient deficiencies [9]. During that period the focus was on consuming certain foods to reverse nutrient deficiencies, whereas since the 1990s, with the problem of overnutrition manifested in the so-called obesity epi-

demic, the use of food as medicine has shifted to optimizing health and reducing chronic disease risks [10, 11].

Food as medicine currently is widely accepted in disease prevention and the movement has been termed "functional foods" [12]. The term "functional foods" has regulatory implications in Japan, where the term is thought to have originated. Conversely, the United States food labeling agency, The Food and Drug Administration (FDA), has yet to establish criteria for this ever-increasingly used term [13]. The nationally accepted definition created by the Institute of Medicine's Food and Nutrition Board regarding functional foods is: any food or food ingredient that may provide

Chocolate: History, Culture, and Heritage. Edited by Grivetti and Shapiro
Copyright © 2009 John Wiley & Sons, Inc.

a health benefit beyond the traditional nutrients it contains [14–16].

During the American Colonial era, the vasodilatation, heart-healthy properties of cocoa would not have been recognized. Nevertheless, chocolate played a prominent medicinal role in disease treatment, especially to counter cholera, smallpox, and yellow fever (see Chapter 6). Revisiting the century just prior to American colonization, the medicinal use of chocolate for disease prevention and cure was chronicled in the book by Sophie and Michael Coe, *The True History of Chocolate* [17]. Still, with few exceptions, little has been written of the medical use of chocolate during early North American history. Moreover, investigations designed to determine contemporary consumer understanding of the concept of chocolate as a functional food are lacking. Conversely, the broad category of consumer attitudes toward functional foods has received increasingly more attention [18].

The International Food Information Council Foundation (IFIC) in 2005 conducted its fourth survey to measure Americans' attitudes toward functional foods [19]. Foods identified in the survey that consumers thought provided health benefits beyond basic nutrition included broccoli, dietary supplements, fortified cereals and juices, omega-3 rich fish, and soy-based products among others [20, 21]. Moreover, in 2006, the IFIC commissioned the consumer research firm Cogent Research of Cambridge, Massachusetts, to conduct a 20-minute Web-based quantitative survey of 1012 adults, ages 18 and older, to measure consumer attitudes toward food, nutrition, and health. This benchmark survey was developed by considering census factors such as age, education, and ethnicity patterns in the United States so that findings could be extrapolated to the American public in general. The survey was designed to provide insights to the barriers or catalysts Americans perceived to be associated with acquiring or maintaining a healthful eating pattern [22]. Taken together, the surveys delineated Americans' views on health in relationship to eating specific foods, and to what extent respondents sought information on the nutritional properties of food.

Chocolate is viewed by some researchers as a food that provides health-promoting benefits [23–25]. The IFIC 2005 Functional Foods survey measured consumer awareness of the relationship between dark chocolate and protection against free radical damage.[1] Significantly, 79 percent of those surveyed were aware of the relationship between dark chocolate and free radical damage implicated in chronic disease [19]. The survey, however, did not investigate attitudes toward using chocolate as a functional food.

Objectives

Our study sought to discover consumers' *knowledge*, *attitudes*, and *behaviors* (KAB) toward chocolate consumption and health implications. Given that certain populations, for example, those residing within the Oaxaca region of Mexico, are known to continue to use chocolate medicinally [8], a goal of the present study was to measure Americans' (1) understanding of the relationship between chocolate consumption and disease attenuation, (2) attitudes toward utilizing chocolate medicinally, and (3) chocolate consumption patterns to treat illness. A second research objective was to determine whether or not the data gathered on the aforementioned research questions differed by age group, culture (as measured by self-described ethnicity), education level, and income. Approximately one-fourth of the survey questions were designed to measure consumers' knowledge regarding chocolate's reported health-promoting properties. However, we believed that by narrowing the focus to consumers' attitudes and behaviors (traits that have a reciprocal relationship) toward chocolate consumption, we could better elucidate salient findings that otherwise would be diluted.

Methods

This project was approved by the University of California, Davis Institutional Review Board (IRB) University Committee on Human Subjects (UCHS). Study participants gave informed consent in their native language (either English or Spanish).

QUESTIONNAIRE DEVELOPMENT PROCEDURES

Focus groups were formed in two "Current Topics and Controversies in Nutrition" (NUT11) classroom discussion sections at the University of California, Davis, campus (UCD). The 45-minute focus group interviews were conducted one week apart. Informed consent was received from 39 students. One student declined to participate and was given an optional reading assignment while the interviews took place. The survey research objectives were explained to the students and sample questions were recited to generate interest in the topic and understandability of the survey questions. Responses to students' attitudes and understanding toward the sample questions and overall interest to the project were recorded. The focus group interviewer encouraged participation from all interviewees and elicited students' opinions by calling on individuals who remained silent. Overall, students displayed interest in the topic, and a synergy developed within the two groups that resulted in a total of eight pages of extensive handwritten notes.

From these results a questionnaire was developed and translated into Spanish. The questionnaire was then back-translated into English using a second bilingual translator who had not seen the original English version [26]. Both translators had experience with the dialects and culture of the subpopulation of Mexican-Americans the survey was designed to reach. Spanish-version iterations were compared and discussed, resulting in a final copy. A bilingual translator then compared the Spanish and English copies to ensure compatible concepts transferred across both languages. The final copies were tested for readability set at the sixth-grade level of comprehension[2] [27].

A computer programmer was hired to format a Web version of the survey. All questions and IRB documents, respondents' Experimental Bill of Rights and Consent to Participate in a Research Study, were converted into PDF documents and uploaded into the SurveyMonkey® software program. The developed survey was then entered into an on-line survey computer program: SurveyMonkey.com® [28]. SurveyMonkey.com software allows identification numbers to be included, which the principal investigator had access to, but which were kept confidential and separate from those participating. Three of the four groups in the study utilized the on-line survey; the fourth group (Mexican-Americans) completed the paper version and the bilingual field assistant administering the survey entered the answers into the on-line software program. A second bilingual assistant checked data entries for accuracy.

SAMPLING PROCEDURES

The three broad variables of interest were what KAB Americans held in relation to chocolate consumption and health outcomes. The sample was stratified among students (nutrition and nonnutrition classes at University of California, Davis), Mexican-Americans (self-defined, residing in Madera, CA), and middle-class, middle-aged adults (residing in greater Sacramento, CA). Within the behavior (B) category we hypothesized that the Mexican-American subgroup would have an 8 percent greater medicinal chocolate usage over the other groups surveyed. Our conservative estimate was based on the 2002 National Health Interview Survey, whereby Mexican-Americans reported a 27 percent Complementary and Alternative Medicine use, due in part to the high cost of medical care [29]. A one-sided 0.05 level of significance and 0.80 level of power were incorporated into our sample size calculation. To achieve significance we calculated that 48 Mexican-Americans and 275 non-Mexican-Americans needed to be enrolled in the survey. We increased the estimated number of Mexican-Americans to 60 to account for project noncompliance and/or dropout.

POPULATION DESCRIPTION AND RECRUITMENT

The study population was divided into four subgroups: (1) Mexican-Americans, (2) UCD students enrolled in a basic nutrition course, (3) UCD students enrolled in a basic world geography course, and (4) middle-class,

middle-aged Americans. Recruitment for Mexican-Americans took place in Madera, California, a moderate-sized[3] central valley farming community with a significant number of Mexican-American citizens with origins from the Mexican state of Oaxaca. In 2000, the United States Census Bureau estimated a 67.8 percent Hispanic or other Latino population [30]. A bilingual field assistant recruited participants via verbal announcements spoken in Spanish at church gatherings, to customers at local, family-owned grocery stores, and at neighborhood Mexican restaurants. Inclusion requirements were self-reported completion of eight years of school as a proxy for sixth-grade literacy level, and a minimum age of 18.

Recruitment at the University of California, Davis, campus required instructor permission.[4] Announcements were made in Nutrition 10, an introductory nutrition class, and Geography 10, an introductory world geography course. With the instructors' permission, students were invited to participate in an on-line survey designed to measure KAB toward chocolate consumption and health implications. The instructor for Nutrition 10 offered students five bonus points toward their total class achievement score if they completed the survey; an alternative extra-credit assignment was offered to students who declined to participate. The instructor for Geography 10 encouraged his students to participate in the study via an announcement during class; no extra class credit was awarded for completion. Students participating were required to be at least 18 years old.

Middle-class, middle-aged participants were recruited from public sites in the greater Sacramento, California, area. Public announcements were generated at community centers, churches, and community-based civic organizations. Inclusion criteria for this subpopulation included a minimum age of 30 and a maximum age of 60 years. For the purpose of this study, self-reported middle-class was defined as a minimum income of $43,000 per two-person household[5] and a maximum income of $250,000 [31].

MEASURES

The survey consisted of closed-ended, multiple choice and opened-ended questions that asked about knowledge, attitudes, and behaviors toward chocolate consumption and health implications. The on-line survey was programmed so that completion of a question was required; the user was then automatically forwarded to the next question. To reduce response burden the survey allowed participates to select "I do not know" or "prefer not to answer" options [32]. The final survey consisted of 69 questions; six closed-ended questions were linked to additional open-ended, fill in the blank questions to further probe participant knowledge of nutrition concepts, or to gain deeper insight regarding eating behaviors.

A codebook was prepared and data were summarized. Data were exported from SurveyMonkey® and imported into Microsoft Excel 2003 program. The primary investigator reviewed the data entries to ensure that responses fell within the appropriate range. Blank or outlier cells were checked against the original survey and corrected.

DATA ANALYSIS

Responses to dependent measures (KAB toward chocolate consumption and health implications) were tallied and separated by subpopulations (nutrition students, nonnutrition students, Mexican-Americans, and middle-aged Americans) and by gender. Demographics, attitudes toward chocolate use and health outcomes, and chocolate consumption behaviors were analyzed within each group using χ^2 (chi-square) and t-tests (Student t-test), and between each group and the total study population to examine gender and group differences. Differences between students (nutrition students and nonnutrition students) and nonstudents (Mexican-Americans and middle-aged Americans) were also examined using bivariate tests. Responses were considered significant at $p < 0.05$.

Results

DEMOGRAPHICS

Demographic traits differed by group and age, but not by gender (Table 50.1). There was a two-thirds majority of females. Participants in the Mexican-American (Mex-Am) and middle-class, middle-aged (MMA) groups on average were 20 years older than the students. No significant differences were found in regard to student age (nutrition or nonnutrition) and nonstudents (Mex-Am or MMA). Likewise, reported Body Mass Index (BMI) differed between groups, with students' BMI being significantly lower than nonstudents'. The older nonstudents had a BMI that would classify them as overweight, while the BMI of the younger, student groups were within normal range[6] [33]. When the groups were collapsed between the two age cohorts, BMI was not significantly different (data not shown). Ninety percent of the middle-class, middle-aged Americans reported completion of at least two years of college, whereas Mexican-Americans reported a 36 percent completion rate.

The nutrition students reported a 90 percent completion of at least some college, while the nonnutrition students only reported a 46 percent completion rate. Seventy-five percent of the nutrition students were 20–22 years of age, and 18.3 percent were 18–19 years of age compared to 25.8 percent and 66.1 percent in the nonnutrition group, respectively (data not shown). This difference in age can be explained by the difficulty of registering for Nutrition 10, a highly sought-after class that continuously exceeds enrollment quotas. As a result, students commonly entered Nutrition 10 as juniors and seniors. Although age affected years of school completed, in terms of absolute years in age it was nonsignificant between the groups.

Ethnicity differed between groups. By study design Mexican-Americans were selected specifically for their Mexican heritage. Conversely, middle-class, middle-aged Americans were not. The University of California, Davis, attracts a large Asian student population and this is reflected in the Nutrition students (NUT) data results (45.0 percent). The geography class (non-NUT), however, reported a higher percentage (54.0 percent) of white/non-Hispanic than all other groups except the MMA group (92.5 percent).

ATTITUDES

The top three health concerns ranked highest within our study population were obesity/weight, heart/circulatory issues, and cancer. These three concerns mirrored the IFIC's benchmark study on *Consumer Attitudes Toward Food, Nutrition & Health* [22]. Whereas 34 percent of both the KAB and IFIC's participants' ranked obesity/weight as their top health concern in terms of absolute ranking (Fig. 50.1), KAB respondents considered obesity/overweight a greater concern than heart/circulatory issues, while IFIC respondents had the aforementioned health concerns ranked inversely.

Consistent with prior surveys [18], participants believed they had a moderate to great amount of control over their health (Table 50.2). Both Mexican-American and MMA males (2.88 ± 0.78 and 2.88 ± 0.34, respectively) reported a higher level of control over females (2.13 ± 0.86 and 2.71 ± 0.46, respectively) than both student groups; however, the differences were nonsignificant.

Seventy percent of participants either "strongly" or "somewhat" agreed that certain foods have health benefits that may reduce the risk of disease (Fig. 50.2). Overall, both female and male participants held similar beliefs that foods provided a role in disease prevention (2.66 ± 0.60 and 2.64 ± 0.63, male and female, respectively). By subgroup, however, Mexican-American females reported a significantly higher positive level of agreement ($p < 0.001$).

When asked specifically about consuming chocolate to *treat* heart disease, 54.4 percent of the female respondents responded positively to utilizing chocolate as a functional food versus 36.6 percent for males (Table 50.2). When subgroups were compared, Mexican-American males reported greater (88.8 percent) medicinal use of chocolate than females (86.1 percent). Compared to using chocolate as a good food source for treating heart disease, fish oil's positive association was reflected in our findings. Males (71.1 percent) more than females

Table 50.1 Demographic Characteristics of Mexican-Americans, Nutrition and Nonnutrition Students, and Middle-Class, Middle-Aged Americans

Characteristics	All Subjects (N = 616)		Mex-Am (N = 56)		NUT (N = 451)		Non-NUT (N = 68)		MMA (N = 40)	
Female	66.5%	N = 410	69.6%	N = 39	65.3%	N = 295	76.2%	N = 52	60.0%	N = 24
Male	33.5%	N = 206	30.4%	N = 17	34.7%	N = 156	23.8%	N = 16	40.0%	N = 16
Age (mean years)	23.5 ± 8.7		35.4 ± 11.6*		20.8 ± 1.7*		19.8 ± 4.6*		44.7 ± 13.7*	
Years in United States										
Six years or less	5.1%		9.3%		5.4%		3.2%		2.5%	
More than 6 years	94.9%		90.7%		94.6%		96.8%		97.5%	
Ethnic background										
White/non-Hispanic	35.9%		0.0%		33.2%		50.8%		92.5%	
Hispanic/Latino	13.8%		98.2%		5.3%		7.9%		2.5%	
Asian	37.1%		0.0%		45.0%		34.9%		2.5%	
Other	13.2%		1.8%		16.5%		6.3%		2.5%	
Education										
High school or less	19.1%		63.6%		9.6%		54.0%		10.0%	
Some college[a]	80.9%		36.4%		90.4%		46.0%		90.0%	
What is your family Income?[b]	604		53		449		62		40	
0–$20,000	10.9%		37.7%		8.9%		9.7%		0.0%	
$20,001–60,000	23.8%		37.7%		23.4%		21.0%		15.0%	
$60,001–100,000	19.2%		7.5%		18.5%		21.0%		40.0%	
$100,001–140,000	12.6%		3.8%		13.1%		9.7%		22.5%	
>$140,001	10.3%		0.0%		11.6%		9.7%		10.0%	
I prefer not to answer	23.2%		13.2%		24.5%		29.0%		12.5%	
Body Mass Index[c]	23.1 ± 4.7		26.8 ± 4.4*		22.3 ± 4.4*		22.7 ± 3.6*		26.9 ± 6.0	

*$p < 0.001$.

[a] Categorical answer: years of school completed either 12–14, 15–16, 17–19, 20–22, >22 years; some college equal to a minimum of 2-years completed.

[b] Not all participants answered this question resulting in a smaller N in three of four subpopulations.

[c] Wt and Ht were self-reported. BMI calculated: $Wt(m)/[Ht(kg)]^2$.

Table 50.2 **Attitudes Toward Chocolate, Disease, and Health**

Survey Questions	All Subjects (610)		Mex-Am (56)		MMA (40)		NUT (451)		Non-NUT (68)	
	Male	Female	Male	Female	Male	Female	Male	Female	Male	Female
	33.4% N = 204	66.6% N = 406	30.4% N = 17	69.6% N = 39	40.0% N = 16	60.0% N = 24	34.7% N = 156	65.3% N = 295	23.8% N = 16	76.2% N = 24
Q7. How much control would you say you have over your health? (0 = No control, 3 = A great amount)	2.48 ± 0.61	2.43 ± 0.65	2.88 ± 0.78	2.13 ± 0.86	2.88 ± 0.34	2.71 ± 0.46	2.45 ± 0.62	2.50 ± 0.60	2.40 ± 0.74	2.46 ± 0.58
Q24. Do you believe that chocolate is a good source of food for treating heart disease? (%Yes)	36.6%**	54.4%**	88.8%	86.1%	42.9%	66.7%	32.9%**	57.0%**	60.0%	41.7%
Q27. Do you believe that fish oil (omega-3) is a good source of food for treating heart disease? (%Yes)	72.1%	68.7%	60.0%	45.5%	60.0%**	83.3%**	75.7%	74.7%	48.3%	50.0%
Q33. Is it possible to develop a weight-reduction diet that permits consumers to eat or drink chocolate everyday? (%Yes)	82.4%	78.8%	57.1%	50.0%	81.8%	90.0%	92.5%	83.1%	90.9%**	73.3%**
Q35. Do you think that overweight people should consume chocolate? (%Yes)	33.3%**	51.8%**	16.7%	25.0%	57.1%	75.0%	32.5%***	55.8%***	42.9%	37.5%
Q13. Do you agree that certain foods have health benefits that may reduce the risk of disease? (0 = Strongly disagree, 3 = Strongly agree)	2.66 ± 0.60	2.64 ± 0.63	2.59 ± 0.71	2.65 ± 0.63	2.63 ± 0.50	2.67 ± 0.56	2.67 ± 0.58	2.63 ± 0.66	2.63 ± 0.71	2.62 ± 0.63
Q14. How interested are you in learning more about foods that have health benefits that may reduce the risk of disease? (0 = Not, 3 = Very)	2.52 ± 0.59*	2.63 ± 0.56*	2.59 ± 0.71	2.65 ± 0.63	2.63 ± 0.50	2.67 ± 0.57	2.67 ± 0.58	2.63 ± 0.68	2.53 ± 0.83	2.62 ± 0.49

Significance of within-group, between gender differences: ln χ^2 or t-test, *$p < 0.05$, **$p < 0.01$, ***$p < 0.001$.

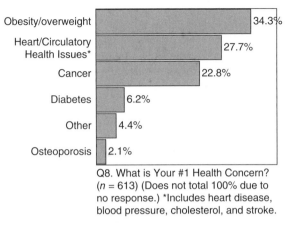

FIGURE 50.1. Consumer top health concerns.

Q8. What is Your #1 Health Concern? (*n* = 613) (Does not total 100% due to no response.) *Includes heart disease, blood pressure, cholesterol, and stroke.

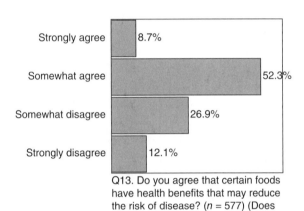

Q13. Do you agree that certain foods have health benefits that may reduce the risk of disease? (*n* = 577) (Does not total 100% due to no response.)

FIGURE 50.2. Consumer attitudes toward functional foods.

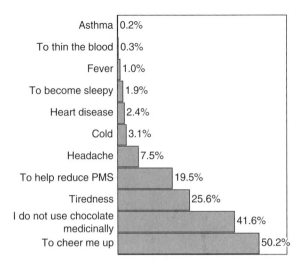

FIGURE 50.3. Using chocolate medicinally.

(68.7 percent) agreed that fish oil (omega-3) was a good source of food for treating heart disease. MMA females reported the highest positive association (83.3 percent) that fish oil was a good source of food for treating heart disease, and the difference between males (60.0 percent) and females was significant ($p < 0.01$).

Attitudes toward weight reduction and menu planning differed between groups. Whereas the majority of participants (82.4 percent male versus 78.8 percent female) believed that a person could develop a weight-reduction diet that permitted consumers to eat or drink chocolate everyday, only 50.0 percent of female Mexican-Americans and 57.1 percent Mexican-American males thought this was possible. Conversely, 90.0 percent of female MMA and 90.9 percent of male non-NUT students agreed that chocolate could be part of an overall weight-reduction diet.

When asked whether overweight people should consume chocolate, a significant reduction of the participants agreed positively—only 33.3 percent of males and 51.8 percent of females thought overweight people should consume chocolate. Female MMA participants differed in that 75.0 percent thought chocolate consumption by overweight people was permissible.

BEHAVIOR

When asked "Do you use food to treat illness?" females had a greater, positive response than males (1.70 ± 0.77 and 1.63 ± 0.83, respectively), except for Mexican-Americans (Table 50.3). Separated by subgroup, the Mexican-American males reported a higher usage than females (1.38 ± 1.02 and 1.24 ± 0.99, respectively) and the difference was significant ($p < 0.001$). Consistent with these results, overall, females also reported a greater usage of food to prevent specific diseases, and when compared to males, the response was significant ($p < 0.001$). To treat specific conditions using chocolate, a little more than 50 percent of the study population reported that they consumed chocolate to "cheer [themselves] up," while 25.6 percent used chocolate to stay alert (Fig. 50.3).

Females reported going on diets more often than males, with the exception of Mexican-Americans, where males had a greater tendency to go on a calorie-restricted diet. Overall, however, none of the groups reported that they dieted frequently. Conversely, 85.9 percent of females and 74.7 percent of males reported making dietary changes to improve their health. The study participants reported eliminating foods high in carbohydrates and focused on removing or eating less snacks, candy, desserts, and foods high in sugar (Fig. 50.4).

Nearly all respondents (94.0 percent) reported that a nutritionist was the most believable source for information on the health benefits of foods (Table 50.4). However, only 37.0 percent of the participants

Table 50.3 Behaviors Toward Chocolate, Disease, and Health

Survey Questions	All Subjects (610)		Mex-Am (56)		MMA (40)		NUT (451)		Non-NUT (68)	
	Male	Female	Male	Female	Male	Female	Male	Female	Male	Female
	33.4% N = 204	66.6% N = 406	30.4% N = 17	69.6% N = 39	40.0% N = 16	60.0% N = 24	34.7% N = 156	65.3% N = 295	23.8% N = 16	76.2% N = 24
Q11. Do you use food (change your diet) to treat illness?[a]	1.63 ± 0.83	1.70 ± 0.77	$1.38 \pm 1.02^{***}$	$1.24 \pm 0.99^{***}$	1.38 ± 1.02	1.78 ± 0.90	1.69 ± 0.78	1.76 ± 0.71	1.62 ± 0.70	1.80 ± 0.69
Q12. Do you use food (change your diet) to prevent specific diseases?[b]	$1.40 \pm 0.88^{***}$	$1.67 \pm 0.76^{***}$	$1.63 \pm 1.03^{***}$	$1.86 \pm 0.99^{***}$	1.13 ± 0.34	1.29 ± 0.46	1.31 ± 0.77	1.25 ± 0.71	1.38 ± 0.87	1.20 ± 0.70
Q34. How often do you go on a calorie-reduction type diet?[c]	$0.50 \pm 0.75^{***}$	$0.86 \pm 0.79^{***}$	0.94 ± 1.09	0.90 ± 0.85	0.50 ± 0.63	0.96 ± 0.69	0.54 ± 0.74	0.86 ± 0.78	0.47 ± 0.83	0.81 ± 0.82
Q41. How often do you eat chocolate?[d]	3.49 ± 1.07	3.68 ± 1.02	2.44 ± 1.59	2.97 ± 1.60	3.94 ± 1.24	4.21 ± 0.72	3.52 ± 0.94	3.70 ± 0.92	3.93 ± 0.62	3.81 ± 0.95
Q54. Have you tried to change your diet to improve your health? (%Yes)	$84.7\%^{***}$	$85.9\%^{***}$	$43.8\%^{***}$	$65.8\%^{***}$	75.0%	87.5%	$81.3\%^{***}$	$90\%^{***}$	$57.1\%^{**}$	$81.3\%^{**}$

Significance of within-group, between gender differences: In χ^2 or t-test, $^*p < 0.05$, $^{**}p < 0.01$, $^{***}p < 0.001$.

[a]0 = I never change my diet; 1 = I usually don't change my diet; 2 = I sometimes change my diet; 3 = I always change my diet.

[b]0 = I never use food; 1 = I usually don't use food; 2 = I sometimes use food; 3 = I always use food.

[c]0 = Never; 1 = Once in a while; 2 = Very often; 3 = Always.

[d]0 = Never; 1 = Holidays only; 2 = Yearly; 3 = Monthly; 4 = Weekly; 5 = Daily.

obtained information from a nutritionist. The next most credible source for seeking information on the health benefits of foods was doctors (80.0 percent), followed closely by dietitians (74.0 percent). The top three sources where respondents actually acquired nutrition information were doctors, teachers/instructors, and mothers/grandmothers (49.0 percent, 48.0 percent, and 43.0 percent, respectively). Although only 4.0 percent of the respondents thought that magazines offered believable nutrition advice, a quarter of the participants acknowledged that they obtained infor-

mation or advice from this source. This same juxtaposition is seen in other media outlets: Seven percent reported that television programming was a believable resource for nutrition advice while 19.0 percent reported that newspapers were credible sources. Yet, the respondents acknowledged that they obtained their nutrition advice from television and newspapers (23.0 percent and 35.0 percent, respectively).

DISCUSSION

Renewed interest in the medicinal use of chocolate (cocoa) began approximately a decade ago (late 20th century) with laboratory investigations focusing on a subgroup of phytochemicals known as flavonoids—more specifically flavan-3-ols, which are found in chocolate. The FDA does not regulate the word "phytochemical," but researchers have reached consensus to define the term as "a non-nutritive plant constituent able to confer a health benefit" [34]. The purported linkage between chocolate consumption and beneficial heart health is related to flavonoids' supposed role in promoting healthy blood vessel function and in decreased platelet activity [35, 36]. From biological and physiological research laboratory findings, a significant body of literature has delineated the relationship between chocolate and health. Still, much less attention has been given to environmental determinants that modulate the medicinal intake of chocolate. Surveys have been conducted on chocolate's perceived sensory attributes [37] and psychological cravings; however,

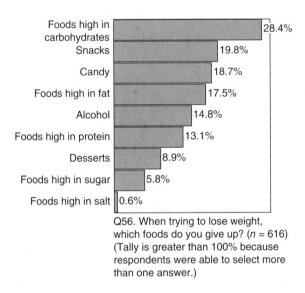

Q56. When trying to lose weight, which foods do you give up? (n = 616) (Tally is greater than 100% because respondents were able to select more than one answer.)

FIGURE 50.4. Foods eliminated when dieting.

Table 50.4 **Believable Sources of Information Versus Where Information Is Obtained**

	Q36. Who/what would you say is the most believable source for information on the health benefits of foods?	Q37. Who or where do you get information on nutrition?
Nutritionist	94%	37%
Your doctor	80%	49%
Dietitian	74%	14%
Friends	38%	26%
Scientist	38%	8%
University health newsletter	37%	14%
Nurse	34%	9%
Teacher/instructor	33%	48%
Health food store	29%	38%
Mother or grandmother	20%	43%
Article in newspaper	19%	35%
Other	19%	2%
Food label	13%	34%
Television show	7%	23%
Magazine	4%	25%
Radio talk show	3%	5%
Spouse	3%	4%

$N = 616$. Tally is greater than 100% because respondents were able to select all answers that applied. Percentage rounded to whole percents.

investigations exploring the medicinal use of chocolate are lacking [38, 39]. With 150 million Americans spending over $20.5 billion annually on functional foods [40], it was timely to investigate Americans' attitudes toward utilizing chocolate medicinally.

To extrapolate our findings to a larger population, selected questions from the *IFIC Survey on Functional Foods 2002* [41] that utilized a national, simple random sample ($n = 1000$) were included in our questionnaire. Both the IFIC survey and the present study results demonstrated that Americans believe they have a moderate to great amount of efficacy toward controlling their health.

The association between dietary intake and health outcomes is a well-established area of study. Researchers have investigated how people view health in relationship to their food selection and how their health perspectives influence dietary patterns [42–44]. The determinants to food choice are multifaceted and evolve over time [45]. These decisions are influenced by cultural, environmental, psychological, and socioeconomic status factors, among other influences [46].

Knowledge of specific, nutrient/health-benefit synergy potentially lowers barriers to consuming food products that have been demonstrated to be health protective [47]. Our study demonstrates an overall greater self-efficacy among men in internal loci over their health. When separated by group and age, older men (Mexican-American and MMA males) retained a higher level of control than females (see Table 50.2). Interestingly, there is incongruity between what females believe regarding the amount of control over their health and reported dietary behavior. Whereas women scored lower on self-efficacy in controlling their health, nearly 86 percent reported changing their diets to improve their health ($p < 0.001$). Our female study participants were 1.5 times more likely to make dietary changes to improve health than the 2006 IFIC's *Food and Health Survey* participants [48].

The present study compared differences in knowledge (not reported), attitudes, and behaviors with respect to utilizing chocolate as a functional food. The hypothesis that men would report consuming chocolate in greater quantities was based on previous research findings [39–49]; only one subgroup (non-nutrition students) within our data reflected this pattern. Seasonal influences were controlled by survey design since all surveys were administered prior to the winter holiday season. Unlike prior studies, females overall and within three subgroups (Mexican-Americans, MMA, and nutrition students) consumed chocolate in greater quantities (3.68 ± 1.02, i.e., almost weekly), but the difference between genders was nonsignificant.

Cocoa mass by weight contains slightly greater than 50 percent lipids[7] [50]. Research has indicated a positive or neutral relationship between certain lipids and cardiovascular health [51, 52]. With the high level of media attention focused on the so-called national obesity epidemic and chocolate's high fat content, we hypothesized that our study participants would have a negative perception between chocolate consumption and heart disease attenuation. Indeed, when asked "Do you believe that chocolate is a good source of food for treating heart disease?" 54.4 percent of females but only 36.6 percent of males ($p < 0.01$) responded positively. A greater percentage, however, 68.7 percent and 72.1 percent (female and male, respectively) thought fish oil was a good source of food for treating heart disease. The positive association between fish oil and heart disease attenuation may be related to better media and governmental promotional messages [53–55]. Further support that the *message* on the relationship between chocolate consumption and heart health promotion has not been heard is demonstrated in our survey results: Only 43 percent of participants (data not shown) responded that they heard that chocolate may be beneficial in treating cardiovascular disease (CVD).

The UCD instructor in Nutrition 10, Dr. Elizabeth Applegate, lectures on the chocolate consumption/CVD attenuation relationship in her nutrition class; since three-fourths of the study participants had access to her chocolate and CVD attenuation lecture prior to completing our survey, it highlights the need to dispense health messages multiple times with the desired effect that consumers will retain the information to memory. Knowledge alone does not change behavior. However, as a basic requirement, the need for consumer awareness of the association between a food and its health benefit is documented in the IFIC's 2006 benchmark survey, where among participants who could identify an association between a nutrient–function relationship, nearly the same percentage consumed the product to improve their health [48].

Aligned with the United States Department of Agriculture's (USDA) (*Dietary Guidelines for Americans 2005*) key recommendation: "Meet recommended intakes within energy needs by adopting a balanced eating pattern," [56] participants in our study overwhelmingly (82.4 percent male, 78.8 percent female) believed that it was possible to develop a weight-reduction diet that permitted consumers to eat or drink chocolate everyday. Our findings revealed, nonetheless, a negative bias against overweight people who consumed chocolate.

Within the subgroup Mexican-Americans, only 25.0 percent of females and 16.7 percent of males thought overweight people should consume chocolate, while this same population thought (50.0 percent female, 57.1 percent male) that it was possible to eat chocolate everyday and still lose weight. This negative bias transcends culture, however. There was a significant difference ($p < 0.01$) between genders, with females more permissive toward overweight people consuming chocolate; this positive difference was maintained throughout the groups with the exception of the nonnutrition students. MMA females had the highest acceptance of overweight people consuming chocolate (75.0 percent). Older, middle-class women also self-reported weight levels that resulted in BMIs classified as overweight (26.9 ± 6.0).

Participants responded that they were "somewhat to very" interested in learning about foods that have health benefits that may reduce the risk of disease. Health messages come from a variety of sources (doctors, the media, and family) and consumers filter these messages through a complex set of variables including—but not limited to—age, gender, income level, personal health, and socioeconomic status [57]. Research questions that our study aimed to examine were if the phytochemicals inherent in chocolate have been determined to be positively associated with improved health outcomes: (1) Who or where did Americans seek out information on the health benefits of foods? and (2) Did they believe these sources to be credible?

Not surprisingly, 94.0 percent of our respondents believed that nutritionists were a believable source for information on the health benefits of food since nearly 74.0 percent ($n = 451$) of our population were nutrition students. Yet, only 37.0 percent of the students actually obtained nutrition-related information from this source. This may be due to income; since they were students, they probably would lack sufficient income to visit a professional nutritionist. Doctors also received a high percentage rating (80 percent) that they would be able to provide dietary advice; however, only half of those responding (49 percent) actually sought a doctor's guidance.

Mass media plays a significant role in dispensing health information. The IFIC's 2006 benchmark survey queried participants on where they sourced health information and their subsequent belief in those messages. Their survey revealed that 72 percent of respondents received health messages from the mass media, while the present study participants ranked magazines, television broadcasts, and newspaper articles as minor resources (see Table 50.4). Our study participants sought advice from mothers and grandmothers (43 percent), although only 20 percent of said group believed that their older relatives were a reliable source regarding nutrition information. This result may be explained given that our population was skewed toward a younger demographic (23.5 ± 8.7 mean years), and therefore this cohort continues to rely on relatives for multiple levels of guidance including health advice.

The research community has published findings in peer-reviewed journals that specific nutrients in chocolate may be health promoting. We also measured public attitudes toward these messages using as a proxy the question: "Do you believe that research

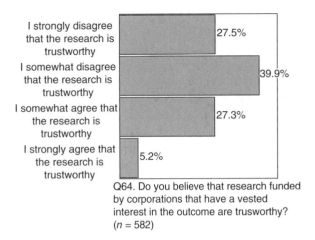

FIGURE 50.5. Attitudes toward research findings funded by corporations.

funded by corporations that have a vested interest in the outcome are trustworthy?" Two-thirds of our respondents ($n = 582$) either strongly or somewhat disagreed that research funded by corporations was trustworthy (Fig. 50.5). Interestingly, respondents also reported that they believed instructors/teachers, scientists, and university health newsletters (33 percent, 38 percent, and 37 percent, respectively) were believable sources for nutrition information. Often the nutrition research these sources espoused was funded by corporations.

Future Research

Our sample size distribution favored young, educated, middle-class participants, which may have biased some of our survey results. When comparing our results with a nationally surveyed population, however, our research predominantly matched previous national data findings. Americans are very interested in learning about food's health promoting properties. Moreover, the present study points out that the majority of participants made dietary changes to improve personal health. Chocolate, a product high in antioxidants and other cardiovascular attenuating nutrients, has the potential to support good health. The disconnect between where people seek nutrition advice—and who or what sources are believable—is an area of study that warrants more attention.

Conclusion

The present study measured consumers' knowledge, attitudes, and beliefs toward utilizing chocolate medicinally. Currently, the message that specifically manufactured cocoa high in polyphenols may attenuate CVD has not been heard by the majority of our study participants. Bias against chocolate as a health-promoting food—due to its high caloric content—needs to be addressed so that consumers can appreciate that chocolate can be part of a healthy dietary pattern. Our study demonstrated that participants consumed chocolate almost weekly, but rarely used the product to treat illness or prevent disease. This result may be attributed to our population's young age and as a result they may not have experienced negative health consequences. Further research is needed to measure barriers to consuming chocolate medicinally.[8]

Acknowledgments

We wish to thank the following who provided assistance during the research component of our work: Elizabeth Applegate, Department of Nutrition, University of California, Davis; Christine Bruhn, Department of Food Science and Technology, University of California, Davis; Dennis Dingemans, Lecturer in Geography, Social Science Program, University of California, Davis; Claudia Hernandez, State Center Community College District, Madera, California; Margaret Rucker, Department of Textiles and Clothing, University of California, Davis; and Sandra Samarron, Department of Nutrition, University of California, Davis. Finally, we especially wish to thank the students in Dr. Applegate's Nutrition 10 course, Fall Quarter, 2005, and the students in Dr. Dingeman's Geography 10 course, Fall Quarter, 2005.

Endnotes

1. Free radicals are molecular species with unpaired electrons in the outer ring. These unpaired electrons are usually highly reactive, so radicals are likely to take part in chemical reactions. Free radicals have been shown to pair with hydrogen molecules in cell membranes and cause damage to the cell.

2. Flesch–Kincaid Grade Level Formula = 0.39 (total words/total sentences) + 11.8 (total syllables/total words) − 15.59.

3. United States Census Bureau, 2000 Census, calculated ~43,000 residents in Madera, California.

4. Instructor permissions were received from Dr. Elizabeth Applegate and Dr. Dennis Dingmans.

5. United States Census Bureau, 2003, U.S. median household income, $43,318; United States Census Bureau, 2000, median persons per household: 2.58.

6. Body Mass Index (BMI) calculated as: weight (kg) / height (m)^2. Ranges for BMI: Underweight ≤ 18.5; Normal weight ≥ 18.5–24.9; Overweight ≥ 25–29.9; Obese ≥ 30.

7. Lipid is a scientific term used to designate fat.

8. Funding for this study was supported through the Department of Nutrition, University of California, Davis.

References

1. Engler, M. B., and Engler, M. M. The emerging role of flavonoid-rich cocoa and chocolate in cardiovascular health and disease. *Nutrition Reviews* 2006;64(3): 109–118.

2. Hollenberg, N. K. Vascular action of cocoa flavanols in humans: the roots of the story. *J. Cardiovascular Pharmacology* 2006;47(Supplement 2):S99–S102.

3. Applegate, E. *Eat Your Way to a Healthy Heart Features Chocolate and 99 Other Foods That Help Your Heart.* Paramus, NJ: Prentice Hall, 1999.

4. Scalbert, A., Manach, C., Morand, C., and Remesy, C. Dietary polyphenols and the prevention of diseases. *Critical Reviews in Food Science & Nutrition* 2005; 45(4):287–306.

5. Buijsse, B., Feskens, E. J., Kok, F. J., and Kromhout, D. Cocoa intake, blood pressure, and cardiovascular mortality: the Zutphen elderly study. *Archives of Internal Medicine* 2006;166(4):411–417.

6. Fisher, N., and Hollenberg, N. K. Flavanols for cardiovascular health: the science behind the sweetness. *Journal of Hypertension* 2005;23(8):1453–1459.

7. Rafter, J. J. Scientific basis of biomarkers and benefits of functional foods for reduction of disease risk: cancer. *British Journal of Nutrition* 2002;88: 219–224.

8. Dillinger, T. L., Barriga, P., Escarcega, S., Jimenez, M., Lowe, D., and Grivetti, L. E. Food of the gods: cure for humanity? A cultural history of the medicinal and ritual use of chocolate. *Journal of Nutrition* 2000;130(8): S2057–S2072.

9. Hasler, C. M. Functional foods: benefits, concerns and challenges—a position paper from the American Council on Science and Health. *Journal of Nutrition* 2002;132(12):3772–3781.

10. Hasler, C. M. The changing face of functional foods. *Journal of the American College of Nutrition* 2000; 19(Supplement 5):S499–S506.

11. Bagchi, D. Nutraceuticals and functional foods regulations in the United States and around the world. *Toxicology* 2006;221(1):1–3.

12. Verschuren, P. M. Functional foods: scientific and global perspectives. *British Journal of Nutrition* 2002;88: 125–131.

13. Hasler, C. Functional foods: their role in disease prevention and health promotion. *Food Technology* 1998;52(2): 57–62.

14. Editorial. Safety assessment and potential health benefits of food components based on selected scientific criteria. *Critical Reviews in Food Science and Nutrition* 1999; 39(3):203–206.

15. Editorial. Position of the American Dietetic Association: functional foods. *Journal of the American Dietetic Association* 2004;104(5):814–826.

16. IOM/NAS. In: Thomas, P. R., and Earl, R., editors. *Opportunities in the Nutrition and Food Sciences.* Washington, DC: Institute of Medicine/National Academy of Sciences, National Academy of Press, 1994; p. 109.

17. Coe, S., and Coe, M. *The True History of Chocolate.* London, England: Thames and Hudson, 1996.

18. International Food Information Council (IFIC). The consumer view on functional foods: yesterday and today. *Food Insight* May/June 2002. Available at http://www.ific.org/foodinsight/2002/mj/funcfdsfi302.cfm. (Accessed January 2, 2007.)

19. IFIC. *2005 Consumer Attitudes Toward Functional Foods/ Foods for Health, Executive Summary.* Washington, DC: International Food Information Council, 2005; p. 16.

20. Webb, G. *Dietary Supplements and Functional Foods.* Oxford, England: Blackwell Publishing, 2006.

21. Hasler, C. M. *Regulation of Functional Foods and Nutraceuticals: A Global Perspective.* Oxford, England: Blackwell Publishing, 2005.

22. IFIC. *Food and Health Survey: Consumer Attitudes Toward Food, Nutrition & Health, A Benchmark Survey 2006.* Washington, DC: International Food Information Council, 2006; p. 88.

23. Keen, C. L., Holt, R. R., Oteiza, P. I., Fraga, C. G., and Schmitz, H. H. Cocoa antioxidants and cardiovascular health. *American Journal of Clinical Nutrition* 2005; 81(1):S298–S303.

24. Arts, I. C., and Hollman, P. C. Polyphenols and disease risk in epidemiologic studies. *American Journal of Clinical Nutrition* 2005;81(1):S317–S325.

25. Fraga, C. G. Cocoa, diabetes, and hypertension: Should we eat more chocolate? *American Journal of Clinical Nutrition* 2005;81(3):541–542.

26. Harkness, J., Van de Vijver, F., and Mohler, P. *Cross-Cultural Survey Methods.* Hoboken, NJ: Wiley, 2003; pp. 36–41.

27. Morrison, G. R., Ross, S. M., and Kemp, J. E. *Designing Effective Instruction.* Hoboken, NJ: Wiley, 2004; p. 210.

28. Finley, R. *SurveyMonkey.com.* Portland, OR: SurveyMonkey.com, 1999–2006.

29. Graham, R., Ahn, A. C., Davis, R. B., O'Conner, B. B., Eisenberg, D. M., and Phillips, R. S. Use of complementary and alternative medical therapies among racial and ethnic minority adults: results from the 2002 National Health Interview Survey. *Journal of the National Medical Association* 2005;97(4):535–545.

30. U.S. Census Bureau; Census 2000 Demographic Profile Highlights: Madera City, CA. Available at http://factfinder.census.gov. (Accessed January 15, 2007.)

31. U.S. Census Bureau; US Census Quick Facts: USA People QuickFacts. Available at http://quickfacts.census.gov. (Accessed January 18, 2007.)

32. Fowler, F. *Improving Survey Questions: Design and Evaluation.* Thousand Oaks, CA: SAGE Publications, 1995; p. 152.

33. National Institutes of Health (NIH). *Clinical Guidelines on the Identification, Evaluation, and Treatment of Overweight and Obesity in Adults.* Washington, DC: NIH Publication, 1998; p. 139.

34. Jardine, N. J. Phytochemicals and phenolics. In: Knight, I., editor. *Chocolate and Cocoa: Health and Nutrition.* London, England: Blackwell Science Ltd., 1999; p. 122.

35. Erdman, J. W., Balentine, D., Arab, L., Beecher, G., Dwyer, J. T., Folts, J., Harnly, J., Hollman, P., Keen, C. L., Mazza, G., Messina, M., Scalbert, A., Vita, J., Williamson, G., and Burrowes, J. Flavonoids and Heart Health: Proceedings of the ILSI North America Flavonoids Workshop, May 31–June 1, 2005, Washington, DC. *Journal of Nutriturm* 2007;37(3), pp. S718 S737.

36. Vlachopoulos, C., Aznaouridis, K., Alexopoulos, N., Economou, E., andreadou, I., and Stefanadis, C. Effect of dark chocolate on arterial function in healthy individuals. *American Journal of Hypertension* 2005;18(6): 785–791.

37. Urala, N., and Lahteenmaki, L. Hedonic ratings and perceived healthiness in experimental functional food choices. *Appetite* 2006;47(3):302–314.

38. Yuker, H. E. Perceived attributes of chocolate. In: Szogyi, A., editor. *Chocolate—Food of the Gods.* Westport, CT: Greenwood Press, 1997; pp. 35–43.

39. Osman, J. L., and Sobal, J. Chocolate cravings in American and Spanish individuals: biological and cultural influences. *Appetite* 2006;47(3):290–301.

40. Burdock, G. A., Carabin, I. G., and Griffiths, J. C. The Importance of GRAS to the functional food and nutraceutical industries. *Toxicology* 2006;221(1): 17–27.

41. *IFIC Survey on Functional Foods 2002.* Cambridge, MA: Cogent Research, 2002; p. 16.

42. Falk, L. W., Sobal, J., Bisogni, C. A., Connors, M., and Devine, C. M. Managing healthy eating: definitions, classifications, and strategies. *Health Education and Behavior* 2001;28(4):425–439.

43. Glanz, K., Basil, M., Maibach, E., Goldberg, J., and Synder, D. Why Americans eat what they do: taste, nutrition, cost, convenience, and weight control concerns as influences in food consumption. *Journal of the Dietetic Association* 1998;98(10):1118–1126.

44. Contento, I., and Murphy, B. Psycho-social factors differentiating people who reported making desirable changes in their diets from those who did not. *Journal of Nutrition Education* 1990;22:6–14.

45. Van den Heuvel, T., Van Trijp, H., Gremmen, B., Jan Renes, R., and Van Woerkum, C. Why preferences change: beliefs become more salient through provided (genomics) information. *Appetite* 2006;47(3): 343–351.

46. Santich, B. Good for you: beliefs about food and their relation to eating habits. *Australian Journal of Nutrition and Dietetics* 1994;51:68–73.

47. Toner, C. Consumer perspectives about antioxidants. *Journal of Nutrition* 2004;134:S3192–S3193.

48. IFIC. *Food and Health Survey: Consumer Attitudes Toward Food, Nutrition & Health, A Benchmark Survey 2006.* Washington, DC: International Food Information Council Foundation, 2006; Appendix I.

49. Douglass, J. S., and Amann, M. M. Chocolate consumption patterns. In: Knight, I., editor. *Chocolate and Cocoa,* London, England: Blackwell Science Ltd., 1999; p. 301.

50. Borchers, A. T., Keen, C. L., Hannum, S. M., and Gershwin, M. E. Cocoa and chocolate: composition, bioavailability, and health implications. *Journal of Medicinal Food* 2000;3(2):77–105.

51. Wang-Polagruto, J. F., Villablanca, A. C., Polagruto, J. A., Lee, L., Holt, R. R., Schrader, H. R., Ensunsa, J. L., Steinberg, F. M., Schmitz, H. H., and Keen, C. L. Chronic consumption of flavanol-rich cocoa improves endothelial function and decreases vascular cell adhesion molecule in hypercholesterolemic postmenopausal women. *Journal of Cardiovascular Pharmacology* 2006; 47(Supplement 2):S177–S186.

52. Heptinstall, S. P., May, J., Fox, S., Kwik-Uribe, C., and Zhao, L. Cocoa flavanols and platelet and leukocyte function: recent *in vitro* and *ex vivo* studies in healthy adults. *Journal of Cardiovascular Pharmacology* 2006; 47(Supplement 2):S197–S205.

53. Kalyanam, K., and Zweben, M. The perfect message at the perfect moment. *Harvard Business Review* 2005;83(11):112–120.

54. Marshall, R. J., Bryant, C., Keller, H., and Fridinger, F. Marketing social marketing: Getting inside those "big dogs' heads" and other challenges. *Heath Promotion and Practice* 2006; pp. 206–212.

55. Korhonen, T., Uutela, A., Korhonen, H. J., and Puska, P. Impact of mass media and interpersonal health. *Journal of Health Communication* 1998;3(2):105.

56. *Dietary Guidelines for Americans 2005.* U.S. Department of Health and Human Services and U.S. Department of Agriculture. Available at http://www.health.gov/dietaryguidelines/dga2005/document/. (Accessed January 29, 2007.)

57. Britten, P., Haven, J., and Davis, C. Consumer research for development of educational messages for the MyPyramid Food Guidance System. *Journal of Nutrition Education and Behavior* 2006;38(6) Supplement (1): S108–S123.

PART

XI

Fieldwork, Methodology, and Interpretation

Chapter 51 (Cabezon and Grivetti) describes the techniques used by paleographers to identify documents by date and relates the skills used to read 16th–18th century handwritten Spanish documents. Their chapter, based on a suite of 1641–1642 St. Augustine chocolate-related documents, reveals the nuances and skills used when translating early Spanish manuscripts. Chapter 52 (Brindle and Olsen) reviews the types and kinds of cacao- and chocolate-associated resources available for inspection in Charleston, South Carolina, and Savannah, Georgia, with specific attention to port records, account ledgers, diaries, and local advertising. Chapter 53 (Lange) identifies and describes the logical processes and decisions taken during the design, development, construction, and refinement of the Chocolate History Portal developed during the course of team activities (2004–2007). His chapter includes a basic overview of the Portal used by team researchers, its content, and navigation. Chapter 54 (Dunning and Fox) examines the structure, construction techniques, and forms of base-metal chocolate pots used in North America during the 18th and 19th centuries. Chapter 55 (Grivetti) considers chocolate during the period of the American Civil War (1860–1865) as recorded through military records and diaries written by soldiers on both sides of the conflict, as well as diaries written by women during this terrible period of American history. Chapter 56 (Grivetti and Shapiro) identifies a broad sweep of chocolate-related historical themes and topics where further research would be rewarding and concludes with a short epilogue and identifies how scholars representing different fields might take part in future activities.

CHAPTER

Symbols from Ancient Times

Paleography and the St. Augustine Chocolate Saga

Beatriz Cabezon and Louis Evan Grivetti

Introduction

A language is a universe of concepts, images, and symbols, and like a universe, languages constantly expand and change. Languages are the products of lengthy evolution through time and space. In this sense, languages are live entities and as long as they are spoken, they grow, change, and adapt to new environments determined by the geographic characteristics of different regions, ecosystems, and peoples. Human societies living in variable environments have created different ways to communicate and express ideas. The world and language of the Inuit people—living in the bitter cold north latitudes of North America—are different from those of the Mayans in hot and humid Central America or those of the Kalahari Desert people in dry and arid Africa.

Just as animals and plants have been classified into phylogenetic groupings, languages have been grouped and clustered into families. Language historians, linguists, and paleographers who study the origin and evolution of languages cluster them into groups that reflect common ancestors and similarities, whether Anglo-Saxon, Indo-European, Latin/Romance, or Xhoi-San. The Danish scholar Rasmus Rask (1787–1832) defined language as "a natural object" and that "its study resembles the study of natural history" [1]. Noam Chomski, widely known for his significant contributions to the field of linguistics during the 20th century, observed that language is part of the human essence and reflects a particular perspective of the world [2]. For Otto Jespersen,

founder of the Linguistic Society of America, the essence of language is determined by the human activity and the need to communicate with and understand others [3].

Language also can be nonverbal or written communication systems used to express thoughts or feelings, as with body language, the language of flowers, even the language of chocolate, the latter reflecting friendship, hospitality, and love. But what were the languages of the Native Americans in Central America who first drank chocolate: Were they Olmecs, Mayans, perhaps Mixtecs or Zapotecs (see Chapter 2)?

Chocolate was offered to Hernán Cortés and his officers at the court of Motecuhzoma Xocoyotzin (sometimes spelled Moctezuma or Montezuma)—the Mexica/Aztec emperor. Was he the first European to taste and drink chocolate (see Chapter 8)? The language spoken by Cortés was Spanish, and so it was that the

Chocolate: History, Culture, and Heritage. Edited by Grivetti and Shapiro
Copyright © 2009 John Wiley & Sons, Inc.

Nahuatl[1] word *xocoatl* (translated as "bitter-water") was incorporated into the Spanish language through a phonetic transformation into the word "chocolate." The history of chocolate, therefore, has been firmly linked with the Spanish language and Spanish was the first non-Amerindian language that expressed the existence and properties of cacao and chocolate in subsequent publications. And it has been 16th–18th century Spanish that has been the language of many documents used as primary sources of information in the present book.

Languages change through time. The Spanish language of the late 15th century at the time of the voyages of Columbus is not the same as 21st century Spanish, a fact obviously true for English and other languages as well. In order to read and understand the original Spanish language manuscripts, researchers (in this sense paleographers) need to rely on various linguistic tools.[2] While Spanish paleography in the strict sense mainly considers manuscripts written before the 16th century, this chapter includes examples from the 16th and 17th centuries. During the period of our efforts, Castilian Spanish grammar[3] was evolving and becoming a new discipline and taught in Spanish universities to a privileged group of scholars. Still, many of the old linguistic ways persisted into the 18th century and sometimes can be seen even in 19th century documents.

Paleography as an auxiliary science of history provides a useful tool to help identify when manuscripts were written. It is not uncommon that the edges or corners of old manuscripts are lost or so damaged that it is impossible to read dates; in other cases dates were not provided as part of the texts. An experienced paleographer, however, frequently can identify the specific time period and often the geographical location where specific manuscripts were written based on regional stylistic conventions [4]. Reading and interpreting archive and library documents provides insights about people/cultures and their ways of life and the information—either obvious or deduced—permits scholars to understand what people ate and drank, and in certain instances the diseases that were prevalent and their treatments. In the specific instance of cacao and chocolate, with a long history of medical and social uses, original archive/ library documents allow researchers to trace the history and cultural aspects of this food and note new innovations and changes through time. But to read and understand these documents from centuries past also requires knowledge of both linguistics[4] and philology.[5]

In the same way that the cacao tree (*Theobroma cacao*) originated from an ancient botanical lineage, the languages spoken today reflect ancient tongues, and Spanish has a unique evolutionary history. The strategic geographic enclave of the Iberian Peninsula, being a bridge between North Africa and the European continent, made Iberia a highly desirable land for conquest and colonization. Since prehistoric times Spain was regularly traversed and settled by a broad diversity of people who left their linguistic and social marks on the land, whether Berbers and Arabic-speaking Moors from North Africa, the Iberians,[6] Celts from the central European Danube plains, or even earlier Phoenicians, Greeks, Romans, and subsequent Visigoths. With each new invasion, other settlers also arrived with still other languages; for example, the Jews, after the Diaspora in the first century CE, reached Iberia and ultimately contributed to a pluralistic and multilingual Spain [5]. Such was the general history of the people and cultures that contributed to the basis of the language known today as Spanish.

Turning to the New World, Amerindian (Native American) tongues enriched not only the Spanish language but English and French as well. Francisco Ximénez, the remarkable Dominican friar[7] who in the early 18th century translated the *Popol Vuh* (see Chapter 3) from the Quiché language to Castilian Spanish, was a passionate student of the Quiché when he wrote:

> Far from being a barbaric the Quiché language is so orderly, harmonious, and exact, and so consistent in character with the nature and properties of things. [6]

In his effort to convince the king of Spain that Christian missionaries should speak Amerindian languages, Ximénez wrote a book regarding the grammar principles of three of the primary Mayan languages spoken in the region of Guatemala: *A Treasure of the Cachiquel, Quiché and Tzutuhil Tongues* (*Tesoro de las Lenguas Cachiquel, Quiché, y Tzutuhil*) [7]. In a subsequent book, *History of the Province of Guatemala*, Ximénez fearlessly demonstrated his appreciation for the Amerindian languages and at the same time revealed his contempt toward the Spanish military conquistadors (and indirectly the Spanish crown) when he defiantly wrote:

> The seeds [of cacao] are born inside of a cob [mazorca] in very green and fresh trees in wet lands; this [tree] is called cacuatl, but because the Spanish that have corrupted everything in this land, have also corrupted the language, and they called [it] cacao; of this [cacao] there is much along this river [i.e., the Tabasco River] shores, and with this [the cacao] they pay in those lands the main tribute to the Spanish. The Indians grew it in all those towns and it serves them well. [8]

In order to understand the many variants of the Castilian Spanish used during the conquest and colonization periods, it is appropriate to examine briefly its linguistic origins. Linguistic scholars have established connections among many different groups of languages and have traced them to ancestral relatives, thus forming a genealogical language tree. One of these

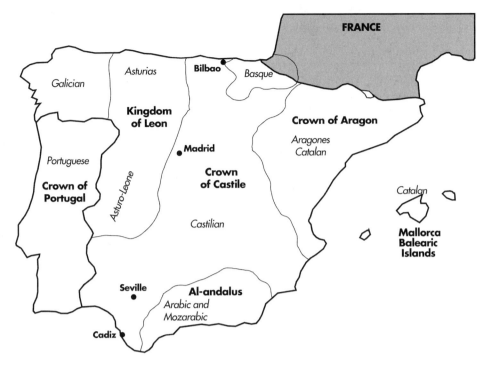

FIGURE 51.1. Iberian Peninsula: kingdoms and linguistic areas ca. 1300 CE.

early groups of languages is called Indo-European.[8] The Italic branch of Indo-European then leads to Latin and from there Latin diverges into the so-called Romance languages: Catalan, Castilian Spanish, Galician, Italian, French, Portuguese, Provençal, and Rumanian [9].

The evolutionary history of a language is often linked with specific historical events, when after military conquest the defeated are forced to adopt the language of the conqueror [10]. The Indo-Europeans[9] through successive waves of migrations settled throughout the European landscape, then separated into different groups, the Italics and the Celts among them, and shared the Danube plains. Through time, the Italics migrated south across the Alps while the Celts began their journey toward the west [11]. When the Iberian Peninsula was reached during the first millennium CE, the Celtic branch encountered the Iberians, a non-Indo-European group, and both cultures formed the so-called Celtiberians, who traded extensively within the Mediterranean with the Phoenicians and Greeks [12]. In 218 CE, the Romans began their conquest of the Iberian Peninsula and imposed both Latin and Christianity. With the decline of the Roman Empire, Iberia was subject to a series of military invasions—first the Vandals and then the Visigoths [13]. In 711, Arabic-speaking North Africans, a mixture of Arabs and Berbers, overran the Visigoth kingdom of Iberia and settled in the south of Spain and there they remained for almost eight centuries until their defeat by the Catholic monarchs (Ferdinand King of Aragon and Isabella Queen of Castile) in 1492 [14]. This historic

kaleidoscope of people and languages also is reflected in the languages of Spain, among them Basque, Catalan, and Galician, and particularly Castilian Spanish (Fig. 51.1).

The legacy of the Roman Empire was Latin and from it the Romances languages evolved.[10] The earliest known example of written Castilian Spanish (*castellano*) is the Emilianense Codex (*Códice Emilianense # 60*) housed in the Monastery of San Millán de la Cogolla in La Rioja, Spain. This priceless document, dated to 1040, contains a small explanatory note (glosa) on the margin written in *castellano* by a monk [15]. Castilian Spanish also developed with influences from the Basque language[11] and Arabic as used by the Moors of *Al-andalus* (modern Andalusia in southern Spain). It had already incorporated through the centuries the lexicon inherited from the Celts, the Iberians, the Greeks, and the Visigoths (Germanic people), becoming in this way a reservoir of diverse cultures and ancient tongues [16].

Oral transmission of knowledge via the spoken language relies on the memory of the individual and the faithful transmission of the information, and it also requires the simultaneous presence of at least two individuals. Written language has the advantage of storing large amounts of information without relying on the memory or the physical presence of the one who possesses the information[12] [17]. Just as the Mayans envisioned cacao as the gift from the gods (see Chapter 3), the Babylonians[13] attributed the invention of letters to the god Nebo, the Egyptians to Thoth, and the Greeks

Table 51.1 Evolution of Spanish Alphabet: 12th through 17th centuries

...tury	14th Century	15th Century	16th Century	17th Century

(The table presents handwritten paleographic letterforms for successive centuries. The left-hand column gives the letter equivalents — a, b, c, d, e, f, g, h, i, k, l, m, n, o, p, q, r, s, t, u, v, x, y, z — followed by their cursive manuscript forms across the 12th/13th through 17th centuries, which cannot be rendered as plain text.)

Source: Muñoz y Rivero, D. J. *Manual de Paleografía Diptomática Española de los siglos XII al XVIII.* Madrid, Spain: Daniel Jorro, editor, 1917; pp. 47 and 58.

to Hermes [18]. The alphabet[14] as a system of letters used today was developed from an early Semitic script.[15] The North Semitic provided the blueprint for Phoenician, which in turn was the origin for the Greek alphabet. Perhaps during the 8th century BCE the Greeks brought their alphabet to Italy (then known as Romania), where it was modified by the Etruscans[16] and later adopted by the Romans, thus becoming the contemporary European script that ultimately evolved into written Spanish [19]. How written Spanish changed through the centuries is another exercise in decipherment by paleographers (Table 51.1).

Methods

The documents presented in this chapter were written during the mid-16th to 17th centuries and were located in the *Archivo General de las Indias*, Seville, Spain (Archive of Indies[17]) by several tenacious archivists who brought to light documents that reveal an astonishing story and show that in the midst of toil and sorrow sometimes a piece of chocolate sweetened personal and financial hardships and uplifted the human spirit.

For the technical aspects of the interpretation and transcription of the manuscripts, it was very useful to have at hand a number of important paleography texts, among them *Album de Paleografía Hispano-Americana de los Siglos XVI y XVII* by Millares Carlo and Mantecón [20]; *Manual de Paleografía Diplomática Española de los siglos XII al XVII* by Jesús Muñoz y Rivero [21]; Rivero's *Paleografía Española* [22]; and the *Lexicon abbreviaturarum* by Adriano Capelli [23]. Other texts that facilitated our navigation through the syntax labyrinth of earlier centuries included *Evolución Sintáctica del Español* by Martín Alonso [24] and *Estudios de Historia*

Lingüística Española by Rafael Lapesa [25]. Two additional volumes, the *Orígenes del Español* and *Manual de Gramática Histórica Española*, both by Ramón Menéndez Pidal [26, 27], the highly respected historian, medievalist, and philologist, were invaluable guides for our understanding of the historical context of the evolution of Castilian Spanish grammar. Finally, the anonymous but comprehensive *Novísimo Diccionario de la Lengua Castellana*, edited by the Spanish Academy [28], the classic treatise *Diccionario de los Idiomas Inglés y Español* compiled by Mariano Velázquez [29], and the medieval Castilian lexicon *Vocabulario Medieval Castellano* by Cejador y Frauca [30] provided access to reliable sources of information instrumental in resolving the puzzles of so many words whose meanings have changed through time.

One of the main difficulties in the process of reading and understanding old texts is that the grammar, syntax, and spelling have changed as well as word meanings. In other instances, some words have basically vanished through the centuries and it is difficult to locate them even in specialized dictionaries. In such cases, it is through comparison of different documents that sometimes it is possible to decipher particular words and deduce or infer their meanings. Furthermore, it is important to know the place of origin (town, region, country) of the scribe or person who wrote the manuscript. The complexity of Spanish languages and regional dialects are important aspects to consider since origin reflects the way a person writes.[18]

Background to the St. Augustine Documents

During the 16th century, enterprises associated with the Spanish crown expanded and turned to the North American Atlantic coast. The Treaty of Tordesillas (1494) that had granted exclusive colonizing–economic privileges to Spain and Portugal did not deter France or England in their efforts to profit also from "discovery" of the new lands. By the mid-16th century, the Spanish crown had organized a combined fleet of merchant vessels and warships collectively known as the *Carrera* of the Indies.[19] Such warships were destined to protect the Spanish galleons and trading vessels loaded with the riches of the Indies from attacks by pirates and commerce raiders who sailed the Mediterranean and Spanish kingdom coasts. King Philippe II of Spain, aware that France would raid the Indies and attempt to settle the North American Atlantic coast, commissioned Pedro Menéndez de Avilés, assigned him the task of establishing a permanent Spanish settlement in Florida, and granted him the title Captain General of the Armada.

Captain General Avilés, a Hidalgo (knight) from the city of Avilés, Asturias, on the northern coast of Spain, was an experienced sea captain who previously had fought French pirates along the north coast of Spain. He also owned a private merchant fleet. On June 27, 1565, the Avilés Indies fleet consisted of at least 12 vessels and included his flagship, *San Pelayo*, a large galleon that he personally owned. By one estimate, the fleet had a total crew of approximately 1500 sailors and officers. The ships left the Spanish port of Cadiz on June 27, 1565. On July 4, having sailed more than 750 miles, the fleet entered the port of Las Palmas on the Great Canary Island to replenenish their water tanks and take on additional supplies. Avilés then set sail for Florida (see Chapter 33) [31, 32].

Given the implications of the voyage, it is worthy to consider one of the first letters of the correspondence maintained between Pedro Menéndez de Avilés—the founder of the San Agustin Presidio in Florida in 1565—and the king of Spain, Philippe II. In the document presented here it is interesting to note that the Caribbean hurricane that almost sank Avilés's ships and forced him to find safe haven in Puerto Rico was the same type of storm that seven decades later brought a substantial cargo of chocolate to the shores of Florida, chocolate that ultimately reached the hands of the San Agustin Presidio soldiers. On August 7, 1565, Pedro Menéndez de Avilés wrote the king from Puerto Rico:

DOCUMENT A: AUGUST 7, 1565

I left from the doldrums of the Gran Canaria[20] [Great Canary Island] on the eighth of the past month in a night with good breeze and [when we were] at two leagues from the land the doldrums caught the [other ships] that were departing; so I waited as much as I could for them; at dawn we were at 8 leagues from the land [still] without seeing none of the ships but one [ship] that seeing me alone and that I could not return he came along with me. I sailed with good weather and when we were at 350 leagues from Florida we were caught in a hurricane and it was a miracle that we did not sink because we were carrying so much artillery and the boat was prepared for war because we were expecting to encounter the French Navy . . . [we had to close] the gunports [and because of] the heavy seas [and because] the wind was so strong it was necessary to throw overboard much of [our] artillery although none of those belonging to Your Majesty was cast away. The great strength of the wind and the fury of the storm blew out all the masts and sails but the main mast, and it lasted two nights and one day, but because the ship was a good seafaring vessel and because it was God's Will that we escaped [from the storm]; then when the weather improved I replaced the [lost] masts with the spare mast I had carried, and with the diligence of the ship that was sailing relatively well I was forced to set sail to Puerto Rico where I arrived on the eighth of this month without have seen land since I departed from the Canary Islands . . . To wait here [for the new masts and riggings to arrive] could result in a great inconvenience because if the French have not yet reached Florida they could do it easily and fortify themselves at their will; this is why I decided to leave from here and sail to Havana or Matanzas [Cuba] and notify the Audiencia [Royal Court of Justice] de Santo Domingo of the best, briefest and safest route to take [to Florida] since they have told me that people in the town [of Santo Domingo] are very fearful to navigate to Florida during the hurricane season. [33]

Yo parti de las calmas degran canaria [de Gran Canaria] a ocho delpasado [mes] y aquella noche de mucha brisa dos leguas de trra [tierra] dio calma a los nauios [navíos] q'iban [que iban] saliendo de manera q'ai [que ahí] repare por los aguardar [aguardarlos] y [e] hize [hice] lo posible al amanecer estaua estaua [estaba] a ocho leguas de trra sin uer [ver] los nauios [navíos] sino uno q'conmigo se alargo y biendome [viéndome] solo y q'no eraposible [era posible] voluer [volver] a tierra nauegue mi viaje con prospero tiempo y a tresientas [tresceintas] leguas de la Florida dio furacan [huracán] q fue milagro no nos hundi. [hundimos] que como traiamos tanta artilleria y zallada puesta a punto de Guerra sperando [esperando] encontrarme con la armada françesa . . . [tuvimos que] cerrar las portas y el viento fue tanto y la mar tan gruesa q'nos conuino [convino] hechar [echar] mucha artilleria alamar [a la mar] aunq'[aunque] de V.M^d [Vuestra Majestad] no se hecho [echó] ningunay la gran fuerza de la tormenta y viento fue tanto que nos lleuo [llevó] todos los arboles y velas sino solo el mastil mayor y duro dos noches y un dia y como el nauio [navío] estaua [estaba] tan estanco y [era] buen marinero fue n^tro [nuestro] señor seruuio [servido] escaparnos abonançando [mejorando] el tiempo volvi [puse de nuevo] lo mejor que pude de algunos mastareos que traia de repuesto y a lo mas que hize [hice] arbol y con la diligencia del nauio [navío] nauegaria [navegaría] mediananam^te [medianamente] y me fue forçoso [forzoso]nauegar [navegar]a este Puerto [Puerto Rico] donde entre [entré] a ocho de este mes sin auer [haber] visto tierra desde canaria [Canaria] . . . la dilacion [dilación] de aguardar aqui [aquí][por mástiles y jarcias] seria [sería] grande y esto podria [podría] resultar grande inconveniente porq' [porque] si los françeses no son llegados a la Florida podrian [podrían] llegar y fortificarse muy a su saluo [salvo] y por esto acuerdo dentro de tres dias [días] despacharme de aqui [aquí] y [e] yrme [irme] a la habana [Habana]o Matanças [Matanzas] y doy auiso [aviso]a la audiencia de Sancto Domingo del biaje [viaje] açertado [acertado] y breue [breve] y mas seguro q' [que] podemos hacer porque me çertificaron [cetificaron] en el pueblo que estan [están] en Sancto Domingo [Santo Domingo]muy temerosos de nauegar [navegar]a la Florida en tiempo de furacanes [huracanes]. [33]

PALEOGRAPHIC ANALYSIS

Spanish *escritura* (script) during the 16th century—of the type presented in the above letter—was characterized by lack of punctuation, irregular separation between words, ligation of several words (words written together), deviation from the initial sentence to intertwine with another sentence, lack of diacritical marks (i.e., accents and tilde), and irregular use of abbrevia-

tions. Writing of the names and surnames of people mentioned in the document, as well as geographical place names, was performed using lowercase letters, as this style is characteristic of 16th century Spanish.

The following examples also illustrate some of the characteristics of the 16th century Spanish writing seen here, where letters have shifted and changed through time, as well as the changes in syntax and the meaning of words. One of the more striking examples is use of the letter **c** with a cedilla, **ç → c**, to produce the soft **s** sound. The author of the document used this form not only before the vowels **a**, **o**, or **u** as would be common in 21st century French (i.e., *ça* (this), *leçon* (lesson) *or reçu* (received as past participle)) but also before the letters **e** and **i**. An excellent example of regionalism is seen in the document as well. Avilés used the letter **f** instead of **h**. Ramón Menéndez Pidal, the renowned Spanish philologist and historian, devoted 21 pages and two linguistic maps to the geographic distribution of the letter **f** as an initial consonant in Spain [34]. It suffices here to say that the shift from the the letter **f** to **h** began with Italic dialects, but by the 13th century the letter **h** had displaced the letter **f** in the region of Castilla–La Rioja (i.e., *fanegas*, a Spanish unit of volume → *hanegas* or *febrero*; the month of February → *hebrero*, to be reversed later during the 18th century). But Captain Menéndez de Avilés, as his last name indicates, was born in Avilés, an Asturian town on the Cantabric coast of Spain, and in Asturias (kingdom of Asturias) the **f** prevailed and still prevails even today in the 21st century. Paleographic examples from this document reveal interesting letter shifts:

b → v	*biendome → viéndome* (seeing me)
	biaje → viaje (journey or trip)
ç → c	*forçoso → forzoso* (needful, indispensable)
	Matanças → Matanzas (name of a Cuban port)
	bonança → bonanza (in *abonançando*) (good weather)
	çertificaron → certificaron (assert, certified)
f → h	*furacan → huracán* (hurricane)
u → v	*uer → ver* (to see)
	nauios → navíos (ships)
	voluer → volver (return)
	saluo → salvo [*a salvo*] (safe, rescued)
	breue → breve (brief)
	nauegar → navegar (to sail)
u → v → b	*estaua → estava → estaba* (I was)
	auer → aver → haber (to have)
y → I	*yrme → irme* (to leave)
z → c	*hize → hice* (I did /made)
	treszientas → trescientas (three hundred)

It is also a characteristic of early 16th century Spanish that the position of the direct object, in this case *los*, precedes the infinitive and it is separated from it:

los aguardar → *aguardarlos* (to wait for them)

In the same way that the use of the letter **c** with a cedilla attached, **ç**, reflects the common origin of Spanish and French as Latin languages, the absence of the initial vowel letter **e** also reveals kinship with Italian, another prominent member of the Latin language family:

sperando → *esperando* (hoping)

sperando → Italian *sperare*, 21st century Spanish → *esperar* (to hope)

Avilés's use of conjunctions also reflects regionalism in his writing. He uses **y** instead of **e** before a word beginning with the letter **e**:

y *yrme* → **e** *irme* (and to leave)

Documents of this era commonly contain abbreviations that are formed in three ways: (1) truncation, (2) syncope, and (3) superscript. An abbreviation by truncation is a method used to shorten words by writing only the initial portion of the word. In most cases, abbreviation marks are added to denote the end of the truncation. An abbreviation by syncope refers to making a contraction of a specific word by removing one or more of the medial (middle) letters, and adding a bar written above the words over the place where the letters were removed. An abbreviation by superscript is seen when superscript letters appear at the end of the word. Examples seen in the above document include the following.

Truncation

hundi. → *hundimos* (we sank)

q' → *que* (what, that) and its derivatives as follows . . .

aunq' → *aunque* (although, even though)

porq' → *porque* (because)

q'ai → *que ahí* (that there)

q'iban → *que iban* (that were going)

Syncope

$t\bar{r}ra$ → *tierra* (land)

Superscript

medianam^{te} → *medianamente* (not too bad—not too good)

This document also is interesting as it contains several words that have changed meaning through the centuries, or have almost vanished from 21st century dictionaries. Several examples include:

Zallada → war supplies (ammunition, arms, and other war-like stores aboard ship)

Abonançando → *bonança* → *bonanza* (bonanza)

Sancto Domingo → *Santo Domingo* (in this case *Sancto* from Latin *Sanctus*)

Introduction to the St. Augustine Documents: Summation of Events

Seventy-six years after the Pedro Menéndez de Avilés document (presented earlier), a manuscript dated September 29, 1641 reported that a ship named the *Nuestra Señora del Rosario y el Carmen*, while navigating across the Bahamas channel[21] on her way to Spain, was caught in the midst of one of the feared Caribbean hurricanes. The captain, Hermenexildo López, saved his ship by casting overboard much of the ship's riggings and crated merchandise in order to lighten his vessel . . . but he did not throw overboard barrels or crates that contained his most precious cargo—chocolate. Captain López was a good navigator and turned his ship northward along coastal Florida and reached safe haven at the port of St. Augustine. And with this event so begins the account of the earliest set of records we have identified—thus far—that document the arrival of cacao and chocolate to the Atlantic coast region of North America.[22]

Once the storm-damaged *Nuestra Señora del Rosario y el Carmen* arrived safely in St. Augustine on September 29, 1641, decisions were made regarding what to do with the cargo that had been salvaged. The highly valuable cargo needed to be stored and protected, but the barrels and crates containing the chocolate and other goods were wet; contents would rot if the crates remained unopened for a long period. The first document of this series, dated December 20, 1641, is a list of the barrels and crates. These containers were unloaded and stored at port facilities. Captain López also took under his personal protection 16 crates and barrels at his house (temporary housing?)[23]

After the decision to store the barrels and crates, a period of nearly three months passes with no reports. We deduce that during this time information was passed from St. Augustine to the owners of the cargo regarding the storm and the decision to store the cargo at the Florida settlement. This three-month period would have allowed time for the information to be passed to the owners—or their agents—and for the gentlemen or their representatives to arrive in St. Augustine.

The second document, dated December 23, 1641, provides the details of a request made by Alfonso Fernández de Cendrera, who represented the merchants whose goods had been damaged at sea. As their legal defender, he specifically asked the officials and judges of the Royal Treasury (in St. Augustine) to grant permission to open the damaged crates, to identify their contents, and to organize an auction of the merchandise in order to recover some of the costs. This initial request, as the story unfolded, ultimately became complicated due to various claims from additional merchants who also requested compensation for their losses.

The third document, dated January 18, 1642, is a petition from the official judges of the Royal Treasury of the province of Florida, Francisco Menéndez Márquez y Nicolás Ponce de León,[24] to the governor of Florida, Damián de la Vega Castro y Pardo. This document presents a charge against Captain Hermenexildo López that he had loaded merchandise on his ship without registration (i.e., that he smuggled goods) with the intent to bring them to Spain in contravention of royal ordinances. Among the items Captain López was accused of smuggling were 32 crates of chocolate and nine chocolate grinding stones.

The fourth document, dated March 6, 1642, is an official authorization to the governor of Florida and the official judges of the Royal Treasury to proclaim publicly the auction of merchandise from the *Nuestra Señora del Rosario y el Carmen,* mandating that the auction period would last nine days. The document also reveals that a total of nine auctions were held in St. Augustine, and the monies received from the sales went to the Royal Treasury to be apportioned later to the owners of the merchandise.

The fifth document, dated March 12, 1642, reveals the legal nightmare of conflicting interests and describes the complexities of the case. Three separate interested parties are identified: the merchants who owned the goods, represented by Alfonso Fernandez de Cendrera, claimed compensation for their losses; Hermenexildo López, captain of the *Nuestra Señora del Rosario y el Carmen*, who under the rights and in compliance with the "*derecho de avería*"[25] asks the merchants/freighters to pay for all the things he had to throw overboard in order to salvage the rest; and the royal officials in the smuggling case pending against Captain López where goods were found aboard ship that had not been registered at the time they were loaded. The document also reveals that the same royal officials petitioned the king of Spain, in the person of the governor of Florida, to declare general damage rights (*derecho de avería gruesa*) from Captain López.

The sixth document, dated May 2, 1642, reveals that the merchant defender, Alfonso Fernandez de Cendrera, has demanded that Captain López should be mandated—under law—to declare the contents of his personal ship's chest, and to explain the presence of seven crates of sugar and two crates of chocolate (a total of 132 small boxes of chocolate weighing 190 pounds) at his residence in St. Augustine (temporary quarters).

The seventh document, dated June 14, 1642, is a letter written by Captain Hermenexildo López to the governor of Florida, wherein he defends himself and explains the circumstances under which the initial ship's inventory was conducted. According to Captain López, the inventory was made at different times by different persons and resulted in duplication of the amount/quantity of merchandise that was loaded aboard the *Nuestra Señora del Rosario y el Carmen*. He also wrote in his defense that the sugar actually belonged to the sailors, but considering that the amount was small he had stored it in his apartment (temporary housing) and that it was his intention to give it back to the sailors at the time he would pay them their wages. The two crates of chocolate are not mentioned.

The eighth document, dated March 1642, imprecisely dated for reasons that will be discussed later, represents the final results of these claims, petitions, and suits: The decision was made that the chocolate and cacao would be auctioned. The chocolate was sold and ultimately distributed to men and boys and to one woman who were administrators, religious persons, and presidio personnel in the city of St. Augustine. A total of 1167 pounds of chocolate and 75 small boxes of gift chocolate were distributed among 222 persons—each person identified by name, occupation, and quantity received. And so begins the St. Augustine chocolate saga.

DOCUMENT 1: DECEMBER 20, 1641

This document is a list of all the barrels, boxes, and crates that Captain Hermenexildo López delivered to the authorities of the Real Hacienda (Royal Treasure) upon his arrival at the port of St. Augustine. The document consists of a long list of 86 individual crates, boxes, or bundles (i.e., water-repellant canvas called *brin*), each with a mark on the outside that consisted of one or two capital letters joined together (as a seal) and with a mark that identified the owner. Several of the containers also exhibited outside identification labels that described the contents therein. Cacao and chocolate were among the few items specified with an outside sign on the crate or bundle. Interestingly, one of the crates had external signs that designated destination to the Castle of Camargo[26] and another to the *Caballero de la Orden de Santiago* (Knight of Saint James), probably the most important order of knights in Spain.[27] By deduction—based on these external signs—we may conclude that the ultimate destination of the *Nuestra Señora del Rosario y el Carmen* was Spain. The document reads as follows:

On December 20, 1641, in the city of San Agustin in the Province of Florida, the Royal Official Judges Lords of these provinces by the grace of Your Majesty, Francisco Menendez Marques, Store Keeper of the Navy [tenedor the bastimentos] and Nicolas Ponce de Leon, Auditor of the Royal Treasure, in my presence, went to the house where Captain Hermenegildo Lopez dwells and found the merchandise that the aforesaid had brought in the ship named Nuestra Señora del Rosario y El Carmen, that arrived to this port because the bad weather with a great damage, dismasted [without masts] and a great damage and with water entering through the cracks of the hull. And what the abovementioned captain delivered to them was:

En la ciudad de San Agustin de la prouincia de la Florida a los veynte dias del mes de diziembre de mill seiscientos quarenta y un, los senores jueces offciales reales de estas dhas prouincias por V. Magest. Nuestro Rey, Franco Menendes marques tenedor de bastimentos y Nicolas Ponçe de Leon, contador por antemi fueron a las casas de la morada del capitan Hermenegildo Lopez y dijeron que encontraron las mercaderias que el susodicho habia traido en el navio nombrado Nuestra Señora del Rosario y el Carmen, que entro en este puerto de arribada desarbolado con mucho daño y haciendo agua y lo que entrego el dho capitan es lo siguiente:

First, eight crates—five of them sheated [i.e., lined] with leather and the [other] three with canvas and [all of the them with] outside marks;

Primera mente ocho caxones los cinco afoxxados en cuexo y los [otros] tres en bxin,[28] *con la maxca de afuexa;*

Another crate sheathed with straw with a sign of remitted [rtt'] to the Castle of Camargo [Castillo de Camargo], Knight of the Order of Santiago of Saint James [Caballero de la Orden de Santiago] [and a sign saying] chocolate;

Otro caxon aforrado en paja con letrero a rtt' del Castillo Camargo, Cavallero dela Orden de Santiago [con un letrero que dice] chocolate;

Another [crate] bound with maguey rope with a sign that says: remitted to Angel Diaz, the administrator of the Royal Treasure [almojarifazgo[29]*] in Seville [with a sign that says] chocolate;*

Otro [caxon] liado con mecate con un letrero que dize: a rrt' diaz angel administrador del almoxaxifazgo chocolate en Seuilla;

Another crate without number or mark on the outside and has nine chocolate grinding stones and 18 pestles;

Otro caxon sin numero o marca de afuera y ten nuebe piedxas de moler chocolate con diez y ocho manos;

Another crate with sugar with a cross carved on the wood;

Otro caxon con açucar con una cruz cabada en la Madera;

And five barrels of sugar with six iron arcs [hoops], each [barrel] with a mark outside.

Mas cinco barriles de açucar con seis arcos de fierro cada uno [con] la marca de afuera.

Four rolls of tobacco bundled in petates [mat made of palms] without mark.

Quatro rollos de tabaco en unos petates liados sin marca. [35]

What we see in this document, as well as in all the others of this series, is the tendency of mid-17th century writing that may be called the "individual factor," that is, how the writing style of a person is determined by cultural background, and particularly by the geographical region of Iberia where the writer was born. Throughout the centuries, Spain had been (and remains) a country with several languages and distinctive dialects. Although Castilian Spanish was the predominant language during the conquest of the Americas, individuals born in different regions reflected regional linguistic characteristics. It is difficult to assess whether or not the St. Augustine manuscripts reflect the writing style of the specific period and location (whether Iberia or in the New World), or personal idiosyncrasies. It would be wise, however, to consider both perspectives.

Consider, for example, punctuation marks and accents versus orthography and abbreviations. During the 12th century, the use of punctuation marks in Spanish had become standardized, but after the 14th century such marks[30] fell into disuse [36]. The set of documents considered here were written by at least seven persons and the total lack of punctuation marks constitutes a constant feature in all. Spelling and abbreviations, however, reflect high variability depending on who wrote the text—and even within the same text. For example, in the December 20 document, the introductory pages and final paragraphs were written by Nicolás Ponce de León, an auditor representing the Royal Treasury, who inspected the containers and at the time most likely dictated (?) remarks to an assistant, as the inventory component of the document is prepared in a separate hand and most likely recorded by a secretary. The absence of accents is also remarkable [37].

In the case of orthography accent is also important to differentiate between variations due to the lack of knowledge of the proper spelling of a word and the variations due to other causes. A comparative series of capital and lowercase letter variations that characterized Spanish script from the 12th through 17th centuries already have been presented in Table 51.1 [38]. The lowercase letter *r*, for example, can take several forms: During the 15th century it appears to be a letter t, while during the 16th century the same letter evolves toward the shape of a z, a symbol that resembles the number 2, and only reaches its definite z shape during the 17th century. The author of Document 1, however, when writing the letter *r*, used a form that resembled the letter *z* but in some cases due to the tilde added to the letter *r*, the form changed again to resemble the letter *X*.

In Document 1 the form of the letter *v* resembles the letter *u* and the author interchanged both *v* and *b*, reflecting individual choice.[31] Use of an open *o*, a symbol that resembles the letter *u* or an *i* attached to a consonant like **d** in **di**, also is characteristic of this document. It is interesting to note that the word chocolate was always spelled in the same way—with a round, closed *o*, not only in this document, but in all the St. Augustine documents as well (and even hundreds of others reviewed during the course of the project). The ligature[32] of letters (i.e., when two letters are combined to represent one) still used during the 17th century can be seen in the words *Order de Santiago* (Order of Saint James), where in the word **Order** the initial *O* is joined to the *R* as in the case of the **ORder** (Order). The use of diacritical marks also called accent marks and abbreviations are discussed in the following documents of this series, but it is important to mention here that accents as well as punctuation signs were almost absent during the 17th century.

DOCUMENT 2: DECEMBER 23, 1641

Three days after Nicolás Ponce de León and his assistant drew up the inventory of the barrels and crates held by Captain López, a second document, dated December 23, 1641, described the petition made by Alonso Fernández de Cendrera, overseer of the Royal Warehouses, who represented the merchant's interests. The document orders the St. Augustine authorities to open and auction the merchandise contained in an initial group of 28 crates and boxes (a total of 86 containers were listed separately from those under the supervision of Captain López). It is not clear from the document why only 28 were selected for the subsequent auction (that lasted over a period of nine days). Perhaps those chosen were the most damaged? Still, this would leave 58 containers —either in the St. Augustine port warehouse or under López's control. It is uncertain whether or not these also contained chocolate, and whether or not they ultimately were forwarded to their eventual destination, Spain.

All the 28 opened crates and boxes contained chocolate or cacao and/or delicate tableware from China associated with making and serving chocolate. Specifically identified were small boxes of chocolate, cacao, chocolate buns,[33] chocolate cups with handles and saucers from China, ordinary chocolate cups, chocolate cups with lids made of coconut shells (*cocos*,[34] or small cups made of coconuts and sometimes ornamented with silver), *jícaras* (chocolate cups made of gourds from the *jícara* tree), either ordinary ones or others finely painted, some of these identified from La Guaira,[35] Venezuela, silver stands to hold the *jícaras*,

fine cotton *pañitos chocolateros* (doilies or napkins used when serving chocolate) identified from Campeche (east coast of Mexico) and fine silk doilies/napkins from China, *molinillos* (used to froth the chocolate), and a variety of fine dishes, dinner, and soup plates from China. Also identified as being shipped in the crates was the highly appreciated red dye made from the cochineal insect.[36]

What accounts for the presence of the goods from China loaded aboard the *Nuestra Señora del Rosario y el Carmen* as cargo? The Manila galleon already had begun its Manila–Acapulco route by this date. Chinese goods reached the Caribbean most likely after being transported—across Mexico—from Acapulco on the west coast to Veracruz on the east coast. Chinese origin goods then could have been loaded at Veracruz, alternatively transshipped to Cuba for subsequent transport to Spain.

The more interesting question, however, must be asked, one that considers and reviews the original port of debarkation for the *Nuestra Señora del Rosario y el Carmen* and its ultimate destination. Clues appear in the documents that describe the goods being transported. Listed on the crate content description (Crate 7) is the term *cocos de Guaira* (coconut chocolate serving cups from Guaira). Guaira was a prominent 17th century port in Venezuela. While it could be that these tropical chocolate cups had been imported northward into Mexico, one cannot discount the possibility that the *Nuestra Señora del Rosario y el Carmen*—as part of the Spanish merchant fleet described earlier in this chapter—began its trading voyage in South America (Venezuela), sailed northward to the Mexican east coast port of Campeche (northwestern coast of the Yucatan peninsula), and then on to Veracruz, where goods from China were loaded along with the unusually high quantities of prepared chocolate and special gift boxes of chocolate that ultimately were destined for Spain.

The *Nuestra Señora del Rosario y el Carmen* then could have sailed from Veracruz, put into Havanna (as noted in the deposition reported on March 12, 1642, as events were being reconstructed), then departed for Spain and passed through what was then called the Bahamas channel (today the Florida Straits), where the ship encountered the destructive hurricane and nearly foundered, but was saved by Captain López throwing overboard much of the cargo, except the containers that were highest in commercial value—chocolate. The arrival in St. Augustine on or before September 29, 1641 was unplanned but had enormous historical implications.

[1] Alonso Fernandes de çendrera [Cendrera], overseer of Your Majesty's warehouses, I appear before Your Lordship as the defender [of the absent merchants] and representing my peers, and said that it came to my knowledge that there are many crates and boxes [that are] at risk of rotting and [if this were

to happen] will not be of any good because it says [on the boxes] for gifts. Because of this I beg you to give permission to open those crates and boxes, and that to declare what they have inside and to auction them in public [se venda a pregón] to those who would give the highest amount of money [for them]. And that the money for all this can be obtained and be deposited in the Royal Chests of Your Majesty until the owners of the merchandise or the persons that they have delegated the power [to represent them] appear. In this way, Your Majesty, it would be a good deed for the parties involved. Duly I am asking Your Majesty to authorize the Royal Official Judges of these provinces [of lorida] that they mandate to open those crates and boxes and declare what is inside them.

Alonso Hernandes de çendrera sobrestante de las fabricas de Su Mag. Aparezco ante Vm^ como defensor de los [sic] y mi spares y digo que por quanto tengo noticia que muchos caxones y caxas de los que se han metido en los almacenes de Su Mag. tienen muchas cosas que corren riesgo de podrir se y no çer de ningun buen uso despues pues por decirse son de regalo y destarse asi tien mucho daño a las dhas m's partes por lo que pido y suplico que autorice a dar licencia a que çe abran esos caxones y caxas y çe man'fieste lo que dentro truvieren y que çeven da a pregon, rrematandose laquienes diese mas y que el dinero que dello rresultase se depoçite en las rreales caxas de Su Mag. asta tanto que aparezcan sus dueños de dha jacienda o persona que tenga sus poderes que en mandallo Vm^ açi çelearabuenaobra a las dhas m's partes Con justiçia que pido Vm^ que los jueces officiales rreales destas provincias [de la Florida] agan abrir los dhos caxones y caxas y abra y man'fieste lo que en ello ai.

———————————

Inventory of the crates' contents

[Crate 1]—First, a crate was opened with an outside mark [that can be seen] in the margin[of this page], and there were 50 small boxes of chocolate, 12 chocolate cups from China with handles and a silver saucer, 3 of them broken, 25 rosaries; each of the small chocolate boxes weighed one pound and four ounces;

[Caxon 1]—Primera mente se abrio un caxon con la marca del margen que hubo cinquenta cajetas de chocolate, doçe taças de China con asas y asiento de plata, las tres dhas taças quebradas y beynte y cinco rrosarios y las cajetas de chocolate peso cada una libra y quarto;

———————————

Crate 2—Another crate was opened with an outside mark [that can be seen] in the margin[of this page], and there were 51 small boxes of chocolate, 6 small cocos ornate with silver and 21 rosaries; each of the small chocolate boxes weighed one and a half pounds;

Caxon 2—Mas se abrio otro caxon con la marca al margen y hubo cinquenta [cincuenta] y una cajetas d chocolate seis coquitos guarnecidos de plata y beyntey quarto rrosarios y las dhas cajetas de chocolate peso cada una una libra y media.

———————————

Crate 3—Another crate was opened with an outside mark, and there were 21 small boxes of chocolate, 12 ordinary cocos, 12 rosaries, and 12 doilies from Campeche; each of the small chocolate boxes weighed one pound and a half;

Caxon 3—Mas se abrio otro caxon de la marca de afuera que hubo beyntenueve cajetas de chocolate doce cocos ordinaries doce rrosarios y doce pañitos de campeche y dhas cajetas hubieron a una libra y m^a cada una;

———————————

Crate 4—Another crate was opened with an outside mark, and there were 20 small boxes of chocolate, 2 broken cocos, 3 intact cocos ornate with silver, 1 dozen of golden jícaras, 4 of them broken, 2 chocolate doilies[37] from China, 2 cotton and silk doilies, 14 pieces [i.e., bolt?] of cloth from Campeche, and another 21 small boxes of chocolate; each of the small chocolate boxes weighed one pound and a half;

Caxon 4—Mas se abrio otro caxon con la marca de afuera q' hubo veynte cajetas de chocolate dos cocos qu = brados tres cocossanos tres coquitos guarnecidos de plata = una Dna de jicras doradas quarto quebradas dos pañitos chocolateros de China = dos pañitos de seda y algodon = catorce paños de campeche de seda y algodon y beyteyu nno [veinte y uno; veintiuno] varias cajetas y dhas cajetas hubieron a libra y media de chocolate;

———————————

Crate 5—Another crate was opened with an outside mark, and there were 45 pounds of tortillas and chocolate buns [bollos de chocolate];

Caxon 5—Mas se abrio otro caxon de la marca de afuera que hubo quarenta y cinco libras detortillas [de tortillas] y bollos de chocolate;

———————————

Crate 6—Another crate was opened with an outside mark, and there were another 58 pounds of the above mentioned tortillas and chocolate buns;

Caxon 6—Mas se abrio otro caxon de la marca de afuera que hubocinquenta y ochco libras de dhas tortillas y bolos de chocolate;

———————————

Crate 7—Another crate was opened with an outside mark, and there were an embroidery bed canopy and a coverlet lined with taffeta and with silk and gold edges, plus another piece of cloth of the same, 6 molinillos, 12 cocos adorned with silver, a large coco ornate with silver and plated with gold, 21 jícaras from Guayra [Guaira-Caracas],10 medium size tecomates,[38] 11 jícaras, small and large, another painted jícara, a small flask with its lid from China, another small flask with its lid, looks like an ink well, several bundles of thin wax tapered candles [cerillas[39]] from China, two white cocos adorned with plated gold silver [golden silver], a silver cup with a stand made of gold plated silver, an ivory crucifix, and 12 fine silk and cotton woven pieces of cloth;

Caxon 7—Mas se abrio otro caxon de la marca de afuera que hubo un dosel y sobrecama de mengue labrado aforrado de

tafetan con punats de seda y oro = mas otro
paño menor de lo mismo — seis molinillos, doce cocos
guarnecidos en plata un coco grande guarneçido enplta
sobredorada = benytiuna jicaras pintadas = tres cocos
ordinaros = seis jicaras de Guayra + diez tocomates
[tecomates] — los dhos medianos once jicaras chicas y
grandes = otra J' [jícara] pintada lo mismo = una cajetilla
de China con su tapadera = otra cajetilla [de China] con su
tapadera q' pareçe tintero = quatro fajos de serillas [cerillas]
de China = dos cocos blancos guarneçidos de plata
sobredorada = una taça de plata de pie sobredorado = un
crucifijo de marfil = doçe paños de seda y algodon = y una
longa q'taya dho caxon por fondo;

Crate 8—Another crate was opened with an outside mark, and
there were 26 small boxes of chocolate each with one and a
half pounds;

Caxon 8—Mas se abrio otro caxon de la marca de afuera que
hubo beynteyseis cajetas de chocolate que peso a una libra y
media cada una;

Crate 9—Another crate was opened with an outside mark, and
there were 47 small boxes of chocolate each one with one
pound and 12 ounces;

Caxon 9—Otro caxon con la marca del margen que hubo
quarenta y siete cajetas de chocolate que peso cada una una
libra y doçe onças;

Crate 10—Another crate was opened with an outside mark,
and there were 96 small boxes of chocolate each with one and
a half pounds;

Caxon 10—Mas se abrio otro caxon de la marca de afuera
que hubo nobenta y dos cajetas de dho chocolate que peso
cada una libra y media;

Crate 11—Another crate was opened with an outside mark,
and there were 46 small dishes from China 1 is broken, 72
crocks [soup plates][escudillas] from China, 3 of them are
broken, 2 large platters [Fuentes], 6 of them large from China,
and 1 very fine cloak [capa[40]] with its hood;

Caxon 11—Mas se abrio otro caxon de la marca de afuera
que hubo quarenta y seis platos chicos de China, el uno
quebrado setenta y dos escudillas de China = tres
quebradas = dos Fuentes grandes de China y una capa con su
capilla;

Crate 12—Another crate was opened with an outside mark,
and there were 83 small boxes of chocolate and 7 boxes of
sweets; 6 of them were large boxes with 42 pounds each and
the others with one pound each; the small chocolate boxes had
one pound each;

Caxon 12—Otro caxon con la marca del margen que
hubo ochenta y tres cajetas de chocolate y siete cajetas de
conserva las seis dellas grandes quehubieron quarenta y dos

libras de dha conserva y las cajetas de chocolate hubo cada
una libra;

Crate 13—Another crate was opened with an outside mark,
and there were 91 small boxes of chocolate each one with a
weight of 2 pounds;

Caxon 13—Mas se abrio otro caxon de la marca de afuera
que hubo nobenta y una cajetas de chocolate que peso cada
Una a dos libras;

Crates 14–15—Another two crates were opened with outside
marks, and there were 50 small boxes of chocolate in each
crate; a total of 100 chocolate boxes and each one with
different weights that amounted to 137 and a half pound;

Caxones 14–15—Mas se abrieron dos caJones de la marca
de afuera que hubo cada caxon a cinquenta cajetas de
chocolate a diferentes pesos, que todas cien cajetas de dhos dos
caJones pesaron ciento y treynta y siete libras y media;

Crate 16—Another crate was opened and we could not find
any outside mark, and there were 4 silver stands [holders] for
jícaras, 6 gilded jícaras, 8 small cocos ornate with silver their
silver lids, 3 painted jícaras, 13 fine pieces of cloth, 3 of them
with their corners made with silk and gold and the other 10
with sides and corners in white color [tablecloths or church
altar cloths?] 4 pieces of embroidered fine cloth [tapestry?] 5
chocolate doilies from China [for the tray used to serve
chocolate], 7 doilies made of cotton, a piece of cloth made of
cotton and silk, and 2 molinillos;

Caxon 16—Mas se abrio otro caJon q' nose le allo marca
ninguna y hubo quatro tenedores de plata de Jicaras = seis
J'caras [jicaras] doradas = diez y ocho coquitos guarnecidos
de plata = treçe paños los tres dellos de puntas de seda y oro
y los diez rrestantes decostados y untas blancas = quatro
paños labrados de China = siete pañitos de algodon = un
paño de algodon y seda = y dos molinillos;

Crate 17—Another crate was opened, also with no outside
mark, and there were 59 small boxes of chocolate each one
with different weights; all the 59 boxes with a weight of 91
pounds 4 ounces;

Caxon 17—Mas se abrio otro caJon que assimismo nose allo
marca alguna que hubo cinquenta y nuebe cajetas de
chocolate que por ser las cajetas de diferentes pesos todas
cinquenta y nuebe cajetas pesaron nobenta y una libra y
quarto onças;

Crate 18—Another crate was opened with an outside mark,
and there were 59 small boxes of chocolate each with one and
a half pounds each;

Caxon 18—Mas se abrio otro caxon de la marca de afuera y
hubo cinquenta y nuebe cajetas de chocolate que cada cajeta
peso una libray media;

Crate 19—Another crate was opened with an outside mark [that can be seen] in the margin [of this page], and there were 24 small boxes of chocolate each with one and a half pounds;

Caxon 19—Mas se abrio otro caxon de la marca de afuera q' hubo beintey quatro cajetas de chocolate a libra y m^a cada cajeta;

———————————————

Crate 20—Another crate was opened with an outside mark, and there were 60 soup plates of China, 60 jícaras, and 5 molinillos;

Caxon 20—Mas se abrio otro caxon de la marca de afuera que hubo sesenta escudillas de China = sesenta Jicaras = y cinco molinillos

———————————————

Crate 21—Another crate was opened with an outside mark, and there were 137 small boxes of chocolate of one pound each;

Caxon 21—Mas se abrio otro caxon de la marca de afuera que hubo ciento y treynta y siete cajetas de chocolate que cada cajeta peso una libra;

———————————————

Crate 22—Another crate was opened with an outside mark, and there were 72 small boxes of chocolate of different types, sometimes two pounds each, and all together 93 pounds and 9 ounces;

Caxon 22—Mas se abrio otro caxon de la marca de afuera que hubo setenta y dos cajetas de chocolate que fueron de diferetnes pesos y psaroon todas las dhas sesenta y dos cajetas [de chocolate] nobenta y tres libras y nuebe onças y media;

———————————————

Crate 23—Another crate was opened with an outside mark, and there were 48 small boxes of chocolate each with one and a half pounds, 4 jícaras, 2 molinillos, 6 chocolate dolies from China [to serve chocolate];

Caxon 23—Mas se abrio otro caxon de la marca de afuera que hubo quarenta y ocho cajetas de chocolate = quatro jicaras = dos molinillos = y seis pannitos chocolateros de China y las dhas cajetas de chocolate peso cada una una libra y media;

———————————————

Crate 24—Another crate was opened that had an outside mark in the margin, and it had 12 molinillos, 24 painted jícaras, and 24 doilies [to serve chocolate] from Campeche;

Caxon 24—Mas se abrio otro caxon de la marca de afuera que tuvo por marca el letrero del margen que tubo doce molinillos = beyntey quatro jicaras pintadas y beite y quatro pañitos de campeche;

———————————————

Crate 25—Another crate was opened with an outside mark, and it had one painted pack cloth sack that was not inside a crate but came as a bale with a total weight of 7 arroba and 12 pounds [i.e., 187 pounds] of wild cochineal; the petate

[palm mat] and the leather [of the bale] had a net weight of 21 pounds, so the net weight of the cochineal was with 6 arrobas and 16 pounds [i.e., 166 pounds] of cochineal;

Caxon 25—Mas se abrio otro caxon de la marca de afuera que tubo un costal de arpillera pintada con cochinillasilvestre q' no fue caxon y el otro tercio peso con la cochinilla siete @ y doce libras bruto y el cuero del petate en que benia aforrado peso beyntey una libras que neta la dicha cochinilla hubo seis @ y diez y seis libras;

———————————————

Crates 26–27—Another two crates were opened with outside marks and lined with leather, and there was fine cochineal; one of the crates had a net weight of 5 arrobas and 5 pounds [i.e., 130 pounds]; the crate and the leather [with which it was lined] weighed 19 pounds, and the sackcloth weighed three pounds. The other crate had a net weight of 5 arrobas and one pound of fine cochineal [i.e., 126 pounds];

Caxones 25–27—Mas se abrieron otros dis [dos] caxones con la marca de afuera que hubo aff^{dos} en cuero que tubieron cochinillafina y el uno de dhos caxones peso Cinco @ y çinco libras netas dos @ y diez y nueb libras que peso el caxon y el cuero en que benia y otras tres libras de costal. El otro caxon peso inco @ y una libra de cochinilla neta bajada de la tara de todo

———————————————

Crate 28—Another crate was opened with an outside mark, and there was a net weight of 4 arrobas and 22 pounds of cacao [i.e., 122 pounds].

Caxon 28—Mas se abrio un tercio [fardo] la marca de afuera que hubo 4 @ y beynte y dos libras de cacao neto.

———————————————

Such inventory was made by the above mentioned sirs, the Royal Officials, in my presence and [in the presence] of Alonso Hernandes de Cendrera, defender of the goods [merchandise]. This I attest signed December 23, 1641.

El qual dho [dicho] inbentario se hiço por los dhos [dichos] ss^{es} [señores] off'' [oficiales] R' [reales] en mi pr' [presencia] y de Al^o H. de çendrera defensor de los dhos [dichos] bienes de que doy fe ff^{do} [fechado] 23 diz$^^$ [diciembre] 1641. [39]

The first paragraph of document 2 is rich in paleographic examples of mid-17th century Spanish writing: absence of punctuation and accents, use of the *c* with the cedilla (or *ç*) not only in front of the vowels *a*, *o*, *u* to produce a soft *z* or *s* sound still maintained today in modern French as explained earlier, but also in front of *e* and *i* as in *çer* → *ser*, or *açi* → *así*. During the 18th century use of the *ç* began to disappear and by the 19th century would be a rare finding. Use of the double *r* or *rr* at the beginning of the word later was replaced by a single *r*, to produce the vibrant strong *r* sound of Spanish, like in *rreales* later transformed into *reales* (the plural for royal in this sense, but also the name of a coin) or *rrematandose* that would become *rematándose* (auctioning; to finish the kill—in English the sense of "making a killing"). Use of the double *rr* at the beginning of a word like use of the ç also began to disappear during the 18th century. Use of the double *f* or *ff* as in *official* → *oficial* (official) was abandoned at the same time as the ç. Common use in the 17th century was also the letter *x* instead of *j*, as in *caxon* → *cajón* (crate or wooden box) or *caxa* → *caja* (box). Use of *j* instead of *h* as in *jacienda* → *hacienda* reflects, too, an important regional variation in the phonetics with the aspiration[41] of the otherwise soundless letter **h** characteristic of Andalusia in the south of Spain. This sharp Andalusian aspiration makes the silent **h** sound like a **j**; starting as a phonetic variation, it finally was adopted in spelling. In the same way that Flamenco dancers and singers vocalize their deep throat sound *Cante Hondo* (Deep Song)—the epitome of Andalusia—so is the way they still pronounce it and the way it was written: *Cante Jondo* → *Cante Hondo* [40].

Another regional variant that became popular in the first half of the 16th century can be seen in the use of the double **ll** instead of **rl** in cases where an infinitive—or in the **dl**, an imperative for the second person plural, *vosotros*—form is followed by the direct object, as in the case of *mandallo* → *mandarlo* (to command, to send) or *mandallo* → *mandadlo* (command it, send it). These are writing forms of phonetic variants that characterize even today the region of Andalusia in southern Spain.[42] The absence of the **h** at the beginning of words, as in *aga* → *haga* (subjunctive form of the verb to do or to make), use of the Latin **i** instead of the Greek **y**, as in *ai* → *hay* (there is, there are), as well as the various abbreviations expressed in the document, like *Su Mag.* → *Su Majestad* (Your Majesty), *Vm* → *Vuestra Merced* (Your Lordship), *dho* → *dicho* or the feminine form *dha* → *dicha* (such, the above mentioned) or *man'fieste* → *manifieste* (show, manifest) and the extensive use of linkage among three or four words that make difficult the reading and understanding of the documents of this century, as in *çelearabuenaobra* → *çe le ara buena obra* → *se le hará buena obra* (it will be good for them), also are seen in other documents of the period. Some examples from the crate inventory list are enlightening. In the inventory we find again the use of **ç**, as in *doçe* → *doce* (twelve), *taça* → *taza* (cup), or *onça* → *onza* (ounce), and abbreviations like *q'* → *que* (that), *m'a* → *media* (half), and *J'cara* → *jícara* (gourd cup). It is important to remember, too, that in the original Mexica/Aztec language, these chocolate cups made out of gourds are written with an **x** → *xicara*.[43]

There are other important, interesting forms in these documents as well. One is the use of **qu** instead of **cu**, as in *quatro* → *cuatro* (four), *quarenta* → *cuarenta* (forty), or *cinquenta* → *cincuenta* (fifty), very similar to French forms as in *quatre, quarante*, and *cinquante* (four, forty, and fifty), respectively. The contraction *dellos* → *de ellos* (of them or from them) or in its feminine form *dellas* → *de ellas* (and in their singular della) arose as a consequence of the natural elision[44] (liaison) between the preposition *de* and the definite article *ellos* or *ellas*: *de + ellas* → *deellas* → *dellas*; this form was used well into the 17th century but disappeared toward the 18th century. The only contractive form that remains in Castilian Spanish in the 21st century is the contraction *de + el* → *del*, but not for the feminine form *de + la*, where the preposition and definite article maintain their individuality. It is easy to see the close relationships among the Romance languages as the Italian *della* commonly is used in surnames. In this instance the contraction between **de** and **la**) is used to indicate place of origin as in the Italian Renaissance philosopher Giovanni Pico della Mirandola[45] [41].

The inventory of crates also provides a good example of the interchangeable use of two important verbs: *haber* and *tener*. Both verbs translate into English as "to have" and their differentiation is a difficult topic for beginning Spanish language students. In the 21st century, *haber* has the meaning of "to have" and is used as an auxiliary in the perfect tenses, as with "I have written" or "I had written." But in a more restricted sense *haber* has the meaning of "to have existence." This is in contrast with the verb *tener* that always implies the possession of something. Castilian Spanish only began to standardize rules in the use of verbs during the 17th century and into the 18th century. Earlier in the the 16th century distinctions between *haber* and *tener* were not well defined; the verb *tener* (to have, to hold) was used to indicate possession, while *haber* was used to mean "to have existence" or "to obtain." During the 17th century when the present document was written, use of *haber* was more confined to the auxiliary of perfect tenses or in structures that indicate existence, like *hay* (there is) or *había* or *hubo* (there was), but the use of both verbs continued to be a mixture of both meanings [42].

Consider Crate 24, with one dozen *molinillos*, two dozen *jícaras*, and two dozen chocolate dolies, and Crate 25, with cochineal. Whoever transcribed the inventory of their contents used the form *tener* when he wrote *tubo* (tuvo), the preterite indicative form of the verb *tener*. For the rest of the crates, however, the verb choice was *hubo*, the preterite form of the verb *haber* used to show existence, as in "there was." It is possible (and is our speculation) that different portions of the crate inventories were written by different persons. Diplomatic paleography (i.e., the paleography of documents) originated due to the need to verify authenticity of medieval documents that required veracity of accounts, so much so that the artistically elaborated signatures of the 15th, 16th, and 17th centuries were not only meant to be beautiful but to make difficult any fraudulent imitations. It is not easy, therefore, to account for the variable use of these verbs in the present document [43].

The crate inventory also reveals use of a very common abbreviation, @ for *arroba*, seen in most of the documents we have read that listed provisions and their weights. It is an abbreviation sign that in our current cyberspace communication system is used to indicate the electronic address of a person. For many readers this is a symbol of the new era of 21st century communications, but in reality it has been used since ancient times as a weight measure [44].[46] It is easy to see that the abbreviation @ consists of a calligraphic form of the letter $a \rightarrow a$, surrounded by a curved abbreviation sign that encircles it. An *arroba* was the equivalent of 25 pounds; thus in Crates 26 and 27, where it was written 5 *arrobas* and 5 pounds in one crate and 5 *arrobas* and one pound of cochineal in the other crate, we may calculate a total quantity of 256 pounds of cochineal; Crate 28, with 4 @ and 22 pounds of cacao, would yield 122 pounds of cacao.

The St. Augustine chocolate story continues with the next document. The royal officials were suspicious of Captain López's decisions and his integrity in handling the salvage process and accused him of loading merchandise aboard his ship without registering the contents (i.e., smuggling).

DOCUMENT 3: JANUARY 18, 1642

The Official Judges of the Royal Treasury of this province of la Florida, the most excellent sirs Francisco Menendez Marquez y Nicolas Ponce de Leon, auditor of the kingdom [of Spain] say that in conformity with the inventory that has been made of the goods and merchandises that came in the ship named Nuestra Señora del Carmen y el Rosario *that moved away from its original course due to the weather[derrotado] arrived at this port of San Agustin and whose owner is Captain Hermenexildo Lopez it seems that there are merchandise and goods that were not registered; and these are: 148 crates of sugar; 32 crates of chocolate; 13 boxes with bundles of sarsaparrilla 29[47]; 4 bundles [petaca][48] of copal[49]; 4 bundles of tobacco; 1 crate of tableware from China; 1 crate of tamarind[50] fruits; 2 crates of fine cochineal; 1 bundle of not so fine cochineal; a small crate with 24 painted jicaras, 24 molinillos and 24 pañitos de chocolate [doilies to serve chocolate]; another crate of chocolate cups and soup plates from China; 1 tercio[51] [i.e., 220 pounds] of cacao; 9 chocolate grinding stones; Some jícaras adorned with silver; 1 cup of gilded silver. All this was confirmed when all the crates were opened in the presence of Your Lordship and in our presence, and also in the presence of the aforementioned captain Ermenexildo Lopez; and these crate carried all the merchandise and goods to the Kingdom of Spain without having register them, as it is mandated by the Royal Ordinances that must be obeyed Your Majesty, [but carrying them] secretly and as goods that have been smuggled, and charge Captain Hermenexildo Lopez, since there is no grand master[maestre 31[52]] on the ship for having carrying them without register in contravention of the aforesaid ordinances. For the reason expressed we ask and we beg Your Lordship that*

you declare, and that this also hold for the palo de campeche[53] [Campeche wood], because the amount stored was more than what had been registered, and charge him for having loaded in the ship without registering it; and [we ask] that in all [these matters] may be prosecuted with justice and in the way that most suits the Royal Treasury of Your Majesty; and on this we ask for justice in the year 1642. Signed by Francisco Menendez Marquez and Nicolas Ponce de Leon.

[Sent to] Damian de la Vega Castro y Pardo Governor and General Captain of these provinces of Florida, in the name of the King Our Lord on January 18, 1642. [45]

Henero 18, 1642

Los jueces off⁵ [>oficiales] de la R′ [Real] Haz^da [Hacienda] desta [de esta] provincia de la Florida, los E ismo [excelentísimos] ss {Señores} fran^co menenz marq⁵ y Nicolas Ponce de Leon, contador del Reino (decimos) que conforme al inbentario [inventario] quesehiço [que se hizo] de la hacienda y mercaderias que vinieron enel [en el] nauio [navío] nnomdo [nombrado] ntra Sra del Carmen y el Rosario quetro [que entro] de Rotado [derrotado] aes tepuerto [a este puerto] (de San Agustin) y quees [que es] del Capp^n Ermenexildo Lopez pareçe que ay mucha mercaderia y xeneros [generos] quenobinieron [que no binieron] rexistradas que son lassiguientes[las siguientes] = ciento quarenta [cuarenta] y ocho cajones de asucar [azúcar]-con varriles [barriles] de dha [dha] asucar [azúcar] = treynta y dos cajones de chocolate = trece cajones de sarçaparrilla catorce tercios [bundles] de dha sarça [sarçaparrilla]-quatro petacasde copal quatro tercios de tabaco en rollo = un cajon de losa [loza] de China = un cajon de tamarintos [tamarindos] dos cajones de grana fina un tercio de grana no tan fina, un cajoncillo con 24 xicaras pintadas y veynteyquatro [veinte y cuatro or veinticuatro] molinillos y veynteyquatro pañitos de chocolate, otro cajon de xicaras y escudillas de China, un tercio de cacao, nuebe [nueve] piedras con dieziocho[dieciocho] manos de moler chocolate, algunas xicaras guarnecidas de plata y una taça de plata sobredorada-Como mas por esto constara por el embarco que de todo se abrio quando seabrieron [se abrieron] los dhos [dichos] cajones en presençia de V^m [Vuestra Merced], y la nuestra estando presente el dho capn Ermenexildo Lopez . . . porque llebababa todas las mercaderias y xeneros a los Reynos de Espanna [España]sin averlos [haberlos] rexistrado como sumag^ lo tiene mandado porque v^tras R⁵ ordenanças se deben aplicar llevandolos escondidos como cosa descaminda que hacerle cargo al capptan Ermenexildo Lopez como dueño de dho [dicho] navio por no aver maestre por averlos [haberlos] traydo [traído] sin rexistro contrabiniendo a las dhas R′ hordenanças = Por tanto a Vmd pedimos y suplicamos lo dec′are [declare] assi [así] y lo mismo sirviere de palo de campeche[Palo de Campeche] y questa [que está] almacenado mas de lo que benia[venía] rexistrado [registrado] y que sehaga [se haga] al dho[dicho] cargo a dho [dicho] capitan Ermenexildo Lopez por averlo embarcado sin rexistro y en todo se (sirva) de proçeder lo que fuera justicia y mas conviene a la R′ Hazienda se Su Mag′ [Su Majestad] sobrequepedimos[sobre lo que pedimos] justicia y en milseiseint/′ [mil seiscientos cuarenta y dos] Signed by Francisco Menendez Marquez and Nicolas Ponce de Leon.

[Sent to] Damian de la Vega Castro Y pardo gobernador y Cappan gen´ destas provyas de la Florida –Por el rrey ntro⁰ seno⁰ [Rey Nuestro Señor] en dieziocho dias del mes de henero [enero] de mil y ∫ᵗᵒˢ quarenta y dos [de mil seiscientos cuarenta y dos]. [45]

In addition to listing the deeds of this daring captain who cast away into the sea many objects and crates to save his ship from the fury of the hurricane and to save his cargo of chocolate, the zealous royal officials, through their calligraphy and syntax provide interesting examples of abbreviations, erratic word separation, juxtaposition (where two or more words are written without separation as a single long word), and a wide array of interchangeable letters, some already encountered in Documents 1 and 2.

Abbreviations have been used since the Roman Empire in order to write more quickly and to use less space. Their use expanded during the last centuries of the Empire as the Romans developed a system of *siglas* (acronyms[54]), for example, AC for Augustus Caesar. Acronyms also served as a type of shorthand—called Tironian notes—used to record formal public speechs [46].[55] Some abbreviations, particularly those related to ecclesiastic or royal matters, became standardized, while others appear to be individual decisions made by writers. Generally, abbreviations can be divided into several categories. In the present documents abbreviations are characterized by suspension or truncation (apocope or apocopation), where the final letters of the word are eliminated and replaced by an abbreviation sign that can be simply a dot—still used in the 21st century as with the abbreviation for the English word *street → st*. In the St. Augustine documents several types of abbreviations may be noted, among them, a slight diagonal segment ⌡, a mark like an apostrophe, or other more elaborated signs:

R´ Hazienda → Real Hacienda [Royal Treasury];

ss. → señores (sirs)

sarça. → sarçaparrilla (sarsaparilla)

gen' → general (military general)

sumag' → Su Majestad (Your Majesty)

Other abbreviations in the documents are formed by syncope (contraction), where the middle letters of the word are removed. Such instances also carry a horizontal line over the abbreviated word, for example:

n̄nom̄do → nombrado (mentioned, above mentioned)

p̄rovyas → provincias (provinces)

p̄cias → provincias (provinces)

d̄h̄a → dicha (said, the aforesaid)

E ismo → excelentísimo (most excellent (sir))

N̄tra S̄ra del Carmen → Nuestra Señora del Carmen (Our Lady of Carmen)

Another type of abbreviation used in this letter is noted by superscripts, either vowels or consonants, commonly used to shorten personal names. It is not uncommon to find a combination of various types, as *in the name and surname abbreviations: franᶜᵒ menenz marqˢ → Francisco Menéndez Márquez* or different forms used to abbreviate titles even in the same document: *Cappⁿ → Capitán* (Captain) but also in a nonsuperscript abbreviation, *Cappan*; or in governmental institutions: *Real Hazᵈᵃ → Hacienda* (Royal Treasury) in contrast with the above mentioned *R´ Hazienda*, and *offˢ → oficiales* (officials); in very formal protocol formulas to address high rank officials or the king as in: *Vᵐ → Vuestra Merced* (Your Lordship) versus the second document *Vm* with no superscript; in edicts, decrees, or ordinances: *vᵗʳᵃˢ Rˢ ordenanças → Vuestras Reales Ordenanzas* (Your Royal Ordinances).

It was common usage in the 17th century to write the year with letters instead of numbers, and in official documents it was normally not abbreviated; this document, however, shows the following abbreviation: *mil y ∫ᵗᵒˢ y quarenta y dos → mil seiscientos cuarenta y dos* (i.e., 1642). The presence of all these variants in the same document shows the lack of regulation of abbreviations at least in the governmental and administrative records of the 17th century. It is interesting to note that sometimes the use of abbreviations did not actually fulfill its original intention (to write faster and use less space); in some instances specific abbreviations became more elaborate than the word itself that the author tried to shorten, and took the same amount of time and space (or more) to write the complete word. We can see this with the following:

de⌡Rotado → derrotado (forced to deviate from the normal ship course)

dec'are → declare (declare)

Another feature of this document from the paleographic point of view is the profusion of examples of letters no longer in use or the use of different spelling variants. In the third document they are pervasive. For the **ç**: *hiço → hizo* (did, made); *pareçe → parece* (seems); *proçeder → proceder* (proceed); *sarçaparrilla → sarzaparrilla* (sarsaparilla); *ordenanças → ordenanzas* (ordinances); *hordenanças → ordenanzas* (ordinances); *taça → taza* (cup). But surprisingly, an **s** instead of a **ç** was used as in *asucar* instead of *açucar → azúcar* (sugar), and **qu** in lieu of **cu**, *quando → cuando* (when), *quatro → cuatro* (four), and *quarenta → cuarenta* (forty). With these examples, as mentioned before, it is not difficult to see a strong resemblance with the French words: *quand, quatre, quarante.*

Additional interesting variants also may be identified such as the substitution of **b** for **v**: *averlos → haberlos* (have them); *benia → venía* (came); *contrabieniendo → contraviniendo* (in contravention); *inbentario → inventario* (inventory), *nuebe → nueve* (nine); and *varriles → barriles* (barrels); the use of the Greek **y** instead of the Latin **i**: *reynos → reinos* (kingdoms); *traydo → traído* (past participle of brought); and predominantly in numerals, especially when writing numbers in the twenties and thirties: *veynte → veinte* (twenty) and *treynta → treinta* (thirty).

Use of the letter **x** as well as the letter **h** (or the absence of **h**) also deserves special comment. Use of the letter **x** is associated with a mixed **s** and **z** sound. The sound of the Castilian speech was more like the English **sh** (as in the word she) or the French **ch** (like Chateaubriand). Toward the end of the 16th century, the **x** sound began to be written as an aspirated **h** or as a **j**, but use of the **x** continued in the St. Augustine documents of the 17th century as a substitute for modern **j** or **g**, as in *xeneros → géneros* (goods, items, merchandise); *rexistradas → registradas* (past participle of register); and *rexistro → registro* (register, record). It is important to mention, too, that in the case of Mexica/Aztec Nahuatl language, words like *xicara* (gourd cup used to serve cacao beverages) transcribed often by the Spanish as *jícara*; this would reflect the Castilian intent to associate the Nahuatl phonetics to what was known to them at the time: The **x** (with the sound of **j** or **g**) is not characteristic of the Nahuatl language. Use of the **x** in words of Nahuatl origin, however, can be seen today fully used in Mexico, like in Xochimilco (floating garden district in Mexico City), Oaxaca (city and state in southern Mexico), or *xicaras*.

The absence of the initial **h**, as in *ay → hay* or *aver → haber* (to have), can be explained as a form that evolved as a consequence of its soundless nature. In Castilian Spanish the presence or absence of the initial **h** does not alter the sound and therefore might have been considered a superfluous letter that could be eliminated. However, as pointed out by Nebrija [47], words that begin with a **u** like *ueso → hueso* from Latin *ossum* (bone) or *uevo → huevo* from Latin *ovum* (egg) had to carry an **h** because in those centuries the **u** and **v** were used indistinctly to represent the vowel sound of the **u** or the consonant sound of the **v**. Use of the **h** preceding a word beginning with **u** indicated that it was the vowel sound that should be used. On the other hand, we also see in this document the use of an initial **h** preceding a vowel (but not a **u**), as in *henero → enero* (January), a usage inherited from Latin that later would disappear.

The juxtaposition of words seen so often in manuscripts from the 16th century, continued during the 17th century, and then fell into disuse during the 18th century. This concatenation of words and sometimes the erratic separation of words or syllables in combination with abbreviations and spelling variants add an extra challenge to reading and transcribing these early Spanish manuscripts. Examples are abundant in Document 3 as seen here:

quesehiço → que se hizo (that was made)

enel → en el (on it)

quetro → que entró (that arrived)

aes tepuerto → a este puerto (at this port)

quenobinieron → que no binieron → que no vinieron (that did not arrive)

lassiguientes → las siguientes (the following)

quees → que es (that is)

seabrieron → se abrieron (were opened)

questa → que está (what is)

sehaga → se haga (be made)

sobrequepedimos → sobre lo que pedimos (on this/what we ask)

Continuing with the St. Augustine saga, three months later while the royal officials were still debating the actions of Captain Hermenexildo López, the auction of the merchandise finally was authorized. In Document 4, it is interesting to observe the emphasis on the chocolate boxes, items always listed first.

DOCUMENT 4: MARCH 6, 1642

On March 6, 1642, in the city of San Agustin Province of Florida, the Castilian Lord and Governor of this City Don Damian de la Vega Castro y Pardo, Governor and Captain of these provinces for His Highness, and the Official Judges of the Real Audiencia [Royal Superior Court in the Americas] Don Francisco Menendez Marques, for His majesty and Sacred Monarch, and the Accountant Adrian Canizares, in the absence of the merchandise owners, say that: according to the edict prepared under the supervision of the notary Juan Ximenez it is mandated to proclaim in public places and to auction all the goods that arrived as cargo in the ship named Nuestra Señora del Rosario y el Carmen, *whose owner and captain is Don Ermenexildo Lopez; and in obeying the above mentioned edict, the chocolate and the other items have been auctioned along with the other things that arrived to the Royal Warehouses have to be auctioned because it is considered [because it is salvage merchandise] unfit to be sold: chocolate, sugar, sarsaparilla, Palo de Campeche, hides, tanned leather, fine cochineal, wild cochineal, indigo blue, tobacco, chocolate grinding stones and other small things that are ordered to be auctioned and that all these goods be publicly proclaimed for nine days. So, then can be auctioned with the assistance of those that represent the merchandise owners and the Royal hacienda officials mandate that Alonso Fernandez Çendrera be present. [48]*

*En la Ciu' de san Agustin Proy^a de la Florida en seis del mes
de março de mill seiscientos quarenta y dos años el señor
castellano damian debega Castro y Pardo Goⁿ y Capp^{an} Gn en
destas dhas Provy^{as} Por el rrey mo^ señor y los jueces ficiales
de la R hacienda ante Don Fran^{co} Menendes marques, por Su
Mag. y sar^ m^a adrian canizares por ausencia del
propietario = dixeron que por quanto paso ante el escibano
Juan Ximenez esta mandado regonar y hacer la almoneda de
los bienes y hacienda que bino de arribada por caso fortuitose
en el nabio llamado Nuestra Señora del Rosario y el Carmen,
de que es duen° el cappn ermenegildo Lopez y en
cumplimiento de dho acto se hiso Remate de chocolate y de
otras menudencias que entraron en los reales almacenes y Por
quanto los generos de mas cantidad y consideracion estan por
venderse y Rematar que son asucar = sarcaparrilla = Palo de
campeche = cueros pelo = cueros curtidos = grana
fina = grana Silvestre = annil = tabaco = piedras de moler
chocolate y otras menudencias mandaban y mandaronse aga
almoneda de los dhos generos y le haygan en Pregones nuebe
dias Para queal cabo dellos Rematar con asistencia de [los
representantes]de los bienes y de los fiscals de la R hacienda
en la Persona que mas sirva Por ellos y echo acordaron
[illegible] mandaron desir sus nombres = a cuyo remate y
almoneda Mandan este presente Alonso fernandez de Sendrera.*
[48]

Here again the author of Document 4 shows irregular use of the letter **h** sometimes replaced by the letter **f**. In Documents 1 and 2, the last name of the government official in charge of the defense of the merchants' interests (Alonso Hernandez/Fernandez de Cendrera) was written with an **H**: *Hernandes*, while in the present document his name is written with the letter **F**: *Fernandez*. This was a common characteristic of the medieval writing that also appears in *Don Quijote*: as in *fermosa → hermosa* (beautiful). The first name of the captain also is spelled differently, in this document it is *Ermenexildo* but *Hermenexildo* in the earlier documents. This change shows once again the individual factor in the variants of the written language in a time when there still were not well-defined rules, and place of origin would determine writing style.

The remainder of Document 4 describes a series of *pregones* (public proclamations) announcing auction of the merchandise, specifying what was to be sold and prices. Between March 7 and March 20, 1642, nine auctions took place in St. Augustine. During the last auction held on March 20, nine chocolate grinding stones (*metates*) were sold at the price of 21 *reales* each. Included with each metate were its "two hands." [56] This important clue provides a very clear indication that the people of St. Augustine already were familiar with chocolate, and the date(s) for introduction of chocolate to the Atlantic coastal region of North America *must precede 1642!*

Although the auctions authorized by the Royal Supreme Court of Justice allowed the merchants to recover some of the money invested, the merchants were determined to obtain a much larger compensation for their losses. Then began a series of complaints and accusations issued against Captain Hermenexildo López. The next two documents, numbers 5 and 6, describe an intricate series of claims and demands from the royal officials against Captain López, and counterclaims/demands from Captain López against the merchants. The captain, as was his right, requested damages himself in accord with laws-of-the-sea, and he sought compensation for all the items that he had to throw overboard in order to save the ship from sinking—and in order to save the rest of the valuable cacao/chocolate merchandise. As part of these claims and charges, the royal officials were particularly interested in knowing the contents of the locked chest that belonged to Captain López and why he held 1050 pounds of sugar and 190 pounds of chocolate at his quarters in St. Augustine.

DOCUMENT 5: MARCH 12, 1642

In the city of Havana on March 12, 1642 appeared in front of me, the notary, and [in front of] the witnesses, here underwritten, Fernando Phelippe de Abovar, lawyer and resident of this city, I attest I know him, and said that the Castilian Lord Don Damian de Vega Castro y Pardo, Governor and General Captain of the Presidio [of San Agustin] and Provinces of Florida, remitted through Captian Andres Gonzales a testimony signed by Pedro de Arisgoechea—notary of the government of these provinces—over a litigation undertaken by Ermenexildo Lopez as the owner and lord of the ship named Nuestra Señora del Rosario y el Carmen, one of the ships of the war and merchant fleet [conserva[57]] under the command of Admiral Juan de Campos. [This ship] arrived to the port of San Agustin of the above mentioned provinces of Florida because of the weather [derrotado[58]]. This litigation involves the freighters [i.e., the merchants who ship goods for other markets] and those concerned with the load carried on the ship, and representing them is Alonso Fernandez de Cendrera—defender [of the merchants] named by the above mentioned Lord Governor; [and refers] to the damage sustained by the merchandise [while at sea]; and the above mentioned Ermenexildo Lopez is asking the freighters to pay for the masts, yards [lugsails] sails, rigging and cordage that he had to cut and threw into the sea, as well as other instruments, furniture [ropa[59]] and other goods that he tossed into the sea during the storm that on September 29 of last year [i.e., 1641] overcame the above mentioned ship and the war fleet [armada de flota] and sailed out of the Bahamas channel at 30 degrees of latitude to shelter and save the ship and the rest of goods that were loaded on it. Attached to this testimony it is also the cause [the lawsuit] that the Royal Officials of the Royal Treasury of Your Majesty in these provinces, Don Francisco Menendez Marques, treasurer, and Nicolás Ponce de León, public auditor, have undertaken against the ship's freighters [the merchants] and those concerned are undertaken—through their defender Alonso Fernandez Cendrera—and against the above mentioned

Ermenexildo Lopez, in relation to the merchandise without register [nondeclared merchandise] that was found aboard of the ship—as detailed in the report that the Royal Officials presented to the Lord Governor on January 18 of this year [1642]. . . . [In relation with the Ermenexildo Lopez cause] We ask VM [Your Lorship] to declare as rightful the general damage rights [derecho de avería gruesa] petitioned for Ermenexildo Lopez and in consequence condemn the freighters of that ship and the load that has been salvaged to apportion the merchandises—through a third party accountant, and that during the 10 days following the auction of the ship, Ermenexildo Lopez must obtain proof of the value of the yards, masts, sails and rigging he threw into the sea, as well as the furniture and objects that belonged to him . . . and also that the salvage goods be apportioned along with the amount in which the forementioned ship was auctioned according to the estimations of Alonso Fernandez de Cendrera, the defender of the merchants, of the merchandise that was cast overboard: Palo de Campeche, boxes, and other goods belonging to private persons and interested owners, and that the cost of the damage merchandises should be paid to the aforementioned ermenegildo Lopez based on the monies that their owners receive. [49]

En la çiudad de La Havana en doze [doce] dias del mes de março [marzo] demil Y seis cientos [mil seiscientos] Y quarenta Y dos [cuarenta y dos] ante mi el escriuano [escribano] y testigos Y uso escriptos [subscritos] parecio [compareció] Fernando Phelippe [Felipe] de Abouar [Abovar] abogado y vecino desta ciudad a quien doy fe que conozco y dijo que el Señor Castellano damian de vega Castro y pardo [Damián de Vega Castro Y Pardo] Gouer^{or} [Governador] y Capitan General del Presidio y prouincias dela florida Por sumagestad [Su Magestad] ¢ [se] rremitio [remitió] con elcapitan [el capitán] Andres Goncales [Andrés Gónzales] un testimonio que Parece estar firmado depedro de aris gochea [de Pedro de Arisgoechea escriuano [escribano] de gouernacion [gobernación] de las dhas prouincias [provincias] deun pleyto [pleito] queseaseguido [que se ha seguido] Por parte de ermegildo LopeB Como dueño y señor que fue del nauio [navío] nombrado nuestra señora del rrosario y el carmen [Nuestra Señora del Rosario y el Carmen] uno [de los barcos] de la conserua [conserva] dela flota de nueva españa {Nueva España] del cargo del almirante Juan de campos {Juan de Campos] que derrotado arribo al puerto de san agustin [San Agustin] De dhas prouincias de la florida Co los Cargadores eiynteresados [e interesados] en lacarga [la carga] del dho nauio Y por los suso dhos [susodichos] en su nombre con Alonso fernandeB deçendrera [Alonso Fernández de Cendrera] Defensor que fue nombrado Por el dho señor Gouer^{or}—[Gobernador] alos [a los] dhos cargadores sobre la aberia [avería] Gruesa que el dho ermegildo LopeB a Pedido selepague [se le pague] Por los otros cargadore ∫De lo aruoles [Earboles] velas y xareza [jarcia] que corto Y echo a la mar y otros pertrechos del dho nauio rropa [ropa] yacienda [y hacienda]abento [aventó] enla tormenta que al dho nauio Y armada de la flota sobreuino aveinte [a veinte] y nueue [nueve] de septiembre del

año proximo pasado de mil yseisceintos y quarenta y uno aviendo [habiendo] desembocado[en] el canal de vahama [Bahama] en altura detreinta [de treinta] Grados Para segurar y saluar [salvar] rl dho nauio Y demas acienda queenel [que en el] venia cargada = En cuyo testimonio viene agregada tra causaquelos [que los] oficales rreales [reales] delarreal [de la Real] acienda [Hacienda] desumag′ [de Su Magestad] de dhas prounicas Don fran^{co} menendeB marqueB[Don Francisco Menéndez Márques] tessorero [tesorero] y nicolaspponce Deleon [Nicolás Ponce de León] contador an [han] seguido y siguen contra los cargadores eYnteresados del dho nauio Y Por ellos contra el dho Alonso fernandes de çendrera como defensor dellos [de ellos] Y contra el dho ermegildo LopeB sobre la carga Y mercaderias que se allaron [hallaron] venir sin rreJistro [registro] en dho nauio contenidas en el escrpto [escrito] de dhos oficiales rreales Presenatdo abte el dho señor goueror endiez y ocho [en dieciocho O] dias del mes de enero deste Presente año. . . . [Y en relacion con la causa de Ermenegildo Lopez] le pedimos a dho señor gouern^{or} declare auer [haber] lugar de derecho laaueria [la avería] gruesa pedida por aprte del dho ermegildo LopeB y que en consecuqncia deello [de ello] condene a los cargadores del dho nauio Ya [y a] la carga que delsesaluo [de él (del navío)] se salvó] a que a precisandosse [apreciandose] toda ella con particularidad la que a cada uno pertenece se repartan por terceros contadores que para ello se nombren en las dihas mercaderias que estan en salvo y en lo que consta por los autos averse [haberse] rrematado [rematado] en dho nauio lo que dentro de 10 dias d'e la notyficacion dela [de la sentencia] prouarese al dho ermegildo LopeB [lo que] valiant las vergas y jarcia que se corto y echo a la amr, del dho nauio, vastimentos [bastimentos] y anelas. Rropa que consta el susodiho echo alamar [a la mar] Y alexo [alegó] perteneciente a dho ermenegildo LopeB. . . . y assemesmo [asimismo] se rratee [proratee] ente la dha acienda que esta en salvo y en lo que se rremato [remató] el dho nauio lo que el dho Alonso Fernandez de Cendrera defensor de dho cargadores prouare [provae] dentro de dhos terminus [lo que] se alexo [alejó/echo] a la amr de palo [Palo de Campeche] y mercaderias caxas [cajas] y otras cosas de personas paricualres eYnteresados del dhoo nauio y lo que prouase valian y assi [así] rrepartido [repartido] lo que le tocase a cada uno ratee por la cantidad de lo alexado [echado] a lo queesta [que está] en salvo y paguen las dhas mercaderias que estan en salvo al dho ermengildo LopeB. [49]

DOCUMENT 6: MAY 2, 1642

Alonso de Cendrera, comptroller of the Treasury of Her Majesty in this city [of San Agustin] and defender of the goods that arrived, moved away from its original course [bienes de arribada[60]], in the ship named Nuestra Señora del Rosario y el Carmen, whose owner and lord, Captain Hermenexildo Lopez presented a cause on the damaged merchandise [avería gruesa] and other issues on the liquidation and apportionment that due to the sentence pronounced by Your Lordship, I said: Your Honor, I ask and I beg you: I have sent you the interrogatory of questions and my witnesses should be examined and having obtained the proof that my conscience tells [me] I am asking for your justice; also I say that the seven crates of sugar that the Captain opened to pay his sailors and that he affirms is his and that he bought them on his account had 42 arrobas [i.e., 1050 pounds] and the two crates of chocolate, one had 52 small boxes of chocolate of 1 pound and a half each [i.e., 70 pounds of chocolate] and the other had 80 small boxes of one pound and a half each [120 pounds of chocolate]; and I do not know what was inside of his chest [escritorio[61]] I am asking, as it is the right of those I represent, to order an appraisement and that the Captain Hermenexildo Lopez declare under oath clearly and openly either confessing or denying the amount of pounds of chocolate that the first crate had and which goods were in his chest [that he opened]; because until now we have no exact proof of what it carried and I cannot find it either; and so we need that the aforesaid be declared so an appraisal can be made and [determine] what is in favor of the ones I represent. To Your Honor, I ask and I beg that you so proceed and mandate the justice I request. Signed by Alonso de Cendrera. [50]

――――――――――――――――

Alonso de Cendrera sobrestante de las fabricas de sumag′ desta ciudad [de San Agustin] defensor de los bienes que de arribada llegaron derrotados al Puerto desta çiudad en el nauio [navío] nombrado ntra [Nuestra] Senora [Señora] del Rosario y Carmen dueño y señor el capitan ermenegildo Lop? en la causa que el susodho çize [hiciese] sobre la avexia [avería] gruesa y demas rreçados que preçenta paxavenirs′ [para avenirse] a la liquidacion y rrateo que pox [por] la çentençia de Vmd [Vuestra Merced] pronunçiada que çeme [se me] notifico esta mandado açer [hacer] digo que paxapro′uar [para aprobar] y liquidar lo que me toca conforme alo [a lo] que demi [de mi] rresa [reza] la dha Centençia, quesaserca [que es acerca] de lo quebalen [que valen] las cosas que çe le embargaron del dho capitan ermengildo lopes y lodemas [lo demás] queladha [que la dicha] çentençia rrefiere ago preçentaçion deste ynterrogatorio [interrogatorio] de preguntas para queporsu [que por su] tenor los testigos quepor [que por] mi parte çe presentaron çe examinen = A Vmd [Vuestra Merced] piso perdony suplico aya [haya] por preçentado dho ynterrogatorio [interrogatorio] y mande queporsu [que por su tenor] çe examinen mis testigos que dada la dha probança [probanza] pedrie loq′ [lo que] mi conuenga [convenga] y pido justicia = Otroçi, [además] digo que los siete caxones [cajones] de asucar que el dho capitan ermengildo lopes abxio [abrió] con quepago [que pagó] la gente de mar que alega çer [ser] suyos y uenia [y que venían] por su quanta [cuenta] de [la cuenta] de su ermano [hermano] tenian quarenta y dos arrobas y los dos caxones [cajones] de chocolate, el uno tenia çinquenta [cincuenta] y dos caxetas [cajetas] de a libra y media ca d auna [cada una] y el otro ochenta caxetas a libra y media cad auna y el dho escritorio no çe [se] sabe liquidamente lo que tenia para que çe aiga [se haga] tasaçion enconformmi dad [en conformidad] dela [de la] facul tad [facultad] que parello [para ello] çe dap or la dha çentençia conuiene [conviene] del dexecho [derecho] de mis partes que el dho capitan ermenegildo

lopes con juramento declaxe [declare] claryad uietamente [clara y abiertamente] confesando negando en conformi dad [conformidad] de haber que cantidad de libras de chocolate tenia el pximer [primer] caxon [cajón] de a libra y de a libra u media cada caxeta y que cosas y xeneros [g′neros] uenian [venían] en el exitorio [escritorio] que abxio [abrió] porque astaoxa [hasta ahora] no consta liquidamente delo [de lo] queuenia [que venía] yauiaen [y había en] el porque yo no lo puedo aueriguaxme [avexiguarme] delo [de lo] por que [porque] el susodho declaxase [declarase] parauenir [para avenir] a la tasaçion y pxoxxata con protestaçion [declaración] deno [de no] estar por su declaraçion mas delo [de lo] que fuexe [fuere] en fabor [favor] de mis partes = A Vmd [Vuestra Merced] pido y suplico açi lo prouea [provea] y mande ques [que es] justicia que pido. [50]

    This letter must have irritated Captain López, who already had lost his damaged ship.[62] In response to these charges he wrote the governor of Florida to explain the confusion and to defend himself from the accusations. While he explained in detail the reason why the inventory of the merchandise differed from what was registered aboard, and why he held the crates with sugar under his personal protection, he remined silent about the two crates of chocolate that the royal officials also found in his possession. At one point in his testimony, Captain López mentioned that he had kept the sugar in his quarters to give it to the sailors, and that he would give it to them when the time came (after the money arrived) to pay their salary. It is not clear from the document, however, if he planned to use the sugar to have chocolate made, or to give the sugar and chocolate to his crew as an added bonus for their hard journey at sea.

DOCUMENT 7: JUNE 14, 1642

The Captain Hermenexildo Lopez in the case with Alonso Fernandez de Cendrera defender of the goods that were lost in my ship, named Nuestra Señora del Rosario y el Carmen, whose owner I was: I said that the aforementioned defender has presented a certification [a document attesting that what is said on it is true] to the Royal Officials of these provinces in which it appears what I have sold to Your Majesty for the service of the hands [the people] of this presidio, as cable [ropes] and anchors as well as the large amount of goods that I delivered to the Royal Dock-Yard / Warehouses [almacenes reales] following the orders of Your Honor. The inventory was taken by the adjutant Martin Ruis when [the merchandise] was about to be disembarked and the inventory was done summarily [without any detail, by the lump]. Thereafter the inventory was done [again] minutely [very thoroughly and in great detail] and the aforesaid defender pretends to adjust the memoir of the aforementioned adjutant with what is registered in the ship's book [libro de abordo] and according to that certification it seems that there is a difference of 21 crates [21 fewer crates] in the inventory of that adjutant to whom I delivered the crates and it also was the cause [of confusion] of the nine petacas de tacamaca[63] [boxes of tacamaca] I delivered [to them]. It also seems that the aforesaid adjutant wrote in the inventory two crates with more than 2 tercios [440 pounds] one of cochineal dye [grana] and the other of cacao that I also delivered. From that 21 crates said to be missing should be subtracted 10 crates in the following manner: 3 crates that had 11 chocolate grinding stones that because they were all broken [broken into pieces], the chocolate grinding stones were delivered as they were [without the crate] and one crate of chalk I had at my house and does not have value; another three crates that carried ammunitions and war supplies, and the remaining three were 2 arrobas of sugar [i.e., 50 pounds] that belonged to different sailors, that because it was a small quantity and did not have [paid] freight, I kept it and gave it to them when I paid them their wages, and for this reason should not be considered in the inventory, because the 9 boxes of tacamaca, and the two tercios [i.e., 440 pounds] [cochineal and cocoa] are already incorporated in the memory account, and should not enter a second time. And about the hides and tanned leather, that according with this memory seems to be more than the one I delivered, they should not be part of the inventory because they were wet and rotten . . . and those concerned [with the merchandise] have the obligation to pay me for the damage according to the sentence pronounced by Your Lordship in this cause . . . and regarding the cables and anchors they should not enter in this apportionment because they belong to the ship that was auctioned. . . . To Your Lorship I ask and I beg that you mandate that all the aforementioned does not enter in the apportionment due to the causes and reasons aforesaid. These [the reasons] I present under the required oath upon which I request justice, Signature: Hermenexildo Lopez for His Majesty in this date of June 14 th., 1642. [51]

El Capp[an] Hermenexildo Lopez en la causa con Alonso Fernandez de Sendrera defençor de la hacienda que se escapo en mi navio n[do] Nuestra Señora del Rosario y el Carmen de q' fui dueño = digo que el dho defençor apreçentado [ha presentado] ſertificaçion [certificación] de los Jueces oficiales R' destas [de estas] provincias por donde consta lo que yo e [he] vendido a Su Magestad [Majestad] para el servicio de las manos del precidio [presidio] de cables y anclas y ansimismo [asimismo] la cantidad de hacienda que yo entregue en los almacenes R[s/] por orden de V[m] asi lo que por inbentario y memoria por mayor tomo mi ayudante Martin Ruis cuando se fue desembarcar como el entrego que despues hizo por menor con la qual ſertificaçion [certificación] pretende dho defençor ajsutar la memoria del dho ayudante con el Rexistro del libro de abordo y como pareçe [parece] ay [hay] diferencia de veintey un [veinte y uno, veintiun] caxones de la memoriade dho ayudante a el [le] entrego y fue la causa que nueve petacas de tacamaca que yo entregue y es tambien demas. Parece [también]que dho ay[te] [ayudante] puso caxones con mas de dos tercios uno de grana y otro de cacao queasimismo [que asimismo] entregue demas de la memoria del dho ayudante y asi estos vajaron [bajaron] de los veinte y uno

caxones que ai [hay] de diferencia, restan diez caxones los quales [los cuales] fueron en esta manera; tres caxones en que binieron [vinieron] once piedras de moler chocolate que por estar hechas pedaços [pedazos] çe [se] entregaron las piedras deporçi [de por si] y un caxon de tissa [tiza] que yo ten[ia] en mi casa y no es de ningun valor y otros tres caxones en que binieron [vinieron] las municiones y pertrechos de Guerra y los tres restantes eran de a dos arrobas de aſucar [azúcar] de diferente marineros que por ser poca cosa y no llevar fletes se lo entregue cuando les pague sus soldadas por lo qual [cual] no deuen [deven deben] entrar lo sussodho [susodicho] en dha [dicha] quanta [cuenta] por quanto [cuanto] las nuebe [nueve] petacas y dos tercios y piedras de moler (chocolate) y municiones y pertrechos estan ya metidos dho [dicha] quanta [cuenta[y demas no deuen [deven → deben] entrar por no ser cossa [cosa] de balor [valor] y en quanto [cuanto] a los cueros que pareçe por dha [dicha] memoria aver [haber] mas de los que yo entregue tampoco an [han] de entrar en dho [dicho] rateo por quanto no ser de ningun prouecho [provecho] y aver [haber] salido moxado [mojado] y podridos del dho [dicho] nauio [navío] . . . y los interesados (en las mercaderias) (tienen) la obligación de pagarmeel daño dellas [de ellas] conforme la sentençia [sentencia]por V[m] pronunçiada [pronunciada] enesta [en esta] caussa [causa] . . . y en quanto a los cables y anclas tampoco deuen [deven deben] entrar en dho rateo por quanto [cuanto] es cossa [cosa] anega [aneja] a el costo del nauio [navío] que se remato AVm pido y suplico mande que todo lo sussodiho [susodicho] no entre en el rateo por las caussas [causas] y raçones [razones] dhas [dichas] la qual [cual] preçento [presento] con el juramento neçeçario [necesario] sobre que pido justicia y esta ſſ[a]

A vuestra Merced le pido y ruego que ordene que todo lo arriba mencionado no se incluya en la distribucion por las causas arriba mencionadas. Estas razones las presento bajo el requerido juramento sobre el que pido justicia [Signature]: Hermenexildo Lopez, autirozado: Pedro de As . . . [illegible] por su majestad. Junio 14 th., 1642. [51]

This letter contains more of the common abbreviations seen before in the earlier documents and previously described. There are abundant examples of the use of the ç and **qu**. For the use of ç instead of **s** or **z**, we can select defençor → defensor (defender) and preçentado → presentado (presented). Regarding the shift of **çe → se**, this is used as a reflexive pronoun in a simplified form of the passive voice: çe entregaron → se entregaron (were given) and pedaços → pedazos (pieces). To illustrate the use of **qu** instead of **cu**, the following examples may be seen: quales → cuales (which); quenta → cuenta (count, account); and por quanto → por cuanto (as a consequence). And once again the document uses the letter **x** instead of **j**, as with in moxado → mojado (wet) and caxon → cajón (crate).

An uncommon variation is the use of the letter **s** in the form of a large Greek sigma, a variant that originated during the 14th century and began to disappear in the 17th century [52]. This variant is seen in ſertificaçion (certification) or appears as an elongate letter **f** in the superscript abbreviation of *fecha* (date), as in the last word of the letter text: ∫∫ᵃ. The author of this document is the only person in all the correspondence we examined that used the double **s**. The double **s** (**ss**) also characterized the dawn of Castilian Spanish [53]. Its letter/sound began as **cz** then evolved into **sz** and finally became **ss** by the end of the 17th century. While an archaic form, it was used in the present document as in *tissa* → *tiza* (chalk) or instead of a single **s**, as in *cossa* → *cosa* (thing) or *caussa* → *causa* (cause). Although this letter form disappeared from modern Spanish, it is widely used in contemporary 21st century Italian, as in *rosso* (red), or in French— *professeur* (professor).

DOCUMENT 8: MARCH 1642

The final St. Augustine document of the series captivated us from first discovery and review: It is an extraordinary list of names and occupations of males (adults and boys) and one woman living at the settlement in March 1642 … an astonishing genealogical and historical record (Fig. 51.2)! After much toil, accusations, explanations, and recriminations, the merchandise had been auctioned and in March 1642,[64] a total of 1167 pounds of chocolate and 75 small gift boxes of chocolate were ordered to be distributed by the governor of the province of Florida. Receiving these gifts were 222 persons living at the presidio and the city of St. Augustine on this date. For the complete list of specific names and quantities of chocolate received see Appendix 5.

We consider this document an important historical document for two reasons. It provides very specific information regarding chocolate distribution and consumption in the first European settlement on the Atlantic coast of North America. More important, however, the document, while rich from a paleographic point of view, represents an extraordinary genealogical treasure by identifying all the names and most of the

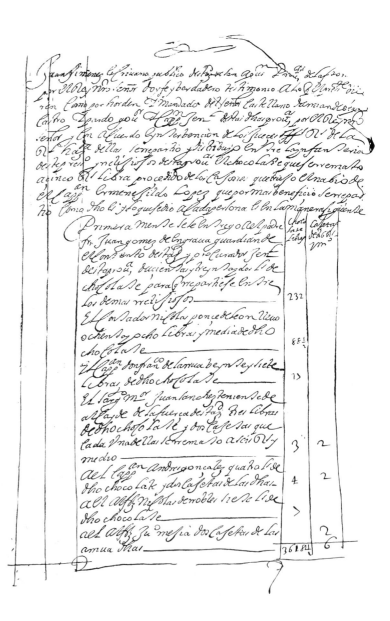

FIGURE 51.2. First page of chocolate distribution document, St. Augustine, Florida, August 1642. *Source:* Courtesy of the Ministry of Culture, Spain, and Archivo General de Indias (Archive of the Indies), Seville, Spain. (Used with permission.)

occupations of 222 persons living at St. Augustine during the years 1641–1642. This long list begins with a paragraph and concludes with a list of 222 names, with professions, and quantities of chocolate received:

I, Juan Jimenez, the Public Commisioner of Peace, in this city of San Agustin, in the Province of Florida, in the name of God I attest and give true testimony that by order of the Castilian Lord Damian de la Vega Castro y Pardo, Governor and General Captain [Capitán General] of these provinces, and in agreement and with the assistance of Official Judges of the Royal Supreme Court [Audiencia Real], I say that the chocolate that came in the ship Nuestra Señora del Rosario, *whose Captain is Ermenexildo Lopez, and [that] was auctioned at 5 reales per pound, has been apportioned and distributed for a maximum benefit [to obtain a maximum benefit] among the infantry [foot soldiers] of this presidio [of San Agustin] and among the religious [priests] of this province [of Florida] in the following way. . . .* [54]

(Then follows a list of 222 names, with profession, and the quantities of chocolate received.)

Juan Jimenez, el comno^ [comisionado] publico de Paz de san Agus^n Proui^as de la Florida por el Rey nro^ Señor doy fe y berdadero testimonio a los Rsentantes [representantes] de los bienes como por horden y mandato del Señor Castellano Damian de Veag Castro y Pardo Gov^or y Capp^n Gen^ destas dhas proui^as por el Rey nro^ Señor con acuerdo e interbencion de los Jueces off^r R^′ de la R^′ haz^da dellas se rrepartio y distribuyo entre la ynfanteria deste press° y rreligiosos desta pro^ua el chocolate que se rremato a cinco R^′ [por] libra procedido de los cajones que trujo el nabio de el Capp^an Ermenegildo Lopez que por mas beneficio se rrepartio como dho [dicho] es y lo que se dio a cada persona en la manera siguiente . . . [list of names . . .]. [54]

In this introductory paragraph, we find the already familiar superscript abbreviations, in this case mainly for the titles of *Capp^n* or *Capp^an* → *Capitán* (Captain) or *Gov^or* → *Gobernador* (Governor). The word chocolate was seldom abbreviated. In the few cases where it was, it received a superscript abbreviation: *choco^te* → *chocolate*. It is important to mention here the use of the double **rr** at the beginning of a word, as in *rreligiosos* → *religiosos* (religious persons/priests), *rremato* → *remató* (auctioned), and *rrepartio* → *repartió* (distributed). The archaic form of *trujo* instead of *trajo* for the preterite indicative of the verb *traer* (to bring) is a derivation of the 13th century Spanish *traxo*. *Trujo* is the derivative of the more literary *traxo* as variably used in speech.

The consistent lack of accents (diacritic marks) in all the documents of this set, but more strikingly in this long list of names and surnames, reflects an important characteristic of the Spanish language before the 18th century. Our transcription of the names shows the names and surnames of the persons as exactly written in the manuscript as abbreviated, some names in lowercase, and without accents. This transcription is a faithful

record and useful guide for those interested in the genealogy of the descendants of these early Spanish settlers on the Atlantic coast of North America. In the second column the names are transcribed as they would have been written today, in the 21st century (see Appendix 5).

The use of orthographic accents may be traced to ancient Greece but during the Middle Ages were rarely used in Castilian Spanish. Nebrija [55] identified and discussed three types of orthographic accents: acute [á], grave [à], and circumflex [â]. While the three are maintained in modern French, only the acute is used today in modern Castilian Spanish. The lack of any accents in the St. Augustine documents confirms once more that orthographic accents were absent during the 16th and 17th centuries or at least highly inconsistent until the 18th century. It was only toward the end of the 19th century that the Royal Spanish Academy established the orthographic accent rules of use [56, 57].

The first or given names of the 222 individuals who received chocolate provide a profusion of examples of superscript abbreviation, among them:

Fran^co Gonçales de Villagarcia → Francisco González de Villagarcía

Bar^me Lopez → Bartolomé López Gabiro

Ju^o de Carbajal → Julio de Carbajal

Ju^o Fran^co M^o → Julio Francisco Moreno

Pedro R^a → Pedro Ribera

The only woman on the list, unfortunately, was not specifically named and was referenced only as *mujer de M. L. Mateo*: wife of M. L. Mateo (see Appendix 5; line number 175).

Surnames can provide historians and genealogists with important clues regarding the early European (in this case Spanish) settlers of the Atlantic coast of North America. This is possible because in Spain as well as in many other places the surname of a person reflects the name of a town, city, or region where the family originated. In the Spanish language, such geographical designations are preceded by the preposition *de* meaning *from*. In the case of Captain Domingo de Almeyda (see Appendix 5; line 218), the surname Almeyda = Almeida could reflect two potential geographical origins: (1) a municipality in east-central Portugal along the border with Spain, or (2) a place now called Vigo, located in the Spanish region of Galicia, near Portugal. For Don Luis Hernández de Viana (see Appendix 5; line 168), Viana do Castelo is a city and a municipality in northern Portugal founded by King Afonso III in the 13th century. In the 16th century, its port gained great importance due to the discoveries. And for Adrián de Cañizares, Sergeant Major, Cañizares is a town in Cuenca, Castilla, La Mancha, Spain (see Appendix 5; line 138).

The document lists 17 different occupations and with a little imagination 21st century readers of this chapter can visualize these individuals going about their daily chores, whatever they were: adjutant aide-de-camp, Agustin friar, auditor, captain, corporal, corporal of the stables, ensign, lawyer, lieutenant, page, parish priest, peace notary, second lieutenant, sergeant, sergeant major, ship boy, superior and guardian of the convent. The first name on the list was Father Juan Gómez, Guardian of the Convent of St. Augustine and General Attorney of the Province of Florida. According to information contained in the California Mission Archives, Santa Barbara, California, the full name of Father Juan Gómez was Juan Gómez de Palma, and he arrived as a missionary in Florida in the year 1609, subsequently became Father Guardian of the Convent of St. Agustine, and held this position from 1631 to 1645. Also on the list are the two persons at the heart of the St. Augustine chocolate saga: Captain Hermenexildo López, he who saved the chocolate, and Nicolás Ponce de León, the zealous royal official who prosecuted him. Although all on the list are identified by name—save one—it may be possible to learn more about specific individuals after further in-depth genealogical research in the Florida archives.

Conclusion

Languages reveal the history of those who speak their words. Languages tell the story of both the conqueror and the conquered. The Spanish language reflects the legacy of Eurasian ancient civilizations, a legacy enriched through the centuries by movements of peoples and groups into and out of the Iberian Peninsula. Spanish also is reflected in geographical names on the landscapes of the New World. From the first moment Spanish conquistadors set foot on the American mainland, they left Spanish language place names on the land (toponyms). These names have enriched the history of North America: Alcatraz, Baja California, Buena Vista, Colorado, El Camino Real, Florida, Los Angeles, Merced, Montana, Monterey, Nevada, Palo Alto, Sacramento, San Francisco, Santa Barbara, Tiburon, Ventura—to name only a few—reveal Spanish linguistic origins.

It is appropriate to end this chapter with a comment on the surname of Captain López. Spanish surnames ending in **ez** (Spanish) or **es** (Portuguese) are most frequently encountered with Spanish and Portuguese genealogies. This ancient "suffix" tradition was brought to the Iberian Peninsula by the Visigoths and means *son of*: Martínez son of Martin, Sánchez son of Sancho, and López son of Lope. The name Lope, Captain López's father, is derived from the Spanish form of *Lupus*, Latin for wolf; thus, López translates as son of a wolf. It is a curious linguistic coincidence, too, that the Spanish language defines an experienced sea captain as a "*lobo de mar*" (literally sea wolf; in the sense of the Jack London novel, *Sea Wolf*). And it was due to this daring *lobo de mar* that in March 1642, the men and boys and one woman of St. Augustine received the most unexpected gift—chocolate.

The period after 1492 changed Castilian Spanish with incorporation of Amerindian tongues and place names, as well as adoption of New World animal and plant terms especially linked with food. Cacao and chocolate—derived from Mexica/Aztec terms—were incorporated into 16th century Spanish, terms that ultimately found their way into global linguistics. Whether spelled in Arabic, French, Greek, Russian, or hundreds of other global languages, the words cacao and chocolate are easily recognized. While the Mexica/Aztecs were conquered by Spain, it was Mexica/Aztec chocolate that ultimately conquered the world.

Finally, what about the one person not identified by name and listed only on the St. Augustine chocolate distribution list as *Mujer de M. L Mateo* (wife of Mateo)? We pose three questions that remain for future research:

1. Who was she?

2. Why was she the only woman on the list?

3. If she had children, are her descendants living today in Florida?

Acknowledgments

We would like to thank Dr. Sylvia Escárcega for her assistance and efforts, working in the Archivo General de las Indias (Archive of the Indies), Seville, Spain, who brought these documents to our attention.

Endnotes

1. Nahuatl is a group of related languages and dialects of the Aztecan branch of the Uto-Aztecan language family. Under the Mexican Law of Linguistic Rights, Nahuatl is recognized as a "national language" with the same validity as Spanish and Mexico's other indigenous languages.

2. Paleography: Muñoz y Rivero begins his *Manual of Spanish Paleography* by explaining that the word paleography derives from the Greek words *palaios*, meaning ancient, and the word *graphos*, meaning to write. It designates the science of old writings and from this point of view encompasses the study of writing through the ages regardless of topic.

3. Antonio de Nebrija (or Lebrija) (1441–1522) was a Spanish scholar born at Lebrija in the province of Seville. He studied in Salamanca and completed his education in Bologna, Italy. In 1492, he published the first grammar of the Spanish language (*Gramática Castellana*), which was the first such grammatical text in any Romance

language. At this time, Castilian became Spanish, the predominant language of Spain, replacing Latin, and in this same year Columbus reached the Americas, carrying with him the Castilian language to the New World.

4. Linguistics is the science of language that includes elements of phonology (evolution of speech sounds), morphology (internal structure and word form), syntax (arrangement of words and sentence structure), and semantics (development of and changes in words and speech forms).

5. Philology is the study of written records, especially literary texts to determine meaning and authenticity.

6. Iberians were an ancient people who spoke a non-Indo-European language. They inhabited the Mediterranean region of what is now called the Iberian Peninsula (today Spain and Portugal). One theory suggests that they arrived in Iberia during the Neolithic period as early as the fourth millennium BCE from a region to the east. A second theory suggests that they may have originated in North Africa.

7. Fray Francisco Ximénez (1666–1729) was a chronicler, philologist, and biographer. He was born in Ecija, Andalusia, on November 23, 1666. He was fluent in several Mayan languages, especially Cakchiquel, Quiché, and Tzutuhil. He discovered and translated to Castilian Spanish the *Popol Vuh*—the Sacred Book of the Mayans (see Chapter 3).

8. The Indo-European language family consists of approximately 100 languages subdivided into subgroups or lineages, among them Albanian, Armenian, Balto-Slavic, Celtic, Germanic, Hellenic (Greek), Indo-Iranian, and Italic groups.

9. By the end of the third millennium BCE, the inhabitants of southeastern Europe and western Asia, vaguely known as the Indo-Europeans, began their migrations.

10. Roman Emperor Caracalla decreed in 212 CE that all those born inside the limits of the Roman Empire should be Roman citizens and called all those outside its limits *Barbari*—meaning unintelligible speech. Two types of Latin may be identified: vernacular (common Latin, sometimes called vulgar Latin) spoken generally (i.e., colloquial) and classical Latin, the literary language of the learned. Romance languages evolved from vernacular Latin.

11. A pre-Roman, non-Indo-European language spoken in the Basque Country (border between Spain and France).

12. Paper was invented in China and subsequently spread globally, initially through trade between China and the Middle East. This product ultimately would replace papyrus as the medium of writing in the Middle East (i.e., paper, from the linguistic root papyrus).

13. Named for their capital city, Babylon, in southern Mesopotamia (modern Iraq).

14. The word alphabet is formed using the first two letters of the Greek alphabet: alpha (α) and beta (β).

15. Semites, both an ethnic and linguistic classification, constitute a large group, among them Acadians, Amorritans, Arabs, Arameans, Ethiopians, Hebrews, and Phoenicians.

16. The Etruscans inhabited the middle portion of the Italian peninsula (modern Tuscany). Their language was non-Indo-European in origin.

17. Indias (Indies) was the official name given to the Spanish possessions in America and Asia, based on the erroneous cosmographic theories of Columbus. The *Archivo General de las Indias* was founded in Seville in 1781 by the Spanish King Carlos III. This archive constitutes the largest and oldest repository of the documents of Spanish America from 1492 until Mexican independence. The *Archivo de Simancas*, founded in Madrid in the 16th century, contains documents of the Castilian crown; the *Archivo Histórico Nacional*, founded in Madrid in 1886, and the *Archivo de Alcalá de Henares* are other important national archives of Spain.

18. Father Junípero Serra, born in Mallorca (largest of the Balearic Islands), spoke and wrote Catalan. Indeed, his first letters upon arrival in New Spain in 1749 were written in Catalan (see Chapter 34).

19. During the 16th century, Spain's control over its overseas possessions depended on the maintenance of communications between the Iberian Peninsula and Spanish America. Merchant vessels and armed ships that formed the fleets that sailed back and forth across the great trajectory collectively were called the *Carrera de Indias* (Route of Indies) and represented a "chain" that held the empire together.

20. *Gran Canaria* (Great Canary), largest island of the seven Canary Islands in the Atlantic Ocean, was colonized by the Spain ca. 1525. *Gran Canaria* became a strategic place for the Spanish fleets to replenish their water tanks and to take on supplies.

21. The ship *Nuestra Señora del Rosario y el Carmen* would have been part of the merchant fleet that traversed the Atlantic between Spain and Nueva España (New Spain: i.e., Mexico). A standard voyage for Spanish trade ships would be to depart from Cadiz, sail south down the coast of Africa, then west to the Canary Island, then after taking on water and supplies, depart westward again until the Caribbean was reached. At this point, the ships would have two primary destinations: Vera Cruz (Veracruz), along the coast of eastern Mexico, or the South American mainland ports of Cartagena, Guaira, and Porto Bello (Portobelo, on the Caribbean coast of Colombia and Venezuela). After loading the ships with the wealth of the Indies, the fleets would gather in Cuba for the return voyage. There, the ships could be refitted and replenished and then the combined flotilla (fleet) would depart Cuba, sailing north through the Straits of Florida and Bermuda. Once the Gulf Stream was reached, the ships passed the Bahamas and set their course for the Azores and Spain. It was a dangerous voyage particularly during the hurricane season.

22. Associations between this mid-17th century chocolate trade, the Bay of Biscay, and the North America Atlantic coast might have originated a century before through connections established by Pedro Menéndez de Avilés, the founder of St. Augustine, Florida.

23. Careful reading of the documents does not reveal what happened. Among the crates (containers) he brought to his headquarters were two that were destined to be delivered to two notables in Spain: one crate was assigned/registered to a castle on the Cantabric coast (Basque coastal region) of Spain, occupied by the Knights of the Order of Saint James, and the second was assigned/registered to the administrator of the Royal Treasury in Seville. Might we speculate—in contrast to the suspicious Spanish officials—that the captain in doing so was attempting to protect these two chocolate-filled crates to assure that they would be intact during his return to Spain?

24. Nicolás Ponce de León identified here was probably a descendant of Juan Ponce de León II, the grandson of the Spanish explorer Juan Ponce de León, who in search of the Fountain of Youth on the island of Bimini was the first European to set foot in Florida. The sacking and burning of St. Augustine in 1688 by the British pirate Robert Searles convinced the Spanish administration to build a strong masonry fort to protect the city. This fort today is known as the San Marcos Castle. Nicolás Ponce de León was among the men who contributed their effort and determination to build this fortification.

25. The word *avería* was the Spanish term given to damage sustained by merchandise aboard ship during transit. The term *avería gruesa* is defined as the damage caused deliberately aboard ship. In this instance, the damage was deliberate since the goods were tossed into the sea in order to lighten and save the ship. Under such conditions, *avería gruesa* normally would have been paid by concerned individuals who had an interest in the salvage of goods. The term *derecho de avería* as used in overseas commerce represented a certain tax imposed on merchants or on merchandise collected to cover damage during shipment.

26. Camargo. This refers to Revilla de Camargo, a locality in Cantabria in northern Spain, near Santander on the Bay of Biscay. Historically, Camargo is linked with Santiago de Compostela, the capital city of Galicia on the northwestern coast of Spain, through an ancient Christian pilgrimage route known as *El Camino de Santiago* (The Way of St. James). Camargo is located on the coastal branch of this ancient pilgrimage route. Walking this route was considered one of three pilgrimages a Christian could make, whereby all sins could be forgiven, the others being the Via Francigena to Rome and the pilgrimage to Jerusalem.

27. *Orden de Santiago de Compostela* (Order of Saint James of Compostela) was a military and religious order of medieval Spain. In 1170, the king of Leon organized a group of knights to defend their land from the Moors and three years later, the king of Castile gave the order a religious character granted by Papal Bull. The order owed its name to the national patron of Spain, St. James the Greater, under whose banner in the 9th century the Christians of Galicia and Asturias drove the Muslim attackers back to Africa.

28. *Bxin = brin*: From the French *brin* and probably from the Celtic *brinos* meaning fiber or filament. In this instance, it represented an ordinary thick form of linen cloth used for lining crates (i.e., similar to canvas).

29. *Almojarifazgo*: The institution in Muslim Spain in charge of tax collection. After the Muslims were expelled from Spain, this administrative structure was incorporated into the Royal Treasury of Castile. Seville was the trade center of medieval Spain and continued to be an important commercial center through the 18th century. Seville at one time was an important port city—although not today in the 21st century. Its decline came about after the silting-up of the Guadalquivir River. It was from Seville that Ferdinand Magellan obtained the ships he used for his circumnavigation of the earth. Indeed, much of the Spanish empire's silver obtained from the New World entered Europe after the Spanish treasure fleet had landed in Seville. The city was home to the *Casa de Contratación* (House of Trade), the Spanish administrative agency that controlled all overseas trade. Seville, today, holds the most important archive of the Spanish administration in the Americas (*Archivo General de Indias*). Other treasures of the Americas passed first through Seville. Some suggest that the first commercial shipment of chocolate from Veracruz to Spain arrived in Seville in 1585.

30. Punctuation marks are ancient signs used to divide and subdivide clauses. The use of punctuation marks first arrived in Iberia through Latin, which had adopted them from the Greek. Between the 12th and 14th centuries, new grammar/writing rules were established and punctuation marks were restricted: a single point (i.e., a period) and a point with a superscript comma (like the inverse of a modern semicolon). During the next centuries, punctuation fell into disuse. Contemporary punctuation marks as known and recognized today were gradually introduced to languages through Italic (archaic Italian).

31. In Old Spanish the letter **b** and the letter **v** constantly have been interchanged. Even the names given to these specific letters in Spanish during the last part of the 19th century and beginning of the 20th century reflect this confusion. The **b** was called/vocalized as **be**, whereas the **v** was called/vocalized as **ube**, meaning "like a **u** with the sound of a **be**."

32. A ligature is the fusion of two ordinary letters to form a new glyph or typographical character. Before printing had been invented, scribes used ligatures to save space, for example, the ligature ου resembles a **V** above an **O** and still is used today in Greek. In the Latin alphabet

examples of ligatures include Æ from **AE** and Œ from **OE**.

33. *Bollos de chocolate*: chocolate buns. This would be an intriguing food item to carry aboard ship. The definition of *bollo* is a round shaped pastry (a bun) made with flour, water, and frequently milk, butter, and eggs. Therefore, a *bollo de chocolate* would be a bun made of all the above mentioned ingredients plus the chocolate. However, this does not make sense to us. Could it be that the term represented large round, bun-shaped molded chocolates, especially made as gifts for presentation? The term also could be a "slang" representation for something else, with the specific meaning lost through the centuries.

34. *Cocos* in this context were cups for serving chocolate made from coconuts. Their inclusion in the crates is especially interesting since these *cocos* stand in sharp contrast to the common, everyday use of *jícaras* or gourd cups also used to serve chocolate drinks. Other documents that we inspected from the 18th century did not refer to *cocos*, only to *jícaras*. The mention of *cocos* in the document also leads us to consider the possibility that the *Nuestra Señora del Rosario y el Carmen* was part of the southern fleet that would have started its return to Spain from a port located along the northern coast of South America before reaching Vera Cruz, where it then loaded the Chinese pottery.

35. La Guaira is the capital of the Venezuelan state of Vargas and the country's chief port. It was founded in 1577 as an outlet for Caracas, located west of the city of Caracas.

36. Cochineal is the name of both crimson/carmine dye obtained from the cochineal insect (*Dactylopius coccus*) that produces it. Cochineal is a domesticated insect cultivated on *Opuntia* spp. cacti by both the Mexica/Aztecs and Incas. The Spaniards upon arrival in Tenochtitlan (now Mexico City) found bags of dried cochineal, sent as tribute to Montezuma, and after the conquest were promptly shipped to Spain. Because of its expense and scarcity, scarlet-dyed cloth quickly became associated with wealth and power: the robes of Roman Catholic cardinals were colored with cochineal, as were the jackets of the British military (i.e., red coats).

37. *Pañito* (doily), piece of fine cloth made out of linen, cotton, or silk, embroidered or made with lace or crochet and used to embellish serving trays, tables, sofa's arms, and other objects. The text here specifies *pañitos chocolateros* presumably used to cover trays or as fine napkins when serving chocolate to guests.

38. *Tecomate*, from the Nahuatl or Mexica/Aztec word *tecomatl*. This is a type of gourd with a narrow neck and hard skin traditionally used to manufacture cups or large containers often used to prepare chocolate.

39. In modern Spanish, *cerillas* are matches used to light candles or stoves, but matches were not yet invented at the time the document was written. We suspect, therefore, that *cerillas* were very thin long wax candles used

to light other candles, that is, with the same purpose as matches in the 21st century.

40. *Capa* is a general word for cloak but the present context seems to suggest the meaning of a special choir cape worn by prelates during special masses or religious processions.

41. Aspiration in phonetics is the strong burst of air that accompanies the release of a phoneme (sound). In English, the aspiration of **h**—as in the word "have"—would be much softer.

42. Transformation of the last letter of a verbal infinitive—in this case **r** or **d** in the command form of the second person plural, *vosotros*—was replaced by the letter **l** when the infinitive or the command was followed by a direct object pronoun like **lo** or **la** (it) or their plural forms. This transformation was used in the first epic saga written in Castilian Spanish ca. 1142, the *Cantar del Mío Cid (Song of the Cid)*.

43. The **x** of the Mexica/Aztec Nahuatl tongue has a different resonance than the Spanish **x**. During the 20th century, the Spanish Royal Academy of Language made the long due amendments to the spelling and pronunciation of words of Nahuatl or Mayan origin when they reported that "the voices of other tongues not adapted to the Spanish language should keep their original orthography. We recommend to restrict the written variants of *Méjico* or *mejicano* to honor the orthographic tradition of the country." In Mexico the letter **x** still represents the palatal fricative phoneme as in the words Oaxaca (city and state of south-central Mexico) or Xochimilco, the area of southern Mexico City and location of the "floating gardens."

44. Elision is the term to define the omission of one or more sounds (such as a vowel, a consonant, or a whole syllable) in a word or phrase, that produces a result easier for speakers to pronounce.

45. Pico della Mirandola was an Italian Renaissance philosopher, born in the city of Mirandola, province of Modena, Italy. He became famous for his writings on religion, philosophy, natural philosophy, and magic. He wrote *Oration on the Dignity of Man*, which has been called the "Manifesto of the Renaissance" and a key text of Renaissance humanism.

46. It is believed that the word *arroba* is a derivative of the word *amphora*, a word referring to a weight and volume/storage vase unit used by ancient Greeks and Romans. *The Novísimo Diccionario de la Lengua Castellana* published in 1878 in Paris defined *arroba* as the "unit of weight of 25 pounds of 16 ounces each pound; also a measurement for liquid substances: Amphora." The *amphora* was used as a measuring unit in Venice and along Mediterranean trade routes since ancient times.

47. *Zarzaparrrila = sarsaparilla: Smilax medical*. This medicinal shrub grew from Mexico south to Brazil in humid forest regions. When the Spaniards arrived in America, they noticed that the natives used this plant to cure

illnesses and named it *zarzaparrilla* due to its similarity to an existing plant in Spain.

48. *Petaca*: from the Mexica/Aztec language, Nahuatl: *petlacalli*—a box made of palm fibers.

49. *Copal* is from the Mexica/Aztec language, Nahuatl: *copalli,* meaning incense. Copal is aromatic tree resin used by the cultures of pre-Columbian Mesoamerica as incense.

50. Native to tropical Africa, the tamarind tree spread early to tropical America and the West Indies. Tamarinds in Mexico grow primarily in the states of Chiapas, Colima, Guerrero, Jalisco, Oaxaca, and Veracruz. Medicinal uses of the tamarind are numerous.

51. A *tercio* is a highly variable weight measure. It sometimes is defined as "half a load" and also can have the meaning of a bundle. For cacao, generally, 1 *tercio* is equal to 2 *fanegas;* each *fanega* being 110 pounds; therefore, 1 *tercio* of cacao would be equivalent to 220 pounds (see Appendix 1).

52. *Maestre*: In this context the word *maestre* refers to the person—after the captain of a ship—who was responsible for the economic administration of the ship: grand master or mate of a merchant ship. Not equivalent to the English "first mate."

53. *Palo de Campeche: Hematoxylon campechianum.* Campeche was a port of call for Caribbean sea trade, due to its geographic location between the Sea of Antilles and the Gulf of Mexico. One of the main products exported from Campeche was *palo de Campeche*, also known as *palo de tinte* a hardwood from which dye could be extracted. This product was in great demand in Europe where textile producers were willing to pay top prices for the product.

54. Acronyms are abbreviations formed using the initial letters of words or word parts in a phrase or name, like UN (United Nations) or UNICEF (United Nations International Children's Emergency Fund).

55. Tironian notes (*notae Tironianae*) represent a system of shorthand said to have been invented by the scribe used by Cato, the Roman commentator, philosopher, and statesman. The name of his scribe was Marcus Tullius Tiro; thereafter, this type of shorthand was referred to as Tironian.

56. *Manos* means hands, but in this context the word should be translated as pestles. The document, therefore, records that two smaller stones (i.e., pestles) were used to grind upon the larger stone.

57. *Conserva* was the name given to a group of vessels that sailed together for common defense, commonly, a mix of merchant ships escorted by war vessels as the fleet departed Spain or, conversely, upon their return journey loaded with the wealth in gold, silver, cacao, and other commodities.

58. *Derrotado* is a seafaring word used to describe a vessel that has been unwillingly blown off course due to wind or stormy weather.

59. Generally, the word *ropa* means clothing but in the present context it is used more widely to include a range of other consumable goods and furniture.

60. *Bienes de arribada*, in seafaring language, refers to goods that arrived at a port (safe haven) in a vessel that has been blown off its original course due to bad weather.

61. In 21st century Spanish, *escritorio* means desk. In the 17th century, however, the term also referred to a large chest made of wood and inlaid with ivory and ebony, with drawers and lockers used to keep jewels and precious items.

62. The hull and primary ship structures of the *Nuestra Señora del Rosario y el Carmen* were auctioned. None of the documents, however, specifies why it was auctioned, any specific damage, or why the ship met this unusual end. If the ship had been demasted during the hurricane, it could have been possible to rig temporary masts that could have allowed it to "hop-scotch" down the Florida coast, then sail southward to Cuba, where it could have been completely refitted. If the *Nuestra Señora del Rosario y el Carmen* was auctioned—and if it had been badly damaged—who would have purchased it and for what purpose? Perhaps it was to salvage the keel and primary ships' timbers? Alternatively, could smaller, lighter, and faster ships have been "created" from the timbers of the *Nuestra Señora del Rosario y el Carmen*? The record—at present—is silent. It would be worth the effort to search the archives in Seville for further documentation.

63. *Tacamaca* (*Copaibacopaitera officinalis*) is also known as the balsam tree (sometimes given as Amacey, Cabimbo, Camíbar, Copayero, Currucay, Marano). This is a South American, tropical species common to the Amazon and the Orinoco, and to Brazil, Colombia, and Venezuela. The tree secretes a resin from which medicinal oil is extracted (i.e., balsam of copaiba). This product (balsam) was used in the 17th century in the Americas as a remedy against venereal disease.

64. The original manuscript was seriously deteriorated at the time of our inspection. The lower right-hand corner of the last page of the document, where dates commonly are written, was missing. Most of the corners of the pages in this document were damaged or missing as well. Another complication can be seen in Document 2 (AGI. Escribanía 155 A. pp. 69r.–70r.). The first page of the original document was given as number 69, the second page had no number, and the third page was inscribed with a repeat number 69. Hence, when we referred to specific pages of this document, we have listed the range as 69r.–70r. and have cited accordingly.

References

1. Stevenson, V. *Words. The Evolution of Western Languages.* New York: Van Nostrand Reinhold Company, 1983; p. 8.

2. Chomski, N. *Language and Mind*. Cambridge, England: Cambridge University Press, 2006; pp. 88–101.

3. Chomski, N. *Essais sur la forme et les sens*. Original title: *Essays on Form and Interpretation*. Translated from English by Joëlle Sampy. Paris, France: Éditions du Seuil, 1977; pp. 36–37.

4. García Villada, Z. *Paleografía Española precedida de una introducción sobre la Paleografía Latina e ilustrada con veintinueve grabados en el texto y ciento diez y seis facsímiles en un album aparte*. Madrid, Spain: Revista de filología española, 1923; pp. 1–5.

5. Sachar, L. A. *A History of the Jews: New York*. New York: Alfred A. Knopf, 1965; pp. 168–183, 249–260.

6. Recinos, A., Goetz, D., and Griswold Morley, S. *The Popol Vuh*. Norman: University of Oklahoma Press, 1950; p. xxi.

7. Ximénez, F. *Historia de la Provincia de San Vicente de Chiapas y Guatemala, de la Orden de Predicadores. Compuesta por el R. P. Pred. Gen. Fray Francisco Ximenez, Hijo de la Misma Provincia*, 3 Volumes. *Biblioteca Goathemala de la Sociedad de Geografía e Historia*. Guatemala City, Guatemala: Centro América, 1929 (Volume I), 1930 (Volume II), 1931 (Volume III); Volume I, Preface.

8. Ximenez, F. *Historia de la Provincia de San Vicente de Chiapas y Guatemala, de la Orden de Predicadores. Compuesta por el R. P. Pred. Gen. Fray Francisco Ximenez, Hijo de la Misma Provincia*, 3 Volumes. *Bioliteca Goathemala de la Sociedad de Geografía e Historia*. Guatemala City, Guatemala: Centro América, 1929; Volume I, p. 322.

9. Stevenson, V. *Words. The Evolution of Western Languages*. New York: Van Nostrand Reinhold Company, 1983; pp. 10–17.

10. Ostler, N. *Empires of the Word. A Language History of the World*. New York: HarperCollins, 2005; passim.

11. Mallory, J. P. *In Search of the Indo-Europeans: Language, Archaeology, and Myth*. New York: Thames and Hudson, 1989; pp. 24–35, 66–94.

12. Twist, C., and Rafterey, B. *Atlas of the Celts*. Buffalo, NY: Firefly Books, 2001; pp. 40, 48–49.

13. Twist, C., and Rafterey B. *Atlas of the Celts*. Buffalo, NY: Firefly Book, 2001; pp. 56–57, 108–109.

14. Elliot, J. H. *Imperial Spain*. New York: St. Martin's Press, 1964; pp. 12–23, 31–40.

15. Anonymous. *Las Glosas Emilianenses*. Madrid, Spain: Ministerio de Educación y Ciencia, 1977; pp. 13–20.

16. Lapesa, R. *Historia de la Lengua Española*. Madrid, Spain: Escelicer S. A., 1962/1959; pp. 11–38, 95–110.

17. Gaur, A. *A History of Writing*. New York: Cross River Press, 1992; pp. 18–52.

18. Firmage, R. *The Alphabet Abecedarium: Some Notes on Letters*. Boston: D. A. Godine, 1993; p. 8.

19. Gaur, A. *A History of Writing*. New York: Cross River Press, 1992; pp. 118–129.

20. Millares Carlo, A., and Mantecón, J. I. *Album de Paleografía Hispano-Americana de los Siglos XVI y XVII. Introducción y Transcripciones*. Barcelona, Spain: Ediciones El Albir, 1975.

21. Muñoz y Rivero, J. *Manual de Paleografía Diplomática Española de los Siglos XII al XVII*. Madrid, Spain: Daniel Jorro, editor, 1917.

22. Millares Cario, A. *Paleografía Española. Ensayo de una Historia de la Escritura en España desde el siglo VIII al XVIII*. Barcelona, Spain: Editorial Labor S. A., 1929.

23. Cappelli, A. *Lexicon abbreviaturarum: dizionario di abbreviature latine ed italiane, usate nelle carte e codici, specialmente del medio-evo*. Milan, Italy: Brachigrafia Medievale Hoepli, 1929; pp. xi–lvi.

24. Alonso, M. *Evolución Sintáctica del Español. Sintaxis Histórica del Español desde el Iberorromano hasta Nuestros Días*. Madrid, Spain: Aguilar, 1964.

25. Lapesa, R. *Estudios de Historia Lingüística Española*. Madrid, Spain: Colección Filológica Paraninfo S.A., 1985.

26. Menéndez Pidal, R. *Orígenes del Español. Estado lingüístico de la Península Ibérica hasta el Siglo XI*. Madrid, Spain: Imprenta, de la librería y casa editorial Hernando (S. A.), 1929; Volume 1.

27. Menéndez Pidal, R. *Manual de Gramática Histórica Española*. Madrid, Spain: V. Suárez, 1934.

28. *Novísimo Diccionario de la Lengua Castellana que comprende la última edición íntegra del publicado por la Academia Española y cerca de cien mil voces, acepciones, frases y locuciones añadidas por una Sociedad de Literatos aumentado con un suplemento de voces de Ciencias, Artes y Oficios, Comercio, Industria, etc. y seguido del Diccionario de Sinónimos de D. Pedro M. de Olive y del Diccionario de la Rima de D. Juan Peñalver*. Paris, France: Librería de Garnier Hermanos, 1878.

29. *A New Pronouncing Dictionary of the Spanish and English Languages*. Compiled by Mariano Velázquez de la Cadena with Edward Gray and Juan L. Iribas. Englewood Cliffs, NJ: Prentice-Hall, 1973.

30. Cejador y Frauca, J. *Vocabulario Medieval Castellano*. New York: Las Americas Publishing Company, 1968.

31. Lyon, E. *The Enterprise of Florida. Pedro Menéndez de Avilés and the Spanish Conquest of 1565–1568*. Gainsville: University Press of Florida, 1976; pp. 1–37, 93–99.

32. Rust. R. L. A brief history of chocolate, food of the gods. *Athena Review Quarterly Journal of Archaeology, History, and Exploration* 1999;2(2):30–34.

33. Archivo General de Indias. AGI. Santo Domingo, 224, R., N.1/2/2r.–N.1/2/2v.

34. Menéndez Pidal, R. *Orígenes del Español. Estado lingüístico de la Península Ibérica hasta el Siglo XI*. Madrid, Spain: 1929; pp. 219–240.

35. AGI. Escribanía 155, pp. 45r–49.

36. Muñoz y Rivero, D. J. *Manual de Paleografía Diplomática Española de los Siglos XII al XVIII*. Madrid, Spain: D. Jorro, 1917; pp. 111–113.

37. Muñoz y Rivero, D. J. *Manual de Paleografía Diplomática Española de los Siglos XII al XVIII*. Madrid, Spain: 1917; p. 113.

38. Muñoz y Rivero, D. J. *Manual de Paleografía Diplomática Española de los Siglos XII al XVIII*. Madrid, Spain: 1917; pp. 47, 58.

39. AGI. Escribanía 155 A. pp. 274r.–276v.

40. Spaulding, R. K. *How Spanish Grew*. Berkeley: University of California Press, 1943; p. 232.

41. Spaulding R. K. *How Spanish Grew*. Berkeley: University of California Press, 1943; p. 166.

42. Alonso. M. *Evolución sintáctica del español. Sintaxis histórica del español desde el iberorromano hasta nuestros días*. Madrid, Spain: Editorial Aguilar, 1964; pp. 303–304.

43. García Villada, Z. *Paleografía Española precedida de una introducción sobre la Paleografía Latina e ilustrada con veintinueve grabados en el texto y ciento diez y seis facsímiles en un album aparte*. Madrid, Spain: Sociedad Anónima Tipográfica, Núñez de Balboa, 1923; pp. 3–8.

44. *Novísimo Diccionario de la Lengua Castellana que comprende la última edición íntegra del publicado por la Academia Española y cerca de cien mil voces, acepciones, frases y locuciones añadidas por una Sociedad de Literatos aumentado con un suplemento de voces de Ciencias, Artes y Oficios, Comercio, Industria, etc. y seguido del Diccionario de Sinónimos de D. Pedro M. de Olive y del Diccionario de la Rima de D. Juan Peñalver*. Paris, France: Librería de Garnier Hermanos, 1878; p. 88.

45. AGI. Escribanía 155 A, pp. 69r.–70r.

46. Capelli, A. *The Elements of Abbreviation is Medieval Latin Paleography*. Translated by David Heigmann and Richard Kay. Lawrence, Kansas: University of Kansas Libraries, 1982.

47. Nebrija, A. *Gramática Castellana*, texto establecido sobre la ed. "princeps" de 1492, por Pascual Galindo Romeo y Luis Ortiz Muñoz, con una introd., notas y facsímil. Prólogo del Sr. D. José Ibáñez Martín. Madrid, Spain: Edición de la Junta del Centenario (facsimile published in 1943); pp. 23–24, 40–42.

48. AGI. Escribanía 155 A, pp. 163r.–163v.

49. AGI. Escribanía 155 A, pp. 184r.–185v.

50. AGI. Escribanía 155 A, pp. 226r–226v.

51. AGI. Escribanía 155 A, pp. 251r.–252v.

52. García Villada, Z. *Paleografía Española precedida de una introducción sobre la Paleografía Latina e ilustrada con veintinueve grabados en el texto y ciento diez y seis facsímiles en un album aparte*. Madrid, Spain: Soc. An. Tipográfica, Nuñez de Balboa, 1923 p. 333.

53. Menéndez Pidal, R. *Orígenes del Español. Estado lingüístico de la Península Ibérica hasta el Siglo XI*. Madrid, Spain: Imprenta de la librería y casa editorial Hernamdo (S. A.), 1929; Volume 1, pp. 62–63.

54. AGI. Escribanía 155 A, pp. 72r.–76r.

55. Nebrija, A. *Gramática castellana*, texto establecido sobre la ed. "princeps" de 1492, por Pascual Galindo Romeo y Luis Ortiz Muñoz, con una introd., notas y facsímil. Prólogo del Sr. D. José Ibáñez Martín. Madrid, Spain: Edición de la Junta del Contenario, 1943. Libro II, cap. ii. (This text was established on the first edition of 1492 by Pascual Galindo Romeo y Luis Ortiz Muñoz with an introduction, notes, and facsimile. Preface by José Ibañez Martín.) Madrid. Spain: Edición de la Junta del Centenario, 1946; pp. 36–41.

56. Muñoz y Rivero, D. J. *Manual de Paleografía Diplomática Española de los Siglos XII al XVIII*. Madrid, Spain: Daniel Jorro, editor, 1917; pp. 113–114.

57. Spaulding R. K. *How Spanish Grew*. Berkeley: University of California Press, 1943; pp. 207–208.

52

Digging for Chocolate in Charleston and Savannah

Laura Pallas Brindle and Bradley Foliart Olson

Introduction

The search for chocolate in South Carolina was conducted at the South Carolina Historical Society (SCHS), Charleston, founded in 1855. This repository represents the oldest and largest collection of South Carolina historical documents and contains an impressive collection of colonial, antebellum, and American Civil War materials. The search for chocolate in Georgia focused on the collection housed by the Georgia Historical Society (GHS) located in Savannah. The GHS was chartered in 1839 and exists as a private, nonprofit organization whose mission is to collect, preserve, and share Georgia history. It also is the oldest cultural institution in Georgia, and one of the oldest historical organizations in the United States.

Data Collection

Research at both archives was facilitated by the availability of on-line keyword searches and card catalog data. These produced lists of manuscript collections and rare books that potentially contained references or information on chocolate and cacao prior to 1900. Keywords used to identify the archive documents included account (records), cacao, chocolate, cocoa, commonplace book, cookery, cooking (manuscripts), export (documents), food, grocer, import (documents), ledger, medicine, menu, newspaper (advertisements), port (records), provision, ration, recipe, ship (import and manifest), and sundries.

Collections matching these keywords were requested for viewing. Documents were scanned for mention of cacao, chocolate, or cocoa. Upon finding a reference the following data were entered into an Excel spreadsheet for subsequent sorting: archive name, collection information (i.e., box/folder number as assigned by the respective archive), document type, title, author, date, and location. Regarding specific ledgers where cacao/chocolate was identified, we recorded additional data as well: account name and number, quantity of chocolate sold, unit, price per unit, and total cost of chocolate purchase.

Chocolate in Charleston
IMPORT AND EXPORT RECORDS

Historically, Charleston was one of the main southeastern ports. A majority of the chocolate- and cacao-related findings appeared in ship manifests and port documents linked to trade between Charleston and England. Microfilm copies of British Public Records Office (BPRO) were searched to locate import/export records [1]. This search identified a suite of documents on microfilm dated between 1709 and 1767 that revealed records for chocolate import/export along

Chocolate: History, Culture, and Heritage. Edited by Grivetti and Shapiro
Copyright © 2009 John Wiley & Sons, Inc.

with other commodities, among them blubber, coffee, sugar, and tar. Summaries of these chocolate-related data from Charleston are presented in Tables 52.1–52.5. Information is displayed by year, port of debarkation, next destination, form/type of chocolate, and quantity.

Cocoa and chocolate were imported to Charleston from the following seven North American ports: Boston; Georgia (specific ports not identified); New Providence, Rhode Island; New York City; Philadelphia; Salem, Massachusetts; and St. Augustine, Florida. Cocoa and chocolate were imported from the following 18 African, European, Central American, and Caribbean ports: Angola (Portuguese Africa), British Isles (specific ports not identified), Anguilla, Antigua, Bahamas, Barbados, Bermuda, Bonaire, Campeche, Cape Français, Curaçao, Grenada, Guadeloupe, Jamaica, Montserrat, St. Vincent, St. Kitts, and St. Lucia.

Cocoa and chocolate were exported from Charleston to the following North American, Caribbean, and European ports: Bristol and London, England; New York City; Marblehead and Salem, Massachusetts; Philadelphia; Providence, Rhode Island; Virginia (specific ports not identified); and the Caribbean ports of Barbados, Bermuda, Jamaica, and St. Vincent.

"Prize cocoa" was obtained from captured or seized ships, with some quantities arriving in Charleston from Philadelphia in 1758. Thirty bags of "prize

Table 52.1 **BPRO Ship Manifest Information on Chocolate/Cocoa Importation to and Exportation from Charleston, South Carolina, from 1709 to 1719**

Year	Import (amount[a])	Export (amount[a])	Form	Port Name
1709		50 lb	Cocoa	Kingston, Jamaica
		1 bbl	Cocoa	Kingston, Jamaica
		13 seroons	Cocoa	Kingston, Jamaica
		10 seroons	Cocoa	Port Royal, Jamaica
		39 seroons	Cocoa	Port Royal, Jamaica
		1 hhd	Cocoa	Port Royal, Jamaica
1711	4542 lb		Cocoa	Port Royal, Jamaica
	594 bbl		Cocoa	Kingston, Jamaica
	20 bags		Cocoa, prize	Port Royal, Jamaica
	18 bags		Cocoa, prize	Port Royal, Jamaica
	20100 lb		Cocoa	Port Royal, Jamaica
	26000 lb		Cocoa	Port Royal, Jamaica
1717	200 lb		Chocolate	Boston
	50 lb		Cocoa	Boston
	4 bags		Cocoa	Boston
	3 bags		Cocoa	Bahamas
	236 bags		Cocoa	Barbados
1718		3 bbl	Cocoa	Jamaica
	47 bbl		Chocolate, store	New York
	6 boxes		Chocolate, store	New York
		7 boxes	Chocolate	New York
1719	1000 wt		Cocoa	Topsham[b] and Barbados
	29 bags		Cocoa	Barbados
	11 bbl		Cocoa	Barbados
	10 bags		Cocoa	Jamaica
	1500 wt		Cocoa	Jamaica
	4 tons		Cocoa, prize	New Providence[c]
	5 cwt		Cocoa, prize	New Providence[c]
	5032 lb		Cocoa, prize	New Providence[c]
		3 bbl	Cocoa	Bristol
		4632 lb	Cocoa	New York
		11 bags	Cocoa	Philadelphia
		1000 lb	Cocoa	—[d]

[a]hhd = hogshead; lb = pound; bbl = barrel.
[b]British Isles.
[c]British Caribbean region.
[d]Location not specified.

Table 52.2 BPRO Ship Manifest Information on Chocolate/Cocoa Importation to and Exportation from Charleston, South Carolina, from 1720 to 1728

Year	Import (amount[a])	Export (amount[a])	Form	Port Name
1720	3 bbl		Cocoa, prize	New York
	27 bags		Cocoa	Port Royal
1722	5300 lb		Cocoa	Barbados
1723	5 kegs		Cocoa	Cape Francais[b]
	1200 wt		Cocoa	St. Lucia[b]
1724	1 box		Chocolate	Boston
	4 bags		Cocoa	Boston
	312 cwt		Cocoa	Barbados
	3960 lb		Cocoa	Barbados
	1900 wt		Cocoa	Barbados
	400 lb		Cocoa	Bermuda
	1800 lb		Cocoa	Jamaica
	500 wt		Cocoa	Jamaica
	2019 lb		Cocoa	St. Thomas[c]
		40 lb	Chocolate	Providence[d]
1725		2 bags	Cocoa	Providence[d]
	109 cwt		Cocoa	Barbados
	12 tierces		Cocoa	Bermuda
	2 cwt		Cocoa	Bermuda
	4 cwt		Cocoa	Jamaica
	5000 lb		Cocoa	Curaçao[c]
1727	28 bundles		Cocoa	Barbados
	1 bundle		Cocoa	Bermuda
1728		1 box	Chocolate	New York

[a]hhd = hogshead; lb = pound; bbl = barrel.
[b]Foreign Caribbean region, not British owned.
[c]Not British owned/claimed.
[d]British Caribbean region.

Table 52.3 BPRO Ship Manifest Information on Chocolate/Cocoa Importation to and Exportation from Charleston, South Carolina, from 1731 to 1739

Year	Import (amount[a])	Export (amount[a])	Form	Port Name
1731		55 bbl	Cocoa	Barbados
	6 bbl		Cocoa, foreign	Curaçao[b]
	10 bags		Cocoa	Antigua
	2 bags		Cocoa	Barbados
	350 lb		Cocoa	Bermuda and Curaçao
	68 bbl		Cocoa	Curaçao
1732	1 bbl		Cocoa	Philadelphia
	2 seroons		Cocoa	Philadelphia
	1251 lb		Cocoa	Anguilla
	20 bags		Cocoa	Bermuda
	8 bbl		Cocoa	Jamaica
	760 lb		Cocoa	Bermuda
		20 bags	Cocoa	Bermuda
1733	2 bbl		Cocoa	Providence[c]
	1500 lb		Cocoa	Curaçao and Bonaire
		9 bags	Cocoa	London
		3 seroons	Cocoa	London

Table 52.3 Continued

Year	Import (amount[a])	Export (amount[a])	Form	Port Name
1734	7 bbl		Cocoa	Jamaica
	3 tierces		Cocoa	Jamaica
	4 half-bbl		Cocoa	Curacoa
	2 quarter-bbl		Cocoa	Curacoa
1735	1 bbl		Cocoa	Jamaica
	2 casks		Cocoa	Jamaica
	25 cwt		Cocoa	Jamaica
	2 cwt		Cocoa	Curacoa
	10 lb		Cocoa	Curacoa
		6 bbl	Cocoa	Bristol
		3 bags	Cocoa	London
		1 hhd	Cocoa	London
1736		115 bags	Cocoa	London
		5 seroons	Cocoa	New York
		5 bags	Cocoa	Barbados
		4 bags	Cocoa and coffee	Port Royal and West Indies
	5 seroons		Cocoa	Georgia
	897 lb		Cocoa	Jamaica
	166 cwt		Cocoa	Montserrat
	18 bbl		Cocoa	New Providence[c]
	1 jar		Cocoa	New Providence[c]
	107 bales		Cocoa	Campeche[d]
	1 jar		Cocoa	Campeche[d]
	1 bag		Cocoa, returned	St. Augustine
1737	5 bags		Cocoa	Barbados
	4 bbl		Cocoa	Jamaica
	8 casks		Cocoa	St. Kitts
		1 bag	Cocoa	New York
		3 bbl	Cocoa	New York
1738		1 bag	Cocoa	Boston
		1 bbl	Cocoa	New England
		4 bbl	Cocoa	New York
		9 bbl	Cocoa	Philadelphia
		4 bbl	Cocoa	Virginia
	20 bbl		Cocoa	Barbados
	1 small bag		Cocoa	Barbados
	2 bags		Cocoa	Jamaica
	1 bbl		Cocoa	Jamaica
	6 casks		Cocoa	Jamaica
1739		1 bbl	Cocoa	Plymouth[e]
		21 seroons	Cocoa	Plymouth[e]
		1 bbl	Cocoa	Boston
		15 seroons	Cocoa	Boston
		10 bags	Cocoa	New York
	3 bags		Cocoa	Jamaica
	3 casks		Cocoa	Jamaica
	1 small parcel		Cocoa	Angola
	30 skins		Cocoa	St. Augustine

[a]hhd = hogshead; lb = pound; bbl = barrel.
[b]Not British owned/claimed.
[c]British Caribbean region.
[d]Foreign Caribbean region.
[e]British Isles.

Table 52.4 BPRO Ship Manifest Information on Chocolate/Cocoa Importation to and Exportation from Charleston, South Carolina, from 1744 to 1759

Year	Import (amount[a])	Export (amount[a])	Form	Port Name
1744	3 tierces		Cocoa, returned	Kingston, Jamaica
	1 bbl		Cocoa, returned	Kingston, Jamaica
		3 seroons	Cocoa	Kingston, Jamaica
		1 bbl	Cocoa	Kingston, Jamaica
1752		32 bbl	Cocoa, foreign	New York
	7 boxes		Chocolate	New York
1753	5 boxes		Chocolate	Philadelphia
	112 lb		Chocolate	Providence[b]
1758		23 bags	Cocoa, prize	Philadelphia
		5 bbl	Cocoa, prize	Philadelphia
	2 boxes		Chocolate	New York
	1 box		Chocolate	Philadelphia
	600 lb		Chocolate	Philadelphia
	403 lb		Chocolate	Rhode Island
	62 bags		Cocoa	Antigua
	10 bbl		Cocoa	Antigua
	1 seroon		Cocoa	Jamaica
		50 lb	Chocolate	Salem + Marblehead
		30 lb	Chocolate	Salem + Marblehead
1759		1 shipment	Chocolate	New Providence[c]
	2 boxes		Chocolate	Boston
	1 box		Chocolate	New York
	2 boxes		Chocolate	Philadelphia
	20 bags		Cocoa	Guadeloupe
		58 lb	Chocolate	Salem + Marblehead

[a]hhd = hogshead; lb = pound; bbl = barrel.
[b]British Isles.
[c]British Caribbean region.

Table 52.5 BPRO Ship Manifest Information on Chocolate/Cocoa Importation to and Exportation from Charleston, South Carolina, from 1760 to 1769

Year	Import (amount[a])	Export (amount[a])	Form	Port Name
1760	7 boxes		Chocolate	Rhode Island
	2 bags		Cocoa	Philadelphia
	34 bags		Cocoa	Antigua
1761		100 wt	Chocolate	Salem + Marblehead
1762		12 bbl	Cocoa	New York
		1 tierce	Cocoa	New York
		6 bbl	Cocoa	Philadelphia
	2 boxes		Chocolate	New York
	31 boxes		Chocolate	Philadelphia
	6 boxes		Chocolate	Rhode Island
	300 lb		Chocolate	Salem
	half ton		Chocolate	Providence[b]
	12 tierces		Cocoa	New York
	36 bags		Cocoa	Antigua
	150 lb		Cocoa	Bermuda
	12 bbl		Cocoa, foreign	Bermuda
	30 bags		Cocoa, prize	St. Kitts
		150 lb	Chocolate	Salem + Marblehead
		20 lb	Chocolate	Salem + Marblehead
		80 lb	Chocolate	Salem + Marblehead

Table 52.5 Continued

Year	Import (amount[a])	Export (amount[a])	Form	Port Name
1763		50 lb	Chocolate	Rhode Island
	4 casks		Cocoa	Grenada
	700 lb		Chocolate	Boston
	8 boxes		Chocolate	New York
	19 boxes		Chocolate	Philadelphia
	7 boxes		Chocolate	Rhode Island
	50 lb		Chocolate	Rhode Island
	400 lb		Chocolate	Salem
		200 wt	Chocolate	Salem + Marblehead
		92 lb	Chocolate	Salem + Marblehead
		100 lb	Chocolate	Salem + Marblehead
		200 lb	Chocolate	Salem + Marblehead
1764		700 lb	Chocolate	Gosport[c]
	1 box		Chocolate	New York
	17 bags		Cocoa	Barbados
	22 bbl		Cocoa	Barbados
		1 box	Chocolate	New York
	8 bbl		Cocoa	Kingston, Jamaica
	810 bbl		Cocoa	Kingston, Jamaica
	12 bags		Cocoa	Kingston, Jamaica
		300 lb	Chocolate	Salem + Marblehead
		50 lb	Chocolate	Salem + Marblehead
		150 lb	Chocolate	Salem + Marblehead
		9 boxes	Chocolate	Boston
1765	13 boxes		Chocolate	Rhode Island
		5 bbl	Cocoa	St. Vincent
		2 boxes	Chocolate	Boston
	140 lb		Chocolate, returned	Salem + Marblehead
1767		600 lb	Cocoa	Jamaica
	5 bbl		Cocoa	Jamaica
	1 keg		Cocoa	Jamaica
	12095 lb		Cocoa	St. Kitts
	5 bbl		Cocoa	Kingston, Jamaica
	1 keg		Cocoa	Kingston, Jamaica
		5 bbl	Cocoa, returned	Kingston, Jamaica
		1 keg	Cocoa, returned	Kingston, Jamaica
1768	860 lb		Chocolate	—[d]
		253 lb	Chocolate	—
1769		4500 lb	Cocoa	Great Britain
	1260 lb		Chocolate	—[d]
		250 lb	Cocoa	—[d]

[a]hhd = hogshead; lb = pound; bbl = barrel.

[b]British Caribbean region.

[c]British Isles.

[d]Location not specified.

cocoa" were imported to Charleston from St. Kitts in 1762. Sometimes the cocoa or chocolate arriving in port did not sell, and merchants would then return the unsold or unused portion. Such transactions are listed in the records as "returned" cocoa or chocolate. Records surveyed showed one bag of cocoa "returned" to St. Augustine from Charleston in 1736. An interesting situation occurred, too, in 1744 when all the cocoa exported from Charleston specifically to Jamaica was returned. This was a curious document and a likely interpretation could be that the cocoa had been placed aboard the wrong ship. Other instances of returned chocolate and cocoa were located: 140 pounds of chocolate was returned to Charleston from Marblehead and

Salem, Massachusetts, and one keg and five barrels were returned to Jamaica from Charleston in 1767. We also encountered the term "store chocolate" referring to chocolate sold under a general company label. Summarizing, more cocoa was imported/exported to/from Charleston than chocolate. Shipments of chocolate were measured in terms of boxes and pounds; cocoa was measured using a variety of terms including bags, barrels, bundles, casks, counterweight, hogsheads, jars, pounds, seroons, skins, and tierces (see Appendix 1).

Beginning in 1768 through 1772, other ship manifest records available on BPRO microfilm included chocolate and cocoa imports/exports within North America with the quantities and types of chocolate designated four ways: chocolate, cocoa, British cocoa, and foreign cocoa [2]. The following ports, grouped by region/colony in a general north to south order, were identified with chocolate and cocoa imports and exports: Newfoundland (no specific port identified); Nova Scotia—Halifax; Quebec—Quebec City; New Hampshire—Piscatagua; Massachusetts—Boston, Falmouth, Salem/Marblehead; Rhode Island (no specific port identified); Connecticut—New Haven, New London; New York (no specific port identified); New Jersey—Perth Amboy; Pennsylvania—New Castle, Philadelphia; Delaware—Lewes; Maryland—Chester, North Potomac, Patuxent, Pocomoke; Virginia—Accomack, James River (Lower), James River (Upper), Rappahanock, Roanoke, South Potomack (sic), York River; North Carolina—Bathtown, Beaufort, Brunswick, Currituck; South Carolina—Charlestown (sic), Port Royal, Wynyaw; Georgia—Savannah; Florida—Pensacola, St. Augustine; Caribbean islands—Bahamas, Bermuda, and St. John's Island. The quantity of imported and exported chocolate and cocoa for Charleston, South Carolina, are summarized in Table 52.6. The origin or destination of the chocolate or cocoa was only specified if exported from Great Britain.

LETTERS AND LETTER BOOKS

Chocolate and cocoa were mentioned in documents known as letter books[1] and in specific letters. A search for ship import records uncovered the Robert Pringle Letter Book [3]. These letters provide insights on the supply and demand for cocoa, as some correspondence indicated low demand and low selling prices of cocoa while other correspondence letters indicated higher demand and higher selling prices. Cocoa was mentioned, generally, in the following letters:

May 4, 1739 letter to Henry Collins (Newport, Rhode Island) [4]

April 19, 1740 letter to William Pringle (Antigua) [5]

Table 52.6 BPRO Ship Manifest Information on Chocolate/Cocoa Importation to and Exportation from Charleston, South Carolina, from 1770 to 1772

Year	Import (amount[a])	Export (amount[a])	Form	Port Name
1770	400 lb		Cocoa, British	—[b]
		600 lb	Cocoa	Great Britain
	2982 lb		Chocolate	—[b]
		50 lb	Cocoa	—[b]
1771	8500 lb		Cocoa, British	—[b]
		6143 lb	Cocoa	Great Britain
	3685 lb		Chocolate	—[b]
		120 lb	Chocolate	—[b]
		1400 lb	Cocoa	—[b]
1772	3050 lb	—[c]	Cocoa, British	—[b]
	950 lb		Chocolate	—[b]
		650 lb	Chocolate	—[b]
		5500 lb	Cocoa	—[b]

[a]lb = pound.
[b]Location not specified.
[c]Import from British and foreign West Indies and ports Southward of Cape Finisterre in Europe and Africa.

October 24, 1743 letter to Andrew Lessly (Antigua) [6]

October 24, 1743 letter to Michael Lovell (Antigua) [7]

More interesting, however, was Pringle's correspondence with Thomas Hutchinson and Thomas Goldthwait of Boston, Massachusetts, where Pringle wrote that cocoa was not selling in Charleston, and therefore, no more should be sent:

February 11, 1744.

I am to take Notice to you that the Cocoa & Blubber per Capt. Blunt happen to be the Two worst Commodities you could possibly have sent here. . . . And there happens to be a Great Deal of Cocoa in Town at present lately Imported from Difernt [sic] Parts which makes it a Drugg [sic]. Shall be Glad to have your Orders to Sell both off at Publick [sic] Vendue,[2] otherways [sic] may Lay a Long time on Hand before they Can be Dispos'd of & Cannot Encourage your sending any Commodity from your Place excepting Rum & Loaf Sugar, & as the Crops fall short in the West Indies your Rum is Likely to be in Demand here. [8]

In a follow-up letter dated March 31, 1744, Robert Pringle asked Hutchinson and Goldthwait whether or not the unsold blubber and cocoa should be auctioned or returned:

The Blubber and Coca [sic] per Capt. Blunt Remains still unsold. Shall be Glad to have your directions to put both up at Publick [sic] Vendue or to Return you the Cocoa. A Great deal of Cocoa has been lately Sold here at Vendue. [9]

A month later, on April 24, 1744, Pringle wrote again to Hutchinson and Goldthwait alerting the exporters that warm weather had arrived and the cocoa and blubber remained unsold:

Your Cocoa & Blubber still Remains on hand unsold, & as our hot Season now begins to Come in, the Blubber wont [sic] keep, so must be Oblidg'd to expose to Publick Vendue. Pray never send any more of it. [10]

On June 30, 1744, Pringle informed Hutchinson and Goldthwait that he returned the unsold cocoa and why there was an oversupply in Charleston:

I have Return'd you per the Sloop Abigail *the Cocoa you sent per Capt. Blunt which is included in the Bill of Lading & put in Seven bbls. for the Conveniency of Stowage. It is a perfect Drugg [sic] here. A large Ship Loaded with Cocoa has been Carried [sic] in a Prize to Cape Fear, which has Glutted this Place with it.* [11]

Pringle also wrote to David Chesebrough of Newport, Rhode Island, on December 17, 1744, and explained that the cocoa market in Charleston was flooded and selling for a low profit margin [12]. Another of Robert Pringle's letters, one written to James Henderson in Kingston, Jamaica, February 9, 1745, reconfirmed the low demand for cocoa:

Neither any thing else from your Island will answer here excepting Good Muscovado Sugar & a Small Quantity of Coffee if Cheap with you, & Sometimes Molasss [sic], No Cocoa, being very Cheap & plenty here. [13]

Several letters in the Pringle collection pertained to prize cocoa obtained from seized ships. On January 19, 1745, Robert Pringle wrote to Andrew Pringle (brother or other family member?) in London about the possibility of purchasing prize cocoa brought into the port at Charleston:

The prize brought in by the Rose *is Condemn'd [sic] here & Valued at One hundred thousand pounds Sterling. The Bulk of Her Cargoe [sic] (besides Gold & Silver which is the Chief) is Cocoa & is said to be very Good. If it sells a Bargain at Vendue [I] intend to be a purchaser especially as you Encourage me to draw on you upon any Good Adventures. I wish you had been so Good as to have mention'd the prices of Indigo, Cocoa, Chocolate, Coffee, Sugar, &c. & how far I might Venture to goe [sic] in the purchase of those Commodity's, we being in the dark here as to the prices at home, which am again to pray you may never omit to advise me of.* [14]

A subsequent letter, of February 22, 1745, informed Andrew Pringle in London of the high cost of cocoa from the *Rose:*

The Cocoa of the Roses *[sic] prize Sells of so dear that it wont [sic] answer to purchase any as the prices you advise are in London.* [15]

Since the prize cocoa on the *Rose* was so expensive, a few days later, on February 25, 1745, Robert Pringle wrote to Robert Ellis in Philadelphia, Pennsylvania, that cheaper cocoa would sell well in Charleston:

I understand that Good Chocolate is Sold at 20/ & 20/ per lb. your Currency. If so, a Quantity would answer well here at that price. [16]

A letter book belonging to Robert Raper contained several general records that mentioned cacao or chocolate but only listed purchases and quantities:

August 22nd, 1759: 1$\frac{1}{2}$ doz pound choacalate. [17]

October 10th, 1767: 1 cacao [meaning unclear]. [18]

DIARIES, JOURNALS, AND LETTERS

Mary Reed Eastman's diary contained notations and descriptions of her honeymoon travels during the years 1832–1833 [19]. She mentioned having "hot cocolate [*sic*]" on January 15, 1833, in Augusta, Georgia [20], and noted that the prisoners in the state penitentiary at Milledgeville, Georgia, were served chocolate [21].

Another document was William Cobbett's account of traveling in America. It was published in 1819 and included information on the accessibility of chocolate:

Groceries, as they are called, are, upon an average, at far less than half the English price. Tea, sugar, coffee, spices, chocolate, cocoa, salt, sweet oil: all free of the boroughmongers' [sic] taxes and their spawn, are so cheap as to be within the reach of every one. Chocolate, which is a treat to the rich, in England, is here used even by the negroes. [22]

Chocolate was not always accessible, especially during times of conflict. We identified only one reference to chocolate or cocoa during the Civil War era (1861–1865) from the Charleston archives, a letter from a Confederate officer. William Gildersleeve Vardell was a major and quartermaster in the C.S.A. On April 21, 1863, he wrote to his wife, Jennie, who was living in Charleston at the time:

I wish you had those nice things the young man brought me especially the chocolate. [23]

Vardell's letter was written during the siege of Vicksburg, Mississippi, when he served with the 23rd South Carolina Infantry Regiment. Although we searched specifically for chocolate and cocoa records during the Civil War era (1861–1865), we found no mention in soldiers' rations or from ledgers of this time period. We surmise, therefore, that chocolate was scarce in Charleston during the war and that money and goods were concentrated more toward war-related items (see also Chapter 55).

SALES AND PURCHASES

Records of sales and purchases yield insight into availability and price fluctuations of the chocolate and cocoa market. James Poyas maintained a daybook for the years 1764–1766 that listed all items sold daily at his dry goods store in Charleston [24]. Among the merchandise itemized were books, buttons, chocolate, cloth, clothing, gunpowder, nails, paper, rum, shoes, ribbons, and tea. Entries detailed customer names along with the amount and selling price of the respective items. Poyas sold chocolate ten times in 1764. Four of these sales were in December, which would have been consumed as a warm beverage during winter [25–28]. The remaining dates for chocolate sales in 1764 may reflect import/supply considerations: May 3 [29], June 18 [30], August 27 [31], September 13 [32], October 18 [33], and November 26 [34].

The supply of cacao and chocolate in Charleston clearly was an issue in 1765 when Poyas sold these items only three times: January 25 [35], February 23 [36], and March 21 [37]. Sales increased during 1766 when he made seven sales that reflected the seasonal nature of supply: January 13 [38] and February 10 [39], then a three-month gap until his next sale on June 20 [40], followed by July 25 [41], then a four-month gap, with sales resuming on November 3 [42] and November 7 [43], then again on December 16 [44].

Family account records for John Chesnut, plantation owner and state legislator from the Camden District of South Carolina,[3] reveal that he made two purchases of chocolate—some 19 years apart! The first was recorded on September 2, 1771, when he noted purchase of "one chocolate" (quantity unknown) at 17 shillings 6 pence [45], and again on January 21, 1790, when he recorded the purchase of "three chocolates" (quantity unknown) that totaled 3 shillings, 9 pence [46]. The two documents are interesting not only because of his minimal purchases, but because of the discrepancy in prices between 1771 and 1790.

Another account and letter book for the merchants Coggeshall & Evans spanned the years 1804–1817 [47]. According to file information, Nathaniel Coggeshall was born in Rhode Island and partnered with his brother, Peter Coggeshall, in the early 1790s. It is not clear which Coggeshall brother then formed the partnership with Evans (first name not known). Ledgers for Coggeshall & Evans contained a very limited number of chocolate purchases, indeed, only four times during 1804–1807: specifically, May 17, 1804 [48], July 23, 1804 [49], November 22, 1805 [50], and January 9, 1807 [51].

The 1846–1849 ledger/account book from Heyward & Deas, grocers in Charleston, record only one sale of chocolate for the three-year period: three units of chocolate (not further defined) on November 17, 1847, at the price of $0.37 cents each [52].

A grocery store account book we identified contained records for monthly rent payments and groceries. The grocery store was located at the corner of King and Calhoun Streets, Charleston, and possibly was owned by O. Von Halden. Most of the ledger entries were written by the clerk, D. W. Ohlandt. The importance of this document is due to the dates covered: 1860–1865, during the American Civil War. Included in the ledger was a single entry on November 20, 1860, for "one box of 24 saved chocolate pieces @ $0.20 cents each" [53]. We suggest that the "saved chocolate" possibly referred to a ration, indicating a scarcity of chocolate during this time.

The rise and fall of chocolate prices in Charleston were recorded in the Bernard O'Neill & Sons ledger covering the period 1870–1871 [54]. This ledger noted 13 sales of boxed chocolate[4] containing between 6 and 25 pieces of caked chocolate that were sold to various customers during this period: two dates in 1870—December 28 [55] and December 29 [56], which would have coincided with the post-Christmas/New Year celebration period, followed by sales in 1871 on January 10 [57], February 8 [58], March 7 [59], March 17 [60], April 4 [61], April 25 [62], no sales in May until June 6 [63], June 14 [64], June 24 [65], June 26 [66], and then a last sale in 1871 on September 19 [67]. The Bernard O'Neill & Sons ledger revealed that the price of chocolate in 1871 increased significantly per box compared to 1870 prices (from $0.25 cents per chocolate box to $0.40 cents/unit by June 14, 1871). Interestingly, the next sale price on June 26—only 12 days later—saw a drop in price to 0.25 cents per box, identical to the list price in 1870. But three months later at the time of the September 19 sale, the price had risen sharply to $0.42 cents/chocolate box, most likely reflecting issues of availability, quality, and quantity on hand.

Since chocolate in Charleston (like elsewhere) primarily was served as a beverage until the mid- to late 19th century, this commodity often had its own serving accoutrements, whether chocolate cups, mills (stirring rods), pots, saucers, or spoons. Mrs. George W. Williams, Sr., kept a scrapbook between 1855 and 1891, now deposited in the South Carolina Historical Society Archives. This interesting collection of clippings contained an advertisement by S. Thomas & Brothers located at 257 King Street in Charleston. The merchants advertised chocolate pots that ranged in price from $1.25 to $4.00. The same advertisement also listed chocolate cups and saucers selling for $1.00 to $5.00 per dozen [68].

PLANTATION RECORDS

Willam Harleston, South Carolina state representative and owner of Bluff Plantation, maintained an account book that recorded purchases for food and household goods between 1816 and 1818. Chocolate was recorded only once, on December 5, 1816 [69]. This entry noted two counterweights[5] of chocolate at 0.375 cents/unit for a total of $0.75 cents. The lack of more chocolate entries may be due to limited availability during this time.

Henry Augustus Middleton, Jr., graduated from Harvard University with a degree in engineering. Between the years 1855 and 1861, he managed the family WeeHaw Plantation for his father, prior to joining the Confederate army during the American Civil War. The WeeHaw Plantation journal regularly listed chocolate under household groceries with regularity during this period [70].

Julian Augustus Mitchell Julian was a lawyer in Charleston and owned the Swallow Bluff Plantation located on Edisto Island, 50 miles south of Charleston. He also was a partner in a grocery store venture with Samuel H. Wilson [71]. The A notation in their account book, dated July 24, 1895, recorded a sale of two pounds of chocolate (Baker's Chocolate) for the sum of $0.80 cents [72].

RECIPES

Chocolate recipes identified in the South Carolina Historical Archives fit two categories: (1) those available from 19th century published cookbooks that form part of the collection, and (2) family recipes, either handwritten or collected/clipped from newspaper/magazines and pasted into commonplace books or family scrapbooks. Examples of the first kind would be those contained in the cookbook by Colin Mackenzie, published in 1825, entitled *Five Thousand Receipts in All the Useful and Domestic Arts*. The Mackenzie cookbook contained a recipe for a chocolate liqueur (*ratafia de chocolat*) made from roasted cocoa nuts, alcohol, sugar, and vanilla [73]. One interesting beverage was called "sassafras cocoa" [74]. The primary value of Mackenzie's book, however, are the two descriptions for making chocolate:

> To Make Chocolate. Roast the cocoa in a frying pan, placed on a clear fire; and having afterwards cleared them of the husks, the nuts must be first powdered coarsely, and afterwards beaten in an iron mortar, the bottom of which is made pretty hot, by placing it on the fire, till the whole runs into a thick kind of oil. In this state it must be poured into thin moulds of any size or shape that is agreeable; and when cold, the cakes may be taken out for use. The Spaniards mix with their cocoa nuts too great a quantity of cloves and cinnamon, besides other drugs without number, as musk, ambergris, &c. The Parisians use few or none of these ingredients; they only choose the best nuts, which are called caracca, from the place from whence they are brought; and with these they mix a very small quantity of cinnamon, the freshest vanilla, and the finest sugar, but very seldom any cloves. Chocolate, fresh from the mill, as it cools in the tin pans into which it is received, becomes strongly electrical: and retains this property for some time after it has been turned out of the pans, but soon loses it by handling. The power may be once or twice renewed by melting it again in an iron ladle, and pouring it into the tin pans as at first; out when it becomes dry and powdery, the power is not capable of being revived by simple melting: but, if a small quantity of olive oil be added, and well mixed with the chocolate in the ladle, its electricity will be completely restored by cooling it in the tin pan as before. . . .
>
> Another Method [to make chocolate]. As the pleasantness of chocolate depends, in a great measure, on the method of preparing it for the table, it is necessary that the strictest attention be made to the following simple direction. To make this chocolate, put the milk and water on to boil, then scrape the chocolate fine, from one to two squares to a pint, to suit the stomach; when the milk and water boils, take it off the fire throw in the chocolate; mill it well, and serve it up with the froth; which process will not take 5 minutes. The sugar may either by put in with the scraped chocolate or added afterward. It should never be made before it is wanted; because heating again injures the flavor, destroys the froth, and separates the body of the chocolate; the oil of the nut being observed, after a few minutes' boiling, or even standing long by the fire, to rise to the top, which is the only cause why this chocolate can offend the most delicate stomach. [75]

Mary M. Pringle clipped recipes and stored them in her personal scrapbook (commonplace book), dated generally to the 1860s. She provides clippings for chocolate cakes and chocolate cream [76] but the recipes are rather common and reflective of mid-19th century cookbooks (see Chapter 9).

The Gibbes family recipe book, actually a commonplace book, contained a range of both culinary and personal interest clippings, observations, and handwritten notes [77]. The Gibbes family lived on the Peaceful Retreat Plantation located on Johns Island, opposite Charleston. The recipe/scrapbook can be dated imprecisely to ca. 1890 and included recipes for chocolate caramels (p. 25), chocolate cake (pp. 27–29, 147–148), chocolate marble cake (p. 33), chocolate moss (p. 40), chocolate pudding (pp. 51, 142), and chocolate icing (pp. 85, 103). This scrapbook represented the largest collection of chocolate recipes that we located in the South Carolina Historical Archives in Charleston.

Search for Chocolate in Savannah

IMPORT AND EXPORT RECORDS

The only Savannah Port Records available listed ships entering and leaving Savannah in the years 1765–1767 [78]. The ledger included the vessel name, captain, quantity of chocolate, and the origin or destination of the chocolate. On March 18, 1765, two boxes of chocolate were headed to Boston in the *Three Friends* under the command of Abraham Gybuat [79]. Dan Waters commanded the *Peggy & Hannaa* on March 27, 1765, headed to Antigua with two boxes of chocolate [80]. One bag of cocoa was noted as being "legally imported" and was heading toward South Carolina in the *Culloden* on April 11, 1765 [81]. John Stanton brought two boxes of chocolate from Boston to Savannah on August 27, 1765 [82]. In 1766, a total of seven boxes of chocolate were registered from Philadelphia and four of these arrived in Savannah on April 16 aboard the *Defiance*, and three boxes arrived May 19 on the *Ogeeche* [83, 84]. Seasonality and/or supply is reflected in the ledger as no additional imports or exports of chocolate were recorded after May 19 until January 12, 1767, when the *Seahorse* commanded by Jon Fredwell brought two boxes of chocolate into Savannah from Rhode Island [85]. An unspecified amount of cocoa arrived from Grenados (i.e., Grenada) aboard the *Laleah & Susannah* captained by James Walden on February 25, 1767 [86]. Two bags of cocoa were sent from Savannah to South Carolina aboard the *Georgia Packer* on March 21, 1767 [87]. The last notation in the ledger for either import or export of chocolate was an entry that one box of chocolate had arrived in Savannah from Philadelphia on May 5, 1767, aboard the *Speedwell* [88]. Earlier and subsequent port records were not available in the Georgia Historical Society Archives, Savannah.

CHANDLER ACCOUNTS

The ledger of an unidentified ship chandler deposited in the archive listed ship names, quantities of chocolate taken aboard, and price information. The following ships were identified as carrying quantities of chocolate: Brig *Casket*, December 15, 1815, 4 pounds chocolate, 2 currency units[6] [89]; Brig *Hollon*, January 5, 1816, 1 counterweight of chocolate; 50 currency units [90]; and *Favorite* (type of ship not recorded), March 25, 1816, 2 units of "choch," 4 currency units [91]. The accounts in this chandler's record do not indicate the respective ships' onward destinations.

DIARIES, JOURNALS, AND LETTERS

While the collection in the Savannah archives was rich in diaries, journals, and letters, the documents were notably limited in references to cacao or chocolate, and we cite only one here. The earliest document we identified was a letter from Isaac Deylon of Savannah to Bernard Gartz of Philadelphia, written on September 24, 1760, that mentioned *chokolet* in the postscript where Deylon listed an inventory of goods with descriptions of how to properly store them [92]. No other letters, journals, or diaries we surveyed contained passages that pertained to the purchase or sale of chocolate, or contained chocolate recipes.

LEDGERS, INVENTORY RECORDS, AND HOUSEHOLD ACCOUNT BOOKS

James Read and James Mossman were business partners in Savannah, and both were active in the community and local politics. The Read–Mossman Ledger, dated 1765–1766, contained 657 pages of detailed sundry accounts listing the account name and number along with the quantity and price for food and goods [93]. Account information for 11 purchases of chocolate is recorded in the ledger (Table 52.7). The price of chocolate increased between September 1765 and October 1766 from 2 £ to 2.6 £ per unit. The ledger entries clearly reveal seasonality, as all purchases were made between June and October with most purchases occurring in September 1766.

An inventory record for the Savannah company of William and Edward Telfair contained an entry for three-dozen chocolate cups that sold for four currency units per dozen cups on January 1, 1775 [94].

Table 52.7 **Information from the Read–Mossman Ledger [2] for Accounts Purchasing Chocolate**

Year	Month	Day	Account Name and Number	Quantity	Unit Price	Total Price (£)
1765	September	4	John Smith #221	2	—	4
1766	June	27	Inglis & Hall #392	6	2	12
1766	August	30	To Daniel Rundles Sales No. 24 #402	1 box	—	3.15
1766	August	30	Kelsall Darling & Munro #337	2	2.6	5
1766	September	2	Kelsall Darling & Munro #337	2	2.6	5
1766	September	6	Gordon & Neitherclift #400	1	—	2.6
1766	September	10	James Read #393	$\frac{1}{2}$ dozen	20.6	10.3
1766	September	18	Clay & Habersham #238	2	2.6	5
1766	September	24	Kelsall Darling & Munro #337	5	2.6	12.6
1766	September	30	Johnson & Wylly #312	1	—	2
1766	October	13	Clay & Habersham #238	1	—	2.6

This entry was the only one found in the Savannah archives that specifically itemized chocolate dishes. For comparison, in the latter portion of the 19th century, chocolate cups sold for between $1 and $5 per dozen in Charleston [95].

The George Galphin collection contained an interesting account book dated between 1767 and 1772. Galphin maintained an Indian trading post located at Silver Bluff, South Carolina, on the Georgia–South Carolina border. The ledger lists that a Robert Toole purchased five "chocolath" in 1767 (month/day not recorded), for the price of 4.7.6, that is, 4 £, 7 shillings, 6 pence [96].

An account book dated to 1820 and kept by the Edward Varner General Store in Putnam County, Georgia, listed several purchases of chocolate, with highly variable price fluctuations, due most likely to different currencies available and used. On May 29, for example, Samuel Holloway purchased one pound of *chockolate* for 3.5 currency units [97], while on the same date John Jones purchased one pound of *chockolate* for 37.5 currency units [98]. On July 29, 1820, C. B. Strong purchased two cakes of chocolate for 50 currency units [99].

The Robert S. Goff daybook dated 1821–1823 contained account/inventory information for Pettingill & Goff, commission merchants located in Savannah. The following selected notations from the daybook reveal erratic sales concentrated between February and March in 1821, then scarcity of product throughout most of 1822–1823. The document reveals, however, that chocolate was sold in boxes that contained 50 pieces/cakes: February 1, 1821—chocolate merchandise at 58.5 currency units [100]; February 3, 1821—chocolate candy, sales # 16 at 15 currency units [101]; and March 14, 1821—two boxes of chocolate merchandise at 4.75 currency units [102]. No mention of chocolate is recorded from mid-March through early August; then on August 7, 1821—nine boxes of chocolate sold to William K. Barnard (price not given) [103]. Then there is no mention of chocolate in the Goff daybook after early August of 1821, through almost all of 1822, and only in late December with a record on December 27, 1822, for three sales of boxes of chocolate (total 7 boxes), each box containing 50 pieces (of chocolate) [104]. The last record of chocolate in the Goff daybook is dated to April 18, 1823, where one box of chocolate containing 50 pieces was sold to a Mr. P. Howard and noted to his account [105].

Two housekeeping accounts were identified that contained references to chocolate purchases/uses. One account was maintained by the Varner family of Butts County, Georgia, and recorded on January 9, 1891, that $0.50 had been spent for *chochlate* [106]. A second account book was located in the Heyward–Howkins Family Papers that listed several dates for chocolate purchases: the family spent $0.25 for cocoa on February 5, 1899 [107], and on April 13, 1900, spent $0.20 for chocolate [108].

RECIPES

On occasion, some documents such as account books and ledgers contained recipes, which ordinarily would not have been expected. One example of such an instance was the Telfair account book of housekeeping expenses we examined, dated to 1850, where recipes were "slipped" into the expense records. Records of the Telfair family from Savannah, in fact, reflected the earliest recipe for chocolate we found in the Georgia Historical Society archives. The recipe was for *chocolate blanchmange* intended to be a warm, rich dessert, one that contained heated milk with grated chocolate and white sugar, thickened with cornstarch [109]. This recipe, in fact, was the earliest we located in the Georgia Historical Society archive.

A second document, also within the Telfair family records, was a notebook that consisted of recipes, including one identified as *fromage of chocolate*. The recipe called for mixing a small quantity of water with melted chocolate, then adding isinglass, sugar, and whipped cream [110].

A cocoa medical recipe appeared in *A Pocket Formulary and Physician's Manual*, published by W. Thorne Williams in Savannah in 1855. Williams recommended boiling two ounces of cocoa in four cups of water followed by simmering an hour over coals. The drink was to be served hot to patients [111].

Other recipes that used chocolate for confections date to the late 19th century and were located in collections of family papers. Mrs. Leila Habersham authored the Woodbridge Family Cookbook, dated to 1895, which contained recipes for chocolate soufflés, chocolate pudding, and chocolate sauce [112]. The Owens–Thomas family collection also contained chocolate recipes that itemized two specific types of sugar (boiled and unboiled). These two recipes, undated, probably could be assigned to the late 19th century as well [113].

Conclusion

We searched the South Carolina Historical Society Archives, Charleston, and the Georgia Historical Society Archives, Savannah, for cacao- and chocolate-related documents. We had more success at the SCHS Archives. This difference mainly can be attributed to the paucity of port records for Savannah. Most of the Charleston records were shipping manifest documents; the majority of information identified for Savannah appeared in account books/ledgers. Both archives exhibited a general lack of chocolate-associated documents from the era of the American Civil War. Further research in southern archives is encouraged.

Acknowledgments

Our search would not have been possible without the support and aid of staff at the Georgia Historical Society (GHS) in Savannah, Georgia, and at the South Carolina Historical Society (SCHS) in Charleston, South Carolina. We would like to thank Stephany Kretchmar, Library Assistant for the GHS, for her outstanding contribution in locating requested collections. Other staff at the GHS deserving thanks for their aid in finding requested collections include Robert Webber, Special Collections Librarian; Nora Galler, Director of the Library and Archives; and Lynette Stoudt, Senior Archivist. We thank Lisa Hayes, SCHS Librarian, for her exceptional help in finding requested documents and for her assistance with the microfilms and ship records. Thanks are also due to Laura Koser, SCHS Archivist, and Mike Coker, SCHS Visual Materials Curator, for their assistance in obtaining requested collections.

Endnotes

1. Letter books contain copies of letters and memos related to merchant correspondence, items of importance to family members, and documents dealing with other merchants in distant cities.

2. Public Vendue = public auction (see Appendix 1).

3. The settlement of Camden, South Carolina, located in Kershaw County, was founded in 1730. Camden is considered the oldest inland city in South Carolina.

4. The ledger term "boxed chocolate" is ambiguous. By 1870–1871 (ledger date), specialty chocolates for "eating" could have been sold by Bernard O'Neill & Sons. It also is possible that "boxed chocolate" were specialty types presented in gift boxes, but with the intent that the contents—bars of chocolate—were to be grated and prepared as chocolate beverages and not eaten directly.

5. Counterweight, that is, weighed at the counter. This is an archaic term used to signify when merchants weighed goods out in front of customers using small beam-balance scales situated on the store countertops near the cash register.

6. Quantity given in the document was four pounds of chocolate at $0.50 cents/pound or $2.00.

References

1. South Carolina Historical Society (SCHS). British Public Records Office ship manifests, microfilm, 387.1.C56M4, Box 199C.

2. SCHS. British Public Records Office ship manifest, 387.1.C56M4, Box 199A.

3. SCHS. Robert Pringle, Letter Book, 1737–1745. (34/1 OvrSz), General citation.

4. SCHS. Robert Pringle, Letter Book, 1737–1745. (34/1 OvrSz), entry for May 4, 1739, p. 86.

5. SCHS. Robert Pringle, Letter Book, 1737–1745. (34/1 OvrSz), entry for April 19, 1740, pp. 182–183.

6. SCHS. Robert Pringle, Letter Book, 1737–1745. (34/1 OvrSz), entry for October 24, 1743, p. 590.

7. SCHS. Robert Pringle, Letter Book, 1737–1745. (34/1 OvrSz), entry for October 24, 1743, p. 592.

8. SCHS. Robert Pringle, Letter Book, 1737–1745. (34/1 OvrSz), entry for February 11, 1744, p. 645.

9. SCHS. Robert Pringle, Letter Book, 1737–1745. (34/1 OvrSz), entry for March 31, 1744, p. 669.

10. SCHS. Robert Pringle, Letter Book, 1737–1745. (34/1 OvrSz), entry for April 24, 1744, p. 676.

11. SCHS. Robert Pringle, Letter Book, 1737–1745. (34/1 OvrSz), entry for June 30, 1744, p. 720.

12. SCHS. Robert Pringle, Letter Book, 1737–1745. (34/1 OvrSz), entry for December 17, 1744, pp. 779–780.

13. SCHS. Robert Pringle, Letter Book, 1737–1745. (34/1 OvrSz), entry for February 9, 1745, p. 810.

14. SCHS. Robert Pringle, Letter Book, 1737–1745. (34/1 OvrSz), entry for January 19, 1745, p. 804.

15. SCHS. Robert Pringle, Letter Book, 1737–1745. (34/1 OvrSz), entry for February 22, 1745, p. 808.

16. SCHS. Robert Pringle, Letter Book, 1737–1745. (34/1 OvrSz), entry for February 25, 1745, p. 820.

17. SCHS. Robert Raper, Letter Book, 1759–1770. (34/511), entry for August 22, 1759.

18. SCHS. Robert Raper, Letter Book, 1759–1770. (34/511), entry for October 10, 1767.

19. SCHS. Mary Reed Eastman, Diary excerpts, 1832–1833. (43/2204).

20. SCHS. Mary Reed Eastman, Diary excerpts, 1832–1833. (43/2204), entry for January 15, 1833, p. 30.

21. SCHS. Mary Reed Eastman, Diary excerpts, 1832–1833. (43/2204), entry for January 19, 1833, p. 37.

22. SCHS. Cobbett, W. *A year's residence in the United States of America: treating of the face of the country, the climate, the soil, the products, the mode of cultivating the land, the prices of land, of labour, of food, of raiment; of the expenses of house-keeping and of the usual manner of living; of the manners and customs of the people; and, or the institutions of the country, civil, political and religious*; in three parts. New York: Clayton and Kingsland, 1819; Part 1, p. 235, Section 335. SCHS accession number R 917.3.C63Y43 1819.

23. SCHS. William G. Vardell, Letters 1862–1863 (271.00). Letter dated April 21, 1863.

24. SCHS. James Poyas, Daybook, 1764–1766. (34/325 OvrSz), General citation.

25. SCHS. James Poyas, Daybook, 1764–1766. (34/325 OvrSz), entry for December 3, 1764, p. 140.

26. SCHS. James Poyas, Daybook, 1764–1766. (34/325 OvrSz), entry for December 6, 1764, p. 143.

27. SCHS. James Poyas, Daybook, 1764–1766. (34/325 OvrSz), first entry for December 11, 1764, p. 145.

28. SCHS. James Poyas, Daybook, 1764–1766. (34/325 OvrSz), second entry for December 11, 1764, p. 145.

29. SCHS. James Poyas, Daybook, 1764–1766. (34/325 OvrSz), entry for May 3, 1764, p. 36.

30. SCHS. James Poyas, Daybook, 1764–1766. (34/325 OvrSz), entry for June 18, 1764, p. 57.

31. SCHS. James Poyas, Daybook, 1764–1766. (34/325 OvrSz), entry for August 27, 1764, p. 84.

32. SCHS. James Poyas, Daybook, 1764–1766. (34/325 OvrSz), entry for September 13, 1764, p. 95.

33. SCHS. James Poyas, Daybook, 1764–1766. (34/325 OvrSz), entry for October 18, 1764, p. 110.

34. SCHS. James Poyas, Daybook, 1764–1766. (34/325 OvrSz), entry for November 26, 1764, p. 132.

35. SCHS. James Poyas, Daybook, 1764–1766. (34/325 OvrSz), entry for January 25, 1766, p. 167.

36. SCHS. James Poyas, Daybook, 1764–1766. (34/325 OvrSz), entry for February 23, 1765, p. 177.

37. SCHS. James Poyas, Daybook, 1764–1766. (34/325 OvrSz), entry for March 21, 1765, p. 190.

38. SCHS. James Poyas, Daybook, 1764–1766. (34/325 OvrSz), entry for January 13, 1766, p. 330.

39. SCHS. James Poyas, Daybook, 1764–1766. (34/325 OvrSz), entry for February 10, 1766, p. 342.

40. SCHS. James Poyas, Daybook, 1764–1766. (34/325 OvrSz), entry for June 20, 1766, p. 392.

41. SCHS. James Poyas, Daybook, 1764–1766. (34/325 OvrSz), entry for July 25, 1766, p. 408.

42. SCHS. James Poyas, Daybook, 1764–1766. (34/325 OvrSz), entry for November 3, 1766, p. 447.

43. SCHS. James Poyas, Daybook, 1764–1766. (34/325 OvrSz), entry for November 7, 1766, p. 448.

44. SCHS. James Poyas, Daybook, 1764–1766. (34/325 OvrSz), entry for December 16, 1766, p. 464.

45. SCHS. Chesnut Family Papers, 1741–1900. (1165.01.01; 12–33), entry for September 2, 1771.

46. SCHS. Chesnut Family Papers, 1741–1900. (1165.01.01; 12–33), entry for January 21, 1790.

47. SCHS. N. Coggeshall & Son, Merchants. Account and Letter Book, 1804–1817. (34/705), General citation.

48. SCHS. N. Coggeshall & Son, Merchants. Account and Letter Book, 1804–1817. (34/705), entry for May 17, 1804.

49. SCHS. N. Coggeshall & Son, Merchants. Account and Letter Book, 1804–1817. (34/705), entry for July 23, 1804.

50. SCHS. N. Coggeshall & Son, Merchants. Account and Letter Book, 1804–1817. (34/705), entry for November 22, 1805.

51. SCHS. N. Coggeshall & Son, Merchants. Account and Letter Book, 1804–1817. (34/705), entry for January 9, 1807.

52. SCHS. James Barnwell Heyward Papers, 1845–1909. (1305.00). Account Book, entry for November 17, 1847.

53. SCHS. Anonymous. Grocery Store Account Book, 1860–1865. (34/706 OvrSz), General citation.

54. SCHS. Bernard O'Neill & Sons. Ledger, 1861–1899. (1189.00), General citation.

55. SCHS. Bernard O'Neill & Sons. Ledger, 1861–1899. (1189.00), entry for December 28, 1870, p. 153.

56. SCHS. Bernard O'Neill & Sons. Ledger, 1861–1899. (1189.00), entry for December 29, 1870, p. 154.

57. SCHS. Bernard O'Neill & Sons. Ledger, 1861–1899. (1189.00), entry for January 10, 1871, p. 171.

58. SCHS. Bernard O'Neill & Sons. Ledger, 1861–1899. (1189.00), entry for February 8, 1871, p. 215.

59. SCHS. Bernard O'Neill & Sons. Ledger, 1861–1899. (1189.00), entry for March 7, 1871, p. 250.

60. SCHS. Bernard O'Neill & Sons. Ledger, 1861–1899. (1189.00), entry for March 17, 1871, p. 265.

61. SCHS. Bernard O'Neill & Sons. Ledger, 1861–1899. (1189.00), entry for April 4, 1871, p. 288.

62. SCHS. Bernard O'Neill & Sons. Ledger, 1861–1899. (1189.00), entry for April 25, 1871, p. 307.

63. SCHS. Bernard O'Neill & Sons. Ledger, 1861–1899. (1189.00), entry for June 6, 1871, p. 346.

64. SCHS. Bernard O'Neill & Sons. Ledger, 1861–1899. (1189.00), entry for June 14, 1871, p. 357.

65. SCHS. Bernard O'Neill & Sons. Ledger, 1861–1899. (1189.00), entry for June 24, 1871, p. 366.

66. SCHS. Bernard O'Neill & Sons. Ledger, 1861–1899. (1189.00), entry for June 26, 1871, p. 367.

67. SCHS. Bernard O'Neill & Sons. Ledger, 1861–1899. (1189.00), entry for September 19, 1871, p. 449.

68. SCHS. Mrs. George W. Williams, Sr., Scrapbook (1855–1891), George Walton Williams Family Papers, 1849–1959 (bulk 1870s–1890s). No page numbers. (1306.00 28-724-4).

69. SCHS. Sarah Quash Harleston, Account Book, 1817–1824. (43/128), entry for December 5, 1816.

70. SCHS. Henry Augustus Middleton, Papers, 1842–1861. (1168.02.07), p. 120.

71. SCHS. Julian A(ugustus) Mitchell, Account Book (1885–1896). Papers, 1870–1902. (154.00), General citation.

72. SCHS. Julian A(ugustus) Mitchell, Account Book (1885–1896). Papers, 1870–1902. (154.00), entry for July 24, 1895.

73. SCHS. Mackenzie, C. *Five Thousand Receipts in All the Useful and Domestic Arts: Constituting a Complete and Universal Practical Library, and Operative Cyclopaedia.* (R 640.M33F58 1825). Philadelphia: Abraham Small, 1825; p. 220.

74. SCHS. Mackenzie, C. *Five Thousand Receipts in All the Useful and Domestic Arts: Constituting a Complete and Universal Practical Library, and Operative Cyclopaedia.* (R 640. M33F58 1825). Philadelphia: Abraham Small, 1825; pp. 239–240.

75. SCHS. Mackenzie, C. *Five Thousand Receipts in All the Useful and Domestic Arts: Constituting a Complete and Universal Practical Library, and Operative Cyclopaedia.* (R 640. M33F58 1825). Philadelphia: Abraham Small, 1825; p. 240.

76. SCHS. Mary M. Pringle, Recipe Book. Alston–Pringle–Frost Papers, 1693–1990, bulk 1780–1958. (1285.00 Box 640, Volume 13). (General citation). Item not numbered.

77. SCHS. Gibbes Recipe Book. Gibbes Family Papers, 1769–ca. 1935. (1035.00 11-152-1). (General citation). Item not numbered.

78. Georgia Historical Society (GHS). Savannah Port Records, MS 704. (General citation).

79. GHS. Savannah Port Records, MS 704, ship ledger entry for March 18, 1765.

80. GHS. Savannah Port Records, MS 704, ship ledger entry for March 27, 1765.

81. GHS. Savannah Port Records, MS 704, ship ledger entry for April 11, 1765.

82. GHS. Savannah Port Records, MS 704, ship ledger entry for August 27, 1765.

83. GHS. Savannah Port Records, MS 704, ship ledger entry for April 16, 1766.

84. GHS. Savannah Port Records, MS 704, ship ledger entry for May 19, 1766.

85. GHS. Savannah Port Records, MS 704, ship ledger entry for January 12, 1767.

86. GHS. Savannah Port Records, MS 704, ship ledger entry for February 25, 1767.

87. GHS. Savannah Port Records, MS 704, ship ledger entry for March 21, 1767.

88. GHS. Savannah Port Records, MS 704, ship ledger entry for May 5, 1767.

89. GHS. Author unknown. Account Book, unidentified, MS 1194, ship chandler entry for December 15, 1815.

90. GHS. Author unknown. Account Book, unidentified, MS 1194, ship chandler entry January 5, 1816.

91. GHS. Author unknown. Account Book, unidentified, MS 1194, ship chandler entry for March 25, 1816.

92. GHS. Isaac Deylon Papers, MS 210.

93. GHS. Read–Mossman Ledger, MS 1635.

94. GHS. William & Edward Telfair & Company. Inventory Book. Telfair Family Papers, MS 793, Box 2, Folder 10, entry for January 1, 1775.

95. SCHS. Mrs. George Williams, Sr., Scrapbook (1855–1891). George Walton Williams Family Papers, 1849–1959 (bulk 1870s–1890s). No page numbers. (1306.00 28-724-4).

96. GHS. George Galphin, Account Books, MS 269, entry for 1767 (no precise date).

97. GHS. Edward Varner General Store, Account Book. Varner Family Papers, MS 1256, Volume 2, entry for May 29, 1820, p. 184.

98. GHS. Edward Varner General Store, Account Book. Varner Family Papers, MS 1256, Volume 2, entry for May 29, 1820, p. 183.

99. GHS. Edward Varner General Store, Account Book. Varner Family Papers, MS 1 256, Volume 2, entry for July 29, 1820, p. 247.

100. GHS. Robert S. Goff, Daybook. Pettingill & Goff Business Records, MS 315, entry for February 1, 1821, p. 44.

101. GHS. Robert S. Goff, Daybook. Pettingill & Goff Business Records, MS 315, entry for February 3, 1821, p. 44.

102. GHS. Robert S. Goff, Daybook. Pettingill & Goff Business Records, MS 315, entry for March 14, 1821, p. 49.

103. GHS. Robert S. Goff, Daybook. Pettingill & Goff Business Records, MS 315, entry for August 7, 1821, p. 64.

104. GHS. Robert S. Goff, Daybook. Pettingill & Goff Business Records, MS 315, entry for December 27, 1822, p. 119.

105. GHS. Robert S. Goff, Daybook. Pettingill & Goff Business Records, MS 315, entry for April 18, 1823, p. 135.

106. GHS. Varner House, Account Book. Varner Family Papers, MS 1256, Volume 9, p. 3.

107. GHS. Heyward–Howkins Family Papers, MS 1278, Box 4, entry for February 25, 1899.

108. GHS. Heyward–Howkins Family Papers, MS 1278, Box 4, entry for April 13, 1900.

109. GHS. Telfair Family Papers, MS 793, Box 8, Folder 74, Item 298.

110. GHS. Telfair Family Papers, MS 793, Box 8, Folder 73, Item 291.

111. GHS. Thomas S. Powell, M.D., 1855. *A Pocket Formulary and Physicians Manual.* RS 125.P68, Savannah, Georgia, p. 126.

112. GHS. Woodbridge Family Cookbook, Caroline Lamar Woodbridge Papers, MS 878, Item 4.

113. GHS. Owens–Thomas Family, MS 602, Box 3, Folder 23.

53

Management of Cacao and Chocolate Data

Design and Development of a Chocolate Research Portal

Matthew Lange

Introduction

The Chocolate History Project, headquartered at the University of California, Davis, Department of Nutrition, is a consortium of more than 100 scholars and technical researchers from all around the world. The primary goal has been to document and determine: how, when, and where chocolate products were introduced into, and dispersed throughout the Americas and globally. The Chocolate Research Portal (CRP)[1] was developed in an effort to aid the collection, categorization, curation, and display of chocolate-related artifacts and documents identified in archives, libraries, and museums, and during field work.

The present chapter is divided into three sections. The first section describes the organization and types of information housed within the CRP and highlights the key advantages and benefits of use for researchers, reporters/editors, and interested members of the lay public. The second section outlines the original motivations, guiding principles, and design challenges faced when constructing the CRP. The third section describes the techniques used to build the CRP. The chapter concludes with suggestions for building similar Web portals, and identifies issues that may be encountered when building such systems.

Organization and Types of Information Within the CRP

The initial development of the CRP was motivated by two driving factors—need for (1) a centralized on-line repository of images and texts collected for the Chocolate History Project, and (2) a centralized information access point where documents could be shared with project-affiliated personnel. As the project grew and expanded during the years 2004–2005, we realized that other types of information could be added to the repository as well. Currently, the CRP contains five basic types of information related to research efforts:

1. Information and images found by researchers during on-line research.

Chocolate: History, Culture, and Heritage. Edited by Grivetti and Shapiro
Copyright © 2009 John Wiley & Sons, Inc.

2. Information and images found by researchers during field work in archives, libraries, and museums, or when conducting interviews in the field.

3. Field notes and annual reports prepared by researchers that summarize the work undertaken, as well as postproject reports that summarize specific research themes.

4. Texts and PowerPoint presentations and project-related publications produced by team members.

5. Concept Maps, Data Folders, Keywords, Smart Folders, and other Metadata (data/content attributes) that describe and classify the above types of information and that also serve as placeholders for future information to be collected.

The Chocolate History Project at the University of California, Davis (UCD), started research activities in 1998 well before construction of the CRP. In September 2004, discussions were initiated regarding the need to create a Web portal that would serve as a repository both for past and present project-related information and for information analysis. Planning for and construction of the CRP began at this time. Cacao- and chocolate-associated information and images obtained through on-line digitized data sources were among the easiest to collect and were the first types of data entered into the CRP. Details of the construction and data management process are presented later in this chapter.

The vast majority of information collected on-line was in the form of images. Images were captured as PDFs, then clustered into folders. The term "image gallery" was used to identify our collection of folders that contained the following primary and secondary groupings:

Advertising
   Boxes and tins
   Posters
   Postcards
   Print media (magazines and newspapers)
   Signs
   Toys
   Trade cards
   Miscellaneous objects used in advertising
Art
   Archaeological materials (i.e., pre-Columbian artifacts)
   Botanical prints
   Fine art
   Posters

Culinary
   Cups and saucers
   Molds (used to shape/form chocolate)
   Pots and vessels
   Spoons and stirring devices
Manufacturing (i.e., images of chocolate production)
Patent documents
Miscellaneous images

When users open and navigate one of the folders located in the image galleries, the CRP automatically selects a random image from the underlying gallery to display. A different image is selected and presented automatically each time the folder is opened.

Users can navigate the image gallery folders by clicking on the representative picture icon. This action directs users to the full suite of images displayed within the gallery. Individual images within the gallery folder can be viewed at a larger scale by clicking directly on the image. When the image appears on screen, additional metadata that describe the item are displayed either to the side or below. Included in the metadata are links to all original URLs where images are located.

As collection of archive, library, and museum information progressed, and copies of cacao- and chocolate-related documents were returned to the central office at UCD. These were digitized and placed in the portal. These documents are clustered within a folder entitled "Archive Documents" and organized by country, then by name of the specific archive.

Just as images of pots, posters, and other material goods were associated with particular metadata, archive documents also were linked with specific, descriptive metadata. The CRP, therefore, can be viewed as a constantly growing/changing product, and efforts currently are underway to transcribe digitized handwritten documents into machine readable (and thus searchable) metadata fields that can be associated with each specific document. This task will be accomplished through Optical Character Recognition (OCR) software. By placing original documents into the CRP—with permission from the respective archives and through what is known as the "fair-use" doctrine—researchers can view the original document and gain new insights on the people, places, and circumstances surrounding the rich history of chocolate in North America.

Oftentimes in the course of archival and anthropological research, notes taken by investigators are as important as the artifacts/documents found or identified. This issue also was taken into consideration during the design phase of the CRP, and a special folder was developed to contain team member field reports. Merely clicking on the links within this folder allows users to inspect and read these important field reports.

Augmenting the field reports by project investigators, high importance also was placed on planning documents and external presentations. PowerPoint® presentations, texts, and posters of presentations made at professional and public meetings, and copies of formal academic publications have been uploaded into the CRP and linked via concept maps to other relevant materials for easy navigation by users.

In addition to the folder hierarchies and metadata fields associated with specific types of information, we embedded what are known as *Smart Folders* and *Concept Maps*. These are two types of content that allow researchers to navigate, organize, and share information on the CRP. The technique used to produce *Smart Folders* is relatively easy or "user-friendly." Researchers can open the chocolate pot image gallery, use the CRP search engine function, and type in the word "pewter;" then all images of pewter chocolate pots already loaded in the portal can be sorted and displayed. Such a search can be saved—to become another part of the CRP—or "trashed" after the researcher has finished.

Part of the pleasure (and one of the challenges) of conducting historical inquiries is searching for relationships between specific concepts, objects, and people. While *Smart Folders* are useful for categorizing similar items based on specific keywords, they do not demonstrate relationships within or among items. *Concept Maps*, on the other hand, are designed specifically for this purpose. Users can open links to a variety of *Smart Folders*, then query the CRP for images, whereupon the various elements are displayed and relationships seen.

Motivations, Guiding Principles, and Design Challenges

One of the most difficult challenges faced by computer programmers and Web designers in the 21st century is how to develop systems that are easy for neophytes or beginning computer users to use and embrace. At the same time, designers also are faced with the daunting task of building data systems that simultaneously allow users robust, queriable access to their own and other sets of complex information and technical content—all in an efficient manner. The main problem to be solved when constructing the CRP was how to design a system with what is called a "low technical barrier," meaning, easy to use, yet able to guide both nontechnical and experienced computer

users toward sophisticated data storage and analysis. The system also has to allow for potential future expansion and be designed in such a way that nontechnical users do not become discouraged. A third issue that had to be considered when constructing the CRP was that project team members would collect data and write their reports off-line (i.e., not connected to the Internet). The design system, therefore, needed to be constructed in such a way that users—both nontechnical and experienced—could easily transmit their findings stored in their own computers and upload the data directly into the CRP or, alternatively, transmit their documents to the central office at UCD, where the documents, files, and spreadsheets could be imported to the CRP with as little additional effort as possible.

Once these basic problems had been identified and the design parameters established, the next step was to develop a set of guiding principles on which the Chocolate Research Portal would be designed (and continues to evolve):

1. Easy operation during periods of field work data collection.

2. Easy tracking and sorting of chocolate-related historical data by date, document type, geographical area, topical theme, and specialized versus general interest.

3. Ability to search documents by keywords and to save searches that could be shared with other researchers and CRP users.

4. Nonreliance on proprietary technology/commercial software licensure.

5. Easy access to students, general public, press corps, and scholars/scientists on the World Wide Web.

6. Eye-catching appeal.

7. Granular security model (GSM) that allows assignments of permission for users, groups of users, documents, and groups of documents, and publication workflow (i.e., the approval process whereby items can be posted to the portal) for storage and content distribution.

8. Ability to aggregate, quantify, and display tabular numerical data from different sources.

9. Ability to annotate images of graphics, artifacts, and documents.

10. Ability to integrate technologies used to construct the CRP with new and maturing Web technologies.

In addition to the problems, issues, and guiding principles identified here, we faced still another design challenge: members of the Chocolate History Research Project—while scholars and experts in their respective fields—were highly heterogeneous in terms of their level of technical skill, network access, and access to computer software. Some team members were quick to recognize and utilize the advantages of the CRP, while others never logged-on or used the system. Some team members worked in geographical areas with ready access to the Internet and could send their data directly to the CRP or to the project office at UCD; others worked in situations and environments where computer access was limited or nonexistent. Some team members, for example, those associated with large academic institutions or companies, had full access to a wide range of software programs; other team members in some locations had only limited access to diversified computer software.

While these principles and guiding questions presented technical challenges, perhaps the biggest challenge of all was convincing historians, many of whose lives revolve around reliving the past, that their participation in cutting-edge information technologies and the Internet could help them delve further into their historical inquiries. Some readily accepted the challenge; others remained reluctant, while still others remained "computer-phobic."

Techniques Used to Build the Chocolate Research Portal

Developing a computer system for use by a team of scholars like those affiliated with the Chocolate History Research Group involved engineering hardware, software, and technical personnel. When attempting to obtain "buy-in" from project participants, "social engineering" was required in order to challenge some project researchers and to change the culture and work habits of some participants so they would embrace and use the newly designed system. This commonly is encountered in a consortium situation (like the current project) where a top–down approach mandating use of a system as in a government or corporate environment would not be feasible. One way of reducing "reluctance" on the part of some team members was our decision to make the system as intuitive and easy to use as possible. This section, therefore, outlines the software tools used to develop the CRP, as well as the workflow design used to retrieve information from the researchers' computers into the portal. In late 2004, at the time design on the CRP was initiated, we were unable to locate a comprehensive free or "open source" solution that would satisfy project needs. The resulting workflow system we designed—so information could flow to the CRP—represented a compilation of multiple technologies of which the Internet-available CRP was only the visible component. In this portion of the chapter we consider the various software components.

CONTENT MANAGEMENT SYSTEM, CONCEPT MAPS, DATABASES, AND FILES

In constructing the portal, we integrated four different information technologies: (1) a concept mapping server, (2) a content management system, (3) a relational database management system, and (4) a basic file system. In doing so, we were able to leverage the power that each technology offered and create a knowledge environment that allowed for easy uploading and synchronization of content resources.

Previous examples in this chapter have shown that concept maps serve several functions inside the Chocolate Research Portal, among them: (1) concept maps allow viewers to visualize information in ways not possible when traditional folder hierarchies are used; (2) concept maps provide intuitive access to content items that are related to search results that have been saved; and (3) concept maps guide users to content items that are related to their interests and alert users to related search terms.

The Plone[2] content management framework (CMF) is the heart of the Chocolate Research Portal. We used Plone technology to produce the basic portal screen layout and file structure. Plone also provides:

1. Built-in granular security, workflow, and document-sharing capabilities.

2. Built-in ability to create and reuse customized content types and their attributes.[3]

3. Integrated relational database access.[4]

4. An integrated search engine that indexes all content/content types that exist within the portal.

5. Customizable *visual skins* that create a dramatic "look-and-feel" within the portal.[5]

Sometimes it was necessary for researchers to collect data off-line. We recommended that data collected should be kept and maintained in tabular or table format using either Microsoft Excel® or Microsoft Word®. When data were kept in this fashion, it was then possible to append these tables to similar tables stored within a relational database management system (RDBMS), in this case PostgreSQL.[6] The RDBMS, which really is just a catalog of related tables of information, could then be synchronized with the Plone content management system. In addition, the PostgreSQL database system allowed for the storage, aggregation, query, and quantification of data that could be fit easily inside table formats. Large files such as images, sound recordings, and videos could be stored in the main file system and managed through the content management framework.

PUTTING EVERYTHING INTO PRACTICE

When project researchers worked in the field (i.e., in archives, libraries, or museums) and were disconnected from the Web, they kept their notes in spreadsheets. When researchers returned from the field, images of artifacts/documents combined with spreadsheets of notes about individual images then could be uploaded to the Web portal directly through a drag-and-drop Web utility.[7]

Images and spreadsheets are stored on the file system where they are cataloged, indexed, and secured through Plone's content management framework. Once the images and spreadsheets are uploaded, the spreadsheets are appended to appropriate tables in the relational database.

Once field data are uploaded into the relational database, the information is associated with the corresponding images to create metadata keywords about the images that were just added to the site. Each row in the spreadsheet is then correlated with one image document; each column represents one metadata element.

Once content is aggregated and cataloged in Plone, it is possible to execute and save searches with specific criteria, forming what are called *Smart Folders*. These *Smart Folders* look and act like regular physical folders, except that they store the results of queries. The actual files listed in the *Smart Folder*, however, are stored in different locations in the portal. It is important to note that content is not moved into a *Smart Folder*—it is only visualized there. The advantage is that the same content can appear in multiple Smart Folders, depending on the folder's search criteria. To reiterate, Concept Maps provide a way to relate multiple Smart Folders to other content items.

Conclusion

Plone and PostgreSQL provide the Chocolate Research Portal (CRP) with many built-in features that facilitate data integration. The current use of *Concept Maps* allows users to explore possible relationships between and among concepts related to the history of chocolate. All of these technologies are new and still maturing. In the early 21st century world of inventive and creative Web-based technologies, we would expect that within a few short years they will integrate and allow Web developers and users even more exciting opportunities and possibilities for evaluating and analyzing historical knowledge.

Future content management projects delving into historical and/or anthropological inquiry would do well to think carefully about the types of content that they will be collecting, and develop appropriate metadata standards and their associated custom content types ahead of time. In addition to developing easy to use and intuitive technologies, project administrators should devote time at the beginning of all projects to consider key components related to portal structure. Administrators would be advised to form focus groups that consist of potential portal users in order to define the necessary components that would ensure and enhance researcher participation and reduce technical jargon.

Acknowledgments

The design and construction of the Chocolate Research Portal was the product of a union of two distinct groups of people who never met each other: those who wrote and created the system software, and those whose research findings provided the content. Constructing the Chocolate Research Portal required hours of customization and programming. I would like to acknowledge the many users in the Plone community who regularly made (and continue to make) themselves available to programmers by answering questions via email lists and IRC channels, especially David Siedband. Thanks also go to Eric Frahm and the systems administration staff at the University of California, Davis, Data Center for their vigilance and professional efforts in keeping the portal servers and network up and running.

The Chocolate Research Portal, of course, would be of little use without end-users who provided content. Special thanks go out to Kurt Richter, Nghiem Ta, and Deanna Pucciarelli. Their suggestions ultimately helped shape the design of the Chocolate Research Portal and the information models contained within. As alpha and beta testers, they were also the first contributors to the portal's contents.

Finally, I would like to thank my children, Ashlyn and Zanevan Lange, whose encouragement for me to construct a "cool site" has never waned.

Endnotes

1. Website address for the Chocolate Research Portal: http://cocoaknow.ucdavis.edu. (Accessed January 10, 2007.)

2. Plone website address: http://www.plone.org. (Accessed January 10, 2007.)

3. Historical persons, for example, would be a *content type* whereas an *attribute* of this type would be the person's occupation, whether auctioneer, broker, manufacturer, ship captain, or other. Another *content type* would be historical chocolate-related utensils; *attributes* of this type would be composition, era, and function—as applied to specific types of equipment (i.e., chocolate cups, pots, and spoons).

4. Relational databases are sets of information tables within the portal that contain data collected by researchers that easily can be searched.

5. *Visual skins* refer to a term used by Plone software developers to describe a separation of the visual display from the content inside a portal. This allows for the same content or information to be "visualized" in multiple, different forms.

6. PostgreSQL is an open source relational database management system. More information can be found at http://www.postgresql.org/. (Accessed January 10, 2007.)

7. A drag-and-drop Web utility allows appropriately credentialed users to add content to the portal using a familiar Windows ©-like interface.

Recommended Readings

GENERAL BACKGROUND

Readers interested in learning more about the problems associated with information overload, and why scholars need new ways to discover, visualize, and make sense of complex data are directed to:

Keyes, J. The new intelligence: the birth of the knowledge management industry. In: *Knowledge Management, Business Intelligence, and Content Management: The IT Practitioner's Guide.* Boca Raton, FL: Taylor and Francis Group, 2006; pp. 1–15.

Readers interested in ways to further integrate different types of information and content into larger knowledge systems regarding agriculture, food, and health are encouraged to explore:

Lange, M., Lemay, D., and German, B. A multi-ontology framework to guide agriculture and food towards diet and health. *Journal of the Science of Food and Agriculture* 2007;87:1427–1434.

KNOWLEDGE MANAGEMENT AND CONTENT MANAGEMENT THEORY

Readers who would like to learn more about how content and knowledge management tools work are encouraged to explore:

Keyes, J. Knowledge engineering techniques. In: *Knowledge Management, Business Intelligence, and Content Management: The IT Practitioner's Guide.* Boca Raton, FL: Taylor and Francis Group, 2006; pp. 45–66.

Keyes, J. Content management. In: *Knowledge Management, Business Intelligence, and Content Management: The IT Practitioner's Guide.* Boca Raton, FL: Taylor and Francis Group, 2006; pp. 129–154.

Rockley, A. Information modeling. In: *Managing Enterprise Content: A Unified Content Strategy.* Toronto, Canada: New Riders Publishing, 2003; pp. 159–181.

Rockley, A. Designing metadata. In: *Managing Enterprise Content: A Unified Content Strategy.* Toronto, Canada: New Riders Publishing, 2003; pp. 183–201.

CONCEPT MAPPING

Readers interested in examining the theoretical foundations underlying concept mapping as a means to capture and archive expert knowledge and as a way to understand and explore new knowledge, may wish to examine the following articles written by early developers of Concept Maps as a learning/teaching tool:

Novak, J., and Cañas, A. *Technical Report IHMC CmapTools.* Boca Raton, FL: Florida Institute for Human and Machine Cognition, 2006; pp. 1–31.

Novak, J., and Cañas, A. The origins of the concept mapping tool and the continuing evolution of the tool. *Information Visualization Journal* 2006;5:175–184.

Additional information on mapping tools is also available at the following web-sites:

http://cmap.ihmc.us/Publications/Research Papers/TheoryCmaps/TheoryUnderlying ConceptMaps.htm. (Accessed January 10, 2007.)

http://cmap.ihmc.us/Publications/Research Papers/OriginsOfConceptMappingTool.pdf. (Accessed January 10, 2007.)

Readers interested in concept maps and how they can be incorporated electronically into a broader knowledge exchange architecture or structure may wish to examine:

Cañas, A., Hill, G., Bunch, L., Carff, R., Eskridge, T., and Pérez, C. Concept Maps: integrating knowledge and information visualization. In: Cañas, A. J., and Novak, J. D., editors. *Concept Maps: Theory, Methodology, Technology. Proceedings of the Second International Conference on Concept Mapping,* 2 Volumes. San Jose, Costa Rica: Universidad de Costa Rica, 2006; Volume 1, pp. 304–310.

Lange, M., and Grivetti, L. Have your chocolate and eat it too. Integrating Concept Maps into a content management framework with relational database connectivity. In: Cañas, A. J., and Novak, J. D., editors. *Concept Maps: Theory, Methodology, Technology. Proceedings of the Second International Conference on Concept Mapping,* 2 Volumes. San Jose, Costa Rica: Universidad de Costa Rica, 2006; Volume 2, pp. 126–129.

Those readers interested in obtaining and using free (open source) concept mapping software, should investigate the concept map sites located at:

http://cmc.ihmc.us/cmc2006Papers/cmc2006-p234.pdf (Accessed January 10, 2007.)

http://cmap.ihmc.us/. (Accessed January 27, 2007.)

PLONE

Readers interested in finding out more about Plone programs and content management systems may examine the following three resources:

McKay, A. *The Definitive Guide to Plone*. Berkeley, CA: Apress, 2004.

Cooper, J. *Building Websites With Plone*. Birmingham, England: Packt Publishing, 2004.

Readers interested in receiving the most up-to-date information regarding Plone, or who are interested in downloading Plone in order to build their own content management Web server, may wish to navigate the following website:

http://www.plone.org. (Accessed January 27, 2007.)

Base Metal Chocolate Pots In North America

Context and Interpretation

Phil Dunning and Christopher D. Fox

Introduction

Chocolate may have begun as an elegant beverage for the affluent, but as the 18th century progressed it became a drink for all classes. In 1701, Sarah Tailer, niece of the lieutenant governor of Massachusetts, had an elegant silver pot made to serve chocolate to her friends [1]. By the early 1770s, frequenters of the Swan Tavern in Yorktown, Virginia, took chocolate from one of "4 copper coffee and chocolate pots" [2]. Chocolate was popular, whether one was at home or under rough conditions. Captain Moses Greenleaf kept a diary of his service in the Continental army during the American Revolution (see Chapter 31). While stationed at Fort Ticonderoga in 1777, he regularly recorded that he "breakfasted on chocolate" [3]. Pelatiah Marek of Falmouth, Massachusetts, wrote to storekeeper John Quimby on January 28, 1783:

I should be glad if you would send me two gallons of Rum & six Pound of Sugar & two # [i.e., pounds] of Coffee and two Pound of Choclate [sic] for I am Bound into the Woods a logging to Morrow morning. [4]

Chocolate could be made in a variety of vessels such as a coffeepot without the lid. Even a small kettle would do in a pinch [5]. However, there exist a number of base metal (copper and brass) pots made specifically for chocolate. They are identified, as are silver examples, by the hole in the center of the lid. This hole was for inserting the stirring stick or mill, as it was called in the 18th century (*moulinet* in French).

Copper and brass chocolate pots present challenges for dating that are absent in silver pieces. They do not carry maker's marks or hallmarks, and few can be dated unless the provenance or other means of identification are known. Archaeological evidence is scarce, and few period pictures show identifiable base metal pots. Only a minority of pots are related in shape to datable silver pieces. Identifying a place of origin equally is difficult. None of the pots studied during the preparation of this chapter have documented provenances. Metal objects made in one country were often sold abroad, and metalworkers in England, France, and Holland regularly copied each others' work if a style proved to be popular. We will see here that one of the simplest forms of pot, the footed baluster, even continues to be made in the Mediterranean today.

To prepare the present chapter, we studied and analyzed the form and construction methods of 30 chocolate pots made available to us in public and private

Chocolate: History, Culture, and Heritage. Edited by Grivetti and Shapiro
Copyright © 2009 John Wiley & Sons, Inc.

collections. This process also allowed assumptions to be made regarding the forms of early mills (i.e., stirring rods) used with these pots.

Base Metal Chocolate Pots: Description and Analysis

Shapes of copper and brass chocolate pots may be clustered into two broad categories: the baluster (with numerous variations) and the tapered cylinder. Most of the shapes exhibit similarities in basic construction. It is useful to understand these construction techniques before examining the forms themselves.

CONSTRUCTION

The sides of almost all pots we examined are made from a single sheet of metal that had been rolled (i.e., shaped or raised) into the desired form. The edges were joined and the joint strengthened by overlapping tabs called dovetails (Fig. 54.1). Next, the dovetails were brazed together, using brass on copper pots, and copper on brass pots. Bottoms of the pots were made of separate pieces of sheet metal, cut to fit, then set into the bottom of the pot. The side of the foot then was

crimped over the bottom sheet to secure the bottom piece in place. Handles for the pots were attached at the side seam. Because copper and brass affect the flavor of liquids—and in some instances can be toxic—these chocolate pots always were coated on the inside with tin (i.e., tinned).

FORMS

The baluster shape appears in some of the earliest illustrations of chocolate pots and was the most common form of those we examined (Fig. 54.2). The bulbous sides of these pots may be softly rounded (Fig. 54.3) or have a sharp shoulder (see Fig. 31.5). A flared foot is integral with the body of the pot. At its simplest, these probably were the least expensive pots. A strap handle constructed of wrought iron, usually ending in a closed loop, was splayed into a flange that permitted attachment to the body of the chocolate pot. The flange was attached to the body with two or more copper rivets (Fig. 54.4).

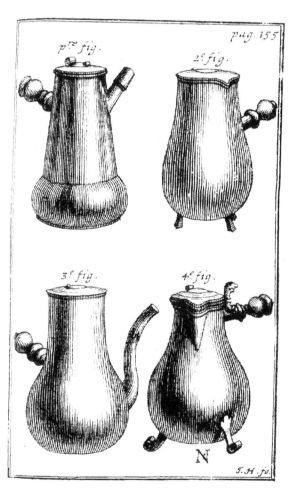

FIGURE 54.2. Chocolate pot construction: earliest known baluster form. *Source*: N. Blegny, *Le bon Usage du Thé, coffe et chocolat*, 1687. Courtesy of the Peter J. Shields Library, University of California, Davis. (Used with permission.)

FIGURE 54.1. Chocolate pot construction: 3–4 tabs visible on seam. *Source*: Courtesy of Christopher D. Fox. (Used with permission.)

FIGURE 54.3. Chocolate pot construction: example of smooth, rounded shoulder on baluster form. *Source*: Courtesy of Winterthur Library and Museum, Winterthur, Deleware. (Used with permission.) (See color insert.)

FIGURE 54.5. Chocolate pot construction: handle attached directly over brazed side construction seam. *Source*: Private collection of Phil Dunning. (Used with permission.)

FIGURE 54.4. Chocolate pot construction: example of simple flange handle. *Source*: Courtesy of Christopher D. Fox. (Used with permission.)

FIGURE 54.6. Chocolate pot construction: comparative European mass produced 19th–20th century type. Construction material is spun brass, identifiable by uniform shape and smooth body form, with heavy cast brass handle attachment panel. *Source*: Courtesy of Eric and Judy Whitacre. (Used with permission.) (See color insert.)

Somewhat more elaborate versions of the footed baluster forms have a turned-wood handle. In these cases, the wrought iron strap was shaped into a tang that was inserted into the wood handle to hold it in place. On another example of footed baluster, the wrought iron strap was shaped into a socket to receive the wood handle. In this case the socket had a small hole in the side for a pin or tack to hold the handle firmly in place.

The footed baluster form appears to have been popular through several centuries and therefore can be difficult to date. The wholesale importation to North America of copperware from North Africa and the Middle East during the late 19th and throughout the 20th and 21st centuries prompts calls for caution when attempting to date pots based only on similarities in form (Figure 54.5). Pots of the baluster form also were produced in Europe during the late 19th and 20th centuries. These pots have brass bodies shaped by spinning on a lathe, rather than being raised by hand; furthermore, the construction seam appears as a straight brazed line without tabs or dovetails (Figure 54.6).

FIGURE 54.7. Chocolate pot construction: unusual baluster form lacking distinctive flared foot that commonly characterizes this style. *Source*: Courtesy of Eric and Judy Whitacre. (Used with permission.) (See color insert.)

FIGURE 54.8. Chocolate pot construction: example of footed-baluster form with spout. *Source*: Courtesy of Phil Dunning. (Used with permission.) (See color insert.)

FIGURE 54.9. Chocolate pot construction: example of baluster form with legs. *Source*: Courtesy of the Stewart Museum, Montreal, Canada. (Used with permission.) (See color insert.)

A few baluster pots were not footed (Fig. 54.7). The bodies of these examined pots were constructed using two sheets of copper. One sheet formed the upper half of the body, while the other was raised into a cup shape and brazed via tabs or dovetails to the body at the widest part of the bulge. These pots tended to have plain wrought iron strap handles.

None of the pots described thus far has a spout or beak to facilitate pouring. The addition of this feature to a chocolate pot brings it closer in design to coffeepots, but the hole in the lid for insertion of the mill remains the important diagnostic feature. One pot with a spout that we examined closely paralleled an example in a 1687 engraving (Fig. 54.2, lower left). The spout was formed from a single sheet, straight-seamed up the back. The turned-wood handle was attached to a copper socket, with a reinforcing panel or shield at the juncture with the body. The lid had a pivoting cover for closing the mill hole. All of these features are found on pots of better quality. A group of footed pots we examined were closely related in form: They shared the same well-made spouts, turned-wood handles, and more complex raised lids with covers for the hole (Fig. 54.8).

An alternative to the spout for pouring is a beak. Like spouts, beaks were made from separate sheets of metal, then brazed to the body. Beaks were found on two types of balusters, those that were "footed" and those with "legs."

The baluster type of base metal chocolate pot with legs is parallel in form/shape with silver chocolate pots of the era (Fig. 54.9). These pots stand on three legs of brass, copper, or iron. Unlike the other chocolate pots described earlier, the bottoms and sides of the balusters with legs were raised from a single sheet of copper with no seams. Beaks were attached separately and lids of this form usually were hinged to the body as was done with silver pots. It is significant that silver pots of this form almost universally are French.

Handles on the spouted and beaked pots invariably are placed on the body at a 90 degree angle to the spout or beak. The shape of these wood handles appears to be of some help in dating. The pots in the 1687 engraving (see Fig. 54.2) have bulbous, baluster-shaped handles with a ball or *knop* on the end. This is paralleled by the handle exhibited in Figure 54.8. If a comparison is made with dated silver chocolate and coffeepots, it can be seen that earlier examples also have similar handles. As the 18th century progressed, a simpler, elongated oval handle became more fashionable.

The other primary type of chocolate pot is the tapered cylinder (Fig. 54.10). These, like the majority

FIGURE 54.10. Chocolate pot construction: example of tapered-cylinder form, perhaps early 18th century. *Source*: Courtesy of Eric and Judy Whitacre. (Used with permission.) (See color insert.)

FIGURE 54.11. Chocolate pot construction: design of lid shape. This is an example of a complex form, fitted over the exterior of the rim with a pivoting cover for the mill hole. *Source*: Courtesy of Eric and Judy Whitacre. (Used with permission.)

of the baluster types just described, were made from one piece of metal for the body, seamed along the side, with a separate inset bottom. The handles typically were turned wood fitted over an iron tang. The iron strap splayed into a decoratively shaped flange. One example of this type/form that we examined at Winterthur, Delaware was engraved with the date 1703.

LIDS

The simplest form of lid was raised from a sheet of metal and fitted over the outside of the pot's rim. It has a hole in the center for the mill. Lids were not usually tinned, given they would not have as much contact with the chocolate liquid as the pot itself. This type may have had a pivoting cover of sheet metal for the central hole. The cover was secured to the lid with a small copper rivet and often had a curled end and a small boss for a secure grip. (Fig. 54.11).

Better quality lids were of more complex construction. These usually were found on pots with spouts or beaks. In such examples the lids were formed of two pieces of copper (Fig. 54.12). The bottom portion of the lid took the form of a copper hoop, whereas the top was raised into a low stepped dome that was crimped over a flange at the top edge of the hoop. Using friction, the hoop could then be fitted into the mouth of the pot. This form, too, often had a cover for the hole. Hole covers on lids of this form were made in either one or two pieces. The one-piece examples were cast from copper and had lathe-finished surfaces. The second type had a cast-copper finial attached to the center of a shaped sheet of copper. These lids invariably are tinned on the inside.

Lids and parts of lids have been recovered archaeologically from several 18th century sites in

FIGURE 54.12. Chocolate pot construction: design of lid shape. This example is characteristic of many pots with spouts or beaks where the lid fits inside the rim of the pot, rather than over the outside. This style commonly has a pivoting cover to close the mill hole. *Source*: Courtesy of Eric and Judy Whitacre. (Used with permission.)

North America. Several lids and pivoting covers were excavated at Fortress Louisbourg in Nova Scotia (see Chapter 25) [8]. One of these was shaped for a beaked pot. Pivoting covers also have been found in excavations at Williamsburg, Virginia (Fig. 54.13).

CHOCOLATE MILLS

Early chocolate beverages, unlike coffee and tea, had to be stirred regularly to keep sediment in suspension and retain a full chocolate flavor. To accomplish this, a

FIGURE 54.13. Chocolate pot mill hole cover, excavated at the James site, Williamsburg, Virginia. *Source*: Courtesy of The Colonial Williamsburg Foundation, Williamsburg, Virginia. (Used with permission.)

stirring stick, or mill, was used. Mills for silver pots usually were made of silver, with ornately turned handles of exotic woods. Those used in base metal pots typically were made entirely of local wood and much less elaborate in design.

Because there are no datable features such as hallmarks on all-wood mills, determining their age is difficult. When they appear in period pictures, all that is visible is the handle protruding through the lid of the pot. A few mills have been found with pots of early forms, and if original to the pot may provide some clues to researchers.

Any implement with a long handle could be used as a mill if necessary; a mixing spoon would do; twigs or sticks, if cut properly, also would work. The most common form of mills seen in antique shops has rounded ends, cut or notched in a star shape (Fig. 54.14). These mills often are found today with baluster-form pots like those in Figure 54.2. Since this form of pot was made over a long period of time, such mills are equally difficult to date.

Several mills have handles with slightly more decorative turnings and have vertical notches or blades. Of the various forms located during the course of our research, these most closely resemble mills depicted in an engraving dated to 1687 (see Fig. 10.1). A third form of mill had horizontal notches or grooves (Fig. 54.15). These mills could fit the description in an English publication of 1675 by John Worlidge: "The mill is only a knop at the end of a slender handle or stick, turned on a turner's lathe, and cut in notches" [9].

Finally, many modern mills made in Mexico have separate, loose rings around the end and usually are cut with decorative designs (Fig. 54.16). The example illustrated appears to be aged; it is especially fine and has bone or ivory inlay. Further research is necessary to determine whether this shape is a traditional or early Spanish form.

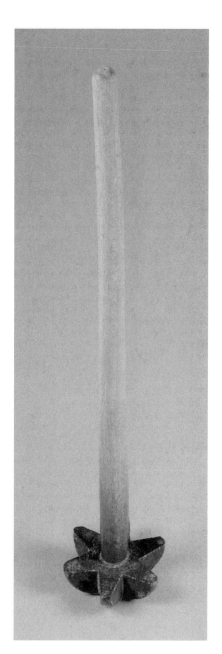

FIGURE 54.14. Chocolate mill: example of a typical star-shaped bottom type (undated). *Source*: Courtesy of Phil Dunning. (Used with permission.)

Conclusion

Archaeological and historical evidence reveals that base metal chocolate pots were used in both English and French North America during the 18th century. By studying the forms and construction of base metal chocolate pots, along with the available archaeological, pictorial, and historical evidence, some general patterns have emerged.

The baluster-shaped pot, in its many varieties, was popular by the late 17th century and continued in use into the 20th century. The footed baluster with no spout or beak was the most common form found

FIGURE 54.15. Chocolate mill: undated form. *Source:* Courtesy of Eric and Judy Whitacre. (Used with permission.)

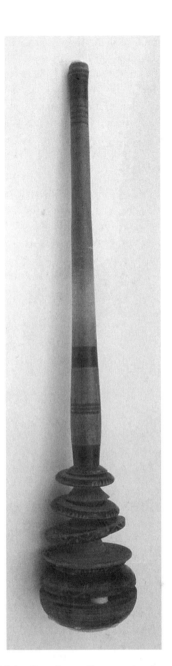

FIGURE 54.16. Chocolate mill: example of modern form with separate loose rings at the base, common to Mexico and Central America in the 20th and 21st centuries. *Source:* Courtesy of Eric and Judy Whitacre. (Used with permission.)

during the course of our research, and the design probably has a longer time span than others. It is still in use in the Mediterranean today. Pictorial evidence shows late 17th century pots with spouts and their popularity continued through the 18th century. The form of the wood handle can help date individual examples within this period. Comparison with hallmarked silver examples suggests that baluster pots with feet are almost universally of French origin. Based on our examination, the tapered cylinder form of pot was much less common than the baluster and, from the single dated example known (1703), may have been an early, short-lived form.

Silver chocolate pots have been studied for decades by connoisseurs of the decorative arts. Base metal pots, in contrast, have received almost no attention. It is our hope that this preliminary study will foster interest in these everyday vessels that have potential for understanding a broader picture of chocolate usage and social behavior.

Acknowledgments

We wish to thank the following for their assistance during the course of research on this project: Christina

Bates, Curator, Canadian Museum of Civilization; John D. Davis, Curator of Metalwork, The Colonial Williamsburg Foundation; Steve Delisle; Ruby Fougère, Collections Technician, Fortress Louisbourg National Historic Site; Sylvie Dauphin, Curator at the Stewart Museum, Montreal; and Eric and Judy Whitacre.

References

1. Ward, G. W. R. *The Silver Chocolate Pots of Colonial Boston: New England Silver & Silversmithing 1620–1815*. Boston: The Colonial Society of Massachusetts, 2001; p. 66.

2. Rice, K. S. *Early American Taverns: For the Entertainment of Friends and Strangers*. Chicago: Regnery Gateway in association with Fraunces Tavern Museum of New York, 1983; p. 165.

3. Wickman, D. H. "Breakfast on Chocolate:" The Diary of Moses Greenleaf, 1777. *Bulletin of the Fort Ticonderoga Museum* 1997;15(6):483–506.

4. Marek, P. Manuscript note, January 28, 1783. Private Collection.

5. Ward, G. W. R. *The Silver Chocolate Pots of Colonial Boston: New England Silver & Silversmithing 1620–1815*. Boston: The Colonial Society of Massachusetts, 2001; p. 80.

6. Fortress Louisbourg National Historic Park archeological collection: Artifact numbers 1B16G1.1764, 16L94H13.5, 1B18K7.28, 1B4E2.32.

7. Ward, G. W. R. *The Silver Chocolate Pots of Colonial Boston: New England Silver & Silversmithing 1620–1815*. Boston: The Colonial Society of Massachusetts, 2001; p. 65.

Blue and Gray Chocolate

Searching for American Civil War Chocolate References

Louis Evan Grivetti

Introduction

It is easily demonstrated that chocolate and cocoa have had long associations with military activity in the Americas and elsewhere. As revealed in several chapters in this anthology, numerous accounts document uses of chocolate as military rations during the French–Indian Wars, the War of the American Revolution, and the period of Native American pacification in the plains and southwestern regions of North America (see Chapters 31 and 33). It would seem logical, therefore, that abundant chocolate-related materials would be associated with the Civil War era. Curiously, this is not the case.

While information on chocolate use during the Civil War era can be identified, the quantity of resources is significantly less when compared to earlier periods of North American history. Whereas chocolate easily is identified as a specific, itemized military ration in Revolutionary War accounts [1–4], chocolate was not a specified military ration during the American Civil War. And while chocolate was not a formal military ration, a diversity of documentation can be identified from the period 1860–1865 that provide culinary and medical uses and consumption by civilians and soldiers, and by adults and children.

The chocolate/cacao-related sources identified in the present chapter are grouped into two clusters, then arranged chronologically. Presented first are reports with specific dates, whether notations in diaries,

letters, or government documents. Following these are interviews and memoirs published subsequent to the war years, sometimes within five years but in other instances almost 75 years after events at Appomattox Courthouse. Within each cluster the documents are arranged chronologically, to allow further chocolate-related findings of the Civil War era to be placed in context.

Eyewitness and Firsthand Accounts, 1860–1865

MAY 20, 1860: DIETARY ADVICE TO RECRUITS

The bombardment of Fort Sumter[1] in Charleston Harbor, on April 12, 1860, initiated hostilities between the Northern and Southern states. Recruitment and training of soldiers on both sides subsequently proceeded in earnest. A little more than one month after events at Fort Sumter, a list of

Chocolate: History, Culture, and Heritage. Edited by Grivetti and Shapiro
Copyright © 2009 John Wiley & Sons, Inc.

recommendations was published in the *New York Times*, on May 20, 1860, in regard to the diet and health of new Federal army volunteers. Since cold water was considered unhealthy at the time, one recommendation encouraged recruits to be cautious of beverage temperature, and drinking chocolate with meals was encouraged:

> *Avoid drinking freely of very cold water, especially when hot or fatigued, or directly after meals. Water quenches thirst better when not very cold and sipped in moderate quantities slowly—though less agreeable. At meals tea, coffee and chocolate are best. Between meals, the less the better. The safest [beverage during] hot weather is molasses and water with ginger or small beer.[2] [5]*

1861: ASSEMBLING MEDICAL SUPPLIES

Volunteers were quick to assist in the war effort as noted in a letter penned by Abby Howland Woolsey to Eliza Newton Woolsey Howland. Her correspondence identified that a wide range of items, including chocolate, was being collected for distribution to hospitals:

> *Dear Eliza: We got off our first trunk of Hospital supplies for Colonel Mansfield Davies' Regiment yesterday. . . . [It was sent to] his headquarters at 564 Broadway [New York] and thence by steamer to Fort Schuyler for the sick soldiers there[3] . . . and contained: 42 shirts, 12 drawers, 6 calico gowns, 24 pairs woolen socks, 24 pairs slippers, 18 pillow sacks, 36 pillow-cases, 18 damask napkins, 36 towels, 24 sponges, 4 boxes of lint, beside old linen, oiled silk, tape, thread, pins, scissors, wax, books . . . ribbon, cloth, etc., and fifty bandages. This morning Mother has been putting up a tin box of stores for Mr. Davies—sardines, potted meats, arrow root, chocolate, guava, and the like with a box of cologne, a jar of prunes and a morocco case with knife, fork, and spoon, fine steel and double plated. [6]*

MAY 20, 1861: REQUEST THAT CHOCOLATE BE SENT TO TROOPS

This letter from John Albion Andrew to John Murray Forbes, representative of the Commonwealth of Massachusetts Executive Department, was written in Boston and specifically requested that Baker's chocolate be sent to Federal soldiers:

> *My dear Sir, I wish you would have some of Baker's chocolate sent on the Pembroke for our troops at Fort Monroe.[4] Some ask for it, and would regard this change [of diet] as a luxury and an advantage.*
>
> *Yours faithfully,*
>
> *J. A. Andrew*
>
> *And [please send] a few dried apples. [7]*

NOVEMBER 1861: THEFT OF PRISONERS' FOOD BY NORTHERN JAILERS

The first Confederate States of America (CSA) officers captured by Federal troops were shipped north and incarcerated at Fort Lafayette,[5] located in New York Harbor. Typically, all goods and money brought with the prisoners were confiscated by the jailers. The confinement quarters (not really cells in the strict sense) contained no furniture and only a few beds. The following account from the diary of Lawrence Sangston reveals his displeasure over treatment by his jailers, who plundered and stole food in his possession. This short account is valuable in that he indirectly reveals that chocolate was available to CSA officers at this early time in the war:

> *Arose at daylight, and went out on deck [of the prison ship] to view the Fort and its surroundings. . . . I was placed in a room with seven others . . . a room sixteen by eighteen feet, lighted by three slits in the wall . . . not a particle of furniture in the room. Shortly afterwards the baggage arrived in carts from the boat, and found all my packages but one trunk and one box had been opened a rifled of more or less of their contents; everything gone that was worth stealing except my cooking apparatus. All my pickles, preserves, bologna sausage, crackers, ten pounds sperm candles, spirits of wine, soap, tobacco, towels, tea, coffee, sugar, lemons, sardines, chocolate, cold tongues, &c., &c.; in fact I was entirely cleaned out of stores of all kinds. [8]*

APRIL 16, 1862: PROVISIONS AUTHORIZED FOR FEDERAL NAVAL FORCES AT PORT ROYAL, SOUTH CAROLINA

The following official Navy Department document from the secretary of the navy to the secretary of the treasury requested provisions be sent to troops of the United States Federal Naval Forces stationed at Port Royal, South Carolina,[6] a request that included chocolate.

> *SIR: I have the honor to inform you that the naval forces at Port Royal, S. C., are in need of live stock [sic], fresh vegetables, and supplies, and other articles of minor importance essential to their comfort, and that an application has been made to the Department by mr. [sic] H. S. Sedgwick, of New York, for permission to send a cargo of such articles to them. He proposes to load his vessel with cattle, potatoes, onions, turnips, apples, and other vegetables, ice, fresh meats, flour, hams, bacon, preserved meats, milk, molasses, sugar, butter, tobacco, cheese, clothing, hats, boots, and shoes, cigars, pipes, coffee, tea, and chocolate. I have, therefore, to request that Mr. Sedgwick may be permitted to clear a vessel from New York for Port Royal, with such articles as are enumerated above, with restrictions not to trade with or supply any portion of his cargo to any other persons than those in the service of the Army or Navy of the United States. [9]*

MAY 1, 1862: LETTER FROM A SOUTHERN SCHOOLGIRL DESCRIBING DAILY LIFE

A year into the war, deprivations and personal sacrifice had not yet reached the small settlement of Pine Grove, South Carolina, as noted by young Miss Harriet Ware. In her letter she describes the monotony of her rural daily life: Breakfast included chocolate, she then walked to school, returned home for lunch (ate waffles, hominy, crackers, sardines, and blackberries, with tea or coffee), then returned to school. She mentioned that her family sold clothing to slave children. She comments elsewhere in her letter that she has been infested with fleas:

After we get downstairs it takes the united efforts of most of the family to get the breakfast on the table, and we are fortunate if we get up from that meal by half-past eight. It generally consists of hominy, very delicious eaten with either milk, butter, or molasses, corn-cakes, or waffles of corn-flour—the best of their kind—concentrated coffee, chocolate, or tea, army bread—when we can get it—crackers, when we can't, and boiled eggs or fried fish, as the case may be. . . . [I go to school and then return home for lunch] at half-past twelve, a cold one generally, sometimes a few waffles or some hominy for variety, but crackers, sardines, and blackberries which we have in abundance now, make a refreshing meal, with tea or coffee when we please. [10]

MAY 30, 1862: LETTER FROM KATHARINE PRESCOTT WORMELEY TO HER MOTHER, WRITTEN ABOARD THE HOSPITAL SHIP *KNICKERBOCKER*

Katharine Wormeley was a volunteer with the United States Sanitary Commission (USSC)[7] during the Peninsular Campaign in Virginia. She wrote to her mother and described how food supplies were distributed and also revealed her concern for gravely wounded soldiers. She described the condition of Confederate prisoners arriving from North Carolina, who appeared to be fit and well fed. She lamented the inefficiency of medical treatment in the camps and commented on the "tension" between the surgeons and members of the Sanitary Commission. At the close of her letter she mentioned that her evenings were "the pleasantest hours of the day" when members of the commission gathered to relate events of the day and they shared whatever food items they could assemble:

I found the condition of things far better than I expected, and infinitely better than it was a week ago. We visited nearly all the [hospital] tents, and gave supplies of beef-tea,[8] milk-punch,[9] arrowroot, and eggs for the worst cases, of which there were comparatively few, for such cases are put on the Commission boats. I found four or five men for whom nothing could be done but to help them to die in peace. . . . The ordinary diet seemed good and plentiful, and quite suited to the majority of the cases. . . . Our evenings are the

pleasantest hours of the day. The Chief and Mr. Knapp and the staff collect on a broken chair, a bed-sack, and sundry carpet-bags, and have their modicum of fun and uinine [unity?]. The person who possesses a dainty—chocolate or gingerbread, for instance—is the hero for the time being. . . . Tell [The Women's Aid Society] . . . that flannel shirts are never in sufficient quantity . . . if anybody proposes to send me anything, say: Good brandy . . . nutmegs . . . Muringer's beef-extract . . . send Muringer's beef-extract. It comes in small cakes looking like dark glue. Send also condensed milk, lemons, and sherry. [11]

MAY 31, 1862: DESCRIPTION OF THE SANITARY COMMISSION FLOATING HOSPITAL ANCHORED IN THE PAMUNKEY RIVER, OFF WHITE HOUSE, VIRGINIA[10]

Charles William Woolsey wrote an informative letter for publication in the *New York Evening Post*. His letter is valuable as it related the daily activities of commission members, both day and night, explained the structure and organization of the hospital ship staff, and described the living conditions of the 500 sick and wounded men aboard ship. He related the difficulties of bringing the sick and wounded men aboard and the problems posed by being anchored in mid-river. He described cases of camp diseases such as typhoid fever, and the "ugliness" of gunshot wounds. Elsewhere in the letter he commented on the inability of men to prepare food aboard ship, and praised the role of women volunteers:

The work of the Sanitory [sic] Commission, as connected [to] the army of the Potomac . . . is a most important and indispensable one. More than two thousand sick and wounded men have been shipped by the Commission to New York, Washington and Boston during the past month. . . . Creature comforts abound in the presence of lady nurses, and from their culinary retreat between decks come forth at all hours of the day a sizzling sound as of cooking arrow-root; armsful [sic] of clean white clothing for the newly washed, and delicacies for the sick without number, sometimes in the shape of milk punch, or lemonade squeezed from real lemons, sometimes a pile of snowy handkerchiefs that leave an odorous wake through the wards. Again, a cooling decoction of currants for the fever case nearest the hatchway . . . or oranges, cups of chocolate and many a novelty, but never a crumb of hard tack . . . or even so faint a suggestion of too familiar salt pork. Suffice it to say that the services of the ladies who are here as nurses of the sick are invaluable to the Commission and duly appreciated by the battle-tried and camp-worn soldiers. [12]

JULY (NO DATE BUT AFTER JULY 9) 1862: MEDICAL SUPPLIES RANSACKED BY NORTHERN TROOPS

Abigail Hopper Gibbons was a prominent member of the Sanitary Commission who helped coordinate the distribution of medical supplies provided by the New York War Relief Agency. She also established two field hospitals in Virginia [13]. While in Washington, DC, she was informed that badly needed medical supplies had arrived and were stored in the Relief Rooms on Market Street, but that these goods had been ransacked by Secessionists living in the city. Gibbons, never faint of heart, demanded a military escort to the home of Atwell Schell, suspected of the theft, who then "surrendered the stores." She described the condition of the supplies:

> Our goods had been packed with much neatness and care, and covered with their own quilts. Everything was turned out [i.e., removed from cases and thrown into a jumble] and package upon package rolled down stairs, until a high stack was formed in the centre of the parlor. There was every variety of garment, bedclothes, delicacies for the sick—such as sugar, tea, chocolate, farina, arrowroot, gelatine [sic], and corn-flour and barley in large packages. [14]

SEPTEMBER (NO DATE) 1862: CHOCOLATE SERVED AT MILITARY HOSPITAL AND GENERAL SCARCITY OF FOOD

Abigail Hopper Gibbons continued her description of the Sanitary Commission work:

> There seems to be an excitement on our waters. Gunboats are passing towards Washington and Baltimore and Fortress Monroe. Six gunboats passed yesterday. We are hourly expecting three hundred patients, which, in addition to what we have already, will keep us active. But for our stores, the soldiers would have suffered much. . . . Thank Mr. Langdell, in strong language, for the solidified milk—the best I ever saw. The grains are finer, and it is free from the flavor peculiar to it. We fix it in a variety of ways for the sick, and they like it very much. This afternoon, they have cocoa for supper. We are kept very busy, with two stoves at our command. . . . Miss Dix has sent us a large supply of currant jelly, which is especially grateful to scurvy patients. . . . The sight of vegetables is a rare luxury here. Last evening a boat brought us, among other things for the hospital, three barrels of cabbages, as many onions, and a few beets. What are they for five hundred men. [15]

SEPTEMBER 13, 1862: ARRIVAL OF FOOD SUPPLIES AT FREDERICK CITY, MARYLAND

Richard Derby enlisted with the Union. He was a first lieutenant and 27 years of age at the time he wrote the following passage in his diary. He was killed at Sharpsville, Maryland, on September 17 just four days later:

> We have just marched through the city, and are bivouacking in the clover-fields near by. . . . We saw the smoke of the cannonading on the mountains, across the valley, as we came down into Frederick. . . . What the rebels mean [by this bombardment] is a mystery. . . . The "box" has come to hand as last! The lemons were so decayed that you could scarcely tell

what they were. The can of raspberry smelt like a bottle of ammonia, and had leaked out a little. . . . The little crackers were all musty, but the cake was still nice, and the sugar, but probably the tea is infected. You cannot send tea with other articles, unless put in air-tight packages. . . . The raisins are nice and very palatable. I have not tried the "corn-starch," but the jelly was nearly eaten at the first opening. The ginger wine was terrible stuff. . . . I had got tired of cocoa. It is too heavy for hot weather; but now the mornings are getting cool, I can make good use of two boxes. . . . I haven't time to speak of the prospects of the war and the country. . . . The Northern people labor under a vital mistake as to the management of the war, and are thoroughly deceived as to the state of affairs. [16]

NOVEMBER 15, 1862: EVENTS AT NEWBERN, NORTH CAROLINA

Zenas T. Haines, a Union soldier, described military encounters in North Carolina at Little Creek, Williamston, and Hamilton. While at Newbern, North Carolina, he wrote the following description of camp food:

> Our troops pushed on to Williamston and Hamilton, where they executed a flank movement, with a fair prospect of bagging the whole rebel force. . . . Newbern has become quite a jolly place to live in. . . . The market supplies splendid Northern apples, southern ditto [i.e., southern apples], cider, honey, ginger cakes, crackers, fish, preserved meats and fruits, oyster, pickles, condensed milk, chocolate, sugar, tea, coffee, military goods, &c. . . . Gingerbread, pies, and even apple-dumplings, are brought to us by the negroes [sic] in profusion, while the sutlers furnish us with butter, cheese, sardines, and all the main essentials of luxurious living. Our regular rations are not to be sneezed at, although at present a scarcity of hops has thrown us back upon hard tack. We are treated to beef steaks, excellent rice soups, fish, has, &c. The general health of the regiment continues good, although a few in each company are weakened by diarrhea, and some few are yet suffering from colds and coughs contracted by our late exposures. [17]

FEBRUARY 3, 1863: FOOD SUPPLIES AT THE LIBBY PRISON HOSPITAL, FORT MONROE

The United States Sanitary Commission not only procured food and medical supplies to be sent to Union field hospitals, but also provided needed supplies to Confederate soldiers held in Union prisons:

> Inventory of six boxes marked "Thos. T. May, Libby Prison Hospital, care Agent of Exchange, Fort Monroe," sent on by the Sanitary Committee of New York for the use of the sick and wounded in this hospital, received February 3rd, 1863: Cotton shirts, 83; condensed milk, 20 cans; ink bottles, 3; cotton drawers, 55 pairs; corn starch, farina, &c.; tapers, 30; lead pencils, 4; woolen shirts, 412; sponges, 9 pieces; chocolate cakes, 5; woolen drawers, 40 pairs; tin plates, 1 dozen; assorted

pickles, 5 gallons; socks, 35 pairs; tin cups, 1 and 9/12th
dozen; pickled peaches, 5 gallons; slippers, 42 pairs;
writing paper, 2 reams; 1 lot of rags and 2 cloth coats;
envelopes, 10 packages; 1 lot of assorted dried fruit; vests, 2;
penholders, 2 dozen; towels, 9 dozen; steel pens, 4 dozen;
assorted soap, 2 bags; combs, 3; cans soup, 1 dozen; [and]
scissors, 6 pair.

JOHN WILKINS

Surgeon in Charge. [18]

APRIL 21, 1863: SHORT NOTE ON CHOCOLATE DEPRIVATION: HUSBAND'S LETTER TO WIFE

At other times and places during the American Civil War chocolate was not always accessible, especially during times of siege. Researchers with our project identified only one reference to chocolate or cocoa during the Civil War era (1861–1865) from the Charleston archives, a letter from a Confederate officer (see Chapter 52). William Gildersleeve Vardell was a major and quartermaster in the C.S.A. Vardell's letter was written during the siege of Vicksburg, Mississippi, when he served with the 23rd South Carolina Infantry Regiment. It may be surmised from his letter that chocolate was scarce in Charleston during the war. On April 21, 1863, Vardell wrote to his wife, Jennie, who lived in Charleston at the time:

I wish you had those nice things the young man brought me especially the chocolate. [19]

JULY 16, 1863: DIARY OF A NURSE VOLUNTEER AT GETTYSBURG

Emily Bliss Thacher Souder wrote to a Mrs. J. Heulings of Moorestown, New Jersey, of her experiences after the decisive battle of Gettysburg (July 1–3, 1863). As a volunteer she had helped arrange the collecting and packing of stores for transport to the field. She arrived at Gettysburg on July 16 and described in detail the terrible conditions, especially the smell from the battlefield:

The atmosphere is truly horrible, and camphor and cologne or smelling salts are prime necessaries for most persons, certainly for the ladies. . . . We rode in an ambulance to the hospital of the Second Corps. The sights and sounds beggar description. There is great need of bandages. Almost every man has lost either an arm or a leg. The groans, the cries, the shrieks of anguish, are awful indeed to hear . . . it requires strong effort to be able to attend to the various calls for aid. The condensed milk is invaluable. The corn-starch, farina, and milk punch are eagerly partaken of, and a cup of chocolate is greatly relished. . . . The Union soldiers and the rebels, so long at variance, are here [inside the hospital] are quite friendly. They have fought their last battle, and vast numbers are going daily to meet the King of Terrors.[11] [20]

JANUARY (NO DATE), JANUARY 23, AND FEBRUARY (NO DATE) 1864: A SUITE OF THREE LETTERS—ATTEMPTS BY SOUTHERN WOMEN TO MAINTAIN SOCIAL ACTIVITIES DURING TIMES OF HARDSHIP

How are civilians during wartime able to keep up "appearances," knowing that events preclude normalcy, but still appearances had to be maintained since one's place continued to be measured on parameters of generosity and hospitality—even as events during the war worsened? A suite of three diary entries penned by Mary Boykin Chesnut, between late January and early February 1864, revealed just these difficulties when she wrote:

People have no variety in war times, but they make up for that lack in exquisite cooking. . . . [21]

On January 23, 1864, Mrs. Chesnut described a party attended by society notables where chocolate, oysters, and special cakes were served:

My luncheon [the one I hosted] was a female affair exclusively. Mrs. Davis [Mrs. Jefferson Davis][12] came early and found Annie and Trudie making the chocolate. Lawrence [her slave] had gone South with my husband; so we had only Molly [a slave] for cooking and parlor-maid. [Food plates were removed from the parlor and taken to the kitchen, where a voice was heard chastising the slave]:"If you eat many more of those fried oysters, they will be missed. Heavens! She is running away with a plug, a palpable plug, out of that jelly cake!" [22]

In her diary entry for February (no date), Ms. Chesnut also mentioned another luncheon for local notables:

Mrs. Davis [Mrs. Jefferson Davis] gave her Luncheon to Ladies Only on Saturday. Many more persons there than at any of these luncheons which we have gone to before. Gumbo, ducks and olives, chickens in jelly, oysters, lettuce salad, chocolate cream, jelly cake, claret, champagne, etc. were the good things set before us. [23]

MARCH 15, 1864: CHOCOLATE IN A NEW FORM

Percival Drayton wrote to his friend, Alexander Hamilton, while aboard the flagship Hartford, off Pensacola, Florida. The account is unusual in that it records that the Hartford had visited Havana, Cuba and, upon return, was ordered to Mobile Bay to await engagement with the Southern ship Tennessee. His letter is unique, however, in his description of a chocolate-related event that took place on shore at Pensacola where he and the admiral dined with the post commander, a former Hungarian refugee and distinguished officer, who served chocolate in an unusual and unforgettable manner:

The first dish was chocolate soup, nothing more or less than such chocolate as you drink served in a tureen and ladled out *like soup. I have seen a good many strange customs but both the Admiral and self agreed that this was beyond both of our experiences.* [24]

MAY 17, 1864: MEDICAL SUPPLIES FOR MILITARY OPERATIONS IN SOUTHEAST VIRGINIA AND NORTH CAROLINA

The following document itemized the supplies needed to establish a field hospital. The goods identified were obtained in Alexandria, Virginia, loaded on barges and a steamer in accord with a requisition by Assistant Surgeon General J. B. Brinton, medical purveyor with the Army of the Potomac. The destination was Port Royal, on the Rappahannock River, Virginia:

It is also desirable that he procure a barge and load it with 3,000 iron bedsteads or wooden cots, 3000 mattresses, 10,000 sheets, 7000 pillows, and 100 brooms, 400 rubber cushions with open center[s], 10,000 pillow-cases, 5000 suits [of] hospital clothing, 2000 blankets, 3000 counterpanes, 500 wooden buckets, 20 cauldrons, 12 cooking stoves with furniture complete, 10 barrels of ferri sulphas[13] for disinfectant purposes, 200 pounds cocoa or chocolate, 200 pounds corn starch, 100 dozen bottles porter, 600 pounds oakum,[14] and 1000 bed-sacks. The above supply to be on a separate boat, and not to be used unless the emergency requires, which emergency will be the establishment of hospitals. [25]

JULY 4, 1864: DIARY OF A NURSE WITH THE SANITARY COMMISSION

Amanda Akin Stearns served as a volunteer with the United States Sanitary Commission and described a special holiday (Fourth of July) treat where chocolate was served to the patients:

A beautiful day, with a pleasant breeze. Gave each of my [hospital] attendants a dollar for a holiday treat, which seemed to afford them much pleasure. Wrote out my list of patients since the last engagements. I remained in the ward to let No. 6, my orderly, and as many of the attendants as possible go out. The hospital gave them an extra dinner, and I made chocolate and treated them in the afternoon, and stewed my last jar of dried cherries for their tea. [26]

DECEMBER 3, 1864: TREATMENT OF SOUTHERN PRISONERS

One would not normally consider that chocolate—an expensive item—would be available at Union prison camps, whether as a ration or as provided for purchase by a sutler. Confederate soldier Henry Clay Dickinson, however, described just such an event in his diary:

The weather continues very mild and pleasant. . . . Water has given out in cisterns at my end of the prison. . . . I hope they may get rid of some of the lice. Mosquitoes continue to annoy us. . . . We get moderate rations issued to us, but through the

sutlery we are living better than at any previous time. Board & Co. and Jones & Co. have on hand a barrel of molasses, three barrels of apples, grapes, sugar, herring, onions, potatoes, nuts, candies, gingerbread, sweet cakes, beer, etc., and many of us never think of touching the rations issued. I have four cooks detailed for my division, who, while the wood lasts, cook pies, puddings, soups, cakes, coffee, etc. Under my bed is flour, butter, coffee, sugar, meat, potatoes, onions, tea, chocolate, etc. enough to last a month. The Yanks generally let us alone and we let them alone. Many spend the days and nights till taps in gambling; others abuse the Yanks; others spend their time in picking lice; others cook and eat and others again grumble over their hard fate. . . . We received a Dixie mail today (no letter for me), from which we learn that the Andersonville prisons are in Savannah. . . . Sherman will strike for Darien and Port Royal and we think in either case must fight not far from Savannah. God grant that he may be overwhelmed. [27]

1864 (NO MONTH OR DATE): DESCRIPTION OF SUPPLIES ORGANIZED BY THE UNITED STATES SANITARY COMMISSION (WESTERN BRANCH)

The following description by Jacob Gilbert Forman details a list of supplies sent to General Grant's army by the Western Commission (branch of the United States Sanitary Commission) during June 1863, prior to the surrender of Vicksburg, Mississippi (July 4, 1863):

No parched and thirsty soil ever drank the dews of heaven, with more avidity, than did those wounded men receive the beneficent gifts and comforts, sent to them through this Commission. The number of articles sent to Gen. Grant's army from the Western Commission during the month of June, preceding the fall of Vicksburg, was 114,697, consisting of . . . 2412 bottles of Catawba wine, 1337 cans of fresh fruit, 1976 cans of condensed milk, 10,000 lemons, 1600 gallons of lager beer, 5477 lbs. dried applies, 2400 lbs. dried peaches, 2088 lbs codfish, 1850 lbs. herring, 11,710 lbs. crackers, 23,060 lbs. ice, 1800 chickens, 3171 dozen eggs, 3068 lbs. butter, 1840 lbs. corn meal, 3145 bushels potatoes . . . bottles of ginger wine, bottles of cassia syrup, bottles of blackberry syrup . . . lbs. of arrowroot, lbs. of tapioca, lbs. of sago, lbs. of pinola, lbs. of flaxseed, lbs. of cassia, lbs. of allspice, lbs. of mustard, lbs. of nutmegs, lbs. of pepper, bottles of pepper sauce, bottles of horseradish, bottles of tomato catsup, bottles of cranberry sauce . . . cans of clams and oysters, cans of spiced tripe, cans of jelly, cans of condensed soup, cans of cocoa paste, lbs. of chocolate, cans of portable lemonade, gallons of ale . . . bottles of extract ginger . . . lbs. of dried beef . . . lbs. of mackerel, lbs. of cheese, lbs. of bread . . . lbs. of coffee, lbs. of tea, lbs. of sugar, lbs. of sour krout [sic], gallons of pickles, gallons of vinegar. . . . Fortunate it was for these brave men that so much preparation and provision had been made for their comfort, and that loving hearts and kind hands had labored for them at home, sending contributions and agents, and volunteer surgeons and nurses, after them, wherever the fortunes of war had led them, to assist in binding up their wounds, in nursing them when sick, and in making them whole. On the fall of Vicksburg, on the following 4th of July, none rejoiced more than these untitled heroes. [28]

FEBRUARY 7, 1865: JOY UPON RECEIVING FOOD PARCELS

Mary Phinney von Olnhausen was an army nurse during the American Civil War. While in Morehead City, North Carolina, she wrote an unidentified friend regarding her experiences and how she and her colleagues were thankful for the arrival of food parcels:

We all sat on the floor and commenced eating. Doctor would taste of everything, and how we did eat! We had to open just a bottle of that currant wine; it was so good, and we felt pretty tired, if we were not sick. After all, we could not decide which cake was best, or whether the cake was better than the cheese, or the pie best of all; and you will no doubt be disgusted to know that we ate a whole one between us. Monday evening I gave a little rest in my room, had the Palmers and some others, and myself. I made chocolate, and Mrs. Palmer made some hot biscuits. . . . C. had two chickens roasted and shot six robins. I never saw people enjoy a supper so much . . . we had celery too, and the pies and cakes and olives and cheese and pickles could n't [sic] be beat—but, oh, that bread and butter. . . . My ward is really doing splendidly. . . . All the amputations are nearly healed. These barracks are so good on account of ventilation. [29]

Postwar Memories

1866: VOLUNTEERS ASSISTING WITH WOUNDED AFTER THE BATTLE OF GETTYSBURG

As noted earlier in this chapter, the Battle of Gettysburg in Pennsylvania occurred July 1–3, 1863. A woman volunteer known only as Mrs. Brady recalled that "for three fearful days piled the ground with bleeding wrecks of manhood" [30]. She described how the field hospital was erected and the available food supply:

Operating in her usual homely but effective and most practical manner, she at once sought a camping ground near as great field hospital, reported for duty to the division surgeon, and had a squad of convalescents assigned to assist her. Her tents were erected, the empty boxes piled so as to wall her in on three sides, and the stoves set up and fuel prepared; so that in two or three hours after reaching Gettysburg, the brigade and division surgeons were pouring in their "requisitions," and the nurses were soon passing from her tent with tubs of lemonade, milk punch, green tea by the bucketful, chocolate, milk toast, arrowroot, rice puddings, and beef tea,—all of which were systematically dispersed in strict obedience to the instructions of the medical men. [31]

1895: DEVELOPMENT OF SPECIAL DIET KITCHENS DURING THE WAR

One of the most influential women during the American Civil War was Annie Wittenmyer. She was instrumental in saving thousands of lives through her careful consideration and ability to implement a proper cooking and food distribution system in field hospitals. The United States Sanitary Commission (USSC) was successful in mobilizing volunteers and collecting tens of thousands of pounds of clothing and food supplies. But once the supplies reached the field, how were they to be distributed, given different dietary requirements for those suffering from camp diseases (scurvy and typhoid) versus simple wounds, versus complicated, abdomen- or thorax-penetrating wounds? Wittenmyer was successful in implementing a very simple plan and had the political acumen to convince the attending physicians that, in fact, it basically was their idea:

No part of the army service was so defective, during the first two years of the war, as the cooking department in the United States government hospitals . . . the supplies coming from the generous people of the North occasioned great anxiety. The surgeons forbade their distribution at the bedside of the patients, on the ground that something might be given which would endanger their lives or retard their recovery, and ordered them turned over to the commissary. Often supplies thus turned over failed to reach the sick or wounded. It was under these trying circumstances that the plan of a system of special-diet kitchens came to me. [She implemented a simple but effective idea whereby doctors prescribed foods for the ill and wounded; the list posted on the hospital bed; and the foods then were cooked in the hospital kitchens by volunteers.]. . . . The first kitchen was opened at Cumberland Hospital, Nashville, Tennessee. . . . There was no opposition to this work. Mr. Lincoln, Secretary Stanton, Surgeon-General Barnes, and Assistant-Surgeon-General Wood, gave me their endorsement and all the aid I needed. [32]

Perhaps the most telling section of her memoir, however, was a list of 67 different foods sent to the special-diet kitchens that Wittenmyer established. It is an astonishing record for several reasons. First, the information is abundant and supplied for only one month (February 1865). Second, it reflects the great diversity of foods believed to be nourishing and helpful for patients suffering from camp diseases and wounds (among the items included were oyster soup, hash, turnips, sauerkraut, and pickles). The Wittenmyer list was expressed as individual foods, and then by number of servings or rations: tea (100,350 rations), coffee (54,818), milk/cold (12,194), milk/boiled (9860), and cocoa (4770) [33].

1897: MEMORY OF POSTWAR DEPRIVATIONS AND LIMITED ACCESS TO FOOD

Phoebe Yates Levy Pember was born in Charleston, South Carolina. She married Thomas Noyes Pember, a respected Bostonian, who died in 1861. Pember then returned to Georgia, where ultimately she became the matron at Chimborazo Military Hospital in Richmond, Virginia. After the fall of Richmond in early April 1865, she "made her way" back to Georgia and ultimately published her memoirs in 1897:

> Whatever food had been provided for the sick since the Federal occupation had served for my small needs, but when my duties [at the hospital] ceased I found myself with a box full of Confederate money and a silver ten-cent piece . . . which puzzled me how to expend. It was all I had for a support, so I bought a box of matches and five cocoa-nut cakes [bars of chocolate]. The wisdom of the purchase there is no need of defending . . . but of what importance was the fact that I was houseless, homeless, and moneyless, in Richmond, the heart of Virginia. [34]

1905: MEMORY OF HOW TO MAKE A CHOCOLATE SUBSTITUTE

A woman identified only as Mrs. Clay (from Alabama) described deprivations during the war. Forty years after the end of hostilities, she published her memoirs. Of interest to the theme of this chapter and anthology is her recipe of how to prepare "substitute" chocolate when "real" chocolate was unavailable:

> Chocolate substitute from peanuts: peanuts roasted, skinned, and pounded in a mortar, then blended with boiled milk and sugar. [35]

1910: RECOLLECTION OF UNSANITARY CONDITIONS

Sometimes the difficulties of war were compounded by the daily grind of dust, dirt, mud, and unsanitary drinking water. On occasion, soldiers compared the color of drinking water to chocolate, as in this letter written from memory by Charles William Bardeen published 45 years after conclusion of hostilities:

> I was born August 28th, 1847. . . . When [Fort] Sumter was fired on April 12th, 1861, I was excited . . . I joined the military company at the Orange country grammar school and took fencing lessons. . . . On July 21st, 1862, I became a Massachusetts soldier, assigned as musician [i.e., drummer boy] to Company D of the 1st Massachusetts Infantry. . . . Often while marching I drank water out of the mud of the road where the troops were treading, and was glad to get it. If chocolate had been made of most of the water we drank while marching in Virginia it would not have changed the color. I don't remember that we ever examined water very closely if it was wet. [36]

Conclusion

During the years prior to and after the American Revolution into the mid-19th century, cocoa and chocolate

regularly were mentioned in documents as rations, as components supplied to state militias, and as foods and medical items appearing in fort and hospital inventories. Cocoa and chocolate are mentioned in innumerable diaries, letters, and political documents throughout the period. Then came the era of the American Civil War when, curiously, chocolate was not identified as a formal ration for soldiers of either side of the conflict.

It is simplistic to conclude that military blockades and civil strife limited access to and distribution of cacao beans, cocoa powder, or prepared chocolate bars. Evidence based on *Shipping News* documents confirms that trade continued between the East Coast ports and cacao supply areas in the Americas [37, 38]. But while our survey documented chocolate supplies sent to hospitals as part of the efforts of Sanitary Commission activities, and chocolate is mentioned in numerous diaries of the period, few records exist when compared to the extensive documentation for the Colonial and early Federal periods. Indeed, project team members (Laura Brindle and Bradley Olsen) working in the state archives of Georgia and South Carolina found extensive mention and use of chocolate prior to and after the war between the states, but only one document dated to the period of hostilities (the brief comment by Major William Gildersleeve Vardell [19]).

That chocolate was available to Federal troops, however, is detailed in records of the Walter Baker & Company in Massachusetts. After the death of Walter Baker in 1852, Henry L. Pierce became the manager of the company. As Amanda Blaschke noted in Chapter 26, Pierce became concerned that camp diseases claimed more lives than actual battlefield casualties. Blaschke's research confirmed that Pierce sold the United States Sanitary Commission more than 20,000 pounds of Baker's chocolate during the period 1862–1864 [39, 40]. Furthermore, Blaschke located an important letter by a physician named E. Donnelly, who suggested that chocolate would be a healthful addition to hospital diet because of its "purity" and strength-giving properties:

> I have just returned from serving my country as Surgeon for over 3 years, and expect to return to the field again. . . . Why do you not make an effort to introduce some of your preparations in to the army, it would be much better than the adulterated coffee the soldiers now drink: a chocolate should be made to keep in a powdered condition, not too sweet, and free from all husks or other irritating substances. Chocolate . . . would be much more nutritious than coffee, not so irritating to the bowels, and it is an excellent antiscorbittic [sic]. Diarrhea in the army is often the result of continually drinking impure and imperfectly made coffee: it would not be so with chocolate. [41]

Despite the paucity of information, the diaries and letters of the period still ring out with interesting chocolate-related information. Men and women, both Northerners and Southerners, longed for chocolate, relished and drank it as a beverage whenever opportunities arose. John Robertson of Marlin, Texas, was a young teenager when he enlisted. Robertson was interviewed late in life while in his 90s—nearly 70 years after he had fought in the American Civil War—and recalled how eating too much chocolate had changed his food-intake behavior:

> I was born near Quincy Florida on March 31st, 1845. I was a soldier in the Confederate Army and served under Maury's division of the Army of Tennessee. I was captured at the battle of Gettysburg in Longstreets charge and was taken to Fort Delaware, an island of 90 acres of land where the Union prisoners were kept. We were detailed to work in the fields and our rations was [sic] corn bread and pickled beef. However, I fared better than some of the prisoner for I was given the privilege of making jewelry for the use of the Union soldiers. I made rings from the buttons from their overcoats and when they were polished the brass made very nice looking rings. These I sold to the soldiers of the Union Army who were our guards and with the money thus obtained I could buy food and clothing. The Union guards kept a commissary and they had a big supply of chocolate. I ate chocolate candy and drank hot chocolate in place of coffee until I have never wanted any chocolate since. [42]

One can have too much of a good thing!

Acknowledgments

A number of team members were attracted to the theme of chocolate and cacao during the era of the American Civil War. Those assisting with archive and library research on this topic were Nicole Guerin, Jennifer Follett, Madeleine Nguyen, Bradley Olsen, and Laura Pallas Brindle. Their long hours of effort in identifying the sources presented in this chapter were greatly appreciated.

Endnotes

1. Construction of Fort Sumter in the harbor at Charleston, South Carolina, began after the War of 1812 in 1829. Bombardment of the fort on April 12, 1861 is considered to be the start of the American Civil War. Today, the fort is administered by the United States Park Service.

2. Small beer, weakly alcoholic, is the beverage produced by pressing the grain twice.

3. The chest described in this passage had been purchased in Beyrut, Syria (sic).

4. Fort Monroe is located on Chesapeake Bay at the eastern tip of the Virginia peninsula, near the James River. Construction began after the War of 1812 and was completed in 1834.

5. Construction of Fort Lafayette, located in New York Harbor, began after the War of 1812 and originally was named Fort Diamond. It served as a prison for captured Confederate soldiers during the American Civil War.

6. The Spanish first landed at what is now Port Royal, South Carolina, in 1520, but the attempt to establish a fort/settlement failed.

7. President Abraham Lincoln established the United States Sanitary Commission (USSC) on June 18, 1861. The objective was to coordinate the efforts of women who volunteered for the war effort.

8. Essentially beef extract, or beef bouillon, commonly served to the sick/infirmed.

9. A beverage not dissimilar to festive eggnog, but commonly served to the sick/infirmed.

10. Jim Gay from The Colonial Williamsburg Foundation, Virginia, reported to us that White House, Virginia, is an unincorporated community across the Pamunkey River from the Pamunkey Indian Reservation and that Martha Washington lived on White House Plantation. It was also the site of a railroad crossing and a Union supply base during the 1862 Peninsula Campaign, which would explain why there would have been a hospital ship and chocolate supplies at this site.

11. King of Terrors—that is the devil.

12. Wife of Jefferson Davis, president of the Confederacy.

13. Ferri sulphas = ferrous sulfate; used as a medicinal tonic.

14. Requisitioned in anticipation that the supply ship might leak, since oakum consisted of unraveled pieces of old ropes soaked with creosote or tar used to calk the seams of wooden boats.

References

1. Massachusetts. General Court. House of Representatives. *Journal of the Honourable House of Representatives, of His Majesty's Province of the Massachusetts-Bay, in New-England, begun and held at Boston, in the County of Suffolk, on Wednesday the thirty-first day of May, Annoque* [sic] *Domini, 1758.* Boston: Samuel Kneeland, 1758; p. 312.

2. Massachusetts. General Court. *Resolves of the General Assembly of the state of Massachusetts-Bay. Begun and held at Watertown, in the county of Middlesex, on Wednesday the 29th day of May, anno domini 1776, and thence continued by one prorogation and three adjournments to Tuesday the the* [sic] *24th day of December following, and then met at Boston in the county of Suffolk.* Boston: Benjamin Edes, 1776; p. 33.

3. *Proceedings of a general court martial, held at White Plains, in the state of New-York, by order of His Excellency General Washington, commander in chief of the army of the United States of America, for the trial of Major General St. Clair, August 25, 1778. Major General Lincoln, president.* Philadelphia: Hall and Sellers, 1778; p. 10.

4. Maryland. General Assembly. House of Delegates. *Votes and proceedings of the House of Delegates of the state of Maryland. Votes and Proceedings, July, 1779*. Annapolis, MD: Frederick Green, 1779; p. 137.

5. *New York Times*, May 20, 1861, p. 2.

6. Letter from Abby Howland Woolsey to Eliza Newton Woolsey Howland, dated imprecisely to 1861. In: *Letters of a Family during the War for the Union 1861–1865*, Volume 1. Edited by Georgeanna Woolsey Bacon and Eliza Woolsey Howland. No city: privately published, 1899; pp. 65–66.

7. Letter from John Albion Andrew to John Murray Forbes, dated May 20, 1861. In: *Letters and Recollections of John Murray Forbes*, Volume 1. Edited by Sarah Forbes Hughes. Boston: Houghton, Mifflin & Company, 1900; p. 214.

8. Diary of Lawrence Sangston, entry for November 1861, In: *The Bastilles of the North*. Baltimore, MD: Kelly, Hedian, and Piet, 1863; pp. 67–68.

9. United States Naval War Records Office, and the United States Office of Naval Records and Library. *Official Records of the Union and Confederate Navies in the War of the Rebellion*. Series I. Volume 12: North Atlantic Blockading Squadron (February 2, 1865–August 3, 1865); South Atlantic Blockading Squadron (October 29, 1861–May 13, 1862). Washington, DC: Government Printing Office, 1901; p. 746.

10. Letter from Harriet Ware, dated May 1, 1862. In: *Letters from Port Royal 1862–1868*. Edited by Elizabeth Ware Pearson. Boston: W. B. Clarke Company, 1906; pp. 23–25.

11. Letter from Katharine Prescott Wormeley, dated May 30, 1862. In: *The Other Side of War. With the Army of the Potomac: Letters From the Headquarters of the United States Sanitary Commission During the Peninsular Campaign in Virginia in 1862*. Boston: Ticknor and Company, 1889; pp. 87–92.

12. Letter from Charles William Woolsey, dated May 31, 1862. In: *Letters of a Family During the War for the Union 1861–1865*, Volume 2. Edited by Georgeanna Woolsey and Eliza Woolsey Howland. No city: privately published, 1899; pp. 397–404.

13. http://www.wtv-zone.com/civilwar/gibbons.html. (Accessed August 28, 2007.)

14. Diary of Abigail Hopper Gibbons, dated July 1862. In: *Life of Abby Hopper Gibbons. Told Chiefly Through Her Correspondence*, Volume 1. Edited by Sarah Hopper Emerson. New York: G. P. Putnam's Sons, 1897; pp. 340–342.

15. Diary of Abigail Hopper Gibbons, dated September 1862. In: *Life of Abby Hopper Gibbons. Told Chiefly Through Her Correspondence*, Volume 1, Edited by Sarah Hopper Emerson. New York: G. P. Putnam's Sons, 1897; pp. 372–383.

16. Letter from Richard Derby, dated September 13, 1862. In: *Soldiers' Letters, from Camp, Battle-field and Prison*. New York: Bunce and Huntington, 1865; pp. 158–159.

17. Letter from Zenas T. Haines, dated November 15, 1862. In: *Letters from the Forty-fourth Regiment M.V.M.: A Record of the Experience of a Nine Month's Regiment in the Department of North Carolina in 1862–3*. Boston: Herald Job Office, 1863; pp. 47–50.

18. United States War Department, United States Record and Pension Office, United States War Records Office . . . et al. *The War of the Rebellion. A Compilation of the Official Records of the Union and Confederate Armies*. Series 2, Volume 5. Washington, DC: Government Printing Office, 1899; p. 833.

19. South Carolina Historical Society. William Gildersleeve Vardell, Letters 1862–1863 (Accession Number: 271.00). Letter dated April 21, 1863.

20. Letter from Emily Bliss Thacher Souder to Mrs. J. Heulings, dated July 16, 1863. In: *Leaves from the Battlefield of Gettysburg: A Series of Letters from a Field Hospital: and National Poems*. Philadelphia: C. Sherman Son and Company, 1864; pp. 21–26.

21. Diary of Mary Boykin Chestnut, dated January 22, 1864. In: *A Diary from Dixie*. Edited by Isabella D. Martin and Myrta L. Avary. New York: D. Appleton and Company, 1905; p. 282.

22. Diary of Mary Boykin Chestnut, dated January 23, 1864. In: *A Diary from Dixie*. Edited by Isabella D. Martin and Myrta L. Avary. New York: D. Appleton and Company, 1905; pp. 282–283.

23. Diary of Mary Boykin Chestnut, dated February 1864. In: *A Diary from Dixie*. Edited by Isabella D. Martin and Myrta L. Avary. New York: D. Appleton and Company, 1905; pp. 282–283.

24. Letter from Percival Drayton To Alexander Hamilton, dated March 15, 1864. In: *Naval Letters from Captain Percival Drayton, 1861–1865*. New York: privately published, 1906; p. 46.

25. United States War Department, United States Record and Pension Office, United States War Records Office . . . et al. *The War of the Rebellion: A Compilation of the Official Records of the Union and Confederate Armies*. Series 1, Volume 36 (part 1). Washington, DC: Government Printing Office, 1891; p. 274.

26. Diary of Amanda Akin Stearns, dated July 1864. In: *The Lady Nurse of Ward E*. New York: Baker and Taylor Company, 1909.

27. Diary of Henry Clay Dickinson, dated December 3, 1864. In: *Diary of Capt. Henry C. Dickinson, C.S.A. Morris Island, 1864–1865*. No city given: Williamson-Haffner, Company, 1910; pp. 132–133.

28. Forman, J. G. *The Western Sanitary Commission; a Sketch of its Origin, History, Labors for the Sick and Wounded of the Western Armies, and Aid Given to Freedmen and Union Refugees, with Incidents of Hospital Life*. St. Louis, MO: Mississippi Valley Sanitary Fair, 1864; pp. 78–79.

29. Letter from Mary Phinney von Olhausen, dated February 7, 1865. In: *Mary Phinney von Olnhausen, Adventures of an Army Nurse in Two Wars*. Edited by James Phinney

Munroe. Boston: Little, Brown, and Company, 1903; pp. 170–173.

30. Moore, F. *Women of the War. Their Heroism and Self-Sacrifice*. Hartford, CT: S.S. Scranton, 1866; p. 48.

31. Moore, F. *Women of the War. Their Heroism and Self-Sacrifice*. Hartford, CT: S.S. Scranton, 1866; p. 48.

32. Memoire of Annie Wittenmyer. In: *Under the Guns. A Woman's Reminiscences of the Civil War*. Introduction by Mrs. U.S. Grant. Boston: E. B. Stillings and Company, 1895; pp. 259–262.

33. Memoire of Annie Wittenmyer. In: *Under the Guns. A Woman's Reminiscences of the Civil War*. Introduction by Mrs. U.S. Grant. Boston: E. B. Stillings and Company, 1895; pp. 264–266.

34. Memoir of Phoebe Yates Pember. In: *Southern Woman's Story: Life in Confederate Richmond*. Jackson, TN: McCowat-Mercer Press, 1959; p. 144.

35. *A Belle of the Fifties. Memoirs of Mrs. Clay of Alabama, Covering Social and Political Life in Washington and the South, 1853–66*. New York: Doubleday and Page. Reprinted in: Spaulding, L. M. and Spaulding, J. *Civil War Recipes. Receipts from the Pages of Godey's Lady's Book*. Lexington: University of Kentucky Press, 1999; p. 24.

36. Memoir of Charles William Bardeen. In: *A Little Fifer's War Diary*. No publication data. Document dated 1910, pp. 17–22.

37. *New York Times*. Arrived (*Shipping News*), August 2, 1861, p. 9.

38. *New York Times*. Arrived (*Shipping News*), November 28, 1861, p. 8.

39. Baker Business Library, Harvard Business School (BBL). A-4, 1863–1868, p. 443.

40. BBL. A-3, 1857–1863, p. 569.

41. BBL. K-1, Folder Three: Incoming Correspondence, 1863–1927. Philadelphia, Pennsylvania, letter dated July 19th, 1864.

42. Interview with John Robertson. Library of Congress, Washington, DC: American Memory Project, American Life Histories. Manuscripts from the Federal Writers' Project, 1936–1940.

Chocolate Futures

Promising Areas for Further Research

Louis Evan Grivetti and Howard-Yana Shapiro

Introduction

From 1998 through 2007, researchers in the Department of Nutrition, University of California, Davis, were awarded substantial grants from Mars, Incorporated, to conduct historical research on chocolate. The chocolate history project was part of long-term funding provided by Mars to several departments and research units on campus (collectively called the MDRU or Multi-Disciplinary Research Unit). Our first chocolate history project (1998–2002) objectives focused on documenting the history of chocolate in Central America (pre-Columbian and Spanish Colonial era), especially development of and transfer to Europe of chocolate-associated culinary and medical information. These goals were accomplished through archive/library research conducted in North, Central, and South America, and the Caribbean, coupled with on-site field interviews with farmers, traditional healers, merchants, and vendors in Cuba, Dominican Republic, Guatemala, Mexico, and Panama. The information obtained was presented at national and international professional venues [1–3] and was published in the *Journal of Nutrition* (United States) and the *Karger Newsletter* (Switzerland) [4, 5].

The second research period of the chocolate history project (2004–2008) built on the first; a team of colleagues was assembled who were affiliated with four institutions: Colonial Williamsburg (Williamsburg, Virginia), Fort Ticonderoga (Ticonderoga, New York), Mars, Incorporated (Elizabethtown, Pennsylvania; Hackettstown, New Jersey; and McLean, Virginia), and the University of California, Davis. Our objectives included the following:

1. Determine when, where, and how chocolate products were introduced and dispersed throughout North America.

2. Trace the development and evolution of chocolate-related technology in North America.

3. Identify culinary, cultural, economic, and dietary–medical uses of chocolate in North America from the Colonial era into the early 20th century.

4. Develop a state-of-the-art database and Web portal containing history of chocolate information for use by students, the general public, and scholars/scientists.

5. Publish chocolate-related findings via popular and scholarly books and journals, and relate the findings via local, national, and international symposia.

A total of 113 team members participated in data collection, analysis, or writing and conducted

Chocolate: History, Culture, and Heritage. Edited by Grivetti and Shapiro
Copyright © 2009 John Wiley & Sons, Inc.

archive/library/museum-based studies at more than 200 locations. Team members included academics, independent contractors, and volunteers with familiarity and experience working in specific archives, libraries, or museum depositories within North, Central, and South America, and the Caribbean, and in Western Africa and Western Europe (Appendices 2, 3, and 4).

Since we could not conduct research in all countries where chocolate has had a strong historical presence, we emphasized the scholarly strengths of our team members and their professional contacts and focused on chocolate-associated information from the following countries: Barbados, Brazil, Canada, Colombia, Cuba, Costa Rica, Dominican Republic, England, Equatorial Guinea, France, Greece, Guatemala, India, Jamaica, Martinique, Mexico, Panama, Portugal, Sao Tomé and Principe, Spain, Trinidad, and the United States.

This decision, predicated upon our existing network of professional contacts, meant that some geographical regions and nations—some with extensive history in regard to chocolate—received only limited or no coverage by our team of researchers. The present anthology reflects these constraints and the present work was not intended to be all-inclusive. It is our hope that other academics, agencies, commodity groups, and independent scholars will be excited and intrigued by our findings and seize further opportunities for chocolate history research. It is in this spirit that we encourage readers of this volume, whether novice or trained archivist/historian, in North America or abroad, to embark upon additional efforts to further the scholarship associated with this most interesting food.

Chocolate Futures

Only two foods—wine and chocolate—have such a rich history and depth of resources available for inspection. The chocolate-associated evidence that our team drew upon spanned almost 3000 years of archaeological data, written accounts, and oral traditions. Our efforts stretched from the period of early domestication of *Theobroma cacao* in South America, through ancient Olmec, Mayan, and Mexica/Aztec stele and codex documents, to letters and books written by famous and obscure authors, to oral histories that team members collected from respondents who live in Guatemalan and Costa Rican rain forests and elsewhere where cocoa trees are tended in the 21st century.

Members of our research team recognized that work on the history of chocolate as food and as medicine could characterize a lifetime of intensive activities and efforts. The kind of chocolate-associated information available for collection was daunting (Table 56.1). Each archive, library, and museum visited brought to light new documents and further insights on chocolate-related historical questions. Sometimes identified documents appeared rather ordinary when first inspected, but aroused new attention as we gathered additional information during the course of the project. This was the case when we reviewed food and medicine procurement lists associated with Colonial hospitals in New Spain, where quantities of medicinal chocolate beverages provided to patients during September 1748 required 10,863 tablets of chocolate [6]! At other times, the documents were intriguing from the start, as when a team member found a manu-

Table 56.1 Sources of Potential Chocolate-Related Information

Advertisements (magazines, newspapers)
Archaeological data
    Analysis of chocolate residues
    Excavation descriptions
    Murals, paintings, vases
Art
    Fine art
    Murals
    Paintings and lithographs
    Popular art
        Posters
        Trade cards
Collectibles
    Chocolate serving equipment (pots and cups)
    Molds
Culinary
    Equipment
    Menus
    Recipes
Government documents (local, state, national, international)
Hospital records
Legal
    Patents
    Trial accounts
Literature
    Autobiographies
    Diaries
    Fiction
    Poetry
    Travel accounts
Magazines (articles)
Manufacturing
    Deeds
    Equipment
    Invoices
Military records
Newspapers (articles)
Price Current documents
Religious documents
Shipping News documents

script in the Archive of the Indies, Seville, Spain, that listed the names and quantities of chocolate assigned to each adult male stationed at the Spanish fort in S. Augustine, Florida, in 1642 [7]. Other documents answered some questions but posed others, as when a press release, dated May 21, 1927, reported that Charles Lindbergh drank chocolate during his historic solo transatlantic flight [8], but our inspection of Lindbergh's published diaries made no mention of this beverage . . . only water [9].

We acknowledge that there remain thousands of archives, libraries, and museums still to be searched for chocolate-related documents, a sobering but exciting thought. What this means, however, is that many additional historical research teams can and should be formed to examine these resources. It means, too, that new annual forums should be organized where chocolate-related historical information can be shared, where both recognized and novice scholars can meet to report their interesting findings.

The purpose of the present chapter, therefore, is to identify a variety of themes and topics where in our opinion future research would be rewarding. It was appropriate in our view to call this chapter "chocolate futures," where we have identified gaps of knowledge where further future efforts might be concentrated. We also present in this chapter selected examples of documents representing themes that our team did not develop fully into chapters. This chapter, therefore, illustrates the variety and richness of the work still ahead.

Expanded Geographical Coverage

First, we identify here the need for expanded geographical coverage. While our team worked within the archives and libraries of many countries, an integrated history of chocolate cannot be written until further work is completed in numerous countries and different global geographical regions. Having identified previously the geographical regions and nations where we concentrated our team's efforts, we identify here the geographical areas and countries that received only limited or no coverage from members of our team where efforts should be concentrated:

Central America, South America, and the Caribbean: Argentina, Aruba, Belize, Chile, Costa Rica, Curaçao, Ecuador, Haiti, Honduras, Nicaragua, Paraguay, Peru, Puerto Rico, Surinam, Uruguay, and Venezuela

Europe: Albania, Austria, Belgium, Bulgaria, Bosnia, Croatia, Czech Republic, Germany, Hungary, Italy,

Macedonia, Netherlands, Romania, Serbia, Sweden, and Switzerland

Africa: Cameroons, Ghana, and Ivory Coast

Asia: Indonesia, Japan, Philippines, Sri Lanka, Thailand, and Vietnam

Pacific Region: Australia, New Zealand, and the Pacific Islands

Potential Thematic Coverage Opportunities

ART AND LITERATURE

There exist tens of thousands of accounts in art and literature that describe cacao harvesting, processing, and manufacturing, and the preparation and serving of chocolate, whether framed and frozen in time as paintings and sculpture, enlivening the pages of novels and poetry, or captured stylistically as themes in dance, music, and theater. Each of these topics represents promising avenues for further enquiry.

Cacao and chocolate encompass much of the world of art: fine art paintings and lithographs, commercial chocolate-related posters, signs, and trading cards, even unusual chocolate-associated signs. These are themes where researchers could make significant contributions to the history of chocolate. A number of fine art paintings are relatively well known and through the centuries have received considerable attention through publications in various chocolate-related books [10–12]. Less well known—but valuable for understanding the historical role of chocolate and its place within various cultures of the world—are the hundreds of chocolate-related posters and trading cards identified during the course of our research (Figs. 56.1– 56.5). While examples our team members identified during the course of research have been summarized and analyzed (see Chapters 14 and 15), we expect that an expanded systematic investigation with analysis and discussion of themes portrayed in such materials would be worthwhile.

Chocolate appears as a theme in thousands of works of literature and poetry. Efforts to focus exclusively on chocolate in literature and poetry in our opinion could serve as the basis for several books. The following selected examples reflect the diverse content and rich potential that such a concentrated effort might bring.

Samuel Pepys, the 17th century English diarist, for example, documented everyday life in London during the mid-17th century. His diary entry for April 24, 1661 reported that he suffered a hangover from overindulging the previous night. He wrote that he felt ill, and that his cure required drinking a specific beverage:

FIGURE 56.1. Chocolate trade card (Suchard), date unknown. *Source:* Courtesy of the Grivetti Family Trust. (Used with permission.) (See color insert.)

Rose [from bed] and went out with Mr. Creed to drink our morning draft, which he did give me in chocolate to settle my stomach. [13]

Alphonse François, Marquis de Sade, primarily known for "other activities," nevertheless, mentioned chocolate in a series of letters written from prison addressed to his wife. In a most interesting letter dated May 16, 1779, he tersely complained of the quality of a food package she previously had sent to him and enumerated his complaints:

The sponge cake is not at all what I asked for: 1) I wanted it iced everywhere, both on top and underneath, with the same icing used on the little cookies; 2) I wanted it to be chocolate inside, of which it contains not the slightest hint; they have colored it with some sort of dark herb, but there is not what one could call the slightest suspicion of chocolate. The next time you send me a package, please have it made for me, and try to have some trustworthy person there to see for themselves that some chocolate is put inside. The cookies must smell of chocolate, as if one were biting into a chocolate bar. [14]

This specific de Sade letter reveals several important pieces of information with direct application to chocolate history. First, the letter hints that the so-called chocolate cookies were prepared from adulterated chocolate (see Chapter 47). Second, the concluding sentence suggests that bars of chocolate for eating pleasure were available in Paris nearly 50 years before Van

Houten's invention of the cocoa press, an invention that some have interpreted as a necessary "tipping-point" required before the development of confectionary chocolate. In 18th century Europe and elsewhere, consumers did not "bite into" standard chocolate tablets. These tablets whether circular, rectangular, or appearing as "globs" were not eaten like 20th and 21st century candy bars; these tablets were grated and used to prepare chocolate beverages. Thus, the phrase "biting into" reveals the probability that bars of confectionary chocolate circulated in France by this early date.

Charles Dickens, the respected English novelist, commonly alluded to chocolate and mentioned the beverage frequently in his works, as in his novel *Bleak House*:

I have lived here many years. I pass my days in court; my evenings and my nights here. I find the nights long, for I sleep but little, and think much. That is, of course, unavoidable; being in the Chancery. I am sorry I cannot offer chocolate. I expect a judgment shortly, and shall then place my establishment on a superior footing. [15]

Chocolate also is mentioned in Chapter II of his more popular work, *A Tale of Two Cities*:

It took four men, all four ablaze with gorgeous decoration, and the Chief of them unable to exist with fewer than two gold watches in his pocket, emulative of the noble and chaste fashion set by Monseigneur, to conduct the happy chocolate to

BITTERN

ONE OF THESE REPRODUCTIONS OF
NORTH AMERICAN WILD-LIFE
ISSUED IN THE INTEREST OF CONSERVATION
FOUND IN EVERY PACKAGE OF

GHIRARDELLI'S
MILK CHOCOLATE

FIGURE 56.2. Chocolate trade card (Ghirardelli—birds of North America), date unknown. *Source:* Courtesy of the Grivetti Family Trust. (Used with permission.) (See color insert.)

Monseigneur's lips. One lacquey [sic] carried the chocolate-pot into the sacred presence; a second, milled and frothed the chocolate with the little instrument he bore for that function; a third, presented the favoured [sic] napkin; a fourth (he of the two gold watches), poured the chocolate out. It was impossible for Monseigneur to dispense with one of these attendants on the chocolate and hold his high place under the admiring Heavens. Deep would have been the blot upon his escutcheon if his chocolate had been ignobly waited on by only three men; he must have died of two. [16]

Newspaper editor Horace Greeley, known for his exhortation—"Go West, young man"—published an account of an overland journey he made from New York to San Francisco during the summer of 1859. Greeley, well into his journey and crossing the Great Plains, reflected at a rest stop and worried that he might be "intercepted" by Cheyenne Indians. He wrote

FIGURE 56.3. Chocolate trade card (Poulain), date unknown. *Source:* Courtesy of the Grivetti Family Trust. (Used with permission.) (See color insert.)

how he dreaded the "simplicity of human existence" during his journey as he cataloged the disappearance of social niceties he had grown used to:

May 12th—Chicago [Illinois]—Chocolate and morning newspapers last seen on the breakfast-table;

May 23rd—Leavenworth [Kansas]—Room-bells and baths make their final appearance;

May 24th—Topeka [Kansas]—Beef-steak and wash-bowls (other than tin) last visible. Barber ditto;

May 26th—Manhattan [Kansas]—Potatoes and eggs last recognized . . . Chairs ditto. [17]

Searching general literature for chocolate-related passages sometimes sheds important light on social customs and events of specific places and eras. The following passage written by John Reed, that appeared in his *Ten Days That Shook the World*, described the severe winter and food shortages in Russia shortly after the 1917 revolution. Reed related how the specter of inflation became rampant in subsequent months, and how the social and personal problems of declining rations were linked with inflationary chocolate prices:

September and October are the worst months of the Russian year—especially the Petrograd year. Under dull grey skies, in

FIGURE 56.5. Chocolate trade card (Bensdorp), date unknown. *Source:* Courtesy of the Grivetti Family Trust. (Used with permission.) (See color insert.)

FIGURE 56.4. Chocolate trade card (Ghirardelli—general advertising), date unknown. *Source:* Courtesy of the Grivetti Family Trust. (Used with permission.) (See color insert.)

the shortening days, the rain fell drenching, incessant. The mud underfoot was deep, slippery and clinging, tracked everywhere by heavy boots, and worse than usual because of the complete break-down of the Municipal administration. . . . Week by week food became scarcer. The daily allowance of bread fell from a pound and a half to a pound, then three quarters, half, and a quarter-pound. Toward the end there was a week without any bread at all. Sugar one was entitled to at the rate of two pounds a month—if one could get it at all, which was seldom. A bar of chocolate or a pound of tasteless candy cost anywhere from seven to ten rubles—at least a dollar. There was milk for about half the babies in the city; most hotels and private houses never saw it for months. [18]

The theme of chocolate and poisoning has been linked for centuries and the image of "death by chocolate" has remained popular into contemporary 21st century literature (see Chapter 21). In her fantasy novel *Poison Study*, Maria Snyder's protagonist is Yelena, an assertive, masterful woman, and designated food taster to the Commander of Ixia. During the course of her training, Yelena is introduced to culinary herbs, medicinal plants, and poison antidotes. In one passage Yelena inspects the seeds contained inside the pod of an unusual tree and contemplates the sweetness of the pulp and bitterness of the beans (that we readers today know to be cacao):

Cutting one of the yellow pods in half, I discovered it was filled with a white mucilaginous pulp. A taste of the pulp revealed it to have a sweet and citrus flavor with a taint of sour, as if it was starting to rot. The white flesh contained seed. I cleaned the pulp from the seeds and uncovered thirty-six of them. They resembled the beans from the caravan. My excitement diminished as I compared seed against bean in the sunlight. The pod seed was purple instead of brown, and when I bit into the seed, I spit it out as a strong bitter and astringent taste filled my mouth. [19]

Chocolate as a component or primary element in poetry is another theme that rings through the ages. Some examples of chocolate-associated poems dated to the 17th century have a risqué texture as revealed in the lines of the following poem dated to 1660:

Nor need the Women longer grieve,
Who spend their oyle [sic] yet not Conceive,
For 'tis a Help Immediate,
If such but Lick of Chocolate.

The Nut-Browne Lasses of the Land,
Whom Nature vail'd [sic] in Face and hand,
Are quickly Beauties of High-Rate,
By one small Draught of Chocolate. [20]

Alexander Pope's well-known masterpiece, *The Rape of the Lock*, published in 1712, also contained a prominent mention of chocolate:

Or as Ixion fix'd, the wretch shall feel
The giddy motion of the whirling Mill,
In fumes of burning Chocolate shall glow,
And tremble at the sea that froaths [sic] below. [21]

Other verses caution readers to resist gluttony and self-indulgence, as with this chocolate-associated poem dated to 1746:

Next day obedient to his word,
The dish appear'd at course the third;
But matters now were alter'd quite,
In bed till noon he'd stretch'd the night.
Took chocolate at ev'ry dose,
And just at twelve his worship rose.
Then eat a toast and sip'd bohea [tea],
'Till one, and fat to dine at three;
And having tasted some half-score,
Of costly things he loath'd before,
He hop'd [sic] his dish of sav'ry meat,
Wou'd prove that still 'twas bliss to eat. [22]

The English poet George Gordon Byron, perhaps better known throughout the West as Lord Byron, died fighting for Greek independence from Ottoman rule. One of Byron's more noted works, *Don Juan*, published in 1821 (but unfinished at the time of the poet's death), also contained mention of chocolate:

Lord Henry, who had not discuss'd his chocolate,
Also the muffin whereof he complain'd,
Said, Juan had not got his usual look elate,
At which he marvell'd, since it had not rain'd;
Then ask'd her Grace what news were of the Duke of late:
Her Grace replied, his Grace was rather pain'd
With some slight, light, hereditary twinges
Of gout, which rusts aristocratic hinges. [23]

COOKBOOK AND COMMONPLACE BOOKS

Major cookbook collections located in North America inspected by team members included the Bancroft Library, Berkeley, California; Huntington Library, San Marino, California; Schlesinger and Houghton Libraries, Harvard University, Boston, Massachusetts; Butler Library, Rare Books Division, Columbia University, New York; New York Public Library; and Library of Congress, Washington, DC. Still, numerous depositories remain untapped. While chocolate-associated information from cookbooks has been considered, we suggest that much further work remains.

Complementing searches through North American cookbooks, we also recommend inspection of what are known as *commonplace books*. These are essentially unpublished idiosyncratic manuscripts pre-

pared by individuals who wrote and preserved their thoughts for the day and commented on daily events. Sometimes the compilers clipped examples of published poetry and pasted these onto the pages as well. One commonplace book reported by Jim Gay, a member of our research team, contained the earliest recipe for chocolate thus far identified in North America, a recipe dating to ca. 1700 (see Chapter 23).

Hundreds if not thousands of commonplace books are kept as family heirlooms. Too often they rest—unread—on dusty home bookshelves. Others, fortunately, have been donated and deposited with local, county, or state historical societies, where, too, they are examined infrequently. We suspect that such handwritten accounts may contain chocolate-related recipes even earlier than 1700 that would push the culinary history of chocolate in North America into the mid- to late 17th century. This is an area where future researchers—and readers of this chapter—could make a critical, substantial contribution to chocolate history.

DIARIES AND FAMILY HISTORIES

We searched hundreds of diaries for chocolate-associated information, while literally thousands more are available for inspection whether in archives, libraries, historical society collections, or through digitized online collections. The excitement of holding a family diary treasure is that one never knows from one to the next when a new chocolate-related citation will be found, when a diary reference might be the first chocolate-associated account for a state or geographical region of North America. Diaries produced by individuals and families who crossed North America during the 19th and early 20th centuries are vast in numbers and were not searched systematically due to time considerations. Family diaries, therefore, offer great potential for finding new chocolate-related citations. Two examples presented here show the important role of chocolate in isolated regions of mid-19th century North America, and document that use of chocolate had spread far and wide and was readily available at relatively obscure and isolated localities.

Fort Hall, Oregon Territory (currently in the state of Idaho), was established in 1834. A ledger kept by Osborne Russell at the fort, dated 1835, documented several items sold along with prices:

2/14/1835	2	Nine inch butcher knives	$2.50
2/14/1835	$\frac{1}{2}$ pt	Red pepper	$0.75
2/14/1835	1 hank	Thread	$0.40
2/14/1835	—	Needles	$0.60
2/16/1835	20 loads	Ammunition for trade	$2.00
2/16/1835	1 bunch	White seed beans	$0.75
2/16/1835	$\frac{1}{2}$ doz.	Flints	$0.25
2/23/1835	2 lbs.	Chocolate	$3.00 [24]

Fur trappers and "mountain men" were used to hardship, living off the land, and loneliness. An annual meeting of fur trappers—called *the rendezvous*—allowed for socialization with fellow "mountain men" who spoke English, French, Spanish, and other languages. These were times to swap stories about regional geography, survival, and unusual experiences. A traveler to the Green River region of the Rocky Mountains in July 1839 captured the essence of one *rendezvous*, and documented for posterity that chocolate was among the treasured items drunk at this event:

> A pound of beaver skins is usually paid for with four dollars worth of goods; but the goods themselves are sold at enormous prices, so-called mountain prices. . . . A pint of meal, for instance, costs from half a dollar to a dollar; a pint of coffee-beans, cocoa beans or sugar, two dollars each; a pint of diluted alcohol (the only spirituous [sic] liquor to be had), four dollars; a piece of chewing tobacco of the commonest sort, which is usually smoked, Indian fashion, mixed with herbs, one to two dollars. . . . With their hairy bank notes, the beaver skins, they can obtain all the luxuries of the mountains, and live for a few days like lords. Coffee and chocolate is cooked; the pipe is kept aglow day and night; the spirits circulate; and whatever is not spent in such ways the squaws coax out of them, or else it is squandered at cards. [25]

EXPEDITIONS AND EXTREME ENVIRONMENTS

Documented use of chocolate as a travel or emergency ration has a long history in North America and elsewhere. One of the more interesting examples where explorers relied on chocolate to provide energy was recorded by Alexander Henry during his 1776 expedition through central Canada:

> On the twentieth [of January, 1776], the last remains of our provisions were expended; but, I had taken the precaution to conceal a cake of chocolate, in reserve for an occasion like that which was now arrived. Toward evening, my men, after walking the whole day, began to lose their strength; but, we nevertheless kept on our feet till it was late; and, when we encamped, I informed them of the treasure which was still in store. I desired them to fill the kettle with snow, and argued with them the while, that the chocolate would keep us alive, for five days at least; an interval in which we should surely meet with some Indian at the chase. Their spirits revived at the suggestion; and, the kettle being filled with two gallons of water, I put into it one square of the chocolate. The quantity was scarcely sufficient to alter the colour [sic] of the water; but, each of us drank half a gallon of the warm liquor, by which we were much refreshed, and in its enjoyment felt no more of the fatigues of the day. In the morning, we allowed ourselves a similar repast, after finishing which, we marched vigorously for six. [26]

An expedition led by Zebulon Montgomery Pike through Louisiana Territory and into New Spain during 1805 through 1807 documented the use of chocolate within the Southwest borderlands (see Chapter 33):

> The house-tops of St. John's were crowded, as well as the streets, when we entered, and at the door of the public quarters we were met by the President Priest. When my companion, who commanded the escort, received him in a street and embraced him, all the poor creatures who stood round stove to kiss the ring or hand of the holy father; for myself, I saluted him in the usual style. My men were conducted into the quarter, and I went to the house of the priest, where we were treated with politeness. He offered us coffee, chocolate, or whatever we thought proper, and desired me to consider myself at home in his house. [27]

Further documentation of chocolate-related information associated with contemporary 20th and 21st century expeditions also would yield good results. Is it not astonishing to learn that the Franklin Expedition of 1845, that intrepid group of men who sought the fabled Northwest Passage but found instead ice-bound, freezing deaths, was provisioned with more than 100,000 pounds of canned food (enough for 67 men for three years) and that one component of this food hoard was 4573 pounds of chocolate [28]?

Survival in harsh environments like the Yukon during the gold rush of 1898 required considerable planning (see Chapter 24). Documents of the era identify items appropriate for the well-supplied miner that included expenses for clothing, mining equipment, and food items, among these beef extract, "condensed" onions, "evaporated" spuds (potatoes), pepper, raisins, yeast, and chocolate [29].

Systematic examination of exploration accounts of Arctic and Antarctic expeditions for chocolate-associated information would be rewarding. Scott's Antarctic expedition in 1901–1904, for example, included 3500 pounds of chocolate and cocoa donated by Cadbury in England. It can be noted, too, that in 2001 a bar of chocolate from this expedition—left behind by Scott in Antarctica—sold at auction for $686.00 [30, 31].

Rations for the subsequent ill-fated Scott expedition of 1911–1912 initially consisted of enough food for 47 men and also contained chocolate:

> 150 tons of roast pheasant, 500 of roast turkey, whole roast partridges, jugged hare, duck and green peas, rump steak, wild cherry sauce, celery seed, black currant vinegar, candied orange peel, Stilton and Double Gloucester cheese, 27 gallons of brandy, 27 gallons of whiskey, 60 cases of port, 36 cases of sherry, 28 cases of champagne, lime juice, 1800 pounds of tobacco, pemmican, raisins, chocolate and onion powder. [32]

After suffering setbacks, the explorers considered their Christmas meal (December 25, 1911) and prepared the following items that included chocolate and horse meat (butchered from the expedition horses that had died):

FIGURE 56.6. Shackleton's hut in Antarctica: exterior view. *Source:* Scott Borg. (Used with permission.) (See color insert.)

[Our] Christmas meal consisted of pony hoosh [a one-pot meal] ground biscuit, a chocolate hoosh made from cocoa, sugar, biscuit and raisins thickened with arrowroot, two-and-a-half square inches each of plum-duff, a pannikin [small tin cup] of cocoa, four caramels each and four pieces of crystallized ginger. [33]

The Norwegian Antarctic South Pole expedition of 1911–1912 led by Roland Amundsen also made extensive use of chocolate:

Alcohol was rationed aboard the voyage on the Fram, *one dram and fifteen drops of spirits were given at dinner on Wednesdays and Sundays. A glass of "toddy" was given on Saturday evenings. The expedition was treated with "bonbons and drops" and "Gala Peter," which was a chocolate bar made by Swiss Peter Chocolate Company, and fruit syrup they could drink.... During a stop in Buenos Aires, they bought pigs, sheep, and some fowl to have fresh meat on the voyage south. The animals were penned in on the top deck.* [34]

Further documentation of chocolate use in the Arctic and Antarctica would be rewarding—for example, more detailed review of the Scott and Shackleton expeditions revealed chocolate was regularly stored at their huts (Figs. 56.6–56.9) and appeared on the Shackleton Christmas dinner menu in 1902 (Fig. 56.10).

Exploration in hot, humid climes, in contrast, posed other types of supply problems. Theodore Roosevelt trekked through the Brazilian wilderness in 1914 and in his account of the expedition identified the components of daily rations for five men that consisted of 26 foods. A representative selection of the Roosevelt expedition foods (weights identified in avoirdupois ounces) is presented here:

Bread	*100 ounces each day*
Corned beef	*70 ounces on Wednesday*

FIGURE 56.7. Shackleton's hut in Antarctica: interior view. Rowntree's chocolate canister. *Source:* Scott Borg. (Used with permission.) (See color insert.)

Sugar	*32 ounces daily*
Baked beans	*25 ounces on Thursday and Saturday*
Figs	*20 ounces on Sunday and Tuesday*
Condensed milk	*17 ounces daily*
Sweet chocolate	*16 ounces each Wednesday*
Coffee	*$10\frac{1}{2}$; ounces daily*
Roast beef	*$6\frac{1}{2}$; ounces on Monday*
Rice	*6 ounces every Sunday, Tuesday, and Friday*
Bacon	*4 ounces every day*
Curry chicken	*$1\frac{3}{4}$ ounces on Friday* [35]

Chocolate also has played important roles in the ascents of Mount Everest and other high peaks on different continents. Edmund Hillary, who in 1953 became the first European to successfully climb Mount Everest together with his Sherpa guide, Tenzing Norgay, once was asked during an interview:

FIGURE 56.8. Scott's hut exterior. *Source:* Scott Borg. (Used with permission.) (See color insert.)

FIGURE 56.9. Scott's hut at Cape Evans, Ross Island, Antarctica, showing Fry's chocolate box and canister. *Source:* Scott Borg. (Used with permission.) (See color insert.)

What did you do when you got to the top? I read somewhere that you ate a chocolate bar? Answer: No. We didn't eat anything on top, but Tenzing buried a little bit of chocolate and some sweets in the snow, which are really a gesture to the gods which the Sherpas believe flit around Everest on all occasions. [36]

Documentation of chocolate use during space flight also would be rewarding for further work, given that both American and Soviet astronauts/cosmonauts included chocolate among their rations. Chocolate pudding, for example, was among the menu items on *Apollo 11* [37], and chocolate bars were served aboard the USSR *Mir* space station in 1988 [38].

MILITARY ACCOUNTS

Chocolate and cacao during the American Revolutionary War have been considered here (see Chapter 31), but additional work on this theme would be rewarding.

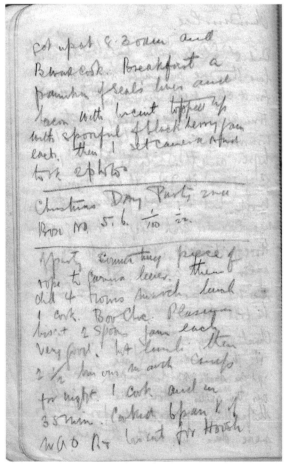

FIGURE 56.10. Shackleton Christmas dinner menu. *Source:* Shackleton Journal, dated December 25, 1902. *Source:* Courtesy of the Polar Research Institute, University of Cambridge, Cambridge, England. (Used with permission.)

Numerous newspaper accounts of the 18th and 19th century, for example, consider the capture of enemy ships (i.e., taking of prizes) and many of these document cacao and chocolate as articles of high value:

Letter from Capt. John Sibbald, Commander of the Wilmington Privateer of this Place, dated on board, the 29th of August 1743, at Cape Nicholas. August the 18th, about 12 Leagues to the Southward of St. Jago de Cuba, we took this Ship coming from Maricabo, loaded with Cocoa, and some Cash on board, which comes home in Capt. Dowel. She had 63 able Men on board, 16 Carriage Guns, and 4 Swivels. The Men I set ashore here. They did not fight. The Captain comes in Capt. Dowel. He is much of a Gentleman, & I desire it as a Favour [sic], that he may be sent to Lisbon by the first Opportunity. The Officers are all in good Health and Spirits, and join in sending their best Respects to you and all the Gentlemen. I am &c. JOHN SIBBALD. [39]

Prior to the outbreak of hostilities during the American Revolution, British and Americans fought against French and Native American troops in what generally are called the French and Indian Wars. The following chocolate-associated document stems from this era and was published in 1747. The letter also documented the "unorthodox" fighting methods of the Native Americans, and how the British/Americans of the time thought such behaviors to be "unmanly." Three decades after this document was written, American revolutionaries, themselves, would adopt "hit-and-run tactics" in contrast to standing and confronting British "battle square" methods:

Extract of a Letter from an Officer at Saraghtoga [sic]. [April 7th] On our part we had 9 Men killed, and 9 wounded, and 6 more missing, but took from the Enemy one Scalp of our own People which they dropt [sic], and 5 kettles, 13 deerskins, 12 blankets, 19 Pair of Indian Shoes, 15 Knives, 3 Looking Glasses, 1 Gun, 1 pistol, 5 lb. of powder, and 4 of Ball, besides several Bags of meal, Bread, Chocolate, paint, &c. which I think is but a small Recompence [sic] for our Loss; but by this you may however see our Diligence, and that whenever the Enemy escapes, it is more owing to their lurking, unmanly Way of fighting, than our Inactivity. [40]

Colonel Ethan Allen with his Green Mountain Boys attacked and captured Fort Ticonderoga, New York, on May 10, 1775. Allen subsequently became a prisoner himself after a failed attack on Montreal. He and others were placed aboard a British prison ship and sent to England where he was incarcerated at Pendennis Castle, Cornwall. Upon his release, Allen described the return voyage to North America and wrote that while aboard the British frigate Solebay, Irish sympathizers for the American cause met the ship at Cork (Ireland) where the Irish made a request to deliver food parcels to the Americans aboard. The British officer-of-the-day agreed in what at the time was a rather nice, civil gesture, and allowed the food distribution as reported in The Pennsylvania Gazette, May 8, 1776:

When Col. Ethan Allen, with about 50 other prisoners, arrived at the Solebay, two gentlemen went on board to enquire into their situation. . . . His treatment on board the Solebay is far different from the barbarous and cruel usage he experienced in his passage from Quebec, being then handcuffed and ironed in the most dreary part of the vessel. . . . A subscription was begun this morning among some friends of the cause, and near fifty guineas collected to buy cloaths [sic] for his men and necessaries for himself. . . . We this day sent a hamper of wine, sugar, fruit, chocolate, &c. on board, for his immediate use. [41]

Images of the American Revolution as a "gentleman's war" is one fostered by Hollywood cinema and fictional literature. Such images seldom portray British soldiers as looters. But as British soldiers evacuated Philadelphia on June 18, 1778, looters took with them vast quantities of "swag" that included chocolate-associated items. Upon reaching the docks and before boarding, some ethical British officers demanded that the stolen goods be left behind on the dock. Imagine the chaos two months later when responsible Philadelphians attempted to match these abandoned goods with their rightful owners:

STOLEN Out of a store whilst the British troops were in this city, and supposed to be sold at the City Vendue [public auction site] THE following Goods . . . Sundry iron pots and kettles; a steel chocolate mill, and large chocolate roaster; several dozens of reedles [?] and reying [?] sieves, iron bound half bushels, and sundry other articles. Any persons who may have purchased the above articles are requested to inform the Printer hereof [i.e., Benjamin Franklin], otherwise if they should be found in their possession by any other means, they may expect, as far as the law directs, to be treated as purchasers of stolen goods. [42]

Other potential areas for further detailed study include the role of cacao/chocolate as a military ration assigned to soldiers after the Revolution but before the American Civil War, and during the post-Civil War era period of attempted Native American pacification. One identified document provided a list of recommendations for new Federal military recruits and cautioned them to drink chocolate over other beverages:

IMPORTANT DIRECTIONS TO VOLUNTEERS:

[Recommendation 2] Avoid drinking freely of very cold water, especially when hot or fatigued or directly after meals. Water quenches thirst better when not very cold and sipped in moderate quantities slowly—though less agreeable. At meals tea, coffee and chocolate are best. Between meals, the less the better. The safest in hot weather is molasses and water with ginger or small beer. [43]

A search of military records associated with activities and campaigns in the central states—especially documentation of possible chocolate use by Custer's 7th Calvary troops—would be interesting.

Interviews with former Civil War veterans were conducted during the American Memory Project instituted by President Roosevelt. Many of these interviews reveal that food and chocolate commonly were among the strongest memories of battle. Such an account by John H. Robertson, a Confederate veteran, is presented here:

> I was captured at the battle of Gettysburg in Longstreet's charge and was taken to Fort Delaware, an island of 90 acres of land where the Union prisoners were kept. We were detailed to work in the fields and our rations was [sic] corn bread and pickled beef. However, I fared better than some of the prisoners for I was given the privilege of making jewelry for the use of the Union soldiers. I made rings from the buttons from their overcoats and when they were polished the brass made very nice looking rings. These I sold to the soldiers of the Union Army who were our guards and with the money thus obtained I could buy food and clothing. The Union guards kept a commissary and they had a big supply of chocolate. I ate chocolate candy and drank hot chocolate in place of coffee until I have never wanted any chocolate since. [44]

An unusual document published in 1881 recommended to the British Admiralty that sailors' daily ration of grog be halted and a daily ration of chocolate substituted. The author of the recommendation was so enthusiastic that a recipe was provided. The document argued further that chocolate was superior to both coffee and tea because it would lessen the possibility of fire aboard ship:

> The corn should be first parched, and then ground very fine, and mixed with as much brown sugar as will make it agreeable—the usual mode here is, 1 lb. Of sugar to the gallon measure, but many put more. A table spoonful of this mixture in a pint of cold water, is much better than a pint of coffee or chocolate, and would furnish an excellent beverage in stormy weather, when it is difficult to make coffee or chocolate. [45]

Another rewarding effort would be summarizing accounts from diaries and interviews of World War II prison camp survivors. Several examples we identified reveal the breadth and depth of chocolate-associated content. The example presented here is the testimony from British airman John Bremner, shot down over Germany, then captured and taken to Titmoning (in southern Bavaria). After the war he recounted the provisions that the Germans gave him and other prisoners:

> A quarter of a pint of very watery soup for lunch and 1 loaf of "black bread" [around] 5:00 in the evening, and large amounts of ersatz [substitute/fake] coffee. . . . [A year after being captured the British Red Cross parcels and] each parcel contained 2 ounces of tea, $1\frac{1}{2}$ pound of sugar, sweets, chocolate, and condensed milk. . . . [A year later Canadian Red Cross parcels arrived] contained 1 pound of butter, $\frac{1}{2}$ pound of coffee, 1 pound of sugar, a large bar of chocolate, sweets, one tin of powdered milk, a large packet of biscuits, and 50 cigarettes. [46]

World War II Red Cross food parcels for Allied prisoners of war actually contained the following items:

> Tea, cocoa, sugar, chocolate, oatmeal, biscuits, sardines, dried fruit, condensed milk, jam, corned beef, margarine, cigarettes/ tobacco, and soap. [47]

RUSSIAN OUTPOSTS IN NORTH AMERICA

We identified the presence and use of chocolate aboard Russian ships that visited the coast of North America during the late 18th and early 19th centuries [48, 49]. Documents from Fort Ross (an English language corruption of the words Fort Russia) in north coastal California, however, were not located and inspected. Colleagues familiar with these documents have suggested to us that chocolate was not among the foods imported to the fort.[1] Our colleagues also mention that the Russians at Fort Ross traded with Mexican suppliers, logically via San Francisco. Hence, further work on Russian–Mexican trade relations in early 19th century California could be rewarding. We offer here, too, the supposition that if chocolate were served at Fort Ross, the beverage could have been in china cups. Examination of the Fort Ross excavation documents and hand inspection of retrieved artifacts especially for "two-eared" (i.e., two-handled) cups commonly used to serve chocolate should be a priority (see Chapter 34).

ADVERTISEMENTS, PRICE CURRENTS, AND SHIPPING NEWS DOCUMENTS

During the past 10 years the Readex Company (Naples, Florida) has digitized and placed on the Web for research use collections of Colonial era newspapers with search-engine capabilities [50]. The volume of chocolate-related information in this rich collection of newspapers exceeds several hundred thousand "hits" and full inspection and documentation of this archive awaits teams of future researchers. While we focused our attention on Colonial era port cities, Boston, Charleston, Philadelphia, and Savannah, much work remains to be completed (see Chapters 27, 30, and 54). During our work we identified more than 900 merchants in Boston associated with the chocolate trade, whether as auctioneers, brokers, manufacturers, or vendors (see Appendix 6). Substantial additional information awaits discovery by searching digitized accounts via Readex from Baltimore, Halifax, Hartford, New York, Philadelphia, Savannah, and elsewhere.

Colonial and Federal era newspapers published weekly or monthly acceptable prices or *Price Currents* for basic commodities, a process that stabilized local economies and alerted consumers to potential price gouging (see Chapter 18). Systematic recording of all available price currents and analysis by port city, season of the year, and historical era through five time periods (i.e., before the American Revolution; during the Revolution; between the Revolution and War of 1812; events during 1812 and the first three months of 1813; and post-War of 1812 to the eve of the American Civil War) would be fruitful. Such an analysis may allow identification of economic trends in 17th–19th century cocoa and chocolate prices that reflect transport costs (e.g., cheaper in Savannah; more expensive in Boston; most expensive at Halifax); summer month hurricane season; and various cultural, economic, and political factors that might correlate with changing values for chocolate of different grades and qualities.

Other documents published in Colonial and Federal era newspapers include port arrival and departure information or *Shipping News* (see Chapter 17). Such documents list the type and name of ship, captain/master, cargo, and ports of debarkation and embarkation. Further research with *Shipping News* documents, coupled with sophisticated cluster analysis for key ships, ports, personnel, and associated cargo, could yield important information whereby specific cacao-carrying ships could be tracked seasonally and by brokers/merchants, and quantities of cacao and chocolate imported to East Coast ports might be quantified.

SOCIAL CUSTOMS/TRADITIONS

Chocolate can be perceived as a "social glue" associated with numerous customs and traditions from remote antiquity to the 21st century. Consider how the continued addition of *achiote* to chocolate in subsequent decades after colonization of Mexico by the Spanish maintained the deep red-brown color of chocolate beverages (but without human blood) [51]; consider the numerous chocolate sculpture contests held frequently in the Americas and Europe and aboard cruise liners plying the west coast of California and Mexico (personal observations); consider the chocolate skulls, death masks, and coffins—all in chocolate—used to commemorate Day of the Dead celebrations in Mexico and North America [52]. During the course of our chocolate research we identified more than 30 different cultural themes in North America, including chocolate-associated contemporary music, games/toys, and bath and body ointments, to the thousands of chocolate-related collectibles available through eBay, Google, and other on-line auction houses and private antique shops (Figs. 56.11–56.13). Further work in this area clearly is warranted (see Table 56.2).

TRANSPORTATION

Before airplanes, automobiles, trains, and transcontinental telegraph and telephones, the west and east coasts of America were linked by a hardy group of horseback riders collectively known as the Pony

FIGURE 56.11. Chocolate mold in the form of fish, dated ca. late 19th century. *Source:* Courtesy of the Grivetti Family Trust. (Used with permission.)

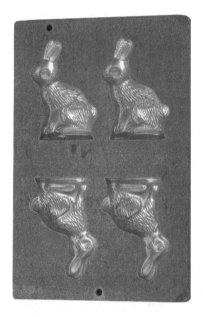

FIGURE 56.12. Chocolate mold in the form of rabbits, dated ca. late 19th century. *Source:* Courtesy of the Grivetti Family Trust. (Used with permission.)

FIGURE 56.13. Baker's chocolate box, dated late 19th century. *Source:* Courtesy of the Grivetti Family Trust. (Used with permission.)

Table 56.2 Chocolate: Cultural Themes

Antiques	Humor
Art	Literature
Clothing	Longevity
Contests	Manners
Crazes	Music
Customs	Needlepoint
Dance	Personal products
Death/executions	Play/recreation
Decorum	Poetry
Dress	Rages
Fads	Religion
Fashions	Sexuality
Festivals	Social traditions
Film/cinema	Spaceflight
Folklore	Style
Gender issues	Tourism
Geography	Vogues
Glamour	World records
Health	Youthfulness

FIGURE 56.14. H.M.S. *Titanic:* chocolate service (White Star Line), dated 1912. *Source:* Courtesy of Brian Hawley at www.luxurylinerrow.com. (Used with permission.) (See color insert.)

Express. Riders began their journey in Sacramento, California, with the terminus at St. Joseph, Missouri. The Pony Express lasted approximately 19 months (early April 1860 through the end of October 1861) before it became outdated by completion of the transcontinental telegraph [53]. Riding day and night throughout the year, teams of horsemen experienced danger and fatigue, and it would have been logical for them to have drunk chocolate along the route.

Several of our project members searched Pony Express records in California and in Missouri. We were puzzled, however, to find no mention of chocolate in rider diaries or Pony Express records, whether in Sacramento, California, or in St. Joseph, Missouri, given that the stimulative properties of cacao were well known by the mid-19th century era of the Pony Express—information that would have benefited tired riders. Chunks of chocolate—already available—could have been carried easily. Despite our initial negative finding, we suggest that Pony Express accounts be searched more diligently within state historical depositories along the route that linked California with Missouri, as chocolate-related information may be present in some more obscure accounts.

Regarding other transportation-related themes, we also suggest that a systematic investigation be undertaken to search for cacao/chocolate-related associations with early airplane, dirigible/zeppelin, railroad, and stagecoach menus. Our brief examination of this topic—especially review of transoceanic dinner menus—revealed interesting examples such as serving chocolate aboard the *Titanic* (Fig. 56.14) and we are certain other information awaits discovery.

Aboard the *Hindenburg* zeppelin—prior to its destructive fire and crash—the items listed in Figure 56.15 were part of the general food service [54].

As briefly mentioned earlier, a controversy exists regarding whether or not Charles Lindbergh drank chocolate in 1927 while piloting his airplane, *Sprit of St. Louis*, from North America to Le Bourget Airport, outside Paris. A United Press International press release dated May 21, 1927, written by Ralph Heinzen, provided the following details:

Lindbergh had been flying the great circle route alone in the cabin of his machine, with sandwiches and hot chocolate, and two bottles of water to sustain his strength. [55]

Lindbergh, however, writing in his diary and ultimately published as *The Spirit of St. Louis*, makes no mention of chocolate, only that he carried with him "meat sandwiches" and "two canteens of water" [56].

Amelia Earhart was the first person to fly solo across the Pacific from Honolulu, Hawaii, to Oakland, California, a distance of more than 2400 miles. During her memorable flight (January 11, 1935), Earhart drank hot chocolate from a thermos and then, after arriving, sat for an interview and provided the following information:

That was the most interesting cup of chocolate I have ever had, sitting up eight thousand feet over the middle of the Pacific Ocean, quite alone. [57]

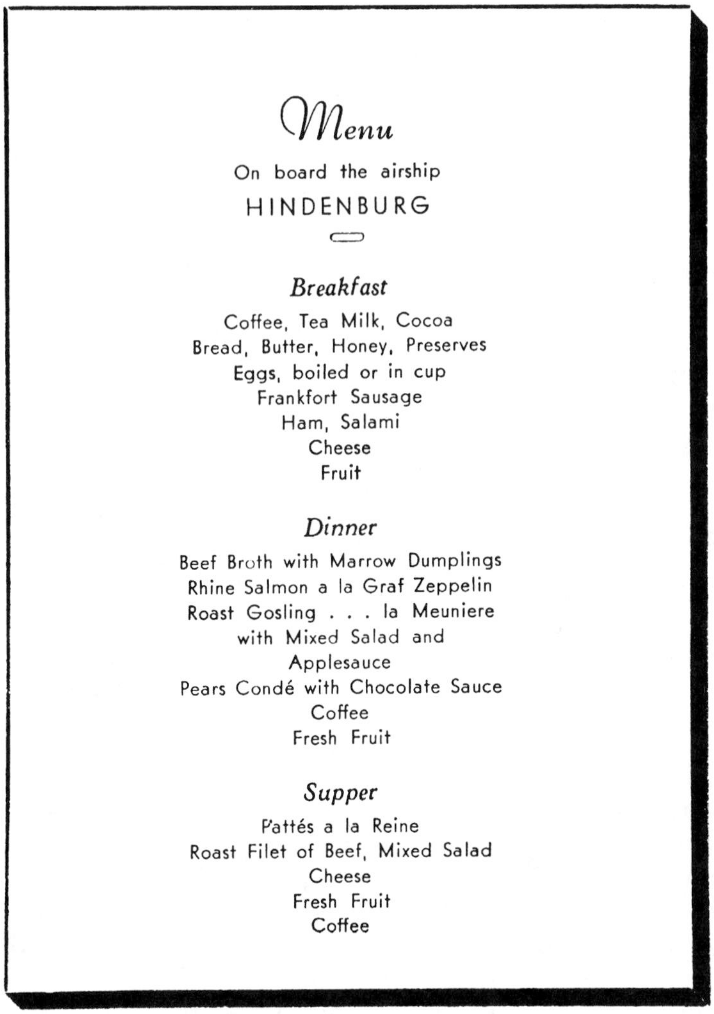

Menu

On board the airship
HINDENBURG

⊂⊃

Breakfast

Coffee, Tea Milk, Cocoa
Bread, Butter, Honey, Preserves
Eggs, boiled or in cup
Frankfort Sausage
Ham, Salami
Cheese
Fruit

Dinner

Beef Broth with Marrow Dumplings
Rhine Salmon a la Graf Zeppelin
Roast Gosling . . . la Meuniere
with Mixed Salad and
Applesauce
Pears Condé with Chocolate Sauce
Coffee
Fresh Fruit

Supper

Pattés a la Reine
Roast Filet of Beef, Mixed Salad
Cheese
Fresh Fruit
Coffee

FIGURE 56.15. *Hindenburg* airship menu; brochure prepared and edited by Luftschiffbau Zeppelin GmbH Friedrichshafen, dated 1937. *Source:* Aeronautical And Space Museum, Smithsonian Institution, Washington DC. Courtesy of Luftschiffbau Zeppelin GmbH. (Used with permission.)

Douglas Groce Corrigan, subsequently known as "Wrong-Way Corrigan," had requested permission several times from New York "authorities" to fly solo from New York to Dublin, Ireland. His requests, however, always had been denied. On July 17, 1938, however, Corrigan received permission to fly nonstop from New York to California and took advantage of this opportunity to make aviation history. Corrigan departed Floyd Bennett Field in Brooklyn but contrary to his presumed flight plan, flew east—not west—and arrived in Dublin 28 hours later, whereupon he became an instant national hero. During his flight, Corrigan reported that he carried two chocolate bars, two boxes of fig bars, and a quart of water [58].

A 1933 trans-Saharan aviation flight to break the England–Cape Town speed record ended in tragedy when William Lancaster crashed. His mummified body was discovered 29 years later on February 12, 1962. Before perishing Lancaster wrote that he had suffered deep lacerations from the crash, felt weak from loss of blood, had two gallons of water, several chicken sandwiches, a thermos of coffee, and a bar of chocolate. While alert after his crash, he jotted notes and calcu-

lated that with these items on hand he could survive for seven days. But he died of thirst when not rescued [59].

While the examples cited here are dated to the 19th or 20th century, there remains an important transportation aspect associated with cacao that needs further, in-depth examination. It is not readily recognized that once the cacao pod is harvested and the beans extracted, the seeds have a very narrow window of opportunity during which they remain viable: if cacao beans are planted after 10 days, the beans are no longer viable. This physiological characteristic raises several important geographical, transportation issues, since it means that transoceanic transport of cacao between the Americas and Africa or between the Americas and Asia required that cacao be transported as seedlings. To this end, we have continued to work on documenting both transatlantic and transpacific shipment of cacao trees. While we have located several critical documents, more effort by international colleagues on this topic would be worthwhile.

Other related topics include the transatlantic and transpacific trade in chocolate-related items. One document we located mentioned transpacific trade in chocolate cups that probably originated in China, via the Manila galleon. Father Francisco Pangua, Apostolic Guardian of the College of San Fernando in Mexico City, described a trip to the Far East made by Father Pedro Bonito Gambon who sailed to Manila in 1780 as the royal chaplain on the frigate *La Princesa*. While in Manila, Father Gambon obtained a large number of items for transshipment to California, including items from China. Gambon returned aboard *The San Carlos*, alias *The Philippino* (i.e., one of the Manila galleons) and arrived in San Diego (original California destination was Monterey) with a consignment of 18 large cases, of which case number 13 contained the following:

> *2 rolls of cinnamon; a parcel of tim sin [?]; 4 rolls of woven Manila hemp for sifting flour; 3 small red boxes of tea; 2 papers of ink; 154 red chocolate cups; 82 blue chocolate cups; 129 small plates; 4 black paper parasols.* [60]

A second set of documents dated to 1802–1803 report that the ship *Nuestra Señora de Guía*, alias *La Casualidad*, arrived in Acapulco from Manila. Upon arrival, the ship was greeted by merchants asking permission to load cacao from Guayaquil (Ecuador) for the return trip to Manila. The amount requested to be loaded was 100 *tercios* (1 *tercio* = 220 pounds for a total of 22,000 pounds). The request was approved by the Royal Aduana (head of the Customs Department for Mexico). In addition to the 22,000 pounds of cacao sent to Manila, other goods sent west aboard *La Casualidad* included:

> *80 loads of chickpeas; 50 boxes of candy; 30 loads of beans; 12 loads of lentils; 8 boxes of books; 300 quintales of copper plate [1 quintale = 100 kilograms]; 2 boxes of feathers; 90 cow hides; 6 "dressed" sheep-skin; 19 boxes of soap; 4 boxes with long and thick wax candles [for religious purposes]; 10 barrels of wine; 20 boxes of hams. Signed in Mexico, December 16th, 1802 by Domingo Goyenochea.*

The total value of the cacao sent to Manila on this ship was 8625 pesos, 4 reales [61].

We expect that many more such documents can be found—especially documents for the transportation of cacao tree seedlings between the Americas and Africa, and the Americas and the Far East (Fig. 56.16).

CHOCOLATE AND FEMININE BEAUTY

At first glance, chocolate and feminine beauty would not be terms readily associated. But this is not the case. Chocolate as a recommended item associated with feminine beauty appeared at least by 1830 in North America with the introduction of "beauty rules" that were published for general public distribution:

> *Rules for a Young Lady: Her breakfast should be something more substantial than a cup of slops, whether denominated tea or coffee, and a thin slice of bread and butter. She should take*

> *a soft boiled egg or two, a little cold meat, a draught of milk, or a cup or two of pure chocolate.* [62]

This initial article apparently struck a nerve among readers and circulated widely within North America. A second set of rules, expanding on the first, were published in 1831. But in this case the anonymous author cautioned against drinking chocolate:

> *The secret of preserving beauty lies in three things— temperance, exercise, and cleanliness. . . . I do not mean feasting like a glutton, or drinking to intoxication. My objection is more against the quantity than the quality of the dishes which constitute the usual repasts of women of fashion. Their breakfasts not only set forth tea and coffee, but chocolate, and hot bread and butter. Both of these latter articles, when taken constantly, are hostile to health and female delicacy. The heated grease, which is their principal ingredient, deranges the stomach; and, by creating or increasing bilious disorders, gradually overspreads the fair skin with a wan or yellow hue.* [63]

While we did not research this theme systematically, the several interesting documents identified here suggest further work on this topic would be warranted.

SHIPWRECK SURVIVORS AND CHOCOLATE ASSOCIATIONS

During the course of archival research, we encountered both factual and fictional accounts of shipwreck survivors and their use of chocolate (or chocolate-related equipment). We include three examples here because they represent unusual literary cases and a topic not commonly associated with chocolate.

The case of Jonathan Dickenson, shipwrecked survivor saved from the "cruelly devouring jawes [sic] of the inhumane cannibals of Florida," is an interesting account not only for its content, but also for its documentation of chocolate aboard ship in North American waters at the end of the 17th century (1699):

> *Our boat was very Leaky, so we gott [sic] her into a Creek to sink her, that the water might swell her. This morning we waited an Opportunity to get leave to depart, which was granted us: Whereupon we asked for such things as they did not make use of, viz. A great Glass, wherein was five or six pound [sic] of butter; some Sugar; the Rundlett of wine [small cask of undefined volume]: and some Balls of Chocolate: All which was granted us; also a Bowle [sic] to heave Water out of the Boat.* [64]

The fictional case of Robinson Crusoe by Daniel Defoe is a literary classic, although less read today than during the 19th and 20th centuries. Still, Defoe's account of perseverance and survival has lessons for readers even in our complicated 21st century:

> *The sea was now very calm, which tempted me to venture to the wreck, not only to get something I wanted, but likewise if there were any body left alive in the ship, to endeavour [sic]*

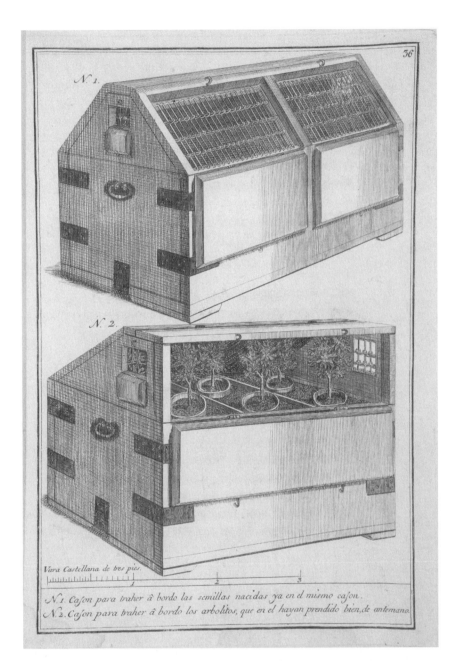

FIGURE 56.16.

Transportation of cacao tree seedlings aboard ship, dated 1779. *Source:* Avis Pour le transport par mer des arbres, des plantes vivaces. Courtesy of Real Jardin Botanico (Royal Botanical Garden), Madrid, Spain. (Used with permission.) (See color insert.)

to save their lives. . . . As I was rummaging about her, I found several things I wanted, viz. a fire shovel and tongs, two brass kettles, a pot to make chocolate, some horns of fine glazed powder, a grid-iron, and several other necessaries. These I put on board my boat, together with two chests, and a cask of rum, and after a great deal of toil and difficulty, got safe back to the island. [Daniel Defoe, 1789. The life and most surprizing [sic] adventures of Robinson Crusoe of York, mariner. Containing a full and particular account of how he lived twenty-eight years in an uninhabited island on the coast of America]. [65]

English, Spanish, and Russian geopolitical competition in the Nootka Sound region of North America in the early 19th century was introduced earlier in this volume (see Chapter 34). A shipwreck account by John Jewett, dated to 1824, described real life adventure in

this region of the northwest Pacific region, with challenges as difficult—or more so—as those experienced by his fictional counterpart, Robinson Crusoe:

[Burning ship abandoned] . . . To my companion and myself it was a most melancholy sight; for with her disappeared from our eyes every trace of a civilized country: but the disappointment we experienced was still more severely felt; for we had calculated on having the provisions to ourselves, which would have furnished us with a stock for years, as whatever is cured with salt, together with most of our other articles of food, are never eaten by these people. I had luckily saved all my tools, excepting the anvil and the bellows, which was attached to the forge, and, from their weight, had not been brought on shore. We had also the good fortune, in looking over what had been taken from the ship, to discover a box of chocolate and a case of Port wine, which, as the Indians were

not fond of it, proved a great comfort to us for some time. . . . About two days after, on examining their booty, the savages found a tierce of rum, with which they were highly delighted, as they have become very fond of spirituous liquors since their intercourse with the whites. [66]

Despite these real or fictional examples of shipwrecks, ocean transportation continued to rise in the decades that followed. There came a time when shipboard etiquette required standardized procedural manuals be developed that dictated behavior when abandoning ship. *The Kedge Anchor, or Young Sailors' Assistant,* published in 1852 and authored by William Brady, was one such manual. Section 387 of this text, entitled Taking to the Boats, itemized responsibilities of ship captains and crew and what items should be carried before abandoning ship and entering the life boats:

Captain: Compass, Maury on Navigation, sextant, spyglass, Nautical Almanac, pencils and writing paper, general chart, pocket watch, pair of compasses.

First Mate: Oars, masts, sails, boat-hooks, bolt of canvass, boat's compass, Bowditch's chart, ensign.

Second Mate: Two or three bags of biscuits, some beakers of water, quadrant, pencils and writing paper, half-gill measure, musket, box of cartridges, and flints or caps.

Cook / Steward: Tinder-box, flints and tinder, small box, lantern and candles, cheese, cabin biscuit, chocolate. [67]

CHOCOLATE AND MANUFACTURING

The history and development of chocolate manufacturing in North America is reviewed in the present anthology (see Chapters 23, 26, and 46). Still, a number of issues associated with chocolate manufacturing and production could be further explored. One, for example, is the potential fire danger posed by the presence of a chocolate mill or factory within a densely populated area. As mills became larger, the risk for fire increased and more than one North American city passed ordinances restricting the location of chocolate factories as seen in this representative document dated to 1786:

An Act to prevent Damage from Fire being communicated from Chocolate Mills and Machines for roasting Cocoa in the Town of Boston. . . . Whereas Chocolate Mills and machines for roasting Cocoa have been erected in the Town of Boston, near to other buildings, to the great hazard of the lives and property of the Inhabitants of the said Town. Be it enacted . . . That any person, from and after the fifth day of July next, shall, within the said Town, roast or cause to be roasted any cocoa, for the purpose of manufacturing the same into chocolate, in any building whatever exception such as may or shall be licenced [sic] for that purpose . . . Shall forfeit and pay, for every such offence, a sum not exceeding—one hundred nor less than fifty pounds. [68]

The second theme that warrants further investigation is identification of documents that show how the quality of chocolate improved through time. The following two examples are representative of many published to inform the public how to differentiate between good quality and adulterated chocolates:

The goodness of chocolate depends first, upon the quality of the cocoa. Of this there are three principal species: caracas, quayaquil, and that from the islands of St. Domingo, Martinique, Curracoa, etc. To make the chocolate the caracas is mixed with the quayaquil; two parts of the caracas and one of the quayaquil. . . . The goodness of chocolate depends, in the second place, on the care with which it is ground and roasted, on the proper proportion of the cocoa, the sugar, and the different aromatics. . . . The characteristics of a good, unadulterated chocolate, are the following: a deep fresh colour; a fine, close, shining grain; small white streaks; an aromatic odour; a facility of dissolving in the mouth, with a sensation of freshness, to produce no appearance of glue in cooling, and to shew [sic] an olily [sic] cream on the top. [69]

In preparing chocolate the cacao beans are roasted in a cylinder similar to those employed for roasting coffee. In this operation the aroma is developed, the bitterness diminished, and the beans are rendered fragile. They are broken under a wooden roller, and winnowed to remove the husk entirely. They may then be reduced to a soft paste in a machine consisting of an annular trough of granite, in which two speroidal [sic] granite millstones are turned by machinery, with knives attached to return the ingredients under the rubbing surface. An equal weight of sugar is here added to the paste, which is finally rendered quite smooth by being ground under horizontal rollers on a plate of iron, heated to about 140 °F. [70]

CHOCOLATE AND UNIVERSITY STUDENTS

Chocolate as part of college and university food settings has a long history. The first university founded in North America was Harvard, established in 1635. Beef, beer, and bread characterized the dining patterns of these early Colonial era students [71]. Other universities—William and Mary, Princeton, Yale—would follow. Food patterns improved during the 18th century, although students still complained. Early in our research, we identified chocolate as a component of student dining foods. We present two examples here, one from Princeton and one from Harvard, and comment that it would be worthwhile to pursue this topic further, even into the 21st century.

Princeton student Peter Elmendorf maintained a diary and recorded his impression of foods offered in the dining commons:

We eat rye bread, half dough, and as black as it possibly can be, and oniony [sic] butter, and some times dry bread and thick coffee for breakfast, a little milk or cyder [sic] and bread, and sometimes meagre [sic] chocolate for supper, very indifferent dinners, such as lean tough boiled fresh beef with dry potatoes. [72]

The Laws of Harvard College, published in 1777 during the Revolutionary War era, included Law Number XXXI that identified specific foods to be supplied to students, among them chocolate:

> Whereas by law 9th of Chapt.VI. It is provided, "that there shall always be chocolate, tea, coffee, and milk for breakfast, with bread and biscuit and butter," and whereas the foreign articles above mentioned are now not to be procured without difficulty, and at a very exorbitant price; therefore, that the charge of commons may be kept as low as possible, Voted, That the Steward shall provide at the common charge only bread or biscuit and milk for breakfast; and, if any of the scholars choose tea, coffee, or chocolate for breakfast, they shall procure those articles for themselves, and likewise the sugar and butter to be used with them. [73]

GENDER AND MINORITIES

Some of the early 16th century documents that we reviewed provide insights on attitudes toward women and native peoples. As we did not conduct a systematic search, we suggest further research on this topic would be highly worthwhile. Presented here for consideration are four documents that provide an astonishing amount of information. The first document dated to 1591 grants permission to Indians allowing them to sell cocoa on the streets of Mexico City:

> In Mexico, August 3rd of the year 1591, a permission is granted to Gabriwl Xuarez and his wife Mariana and other "naturals" [i.e., natives] of this city in order to allow them to sell chocolate, firewood, and [illegible] of cacao and fruits. [74]

Approximately 50 years later, in 1645, the king of Spain issued a royal decree that rescinded all such vending permissions. The document translated here lists five "types" of cacao street peddlers—and one of these types was identified as Chinese! The impact of this royal decree was that sale of "street" cacao was placed under the authority of the king of Spain within all his dominions:

> And such people that are Spanish, Mestizo, Mulatto, Black and Chinese are selling [cocoa and chocolate] on the streets, convents and private. . . . Those vendors that sell in small amounts, the peddlers or hawkers [buhoneros], are clandestine, furtive and they hide, and they sell it [cocoa and chocolate] very cheap, deceiving the neighbors. They have to be punished and the neighbors be warned . . . the most important thing that has to be taken into account in matters of supplying a town with provisions and particular those provisions that are a primary food is that the supply has to be comparable. The frauds and deceits that malicious people are seeking are well known and causes detriment to the safety of the merchandises . . . and they commit a criminal act, a delinquency, selling "old" cocoa [cacao añejo], of an inferior quality and noxious

to the people's health, selling what is bad for good. . . . Therefore it has been prohibited to sell such cocoa because it is to pestilent and because is too harmful that has been the cause of death. . . . It is then ordained and enacted by the Royal Law that the peddlers cannot wander on the streets of the cities, on the towns, villages, and other places of the Kingdom of Her Majesty and selling to the private houses the merchandise they carry in their peddler. [75]

In 1695, a vendor controversy developed when an adult male (unnamed) challenged the right of a woman, Maria de Iribarne, to maintain a vendor's stand adjacent to a road close to the El Camino Real (the "King's Road" in Mexico) in the town of Aztapalapa, Mexico. The petition argued that she be removed and be ordered to stop selling food for travelers:

> Maria has built four walls of adobe and she sells chocolate, bread, brown sugar candy, and milk custard. [76]

Maria countered the accusation with a letter directed to the Justice of Peace of Mexico (the Alcade) that outlined her defense. She requested of the Alcade that she be permitted to continue to sell: "because I need to feed my children and do not have anything else" [77]. Our search through the Mexican National Archives did not reveal the outcome of this dispute.

Turning to North America, examination of Colonial and early Federal era newspaper advertisements reveals names of a large number of adult women associated with chocolate manufacturing or serving as proprietors of chocolate shops. Sometimes the document reveals the woman was a widow; other times the woman was a close family relation. In such cases, researchers interested in gender issues and how women fared in 18th and 19th century North America would find ample evidence that women played important roles in the chocolate trade. Two examples from the pre-Revolutionary War era are presented here:

> To be sold by ELEANOR DEXTER, Living in White horse alley, near the Presbyterian Meeting house, RUM, sugar, molasses, tea, coffee, chocolate, rice, indigo, allum [sic], copperas, brimstone, rosin, all sorts of spices, cinnamon, aniseed, clove, tansey, snake root waters, and sundry other cordials, brandy, Stoughton elixir, spirits of calamus [i.e., from Acorus calamus], Hungary water; coarse and fine salt; all sorts of earthen ware, glasses and china. [78]

> MARY CRATHORN, Begs leave to inform the public (and particularly those that were her late husband customers) that she has removed from the house she lately occupied in Laetitia Court, to the house lately occupied by Mrs. Aris, at the corner of the said court, in Market street, where she continues to sell by wholesale and retail, THE genuine FLOUR of MUSTARD, of different degrees of fineness; chocolate, well manufactured, and genuine raw and ground coffee, tea, race [mace?] and

ground ginger, whole and ground pepper, alspice [sic], London fig blue, oat groats, oatmeal, barley, rice, corks; a fresh assortment of spices, domestic pickles, London loaf sugar, by the loaf or hundred weight, Muscovado sugars, choice raisins by the keg or less quantity, best thin shell almonds, olives and capers, with sundry other articles. [79]

POLITICS

In the skeptical political climate of 21st century America, it was refreshing to discover that citizens of an independent America sometimes celebrated the Fourth of July with jubilation and events that sometimes began at 6:00 a.m. with a full day of musical performances and singing, food, and revelry. Would Americans in 2009 (and subsequently) arise at dawn just to celebrate the birth of freedom in the United States? This document dated to 1791 reveals how proud these early Americans were of their newly won freedom:

> *[To be held at] GRAYS GARDENS. A CONCERT of Vocal and Instrumental Music will begin on Monday, the glorious 4th of July, at six o'clock in the morning, and conclude at ten at night, should the day be fair, to celebrate American Independence. Songs, with harmony and martial music, in honor of the day, will be performed. In order to furnish the public with refreshments, tea, coffee and chocolate, and fruits of the season, will be ready for breakfast. Three tables, with 100 covers each, furnished with roast beef, rounds, hams, &c. &c. ready to cut-and-come-again, from morning until night.* [80]

Coronations of kings and the installations of presidents are political events that have been cause for celebrations through the centuries. Commonly associated with pomp and circumstance, coronation and inaugural banquets typically exhibited foods in enormous varieties and quantities. A systematic investigation of the role of chocolate at such events would be worthwhile. Two such examples are presented here to show richness of content: the coronation of King Louis XV and the inaugural banquet hosted by President Abraham Lincoln when he was elected for his second term of office.

> *[Dinner hosted by the Duke of Orleans after the coronation of King Louis 15th, August, 1810] . . . 29,045 heads of game and poultry; 36,464 eggs; 80,000 bottles of Burgundy and Champain [sic]; 1400 bottles of English beer and cider; 8000 pounds of sugar; 2000 pounds of coffee, besides tea; 1500 pounds of chocolate; 65,000 lemons and oranges (sweet and sour); 800 pomegranates; 150,000 apples and pears; 30,000 China plates and dishes for dessert; 115,00 decanters and glasses; 3300 table cloths; 900 dozen of napkins; and 2000 dozen of aprons were used by the cooks and others.* [81]

President Abraham Lincoln's second inauguration in 1865 was a highly anticipated event in

FIGURE 56.17. Menu: Abraham Lincoln second inaugural ball and dinner. *Source:* Personal Collection of Louis Szathmary. Courtesy of Johnson and Whales University, Culinary Museum. (Used with permission.)

Washington, DC. In less than a month, the American Civil War would be over. Festivities to commemorate the inauguration included preparations for a ball with a formal meal. The menu for this event is presented here (Fig. 56.17). An expanded list of the various foods served was published as a front-page article by the *New York Times* and contained numerous chocolate dishes. Imagine the gaiety and festivities: Who would have imagined that a mere six weeks later President Lincoln would be assassinated?

> *[Bill of Fare: Inaugural Ball of President Abraham Lincoln]*
>
> *Oyster stew, terrapin stew, oysters pickled; roast beef; fillet de beef; beef al la mode; turkey; roast chicken; boned grouse; roast grouse; pheasant; quail; venison; patete of duck en gelee [sic]; patete de foie gras [sic]; smoked ham; tongue plain; salads; lobsters; ornamental pyramids—caramel with fancy cream*

candy, cocoanut [sic], macaroon, chocolate; pound cake; sponge cake; ice cream (vanilla, lemon, coffee, chocolate, burnt almond), maraschino fruit ices, strawberry, orange, and lemon ices; for dessert—grapes, almonds, raisins, &c, coffee and chocolate. [82]

ETHICS, SLAVERY, AND INDENTURED SERVITUDE

Numerous accounts identified during the course of our research revealed use of slaves and indentured servants in chocolate manufacturing during the Colonial and early Federal period, up to the eve of the American Civil War. The documents represent several types of information. Commonly, the documents appear as newspaper advertisements that announced rewards for runaway slaves or servants. Descriptions in the texts mentioned that the men (or women) had specific chocolate-associated training. A second theme for further investigation is the relocation of free blacks (coloreds as the term was used in the mid-19th century) or the relocation of slaves to geographical locations outside North America, specifically, to western Africa, where they would become involved in the foundation of the Republic of Liberia, or to Nicaragua, where southern slave-holders hoped to develop plantations there. A study of chocolate and slavery as it existed in North America would be enlightening and might show whether or not the chocolate beverages drunk by the Founding Fathers, in fact, were prepared by slaves:

[1749] To be SOLD, A Likely young Negroe [sic] man, by trade a painter, and fit to wait upon a gentleman. Enquire of Thomas White, chocolate maker, opposite to Norris's Alley, in Front street. [83]

[1750] TO BE SOLD, Two likely Negro men, one understands country business, and is a very good mason; the other is a good chocolate grinder. Enquire of Thomas Hart, over against the church, in Second street, Philadelphia. [84]

[1763] Ran away from his Master. Daniel Sharley, of Boston, on Monday last, a Negro Fellow about 35 Years of Age, named Sam, has a large Bump over his right Eye. . . . He has been used to grind Chocolate, and carry it about for Sale. Whoever takes up said Negro, and will bring him to his Master, shall have TWO DOLLARS Reward, and all necessary Charges paid by DANIEL SHARLEY. [85]

[1772] RUN away from the Subscriber, living at Litchfield, in Connecticut, on the 9ᵗʰ of May, last, a Negro Man, named Guinea, a short thick Fellow, about 40 Years of Age; he pretends to be free, is a Clothier by Trade, and sometimes a Pedler [sic] of Chocolate, Gingerbread, Indigo, and Sleeve-buttons. Whoever takes up said Fellow, and returns him to me, shall have Fifteen Dollars Reward, and reasonable Charges, paid by HEZEKIAH ORTON. [86]

[1788] THREE DOLLARS REWARD. RUN AWAY [sic] from the subscriber, on Thursday, the first of May last, an indented [sic] servant BOY, sixteen years old, born in the Highlands of Scotland, named DUNCAN CHESHOLME, but has changed it; is very short and thick set, carries his head very erect, fair and florid countenance, but now may be tanned; and when laughs his cheeks and chin have dimples; he speaks very good English, and calls himself an Englishman, writes and ciphers, used to chocolate making and attending a grocery store, four years of his servitude unexpired; he is a cunning, artful fellow, and is capable of deceiving any one, having defrauded his master considerably . . . MOSES GOMEZ, Jr. At No. 5, Duke-Street, New-York. N. B. It is reported that he is bursted [sic] by being thrown from a horse. All persons are forbid harbouring [sic] this rascal, as they shall answer it at their peril. [87]

By the mid-18th century, trade in African slaves had been thoroughly established and integrated throughout Brazil, the Caribbean, and east coast portions of North America. On the British island of Barbados, human lives were bought and sold—not for currency, but for quantities of cacao beans, as documented in this account:

But as to St. Lucia, St. Vincent, and Dominico, these are already fine Islands, have Governors, Large Towns, many Settlements, with Numbers of Plantations. The People from this Island [Barbados] supply the islands just mention'd with Negroes [sic]; and take their Commodities, such as Coffee, Cocoa, &c in Return. Hence these Islands cannot fail [to prosper] in a few Years, of being as good as any in the West Indies. [88]

During the early 19th century, educated free blacks (known then as coloreds) were being encouraged to leave continental North America to take up residence on the Caribbean island of Trinidad. Articles published described the economic advantages of emigration and commonly touted the advantages of leaving America. This 1839 document, for example, informed the readership that newly arrived North American blacks would find "happy people" in Trinidad:

[Advantages to be derived from emigration to Trinidad.] The present inhabitants attend to little but the cultivation of sugar, cocoa and coffee. The corn consumed by the population of Port of Spain is, much of it, brought from Connecticut, and great part of the remainder from the South American continent. . . . An emigrant in Trinidad, would not find himself shut up in a distant part of the globe, in the midst of barbarians; but surrounded by a free happy and prosperous population. [89]

After more than 25 years of struggle by blacks (coloreds) who had emigrated from North America, the Republic of Liberia was founded in 1847 in West Africa. The fledgling colony (and subsequent state) needed immigrants, and these were recruited by Liberian representatives who came to

North America to present public forums that outlined advantages of blacks returning to Africa. In the document presented here, a representative emphasized that raising cacao in Liberia would be a profitable enterprise:

> Upon arrival, emigrants, as soon as their acclimation permitted them, performed their own labor; as an excellent substitute for bread, cassada [cassava meant] was first planted; sweet potatoes; yams; lima beans; ginger; arrow root; Indian corn; rice; coffee; cocoa of a quality vastly better than your chocolate is raised abundantly; ground nuts; cabbages; radishes; tomatoes; bananas; plantains. . . . Liberia offers us, as its greatest gift, a free country. Our own race are in power and honor. We are a free and independent State, having a Constitution and Bill of Rights, like that of the United States. We do our own voting. [90]

Economic conditions in the American South on the eve of the Civil War reflected future uncertainty, especially if slavery could not be maintained. This document discusses a plan to remove slaves in Louisiana to Nicaragua, where economic conditions—and slavery—could be maintained.

> Numbers of slaveholders have already written to us to know if they could safely take their slaves into Nicaragua, to cultivate sugar, coffee, rice, indigo, or chocolate plantations, as the case might be. We have always assured our correspondents that though slaves were not recognized by law in Nicaragua, we have no doubt they would be secured to their owners during Walker's administration, and that ultimately slavery would have an existence their of [sic] law as a well as a fact. [91]

CHOCOLATE AND RELIGION

A full search for chocolate in religious documents is yet to be undertaken. In our present research we focused on chocolate and cacao as it existed as a theme in Spanish Inquisition accounts (see Chapter 4). The theme of religious fasting and whether or not chocolate would be permitted or forbidden has a long history [92]. The debate continued in North America into the 19th century as revealed in the following document published by the Archdiocese of Baltimore, which in 1848 still needed to resolve the issue and educate its parishioners:

> [Lent Regulation 7] General usage has made it lawful to drink in the morning some warm liquid; as tea, coffee, or thin chocolate, made with water, to which a few drops of milk may be added, serving rather to color the liquids, than to make them substantial food. [93]

Another example of a religious document associated with chocolate is the following account that identifies chocolate as a beverage commonly drunk at breakfast by Pope Pius IX:

> Every day he rises at half-past six in the morning. . . . At seven o'clock the Pope says a Mass alone in his oratory, and he hears one afterwards. . . . At half-past eight, the Pope has accomplished his pontifical duties. . . . He goes out of the chapel, and makes a light collation, by dipping a few biscuits in a mixture of coffee and chocolate, a usage common in Italy. [94]

Another potential theme that links religion with chocolate is the association with specific festivals, whether Day of the Dead in Mexico or chocolate festivals elsewhere. A review of early commercial advertising would be worthwhile to determine when chocolate in North America—and elsewhere—was first associated with Christian Christmas and Easter celebrations.

ECONOMICS

Three distinct potential lines for further enquiry are associated with chocolate and economics. First is the continued search for original documents that describe the use of cacao beans as money or currency.[2] While some of these documents date to the pre-Columbian era, or immediate decades after Spanish contact, others identified during the course of our work revealed that cacao beans continued to be used as a form of money into the late 18th century.

The highly respected archaeologist Rene Millon reviewed the history of cacao beans as currency in his Ph.D. dissertation completed in 1955, When Money Grew on Trees, where he summarized much of the information related to pre-Columbian uses [95]. Prior to the Spanish conquest of Mexico, Fernandez de Oviedo accompanied the fleet captained by Pedro Arias de Avila in his expedition along the coast of eastern Central America in 1513 and subsequently wrote a book of his experiences in 1526 entitled Historia General y Natural de las Indias. Oviedo observed the inhabitants of Nicaragua and Nicoya (vicinity of the Nicoya Peninsula in Costa Rica) and related important chocolate-related information regarding the introduction of cacao trees and use of cacao beans as currency:

> And because those [people] of Nicaragua and their tongue, wherever they have come from, are the ones that brought to the land [i.e., Costa Rica] the cacao or cacao beans that are used as money in those places, and [because] they have in their inheritance the trees that carry that fruit, there is not a single tree in the hands of the chorotegas [i.e., Costa Rican natives]. [96]

> So that in that province . . . a rabbit cost 10 of these cacao beans; and one slave cost hundreds [of them] . . . and because in that land there are women that give their bodies for a price . . . those who wanted for their libidinous use, gave [to the women] 8 to 10 cacao beans for a ride [por una carrera]. [97]

Another early European account dated to 1524 was written by Toribio de Benavente, a Franciscan priest, one of "the twelve" who arrived in New Spain shortly after the defeat of the Mexica/Aztecs by Cortés. Benavente's account also described use of cacao beans as currency:

> The more general use of these grains [cacao beans] is as a coin. This form of currency is used all over the land. A "load" has three times the number 8000. The Indians called it xiquipile. A "load" is 24,000 [i.e., 8000 × 3] cocoa seeds. A "load" has a value of 5 or 6 gold pesos in the areas where it is harvested. And the price increases when it is brought inland. The price also varies according with the year. If the year is good [warm] it produces a lot; if the year is cold there will be less cocoa, because it is a very delicate tree. This cocoa is a very general drink; they grind it and mix it with corn and other ground seeds; and this is also a major use of the cocoa seeds. It is good, it is good and it is considered as a nutritious drink. [98]

Francis Pretty was aboard Thomas Candish's English expedition that circumnavigated the globe between 1586 and 1588. As they sailed up the coast of South America they looted and burned Spanish towns. They passed Guaiaquil (Guayaquil in modern Ecuador), the Island of Puna, Rio Dolce, Rio Copalita, and reached Aquatulco. His 16th century English spelling has been maintained in the following account:

> Wee landed there, and burnt their towne, with the church and custome-house which was very faire and large: in which house [there were] 600 bags of anile to dye cloth; every bag whereof was worth 40 crownes, and 400 bags of cacaos: every bag whereof is worth ten crownes. . . . These cacos goe among them for meate and money. For 150 of them are in value one rial of plate [silver] in ready payment. They are very like unto an almond, but are nothing so pleasant in taste: they eat them, and make drinke of them. . . . After we had spoyled and burnt the towne, wherein there were some hundred houses . . . we set saile. [99]

Samuel de Champlain traveled to the West Indies and Mexico during 1599–1602 and described use of cacao beans as currency (see Chapter 24). His manuscript also described the cacao, used as an appetite suppressant, and was among the first to document medicinal uses for cacao bark, pith, and leaves:

> There is another tree, which is called cacou [sic], the fruit of which is very good and useful for many things, and even serves for money among the Indians, who give sixty for one [Spanish] real. Each fruit is of the size of a pine-seed, and of the same shape, but the shell is not so hard. The older it is the better and to buy provisions, such as bread, meat, fish, or herbs. This money may serve for five or six objects. . . . From the bark of this tree vinegar is made, as strong as that from wine, and taking the heart of this tree [i.e., pith] and pressing it, there comes out very good honey. Then drying the pith thus pressed in the sun, it serves to light fires. Moreover, in pressing the leaves of this tree, which are like those of the olive tree, there proceeds from them a juice, of which the Indians make a beverage. [100]

One of our team members, Patricia Barriga, discovered a document in the Mexican Archives, whereby the king of Spain ordered new coinage be developed for use in Spanish provinces overseas. When the new coins arrived in Mexico, the main problem was lack of enough coins of small value (i.e., half *reals*) needed for everyday purchases throughout the kingdom. This monetary discrepancy posed a problem to the governor; the captain general of the province of Yucatan wrote to officials at the royal mint in Mexico City that he had been forced to reinstitute the practice of using cacao beans as currency in order for customers to receive proper "change" from merchants/vendors:

> February 3rd, 1773: Very Honorable Sir. In my [previous] letter dated on July 5th, 1772 I expressed my concern about the bad consequences that would result with the re-collection [removal from circulation] of the "macuquina" coin, the most abundant coin used in this province. Of the 99,788 pesos and half pesos of new coins sent to this province, there are only 80,000 in pieces of 2 and 1 real and none of the half real; and these half reals are the ones that are needed the most. It is important to replace the old coin for the same amount of the new coin. Today it is a common practice to use as exchange cocoa grains [beans] which have a value regulated as follows: 90 cocoa grains equal to a half real coin. If there are not enough new coins to replace the old half real coins it will become necessary to double the amount of cocoa that is stored in the marketplaces to facilitate the purchases that have a price lower than a silver real. There are 860,336 Indians in all the province [of Yucatan] and all of them prefer in their capricious and obstinate idea to have the "macuquina" coin instead of the circular coin regardless that this circular coin is made of gold. [101]

The second line of inquiry for further research and assessment associated with economics is the linkage between epidemics, decline of Native American health, and the impact of labor shortages on cacao production. As a consequence of this dual problem, the production of cacao became the responsibility of local authorities—and caused an increased demand for the importation of slaves to work the plantations.

A document from Guatemala dated 1623 related that a plague (epidemic) resulted in the deaths of many children and young Native Americans, and as a result the tribute in cacao that normally would be sent to the authorities was low.[3] This document also commented that the plague ended at the beginning of August the same year [102].

In 1626, a Guatemalan government official, Alonso Diaz, wrote authorities in New Spain (Mexico City) to permit the importation of cacao from Peru—not for food, but for use as money because of scarcity of local beans caused by an epidemic:

The cacao from Peru is the coin that is more used; the silver coin is of a less use. With that cacao they buy food in the towns of Indians as well as in the towns of the Spanish. They use it in the Plazas, butcher shops, and also they buy chocolate with it. It serves as coin and as sustenance and because of this it is very necessary. Without it people cannot live and if they were to lack cacao, particularly the poor people, they will perish. [103]

In 1632, a representative from the province of Nicaragua requested permission to import cacao beans from Peru due to the scarcity of this crop within his region [104].

Shortages of available cacao continued throughout much of Central America and into Mexico, so much so that in 1638 another order was issued by the government of New Spain that all persons who had stored cacao had to relinquish their supplies and the cacao had to be brought to the public granary (*alhondiga*) within the various towns. The order specified that it was forbidden to hide cacao in private homes or in "any other forbidden places." The document ended ominously:

Any person that knows where there is cacao hidden, must declare under the penalty to proceed against them [i.e., turn in or denounce the hoarders] if they will not comply. [105]

As a consequence of cacao shortages, new avenues of commerce developed that linked New Spain and Spanish provinces in Central and South America. Numerous documents report importation of cacao to Veracruz from Caracas, and from Callao (Peru) to Acapulco. This thriving trade blossomed and resulted in numerous governmental attempts to control imports and to levee tariffs and import taxes. As a result, smuggling rose exponentially and such illicit imports and associated crimes soon plagued New Spain. The following example revealed that cacao-related piracy had a multireligious nature (Fig. 56.18):

1763. On June 8th of the last year [1762] a sloop carrying 321 fanegas of cacao from private people [i.e., merchants] left Ocumare [today the Port of Cumana, in northern Venezuela]—sailed down the coast with the intent to reach the Port of Guaira [northern Venezuela] but on June 10th was attacked by a Dutch sloop from Curacao, the ship captain [being] Phelipe Christiano, [other passengers being] a Jewish merchant named Mich [?] and another [Jew]—the owner [of the ship] named Abraham—and they took with violence that cacao. [106]

The letter continued that the governor of Caracas requested that the governor of Curacao return the cacao that was stolen, or pay for it. He also asked the Curacao governor to stop such pirate attacks that were so common along these coasts [107]. But there was no reply from the governor of Curacao.

A full study and analysis of the Mexican cacao trade remains an exciting potential avenue for further research, for example, importation of cacao beans to the west coast port of Acapulco, Mexico, from Guayquil, Eduador, and from Callao, Peru, and to the east coast port of Veracruz from Caracas and Maracaibo, Venezuela. Documents we perused dated from 1755 through 1796 and were the consequence of the numerous decades when Native Americans who worked the cacao plantations died from various epidemics [108, 109].

The third line of economic-related research might consider the development and evolution of cacao- and chocolate-associated tariffs and economic trade issues within North America. These documents are numerous and were not systematically investigated by our team. We noted, however, that with the conclusion of the American Revolution, the new country focused on development of its economic base. Economic documents published during the early years after independence identified the basic commodities available for export as noted here in 1787:

A list of raw materials and natural productions which now are or may readily be, furnished by the united states of America [sic] . . . for home consumption and use, and for exportation: Cod-fish (cured), Cheese Cyder [sic], Chocolate, Candles and candle-wick, Cottons (printed). [110]

Adam Smith, considered by many to be the founder of modern economics, published his *Inquiry into the Nature and Causes of the Wealth of Nations* in 1789. He considered wealth in the sense of fixed assets and the mechanisms whereby assets were generated in order to maintain capital. His thinking clearly differentiated between commodities produced domestically versus those necessary for import, and the interplay between colonies and the "mother country." One underlying theme of his text, however, considered how to reduce economic competition, and his view had a direct impact on the import/export possibilities of North American chocolate manufacturers:

The enumerated commodities are of two sorts: first, such as are either the peculiar produce of America, or as cannot be produced, or at least are not produced, in the mother country [i.e., England]. Of this kind are melasses [sic], coffee, cacao-nuts, tobacco, pimento, ginger, whale-fins, raw silk, cotton-wool, beaver and other peltry of America. . . . The largest important of commodities of the first kind [i.e., those identified here] should not discourage the growth or interfere with the sale of any part of the produce of the mother country. [111]

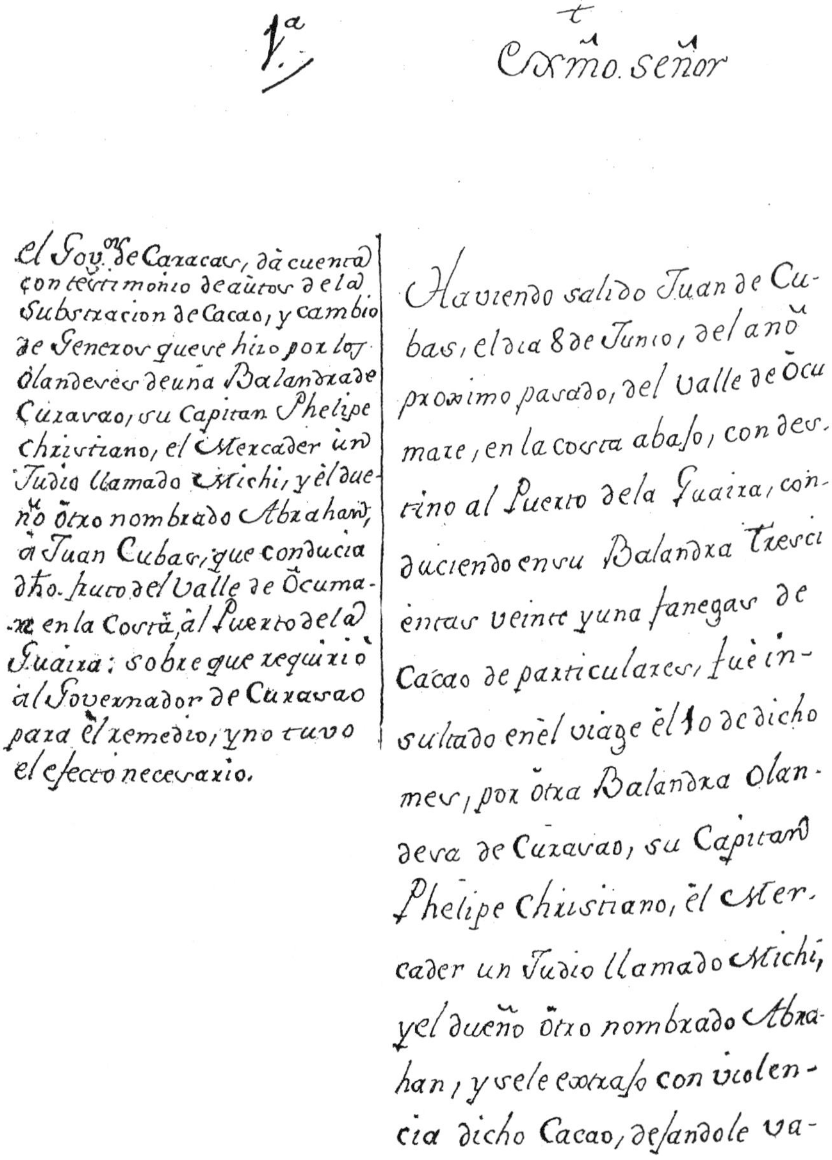

1.ª

Ex.mo señor

El Gov.or de Caracas, dà cuenta con testimonio de autos de la Substracion de Cacao, y cambio de Generos que se hizo por los olandeses de una Balandra de Curavao, su Capitan Phelipe Chistiano, el Mercader un Judio llamado Michi, y el dueño otro nombrado Abraham, à Juan Cubas, que conducia dho. fruto del Valle de Ocumare en la Costa, al Puerto de la Guaira: sobre que requirio al Governador de Curavao para el remedio, y no tuvo el efecto necevario.

Haviendo salido Juan de Cubas, el dia 8 de Junio, del año proximo pasado, del Valle de Ocumare, en la costa abajo, con destino al Puerto de la Guaira, conduciendo en su Balandra Trescientas veinte y una fanegas de Cacao de particulares, fue insultado en el viage el 10 de dicho mes, por otra Balandra olandesa de Curavao, su Capitan Phelipe Chistiano, el Mercader un Judio llamado Michi, y el dueño otro nombrado Abraham, y se le extrajo con violencia dicho Cacao, dejandole va-

FIGURE 56.18. Report of Jewish pirates seizing cacao destined for the Port of Guaira, northern Venezuela; dated June 8, 1763. *Source:* Caracas, Number 438. Courtesy of Archivo General de las Indias (Archive of the Indies), Seville, Spain. (Used with permission.)

The reality of dealing with day-to-day political issues and the need to extract income (i.e., taxes) from both imported and exported goods was a common theme before, during, and after the American Revolution. A law passed by the General Assembly of the Colony of New York in July of 1715 regulated the import and taxation of cacao:

> For every Hundred Weight of Cocoa, imported directly from the Place of its Growth, or from any Island, part or Place of the West-Indies, One ounce of [silver] Plate aforesaid. [112]

During the American Revolution numerous documents produced by the colonies commented upon the difficulty of maintaining imports of food and basic commodities and restricted the export of essential goods. The House of Representatives of Massachusetts-Bay State passed the following act to prevent "Monopoly and Oppression" on March 15, 1777:

> Resolved, That [sic] no Beef whether the Produce of America or elsewhere shall be Sold for more than Three Pounds fourteen Shillings per Barrel. And also, Resolved, That the Price stipulated in the above said Act for Peas, Beans, Pork, Wheat, Rye, Hides, Loaf Sugar, Chocolate, Butter, Potatoes, Oats, Tallow, Flour, and Bar Iron, the Produce of America, shall be the Price of those Articles already or that may hereafter be Imported or Captured. [113]

John Hancock, Speaker of the House for the state of Massachusetts-Bay, signed a bill on September 23, 1779, forbidding the export to neighboring states of certain foods and essential goods subject to confiscation and forfeit of the value of any such goods:

Be it therefore enacted by Council and House of Representatives in General Court assembled, and by the authority of the same, That no exportation be permitted of rum, wine, or any kind of spirits, molasses, sugar, cotton wool, sheep's wool, wool cards, flax, salt, coffee, cocoa, chocolate, linnen [sic], woolen and cotton goods of all kinds; provisions of all and every sort; live-stock, shoes, skins, and leather of all kinds; either by land or by water from any part of this State. [114]

Literally hundreds of such economic-related documents exist and further detailed examination and analysis is warranted.

Conclusion

In this chapter we identified numerous themes and topics where we believe additional scholarly investigation would be productive and lead to new insights on chocolate history. At the start of our research we could not have conceived how our work would lead to such a diversity of records, some extremely unusual. We counted the following examples as among the more interesting and unusual.

1. King Philip IV of Spain, clearly appreciated chocolate. In a letter written in 1637 and addressed to the judge and officials of the Casa de Contracion[4] in the city of Seville—personally signed by the king—he instructed Pedro de Vibanco, president of the Contracion, to pay 12,300 reales for 41 arrobas of chocolate (the equivalent of 1025 pounds), and that it was "his will" that the persons who brought the chocolate to Madrid would be paid. The document is signed by the royal hand—*Yo, el Rey* [115].

2. In 1726, upon her death, Lady Palmerston bequeathed to her husband the following items:

As a remembrance of death and also of the fondest and faithfulest [sic] friend he ever had—two gold chocolate cups made out of mourning rings, and used by her daily as a memorial of her departed friends and of eternity. [116]

3. There is an account dated to the mid-18th century that George Frederick Handel regularly drank chocolate when he was composing *The Messiah*:

Being once inquired of as to his ideas and emotions when writing the Hallelujah Chorus, he replied, in the best English he could command, "I did think I did see heaven all before me, and the great God himself" . . . *when Handel's servant used to bring him in his chocolate in the morning he often stood in silent astonishment to see his master's tears mingling with his ink as he wrote his masterly works. Indeed, it appears to have been usually the case that during his compositions his face would be bathed in tears.* [117]

4. There was an announcement of the tragic suicide in 1772 of Boston chocolate grinder Dunn (first name not provided), who became delirious, defenes-

trated himself, and suffered terribly as he lay on the ground throughout the night. The account of Dunn's death was that he:

Threw himself out of a garret window into the street, where he was found the next morning with some signs of life but then expired. [118]

5. There is a report of the arrival of a ship captained by a Mr. Kennedy, to the port of Boston, which carried as cargo an elephant from India. The document related the difficulties of unloading the beast. More interesting, however, were the authors' comments regarding the voyage to and from India where he claimed:

Not a drop of ardent spirit was drunk on board Captain Kennedy's ship, from the day of her departure to her return. Plenty of hot coffee and chocolate supplied its place in cold weather, and the yankee switchell [a rum-based cocktail] preserved the health of the men in Calcutta, while half the rum-run-drinking crews there were in the hospital. [119]

6. Finally, there is an account that describes an insane gentleman identified as Max Sefelege, a native of Potsdam, Germany, who in 1850 claimed that he had "invented" chocolate [120].

The present anthology has integrated chocolate-associated information from the pre-Columbian era in Central America, through recent 20th and 21st century discoveries, but there is much more to cover, new archives and libraries to search, and museums to visit. It is our hope that readers of our work will continue these efforts in the months and years to come.

History, however, is but one facet of the chocolate story. Recent scientific discoveries confirm a range of positive health-related outcomes associated with drinking or eating certain types of chocolate. In 2006, Wiley-Interscience contracted with University of California, Davis, researchers to produce two books that focused on chocolate: the present anthology on chocolate history and a second on recent scientific findings related to chocolate. This companion volume, edited by Alan Bennett, Carl Keen, and Howard Shapiro, is entitled *Theobroma cacao. Biology, Chemistry, and Human Health* [121]. Together, both books provide complementary reading experiences that blend history with cutting-edge 21st century scientific research. We hope that readers interested in the most recent chocolate-associated science, who wish to understand the potential role of chocolate as part of a healthy diet, will consider their work as a companion volume to our historical coverage.

Epilogue

During the dawn of prehistory in a geographical region located in what is now northeastern Ecuador, early humans ate the sweet whitish pulp contained within the attractive, colorful pods gleaned from wild cacao trees. This sweet pulp probably was the sole dietary use people made of these pods at this time in prehistory; they would have tasted and then discarded the dark, bitter beans within the pods. Others, however, perhaps healers or shamans, experimented with the cacao beans and in so doing determined that chewing the bitter beans altered consciousness and inability to sleep. This dual contribution from *Theobroma cacao* pods, the sweet edible pulp and the bitter beans, ultimately led others to select and tend wild trees that bore pods of unusual color or shape, a process that ultimately led to domestication or genetic differentiation from the wild progenitor. Scientists expect that this process occurred perhaps by 4000 BCE. While pods from *Theobroma cacao* were used by these early humans, they did not, in turn, produce chocolate whether as a beverage or as a solid. This subsequent step took place several thousand years later north of the equator in Central America within a geographical region located in the modern states of southern Mexico, Guatemala, and Beliz. It is to these unknown early Central Americans we should offer thanks and blessings, for developing the technical skills that ultimately led to the production of chocolate as we know it today.

The historical, cultural, and culinary saga of chocolate thus extends from ca. 4000 BCE to 2009 CE a span of at least 6000 years. Chocolate in the 21st century is celebrated on every continent, even Antarctica. Is there any country where chocolate is unknown? Chocolate history past and present can be visualized as an enormous three-dimensional global jigsaw puzzle consisting of a myriad of geographical points in time and space. Our project team members have searched for initial beginnings and introduction of chocolate to North American and elsewhere, and how chocolate products and varieties were developed and dispersed globally. The chocolate preferred by North American colonists before the Revolutionary War was not the same as that mass-produced today for local, regional, national, and international markets: Ingredients have changed, processing has been refined, new chocolate-related inventions/ patents developed, and advertising by major, minor, and artisan manufacturers has created different taste, flavor, and texture demands among chocolate connoisseurs.

Despite these changes and development in chocolate manufacturing through the ages, the fact remains that the bitter dark beans from trees initially domesticated in South America have been transformed into pleasurable taste sensations admired and craved throughout the world. The near global attraction to chocolate is a phenomenon unlike any other food. The volumes written about chocolate outrival all other foods and beverages, with perhaps the exception of

wine produced from *vinifera* grapes. Chocolate is, as one of our project team members once aptly stated, *the social glue that binds peoples and cultures throughout the world*. In our research we explored and surveyed chocolate-associated antiques and art, culinary and medical uses, dance, literature and poetry, music, and theater. We explored how chocolate seeps, melts, and blends with human intimacy and sexuality as evidenced by the broad range of chocolate-related personal products such as body paint. We found chocolate use beyond earth, where it characterized many meals consumed by American and Soviet astronauts and cosmonauts. Chocolate is the true global, universal food product.

The material presented in this anthology reflects but a sampling of the hundreds of chocolate-related topics that could be developed into full-length topics and explored more in depth. Generous funding from Mars, Incorporated, enabled us to complete the work that resulted in the present publication. We suggest that with many hundreds of topics remaining to be explored, with literally thousands of additional archives, libraries, museums, and private research collections to be searched for chocolate-related information, other companies would see the benefit of initiating their own contributions to chocolate history work. Should such contributions be forthcoming, would it not be logical to plan for national and international meetings where new, exciting chocolate history findings could be reported four or five times each decade?

If you have enjoyed the material contained within the present anthology, why not join together in future chocolate history research ... and let us plan together for a national chocolate history meeting in 2010 to share information.

Acknowledgments

We wish to thank the following colleagues for their suggestions during the preparation of this chapter: Bertram Gordon, Mills College, Oakland, California; Carl Keen, Department of Nutrition, University of California, Davis; Deanna Pucciarelli, Department of Nutrition, University of California, Davis; and Kurt Richter, Graduate Group in Geography, University of California, Davis.

Endnotes

1. Personal communications from Kenneth Owens (December 11, 2006) and Glen Farris (December 14, 2006).

2. A number of accounts continue to circulate, supposedly based on original accounts, reportedly documenting the specific value of cocoa beans in respect to barter and purchasing. Oviedo reportedly stated: 1 rabbit = 10

cocoa grains [i.e., beans], 8 zapote fruits = 4 cocoa beans; favors of a prostitute = up to 10 cocoa beans. Alexander Humbolt wrote: "1151 cocoa grains = one silver peso". In 1911, in Chiapas, southern Mexico, it was reported that 1000 cocoa beans were equivalent to a Mexican silver peso (from Curado, Anselmo J. Chocolate. In: *Liquid Gold*. Edited in Barcelona, Spain: 1996; Chapter 3).

3. It is likely that the decline of the Native American population due to diseases introduced from Europe resulted in the need for more labor imports (slaves and indentured persons) from Africa and from the Philippines. Those from the Philippines commonly were called Chinois. They should not automatically be identified as Chinese.

4. The Casa de Contración was an institution created in 1503, in Seville, to promote and control commerce with New Spain.

References

1. Grivetti, L. Chocolate: food of the gods; medicine for humans. Presented at the British Association for the Advancement of Science, Festival of Science, September 3–7, 2001, Glasgow, Scotland.

2. Grivetti, L. *Medical, Cultural, and Culinary Uses of Chocolate. A 450 Year Perspective*. Washington, DC: National Academy of Science, 2004.

3. Grivetti, L. *Cocoa and Chocolate during the American Revolution Era*. Washington, DC: National Academy of Science, 2004.

4. Dillinger, T. L., Barriga, P., Escarcega, S., Jimenez, M., Salazar Lowe, D., and Grivetti, L. E. Food of the gods: cure for humanity? A cultural history of the medicinal and ritual use of chocolate. *Journal of Nutrition* 2000; 130(Supplement):2057S–2072S.

5. Grivetti, L. E. From aphrodisiac to health food; a cultural history of chocolate. In: *Karger Gazette*, Issue 68 (Chocolate). Basel, Switzerland: S. Karger, 2005.

6. AGN. Archive Document dated 1748–1751; Hospital de Jesus, September 1748.

7. AGI. Escribanía 155 A. (Undetermined date between March 6 and March 20, 1642. Document is attached to the back of a document dated March 6, 1642. Document consists of nine pages; corners are damaged or missing.)

8. http://asbout.upi.com/AboutUs/index.php?ContentID=20051018121523-29779&SectionName=About Us. (Accessed, January 10, 2007.)

9. Lindbergh, C. A. *The Spirit of St. Louis*. New York: Charles Scribner's Sons, 1953; pp. 209–210, 484–485.

10. Bailleux, N., Bizeul, H., Feltwell, J., Kopp, R., Kummer, C., Labanne, P., Pauly, C., Perrard, O., and Schiaffino, M. *The Book of Chocolate*. New York, Flammarion, 1996.

11. Schuhmacher, K., Forsthofer, L., Rizzi, S., and Teubner, C. *El gran libro del Chocolate*. (No date, no publishing information; purchased in San Jose, Costa Rica, 2006.) Available at http://www.paradigmalibros.com/libros/4/842419204.html#. (Accessed January 10, 2007.)

12. Coe, S. D., and Coe, M. D. *The True History of Chocolate*. London, England: Thames and Hudson, 2000.

13. Peyps, S. Diary Entry for April 24, 1661. Available at http://www.pepysdiary.com/archive/1661/04/. (Accessed January 10, 2007.)

14. de Sade, A. F. (Marquis de Sade). *Letters from Prison*. Translated by R. Seaver. New York: Arcade Publishing, 1999. Letter to his wife, dated May 16, 1779.

15. Dickens, C. 1852–1853. *Bleak House*. The Project Gutenberg E-Text. Release date August 1997 (E-Text 1023). Recently updated January 30, 2006. Chapter V.

16. Dickens, C. 1859. *A Tale of Two Cities. A Story of the French Revolution*. The Project Gutenberg E-Text. Release date September 25, 2004 (E-Text #98). Book 2, Chapter 7.

17. Greeley, H. *An Overland Journey, New York to San Francisco, the Summer of 1859*. New York: C. M. Saxton, Barker, and Company, 1860; p. 78.

18. Reed, J. 1919. *Ten Days That Shook the World*. The Project Gutenberg E-Book. Release date December 16, 2000 (E-Text #3076). Chapter 1.

19. Snyder, M. *Poison Study*. New York: Luna Books, 2005; p. 159.

20. Anonymous. *The Vertues [sic] of Chocolate [and] The Properties of Cavee*. Oxford, England: Henry Hall, 1660. Available at http://homepage.univie.ac.at/thomas.gloning/tx/vertues.htm. (Accessed January 10, 2007.)

21. Pope, A. *The Rape of the Lock*, 1712; CANTO II: lines 133–136.

22. Greville, H. The Miserable Glutton, or, The Pleasures of Sense, Dependent on Virtue. *The American Magazine and Historical Chronicle* July 1746:326.

23. Byron, G. G. 1824. *Don Juan*. Canto xvi: Verse xxxiv. Available at http://www.geocities.com/~bblair/ednote.htm. (Accessed January 10, 2007.)

24. http://roxen.xmission.com/~drudy/mtman/RuslPaper.html. (Accessed January 10, 2007.)

25. Wislizenus, F. A. *A Journey to the Rocky Mountains in 1839*. St. Louis: Missouri Historical Society, 1912; Chapter 11, no page numbers.

26. Henry, A. *Travels & Adventures in Canada and the Indian Territories Between the Years 1760 and 1776*. Edited by Elliot Couse. Boston: Little, Brown and Company, 1901; p. 272.

27. Pike, Z. M. *The Expeditions of Zebulon Montgomery Pike, to the Headwaters of the Mississippi River, Through Louisiana Territory, and in New Spain, During the Years 1805-6-7*, 2 Volumes. Edited by Elliot Coues. New York: F. P. Harper, 1895; Volume 2, p. 601.

28. http://www.arcticwebsite.com/Franklin1845 provisions.html. (Accessed January 10, 2007.)

29. http://www.arcticwebsite.com/goldrushsupplies 1898-2.html. (Accessed January 10, 2007.)

30. http://www.mariner.org/exploration/index.php?type=explorersection&id=14. (Accessed January 10, 2007.)

31. http://archives.cnn.com/2001/WORLD/Europe/09/26/antartica.chocolate/index.html. (Accessed January 10, 2007.)

32. http://www.solarnavigator.net/history/scott_of_the_antarctic_explorer_captain_robert_falcon.htm. (Accessed January 10, 2007.)

33. http://www.solarnavigator.net/history/scott_of_the_antarctic_explorer_captain_robert_falcon.htm. (Accessed January 10, 2007.)

34. http://www.mariner.org/exploration/index.php?type=webpage&id=52. (Accessed January 10, 2007.)

35. Roosevelt, T. *Through the Brazilian Wilderness*. London, England: John Murray, 1914; Appendix B.

36. http://www.achievement.org/autodoc/page/hil0int-8. (Accessed January 10, 2007.)

37. http://www.nasm.si.edu/exhibitions/attm/nojs/a11.jo.es.1.html. (Accessed January 10, 2007.)

38. http://www.spacetoday.org/SpcShtls/AstronautsEat.html. (Accessed January 10, 2007.)

39. *The Pennsylvania Gazette*, October 6, 1743.

40. *The Pennsylvania Gazette*, April 16, 1747.

41. *The Pennsylvania Gazette*, May 8, 1776.

42. *The Pennsylvania Gazette*, August 20, 1778.

43. *New York Times*, May 20, 1861; p. 2.

44. http://memory.loc.gov/ammem/today///ammem/wpaintro/wpahome.html. (Accessed January 10, 2007.)

45. *The National Recorder*, February 19, 1820; p. 124.

46. http://bremnerhome.com/. (Accessed January 10, 2007.)

47. http://www.diggerhistory.info/pages-discipline/../images/asstd/red-cross-parcel.jpg. (Accessed January 10, 2007.)

48. Langsdorff, G. H. von. *Voyages and Travels in Various Parts of the World, during the Years 1803, 1804, 1805, 1806, and 1807*. Illustrated by engravings from original drawings. London, England: Printed for Henry Colburn and Sold by George Goldie, Edinburgh; and John Cumming, Dublin.

49. Langsdorff, F. von. *G. H. Langsdorff's Narrative of the Rezanov Voyage to Neuva California in 1806*. San Francisco, CA: Thomas C. Russell, 1927.

50. http://www.readex.com/readex/. (Accessed January 10, 2007.)

51. Cárdenas, Juan De. *Problemas y Secretos Maravillosos de las Indias*. Madrid, Spain: Ediciones Cultura Hispanica, 1591; Chapter 7, pp. 112f–112v.

52. Salvador, R. J. What Do Mexicans celebrate on the Day of the Dead? In: Morgan, J. D., and Laungani, P., editors. *Death and Bereavement in the Americas*. Death, Value and Meaning Series, Volume II. Amityville, NY: Baywood Publishing Company, 2003.

53. http://www.ponyexpress.org/. (Accessed January 10, 2007.)

54. http://www.hindenburg.net/background.htm. (Accessed January 10, 2007.)

55. http://asbout.upi.com/AboutUs/index.php?ContentID=20051018121523-29779&SectionName=About Us. (Accessed January 10, 2007.)

56. Lindbergh, C. A. *The Spirit of St. Louis*. New York: Charles Scribner's Sons, 1953; pp. 484–485.

57. http://www.californiamuseum.org/halloffame/inducties/amelia_earhart/. (Accessed January 10, 2007.)

58. http://www.historynet.com/exploration/adventurers/3032866.html?featured=y&c=y. (Accessed January 10, 2007.)

59. http://www.historynet.com/air_sea/aviation?history/3029921.html?page+1&c=y. (Accessed January 10, 2007.)

60. Father Francisco Pangua, Apostolic Guardian of the College of San Fernando in Mexico City: Santa Barbara Mission Archive, Father Serra Collection, Document #885a.

61. AGN. Marina. Volume 86, fs. 231–241.

62. Anonymous. Rules for a young lady. *Ladies' Magazine and Literary Gazette* 1830;3:239.

63. Anonymous. On the female form. *Lady's Book*. November 1831:294.

64. Dickinson, J. 1699. *Gods protecting providence man's surest help and defence [sic] in the times of the greatest difficulty and most imminent danger; evidenced in the remarkable deliverance of divers persons, from the devouring waves of the sea, amongst which they suffered shipwreck [sic]. And also from the more cruelly devouring jawes [sic] of the inhumane cannibals [sic] of Florida. Faithfully related by one of the persons concerned therein, Jonathan Dickenson*. Philadelphia, Pennsylvania: Reinier Jansen, p. 18.

65. Defoe, D. 1808. *The Life and Adventures of Robinson Crusoe*. Project Gutenburg. Release date June 15, 2004 (E-Text #12623). (Accessed January 10, 2007.)

66. Jewitt, J. R. *The Adventures and Sufferings of John R. Jewitt. Only Survivor of the Ship Boston, During a Captivity of*

Nearly Three Years Among the Savages of Nootka Sound; With an Account of the Manners, Mode of Living, and Religious Opinions of the Natives. Edinburgh, Scotland: Archibald Constable, 1824; pp. 50–51.

67. Brady, W. N. *The Kedge Anchor, or Young Sailors' Assistant. Appertaining to the Practical Evolutions of Modern Seamanship, Rigging, Knotting ... and Other Miscellaneous Matters, Applicable to Ships of War and Others*, 6th edition. New York: Published by the author, 1852; p. 208.

68. *The by-laws and town-orders of the town of Boston, made and passed at several meetings in 1785 and 1786. And duly approved by the Court of Sessions.*

69. *The Literary Magazine and American Register* 1804; March Issue:427.

70. *Scientific American* 1852; 8(1): p. 3.

71. Grivetti, L. E. Harding Distinguished Lecture. Beer–Beef–Bread. A Perspective on Food and Health of University-Age Students. In: *Nutrition in Action IV. The Relation of Nutrition to Health in Young Adults. Proceedings, Fourth Ethel Austin Martin Visiting Professorship in Human Nutrition at South Dakota State University.* Brookings, SD: Human Nutrition Fund Committee, 1985; pp. 1–16.

72. http://etcweb.princeton.edu/Campus/text_revolution.html. (Accessed January 10, 2007.)

73. Quincy, J. *The History of Harvard University*, 2 Volumes. Boston: Crosby, Nichols, Lee, and Company, 1840; Volume 2, p. 541.

74. AGN. Indios. Volume 3, exp. 860, f. 207v. Ano 1591.

75. AGN. Reales Cedulas Duplicadas, July 28, 1645. Volume l 15, exp. 176, fs. 136r–139r.

76. AGN. Tierras. Volume 2920, exp. 2, fs. 196r–203v.

77. AGN. Tierras. Volume 2920, exp. 2, fs. 196r–203v.

78. *The Pennsylvania Gazette*, March 26, 1754.

79. *The Pennsylvania Gazette*, February 11, 1768.

80. *The Gazette of the United States*, July 2, 1791; p. 72.

81. Anonymous. Dinner hosted by the Duke of Orleans after the coronation of King Louis XV. *Select Reviews and Spirit of the Foreign Magazines* 1810; August Issue:138.

82. *New York Times*, March 8, 1865; p. 1.

83. *The Pennsylvania Gazette*, August 31, 1749.

84. *The Pennsylvania Gazette*, January 16, 1750.

85. *The Boston News-Letter and New-England Chronicle*, June 2, 1763; p. 3.

86. *The Providence Gazette and Country Journal*, November 7, 1772; p. 3.

87. *New Jersey Journal*, August 20, 1788; p. 3.

88. *The Pennsylvania Gazette*, November 15, 1750.

89. *The Colored American*, August 31, 1839.

90. *The African Repository* 1850; August Issue:241–243.

91. *The National Era* (Washington DC) 1856; X(516): 188.

92. Pinelo, A. de León. *Question Moral. Si el Chocolate Quebranta el Ayuno Eclesiastico.* Facsímile de la primera edición (Madrid, Spain, 1636). Mexico City, Mexico: Centro de Estudios de Historia de Meico [CONDU-MEX], 1994; passim.

93. *The United States Catholic Magazine and Monthly Review* 1848; March Issue:151.

94. *The United States Catholic Magazine* 1849; February 17:97.

95. Millon, R. F. *When Money Grew on Trees: A Study of Cacao in Ancient Mesoamerica.* Ph.D. dissertation. New York: Columbia University, 1955.

96. Oviedo, F. de. *Historia General y Natural de las Indias.* 1526. XI, p. 109.

97. Oviedo, F. de. *Historia General y Natural de las Indias.* 1526. II, p. 245.

98. Fray Toribio Benavente [Motolinía]. 1524. *Historia de los indios de la Nueva España.* Historia 16. Madrid, Spain: Edición de Claudio Esteva, 1985; p. 240.

99. Pretty, F. Account by Master Francis Pretty. The Admirable and Prosperous Voyage of the Worshipfull Master Thomas Candish of Trimley in the Countrie of Suffolke Esquire, into the South Sea, and from Thence Round About the Circumference of the Whole Earth, Begun in the Yeere [sic] of Our Lord 1586, and Finished 1588. In: Hakluyt, R., editor. *The Principal Navigations, Voyages, Traffiques [sic], and Discoveries of the English Nation Made by Sea or Over-land to the Remote and Farthest Distant Quarters of the Earth at Any Time Within the Compasse [sic] of These 1600 Yeeres [sic].* Glasgow, Scotland: James MacLehose and Sons, 1904; Volume 11, pp. 315–320.

100. Champlain, S. (manuscript dated ca. 1603). *Narrative of a Voyage to the West Indies and Mexico in the Years 1599–1602.* Translated by A. Wilmere; edited by N. Shaw. Series 1, Number 23. London, England: The Hakluyt Society, 1859; pp. 26–28.

101. AGN. Case de Moneda, 1772. Volume 90, exp. 4/5, fols. 313r–357r.

102. AGI. Guatemala. 42, N.22/1/1.

103. AGI. Guatemala. 43, N. 36/3/1.

104. AGI. Guatemala. 43, N. 36/1/1.

105. AGI. File identified as Ministerio de Educación. AHN (Archivo Histórico Nacional). Diversos, 31, Doc. 70.

106. AGI. Caracas. Number 438.

107. AGI. Caracas. Number 438.

108. AGN. Marina. Volume 11, exp. 13, fs. 35–38b.

109. AGN. Marina. Volume 56, fs. 265–269v.

110. *The American Museum* 1787; September Issue:258.

111. Smith, A. *Inquiry into the Nature and Causes of the Wealth of Nations.* Edited by E. Carnan, 5th edition. London, England: Methuen and Company, 1904; Volume 1, p. 315.

112. *Laws, Session, May, 1715. Acts passed by the General Assembly of the Colony of New-York in July, 1715.* New York: William Bradford, p. 207.

113. *The Boston Gazette and Country Journal*, March 17, 1777, p. 3.

114. *Continental Journal*, October 7, 1779, p. 4.

115. AGI. Indiferente. 434, L.7/1/783.

116. *The Dollar Magazine* 1851; September Issue:138.

117. Belcher, J. *Historical Sketches of Hymns, Their Writers, and Their Influence*. Philadelphia: Lindsay and Blakiston, 1859; p. 368.

118. *The Massachusetts Spy*, February 27, 1772, p. 207.

119. *The Friend. A Religious and Literary Journal*, January 14, 1832, p. 106.

120. *The Albion. A Journal of News, Politics, and Literature*, June 15, 1850, p. 286.

121. Bennett, A., Keen, C., and Shapiro, H.-Y. *Theobroma cacao. Biology, Chemistry, and Human Health*. Hoboken, NJ: Wiley-Interscience, 2009.

Appendices

Lexicon and Abbreviations

Term	Definition
&c.	Archaic English abbreviation for *et cetera*.
Achiote	Red paste from *Bixa orellana*, commonly added to chocolate (see annato, rocoa).
Acompting-house	Archaic English term for bookkeeping; an accounting office.
ADCM	Archives départementales de la Charente-Maritime (Canadian Archive).
Adulteration	Addition of impure substances for the purpose of defrauding customers (e.g., adulterated chocolate).
Aduna Royal	Port authority; head of the Royal Customs Department.
Aga	Turkish civil leader.
AGI	Archivo General de las Indias (Archive of the Indies, Seville, Spain).
AGN	Archivo General de la Nación (General Archive of the Nation), Mexico City, Mexico.
Aguardiente	Commonly translated as brandy.
AHRCIL	Archivo Histórico del Real Colegio de San Ignacio de Loyola.
AHU	Historical Overseas Colonial Archive, Lisbon, Portugal.
Aiguillette	In the sense used in the present book, a long strip of chocolate candy.
AJHC	American Jewish Historical Society.
Ajiaco	Hot, spicy stew of various meats.
AKA	Abbreviation meaning "also known as;" commonly used in criminal cases and trial documents.
Alcade	Literally, an administrative major of a town with legal and judicial powers.
Alhondiga	Spanish term for public granary.
Almud	Spanish unit of weight: 1 *almud* $= \frac{1}{12}$ to $\frac{1}{2}$ of a *fanega* (1 *fanega* $= 110$ pounds).
Alta	Used in the sense of *Alta* California (i.e., upper California) or the geographical area of modern day California.
Ambergrease	Spelling variant of ambergris.

Chocolate: History, Culture, and Heritage. Edited by Grivetti and Shapiro
Copyright © 2009 John Wiley & Sons, Inc.

Term	Definition
Ambergris	Waxy vomit of sperm whales, sometimes added to chocolate.
Ambre gris	Spelling variant of ambergris, literally translated as "gray amber."
Amenolado	Highly aromatic variety of forestero cacao; local varieties of amenolado cacao in Ecuador called arriba or nacional.
Analeptique	Something that restores health; in the sense used in the present book, a type of medicinal chocolate.
Anana	Common term for pineapple.
Anatto	Spelling variant of annotto, or achiote (*Bixa orellana*); red powder or paste commonly added to chocolate.
Anker	English liquid measure equal to approximately 10 gallons.
Annotto	Red colored pulp covering seeds of *Bixa orellana*. (See Achiote and Rocoa.)
Antebellum	In the sense used in the present book, prior to the American Civil War.
Antiscorbittic	Variant spelling for antiscorbutic, a medicine or food used to treat scurvy.
Apastle	Clay pot used to serve food or chocolate.
Apothecary	Merchant who sells drugs and medicines.
APS	Public Archive of Bahia/Salvador, Brazil.
Aqua vita	Literally, "water of life," strong drink, or distilled spirits.
Arrateis	Spanish unit of weight: 1/32nd part of an *arroba*.
Arroba	Spanish unit of weight; commonly 1 *arroba* = 25 pounds = 32 *arrateis*.
Arrowroot	Nutritious starch prepared from the roots of *Maranta arundinacea*.
Askos	In the sense used in the present book, a bronze ewer.
Atole	Cornmeal gruel; sometimes prepared with chocolate.
Aubergiste	Term for innkeeper in New France.
Auto de fé	Public procession of persons convicted of heresy, followed by burning at the stake.
Avocado bone	Avocado pits sometimes used by unscrupulous merchants to substitute for cacao beans.
Azulejos	Glazed tiles, commonly formed into a mural.
BADM	British Admiralty Office Archive.
Baguette	French term for stick used to beat (froth) chocolate, sometimes elaborately carved.
Baja	Used in the sense of *Baja* California (i.e., lower California); part of modern Mexico.
Bandeirantes	Groups of roving slave raiders.
Banilla	Spelling variant of vanilla.
Bark	Three-mast sailing ship commonly used to transport cacao.
Barque	Alternative spelling of bark, a three-mast sailing ship.
Bason	Archaic English spelling of the word basin, a large cooking pot.
Bateau	French term for small, flat-bottomed rowboat used to transport goods on rivers and lakes in Colonial America.
Batedor	Portuguese term for a stick to froth chocolate (see also *molinillo*).
Bâton	French term for stick used to beat (froth) chocolate, sometimes elaborately carved.
Bbls	Abbreviation for barrels (undefined capacity).
BCE	Before the Common Era.

Term	Definition
BCO	British Colonial Office Archive.
Beef (tea)	Beef extract (i.e., beef bouillon) served in field hospitals.
Beer (small)	Beer prepared from the third mashing of grain, weaker than ale and second-mashed beer.
Blast	Colloquial English term used for destruction of cocoa plantations in the Caribbean due to disease.
Bloom	In the sense used in the present book, the dull efflorescence of cacao butter that sometimes appears on the surface of chocolate bars.
BMI	Body Mass Index (a relative measure of obesity).
BNRJ	Biblioteca Nacional, Rio de Janeiro, Brazil.
Boel	In the sense used in the present book, an archaic spelling for bowl (i.e., a chocolate bowl).
Bohea tea	Variety of tea from China, grown along the border of Jiangxi and Fujian Provinces (Bohea Hills). Pronounced "bu-ee."
Boroughmonger	Nineteenth century English slang term for town administrator.
Botica	Portuguese word for pharmacy.
BPRO	British Public Records Office (see also PRO).
Breviaries	Prayer books used by priests.
Bribri	Indigenous culture of southeastern Costa Rica.
Brick (dust)	Sometimes added to chocolate as an adulterant.
Brigantine	Two-mast sailing ship commonly used to transport cacao.
BT	British Board of Trade Archive.
Buhoneros	Spanish term for a street merchant, a peddler or hawker (of chocolate).
Bumbo	Variety of Colonial era punch commonly prepared using rum, sugar, and water.
C.wt (or cwt)	English unit of weight = 100 pounds; more commonly abbreviated cwt.
Caballería	Unit of measure used in pre- and post-Colonial Cuba; usually defined as 42 hectares or approximately 84 acres.
Cacahoacentli	Nahuatl word for the pod of the cacao tree (*Theobroma cacao*).
Cacahoatl	Nahuatl word for land with cacao trees (spelling variants include *cacaotal* and *cacaguatal*).
Cacao (green)	Unroasted or green cacao; chocolate beverages prepared from green cacao thought to intoxicate consumers.
Cacao añejo	Literally, old cocoa; a product of inferior quality and reportedly harmful to the health of consumers.
Cacao costeño	Term applied to cacao growing along the Caribbean coast of northern Colombia.
Cacaoeiro	Portuguese word for cacao farmer.
Cacaotales	Groves of cacao trees.
Cacap	Archaic French term used to describe a "fruit" similar to almonds; commonly applied to cacao beans.
Cacaueiros	Portuguese word for cacao trees.
Café con leche	French term for coffee with milk.
Cafetière	French term for coffeepot; a cafeteria where coffee is served.
Caisson	Large chest used to transport goods on mule back.
Calamus	Aromatic root of sweet flag (*Acorus calamus*).

Term	Definition
Calipha	Fictional Amazon queen; ruler of the island of California.
Camino Real, El	Literally, Spanish for "The King's Road."
Canella	Cinnamon.
Canon	Collection of books accepted as Holy Scripture (i.e., Bible); word also applied to a list of saints or a list of religious rules.
Canónigo	Member of the clergy belonging to a cathedral or collegiate church.
CAOM COL	Centre des archives d'outre-mer, Colonies (Canadian Archive).
CAOM DPPC	Centre des archives d'outre-mer, Depôt des papiers publics des colonies (Canadian Archive).
Capitanéa	Portuguese colonial administrative zone (in Brazil).
Caracas (cacao)	Variety of cacao beans from Venezuela shipped from the port of La Guaira.
Carga	Spanish unit of weight: one *carga* of cacao is equivalent to two *tercios*, or 440 pounds.
Carne asada	Spanish term for meat broiled on a spit.
Carnival	Pre-Lenten festival commonly held in Christian countries.
Carta Régia	Portuguese royal directive.
Caserío	Simple dwellings occupied by laborers working on cacao plantations.
Cassada	Alternative spelling for cassava.
Cassia	Sometimes called Chinese cinnamon, a member of the genus *Cassia*; pulp used medicinally and as a flavoring.
Cassonade	Raw, unrefined sugar commonly transported in a *caisson* (large chest).
Catalan	Pre-Castilian Spanish language of the Iberian Peninsula.
Catarrh	Inflammation of the mucous membranes of the nose and throat.
CE	Common Era.
Cedula of 1783	Spanish edict based on race and slave-holdings that offered land in Trinidad to white and free black settlers.
Cenote	Limestone sinkhole commonly filled with vegetation and sometimes wild cacao trees; when dry, cenotes sometimes are used for storage. (See rejollata.)
Censo	Obligatory fee paid to the Catholic Church.
CFR	Code of Federal Regulations.
Chaak	Mayan god of rain.
Chacha	Variety of medicinal chocolate.
Chamal	Cotton or woolen blankets worn by Native American women.
Champurrado	Traditional chocolate beverage of Mexico.
Chandler	Commonly, port-side merchants who provided ship supplies (e.g., chocolate to whaling ships).
Charibbee Islands	Spelling variant for Caribbean Islands.
Chimera	Fire-breathing monster or grotesque product of ones imagination.
China	In the sense used in the present book either (1) the country of China or (2) porcelain.
Chinois	Although Chinois is the Spanish term for Chinese, in early documents from New Spain the word refers to Asians in general (i.e., perhaps Chinese or perhaps Philippinos).
Chlorosis	The "green sickness."

Term	Definition
Chocholá	Mayan pottery style.
Chocolat fondant	Chocolate with cocoa butter added; sometimes identified as melting chocolate and used to prepare "eating" chocolate or various varieties of chocolate candy.
Chocolat ouvragé	Variety of hard chocolate used by sculptures to produce chocolate art.
Chocolate (amber)	Chocolate with ambergris added (not to be confused with fossilized amber).
Chocolate (de salud)	Literally, health chocolate; any of several varieties prepared as medicine to aid in the recovery of patients.
Chocolate (fake)	Various forms of "pseudo-chocolate" prepared using ground almonds and grape sugar.
Chocolate (French)	Combination of nibs, sugar, and vanilla.
Chocolate (labrado)	Literally, "good chocolate."
Chocolate (ordinario)	Literally, "ordinary chocolate" for everyday use.
Chocolate (pâte)	Chocolate paste.
Chocolate (quemado)	Literally, "burned chocolate;" a variety of very dark chocolate.
Chocolate (regalo)	Literally, "gift chocolate," especially fine varieties of chocolate commonly presented in gift chests.
Chocolate (rice)	A pudding-like dessert prepared from chocolate, boiled rice, and molasses.
Chocolate (Spanish)	Combination of almonds, cacao from Curacao, cinnamon, cloves, and sugar.
Chocolate (stone)	American English term for a saddle quern or grinding stone.
Chocolate (vanilla)	Combination of cacao from Caracas, cinnamon, cloves, and vanilla from Mexico.
Chocolate (wet)	Chocolate that would not harden (or set) due to high summer temperatures and humidity.
Chocolate (white)	Cocoa beans placed into a bag of water, left to soak in a pit for months; product becomes defatted and turns white. Feather-like tool used to remove the bean shells; small white beads that result from the process are retrieved. Regular cacao prepared and poured into a bowl; the white beads are whipped into a froth and poured on top of the chocolate (like the top of a cup of cappuccino coffee).
Chocolate (wild)	Beverage prepared from crushed, dried bloodroot; powder boiled then sugar and milk added.
Chocolateiro	Stone or ceramic grinding device used to pulverize cacao beans and spices.
Chocolatephilos	Lovers of chocolate.
Chocolatería	Place where chocolate is served to the public, also where chocolate is made and sold.
Chocolatero	Pot used to serve chocolate.
Chocolatero/a	Person (male or female) who makes or sells chocolate; alternatively, a person who likes to drink chocolate very often or a pot used to serve chocolate.
Chocolatière	Chocolate pot.
Chocolattexos	Archaic 18th century spelling of chocolatero.
Chorotegas	Early Spanish slang for the natives of Costa Rica.
Churro	Spiral or cylindrical sweet dough, deep-fried, coated with cinnamon and sugar usually served with hot chocolate.
Cioccolateiri	Master chocolate-maker or chocolate chef.
Cloister	Covered walk in a religious institution usually opening out onto a courtyard.
Clyster	Medicinal enema.

Term	Definition
Cochineal	Domesticated insect from which red dye is obtained.
Cocoa (essence)	Pure cocoa after 60–70 percent of fat has been extracted.
Cocoa (flakes)	Whole nibs and husks ground together to a flake-like appearance.
Cocoa (homeopathic)	Same as rock cocoa, but without sugar.
Cocoa (maravilla)	Combination of cacao, sago flour, and sugar.
Cocoa (nibs)	Bruised, roasted cacao beans deprived of their covering.
Cocoa (pressed)	Cocoa nibs containing only 30 percent residual cocoa fat.
Cocoa (rock)	Same as flake cocoa, with addition of arrowroot and sugar.
Cocoa (spoiled)	Cocoa adulterated accidentally or on purpose with salt (usually sea water during ocean transit).
Cocoa (walk)	English term for cocoa plantations in the Caribbean, especially Jamaica.
Cocos	Generally signifies coconuts; term also applied to small cups made of coconuts, sometimes ribbed/girded with silver.
Code Noir	Literally, Black Code. The French *Code Noir* of 1685 expelled Jews from Martinique and other French Caribbean islands.
Codex	Hand-painted, decorative pre-Columbian era text.
Coffee (crust)	Common beverage served to the sick: prepared by toasting bread, pouring boiling water over the well-done toast, then straining the water; cream, sugar, and sometimes nutmeg added.
Colera morbus	Archaic English term for dysentery (derived from Latin).
Collation	A light meal taken during days of general fasting.
Comal	Mexican-style clay cooking griddle.
Comistrajo	Generally, food of low quality; alternatively, a one-pot stew or hodgepodge.
Commandante	Military commander; garrison commander.
Commissary of the Holy Office	Another term for Commissary of the Inquisition.
Commissary of the Inquisition	Name given to each of the priests who represented the ecclesiastic tribunal in the main cities of the Kingdom of Spain (including all Spanish possessions in the Americas); each of the priests who represented the ecclesiastic tribunal in the main cities of the Kingdom of Spain (including all Spanish possessions in the Americas).
Commission of the Holy Office	Office of the Inquisition.
Commonplace book	Personal handwritten journal containing thoughts, recipes, quotes, and comments.
Companhia do Grão-Pará e Maranhão	Portuguese state-regulated trading company operating in Brazil.
Compania volante	Literally, "road troops," or Spanish soldiers who accompanied expeditions and supply trains.
Conching	Grinding, heating process used in the manufacture of chocolate.
Confection	Original meaning was a medicinal preparation mixed/bound with honey or sugar syrup.
Constirutior	Constitution, in the sense of general body health.
Convent	House or monastery where religious persons live under the rules of their institution. Convents can be for either priests or nuns.
Converso	A Jew converted to Christianity.

Term	Definition
Coreligionist	Term commonly used by specific religious groups, whether Christians, Jews, or Muslims, to define themselves.
Corregedor	Portuguese customs official.
Corsair	In the sense used in the present book, a pirate or a pirate ship.
Costa Firme	Literally, "firm coast;" term used in colonial times to define the shores of the American continent (North, Central, and South) in contrast to shores of the various respective Caribbean islands.
Costeño	General term for persons living near/on the Caribbean coast in northern Colombia.
Counterweight	In the sense used in the present book, a quantity of chocolate weighed on the merchant's scale (where the scale rested on the counter near the cash till).
Courlander	Person from the Duchy of Courland, a fief of the United Kingdom of Poland and Lithuania.
Couvercle	Unattached, hinged cover of a chocolate pot.
Crater	Large packing crate used by muleteers to transport goods (commonly chocolate).
Creole	Definition depends on geography: sometimes the designation for a person of European descent born in the West Indies or Spanish America; alternatively, a person culturally related to Spanish or Portuguese settlers in the American Gulf states; alternatively, a person of mixed black and European ancestry; alternatively a black slave born in the Americas.
Criollo	Variety of cacao beans grown in Mesoamerica; term also used to designate a white person born in the New World, specifically of white, Spanish parents.
Crust coffee	Beverage for the sick: prepared by toasting bread, pouring boiling water over the well-done toast, then straining the water; cream, sugar, and sometimes nutmeg added.
Crypto-Jew	Person of Jewish faith who practiced his/her religion in secret. Crypto-Jews were primary targets of the Spanish Inquisition.
CSA	Confederate States of America.
Cuartillo	Spanish coin with the value of $\frac{1}{4}$ real, minted by Enrique IV, king of Castile (1454–1474).
Cudgeon	Baton or large stick used for crushing.
Cupping glass	Globular glass used during bleeding operations (to remove bad humors).
Curandero	Traditional healer.
CWT or cwt	English unit of weight = 100 pounds; sometimes abbreviated C.wt.
d	English unit of currency; abbreviation for pence or cent.
Damask	Reversible fabric of cotton, silk, or wool, woven with patterns.
Damnified	Colonial era spelling of indemnified, to protect against damage or loss.
Defatted	In the sense used in the present book, defatted means to press out or remove cocoa butter during chocolate manufacturing.
Déjeuner	Breakfast service or meal; chocolate commonly a component.
Denizen	A foreigner granted rights of residence, sometimes citizenship.
Department	In the sense used in the present book, an administrative district.
Dia de la Muertos	Day of the Dead.
Disease (new)	The "new disease" (i.e., syphilis).
Dog days	Early July through early September; the hottest part of summer.

Term	Definition
Doily	Small napkin used during a dessert course, especially when serving chocolate.
Domine	A clergyman of the Dutch Reformed Church.
Drachm	An apothecary weight equal to 1 dram or $\frac{1}{8}$ ounce; also equal to 60 grains.
Dragées	French term for chocolate bonbons.
Draught	A drink or a sip.
Dropsy	Archaic medical term for swelling of soft tissue in the body due to water accumulation.
DRWW	Center for Documentary Research of the Southwest; archive located at the Arizona State Museum, University of Arizona, Tucson.
Duff	Molasses-sweetened fruitcake.
Dutch East India Company	Founded in 1602 as *Vereenigde Oostindische Compagnie* (United East Indies Company) with monopoly to conduct trade with eastern Asia.
Dutch pink	Form of dyed chalk used to stain paper; sometimes added as an ingredient to tea.
Dutching	Addition of alkali-potash to cacao nibs before roasting.
Earspool	Mayan decorative device; wood or stone spools inserted into slits in the wearer's earlobes.
Ek' Chuwah	Mayan god of cacao and merchants.
English East India Company	Chartered by Queen Elizabeth I in 1600 for trade with India and the East Indies.
Entremets	Side dishes.
Entrepôt	Trading or market center; distribution point for goods.
Ephemera	Printed material of general interest (exclusive of books or manuscripts).
Erido	Literally, wounded in battle.
Ersatz	Literally, substitute or false.
Espanto	Disease thought to be caused by fright.
exp.	Abbreviation for Spanish term *expedient*; used in archive research to represent a set of papers related to a single issue.
f. or fs.	Abbreviation for Spanish term *foja* (folio); used in archive research to represent all the pages of a single, official document.
Faience	Glazed pottery.
Familiares	Secular police of the Inquisition; responsible for conducting secret investigations.
Fanega	Spanish unit of volume; commonly 1 fanega = 1 English or 1 Spanish bushel; commonly 36 liters, 8 gallons, or 110 pounds.
Farina	Finely ground meal from various cereals sometimes added to chocolate as an adulterant.
Fazenda	Portuguese term for estate.
FDA	Food and Drug Administration.
Fecula	In the sense used in the present book, starch obtained from different plants (i.e., arrowroot, potatoes) sometimes added to chocolate as an adulterant.
Femme covert	A wife's legal identity as defined by English Common Law.
Femme sole	In English Common Law, a widow with rights to make contracts, run businesses, buy and sell property, and subject to taxation.
Finial	An ornamental post (commonly hinged) in the center of a chocolate pot lid.
Firkin	English unit of capacity; commonly 1 *ferkin* = $\frac{1}{4}$ barrel or 9 gallons.

Term	Definition
Fisco de la Inquisición	Inquisition Treasury: funds obtained from the auction of goods confiscated from those accused and convicted of heresy.
Flakes	Literally, windblown ashes/embers.
Flavanol	A class of plant compounds shown to be health-promoting.
Florin	Another term for a guilder, basic monetary unit of the Netherlands and Dutch colonies.
Fo'c'sle	Common English abbreviation of forecastle, the superstructure at the bow of a merchant ship (crew's quarters).
Foodways	Culinary practices of individuals, cultures, regions, and historic periods.
Forestero	Variety of cacao beans from the Amazon basin.
Formulary	Book of pharmaceutical formulas giving uses and preparation methods.
Franciscans	Order of friars founded by St. Francis of Assisi in 13th century Italy; also known as the Friars Minor, or Grey Friars.
Friandises	French-style sweet desserts that sometimes contained chocolate.
Friar	English religious term meaning Brother; member of one of the mendicant orders (i.e., those taking vows of poverty and living from alms)—defined as Augustine, Carmelite, Dominican, and Franciscan Orders.
Frigate	In the sense used in the present book, a fast, medium-size sailing ship.
Fritzie	Shortbread cookie with nutmeg added.
Fulling mill	Mill used to clean cloth.
Functional food	Any food or food ingredient that may provide a health benefit beyond the nutrients contained.
Furrio	Archaic Cuban Spanish word for chocolate.
Fyal (also Faial, Fyall)	An island in the Azores known for its wine.
Gadrooning	Ornamental band on an object where the silverwork is embellished with fluting or another continuous pattern.
Galactagogue	Item or beverage consumed by women in the belief such products will improve or increase lactation performance (increase volume of breast milk).
Galenical medicine	Humoral medicine as used by the ancient Greek physician Galen.
Ganache	Frosting or icing made from semisweet chocolate used as a filling for cakes and pastries.
Ganta	Spanish measure of volume; one *ganta* = 3 liters.
Gauchupin	Person born in Spain, but residing in New Spain.
Germ theory	Concept proposed by Louis Pasteur that the unseen world of "germs" could cause illness and diseases, and could be transmitted through air, water, and human-to-human personal contact.
Ghetto	Originally, a section of a European city restricted for Jews.
GHS	Georgia Historical Society.
Gill	Liquid measure equal to $\frac{1}{4}$ pint.
Goal	Archaic English spelling for jail.
Graino(s)	Spanish unit of currency, subdivision of a Spanish *real*; specific meaning of a small coin equivalent to a twelfth part of a *tomín*; sometimes used to designate cacao beans.
Grand Banks	Shoals in the western Atlantic off the coast of Newfoundland; major fishing location.

Term	Definition
Great	In this sense a verb: spelling variant for grate (i.e., to grate a bar of chocolate).
Great Awakening	Period of dramatic religious revival that swept the American colonies during the 1730s and 1740s.
Greenhand	Term applied to ordinary sailor.
Grog	Rum mixed with water; commonly served daily to sailors when aboard ship.
Gruit	Mixture of spices added to ale during the brewing process.
Guaraná	Stimulating beverage prepared from the Amazonian plant *Paullinia cupana.*
Guarapo	Spirit beverage made from ground sugar cane.
Guardian (Father Guardian)	Head of a Franciscan community of friars.
Guaxnicion	Military garrison.
Guilder	Basic monetary unit of the Netherlands and Dutch colonies.
Gum tragacanth	Medicinal gum prepared from the herb *Astragalus gumifer.*
Guman	Tree resin.
Gumdragon (see Tragacanth)	Alternative spelling for gum tragacanth.
H.M.	His/Her Majesty.
Habit	Ordinary clothing used by priests.
Hacendado	Landowner.
Hadienca	In the sense used in the present book, a Spanish cacao plantation in the New World.
Haham	Hebrew term for wise man or sage.
Hardtack	Hard biscuit, commonly a military ration, made of flour, salt, and water.
Hechura	Wooden painted religious statue.
Hhd	English abbreviation for a hogshead cask.
Hidalgo	Spanish knight.
Hispederia	Travelers lodge, often within a mission or presidio.
Hogshead	English unit of volume: a barrel with the capacity of 63 gallons.
Holli	Tree resin.
Holy Thursday	Thursday before Easter Sunday.
Hoosh	Literally, one-pot meal or stew.
Hops	Leaves of *Humulus lupulus* added during the brewing process to make beer.
Houacaca powder	Archaic spelling for ground cocoa powder from Oaxaca.
Huautli	Mexica/Aztec word for amaranth (*Amaranthus* spp.), sometimes used in the 19th century as a chocolate adulterant.
Hudson's Bay Company	English fur-trading company chartered by King Charles II, May 1670, granted to "the Governor and Company of Adventurers of England trading into Hudson Bay."
Humors	The four body fluids (blood, phlegm, black bile, yellow bile); humoral balance believed to determine person's health.
Hydrophilic	Water-loving; term sometimes applied to sugar.
Hydrophobic	Water-fearing; term sometimes applied to cacao butter.
Hypothermia	Dangerously low body temperature, sometimes relieved by drinking hot chocolate.

Term	Definition
Iberia	Western European peninsula consisting of the countries of Spain and Portugal.
IFIC	International Food Information Council.
Île Royale	Site of Fortress Louisbourg in New France (Canada).
Île St.-Jean	Prince Edward Island, Canada.
Indenture	A contract whereby a person was obligated to the service of another (i.e., indentured servant).
Inquisition	Roman Catholic tribunal for the suppression of heresy.
IRB	Institutional Review Board, University Committee on Use of Human Subjects.
Iricia	Bribri goddess; niece of Sibö.
Isinglass	In the context used in the present book, a transparent gelatin prepared from the air bladder of sturgeons, used as a clarifying agent when cooking.
Itzamna	Mayan goddess, primordial creator of the earth, flowers, and wind.
Ixik Kaab'	Mayan earth goddess.
Jacobite	Supporter of James II or Stuart pretenders to the throne of England after 1688.
Jakaltak	Indigenous Mayan ethnic group of Guatemala.
Jarro	Simple mug used to serve chocolate.
Jigger	Small glass.
Jocolato	Early Colonial New England spelling for chocolate.
Jordan Almonds	Roasted almonds coated with a thin candy shell.
Jurisdiction	In the sense used in the present book, a province.
K'awil	Mayan god of sustenance represented by lightning bolt.
K'iche	Indigenous Mayan ethnic group of Guatemala (alternative spelling Quiché).
KAB	Acronym for knowledge, attitudes, behaviors.
Kakaw yal ixim	Mayan beverage prepared from a mixture of maize and cacao, same as pozole.
Kedge	Small anchor.
Kimil	Mayan god of death.
Kisin	Mayan god of the underworld.
Knop	Small decorative knob or boss.
L (£)	English unit of currency; abbreviation for 1 English pound.
Labor	Variable interpretations. In this book, the word "labor" is used in two specific contexts: (1) a term meaning building supplies (i.e., bricks and tiles, where 1 *labor* = 1000 bricks/tiles) and (2) a term meaning a quantity of chocolate produced by a laborer during an official day of work, commonly established at 48 pounds.
Ladino	Judeo-Spanish language primarily spoken by Sephardic Jews.
Laudanum	Opiate derivative sometimes used as an adulterant of chocolate.
lb	Abbreviation for pound (weight).
League	A land distance measurement = 4.2 kilometers or 2.6 miles. A maritime measurement of distance, highly variable: commonly, 1 league = 3 miles.
Lett	Archaic spelling for let (i.e., to rent).
Licenciate/Licenciado	In the context used in the present book, a university graduate.
Lignum vitae	Literally, "tree of life;" a commercial hardwood from the tropical tree *Guaiacum officinale* or *G. sanctum,* with various medicinal uses.
Limonadier	French for a member of the café-owners guild.

Term	Definition
Liquor	In the context used in the present book, the liquid extracted from dried cocoa nibs during the grinding process.
Load	In the context used in the present book, a designated quantity of cacao beans: 1 load = 2 tierces or 4 fanegas. One load of cacao was equal to 440 pounds, commonly 8000 beans (= 1 *xiquipile*). Spanish term for load is *carga*.
Lobscouse	Stew of salted meat, potatoes, and vegetables.
Lung sien hiang	Chinese term translated as "dragon spittle perfume" (i.e., ambergris), mixed with chocolate.
Macuquina	Slang for small coin, subdivision of a real; used in the province of Yucatan.
Madrerías	Spanish term for "mother trees," planted first to provide shade and protection to the more delicate, younger cacao seedlings.
Maize flower	Archaic spelling of maize flour or cornmeal.
Mam	Indigenous Mayan ethnic group of Guatemala.
Mamarrachos	Cuban summer festivals.
Mambí	Afro-Antillan word applied to revolutionaries from Cuba and Santo Domingo who fought against Spain in the War of Independence (War of the 10 Years).
Mancerina	Chocolate stand or *trembluese*.
Manchet	Loaf of bread made from fine, white flour.
Manipulador	In the context used in the present book, Portuguese term for "handler," or person providing medicinal cacao supplies.
Mano	Spanish term for "hand;" In the context used in the present book, a stone roller used for grinding on a *metate*.
Manteiga de cacão	Cocoa butter.
Maracaibo (cacao)	Variety of cacao beans from the Venezuelan Andes.
Marrano	Person who professed to be Catholic, but secretly practiced Jewish rites.
Masa	Corn flour prepared as dough.
MDRU	Multi-Disciplinary Research Unit located at the University of California, Davis.
Mecacacahoatl	Nahuatl term for a type of cacao bean.
Mecate	Spanish term for a rope made from maguey fiber.
Mecaxuchil	Nahuatl term for vanilla: *Vanilla planifolia*.
Mélangeur	Two heavy round millstones supported on a granite floor and used to grind chocolate.
Merienda	Spanish term generally translated as "snack."
Mestizo/a	Person of Spanish-Indian heritage.
Metate	Grinding stone.
Metayer	As used in the Caribbean, an emancipated former slave who worked as a sharecropper on cacao plantations.
Metropole	Used in the sense of an urban area.
Mexica	Preferred term for Aztec.
Milk-punch	Medicinal beverage prepared using milk mixed with liquor, sugar, and flavoring agents.
Mill	One-tenth of a cent.
Miniestora	Ration of dry, cooked legumes.
Mixtec	Indigenous ethnic group of south-central Mexico, region of Oaxaca; Mixtec communities also prominent in the town of Madera, California.

Term	Definition
MOA	Making of America (Digital Archive).
Molinet	Term for chocolate stirring rod.
Molinillo	Stirring rod used to beat (froth) chocolate, sometimes elaborately carved.
Monoculture	Agricultural practice of planting only one crop (i.e., all one's land used to produce only cacao, coffee, or other crop).
Monteith	Large punch bowl, usually silver, having a notched rim for suspending punch cups.
Moors	In the context used in the present book, Muslims from North Africa who invaded Iberia.
Morocco	Soft, fine goatskin leather tanned with sumac; commonly used for book bindings and shoes.
Moulding	In the sense used in the present book, manufacturing process used to shape or mold chocolate.
Moussoir	French term for chocolate stirring rod.
Mouthfeel	Term used by food scientists to describe the interactions of food taste and texture. In reference to chocolate, mouthfeel is used to describe how the product melts and flows inside the mouth, and whether or not the product is smooth or contains gritty particles.
Muddler	English term for chocolate stirring rod.
Muscovado	Unrefined, raw sugar.
N.B.	Abbreviation of the Latin, *nota bene*, meaning pay special attention.
NAC	National Archives of Canada.
Nahuatl	Mexica/Aztec language.
Nal	Mayan god of maize.
NAS	National Archive of Scotland.
Nation, The	Term sometimes used to identify Jews in Colonial era documents; derived from the portuguese phrase, *os da nação*.
Natural	In the sense used historically in the present book, a Native American.
NBWM	New Bedford Whaling Museum, New Bedford, Massachusetts (Archive).
New France	In the broadest sense, French colonies in North America.
New Netherlands	In the broadest sense, the region of North America settled by the Dutch that included portions of the present states of Connecticut, Delaware, New Jersey, New York, and Pennsylvania.
New Spain	In the broadest sense, Spanish colonies located in North and Central America.
Nibs (cocoa)	Shelled, roasted cacao beans.
Nigh	Archaic English for nearby.
Nik	Mayan god of wind, life, and flowers.
Nog	Variety of Colonial era punch prepared from milk, rum, and various spices.
NWC	North West Company, fur-trading company based in Montreal.
NYWRA	New York War Relief Agency (American Civil War era).
Oakum	Loose fiber prepared by untwisting and separating strands of old rope; used to caulk ship seams and pack pipe joints.
OBP	Old Bailey Proceedings (English trial archives).
Olmec	Mesoamerican civilization, ca. 1000–400 BCE, that inhabited the southern Gulf Coast of Mexico.
Otomí	Indigenous ethnic group of central Mexico.

Term	Definition
Oyl	Oil; alternative spelling, *oyle*; archaic 18th century vulgar English slang for female sexual lubrication.
Pacotilles	Private cargo aboard ship that sometimes consisted of personal chocolate stores.
Panada	Bread crumb paste or gruel.
Pancit	Chinese-style noodles.
Panda/Pandao	Spelling variants of panada, bread crumb paste or gruel.
Pannikin	Small tin cup.
Panocha	Depending on context, can be a solid cone of brown sugar (see Piloncillo), or a brown sugar fudge-like confection made with butter, milk, and nuts.
Pañuelo	Woman's scarf, used in the sense of a fine silk scarf imported to New Spain from China via the Manila galleon.
Panyols	Poor Venezuelan farmers who immigrated to Trinidad in the 19th century to work cacao plantations.
Passions	In the context used in the present book, religious chanting during Holy Week (i.e., the week before Easter).
PCSSCO	Pacific Coast Steam Ship Company.
PEI	Prince Edward Island, Canada.
Pemmican	Native American travel ration prepared using dried meat pounded into powder, mixed with fat and dried fruits/berries, then pressed into small cakes.
Peso (silver)	Spanish unit of currency.
Pesos fuertes	Literally, "strong pesos," meaning old silver Spanish coins of one ounce, equivalent to 8 reales.
Petate	Mat made from palm fiber.
Phylloxera	*Daktulosphaira vitifoliae*; parasite to *Vitis vinifera,* the domesticated grapes used to make wine.
Physicke	Archaic term for physician.
Phytochemical	A nonnutritive plant constituent that confers a health benefit to the consumer.
Pilón/Piloncillo	Solid cone of brown sugar. (See also Panocha.)
Pinol	Archaic spelling: toasted corn flour sometimes prepared with chocolate.
Pinole	Contemporary spelling: toasted corn flour sometimes prepared with chocolate.
Pipe	English unit of volume; a type of barrel with the capacity of 126 gallons = 2 hogsheads.
Pirogue	Dugout canoe.
Pistoles	Chocolate drops or large chips/chunks of chocolate.
Plaza	Central square of a Spanish settlement.
Podagra	Archaic term for gout.
Pompholix	Unrefined zinc oxide sometimes added to chocolate as an adulterant.
Popol Vuh	Mayan creation text.
Porcelain	Hard, white, translucent ceramic made of pure clay, glazed with various colored materials (slips). (See China.)
Porringer	Shallow cup or bowl with a handle.
Portable soup	Kind of dehydrated food used in the 18th and 19th centuries, similar to meat extract and bouillon cubes; also known as *pocket soop* (sic) or *veal glew* (sic); staple of seamen and explorers.

Term	Definition
Portage	Carrying goods and supplies overland from one navigable water (lake, river) to another.
Porteria	Porter's hut usually adjacent to the mission or fort entrance gate.
Potosí	A city of Mexico founded in 1545 and famous for its silver mines during the 16th–17th centuries. The term *tener un Potosí* (to have a Potosí) was widely used to signify great wealth.
Pozole	Beverage prepared from a mixture of maize and cacao. (See also Kakaw yal ixim.)
Pre-Columbian	Generally applied to activities and events in what is now North, Central, and South America, and the Caribbean before arrival of Christopher Columbus in 1492.
Presidio	Spanish fort and administrative center.
Price Currents	Newspaper announcement of local/regional fair prices for selected commodities.
Privateer	Privately owned ship under government authorization to attack and capture enemy vessels during wartime; also the commander of such a ship.
Prize	In the context used in the present book, a captured enemy ship.
PRO	Public Records Office, London (Archive).
Procurador	Priest of a religious community in charge of economic decisions.
Prussian Blue	Insoluble dark blue pigment composed of ferric ferrocyanide.
ps.fs	Spanish abbreviation for the monetary term, *peso fuertes* (i.e., strong peso), that is silver backed.
Punch	Strong alcoholic beverage prepared from spirits, spices, lemon juice, sugar, and water—hence the terms "punch-drunk" and "punch-bowl."
Puyomate	A plant indigenous to Mexico widely used by traditional healers and others for the purposes of attraction and repulsion (i.e., a "love" plant).
Pye	Archaic English spelling of pie.
Quarrée	French unit of land measurement in the Caribbean, equal to 3.2 acres.
Quartillo	Spanish unit of volume; commonly, 1 quartillo = $\frac{1}{4}$ liter.
Quauhcacahoatl	Nahuatl term for a type of cacao bean.
Quern (saddle quern)	Primitive saddle-shaped stone used to grind grain (and cacao beans).
Quetzalcóatl	Toltec/Mexica/Aztec god who brought the cacao tree to humans.
Quiché	Indigenous Mayan ethnic group of Guatemala (alternative spelling K'iche).
Quintal	Spanish unit of weight: 1 *quintal* = 100 kilograms.
Ratsbane	Type of poison (arsenic trioxide).
Real	*Real de plata* (Real of silver): Spanish unit of currency; 10 *reales* = 1 silver *peso*.
Reconciliado(a)	A heretic, first offender, but one who returned to the church. During Inquisition trials the property of a reconciliado/a was confiscated and the person usually required to wear a *sanbenito* in public.
Recto	Right or front side of an unnumbered document.
Red scutcheneel	Alternative term for cochineal; red dye obtained from a domesticated insect.
Reddle	Iron ore used in dying and marking, sometimes added to chocolate as an adulterant.
Reducciones	Term applied to California Indian Christian converts living on mission land.
Refectory	Dining hall in a religious house or college.
Refiner	Stage of chocolate production after chocolate paste has been removed from a mélangeur; a finer grinding machine.

Term	Definition
Rejollada	Limestone sinkhole (dry); sometimes used for storage by the Maya. (See Cenote.)
Relación	An account of an event.
Relajado	Person who is to be "relaxed;" person burned at the stake after being convicted of heresy (i.e., the one being relaxed).
Relaxation	An Inquisition procedure to deliver a condemned heretic to the hands of the secular authorities for execution. Being an ecclesiastical organization, the Inquisition could not itself carry out the execution since one of the ten commandments is—*thou shall not kill.*
Religious	In the sense used in the present book, a plural noun signifying priests and nuns.
Rendezvous	In the sense used in the present book, an annual meeting of fur trappers in the Green River region of the Rocky Mountains.
Replacement	In the context used in the present book, a fat substitute added to chocolate after the natural cocoa butter has been removed (i.e., lard, suet, tallow).
Repoussé work	Metalworking design where patterns are formed in relief by hammering and pressing on the reverse side.
Resma	Spanish measure of paper weight and quantity = 500 pages (i.e., ream).
Retablos	Small altar ornaments.
Revista de Agricultura	Journal of Agriculture.
Rheum	Discharge from the mucous membranes, especially from the eyes, nose, or respiratory tract.
Roças	Portuguese term for cacao plantation.
Rocoa	Red colored pulp covering seeds of *Bixa orellana*, from which *annotto* is prepared. (See Achiote.)
Ropa-vieja	Boiled meat subsequently pan-fried.
Rose of Alexaxndria	Old World rose petals/essence sometimes added to chocolate.
Rowl	Archaic spelling variant for roll.
Rugg	Archaic spelling for a heavy blanket.
Rundlett	Small wine cask of undefined volume.
S.J.	*Sacerdote jesuita*: Society of Jesus (a Jesuit).
Sack	English term for sherry.
Sacristy	Room in a religious house used to store religious objects and sacred vessels.
Sagoe	Spelling variant of sago; starch obtained from sago palm.
Saint Anthony's fire	Medieval term for ergot poisoning.
Saint Domingue	Original term for the western portion of the island of Hispaniola (i.e., Haiti).
Salep	Starchy food paste prepared from dried roots/tubers and orchids.
SAMCC	Sacramento Archives and Museum Collection Center (Sacramento, California).
Sanbenito	A penitential garment (i.e., dunce cap and cape) worn over the outer clothing of the guilty convicted of heresy for a period of time determined by the inquisitors.
Sangria	Spanish fruit and wine-based beverage.
Santonin	Compound obtained from wormwood, a type of antihelmenthic (medicine to get rid of intestinal worms).
Save	English language variant for the town, Sèvres, the French city known for its manufacture of china.
SBMA	Santa Barbara Mission Archives.

Term	Definition
SBPA	Santa Barbara Presidio Archive.
Schooner	Two-mast sailing ship; commonly used to transport cacao.
SCHS	South Carolina Historical Society.
Scruple	Apothecary weight = 20 grains.
Sea pie	English slang term for fish or meat dumpling.
Searce	Archaic English word meaning "to sieve" or "to sift."
Selectmen	English term for elected city officials.
Sephardim	Jews from the Iberian Peninsula.
Serch	Spelling variant for "searce," archaic English term meaning "to sift."
Seroon	Bale or package.
Serviçais	In the sense used in the present book, new laborers (not necessarily servants).
Service à la Française	Courses served buffet style in contrast to Service à la Russe, where courses are brought sequentially to the table.
Service à la Russe	Courses brought sequentially to the table in contrast to Service à la Française or more buffet style dining.
Shallop	Large, heavy boat with two masts and fore- and aft-sails; alternatively, a small, open boat with oars or sails used in shallow water.
Shells	In the sense used in the present book, the shells of cacao beans; also brewed to make a chocolate-shell tea.
Shilling	Predecimal monetary system of England; 1 shilling = 10 pence; 12 shillings = 1 English pound (£).
Shipping News	Newspaper column identifying the arrival and departure of ships from various North American ports.
Sibö	Primary god of the Bribri of Costa Rica.
Sickness (green)	General term for chlorosis.
Sloop	In the sense used in the present book, a single-mast vessel commonly used to transport cacao.
Slop chest	Sea captain's chest containing nonessential items (sometimes chocolate).
Slops	Archaic English slang for unappetizing watery food or soup.
Small beer	Lightly alcoholic beer prepared from the third pressing of the barley grain.
Soconosco (cacao)	Variety of cacao beans from southern Mexico, region of Chiapas.
SOI	Abbreviation for Standard of Identity; used by the food industry to define foods and purity standards.
Soldados de cuera	Literally, leather-jacket soldiers who commonly were assigned for escort and protection duties in New Spain.
Solitaire	A beverage service for one person (commonly chocolate).
Sotana	Spanish term for priest's cassock.
Souchong tea	Black tea from Fujian Province, China; sometimes called Russian caravan tea.
Soupçon	French term to describe quantity (i.e., a very small amount), a hint, or a trace.
SP	State Paper Office, London (archive).
Spanish Main	Geographical term with two meanings: (1) Caribbean mainland of eastern Central America including the coast of Mexico south to the isthmus of Panama; (2) romantic term for the Caribbean Sea during the period of the Spanish treasure ships and piracy.
Spend	Archaic English slang for orgasm.

Term	Definition
Spermaceti	Waxy, highly valuable amber-colored oil contained in the head of the sperm whale.
Spiked (cannon)	Act of driving a spike into the touch-hole (fire-hole) of a cannon, rendering it useless.
Spinage	Archaic spelling variant for spinach.
Stiver	Small coin used in the Netherlands and Dutch colonies, worth 1/20th of a guilder.
Store chocolate	Chocolate sold under a company label.
Subaltern officer	Commissioned officer below the rank of captain.
Sufi	Follower of mystic traditions within Islam.
Sugar Trust	American organization of eight major refining companies formed in 1887.
Sulpicien	Society of secular French priests founded in 1642 to train men to teach in seminaries.
Susto	Disease thought to be caused by fright.
Sutler	Person who followed an army or maintained a store on an army post to sell provisions to soldiers.
Sware	Archaic English spelling for to swear, as to swear in court.
Switchell	A rum-based cocktail.
Syllabub	Beverage prepared from sweetened milk or cream curdled with wine or spirits.
Syndic	Religious magistrate or administrative representative.
Tabililla	Commonly, a chocolate tablet.
Tansy tea	Herbal tea prepared from the herb *Tanacetum vulgare*.
Teamster	One who drives a team of horses for hauling goods.
Tecomate	A type of squash with a narrow neck and hard shell (when dried) used to manufacture large vessels and drinking cups.
Temperance Movement	Antialcohol movement that commonly recommended cacao as a beverage instead of beer, wine, and spirits.
Tempering	In the sense used in the present book, a chocolate production process employed to reduce bloom.
Tenor	Archaic English term for money.
Terra alba	Literally, "white earth;" a type of generic clay used as a chocolate adulterant.
Terrine	Large clay pot used to serve food, sometimes chocolate.
Tête-à-tête	In the sense used in the present book, a beverage service for two persons (commonly chocolate).
The Twelve	Term used to identify the group of twelve priests who arrived in Mexico after the Conquest.
Théière	French term for teapot.
Tibores de chocolate	Chocolate jars.
Tierce	A variable measure of liquid or weight. A liquid *tierce* commonly equaled 1/3 of a *pipe* (42 gallons). As a unit of weight when measuring cacao, it was equal to 220 pounds, sometimes described in documents as half-a-load.
Tiffin	English term for a "light lunch."
Tlalcacahoatl	Nahuatl term for a type of cacao bean.
Tlaloc	Toltec/Mexica/Aztec god of rain.
Tlilixochitl	Nahuatl term for vanilla: *Vanilla planifolia*.

Term	Definition
TO	Treasury Office, London (archive).
Tomín	A silver coin used in certain parts of Spanish America during colonial times; also used as a weight measure in pharmacy.
Tonalamatl	Toltec book of prophecies.
Tory	American who sided with the British during the Revolutionary War.
Tragacanth (see Gumdragon)	Gum used in adhesives, cooking, pharmacy, and textile printing and obtained from *Astragalus gummifer*.
Transport	In the sense used in the present book, an English court verdict; to "transport" (deport) criminals out of England to the Americas or to Australia (Botany Bay).
Trembluese	Chocolate stand shaped like a shell with attached cup holder.
Trinitario	Variety of cacao produced on the island of Trinidad through genetic crosses of forestero and criollo varieties.
Trunnion	Cylindrical projections on a cannon, one on each side for supporting the cannon on its carriage.
Try Works	Large iron cauldrons used to process whale blubber into oil while at sea.
Tsura	Bribri goddess; sister and wife of Sibö.
Twelve (The)	Term used to identify the group of twelve priests who arrived in Mexico after the Conquest.
UCHS	University Committee on Use of Human Subjects.
Umber	Earth rich in iron and manganese oxide that exists as a brown pigment (raw umber); when heated turns red-brown (burnt umber); sometimes added to chocolate as an adulterant.
USAID	United States Agency for International Development.
USSC	United States Sanitary Commission.
Vacaca Chinorum	Archaic term for Chinese chocolate.
Vauban style	Marquis Sébastien le Prestre de Vauban was a French military engineer who revolutionized fortification and siege strategies during the reign of King Louis XIV.
Vendue	Auction house or public sale.
Venetian red	A light, warm, unsaturated scarlet pigment derived from nearly pure ferric oxide; sometimes added to chocolate as an adulterant.
Verdigris	Green patina or crust of copper sulfate.
Vermifuge	Medicine to remove intestinal worms.
Vermillion	A bright red pigment prepared from mercuric sulfide; sometimes added to chocolate as an adulterant.
Verso	Left or back side of an unnumbered document.
Viceroy	Governor of a province.
Vi-Cocoa	Cocoa with addition of kola nut extract.
Victuler	Supplier of food items (i.e., grocer).
Walk	In the sense used in the present book, an English term for cocoa plantation in the Caribbean, especially Jamaica (i.e., a cocoa walk).
Walloon	French-speaker of southeast Belgium.
Wax	In the sense used in the present book, religious candles.
wch	Archaic abbreviation for "which."
wd	Archaic abbreviation for "would."

Term	Definition
Wharfage	Fee charged to dock a vessel while in port.
Whig	In the context used in the present book, a supporter of the war against England during the American Revolution; during the 19th century, a political party that favored high tariffs and loose interpretation of the American Constitution.
Whisk	Term for chocolate stirring rod.
Windward Islands	Collective term for the Caribbean islands of Dominica, Grenada, St. Lucia, and St. Vincent.
Wine (Rennish)	Early English term for German wines.
Witches broom	Disease of cacao trees caused by the parasitic fungus *Crinipellis perniciosa*, which produces abnormal brush-like growth of small thin branches on cacao trees.
Wort	An infusion of grain and water; a step in the production of beer.
WTO	World Trade Organization.
Xiquipile	Measure of cacao bean quantity; 1 *xiquipile* = 8000 cacao beans. 1 load = 5–6 gold *pesos*.
Xmucane	Mayan creation god.
Xochicacahoatl	Nahuatl term for a type of cacao bean.
Xochiquetzatl	Toltec goddess.
Xocolatl	Nahuatl term for chocolate, commonly translated as "bitter water."
Yacatecuhtli	Mexica/Aztec god of merchants and travelers.
Yarrow	Leaves of *Achillea millefolium* sometimes added during the brewing process to make beer.
Yucca	Synonym for cassava.
Yuta	Alternate spelling for Ute Indians.
Zapotec	Indigenous ethnic group of southern Mexico.
Zontle	Measure of quantity; 1 *zontlea* = 400 cacao beans.

Archives, Institutions, Libraries, and Museums Consulted

Aeronautical and Space Museum, Smithsonian Institution, Washington, DC

Albany Institute of History and Art, Albany, New York

American Antiquarian Society, Worcester, Massachusetts

American Jewish Archives, Hebrew Union College, Cincinnati, Ohio

American Jewish Historical Society, New York, New York

Annenberg Rare Book & Manuscript Library, University of Pennsylvania, Philadelphia, Pennsylvania

Archiv der Luftschiffbau Zeppelin GmbH, Zeppelin Museum, Friedrichshafen, Germany

Archives Communales de Bayonne, Bayonne, France

Archives de la Marine, Paris, France

Archives départmentales, de la Charente-Maritime, La Rochelle, France

Archives des colonies, Paris, France

Archives Municipales de Bayonne, Bayonne, France

Archives Nationales de France Outre-Mer, Aix-en-Provence, France

Archives Nationales, Paris, France

Archivo General de Indias (AGI), Seville, Spain

Archivo General de la Nación (AGN), Mexico City, Mexico

Arizona Historical Society, Tucson, Arizona

Arizona State Museum, University of Arizona, Tucson, Arizona

Arquivo Historica Ultamarino, Lisbon, Portugal

Arquivo Historico do São Tome, São Tomé & Príncipe, West Africa

Arquivo Nacional da Torre do Tombo, Lisbon, Portugal

Art Institute of Chicago, Chicago, Illinois

Baker Business Library, Harvard Business School, Boston, Massachusetts

Bancroft Library, University of California, Berkeley, California

Barbados National Archives and Museum, Garrison, Barbados

Basil Bartley Archive, Shields Library, University of California, Davis, California

Beinecke Archives, Yale University, New Haven, Connecticut

Biblioteca Manuel Orozco y Berra, Mexico City, Mexico

Biblioteca Nacional do Rio de Janeiro, Brazil

Biblioteca Nacional Fondo Reservado, Mexico City, Mexico

Biblioteca Portuguesa do Rio de Janeiro, Brazil

Chocolate: History, Culture, and Heritage. Edited by Grivetti and Shapiro
Copyright © 2009 John Wiley & Sons, Inc.

Bibliothèque Fornay, Paris, France

Bibliothèque Historique de la Ville de Paris, Paris, France

Bibliotheque National de France, Paris, France

Bodleian Library of Commonwealth and African Studies at Rhodes House, Oxford, England

Bodleian Library, Oxford University, Oxford, England

Boston Public Library, Rare Books Collection, Boston, Massachusetts

Botanical Collections, Royal Botanical Gardens, Kew, Richmond, England

Botany Library, Natural History Museum, London, England

Braun Research Library, Southwest Museum, Los Angeles, California

British Library and Museum, London, England

Butler Library, Rare Books Division, Columbia University, New York, New York

Cadbury's Archives, Bournville, Birmingham, England

California Mission Archive, Santa Barbara, California

California State Historical Society, San Francisco, California

California State Library, Sacramento, California

California State Railroad Museum, Sacramento, California

Cambridge University Library, Cambridge, England

Caribbean Agricultural Research and Development Institute (CARDI), St. Augustine, Trinidad and Tobago

Centro de Estudios de Historia de Mexico (Condumex), Mexico City, Mexico

Chantal de Chouchet, Art & Gastronomie, Versailles, France

Chinese Historical Society, San Francisco, California

Christie's, New York, New York

Christie's, Paris, France

Clark Art Institute, Williamstown, Massachusetts

Cocoa Museum, Martinique

Cocoa Research Centre Library, University of the West Indies, Trinidad

College of Physicians, Philadelphia, Pennsylvania

Colonial Williamsburg Foundation, Williamsburg, Virginia

Connecticut Historical Society, Hartford, Connecticut

David Library of the American Revolution, Washington's Crossing, Pennsylvania

Debauve et Gallais, Paris, France

Department of Archives Black Rock, St. James, Barbados

Deutsche Zeppelin-Reederei GmbH, Friedrichshafen, Germany

Dunbarton Oaks Library, Washington, DC

DuPont Hagley Museum and Library, Wilmington, Delaware

Earl Gregg Swem Library, College of William and Mary, Williamsburg, Virginia

Edward Kiev Judaica Collection, The Gelman Library, George Washington University, Washington, DC

Escuela de Estudios Hispanoamericanos, Seville, Spain

Fort Ticonderoga, Ticonderoga, New York

Fortress of Louisbourg Archaeological Active Research Collection, Sydney, Canada

Fortress of Louisbourg Archives, Sydney, Canada

Fortress of Louisbourg Curatorial Collection, Sydney, Canada

Free Library of Philadelphia, Philadelphia, Pennsylvania

Frick Art Reference Library, New York, New York

Frost Library, Amherst College, Amherst, Massachusetts

Fundacao Oswaldo Cruz Library, Rio de Janeiro, Brazil

Georgia Historical Society, Savannah, Georgia

Getty Institute, Los Angeles, California

Glenbow Museum, Alberta, Canada

Glendale Public Library Special Collections, Glendale, California

Harvard University, Cambridge, Massachusetts

Hayden Library, Arizona State University, Tempe, Arizona

Hellenic Literary and Historical Archives, Athens, Greece

Henriet Chocolatier, Biarritz, France

Hispanic Society of America, New York, New York

Historic American Cookbook Project, Michigan State University, East Lansing, Michigan

Historic Deerfield Archives and Library, Deerfield, Massachusetts

Historic Hudson Valley Association, Tarrytown, New York

Historical Archives of Cartagena, Colombia

Historical Archives of Samos, Greece

Historical Overseas Colonial Archive, Lisbon, Portugal

Historical Society of Philadelphia, Philadelphia, Pennsylvania

Houghton Library, Harvard University, Boston, Massachusetts

Hudson's Bay Company Archives, Winnipeg, Canada

Huguenot Historical Society, New Paltz, New York

Huntington Library, San Marino, California

Instituto de Agroquimica y Tecnologia de Alimentos, Valencia, Spain

Instituto de Estudios Documentales e Historicos Sobre la Ciencia, University of Valencia, Spain

Instituto do Valle Flor, Lisbon, Portugal

Instituto Mora Library, Mexico City, Mexico

Instituto Nacional de Antropologia e Historia (INAH), Mexico City, Mexico

Instituto Português dos Museus, Lisbon, Portugal

John Carter Brown Library, Brown University, Boston, Massachusetts

John D. Rockefeller Library, Colonial Williamsburg, Williamsburg, Virginia

John F. Kennedy Library, California State University, Los Angeles, California

John Grossman Collection of Antique Images, Point Richmond, California (now located in Tucson, Arizona)

Kiev Collection, George Washington University, Washington, DC

Librairie Gourmande, Paris, France

Library and Archives of Canada, Ottawa, Canada

Library of Congress, Washington, DC

Library, School of Geography, Oxford, England

Los Angeles Public Library, Los Angeles, California

Luis Angel Arango Library, Bogota, Colombia

Manufacture Nationale de Céramique, Sevres, France

Manufacture Nationale de Sevres Archives, Sevres, France

Maritime Museum, San Francisco, California

Mary Norton Clapp Library, Occidental College, Los Angeles, California

Maryland State Archives, Annapolis, Maryland

Massachusetts Historical Society, Boston, Massachusetts

McCord Museum of Canadian History, Montreal, Canada

Metropolitan Museum of Art, Collections, Records, and Archives, American Wing, New York, New York

Metropolitan Museum of Art, New York, New York

Metropolitan Museum of Art, Thomas J. Watson Library, New York, New York

Michel Chaudun, Chocolatier, Paris, France

Minneapolis Institute of Arts, Minneapolis, Minnesota

Mount Holyoke College Library, Special Collections, South Hadley, Massachusetts

Mount Vernon Collection, Mount Vernon, Virginia

Musée Basque, Bayonne, France

Musée Carnavalet, Paris, France

Musée de la Publicité, Paris, France

Musée des Arts Décoratifs, Paris, France

Musée du Chocolate, Biarritz, France

Musée du Louvre, Département des Objets d'Art, Paris, France

Musée National Adrien-Bubouché (Ceramics), Limoges, France

Musée National des Arts et Traditions Populaires, Paris, France

Musée National du Château et des Trianons, Versailles, France

Musée Picasso, Paris, France

Museo de América, Madrid, Spain

Museo delle Porcellane, Palazzo Pitti, Florence, Italy

Museo di Doccia/Museo Richard Ginori, Sesto Fiorentino, Italy

Museo Duca di Martina, Villa Floridiana, Naples, Italy

Museo National del Prado, Madrid, Spain

Museo Regional de Yucatán Palacio Cantón, Mérida, Yucatán, Mexico

Museu do Arte Antiga, Lisbon, Portugal

Museu do Azulejo, Lisbon, Portugal

Museu do Indio Library, Rio de Janeiro, Brazil

Museum of Fine Arts, American Decorative Arts and Sculpture Department, Boston, Massachusetts

Museum of Fine Arts, Boston, Massachusetts

National Archives of Canada, Ottawa, Canada

National Archives, Kew Gardens, Richmond, England

National Archives, Washington, DC

National Gallery, Trafalgar Square, London, England

National Library of Medicine, National Institutes of Health, Bethesda, Maryland

National Library of Scotland, Edinburgh, Scotland

New Orleans Auction Galleries, Incorporated, New Orleans, Louisiana

New York Historical Society Library, New York, New York

New York Public Library, New York, New York

New York State Library and Archives, Albany, New York

Newberry Library, Chicago, Illinois

Newport Historical Society, Aaron Lopez Collection, Newport, Rhode Island

Oakland Museum, Oakland, California

Oakland Public Library, Oakland History Room, Oakland, California

Oxford Brookes University Library, Oxford, England

Oxfordshire Record Office, Oxford, England

Palazzo Ginori, Florence, Italy

Pasadena Public Library (Centennial Room), Pasadena, California

Peabody Essex Museum, Salem, Massachusetts

Peabody Museum of Archaeology and Ethnology, Harvard University, Cambridge, Massachusetts

Peter J. Shields Library, University of California, Davis, California

Pierpont Morgan Library and Museum, New York, New York

Pocumtuck Valley Memorial Association, Deerfield, Massachusetts

Portsmouth Athenaeum, Portsmouth, New Hampshire (virtual site)

Public Archive of Bahia/Salvador, Brazil

Public Record Office, National Archives, Kew Gardens, Richmond, England

Rare Book & Manuscript Library, Columbia University, New York, New York

Real Jardin Botanico, Madrid, Spain

Rhode Island Historical Society Library, Providence, Rhode Island

Rhode Island State Archive, Providence, Rhode Island

Rhodes House Library, Oxford, England

Robert E. Kennedy Library, California State Polytechnic University, San Luis Obispo, California

Rodney Snyder Family Private Collection, Elizabethtown, Pennsylvania

Royal Collections, London, England

Royal Geographical Society, London, England

Sacramento Archives and Museum Collection Center, Sacramento, California

Salon du Chocolat, Paris, France

Santa Barbara Mission Archive and Library, Santa Barbara, California

Santa Barbara Presidio Research Library, Santa Barbara, California

Schlesinger Library, Harvard University, Boston, Massachusetts

Seaver Center for Western History, Natural History Museum, Los Angeles, California

Sotheby's, New York, New York

Sotheby's, Paris, France

South Carolina Historical Society, Charleston, South Carolina

Spadina Museum, Toronto, Canada

State Hermitage Museum, Saint Petersburg, Russia

Sterling and Francine Clark Art Institute, Williamstown, Massachusetts

Steward Museum, Montreal, Canada

Sutro Library, San Francisco, California

Sutter's Fort Archive, Sacramento, California

The Toledo Museum of Art, Toledo, Ohio

Thomas H. Manning Polar Archives, Scott Polar Research Institute, University of Cambridge, England

Thompson-Pell Research Center, Fort Ticonderoga, Ticonderoga, New York

Toledo Museum of Art, Toledo, Ohio

University Library, Cambridge, England

University Library, Mona, Jamaica

University of Arizona Library, Tucson, Arizona

University of California Library System, Berkeley, California

University of California Library System, Davis, California

University of California Library System, Los Angeles, California

University of the West Indies Library, Cave Hill Campus, Bridgetown, Barbados

Van Pelt Library, Rare Books Division, University of Pennsylvania, Philadelphia, Pennsylvania

Victoria & Albert Museum, South Kensington, London, England

W. E. B. Du Bois Library, University of Massachusetts, Amherst, Massachusetts

Wallace Collection, London, England

Wethersfield Historical Society, Wethersfield, Connecticut

Widener Library, Harvard University, Cambridge, Massachusetts

William L. Clements Library, University of Michigan, Ann Arbor, Michigan

Winslow Papers Collection, University of New Brunswick, Canada

Winterthur Library and Museum, Winterthur, Delaware

Yale University Art Gallery, New Haven, Connecticut

Yale University, The Lewis Walpole Library of Yale University, Farmington, Connecticut

Commonsense Rules for Working in Archives, Libraries, and Museums

Bertram M. Gordon and Louis Evan Grivetti

Why should there be a need for an appendix with this title? Is it not logical that common sense and good manners should dictate research behavior at such locations? Both of us, collectively, have spent more than 75 years working in archives and libraries. During our work we have practiced what might be called the "golden rule" and, by doing so, always have been invited back and have developed strong, professional associations with colleagues responsible for private or public document collections. At the same time, we both have observed other researchers who did not follow basic common sense and whose activities were disgraceful.

We have observed researchers who were boorish, critical, and pompous. We have watched in amazement as some researchers became angry, argumentative, loud and verbally abusive to hard-working staff. We have observed some who demanded immediate attention—as other researchers waited patiently for their requests to be fulfilled. On more than one occasion, we have overheard conversations between archive, library, or museum officials who banned specific researchers from returning—because of poor behavior and because basic commonsense rules were not followed. The basic rules that govern all work in archives and libraries are based on common sense and polite behavior. Remember, too, the axiom—one never has a second chance to make a first impression. Behave well and good things follow.

1. *Inspection of Archive/Library/Museum Documents Is a Privilege, Not a Right.* Archive, library, and museum collections, whether public or private, are invaluable pieces of national, regional, or local historical records. Most materials in these collections are irreplaceable. The primary duty of archivists and librarians is to preserve the documents under their authority, a duty more important than the secondary function of making the documents available to researchers.

2. *Make an Appointment in Advance.* It may not be necessary to have an appointment, but establishing an initial contact with the archive administration is a good idea. This type of behavior establishes researchers as legitimate and serious. Take time as appropriate to speak with the head archivist, explain your work, objectives, and basic research questions. Explain how and why documents in their respective archive can help answer the questions critical to your research.

Chocolate: History, Culture, and Heritage. Edited by Grivetti and Shapiro
Copyright © 2009 John Wiley & Sons, Inc.

3. *Know Your Research Priorities in Advance.* Most archives will allow researchers to inspect only one or two documents or files at a time (i.e., per request). In order to save time and to avoid misunderstandings, have your priorities in order before arriving at the archive. While you may consider filling out request forms to be tedious and tiresome, and while you may grumble that "certain archives/libraries" should be more up-to-date with computerized request facilities, recognize that there are longstanding, established identification and viewing protocols for each archive. How specific archives, libraries, or museums are managed is not your choice. In other instances, some archives and libraries may have their holdings already cataloged and placed on-line for ease of computer access. It is helpful, therefore, to have completed as much of your research as possible on-line before visiting a respective archive, library, or museum, as this advance work will maximize your time and effort spent. Furthermore, once in the research library or archive, familiarize yourself with its general catalogs as these may contain material not listed in the on-line catalogs you consulted prior to your visit.

4. *When Entering and Leaving, Submit Graciously to Inspection.* Some but not all archives and libraries are equipped with scanning devices. Entrances to the document depositories also may be coupled with security officers who have as their primary responsibility the integrity of the research collection. Scanning and personal inspection is routine to prevent document theft, and to identify objects such as markers, pens, staplers, food, and beverages that could damage or destroy documents. Know in advance, therefore, what you can or cannot take into the document depository. Submit pleasantly and graciously to inspection of your briefcase, camera, bulky clothing, purse, tape recorder, and umbrella. Upon entering, turn off your mobile telephone! Upon leaving, do not complain if asked to step aside to have your long shirt sleeves "patted-down." If asked to remove your boots or shoes do not complain. Do your best to comply with the requests of the inspecting officers.

5. *Be Courteous to Archivists and Librarians.* Archivists and librarians are knowledgeable about the structure and organization of their respective collections. While you may be an expert in a specific field, the chief archivist and/or his or her assistants most likely can provide you with a number of research leads of which you probably were unaware. Send each person who assisted you a thank you note once you have returned home. Small social niceties, coupled with proper professional behavior, can develop friendships and mutual respect through the years. Small efforts and being polite yield great benefits on return visits.

6. *Make Use of Reference Works in the Archive or Library.* Archives and libraries, especially larger ones, often have reference books and inventories related to their collections on open shelves, either in the readers' room or an anteroom nearby. Inventories of materials held in the archive or library collections frequently contain older information not on-line even after collection catalogs have been digitized. These books can be excellent sources for material of which you may have been unaware. Archivists and librarians may be able to point you in the direction of the reference books/inventories that will be most useful to you. An excellent way of using the time waiting for your documents to be brought to you is to walk along the walls where the open stacks are located and glance at the holdings. You may be pleasantly surprised at what you find.

7. *Respect Camera, Copying, and Laptop Rules.* Inquire about use of laptops before your visit. Most archives and research libraries will allow laptop use but some smaller ones may not. Most archives and libraries do not allow use of pens or marking devices, and require that handwritten notes be made in pencil. Although research libraries and archives usually will have pencils available for use, bringing your own supply signals your professionalism. Regarding copying regulations, each archive has its own and regulations commonly are highly variable. Some allow digital cameras (providing flash is not used); others do not. Some archives—after inspection of a document by the archivist—will allow researchers themselves to make photo/Xerox copies. Other archives require that any photo/Xerox copy be made by the archivist; after inspection of a document the archivist/librarian may deem it too fragile to copy, whereupon only handwritten notes may be taken. Other archives will provide copies to researchers on a fee basis, with a maximum number of copies prepared daily (commonly under 30), and in some instances digitized copies can be supplied on CDs based on a posted fee schedule. It makes sense, then, always to inquire in advance regarding laptops, digital cameras, and copying procedures so as to avoid potential problems upon arrival.

8. *Respect Opening and Closing Time Rules.* Many archives and libraries are opened for staff administrative and research activities in the early morning. While these employees are working, that does not mean you have the right to impose on their time and request an "early admission exception." Furthermore, closing time commonly means that the building must be empty and locked by a specific hour. This means that guest researchers may be asked to vacate 15 to 30 minutes prior to the listed closing time. Be prepared to honor these rules and take them to heart; do not behave in such a way that a guard needs to be called to evict you, which may lead to loss of archive/library privileges.

9. *Take Breaks During the Workday.* Researchers working with limited time in a distant archive or library may be tempted to work through the entire day, from opening to closing, without a break. A lunch break may involve an hour's interruption, especially if there is no cafeteria or restaurant in the building or nearby. Many archives and libraries, however, have vending machine areas with tables where researchers can obtain a quick snack or beverage and eat without bringing food into the reading rooms. Use these short breaks to refresh yourself during the workday. Why? You never know whom you might meet at the vending machines. Other researchers you meet may be working on similar topics and have interesting and helpful tips to share with you. Also, as you walk up and down the aisles or pass the tables where other researchers are working in the reading room, be alert: You might spot someone working on a related topic and you could make useful professional contacts and lifelong friends.

10. *Never Steal a Document!* This last "commandment" should be obvious; it should never cross your mind to remove a document from an archive or library—for whatever purpose. Beyond the embarrassment of being challenged by an archivist or guard and being caught, is the probability that theft (alternatively even the suspicion of theft) at the very least will result in losing access privileges for life and commonly leads to prosecution, fines, and prison time.

Yes, there is great joy in holding a rare map, manuscript, or letter from a notable from the 17th century . . . but the joy should not turn into possessive behavior. Archive, library, and museum holdings are for all to enjoy—hopefully—forever!

Acknowledgments

We wish to thank Jennifer Follett, Deanna Pucciarelli, Peter Rose, Celia Shapiro, and Timothy Walker who offered suggestions during the early stages of manuscript preparation.

Digitized Resources Consulted

African American Newspaper Collection (via Accessible Archives)

American Civil War Letters and Diaries

American Periodical Series Online (via Proquest)

Calisphere

Early American Imprint Series II 1801–1819 (Shaw-Shoemaker)

Early American Newspapers (via NewsBank)

Early Americas Digital Archives

Early Encounters in North America (EENA): Peoples, Cultures, and the Environment

Evans Digital Collection (1639–1800)

Gerritsen Collection—Women's History Online

HarpWeek (Harpers' Weekly) 1857–1912

In the First Person: An Index to Letters, Diaries, Oral Histories, and Other Personal Narratives

Journals of the Lewis and Clark Expedition Online

Library of Congress: American Memory Database: American Life Histories. Manuscripts from the Federal Writers' Project, 1936–1940

Library of Congress: American Memory Database: American Notes: Travels in America, 1750–1920

Library of Congress: American Memory Database: An American Time Capsule. Three Centuries of Broadsides and Other Printed Ephemera

Library of Congress: American Memory Database: California as I Saw It. First-Person Narratives of California's Early Years, 1849–1900

Library of Congress: American Memory Database: George Washington Papers at the Library of Congress, 1741–1799

Library of Congress: American Memory Database: Pioneering the Upper Midwest. Books from Michigan, Minnesota, and Wisconsin, ca. 1820–1910

Library of Congress: American Memory Database: The Capital and the Bay. Narratives of Washington and the Chesapeake Bay Region, ca. 1600–1925

Library of Congress: American Memory Database: The First American West: The Ohio River Valley, 1750–1820

Library of Congress: American Memory Database: The Nineteenth Century in Print. The Making of America in Books and Periodicals: Books

Library of Congress: American Memory Database: The Nineteenth Century in Print. The Making of America in Books and Periodicals: Periodicals

Library of Congress: American Memory Database: The Stars and Stripes. The American Soldiers' Newspaper of World War I, 1918–1919

Library of Congress: American Memory Database: Trails to Utah and the Pacific. Diaries and Letters, 1846–1869

Library of Congress: American Memory Database: Westward by Sea. A Maritime Perspective on American Expansion, 1820–1890

Making of America Books

North American Women's Letters and Diaries (NAWLD)

Online Archive of California

Pennsylvania Gazette (via Accessible Archives)

ProQuest Historical Newspapers

Readex/NewsBank

Chocolate: History, Culture, and Heritage. Edited by Grivetti and Shapiro
Copyright © 2009 John Wiley & Sons, Inc.

APPENDIX

Saint Augustine, Florida, 1642

Chocolate Distribution List by Name, Occupation, and Quantities Received

Beatriz Cabezon, translator

List Position	Name: 17th Century Spelling	Name: 21st Century Spelling	Occupation	Chocolate Weight in Pounds	Chocolate Number of Boxes
1	Father Juan Gomez	Father Juan Gomez	Superior and Guardian of the Convent of this city, and Attorney General of these provinces	232	0
2	Nicolas Ponce de Leon	Nicolás Ponce de León	Auditor	88.5	0
3	Francisco de larrua	Francisco de la Rua	Captain	27	0
4	Juan Sanchez	Juan Sánchez	Second Lieutenant; regiment of this city	3	2
5	Andres Gonçales	Andrés González	Captain	4	2
6	Nicolas de Robles	Nicolás de Robles	Second Lieutenant	7	0
7	Julio Mejia	Julio Mejía	Second Lieutenant	0	2
8	Alvaro Sanchez	Álvaro Sánchez	Sergeant	4.6	0
9	Andres Amador	Andrés Amador	Sergeant	3.5	0
10	Pedro Falcon	Pedro Falcón	Adjutant aide-de-camp	58	0
11	Ju^{lio} de Mercado	Julio de Mercado	Adjutant aide-de-camp	9	0

Chocolate: History, Culture, and Heritage. Edited by Grivetti and Shapiro
Copyright © 2009 John Wiley & Sons, Inc.

List Position	Name: 17th Century Spelling	Name: 21st Century Spelling	Occupation	Chocolate Weight in Pounds	Chocolate Number of Boxes
12	Fran*co* Man. Montes	Francisco Manuel Montes	Sergeant Mayor	3	2
13	Ant*o* de Enera	Antonio de Enera	Sergeant Major	7	0
14	Julio de Monzon	Julio de Monzón	Sergeant Major	16.5	0
15	Adrian de Canizares	Adrián de Cañizares	Sergeant Major	5	0
16	Pedro Martinez	Pedro Martínez	Sergeant Major	6	0
17	Francisco Salvador	Francisco Salvador	Sergeant Major	3	0
18	Pedro Muñoz	Pedro Muñoz	Sergeant Major	3	0
19	Manuel Ruiz	Manuel Ruiz	Adjutant aide-de-camp	9	0
20	Pedro de San m m	Pedro de San m m ?	Adjutant aide-de-camp	4	0
21	Antonio de Cuevas	Antonio de Cuevas	Adjutant aide-de-camp	3	0
22	Marcos Estivel	Marcos Estivel	Second Lieutenant	0	2
23	Don Francisco de labanera	Don Francisco de la Banera	Second Lieutenant	24.6	0
24	Pedro de sorres	Pedro de Sorres	Second Lieutenant	3	0
25	Andres del Cobedo	Andrés del Cobedo	Second Lieutenant	3	0
26	Antonio Sartucho	Antonio Sartucho	Second Lieutenant	3	0
27	Mn Ascay	Manuel Ascay	Second Lieutenant	3	0
28	Alfonso Solana	Alfonso Solana	Second Lieutenant	6	0
29	Patricio de florencia	Patricio de Florencia	Second Lieutenant	5	0
30	Fran*co* Castellanos	Francisco Castellanos	Second Lieutenant	0	2
31	Fran*co* de Carnuebo	Francisco de Carnuebo	Second Lieutenant	0	2
32	Pedro de Brana	Pedro de Brana	Peace Notary (Escribano de Paz)	2	2
33	M*mo* Lopez de Mesa	Máximo López de Mesa	Second Lieutenant	13	0
34	Matias de los Rios	Matías de los Ríos	Second Lieutenant	5	0
35	Diego Diaz	Diego Díaz	Second Lieutenant	5.12	0
36	Felipe de Caluya	Felipe de Caluya	Sergeant	3	0
37	Bar*me* Lopez Gabiro	Bartolomé López Gabiro	Sergeant	6	0
38	Pedro de Azpiolea	Pedro de Azpiolea	Not specified	7	0
39	Fran*co* Gonçales de Villagarcia	Francisco Gonzalez de Villagarcía	Not specified	13.14	0
40	Ant*o* Mon	Antonio Montero ?	Not specified	2	0
41	Juan Ant*o* Marquez	Juan Antonio Márquez	Not specified	3	0
42	Juan Perez de Craso	Juan Pérez de Craso	Not specified	21	2
43	A Pz de Çendrera	Antonio? Pérez de Cendrera	Sergeant	26.05	0
44	Juan entonado	Juan Entonado	Not specified	8	0
45	Fran*co* Albarez	Francisco Álvarez	Corporal of the Stables (*Cabo de cuadra*)	8	0
46	Ju*o* de Carbajal	Julio de Carbajal	Corporal	0	2

List Position	Name: 17th Century Spelling	Name: 21st Century Spelling	Occupation	Chocolate Weight in Pounds	Chocolate Number of Boxes
47	Fran^{co} Pontevedra	Francisco Pontevedra	Corporal	3	0
48	Jayme de Torres	Jaime de Torres	Corporal	2	0
49	Luis Diaz	Luis Díaz	Corporal	3	0
50	Bar^{me} de Castañeda	Bartolomé de Castañeda	Corporal	2.5	0
51	Nicolas de Carmenales	Nicolás de Carmenales	Corporal	3	0
52	Bar^{me} R.	Bartolomé R. (Rodríguez?)	Corporal	0	2
53	Ant^{o},Menendez de Posada	Don Antonio Menéndez de Posada	Not identified	13.5	0
54	Andres ceron	Andrés Cerón	Not identified	2	0
55	Ant^{o} M^{n} Crismondo	Antonio Menéndez (Mendez?) Crismondo	Not identified	0	2
56	Ant^{o} Lopez	Antonio López	Not identified	0	2
57	Andres de Junco	Andrés de Junco	Not identified	0	2
58	Ant^{o} Cuello	Antonio Cuello	Not identified	0	2
59	Andres de Binçes	Andrés de Vinces (?)	Not identified	0	1
60	Domingo . . .ez . . . .de Bran..(illegible)	Domingo . . .ez . . . .de Bran.. (illegible)	Not identified	2	0
61	Diego de Çavallos	Diego de (Zaballos Ceballos)	Not identified	2	0
62	Diego de Rivas	Diego de Rivas	Not identified	2	0
63	Diego de Carrasco	Diego de Carrasco	Not identified	2	0
64	Diego de Lopez Dueñas	Diego de López Dueñas	Not identified	2	0
65	Dom^{o} Muñoz	Domingo Muñoz	Not identified	0	1
66	Diego Jorje	Diego Jorge	Not identified	0	2
67	Diego Diaz Alfran	Diego Díaz Alfrán	Not identified	1.5	1
68	Diego de Monçon	Diego de Monzón	Not identified	6	0
69	Diego ernandez	Diego Hernández	Not identified	6.6	0
70	Fran^{co} Lpez de Castro el rio	Francisco López de Castro del Río	Not identified	24	2
71	Fran^{co} trivaldos	Francisco Trivaldos	Not identified	2	0
72	Franco Jimenez	Francisco Jiménez	Not identified	2	0
73	Fernando Moreno	Fernando Moreno	Not identified	3	0
74	Franco R^{a} de labera	Francisco Ribera de la Vera (Bera)	Not identified	6	0
75	Franco Sanchez	Francisco Sánchez	Not identified	1	0
76	Felipe de santiago	Felipe de Santiago	Not identified	1	0
77	Franco de los Santos	Francisco de los Santos	Not identified	1	0

List Position	Name: 17th Century Spelling	Name: 21st Century Spelling	Occupation	Chocolate Weight in Pounds	Chocolate Number of Boxes
78	Franco Albarez de san Agustin	Francisco Álvarez de San Agustín	Not identified	3	0
79	Fran^{co} Maroto	Francisco Maroto	Not identified	1	1
80	Fran^{co} de la Cruz	Francisco de la Cruz	Not identified	3	0
81	Fernando Lopez	Fernando López	Not identified	2	0
82	Fran^{co} de rrivas	Francisco de Rivas	Not identified	1	0
83	Fran^{co} gomez dueñas	Francisco Gómez Dueñas	Not identified	2	0
84	Fran^{co} de florencia	Francisco de Florencia	Not identified	1.5	0
85	Fran^{co} bernal	Francisco Bernal	Ship's Boy	1	0
86	Fran^{co} Mendez	Francisco Méndez	Ship's Boy	1	0
87	Andres de (illegible)	Andrés de (illegible)	Not identified	2	0
88	Ant^o rroebas	Antonio Roebas	Not identified	1	0
89	Andres botin	Andrés Botin (Botín)	Not identified	1.5	0
90	Ant^o Serrano	Antonio Serrano	Not identified	3	0
91	Ant^o Arguelles	Antonio Argüelles	Not identified	3	0
92	Ant^o Monçon	Antonio Monzón	Not identified	6	0
93	Ant^o de Dueñas	Antonio de Dueñas	Not identified	3	0
94	Ant^o de Florencia	Antonio de Florencia	Not identified	1.5	0
95	Al^o Solana	Alfonso Solana	Not identified	1.5	0
96	Ant^o Fran^{co}	Antonio Francisco	Not identified	1.5	0
97	de los rreyes	Antonio de los Reyes	Not identified	1.5	0
98	Andres de gonçalez	Andrés de González	Not identified	1.5	0
99	borondon	Antonio Borondón	Not identified	0	1
100	Ant^o m. L.	Antonio (Moreno L.?)	Not identified	1	0
101	Ant^o Ribera paje	Antonio Ribera	Page	1	0
102	Andres Gomez m^o	Andrés Gómez Moreno	Not identified	1	0
103	Ant^o de ensonado	Antonio de Ensonado	Page	2	0
104	Ant^o de flor^a	Antonio de Florencia	Page	3	0
105	Ant^o Cuello M^o	Antonio Cuello Moreno	Not identified	0	1
106	Benito Diaz	Benito Díaz	Not identified	2	0
107	bar^{me} Jimenez	Bartolomé Jiménez	Not identified	2	0
108	bernabe Lopez	Bernabé López	Not identified	2	0
109	blas de Cabrera	Blas de Cabrera	Not identified	2	0
110	bernabe frayofo (?)	Bernabé Frayofo (?)	Not identified	1.5	0
111	bar^{me} belazquez	Bartolomé Velázquez	Not identified	1.5	0
112	bernardino Jimenez	Bernardino Jiménez	Not identified	1.5	0

List Position	Name: 17th Century Spelling	Name: 21st Century Spelling	Occupation	Chocolate Weight in Pounds	Chocolate Number of Boxes
113	Bernardo de R^a	Bernardo de Rivera	Not identified	1.5	0
114	bar^{me} perez	Bartolomé Pérez	Page	1.5	0
115	bar^{me} de frias	Bartolomé de Frías	Not identified	1	0
116	Claudio de Florencia	Claudio de Florencia	Not identified	6	0
117	Domingo de bergança	Domingo de Berganza	Not identified	3	0
118	Fran^{co} fernanz	Francisco Fernández	Not identified	2	0
119	Fran^{co} diaz	Francisco Díaz	Ship's Boy	0	1
120	Gonçalo el almendral	Gonzalo del Almendral	Not identified	3	0
121	R^{mo} ernandez	Romero(?) Hernández	Not identified	1.5	0
122	gabriel de monçon	Gabriel de Monzón	Not identified	12.5	0
123	R^{mo} de cabrera	Romero(?) de Cabrera	Not identified	2.5	1
124	don gaspar de çea	Don Gaspar de Zea	Not identified	1.5	0
125	gaspar gonçales M^o	Gaspar González Moreno	Not identified	0	1
126	Juan Sanchez de Sevilla	Juan Sánchez de Sevilla	Not identified	0	0
127	Ju^o mallo	Julio Mallo (Mayo)	Not identified	0	2
128	Juan de Laguardia	Juan de la Guardia	Not identified	0	2
129	Ju^o descobedo	Julio de Escobedo	Not identified	0	2
130	Juan Luis	Juan Luis	Not identified	0	2
131	Jil gonçales	Gil González	Not identified	0	1
132	Juan Fran^{co} de los Santos	Juan Francisco de los Santos	Not identified	0	1
133	Juan albarez feran	Juan Álvarez Feran (?)	Not identified	0	1
134	Juan de Medina	Juan de Medina	Not identified	1.5	0
135	Ju^o Lopez de cerpa	Julio López de Serpa	Not identified	1.5	0
136	Jacinto de los rreyes	Jacinto de los Reyes	Not identified	1.5	0
137	Juan mellado	Juan Mellado	Not identified	1.5	0
138	Joseph de Cañizares	José de Cañizares	Not identified	1.5	0
139	Juan Alonso	Juan Alonso	Not identified	1.5	0
140	Juan rrodriguez	Juan Rodríguez	Not identified	1.5	0
141	Ju^o de Contreras	Julio de Contreras	Not identified	1.5	0
142	Juan Ruiz Salguero	Juan Ruiz Salguero	Not identified	12	0
143	Juan Sanchez de nabalque	Juan Sánchez de Navalque	Not identified	7.5	0
144	Juan de Soto	Juan de Soto	Not identified	2	0
145	Jacinto de Mendoça	Jacinto de Mendoza	Not identified	6.6	2
146	Juan Sanchez Arriaga	Juan Sánchez Arriaga	Not identified	0	2
147	Juan Diaz mejia	Juan Díaz Mejía	Not identified	3	0

List Position	Name: 17th Century Spelling	Name: 21st Century Spelling	Occupation	Chocolate Weight in Pounds	Chocolate Number of Boxes
148	Juan Cartaya	Juan Cartaya	Not identified	5.12 (?)	0
149	Juan Moreno	Juan Moreno	Not identified	4.12 (?)	0
150	Juan Muñoz	Juan Muñoz	Not identified	14.4	0
151	Juan rroldan	Juan Roldán	Not identified	3	0
152	Juan belarde	Juan Belarde (Velarde?)	Not identified	3	0
153	Juan melendez	Juan Meléndez (Menéndez?)	Page	2.5	0
154	Juan de Acosta	Juan de Acosta	Not identified	3	0
155	Ju^o de Contreras, ag^n	Julio de Contreras	Augustine Friar	3.5	0
156	Juan Gomez, agus^n	Juan Gómez	Augustine Friar	3.5	0
157	Juan barbosa (de) lisboa	Juan Barbosa de Lisboa	Not identified	3	0
158	Juan de armenteros	Juan de Armenteros	Not identified	1	0
159	Joseph de la Cruz	José de la Cruz	Not identified	1	0
160	Juan de rrivas	Juan de Rivas	Not identified	2	0
161	Ju^o descalona	Julio de Escalona	Not identified	3	0
162	Ju^o Dominguez	Julio Domínguez	Not identified	1	0
163	Ju^o de rribas	Julio de Rivas,	Augustine Friar	1	0
164	Ju^o Fran^{co} M^o	Julio Francisco Moreno	Not identified	2	0
165	Juan de errera	Juan de Herrera	Sailor	0	1
166	Luis Mn	Luis Menéndez (?)	Not identified	2	0
167	Lorenço Santiago	Lorenzo Santiago	Not identified	3	0
168	Luis ernandez de biana	Luis Hernández de Viana	Not identified	1	0
169	Luis noe	Luis Noé	Not identified	2	0
170	Laçaro descalona	Lázaro de Escalona	Not identified	1.5	0
171	Mateo Luis de floza	Mateo Luis de Florencia	Not identified	3	0
172	m. l. borondon	Mateo Luis (?) Borondón	Not identified	0	2
173	matias cabeças	Matías Cabezas	Not identified	4	0
174	Mrn(illegible) gonç..z	Mariano....(?) González	Not identified	5	?
175	Mujer de M.L Mateo	Wife of M..L.. Mateo	Not identified	2.12	0
176	M-de banios	Mateo(?) de Banios	Not identified	1.0	0
177	marcos de Lujan	Marcos de Luján	Not identified	2.0	0
178	mateo R^a	Mateo Ribera (?)	Not identified	44.2	0
179	m. albarez	Mateo Álvarez	Not identified	3.0	0
180	m. L. Carballo	Mateo Luis Carballo	Not identified	1.0	0
181	mateo Luis de san agus^n	Mateo Luis de San Agustín	Not identified	1.5	0
182	m. L Jacinto	Mateo Luis Jacinto	Not identified	2.0	0
183	marcos Martin	Marcos Martín	Not identified	3.0	0
184	mateo ernandez	Mateo Hernández	Not identified	1.0	0
185	melchor de los rreyes	Melchor de los Reyes	Not identified	1.0	0
186	matias rromero	Matías Romero	Not identified	1.5	0

List Position	Name: 17th Century Spelling	Name: 21st Century Spelling	Occupation	Chocolate Weight in Pounds	Chocolate Number of Boxes
187	miguel albarez	Miguel Álvarez	Not identified	1.5	0
188	miguel de guebara	Miguel de Guevara	Not identified	4.0	0
189	Don matias de batero	Don Matías de Batero	Not identified	6.0	0
190	mateo de Faxtecha	Mateo de Faxtecha (Fantecha)	Not identified	2.0	0
191	mrn. de toro M°.	Mariano (Menéndez) de Toro Moreno	Not identified	2.0	0
193	Nicolas R. m°	Nicolás Ribera Moreno	Not identified	2.5	0
194	Onorato mrn	Honorato Menéndez	Not identified	1.5	0
195	pedro delgado	Pedro Delgado	Not identified	1.5	0
196	pedro enrrique	Pedro Enrique	Not identified	1.5	0
197	pedro Nuñez	Pedro Núñez	Not identified	2.0	2
198	pedro osorio	Pedro Osorio	Not identified	2.0	0
199	pedro gonçalez	Pedro González	Not identified	2.0	0
200	pedro R^a	Pedro Ribera	Not identified	0.0	1
201	pedro de la Cruz	Pedro de la Cruz	Not identified	19.4	0
202	pedro Antonio	Pedro Antonio	Not identified	2.0	0
203	pedro ernandez Leon	Pedro Hernández León	Not identified	2.0	0
204	rroque Lopez	Roque López	Not identified	4.5	0
205	salvador de pedrosa	Salvador de Pedrosa	Not identified	2.0	0
206	fe . . . de bastiano	Felipe (?) de Bastiano	Not identified	0.0	2
207	salvador gomez	Salvador Gómez	Not identified	2.0	0
208	Sebastian Rs	Sebastián Rivas	Not identified	3.0	0
209	Sebastian rrico	Sebastián Rico	Not identified	4.6	0
210	salvador Vicente	Salvador Vicente	Not identified	0.0	1
211	tomas de lamon	Tomás de Lamon	Not identified	0.0	1
212	tomas Fran^{co}	Tomás Francisco	Not identified	3.0	0
213	Xpoval Muñoz	Xpoval Muñoz	Not identified	2.0	0
214	Xpoval Sanchez	Xpoval Sánchez	Not identified	3.0	0
215	Xpoval Ortiz	Xpoval Ortiz	Not identified	1.5	0
216	Xpoval rrodriguez	Xpoval Rodríguez	Not identified	2.5	0
217	Xpoval Lopez montero	Xpoval López Montero	Not identified	6.0	0
218	Domingo de Almeyda	Domingo de Almeyda	Captain	21.0	0
219	Ju° Gomez nabarro	Julio Gómez Navarro	Ensign (?)	2.0	4
220	pedro berdugo de la silbeyra	Pedro Verdugo de la Sylbeira	Parish priest	9.0	0
221	Antonio Calvo	Antonio Calvo	Licenciate lawyer	10.0	6
222	Ermenegildo Lopez	Hermenegildo López	Captain	12.5	0
Total				*1,167.22*	78

Boston Chocolate, 1700–1825

People, Occupations, and Addresses

Louis Evan Grivetti

Name	Profession	Date	Street Address
Abbot, Samuel	Merchant	1755	Shop located near the Mill Bridge, opposite to Mr. Joseph H [?] Tinman, in Boston
Abbot, Samuel	Merchant	1758	Shop located at the head of the Town-Dock, at the Corner of Wing's Lane
Abraham, Nathaniel	Merchant	1760	Shop located at the sign of the Golden Key in Ann Street
Adams, John	Merchant	1772	Shop located in the establishment lately improv'd by Mr. Nathaniel Tucker, Apothecary, opposite the Old South Meeting House, Boston
Adams, John & Comy.	Merchant(s)	1771	Shop located nearly opposite the Old South Meeting-House, lately improved by Mrs. Bethiah Oliver, deceased [she had been a chocolate merchant]
Allen, Jolley	Merchant	1768	Shop located opposite the Heart and Crown in Cornhill, Boston
Alline, William	Auctioneer	1795	Auction located at his office [no address provided]
Ammidon and Parker	Merchant(s)	1801	Shop located at Number 8, Central Wf [i.e., wharf]
Amory, Francis	Auctioneer	1815	Auction held at shop Number 41, Marlboro' Street
Amory, Jonathan & John	Merchant(s)	1757	Shop located at the Sign of the Horse, at the head of Dock Square, Boston
Amory, Rebeckah [Rebekah]	Merchant	1732	Shop located near Mr. Colman's Meeting House
Andrews & Domett	Merchant(s)	1764	Shop located at Number 10, Hancock's Wharff [sic]
Andrews & Domett	Merchant(s)	1765	Shop located at Number 1, South side of the Town Dock (Opposite the Swing Bridge)

Chocolate: History, Culture, and Heritage. Edited by Grivetti and Shapiro
Copyright © 2009 John Wiley & Sons, Inc.

Name	Profession	Date	Street Address
Andrews, Benjamin	Merchant	1776	Warehouse located at the site lately occupied by Mr. Joshua Blanchard, Jr. At the head of the Town Dock, Boston
Andrews, Benjamin Junior	Merchant	1769	Shop located opposite the Swing Bridge
Andrews, James	Merchant	1796	Shop located at Number 50, Long Wharf
Andrews, James	Merchant	1797	Shop located at Number 39, Long Wharf
Andrews, James	Merchant	1823	Shop located at Number 8, Central Wharf
Andrews, James & Co.	Merchant(s)	1800	Shop located at Number 39 Long Wharf
Appleton, Nathaniel	Merchant	1755	Shop located near the Town house in Cornhill, Boston
Arms, Stephen	Paper-dealer	1818	Shop located at Number 39, Ann Street
Austin & Thayer	Merchant(s)	1809	Shop located at number 17, India Street
Ayres, Thomas	Merchant	1747	Shop located over against the [Red?] Lyon
Bailey, Calvin	Merchant	1815	Shop located at Number 10, Foster's Wharf
Baker & Bridgham	Merchant(s)	1764	Shop located in Union Street
Baker & Brown	Merchant(s)	1815	Shop located at Number 40, India Street
Baker & Hodges	Merchant(s)	1817	Shop located at Number 30, India Street, the store recently occupied by Messrs. Stanton and Spellman
Baker, Brown & Co.	Merchant(s)	1816	Shop located at Number 40, India Street
Baker, Eben. and Co.	Merchant(s)	1803	Shop located at Number 4, Court Street
Baker, Ephraim	Auctioneer	1737	Auction held at the Crown Tavern, at the Head of Scarlet's Wharff, North End of Boston
Baker, Ephraim	Auctioneer	1738	Auction held at the house in which Mr. John Darrill lately dwelt in Union Street, at the Head of the Town Dock, Boston
Baker, John	Merchant	1749	Shop located near the Baptist Meeting House
Baker, John	Sugar baker	1750	Shop located in Back Street
Baker, John	Merchant	1753	Warehouse located in Back Street
Baker, John and Co.	Merchant(s)	1801	Shop located at Number 47 State Street (formerly occupied by Messer's Dorr's)
Baker, T. M.	Auctioneer	1823	Main Street near the Bridge in Charlestown
Balch, Nathaniel	Merchant	1766	Shop located in Mr. Bligh's [establishment] in Marlborough Street
Balch, Nathaniel	Merchant	1766	Shop located in Mr. John Langdon's [establishment] in Fore Street, near the Draw Bridge, next door to Messieurs Cushing and Newman's
Baldwin, Cyrus	Merchant	1772	Shop located nearly opposite the Pump in Cornhill, Boston
Baldwin, Luke & Co.	Merchant(s)	1806	Shop located at Number 44, Long Wharf
Ballard, John Jr. & Co.	Auctioneer(s)	1809	Auction located at Number 32 Marlboro' Street
Ballard, Joseph	Merchant	1764	Shop located at the South End, near the Hay Market
Ballard, Mary	Coffee house manager	1755	Coffee Shop: located on King's Street
Bant, William	Merchant	1771	Shop located fronting Dock Square, Boston
Barker & Bridge	Auctioneer(s)	1805	Auction held on Robbins' Wharf, Washington Street, South End
Barnes, Henry	Merchant	1760	Shop located a quarter of a mile past of the Meeting House in Marlborough [Street]
Barns, Henry	Merchant	1750	Shop located next door to Mr. Dowse's Insurance Office at the lower end of King Street
Barrell, Joseph	Merchant	1768	Shop located at Number 3, on the South Side of the Town Dock
Bartlett & Gay	Merchant(s)	1809	Shop located under the Exchange Building Keep
Bartlett, N. H.	Merchant	1806	Shop located at Number 30, Marlboro' Street
Baxter, Charles	Merchant	1819	Shop located at Number 54, Long Wharf
Baylies and Howard	Merchant(s)	1801	Shop located at Number 54, State Street
Baylies and Howard	Merchant(s)	1803	Shop located at Number 28, Long Wharf
Baylis, William	Auctioneer	1796	Auction located at his office on State Street
Beck, F.	Merchant	1815	Shop located at Number 11, India Street

Name	Profession	Date	Street Address
Beckford & Bates	Merchant(s)	1812	Shop located at Number 39, Long Wharf
Bell, Daniel	Merchant	1773	Shop located on the Town-Dock, lately improved by Mr. John Williams, and the next door southward to Mr. Samuel Eliot, Jr.
Bell, Daniel	Merchant	1775	Shop located directly opposite the East end of Faneuil Hall, Boston
Bell, S.	Deputy sheriff	1817	Sheriff's Sale held at the corner of Broad and Milk Streets
Beman, D.	Auctioneer	1810	Auction at Number 9, Kilby Street
Beman, David	Merchant	1806	Shop located at Number 1, Sumner Street, within fifteen rods of the new State House
Bemis & Eddy	Merchant(s)	1809	Shop located at Number 12, Long Wharf
Bernard, Mr.	Merchant	1728	Shop located next to the Town-House
Biglow and Beale	Merchant(s)	1798	Shop located at Number 17, Kilby Street
Biglow and Beale	Merchant(s)	1800	Shop located at Number 37, State Street
Billings, Joseph	Tailor	1774	No address provided
Billings, Mary	Merchant	1760	Shop located opposite the Province-House
Binney & Ludlow	Merchant(s)	1819	Shop located at Number 33, Long Wharf
Blake & Cunningham	Auctioneer(s)	1818	Auction held at their office, Number 5, Kilby Street
Blanchard, Caleb	Merchant	1754	Shop located in Union Street, adjoining to Mr. Joseph Scott's
Blanchard, Joshua	Merchant	1776	Shop located at the old store and wine cellar on Dock Square, Boston
Blanchard, Joshua	Merchant	1796	Shop located at Number 4, Dock Square
Blanchard, Joshua Jr.	Merchant	1772	Shop located over Blanchard's Wine Cellar, Dock Square
Blanchard, Joshua jun.	Merchant	1774	Shop located in establishment lately occupied by Mr. Joseph Tyler, opposite to Mr. Scott's Irish Linen Store, in Ann-Street, and fronting the North side of Faneuil-Hall Market, Boston
Blanchard, Joshua, Jr.	Merchant	1776	Shop located next to Messrs. Breek and Hammatt's, near the Golden Ball, Boston
Blodgett, Samuel	Merchant	1761	Shop located in Marlboro Street, at the corner of Bromfield's Lane, almost opposite to the late Dr. Gibbons's
Boies & M'Lean	Merchant(s)	1788	No address provided; announcement of a fire at their chocolate mill
Boit and Hunt	Merchant(s)	1785	No address provided
Boit's Cheap Fruit Store	Merchant	1783	Shop located on the South side of the Market
Boit's Store	Merchant	1791	Shop located on the South side of the Market
Bond, Nathan	Merchant	1794	Shop located at Mr. Benjamin Hammatt's Store, Number 8, Green's Wharf
Bond, Nathan	Merchant	1795	Shop located at Number 9, Green's Wharf
Bongard, Philip	Merchant	1746	Shop located in King's Street
Borland, J. B.	Merchant	1811	Shop located at Number 45, Long Wharf
Borland, John	Merchant	1768	Shop located North side of the Swing Bridge
Bowen, Penuel	Merchant	1772	Shop located in the West-India and Grocery Way of Business, at the house lately improved as a Public House by Capt. Daniel Jones, a little to the Southward of Deacon Church's Vendue Room
Bowen, Penuel	Merchant	1773	Shop located in Newbury Street
Bowen, Penuel	Merchant	1773	Shop located to the establishment lately improved by Messi'rs Joseph Palmer & Son, a little to the Southward of the Market, & directly opposite the Golden Ball—where he has to dispose of almost every article in the West India & Grocery Business
Bowen, Penuel	Merchant	1784	Shop located on Dock Square, opposite to Deacon Timothy Newell's
Boydell, Hannah	Merchant	1731	Shop located next to the Naval Office and over against the Bunch of Grapes Tavern in King's Street, Boston

Name	Profession	Date	Street Address
Boydell, John	Merchant	1736	Shop located at the Sign of the Pine Tree and Two Sugar Loaves, in King's Street, Boston
Boyles, John	Merchant	1769	Shop located a little Northward of Dr. Gardiner's in Marlborough Street, Boston
Boylston, Zabdiel	Apothecary	1711	Shop located at Dock Square, Boston
Bradford, John R. & Co.	Merchant(s)	1811	Shop located at Number 16, Dock Square at the establishment lately improved by Mr. Rowland Freeman
Bradford, Richard	Merchant	1763	Shop located near the Conduit in Union Street
Bradford, Richard	Merchant	1765	Shop located near the Conduit, in Union Street, Boston
Bradford, S.	Auctioneer	1801	Auction held opposite Number 3, Codman's Wharf
Bradford, S.	Auctioneer	1802	Auction held at Number 26, Long Wharf
Bradford, S.	Auctioneer	1802	Auction held at Shop Number 7, Doan's Wharf
Bradford, S.	Auctioneer	1802	Auction held at the large store, lately Gridley's, Battery March; sale of sundry effects of Henry Hastings, bankrupt
Bradford, S.	Auctioneer	1802	Auction held opposite Number 4, Butler's Row
Bradford, S.	Auctioneer	1802	Auction held opposite Shop Number 10, Long Wharf
Bradford, S.	Auctioneer	1803	Auction held at the store of Mr. Thomas Mayo, situated in Roxbury Street
Bradford, William B.	Merchant	1811	Shop located at Number 4, South Side Market House
Bradford, William B., Jr.	Merchant	1814	Shop located at Number 4, South side of Faneuil Hall
Bradlee, Josiah	Merchant	1802	Shop located at Number 12, Long Wharf
Bradlee, Josiah	Merchant	1810	Shop located at Number 29, India Street Wharf
Bradlee, Josiah	Merchant	1810	Shop located at Number 6, Long Wharf
Brazer, Samuel and Son	Merchant(s)	1799	Shop located at Number 9, Market Square
Breck and Hammatt	Merchant(s)	1773	Shop located at the corner of Greene's Wharf
Breck, Samuel	Merchant	1778	Shop located opposite the Golden Ball
Brewster, John	Merchant	1730	Shop located in the North End in Boston, at the Sign of the Boot
Brewster, John	Chocolate maker	1732	Shop located at the Black Boy, North End of Boston
Brewster, John	Chocolate maker	1738	Shop located at the Sign of the Boot in [?] Street
Brewster, John	Chocolate maker	1743	Shop located at the North End of Boston near the Red Cross
Brewster, John	Chocolate maker	1766	Auction held at his estate located in Fish Street, Boston
Bridge and Kidder	Merchant(s)	1793	Shop located in Court Street, near the Court House
Brigham and Bigelow	Merchant(s)	1809	Shop located at Number 18, Long Wharf
Brigham and Bigelow	Merchant(s)	1809	Shop located at Number 23, Long Wharf
Brimmer, Herman	Merchant	1764	Shop located at the House of Mr. Samuel Woodburn, the Sign of the Lion, in Waltham; relocated due to the smallpox outbreak
Brimmer, Martin	Merchant	1776	Shop located at Number 25, Long Wharf
Brinley, F. and E. Jr.	Merchant(s)	1800	Shop located at Number 5, Town Dock
Brown, John	Chocolate maker	1771	[No address provided]
Brown, John	Merchant	1817	Shop located at Number 40, India Street
Brown, Stephen	Merchant	1803	Shop located at Number 11, South side of the Town Dock
Brown, William & Son	Merchant(s)	1815	Shop located at Number 13 Broad Street
Bryant, James	Merchant	1775	Shop located next to the seat [?] of John Dennis, Esquire, at Little Cambridge
Bryant, James	Merchant	1809	Shop located at Number 66, Long Wharf
Bryant, James	Merchant	1809	Shop located at Number 67, Long Wharf
Bull, Fred. G.	Merchant	1811	Shop located at Number 59, Long Wharf
Bullard & Parker [Silas Bullard and Isaac Parker]	Merchant(s)	1817	Shop located at Number 32, Central Wharf
Bullard, Nath'l	Deputy sheriff	1810	Auction held at the house of Mr. Jeremiah Daniels
Burnham & Burt	Merchant(s)	1811	Shop located at Number 6, Long Wharf

Name	Profession	Date	Street Address
Burroughs, G. Jr. & Co.	Auctioneer(s)	1807	Shop located at Number 75, State Street
Burroughs, G. Jun'	Auctioneer	1811	Auction located at his office on India Street
Cabot, Samuel Jun.	Apothecary	1812	Shop located at Number 39, India Street
Callahan & Austen	Merchant(s)	1807	Shop located at Number 70, Long Wharf
Callender, Charles	Librarian/merchant	1814	Combination Library/Tea shop located at Number 11, Marlboro' Street
Callender, Joseph	Merchant	1803	Shop located at Number 40, Marlboro' Street, in front of the brick building lately improved by the Hon. Benjamin Hichborn, Esq. and one door North of Messrs. S. & S. Salisbury's Store
Callender, Joseph & Son	Merchant(s)	1812	Shop located at Number 40, Marlboro' Street, near the Old South Meeting House
Callender, Joseph Jr.	Merchant	1791	Shop located on the South side [of] the Market House
Callender, Wm. B.	Merchant	1817	Shop located at Number 24, Court Street
Carleton, Jonathan	Merchant	1813	Shop located at Number 7, Custom House Street
Carter, G.	Auctioneer	1806	Auction held near Leach's Wharf, North End
Carter, G.	Auctioneer	1808	Auction held at his office [no address given]
Catton, Richard	Chocolate maker	1769	Shop located at the house of Mr. Moses Peck in Williams' Court
Chandler & Howard	Merchant(s)	1818	Shop located at Number 33, Central Wharf
Chandler & Howard	Merchant(s)	1822	Shop located at Number 25, Central Wharf
Chandler & Howard	Merchant(s)	1823	Shop located at Number 16, Central Wharf
Channing, George G.	Auctioneer	1825	Shop located at Number 6, Lindall Street
Chase & Speakman	Merchant(s)	1764	Shop located at their Distill-House at the South End
Chase, Hezekiah	Merchant	1821	Shop located at Number 26, Long Wharf
Checkley, John Webb	Merchant	1792	Shop located in establishment late occupied by Doctor Winship, near the Mill Bridge
Checkley, John Webb	Merchant	1795	Shop located at Number 3, near the Mill Bridge
Church, Benjamin	Auctioneer	1743	Auction held at his house, at the South End of Boston
Clapp, Derastus	Auctioneer	1825	Shop located on Sea Street
Clark & Calender	Merchant(s)	1817	Shop located at Number 4, Boylston Square, South side New Market
Clark and Eustis	Auctioneer(s)	1797	Auction held at their office [address not provided]
Clark, John	Merchant	1825	Shop located opposite the Northeast corner of the Old Market
Clark, Thomas	Auctioneer	1802	Auction held at his store in Kilby Street, lately occupied by Messrs. Hathorne's
Clark, Thomas	Auctioneer	1817	Auction held at the house of Mrs. Hatch, Columbia Street
Clarke, Edward	Merchant	1761	Shop located opposite Green's Wharff, near the Golden Ball
Cleverly, Stephen & Company	Merchant(s)	1764	Shop located in Dedham, the next house above Samuel Dexter, Esq.; near the Meeting House
Coates, Benjamin	Merchant	1803	Shop located at Number 45, Ann Street
Coffin, Thomas & Co.	Merchant(s)	1808	Shop located at Number 7, Long Wharf
Coit, William	Merchant	1782	Shop located opposite the Court House [advertisement ran in the Boston Gazette and the Country Journal, August 12, 1782]
Collier, E. H.	Merchant	1813	Shop located at Number 104, Broad Street
Comerin, John	Merchant	1741	Shop located in Royal Exchange Lane
Coo[kson], Obadiah	Merchant	1742	Shop located at the Cross'd Pistols in Fish Street
Cooke, J.	Chocolate maker	1771	Shop located adjoining to Mr. Sweetfer's on the Town Dock
Cooke, J.	Chocolate maker	1771	Shop located at the Slitting Mills in Milton
Cooke, James	Chocolate maker	1770	Shop located at the new Mills, Boston
Cookson, Obadiah	Merchant	1746	Shop located at the [Sign of the] Cross Pistols, Fish Street, at the North End
Coolidge, Cornelius	Merchant	1800	Shop located at Number 7, Long Wharf
Coolidge, Cornelius	Merchant	1801	Shop located at Number 32, Long Wharf

Name	Profession	Date	Street Address
Coolidge, S. F.	Auctioneer	1822	Shop located at Number 38, Brattle Street
Coolidge, S. F.	Auctioneer	1823	Shop located at Number 9, Merchant's Row
Coolidge, Samuel B	Merchant	1824	Shop located at the head of Poplar Street
Copeland, E. Jr.	Merchant	1823	Shop located at Number 69, Broad Street
Cordis, Cord	Merchant	1753	Shop located in his house at New Boston
Coverly, Thomas	Merchant	1776	Shop located at the establishment formerly improved by Henry Barns, next to the Meeting House in Marlborough
Cudworth, Nathaniel	Merchant	1767	Shop located in King's Street, opposite the Sign of Admiral Vernon
Cudworth, Nathaniel	Merchant	1770	Shop located to the South End, three doors above the Sign of the Lamb
Cunningham, Archbald	Merchant	1769	Shop located near the Draw Bridge, in Fore Street, Boston
Cunningham, Archbald	Merchant	1773	Shop located on Ann Street near the Draw Bridge
Cushing, John	Merchant	1776	Shop located nearly opposite East end of the Market, Boston
Cushman & Topliff	Merchant(s)	1806	Shop located at Numbers 17 and 18, Green's Wharf, opposite Parkman's Buildings
Dabney, Charles	Stay-maker	1753	Shop located in the front of Captain Tyng's house in Milk Street
Damon, Nathaniel L.	Broker	1819	Computing room located at Number 9, Long Wharf
Dana, Dexter	Merchant	1796	Shop located in the new brick store, on the spot formerly kept by Mr. Joseph Callender, Jr. South side [of] the market
Dana, Dexter	Merchant	1805	Shop located at Number 4, South Side Market House
Dana, Dexter	Merchant	1812	Shop located at Number 15, Exchange Street
Davenport & Dana	Merchant(s)	1810	Shop located in the New England Buildings, Exchange Street
Davenport & Dana	Merchant(s)	1811	Shop located in [the] New England Buildings, Exchange Street
Davenport and Dana	Merchant(s)	1812	Shop located on Exchange Street
Davenport and Tucker	Merchant(s)	1799	Shop located at Number 23 and Number 24, Long Wharf
Davenport, Elijah and Samuel		1801	Shop located at Number 8, Long Wharf
Davenport, Elijah and Samuel	Merchants	1802	Shop located at Number 17, Long Wharf
Davenport, R. & E.	Merchant(s)	1817	Shop located at Number 14, Central Wharf
Davenport, R. & E.	Merchant(s)	1817	Shop located at Number 31, Long Wharf
Davenport, Rufus	Merchant	1797	Shop located at Number 23 and Number 24, Long Wharf
Davenport, Rufus	Merchant	1816	Shop located at Number 31, Long Wharf
Davis, Deacon	Chocolate maker	1772	No address provided
Deblois, Stephen Jr.	Merchant	1761	Shop located at the Golden Eagle, Dock Square, Boston
Degrand, P. P. F.	Merchant	1812	Shop located at Number 20, India Street
Dennie, John	Merchant	1776	Shop located in Little Cambridge
Dennie, Thomas	Merchant	1776	Shop located in Little Cambridge
Deshon, Moses	Auctioneer	1763	Auction held at the House of the Widow Stevens, deceased, at the North End, near Thomas Hancock, Esqrs. Wharf
Deshon, Moses	Auctioneer	1765	Auction held at the newest auction room, opposite the West End of Faneuil Hall, Dock Square
Dexter, Knight	Merchant	1759	Shop located at the Sign of the Boy with a Book in his Hand
Dexter, Samuel	Merchant	1749	Shop located in Hanover Street, near the Mill Bridge
Dexter, Samuel	Merchant	1752	Shop located in his house, near the Mill Bridge
Dexter, Samuel	Merchant	1753	Shop located near the Mill Bridge

Name	Profession	Date	Street Address
Dexter, Samuel	Merchant	1755	Shop located in the establishment where Mr. Nathaniel Holmes lately lived, where Mr. Maverick used to keep the Sign of the Horse & Plough, just by the corner of Cross Street, a little to the northward of said Mill Bridge, where his name is wrote [*sic*] over the door
Dillaway, Samuel jun.	Merchant	1799	Shop located at Number 10, Long Wharf
Dimond, Peter	Merchant	1796	Shop located at Number 4, Dock Square, North side of the Market [relocated his shop due to a fire on State Street—did he have his shop there?]
Dimond, Peter	Merchant	1797	Shop located at Number 9, Market Square
Dobel, John [is he the same as Doble, and Dobell?]	Merchant	1749	Shop located at the Sign of the Three Sugar Loafs, in Dock Square, near the Market
Dobel, John [is he the same as Doble, and Dobell?]	Merchant	1759	Shop located in King's Street at the Sign of the Black Boy and Two Sugar Loaves
Dobell, John [in subsequent advertisements spelled John Doble]	Merchant	1748	Shop located at the North End, over against the Red Lion
Doggett, Jesse	Auctioneer	1810	Auction held at the store of Seaver and Woods, in Roxbury
Domett, Joseph	Merchant	1768	Shop located at the corner of Green's Wharff, opposite [the] Golden Ball
Domett, Joseph	Merchant	1769	Shop located directly opposite the east door of Faneuil Hall Market
Dow, Benjamin	Merchant	1823	Shop located at Number 7, Kilby Street
Dowrick, William	Merchant	1748	Shop located near the North Latin School
Dowse, Samuel	Chocolate maker	1763	Chocolate mill: located in Charlestown
Draper, Jeremiah	Auctioneer	1814	Auction held at his office, Corner of Kilby and Central Street
Draper, Jeremiah	Merchant	1814	Shop located at Number 13, Kilby Street
Dupee, Elias	Auctioneer	1766	Auction held at the New Auction Room in Royal Exchange Lane
Durant, Thomas	Merchant	1749	Shop located at the sign of the Blue Boar, opposite to the Ferry-House at the north side of Portsmouth
Dyde, Robert	Merchant	1813	Massachusetts Tavern and Hotel—a New Establishment—located at School Street, Boston, establishment formerly occupied by Mrs. Vose
Dyre, Mrs.	Merchant	1736	Shop located at the North end of Boston near the Old Winnisimut Ferry
Eaton, Israel	Chocolate maker	1752	Shop located in his house near Mill Bridge
Eaton, Israel	Chocolate maker	1760	Shop located near Mill Bridge
Eaton, J. And D. M.	Auctioneer(s)	1802	Auction to be held at their shop [no address given]
Edwards and Norwood	Merchant(s)	1798	Shop located at the establishment formerly occupied by Messrs. Martin Bieker and Son, [at] Number 3, Ann Street, leading from thence to the Market
Edwards and Richardson	Merchant(s)	1794	Shop located at Number 6, Marlborough Street, near Seven Star Lane
Edwards, J.	Merchant	1748	Shop located in Cornhill
Edwards, R.	Auctioneer	1804	Auction held at the shop of Mr. John D. Dyer, in Marshal's Lane, near the Boston Stone [?]
Edwards, R.	Auctioneer	1810	Auction held at Number 63, State Street
Edwards, Richard	Merchant	1806	Shop located at Number 32, Long-Wharf
Edwards, Richard	Merchant	1809	Shop located at Number 47, Long Wharf
Eliot, Josiah	Merchant	1770	Shop located at the establishment lately improved by Mr. John Head, on Treat's Wharf
Erving & Callender's Store	Merchant(s)	1786	Shop located next door to the Golden Ball, Merchants Row
Erving, John	Merchant	1791	Shop located at Number 42, Marlborough Street

Name	Profession	Date	Street Address
Eustis, William	Auctioneer	1800	Auction held at his office [no address given]
Fairfield & Choate	Auctioneer(s)	1821	Shop located at Number 14, Central Street
Fairfield & Choate	Auctioneer(s)	1821	Shop located at the North side [of] India Wharf
Fairfield, John	Auctioneer	1824	Shop located at Number 14, Central Street
Fales, Stephen	Merchant	1792	Shop located at the establishment improved by Mr. Samuel Blagge, opposite the East End of the Market
Fanuel, Andres	Merchant	1729	Warehouse located at the lower end of King Street
Farmer, William	Merchant	1797	Shop located at the establishment lately improved by Mr. Hager, in Ann Street, opposite Messrs. Gridley and Nolen's
Faxon, Richard Jr.	Merchant(s)	1803	Shop located at Number 28, State Street
Fay, Otis	Merchant	1812	Shop located at Number 40, Middle Street
Featherstone, George	Merchant	1743	Shop located at the Two Blue Sugar-Loaves and Green Cannister [sic], over against Dr. Gibbon's in Marlborough Street, Boston
Fechem, George	Chocolate maker	1776	Shop located near Watertown Bridge
Fessenden, M. Jr.	Merchant	1810	Shop located at Number 57, India Wharf
Fetton, Nathaniel	Scythe-maker	1764	Shop located in his house in Roxbury. Shop goods transported out of Boston due to the smallpox outbreak; the goods belonged to Rebecca Walker [a relative?]
Fiske, Oliver	Merchant	1798	Shop located opposite Mr. Barker's Tavern
Flagg, Samuel H.	Merchant	1807	Shop located at Number 15, Broad Street
Flagg, Samuel H.	Merchant	1807	Shop located at Number 9, Broad Street
Fletcher, Samuel	Merchant	1759	Shop located near the Draw Bridge, Boston
Fletcher, William A	Merchant	1813	Shop located at Number 36, India Street
Floyd, Richard	Chocolate maker	1763	Shop located at the Sign of the Tea Kettle, South End
Forsyth, Alexander	Merchant	1732	Shop located at the Sign of the Two Jarrs [sic], and Four Sugar Loaves in Prince Street alias Black-Horse Lane (near the Charlestown Ferry)
Foster, Benjamin W.	Merchant	1801	Shop located at Number 64, Long Wharf
Foster, Benjamin W.	Merchant	1809	Shop located at Number 3, Long Wharf
Foster, Bossenger & William	Merchant(s)	1769	Shop located on Spear's Wharf
French, Leonard	Merchant	1813	Shop located at Number 11, Sudbury Street
Fuller, Josiah	Ship master	1757	At the Vice Admiralty Court, Boston
Furness & Walley's Store	Merchant(s)	1792	Shop located opposite the East end of the Market House
Gamage & Mason	Merchant(s)	1813	Shop located at Number 4, Central Street
Gardner and Downer	Merchant(s)	1809	Shop located at Number 19, Merchant's Row
Gardner, Joshua	Merchant	1800	Shop located at Number 5, Butler's Row
Gardner, R.	Auctioneer	1801	Auction held at Central Wharf
Gardner, R.	Auctioneer	1801	Auction held on Scott's Wharf (formerly Hancock's Wharf)
Gardner, R.	Auctioneer	1803	Auction held at the head of Treat and Lane's Wharves, Fore Street, Bottom of North Square
Gay & Eaton	Merchant(s)	1807	Shop located at Number 63, State Street
Gerrish, J.	Auctioneer	1767	Auction held at the Public Vendue-Office, North End
Gerrish, John	Auctioneer	1741	Auction held at the house of Mr. Joseph Lewis, Tobacconist
Gerrish, John	Auctioneer	1746	Auction held at his Public Vendue House, Dock Square
Gibbs, C. [Gibb also Gibbs in subsequent advertisements]	Merchant	1793	Shop located at Number 9, Butler's Row
Goldsmith, Hepzibah	Merchant	1773	[No address provided]
Goldsmith, John	Chocolate maker	1768	Shop located on the corner leading down [to] John Hancock, Esq'rs Wharff
Goldsmith, John	Merchant	1772	Shop located at the establishment of the late Mr. John Brewster [chocolate maker]

Name	Profession	Date	Street Address
Goldsmith, John	Chocolate maker	1772	Shop located at the head of John Hancock Esq'rs Wharff
Gould, John	Merchant	1752	Shop located about a mile from Watertown Bridge
Gould, John	Merchant	1753	Shop located near the Mill Bridge
Gould, Robert	Merchant	1764	Shop located in the establishment several years improv'd by Mr. James Tifer in Sudbury; he relocated: outside the city "that the Small-Pox being rife in Town, has occasioned his Remove into the Country"
Graham, John	Merchant	1774	Shop located on Wendel's Wharf, near the South Battery, Fort-Hill
Grant, John Jr.	Merchant	1769	Shop located near the Draw Bridge
Grant, Moses	Paper-dealer	1818	Shop located at Number 7, Union Street
Grant, Moses	Paper-dealer	1818	Shop located at the mills, Newton Lower Falls
Gray, George	Coffee house manager	1738	Coffee House: located at the head of the Long Wharff
Gray, Lewis	Merchant	1764	Shop located at Number 1, Butler's Row, near Faneuil Hall Market
Gray, Samuel	Merchant	1773	Shop located near the Heart and Crown in Cornhill district of Boston
Greaton, John	Merchant	1763	Shop located opposite Mr. Harris the Baker, at the South End of Boston
Greaton, John Jun.	Merchant	1764	Shop located in Roxbury
Green & Russell	Merchant(s)	1764	No address provided
Green & Walker	Merchant(s)	1762	Shop located at the North corner of Queen Street, near the Town House, Boston
Green, Edward	Merchant	1755	Shop located over against Messrs. Ebenezer Storer and Son, in Union Street
Green, James	Merchant	1760	Shop located opposite John Angell, Esq.; at the sign of the Elephant [Rhode Island merchant advertising in the *Boston Post Boy*, September 29, 1760; advertisement first placed in Providence, September 13, 1760]
Green, Joshua	Merchant	1765	Shop located at the corner of Queen Street, near the Town House, in Boston
Greenleaf, Oliver	Merchant	1764	Shop located between the White Horse Tavern and the Great Elm Trees, South End of Boston
Greenleaf, William	Merchant	1783	Shop located in Cornhill
Haislup, William	Merchant	1741	Shop located over against Corn-Fields in Union Street, Boston
Hall, Thacher & Co.	Merchant(s)	1810	Shop located at Number 10, Bray's Wharf
Hallet, Geo.	Merchant	1815	Shop located at Number 10, India Street
Hallet, Geo.	Merchant	1815	Shop located at Number 3, India Street
Hallowell, Benjamin	Merchant	1750	Warehouse located next door but one below Dowse's Insurance Office, the lower end of King Street
Hammatt & Brown	Merchant(s)	1769	Shop located at the establishment lately improved by Messrs Baker and Bridgham, near the Sign of the Cornfield, in Union Street, Boston
Hammatt, [?]	Merchant	1804	Shop located near [the] Concert Hall
Hammatt, Benjamin Jr.	Merchant	1791	Shop located near the Market
Hammatt, Benjamin Junior	Merchant	1778	Shop located near the Golden Ball
Hammatt, Benjamin Junior	Merchant	1797	Shop located next to his Dwelling House, in Southak's Court near Concert Hall
Hammond & Coolidge	Merchant(s)	1810	Shop located at Number 42, Long Wharf
Hancock, John	Merchant	1798	Shop located at Number 8, Merchant's Row
Hancock, Thomas and John	Merchant(s)	1795	Shop located at the East End of the Market
Hancock, Thomas and John	Merchant(s)	1796	Shop located at Number 8, Merchant's Row

Name	Profession	Date	Street Address
Harris, Samuel	Merchant	1794	Shop located next to Bridge and Kidder's in Court Street
Haskins, John jun.	Merchant	1800	Shop located at Number 35, State Street
Hastings, Samuel	Merchant	1778	Shop located near the Stump of Liberty Tree
Hastings, Samuel	Merchant	1794	Shop located next to [the] Liberty Pole, South End, Boston
Haven, Calvin & Co.	Merchant(s)	1817	Shop located at Number 8, India Street
Hawes, Prince and Co.	Merchant(s)	1824	Shop located at Number 25, India Street
Hawkes, Samuel	Merchant	1752	Shop located at the North End
Hayt, Lewis	Auctioneer	1790	Auction held on State Street
Hayt, Lewis	Auctioneer	1802	Auction held at number 13 Orange Street, South End
Hayward, C.	Auctioneer	1811	Auction held at his office at Number 71, State Street
Hayward, C.	Auctioneer	1814	Auction held at Numbers 24 and 25, Long Wharf
Head, John	Merchant	1766	Shop located in Roxbury in the establishment formerly improved by Mr. Samuel Williams, opposite the Burying Ground, and near the Entrance of Boston Neck
Head, John	Merchant	1770	Shop located at the establishment lately improved by Mr. Peter Hughes, next door below Mr. Robert Moodie's, near the East End of Faneuil Hall Market
Head, John	Merchant	1771	Shop located at Number 5 on the South-side [of the] Town-Dock, just below the Swing Bridge
Hersey & Durell	Auctioneer(s)	1823	Shop located at Number 11, Bowdoin Square
Hersey & Durell	Auctioneer(s)	1823	Shop located at Number 16, Custom House Street
Hersey & Durell	Auctioneer(s)	1823	Shop located at Number 38, Brattle Street
Hersey & Durell	Auctioneer(s)	1823	Shop located at the corner of Franklin and Hawley Streets
Hersey, Daniel	Auctioneer	1821	Shop located at Number 9, Washington Street
Hersey, Daniel	Auctioneer	1822	Auction held at Number 13, Brattle Street
Hersey, Daniel	Auctioneer	1822	Shop located at Number 22, Brattle Street
Hersey, Daniel	Auctioneer	1822	Shop located at the corner of Salem and Tileston Streets
Hersey, Daniel	Auctioneer	1822	Shop located in Hatter's Square
Hersey, Daniel	Auctioneer	1823	Shop located at Number 9, Charles Street
Hersey, Daniel	Auctioneer	1824	Shop located at Number 103, Ann Street
Hersey, Daniel	Auctioneer	1824	Shop located at the corner of Hancock's Wharf
Hersey, Daniel	Auctioneer	1825	Auction held at Number 115 Hanover-Street
Hersey, Daniel	Auctioneer	1825	Shop located at the corner of High and Summer Streets
Hersey, Daniel	Auctioneer	1825	Shop located on Leveret Street
Hewes, N. P.	Auctioneer	1806	Auction held at Number 1, Cambridge Street, near Concert Hall
Hill & Chase [Noah Hill and Hezekiah Chase]	Merchant(s)	1813	Shop located at Number 17, Milk Street, three doors from Broad Street
Hill, T. Q. & W.	Merchant(s)	1814	Shop located at Number 24, Long Wharf
Hill, T. Q. & W.	Merchant(s)	1818	Shop located at Number 20, India Wharf
Hinckley, David	Merchant	1818	Shop located at Number 46, Central Wharf
Hinkley and Woods	Auctioneer(s)	1793	Auction held at their store on Boston Neck
Holden, Edward	Merchant	1798	Shop located at Number 54, State Street
Holden, Edward	Merchant	1800	Shop located at Number 39, State Street
Holden, Edward	Merchant	1808	Shop located at Number 22, Long Wharf
Holmes & Rogers	Merchant(s)	1822	Shop located at Number 54, State Street
Holyoke & Soren	Merchant(s)	1792	Shop located at Number 44, Cornhill, near the Market
Homes, William	Merchant	1764	Shop located at Number 2, South side of the Town Dock, just below the Swing Bridge
Hoskins, William	Merchant	1773	Shop located near Mr. Harrod's, Black Horse Lane
How, Rebecca	Merchant	1767	Shop located at the bottom of Peter Chardon, Esq'rs Lane, adjoining to Mr. Thomas Jackson's Distill-House, at New Boston

Name	Profession	Date	Street Address
Howard, J. & B	Merchant(s)	1817	Shop located at Number 17, Commercial Street
Howard, J. & B	Merchant(s)	1817	Shop located at Number 6, Central Wharf
Howard, J. & B	Merchant(s)	1818	Shop located at Number 33, Central Wharf
Howe & Fletcher	Merchant(s)	1815	Shop located at Number 33, India Street
Howe, Abraham F. & Co.	Merchant(s)	1815	Shop located at Number 33, India Street
Hunt & Torrey	Merchant(s)	1756	Shop located near the Prison, Boston
Hunt, Samuel W.	Merchant	1785	Shop located at the establishment lately improved by Boit and Hunt
Huse, Enoch	Merchant	1804	Shop located at Number 42, State Street
Hussey & Barker	Merchant(s)	1809	Shop located at Number 7, Long Wharf
Hussey, Timothy	Merchant	1811	Shop located at Number 7, Long Wharf
Hutchinson & Brinley	Merchant(s)	1753	Shop located at the Three Sugar Loaves and Cannister [sic] in King's Street, Boston
Hutchinson, Shrimpton	Merchant	1749	Shop located at the Three Sugar Loaves and Canister in King's Street, near the Town House, Boston
Inches, Thomas	Merchant	1741	Shop located on Marlborough Street, near the Old South Meeting Hall
Ingraham & West	Merchant(s)	1810	Shop located at Number 14, India Wharf
Ingraham, Duncan Jr.	Merchant	1776	Shop located at Captain Joseph Adams's address in Lincoln
Inman, Ralph	Merchant	1764	Shop located at the house of Mr. Samuel Woodburn, at the Sign of the Lion, in Waltham; removed his goods from previous location to keep them free from infection [smallpox]
Jackson, James	Merchant	1765	Shop located in Union Street, Boston
Jackson, Newark	Chocolate maker	1740	Shop located near Mr. Clark's Shipyard at the North End of Boston
Jackson, William & James	Merchant(s)	1764	Shop located at the Sign of the White Horse in Waltham because of the "spreading of the Small Pox in Boston"
Jacobs, Daniel	Not identified	1772	[No address provided; announcement of a house fire]
Jenkins, William Wainright	Merchant	1796	Shop located in Milk Street, at the bottom of Federal Street, next door to Mons. Julien's Restorator [sic]
Johnson, Mary	Merchant	1750	Shop located at the Sign of the Blue Boar, South End
Jones and Bass	Auctioneer(s)	1794	Auction held on State Street
Jones, Daniel	Merchant	1759	Shop located at the Hatt [sic] & Helmet, South End, Boston
Jones, Elnathan	Merchant	1772	Shop located in Concord [advertised in the *Boston Gazette and Country Journal*, May 11, 1772]
Jones, T. K.	Auctioneer	1791	Auction held on State Street
Jones, T. K.	Auctioneer	1801	Auction held at the store back of Messrs. Bradford & Palfrey, Lendalt's Lane
Jones, T. K. and Company	Auctioneer(s)	1825	Auction held at Number 8, Kilby Street
Jones, T. K. and Company	Auctioneer(s)	1811	Auction held at their office on Kilby Street
Jones, Thomas. K.	Auctioneer	1790	Auction held on State Street
Judkins and Ford [Judkins, Moses S. and Ford, Samuel]	Merchant(s)	1807	Shop located at Number 65, Broad Street
Kelley and Clark	Merchant(s)	1795	Shop located on Mr. Nathan Spear's Wharf, near the Market
Kennedy, John	Merchant	1797	Shop located at Number 46, Long Wharf
Kneeland, John	Merchant	1742	Shop located at the King's Arms, adjoining Mill Bridge
Kneeland, John	Merchant	1774	Shop located at the head of Green's Wharf, opposite to John Rowe, Esquire
Kneeland, John	Merchant	1774	Shop located at the head of Green's Wharf, opposite to John Rowe, Esquire, near the Market
Ladd, William	Merchant	1794	Shop located at Number 14, Butler's Row

Name	Profession	Date	Street Address
Ladd, William	Merchant	1796	Shop located at Number 13, Codman's Wharf
Lane & Cutter	Merchant(s)	1798	Shop located at Number 62, Long Wharf
Larkin, E.	Real estate agent?	1807	Shop located at Number 47, Cornhill, Boston
Leach, James	Merchant	1793	Shop located at the Sign of the Grand Turk, Newbury Street
Learned, Lydia [pseudonym?]	Merchant	1771	Shop located in Brookline, near the Sign of the Punch Bowl
Ledain, Mary	Merchant	1744	Shop located next door to Captain Sigourney's, in the Lane leading to the Charlestown Ferry
Lefebure, Rebecca	Merchant	1770	Shop located opposite the West End of the Town House
Leigh, Benjamin	Chocolate maker	1771	Shop located [near?] the White Horse Inn
Leland & Bailey	Merchant(s)	1812	Shop located at Number 10, Foster's Wharf
Leland & Robinson	Merchant(s)	1816	Shop located at Number 7, Long Wharf
Leonard, George	Chocolate maker	1771	Shop located at the New Mills, near the Mill Bridge
Leverett, Benjamin	Merchant	1806	Shop located at Number 48, Broad Street, bottom of Milk Street
Leverett, John	Merchant	1759	Shop located on the North side of the Town Dock, below the Swing Bridge, where the late Deacon Bridgham lived
Leverett, John	Merchant	1758	Shop located at the house late in the occupation of Mr. William White, near the Conduit, at the head of the Town Dock
Leverett, John and Thomas	Merchant(s)	1752	Shop located opposite to the Stationer's Arms in Corn[hill]
Leverett, Thomas	Merchant	1758	Shop located in Cornhill
Lewis, Ezekiel Junior	Merchant	1764	Shop located to the upper end of Roxbury (commonly called Spring-Street) in the House of Mr. Ebenezer Whiting, due to the smallpox outbreak
Lillie, Theophilus	Merchant	1764	Shop located near the Milton Meeting House
Lincoln, Francis	Merchant	1806	Shop located at Number 33, Long Wharf
Little, William	Merchant	1813	Shop located at Number 2, Long Wharf
Logan, Walter	Merchant	1761	Warehouse located next to Messrs. John and William Powell's, on Dyer's Wharff
Lord and Kimball	Merchant(s)	1797	Shop located at Number 54, Long Wharf
Loring, Nathan	Merchant	1751	Shop located near Faneuil Hall
Low, Caleb	Auctioneer	1777	[No address provided]
Lowell, Ebenezer	Merchant	1748	Shop located in King Street
Lowell, John & Company	Merchant(s)	1768	Shop located on Hancock's Wharf
Lowell, Michael	Merchant	1750	Shop located on the corner next to Mr. James Davenport, Baker, near the Sign of the Cornfields
Lubbuck, James	Chocolate maker	1727	Shop located near the Reverend Mr. Colman's Meeting House
Mann, Joseph	Merchant	1762	Shop located under the New Auction Room, Dock Square
Mann, Joseph	Merchant	1763	Shop located in Hanover-Street, near Mr. John Smith, Merchant, opposite the Head of Wing's Lane
Mann, Joseph	Chocolate maker	1769	Shop located in his house in Water-Street, next door to Mr. Ebenezer Torrey-Baker, at the Sign of the Wheat Sheaff [sic] near Oliver's Dock, Boston
Mariott, Powers	Merchant	1753	Shop located opposite to the Sign of the Heart and Crown in Cornhill
Marston & Thayer	Auctioneer(s)	1819	Auction held at store Number 24, Long Wharf
Marston, David	Merchant	1807	Shop located at Number 41, Long Wharf
Marston, David	Merchant	1807	Shop located at Number 58, Long Wharf
Marston, David	Merchant	1808	Shop located at Number 31, Long Wharf
Marston, David	Merchant	1809	Shop located at Number 42, Long Wharf
Marston, William	Merchant	1805	Shop located at Number 4, Central Wharf
Marston, William	Merchant	1817	Shop located at Number 35, Central Wharf

Name	Profession	Date	Street Address
Maxwell, John	Merchant	1733	Shop located in his house in Long Lane, Boston
Maxwell, Rosanna	Merchant	1775	Shop located next door to the British Coffee-House and opposite the Bunch of Grapes Tavern, in King's Street, Boston
Maxwell, William	Merchant	1767	Shop located in King's Street, next door to the British Coffee House
M'Carthy, Daniel	Merchant	1767	Shop located in the street leading from the Cornfields to the Mill Bridge
McLellan, Isaac & Co.	Merchant(s)	1817	Shop located at Number 40, Central Wharf
Melvil, Allan and Melvil, John	Merchant(s)	1759	Warehouse located next to Messrs. John & William Powell's, on Dyer's Wharf
Merchant, J.	Auctioneer	1807	Auction held at Number 37, Fish Street
Merrett, John [sometimes spelled Merrit in subsequent advertisements]	Merchant	1732	Shop located at the Three Sugar Loaves and Canister, near the Town House, Boston
Merrill, Ezekiel	Merchant	1819	Shop located at Number 35, Long Wharf
Merritt and Fletcher	Merchant(s)	1731	Shop located at the Three Sugar Loaves and Cannister [sic], in King's Street, near the Town House, Boston
Merritt and Fletcher	Merchant(s)	1731	Warehouse located at the Letter "C" in Elis Esqu's buildings [?], near the Long Wharf
Merritt, John	Merchant	1739	Shop located at the Three Sugar Loaves and Cannister [sic], King Street, near the Town House, Boston
Miller, Charles	Merchant	1767	Shop located on Minot's "T," Boston
Miller, Charles	Merchant	1771	Shop located in Mr. Palmer's Store, in King Street
Miller, Joseph	Merchant	1759	Shop located in his house, opposite the South-East corner of the Town House
Miller, Joseph	Merchant	1760	Shop located in his house, opposite the South-East corner of the Town House, Boston
Minchin, Edward	Merchant	1798	Shop located at Number 10, Kilby Street
Minchin, Edward	Merchant	1798	Shop located at Number 40, State Street
Minot, Mary	Merchant	1754	Shop located at her house in Cole Lane
Minott, Samuel	Merchant	1772	Shop located opposite Williams's Court, Cornhill in Boston
Minott, Samuel	Merchant	1773	Shop located Southward of the Town House, next shop to Sign of Buck & Glove in Cornhill, Boston
M'Lean, John	Auctioneer/real estate agent?	1804	[No address provided]
Morrill & Baker	Merchant(s)	1806	Shop located at Number 47, State Street
Morse, Gilead	Merchant	1807	Shop located at Number 5, Spear's Wharf
Moseley & Babbidge	Merchant(s)	1814	Shop located at Number 17, Milk Street, corner of Broad Street
Moseley & Babbidge	Merchant(s)	1815	Shop located at Number 39, India Street
M'Tagart, Peter	Merchant	1750	Shop located in his house, near Doctor Cutler's Church, and the North End of Boston
Mumford, Peter	Merchant	1760	Shop located opposite to Captain John Brown's in the Main Street in Newport
Munson & Barnard	Merchant(s)	1811	Shop located at Number 27, India Street
Munson, Israel	Merchant	1802	Shop located at Number 5, Long Wharf
Murdock, Artemas	Auctioneer/real estate agent?	1803	[No address provided]
Murdock, George	Merchant	1806	Shop located at Number 14, Market Square, directly opposite Messrs. Braistreet & Story
Murdock, Isaac	Merchant	1812	Shop located in Common Street, at the head of Bromfield's Lane, in the establishment formerly occupied by Mr. Fairbanks
Nazro, John	Merchant	1768	Shop located at the establishment lately improved by Dr. William Greenleaf in Cornhill, just below the Court House
Nazro, John	Merchant	1787	Shop located at the corner of the New County Road, near the Meeting-House in Worcester

Name	Profession	Date	Street Address
Nazro, John	Merchant	1798	Shop located in Worcester
Newell and Niles	Merchant(s)	1802	Shop located at Number 9, Doane's Wharf
Newell, Joseph	Merchant	1801	Shop located at Number 9, Doane's Wharf
Nichols & Gerrish	Auctioneer(s)	1739	Auction held at Mr. Weekes's the Tobacconist
Nichols, William	Auctioneer	1738	Auction held at the Royal Exchange Tavern, King's Street, Boston
Nolen & Gridley	Auctioneer(s)	1813	Auction held at their office, Numbers 27, 28, and 29, Cornhill
Norton and Holyoke	Merchant(s)	1794	Shop located at Number 44, Cornhill, formerly the Post Office
Noyes, Joseph	Merchant	1815	Shop located at Number 5, Long Wharf
Oliver and Proctor	Merchant(s)	1799	Shop located at Number 3, Long Wharf
Oliver, Bethiah	Merchant	1765	Shop located opposite Rev. Dr. Sewall's Meeting House, Boston
Oliver, Borland & Abbot	Auctioneer(s)	1814	Auction held at Number 3, Kilby Street
Oliver, Borland & Abbot	Auctioneer(s)	1817	Auction held at their new store, located at Number 21, Central Wharf
Oliver, Ebenezer	Merchant	1773	Shop located nearly opposite the Old South Meeting-House, Boston
Oliver, James	Merchant	1734	[No address provided]
Osborne, John	Chocolate maker	1763	Chocolate mill: located in Charlestown
Otis, Samuel Allyne	Merchant	1767	Shop located at Number 5, South side of the Town Dock
Otis, Samuel Allyne	Merchant	1770	Shop located at Number 1, Butler's Row
Otis, Samuel Allyne	Broker [?]	1774	Shop located next door to Mr. Thomas H. Peck's
Otis's Grocery Store	Merchant	1775	Shop located between Mr. Tho's H. Peck's and the Vernon Corner
Pailhes, Peter	Merchant	1807	Shop located at the store opposite Wm. Smith, Esq., near Concert Hall, Court Street
Paine, Timothy	Not identified	1766	[No address provided]
Palmer & Company Store	Merchant(s)	1767	Shop located on Minot's T
Palmer and Cranch	Chocolate maker(s)	1751	Shop located in School Street, Boston
Palmer, Joseph	Merchant	1769	Shop located on Minot's T
Palmer, Joseph and Cranch, Richard	Chocolate maker(s)	1750	Shop located in their house in South School Street
Palmer, Joseph and Son	Chocolate makers	1773	Shop located opposite the Golden-Ball
Palmer, Joseph P.	Merchant	1773	Shop located at Number 18, South side of the Town Dock
Palmes, Richard	Merchant	1765	Shop located next door to the Sign of the Fan, in Marlborough Street, Boston
Parker & Stevens	Merchant(s)	1816	Shop located at Number 19, Long Wharf
Parker and Warner	Merchant(s)	1805	Shop located at Number 2, Doane's Wharf
Parker, Elizabeth	Merchant	1748	Shop located at the corner house, nest to Dr. Cutler's church
Parks, E. & L.	Auctioneer(s)	1811	Auction held at Number 19, Kilby [Street]
Parks, E. & L.	Auctioneer(s)	1810	Auction held at Number 1, Kilby Street
Parks, E. & L.	Auctioneer	1814	Auction held at Number 40, India Street
Parks, E. & L.	Auctioneer	1814	Auction held at Number 56, India Wharf
Parks, E. & L.	Auctioneer	1814	Auction held at Number 8, Butler's Row
Parks, Elisha	Auctioneer	1810	Auction held at Number 22, State Street
Parks, Elisha	Auctioneer	1817	Auction held at his office, Number 19, Kilby Street
Parks, Elisha	Auctioneer	1819	Auction held at "a store in Sudbury Street" [going out of business auction]
Parks, Elisha	Auctioneer	1819	Auction held at Number 5, Long Wharf
Parks, Elisha	Auctioneer	1821	Shop located at Number 19, Central Street
Parks, Luther	Auctioneer	1815	Auction held at his office, Number 6, Kilby Street
Parks, Luther	Merchant	1815	Shop located at Number 6, Kilby Street
Parks, Luther	Auctioneer	1819	Auction held at Number 2, Lock's Wharf; Sheriff's Sale; Luke Baldwin, Deputy Sheriff

Name	Profession	Date	Street Address
Pattin, William	Merchant	1740	Shop located at the Two Sugar Loaves in Cornhil [sic], Boston
Pattin, William	Merchant	1767	Shop located at the sign of the Two Sugar Loaves, next to the Golden Lyon in Cornhill
Peabody, Jacob and Company	Merchant(s)	1818	Shop located at Number 12, Kilby Street
Peabody, Jacob and Company	Auctioneer(s)	1823	Shop located at Number 4, Long Wharf
Peabody, Jacob and Company	Auctioneer(s)	1823	Shop located at Number 54, State Street
Peabody, Jacob and Company	Auctioneer(s)	1823	Shop located at Number 6, Washington Street
Peabody, Jacob and Company	Auctioneer(s)	1824	Shop located at Number 38, Long Wharf
Pearce, David, Jr.	Merchant	1801	Shop located at Number 35, Long Wharf
Peirce, Joseph, Jr.	Merchant	1792	Shop located at Number 69, Long Wharf
Penniman, S. & A.	Merchant(s)	1806	Shop located at Number 3, Long Wharf
Penniman, Silas	Merchant	1810	Shop located at Number 104, Broad Street
Penniman, Silas	Merchant	1810	Shop located at Number 25, Long Wharf
Perkins, Elizabeth	Merchant	1774	Shop located two doors below the British Coffee-House, North Side of King's Street
Perkins, James	Merchant	1771	Shop located two doors below the British Coffee-House, North Side of King's Street
Peters & Pond	Merchant(s)	1812	Sale taking place at the corner of Milk and India Street, aboard the ship *Magnet*.
Peters & Pond	Merchant(s)	1812	Shop located at the corner of Middle and India Streets
Peters & Pond	Merchant(s)	1813	Shop located at the corner of Milk and India Streets
Peters, Pond & Co.	Merchant(s)	1813	Shop located at Number 7, Milk Street
Phelps and Rand	Merchant(s)	1800	Shop located at Codman's Wh'f
Phelps and Rand	Merchant(s)	1800	Shop located at Number 28, Long Wharf
Pickney, John	Merchant	1736	Shop located near the Red Lion at the North End of Boston
Pigeon, John	Merchant	1754	Shop located near Doctor Clark's [?] in Fish Street
Pigeon, John	Merchant	1761	Shop located near Notomy Meeting House
Pinkney, John [spelled Pinkny or Pinckny in subsequent advertisements]	Merchant	1732	Shop located near the Red Lyon at the North End
Plympton & Marett	Auctioneer(s)	1815	Auction held at Battery Wharf; sale is for the ship *Globe* and its "appurtenances"
Pratt, John & Son	Merchant(s)	1818	Shop located at Number 19, India Wharf
Prentiss & Co.	Auctioneer(s)	1801	Auction held at their office on Kilby Street
Prentiss, Appleton	Auctioneer	1799	Auction held at his office, North side [of] the Market
Prentiss, Appleton	Broker	1801	Shop located at Number 42, State Street, near the head of the Long Wharf
Prentiss, Henry	Merchant	1770	Shop located on the South Side of Faneuil Hall Market
Prentiss, Henry	Merchant	1772	Shop located at the head of Green's Wharf
Prince, Job	Merchant	1769	Shop located at the establishment lately improved by Captain Job Prince, just below Minot's T, on the Long Wharf
Prince, Samuel	Merchant	1799	Shop located at Number 38, Long Wharf
Procter, Thomas	Merchant	1759	Shop located on Queen Street, next door to the Concert Hall
Proctor, Thomas	Merchant	1766	Shop located next door to the Concert Hall
Quincy, John W.	Auctioneer	1797	Auction held at his office, Number 54, State Street
Read, Mrs. [first name not provided]	Chocolate house manager	1731	Chocolate House: located on King's Street, on the North side of the Town House
Renken, Susanna	Merchant	1764	Shop located near the Draw Bridge in Fore Street, Boston

Name	Profession	Date	Street Address
Rhodes, B. [and] Harris, J.	Merchant(s)	1797	Shop located at Number 44, North Water Street, Philadelphia [advertised in the *Boston Columbian Centinel* [*sic*] August 12, 1797]
Richards, H. W.	Merchant	1812	Shop located at Number 9 and Number 10, Central Street
Righton, Francis	Merchant	1743	Shop located in Ann Street near the Draw Bridge
Ripley, Joseph	Merchant	1800	Shop located at Number 8, D. Spear's Wharf
Ripley, Joseph	Merchant	1800	Shop located on D. Spear's Wharf
Ripley, Joseph	Auctioneer	1802	Auction held at D. Spear's Wharf
Ripley, Joseph	Auctioneer	1810	Auction held at his office on India Street
Ripley, Joseph	Auctioneer	1812	Auction held at his office, Number 3, India Street
Ripley, Joseph	Auctioneer	1812	Auction held opposite Number 4, India Street
Ripley, Joseph	Auctioneer	1812	Auction held opposite Number 5, India Street
Risbrough, John	Merchant	1802	Shop located adjoining his former stand, head of Hancock's Wharf [he identifies himself in the advertisement as a grocer]
Ritchie, Robert	Merchant	1757	Warehouse located next to Messrs. Melvill's, on Treat's Wharff [*sic*].
Rogers, J.	Auctioneer	1812	Auction held at his office, Number 7, Kilby Street
Ropes & Pickman	Merchant(s)	1811	Shop located at Number 41, India Wharf
Rowe, John	Auctioneer	1765	Auction held at John Rowe's Wharf
Ruddock, Abiel	Merchant	1772	Shop located near the Winnisnnet Ferry
Rumney, Edward	Merchant	1784	Shop located opposite the East corner of the Market
Rumney, Edward	Chocolate maker	1793	Shop located at his Mills in Malden, Massachusetts
Rumney, Edward	Chocolate maker	1797	Shop located at Capt. Joseph Howard's Store, Fore Street
Rumney, Edward	Chocolate maker	1797	Shop located in Back Street
Rumney, Edward	Merchant	1805	Shop located at Number 57, Back Street
Russell, Benjamin	Merchant	1782	Shop located opposite the Old Meeting House in Marblehead
Russell, J.	Auctioneer	1768	Auction held at the Auction Room in Queen Street
Russell, J.	Auctioneer	1771	Auction held in Queen Street
Salter, Richard Junior	Merchant	1798	Shop located at Number 4, Court Street
Salter, Richard Junior	Merchant	1799	Shop located at Number 2, Court Street
Savage, Arthur	Merchant	1728	Shop located at his house in Brattle Street
Savage, Arthur and Company	Merchant(s)	1761	Shop located on the Town Dock
Savage, Arthur and Company	Merchant(s)	1763	Shop located at Number 11, Long Wharf
Savage, Arthur and Company	Merchant(s)	1763	Shop located on the South side of the Town Dock
Savage, Arthur and Company	Merchant(s)	1764	Shop located at the West Parish in Cambridge (commonly called Menotomy), near the Meeting-House
Savage, Arthur Junior & Company	Merchant(s)	1759	Shop located at Number 2, near the Swing Bridge on the Town Dock, Boston
Savage, Habijah	Merchant	1764	Shop located at Number 16, Long Wharff
Savage, John	Merchant	1762	Shop located at Number 13, Long Wharf
Savage, Thomas	Merchant	1757	Shop located at Number 14, Long Wharff
Savage, Thomas	Merchant	1764	Shop located at Number 16, Long Wharff
Scott, William	Merchant	1771	Shop located in Ann Street, near the Draw Bridge
Scudder, Daniel	Merchant	1798	Shop located at Number 8, Codman's Wharf
Sea[?], Reynolds	Merchant	1751	Shop located at the North end, near Mr. Clark's Building Yard
Sewall, Mary	Merchant	1732	Shop located next door to Mr. Sa[?] Waldo's near the Orange Tree
Seward & Loring	Merchant(s)	1811	Shop located at Number 7, Custom House Street
Sharley, Daniel	Merchant ?	1763	[No address provided]
Shaw, R. G. and Co.	Auctioneer(s)	1801	Auction held at the store lately occupied by Mr. N. Copeland, near Proctor's Lane, Fore Street

Name	Profession	Date	Street Address
Shaw, R. G. and Co.	Auctioneer(s)	1802	Auction held at Number 55 Marlborough Street
Shaw, Robert G.	Merchant	1812	Shop located at Number 53, State Street
Sheaffe, Mrs.	Merchant	1772	Shop located at the North corner of Queen Street, lately occupied by Mr. Joshua Green
Sheaffe, Mrs.	Merchant	1776	Shop located at the establishment formerly occupied by Messieurs Amory, Taylor, and Rogers, in King's Street
Sheaffe, Mrs.	Merchant	1776	Shop located at the establishment occupied last by Richard Salter, Cornhill
Shed, Samuel A.	Merchant	1814	[No address given]
Sherman, Oliver	Merchant	1810	Shop located at Number 16, Long Wharf
Silsbe, Daniel	Merchant	1772	Shop located opposite the South Side of Faneuil Hall
Simpkins, John	Merchant	1770	Shop located in Cornhill, near the Town House, next Door to Mr. Bagnall's the Watch Maker
Simpson, John	Merchant	1815	Shop located at Number 17, Exchange Street
Smith, Eben.	Merchant	1812	Shop located at Craigie's Bridge
Smith, Eben.	Merchant	1812	Shop located at Pond Street, that leads from [the] Charles River Bridge through the Mill Pond, the store formerly occupied by Mr. Wait
Smith, Henry	Merchant	1792	Shop located at Number 1, Town Dock
Smith, Samuel	Auctioneer	1758	Auction held at Smith's Vendue Room, Chardon's Wharff, at the North End
Smith, William	Merchant	1779	Shop located opposite the Golden Ball
Snow, Henry	Chocolate maker	1767	Shop located on Temple Street, New Boston
Spooner & Avery	Merchant(s)	1768	Shop located on Treat's Wharf, just to the Southward of the Town Dock
Stanton & Spelman	Merchant(s)	1806	Shop located at Number 10, Colman's Wharf
Stickney, John	Merchant	1797	Shop located nearly opposite the late Capt. John Stanton's in Worcester
Stickney, Thomas	Merchant	1796	Shop located in Worcester
Stickney, Thomas	Merchant	1796	Shop located opposite Mr. Barker's Tavern
Stickney, Thomas	Merchant	1798	Shop located opposite Capt. Hitchcock's Tavern, in Brookfield, West Parish
Swan, James	Merchant	1772	Shop located at Number 6, next to Mr. Ellis Gray's, opposite the East end of Faneuil Hall, and leading down to Treat's and Spear's Wharfs
Sweetser, Benjamin	Merchant	1797	Shop located at the establishment lately occupied by Nicholas Brown in Back Street
Sweetser, Thomas W.	Merchant	1816	Shop located at Number 15, Milk Street
Taylor & Gamage	Merchant(s)	1812	Shop located at Number 15, Central Street
Taylor, William	Merchant	1769	Shop located in King Street, opposite the Sign of Admiral Vernon
Taylor, William	Merchant	1769	Shop located on the Long Wharf
Taylor, William and Company	Merchant(s)	1765	Shop located at Number 18, Long Wharf
Taylor, Winslow	Merchant	1767	Shop located on King Street
Taylor, Winslow	Merchant	1767	Shop located opposite the Sign of Admiral Vernon in King Street
Thayer, Jeconias	Merchant	1818	Shop located in the Phillips' Buildings, Water Street, East of the Post Office
Ti[nkum], John	Chocolate maker	1738	Shop located next door to Captain Alden's in Milk Street
Ticknor & Marsh	Merchant(s)	1813	Shop located at the store recently improved by Mr. Elisha Ticknor, at the Sign of the Bee Hive, Number 42, Marlboro' Street
Ticknor, Elisha	Merchant	1795	Shop located at Number 42, Marlborough Street, at the establishment lately occupied by Mr. John Erving [identified as a grocery store]
Ticknor, Elisha	Merchant	1806	Shop located at the Sign of the Bee Hive, Number 42, Marlboro' Street

Name	Profession	Date	Street Address
Ticknor, Elisha and Company	Merchant(s)	1803	Shop located at the Sign of the Bee Hive, Number 42, Marlboro' Street
Tillinghast, Daniel	Merchant	1796	[No address provided]
Tinkum, John	Chocolate maker	1736	Shop located opposite to the house [where] the late Mrs. Smith dwelt, in Milk-Street, below the South Meeting House in Boston
Titcomb & Clark	Merchant(s)	1813	Shop located at Number 25, India Street
Tolmam, Robert P. and Co.	Merchant(s)	1818	Shop located at Number 15, Commercial Street, (formerly Milk Street)
Tolman, Robert P.	Merchant	1812	Shop located at Number 9, India Street
Traill, Robert	Merchant	1749	Warehouse located at Number 4, Butler's Row
Tucker, B.	Auctioneer	1807	Auction held at Number 66, Long Wharf
Tucker, B.	Auctioneer	1812	Auction held at his office, Number 62, State Street
Tucker, Benjamin	Auctioneer	1808	Auction held at Number 66, State Street
Tucker, William	Merchant	1814	Shop located at Number 22, Long Wharf
Tufts, Simon, Jr.	Merchant	1774	Shop located near Faneuil Hall Market
Turell, Joseph	Merchant	1770	Shop located at Number 7, South side [of] Swing Bridge
Turell, Joseph	Merchant	1770	Shop located at the corner of Kilby Street
Turner, Robert	Merchant	1801	Shop located at Number 1, West Row, Cambridge Street, near [the] Concert Hall
Tyler, John	Auctioneer	1818	Auction held at Number 72, Orange Street, Corner of Essez [Essex?] Street
Tyng, Edward	Merchant	1737	Warehouse located in Milk Street, near the old South Meeting House, Boston
Vaughan, Daniel	Merchant	1812	Shop located at Number 63, Cambridge Street, corner of Beknap Street [states he is of the late firm of Wright & Vaughan and solicits the patronage of the public, and his friends]
Wait, John	Brick-maker	1800	Brickyard located in Charlestown; he sells out of his house on Prince Street near the Charlestown Bridge
Wait, John Jr.	Chocolate maker	1816	[No address provided]
Waldo, Daniel and Son	Merchant(s)	1787	Shop located opposite the prison in Worcester
Waldo, John	Merchant	1759	Shop located at Number 17, Long Wharf opposite the first crane
Waldo, John	Merchant	1771	Shop located at his house and store, South part of the Town, third door from the Sign of the Lamb
Waldron, [?]	Merchant	1797	Shop located at the head of Hooton's Wharf, [at] Fish Street
Wales & Beale	Merchant(s)	1807	Shop located at Number 54, Long Wharf
Walker, Edward	Merchant	1765	Shop located in the establishment lately improved by Mr. Christopher Clarke, next door to Messiers Fleets Printing-Office, in Cornhill, Boston
Walker, J.	Merchant	1803	Shop located at Number 14, Cambridge Street, entrance of Bowdoin Square
Walker, Rebecca	Merchant	1764	Shop located opposite the Blue-Ball near Mill Bridge
Walley, Samuel H.	Merchant	1801	Shop located at Number 34, Long Wharf
Walley, Thomas	Auctioneer	1762	Auction held under [?] the New Auction Room, Dock Square
Walley, Thomas	Merchant	1765	Shop located on Dock Square
Wallis, Samuel	Merchant	1787	Shop located on the North side of the Town Dock
Ward & Fenno	Merchant(s)	1774	Shop located at the head of Dock Square
Warren & Eustis	Auctioneer(s)	1799	Auction held at their office [no address provided]
Waters, Josiah	Merchant	1765	Shop located in Ann Street, next door to the Sign of [the] Seven Stars
Waters, Josiah and Son	Merchant(s)	1772	Shop located on Ann Street near the Draw Bridge
Watts, Mary	Merchant	1751	Shop located on Middle Street
Watts, Samuel	Chocolate maker	1751	Shop located up [by] the Prison Yard
Watts, Samuel	Chocolate maker	1752	Shop located at the bottom of Mr. Charden's Garden, at New Boston

Name	Profession	Date	Street Address
Webb, Joseph	Merchant	1769	Shop located at the IRON WARE STORE, between Philips and Moore's Wharves, and leading down from Kilby Street, near Oliver's Dock, Boston
Webb, Thomas	Merchant	1768	Shop located at the corner near the south east end of Faneuil Hall Market
Webster, Grant	Merchant	1773	Shop located at the Intelligence Office, on the way leading down to Treat's Wharf
Welch, Francis	Merchant	1767	Shop located on the South Side of Faneuil Hall Market
Welsh, Jonas	Chocolate maker	1792	Chocolate Mill: located near the Charles River Bridge, Boston
Welsh, Mr.	Chocolate maker	1782	Shop located at the North Mills in Boston
Welsh, Mr.	Chocolate maker	1785	Shop located in the North End
Wheatley, Nathaniel	Merchant	1764	Shop located in King's Street
White, Thomas & Co.	Merchant(s)	1817	Shop located at Number 17, India Street
White, William	Merchant	1792	Shop located at Numbers 55 and 56, Long Wharf
Whitewell & Bond	Auctioneer(s)	1809	Shop located on Kilby Street
Whitewell & Bond	Auctioneer(s)	1813	Auction held in Brookline, at the late residence of William Marshall, Sen. of Brookline
Whitney and Lobre	Merchant(s)	1795	Shop located at Number 27, Union Street
Whitney, Cutler & Hammond	Merchant(s)	1811	Shop located at Number 34, Broad Street
Whittmore, W. & Co.	Auctioneer/real estate agent?	1802	Business [?] located on Dock Square
Whitwell & Bond	Auctioneer(s)	1811	Auction held opposite Number 10, India Street
Whitwell & Bond	Auctioneer(s)	1812	Auction held at Number 56, State Street
Whitwell, Bond & Co.	Auctioneer(s)	1818	Auction held at their store, Number 8, Merchant's Row
Whitwell, Bond & Co.	Auctioneer(s)	1819	Auction held at their office, Number 2, Kilby Street
Whitwell, Bond & Company	Auctioneer(s)	1822	Shop located in Doane Street
Wilby, Francis & Co.	Auctioneer(s)	1817	Auction held in the Assembly Room, at the Exchange Coffee House, in Boston
Wild, Daniel	Auctioneer	1801	Auction held at his office on State Street
Williams and Vincent	Merchant(s)	1776	Shop located one door above the American Coffee House, King Street [advertisement dated July 11, 1776]
Williams, Jonathan	Merchant	1751	Shop located upon the Mill Bridge
Williams, Jonathan Jun'r.	Merchant	1772	Shop located in Ann Street, below the Draw Bridge, and directly opposite the Drum-makers at the head of Barrett's Wharf
Williams, Mary	Merchant	1752	Shop located at the Two Sugar Loaves in Fleet Street, leading from Scarlet's Wharff to Bennet Street, near Mr. Mather's Meeting House at the North End
Williams, Samuel	Merchant	1801	Shop located at Number 12, Butler's Row
Winflow and Winniett	Merchant(s)	1752	Shop located in the former establishment of Mr. Caleb Lymann, deceased, over against the Sun Tavern in Cornhill
Winslow, Channing & Company	Auctioneer(s)	1821	Shop located at Number 8, Kilby Street
Winslow, Channing & Company	Auctioneer(s)	1821	Shop located opposite Number 14, Central Wharf, North Side
Winter, William	Auctioneer	1757	Auction held in his Vendue Room in Wing's Lane
Winter, William Jr.	Merchant	1737	Shop located at the Five Sugar Loaves in Fish Street, Boston
Winthrop, Adam	Merchant	1743	Shop located at his house on Brattle Street, next to the Rev. Dr. Colman's
Wood, Abial	Merchant	1799	Shop located at Number 39, Long Wharf
Woodward, A. & J.	Merchant(s)	1816	Shop located at Number 10, Newbury Street
Wright, Winslow & Co.	Merchant(s)	1817	Shop located at Number 53, State Street
Young, William	Merchant	1743	Shop located at the North End

7

The Ninety and Nine

Notable Chocolate-Associated Quotations, 1502–1953

Person	Date	Quotation	Source
Hillary, Edmund	1953	[Interview in 1973 with first person to reach the summit of Mount Everest]: What did you do when you got to the top? I read somewhere that you ate a chocolate bar? Answer: No. We didn't eat anything on top, but Tenzing buried a little bit of chocolate and some sweets in the snow, which are really a gesture to the gods which the Sherpas believe flit around Everest on all occasions.	Academy of Achievement Interview, Washington, DC. Available at http://www.achievement.org/ autodoc/page/hil0int-8. (Accessed October 2, 2007.)
Earhart, Amelia	1935	[Interview after her record-setting Hawaii to U.S. mainland flight]: That was the most interesting cup of chocolate I have ever had, sitting up eight thousand feet over the middle of the Pacific Ocean, quite alone.	Available at http://www. ameliaearhart.com/about/ biography2.html. (Accessed October 2, 2007.)
Reed, John	1917	[Description of harsh winter and food shortages during Russian Revolution]: Toward the end there was a week without any bread at all. Sugar one was entitled to at the rate of two pounds a month—if one could get it at all, which was seldom. A bar of chocolate or a pound of tasteless candy cost anywhere from seven to ten rubles—at least a dollar. There was milk for about half the babies in the city; most hotels and private houses never sat it for months.	Reed, J. *Ten Days That Shook the World*. The Project Gutenberg E-Book. Release date December 16, 2000. (E-Text #3076). Chapter 1.

Chocolate: History, Culture, and Heritage. Edited by Grivetti and Shapiro
Copyright © 2009 John Wiley & Sons, Inc.

Person	Date	Quotation	Source
Roosevelt, Theodore	1914	[What to take on a South American jungle expedition]: Bread 100 ounces each day; Corned beef 70 ounces on Wednesday; Sugar 32 ounces daily; Baked beans 25 ounces on Thursday and Saturday; Figs 20 ounces on Sunday and Tuesday; Condensed milk 17 ounces daily; Sweet chocolate 16 ounces each Wednesday; Coffee $10\frac{1}{2}$ ounces daily; Roast beef; $6\frac{1}{2}$ ounces on Monday; Rice 6 ounces every Sunday, Tuesday, and Friday; Bacon 4 ounces every day; Curry chicken $1\frac{3}{4}$ ounces on Friday.	Roosevelt, T. *Through the Brazilian Wilderness.* London, England: John Murray, 1914; Appendix B.
Lincoln, Abraham	1865	[Newspaper announcement, President Lincoln's second inaugural ball bill of fare]: Oyster stew, terrapin stew, oysters pickled; roast beef; turkey; roast chicken; roast grouse; quail; venison; smoked ham; salads; lobsters; chocolate; ice cream; fruit ices; for dessert—grapes, almonds, raisins, &c, coffee and chocolate.	*New York Times*, March 8, 1865, p. 1.
Robertson, John H.	1863	[Memory from the American Civil War]: I was captured at the battle of Gettysburg in Longstreet's charge and was taken to Fort Delaware, an island of 90 acres of land where the Union prisoners were kept. We were detailed to work in the fields and our rations was [*sic*] corn bread and pickled beef. However, I fared better than some of the prisoners for I was given the privilege of making jewelry for the use of the Union soldiers. I made rings from the buttons from their overcoats and when they were polished the brass made very nice looking rings. These I sold to the soldiers of the Union Army who were our guards and with the money thus obtained I could buy food and clothing. The Union guards kept a commissary and they had a big supply of chocolate. I ate chocolate candy and drank hot chocolate in place of coffee until I have never wanted any chocolate since.	Library of Congress, Manuscript Division, WPA Federal Writers' Project Collection.
Anonymous	1861	[Newspaper announcement, instructions for Federal military volunteers]: Recommendation 2. Avoid drinking freely of very cold water, especially when hot or fatigued, or directly after meals. At meals tea, coffee, and chocolate are best. The safest [beverage] in hot weather is molasses and water with ginger or small beer.	*New York Times*, May 20, 1861, p. 2.
Dickens, Charles	1859	[Passage from his novel]: It took four men, all four ablaze with gorgeous decoration, and the Chief of them unable to exist with fewer than two gold watches in his pocket, emulative of the noble and chaste fashion set by Monseigneur, to conduct the happy chocolate to Monseigneur's lips. One lacquey carried the chocolate-pot into the sacred presence; a second, milled and frothed the chocolate with the little instrument he bore for that function; a third, presented the favoured napkin; a fourth (he of the two gold watches), poured the chocolate out. It was impossible for Monseigneur to dispense with one of these attendants on the chocolate and hold his high place under the admiring Heavens. Deep would have been the blot upon his escutcheon if his chocolate had been ignobly waited on by only three men.	Dickens, C. *A Tale of Two Cities. A Story of the French Revolution.* The Project Gutenberg E-Text. Release date September 25, 2004 (E-Text #98). Book 2, Chapter 7.

Person	Date	Quotation	Source
Loucks, Andrew	1849	[Published interview, exploring the keys to longevity]: We ate lighter food when I was a boy than at present—such as soups; used a great deal of milk, and but little tea and coffee. We sometimes made chocolate by roasting wheat flour in a pot.	*Scientific American*, Vol. 4, No. 31, April 21, 1848.
Thoreau, Henry David	1848	[Essay, including passage on breakfast in prison]: In the morning our breakfasts were put through the hole in the door, in small oblong-square tin pans, made to fit, and holding a pint of chocolate, with brown bread, and an iron spoon.	Thoreau, Henry David. *Resistance to Civil Government*.
Cooper, James Fenimore	1843	[Book, selected passage]: Take America, in a better-class house in the country, and you reach the *ne plus ultra*, in that sort of thing. Tea, coffee, and chocolate, of which the first and last were excellent and the second respectable; ham, fish, eggs, toast, cakes, rolls, marmalades, &c. &c. &c. were thrown together in noble confusion.	Cooper, James Fenimore. *Wyandotte; or, the Hutted Knoll*, Vol. 1. Philadelphia: Lea and Blanchard, 1843; p. 88.
Wislizenus, F. A.	1839	[Diary entry regarding Rocky Mountain Green River fur trapper rendezvous]: A pound of beaver skins is usually paid for with four dollars worth of goods; but the goods themselves are sold at enormous prices, so-called mountain prices. . . . A pint of meal, for instance, costs from half a dollar to a dollar; a pint of coffee-beans, cocoa beans or sugar, two dollars each; a pint of diluted alcohol (the only spiritous liquor to be had), four dollars; a piece of chewing tobacco of the commonest sort, which is usually smoked, Indian fashion, mixed with herbs, one to two dollars. . . . With their hairy bank notes, the beaver skins, they can obtain all the luxuries of the mountains, and live for a few days like lords. Coffee and chocolate is cooked; the pipe is kept aglow day and night; the spirits circulate and whatever is not spent in such ways the squaws coax out of them, or else it is squandered at cards.	Wislizenus, F. A. *A Journey to the Rocky Mountains in 1839*. St. Louis, Missouri: Missouri Historical Society, 1912; Chapter 11, no page numbers.
Santa Anna, Antonio López de	1833	[General Order, issued by president and commander in chief of the Mexican Army]: November 18th, for the Officers Mess the following items were to be procured: For the breadmaker—50.00 pesos; lard 10.47; Chocolate 43; Beans 17; For the butcher 240.00; Chickpeas 27; coal [for cooking] 30.	Centro de Estudios de Historia de México. CONDUMEX. 355.07 MEX. Núm.33825. Secretaria de Guerra y Marina, Secion 7a. Reglamento para el Colegio Militar.
Peixotto, Daniel L. M.	1832	[Announcement, concerning cholera, fasting, and Jewish Holy Days]: Allow me to suggest, that on the present occasion a slight meal, say of coffee, tea, or cocoa, with dry toast, be allowed at early rising, and a few draughts through the day of toast-water, or tea. This will obviate any mischief which might otherwise result from severe abstemiousness in the first place.	No Author. Miscellaneous Items Relating to Jews in New York. *Publications of the American Jewish Historical Society (1893–1961)*, Vol. 27, p. 150, 1920.

Person	Date	Quotation	Source
Anonymous	1830	[Published rules for young ladies]: Her breakfast should be something more substantial than a cup of slops, whether denominated tea or coffee, and a thin slice of bread and butter. She should take a soft boiled egg or two, a little cold meat, a draught of milk, or a cup or two of pure chocolate.	Anonymous. Rules for a Young Lady. *Ladies' Magazine and Literary Gazette*, Vol. 3, p. 239, 1830.
Graham, Thomas J.	1828	[For the treatment of asthma]: This diet should be uniformly light and easy to digest, consisting mainly of a fresh food of animal origin, such as eggs, as well as bread, tea and chocolate.	Graham, T. J. 1828. *Medicina Moderna Casera. O Tratado Popular, en el que se ilustran el caracter, sintomas, causas, distincion, y plan curativo correcto, de todas las enfermedades incidentales al cuerpo humano.* London, England: Juan Davy; p. 231.
Brillat-Savarin, Anthelme	1825	[Commentary on the qualities of chocolate]: Chocolate, when properly prepared, is a food as wholesome as it is agreeable . . . it is nourishing, easily digested . . . and is an antidote to the [inconveniences] . . . ascribed to coffee.	Brillat-Savarin, J. A. *The Physiology of Taste.* New York: Houghton Mifflin, 1825; p. 95.
Byron (Lord), George Gordon	1821	[Poem]: Lord Henry, who had not discuss'd his chocolate, Also the muffin whereof he complain'd, Said, Juan had not got his usual look elate, At which he marvell'd, since it had not rain'd;	Byron, G. G. 1824. *Don Juan.* Canto xvi: Verse xxxiv.
Clark, William	1806	[Diary, account of Clark's illness toward the end of the expedition]: I felt my self [*sic*] very unwell and derected [*sic*] a little Chocolate which Mr. McClellin gave us, prepared of which I drank about a pint and found great relief.	*The Journals of Lewis and Clark.* Entry for September 13, 1806. (Edited by Bernard DeVoto. The American Heritage Library Edition. Boston: Houghton Mifflin; p. 473.)
Anonymous	1804	[Discussion on the quality of chocolate]: The goodness of chocolate depends first, upon the quality of the cocoa. Of this there are three principal species: caracas, quayaquil, and that from the islands of St. Domingo, Martinique, Curracoa, etc. To make the chocolate the caracas is mixed with the quayaquil; two parts of the caracas and one of the quayaquil. . . . The goodness of chocolate depends, in the second place, on the care with which it is ground and roasted, on the proper proportion of the cocoa, the sugar, and the different aromatics. . . . The characteristics of a good, unadulterated chocolate, are the following: a deep fresh colour; a fine, close, shining grain; small white streaks; an aromatic odour; a facility of dissolving in the mouth, with a sensation of freshness, to produce no appearance of glue in cooling, and to shew [*sic*] an olily [*sic*] cream on the top.	*The Literary Magazine and American Register*, March 1804, p. 427.
Daboll, Nathan	1800	[Arithmetic book, sample question]: If 40 pounds of chocolate be sold at 25 cents per pound, and I gain 9%. . . . What did the whole cost me? [Answer: 9 dollars 17 cents and 4 mills.]	Nathan Daboll. 1800. *Daboll's Schoolmaster's Assistant. Being a Plain Practical System of Arithmetic Adapted to the United States*; p. 122.
Mather, Cotton	1796	[Sermon, describing Milton, Massachusetts]: There are seven mills upon the Naponset River, in three of which the manufacture of Paper is carried on; chocolate, slitting, saw and grist mill make up the rest. The inhabitants subsist chiefly by agriculture.	Mather, Cotton. 1796. Sermon: *The Comfortable Chambers*; p. 38.

Person	Date	Quotation	Source
Allen, Israel	1796	[Medical tract, treatment for dysentery]: The diet should contain chocolate, coffee, tea, milk, rice, &c. I have always observed children at the breast, bear the disease better than those who are weaned.	Allen, Israel. 1796. *A Treatise on the Scarlatina Anginosa, and Dysentery*. Leominster, MA: Charles Prentiss.
Hancock, Jr., John	1795	[Advertisement, maintaining the family tradition]: Fresh Philadelphia Flour, Just arrived, and for Sale by THOMAS and JOHN HANCOCK, At their Store, near the Market; Also, French Brandy, good Brown Sugars, Nails all sizes, Gun-Powder, Anchors, Coffee, Chocolate, Bar Soap in Boxes, Lemons, Prime Beef, and a few Bales [of] India Cotton, &c.	*Columbian Centinel* [sic] [Boston], June 27, 1795, p. 3.
Franklin, Benjamin	1794	[Essay, on what to take on a long sea voyage]: You ought also to carry with you good tea, ground coffee, chocolate, wine of the sort you like best, cyder [sic], dried raisins, almonds, sugar, capillaire, citrons.	*The Works of the Late Dr. Benjamin Franklin consisting of his Life written by himself together with Essays*. New York: Tiebout and Obrian; p. 64.
Woodbridge, Samuel	1792	[Advertisement, Christmas foods]: Green Teas, Coffee, Chocolate, Pepper, Spices, Codfish, Molasses . . . And many other good things for Christmas.	*Connecticut Gazette*, December 13, 1792, p. 4.
Anonymous	1792	[A small fleet arrived at Cadiz, Spain, carrying Jesuit priests who had been expelled from Mexico]: In unloading the vessels, eight large cases of chocolate were said to have been found. . . . These cases threatening to break the backs of the porters . . . [customs] officers became curious to know the cause. They opened one amongst themselves, and found nothing but very large cakes of chocolate, piled on each other. They were all equally heavy, and the weight of each surprising [sic]. Attempting to break one, the cake resisted, but the chocolate shivering off, [revealed] an inside of gold covered round with chocolate to the thickness of an inch.	*Massachusetts Magazine / Monthly Museum*, 1792, p. 416.
Anonymous	1791	[Announcement of a 4th of July celebration to be held at]: GRAYS GARDENS. A CONCERT of Vocal and Instrumental Music will begin on Monday, the glorious 4th of July, at six o'clock in the morning, and conclude at ten at night, should the day be fair, to celebrate American Independence. Songs, with harmony and martial music, in honor of the day, will be performed. In order to furnish the public with refreshments, tea, coffee and chocolate, and fruits of the season, will be ready for breakfast. Three tables, with 100 covers each, furnished with roast beef, rounds, hams, &c. &c. ready to cut-and-come-again, from morning until night.	*The Gazette of the United States*, July 2, 1791, p. 72.
Rush, Benjamin	1789	[Medical essay, treatment of smallpox]: The diet should consist chiefly of vegetables. Tea, coffee, and even weak chocolate, with biscuit or dry toast.	Rush, Benjamin. *Medical Inquiries and Observations on the Treatment of Smallpox*, Vol. 1. Philadelphia: Prichard & Hall, 1789.
Wadsworth, Jeremiah	1788	[Letter; chocolate, politics, and headaches]: My dear friend . . . I received several letters from Harriet and hear you are gone to Litchfield. I am well and have no head acch [sic] except ones on eating chocolate at an Assembly. New York is more disagreeable than ever.	Jeremiah Wadsworth to Mehitable Wadsworth, dated February 10, 1788. *Letters of Delegates to Congress*: Volume 24, November 6, 1786–February 29, 1788. Washington, DC: Library of Congress.

Person	Date	Quotation	Source
Anonymous	1787	[Report of negligence leading to death]: Saturday sennight [*sic*], master Robert Clough, son of Mr. Robert Clough, of this town [Worcester], being at play with another child, in Mr. Makepeace's chocolate mills, north end, by some accident the former got entangled in the machinary [*sic*] of the works, while the horses were on the go, and was so shockingly mangled in his bowels, &c. that he died at two o'clock, Sunday morning.	*Worcester Magazine*, September 1, 1787, p. 307.
Anza, Juan Bautista de	1786	[Receipt, military expenses]: Account of what was received and of the expenditures, for the extraordinary expenditures of the Peace and War of this Province of New Mexico, according to the orders that were given to me [José Maldonado] the Lieutenant and Paymaster of the Presidio Santa Fe in the above mentioned Province in the year 1787: 3 arrobas [=75 pounds] of ordinary chocolate at 3 reales and 6 grains [?] 32 pesos 6 reales 6 grains.	AGN. Provincias Internas, Vol. 67, exp. 1, f.1, 1786, p.114v.
Jefferson, Thomas	1785	[Letter; Jefferson's vision of the future]: Chocolate . . . the superiority of the article both for health and nourishment will soon give it the same preference over tea and coffee in America.	Thomas Jefferson to John Adams. Paris, dated November 27, 1785.
Arnold, Benedict	1785	[Advertisement; subsequent use of Arnold's abandoned store]: William & Samuel Helms [have] taken Possession of that range of Stores in Water-Street [New-Haven, Connecticut], a few Rods East of Union-Street, formerly occupied by Sheriff Mansfield and Benedict Arnold, where they continue to sell Rum, Sugar, Tea, Coffee, Chocolate, Indigo, Pepper, Allspice, Ginger, Alum, &c. &c.	*New-Haven Gazette*, August 4, 1785, p. 4.
Webster, Noah	1783	[Article; training children in the art of manners]: Will you drink a dish of tea? I choose not to drink any. Perhaps you like coffee better. No, Sir, I like chocolate.	Webster, Noah. *A Grammatical Institute of the English language.* 1st edition. Hartford: Hudson & Goodwin, 1783.
Elmendorf, Peter	1782	[Diary; student food at Princeton University]: We eat rye bread, half dough, and as black as it possibly can be, and oniony [*sic*] butter, and some times dry bread and thick coffee for breakfast, a little milk or cyder [*sic*] and bread, and sometimes meagre [*sic*] chocolate for supper, very indifferent dinners, such as lean tough boiled fresh beef with dry potatoes.	Peter Elmendorf's Diary. 1782. Available at http://etcweb1. princeton.edu/Campus/text_ revolution.html (Accessed October 2, 2007.)
Huntington, Samuel	1780	[Letter; difficulties facing American soldiers during Revolution]: We have omitted observing that the Medical department are destitute of those necessaries, which are indispensable for the sick. They have neither wine, Tea, sugar, Coffee, Chocolate, or spirits. We wish orders may be given for an immediate supply, as the army grow more sickly every hour.	Letter from P. Schuyler, Jno. Matthews, and Nathl. Peabody to Samuel Huntington, dated May 10, 1780. *Letters of Delegates to Congress*, Vol. 15, April 1, 1780– August 31, 1780. Washington, DC: Library of Congress.

Person	Date	Quotation	Source
Continental Congress	1779	[Continental Congress Act; general military provisions]: Motion by Mr. Gouverneur Morris, and seconded by Mr. Thomas Burke: A Colonel, Lieutenant Colonel, or Chaplain of a Brigade shall be entitled to draw only six gallons of Rum, either four pounds of Coffee or Chocolate, or one pound of tea and twelve pounds of Sugar monthly; a Major or Regimental Surgeon, only four gallons of Rum, either three pounds of Coffee or Chocolate, or three quarters of a pound of tea and eight pounds of Sugar monthly.	*Journals of the Continental Congress, 1774–1789.* Entry for Wednesday, April 17, 1779. Washington, DC: Library of Congress.
Adams, John	1779	[Diary, observations in Spain]: The servant then brought in another salver [serving tray] of cups and hot chocolate. Each lady took a cup and drank it, and then cakes and bread and butter were served; then each lady took another cup of cold water, and here ended the repast.	Diary of John Adams. (Excerpted in *The Dollar Magazine*, December 1851, pp. 241–244.)
de Sade, Donatien Alphonse François Marquis	1779	[Early mention of chocolate bars]: The next time you send me a package, please have it made for me, and try to have some trustworthy person there to see for themselves that some chocolate is put inside. The cookies must smell of chocolate, as if one were biting into a chocolate bar.	de Sade, A. F. (Marquis. Marquis de Sade). *Letters from Prison.* Translated by R. Seaver. New York: Arcade Publishing. Letter to his wife, dated May 16, 1779.
Adams, Abigail	1777	[Letter; complaint over high prices during wartime]: New England rum, 8 shillings/gallon; coffee 2 and sixpence/pound; chocolate, three shillings/pound. What can be done/ Will gold and silver remedy this evil?	Abigail Smith Adams to John Adams, dated April 20, 1777.
Wayne, Anthony	1776	[Diary/Day book, Military Hospital, Fort Ticonderoga]: Doctor Johnson will order Chocolate and Sugar for the Sick in the Hospital and such other suitable Refreshments as may be had on the Ground.	*Orderly Book, Fort Ticonderoga.* Colonel Anthony Wayne. Entry for December 19, 1776.
Pennsylvania District Committee	1776	[Advertisement; condemnation of hording food and commodities during wartime]: Salt, Rum, Sugar, Spice, Pepper, Molasses, Cocoa, and Coffee. . . . The scarcity of those articles is artificial; and several persons whose names are returned to this Committee have formed a cruel design to add to the distresses of their suffering fellow citizens and country, by collecting great quantities of, and exacting exorbitant prices for, the above articles. [Such] conduct at any time [is] shameful, but at a period of public calamity most barbarous and oppressive, more especially on the poor and middling ranks of life.	*Pennsylvania Evening Post*, March 7, 1776, p. 119.
Avery, Elisha	1776	[Inventory list; military provisions, Fort Ticonderoga]: 565 barrels Pork; 30 barrels Sugar; 60 barrels Rum; 300 pounds Coffee; 800 pounds Chocolate; 50 bushels Salt.	*Report of the Deputy Commissary Officer,* dated August 23, 1776.

Person	Date	Quotation	Source
Allan, Ethan	1776	[Newspaper article, description of reatment of American Revolutionary hero]: When Col. Ethan Allen, with about 50 other prisoners arrived [in Ireland aboard a British prison ship] . . . a subscription was begun among some friends of the cause. . . . We this day sent a hamper of wine, sugar, fruit, chocolate, &c. on board, for his immediate use.	*New-Hampshire Gazette and Historical Chronicle*, January 9, 1776, p. 1.
Font, Pedro	1776	[Chocolate used as an engagement gift by members of the 2nd Anza California expedition]: While we were in San Diego the corporal of the guard of San Gabriel fell in love with a girl of our expedition; and since he had nothing to give her as a means of getting into her good graces, he urged the muleteers to give him something of what came in their charge, and, condescending, they gave him chocolate and other things.	Font, P. Diary. Entry for March 7, 1776.
Alexander, Henry	1776	[Use of chocolate during expeditions]: On the twentieth [of January 1776], the last remains of our provisions were expended; but, I had taken the precaution to conceal a cake of chocolate, in reserve for an occasion like that which was now arrived. Toward evening, my men, after walking the whole day, began to lose their strength; but, we nevertheless kept on our feet till it was late; and, when we encamped, I informed them of the treasure which was still in store. I desired them to fill the kettle with snow, and argued with them the while, that the chocolate would keep us alive, for five days at least; an interval in which we should surely meet with some Indian at the chase . . . the kettle being filled with two gallons of water, I put into it one square of the chocolate. The quantity was scarcely sufficient to alter the colour of the water; but, each of us drank half a gallon of the warm liquor . . . and in its enjoyment felt no more of the fatigues of the day. In the morning, we allowed ourselves a similar repast, after finishing which, we marched vigorously for six.	Henry, A. *Travels & Adventures in Canada and the Indian Territories Between the Years 1760 and 1776.* Edited by Elliot Couse. Boston, MA: Little, Brown and Company, 1901; p. 272.
Maxwell, Rosanna	1775	[Advertisement; Loyalist merchant]: Choice cane and Jamaica spirits, West India and New England Rum, Port, Malaga and Sherry Wines, Loaf and Brown Sugar, Coffee, Chocolate, Spices, English and Rhode-Island Cheeses.	*Boston Gazette and Country Journal,* February 13, 1775, p. 4.
General Assembly, Colony of Connecticut	1775	[Government of Connecticut; act to procure military provisions]: Three-quarters of a pound of Pork, or one pound of Beef, per diem; Fish three times per week; Three pints of Beer per diem; Half a pint of Rice or one pint of Meal; six ounces of Butter; three pints of Peas, or Beans per week; Chocolate, six pounds per Company per week; One gill of Rum per man, on fatigue days only. The Rations will cost— when Pork is issued, eleven pence per diem; when fresh Beef, ten pence.	*Commissary Order for Troop Provisions,* dated May 1775, General Assembly, Colony of Connecticut.

Person	Date	Quotation	Source
Santa María, Vicente	1775	[Diary, discovery of San Francisco Bay: As we came to the shore, we wondered much to see Indians , lords of these coasts, quite weaponless and obedient to our least sign to them to sit down, doing just as they were bid. . . . The Indians made various signs of good will by way of salutations. . . . One of the sailors had brought some chocolate. He gave it to an Indian who finding it sweet, made signs that he would get something of similar flavor. He did so, bringing back to him a small tamale that had a fairly sweet taste and is made from a seed resembling polilla.	Santa María, V. 1971. *The First Spanish Entry into San Francisco Bay, 1775. The Original Narrative, Hitherto Unpublished by Fr. Vicente Santa María, and Further Details by Participants in the First Explorations of the Bay's Waters*. Edited by J. Galvin. San Francisco, CA: John Howell-Books; p. 45.
Hill, J.	1775	[Commentary on the importance of breakfast]: Breakfasts are as necessary as suppers; only those who are troubled with phlegm should eat less at this meal than others. A cup of chocolate, not made too strong, is a good breakfast. Coffee I cannot advise generally: But the exceptions against tea are in a great measure groundless.	Hill, J. 1775. *The Old Man's Guide to Health and Longer Life. With Rules for Diet, Exercise, and Physic. For Preserving a Good Constitution, and Preventing Disorders in a Bad One*. Philadelphia, PA: John Dunlap; p. 12.
Schuyler, Philip	1775	[Letter describing soldier's tricks to avoid military service during American Revolutionary War]: It is certain, however, that some [of the recruits] have feigned sickness; for Dr. Stringer informs me, that on his way up here, about the 6th of September [1775], he met many men that looked very well; and, upon inquiry, some acknowledged they had procured their discharges by swallowing tobacco juice, to make them sick. Others had scorched their tongues with hot chocolate, to induce a believe [sic] that they had a fever, &c.	Letter: Philip Schuyler to Governor Jonathan Trumbull, written at Fort Ticonderoga, October 12, 1775. *American Archives*, Series 4, Vol. 3, p. 1035.
Hancock, John	1770	[Advertisement]: Choice chocolate made and Sold by John Goldsmith, at the Corner Shop leading down [to] John Hancock, Esq'rs, Wharff [sic], by the large or small Quantities. The Chocolate will be warranted good, and sold at the cheapest Rates—Cash given for Cocoa. Cocoa manufactured for Gentlemen in the best Manner—ALSO, Choice Cocoa and Cocoa Nuts Shells.	*Boston News-Letter*, March 8, 1770, p. 2.
Serra, Junipero	1770	[Letter describing supply needs for California missions]: What we really need is wax for the masses, underwear robes, because, as I said it in other occasion is very cold here, also two good blankets for each of the new missions [San Diego and Monterey] and for the next one I hope soon will be established [Buenaventura] with a label each of them [2 for San Diego, 2 for Monterey and 2 for Buenaventura] so they will remain where they belong; we also need some chocolate, that Thanks God, we were not lacking up to now, due to the care of the Illustrious General Surveyor; also many heavy wool clothing and flannels to cover so many naked poor people that here are peaceful and docile.	Santa Barbara Mission Archives. SBMA. Father Serra Collection, Document 217. Letter of Father Serra to Father Guardian Juán Andrés, dated June 12, 1770.

Person	Date	Quotation	Source
Washington, George	1769	[Letter; written at Mount Vernon]: I should be obliged to you for bringing me the following Articles [after] first deducting the Freight and Commissions: 1 barl. Of very best bro[wn] Sugar; 200Wt. of Loaf refined Sugar if good and Cheap; 1 Pot, about 5 pounds preserved Green Sweetmeats; 2 or 3 doz. Sweet Oranges; 1 dozn. Cocoa Nuts; A few Pine Apples if in Season; and the residue of the money, be it little or much, to be laid out in good Spirits.	George Washington to Lawrence Sanford, dated September 26, 1769, Account Book 2. Washington, DC: Library of Congress.
Anonymous	1769	[Items necessary for a whaling voyage]: Pork, beef, bread, corn and meal, as may be thought sufficient, 1 barrel of beans, 1 barrel of salt, 2 barrels of flour, 1 barrel of rum, 2 barrels of molasses, 12 barrels of potatoes, 14 pounds either coffee or chocolate, 1 pound pepper, 56 pounds sugar, 50 pounds either butter or cheese, 75 pounds fat, 112 pounds rice. Dry fish if they go to the Western Islands.	Articles of Agreement Relative to the Whale-fishery. Boston, February 10, 1769 (Broadside).
Crathorn, Mary	1768	[Advertisement; a widow's plea]: [I beg] leave to inform my late husband's customers that [I] continue to sell chocolate well manufactured . . . with sundry other articles.	*Pennsylvania Gazette*, February 11, 1768, no page number.
Lopez, Aaron	1766	[Description of work effort]: Prince Updike who prepared 2000 pounds of chocolate from 2500 pounds of cocoa in 1766.	Ledger Entry: Aaron Lopez Ledgers for 1766, pp. 17, 61, 71, 77, 107.
Arnold, Benedict	1765	[Advertisement; prior to his disgrace for treason]: Benedict Arnold, Has just imported (via New-York) and sells at his Store in New-Haven, A very large and fresh Assortment Of Drugs and Chymical [*sic*] Preparations: Essesnce [*sic*] Water, Dock, Essence Balm Gilded, Birgamot . . . TEA, Rum, Sugar, m [*sic*], Fine Durham Flower, Mustard, & choc . . . many other Articles, very cheap, for Cash or Short Credit.	Massachusetts Historical Society, Amory Fiche, 41515, Broadside. Benedict Arnold, 1741–1801 (New Haven, 1765).
Anonymous	1765	[Newspaper description of a terrible accident]: One morning last week [last week of November, 1765] a child about 3 years old accidentally overset a large sauce-pan of hot chocolate into its bosom, whereby it was so terribly scalded that it died soon after.	*Boston Evening Post*, December 2, 1765, p. 3.
Walker, Rebecca	1764	[Advertisement; smallpox epidemic]: All sorts of garden seeds imported in the last ship from London . . . Peas, beans, red and white clover; hemp seed; flour of mustard; sugar and chocolate . . . But now removed out of Boston because of the Small-Pox.	*Boston Gazette and Country Journal*, February 20, 1764, p. 4.
Palmes, Richard	1763	[Advertisement; chocolate as medicine]: At his Shop in New-London; a complete assortment of Medicines, Chymical and Galenical, among which are, Turlington's Balsom of Life, Bateman's Drops. ALSO, All Kinds of Surgeon's Instruments. Likewise Groceries, consisting of Raisins, Figgs, Currants, Pepper, Coffee, Almonds in Shells, Mustard, Chocolate, Salt Petre, Choice new Rice—All of which he will Sell very cheap.	*New-London Gazette*, December 9, 1763, p. 3.

Person	Date	Quotation	Source
Don José Solano y Bote, Governor and Captain-General, Caracas	1763	[Letter, description of Jewish pirates raiding Venezuelan cacao ships]: On June 8th of the last year [1762] a sloop carrying 321 fanegas of cacao from private people [i.e., merchants] left Ocumare [today the Port of Cumana, in northern Venezuela]— sailed down the coast with the intent to reach the Port of Guaira [northern Venezuela] but on June 10th was attacked by a Dutch sloop from Curacao, the ship captain [being] Phelipe Christiano, [other passengers being] a Jewish merchant named Mich [?] and another [Jew]—the owner [of the ship] named Abraham—and they took with violence that cacao.	AGI. Caracas, Number 438.
Jones, Daniel	1760	[Advertisement; Loyalist merchant]: Loaf and Brown sugar, tea, coffee, chocolate and spices. Notice: Officers and Soldiers who have been in this Province service, may be supply'd with any of the above Articles on Credit.	*Boston Evening Post*, December 1, 1760, p. 4.
Traill, Robert	1757	[Advertisement, announcing importation of chocolate grinding stones from England]: Chocolate mills imported by Robert Traill, from the ship *Snow Perks*, out of Bristol . . . Sold at wholesale at his stores on the Long Wharff [*sic*].	*New Hampshire Gazette*, November 18, 1757, no page number.
Anonymous	1752	[Monastery rules regarding gender interactions]: The conversations with women no matter how brief and light they are always are distracting and so it is commanded that the Missionaries should avoid them. As a consequence chocolate should not be administered or given to any women even if she is a Syndic during breakfast, lunch, meal or afternoon merienda [snack] at the porteria or talk to her in the said place, in the Church, or in the Sacristy, or in any other part that belongs to the convent.	Santa Barbara Mission Archives (SBMA). Father Serra Collection, Document 53 dated 1751–1752.
Watts, Samuel	1751	[Summer chocolate supply problems]: By reason of the heat of the weather, [he] has been unable to supply his customers with chocolate for these few weeks past, [and] fears to the loss of many of them.	*The Boston Gazette or Weekly Journal*, October 8, 1751, p. 2.
Serra, Junipero	1749	[While aboard ship en route to Mexico]: On the feast of "Our Lady of the Holy Rosary" in the fear that there was not enough water, rationing was introduced for all and the ration was so small that we could not drink the chocolate we brought with us because there was not water to boil it with; and this [rationing] lasted for all the 15 days that took us to reach the port of Puerto Rico that is 1200 to 1300 leagues from Cádiz.	Santa Barbara Mission Archives (SBMA). Father Serra Collection, Document 37, dated May 14, 1769.
Navarro, Isaac	1748	[Advertisement; Jewish merchant]: Notice is hereby given the Subscriber makes and sells as good Chocolate as was ever made in England at 4s [shillings] 6d [pence] per pound. Constant attention is given at said House by Isaac Navarro.	*Maryland Gazette,* November 9, 1748, p. 3.

Person	Date	Quotation	Source
Hamilton, Alexander	1744	[Diary: notice of dirty chocolate]: In 1744, Dr. Alexander Hamilton traveling in Maryland noted in his diary, I breakfasted upon some dirty chocolate, but the best that the house could afford.	Hamilton, Dr. Alexander. *Hamilton's Itinerarium; being a narrative of a journey from Annapolis, Maryland, through Delaware, Pennsylvania, New York, New Jersey, Connecticut, Rhode Island, Massachusetts and New Hampshire, from May to September, 1744.* Edited by Albert Bushnell Hart; p. 16.
Brewster, John	1738	[Advertisement; chocolate being branded/marked]: At the sign of the Boot in Ann-Street, sells his best Chocolate for Thirteen Shillings per single Pound, and Twelve Shillings and six Pence by the Dozen, Marked I.B.	*Boston Weekly News-Letter*, Thursday, February 16 to Thursday, February 23, 1738, p. 2.
Stiger, Gaspar	1737	[Letter, request for chocolate]: If only Your Reverence could send the mule train driver earlier every year because the chocolate melts when the alms come in June. But I see that Your Reverence is not at fault for the late arrival of the mule trains in Mexico.	Personal communication from Professor Bernard Fontana, Historian, University of Arizona Museum, Tucson, June 2005. Manuscript deposited in the archives, Documentary Research of the Southwest (DRSW), at the Arizona State Museum, University of Arizona, Tucson.
Anonymous	1737	[Newspaper announcement of a new chocolate manufacturing invention]: By a Gentleman of this Town [Salem] is this Day bro't [*sic*] to Perfection, an Engine to Grind Cocoa; it is a Contrivance that cost much less than any commonly used; and will effect all that which the Chocolate Grinders do with their Mills and Stones without any or with very Inconsiderable Labour; and it may be depended on for Truth, that will in less than Six Hours bring one Hundred weight of Nuts to a consistence fit for the Mold. And the Chocolate made by it, is finer and better, the Oyly [*sic*] Spirit of the Nut being almost altogether preserved. And there is little or no need of Fire in the making.	*New England Weekly Journal*, September 13, 1737, p. 1.
Anonymous	1735	[Newspaper announcement of an attempted murder using chocolate]: Last Tuesday [August 4, 1735] a horrid Attempt was made here to poison Mr. Humphrey Scarlet of this Town, Victuller [*sic*], his Wife and 2 Children; some Arsenick [*sic*] or Ratsbane having been put into a Skillet of Chocolate when they eat for Breakfast; Finding themselves soon after all disordered thereby, immediately applied to a Physician, and upon using proper Means, we hear, they are in a likely Way to do well.	*Pennsylvania Gazette,* August 21, 1735.
Reed, Mrs.	1731	[Advertisement, announcing the opening of a chocolate house]: This may serve to Inform Gentlemen in the Town of Boston, that Mrs. Reed has opened a Chocolate House in King-street, on the North side of the Town House, where they may Read the News, and have Chocolate, Coffee or Tea ready made any time of the Day.	*Boston Gazette,* Monday, September 6, to Monday, September 13, 1731, p. 2.
			Hamilton, Dr. Alexander. *Hamilton's*

Person	Date	Quotation	Source
Quélus, [?] de	1718	[Commentary on chocolate and longevity]: There lately died at Martinico [Caribbean Island of Martinique meant] a councilor about a hundred years old, who, for thirty years past, lived on nothing but chocolate and biscuit. He sometimes, indeed, had a little soup at dinner, but never any fish, flesh, or other victuals: he was, nevertheless, so vigorous and nimble, that at fourscore and five, he could get on horseback without stirrups.	Quélus, de. *The Natural History of Chocolate: Being a Distinct and Particular Account of the Cocoa-Tree . . . the Best Way of Making Chocolate is Explain'd; and Several Uncommon Medicines Drawn from It, Are Communicated.* London, England: J. Roberts, 1730; p. 58.
Simpson, Nathan	1714	[Letter; from Jewish merchant]: I understand that a Certain Person at N. York has yet a Quantity of Cocoa of about 20 Tons and there are orders Sent to Curasow [*sic*] for more which made me willing to part with Yours.	Richard Janeway to Nathan Simpson, dated June 15, 1714.
Pope, Alexander	1712	[Poem]: Or as Ixion fix'd, the wretch shall feel, The giddy motion of the whirling Mill, In fumes of burning Chocolate shall glow, And tremble at the sea that froaths below.	Pope, A. Rape of the Lock, 1712, CANTO II: lines 133–136.
Boylston, Zebadiah	1711	[Advertisement, chocolate as medicine]: To be sold at his Apothecary's Shop in dock-Square, Boston . . . Tea, sweetmeats, rice, chocolate . . . excellent perfumes, good against deafness and to make Hair grow.	*Boston News-Letter*, March 17, 1711, p. 2.
Leblond, James	1705	[Advertisement; among earliest in North America]: To be Sold in Boston at the Ware-house of Mr. James Leblond on the Long Wharf near the Swing-Bridge, new Lisbon Salt at 2 [illegible] s. [shillings] per Hogshead, & 4 s. per Bushel; also Rum, Sugar, Molasses, Wine, Brandy, sweet Oyl [*sic*], Indigo, Brasilet [?], Cocoa, Chocolate, with all sorts of Spice, either by Wholesale or Retail, at reasonable Rates.	*The Boston News-Letter*, December 3, 1705, p. 4.
Anonymous	1700	[Earliest chocolate recipe from North America]: 1700 . . . To Make Chocolate Almonds: Take your Sugar & beat it & Serch [sift] it then: great youre [*sic*] Chocolatt: take to 1 lb Sugar: 5 oz. of Chocolatt mix well together put in 2 Spoonfull Gumdragon Soaked in rosewater & a grn musk & ambergrease & beat all well together in mortar rowl out and mark wt: ye molds & lay on tin plats [*sic*] to dry turn everyday.	Harbury, Katherine E. *Colonial Virginians Cooking Dynasty.* Columbia: University of South Carolina Press, 2004; p. 189.
Dickenson, Johathan	1699	[Letter]: This morning We waited an Opportunity to gett [*sic*] leave to depart, which was granted. Whereupon we asked for such things as they not make use of, viz. A great Glass, wherein was five or six pounds of Butter, some Sugar, And some Balls of Chocolate. All which was granted us.	Dickenson, Johathan. *Gods [sic] Protecting Providence . . . In the Times of Greatest Difficulty.* Philadelphia: Reinier Jansen, 1699.
Kino, Eusebio Francisco	1697	[Diary, southwestern exploration]: When the Pimas and Pima-Sobaipurís struck the first blow against the enemy Jocomes, we had a fiesta for this Great Lady and Sovereign Quenn of the Angels. . . . We killed some cattle, for our native Pima, and we gave chocolate to some of them and told them goodbye—and in saying goodbye to these Pimas of the North and Northwest I told them how, within ten days I had to go to the Rio Grande of the Sea of California.	Bolton, H. E. *Rim of Christendom. A Biography of Eusebio Francisco Kino. Pacific Coast Pioneer.* Tucson, AZ: The University of Arizona Press, 1936; p. 392.

Person	Date	Quotation	Source
Sewall, Samuel	1697	[Diary: Samuel Sewall]: On October 20th Reverend Samuel Sewall wrote in his diary, I wait on the Lieut. Governor at Dorchester, and there meet with Mr. Torry, breakfast together on Venison and Chockalatte: I said Massachutset and Mexico met at his Honours Table.	Thomas, M. Halsey. *The Diary of Samuel Sewall, Volume 1, 1674–1708*. New York: Farrar, Straus and Giroux, 1973; p. 380.
Massanet, Damián	1689	[Letter of Father Damian Massanet to Don Carlos de Siguenza]: Captain Leon had a compadre along, Captain [?], so honorable that he never failed to play the tale-bearer and excite quarrels; so kind-hearted that only his friend Leon drank chocolate, and the others the lukewarm water; so considerate of others that he got up early in the morning to drink chocolate, and would afterward drink again with the rest; so vigilant that he would keep awake and go at midnight to steal the chocolate out of the boxes; perhaps this vigilance was the reason why, while, by order of His Excellency, Captain Leon should have left for the priests three hundredweight of chocolate and the same quantity of sugar, but he left only one and one-half hundredweight of each.	Massanet, D. Letter of Fray Damián Massanet to Don Carlos de Sigüenza. In: *Spanish Exploration in the Southwest, 1542–1706*. New York: Charles Scribner's Sons, 1916; p. 386.
Ortiz, Julio	1688	[Chocolate and Easter celebrations]: May my Brother have a very happy Easter beginning and a very happy Easter end, and may you find consolation of the absence of the Brother Bartolome Gonzales. I am writing to you, my Brother, asking your kindness and benevolence, to send me 150 pesos to pay the Sunday's people and 2 arrobas of chocolate, one arroba of one and another arroba of the other and one arroba of sugar. May God guard my Brother for many years Signed, April 1688 Julio Ortiz.	Archivo General de la Nación (AGN). Jesuitas, Vol. III-10, year 1688.
Pepys, Samuel	1661	[Diary, use of chocolate to cure hangover]: Rose [from bed] and went out with Mr. Creed to drink our morning draft, which he did give me in chocolate to settle my stomach.	Peyps, S. Diary entry for April 24, 1661.
Anonymous	1660	[Poem]: Nor need the Women longer grieve, Who spend their oyle [sic] yet not Conceive, For 'tis a Help Immediate, If such but Lick of Chocolate. The Nut-Browne Lasses of the Land, Whom Nature vail'd [sic] in Face and hand, Are quickly Beauties of High-Rate, By one small Draught of Chocolate.	Anonymous. *The Vertues [sic] of Chocolate [and] The Properties of Cavee*. Oxford, England: Henry Hall, 1660.

Person	Date	Quotation	Source
Philip II (Felipe II), king of Spain	1645	[Royal edict]: And such people that are Spanish, Mestizo, Mulatto, Black and Chinese are selling [cocoa and chocolate] on the streets, convents and private. . . . Those vendors that sell in small amounts, the peddlers or hawkers [buhoneros], are clandestine, furtive and they hide, and they sell it [cocoa and chocolate] very cheap, deceiving the neighbors. They have to be punished and the neighbors be warned. . . . The frauds and deceits that malicious people are seeking are well known and causes detriment to the safety of the merchandises . . . and they commit a criminal act, a delinquency, selling "old" cocoa [cacao añejo], of an inferior quality and noxious to the people's health, selling what is bad for good. . . . Therefore it has been prohibited to sell such cocoa because it is too pestilent and because is too harmful that has been the cause of death. . . . It is then ordained and enacted by the Royal Law that the peddlers cannot wander on the streets of the cities, on the towns, [and] villages.	AGN. Indios, Vol. 3, exp. 860, f. 207v. Ano 1591.
Jimenez, Juan	1642	[Early mention of chocolate in North America, the St. Augustine shipwreck document]: I, Juan Jimenez, Public Auditor [Notario Público] Pardo, in this city of San Agustin, in the Province of La Florida, in the name of God I attest and give true testimony that by order of the Castilian Lord Damian de la Vega Castro y Pardo, Governor and General Captain [Capitán General] of these provinces, and in agreement and with the assistance of Official Judges of the Royal Supreme Court [Audiencia Real]. I say that the chocolate that came in the ship *Nuestra Señora del Rosario*, whose Captain is Ermenexildo Lopez, and [that] was auctioned at 5 reales per pound, has been apportioned and distributed for a maximum benefit among the infantry [foot soldiers] of this presidio [of San Agustín] and among the religious [priests] of this province [of Florida] in the following way [then the people are named in the document].	Archivo General de Indias/Escriban (Seville, Spain). Document #155A/1643-1652/Residencias Florida
Diaz, Alonso	1626	The cacao from Peru is the coin that is more used; the silver coin is of a less use. With that cacao they buy food in the towns of Indians as well as in the towns of the Spanish. They use it in the Plazas, butcher shops, and also they buy chocolate with it. It serves as coin and as sustenance and because of this it is very necessary. Without it people cannot live and if they were to lack cacao, particularly the poor people, they will perish.	AGI. Guatemala, 43, N. 36/3/1.

Person	Date	Quotation	Source
Acosta, José de	1604	[Spices added to chocolate]: The Spaniards, both men and women, are very greedy of this chocolaté. They say they make diverse sortes [sic] of it, some hote [sic], some colde [sic], and put therein much chili . . . that the chocolate paste is good for the stomach.	Acosta, J. de. *The Naturall and Morall Historie of the East and West Indies. Intreating of the Remarkable Things of Heaven, of the Elements, Metals, Plants and Beasts.* London, England: Val. Sims for Edward Blount and William Aspley, 1604; p. 271.
Cárdenas, Juan de	1591	[Chocolate and digestion]: The most appropriate time to [drink chocolate] is in the morning at seven, or at eight, and before breakfast, because then the heat of this substantial drink helps to spend all that phlegm which has remained in the stomach from the [previous] dinner and supper. . . . The second time for using it is at five or six in the afternoon, when it is presumed that digestion is completed of what was eaten at noon, and then it helps him to distribute [that] heat throughout the whole body.	Cárdenas, Juan de. *Problemas y Secretos Maravillosos de las Indias.* Madrid, Spain: Ediciones Cultura Hispanica, 1591; Chapter 8, pp. 117v–118f.
Anonymous	1591	[Permission for Mexican Indians to sell chocolate]: In Mexico, August 3rd of the year 1591, a permission is granted to Gabriwl Xuarez and his wife Mariana and other "naturals" [i.e., natives] of this city in order to allow them to sell chocolate, firewood, and [illegible] of cacao and fruits.	AGN. Indios, Vol. 3, exp. 860, f. 207v. Ano 1591.
Hernández, Francisco	1577	[Botanical description of cacao tree]: The cacahoaquahuitl is a tree of a size and leaves like the citron-tree, but the leaves are much bigger and wider, with an oblong fruit similar to a large melon, but striated and of a red color, called cacahoacentli, which is full of the seed cacahoatl, which, as we have said, served the Mexicans as coin and to make a very agreeable beverage. It is formed of a blackish substance divided into unequal particles, but very tightly fit among themselves, tender, of much nutrition, somewhat bitter, a bit sweet and of a temperate nature or a bit cold and humid.	Hernández, F. *Historia de las Plantas de la Neuva Espana.* Mexico City, Mexico: Imprenta Universitaria, 1577; p. 304.
Oviedo, Fernadez de	1526	[Discussion of Costa Rica and Nicaragua]: And because those [people] of Nicaragua and their tongue, wherever they have come from, are the ones that brought to the land [i.e., Costa Rica] the cacao or cacao beans that are used as money in those places, and [because] they have in their inheritance the trees that carry that fruit, there is not a single tree in the hands of the chorotegas' [i.e., Costa Rican natives].	Oviedo, F. de. *Historia General y Natural de las Indias*; 1526. XI, p. 109.
López-Gómara, Francisco	1520	[Discussion of the marketplace at Tlatelolco]: The most important of all, which is used for money, is one that resembles the almond, which they call cacahuatl, and we cacao, as we knew it in the islands of Cuba and Haiti.	López-Gómara, F. 1552. *Cortés: The Life of the Conqueror by His Secretary (from the Istoria de la Conquista de Mexico).* Berkeley, CA: University of California Press, 1964; p. 162.

Person	Date	Quotation	Source
Díaz del Castillo, Bernal	1520	[Spanish conquistadors dining with Montezuma]: From time to time the men of Montezuma's guard brought him, in cups of pure gold a drink made from the cacao-plant, which they said he took before visiting his wives. I saw them bring in a good fifty large jugs of chocolate, all frothed up, of which he would drink a little. I think more than a thousand plates of food must have been brought in for them, and more than two thousand jugs of chocolate frothed up in the Mexican style.	Díaz del Castillo, B. 1560–1568. *The Discovery and Conquest of Mexico*; pp. 226–227.
Columbus, Ferdinand	1502	[Christopher Columbus's son, Ferdinand's description of events related to the fourth voyage of Columbus to Honduras and Panama]: Our men brought the canoe alongside the flagship. For provisions they had such roots and grains as the Indians of Española eat, also a wine made of maize [i.e., chicha] that tasted like English beer. They had as well many of the "almonds" which the Indians of New Spain use as currency; and these the Indians in the canoe valued greatly, for I noticed that when they were brought aboard with the other goods, and some fell to the floor [of the ship], all the Indians squatted down to pick them up as if they had lost something of great value.	Keen, Benjamin. *The Life of the Admiral Christopher Columbus by His Son Ferdinand*. New Brunswick, NJ: Rutgers University Press, 1959; pp. 230–233.

Chocolate Timeline

Chocolate-Related Events in North America	Date ←	Decade ↔	Date →	Chocolate-Related Events Elsewhere and Historical Context
First systematic review of chocolate and its effects on learning and memory presented at AAAS meeting in San Francisco.	2007	2000–2009	2007	British report that eating chocolate causes human heart rate to rise faster than kissing loved one.
Development of Mars American Heritage chocolate (Colonial style).	2006	2000–2009	2006	250th anniversary of the birth of Wolfgang Amadeus Mozart.
Emergency personnel responded to the Debelis Company chocolate factory when a 21-year-old worker became stuck in a vat of 110-degree chocolate after slipping into it while trying to shut it down. He was treated for minor ankle injuries.	2006	2000–2009	2006	100th anniversary celebrations/festivities commemorating the 1906 San Francisco earthquake and fire.
Mars Incorporated offers 2 million dark chocolate M&Ms for the return of Edvard Much's painting, *The Scream*, stolen from the Munch Museum in Oslo, Norway, in 2004.	2006	2000–2009	2006	Global worry over avian flu virus.
Adam Smith, owner of *Fog City News* in San Francisco, and his staff have tasted, evaluated, and documented more than 2500 types of chocolate.	2005	2000–2009	2005	World population reached 6,453,628,000.
Hershey Corporation announced acquisition of Scharffen Berger Chocolate Maker Incorporated, Berkeley, California.	2005	2000–2009	2005	Steve Fossett completes first nonstop, nonrefueled solo flight around the world.
Jacques Torres opens his Chocolate Haven factory and retail store in New York City.	2004	2000–2009	2004	China's first Salon du Chocolat, an international exposition of fine chocolates, is held in Beijing.
	2003	2000–2009	2003	A chocolate-scented mousetrap has been developed by U.K. scientists to catch the pests without the need for bait.
United States Food and Drug Administration establishes Standard of Identification for White Chocolate.	2002	2000–2009	2002	NASA's *Odyssey* space probe begins mapping the surface of Mars.

Chocolate: History, Culture, and Heritage. Edited by Grivetti and Shapiro
Copyright © 2009 John Wiley & Sons, Inc.

Chocolate-Related Events in North America	Date ←	Decade ↔	Date →	Chocolate-Related Events Elsewhere and Historical Context
Debut of M&Ms Dulce de Leche Caramel Chocolate Candies.	2001	2000–2009	2001	Bar of chocolate from Scott's ill-fated Antarctica expedition auctioned for £470 US$686; September 11, 2001 terrorist events.
Name of Milky Way Dark bar changed to Milky Way Midnight.	2000	2000–2009	2000	Millennium celebrations throughout the world.
	1998	1990–1999	1998	Divine, a fair trade chocolate bar, launched in England by Day Chocolate.
Scharffen Berger chocolate founded in Berkeley, California.	1997	1990–1999	1997	Sheep cloned; Serge Couzigou establishes Musée du Chocolat in Biarritz, France.
Tan M&Ms replaced by blue.	1995	1990–1999	1995	First introduction of DVDs.
Kudos Cookies N Cream bar introduced.	1994	1990–1999	1994	First genetically engineered vegetable, *Flavr Savr* tomato.
Debut of Dove dark chocolate bar.	1992	1990–1999	1992	European Economic Union (EEC) founded.
Milky Way Dark bar introduced.	1989	1980–1989	1989	*Exxon Valdez* oil spill in Alaska.
Symphony milk chocolate bar introduced.	1989	1980–1989	1989	Fall of the Berlin Wall; symbolic end of the Cold War.
	1988	1980–1989	1988	Chocolate bars among the menu items served aboard the USSR *Mir* space station.
	1986	1980–1989	1986	Michel Chaudon establishes chocolate shop in Paris.
Ghirardelli Square, San Francisco, named a historic site by the National Park Service.	1982	1980–1989	1982	*Cats*, musical by Andrew Webber, opened in New York City.
Marathon bar introduced.	1974	1970–1979	1974	First implementation of government mandated food labeling.
	1955	1950–1959	1955	Bernachon Chocolates established in Lyon.
	1953	1950–1959	1953	Edmond Hillary and Tenzing Norgay reach summit of Mount Everest: expedition supplies include chocolate, sardines, canned meat, soup, crackers, oatmeal, raisins, nuts, sugar, tea, coffee, and lemonade; Hillary recalled later that Norgay buried a little bit of chocolate at the top in the snow, as a gesture to the gods.
	1947	1940–1949	1947	Cluizel Chocolates established in Paris.
Forrest Mars Sr. and Bruce Murrie introduce M&Ms.	1940	1940–1949	1940	Color television invented.
Debut of Hershey Miniatures.	1939	1930–1939	1939	Pan American Airlines inaugurates first transatlantic passenger service between New York and Marseilles, France.
Hindenburg menu lists dessert of pears condé with chocolate.	1937	1930–1939	1937	Crash of the zeppelin *Hindenburg* at Lakehurst, New Jersey.
Forever Yours bar introduced; debut of 5th Avenue bar (peanut butter crunch and chocolate).	1936	1930–1939	1936	Construction of Hoover Dam completed; first scheduled transatlantic zeppelin flight from Germany to Lakehurst, New Jersey.
PayDay bar introduced; debut of Three Musketeers bar; debut of Mars Bar (almonds, chocolate).	1932	1930–1939	1932	Franklin D. Roosevelt elected president of the United States; Charles King isolates vitamin C; construction begins on the San Francisco–Oakland Bay Bridge.

Chocolate-Related Events in North America	Date ←	Decade ↔	Date →	Chocolate-Related Events Elsewhere and Historical Context
Chocolate tempering process developed to control bloom in chocolate molding.	1931	1930–1939	1931	Canada declares independence from Great Britain.
Commonly accepted date for first recipe for Toll House Cookies (i.e., chocolate chip cookies); debut of Snickers bar.	1930	1930–1939	1930	First appearance of Mickey Mouse in a comic strip; Max Theiler develops yellow fever vaccine.
	1929	1920–1929	1929	Jose Rafael Zozaya and Carmelo Tuozzo develop El Rey chocolate company in Venezuela.
Dreyer's ice cream established in 1928 in Oakland. Helps popularize chocolate through ice cream.	1928	1920–1929	1928	School Hygiene in Athens, Greece, recommends milk with coffee or chocolate for children; Alexander Fleming discovers penicillin.
Welch's Fudge bar introduced.	1927	1920–1929	1927	Greek physicians recommend chocolate served to tuberculosis patients; also "[Breakfast] milk, plain or with chocolate, or with cocoa, jam, two eggs, butter and bread."
United Press International report that Lindbergh drank hot chocolate aboard the *Spirit of St. Louis* (Lindbergh's diary does not mention chocolate).	1927	1920–1929	1927	Charles Lindbergh completes first solo transatlantic flight; pilots the *Spirit of St. Louis* from New York to Paris.
Mr. Goodbar introduced.	1925	1920–1929	1925	Scopes trial in Dayton, Tennessee.
	1924	1920–1929	1924	Valrhona Chocolate established in Tain L'Hermitage near Lyon.
Debut of Milky Way bar.	1923	1920–1929	1923	Tomb of King Tutankhamen discovered by Howard Carter in Egypt.
Charles See and his mother, Mary See, create See's Candies in Los Angeles, California.	1921	1920–1929	1921	Albert Einstein awarded Nobel Prize.
	1920	1920–1929	1920	Quantity of cacao exported from Trinidad, 62 million pounds.
	1917	1910–1919	1917	John Reed in his *Ten Days That Shook the World*, described food shortages in Russia.
Mountain Tacoma bar introduced.	1915	1910–1919	1915	Albert Einstein postulates general theory of relativity.
Chocolate serving jugs part of dining items aboard the *Lusitania*.	1915	1910–1919	1915	Sinking of the *Lusitania*.
	1914	1910–1919	1914	Panama Canal opened. Theodore Roosevelt treks through the Brazilian wilderness carrying 26 foods, among his supplies sweet chocolate (16 ounces served each Wednesday to men).
Debut of Goo Goo Clusters (caramel, chocolate, marshmallow, peanuts); Klein Chocolate Company founded in Elizabethtown, Pennsylvania.	1913	1910–1919	1913	Jules Suchard invents filled chocolate bonbon. Ford Motor Company institutes revolutionary idea—assembly line production of automobiles.
Debut of Whitman's Sampler boxed chocolates; chocolate listed on *Titanic* menu.	1912	1910–1919	1912	Bones of Piltdown Man shown in exhibition (later proved to be a hoax). *Titanic* strikes iceberg and sinks with great loss of life.
Jane Earyre Fryer publishes *The Mary Frances Cook Book; Or, Adventures Among the Kitchen People*. Provides recipe for fudge.	1912	1910–1919	1912	Arizona admitted to the Union as 48th state.
Foundation of Mars chocolate in Tacoma, Washington.	1911	1910–1919	1911	First North American transcontinental airplane flight. Ronald Amundsen reaches the South Pole, with chocolate among supplies carried. Scott and his team of Antarctic explorers celebrate their Christmas meal, which includes chocolate.

Chocolate-Related Events in North America	Date ←	Decade ↔	Date →	Chocolate-Related Events Elsewhere and Historical Context
Commonly accepted date when Arthur Ganong introduced first 5 cent chocolate bar to Canada. Alida Frances Pattee publishes *Practical Dietetics with Reference to Diet in Disease*: provides food values for cocoa and discusses breakfast cocoa; cocoa or chocolate recommended during convalescence from constipation, chronic rheumatism, tuberculosis, and typhoid fever.	1910	1910–1919	1910	A menu from the Canton Hotel, Canton, China (in English), listing cocoa as an item along with coffee and several varieties of tea. Start of the Mexican Revolution. Chocolate pudding served at a banquet on the Greek island of Samos to honor a visit by the Admiral of the Austrian Navy.
Julius Friedenwald publishes *Diet in Health and Disease*. Identifies "good uses of chocolate" for children and convalescents; for singers' voices; carcinoma of the stomach; relief of leanness; tuberculosis, epilepsy, debility, and anemia.	1909	1900–1909	1909	Plastic invented.
The Hotel Alexandria in downtown Los Angeles lists chocolate ice cream, chocolate éclairs, as well as both chocolate and cocoa as beverages, in proliferation of chocolate usage.	1908	1900–1909	1908	Women compete in modern Olympic games for the first time.
Hints to health and beauty by Ethel Lynne: "Gain in flesh by the following rules: get ten to twelve hours' sleep out of every 24, abstain from tobacco, and eat plenty of well cooked foods such as chocolate, cocoa, sugar, butter, and fruits."	1908	1900–1909	1908	Theodore Tobler develops triangular nougat-filled chocolate bar.
Nellier Eward publishes *Daily Living*. Provides numerous chocolate recipes. Writes that chocolate is "better prepared several hours before serving time and then reheated."	1908	1900–1909	1908	Model T Ford introduced to American public.
Amended sections of the United States Food Law: "Candies and chocolates shall be deemed to be adulterated if they contain terra alba, barytes, talc, chrome yellow, or other mineral substances, or poisonous colors or flavors, or other ingredients deleterious or detrimental to health."	1907	1900–1909	1907	Color photography invented.
Advertisement for Huyler's Cocoa: "Your grocer or druggist sells both a perfect health food, a satisfying lunch biscuit, Huyler's CHOCOLATE DIPPED Triscuit is the highest achievement in the science of food production. The natural Food Co's [Company] shredded wheat Triscuit dipped by Huyler's in their World Famed Chocolate, making it doubly Strengthening and Invigorating. The best children's biscuit ever produced."	1907	1900–1909	1907	Polio epidemic in New York City.
Report by Major General Adolphus Greely: *Earthquake in California, April 18th*: "Food stocks available for the sick: cocoa, chocolate, tea, soup, fancy canned meat, conned and evaporated fruits, preserves, canned vegetables, special foods for infants and invalids, cereals, crackers. Purchases were made of ice, fresh meat, vegetables, oranges, lemons, eggs, butter, milk, bread."	1906	1900–1909	1906	San Francisco earthquake and fire.

Chocolate-Related Events in North America	Date ←	Decade ↔	Date →	Chocolate-Related Events Elsewhere and Historical Context
The Pacific Coast Steamship Company (P.C.S.S.CO.), established in 1875, plies routes from Panama to San Francisco from California to Japan and China. One of their menus, of December 1906, shows coffee, tea, and cocoa served as beverages on the S.S. *Spokane*.	1906	1900–1909	1906	Cadbury introduces Bournville alkalized cocoa powder.
Debut of Hershey's Kisses.	1906	1900–1909	1906	Passage of the United States Pure Food and Drug Act.
	1905	1900–1909	1905	Legality of compulsory vaccination upheld by the United States Supreme Court. Cacao planted in the Côte d'Ivoire. Cadbury Dairy Milk bar launched. Albert Einstein develops special theory of relativity.
Sarah Tyson Rorer publishes *World's Fair Souvenir Cookbook*. Provides recipes for chocolate mousse, iced chocolate, cake, ice cream, icing, and pudding.	1904	1900–1909	1904	Construction begins on Panama Canal.
	1903	1900–1909	1903	Henry Ford establishes Ford Motor Company, Detroit, Michigan. First successful heavier than air flight by Orville and Wilber Wright.
Ellen Duff publishes *A Course in Household Arts*. Extensive information on cocoa and chocolate. "Travelers who may be obliged to go where food cannot be easily obtained, frequently carry a supply of chocolate, which has the advantage of containing much nutriment in small space."	1902	1900–1909	1902	End of the Boer War.
Helena Viola Sachse publishes *How to Cook for the Sick and Convalescent*. Identifies sugar as the cause for gout, indigestion, hepatic disorders, and diabetes. Includes numerous chocolate recipes.	1901	1900–1909	1901	Walter Reed identifies mosquito vector as the source of transmission of the yellow-fever virus.
Students at Mills College in Oakland, California, recommend, as part of a facetious weight gain diet, a "Splendid Flesh-Producing Diet," to help "thin women gain flesh." This includes a breakfast of porridge and milk, followed by cocoa and Fritzie (a shortbread cookie with nutmeg, popular in the early 20th century), weak tea and coffee or milk, with fat bacon, fish, and jam. Afternoon tea was to be replaced by cocoa.	1900	1900–1909	1900	Boxer Rebellion in China.
Debut of Hershey's chocolate bar.	1900	1900–1909	1900	Life expectancy in the United States: 47.3 years.
Edward Munson, Assistant Surgeon U.S. Army, prepares for publication: *Emergency Diet for the Sick in the Military Service*. Identifies importance of medical stores, including chocolate.	1899	1890–1899	1899	Scott Joplin composes *Maple Leaf Rag*.
First production of coco-Quinine, a liquid medicine that contained yerba santa and chocolate.	1899	1890–1899	1899	Cuba liberated from Spain by the United States.
Common North American restaurant prices: hamburger steak $1.50; Coffee, with bread and butter, $1.00; Coffee, tea, or chocolate, $0.25 a cup.	1899	1890–1899	1899	Beginning of the Boer War

Chocolate-Related Events in North America	Date ←	Decade ↔	Date →	Chocolate-Related Events Elsewhere and Historical Context
Horace Clark, Charles Heitman, and Charles Consaul publish *The Miner's Manual*; list Klondike food supply prices at Sitka, Alaska: "1 package chocolate $0.30."	1898	1890–1899	1898	Klondike diary of Robert Hunter McCreary of Leechburg, Pennsylvania: "Entry Thursday, March 24th, Broke camp and moved to within three miles of the line. Traveled 14 miles, all very tired, wind broke center pole, got no breakfast, eat cold beans for dinner, had a good hot supper at 7:30 PM corn cakes, bacon, stewed onions, rolled oats, chocolate."
U.S. Supreme Court case: *US v. Salambrier* regarding duty on imported chocolate identified as "sweetened."	1898	1890–1899	1898	Beginning and end of Spanish–American War.
	1897	1890–1899	1897	Report that 22,400,000 pounds of raw cocoa or "nibs" consumed in united Kingdom yearly; 3,200,00 pounds in London; 8800 pounds daily; 4400 pounds drunk as cocoa, with remainder used in manufacture of candy.
Publication of *Manual for Army Cooks*. Contains recipe for preparing chocolate ice cream.	1896	1890–1899	1896	Gold discovered in the Klondike region of North America. Menu from the Hotel Sovereignty of Samos (Greek island of Samos): lists chocolate.
Advertisement in the *Los Angeles Times*: "Free to all sufferers from Asthma, Bronchial or Lung Trouble, regular size bottle of Dr. Gordia's Chocolate Emulsion of Cod Liver Oil with Hypophosphites, 'As palatable as milk or honey' Thomas & Ellington Cut-rate Druggists, Corner Temple and Spring Streets."	1896	1890–1899	1896	Revival of the Olympic Games in Athens, Greece.
Savannah resident Caroline Lamar Woodbridge prepares handwritten recipes attending Ms. Leila Habersham's cooking class. Recipes include how to prepare terrapin (turtle), chocolate soufflé, chocolate pudding, and chocolate sauce.	1895	1890–1899	1895	Greek advertisements for *Loukoumia Sokolatas* (chocolate cookies). Letter from Mabel Hubbard Bell to Alexander Graham Bell, dated May 20th, written in Paris at the Hotel Vendome: "I have cooked the children's nightly cup of cocoa over my alcohol lamp, tucked them in bed, and now come back to finish my letter."
Chicago World's Fair: Poster advertising Buffalo Bill's Wild West show with notation: "Coffee, tea, and cocoa can be had in tin pails with cup attached by applying in the Lunch Room, Ground Floor."	1893	1890–1899	1893	Henry Ford invents successful gasoline engine.
Mary Boland publishes *A Handbook of Invalid Cooking for the Use of Nurses n Training Schools, Nurses in Private Practice, and Others Who Care for the Sick*. Stresses importance of boiling water and serving hot beverages to patients. Identifies cocoa as an easily digested food; provides recipes.	1893	1890–1899	1893	Polio epidemic in Boston spreading into New England (Vermont).
	1892	1890–1899	1892	Clémentine and Auguste Rouzaud begin making fine chocolates in Royat (France). George Miller Sternberg publishes *A Manual of Bacteriology*.
Mary L. Clarke publishes *Cooking for the Sick and Convalescent*. Provides recipes for cocoa and chocolate.	1892	1890–1899	1892	Historicus (Richard Cadbury) publishes *Cocoa: All About it*.

Chocolate-Related Events in North America	Date ←	Decade ↔	Date →	Chocolate-Related Events Elsewhere and Historical Context
Ledger from Coloma, California (boom-town where gold was discovered), entry for September 3: "Martha Bayne purchased: sugar $1.00; corn $0.50; chocolate $0.25; and lard $0.65."	1891	1890–1899	1891	Thomas Edison issued patent for "transmission of signals electrically" (i.e. radio).
The Weinstock and Lubin mail order house sells chocolate in several variations that include chocolate cream candies, along with chocolate bars wrapped in paper by Ghirardelli, Baker, and other companies. Also available are cocoa in cans by Ghirardelli, breakfast cocoa by Baker, half-pound cans of Epp's homeopathic cocoa, and cocoa shells, the husks removed from roasted cacao beans and used as mulch.	1891	1890–1899	1891	Game of basketball created by James Naismith.
Varner House Hotel Records kept by Edward Varner, Butts County, Georgia: entry for June 9th "[illegible] and wood—$1.40; 3 chickens and chocolate—$1.05."	1891	1890–1899	1891	Work on trans-Siberian railroad begins.
Summary report on adulterants added to chocolate.	1891	1890–1899	1891	Thomas Edison issued patent for wireless telegraphy.
Walter Baker and Company publish *The Chocolate Plant and Its Products*.	1891	1890–1899	1891	Thomas Edison issued patent for motion picture camera.
Hattie Burr publishes *The Woman [sic] Suffrage Cook Book*; includes recipe for chocolate cake.	1890	1890–1899	1890	Massacre of Native Americans at Wounded Knee, South Dakota.
Summary review of cacao, identified as one of America's most important tropical products.	1890	1890–1899	1890	Influenza pandemic.
	1889	1880–1889	1889	Freia, Norwegian chocolate company, founded.
The Santa Fe Railroad, serving California and the West, offers an unusual entrée listed as "Sweetbreads, Saute [sic] with Mushrooms, Spanish Puffs, Chocolate Glace."	1888	1880–1889	1888	*National Geographic* publishes first issue.
Harriet Harper writes a description of the various foods available in Los Angeles: "To saunter about the streets of Los Angeles is to conceive a wonderful idea of the Los Angeleans' hospitality. . . . It is not at all the thing to sleep and eat in the same house. Everyone is a Bohemian, a civilized nomad, and finds his meals whenever fate lands him. If he discovers the aching void while he is in the adobe town, he asks the handsome, dark-eyed senorita for hot tamales and chocolate. If he is in Chinatown he gets a cup of delicious tea and a bird's nest. If he feels flush he crosses the elegant threshold of Koster's; if poor he tries the Silver Moon, where for twenty-five cents you get soup, several kinds of meat, all vegetables in the market, ice-cream, pie and coffee. You can live like a prince, or you can dine luxuriously as a pauper."	1888	1880–1889	1888	Yellow fever epidemic ravages Florida.

Chocolate-Related Events in North America	Date ←	Decade ↔	Date →	Chocolate-Related Events Elsewhere and Historical Context
Mrs. John Sherwood, in *The Modern Dinner Table. Manners and Social Usages*, lists duties/expectations of a lady's maid. "The maid, if she does not accompany her mistress to a party and wait for her in the dressing room, should await her arrival at home, assist her to undress, comb and brush her hair, and get ready the bath. She should also have a cup of hot tea or chocolate in readiness for her."	1887	1880–1889	1887	Russell Hodge invents barbed wire; will change forever cattle ranching in the American West.
New York Times editorial on diet and dyspepsia summarizes the opinions of Henry Thompson: "Chocolate, thick cocoa, or even milk, are not so efficacious in allaying thirst as water."	1886	1880–1889	1886	The Debauve et Gallais Company of France sells *pistoles* (chocolate chips) and chocolate bars on the Nanjing Lu Road in Shanghai.
Hattie A. Burr publishes *The Woman Suffrage Cook Book. Containing Thoroughly Tested and Reliable Recipes for Cooking, Directions for Care of the Sick and Practical Suggestions.* Provides recipe for chocolate cake (it is called chocolate cake because chocolate icing is spread between the cake layers).	1886	1880–1889	1886	Yellow fever epidemic sweeps Jacksonville, Florida.
Henry George Moran acquitted of attempting to poison Alfred Calvert by putting corrosive sublimate into chocolate.	1886	1880–1889	1886	Pharmacist John Stith Pemberton invents a carbonated beverage ultimately named "Coca-Cola."
W. Baker and Company publish *Cocoa and Chocolate*.	1886	1880–1889	1886	Statue of Liberty, French gift to the United States, dedicated in New York Harbor.
United States commerce data reports imports of 12,235,304 pounds of chocolate/cacao beans.	1885	1880–1889	1885	Cholera epidemic in Spain.
Report of vandals in New York inserting pins into chocolate candy distributed to local children.	1885	1880–1889	1885	Louis Pasteur develops vaccine against rabies.
Emilie Sichel tried and acquitted for smuggling chocolate, corsets, and lace into United States.	1885	1880–1889	1885	Canadian Pacific Railroad completed.
New York Times editorial chastisement of uneducated American diplomats who entertain foreign dignitaries by serving spareribs, sausages and chicken pie, garnished with fried potatoes, canned corn, tomatoes, and current jelly, and crowned with ice cream and chocolate meringues: "Crimes of this magnitude are now, it is true, rarely perpetuated."	1885	1880–1889	1885	Typhoid fever outbreak in Plymouth, Pennsylvania.
Susanna Way Dodds publishes *Health in the Household*. Claims that pepper and condiments are vegetable poisons, and there is no need to drink at meals since enough water is found in fruits and vegetables. "No one would care for chocolate, if it were not for the quantities of milk and sugar that are used in it as seasonings. Moreover, it is prepared from the oily seeds of the *Theobroma Cacao* [sic] and is, therefore, a greasy substance, not fit to moisten the food preparatory to its being received into the stomach."	1885	1880–1889	1885	Bechuanaland Protectorate established in southern Africa.

Chocolate-Related Events in North America	Date ←	Decade ↔	Date →	Chocolate-Related Events Elsewhere and Historical Context
New York Times announcement of the burning of the Runkle Brothers' chocolate factory: "Although the police believe that the building was fired by thieves, it is not at all impossible that it may have been accidentally started by policemen engaged in searching the building incautiously throwing away burning matches."	1885	1880–1889	1885	Publication of *Adventures of Hucklebery Finn* by Mark Twain.
Henry Thompson condemns the practice of mixing chocolate and thick cocoa with water.	1885	1880–1889	1885	Mahdist uprising in Sudan; General Charles "Chinese" Gordon killed.
A. W. Nicholson publishes *Food and Drink for Invalids*. Claims that chocolate is hard to digest.	1884	1880–1889	1884	United States adopts Standard Time.
Recipe for chocolate macaroons published in the *Los Angeles Times*.	1883	1880–1889	1883	Publication of "Cousin Cocoa" in the *Chautauquan Weekly Newsmagazine*.
Miss Parloa's New Cook Book provides recipe for chocolate eclairs.	1882	1880–1889	1882	British invade Egypt and occupy Cairo.
The Pico House in Los Angeles serves chocolate cream cake as a dessert and chocolate as a beverage.	1881	1880–1889	1881	Guerrilla war led by Apache chief, Geronimo.
Professor Daniel C. Eaton presents lecture on chocolate to the Sheffield Scientific School, New Haven, Connecticut.	1881	1880–1889	1881	Clara Barton establishes American branch of the Red Cross.
Madeleine Vinton Dahlgren, in *Etiquette of Social Life in Washington [DC]*, writes: "As to the refreshments proper to provide at a morning reception, the choice is quite optional here, as in other of our cities. A cup of chocolate is, however usually offered, and many still preserve the old custom, and add other refreshing drinks and many tempting comfits."	1881	1880–1889	1881	British naval recommendation that daily ration of rum (grog) be replaced with daily ration of soluble chocolate and sugar (recommendation failed). Valor, Spanish chocolate company, founded. W. J. Sinclair publishes *Beverages, Tea, Cocoa, etc.*
J. M. Fothergill publishes *Food for the Invalid*. Provides recipes using cocoa nibs and for the preparation of chocolate cream.	1880	1880–1889	1880	University of California, Los Angeles, founded.
Julia A. Pye publishes *Invalid Cookery. A Manual of Recipes for the Preparation of Food for the Sick and Convalescent*. Provides recipes for chocolate, cocoa, and broma.	1880	1880–1889	1880	Thomas Edison patents his incandescent lamp.
James William Holland publishes *Diet for the Sick. Notes: Medical and Culinary*. Writes that chocolate is an irritant to dyspeptics.	1880	1880–1889	1880	Safety razor and film-roll camera developed.
Mari Parola publishes *Miss Parola's Kitchen Companion*. Provides chocolate recipes for the sick.	1880	1880–1889	1880	Start of Alaska gold rush.
Report by George T. Angell regarding adulteration of food: "Cocoa and chocolate are adulterated with various mineral substances. Several mills in New England are now engaged in grinding white stone into a fine powder for purpose of adulteration."	1879	1870–1879	1879	Commonly accepted date when Rudolphe Lindt invents chocolate "conching" machine. Daniel Peter and Henri Nestlé form partnership.

Chocolate-Related Events in North America	Date ←	Decade ↔	Date →	Chocolate-Related Events Elsewhere and Historical Context
National Police Gazette report that a Mr. Wynkoop put poison in chocolate that was drunk by Mary Kiehl; Wynkoop was not a relative of Kiehl, but was her sole heir.	1879	1870–1879	1879	Thomas Edison demonstrates first practical electric light bulb.
Thomas Chambers publishes *Lesson's [sic] in Cookery* (American edition). Lists rations for convicts and mentions cocoa.	1878	1870–1879	1878	Yellow fever ravages Alabama, Louisiana, Mississippi, Ohio, and Tennessee.
U.S. Supreme Court *Arthur v. Stephani*, 96 U.S. 125 (1877): alleged excess of duty paid upon certain chocolate imported by them from Liverpool, in 1873, and upon which the collector of the port of New York, holding it to be "confectionery," exacted a duty of fifty cents per pound ad valorem, under the first section of the act of June 30, 1864 (13 Stat. 202). The importers claimed it to be dutiable as "chocolate," under the first section of the act of June 6, 1872 (17 id. 231), which imposes a duty of five cents per pound.	1877	1870–1879	1877	Root beer first marketed by Elmer Hires.
Warning to merchants and general public that popular green paper used to wrap chocolate contains arsenic.	1877	1870–1879	1877	Thomas Edison invents the phonograph.
Juliet Corson publishes *Fifteen Cent Dinners for Families of Six*: "Cocoa makes a wholesome and nourishing drink . . . The cost of tea is .03; coffee .06 and cocoa .06 using cocoa shells."	1877	1870–1879	1877	Nez Perce, Chief Joseph, crosses the Musselshell River east of Harlowton, Montana.
Anonymous publication of *Every Days Need. A Collection of Well Proven Recipes Furnished by the Ladies of the Business Woman's Union*. Provides numerous recipes using chocolate.	1876	1870–1879	1876	Daniel Peter introduces a line of milk chocolate designed for eating. Alexander Graham Bell patents the telephone.
Fannie Fitch publishes *A Gem Cookbook*. "This [chocolate] is the most wholesome and invigorating of all beverages, and is recommended by all physicians, specially [sic] to invalids, children, and adults of weak digestion. It is invaluable to persons debilitated by excessive brain work or violent exercise."	1875	1870–1879	1875	Beginning of the Black Hills gold rush. Daniel Peter produces hardened milk chocolate in tablets and bars.
The 25th anniversary of the 1849 arrival of the steamship *California* in San Francisco is celebrated in 1874 with coffee, tea, chocolate, lemonade, and wine.	1874	1870–1879	1874	Levi Strauss starts to sell blue-jeans with copper rivets.
Catherine Beecher publishes *Miss Beecher's Housekeeper and Healthkeeper, Containing Five Hundred Recipes for Economical and Healthful Cooking*. Identifies use of cocoa and chocolate for sick and young children.	1873	1870–1879	1873	Alexandre Dumas's *Grand dictionnaire de cuisine* refers to Chinese chocolate. Alfred Wanklyn publishes *A Practical Treatise on the Analysis of Tea, Coffee, Cocoa, Chocolate*. Epidemics of cholera, smallpox, and yellow fever sweep American South.
Article in the *Manufacturer and Builder* describes rolled tin (i.e., tin foil) as protective against oxidation and is widely used for chocolate and tobacco products.	1872	1870–1879	1872	Great Boston fire.

Chocolate-Related Events in North America	Date ←	Decade ↔	Date →	Chocolate-Related Events Elsewhere and Historical Context
Charles Nordhoff travels from Chicago to Omaha by train and lists the menu and food prices: "French Coffee, English Breakfast Tea, and Chocolate $0.15; French Coffee, Tea, Chocolate, without an order $0.25."	1872	1870–1879	1872	Jules Verne writes *Around the World in 80 Days*.
S. D. Farrar publishes *The Homekeeper. Containing Numerous Recipes for Cooking and Preparing Food in a Manner Most Conductive to Health, Directions for Preserving Health and Beauty, and for Nursing the Sick*. "Chocolate, for instance, may be made thick or thin, sweet or not, just as a person fancies, and still be good and wholesome."	1872	1870–1879	1872	Yellowstone Park established as world's first national park.
Herald of Health magazine identifies chocolate as nutritious but difficult to digest.	1871	1870–1879	1871	Announcement that, in the neighborhood of Bingen-on-the-Rhine, adulteration of chocolate was being carried out. Siege of Paris (105th day): Ebenezer Washburne reports "no change in the price of coffee, chocolate, wine, liqueurs." Deranged English woman arrested for lacing chocolate with poison, and lefting children eat it.
Marion Harland publishes *Common Sense in the Household: A Manual of Practical Housewifery*. Provides recipes using cocoa nibs, shells, cocoa to prepare chocolate cake, caramels, icing, ice cream.	1871	1870–1879	1871	Great Chicago fire.
American consumers warned that some manufacturers add paraffin to their chocolate.	1870	1870–1879	1870	Quantity of cacao exported from Trinidad—8.5 million pounds. John D. Rockefeller establishes Standard Oil Company.
Emily Edson Briggs describes a Washington DC social party: "For hours this distinguished sea of humanity whirled and surged through the mansion. Waiters managed, by some secret known only to themselves, to wedge their way through the dense throng and refresh the guests with cakes and ices. A room was provided where coffee and chocolate were served, but no costly wine or any other beverage that intoxicates was seen at the reception of the Secretary of War."	1870	1870–1879	1870	Louis Pasteur and Robert Koch establish germ theory of disease transmission.
	1869	1860–1869	1869	Completion of Union Pacific transcontinental railroad; *Golden Spike* driven at Premonitory Point, Utah.
Etienne Guittard arrives in San Francisco; starts chocolate business.	1868	1860–1869	1868	University of California chartered in Berkeley, California. Eight-hour work day passed by Congress.
	1867	1860–1869	1867	Anglo-Swiss Company develops condensed milk.
Chocolate served at President Lincoln's second inaugural banquet.	1865	1860–1869	1865	Abraham Lincoln assassinated on Good Friday, while attending the theater. Mexican proverb: "Beware of tasting Chiapa[s] chocolate." Article in *Harper's Weekly* repeats oft-told story of how the Bishop of Chiapa[s] was poisoned by angry ladies of the region.

Chocolate-Related Events in North America	Date ←	Decade ↔	Date →	Chocolate-Related Events Elsewhere and Historical Context
Medical and Surgical Report: "The Medical Supply Table authorizes the issue, on proper requisition, to all hospitals, of barley, extract of beef, cinnamon, cocoa, extract of coffee, corn-starch, farina, gelatin, ginger, concentrated milk, nutmegs, pepper, porter, white sugar, black tea, and tapioca; together with whisky, brandy, port, and Taitagona [?] wine."	1864	1860–1869	1864	Abraham Lincoln reelected president of the United States.
Army of the Potomac, supply requisition, dated May 17: "200 pounds cocoa or chocolate; 200 pounds corn starch; 100 dozen bottles porter; 3000 iron bedsteads; 3000 mattresses, 10,000 sheets, 7000 pillows, 2000 blankets; 12 cooking stores."	1864	1860–1869	1864	International Red Cross founded.
Percival Drayton writes to his friend Alexander Hamilton describing chocolate soup as something unexpected "beyond his experience."	1864	1860–1869	1864	Auguste Debay identifies medicinal uses of chocolate.
Request to Walter Baker that his company supply Federal troops with chocolate.	1864	1860–1869	1864	American Civil War battles: Atlanta and the "march to the sea."
Emily Bliss Thacher Souder, nurse to wounded soldiers, writes: "A cup of chocolate is greatly relished."	1863	1860–1869	1863	American Civil War battles: Gettysburg (Pennsylvania), Vicksburg (Mississippi); and Chattanooga (Tennessee).
Confirmation from the Libby Prison Hospital, Fort Monroe, to the Sanitary Committee of New York acknowledging receipt of "20 cans condensed milk; 5 chocolate cakes; 5 gallons assorted pickles; 5 gallons pickled peaches."	1863	1860–1869	1863	French capture Mexico City; Archduke Maximilian of Austria established as ruler.
Letter from the Secretary of the Navy to the Secretary of the Treasury, requesting provisions for Federal naval forces at Port Royal, South Carolina: among the items requested—cattle, potatoes, onions, turnips, apples, ice, fresh meat, flour, hams, bacon, milk, molasses, sugar, butter, tobacco, cheese, clothing, boots and shoes, cigars, coffee, tea, and chocolate.	1862	1860–1869	1862	Mexican army defeats French at battle of Puebla (May 5: Cinco de Mayo).
	1862	1860–1869	1862	A. Mangin publishes *Le Cacao et la Chocolat*.
Letter from John Albion Andrew to John Murray Forbes requesting that Baker's chocolate be sent to Federal troops at Fort Monroe.	1861	1860–1869	1861	Louis Pasteur publishes his germ theory of fermentation.
Powell Lopell and Ellis Elting counterfeit coins to purchase chocolate (tried and acquitted).	1861	1860–1869	1861	First transcontinental telegraph message ultimately forces end of Pony Express.
Chocolate recommended as a beverage to Federal military volunteers: "Avoid drinking freely of very cold water, especially when hot or fatigued, or directly after meals. Water quenches thirst better when not very cold and sipped in moderate quantities slowly—though less agreeable. At meals tea, coffee and chocolate are best. Between meals, the less the better. The safest in hot weather is molasses and water with ginger or small beer."	1861	1860–1869	1861	Federal troops fire on Fort Sumter, South Carolina; start of American Civil War.

Chocolate-Related Events in North America	Date ←	Decade ↔	Date →	Chocolate-Related Events Elsewhere and Historical Context
Army of Virginia and Florence Nightingale publish *Directions for Cooking by Troops in Camp and Hospital*. "Cocoa is often recommended to the sick in lieu of tea or coffee … It is an oily starchy nut, having no restorative [*sic*] at all, but simply increasing fat. It is pure mockery of the sick, therefore, to call it a substitute for tea. For any renovating stimulus it has, you might just offer them [the wounded/sick] chestnuts instead of tea."	1861	1860–1869	1861	United States Sanitary Commission established by Simon Cameron to provide food and medical care to soldiers and to provide education to combat the spread of disease during the American Civil War.
Posted prices at Mrs. S. L. Skilton's Eating House, #6 Lindall Street, Boston: "cup of cocoa, $0.04."	1860	1860–1869	1860	Pony Express riders carry mail between Sacramento, California, and St. Joseph, Missouri.
New York Times advertisement for chocolate: "Chocolate is healthy, agreeable to the taste and nutritive. It is used everywhere as a pectoral, as a strengthening agent. It is acknowledged by every Medical Institute and recommended for the benefit of feeble constitutions and sickly persons. Use Chocolate Bonbons for your children, instead of any other bonbons or candies. Use Chocolate for your LUNCH. Every business man ought to keep in his desk a tablet of MENDES Health Chocolate. It is used in Paris and in London by all persons whose business does not allow their going home at mid-day as the most nutritive lunch, as the most agreeable food. The best lunch is a cake of good Chocolate with a roll of bread and a glass of water for beverage."	1860	1860–1869	1860	A. Gosselin publishes *Manuel des Chocolatiers*.
Newspaper editor Horace Greeley writes an account of his overland journey from New York to San Francisco: "May 12th—Chicago: chocolate and morning newspapers last seen on the breakfast-table."	1859	1850–1859	1859	Charles Dickens writes in *A Tale of Two Cities*. John Brown leads raid on the Federal Arsenal at Harper's Ferry.
Julie Andrews publishes *Breakfast, Dinner, and Tea Viewed Classically, Poetically, and Practically*. Discusses chocolate and reports a chocolate-related poem by Francisco Redi (1626–1698), who disliked chocolate: "Talk of chocolate! Talk of tea! Medicines made—ye gods! Are no medicines made for me."	1859	1850–1859	1859	A. Mitscherlich publishes *Der Kakao und die Schokolade*.
Elizabeth Emma Stuart writes that chocolate predisposes to headache but is excellent nourishment for mother and baby.	1858	1850–1859	1858	First transcontinental overland mail service by stagecoach connects East Coast and West Coast.
The Great Western Cook Book, or Table Receipts, Adapted to Western Housewifery (Anonymous author). Recipe for chocolate as a beverage: "Put the milk and water on to boil, then scrape from one to two squares of chocolate to a pint of milk and water mixed. When the milk and water boil, take it off the fire; throw in the chocolate, beat it up well, and serve it up with the froth."	1857	1850–1859	1857	Traditional date for the introduction of cacao trees into Ghana.

Chocolate-Related Events in North America	Date ←	Decade ↔	Date →	Chocolate-Related Events Elsewhere and Historical Context
Proposal circulates in Washington DC to send Louisiana slaves to Nicaragua to work on cacao plantations.	1856	1850–1859	1856	Native Americans and settlers skirmish at Seattle settlement
The Ghirardelli Company advertises in the newspaper *Alta California* as making chocolate, as well as syrups, liquors, and ground coffee.	1856	1850–1859	1856	Vigilance Committee founded in San Francisco to protect citizens.
Six thousand pounds of cocoa are exported through the Port of San Francisco, with an ad valorum duty paid of $447.	1856	1850–1859	1856	George Mendel starts research on variation in garden peas (considered father of modern genetics).
Thomas S. Powell publishes *A Pocket Formulary and Physician's Manual, Embracing the Art of Combining and Prescribing Medicines to the Best Advantage; with Many Valuable Recipes, Tables, etc.* "Cocoa. Boil two ounces of good Cocoa in a quart of water, and, as soon as it boils, set it on coals to simmer gently for an hour or more. To be used hot."	1855	1850–1859	1855	Patent developed in England for medicinal gluten chocolate: two parts cocoa, two parts sugar, one part gluten-bread reduced to a fine powder.
	1855	1850–1859	1855	Letter from the Crimea: "Sir, Our hapless soldiers have been supplied with raw coffee and found it almost useless, for want of fuel, roasters, and mills. Being familiar with foreign lands and rough travel, I venture to suggest the use of soluble chocolate, packed in pound parcels (for retail purchases), containing 36 square bits, two of which make an ample breakfast, and one of which, eaten even without any preparation, will stay hunger for four or five hours."
Frederick Douglass claims betel-nut, chocolate, coffee, hemp, opium, tea, and tobacco are poisons.	1855	1850–1859	1855	Florence Nightingale applies hygienic standards in military hospitals during Crimean War.
Godey's Lady's Book publishes recipe for chocolate custard.	1854	1850–1859	1854	United States establishes trade with Japan.
Debut of Whitman's boxed chocolates.	1854	1850–1859	1854	John Snow identifies Broad Street water pump handle as source of cholera contamination; cholera confirmed as a water-borne disease associated with poor sanitation (beginning of the field of epidemiology).
Frank Lecouvreur wrote about the reconstruction of San Francisco after the 1849 fire: "San Francisco, too, has changed wonderfully in this one year since I first saw it; one can hardly believe his own eyes. Not only have frame and sheet-iron buildings been torn down and replaced by magnificent brick buildings, and new ones been added; but whole streets have been opened and built up; and where as late as last January large ships have discharged or taken in cargo, there you may see today buildings two and three stories high on solid foundations; and on the very spot where we passengers of the *Aurora* managed to land by climbing the narrow ladder on the California wharf—one may now take his cup of chocolate in a beautifully furnished establishment and may indulge in finest confectionery."	1853	1850–1859	1853	Commodore Matthew Perry sails into Tokyo Bay.

Chocolate-Related Events in North America	Date ←	Decade ↔	Date →	Chocolate-Related Events Elsewhere and Historical Context
Godey's Lady's Book describes chocolate at breakfast as: "Strangely suggestive of Spain, and the bright eyes of the Andalusian maidens, and the Alhambra, with its wealth of Moorish lore, and the muleteers, singing gaily over the mountains, as they wend their way homewards amid the melodious jangling of the pendent bells."	1853	1850–1859	1853	Inauguration of railroad service between New York and Chicago.
Lewis Carstairs Gunn writes regarding his experiences in the California gold rush: "Before we reached our house, we stopped at the stage office to leave some passengers, and such lots of men came out to see us! I was introduced to I don't know how many—all coming to see us soon. Well, we got home, and went up stairs and washed, and had supper, and after a while went to bed. Our supper was baker's bread and butter and sardines and chocolate. And we all slept on the floor."	1852	1850–1859	1852	Formation of California Historical Society.
Letter from Jasper Boush to William H. Starr lauding immigration to Liberia, where immigrants can raise cacao and reap profits.	1852	1850–1859	1852	*Uncle Tom's Cabin* published by Harriet Beecher Stowe.
Luch Rutledge Cooke diary: "[At Fort Laramie] My dear husband has just been over to the store there to see if he could get anything to benefit me. . . . He bought two bottles of lemon syrup at $1.25 each, a can of preserved quinces, chocolate, a box of seidlitz powders, a big packet of nice candy sticks, just the thing for me to keep in my mouth, and several other goodies. Oh, it all seemed a Godsend to me."	1852	1850–1859	1852	Sanitary Commission of London report on adulterants added to chocolate.
Louise Amelia Knapp Smith Clappe writes: "[Living at the mines] When we arrived at the little oak-opening, we were, of course, in duty bound to take a draft from the spring, which its admirers declare is the best water in all California. When it came to my turn, I complacently touched the rusty tin cup, though I never did care much for water, in the abstract, as water. Though I think it very useful to make coffee, tea, chocolate, and other good drinks. I could never detect any other flavor in it than that of cold, and have often wondered whether there was any truth in the remark of a character in some play, that, ever since the world was drowned in it, it had tasted of sinners."	1852	1850–1859	1852	William Brady, author of *The Kedge-Anchor, or Young Sailor's Assistant*, writes that it was the responsibility—when "taking to the boats, i.e., abandoning ship, that the Cook/Steward be responsible for carrying into the lifeboat the following items: tinder-box, flints and tinder, a lantern and candles, cheese, cabin biscuits, and chocolate.

Chocolate-Related Events in North America	Date ←	Decade ↔	Date →	Chocolate-Related Events Elsewhere and Historical Context
New York Times announcement for the treatment of cholera: "Immediately on ascertaining the existence of the foregoing symptoms the patient should be put to bed, and covered up warm; then, if the patient be an adult, give immediately, or as soon as can be obtained from the nearest apothecary, a pill composed of the following ingredients, vis: Opium 12 grain; Camphor 2 grains; Calomel 5 grains; Cayenne 1 grain. Six hours after the pill has been given, the patient should take one ounce of Castor Oil, with twenty-five drops of Laudanum. This, with sufficient rest, and the most digestible diet, as gruel, rice, light soup, or chocolate, is all that is required to cure either disease in their incipient stages."	1852	1850–1859	1852	Wells, Fargo & Company founded.
A. L. Webster publishes *The Improved Housewife, or Book of Receipts*. Contains recipe where rice flour is added to chocolate. Mentions "cocoa shells: soak overnight, boil in morning. They are healthful and cheap."	1852	1850–1859	1852	Yellow fever epidemic in Louisiana; New Orleans especially hard hit.
Godey's Lady's Book recommends chocolate to improve the appetite of invalids.	1851	1850–1859	1851	Great San Francisco fire.
Publication of *The American Matron, or, Practical and Scientific Cookery* (anonymous author) suggests that homemakers should have "scientific knowledge associated with the methodology in cooking." Provides recipes using chocolate, cocoa, and cocoa nibs.	1851	1850–1859	1851	Western Union telegraph company founded.
Shipping invoices from Sutter's Fort, Sacramento, California: chocolate imported to San Francisco (British vessel, *Dorset*, out of Hobart, Tasmania, to San Francisco; British vessel, *Erin's Queen*, out of Liverpool to San Francisco; French vessel, *Antoinette*, Bordeaux to San Francisco; British vessel, *Archibald Gracie*, Mansanillo, Mexico, to San Francisco).	1851	1850–1859	1851	Cholera epidemic in Illinois, Great Plains, and Missouri region.
William Alcott publishes *The Young Mother, or, Management of Children with Regard to Health*: "Chocolate can quench thirst, but not as well as water due to its mucilage and nutriment properties."	1851	1850–1859	1851	Isaac Singer patents his sewing machine.
Chocolate advertised as available in the Miners' Variety Store in Sacramento.	1851	1850–1859	1851	First settlement at Alki Point, later would become Seattle, Washington.
J. B. Starr ledger in Sacramento records some 25 or more chocolate or cocoa transactions, beginning October 18, 1850 and extending through August 13, 1851. The sales include tins and boxes (which appear to have been identical) of chocolate and cocoa, cases of chocolate, and boxes of cocoa paste.	1850	1850–1859	1850	Max Sefelege, native of Potsdam, confined in the hospital (insane asylum) at Spandau." He pretended, for some time past, that he was the inventor of chocolate."

Chocolate-Related Events in North America	Date ←	Decade ↔	Date →	Chocolate-Related Events Elsewhere and Historical Context
Freidrich Wilhelm Christian Gerstäcker writes: "We are in San Francisco . . . [Description of the gaming rooms]. On the right of the room, behind a long bench stands a young and very beautiful girl; she is coquettishly dressed in a gown of black silk; with her slender white hands, fitted with rings, she serves to some tea, to others coffee, and chocolate with cake or jam, while in the opposite corner of the room a man sells wine and liquors. Before the tea table two or three tall figures, rough and little refined, stand as in ecstasy before the young girl; in order to have a pretext to say where they are, these persons make a purchase of tea at 1 frank 25 a cup."	1850	1850–1859	1850	Oregon gold rush.
Enos Christman writes: "Soon after dinner [mid-day] we fell in with four or five broken-down-looking fellows on foot who were returning from the mines, and they gave us some rather discouraging accounts. . . . About four in the afternoon we came to a small stream crossing the prairie, making the only camping ground for several miles; here we are not encamped for the night. We have just concluded supper, cooking nothing but a cup of chocolate and soaking pilot bread in it. Here are three tired boys! Oh, how my poor legs ache! I think I could almost rest forever."	1850	1850–1859	1850	First transatlantic steamship schedule links England and United States.
Sahal ben Haroun writes in *Holdens Dollar Magazine of Criticism* that "the great lexicographer, Sam[uel] Johnson put sugar in his port wine, and not only mixed his chocolate with cream, but poured in melted butter as an additional dulcifier."	1850	1850–1859	1850	Descriptive account of travel by stagecoach in Mexico: passengers awakened at 3:00 a.m. and served chocolate.
Chocolate produced in Liberia, West Africa, claimed to be better than chocolate in North America.	1850	1850–1859	1850	Cholera epidemic sweeps through Midwest.
Saturday Evening Post publishes recipe for chocolate ice cream.	1850	1850–1859	1850	Great San Francisco fire.
Scientific American report that individuals habitually drinking chocolate do not experience attacks of cholera or "dysenteric affections" but other family members who drink coffee, tea, or cold water experience these diseases.	1849	1840–1849	1849	Fort Worth, Texas, founded.
John Frémont stays at the home of Kit Carson in Taos, New Mexico; served his morning cup of chocolate while still in bed.	1849	1840–1849	1849	Cholera epidemic in St. Louis, Missouri.

Chocolate-Related Events in North America	Date ←	Decade ↔	Date →	Chocolate-Related Events Elsewhere and Historical Context
Cholera epidemic in Cincinnati, Ohio; chocolate among the dietary recommendations.	1849	1840–1849	1849	Daily schedule of Pope Pius IX (after 8:30 a.m.): The Pope "dips biscuits in a mixture of coffee and chocolate" and begins his daily business affairs.
Caution issued against drinking chocolate. Argument is based on concept that water-drinking in childhood leads to good health whereas "the nerves and blood system are over excited by taking viands, spices, beer, wine, chocolate, coffee, &c. and thus a constant, artificial state of fever is maintained, and the process of life is so much accelerated by it, that children fed in this manner, do not attain perhaps half the age ordained by nature."	1849	1840–1849	1849	Cadbury introduces a line of milk chocolate prepared after the "Sloane Recipe" (after Hans Sloane's recipe of 1687).
Bayard Taylor describes San Francisco at the beginning of September: "At the United States and California [restaurants] on the plaza, you may get an excellent beefsteak, scantily garnished with potatoes, and a cup of good coffee or chocolate, for $1.00. Fresh beef, bread, potatoes, and all provisions which will bear importation, are plenty; but milk, fruit and vegetables are classed as luxuries, and fresh butter is rarely heard of."	1849	1840–1849	1849	Influx of emigrants to California in search for gold (49ers).
William Z. Walker writes of crossing the plains to reach Utah as part of the Mormon migration: "June 9th, Started at 5 o'clk [sic] without our breakfast, there being no wood for fuel and the buffalo chips being to [sic] damp to cook with. We traveled till dark and encamped near a fine Spring of cool water of which we drank very plentifully. We luxuriated [sic] on a fine dish of chocolate at supper."	1849	1840–1849	1849	St. Louis destroyed by fire.
J. M. Letts describes crossing the isthmus of Panama on the way to California: "We usually boiled our coffeewater in the camp-kettle, but this being full of game. . . . The ham was frying briskly by the fire, our chocolate dissolving."	1849	1840–1849	1849	Elizabeth Blackwell becomes first woman physician in the United States.
Andrew Loucks (97 years old) was interviewed and asked to what he owes his health and longevity. He replies: "We ate lighter food when I was a boy than at present—such as soups; used a great deal of milk, and but little tea and coffee. We sometimes made chocolate by roasting wheat flour in a pot, though not often . . . young people are now up late at nights—to run about evenings is not good."	1849	1840–1849	1849	Cholera epidemic across the Great Plains region.

Chocolate-Related Events in North America	Date ←	Decade ↔	Date →	Chocolate-Related Events Elsewhere and Historical Context
Stephen Chapin Davis provides list of provisions with San Francisco prices: "Chocolate at $0.40 dozen."	1849	1840–1849	1849	Continued famine in Ireland.
Felix Paul Wierzbicki, in his *Guide to the Gold Region*, writes: "As a substitute for tea or coffee, chocolate may be used advantageously; it is not a drug; it possesses no remedial powers, in the proper sense of the word; it is only alimentative, nourishing by its natural oil and substance. And as an aliment, an excessive quantity must be guarded against, as it is the case with any other article of food." He also lists chocolate among the provisions commonly used by miners.	1849	1840–1849	1849	Steamboat service connects west and east coasts of North America.
Enos Christman writes: "It was with the greatest difficulty we could get the fire to burn sufficient to boil our beer, potatoes and chocolate. We bought a pound of chocolate for $1.00."	1849	1840–1849	1849	Aboard ship—going around the Horn to California—Enos Christman writes: "Today I stood nearly an hour in the rain and cold until my fingers were quite benumbed, holding a tin cup to catch water as it dripped off one of the small [life] boats. This was free from salt and made a good cup of chocolate."
William Redmond Ryan describes pearl divers in Baja California: "Dealers on shore, who always accompany these expeditions with spirituous liquors, chocolate, sugar, cigars, and other articles of which these Indian divers are especially fond."	1848	1840–1849	1848	Cholera epidemic across the Great Plains region.
	1848	1840–1849	1848	Poulain Chocolate Company created at Blois, France.
Fasting regulations for Lent issued by the Baltimore, Maryland, Catholic Diocese: "Lawful to drink thin chocolate, made with water, to which a few drops of milk may be added."	1848	1840–1849	1848	Continued famine in Ireland.
Henry David Thoreau was served chocolate for breakfast while in prison.	1848	1840–1849	1848	American Association for the Advancement of Science founded.
William Radde, agent for the Leipsic (*sic*) Central Homoeopathic Pharmacy, sells pure spirits of wine, corks, diet papers, and homoeopathic chocolate.	1848	1840–1849	1848	Treaty of Guadalupe Hidalgo ends United States war with Mexico; United States receives Texas, New Mexico, California, Utah, Nevada, Arizona, and portions of Colorado and Wyoming from Mexico.
Mrs. T. J. Crowen publishes *Every Lady's Book*. Recipe for chocolate: "Make paste with chocolate and milk. Then boil water, add paste, boil or simmer longer 30 minutes, add cream or milk to mixture and sugar to taste."	1848	1840–1849	1848	Gold discovered at Sutter's Mill on the American River east of Sacramento, California.
Eliza Leslie publishes her *Lady's Receipt-Book* and includes a recipe for chocolate pudding.	1847	1840–1849	1847	Smithsonian Institution dedicated in Washington, DC.
	1847	1840–1849	1847	Salt Lake City founded by Brigham Young.

Chocolate-Related Events in North America	Date ←	Decade ↔	Date →	Chocolate-Related Events Elsewhere and Historical Context
Handwritten document listing relief supplies that include 100 pounds of cacao, assembled for "those stranded in the California mountains."	1847	1840–1849	1847	Donner Party disaster in the California mountains.
	1847	1840–1849	1847	Commonly accepted year when Joseph Fry and Sons prepared first chocolate candy bar by mixing melted cacao butter back into defatted cocoa powder; This could be pressed into a mold.
Bidwell, quartermaster ledger entries include purchases of cocoa on December 18.	1846	1840–1849	1846	Start of United States war with Mexico.
Heinrich Lienhard writes: "Having completed my business in Sacramento, I decided to return to Eliza City, where, before I had erected the hotel, I had purchased a tent. Instead of incurring expenses for board, I now decided to live the life of a bachelor as I had done before. Purchasing a large amount of supplies, such as sardines, soups, sausages, peas, etc., as well as a six-gallon keg of Chinese honey cakes, chocolates, sweetmeats, sugar, and a whole box of Chinese tea, I began to keep house. Meanwhile, I was looking eagerly forward to leaving this land of gold, which was also the land of murderers, drunkards, and cut-throats, within a short time."	1846	1840–1849	1846	Irish potato famine.
Publication (author unknown) of *A Manual of Homoeopathic Cookery*. Recipe: "Take two ounces of cocoa nibs and put them in a coffee boiler (but which has never been used for making coffee) with two pints of water; allow it to simmer for eight hours by the side f the fire, and then pour it gently off for use, leaving the nibs in the boiler."	1846	1840–1849	1846	Auguste Saint-Arroman identifies medicinal uses of chocolate.
James Fennimore Cooper publishes *Satanstoe*; chocolate mentioned in Volume 1, Chapter 8.	1845	1840–1849	1845	John Frémont arrives in California.
Franklin expedition departs England to search for the Northwest Passage; supplies include 4573 pounds of chocolate.	1845	1840–1849	1845	Texas annexed by the United States.
James Fennimore Cooper publishes *Wyandotte*; chocolate mentioned in Volume 1, Chapter 6.	1843	1840–1849	1843	Yellow fever epidemic in Mississippi Valley.
William Heath Davis writes: "The Californians were fond of cocoa and chocolate; the manufacture of the latter from the cocoa was done by women, who prepared a choice article with the hand-mill or *metate*."	1841	1840–1849	1841	Woskresensky identifies theobromine in cacao beans.
Black women form trading association to sell foods and household items, among them chocolate, cocoa, coffee, and tea.	1841	1840–1849	1841	Russians sell Fort Ross to John Sutter.

Chocolate-Related Events in North America	Date ←	Decade ↔	Date →	Chocolate-Related Events Elsewhere and Historical Context
F. A. Wislizenus diary discussing Rocky Mountain Green River fur trapper rendezvous: "A pound of beaver skins is usually paid for with four dollars worth of goods; but the goods themselves are sold at enormous prices, so-called mountain prices. . . . A pint of meal, for instance, costs from half a dollar to a dollar; a pint of coffee-beans, cocoa beans or sugar, two dollars each; a pint of diluted alcohol (the only spirituous liquor to be had), four dollars; a piece of chewing tobacco of the commonest sort, which is usually smoked, Indian fashion, mixed with herbs, one to two dollars. . . . With their hairy bank notes, the beaver skins, they can obtain all the luxuries of the mountains, and live for a few days like lords. Coffee and chocolate is cooked; the pipe is kept aglow day and night; the spirits circulate and whatever is not spent in such ways the squaws coax out of them, or else it is squandered at cards."	1839	1830–1839	1839	Charles Goodyear invents vulcanized rubber.
Account by William Heath Davis of California: "The Californians were early risers. The ranchero would frequently receive a cup of coffee or chocolate in bed . . . breakfast was a solid meal, consisting of *carne asada* [meat broiled on a spit], beefsteak with rich gravy or with onions, eggs, beans, tortillas, sometimes bread and coffee, the latter often made of peas."	1839	1830–1839	1839	Introduction of daguerreotype photographic equipment to United States.
American blacks encouraged to emigrate to Trinidad and grow cacao.	1839	1830–1839	1839	Slaves mutiny aboard the Spanish ship *Amistad.*
James Fennimore Cooper publishes *Homeward Bound*; chocolate mentioned in Chapter 7.	1838	1830–1839	1838	Cherokee forced removal (Trail of Tears).
Sylvester Graham condemns chocolate, claims it to be a vile indigestible substance.	1837	1830–1839	1837	Samuel Morse perfects the telegraph.
Don Agustin Janssens, a Belgian who had joined the Hijar-Padres expedition to California, noted atrocities in his diary and wrote: "That for Sergeant Macedonio Gonzalez of San Diego, shooting an Indian is as easy as taking a cup of chocolate."	1836	1830–1839	1836	Siege and defense of The Alamo; subsequent Texas independence from Mexico.
Osborne Russell ledger, Fort Hall, Oregon Territory: "2 nine inch butcher knives, $2.50; 1/2 pt red pepper, $0.75; 1 hank thread, $0.40; [quantity not given] Needles, $0.60; 20 loads Ammunition for trade, $2.00; 1 bunch white seed beans, $0.75; 1/2 dozen flints, $0.25; 2 pounds chocolate, $3.00."	1835	1830–1839	1835	Samuel Morse demonstrates potential of the telegraph (range approximately 40 feet).

Chocolate-Related Events in North America	Date ←	Decade ↔	Date →	Chocolate-Related Events Elsewhere and Historical Context
William Henry Dana comments on chocolate en route to California (*Two Years Before the Mast*): "I never knew a sailor, who had been a month away from the grog shops, who would not prefer a pot of hot coffee or chocolate, in a cold night, to all the rum afloat. . . . There was not a man on board who would not have pitched [his] rum to the dogs for a pot of coffee or chocolate."	1835	1830–1839	1835	Marcus Whitman wagon train reaches Oregon territory (Narcissa Prentise Whitman is the first European women to cross the continent).
	1834	1830–1839	1834	Directive from José Maria Tornel, Secretary of War, Mexico City, identifying chocolate among the items served at the officer's school.
Wilkes-Barre merchant advertises using poetry: "Chocolate and Poland starch, and salmon, Saltpetre'd well, and well-smoked gammon; From Arab's land the best of dates, And Cocoa-nuts as big as pates!"	1833	1830–1839	1833	City of Chicago founded.
Description of meals served to children at the Green Bay, Michigan Territory Mission: "Breakfast . . . coffee or tea, or chocolate, or milk; mush; potatoes; bread and butter, or molasses, or meat; Dinner . . . potatoes or other vegetables, soup, or beef, or pork, or fish, or venison, bread; Supper . . . tea or milk, bread and butter, or molasses."	1832	1830–1839	1832	Treaty of London creates an independent Kingdom of Greece.
Notice: "There can be no doubt as to the nutritive and wholesome qualities of the cocoa nut. . . . To the traveler, a bowl of good chocolate, and a slice of bread, before setting forth on his journey of a cold winter morning, will really produce all the good effects which have been erroneously attributed to ardent spirits, or to wine. . . . The best way to prepare chocolate for persons of delicate habits, or valetudinarians, is to boil it in water, and allow it to grow cold. . . . Some persons with very weak digestion, make use advantageously of the shells of the cocoa, boiled in milk, or even in water."	1832	1830–1839	1832	Cholera epidemic in London, ultimately spreads across the Atlantic to sweep Canada and the United States along the Mississippi River.
Captain Kennedy of the ship *Rome* transports an elephant from India to Boston and return: "It can do no harm to mention, that not a drop of ardent spirit was drunk on board Captain Kennedy's ship, from the day of her departure to her return. Plenty of hot coffee and chocolate supplied its place in cold weather, and the Yankee 'switchell' preserved the health of the men in Calcutta, while half the rum-drinking crews there were in the hospital."	1832	1830–1839	1832	Benjamin de Bonneville explores Rocky Mountains, Columbia River, and Northwest Territory.

Chocolate-Related Events in North America	Date ←	Decade ↔	Date →	Chocolate-Related Events Elsewhere and Historical Context
Synonymous sounds (essay to reduce confusion over the words): "Cocoa: name of the root and palm tree upon which it grows, and of an oil produced from it; Cacao: name of a fruit and tree, and of the shells of the kernels of the fruit; and it is from this fruit that chocolate is manufactured; Coco: does not enter into our commerce; it is the name of a creeping plant about the size and shape of the leaf of the kidney bean; it has an aromatic flavor, and is used in South America and in Hindoostan [sic], in combination with other ingredients, as a luxury."	1832	1830–1839	1832	Jacksonville, Florida, receives charter.
Washington Irving publishes *The Alhambra*; chocolate mentioned in Volume 2.	1832	1830–1839	1832	Greece becomes a monarchy.
Letter to the editor of the *New Hampshire Gazette*, relating the objective of Andrew Jackson as president was to lower taxes on chocolate, coffee, molasses, salt, and tea, and thereby save every family in the country "an average tax of several dollars a year."	1831	1830–1839	1831	José Antonio Alzate describes cocoa production in Mexico: "Two main harvests each year. One on Christmas Eve, called the Christmas harvest, and the other on [the Feast Day of] San Juan."
Account in the *Journal of Health* concerning breakfast: "In many instances, good chocolate, properly prepared, will constitute an excellent substitute for coffee."	1831	1830–1839	1831	Cyrus McCormick demonstrates grain reaper.
Discussion of the female body: "The secret of preserving beauty lies in three things—temperance, exercise, and cleanliness. . . . Breakfast not only set forth tea and coffee, but chocolate, and hot bread and butter. Both of these latter articles, when taken constantly, are hostile to health and female delicacy."	1831	1830–1839	1831	Nat Turner slave revolt in Virginia.
H. L. Barnum publishes *Family Receipts*. Recipe calls for use of a chocolate mill; also recipe for chocolate drops: "Chocolate nourishes greatly, strengthens the stomach, helps digestions, and softens humours [sic]; whence it is proper for weak stomachs and consumptive persons."	1831	1830–1839	1831	John Cadbury starts to manufacture drinking chocolate and cocoa.
Rules for a young lady (revised): "Her breakfast should be something more substantial than a cup of slops, whether denominated tea or coffee, and a thin slice of bread and butter. She should take a soft boiled egg or two, a little cold meat, a draught of milk or a cup or two of pure chocolate."	1830	1830–1839	1830	Greece becomes independent from Turkey.
	1830	1830–1839	1830	Charles-Amadée Kohler makes *noisettes au chocolat* (chocolate with nuts) in Lausanne, Switzerland.
	1830	1830–1839	1830	Chocolate production at Lynn, Massachusetts, reaches 60 tons/year.

Chocolate-Related Events in North America	Date ←	Decade ↔	Date →	Chocolate-Related Events Elsewhere and Historical Context
Instructions for breakfast: "If the maid has prepared bread and butter, always ask for toast. If she has given you toast, prefer buckwheat cakes, and be very much astonished that she has not yet discovered your liking. When she has provided cakes, desire her to bring you muffins. Should she prepare coffee, be sure you want tea; and if tea, order chocolate forthwith. always make yourself very difficult to suit."	1829	1820–1829	1829	James Smithson, British chemist, bequeaths funds for the foundation of the Smithsonian Institution, Washington, DC.
	1828	1820–1829	1828	*Saturday Evening Post* account from Paris; "Among the novelties offered for sale in Paris, is the article of white chocolate."
	1828	1820–1829	1828	Patent entitled "Method for pressing the fat from cocoa beans," filed by Caspar Van Houten, April 4.
	1828	1820–1829	1828	Thomas Graham identifies medicinal uses of chocolate.
Auguste Bernard Duhaut-Cilly, a French sea captain seeking to trade for furs on the Pacific coast of North America, is served chocolate at Mission San Luis Rey.	1827	1820–1829	1827	Antoine Gallais publishes his *Monographie du Cacao, ou Manuel de l'Amateur de Chocolat* in Paris. He writes that the Chinese take much chocolate but as their cacao arrive in paste (form), without spices, they have, in tin or porcelain boxes, a powder composed of vanilla, cinnamon, and ambergris.
	1826	1820–1829	1826	Philippe Suchard establishes his chocolate factory in Serrieres, Switzerland.
	1826	1820–1829	1826	Earliest advertisement for chocolate candy in England (Fry's chocolate lozenges).
	1826	1820–1829	1826	Report of papal benediction at Rome: "We found a brilliant assemblage of foreigners, in magnificent dresses, mixed with a large party of our own countrymen; who were regaling themselves with chocolate, ices, lemonade, and a profusion of other refreshments."
	1826	1820–1829	1826	Trinidad cacao exports (total) were 3,011,091 pounds.
	1825	1820–1829	1825	Anthelme Brillat-Savarin identifies medicinal uses of chocolate.
	1825	1820–1829	1825	Jean-Antoine Brutus Menier begins producing chocolate at Noisiel.
	1825	1820–1829	1825	Description of Lisbon, Portugal: "In Lisbon there are very few boarding houses such as we have in America. . . . For lodging and meals it costs $2.25 cts. Per diem. The coffee-houses here are also different fro those in our country; they take no lodgers and give no dinners; they are appropriated entirely to the selling of coffee, chocolate, &c."
Thomas Cooper publishes *Treatise of Domestic Medicine, to which is added, A Practical System of Domestic Cookery.* Provides chocolate recipes.	1824	1820–1829	1824	Sequoyah, Cherokee scholar, develops first written American Indian language.

Chocolate-Related Events in North America	Date ←	Decade ↔	Date →	Chocolate-Related Events Elsewhere and Historical Context
Notice: Christmas Holidays. Late additions and donations to the Peale's Museum, Baltimore: "Several water vessels, which were disinterred from one of the Inca Cities, near Traxillo, S. A.; Modern water vessels, from South America; Cloth made of bark, by the descendants of Christian, the mutineer of the ship *Bounty*; Cloth made of bark by the natives of the Sandwich Islands; Pods of the Cocoa, of which chocolate is made. Presented by Captain W. H. Conkling."	1824	1820–1829	1824	Philippe Suchard opens chocolate factory in Neuchâtel, Switzerland.
Cincinnati Literary Gazette publishes the poem "School for Politeness", and the lines: "The liquor drank, the garments chang'd; The family round the fire arrang'd; The mistress begg'd to know if he Chose coffee, chocolate, or tea?"	1824	1820–1829	1824	John Cadbury, a young Quaker, announces on March 1st the opening of a shop in Bull Street in Birmingham in the *Birmingham Gazette*: "John Cadbury is desirous of introducing to particular notice 'Cocoa Nibs', prepared by himself, an article affording a most nutritious beverage for breakfast."
Rhode Island announcement: "Public Auction. To be sold at Public Auction at Olneyville, in the Mill Brown George, on MONDAY, the 15th [November] instant, at 11o-clock, a. m. All the Machinery, Apparatus, Utensils, and Implements generally used for manufacturing Chocolate, consisting of the following articles, which were mortgaged as security for the payment of the rents of the said Brown George Mill, and the mansion-house of the late Colonel Christopher Olney, viz: 1 pair Millstones for grinding Chocolate, with the whole apparatus there-to belonging, Frame, Gearing, &c attached to the same; 3 Kettles, with Stirrers, Shafts and Cogs; 1 other Kettle for melting Chocolate; 1 Cracking Machine and Apparatus; 1 Fanner and Apparatus; 1 Roaster with Crane and Apparatus; 6000 Tin Chocolate Pans; 2 small Scale Beams and Weights; Set of Press Screws with Bolts; Box, containing Stamps and four Plates Sundry other small articles; 1 large Water Wheel."	1824	1820–1829	1824	Rocky Mountains explored by Jedediah Smith.
Report in *New York Mirror* on the preservation of beauty of young ladies: "3rd Recommendation: Her breakfast should be something more solid than a cup of transy tea and a thin slice of bread and butter. She should take an egg or two, a little cold meat, or a cup of chocolate."	1823	1820–1829	1823	Letter from Colombia: "It is the custom here, on first rising in the morning, to take a dish of strong coffee, the delicious flavour of which and its gently stimulating effects upon the nerves, is little recognized at home ... In the intervals between the meals, the small *fondas*, or coffee-rooms, afford for a *medio*, or five-penny bit, an excellent cup of chocolate and a small roll of exquisite bread."
E. S. Thomas, Baltimore merchant, advertises "Good Things for Christmas ... CHOCOLATE."	1822	1820–1829	1822	English merchants prohibited to export chocolate to Russia.
	1821	1820–1829	1821	English poet, George Gordon Byron writes of the exploits of *Don Juan* and mentions chocolate.
	1821	1820–1829	1821	Breakfast of Cuban *caballeros* identified only as a cup of chocolate.

Chocolate-Related Events in North America	Date ←	Decade ↔	Date →	Chocolate-Related Events Elsewhere and Historical Context
William Hewes sells chocolate on Conti Street, New Orleans, Louisiana.	1820	1820–1829	1820	Stephen Long explores headwaters of the Platte and Arkansas rivers.
	1819	1810–1819	1819	François-Louis Cailler opens first Swiss chocolate factory at Corsier.
Hayt & Sherman, merchants at Claiborne, Alabama, advertise: "Rat Traps; Flunch Coffee Mills; American Rum; Chocolate; Rice; Imperial Tea; Starch, Vinegar, &c."	1819	1810–1819	1819	Commonly accepted date for first transportation of live cocoa trees from Brazil to São Tomé and Príncipe in Portuguese Africa.
	1818	1810–1819	1818	Englishman Peter Durand arrives in the United States with the idea that tin cans could be used to seal food.
Research identifies *Holcus bicolor* as a potential substitute for chocolate.	1818	1810–1819	1818	Anglo-American Convention fixes 49th parallel as the border with Canada.
Smugglers caught with fraudulent invoices; so-called boxes of chocolate actually contain ribbons, silk stockings, boots, and shoes.	1817	1810–1819	1817	Construction begins on Erie Canal.
Death of John Juhel, New York merchant and cocoa importer.	1817	1810–1819	1817	J.-J. Machet writes *Le Confiseur Moderne, ou l'Art du Confiseur et du Distillateur and* identifies "Chinese chocolate."
Sons of John Wait, Boston chocolate maker, scalded to death when ship boiler explodes.	1816	1810–1819	1816	Elgin Marbles purchased for the British Museum, London.
Oliver Prescott, in *The New England Journal of Medicine*, describes a fatal attack of colic in a patient who swallowed a cacao nut that lodged in the "appendicula vermiformis" (i.e., appendix).	1815	1810–1819	1815	Samuel Mitchell letter to James Monroe, Secretary of War regarding how to preserve the health of soldiers while in camp: "Chocolate, although in my judgment very good when taken alone, and in its dry state, is nevertheless much more efficacious when dissolved in water, and received hot into the stomach . . . the chief effect[s] of chocolate … can rarefy the blood, enlarge the veins, render a man more lively, and make him more prompt to perform any given [military] service."
	1815	1810–1819	1815	Caspar van Houten builds a chocolate mill in Amsterdam, Holland.
Boston merchants Mosley and Babbidge import 10,000 pounds of cocoa from the Portuguese territories of South America.	1815	1810–1819	1815	Andrew Jackson defeats British in the battle of New Orleans.
Letter to President Madison regarding wartime economic problems: "Before these times we could buy eight pounds of good sugar for a dollar: now we can buy the same quantity, but of the worst kind, for fourteen shillings. Tea, coffee, rice, a chocolate, &c. have risen in nearly the same degree . . . seeing there is no hope of a speedy end of this war . . . the non-intercourses [i.e., trade restrictions], and war, have cost us a great deal, and done no good whatever. . . . Return then, I implore you to the principles and the disciples of Washington. Under these you prospered formerly; under these you may prosper again, if you will."	1813	1810–1819	1813	McGill University founded in Montreal.
Samuel Smith sells chocolate in Clinton, Ohio.	1813	1810–1819	1813	Battle of Lake Erie.

Chocolate-Related Events in North America	Date ←	Decade ↔	Date →	Chocolate-Related Events Elsewhere and Historical Context
Advertisement: Robert Dyde opens the Massachusetts Tavern and Hotel in Boston; serves chocolate in the tavern bar every day from 11:00 to 2:00.	1813	1810–1819	1813	Mexico declares independence from Spain.
F. Beck, Boston merchant, advertises chocolate and other goods as a "small assortment of good things for Thanksgiving."	1813	1810–1819	1813	American forces capture York (Toronto) and Fort St. George.
Robert Tolman opens chocolate store at #9 India Street, Boston: "All chocolate warranted with money refunded guarantee."	1812	1810–1819	1812	Napoleon retreats from Moscow.
Gardner and Downer, Boston merchants, announce they have imported 10,000 pounds of cocoa, suitable for chocolate making.	1812	1810–1819	1812	American Antiquarian Society founded in Worcester, Massachusetts.
Joseph Ripley, Boston auctioneer, advertises for sale 1000 pounds of chocolate shells and 75 bundles of wrapping paper.	1812	1810–1819	1812	United States declares war against England (War of 1812).
F. Beck, Boston merchant, advertises that he has imported 20,000 pounds of first quality cocoa.	1812	1810–1819	1812	Russians construct Fort Rossiya (i.e., Fort Ross) in northern California.
Welsh's chocolate brokered through the shop of Davenport & Dana in Boston.	1811	1810–1819	1811	President Madison prohibits trade with England.
Edward Cutbush describes treatment of typhus among sailors aboard the frigate *Constellation*: "Barley water acidulated with lemon juice, also weak gentian or snake-root tea, were given as common drink; very little wine was used until the convalescent state commenced, which I dated from the time the gums became sore, and then, it was given with a decoct, cort. peruv. or with sago, tapioca and gruel: rice, chocolate, barley, and portable soup, composed the chief diet of the sick."	1811	1810–1819	1811	"The debility and lowness with which his Majesty is frequently afflicted, often proceeds from his excessive propensity to talking; within these few days he talked nearly twenty two hours, with little cessation. It is also increased by frequent violent attacks of his malady. On Wednesday morning he took some chocolate for his breakfast and on the same day he partook of a [?]ude dish and some mutton for his dinner. His appetite is frequently as good as ever it was."
Welsh's chocolate brokered through the shop of Elisha Ticknor in Boston.	1811	1810–1819	1811	"The King of England takes for his breakfast chocolate, and eats very heartily; he also takes a hearty dinner; his refreshment in the course of the day, besides, is very little, except some coffee, which is made very strong; he often drinks, in the course of the day, lemonade, which has always been a favorite beverage with him."
Lopez & Dexter, Newport, Rhode Island, auctioneers, announce sale of the Progressive Company entire stock in trade. Among items to be auctioned "20 boxes of chocolate."	1810	1810–1819	1810	Coronation dinner for King Louis XV of France includes 1500 pounds of chocolate.
Jacob & Joshua Lopez, Newport, Rhode Island, merchants, advertise for sale: "Best Cognac Brandy, Mamaica Spirits, St. Croix Rum, Bohea Tea, Loaf Sugar, Batavia Coffee, Pimento, Pepper, Currants, Raisins, Chocolate, Almonds, Ginger, London Mustard."	1809	1800–1809	1809	Method developed to preserve food by canning (Nicolas Appert).

Chocolate-Related Events in North America	Date ←	Decade ↔	Date →	Chocolate-Related Events Elsewhere and Historical Context
Announcement of theft of chocolate marked with the name "A. Child" along with other goods from Juno Kennedy's store, Woburn, Massachusetts: $50.00 reward.	1808	1800–1809	1808	Russia establishes colony of Noviiy Rossiya in California.
New York Shipping News announces arrival of brigantine *Delight*, from Cayenne, via Antigua, with cocoa consigned to John Juhel, New York merchant.	1808	1800–1809	1808	Account of chocolate purchased for patients at the San Lorenzo Hospital, Mexico: "750 pesos have been paid to Felipe de Mendoza for 120 arrobas of chocolate to be given to the sick [1 arroba = 25 pounds, therefore, quantity identified as 3000 pounds].
Announcement of theft of chocolate branded "E. Baker Dorchester, Mass." from # 22 Long Wharf, Boston: $5.00 reward for recovery.	1808	1800–1809	1808	Announcement that Bayonne, France, is famous for hams, chocolate, and bayonets.
Lopez & Dexter, Newport, Rhode Island, merchants, advertise for sale: "West-India and Batavia Coffee, Jamaica Rum, Havana brown and white Sugars, Rice, and chocolate."	1808	1800–1809	1808	Importation of slaves into United States forbidden.
Peter Pailhes opens ice cream store in Boston and sells rose-, lemon-, raspberry-, currant-, coffee-, and chocolate-flavors.	1807	1800–1809	1807	Robert Fulton develops first commercially successful steamboat in North America for river traffic.
Lopez & Dexter, Newport, Rhode Island, merchants, advertise for sale: "Coffee and Chocolate. Three Thousand wt. Surinam Coffee, 35 boxes Chocolate."	1807	1800–1809	1807	First railroad passenger service begins in England.
Lopez & Dexter, Newport, Rhode Island, merchants, advertise for sale: "Muscavado SUGARS, MOLASSES, Jamaica RUM, St. Croix RUM, Malaga BRANDY, Lisbon WINE, Surinam COFFEE, Boxes CHOCOLATE, real Spanish CIGARS."	1807	1800–1809	1807	Widely circulated English propaganda report that Pauline Riotti, a former mistress servant of Napoleon Bonaparte, attempted to murder him by putting poison in his chocolate (apocryphal).
Caspar Wistar describes a patient who had diarrhea but then followed a diet of only rice, chocolate (beverage), wheat bread, and butter, with a small quantity of fresh meat. Subsequently, she reduced the quantity of her animal food and increased the quantity of chocolate. She lived on this dietary plan for a year without any visible ill effects. Subsequently, she became scorbutic. Wistar concluded that the scurvy was not caused by lack of fluids, or for "want of nourishment," and not from deficiency in oxygen since she lived in the countryside. He concluded that scurvy was due to a deficiency in the diet, something lacking in the juices of vegetables.	1806	1800–1809	1806	Noah Webster publishes first American English dictionary.

Chocolate-Related Events in North America	Date ←	Decade ↔	Date →	Chocolate-Related Events Elsewhere and Historical Context
Freiherr Langsdorff, passenger on the *Rezanov*, on a voyage to Neuva California, reports scarcity of chocolate, sugar, and iron tools at the Russian settlements from Okhotsk to the northwest coast of America.	1806	1800–1809	1806	Development of first federal highway system in United States.
Lopez & Dexter, Newport, Rhode Island, merchants, advertise for sale: "Cognac Brandy, London Stout, good retailing Sugar, Chocolate, Pilot Bread, Powder Horns, sundry Muskets, boarding Pikes, Tomahawks, Musket Balls and Canister Shot."	1806	1800–1809	1806	British invade South Africa; Dutch in Cape Town surrender to the British.
From the diary of Meriwether Lewis, September 13: "I felt my self [sic] very unwell and derected [sic] a little chocolate which Mr. McClellin gave us, prepared of which I drank about a pint and found great relief."	1806	1800–1809	1806	Meriwether Lewis and William Clark reach Pacific Coast; begin return trip to St. Louis.
Gideon Foster, Boston chocolate manufacturer, apologizes to customers for deficiency in weights in some of the chocolate he recently sold and blames an incorrect balance scale.	1806	1800–1809	1806	Zebulon Pike explores the American Southwest.
Yellow fever epidemic at Baltimore: "This was not infrequently the case in the yellow fever. To use stimulating or tonic medicines, was dangerous. The cure, therefore, was submitted to something a little nutritious—as weak broth in small quantities; and the symptoms soon indicated, whether stimulating soups, oysters, chocolate, mush and milk, &c. might be admitted. In this stage the cure was almost totally relinquished to such means, and to the attentions of the nurse."	1805	1800–1809	1805	Nelson defeats the French fleet at the battle of Trafalgar.
Boston advertisement provided in verse: "Biscuit and Butter, Eggs and Fishes, Molasses, Beer and Earthen Dishes; hard Soap & Candles, Tea & Snuff, Tobacco pipes perhaps enough; Shells, Chocolate & Stetson's Hoes, As good as can be (I suppose)."	1805	1800–1809	1805	United States and England break diplomatic relations over West Indies.
The descriptive qualities of pure, unadulterated chocolate defined in the *American Register*.	1804	1800–1809	1804	Start of Lewis and Clark expedition.
John Sibley suggests it might be possible to grow cacao in the western limits of Louisiana toward the River Grande (Rio Grande in Texas Territory).	1804	1800–1809	1804	Total Trinidad cacao exports to United States this year were 122,211 pounds.
New York Shipping News announces arrival of the schooner *Gardner* from St. Pierre's, Martinique, carrying 112 bags of cocoa consigned to John Juhel, New York merchant.	1804	1800–1809	1804	Alexander Hamilton killed in duel with Aaron Burr.

Chocolate-Related Events in North America	Date ←	Decade ↔	Date →	Chocolate-Related Events Elsewhere and Historical Context
New York Shipping News announces arrival of the schooners *Hannah&Betsy*, from Martinique, via Newport: sugar, coffee, and cocoa consigned to John Juhel, New York merchant.	1804	1800–1809	1804	Trinidad shipping records report: January 7th ship called *Trial*, built in Salem in 1794, bound for Salem with 22 casks of cocoa onboard; Feb 3rd ship called *Sally* bound for Philadelphia with 163 casks of cocoa, 271 bags, 144 bushels; From April 7th to July 4th, destination of ships leaving the Port of Spain included Philadelphia, Savannah, Washington, Norfolk, Baltimore, Boston and New York, exporting a total of 292 bags and 67 casks and cocoa; July 13th – Oct 6th destinations included New York, Boston, New London, and Philadelphia, for a total of 122,211 lbs of cocoa exported to the United States during this period; and July 13th, 3500 pounds of cocoa departed from Trinidad bound for New York.
Legislature of Massachusetts, House of Representatives: "Bill to regulate the manufacture of Chocolate in this Commonwealth, and to prevent deception in the quality and exportation thereof."	1803	1800–1809	1803	Yellow fever epidemic in New York.
Detailed description of the cacao tree, or chocolate-nut, in a juvenile magazine.	1803	1800–1809	1803	Louisiana Purchase.
Chocolate-related poem published in the *Boston Weekly Magazine*, devoted to "Morality, Literature, Biography, and History."	1803	1800–1809	1803	Fort Dearborn established near Lake Michigan (site later will become Chicago).
John Rodgers Jewitt, held captive by Nootka Native Americans, along the Northwest coast: Native Americans boarded his ship and set it afire; later, he made an inventory of the loss: "We had also the good fortune in looking over what had been taken from the ship to discover a box of chocolate and a case of Port wine, which as the Indians were not fond of it proves a great comfort to us, for some time."	1803	1800–1809	1803	Parmentier publishes *On the Composition and Use of Chocolate*.
	1802	1800–1809	1802	Mary Hassal letter to Aaron Burr describes her breakfast at Barracoa (in Cuba) that consisted of chocolate served in little calabashes.
	1801	1800–1809	1801	Jeanne Dunatte opens a chocolate shop in Ustaritz, near Bayonne; possibly France's the first female chocolate maker.
Samuel Brown describes diet for yellow fever patients: "FOOD—None, till after the crisis. Then begin with the lightest and mildest kind, and such as is easiest of digestion, and such as can be taken in a liquid form. Weak tea and coffee, milk or water porridge, milk in water, roasted or baked fruits, chocolate, sago, weak chicken or veal broth; from these gradually advance in the use of the more substantial foods, until the powers of digestion are fully and permanently restored."	1880	1800–1809	1800	Louisiana Territory ceded by Spain to France.
Blue paper suitable for wrapping candles or chocolate sold for cash in Boston.	1800	1800–1809	1800	Library of Congress established.

Chocolate-Related Events in North America	Date ←	Decade ↔	Date →	Chocolate-Related Events Elsewhere and Historical Context
Nathan Daboll uses examples of chocolate quantities to teach arithmetic.	1800	1800–1809	1800	Washington, DC established as nation's capital.
	1800	1800–1809	1800	Sulpice Debauve, a pharmacist, establishes a chocolate shop in Paris. In 1827, his nephew, Antoine Gallais, also a pharmacist, joins him and the shop is known as Debauve and Gallais. It still exists as a center for fine chocolate in Paris.
Risque chocolate poem appears in book published for the "improvement of young minds."	1799	1790–1799	1799	United States Congress standardizes weights and measures.
Announcement: "TEN DOLLARS REWARD. STOLEN from #38 Long-Wharf, Boston . . . one box chocolate, containing 50 wt. Box branded 'Watertown.' Any person who shall bring the Thief to justice shall be entitled to the above reward—or the Chocolate, for their trouble."	1799	1790–1799	1799	Widespread introduction of smallpox vaccination in England.
Priscilla Wakefield uses chocolate-related dialogue to teach mental improvement.	1799	1790–1799	1799	Cocoa Tree Chocolate House relocated to Number 64, St. James Street, London.
James Tyler publishes treatise on the plague and yellow fever; argues that chocolate possesses no exhilarating virtue, or only in a small degree, but that chocolate is more "nutritive" than either coffee or tea.	1799	1790–1799	1799	Rosetta Stone, a trilingual stele, found near Rosetta, Egypt.
	1798	1790–1799	1798	*Estimate of the Annual Amount and Value of the Depredations committed on Public and Private Property in the Metropolis [London] and its Vicinity, IN ONE YEAR:* "Thefts upon the River and Quays: raw sugar, rum, coffee, chocolate, pimento, ginger, cotton, dying woods, and every other article of West-India produce £150,000."
John Juhel, New York merchant, Washington Street between Cedar and Liberty Streets, advertises the importation of 60,000 pounds of cocoa.	1798	1790–1799	1798	Napoleon invades Egypt.
Thomas Condie and Richard Folwell recommend a diet for yellow fever patients: "Weak tea and coffee may be taken in the beginning of the fever. In its second stage, the patient may eat bread and milk, with roasted apples, or soft peaches, chocolate, sago, tapioca, ripe fruits, weak chicken or veal broth, and a little boiled chicken."	1798	1790–1799	1798	Yellow fever epidemic in New York City.
Hannah Foster publishes *The Coquette*; chocolate mentioned.	1797	1790–1799	1797	Albany replaces New York City as the capital of New York.
American House of Representatives response to hardship petition experienced by manufacturers of chocolate, support additional duty of two cents per pound on cocoa (petition honored and duty repealed).	1797	1790–1799	1797	John Adams succeeds George Washington as president of the United States.
Broadside published in Philadelphia (Dr. Giffts) recommends chocolate in the diet of yellow fever patients.	1797	1790–1799	1797	Erasmus Darwin suffers gout; treats himself with a diet that contains chocolate.

Chocolate-Related Events in North America	Date ←	Decade ↔	Date →	Chocolate-Related Events Elsewhere and Historical Context
Sarah Trimmer describes cacao trees in her children's nature book.	1796	1790–1799	1796	John Hunter, physician, describes how he bled a pregnant patient who had eaten only dry toast with a cup of chocolate for breakfast, and found the blood "inflamed."
	1796	1790–1799	1796	Thomas Hayes recommends chocolate in the diet of asthmatic patients.
Cotton Mather describes seven mills on the Naponset River, Massachusetts, one a chocolate mill.	1796	1790–1799	1796	United States takes possession of Detroit from British.
Report of foods distributed to Spanish officers, sergeants, petty officers, and soldiers stationed at the presidio in Santa Barbara. Among the items identified: 75 pounds of Caracas cacao, fresh and clean; 79 pounds of Colorado (green) cacao, fresh and clean; 72 *molinillos* (stirring rods) for chocolate; 72 copper chocolate mugs; 1050 pounds of ordinary chocolate; 300 pounds of good, gift chocolate; 175 pounds of superior, gift chocolate.	1796	1790–1799	1796	Edward Jenner develops vaccination for smallpox.
Israel Allen recommends a diet to treat dysentery that consists of chocolate, coffee, tea, milk, and rice.	1796	1790–1799	1796	Antonio Lavadan identifies medicinal uses of chocolate.
Christian Salzmann uses chocolate-related dialogue to teach moral values.	1795	1790–1799	1795	Metric system adopted by France.
American schooner *Hibernia*, under Captain Merry of Boston, captured by British (pirates) sold in Halifax; cargo consists of wine, rum, molasses, tea, tobacco, raisins, gin, and chocolate.	1795	1790–1799	1795	François Appert experiments with food preservation jars.
	1795	1790–1799	1795	Joseph Fry installs a steam engine in his chocolate manufactury in Bristol, England.
Thomas and John Hancock, Jr. advertise chocolate for sale in Boston.	1795	1790–1799	1795	Treaty with Spain opens navigation on Mississippi River.
Letter to George Washington, from his cousin: "I wd. [would] take the liberty of requesting you'll be so good as to procure and send me 2 or 3 Bush[els]; of the Chocolate Shells such as we're frequently drank Chocolate of at Mt. Vernon, as my Wife thinks it agreed with her better than any other Breakfast."	1794	1790–1799	1794	Continued executions in France: Antoine Laurent Lavoisier (considered to be the founding father of nutrition) among those killed.
Mr. Valette & Co. sweet meat makers and distillers of all sorts of cordials advertise: "Chocolate de la Vanille & of Health. . . . They make also Ice Cream, of all kinds and tastes, and will send it where wanted at 1 shilling 4 1/4 cents per glass."	1794	1790–1799	1794	Eli Whitney receives patent for cotton gin.
Benjamin Rush identifies diet to be followed by those considering inoculation for smallpox: "Tea, coffee, and even weak chocolate, with biscuit or dry toast, may be used as usual."	1794	1790–1799	1794	Hugh Smith publishes *An Essay on Foreign Teas, with Observations of Mineral Waters, Coffee, Chocolate, Etc.*
Benjamin Rush of Philadelphia recommends chocolate in the diet of yellow fever patients: "The food should consist of gruel, sago, panada, tapioca, tea, coffee, weak chocolate, wine whey, chicken broth, and the white meats, according to the weak or active state of the system."	1794	1790–1799	1794	Reign of Terror: mass executions in Paris continue.

Chocolate-Related Events in North America	Date ←	Decade ↔	Date →	Chocolate-Related Events Elsewhere and Historical Context
Daniel Defoe writes of fathers and daughters, sitting together and drinking chocolate while discussing courtship and marriage.	1793	1790–1799	1793	Report dated May 31, cautioning American manufacturers importing chocolate into India to be ethical in their trade dealings: "If your chocolate, instead of being made of Cocoa, should be found to be little else but a greasy mixture, no one will touch it twice."
Jonas Welsh advises public he sells his best chocolate at his mill along the Charles River, and that "NO Chocolate made as the Mills in Boston," is sold at William White's store on Long Wharf.	1793	1790–1799	1793	France declares war on England and Holland.
Samuel West publishes *Essays on Liberty and Necessity*; includes passage how gentlemen select from among chocolate, coffee, or tea when each beverage is liked equally well.	1793	1790–1799	1793	The Louvre, Paris, developed as the French national art gallery.
William Cogswell, New York merchant, advertises: "113 boxes Welsh's Boston made CHOCOLATE Very fresh."	1793	1790–1799	1793	Alexander Mackenzie becomes first European to cross Canada from coast to coast.
William Duncan sells chocolate at his store in Concord, Vermont.	1793	1790–1799	1793	King Louis XIV of France and Queen Marie Antoinette executed.
Richard Briggs publishes *The New Art of Cookery*; includes a recipe for chocolate puffs.	1792	1790–1799	1792	Dollar coins first minted in United States.
Announcement of Kentucky prices: "Chocolate from 1 shilling 6 pence to 1 shilling 8 pence; spices [here] 25% higher than in Philadelphia or Baltimore."	1792	1790–1799	1792	Republican Party founded by Thomas Jefferson; Federalist Party founded by Alexander Hamilton and John Adams.
George Riley uses chocolate-related dialogue to teach moral values.	1792	1790–1799	1792	Smallpox epidemic in Boston.
Alexander Hamilton establishes American tariffs for chocolate and sugar.	1792	1790–1799	1792	Gray and Vancouver discover Columbia River (river named after Gray's ship).
Samuel Woodbridge, Norwich, Connecticut, merchant, advertises: "West-India and New-England Rum, Madeira and Malaga Wines, Bohea and Green Teas, Coffee, Chocolate, Pepper, Spices, Codfish, Molasses; and many other good things, for CHRISTMAS."	1792	1790–1799	1792	Jesuits evicted from Mexico caught smuggling gold into Spain; gold discovered and hidden inside "globs" of chocolate.
Gordon Johnson uses chocolate-related examples in his general education arithmetic textbook.	1792	1790–1799	1792	Yellow fever epidemic along east coast of North America (Charleston, Philadelphia, New Haven, New York, and Baltimore).
Pennsylvania Hospital account ledger: "Paid this Year: Sugar, tea, coffee, and chocolate—£116–0 [shillings]–3 pence."	1791	1790–1799	1791	All French guilds are dissolved; Jews granted legal citizenship, ending Bayonne chocolate guild controversy.
Announcement: "Events start at 6:00 AM and in celebration of American Independence, public will be furnished with chocolate and fruits for breakfast."	1791	1790–1799	1791	First American ships visit Japan.

Chocolate-Related Events in North America	Date ←	Decade ↔	Date →	Chocolate-Related Events Elsewhere and Historical Context
Samuel Sterns recommends chocolate to patients at the Saratoga Spa, New York.	1791	1790–1799	1791	Bill of Rights ratified (first ten amendments to U.S. Constitution).
	1791	1790–1799	1791	Marie Antoinette, hoping to flee the French Revolution with her family, orders a *nécessaire de voyage* [portable case holding implements to make tea, coffee, and chocolate] to be made. Case contains a silver chocolatière. [This *nécessaire de voyage* (and chocolatière) appear as inventory items 1794, a year after her execution].
Anonymous physician recommends reducing appetite by drinking a cup of coffee or chocolate, followed by eating raisins or an apple.	1791	1790–1799	1791	Mission Santa Cruz founded by Father Fermín Francisco de Lasuén.
Chocolate, cigars, and sugar distributed by Spanish troops to pacify Apache Indians in American Southwest.	1791	1790–1799	1791	References to chocolate appear frequently in the writings of the Marques de Sade.
Convicted thief, Thomas Mount, confesses to sealing chocolate, rum, and tobacco before he is hanged.	1791	1790–1799	1791	Thomas Jefferson and James Madison visit Fort Ticonderoga and Lake Erie region.
Estate sale: "Household Goods Belonging to Mr. Joseph Mass, deceased, Philadelphia Chocolate maker and manufacturer, to be sold for cash; Also all the articles belonging to the Chocolate making business."	1790	1790–1799	1790	First patent law written for the United States.
Harvard University rules state students to be supplied with beer, chocolate, coffee, cider, and tea.	1790	1790–1799	1790	Philadelphia becomes national capital of United States.
Announcement that "chocolate manufactured or made in the states of North-Carolina, or Rhode-Island and Providence Plantations, and imported or brought into the United States, shall be deemed and taken to be, subject to the like duties, as goods of the like kinds, imported from any foreign state, kingdom, or country."	1790	1790–1799	1790	Attempt to assassinate King Fredrick of Prussia by putting poison in his chocolate: "From that period, Frederick, before he took his chocolate, always gave a little to his dogs."
Jebidiah Morse discusses production of cacao in his *History of America*.	1790	1790–1799	1790	United States Congress authorizes first national census.
Concise Observations on the Nature of Our Common Food: "CHOCOLATE, Which is the cocoa-nut mixed with flour or sugar, is very wholesome."	1790	1790–1799	1790	American Captain Robert Gray circumnavigates the world; trade opened between China and New England.
A document dated 1789 signed in Chihuahua, Mexico, identifies the foods supplied by Don Francisco Guisarnotegui (i.e., Francisco Guizarnotegui) to the presidios of New Mexico and in Chihuahua, and "road troops" (*compañia volante*) by the muleteer Don Rafael Villagran "a neighbor of this city" but purchased in Mexico City. The repetitive use of boxes, cuts, and weights implies that at least 8 mules would be used: 8 boxes of gift chocolate cut in 12 (pieces) each box with 7 arrobas—a total of 1400 pounds.	1789	1780–1789	1789	Publication in London of the *New Etiquette [sic] for Mourners*. "A Husband Losing His Wife. Must weep, or seem to weep at the funeral. Should not appear at the chocolate-house the first week; should vent a proper sigh whenever good wives, or even matrimony is mentioned. May take a mistress into keeping the third week, provided he had not one before. May appear with her in public at the end of the month; and as he probably may not chuse [sic] to marry again, he may, at the close of the second month, be allowed a couple of mistresses to solace him in his melan choly [sic]."

Chocolate-Related Events in North America	Date ←	Decade ↔	Date →	Chocolate-Related Events Elsewhere and Historical Context
George Washington writes his agent regarding the wish of his wife, Martha: "She will . . . thank you to get 20 pounds of the shells of Cocoa nuts, if they can be had of the Chocolate makers."	1789	1780–1789	1789	Thanksgiving Day celebrated in United States as national holiday.
Chocolate recommended for coughs caused by "obstructed perspiration."	1789	1780–1789	1789	Fletcher Christian leads mutiny aboard the ship *Bounty*, captained by William Bligh.
Chocolate mill owned by Boiers and M'Lean in Milton, Massachusetts, burns to the ground.	1789	1780–1789	1789	Daniel Defoe writes *Robinson Crusoe*; Crusoe salvages a chocolate pot from a wrecked ship.
Benjamin Rush recommends chocolate in the diet of smallpox patients.	1789	1780–1789	1789	Adam Smith argues that it is illogical for England to import manufactured goods from America.
Father Francisco Palou writes regarding chocolate in California: "We were allowed to make chocolate . . . a bag of hard-tack, chocolate, figs, and raisins so that he would have something to give as presents to the Indians. . . . He told him of the meager supply of food at Monterey, how there had not been even a little cake of chocolate with which to honor their presence at their breakfast."	1789	1780–1789	1789	George Washington becomes first president of United States.
William Darton uses chocolate-related dialogue to teach moral values.	1789	1780–1789	1789	Beginning of the French Revolution.
Announcement by Moses Gomez, in the *New Jersey Journal*, that his indentured servant, Duncan Chesholme, born in the Highlands of Scotland, has fled. Chesholme identified as able to read and write, and "used to chocolate making and attending a grocery store."	1788	1780–1789	1788	Fire in New Orleans kills 25 percent of the population.
Jeremiah Wadsworth claims that drinking chocolate leads to headache.	1788	1780–1789	1788	Measles epidemic in Philadelphia and New York.
Broadside (large printed handbill) announcing a parade to honor the Constitution of the United States; chocolate makers included in the "order of procession."	1788	1780–1789	1788	Hannah Glasse publishes *The Art of Cookery, Made Plain and Easy*, with recipes for chocolate.
Nicolas Pike uses chocolate-related sample questions to teach arithmetic.	1788	1780–1789	1788	Cincinnati, Ohio, founded.
Ashur Shephard sells chocolate at his medicinal store in Bennington, Vermont.	1788	1780–1789	1788	Great New Orleans fire.
Document dated December 19, signed by Governor Juan Bautista de Anza, authorizing payment of chocolate to the Comanche Indians for their assistance in the Spanish–Apache wars: "Merchandise that has been delivered to the Lieutenant Jose Maldonado, Pay Master of the Presidio of Santa Fe—New Mexico—for the reward of the Comanches: un cajon de buen chocolate [A large box of good chocolate]. Signed, Jose Maldonado, Pay Master . . . Juan Batista de Anza, Governor Colonel."	1788	1780–1789	1788	American Constitution ratified by New Hampshire; Constitution now in force.

Chocolate-Related Events in North America	Date ←	Decade ↔	Date →	Chocolate-Related Events Elsewhere and Historical Context
Document dated June 16, signed by Governor Juan Bautista de Anza authorizing chocolate and sugar be given to "explorers of the road" on an expedition to Louisiana: Exchequer bills for the extraordinary expenditures to equip the individuals who will march with Pedro Vial, for the purpose of finding a direct road to Nachitoches (*el camino derecho para Nachitoches*): 6 lb piloncillo (brown solid sugar cone); 6 lb of chocolate; 10 lb of sugar.	1788	1780–1789	1788	Adam Smith publishes *An Inquiry into the Nature and Causes of the Wealth of Nations*. Identifies cacao-nuts among the "enumerated commodities of two sorts: those produced in America, or not produced in the mother country."
Announcement of raw materials and natural products available for export by the United States of America: among items identified—bear-skins and chocolate.	1787	1780–1789	1787	Constitutional Convention meets in Philadelphia.
"Robert Clough, being at play with another child, in Mr. Makepeace's chocolate mills, north-end [of Boston], by some accident the former got entangled in the machinery of the works, while the horses were on the go, and was so shockingly mangled in his bowels, &c. that he died."	1787	1780–1789	1787	Dollar established as unit of American currency.
Robert Gray expedition to the northwest coast of North America includes chocolate among the provisions.	1787	1780–1789	1787	Establishment of the Northwest Ordinance.
Document dated April 5, and signed by the paymaster of the presidio at Santa Fe, New Mexico, and countersigned by Governor Juan Batista de Anza: "Account of what was received and of the expenditures, for the extraordinary expenditures of the Peace and War of this Province of New Mexico, according to the orders that were given to me [José Maldonado] the Lieutenant and Paymaster of the Presidio Santa Fe in the above mentioned Province in the year 1787: 3 arrobas [=75 pounds] of ordinary chocolate at 3 reales and 6 grains [?]."	1787	1780–1789	1787	Additional prison ships arrive and England establishes formal prison colony in New South Wales and first permanent settlement in Australia.
Town meeting in Boston to receive Town-Orders and By-Laws: "CHOCOLATE-MILLS. An Order, that the Law of this Commonwealth, entitled, 'An Act to prevent damage by fire from Chocolate-Mills and Machines for roasting Cocoa in the town of Boston,' be inserted and published with the By-Laws, for the future good government of the town."	1786	1780–1789	1786	First ships carrying convicts sail from England to Botany Bay (Australia).
John Bonnycastle uses chocolate-related sample questions to teach economics of barter.	1786	1780–1789	1786	Start of Shays' Rebellion in Massachusetts.

Chocolate-Related Events in North America	Date ←	Decade ↔	Date →	Chocolate-Related Events Elsewhere and Historical Context
Letter from Benjamin Franklin to Alphonsus Le Roy regarding what to take along during long sea voyages: "Whatever right you may have by agreement in the mass of stores laid in by him [the captain] for the passengers, it is good to have some particular things in your own possession, so as to be always at your own command. 1. Good water, that of the ship being often bad. 2. Good tea, 3. Coffee ground, 4. Chocolate, 5. Wine of the sort you particularly like, and cyder, 6. Raisins, 7. Almonds, 8. Sugar, 9. Capillaire, 10. Lemons, 11. Jamaica spirits, 12. Eggs greas'd, 13. Diet bread, 14. Portable soup, 15. Rusks. As to fowls, it is not worth while to have any called yours."	1786	1780–1789	1786	Mission Santa Barbara founded by Father Fermín Francisco de Lasuén.
Robison & Hale, Albany merchants, located opposite the Dutch Church, "have for sale Tea, Coffee and Chocolate."	1785	1780–1789	1785	First medical dispensary established in America (Philadelphia Dispensary) by Benjamin Rush.
James Caldwell advertises his store as "near the Low Dutch Church in Albany" and announces that "He has at his CHOCOLATE MANUFACTORY, A large quantity of Excellent CHOCOLATE, Made from the best CAYENNE COCOA Which he will sell as low as can be purchased in NEW-YORK."	1785	1780–1789	1785	University of Georgia founded.
Abigail Adams, writing to John Quincy Adams, describes drinking chocolate for breakfast while in London, living at Grosvenor Square.	1785	1780–1789	1785	Jacques Casanova includes many references to chocolate in his memoirs.
Benedict Arnold's abandoned store reopens with chocolate among the items for sale.	1785	1780–1789	1785	New York becomes temporary capital of United States.
Thomas Jefferson predicts that chocolate will become the favorite beverage in North America over coffee and tea.	1785	1780–1789	1785	Early attempt to establish dollar system as unit of American currency.
Geographical Gazetteer of Massachusetts identifies two chocolate mills at Dorchester.	1784	1780–1789	1784	China trade via Cape Horn links Asia with east coast of North America.
Robison and Hale sell chocolate at Albany, New York.	1784	1780–1789	1784	American independence from Britain recognized by Treaty of Paris.
Penuel Bowen sells drugs, medicines, chocolate, coffee, and tea at his Boston shop.	1784	1780–1789	1784	Spanish land grants authorized in *Alta California*.
Aaron Hastings sells chocolate, drugs, and medicines in Bennington, Vermont.	1784	1780–1789	1784	Russia establishes first settlements in Alaska.
John Adams writes of diplomatic negotiations with the Portuguese in order to "secure an admission to all the Azores, and to have these islands, or some of them, made a depot for the sugars, coffee, cotton, and cocoa, &c., of the Brazils [sic]" and that the ambassador agreed to sell these commodities to the American Government at 15% less than if purchased "from the English."	1783	1780–1789	1783	First daily newspaper published in United States (*Pennsylvania Evening Post*).

Chocolate-Related Events in North America	Date ←	Decade ↔	Date →	Chocolate-Related Events Elsewhere and Historical Context
Noah Webster employs a chocolate-related dialogue to teach English.	1783	1780–1789	1783	Land given to loyalists who want to move from the United States to Canada.
James Caldwell sells chocolate at grocery store in Albany, New York.	1783	1780–1789	1783	Treaty of Paris formalizes American independence and establishes geographical boundaries of the nation.
Chocolate identified among the foods served to Princeton University students.	1782	1780–1789	1782	Congress declares a day of Thanksgiving.
While recovering from an injury, Father Serra writes Father Lasuen at the Santa Clara Mission that he has no more chocolate.	1781	1780–1789	1781	Bank of North America established by Congress.
Benjamin Rush reaffirms that "tea, coffee, and even weak chocolate with biscuit" would help those to be inoculated for smallpox.	1781	1780–1789	1781	El Pueblo de Nuestra Señora la Reina de los Angeles de Porciuncula (Los Angeles, California) founded.
Thomas Dilworth uses examples of chocolate quantities to teach economics of barter.	1781	1780–1789	1781	British surrender at Yorktown, Virginia.
United States Congressional Medical Committee directive to Jonathan Potts: "The Medical Committee direct that D. Potts purveyor of the Hospitals in the middle district dispose of the money granted by Congress . . . as follows: Sixty thousand dollars for purchasing & laying in hospitals stores, such as, Coffee, Tea, Chocolate, Rice or Barley; but the Committee do not conceive wine to be absolutely necessary therefore none is to be purchased at present."	1780	1780–1789	1780	Benedict Arnold meets with John André to surrender West Point to the British; André captured with map of West Point fortifications in his boots. André later was executed; Arnold fled to British-occupied New York.
Rebecca Gomez advertises her Chocolate Manufactory at the upper end of Nassau-Street, New York, selling, "SUPERFINE warranted CHOCOLATE, wholesale and retail" along with vinegar, whale oil, soap, and playing cards.	1780	1780–1789	1780	Completion of Diderot's *Encyclopedia* (containing images of chocolate production).
Father Palou documents the standard provision list that Father Junipero Serra obtained for the California missions: 10 boxes of ham; 6 crates of ordinary chocolate at a weight of 155 pounds each; 4 barrels of wine from Castile; 960 pounds of good flour; 900 bushels of corn; 250 bushels of beans; and 16 boxes of sugar.	1780	1780–1789	1780	Hurricane destruction in Barbados and Martinique, destroying crops and cacao production.
Samuel Huntington petitions Continental Congress for more rations: "They have neither wine, Tea, sugar, Coffee, Chocolate, or spirits. We wish orders may be given for an immediate supply, as the army grow more sickly every hour."	1780	1780–1789	1780	Benedict Arnold appointed commandant at West Point, New York.
Sundries Bot [sic] in Philad[elphia] by J. Randall for the Maryland Troops: "4 lbs Bohea Tea £90; 4 lbs Chocolate £21."	1780	1780–1789	1780	Charleston, South Carolina, captured by British troops.

Chocolate-Related Events in North America	Date ←	Decade ↔	Date →	Chocolate-Related Events Elsewhere and Historical Context
Poor Will's Almanac for the year 1779 records a "Receipt for curing the Bite of a Mad-dog, &c." whereby the patient is bled using a cupping-glass, but if a cupping-glass is not available, then a chocolate cup may be used.	1779	1770–1779	1779	Establishment of Fort Nashborough by James Robertson (later Nashville, Tennessee)
Continental Congress includes chocolate among rations.	1779	1770–1779	1779	John Adams, writing from Spain, mentions chocolate in his diaries.
Massachusetts-Bay Colony reaffirms prohibition on exportation of cocoa and chocolate.	1779	1770–1779	1779	Death of Captain James Cook in Hawaii.
Continental Congress establishes that colonels, lieutenant colonels, and chaplains of brigades "shall be entitled to draw only six gallons of rum, either four pounds of coffee or chocolate, or one pound of tea and twelve pounds of sugar monthly."	1779	1770–1779	1779	Alphonse François, Marquis de Sade, writes on May 16: "The sponge cake is not at all what I asked for: 1) I wanted it iced everywhere, both on top and underneath, with the same icing used on the little cookies; 2) I wanted it to be chocolate inside, of which it contains not the slightest hint; they have colored it with some sort of dark herb, but there is not what one could call the slightest suspicion of chocolate. The next time you send me a package, please have it made for me, and try to have some trustworthy person there to see for themselves that some chocolate is put inside. The cookies must smell of chocolate, as if one were biting into a chocolate bar."
Continental Congress establishes prices for basic commodities: chocolate not to exceed 20 shillings per box; 22 shillings per dozen, and 24 shillings per pound.	1779	1770–1779	1779	John Hannon, partner with James Baker, reported lost at sea.
Pennsylvania General Assembly resolves to supply officers and soldiers with "West-India rum at 5 shillings per gallon; Muscovado sugar at 3/9 per pound; coffee at 3/9 [i.e, three shillings; nine pence] ditto; tea at 12 shillings ditto; chocolate at 3/9 ditto."	1779	1770–1779	1779	English law passed that chocolate smugglers found guilty a second time will be banished to the East Indies or the African Coast colonies for life.
Massachusetts-Bay Colony establishes price for chocolate: 20 shillings per pound per box.	1779	1770–1779	1779	John Paul Jones victory over the *HMS Serapis*.
William Buchan recommends chocolate in the treatment of "fainting fits."	1778	1770–1779	1778	Voltaire writes that imported chocolate is among the items causing poverty in France.
Proceedings of the General Court Martial of Major General St. Clair, by order of his Excellency General Washington, Commander in Chief of the Army of the United States of America: report identifies 12 boxes of chocolate had been abandoned at Ticonderoga.	1778	1770–1779	1778	James Cook explores Hawaii.

Chocolate-Related Events in North America	Date ←	Decade ↔	Date →	Chocolate-Related Events Elsewhere and Historical Context
Advertisement by Moses Gomez, Junior, New York merchant: "THREE DOLLARS REWARD. RUNAWY [sic] from the subscriber [Gomez], on Thursday, the first of May last, an indentured servant BOY, sixteen years old, born in the Highlands of Scotland, named DUNCAN CHESHOLME, but has changed it; is very short and thick set, carries his head very erect, fair and florid countenance, but now may be tanned; and when laughs his cheeks and chin have dimples; he speaks very good English, and calls himself an Englishman, writes and cyphers, used to chocolate making and attending a grocery store, four years of his servitude unexpired; he is a cunning, artful fellow, and is capable of deceiving any one, having defrauded his master considerably. He was lately at Mr. Scenck's house, Pleasant Valley, and is either lurking in Jersey or gone to Philadelphia. . . . It is reported that he is bursted by being thrown from a horse. All persons are forbid harbouring this rascal, as they shall answer it at their peril."	1778	1770–1779	1778	Daniel Boone attacks British outposts in Kentucky.
	1778	1770–1779	1778	A Monsieur Doret is given an award for developing a machine to mix ground chocolate.
Chocolate listed among the supplies taken on board by John Adams as he debarked for France.	1778	1770–1779	1778	George Rogers Clark attacks British outposts in Illinois.
British loot Philadelphia, leave behind on dock piles of stolen goods including iron pots and a steel chocolate mill.	1778	1770–1779	1778	British evacuate Philadelphia.
Abigail Adams complains about the high cost of chocolate.	1777	1770–1779	1777	Washington's troops spend dreadful winter at Valley Forge, Pennsylvania.
State of Rhode Island, War Council Declaration: "It is absolutely necessary that the continental forces raised within this state should be immediately supplied with ten thousand weight of sugar, one thousand gallons of rum, one thousand weight of coffee, and five hundred weight of chocolate."	1777	1770–1779	1777	Settlement of San Jose founded in California.
Massachusetts-Bay Colony establishes price for cocoa: best cocoa at 6 £ 10 shillings a cwt (hundred pounds). American manufactured chocolate at 1 shilling 8 cents per pound.	1777	1770–1779	1777	*Articles of Confederation* submitted to states for ratification.
Inventory of chocolate supplies on hand at Fort Ticonderoga—16 boxes; at Fort George—10 boxes; at Albany—6 boxes.	1777	1770–1779	1777	Battle of Saratoga.
Moses Greenleaf breakfasts on chocolate while serving at Fort Ticonderoga.	1777	1770–1779	1777	General Arthur St. Clair abandons Fort Ticonderoga to British.
Continental Congress imposes price controls during wartime: "The following Articles be hereafter Sold at the Prices following, or not exceeding such Prices: cocoa at seven £ per Hundred Weight; Chocolate at two Shillings per Pound by Wholesale, and two Shillings and two Pence per Pound by Retail."	1777	1770–1779	1777	Congress authorizes flag consisting of 13 stripes (red and white) with 13 stars.

Chocolate-Related Events in North America	Date ←	Decade ↔	Date →	Chocolate-Related Events Elsewhere and Historical Context
Massachusetts-Bay Colony prohibits the export of cocoa and chocolate.	1777	1770–1779	1777	Marquis de Lafayette of France assists in training Continental Army.
Massachusetts-Bay Colony freezes prices of "Peas, Beans, Pork, Wheat, Rye, Hides, Loaf Sugar, Chocolate, Butter, Potatoes, Oats, Tallow, Flour, and Bar Iron."	1777	1770–1779	1777	Philadelphia occupied by British.
Francis Daymon, Philadelphia merchant, has "a very good assortment of MEDICINES, and the very best chocolate, already sweetened fit for the gentlemen of the army."	1777	1770–1779	1777	Battle of Brandywine.
Announcement: "A part of General Washington's army occupying the houses and stores belonging to William Richards, at Lamberton, near Trenton, for barracks, hospitals, and slaughter houses, on Friday the third instant, the dwelling-house was burnt down [supposed by accident] with a large quantity of mustard seed, some household goods, and a chocolate, mill, &c."	1777	1770–1779	1777	Vermont declares independence from New York.
Moses M. Hayes, Newport, Rhode Island, merchant, advertises for sale: "Best Turkey Rhubarb, Jesuit's Bark, Nutmegs, Cinnamon, Wine Glasses, &c. in Crates, Vinegar, Linens, Chocolate by the Box, &c."	1776	1770–1779	1776	Thoms Payne publishes *Common Sense*.
Anthony Wayne at Fort Ticonderoga identifies chocolate as a medicinal food.	1776	1770–1779	1776	Phi Beta Kappa Society founded at College of William & Mary, Williamsburg, Virginia.
Letter from Charles Cushing to his brother written July 8: "We have now been at Crown Point eight days, and the sick considerably longer; and since their arrival, we have buried great numbers. . . . But few have died [this statement does not follow from the previous sentence], except with the small pox. Some regiments which did not inoculate have lost many. . . . Here are likewise sutlers who have spirits of all kinds—wines, sugar, chocolate, &c. to sell, though at a very dear rate—sugar three shillings, lawful money, per pound, &c."	1776	1770–1779	1776	Nathan Hale executed. ("Give me liberty, or give me death.")
Committee of Newark, New Jersey acting to prevent: "Undue advantages being taken by reason of the scarcity of sundry articles . . . Have resolved to regulate the prices of West-India product[s] to be sold in this Township, as follows, to wit: Chocolate, 2 shillings per pound. . . . On proof being made to this Committee of any person having, after the publication hereof, contravened, or in anywise [sic] acted in defiance of the said recommendation, the delinquent shall be exposed by name to publick [sic] view, and as an enemy to his country."	1776	1770–1779	1776	American Declaration of Independence signed.

Chocolate-Related Events in North America	Date ←	Decade ↔	Date →	Chocolate-Related Events Elsewhere and Historical Context
Philadelphia Committee acting in the public good to check mercenary practices: "Declare that they will expose such persons, by name, to publick [sic] view, as sordid vultures, who arte [sic] preying on the vitals of their country in a time of general distress." Prices established include "cocoa, 5 £ per hundred [weight]; chocolate, 16 pence per pound."	1776	1770–1779	1776	Thomas Payne publishes *Common Sense*.
Announcement in New Haven, Connecticut, dated July 10: "To morrow [sic], will be ready for sale, The Resolves of the Congress, declaring the United Colonies, FREE, and INDEPENDENT STATES." Advertisement also lists "PEPPER by the 100Wt. Or smaller Quantity, a few KEGS of RAISINS, INDIGO, COFFEE, ALSPICE [sic] and CHOCOLATE, to be sold by Elijah and Archibald Austin."	1776	1770–1779	1776	Joseph Fry predicts a vast export market for cacao if trade duties be favorably aligned: "As it is certain that good and beautiful Chocolate cannot be made in warm Climates, and that abroad that English Chocolate is prefferr'd to most if not any other sort, doubtless very large Quantities would be exported to the East and West Indies and many other Places, if a suitable Drawback were to be allowed in Exportation."
George Fechem, Boston merchant, advertises to sell: "A good Chocolate Mill, which will go with a Horse and grind 120 wt. Of Chocolate in a Day. Said Mill consists of three good Kettles, with twelve Pestles, in a Kettle well Leaded, nine Dozen Pans, one good Nut cracker, a good Muslin for to clean the shells from nuts."	1776	1770–1779	1776	Settlement of Yerba Buena (San Francisco, California) established by Spanish Franciscans. Francisco Palou establishes mission San Francisco de Asis
Henry Alexander explores central Canada; when rations depleted he survives on chocolate, which provides him energy to keep walking.	1776	1770–1779	1776	British evacuate Boston.
Beginning of five-year supply system between Mexico City and California missions that include chocolate, lentils, rice, and barrels of Castilian wine.	1776	1770–1779	1776	Complaint by Joseph Fry, chocolate manufacturer of Bristol, regarding loss of trade due to chocolate being smuggled into England and Wales from Ireland.
Pedro Font writes in his diary that chocolate was offered as an engagement present to a couple in San Diego.	1776	1770–1779	1776	Basket of supplies sent by "friends of the suffering America" to English prisoner, Ethan Allen: "We this day sent a hamper of wine, sugar, fruit, chocolate, &c. on board for his immediate use."
"Inventory by Elisha Avery, Deputy Commissary: Provisions on hand in the Magazine at Ticonderoga, August 23rd, 1776: 565 barrels Pork; 30 barrels Sugar; 6 tierces Peas; 46 barrels Molasses; 60 barrels Rum; 10 hogsheads Rum; 50 barrels Spirits, of different kinds; 509 barrels Flour; 300 pounds Coffee; 800 pounds Chocolate; 300 pounds Loaf Sugar; 7 boxes mould Candles; 50 bushels Salt; 5 tierces Indian Meal; 4 barrels soft Soap; 1 barrel Pearlash, and a potash Kettle, cracked."	1776	1770–1779	1776	Doret invents a hydraulic machine to grind cacao seeds into a paste.
Rations for troops raised by Colony of Connecticut: chocolate, 6 pounds per company per week.	1775	1770–1779	1775	Ethan Allen captured by British at battle of Montreal.

Chocolate-Related Events in North America	Date ←	Decade ↔	Date →	Chocolate-Related Events Elsewhere and Historical Context
Juan Bautista de Anza mentions chocolate in his diary.	1775	1770–1779	1775	Battle of Breed's Hill (Bunker Hill) in Boston.
Letter from a gentleman in Philadelphia, dated May 21: "[British] Guards are appointed to examine all trunks, boxes, beds, and every thing else to be carried out; these have produced such extremities, as to take from the poor people a single loaf of bread, and half pound of chocolate; so that no one is allowed to carry out a mouthful of provisions; but all is submitted to quietly. The anxiety indeed is so great to get out of Town, that even were we obliged to go naked, it would not hinder us."	1775	1770–1779	1775	Smallpox epidemic in New England.
Letter from General Schuyler to Governor Trumbull: "It is certain, however, that some [of the recruits] have feigned sickness; for Dr. Stringer informs me, that on his way up here, about the 6th of September, he met many men that looked very well; and, upon inquiry, some acknowledged they had procured their discharges by swallowing tobacco juice, to make them sick. Others had scorched their tongues with hot chocolate, to induce a believe [sic] that they had a fever, &c."	1775	1770–1779	1775	Daniel Boone blazes the Wilderness Road (Shenandoah Valley to the Ohio River).
Rosanna Maxwell sells chocolate in her Boston shop adjacent to the British coffeehouse.	1775	1770–1779	1775	Ethan Allen and Benedict Arnold capture Fort Ticonderoga.
Chocolate identified as a good breakfast food for elderly men.	1775	1770–1779	1775	Armed skirmishes between British and Revolutionaries at Lexington and Concord Bridge.
Committee for County of Baltimore, Maryland: "Captain George Southward, Schooner Hope, arriving from Salem, appeared, and reported, on oath, his cargo, consisting of Ruin [rum meant?], Coffee, and chocolate."	1775	1770–1779	1775	Paul Revere's ride.
Elizabeth Perkins advertises the sale of chocolate at her shop, "two Doors below the British Coffee-House, North-Side of Kingstreet [sic], Boston."	1774	1770–1779	1774	Year of commonly cited story that Pope Clement XIV was murdered by poisoned chocolate prepared by Jesuit priests.
Seth Lee, Farmington merchant, advertises drugs and medicines, including "Turlington's Balsam of Life, Eaton's Styptic, James's universal Fever Powder, Spirits of Scurvy Grass, also Mace, Cinnamon, Cloves, Nutmegs, Ginger, Pepper, Allspice, Raisins, Figgs [sic], Chocolate."	1774	1770–1779	1774	Quartering Act reestablished.
Letter from Father Palou to Father Serra (written at Monterey): "We are in great need. Last year the Indians had pine nuts—but not this year—only in the Mission of San Gabriel they have a crop of corn that survived—a very small one. Here [in Monterey] we only have a small amount of flour and beans; chocolate we have none . . . so our breakfast is only milk."	1774	1770–1779	1774	Great Britain passes Boston Port Act: port of Boston ordered closed.

Chocolate-Related Events in North America	Date ←	Decade ↔	Date →	Chocolate-Related Events Elsewhere and Historical Context
General Assembly of Hartford, Connecticut: "Resolved . . . that the allowance of provisions for said troops be as follows, viz: three-quarters of a pound of pork, or one pound of beef, and also one pound of bread or flour, with three pints of beans to each man per day, the beef to be fresh two days in the week; and also half a pint of rice or a pint of Indian-Meal, and also six ounces of Butter; also three pints of peas or beans to each man per week; also one gill of rum to each man upon fatigue per day, and not at any other time; milk, molasses, candles, soap, vinegar, coffee, chocolate, sugar, tobacco, onions in their season, and vegetables be provided for said troops, at the discretion of the General and Field Officers."	1774	1770–1779	1774	Juan Perez Hernandez explores coast of the Pacific Northwest.
Juan Josef Perez Hernandez explores Pacific Northwest coast; provisions include chocolate.	1774	1770–1779	1774	Passage of the Intolerable Acts.
Rules for the Regulating of Salem Hospital (Massachusetts): "Bohea Tea, Coffee, Chocolate, Cocoa Nut Shells, white and brown Bread, brown Bisket [sic], Milk Bisket [sic], Gruel, Milk Porridge, Rice, Indian and Rye Hasty-Pudding, Milk, Molasses, best brown Sugar, Scotch Barley, Flour, Bread, or Rice Pudding, Turnips, Potatoes, Veal, Lamb, Mutton, Beef, dry Peas and Beans, Vinegar, Salt, Mustard, Pepper, one Pound of Raisins daily for each Room, Spruce Beer."	1773	1770–1779	1773	Boston Tea Party.
Hephzibah Goldsmith, executrix to John Goldsmith the chocolate-grinder of Boston, deceased: "She carries on the Business of Chocolate Grinding, where the former Customers and others may be supply'd at the cheapest Rate."	1773	1770–1779	1773	Tea Act passed by Parliament.
Joseph Mann, Boston manufacturer, advertises: "Choice Chocolate, warranted pure . . . Cocoa taken in to Grind, Where any Gentlemen [sic] may have it done with Fidelity and Dispatch, and their Favours gratefully acknowledged."	1773	1770–1779	1773	British Parliament grants the British East India Company a monopoly on the North American tea trade.
Letter from Father Junipero Serra to the Viceroy of California, Antonio Bucarelli: A Sergeant Periquez had written Father Serra about the abuses of his commandant, Pedro Fages. Chocolate had been ordered by the doctor/surgeon, Pedro Prat, to be given to sick soldiers, but the chocolate was confiscated by Fages who took it for his personal use. Sergeant Periques sent his letter of complaint to Father Serra via Captain Agustin Callis, captain of the presidio at Monterey. Father Serra recopied it and forwarded it to Bucarelli (result: Pedro Fages removed form his position).	1773	1770–1779	1773	Jesuit influence curbed by Pope Clement XIV.

Chocolate-Related Events in North America	Date ←	Decade ↔	Date →	Chocolate-Related Events Elsewhere and Historical Context
Receipt from Edward Preston: "Received from Mr. James Baker amount of one pound thirteen shillings and four pence in full, for my part of the rent of the chocolate mill, until the 11th of December."	1773	1770–1779	1773	Antigua, Guatemala, destroyed by earthquake; capital of Guatemala moved to what is now Guatemala City.
Joseph Barrell, Boston merchant, advertises: "CHOCOLATE, Neither Hannon's much approv'd, Palmer's superior, nor made by Deacon Davis, Yet warranted Pure, and at least, equal to either."	1773	1770–1779	1773	First asylum for the mentally ill in America established at Williamsburg, Virginia.
Annual importation of cocoa beans by English colonists exceeds 320 tons.	1773	1770–1779	1773	Daniel Boone leads settlers into Kentucky.
Father Palou organizes mule train to transport supplies along the California mission chain; he leaves a box containing 112.5 pounds of ordinary chocolate at Mission San Luis Obispo.	1772	1770–1779	1772	Slavery outlawed in England.
Letter from Father Junipero Serra to Captain Pedro Fages, addressing need to use the king's mules to help in transport of supplies to the missions before the beginning the winter rainy season: "Not all the mules will be used to carry flour and beans, because it is necessary that some mules carry chocolate for the Fathers."	1772	1770–1779	1772	Measles epidemic throughout North America.
Suicide report: "Last Monday night one Dunn, chocolate grinder, journeyman to Mr. Wallace, at the north-end [of Boston] being delirious, threw himself out of a garret window into the street, where he was found the next morning."	1772	1770–1779	1772	Correspondence of the governor and general captain of the Province of Yucatan: introduction of too few new silver 1/2 real coins posed a difficulty in daily commerce; since change could not be given by merchants, they resorted to using cacao beans and pegged to 90 "grains" of cocoa to the 1/2 silver real.
James Baker establishes his own chocolate mill.	1772	1770–1779	1772	Chocolate appears as a word in German texts used to teach grammar.
Fire burns house of Mr. Daniel Jacobs in Danvers, Massachusetts. Loss included: "Furniture, Provisions, &c. 3000 pounds of Cocoa, and several 100 pounds of Chocolate, which were nearly or quite all destroyed. The Loss, at a Moderate Computation, is said to amount to Fire Hundred Pounds, lawful Money."	1772	1770–1779	1772	Samuel Adams and Joseph Warren form the first Committee of Correspondence.
Philadelphia slave document: individual identified as "thoroughly acquainted with the Chocolate Manufactory."	1772	1770–1779	1772	Mission San Luis Obispo de Tolosa founded in San Luis Obispo, California.
J. Coke advertises: "CHOCOLATE warranted good . . . In large or small Quantities, at the lowest Prices for Cash, or exchang'd for good Cocoa at the Market Price. Cocoa taken in to manufacture at the above Place, and warranted to be done in the best and neatest Manner, and with the greatest Dispatch, at Ten Shillings Lawful Money per C. Weight."	1771	1770–1779	1771	Epidemic of plague ravages Moscow.

Chocolate-Related Events in North America	Date ←	Decade ↔	Date →	Chocolate-Related Events Elsewhere and Historical Context
Joseph Mann, Boston merchant, announces:"Cocoa taken in to Grind for Gentlemen and done with Fidelity and Dispatch." Also sells cocoa shells and certifies his chocolate pure from any adulteration.	1771	1770–1779	1771	Fathers Pedro Cambon and Angel Somera found Mission San Gabriel Archangel in what is now San Gabriel, California
J. Russell, Boston auctioneer, announces sale of calico, Irish linen, striped Holland (cloth), India damask, and a "very good Mill for grinding of Chocolate."	1771	1770–1779	1771	Report from London: "It is remarkable, that Lord Mansfield rises every morning by three or four o'clock, when he drinks a dish of chocolate, and retires to his study (where there is a fire kept night and day) till about seven or eight o'clock, then he goes to bed again, and lays for an hour or two. He has accustomed himself to this regimen ever since he had the weight of Government conferred on him."
		1770–1779	1771	Old Bailey trial record: "The woman said the chocolate was smuggled chocolate. It has a stamp and our own private mark on it."
Thomas Dimsdale recommends drinking chocolate before undergoing smallpox inoculation.	1771	1770–1779	1771	Great fire in New Orleans.
	1771	1770–1779	1771	Smith's Chocolate House, London, located at the end of the heath near Blackheath Hill.
Joseph Palmer, merchant of Boston, "designing soon to embark for ENGLAND . . . He has to Sell at his Store on Minot's T . . . Chocolate warranted of the best quality."	1770	1770–1779	1770	East coast of New Holland (Australia) claimed for England by James Cook.
Another letter from Father Junipero Serra requesting chocolate specifically for the missions at San Diego, Buenaventura, and Monterey.	1770	1770–1779	1770	Mission at Monterey, California, established.
Letter from Father Junipero Serra to the Viceroy of California, written at the Presidio of San Blas: "The ration for the people is half a quarter of corn, half a pound of beef, one ounce of sugar, two ounces of *Pinole*. Here and there I give a tablet of chocolate to those that labor well and I also give chocolate to the sick whenever they need it."	1770	1770–1779	1770	Marie Antoinette arrives at the French court.
Archibald Cunningham, Boston merchant, advertises: "Choice Capers and Anchovys [*sic*] . . . Chocolate, Loaf and brown Sugar, Mustard per pound or in Bottles. A Quantity of NUTMEGS to be sold cheap at said Shop."	1770	1770–1779	1770	Gaspar de Portola and Father Junipero Serra establish settlement at Monterey, California.
Letter from Father Junipero Serra to Father Guardian Fray Juan Andreas, describes hardships of the overland journey from San Diego to Monterey. Arriving, he writes: "What we much need for the masses, warm undergarments, because it is very cold here . . . and for the new missions [i.e., at Monterey and San Juan Buena Ventura] it is important that they have always chocolate, that thanks God, so far we always had."	1770	1770–1779	1770	British Parliament repeals the Townshend Acts; tax remained on tea.

Chocolate-Related Events in North America	Date ←	Decade ↔	Date →	Chocolate-Related Events Elsewhere and Historical Context
"A Record of the Merchandise and Goods which are Embarked on the Ship *San Jose,* also called the *Discoverer,* with Destination to the West Coast of California Where It Is To Be Delivered to the Surgeon Pedro Prat: 10 barrels of meat, 33 sacks of flour, 6 sacks of rice, 10 sacks of beans, 29 bushels of corn, 2 sacks of barley seed, 10 bushels of brown sugar, 1 box of coarse chocolate, 18 pounds of melon seeds, 10 pounds of watermelon seeds, 10 pounds of pumpkin seeds, crackers [amount unspecified], 1 box of fine chocolate, 1 barrel of California vinegar, and lemon juice [amount unspecified]."	1770	1770–1779	1770	Boston Massacre.
Publication in the *Pennsylvania Chronicle and Universal Advertiser* comparing the economic advantage of drinking chocolate over tea: "A Brekfast [*sic*] of Tea for nine persons 19 1/2 pence; Chocolate for nine Persons 16 1/2 pence: Balance in favor of Chocolate, THREE-PENCE."	1769	1760–1769	1769	Famine in Bengal, India, kills a reported 10 million people, reportedly the worst natural disaster in human lives lost.
Father Junipero Serra establishes first mission in Alta California (upper California) at San Diego; describes his voyage from Loreto (Baja California) to San Diego: "The Governor ordered to be given to the Father, some cattle, Biscuits, flour, chocolate, raisins, and figs."	1769	1760–1769	1769	Captain James Cook arrives in Tahiti.
Letter from Father Palou to Father Guardian Andre; describes the Franciscan takeover of the Jesuit mission at Loreto Presidio, where the priests had a "stash" of chocolate at the mission.	1769	1760–1769	1769	Chocolate sent by Governor Gálvez among the supplies at Velicata Mission, Baja California.
T. Philomath writes that drinking chocolate during Spring should be discouraged.	1769	1760–1769	1769	Nicolas-Joseph Cugnot demonstrates first steam-powered automobile.
Joseph Mann, Boston merchant, announces that he "Makes and Sells Chocolate, which will be warranted free of any Adulteration, likewise New-England Mustard, manufactured by said Mann."	1769	1760–1769	1769	Dartmouth College founded in Hanover, New Hampshire.
George Washington orders rum, sugar, sweet oranges, pineapples, and one dozen cocoa nuts shipped to Mount Vernon.	1769	1760–1769	1769	Daniel Boone begins exploration of Kentucky.
Announcement: "Theft of Two Firkins Butter, about 50 or 60 pounds of Chocolate mark'd 'W. Call, and S. Snow,' half a Barrel of Coffee, nine or ten Pair Lynn Shoes, some Cocoa and Sugar." Reward offered by Bossenger and William Foster, Boston.	1769	1760–1769	1769	Charles III of Spain sends additional missionaries to New Spain.
Contributions to the Pennsylvania Hospital for the year ending 5th Month, 1769, include a horse, butchers meat, and chocolate.	1769	1760–1769	1769	James Watt patents first practical steam engine.

Chocolate-Related Events in North America	Date ←	Decade ↔	Date →	Chocolate-Related Events Elsewhere and Historical Context
Stores as necessary for a whaling voyage: "PORK, Beef, Bread, Corn and Meal, as may be thought sufficient, 1 Barrel of Beans, 1 Barrel of Salt, 2 Barrels of Flour, 1 Barrel of Rum, 2 Barrels of Molasses, 12 Bushels of Potatoes, 14 pounds either Coffee or Chocolate, 1 pound Pepper, 56 pounds Sugar, 50 pounds either Butter or Cheese, 75 pounds Fat, 112 pounds Rice, Dry Fish if they go to the Western Islands."	1769	1760–1769	1769	Moncada and Portolá expeditions arrive in San Miguel (modern San Diego). Settlement of San Diego founded by Spanish Franciscans.
Widow Mary Crathorn of Philadelphia informs customers she sells chocolate; also marks the paper around each pound of chocolate with a stamp.	1768	1760–1769	1768	Captain James Cook, in the ship *Endeavour,* sails from Plymouth to explore south Pacific and collect botanical specimens.
Ledger data show that Prince Updike (slave or free black man?) reduces 5000 pounds of cocoa to 4000 pounds of ground chocolate for Aaron Lopez, Newport, Rhode Island.	1768	1760–1769	1768	Boston citizens refuse to quarter British troops.
John Goldsmith, Boston merchant, advertises chocolate as "warranted good and sold at cheapest rates—cash given for cocoa."	1768	1760–1769	1768	British troops land and occupy Boston.
Charles Miller, Boston merchant, advertises: "CHOCOLATE, Warranted equal to any, and superior to most that's made TO BE SOLD."	1768	1760–1769	1768	Samuel Adams incites opposition to the Townshend Act taxes.
Advertisement: "Theft of silver chocolate ladle, weighing 3 1/2 ounces, marked 'M.S.' on the bottom: One DOLLAR reward. No Questions asked."	1768	1760–1769	1768	First advertisements for mustard being sold in American colonies.
Hutchin's Improved Almanack [sic] Chocolate among the foods recommended when preparing for smallpox inoculation.	1768	1760–1769	1768	Island of Corsica ceded by Genoa to France
Last words of Arthur, identified as a Negro man, executed at Worcester, Massachusetts: "At Weston we [with his accomplice Frasier] stole some butter; we broke into a house belonging to one Mr. Fisk, from whom we took a small sum of money, some chocolate; at Watertown we stole a brass kettle from one Mrs. White of that place . . . I humbly ask forgiveness of all whom I have injured, and desire that they would pray that I may receive forgiveness of God, whom I have most of all offended."	1768	1760–1769	1768	Explorers under Fernando de Rivera y Mondada and Gaspar de Portolá depart Loreto, Baja California, for Alta California.
	1767	1760–1769	1767	Following appeals by local Jewish leaders, the Bordeaux parlement dissolves the Bayonne chocolate makers' guild, handing the Jews a victory in the Bayonne chocolate war.
Boston announcement: "A Machine, the newest that has been made in Boston, to grind Chocolate, and will be warranted. Likewise a Cleaner of Cocoa for Chocolate, fit to grind 500 wt. In Ten Hours. This Mill is warranted to grind 14 wt. In Two Hours. Any Gentleman inclining to purchase the above Machine, may have it a Pennyworth for Cash. The same is to be Sold by Henry Snow in Temple-Street, New-Boston. Choice Chocolate made and Sold by said Snow."	1767	1760–1769	1767	Advertisement from Morlaix, Brittany, lists chocolate at 32 sous per pound, with home delivery.

Chocolate-Related Events in North America	Date ←	Decade ↔	Date →	Chocolate-Related Events Elsewhere and Historical Context
J. Mackenzie recommends chocolate to treat "swooning" and fainting-fits.	1767	1760–1769	1767	New York Assembly suspended for refusing to support quartering of British troops.
Benjamin and Edward Thurber sell chocolate in North Providence, Rhode Island.	1766	1760–1769	1766	Charles Mason and Jeremiah Dixon survey boundary between Maryland and Pennsylvania (Mason–Dixon line).
Thief, Timothy Williams, captured in Boston with "sundry Articles" including "6 pounds Tea; two Cakes Chocolate; one Cake Soap. Any Person or Persons claiming all or any of the aforesaid Articles may apply for them to the said Timothy Paine, Esq.; of Worcester."	1766	1760–1769	1766	Law passed in Stockholm, Sweden, that identifies unnecessary merchandizes [sic] which will be prohibited January 1, 1767: Among the items identified: "Coffee, chocolate, arrack, punch, strong waters, and most sorts of wine; Confectionary; use of tobacco except the person be 21 years of age."
Benjamin Franklin replies to a British criticism of American food saying: "We have oatmeal in plenty. . . . There is every where plenty of milk, butter, and cheese; while the islands yield us [America] plenty of coffee and chocolate. Let the gentleman do us the honour of a visit in America, and I will engage to breakfast him every day in the month with a fresh variety, without offering him either tea or Indian corn."	1766	1760–1769	1766	Townshend Revenue Act passed by British Parliament.
Aaron Lopez employs local black men (free or slaves?) in Newport, Rhode Island, to grind chocolate. Company ledgers note that one black man, named Prince Updike, reduces 2000 pounds of chocolate from 2500 pounds of cocoa during this year and next (1767).	1766	1760–1769	1766	Stamp Act repealed.
John Theobald recommends chocolate as a nourishing, liquid food.	1765	1760–1769	1765	Patrick Henry attacks the Stamp Act in the Virginia House of Burgesses.
Boston advertisement: "To be Sold at Public VENDUE, by Moses Deshon, At the Newest Auction-Room, opposite the West End of Faneuil-Hall, Dock-Square; Next Tuesday, Wednesday, and Friday Evenings; An Assortment of English Goods, a Box of good Indigo, Household Furniture, Men's Great Coats, two Chocolate Stones and Rollers, hand-some [sic] Pictures under Glass, &c."	1765	1760–1769	1765	First school of medicine established in American colonies in Philadelphia.
James Read (Reid) ledger: "Savannah in Georgia September: [sold to John Smith] 1/2 gallon bowls, 2 padlocks, 2 [?] chocolate, 1 oz nutmeg, 1 nutmeg grater, 1 oz thread, 1 pair scruff shoes, 300 [?] bar iron, 3 yards ribbon, 25 [?] small shot, 3 penknives, 16 spikes, 1 [?] barley sugar."	1765	1760–1769	1765	Quartering Act passed by British Parliament.
John Harmon and James Baker begin to manufacture chocolate at the Milton Lower Mills, near Dorchester, Massachusetts.	1765	1760–1769	1765	Stamp Act passed by British Parliament.

Chocolate-Related Events in North America	Date ←	Decade ↔	Date →	Chocolate-Related Events Elsewhere and Historical Context
Ralph Inman, chocolate merchant, relocates to the house of Mr. Samuel Woodburn, in Waltham, due to the smallpox outbreak in Boston, where his customers may be supplied.	1764	1760–1769	1764	Sugar Act passed by British Parliament to help pay costs of the French and Indian War.
Rebecca Walker, chocolate merchant, relocates to Roxbury due to smallpox outbreak in Boston.	1764	1760–1769	1764	Smallpox epidemic in Boston and New England.
Nathaniel Barrel sells chocolate at the Sign of the State House, Portsmouth, New Hampshire.	1764	1760–1769	1764	Foundation of St. Louis (Missouri).
Simson's store, Stone-Street, New York, sells "wrapping paper fit for shopkeepers, tobacconists, chocolate-makers, tallow-chandlers, hatters, &c."	1764	1760–1769	1764	Currency Act passed by British Parliament forbidding American colonies from issuing their own currency.
A document from Sonora provides a list of foods, with chocolate among the items, being sent to different presidios, among them Tubac. Document mentions Apaches taking cattle and chocolate is mentioned on the list of items supplied (but in small amounts). Items supplied, including the chocolate, were sold to the soldiers.	1764	1760–1769	1764	Chief Pontiac surrenders to the British.
	1763	1760–1769	1763	Report of Jewish pirates attacking Dutch sloop out of Ocumare (Cumana), Venezuela destined for Curaçao.
	1763	1760–1769	1763	*Encyclopedia* of Diderot and d'Alembert contains an illustration of a worker grinding chocolate.
Announcement: "CHOCOLATE-MILL in Charlestown [Massachusetts], late in the Occupation of John Osborne, is now improved by Samuel Dowse. All Persons who have NUTS to grind, may depend on their being well ground, with Fidelity and Dispatch."	1763	1760–1769	1763	France cedes Canada to Great Britain; Treaty of Paris.
William Rogers sells chocolate on North side of the Parade grounds, Newport, Rhode Island.	1763	1760–1769	1763	King George III bans American settlements west of the Appalachians.
Richard Palmes sells medicinal chocolate in his Boston shop.	1763	1760–1769	1763	Report of cacao grown in the Philippines.
	1762	1760–1769	1762	Inventory of pieces produced at the Royal Porcelain Manufactory at Sèvres France includes *gobelets à chocolat,* possibly made for Madame de Pompadour.
	1762	1760–1769	1762	Father Junipero Serra reported from Mexico that priests and Indians were suffering from local epidemics and lack of provisions; describes hardships and how they were sustained by chocolate and *atole.*
Knight Dexter, Providence, Rhode Island, merchant, sells textiles, bibles, hammers, chocolate, coffee, and tea.	1762	1760–1769	1762	Report of the Conquest of Martiniaco [sic] and British possession of the Neutral Islands: "Dominico: The French cultivated on it great Quantities of Coffee, Cocoa, and Cotton . . . St. Vincent: It is of considerable Extent, and its present Produce is Coffee, Cocoa and Tobacco."
James Green, Providence, Rhode Island, merchant, sells "Goods made in Brass, Steel, Iron, Pewter, &C Also . . . Coffee, chocolate, tea."	1762	1760–1769	1762	British establish settlement at Maugerville, New Brunswick.

Chocolate-Related Events in North America	Date ←	Decade ↔	Date →	Chocolate-Related Events Elsewhere and Historical Context
Job Bennet, Junior, sells chocolate and nails near Point Bridge, Newport, Rhode Island.	1762	1760–1769	1762	France transfers control of Louisiana to Spain.
Peter Cooke, Newport, Rhode Island, merchant, sells wine, coffee, tea, sugar, and chocolate.	1762	1760–1769	1762	Edward Gibbons reports seeing men at the Cocoa-Tree Coffee Shop in London dining on cold meat and sandwiches.
	1761	1760–1769	1761	Announcement from London: "Perhaps very few princes in the world have ever so many virtues, or much wisdom as his present majesty king George the third . . . His majesty riseth at five o'clock every morning, lights his own candle from the lap which burnt all night in his room, dresseth himself, then calleth a servant to bring him bread, butter and chocolate."
	1761	1760–1769	1761	Death of King George II: "That his Majesty was waited on as usual without any apparent [sic] signs of indisposition, drank his chocolate, and enquired about the wind, as if anxious for the arrival of mails, opened the window of his room and perceiving a fine day, said he would walk in the garden; That his chocolate maker being the last person with his Majesty (who appropriated the early hours of the morning to retirement) observed him give a sigh on quitting his presence, and soon after hearing a noise like the falling of a billet of wood from the fire, he returned and found the King dropt from his seat, as if attempting to ring the bell."
	1761	1760–1769	1761	Non-Jewish chocolatiers of Bayonne organize a guild to prohibit local Jews from selling chocolate in the town.
Letter from Isaac Delyon of Savannah to Bernard Gratz of Philadelphia provides an inventory of goods and relates how the goods should be stored aboard ship: "25 wt. Chokolet [sic]."	1760	1760–1769	1760	First bifocal lenses invented by Benjamin Franklin.
Nathaniel Abraham, Boston merchant, sells chocolate and women's best low-heel shoes.	1760	1760–1769	1760	Smallpox epidemic in Charleston, South Carolina.
Israel Eaton grinds chocolate near the Mill Bridge, Boston.	1760	1760–1769	1760	Botanical gardens at Kew, London, opened.
Richard Saunders in *Poor Richard's Almanack* recommends chocolate to smallpox patients.	1760	1760–1769	1760	Great Boston fire.
Daniel Jones, loyalist merchant, sells chocolate in his Boston shop.	1760	1760–1769	1760	Death of George II; George III becomes king of England.
Marry Billings sells Irish linens and chocolate in her Boston store opposite Province House.	1760	1760–1769	1760	French surrender Detroit to British.
	1759	1750–1759	1759	Several references to chocolate in *Candide* by Voltaire.

Chocolate-Related Events in North America	Date ←	Decade ↔	Date →	Chocolate-Related Events Elsewhere and Historical Context
Benjamin Jackson, Philadelphia merchant, advertises himself as mustard and chocolate maker from London. Advertisement depicts Jackson's crest flanked by a bottle of mustard and bar of chocolate.	1759	1750–1759	1759	Old Bailey trial record: "The [stolen] chocolate was marked with the two initial letters of our names, W. T."
Thomas Proctor sells chocolate in Queen Street, Boston, next door to the Concert Hall.	1759	1750–1759	1759	British capture Quebec.
Recommendation: "Upon the appearance of smallpox eruptions . . . abstain entirely from all liquors stronger than small beer, and from broth and meat of every kind; but may nourish himself with milk, panda, chocolate, sago, gruels of all sorts, bread, biscuits, puddings, tarts, greens, and roots."	1759	1750–1759	1759	Play opens in London entitled *The Chocolate-Makers: or, Mimicry Exposed.*
Privateers bring French sloop into New York Harbor; cargo contains sugar, cotton, coffee, and cocoa.	1759	1750–1759	1759	French abandon Fort Ticonderoga.
Samuel Abbot sells chocolate near the corner of Wing's Lane, Boston.	1758	1750–1759	1758	British attack the French at Fort Duquesne; fort renamed Pittsburgh.
Chocolate-related poetry appears in *Poor Richard's Almanack*.	1758	1750–1759	1758	British attack Fort Ticonderoga and Fortress Louisbourg.
Boston advertisement: "To be sold by William Winter, at his Vendue-Room in Wing's Lane [Boston], a Chocolate Mill with its appurtenances; a gun; a pair of pocket pistols . . . Also . . . A chocolate kettle, pestle, and pans."	1758	1750–1759	1758	French warship *l'Hermoine* (600 ton, 28 gun frigate) captured leaving Louisbourg with 29 English prisoners aboard being sent to France. The galley list includes "Chocola" among many luxuries.
Samuel Smith, Boston merchant, advertises: "A Great Assortment of European Goods . . . A Chocolate Stone & Cooler, Delph Ware, a good Watch, Caps. &c."	1758	1750–1759	1758	Inquisition document (Mexico): support for the banning of chocolate crucifixes and chocolate images of Mary.
Anonymous poet commends for breakfast toast, butter and tea, and disapproves of coffee and chocolate.	1758	1750–1759	1758	Chocolate exported to Virginia from Glasgow, Scotland.
Boston advertisement: "Chocolate mills imported by Robert Traill, from the ship *Snow Perks*, out of Bristol . . . Sold at wholesale at his stores on the Long Wharff [sic], and that adjoining his House at Portsmouth."	1757	1750–1759	1757	French under General Montcalm attack Fort William Henry.
	1757	1750–1759	1757	Forestero cacao introduced to Trinidad from Venezuela.
	1757	1750–1759	1757	Chinese export porcelain chocolate pots via the Dutch East Indies Company.
	1756	1750–1759	1756	First chocolate factory established in Germany.
Benjamin Franklin provides military supplies, including 6 lb loaf sugar, 1 lb good bohea tea, 6 lb good ground coffee, and 6 lb chocolate.	1755	1750–1759	1756	England declares war on France (French–Indian War).
Mary Ballard, Boston merchant, opens a coffeehouse for the "Entertainment of Gentlemen, Benefit of Commerce, and Dispatch of Business . . . Tea, Coffee, or Chocolate and constant Attendance given by their humble Servant."	1755	1750–1759	1755	Great Expulsion—Arcadians forced from their lands; many take refuge in Louisiana (origin of the Cajuns).
Samuel Dexter sells chocolate at Cross Street near the Mill Bridge in Boston.	1755	1750–1759	1755	General Braddock, commander of British forces in North America, arrives in Williamsburg, Virginia.

Chocolate-Related Events in North America	Date ←	Decade ↔	Date →	Chocolate-Related Events Elsewhere and Historical Context
Mary Minot, Boston merchant, sells "GOOD Chocolate by the Quantity or single Pound."	1754	1750–1759	1754	Mark Catesby describes cacao production by Spanish in the Caribbean; recommends British pay attention to Jamaica for economic gain.
Widow Eleanor Dexter resumes selling chocolate at the family shop in White Horse Alley, Philadelphia.	1754	1750–1759	1754	Etienne Bachot publishes dissertation entitled *An chocolatae usus salubris?*
	1753	1750–1759	1753	British government buys the collection of Sir Hans Sloane, a botanist and physician who had traveled to Jamaica. Collection includes Italian-made ceramic chocolate cups now on display at the British Museum.
Early record of chocolate manufacturing in Canada: Schooner *Jolly Robin* leaves Halifax with cargo containg 2 boxes of chocolate identified as "here made."	1752	1750–1759	1752	Benjamin Franklin conducts his kite/key expreiment.
Chocolateria (chocolate shop) established at San Fernando College, Baja California.	1752	1750–1759	1752	The Dutch East Indies Company's ship *Geldermalsen,* wrecked during its return to Amsterdam from Canton, contained a cargo of some 150,000 porcelain items, of which 9735 chocolate cups and saucers were listed.
Obadiah Brown manufactures chocolate in Providence, Rhode Island.	1752	1750–1759	1752	First general hospital in North America opened in Philadelphia.
Report of Boston heat wave: "Whereas Samuel Watts up [by] the Prison Yard, by Reason of the heat of the Weather, has been unable to supply his Customers with Chocolate for these few Weeks past, he fears to the Loss of many of them; humbly begs Leave to inform them, that he has a Quantity very good and very cheap."	1751	1750–1759	1751	Report from Madrid: "The Cargoes of the four Ships lately arrived from Vera Cruz at Cadiz, are as follows: . . . 1986 Chests of Tobacco Stalks; 1,073,950 Pounds of fine Cochineal; 104,719 pounds of Vanillas [*sic*]; 1394 chests of Sugar; 1006 *Serons* of Cocoa [note: 1 seron = 200 pounds]; 1008 Chests of Chocolate."
Joseph Palmer and Richard Cranch manufacture chocolate in Boston.	1751	1750–1759	1751	Thomas Bond establishes Philadelphia Hospital with support of Benjamin Franklin.
Publication of a "Friendly Caution against Drinking Tea, Coffee, Chocolate, &c. Very Hot. Argument that these beverages should be served at temperatures below normal body heat, lest the higher degree of heat coagulate and thicken the Blood to such a Degree as to endanger Life."	1751	1750–1759	1751	Philip Dormer Stanhope, Earl of Chesterfield, advises his son: "Drink no chocolate, take your coffee without cream; you cannot possibly avoid suppers at Paris, unless you avoid company too, which I would by [no] means have you do; but eat as little at supper as you can."
Widow Jones sells chocolate at her Philadelphia coffeehouse.	1750	1750–1759	1750	Discovery of Cumberland Gap and beginning of western expansion into the interior of North America.
Announcement of slave sale in Philadelphia: one man to be sold identified as "a good chocolate grinder."	1750	1750–1759	1750	Barbados account of French selling slaves for quantities of cocoa and coffee.
Aaron Lopez moves to Newport, Rhode Island, with interests in whaling, trade with the West Indies, and chocolate manufacture.	1750	1750–1759	1750	Report of attempt to assassinate the Knights of Rhodes, by the Bashaw [Pasha]: "He caused them to buy a great Quantity of Poison, which they were to deliver to him when those Vessels returned home. This Poison he intended to throw into the great Aqueduct which supplies the Fountain with Water, and to cause some to be mixed in the Knights Coffee and Chocolate."

Chocolate-Related Events in North America	Date ←	Decade ↔	Date →	Chocolate-Related Events Elsewhere and Historical Context
Thomas White, chocolate maker, Philadelphia, announces sale of a slave.	1749	1740–1749	1749	Fortified port of Halifax, Nova Scotia, established.
Swedish botanist Pehr Kalm, traveling in New France, observed that chocolate was a common breakfast, as well as eau-de-vie with bread and coffee with milk.	1749	1740–1749	1749	Letter written by Father Junipero Serra during his voyage from Cadiz, Spain, to Vera Cruz. Letter confirms that Serra brought chocolate with him on the voyage and he laments: "There was not enough water [aboard] and rationing was introduced for all . . . nor could we drink chocolate we brought with us because there was not water to boil it with."
Captured French ship brought into New York with 10 tons of cocoa on board.	1748	1740–1749	1748	Treaty of Aix-la-Chapelle signed on October 18. The Island of Cape Breton returned to France in exchange for the Austrian Netherlands, Maestricht and Bergen-op-Zoom, the two frontier fortresses of Holland, Madras, and with it command of the whole Coromandel Coast in India.
Isaac Navarro, Annapolis, Maryland, merchant, manufactures and sells chocolate.	1748	1740–1749	1748	France regains Fortress Louisbourg by treaty.
Native Americans flee after brief battle, leaving behind a cache of chocolate.	1747	1740–1749	1747	Benjamin Franklin begins experiments with electricity.
	1747	1740–1749	1747	R. Campbell publishes *The London Tradesman*, which contains the account: "Chocolate is made of Cocoa, the Product of the West-Indies. It is stripped of its Shell, or rather Husk, and wrought upon a Stone over Charcoal Fire till it is equally mellow put into Moulds, which shapes it into Cakes. To perfume it they mix it Venello [i.e., vanilla]. It is a hot laborious Business, but does not require much Ingenuity. Journeyman's Wages is from Twelve to Fifteen Shillings a Week, are not employed much in the Summer. They require Heat to work with, but cold Weather is necessary to dry it."
Evan Morgan, Philadelphia merchant, sells Jamaica spirit, cinnamon water, Barbados citron water, chocolate, sallad [sic] oil, and nails.	1747	1740–1749	1747	Simon Lord Lovant requests and is served chocolate on the day of his execution.
Minister's complaint to the editor of the *Boston Evening Post* of high prices for food and essentials; provides monthly budget for essentials with comparative prices for 1746 and 1706.	1747	1740–1749	1747	Juan de la Mata publishes his cookbook, *Arte de Reposteria*, which contains various recipes for chocolate.

Chocolate-Related Events in North America	Date ←	Decade ↔	Date →	Chocolate-Related Events Elsewhere and Historical Context
Brigantine called *The Victory*, en route from Martinique, captured by the New England troops as the ship approached Louisbourg. Originally commanded by Loring and captured by the French in 1744, it was then taken by the English. *The Victory* had on board a cargo of rum, molasses, coffee, syrup, and chocolate.	1746	1740–1749	1746	College of New Jersey (later Princeton University) founded.
Greville publishes *The Miserable Glutton* and writes: "Next day obedient to his word, The dish appear'd at course the third; But matters now were alter'd quite, In bed till noon he'd stretch'd the night. Took chocolate at ev'ry dose, And just at twelve his worship rose. Then eat a toast and sip'd bohea, "Till one, and fat to dine at three."	1746	1740–1749	1746	Simon Paulli writes treatise on chocolate, coffee, tea, and tobacco; argues for the rejection of these beverages as they cause effeminacy, impotency, and sterility.
French ship, *Notre Dame de la Deliverance*, brought to Louisbourg: cargo contained 14,840 double doubloons; 1,320,500 (silver) dollars; 7867 ounces of gold; 283 pounds of silver; and 316 bags of cocoa.	1745	1740–1749	1745	Smallpox epidemic in New York, Delaware, and Maryland.
List of provisions issued to a military expedition from Quebec to march against New England. Chocolate listed as supplies to be issued to Compagnie Franches soldiers, cadets, and officer level volunteers, but not for common soldiers.	1745	1740–1749	1745	Jews expelled from Prague.
Thomas Cadwalader publishes his essay on the West India dry-gripes with "the method of preventing and curing that cruel distemper." Remarks that to stay well in the West Indies one should abstain from strong punch and highly seasoned meats; immoderate exercise, which raises sweat, and "to rise early in the Morning . . . to take Chocolate for Breakfast and Supper."	1745	1740–1749	1745	Fortress Louisbourg, Cape Breton Island, Nova Scotia, captured by English: letter from William Shirley, Esq., Governor of Massachutes Bay to his Grace the Duke of Newcastle. William Shirley, Esq. announcing the suffender of Fortress Louisbourg to the English on June 17th.
Prices of provisions at Louisbourg, November 22: "Fresh Beef, Mutton and Pork, 4 shillings a pound; Dunghill Fowls, 10 shillings apiece; Geese, 30 shillings apiece; Apples and Potatoes, 20 shillings a bushel; Rum, 24 shillings a gallon; Chocolate, 21 shillings a pound."	1745	1740–1749	1745	Saratoga, New York, burned by French and Indian forces.
Alexander Hamilton notes in his diary: "I breakfasted upon some dirty chocolate, but the best that the house could afford."	1744	1740–1749	1744	First brewery established in Baltimore.
Provision list for a sortie party of 120 French and 200 Native Americans leaving Quebec in January to march against the English at Annapolis Royal (former French territory of Acadia). Identified on the list were 84 pounds of chocolate and 800 pounds of sugar.	1744	1740–1749	1744	King Louis XV of France declares war against England.
When the governor of Île Royale died at Louisbourg, his inventory contained 29 pounds (14.2 kg) of chocolate.	1744	1740–1749	1744	Start of King Georges War (Third French–Indian War) at Port Royal, Nova Scotia.

Chocolate-Related Events in North America	Date ←	Decade ↔	Date →	Chocolate-Related Events Elsewhere and Historical Context
Inventory of personal belongings of Bertrand Dargaignanats. He was staying at the DeChevary Home and upon his death his estate was submitted with a bill for food and lodging. The food bill included 3 cartons of chocolate, 1 livre of chocolate, and another $\frac{1}{2}$ deml livre of chocolate.	1744	1740–1749	1744	Alamo mission fortified.
Mordecai Dunbar, merchant of Newport, Rhode Island, sells "good Chocolate by Retail for Eight Shillings a Pound, old Tenor."	1744	1740–1749	1744	Old Bailey trial record: testimony of Charles Gataker apprentice to Mr. Barton: "Question: How do you know this to be your master's chocolate? Gataker: It was marked with our mark; I believe she was a little in liquor when she did it [i.e., stole the chocolate].
David-Bernard Muiron, a contractor for the fortifications on Louisbourg submits written allegations to the Bailage Court that Valerien Louis dit le Bourguignon, a former soldier from France and now working as a stone cutter for Muiron, had been stealing from him over a period of four months iron, steel, chocolate, coffee, and table knives.	1743	1740–1749	1743	Pesthouse used to quarantine immigrants established in Philadelphia.
The ship *Midnight* from New England arrived at Louisbourg with cargo. Items unloaded included one box of chocolate.	1743	1740–1749	1743	Yellow fever epidemic in New York.
Captain Dowell brings captured Spanish ship to Philadelphia loaded with 157.5 tons of cocoa, and one ton of chocolate (value: 10,000 pieces of eight).	1743	1740–1749	1743	Old Bailey trial record: testimony of Josias Taylor: "I had 120 pound weight of chocolate brought from the office, which had been sent there to be stamped [marked]. It was brought in on the Saturday and the Monday before the Tuesday Night on which this [the theft] happened. It was put in the window to dry, because it is damp when it comes from the office."
Bard and Lawrence, Philadelphia merchants, sell Caracas cocoa and chocolate "made of the said Nuts, for 3 shillings per pound."	1743	1740–1749	1743	Old Bailey trial records: John Read, convicted of stealing 26 pounds of chocolate.
Elsie Smith publishes recipe for chocolate almonds in her *Compleat [sic] Housewife*.	1742	1740–1749	1742	British attack Spanish port of Guayra in modern Venezuela; port of Guayra a prominent locale for exportation of cacao.
William Vanderspiegle, broker for male and female Welsh servants, reports he also sells cocoa.	1742	1740–1749	1742	Benjamin Franklin invents the so-called Franklin stove.
George Wilhelm Steller reports chocolate enjoyed on voyage from Russia to America.	1741	1740–1749	1741	Carl von Linné (Linnaeus) identifies medicinal uses of chocolate.
Captured French ship brought to Boston; cargo contains 100 tons of cocoa.	1741	1740–1749	1741	Great Awakening religious revival reaches peak.
Privateer brings booty into Philadelphia containing between 40 and 50 thousand weight of cocoa.	1740	1740–1749	1740	British Admiral Haddock arrives in Portsmouth with Spanish ships from Caracas, with about 350 tons of cocoa.

Chocolate-Related Events in North America	Date ←	Decade ↔	Date →	Chocolate-Related Events Elsewhere and Historical Context
John Gottlieb Ernestus Heckewelder begins Christian mission among the Delaware and Mohegan (sic) Indians: "We also made use of some other vegetables and greens. Beside we had brought along some tea and chocolate, which we drank as we could without milk. . . . We had for our supper chocolate and dry bread . . . by this time the fire was in fully blaze and then the bread was baked and the chocolate or coffee prepared. Each one had his own cup, spoon and knife for chocolate. . . . To Penmaholen we presented some chocolate, sugar &c. He was friendly. . . . We had bread, pork, butter, milch, sugar, chocolate &c . . . A desent [sic] table to a stranger, should be furnished with bread, meat, cheese, milch [sic], tea, coffee or chocolate, whatever he chooses."	1740	1740–1749	1740	First breweries established in Georgia to provide beer for Colonial soldiers.
New York advertisement: Auction House of Mr. Benjamin D'harriett, opposite the New-Dutch Church—for auction—"a chocolate stone and rowler [sic]."	1740	1740–1749	1740	University of Pennsylvania founded.
Benjamin Franklin sells bibles and other books "too tedious to mention," also pencils, ink, writing paper, and "very good Chocolate."	1739	1730–1739	1739	H. T. Baron publishes *An Senibus Chocolatae Putus?*
French explorers Pierre and Paul Mallet pack and carry a handmill and three chocolate cups during their expedition to Sante Fe (New Mexico).	1739	1730–1739	1739	French explorers Pierre and Paul Mallet reach the Rocky Mountains.
Boston advertisement: "THIS is to give Notice that all Gentlemen may be Daily supply'd with Coffee and Chocolate ready made, by George Gray, at the Crown Coffee-House, at the head of the Long Wharff [sic]."	1738	1730–1739	1738	F. E. Bruckman publishes *Relatio de Cacao.*
Joseph de Goxxaes (Gonzales) from the Mission La Punta de Lamparo dated to 1738 and written in the Kingdom of Leon (Mexico) denounced the ill treatment of the Chichimecos Indians by missionaries and soldiers at the sister mission of Santa Maria de Los Dolores (Kingdom of Leon). The document primarily explored the problem and concluded with a list of provisions provided to the Santa Maria mission by Father Gabriel de Vergara during 1738: 3 frac12; arrobas of ordinary chocolate cut in 12 parts, and 1 arroba of ordinary chocolate (no further designation).	1738	1730–1739	1738	Smallpox epidemic among Cherokee Native Americans in Georgia; smallpox epidemic in South Carolina.
John Brewster, Boston merchant, marks chocolate with his initials "I.B." to certify product source.	1738	1730–1739	1738	English and French fur traders compete in vicinity of Lake Erie.

Chocolate-Related Events in North America	Date ←	Decade ↔	Date →	Chocolate-Related Events Elsewhere and Historical Context
Inventor in Salem, Massachusetts, develops an engine to grind cocoa, that costs less to run than any commonly in use; can produce 100 weight of chocolate in six hours, and chocolate made is "finer and better, the Oyly Spirit of the Nut being almost altogether preserved . . . and there is little or no need of fire in the making."	1737	1730–1739	1737	Letter from Governor William Mathew to Council of Trade and Plantations: describes passage of an Act of Montserrat that restricted slaves and prohibited them from "planting indigo, cotton, ginger, coffee, or cocoa, and from keeping a public market on Sundays."
Abraham Wendell produces chocolate in the Hudson Valley.	1737	1730–1739	1737	Richmond, Virginia, founded.
William Graham, Philadelphia merchant, sells Antigua rum, St. Kitts molasses, chocolate, and pepper.	1735	1730–1739	1735	Carl von Linné (Linnaeus) writes *Systema naturae*.
Servants attempt to poison Humphrey Scarlet, his wife and children, using chocolate laced with ratsbane.	1735	1730–1739	1735	Augusta, Georgia, founded.
"To be SOLD, BY the Printer hereof, very good Chocolate at 4 shillings per Pound by the half Dozen, and 4 shillings 6 pence by the single Pound." (Note: printer of the *Pennsylvania Gazette* was Benjamin Franklin.)	1735	1730–1739	1735	Epidemics of diphtheria and scarlet fever throughout New England.
A sloop belonging to Rodrigo Pacheo carries a cargo to Curaçao, where it is loaded with cocoa and lime, to be exchanged in Charleston for rice.	1734	1730–1739	1734	The text *A Nouvelle Instruction pour les confitures, les liqueurs, et les fruits*, notes that chocolate can be made as a beverage, or in solid form for candies and biscuits.
John Pinkney, Boston merchant, sells "good Chocolate, at Eleven Shillings per Pound."	1734	1730–1739	1734	John Winthrop describes decimation of Native Americans by smallpox.
William Byrd, describing his journey through Carolina: "I filed off with my honest friend, Mr. Banister, to his habitation on Hatcher's run. . . . In the gayety [*sic*] of our hearts we drank our bottle a little too freely, which had an unusual effect on persons so long accustomed to simple element. . . . [Next morning] After pouring down a basin of chocolate, I wished peace to that house, and departed."	1733	1730–1739	1733	James Oglethorpe establishes settlement of Savannah, Georgia.
John Merrett, grocer of Boston, sells "Cocoa, Chocolate, Tea Bohea & Green Coffee Raw & Roasted, all sorts of Loaf, Powder & Muscovado Sugar."	1732	1730–1739	1732	Port of Amsterdam receives 13,240 chocolate cups and 6493 chocolate cups with handles from Canton, China.
	1732	1730–1739	1732	A Monsieur de Buisson devises a table for the "chocolate-worker," the person grinding the beans, to stand upright rather than have to bend over the grinding stone.
	1732	1730–1739	1732	Additional English patent issued to Walter Churchman for "New invention of making chocolate without fire, to greater perfection, in all respects, than by the common method."
Mrs. Hannah Boydell, wife of Mr. John Boydell and merchant of Boston, "Sells Sugar, Coffee, Chocolate, Starch, Indigo, Spices and Grocery Ware[s], reasonably."	1731	1730–1739	1731	Benjamin Franklin establishes first circulating library in the New World at Philadelphia.

Chocolate-Related Events in North America	Date ←	Decade ↔	Date →	Chocolate-Related Events Elsewhere and Historical Context
Mrs. Reed (first name unknown) opens chocolate house in Boston, has chocolate, coffee, or tea ready made any time of the day.	1731	1730–1739	1731	Smallpox epidemic in Philadelphia and New York.
Governor of Texas, Don Juan Antonio de Bustillo y Lerna, reported the immediate need to receive the following supplies: "90 men, that as helpers, I invited from the Kingdom of Leon [in Mexico; not Spain] with some neighbors of this presidio in the town of San Antonio de Valero—60 friendly Indians of this town and also from the nations Pampopas and Sanes: 44 loads of flour; 50 fanegas of corn; 2 grind loads of corn for [making] pinole for the Indians; 250 pounds of 'chocolate de regalo' [gift chocolate]."	1731	1730–1739	1731	French besiege Natchez, Mississippi.
	1730	1730–1739	1730	Dutch East India Company ship, *Coxhoorn*, leaves Canton and arrives in Amsterdam with 9457 Chinese chocolate cups and pots.
	1730	1730–1739	1730	Richard Brookes comments that chocolate can be eaten in solid form (early mention of chocolate consumed other than as a beverage).
	1730	1730–1739	1730	De Quelus identifies medicinal uses of chocolate.
Andrew Fanueil, Boston merchant, advertises: "Good New York Flower [sic] To be Sold at 35 shillings per Hundred [weight]; as also good Chocolate, just imported."	1729	1720–1729	1729	A chocolatière, of gilded silver (gold on silver), Rococo style, by Henry-Nicolas Cousinet, is given by Louis XV to his queen, Maria Leczinska, on the occasion of the birth of their son, the Dauphin.
	1729	1720–1729	1729	Letter dated April 20th from Franciscan priest and medical surgeon, Fr. Antonio de la Concepción, living in southern China, requesting approval for one of the Franciscan brothers to make 130 pesos of chocolate to give to the Jesuit fathers at the Chinese court.
	1729	1720–1729	1729	Walter Churchman petitions king of England for patent and sole use of an invention for the "expeditious, fine and clean making of chocolate by an engine."
Letter to the *New England Weekly Journal*, Boston: "Necessary Expenses in a Family of but middling Figure, and no more than Eight Persons: Breakfast. Bread and 1 pint of milk, 3 cents; Dinner. Meat, roots, salt, and Vinegar, 10 cents; Supper as the Breakfast, 3 cents; Small Beer for the whole Day, Winter & Summer, 2 cents. For one Person, one Week i.e. 7 days = 10 shillings 6 cents, for Eight Persons for one Day = 12 shillings. But . . . No Butter, Cheese, Sugar, Coffee, Tea, nor Chocolate; No Wine, nor Cyder, nor any other Spirituous Liquors not so much as Strong Beer; No Tobacco, Fruit, Spices, nor Sweet-meats; No Hospitality, or Occasional Entertaining, either Gentlemen, Strangers, Friends, or Relatives [long list of other things/activities not possible]."	1728	1720–1729	1728	Fry and Sons begin chocolate production in Bristol, England.
	1728	1720–1729	1728	Walter Churchman starts to manufacture chocolate in Bristol, England.
	1728	1720–1729	1728	Benjamin Franklin opens his printing shop in Philadelphia.

Chocolate-Related Events in North America	Date ←	Decade ↔	Date →	Chocolate-Related Events Elsewhere and Historical Context
John Lubbuck, Boston chocolatemaker, flees his creditors.	1728	1720–1729	1728	Arrival of "casket girls" in Louisiana.
James Lubbuck, chocolategrinder of Boston, "sells best Chocolate at the lowest prices; takes in Cocoa-Nuts to grind with expedition at 6 pence per pound."	1727	1720–1729	1727	Earthquake destroys Martinique's cacao plantations; will never fully recover.
	1727	1720–1729	1727	Announcement from London of cargo from the Spanish fleet: "12,474,00 pieces of eight; 3700 marks of silver in pigs [ingots], 17,230 marks of wrought silver; 74,200 pistoles (coins) in gold . . . 585 arrobas [1 arroba = 26 pounds] of chocolate."
	1727	1720–1729	1727	Jamaica cacao plantings destroyed by the "blast" (plant disease), perhaps destruction by *Ceratocystis* wilt or *Phytophthora* (bark canker).
French soldiers diving off the coast of Cape Breton to salvage the royal transport vessel *Le chameau* were fed fresh meat and chocolate.	1726	1720–1729	1726	Jonathan Swift publishes *Gulliver's Travels.*
First documented mention of chocolate in Baja California.	1726	1720–1729	1726	Poor residents of Philadelphia riot; uprising put down harshly.
Peter Zenger writes on duties assigned to cocoa and chocolate.	1726	1720–1729	1726	Upon her death, Lady Palmerston bequeathed to her husband: "As a remembrance of death and also of the fondest and faithfulest friend he ever had--two gold chocolate cups made out of mourning rings, and used by her daily as a memorial of her departed friends and of eternity."
Cotton Mather writes *Manuductio ad Ministerium [Directions for a Candidate of the Ministry]* giving dietary recommendations: "To feed much on Salt-Meats, won't be for your Safety. Indeed, if less Flesh were eaten, and more of the Vegetable and Farinaceous Food were used, it were better. The Milk-Diet is for the most part some of the wholesomest [sic] in the World! And not the less wholesome, for the Cocoa-Nutt giving a little Tincture to it."	1726	1720–1729	1726	Petition of Charles Dunbar, Surveyor General of the Customs, Antigua, to the King: "In Feb. 1726/7 [February 7th] the Spanish ship *Santa Reta* was cast away on the shoals of Barbuda. About 70,000 weight of damaged cocoa, 1600 weight of snuff and a parcel of Spanish coin said to amount to 60,000 pieces of eight, was brought to Antigua, and the snuff and cocoa, was lodged in the Custom House by petitioner to prevent clandestine trade or embezzlement."
	1724	1720–1729	1724	R. Brooks translates and publishes the work of De Chélus: *Natural History of Chocolate.*
Duties computed in South-Carolina by an Act in 1722–3: "Cwt [hundred weight] of Cocoa, 15 shillings; Pound of Chocolate, 1 shilling."	1722	1720–1729	1722	A letter by Jesuit Father François Xavier d'Entrecolles indicates that Chinese chocolate cups were being produced in Jingdezhen.
Letter from Isaac Bobon to George Clarke, Secretary of New York Colony: "A list of Sundarys [sic] sent by Riche March: Three pound [sic] of Chocolate £ 0.7.6. . . . If ye prices of any of the above particulars are not liked they will be taken again."	1722	1720–1729	1722	Initial construction at the Alamo Mission in San Antonio, Texas.
	1721	1720–1729	1721	Ann Tompkins tried in Old Bailey, London, for felonious theft of 80 pounds of cocoa-Nuts and 10 pounds of chocolate (acquitted).

Chocolate-Related Events in North America	Date ←	Decade ↔	Date →	Chocolate-Related Events Elsewhere and Historical Context
Charles Read, Philadelphia merchant, advertises to grind cocoa and sells chocolate, cheap.	1720	1720–1729	1720	Account of Captain Sebastia de Menthata of Portugal, accused of smuggling 8800 pounds or 70 bags of St. Domingo cocoa nuts into England aboard the ship, *St. Augustine.*
Letter from Isaac Bobon to George Clarke, Secretary of New York Colony: "Prices as follows: a Doz. pounds of chocolate 1£ 2 shillings."	1720	1720–1729	1720	Construction completed on Governor's Palace, Williamsburg, Virginia.
	1720	1720–1729	1720	Letter to the British Council of Trade and Plantations recommending settlement of the island of St. Lucia: "The soils of the same nature as Martinique, and very proper for producing cocoa; a species of merchandize which the English have never yet produced, and which they consume great quantities of, especially in our Colonies, where every mechanic drinks chocolate for breakfast and supper."
	1719	1710–1719	1719	De Chélus publishes *Histoire Naturelle du Cacao et du Sucre.*
French suggest planting cacao trees in southern Mississippi River region could stimulate trade with Havana, Cuba.	1716	1710–1719	1716	First black slaves brought to Louisiana Territory.
General Assembly, Colony of New York, passes act to support the king's government expenses in New York: "For every Hundred Weight of Cocoa, imported directly from the Place of its Growth, or from any Island, part or Place of the West-Indies, One Ounce of [silver] Plate [is levied]."	1715	1710–1719	1715	Death of King Louis XIV of France.
Nathan Simpson, New York merchant, trades in cacao and sells chocolate.	1714	1710–1719	1714	Tea first introduced to North America.
Massachusetts Charter: "And be it further Enacted by the Authority aforesaid, That no Person shall presume to Keep a Coffee-House for the Selling of Coffee, Tea, Chocolate, or any Distilled Liquors, but how shall be Licensed in such manner as is provided in and by the Law.	1714	1710–1719	1714	Louis Juchereau de St. Denis establishes Fort St. Jean Baptiste at present day Natchitoches, Louisiana (first permanent European settlement in the Louisiana Territory).
	1713	1710–1719	1713	Correspondence between Nathan Simpson and Richard Janeway: "[When price of] sugar falls, cocoa will rise."
	1712	1710–1719	1712	Letter, dated February 16, from Franciscan priest and medical surgeon, Fr. Antonio de la Concepción, living in southern China, thanks his superior in Canton for the receipt of eight gantas [1 ganta = 3 liters] of chocolate to be distributed as instructed.

Chocolate-Related Events in North America	Date ←	Decade ↔	Date →	Chocolate-Related Events Elsewhere and Historical Context
Governor Spotswood of Virginia writes to the English Council of Trade and Plantations, on May 15, that "the Bedford galley arrived here the other day, and brought in a French merchant ship loaded with sugar, indico and cocoa, and I hear Capt. Pudner in the Severn, one of the convoys to the Virginia Fleet, has taken and carried into New York a French privateer of 180 men, wch. Very much infested this coast."	1712	1710–1719	1712	Alexander Pope, in *The Rape of the Lock*, writes: "In fumes of burning chocolate shall glow, and tremble at the sea that froaths [*sic*] below!"
	1711	1710–1719	1711	A dispute over cacao booty taken in Nantes, France, from three captured enemy ships leads to a decree by Louis XIV's government, that cacao taken as a prize from captured ships is subject to special royal taxes.
	1711	1710–1719	1711	Old Bailey (London prison) trial records: Charles Goodale convicted of stealing 1 silver chocolate pot, value 12£ 1 shilling.
	1711	1710–1719	1711	Exports from Jamaica: "Cocoa, to England, 41 pipes, 189 hhds., 42 tierces, 100 barrels, 557 casks, 1804 seroons, 440 bags, 8000 pounds."
Zebadiah Boylston sells chocolate in his Boston apothecary shop.	1711	1710–1719	1711	Letter, dated February 22, from Franciscan priest and medical surgeon, Fr. Antonio de la Concepción, living in southern China, mentions receipt of chocolate sent as alms to the missionaries in China, and had been shared with those present.
Governor Hunter writes to the Commissioners of Customs in New York, on May 11, reporting: "In Sept. last [i.e., 1710] the captains of two privateers brought in hither a large ship loaden with cacoa upon my promise that no injury or hardship should be offered them; the collector agreed to the unloading of the vessell after condemnation putting the effects into safe store-houses under lock and key in his possession for securing Queen's dutys, condescending to let them sell from time to time what they could paying the dutys as they sold."	1711	1710–1719	1711	Trade in cacao between Lisbon and London.
	1710	1700–1709	1710	A drinking cup from the Kangxi period (1661–1722) described in the period records of the Dutch East Indies Company listed as a chocolate cup.
William Byrd records in his diary: "I rose at 6 o-clock this morning and read a chapter in Hebrew and 200 verses in Homer's *Odyssey*. I said my [prayers] and ate chocolate for breakfast with Mr. Isham Randolph, who went away immediately after."	1709	1700–1709	1709	Swiss and German immigrants settle the Carolinas.
	1708	1700–1709	1708	Missionaries on Trinidad who had organized Indians to work cacao orchards forced to leave the island.
	1707	1700–1709	1707	Chocolate as a pharmaceutical appears frequently in the requests of the Franciscans priest and medical surgeon, Fr. Antonio de la Concepción, living in southern China. Letter, dated November 17, notes that the chocolate requested earlier, has not arrived.
	1706	1700–1709	1706	(Dr.) Duncan publishes: *Wholesome Advice Against the Abuse of Hot Liquors, Particularly of Coffee, Tea, Chocolate, etc.*

Chocolate-Related Events in North America	Date ←	Decade ↔	Date →	Chocolate-Related Events Elsewhere and Historical Context
Earliest cacao/chocolate/cocoa newspaper advertisement thus far located: "To be Sold in Boston at the Ware-house of Mr. James Leblond on the Long Wharf near the Swing-Bridge . . . Rum, Sugar, Molasses, Wine, Brandy, sweet Oyl, Indigo, Brasilet [?], Cocoa, Chocolate, with all sorts of Spice, either by Wholesale or Retail, at reasonable Rates" (*The Boston News-Letter,* December 3 to December 10, 1705, Issue 86, p. 4).	1705	1700–1709	1705	Virginia slavery act states all imported slaves remain in lifelong servitude.
Lewis Moses Gomez, chocolate maker, arrives in New York.	1702	1700–1709	1702	First regular newspaper published in England (*Daily Courant*).
English report of Dutch parties at New Amsterdam: "Among wealthier families chocolate parties [are] much in vogue, which a *domine* objected to as keeping people up till nine o'clock at night."	1700	1700–1709	1700	Mission San Xavier del Bac founded in New Spain near Tucson.
Mission at Loreto (Baja California) reports no chocolate supplies received from Mexico for 14 months.	1700	1700–1709	1700	Bartolomeo Cristofori of Padua, Italy, invents the piano.
Anonymous commonplace book recipe: "To Make Chocolate Almonds. Take your Sugar & beat it & Serch [sift] it then: great youre [sic] Chocolatt [sic]: take to 1 lb Sugar: 5 oz. Of Chocolatt [sic] mix well together put in 2 Spoonful Gundragon Soaked in rosewater & a grm musk & ambergrease & beat all well together in mortar. Rowl [sic] out and markwt: ye molds & lay on tin plates to dry turn everyday."	1700	1700–1709	1700	Samuel Sewell publishes first American protest against slavery.
Jonathan Dickenson, shipwrecked along the coast of Florida, survives on butter, chocolate, sugar, and wine.	1699	1690–1699	1699	Colonial capital moved from Jamestown to Williamsburg, Virginia.
	1699	1690–1699	1699	Reported attempt to assassinate King Charles II of England with poisoned chocolate (apocryphal?).
	1698	1690–1699	1698	Chocolate served during Easter vigils at the Real Colegio de San Ignacio de Loyola, Mexico City.
	1698	1690–1699	1698	Cocoa Tree Chocolate House, Pall Mall, London, established at least by this date (but possibly earlier in 1692).
Samuel Sewell breakfasts on venison and chocolate.	1697	1690–1699	1697	Chocolate served to priests who officiated at the burial of nuns at the Santa Clara Convent, Mexico City.
	1697	1690–1699	1697	Smallpox ravages Charleston, South Carolina.
	1695	1690–1699	1695	James Lightbody publishes *Every Man His Own Gauger* and identifies how to make chocolate "cakes" and "rowles."
	1695	1690–1699	1695	Unnamed man challenges right of Maria de Iribarne to maintain a vendng stand in Aztapalapa, Mexico, where she sells bread, brown sugar candy, chocolate, and milk custard. She counters his accusation and informs the Alcade that she sells "because I need to feed my children and do not have anything else." Disposition of the case uncertain.

Chocolate-Related Events in North America	Date ←	Decade ↔	Date →	Chocolate-Related Events Elsewhere and Historical Context
Mention of "le llebaron refresco de chocolate, tabaco y papel [Chocolate refreshments [cold drinks], tobacco, and paper]" (probably to roll cigarettes) sent to Presidio San Joseph in what is now New Mexico.	1695	1690–1699	1695	Marcus Mappus publishes *Dissertationes Medicae Tres de Receptis Hodie Etiam in Europa, Potus Calidi Generibus Thée, Care, Chocolata.*
	1693	1690–1699	1693	French edict gives all grocers and merchants in France the freedom to buy and sell chocolate, cacao, and tea; edict remains in force until 1790.
	1693	1690–1699	1693	Chocolate served to a Turkish Aga in Smyrna (Izmir) by Gemelli Careri.
	1693	1690–1699	1693	White's Chocolate House, St. James' Street, founded by the Italian, Francis White.
	1691	1690–1699	1691	Bayonne non-Jewish chocolatiers gain passage of a municipal ordinance prohibiting Jews from selling chocolate within the city gates.
	1691	1690–1699	1691	François Massaliot's *Cuisinier Roial et Bourgeois*, a guide to organizing meals for the French upper classes, offers recipes for *Macreuse en ragoût au Chocolat*, a savory poultry dish with a chocolate sauce, and *Crême de Chocolat,* using milk, sugar, and egg yolk, in addition to chocolate.
Reference to a "jocolato pot" appears in an inventory of Chinese export ware belonging to William Pleay in New York.	1690	1690–1699	1690	Yellow fever epidemic in New York.
Letter of Father Damian Massanet to Don Carlos de Siguenza: "Captain Leon had a compadre along, Captain [?], so honorable that he never failed to play the tale-bearer and excite quarrels; so kind-hearted that only his friend Leon drank chocolate, and the others the lukewarm water; so considerate of others that he got up early in the morning to drink chocolate, and would afterward drink again with the rest; so vigilant that he would keep awake and go at midnight to steal the chocolate out of the boxes; perhaps this vigilance was the reason why, while, by order of His Excellency, Captain Leon should have left for the priests three hundredweight of chocolate and the same quantity of sugar, but he left only one and one-half hundredweight of each."	1690	1690–1699	1690	First colonial newspaper published in Boston (*Publik Occurrences*); article mentions smallpox epidemic in Boston.
	1690	1690–1699	1690	Hans Sloane adds milk to chocolate to make it more palatable.

Chocolate-Related Events in North America	Date ←	Decade ↔	Date →	Chocolate-Related Events Elsewhere and Historical Context
	1690	1690–1699	1690	Inquisition document (Mexico): declaration related to witchcraft. Denunciation of Simón Hernán, Spanish, 35 years old: "During the religious office at the Nuestra Señora de la Merced in this Province of Mexico he [Simón] give some powders to a Spanish woman named Micaela, and he gave them to her mixed with chocolate. Eight days later and without telling anything to his wife, in another religious office, he gave her again those powders mixed with chocolate. He [Simón] declared that those powders were meant to seduce Micaela and in doing so be able to enjoy her favors, but that the powders had no effect on Micaela. "This, he [Simón] declares as true and under oath."
	1689	1680–1689	1689	Parliament announcement to change the excise tax on imported coffee, chocolate, and tea based on volume of a beverage serving (a pound of chocolate makes four gallons of liquor); a pound of cocoa-nut makes two pounds of chocolate).
	1689	1680–1689	1689	The Duc d'Orléans stages a lottery at Saint-Cloud to give away "jewels for women." One woman, Madame de Maré, wins two chocolatières, one of silver and one of "porcelain," together with seven chocolate beating sticks [bastons], and a tea box. The "porcelain" chocolatière may have been made of faïence.
	1687	1680–1689	1687	Nicolas de Blégny identifies medicinal uses of chocolate: Le bon Usage de Thé, du Caffé, et du Chocolat pour la Preservation et pour la Guerison des Malades.
	1687	1680–1689	1687	Sir Hans Sloane traveling in Jamaica records a native recipe that he brings back to Europe, the so-called Sloane Recipe.
	1687	1680–1689	1687	First mention of chocolate production in Bayonne, France.
	1686	1680–1689	1686	Father Buenaventura Ibañez, in a letter dated February 24 (written in China) refers to the use of chocolate as a pharmaceutical. He complains that he could eat neither meat nor chicken and only a little fish, and found, however, that a good bowl of chocolate in the mornings was helpful.
	1686	1680–1689	1686	King of Siam sends gift of nine chocolatières (chocolate pots) to King Louis XIV of France.
	1685	1680–1689	1685	J. Chamberlaine publishes translation of Sylvestre Dufour's book: The Manner of Making Coffee, Tea and Chocolate.

Chocolate-Related Events in North America	Date ←	Decade ↔	Date →	Chocolate-Related Events Elsewhere and Historical Context
	1685	1680–1689	1685	Sylvestre Dufor identifies medicinal uses of chocolate. *Traitez Nouveaux et Curieux du Café, du Thé et du Chocolat.*
	1685	1680–1689	1685	Henry Mundy publishes *Opera Omnia Medico-Physica de Aere Vitali, Esculentis et Potulentis cum Appendice de Parergis in Victu et Chocolatu, Thea, Caffea, Tobacco.*
	1684	1680–1689	1684	François Foucault, in his dissertation *An Chocolatae usus salubris?*, argues for the medicinal benefits of chocolate.
Henri Joutel explores Mississippi River; mentions chocolate in his diary.	1684	1680–1689	1684	Charles II annuls the 1629 charter of Massachusetts Colony.
	1693	1680–1689	1683	Father Buenaventura Ibañez, in a letter dated February 15 (written in China), contains a reference to the Franciscans using chocolate during fast days.
British report of cocoa exports from Jamaica to Boston.	1682	1680–1689	1682	Spanish settlement of Yselta founded near modern El Paso, Texas.
	1691	1680–1689	1681	Father Buenaventura Ibañez, in a letter dated March 4 (written in China), mentions chocolate received as annual alms to the Franciscan mission in China.
	1680	1680–1689	1680	Joseph del Olmo, describing the Inquisition executions in Madrid, suggests that "High-ranking officials attending the spectacle of an auto-de-fé were served biscuits and chocolate, sweets and sweet drinks."
	1680	1680–1689	1680	The word chocolate first appears in Pierre Richelet's French dictionary.
The ship *Charles*, out of New York and owned by Frederick Philipse, pays taxes on "2 hhds [hogsheads] cocoa vallued [sic] at 114.10 English pounds."	1679	1670–1679	1679	Great Boston fire.
	1679	1670–1679	1679	The ship *Triomphant* arrives in the French port of Brest from Martinique with a cargo of cacao, marking the opening of official trade in cacao between France and the French Caribbean.
	1678	1670–1679	1678	Writing from Canton, China, to his minister provincial on March 5, Father Buenaventura Ibañez noted that he had requested a cargo of cacao to be sent from New Spain with all the accoutrements necessary to make chocolate as a present for the "king of Canton." This is one of the rare sources where the intent is to give chocolate to the Chinese rather than keeping it among the missionaries.
	1677	1670–1679	1677	Commonly accepted year when cacao plantations were established in the Brazilian state of Para.
	1676	1670–1679	1676	French archives report: Jean de Herrera appointed chocolatier to the queen of France.

Chocolate-Related Events in North America	Date ←	Decade ↔	Date →	Chocolate-Related Events Elsewhere and Historical Context
	1675	1670–1679	1675	Royal edict: a proclamation for the suppression of coffeehouses: "After the Tenth Day of January next ensuing, to keep any Public Coffee House, or to utter or sell by retail, in his, her or their house or houses any Coffee, Chocolet [sic], Sherbett [sic] or Tea, as they will answer the contrary at their utmost perils."
	1674	1670–1679	1674	Writing from Macao on March 27, Fr. Buenaventura Ibañez, a Franciscan missionary, refers to himself as an old man aged sixty-five, with difficulty consuming anything other than rice and liquids. He asks a visiting colleague to bring cacao with vanilla, that he might then eat in the mornings from a chocolate bowl (*una escudilla de chocolate*), which could be prepared locally.
	1673	1670–1679	1673	English Parliament imposes plantation duties upon certain goods: Coco Nuts [sic] levied at 1 cent [quantity of cocoa nuts not identified].
	1672	1670–1679	1672	Letter from Thomas Lynch to H. Slingesby discusses the "blasts" (plant disease) that ruined the cocoa trees on the islands of Hispaniola and Cuba.
	1672	1670–1679	1672	Letter from Thomas Lynch to Joseph Williamson mentions that he has sent the king some cocoa and vanillas to Mr. Chiffins.
	1672	1670–1679	1672	Letter from Thomas Lynch to Lord Arlington mentions that he has sent Sir C. L[yttleton] a package of excellent wild cocoa, containing 125 pounds; 50 for the King, as much for his Lordship, and the rest for Lady Herbert.
	1672	1670–1679	1672	William Hughes identifies medicinal uses of chocolate, with specific comments on asthma: *The American Physitian [sic] . . . Whereunto is added a Discourse on the Cacao-Nut-Tree, and the Use of Its Fruit, with All the Ways of Making Chocolate.* "[Cacao] *being very nourishing . . . and keepeth the body fat and plump* . . . Chocolate is good against all coughs shortness of breath, opening and making the roughness of the artery smooth . . . [chocolate] strengthens the vitals and is good against fevers, catarrhs, asthmaes [sic], and consumptions of all sorts."
	1671	1670–1679	1671	Letter from Thomas Lynch to Joseph Williamson describes a nearly destitute island of Jamaica; "There is no money in the treasury, a dry season has blasted all the cocoa and sugar [plantings]." Begs for God's sake for frequent letters and directions.

Chocolate-Related Events in North America	Date ←	Decade ↔	Date →	Chocolate-Related Events Elsewhere and Historical Context
	1671	1670–1679	1671	Sylvestre Dufour reviews publication by Moreau and publishes *De l'Usage du Caphé, du Thé, et du Chocolat*.
	1671	1670–1679	1671	Nicephorus Sebastus Melissenus reviews the ecclesiastical views regarding chocolate in *De Chocolatis Potione Resolutio Moralis*.
	1671	1670–1679	1671	First French chocolatière reference is in a letter by the Marquise de Sévigné to her adult daughter, Madame de Grignan, whom she advises to take chocolate as a medical restorative for fatigue. The Marquise writes: "you have no chocolatière—I've thought about it a thousand times! How will you make it [chocolate]?"
Dorothy Jones and Jane Barnard open a public house in Boston and sell chocolate.	1670	1670–1679	1670	Hudson's Bay Company founded.
	1670	1670–1679	1670	Portuguese report presence of cacao beans during exploration of Amazon River headwaters.
	1670	1670–1679	1670	Commonly accepted year when cacao plantations began to be developed in the Philippines by Pedro Bravo do los Camerinos.
	1670	1670–1679	1670	First mention of chocolate in the Bayonne Municipal Archive. Used by town officials as gifts for persons of "consequence." Chocolate brought to France from Spain.
	1670	1670–1679	1670	Statement by Sir Thomas Modyford regarding Jamaica: argues that the island should be populated as quickly as possible and that owners of ships will willingly transport slaves, "the price being males 12£ to 15£, females 10£ to 12£ ready money, with which they buy cocoa which nearly doubles at their return, so that many [slaves] have been brought hither [to Jamaica] within these ten months."
	1667	1660–1669	1667	A. Vitrioli translates Moreau's French translation of Colmenaro de Ledesma's text into Italian: *Della Cioccolata Discorso*.
	1665	1660–1669	1665	Simon Paulli writes the uses/abuses of tobacco with commentaries on chocolate and tea: *Commentarius de Abusu Tabaci*.
	1664	1660–1669	1664	Colbert creates the West Indies Company and gives it trading monopoly in the French Antilles. D'Acosta and other Jews in Martinique lose the right to trade in cacao.
	1664	1660–1669	1664	Franciscus Maria Brancatius discusses chocolate: *De Chocalatis Potu Diatribe*.
	1662	1660–1669	1662	Henry Stubbe identifies medicinal uses of chocolate: *The Indian Nectar or a Discourse Concerning Chocolata*.

Chocolate-Related Events in North America	Date ←	Decade ↔	Date →	Chocolate-Related Events Elsewhere and Historical Context
	1661	1660–1669	1661	Samuel Pepys writes: "Went to Mr Bland's and there drank my morning draft of chocollatte."
	1661	1660–1669	1661	On orders of the Governor and Council of Jamaica, the price of cocoa in Jamaica fixed at 4 cents per pound.
	1660	1660–1669	1660	King Louis XIV of France marries Thérèse, another Spanish Infanta; she brings chocolate from Spain to France. This event, authenticated in memoirs of Madame de Montespan.
	1660	1660–1669	1660	Several sources indicate that the first cacao trees are cultivated in Martinique by Benjamin Andreade (or d'Andrade) D'Acosta, a Marrano Jew, fleeing possible Portuguese encroachment upon Curaçao.
	1660	1660–1669	1660	Anonymous author publishes erotic chocolate poem in *The Vertues [sic] of Chocolate, East-India Drink.*
	1660	1660–1669	1660	English Parliament imposes duty of 8 pence on every gallon of chocolate, sherbet, and tea, made for sale; tax on coffee and foreign spirits 5 pence.
	1659	1650–1659	1659	King Louis XIV of France grants David Chaliou royal patent to manufacture liquid, pastille (lozenge) or box form chocolate.
	1659	1650–1659	1659	Letter from William Dalyson to his cousin, Robert Blackborne, Secretary to the Admiralty Commissioners: he has sent cocoa in anticipation that his debts will be discharged; "hopes to get a dividend [sic] of a small prize taken upon the Main laden with cocoa."
	1657	1650–1659	1657	Chocolate as a beverage introduced in London: "In Bishopsgate Street, in Queen's Head Alley, at a Frenchman's house, an excellent West Indian drink called chocolate to be sold, where you may have it ready at any time, and also unmade at reasonable rates."
	1652	1650–1659	1652	J. Wadsworth translates Colmenero de Ledesma's text into English under his own name as: *Chocolate. Or an Indian Drinke.*
	1651	1650–1659	1651	Benjamin d'Acosta establishes first cacao-producing plant in New World on Martinique.
	1650	1650–1659	1650	First coffeehouse in England opens at Oxford; chocolate beverages sold: "Cirques Jobson, a Jew and Jacobite, borne neare Mount-Libanus, sold coffee in Oxon [Oxford] in an house between Edmund hall and Queen Coll. [Queen College] corner. Coffey, which had been drank by some persons in Oxon [Oxford] 1650, was this yeare publickly sold at or neare the Angle within the east gate of Oxon [Oxford]; as also chocolate, by an outlander or a Jew."

Chocolate-Related Events in North America	Date ←	Decade ↔	Date →	Chocolate-Related Events Elsewhere and Historical Context
	1649	1640–1649	1649	Cacao introduced into the Antilles.
	1648	1640–1649	1648	Thomas Gage identifies medicinal uses of chocolate.
	1647	1640–1649	1647	Inquisition document (Mexico) identifies loaves of chocolate belonging to Manuel Acosta.
	1645	1640–1649	1645	Lent celebrated with chocolate gifts at the Jesus Maria Convent, Mexico City.
	1645	1640–1649	1645	King of Spain issues royal decree recinding permission for vendors (defined as mestizo, mulatto, black, and Chinese) to sell cocoa and chocolate within all his dominions.
	1644	1640–1649	1644	J. G. Volckamer translates Colmenero de Ledesma's text into Latin: *Chocolata Inda, Opusculum de Qualitate et Natura Chocolatae.*
	1643	1640–1649	1643	Barthélémy Marradon publishes *Dialogue touchant le même chocolat,* as part of René Moreau's, *Du chocolat, discours curieux divisé en quatre parties.*
	1643	1640–1649	1643	René Moreau translates Colmenero de Ledesma's text into French: *Du Chocolat Discours Curieux.* He dedicates his work to Cardinal Alphonse de Richelieu of Lyon (brother of the more famous Richelieu), who reportedly used chocolate as a medicine for his spleen.
Spanish ship, *Nuestra Senora del Rosario del Carmen,* arrives in St. Augustine, Florida, with crates of chocolate.	1641	1640–1649	1641	*Body of Liberties* adopted in Massachusetts (presages the *Bill of Rights*).
	1641	1640–1649	1641	Don Diego de Vades-Forte (pseudonym of J. Wadsworth) translates Colmenero de Ledesma's text into English: *A Curious Treatise of the Nature and Quality of Chocolate.*
	1638	1630–1639	1638	Cacao trees introduced to Jamaica.
	1637	1630–1639	1637	Letter signed by King Philip IV of Spain instructs payment of 12,300 reales for 41 arrobas of chocolate (1025 pounds). Letter states that it was "his will that the persons who brought the chocolate to Madrid, would be paid.
	1636	1630–1639	1636	Antonio de Leon Peinelo publishes *Question Moral se el Chocolate Quebranta el Ayuno Eclesiastico:* moral issues regarding whether or not chocolate is a food, can be consumed during Lent and so on.
	1636	1630–1639	1636	Cacao introduced to Puerto Rico.

Chocolate-Related Events in North America	Date ←	Decade ↔	Date →	Chocolate-Related Events Elsewhere and Historical Context
	1631	1630–1639	1631	Colmenero de Ledesma publishes first book dedicated entirely to chocolate: *Curioso Tradado de la Naturaleza y Calidad del Chocolate, Dividido en Quatro Puntos.*
	1629	1620–1629	1629	Inquisition document (Mexico): declaration related to witchcraft. "Magdalena Mendes the half woman of Francisco Palacios, a Spanish man, said that a mulatto woman named María gave her a herb for her to give it to the man that she was with, so he will forget the other woman. And she told her that she should dissolve that herb in chocolate and give it to him four times. She also told her, that another thing she could do in order for him to forget the other woman, was to find the heart of a crow, mixed with the excrements of that woman, and give this potion to him mixed with his chocolate drink."
	1626	1620–1629	1626	Inquisition document (Mexico): declaration related to witchcraft. "In the city of Tepeaca . . . Before Mr. Domingo Carvajal y Sosa, Commissioner of the Holy Office, appeared, without being called and swearing to tell the truth, a woman who said her name was Ana Perdomo, widow of Ardiga, of this city, who is 28 years old. [A long, complicated story is told, basically: her husband was in love with a woman named Augustina de Vergara and Augustina] "gave him menstrual blood to drink in his chocolate."
	1624	1620–1629	1624	Santiago de Valverde Turices identifies medicinal uses of chocolate.
	1624	1620–1629	1624	Rauch publishes his *Disputatio Medico Dioetetica* and condemns cocoa as a beverage that inflames human passion.
	1621	1620–1629	1621	Inquisition document from Mexico, related to witchcraft, contains testimony denouncing a women whereby it was found "under her bed a piece of flesh from one of the quarters of a man who had been hanged, and that she had roasted it and mixed with her chocolate."
	1621	1620–1629	1621	Inquisition document from Mexico, related to witchcraft, contains a spell using dirt from a woman's grave mixed with chocolate to make a violent husband harmless.
	1616	1610–1619	1616	Report of wild cocoa trees growing in Trinidad.

Chocolate-Related Events in North America	Date ←	Decade ↔	Date →	Chocolate-Related Events Elsewhere and Historical Context
	1615	1610–1619	1615	Marriage of Anne of Austria, Infanta of Spain, daughter of Phillip III, with Louis XIII, commonly reported to be occasion for introduction of chocolate to France. Story cannot be confirmed and is considered apocryphal.
	1615	1610–1619	1615	Francisco Hernández publishes *Quatro Libros de la Naturaleza.* Provides sketch of cacao tree and discusses uses of cacao.
	1606	1600–1609	1606	Commonly cited year for the introduction of chocolate into Italy.
	1604	1600–1609	1604	English edition of José de Acosta's book on chocolate.
	1602	1600–1609	1602	Henry IV charges the Gramont family, who govern the city of Bayonne in the French Basque region, to offer protection to Jews fleeing the Inquisition in Spain. It is believed that some of these refugees were chocolate makers.
	1602	1600–1609	1602	Samuel de Champlain travels in the Caribbean; describes cacao trees and their uses: "there is a tree, which is called cacao . . . When this fruit is desired to be made use of, it is reduced to powder, then a paste is made, which is steeped in hot water, in which honey . . . is mixed and a little spice; then the whole being boiled together, it is drunk in the morning, warm . . . and they find themselves so well after having drunk of it that they can pass a whole day without eating or having great appetite."
	1598	1590–1599	1598	First mention of chocolate in France (French translation of Jose de Acosta's *Histoire naturelle et morale des Indes tant occidentales qu'orientales).*
	1592	1590–1599	1592	Agustin Farfan identifies medicinal uses of chocolate.
	1591	1590–1599	1591	Indians, with permits, granted right to sell chocolate on the streets of Mexico City.
	1591	1590–1599	1591	Juan de Cárdenas identifies medicinal uses of chocolate.
	1590	1590–1599	1590	Florentine Codex identifies medicinal uses of chocolate.
	1590	1590–1599	1590	José de Acosta identifies medicinal uses of chocolate.
	1575	1570–1579	1575	Giralimo Benzoni writes in his *History of the New World,* that chocolate is more a drink for pigs than for humans.
	1552	1550–1559	1552	Martin de la Cruz (Juan Badianus) writes and illustrates the Mexica herbal today known as the Badianus Codex. Contains image of a cacao tree.

Chocolate-Related Events in North America	Date ←	Decade ↔	Date →	Chocolate-Related Events Elsewhere and Historical Context
	1544	1540–1549	1544	Commonly cited year for the introduction of chocolate to the Spanish court by Kekchi Mayans brought to Spain by Dominican priests.
	1544	1540–1549	1544	Melchor and Alonso Pacheco establish small cacao farms in northern Belize.
Hypothetical: Cabrillo expedition to explore the Pacific coast of North America. Chocolate is not mentioned in the document but these are several tantalizing passages: "[Latitude 30 degrees 24 minutes North; 27 miles northwest of Punta Baja, and five or six miles southwest of the village of San Quentín]. On Thursday they saw some smokes and, going to them with the boat, they found some thirty Indian fishermen, who remained where they were. They brought to the ship a boy and two women, gave them clothing and presents, and let them go." Was chocolate among the presents?	1542	1540–1549	1542	Dutch War of Independence from Spain establishes Dutch Republic.
	1526	1520–1529	1526	Gonzalo Fernández de Oviedo y Valdéz publishes La historia general y natural de las Indias; identifies chocolate as healthy.
	1524	1520–1529	1524	Account by Father Toribio de Benavente, one of "The Twelve" Franciscan priests sent to Mexico, describes use of cacao beans as money, and as a general beverage: "This cocoa is a very general drink; they grind it and mix it with corn and other grund seeds; and this is also a major use of the cocoa seeds. It is good and it is considered as a nutritious drink."
	1519	1510–1519	1519	Hernán Cortés marches overland to Tenochtitlan; enters the Mexica/Aztec city; drinks chocolate with Montezuma.
	1502	1500–1509	1502	Christopher Columbus's son, Ferdinand, describes events related to the fourth voyage of his father to Honduras and Panama: first mention of cacao/chocolate in a European language. "[Our men brought the canoe alongside the flagship] For provisions they had such roots and grains as the Indians of Española eat, also a wine made of maize [i.e. chicha] that tasted like English beer. They had as well many of the almonds which the Indians of New Spain use as currency; and these the Indians in the canoe valued greatly, for I noticed that when they were brought aboard with the other goods, and some fell to the floor [of the ship], all the Indians squatted down to pick them up as if they had lost something of great value."
	1492	1490–1499	1492	Columbus reaches Caribbean landfall.
		1490–1499	1492	Ferdinand and Isabella finance voyage of Christopher Columbus. Expulsion of Jews and Muslims from Spain.
		1480–1489	1480	Start of Spanish Inquisition.
		1470–1479	1479	Union of the kingdoms of Aragon and Castile under Ferdinand and Isabella.
		1450–1459	ca. 1450	Chocolate mentioned in poem by King of Texcoco, Netzahualcoyotl.
		1440–1449	1447	Johann Gutenberg develops movable type printing press in Europe.

Chocolate-Related Events in North America	Date ←	Decade ↔	Date →	Chocolate-Related Events Elsewhere and Historical Context
		Uncertain	15th century	Nuttal Codex from Oaxaca, Mexico, illustrating exchange of cacao during a wedding ceremony.
		Uncertain	15th century	Madrid Codex, pages showing chocolate and four deities piercing their ears with obsidian blades; flower god Nik in association with a cacao tree; wedding ceremony featuring honeycomb and cacao beverage.
		Uncertain	Late 14th, early 15th century	Dresden Codex depiction of Mayan god of sustenance K'awil holding a bowl containing cacao.
		1350–1359	1350	Beginning of the Inca empire in modern Peru.
		1340–1349	1340	Start of the Black plague epidemic in Europe that ultimately will kill one-third of the European population.
		1320–1329	ca. 1325	Tenochtitlán (modern day Mexico City) founded by Mexica/Aztecs.
		1230–1239	1234	Chae Yun-eui invents printing press.
		1210–1219	1215	Magna Carta signed.
		1040–1049	ca. 1041	Bi Sheng invents movable type.
		1000–1009	ca. 1004	Viking settlements established in Labrador and Newfoundland by Thorvald Erikson and Thorfinn Karlseffin.
		1000–1009	ca. 1000	Leif Erickson crosses Atlantic, reaches North America.
		900–1000	900	Red Temple at Cacaxtla, Tlaxcala, Mexico with mural of Mayan merchant deity, God L, pictured in front of a cacao tree.
		880–899	881	Painted capstone from the Temple of the Owls, Chichén Itzá, Mexico, showing emergence of the deity K'awil from the Underworld.
		Uncertain	600–900	Bowl with cacao beans carved out of shells, from royal tomb at Ek' Balam, Yucatán, Mexico.
		460–469	ca. 450	Vessel from Tomb 19, Rio Azul, Guatemala, contains chocolate residue.
		600–609 BCE	ca. 600 BCE	Excavated Mayan pot from Colha, Belize, contains chocolate residue.
		1000–1009 BCE	ca. 1000 BCE	First pottery appears along the Atlantic coast of North America.
		1100–1109 BCE	ca. 1100 BCE	Analysis of pottery fragments from Rio Ulúa Valley, northern Honduras, reveal presence of theobromine and suggested use of a "chocolate beer."
		1500–1509 BCE	ca. 1500 BCE	Suggested date for the domestication of *Theobroma cacao* in northeastern Ecuador.
		ca. 4000 BCE	ca. 4000 BCE	First documented human settlements along the Pacific coast of North America.

Early Works on Chocolate

A Checklist

Axel Borg and Adam Siegel

*C*hocolate and books are a wonderful combination. In this book about chocolate, we offer a preliminary list of early books about chocolate. In examining this list, interesting patterns and pathways can be seen as chocolate originally was brought to Spain from the Americas and from there spread across Europe.

The first substantial book on chocolate was written by Antonio Colmenero de Ledesma and published in Madrid in 1631. Like chocolate, this book spread across Europe. A French translation was printed in Paris in 1634 and followed by an English translation in 1640 and a Latin translation printed in Nurnberg in 1640. In addition to his book, Colmenero de Ledesma's work was incorporated into other books about chocolate, sometimes with credit and at other times without. Little is known about Antonio Colmenero de Ledesma other than he appears to have been an Andalusian physician. Perhaps he was a Morisco, a Jew forcibly converted to Christianity. Colmenero de Ledesma's book is on the health aspects of chocolate, a theme that continues today.

One very significant pattern to note is the coupling of three beverages: chocolate, coffee, and tea. When Spanish conquistadors first encountered chocolate, it was a beverage, and with its introduction to Europe, it remained a beverage for many years. This triumvirate of stimulants proved very popular with all social classes, as is documented both by the items in this list and the texts in this book.

Another early work on chocolate by Antonio de León Pinelo, published in Madrid in 1636, addressed its theological aspects. Today, we have a good laugh about the "sinfulness" of chocolate. However, when Europeans first encountered chocolate, issues of its morality were of great concern. After all, it was a new food, from a new world, and not part of accepted, and understood, Christian culture (Diaz del Castillo believed the chocolate drink to be an aphrodisiac, a reputation never quite dispelled). In 1645, Tomás Hurtado, a professor of theology at the University of Salamanca, addressed both chocolate and tobacco. Both of these plants have played a significant role in human nutrition and health. Later, *De chocolates potione resolution moralis* . . . (3rd edition, 1671) by Sebasto Nicéforo continues to raise the issue of morality.

Pre-19th century publication patterns are in their own right revealing. While initial accounts of chocolate and cacao are unsurprisingly found in primary sources—Cortés, Columbus, Diaz del Castillo, and so on—the *Urtext*, more or less, for much of the 17th century's publicaton history is Colmenero de Ledesma. By the late 1600s, we see a print culture in which simultaneous publications, pirated editions, misattributions, and murky attributions are commonplace, with a handful of works—Chamberlayne, Colmenero de Lesdesma, Dufour, Buc'hoz, and so on—appearing in multiple editions and translations throughout Europe.

Chocolate: History, Culture, and Heritage. Edited by Grivetti and Shapiro
Copyright © 2009 John Wiley & Sons, Inc.

The locus of this activity is primarily northern European centers of trade and scholarship.

The following on-line union catalogs were consulted in the compilation of this checklist, which aspires to comprehensiveness: Worldcat (OCLC), the Karlsruhe Virtueller Katalog (including all represented German union catalogs and OPACs), the *Verzeichnisse der deutschsprachigen Schrifttums der 16. und 17. Jahrhunderten*, national union catalogs (France, England, Austria, Italy, Spain, Portugal), and, in print, Wolf Mueller's *Bibliographie des Kaffee, des Kakao, der Schokolade, des Tee und deren Surrogate bis zum Jahre 1900* (Bad Bocklet: Walter Krieg Verlag, 1960). All research was done at the Peter J. Shields Library in Davis, California. Items are listed in reverse chronological order and include duplicate and fugitive editions. As a preliminary list, we hope that this will prompt other bibliographers and scholars to add to our initial effort. We are confident that there are variant editions as well as books in other languages, just waiting to be rediscovered and enjoyed anew.

1700–1800

Effeitos do chocolate!—:cançoneta original do auctor e representada em familia por um irmão do Senhor dos Passos. Lisboa: Typ. Rua do Telhal, 1800–1899? 8 pp. 15 cm.

Marschall, C. F. *Surrogate des Kaffees, Zuckers und Thees, un schädliche Schminken, Zahnschmerzstill. Pulver, Liqueurs, Tuschen, Kräuterbiere und -Weine, Chocoladen, Marzipans, Syrup.* Leipzig, 1799. 8o. VI, 185, (16) S.

United States Congress. House. Committee of Commerce and Manufactures. *Report of the Committee of Commerce and Manufactures, on the memorials of sundry manufacturers of chocolate 8th February 1797, referred to the committee of the whole House, to whom is committed the report of the Committee of Ways and Means, of the 3d ultimo, on the subject of further revenues.* Published by order of the House of Representatives. Philadelphia: Printed by William Ross?, 1797. 4 pp. 23 cm (8vo).

O Fazendeiro do Brazil melhorado na economia rural dos generos já cultivados, e de outros, que se podem introduzir; e nas fabricas, que lhe são proprias, segundo o melhor, que se tem escrito a este assumpto ... / colligido de memorias estrangeiras por Fr. José Mariano da Conceição Velloso; [il.] Manuel Luís Rodrigues Vianna por Veloso, José Mariano da Conceição, 1742–1811, O.F.M., compil.; Viana, Manuel Luís Rodrigues, 1770–?, il.; Ferreira, Simão Tadeu, fl. 178– 182-, impr.; Silva, João Procópio Correia da, fl. 1798– 1800, impr.; Régia Oficina Tipográfica, impr. Lisboa: na Regia Officina Typografica, 1798–1806. 10 vol.: il.; 8o (19 cm). A Casa Literária do Arco do Cego (1799–1801) 137. Tomo III–Parte 3: Bebidas Alimentosas. Cacao. Lisboa: na Impressam Regia, 1806. [12], 349, [3] p. 1 grav. desdobr.

Lavedan, Antonio. *Tratado de los usos, abusos, propiedades y virtudes del tabaco, café, té y chocolate: extractado de los mejores autores que han tratado de esta materia, á fin de que su uso no perjudique á la salud, antes bien pueda servir de alivio y curación de muchos males.* Madrid: En la Imprenta Real, 1796. [10], 237, [1] p.; 20 cm.

Smith, H., fl. 1794–1830. *An essay on the nerves, illustrating their efficient, formal, material, and final causes; with a copperplate, ... To which is added an essay on foreign teas; ... to demonstrate their pernicious consequences on the nerves, ... With observations on mineral waters, coffee, chocolate, &c.* London: printed by P. Norman; for G. G. and J. Robinson, 1795. Plate; 80 pp.

La manovra della cioccolata e del caffé trattata per principi da F. Vincenzo Corrado, 2nd edition. In Napoli: nella stamperia di Nicola Russo, 1794. 63 pp.; 19 cm (8vo).

Seidenberg, John Gottlieb. *Anweisung für Frauenzimmer die ihrer Wirthschaft selbst vorstehen wollen Zweytes Stück: vom Mariniren, Einsalzen und Räuchern der Fische, vom Hausschlachten, vom Kaffee, vom Thee, und von der Chocolate.* Berlin: Wever, 1790. 126 pp.

Tea, coffee and chocolate, at T. Boot's warehouse, no. 212, in Piccadilly, the corner of Eagle-Street; where the nobility, gentry, and others, may be supplied with some fine fresh teas, coffee and chocolate. London: T. Boot, 1790. Broadside.

Osasunasco, Desiderio de. *Observaciones sobre la preparacion y usos del chocolate.* Mexico: D. Felipe de Zuñiga y Ontiveros, 1789. [14] pp. 15 cm.

Sbalbi, Giambattista ital. et lat. *La cioccolata. Versione dal Latine in rime Toscane (da) Giambatista Sbalbi. (Text. lat. et ital.).* Piacenza, Dr.: Orcesi, 1789. XVI pp.

Lardizábal, Vicente. *Memoria sobre las utilidades de el chocolate Para precaber las incomodidades, que resultan del uso de las aguas minerales, y promover sus buenos efectos, como los de los purgantes, y otros remedios: y para curar ciertas dolencias.* Pamplona: Por Antonio Castilla, impresor, 1788. 31 (i.e., 21) pp. 23 cm.

Buc'hoz, Pierre-Joseph, 1731–1807. *Dissertations sur l'utilité, et les bons et mauvais effets du tabac, du café, du cacao et du thé, ornées de quatre planches en taille-douce par M. Buc'hoz, médecin de Monsieur, frère du Roi, ancien médecin de Monsieur Comte d'Artois, & de seu Sa Majesté le Roi de Pologne, Duc de Lorraine & de Bar, membre de plusieurs académies, tant etrangeres que nationales,* Seconde édition. Paris: chez l'auteur, rue de la Harpe, no. 109. De Bure, l'ainé, libraire, rue Serpente, et la veuve Tilliard, & fils, libraires, rue de la Harpe, 1788. [4], 185, [1] p., [4] folded leaves of plates: ill.; 22 cm (8vo). *Note:* A Paris and Liège edition printed in 1787 is dismissed as a pirated edition by the author in the preface (see below); a Liège edition was reprinted in 1789.

Buc'hoz, Pierre-Joseph. *Dissertation sur le cacao, sur sa culture, et sur les différentes des préparations de chocolat.* Paris: self-published, 1787. 12 pp. 2 tables; 2″.

Anna Fry and Son. *Patent cocoa, genuine and unadulterated, made by Anna Fry and Son, patentees of Churchman's chocolate, in Bristol.* (Bristol, 1787–1803?) 1 sheet; 1/40.

Parmentier, A. A. *Abus dans la fabrication du chocolat.* Paris, 1786. 6 pp.

Bergius, Bengt. *Tal om laeckerheter, både i sig sjelfva sådana, och för sådana ansedda genom Folkslags bruk och inbillning.* Stockholm: Lange, 1785. [5], 272, [1] p.; 8.

Proposals for opening a Scotch eating house, or north country ordinary, and Scotch chocolate house, in the neighbourhood of St. James's. London: R. Bassam, 1785. Broadside. *Note:* Broadside–satire.

Tissot, Wilhelm. *Die Weinflasche, Chocolaten- und Theekanne, ein curiöses Gespräch eines Herren mit einer Dame von Wein, Chocolate, Thee, Branntwein, Bier, Wasser, Schnupf- und Rauchtabak.* [s. n.]. 1782. 8o. 48 pp.

Anfossi, Giovanni Battista. *Dell'uso ed abuso della cioccolata.* In Venezia: Appresso Francesco Locatelli, 1779. 100 pp. 21 cm.

Navier, Pierre Toussaint, 1712–1779. *Bemerkungen über den Cacao und die Chokolate: worinnen der Nutzen und Schaden untersuchet wird, der aus dem Genusse dieser nahrhaften Dinge entstehen kann. Alles auf Erfahrung und zergliedernde Versuche mit der Cacao-Mandel gebauet. Nebst einigen Erinnerungen über das System des Herrn de la Mure, betreffend das Schlagen von Puls-Adern. Aus dem Französischen übersetzt. Nebst einer Vorrede D. Carl Christian Krausens.* Naumburg und Zeitz: Bey Heinr. Wilh. Friedr. Flittner, 1776. [10] leaves, 164 pp.; 17 cm (8vo).

Navier, Pierre Toussaint. *Bemerkungen über den Cacao und die Chocolate, worinnen der Nutzen und Schaden untersuchet wird, der aus dem Genusse dieser nahrhaften Dinge entstehen kann: Alles auf Erfahrung und zergliedernde Versuche mit der Cacao-Mandel gebauet; Nebst einigen Erinnerungen über das System des Hrn. De-La-Müre, betreffend das Schlagen der Puls-Adern. Aus dem Französischen übersetzt / Nebst einer Vorrede. D. Carl Christian Krausens.* Leipzig: Saalbach, 1775. Drucker: Saalbach, Ulrich Christian. Umfang: [10] s., 164 pp. Anmerkung: Die Vorlage enth. insgesamt. Werke Vorlageform des Erscheinungsvermerks: Leipzig, bey Ulrich Christian Saalbach, 1775.

El bo. frei d. Antonio Maria Bucareli, y Ursúa, Henestrósa, Laso de la Vega, Villacis y Córdova . . . *Por quanto conducido el Real tribunal del consulado de esta capital del notorio zelo y amor que siempre há demostrado por el bien del público y de su continua vigilancia en el arreglo del comercio . . . me representó los engaños y perjuicios que sufria el reyno en la provision del cacao.* Mexico, 1775. [3] pp.

Muzzarelli, A. *Il tempio della fedeltà, La cioccolata, La bottega del caffè; poemetti.* Bologna, 1774. [xciv]. 4o.

Navier, Pierre Toussaint. *Observations sur le cacao et sur le chocolat: où l'on examine les avantages et les inconvéniens qui peuvent résulter de l'usage de ces substances nourricières. Suivies de réflexions sur le systême de M. de Lamure, touchant le battement des artères.* Paris: chez P. Fr. Didot jeune, libraire Quai des Augustins, a Saint Augustin, 1772. [4], 152 pp. 12o.

Merli, Francesco. *Guida medica, intorno all'uso del the, caffè, e cioccolata.* Napoli, Vincenzo Flauto, 1768. [4], 111 pp. 20 cm.

Buc'hoz, Pierre-Joseph. *Dissertations 1) sur le tabac, 2) le café. 3) le cacao, 4) l'ipo, 5) e quassi . . . : Ces dissertations formant* une nouvelle édition de l'Histoire générale et économique des trois regnes. Paris, 1767. Colored tables; 2″.

Linné, Carl von, 1707–1778, praeses. *Dissertatio medica inauguralis de potu chocolatae . . .* Holmiae: Literis direct. L. Salvii, 1765. 10 pp. 18 × 15 cm. *Note:* Hoffman, Anton, 1739–1782; respondent.

Maria Theresia, Kaiserin von Österreich. *Über den Zoll auf Kaffee, Thee und Schokolade.* Wien, 30 März 1763. Folio. 2 pp.

Stayley, George, 1727–1779? *The rival theatres, or, A play-house to be let a farce; to which is added The chocolate-makers, or, Mimickry exposed: an interlude: with a preface and notes, commentary and explanatory by Mr. George Stayley, comedian.* London: Dublin printed and London reprinted for W. Reeve, 1759. 4 pp. l., 44 pp. 21 cm.

Der geschickte Wein- und Bier-Künstler: mit einem nützlichen Unterricht Thee, Coffe, Chocolate, und Mandelate zu machen: wie auch allerhand Brandtewein, Essig und andere gebrandte Wasser zu verfertigen: allen Hausvätern und Hausmüttern als ein Handbuch zum täglichen Gebrauch herausgegeben. Frankfurt und Leipzig: Bey Johann Georg Cotta, 1755. [8], 158, [10] pp. 17 cm (8vo).

Young, Edward, 1683–1765. *A dish of chocolate for the times addressed to the Reverend Edward Young, LL.D.* Dublin: [s.n.], 1754. 14 pp. 20 cm.

The female parliament: a seri-tragi-comi-farcical entertainment, nver acted in Eutopia before: wherein are occasionally exhibited, the humours of Fanny Bloom and Lady Niceairs: together with the amours of Sir Timothy Fopwell and Justice Vainlove; with several pieces of chocolate-scandal, tea-calumny, and midnight-slander, entirely new. London: Printed next door to the Saddle on the Right Horse, 1754. 57 pp. folded front; 18 cm.

Díaz Bravo, José Vicente, d. 1771. *El ayuno reformado, segun practica de la primitiva iglesia, por los cinco breves de . . . Benedicto XIV: Obra historica, canonico-medica . . . Con noticia particular de los privilegios, que aun despues de los breves, gozan en España los soldados. Y una disertacion historica, medico-chymica . . . de el chocolate, y su uso, despues de los nuevos preceptos.* Pamplona: En la oficina de Pasqual Ibañez, 1753. [32] leaves, 347 pp. [8] leaves; 21 cm (4to). *Note:* Palau; 72311.

Gómez Arias. *Tratado physico medico, de las virtudes, qualidades, provechos, uso, y abuso, del cafè, del the, del chocolate, y del tabaco: su autor don Gomez Arias.* Madrid: En la Imprenta de los Herederos de Francisco del Hierro, 1752. [4], 11 pp. 20 cm.

Navas de Carrera, Manuel. *Dissertacion histórica phisico-Chimica, y analysis del cacao, su uso, y dossis que para beneficio comun da al publico Don Manuel Navas de Carrera.* Zaragoza: por Francisco Moreno, 1751. Sign. a4–c4, 62 pp.

Schreiben eines Cavalliers an ein Fräulein betreffend den neuerfunden Damen-Thee, Caffee und Chocolate, oder eigentlich Mandelate. 2. Auflage. Erfurt, 1749 in dem Loberischen Buchladen. 8o. 24 pp.

Concina, Daniele, 1687–1756. *Memorie storiche sopra l'uso della cioccolata in tempo di digiuno: esposte in una lettera a Monsig. illustriss., e reverendiss. Arcivescovo N.N.* Venezia: Appresso S. Occhi, 1748. [4], cxcvi pp. 20 cm.

Paulli, Simon, 1603–1680. *A treatise on tobacco, tea, coffee, and chocolate: in which i. The advantages and disadvantages attending the use of these commodities, are not only impartialy considered, upon the principles of medicine and chymistry, but also ascertained by observation and experience. ii. Full and distinct directions laid down for knowing in what cases, and for what particular constitutions, these substances are either beneficial, or hurtful. iii. The Chinese or Asiatic tea, shewn to be the same with the European chamelæagnus, or myrtus brabantica: the whole illustrated with copper plates, exhibiting the tea utensils of the Chnese and Persians written originally by Simon Pauli, and now translated by Dr. James.* London: Printed for T. Osborne . . . [et al.], 1746. [1], 171, [1] pp. [2] folded leaves of plates: ill.; 20 cm. *Note*: Translation of: *Commentarius de abusu tabaci americanorum veteri, et herbae thee asiaticorum in Europa novo.*

Milhau, Mr. (Jean Louis). *Dissertation sur le cacaoyer: presentée a messieurs de la Societé-Royale des Sciences do Montpellier par Mr. Milhau.* A Montpellier: De l'Imprimerie de Jean Martel, Imprimeur du Roi, 1746. 32 pp. 18 cm (8vo).

Bianchini, Giuseppe Maria, 1685–1749. *Saggi di poesie diverse dell'illustrissimo, e clarissimo sig.senatore Marcello Malaspina de'marchesi di Filattiera e Terra Rossa.* Firenze: Nella stamperia di Bernardo Paperini, 1741. 2 pts. in 1 v. (xxxi, [1], 131, [1]; viii, 42, [2] pp.), [1] leaf of plates: port.; 27 cm (4to). *Note*: Second part is: Bacco in America: componimento ditirambico in lode della cioccolata / colle note fatte al medesimo dal dottore Giuseppe Bianchini di Prato.

Baron, Hyacinth Théodore et Le Monnier, Ludov. Guil. *An Senibus Chocolatae potus?* Paris, 1739. 4o.

Boeckler, Joh. Philipp. *Dissertatio de Chocolate Indorum.* Argentorati, 1736.

Disputatio Medica Inauguralis De Balsamo Cacao. Qvam ~D. Sub Praesidio Andreae Ottomari Goelicke, ~D. D. VI. Jul. Ann. MDCCXXXVI. ~D. publice ~D. suscipit Carolus Fridericus Semprecht, Wolavia-Silesius. Francofvrti Cis Viadrum: Schwartze, 1736. 18 pp.

Eschweiler, Joh. Mart. Jos. *Disertatio inauguralis medica de Chocolata Indorum . . . Ejusque viribvs medicis . . . Praeside Dn. D. Ivone Joanne Stahl, . . . Pvblico ervditorvm examini svbmittet. Jo Martin. Joseph Eschweiler.* Erffurtensis, 1736.

Mauchart, Burchard David, 1696–1751. *Butyrum cacao novum atque commendatissimum medicamentum praeside Burcardo Davide Maucharto . . . pro licentia honores et privilegia doctoris medici rite capessendi . . . publico submittit examini Theophilus Hoffmann.* Tubingensis. Die Maj 1735. 24 pp. 4o.

Raminasius, Georg. *Chocolata, sive in laudem potionis Indicae quam appellant Chocolate.* [s.n.], 1733. 11 pp. 8o.

Waltenhofen, F. Xaver von. *Dissertatio de usu et abusus potus Chocolatae.* Oeniponti, 1733. 4o. 22 pp.

Great Britain. Laws, statutes, etc., 1727–1760 (George II). *Anno regni Georgii II. regis Magnæ Britanniæ, Franciæ, & Hiberniæ, quatro. At the Parliament begun and holden at Westminster, the twenty third day of January, anno Dom. 1727. In the first year of reign of Our Sovereign Lord George II. . . . And from thence continued by several prorogations to the 21st day of January, 1730 . . .* [An act to prevent fraud in the revenue of excise, with respect to starch, coffee, tea, and chocolate.] London: Printed by the assigns of His Majesty's printer, and of H. Hills deceas'd, 1731. 24 pp. 15 cm.

Discorso apologetico por el Chocolate. Madrid, 1730.

Quélus, D. *The natural history of chocolate being a distinct and particular account of the cocoa-tree . . . : wherein the errors of those who have wrote upon this subject are discover'd, the best way of making chocolate is explain'd . . . translated from the last edition of the French by R. Brookes, M.D.* The 2nd edition (reissue) London: Printed for J. Roberts, 1730. viii, 95 pp. 21 cm.

T., J. G. *Ein vor alle Menschen wohleingerichtetes Gesundheits-Büchlein: von den Verhalten der Gesunden und Krancken, wie sich die meisten übel bey ihrer Gesundheit verhalten, und wie sie sich besser verhalten könten, wobey von vielen Dingen, als von glücklichen Kinderzeugen, Erziehung der Kinder, von den Ammen, von Speisen, was, wenn, wie viel, ein ieder Mensch essen soll, vom Trincken, vom Toback, Wein, Thee, Coffee, Chocolate gehandelt: item, von der Diaet der Krancken, von Medicamenten, Haussmitteln u. ein Sentiment gefället wird.* [S.l.: s.n.], 1730. 80 pp. 18 cm.

Cortijo Herraiz, Tomás. *Discurso apologetico, medico astronomico pruebase la real influencia de los cuerpos celestes en estos sublunares, y la necessidad de la observancia de sus aspectos, para el mas recto uso, y exercicio de la medicina: Con un examen sobre el uso de el chocolate en las enfermedades: Escrito por el Doct. Don Thomas Cortijo Herraiz . . . Dedicado al maestro don Diego de Torres . . . Con licencia.* Salamanca: Impr. E. Garcia de Honorato y San Miguèl, 1729. 24 p. l., 120 pp. 16vo.

Cornaro, Luigi, 1475–1566. *Herrn Ludovici Cornelii, Eines Edlen Venetianers Goldener Tractat, Von der Nutzbarkeit eines nüchtern Lebens . . . von dem . . . Patre Lessio . . . ausz der welschen Sprach in die Lateinische übersetzt; und endlich von . . . Bernardino Ramazzini . . . mit . . . Anmerckungen erkläret . . . in die Teutsche Sprach übersetzt worden; Dann folget auch . . . Gio: Battista Felici, Parere über den Gebrauch der Ciocolata . . . in die Teutsche Sprach übertragen worden.* Brünn: bey J.M. Swoboda, 1729. [2], 131 pp. 17 cm (8vo).

Avanzini, Giuseppe, 18th century. *Lezione accademica in lode della cioccolata del dottore Giuseppe Avanzini . . . ; recitata nell'Accademia degli apatisti il dì ultimo di maggio dell'anno 1728.* Firenze: Nella stampería di Bernardo Paperini, 1728. 32 pp. 22 cm (4to).

Altro parere intorno alla natura, ed all'uso della cioccolata disteso in forma di lettera indirizzata all'illustrissimo signor conte

Armando di Woltsfeitt. Contiene anche: Scherzo ditirambico in lode della cioccolata di Girolamo Giuntini; In lode dell'albero che produce il cacao di B.P.S.E Dedicatoria di Francesco Zeti Iniziali e fregi xil. Opera attribuita al Giuntini (cfr. G. Melzi, *Dizionario di opere anonime e pseudonime* . . . , v. 1, p. 40 Segn.: A-D4 Vignetta calcogr. sul front.) Firenze: si vende allato alla chiesa di Sant'Apollinare, 1728. 32 pp. 4o.

Brückmann, Franz Ernst, 1697–1753. *Relatio brevis historico-botanico-medica de avellana Mexicana, vvlgo cacao. / Dicta, rec. et delineata a Fr. Ernest. Bruckmann. Ed. 2., priori auctior.* Brunsvigae: Schroeder, 1728. Drucker: Schroeder, Ludolph. 29 pp. [1] leafl.: Ill.; 8° (4°).

Felici, Giovanni Batista. *Parere intorno all'uso della cioccolata scritto in una lettera dal conte dottor Gio Batista Felici all'illustriss. signora Lisabetta Girolami d'Ambra.* Firenze: Appresso G. Manni, 1728. xi, [1], 88, 32, [8], 22 pp. 22 cm.

Mutain (Le Fevre, Jean François). *De natura, usu et abusu Caffé, Thé, Chocolatae et Tabaci. Thèse Medical.* Vesontione, Couché, 1726. 40 pp. 4o. (Attribution per Mueller.)

Quélus, D. *The natural history of chocolate: being a distinct and particular account of the cocoa-tree, its growth and culture, and the preparation, excellent properties, and medicinal vertues of its fruit. Wherein the errors of those who have wrote upon this subject are discover'd; the best way of making chocolate is explain'd; and several uncommon medicines drawn from it, are communicated.* The 2nd edition. London: Printed for D. Browne, junr., 1725. viii, 95, [1] p.; 21 cm (8vo).

Ordonantie, na welke ridderschap en steden de Staten van Overyssel, hebben goedgevonden te introduceren een impositie op de consumptie van coffy, thee, chocolate en andere diergelyke specien met water gemengt, by forme van taxatie. Te Zwolle: gedrukt bij D. Rampen en F. Clement, 1725. v.; in-4.

Wir Burgermeistere und Rath, dieser des Heiligen Reichs Stadt Franckfurt, fügen hiemit zu wissen: Demnach uns die hiesige Caffè-Schencken, mittels überreichten Memorialis, wehmüthig zu vernehmen gegeben, wie ob sie wohlen allein berechtiget wären, allhier einen öffentlichen Caffè-Schanck zu treiben, und dafür der hiesigen Stadt Aerario ein nahmhafftes Stück Geldes beyzutragen hätten, sich dannoch seith einiger Zeit vielerley ihnen in solcher ihrer Profession und Nahrung höchst schädliche und nachtheilige Beeinträchtigungen darinnen geäussert und hervor gethan hätten, daß auch die Gast- und Baum-Wirthe, Zuckerbeckere, Parfumeurs, Bierbräuere, die Köche und Speißmeistere bey denen Hochzeiten . . . Caffè, Thé, Chocolate, Rossoli und dergleichen, umbs Geld ausschenckten, . . . ; Daß Wir dannenhero keinen Umbgang nehmen können, hierinnefalls ein ernstlich Einsehen zu haben, und solchem Unwesen mit Nachtruck zu steuren und abzuhelffen: Es wird also hiemit und in Krafft dieses, sowol denen öffentlichen Gast-Wirthen, . . . die Ausschenckung des Caffè, Thé, Chocolate, Rossoli, und alle dergleichen Geträncks umbs Geld, hiemit ein vor alle mahl . . . verbotten . . . : Geschlossen bey Rath, Donnerstags, den 9. Augusti 1725. Frankfurt (Main): [S.l.], 1725.

The Case of the drugists, grocers, and other dealers in coffee, tea, chocolate &c. in relation to the bill now depending concerning those commodities. London: [s.n.], 1724. Broadside.

An act for repealing certain duties therein mentioned, payable upon coffee, tea, cocoa nuts, chocolate, and cocoa paste imported: and for granting certain inland duties in lieu thereof: and for prohibiting the importation of chocolate ready made, and cocoa paste: and for better ascertaining the duties payable upon coffee, tea, and cocoa nuts imported. London: Printed by John Basket . . . and by the Assigns of Henry Hills, 1724. 114 pp. 16 cm (12mo).

The Case of the wholesale and retale dealers in coffee, tea, chocolate, &c., in relation to the bill now depending. S.l.: s.n., 1724. Broadside.

Great Britain. Commissioners of Excise. Instructions for officers in the country, who take account of coffee, tea, cocoa nuts, and chocolate, and survey chocolate-makers. London: Printed by J. Tonson, 1724. 23 pp. 23 × 10 cm.

Quélus, D. *The natural history of chocolate: Being a distinct and particular account of the cocao-tree, its growth and culture, and the preparation, excellent properties, and medicinal vertues of its fruit. Wherein the errors of those who have wrote upon this subject are discover'd; the best way of making chocolate is explain'd; and several uncommon medicines drawn from it, are communicated being published with the particular approbation of Mons. Andry, counsellor, lecturer, and Regal Professor, Doctor, Regent of the Faculty of Medicine at Paris, and Censor-Royal of books; Translated from the last edition of the French, by a physician.* London: Printed for J. Roberts, near the Oxford-Arms in Warwick-Lane, 1724. viii, 95 [1] p.; 19 cm.

Great Britain. Commissioners of Excise. *Instructions for officers in the country, who take account of coffee, tea, cocoa nuts, and chocolate, and survey chocolate-makers.* London: Printed by J. Tonson, 1724. 23 pp. 23 × 10 cm.

A Table of the matters contained in the act for laying inland duties on coffee, tea, and chocolate. London: [s.n.], 1724. 41 pp. 16 cm (12mo).

Anno regni Georgii regis Magnæ Britanniæ, Franciæ, & Hiberniæ, decimo. At the Parliament begun and holden at Westminster, the ninth day of October, anno Dom. 1722. *In the ninth year of the reign of Our Sovereign Lord George . . . And from thence continued by several prorogations to the ninth day of January, 1723 . . . [An act for repealing certain duties therein mentioned, payable upon coffee, tea, cocoa nuts, chocolate, and cocoa paste imported; and for granting certain inland duties in lieu thereof; and for prohibiting the importation of chocolate ready made, and cocoa paste, and for better ascertaining the duties payable upon coffee, tea, and cocoa nuts Imported.]* London: Printed by J. Baskett, printer to the King's Most Excellent Majesty, and by the assigns of H. Hills, deceas'd, 1724. 114 pp. 2 l., 41 pp. 15 cm.

The Case of the drugists, grocers, and other dealers in coffee, tea, &c. in relation to the bill now depending concerning those comodities. London, 1723. 1 l.; 34 cm.

Recueil de chansons sur l'usage du Caffé, du Chocolat et du Ratafiat, avec leurs propriété et la manière de les bien préparer. Paris, 1723. 24 pp. 12o.

Of the use of tobacco, tea, coffee, chocolate, and drams, under the following heads: I. Of smoking tobacco. II. Of chewing. II. Of snuff. IV. Of coffee & its grounds. V. Of tea. VI. Of chocolate. VII. Of drams. London: Printed by H. Parker, 1722. 15, [1] p.: ill.; 18 cm (8vo).

Broadbent, Humphrey. *The domestick coffee-man, shewing the true way of preparing and making of chocolate, coffee and tea by Humphrey Broadbent.* London: Printed for E. Curll . . . and T. Bickerton, 1722. [4], 26, [2] pp. 20 cm.

Meisner, Leonhard Ferdinand. *Leonh. Ferdinand. Meisneri, Med. Doct. & Prof. Regii, De caffe, chocolatae, herbae thee ac nicotianae natura, usu, et abusu anacrisis, medico-historico-diaetetica.* Norimbergae: Sumpt. Joh. Frider. Rudigeri, 1721. [2], 124 pp. 4 leaves of plates (2 folded): ill. (engravings); 17 cm (8vo). *Note:* Originally published as a dissertation (L. F. Meisner, praeses) under the title: *Anacrisis medico-historico diaetetica, seu dissertations quadripartitae, de caffe et chocolatae, nec non de herbae thee ac nicotianae natura, usu, & abusu. Vetero-Pragae: In typis Georgii Labaun, 1720.*

Spies, Johann Carl. *Dissertatio botanico-medica inauguralis de Avellana Mexicana / quam praeside Jo. Carolo Spies . . . ; disquisitioni submittit Franciscus Ernestus Brückmann.* Helmstadii, 1721; 4'.

Meisner, Leonhard Ferdinand. *Anacrisis medico-historico diaetetica, seu dissertations quadripartitae, de caffe et chocolatae, nec non de herbae thee ac nicotianae natura, usu, & abusu.* Vetero-Pragae: In typis Georgii Labaun, 1720.

Histoire Naturelle Du Cacao Et Du Sucre: Divisée En Deux Traitez, Qui contiennent plusieurs faits nouveaux . . . / [D. de Cailus]. Seconde Édition, Revue & corrigée par l'Auteur. Amsterdam: Strik, 1720. [6] l., 228 pp. Ill.; 8°.

Quélus, D. *Histoire naturelle du cacao, et du sucre, divisée en deux traités, qui contiennent plusieurs faits nouveaux, & beaucoup d'observations également curieuses & utiles.* Paris: L. d'Houry, 1719. 4 p.l., 227, [10] pp. 6 pl. (4 fold.); 17 cm.

Kühne, Joh. Gottfr. *Nachricht von der Chocolate.* 2. Aufl. Nürnberg, 1719.

Durante, Castore. *Herbario nuovo che nascono in tutta Europa, & nell'Indie orientali, & occidentali, con versi latini, che comprendono le facoltà di i semplici medicamenti, e con discorsi che dimostrano i nomi, le spetie, la forma, il loco, il tempo, le qualità, & le virtù mirabili dell'herbe . . . ; con aggionta de i discorsi à quelle figure, che erano nell'appendice, fatti da Gio. Maria Ferro . . . & hora in questa novissima impressione vi si è posto in fine l'herbe thè, caffè, ribes de gli Arabi, e cioccolata.* Venetia: Presso Michele Hertz, 1718. [8], 480, [27] pp. ill.; 33 cm (folio).

Kühne, Johann Gottfried. *Vollständige Nachricht von der Chocolate.* Nürnberg, 1717. 79 pp.

Paradisi, Agustín. *Due lettere dell' illustrissimo sig. Agostino Paradisi . . . scritte all' illustrissimo Fr. Cappone Capponi . . . nella prima di dette due lettere si discorre della natura, delle propieta, e della virtú del Cioccolato, nell' altra si cerca se il cioccolato in bevanda Rompa il diviuno Ecclesiastico.* Modena: Per il Soliani Stampatore Ducale, 1715. xv, 48 pp. 23 cm.

France. Conseil d'Etat. *Arrest du conseil d'estat du roy, par lequel Sa Majesté declare n'avoir entendu comprendre dans la décharge des droits, accordée par l'arrêt . . . du 12. may 1693. en faveur du cacao declaré, pour estre mis en entrepôt & transporté à l'etranger, celui de trois pour cent, dont le fermier du domaine d'occident a droit de joüir sur toutes les marchandises & denrées du crû des isles de l'Amerique, arrivant dans les ports du royaume; comme aussi ordonne Sa Majesté que les negocians de la ville de Bordeaux, payeront . . . le droit de trois pour cent sur le cacao du crû desdites isles . . . Du vingt-cinquiéme juin 1715. Extrait des registres du Conseil d'etat.* Paris: Chez la veuve Saugrain, 1715. 8 pp. 27 cm (4to).

Tr. des Jeux de Hazard, defendu contre les objections de Mr de Joncourt: Histoire naturelle du Cacao et du sucre. Jean LaPlacette. La Haye, Amsterdam, 1714, 1720. Second, edition.

Andry de Bois-Regard, Nicolas, 1658–1742. *Traite des alimens de caresme, où, L'on explique les differentes qualitez des légumes, des herbages, des racines, des fruits, des poissons, des amphibies, des assaisonnemens: des boissons mêmes les plus en usage, comme de l'eau, du vin, de la bierre, du cidre, du thé, du caffé, du chocolat: et où l'on éclaircit plusieurs questions importantes sur l'abstinence et sur le jeûne, tant par rapport au Carême, que par rapport à la santé par Me Nicolas Andry.* Paris: Chez Jean-Baptiste Coignard, imprimeur ordinaire du Roi, rüe S. Jacques, à la Bible d'or, 1713. 2 v.; 17 (12mo).

Duncan, Daniel. *Von dem Mißbrauch heißer und hitziger Speisen und Geträncke, sonderlich aber des Caffes, Schockolate und Thees.* Leipzig: Gleditsch, 1707.

Duncan, M. (Daniel), 1649–1735. *Wholesome advice against the abuse of hot liquors, particularly of coffee, chocolate, tea, brandy, and strong-waters with directions to know what constitutions they suit, and when the use of them may be profitable or hurtful by Dr. Duncan . . . ; done out of French.* London: Printed for H. Rhodes . . . and A. Bell . . . , 1706. [8], 280 pp. 20 cm (8vo).

Dufour, Philippe Sylvestre, 1622–1687. *Jacobi Sponii medici Lugdunensis Bevanda asiatica, hoc est, Physiologia potus café a D. Dre Manget notis & seorsim a Constantinopoli plantae iconismis recens illustrata.* Lipsiae: [s.n.], 1705. [8], 56 pp., 6 folded leaves of plates: ill.; 22 cm.

Blankaart, Steven, 1650–1702. *Stephani Blancardi, phil. & med. doct. und practici ordinarii zu Amsterdam, Haustus Polychresti, oder, Zuverlässige Gedancken vom Theé, Coffeé, Chocolate und Taback: mit welchen der grosse Nutze dieser ausländischen Wahren so wol in gesunden als krancken Tagen gründlich und umständlich gelehret wird.* Hamburg: Zu finden bey Samuel Heyl, und Johann Gottfried Liebezeit . . . : Gedruckt bey Friedr. Conr. Greflinger, 1705. [14], 222 pp., [1] leaf of plates: 1 ill.; 17 cm (8vo).

Blankaart, Stephen, 1650–1702. *Haustus polychresti; oder, Zuverlässige gedancken vom theé, coffeé, chocolate, und taback, mit welchen der grosse nutze dieser ausländischen wahren so wol in gesunden als krancken tagen gründlich und umständlich gelehret wird. Stephanus Blancardus.* Hamburg: Heyl–Liebezeit, 1705. [14], [222] l.; 16mo.

Bontekoe, Cornelis, 1640?–1685. *Kurtze Abhandlung von dem menschlichen Leben, Gesundheit, Kranckheit und Tod: in drey unterschiedenen Theilen verfasset. Davon das I. Unterricht giebet von dem Leibe und desselben zur Gesundheit dienlichen Berrichtungen. II. Von der Kranckheit und derselben Ursachen. III. Von denen Mitteln, das Leben und die Gesundheit zu unterhalten und zu verlängern; die meisten Kranckheiten aber und ein daraus entstehendes beschwerliches Alter, durch Speise, Drank, Schlassen, Thee, Coffee, Chocolate, Toback und anderer vergleichen zur Gesundheit diensiche [illegible] eine gerauenteeit zu verhüten. Wobey noch angehänget drei kleine Tractätlein, I. Von der Natur. II. Von der Experienz oder Erfahrung. III. Von der Gewissheit der Medicin oder Heilkunst. Erstlich in holländischer Sprache beschrieben durch Cornelium Bontekoe, Med. D. Churfl. Durchl. zu Brandenb. Rath und Leib-Medicus. Itzo aber in die Hochteutsche Sprache versetzet.* Budisin: verlegts Johann Wilisch, 1701. [12], 353 (i.e., 553), [7] p.: port.; 17 cm.

Dufour, Philippe Sylvestre, 1622–1687. *Drey neue curieuse Tractätgen, von dem Trancke Cafe, sinesischen The, und der Chocolata, welche nach ihren Eigenschafften, Gewächs, Fortpflantzung, Praeparirung, Tugenden und herlichen Nutzen sehr curieus beschrieben, und nunmehro in die hochteutsche Sprache übersetzet von dem, welcher sich jederzeit nennet Theae potum maxime colentem.* Budissin: In Verlegts Johann Wilisch, 1701. [6], 247 (i.e., 245), [3] pp., [4] leaves of plates: ill. (woodcuts); 17 cm (12mo).

Lightbody, James. *Every man his own gauger: wherein not only the artist is shown a more ready and exact method of gauging than any hitherto extant . . . Also, What a pip, hogshead, &c. amounts to at the common rate and measure . . . To which is added, the art of brewing beer, ale, mum . . . The wintners art if fining, curing, preserving and recifying all sorts of wines . . . Together with the compleat coffee-man teaching how to make coffee, tea, chocolate, content . . . by J. Lightbody.* London: Printed for G.C. at the Ring in Little-Brittain, 1700s. 68 pp. 15 cm (12mo).

[Acte royal. 1692-01-00. Versailles]. *Édit . . . portant règlement pour la vente et débit du caffé, thé, sorbec, chocolat, du cacao et vanille. Publié en audiance publique [du Parlement de Dauphiné] le 5 may 1692* (Texte imprimé). Grenoble: A. Giroud, 1700. In-4°, 8 pp.

Man, Alexander. *To the honorable the Commons of England in Parliament assembled the coffee men dwelling within the bills of mortality, humbly present this their answer to a printed paper lately published, entituled, Propositions for changing the excise now laid upon coffee, chocolet, and tea inno [!] an imposition upon those commodities at their importation.* S.l.: [s.n]., 1700. 1 Broadside; 30 × 40 cm.

Collège des quatre-nations.; Laboratoire royal. Le Bon usage du chocolat dégraissé preparé à Paris, par les artistes du Laboratoire royal du College des quatre nations. Paris: s.n., 1700–1799? [36] p., [1] leaf of plates: ill. (Credited in Mueller to Leonard Biet, 17th century.)

Imperial Chocolate made by a German lately come to England. London, ca. 1700; folio, 1 sheet.

1600–1699

Dufour, Philippe Sylvestre, 1622–1687. *Novi tractatus de potu caphe, de Chinensium thé, et de chocolate à D.M., notis illustrati.* Genevae: Apud Cramer & Perachon, 1699. [4], 188 p., [3] leaves of plates: ill.; 15 cm. *Note:* "First published in 1671. It is probable that the original was compiled by Dufour and was translated by Spon. The treatise on chocolate on pp. 132–178 is mainly from the Spanish of Antonio Colmenero de Ledesma.—NUC pre-1956. Published in 1685 as *Tractatus novi . . .* Translation and abridgement by Jacob Spon of Dufour's *Traitez nouveaux & curieux du café, du thé, et du chocolate*, 1685." This note appear in the bibliographic record. It's very intermatine, so we chose to keep it. However, it was written by Anciker.

Ordonantie van den impost op de consumptie von Coffij, Thee, Chocolade. . . . s'Gravenhage, 1699. 4o.

Theophilotechnus. *Vier Sonderbare und allgemeine Gesundheits-Mittel: Wordurch Allen und jeden Kranckheiten vorzukom[m]en/ bey Weibs und Manns-Persohnen/ auch selbige der mehrere Theil können curiert werden. Oder Ausführliche Beschreibung/ Ursprung/ Praeparierung . . . des Chineser-Thee, Türckischen Coffe, Spanischen Chocolatho, und Teutschen Taback / Allen Die ihr Gesundheit und langes Leben lieben/ zu lieb aufgesetzet. Von Theeophilotechno.* Augspurg: Stretter, 1699.

Great Britain. Laws, statutes, etc. 1694 (William III). *Abstracts. Anno 9 & 10 W. III.* London: Printed by C. Bill, and the executrix of Thomas Newcomb, deceas'd, Printers to the Kings most Ex elent [sic] Majesty, 1698. 4 pp. 29 × 18 cm. Contents: An abstract of the act for granting to His Majesty several duties upon coals and culm.—An abstract of the act for continuing the duties upon coffe, tea and chocolate and spices, towards satisfaction of the debt due for transport service for the reduction of Ireland.—An abstract of the act for determining differences by arbitration.—An abstract of the act to execute judgements and decrees saved in a clause in an act of the first year of the reign of King William and Queen Mary, intituled, An act for taking away the court holden before the president and council of the marches of Wales.—An abstract of the act for the better payment of inland bills of exchange.—An abstract of the act to naturalize the children of such officers and soldiers, and others the natural born subjects of this realm, who have been born abroad during the war, the parents of such children having been in the service of this government.—An abstract of the act for the better preventing the counterfeiting, clipping, and other diminishing the coin of this kingdom.

Chamberlayne, John, 1666–1723. *Schatz-Kammer rarer und neuer Curiositäten in den aller-wunderbahresten Würckungen der Natur und Kunst: darinnen allerhand seltzame und ungemeine Geheimnisse, bewehrte Artzneyen, Wissenschafften und Kunst-Stücke zu finden* . . . *der vierte Druck, jetzo mit dem dritten Theil von vielen chymischen Experimenten und andern Künsten vermehrt: deme angehänget ist ein Tractat, Naturgemässer Beschreibung der Coffeé, Theé, Chocolate, Tobacks und dergleichen.* Hamburg: bey Benjamin Schillern, 1697. 6, 656, [24] pp. ill.; 17 cm (8vo).

Kurtze/ jedoch sehr Accurat und Nützliche Beschreibung/ Der Thee Coffe und Choccolate: wie auch Deß in Teutschland bisanhero ziemlich unbekannten/ iedoch überaus herrlichen und nützlichen Chinesischen Saamens/ genant Padian oder Anisum Stellatum Chinense / Aus unterschiedenen Schrifften berühmter Medicorum, und den neuesten courieusen Reise-Beschreibungen zusammen gezogen/ Von dem Italiener Bartholomaeo Belli aber in Auerbachs Hoffe/ bey welchen dieses alles recht frisch/ aufrichtig und unverfälscht zu bekommen/ auff seine eigene Unkosten zum druck befördert. Leipzig, 1695. [8] sheets. 4°.

Lightbody, James. *Every man his own gauger wherein not only the artist is shown a more ready and exact method of gauging than any hitherto extant, but the most ignorant, who can but read English, and tell twenty in figures, is taught to find the content of any sort of cask or vessel, either full, or in part full, and to know if they be right siz'd. Also what a pipe, hogshead, &c. amounts to at the common rate and measure they buy or sell at. With several useful tables to know the content of any vessel by likewise a table shewing the price of any commodity, from one pound to an hundred weight, and the contrary. To which is added, the art of brewing beer, ale, mum, of fining, preserving and botling brew'd liquors, of making the most common physical ales now in use, of making several fine English wines. The vintners art of fining, curing, preserving all sorts of wines* . . . *together with the compleat coffee-man, teaching how to make coffee, tea, chocolate* . . . *Of great use for common brewers, victuallers,* . . . *and all other traders by James Lightbody, philomath.* London: Printed for Hugh Newman, at the Grasshopper near the Rose Tavern in the Poultrey, 1695. [14], 115 pp. tables; 12mo.

Mappus, Marcus. *Dissertationes medicae tres de recentis hodie etiam in Europa potus calidi generibus, Thee, Café, Chocolata.* Argentorati: Literis Joh. Friderici Spoor, 1695. 4o.

Colmenero de Ledesma, Antonio. *Della cioccolata. Discorso.* Bologna, 1694. 72 pp. 24to.

Worlidge, John, fl. 1669–1698. *Mr. Worlidge's two treatises: the first, Of improvement of husbandry, and advantage of enclosing lands, of meadows and pastures, of arable land and tillage, and the benefit of hempseed and flax: with a description of the engine for dressing hemp and flax. Of the manuring, dunging, and soiling of lands; and the benefit of raising, planting, and propogating of woods. The second, A treatise of cyder, and of the cyder-mill, and a new sort of press. Of currant-wine, apricock-wine, rasberry-wine: of making chocolate, coffee, tea: of the extract of juniper-berries, and of mum. To which is added, An essay towards the discovery of the original of fountains, and springs.* London: Printed for M. Wotton, 1694. 1 p. sh., xlix, [3], 191, [1] p.; plates; 20 cm.

Dufour, Philippe Sylvestre, 1622–1687. *Traitez nouveaux & curieux du café, du thé et du chocolate ouvrage également neces-saire aux medecins, & à tous ceux qui aiment leur santé par Philippe Sylvestre Dufour; à quoy on a adjouté dans cette édition, la meilleure de toutes les mèthodes, qui manquoit à ce livre, pour composer l'excellent chocolate, par Mr. St. Disdier. Troisiéme ed.* A La Haye: Chez Adrian Moetjens, 1693. [2], 404, [4] pp., [3] leaves of plates: ill.; 14 cm. *Note*: An enlarged edition of *De l'usage du caphé, du thé, et du chocolate* (1671), compiled from various authors, among them A. F. Naironi, Alexandre de Rhodes, and Thomas Gage. The section on chocolate is from the Spanish of Antonio Colmenero de Ledesma.

Tozzi, Guglielmo. *Tractatus novus de potu cophe, de sinensium thee et de chocolate.* Francofurti, 1693. 496 pp. + index. 14 cm. (Mueller: "Der Inhalt entspricht wörtlich dem Werke Mundays." See below: Munday, 1685.)

Bontekoe, Cornelis, 1648–1685. *Kurtze Abhandlung Von dem Menschlichen Leben/ Gesundheit/ Kranckheit und Tod: In Drey unterschiedenen Theilen verfasset/ Davon das I. Unterricht giebet von dem Leibe/ und desselben zur Gesundheit dienlichen Verrichtungen. II. Von der Kranckheit / und derselben Ursachen. III. Von denen Mitteln/ das Leben und die Gesundheit zu unterhalten und zu verlängern/ die meisten Kranckheite[n] aber* . . . *durch Speise/ Tranck/ Schlaffen/ Thee, Coffee, Chocolate, Taback/ und andere* . . . *Mittel/ eine geraume Zeit zu verhüten Wobey noch angehänget/ Drey kleine Tractätlein/ I. Von der Natur/ II. Von der Experienz oder Erfahrung/ III. Von der Gewißheit der Medicin, oder Heilkunst / Erstlich in Holländischer Sprach beschrieben/ durch Cornelium Bontekoe, Med. D. . . . Itzo aber in die Hochteutsche Sprach versetzt.* Budißin: Arnst; Rudolstadt: Löwe, 1692. [8] s., 553 pp. [3] l.: Frontisp. (Portr.), Tbl. r&s.; 8°. Main title: *Korte verhandling van's menschen leven, gesondheid, siekte en dood.*

Dufour, Philippe Sylvestre, 1622–1687. *Drey Neue Curieuse Tractätgen, von Dem Tranck Cafe, Sinesischen The, und der Chocolata, Welche Nach ihren Eigenschafften, Gewächs, Fortpflantzung, Praeparirung, Tugenden und herrlichen Nutzen sehr curieus beschrieben, Und nunmehro in die hochteutsche Sprache übersetzet Von dem, Welcher sich iederzeit nennet Theæ Potum Maxime Colentem.* Budissin: In Verlegung Friedrich Arnsts, 1692. [6], 247 (i.e., 245), [3] pp. 3 plates; 17 cm.

[Acte royal. 1692–01–00. Versailles]. *Édit* . . . *portant établissement de droits sur le caffé, thé, sorbec et chocolat* . . . *Vérifié en Parlement le 26 février 1692* (Texte imprimé). Paris: G. Desprez, 1692. In-4°, 8 pp.

[Acte royal. 1692–01–00. Versailles]. *Édit* . . . *portant règlement pour la vente et distribution du caffé, thé, chocolat, cacao et vanille* . . . *Registré en Parlement* (Texte imprimé). Paris: E. Michallet, 1692. In-4°, 8 pp.

Dufour, Philippe Sylvestre, 1622–1687. *Drey Neue Curieuse Tractätgen, von Dem Tranck Cafe, Sinesischen The, und der Chocolata, Welche Nach ihren Eigenschafften, Gewächs, Fortpflantzung, Praeparirung, Tugenden und herrlichen Nutzen sehr curieus beschrieben, Und nunmehro in die hochteutsche Sprache übersetzet Von dem, Welcher sich iederzeit nennet Theæ Potum Maxime Colentem.* Budissin: In Verlegung Friedrich Arnsts, 1692. [6], 247 (i.e., 245), [3] pp. 3 plates; 17 cm.

Price, Samuel, in Christ-Church Hospital. *The virtues of coffee, chocolette, and thee or tea, experimentally known in this our climate.* (London): [s.n.], 1690. Broadside.

England and Wales; Parliament; House of Commons. *An Answer to a paper set forth by the coffee-men directed to the Honourable, the Commons in Parliament assembled being reflections upon some propositions that were exhibited to the Parliament for the changing the excise of coffee, tea, and chocolate into a custom upon the commodities.* (London), 1680–1689? Broadside.

England and Wales; Parliament; House of Commons. *To the honourable, the knights, citizens, and burgesses in Parliament assembled, propositions for changing the excise, now laid upon coffee, chacholet, and tea, into an imposition upon those commodities at their importation.* (London?): [s.n.], 1689. Broadside, 44 cm.

Chamberlayne, John, 1666–1723. *Schatzkammer rarer und neuer Curiositäten in den aller-wunderbahresten Würckungen der Natur und Kunst darinnen allerhand seltzame und ungemeine Geheimnüsse, bewehrte Artzneyen, Wissenschafften und Kunst-Stücke zu finden. . . . 3. Druck.* Hamburg: auff Gottfried Schultzens Kosten, 1689. [4], 592, [24] p., [1] leaf of plates: ill.; 17 cm.

Schatzkammer Rarer und neuer Curiositäten/ In den aller-wunderbahresten Würckungen der Natur und Kunst: Darinnen allerhand seltzame und ungemeine Geheimnüsse/ bewehrte Artzneyen/ Wissenschafften und Kunst-Stücke zu finden . . . Deme angehenget ist Ein Tractat, Naturgemässer Beschreibung Der Coffee, Thee, Chocolate, Tabacks/ und dergleichen Ausgabe: Der Dritte Druck/ Jetzo mit dem dritten Theil von vielen Chymischen Experiementen und andern Künsten vermehret. Hamburg: Schultze, 1689. [4] Bl., 592 S., [12] Bl.; Kupfert.; 8°.

Schatz Kammer neuer Curiositaeten Bibliogr. Vorlageform des Erscheinungsvermerks: Hamburg/ Auff Gottfried Schultzens Kosten/ 1689. Enth. ausserdem: Naturgemässe Beschreibung Der Coffee, Thee, Chocolate und Tabacks . . . Mit einem Tracktätlein.Von Hollunder- und Wachholder-Beeren . . . / Auß der Englischen in die Hochteutsche Sprache übersetzet/ durch J. L. M. C. Nachweis: Weiss, Hans U.: Gastronomia, Nr. 3345.

Schatzkammer Rarer und neuer Curiositäten, Jn den aller-wunderbahresten Würckungen der Natur und Kunst: Darinnen allerhand seltzame und ungemeine Geheimnüsse, bewehrte Artzneyen, Wissenschafften und Kunst-Stücke zu finden Dessen Jnhalt auff folgenden Blat zu sehen ist; Ein Werck, so jedermänniglich, wes Standes, Geschlechtes und Alter er ist, nützlich und ergetzlich seyn wird / Lange, Johann.—Der Dritte Druck, Jetzo mit dem dritten Theil von vielen Chymischen Experimenten und anderen Künsten vermehret. Hamburg, 1689.

Dufour, Philippe Sylvestre, 1622–1687. *Traitez nouveaux & curieux du café, du thé et du chocolate*, 2nd edition. Lyon: Chez Jean Baptiste Deville, 1688. [24], 444, [10] pp. [1] leaf of plates: ill.; 16 cm.

Dufour, Philippe Sylvestre, 1622–1687. *Drey Neue Curieuse Tractätgen/ Von Dem Trancke Cafe, Sinesischen The, und der Chocolata:Welche Nach ihren Eigenschafften/ Gewächs/ Fortpflantzung/ Praeparirung/ Tugenden und herrlichen Nutzen/ sehr curieus beschrieben/ Und nunmehro in die Hoch-teutsche Sprache übersetzet / Von dem/ Welcher sich jederzeit nennet Theae Potum Maxime Colentem.* Budissin: Arnst; Budissin: Richter, 1688. [4] Bl., 247 (i.e. 245) S., [1] Bl., [3] Bl.: Frontisp., 3 Ill. (Kupferst.); 8°. its- ine. r-it weun 3 1688A.

Chamberlayne, John, 1666–1723. *Schatzkammer rarer und neuer Curiositäten: in den aller-wunderbahresten Würckungen der Natur und Kunst, darinnen allerhand seltsame und ungemeine Geheimnüsse, bewehrte Artzneyen, Wissenschafften und Kunst-Stücke zu finden, dessen Inhalt auf folgenden Blat zu sehen ist. Ein Werck, so jedermänniglich, wes Standes, Geschlechtes und Alter er ist, nützlich und ergetzlich seyn wird. Der andere Druck, jetzo mit dem dritten Theil von vielen chymischen Experimenten und anderen Künsten vermehret, deme angehenget ist ein Tractat naturgemasser Beschreibung der Coffee, Thee, Chocolate, Tabacks, und vergleichen. Mit Thur-Sächsicher Gnäd. Befreyung nicht nachzudrucken.* Hamburg: Auff Gottfried Schultzens Kosten, 1686. 592, [23] pp. 17 cm.

Dufour, Philippe Sylvestre, 1622–1687. *Drey Neue Curieuse Tractätgen, Von Dem Trancke Cafe, Sinesischen The, und der Chocolata,Welche Nach ihren Eigenschafften, Gewächs, Fortpflantzung, Præparirung, Tugenden und herlichen Nutzen, sehr curieus beschrieben, Und nunmehro in die Hoch-teutsche Sprache übersetzet Von dem,Welcher sich jederzeit nennet Theæ Potum Maxime Colentem.* Budissin: In Verlegung Friedrich Arnsts, drucks Andreas Richter, 1686. [6], 247 (i.e., 245), [3] pp. 4 plates; 16 cm. *Note:* This translation, by J. Spon, also includes Marradón's "Dialogue" and St. Disdier's "Method of preparing chocolate."

Dufour, Philippe Sylvestre, 1622–1687. *The manner of making of coffee, tea, and chocolate as it is used in most parts of Europe, Asia, Africa, and America, with their vertues newly done out of French and Spanish.* London: Printed for William Crook, 1685. [12], 116 pp.: ill.; 15 cm.

England and Wales; Parliament; House of Commons. *To the honourable the knights, citizens, and burgesses in Parliament assembled. Propositions for changing the excise now laid upon coffee, chacholet, and tea into an imposition upon those commodities at their importation.* S.l.: [s.n.], 1689. 1 Broadside; 44 cm.

Dufour, Philippe Sylvestre, 1622–1687. *Traitez nouveaux & curieux du café, du thé et du chocolate: ouvrage également necessaire aux medecins, & à tous ceux qui aiment leur santé par Philippe Sylvestre Dufour*, 2nd edition. Lyon: Chez Jean Baptiste Deville, 1688. [24], 444, [10] pp. [1] leaf of plates: ill.; 16 cm.

Blégny, (Nicolas), Monsieur de, 1652–1722. *Le bon usage du thé, du caffé et du chocolat pour la préservation & pour la guérison des maladies*, 2nd edition. Paris, L'auteur (etc.), 1687. [20], 358, [4] pp. illus. 15 cm.

Blégny, (Nicolas), Monsieur de, 1652–1722. *Abregé des traitez du caffé, du thé et du chocolat: pour la preservation, & pour la guerison des maladies.* Lyon: Chez Esprit Vitalis, 1687. 70, [2] pp. 14 cm (12mo).

Blégny, (Nicolas), Monsieur de, 1652–1722. *Le bon usage du thé du caffé et du chocolat: pour la preservation & pour la guerison des maladies par Mr de Blegny.* Lyon, & se vend a Paris: Chez Jacques Collombat, 1687. [24], 120, 131–226, 219–357, [5] p.: ill.; 16 cm (12mo).

Blégny, (Nicolas), Monsieur de, 1652–1722. *Le bon usage du thé, du caffé, et du chocolat pour la preservation & pour la guerison des maladies par Mr. de Blegny.* Paris: Chez Estienne Michallet, 1687. [23], 357, [5] p.; ill.; 15 cm. Series: Goldsmiths'–Kress Library of Economic Literature; no. 2675.1. *Note:* another edition; Lyon: Chez Thomas Amaulry, 1687.

Chamberlayne, John, 1666–1723. *Schatzkammer rarer und neuer Curiositäten: in den aller-wunderbahresten Würckungen der Natur und Kunst, darinnen allerhand seltsame und ungemeine Geheimnüsse, bewehrte Artzneyen, Wissenschafften und Kunst-Stücke zu finden, dessen Inhalt auf folgenden Blat zu sehen ist. Ein Werck, so jedermänniglich, wes Standes, Geschlechtes und Alter er ist, nützlich und ergetzlich seyn wird. Der andere Druck, jetzo mit dem dritten Theil von vielen chymischen Experimenten und anderen Künsten vermehret, deme angehenget ist ein Tractat naturgemasser Beschreibung der Coffee, Thee, Chocolate, Tabacks, und vergleichen. Mit Thur-Sächsicher Gnäd. Befreyung nicht nachzudrucken.* Hamburg: Auff Gottfried Schultzens Kosten, 1686. 592, [23] pp. 17 cm. *Note:* "Sometimes attributed to John Chamberlayne. Chamberlayne was born in 1666, making him twenty at the time of publication (and younger when the English original would have been written or translated); thus this attribution is by no means a certainty. The books also attributed to Chamberlayne on which this German edition is based are *The Natural history of coffee, chocolate, thee, tobacco, in four several sections* (London: Christopher Wilkinson, 1682), itself a translation of Philippe Sylvestre Dufour's *Traitez nouveaux & curieux du café, du thé et du chocolate*, and *A treasure of health* (London: William Crook, 1686), a translation allegedly by Chamberlayne of Castore Durante's *De bonitate et vitio alimentorum centuria* (Translator unknown)." This note appear in the bibliographic record. It's very intermatine, so we chose to keep it. However, it was written by Anciker.

Dufour, Philippe Sylvestre, 1622–1687. *Drey Neue Curieuse Tractätgen, Von Dem Trancke Cafe, Sinesischen The, und der Chocolata, Welche Nach ihren Eigenschafften, Gewächs, Fortpflantzung, Præparirung, Tugenden und herlichen Nutzen, sehr curieus beschrieben, Und nunmehr in die Hoch-teutsche Sprache übersetzet Von dem, Welcher sich jederzeit nennet Theca Potum Maxime Colentem.* Budissin: In Verlegung Friedrich Arnsts, drucks Andreas Richter, 1686. [6], 247 (i.e., 245), [3] pp. 4 plates; 16 cm.

Dufour, Philippe Sylvestre, 1622–1687. *Drey Neue Curieuse Tractätgen / Von Dem Trancke Cafe, Sinesischen The, und der Chocolata: Welche Nach ihren Eigenschafften / Gewächs / Fortpflantzung / Praeparirung / Tugenden und herrlichen Nutzen / sehr curieus beschrieben / Und nunmehr in die Hoch-teutsche Sprache übersetzet / Von dem / Welcher sich jederzeit nennet Theae Potum Maxime Colentem. [Philippe Sylvestre Dufour].*

Budissin: Arnst; [Leipzig]: Krüger, 1686. [4] Bl., 247 (i.e., 245) pp., [1] s., [3] l.: Frontispiece, 3 Ill. (Kupferst.); 8°.

Bontekoe, Cornelis, 1640–1685. *Tractaat van het excellenste kruyd thee: 't Welk vertoond het regte gebruyk, en de groote kragten van 't selve in gesondheyt, en siekten: Benevens een kort discours op het leven, de siekte, en de dood: mitsgaders op de medicijne van dese tijd. Ten dienste van die gene, die lust hebben, om langer, gesonder, en wijser te leven / door Cornelis Bontekoe. Den derden druk vermeerdert, en vergroot met byvoeginge van noch twee korte verhandelingen, I. Van de coffi; II. Van de chocolate; Mitsgaders van een apologie van den autheur negens sijne lasteraars.* 'sGravenhage: Pieter Hagen, 1685. [56], 344 pp. 18 cm.

Bontekoe, Cornelis, 1648–1685. *Kurtze Abhandlung Von dem Menschlichen Leben / Gesundheit / Kranckheit / und Tod: In Drey unterschiedenen Theilen verfasset / Davon das I. Unterricht giebet von dem Leibe / und desselben zur Gesundheit dienlichen Verrichtungen. II. Von der Kranckheit / und derselben Ursachen. III. Von denen Mitteln / das Leben und die Gesundheit zu unterhalten und zu verlängern / die meisten Kranckheite[n] aber . . . durch Speise / Tranck / Schlaffen / Thee, Coffee, Chocolate, Taback / und andere . . . Mittel / eine geraume Zeit zu verhüten Wobey noch angehänget / Drey kleine Tractätlein / I. Von der Natur / II. Von der Experienz oder Erfahrung / III. Von der Gewißheit der Medicin, oder Heil-Kunst / Erstlich in Holländischer Sprache beschrieben / durch Cornelium Bontekoe, Med. D. . . . Anitzo aber in die Hochteutsche Sprache versetzet / Von R. J. H.* Budissin: Arnst; Budissin: Richter, 1685. [11] Bl., 553 S., [4] Bl.; 8°.

Chamberlayne, John, 1666–1723. *The manner of making coffee, tea, and chocolate. As it is used in most parts of Europe, Asia, Africa, and America. With their vertues.* London: Printed for William Crook at the Green Dragon without Temple Bar near Devereux Court, 1685. 6 pp. leaf, 116 (i.e., 108) pp. illus. 15 cm.

Dufour, Philippe Sylvestre, 1622–1687. *Traitez nouveaux & curieux du café, du thé et du chocolate ouvrage également necessaire aux medecins, & à tous ceux qui aiment leur santé par Philippe Sylvestre Dufour; à quoy on a adjouté dans cette édition, la meilleure de toutes les mèthodes, qui manquoit à ce livre, pour composer l'excellent chocolate.* La Haye: Chez Adrian Moetjens, 1685. [4], 403, [5] pp., [3] leaves of plates: ill.; 14 cm. Series: Goldsmiths'–Kress Library of Economic Literature; no. 2582.6 suppl. *Note:* An enlarged edition of *De l'usage du caphé, du thé, et du chocolate* (1671), compiled from various authors, among them A. F. Nairoro, Alexandre de Rhodes, and Thomas Gage. The section on chocolate is from the Spanish of Antonio Colmenero de Ledesma, *Suivant la copie de Lyon.* Includes *Dialogue du chocolate, entre un medecin, un Indien, et un bourgeois* from the Spanish of Bartolomé Marradón (pp. 381–403).

Dufour, Philippe Sylvestre, 1622–1687. *The manner of making of coffee, tea, and chocolate as it is used in most parts of Europe, Asia, Africa, and America, with their vertues newly done out of French and Spanish.* London: Printed for William Crook, 1685. [10], 116 p.; ill. *Note:* The tracts tea and chocolate have special title pages. Those on tea and coffee are translated by John Chamberlayne from the French of Philippe Sylvestre Dufour; that on chocolate from the Spanish of Antonio Colmenero de Ledesma (cf. British Museum catalog).

Dufour, Philippe Sylvestre, 1622–1687. *Traitez nouveaux & curieux du café, du thé, et du chocolate: ouvrage également necessaire aux medecins, & à tous ceux qui aiment leur sante par Philippe Sylvestre Dufour*, 2nd edition. Lyon: Chez Jean Baptieste Deville, 1685. [18], 444, [8] p., [3] leaves of plates: ill.; 16 cm (12mo).

Dufour, Philippe Sylvestre, 1622–1687. *Traitez nouveaux & curieux dv café, dv thé et dv chocolate. Ouvrage également necessire aux medecins, & à tous ceux qui aiment leur santé. Par Philippe Sylvestre Dvfovr.* Lyon: I. Girin & B. Riviere, 1685. 11 p.l., 445, [5] p. illus., plates; 16 cm. *Note: Dialogue du chocolate. Entre un medecin, un indien & un bourgeois* (pp. 423–445) is from the Spanish of Bartolomé Marradón.

Dufour, Philippe Sylvestre, 1622–1687. *Tractatvs novi de potv caphé; de chinensivm thé; et de chocolata.* Paris: Apud Petrum Muguet, 1685. [8], 202, [4] p., 3 leaves of plates: ill. (engravings); (12mo). *Note: Dialogus de chocolata* is from the Spanish of Bartolomé Marradón.

Munday, Henry (Mundius). *Opera omnia Medico-Physica, Tractatibus tribus comprehensa. De Aera vitalii; De esculentis, De potulentis. Una cum appendice de Paerergis in Victu ut Chocolata, Coffe, Thea, Tabao &c. . . . Henri Mundii. Medic. Doct. Londinens.* Lugdunum Batavorum, Apud Petrum van der Aa, 1685. 8o. (22), 362 pp.

Spon, Jacques lat. *Tractatus novi de potu Caphe, de Chinensium The, et de Chocolata.* Paris, Muguet, 1685.

Bachot, Jos., Praes. Et Fr. Fourcault, Resp. *An Chocolatae usus salutaris?* Paris, 1684. apud Franciscus Muguet. 4o.

C., J. L. M. *Natur-gemässe Beschreibung der Coffee, Thee, Chocolate, Tabacks . . . : Mit einem Tractätlein von Hollunderund Wacholder-Beeren, welches anweiset wie nützlich dieselben seyn könten in unsern Coffee-Häusern wie auch den Weg die Mumme zu bereiten . . . auss der Englischen.* Hamburg: Auff G. Schulzens Kosten, 1684. 71 pp. 12.

J.L.M.C. *Natur-gemässe Beschreibung der Coffee, Thee, Chocolate, Tabacks: In vier unterschiedlichen Abtheilungen, mit einem Tractätlein von Hollunder- und Wacholder-Beeren, welches anweiset wie nützlich dieselben seyn könten in unsern Coffee-Häusern . . . Aus den Schriften der besten medicorum . . . zusammen gelesen, und aus der engl. in die hoch-teutsche Sprache übers. durch J.L.M.C.* Hamburg, 1684. in-12. *Note:* J.L.M.C. is Johann Lange.

Herbario nvovo di Castore Durante. *Con figure, che rappresentano le viue piante, che nascono in tutta Europa, & nell'Indie Orientali, & Occidentali, con versi Latini, che comprendono le facoltà de i semplici medicamenti, e con discorsi che dimostrano i nomi, le spetie, la forma, il loco, il tempo, le qualità, & le virtù mirabili dell'herbe, insieme col peso, & ordine da vsarle, scoprendosi rari secreti, & singolari rimedij da sanar le più difficili infirmità del corpo humano. Con dve tavole copiossime, l'una delle herbe, & l'altra delle infirmità, & di tutto quello che nell'opera si contiene. Con aggiunta de i discorsi à quelle figure, che erano nell'appendice, fatti da Gio. Maria Ferro spetiale alla sanità; & hora in questa nouissima impressione vi si è posto in fine l'herbe thè, caffè, ribes de gli Arabi, e cioccolata.* Venetia: Presso Gian Giacomo Hertz, 1684. 6 p. l., 480, [27] pp. illus. 36 cm (fol.).

Bontekoe, Cornelis, 1640–1685. *Korte verhandeling van 'smenschen leven, gesondheid, siekte, en dood.* 'sGravenhage: P. Hagen bookseller, 1684 (title page χ1: 1685 A1, l. 5, zyt. Incorporates: C. Bontekoe, Drie verhandelingen, 1685. 8o.

Chamberlayne, John, 1666–1723. *The natural history of coffee, thee, chocolate, tobacco, in four several sections: with a tract of elder and juniper-berries, shewing how useful they may be in our coffee-houses: and also the way of making mum, with some remarks on that liquor collected from the writings of the best physicians, and modern travellers.* London: Printed for Christopher Wilkinson, 1682. 36, [4] p.; 20 cm (4to).

Paulli, Simon, 1603–1680. *Simonis Paulli, D. Medici Regii, ac Prælati Aarhusiensis Commentarius de abusu tabaci americanorum veteri, et herbæ thee asiaticorum in Europa novo, quæ ipsissima est chamæleagnos Dodonæi. Editio secunda priori auctior & correctior.* Argentorati (Strasbourg): Sumptibus B. authoris filii Simonis Paulli, 1681. [60], 87, [13] p., [2] folded leaves of plates: ill., port.; 22 cm. (4to).

Bontekoe, Cornelis, 1640–1685. *Tractaat van het excellenste kruyd thee: 't welk vertoond het regte gebruyk, en de groote kragten van 't selve in gesondheyd, en siekten: benevens een kort discours op het leven, de siekte, en de dood: mitsgaders op de medicijne van dese tijd: ten dienste van die gene, die lust hebben, om langer, gesonder, en wijser te leven Cornelis Bontekoe. Den tweeden druk vermeerdert, en vergroot met byvoeginge van noch twee korte verhandelingen, I. Van de coffi; II. Van de chocolate, mitsgaders van Een apologie van den autheur tegens sijne lasteraars.* 'sGravenhage: Pieter Hagen, 1679. [31], 367 (i.e., 341) pp. 16 cm.

Bontekoe, Cornelis, 1640–1685. *Tractaat van het excellenste kruyd thee: 't Welk vertoond het regte gebruyk, en de grote kragten van 't selve in gesondheid, en siekten: benevens een kort discours op het leven, de siekte, en de dood: mitsgaders op de medicijne, en de medicijns van dese tijd, en speciaal van ons land. Ten dienste van die gene, die lust hebben, om langer, gesonder, en wijser te leven, als de meeste menschen nu in 't gemeen doen; Cornelis Bontekoe.* 'sGravenhage: by Pieter Hagen op de Hoogstraat, in de Koning van Engeland, 1678. [36], 76, 324, [4] pp. in-12.

Colmenero de Ledesma, Antonio. *Della cioccolata.* Venetia: Per il Valvasense, 1678. 80 pp. 15 cm. *Note:* A translation of his *Curioso tratado de la naturaleza y calidad del chocolate.*

Hughes, William, fl. 1665–1683. *The flower garden enlarged: Shewing how to order and increase all sorts of flowers, whether by layers slips off-sets. Cuttings. Seeds &c. Also how to draw a horizontal dial as a knot in a garden. To which is now added a* treatise of all the roots, plants, trees, shrubs, fruits, herbs, &c. Growing in His Majesties plantations. By W. Hughe's. London: Printed for C. Wall in the Strand, 1677. [12], 100, [30], 159, [7] p., [4] leaves of folded plates: ill.; 15 cm (12mo).

Vieheuer, Christoph. Gründliche Beschreibung fremder Materialien und specereyen, Ursprung, Wachstum. Leipzig: Johann Fritzsche, 1676. 4o. XII, (16), 204 pp.

Brancaccio, Francesco Maria, Cardinal, 1592–1675. Francisci Mariae, episcopi Portvensis, cardinalis Brancatii, dissertationes. Romae, ex typ. N.A. Tinassij, 1672. 360 pp. plates.

Hughes, William, fl. 1665–1683. The American physitian; or, a treatise of the roots, plants, trees, shrubs, fruit, herbs, &c. growing in the English plantations in America. Describing the place, time, names, kindes, temperature, vertues and uses of them, either for diet, physick, &c. Whereunto is added a discourse of the cacaonut-tree, and the use of its fruit; with all of the ways of making chocolate. The like never extant before. By W. Hughes. London: Printed by J.C. for William Crook, at the Green Dragon, without Temple-Bar, 1672. 11 p. l., 159 p., 3 sh. 15 cm. (12mo).

Dufour, Philippe Sylvestre, 1622–1687. De l'vsage du caphé, dv thé et dv chocolate. Lyon: Chez Iean Girin, & Barthelemy Riviere, 1671. [24], 188 p.; 1 ill. (woodcut); 14 cm (12mo). Note: "The sections on coffee and tea are compiled from various authors, among them A. F. Naironi and Alexandre de Rhodes; the section on chocolate is from the Spanish of Antonio Colmenero de Ledesma. An additional part, Du chocolate: dialogue entre un medecin, un Indien, et un bourgeois, was composed by B. Marradón. Jacob Spon, who translated a later edition into Latin, may have assisted Dufour with the original compilation; the work is sometimes attributed to him." This note appear in the bibliographic record. It's very intermatine, so we chose to keep it. However, it was written by Anciker.

Du Four de La Crespelière, Jacques. Commentaire en vers françois, sur l'Ecole de Salerne, contenant les moyens de se passer de medecin, & de viure longtemps en santé, avec une infinité de remedes contre toutes sortes de maladies, & un traitté des humeurs, & de la saignée, où sont adjoustez. La sanguification, circulation, et transfusion du sang, la poudre & l'onguent de sympathie, le thé, le caphé, le chocolate, et le grand secret de la pierre philosophale, ou la veritable maniere de faire de l'or, aussi en vers françois, et l'ouromantie, scatomantie, et hydromantie en prose. Par monsieur D.F.C., docteur en la Faculté de medecine. Paris: Chez Gilles Alliot, 1671. 30 p.l., 714 (i.e., 696) pp. 15 cm.

Sebasto, F. (Nicephoro). De chocolatis potione resolvtio moralis avthore F. Nicephoro Sebasto Melisseno Or. Er. S. Aug. . . . 3a ed. Neapoli: Typis Io: Francisci Paci, 1671. 36, [2] p., [1] leaf of plates: ill.; 14 cm (18mo). Note: No earlier edition found.

Colmenero de Ledesma, Antonio. Della cioccolata: discorso diuiso in quattro parti. Roma: nella stamparia della R[everendissima]. C[amerata]. A[postolica]., 1667. 72, 71–94 p.; 15 cm (12mo).

Paulli, Simon, 1603–1680. Simonis Paulli . . . Commentarius de abusu tabaci Americanorum veteri, et herbæ theé Asiaticorum in Europa novo: quæ ipsissima est chamæleagnos Dodonæi, aliàs myrtus brabantica, danicè porss, german. post, gallicè piment royal, belgicè gagel dicta; cum figuris æneis, utensilia quædam Chinensivm eaq pretiosissima repræsentantibus. Argentorati: Sumptibus authoris filij S. Paulli, 1665. 11 p. l., 61 numb. l. incl. port. fold. pl. 20 × 16 cm.

Brancaccio, Francesco Maria, Cardinal, 1592–1675. De chocolatis potu diatribe Francisci Mariae Cardinalis Brancatii. Romae: Per Zachariam Dominicum Acsamitek à Kronenfeld, Vetero Pragensem, 1664. [2], 37, [4] p., [1] leaf of plates: ill.; 21 cm. Note: Engraved allegorical plate depicts Neptune receiving chocolate from a personified America.

Vesling, Johannes. Epistola tres de Balsamo, duae de Taxo, de Coffea, de Saccharo duae de Corto de Calamo et Cacao in epist. Posthumis a Bartholina. Hafniae, 1664. 8o.

Armuthaz, Bollicosgo. The coffee-mans granado discharged upon the Maidens complaint against coffee, in a dialogue between Mr. Black-burnt and Democritus, wherein is discovered severall strange, wonderful, and miraculous cures performed by coffee . . . : also some merry passages between Peg and Cis, two merry milk-maids of Islington, touching the rare vertues of chocolate written by Don Bollicosgo Armuthaz, to confute the author of that lying pamphlet. London: Printed for J. Johnson, 1663. 8 p., [1] folded leaf of plates: ill.; 19 cm.

Stubbe, Henry, 1632–1676. The Indian nectar, or A discourse concerning chocolata: wherein the nature of cacao-nut, and the other ingredients of that composition, is examined, and stated according to the judgment and experience of the Indians, and Spanish writers, who lived in the Indies, and others; with sundry additional observations made in England: the ways of compounding and preparing chocolata are enquired into; its effects, as to its alimental and venereal quality, as well as medicinal (especially in hypochondriacal melancholy) are fully debated. Together with a spagyrical analysis of the cacao-nut, performed by that excellent chymist, Monsieur le Febure, chymist to His Majesty. By Henry Stubbe formerly of Ch. Ch. in Oxon. physician for His Majesty, and the right honourable Thomas Lord Windsor in the island of Jamaica in the West-Indies. London: Printed by J[ames]. C[ottrell]. for Andrew Crook at the sign of the Green Dragon in St. Paul's Church-yard, 1662. [16], 184 pp. 8vo.

Dupont, Michel et Car. Brisset. An Salubris usus chocolatae. Thèse. Paris, 1661. 4o.

The vertues of chocolate East-India drink. (Oxford): Henry Hall, 1660. Broadside.

Louis XIV. Lettre patente du Roi accordant pour une durée de 29 ans le privilège exclusif de fairefaire, vendre et débiter une certaine composition, qui se nomme chocolate . . . à David Chalious, officier de la Reine. Toulouse, 28. V. 1659.

Colmenero de Ledesma, Antonio. Chocolate: or, An Indian drinke: By the wise and moderate use whereof, health is preserved, sickness diverted, and cured, especially the plague of the guts; vulgarly called the new disease; fluxes, consumptions, & coughs of the lungs, with sundry other desperate diseases. By it also, conception is caused, the birth hastened and facilitated, beauty gain'd and continued. Written originally in Spanish, by Antonio Colmenero of Ledesma, Doctor in Physicke, and faithfully rendred in the English, by Capt. James Wadsworth. London: Printed by J.G. for Iohn Dakins, dwelling neare the Vine Taverne in Holborne, where this tract, together with the chocolate it selfe, may be had at reasonable rates, 1652. [14], 40, [4] pp.

Hurtado, Tomás, *Chocolate y tobaco ayuno eclesiastico y natural: si este le que brante el chocolate: y el tabaco al natural, para la sagrada comunion. Al ilustrissimo y excelentissimo Señor don Fr. Domingo Pimentel . . . Consagra el Padre Maestro Tomas Hurtado de los C.R.M. Toledano.* Madrid: Francisco García, 1645. [11], 144, [8] l. *Note:* Hurtado Professor of Theology at Salamanca.

Colmenero de Ledesma, Antonio. *Chocolata inda. Opusculum de qualitate & natura chocolatae . . . Hispanico antehac idiomate editum: nunc vero curante Marco Aurelio Severino . . . in Latinum translatum.* Norimbergae: Typis Wolfgangi Enderi, 1644. [20], 73, [6] pp. plate; 13 cm.

Colmenero de Ledesma, Antonio. *Du chocolate discours curieux, divisé en quatre parties . . . Traduit d'espagnol en françois sur l'impression faite à Madrid l'an 1631. & esclaircy de quelques annotations. Par René Moreau . . . Plus est adjousté un dialogue touchant le mesme chocolate.* Paris: Sebastien Cramoisy, 1643. [8], 59 pp. 21 cm. *Note: Du chocolate. Dialogue entre un medecin, un indien, & un bourgeois. Composé par Barthelemy Marradon . . . imprimé à Seville l'an 1618. Tourné à present de l'Espagnol, & accommodé à la françoise,* pp. 45–59.

Castro de Torres. *Panegirico del Chocolate.* Segovia, 1640.

Colmenero de Ledesma, Antonio. *A curious treatise of the nature and quality of chocolate. VVritten in Spanish by Antonio Colmenero, doctor in physicke and chirurgery. And put into English by Don Diego de Vades-forte.* Imprinted at London: By I. Okes, dwelling in Little St. Bartholomewes, 1640. [6], 21, [1] p. *Note:* Don Diego de Vades-forte = James Wadsworth; a translation of *Curioso tratado de la naturaleza y calidad del chocolate.*

León Pinelo, Antonio de, 1590 or 1591–1660. *Question moral si el chocolate quebranta el ayuno eclesiastico: tratase de otras bebidas j confecciones que se usan en varias provincias, a D. Garcia de Avellaneda y Haro . . . por el licdo. Antonio de Leon Pinelo, relator del mismo consejo.* Madrid: Por la viuda de Iuan Gonçalez, 1636. [6], 122, [12] leaves; 19 cm (4to).

Colmenero de Ledesma, Antonio. *Du chocolate. Discours curieux . . . Traduit d'Espagnol en François sur l'impression faite à Madrid l'an 1631 & esclaircy de quelques annotations. Par René Moreau . . . Plus est adjousté vn dialogue touchant le mesme chocolate (composé par Barthelemy Marradon . . . Tourné à present de l'Espagnol, & accommodé à la Françoise), etc.* Paris: S. Cramoisy, 1634.

Colmenero de Ledesma, Antonio. *Curioso tratado de la naturaleza y calidad del chocolate, dividido en quatro puntos. En el primero se trata, que sea chocolate; y que calidad tenga el Cacao, y los demas ingredientes. En el segundo, se trata la calidad que resulta de todos ellos. En el tercero se trata el modo de hazerlo, y de quantas maneras se toma en las Indias, y qual dellas es mas saludable. El ultimo punto trata de la quantidad, y como se ha de tomar, y en quetiempo y que personas. por Antonio Colmenero de Ledesma.* Madrid: F. Martinez, 1631. [12] leaves.

Schröder, Johann. *Pharmacopoeia Medico, sive Thesaurus Pharmacologicus, quo composite quaeq celebriora hinc Mineralia, Vegetabili & Animalia Chymico-Medicè describuntur. atq insuper Principia Physicae Hermetico-Hippocraticae candied exhibentur; Opus Non minus utile Physicis quàm Medicis: Authore Johanne Schroedero, M. D. Medico Reip.* Moeno-Francofurtanae: Ulmae, Sumptibus johannis Gerlini Bibliopolae. 1631. 4o. (Mueller notes that "Succolada" is covered, its property *"quod Venerem proritare dicatur."* A second edition expands on this, and corrects the spelling.)

Valverde Turices, Santiago de, fl. 1625. *Discurso de una bevida [aloxa], que aunque en las Indias aya fido antigua, en este lugar es mas nueva, como es el chocolate.* Seville: J. Cabrera, 1625. [8] leaves: t.p. with engr. coat-of-arms (4to).

Valverde Turices, Santiago de, fl. 1625. *Un discurso del chocolate.* Seville: J. Cabrera, 1624. [32] p.; 21 cm (4to).

Marradón, Bartolomeo. *Del Tabago, los daños que causa y del Chocolota.* Sevilla: Gabriele Ramos, 1616 [1618]. 8o.

Hernandez, Francisco. *Qvatro Libros de la Natvraleza, y Virtudes de Las Plantas y animales que estan receuidos en el usu de Medicina en la Neuea-Espana, y la Methodo, y corrección, y preparación, que para a deministrallas se requiere con lo que el Doctor Francisco Hernandez escriuió en la lengua Latina. Traducido y aumentados muches simples, y compuestos y otros muchos secretos curativos, por Francisco Ximenez.* Mexico: en casa de la Viuda de Diego Lopez Daualos, 1615. 4o. 5, 205, 7 Bl. (Mueller includes a section on *"Del Arbol de Cacao que llaman cacahuaquahuitl."*)

Barrios, Juan de. *Libro en el cual se trata del chocolate, y que provecho haga, y se es bebida saludable ó no, y en particular de todas las cosas que lleva, y que receta conviene para cada persona, y como se conocerá cada uno de que complexión sea, para que pueda beber el Chocolate de suerte que no la haga mal.* Mexico: Fernando Balbi, 1609.

1500—1599

Cardenas, Juan de. *Primera Parte de los Problemas y Segretos Maravillosos de las Indias. Compuesta por el Doctor Iuan de Cardenas, Medico.* Mexico: en casa de Pedro Ocharte, 1591, 8o. 8 s., 246 l.

Acosta, José de. *Historia natural y moral de las Indias: en que se tratan las cosas notables del cielo, y elementos, metales, plantas, y animales dellas: y los ritos, y ceremonias, leyes, y gouierno, y guerras de los Indios. Compuesta por el Padre Joseph de Acosta. Religiose de la Compañia de Jesus.* Impresso en Seuilla: en casa de Iuan de Leon, 1590. 535, [37] pp. 21 cm (8vo).

Clusius, Carolus (Caroli Clusii Atreb). *Aliquot notæ in Garciæ Aromatum historiam.: Eiusdem Descriptiones nonnul-*

larum stirpium, & aliarum exoticarum rerum, qu[a]e à generoso viro Francisco Drake equite Anglo, & his obseruatæ sunt, qui eum in longa illa nauigatione, qua proximis annis vniuersum orbem circumiuit, comitati sunt: & quorundam peregrinorum fructuum quos Londini ab amicis accepit. Antuerpiæ: Ex officina Christophori Plantini, 1582. 43, [5] p.; ill.; 80. (Mueller includes description and image of cacao.)

Hurtado, Tomaso. *Tractatus varii Resolutionum Moralium in quibus multiplices casus et principiis Theologia moralis S. Thomae &c Eminentiss. Caietani, methodo brevi, resoluta & clara enucleantur.* Lugduni: Sumptibus Laurentii Anisson & Soc., 1651. Folio. (Mueller: Part Two discusses *"De Potiones Cocolatica Sumenda, vel non sumenda in die jeiunii ecclesiastici."*)

Benzoni, Girolamo. *La Historia del Mondo Nvovo di M. Girolamo Benzoni, Milanese. Laquel tratta dell'Isole, & Mari nuouamente ritrouati, & delle nuoue Città da lui proprio vedute, per acqua & per terra in quattordeci anni. Con Priuilegio della Illustrissima Signoria di Venetia, Per anni XX.* In fine: Venezia: Fr. Rampazetto, 1565. 8o. 175 l. (Includes early printed images of the cacao tree, as well as a description of its cultivation and preparation.)

Diaz del Castillo, Bernal. *Historia Verdadera de la Conquista de la Nueva-España. Escrita Por el Capitan Bernal Diaz del Castillo, vno de sus Conquistadores. Sacada a Lvz Por el P. M. Fr. Alonso Remon, Predicador y Coronista General del Orden de Nuestra Senora de la Merced Redempcion de Cautivos.* Madrid: en la Imprenta del Reyno, 1632. Folio. (12), 254, 6 Bl. (Moctezuma being served chocolate.)

Martyr de Angieria, Petrus. *De Orbe novo Petri Martyris ab Angiera mediolanensis protonotario Caesaris Senatoris Decades VIII. Cum privilegio imperiali.* Compluti Alcala apud Michaelem Eguia, 1530. Folio. 2 Bl., III, CXIII Bl. 3 Bl. (Earliest account of cacao bean as currency and delicacy.)

Oviedo y Valdes, Gonzalo Hernandez de. *Natural y general Historia de las Indias . . . Por industria de maestro Remon de Petras.* Toledo, 1526. Folio.

Cortés, Hernán. *Carta. . . . 30.10.1520.* Sevilla: J. Cromberger, 1522. Folio. 28 s. (First print reference to chocolate in Europe.)

Colombo, Fernando. *Historia del S. D. Fernando Colombo, nelle quali s'ha particolare e vera relatione della vita, & deit Fatti ch'egli fece dell'Indie occidentali dello Mondo Nuovo, hora posseduto dal S. re Cattolico, nuovamenta di lingua spagnuola tradotte nell'Italiana dal Sig. Alfonso Ulloa.* Venetia: Apresso Francesco de'Francheschi, Senese, 1521. 19 s. 247 l. (Europeans first encounter the cacao bean, July 30, 1502, Isla de Piños.)

Nutritional Properties of Cocoa

Robert Rucker

Relative to other seeds, the nutritive value of cacao may be viewed as very good to excellent based on its overall composition. Indeed, its ability to serve as a relatively complete food in part provides rationale as to why cocoa evolved as a substantial component of the diets in southern Mexico and Central and South America [1].

The cacao fruit or pod is melon-shaped (12–30 cm long and 8–12 cm wide) with an average weight of 400–600 g. Each pod usually contains 30–40 seeds or beans imbedded in a white pulp that is sweet when the pods are fresh. The seeds have a reddish brown external color and a dark brown interior owing to their rich polyphenolic content. The pulp is aromatic and mucilaginous, composed of spongy parenchymatous cells that contain a sap rich in carbohydrates. The beans and pulp together amount to about one-third to one-half the total weight of the pod (125–200 g). Although it takes 20 or more pods to harvest 1 kilogram of processed commercial cocoa powder, from a food perspective the beans and pulp from a single pod are sufficient to provide half the daily need of a small person (450–700 calories).

With regard to cocoa powder, 100 hundred grams (about $\frac{1}{4}$ pound) can provide 10–20 percent of the energy needs of a young adult [2] and contribute substantial amounts of a person's vitamin and mineral needs (Table A10.1) [3].

Cacao seeds contain about 40–50 percent fat and 14–18 percent protein depending on the time of harvest. Although cocoa is not an optimal vitamin source as a single food, on a per unit energy basis (e.g., expressed as an amount per calories), cocoa is adequate. As a single source of food, however, cocoa is an excellent source of essential minerals. For example, 50–100 grams of processed cocoa powder can easily meet the daily human needs of most of the essential so-called trace minerals or elements (Table A10.1). Regarding protein, its content in cacao is adequate on a caloric or energy basis, but the quality of the protein is somewhat marginal because the concentration of two essential amino acids, methionine and cysteine, is too low to meet independently human requirements (see section regarding Protein and Amino Acids and Fig. A10.1).

Lipids

The fat in cocoa (cocoa butter) is neutral in flavor. Of the caloric content in typical whole beans, about 50–60 percent comes from cocoa butter. The fatty acids that predominate the triacylglycerol (triglyceride) fraction of cocoa butter are oleic, palmitic, and stearic acids (>90 percent) in combination with several polyunsaturated fatty acids, mostly arachidonic and linolinic acids (5–10 percent). Although concern has been expressed about the high saturated fat content of cocoa butter, the usual health assumptions made regarding saturated fats

Chocolate: History, Culture, and Heritage. Edited by Grivetti and Shapiro
Copyright © 2009 John Wiley & Sons, Inc.

Table A10.1 Nutritional Properties of Cocoa: Cocoa as a Food Source[a]

Nutrient or Component RDA[b]	Amount per 100 grams (Dry Powder)	Percentage of U.S. (grams per 100 grams Dry Powder)
Food energy	225–325 kcal (1000–1500 kJ)	See text
Protein (g)	18–19	See text
Fat (g)	20–25	See text
Carbohydrate (g)	45–55	See text
Fiber (g)	25–35	
Digestible (g)	10–20	
Simple sugars (g)	1–2	
Ash (minerals) (g):	5–6	
Calcium	100–180 mg	10–20
Phosphorus	750–1000 mg	100
Iron	10–15 mg	60–80
Magnesium	500–600 mg	100
Copper	4–6 mg	100
Zinc	5–10 mg	80–100
Potassium	1500–2000 mg	100
Manganese	3–5 mg	100
Selenium	15–20 μg	~40
Vitamins		
Vitamin A equivalents	10–20 RE (μg)	<1
Vitamin E	0.1–0.2 mg as αTE	<5
Vitamin K1	2.5 μg	<5
Vitamin B1, thiamin	0.1 mg	<10
Vitamin B2, riboflavin	0.2–0.3 mg	<20
Niacin	2–4 mg	<20
Vitamin B6, pyridoxine	0.1–0.2 mg	<10
Folate, total	32–40 mcg	10
Pantothenate	0.2–0.4 mg	<10
Other		
Polyphenols	7–18 g	
Theobromine	2–3 g	
Caffeine	0.1 g	
Cholesterol	0	

[a]Based on a composite of compositions for cocoa dry powder [3].
[b]Based on the needs of a small adult [2].

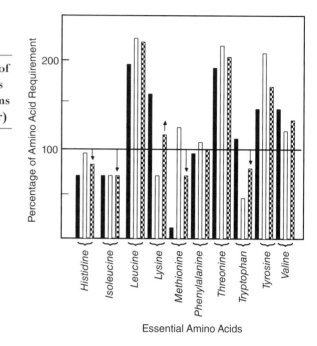

FIGURE A10.1. Amino acid requirements, cocoa and corn protein isolates and their combination. The ability of cocoa (solid bar), corn (open bar), or a combination (50:50 mixture, checkered bar) to meet human amino acid requirements is represented when fed these protein or the combination is fed at the equivalent of 25 grams of protein (i.e., about an ounce). The requirement for each amino acid is arbitrarily set at 100 percent [2]. By complementing corn protein, cocoa improves biological quality so that the requirement for lysine is met and the requirement for tryptophan is almost met. By complementing cocoa protein, corn improves biological quality so that the requirements for histidine and methionine are almost met. Either corn or cocoa when consumed at the equivalent of 25 grams of protein contain sufficient isoleucine to meet ~75 percent of the requirement.

may not apply to cocoa. The lipids in cocoa butter, for example, are not as efficiently or as rapidly absorbed as those in other vegetable fat because of the rather complex molecular configuration of these fatty acids in cocoa triglycerides, which slows the time for overall metabolism and, in particular, slows intestinal absorption [4, 5].

With respect to the consumption of cocoa butter and its effects on serum lipids, the slower rate of absorption of the triacylglycerides in cocoa butter and the high polyphenolic content in cocoa actually have a salutary effect. In humans fed cocoa, plasma LDL cholesterol, oxidized LDL (so-called bad cholesterol components), and apo B (an important lipoprotein polypeptide critical to LDL function) are decreased, while plasma HDL cholesterol (so-called good cholesterol) is increased [6–8].

Protein and Amino Acids

The protein content in cocoa beans is typical of many seeds but, as noted previously, protein quality—based on the distribution of certain essential amino acids—is relatively poor [3, 9]. Compared to a typical reference protein, for example, egg white, whose biological value and chemical score of 100 serves as a comparative baseline, the chemical score for cocoa protein is less than 25, due to the low amount of methionine, an essential amino acid needed to make a "complete" protein. The chemical score is a method for rating proteins based on their chemical composition (more specifically the essential amino acid levels relative to a chosen reference). To determine the chemical score, a protein is picked as a reference and other proteins are rated relative to that reference protein. Typically, egg protein is used with the assumption that the amino acid profile of egg is the ideal for humans; that is, the essential amino acid requirements will be met when fed at the amount needed to meet the protein requirement [10].

For additional comparisons, the biological values and chemical scores for two relative "good" protein sources, soybean protein isolate and casein, are 65 and 50, respectively. The scores of 65 and 50 mean that when one of them is consumed as the sole source of protein, there is one essential amino acid whose content relative to the requirement is only 50–65 percent. However, many types of proteins are consumed through the course of a day. Cocoa was often coconsumed with corn, a very common practice in Central America and Mexico.

When proteins are consumed together, the chemical score of the mixture can markedly improve along with the biological quality depending on the relative distribution of different proteins in these two sources. When an improvement is achieved, the proteins are designated as complementary. Cocoa and maize are complementary in that the amino acid content of the mixture is improved. For example, methionine is higher in corn and the amino acid composition of cocoa improves the low lysine and tryptophan content of corn (Fig. A10.1). When human diets are sufficient in calories, the protein requirement needed to maintain positive nitrogen balance in a human is about 0.8 g per kg of body weight or 12–16 percent of total daily calories. Accordingly, with the appropriate complementation brought about by consuming cocoa and corn in combination, as little as $\frac{3}{4}$ ounces (about 25 grams) of crude protein from a 50:50 mix of cocoa powder and corn protein almost meets a person's essential amino acid needs.

Carbohydrate

Freshly harvested cocoa beans contain about 12–14 percent potentially digestible carbohydrate and considerable quantities of nondigestible carbohydrate or fiber. In fresh pulp there is sufficient simple sugar to have a sweet taste. The pulp also contains pentosans (2–3 percent), citric acid (1–2 percent), and salts (8–10 percent), mainly potassium complexes. However, after six or more days of bacterial fermentation, as per the traditional method of processing cacao beans, the digestible carbohydrate content and composition of the cocoa pod (pulp and beans) is altered significantly [11].

During this process the pulp is digested and a number of complex products are produced, among them various short-chain sugars and ketoacids, and eventually acetic acid and alcohol. The fermentation process also reduces the carbohydrate content of beans to 5–6 percent, owing to the production of fermentation products. In this regard, it is important to note that cacao beans may be prepared by simply drying them or curing them by fermentation, often referred to as "sweating," because of the heat and moisture produced. Drying usually produces a product that is more bitter than that prepared by "sweating" or extensive fermentation.

Vitamins and Minerals

Expressed as amounts per kilocalorie or other units of food-derived energy, cocoa is an excellent source of most essential minerals, especially calcium, copper, iron, manganese, magnesium, phosphorus, potassium, and zinc. For copper, iron, and magnesium, the human requirement on a caloric basis is often exceeded severalfold [2].

Although cocoa is not high in certain vitamins, when consumed with maize and with condiments and spices, especially pigmented fruits and vegetables, these complementary additions provide vitamins whose amounts are missing in cocoa. Noteworthy, the cacao shell has been reported to contain 20–30 International Units of vitamin D_2 (ergocalciferol) per gram or 300 International Units per gram of extractable fat. Consequently, by-products of cocoa—such as the shells—have been used successfully as components of animal feeds [12–15] to achieve maximal rates of growth or milk production.

Other Components

Some of the health effects of the other components in cocoa have been discussed in considerable detail elsewhere [16]. The amounts of nonnutritive compounds (only in the sense that they do not provide energy or may be classed as traditional vitamin) contained in 100 grams of cocoa powder are significant and may account for 20 percent of powder weight: polyphenolics, 7–18 grams (mostly anthocyanidins, proanthocyanins, and other catechin derivatives); theobromine, 2–3 grams; and caffeine, 0.1 gram per 100 g of powder.

Conclusion

For hundreds of years, chocolate in its many forms has been used as a food component. According to legend, the Mexican god Quetzalcoatl left the cocoa tree for the people, that is, as "the food of the gods." As a bean or seed, it is clearly superior to most beans from a nutritional perspective. It is almost a complete food when the nutrient content is expressed per food Calorie. With appropriate complementation, cocoa is capable of serving as the central part of even a simple diet.

References

1. Minifie, B. W. *Chocolate, Cocoa, and Confectionery: Science and Technology*, 3rd ed. New York: Van Nostrand Reinhold, 1989.

2. *Dietary Reference Intakes: Guiding Principles for Nutrition Labeling and Fortification*. Washington DC, Food and Nutrition Board, National Academy of Sciences, 2003.

3. *Composition of Foods: Raw, Processed, Prepared USDA Nutrient Database for Standard Reference*. Beltsville, MD: U.S. Department of Agriculture, Agricultural Research Service, Beltsville Human Nutrition Research Center, Nutrient Data Laboratory, 1999.

4. Porsgaard, T., and Hoy, C. E. Lymphatic transport in rats of several dietary fats differing in fatty acid profile and triacylglycerol structure. *Journal of Nutrition* 2000;130:1619–1624.

5. Sanders, T. A., Berry, S. E., and Miller, G. J. Influence of triacylglycerol structure on the postprandial response of factor VII to stearic acid-rich fats. *American Journal of Clinical Nutrition* 2003;77:777–7782.

6. Mursu, J., Voutilainen, S., Nurmi, T., Rissanen, T. H, Virtanen, J. K, Kaikkonen, J., Nyyssonen, K., and Salonen, J. T. Dark chocolate consumption increases HDL cholesterol concentration and chocolate fatty acids may inhibit lipid peroxidation in healthy humans. *Free Radical Biology Medicine* 2004;1:1351–1359.

7. Wan, Y., Vinson, J. A., Etherton, T. D., Proch, J., Lazarus, S. A., and Kris-Etherton, P. M. Effects of cocoa powder and dark chocolate on LDL oxidative susceptibility and prostaglandin concentrations in humans. *American Journal of Clinical Nutrition* 2001;74:596–602.

8. Denke, M. A. Effects of cocoa butter on serum lipids in humans: historical highlights. *American Journal of Clinical Nutrition* 1994;60(Suppl):1014S–1016S.

9. Timbie, D. J., and Keeney, P. G. Extraction, fractionation, and amino acid composition of Brazilian cacao proteins. *Journal of Agriculture and Food Chemistry* 1977;25:424–426.

10. Young, V. R., and Pellett, P. L. Protein evaluation, amino acid scoring and the Food and Drug Administration's proposed food labeling regulations. *Journal of Nutrition* 1991;121:145–150.

11. Bucheli, P., Rousseau, G., Alvarez, M., Laloi, M., and McCarthy, J. Developmental variation of sugars, carboxylic acids, purine alkaloids, fatty acids, and endoproteinase activity during maturation of *Theobroma cacao* L. seeds. *Journal of Agriculture and Food Chemistry* 2001;49:5046–5051.

12. Dei, H. K., Rose, S. P., and Mackenzie, A. M. Apparent metabolisable energy and digestibility of shea (*Vitellaria paradoxa*) fat, cocoa (*Theobroma cacao*) fat and soybean oil in broiler chicks. *British Poultry Science* 2006;47:607–612.

13. Payne, R. L., Lirette, R. D., Bidner, T. D., and Southern, L. L. Effects of a novel carbohydrate and protein source on sow performance during lactation. *Journal of Animal Science* 2004;82:2392–2396.

14. Johnston, L. J., Pettigrew, J. E., Baidoo, S. K., Shurson, G. C., and Walker, R. D. Efficacy of sucrose and milk chocolate product or dried porcine solubles to increase feed intake and improve performance of lactating sows. *Journal of Animal Science* 2003;81:2475–2481.

15. Aregheore, E. M. Chemical evaluation and digestibility of cocoa (*Theobroma cacao*) byproducts fed to goats. *Tropical Animal Health Production* 2002;34:339–348.

16. Rice-Evans, C. A., and Packer, L. *Flavonoids in Health and Disease*, 2nd edition. Antioxidants in Health and Disease Series. New York: Marcel Dekker, 2003.

Illustration Credits

Cover: Peter J. Shields Library, University of California, Davis. Used with permission.

Title page: Badianus Codex. Manuscript page 38 verso, top register. Courtesy of the National Institute of Anthropology and History, Mexico City D.F., Mexico. Used with permission.

Figure 1.1: Museo Nacional de Arqueología y Etnología del Ministerio de Cultura y Deportes de Guatemala, Guatemala City. Used with permission.

Figure 1.2: The Codex Nuttall. A Picture Manuscript from Ancient Mexico, Edited by Zelia Nuttall. Dover Publications, New York, New York. Used with permission.

Figure 1.3: Ernest Förstemann. Die Maya-handschrift der Königlichen offentlichen Bibliothek zu Dresden. Plate 12a. Dresden, Germany: R. Bertling, 1892. E. Wyllys Andrews V. Director, Middle American Research Institute, Tulane University, New Orleans, Louisiana. Used with permission.

Figure 1.4: Códice Trocortesiano (Madrid Codex). Museo de América, Madrid, Spain. Used with permission.

Figure 1.5: Códice Trocortesiano (Madrid Codex). Museo de América, Madrid, Spain. Used with permission.

Figure 1.6: God "L" and Cacao Tree. From the Red Temple at Cacaxtla, Tlaxcala, Mexico. David R. Hixson © 1997–2001. Used with permission.

Figure 1.7: Códice Trocortesiano (Madrid Codex). Museo de América, Madrid, Spain. Used with permission.

Figure 1.8: Códice Trocortesiano (Madrid Codex). Museo de América, Madrid, Spain. Used with permission.

Figure 1.9: The Peabody Museum of Archaeology and Ethnology, Harvard University, Cambridge, Massachusetts. Used with permission.

Figure 1.10: Ernest Förstemann. Die Maya-handschrift der Königlichen offentlichen Bibliothek zu Dresden. Plate 25a. Dresden, Germany: R. Bertling, 1892. E. Wyllys Andrews V. Director, Middle American Research Institute, Tulane University, New Orleans, Louisiana. Used with permission.

Figure 1.11: Ernest Förstemann. Die Maya-handschrift der Königlichen offentlichen Bibliothek zu Dresden. Plate 33b. Dresden, Germany: R. Bertling, 1892. E. Wyllys Andrews V. Director, Middle American Research Institute, Tulane University, New Orleans, Louisiana. Used with permission.

Figure 1.12: Códice Trocortesiano (Madrid Codex). Museo de América, Madrid, Spain. Used with permission.

Figure 2.1: Project Work Product. Mediaworks Group, University of California, Davis.

Figure 2.2: Matthew G. Looper and Ian Graham. Used with permission.

Chocolate: History, Culture, and Heritage. Edited by Grivetti and Shapiro
Copyright © 2009 John Wiley & Sons, Inc.

Figure 2.3. Matthew G. Looper. Used with permission.

Figure 2.4. Matthew G. Looper. Used with permission.

Figure 3.1. Religious procession in Oaxaca, Mexico. Participants carrying cacao pods. Copyrigt 2008, Guillermo Aldana. Used with permission.

Figure 3.2. Santo image. Day of the Dead Ofrenda Image. Copyright 2008, Howard Shapiro.

Figure 3.3. Day of the Dead Ofrenda Table. Copyright 2008, Howard Shapiro.

Figure 3.4. Private Collection. Grivetti Family Trust. Used with permission.

Figure 3.5. Private Collection. Mari Montejo. Used with permission.

Figure 5.1. Museo Nacional del Prado, Madrid, Spain. Used with permission.

Figure 5.2. Aaron Lopez Collection (P-11), Box 12, Miscellaneous. Accounts–Household. American Jewish Historical Society, New York, New York. Used with permission.

Figure 5.3. *Royal Gazette,* February 12th, 1780, Issue 436, Page 3. Early American Newspapers, Series I: 1690–1876, an Archive of Americana Collection published by Readex (Readex.com) a division of NewsBank, and in cooperation with the American Antiquarian Society. Used with permission.

Figure 5.4. Aaron Lopez Ledger for the year 1766. Aaron Lopez Collection, Lopez Account Book 715, page 37, left. Newport Historical Society, Newport, Rhode Island. Used with permission.

Figure 6.1. Badianus Codex. Manuscript page 38 verso, top register. National Institute of Anthropology and History, Mexico City. Used with permission.

Figure 6.2. Peter J. Shields Library, University of California, Davis. Used with permission.

Figure 6.3. Early American Newspapers, Series I: 1690–1876, an Archive of Americana Collection published by Readex (Readex.com) a division of NewsBank, and in cooperation with the American Antiquarian Society. Used with permission.

Figure 6.4. Early American Imprints, Series I: Evans 1639–1800, an Archive of Americana Collection, published by Readex (Readex.com) a division of Newsbank, and in cooperation with the American Antiquarian Society. Used with permission.

Figure 6.5. Early American Imprints, Series I: Evans 1639–1800, an Archive of Americana Collection, published by Readex (Readex.com) a division of Newsbank, and in cooperation with the American Antiquarian Society. Used with permission.

Figure 7.1. Early American Imprints, Series I: Evans 1639–1800, an Archive of Americana Collection, published by Readex (Readex.com) a division of Newsbank, and in cooperation with the American Antiquarian Society. Used with permission.

Figure 7.2. Early American Newspapers, Series I: 1690–1876, an Archive of Americana Collection published by Readex (Readex.com) a division of NewsBank, and in cooperation with the American Antiquarian Society (used with permission).

Figure 7.3. Early American Newspapers, Series I: 1690–1876, an Archive of Americana Collection published by Readex (Readex.com) a division of NewsBank, and in cooperation with the American Antiquarian Society. Used with permission.

Figure 7.4. Early American Newspapers, Series I: 1690–1876, an Archive of Americana Collection published by Readex (Readex.com) a division of NewsBank, and in cooperation with the American Antiquarian Society. Used with permission.

Figure 7.5. Early American Newspapers, Series I: 1690–1876, an Archive of Americana Collection published by Readex (Readex.com) a division of NewsBank, and in cooperation with the American Antiquarian Society. Used with permission.

Figure 7.6. Early American Newspapers, Series I: 1690–1876, an Archive of Americana Collection published by Readex (Readex.com) a division of NewsBank, and in cooperation with the American Antiquarian Society. Used with permission.

Figure 7.7. Early American Newspapers, Series I: 1690–1876, an Archive of Americana Collection published by Readex (Readex.com) a division of NewsBank, and in cooperation with the American Antiquarian Society. Used with permission.

Figure 7.8. Early American Newspapers, Series I: 1690–1876, an Archive of Americana Collection published by Readex (Readex.com) a division of NewsBank, and in cooperation with the American Antiquarian Society. Used with permission.

Figure 9.1. Peter J. Shields Library, University of California, Davis. Used with permission.

Figure 9.2: American Antiquarian Society, Worcester, Massachusetts. Used with permission.

Figure 9.3: Private Collection. The John Grossman Collection of Antique Images, Tucson, Arizona. Used with permission.

Figure 9.4: Peter J. Shields Library, University of California, Davis. Used with permission.

Figure 9.5: Peter J. Shields Library, University of California, Davis. Used with permission.

Figure 9.6: Peter J. Shields Library, University of California, Davis. Used with permission.

Figure 9.7: Huntington Library, San Marino, California. Used with permission.

Figure 10.1: Peter J. Shields Library, University of California, Davis. Used with permission.

Figure 10.2: Historic Deerfield, Deerfield, Massachusetts. Used with permission.

Figure 10.3: Historic Deerfield, Deerfield, Massachusetts. Used with permission.

Figure 10.4: Monticello and Thomas Jefferson Foundation, Charlottesville, Virginia. Used with permission.

Figure 10.5: The Winterthur Museum, Winterthur, Delaware. Used with permission.

Figure 10.6: Historic Deerfield, Deerfield, Massachusetts. Used with permission.

Figure 10.7: The Winterthur Library. Joseph Downs Collection of Manuscripts and Printed Ephemera. Winterthur, Delaware. Used with permission.

Figure 10.8: Historic Deerfield, Deerfield, Massachusetts. Used with permission.

Figure 10.9: Mount Vernon Porcelain Collection. Mount Vernon Ladies' Association, Mount Vernon, Virginia. Used with permission.

Figure 10.10: Mount Vernon Porcelain Collection. Courtesy: Mount Vernon Ladies' Association, Mount Vernon, Virginia. Used with permission.

Figure 10.11: Historic Deerfield, Deerfield, Massachusetts. Used with permission.

Figure 11.1: Judge Samuel Sewall, Portrait by John Smibert (1688–1751). Date: 1729. Bequest of William L. Barnard by Exchange, and Emily L. Ainsley Fund. Accession Number 58.358. Museum of Fine Arts, Boston, Massachusetts. Used with permission.

Figure 11.2: Museo Nacional del Prado, Madrid, Spain. Used with permission.

Figure 11.3: Chocolate Pot, Manufactured by John Coney (1665–1722). Date: 1701. Gift of Edwin Jackson Holmes. Accession Number 29.1091. Museum of Fine Arts, Boston, Massachusetts. Used with permission.

Figure 11.4: The Metropolitan Museum of Art, Gift of George O. May, 1943 (43.108). Photograph © 1989. The Metropolitan Museum of Art, New York, New York. Used with permission.

Figure 11.5: Chocolate Pot, Manufactured by John Coney. Date: c. 1710–1722. Partial Gift of Dr. Lamar Scutter, Theodora Wilbour Fund in Memory of Charlotte Beebe Wilbour, and the Marion E. Davis Fund. Accession Number 1976.771. Museum of Fine Arts, Boston, Massachusetts. Used with permission.

Figure 11.6: Chocolate Pot, Manufactured by Edward Winslow (1669–1753), Boston, Massachusetts. Date: c. 1705. Courtesy: The Metropolitan Museum of Art, Bequest of Alphonso T. Clearwater, 1933 (33.120.221). Photograph © 1980. The Metropolitan Museum of Art, New York, New York. Used with permission.

Figure 11.7: Chocolate Pot, Manufactured by Peter Oliver (ca. 1682–1712), Boston, Massachusetts. Date: c. 1705. Private Collection. Used with permission.

Figure 11.8: Chocolate Pot, Manufactured by Edward Webb (1666–1718), Boston, Massachusetts. Date: c. 1710. Gift of a Friend of the Department of American Decorative Arts and Sculpture, and Marion E. Davis Fund. Accession Number 1993.61. Museum of Fine Arts, Boston, Massachusetts. Used with permission.

Figure 11.9: Monteith [large silver punch bowl), manufactured by William Denny, London, England. Date 1702. Christie's Images. Used with permission.

Figure 11.10: Chocolate Pot, Manufactured by Zachariah Brigden (1734–1787), Boston, Massachusetts. Date: c. 1755–1760. Gift of Misses Rose and Elizabeth Townsend. Accession Number 56.676. Museum of Fine Arts, Boston, Massachusetts. Used with permission.

Figure 11.11: Chest-on-Chest, Black Walnut, Burl Walnut Veneer, and Eastern White Pine. Manufacturer Unknown. Boston, Massachusetts. Date: c. 1715–1725. Gift of a Friend of the Department of American Decorative Arts and Sculpture and Otis Norcross Fund. Accession Number 1986.240. Museum of Fine Arts, Boston, Massachusetts. Used with permission.

Figure 11.12: Desk and Bookcase, Walnut, Walnut Veneer, Eastern White Pine. Manufacturer Unknown. Boston, Massachusetts. Date: c. 1715–1720. The M. and M. Karolik Collection of Eighteenth-Century American Arts. Accession Number 39.176. Museum of Fine Arts, Boston, Massachusetts. Used with permission.

Figure 11.13: Silver Sugar Box, Manufactured by John Coney, Boston, Massachusetts. Date: 1680–1690. Gift of Mrs. Joseph Richmond Churchill. Accession Number 13.421. Museum of Fine Arts, Boston, Massachusetts. Used with permission.

Figure 11.14: Silver Sugar Box, Manufactured by Edward Winslow (1669–1753), Boston, Massachusetts. Date: c. 1700. The Philip Leffingwell Spalding Collection. Given in His Memory by Katherine Ames Spalding and Philip Spalding, Oakes Ames Spalding, and Hobart Ames Spalding. Accession Number 42.251. Museum of Fine Arts, Boston, Massachusetts. Used with permission.

Figure 11.15: Un Cavalier et une Dame beuvant du chocolat (Cavalier and Lady Drinking Chocolate), Engraving on Paper by Robert Bonnart. Date: c. 1690–1710. Pierpont Morgan Library and Museum, New York. Used with permission.

Figure 11.16: La Jolie Visiteuse (The Pleasant Houseguest) by Jean-Baptiste Mallet (1759–1835). Date: c. 1750. Forsyth Wickes Collection. Accession Number 65.2585. Museum of Fine Arts, Boston, Massachusetts. Used with permission.

Figure 11.17: Jean-Baptiste Le Prince (French, 1734–1781), La Crainte (Fear) 1769, oil on canvas, 50 x 64 cm, Toledo Museum of Art, Purchased with funds from the Libbey Endowment, Gift of Edward Drummond Libbey, 1970.444. The Toledo Museum of Art, Toledo, Ohio. Used with permission.

Figure 12.1: Engraving by Antoine Trouvain, Date: 1694–1698. Musée National du Château et des Trianons, Versailles, France. © Réunion des Musées Nationaux, France, and Art Resource, New York. Used with permission.

Figure 12.2: Chocolatière in the Louis XV Silver-Gilt Rococo Style, A Gift from the King to the Queen on the Occasion of the Birth of the Dauphin. Manufactured by Henry-Nicolas Cousinet. Date: c. 1729–1730. Musée du Louvre. © Réunion des Musées Nationaux, France, and Art Resource, New York. Used with permission.

Figure 12.3: Chocolatière, Silver Spirally Fluted, Pear-Shaped Style, Manufactured by François-Thomas Germain, Paris, France, Date: c. 1765–1766. The Metropolitan Museum of Art, Bequest of Catherine D. Wentworth, 1948 (48.187.407). Photograph by Bobby Hansson. Photograph © 1988. The Metropolitan Museum of Art, New York, New York. Used with permission.

Figure 12.4: Chocolatière, Silver Three-Footed, Pear-Shaped Style, Manufactured by Pierre Vallières, Paris, France. Date: 1781. The Metropolitan Museum of Art, Rogers Fund, 1928 (28.156). Photograph © The Metropolitan Museum of Art, New York, New York. Used with permission.

Figure 12.5: Chocolatière, Silver, Commissioned by Catherine the Great of Russia as a Gift for Count Gregorii Orlofv. Manufactured by Jacques-Nicolas Roettiers, Paris, France. Date: early 1770s. Christie's Sales Catalogue, May 25th, 1993, Object Number 296. Christie's Auction House, New York, New York. Used with permission.

Figure 12.6: Chocolatière, Faïence Fine (Earthenware), Manufactured by the Rue de Charenton Factory, Paris, France. Date: c. 1750s-1760s. Accession Number MNC 4007, Musée National de Céramique, Sèvres, France. © Reunion des Musees Nationaux, France, and Art Resource, New York. Used with permission.

Figure 12.7: Chocolatière, Sèvres Porcelain, Owned by Prince Yusupov. Chinoiserie Style. Date: 1781. Accession Number 20549. State Hermitage Museum, Saint Petersburg, Russia. Used with permission.

Figure 12.8: Chocolatière, Sèvres Porcelain With White Background. Part of a Breakfast Service (déjeuner) Belonging to Count Shuvalov. Date 1780s. Accession Number 24984. State Hermitage Museum, Saint Petersburg, Russia. Used with permission.

Figure 12.9: Chocolatière, Sèvres Porcelain, Owned by Prince Yusupov. Chinoiserie Style. Date: 1781. Detail Lid Top. Accession Number 20549. Courtesy: State Hermitage Museum, Saint Petersburg, Russia. Used with permission.

Figure 12.10: Chocolatière, Sèvres Porcelain, Owned by Prince Yusupov. Chinoiserie Style. Date: 1781. Accession Number 20549. Detail Lid Bottom. Accession Number 20549. Courtesy: State Hermitage Museum, Saint Petersburg, Russia. Used with permission.

Figure 12.11: Chocolatière, Sèvres Porcelain, Owned by Prince Yusupov. Chinoiserie Style. Date: 1781. Accession Number 20549. Detail of Chocolatière Bottom. Accession Number 20549. Courtesy: State Hermitage Museum, Saint Petersburg, Russia. Used with permission.

Figure 12.12: Chocolatière, Sèvres Porcelain Off-white With Gold Decoration. Manufactured for Madame Victoire (Daughter of King Louis XV). Date: 1786. Musée National Adrien-Bubouché [Ceramic Museum], Limoges, France. © Reunion des Musees Nationaux, France, and Art Resource, New York. Used with permission.

Figure 12.13: Déjeuner (Breakfast), by François Boucher. Informal Scene of Drinking Chocolate and Children. Date: 1739. Musée du Louvre, Paris, France. Courtesy: © Réunion des Musées Nationaux, France, and Art Resource, New York. Used with permission.

Figure 12.14: La Tasse de chocolat (The Cup of Chocolate), by Jean-Baptiste Charpentier. Date: 1768. Accession Number MV 7716, Musée National du Château et des Trianons, Versailles, France. © Réunion des Musées Nationaux, France, and Art Resource, New York. Used with permission.

Figure 12.15: Portrait de Mme. du Barry à sa Toilette (Madame du Barry at her Dressing Table), by Jean-Baptiste André Gautier d'Agoty. Date: between 1769–1786. Musée National du Château et des Trianons, Versailles, France. © Réunion des Musées Nationaux, France, and Art Resource, New York. Used with permission.

Figure 12.16: La Tasse de Thé (The Cup of Tea), by Nicholas Lancret. Detail Showing a Family Taking Chocolate Outdoors. Date: c. 1740. Courtesy: Trustees of the National Gallery, London, England. Used with permission.

Figure 12.17: Hand-painted Meissen Porcelain Figure by Johann Joachim Kändler. Date: c. 1744. The Metropolitan Museum of Art, The Jack and Belle Linsky Collection, 1982 (1982.60.326). Photograph © The Metropolitan Museum of Art, New York, New York. Used with permission.

Figure 12.18: Chocolatière, Silver, Owned by Marie Antoinette. Manufactured by Jean-Pierre Charpenat, Paris. Date: 1787–1789. Musée du Louvre Accession Number, OA 9594. © Reunion des Musees Nationaux, France, and Art Resource, New York. Used with permission.

Figure 12.19: Verseuse (Pitcher), Silver, Owned by Napoleon, Converted into Chocolatière. Manufactured by Martin-Guillaume Biennais. Date: 1798–1809. Musée du Louvre, Accession Number, OA 10270. © Reunion des Musees Nationaux, France, and Art Resource, New York. Used with permission.

Figure 12.20: Verseuse (Pitcher), Silver, Owned by Napoleon, Converted into Chocolatière. Detail of Imperial Arms. Manufactured by Martin-Guillaume Biennais. Date: 1798–1809. Source: Museum Accession Number: OA 10270, Musée du Louvre, Paris, France. © Réunion des Musées Nationaux, France, and Art Resource, New York. Used with permission.

Figure 12.21: Painting on Sèvres Porcelain Tray by Charles Develly. Date: 1836. The Metropolitan Museum of Art, Purchase, The Charles E. Sampson Memorial Fund and Gift of Irwin Untermyer, by Exchange, 1986 (1986.281.41). Photograph © 1997 The Metropolitan Museum of Art, New York, New York. Used with permission.

Figure 12.22: Painting on Sèvres Porcelain Tray by Charles Develly. Detail. Date: 1836. The Metropolitan Museum of Art, Purchase, The Charles E. Sampson Memorial Fund and Gift of Irwin Untermeyer, by Exchange, 1986 (1986.281.41). Photograph © 1997 The Metropolitan Museum of Art, New York, New York. Used with permission.

Figure 12.23: Coffeepot. One of four pieces from a Coffee set. Factory: Sevres Manufactory 1836. The Metropolitan Museum of Art, Purchase, The Charles E. Sampson Memorial Fund and Gift of Irwin Untermyer, by Exchange, 1986 (1986.281.1ab). Photograph © 1997 The Metropolitan Museum of Art, New York, New York. Used with permission.

Figure 13.1: Private Collection. Previous owner Robert Elliott & Charles Wamsley Jr. Used with permission.

Figure 13.2: Private Collection. Previous owner Michael Tese. Used with permission.

Figure 13.3: Private Collection. Previous owner Wen Chen. Used with permission.

Figure 13.4: Private Collection. Previous owner Gareth and Amy Jones. Used with permission.

Figure 13.5: Private Collection. Previous owner Wendy and Tim Burri. Used with permission.

Figure 13.6: Private Collection. Previous owner Robert Brenner. Used with permission.

Figure 13.7: Private Collection. Previous owner Michael Moore. Used with permission.

Figure 14.1: Private Collection. The John Grossman Collection of Antique Images, Tucson, Arizona. Used with permission.

Figure 14.2. Private Collection. The John Grossman Collection of Antique Images, Tucson, Arizona. Used with permission.

Figure 14.3. American Antiquarian Society, Worcester, Massachusetts. Used with permission.

Figure 14.4. American Antiquarian Society, Worcester, Massachusetts. Used with permission.

Figure 14.5. American Antiquarian Society, Worcester, Massachusetts. Used with permission.

Figure 14.6. American Antiquarian Society, Worcester, Massachusetts. Used with permission.

Figure 14.7. American Antiquarian Society, Worcester, Massachusetts. Used with permission.

Figure 14.8. American Antiquarian Society, Worcester, Massachusetts. Used with permission.

Figure 14.9. Private Collection. The John Grossman Collection of Antique Images, Tucson, Arizona. Used with permission.

Figure 14.10. Private Collection. The John Grossman Collection of Antique Images, Tucson, Arizona. Used with permission.

Figure 15.1. Compagnie Francaise des Chocolats et des Thes Color Lithograph. Theophile Alexandre Steinlen. Date: c. 1898. Bibliotheque des Arts Decoratifs, Paris, France. The Bridgeman Art Library, London, England. Used with permission.

Figure 15.2. Gala-Peter Color Lithograph. Date: Late 19th Early 20th Century. The Bridgeman Art Library, London, England. Used with permission.

Figure 15.3. Suchard Color Lithograph. Date: Late 19th Century. The Bridgeman Art Library, London, England. Used with permission.

Figure 15.4. Kasseler Color Lithograph. Date: Early 20th Century. Bibliotheque des Arts Decoratifs, Paris, France, Archives Charmet. The Bridgeman Art Library, London, England. Used with permission.

Figure 15.5. Vinay Color Lithograph. Date: Late 19th early 20th Century. The Bridgeman Art Library, London, England. Used with permission.

Figure 16.1. Bunte Christmas Advertisement. Date: 1930. The Bridgeman Art Library, London, England. Used with permission.

Figure 16.2. J. S. Fry of Bristol and London, Agricultural Building, Centennial Exposition, Philadelphia, 1876. Image C020428. Print and Picture Collection, The Free Library of Philadelphia. Used with permission.

Figure 16.3. Stephen F. Whitman and Son of Philadelphia, Machinery Hall, Centennial Exposition, Philadelphia, 1876. Image C021232. Print and Picture Collection, The Free Library of Philadelphia. Used with permission.

Figure 16.4. Chocolate Exhibit: Henry Maillard, New York, Machinery Hall, Centennial Exposition, Philadelphia, 1876. Source: Image C02182. Courtesy: Print and Picture Collection, The Free Library of Philadelphia (used with permission).

FIGIURE 16.4: Peter J. Shields Library, University of California, Davis. Used with permission.

Figure 16.5. Private Collection. Nick Westbrook. Used with permission.

Figure 16.6. Private Collection. Nick Westbrook. Used with permission.

Figure 16.7. Private Collection. Nick Westbrook. Used with permission.

Figure 16.8. Private Collection. Nick Westbrook. Used with permission.

Figure 16.9. Photograph by Amy Houghton, 2008. Used with permission.

Figure 16.10. American Antiquarian Society, Worcester, Massachusetts. Used with permission.

Figure 16.11. Walter Baker & Company Pavilion, Pan-American Exposition, Buffalo, 1901. Image Number 475. Earl Taylor, Dorchester Atheneum, Dorchester, Massachusetts. Used with permission.

Figure 17.1. Early American Newspapers, Series I: 1690–1876, an Archive of Americana Collection published by Readex (Readex.com) a division of NewsBank, and in cooperation with the American Antiquarian Society. Used with permission.

Figure 18.1. Project Work Product. New York City Cocoa Prices, 1797–1820.

Figure 18.2. Project Work Product. New York City, Boston and Philadelphia Cocoa Prices, 1797–1820.

Figure 18.3. Project Work Product. London Cocoa Prices by Variety, 1785–1817.

Figure 18.4. Project Work Product. Liverpool Cocoa Prices, 1799–1818.

Figure 19.1. Early American Reprints, Series I: Evans 1639–1800, an Archive of Americana Collection published by Readex (Readex.com) a division of NewsBank, and in cooperation with the American Antiquarian Society. Used with permission.

Figure 19.2: Early American Reprints, Series I: Evans 1639–1800, an Archive of Americana Collection published by Readex (Readex.com) a division of NewsBank, and in cooperation with the American Antiquarian Society. Used with permission.

Figure 20.1: Peter J. Shields Library, University of California, Davis. Used with permission.

Figure 21.1: Early American Newspapers, Series I: 1690–1876, an Archive of Americana Collection published by Readex (Readex.com) a division of NewsBank, and in cooperation with the American Antiquarian Society. Used with permission.

Figure 21.2: Early American Newspapers, Series I: 1690–1876, an Archive of Americana Collection published by Readex (Readex.com) a division of NewsBank, and in cooperation with the American Antiquarian Society. Used with permission.

Figure 21.3: Early American Newspapers, Series I: 1690–1876, an Archive of Americana Collection published by Readex (Readex.com) a division of NewsBank, and in cooperation with the American Antiquarian Society. Used with permission.

Figure 22.1: Toby Fillpot. Accession Number 1971-480. The Colonial Williamsburg Foundation, Williamsburg, Virginia. Used with Permission.

Figure 22.2: *A Midnight Modern Conversation*, by William Hogarth. Date: 1732–1733. Accession Number 1966-498. The Colonial Williamsburg Foundation, Williamsburg, Virginia. Used with permission.

Figure 22.3: *Christmas in the Country*, by Barlow. Date: January 7th, 1791. Accession Number 1975-286. The Colonial Williamsburg Foundation, Williamsburg, Virginia. Used with permission.

Figure 22.4: *An Election Entertainment*, by William Hogarth. Date: February 24th, 1755. Accession Number 1966-499. The Colonial Williamsburg Foundation, Williamsburg, Virginia. Used with permission.

Figure 22.5: *Beer Street*, by William Hogarth. Date: 1761. Accession Number 1972-409,93. The Colonial Williamsburg Foundation, Williamsburg, Virginia. Used with permission.

Figure 22.6: *Gin Lane*, by William Hogarth. Date: 1761. Accession Number 1972-409,94. The Colonial Williamsburg Foundation, Williamsburg, Virginia. Used with permission.

Figure 22.7: Reproduction of Benjamin Rush's *Moral and Physical Thermometer*, 1790. The Colonial Williamsburg Foundation, Williamsburg, Virginia. Used with permission.

Figure 22.8: *Un Caffetier*, by Martin Englebrecht, Germany c. 1720. Accession Number 1955-150, 1A. The Colonial Williamsburg Foundation, Williamsburg, Virginia. Used with permission.

Figure 24.1: Fortress of Louisbourg, National Historic Site of Canada, Parks Canada. Used with permission.

Figure 24.2: L'arbe qui s'appelle cacao (The Tree That is Called Cacao). Manuscript Version: *Brief discours des choses plus remarquables que Samuel Champlain de Brouage a reconnues aux Indes occidentals.* The John Carter Brown Library At Brown University. Used with permission.

Figure 24.3: L'arbe qui s'appelle cacao (The Tree That is Called Cacao). Published Version: *Brief discours des choses plus remarquables que Samuel Champlain de Brouage a reconnues aux Indes occidentals.* The John Carter Brown Library At Brown University. Used with permission.

Figure 24.4: Mather Byles' Letter Books. Date: 1785. The Winslow Family Papers, Volume 25-4, p. 16–17. Archives and Special Collections, Harriet Irving Library, University of New Brunswick. Used with permission.

Figure 24.5: Stephen Miller's Letter Books, Number 1, Page 5. Date: 1759. The Winslow Family Papers, Vol. 28-1, p. 5. Archives and Special Collections, Harriet Irving Library, University of New Brunswick. Used with permission.

Figure 24.6: The Cowan Cocoa Company of Toronto, 1915. Nestlé Canada, North York, Ontario. Used with permission.

Figure 24.7: Laura Secord Chocolate Boxes. Accession Number M996x.2.512.1-2. McCord Museum, Montreal, Canada. Used with permission.

Figure 25.1: *View from a Warship* by Lewis Parker. Photo by Ruby Fougère. Image Accession Number E-00-29. Fortress of Louisbourg, National Historic Site of Canada, Parks Canada. Used with permission.

Figure 25.2: Atlantic Service Centre, Parks Canada. Photo by Ron Garnett. Image Accession Number NS-FL-1999-14. Fortress of Louisbourg, National Historic Site of Canada, Parks Canada. Used with permission.

Figure 25.3: Archaeological Collection of the Fortress of Louisbourg. Accession Numbers: Lids 1B18K7-28, 1B16G1-1764, 16L94H13-5. Swivels 1B4E2-32, 4L54K9-4. Photograph by Heidi Moses. Image Accession Number 6796E. Fortress of Louisbourg, National Historic Site of Canada, Parks Canada. Used with permission.

Figure 25.4: Archaeological Collection of the Fortress of Louisbourg. Accession Number: 1B18K7-28. Photograph by Heidi Moses. Image Accession Number 6652E. Fortress of Louisbourg, National Historic Site of Canada, Parks Canada. Used with permission.

Figure 25.5: Archaeological Collection of the Fortress of Louisbourg. Accession Number 1B18K7-28. Photograph by Heidi Moses. Image Accession Number 6652E. Fortress of Louisbourg, National Historic Site of Canada, Parks Canada. Used with permission.

Figure 25.6: Archaeological Collection of the Fortress of Louisbourg. Accession Number 1B16G1-912. Photograph by Heidi Moses. Image Accession Number 6629E. Fortress of Louisbourg, National Historic Site of Canada, Parks Canada. Used with permission.

Figure 25.7: Archaeological Collection of the Fortress of Louisbourg. Accession Numbers 4L56C3-11, 1B18H2-20, 1B18H2-41, 4L53F2-29. Photo by Heidi Moses. Image Accession Number 6792E. Fortress of Louisbourg, National Historic Site of Canada, Parks Canada. Used with permission.

Figure 25.8: Archaeological Collection of the Fortress of Louisbourg. Accession Numbers: 3L19A3-10, 4L52M21-1, 1B5A7-4. Photo by Heidi Moses. Image Accession Number 6791E. Fortress of Louisbourg, National Historic Site of Canada, Parks Canada. Used with permission.

Figure 25.9: Archaeological Collection of the Fortress of Louisbourg. Accession Number 3L1Q5-24. Photo by Heidi Moses. Image Accession Number 6806E. Fortress of Louisbourg, National Historic Site of Canada, Parks Canada. Used with permission.

Figure 25.10: Archaeological Collection of the Fortress of Louisbourg. Accession Number 1B18H2-41. Photo by Heidi Moses. Image Accession Number 6613E. Fortress of Louisbourg, National Historic Site of Canada, Parks Canada. Used with permission.

Figure 25.11: Archaeological Collection of the Fortress of Louisbourg. Accession Number 17L31C3-1. Photo by Heidi Moses. Image Accession Number 6799E. Fortress of Louisbourg, National Historic Site of Canada, Parks Canada. Used with permission.

Figure 25.12: Archaeological Collection of the Fortress of Louisbourg. Accession Number 6L92N12-3. Photo by Heidi Moses. Image Accession Number 6798E. Fortress of Louisbourg, National Historic Site of Canada, Parks Canada. Used with permission.

Figure 25.13: Curatorial Collection of the Fortress of Louisbourg. Photograph by Ruby Fougère. Image Accession Number BL.68.1.117. Fortress of Louisbourg, National Historic Site of Canada, Parks Canada. Used with permission.

Figure 25.14: Archaeological Collection of the Fortress of Louisbourg. Accession Numbers 4L52M14-4 and 3L1Q5-4. Photo by Heidi Moses. Image Accession Number 6794E. Fortress of Louisbourg, National Historic Site of Canada, Parks Canada. Used with permission.

Figure 25.15: Curatorial Collection of the Fortress of Louisbourg. Photo by Ruby Fougère. Image Accession Number BL.68.1.122. Courtesy: Fortress of Louisbourg, National Historic Site of Canada, Parks Canada. Used with permission.

Figure 25.16: Archaeological Collection of the Fortress of Louisbourg. Accession Number 4L52M12-19. Photo by Heidi Moses. Image Accession Number 6793E. Fortress of Louisbourg, National Historic Site of Canada, Parks Canada. Used with permission.

Figure 25.17: Dossier Relating to the Succession of Jean-Baptiste-Louis LePrevost Duquesnel. Dated 1744. Page 47 of the Inventory Itemizes Objects in His Cabinet, Including a Gold-Framed Mirror, an English Desk, and Chocolate from Manila. CAOM DPPC GR 2124. Archives nationales d'outre-mer (Aix-en-Provence), France. Used with permission.

Figure 25.18: Private Cabinet. Governor Duquesnel's Reconstructed Apartments at the Fortress of Louisbourg. Photograph Depicts Antique Gold-Framed Mirror, Antique English Desk, Reproduction Porcelain, and Antique Chocolate/Coffee Pot as Listed in His Inventory (Figure 25:17). Photograph by Ruby Fougère. Image Accession Number 2007-21S-1. Fortress of Louisbourg, National Historic Site of Canada, Parks Canada. Used with permission.

Figure 26.1: Benedict Arnold Advertisement Date 1765. Accession Number 1923.42. Toledo Museum of Art. Used with permission.

Figure 26.2: Walter Baker & Company, Limited, Factory Complex, Late 19th Century. Image 186. Earl Taylor, Dorchester Atheneum, Dorchester, Massachusetts. Used with permission.

Figure 26.3: Baker's Breakfast Cocoa Advertisement, 1878. Image 2511. Earl Taylor, Dorchester Atheneum, Dorchester, Massachusetts. Used with permission.

Figure 26.4: Walter Baker & Co.'s Caracas Sweet Chocolate Advertisement. Earl Taylor, Dorchester Atheneum, Dorchester, Massachusetts. Used with permission.

Figure 27.1: Early American Newspapers, Series I: 1690–1876, an Archive of Americana Collection published by Readex (Readex.com) a division of NewsBank, and in cooperation with the American Antiquarian Society. Used with permission.

Figure 27.2: Private Collection. Grivetti Family Trust. Used with permission.

Figure 27.3: Early American Newspapers, Series I: 1690–1876, an Archive of Americana Collection published by Readex (Readex.com) a division of NewsBank, and in cooperation with the American Antiquarian Society. Used with permission.

Figure 28.1: Rochester Museum and Science Center. Accession Number RFC 5283/28. Rochester, New York. Used with permission.

Figure 29.1: Abraham Wendell (1715-?), Attributed to John Heaten. Date: 1730–1745. Accession Number 1962.47. Gift of Governor and Mrs. W. Averell Harriman, Dorothy Treat Arnold (Mrs. Ledyard, Jr.) Cogswell Gates, and Bessie De Beer Aufsesser. Albany Institute of History and Art, Albany, New York. Used with permission.

Figure 29.2: View of Rensselaerville Manufactory, Artist Unidentified. Date: 1792. Caldwell Family Papers, GQ 78-14. Box 1/F-2. Albany Institute of History and Art, Albany, New York. Used with permission.

Figure 29.3: Copper Engraving Plate Inscribed Caldwell's Super Fine Chocolate, Albany. Date: c. 1790–1820. Accession Number 1959.123.20. Gift of Mrs. Jean Mason Browne. Albany Institute of History and Art, Albany, New York. Used with permission.

Figure 29.4: Chocolate Receipt Prepared by Captain Marte G. Van Bergon for William DePeyster. Date: 1759. Manuscript Number 2229. Albany Institute of History and Art, Albany, New York. Used with permission.

Figure 29.5: Early American Newspapers, Series I: 1690–1876, an Archive of Americana Collection published by Readex (Readex.com) a division of NewsBank, and in cooperation with the American Antiquarian Society. Used with permission.

Figure 30.1: Wythe Kitchen (Contemporary Photograph). The Colonial Williamsburg Foundation, Williamsburg, Virginia. Used with permission.

Figure 30.2: Early American Newspapers, Series I: 1690–1876, an Archive of Americana Collection published by Readex (Readex.com) a division of NewsBank, and in cooperation with the American Antiquarian Society. Used with permission.

Figure 30.3: Hand Made Chocolate (Contemporary Photograph). The Colonial Williamsburg Foundation, Williamsburg, Virginia. Used with permission.

Figure 30.4: Early American Newspapers, Series I: 1690–1876, an Archive of Americana Collection published by Readex (Readex.com) a division of NewsBank, and in cooperation with the American Antiquarian Society. Used with permission.

Figure 30.5: Early American Newspapers, Series I: 1690–1876, an Archive of Americana Collection published by Readex (Readex.com) a division of NewsBank, and in cooperation with the American Antiquarian Society. Used with permission.

Figure 31.1: Johnson Hall. Sir William Johnson Presenting Medals to the Indian Chiefs of the Six Nations at Johnstown, New York, 1772. Art Purchase 1993.44. Albany Institute of History and Art, Albany, New York. Used with permission.

Figure 31.2: Collection of the Fort Ticonderoga Museum. Fort Ticonderoga, Ticonderoga, New York. Used with permission.

Figure 31.3: Library and Archives of Canada and Brechin Imaging Services, Ottawa, Ontario. Used with permission.

Figure 31.4: Chocolate Purchases by George Prince Eugane. Date: 1762–1763. Tailer & Blodgett Account Book, Manuscript Number M-2178. Collection of the Fort Ticonderoga Museum. Fort Ticonderoga, Ticonderoga, New York. Used with permission.

Figure 31.5: Collection of the Fort Ticonderoga Museum. Fort Ticonderoga, Ticonderoga, New York. Used with permission.

Figure 31.6: Fort Ticonderoga (Contemporary Photograph, June, 2004). Fort Ticonderoga, Ticonderoga, New York. Used with permission.

Figure 31.7. Fort Ticonderoga (Contemporary Photograph, June, 2004). Fort Ticonderoga, Ticonderoga, New York. Used with permission.

Figure 31.8. Project Work Product. Summarized from Tailer & Blodgett Account Book, Manuscript Number M-2178, Collection of the Fort Ticonderoga Museum. Fort Ticonderoga, Ticonderoga, New York. Used with permission.

Figure 32.1. Early American Imprints, Series I: Evans 1639–1800, an Archive of Americana Collection, published by Readex (Readex.com) a division of Newsbank, and in cooperation with the American Antiquarian Society. Used with permission.

Figure 33.1. Project Work Product. Mediaworks Group, University of California, Davis.

Figure 34.1. Project Work Product. Mediaworks Group, University of California, Davis.

Figure 34.2. Project Work Product. Mediaworks Group, University of California, Davis.

Figure 34.3. Father Serra Collection, Document 332. Santa Barbara Mission Archive-Library, Santa Barbara, California. Used with permission.

Figure 35.1. John Bidwell Donner Party Rescue Document. Date: December 1846. California State Parks, Sutter's Fort Archives, Patty Reed Lewis Collection. Used with permission.

Figure 35.2. Ghirardelli Chocolate Company, San Leandro, California. Used with permission.

Figure 35.3. Private Collection. Ghirardelli Chocolate Company, San Leandro, California, and Grivetti Family Trust. Used with permission.

Figure 35.4. Ghirardelli Chocolate Company, San Leandro, California. Used with permission.

Figure 35.5. Menu (Carte du Jour) Pico House, Los Angeles. Date: Tuesday, February 22nd, 1881. Ephemera, E38-136. Huntington Library, San Marino, California. Used with permission.

Figure 35.6. Menu The Raymond, South Pasadena, California. Date: November 13th, 1886. Ephemera, E38-187. Huntington Library, San Marino, California. Used with permission.

Figure 35.7. Menu The Raymond, South Pasadena, California. Date: November 20th, 1886. Ephemera, E38-188. Huntington Library, San Marino, California. Used with permission.

Figure 35.8. Menu The Arlington, Santa Barbara, California. Date: Christmas, 1886. Ephemera, E38-212. Huntington Library, San Marino, California. Used with permission.

Figure 35.9. Menu Santa Fe Railroad. Date: May 18th, 1888. Ephemera, E38-196. Huntington Library, San Marino, California. Used with permission.

Figure 36.1. Project Work Product. Jamaica: Cocoa Production (19th Century).

Figure 36.2. Project Work Product. St. Lucia: Cocoa Exports (1831–1910).

Figure 36.3. Project Work Product. St. Lucia: Cocoa Production (1831–1866).

Figure 36.4. Project Work Product. St. Lucia: Cocoa Prices (1831–1881).

Figure 36.5. Project Work Product. Grenada: Cocoa Production (19th Century).

Figure 36.6. Project Work Product. Grenada: Cocoa Production, Pounds Produced (19th Century).

Figure 36.7. Project Work Product. Trinidad: Cocoa Acreage, Production, and Export (19th Century).

Figure 40.1. Project Work Product. Redrawn form Russell-Wood, A. J. R. *A World on the Move*, 1993. Mediaworks Group, University of California, Davis.

Figure 41.1. Chocolate Pot of Portuguese King João V. Date: c. 1720–1722. Museu do Arte Antiga Inventory Number 1872. Instituto Português dos Museus, Lisbon, Portugal. Used with permission.

Figure 41.2. Portuguese King João V With Chocolate. Date: 1720. Museu do Arte Antiga Inventory Number, 58 Min. Instituto Português dos Museus, Lisbon, Portugal. Used with permission.

Figure 41.3. Museo Nacional de Cerámica y Artes Suntuarias Gonzalez Marti. Valéncia, Spain, Inventory Number 1/803-804. Used with permission.

Figure 42.1. Peter J. Shields Library, University of California, Davis. Used with permission.

Figure 42.2. Request of Bayonne City Authorities to Purchase Chocolate from Spain. Date: April 6th, 1670. Source: CC 314, #105 [Document Entitled] Pour le Sieur Martin de Lafourcade bourgeois et trésorier de la ville de cette présente année 1670," 6 April 1670. Archives Communales de Bayonne, France. Used with permission.

Figure 42.3. Peter J. Shields Library, University of California, Davis. Used with permission.

Figure 42.4. Peter J. Shields Library, University of California, Davis. Used with permission.

Figure 42.5. Private Collection. Mary and Philip Hyman. Used with permission.

Figure 42.6. Private Collection. Mary and Philip Hyman. Used with permission.

Figure 42.7: Peter J. Shields Library, University of California, Davis. Used with permission.

Figure 42.8: Peter J. Shields Library, University of California, Davis. Used with permission.

Figure 42.9: Gazetin du comestible, No. 12. December 1767. Biblothèque nationale de France [BnF], online Gallica site. Used with permission.

Figure 43.1: Peter J. Shields Library, University of California, Davis. Used with permission.

Figure 43.2: Minneapolis Institute of Arts, Minneapolis, Minnesota. Used with permission.

Figure 43.3: Temple Newsam House, Leeds, England, and The Bridgeman Art Library International, New York, New York. Used with permission.

Figure 43.4: The Trustees of the British Museum, London, England. Used with permission.

Figure 44.1: Chinese Chocolate Cup and Saucer, 1729–1735; Decoration by Ignatius Preissler, 1729–1735. Accession Number AE 85-558. Peabody Essex Museum, Salem, Massachusetts. Used with permission.

Figure 44.2: Museo Nazionale della Ceramica Duca di Martina, Naples, Italy. Used with permission.

Figure 44.3: Menu: Canton Hotel, Dinner Bill of Fare. Date: January 9th, 1910. New York Public Library Menus Collection, Buttolph Collection. New York Public Library, New York, New York. Used with permission.

Figure 44.4: Menu: Canton Hotel, Tiffin Bill of Fare. Date: January 10th, 1910. New York Public Library Menus Collection, Buttolph Collection. New York Public Library, New York, New York. Used with permission.

Figure 45.1: Ghirardelli Chocolate Company, San Leandro, California. Used with permission.

Figure 46.1: Public Domain. Private Collection. Rodney Snyder. Used with permission.

Figure 46.2: Public Domain. Private Collection. Rodney Snyder. Used with permission.

Figure 46.3: Public Domain. Private Collection. Rodney Snyder. Used with permission.

Figure 46.4: Public Domain. Private Collection. Rodney Snyder. Used with permission.

Figure 46.5: Public Domain. Private Collection. Rodney Snyder. Used with permission.

Figure 46.6: Public Domain. Private Collection. Rodney Snyder. Used with permission.

Figure 46.7: Public Domain. Private Collection. Rodney Snyder. Used with permission.

Figure 46.8: Public Domain. Private Collection. Rodney Snyder. Used with permission.

Figure 46.9: Public Domain. Private Collection. Rodney Snyder. Used with permission.

Figure 46.10: Public Domain. Private Collection. Rodney Snyder. Used with permission.

Figure 46.11: Public Domain. Private Collection. Rodney Snyder. Used with permission.

Figure 48.1: Chocolate Cream and Pot (Contemporary Photograph). The Colonial Williamsburg Foundation, Williamsburg, Virginia. Used with permission.

Figure 48.2: Roasting Cacao Beans (Contemporary Photograph). The Colonial Williamsburg Foundation, Williamsburg, Virginia. Used with permission.

Figure 48.3: Cracking Roasted Cacao Beans to Remove Nibs (Contemporary Photograph). The Colonial Williamsburg Foundation, Williamsburg, Virginia. Used with permission.

Figure 48.4: Winnowing Roasted Cacao Beans (Contemporary Photograph). The Colonial Williamsburg Foundation, Williamsburg, Virginia. Used with permission.

Figure 48.5: Crushing Nibs (Contemporary Photograph), The Colonial Williamsburg Foundation, Williamsburg, Virginia. Used with permission.

Figure 48.6: Grinding Cacao Beans (Contemporary Photograph). The Colonial Williamsburg Foundation, Williamsburg, Virginia. Used with permission.

Figure 48.7: Pouring Chocolate (Contemporary Photograph). The Colonial Williamsburg Foundation, Williamsburg, Virginia (used with permission).

Figure 50.1: Project Work Product. Consumer Top Health Concerns.

Figure 50.2: Project Work Product. Consumer Attitudes Towards Functional Foods.

Figure 50.3: Project Work Product. Using Chocolate Medicinally.

Figure 50.4: Project Work Product. Foods Eliminated When Dieting.

Figure 50.5: Project Work Product. Attitudes Towards Corporation Funding of Research.

Figure 51.1: Project Work Product. Mediaworks Group, University of California, Davis.

Figure 51.2: Ministry of Culture, Spain, and Archivo General de Indias (Archive of the Indies), Seville, Spain. Used with permission.

Figure 54.1: Chocolate Pot Construction: 3–4 Tabs Visible on Seam. Private Collection. Christopher D. Fox (used with permission).

Figure 54.2: Peter J. Shields Library, University of California, Davis. Used with permission.

Figure 54.3: Winterthur Library and Museum, Winterthur, Delaware. Used with permission.

Figure 54.4: Chocolate Pot Construction: Handle attached directly over brazed side construction seam. Private Collection: Phil Dunning. Used with permission.

Figure 54.5: Chocolate Pot Construction: Comparative North American 19th-20th Century Type. Construction Material Spun Brass, Identifiable by Uniform Shape and Smooth Body Form, with Heavy Cast Brass Handle Attachment Panel. Private Collection. Eric and Judy Whitacre. Used with permission.

Figure 54.6: Chocolate Pot Construction: Unusual Baluster-Form Lacking Distinctive Flared Foot That Commonly Characterizes This Style. Private Collection. Eric and Judy Whitacre. Used with permission.

Figure 54.7: Chocolate Pot Construction: Example of Footed-Baluster Form With Spout. Private Collection. Phil Dunning. Used with permission.

Figure 54.8: Chocolate Pot Construction: Example of Baluster Form with Legs. Stewart Museum, Montreal, Canada. Used with permission.

Figure 54.9: Chocolate Pot Construction: Example of Tapered-Cylinder Form, Perhaps Early 18th Century. Private Collection. Eric and Judy Whitacre. Used with permission.

Figure 54.10: Chocolate Pot Construction: Design of Lid Shape. Example of Complex Form, Fitted Over the Exterior of the Rim With a Pivoting Cover for the Mill Hole. Parks Canada, Fortress Louisbourg National Historic Park, Louisbourg, Nova Scotia. Used with Permission.

Figure 54.11: Chocolate Pot Construction: Design of Lid Shape. Example Characteristic of Many Pots with Spouts or Beaks Where Lid Fits Inside the Rim of the Pot, Rather than Over the Outside. This Style Commonly Has a Pivoting Cover to Close the Mill Hole. Private Collection. Eric and Judy Whitacre. Used with permission.

Figure 54.12: Chocolate Pot Mill Hole Cover Excavated at the Geddy House Site. Object # 00279-19BB; Context number: E. R. 1346G-19.B. The Colonial Williamsburg Foundation, Williamsburg, Virginia. Used with permission.

Figure 54.13: Chocolate Mill: Example of a Typical Star-Shaped Bottom Type (undated). Private Collection. Phil Dunning. Used with permission.

Figure 54.14: Chocolate Mill: Vertically-notched Mill with Decoratively-Turned Handle Stylistically Similar to Some Early Dated Illustrations. Associated with Chocolate Pot in Figure 54.09. Private Collection. Eric and Judy Whitacre. Used with permission.

Figure 54.15: Chocolate Mill: Examples of Modern Forms With Separate Loose Rings at the Base, Common to Mexico and Central America in the 20th and 21st Century. Privates Collection. Eric and Judy Whitacre. Used with permission.

Figure 56.1: Private Collection. Grivetti Family Trust. Used with permission.

Figure 56.2: Private Collection. Ghirardelli Chocolate Company, San Leandro, California, and Grivetti Family Trust. Used with permission.

Figure 56.3: Private Collection. Grivetti Family Trust. Used with permission.

Figure 56.4: Private Collection. Ghirardelli Chocolate Company, San Leandro, California, and Grivetti Family Trust. Used with permission.

Figure 56.5: Private Collection. Grivetti Family Trust. Used with permission.

Figure 56.6: Photograph by Scott Borg (used with permission).

Figure 56.7: Photograph by Scott Borg. Used with permission.

Figure 56.8: Photograph by Scott Borg. Used with permission.

Figure 56.9: Photograph by Scott Borg. Used with permission.

Figure 56.10: The Thomas H. Manning Polar Archives. Scott Polar Research Institute, University of Cambridge, Cambridge, England. Used with permission.

Figure 56.11: Private Collection. Grivetti Family Trust. Used with permission.

Figure 56.12: Private Collection. Grivetti Family Trust. Used with permission.

Figure 56.13: Private Collection. Grivetti Family Trust. Used with permission.

Figure 56.14: Brian Hawley at <www.luxurylinerrow.com>. Used with permission.

Figure 56.15: National Air and Space Museum, LTA File A3G-309329-01, Smithsonian Institution, Washington DC. and Luftschiffbau Zeppelin GmbH, Friedrichshafen, Germany. Used with permission.

Figure 56.16: Real Jardin Botanico (Royal Botanical Garden) Library, Madrid, Spain. Used with permission.

Figure 56.17: Johnson and Whales University and Culinary Museum. Personal Collection of Louis Szathmary. Copy and Reuse Restrictions Apply [www.culinary.org], Providence, Rhode Island. Used with permission.

Figure 56.18: Ministry of Culture, Madrid, Spain, and Archivo General de las Indias (Archive of the Indies), Seville, Spain. Used with permission.

Index

Chocolate: History, Culture, and Heritage. Edited by Grivetti and Shapiro

Copyright © 2009 John Wiley & Sons, Inc.